WORKSHOP MANUAL

VOLKSWAGEN 1200, Type 11, 14 and 15

Selected Books and Repair Information From Bentley Publishers

Motorsports

Alex Zanardi: My Sweetest Victory
Alex Zanardi and Gianluca Gasparini
ISBN 978-0-8376-1249-2

The Unfair Advantage *Mark Donohue and Paul van Valkenburgh*
ISBN 978-0-8376-0069-7

Equations of Motion - Adventure, Risk and Innovation
William F. Milliken
ISBN 978-0-8376-1570-7

Engineering

The Hack Mechanic Guide to European Automotive Electrical Systems *Rob Siegel*
ISBN 978-0-8376-1751-0

Physics for Gearheads
Randy Beikmann ISBN 978-0-8376-1615-5

Bosch Automotive Handbook
Updated 9th Edition *Robert Bosch GmbH*
ISBN 978-0-8376-1732-9

Bosch Fuel Injection and Engine Management *Charles O. Probst, SAE*
ISBN 978-0-8376-0300-1

Maximum Boost: Designing, Testing, and Installing Turbocharger Systems
Corky Bell ISBN 978-0-8376-0160-1

Supercharged! Design, Testing and Installation of Supercharger Systems
Corky Bell ISBN 978-0-8376-0168-7

Audi

Audi A4 Service Manual: 2002-2008, 1.8L Turbo, 2.0L Turbo, 3.0L, 3.2L
Bentley Publishers
ISBN 978-0-8376-1574-5

Audi TT Service Manual: 2000-2006, 1.8L turbo, 3.2 L *Bentley Publishers*
ISBN 978-0-8376-1625-4

Audi A6 (C5 platform) Service Manual: 1998-2004 *Bentley Publishers*
ISBN 978-0-8376-1670-4

BMW

Memoirs of a Hack Mechanic
Rob Siegel ISBN 978-0-8376-1720-6

BMW 3 Series (F30, F31, F34) Service Manual: 2012-2015 *Bentley Publishers*
ISBN 978-0-8376-1752-7

BMW X3 (E83) Service Manual: 2004-2010 *Bentley Publishers*
ISBN 978-0-8376-1731-2

BMW X5 (53) Service Manual: 2000-2006 *Bentley Publishers*
ISBN 978-0-8376-1643-8

BMW 3 Series (E90, E91, E92, E93) Service Manual: 2006-2011
Bentley Publishers ISBN 978-0-8376-1723-7

BMW 3 Series (E46) Service Manual: 1999-2005 *Bentley Publishers*
ISBN 978-0-8376-1657-5

BMW 3 Series (E36) Service Manual: 1992-1998 *Bentley Publishers*
ISBN 978-0-8376-1709-1

BMW 3 Series (E30) Service Manual: 1984-1990 *Bentley Publishers*
ISBN 978-0-8376-1647-6

BMW Z3 (E36/7) Service Manual: 1996-2002 *Bentley Publishers*
ISBN 978-0-8376-1617-9

BMW 5 Series (E60, E61) Service Manual: 2004-2010 *Bentley Publishers*
ISBN 978-0-8376-1689-6

BMW 5 Series (E39) Service Manual: 1997-2003 *Bentley Publishers*
ISBN 978-0-8376-1672-8

Corvette

Corvette: America's Star-Spangled Sports Car *Karl Ludvigsen*
ISBN 978-0-8376-1659-9

Zora Arkus-Duntov: The Legend Behind Corvette *Jerry Burton*
ISBN 978-08376-0858-7

Porsche

Porsche 911 (996) Service Manual: 1999-2005 *Bentley Publishers*
ISBN 978-0-8376-1710-7

Porsche 911 (993) Service Manual: 1995-1998 *Bentley Publishers*
ISBN 978-0-8376-1719-0

Porsche Boxster Service Manual: 1997-2004 *Bentley Publishers*
ISBN 978-0-8376-1645-2

Porsche 911 Carrera Service Manual: 1984-1989 *Bentley Publishers*
ISBN 978-0-8376-1696-4

Porsche 911 SC Service Manual: 1978-1983 *Bentley Publishers*
ISBN 978-0-8376-1705-3

Porsche: Excellence Was Expected
Karl Ludvigsen ISBN 978-0-8376-0235-6

Porsche — Origin of the Species
Karl Ludvigsen ISBN 978-0-8376-1331-4

Volkswagen

Volkswagen Rabbit, GTI Service Manual: 2006-2009 *Bentley Publishers*
ISBN 978-0-8376-1664-3

Volkswagen Jetta Service Manual: 2005-2010 *Bentley Publishers*
ISBN 978-0-8376-1616-2

Volkswagen Jetta, Golf, GTI Service Manual: 1999-2005 *Bentley Publishers*
ISBN 978-0-8376-1678-0

Volkswagen Jetta, Golf, GTI: 1993-1999, Cabrio: 1995-2002 Service Manual *Bentley Publishers*
ISBN 978-0-8376-1660-5

Volkswagen GTI, Golf, Jetta Service Manual: 1985-1992 *Bentley Publishers*
ISBN 978-0-8376-1637-7

Volkswagen Corrado Repair Manual: 1990-1994 *Bentley Publishers*
ISBN 978-0-8376-1699-5

Volkswagen Passat, Passat Wagon Service Manual: 1998-2005
Bentley Publishers
ISBN 978-0-8376-1669-8

MINI Repair Manuals

MINI Cooper Service Manual: 2007-2013 *Bentley Publishers*
ISBN 978-0-8376-1730-5

MINI Cooper Service Manual: 2002-2006 *Bentley Publishers*
ISBN 978-0-8376-1639-1

Mercedes-Benz

Mercedes-Benz C-Class (W202) Service Manual 1994-2000
Bentley Publishers
ISBN 978-0-8376-1692-6

Mercedes-Benz Technical Companion
Staff of The Star and members of Mercedes-Benz Club of America
ISBN 978-0-8376-1033-7

WORKSHOP MANUAL

VOLKSWAGEN 1200, Type 11, 14 and 15

BENTLEY PUBLISHERS · CAMBRIDGE, MASSACHUSETTS

BENTLEY PUBLISHERS® | Automotive Reference™

Bentley Publishers, a division of Robert Bentley, Inc.
1734 Massachusetts Avenue
Cambridge, MA 02138 USA
800-423-4595 / 617-547-4170

Information that makes the difference®

BentleyPublishers®.com

Technical contact information

We welcome your feedback. Please submit corrections and additions to our Volkswagen technical discussion forum at:

`http://www.BentleyPublishers.com`

Updates and corrections

We will evaluate submissions and post appropriate editorial changes online as updates or tech discussion. Appropriate updates and corrections will be added to the book in future printings. Check for updates and corrections for this manual before beginning work on your vehicle. See the following web address for additional information:

`http://www.BentleyPublishers.com/updates/`

WARNING—Important Safety Notice

This manual is a reproduction of original, dated material and is for historical reference only. This manual has not been updated or revised to include safety precautions, cautions or warnings, nor does it include technical changes, corrections or updates that may have been made to this information since its original publication. Therefore, procedures, specifications, part numbers, special tools and safety equipment are for historical reference only and may be obsolete or unacceptably hazardous by current standards.

Do not use this manual unless you are familiar with automotive repair procedures and safe workshop practices. This manual illustrates the workshop procedures required for some maintenance and service work. It is not a substitute for full and up-to-date information from the vehicle manufacturer or for proper training as an automotive technician. Note that it is not possible for us to anticipate all of the ways or conditions under which vehicles may be serviced or to provide cautions as to all of the possible hazards that may result.

FOR THESE REASONS, NEITHER THE PUBLISHER NOR THE AUTHOR MAKES ANY WARRANTIES, EXPRESS OR IMPLIED, THAT THE INFORMATION IN THIS BOOK IS FREE OF ERRORS OR OMISSIONS AND WE EXPRESSLY DISCLAIM THE IMPLIED WARRANTIES OF MERCHANTABILITY AND OF FITNESS FOR A PARTICULAR PURPOSE, EVEN IF THE PUBLISHER OR AUTHOR HAVE BEEN ADVISED OF A PARTICULAR PURPOSE, AND EVEN IF A PARTICULAR PURPOSE IS INDICATED IN THE MANUAL. THE PUBLISHER AND AUTHOR ALSO DISCLAIM ALL LIABILITY FOR DIRECT, INDIRECT, INCIDENTAL OR CONSEQUENTIAL DAMAGES THAT RESULT FROM ANY USE OF THE EXAMPLES, INSTRUCTIONS OR OTHER INFORMATION IN THIS BOOK. IN NO EVENT SHALL OUR LIABILITY WHETHER IN TORT, CONTRACT OR OTHERWISE EXCEED THE COST OF THIS MANUAL.

Before attempting any work on your Volkswagen, read the **Warnings and Cautions** on page **vi** and any **Warning** or **Caution** that accompanies a procedure in the service manual. Review the **Warnings and Cautions** on page **vi** each time you prepare to work on your Volkswagen.

Your common sense and good judgment are crucial to safe and successful service work. Read procedures through before starting them. Think about whether the condition of your car, your level of mechanical skill, or your level of reading comprehension might result in or contribute in some way to an occurrence which might cause you injury, damage your car, or result in an unsafe repair. If you have doubts for these or other reasons about your ability to perform safe repair work on your car, have the work done at an authorized Volkswagen dealer or other qualified shop.

This publication is a reproduction of the original Volkswagen service and repair information covering the original removal, installation and adjustment procedures for the 1961 through 1965 Volkswagen Type 1 models: Sedan and Convertible, and Karmann Ghia Coupe and Convertible. Because this manual is a reprint of historic, dated, information it is reprinted here for historical interest only. The publisher cannot be responsible for the result of any error in this manual.

This manual is prepared, published and distributed by Bentley Publishers, 1734 Massachusetts Avenue, Cambridge, Massachusetts 02138 USA. Volkswagen has not reviewed and does not vouch for the accuracy of the technical specifications and procedures described in this manual. The publisher would like to thank Volkswagen of America, Inc., for its support of this project.

Library of Congress Catalog Card No. 90-82469
ISBN 978-0-8376-1702-2
VWoA Inc. Part No. LPV 800 121
VWC Inc. Part No. MAN 101 113
Job code: V165-11

The paper used in this publication is acid free and meets the requirements of the National Standard for Information Sciences—Permanence of Paper for Printed Library Materials. ∞

Manufactured in the United States of America.

202506R002

CONTENTS

Index to Sections:

Please read these Warnings and Cautions before proceeding with any maintenance and repair work.

WARNING—

- This Volkswagen Type 1 Workshop Manual is the original service and repair information which was available in 1965. Volkswagen has constantly improved its cars, and these changes, both in parts and specifications, are often applicable to earlier models. This manual has not been updated or revised to include safety precautions, cautions or warnings, not does it include technical changes, corrections or updates that may have been made to this information since its original publication. Therefore, procedures, specifications and part numbers are for historical reference only and may be obsolete or unacceptably hazardous by current standards.

- Do not re-use any fasteners that are worn or deformed in normal use. Many fasteners are designed to be used only once and become unreliable and may fail when used a second time. This includes, but is not limited to, nuts, bolts, washers, self-locking nuts or bolts, circlips, cotter pins. Always replace these fasteners with new parts.

- Never work under a lifted car unless it is solidly supported on stands designed for the purpose. Do not support a car on cinder blocks, hollow tiles, or other props that may crumble under continuous load. Never work under a car that is supported solely by a jack. Never work under the car while the engine is running.

- If you are going to work under a car on the ground, make sure that the ground is level. Block the wheels to keep the car from rolling. Disconnect the battery negative (-) terminal (ground strap) to prevent others from starting the car while you are under it.

- Never run the engine unless the work area is well ventilated. Carbon monoxide kills.

- Tie long hair behind your head. Do not wear a necktie, a scarf, loose clothing, or a necklace when you work near machine tools or running engines. If your hair, clothing, or jewelry were to get caught in the machinery, severe injury could result.

- Finger rings, bracelets and other jewelry should be removed so that they cannot cause electrical shorts, get caught in running machinery, or be crushed by heavy parts.

- Do not attempt to work on your car if you do not feel well. You increase the danger of injury to yourself and others if you are tired, upset or have taken medicine or any other substance that may impair you from being fully alert.

- Illuminate your work area adequately but safely. Use a portable safety light for working inside or under the car. Make sure the bulb is enclosed by a wire cage. The hot filament of an accidentally broken bulb can ignite spilled fuel or oil.

- Catch draining fuel, oil, or brake fluid in suitable containers. Do not use food or beverage containers that might mislead someone into drinking from them. Store flammable fluids away from fire hazards. Wipe up spills at once, but do not store the oily rags, which can ignite and burn spontaneously.

- Always observe good workshop practices. Wear goggles when you operate machine tools or work with battery acid. Gloves or other protective clothing should be worn whenever the job requires working with harmful substances.

- Friction materials such as brake and clutch discs may conta asbestos fibers. Do not create dust by grinding, sanding or cleaning with compressed air. Avoid breathing asbestos fibers ar asbestos dust. Breathing asbestos can cause serious diseases su as asbestosis or cancer, and may result in death.

- Disconnect the battery negative (-) terminal (ground strap) whe ever you work on the fuel system or the electrical system. Do n smoke or work near heaters or other fire hazards. Keep approved fire extinguisher handy.

- Batteries give off explosive hydrogen gas during charging. Ke sparks, lighted matches and open flame away from the top of t battery. If hydrogen gas escaping from the cap vents is ignited, will ignite gas trapped in the cells and cause the battery to explod

- Connect and disconnect battery cables, jumper cables or a b tery charger only with the ignition switched off, to prevent spark Do not disconnect the battery while the engine is running.

- Do not quick-charge the battery (for boost starting) for long than one minute. Wait at least one minute before boosting the b tery a second time.

- Do not use excessive battery charging voltage. If the batte begins producing gas or boiling violently, reduce the charging ra Boosting a sulfated battery at a high charging rate can cause explosion.

- Some aerosol tire inflators are highly flammable. Be extrem cautious when repairing a tire that may have been inflated using aerosol tire inflator. Keep sparks, open flame or other sources ignition away from the tire repair area. Inflate and deflate the t at least four times before breaking the bead from the rim. Co pletely remove the tire from the rim before attempting any repai

- Greases, lubricants and other automotive chemicals contain to substances, many of which are absorbed directly through the sk Read manufacturer's instructions and warnings carefully. Use ha and eye protection. Avoid direct skin contact.

CAUTION—

- Before starting a job, make certain that you have all the necess tools and parts on hand. Read all the instructions thoroughly, do attempt shortcuts. Use tools appropriate to the work and use o replacement parts meeting Volkswagen specifications. Makes tools, parts and procedures will not make good repairs.

- Use pneumatic and electric tools only to loosen threaded pa and fasteners. Never use these tools to tighten fasteners, especi on light alloy parts. Always use a torque wrench to tighten fasten to the tightening torque specification listed.

- Be mindful of the environment and ecology. Before you drain crankcase, find out the proper way to dispose of the oil. Do not p oil onto the ground, down a drain, or into a stream, pond, or la Consult local ordinances that govern the disposal of wastes.

Foreword

The Volkswagen Beetle's sales success and its dramatic impact on the automotive world is legendary. From its pre-war beginnings through its remarkable ascent from industrially ravaged post-war Germany, the Type 1, or Beetle, went on to sell over twenty million units of production over a span of more than three decades, becoming one of the most significant and influential automobiles of all time.

Volkswagen's elegantly simple, functional design and outstanding sales success produced a vast and loyal following of owners and enthusiasts. More than any of its contemporaries, the Beetle was fun to drive, inexpensive to own and operate, and easy to maintain. These same virtues are the basis for the enthusiasm that continues to surround the cars to this day. Interest in the preservation and restoration of older Volkswagens has never been greater. This manual provides fascinating historical perspective on these early cars, and is a comprehensive source of original technical information.

This manual is the original Volkswagen service and repair information available to the authorized Volkswagen dealer technician in 1965. This manual covers the following Volkswagen Type 1 models produced between 1961 and 1965:

1200 Sedan
1200 Convertible
1200 Karmann Ghia Coupe
1200 Karmann Ghia Convertible

Users of this manual are cautioned to be particularly aware of its limitations as a present-day guide to service and repair. Because this manual is a reproduction of dated information, no attempt has been made to revise or update the information to reflect changes in safety precautions, procedures, materials or specifications which may have been made by Volkswagen at a later time. References to procedures, specifications and part numbers are historical and may not accurately represent today's knowledge of workshop practices, assembly materials or repair methods, fluid and lubricant specifications, or important health and safety issues.

Volkswagen improved their cars many times throughout the production cycle. Sometimes these improvements did not coincide with a new model year but were noted by the date of production and the engine or chassis number of the first car to carry the change. Where the point of introduction of an improvement is less clear, the repair information in this manual has been amended to note that it covers "Early" or "Late" models within the 1961 to 1965 range of production.

Robert Bentley, Inc.

VW SERVICE

Engine and Clutch

Contents: M

Engine

Clutch

Automatic Clutch

Important — Friction materials such as brake or clutch discs may contain asbestos fibers. Do not create dust by grinding, sanding, or by cleaning with compressed air. Avoid breathing asbestos fibers and asbestos dust. Breathing asbestos can cause serious diseases such as asbestosis or cancer, and may result in death.

THIS PAGE INTENTIONALLY
LEFT BLANK

Description of Engine

General Description

The air-cooled Volkswagen engine is of the 4-cylinder, 4-cycle, O. H. V. type with two pairs of cylinders horizontally opposed. It is mounted on the flange of the rubber-cushioned transmission case by means of four bolts.

Crankcase

The bipartite crankcase is a light-metal die cast. The crankcase halves are machined in pairs to very close limits and in consequence replacements must be made in pairs.

Crankshaft

The crankshaft is heat-treated at all bearing points. The crankshaft has 4 special light metal bearings. No. 2 bearing-seen from the clutch-is of the split type. No. 1 bearing is lead coated and takes up the crankshaft end thrust. The flywheel with starter gear ring is held by a gland nut and additionally secured to the crankshaft by four dowel pins. Timing gear and distributor drive gear are secured in place by Woodruff keys. The fan pulley is bolted to the crankshaft. An oil seal is fitted to the clutch side of the crankshaft and an oil thrower and oil return thread to the pulley side.

Connecting Rods

The crank ends of the four connecting rods contain replaceable lead-bronze bearings. The piston ends are provided with bronze piston pin bushings.

Pistons

The pistons are of light-metal alloy and have three rings, of which the bottom one is the oil scraper ring. The piston pins are fully floating and held in place by means of circlips.

Cylinders

The four cylinders of special cylinder casting are interchangeable and can be replaced separately together with the corresponding piston. The cylinders are provided with fins for more efficient cooling.

Cylinder Head

Each pair of cylinders has one mutual detachable cylinder head of light alloy casting. The cylinder head is also provided with cooling fins and it incorporates shrunk-in valve seat inserts and valve guides. The valves are of the overhead type. No gasket is used between the jointing faces of cylinder and cylinder head.

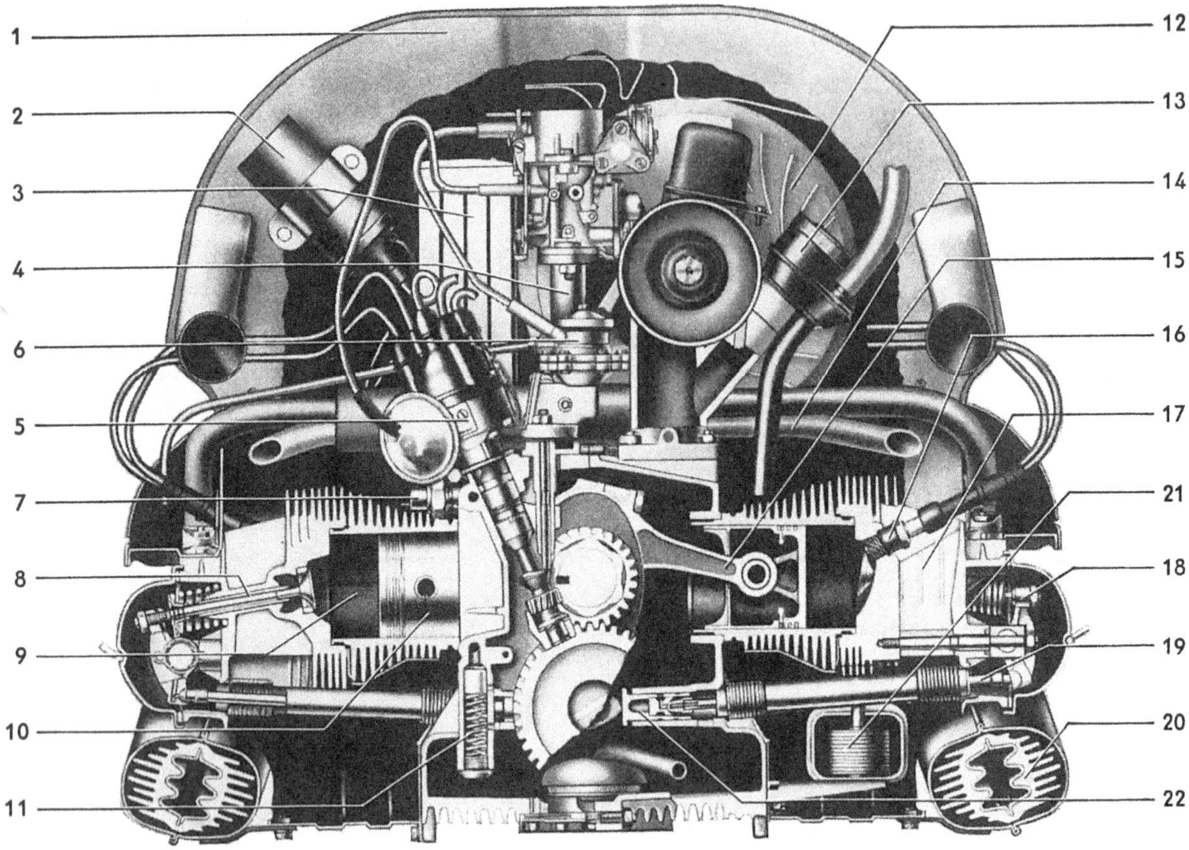
1
2
3
4
6
5
7
8
9
10
11
12
13
14
15
16
17
21
18
19
20
22

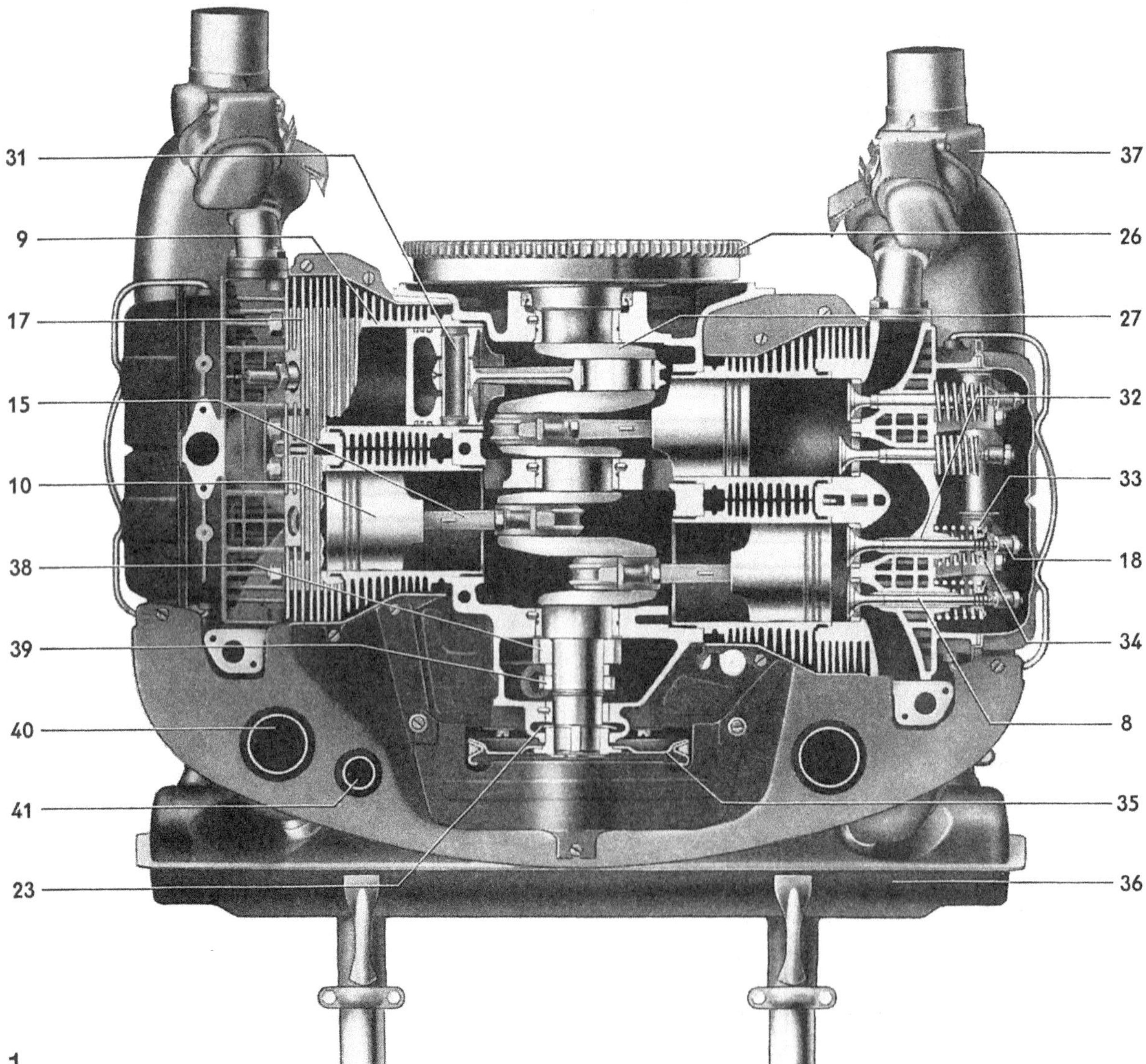
31
9
17
15
10
38
39
40
41
23
37
26
27
32
33
18
34
8
35
36

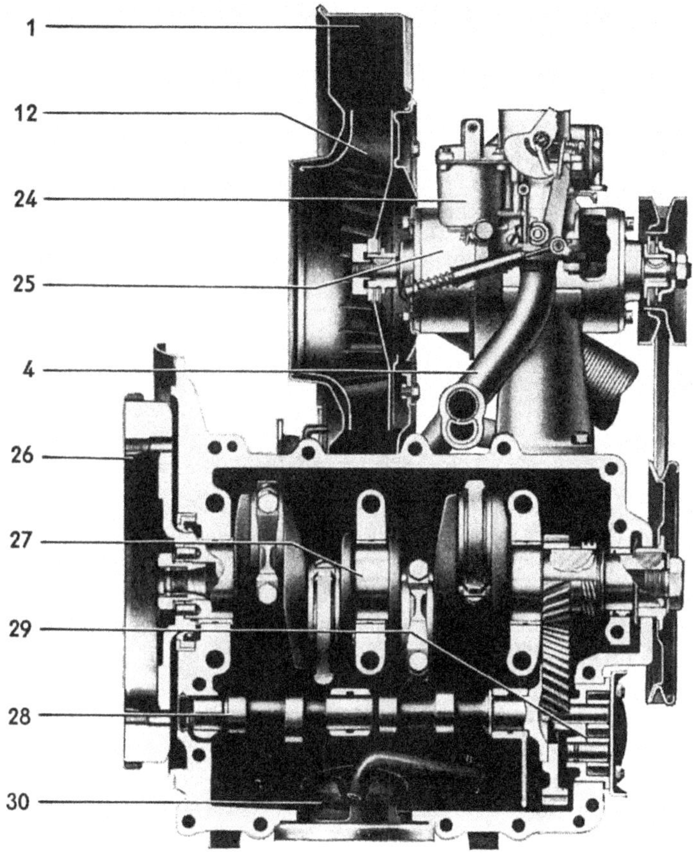

The Volkswagen Engine

1 - Fan housing
2 - Ignition coil
3 - Oil cooler
4 - Intake manifold
5 - Distributor
6 - Fuel pump
7 - Oil pressure switch
8 - Valve
9 - Cylinder
10 - Piston
11 - Oil pressure relief valve
12 - Fan
13 - Oil filler neck
14 - Pre-heater pipe
15 - Conecting rod
16 - Spark plug
17 - Cylinder head
18 - Rocker arm
19 - Push rod
20 - Heat exchanger
21 - Thermostat
22 - Cam follower
23 - Oil deflector washer
24 - Carburetor
25 - Generator
26 - Flywheel
27 - Crankshaft
28 - Camshaft
29 - Oil pump
30 -Oil strainer
31 - Piston pin
32 - Valve guide
33 - Valve spring
34 - Valve spring cap
35 - Pulley
36 - Muffler
37 - Junction box
38 - Crankshaft gear
39 - Distributor drive gear
40 - Heating air hose
41 - Carburetor pre-heater hose

Valve Actuating Mechanism

The camshaft is carried in three bearings machined in the crankcase. It is driven from the crankshaft by helical gears. The camshaft timing gear is of light metal. The valves are operated by the cams via cam followers, push rods and rocker arms. Each cam operates in turn one of the valves of two opposed cylinders. The exhaust valves are plated with high quality chrome-nickel steel.

Cooling System

The air cooling is done by means of a fan, which is attached to the extended generator shaft. It is driven from the crankshaft by an adjustable V-belt at approx. twice the engine revolutions. The fan sucks in air through an opening in the fan housing, and the air cools the engine by being forced through the fins of the cylinders and cylinder heads. The air flow is directed by air deflector plates; some of them are situated in the fan housing, and the others cover the cylinders. The throttle ring at the air intake opening of the fan housing is thermostat-controlled which ensures that the operating temperature is quickly attained and uniformly maintained.

Oil Circulation

The pressure feed lubrication system includes a special oil cooler. The gear-type oil pump is situated at the gear side of the camshaft, from which it receives its drive. Oil is drawn from the lowest point of the crankcase and forced into the oil passages via the oil cooler. Part of the oil is fed via the crankshaft main bearings through the drilled passages in the crankshaft to the connecting rod bearings. Another part lubricates the camshaft bearings, and the rest is fed through the hollow push rods to the rocker arms, lubricating their bearings and the valve stems. Cylinder walls, pistons and piston pins are lubricated by splash. The oil returns to the crankcase bottom, where it is filtered by a gauze strainer at the lowest point before again entering the circulation.

The oil cooler on the crankcase is positioned in the ducted air flow. The situation of the oil cooler is such that the oil forced up by the pump must pass through it before reaching the lubrication points. The drop in temperature in the oil cooler amounts to 20° C (36° F). This enables the oil to maintain its lubricating qualities even at high outside temperatures and at sustained high engine speeds. In cold weather, when the oil is of higher viscosity, an oil pressure relief valve makes it possible for the engine to be lubricated directly, that is, by by-passing the oil cooler.

An automatic oil pressure switch for the oil pressure warning light is fitted to the oil pipe between pump and cooler. This switch opens an electric contact at a pressure of 0.15—0.4 kg/cm² (2.1—6 psi) interrupting the supply of electric current to the warning light. The lamp lights up when switching on the ignition and when the oil pressure is insufficient.

Note:

From Chassis No. 4846836 (Engine No. 6916251), some parts of the 1200 cc. engine (40 bhp) were modified.

Part Numbers:

	new	previous
Cylinder head	113 101 351 C	113 101 351 B
Oil cooler	111 117 021 D	111 117 021 C
Intake manifold	113 129 701 D	113 129 701 B
Fan housing	113 119 025 C	113 119 025
Fan	113 119 031 A	113 119 031

The parts listed can be service installed in all 40 bhp engines. The previous type cylinder head and oil cooler will no longer be available after stocks have been used up.

Main Modifications to the Engine of the 1200 Volkswagen

The following modifications were made to the 1.2 liter engine, from Chassis No. 116000001 (Engine No. D0000001) in August 1965:

Part	Part Number	Remarks
Crankcase	113101025 B	With camshaft bores for bearing shells Oil intake pipe: 12 mm diameter (was 10 mm)
Cylinder	113101311 C	18 cooling fins. Can be used as replacement for or be paired with previous cylinders.
Camshaft bearing No. 1 Camshaft bearing No. 2 Camshaft bearing No. 3, left (with guide shoulder) Camshaft bearing No. 3, right	113101501 113101511 113101521 113101522	The camshaft bearings can be obtained individually or in SP tets (SP 50). Cannot be service installed in previous crankcase.
Crankshaft bearing No. 2	131105531	Steel-backed bearings. Can be service installed in previous engines.
Push rods	113109301 D	Ball head inside rod. Can be interchanged with previous design.
Oil pump housing Oil pump drive shaft Oil pump gear Oil strainer	311115107 311115115 311115123 A 311115175 Mark: Annular groove	Cannot be service installed in previous 1.2 liter engines. Previous parts still available.
Engine front cover plate	211119517 A	Can be service installed in previous engines.
Clutch operating shaft Left bearing bush for clutch operating shaft	113141701 C 113141707	Can be service installed in previous crank-case.

THIS PAGE INTENTIONALLY
LEFT BLANK

Engine Removal and Installation

Following is a list of the facilities which are available for removing and installing the engine:

To lift car

Gantry VW 605
Free-wheel lift
Roll-on lift

To keep car in raised position

Vehicle trolley VW 603/1
Trestle VW 633
Front end lifter VW 606 (with VW 633)

To remove engine

Engine trolley VW 600
Gantry cross tube
Hydraulic or mechanical trolley jack

Depending on the equipment of the workshop various possibilities exist. The following pictures show the combination.

a - Trolley with Gantry and cross tube

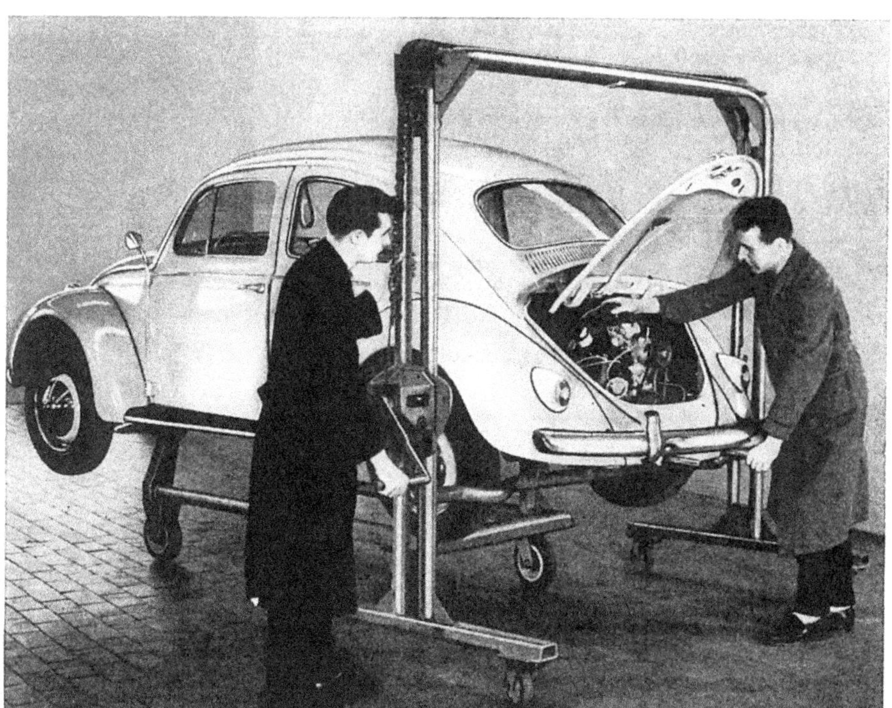

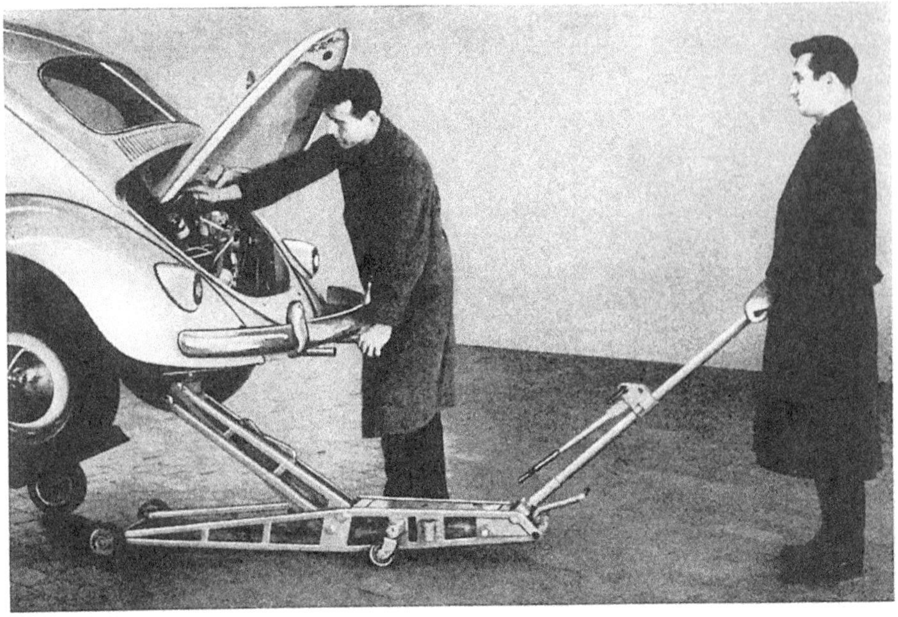

b - Trolley and Hydraulic Jack

When jacking up, care should be taken that the rear end of the vehicle clears the ground by about one yard prior to removing the engine.

Engine Removal

1 - Disconnect ground strap from battery.

2 - Turn fuel tap to "off" position.

3 - Open rear hood.

4 - Remove air cleaner and engine rear cover plate.

5 - Disconnect: cables from terminals 51 and 61 on generator, from carburetor, cable from terminal 15 on ignition coil, and cable from oil pressure switch.

6 - Disconnect accelerator cable from carburetor.

7 - Loosen mounting screw on the distributor support and turn distributor far enough to allow vacuum chamber to clear rear cover plate when the engine is removed.

8 - Raise vehicle at rear or lift it on trolley stand or trestle.

9 - Disconnect both heating control cables and loosen flexible heater pipes from engine.

10 - Disconnect fuel hose from engine.

11 - Unscrew two nuts of the lower engine mounting bolts.

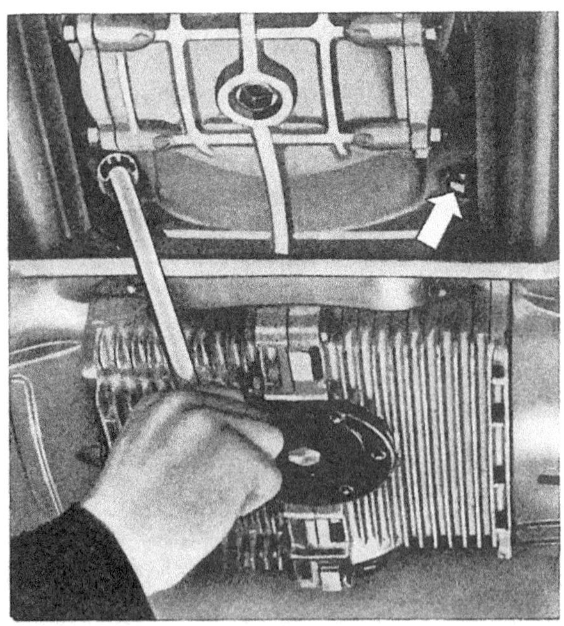

12 - Withdraw accelerator cable from guide tube.

13 - Place engine trolley, or gantry cross tube, or trolley jack under engine.

14 - Have an assistant remove the nuts, while you hold the two upper engine mounting bolts.

15 - Engine trolley:

Lower car until engine rests on trolley.

Trolley jack:

Raise jack until platform contacts engine.

16 - Withdraw engine until clutch release plate clears main drive shaft.

17 - Engine trolley: Lift up car, —

Trolley jack: Lower jack, —

and tilt the engine down at its rear end until it can be withdrawn.

When carrying out this operation, be sure not to distort or damage the clutch release plate or main drive shaft.

Engine Installation

This is a reversal of the preceding operations, but the following points should be observed:

1 - Install engine only with engine rear cover plate removed.

2 - Loosen mounting screw on the distributor and turn distributor.

3 - Check central position of clutch plate, using special tool VW 219. If necessary, depress clutch release plate using VW tool 657 (local manufacture) and center the clutch plate.

4 - Check clutch release bearing and clutch release plate for wear and cracks, renew if necessary.

If a clutch release bearing with plastic ring is fitted, roughen the ring up with coarse emery and rub in a molybdenum disulphide lubricant such as

Molykote Paste G or
Molykote Powder Z or
Liqui Moly LM 11

5 - Examine needle bearing in flywheel gland nut for wear and pack it with 10 g (0.35 oz.) Universal Grease.

6 - Lubricate starter shaft bush, starter drive pinion, and flywheel gear ring with Universal Grease.

7 - Lubricate main drive shaft splines and pilot with transmission oil. Apply with a clean dry cloth.

8 - Thoroughly clean transmission case and engine flange.

9 - When replacing engine, care must be taken to prevent damage to gland nut needle bearing and clutch release bearing and to avoid bending the main drive shaft.

To facilitate entry of main drive shaft into clutch plate and gland nut needle bearing rotate engine at V-belt (engage a gear to steady main drive shaft).

Note:

To avoid interference with the function of the automatic clutch, take care to route the connecting hoses so that they are not kinked or jammed when installing the engine. This applies particularly to the small diameter pipe from the control valve to the carburetor venturi which will only work properly if routed in the original production manner.

10 - When installing engine, first insert the lower engine mounting bolts into their corresponding holes in transmission case flange. Press engine firmly against flange until it is contacting properly all round. First slightly tighten the upper mounting bolt nuts and then the lower ones. After that, fully tighten in the same order.

11 - Adjust accelerator cable correctly.

12 - Adjust ignition timing.

THIS PAGE INTENTIONALLY
LEFT BLANK

Engine Disassembly and Assembly

It is recommended to adopt the following sequence of operations when disassembling and assembling the engine:

Disassembly

1 - Drain engine oil.
2 - Remove front engine cover plate.
3 - Remove fan belt.
4 - Disconnect cable between distributor and ignition coil.
5 - Remove fan housing and generator as a unit.
6 - Remove crankshaft pulley.
7 - Remove fuel lines.
8 - Remove intake manifold and pre-heating pipe.
9 - Remove muffler (silencer) and heater assy.
10 - Remove heating channels and both cylinder cover plates.
11 - Remove clutch.
12 - Remove cylinder head covers.
13 - Remove rocker arm shafts.
14 - Remove cylinder heads.
15 - Remove valve push rod tubes and valve push rods.
16 - Remove deflector plates below the cylinders.
17 - Remove cylinders.
18 - Remove pistons.
19 - Remove oil cooler.
20 - Remove oil pump.
21 - Remove oil strainer
22 - Remove fuel pump.
23 - Remove distributor and distributor drive pinion.
24 - Remove flywheel.
25 - Disassemble crankcase.
26 - Remove camshaft and crankshaft.

Assembly

Assembling the engine is a reversal of the above operations. Special hints concerning the installation are given on the following pages.

It is recommended to disassemble or assemble the engine on the Stand VW 643 (Local Manufacture) in conjunction with VW 307 or Fixture VW 313 in conjunction with VW 307. To catch any oil drips the Oil tray VW 631 (Local Manufacture) is used. The engine and its components should be cleaned in a wash plant similar to Wash Plant VW 630 (Local Manufacture). Disassembled engine components should be kept in the Disassembly Trolley VW 651 (Local Manufacture) and the Disassembly Rack VW 652/3 (Local Manufacture) so as to avoid confusion during assembly. Small parts such as screws, nuts, washers etc. should be left in a strainer in which they are cleaned in the wash plant.

THIS PAGE INTENTIONALLY
LEFT BLANK

Removing and Installing Cover Plates

Removal

When installing and removing cover and deflector plates, the following sequence of operations should be adopted:

1 - Take off engine rear cover plate prior to removing engine.

2 - Remove engine front cover plate.

3 - Remove fan housing and generator as a unit.

4 - Remove heating channels after having removed the exhaust system.

5 - Remove both cylinder cover plates.

6 - Remove crankshaft pulley cover after having removed the pulley.

7 - Remove deflector plates after having taken off valve push rod tubes.

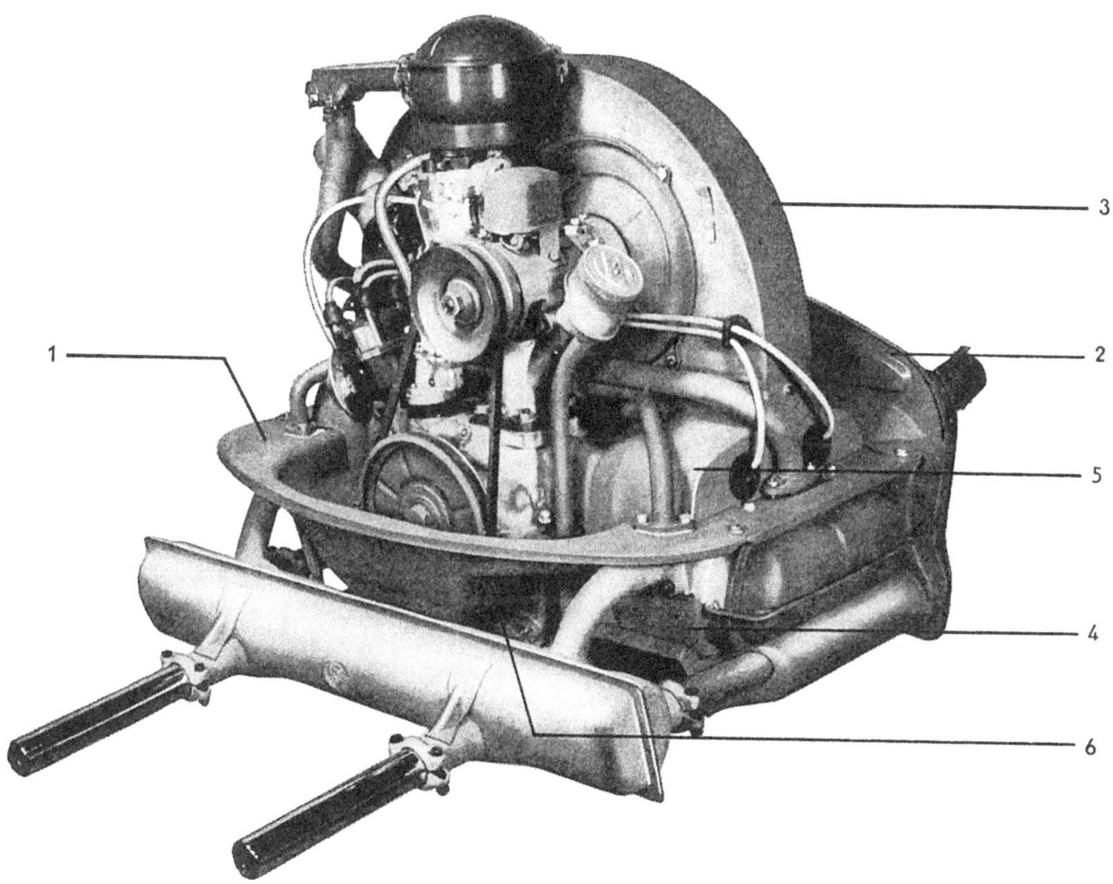

1 - Engine rear cover plate
2 - Engine front cover plate
3 - Fan housing
4 - Heating channel
5 - Cylinder cover plate
6 - Cover plate below crankshaft pulley

2 - When replacing cylinder cover plates, attention should be paid to condition and sealing of spark plug rubber caps.

3 - The cylinder cover plates should fit snugly on the exterior of the fan housing to prevent loss of cooling air.

4 - Prior to installing the engine front cover plate, check condition of weatherstrip.

Installation

When installing, the following points should be observed:

1 - The deflector plates below the cylinders must be installed prior to the push rods and push rod tubes and checked for correct seating. If the necessity should arise, bend the plates until they tightly bear on the ,ylinder head studs to prevent them from rattling or working loose.

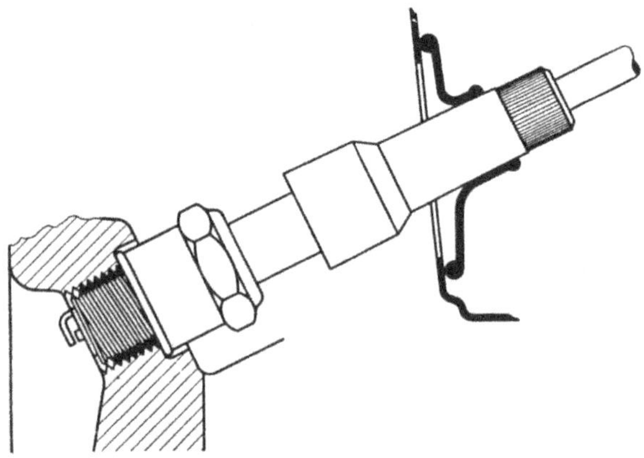

Removing and Installing Engine Rear Cover Plate

(Engine in situ)

Removal

1 - Remove slotted screws of engine cover plate.

2 - Remove engine cover plate, by lifting it off backwards.

Installation

When installing, the following points should be observed:

1 - Do not forget to reinstall the washers for the slotted screws.

2 - After engine has been installed, the weatherstrip lips should be positioned as shown in the illustration.

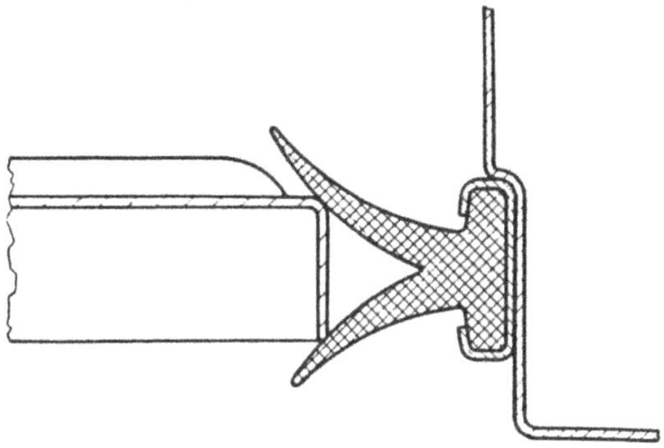

Damaged weatherstrips are to be renewed.

3 - The breather pipe grommet must squarely bear on the cover plate.

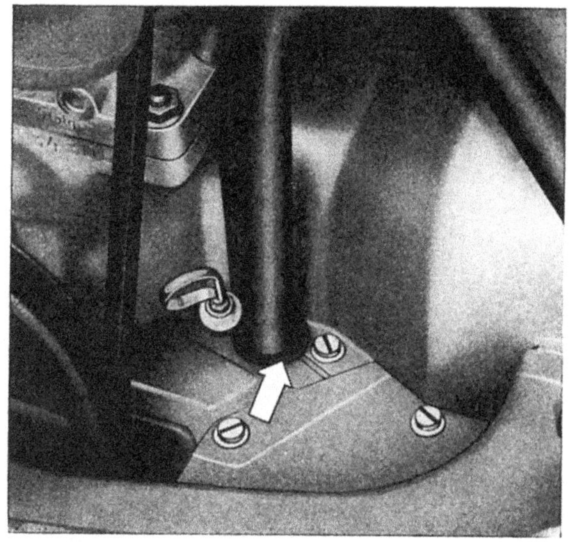

Removing and Installing Fan Housing

Removal

1 - Remove fan belt.

2 - Remove generator strap and pull off cable from ignition coil.

3 - Remove rubber holders for ignition cables from fan housing.

4 - Remove slotted screws on both sides of fan housing.

5 - Detach spring of automatic cooling air control and remove throttle ring screws.

6 - Lift off fan housing and generator as a unit.

Installation

When installing, the following points should be observed:

1 - Examine fan housing for damage and loose air deflector plates.

2 - There must be no "blow-past" between fan housing and cover plates of cylinders. If necessary, bend plates into correct position.

3 - Insert throttle ring and screw it to the holding plate on the operating shaft, taking care that the throttle ring is not off-centered from the intake flange. The throttle ring is at an angle to the intake flange - as seen from the side and above - and no attempt should be made to bend the holding plate as this would have a detrimental effect on the cooling system.

The throttle ring is centered to the intake flange by moving it in its clearance holes.

4 - Connect return spring.

5 - Adjust throttle ring.

6 - Install rubber holders for ignition cables at fan housing.

Removing and Installing Fan Housing

(Engine in situ)

Removal

1 - Disconnect battery.

2 - Remove rear hood together with hinge bracket.

3 - Disconnect cables from generator, carburetor and ignition coil as well as cable from oil pressure switch.

4 - Remove vacuum line from ignition distributor and carburetor and fuel hose.

5 - Pull out accelerator cable and conduit tube.

6 - Remove carburetor mounting nuts and take off carburator.

The other operations for removing and installing fan housing are the same as with the engine removed.

Removal

1 - Remove the four screws on fan cover with T-wrench VW 106.

2 - Remove generator and fan.

3 - Remove special nut on fan with special wrench VW 112. A second mechanic should hold the fan.

4 - Remove fan, spacer washers, and hub.

Installation

1 - Place hub on generator shaft. Make sure the woodruff key is properly seated.

2 - Insert spacer washers.

3 - Place fan in position.

4 - Tighten special nut with torque wrench and socket to 5.5—6.5 mkg (40—47 ft. lbs.).

5 - Check distance between fan and cover (a = 1.5—1.8 mm/0.06—0.07"). To obtain correct distance, spacer washers can be inserted between hub and carrier plate. If only one is used, the other two should be placed between lock washer and fan.

6 - Insert generator in fan housing.

7 - Tighten the four screws on fan cover with T-wrench VW 106.

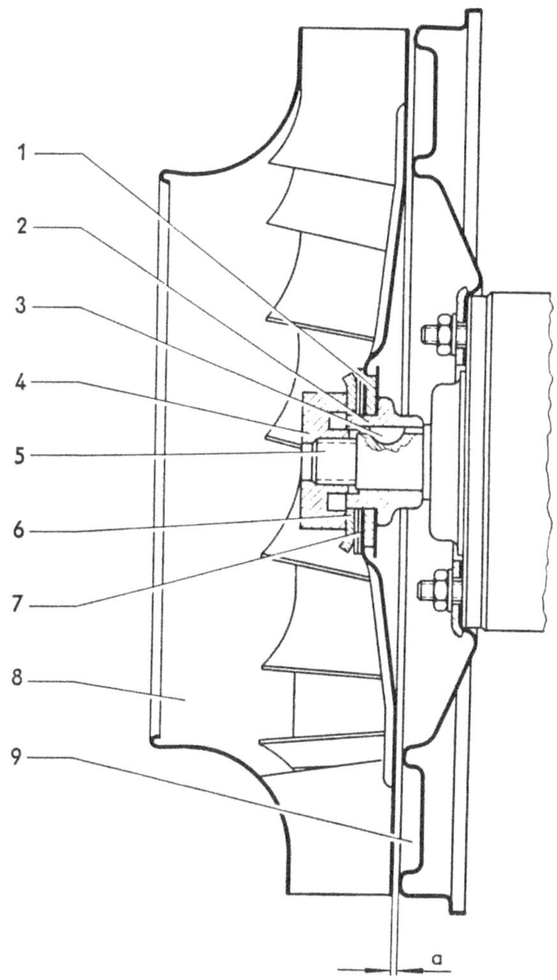

a = 1.5—1.8 mm/0.06—0.07"

1 - Hub carrier plate
2 - Fan hub
3 - Woodruff key
4 - Special nut
5 - Generator shaft
6 - Lock washer
7 - Spacer washer
8 - Fan
9 - Fan cover

Automatic Air Intake Control

Inspection and Adjustment

The adjustment of the automatic air intake control should be checked during inspections and at the beginning of the cold and warm seasons, and adjusted if necessary.

When adjusting, bear in mind that if the throttle ring opens too soon or remains open, the engine attains its operating temperature too slowly and in some cases leads to continuous spitting and excessive fuel consumption. If the throttle ring opens too far, it may foul the fan and cause considerable noise. If it opens too slowly in the warm seasons the engine will overheat when under sustained high loads.

If the throttle ring remains open whilst the engine is cold, the thermostat may be defective. To prevent the engine from overheating, the throttle ring automatically opens fully if the cooling system should go out of order.

Note:

From Chassis No. 3 866 883 (Engine No. 5 773 144), an 8–9 mm high rubber plug was installed on the fan housing as a stop for the throttle ring. The adjustment of the ring is not affected. The plug can be service installed.

Note:

From Chassis No. 4 166 056 (Engine No. 6 120 731), the opening temperature of the thermostat was reduced from 75–80° C to 65–70° C (149–150° F). The bracket has been reinforced for this reason.

The previous type of bracket will be discontinued when existing stocks are exhausted. The modified bracket can be used on all engines. When an old thermostat is replaced by the new pattern, the modified bracket must be installed at the same time. The figures "65°" are stamped on the bottom of the modified thermostat.

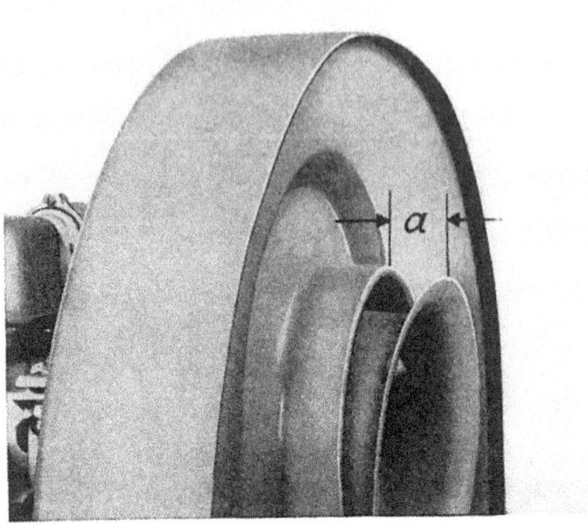

Inspection

1 - When the engine is cold, the throttle ring should rest slightly pre-loaded against the air intake flange.

2 - The engine should be allowed to warm up until the upper end of the thermostat touches the upper stop of the support.

The distance (a) from the middle of the intake flange to the edge of the throttle ring should measure 25—30 mm (0.98—1.18″) in this position.

Adjustment when Assembling Engine

1 - Lift thermostat to the upper stop of its support.

2 - Adjust throttle ring so that it opens 20 mm (0.79″).

3 - Tighten operating lever.

4 - Tighten thermostat. Be sure the faces milled in the tapped boss of the thermostat fit properly in the guide hole of the support. For this purpose it may be necessary to turn the thermostat back by up to half a turn. When the thermostat has been tightened, the throttle ring rests slightly pre-loaded against the intake flange.

5 - Connect return spring.

6 - Install right heating channel.

Adjustment with Engine in Situ

1 - Detach return spring.

2 - Release throttle ring operating lever.

3 - Allow engine to warm up until the upper end of thermostat touches the upper stop of the support.

4 - Adjust throttle ring so that it opens 25—30 mm (0.98—1.18″).

5 - Tighten operating lever.

6 - Connect return spring.

Make sure that throttle ring and linkage move freely in all positions.

General

Generator and cooling system are driven by the fan belt. Due to the power absorbed by these two units, the fan belt is subjected to considerable stress, especially at high speed and when shifting down. To insure adequate cooling and long service life of the belt, it is of utmost importance to maintain the belt at correct tension. If the belt is too slack, it will slip and lead to an overheating of the engine. Excessive tension creates undue stress and is liable to cause breakage of the belt and damage to the generator bearings.

Examination

The belt, when firmly pressed with the thumb at midpoint, must yield approximately 15 mm a = 15 mm (0.6″).

No traces of excess wear, such as frayed edges and cracks, should be perceptible.

It is important to prevent oil getting on to the fan belt when lubricating the engine. Oily fan belts can in many cases be reserviced by washing them in an alkaline degreasing solution. Fuel should not be used.

Fan belts which have become impregnated with oil are generally unserviceable and must be renewed.

Adjusting Fan Belt Tension

1 - Remove nut from generator shaft pulley. When loosening or tightening nut, insert a screwdriver in the slot cut into the inner half of the pulley, and support it against upper generator housing bolt.

2 - Remove outer pulley half.

3 - Arrange spacer washers as required to correct the fan belt tension.

The tension of the fan belt is adjusted by fitting spacer washers—the number varies—between

the two pulley halves until the belt yields approximately 15 mm (0.6″) by a firm thumb pressure. Belt slackness is taken up by removing one or more washers, and if the belt is too tight, one or more washers should be added.

When the belt has stretched, or is worn, to an extent where no washers remain between the two pulley halves to obtain the correct tension, the belt should renewed, as the amount of cooling air then becomes inadequate due to the reduced number of fan revolutions. It is also important that the belt does not bear on the base of the pulley well, that is, on the washers.

4 - Install outer pulley half.

5 - Place all surplus washers between outer pulley half and pulley nut so that all the spacer washers are retained on the pulley hub.

6 - Tighten pulley nut.

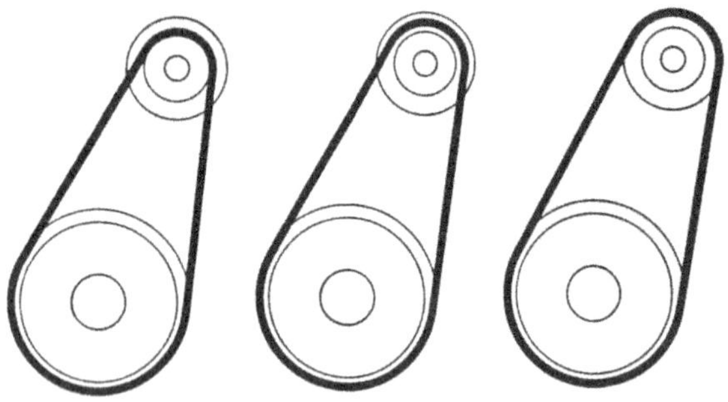

Wrong Correct Wrong

Important

New belts stretch after a short period of operation approximately after 50 to 100 km (30 to 60 miles) and cause belt slackness. The belt should, therefore, be checked in time, and adjusted if necessary.

No attempt should be made to remove the fan belt by means of a screwdriver without backing off the pulley nut, as such practice will destroy the belt and damage the pulley.

Removing and Installing Crankshaft Pulley

Removal

1 - Take off fan belt.

2 - Take out bolt on crankshaft pulley.

3 - Remove crankshaft pulley by using Fan Pulley Extractor VW 203b and Fan Pulley Thrust Pad VW 203d.

4 - Take out screws of cover plate below crankshaft pulley.

5 - Take off cover plate.

Installation

When assembling a completely disassembled engine, the cover plates and the crankshaft pulley should be installed before the ignition distributor drive pinion has been inserted and after the crankcase has been assembled. Before that the oil pump has to be installed.

1 - Check crankshaft pulley before installation for proper seating and belt contact surface. Clean oil return thread.

2 - Make sure crankshaft pulley has no run-out.

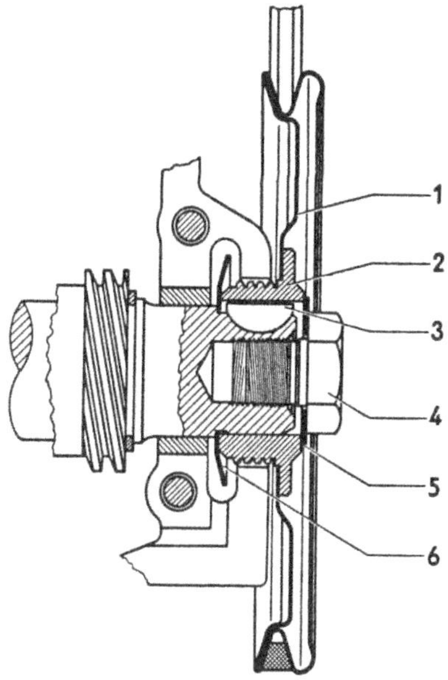

1 - Crankshaft pulley
2 - Oil return thread
3 - Spring plate
4 - Mounting bolt for crankshaft pulley
5 - Spring washer
6 - Oil thrower (crankshaft)

Note:

From Chassis No. 3200001 (Engine No. 5016001), the thickness of the material for the rear pulley disc (Part No. complete 113903109) was increased from 2 ± 0.13 to 2.5 ± 0.15 mm.

The material thickness of the front pulley disc remains 2.0 ± 0.13 mm.

Note:

From Chassis No. 3627442 (Engine No. 5507794), the ratio was changed from 1 : 1.8 to 1 : 1.75 and the number of washers from 8—11 to 3—7.

Crankshaft pulley,

new	113105251 A	(Outside dia. 170 mm)
formerly	113105251	(Outside dia. 177.5 mm)

Pulley with hub, unchanged
113903109 (Outside dia. 108 mm)

The modified crankshaft pulley can be service installed. The belt length remains unchanged. The Part No. 113105251 pulley is no longer supplied as a spare part.

Note:

From Chassis No. 5199980 (Engine No. 7336420), the heating of the Volkswagen 1200 was modified. The air for the heating is now taken from the fan housing near the fan and fed through hoses to the heat exchangers as on the Volkswagen 1500.

The following new parts are fitted on the 34 bhp (40 SAE bhp) engine with fresh air heating:

Part No.	Designation	Qty.
113 119 025 B	Fan housing	1
113 119 303 A	Cylinder cover plate, left	1
113 119 304 B	Cylinder cover plate, right	1
113 119 319	Screening plate on cylinder cover plate	2
113 119 351 B	Lower heater channel, left	1
113 119 352 B	Lower heater channel, right	1
113 119 357 A	Rear air deflector plate, left	1
113 119 358	Rear air deflector plate, right	1
113 119 523 A	Rear engine cover plate	1
113 119 533 B	Cover plate under fan pulley	1
113 119 551	Fan pulley cover	1
113 119 577	Pre-heater pipe sealing plate, left	1
113 119 578	Pre-heater pipe sealing plate, right	1
113 119 585 A	Grommet for heater hose	2
113 119 597	Rear half of pre-heater pipe gasket	2
113 119 599	Front half of pre-heater pipe gasket	2
113 129 511 A	Pipe for carburetor pre-heater	1
113 251 053 D	Muffler	1
211 251 053 F	Muffler	1
113 255 105 A	Heat exchanger, left	1
113 255 106 A	Heat exchanger, right	1
113 255 165	Connecting pipe for heater hose	2
113 255 291 A	Heater hose, left	1
113 255 292 A	Heater hose, right	1
113 255 341 A	Clip, 25 mm wide, for connecting pipe	2

THIS PAGE INTENTIONALLY
LEFT BLANK

Removal and Installation of Cover Plates
(Engine with fresh air heating)

Removal

It is advisable to remove and install the cover plates in the following sequence:

1 - Remove front engine cover plate.

2 - Take off hoses between fan housing and heat exchangers.

3 - Remove hose between warm air adaptor and air cleaner.

4 - Remove fan pulley cover.

5 - Remove pre-heater pipe sealing plate screws.

6 - Remove rear engine cover plate.

7 - Remove fan housing with generator and air control flaps.

8 - Remove intake manifold with pre-heater pipe.

9 - Remove air deflector plate screws at rear and warm air duct lower part screws on both sides. On the left-hand side the adaptor pipe for the carburetor pre-heating must also be removed.

10 - Lift cylinder cover plates off.

11 - Remove plate underneath fan pulley after taking fan pulley off.

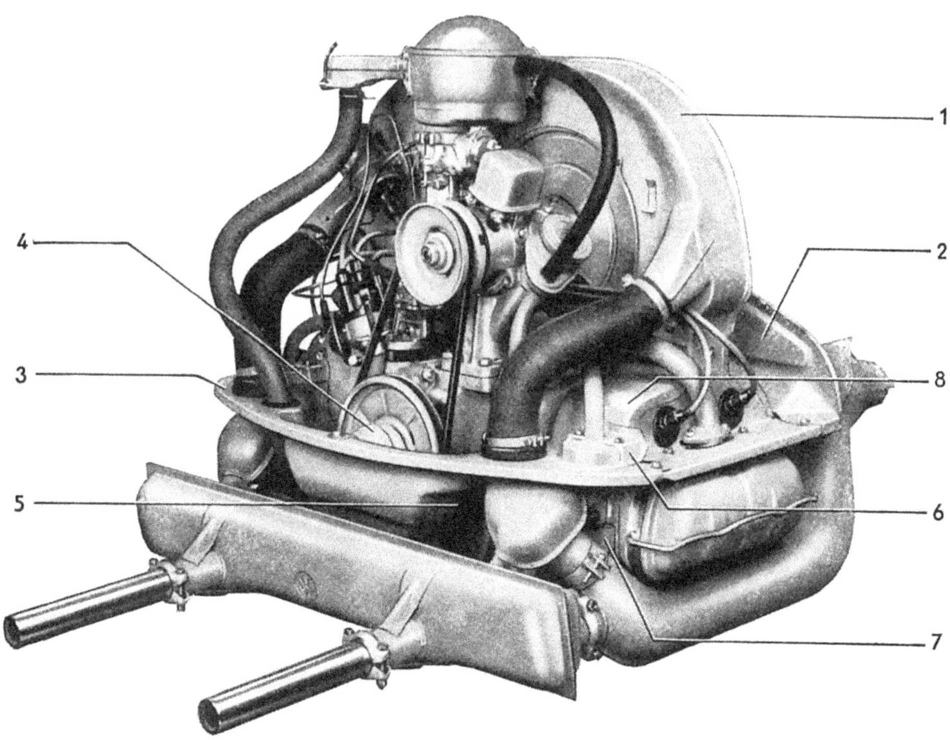

1 - Fan housing
2 - Front engine cover plate
3 - Rear engine cover plate
4 - Fan pulley cover
5 - Fan pulley lower plate
6 - Pre-heater pipe sealing plate
7 - Air deflector plate
8 - Cylinder cover plate

Installation

Note the following points when installing:

1 - The deflector plates *underneath the cylinde* must be located before the push rods and pu rod tubes are installed. Check that the plat are correctly positioned. If necessary, bend t plates so that they press firmly against the sha of the cylinder head studs and cannot rattle fall down when the vehicle is in motion.

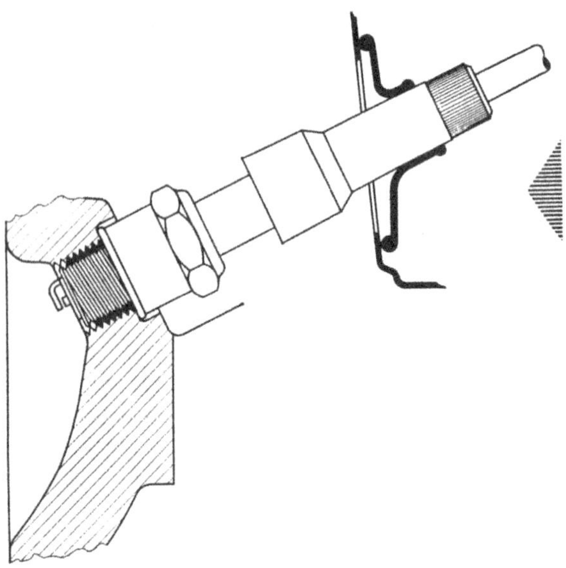

2 - When installing the cylinder cover plat ensure that the rubber caps on the spark plu are tight and in good condition.

3 - When the fan housing is installed, the cyl der cover plates must seal properly on outside of the fan housing.

4 - Before installing the front engine cover pla check the condition of the rubber seal.

Removal and Installation of Rear Engine Cover Plate

(Engine installed)

Removal

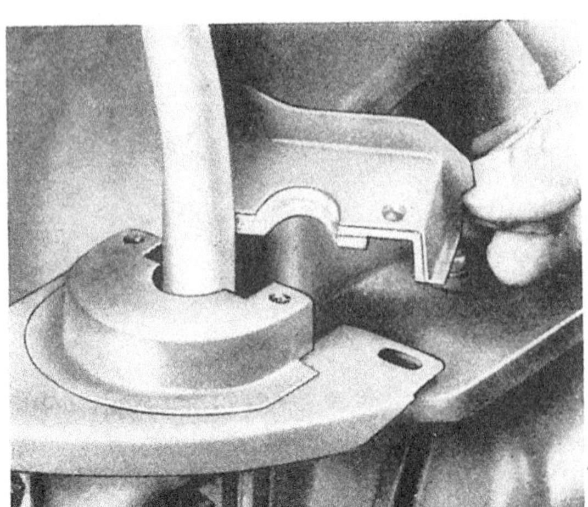

1 - Take off hoses between fan housing and h exchangers.

2 - Remove carburetor pre-heater hose.

3 - Remove fan pulley cover.

4 - Remove manifold pre-heater pipe sea plate.

5 - Remove engine cover plate.

Installation

Note the following points when installing:

1 - Check the asbestos seal in the cover plate and in the manifold pre-heater pipe sealing plate.

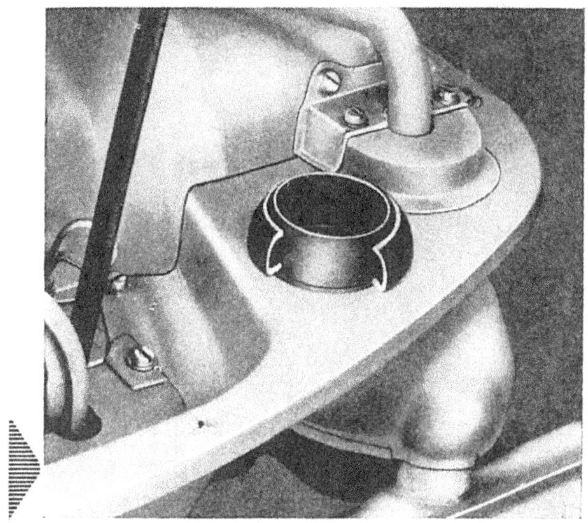

2 - The cover plate must not touch the connections on the heat exchangers.

3 - The flat sides of the two rubber grommets must be on the engine cover plate.

4 - When the engine cover plate is installed, the upper lip of the rubber seal must be over the cover plate and the lower lip on the lower edge.

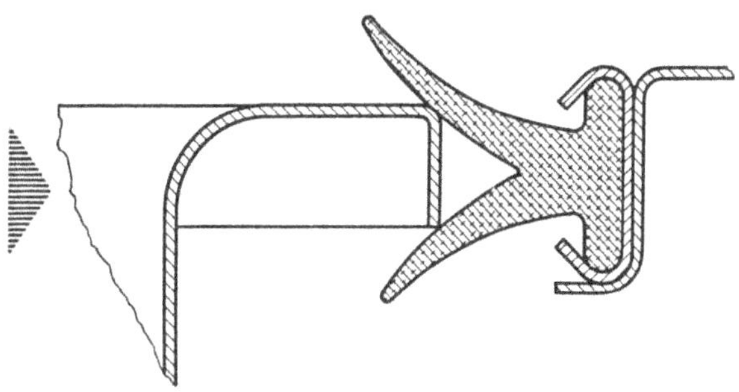

Removing and Installing Fan Housing

Removal

1 - Remove heater hoses.

2 - Take fan belt off.

3 - Remove generator strap and pull cable out of coil.

4 - Take distributor cap off and pull spark plug connectors off.

5 - Remove screws on both sides of fan housing.

6 - Remove thermostat securing screws through holes in lower part of warm air duct.

7 - Screw thermostat off connecting rod and take it out.

8 - Take off fan housing with generator.

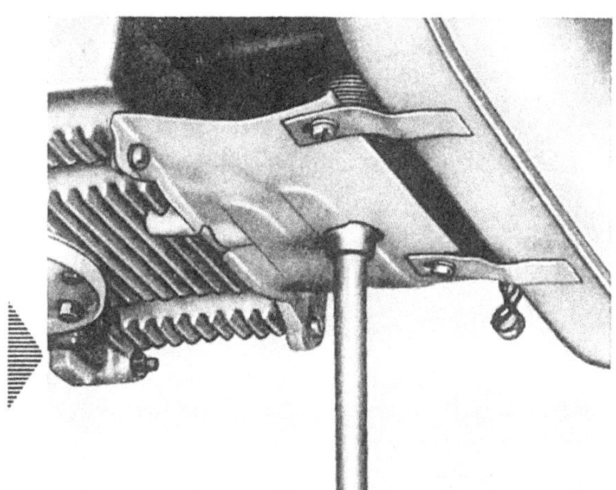

Installation

Note the following points:

1 - Check fan housing for damage and loose air deflector plates.

2 - Install flap assemblies in fan housing.

3 - Insert thermostat connecting rod into hole in cylinder head and lower fan housing.

4 - The fan housing must fit properly on the cylinder cover plates to prevent loss of cooling air. If necessary, the cover plates must be bent slightly.

5 - Install generator strap.

6 - Install screws on both sides of fan housing.

7 - Screw thermostat on to connecting rod and secure it to the thermostat bracket.

Removing and Installing Fan

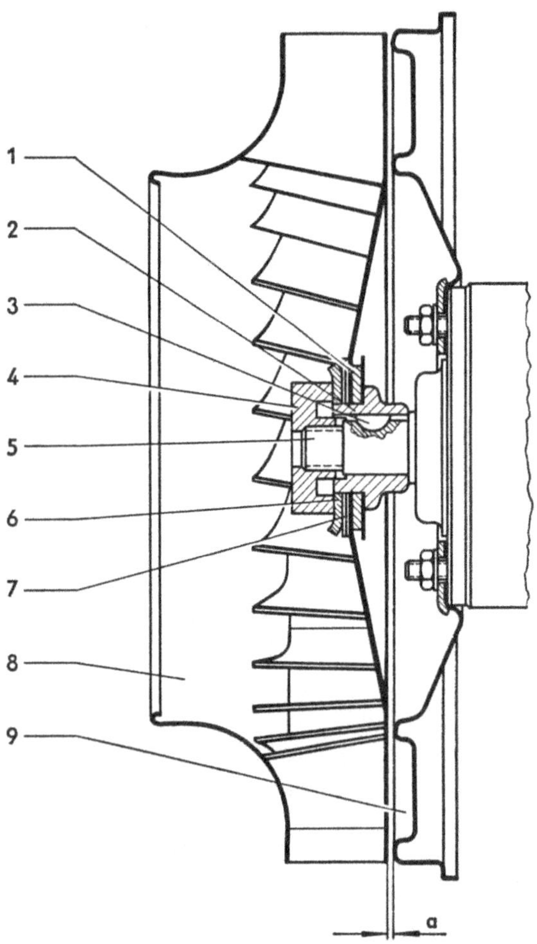

a = approx. 2.0 mm

1 - Thrust washer
2 - Fan hub
3 - Woodruff key
4 - Special nut
5 - Generator shaft
6 - Lock washer
7 - Spacer washers
8 - Fan
9 - Fan cover

Removal

1 - Remove the four screws on fan cover with T-wrench VW 106.

2 - Remove generator and fan.

3 - Remove special nut on fan (36 mm across flats). A second mechanic should hold the fan.

4 - Remove fan, spacer washers, and hub.

Installation

1 - Place hub on generator shaft. Make sure the woodruff key is properly seated.

2 - Insert spacer washers.

3 - Place fan in position.

4 - Tighten special nut to 5.5—6.5 mkg (40 to 47 lb. ft.).

5 - Check distance from fan to cover (a = 2.0 mm). This spacing is achieved by inserting spacer washers between hub and thrust washer. If only one is used, the other two should be placed between lock washer and fan.

6 - Insert generator in fan housing.

7 - Tighten the four screws on cover with T-wrench VW 106.

Note:

The dimensions of about 150,000 fan hubs, Part No. 111 119 123, were changed. For this reason, up to nine shims, Part No. 111 119 133 B, instead of up to three shims are now installed between hub and pulley when setting the distance between fan and fan cover (approximately 2 mm). *If fewer than nine shims are required,* the rest must be inserted between the spring washer and the fan.

Removing and Installing Air Control Flaps

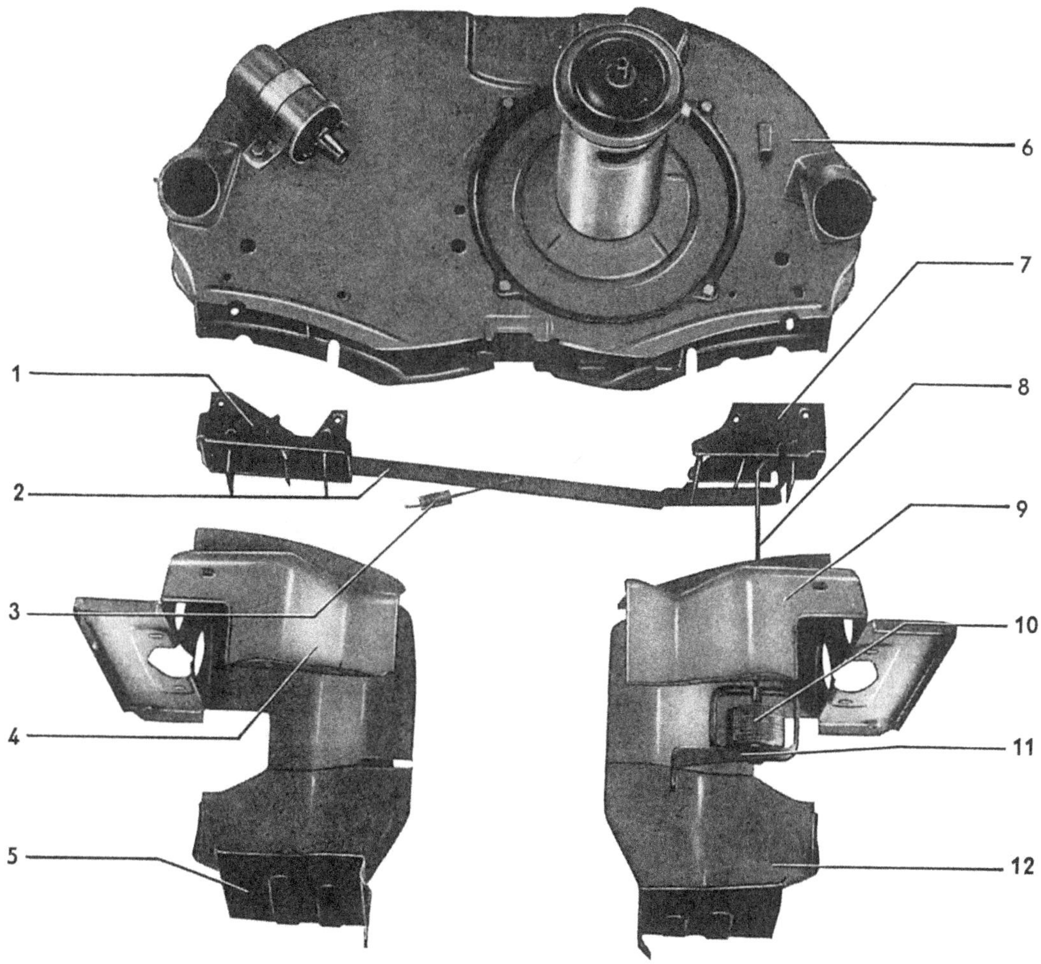

1 - Control flaps, left
2 - Connecting link
3 - Return spring for link
4 - Cylinder cover plate, left
5 - Air duct, lower part, left
6 - Fan housing
7 - Control flaps, right
8 - Thermostat rod
9 - Cylinder cover plate, right
10 - Thermostat
11 - Thermostat bracket
12 - Air duct, lower part, right

Removal

1 - Unhook connecting link return spring.

2 - Remove eight screws securing the control flaps.

3 - Take both flap housings out.

Installation

Installation takes place in the reverse order. Ensure that the rubber stop is fitted in the right-hand flap housing.

Adjustment

1 - Install cylinder cover plates and fan housing complete with control flaps.

2 - Screw thermostat on to connecting rod.

3 - Loosen nuts securing thermostat bracket.

4 - Press the flaps into the open position. Move the thermostat bracket in the slot until the thermostat is touching the upper part of the bracket.

5 - Tighten bracket screws.

6 - Check operation of flaps by moving the thermostat to and fro.

7 - Tighten thermostat securing screw.

Checking Fan Belt Tension

General

Generator and cooling system are driven by the fan belt. Due to the power absorbed by these two units, the fan belt is subjected to considerable stress, especially at high speed and when shifting down. To ensure adequate cooling and long belt life, it is of utmost importance to maintain the belt at correct tension. If the belt is too slack, it will slip and cause the engine to overheat. Excessive tension creates undue stress and is liable to cause breakage of the belt and damage to the generator bearings.

Examination

The belt, when firmly pressed with the thumb at midpoint, must yield approximately 15 mm a = 15 mm (.6″).

No traces of excess wear, such as frayed edges and cracks, should be perceptible.

It is important to prevent oil getting on to the fan belt when lubricating the engine. Oily fan belts can in many cases be reserviced by washing them in an alkaline degreasing solution. Fuel should not be used.

Fan belts which have become impregnated with oil are generally unserviceable and must be renewed.

Adjusting Fan Belt Tension

1 - Remove nut from generator shaft pulley. When loosening or tightening nut, insert a screwdriver in the slot cut into the inner half of the pulley. and support it against upper generator housing bolt.

2 - Remove outer pulley half.

3 - Arrange spacer washers as required to correct the fan belt tension.

The tension of the fan belt is adjusted by fitting spacer washers — the number varies — between the two pulley halves until the belt yields approximately 15 mm (.6″) when pressed firmly with the thumb. Belt slackness is taken up by removing washers, and if the belt is too tight, washers should be added.

When the belt has stretched, or is worn, to such an extent that no washers remain between the two pulley halves when the tension is correct, the belt should be renewed.

4 - Install outer pulley half.

5 - Place all surplus washers between outer pulley half and pulley nut so that all the spacer washers are retained on the pulley hub.

6 - Tighten pulley nut.

Do not attempt to remove the fan belt with a screwdriver without backing off the pulley nut, as this will destroy the belt and damage the pulley.

Removing and Installing Crankshaft Pulley

Removal

1 - Take off fan belt.

2 - Take out bolt on crankshaft pulley.

3 - Remove crankshaft pulley by using fan pulley extractor VW 203b and fan pulley thrust pad VW 203d.

4 - Remove screws in cover plate below crankshaft pulley.

5 - Take off cover plate.

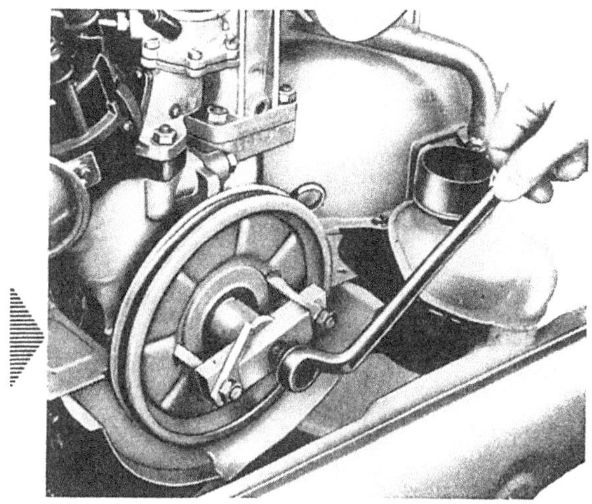

Installation

A fully dismantled engine should be assembled in the following sequence: Oil pump, cover plate under crankshaft pulley, then the pulley and finally the distributor drive shaft.

1 - Check crankshaft pulley before installation for proper seating and belt contact surface. Clean the oil return thread and lubricate it with oil with a molybdenum-disulphide additive.

2 - Make sure that the crankshaft pulley has no run-out.

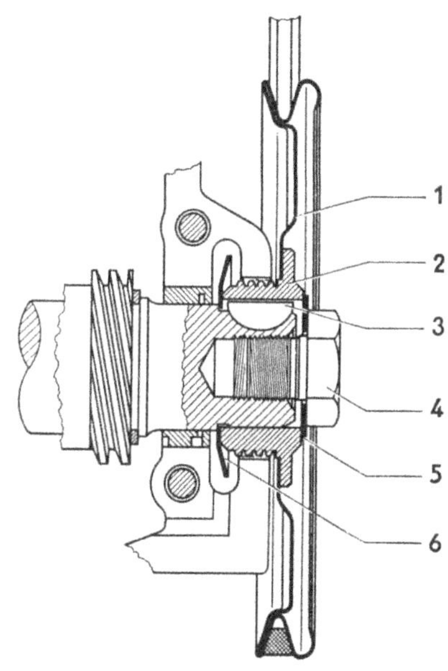

1 - Crankshaft pulley
2 - Oil return thread
3 - Woodruff key
4 - Mounting bolt for crankshaft pulley
5 - Spring washer
6 - Oil thrower for crankshaft

Note:

The crankshaft pulley can also be removed with the engine in situ after the rear cover plate has been taken off.

THIS PAGE INTENTIONALLY
LEFT BLANK

Removing and Installing Intake Manifold with Preheater Pipe

Removal

1 - Remove fan housing.

2 - Remove carburetor.

3 - Screw off fuel line to pump.

4 - Remove nuts and screws at intake manifold flanges with T-wrench 10 mm (VW 106).

5 - Take off manifold.

6 - Remove manifold gaskets.

Installation

When installing, the following points should be observed:

1 - The contact surfaces of the flanges must be clean and smooth.

2 - Use new manifold gaskets.

3 - When installing the manifold, make sure that it is not warped (alignment of flange holes with studs and tapped holes).

If necessary, straighten the manifold. If it has been heated, make sure that no scale remains in the interior.

Note:

From Chassis No. 4 205 477 (Engine No. 6 161 125), the material thickness of the pre-heating pipe on the intake manifold was increased from 20 x 2.0 mm (.787 x .080") to 20 x 2.5 mm (.787 x .098").

4 - Tighten screws and nuts evenly to avoid leakage.

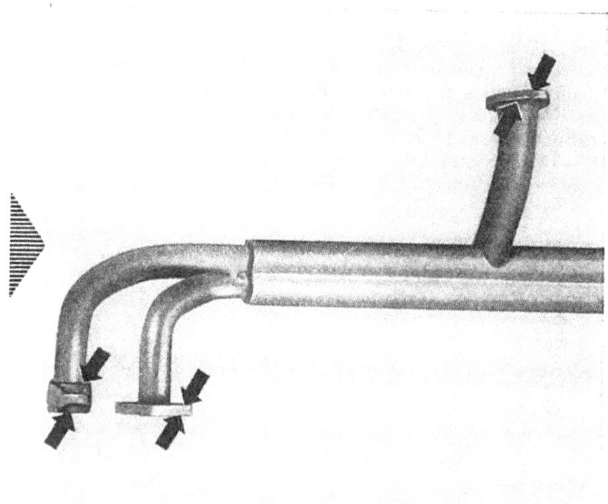

Note:

From Chassis No. 4 846 836 (Engine No. 6 916 251), a modified cylinder head was installed in all 41.5 SAE bhp engines.

Part No. **new** 113 101 351 C
previously 113 101 351 B

The diameter of the intake port and the outer diameter of the sealing surface have been increased. The sealing washer N 13 823 1 used in the 54 bhp (SAE) engine serves as a seal between the intake manifold and the cylinder head.

Service Installation

The modified cylinder head (113 101 351 C) together with the sealing washer N 13 823 1 can be service installed in all 41.5 bhp engines.

After stocks have been used up, the previous type cylinder head (113 101 351 B) will no longer be supplied.

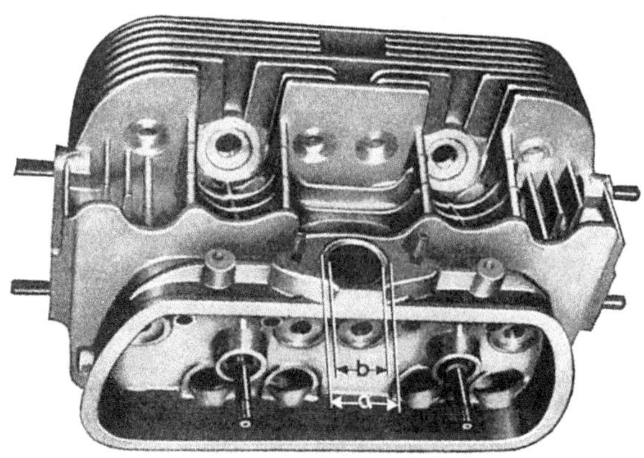

a - **new** 34 mm (1.33") dia.
previously 32 mm (1.25") dia.
b - **new** 27 mm (1.06") dia.
previously 25 mm (.98") dia.

Note:

The intake manifold for 34 bhp (41.5 SAE bhp) engines produced up to August 1962 will no longer be supplied as a spare part. Only the larger diameter intake manifold manufactured since August 1962 will be supplied.

If the new intake manifold 113 129 701 D is fitted to an old cylinder head 113 101 351 B, the compensator sealing ring 113 129 707 should be used.

The wide face of the ring must contact the head and the narrow face the intake pipe.

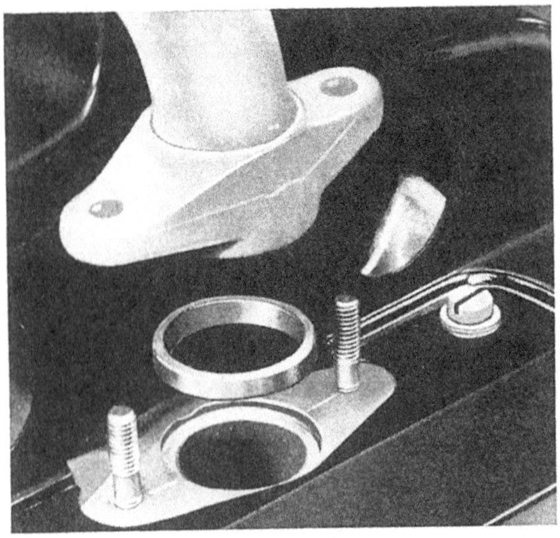

Note:

From Chassis No. 3 223 145 (Engine No. 5 042 363), a modified gasket is installed on the left connection flange of the pre-heater pipe. The internal diameter of this gasket (Part No. 113 251 265 **new**) has been reduced from 16 to 6 mm/.629 to .236". The gasket on the right connection flange (Part No. 113 251 263) remains unchanged.

Hints:

When back-firing occurs, the following checks should be carried out in the sequence given.

1 - Check that the modified gasket has been installed.

2 - Check idling adjustment. It may be necessary to turn the volume control screw out about 1/8 to 1/4 of a turn to richen the idling mixture slightly.

3 - Check that the spark plug gaps are set correctly to 0.7 mm/.28". The tendency to back-fire is stronger if the gaps are too small.

4 - Check to leakage in the exhaust system by blocking the tail pipes. If the system is in order, the idling engine will stop in a very short time. A twittering noise indicates leakage.

5 - Check ignition cables for signs of burning or damage.

6 - Check intake manifold, connecting flanges and the gaskets between cylinder head and intake pipes. If the nuts are too tight the flanges become distorted and the gaskets squeezed so that air can enter and weaken the carburetor mixture. The joints between intake pipes and flanges should be carefully inspected as leakage in the intake pipes can also cause back-firing.

Removing and Installing Muffler

Removal

1 - Release clamps at connection of front exhaust pipes.

2 - Remove four nuts and four screws from muffler flanges.

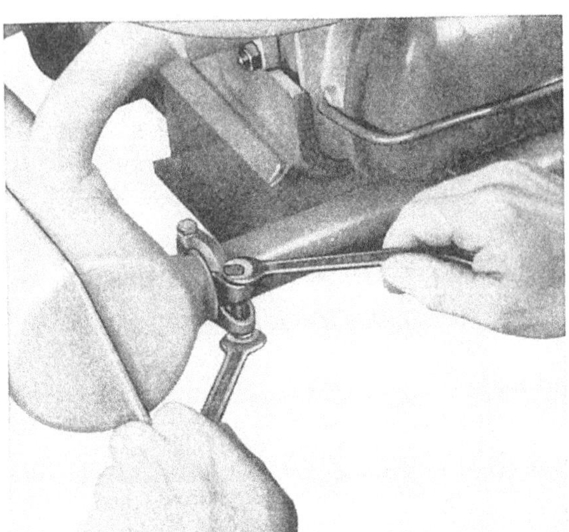

3 - Take off muffler. Remove gaskets from cylinder head flanges and muffler.

Installation

When installing, the following points should be observed:

1 - Check muffler and exhaust pipes for cracks and damage. If necessary, the pipes can be straightened.

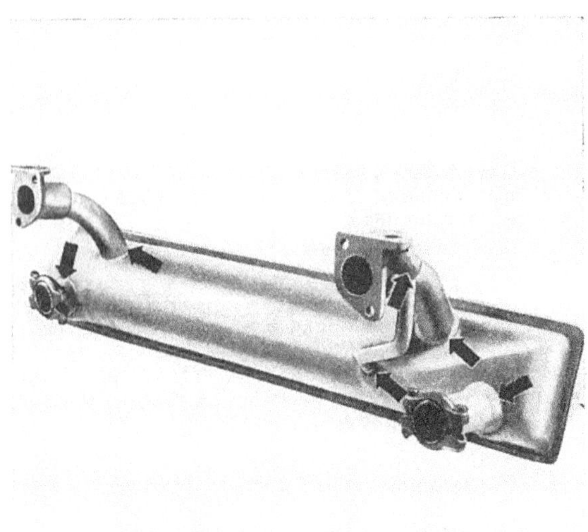

The welded joint of the muffler and the tail pipe is particularly susceptible to damage by impacts. Leaks may result in the exhaust fumes entering the engine compartment and the interior of the car when the heating is turned on.

Always replace all bent or out-of-round tail pipes. If the cartridges are no longer serviceable the tail pipes have to be replaced.

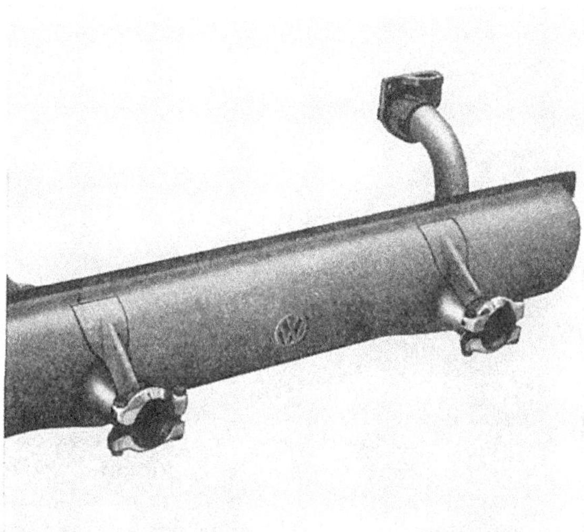

2 - Use new gaskets.

3 - There should be a perfect seal at connection to front exhaust pipes.

4 - Push tail pipes into the exhaust pipes and make sure there is a perfect seal at the connection with the exhaust pipes. The tail pipes should protrude approximately 190 mm (7.5") out of the muffler.

5 - With the engine in situ, the tail pipes must not touch the lower edge of the body. If necessary, remove tail pipes and heat exhaust pipes prior to bending them.

Removing and Installing Muffler

(Engine in Situ)

Removal

1 - Raise car at rear and support on trestles.

2 - Remove rear cover plate.

3 - Take off four nuts at flanges of preheating pipe.

4 - Loosen clamps on the tail pipes and take pipes out.

5 - Loosen clamps at front exhaust pipes.

6 - Take off four nuts at flanges of muffler.

7 - Draw back muffler and take it off from below. Remove gaskets from flanges of cylinder heads, muffler and preheating pipe.

Installation

When installing, the same points apply as with engine removed.

Removing and Installing Heating Junction Box and Exhaust Pipe

Removal

1 - Remove nuts from exhaust flange.

2 - Loosen the clamp at the muffler.

3 - Remove slotted screw at the junction box.

4 - Remove the spring clip and unhook the connector rod from the heat control flap.

5 - Take off heating junction box and exhaust pipe.

Installation

When installing, the following points should be observed:

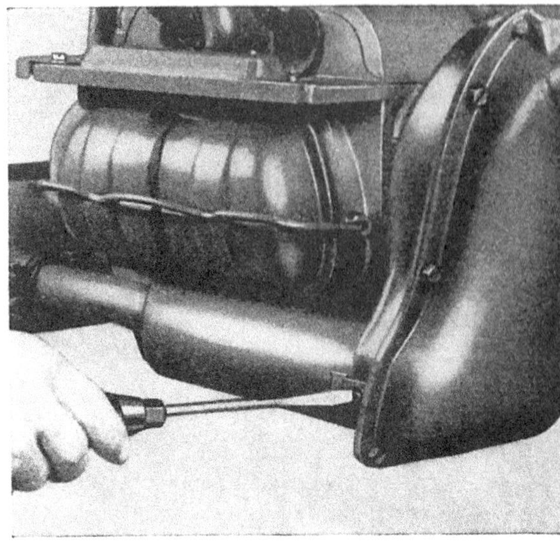

1 - Important. Check heating junction box and exhaust pipe for leaks and damage. Leaks at this point may lead to exhaust fumes entering the interior of the car through the warm air heating system.

Note:

From Chassis No. 3 915 597 (Engine No. 5 840 597), the wall thickness of the exhaust pipe for cylinders 1 and 3 was increased from 1.5 to 2.0 mm (outside diameter remains: 32 mm). The Part Nos. remain unchanged.

2 - The contact surfaces of the flanges must be clean and even. Distorted or bent flanges are to be repaired or straightened.

3 - Use new gaskets.

4 - All moving joints are to be lubricated with high melting point graphite grease.

Disassembling and Assembling Heating Junction Box

Disassembly

1 - Remove the spring clip and unhook the connector rod from the heat control flap.

2 - Remove flange screws.

3 - Take off junction box half.

Note:

From Chassis No. 4 011 959 (Engine No. 5 959 575), the method of securing the heater junction box halves together was changed.

The halves which were formerly held together by tapping screws are now held together by bending over metal lugs on the front half.

If the lugs break off when carrying out repairs, the junction boxes can be secured with screws as they were formerly.

Assembly

When assembling, the following points should be observed:

1 - Clean parts and examine for damage.

2 - Check control flaps for correct operation.

Removing and Installing Heating Control Cable

Removal

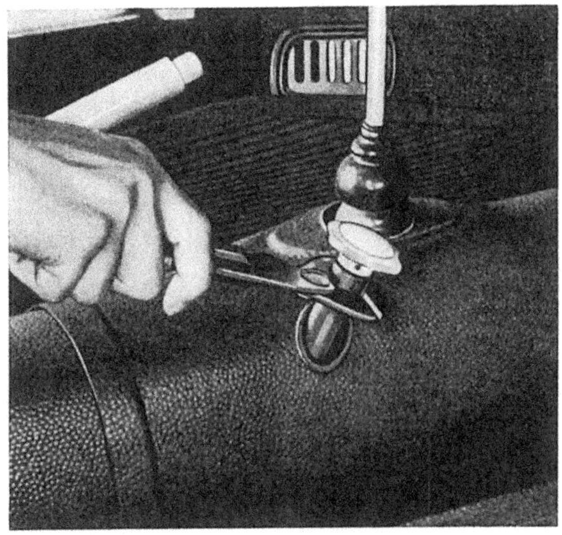

1 - Raise the car.

2 - Release nut of clamping device (use 9 and 10 mm open end wrenches to avoid breakage of cable due to distortion of link).

3 - Disconnect control cable from clamping device.

4 - Remove rubber grommets from conduit tubes and slide them off the cable.

5 - Release the threaded cap and pull out the knob and cable (note correct position of cable ends into conduit tubes).

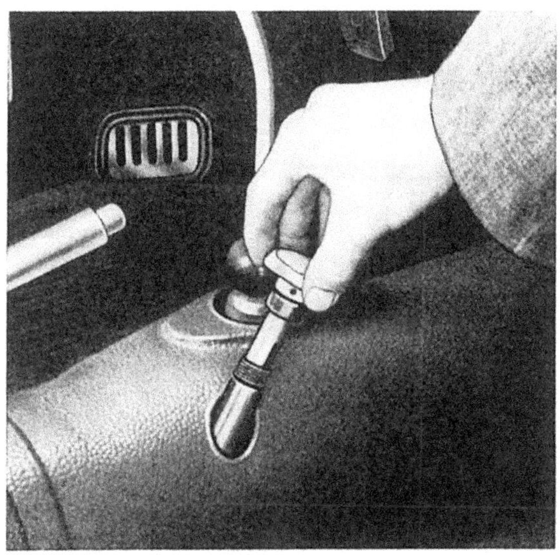

Installation

When installing, the following points should be observed:

1 - Grease cable with Universal Grease.

2 - The longer end of the heating control cable must be inserted into the right-hand conduit tube (as seen in driving direction).

3 - Before installing the heating control knob unit, turn the knob anti-clockwise until stop can be felt and then turn it clockwise three turns.

4 - Insert heating control knob unit, taking care that the guide nose enters the slot in the threaded sleeve. The heating control knob unit is properly installed if the threaded sleeve does not protrude above the edge of the conduit tube.

5 - Check rubber grommets for wear and replace as necessary.

6 - Check heating system for proper functioning.

Heating Control Knob Unit Disassembly and Assembly

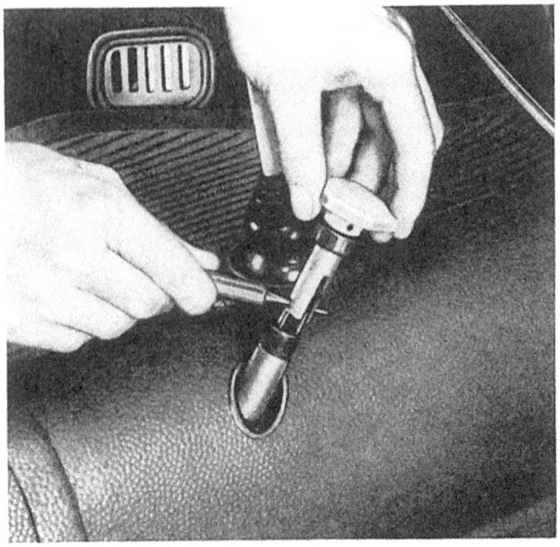

Disassembly

1 - Release clamping device and withdraw the cable.

2 - Release threaded cap and pull out heating control knob unit until the cable can be seen.

3 - Remove the cable pin by means of a drift and take off heating control knob unit.

4 - Screw off the threaded sleeve.

5 - Drive out grooved pin and withdraw knob from the spindle.

6 - Lift off threaded cap and spring washer.

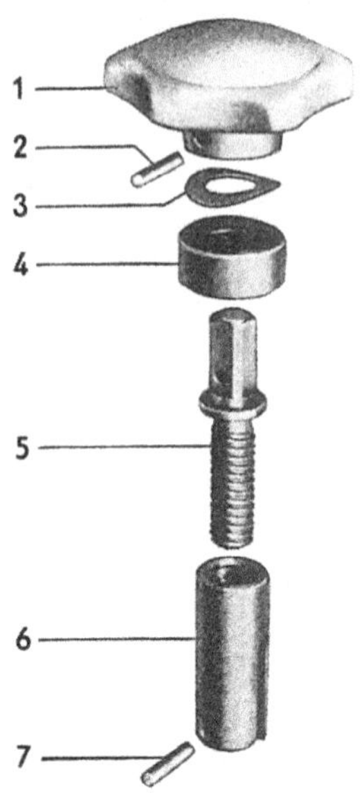

1 - Knob
2 - Grooved pin
3 - Spring washer
4 - Threaded cap
5 - Spindle
6 - Threaded sleeve
7 - Cable pin

Assembly

When assembling, the following points should be observed:

1 - Clean all parts and lubricate them with Universal Grease.

2 - Carefully drive in the grooved pin in order not to damage the knob.

Engine and Heating System

In order to insure satisfactory operation of the heating system, the following points should be observed:

a - Adjust heating control linkage, valves and sheets.

b - To avoid excessive loss of heat, eliminate leaks in heating channels and body.

c - Check for engine oil leaks.

d - Check for oil leaks at oil filler and breather tube.

e - Check for leaks at flanges between exhaust pipes, muffler, and pre-heating pipe.

f - Check for obstructions in cooling air duct to engine.

An unsatisfactory heating system should be carefully examined, especially in the case of unpleasant fumes in the heating air. This can be caused by dirty cylinder and cylinder head cooling fins or a dirty oil cooler. Apart from the fact that a dirty engine causes the operating temperature to rise, which in turn can damage the engine, fumes may arise which can have a more or less unpleasant effect on the passengers in the vehicle.

Removing and Installing Cylinder Cover Plates

Removal

1 - Remove threaded ring in oil filler by means of socket VW 170. Take off oil filler and gasket.

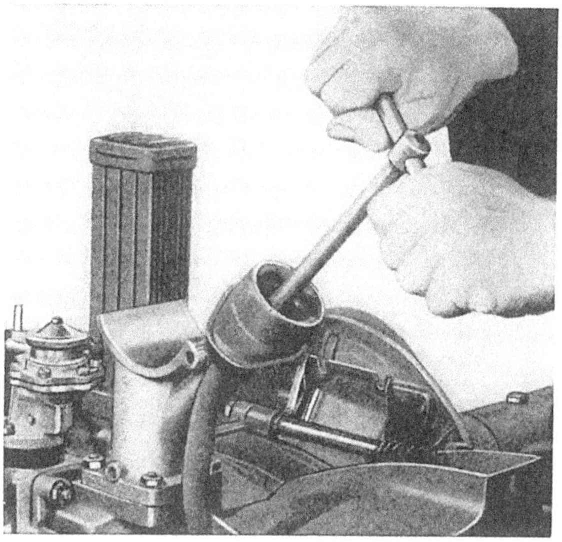

2 - Remove screws that attach cylinder cover plates to heating channels.

3 - Remove both heating channels.

4 - Remove thermostat attaching screw.

5 - Unscrew thermostat from connector rod.

6 - Disconnect connector rod from operating lever.

7 - Remove nut of throttle ring shaft and withdraw the saft.

8 - Remove cover plate screws beside both end of the intake manifold.

9 - Remove both cover plates.

Installation

This is effected by reversing the disassembly pro cedure. However, the right heating channel should only be installed after the fan housing has been placed in position and the throttle ring adjusted.

Removing and Installing Heating Channel

(Engine installed)

Removal

1 - Remove engine rear cover plate.

2 - Remove muffler.

3 - Detach heating control cable.

4 - Remove heating junction box and exhaust pipe.

5 - Remove screws of heating channel.

6 - Remove heating channel.

Installing

When installing, the following points should be observed:

1 - Check heating channel for damage prior to installation.

2 - Both the heat control valve and the control sheet at the rear should freely move and must fully open and close simultaneously to ensure correct heating control.

3 - Adjust heating control cable so that the valve at the front of the heating channel is fully closed when the heating is turned off.

Removal and Installation of Carburetor Pre-heater Pipe

(Engine with fresh air heating)

Removal

1 - Remove carburetor pre-heater hose.

2 - Remove rear engine cover plate.

3 - Remove nut on left exhaust flange.

4 - Pull pipe out to the rear.

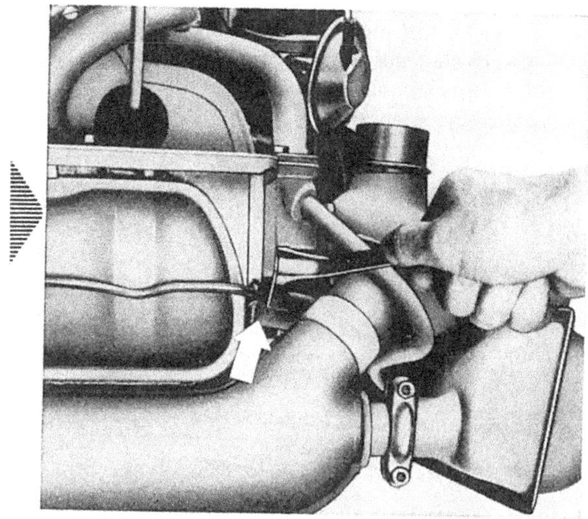

Installation

This takes place in the reverse order. Ensure that the rubber grommets in the engine cover plate are correctly located.

Removal and Installation of Muffler

(Engine with fresh air heating)

Removal

1 - Loosen heat exchanger connection clips.

2 - Take off warm air pipe connection clips.

3 - Remove nuts on flanges and muffler. Take air intake pre-heater pipe off.

4 - Remove four screws securing manifold pre-heater pipe.

5 - Withdraw muffler. Take gaskets off cylinder head flanges or muffler.

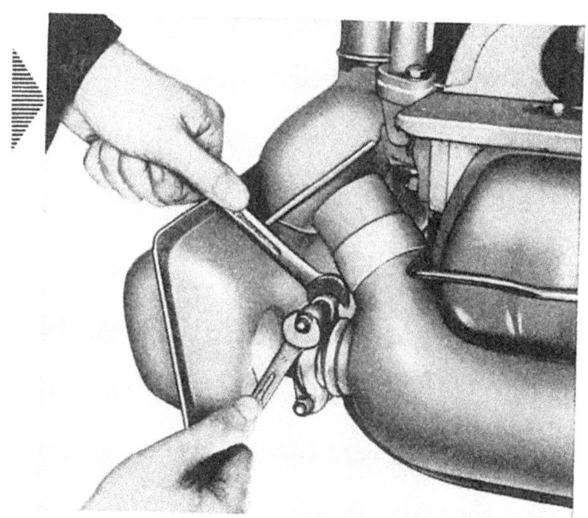

Installation

Note the following points:

1 - Check muffler and pipe for leakage and damage.

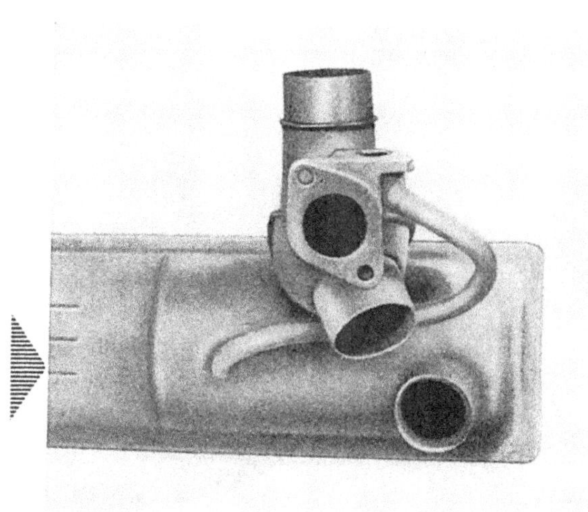

The pipes welded to the muffler can be straightened when necessary. The tapered surfaces must not be dented.

2 - Use new gaskets.

3 - Ensure that the heat exchanger connections seal properly.

4 - Push the tail pipes into the exhaust pipes and ensure that the connections seal properly. The tail pipes should project about 190 mm (7.5″) out of the silencer.

5 - The tail pipes must not touch the body when the engine is installed. If they do, the tail pipes should be removed again and the exhaust pipes heated up and bent slightly.

Note:

We have reason to point out that the heat exchangers and the muffler must be secured to the cylinder heads with **self-locking** M 8 hexagon nuts (311 101 463).

The clamps connecting the heat exchangers or the tail pipes to the muffler must also always be fitted with **self-locking** M 6 nuts (113 251 273).

Other hexagon nuts, even with lock washers, must not be used.

Note:

Two different sealing rings must always be used to seal the exhaust pipes.

1 - A new seal (111 251 **241**) should be fitted between muffler and heat exchangers. The inside diameter of this seal has been reduced by 0.4 mm (.016″) to 35.1 mm (1.38″) and it is marked with **a white paint spot or line.**

2 - The seal (111 251 **231**) should still be used between muffler and tail pipes.

The new seal will be included in the SP sets of engine seals.

Removal and Installation of Heat Exchangers

(Engine with fresh air heating)

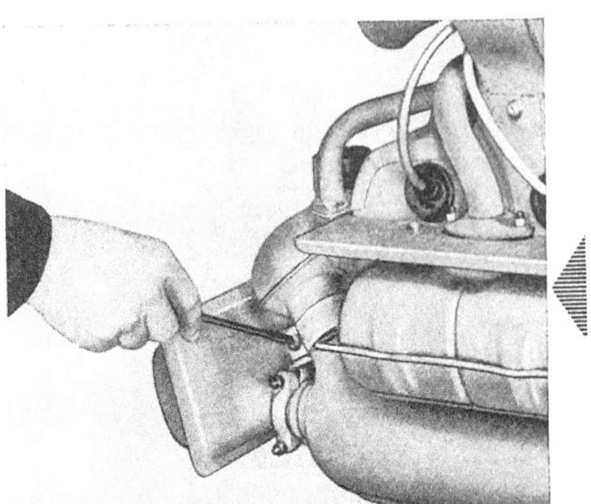

Removal

1 - Take off hoses between fan housing and heat exchangers.

2 - Remove rear engine cover plate.

3 - Remove front nuts on cylinder head and warm air pipe connecting clips.

4 - Remove screws in cover plate below fan pulley.

5 - Remove exhaust pipe clips.

6 - Take heat exchanger off to the front.

Installation

Note the following points when installing:

1 - Check outer shell and particularly the exhaust pipes for leakage and damage. If the heat exchangers leak, the poisonous exhaust gases can enter the heating system.

2 - The flange sealing surfaces must be clean and smooth. Flanges which are distorted or have been bent by excessive tightening should be straightened or machined.

Use new gaskets.

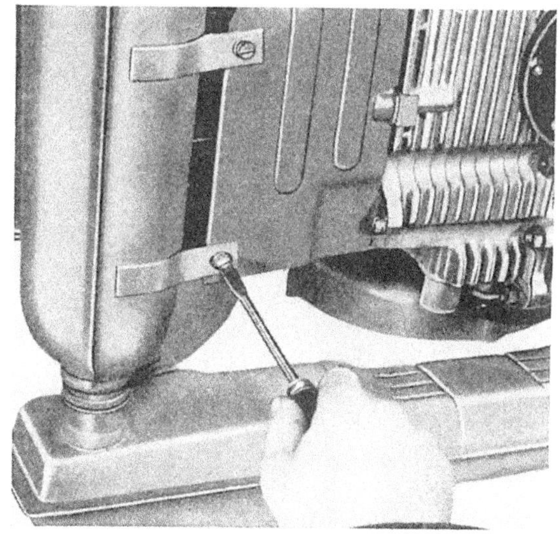

Note:

From June 1964, Chassis No. 6379903 (Engine No. 8678999) the heat exchangers are fitted with 60 mm diameter outlets.

From August 1964, Chassis No. 115000001, the heater pipes in the body were increased in size from 50 to 62.5 mm.

The following parts are new:		Part No. new	old
Heat exchanger, left	for 34 bhp engines	113 255 105 B	113 255 105 A
Heat exchanger, right		113 255 106 B	113 255 106 A
Heat exchanger, left	for 30 bhp engines	111 255 105 B	111 255 105 A
Heat exchanger, right		111 255 106 B	111 255 106 A
Heater flap lever for left heat exchanger		113 255 147 B	113 255 147 A
Heater flap lever for right heat exchanger		113 255 148 B	113 255 148 A
Warm air hose 60/62.5 mm		113 255 355 B	—
Warm air hose 60/50 mm		113 255 355 C	111 255 355
Warm air hose 62.5/50 mm		113 255 355 D	—
Adaptor ring 60/50 mm		211 255 427	—

The previous type of heat exchanger will be discontinued when stocks are used up.

The old and new parts can be combined as follows:

*) If the existing hose (111 255 355) is to be used again, an adaptor (211 255 427) 60/50 mm ends must be fitted on the heat exchanger.

Combination	**former** heat exchanger 50 mm end	**new** heat exchanger 60 mm end
former body 50 mm end	warm air hose 50/50 mm ends 111 255 355	warm air hose*) 60/50 mm ends 113 255 355 C
new body 62.5 mm end	warm air hose 62.5/50 mm ends 113 255 355 D	warm air hose 60/62.5 mm ends 113 255 355 B

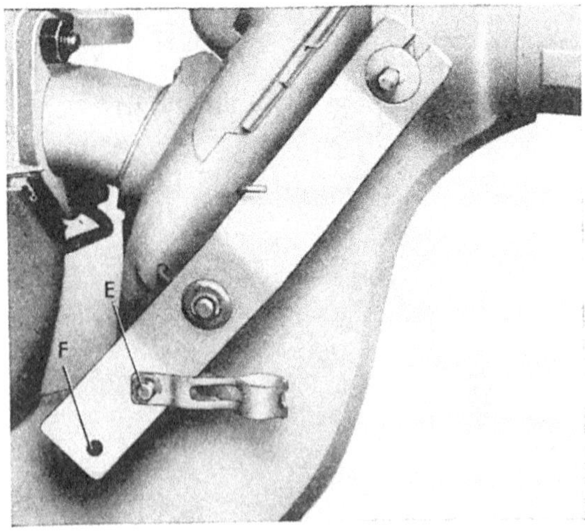

When service installing heat exchangers with the 60 mm outlets the cable link (111 255 309) on the heater flap lever of the heat exchanger (113 255 147 B/148 B) must be connected to the upper hole.

E = Upper hole for the former rotary knob heater control.

F = Lower hole for new lever heater control.

Disassembling and Assembling Heater Control Boxes

(Engine with fresh air heating)

Disassembly

1 - Remove locking washer for heater flap operating lever.

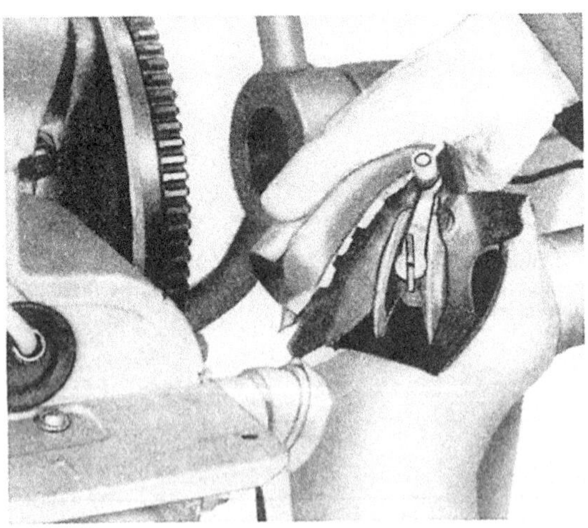

2 - Take lever off.

3 - Bend up metal tabs on heater control box.

4 - Hook flap housing off inwards.

Assembly

When assembling note the following points:

1 - The control flaps must contact all round.

2 - Broken off metal tabs can be replaced with screws.

3 - All bearing points should be lubricated with high melting point graphite grease.

Note:

Since September 1964, from Chassis No. **115084567** (Engine No. 8841279) the heat exchangers have been fitted with internally ribbed exhaust pipes. The diameter of the warm air ducts has not changed.

Part Nos.	**new**	old
Heat exchanger, left	113255105C	113255105A
Heat exchanger, right	113255106C	113255106A

The new heat exchangers can be subsequently installed. Stocks of the old type should be used up.

Note:

Since May 1965, Chassis No. 115855772 (Engine No. 9623350) all heat exchangers have been fitted with galvanized heater flap shafts. This prevents the shafts from sticking as has occurred occasionally.

Type	**Chassis No.**	**Engine No.**
1/1200	116102780	D0016999
1/1300	116095979	F0097227

In addition, since August 1965, the hole in the heat exchanger casing for the heater flap shafts has been enlarged from 6.5 mm to 7.5 mm.

Heat exchangers on which the heater flap shafts no longer work properly should now be repaired and not replaced. Outlet pipes with flaps are now available for this purpose. It is advisable to carry out the repair on both heat exchangers at the same time on the vehicles concerned.

Repair instructions:

1 - Remove engine and place it in the repair stand.

2 - Remove front engine cover plate.

3 - Bend up the edges of the cover plate for the warm air opening and take plate off.

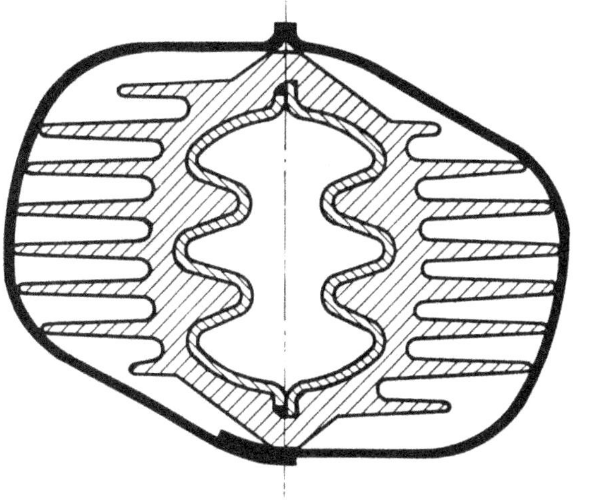

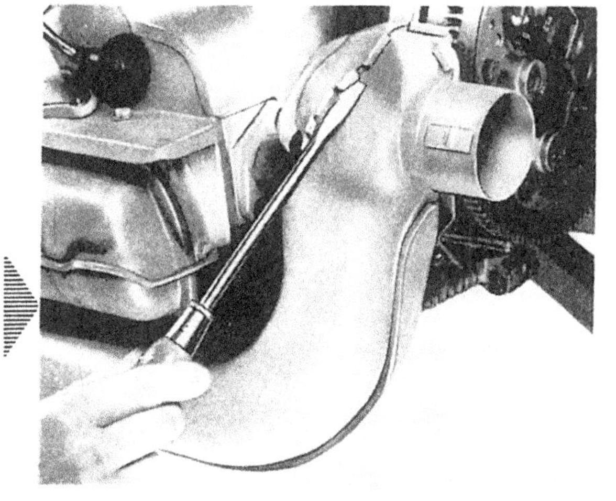

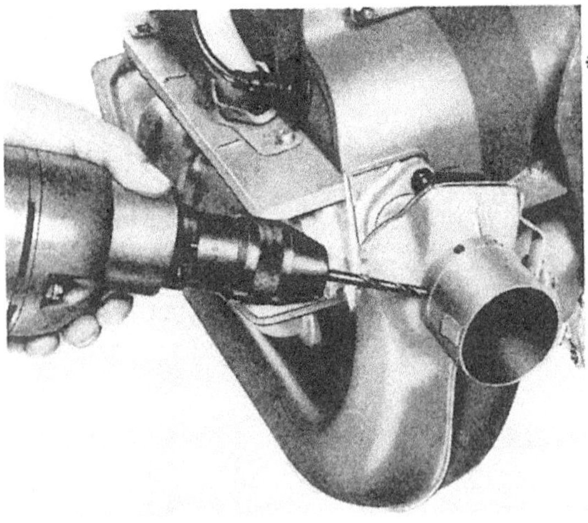

4 - Drill out the spot welds which secure the outlet pipe to the heat exchanger casing with a .24 in. (6 mm) drill.

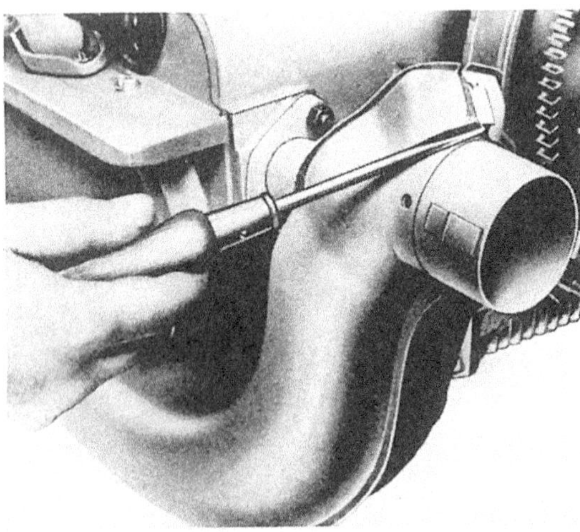

5 - Bend up the lip which holds the heat exchanger casing together.

Note:

Since May 1966 all vehicle engines have been provided with modified parts for the pre-heating. The hot air is drawn from the heat exchanger.

Service installation

With the introduction of the improved carburetor pre-heating, only heat exchangers with the new hot air withdrawal system will now be supplied, but without the heater flap lever. When installing a new heat exchanger in older vehicles, on which the hot air is still taken from the cylinder head, the hot air outlet on the heat exchanger must be closed with plug 113255117.

If, when carrying out repairs, a heat exchanger of new design is installed and there is a stubborn case of carburetor icing, the increased pre-heating of the carburetor can be utilized. The parts for conducting the hot air from the heat exchanger to the carburetor intake, as at present installed in production, should be installed. In addition, the opening for the previous hot air hose in the engine rear cover plate should be closed with plug 113119581.

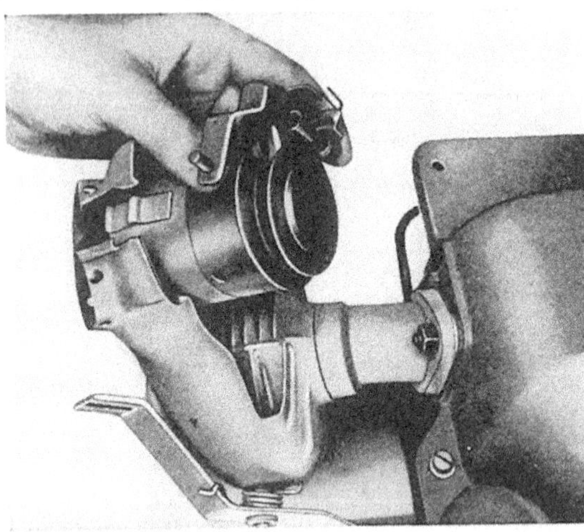

6 - Bend casing apart slightly and take outlet pipe out.

7 - If necessary, rework the heater flap shaft hole in the casing and cover plate.

8 - Install new outlet pipe and bend back the lip of the casing.

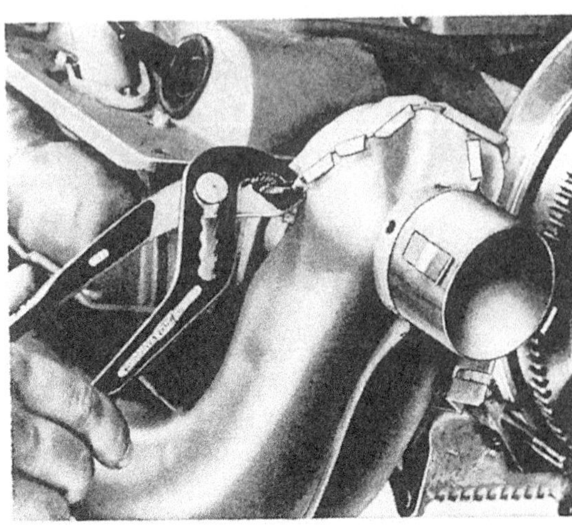

9 - Install cover plate and press the edges together.

10 - Tack weld the outlet pipes to the heat exchanger casing. The heater flaps should be opened when doing this.

11 - Check that flaps work easily.

12 - Paint new outlet pipes with cold zinc paint.

13 - Install engine.

There are no replacement parts available for heat exchangers with 50 mm diameter warm air outlets.

Removing and Installing Valve Rocker Mechanism

Removal

1 - Remove cylinder head cover.

Note:

If the cylinder head covers are not removed carefully, the accumulated dirt on the cylinder heads and covers can fall on to the valve springs and rocker arms. If this dirt is not removed quickly and thoroughly it can cause considerable damage and eventually lead to premature engine failure. To avoid this, clean dirt-encrusted cylinder heads and engine cover plates carefully before removing the cylinder head covers and ensure that all exposed valve gear is scrupulously clean before installing the cylinder head covers.

2 - Remove rocker arm shaft retaining nuts.

3 - Remove rocker arm shaft and rocker arms.

4 - Remove the stud seals.

Installation

When installing, the following points should be observed:

1 - Install the stud seals.

Note:

From Engine No. 5979933, the contact surface of the rocker shaft stud was increased from 10.8 to 11.3 mm dia. and the sealing shoulder from 11.5 to 12.5 mm dia. The cylinder head and the sealing ring have also been altered to suit.

Rocker shaft stud	**new:**	113101397A
	previously:	113101397
Sealing ring for stud	**new:**	113109449A
	previously:	113109449

The Part No. of the cylinder head remains unchanged: 113101351B.

Service Installation

New and old type parts must not be installed together. The modified studs (Part No. 113101397A) cannot be installed in the old type cylinder heads with normal workshop tools.

Both types of sealing ring are supplied together in the gasket set SP 1 GG but the packets containing the rings are marked to avoid confusion.

Note:

From December 1962, Chassis No. 5208482 (Engine No. 7366315) the thread end at the inner shoulder of the rocker shaft stud (Part No. 113101397A) was increased in diameter from 6.2 mm to 7.0 mm. This makes the stud stronger at this point.

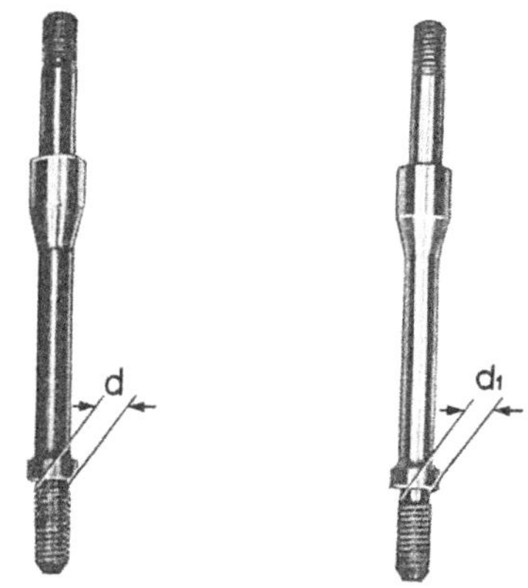

new: $d = 7$ mm−0.2 mm
formerly: $d_1 = 6.3$ mm−0.2 mm

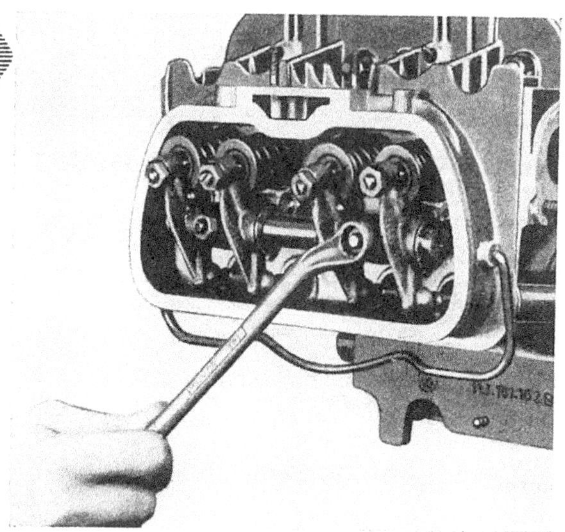

Note:

When assembling the engine, hexagon M 8 nuts of various quality grades are used. The rocker shaft must only be secured with M 8 nuts of the 8 G grade (Part No. N 110085). These nuts can be distinguished from the others by their copper color.

2 - The ball ends of the valve push rods must be central in the sockets of the rocker arms. Tighten the nuts to 14 lb. ft. (2 mkg).

3 - To make the valves rotate during operation, the rocker arm adjusting screws should contact the valve stem slightly offset (to the right). Wear on the stem face and deposit formation on the valve seating faces are thus reduced by the gradual rotation.

4 - Adjust the valves.

5 - Stick new gaskets in cylinder head covers with multi-purpose grease.

6 - Install cylinder head covers.

Note:

From March 1964, Chassis No. 6192906 (Engine No. 8433871), the guide shoulder on the rocker shaft stud was lengthened from 12 to 22 mm (a). The hole in the cylinder head (Part No. unchanged) is now a clearance hole.

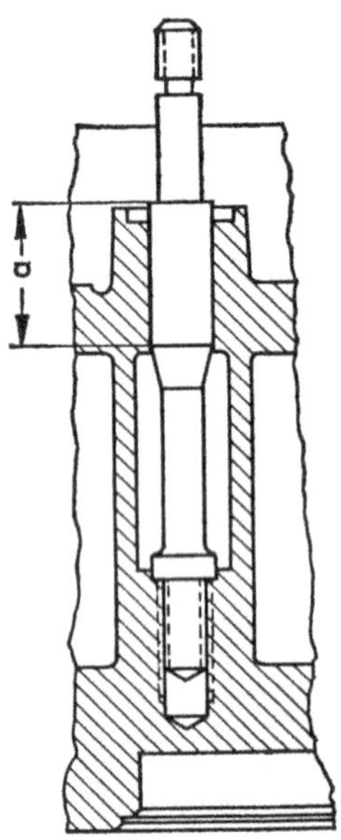

new

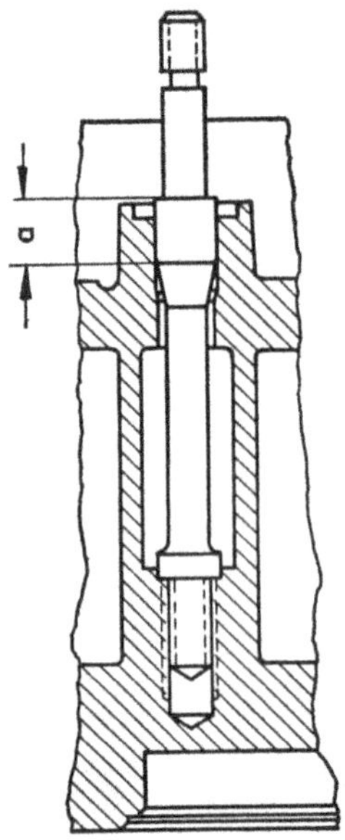

old

Part. No.	new	old
Rocker shaft stud	113101397 B	113101397 A

The 113101397 A stud will remain available. The new stud cannot be service installed in place of the previous type.

Note:

From Chassis No. 4242951 (Engine No. 6210000), the sealing between cylinder head and cylinder head cover was modified.

The sealing surface on the cylinder head below the intake manifold now runs in a straight line and the cylinder head cover has been modified to suit.

Cylinder head cover formerly: 113101475
new 113101475 A

formerly

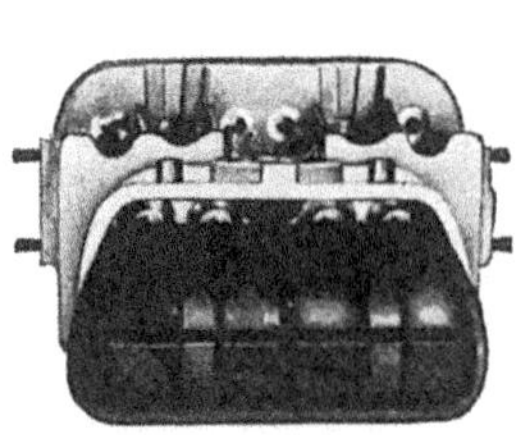

new

The Part. Nos. of cylinder head and cork gasket remain unchanged.

Note:

1 - The older version cylinder head cover and cork gasket (Part. No. 113101475 and 113101481) must **not** be used for engines with modified cylinder heads. If these parts are used the sealing near the intake port will be defective.

2 - Modified cylinder head covers (Part No. **new** 113101475 A) must not be used for engines with cylinder heads of previous design as the sealing in the region of the intake port would not be sufficient.

> **Important**
> If cylinder head covers are used during Inspections which do not belong to the particular engine and have been cleaned in advance, care must be taken to store the two different types separately in order to avoid confusion.

Note:

From December 1963, Chassis No. 6009513 (Engine No. 8250020) the rocker arm mechanism was improved. As a result, some parts of the valve gear have been modified. The angle of the valves was increased from 9° to 9° 30'.

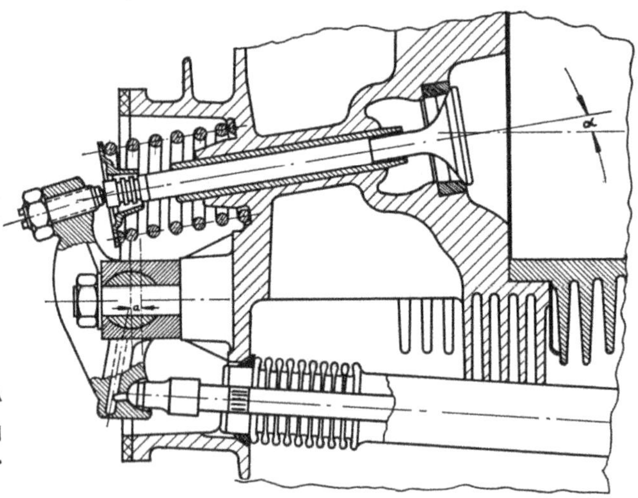

The modified cylinder head can be identified externally by the letter "C" behind the embossed casting number between the rocker shaft studs.

Note:

Replacement M 15 × 1.5 studs (Part No. 113101398) are now available for the repair of old type cylinder heads (with round boss) in which the rocker shaft studs are loose or have pulled out of the threads. The cylinder heads should be removed to carry out the repair and both heads must be repaired at the same time.

Important

The valves in engines which have been fitted with these studs should be set to **0.1 mm (.004 in.)** and the engine marked with the valve clearance sticker as has been used in the past. The avoid confusion, a metal tag marked 0.1 mm (.004 in.) should also be fitted to one of the rocker shaft studs.

Repair instructions

1 - Unscrew loose studs.

2 - Align cylinder head on the drill table and clamp it in position. The clamp and the flat plate for the cylinder head from the local manufacture tool VW 698 are suitable for this purpose.

3 - Enlarge hole to .512 in. (13 mm).

4 - Without moving the cylinder head, tap an M 15 × 1.5 thread in the hole. A suitable lathe center can be used to guide the tap.

Do not forget lubrication.

5 - Rework second hole in the same way.

6 - Coat the shoulder of the special stud with D 2 sealing compound and screw stud in, using a cap nut.

Note:

Hand taps (1 pilot and 1 plug tap) suitable for this work can be obtained under the designation "M 15 × 1.5 nach DIN 2181, Festsitz nach NB 20—208 des Volkswagenwerkes" from Messrs. Prototyp-Werke GmbH, Zell-Harmersbach/Schwarzwald, Germany.

The rocker shaft has been moved outwards 1 mm (a) and 0.5 mm longitudinally. The bores for the rocker shaft are now off center in the supports. The support is marked "Top" to facilitate installation; the marked side must be outwards.

The length of the push rod has been increased by 2 mm. The shaft of the longer rod is knurled.

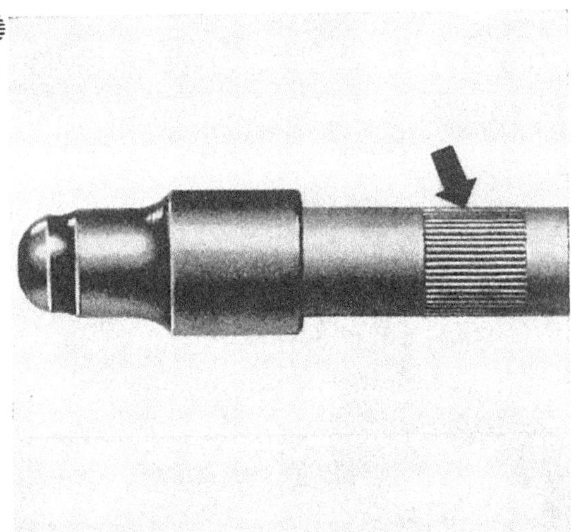

Part No.	new	formerly
Cylinder head	113101371D	113101371C
Rocker shaft support	113109427A	113109427
Push rod	113109301B	113109301A

The old type cylinder head will be discontinued when stocks are exhausted. New and old type cylinder heads can be installed on one engine if necessary.

The new push rods may only be fitted together with the new rocker shaft supports.

Note:

From February 1964, Chassis No. 6092111 (Engine No. 8339393), the rocker arms of all engines are provided with a drilling which runs from the rocker arm bearing to the adjusting screw thread. The engine oil is pumped from the rocker arm bearing through the drilling to the threaded hole for the adjusting screw. From here it passes through the thread to the point where the adjusting screw contacts the valve stem. The rib through which the drilling passes has been enlarged to 5 mm (.2″).

This modification will increase the wear resistance of the valve stem ends and reduce the valve gear noise and the adjusting screw wear.

An oil deflector ring (Part No. 113109619) on the valve stem limits the ingress of oil into the valve guides. The valve guides have been shortened 1 mm in this connection. The measurement "a" from the end of the valve guide to the contact surface for the valve spring is now 20—0.5 mm (formerly 21—0.5 mm).

Fitting instructions

The special sleeve VW 699 (local manufacture) should be used to facilitate fitting the oil deflector ring.

Service installation

When the new rocker arms are fitted, the oil deflector rings (Part No. 113109619) must always be installed as well. To do this, all the valve guides must be shortened until measurement "a" = 20–0.5 mm.

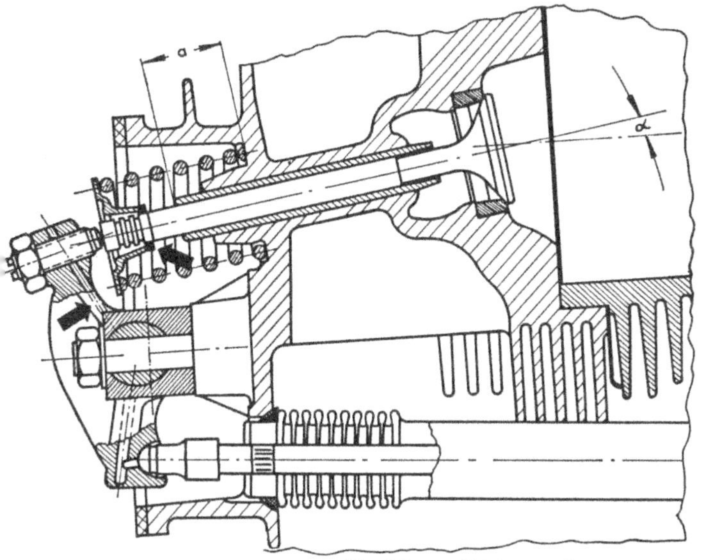

a = 20–0.5 mm (.787–.020")
formerly a = 21–0.5 mm (.826–.020")

The valves guides can be shortened with a normal M 8×8.4 end cutter on which the pilot has been ground down to 7.95 ± 0.02 mm (b). The surface must be level. The inner edge of the valve guide bore must not be chamfered during the cutting process.

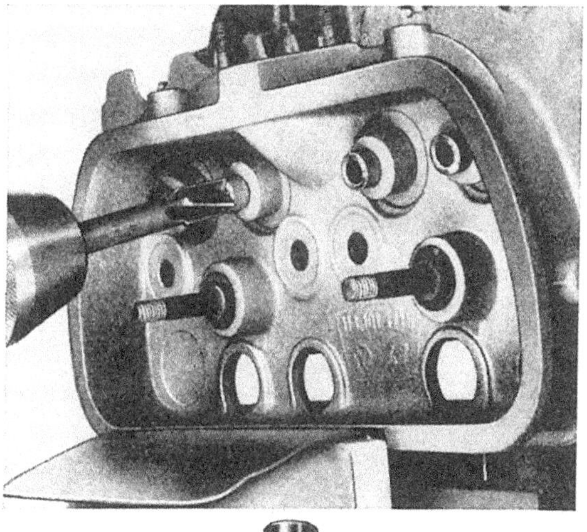

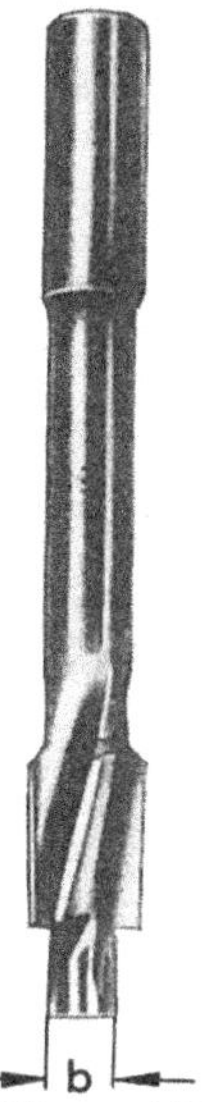

b = 7.95 ± 0.02 mm (.312 ± .0007")

Part. No.

	new	old
Rocker arm	113109443B	113109443A
Oil deflector ring	113109619	

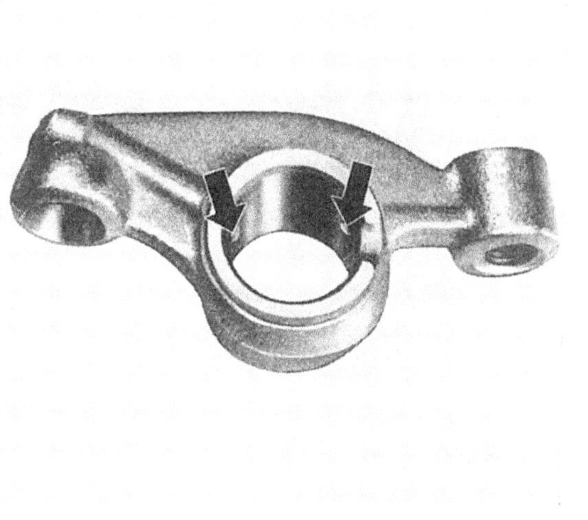

The previous type of rocker arm remains available.

The new type rocker arms have two oil drillings.

Note

When driving hard for long periods under certain conditions (e. g. strong head winds and high air temperatures) the former type of cylinder heads (round boss) can be stressed thermally to such an extent that the valves may be damaged.

In all cases of damage where installation of the new type cylinder head (square boss) is not justified, three 10 mm diameter holes should be drilled in the two ribs near the intake port when the head has been repaired. This measure should also be carried out when using exchange heads of the previous type (round boss).

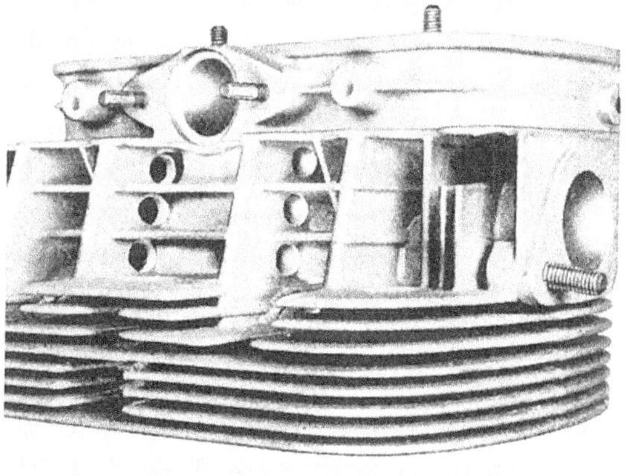

This will reduce the cylinder head temperature at full load, particularly near the exhaust valve seats and guides. It will also reduce the tendency to pink in the part-load range and on acceleration. It is also possible to use a crankshaft pulley of slightly larger outside diameter (211105251). This will increase the amount of cooling air at high engine speeds.

Note:

Cylinder heads in which the ribs have been drilled as instructed here will be accepted by the Exchange Service of the Parts Department.

Disassembly and Assembly of Rocker Arm Mechanism (up to 1965)

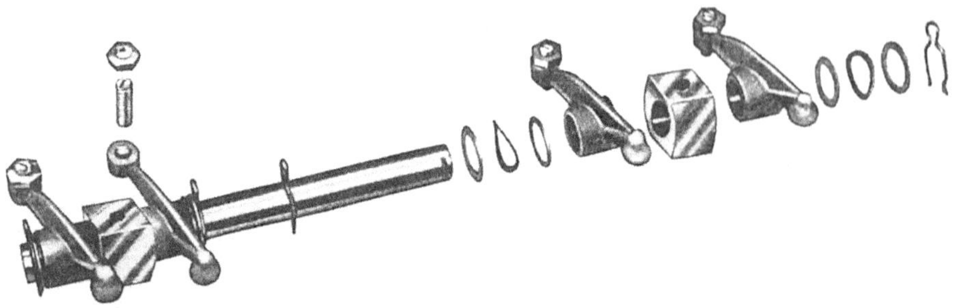

Disassembly

1 - Remove spring clips from rocker arm shaft.

2 - Remove washers, rocker arms, and bearing supports.

Note:

If the valves tend to be noisy even though the clearance is correct and the adjusting screws and valve stem ends are in good condition, the rocker mechanism should be checked for ease of movement. In particular, the side thrust surfaces of the rocker arms and the rocker shaft, supports should be inspected for scoring and roughness. Any signs of unevenness must be removed with fine emery cloth.

Assembly

When installing, the following points should be observed:

1 - Check rocker arm shaft for wear.

2 - Examine seats and ball sockets of rocker arms and valve adjusting screws for wear.

3 - Loosen adjusting screws prior to installing rocker arms.

Note:

In February 1964, a large number of cylinder heads were fitted with slotted rocker arm supports (113109427 B). The support without slot (113109427 A) will be supplied as a replacement part for these engines.

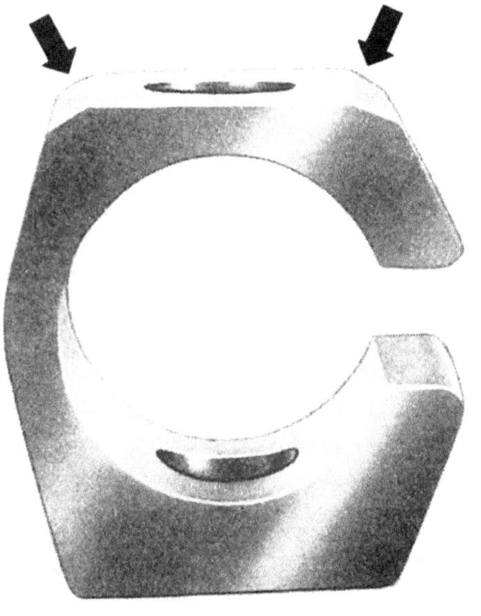

Note:

Both supports on one rocker shaft must be of the same type. Ensure also that the chamfered edges always face outwards and the slot upwards.

Removing and Installing Cylinder Head (up to 1965)

Removal

1 - Remove cylinder head nuts using a 15 mm wrench (VW 165).

2 - Lift off cylinder head.

If it is intended to remove the cylinder heads only, the cylinder retainer VW 650/1 (local manufacture) should be used to prevent the cylinders from being withdraw together with the head, thus avoiding ingress of dirt.

Installation

When installing, the following points should be observed:

1 - Check cylinder head for cracks in combustion chamber and exhaust ports. Cracked cylinder heads must be replaced.

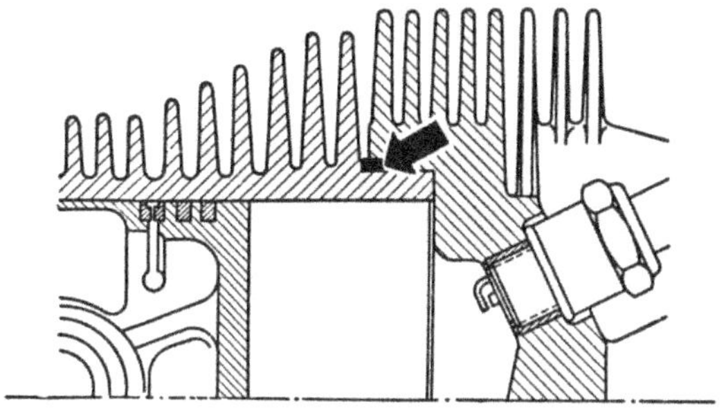

2 - Check studs for security. If necessary use Heli-Coil threaded inserts.

3 - There is no gasket between the upper edge of cylinder and the corresponding contact surface in cylinder head.

4 - Renew gasket between shoulder of cylinder and cylinder head. The slotted side of the gasket must be towards the cylinder head.

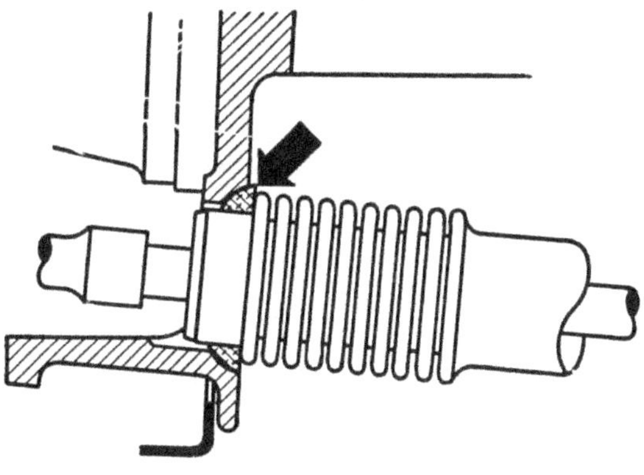

5 - When installing cylinder head, make sure that the oil seals at the ends of the push rod tube are properly seated.

6 - Insert cylinder head nut washers.

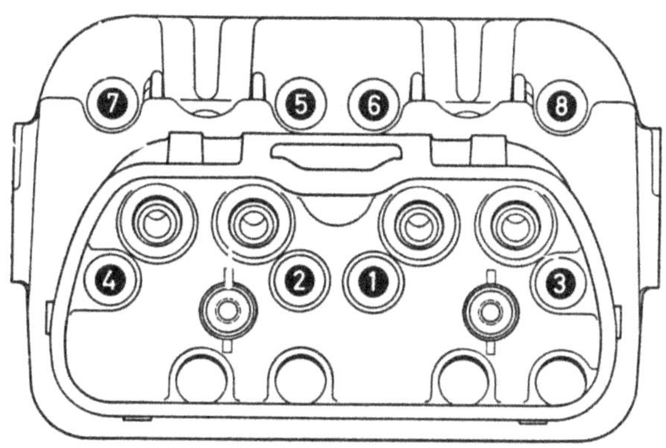

7 - Coat cylinder head nuts with graphite paste and screw them down until resistance can be felt. Tighten with a torque wrench to 1 mkg (7 ft. lbs.) in the order indicated on the left.

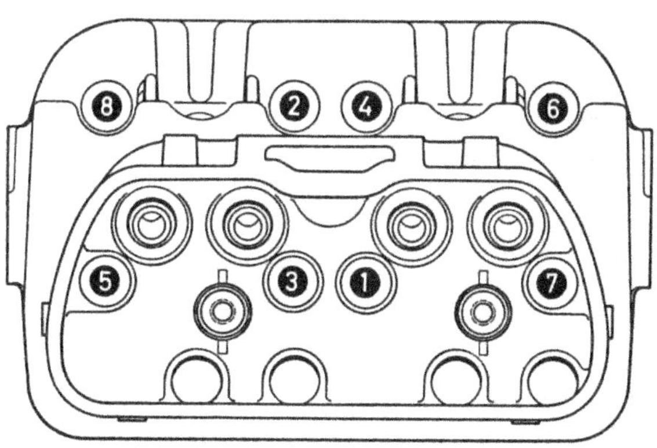

8 - Fully tighten to a torque of between 3.0 and 3.2 mkg (22 and 23 ft. lbs.) in the order indicated on the left.

Disassembling and Assembling Rocker Arm Mechanism (1965)

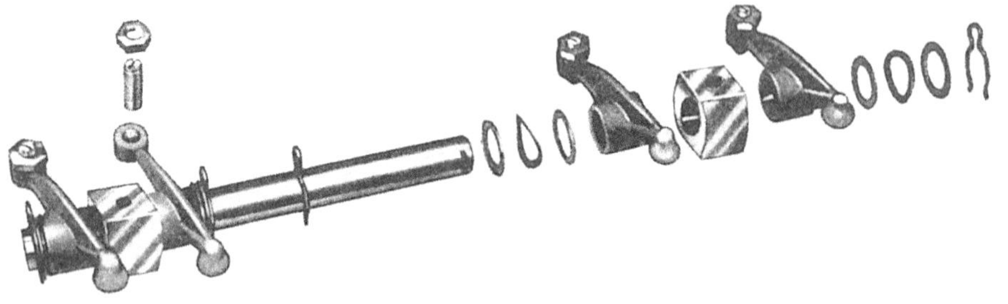

Disassembly

1 - Remove spring clips from rocker arm shaft.

2 - Remove washers, rocker arms, and bearing supports.

Note:

If the valves tend to be noisy even though the clearance is correct and the adjusting screws and valve stem ends are in good condition, the rocker mechanism should be checked for ease of movement. In particular, the side thrust surfaces of the rocker arms and the rocker shaft supports should be inspected for scoring and roughness. Any signs of uneveness must be removed with fine emery cloth.

Note:

From November 1964, Chassis No. 115429385 (Engine No. 9205700), with the introduction of the new cylinder heads — 113101353 — a slotted rocker shaft support (113109427C) was fitted.

Important

Originally the supports were installed with the slot downward. In the meantime the position has been changed.

The slot now points upward

The supports should only be turned round when carrying out repairs to the cylinder head.

Assembly

When installing, the following points should be noted:

1 - Check rocker arm shaft for wear.

2 - Examine seats and ball sockets of rocker arms and valve adjusting screws for wear.

3 - Loosen adjusting screws prior to installing rocker arms.

Removing and Installing Cylinder Head (1965)

Removal

1 - Remove cylinder head nuts using a 15 mm hexagon wrench (VW 165).

2 - Lift off cylinder head.

If it is intended to remove the cylinder heads only, the cylinder retainer VW 650/1 (local manufacture) can be used to prevent the cylinders from being withdrawn unintentionally and avoid the ingress of dirt.

Installation

When installing, the following points should be noted:

Note:

After carrying out all engine repairs involving the removal of the cylinder covers but not the disassembly of the crankcase, the tightness of the crankcase studs and nuts must be checked by applying the correct torque. This check is especially important where older engines are concerned.

Tightening torques:

Nuts M 12 × 1.5	25 lb. ft. (3.5 mkg)
Studs and nuts M 8	14 lb. ft. (2.0 mkg)

1 - Check cylinder head for cracks in combustion chamber and exhaust ports. Cracked cylinder heads must be replaced.

8 - Tighten the nuts to 7 lb. ft. 1 (mkg), first in the order shown.

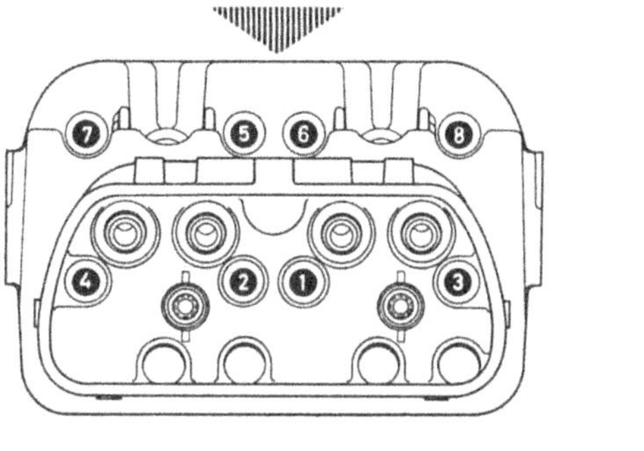

9 - Fully tighten the nuts to a torque of between 22 and 23 lb. ft. (3.0 and 3.2 mkg) in different order.

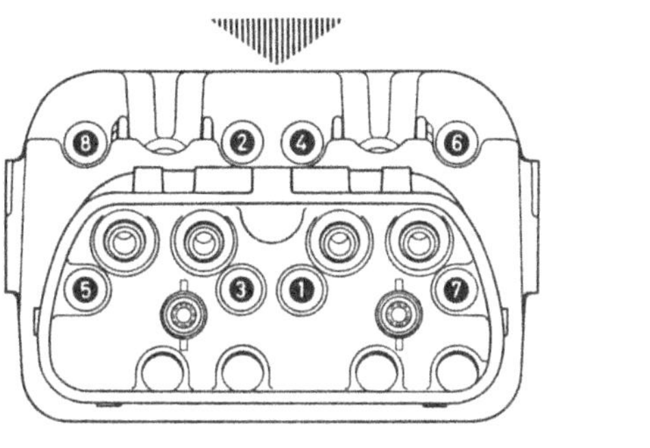

Note:

When driving hard for long periods under certain conditions (e. g., strong head winds and high air temperatures), the former type cylinder heads (round boss) are stressed thermally to such an extent that the valves may be damaged.

In all cases of damage where installation of the new type cylinder head (rectangular boss) is not justified, three 10 mm diameter holes should be drilled in the two ribs near the intake port when the head has been repaired. This measure should also be carried out when using exchange heads of the previous type (round boss).

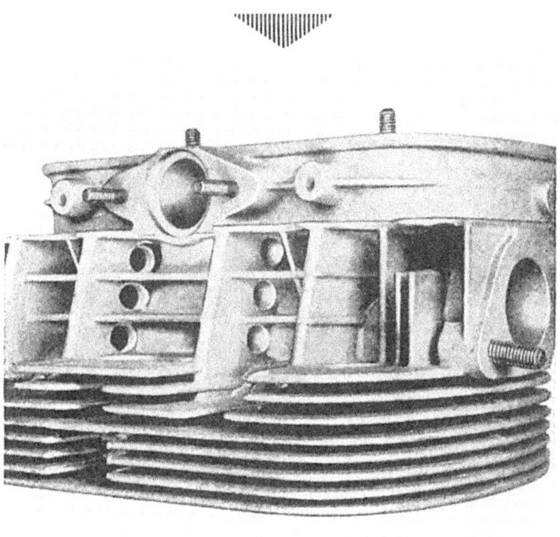

This will reduce the cylinder head temperature at full load, particularly near the exhaust valve seats and guides. It will also reduce the tendency to pink in the part-load range and on acceleration.

It is also possible to use a crankshaft pulley of slightly larger outside diameter, part number 211105251. This will increase the amount of cooling air at high engine speeds.

Note:

Cylinder heads in which the ribs have been drilled as instructed will be accepted by the Exchange Service of the Parts Department.

Note:

From November 1964, Chassis No. 115429385 (Engine No. 9205700) the rocker shaft is secured with studs which are screwed into the two bosses in the valve chamber as in the 30 bhp engine. The studs have a 9 mm dia. shoulder which centers the rocker shaft supports. The bosses are now rectangular in shape and not round as they were formerly. The new heads can be identified by this feature when the cylinder head covers have been removed.

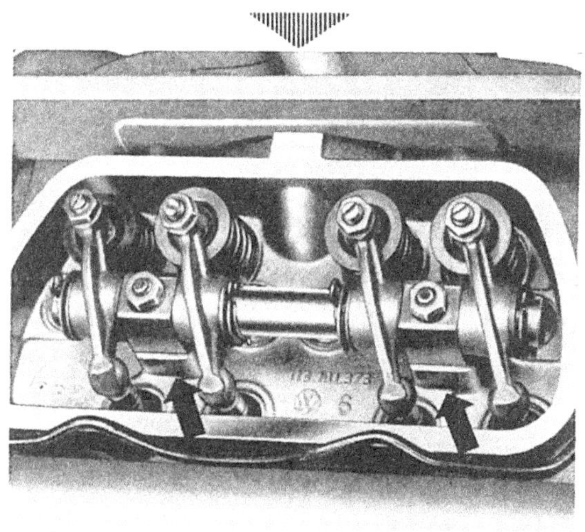

new — rectangular boss

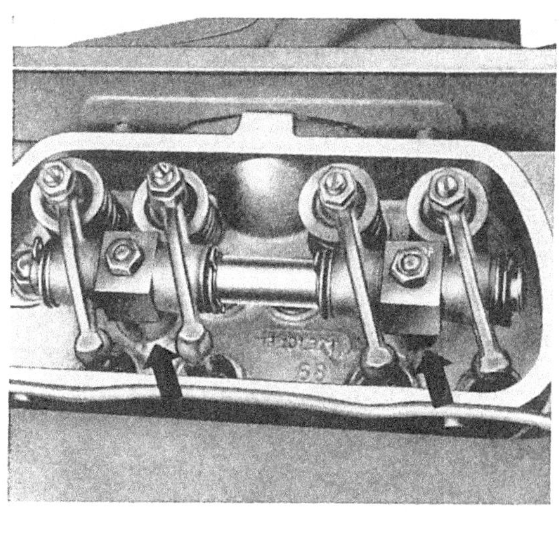

old — round boss

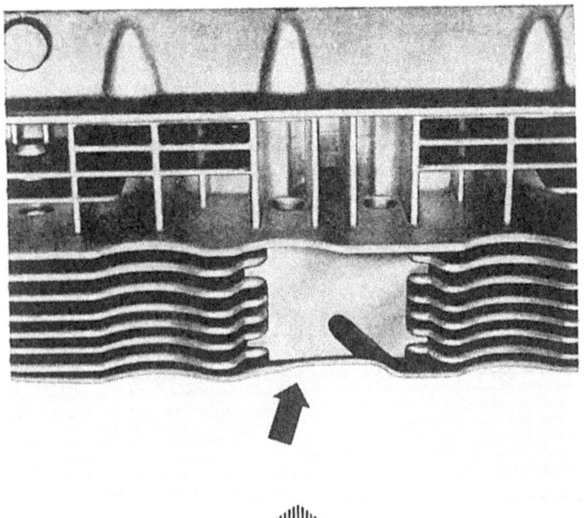

A deflector plate has been fitted on the underside of the cylinder head to improve the cooling air distribution. Furthermore, the four upper cylinder head securing studs have been lengthened.

The following parts were changed:

Part No.	new	old
Cylinder head	113101373	113101371 D
Rocker shaft support	113109427C	113109427A
Rocker shaft stud	113101399B	113101397B
Sealing ring	113109449B	113109449A
Stud, AM 10 AX 193	N 145131	N 145041

Service installation

The new cylinder heads can only be installed if both heads and the rocker shaft supports are replaced at the same time.

The 2 mm longer push rods with knurled shafts (113109301 B) must be installed together with the new cylinder heads. These parts were introduced continuously from Engine No. 8250020.

Drilled out rocker arms (113109443B) can also be installed but only together with oil deflector ring (113109619).

The following should also be noted:

8 studs (N 145041) securing the cylinder heads should be replaced by the longer version (N 145131).

Note.

If the threaded holes in the crankcase are damaged, Heli-Coil inserts should be fitted. To ensure that the studs fit tightly, use only the M 10 × 1.5 tap (009110u).

The previous type of cylinder head will remain available

Removing and Installing Valves

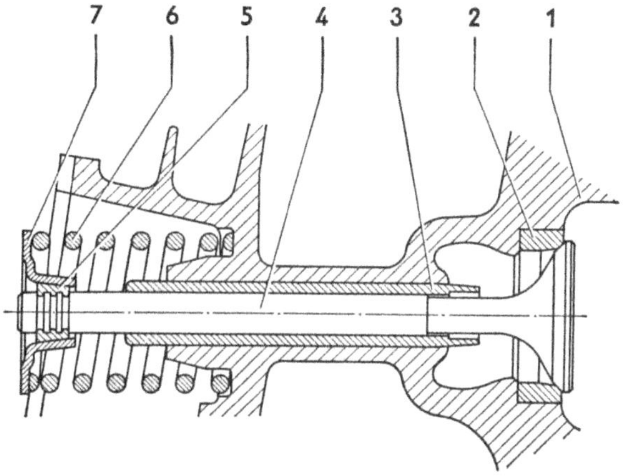

1 - Cylinder head
2 - Valve seat insert
3 - Valve guide
4 - Valve
5 - Valve cotter
6 - Valve spring
7 - Valve spring cap

Removal

1 - Remove cylinder head.

2 - Place cylinder head in valve extractor VW 311 h, press down valve spring seat, remove valve cotters and valve spring caps.

3 - Remove valve springs.

4 - Extract valves.

On older vehicles the seating surface of the cotters on the valve stem may be burred slightly.

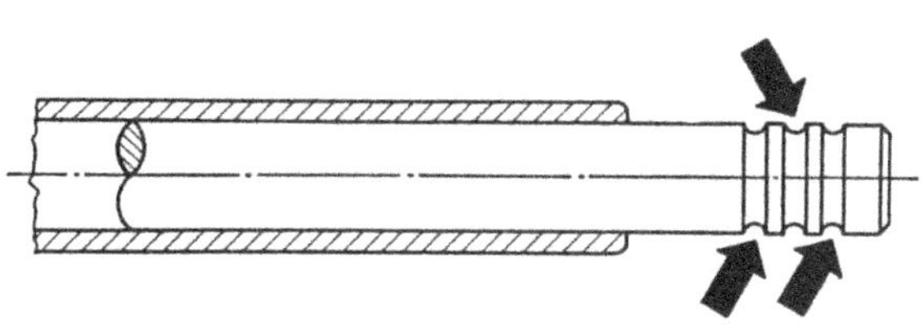

After the burr has been removed with a smoothing file, the valve can be extracted from its guide.

Installation

When installing, the following points should be observed:

1 - Test valve springs.

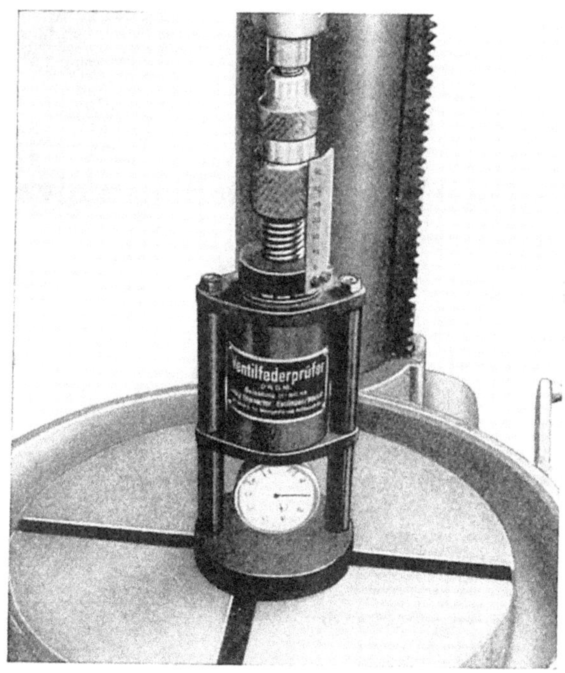

Loaded length	34.3 mm (1.35″)
Load	46.3 ± 3 kg (102 ± 6 lbs.)

Note:

From Chassis No. 4750946 (Engine No. 6850940), all engines are fitted with progressively coiled valve springs.

Part No. **new:** 113109623 A
formerly: 113109623

The progressively coiled spring can be recognised by the difference in the pitch of the coils.

Data:

Loaded Length	Load
33.4 mm (1.314″)	43.8 ± 3 kg (96.5 ± 6 lbs.)

Installation:

Ensure that the spring is installed with the closely coiled end towards the cylinder head.

New and old type springs can be installed together in one engine.

The former valve springs (Part No. 113109623) will be discontinued when stocks are exhausted.

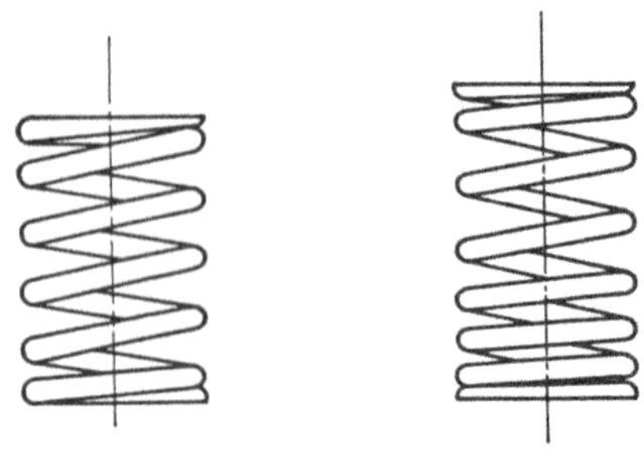

old   new

2 - Check valve cotters for wear prior to assembly.

3 - Test valve stem for run-out (Max. permissible run-out 0.01 mm/0.0004").

4 - Examine valve guides for wear.

5 - Examine valves for wear and leaks. Rough valve stems should be polished carefully with fine emery cloth.

Removing and Installing Valve Springs

(Engine assembled)

Removal

1 - Take off cylinder head cover.

2 - Remove hexagon nuts on rocker arm shaft.

3 - Take off rocker arm shaft with rocker arms and seals.

4 - Install valve spring compressing tool VW 653/1 (local manufacture) in place of rocker arm shaft.

5 - Take out spark plug.

6 - Screw valve holder into spark plug aperture and hold valve.

7 - Depress valve spring seat. Take out valve cotters and remove valve spring cap.

8 - Take out valve spring.

Note:

II push rod breakage occurs, replace all the 113109623A valve springs by the type used on the 1.5 liter engine (113109623C) together with the new valve spring caps (113109641 B).

The springs can be replaced with engine installed, using the spring compressing tool VW 653/2 (local manufacture). Keep the exposed valve gear clean.

Installation

When installing, the following points should be observed:

1 - Check valve springs.

2 - Check valve cotters.

Testing Valve Seats

The valve seat must be concentric with the valve guide. A check is carried out by means of the marking gauge VW 311 k.

1 - Spread a thin film of Prussian blue (engineer's marking) on the contact surface of the gauge.

2 - Insert pilot into valve guide and turn the gauge with light pressure one quarter turn in the valve seat.

3 - Check valve seat. If the blue does not transfer evenly to the valve seat, the seat is to be refaced.

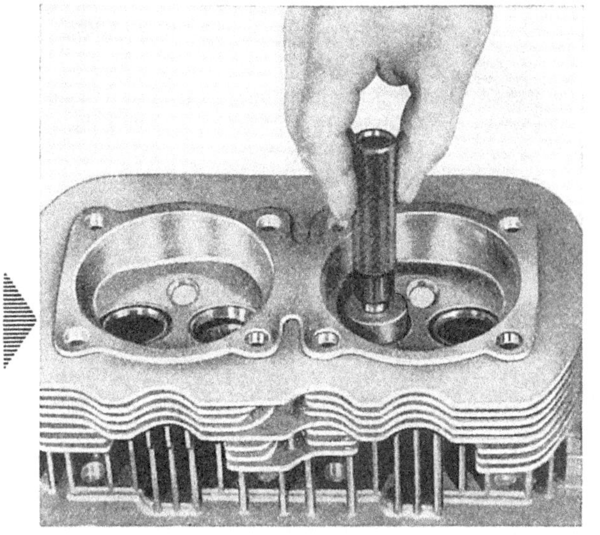

Checking Valve Guides

Replacement of the valve guides is not possible with the tools and appliances which are available in a workshop, as the guides are shrunk (chilled) into position and pressing out the old guides could lead to damage of the cylinder head. Deposits in the valve guides can be removed by using a reamer or broach. The broach is clamped in a press or drill chuck and gradually forced through the guide.

The broach should never be allowed to revolve when passing through the guide.

Valve guide	Dimension
Intake and Exhaust	8.0 dia. H 7 (0.3150″)

The clearance between valve stem and guide is:

Intake 0.050—0.075 mm (0.0020—0.0030″)
Exhaust 0.080—0.105 mm (0.0031—0.0041″)

If the clearance is near the wear limit of 0.16 mm (0.0063″) for both the intake and exhaust valve, the cylinder should be replaced by a new or reconditioned one.

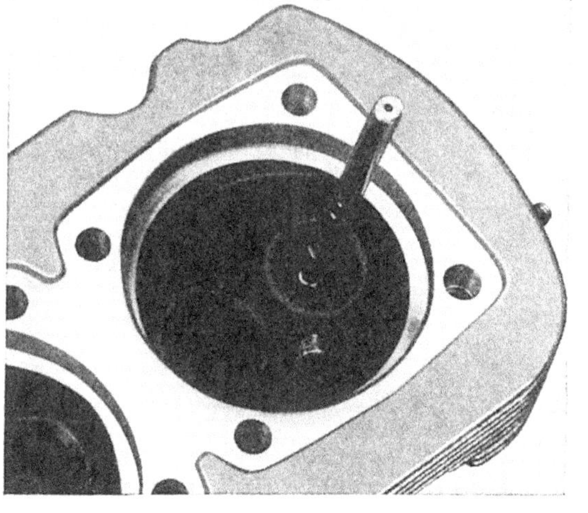

A check is carried out with the Plug Gauge VW 311 k after the valve guide has been cleaned from deposits that may have accumulated.

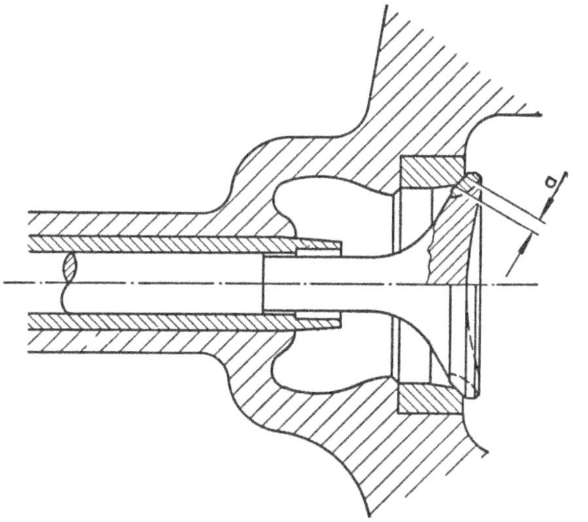

Refacing Valve Seat Inserts

Damaged or burnt inserts may be reconditioned by means of a seat cutter of 45° as long as the specified width of the seat face is maintained and the outer edge of the 15° chamfer does not exceed the outer diameter of the valve seat insert.

Seat width (a):

Intake 1.3—1.6 mm (0.051—0.063″)
Exhaust: 1.7—2,0 mm (0.067—0.079″)

Checking Valve Guides

When repairing engines the valves of which do not seat properly, it is not sufficient to rework the valve seats or to replace the valves. It is also necessary to check the valve guides for wear and replace them if necessary. This check is especially important on engines which have been operating for a long period of time and it must also be carried out on the exhaust valve guides.

Technical Recommendation M 9, 1964 edition, contains information on replacing valve guides.

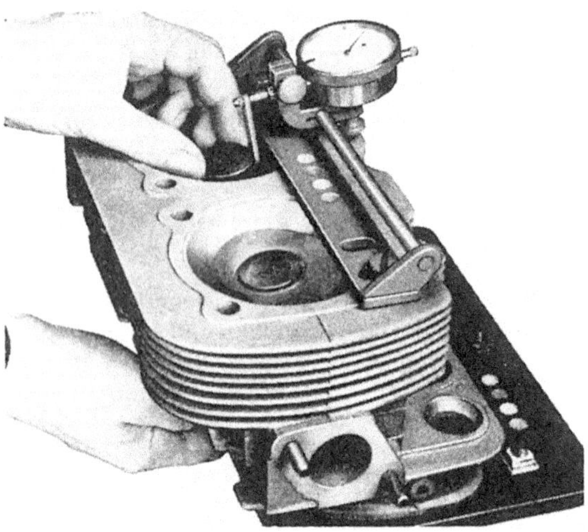

The valve guides must be checked with the "Mounting Plate for Cylinder Head together with Valve Guide Wear Measuring Appliance", VW 689/1 (local manufacture).

Check

1 - Remove deposits with a cleaning broach.

2 - Place cylinder head on mounting plate with the combustion chamber upward and secure it with guide.

3 - Insert dial gauge into bracket and attach dial gauge extension.

4 - Insert new valve into guide to be measured and hold it so that valve stem is flush with end of valve guide.

5 - Set dial gauge and determine amount of rock.

	Intake	Exhaust	Wear Limit
Permissible "rock" of new or repaired valve guides	0.21—0.23 mm (.008 in.—.009 in.)	0.28—0.32 mm (.011 in.—.013 in.)	0.8 mm (.031 in.)

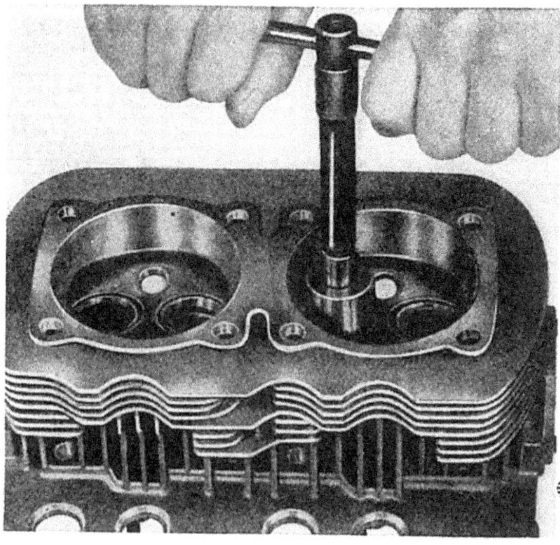

If the chamfer exceeds the outer diameter of the seat insert, the cylinder head is to be replaced by a new or reconditioned one. The inserts cannot be replaced with normal workshop tools as they are shrunk (chilled) into position.

The cutters contained in VW 311 k are required for reconditioning the valve seat inserts.

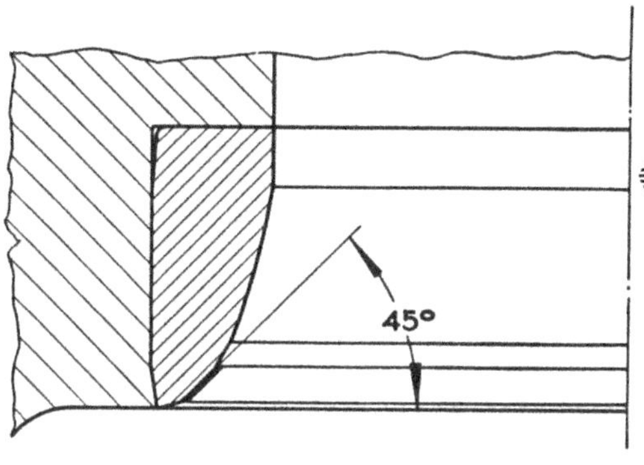

Sequence of operations

1 - Cut the 45° seat face. Considerable care must be taken when cutting to obtain a concentric seating surface. Take off only the minimum of metal as otherwise the life of the inserts will be affected. The operation should cease as soon as the whole surface has been cut.

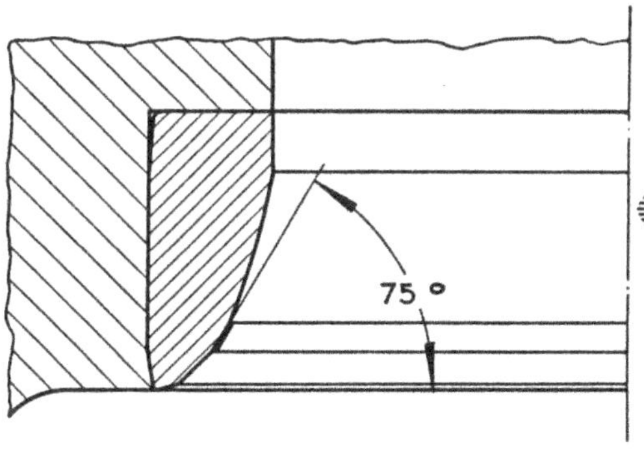

2 - Cut the 75° face: Slightly chamfer the lower edge of the valve seat face with the 75° cutter.

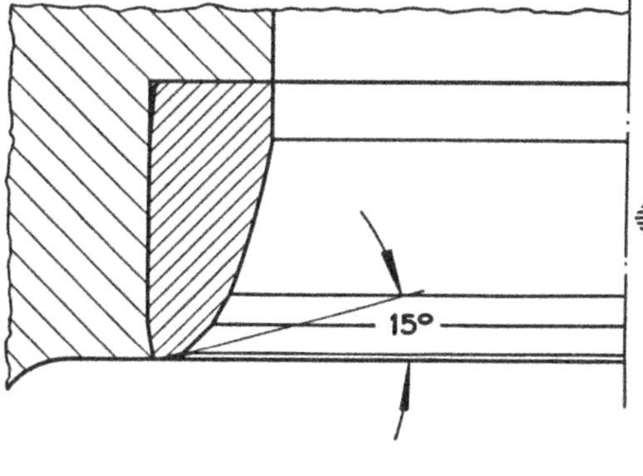

3 - Cut the 15° face: Chamfer the upper edge of the valve seat face with the 15° cutter until the correct seat width is reached.

Inspecting Valves

1 - Thoroughly clean all traces of deposits from the valves with a wire brush.

2 - Examine valve faces for wear and burns. Reface them if necessary.

Considering the high thermal stress imposed on the exhaust valves, it is important not to reduce the dimension "b" of the valve head more than specified.

If the valve face is badly burnt or pitted, the valve should be replaced.

Note:

From Chassis No. 4163317 (Engine No. 6114986), exhaust valves with modified armoured seating surfaces were installed. The new valve can be service installed. The heads of the new valves are flatter so they can easily be distinguished from the old type.

3 - Discard all valves with warped stems, signs of seizure, or damaged valve cotter seats. No attempt should be made to straighten or grind valve stems.

Note:

Valves with worn stem ends can be made usable with the aid of valve caps (Part No. 113109621). The cap is merely placed on the stem before the rocker arms are installed.

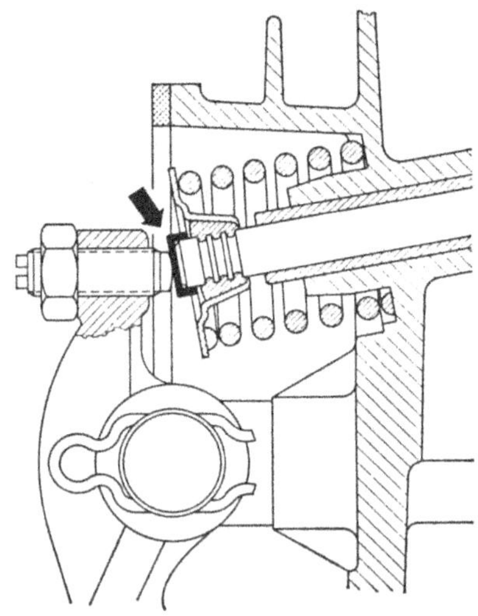

Refacing Valves

Badly burnt or pitted valves which cannot be ground in (lapped) can be refaced with a valve facing machine or valve turning machine.

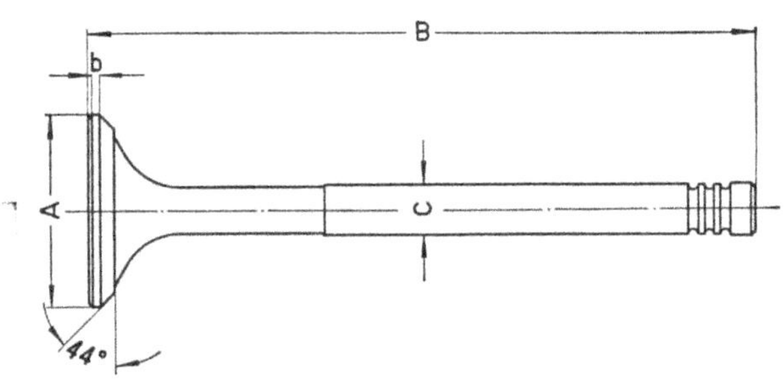

	Intake	Exhaust
A	31.4 mm - 31.6 mm dia. (1.236" - 1.244" dia.)	29.9 mm - 30.1 mm dia. (1.77" - 1.185" dia.)
B	111.4 mm - 112.2 mm (4.386" - 4.417"	111.6 mm - 112.4 mm (4.394" - 4.425")
C	7.94 mm - 7.95 mm dia. (0.3126" - 0.3130" dia.)	7.91 mm - 7.92 mm dia. (0.3114" - 0.3118" dia.)
b	0.8 mm - 1.5 mm (0.031" - 0.059")	1.0 mm - 1.7 mm (0.039" - 0.067")

Note:

From January 1963, Chassis No. 5271918 (Engine No. 7434715) the face angle of the exhaust valve was altered from 44° to 45°. On the inlet valve the angle remains 44° as before.

The angle of the valve seats in the cylinder head remains 45° for inlet and exhaust valves.

The part number of the exhaust valves has not changed. The exhaust valves with the former face angle will be discontinued when stocks are exhausted. When grinding exhaust valves, keep to the new angle.

THIS PAGE INTENTIONALLY
LEFT BLANK

Checking Valve Sealing

With new valves and accurately refaced seats, grinding (lapping) is not always necessary. A simple check can be made by examining the contact pattern.

Contact Pattern

1 - Lightly coat the valve face with Prussian blue.

2 - Insert valve into valve seat and, with light hand pressure, rotate the valve a quarter of a turn.

3 - Lift off the valve. If the valve is seating correctly, the blue should be transferred evenly to the valve seat face. If necessary, grind the valves.

Lapping Valves

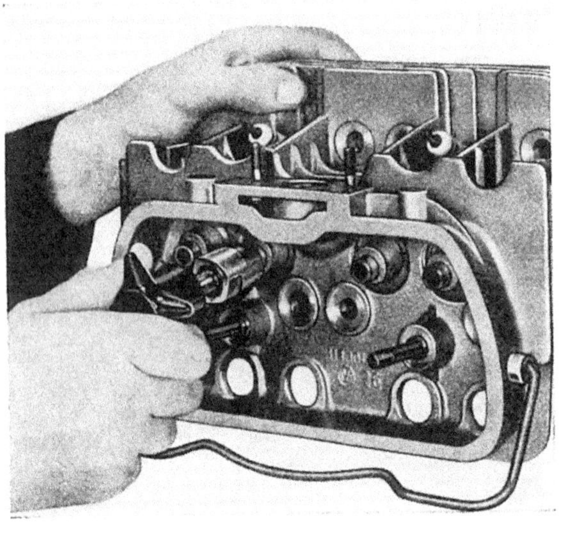

Use the Chuck VW 311 f to grind the valves.

Raise the valve frequently and revolve it evenly to prevent grooves from forming on the faces.

After grinding, thoroughly clean off all traces of grinding compound.

The valve clearance can be checked and adjusted when the engine is cold or up to a maximum oil temperature of 50° C (122° F).

Valve clearance: **Intake** **0.20 mm (.008 in.)**

Exhaust **0.30 mm (.012 in.)**

The valve clearance must be set accurately and with great care. If it is too large, the engine will be noisy. As the clearance **decreases** as the engine warms up, there is a danger that the clearance will disappear altogether if it is set too small when the engine is cold. This then causes exhaust valve damage.

Note:

Since November 1964, Chassis No. 115429385 (Engine No. 9205700) and as a result of the alteration to the rocker shaft securing studs, the valve clearance, has been reduced. On all engines with the new cylinder heads (identified by square bosses) the clearance is **0.10 mm/.004 in.** for exhaust and inlet valves with the engine cold or oil temperature not above 50° C.

The valve clearance for all other engines remains unchanged.

Engine identification

Engines with the modified cylinder heads are marked with a sticker on the cooling air intake housing.

Valve adjustment gives the desired result only if the valves seal tightly, there is no undue clearance in the valve guides and the valve stem ends are not pitted.

Valve clearance insufficient:

Burning of valves and valve seats.
Distortion of valves.
Poor performance due to low compression.
Uneven engine running.
Altered valve timing.

Valve clearance excessive:

Noisy timing mechanism.
Uneven engine running.
Altered valve timing.
Poor performance resulting from insufficient cylinder charge.

Checking valve timing

1 - Mark the TDC position of the No. 1 cylinder on the crankshaft pulley: .6 in. (15 mm) to the left of the right-hand ignition timing mark.

2 - Make a timing mark .35 in. (9 mm) to the right of the TDC mark (.24 in./6 mm up to Engine No. 6930129).

3 - Set the valve clearance to .04 in. (1 mm) on No. 1 cylinder.

4 - Turn the crankshaft to the right. When the timing mark is in line with the crankcase joint, the inlet valve on No. 1 cylinder should open.

Offsetting the camshaft gear by one tooth alters the opening and closing times by about .87 in. (22 mm).

Adjusting Valves

Valve adjustment is best effected in the following sequence: 1st—2nd—3rd—4th cylinder. Adjust the valves when the piston of the corresponding cylinder is in top dead center position of the compression stroke, as the two valves are then closed.

Starting with the 1st cylinder, crank the engine over slowly to the left at the fan pulley, until both valves are in fully closed position and the timing mark on the pulley is in line with the vertical jointing faces of the crankcase.

1 - Remove cylinder head cover.

2 - Set engine to the firing position of the cylinder to be adjusted.

3 - Check valve clearance with a feeler gauge. 0.20 mm (0.008″) for the inlet valves, 0.30 mm (.012″) for the exchaust valves.

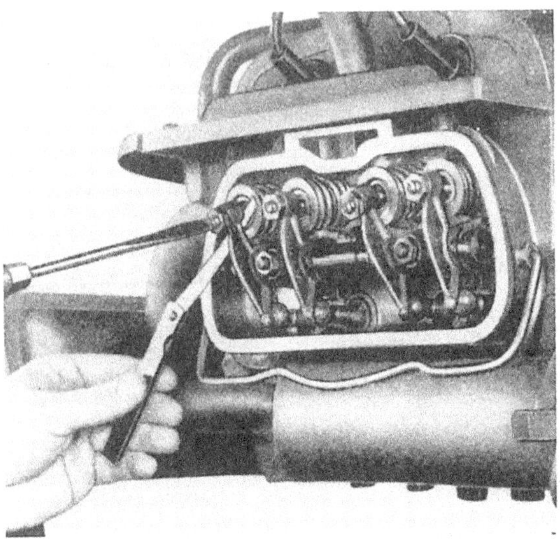

Note:

The valve clearance is set properly if the feeler blade can be pushed between adjusting screw and valve stem easily. It is wrong to pull the feeler blade through with a fair amount of forceas this is bound to lead to insufficient clearance. As the clearance alters when the lock nut is tightened, the clearance should be checked again after tightening the nut.

4 - Loosen lock nut of the adjusting screw.

5 - Turn adjusting screw as required to obtain the proper clearance.

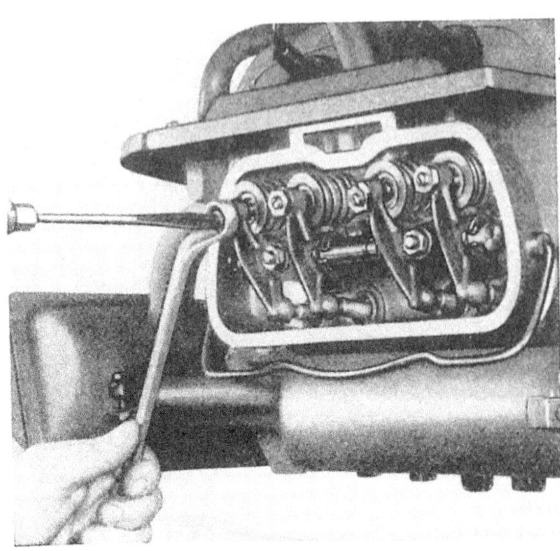

6 - Hold adjusting screw with a screwdriver while tightening lock nut.

7 - Recheck adjustment.

8 - Check and adjust the other valves in the same manner.

9 - Replace cylinder head cover, using a gasket which is in good condition.

Note:

Cylinder heads with round bosses

Exchange engines are no longer being fitted with cylinder heads with round bosses and long studs. The heads on these engines are equipped with short replacement studs which means that the valve clearance is altered to .004 in. (0.1 mm) for inlet and exhaust valves.

Identification

Engines equipped with these heads are marked with the well-known sticker "Valve clearance .004 in./0.1 mm" (Part No. 311 100 175) and with the clip (311 100 177) which was introduced later.

> As the heads with short studs cannot be differentiated from those with long studs when the heads are on the engine, the former means of noting the valve clearance — round boss = 0.2/0.3 mm, square boss = 0.1 mm — is no longer applicable. The proper valve clearance is indicated only by the sticker on the engine and the clip on the rocker shaft.

Exchange parts:

Exchange cylinder heads with round bosses supplied as individual parts will be equipped with long studs or with short studs. The Part Nos. of these heads have the letters X and Y added to them.

Heads with 0.2/0.3 mm valve clearance

Long studs

113 101 061 BX
113 101 061 CX
113 101 061 DX

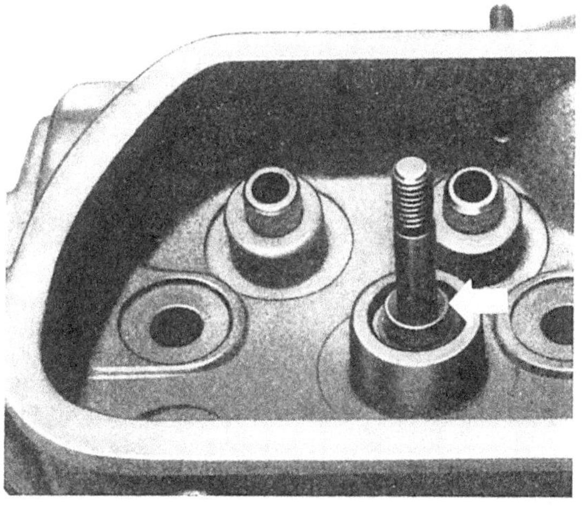

Heads with 0.1/0.1 mm valve clearance *)

Short replacement studs

113 101 061 BY
113 101 061 CY
113 101 061 DY

*) Metal tag is included

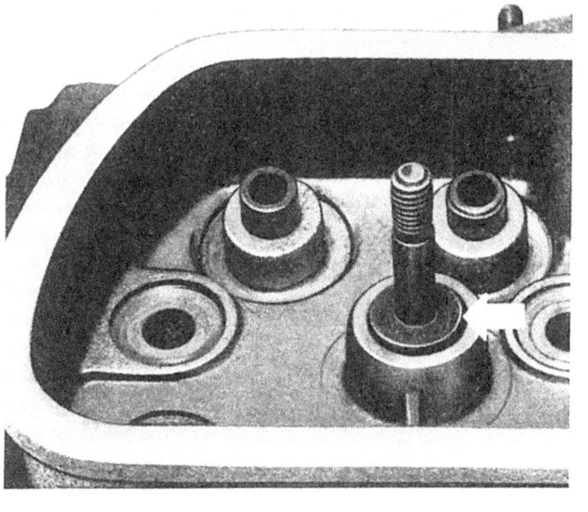

Service installation

The installation of heads with short studs on engines which had heads with long studs is only permissible if both heads are replaced. Engines which are converted in this way should be marked with the valve clearance sticker and clip or by attaching a metal tag showing the clearance .004 in./0.1 mm to one of the studs.

Note:

The short studs can also be fitted in normal workshops or service stations. Repair instructions are given on page M-6/2.

THIS PAGE INTENTIONALLY
LEFT BLANK

Removing and Installing Cylinders

Removal

1 - Remove valve push rods and valve push rod tubes.

2 - Remove deflector plate below the cylinders.

3 - Take off cylinders.

Installation

When installing, the following points should be observed:

1 - Check cylinder for wear and, if necessary, replace by another matched cylinder and piston of the same size.

2 - Cylinder seating surface on crankcase, cylinder shoulder, and gasket must be perfectly clean. Foreign matter at this point may cause distortion of cylinder.

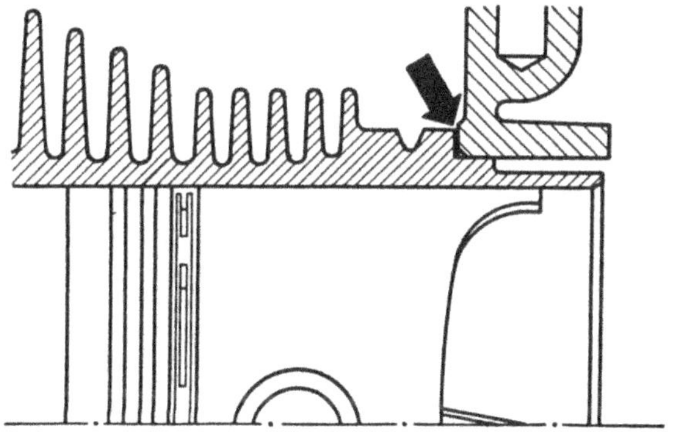

3 - Always use a new gasket between cylinder and crankcase.

4 - Apply some oil to piston and piston pin.

5 - Compress the rings with the compressing tool VW 123a and make sure that the ring gaps are offset. The oil ring gap must always be at the top when the pistons are in their horizontal position in the engine.

6 - Oil the cylinder wall and slide the cylinder over the piston. The crankcase studs must not contact the cylinder cooling fins.

7 - Install deflector plates. Make sure that they are correctly seated. If necessary, bend the plates until they tightly bear on the cylinder head studs to prevent them from rattling and working loose during operation.

8 - Install push rods and push rod tubes. Insert tubes so that the seam is facing upwards. Before installing used tubes, they should be stretched to proper length in order to ensure perfect sealing between tube and crankcase and/or cylinder head.

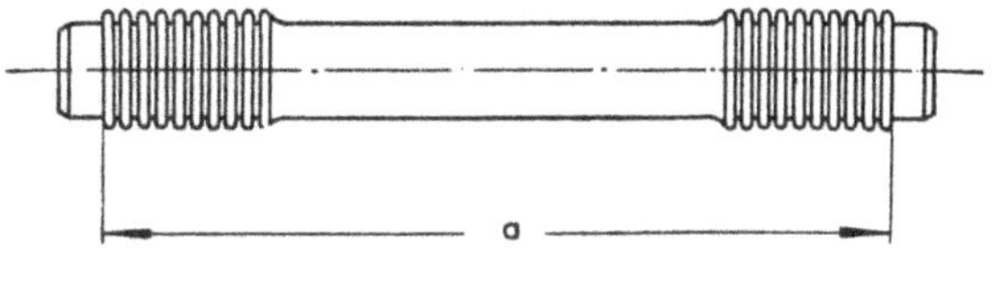

a = 180.5 — 181.5 mm (7.11—7.16")

This operation should be carried out carefully so as to avoid cracking the tubes.

Removing and Installing Pistons

Removal

1 - Remove cylinder.

2 - Mark the piston to make sure that it is reinstalled in its original position.

3 - Remove piston pin circlip, using circlip pliers.

4 - Heat the piston to 80° C (176° F) with electric piston heating tool VW 205 a.

5 - Remove piston pin, using pilot drift VW 207 or tool VW 207 a.

6 - If the piston rings have to be removed use the piston ring tool. The piston rings should, if possible, remain on the piston to avoid damage.

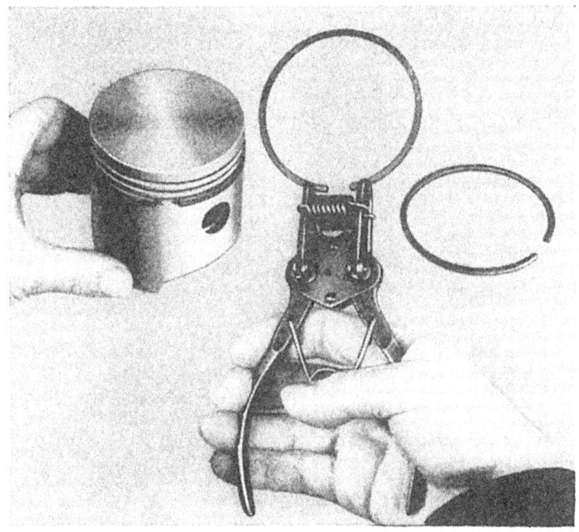

Installation

When installing, the following points should be observed:

1 - Clean piston. Remove carbon from piston tops and piston ring grooves without damaging the piston surface. Do not use emery cloth to remove carbon from piston skirts (if necessary, use a fine corundum stone with oil). A poor wear pattern and uneven formation of carbon deposits on the piston can be caused by a distorted connecting rod.

2 - Check piston for wear, if necessary replace it by one of the corresponding size. The difference in weight between the pistons must not exceed 10 g.

Note:

The distance from piston crown to center of piston pin hole is 39 mm (1.535″). An index letter is stamped on the piston crown to avoid confusion with the pistons in the modified Transporter engine from May 1959.

Marking of pistons

A - Arrow (indented or stamped on) which must point towards the flywheel when piston is installed.

B - Details of piston pin bore size indented or stamped on (s = black, w = white).

C - Paint spot indicating matching size (blue, pink, green).

D - The letter near the arrow corresponds to the index of the part number of the piston concerned. It serves as an identification mark.

E - Details of weight grading (+ or —) indented or stamped on.

F - Paint spot indicating weight grading (brown = — weight, grey = + weight).

G - Details of piston size in mm.

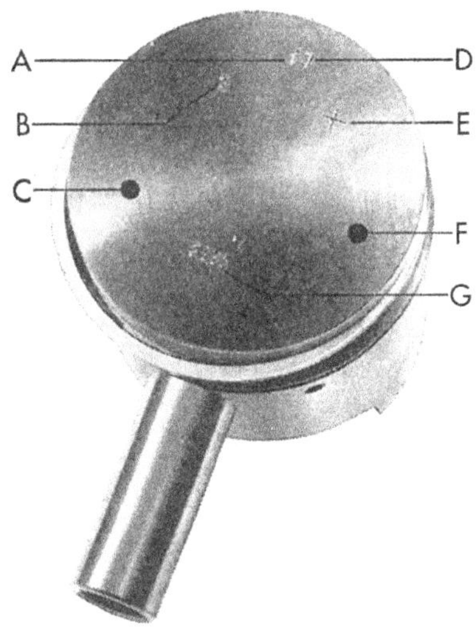

3 - Select piston rings of proper size. Check gap with a feeler gauge after the ring has been inserted in the cylinder and squarely pushed down about 5 mm (.2") with the piston.

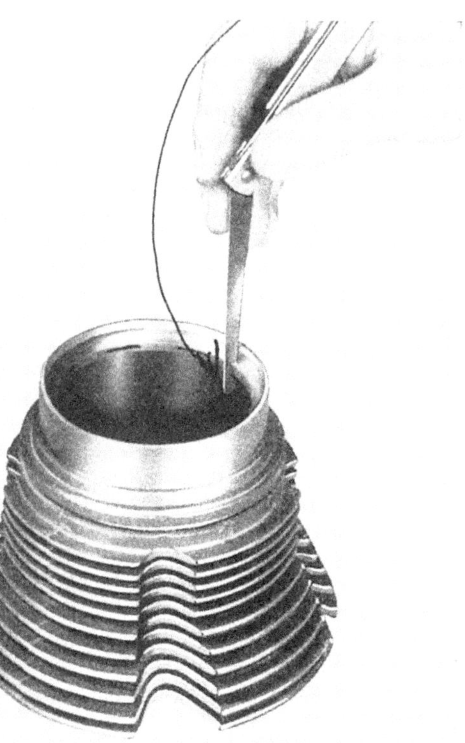

Gap of compression rings: 0.30—0.45 mm (.012—.018"); max. 0.95 mm (.037").

Gap of oil scraper ring:

0.25—0.40 mm (.010—.016");
max. 0.95 mm (.037").

When re-installing the cylinder, make sure that the oil scraper ring gap is at the top and that each piston ring gap is offset by 120°.

Check piston ring side clearance in grooves using a feeler gauge.

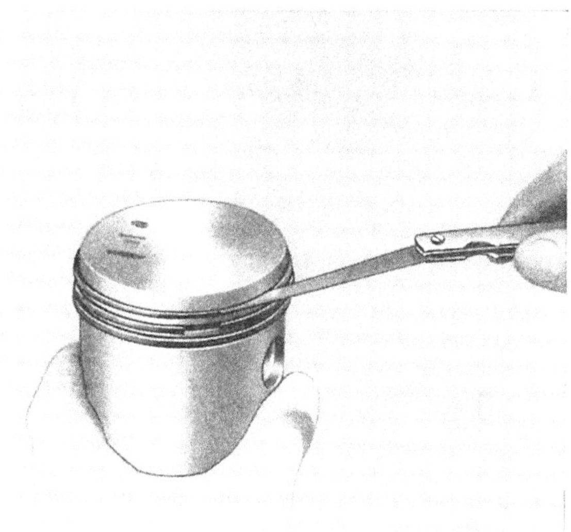

Compression ring, top	0.065—0.092 mm; max. 0.12 mm (.0026—.0036"; max. .0047").	
Compression ring, lower	0.045—0.072 mm; max. 0.01 mm (.0018—.0028"; max. .004").	
Oil ring	0.025—0.052 mm; max. 0.01 mm (.001—.002"; max. .004").	

The rings must only be installed with the piston ring tool to avoid damaging the piston.

Make sure that the compression rings are installed with the marking "Top" or "Oben" toward the top of the piston.

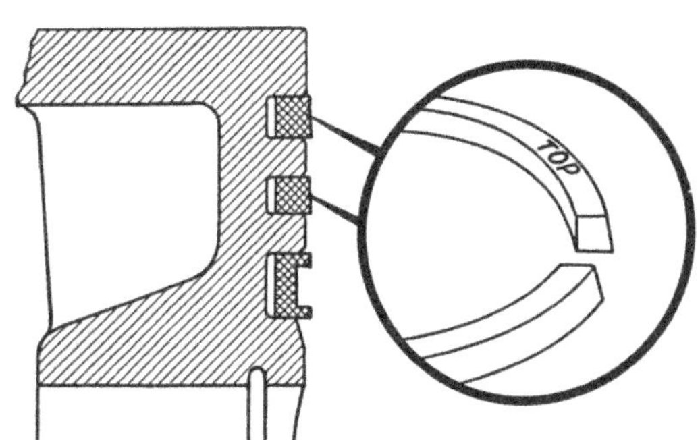

Note:

From Chassis No. 4 988 623 (Engine No. 7 076 057), modified pistons and piston rings are installed in all 41.5 SAE bhp engines. The depth of the two upper piston ring grooves has been reduced and the upper rings chamfered on the inner edges.

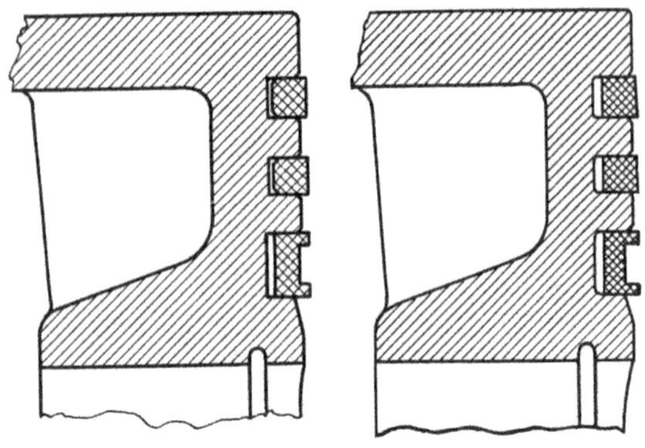

new old

Identification marks on modified pistons

Mahle: Embossed circle inside piston skirt on the right of the VW sign.

KS: Embossed circle inside piston skirt above VW sign.

Nüral: Indented circle inside extended part of piston skirt.

Part Nos.:	**new**	previously
Piston (Mahle)	113 107 111 P	113 107 111 F
Piston (Nüral)	113 107 111 R	113 107 111 M
Compression rings	111 107 351 A	111 107 351
Scraper ring	111 107 361 B	111 107 361 A
Piston and cylinder set	SP 2 F	SP 2 E

Parts of the previous pattern will be discontinued when stocks are exhausted.

Note:

a - The new piston rings can be installed in the old type pistons.

b - The old piston rings must **not** be installed in the new type pistons.

Note:

The following piston ring sets are available for service installation in engines having an excessively high oil consumption:

Bore	Part No.
77 mm	SP 103 A
77.5 mm	SP 107 A
78 mm	SP 111 A

The installation of these special piston ring sets is justified if

1 - the oil consumption is actually found to exceed 1 liter (about 2 pints) per 1000 km (600 miles), or if

2 - the cylinder out-of-round is not in excess of 0.01 mm (.0004"). This refers roughly to engines with a mileage below 30,000 km (19,000 miles).

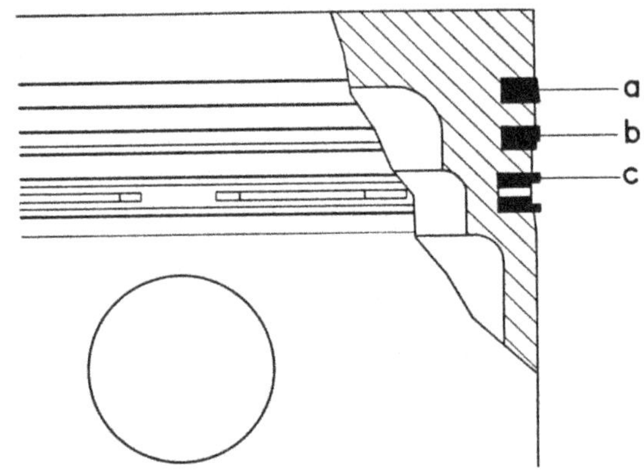

a - Top ring
b - Narrow face contact ring
c - Oil ring

The installation is useless if

1 - it is not definitely known whether oil consumption or an oil leak is involved.

2 - the engine has not covered more than 5000 km (3000 miles). New pistons often have a higher oil consumption than those already worn in, or if

3 - the engine shows excessive wear or has covered considerably more than 30,000 km (19,000 miles).

Check the following when installing:

1 - correct installation of the rings.

2 - the tolerance limits specified for the side clearance and gaps of the rings.

3 - correct clearance between pistons and cylinders.

4 - good piston contact pattern.

Special piston ring sets of other designs should not be used for the following reasons:

1 - Substantially higher cost.

2 - Shorter service life on account of heavy wear on cylinders and rings (pressure against cylinder wall, matching of materials).

3 - Loss of power by higher frictional resistance.

4 - The VW engine does not require breaking-in, even when fitted with narrow face contact rings. Most other special piston rings sets require carefully breaking-in.

4 - First insert the piston pin circlip which faces towards the flywheel.

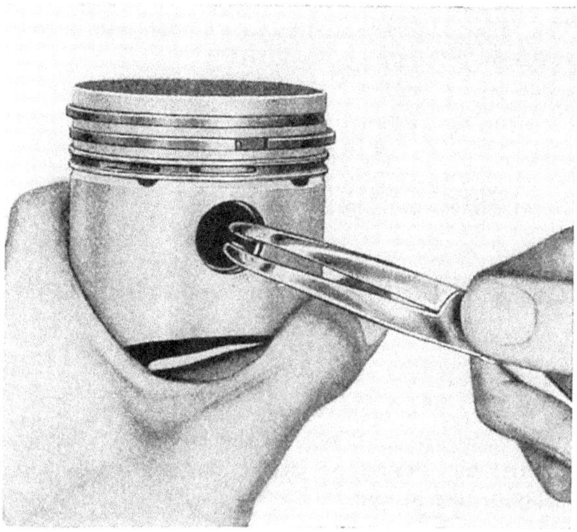

As the piston pin holes are offset make sure that the arrow or word "vorn" points towards the flywheel when installing. Due to the piston pin being offset, the connecting rod angle changes and the piston alters its bearing on the cylinder before reaching TDC. Combustion has not yet taken place and the side stresses are relatively small. The piston tilts gently but not abruptly towards the opposite side of the cylinder wall, thus avoiding the usual slapping wall, thus avoiding the usual slapping noise

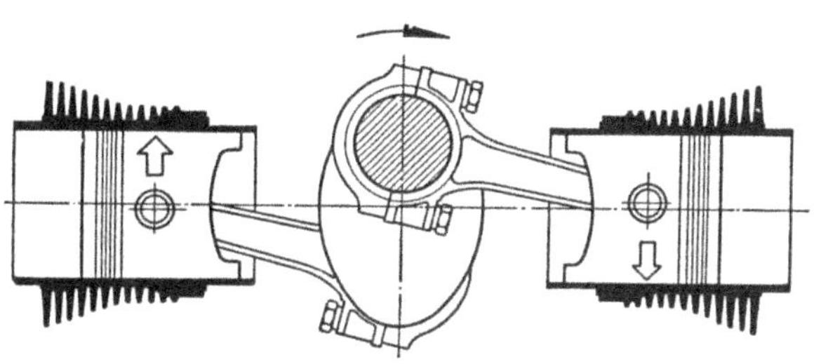

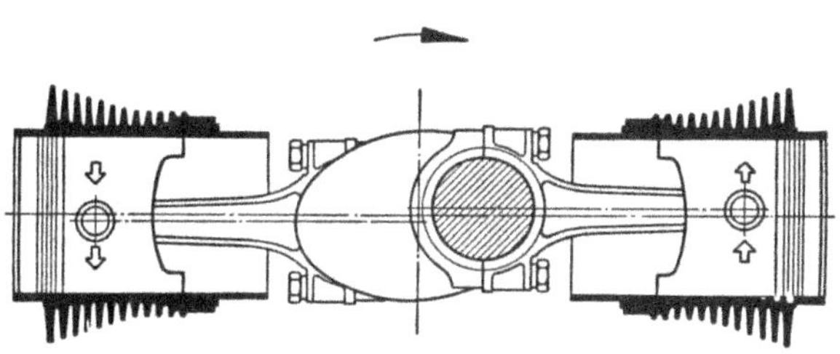

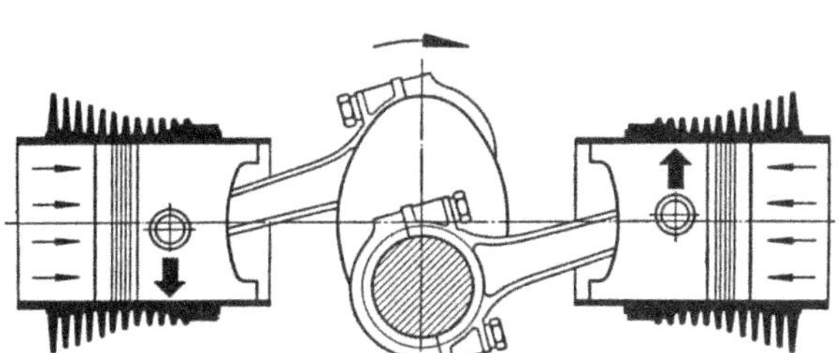

caused by the piston tilting especially when clearance between piston and cylinder is large.

5 - Check and fit piston pin. Depending on the combination of tolerances between piston pin and bearing, the pin may be found to be a light finger-push fit in the piston even when the piston is cold. This condition is quite normal, even when the piston pin slides out of the piston under its own weight. There is no reason whatsoever to replace piston pin, piston, or both, in such cases. To determine the proper size the piston pin and the piston pin hole are color marked. For pistons with a piston pin hole dia. in excess of 20.001 mm/.7874" over-size pins — marked green — are available.

The clearance between piston pin and connecting rod bush is 0.01—0.02 mm (.0004—.0008"). If the clearance is near the wear limit of 0.04 mm (.0016"), renew the piston pin and the connecting rod bush.

It is not permissible to install an oversize pin in this case.

Colour	Piston pin dia. (mm)	Pin bearing dia. (mm)
Black	19.994—19.997	19.996—19.999
White	19.997—20.000	19.999—20.001
Green	20.001—20.004	Pins only

Heat the piston in all cases where the pin is not a light finger push fit in the cold piston. Heat it by means of the piston heating tool VW 205 a or in oil to about 80° C (176° F). The pin is then pushed in with the VW 207 drift in one movement until it contacts the circlip.

6 - Insert the other circlip. It is important that the circlip fit in their grooves perfectly.

Checking the Clearance between Cylinder and Piston

The fitting clearance between cylinder and piston is 0.036—0.54 mm (.0014—.0021″).

Permissible wear limits:

Clearance cylinder/piston max. 0.20 mm (.008″).

The clearance must not be checked with a feeler gauge, but should be determined by measuring cylinder and piston.

The cylinders are measured with a dial gauge for inside diameters which is set with a micrometer according to the cylinder size.

The cylinder should be measured 10—15 mm (.4—.6″) below the upper edge.

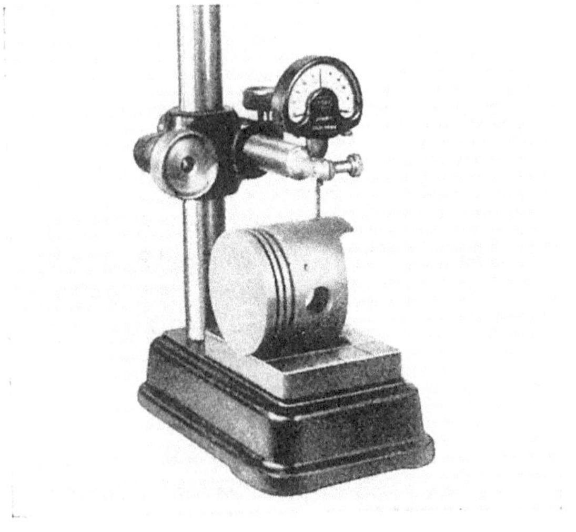

The nominal diameter is stamped on the piston top. The reading must be taken at the bottom end of the skirt at right angles to the piston pin axis.

The size grading of cylinders and corresponding pistons is as indicated below:

	Colour	Cylinder mm ø	Corresponding Piston mm ø
	Blue	76.990—76.999	76.95
Standard Size	Pink	77.000—77.009	76.96
Nominal Dimension 77 mm ø	Green	77.010—77.020	76.97
	Blue	77.490—77.499	77.45
1st Oversize	Pink	77.500—77.509	77.46
Nominal Dimension 77.5 mm ø	Green	77.510—77.520	77.47
	Blue	77.990—77.999	77.95
2nd Oversize	Pink	78.000—78.009	77.96
Nominal Dimension 78 mm ø	Green	78.010—78.020	77.97

Note:

To simplify repairs which involve the replacement of pistons and cylinders, the weight tolerance for the pistons has been raised to 10 g (.35 oz.) for all engines.

Provided that the piston weights are checked, (piston weight = bare weight without piston pin and rings) pistons of different makes and matching sizes (blue, pink, green) can be installed in **one** engine. The former weight grading (grey, brown) has, therefore, lost its significance during repairs. It is still only permissible to install pistons and cylinders of the same size grading (Standard, 1st oversize, 2nd oversize) in **one** engine.

The instruction that a piston and the associated cylinder must have the same color marking for the matching size, still remains valid as this concerns the selective fitting which gives the correct running clearance for the piston in the cylinder.

If measurement of piston and cylinder shows that the running clearance is near 0.2 mm (.008″), the piston and cylinder should be replaced by a set of the same size grading (normal or oversize). The difference in weight between the pistons in **one** engine must not exceed 10 g. Pistons must not be replaced individually if the cylinders to which they belong show signs of wear. If the cylinder of a damaged piston shows no signs of wear it is usually sufficient to install a new piston of the appropriate matching size.

Since the compression ratio should remain the same when installing reconditioned cylinders, the distance from the crown to the boss of the oversize pistons (77.5 or 78 mm dia.) is appropriately reduced.

Important

Beside the wear, the oil consumption of the engine is an important factor in deciding whether or not a new cylinder and piston must be installed. If the oil consumption is in excess of 1 liter per 1000 km (2 pints per 600 miles) the engine needs an overhaul. An oil of higher viscosity, for example SAE 30, can be used during the warmer season with engines having a marked oil consumption.

THIS PAGE INTENTIONALLY
LEFT BLANK

Oil Circulation

General

The VW engine has a full pressure lubrication system with a gear type oil pump.

The gear-type oil pump is situated at the gear side of the camshaft, from which it receives its drive. Oil is drawn from the lowest point of the crankcase and forced into the oil passages via the oil cooler. Some of the oil is fed via the crankshaft main bearings through the drilled passages in the crankshaft to the connecting rod bearings. Oil is also fed to the camshaft bearings and through the hollow push rods to the rocker arms, lubricating their bearings. The splash oil and oil mist lubricates the valve stems and then flows down the push rod tubes to the crankcase. Cylinder walls, pistons and piston pins are also splash-lubricated. The oil returns to the crankcase where it is filtered by a gauze strainer at the lowest point before again entering the circulation.

The oil cooler on the crankcase is positioned in the ducted air flow. This enables the oil to maintain its lubricating qualities even at high outside temperatures and at sustained high engine speeds.

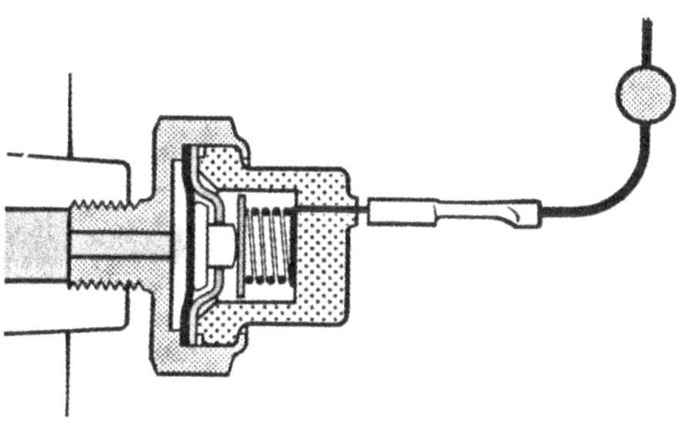

The oil pressure switch gives a check on the oil pressure.

a - Pressure builds up after starting the engine:

Contact opens — 0.3—0.6 kg/sq. cm
(4.3—8.5 lbs./sq. in.)

Warning light goes out.

Note:

From Chassis No. 3924800 (Engine No. 5843201) the operating pressure is 0.15—0.45 kg/sq. cm (2.1—6.3 lbs/sq. in.)

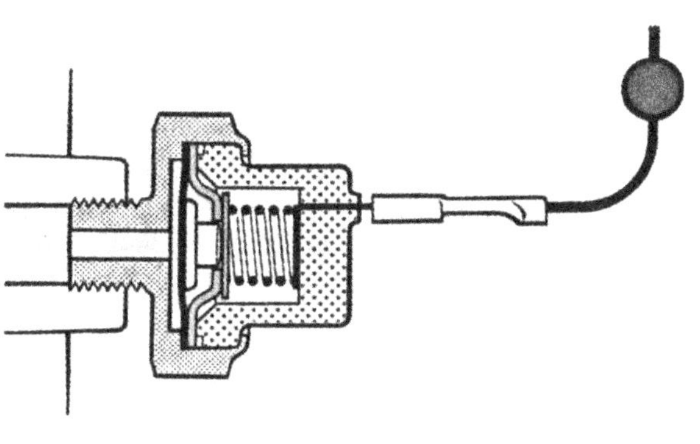

b - Oil pressure is too low:

Contact closes —

Warning light lights up.

The oil strainers must, on no account, be confused with one another as otherwise serious engine damage could arise:

1 - to 54 bhp engine:

due to oil starvation, since the wire mesh will become crushed and part of the oil suction pipe opening will become blocked off.

2 - to 40 bhp engine:

due to oil contamination since some of the oil is drawn up through the space between suction pipe and oil strainer instead of all the oil becoming sucked through the oil strainer.

The oil strainer of the 54 bhp engine (Part No. 311115175) is distinguished by a circular depression (A) in the wire mesh.

Note:

If the oil strainer is incorrectly fitted and the oil intake pipe too low down, damage may be caused to the crankshaft and connecting bearings. When the oil intake pipe is too low, the end of the pipe tends to pierce the bottom of the strainer after it has been removed and installed several times. The engine then draws in unfiltered oil.

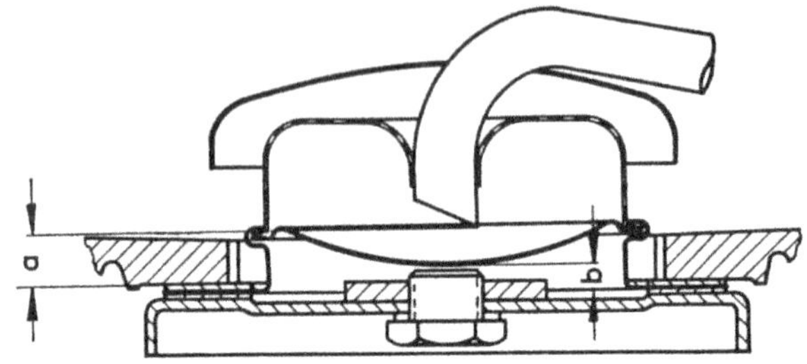

a - Measurement from strainer contact surface on crankcase to tip of pipe.

b - Measurement from strainer flange to bottom of strainer.

Engine	Oil strainer	Measurement "a" in mm	Measurement "b" in mm
30 bhp 36 SAE bhp	111 115 175A	13,5 ± 1	8,5 ± 1
34 bhp 41,5 SAE bhp	113 115 175	10 ± 1	6 ± 1

Important

Before installing the strainer, check if the oil pipe has made marks on the bottom of it. If it has, bend the pipe slightly and fit a new strainer.

Removing and Installing Oil Pressure Relief Valve

Check the oil pressure relief valve when disturbances in the oil circulation occur, especially when the oil cooler leaks. If the plunger sticks at TDC when the oil is thick, there is a danger of the oil cooler leaking. If the plunger sticks at BDC the oil will flow directly back to the sump.

1 - Plunger
2 - Spring
3 - Gasket
4 - Plug

Removal

1 - Remove plug.

2 - Remove spring and plunger. If the plunger has stuck it can be removed by screwing a tap M 10 (10 mm metric thread) into it.

Installation

When installing, the following points should be observed:

1 - Check plunger and bore in crankcase for signs of seizure. Carefully remove signs of seizure, renew plunger if necessary.

2 - Examine spring.

Condition	Length	Load in kg (lbs.)
loaded	23.6 mm (0.93″)	7.75 (17.1)

3 - Make sure that the upper end of the spring does not scratch on the bore wall.

4 - Renew gasket.

Note:

The modified relief valve piston with annular groove (311415411) is a standard fitting in the twin-carburetor engine. This piston can be subsequently installed in all engines (except 30 bhp type) which are used mainly in **hot climates.**

As this measure reduces the engine oil temperature over the entire range, the modified piston must **not** be installed in countries with temperate or cold climates.

Removal

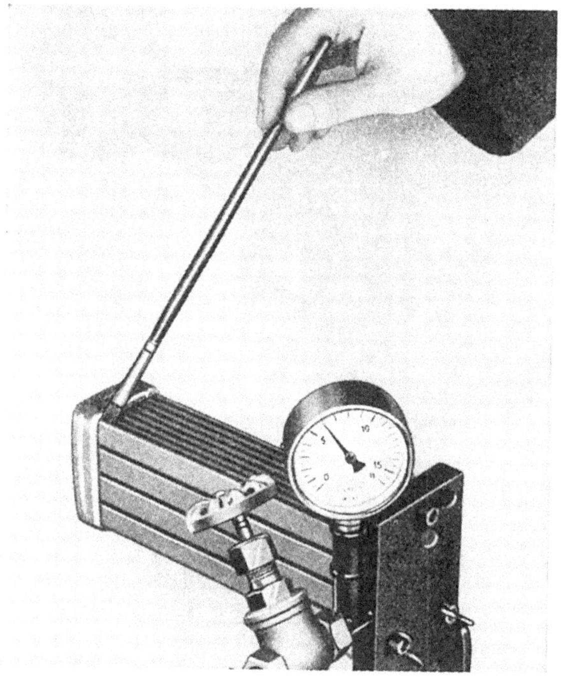

1 - Remove oil cooler retaining nuts.

2 - Remove oil cooler and gaskets.

Installation

When installing, the following points should be observed:

1 - Check studs and bracket on cooler for tightness. Test pressure: 6 kg/sq. cm (85 psi). Oil cooler test appliance: VW 661/2 (local manufacture).

2 - If cooler leaks, check oil pressure relief valve.

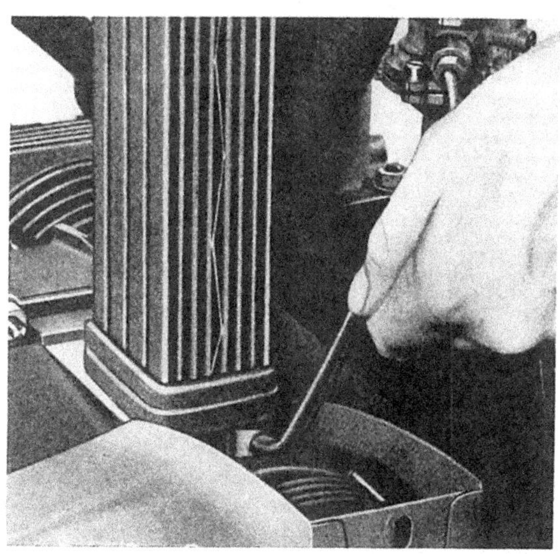

3 - The oil cooler tubes must not touch one another and the partition plate must not be loose.

4 - Fit new gaskets.

Note:

When removing the oil cooler with engine installed or assembled, the fan hausing should be taken off and the cooler nuts loosened with a VW 109 box wrench.

Removing and Installing Oil Pump

Removal

1 - Remove nuts on oil pump cover and take off cover and gasket.

2 - Remove the gears.

3 - Remove oil pump body with extractor VW 201.

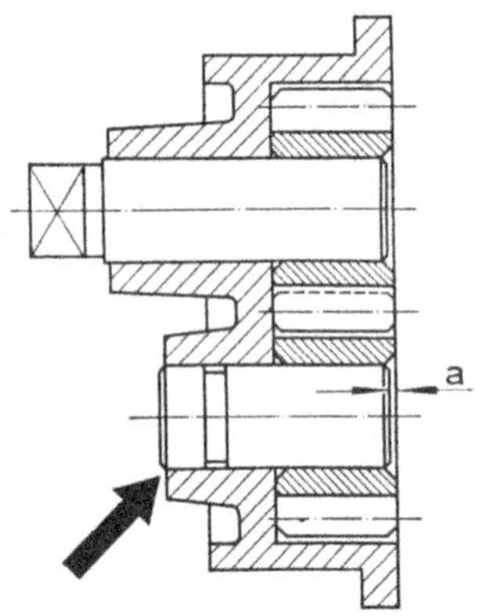

Installation

Observe the following points:

1 - Check oil pump body for wear, especially the gear seating, prior to assembly. Wear in the pump body will result in an excessive loss of pressure.

2 - Check driven gear shaft for tightness. If necessary, peen it or replace pump body (a = 0.5—1.0 mm/.02—.04″).

3 - Check gears for wear.
Backlash 0.03—0.08 mm (.0012—.0031″).
Maximum end play, without gasket 0.1 mm (.004″).

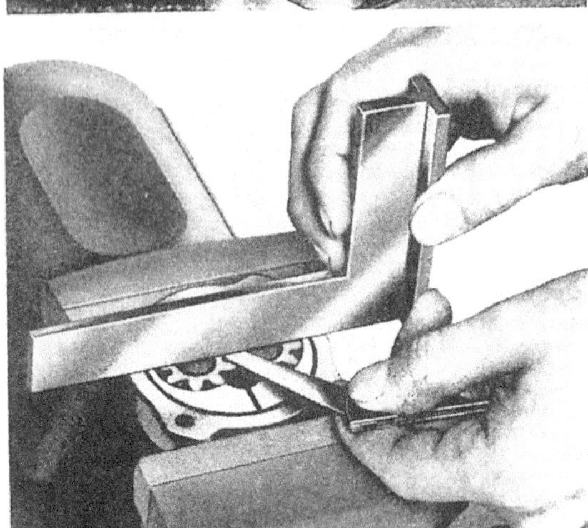

4 - The measurement can also be made with an engineer's square and a feeler gauge.

5 - Check pump body mating surface on crankcase for dirt and damage.

Note:
Plastic-coated gaskets have been used for the oil pump and the oil strainer cover for some time now. If these gaskets adhere firmly, they should not be removed with a scraper, as the sealing surfaces are easily damaged. If they are difficult to remove, the gaskets can be loosened with a 5% ammonia solution and, after a short time, can be easily removed.

6 - Install pump body with gasket and without sealing compound.

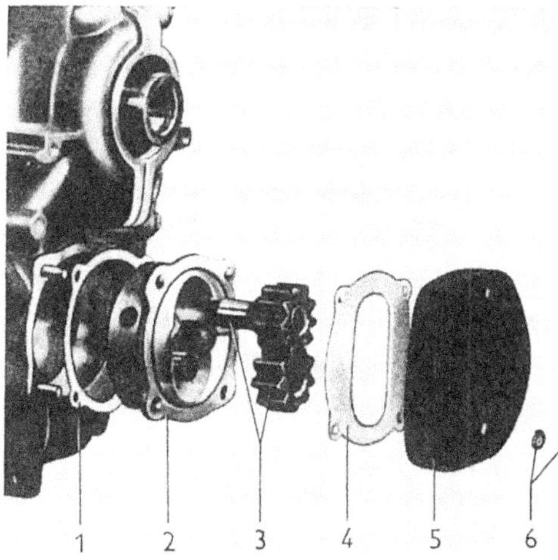

1 - Gasket
2 - Oil pump body
3 - Gears
4 - Gasket
5 - Oil pump cover
6 - Nut and washer

7 - Insert oil pump pilot VW 665 (local manufacture) into body instead of the oil pump drive shaft.

8 - Turn camshaft 360° (one complete turn) = two turns of the crankshaft. This will center the pump body opposite the slot in the camshaft.

9 - Mark the seating position of the pump body on the crankcase so that the position can be checked after the cover has been installed.

10 - Remove pilot and install gears.

11 - Check the cover. Machine or replace worn covers.

12 - Fit new gaskets without sealing compound and install cover. When tightening the nuts, do not disturb the position of the pump body.

Note:

Type	from Chassis No.	from Engine No.
1/1200	118 001 312	D 0 230 001
1/1300	118 000 226	F 1 225 019
1/1500	117 811 483	H 0 822 052

From June 1967 four M 8 studs (were M 6) are used for securing the oil pump to the crankcase. At the same time, the washer, N 11 5071, and the M 6 hexagon nuts were replaced with M 8 sealing nuts with pressed-in plastic rings. The plastic ring must face the oil pump cover.

Note:

Complaints are occasionally received of high oil consumption which is not caused by loss of oil at joints and seals. Often the fault lies in faulty measurement of the oil level at filling stations and the oil consumption is actually quite normal.

Note the following when checking the oil level:

1 - The vehicle must be standing on a level surface. Even a slight angle, as on a hilly street, can cause measuring errors of more than 1 liter (2.1 US pints, 1.75 Imp. pints).

2 - The oil level should not be checked until the engine has been stationary for at least 5 minutes so that the oil can drain down to the sump from the various drillings and the oil cooler.

If an oil consumption of more than 1 liter per 600 miles (1000 km) is established, despite observation of the above points, proceed as follows:

A - **Check if the engine is throwing oil out**

Some engines can throw oil, which has collected in the drainage pipe, out of the crankcase breather. First check whether it really is engine oil and is not condensate that has been mixed with traces of oil to form an emulsion in the rubber valve. Normal condensate which forms mainly in the cold season when driving short distances and which is unavoidable cannot be regarded as loss of oil.

Even though there are only slight traces of oil at the rubber valve the engine can have lost oil via the crankcase breather. The engine can be checked as follows on a roller test stand or an engine test stand:

1 - Replace the hose between oil filler and air cleaner with a transparent hose.

2 - Allow engine to cool down to room temperature.

3 - Start engine and let it run for a few seconds to ensure that the oil pressure is adequate, then open throttle fully and run engine at 5000 rpm under load.

4 - After about 10—15 seconds, oil will be visible in the transparent hose on engines which are throwing oil out of the crankcase breather. When this happens, reduce speed immediately and then switch engine off.

Remedy:

1 - Check oil level in the oil bath air cleaner. If the oil level is too high, the crankcase exhaust fumes collect and promote oil throwing from the breather. The oil must, therefore, be only up to the mark on the oil bath air cleaner.

2 - Measure the compression pressure. If the pressure in one cylinder is considerably lower than that in the others, it is possible that a piston ring is broken. This causes an increase in pressure in the crankcase and promotes oil throwing.

3 - Make an oil deflector from 1 mm sheet metal as shown in the drawing and install it. In addition, a second gasket, Part No. 113 101 219, must be used for the generator support. The oil deflector is installed between crankcase and generator support so that the slots of the tube are opposite the filler opening.

For Type 14/1300 with 105 mm diameter generator and modified generator support, make an oil deflector with dimension a = 1.3 in. (33 mm) and dimension b = 1.54 in. (39 mm) — all other dimensions are unchanged.

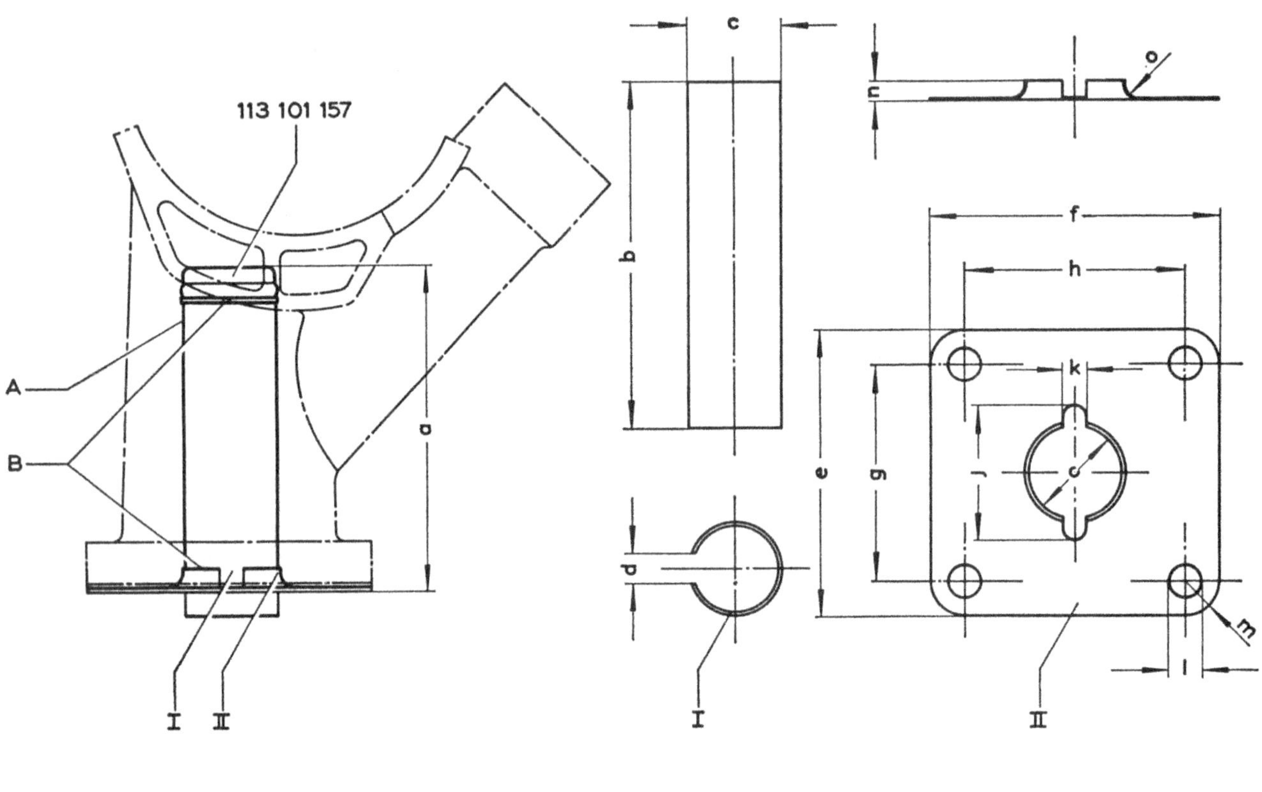

A - The slot of the tube must be opposite the filler opening.

B - Tack sheet metal flange and camshaft end plug, Part No. 113 101 157, to the tube at two locations

a = 3.39 in. (86 mm) or 1.3 in. (33 mm)	h = 2.32 in. (59 mm)
b = 3.62 in. (92 mm) or 1.54 in. (39 mm)	j = 1.42 in. (36 mm)
c = .96 in. (24.5 mm)	k = .24 in. (6 mm)
d = .31 in. (8 mm)	l = .33 in. (8.5 mm)
e = 3 in. (76 mm)	m = .35 in. (9 mm)
f = 3.03 in. (77 mm)	n = .2 in. (5 mm)
g = 2.3 in. (58 mm)	o = .16 in. (4 mm)

Note:

In countries with very low outside temperatures, isolated cases of icing up of the crankcase ventilation can occur if an oil deflector is installed on vehicles which are used mainly for only short distances.

4 - Enlarge hole in generator bracket flange and cut a halfmoon shaped hole in the rib underneath the flange (see page M-9/9).

B - **Increased oil consumption via rocker arms**

In some cases it was noted that oil was drawn through the inlet valve guides after the introduction of the modified rocker arm lubrication. This can be eliminated as follows:

1 - Remove rocker shaft and clean shaft and rocker arms.

2 - Close the outer oil hole on all rocker arms with an electric welding spot (A). The minus electrode should be connected to the rocker arm each time to prevent welding occurring between rocker arm and rocker shaft. Gas welding is not suitable.

3 - Grind spot weld lightly (B).

4 - Check that all valve stems are fitted with oil deflector rings (113 109 619) and that the rings fit tightly.

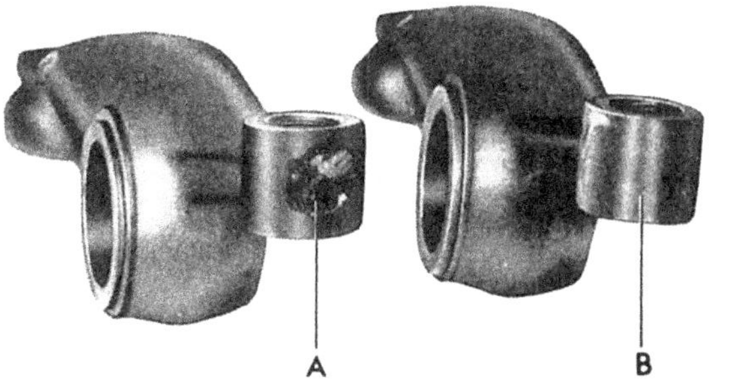

A = Hole welded B = After grinding

C - **Piston ring contact pattern**

If the oil consumption cannot be brought within limits by the measures given, the engine should be dismantled and the piston rings checked.

The two piston rings must contact on the lower edge first. The smooth surfaces show which edge of the ring has contacted. If it is seen that the upper edge has contacted first, even though the "Top" marking is upward, the ring should be replaced.

D - **Oil control rings with spiral expander springs**

Oil control rings with spiral expander springs are available for installation in engines which are using too much oil. Further details are given on page M-7/4.

Note:

The 34 bhp (41.5 SAE bhp) engines are normally fitted with two 0.6 mm thick washers (111 105 235 A) under the distributor drive shaft. On some exchange engines, the drive shaft bore in the crankcase is machined out and the **lower** washer replaced by a 3 mm thick washer (113 105 235 B). When repairing these engines, take care to ensure that the thick washer is always underneath the thin washer.

Note:

Type	From Chassis No.	From Engine No.
1/1200	116628529	D 0060316
1/1300	116625936	F 0588337

Since March 1966 all engines have been fitted with a modified distributor drive shaft, Part No. new 113 105 231 B in production. With this drive shaft, two shims each .024 in. (0.6 mm) thick, Part No. 111 105 235 A, are installed between the thrust shoulder and the bore in the crankcase.

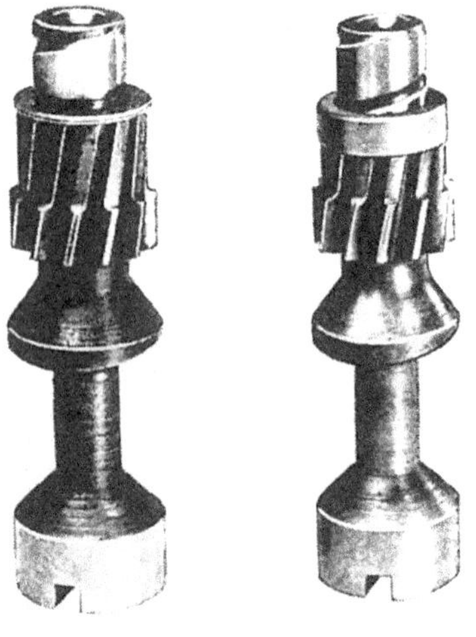

new: 113 105 231 B   previous: 113 105 231

The distributor drive shaft of new design, together with the two shims, can be service installed in all previously manufactured 1.3 and 1.2 liter engines from Engine No. 5497750. The distributor drive shaft, Part No. 113 105 231, for the crankcase of the 1.2 liter engines with higher thrust surface, from Engine No. 5 000 000 to 5 497 750 will still be supplied.

Note:

From Chassis No. 3616527 (Engine No. 5497750), the thrust surface for the distributor drive shaft in the left crankcase half is machined 5 mm (0.196″) deeper. At the same time the distributor drive shaft has been modified and now has a thrust shoulder with a flat machined on one side (see arrow) in place of the reduced diameter toothing. The thrust washer (Part No. 111105235) is still used.

The modified distributor drive shaft can only be removed and installed when No. 1 cylinder is at firing point as the flat on the thrust shoulder is then towards the distributor drive gear. In this position the offset slot in the upper part of the shaft is at right angles to the crankcase joint. Any attempt to remove or install the shaft in another position will damage the distributor drive gear.

Installation Instructions

1 - Modified crankcase and former distributor drive shaft:
Only the modified crankcase will be supplied as spare parts. When installing the former distributor drive shaft in this crankcase, a modified thrust washer (Part No. 113105235A) must be inserted at the same time. This thrust washer (6.3mm/0.248″ thick) will be included with every crankcase for approximately 2 months. The former thrust washer (Part No. 111105235) must not be installed.

2 - Former crankcase and modified distributor drive shaft:

The modified distributor drive shaft cannot be installed in the former crankcase.

Removing and Installing Flywheel (Early)

General Description

The flywheel is attached to the crankshaft by means of a gland nut and dowel-located by 4 dowels. A paper gasket is fitted between flywheel and crankshaft. An oil seal is recessed in the crankcase casting at main bearing No. 1. The oil seal lip fits over the flywheel flange. A needle bearing, which supports the main drive shaft, is situated in the gland nut.

Removal

1 - Remove clutch pressure plate.

2 - Remove clutch driven plate.

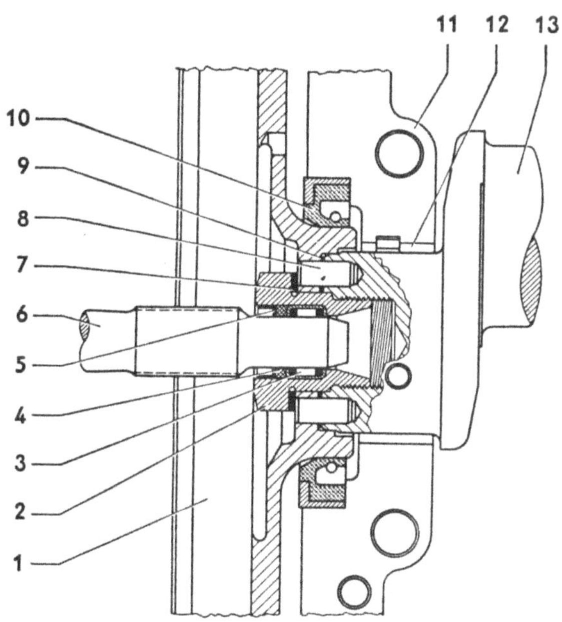

1 - Flywheel
2 - Gland nut
3 - Needle bearing
4 - Gasket
5 - End ring
6 - Main drive shaft
7 - Lock washer
8 - Dowel pin
9 - Gasket
10 - Oil seal
11 - Crankcase
12 - Crankshaft bearing
13 - Crankshaft

3 - Loosen gland nut, using 36 mm special wrench VW 112 and flywheel retainer VW 215b. Remove guide plate of the special wrench.

4 - Remove gland nut.

5 - Withdraw flywheel.

Installation

Installing the flywheel is a reversal of the above, but the following points should be observed:

1 - Check flywheel teeth for wear and damage. Lightly damaged gear rings may be remachined, removing up to 2 mm (0.08″) metal at the clutch side of the gear ring. Rechamfer the teeth to assure proper engagement with the starting motor pinion.

2 - Check dowel holes in flywheel. If they are worn, place drill jig VW 231 c/d on flywheel, drill new holes of 7.8 mm dia. (0.307″) 45° offset from the original holes, and ream them to 8 mm (0.315″). Plug up one of the old holes to avoid confusion.

3 - Inspect dowel holes in crankshaft for wear. If they are worn, remove crankshaft, drill new holes of 7.8 mm dia. (0.307″) with drill jig VW 231 c/d 45° offset from the original holes, and ream them to 8 mm (0.315″).

4 - Renew dowels if necessary.

5 - Adjust crankshaft end play.

6 - Check needle bearing in gland nut for wear.

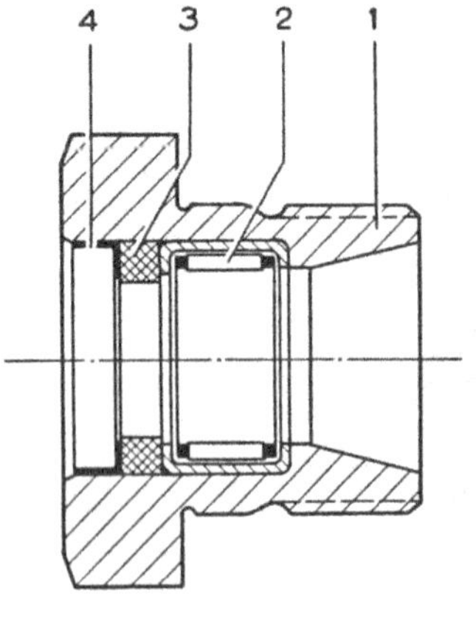

1 - Gland nut
2 - Needle bearing
3 - Oil seal
4 - End ring

The needle bearing is lubricated with about 10 g (0.35 oz.) Universal Grease. Make sure that the needle cage is adequately greased.

7 - Renew flywheel gasket.

8 - In order to counteract the existing permissible unbalance of crankshaft, flywheel, and clutch, the heaviest points of these parts are marked. On assembly, ensure that these marks are 120° offset. If only two of the three parts are marked, the marks should be 180° offset.

Parts	Mark
Crankshaft	Paint, dot in crankshaft tapped hole which accepts flywheel gland nut.
Flywheel	Paint dot and 5 mm dia. (0.2″) hole on the face which is towards the clutch.
Clutch	Paint line at the outer edge of the clutch pressure plate.

9 - Tighten gland nut to 30 mkg (217 ft. lbs.).

10 - Check flywheel for true running:
Lateral run-out max. 0.3 mm (0.012″).

Removing and Installing Flywheel
(Late)

General

The flywheel is attached by a gland nut to the crankshaft and secured by four dowel pins. A metal gasket is fitted between flywheel and crankshaft. An oil seal is fitted into a recess in the crankcase at No. 1 main bearing. The oil seal lip bears on the flywheel shoulder. The main drive shaft pilot is supported by a needle bearing in the gland nut.

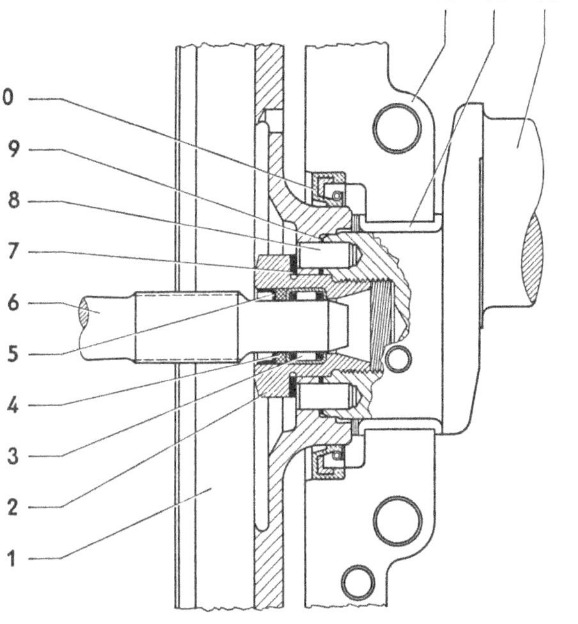

1 - Flywheel
2 - Gland nut
3 - Needle bearing
4 - Felt ring
5 - Retaining ring
6 - Main drive shaft
7 - Lock washer
8 - Dowel pin
9 - Gasket
10 - Oil seal
11 - Crankcase
12 - Crankshaft bearing
13 - Crankshaft

Removal

1 - Remove clutch and clutch plate.

2 - Loosen gland nut with special wrench VW 112a and flywheel retainer VW 215b. Remove guide plate for the special wrench.

3 - Remove gland nut.

4 - Take off flywheel.

Installation

Note:

The following seals are to be used between crankshaft and flywheel:

	Paper gasket 113105279	Metal gasket 113105279 A
Contacting surface for crankshaft in flywheel	straight	0.05—0.06 mm conical
Engines with 180 mm dia. clutch, flywheel 113105271 A	*	
Engines with 180 mm dia. clutch, flywheel 113105271 B Designation: "B" should be stamped on back of flywheel		*
Type 1 with Saxomat up to Chassis No. 4813430 up to Engine No. 6827458	*	
Type 1 with Saxomat from Chassis No. 4813431 from Engine No. 6827459		*
Exchange flywheels Part No. 111105271 x 113105271 B x Designation: "Austausch" stamped on	*	
Exchange flywheels Part No. 113105271 Bx Designation: "Austausch" stamped on, turned-out ring 100 mm dia.		*

If there is any doubt, measure the flywheel surface with a dial gauge.

Note:

Type	from Chassis No.	from Engine No.
1/1200 *	117 000 001	D 0 095 050
1/1300	117 000 002	F 0 940 717
1/1500	117 000 003	H 0 204 001

* Vehicles with Saxomat are excluded from this modification.

From August 1966 the flywheels of the vehicle engines have a different number of teeth and a new starter. The flywheel now has 130 teeth instead of 109 and its outside diameter has been increased by 4 mm to 276 mm.

The transmission housing too, has been changed due to the increased flywheel diameter. The space for the clutch in the housing has been enlarged by machining. In addition, the diameter of the hole for the starter pinion shaft has been reduced from 12.48 mm to 10.98 mm.

> Before replacing an engine always carefully check that the number of teeth of the installed starter corresponds to the number of teeth of the flywheel. If this is not done, damage can occur when starting the engine.
>
> In addition, ensure that the voltage of the engine corresponds with that of the vehicle (6 or 12 volt).

The service installation of a new flywheel is only possible if, at the same time, starter and transmission of new design are used. Previous type flywheels will still be supplied.

Note the following points:

1 - Check flywheel teeth for wear and damage.
A maximum of .08 in. (2 mm) can be removed from a damaged gear ring on the clutch side.
After removing the burr, cut a .08 in. (2 mm) chamfer on the teeth at 45°

2 - Check dowel pin holes in crankshaft for wear.

3 - Renew dowel pins if necessary.

4 - Adjust crankshaft end play.

5 - Check needle bearing for wear.

Note:

After cleaning a used gland nut or after dewaxing a new one, the sealing ring must be lubricated with engine oil and the needle bearing with a small amount of lithium grease. Excess lubricant must be wiped off.

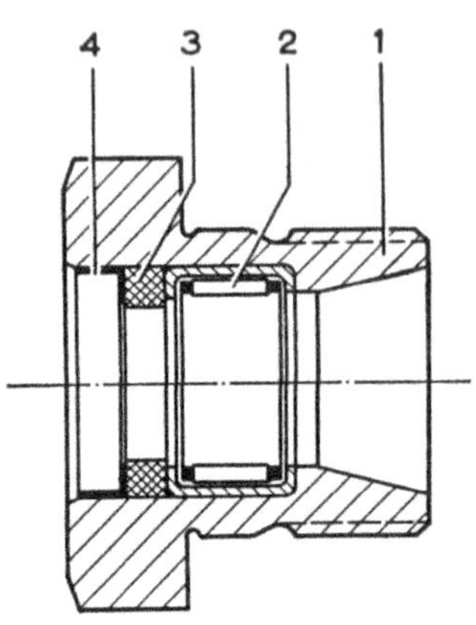

1 - Gland nut
2 - Needle bearing
3 - Oil seal
4 - End ring

6 - Insert metal gasket for flywheel.

7 - Observe the unbalance marks. Taking the mark on the crankshaft into consideration, install the flywheel and clutch so that the marks are approximately 120° offset. If only two parts are marked they should be installed with an offset of 180°.

Parts	Mark
Crankshaft	Paint dot **on side** of hole for gland nut.
Flywheel	Paint dot and (.2 in.) 5 mm dia. hole on the face toward the clutch.
Clutch	Paint line on the outer edge of the clutch pressure plate.

Note:

Type	from Chassis No.	from Engine No.
1/1200	116 979 941	D 0 084 620
1/1300	116 861 446	F 0 774 757

Since May 1966 all engines have been fitted in production with a modified gland nut, Part No. new 111 105 305 E.

This gland nut has been increased in length by 1.5 threads (approximately .08 in./2 mm) to secure the flywheel better. The modified gland nut can be installed in all engines except the 25 and 30 bhp (30 and 36 SAE bhp).

Note:

25 and 30 bhp (30 and 36 SAE bhp) only

When installing crankshaft — 111 105 101 — and flywheel — 111 105 271 — the previous gland nut, Part No. 111 105 305 C must still be installed.

8 - Tighten gland nut to 217 lb. ft. (30 mkg).

9 - Check flywheel for run-out.

Note:

Type	from Chassis No.	from Engine No.
1/1200 *	116 741 602	D 0 076 332
1/1300	116 796 901	F 0 741 385

Since April 1966 the metal seal, Part No. 113 105 279 A, between crankshaft and flywheel has been discontinued. Instead, rubber seal, Part No. 311 105 295 A will be installed.

In addition to the aforementioned change, also the crankshaft, the flywheel and the shims for the end play adjustment have been modified.

* All vehicles with Saxomat are excluded from this modification. Engines with Saxomat will be still provided with a metal seal between crankshaft and flywheel.

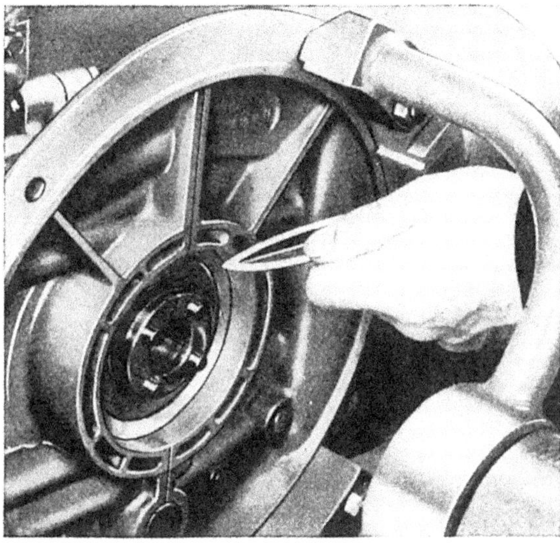

Removal

1 - Remove flywheel. Inspect oil seal lip contact surface on flywheel flange.

2 - Remove old oil seal.

Installation

1 - Clean oil seal recess in crankcase and coat it with a thin film of sealing compound. If necessary, chamfer the outer edge slightly with a scraper. Clean metal chips from the recess.

2 - Install new oil seal using tool VW 204 b. Screw the tool into crankshaft and insert oil seal by tightening the guide piece. The oil seal must bed squarely on the bottom of its recess.

3 - Remove the tool.

4 - Reinstall flywheel. The oil seal lip contact surface is to be lubricated oil.

Note:

From Chassis No. 4 244 394 (Engine No. 6 213 300), flywheels with a modified sealing surface for the crankshaft flange were installed in all engines. The sealing surface in the area of the four dowel pin holes now has a slight inward conical shape. In conjunction with this modification the paper gasket has been replaced by a sheet metal gasket.

The modified parts were installed intermittently in engines between numbers: **6 119 357** and **6 159 289.**

Flywheels with conical sealing surfaces (Part No. 113 105 271 B) must only be installed with the metal gasket (113 105 279 A). The gasket should be installed so that the burr on the outer edge caused during manufacture is towards the crankshaft. The paper gasket (Part No. 113 105 279) is still used for sealing flywheels with level surfaces (Part No. 113 105 271 A). The modified flywheels are marked with a green paint spot or a letter "B".

Note:

In certain cases, the crankshaft oil seal outer diameter expands to a lesser degree than the respective crankcase bore when the engine is at operating temperature. With oil seals in the lower tolerance range, this may, in the course of time, lead to leaks.

Oil losses at the flywheel side, resulting from an leaky oil seal, are quite frequently believed to result from leaks at the crankcase joint and, as a result, unnecessary and rather expensive repairs are carried out to eliminate the trouble. It is, therefore advisable always to check the crankshaft oil seal first whenever oil losses are noticed at the flywheel side. If necessary, fit a new oil seal.

Note:

From September 1964, Chassis No. 115 125 097 (Engine No. 8 908 165) the material and the shape of the lip on the crankshaft oil seal were changed. The new seal can be recognized by the number moulded into the edge — 113 105 245 **C, G or F.** The seal is supplied by three different manufacturers. Only the 113 105 245 **F** seal is supplied as a replacement part.

Oil seals of the previous pattern (113 105 245 B) should be used up.

Note:

In many cases, the problem of deciding whether the oil seals for crankshaft (113 105 245 F) or drive shaft (111 307 113 B) should be replaced due to leakage or not, appears to be causing difficulty.

1 - The fitting of a new oil seal is **justified,**

a - if oil is getting into the clutch and making it slip

b - if so much oil has leaked out that the bottom of the clutch housing is completely or partially covered.

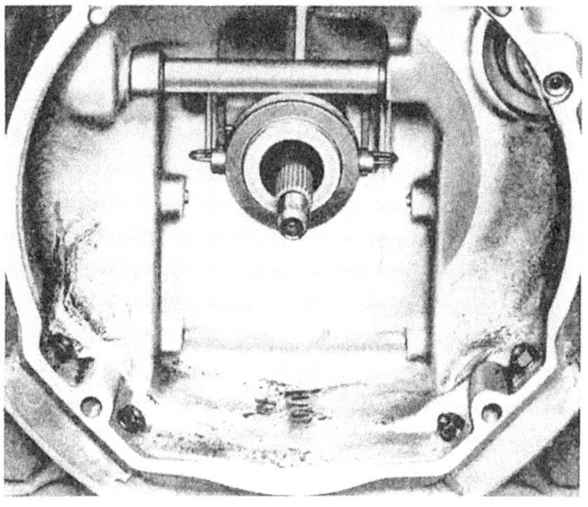

2 - The fitting of a new seal is **not justified,** if slight amounts of oil are found on the outer edge of seals or if there are small splashes of oil or grease on the walls of the clutch housing.

A good oil seal must leak a small quantity of oil to lubricate the sealing lips and prevent them from burning. The thin smear of oil which coats the walls of the housing together with dust from the clutch linings, does not indicate that the oil seal is defective.

Removing and Installing Oil Filler and Breather

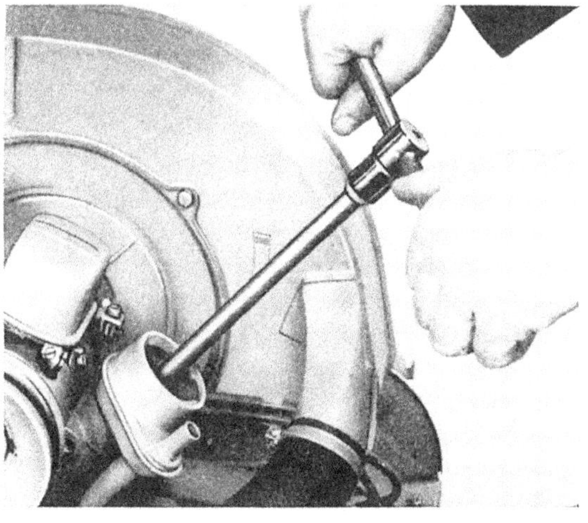

Removal

1 - Pull off connection hose.

2 - Remove threaded ring with VW 170.

3 - Take off oil filler and water drain pipe.

Installation

1 - Put the rubber cap on the water drain pipe.

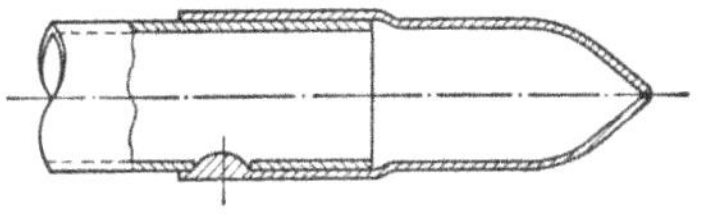

2 - Slide the rubber valve onto the drain pipe until the knob engages in the hole in the pipe.

3 - Do not forget the gasket between generator support and oil filler.

Disassembling and Assembling Crankcase

Disassembly

1 - Remove oil pressure switch with VW 159 a.

2 - Remove oil filler.

3 - Remove crankcase nuts.

4 - Remove oil strainer.

5 - Use a rubber hammer to remove the right-hand crankcase half. Do not insert tools, such as screwdrivers etc., between the joining faces.

6 - Remove crankshaft oil seal.

7 - Remove camshaft end plug.

8 - Lift out camshaft and crankshaft.

9 - Remove cam followers.

10 - Remove No. 2 crankshaft bearing shells.

11 - Remove oil pressure relief valve.

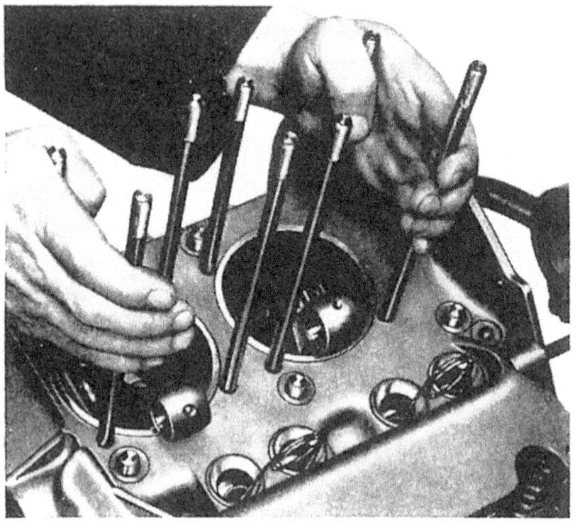

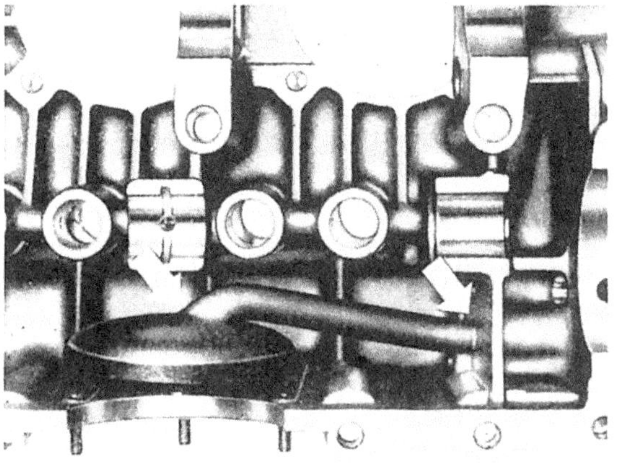

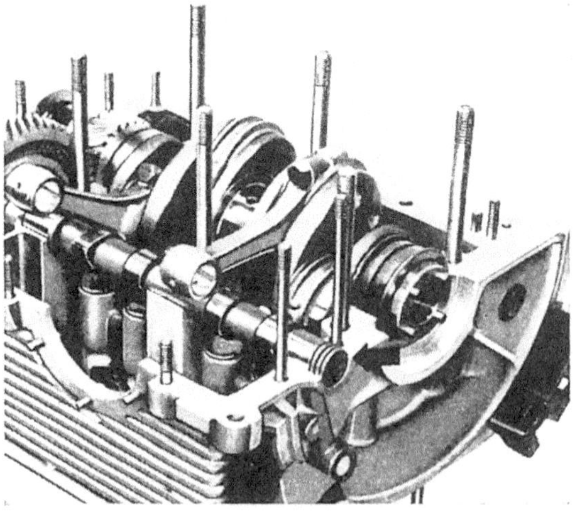

Assembly

Note the following points:

1 - Check crankcase for exterior damage and cracks.

2 - Use a solvent to remove all traces of the old sealing compound from the joining faces.

3 - Make sure that the joining faces are perfectly even and clean.

4 - Join the crankcase halves and tighten the nuts to the prescribed torque. Measure the crankcase bores for the crankshaft bearings with a special dial gauge and micrometer.

5 - If necessary, slightly chamfer the edges of the bearing bores.

6 - Flush out the oil passages and blow them out with compressed air.

7 - Check oil suction pipe for tightness and leaks; if necessary, secure it with a peening tool.

8 - Check studs for tightness. If the tapped holes are worn, Heli-Coil inserts can be installed.

9 - Check cam followers and their bores in the crankcase.

10 - Insert cam followers.

11 - Insert crankshaft bearing dowel pins.

12 - Install crankshaft and camshaft. Note position of timing marks on timing gears.

13 - Install camshaft end plug, using sealing compound.

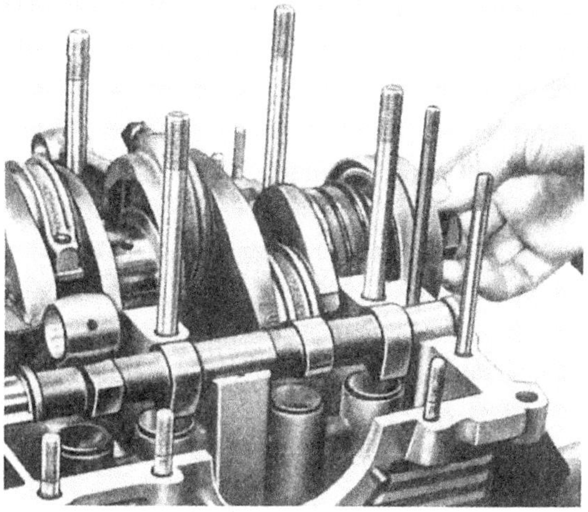

14 - Install thrust washers and crankshaft oil seal with the installing tool VW 204 b. The oil seal must bed squarely in the crankcase recess.

15 - Hold the cam followers of the right-hand crankcase half in position by using the retaining springs VW 171.

16 - Spread an even film of sealing compound on the joining faces of the crankcase halves. On no account must the sealing compound enter the oil passages of crankshaft and camshaft bearings.

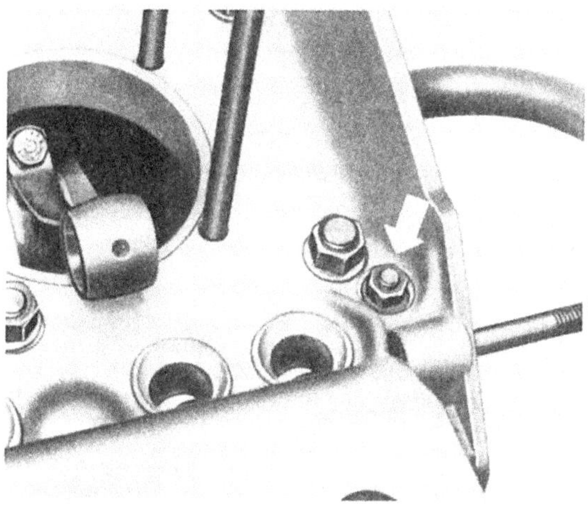

17 - Join the crankcase halves and tighten the nuts.

M 12 nuts = 24—26 lb. ft. (3.4—3.6 mkg).

M 8 nuts = 14 lb. ft. (2 mkg).

First tighten the M 8 nut near the M 12 stud of No. 1 crankshaft bearing. Only then must the M 12 nuts be tightened fully. This tightening sequence must be adhered to.

18 - Turn the crankshaft to check for ease of movement.

19 - Install oil pressure relief valve.

Note:

From August 1964, Chassis No. 115 013 672 (Engine No. 8 803 980), the oil return from generator bracket to crankcase was modified. The upper hole in the bracket flange was enlarged and a half-moon shaped hole made in the rib underneath the flange.

Crankcases in which the holes have not been enlarged in this manner can be reworked subsequently. This measure should be carried out on engines which are throwing oil out of the filler.

Note:

In the case of seizure of parts such as main bearings and connecting rod bearing shells, when metal particles or metal dust is discovered in larger quantities in the crankcase, the crankcase must be especially carefully cleaned and the oil cooler must be replaced to avoid consequential damage.

a = 27 mm (1.06 in.), b = 18 mm (.708 in.)

c = 54 mm (2.125 in.), d = 25 mm (1 in.)

Note:

From December 1964, Chassis No. 115 344 045 (Engine No. 9 141 876) the cam follower (Part No. 113 109 309 C as before) was modified. The diameter of the flange was reduced from 29.5 mm to 28.5 mm and the thickness of the flange increased by 1 mm. New and old cam followers can be installed together in one engine.

Cam followers of the previous type should be used up.

Note:

Type	from Chassis No.	from Engine No.
1/1200	117 204 283	D 0 109 385
1/1300	117 197 986	F 0 991 728
1/1500	117 198 502	H 0 398 526

From September 1966, all six crankcase studs M 12×1.5 are sealed with rubber seals — A —, Part No. 113 101 125. The seals are between the crankcase halves. For this reason, the stud holes in both crankcase halves have been countersunk.

In connection with the above the two sealing nuts, Part No. 113 101 131 A, for the center crankcase studs near bearing No. 2 are being replaced with the formerly used hexagon nuts **with** washers. The sealing nuts for sealing the previous type crankcases will still be supplied.

The crankcase part numbers remain unchanged. A reworking of the previous type crankcases is not being contemplated.

Assembly note:

Prior to assembly of the crankcase halves, slide the seals over the studs until they bear against the crankcase.

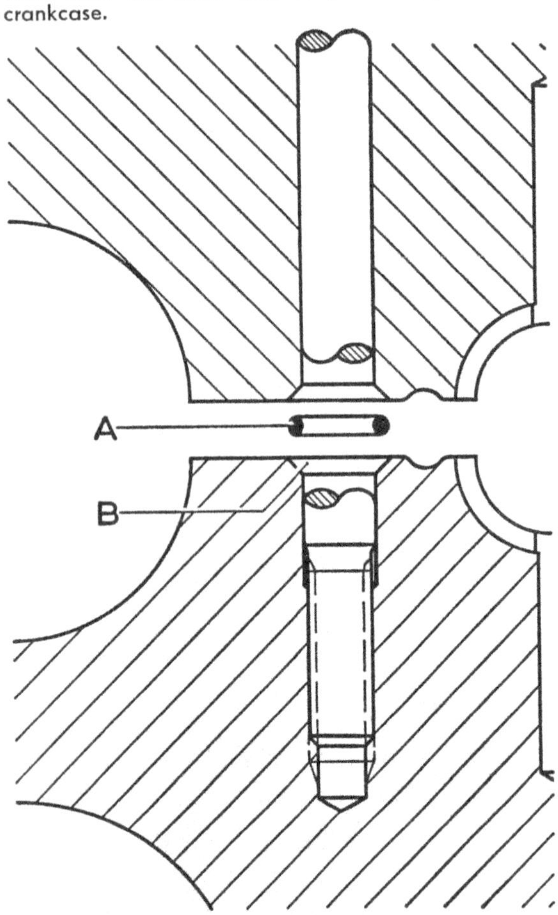

A - Rubber seal B - Countersink

Crankcase for exchange engines and exchange crankcase

These crankcases vary from the standard ones in that they have cylindrical countersinks in the **right** crankcase halves only. The countersinks are so deep, however, that the whole seal fits in them. The seal seals the stud hole by bearing with its outer circumference against the wall and on the inside it rests against the stud.

Note:

Type	from Chassis No.	from Engine No.
1/1200	117 707 100	D 0 182 983
1/1300	117 710 493	F 1 124 726
1/1500 6 v	117 571 250	H 0 646 887
1/1500 12 v	117 377 334	H 0 530 915

From December 1966, the crankcase is sealed with a **light-brown** compound (D 3) instead of the former light-gray compound (D 2). The new compound is supplied in 175 gram tins with a screw cap. A special pump available under Part No. 111 012 361 can be fitted on to the tin and used to apply the sealer.

The **light-gray** compound should no longer be used. The D 1 A compound can still be used to seal transmission cases and steering boxes.

Note:

Type	from Chassis No.	from Engine No.
1/1200	117 018 982	D 0 101 334
1/1300	117 071 626	F 0 966 806
1/1500	117 070 165	H 0 230 323

From August 1966, the center M 12×1.5 crankcase studs near No. 2 bearing are sealed in production with sealing nuts, Part No. 113 101 131 A. Washers are **not** installed. The plastic ring pressed into the sealing nut faces the crankcase.

It is possible to service install these new sealing nuts on older vehicles, but washers are **not** to be installed. If sealing nuts are reused, ensure that the pressed-in plastic ring is in good condition.

These sealing nuts should be torqued to **18 lb. ft. (2.5 mkg)** instead of the previous 25 lb. ft. (3.5 mkg).

Note:

When unscrewing the sealing nuts the sealing ring is sometimes detached from the nut and adheres to the threads of the stud. Such rings can be removed with a locally manufactured sealing ring remover made from piece of 15×1.5 mm tube. The points of the sealing ring remover are hooked into the sealing ring and the remover turned anti-clockwise to remove the ring from the stud.

The sealing ring must not be levered off with a screwdriver or similar tool as this will damage the crankcase.

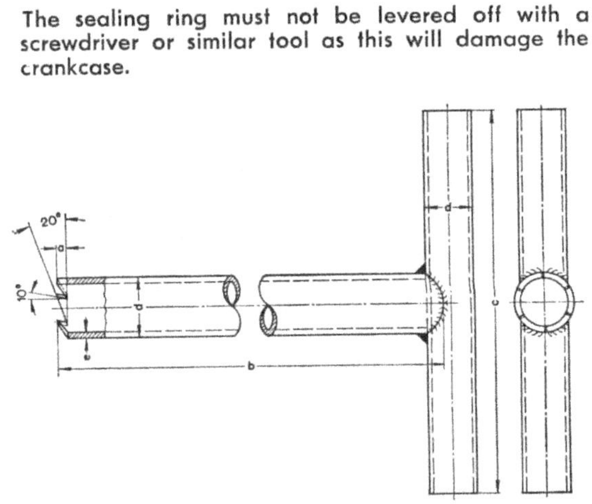

a = .1 in. (2.5 mm)
b = 10.24 in. (260 mm)
c = 3.94 in. (100 mm)
d = .59 in. (15 mm) diameter
e = .06 in. (1.5 mm)

Checking Crankshaft End Play

Crankshaft end play should be within 0.065 mm and 0.125 mm (0.0028 and 0.0047″), the wear limit being 0.15 mm (0.006″). The end play can conveniently be checked with the engine installed or removed. A bracket (local manufacture VW 659) has been designed to hold a dial indicator.

Engine Installed in Vehicle

The end play reading is taken at the pulley. The dial indicator bracket is mounted on the rearmost crankcase stud. An end play reading is obtained by rocking the crankshaft back and forth at the pulley hub.

Engine Assembled

The end play reading is taken at the flywheel. The dial indicator is held by one of the engine mounting bolts.

Adjusting End Play

1 - Force the installed crankshaft towards the flywheel side of the engine (with flywheel removed) so that it contacts the inner thrust surface of No. 1 bearing.

2 - Insert dial gauge VW 292 in flywheel seat so that it contacts the crankshaft and measure distance from crankshaft face to outer face of main bearing No. 1

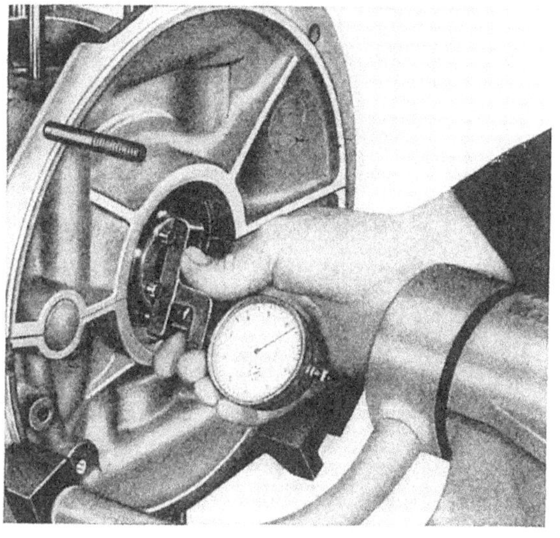

3 - Place gauge VW 292 on flywheel flange and measure depth of crankshaft seat.

4 - The thickness of the shims to be used is decided by the difference in both readings (taking into account the paper gasket). The thickness of the paper gasket is 0.2 mm (0.0078″) and is compressed by 0.05 mm (0.0019″) in assembling, leaving 0.15 mm (0.0059″) to be considered when deciding thickness of shim.

Shims of the following thicknesses are available:

0.24 mm
0.30 mm
0.32 mm
0.34 mm
0.36 mm
0.38 mm

The thickness is etched on each shim. If necessary, measure the thickness with a micrometer.

Three shims of the required thickness are to be installed in each case.

Never use more than one paper gasket.

Example:

Distance crankshaft face/main bearing No. 1	9.070 mm
Depth of crankshaft seat in flywheel	— 8.240 mm
	0.830 mm
Thickness of paper gasket installed	+ 0.150 mm
	0.980 mm
Shims to be used: three shims 0.30 mm each	— 0.900 mm
End play	0.080 mm

Steel Backed Crankshaft Bearings
(for Cold Climates)

From Chassis No. 3 192 507 (Engine No. 5 000 001) engines for countries with arctic climates are equipped with steel backed bearing shells for crankshaft bearings Nos. 1, 2 and 3 (M 173).

Disassembling and Assembling Crankcase

When installing these bearings, observe the following points:

1 - R/H crankcase half.

Insert the dowel pins (Part. No. 111 101 123) for bearings No. 2 and 3 into the crankcase and press the bearing shells in by hand.

M-9

2 - L/H crankcase half

a - Insert all dowel pins in the crankcase and press bearing shells for bearings No. 1, 2 and 3 in by hand. As seen in the picture, the shell for No. 1 bearing is installed at a slight angle to facilitate the centering of the crankcase halves.

b - Place crankshaft with No. 4 ring bearing installed, into position and locate No. 4 bearing on the dowel pin as usual.

c - Place second half of No. 1 bearing on the crankshaft.

> **Important**
> The oil pockets in the bearing shoulder must be towards the flywheel.

3 - Place the R/H crankcase half, with bearings No. 2 and 3 installed, into position.

Service Installation

These bearings can also be service installed from engine No. 3 520 333. A dowel pin hole for No. 3 bearing must be drilled in the R/H crankcase half (see drawing).

To mark the hole, a template made of an old No. 3 ring bearing in which the dowel pin hole has been drilled through, can be used. Cut the bearing in half so that the edges are level with the joining face of the crankcase and the dowel pin hole is exactly vertical.

Important
Place the finished template in the crankcase so that the offset dowel pin hole is towards No. 2 crankshaft bearing.

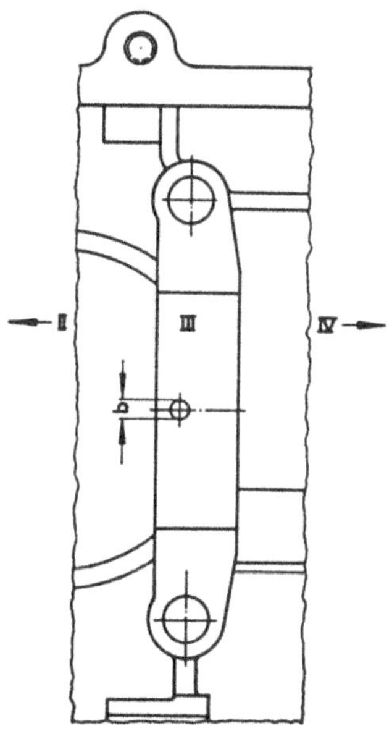

b = 5 + 0.075 mm (.197 + .003")

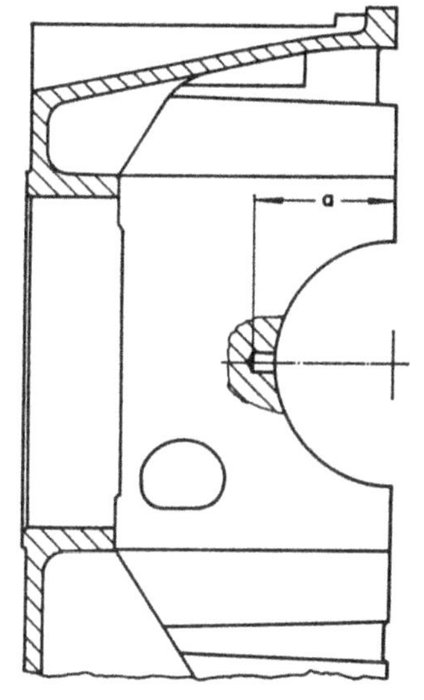

a = 37.9 mm (1.491")

Measuring Crankshaft Bearing Clearance

The crankshaft bearing shell clearance can be measured directly by the method described here. The micrometer and internal measuring gauges previously used are no longer required.

The clearance is measured by means of a plastic wire "Plastigage": A piece of this wire is placed across the bearing journal, the crankshaft is installed and the crankcase halves tightened to the prescribed torque. The nuts and bolts should then be loosened and a crankcase half removed. The flattened wire which is usually stuck to the bearing shell is now compared with the scale on the reverse of the paper container and the clearance ascertained.

Plastigage is manufactured by Messrs. Perfect Circle, Hagerstown, Indiana, U.S.A. It can be obtained in Germany from Messrs.

Motorenteile GmbH. Karlheinz Ern, Duesseldorf, Corneliusstrasse 65—67, Tel. No. 80404.

Note:

Crankcases with worn bearing bores can be re-used by installing crankshaft bearings with a larger outside diameter. The outside diameter of these bearings is 0.5 mm (.020") larger than that of the standard bearings. The crankshaft bearing bores must be enlarged 0.5 mm with the crankcase reconditioning tool. The procedure is described in detail in Technical Recommendation M 5 "Crankcase Reconditioning".

The No.1 crankshaft bearing (Part No. 113 105 503 A) with the larger outside diameter has a 1 mm thicker shoulder on the flywheel side. The bearing seat in the crankcase must be cut 1 mm (a) deeper (b = 20,993 — 0.025 mm/ .8264 — .001").

Crankcases on which the No. 1 bearing bores are only worn axially cannot be repaired by fitting a crankshaft bearing with a thicker shoulder and a normal outside diameter as was formerly the case with the 25 and 30 bhp engines. An oversize bearing must also be used in these instances. As the crankcase bores can only be all reamed out together, all the bearings must always be replaced by the oversize type.

The running surface for the oil return thread on the crankshaft pulley is also enlarged 0.5 mm during the repair so that the standard pulley must be replaced by an oversize pulley.

Engines in which oversize crankshaft bearings have been installed, should be marked with a "0" below the engine number. Exchange engines are also marked in the same manner.

Oversize crankshaft bearings are only available with the normal inside diameter.

Part Nos. and crankcase bore diameters:

	Part No.	Bore dia.	
No. 1 crankshaft bearing	113 105 503 A	D_1 65,500—65,519	(2.5786—2.5793")
No. 2 crankshaft bearing	113 105 533 D	D_2 65,500—65,519	(2.5786—2.5793")
No. 3 crankshaft bearing	113 105 563	D_3 65,500—65,519	(2.5786—2.5793")
No. 4 crankshaft bearing	113 105 593 A	D_4 50,500—65,519	(1.9881—1.9888")
Crankshaft pulley	113 105 253	D_4 50,500—65,519	(1.9881—1.9888")

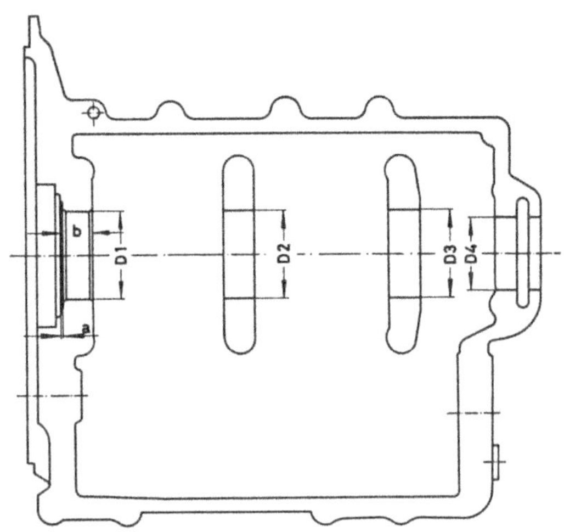

a = 1 mm b = 20.993—0.025 mm

Removing and Installing Camshaft

Removal

1 - Open the crankcase.

2 - Remove camshaft.

Installation

When installing, the following points should be observed:

1 - Examine riveted joint between camshaft timing gear and camshaft.

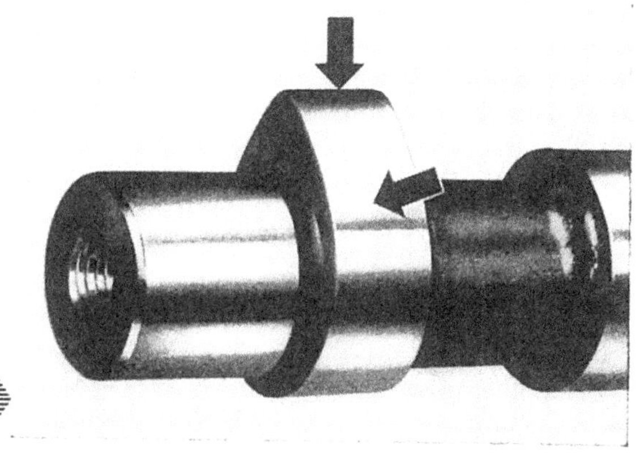

2 - Check camshaft bearing points and cam faces for wear. The cam faces must not be scored and must be perfectly smooth and square. See "List of Tolerances and Wear Limits" for permissible end play.

Check bearing journals, cams, and timing gear for damage. Slight damage may be smoothed down with an oilstone (silicon carbide) — a 100—120 grain stone should be used before polishing with a 280—320 grain stone.

3 - Check camshaft for run-out.

4 - Examine camshaft timing gear for wear and correct tooth contact.

5 - Apply engine oil to bearing journals and cams.

6 - When installing camshaft, take care that the timing gear tooth marked "0" is situated between the two teeth of the crankshaft timing gear marked by a center punch.

7 - Check backlash of camshaft timing gear 0.0—0.052 mm (0.0020″).

To ensure that the timing gears operate silently, the prescribed backlash must be adhered to. A check is carried out by rocking the gears back and forth with both hands while gradually revolving the camshaft timing gear until it has made a complete turn.

To assist in obtaining the specified backlash, the camshafts are available with timing gears in various sizes under different part numbers.

The timing gears are marked —1, 0, +1, +2 etched on their inner face. The digits indicate in $^1/_{100}$ mm how much the pitch radius departs from the standard pitch radius which is denoted by "0".

Important

The mark "0" on the outer face of each camshaft timing gear is to ensure correct installation of the camshaft in relation to the crankshaft timing gear and has nothing to do with the aforementioned size markings.

The crankshaft timing gear is obtainable in one size only.

Note:

From Chassis No. 3192507 (Engine No. 5000001), the thrust surface in the crankcase was widened

new:	38.00 mm
previously:	36.00 mm

From Chassis No. 3195309 (Engine No. 5016085), the camshaft thrust bearing was provided with an annular oil groove.

From Chassis No. 3243061 (Engine No. 5067818), the end play was increased by modifications to the camshaft.

new:	0.06—0.114 mm
previously:	0.03—0.084 mm

The wear limit is 0.14 mm (formerly 0.10 mm).

Note:

1 - From Chassis No. 4810758 (Engine No. 6864207), a camshaft with cams of a different shape was installed.

new Part No. 113109019 D — 035 D

Until a final identification mark is introduced, these camshafts will be marked with green and yellow paint spots between the cams.

2 - From Chassis No. 4868581 (Engine No. 6930129), cam followers made out of one piece of material were installed (**new** Part No. 113109309 C, version 4).

Prior to the above introduction dates, three different versions of cam follower were installed. The individual versions are illustrated in the following drawing.

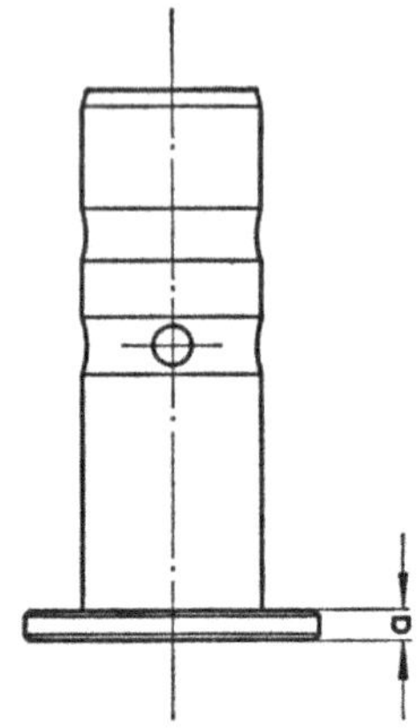

Version 1
113109309 A
a = 3.2 mm

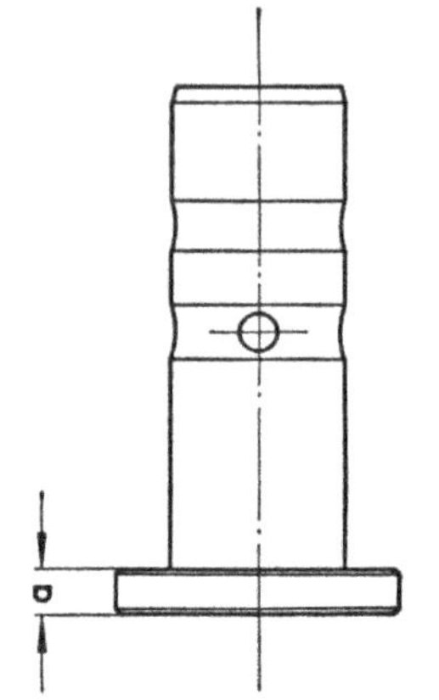

Version 2
113109309 B
a = 4.5 mm

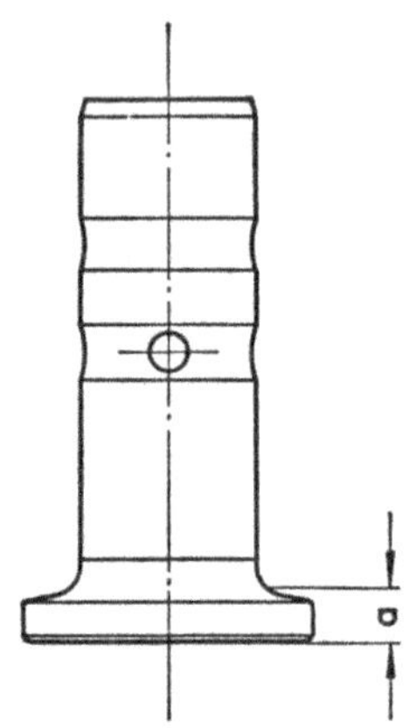

Version 3
113109309 B
a = 5.4 mm
radius between head and stem

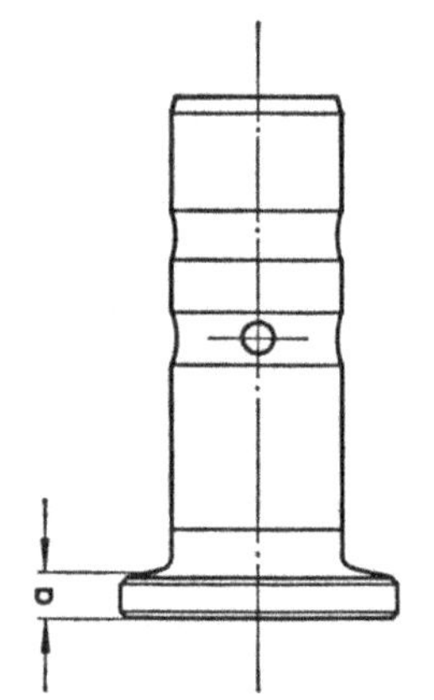

Version 4
113109309 C
a = 3.9 mm
one piece cam follower

3 - Service Installation

If the camshaft or the cam followers have to be renewed during a repair, the camshaft **and** cam followers should always be renewed. Before assembling the engine, the oil drillings in the crankcase and the oil cooler should be thoroughly cleaned.

Only the 113109019 D — 035 D camshafts and the 113109309 C cam followers (version 4) may be used. (See instructions in para. 4). At the same time, the progressively coiled valve springs (113109623 A) should be installed.

Engine marking

Engines in which the modified parts have been installed should be clearly marked with a white cross on the fan housing.

4 - Tools for shortening cam follower guides

Before the cam followers (113 109 309 C, version 4) are installed, measure the distance from the crankcase joining face to the upper edge of the cam follower guides. The measurement "c" should be 30.5 + 0.5 mm. If not, the cam follower guides must be shortened.

Cutter set

The cutter set can be obtained from Messrs. Matra-Werke GmbH., Frankfurt am Main, under Order No. W 70 at a price of DM 49.50. Orders should be sent direct to Messrs. Matra.

If this cutter set is not available, a locally manufactured cutting tool can be used in urgent cases.

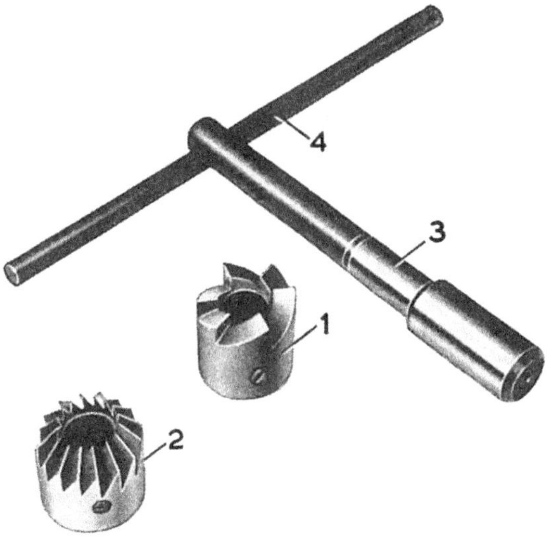

1 - Face milling cutter
2 - 45° chamfer cutter
3 - Pilot
4 - Handle

Locally manufactured cutting tool

A - T handled cutter spindle (1) with taper (2) and threaded portion (3).

1 - Spindle diameter: 16 mm (.63") minimum.

2 - Smallest diameter of taper: 15.4 mm (.61") (lower) taper: 1 : 50.

3 - Threaded portion: 10 mm (.394") dia., 70 mm (2.76") long, M 10 thread.

B - 15° cutter, Matra 32/19 from valve grinding set VW 311 b.

C - Pilot sleeve: Length 50 mm (1.97"), outside dia. 18.94—18.96 mm (.745—.746"), inside diameter 10.1 mm (.40").

D - M 10 nut. The width across corners of nut is to be ground down to 18.5 mm (.73").

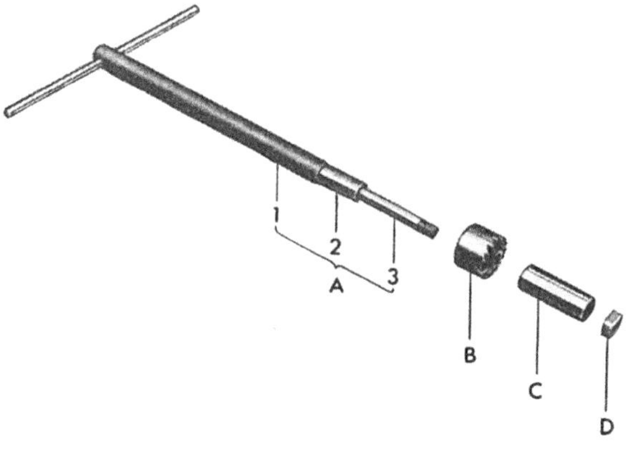

Removing and Installing Crankshaft and Connecting Rods

Removal

1 - Open the crankcase.

2 - Remove camshaft.

3 - Remove crankshaft.

> **Important**
>
> Removed crankshafts must not be stored without being greased or oiled to prevent corosion.

Installation

When installing, the following points should be observed:

1 - The crankshaft bore in the crankcase must not have sharp edges at the joining faces. Slightly chamfer the edges.

2 - Check dowel pins for tightness.

3 - The oil passages in the crankshaft must not have sharp edges. Should foreign matter be embedded in the main bearings, it may be removed with a sharp scraper. Care must be exercised not to remove metal from the bearing shell itself.

4 - Place one half of No. 2 crankshaft bearing in crankcase.

5 - Slide on crankshaft bearing No. 1 so that the dowel pin hole is toward the flywheel.

6 - Install crankshaft. Make sure that the dowel pins are correctly seated in the crankshaft bearings.

7 - Note the marks on the timing gears when installing camshaft.

Note:

From February 1963, Chassis No. 5 301 820 (Engine No. 7 484 424) the piston pin bearing in the connecting rod is offset 1 mm in relation to the connecting rod bearing. The piston pin bearing is thus located centrally in the cylinder bore. The offset connecting rod has a forged mark on the shank which must be upwards when the connecting rod is installed.

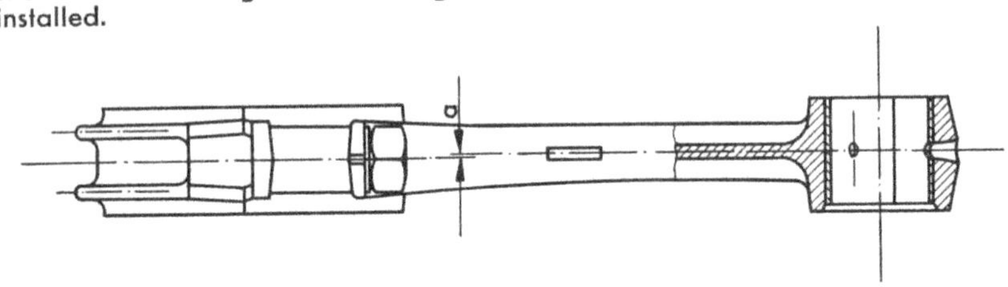

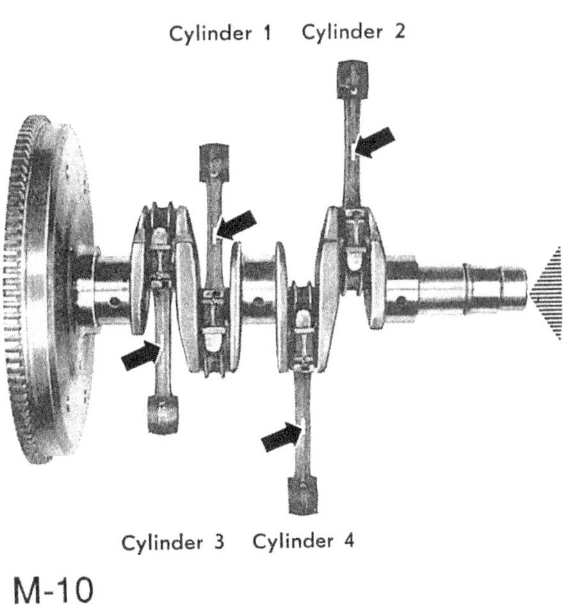

In the illustration, each connecting rod is pointing to its cylinder.

Part No.

Connecting rod

new	old
113 105 401 A	113 105 401

The previous type of connecting rod will be discontinued when stocks are exhausted.

Disassembly

1 - Attach the crankshaft to holding fixture VW 310 a.

2 - Remove woodruff key.

3 - Remove oil thrower.

4 - Remove bearing No. 4.

Note:

From Chassis No. 3 465 332 (Engine No. 5 326 955) a No. 4 crankshaft bearing with an annular oil groove and an oil drain groove was installed.

5 - Remove distributor drive gear retaining ring, using tool VW 161 a.

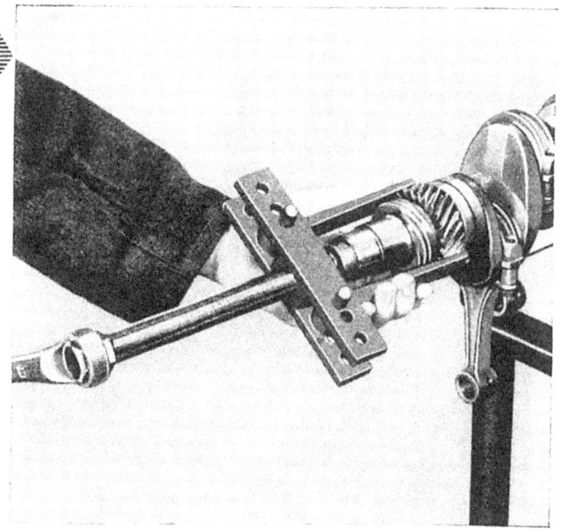

6 - Remove distributor drive gear, spacer, and crankshaft timing gear, using extractor VW 202 in conjunction with VW 202 a and VW 202 f or VW repair press.

Light signs of seizure can be removed, provided that the press fit will not be affected.

7 - Remove No. 3 crankshaft bearing.

8 - Remove connecting rods.

Important

Do not store removed crankshafts without oiling or greasing them to prevent rust.

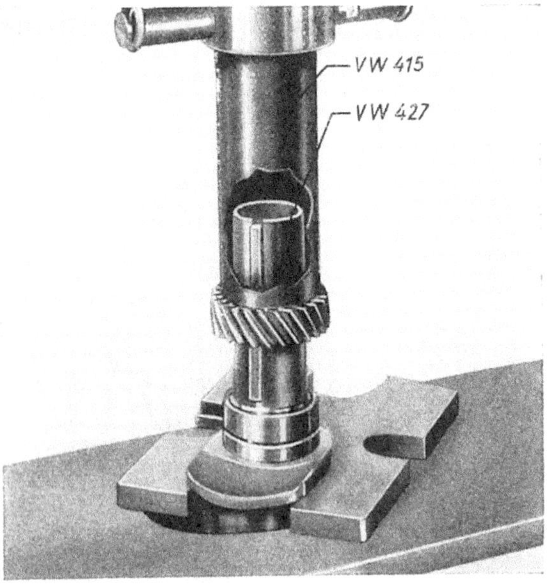

Assembly

When installing, the following points should be observed:

1 - Check crankshaft for run-out, cracks (ringing test), and wear. Should it become necessary, regrind or renew crankshaft and install new main bearings. To ensure correct assembly, first place main bearings Nos. 1, 3, and 4 in left-hand half of crankcase, noting proper position of dowel holes and oil holes which must register with the oil passages in the crankcase. The dowel hole in main bearing 1 must be towards the flywheel.

To facilitate fitting of main bearings on the dowels when installing crankshaft, it is recommended that the bearings are marked at the crankcase joining faces with a pencil.

2 - Check dowel holes in crankshaft for wear. If they are worn insert crankshaft drill jig VW 231 c/d and drill new holes 7.8 mm dia. (.307") 45° offset, and ream them to 8 mm (.315").

3 - Slide main bearing No. 3 into position and insert woodruff key for crankshaft timing gear distributor drive gear.

4 - Check crankshaft timing gear for wear and check tooth contact.

Heat the gear to about 80° C (176° F) in an oil bath and press it into position using guide tube VW 427. Slide spacer on crankshaft.

5 - Check distributor drive gear for wear. Heat the gear to about 80° C (176° F) and press it into position using guide tube VW 427.

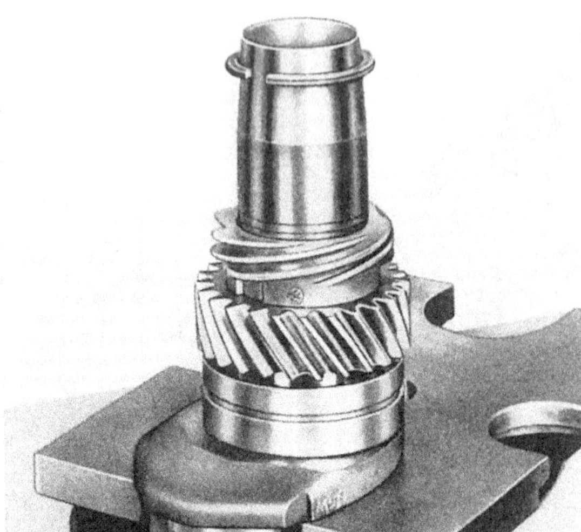

6 - Install retaining ring, using tapered guide tube VW 428 to avoid damage to the crankshaft journal. Check gears for secure seating after they have cooled down.

7 - Clean out oil passages, using compressed air.

8 - Slide main bearing No. 4 on to crankshaft.

9 - Install oil thrower so that the concave surface faces the crankshaft pulley.

10 - Insert woodruff key.

11 - Assemble connecting rods.

Removing and Installing Connecting Rods

Removal

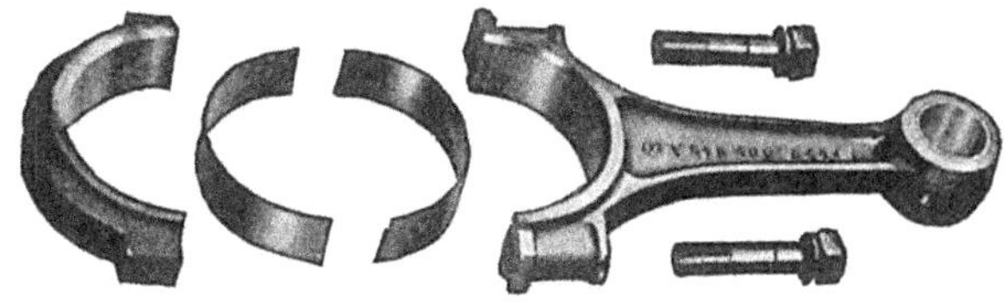

1 - Remove crankshaft and attach it to the Holding Fixture VW 310 a.

2 - Remove connecting rod clamping bolts and remove connecting rods and caps.

Installation

When installing, the following points should be observed:

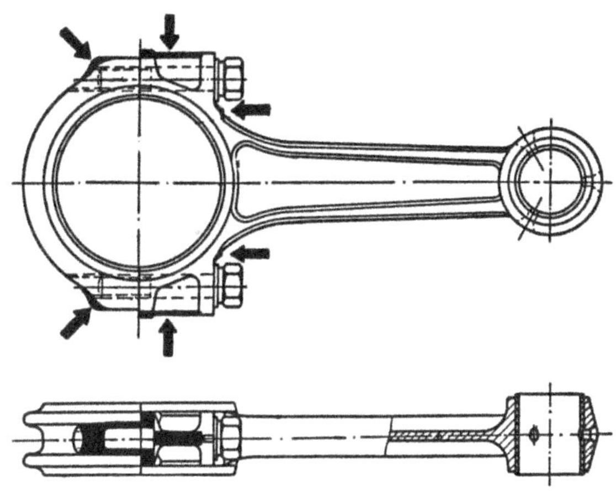

1 - Check weight of connecting rods. The difference in weight of the connecting rods in one engine must not be in excess of 10 grams to maintain proper engine balance. If necessary, metal should be removed from the heavier connecting rods at the points indicated on the right. Thus, a reduction of the weight by about 8 grams can be obtained.

2 - Inspect piston pin bush. With a new bush, the correct clearance is indicated by a light finger push fit of the pin at room temperature.

3 - Check and, if necessary, correct connecting rod alignment.

4 - Reinsert connecting rod bearing shells after all parts have been thoroughly cleaned and assemble connecting rods on crankshaft. The identification numbers stamped on connecting rods and bearing caps must both be on one side.

1961-1964

5 - Tighten connecting rod bolts to a torque of 4.5 ± 0.5 mkg (32 ± 3.6 ft. lbs.). A slight pretension between the bearing halves, which is likely to occur when tightening connecting rod bolts, can be eliminated by light hammer taps.

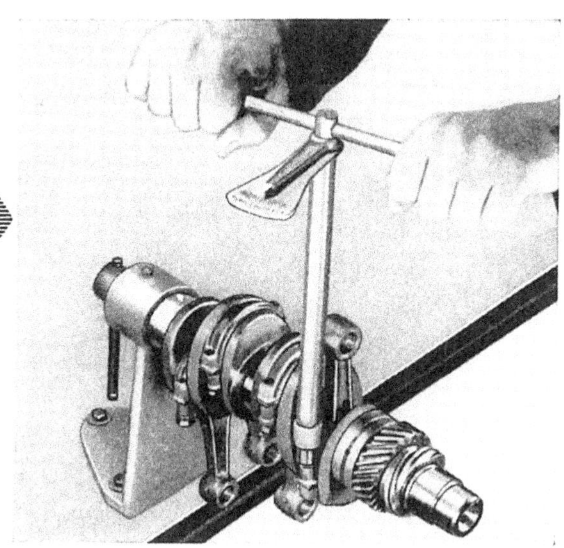

1965

Tighten connecting rod bolts to 3.0—3.5 mkg (21—25 lb. ft.) with bearing surfaces oiled.

> **Important**
>
> Always use new connecting rod screws. The new screws must be de-waxed before installation.

The connecting rods, lubricated with engine oil prior to assembly, must slide on the crank pin by their own weight. The connecting rod bushes must not be scraped, reamed or filed during assembly.

6 - Secure connecting rod bolts in place, using chisel VW 124.

Note:

From June 1964, Chassis No. 6454172 (Engine No. 8706021) the tightening torque for the connecting rod screws was altered:

new	old
4.5 ± 0.5 mkg	5 ± 0.5 mkg
(32 ± 3.6 ft. lbs.)	36 ± 3.6 ft. lbs.)

Important

Every time the connecting rods are removed, new screws must be fitted.

This also applies for connecting rod screws 113105425 until they are used up.

If the connecting rods are slightly bent or have worn bushes, they should be aligned and new bushes installed.

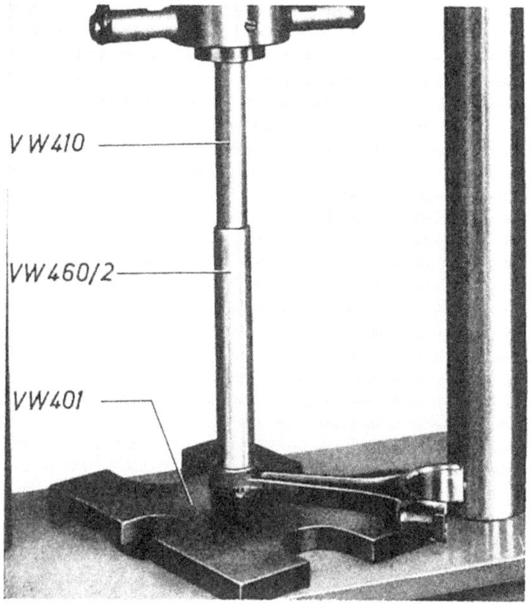

1 - Remove piston pin bush in the VW repair press in conjunction with VW 401, VW 460/2 and VW 410.

2 - Place connecting rod in the device VW 214b.

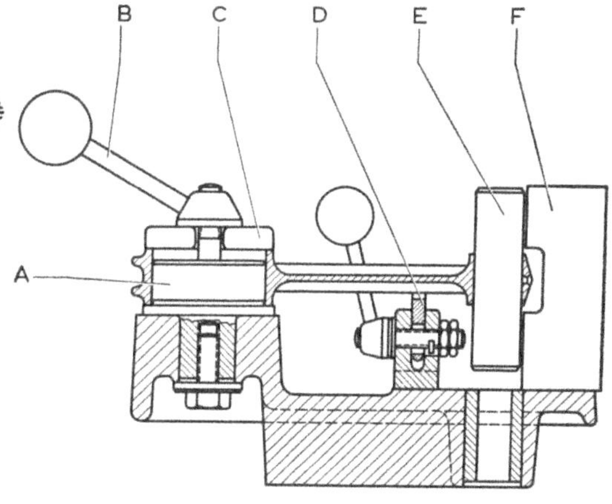

Device VW 214 b

A - Mandrel
B - Locking lever
C - Washer
D - Support
E - Pin
F - Gauge

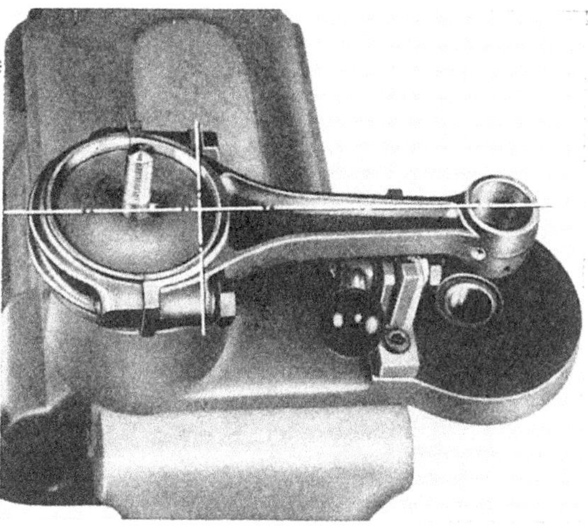

Turn the mandrel so that the flat is at right angles to the centre line of the connecting rod.

3 - After inserting the washer C, tighten the locking lever B until the connecting rod can just be moved in both directions. Support D is not tightened.

4 - Insert pin E into the connecting rod and push it towards the mandrel A with two fingers so as to ensure that there is no tilt between mandrel and big end of connecting rod nor between connecting rod eye and pin.

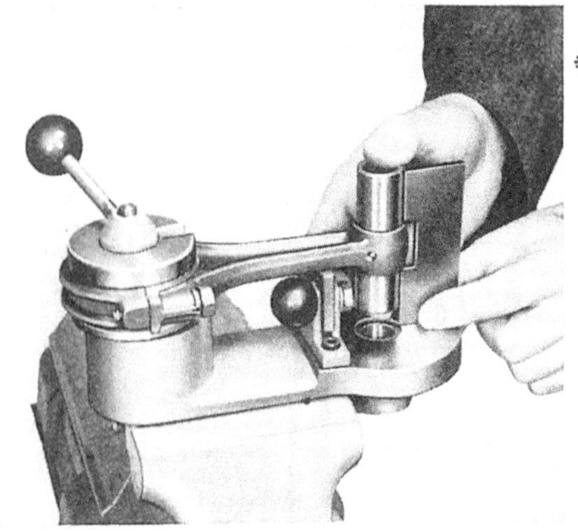

5 - Check connecting rod for twist and parallelism with the gauge F.

(The picture shows the parallelism check.)

Note:

The letters quoted here refer to the drawing on page M-10/7.

If deviations occur, tighten locking lever B and straighten connecting rod with the bar.

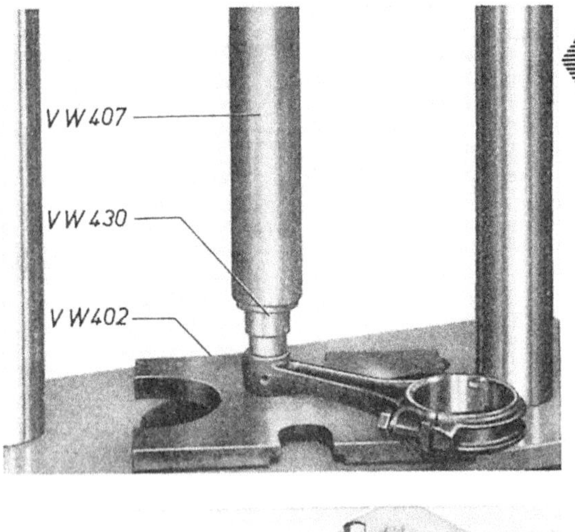

6 - Press in the connecting rod bush on the VW repair press in conjunction with VW 402, VW 430 and VW 407.

7 - Insert reamer through the connecting rod eye and the corresponding hole in the device.

The conical bush ensures correct centering of the piston pin bush. Tighten locking lever B and support D.

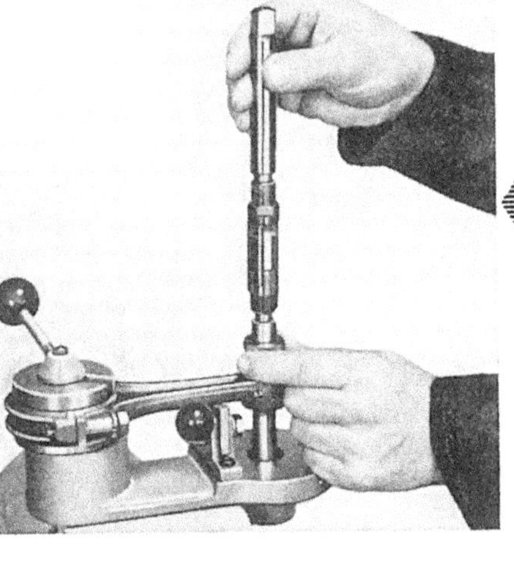

8 - Ream up piston pin bush.

Inner diameter for piston pin:

with black dot 20.000—20.006 mm (.7874—.7876")

with white dot 20.007—20.013 mm (.7877—.7879")

The bush bores must be free from scores and chatter marks after reaming. Without applying oil, the piston pin must be a light push fit. It is not permissible to fit an oversize piston pin in order to eliminate clearance between piston pin and bush. A new bush must be installed and reamed out to size.

9 - Re-check connecting rod for parallelism and twist as mentioned previously, but this time with the piston pin installed. Slight deviations can be corrected by inserting a bar in the piston pin and levering to straighten.

Reconditioning Crankshaft

Crankshafts which require regrinding should, if possible, be sent to the factory, as this process requires first class equipment and skill.

If it is not possible to send in the crankshaft for reconditioning, the following details will be found helpful:

	Main Bearings 1—3 and Connecting Rod Bearings (D)						Main Bearing 4 (d)					
	Nominal Dia.		Ground Dia.		Lapped Dia.		Nominal Dia.		Ground Dia.		Lapped Dia.	
	mm	in.	mm	in.	mm	in.	mm	in.	mm	in.	mm	in.
Standard	55.00	2.1654	—		54.990 54.971	2.1650 2.1642	40.00	1.5748	—		40.000 39.984	1.5748 1.5742
1st Undersize	54.75	2.1555	54.749 54.740	2.1555 2.1551	54.740 54.721	2.1551 2.1544	39.75	1.5650	39.760 39.750	1.5654 1.5650	39.750 39.734	1.5650 1.5643
2nd Undersize	54.50	2.1457	54.499 54.490	2.1456 2.1453	54.490 54.471	2.1453 2.1445	39.50	1.5551	39.510 39.500	1.5555 1.5551	39.500 39.484	1.5551 1.5545

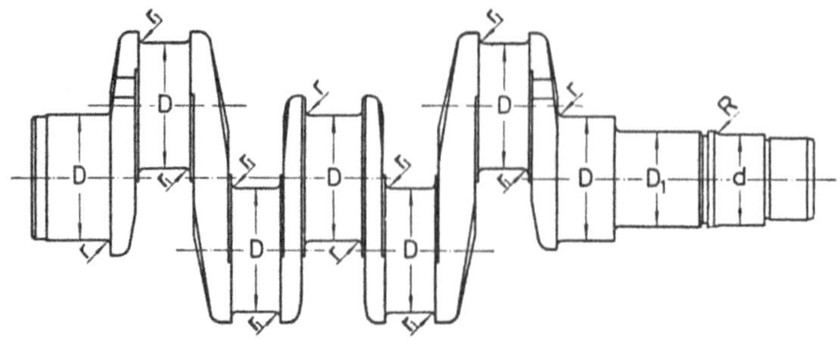

$$D_1 = \frac{42.006}{41.995} \text{ mm dia. } (1.6538''/1.6533'') \quad R = \frac{4.0}{3.5} \text{ mm } (.16''/.14'') \quad r = \frac{2.0}{1.5} \text{ mm } (.08''/.06'')$$

$$r_1 = \frac{2.5}{2.0} \text{ mm } (.09''/.08'')$$

Careful grinding of the radii is of great importance to the life of the crankshaft. Try to reach the values 2.0 mm or 2.5 mm. Repolish the radii.

Regrinding

On no account must the bearing shells be remachined.

After grinding, remove sharp edges on oil passages by slightly chamfering them. Crankshaft timing gear and distributor drive gear must be a press fit on the crankshaft $\frac{42.006}{41.995}$ mm dia. (1.6538″/1.6533″).

Where the gears are worn from being removed and installed several times, the press fit can be restored by chromium plating or metal spraying.

To make sure the crankshaft is free from internal cracks, it is advisable to carry out a "ringing test" prior to installation. The crankshaft must then be checked for run-out.

Bench Test Specifications

Bench testing and inspecting engines

General

The basic rules when bench testing an engine are as follows:

1 - Warm engine up at a medium speed and under a light load.

2 - Check engine for leaks, oil pressure, noise fan operation and performance.

A test stand with a water brake is particularly suitable for carrying out these checks. On this type of stand the engine loading can be regulated to suit the requirements of the performance and fuel consumption tests.

Reconditioned engines

These are engines which have, amongst other things, new pistons and cylinders, new bearings and reconditioned cylinder heads.

Preparation

a - Adjust valve clearance.

b - Adjust contact points and set ignition.

c - Check fan belt tension.

d - Fill engine with 2.5 liters of SAE 10 W or 20 W/20 oil.

Starting the engine

Before starting, crank the engine several revolutions by hand. If the engine has been in storage for a long time, it is good practice to inject a few drops of oil into the carburetor air intake while starting.

When the engine has been started, the green oil pressure warning lamp must go out as soon as the speed increases. Otherwise the oil pump has not drawn in any oil and the bearings and moving parts are not receiving the necessary lubricant.

The red generator warning lamp should also go out as soon as the speed increases.

The test run

The warming up time for the engine on the test bench can generally be limited to 30 minutes which is divided as follows:

10 minutes at 1500 rpm with a 2 kg load (4.4 lbs.)

20 minutes at 2000 rpm with a 4 kg load (8.8 lbs.)

When the engine has been warmed up, the tests can be carried out.

Checks during the bench test

1 - Fuel system

At the beginning of the test, make sure that the fuel pump, fuel lines and carburetor are not leaking and test the fuel pump pressure. Adjust idling speed while the engine is warm.

2 - Generator, regulator and fan

Check that the generator and regulator are functioning properly. The fan should not whine at 3600 engine rpm or foul the housing or throttle ring.

3 - Fuel consumption and performance test.

Towards the end of the 30 minute test run, check the fuel consumption at the specified speeds. When using normal fuel the consumption times should be as given in the "Special Instructions" section.

Finally, the engine output should be tested. The values are given in the performance graphs.

A variation of $\pm 5\%$ is permissible to compensate for manufacturing tolerances and differences in the test conditions.

The measured output is then converted to 760 mm mercury and 20° C with the following formula:

$$\boxed{N_o = N_e \cdot f} \ (BHP_o)$$

$$N_e = \frac{P \cdot n}{1000} \ (BHP_e)$$

$$f = \frac{760}{b} \cdot \sqrt{\frac{273 + t}{293}} \ (—)$$

When

P (kg)	= Test brake loading
n (rpm)	= Engine speed
N_e (BHP_e)	= Effective engine output (measured)
N_o (BHP_o)	= Normal engine output
t (°C)	= Outside temperature
b (mm mercury)	= Atmospheric pressure
f (—)	= Correction factor

4 - Check compression pressure

Check with a suitable tester and engine warm, throttle open, all spark plugs out, using starter to turn engine. The pressure should be between 7.0 and 9.0 kg/sq. cm. (100—128 psi.). The wear limit is 4.5 kg/sq. cm. (65 psi.).

Final Inspection

a - Checking for oil leaks

After the full-load and fuel consumption test, the engine should be checked for oil leaks. Special attention should be paid to valve push rod tubes, oil pump, oil cooler, and crankcase joints.

b - Re-check

Prior to installation in the car, check the valve clearance and fan belt tension. The air cleaner should be clean and filled up with the required amount of oil.

c - Storage of engines

Engines which are to be stored for a prolonged period must be specially treated to prevent damage due to corrosion. All remaining traces of fuel and combustion gases will become chemically active and attack the cylinder walls, valve guides, etc. In order to prevent damage of this kind, it is recommended that anti-corrosion oil be injected through the carburetor air intake during the last engine revolutions, or that the spark plugs be removed and anti-corrosion oil injected directed into each cylinder.

The exterior of the engine should be sprayed with anti-corrosion oil.

If an engine has only been partly reconditioned, e. g., valves renewed and ground-in the full load test must not be carried out until the engine has attained operating temperature (approximately 60—80° C).

The following points also apply to partly reconditioned engines:

1 - Pre-inspection.

2 - Checking of fuel system.

3 - Fuel consumption and performance test.

4 - Checking for oil leaks.

5 - Final inspection.

Important

To avoid blisters and discoloration on tail pipes caused by pushing exhaust extractor pipes over the tail pipes while on testing stands, it is advisable to remove the normal tail pipes and to fit special ones for this purpose.

Note:

As already known, only premium fuels (min. Research O. No. 95) may be used in vehicles with the twin-carburetor engine. The same requirement exists when the engine is being tested on a test stand.

In order to avoid changing the fuel when testing single and twin-carburetor engines on one stand, it is pointed out that all VW engines may be tested for output and fuel consumption with premium fuel.

Differences in engine output are not to be expected if the ignition is set exactly as specified for the individual engines. The quantity of fuel usually employed for the consumption test is 100 cc. With such a small amount, the higher specific weight of the premium fuel will not affect the consumption time noticeably. It is, therefore, also possible to test the fuel consumption of all engines with premium fuel. Only in cases of doubt would it appear advisable to carry out this test with the correct fuel for the engine concerned.

It is not intended to equip the test stands with an additional fuel container in the future.

Power Curves

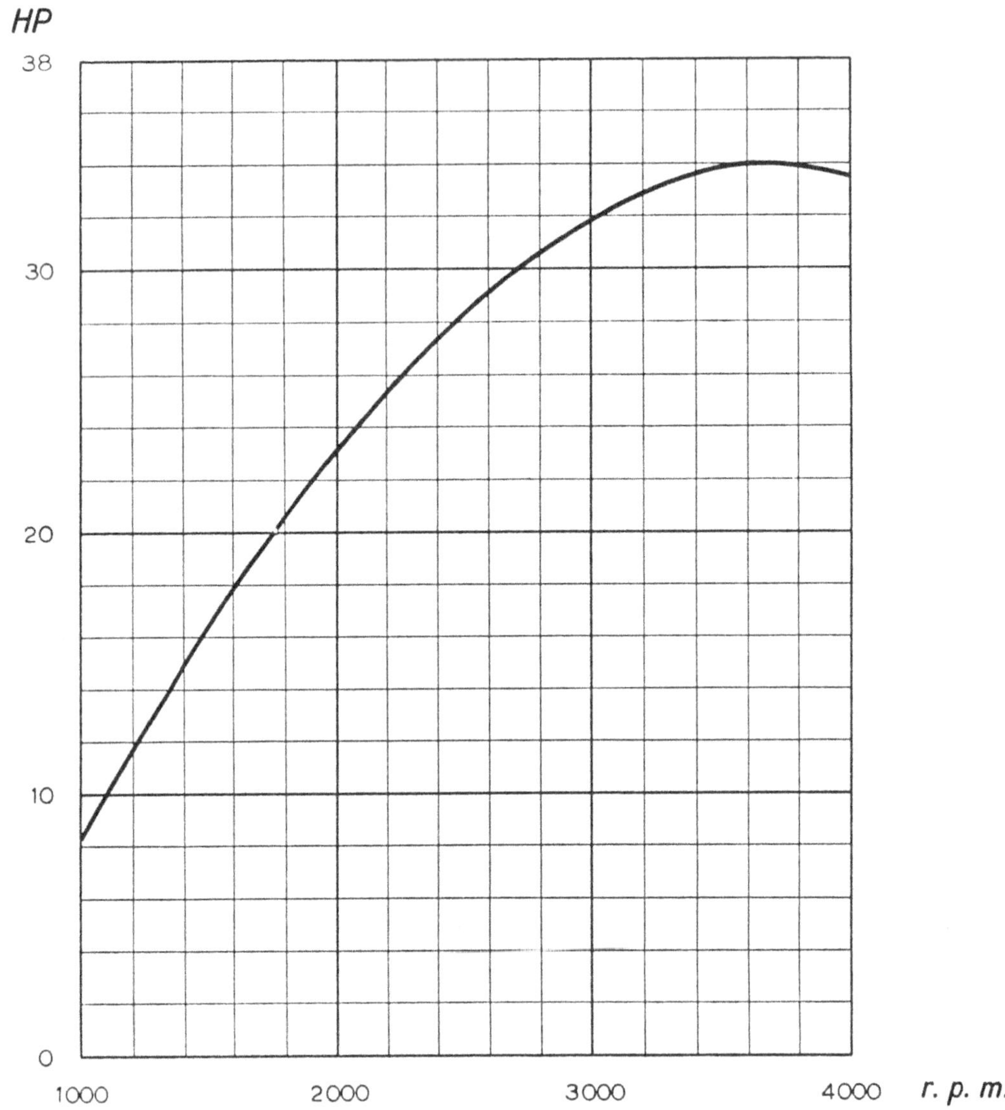

Ne = Performance (HP)

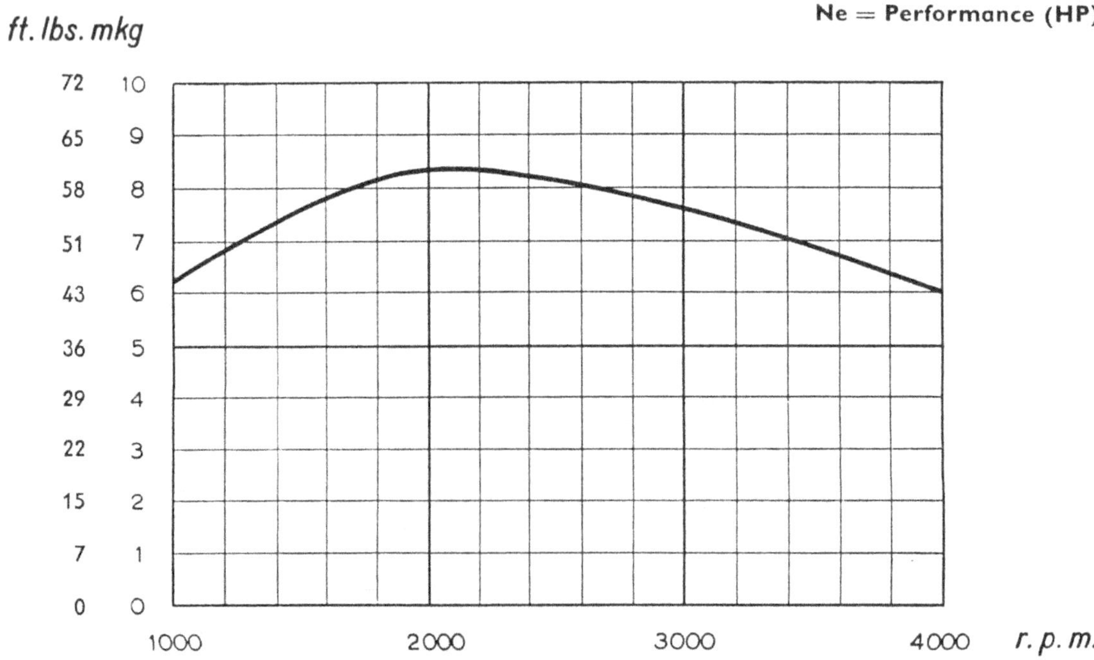

Md = Torque

M-11

Crankcase Ventilation

1 - When oil leaks occur at the oil filler, an insert can be installed in the oil filler passage. This insert can be manufactured locally according to the drawing.

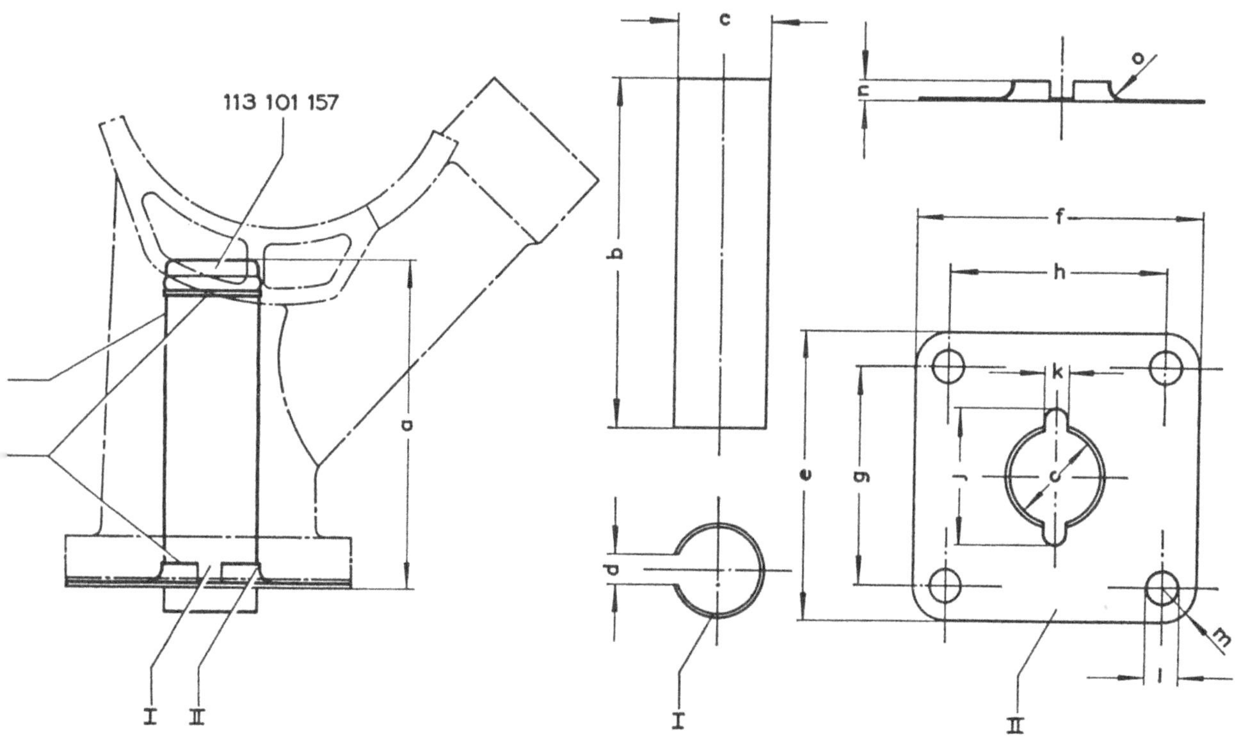

A = Slot on this side

B = Tack-welded twice at base

a = 86 mm (3.39″)	f = 77 mm (3.03″)	l = 8.5 mm (0.33″)
b = 92 mm (3.62″)	g = 58 mm (2.28″)	m = 9 mm (0.35″)
c = 24.5 mm (0.96″)	h = 59 mm (2.32″)	n = 5 mm (0.20″)
d = 8 mm (0.31″)	j = 36 mm (1.42″)	o = 4 mm (0.16″)
e = 76 mm (2.99″)	k = 6 mm (0.25″)	

The insert is positioned between crankcase and generator support so that the slot in the insert is facing away from the oil filler cap. An additional gasket for the generator support (Part. No. 113101 219) has to be used.

2 - The ventilation of the crankcase has been modified.

Formerly the oil vapor passed from the crankcase through the breather pipe into the open air below the engine. These vapors are now lead to the oil bath air cleaner and burned with the fuel air mixture.

Note:

De Luxe Sedan; VW Convertible

From Chassis No. 3806249 (Engine No. 5703138)

1 - A = Oil filler (Part No. 211115451)

B = Connecting hose 330 mm (13″) long.

The connecting hose (Part No. 211129653 A), 905 mm/35.5″ long, can be used as a spare part.

C = Oil bath air cleaner

This version is not available as a spare part.

2 - The parts listed under point **1**, B and C are only available for production in limited quantities. When these parts are used up, the following parts will be installed.

B = Connecting hose 455 mm (18″) long.

The connecting hose (Part No. 211129653 A), 905 mm long, can be used as a spare part.

C = Oil bath air cleaner

The oil bath air cleaner (Part No. 113129613 B) is supplied as a spare part.

Note:

Connecting hoses of a synthetic material are being used temporarily. They are secured with a hose clip at each joint. Engines which have been delivered with synthetic hoses but without hose clips should be subsequently fitted with clips.

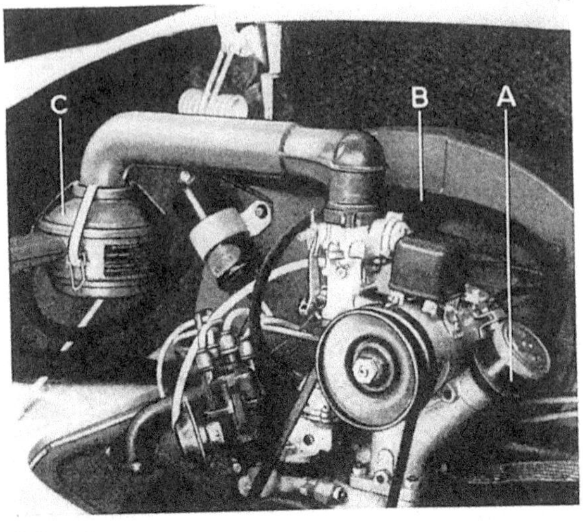

Karmann Ghia Models:

From Chassis No. 3931154 (Engine No. 5855551).

1 - A = Oil filler (Part No. 211115451 A).

B = Connecting hose 905 mm long (Part No. 211129653 A).

C = Oil bath air cleaner

The oil bath air cleaner (Part No. 261129613 A) is supplied as a spare part.

Note:

1 - The crankcase ventilation fitted to models 113, 114, 117, 118 and 15 since April 1961 and model 14 since June 1961 has been modified.

The connecting hose which was formerly attached to the air cleaner intake tube is now connected to the lower part of the air cleaner. The oil filler remains unchanged.

Introduced:

Model	11	14	15
From Chassis No.	4519277	4545642	4547060
From Engine No.	6502426	6502426	6502426

A - Oil filler (Part No. 211 115 451)

B - Connecting hose

Length **new:** 275 mm (11")
formerly: 460 mm (18")

The connecting hose — Part No. 211 129 653 A — 935 mm/37" long, can be used as a spare part.

C - Oil bath air cleaner Part No. 113 129 613 B as before.

Model 11 and 15

new

old

Model 14

new

A - Oil filler (Part No. 211 115 451)

B - Connecting hose

Length **new:** 670 mm (26.5")
formerly: 935 mm

The connecting hose — Part No. 211 129 653 A — 935 mm long, can be used as a spare part.

C - Oil bath air cleaner Part No. 261 129 613 A as before.

2 - Service Installation

The crankcase ventilation can be subsequently installed in the 34 bhp (41.5 SAE bhp) engine. Oil bath air cleaners which have no connection for the ventilation hose can be modified to suit.

Models 11 and 15

Weld or solder a 25 mm (1") long piece of 13 × 0.5 mm (.5" × .020") tube to the lower part of the air cleaner at the location shown in the illustration.

a = 13 mm (.51") b = 20 mm (.8") c = 5 mm (.2")

Model 14

Weld or solder a 25 mm (1") long piece of 13 × 0.5 mm (.5" × .020") tube to the air cleaner opposite the retaining clip marked with an arrow.

a = 15 mm (.6") b = 20 mm (.8")

The tube must project into the air cleaner to prevent oil from entering the connecting hose when the oil level varies.

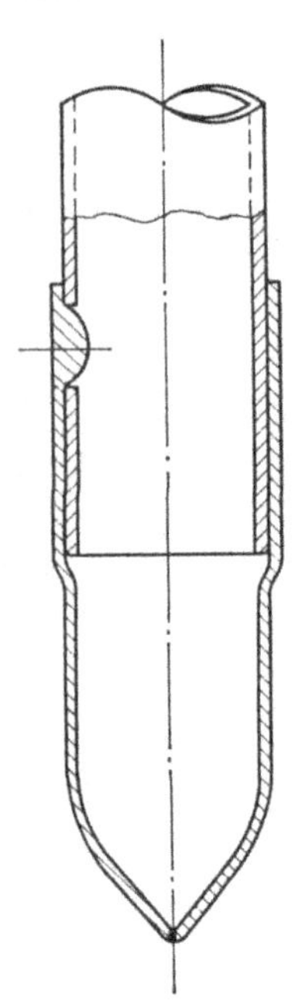

Note:

From October 1963, Chassis No. 5 815 778 (Engine No. 8 046 097) the oil breather is equipped with a condensed water drain pipe. The pipe is fitted in the same way as it was on the old oil breather before the fumes were drawn off through the oil bath air cleaner.

The end of the pipe is sealed with a rubber valve. The valve is secured by means of an internal button which engages in a hole in the drain pipe.

The rubber valve regulates the draining of the condensed water automatically. When a certain amount of water has collected in the pipe, the weight of the water opens a slot in the valve and the water drains off.

Important

At every 5000 km (3000 mile) maintenance service, the rubber valve must be checked. If it is damaged or if the slot no longer closes properly, it must be replaced.

Part. Nos.	new	old
Oil breather Rubber valve	211 115 451 A 311 115 541	212 115 451 —

Service Installation

The new oil breather can be installed in all 34 bhp (41.5 SAE bhp) engines from Engine No. 5 703 138 (June 1961). Proceed as follows:

1 - Remove old oil breather.

2 - Drill a 22 mm hole in the cover plate below the crankshaft pulley and clean up edges. A template for marking off the hole can be made with the aid of the sketch below.

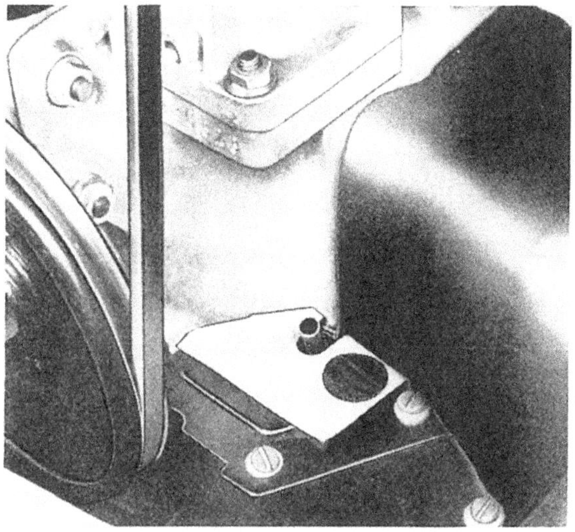

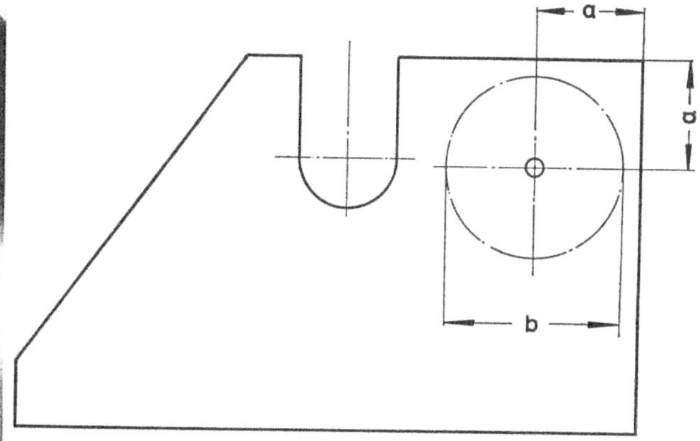

a = 13 mm (.51″) b = 22 mm (.86″) dia.

3 - Push rubber cap (Part No. 111115491) on to the condensed water drain pipe.

4 - Slide rubber valve (Part No. 311115541) on to the drain pipe until the button engages in the 7 mm hole in the pipe.

5 - Install oil filler.

6 - Cut new breather hose (Part No. 211129653A) to length and install.

If the new oil breather (Part No. 211115451 A) is installed in 34 bhp engines on which the crankcase fumes are not conducted to the air cleaner (up to Engine No. 5703137), the upper adaptor for the breather hose must be sealed with a plug (Part No. 111115465). The rubber valve must **not** be installed in these engines as otherwise the crankcase fumes cannot escape.

Note:

From June 1964, Chassis No. 6469413 (Engine No 3970858), the breather on the 30 bhp engine is also fitted with a water drain pipe.

Part No.	new	old
Breather	111115451 C	111115451 B
Rubber valve	311115541	

Service installation

The new oil filler with breather can be service installed in all 30 bhp engines from Engine No. 3942539. The procedure is the same as on the 34 bhp engine.

c = 35 mm

d = 22 mm

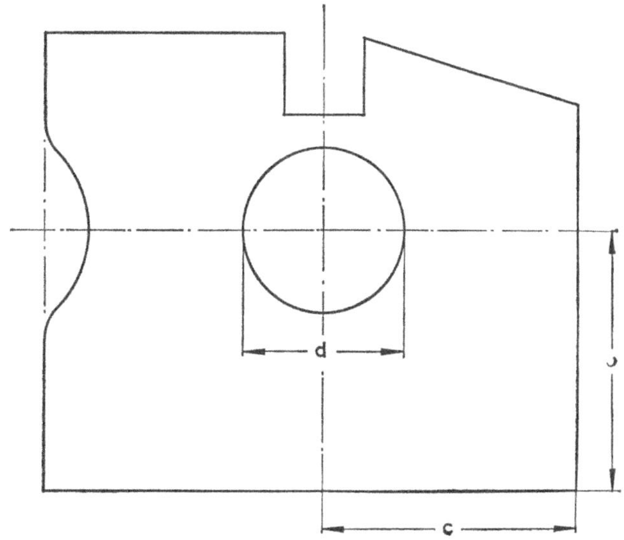

THIS PAGE INTENTIONALLY
LEFT BLANK

Description of Clutch

General Description

The single-plate, dry disc type clutch between engine and transmission is fitted to the flywheel. The driven plate (disc) to which the friction linings are riveted is splined to the main drive shaft. The clutch cover, which carries pressure plate, thrust springs, release levers and release plate, is bolted centrically to the flywheel. When engaged, the driven plate is forced against the flywheel by the spring-loaded clutch pressure plate. Thus the engine power is transmitted to the transmission.

Clutch operating shaft and clutch release bearing are located in the transmission case. The release bearing carries a carbon thrust ring and is maintenance-free.

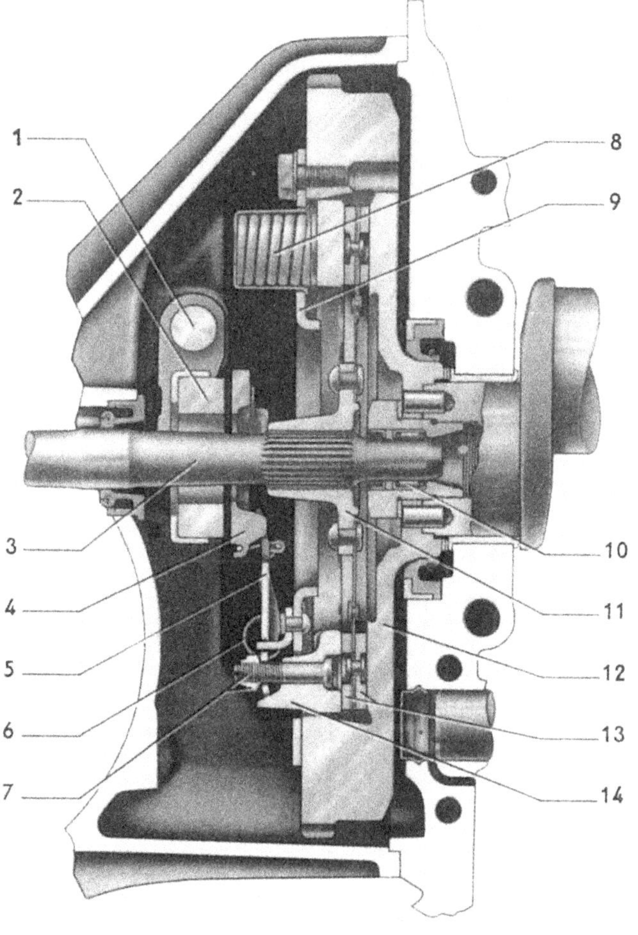

1 - Operating shaft
2 - Carbon thrust ring
3 - Main drive shaft
4 - Release plate
5 - Release lever
6 - Release lever spring
7 - Bolt and special nut
8 - Thrust spring
9 - Cover
10 - Needle bearing for gland nut
11 - Driven plate (disc)
12 - Flywheel
13 - Lining
14 - Pressure plate

Operation

Release is accomplished by depressing the clutch pedal. The pedal movement is transmitted through a cable to the release mechanism. As pressure is applied to the pedal, the release levers are moved inward by the release bearing and the pressure plate is moved away from the driven plate, disengaging the clutch.

Adjustment

Clutch servicing is limited to adjusting the clutch pedal free-play of 10—20 mm (0.4″—0.8″) as the lining wears. An adjustment of the clutch itself is only necessary when the clutch is disassembled for replacement of parts. This adjustment is carried out at the flywheel with the engine removed by using Clutch Adjustment Gauge VW 254a.

Clutch Removal and Installation

Removal

1 - Remove engine.

2 - Evenly release clutch cover securing bolts diametrically opposite in turn, giving each bolt one or two turns at a time to prevent distortion due to the tension of the thrust springs.

3 - Take off clutch cover.

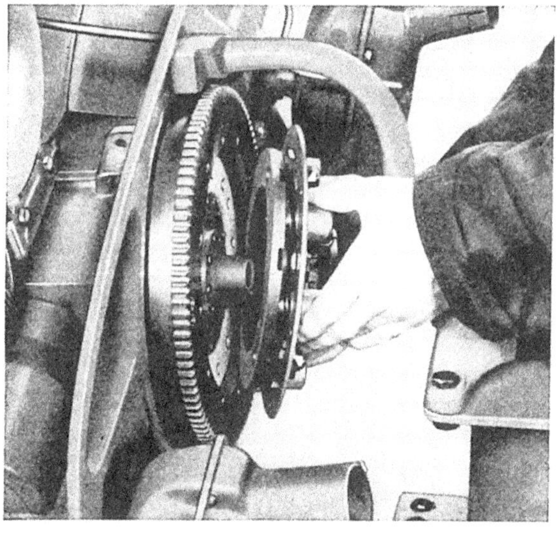

4 - Withdraw clutch driven plate.

Installation

When installing, the following points should be observed:

1 - Clean clutch contact surface in flywheel and inspect it for wear. Regrind as necessary (max. 0.2 mm/0.008″) and polish, using fine emery cloth. Replace flywheel if necessary.

2 - Inspect clutch driven plate for wear of linings, run-out, correct setting of cushion segments, and rivets at hub for tightness. Renew linings or complete clutch driven plate if necessary.

3 - Inspect clutch pressure plate for wear and distortion. An uneven contact surface of the pressure plate causes the clutch to chatter and grab. Regrind contact surface or renew pressure plate as necessary.

4 - Examine release levers and springs, replace if necessary.

5 - Inspect release plate for wear and damage. Renew the release plate if it is damaged or has been affected by heat.

6 - Check carbon thrust ring release bearing for wear. Renew complete release bearing if necessary. Note correct position of retaining springs.

7 - Inspect bearing points of clutch operating shaft in transmission case for wear.

8 - Pack needle bearing in flywheel gland nut with about 10 grams universal grease.

9 - Reinstall clutch driven plate, using Pilot Mandrel VW 219 to ensure correct centering.

Note proper position of clutch cover locating lugs in flywheel.

10 - Evenly tighten clutch cover securing bolts diametrically opposite in turn, giving each bolt one or two turns at a time to prevent distortion due to the tension of the thrust springs.

Tighten screws to 2.5 mkg (18 lb. ft.) evenly and diagonally.

11 - Check proper distance and parallelism between the clutch cover contact face at the flywheel and the clutch release plate by means of the Clutch Adjustment Dial Gage VW 254a. Adjust release plate if necessary.

Distance from flywheel to release ring

26.7—30.0 mm

Release ring run-out

max. 0,6 mm

Note:

The clutch covers and pressure plates of **exchange clutches** are sprayed with a wax-based preservative after assembly. The preservative is only sprayed very thinly and is not easy to see.

Important

Before installation, these clutches must be cleaned with benzine and blown out with compressed air. The friction points of the withdrawal levers should then be lubricated lightly with lithium grease.

Clutch Disassembly

Disassembly

1 - Remove clutch.

2 - Install clutch cover and pressure plate in flywheel (with clutch driven plate) and tighten by means of the six clutch cover securing bolts. Tighten the bolts evenly giving each bolt one or two turns at a time to prevent distortion. Mark all parts so that they can be reassembled in the same position without having to rebalance them.

3 - Remove the special nuts securing the release levers, using a saw to remove the metal peened over. Lift off release together with the springs and the release plate.

4 - Evenly release clutch cover securing bolts and take off clutch cover. Remove thrust springs, spring seats, and pressure plate.

Carefully inspect all clutch components before reassembly.

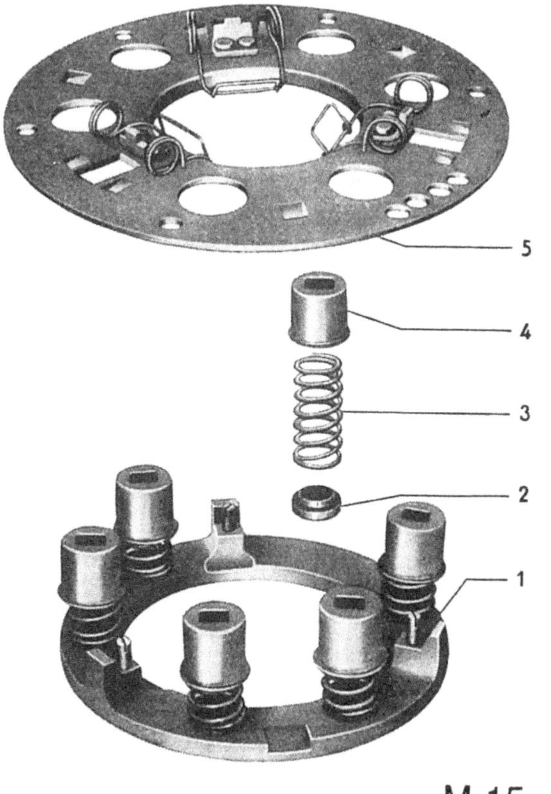

1 - Pressure plate
2 - Spring seat
3 - Thrust spring
4 - Spring cap
5 - Clutch cover

Clutch Cover and Pressure Plate

Inspection

1 - Check clutch cover for distortion and straighten it as necessary. Distortion of the cover is mostly caused by unevenly tightening or releasing the bolts.

2 - Thoroughly clean pressure plate and inspect it for distortion, wear and cracks. The contact surface of the pressure plate must bear evenly all the way round to avoid clutch chattering. In the event of run-out, the pressure plate may be reground. The pressure plate must otherwise be replaced.

3 - Examine thrust springs.

Loaded length in mm (in.)	28.3 (1.11″)
Load in kg (lbs.)	58.0 ± 3.0 (128 ± 6 lbs.)

4 - Inspect release plate for wear and damage. Renew damaged release plates or those which have become discolored due to excessive heat.

Note:

1 - From Chassis No. 4 464 038 (Engine No. 6 430 518), the clutch pressure was increased from 300—325 kg to 315—340 kg by the installation of stronger clutch springs.

Clutch cover with pressure plate
new: 111 141 025 B
formerly: 111 141 025 A

The clutch cover of the new clutch is marked with a letter "B".

111 141 025 A
— B

In addition to this, the springs in this clutch can be distinguished by the color. Three are yellow and three grey-blue.

If the springs have to be renewed on this clutch always use six springs of the 111 141 151 C version.

2 - From Chassis No. 4 683 160 (Engine No. 6 719 146) a clutch with six springs of equal strength is installed. The part number for the clutch cover with pressure plate remains unchanged 111 141 025 B. The clutch pressure is 315—340 kg.

The springs are brown (Part Number **new:** 111 141 151 C).

Data:

Loaded length	Load
28.3 mm (1.113″)	60 ± 3 kg (132 ± 6 lbs.)

These springs (Part No. 111 141 151 C) can be installed in previous clutches. The former springs (Part No. 111 141 151 B) will remain available as spares.

Note:

From July 1963, Chassis No. 5 661 082 (Engine No. 7 860 588) the clutch is equipped with heat-resistant springs. The Part No. of the clutch cover with pressure plate is still 111 141 025 B. The clutch pressure is 315—350 kg (694—771 lbs.).

The springs are painted brown and marked with a gold-bronze line (**New** Part No. 111 141 151 D — formerly 111 141 151 C).

The properties of the heat resistant springs are such that they only lose a certain amount of tension when heated up by excessive strain on the clutch. The springs then settle slightly. When overhauling the clutch, the settled springs can be used again provided that they are still within the test limits.

New clutch springs with the high initial tension must not be installed together with settled springs as the difference in tension between the springs is too large. They must only be installed in complete sets.

Test data:

Compressed length	Load, new	Load, settled
28.3 mm (1.114")	61 ± 2.5 kg (134 ± 5.5 lbs)	53.5 ± 2.5 kg (118 ± 5.5 lbs)

The previous type of springs will be discontinued when stocks are exhausted. The new springs can be subsequently installed in former clutches as a complete set.

Clutch Driven Plate

Inspection

1 - Check the clutch driven plate. Eight segments are riveted to the carrier plate. The cushion segments have a slightly curved setting. It is essential for perfect clutch operation that the segments are equally set.

The hub of the plate must slide freely on the splined main drive shaft without undue radial clearance. Worn parts must be renewed.

2 - Inspect clutch linings. Renew the linings if they are oily, burnt, cracked, or nearly worn down to the rivets.

Important
Only approved clutch linings must be used.

Clutch Linings

Outer dia.	109.0—181.0 mm/7.04—7.09″
Inner dia.	124.0—125.0 mm/4.88—4.92″
Thickness	3.7— 3.9 mm/0.146—0.154″

When riveting clutch linings, it should be noted that every second hole is countersunk. The riveting is carried out on alternate sides.

The linings are attached to the convex side of the cushion segments.

3 - Check clutch driven plate for run-out with linings assembled. Permissible run-out: max. 0.8 mm (0.03″).

Note:

The thickness of the plate 'b' measured when compressed, must not be less than 5.4 mm (.212″).

Note:

From April 1962, Chassis No. 4659008 (Engine No. 6684821), two different types of lining are riveted to the clutch plate in order to obtain the most favourable frictional matching of the clutch linings to the different materials of flywheel and clutch pressure plate.

The following combinations are used:

Combination	1	2
Flywheel side Clutch side	Jurid Textar	Beral Textar

Assembling Clutch

Assembly

When assembling, the following points should be observed:

1 - Place clutch cover and driven plate into flywheel.

2 - Tighten the clutch cover securing bolts evenly, giving each bolt one or two turns at a time to prevent distortion.

3 - Lightly grease the moving joints of the release levers, using special grease.

4 - Renew release lever special nuts and bolts, if these have become unserviceable in disassembly.

5 - Place Clutch Adjustment Dial Gauge VW 254a in position. Tighten special nuts at the release levers until the marking groove of the test plate pin is flush with the edge of the bracket and the test plate makes full contact with the release plate. The distance between release plate and clutch cover bearing surface on flywheel amounts then to 27 mm (1.062″).

6 - The height and parallelism of the clutch release plate may also be checked by using a straight edge and a depth gauge.

7 - Secure special nuts, using peening tool VW 124.

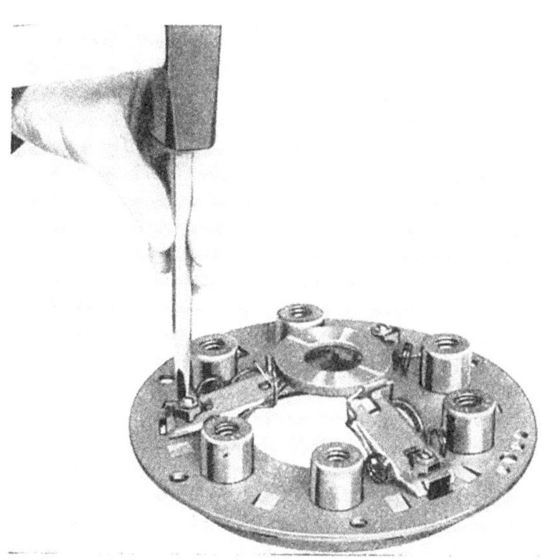

THIS PAGE INTENTIONALLY
LEFT BLANK

Clutch Overhaul
(Late)

Note:

Coil spring clutches can be checked quicker with tool VW 254 b and then, together with bracket VW 782, adjusted more accurately.

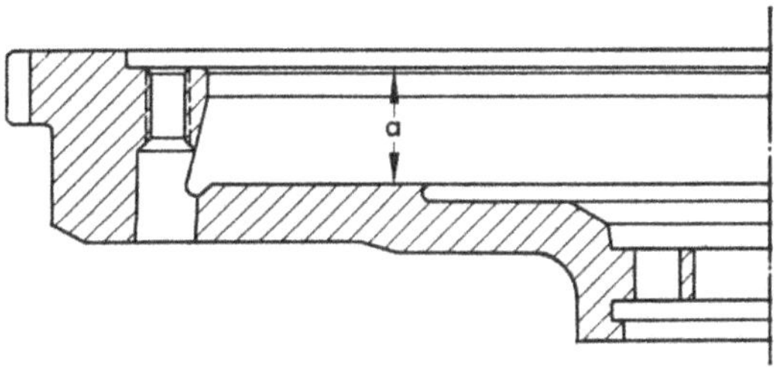

a = 21 mm for 200 mm clutch
a = 24 mm for 180 mm clutch

To supplement the tool a flywheel for each of the 180 and 200 mm diameter clutches is required. For this purpose, however, flywheels must be chosen, the dimension "a" of which must be 24 ± 0 mm for 180 mm diameter clutch and 21 ± 0 mm for the 200 mm diameter clutch. If necessary, the flywheels must be reworked.

Notes on clutch adjustment:

a - For adjustment purposes, torque the six securing screws with installed lock washers to the prescribed value of 18 lb. ft. (2.5 mkg).

b - Depress clutch several times prior to adjusting the clutch release ring.

c - Before checking the adjustment, the release lever and the pivoting bolt must be pressed outward.

Clutch Disassembly

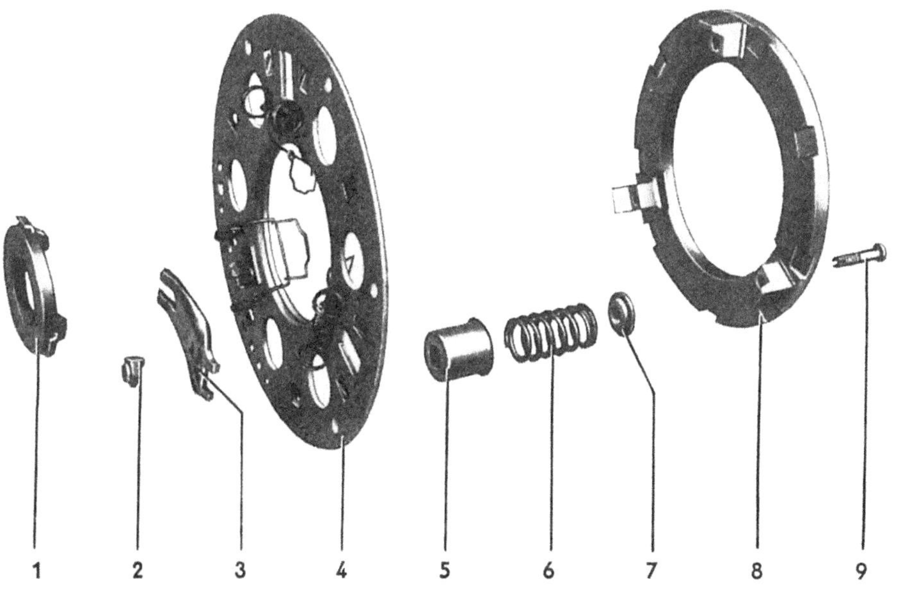

1 - Release ring
2 - Adjusting nut
3 - Release lever
4 - Cover
5 - Spring cap
6 - Spring
7 - Spring seat
8 - Pressure plate
9 - Bolt for release lever

Disassembly

The assembled clutch is balanced dynamically by the manufacturer. Clutches with an unbalance up to 5 cmg are not marked; from 5 to 15 cmg the location of the unbalance is marked with a white paint dot. When disassembling, therefore, the release levers and the position of the clutch cover in relation to the clutch pressure plate must be marked to avoid the risk of considerable unbalance after assembling.

The maximum permissible clutch unbalance is 15 cmg.

3 - Place clutch driven plate and clutch into the flywheel and press down the clutch on the repair press by using a flat piece of wood and three tubes of equal length.

4 - Take off adjusting nuts.

5 - Release the press slowly and disassemble clutch.

Clutch Checking

Checking

1 - Disassemble clutch.

2 - Check clutch pressure plate. Warped or grooved pressure plates should be re-machined, reground or, if necessary, renewed. A pressure plate with an uneven contact pattern will encourage clutch chattering.

3 - Check the thrust springs.

The total pressure should be 315—350 kg (694—771 lbs.).

The clutch is fitted with heat resistant springs.

Note:

The properties of the heat resistant springs are such that they only lose a certain amount of tension when heated up by excessive strain on the clutch. The springs then settle slightly. When overhauling the clutch, the settled springs can be used again provided that they are still within the test limits (53.5 kg to 61 kg) and that the difference between the springs is not more than ± 2.5 kg.

New clutch springs with the high initial tension must not be installed together with settled springs as the difference in tension between the springs is too large. They must only be installed in complete sets.

Test data

Compressed length	Load, **new**	Load, settled
28.3 mm (1.113 in.)	61 ± 2.5 kg (134 ± 5.5 lbs.)	53.5 ± 2.5 kg (118 ± 5.5 lbs.)

4 - Check the clutch cover for warpage or cracks. Straighten warped covers; replace if cracked.

5 - Check release levers. Replace levers if warped, cracked or worn.

6 - Check return springs for tension; replace weak springs.

7 - Check clutch release ring. If the grooves on the release ring for the release levers are worn, the release ring must be replaced.

Assembly and Adjustment

1 - Insert release lever pins into pressure plate. Always use new pins.

2 - Insert driven plate and pressure plate into the flywheel (a new driven plate must always be used when adjusting the clutch). Insert seats, thrust springs and caps and install clutch cover.

Important

The marks made on the release levers, clutch cover and pressure plate before disassembly must coincide. Make sure that the spring seats are correctly seated in the pressure plate.

3 - Turn the clutch until the holes for the securing bolts in the clutch cover are flush with the threaded holes in the flywheel.

4 - Screw in a hexagon bolt loosely (approximately 50 mm/2 in. long).

5 - Compress the clutch carefully on the press. Make sure that the square holes in the clutch cover are not damaged and that the centering lugs fit into the recesses in the flywheel.

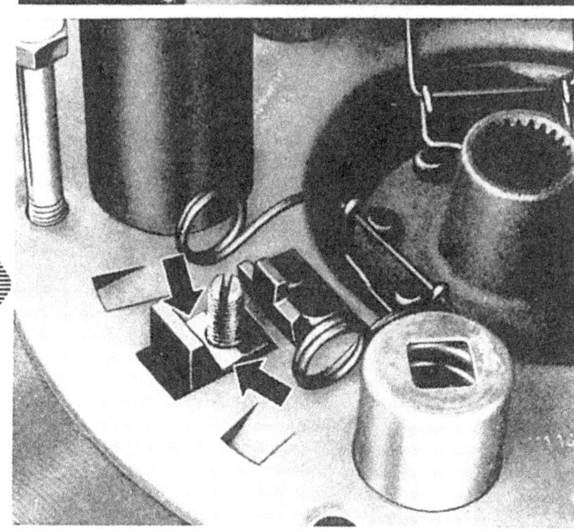

6 - Insert securing bolts into every second hole and tighten them. The guide bolt can now be removed.

7 - Release the pressure and take out the piece of wood and the tubes.

8 - Lightly lubricate the release levers at the pivoting points with lithium grease and install them. Install the adjusting nuts and adjust the release levers provisionally. New adjusting nuts must always be used.

9 - Lubricate the grooves for the release levers on the release ring with lithium grease. Install the release ring and insert the return springs.

10 - Place the clutch adjustment gauge VW 254 a in position and adjust the parallelism and release ring clearance. Tighten the adjusting nuts until the upper edge of the marking groove in the bolts is flush with the shoulder of the bracket.

The maximum permissible run-out is 0.3 mm (.012 in.). The adjustment can also be carried out with a straight edge and depth gauge. The distance between flywheel and release ring is 27 ± 0.3 mm (1.062 ± .012 in.).

11 - After the clutch has been adjusted it should be operated a few times and the adjustment checked.

12 - Unscrew the three mounting bolts and remove the clutch. To prevent distortion of the clutch, the piece of wood and tubes must be used again.

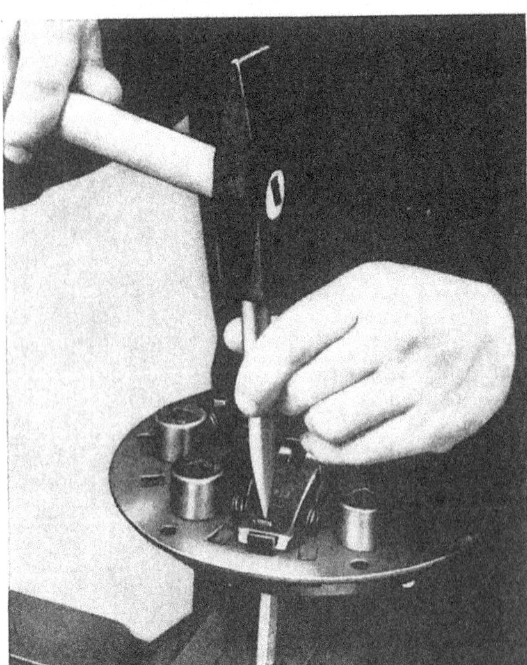

13 - Lock the adjusting nuts.

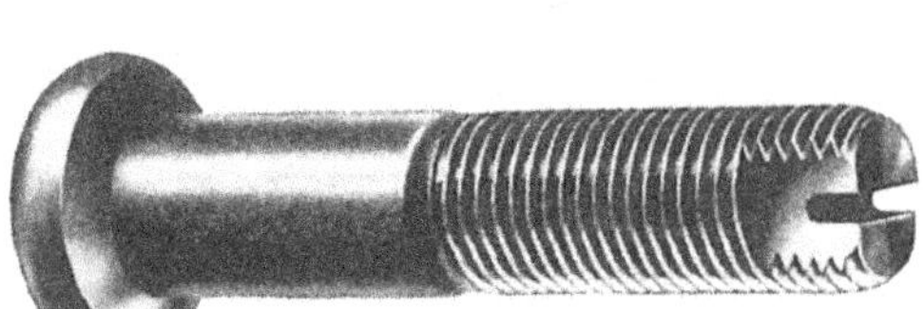

Note:

On the 180 mm diameter clutches — 111 141 025 A/B the adjusting nuts have, up to now, been secured by peening the nut shoulder with chisel VW 124. The clutch lever bolts, new 111 141 133 B, now have two flats on the end of the thread and the adjusting nuts, new 111 141 137 B, have a wider shoulder. These nuts are secured as on the newer type clutches, by squeezing the shoulders.

A number of the clutch lever bolts do not have the slot for adjustment purposes. When installing new bolts, this slot should be sawn max. .08 in. (2 mm) deep at right angles to the two flats.

Note:

From November 1964, Chassis No. 115318171, (Engine No. 9112176) a reinforced clutch was fitted. The clutch pressure is now 320—345 kg (705—760 lb.) instead of 315—350 kg (694—771 lb.).

new

old

This clutch has 3 dark-brown springs and 3 light-brown springs which are fitted alternately.

Test data:

	Loaded length	Load, new	Load, settled
111 141 151 E dark-brown	28.3 mm (1.114 in.)	62.5 + 3 kg (138 + 6.6 lb.)	55 + 3 kg (121 + 6.6 lb.)
111 141 151 F light-brown	28.3 mm	62.5—3 kg	55—3 kg

The other data remains unchanged.

The new clutch can be service installed.

Clutches of the previous type should be used up. The individual parts, apart from the pressure springs, will remain available.

Clutch springs of the previous type will be discontinued when stocks are exhausted. The new springs can be installed in the previous type of clutch as a complete set.

Note:

Type	From Chassis No.	From Engine No.
1/1200	116377154	D0045573
1/1300	116374949	F 0346052

Since November 1965 the clutch — 111 141 025 D — which has been previously installed on 1.3 liter engines has been provided with different release levers and improved mountings. It is also now being installed on 1.2 liter engines. The improved design, the part number remains unchanged, can be recognized by a green paint dot on the clutch cover or by the manufacturer's designation "F & S" stamped on the longitudinal axis of the release lever. The manufacturer's designation was previously stamped in at right angles to the longitudinal axis.

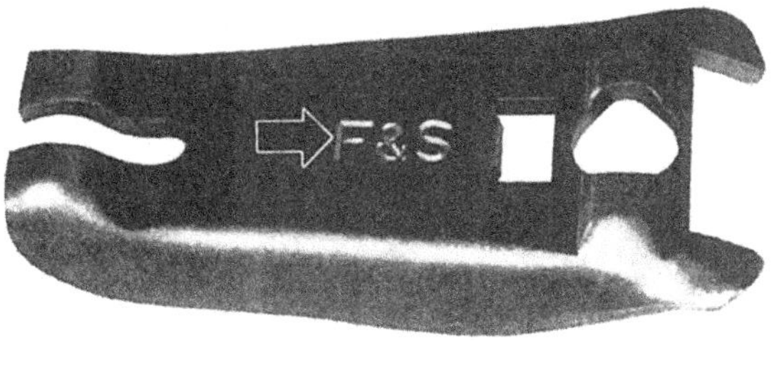

new

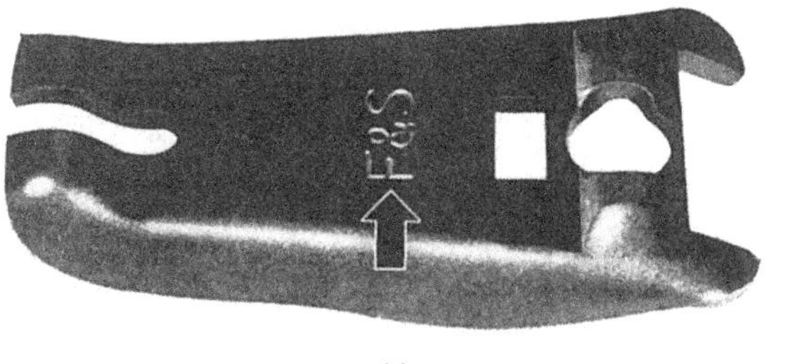

old

Only the improved design will be supplied as replacement part for 1.2 and 1.3 liter engines, under Part No. 111 141 025 D.

Repair Notes

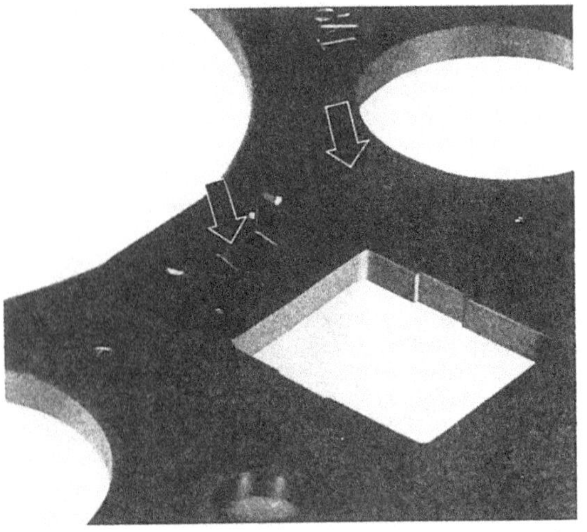

1 - Prior to disassembly, mark the individual parts of the clutch, otherwise the clutch will be out of balance on assembly.

2 - Disassemble clutch.

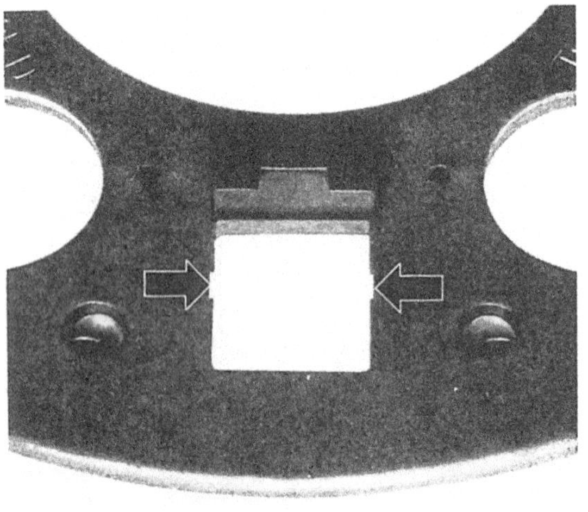

3 - Put a fine chamfer on the inside edges of the release lever supports riveted to the clutch cover, using a carborundum stone. Clutch covers, the edges of whose supports are broken, must be replaced with new ones.

4 - Using a file, remove burrs from the sides of the cutouts in the clutch cover in which the pivot bolt supports are located, then file the grooves as shown in the illustration.

5 - Remove the sharp edges of the pivot pin supports on the pressure plate (black arrows).

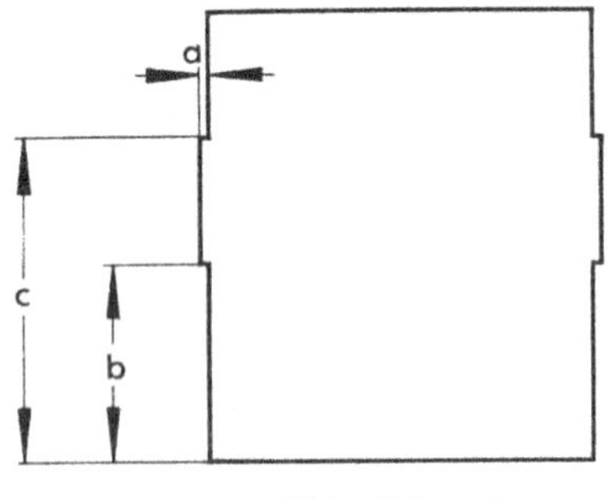

a = .02 in. (0.5 mm)
b = .47 in. (12 mm)
c = .71 in. (18 mm)

6 - Grind the crests of the threads off the pivot pin, on the outside (center arrow) near the edge of the release lever. This has to be done only if the length of the thread is more than .4 in. (10 mm).

7 - Install only dark and light blue springs 111 141 151 G/H, in alternate sequence in clutches 111 141 025 **C and D.**

Note:

Dark and light brown springs, 111 141 151 E/F, are to be used for clutch, 111 141 025 **B.**

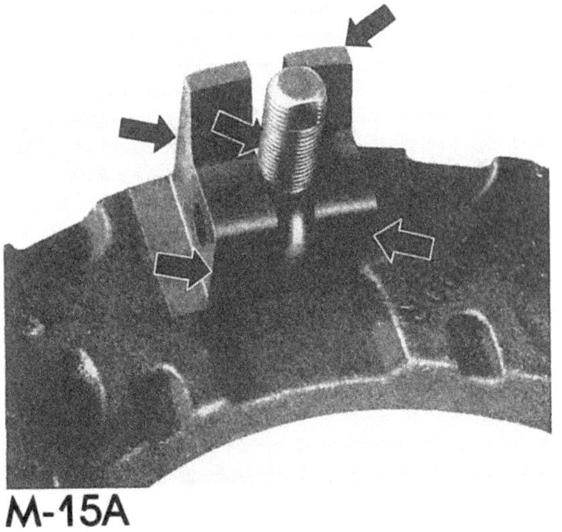

8 - Always replace the release levers with the new type.

9 - Assemble clutch according to markings made on disassembly and to setting measurements.

Checking and Relining Clutch Driven Plate

The eight spring segments are each attached to the driven plate with two rivets. The individual spring segments are curved and the linings are riveted on alternate sides. The lining is thus under spring tension and ensures soft clutch operation.

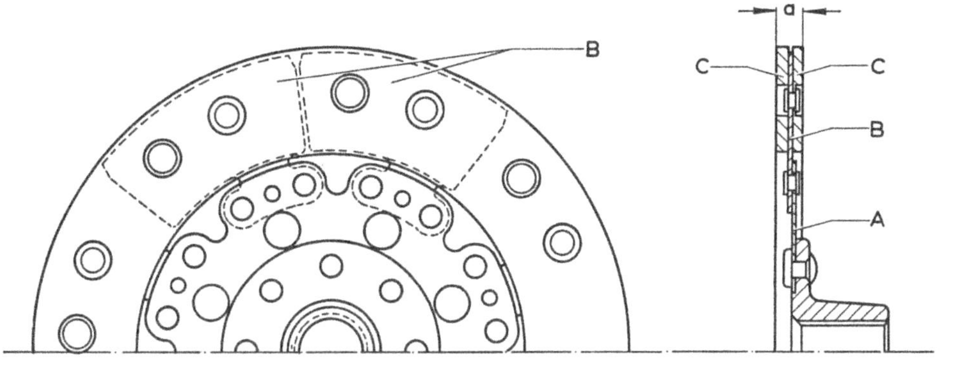

A - Driven plate B - Spring segments C - Clutch linings

Checking

1 - The driven plate must slide freely on the main drive shaft without undue radial clearance. Replace worn parts.

2 - Check the rivets; replace driven plate if necessary.

3 - If the spring segments or carrier plate are cracked, replace the complete driven plate.

4 - Check clutch lining. Replace worn, cracked, oily or burned linings. Maximum permissible wear from the total lining thickness is 2.75 mm (.107 in.) i. e. if either of the linings is worn to 2.35 mm (.091 in.) the driven plate must be relined or replaced.

Relining

1 - Worn linings must not be torn off. Always drill out the rivets.

Important

Only use linings which have been approved by Volkswagenwerk.

Clutch Linings

Outside dia.	179.0—180.0 mm (7.04—7.08 in.)
Inside dia.	124.0—125.0 mm (4.88—4.92 in.)
Thickness	3.7— 3.9 mm (.146—.154 in.)

Important

Two different types of lining are riveted to the clutch plate in order to obtain the most favorable frictional matching of the clutch linings to the different materials of flywheel and clutch pressure plate.

The following combinations are used:

Combination	1	2
Flywheel side	Jurid	Beral
Clutch side	Textar	Textar

2 - When riveting linings, note that every second hole is a clearance hole. Corresponding to the curved surface of the spring segment, the linings are riveted on alternate sides.

3 - After relining, check the driven plate for run-out; if necessary straighten with a suitable tool. On no account machine the linings. Maximum permissible run-out 0.8 mm (.031 in.).

Straighten the driven plate carefully and do not damage the linings.

4 - The distance between the friction faces of the driven plate must not be less than 5.4 mm (.212 in.).

Note:

From November 1964, Chassis No. 115366664 (Engine No. 9099857) rivets with a 2 mm hole in them (311141195) are used to secure the clutch linings.

Clutch plates of the previous pattern should be used up. Before installing these plates, the rivets should be drilled with a 2 mm drill.

From October 1964, Chassis No. 115162787 (Engine No. 8963731) the splines in the hub of the clutch plate have a phosphated sliding finish. The splines of the drive shaft are being treated with a molybdenum disulfide based coating.

Important

Before installing the engine after carrying out repairs, the splines on the drive shaft and in the clutch plate must be cleaned and dried carefully. A molybdenum disulfide based powder is then applied to the splines of the shaft with a brush. Excess powder should be blown off.

From March 1965, Chassis No. 115579323 (Engine No. 9282492) the "Beral" or "Jurid" linings on the flywheel side have radial grooves.

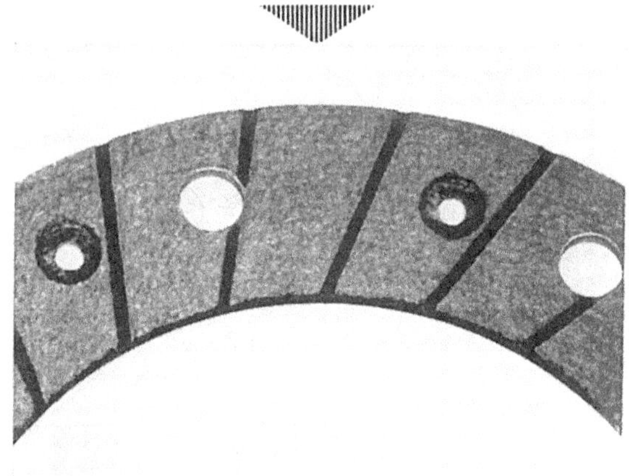

Part No.	unchanged
Clutch plate	111 141 031 B

Clutch plates of the previous type should be used up.

Note:

The following clutch plates with torsion springs are available for service installation:

Part Number	to be combined with clutch
111 141 031 C	111 141 025 B 111 141 025 C*) 111 141 025 D*)

*) Use release lever return spring 111 141 141 A.

These clutch plates should only be installed when it is necessary to eliminate load-change noises or rattling noises when coasting.

Removing and Installing Clutch Release Bearing

Removal

1 - Remove engine.

2 - Remove clutch release bearing retaining springs.

3 - Withdraw release bearing.

Installation

When installing, the following points should be observed:

1 - Inspect carbon thrust ring of release bearing for wear and cracks. Renew release bearing should the carbon thrust ring be damaged. The ring should not be replaced separately as it will definitely suffer damage when pressing it in position.

2 - Note correct position of retaining springs.

3 - Readjust clutch pedal free-play after engine has been installed.

Note:

Clutches which are subjected to undue stress, as in driving school cars, can be equipped with a ball thrust release bearing obtainable as a Service Part. The ball thrust bearing is installed in the same manner as the standard release bearing, except for the two retaining springs (a).

The ball thrust release bearing is already provided with grease at the factory and does not need any further lubrication.

The ball thrust release bearing must never be cleaned with fuel, but should be wiped off with a clean rag. Dirty, and consequently, noisy bearings have to be replaced.

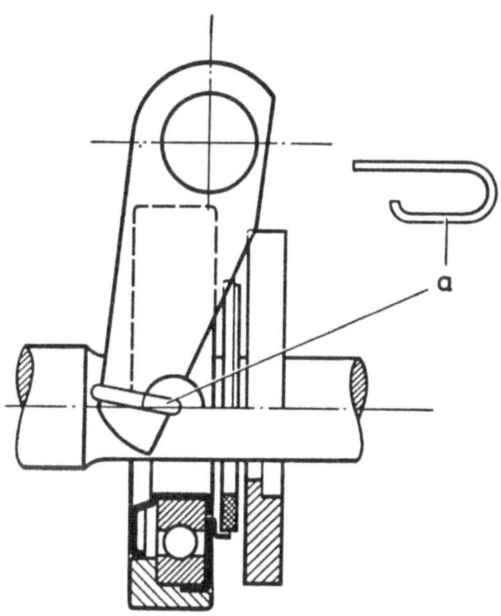

Note:

From October 1964, Chassis No. 115 162 922 (Rear Axle No. 7 256 130) a ball release bearing with a plastic ring (111 141 167 C) is installed in place of the release bearing with graphite ring (111 141 165).

This bearing is attached to the clutch operating shaft with two spring clips (111 141 177 A).

Note:

The release bearing requires no maintenance. It should merely be wiped with a clean cloth and not, in any circumstances, be washed in benzine or other solvents.

> **Important**
>
> The release bearing with graphite ring — 111 141 165 — is still used in the Volkswagen 1200 with Saxomat clutch. It must not be replaced by a ball release bearing.

Note:

Since December 1965, Chassis No. 116 407 142 (Rear Axle No. 8 729 521), the plastic ring of the release bearing, Part No. unchanged 111 141 167 C, has been provided with a surface film of a molybdenum-disulfide base to reduce noises when operating the clutch.

From April 1967, Chassis No. 117 604 080 (Rear Axle No. 0 211 071) the plastic ring of the clutch release bearing, unchanged 111 141 167 C, was given a rough surface by machining so that the lubricant film of a molybdenum-disulfide base can adhere better.

If a clutch release bearing causes noises when operating the clutch, it should **not** be replaced. In such a case, roughen the plastic ring with coarse emery cloth and rub in a molybdenum-disulfide based lubricant such as

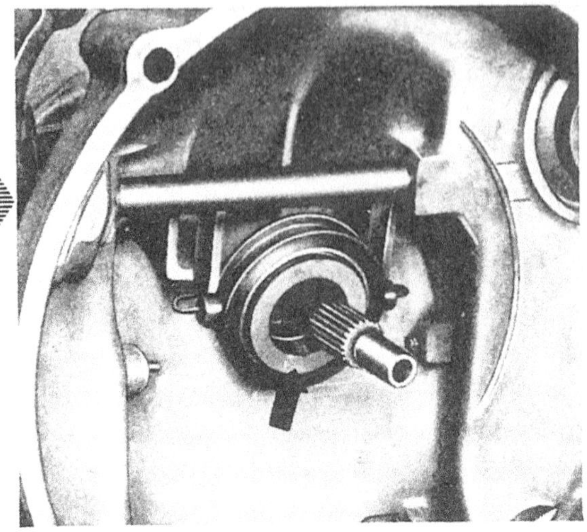

Molykote Paste G or
Molykote Powder Z or
Liqui Moly LM 11.

The film of lubricant should be renewed each time the engine is installed.

Note:

From March 1967, Chassis No. 117 618 945 (Engine No. D 0 164 834) a clutch release bearing with graphite ring (111 141 165) is installed in the Volkswagen 1200. Instead of the clutch release ball bearing (111 141 167 C). This made it necessary to use a larger clutch release ring (61 mm diameter instead of 54 mm) on the 180 mm diameter clutch.

When carrying out repairs, the release bearing with graphite ring can be replaced with the release ball bearing. If a clutch with a large release ring is replaced with a clutch with a small release ring (111 141 025 D), the clutch release ball bearing must be installed with it.

Clutches with large release rings and individual large clutch release rings are not obtainable as replacement parts.

Removing and Installing Clutch Operating Shaft

(transmission installed)

Removal

1 - Remove clutch release bearing.

2 - Remove nut at clutch operating lever and withdraw lever together with return spring and spring seat.

3 - Remove bolt for operating shaft bush.

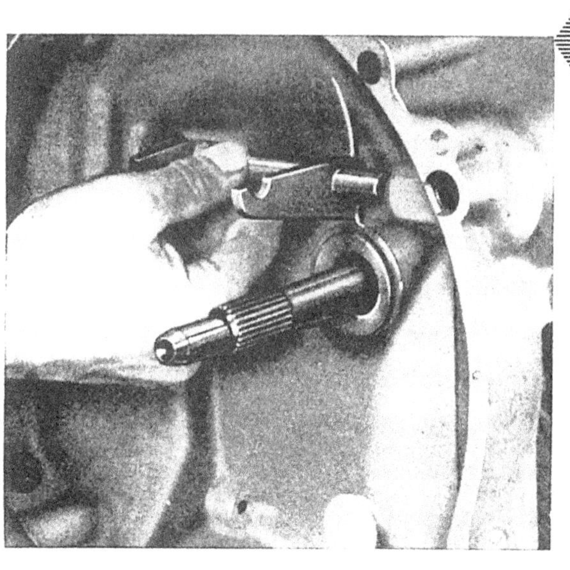

4 - Push operating shaft to the left and take out bush.

5 - Withdraw operating shaft to the right. Remove washer.

Installation

When installing, the following points should be observed:

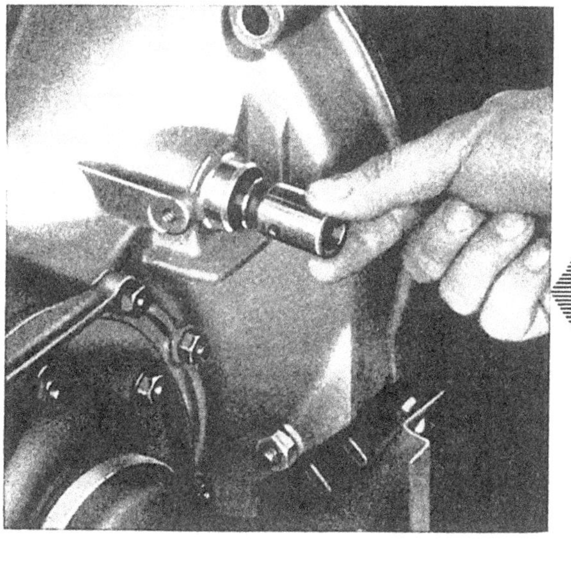

1 - Check right bush in transmission case for wear, replace if necessary.

2 - Apply Special Grease to operating shaft and install it.

3 - Check bush and washer for wear and install them.

4 - Screw bolt for bush into position.

5 - Check return spring and renew it if found to be weak.

6 - With the clutch correctly adjusted, check position of operating lever.

The position of the operating lever is correct, if:

a - **with clutch engaged,** the clutch cable threaded end piece only lightly touches the lower edge (A) of the bevelled hole in the lever;

b - **with carbon thrust ring just touching the release plate,** the inclination of the lever towards the front is not in excess of 2° from the vertical;

c - **with clutch fully released,** the upper edge (B) of the bevelled hole is not forced against the threaded end piece.

Incorrect position of the lever due to the clutch being mal-adjusted, an excessively worn carbon thrust ring, or a restricted movement of the adjusting nut in its concave seating is liable to kink the threaded end piece and may eventually lead to breakage of the cable.

The following advice should be followed to prevent breakage:

a - Renew release bearing if badly worn.

b - Grease the adjusting nut with Universal Grease when carrying out adjustments or repairs.

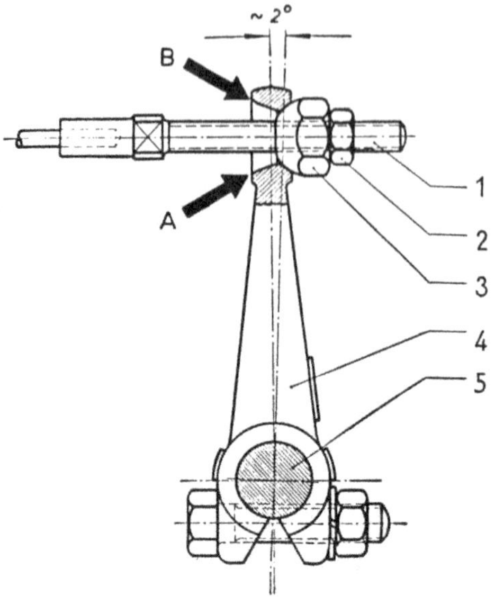

1 - Clutch cable threaded end piece
2 - Lock nut
3 - Adjusting nut
4 - Operating lever
5 - Operating shaft

Note:

From Chassis No. 2765107, the sizes of the threads of the clutch cable end piece, the adjusting nut and the lock nut were increased from M 6 to M 7. At the same time the internal diameter of the cable sleeve has been altered from 7 + 0.5 mm (0.275 + 0.020") to 7.5 + 0.5 mm (0.295 + 0.020").

When replacing a clutch cable with the M 6 threaded end piece by one with the M 7 threaded end piece, the new adjusting nut and lock nut must be installed at the same time. The cable sleeve need not be replaced by the new pattern.

Note:

From Chassis No. 4040690, the clutch operating lever clamp bolt was discontinued and the lever is now secured to the shaft by means of splines and a circlip. A second circlip prevents the operating shaft from moving sideways in the transmission and enables the spacer sleeve to be dispensed with.

Clutch Cable Removal and Installation

Removal

1 - Raise rear end of car and remove left-hand rear wheel.

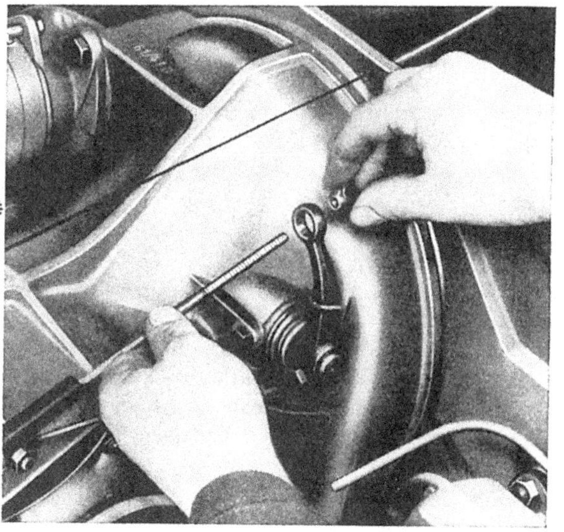

2 - Disconnect clutch cable from clutch operating lever on transmission case.

3 - Withdraw rubber boot from guide tube and pull out clutch cable from rubber boot.

4 - Remove the piston push rod in the master brake cylinder.

5 - Disconnect accelerator cable.

6 - Remove pedal cluster.

7 - Pull out clutch cable through the hole provided for the pedal cluster in the frame tunnel.

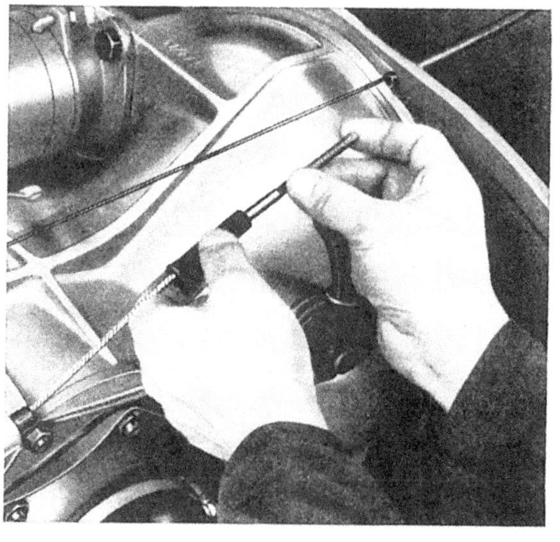

Installation

When installing the following points should be observed:

1 - Grease clutch cable with universal grease.

2 - Insert clutch cable.

Put threaded cable end into your left hand between middle and index finger and pass it through the pedal cluster hole to the guide tube. Put both fingers behind the guide tube and in this position push the cable in. Push it forward with the right hand until the clutch cable has fully entered the guide tube. Then push the cable completely through the tunnel.

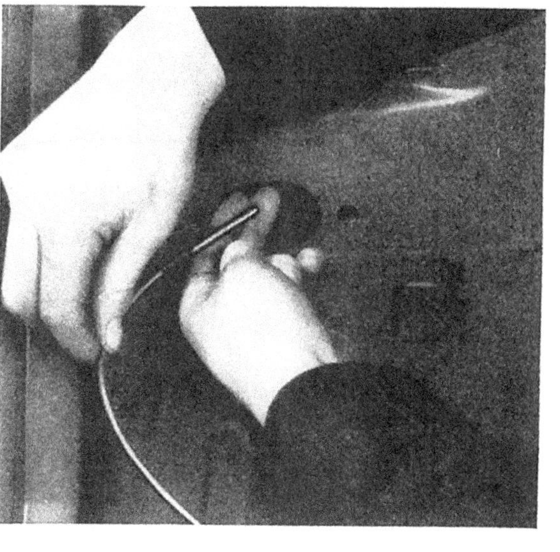

3 - Note correct position of the rubber boot at the end of the guide tube.

4 - Grease clutch cable eye and clutch pedal shaft with universal grease.

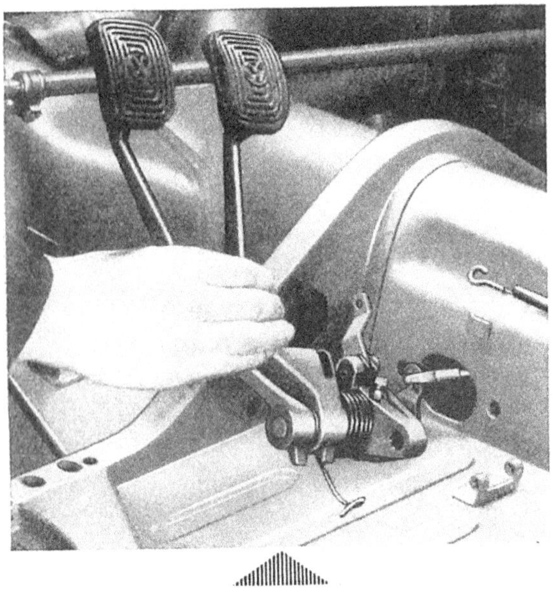

5 - When the cable eye is attached to the hook, the clutch pedal must be held in a vertical position to prevent the cable becoming disconnected. An assistant should keep it under tension at the other end.

6 - Ensure that the pedal stops are in correct position. The piston push rod must have a clearance of 1 mm (0.04″) in the piston.

7 - Grease clutch cable adjusting nut with universal grease.

8 - Adjust clutch pedal free-play.

9 - The clutch cable guide at the end of the frame tunnel should sag 25—45 mm (1—1.8″) (B). This preload is obtained by inserting washers between the bracket on the transmission and the end piece of the guide (A).

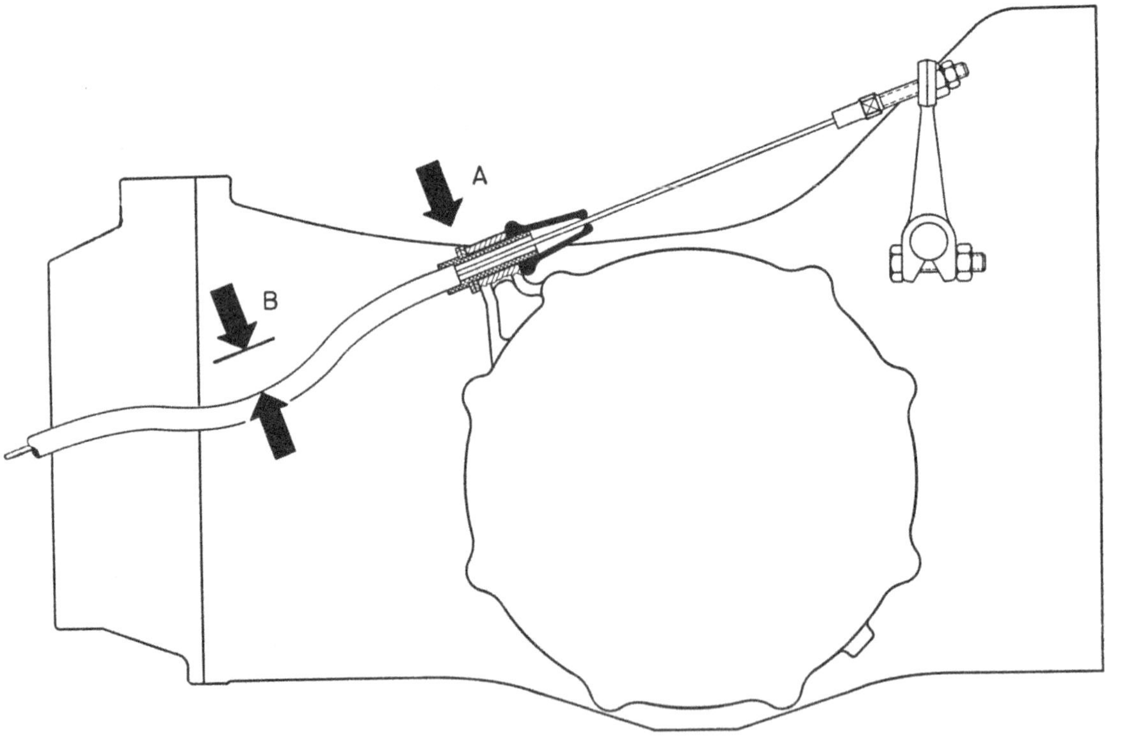

Note:

If the cable guide tube is bent down too much on vehicles with fully synchronized transmissions it will make the cable stiff in operation and can create noises or cause breakage of the cable. This condition can be rectified by shortening the guide tube at the rear end after removing tube and cable.

If the amount of bend is insufficient it can be increased by inserting washers between guide tube and the boss cast on the final drive cover. The amount of bend should be 25—45 mm (1—1.8″).

Note:

1 - Clutch pedal shaft

From January 1963, Chassis No. 5 235 301 the cable hook on the clutch pedal shaft was inclined 3° more to the rear in order to obtain the most favourable fitting position for the clutch cable.

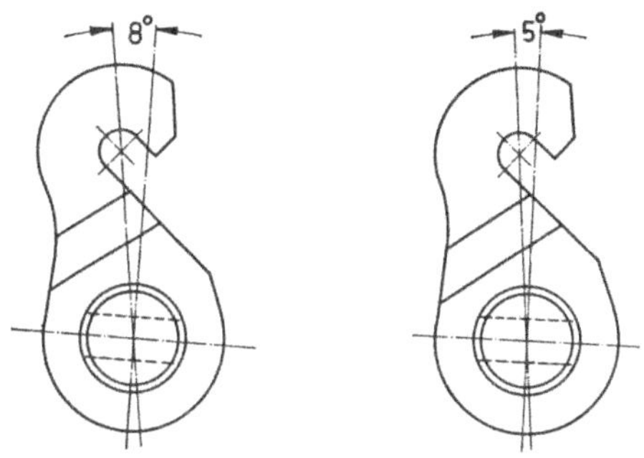

new version previous version

When installing a new clutch pedal shaft, a 10 mm shorter clutch cable (new Part No. 113 721 335) must be used.

2 - Clutch cable

From January 1963, Chassis No. 5 261 830 a 10 mm shorter clutch cable was installed (new Part No. 113 721 335). The 10 mm shorter clutch cables are marked with blue paint on the cable eye. Cables of the previous type (Part No. 111 721 335) will remain available.

Note:

From the 8th to 24th January 1963 (Chassis No. 5 235 301 to 5 261 830) the old type clutch cable (Part No. 111 721 335) was installed together with the modified clutch pedal shaft in production. If difficulty is experienced on these vehicles when adjusting the clutch play, the short cable (Part No. 113 721 335) should be installed.

THIS PAGE INTENTIONALLY
LEFT BLANK

Adjustment of Clutch Pedal Free-Play

The clutch is to be adjusted so that there is a clearance of 1—2 mm (.04″—.08″) between the carbon thrust ring of the release bearing and the clutch release plate with the clutch engaged. Measured at the clutch pedal this clearance is 10—20 mm (.4—.8″). The clearance may be adjusted at the adjusting nut on the cable end.

As the wear on the clutch lining increases, the clearance between the carbon thrust ring and the release plate is reduced until these two parts contact each other. This condition leads to excessive wear or damage. At the same time it reduces the clutch pressure which is liable to result in slippage and burning of the lining.

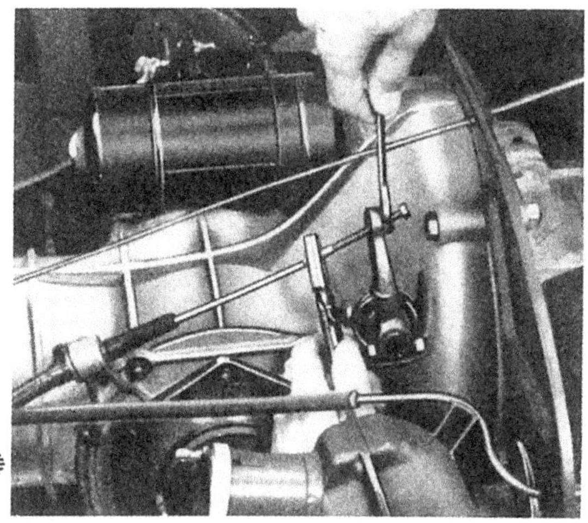

Adjustment

1 - Release lock nut on the threaded cable end.

2 - Adjust clutch clearance by turning the adjusting nut on cable end until the clutch pedal free-play is 10—20 mm (.4″—.8′). Depress clutch pedal several times and recheck pedal free-play.

3 - When the correct adjustment has been reached, hold adjusting nut and tighten lock nut against it.

4 - Grease clutch cable adjusting nut with universal grease.

a = 10—20 mm (.4″—.8″)

Note:

From April 1965, Chassis No. 115685587 (Rear Axle No. 7889618) all Volkswagen are fitted with a modified clutch operating lever (131141719) and a special wing nut (131721349) to facilitate clutch adjustment. It is necessary to ensure that the lugs on the wing nut engage in the grooves in the clutch lever. After making the adjustment, coat the thread at the end of the cable and the wing nut with grease.

The new clutch lever and the wing nut can only be subsequently installed if the transmission case and the left final drive cover are replaced.

Clutch levers of the previous type (111141719B), adjusting nuts (111721349A) and hexagon M7 nuts (N110071) will remain available.

THIS PAGE INTENTIONALLY
LEFT BLANK

Clutch Trouble Shooting

Symptom	Cause	Remedy
1. Noise	a - Needle bearing in flywheel gland nut worn	a - Renew gland nut and fill it with 10 gr. Universal Grease.
	b - Carbon thrust ring excessively worn	b - Renew carbon thrust ring. Insure that release plate and clutch clearance are correctly adjusted. Tell driver not to use clutch pedal as a foot rest
	c - Driven plate fouling pressure plate	c - Renew or straighten driven plate
	d - Weak release lever spring or unequal tension	d - Renew springs
2. Chatter or Grabbing	a - Transmission case not tightly mounted	a - Tighten mounting bolts and nuts
	b - Sag of cable conduit too slight or excessive	b - Correct the sag to 20 or 30 mm (.8″ or 1.2″)
	c - Grease or oil on clutch	c - Renew oil seal. Clean all parts and reline driven plate
	d - Uneven contact of pressure plate, or worn	d - Renew or regrind pressure plate
	e - Release plate not running true	e - Adjust or replace release plate
	f - Unequal tension of thrust springs	f - Renew thrust springs
	g - Cushion segments excessively or unequally set	g - Reset cushion segments or replace clutch driven plate

Symptom	Cause	Remedy
3. Dragging or Incomplete Release	a - Excessive pedal free-play	a - Adjust clutch clearance: 10—20 mm (0.4″—0.8″) at clutch pedal
	b - Sag of cable conduit too great	b - Correct the sag to 20 or 30 mm (.8″ or 1.2″)
	c - Driven plate not running true	c - Straighten or replace driven plate
	d - Cushion segments excessively or unequally set	d - Reset cushion segments or replace clutch driven plate
	e - Plate linings broken	e - Install new linings or replace clutch driven plate
	f - Main drive shaft not running true with gland nut. (Installation tolerances)	f - It is sometimes sufficient to loosen the engine mounting bolts, move the engine slightly and retighten the bolts. Otherwise check gland nut. If the thread is damaged or there is excessive play between inner and outer thread the gland nut cannot be centered correctly
	g - Needle bearing in gland nut defective or insufficiently greased	g - Replace gland nut or grease needle bearing
	h - Splines on main drive shaft or clutch driven plate dirty or burr formation	h - Clean splines. Remove burr
	i - Main drive shaft splines not sufficiently greased	i - Lubricate splines with transmission oil
	k - Sticky clutch linings	k - Wash the linings with fuel
	l - Felt ring in gland nut tight on main drive shaft	l - Replace the gland nut by one which has a better fitting felt ring
	m - Stiffness in the pedal cluster, clutch cable and the operating shaft	m - Grease the parts thoroughly with Universal Grease VW—A 052
	n - Distance between the release plate and clutch cover bearing surface incorrectly adjusted	n - Re-adjust the distance with the clutch adjustment gauge VW 254 a
4. Slippage	a - Lack of pedal free-play due to wear of linings	a - Adjust clutch clearance: 10—20 mm (0.4″—0.8″) at clutch pedal
	b - Grease or oil on clutch linings	b - Replace clutch linings. Replace engine or transmission oil seal if necessary

Description

General

The automatic clutch is a combination of a starting clutch and a gearshift clutch which both work independently of each other. The normal clutch pedal is dispensed with.

When moving off and stopping the vehicle, the clutch operation is carried out solely by means of the starting clutch (centrifugal roller type) in connection with the engine speed.

For gear shifting when the vehicle is in motion the gearshift clutch is disengaged with the assistance of the engine vacuum via a solenoid actuated control valve and a clutch servo. The gears can then be shifted in the normal manner. The re-engagement of the clutch is controlled by means of a two stage vacuum reduction after the gear has been shifted.

1 - Starter ring
2 - Socket head screw
3 - Hexagon head screw
4 - Inner and outer clutch springs
5 - Clutch operating shaft with lever, spring and clamping bolt
6 - Clutch release bearing
7 - Clutch cover
8 - Release plate
9 - Oil seal
10 - Release lever
11 - Centering pin
12 - Centrifugal rollers
13 - Roller housing
14 - Pressure plate
15 - Drive housing
16 - Socket head screw
17 - Socket head screw
18 - Dowel pin
19 - Gland nut with inner serrations
20 - Free wheel
21 - Bearing flange
22 - Clutch plate
23 - Flywheel
24 - Starting clutch plate

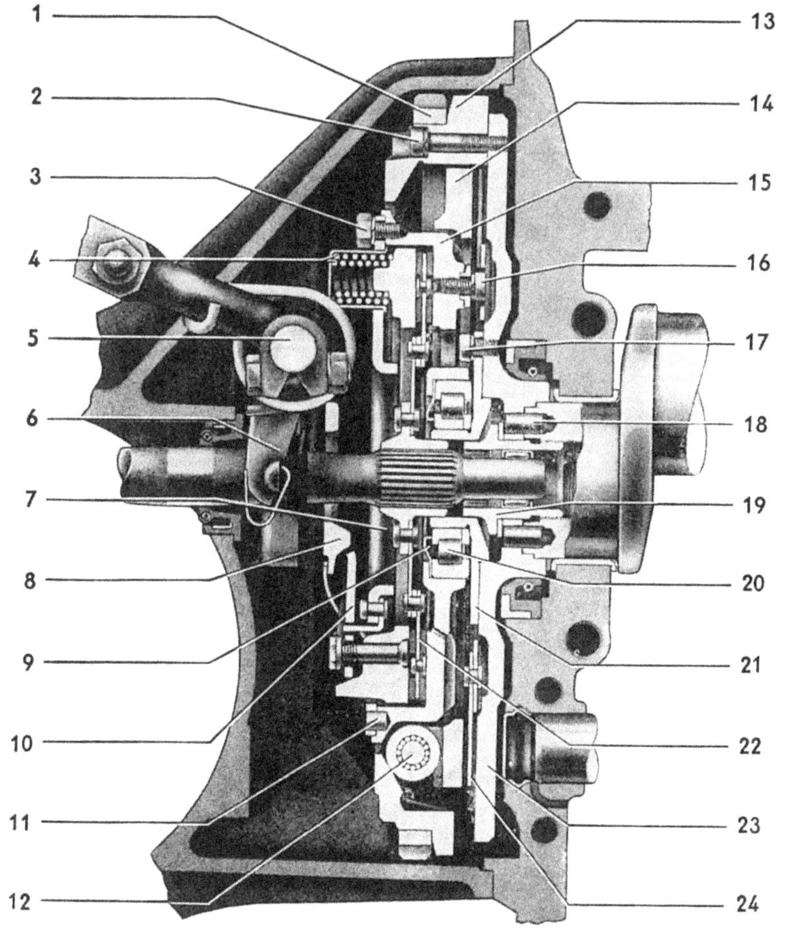

Construction of the System

The starting clutch with is ten centrifugal weights is secured to the engine flywheel. Each weight consists of two small cylindrical rollers which roll on the pressure plate as the engine speed increases and a larger, slightly barrel-shaped roller which is guided by a groove in the pressure plate and forced outwards against a sloping surface in the clutch housing by centrifugal force.

The starting clutch plate is connected to the drive housing of the gearshift clutch by means of spring bands. The plate segments with the sintered-bronze friction linings are riveted to a spring steel carrier ring. The segments are offset axially and when the clutch pressure is applied an undulating elastic deformation of the ring takes place. This action has a considerable influence on the smooth operation of the starting clutch. The drive housing for the gearshift clutch is connected to the engine flywheel by a bearing flange. Between the bearing flange and the drive housing is a freewheel device which provides a constant power connection when the engine is acting as a brake.

The gearshift clutch is attached to the drive housing and is identical in construction and operation to the normal dry single-plate clutch.

The solenoid-operated control valve is attached by three hexagon nuts and bonded rubber mountings to the left side of the engine compartment (looking forward). From the solenoid, one cable goes to terminal 15 of the ignition coil and one to the gearshift lever.

The clutch servo is mounted on the left side of the transmission on a bracket which is secured with the transmission housing screws.

An adjustable rod is screwed into the clutch servo diaphragm and connected to the clutch servo lever by means of an adjusting nut. The adjusting nut is secured with a cotter pin. The clutch servo lever and the clutch operating lever are bolted together with a hexagon head bolt and complete the connection between the clutch servo and the clutch operating shaft. Between the clutch servo lever and the clutch operating lever contact surfaces is an aluminium washer.

Both the clutch servo and the vacuum tank are connected to the control valve by separate pipes. The pipe from clutch servo to control valve and the cable from gearshift lever to control valve pass through a hole, with a grommet, in the front engine cover plate. Another pipe connects the control valve to the intake manifold. A 3.5 mm diameter pipe connects the control valve to the carburetor venturi via a special connection on the carburetor provided for this purpose.

The vacuum tank with welded-on bracket and nuts is attached to the right side of the engine compartment by three hexagon head screws.

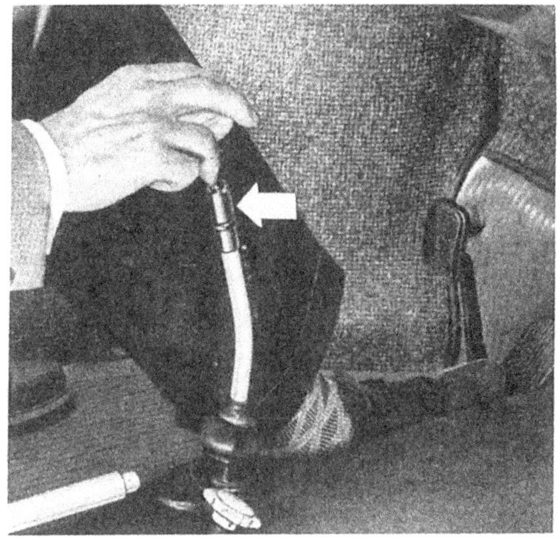

A gearshift lever with a contact piece and connecting cable is installed instead of the normal gearshift lever. The cable is secured with a rubber ring above the hole where it passes through the lever and is then carried along the frame tunnel to the control valve.

Between the control valve and the coil is an 8 amp. fuse.

Maintenance

The following checks and adjustments should be carried out during the Maintenance Service:

1 - Check clutch free play every 5000 km (3000 miles) and adjust if necessary. The play should be checked at the clutch lever and adjusted by means of the adjusting nut on the clutch servo rod.

2 - Polish and adjust gearshift lever contacts every 5000 km. The contact gap should be 0.25 mm (0.010″).

3 - Clean control valve air filter every 25000 km (15000 miles).

The Operation of the Gearshift Clutch

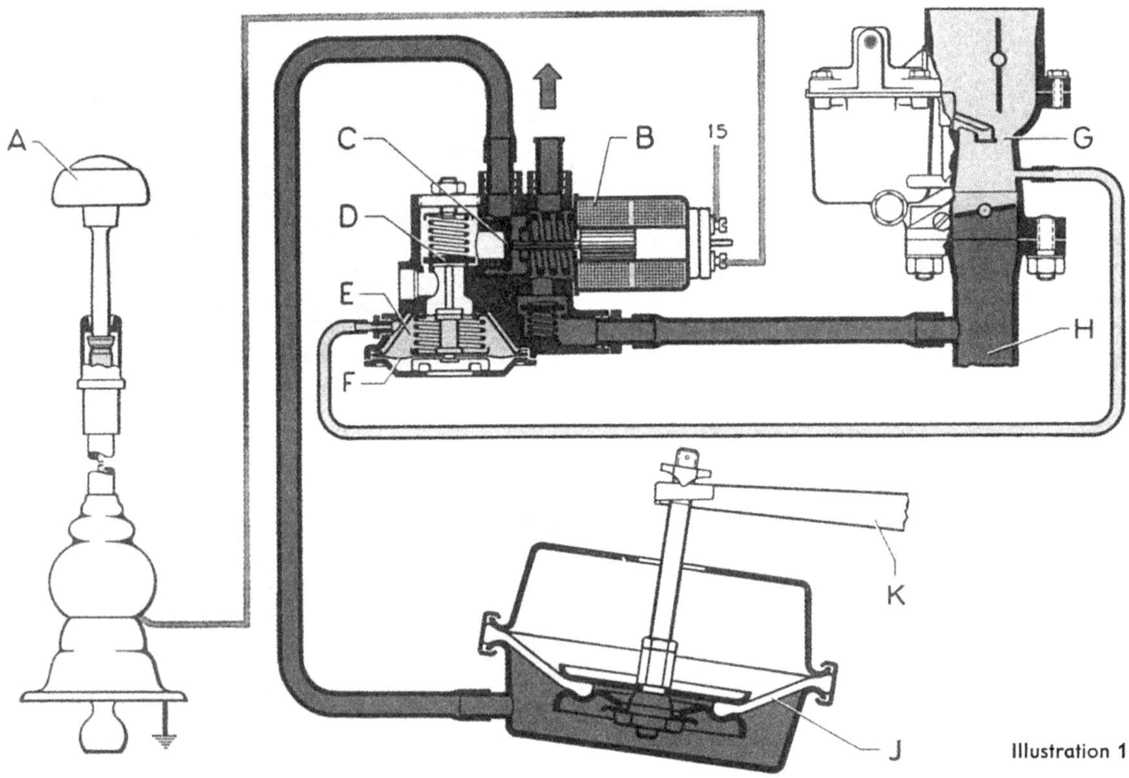

Illustration 1

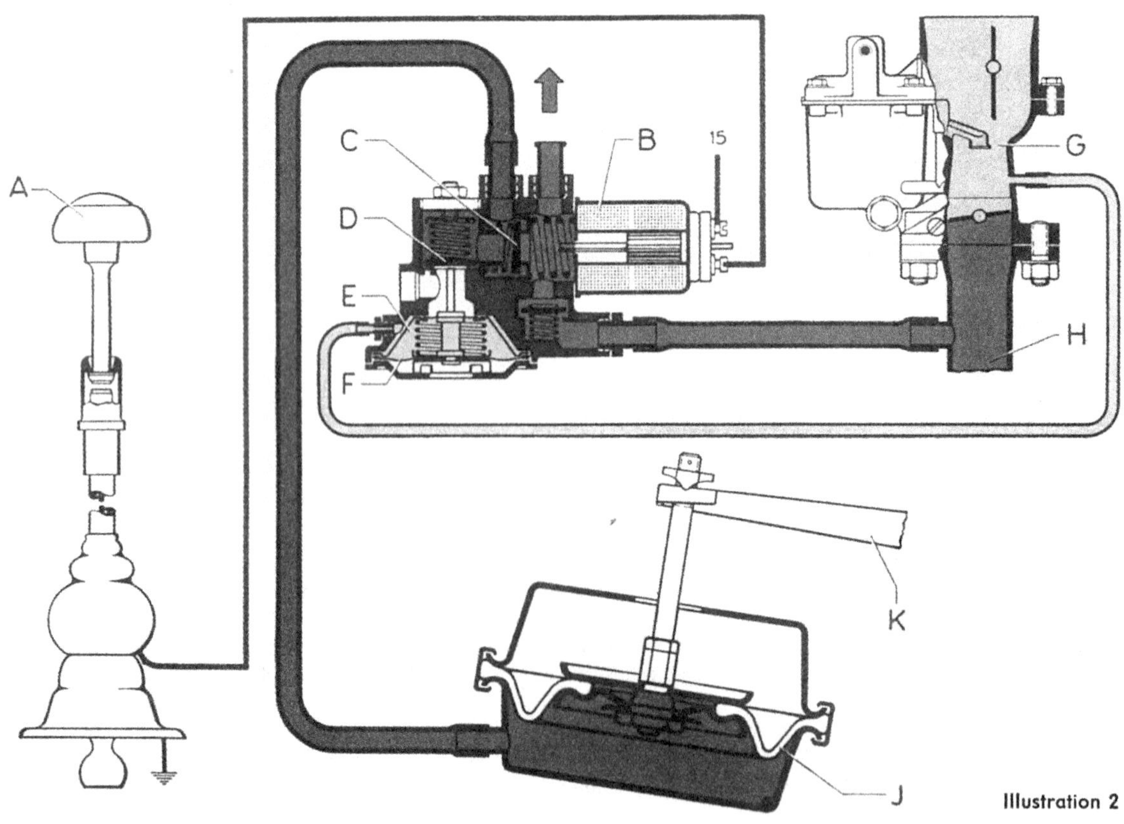

Illustration 2

When the gearshift lever A is touched, current is supplied to the solenoid B (Illustration 1). The solenoid actuates valve C and connects the servo system with the intake manifold H of the engine. Atmospheric pressure then forces the diaphragm J of the clutch servo down and this movement, via the clutch servo lever K, the clutch operating lever and the release bearing, disengages the gearshift clutch. The gearshift clutch is thus disengaged at the beginning of each gear shifting operation. A vacuum tank is installed to build up a reserve of vacuum and ensure that an adequate supply is available under all operating conditions. The re-engagement of the clutch commences as soon as the gearshift lever is released. This cuts the flow of current to the solenoid and allows valve C to return to its initial position and disconnect the servo system from the intake manifold.

The reduction of the vacuum in the clutch servo, and the corresponding increase in the torque to be transmited by the gearshift clutch, takes place in two stages.

In the first stage (Illustration 2), the vacuum is reduced very quickly by means of the reducing valve D until the gearshift clutch begins to engage. The braking effect of the engine, which is only running slowly immediately after the gear shift, is naturally very slight. The degree of vacuum to be reduced in the first stage can be regulated by means of the reducing valve D.

In the second stage the residual vacuum is dissipated gradually via a small jet so that, after a few seconds, the gearshift clutch is fully engaged. This delay only takes full effect if the accelerator is not depressed.

In order to prevent the vacuum which still exists in the servo system causing excessive gearshift clutch slip if the accelerator is depressed immediately after the gear shift has been completed, the reduction of the vacuum is speeded up in connection with the depression in the carburetor venturi. The chamber in the control valve is provided for this purpose and is connected to the carburetor G by a small diameter plastic tube. This chamber is sealed at the bottom by a spring-loaded diaphragm F and is connected to the reducing valve D by means of a plunger. As long as there is no vacuum or a very slight vacuum in the chamber E, the diaphragm F remains in the lower position (Illustration 1 and 2) and the plunger does not contact the reducing valve D. This condition exists when the engine is stopped or running very slowly and the thro tle valve in the carburetor is closed.

When the thrcttle valve is opened and the engine speed starts to rise, the vacuum in chamber E increases, lifts the diaphragm F and opens the reducing valve D. This control of the clutch operation after the gear has been shifted gives an increasingly rapid and positive engagement to the gearshift the higher the engine speed rises and the further the throttle valve is opened. It also ensures the smooth engagement of the gearshift clutch when changing down to use the engine as a brake, even if the engine should stall during the gear shifting operation.

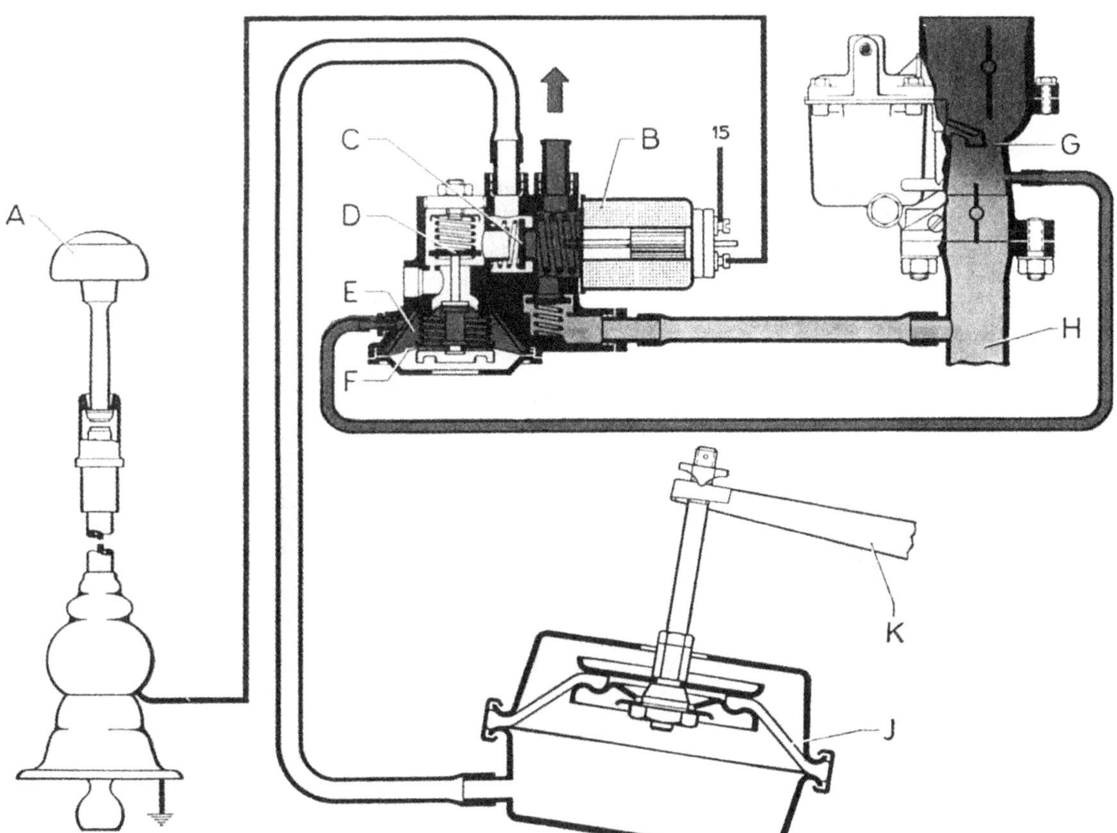

Illustration 3

The Operation of the Gearshift and Starting Clutches

When the accelerator is depressed and the engine speed increases, the rollers are pressed outwards by centrifugal force. This pressure causes the roller pressure plate to overcome the release springs and it bears gradually against the starting clutch plate until the vehicle commences to move. The starting operation begins at approximately 950—1000 r.p.m. and full torque is transmitted to the transmission when the speed has reached about 1500 r.p.m. The slip in the starting clutch thus ceases at approximately 1500 r.p.m. at full throttle and at a correspondingly lower speed at part throttle. The gearshift clutch is always engaged during the starting operation.

When the brakes are applied, the starting clutch remains engaged until the engine speed has dropped to about 1000 r.p.m. and the braking effect of the engine is retained via the starting clutch down to this speed. As soon as the speed drops further and the starting clutch ceases to operate, the freewheel device comes into action and transmits the braking effect of the engine via the gearshift clutch, thus cutting out the starting clutch. In this way the engine braking action is effective without interruption right down to idling speed and, when the ignition is switched off, until the vehicle is stationary.

The freewheel device also provides a connection between the engine and road wheels when the vehicle is parked on a gradient and a gear engaged. It is essential, however, to engage a forward gear if the vehicle is parked facing downhill and reverse gear if facing uphill.

When tow-starting the vehicle, the freewheel device also provides a power connection between the engine and the gearshift clutch.

Illustration 1:

The rollers A rest in the starting position when the engine is at a standstill and idling. The pressure plate is lifted by means of springs thereby freeing the starting clutch plate.

In this condition the starting clutch is separated and the connection between engine and transmission interrupted.

Illustration 2:

When the accelerator is depressed and the engine speed increases, the rollers are pressed outwards by centrifugal force. This causes the roller pressure plate to overcome the release springs and it bears gradually against the starting clutch plate until the vehicle commences to move. The starting operation begins at approximately 950—1000 r.p.m. and full torque is transmitted to the transmission when the speed has reached about 1500 r.p.m.

Illustration 3:

When shifting gear whilst driving, the gearshift clutch is released with the aid of the vacuum in the intake manifold, the clutch plate then runs free so that the shifting of the gear can be carried out in the normal manner.

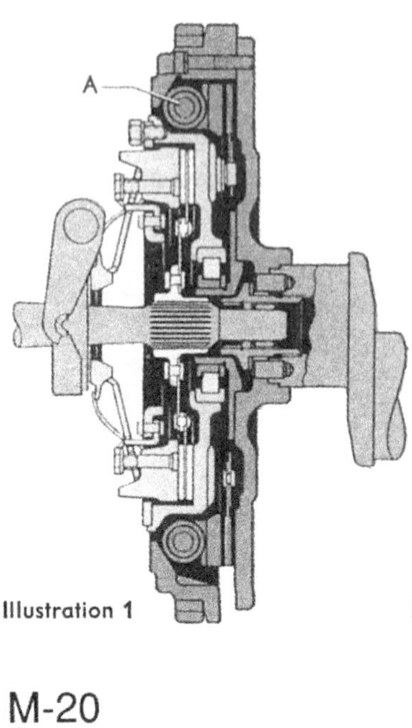

Illustration 1

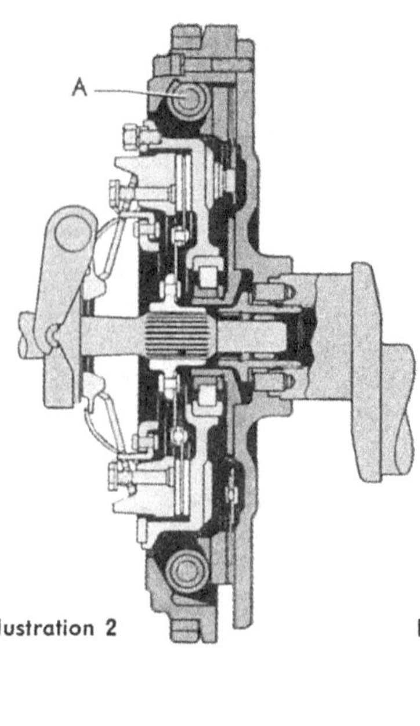

Illustration 2

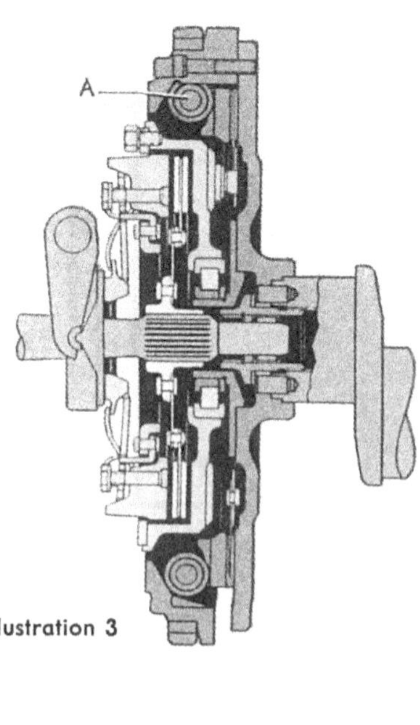

Illustration 3

Automatic Clutch Removal and Installation

Removal

1 - Remove engine, paying attention to the following points:

a - Remove the cable from the ignition coil to the control valve.

b - Remove the pipes from the intake manifold to the control valve and from the control valve to clutch servo.

c - Remove the pipe from carburetor to control valve.

2 - Unscrew the screws securing the clutch cover to the drive housing evenly. Loosen the screws one or two turns alternately in a cross fashion until the spring tension is released, to avoid distorting the clutch cover.

3 - Remove clutch pressure plate and clutch plate.

4 - Remove the gland nut with a special socket VW 173 and a flywheel retaining clip VW 215 b.

5 - Pull flywheel off, complete with starting clutch.

Installation

Installation takes place in the reverse order, paying attention to the following points.

1 - Check condition of flywheel teeth.

2 - Check the dowel pin holes in the flywheel.

3 - Check the dowel pin holes in the crankshaft. If necessary use new dowel pins.

4 - Adjust crankshaft end play.

5 - Check needle bearing for wear. If necessary, clean the bearing and lubricate with Universal Grease.

6 - Replace flywheel gasket if necessary.

7 - When installing the flywheel with starting clutch, the unbalance paint marks on the drive housing edge and the crankshaft should be spaced 180° from each other.

8 - When installing the clutch pressure plate with clutch plate, the unbalance paint marks on the drive housing edge and the clutch pressure plate should be spaced 180° from each other.

Part	Distinguishing Mark
Crankshaft	Paint mark at side in base of gland nut boring
Flywheel with starting clutch	Paint mark on the centrifugal roller housing
Gearshift clutch	Paint marks on gearshift clutch housing and on the outer edge of the clutch cover

9 - Tighten gland nut to a torque of 30 mkg (215 ft. lbs.).

10 - When the flywheel has been installed, the drive housing should turn easily in the freewheel direction.

11 - Check the clutch plate contact surface in the drive housing for wear and polish with polishing cloth if required. If necessary, disassemble the flywheel with starting clutch and replace the drive housing complete with bearing flange and freewheel.

12 - Check the clutch plate for lining wear, run-out (maximum 0.8 mm/0.031″), condition of segments and riveting at the hub. If necessary, replace clutch plate complete.

13 - Check pressure plate for wear and distortion. If the plate shows signs of uneven wear it should be repaired or renewed.

14 - Examine release levers and springs. If necessary, disassemble pressure plate and replace damaged parts.

15 - Check release plate for wear and damage.

16 - Check carbon thrust ring for wear and cracks. If necessary, renew complete release bearing. Note correct position of retaining springs. Under no circumstances should the carbon ring release bearing be replaced by the ball bearing type.

17 - Check the clutch operating shaft bushes in the transmission case for wear.

18 - Install the clutch plate with mandrel VW 219 to facilitate centering. The retaining lever VW 682 can be used to hold the clutch pressure plate when tightening the hexagon head screws. Make sure that the holes in the clutch cover seat uniformly on the dowel pins in the drive housing.

19 - To avoid distortion of the clutch cover the screws should be tightened evenly one or two turns at a time.

Starting Clutch Disassembly

General

On account of the freewheel device the drive housing can only be turned to the right. Each socket head screw must, therefore, be screwed completely out. if only loosened, the heads of the screws would strike the back of the drive housing and render the complete starting clutch useless. Before disassembling the starting clutch, mark all parts so that they can be assembled in the original position.

Disassembly

1 - Remove the nine socket head screws from the starting clutch with a 5 mm Allen wrench.

2 - Turn the drive housing until the two holes are over two of the six socket head screws.

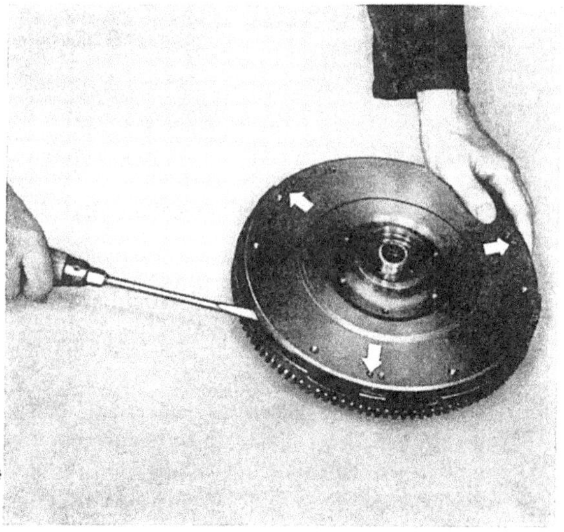

3 - Remove the screws completely with a 5 mm Allen wrench.

4 - Turn the drive housing until all screws are removed.

5 - With a screwdriver, lever the flywheel evenly off the three dowel pins in the pressure plate, after marking the position of the parts.

6 - Remove the gland nut.

7 - Remove the three socket head screws securing the driven plate, after marking the position of the parts.

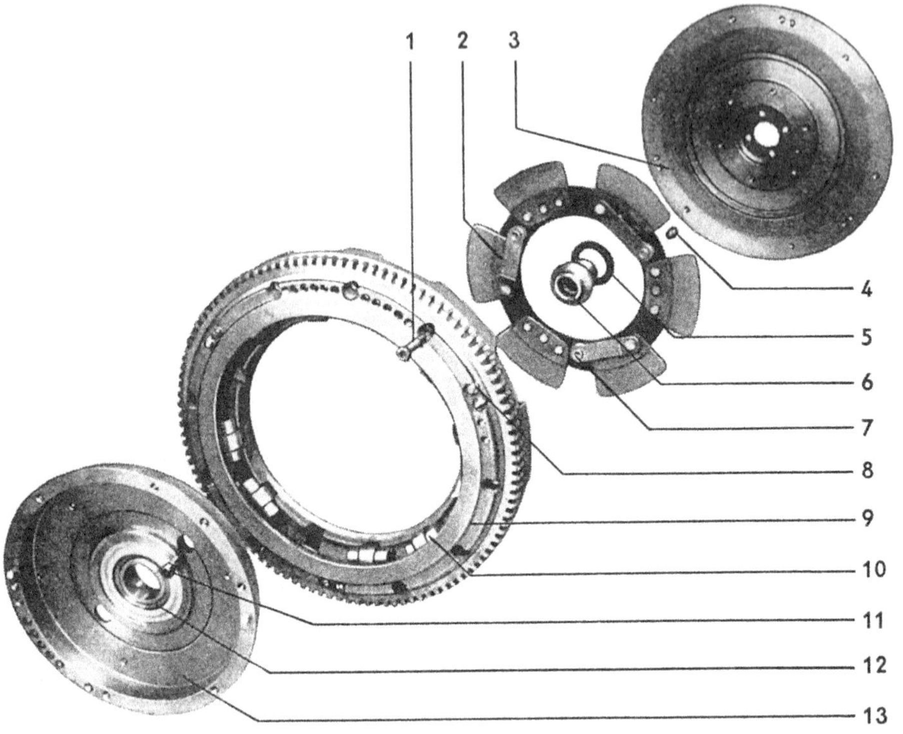

1 - Socket head screw
2 - Starting clutch plate
3 - Flywheel
4 - Socket head screw
5 - Lock washer
6 - Gland nut with inner serrations
7 - Spacer sleeve
8 - Connecting screw
9 - Roller housing
10 - Centrifugal rollers
11 - Socket head screw
12 - Freewheel
13 - Drive housing

8 - Remove drive housing with bearing flange and freewheel.

9 - If necessary, clean centrifugal rollers and working surfaces. The rollers can be removed from their seats if the roller housing is lifted slightly with a screwdriver. Clean rollers and working surfaces with benzine and dry with compressed air. If necessary, polish the working surfaces with polishing cloth.

Note:

The connecting screws of the pressure plate and roller housing are adjusted by the manufacturer and may not be altered.

10 - Check that the freewheel operates correctly.

Note:

The freewheel is adequately lubricated by the manufacturer and cannot be lubricated subsequently. Under no circumstances should the freewheel be washed in benzine or similar cleaning solutions. Defective freewheels must be replaced complete with bearing flange and drive housing.

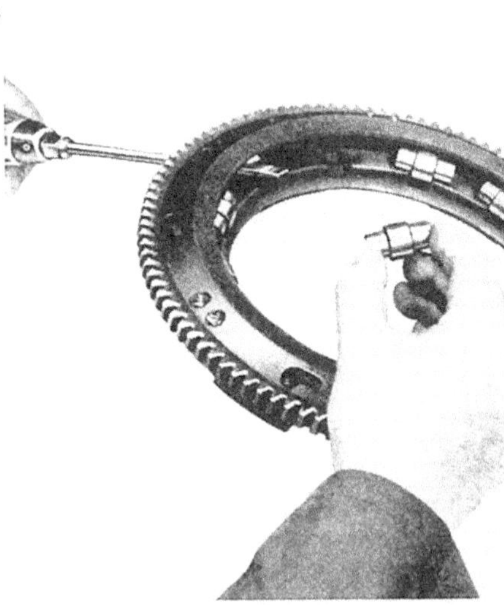

Assembly

The assembly takes place in the reverse order, noting the following points.

1 - Place the flywheel on the three dowel pins in the pressure plate, after aligning the marks.

2 - Tighten all socket head screws in a cross fashion to a torque of 1.5—1.8 mkg (11—13 ft. lbs.).

Clutch Cover with Pressure Plate

Clutch Disassembly

1 - Remove gearshift clutch.

2 - Mark the parts so that it will not be necessary to re-balance the clutch after assembly.

3 - Free the three special release lever nuts by grinding off the spot welds and remove the levers, springs and release plate.

4 - Loosen the clutch cover screws evenly and remove the cover. Take out the clutch springs, spring seats and pressure plate.

Check all individual parts before reassembling clutch.

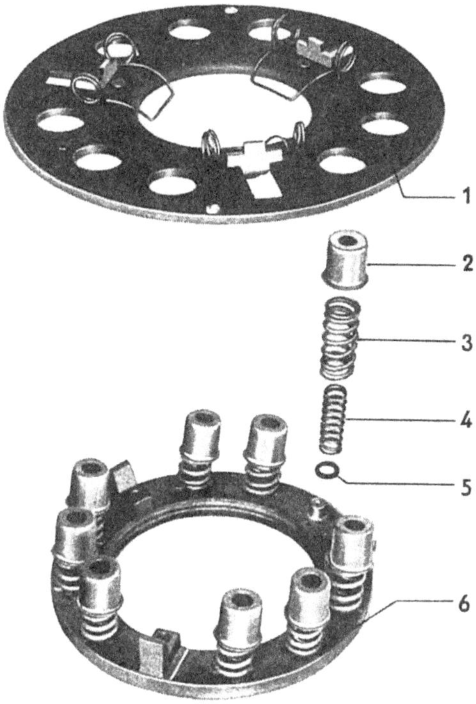

1 - Clutch cover
2 - Spring cup
3 - Clutch spring — outer
4 - Clutch spring — inner
5 - Spring seat
6 - Pressure seat

Checking

1 - Check clutch cover for distortion and straighten if necessary. Clutch cover distortion is usually caused by the screws being loosened or tightened unevenly.

2 - Clean the pressure plate and check for distortion, wear and cracks. If the friction surface of the plate shows an uneven contact pattern, the clutch will have a tendency to grab. If the run-out does not exceed 0.1 mm (0.004″) the pressure plate can be reground and polished with polishing cloth. Otherwise the plate should be replaced.

3 - Check clutch springs.

	Outer spring	Inner spring
Free length in mm	35.5—2 1.396—0.078″	33.5—1.5 1.318—0.058″
Compressed length in mm	21.5/0.845″	19.5/0.767″
Loading in kg	26.0+2 57.2+4.4 lbs.	11.0+1 24.2+1.1 lbs.
Wear limit in mm	2.0—3.0 0.78—0.118″	2.0—3.0

The lengths of the inner and outer springs must not vary by more than 3 mm, as unequal spring pressures can lead to clutch grabbing.

4 - Check release plate for wear and damage. Plates which are damaged or marked by excessive heat should be replaced.

Clutch Plate

Checking

1 - Check clutch plate. The six spring segments riveted to the driven plate are set at a slight curve to give a cushioning effect to the clutch plate operation. To ensure proper clutch operation it is essential that all the segments are set to the same extent.

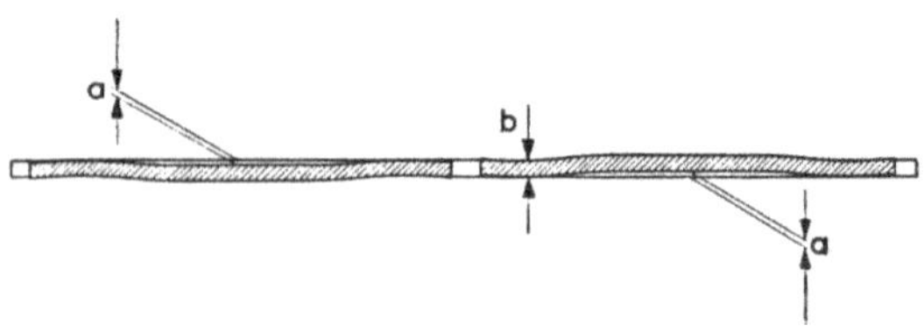

a = 0.5—0.8 mm (0.196"—0.0314")   b = 1.3 mm (0.0512")

The clutch plate should slide easily on the drive shaft splines without excessive side play. Worn parts should be replaced.

2 - Check clutch linings. If they are oily, burnt, cracked or worn almost down to the rivets, they should be replaced. To correspond with the curved setting the linings should be riveted on the convex side of the segments.

3 - Check clutch plate for run-out with the linings installed. The maximum permissible run-out is 0.8 mm (0.0314").

4 - Check the thickness of the clutch plate.

> **Important**
>
> Only clutch linings approved by the Volkswagenwerk should be installed.

Clutch lining

Outside diameter	160 ± 1 mm/6.3 ± 0,040"
Inside diameter	110 ± 1 mm/4.330 ± 0.040"
Thickness	2.75 ± 0.1 mm/0.107 ±0.004"

When riveting the linings to the curved segments make sure that each second hole in the lining is a clearance hole. The riveting should be carried out so that both linings are riveted separately to each segment.

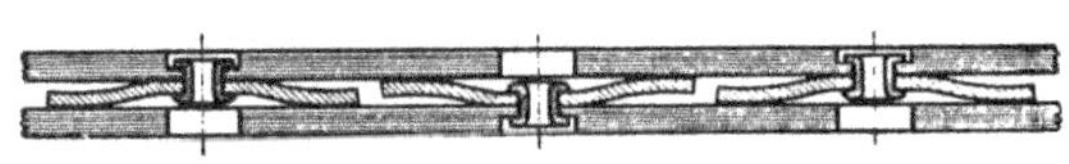

c = 6.3 + 0.3—6.8 + 0.3 mm
(0.248 + 0.0118"—0.267 + 0.0118")

Note the following points when assembling the clutch:

1 - Renew the special nuts and bolts for the release levers.

2 - Place the clutch cover with clutch plate in the drive housing.

3 - Tighten the cover securing screws evenly one or two threads at a time to avoid distorting the cover.

4 - Lubricate the working points of the release levers lightly with Special Grease.

5 - The distance from the contact surface for the clutch cover on the drive housing to the release plate is 24 ± 0.2 mm/0.945 ± 0.008".

6 - Check the height and parallelism of the release plate with a depth gauge.

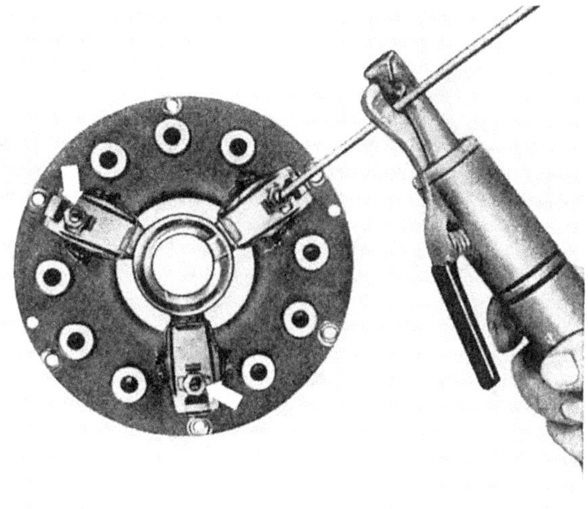

7 - Secure the special nuts by spot welding.

THIS PAGE INTENTIONALLY
LEFT BLANK

Control Valve Removal and Installation

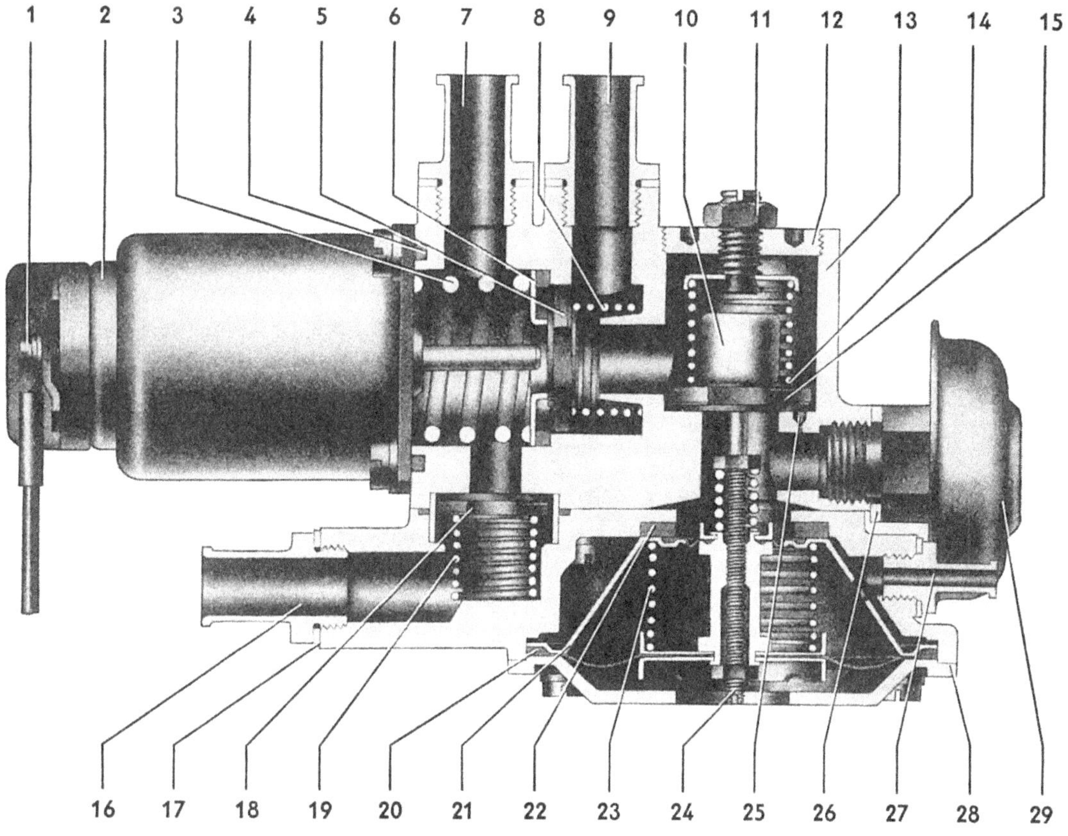

1 - Screw for cable
2 - Solenoid
3 - Distance spring
4 - Sealing ring
5 - Main valve
6 - Main valve seat
7 - Connection to vacuum tank
8 - Main valve spring
9 - Connection to clutch servo
10 - Compensating weight
11 - Screw and lock nut for reducing valve adjustment
12 - Screw cap
13 - Housing for control valve
14 - Reducing valve spring
15 - Reducing valve
16 - Connection to intake manifold
17 - Gasket
18 - Check valve
19 - Check valve spring
20 - Diaphragm
21 - Diaphragm housing
22 - Sealing ring
23 - Diaphragm spring
24 - Diaphragm push rod
25 - Jet
26 - Gasket
27 - Connection to carburetor
28 - Vacuum chamber cover
29 - Air filter

Removal

1 - Place rear of vehicle on blocks and remove left-hand rear wheel.

2 - Disconnect cable from solenoid.

3 - Remove the four pipes from the control valve.

4 - Remove securing nuts and take control valve off side panel.

5 - If necessary, remove screws and detach solenoid.

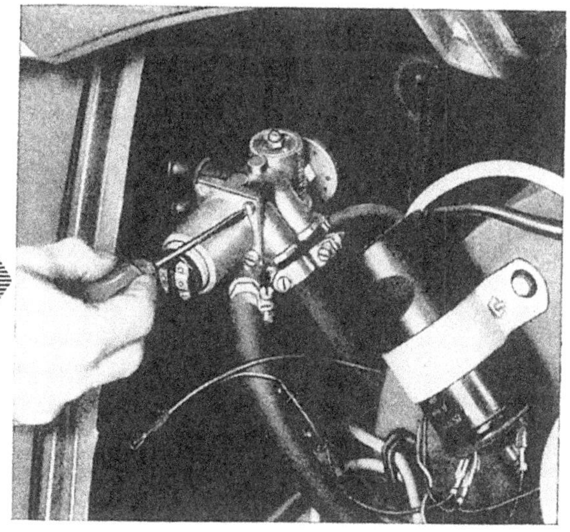

Installation

Installation takes place in the reverse order.

If the control valve does not operate satisfactorily, only the solenoid can be replaced. The valves are set at the factory and require no maintenance.

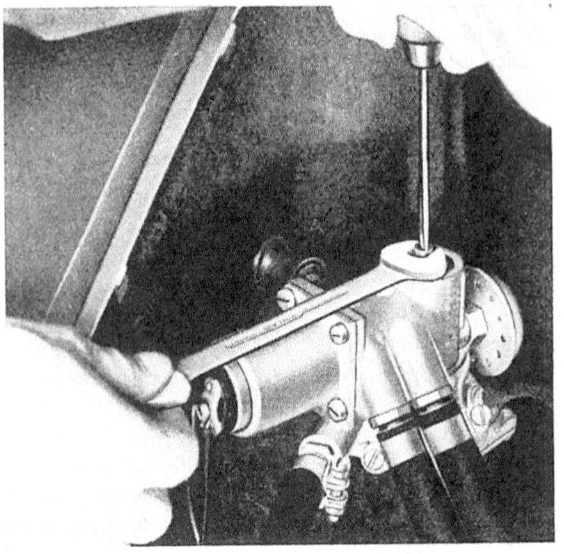

Note:

The control valve of the automatic clutch is adjusted so that the clutch engages smoothly after the gear has been shifted. After being in use for some time the clutch plate contact pattern improves and this may cause the clutch to grab under certain conditions. This alteration can be compensated for by adjusting the reducing valve in the control valve.

Clutch engages too quickly:

1 - Loosen locknut.

2 - Turn adjusting screw 1/4—1/2 a turn to the right.

3 - Hold screw and tighten locknut.

Clutch engages too slowly:

1 - Loosen locknut.

2 - Turn adjusting screw 1/4—1/2 a turn to the left.

3 - Hold screw and tighten locknut.

The adjustment of the reducing valve should be checked during a road test and can be adapted to the customer's individual driving style.

Note:

When fitting new hoses between the control valve and the intake manifold, servo motor or vacuum container on Volkswagen 1200 vehicles with automatic clutches, it is only permissible to use the proper type of hose (Part No. 141142361). This hose is reinforced with coiled wire and does not collapse under the influence of the vacuum.

On no account should other types of hose — such as the hose between crankcase breather and air cleaner for example, which looks very similar — be used for this purpose.

Clutch Servo and Vacum Tank Removal and Installation

Clutch Servo

Removal

1 - Place vehicle on blocks and remove left-hand rear wheel.

1 - Cotter pin
2 - Adjustment nut
3 - Clutch rod
4 - Boot
5 - Clutch servo
6 - Piston rod
7 - Lock nut
8 - Diaphragm
9 - Rubber plug
10 - Connection to control valve

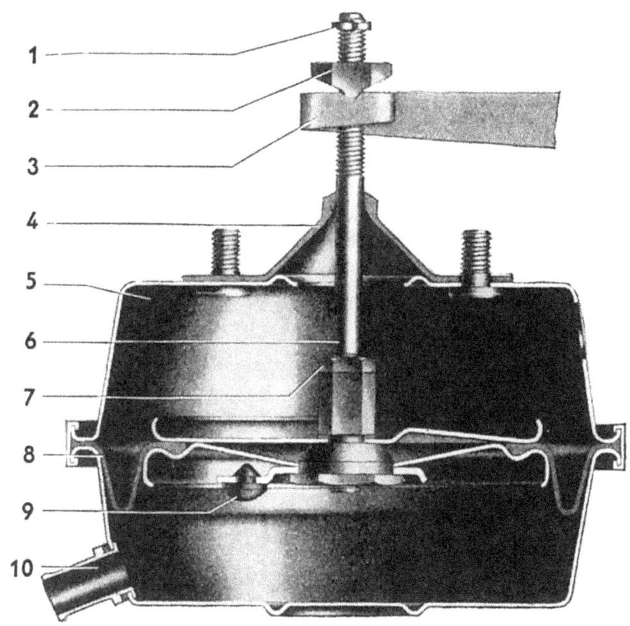

2 - Remove cotter pin from servo rod.

3 - Press clutch lever down and screw adjusting nut off.

4 - Remove connecting pipe.

5 - Unscrew retaining nuts on bracket and remove clutch servo.

6 - Push the rubber boot off the rod and loosen the locknut while holding the threaded portion in the diaphragm with a second wrench. Screw the rod out of the diaphragm.

Installation

Installation takes place in the reverse order, paying attention to the following points:

1 - Screw the rod into the threaded portion of the diaphragm to the last thread and tighten the locknut.

2 - Check that the boot between the clutch servo and bracket is properly seated.

3 - Secure the adjusting nut with a new cotter pin.

4 - The adjustment of the clutch lever is described under the heading "Clutch Adjustment".

Vacuum Tank

Removal

1 - Place rear of vehicle on blocks.

2 - Remove connecting pipe.

3 - Unscrew three securing screws and remove tank.

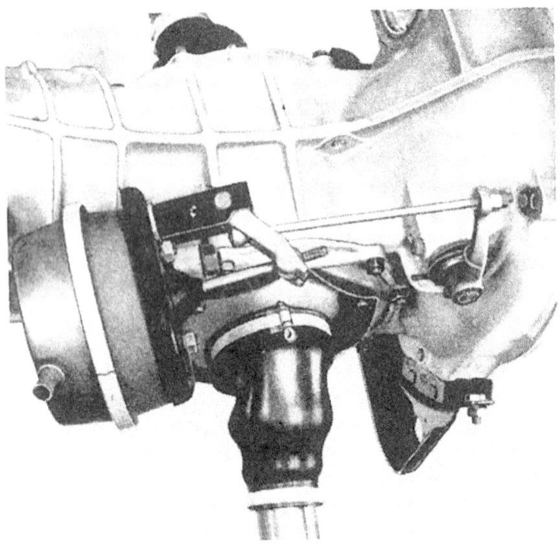

Installation

Installation takes place in the reverse order.

Note:

From December 1962, Chassis No. 5 218 324 when the fresh air heating was introduced, the location of the vacuum tank was changed and the tank made flatter in shape. At the same time, the position of the clutch servo and the operating linkage was altered. The long lever used formerly has been replaced by a double linkage.

Part Numbers:	**new**
Vacuum tank	113 142 061 A
Special bolt for tank	113 142 347 A
Bracket for clutch servo	311 142 303
Lever for clutch rod	311 142 319
Clutch rod for servo	311 142 711
Adjusting nut for piston rod	311 142 721

The previous parts will remain available.

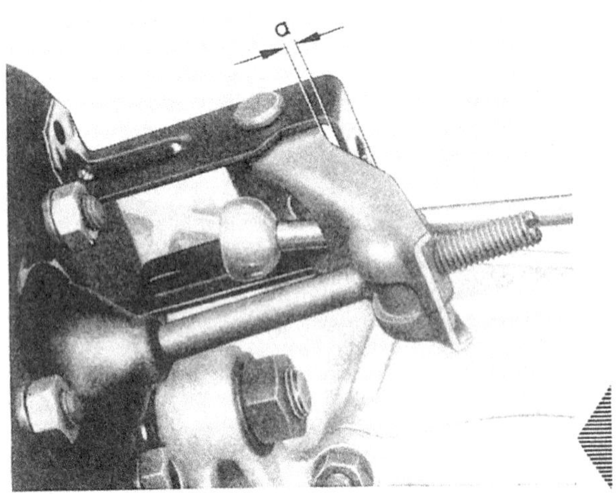

Adjusting clutch play

Basic adjustment

The local manufacture tools VW 683 and VW 684/1 are not required when adjusting the clutch play on vehicles with the new type of linkage.

1 - Remove left rear wheel.

2 - Take vacuum hose off clutch servo.

3 - Loosen lock nut and adjusting nut on clutch rod and screw them back.

4 - Pull piston rod out until piston bottoms in housing.

5 - Adjust piston rod nut so that the lever is 1 mm (a) from the support bracket when in the rest position.

6 - Turn clutch rod adjusting nut until the clutch release bearing touches the release plate.

7 - Slacken adjusting nut until there is about 10 mm (S) play at the clutch rod lever.

8 - Lock adjusting nut.

9 - Connect vacuum hose and tighten clip.

10 - Fit rear wheel.

Adjusting clutch play

1 - Remove left rear wheel.

2 - Disconnect vacuum hose from servo.

3 - Loosen locknut.

4 - Turn clutch rod adjusting nut until the play at the clutch rod lever is 10 mm.

5 - Check basic setting of clutch rod lever on support bracket.

6 - Lock adjusting nut.

7 - Connect vacuum hose and tighten clip.

8 - Fit rear wheel.

Vacuum Tank
(from December 1962)

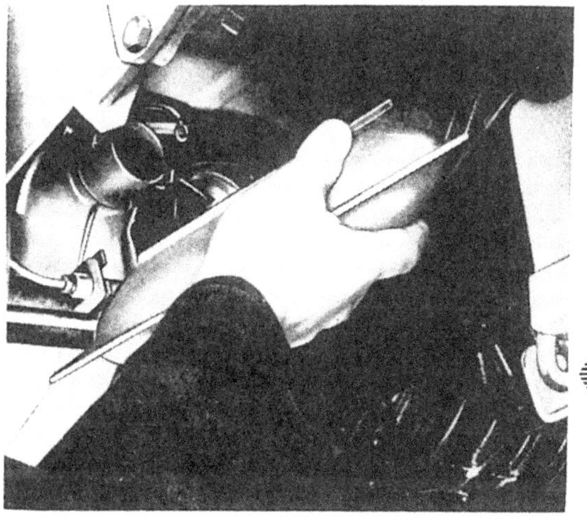

Removal

1 - Place rear of vehicle on blocks.

2 - Remove connecting pipe.

3 - Remove right-hand noise damper.

4 - Remove starter.

5 - Unscrew four nuts on luggage compartment floor.

6 - Take vacuum tank off.

Installation

This takes place in the reverse order.

THIS PAGE INTENTIONALLY
LEFT BLANK

Cleaning and Adjusting Gearshift Lever Contacts

Removal

1 - Bend up gearshift lever sleeve lockplate.

2 - Hold sleeve with an open-end wrench and loosen locknut.

3 - Screw sleeve off and remove lever upper part with spring.

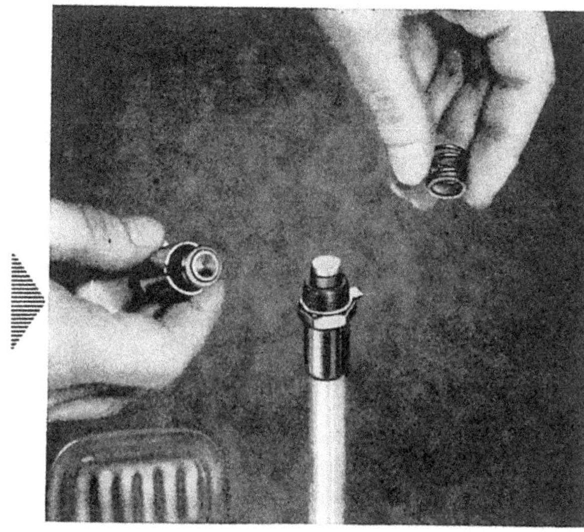

4 - If necessary, clean the contact surfaces with polishing cloth and remove any rough edges with a fine file.

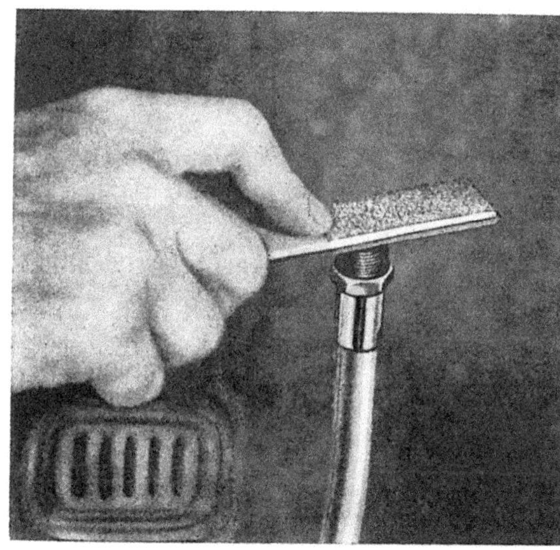

5 - If the contact surfaces are badly worn, the contacts should be replaced. To do this it is necessary to disconnect the cable at the control valve and pull it through to the gearshift lever. Then lift the gearshift lever boot and push the rubber ring off the cable. Pull the contact and cable out of the lever. The installation of a new contact takes place in the reverse order.

Note:

From Chassis No. 3452000, the number of wires in the cable from the gearshift lever to the control valve was trebled without increasing the cross sectional area. The bending strength has been improved by a modification to the insulating material. The new cable is white in color and the Part No. 113142409 remains unchanged.

To improve the routing of the cable the hole in the gearshift lever has been moved 180°.

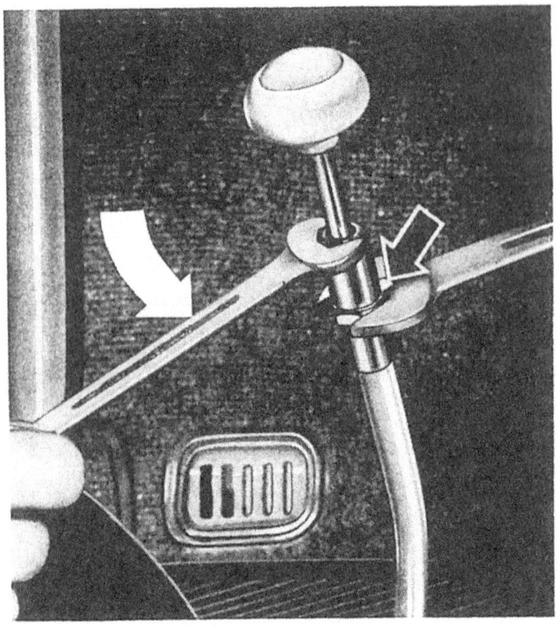

Installation

6 - Place a new lockplate over the locknut on the shift lever with the small raised portion uppermost.

7 - Place the upper part of the lever with sleeve and spring on the shift lever and screw on tightly.

8 - Adjust contact gap. Screw the sleeve on until the contacts are together. Turn the locknut until it touches the sleeve, taking care that the raised portion in the lockplate engages properly in the sleeve. Then turn the sleeve back 1/3 of a turn to give the correct clearance of 0.25 mm (0.010″).

9 - Hold the sleeve with a wrench and tighten the locknut.

10 - Bend the lockplate down over the locknut.

Note:

The contact in the lower part of the gearshift lever is pressed into the insulating sleeve. If there is any looseness at this point the cable will tend to push the contact upwards when the shift lever is moved. This brings the contacts together and causes the gearshift clutch to be operated unintentionally. In such cases the contact support piece should be screwed off the shift lever and replaced complete with contact.

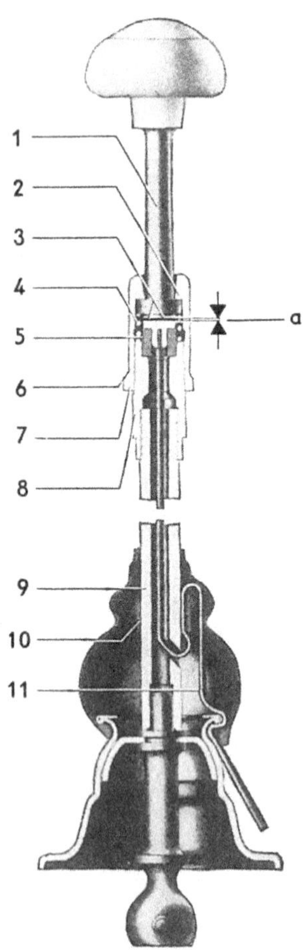

a = 0.25 mm (0.010″)

1 - Shift lever, upper part
2 - Shift lever sleeve
3 - Contact surfaces
4 - Spring
5 - Contact piece
6 - Lock plate
7 - Lock nut
8 - Shift lever extension
9 - Shift lever, lower part
10 - Rubber ring
11 - Cable

Note:

From Chassis No. 3597114 a stronger spring was installed in the upper part of the gearshift lever to eliminate the slight clicking noise which occurred sometimes, particularly when driving on bad roads. The new spring can be service installed.

General

As the clutch linings wear the clearance between the carbon ring and the clutch release plate decreases until these parts are touching each other. When this occurs the adjusting nut should be screwed back to the cotter pin.

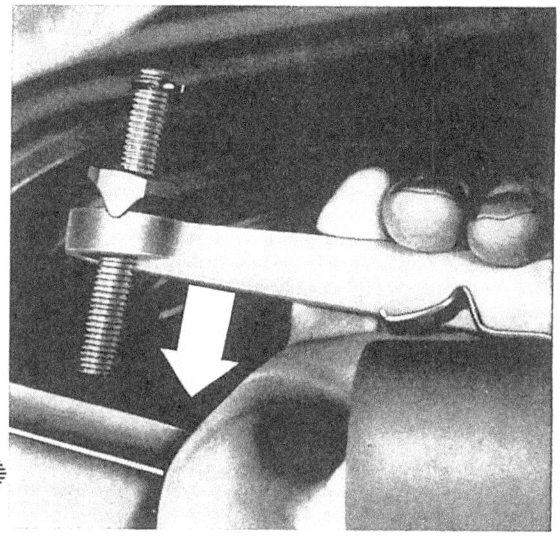

Checking and Adjusting Clutch Clearance

1 - Lift the vehicle.

2 - Pull the clutch lever down until the resistance of the carbon ring contacting the release plate is felt.

3 - If clutch lever can only be moved a small amount (5—10 mm/0.2—0.4″) instead of about 30 mm (1.18″) which is obtained with a newly lined clutch plate, the adjusting nut must be turned back to the cotter pin. The lugs on the adjusting nut must engage the slots in the clutch lever.

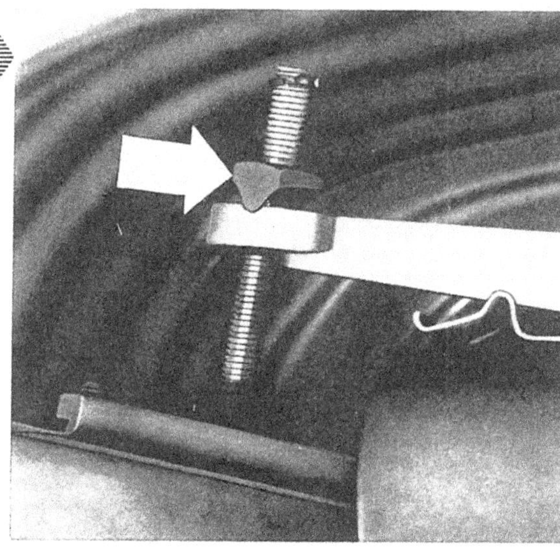

Note:

If the adjusting nut has already been turned back to the cotter pin and the lever cannot be moved downwards, the clutch plate must be relined or replaced.

Adjusting Clutch Clearance (Basic Adjustment)

Every time an engine, transmission and automatic clutch is installed, the basic adjustment must be checked as follows and, if necessary, rectified.

1 - Lift vehicle and remove left-hand rear wheel.

2 - Pull the clutch lever down until resistance is felt and place the retaining device VW 683 (local manufacture) between clutch lever and luggage compartment floor.

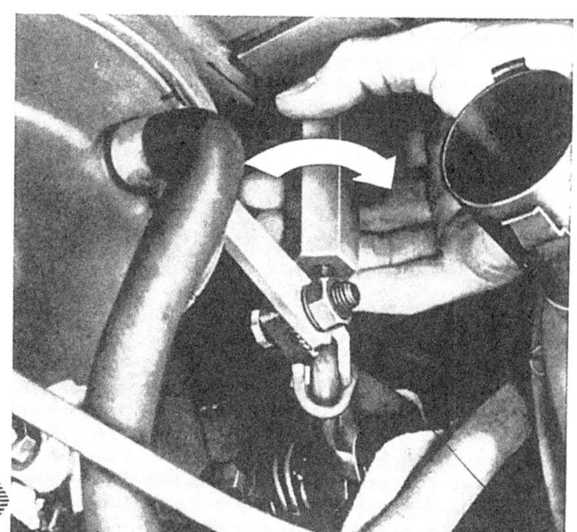

3 - Set the adjusting nut to the measurement a = 17 mm (0.67″) with the help of the gauge VW 684/1 (local manufacture).

4 - Check measurement b = 42 mm (1.65″) between clutch lever and clutch servo bracket with the gauge VW 684. If necessary, loosen the clamp bolt and adjust the clutch lever to the correct measurement of 42 mm. If necessary, replace the aluminium washer. Then retighten clamp bolt to a torque of 6 mkg (43 ft. lbs.).

5 - Remove the retaining device VW 683.

Note:

When a new clutch operating lever or shaft has been installed, take care that the play between these parts is taken up in the working direction of the clutch lever before the clutch shaft clamp bolt is tightened.

6 - Coat the clutch servo rod thread with Molykote Paste "G"

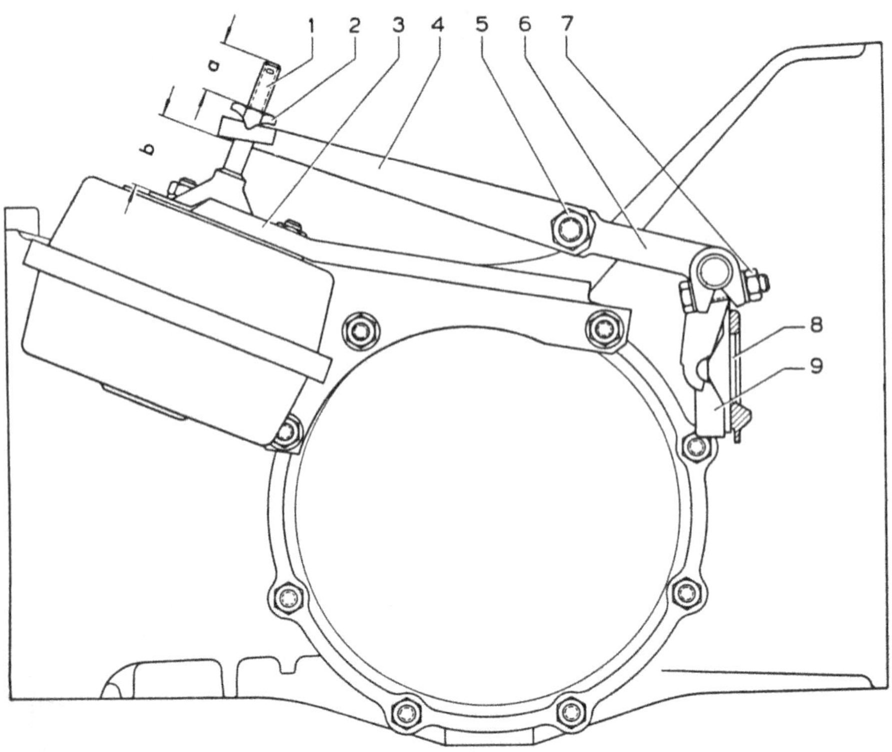

a = 17 ± 2 mm
b = 42 ± 2 mm

1 - Piston rod
2 - Adjustment nut
3 - Bracket
4 - Clutch servo lever
5 - Clamping bolt
6 - Clutch operating lever
7 - Clamping bolt
8 - Clutch release plate
9 - Clutch release bearing

Note:

In isolated instances the clutch lever comes into contact with the edge of the clutch servo bracket when de-clutching. This limits the stroke of the lever and prevents the clutch from disengaging completely. In such cases the lever should be modified slightly according to the measurements given in the drawing.

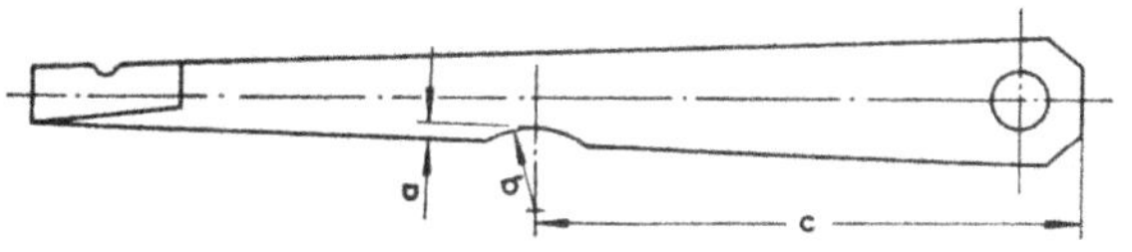

a = 3 mm (0.12″) b = 15 mm (0.59″) r
c = 96 mm (3.78″)

Note:

From Chassis No. 3364161 (Engine No. 5223422), the following modifications were introduced to prevent the clamp bolt between the clutch operating lever and the servo lever from working loose.

a - In place of the 10.5 mm dia. hole, the operating lever now has an M 10 threaded hole which is countersunk. The top part of the lever has been strengthened by 2 mm.

b - The M 10 × 30 mm clamp bolt, the M 10 nut and the spring washer are no longer used.

c - The servo lever is now secured to the clutch operating lever with an M 10 × 20 mm hexagon head screw with spring washer. The tightening torque for the screw remains 6 mkg (42 ft. lbs.). When installing the servo and clutch operating levers, check the aluminium washer and replace if necessary. This modification will be introduced when stocks of present parts are exhausted.

THIS PAGE INTENTIONALLY
LEFT BLANK

Automatic Clutch Trouble Checking

Symptom	Cause	Remedy
Noise in clutch	a - Needle bearing in flywheel gland nut worn	a - Renew gland nut and pack with Universal Grease
	b - Carbon thrust ring badly worn	b - Renew carbon thrust ring. Check that the release plate and clutch clearance are properly adjusted
	c - Clutch plate striking pressure plate	c - Renew or true-up clutch plate
	d - Release lever springs weak or tension uneven	d - Renew springs
Grabbing (Gearshift clutch)	a - Oily clutch plate linings	a - Replace crankcase or transmission case oil seal, clean clutch and re-line clutch plate
	b - Pressure plate worn or contacting unevenly	b - Renew pressure plate
	c - Release plate not running true	c - Renew release plate
	d - Clutch spring tension not uniform	d - Renew springs
	e - Clutch plate segments excessively or unevenly set	e - Renew clutch plate
Grabbing (Starting Clutch)	a - Oily plate linings	a - Replace starting clutch
	b - Friction surfaces of plate or flywheel uneven	b - If necessary, replace the flywheel complete with starting clutch
Clutch drags (vehicle tends to creep)	a - Idling speed too high	a - Adjust idling speed
	b - Starting clutch release springs weak	b - Replace starting clutch
	c - Centrifugal rollers sticking	c - Replace starting clutch
	d - Clutch plate or main drive shaft out of true	d - Straighten or replace clutch plate or main drive shaft
	e - Clutch plate too thick or segments unevenly set	e - Reset segments or replace clutch plate
	f - Clutch linings broken	f - Install new linings or replace clutch plate
	g - Needle bearing in gland nut faulty or insufficiently lubricated	g - Replace gland nut or lubricate needle bearing

Symptom	Cause	Remedy
	h - Splines on main drive shaft or in clutch plate dirty or burred	h - Clean splines or remove burr
	i - Clutch clearance excessive.	i - Adjust clutch clearance
Clutch slip	a - Oily clutch plate	a - Replace clutch linings. If necessary renew engine or transmission oil seal
	b - Gearshift clutch faulty	b - Replace gearshift clutch
	c - Insufficient clutch clearance due to wear in linings	c - Adjust clutch clearance
	d - Sintered coating on starting clutch plate worn	d - Replace starting clutch
Clutch does not disengage when changing gear	a - Solenoid electrical circuit faulty	a - Check solenoid connections and, if necessary, replace fuse
	b - Gearshift lever contacts dirty or burnt	b - Clean contacts and remove burred edges or install new contacts and readjust
	c - Defective control valve solenoid	c - Replace solenoid
	d - Connecting pipes leaking or disconnected	d - Tighten or replace connecting pipes
	e - Clutch servo diaphragm defective	e - Replace clutch servo
Vehicle jerks badly when gearshift lever is released after engaging a gear, with engine at idling speed	a - Starting clutch not freeing properly. Centrifugal rollers sticking or roller plate surfaces dirty	a - Clean rollers and roller plate surfaces. Replace starting clutch if necessary
	b - Starting clutch plate release springs weak	b - Replace starting clutch
Gearshift clutch does not engage after gear has been shifted, even when accelerator is depressed	a - Gearshift lever contacts sticking or being bridged by burred edges	a - Clean contacts and remove burr. If necessary, replace contacts
	b - Short circuit in cable between contact piece and solenoid	b - Eliminate short circuit. If necessary replace contact piece with cable
	c - Control valve solenoid sticking	c - Replace control valve

Symptom	Cause	Remedy
Gearshift clutch slips too long after gear has been shifted	a - Connecting pipe between carburetor and control valve diaphragm leaking or disconnected	a - Tighten or replace pipe
	b - Control valve diaphragm defective	b - Replace control valve
Clutch occasionally fails to engage after shifting a gear, but grips hard when accelerator is depressed	a - Freewheel defective	a - Replace freewheel complete with bearing flange and drive housing

THIS PAGE INTENTIONALLY
LEFT BLANK

1 - VW Special Tools

VW 106	T. Wrench 10 mm socket
VW 109	Box Wrench 10 mm
VW 112	Special Wrench 36 mm with Guide Plate
VW 122 b	Circlip Pliers
VW 123 a	Piston Ring Compressing Tool 77 mm dia.
VW 124	Chisel (Peening)
	Oil Pressure Switch Wrench
VW 161 a	Circlip Pliers
VW 165	Socket Wrench for Cylinder Head Nut
VW 170	Socket for Threaded Ring
VW 171	Retaining Spring
VW 173	Special Socket
VW 174	Special Socket M 6
VW 201	Oil Pump Extractor
VW 202	Extractor
VW 202 a	Extractor Hooks
VW 202 f	Thrust Pad
VW 203 b	Fan Pulley Extractor
VW 203 d	Thrust Pad for Fan Pulley
VW 204 b	Crankshaft Oil Seal Installing Tool
VW 205 a	Electric Piston Heating Tool (75—78 mm dia.)
VW 207	Piston Pin Pilot Drift
VW 207 a	Piston Pin Removing and Installing Tool
VW 212 a	Piston Pin Bearing (little end) Removing and Installing Tool
VW 214 b	Device for Checking, Straightening, and Reaming Connecting Rods
VW 215 b	Flywheel Retainer
VW 219	Clutch Pilot
VW 231 c	Drill Jig for Flywheel
VW 231 d	Drill Jig for Crankshaft
VW 247 a	Gauge Ring for Crankcase Center Bore
VW 252 d, e, f	Master Ring Gages
VW 542 a	Clutch Adjustment Dial Gage
VW 292	Gauge
VW 307	Fixture
VW 310 a	Fixture
VW 311 k	Valve Grinding Kit
VW 311 h	Valve Extractor
VW 311 f	Chuck
VW 313	Fixture
VW 400	Repair Press 15 t
VW 401	Thrust Pad
VW 402	Thrust Pad
VW 407	Punch
VW 408	Punch
VW 411	Punch
VW 415	Tube 75 mm dia.
VW 419	Tube

VW 422	Slotted Tube
VW 423	Punch
VW 427	Guide Tube
VW 428	Guide Tube (tapered)
VW 429	Thrust Ring
VW 435	Mandrel

2 - Workshop Equipment for Local Manufacture

VW 600	Engine Trolley
VW 601	Trolley for Assembles and Units
VW 603/1	Vehicle Trolley
VW 605	Gantry Crane
VW 606	Front End lifter
VW 630	Unit Wash Plant
VW 631	Oil Tray for Repair Stand
VW 633	Trestle
VW 643	Stand
VW 650/1	Cylinder Retainer
VW 651	Disassembly Trolley
VW 652/3	Disassembly Rack
VW 653/1	Valve Spring Compressing Tool
VW 657	Clutch Pressure Plate Compressing Tool
VW 659	Dial Indicator Bracket
VW 661/2	Oil Cooler Tester
VW 665	Oil Pump Pilot
VW 667	Clutch Cable Adjusting Tool
VW 682	Retaining Lever for Clutch Pressure Plate
VW 683	Holding Device for Clutch Operating Lever
VW 684	Adjusting Plate for Clutch Opecating Lever

3 - Normal Hand Tools

Screwdriver
Combination pliers
Waterpump Pliers
Center Punch
Pin punch, 2 mm
Mechanic's hammer, 300 grams
Mechanic's hammer, 500 grams
Rubber hammer, 85 x 50 mm
Aluminum hammer
VW Spark plug wrench
Triangular scraper
Flat scraper
Flat file, 180 mm in length
Socket wrench, 13 mm
Socket wrench, 14 mm
Socket wrench, 17 mm
Socket wrench, 19 mm
Socket wrench, 36 mm
Open-end wrench, 7 mm
Open-end wrench, 9 mm
Open-end wrench, 10 mm
Open-end wrench, 11 mm
Open-end wrench, 13 mm
Open-end wrench, 14 mm

Open-end wrench, 17 mm
Open-end wrench, 19 mm
Open-end wrench, 22 mm
Box wrench, 10 mm
Box wrench, 13 mm
Box wrench, 14 mm
Box wrench, 17 mm
Box wrench, 19 mm
Box wrench, 27 mm
Box wrench, 30 mm
Wire brush
Oil-can
Can for derusting fluid
Grease container
Scriber
Piston Ring Pliers
Dial Indicator
Set of feeler gages 0.05—1.00 mm
Micrometer 0— 25 mm
Micrometer 25— 50 mm
Micrometer 50— 75 mm
Micrometer 75—100 mm
Caliper square, 200 mm long, 1/10 mm scale
Depth gage, 200 mm long, 1/10 mm scale
Engineer's Square 90°
Dial gage for checking 18—100 mm inside diameters, readings ////// mm
Broach 8.0 H 7
Torque wrench
Inspection lamp with cable and plug
Electric drill
Oil funnel
Metal saw

4 - Additional Workshop Equipment

Valve spring tester
Engine test stand
Valve grinding machine or
Valve refacing machine
Compression tester
Run-out test equipment

THIS PAGE INTENTIONALLY
LEFT BLANK

Fuel System

Contents: K

THIS PAGE INTENTIONALLY
LEFT BLANK

Description of Fuel System

(with 28 PICT carburetor)

General Description

The fuel system is composed of the fuel tank with three-way tap, fuel lines, mechanical fuel pump, the downdraft carburetor and oil bath air cleaner with pre-heating connection.

The fuel tank, having a capacity of 40 liters (10.5 U.S. gals.; 8.8. Imp. gals.), is accessible by lifting the front hood. The remote controlled fuel tap, situated underneath the tank, regulates the flow of fuel and can be operated from the front seats. Its three positions are: closed, open, and reserve. The fuel reserve (5 liters 1.3 U.S. gal.; 1.1 Imp. gal.) can only flow out after the tap has been turned to the reserve position.

The mechanical pump serves to draw the fuel from the tank through the fuel line in the frame tunnel and then force it on into the carburetor.

The air drawn in by the carburetor passes through an oil bath type air cleaner with pre-heating connection to prevent dirt, dust and other abrasive particles from entering the engine.

Maintenance

Special care should be taken when filling the tank from cans to avoid dirt and foreign matter entering the tank. It is recommended that the fuel is filtered through a clean piece of chamois when filling the tank.

The air cleaner should be serviced in accordance with the instructions given under a separate heading in this section. Under severe dust conditions, the air cleaner must be serviced more frequently than indicated in the Maintenance Chart, in extreme cases even daily.

The fuel pump requires no service attention apart from regular filter cleaning. The pump pressure should be checked when fuel consumption is excessive or engine looses power at high speeds.

The idling adjustment should be checked at the maintenance service and, if necessary, readjusted to ensure good engine performance under the prevailing operating conditions (climate, altitude, etc.). If complaints are made with regard to excessive fuel consumption, no attempt should be made to effect a remedy by changing the jets. The first step to be taken in analyzing such complaints is to determine the fuel consumption as outlined under the heading "Special Hints" at the end of this section. Driving habits or abnormal operating conditions are often responsible for excessive fuel consumption.

The valve in the oil bath air cleaner intake pipe should also be checked for freedom of movement and set in the correct position to suit the prevailing temperature.

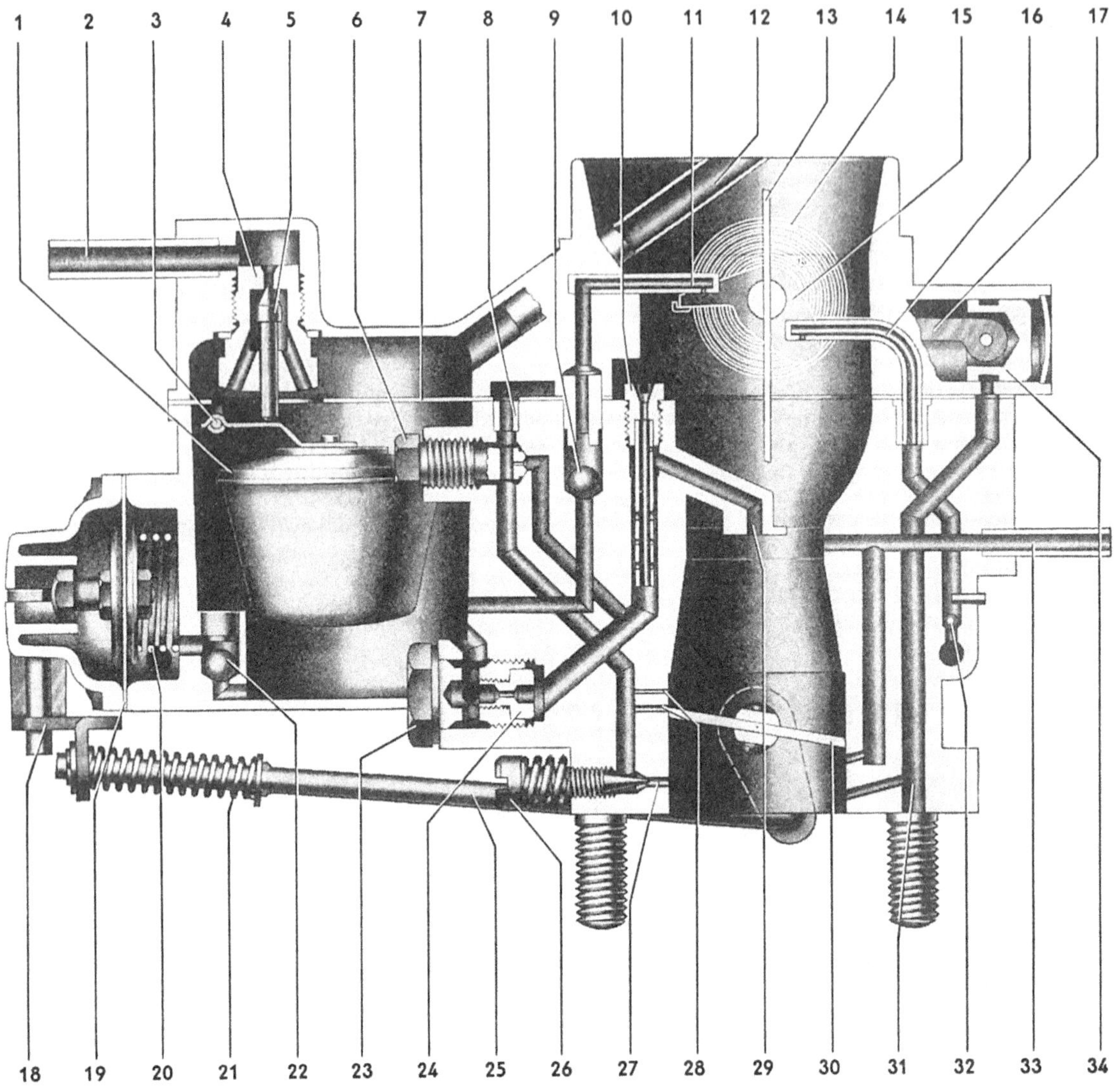

Solex 28 PICT

1 - Float
2 - Fuel line
3 - Float lever
4 - Float needle valve
5 - Float needle
6 - Pilot jet
7 - Gasket
8 - Pilot air drilling
9 - Ball check valve in power fuel system
10 - Air correction jet with emulsion tube
11 - Power fuel tube
12 - Float bowl vent tube
13 - Choke valve
14 - Bi-metal spring
15 - Operating lever
16 - Accelerator pump discharge tube
17 - Piston rod
18 - Pump lever
19 - Pump diaphragm
20 - Diaphragm spring
21 - Spring
22 - Ball check valve for accelerator pump
23 - Main jet carrier
24 - Main jet
25 - Pump connector rod
26 - Volume control screw
27 - Idle port
28 - By-pass port
29 - Discharge arm
30 - Throttle valve
31 - Vacuum drilling
32 - Ball check valve in accelerator pump drilling
33 - Vacuum connection
34 - Vacuum piston

Description of Fuel System

(with 28 PICT-1 carburetor)

The fuel system of the Volkswagen 1200 consists of the fuel tank, the fuel lines, the mechanical pump and the down-draft carburetor with oil bath air cleaner and pre-heater pipe.

The tank holds 40 liters (10.5 US galls,; 8.8 Imp. galls) and is located under the front hood. The tank is vented by means of a breather pipe which ends outside the body. From the tank, the fuel passes to the pump via the line in the frame tunnel and is pumped from there to the carburetor.

The dust and dirt in the air drawn in by the carburetor is extracted by the oil bath air cleaner.

Maintenance

When filling the tank from cans it is advisable to filter the fuel through a leather to prevent dirt from getting into the tank.

The air cleaner should be checked in accordance with the special instructions given here or, at least, at the intervals quoted in the Maintenance Chart. In very dusty conditions, the cleaner must be checked much more frequently, even daily when necessary.

The fuel pump requires no maintenance apart from filter cleaning as laid down in the Maintenance Chart.

During the Maintenance Check the idling setting should be checked and, if necessary, adjusted so that engine idling speed and transfer suit the local conditions (altitude) and climate (summer-winter). The flap in the air cleaner intake pipe should also be checked each time for freedom of movement.

The flap should not normally be fixed.

It is wrong to try to reduce a high fuel consumption by experimenting with the jet sizes. The best way is to make a fuel mileage test under fixed conditions or run a consumption test on a test stand and find out if the engine is actually responsible for the high consumption. If this is not the case, the trouble is due to operating conditions, the weather and the mode of driving.

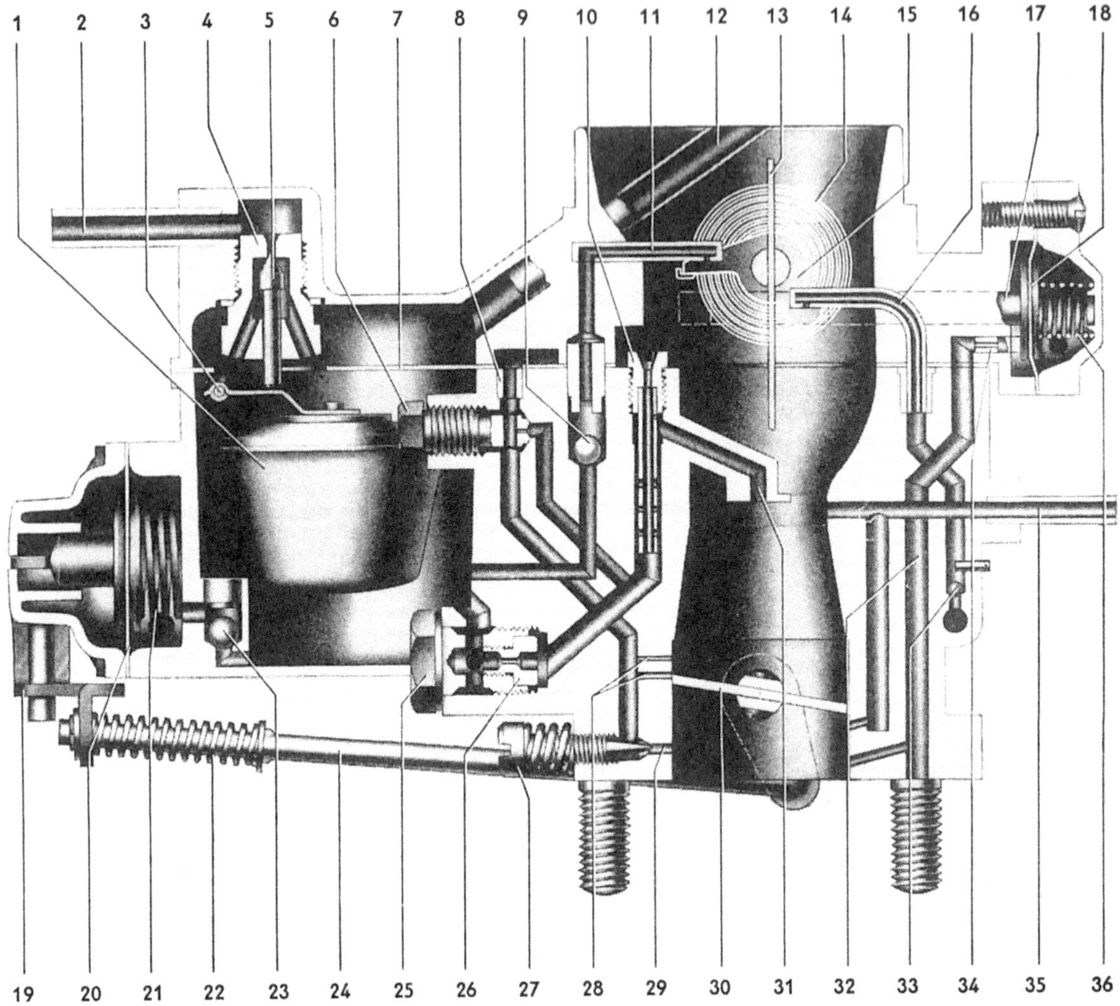

SOLEX 28 PICT-1

1 - Float
2 - Fuel line
3 - Float lever
4 - Float needle valve
5 - Float needle
6 - Pilot jet
7 - Gasket
8 - Pilot air drilling
9 - Ball check valve in power fuel system
10 - Air correction jet with emulsion tube
11 - Power fuel tube
12 - Float bowl vent tube
13 - Choke valve
14 - Bi-metal spring
15 - Operating lever
16 - Accelerator pump discharge tube
17 - Diaphragm rod
18 - Vacuum diaphragm
19 - Pump lever
20 - Pump diaphragm
21 - Pump spring
22 - Spring
23 - Ball check valve for accelerator pump
24 - Pull rod for vacuum diaphragm
25 - Main jet carrier
26 - Main jet
27 - Volume control screw
28 - By-pass port
29 - Idle port
30 - Throttle valve
31 - Discharge arm
32 - Vacuum drilling
33 - Ball check valve in accelerator pump drilling
34 - Jet in vacuum drilling
35 - Vacuum connection
36 - Spring for vacuum diaphragm

Carburetor (Solex 28 PICT)

General

The carburetor consists of two main parts: the upper part and the lower part. A gasket is fitted between the two parts and they are held together by five slotted screws.

The connection pipe for the fuel line and the float needle valve are installed in the upper part of the carburetor. The vent tube for the float chamber and the delivery pipe for the additional fuel are pressed into the upper part which also accomodates the automatic choke. This consists of the choke valve shaft with the choke valve, a fast idle cam and a lever, on one side of the barrel. On the other side is a heater element and bi-metal thermostatic spring embedded in a ceramic plate which is secured by a retaining rind and three screws in a housing cast onto the carburetor. A further cylinder cast on this housing contains a sliding vacuum piston. This cylinder is connected by a passage to the vacuum below the throttle valve.

The lower part of the carburetor contains the mixing and float chambers, the float and all the parts required for the preparation of the fuel-air mixture. The carburetor is secured to the induction manifold by two studs in the lower flange. The throttle valve and shaft are located below the mixing chamber and operated by the throttle lever. The accelerator pump, housed in a casting on the side of the float chamber, is connected to the throttle valve shaft by means of a lever and a connecting rod. The fuel is maintained at a constant level in the carburetor by the float and float needle valve.

The automatic choke arrangement in the upper part of the carburetor facilitates engine starting and controls the mixture preparation for idling and driving until the engine has attained its normal operating temperature.

The fuel-air mixture for the engine is produced in the carburetor. The fuel is atomised and thoroughly mixed with air in the proper proportions for efficient combustion. The principle of the downdraft carburetor makes full use of gravity to improve the cylinder charge and this increases engine flexibility and performance and facilitates the starting procedure.

The carburetor has a central air intake fitted with an oil bath air cleaner which ensures that the air for the mixture and the float chamber ventilation is cleaned under all operating conditions. This arrangement reduces the ingress of dirt into the carburetor to a minimum and prevents the fuel consumption from being influenced by the accumulation of dirt in the air cleaner.

The air drawn in by the engine enters the air cleaner via a tube mounted at the side and fitted with a flap valve with a balance weight. This valve opens and closes automatically according to engine speed and controls the entry of cold and pre-heated air. At idling speed the proportion of warm air is at its largest and this helps to eliminate carburetor icing. At temperatures over 20° C (68° F) the valve should be fixed in the open position with the clip.

The function of the idling circuit can be considered as that of an auxiliary carburetor which takes over the preparation of the mixture when the throttle valve is almost closed. The idling mixture can be enriched or weakened by means of the volume control screw. The engine idling speed is controlled by the idle adjusting screw. The newly designed idling stop segment on the fast idle cam ensures that the idling speed corresponds correctly with the choke valve position from the moment the engine is started until the normal operating temperature is attained.

The pre-heating of the intake not only prevents the condensation of the fuel vapor but also assists the creation of a readily combustible mixture.

Operation

The automatic choke improves engine starting under all weather conditions. The choke valve shaft is tensioned by a spiral shaped bi-metal spring which responds to every change of temperature. When the engine is cold the choke, actuated by the movement of the bi-metal spring as it cools off, is more or less closed according to the outside temperature. When the bi-metal spring is heated it loses its closing effort and the choke valve can open progressively until the normal operating temperature is attained and the air intake is completely clear. The bi-metal spring is heated by the heater element embedded in the ceramic cover on the spring housing The opening of the choke valve is assisted by the off center mounting of the choke valve shaft in the air intake passage. The air stream being drawn in thus tends to open the valve and this effort is further influenced by the fact that the freely pivoted fast idle cam on the choke valve shaft is designed as a balance weight and also works in the opening direction. When the choke valve is closed the throttle valve is simultaneously opened slightly. This is achieved by the fixed lever on the choke valve shaft lifting the fast idle cam when the choke closes and bringing the stop segment, on which the idle adjusting screw on the throttle lever rests, into operation.

When the choke is fully closed the idle adjusting screw is pushed into the outer position thereby opening the throttle valve slightly and allowing the vacuum created when starting the engine to take full effect below the choke valve.

This vacuum lifts the fuel from the emulsion tube into the mixing chamber via the discharge arm. The air required for the mixture is drawn in past the choke valve which then commences to flutter between the open position — under the influence of the vacuum — and the closed position — under the influence of the bi-metal spring tension. In this manner a very rich mixture is created which ensures easy engine starting even at very low temperatures. As the engine warms up the choke begins to open and the proportion of air in the mixture increases and automatically weakens it. This weakening continues until the choke valve reaches the vertical operating position.

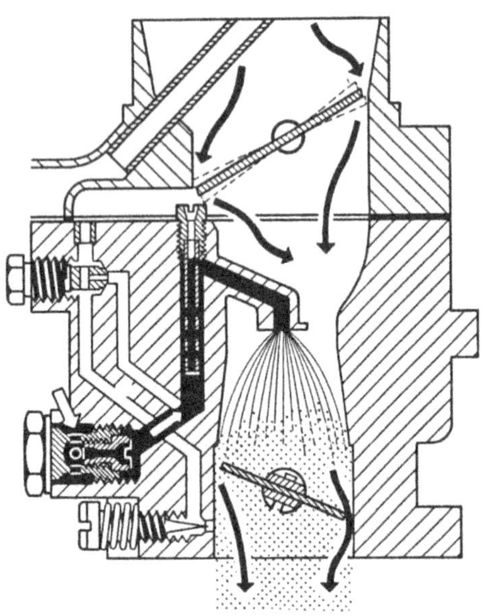

The vacuum piston belonging to the automatic choke is connected to the choke valve shaft by a piston rod and lever. The sealed artight chamber above the piston is connected to the space below the throttle valve by a passage so that the vacuum existing there is also effective at the piston. The function of the vacuum piston is to open the choke valve slightly against the pressure of the bi-metal spring when the engine speed increases after starting, under light loads or when the vehicle is over-running the engine and thus provide an air bleed to prevent the mixture from becoming excessively rich. This is brought about by the piston being moved by the vacuum below the throttle valve increasing as the engine speed rises.

On starting a cold engine, the fuel flows from the fuel line through the float needle valve into the float chamber and on via the main jet, emulsion tube and spraying arm into the mixing chamber where it mixes with the air being drawn past the fluttering choke valve. This very rich mixture then passes to the engine through the slightly opened throttle valve.

At **idling speed**, that is, when the throttle valve is almost closed the micture takes place in an entirely different manner. With the throttle valve in this position the air speed, and consequently the vacuum in the venturi, is so low that the fuel will not be drawn out of the spraying arm. The carburetor is therefore equipped with an idling circuit which provides an approximately correct mixture by means of a pilot jet and a fixed pilot air bleed. This mixture is fed to three small ports below and level with the throttle valve. The idling mixture can be enriched or weakened by means of the volume control screw. Two of the three ports are known as by-pass ports. The idling mixture is also drawn from the port which is exactly in line with the edge of the throttle valve. The other port, which is slightly above the edge of the closed throttle valve only comes into operation when the throttle is opened slightly. Both ports help to improve the transfer from the idling circuit to the main jet system as the mixture is drawn out of the drillings by the increased speed of the air passing the partly opened throttle valve. When the throttle is closed quickly the idling mixture is automatically weakened if the idling has been adjusted properly by means of the idle adjusting screw on the throttle lever.

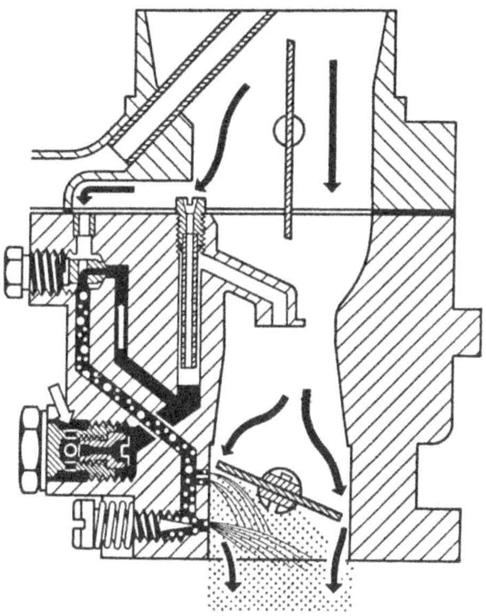

The fuel flows from the fuel line through the float needle valve to the float chamber and then via the main jet to the pilot jet. With the choke valve fully open the air is fed through the fixed pilot air bleed and the mixture passes to the idling and by-pass ports. The air passing the partly opened throttle valve draws the mixture with it. The adjustment of the volume control screw governs the proportions of fuel and air in the mixture.

During normal operation the vacuum in the intake passage draws the fuel out of the spraying arm which is connected to a vertical well containing the air correction jet with the emulsion tube pressed into it. Fuel flows to this well from the float chamber via the main jet. The factors governing the amount of vacuum are the size of the cost-in venturi, the engine speed and the throttle opening. The specially designed restriction of the venturi — where the spraying arm ends — has an accelerating effect on the air being drawn in by the engine. This gives — in connection with the position of the throttle valve — a varying degree of vacuum and draws different quantities of fuel from the spraying arm. As the fuel level in the emulsion tube drops due to the rising vacuum created by the increasing throttle opening, air is simultaneously drawn in through the air correction jet. This air passes out of the holes in the emulsion tube as they are uncovered and mixes with the fuel flowing from the main jet, thus progressively weakening the mixture as the engine speed increases.

In the part-load and particularly in the full-load ranges the fuel flows from the fuel line through the float needle valve to the float chamber and on via the main jet and emulsion tube well to the spraying arm. Here it mixes with the induction air in the mixing chamber — at part-load — with the choke fully open and the throttle partly open. At full-load, additional air enters via the air correction jet and mixes with the fuel in the emulsion tube well. The choke and throttle are both fully open.

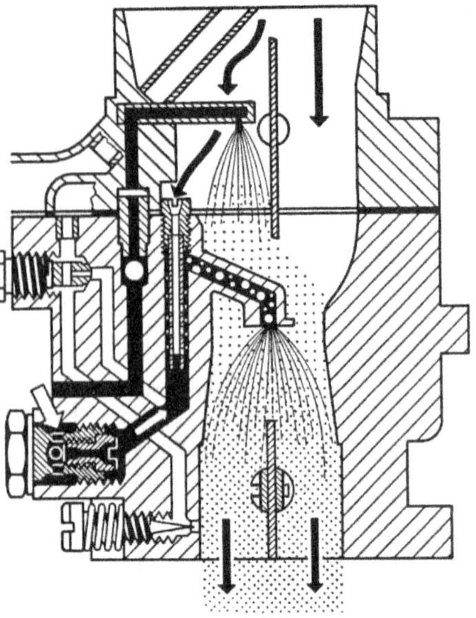

The function of the additional **Power Fuel System** is to provide the richer fuel mixture required at full load conditions and high engine speeds to enable the engine to give maximum performance. A calibrated power fuel tube connected to the float chamber by a passage in the carburetor body is installed in the upper part of the carburetor. The tube ends roughly at the level of the choke valve shaft, in an area where the vacuum is reduced. At low and medium revolutions the vacuum is not high enough to draw fuel out of the power fuel tube as y ball has to be lifted before fuel can flow through the passage. Only when the engine is at full load speeds and the vacuum is high enough to lift the fuel up to the power fuel tube does the additional fuel feed from the power fuel system commence. The feed is progressive and increases in quantity until the maximum engine speed is reached. This ensures that the engine gives full power with the lowest possible part-load consumption and low specific consumption in the lower full-load range.

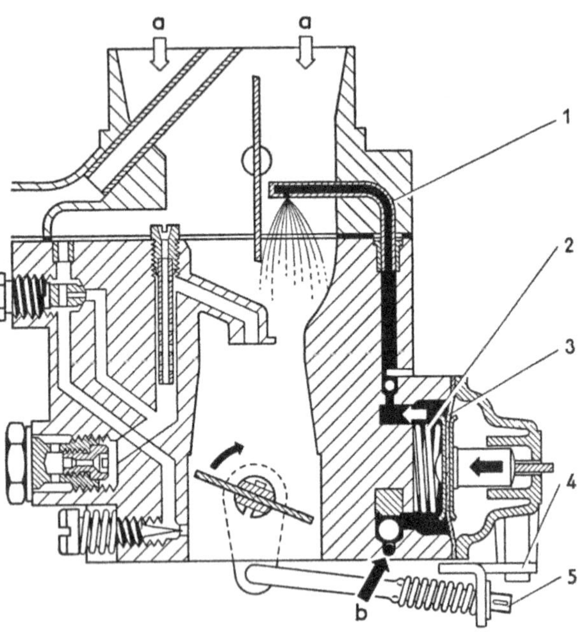

a - Air
b - Fuel from float bowl

1 - Discharge B tube
2 - Diaphragm spring
3 - Pump diaphragm
4 - Pump lever
5 - Pump connector rod

The **accelerator pump** is operated from the throttle shaft by a connector rod and pump lever. When the trottle valve is closed the diaphragm spring presses the pump diaphragm to the rest position and the chamber in front of the diaphragm fills with fuel drawn in from the float chamber via a ball valve. When the throttle valve is opened, the movement is transmitted to the diaphragm through the linkage. The pressure stroke of the pump opens a ball check valve in the delivery passage and fuel is forced through a calibrated discharge tube into the mixing chamber. This additional fuel enriches the mixture and ensures a smooth transfer and rapid acceleration. The amount of fuel delivered is governed by the stroke of the pump. The calibrated discharge tube only controls the amount of fuel delivered in a given time and therefore the duration of the injection.

The accelerator pump enriches the mixture in the low and medium speed range only, as the spring on the connecting rod holds the diaphragm in the depressed position as the throttle is opened further and prevents the pump from refilling with fuel until the throttle is almost closed again.

Adjustment

The carburetor settings have been tested and established at the Volkswagen Factory and should not normally be altered. Excessive fuel consumption and poor engine performance are generally due to other causes. If the idling mixture is too rich the engine will tend to stall when braking sharply. The prescribed jet combination and a correctly adjusted idling speed are essential requirements for proper carburetor functioning. The idling speed should, therefore, be carefully adjusted when the engine is warm. Furthermore, the valve in the intake tube on the air cleaner should only be fixed in the open position when the outside temperature is over 20° C.

Removing and Installing Carburetor

Removal

1 - Detach pre-heater connection on the intake tube of the oil bath air cleaner.

2 - Remove oil bath air cleaner.

3 - Disconnect the fuel and vacuum lines from carburetor.

4 - Detach push-on connector for heater element.

5 - Disconnect accelerator cable at throttle lever and remove spring, washer and cable swivel pin.

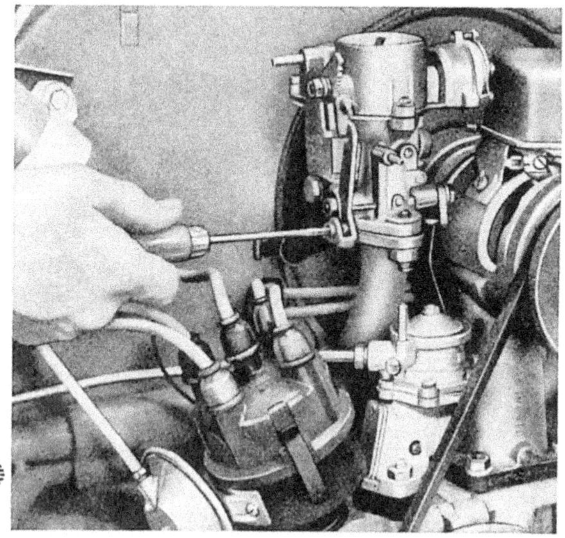

6 - Unscrew carburetor securing nuts with a 13 mm starter spanner and remove carburetor.

Installation

The following points must be observed during installation:

1 - Install new gasket at intake manifold flange.

2 - Secure accelerator cable on the throttle lever as follows: Open throttle so that there is a clearance of about 1 mm (.04") between throttle lever and the stop on the carburetor body. Place the accelerator pedal in the fully depressed position and connect the cable to the throttle lever. All other operating linkage should be installed free of play and stress.

3 - Do not overtighten the intake elbow securing screw.

4 - Do not forget the heater element connection.

5 - Adjust idle speed with engine warm.

Disassembly and Assembly of Carburetor

Disassembly

1 - Remove carburetor.

2 - Remove the five screws securing the upper part and lift it off.

3 - Screw out float needle valve.

4 - Remove the three screws in the automatic choke and take off retaining ring together with the ceramic plate, bi-metal spring and heater element.

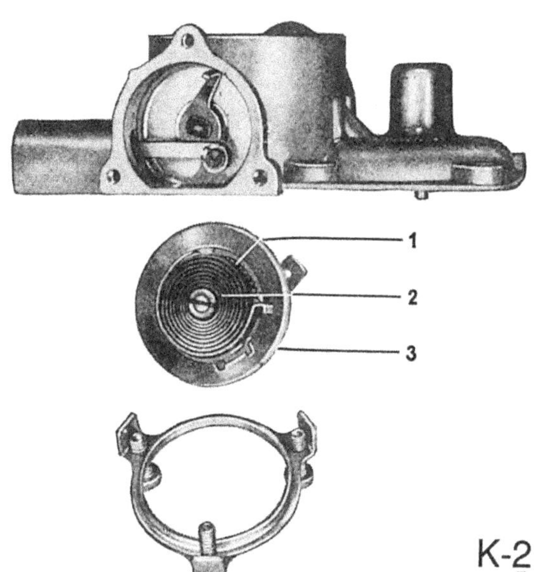

1 - Heater element 2 - Bi-metal spring 3 - Ceramic plate

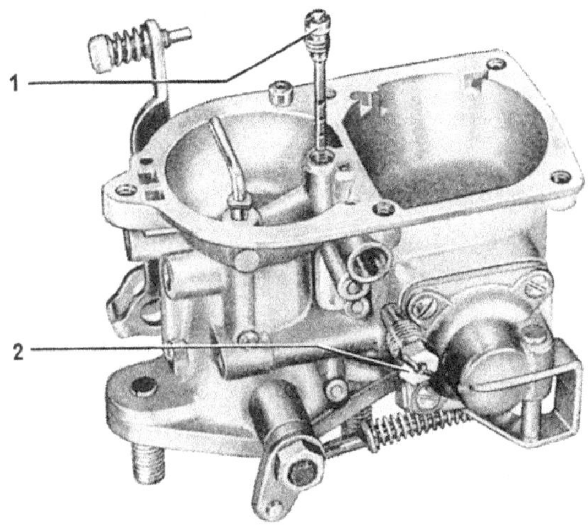

5 - Take out the float.

6 - Screw out the air correction jet with the emulsion tube and the idling jet.

1 - Air correction jet with emulsion tube
2 - Pilot jet

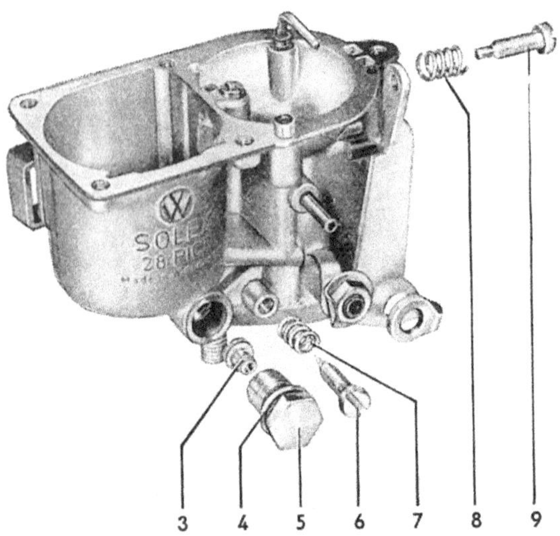

7 - Screw out the main jet carrier with main jet and the volume control screw.

3 - Main jet
4 - Gasket
5 - Main jet carrier
6 - Volume control screw
7 - Spring
8 - Spring
9 - Idle adjusting screw

8 - Remove pump lever cotter pin from the connector rod.

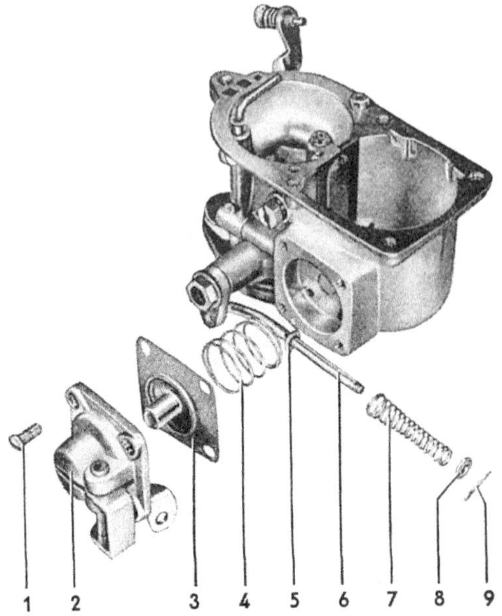

9 - Remove four pump cover retaining screws and take off cover, diaphragm and spring.

1 - Screw
2 - Pump cover
3 - Pump diaphragm
4 - Diaphragm spring
5 - Washer
6 - Pump connector rod
7 - Spring for connector rod
8 - Washer
9 - Cotter pin

Cleaning

Under no circumstances should pins or pieces of wire be used to clean the jets as they will damage and enlarge the calibrated drillings.

Checking and Assembly

The assembly is a reversal of the disassembly procedure. When checking the parts observe the following points:

Upper part

1 - Check needle valve for leaks.

2 - Examine needle valve gasket condition and check correct seat when installed.

3 - Check carburetor body gasket.

4 - Check play in choke valve shaft.

5 - Check heater element and bi-metal spring. If one of the parts is damaged the complete ceramic plate must be replaced.

Important

When installing the ceramic plate with bimetal spring and heater element, take care that the marking on the ceramic plate corresponds with that on the spring housing.

Note:

It may occasionally be found that the automatic chohe becomes very dirty and the vacuum piston seizes in the cylinder due to the formation of deposits from the crankcase ventilation system. The choke valve can then no longer open fully so that difficulty is experienced when starting from cold and the fuel consumption rises considerably.

When this occurs, proceed as follows:

Check the automatic choke

1 - Take ceramic cover off carburetor. The heater element and bi-metal spring must not be encrusted with burnt oil.

2 - Try to turn the choke valve shaft with the operating lever. If there is no sign of movement, the choke valve shaft is seized. If there is a small amount of movement due to the play in the linkage connecting the vacuum piston, the choke shaft is free but the vacuum piston is seized.

Remedy

1 - Remove top of carburetor.

2 - Drill the cylinder end cover with a 4—5 mm drill and pull cover out.

3 - Clean automatic choke housing with acetone.

4 - Squirt acetone into cylinder from both sides and move piston to and fro. Clean cylinder thoroughly.

5 - Blow out the vacuum drilling (A) to the cylinder with compressed air and clear the opening (B) in the cylinder with a piece of thin wire.

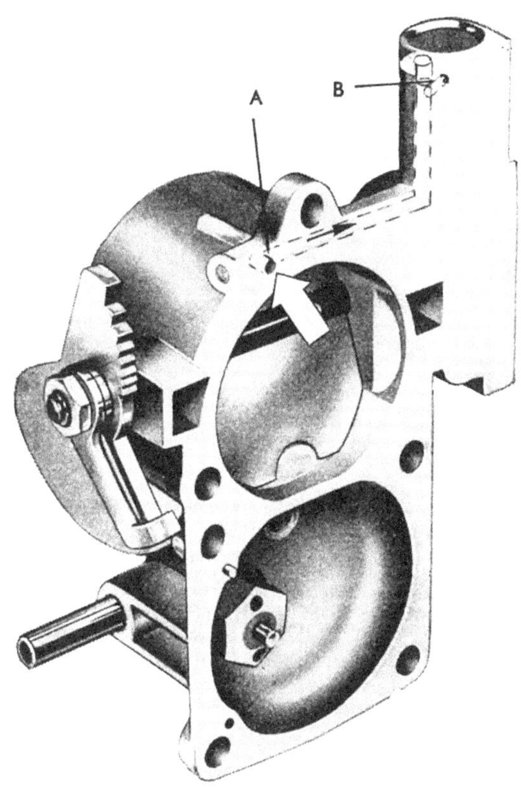

A = Vacuum drilling
B = Opening in cylinder

6 - Place new end cover (C) — Part No. 113 129 119 — in the cylinder with the convex side outwards and tap lightly with a hammer and 11—12 mm diameter punch so that the cover spreads (D). Seal edges with sealing compound.

7 - Install a new ceramic cover (Part No. 113 129 191) if the heater element is encrusted with deposits.

8 - Install carburetor top.

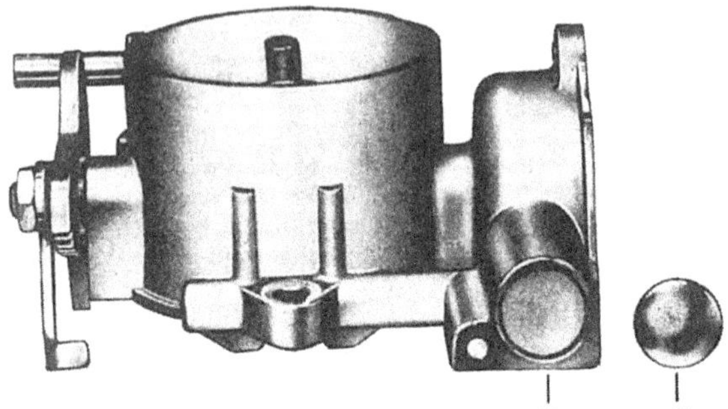

C = New end cover
D = Cover installed and sealed

Note:

From November 1963, Chassis No. 5 909 656 (Engine No. 8 154 031) the carburetor was fitted with a new top part with a modified automatic choke. The designation is now **28 PICT-1.**

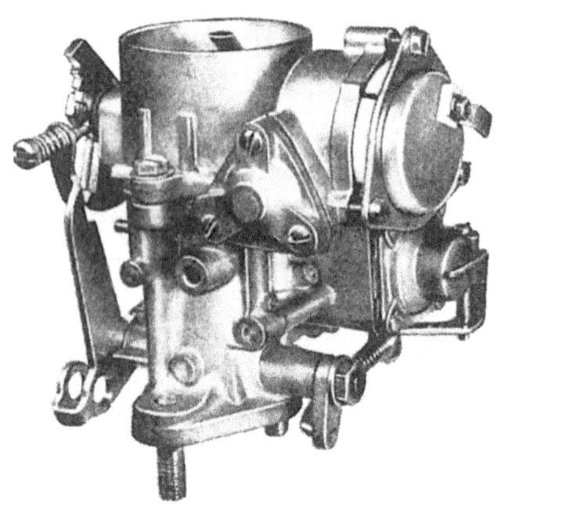

The new automatic choke incorporates the following modifications:

1 - The housing for the automatic choke is larger and the ceramic cover (1) is insulated externally by a metal cap (2). The lateral support and heat insulation of the bimetal spring has been improved by a plastic insert (3) in the housing. A ceramic rod (4) prevents the heater element from coming into contact with the bimetal spring. The plastic insert has a small projection on the outer edge which engages in a slot in the automatic choke housing.

The retaining ring for the cover is supported against the automatic choke housing by small spacer bushes (Part No. 113 129 198).

2 - The vacuum piston used formerly has been replaced by a diaphragm which opens the choke valve as the vacuum increases.

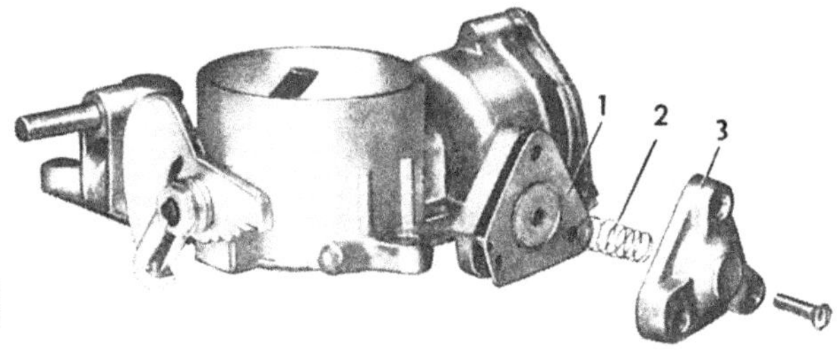

1 - Vacuum diaphragm
2 - Diaphragm spring
3 - Cover for diaphragm

3 - The previous fast idle cam had 9 steps (A), the new one has 6 (B). This means that the idling speed now alters in slightly larger stages.

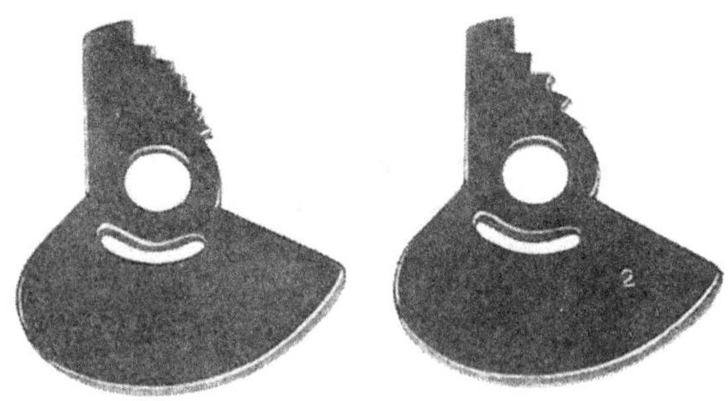

A B

4 - The same jets are used in the carburetor.

5 - The bottom of the oil bath air cleaner (Part No. 113 129 613 B as before) has a small depression to suit larger dimensions of the automatic choke housing.

6 - The two generator mounting holes in the fan housing cover (new Part No. 113 119 261 A) have been turned to 10° 30′ to the right (formerly 5° 30′). The generator is thus inclined more so that the space between the choke housing and the voltage regulator is larger.

The modified carburetor or the modified carburetor upper part can be service installed in all 34 bhp engines.

On installation, ensure that there is sufficient space between the automatic choke housing and the regulator on the generator as otherwise there is a risk of a short circuit at the heater element terminal.

If the space is not sufficient, the generator should be removed and the fan housing cover plate exchanged. When the new cover (Part No. 113 119 261 A) is used, the generator is turned to 10° 30′ to the right.

On the Type 1 models 11 and 15, the bottom of the air cleaner must be depressed 2—3 mm with a hammer in line with the upper screw of the choke retaining ring so that the flange of the air cleaner goes on to the carburetor upper part far enough.

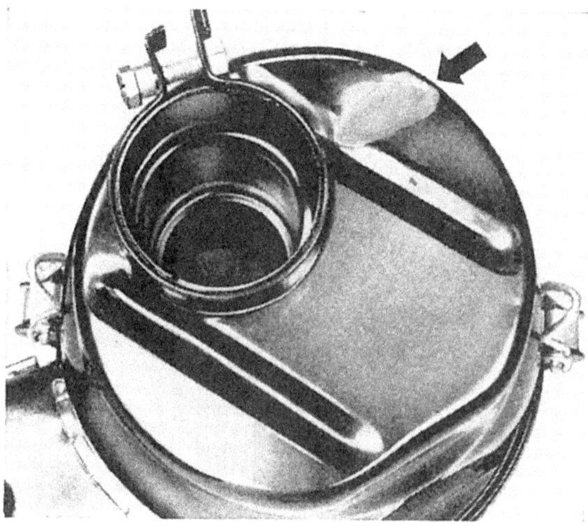

Should it be found necessary to depress the bottom of the air cleaner more than this, it is better to fit a new air cleaner because if depressed too much the cleaner may not seal properly on the carburetor or the sheet metal may be torn.

Note:

In some 28 PICT-1 carburetors the bi-metal spring in the automatic choke is tensioned too much even though the marks are aligned properly. This makes the choke open late and gives a very rich mixture. When starting from cold this, in turn, causes excessive smoke at the exhaust and uneven engine running.

This trouble can be eliminated by turning the ceramic cover of the automatic choke about 4—5 mm to the right (a). This opens the choke earlier and weakens the fuel-air mixture.

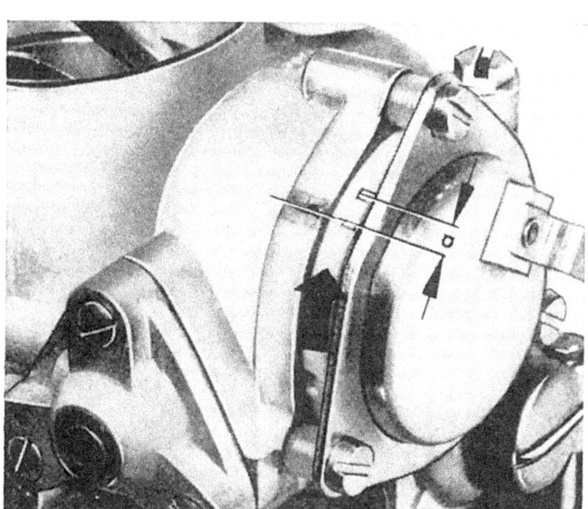

a = 4—5 mm (.16—.2″)

Note:

Spitting in the carburetor, poor and sluggish transfer in the part-load range and low output has been reported recently on some of the 30 bhp engines fitted in the Volkswagen 1200 A with fully synchronized transmission (from November 1964). This trouble is caused by insufficient spark advance in the part-load range and can be remedied as follows:

1 - Remove the carburetor and open out the drilling behind the threaded connection for the vacuum pipe with a 1.7 mm twist drill. The drill should be passed through the connection. Blow out metal chippings carefully with compressed air.

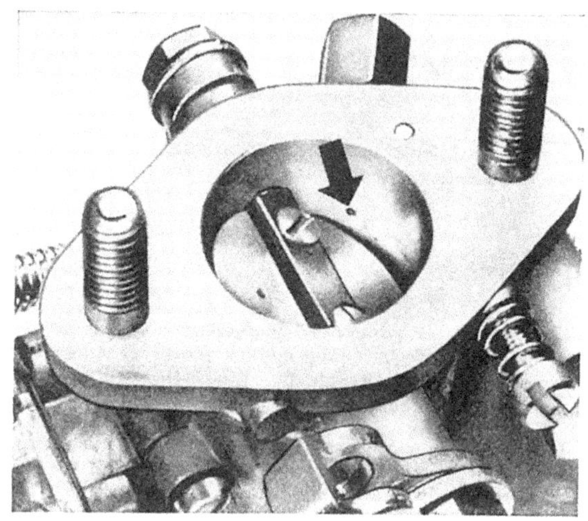

2 - Set ignition timing to 10° before TDC (right-hand mark on crankshaft pulley) with the engine cold.

Lower part

1 - Check pump diaphragm for leaks and replace if necessary.

Note:

Spitting in the carburetor on sudden acceleration generally indicates a leaky pump diaphragm.

2 - Dip the float in hot water. If bubbles appear the float is leaking and must be replaced. (See "Adjustment Data" for correct float weight).

3 - Check all jets for correct sizes as shown in the "Adjustment Data".

When replacing jets or valves only genuine "SOLEX" parts should be used. Only these parts are accurately calibrated and ensure correct adjustment and low fuel consumption.

Use only main jets and carriers with ring grooves.

4 - Check throttle shaft clearance. Excessive clearance encourages the ingress of secondary air and has a detrimental effect on the starting and idling conditions. If necessary, the holes for the throttle shaft must be bushed.

5 - Examine the tapered portion of the volume control screw and replace if the tip is bent or broken. Only brass volume control screws should be used. Check the tapped hole and seat for the screw in the carburetor lower part and remove the tip of the old screw if it has broken off.

6 - Insert the float.

7 - Oil the steps on the fast idle cam.

Note:

On engines with the new 28 PICT carburetor it may sometimes be found that, due to corrosion caused by prolonged delivery or storage times, the fast idle cam sticks and the choke valve does not open fully.

This condition can be remedied by applying a few drops of anti-corrosion oil to the bearings and moving the cam back and forth several times.

Note:

From August 1960, the carburetor lower bodies supplied as spares without the ball in the power fuel system (Part No. 141 129 301 — Karmann Ghia models and Transporter) can be distinguished from the lower bodies with ball (Part No. 113 129 301 A — De Luxe Sedan and VW Convertible) by means of the **yellow spot** on the float housing near the word "SOLEX".

Note:

From Chassis No. 3 510 198 the retaining ring for the ceramic cover was modified and now has stronger lugs.

The lugs on the previous ring tended to bend in some cases and allow the three securing screws to become loose. The ceramic cover could then turn so that the bi-metal spring no longer closed the choke valve properly and starting difficulties ensued. The setting mark on the ceramic cover must always be exactly in line with the mark on the carburetor body (B).

The modified ring (Part No. 113 129 199) as before can be installed without difficulty.

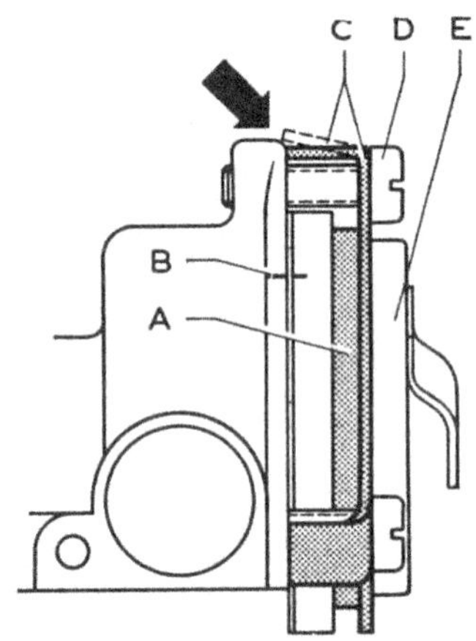

A - Retaining ring
B - Setting mark
C - Lugs on ring
D - Securing screws
E - Ceramic cover

Note:

Deposits from the crankcase ventilation system which build up in side the 28 PICT carburetor can often cause carburation troubles.

The formation of deposits not only make the vacuum piston and choke valve shaft stiff in operation, they also tend to block the pilot air bleed drilling and reduce the diameter of the air correction jet. This results in poor idling and increased fuel consumption.

Remedy

Remove carburetor and take top part off body. Soak both parts in "Speedclene" cleaning solution for several hours and then rinse thoroughly with trichlorethylene. This method is guaranteed to remove all traces of the deposits.

The carburetor can also be cleaned with trichlorethylene or acetone alone.

Manufacturer of Speedclene:
Bendix Products Division
South Bend 20
Indiana/USA

Distributor in Germany:
Messrs. A. Pierburg
Auto- und Luftfahrt-Gerätebau KG
404 Neuss/Rhein
Postfach 867

Note:

From Chassis No. 3 192 507, a different carburetor and ignition distributor are installed in vehicles with the 34 bhp engine (41.5 SAE bhp) which are equipped with a cyclone air filter.

De Luxe Sedan (M 153)

The carburetor (Part No. 113 129 023 F) for engines with the cyclone air filter differs from the standard carburetor (Part No. 113 129 023 D) in that the upper vacuum drilling to the venturi and the power fuel jet are blocked.

The Bosch VJU 5 BR 8 distributor (Part No. 111 905 205 F) is used for this version. The ignition timing point is 12,5° before T.D.C. The right hand mark on the crankshaft pulley should be 4 mm (.16") to the left of the crankcase joint.

Karmann Ghia (M 154)

The carburetor (Part No. 141 129 023 E) which is used on the Karmann Ghia models has a 130 main jet, a 170 Z air correction jet and the vacuum drilling to the venturi is blocked.

The distributor and the ignition timing are the same as on the de Luxe Sedan.

The SP Set No. 80 A is available for service installation in the De Luxe Sedan and Set 81 A for the Karmann Ghia models. The installation is described in the supplement to Technical Bulletin K 7.

Note:

From January 1963, 34 bhp (41.5 SAE bhp) engines with the fresh air heating which was introduced in December 1962 can be fitted with a cyclone air filter as an optional extra. The order numbers are:

M 153 .... Type 1, Model 11
M 154 .... Type 1, Model 14

De Luxe (M 153)

Engines with cyclone filters are fitted with the 113 129 023 F carburetor and the Bosch distributor VJU 4 BR 8 (Part No. 111 905 205 F). The spark advance on this distributor is controlled by vacuum and centrifugal force. The upper vacuum drilling in the body of the above-mentioned carburetor is closed. The ignition timing is set to 12.5° before TDC. The right-hand mark on the crankshaft pulley should be 4 mm to the left of the crankcase joint.

Karmann Ghia models (M 154) and VOLKSWAGEN Transporter (M 155)

Model 14 and Type 2 vehicles are fitted with the 141 129 023 E carburetor.

The instructions on the distributor and the ignition timing are as for the De Luxe Sedan.

Servicing instructions

Note the following points on oil bath air cleaner servicing:

Type 1, Model 11

The air cleaner should be removed for cleaning and oil changing purposes. Proceed as follows:

1 - Loosen screw (A) on the cyclone filter connecting pipe clip.

2 - Loosen screw (B) on the clip for the hose between carburetor and oil bath air cleaner and pull hose (C) off.

3 - Loosen two retaining straps (D) at quick release clips and press apart.

4 - Lift oil bath air cleaner out of engine compartment.

5 - Release clips on lower part of cleaner (E). Clean lower part and fill to level mark with fresh engine oil.

> **Important**
>
> When installing the air cleaner, ensure that all connection screws are tightened properly in order to prevent leakage.

Model 14 and Type 2

1 - Pull crankcase ventilating hose off cyclone filter elbow (F).

2 - Remove cyclone filter by loosening screw (G) in elbow securing clip and pulling the rubber hose (H) out of the lower engine compartment panel.

3 - Release two clips on lower part of air cleaner. Take lower part out of engine compartment complete with insert.

> **Important**
>
> When the lower part has been cleaned, it should be filled to the mark with fresh engine oil. After installing the air cleaner, ensure that the screw in the cyclone filter clip is tightened fully. The groove at the lower edge of the rubber hose should be pressed into the opening in the engine compartment panel all round. If this is not done, the dust extracted by the cyclone filter will fall back into the engine compartment.
>
> A supplement to Technical Bulletin K 7 describing the service installation of the cyclone filter in 41.5 SAE bhp engines with fresh air heating, has been issued.

To ensure satisfactory mixing in the carburetor even at high altitudes, a high altitude corrector can be installed in place of the main jet carrier.

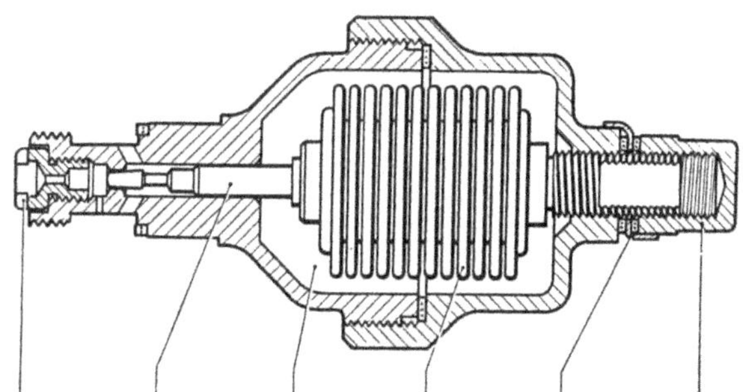

Operation of the high altitude corrector

The main jet carrier is replaced by the high altitude corrector which is mounted in the same manner as the main jet carrier.

The main jet (1) through which the fuel, coming through an annular chamber from the float housing has to pass, is screwed to one end of the high altitude corrector.

The fuel is supplied to the main jet through 4 oblique holes, and through a small hole between the main jet and the 4 oblique holes. The fuel also flows into the pressure unit chamber (3).

The pressure unit (4) is supported axially and fixed in position at one end by means of a pin which is screwed into the housing. Mounted at the other end of the pressure unit the needle (2) slides freely in the passage leading to the main jet.

When the pressure unit (4) expands due to the atmospheric pressure at high altitude, the needle progressively throttles the fuel supply for the main jet, as the change in atmospheric pressure alters the length of the pressure unit and the position of the needle, thus governing the fuel supply.

At sea level, the needle will be in a position which does not affect the normal fuel consumption.

Caution

The pressure unit is set during assembly by means of the adjusting nut (6) which is secured with a lock plate (5). This setting should not be disturbed.

Note:

From January 1963, all high altitude correctors are designed for certain jet sizes and the correct main jet must be installed in the appropriate corrector body in every case. It is not possible to adjust the pressure unit to suit a jet size other than that for which it was intended.

The following list gives the various versions with the correct jet sizes and the carburetors for which the correctors are intended. The corrector type is marked on the end opposite to the jet between two of the reinforcing ribs.

Engine	Main jet	Carburetor type	Type	Part No.
36 SAE bhp	117.5	28 PCI	H 15	111 129 501
41.5 SAE bhp	122.5	28 PICT	H 14	113 129 501
41.5 SAE bhp with cyclone filter	130	28 PICT	H 23	113 129 501 AJ

Carburetor Adjustment

Every carburetor is tested at the factory and adjusted on the engine to suit branded fuels. Any alteration in the settings such as replacing the jets by other than the prescribed sizes is detrimental under normal operating conditions and chould be avoided. When changing from ordinary fuel to a benzine-benzole mixture it is generally only necessary to adjust the idling speed.

Specifications

Venturi (cast-in)	22.5 mm dia.
Main jet	122.5
Air correction jet (with emulsion tube)	130 y (145 y for Ghia models)
Pilot jet	55
Pilot air bleed	2.0 mm dia.
Pump jet	50
Power fuel jet	1.0
Float needle valve	1.5 mm
Float weight	5.7 g
Pump delivery quantity	0.8—1.0 cm^3/stroke

Note:

From July 1963, the solenoid switch was modified on all electro-magnetic pilot jets. There is now only the tab (1) for the cable and a grubscrew (2) on the head of the solenoid switch.

When necessary, the grubscrew can be turned to withdraw the needle valve from the pilot jet.

Turning the screw as far as it will go

to the right — closes the needle valve
to the left — opens the neddle valve

The part numbers of the electro-magnetic pilot jets available for the various engines are given in the following list.

Type	Engine	Carburetor	Pilot jet	Part No.	Voltage
1 and 2	36 SAE bhp	28 PCT	g 50 g 50	113 129 413 A 113 129 413 B	6 V 12 V
	41.5 SAE bhp	28 PICT	g 55 g 55	113 129 413 113 129 413 C	6 V 12 V

Note:

It has been noticed occasionally that the 34 and 42 bhp (41.5 and 51 SAE bhp) engines tend to idle unevenly despite the fact that the carburetor and ignition settings are correct. This trouble is caused by the distributor spark advance coming in too soon.

For this reason the vacuum drilling near the throttle valve has been moved up 0.5 mm on all 28 PICT carburetors (from Carburetor No. 4 096 708 and 8 009 601).

This does not apply to carburetors from No. 8 019 651 to 8 021 498. On these carburetors the vacuum drilling has not been moved.

Idling adjustment

The idling adjustment may require regulating from time to time. This should be carried out when the engine is warm.

> **Important**
>
> Check that the idle adjusting screw is not resting on one of the steps of the fast idle cam.

1 - Set the engine idling speed to approximately 550 rpm. with the idle adjusting screw.

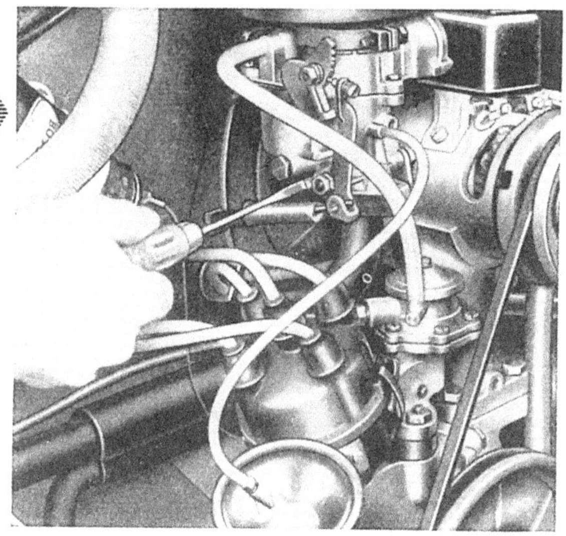

2 - Turn the volume control screw to the right until the engine speed begins to drop, then give it a quarter to a third of a turn to the left.

Note:

Turn the volume control screw carefully to avoid damaging the screw tip or the idle port.

3 - Regulate the idling speed as required.

The normal idling speed is usually attained with the volume control screw $1^1/_4$ to $1^1/_2$ turns from the fully closed position. Take care when closing the screw as careless handling can easily lead to the tip of the screw or the idle port being damaged. Accurate idling adjustment is of great importance as it has a considerable influence on fuel consumption in the low and medium speed ranges. In some cases the consumption can be increased by as much as $^1/_2$ litre over 100 kms (1 Imp. gallon over 560 miles or 1 U.S. gallon over 470 miles). This affect will be particularly noticeable on vehicles which are already operating under unfavourable conditions.

The adjustment is correct if the warm engine continues to run when the throttle is opened and closed suddenly or when the clutch pedal is depressed. If the engine stalls, the idling mixture is too weak and the volume control screw can be screwed out $^1/_8$ of a turn. Finally check that the transfer is smooth by operating the throttle slowly and gradually increasing the engine speed. Should the engine stall when the throttle is closed quickly, for instance when braking sharply, the idling mixture is probably too rich. If the engine idles unevenly when all the adjustment have been checked, the fault may lie in a damaged intake manifold flange gasket, a cracked or loose intake manifold or an improperly adjusted fuel pump.

Faults in the ignition system and excessive variations between the compression pressures in the individual cylinders can also have a detrimental effect on the idling.

Note:

Idling adjustment on vehicles with "Saxomat"

When the engine is started from cold it runs at a fast idling speed for a short time. As the starting clutch begins to grip at engine speeds over 950 rpm it is possible for a vehicle on which the handbrake is not set to start moving of its own accord as soon as a gear is engaged.

2 - This possibility can be reduced considerably by adjusting the idling speed to 500/550 rqm when the engine is warm. It is advisable to use a revolution counter when doing this as it is very difficult to estimate the speed accurately.

When the cold engine has started and been accelerated once, the idle screw will move to the 3rd or 4th step on the fast idle cam. In this position the engine speed is sufficently low to enable a gear to be selected without causing the vehicle to move, if the idling speed is properly adjusted.

3 - The section "Driving" in the "Saxomat" supplement for the Volkswagen Instruction Manual also deals with this subject. The instructions on starting vehicles with the Saxomat should be brought to the attention of all workshop staff.

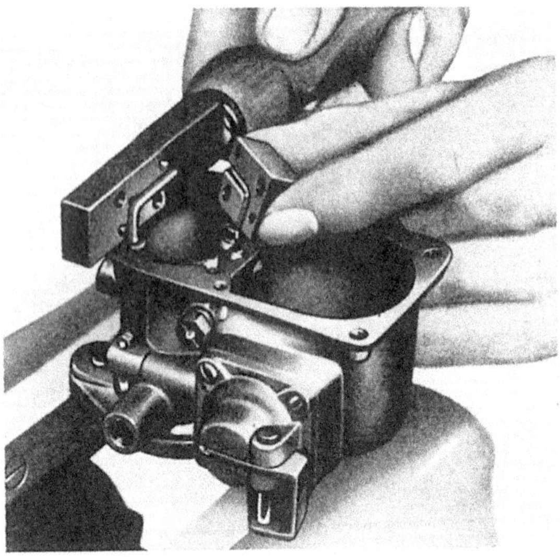

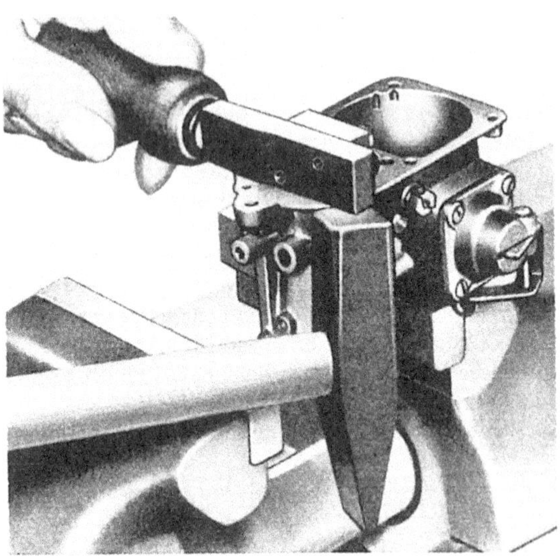

Note:

It may occasionally be noticed that the 28 PICT carburetor has a "flat spot" on acceleration. This can originate from various sources:

1 - **Air cleaner warm air regulating flap in fixed position**

The warm air regulating flap in the intake tube of the air cleaner should be in the working position and move easily.

2 - **Movement of the accelerator pump lever restricted**

This is caused by the lever contacting the generator.

Remedy:

a - Loosen the four nuts on the generator support flange, push the generator to the right as far as possible and tighten the nuts in this position.

b - Move the carburetor in a similar manner as far as possible to the left on the intake manifold flange.

c - Bend the intake manifold support as necessary.

3 - **Accelerator pump discharge tube bent or dirty**

Check the operation of the accelerator pump in the carburetor:

A - When the throttle opens, the fuel should be sprayed straight down through the carburetor without touching the walls of the mixing chamber.

Remedy:

Adjust the discharge tube.

B - If the spray is uneven or varies in direction, the jet is dirty.

Remedy:

Pull out the discharge tube and replace it with a new one (Part No. 113 129 323). A tool for this purpose is being published in a supplement to VW "Local Manufacture of Workshop Equipment" under the number VW 646.

Important

Do not turn the discharge tube when removing or installing it.

Removing the discharge tube with VW 646

a - Loosen wing nuts and take off clamp jaw.

b - Position the tool so that the discharge tube is located in the recess provided.

c - Fit jaw and tighten wing nuts.

d - Pull the tube out by tapping the end of the tool with a hammer.

The new tube is installed in the reverse order.

C - The accelerator pump discharge tube (Part No. **new:** 113 129 323 A) has been provided with a slightly wider shoulder to facilitate removal with a locally manufactured tool. The special tool for the 28 PICT carburetor discharge tube — VW 646 — has been altered and redesignated VW 646/1. New drawings for the publication "Local Manufacture of Workshop Equipment" will be issued in the near future so that existing VW 646 tools can be modified.

The end piece of the discharge tube is now sealed with a grub screw and not with solder as it was formerly. Both modifications make the cleaning of the accelerator pump system easier.

Service installation in older carburetors is not possible.

4 - **Intake valve for accelerator pump leaking**
If insufficient fuel is injected by the accelerator pump, it will invariably be found that the intake valve is leaking. This can be checked by removing the top part of the carburetor body, taking out the float and putting a small quantity of fuel into the float bowl. When the pump is operated, it will be seen clearly that fuel is forced back into the float bowl from the intake opening of the pump.

Remedy:

Replace lower part of carburetor body.

5 - **Accelerator pump delivery valve sticking**

a - If the delivery valve sticks in the closed position it will not be possible to move the pump lever.

b - If the delivery valve does not close, the intake stroke of the pump will draw in air via the valve in the pump housing instead of drawing fuel from the float bowl. Fuel will only be drawn in if the hole in the discharge tube is closed with the finger.

Remedy:

Pull out discharge tube. Remove the retaining pin and take out the delivery valve ball. Clean the drilling, renew the ball and install parts.

Synthetic hose between fuel pump and carburetor

On engines up to No. 5 222 818 which are still equipped with a synthetic hose between fuel pump and carburetor, it may be found that the hose material dissolves and forms a sticky deposit in the carburetor when fuels containing alcohol are used. These carburetors should be soaked in methylated spirits for 24 hours to remove the deposit. The rubber hose — Part No. 111 209 185 A — should then be installed.

Note:

In January 1963, between the following numbers:

	Chassis No.	**Engine No.**
from	5 220 713	7 350 401
to	5 250 430	7 435 393

about 50000 28 PICT carburetors with an accelerator pump capacity of 1.1—1.4 cc per stroke (formerly 0.8—1.0 cc) were fitted. For this purpose, a stronger diaphragm spring new Part No. 113 129 467 A (formerly 113 129 467) is used. The larger amount of fuel injected will eliminate the transfer difficulties which occur occasionally.

After the above alteration, the normal 28 PICT carburetor was fitted again. To increase the amount injected on these carburetors, a 2 mm thick washer is fitted at the connecting rod cotter pin instead of the 1 mm thick washer (Part No. 111 129 147).

From June 1963, Chassis No. 5 578 122 (Engine No. 7 777 338), 28 PICT carburetors with the increased accelerator pump injection amount of 1.1—1.4 cc per stroke were fitted. The following parts were modified:

	Part No. new	Part No. old
Diaphragm spring	113 129 467 A	113 129 467
Connecting rod	113 129 481 A	113 129 481

The modified version begins from carburetor No. 4 028 618 and 7 918 801.

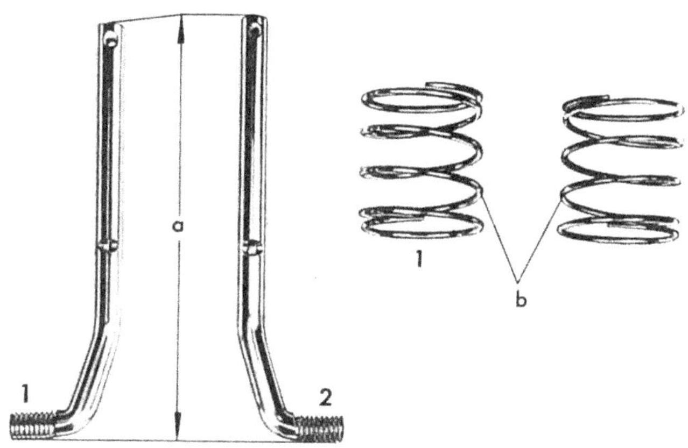

Version	Length a	Wire thickness b
1 = new	61.0 mm (2.401″)	1.1 mm (.043″)
2 = old	62.5 mm (2.460″)	0.9 mm (.035″)

Service installation

a - The stronger spring can be installed together with the former connecting rod.

b - The new connecting rod must only be installed with the stronger spring.

Parts of the previous pattern should be used up.

Air Cleaner

The oil bath air cleaner has an air intake tube which is fitted with a controllable flap valve. In the released position, that is at temperatures under 20° C (68° F) the valve is opened by the air drawn in as the engine speed increases. The valve shaft is equipped with a small weight for compensation purposes. In the fixed position — at temperatures of 20° C (68° F) and over — the air intake tube is opened and the connection for the pre-heater pipe from the left heater junction box is closed by the

Symptoms	Cause	Remedy
7 - Engine stalls when accelerator pedal is released suddenly	Idling mixture too rich	Adjust idling correctly
8 - Engine runs unevenly (surges) with black exhaust smoke at low idling speed and smokes badly as idling speed increases. Spark plugs soot-up quickly and mis-fire	a - Excessive pressure on the float needle valve b - Leaky float c - Float needle valve not closing	a - Check fuel pump pressure and reduce if necessary b - Replace float c - Check needle valve and replace if necessary
9 - Engine runs unevenly at full throttle, misfires and cuts out or loses power	Fuel starvation	a - Clean main jet and power fuel system b - Clean float needle valve c - Check fuel pump pressure and increase if necessary d - Clean fuel tank and tap
10 - Excessive fuel consumption	a - Jet sizes not properly matched b - Excessive pressure at float needle valve c - Leaky float d - Float needle valve not closing e - Automatic choke not working properly	a - Install correct set of jets. Check spark plug condition b - Check fuel pump pressure and reduce if necessary c - Replace float d - Check needle valve and replace if necessary e - Check as at Point 2

Note:

A poor transfer and a tendency to stall when idling can also be caused by insufficient ignition advance, inadequate breaker point gap or dirty spark plugs. Always check ignition system when in doubt.

General

The accelerator cable passes through the frame tunnel and fan housing of the engine in a guide tube. It is attached to the accelerator pedal rod at one end and to a swivel pin in the throttle valve lever at the other.

The spring pushed over the guide tube at the fan housing returns the accelerator cable and closes the throttle. A sleeve prevents the spring from becoming deflected.

Removal

The removal and installation is facilitated by lifting the rear end of the vehicle.

1 - Disconnect accelerator cable from throttle valve lever.

2 - Compress the spring and remove spring seat. Take off sleeve and spring.

3 - Detach bolt from accelerator pedal and disconnect cable from bolt.

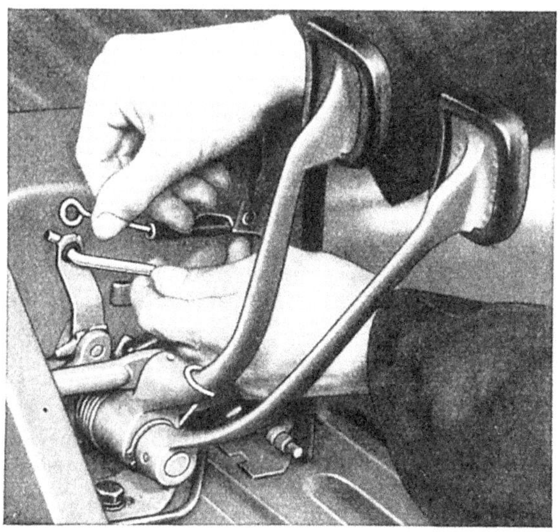

4 - Pull accelerator cable out of the guide tube in the fan housing towards the front.

5 - Take off rubber boot at the end of the cable guide tubes in the frame.

6 - Pull the accelerator cable towards the front out of the guide tube.

Installation

When installing, the following points should be noted:

1 - Grease accelerator cable with universal grease.

2 - The accelerator cable must be laid straight between the guide tubes.

3 - Make sure that the rubber boot is correctly seated to prevent water entering the guide tube.

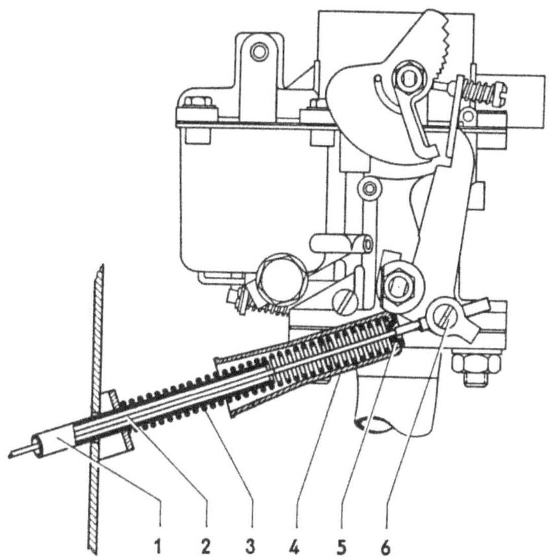

1 - Accelerator cable guide tube
2 - Accelerator cable
3 - Accelerator cable spring
4 - Spring sleeve
5 - Spring seat
6 - Accelerator cable swivel pin

1 - Rubber boot 2 - Accelerator cable 3 - Clutch cable

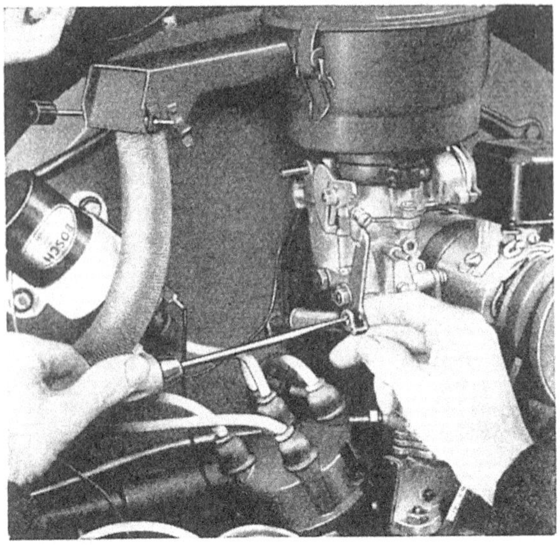

4 - Special care must be taken when attaching the accelerator cable to the throttle valve lever, as otherwise undue tension may occur at full throttle, leading to breakage of the cable. Open throttle valve so that there is a clearance of about 1 mm (.04") between throttle lever and stop at carburetor body. Fully depress accelerator pedal and connect cable to throttle valve.

Accelerator Cable and Accelerator Pedal

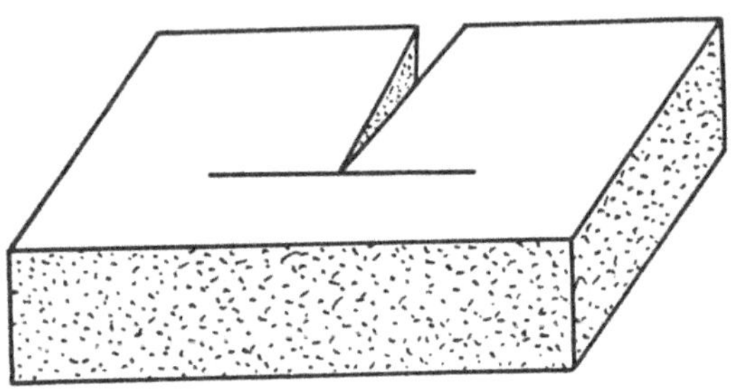

Length: 65 mm (2.6")
Width: 35 mm (1.4")
Height: 15 mm (.6")

Sticking of the accelerator pedal hinge caused by snow and dirt accumulations can generally be avoided by greasing the hinge well. Under severe conditions a more effective remedy can be obtained by covering the hinge with a PVC foam rubber strip.

1 - Cut out a piece of water-repellent PVC foam rubber and provide it with a T-shaped slit.

2 - The portion of the strip situated under the accelerator pedal should be about 12—15 mm (.5"—.6") wide.

3 - Coat the strip with adhesive and position it round the accelerator pedal so that the leg of the T points in the direction of travel and the cross of the T is positioned on and parallel to the hinge pin.

4 - The opening in the floor mat should be trimmed accordingly.

If the accelerator cable has become frozen in the guide tube due to the *formation of condensation*, remove it, blow out the guide tube with compressed air and re-install the cable after greasing it well. Ensure that the carburetor linkage is well lubricated.

Note:

During the winter, the accelerator movement on vehicles from Chassis No. 1 600 440 can become restricted by the build-up of hard snow beneath the pedal hinge. The following modification can be made to prevent this from occurring:

1 - Remove accelerator pedal.

2 - Shorten the pedal plate at the bottom by approximately 13 mm as shown in drawing.

3 - Bend the hinge until the eye is level with the pedal plate and cement a new rubber cover (Part No. 113 721 647 A) into position.

4 - Install pedal and tuck end flap of the new rubber cover underneath the floor covering.

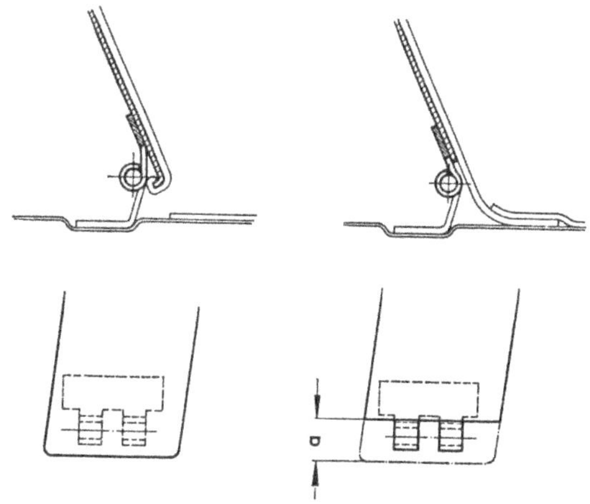

a = 13 mm (.5")

Note:

From Chassis No. 3 405 001 the accelerator pedal hinge was raised and the rubber cover lengthened at the bottom. The new version (Part No. 111 721 507 B — RH drive 113 721 507 B) prevents the pedal hinge from becoming blocked by ice and snow.

It can be installed in vehicles from Chassis No. 1 600 440 (1 st August 1957) as follows:

1 - Remove accelerator pedal.

2 - Drill out the hinge spot welds.

3 - Drill 6.2 mm (.25") holes at the spot weld bosses in the new hinge (Part No. 111 701 535 A) and secure it to the floor plate with two M 6×15 mm screws. Install a rubber packing under the hinge to prevent leaks.

4 - Cement a new rubber cover (Part No. 113 721 647 A to the pedal plate and tuck the extended flap underneath the floor covering.

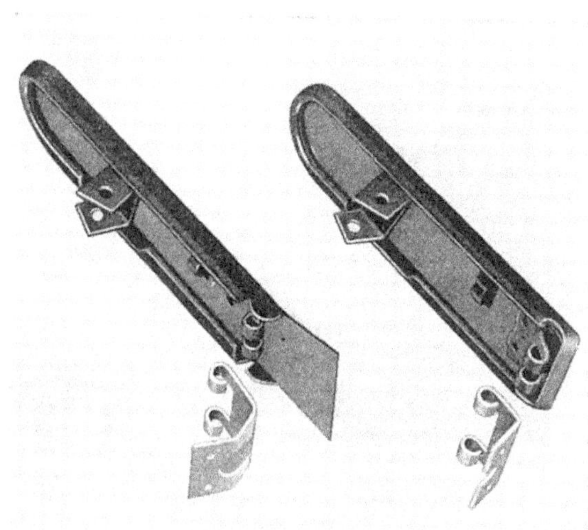

New Old

THIS PAGE INTENTIONALLY
LEFT BLANK

The upper and lower parts are held together by five screws and a gasket is fitted between the two parts.

The connection pipe for the fuel line and the power fuel pipe are pressed into the upper part and the float needle valve is screwed into it. An angled drilling ventilates the float chamber to the air intake flange.

The automatic choke is installed in a small housing cast on to the upper part. The main parts of the automatic choke are the choke valve, choke valve shaft with a two-arm lever riveted to it and the freely mounted fast idle cam with the operating lever.

The lever on the choke valve shaft engages in a recess in the pull rod for the vacuum diaphragm which is connected via a drilling to the vacuum below the throttle valve. As the vacuum increases, the diaphragm opens the choke valve slightly.

The automatic choke is closed with a ceramic cover in which are the heater element and the bi-metal spring.

The lower part of the carburetor contains the mixing and float chambers, the float and all the parts necessary for the preparation of the fuel/air mixture. At the bottom of the mixing chamber is the throttle valve shaft with the throttle valve which is operated by the throttle valve lever.

The accelerator pump is located in a housing cast on to the float chamber and is connected to the throttle valve shaft via the connecting link and the pump lever.

The function of the idling circuit is that of an auxiliary carburetor which takes over the preparation of the mixture when the throttle valve is almost closed. The idling mixture can be enriched or weakened by means of the volume control screw. The engine idling speed is controlled by the idle adjusting screw. The fast idle cam increases the idling speed as the choke valve closes when starting from cold and decreases it as the engine warms up.

Operation

Automatic Choke

The automatic choke provides a rich mixture for the cold engine, dependent on the outside temperature, and weakens the mixture gradually as the engine warms up. Furthermore, it increases the idling speed as long as the engine is not properly warm. This makes it possible to start and run the cold engine satisfactorily until it is thoroughly warmed up. The operation of the choke is fully automatic.

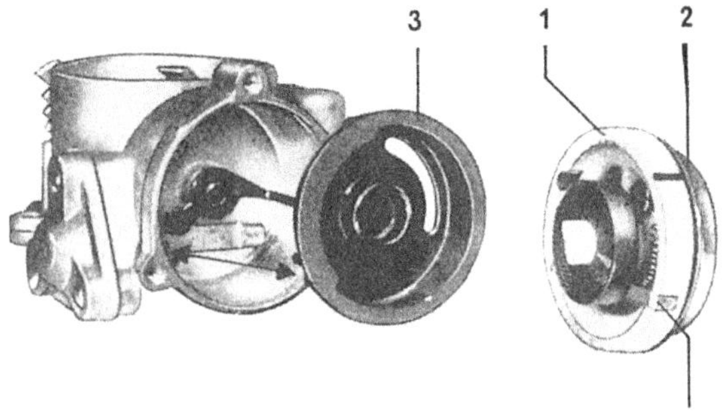

The choke unit (1) is of ceramic material and is insulated externally by a metal cap (2). Inside the unit is a heater element and a bi-metal spring which has a hook-shaped outer end. A plastic insert (3) in the choke housing provides lateral support and heat insulation for the bi-metal spring. A ceramic rod (4) prevents the heater element from coming into contact with the bi-metal spring.

At low temperatures the bi-metal spring commences to uncurl and this movement is transmitted by the hooked end of the spring to the angled arm of the lever riveted to the choke valve shaft. This closes the choke valve and the engine receives the rich mixture required when starting from cold.

The rotary movement of the choke valve shaft is also transmitted to the operating lever which moves the fast idle cam. The closing force of the bi-metal spring thus turns the fast idle cam to a position where the idling screw rests on the highest step of the cam, when the accelerator pedal is released. According to the degree of force exerted by the bi-metal spring, the idling screw may come to rest on one of the six steps on the cam. This opens the throttle slightly and increases the idling speed. As the closing effort decreases due to the increasing heat, the fast idle cam turns so that the idling screw rests on the lower step and the engine runs at the normal idling speed.

Several forces work together to open the choke valve. As a result of the off center mounting of the choke valve shaft in the air intake flange of the carburetor, the flaps of the choke valve are of different sizes. The larger flap moves downwards as it opens so that the flow of air being drawn in tends to force the choke open. At the same time, the closing effort of the bi-metal spring is reduced by the heat from the heater element. When the ignition is switched on, the heater element receives current via a cable from terminal 15 on the coil to the tab connector on the automatic choke cover. The choke valve takes about 2 to 3 minutes to open fully after the ignition has been switched on.

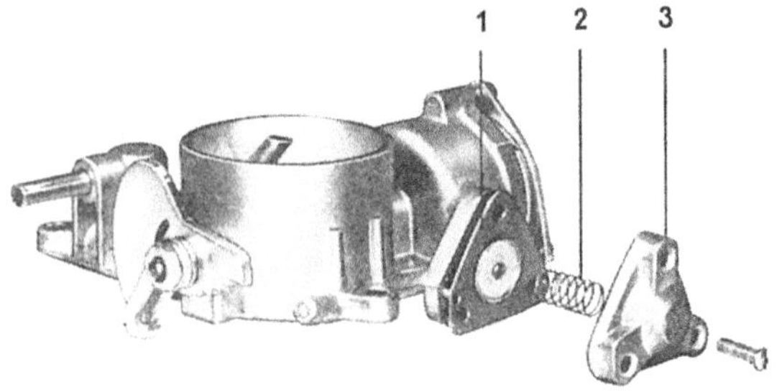

1 - Vacuum diaphragm
2 - Diaphragm spring
3 - Cover for vacuum diaphragm

The choke valve is also open slightly against the tension of the bi-metal spring by the vacuum diaphragm (1). On one side of the vacuum diaphragm is a drilling which leads to below the throttle valve. As soon as the throttle valve is opened slightly there is a high vacuum at this point and this actuates the diaphragm. The pull rod attached to the diaphragm engages the arm of the lever on the choke valve shaft and opens the choke. This ensures that the rich mixture is automatically weakened to suit the engine conditions.

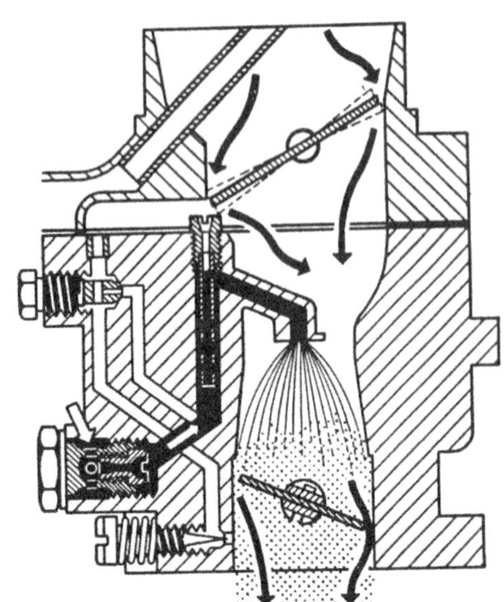

Before starting an engine from cold, the accelerator pedal must be depressed once so that the bi-metal spring can close the choke valve and enable the automatic choke to work properly. The starter must be operated as soon as the ignition is switched on so that the choke valve cannot open prematurely due to the heat from the heater element.

When starting the engine from cold, the vacuum effective under the choke valve lifts fuel from the emulsion tube into the mixing chamber via the discharge arm. The air required for the formation of the mixture is drawn in past the choke valve which is then opened slightly by the vacuum. The tension of the bi-metal spring closes the choke valve so that it alternates between open and closed when starting. This gives a rich starting mixture which ensures easy starting even at very low temperatures. The vacuum building up at the throttle valve then starts to move the diaphragm so that the choke valve opens further.

This prevents the starting mixture from becoming excessively rich. The choke valve is then opened fully by the bi-metal spring as it warms up and the mixture automatically weakens to suit the operating conditions of the engine.

If the accelerator pedal is released when moving off, the idling adjustment screw rests on one of the steps on the fast idle cam and ensures that the engine continues to run at a fast idling speed. The engine does not run at the pre-set idling speed when the accelerator pedal is released until the idling screw is resting on the lowest step.

Idling operation

At idling speed, the mixture takes place in a different manner. When the throttle is almost closed the air speed, and consequently, the vacuum in the venture is so low that no fuel will be drawn out of the discharge arm. The engine now gets the mixture from the idling system.

The fuel flows from the float chamber through the main jet into a drilling leading up to the pilot jet. Air is drawn via the pilot air bleed drilling above this jet. The idling mixture thus created passes through a drilling leading downwards to the volume control screw and the discharge opening below the throttle valve. The air passing by at high speed draws the idling mixture out of the discharge hole. The idling mixture can be enriched or weakened with the volume control screw. The adjustment of this screw is, therefore, very important.

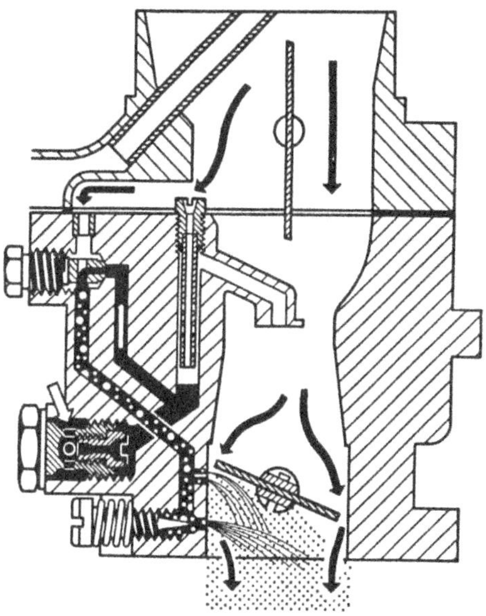

The idling adjustment also includes the idling speed of the engine. This setting must be correct so that the speed is not too high or too low. This part of the adjustment is made with the idling screw which alters the position of the throttle valve at idling.

When the throttle valve is opened slightly and the idling speed thus increased, the engine requires more fuel. The vacuum zone which is effective at the throttle valve gap moves up into the area near the by-pass ports which are situated just above the idle port and are connected to the idling mixture drilling. Additional idling mixture is drawn from the by-pass ports and this helps to improve the transfer from the idling circuit to the main jet system.

Normal operation

During normal operation the vacuum in the intake passage draws the fuel out of the spraying arm which is connected to a vertical well containing the air correction jet with the emulsion tube pressed into it. Fuel flows to this well from the float chamber via the main jet.

The factors governing the amount of vacuum are the size of the venturi, the engine speed and the throttle opening. The specially designed restriction of the venturi — where the spraying arm ends — has an accelerating effect on the air being drawn in by the engine. This gives — in connection with the position of the throttle valve — a varying degree of vacuum and draws different quantities of fuel from the spraying arm. As the fuel level in the emulsion tube drops due to the rising vacuum created by the increasing throttle opening, air is simultaneously drawn in through the air correction jet. This air passes out of the holes in the emulsion tube as they are uncovered and mixes with the fuel flowing from the main jet, thus progressively weakening the mixture as the engine speed increases.

Accelerator Pump

The accelerator pump is operated from the throttle shaft by a connector rod and pump lever. When the throttle valve is closed, the diaphragm spring presses the pump diaphragm to the rest position and the chamber in front of the diaphragm fills with fuel drawn in from the float chamber via a ball valve.

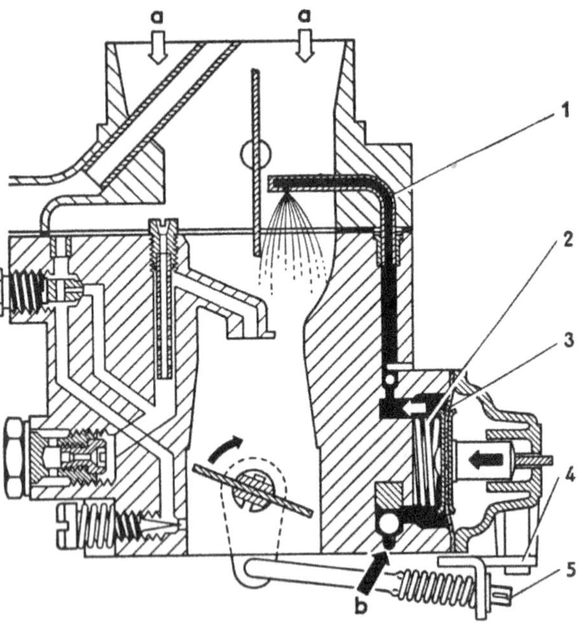

a - Air
b - Fuel from float chamber
1 - Injection tube
2 - Pump spring
3 - Pump diaphragm
4 - Pump lever
5 - Connecting rod

When the throttle valve is opened, the movement is transmitted to the diaphragm through the linkage. The pressure stroke of the pump opens a ball check valve in the delivery passage and fuel is forced through a calibrated discharge tube into the mixing chamber. Injection begins when the throttle is opened and ends at an opening angle of about 30°. This additional fuel enriches the mixture and ensures a smooth transfer and rapid acceleration.

The accelerator pump enriches the mixture in the low and medium speed range only, as the spring on the connecting rod holds the diaphragm in the depressed position as the throttle is opened further and prevents the pump from refilling with fuel until the throttle is almost closed again.

Power Fuel System

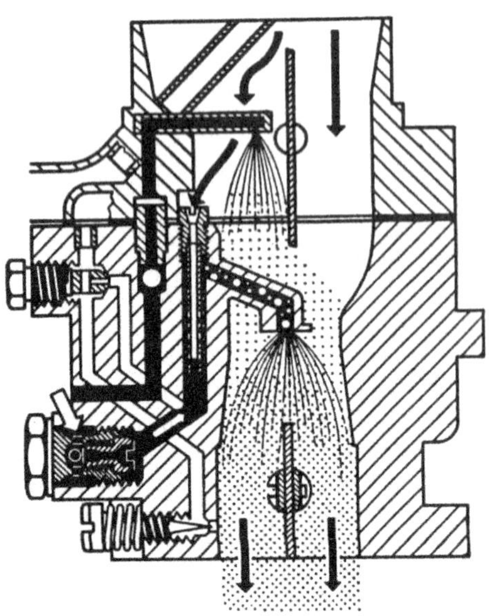

The power fuel system provides a richer mixture at full load and high engine speeds to enable the engine to give its maximum output. A calibrated power fuel tube is pressed into the upper part of the carburetor and connected by a drilling to the float chamber. The tube ends above the discharge arm in an area where the vacuum is reduced. At low and medium revolutions the vacuum is not high enough to draw fuel out of the power fuel tube. The additional fuel supply from the power fuel system only commences when the engine speed and the vacuum is high.

Under these operating conditions, the mixture is enriched by fuel which is supplied through the accelerator circuit. The vacuum at the end of the injection tube is now so large that fuel is drawn from the float chamber through the accelerator pump.

The oil bath air cleaner consists of the upper part with the filter element and the lower part with the air intake tube. The two parts are held together by two quick-release clips.

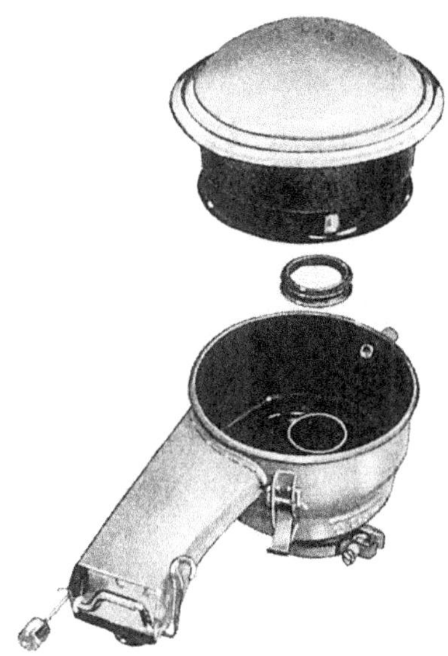

In the air intake tube in the lower part is a small flap valve which controls the flow of fresh or pre-heated air in accordance with the speed of the engine. The flap is fitted with a balance weight so that at low speeds only preheated air can center the carburetor. The warm air is taken from the engine cooling air near the left rear exhaust elbow and passes through a hose to the air intake tube on the cleaner. In this manner, the carburetor receives only pre-heated air at idling and medium speeds and this helps to prevent the formation of ice in the carburetor. As the engine speed increases and more air is drawn in, the air stream presses the flap valve down against the weight and cuts off the intake of pre-heated air so that the carburetor draws in only cold air. Only in exceptional cases when the outside temperature is very high should the flap be fixed so that the intake of warm air is closed continuously.

The lower part of the air cleaner is fitted with an adaptor pipe which is connected with a hose to the oil filler and gives crankcase ventilation.

The air cleaner filters the dust and dirt out of the air drawn in by the engine and also acts as an intake air noise damper. The cleaner must be checked at the intervals given in the lubrication chart and the lower part cleaned as required. The lower part should always be cleaned when there is only 4—5 mm of clear oil above the sludge layer.

In very dusty conditions the cleaner must be checked more frequently, even daily if necessary. The lower part should then be cleaned as required and filled with fresh oil.

Note:

The securing clip on the air cleaner has a red mark which should be in line with the left rib on the upper part of the carburetor. This ensures that the flat part of the cleaner is parallel to the fan housing.

Note:

5/66

From May 1966

Type	Chassis No.	Engine No.
1/1200	116 807 190	D 0 079 454
1/1300	116 852 850	F 0 767 046

all vehicle engines are fitted with a modified carburetor pre-heating system. The warm air is taken from the left heat exchanger except on the 14 Models where it is taken from the right side.

At temperatures below +10° C (+50° F) the balance-weighted flap in the intake of the oil bath air cleaner must be free to operate on its own.

If higher temperatures prevail, the flap should be held in position to cut off the hot air flow. To do this, the flap lever is pushed down past the folded edge of the air intake or the control box.

If these measures are not carried out, the engine can tend to pink if the vehicle is operated for longer periods at high outside temperatures and the engine performance will be reduced.

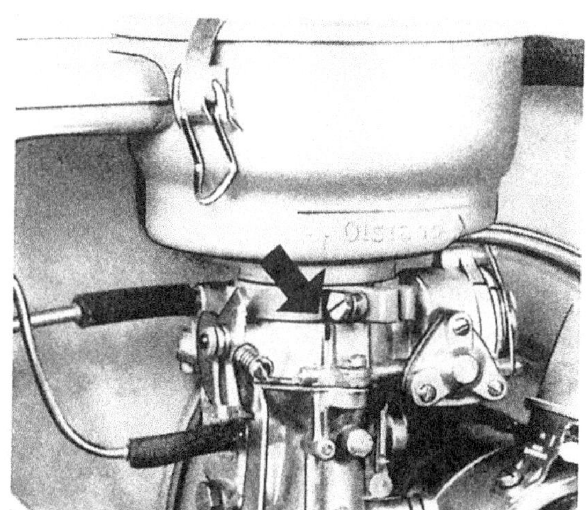

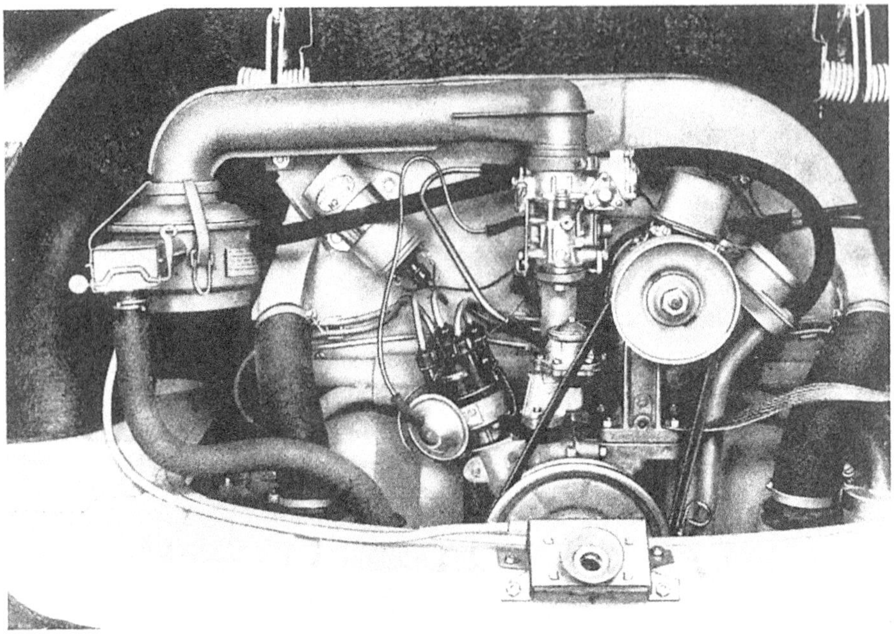

Removal of Intake Elbow

Contrary to the other 1200 Volkswagen Models, the Karmann Ghia models have a suspended cleaner which is connected to the carburetor with an elbow.

The intake elbow must always be removed before taking off the top part of the carburetor or the complete carburetor. The same applies when removing and installing the engine.

Removal

1 - Remove pre-heating pipe from air cleaner intake tube.

2 - Pull crankcase breather pipe off air cleaner.

3 - Release the clips and take air cleaner off elbow.

4 - Loosen elbow securing screw at carburetor.

5 - Remove screw at air cleaner retainer and take cleaner support off.

6 - Lift elbow off.

The air cleaner retainer can be taken off after removing the ignition coil.

Installation

When installing, note the following:

1 - Install intake elbow on carburetor upper part and ensure that it seals properly.

2 - Do not overtighten the screw on the carburetor end of the elbow as this can distort the upper part of the carburetor and make the choke valve shaft stiff in operation.

Removal

1 - Remove pre-heater pipe from air cleaner intake tube.

2 - Pull crankcase breather pipe off air cleaner.

3 - Loosen clamp screw on cleaner and take cleaner off.

4 - Take fuel hose and vacuum hose off carburetor.

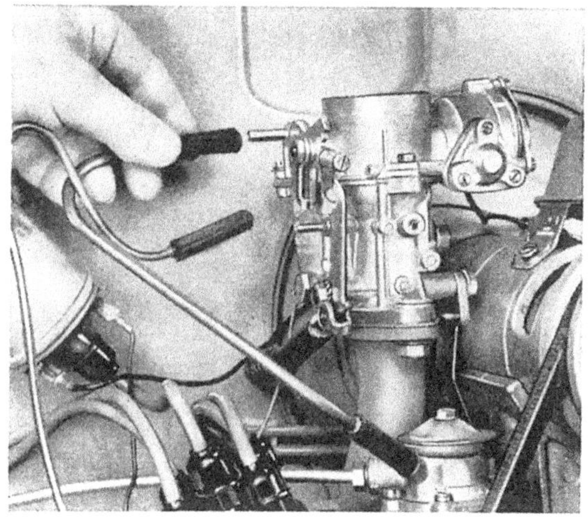

5 - Pull automatic choke cable off tab.

6 - Detach accelerator cable from throttle valve lever and take off spring, spring plate and cable pin.

7 - Remove two nuts and take carburetor off.

Installation

When installing, note the following:

1 - Fit a new gasket on intake manifold flange.

2 - Tighten nuts evenly but do not overtighten.

3 - Secure accelerator cable so that there is 1 mm play between throttle valve lever and the stop on the carburetor body when the accelerator pedal is fully depressed.

4 - Set the idling when the engine is warm.

Disassembly and Assembly of Carburetor

1 - Remove carburetor.

2 - Remove the five screws securing the upper part and lift it off.

3 - Take float out.

4 - Screw float needle valve out of upper part.

5 - Remove the three screws in the automatic choke and take off retaining ring and distance pieces.

6 - Take off the ceramic plate, bi-metal spring, heater element and plastic cap.

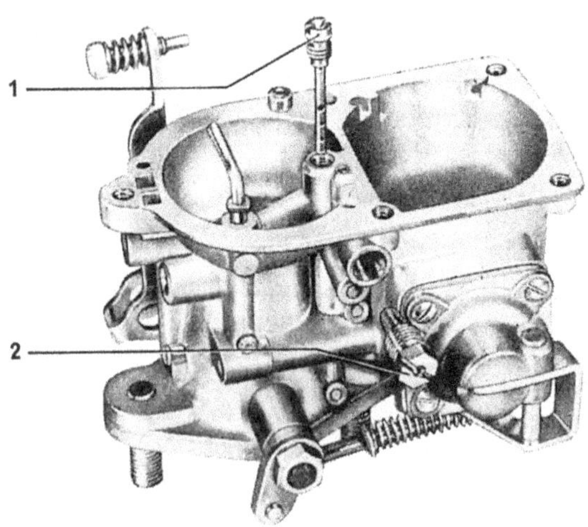

7 - Screw out the air correction jet with the emulsion tube and the pilot jet.

1 - Air correction jet with emulsion tube
2 - Pilot jet

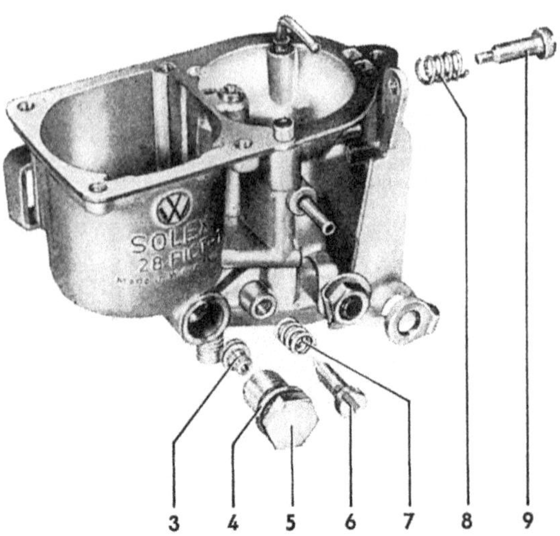

8 - Screw out the main jet carrier with main jet and the volume control screw.

3 - Main jet
4 - Gasket
5 - Main jet carrier
6 - Volume control screw
7 - Spring
8 - Spring
9 - Idle adjusting screw

9 - Remove pump lever cotter pin from the connector rod.

10 - Remove four pump cover retaining screws and take off cover, diaphragm and spring.

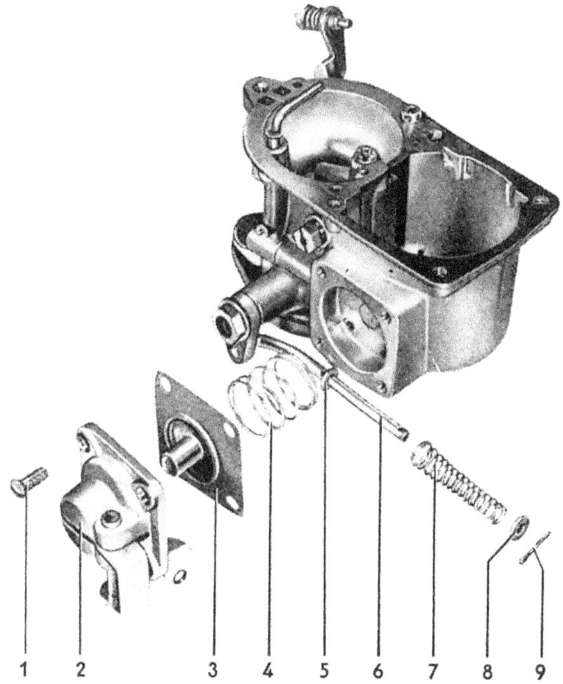

1 - Screw
2 - Pump cover
3 - Pump diaphragm
4 - Diaphragm spring
5 - Washer
6 - Pump connector rod
7 - Spring for connector rod
8 - Washer
9 - Cotter pin

Cleaning

1 - Clean all parts in gasoline with the exception of the cover for the automatic choke.

2 - Clean jets, valves in accelerator system and drillings with compressed air.

Under no circumstances should pins or pieces of wire be used to clean the jets as they will damage or enlarge the calibrated drillings.

Removal of accelerator pump discharge tube

1 - Loosen wing nuts of VW 646/1 clamp (local manufacture) and take off clamp jaw.

2 - Position the tool so that the discharge tube is located in the recess provided.

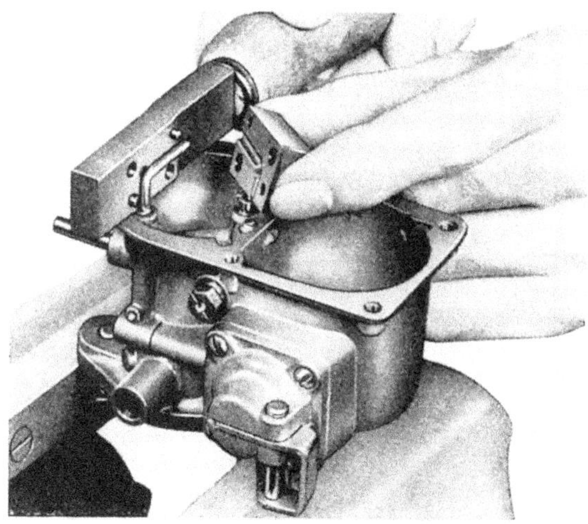

3 - Fit jaw and tighten wing nuts.

4 - Pull the tube out by tapping the end of the clamp with a hammer. Do not turn the tube when removing or installing.

5 - Install a new discharge tube and check direction of spray. When the throttle opens, the fuel should be sprayed straight down past the throttle valve. If this is not so the tube should be aligned properly.

Checking and Assembling

Note the following points when checking the various parts:

Upper part:

1 - Check float needle valve for leakage. When the needle is pressed lightly on to the seat, it should not be possible to blow air through the valve.

2 - Check condition of needle valve gasket and location when installed.

3 - Check gasket between upper and lower parts.

4 - Check choke valve shaft and fast idle cam for freedom of movement.

5 - Check heater element and bi-metal spring. If one of these parts is damaged the complete cover must be replaced.

6 - The lug on the plastic insert must engage in the notch in the automatic choke housing.

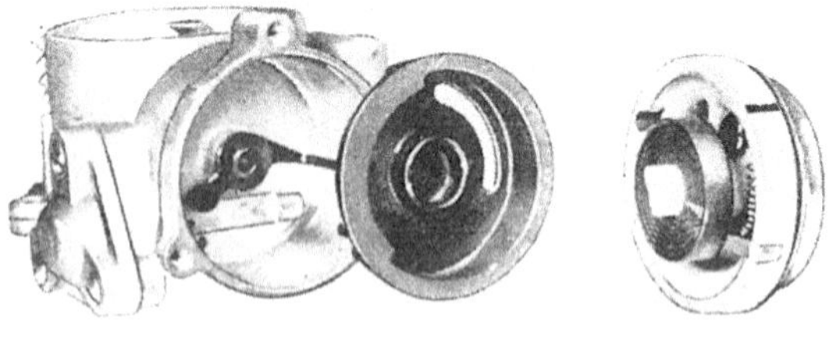

7 - When fitting the ceramic cover, ensure that the ceramic rod between heater element and bi-metal spring is located properly.

8 - The operating lever must engage the hooked end of the bi-metal spring when the cover is installed.

9 - Install outer cap and retaining ring with three screws and distance pieces.

10 - Turn the cap until the mark on the ceramic cover is in line with the lug on the automatic choke housing. Do not overtighten the screws.

11 - Oil the fast idle cam on the choke valve shaft lightly.

Lower part:

1 - Check pump diaphragm for leakage and replace if necessary.

2 - When tightening the screws for the pump cover, the pump lever should be pressed away from the float chamber so that the diaphragm is secured in the pressure stroke position.

3 - Test float for leakage by immersing it in hot water. If bubbles appear, the float is leaking and must be replaced.

4 - Check jet sizes and float weight with the data list.

When replacing jets, floats or valves, use only parts marked "SOLEX".

5 - Check throttle shaft clearance.

Excessive clearance encourages the ingress of secondary air and has a detrimental affect on starting and idling. If necessary, the holes for the throttle shaft must be bushed.

6 - Check volume control screw. The tapered portion must not be grooved, bent or pressure marked.

7 - Install float.

8 - Check position of accelerator pump injector tube in the carburetor bore. When the throttle is opened, *the fuel jet must be sprayed straight* down past the throttle valve. If necessary, align the tube properly.

Every carburetor is tested at the factory and adjusted on the engine to suit branded fuels. Any alteration in the settings such as replacing the jets by other than the prescribed sizes is detrimental under normal operating conditions and should be avoided. When changing from ordinary fuel to a premium mixture it is generally only necessary to adjust the idling speed.

Specifications

Venturi	22.5 mm dia.
Main jet	122.5
Air correction jet (with emulsion tube)	130 y (145 y for Ghia models)
Pilot jet	g 55
Pilot air bleed	2.0 mm dia.
Pump jet	0.5
Power fuel jet	1.0
Float needle valve	1.5 mm
Float weight	5.7 g
Pump delivery quantity	1.1—1.4 cc/stroke

Idling adjustment

The idling adjustment requires resetting from time to time. This should only be attempted when the engine is warm, making sure that the idle adjusting screw is not resting on one of the steps on the fast idle cam.

10/65

1 - Set the idling speed to about 700—800 rpm with the idle adjusting screw.

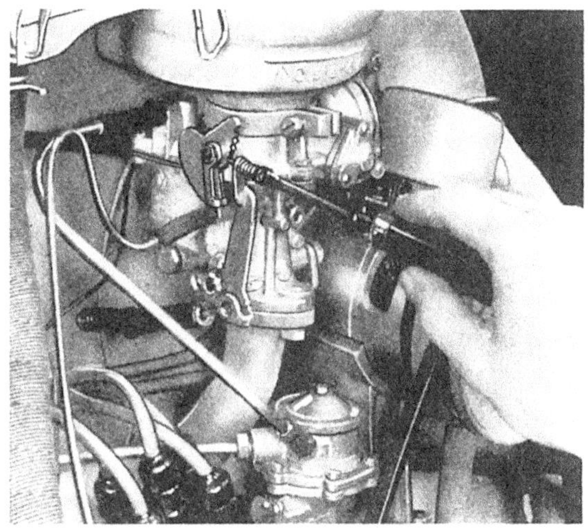

2 - Turn volume control screw in slowly until the engine speed begins to drop, then turn the screw out until the engine runs smoothly. Then turn the volume control screw out about $^{1}/_{4}$ of a turn.

3 - The idling adjustment is correct if the warm engine continues to run after the throttle has been opened and closed suddenly with the clutch pedal depressed.

Idling adjustment on vehicles with "Saxomat"

1 - When the engine is started from cold it runs at a fast idling speed for a short time. As the starting clutch begins to grip at engine speeds over 950 rpm it is possible for a vehicle on which the handbrake is not set, to start moving of its own accord as soon as a gear is engaged.

2 - This possibility can be reduced considerably by adjusting the idling speed to 500/550 rpm when the engine is warm. It is advisable to use a revolution counter when doing this as it is very difficult to estimate the speed accurately.

When the cold engine has started and been acelerated once, the idle screw will move to the 2nd step on the fast idle cam. In this position the engine speed is sufficiently low to enable a gear to be selected without causing the vehicle to move, if the idling speed is properly adjusted.

Carburetor Trouble Shooting

Symptoms	Cause	Remedy
1 - Engine will not start (with fuel in tank and ignition in order)	a - Automatic choke not working properly	a - Check the vacuum diaphragm for freedom of movement
	b - Choke valve sticking	b - Free the choke valve shaft with easing oil
	c - Bi-metal spring unhooked or broken	c - Re-connect spring, or if broken, replace complete ceramic plate
	d - Ceramic plate broken	d - Replace ceramic plate. **Note marks when installing**
	e - Float needle valve sticking and carburetor flooding	e - Clean or replace float needle valve
		Important: If a large quantity of fuel has passed from the flooded carburetor into the engine, switch on the ignition and wait 1 minute before starting and then open throttle fully.
2 - Engine runs continually at a fast idle	a - Automatic choke not switching off	a - Check heater element and both connections
	b - Heater element defective	b - Replace complete ceramic plate
3 - Engine idles unevenly or stalls	a - Idling adjustment incorrect	a - Adjust idling correctly
	b - Pilot jet blocked	b - Clean jet
4 - Engine "runs-on" when ignition is switched off	a - Idling mixture too rich	a - Weaken idling mixture
	b - Idling speed too fast	b - Regulate idling speed
		c - Fit electro-magnetic cut-off valve if necessary
5 - Banging in the exhaust when vehicle is overrunning the engine	Idling mixture slightly weak	Enrich mixture by turning the volume control screw approximately $^1/_8$ of a turn
6 - Poor transfer from idling to normal running	a - Accelerator pump dirty (pump passages blocked, ball sticking)	a - Clean accelerator pump and check action
	b - Torn diaphragm	b - Replace diaphragm
	c - Idling adjustment incorrect	c - Adjust idling correctly

Symptoms	Cause	Remedy
7 - Engine stalls when accelerator pedal is released suddenly	Idling mixture too rich	Adjust idling correctly
8 - Engine runs unevenly (surges) with black exhaust smoke at low idling speed and smokes badly as idling speed increases. Spark plugs soot-up quickly and mis-fire	a - Excessive pressure on the float needle valve b - Float leaking c - Float needle valve not closing	a - Check fuel pump pressure and reduce if necessary b - Replace float c - Check needle valve and replace if necessary
9 - Engine runs unevenly at full throttle, misfires and cuts out or lacks power	Fuel starvation	a - Clean main jet and power fuel system b - Clean float needle valve c - Check fuel pump pressure and increase if necessary d - Clean fuel tank and tap
10 - Excessive fuel consumption	a - Jet sizes not properly matched b - Excessive pressure at float needle valve c - Float leaking d - Float needle valve not closing e - Automatic choke not working properly	a - Install correct set of jets. Check spark plug condition b - Check fuel pump pressure and reduce if necessary c - Replace float d - Check needle valve and replace if necessary e - Check as at Point 2

Note:

A poor transfer and a tendency to stall when idling can also be caused by insufficient ignition advance, inadequate breaker point gap or dirty spark plugs. Always check ignition system when in doubt.

Note:

In some 28 PICT-1 carburetors the bi-metal spring in the automatic choke is tensioned too much even though the marks are aligned properly. This makes the choke open late and gives a very rich mixture. When starting from cold this, in turn, causes excessive smoke at the exhaust and uneven engine running.

This trouble can be eliminated by turning the ceramic cover of the automatic choke about 4—5 mm to the right (a). This opens the choke earlier and weakens the fuel-air mixture.

a = 4—5 mm

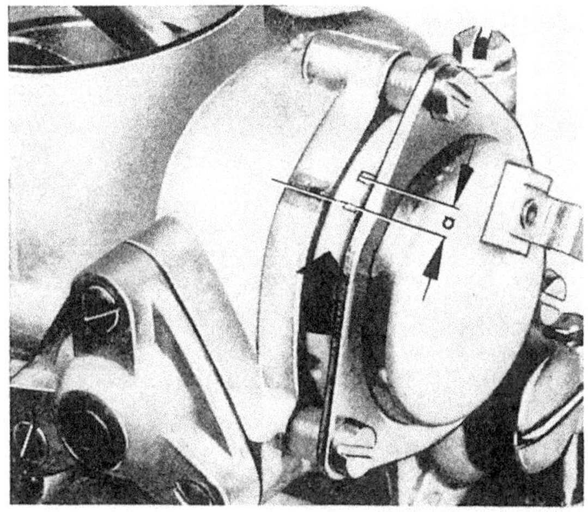

Note:

On some 28 PICT-1 carburetors of a certain manufacturing series (only version 2) it may be found that the power fuel system does not work properly. The result is that on some Volkswagen 1200 De Luxe Sedans and Convertibles, the top speed is considerably lower than the specified 115 kph (72 mph).

When complaints of this nature are received, the first thing to do is to check the engine and fuel system settings and rectify as necessary. If the engine output remains low, the carburetor itself should be checked carefully. Carburetors of version 2 should then be replaced by carburetors of version 1 of the same type — Part No. 113 129 023 J.

The defect has been eliminated from Engine No. 9 074 380.

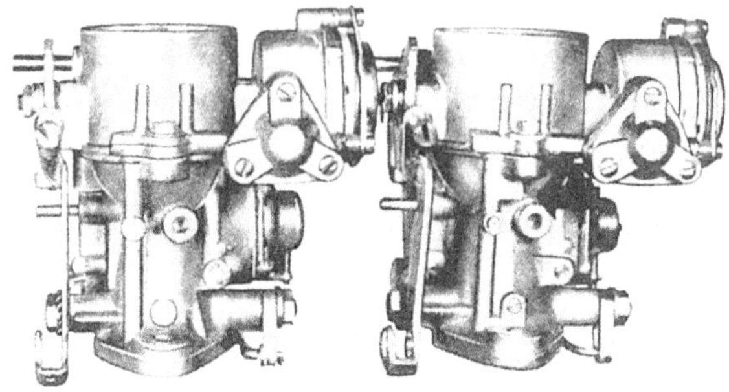

Version 1 Version 2

Removing and Installing Accelerator Cable

The accelerator cable passes through the frame tunnel and fan housing of the engine in guide tubes. It is attached to the accelerator pedal rod at one end and to a swivel pin in the throttle valve lever at the other. Between tunnel and front engine cover plate, the cable passes through a plastic hose.

The spring pushed over the guide tube at the fan housing returns the accelerator cable and closes the throttle A sleeve prevents the spring from becoming deflected.

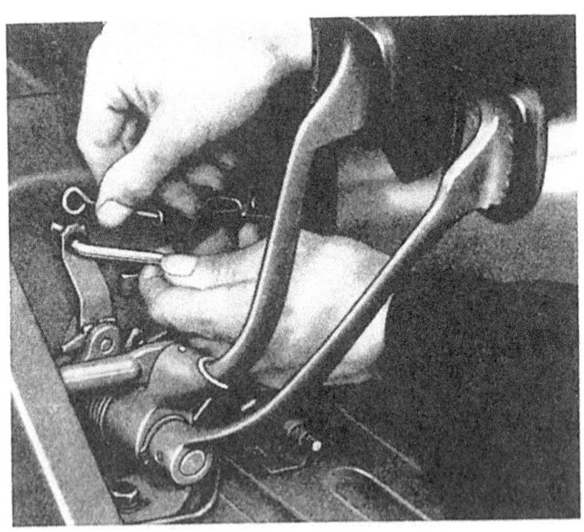

Removal

The removal and installation is facilitated by lifting the rear end of the vehicle.

1 - Disconnect accelerator cable from throttle valve lever.

2 - Compress the spring and remove spring seat. Take off sleeve and spring.

3 - Detach bolt from accelerator pedal and disconnect cable from bolt.

4 - Pull out accelerator cable from its guide tube in the fan housing towards the front.

5 - Pull plastic hose off cable.

6 - Take off rubber boot at the end of the cable guide tubes in the frame.

7 - Pull the accelerator cable towards the front out of the guide tube.

1 - Rubber boot
2 - Accelerator cable with plastic hose
3 - Clutch cable

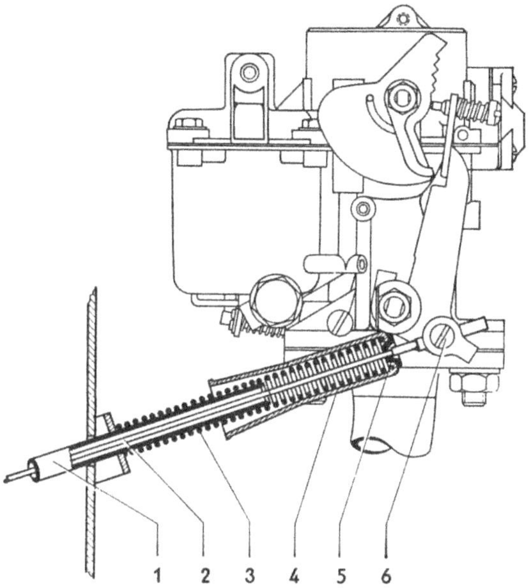

1 - Accelerator cable guide tube
2 - Accelerator cable
3 - Accelerator cable spring
4 - Spring sleeve
5 - Spring seat
6 - Accelerator cable pin

Installation

When installing, the following points should be noted:

1 - Grease accelerator cable with universal grease.

2 - The accelerator cable must be laid straight between the guide tubes.

3 - Make sure that the rubber boot and the plastic hose are correctly seated to prevent water from entering the guide tubes.

4 - Special care must be taken when attaching the accelerator cable to the throttle valve lever as otherwise excessive tension may occur at full throttle and cause breakage of the cable. Open throttle valve

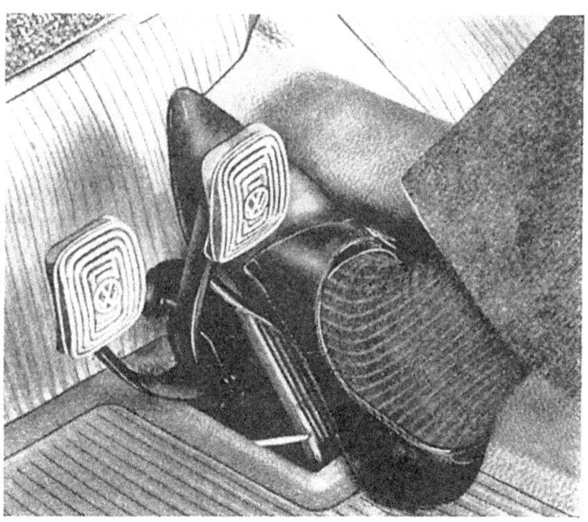

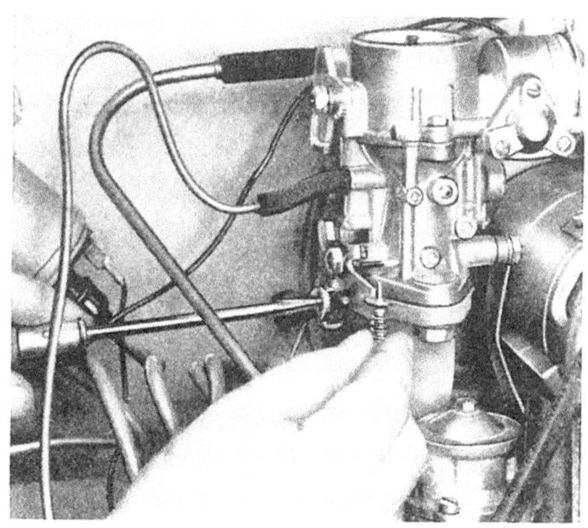

so that there is a clearance of about 1 mm (.04") between throttle lever and stop on carburetor body. Fully depress accelerator pedal and connect cable to throttle valve.

8/66

Note:

From August 1965, Chassis No. 116 000 002 (Engine No. F 0 000 001), all 1.3 liter engines are fitted with a modified SOLEX carburetor 30 PICT-1. Externally, this carburetor is the same as the 28 PICT-1 but the venturi, jets and the diameter of the mounting flange are different.

THIS PAGE INTENTIONALLY
LEFT BLANK

Fuel Pump

Operation

Fuel is fed to the carburetor by a diaphragm pump which is attached to the crankcase. It is operated mechanically from an eccentric on the distributor drive shaft. The flow of fuel delivered by the pump is automatically regulated as the fuel is used up from the float bowl.

The fuel pump consists of the top half, containing suction valve and delivery valve, and the bottom half, incorporating the rocker mechanism. The diaphragm and spring are situated between the cover and the body. The diaphragm consists of several layers of special flexible, clothlike material that is not affected by the fuel and two support washers which are riveted to the diaphragm pull rod.

General Description

As the distributor drive shaft revolves, the eccentric causes the push rod to move against the rocker arm which pulls the diaphragm downward against the diaphragm spring. This movement creates a vacuum above the diaphragm which lifts the suction valve off its seat so that fuel can be drawn in. When the push rod moves downwards, the loaded diaphragm spring pushes the diaphragm upward, forcing the fuel in the pump through the delivery valve into the carburetor. This process is repeated at every turn of the eccentric (once every two revolutions of the engine).

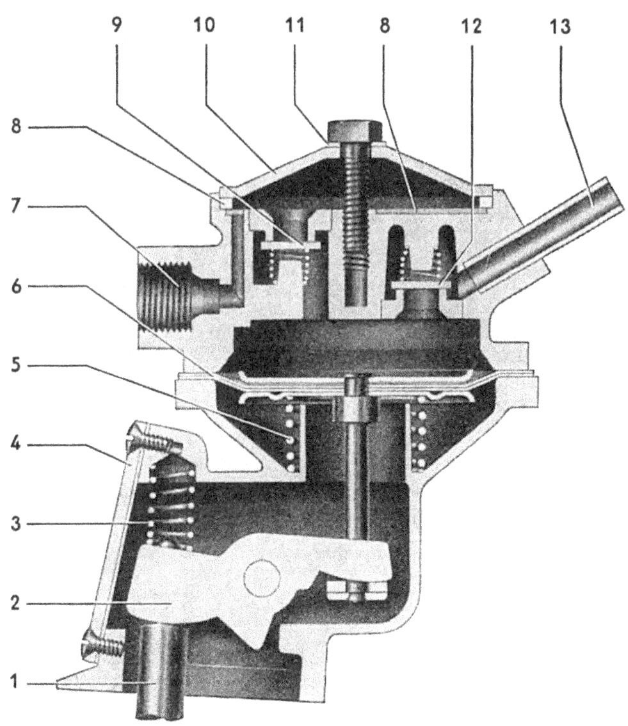

1 - Push rod
2 - Rocker arm
3 - Rocker arm spring
4 - Inspection cover
5 - Diaphragm spring
6 - Diaphragm
7 - Fuel intake
8 - Filter
9 - Suction valve
10 - Fuel pump cover
11 - Gasket
12 - Delivery valve
13 - Fuel outlet

The pump pressure depends on how much the spring is compressed during the pump suction stroke. This pressure is balanced by the buoyancy of the carburetor float, which causes a corresponding pressure at the needle valve seat. As fuel rises in the float bowl, the needle is forced up with greater pressure. Thus the pressure in the fuel line and pump chamber increases, while the pump working stroke decreases. With normal engine operation, the diaphragm is moved only a fraction of a millimeter.

A hole is provided to ventilate the chamber below the diaphragm. This hole also permits the draining of fuel which may have entered the lower chamber.

The fuel pump requires no regular lubrication, as it is lubricated from the crankcase. The filter should be cleaned at the intervals given in the Maintenance Chart.

Note:

From Chassis No. 3 275 700 all engines were equipped with a fuel pump (Part No. 113 127 025) which has a bent delivery pipe in place of the former straight vertical type.

Before this modification was introduced, the throttle lever on the 28 PICT carburetor sometimes came into contact with the pipe between the fuel pump and carburetor when the throttle was opened fully. This should be rectified on all vehicles at the next opportunity by making the following alteration:

Bend the vertical pipe towards the ignition coil with two cranked 6 mm ring wrenches as shown overleaf.

The pipe must be held firmly with the lower wrench and bent with the upper wrench. Care should be taken to avoid loosening the pipe and causing it to leak or breaking the die-cast pump housing at this point.

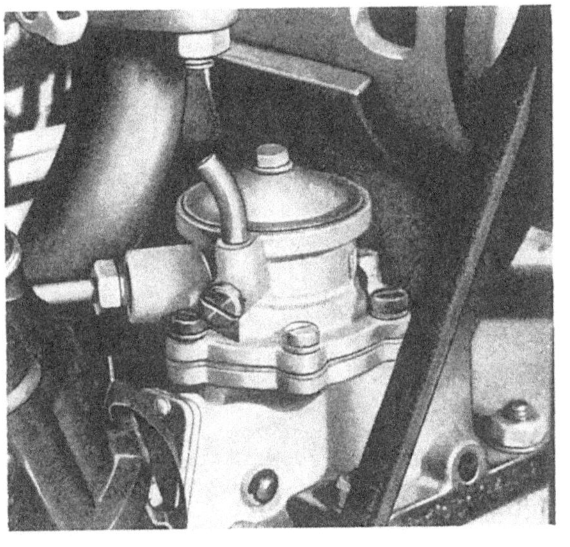

Note:

From Chassis No. 3 606 032 the delivery pipe on the fuel pump (unchanged Part No. 113 127 025) is angled upwards.

At the same time, the filter (Part No. **new** 113 127 177 A) was modified and an additional gasket (Part No. 113 127 183) installed between filter and pump cover.

The new filter and gasket can be installed in previous pumps but the old filter will not fit the new pump upper part.

Note:

From Chassis No. 3 373 469 all engines are equipped with a black rubber hose with a woven outer covering (Part No. **new** 111 209 185 A) between the fuel pump and carburetor in place of the light-colored synthetic hose.

When stocks of Part Nos. 113 127 511 and 111 127 511 B have been used up, the new type (Part No. 111 209 185 A) should be installed.

Note:

From Chassis No. 4 636 869, the black fuel hose (Part No. 111 209 185 A) between fuel pump and carburetor was replaced by a pipe with hose connecting pieces at each end. Only the previous type of hose (Part No. 111 209 185 A) will be supplied as a spare part.

On installation, ensure that the pipe part of the new line does not contact the vacuum line.

Note:

From September 1964, Chassis No. 115 083 659 (Engine No. 8 888 105) the fuel line on the 34 bhp (41.5 SAE bhp) engine is fitted with a valve (Part No. 113 127 405) which stops the flow of fuel to the carburetor when the engine is not running. When the engine is started, the valve is opened by the pressure from the pump.

The fuel lines have been altered slightly to fit the valve. The line between the front engine cover plate and the fuel pump is secured by a clip which is fixed to the screw attaching the fan housing to the cylinder cover plate.

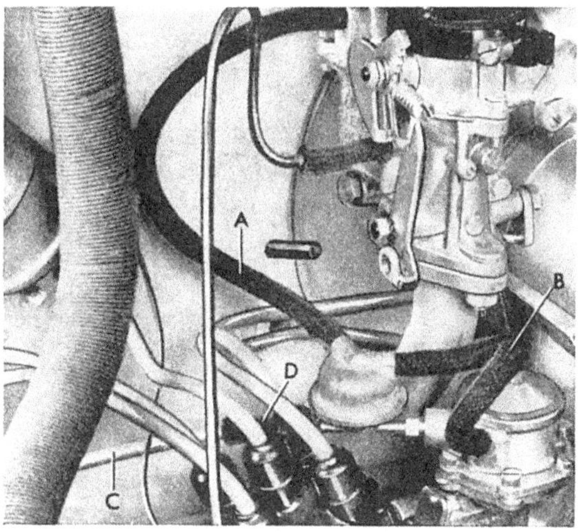

The following parts, which are also required for service installation, have been changed:

	Part No.
Valve for fuel line	113 127 405
Bracket for fuel hose	311 127 525
Nipple	111 127 531 A
Fuel hose (410 mm long)	N 20 355 1

When required, the valve can be service installed as follows:

1 - Seal the fuel line and remove fuel lines and hoses in engine compartment.

2 - Place union nut and new nipple on the pipe in the valve. Connect pipe to fuel pump, align valve properly and fully tighten union nut.

3 - Install a new hose, 250 mm long (A), between valve and carburetor.

4 - Install a new hose, 160 mm long, between pump and valve (B).

5 - Shorten the existing fuel line (C) 90 mm on the nipple side and install it. Cut a 45 mm long piece (D) off an existing hose and install it between line and valve. Install the hose coming from the frame on the fuel line.

6 - Cut a 20 mm long piece of existing hose open lengthwise and fit it on the line to the pump. Install clip (311 127 525) on this piece of hose and secure clip with the screw attaching fan housing to left cylinder cover.

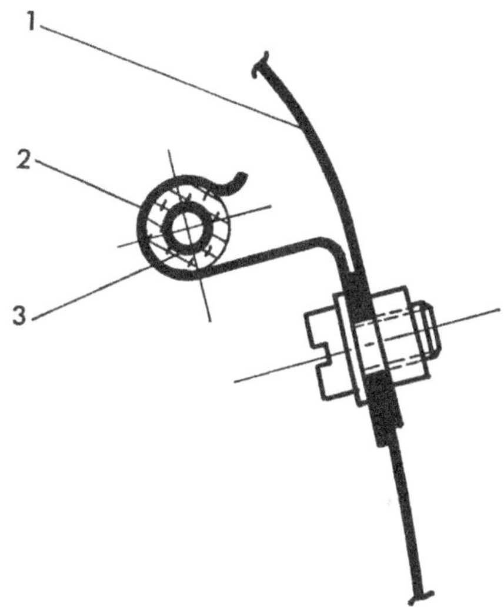

1 - Fan housing
2 - Clip for hose
3 - Fuel line

7 - Start engine and check all lines and hoses for leakage.

THIS PAGE INTENTIONALLY
LEFT BLANK

Checking Pump Pressure

The maximum pump pressure should be 0.2 kg/sq. cm (2.8 lbs./sq. in.) with the needle valve closed and the engine running at 3000 rpm. The minimum amount of fuel delivered is 267 c.c. per minute (16.3 cu. ins.). To check the fuel pump pressure, connect a gauge (range: 0—0.4 kg/sq. cm = 0—5.7 lbs./sq. in.) which is brazed to a fuel test line between the pump and the carburetor by means of a T-piece. The fuel test line is fitted with a fuel tap behind the gauge (drawing VW 663/1 for local manufacture).

The pump pressure is determined by the correct adjustment of the push rod stroke and the diaphragm spring tension.

Adjustment of the push rod stroke is afected by adding or removing flange gaskets as described under the heading "Removing and Installing Fuel Pump".

If the stroke adjustment does not give the desired result, replace the diaphragm spring. If the pump pressure is too high, the intermediate turns of the spring may be brought further together, and if the pump pressure is too low they may be stretched apart as an emergency measure.

If the pressure is too high, flooding, and dilution of the engine oil will be the result.

If it is too low, insufficient fuel will be delivered and faulty engine performance will result.

Removing and Installing Fuel Pump

Removal

1 - Disconnect fuel hose from pump.

2 - Remove the two nuts from the mounting studs at the flange (Wrench VW 126a or 126b).

3 - Take off pump.

4 - Remove push rod, intermediate flange and gasket.

Adjusting Fuel Pump Stroke

1 - Place intermediate flange, push rod and two gaskets, which should be in perfect condition, on fuel mounting studs. The convex end of the push rod must be at the eccentric of the distributor drive pinion.

2 - Attach Gauge VW 328c to the flange and tighten it to the same torque as for the fuel pump in order to compress the gaskets to their usual thickness.

The push rod stroke of about 4 mm (.16") is determined by the eccentric on the distributor drive shaft. The stroke should move within a range of 5 mm (.2") which is marked on the gauge. The marks correspond to a length of 8 mm (.31") and 13 mm (.51") measured from the fuel pump contact flange (incl. gaskets) to the projecting push rod end. Rotate the engine to check the pump stroke. The stroke can be adjusted by fitting an appropriate number of gaskets to the intermediate flange. Do not fit less gaskets than required, as this would have a detrimental effect on the diaphragm and the drive mechanism.

Installation

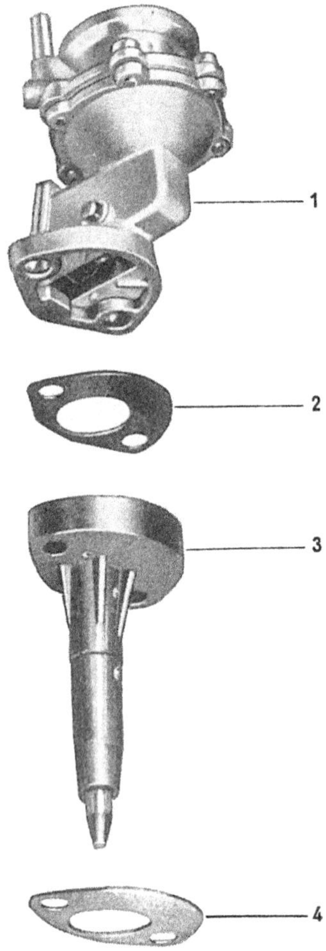

> **Caution**
>
> Do not insert the push rod until the intermediate flange has been installed as otherwise there is a danger that it may slip through the flange into the crankcase.

1 - Fill lower pump chamber with Universal Grease before installation.

2 - Install fuel pump so that the inspection cover is facing towards the left. Retighten nuts when the engine has attained working temperature but do not overtighten.

3 - Connect fuel line and hose.

Check that the fuel line rubber grommet is correctly seated in the engine front cover plate.

1 - Fuel pump
2 - Gasket
3 - Intermediate flange
4 - Gasket

Reconditioning Fuel Pump

Disassembly

1 - Remove fuel pump.

2 - Remove lock ring on rocker arm pin.

3 - Drive out rocker arm pin.

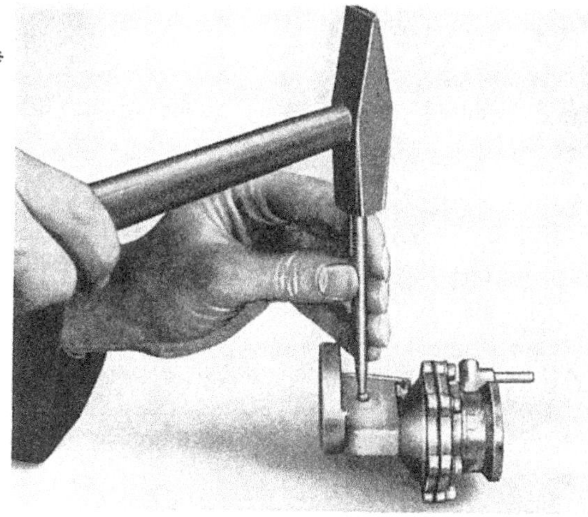

4 - Unscrew the two slotted screws on the inspection cover and remove cover.

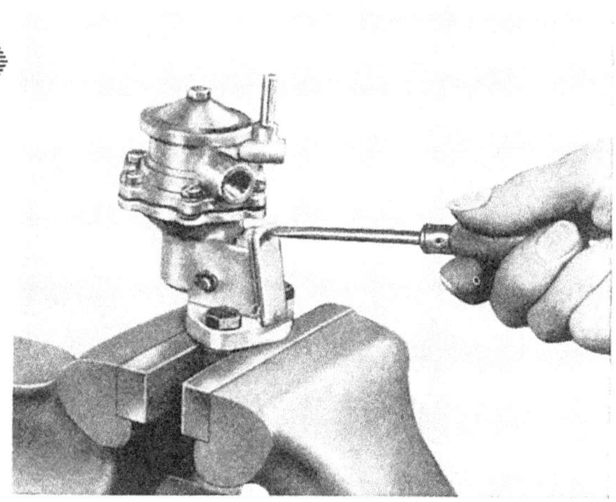

5 - Remove rocker arm return spring with a screwdriver.

6 - Remove cover screw with an 8 mm open-end wrench.

7 - Withdraw filter carefully from top half.

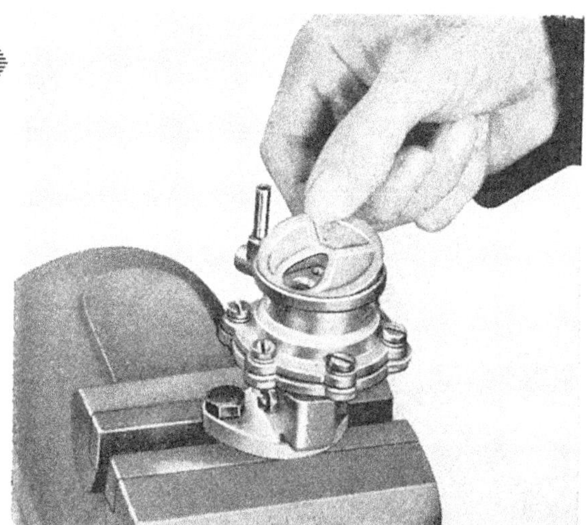

8 - Unscrew the six slotted screws and take off the top half of pump.

9 - Press down diaphragm and remove rocker arm.

10 - Take diaphragm with spring and gasket out of bottom half of pump.

Assembly

When installing, the following points should be observed:

1 - Check valves carefully; replace top half if necessary.

2 - Replace diaphragm and rubber gasket if they shows signs of hardening.

3 - Press down diaphragm and spring and insert rocker arm in diaphragm.

Insert pin and secure by means of lock ring.

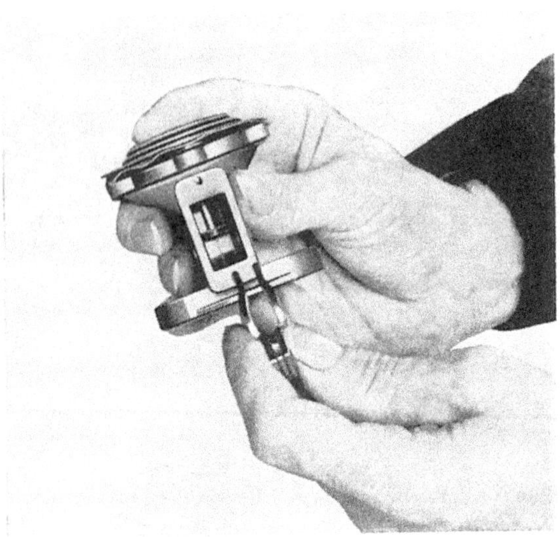

4 - Place lower half of fuel pump in a vice with the Gauge VW 328d inserted. Thus the rocker arm is pressed in 14 mm (from flange jointing face), bringing the diaphragm to the required assembling position.

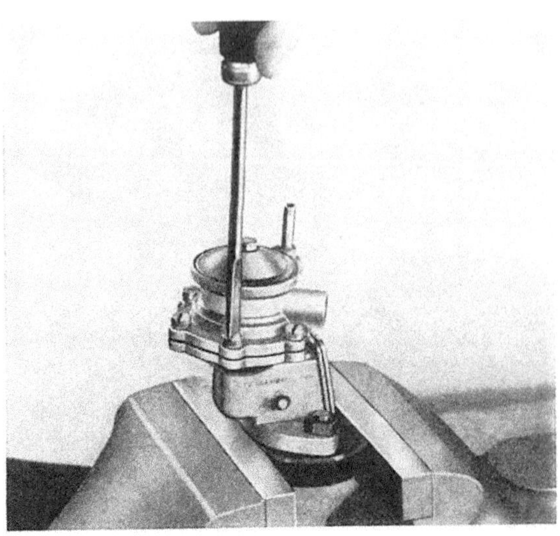

5 - Place the top half of pump in position so that the fuel connections are situated above the inspection cover. Make sure that the diaphragm is not creased. Insert filter so that the flat side is positioned downwards. Do not omit the sealing washer between pump cover and mounting screw.

6 - Fill lower half of pump with Universal Grease (cold-resistant). The grease assumes a liquid condition at operating temperature and lubricates all moving parts. Rocker arm and push rods which are devoid of grease or oil indicate a leaky diaphragm.

7 - Check rocker arm spring for proper seating.

8 - Check inspection cover gasket for wear, replace if necessary.

Note:

When replacing parts, use only the genuine spare parts.

Note:

From Chassis No. 4 432 262, the pump operating lever (Part No. **new:** 113 127 117 A) was modified and is now a one-piece pressing. The lever return spring (Part No. **new:** 113127131 A) has been lengthened at the same time.

The new operating lever may only be installed in previous type pumps together with the new spring. Old type levers and springs can be installed in the modified pumps.

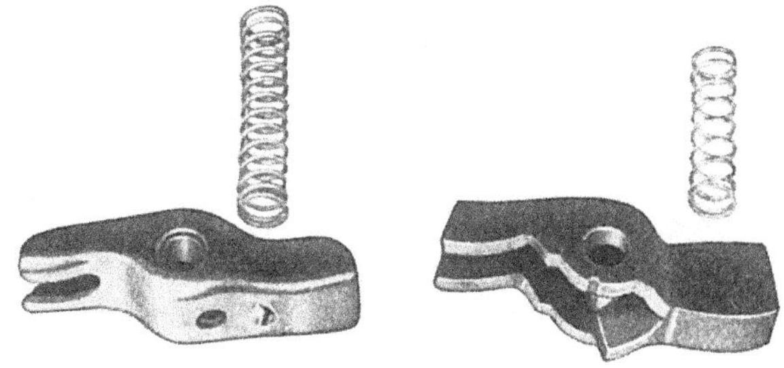

new old

Note:

The diaphragm springs for the fuel pumps of Type 1 and 2 vehicles differ in color and wire thickness from the spring for the Type 3.

If the springs are interchanged, the pump pressure would be either too low or too high. To avoid confusion in the future, the following distinguishing features are used:

	Wire thickness	Color	Part No.
Type 1 and 2	1.5 mm dia.	silver	113 127 147
Type 3	1.6 mm dia.	brass	311 127 147

Note:

From June 1963, Chassis No. 5 578 122, Engine No. 7 777 338 (except 30 bhp engine) a fuel pump with a modified upper part was fitted. This pump has a leaf spring flap valve on the suction side and a plastic disc valve on the pressure valve. The pump is fitted on the 34 bhp engine and is marked VW 3.

Pump	Part No. 211 127 025
Upper part	Part No. 211 127 171
Diaphragm spring (wire thickness 1.5 mm)	Part No. 113 127 147
Delivery output: 5.25 gall/hour = 400 cc/min.	
Maximum pressure 2.8 psi	

The picture shows where the identification marks are on the pump upper and lower parts

Note:

From August 1963, Chassis No. 5 700 147, Engine No. 7 860 830 (except 30 bhp engine) the intermediate flange for the fuel pump is made of a different material (Part No. 113 127 303 as before). The new flange can be distinguished from the former dark-brown version by its red-brown color.

Only the new flange will be supplied as a spare part. The old type parts must not be used.

Important

When installing the new flange, take care that the tightening torque for the securing nuts does not exceed 1.5 mkg (11 ft. lbs.).

Fuel Pump Trouble Shooting

Symptoms	Cause	Remedy
1 - Pump leaking at joining faces: Loss of fuel	a - Slotted screws loose b - Diaphragm cracked	a - Tighten screws b - Renew diaphragm (VW 328d)
2 - Diaphragm leaks at rivets: Loss of fuel	Diaphragm damaged by incorrect assembly	Renew diaphragm (VW 328d)
3 - Diaphragm material leaky: Loss of fuel	Diaphragm material damaged by solvent substance in fuel	Renew diaphragm (VW 328d)
4 - Excessive pump stroke: Overstraining the diaphragm	Pump incorrect installed, gasket too thin	Install pump correctly (VW 328c), check diaphragm, if necessary
5 - Pump pressure low	a - Pump incorrectly installed, gasket too thick b - Spring pressure low	a - Install pump correctly (VW 328c) b - Renew spring or, if necessary stretch coils apart
6 - Pump pressure excess-ive: Float needle valve forced down	a - Pump incorrectly installed, gasket too thin b - Spring pressure excessive	a - Install pump correctly (VW 328c) b - Renew spring or, if necessary, bring intermediate turns closer together
7 - Fuel pump inoperative or insufficient fuel delivery	Valves leaking or sticking	Renew top half of pump

General

The fuel tank has a capacity of 40 liters (10.5 U.S. gals.; 8.8 Imp. gals.). Access is obtained by lifting the rearhinged front hood.

Removing and Installing Fuel Tank

Removal

1 - Lift out spare wheel, jack, and tools

2 - Remove luggage compartment lining.

3 - Shut the fuel tap.

4 - Remove cotter pin at the end of the fuel tap operating rod, after taking the right hand front wheel off, and withdraw operating rod.

5 - Pull the fuel hose off the fuel tap.

6 - Remove fuel tank breather pipe.

7 - Remove the four tank retaining screws and lift off the tank.

8 - Remove fuel tap.

9 - Flush the tank with fuel and blow it out with compressed air.

Installation

When installing, the following points should be observed:

1 - Place fuel tank anti-squeak packing in position; renew packing if damaged.

2 - The fuel tap operating rod must neither be tight nor rattle in its hole in the body. Note. correct seating of grommet. Tight fittings rods should be installed with the use of French chalk or brake fluid. (Caution: Brake fluid will affect paint.)

3 - Replace breather pipe.

Note:

The fuel tank filler cap serves the dual purpose of providing a seal and compensating for any difference in the pressures of the atmosphere and the tank.

If fuel leaks out at the filler cap, proceed as follows to eliminate the source of the trouble:

1 - Remove the cap and make sure that the cork gasket is at least 2.3 mm (.09″) thick and sufficiently supple. Renew hard or damaged gaskets. If no new gasket is available, place old gasket in warm water for a few minutes and reinstall it.

2 - Check filler neck edge for evenness (e.g. with a glass plate and a feeler gauge). Premissible unevenness: 0.1 mm (.004″). Greater deviations and damage, which might have been caused in refueling, are not compensated for by the cap gasket. In such cases, smooth off the filler neck edge with a finishing file. Make sure that no metal chips drop into the tank. If necessary, flush out the tank.

3 - If the above measures do not eliminate the trouble, replace the filler cap, as the sealing in the cap is then likely to be defective.

Fuel Tap

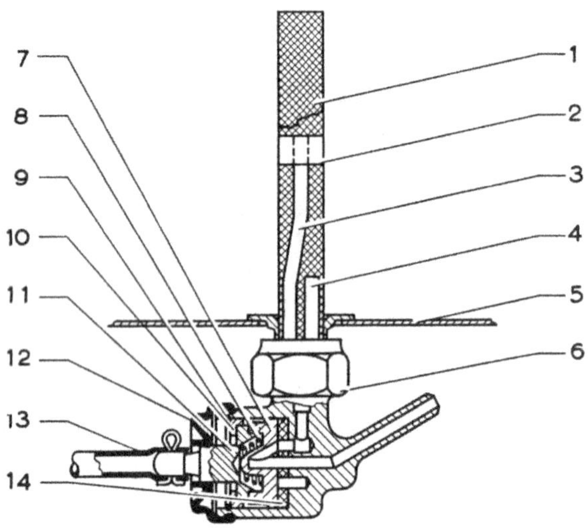

Fuel passes through a filter in the tank via the fuel tap into the fuel line. The tap is operated from the front seat. Its three positions are: "open", "reserve" and "off".

1 - Filter
2 - Cork washer
3 - Fuel pipe with tap at "Open"
4 - Fuel pipe with tap at "Reserve"
5 - Fuel tank
6 - Nut
7 - Plate
8 - Spring
9 - Retaining ring
10 - Stop washer
11 - Tap spindle
12 - Rubber cap
13 - Operating rod
14 - Gasket

Removing and Installing Fuel Tap

To clean the filter and the tank, the tap must be removed.

Removal

1 - Remove and drain fuel tank.

2 - Loosen fuel tap nut and take off the tap.

3 - Clean filter with compressed air.

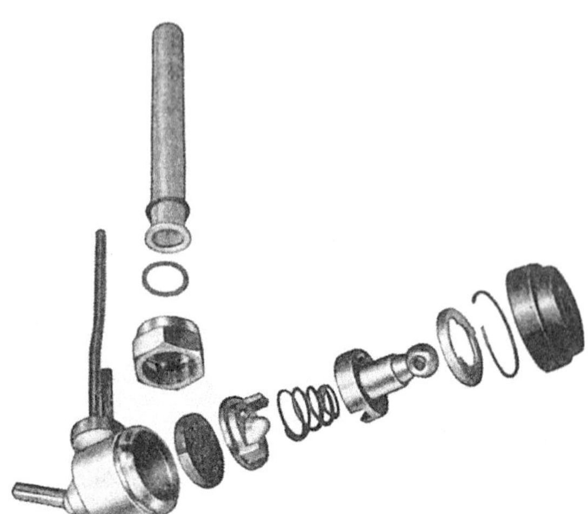

Installation

When installing, the following points should be observed:

1 - Carefully clean fuel tank.

2 - Use new fuel tap gaskets.

> **Note!**
>
> The cork disk between the filter and the long drain pipe should be nearly flush with the upper edge of the pipe.

3 - After installation, check connections for leaks.

Reconditioning Fuel Tap

Disassembly

Remove the fuel tap and proceed as follows:

1 - Remove protective cap.

2 - Press down stop washer with a piece of tube and remove retaining rind and washer.

3 - Remove spindle, spring, plate and thiokol gasket.

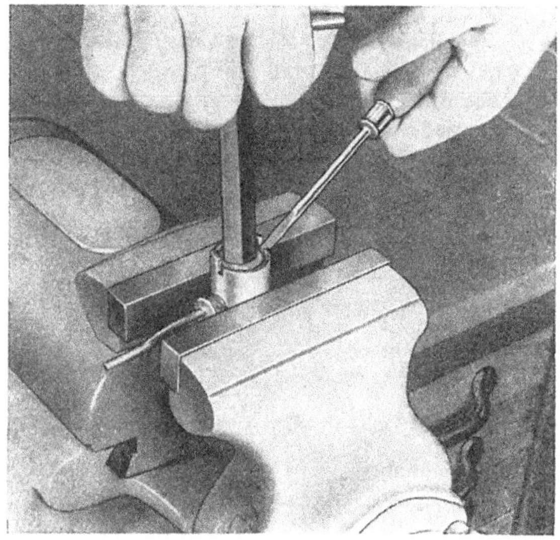

Assembly

When installing, the following points should be observed:

1 - Thoroughly clean all components and dry them with compressed air.

2 - Use a new thiokol gasket for the spindle plate.

Note:

From Chassis No. 3 191 531 the fuel taps on all vehicles are equipped with a Thiokol gasket (unchanged Part No. 111 209 113 A). The increased hardness of this gasket material prevents fuel stoppages due to restricted delivery openings.

Derusting of Fuel Tanks

A fuel tank which shows formation of rust on the bottom due to the content of water in the fuel is liable to lead to constant contamination of the fuel system and resultant engine failure. Such tanks can be cleaned by applying acid type chemicals as detailed below.

There are two methods which may be used, but preference should be given to method "A". It is not so severe, has a good pickling effect and leaves a phosphate film.

In both cases, however, the success of the acid treatment is dependent on the fact that the fuel tank must be rinsed out immediately with soluble oil. If not rinsed, rust will form again.

A - **Derusting Phosphate Agent "Antox Extra M"**

The solution should be mixed as follows:
1 part of Antox Extra M to 10 parts of water.

B - **Hydrochloric acid solution (for industrial use) Specific weight 1.19**

It should be mixed as follows:
20 parts of hydrochloric acid solution to 80 parts of water and 1 part of restrainer.

When rinsing use a mixture of 1 part of coolant (usual mineral oil type) to 20 parts of water.

The fuel tank should be treated in the following manner:

1 - Place the tank horizontally in a suitable rack and replace the tap by a snugly fitting wooden plug.

2 - Pour either of the two solutions, A or B, into the tank. The tank should be filled up to the filler, as otherwise the evaporized acid will adversely affect the tank wall above the fluid level.

3 - A slight accumulation of rust calls for a treatment of 40 or 60 minutes while the time required in more severe cases may vary from 3 to 8 hours. The filled can tank be left standing

overnight to obtain best results without loss of time to the owner.

4 - After the treatment the tank should be washed out immediately with the rinsing solution. About 4—5 litres (0.9—1.1 Imp. gal.) should be used and the tank rocked to and fro so that the complete interior is moistened.

5 - Drain the tank and dry it out with compressed air. Treatment with soluble oil leaves a thin film and consequently further treatment is not necessary.

An adequate amount of acid type cleaner — 45—50 litres (9.9—11.0 Imp. gal.) — should be stored in glass containers ready for use when needed. The same cleaner can be used from 10—15 times.

Available are:

Designation	Manufacturer	Address
Antox Extra M	Metallgesellschaft Frankfurt/Main	Frankfurt, Reuterweg 14
Hydrochloric aced solution (Industrial use)	Usual type	Chemical Dealers, Chemists
Coolant, mineral oil type	Usual type	Mineral oil firms
Original Dr. Vogel's Sparbeize	Messrs. Max Hoeck Chemical Factory	Duesseldorf-Oberkassel Kaiser-Wilhelm-Ring 45/46, Postfach 6 Germany
ACP-Rodine 50 Sparbeize	American Chemical Paint Co.	Ambler, Pa. U.S.A.
Henkel's Sparbeize	Messrs. Henkel & Cie.	Duesseldorf, Germany

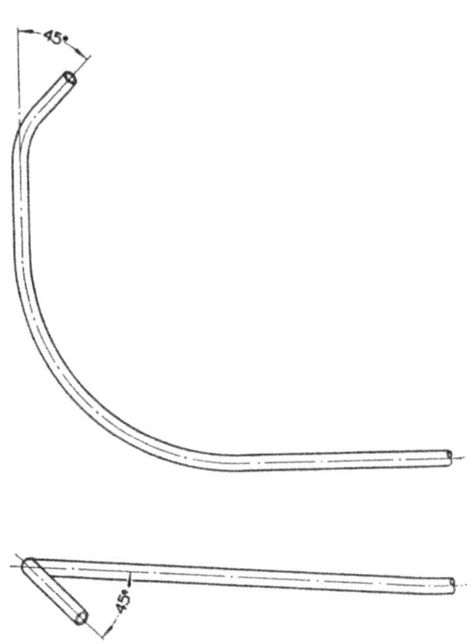

Note:

From Chassis No. 3 631 277 the front end of the fuel line was bent 45° backwards and to the left above where it enters the frame tunnel.

This modification improves the routing of the hose from the fuel tap to the line so that it does not become kinked.

The Part No. of the fuel line remains unchanged — 111 209 211 B.

Note:

If trouble is experienced with the fuel supply due to a kinked fuel hose, the fuel line should be bent carefully at the appropriate place.

Note:

From Chassis No. 4 010 995 all Volkswagens are equipped with a mechanical fuel gauge. The sender unit and the gauge are connected by means of a cable.

The fuel gauge in the instrument panel shows the amount of fuel in the tank. When the needle is on "R" —Reserve— there are about 5 liters (1.1 Imp. galls., 1.3 US galls.) left in the tank and it is time to refuel at the next opportunity.

The fuel tap has been discontinued but will remain available as a spare part.

Adjustment

The gauge can be adjusted by means of a knurled screw on the back of the instrument. It is advisable to do this when the tank is empty. The screw must be turned in the direction of arrow until the needle is resting on its stop. Further calibration to a certain fuel content is unnecessary.

This process can also be used when there is still a certain amount of fuel in the tank. In this case the sender unit cover should be removed and the lever, to which the cable is attached, pressed to the rear until the float is at its lowest position.

This adjustment ensures sufficient accuracy in the lower range and guarantees that at least 5 liters of fuel is left in the tank when the needle is on the reserve mark. There may be slight inaccuracies in the upper range but they are not important.

Troubles

Complaints of inaccurate readings are occasionally received. This is particularly unpleasant when the inaccuracy is in the reserve range and the actual amount left in the tank is considerably less than 5 liters.

These variations are caused by defects at the sender unit and invariably occur when a "VDO" sender unit is paired with a gauge from Messrs. Magura. No trouble is experienced with the other combination or when items from the same manufacturer are used.

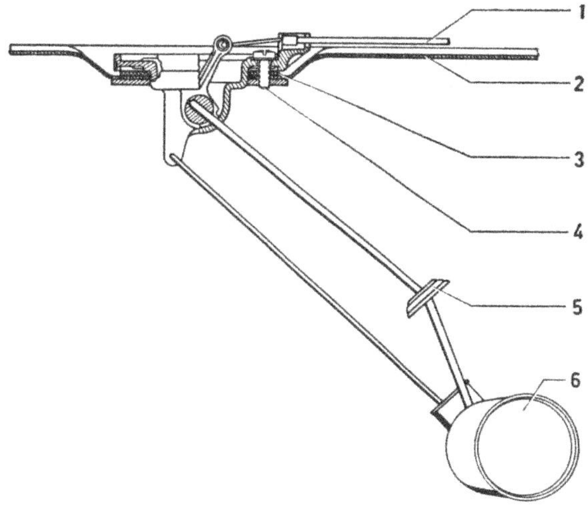

1 - Cable
2 - Fuel tank
3 - Cork gasket
4 - Screw
5 - Stop washer
6 - Float

As a provisional remedy we recommend that the VDO sender unit is replaced by the Magura type when the combination mentioned above is found. Only the Magura sender unit is supplied as a spare part. Efforts are being made in the meantime to modify the parts so that each combination works satisfactorily.

Note:

From June 1962, the plastic float of the VDO tank unit (Part No. 113 919 049) was altered in length **from 105 mm to 145 mm**. It is now possible to combine this float with a Magura fuel gauge.

The VDO fuel gauge (Part No. 113 919 029) has been fitted with a stronger spring. After the introduction of this modification in March 1962, Magura tank units can be paired with VDO fuel gauges without hesitation. Both makes of fuel gauge have the production date marked on the back of the housing. The figures 3 62 for example indicate that the gauge was manufactured in March 1962.

THIS PAGE INTENTIONALLY
LEFT BLANK

Inspection of Fuel System

If the fuel supply to the carburetor is interrupted, check as outlined below.

Sequence of Operations

1 - Check quantity of fuel in tank.

2 - Check position of fuel tap.

3 - Disconnect fuel line at the carburetor. Operate the starter and check if fuel is delivered by the pump.

a - If fuel is delivered by the pump: Check pump pressure. Inspect float needle valve and carburetor jets for dirt and foreign matter.

b - If no fuel is delivered by the pump:

4 - Disconnect fuel line to fuel pump.

a - If fuel comes out: Check fuel pump for leaks. Re-tighten screws and connections; remove and check fuel pump, if necessary.

b - If no fuel comes out:

5 - Remove fuel tank; check fuel tap; clean filter; inspect gaskets.

6 - Blow fuel line out with compressed air.

Fuel Consumption Test

Average consumption test on the road

An accurate measurement of the fuel consumed can be made with a fuel-mileage tester which should be installed near the driver's seat. The fuel-mileage tester is connected to the engine by means of hoses. It should be possible to switch from normal to test equipment.

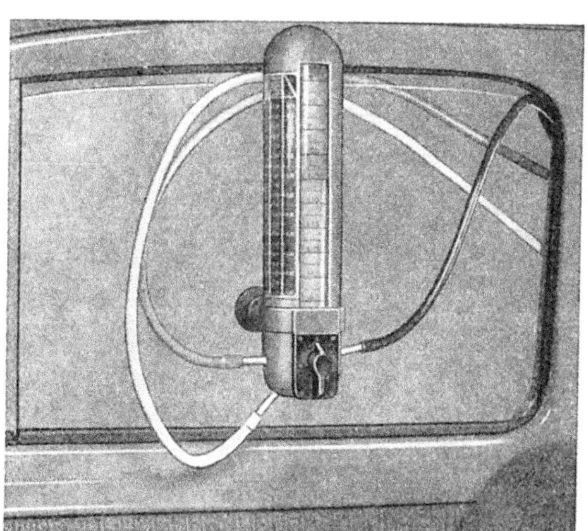

If the test result corresponds to the figures given by the factory, the driving habits or special operating conditions are responsible for high fuel consumption.

When carrying out the test, the following points should be noted:

1 - Adjustment of fuel pump, carburetor and ignition must be correct.

2 - The engine must have attained operating temperature before conducting the test.

3 - The car must be normally loaded.

4 - Conduct the test on a level road and under average driving conditions. Periods of idling, stop and start operation, continuous driving in low gear in flat or hilly districts, rapid acceleration, and driving at top speed have a definite effect on the fuel consumption.

5 - Only normal branded fuel should be used. The consumption should be determined as follows:

Liter per 100 km (metric)

$$= \frac{\text{Fuel consumed (liter)}}{\text{Length of test course (km)}} \times 100$$

6 - The engine and the fuel system should be thoroughly checked if there is a considerable difference between the result of the fuel-mileage test carried out as outlined above and the nominal fuel consumption figures given by the factory.

Note:

A consumption test with a fuel level gauge and the mileage recorder generally gives only an approximate fuel consumption figure and is not, therefore, be recommended.

Fuel Consumption Test on Engine Test Bench (Dynamometer)

The fuel consumption can also be checked on the engine test bench by means of a gauged test container and a stop-watch. The consumption can be determined by subjecting the engine to a given load and speed during a fixed time.

The consumption figures for 100 cc. (6 cu. ins.) are given in the chart below.

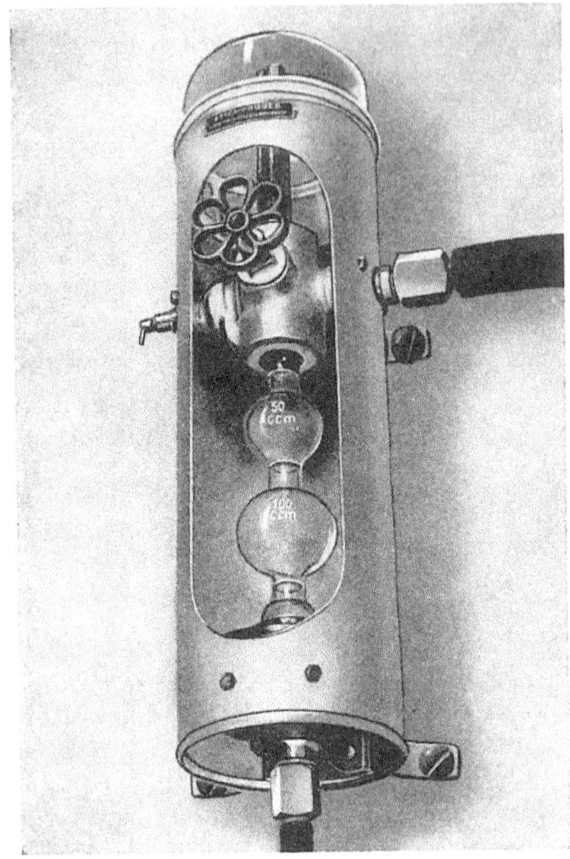

Example:

At a speed of 2180 rpm and a brake load of 5 kg the consumption time for 100 cc. of fuel should be between 84.5 and 91.5 seconds.

The table only applies to engines which are completely broken in (run in) and equipped with the SOLEX 28 PICT carburetor.

The table contains only the normal measurements which correspond to road speeds of 50, 70 and 90 kms and full throttle. Further measurements are unnecessary as the consumption characteristics of the carburetor are uniform throughout the complete load range.

Brake load kg	Brake load lbs.	Engine speed rpm	Corresponding road speed km	Corresponding road speed miles	Consumption time for 100 cc. (6 cu. ins.) seconds
3.87	8.5	1550	50	31	144±6
5.0	11.0	2180	70	44	88±3.5
6.71	14.8	2800	90	57	56±2.2
9.1—9.9 (minimum)	20—22	3600	Full throttle		31.5±1.3

Note:

Generally the brake load at 3600 rpm should be at least 9.1 kg. If a higher loading is reached the consumption time can be shorter. If the given brake loading is not reached the consumption time may increase.

Fuel Consumption Test on Chassis Dynamometer

The fuel consumption can also be checked exactly with the car on a chassis dynamometer.

1 - Remove fuel line union nut from fuel pump and connect fuel hose of dynamometer to pump.

2 - Warm up the engine.

3 - Operate engine at full load until the dynamometer indicates a speed of 110 km/h (68 mph).

4 - Meter the consumption of 100 cc. of fuel by means of the measuring device on the dynamometer and a stop watch.

Note:

If complaints about excessive fuel consumption are received, the following points should be checked on the 28 PICT carburetor:

1 - The flap in the intake tube of the oil bath air cleaner must move freely. It should not be secured at temperatures below 20° C (68° F).

2 - The throttle valve spindle must move freely. If it is still tight after being moved back and forth several times, the lower part of the carburetor should be replaced.

3 - Check the ease of movement of the fast idle cam. If the cam is sticking, lubricate the bearings with a few drops of anti-corrosion oil.

4 - Check the location of the ceramic cover. The mark on the cover should be aligned with the lug in the housing.

5 - Check the operation of the automatic choke. First lift the throttle valve lever out of the steps on the fast idle cam so that the choke valve can move under the influence of the bi-metal spring. The automatic choke is working properly when:

a - the choke valve is completely closed in winter at temperatures of 0° C (32° F) and below or

b - is opened so far in summer at roughly 20° C that the idle adjusting screw rests on the third or fourth step from the top on the fast idle cam.

c - the choke valve is fully open when the engine is at operating temperature and the ignition switched on

6 - The heater element should warm up when the ignition is switched on. This can be checked at the outside of the ceramic cover.

When the engine and carburetor have cooled down to about 20° C, the choke valve should open fully in not more than 5 minutes from the time the ignition is switched on, with engine stationary and rear hood open. If this is not the case, the following defects may be found:

a - Faulty cable connection or loose terminal screw in ceramic cover.

b - Cable to ignition coil or heater element damaged.

c - Bi-metal spring sticking or rubbing in the housing.

d - Ceramic cover or locating tongue on the bi-metal spring broken.

7 - Check the idling adjustment and rectify if necessary.

Note:

The commercial fuels in many countries vary quite considerably in certain characteristics which are not exactly defined in the normal specifications. This also applies to the flash point of the fuel. If the flash point of a fuel is slightly low it can cause an engine to run-on when the ignition has been switched off, especially when assisted by the following conditions:

1 - Incorrect carburetor idling adjustment (too rich or too fast).

2 - Incorrect ignition timing (see instructions in Workshop and Instruction Manuals).

3 - Incorrect cooling air throttle ring adjustment (20 mm when assembling engine; 25—30 mm when engine warm).

   To improve cooling, the rubber plug which acts as a stop for the throttle ring can be replaced by an 8—9 mm high plug (Part No. 111 119 189). See sechon M-4/6.

4 - Excessive carbon deposits in the combustion chambers in conjunction with working conditions, way of driving and type of fuel used.

5 - Engine cooling affected by a loose fan belt or dirt deposits on engine exterior.

In exceptional cases where the pre-ignition cannot be satisfactorily cured by normal workshop means, a "Pilot jet with electro-magnetic shut-off valve" which is available for the 20 SAE hp engine under the Part No. 113 129 413, can be installed.

Installation:

a - Remove standard pilot jet and install the pilot jet with electro-magnetic shut-off valve.

b - Install a cable between shut-off valve terminal and terminal 15 on the ignition (length of cable about 400 mm/16").

Operation

When the ignition is switched off, the electro-magnetically controlled needle closes the pilot jet and stops the flow of fuel. This effectively eliminates pre-ignition. When the ignition is switched on, the jet opening is cleared.

Workshop Equipment

1 - VW Special Service Tools

VW 126a	Fuel Pump Wrench
VW 126b	Fuel Pump Wrench
VW 328c	Fuel Pump Push Rod Gauge
VW 328d	Fuel Pump Diaphragm Gauge

2 - VW Workshop Equipment for Local Manufacture

VW 663/1	Fuel Pump Tester

3 - Normal Hand Tools

Screwdriver,
Carburetor screwdriver
Combination pliers
Waterpump pliers
Mechanic's hammer, 300 grams
Triangular scraper
Open-end wrench, 8 mm
Open-end wrench, 11 mm
Open-end wrench, 13 mm
Open-end wrench, 12 mm
Open-end wrench, 14 mm
Open-end wrench, 17 mm
Open-end wrench, 22 mm
Box-wrench, 14 mm
Box-wrench, 17 mm
Bell cranked box wrench
Wire brush
Hand vice
Oil-can
Grease container
Caliper square, 200 mm in lenght, measuring in 1/10 mm
Tap M 6
Die stock, size 1
Die M 6
Tap wrench, size 1, adjustable

4 - Supplementary Workshop Equipment

Fuel mileage tester
Engine dynamometer
Chassis dynamometer

THIS PAGE INTENTIONALLY
LEFT BLANK

Contents: V

Front Axle

Steering

Front Wheel Alignement

Worm and Roller Steering

Important — Do not re-use any fasteners that are worn or deformed in normal use. Many fasteners are designed to be used only once and become unreliable and may fail when used a second time. This includes, but is not limited to, nuts, bolts, washers, self-locking nuts or bolts, circlips, cotter pins. Always replace these fasteners with new parts.

THIS PAGE INTENTIONALLY
LEFT BLANK

Description of Front Axle

General Description

The front axle consist of two rigidly joined tubes secured to the frame head by means of four bolts. Pivoted in each tube is a torsion bar, anchored at the middle to prevent twisting and lateral movement. The ends of the torsion bars are joined to the torsion arms, each of which is free to pivot in one fibre bushing and needle bearing in the tube. The outer ends of torsion arm are connected to the torsion arm links by adjustable pins, which are free to pivot. The stub axle (steering knuckle) swivels on a king pin, which passes through bushes in the torsion arm link. A thrust bearing is fitted between the upper end of the stub axle and the torsion arm link. Front axle tubes, torsion arms and torsion arm link nearly form a parallelogram. Hydraulic, double-acting telescopic shock absorbers dampen road shocks and take up the return swing. Stops with rubber buffers prevent excessively violent springing action.

The De Luxe Sedan, VW Convertible and Karmann Ghia models are equipped with a stabiliser and steering damper.

Note:

From Chassis No. 3 912 101 (Front Axle No. 3 872 754) the torsion arm stop is stamped with an identification letter and figure which is the production number of the axle. The front axle number itself is marked in white paint on the right-hand end plate. The front axle number is no longer stamped into the torsion arm stop.

Lubrication

Under normal driving conditions, the front axle should be greased at regular intervals as prescribed in the Lubrication Chart. If, however, the car is mainly used on rough roads or if it is driven not more than 12,000 km (7500 miles) in a year, that is a monthly average of below 1000 km (600 miles), it is advisable to grease more frequently than specified in the chart, say every 1250 km (800 miles), particularly at the lubricating points of the torsion arm links and the outer tie (track) rod joints. The front wheel bearings should be thoroughly cleaned and repacked with grease every 50,000 km (30,000 miles) or at least once a year.

Important — Perfect lubrication of the front axle bearing points is only ensured with the front axle raised (unloaded condition). Grease should be injected into the fittings until the excess grease begins to emerge at the edges of the lubrication points. The service life of the front axle depends on skilled lubrication with branded lubricants which conform with the specifications given by the Volkswagenwerk. When lubricating, the grease fittings and grease guns should be perfectly clean.

Maintenance

To maintain the riding qualities and the standard of safety of the car, regular inspection and maintenance of the front axle and steering is absolutely necessary. The following operations are prescribed:

1 - Check and adjust front wheel bearings.
2 - Check and adjust torsion arm link pins.
3 - Check and adjust toe-in of front wheels.
4 - Check and adjust play in the steering.
5 - Check tie rods and steering gear for tightness.
6 - Check operation and mounting of shock absorbers and steering damper.
7 - Check wheel mounting bolts for tightness.
8 - Check and correct tire pressures.

Volkswagen Front Axle
(Sectional View)

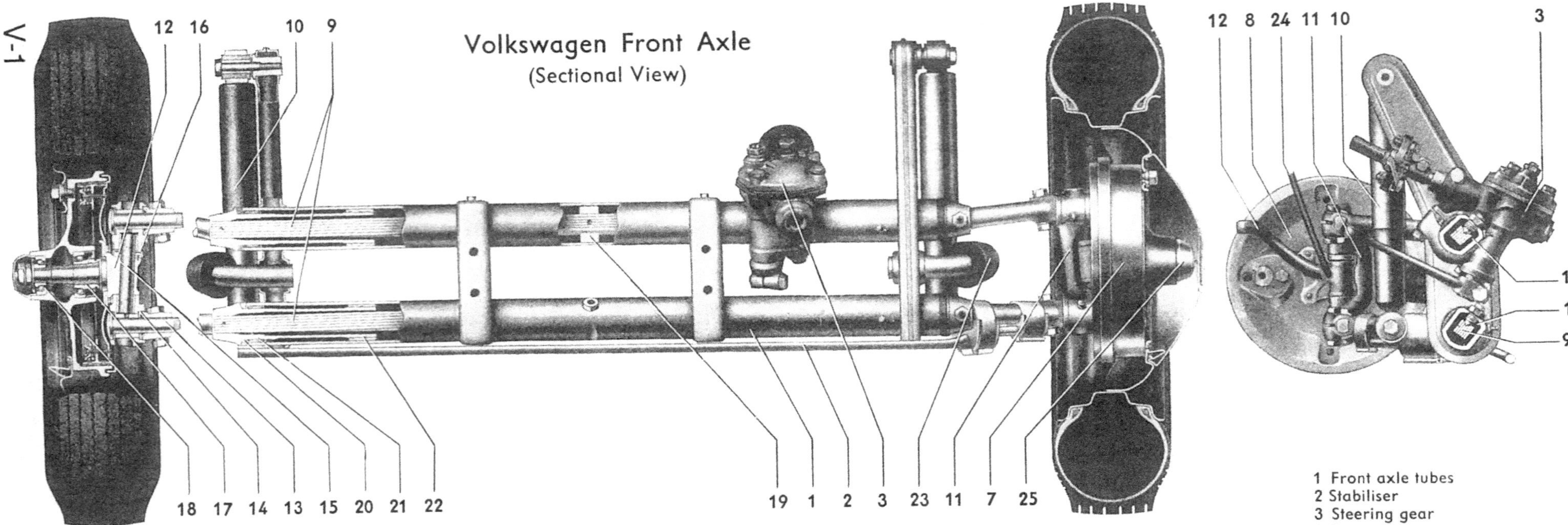

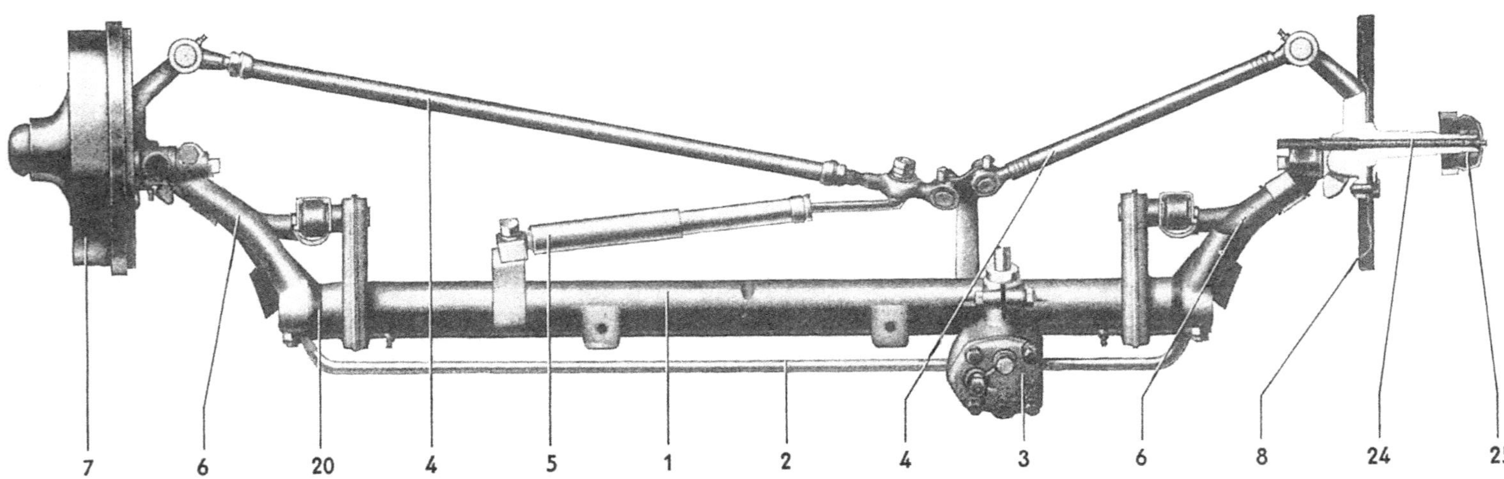

1 Front axle tubes
2 Stabiliser
3 Steering gear
4 Tie rods
5 Steering damper
6 Torsion arm
7 Brake drum
8 Brake back plate
9 Torsion bar
10 Shock absorber
11 Torsion arm link
12 Stub axle (steering knuckle)
13 Torsion arm link pin
14 Dust excluder
15 King pin
16 Thrust bearing
17 Front wheel bearing, inner
18 Front wheel bearing, outer
19 Center anchor
20 Seal
21 Needle bearing for torsion arm
22 Bush for torsion arm
23 Rubber stop
24 Speedometer cable
25 Grease cap

Removing and Installing Front Axle

Removal

1 - Loosen wheel bolts, raise the car and take off front wheels.

2 - Turn off the fuel tap, disconnect operating rod from tap and pull off fuel hose.

3 - Remove protection cover at back of instrument panel and luggage compartment and take out fuel tank.

4 - Jack up the car, and lower it on to trolley or trestles.

For this the following auxiliary equipment may be used:

To lift:

Volkswagen hoists
Front axle jack VW 606
Trolley jack

To lower:

Vehicle trolley VW 603/1

To support:

Trestle VW 633

5 - Disconnect brake hoses at brake back plates and plug them up, using wooden plugs.

Important — When replacing the front axle, the brake cables must remain on the axle.

6 - Take off steering damper from front axle tube bracket.

7 - Remove speedometer cable at front wheel.

8 - Loosen steering column clamp, pull off horn cable and withdraw steering column with steering wheel from steering coupling.

9 - After unlocking and unscrewing the nuts with VW 266f, pull off the tie rod joints and remove the rods.

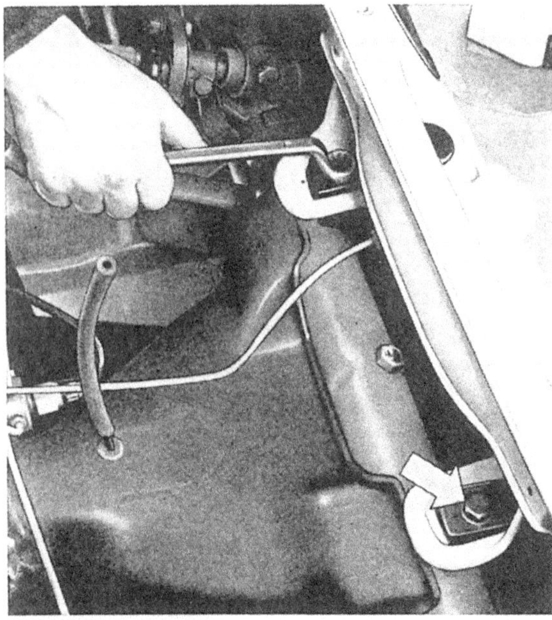

10 - Remove the two body mounting bolts M 10.

11 - Remove the four front axle mounting bolts, after the lock plates have been prised off, and take off the front axle.

The lower front axle mounting bolts become accessible and can be removed if the stabilizer is pressed down slightly. Insert a piece of wood between the axle tube and stabilizer and use it is a lever.

12 - To remove the front axle the best method is to use the front axle cradle in conjunction with the trolley jack VW 610.

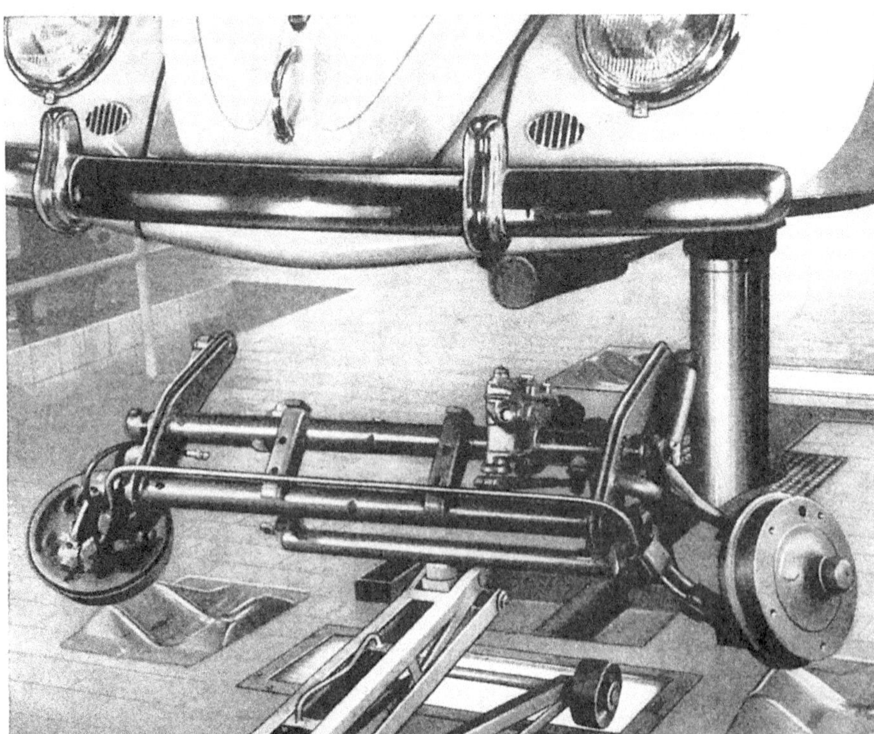

Installation

This is accomplished by reversing the removal procedure, but the following points should be observed:

1 - Before installing the front axle, place a rubber packing on each of the threaded bushes on the axle beam.

2 - Use new lock plates.

3 - Tighten the front axle mounting bolts to 5.0—6.0 mkg (36—43 ft. lbs.).

4 - Tighten two body bolts M 10 to 3.0 mkg (21 ft. lbs.).

5 - Tighten the tie rod end nuts and secure with a cotter pin.

6 - Make sure that there is a good ground connection at the steering column coupling.

7 - When pushing the steering column on the coupling flange pin, make sure that the steering wheel is properly positioned. Secure steering column clamp screw with a new lock plate.

8 - Install the steering damper bolt in the front axle beam bracket with a new lock plate and secure the bolt.

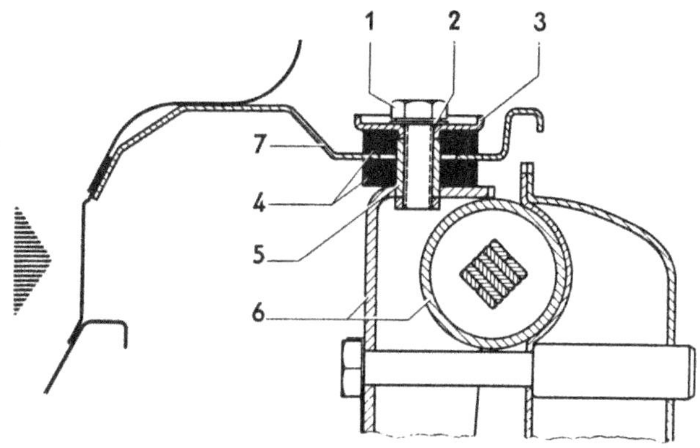

1 - Body mounting bolt
2 - Lockwasher
3 - Spacer
4 - Rubber packings
5 - Threaded bush
6 - Front axle
7 - Body

9 - Take care that the brake hoses are not twisted. Bleed braking system and adjust brakes.

10 - Check toe-in and camber of the wheels and caster of the axle tubes.

THIS PAGE INTENTIONALLY
LEFT BLANK

Disassembly and Assembly of Front Axle

General

As the front axle can also be partly disassembled while in position in the vehicle, repairs can generally by effected without having to remove the axle.

All operations to be carried out on the removed front axle can best be accomplished with the axle mounted on the Holding Fixture VW 643 in conjunction with the plate VW 309.

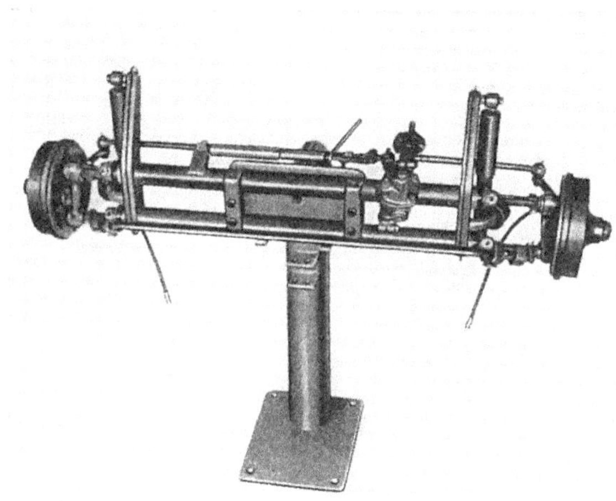

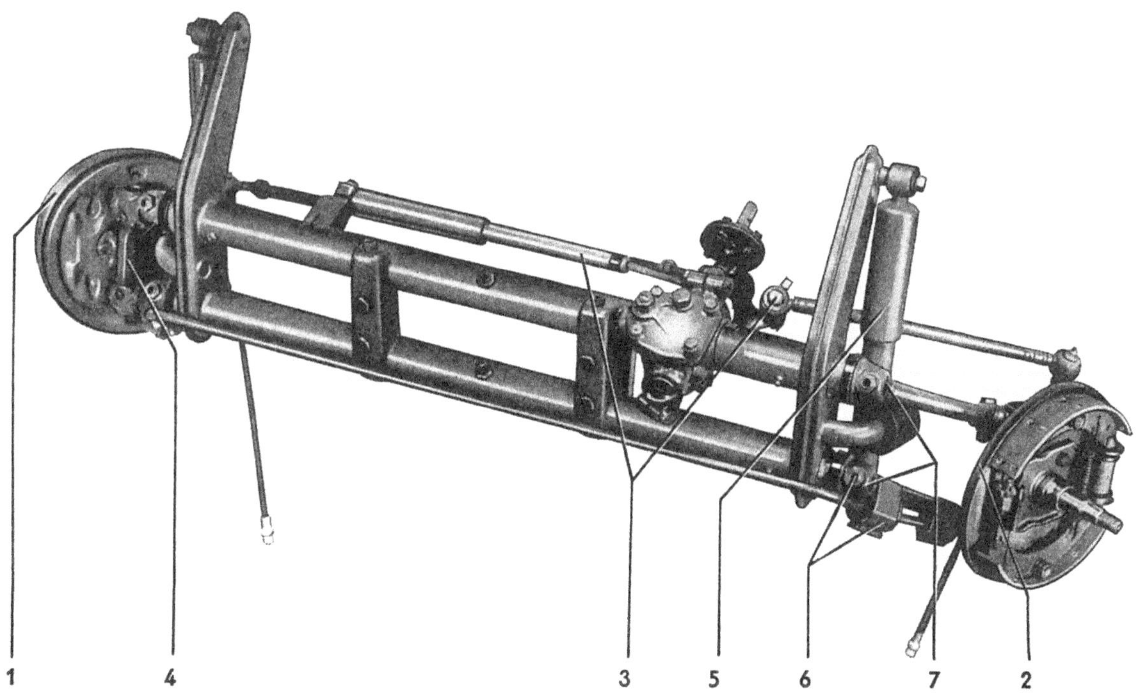

To facilitate disassembly and assembly, the following sequence of operations should be followed:

Disassembly

1 - Remove brake drums.

2 - Remove brake back plates.

3 - Remove tie (track) rods and steering damper.

4 - Remove torsion arm links and stub axles (steering knuckles).

5 - Remove shock absorbers.

6 - Remove torsion arms and stabilizer.

7 - Remove torsion bars.

Assembly

Assembly is accomplished by reversing the above procedure. The hints given on the following pages should, however, be observed.

THIS PAGE INTENTIONALLY
LEFT BLANK

Removing and Installing Brake Drum

(Front axle in situ)

Removal

1 - Remove the hub cap and loosen wheel mounting bolts.

2 - Jack up the car or support on trestles.

3 - Remove the wheel.

4 - On the left-hand front wheel: Remove the cotter pin securing the speedometer drive cable. Pull off inner hub cap with VW 637.

5 - Turn up lock plate and remove nuts and thrust washer (at left stub axle: left-hand thread!).

6 - Pull off the brake drum with the special tool VW 202 in conjunction with ring VW 202c, hooks VW 202d and thrust pad VW 202i. The nose of the thrust pad must engage in the slot cut in the wheel spindle.

Installation

This is a reversal of the above operations, but the following points should be observed:

1 - Check tapped holes of brake drum for damage and make sure that the braking surface is smooth.

2 - Clean brake drum hub and front wheel bearing carefully. Lubriate front wheel bearing and brake drum hub as prescribed under lubricant specifications.

3 - Adjust front wheel bearing as prescribed.

Reconditioning Brake Drum

Re-Tapping Brake Drum Holes

Brake drums with damaged threads for the wheel mounting bolts can be repaired by cutting a thread M 14 × 1.5 in the five tapped holes M 12 × 1.5 of the drum. Oversize wheel bolts M14 × 1.5 are available to suit the enlarged holes.

1 - Bore up the brake drum holes with a drill 12.0—12.1 mm dia. (0.470—0.474").

2 - Cut thread M 14 × 1.5.

Important
Oversize wheel bolts must be tightened to between 11.0 and 13.0 mkg (79—93 ft. lbs.).

Tapped Inserts for Brake Drum Holes

(Part No. 111 405 619)

Another method of reconditioning brake drums having damaged threads is to press tapped inserts M 12 × 1.5 into the brake drum holes.

1 - Bore up the five brake drum holes with drill 17.8 mm dia. (.7008″).

2 - Ream the holes with reamer 18 H 8 = 18.000 — 18.027 mm (a). (.7087—.7097″).

3 - Counterbore the holes with a drill 20.0—21.5 mm dia. (.7874—.8466″) (b) so that a wall thickness of 8.0—8.3 mm .3150—.3268″ (c) remains.

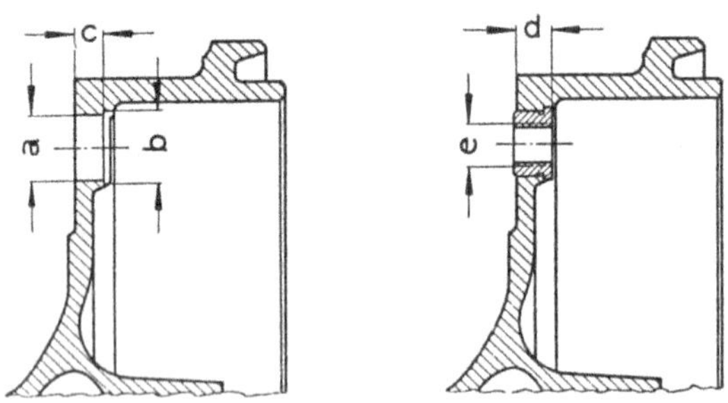

$$a - 18 \text{ dia. H } 8 = \frac{18.027}{18.000} \text{ mm dia.} \quad \left(\frac{.7097}{.7087} \text{ ins. dia.}\right)$$

$$b = \frac{20.0}{21.5} \text{ mm dia.} \quad \left(\frac{.7874}{.8466} \text{ ins. dia.}\right)$$

$$c = \frac{8.3}{8.0} \text{ mm} \quad \left(\frac{.3268}{.3150} \text{ ins. dia.}\right)$$

$$d = \frac{10.3}{9.8} \text{ mm} \quad \left(\frac{.4055}{.3858} \text{ ins. dia.}\right)$$

e = M 12 × 1.5

4 - Press the tapped inserts into the brake drum holes. The dimension 9.8—10.3 mm (d) must be strictly adhered to. The chamfered portion 0.5 × 45° of the insert may project from the brake drum surface, but never the knurled portion.

Use of Heli-Coil Tapped Inserts

The tapped holes can also be reconditioned by using Heil-Coil tapped inserts. The threads have to be recut and M 12 inserts screwed in. In this case new wheel bolts M 12 × 1.5 have to be used.

Refacing Brake Drum

If the braking surface of the drum is scored, out-of-round or bell-mouthed, it may be refaced on a lathe.

1 - Turn the drum to an inside diameter of 231.5 mm (9.11″). The thickness of a refaced drum must never be below 4 mm (0.16″). The braking surface must not be tapered more then 0.1 mm (.004″).

The permissible lateral and radial run-out of the drum is 0.25 mm (.010″).

2 - The brake shoes of refaced drums must be fitted with oversize brake linings to correspond with the radius of the braking surface.

Removing and Installing Front Wheel Bearings

(Front axle in situ)

Removal

1 - Remove brake drum.

2 - Pull off the spacer and the inner race of the inner bearing by means of the puller VW 202 in conjunction with hooks VW 202s and thrust pad VW 202i.

3 - Remove oil seal from inner bearing of drum and take out ball cage.

4 - Remove outer race of the inner bearing (62 mm dia.) on Repair Press VW 400 with thrust plate VW 447 in conjunction with VW 412 and VW 446.

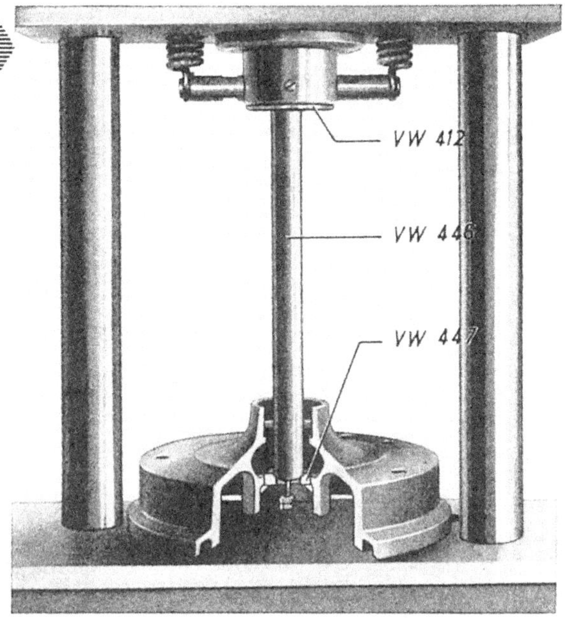

5 - Remove outer race of the bearing (52 mm dia.) on Repair Press VW 400 with punch VW 407 in conjunction with VW 401.

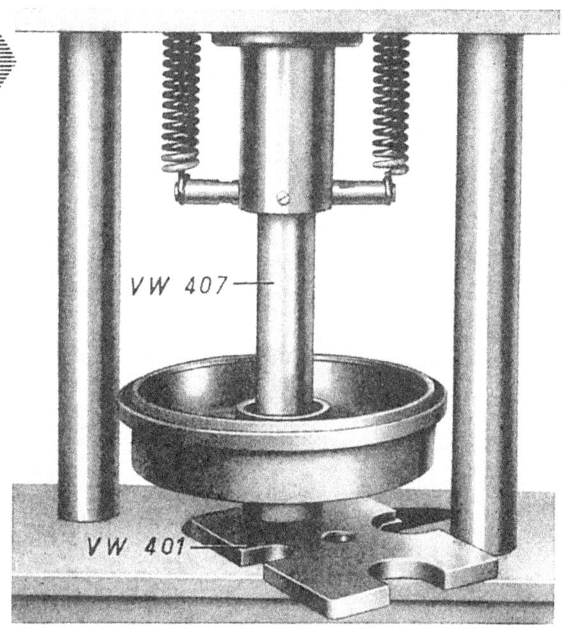

Installation

This is accomplished by reversing the above procedure, but the following points should be observed:

1 - Before installing the brake drum, clean the brake shoes and brake back plate thoroughly with compressed air. Dirt and dust produced by lining wear have an abrasive effect if allowed to enter the bearings and seatings. If the outer bearing race is no longer a press fit in the drum, the latter must be renewed.

2 - Thoroughly clean the wheel spindle and bearing chambers between the seatings of the brake drum with benzine and dry with a clean rag.

3 - Clean old grease from the ball races, cages and balls of the bearings. This grease must be removed completely. The best method is to thoroughly wash out the bearings.

4 - Carefully inspect the bearings for pits, cracks, or other signs of wear. If necessary, use new bearings.

Important — Ball races or balls must never be replaced separately.

5 - Inspect spacer of inner bearing for cracks and wear, replace if necessary, using the sleeve VW 244 to drive it into position. A correct sealing of the brake drum oil seal is only insured, if the surface of the spacer is absolutely clean and smooth.

6 - Install inner ball race of inner bearing with sleeve VW **244**. If the seating surfaces for the inner ball races have become worn by repeated removal and the adhesion or sliding fit is no longer present, a new stub axle must be used.

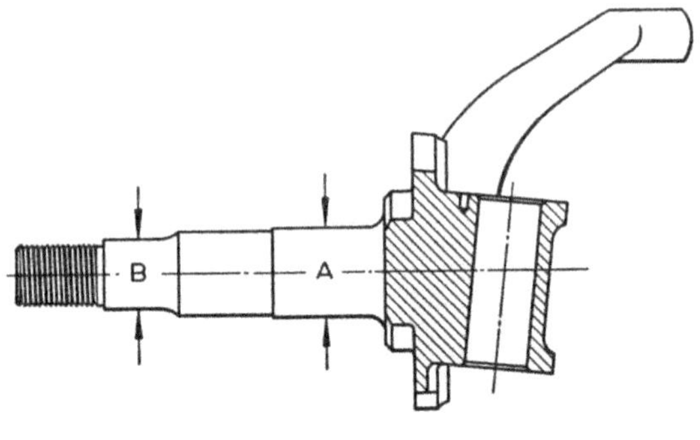

Seating A for inner bearing 25_{h6}	=	25.000 mm dia. / 24.987 mm dia.
		0.9843 in. dia. / 0.9837 in. dia.
Ball bearing inner dia.	=	25.00 mm / 25.01 mm
Seating B for outer bearing 20_{g5}	=	19.993 mm dia. / 19.984 mm dia.
		0.7871 in. dia. / 0.7868 in. dia.
Ball bearing inner dia.	=	20.00 mm / 20.01 mm

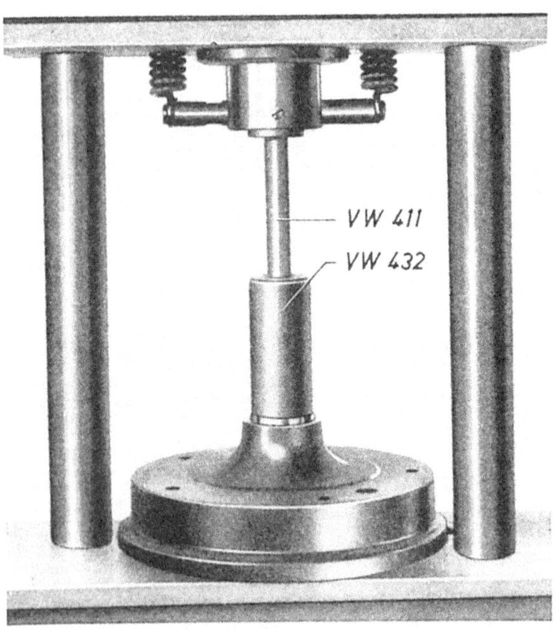

7 - Install outer race of outer bearing on VW Repair Press with thrust mandrel VW 432 in conjunction with VW 411.

8 - Install outer race of inner bearing on VW Repair Press with thrust block VW 433 in

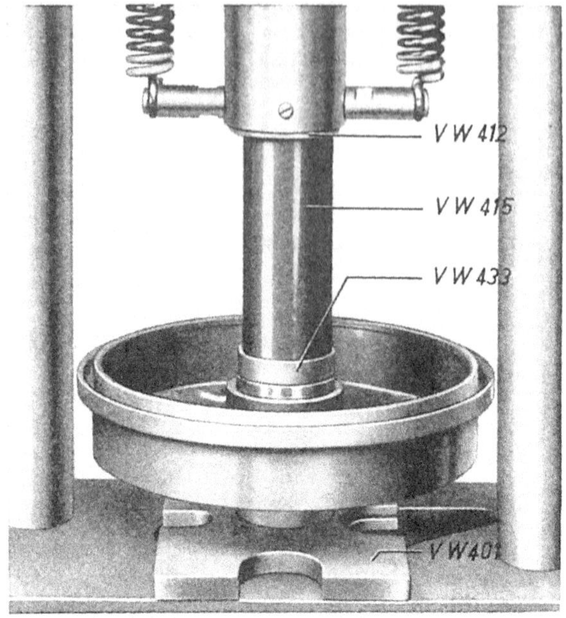

conjunction with VW 401, 412, and 415. Insert the ball cage.

9 - Press in the oil seal on the Repair Press with thrust block VW 433 in conjunction with VW 401,

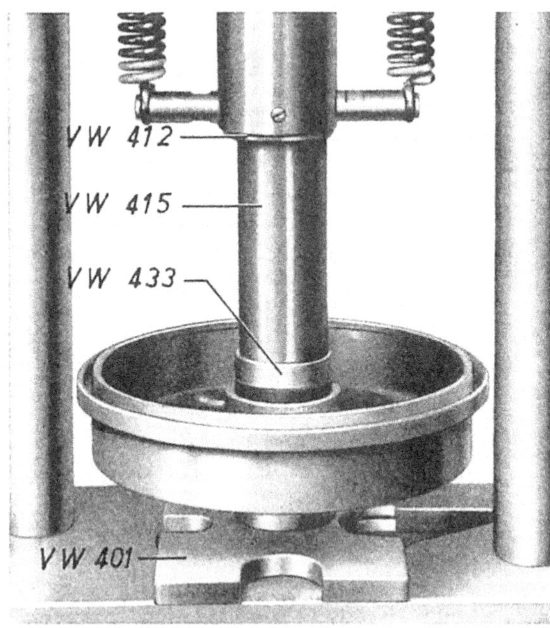

412, and VW 415. Ensure that the oil seal is correctly seated.

Check oil seal for wear. If in doubt, renew.

10 - Lubricate bearings with grease of the specified grade. Press the grease well into the cages and between the balls. Apply a light film to the seatings and to the surfaces of the ball races. Only use grease of proven brands.

Important

Mixing greases of different brands or types can have harmful after-effects and is to be avoided at all costs.

Too much grease can also be harmful.

11 - During lubrication and assembly take care to prevent dirt or foreign matter from entering.

12 - Install brake drum and drive the inner race of the outer bearing into position, using sleeve VW 244.

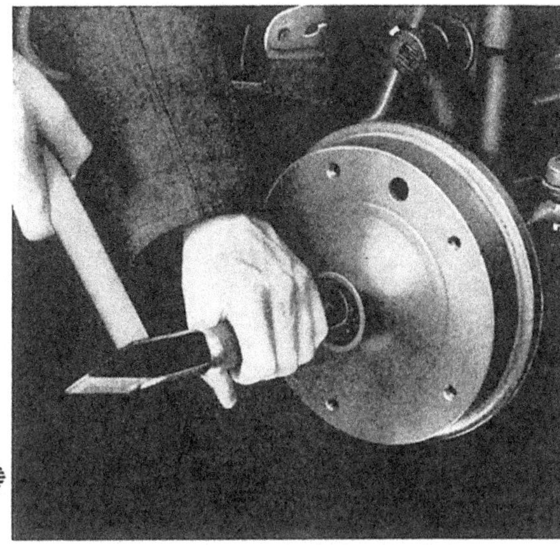

13 - Make sure the thrust washer is correctly positioned. A tilted washer causes maladjustments.

14 - Use new lock plate.

15 - Adjust bearings as specified.

Testing and Adjusting Front Wheel Bearings

Maladjustment of the wheel bearings leads sooner or later to damage or destruction of the bearings. If they are too loose, the whole load of the front axle is taken by only a few balls, resulting in scoring due to fatigue of the ball races and the balls.

If the bearings are too tight, the balls are subjected to excessive pressure and consequently become overloaded until they finally collapse.

Close attention must therefore be paid to the instructions in the section "Testing and Adjusting Front Wheel Bearings".

Checking adjustment

1 - Take off grease cap from brake drum using VW 637.

2 - Adjustment is correct if the thrust washer at the outer bearing can just be moved laterally by a screwdriver and if no axial bearing play can be felt when rocking the brake drum.

Adjustment

1 - Pry up noses of lock plate at stub axle (steering knuckle) nuts.

2 - Loosen nuts by means of the two open end wrenches 27 mm, VW 113.

3 - Use a new lock plate.

4 - Tighten inner nut until the thrust washer at the outer bearing can just be moved laterally by a screwdriver and no axial bearing play can be felt when rocking the brake drum. Failure to observe this instruction could cause the inside nut to be tightened up immediately to the prescribed clearance, so that when the lock-nut is tightened up, it will exert and additional pressure on the inner nut, owing to backlash, thus neutralizing the original bearing play.

5 - Secure nut by alternately bending down nose of lock plate.

The front wheel bearing adjustment is supplemented by the following procedure which is based on exact values for tightening and backing off the inner nut.

For this procedure the following tools are required:

1 - Torque wrench, suitable for 4 mkg (29 ft. lbs.) with 27 mm socket (if necessary with 3/4"—1/2" reducing adapter).

2 - 1 set of open end 27 mm wrenches VW 113.

Sequence of Operations:

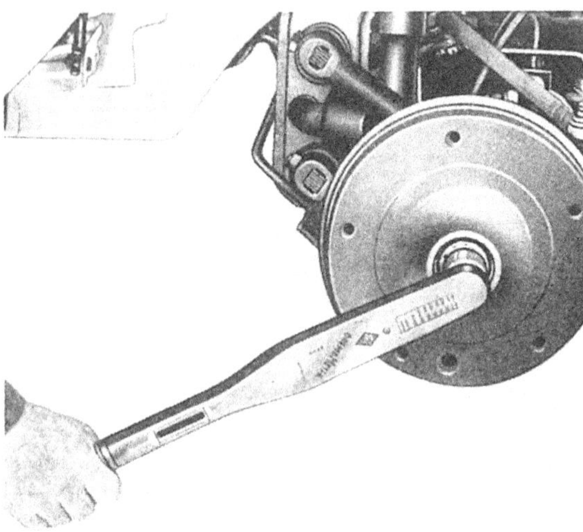

1 - Pry up noses of lock plate and unlock nuts.

2 - Loosen nuts by means of the open and wrenches 27 mm — VW113. Remove outer nut and lock plate.

3 - Tighten inner nut to a torque of 4.0 mkg (29 ft. lbs.).

4 - Install new lock plate.

5 - Loosely screw on outer nut.

6 - Loosen inner nut 72 degrees by means of the offset open end wrench 27 mm — VW 113 as follows:

Put open end wrench onto inner nut. Align one tapped hole with the edge of the open end wrench. Back off the nut until edge of open end wrench is aligned with the next hole in the brake drum.

A = Direction of Loosening α = Angle of Loosening = 72°

7 - Tighten outer nut.

Important

Adjustment is correct if the thrust washer at the outer bearing can just be moved laterally by a big screwdriver and if no bearing play can be felt when rocking the brake drum.

8 - Secure nuts by *alternately bending down nose* of lock plate.

Adjusting Torsion Arm Link Pins

General

The adjustment of the torsion arm link pins must be checked at regular intervals as specified for the maintenance service, and corrected if necessary.

Inspection

1 - Raise front end car.

2 - Rock the wheel by hand to check for end play between torsion arm link and torsion arms. If play is present, adjust torsion arm link pins.

Adjustment

1 - Raise front end of car. To adjust the torsion arm link pins the weight must always be taken of the wheels.

2 - If necessary grease king pins and torsion arm link pins.

3 - Back off pinch bolts at torsion arm eyes.

4 - Fully tighten the link pins and then back off 1/8 of a turn. Retighten the link pins until the first light resistance of the shoulder making contact is felt. This corresponds to a slackening angle of about 10—12°.

If the adjustment has no effect the shims are worn and must be replaced.

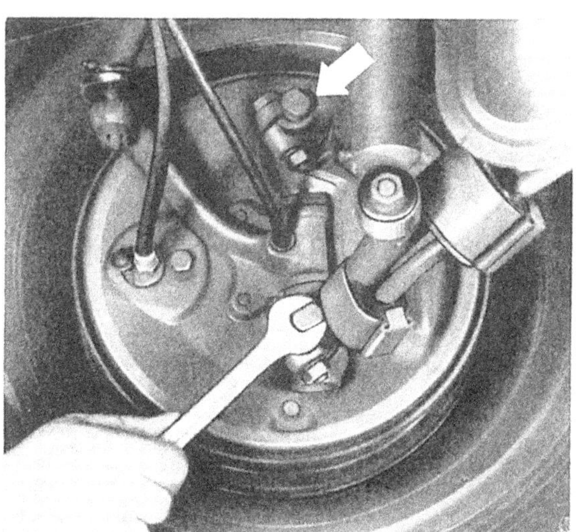

5 - Tap the ends of the link pins lightly to free the tension caused between the parts when the pins were fully tightened.

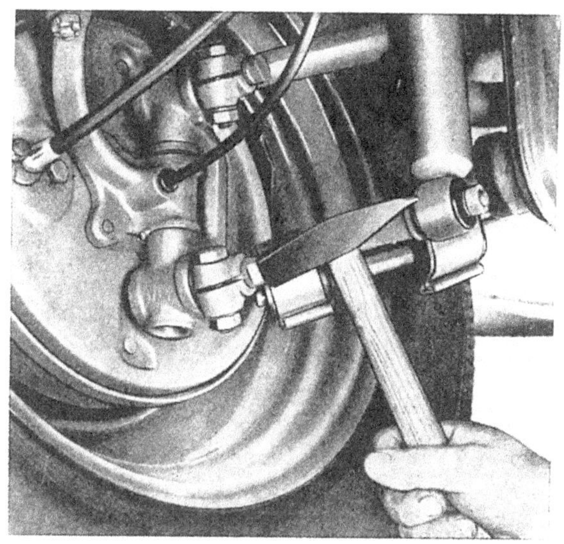

6 - Tighten pinch bolts in the torsion arm eyes.

These adjustments should be followed carefully. If the torsion arm link pins are adjusted too tightly the action of the torsion arm needle bearings and consequently the desired freedom of spring movement will be affected. On the other hand the link pins should not be adjusted too loosely or show any perceptible play.

> **Important**
> After the torsion arm link pins have been adjusted, it is absolutely necessary to check the toe-in.

Removal and Installation of Torsion Arm Link Complete with Stub Axle (Steering Knuckle) (Front axle in position)

Removal

1 - Raise front end of car and if necessary, support it behind the frame head.

2 - Remove front wheels.

3 - Left front wheel: detach speedometer cable.

4 - Remove outer tie rod with tool VW 266f.

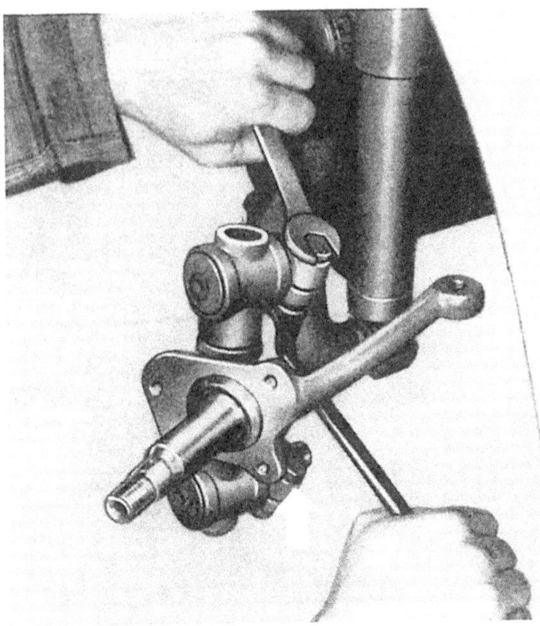

5 - Remove brake drum and brake back plate. It is not necessary to detach the brake hose.

6 - Remove pinch bolts at torsion arm eyes.

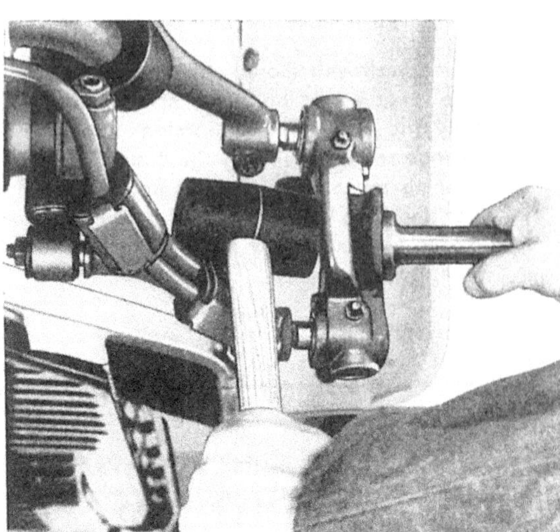

7 - Remove torsion arm link and stub axle by equally driving out both torsion arm link pins.

Installation

This is a reversal of the above procedure, but the following points should be observed:

1 - Check for wear and replace as necessary: torsion arm link pins, torsion arm link pin bushes, shims, and retainer with dust excluder.

2 - Examine faces of torsion arm eyes for wear and roughness. If necessary, reface them with cutter VW 217.

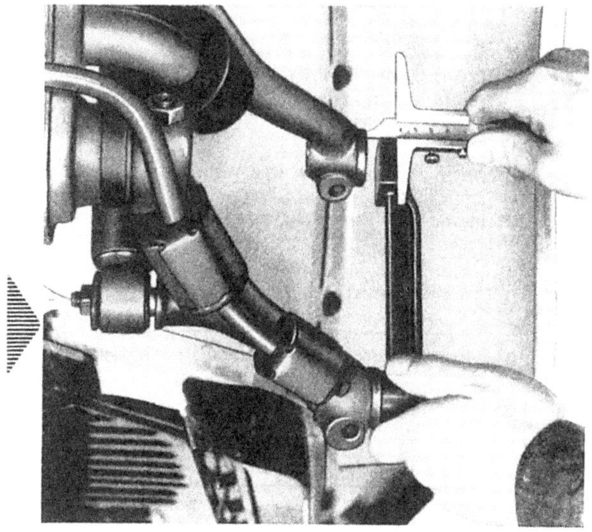

3 - Using gauge VW 270 a, measure offset of eye faces, which should be 7 mm (0.28"). Departures from that dimension must not exceed ± 1.5 mm (0.06"). Correct the offset with shims of 0.5 mm (0.02") thickness.

The installation of the retainer reduces the total number of shims from 10 to 8.

Arrangement of Shims on Torsion Arm Link Pins				
Offset in mm	Number of Shims for			
	Upper Torsion Arm		**Lower Torsion Arm**	
	Inner Shims (A)	Outer Shims (B)	Inner Shims (C)	Outer Shims (D)
5,5	2	6	5	3
6	2	6	4	4
6,5	3	5	4	4
7	**3**	**5**	**3**	**5**
7,5	4	4	3	5
8	4	4	2	6
8,5	5	3	2	6

Note:

a - There must always be 8 shims and one retainer with dust excluder fitted to each torsion arm link pin.

b - If the offset is 7 mm B and D have 5 shims each and A and C 3 each.

c - If the offset is in excess of 7 mm, shims should be added to A and removed from C.

d - If the offset is below 7 mm, shims should be removed from A and added to C.

e - Number of shims at B and D should always be brought up to 8.

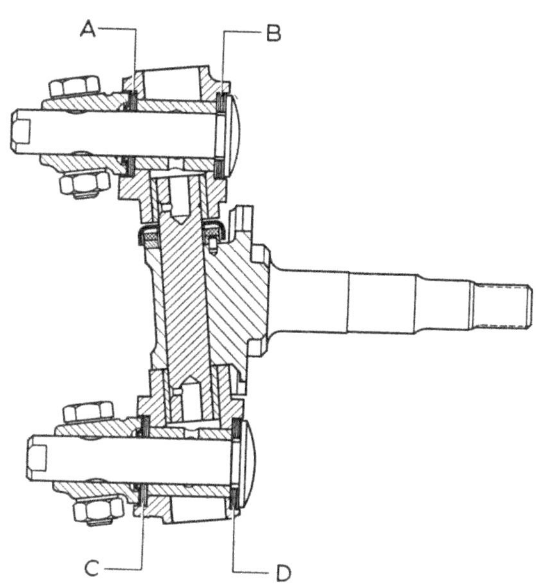

Example:

a - The measurement of the offset is 8.2 mm. The value measured is rounded off, in this case to 8.0 mm.

b - The deviation from the nominal dimension (7 mm) is 8.2—7 = **1.2 mm.** This value corresponds to the thickness of 2 shims of 0.5 mm each.

c - The shims must be arranged as follows:

Upper torsion arm		Lower torsion arm	
Inner Shims (A)	Outer Shims (B)	Inner Shims (C)	Outer Shims (D)
4	4	2	6

If deviation exceeds ± 1.5 mm, it is not permissible to add further shims. In order to locate the cause of the trouble the torsion arms must be removed and checked (gauge VW 282b) and the entire front axle assembly inspected (with front axle tube alignment gauge VW 256a).

Bent torsion arms must always be replaced. It is not permissible to straighten these parts.

Note:

If the torsion arm link offset (correct value: 7.0 mm) departs from the tolerance of ± 0.2 mm, this difference has to be considered when arranging the shims. Such a deviation is present when the torsion arm link offset reading which is measured with a depth gauge from the back of the gauge to the inner shoulder of the torsion arm link, is less than 19.8 mm or more than 20.2 mm. In this case, proceed as follows:

If the torsion arm link offset reading is less than 19.8, the difference between this reading and 20.0 mm has to be subtracted from the torsion arm offset. If the difference is more than 20.2 mm the amount over 20.0 mm is added to the torsion arm offset.

Example:

a - The torsion arm offset as previously measured is 8.2 mm.

b - The reading for the torsion arm link offset is accepted as being 20.3 mm.

c - The amount over 20.0 mm is:

20.3 — 20.00 = 0.3 mm.

d - The torsion arm offset is, therefore:

8.2 + 0.3 = 8.5 mm.

e - The shims should then be arranged as follows:

Upper torsion arm		Lower torsion arm	
Inner Shims (A)	Outer Shims (B)	Inner Shims (C)	Outer Shims (D)
5	3	2	6

4 - Insert torsion arm link pins and shims with universal grease. The dust excluder retainer and the dust excluder should be pushed on to the torsion arm link pin so that the dust excluder seats correctly and is not crushed.

If the offset has been taken into consideration, the torsion arm link pins can be inserted fully into the torsion arm eyes and the dust excluder retainers contact the upper and lower torsion arm eyes at the same time.

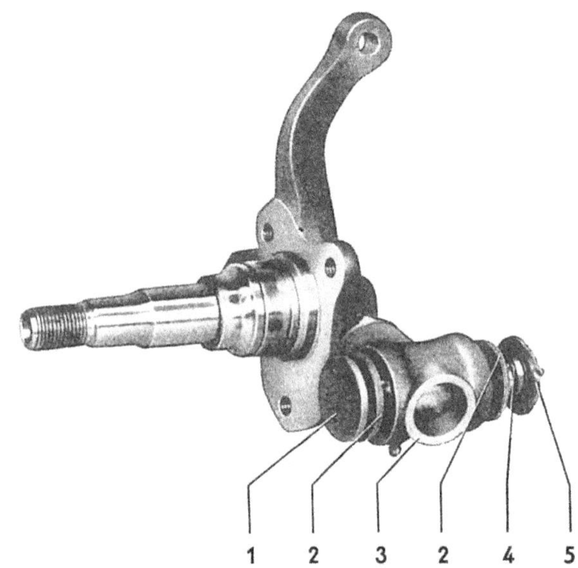

1 - Torsion arm link pin
2 - Shims
3 - Torsion arm link
4 - Dust-excluder
5 - Retainer for dust excluder

5 - When assembling, the dust excluder retainers should be turned so that their noses fit into the slots in the torsion arm eyes.

6 - Adjust torsion arm link pins.

7 - Bleed and adjust brakes.

8 - Adjust front wheel bearings as specified.

9 - Check camber and toe-in of front wheels.

Removal

1 - Remove torsion arm link and stub axle (steering knuckle).

2 - Remove torsion arm link pin bushes on VW Repair Press with thrust mandrel VW 434 in conjunction with VW 408 and VW 401.

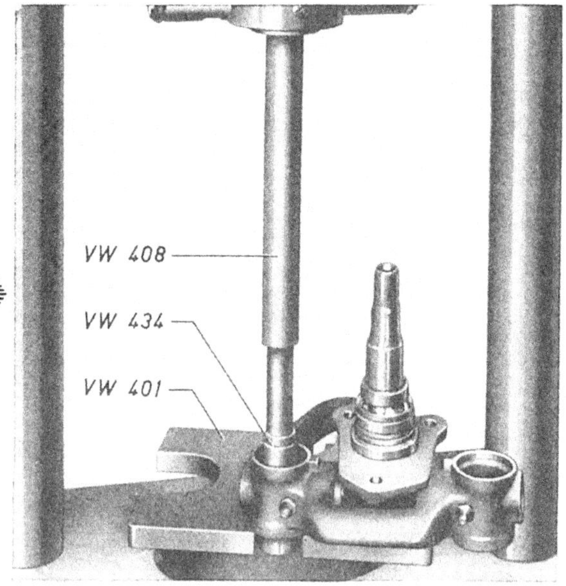

Installation

When installing torsion arm link pin bushes, the following points should be observed:

1 - Check depth of bores for torsion arm link pin bushes using gauge VW 259.

2 - Check torsion arm link for correct offset by means of gauge VW 259.

3 - Check clearance of king pin in the bushes. Replace pin and bushes if necessary.

4 - When inserting the bushes of the torsion arm link pins ensure that the lubrication passages are correctly positioned.

5 - Install bushes for torsion arm link pin on VW Repair Press with punch VW 408 in conjunction with VW 401 and VW 418. The bushes must be a force fit in the torsion arm link.

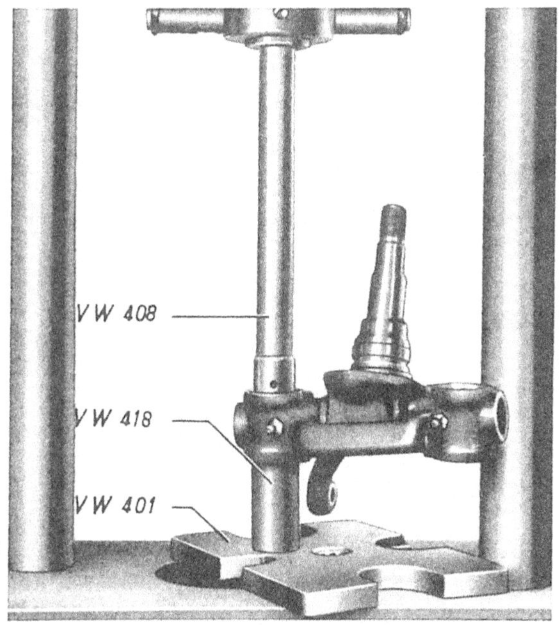

Removing and Installing Stub Axle (Steering Knuckle)

Removal

1 - Remove torsion arm link pin bushes.

2 - Drive out king pin on VW Repair Press with punch VW 411 in conjunction with VW 401 and VW 418. To avoid damage, the stub axle should be heated in an oil bath to 80—90°C (175—195°F).

3 - Remove stub axle and thrust bearing assembly from torsion arm link.

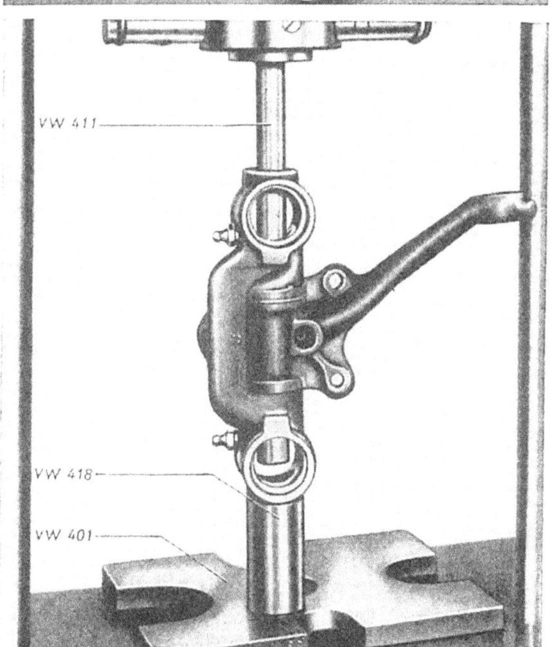

Checking Stub Axle

1 - Pull off front wheel bearing spacer with extractor VW 202 in conjunction with VW 202 S and 202 i.

2 - Check the stub axle with gauge VW 258 a for the following:

A - The face of the stub axle must be completely flush with the side of the gauge.

B - The bore in the stub axle steering arm must align with that in the gauge.

C - The face of the steering arm eye must be parallel to the corresponding surface of the gauge.

D - Condition of king pin bores. If the bore is worn as a result of king pin seizure, the stub axle must be replaced.

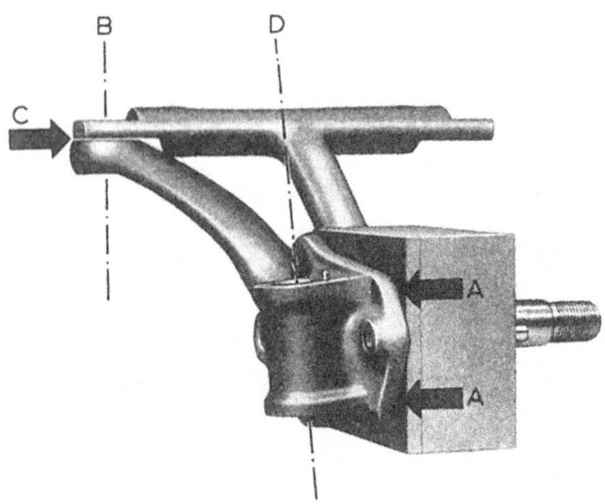

> **Important**
>
> Bent stub axles must not be straightened, but should be replaced.

3 - Examine front wheel bearing seating surfaces on stub axle. If the seating surfaces for the bearings have become worn by repeated removal of the bearings to an extent that the correct seating of the inner races cannot be ensured, the stub-axles must be renewed.

4 - Inspect thrust washer contact surface on stub axle for smoothness and remove any burr.

Checking Torsion Arm Link

1 - Drive out torsion arm link pin bushes.

2 - Inspect torsion arm link with gauge VW 259. The torsion arm link pin shims are positioned at the shoulder. These contact surface wear due to friction. The bores for the torsion arm link pin bushes should, therefore, be checked for sufficient length. The torsion arm link is placed

— first with its upper bore and then with the lower bore — over the mandrel of the offset gauge. The shoulder of the bores should be higher than the face of the mandrel. If this is not the case, install a new torsion arm link.

3 - Check the torsion arm link with gauge VW 259 for correct offset.

Correct offset 7.0 mm

Tolerance ± 0.2 mm

Place gauge and torsion arm link in a vise and measure the offset with a depth gauge. With the depth gauge contacting the back of the gauge, a dimension of 20.0 mm up to the shoulder of the torsion arm link, corresponds to an offset of 7.0 mm.

4 - Important:

a - If the reading exceeds 20.0 mm, the offset is less than 7.0 mm. If the reading is less than 20.0 mm, the offset exceeds 7.0 mm.

b - Readings between 19.8 and 20.2 mm which correspond to the tolerance of from 6.8 to 7.2 mm, are not taken into consideration. The torsion arm link can still be used provided that the bores for the torsion arm link pin bushes are sufficiently long.

c - If the torsion arm link offset reading is less than 19.8 mm or greater than 20.2 mm, such a deviation must be considered when arranging the shims as is described during the assembly of stub axle and torsion arm link.

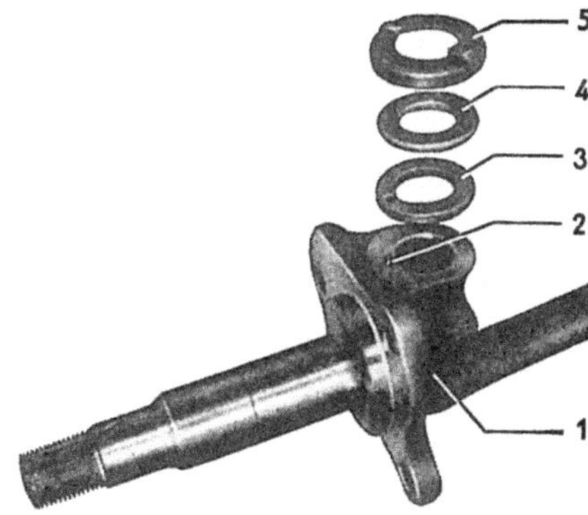

1 - Stub axle (steering knuckle)
2 - Dowel pin
3 - Friction washer
4 - Plastic washer
5 - Cover

Installation

This is a reversal of the removal procedure but the following points should be observed:

1 - Check king pin for wear. King pins must only be replaced together with the bushes in the torsion arm link.

2 - Inspect seating of thrust washer dowel pin in stub axle.

3 - The torsion arm link must be assembled with the stub axle and the thrust bearing (thrust washer, plastic washer and cover) so that no end play is evident. The assembly should take place with a preload of 0.02 ± 0.02 mm.

Any end play felt is eliminated by installing a thicker thrust washer. Steel thrust washer of various thicknesses from 3.56 — 3.63 — 3.68 — 3.73 mm etc. up to 4.08 and 4.30 mm are available.

The thrust washer is dowel-located in the stub axle and the cover is secured in position by slots in the torsion arm link.

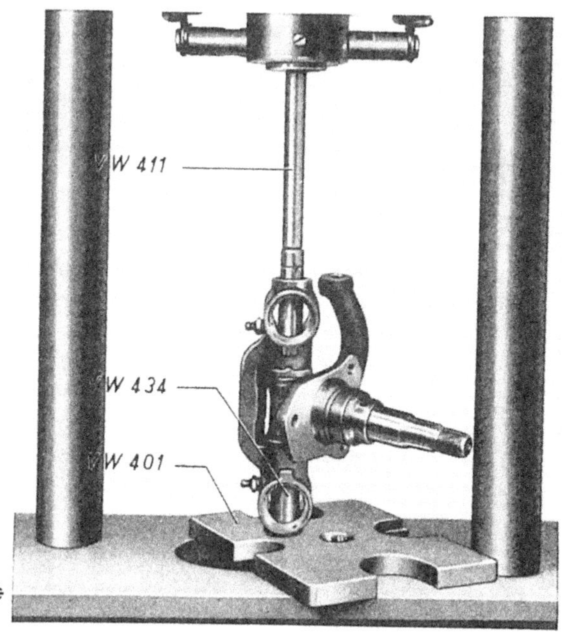

4 - The king pin is a press fit in the stub axle and the stub axle must, therefore, be heated in an oil bath to approximately 80 °C (175 °F). Press in the king pin on the VW Repair Press with arbor VW 434 in conjunction with VW 401 and 411.

The torsion arm link and steering knuckle must be free enough to be turned by hand. If necessary, tap the parts several times with an aluminium mallet.

5 - In connection with the installed steering damper, it is of particular importance that the stub axle moves sufficiently free in the torsion arm link. The movement must be checked each time repairs are carried out, and it can best be done with a spring balance which is hooked into the stub axle arm. The force applied to move the arm should be a maximum of 2 kg (4.4 lbs.).

Removing and Installing King Pin Bushes

Removal

1 - Remove stub axle (steering knuckle).

2 - Drive out king pin bushes on VW Repair Press with punch VW 411 in conjunction with VW 401, 422 and 423.

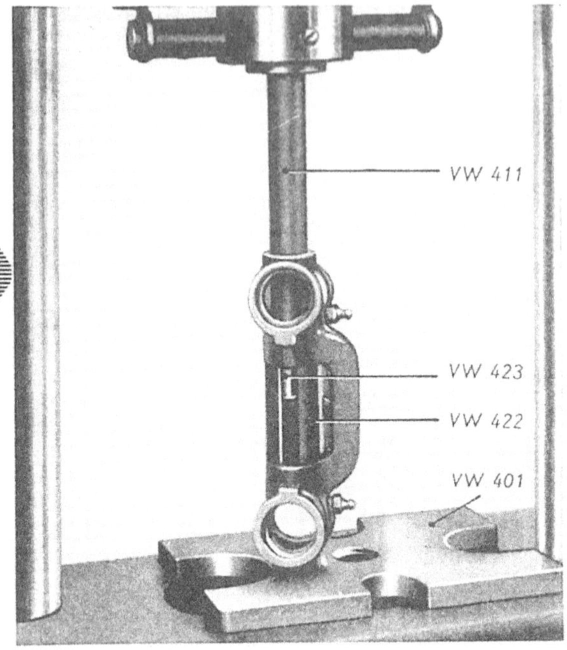

Installation

When installing king pin bushes, the following points should be observed:

1 - Press in new bushes from inside of torsion arm link on VW Repair Press with punch VW 411 in conjunction with VW 431, VW 432 and VW 401.

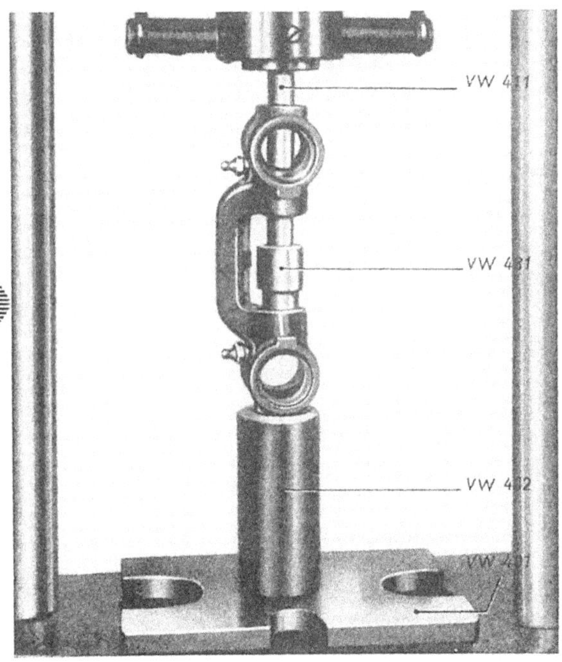

2 - File slots in the upper bush for the thrust bearing locating lugs.

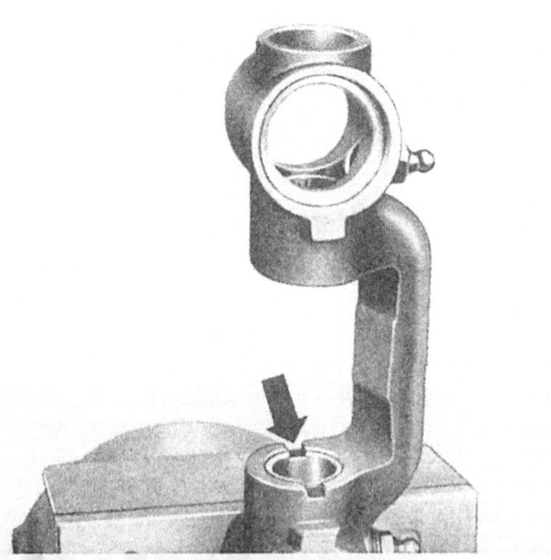

3 - Ream up bushes by means of reamer VW 224 (18—21 mm dia.). The tapered bush of the reamer serves as a guide.

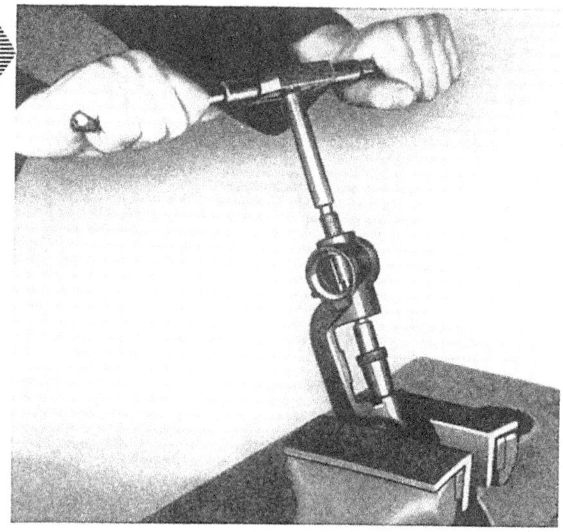

Correct dia. $\frac{18.034\ mm}{18.016\ mm} = \frac{.7099''}{.7092''}$

Correctly reamed up bushes must be free from score and chatter marks, allowing the king pin to be turned by hand, without play being felt.

Note:

When reaming up the rolled and slit type bronze bush there is a possibility of the cutter becoming prematurely blunt.

THIS PAGE INTENTIONALLY
LEFT BLANK

Removing and Installing Stabilizer

(With front axle installed)

Removal

1 - Loosen wheel bolts, raise or support vehicle on trestles and take off front wheels.

2 - Bend up lugs of spring retaining clips and remove clips from inner and outer clamp on both sides.

3 - Expand the clamps and lift them off. Take off shims.

4 - Take off stabilizer and slide off the rubber cushions.

Installation

To install, reverse the preceding operations and observe the following points:

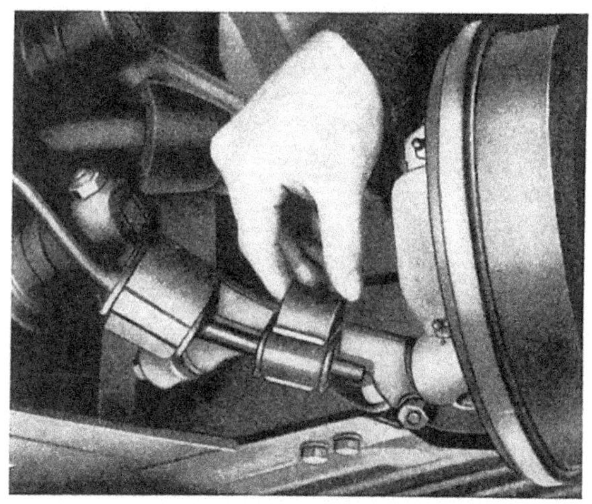

1 - Check stabilizer, rubber cushions, clamps, shims and clamp links for damage, replace as found necessary.

2 - The clamp edges are tapered. When installing, make sure that the edges of both clamps and links taper down toward the wheel.

3 - Make sure the inner clamps occupy a position which prevents them from fouling the shock absorbers during operation.

4 - Tension the clamp with a hand vise and position the retaining clips. The lug of the retaining clips must be towards the front axle tubes.

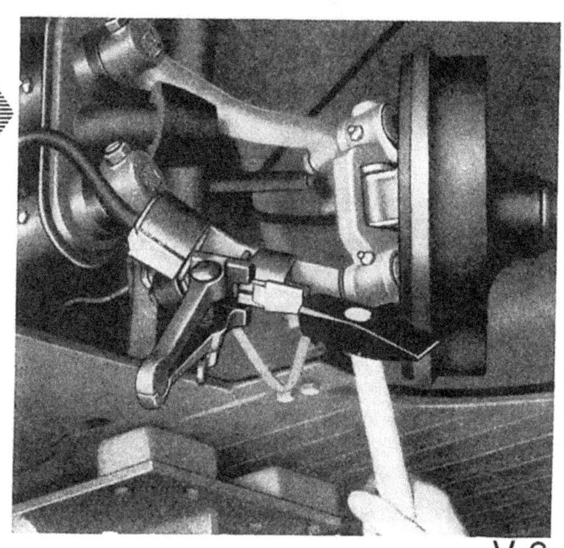

Secure the retaining clips by bending over the lug.

Note:

If the front axle with stabilizer is to be removed, the stabilizer must be pressed down slightly to give access to the lower front axle mounting bolts. It is best to insert a piece of wood between the axle tube and stabilizer and use it as a lever.

Removing and Installing Torsion Arms

(Front axle in situ)

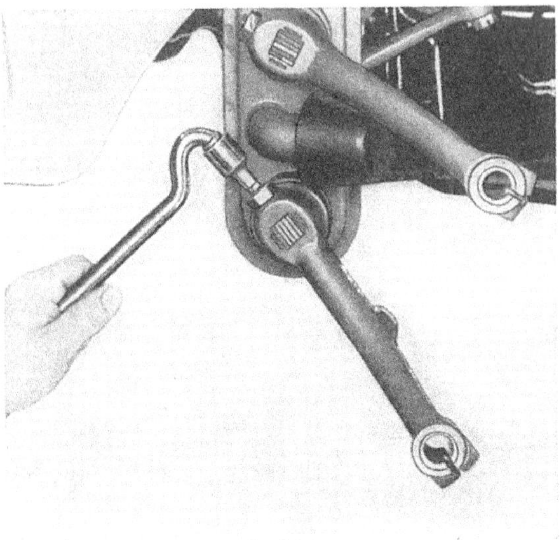

Removal

1 - Remove torsion arm link and steering knuckle.

2 - Remove shock absorber.

3 - Remove the stabilizer.

4 - Remove lock nuts and set screws at torsion arms using offset handle VW 150 in conjunction with 8 mm key (VW 156).

5 - Take off torsion arm with dust seals.

6 - Pull off torsion arm rubber stop and remove upper torsion arm.

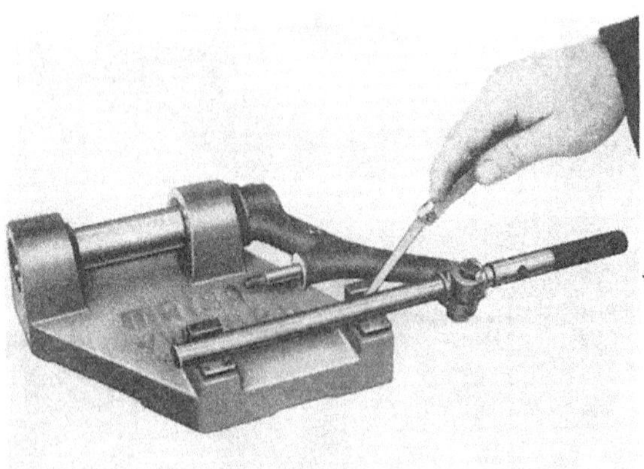

Checking Torsion Arms

1 - Check torsion arms for parallelism and twist. Insert the torsion arm in the bores of the gauge VW 282 b together with the bushes provided. Insert test mandrel in torsion arm eye and tighten it with the clamping screw. The clearance between the mandrel and the test faces can be checked with a feeler gauge. The permissible departure from the parallelism is 0.5 mm (.020").

No attempt should be made to straighten bent torsion arms: they must in all cases be exchanged.

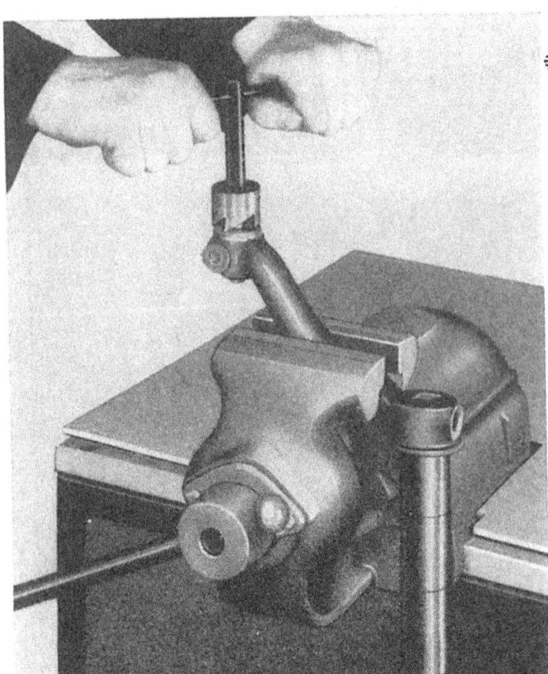

2 - Check contact faces of torsion arm eyes for wear. If worn, reface them by means of cutter VW 217.

3 - Inspect bearing points for wear. If only slightly worn, the torsion arms of both sides can be interchanged. Signs of seizure or increased wear necessitate a removal of the torsion arms.

Note:

Upper and lower torsion arms are not interchangeable (shock absorber mounting).

> **Important**
>
> The outer needle bearing *surface on the* torsion arm is surface hardened.

Installation

This is a reversal of the removal procedure, but the following points should be observed:

1 - Check needle bearings torsion arm bushes in front axle tubes for wear, renew them if necessary.

2 - If necessary, renew the dust seals.

3 - Grease torsion arm with Universal Grease and insert it in the front axle tube until the tapped hole is in line with the recess in the torsion bar. Tighten set screw and secure it by means of the nut.

4 - Lift up upper torsion arm and push rubber stop in position.

Note:

The ends of the 8-leaf torsion bars are not welded. The outer edges of the square holes in the torsion arms are peened as shown in the drawing to prevent the outer split leaves from moving outward. A 14 mm chisel is suitable for peening.

5 - Re-install shock absorber, using new lock-washer, and tighten it to 3.0—3.5 mkg.

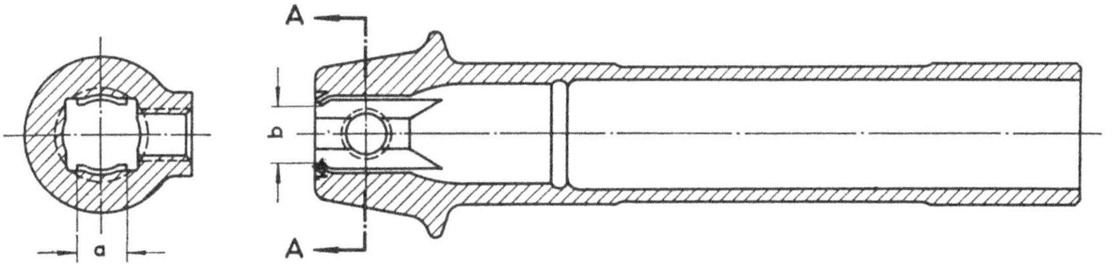

a = 14 ± 1 mm (0.551″ ± 0.039″) b = 17 mm max. (0.669″)

Removing Shock Absorber Mounting Stud from Torsion Arm

General Note

When renewing a shock absorber mounting stud at the lower torsion arm, a 0.5 mm (.02″) oversize stud must be used. The hole in the torsion arm has become larger by the pressing in of the original stud so that another stud of the same size would have a loose fit. It is, therefore, necessary to use the oversize stud D_2 12.489—12.500 mm (.4917″—.4921″) instead of the standard stud 11.989—12.000 mm (.4720″—.4724″).

Removal

1 - Remove torsion arms.

2 - Drive out retaining pin.

3 - Remove stud. In the case of the stud being broken off, mark the piece in the torsion arm by means of a center punch and drill a center hole, using a 3 mm (.12″) drill. Bore out the remaining piece with a 10.75 mm (.23″) drill. The thin shell remaining around the drill will come out during the final drilling revolutions.

Installation

1 - Bore the hole with a 12.3 mm (.48″) drill and ream it with a reamer 12.5 mm (.49″) P 8 = 12.455—12.482 mm D_1 (.4903″—.4914″). If a reamer of the size 12.5 mm P 8 should not be available, make the stud to fit the hole by grinding it to the required diameter. The stud must in all cases be a press fit 0.01—0.05 mm (.0004—0020″) in the torsion arm.

2 - Press oversize stud in place, taking care that the projecting end has a length of a = 45.0—45.5 mm (1.77″—1.79″).

3 - Drill hole d = 4.00—4.08 mm dia. (.157″—.160″) for retaining pin in the stud.

4 - Drive in retaining pin.

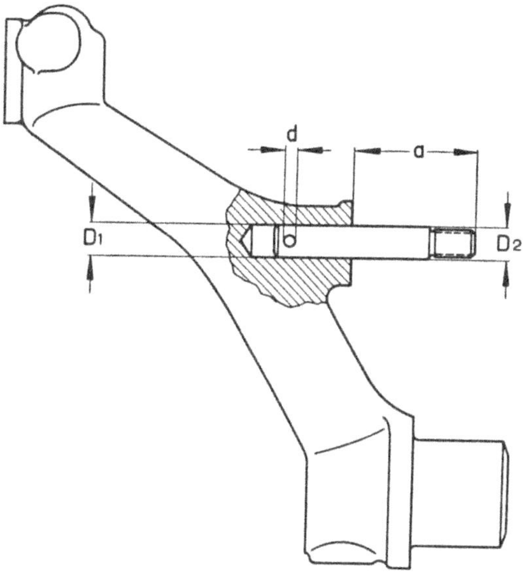

Torsion Bars

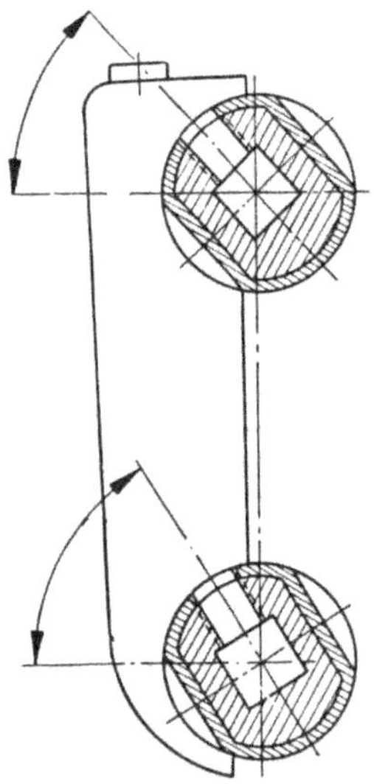

General Description

The torsion bars of the front axle consists of two groups of eight steel leaves each. The torsion bars are anchored in the center of the tubes and secured in position by means of a set-screw and a lock-nut each. The mounting angles are:

Upper torsion bar 49° ± 1°

Lower torsion bar 51° 30′ ± 1°

The device VW 668 is available for checking the mounting angles.

Note:

When replacing a torsion bar, the new bar must have the same number of leaves as the old one. The mounting angle of the center anchor is different for torsion bars with different numbers of leaves. Thus eight-leaf torsion bars with the above setting must only be replaced by eight-leaf bars, otherwise the smooth riding qualities will be adversely affected, causing the suspension to "bottom".

Removing and Installing Torsion Bars

(Front axle in situ)

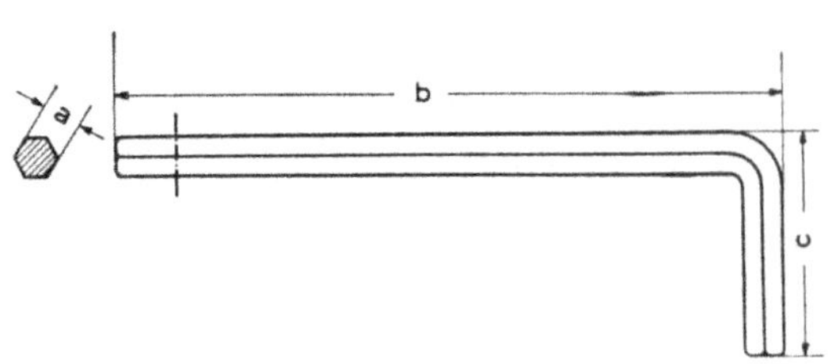

a = 8 mm (.31″)
b = 135 mm (5.31″)
c = 45 mm (1.77″)

Removal

1 - Remove stabilizer and shock absorbers on both sides.

2 - Remove torsion arms on one side.

3 - Release lock nut and setscrew at center anchor.

Note:

The setscrew for the upper torsion bar can no longer be removed by means of VW 156 (8 mm socket wrench) and VW 150 (angular handle) with the front axle installed.

The key for the earlier type oil drain plug (8 mm hex.) can be used for this purpose. If not available, this key can be made locally of 8 mm (.31") hex. section steel.

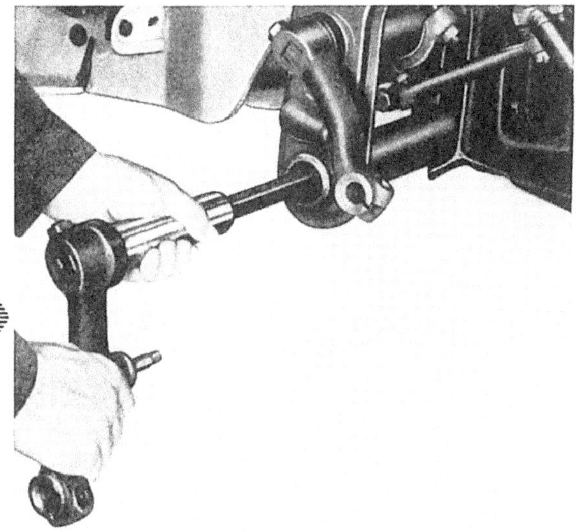

4 - Take off rubber stop on torsion arm.

5 - Pull out torsion arm of the opposite side together with the torsion bar.

Note:

It is not necessary to mark the direction of torsion imposed on the bars, as this has no bearing on the service life of the bars.

Installation

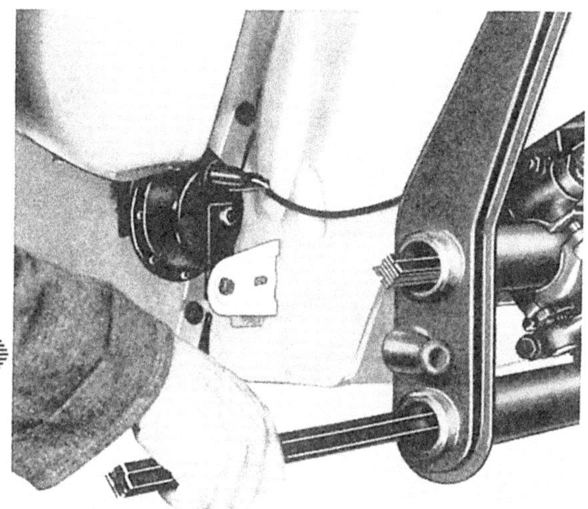

This is accomplished by reversing the removal procedure, but the following points should be observed:

1 - Clean torsion bars, check for cracks, replace if necessary.

2 - Check torsion arms, needle bearings and torsion arm bushes, replace if necessary.

3 - When installing torsion bars, note the number of steel leaves.

4 - Coat torsion bar with Universal Grease prior to its installation.

5 - Bring the recess in the center of the bar in line with the hole for the setscrew; tighten center setscrew and secure with lock nut.

Removing and Installing Torsion Arm Bushes and Bearings

Removal

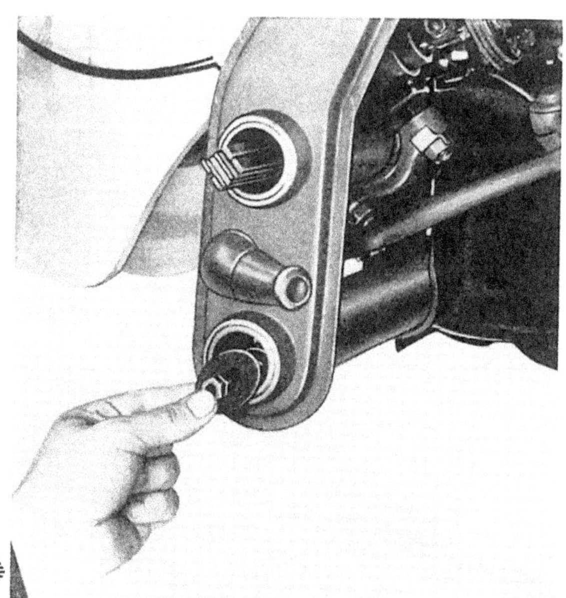

1 - Remove torsion bars.

2 - Remove grease nipples in front axle tube.

3 - Remove the needle bearing and inner bush with torsion arm bush drift VW 638 (local manufacture).

4 - Attach extractor disk for passenger cars to the rod. Insert the rod in the axle tube, guiding it by means of the center anchor.

5 - Insert the extractor disk and move the rod towards the disk until it meets the shoulder of the needle bearing.

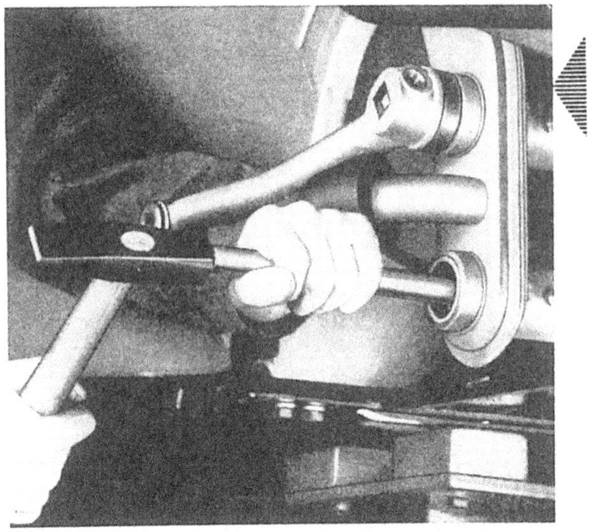

6 - Drive out the needle bearing by hammering the end of the rod.

7 - Insert the disk again and drive out the inner bush in the same manner.

Installation

1 - Clean front axle tubes, especially the seating surfaces of needle bearing and bushes.

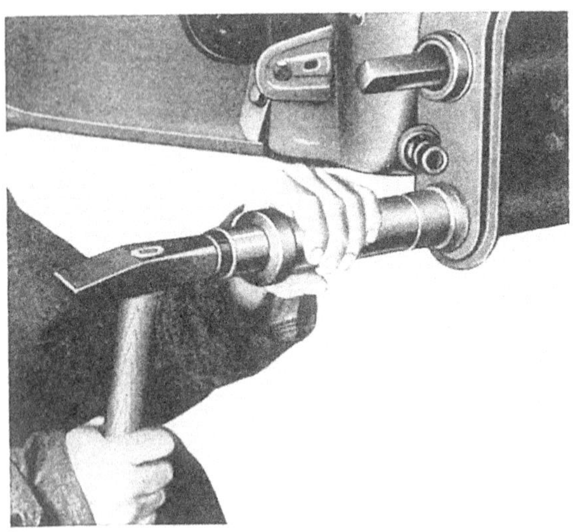

2 - Check the condition and measurements of the needle bearing seating surface. The needle bearings are a press fit in the axle tubes. The diameter of the bearing seat:

46^{N7} = 45.992/45.967 mm (1.8107"—1.8097")

Normally needle bearings with a diameter of 46 mm are installed. In isolated cases the diameter is sometimes larger and then is:

46.2^{N7} = 46.192/46.167 mm (1.81858"—1.81759")

The corresponding needle bearing has an outer diameter of 46.2 mm (1.820").

In certain cases it is possible to eliminate slight signs of wear on the needle bearing seating surface, but in all cases a correct press fit must be ensured. If the diameter of the bearing seat in the axle tube is no longer correct, new front axle tubes must be installed as it is not possible to rework the seating surface.

3 - Clean the needle bearing thoroughly by removing all traces of the old grease. Check the condition of the bearing. If it shows signs of wear or damage it must be replaced.

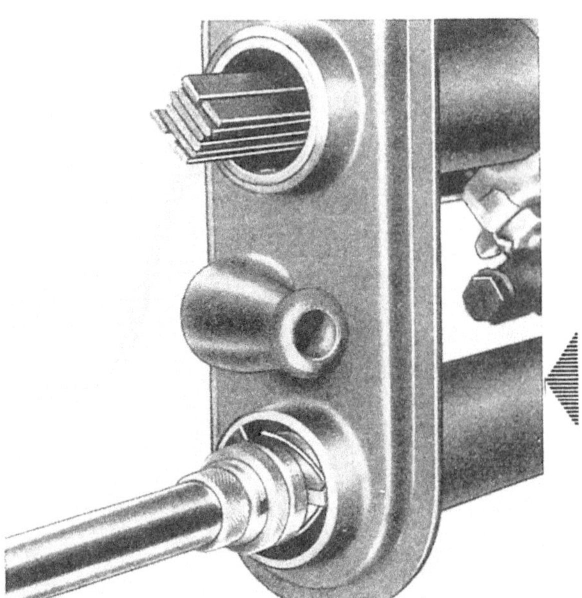

4 - Before installing the needle bearing check the diameter of the bearing seat. Only install bearings with a corresponding outer diameter.

5 - Place inner bush on the longer guide portion of the drift VW 273a so that the open side of the lubrication groove faces the grease nipple. Drive in the bush until the collar of the drift buts against the tube.

6 - Ream up inner bush with reamer VW 274a. The reamer is centered by the guide rod which enters the anchor in the center of the tube. The reamer is supported in the bearing seating surface by the guide bush EL 20.

The fitting clearance between torsion arm and bushes is:

0.20—0.27 mm (.008"—.011")

Ream the bushes to the following dimensions:

Torsion Arm	1 - Bushes 2 - Reaming dimension
From Chassis No. 2921552 36.95—37.00 mm dia. (1.4539"—1.4547")	1 - 37.15—37.20 mm dia. (1.4626"—1.4646") 2 - 37.20—37.25 mm dia.

7 - Blow out the axle tube with compressed air and carefully clean the bearing seats.

8 - Grease the seating surface, needle bearing outer race. Push the bearing on to the short guiding portion of the drift and drive it into the axle tube.

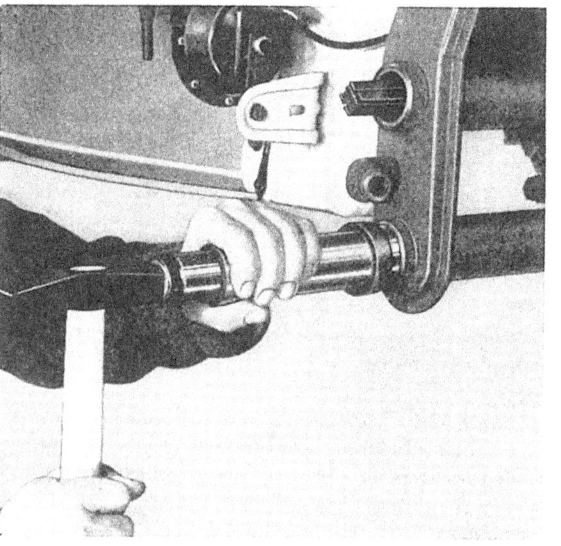

9 - Reinstall grease fitting. Renew damaged grease fittings.

10 - Lubricate the front axle assembly with Universal Grease.

Checking Front Axle Tubes

General

Front axle tubes which are suspected to be out of alignment after an accident should be checked with the gage VW 256 a.

> **Important**
> Bent front axle beams must not be straightened. Apart from the fact that heat treatment or cold straightening would reduce the strength and consequently the safety factor, workshops have not got the means of checking the alignment of the tubes, the angle of the centre anchors and the quality of the welding seams.

Inspection

1 - Check torsion arm bushes and needle bearings for wear. If necessary, install new bushes and bearings and ream them up as otherwise the tube alignment cannot be correctly measured.

2 - Insert the two mandrels of the gage VW 256 a in the front axle tubes as far as they will go.

3 - Contact test plate with the four cylindrical test surfaces of the mandrels and check for twist by means of a feeler gage.

Permissible deviation from the parallel: 0.2 mm (.008").

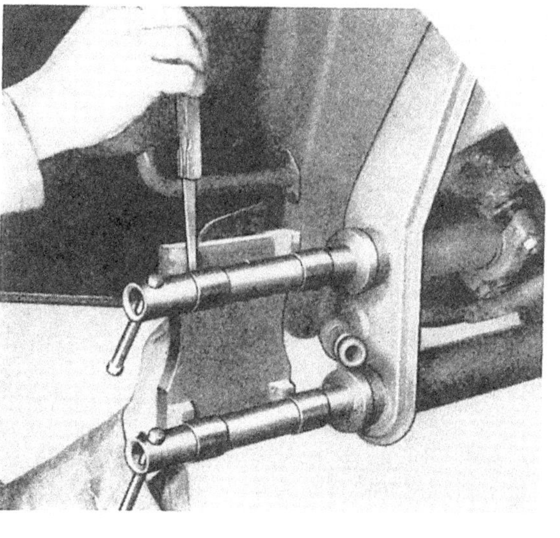

Shock Absorbers

General

The front axle of the Volkswagen is equipped with double-acting, hydraulic shock absorbers of the telescope type which absorb road shock and prevent excessive rebound. The resistance of the shock absorber is progressive, that is, the damping effect increases when large or sudden movements of the wheel take place.

As both the compression and rebound actions of the shock absorbers correspond to the springing characteristics of the car, no attempt should be made to change the adjustment or to fit shock absorbers of other characteristics. The riding qualities would otherwise be seriously affected.

Inspection and Maintenance

Shock absorber action may be roughly checked by bouncing each corner of the car in turn or by riding the car over uneven roads. A more accurate check can be made by means of testing appliances which are generally not available at the workshop.

A simple check of the removed shock absorber can be carried out by holding it vertically and compressing it by hand. But this method will only give an indication whether or not there is a resistance; the degree of efficiency of the compression and rebound strokes cannot be determined.

The equipment for the VW test station contains a testing machine on which the operation of the shock absorbers installed on the car can be properly tested. Further information can be obtained from the Technical Service Department on request. Shock absorbers which do not function correctly or which lose fluid at a high rate must be replaced — preferably by one of the same make.

If only a slight loss of fluid has taken place and the function of the shock absorbers is still satisfactory, there is no need for an exchange, as an adequate fluid reservoir compensates smaller losses. In this respect the shock absorber requires no maintenance attention. Service is, therefore, confined to a check for normal function and a periodical examination of the anchorage at front axle and torsion arm.

Replacing Shock Absorbers

When exchanging shock absorbers it is permissible to use shock absorbers from different manufacturers on one front axle.

Shock absorbers for the front axle are painted black, and for the rear axle red-brown.

Important

When changing front shock absorbers, make sure that only shock absorbers meant for the front axle are used. It is inadmissible to use any other shock absorbers on account of the totally different characteristics.

Removing and Installing Shock Absorber

Removal

1 - Raise car and remove front wheel.

2 - Remove torsion arm stud nut and fixing bolt at front axle beam.

3 - Remove shock absorber.

Installation

This is accomplished by reversing the removal procedure, but the following points should be observed:

1 - Check shock absorber; if necessary exchange shock absorber for a new one of the same type.

2 - Examine shock absorber rubber bushes for wear. Renew worn bushes.

3 - Inspect fixing stud and bolt for wear and replace them if necessary.

4 - Use new lock washers if necessary.

5 - Tighten nut and bolt to 3.0—3.5 mkg (22 to 25 ft. lbs.) so that they bear securely against the rubber bearing bushes, otherwise premature wear and rattling during operation will be the result.

Note:

1 - From Chassis No. 3 250 900 all axles are equipped with softer shock absorbers which improve the springing characteristics and riding qualities.

Part Numbers	front		rear
New:	111 413 031 J 111 413 031 K	intermittently	111 513 031 F 111 513 031 G
Previously:	111 413 031 F 111 413 031 G	intermittently	111 513 031 C 111 513 031 D

The shock absorber which was installed prior to the series introduction of both shock absorbers is marked with a white paint spot on one of the securing eyes. The new shock absorbers can be service installed on all Volkswagen from Chassis No. 2 528 668 (August 1959), but only on all four wheels at the same time. The previous type shock absorbers will remain available as spare parts.

2 - Shock absorbers from Messrs. Hoesch are being installed on the front axle from September 1963, Chassis No. 5 765 471 and the rear axle from October 1963, Chassis No. 5 813 842 in addition to those from Boge or Fichtel and Sachs. The Hoesch shock absorbers differ externally from the other two types in that the protective tube is made of a PVC synthetic material. Contrary to the instructions for the Boge or Fichtel and Sachs types, these shock absorbers are installed with the protective tube at the **bottom.** The Hoesch shock absorbers are not supplied as spare parts.

The following shock absorbers supplied as spare parts can be installed together with the Hoesch type:

	Make	Part No.	Remarks
front axle	Boge Fichtel & Sachs	113 413 031 A 113 413 031 B	
front axle	Boge Fichtel & Sachs	111 413 031 J 111 413 031 K	These shock absorbers will be discontinued when stocks are exhausted.
rear axle	Boge Fichtel & Sachs	111 513 031 P 111 513 031 Q	

The following shock absorbers **must not** be installed together with the Hoesch type:

	Make	Part No.	Remarks
rear axle	Boge Fichtel & Sachs	111 513 031 F 111 513 031 G	These shock absorbers will remain available as spare parts.

Important

On vehicles up to Chassis No. 5 813 841, only Boge shock absorbers — Part No. 111 513 031 F — and Fichtel and Sachs shock absorbers — Part No. 111 513 031 G — are to be used on the rear axle.

3 - Special heavy duty shock absorbers which are not greatly affected by tropical heat, high continuous stress or flying stones are now available. These shock absorbers can be ordered as optional extras under the number "M 103".

The shock absorbers are identified by the following numbers:

Front	Color	Rear	Color
111 413 031 N/M	black	111 513 031 J/K	red-brown

These numbers, with the first letter N and J are the Part Nos.

The shock absorbers differ from the standard version in the following parts:

Outside diameter of outer tube:
54 mm (standard: 46 mm)

Larger diameter piston:
32 mm (standard: 27 mm)

Thickness of outer tube:
2.25 mm (standard: 1.0 mm)

Increased oil capacity.

Always replace all four shock absorbers when carrying out the installation subsequently. On the front axle the following additional parts are required:

Top:	2 spacer rings	111 413 391
	2 screws	111 413 401 B
Bottom:	2 spacer rings	111 413 391
	2 M 10 nuts (8 G)	N 11 010 4

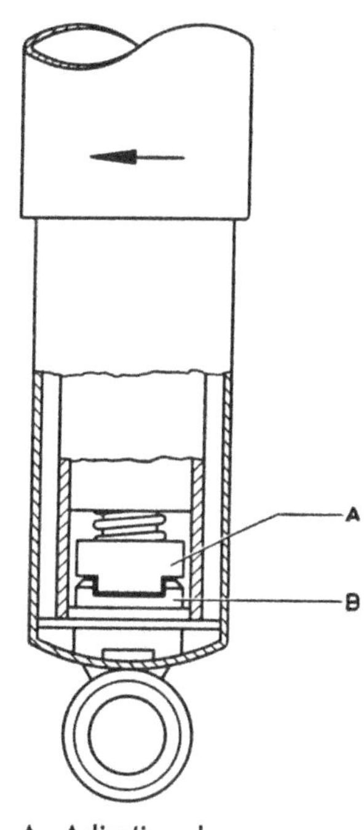

A - Adjusting dog
B - Bottom valve

4 - **Reinforced shock absorbers**

In addition to the standard shock absorbers from Messrs. Boge and Fichtel & Sachs, a reinforced type is available from Messrs. Koni. The Koni shock absorbers Part No. 211 413 031 B (front) and 111 513 031 E (rear) are particularly suitable for use in areas where road conditions are very bad.

Adjustment

1 - Fully compress the shock absorber and turn to the left until the adjusting dog engages in the slot in the bottom valve.

2 - Turn the shock absorber from the zero position (left stop) the prescribed amount to the right in the direction "heavy".

Front shock absorber: $1/2$—1 turn
Rear shock absorber: 1—$1\,1/2$ turns

Note:

Other settings are not recommended on account of the accompanying disadvantages.

5 - From Chassis No. 4 847 723 — Front Axle No. 4 818 287 (1st August 1962 the shock absorbers are secured at the top mounting by a hexagon bolt M 12 × 1.5 (previously M 10 × 1.5), Part. No. 111 413 403 A. As a result the thread in the anchor plate has also been altered.

The hexagon bolts M 12 × 1.5 can also be service installed. To do this the bush in the anchor plate must be re-bored with a 10.2 mm dia. (.40") drill and a M 12 × 1.5 thread cut in the bush.

Removing and Installing Shock Absorber Rubber Bushes

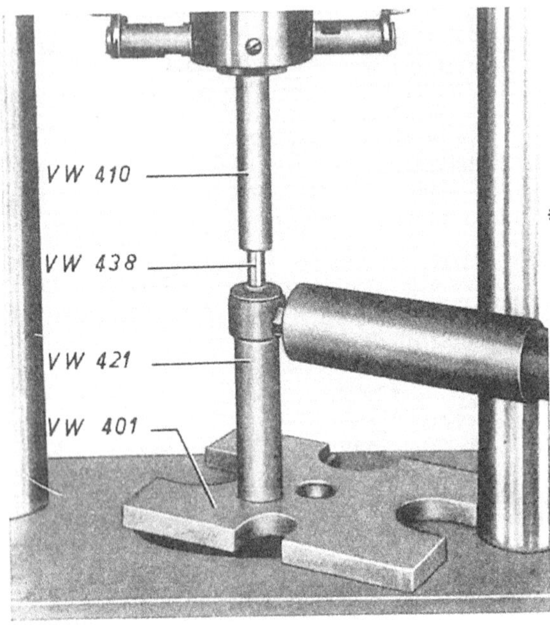

Removal

1 - Remove shock absorber.

2 - Press bush steere out on VW Repair Press with guide pin VW 438 in conjunction with VW 401, VW 410 and VW 421.

3 - Press bush out.

Installation

1 - Press bush in.

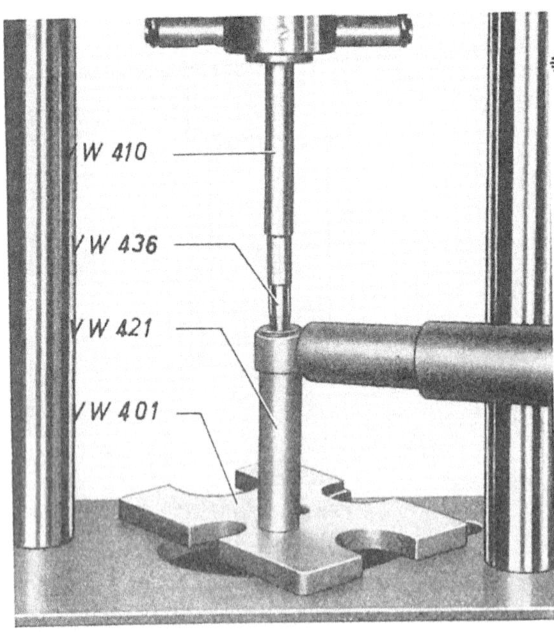

2 - Press bush steere in with guide pin VW 436 in conjunction with VW 401, VW 410 and VW 421.

3 - Install shock absorber.

Note:

Under Part No. SP 131 the Parts Department supplies a set of rubber bushes and sleeves for reinforced adjustable shock absorbers.

Damaged or worn rubber bushes and sleeves can now be replaced. The procedure is as follows:

1 - Press out worn or damaged sleeves on the repair press VW 400 with the help of the punch VW 410, cylindrical guide pin VW 438, tube VW 421 and thrust plate VW 401.

2 - Replace worn or damaged rubber bushes.

3 - Press in the new bushes with the punch VW 410, tapered guide pin, tube VW 421 and thrust plate VW 401. The tapered guide pin must be made in accordance with the sketch.

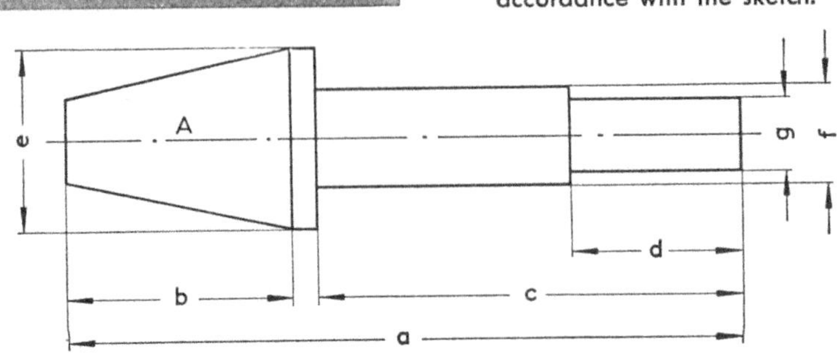

A = taper 1 : 2.1	b = 25 mm (0.98")	d = 20 mm (.78")	f = 11.5 mm (.45") dia.
a = 80 mm (3.14")	c = 50 mm (1.96")	e = 21 mm (.82") dia.	g = 8.5 mm (.33") dia.

Description of Steering

General Description

The rotation of the steering wheel is transmitted to the steering gear through the steering column and the coupling disc. The adjustable steering worm is carried in ball bearings within the steering gear case which is attached to the upper front axle tube. The hemispherical steering sector is freely located in the concave recess of the sector shaft. Tie (track) rods connected to the end of the drop arm (pitman arm) carry the steering motion to the steering arms at the front wheels.

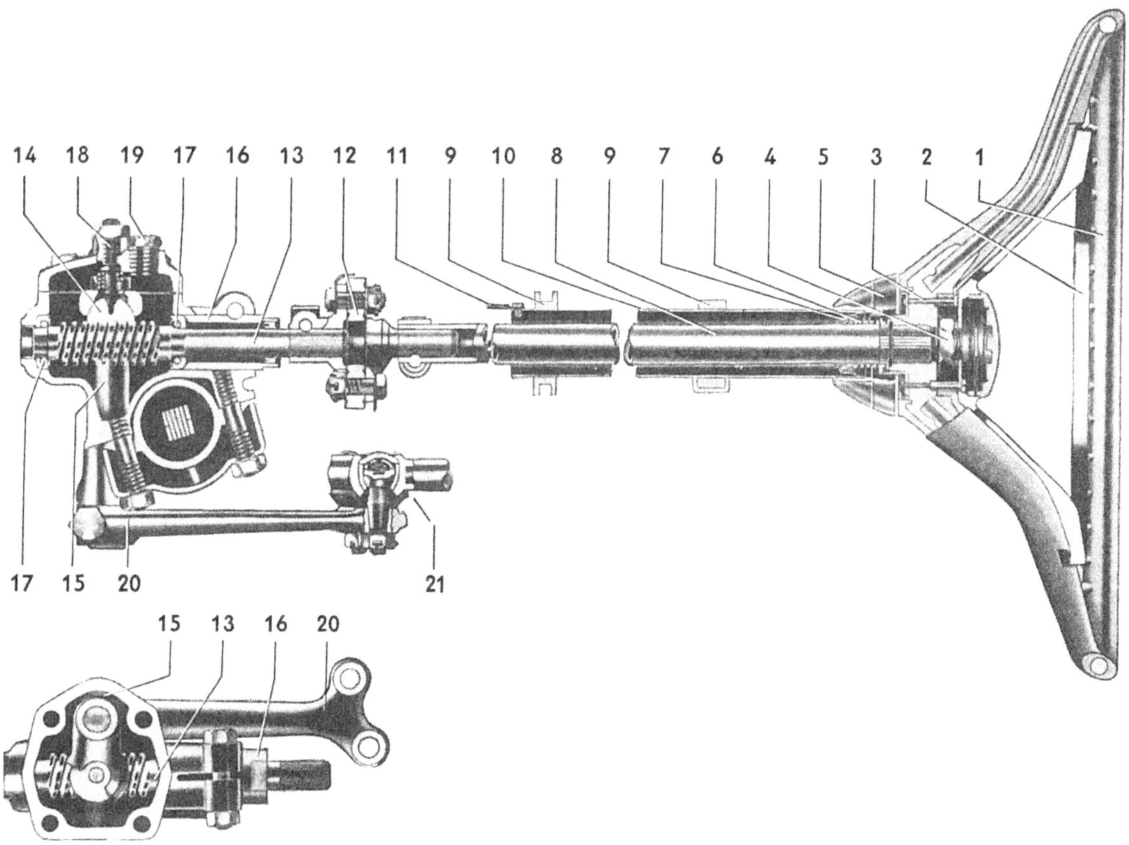

Volkswagen Steering
(Sectional View)

1 - Steering wheel
2 - Horn ring
3 - Contact pins
4 - Steering wheel nut
5 - Flashing indicator switch
6 - Contact spring
7 - Ball bearing
8 - Steering column
9 - Rubber mounting for steering column tube
10 - Steering column tube
11 - Connection for horn cable
12 - Coupling disc
13 - Steering worm } Steering gear set
14 - Sector } Steering gear set
15 - Sector shaft } Steering gear set
16 - Adjusting sleeve (axial play)
17 - Ball bearing
18 - Adjusting screw (radial play)
19 - Oil filler plug
20 - Drop arm (Pitman arm)
21 - Tie (track) rod end

Lubrication

The steering gear should in conformity with VW lubricant specifications be lubricated exclusively with transmission oil of the specification SAE 90, under no circumstances with grease or other oils. It is accessible through a hand-opening underneath the spare wheel. The oil level should be checked regularly in accordance with the lubrication chart. The level should be kept to the lower edge of the threaded hole for the filler plug. An oil change is not required.

The tie (track) rod joints should be lubricated together with the other lubricating points of the front axle at intervals prescribed on the lubrication chart.

Maintenance

The steering gear adjustment should be checked at the intervals specified in the maintenance chart.

Free play in the steering must not be too excessive. The play in the steering system should be kept down to a minimum, but the wheel should automatically return to the straight-ahead position after turning a corner. Regular attention in accordance with the specification after the prescribed mileage considerably influences the free play, while retaining steering ease and extending the service-life of the steering.

A regular check of the toe-in insures that the front wheels remain parallel, preserves the good riding qualities of the car and reduces tire wear to a minimum. If it is suspected that the front axle has been affected by an accident, all steering linkage components should be thoroughly checked functionally whilst at the same time checking the vehicle measurements.

Removing and Installing the Steering Damper

Removal

1 - Raise or support vehicle on trestles and take off front wheels.

2 - Bend up the hexagon head screw locking plate on the bracket at the axle beam and remove screw.

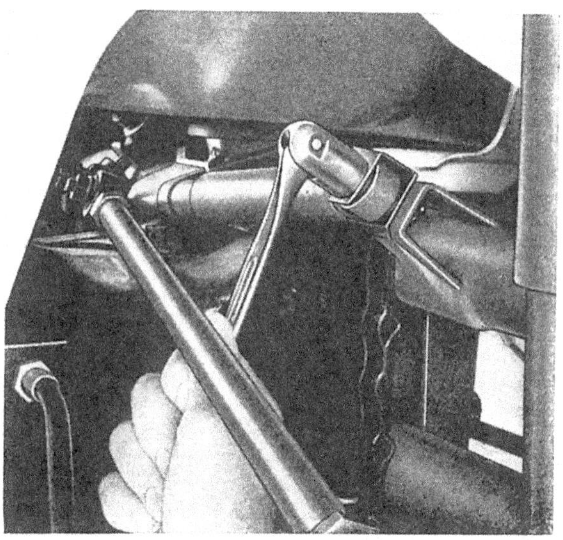

3 - Remove two nuts at tie rod eye and take steering damper off.

Installation

1 - Check the damper by hand. It should move slowly and uniformly. If there is a "flat spot", the damper must be replaced.

Important

There are two different types of steering damper which are installed intermittently and supplied under the same part number. The difference is as follows:

1st type: two telescopic tubes (Stabilus).

2nd type: compensator cylinder at right angles to damper body, and chrome plated piston rod.

Only the 1st type steering damper can be installed in all R.H.D. vehicles.

Note:

From Chassis No. 3 447 652 (1st December 1960), only the following shock absorbers are installed:

LHD: Part No. 113 415 901 B (Boge)

RHD: Part No. 141 415 901 (Hemscheidt)

From Chassis No. 5 123 818 (13th Oct. 1962), left-hand drive vehicles are fitted with shock absorbers from Messrs. Hemscheidt (Part No. 113 415 901 A) and Boge (Part No. 113 415 901 B).

2 - Check bush and rubber mounting on steering damper and tie rod eye for wear, replace if necessary.

3 - Tighten the inner flat hexagon nut at tie rod eye to between 2.5 and 3.0 mkg (18 and 22 ft. lbs.) and then secure with outer nut.

4 - Install the hexagon screw at the axle beam bracket with a new locking plate. The locking plate must be installed so that the opening of the "U" shaped plate points towards the front of the vehicle and the short lip bears on the axle beam bracket. Tighten the screw to between 2.5 and 3.0 mkg (18 and 22 ft. lbs.) and secure it.

Removing and Installing Steering Damper Rubber Bushes

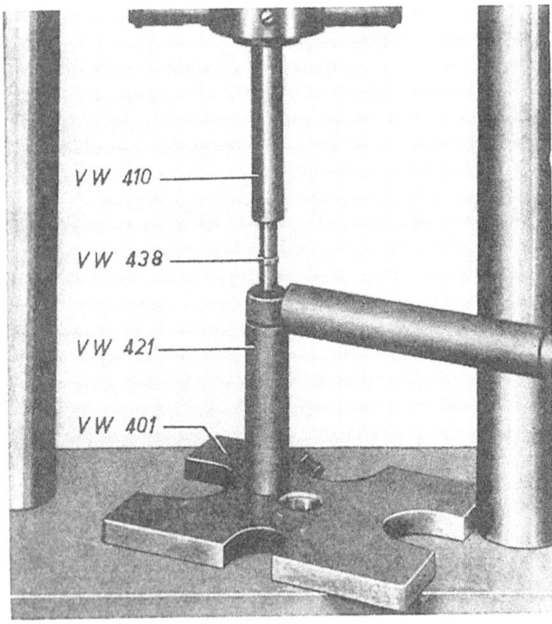

Removal

1 - Remove steering damper.

2 - Press the sleeve out of the bush on the Repair Press VW 400 with the cylindrical guide pin VW 438 in conjunction with VW 401, VW 410 and VW 421.

3 - The rubber bush can usually be pressed out of the eye without tools.

Installation

Reverse the above procedure, but observe the following:

1 - Place the sleeve in the rubber bush and press it into position with a vise.

2 - Tighten the steering damper as prescribed and secure it.

Removing and Installing Tie Rods

Removal

1 - Jack up the car and remove front wheels.

2 - Take off the steering damper at the axle beam bracket.

3 - Remove nuts of ball studs.

4 - Press ball studs out with special tool VW 266 f.

Remove tie rods with steering damper.

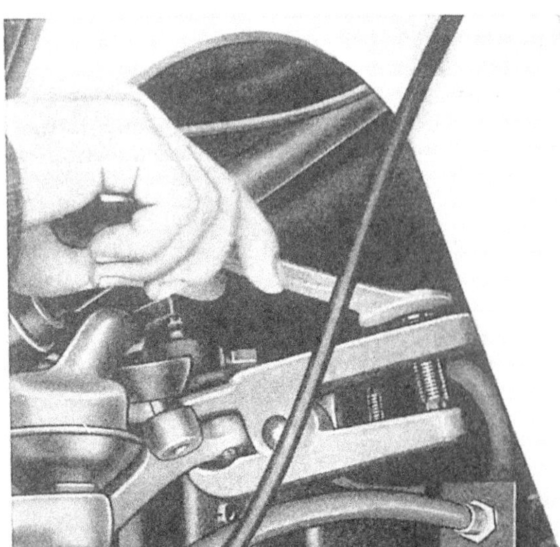

Installation

This a reversal of the above, but the following points should be observed:

1 - Check tie rods for distortion and damage. Bent or damaged tie rods should, for safety reasons, be renewed and not repaired.

2 - Check tie rod ends for wear. If necessary, replace individual tie rod ends or complete tie rods.

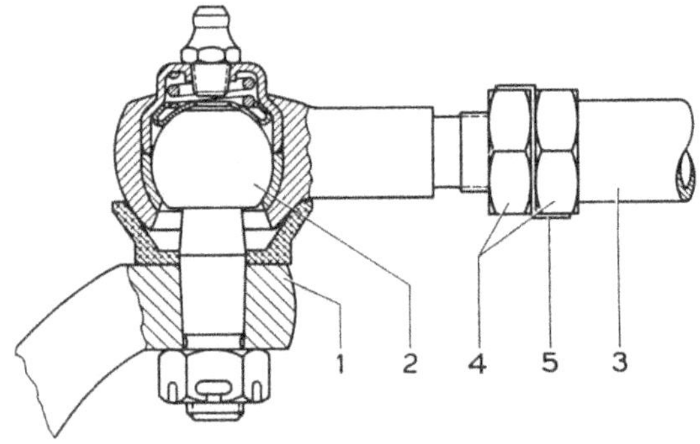

1 - Steering arm at steering knuckle
2 - Tie rod end
3 - Tie rod end
4 - Nut
5 - Locking plate

3 - Damaged or clogged up grease nipples should be exchanged; angle grease nipples are to be fitted to the inner tie rod joints.

4 - Install short (with left-hand steering the left) tie rod so that the bent end is attached to the steering drop arm.

5 - Check the steering damper bushes for wear, replace if necessary.

6 - Attach the steering damper to the tie rod and axle tubes as described.

7 - Tighten ball stud nuts and secure.

8 - Lubricate the joints and adjust toe-in.

Note:

From Chassis No. 4010995, all models of the Volkswagen are equipped with maintenance-free tie rods to simplify servicing.

When existing stocks are exhausted, only the maintenance-free tie rods will be supplied as spares. Installation in previous vehicles is possible.

The Tie Rod End Remover VW 266h is intended for pressing off the maintenance-free tie rod ends. It can also be used for removing all other tie rod ends.

Removing and Installing Steering Wheel

Removal

1 - Remove ornamental horn ring cover.

2 - Screw off steering wheel nut and take off steering wheel.

Note:

From August 1963, Chassis No. 5677119, the steering wheel — Part No. 113415651 EF — was modified on all models except 111, 112, 115 and 116. The two vertical recesses in the spoke for the horn ring have been discontinued. On the new steering wheel—Part No. 113415651AF — the horn is operated by means of a lever — Part No. 311051531.

The previous type of horn ring — Part No. 113951531 D — and the steering wheel — Part No. 113415651 EF — will remain available as spare parts.

Service installation

The new steering wheel and horn lever can be installed in former vehicles without difficulty.

Installation

The installation is a reversal of the removal, but the following points should be observed:

1 - Make sure contact ring is seated correctly. With wheels in straight-ahead position the contact ring notch has to point exactly to the right.

2 - Install steering wheel with front wheels in straight-ahead position, make sure that prong of actuating ring of the self-cancelling switch seats properly in notch of contact ring and that steering wheel spokes are horizontal.

3 - Insert spring washer and tighten steering wheel mounting nut to a torque of 5 mkg (36 ft. lbs.).

Removing and Installing Steering Column

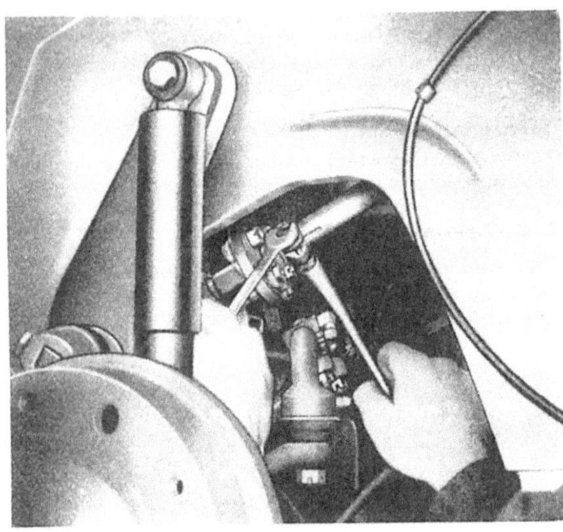

Removal

1 - Unlock hexagon nut on steering column bracket, unscrew it and remove bracket.

2 - Pull off the horn wire from the steering column tube contact.

3 - Slacken the steering wheel nut slightly and withdraw the column with steering wheel out of the column tube. Take off steering wheel.

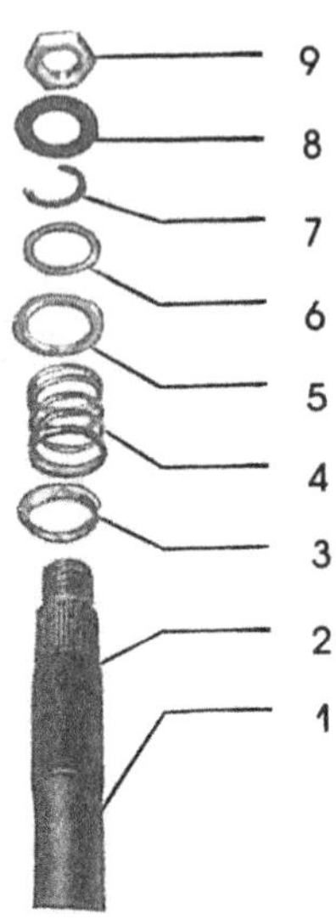

4 - Press out circlip on steering column. Take off plastic washer, contact ring, contact spring and spring seat.

1 - Steering column
2 - Insulating sleeve
3 - Spring seat
4 - Contact spring
5 - Contact ring
6 - Plastic washer
7 - Circlip
8 - Spring washer
9 - Steering wheel nut

Installation

The installation is a reversal of the removal, but the following points should be observed:

1 - Check operational ease of steering column ball bearing in steering column tube, if necessary install new ball bearing.

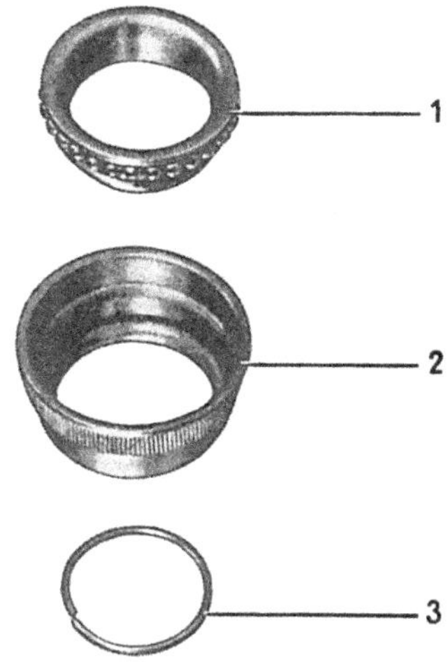

1 - Inner race with balls
2 - Outer race
3 - Circlip

Note:

The steering column ball bearing is filled with a special grease and does not require any maintenance. If the ball bearing has to be exchanged, it can be pushed upward from the lower end of the steering column tube towards the top and out.

2 - Put spring seat, contact spring, contact ring and plastic washer onto steering column and secure with circlip.

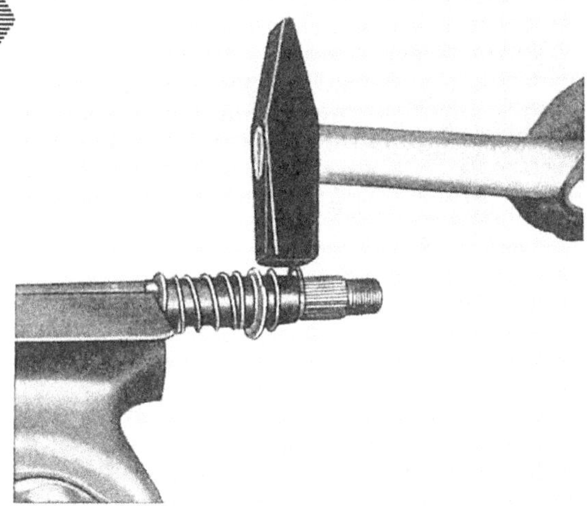

3 - With front wheels in straight-ahead position, turn notch of contact ring so that it points exactly to the right.

4 - Make sure the flashing indicator switch is correctly axially aligned.

5 - Push on the horn wire at the steering column tube.

6 - Install steering column bracket with new lock plate. Tighten and secure clamping screw.

Removing and Installing Steering Column Tube

Removal

1 - Pull off the horn wire from the steering column tube contact.

2 - Take off the steering wheel.

3 - Take off flashing indicator clamping screw. To this end it is necessary to pull off the individual wires at the flasher relay and to pull out the wire leading to the switch from the instrument panel.

4 - Remove circlip and take off washers, contact spring and spring seat.

5 - Remove bracket at the steering column tube.

6 - Withdraw steering column tube from the rubber mounting. The tube should be slightly turned when pulling.

Installation

This is accomplished by reversing the preceding operations, but the following points should be noted:

1 - Check operational ease of ball bearing, replace if necessary.

2 - Check upper and lower rubber mounting of steering column tube and replace, if damaged.

3 - Adjust the self-cancelling flashing indicator switch as follows:

a - Before tightening the mounting screw it should be insured that the self-cancelling switch is inserted right up to the stop on the steering column tube.

b - The distance between cancelling switch and steering wheel hub should not exceed 2 + 0.5 mm. The distance is to be adjusted with the steering wheel installed and the self-cancelling switch properly fixed by pushing the steering column tube further up or down.

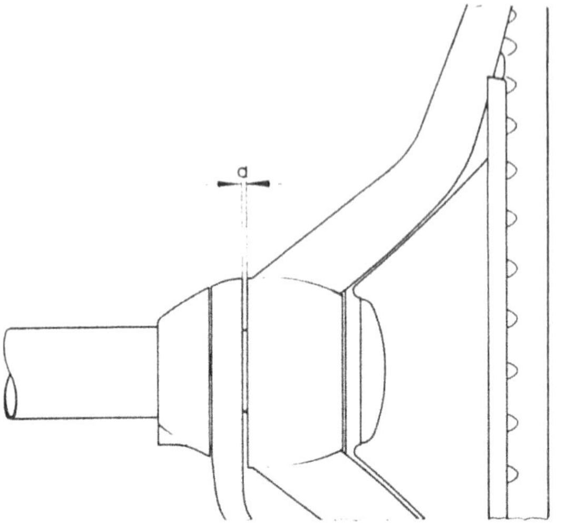

a = 2.0 + 0.5 mm

4 - Slide the spring seat, contact spring and washers on to the steering column in correct sequence and secure with circlip.

5 - Push on the wire at the flasher relay and column tube contact.

Removing and Installing Steering Gear

(Front axle in situ)

Removal

1 - Raise or support vehicle on trestles and remove left front wheel (R. H. D.: right front wheel).

2 - Press out ball studs at steering drop arm (pitman arm), using special tool VW 266f.

3 - Remove clamping screw at the steering column clamp.

4 - Pull off the wire from the steering column tube contact.

5 - Withdraw steering column from the coupling flange.

6 - Remove steering gear case mounting clamp.

7 - Remove steering gear.

Installation

The installation is a reversal of the romoval procedure, but the following points should be observed:

1 - Check and adjust steering gear. If necessary, disassemble steering gear and exchange damaged or worn components. Check seating steering drop arm (pitman arm) on the sector shaft. Observe the clearance 6 = 0.4—1.0 mm (.016"—.04") between drop arm and steering gear case.

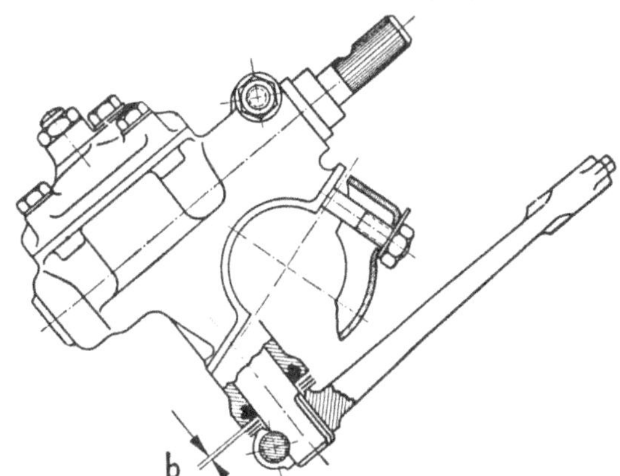

b = 0.4—1.0 mm (.016"—.04")

2 - Place the steering gear between the two shoulders welded on to the axle tube and screw up tight. The shoulders restrict the position of the steering gear so that the distance between the center of the oil filler plug and the mid-point of the front axle assembly amounts to 260 mm (10.24") (a). The inclination of the steering worm is about 25°. Ensure that the worm and steering column are properly aligned, and that the steering column does not foul the steering column tube.

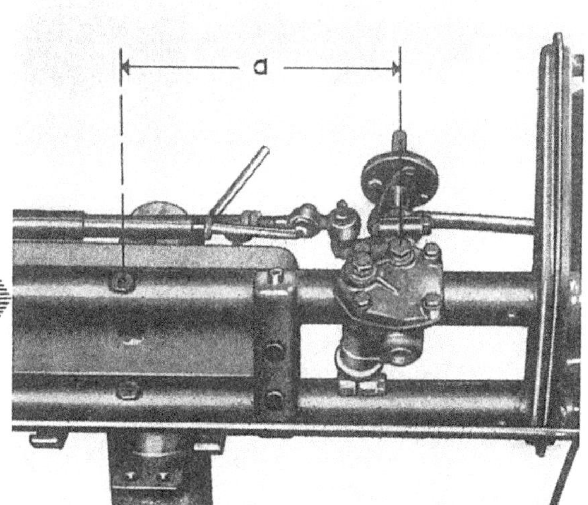

Important
In the case of VW Passenger Cars with right-hand drive the distance a should be 224—226 mm (8.82″ to 8.90″).

3 - Tighten up the nuts on the steering gear case with 2.5—3.0 mkg (18—22 ft. lbs.) and secure with new lock plates.

4 - Check oil level, if necessary top up to near the lower edge of the filler hole with oil of the proper specification.

5 - Fit the clip on the steering column tube with a new lock plate. Tighten up the clamping screw and lock.

6 - Check and, if necessary, adjust toe-in.

Important
Each time the steering gear has been re-installed, or if the position of the steering gear has been altered, it is absolutely necessary to check the toe-in.

Disassembly and Assembly of Steering Gear

To facilitate these operations it is recommended to attach the steering gear to a tube having the size of the front axle tube.

Disassembly

1 - Release clamping screw at steering worm and withdraw complete steering column coupling.

2 - Remove bolt that attaches the steering drop arm (pitman arm) to the sector shaft and pull off steering drop arm.

3 - Release the four screws of the steering gear case cover and take off cover.

4 - Take off thrust pin and spring from sector shaft.

5 - Withdraw sector shaft.

6 - Take off steering sector.

7 - Remove clamping screw at adjusting sleeve.

8 - Withdraw steering worm with adjusting sleeve, oil seal and upper thrust bearing.

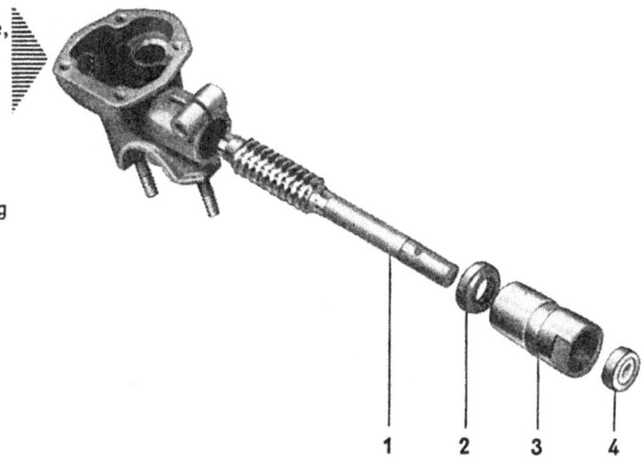

1 - Steering worm
2 - Upper thrust bearing
3 - Adjusting sleeve
4 - Oil seal

9 - Press out end cap at lower thrust bearing.

10 - Press out lower thrust bearing from outside towards the interior of the steering gear case.

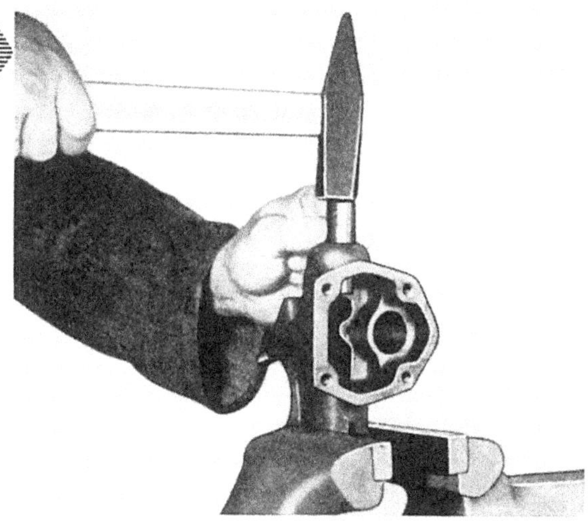

Assembly

Assemble in the reverse sequence, observing the following points:

1 - Thoroughly clean steering gear case and components.

2 - Correct steering operation can only be ensured if the surface condition of the worm, sector and sector shaft are perfect and tooth contact is good. When new, the steering sector bears only on the base and the backlash amounts to 0.01—0.02 mm (.0004"—.0008"). After some time in service, the backlash will disappear and the teeth will also bear at the flanks.

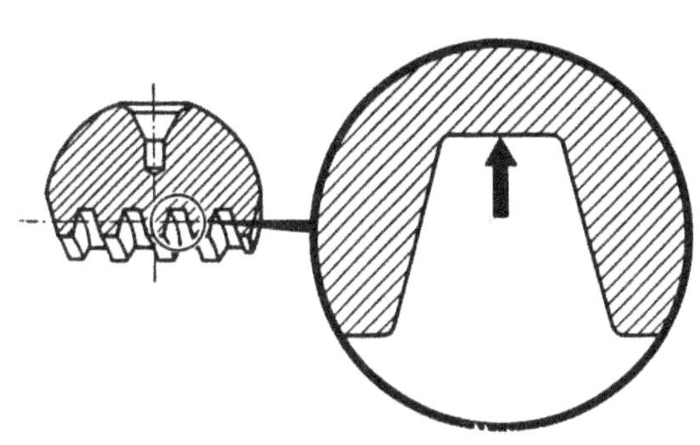

To check the tooth contact is the best way to ascertain if the steering operation is correct. When doing this, however, engineer's blue should be applied lightly. The tooth contact between sector and the recess in the sector shaft should be checked in the same manner.

The contact surfaces on the worm for the thrust bearings should not show sign of excessive wear or damage.

If the worm, sector and recess in the sector shaft show considerable wear or poor tooth contact, the parts must be replaced. Spares of these three parts are only available in sets. However, it is important to note that steering worm and sector must only be replaced in pairs.

3 - Inspect thrust bearings for wear and replace, if necessary. Install lower bearings on the VW Repair Press, using thrust disc VW 412 in conjunction with VW401 and VW417. Never install the bearings by hammer blows. The thrust surface of the tube VW417 for pressing in the lower bearing must only contact the outer race of the bearing.

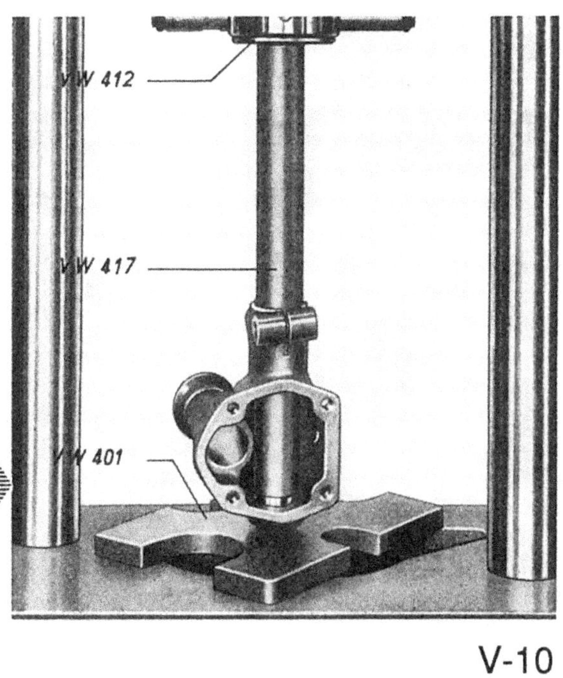

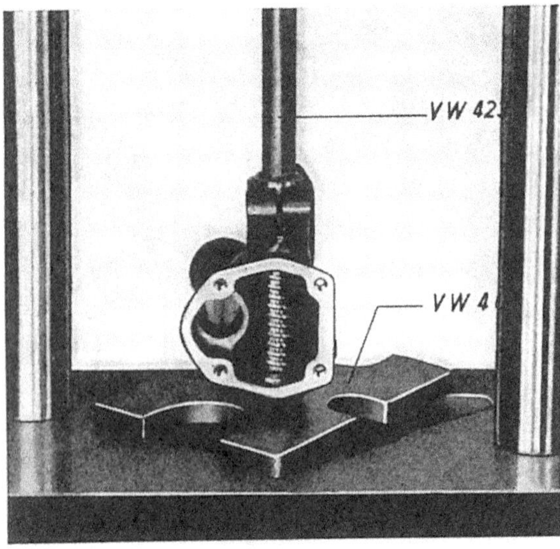

4 - With the steering worm shaft installed, press in upper thrust bearing on the VW Repair Press with tube VW 423 in conjunction with tube VW 412 and VW 401.

5 - Examine steering gear case cover gasket and sector shaft oil seal, renew them if necessary.

6 - Observe the installation direction when installing a new seal in the adjusting sleeve. The seal must be installed to a depth of 10.0—11.5 mm (.393″—.452″) in the adjusting sleeve.

7 - Remove burr at the sector shaft splines before installing it. Moreover, ensure that the contact surface for the lip of the seal is smooth and clean.

Avoid turning the sector shaft when inserting it in the steering gear case as otherwise the lip of the seal may be damaged.

8 - When installing thrust pin and spring for sector shaft, make sure that they are within the dimensions given below:

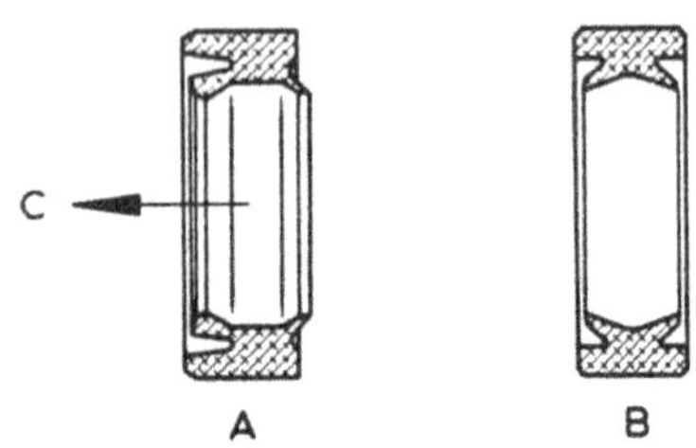

A = 1st type
B = 2nd type
C = Direction of assembly

$$\text{Spring} = \frac{23.8}{23.0}\ \text{mm}\ \frac{.94''}{.90''}$$

Load on
assembled spring = 60—75 kg/132—165 lbs.

$$\text{Pin} = \frac{20.1}{19.9}\ \text{mm}\ \frac{.79''}{.78''}$$

Thrust pins and springs not within the above limits result in mal-adjustment of the steering and excessive wear of the components.

9 - Tighten up the bolt of the drop arm with 6.5—7.5 mkg (47—54 ft. lbs.).

10 - To insure a proper sealing of the steering gear case, sealing compound should be applied as follows:

a - At stud holes in steering gear case. Additionally, dip stud ends in sealing compound.

b - Seating surface of end cap at lower thrust bearing.

c - Steering gear case cover gasket.

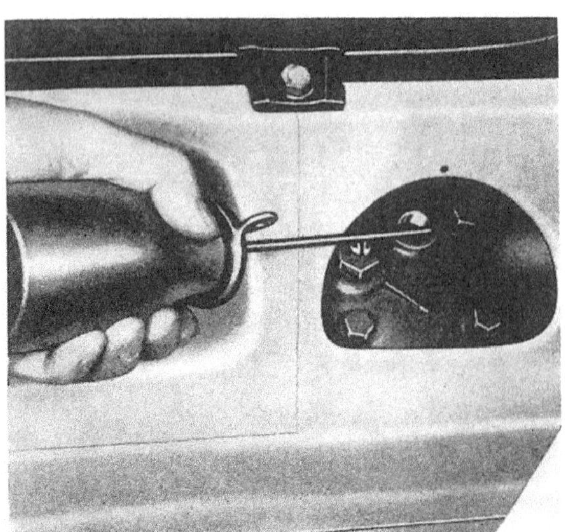

11 - Fill steering gear case with 125 c. c. (.26 U. S. pint; .22 Imp. pint) transmission oil SAE 90.

General Note

The play between steering worm and concave recess in the sector shaft is determined by the play between the thrust pin and sector shaft. This play is adjusted by means of the adjusting sleeve and must not exceed 0.2 mm (.008").

A careful adjustment is absolutely necessary to obtain a proper meshing of the sector with the worm and a good seating of the sector in the sector shaft press.

This will ensure that wear of the steering components is kept at a minimum. The play should be kept at the low limit while taking care that the steering does not bind when turning the steering wheel in both directions as far as it will go. The front wheels should automatically return to the straight-ahead position after having taken a turn.

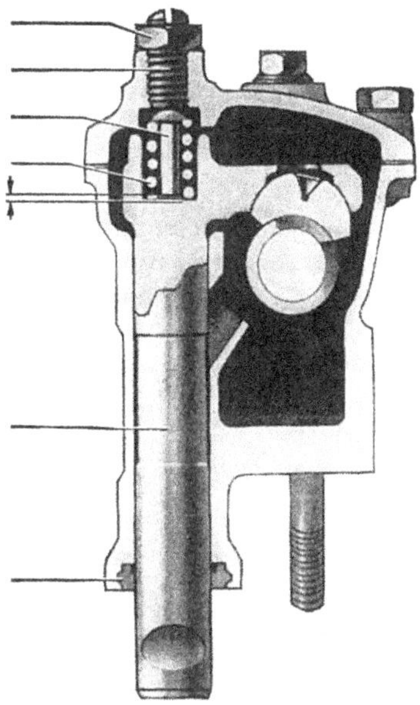

a = max. 0.2 mm (.008")

1 - Lock nut
2 - Adjusting screw
3 - Thrust pin
4 - Spring
5 - Sector shaft
6 - Oil seal

Adjustment (Steering gear removed)

After installing the adjusting sleeve with seal, the adjusting sleeve is turned till the clamping screw can be inserted. The nut is screwed loosely on to the clamping screw.

Worm-Shaft End Play

1 - Turn the steering worm to and fro and tighten adjusting sleeve clockwise till no end play can be felt and the worm can still be turned easily.

2 - The adjusting sleeve presses against the upper bearing. This pressure is also imposed on the lower bearing via the worm. If the adjusting sleeve is overtightened, the bearings will be damaged and stiffness will result in the steering gear.

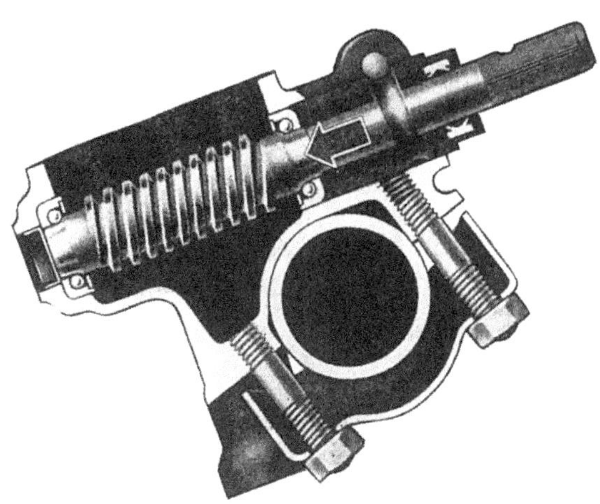

3 - Tighten adjusting sleeve clamping screw with nut.

Sector-Shaft End Play

4 - When adjusting the radial play turn the worm so that the sector shaft arm is at right angles to the axis of the worm. With the steering gear installed the front wheels are then exactly in the straight ahead position.

5 - After installing the thrust spring and pin and tightening the cover, the radial play is adjusted as follows: Tighten the adjusting screw as far as it will go and back it off approximately 1/4 of a turn (90 to 70°).

6 - Hold adjusting screw and tighten lock nut.

> **Important**
> Make sure that the adjusting screw is backed off the proper amount as this ensures that the correct play is obtained between thrust pin and sector shaft.

After the adjustments are completed and the steering gear has been tightened on the axle beam, check the steering lock by turning the steering wheel in both directions whilst the vehicle is still on the trestles.

Adjustment (steering gear installed)

The steering gear adjustment should be checked regularly during the maintenance service. The vehicle should be raised or supported on trestles and inspected as follows:

1 - Back off lock nut and sector shaft adjusting screw.

2 - Back off the adjusting sleeve clamping screw and tighten the adjusting sleeve clockwise until no end play is felt at the worm.

> **Important**
> If tightened excessively the thrust bearings will be damaged.

3 - Tighten the clamping screw and adjusting sleeve nut.

4 - Turn the wheels straight ahead. In this position the sector shaft arm is at right angles to the worm. Tighten the adjusting screw as far as it will go and back it off approximately 1/4 of a turn (90 to 70°).

5 - Hold adjusting screw and tighten lock nut.

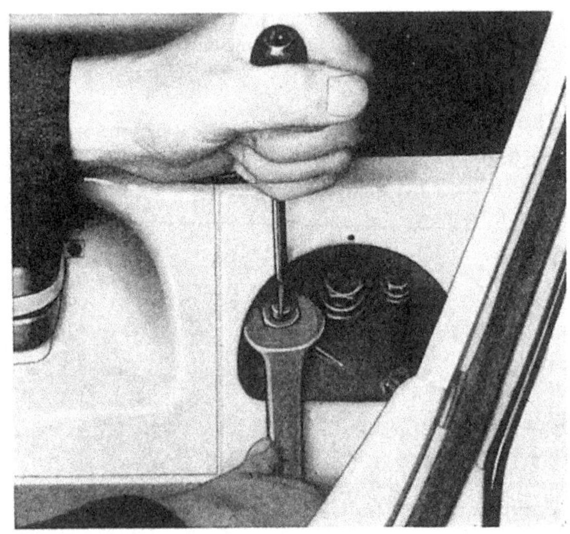

6 - After completing the adjustment, check the steering lock by turning the wheels in both directions.

If the steering feels hard, despite the fact that the adjustments have been carried out as described above, the steering gear should be removed and disassembled for checking all the components.

Steering Trouble Shooting

Symptom	Cause	Remedy
Hard Steering Steering is equally stiff from lock to lock, jams or does not automatically selfcenter	a - Front axle inadequately lubricated	a - Jack up front end of car and thoroughly lubricate front axle
	b - King pins stiff or seized	b - Jack up car and disconnect tie rods. Try to make stiff king pins move freely by thoroughly lubricating them. If necessary, remove steering knuckle and exchange damaged components. Install steering knuckle and thrust bearing with no end play or a preload up to 0.04 mm (0.0016″)
	c - Steering gear maladjusted	c - Check steering gear adjustments. Adjust sector shaft and worm shaft end play as prescribed. If necessary, exchange steering set (sector shaft, sector, and worm). Check oil level
	d - Steering sector does not match the steering worm. The sector does not bear on the root when new, but only on the flanks, allowing it to tilt during operation	d - Check tooth contact between steering worm and sector, if necessary replace steering gear set (worm, sector, and sector shaft). In less severe cases, a remedy may be effected by carefully lapping the sector to the worm. To do this, clamp the worm in the chuck of a lathe and use fine grinding compound. Prior to reassembly, remove all traces of the grinding compound carefully and recheck tooth contact.
	e - Ball-shaped surface of sector or concave recess in sector shaft damaged	e - Renew steering gear set. Check impression of contact before final assembly
	f - Sector shaft seized up in bore of steering gear case	f - If only slight signs of seizure are evident, these should be carefully removed and the shaft polished. Permissible clearance between shaft and bore: 0.040—0.082 mm (.0016″—.0032″). If signs of seizure are severe, replace the steering gear set and case

Symptom	Cause	Remedy
	g - The spring pressure on the sector shaft is too high: Sector shaft end play improperly adjusted, thrust pin too short, thrust spring too long or has too much tension Shaft length of thrust pin 20.1/19.9 mm (.79"/.78") Free length of thrust spring 23.8/23.0 mm (.94"/.91") Length and load of installed spring: 60—75 kg (132—165 lbs.) 20.3 mm (.80")	g - Check parts and replace as found necessary. The spring may be ground off so that its end is only 0.9—1.5 mm (.035"—.059") above the thrust pin face in the unloaded condition
	h - Steering worm thrust bearing cups damaged, restricting the movement of the balls	h - Check thrust bearings for free movement, replace if necessary. Care should be exercised when pressing the end cover and the upper thrust bearing into the steering gear case. Sealing compound should not enter the lower thrust bearing. The outer edge of the upper thrust bearing cup should be free from burrs in order not to scrape metal chips off the wall of the bore when pressing the cup in position. Such metal chips would enter the lower thrust bearing
Steering stiff in every position	a - Ball bearing in steering column tube damaged or stiff	a - Install new ball bearing
	b - Steering column fouling the ball bearing	b - Check if the steering column and steering gear are correctly positioned, correct if necessary. Check toe-in afterwards
Front wheels do not self-center, although there is no binding in the steering system	a - Front wheels improperly adjusted	a - Check and adjust front wheel alignment (caster, camber and toe-in)
	b - Steering arms of stub axles (steering knuckles) bent or twisted	b - Remove stub axles and check them for bends and twisting with the gauge VW 258 a. Replace if necessary
Excessive Play in Steering System Excessive play in steering gear	a - Steering gear improperly adjusted	a - Check steering gear adjustments. Adjust sector shaft worm shaft end play as prescribed
	b - Steering set worn	b - Exchange steering set (sector shaft, sector, and worm)
Excessive play in tie rod joints	Ball studs worn	Exchange worn ball studs
Excessive play in front wheel suspension	Worn bearing points (torsion arms, torsion arm links, stub axle, and front wheel bearings)	Check adjustments of torsion arm *link pins and front wheel bearings.* Renew excessively worn parts

When the steering wheel is turned, the wheels rotate about the king pins until the torsion arm link of the wheel on the outside of the curve strikes the front boss of the stub axle where the bolt of the brake backplate is fixed. The design of the steering parallelogram of the VW passenger car is such that, when the steering wheel is turned, the wheel on the inside of the curve always moves through a bigger angle than the outer wheel. The angles at full lock should be as follows:

Model	Inside wheel	Outside wheel
De Luxe Sedan VW Convertible Karmann Ghia Coupé Karmann Ghia Convertible	34° ± 2°	28° - 1°

Hence the smallest turning circle is about 10.5 meters (34.5 ft.).

Deviations from the specified values of lock angle may be due to various causes and influence the riding properties and cornering of the car. The possible remedies are given in the following table.

Symptom	Cause	Remedy
1 - Turning circle too large at full lock on either side	Lock angle too small on both sides	Grind down the front boss on both stub axles for the mounting bolt of the brake backplate, until the lock angle is correct
2 - Turning circle too small at full lock on either side (wheels fouling fenders)	Lock angle too big on both sides	Substitute a longer mounting bolt M10 × 1.5, tighten up at 4.0—4.5 mkg (29—33 ft. lbs.). Grind off the projecting end of the bolt till the lock angle is correct
3 - Turning circle too large or too small at full lock on one side only	Lock angle of both wheels to big or too small on one side	Apply the remedy under 1 or 2 above to one wheel only. If the defect is only apparent on the left-hand lock, carry out the repair to the right-hand wheel and vice-versa
4 - Lock angle of one wheel to big or too small	Steering parallelogram not functioning properly	Carry out a dimensional survey of the car and check the difference in wheel angularity with the inside wheel turned 20°. Check the steering gear seating for clearance and inclination, also drop arms, tie (track) rods, steering arms of stub axle, for deformation. Take great care that the shorter tie (track) rod is the right length. For further information, see "Vehicle Measurement"

The lock angle can best be measured on a turnable which is on a floating mounting and can be locked, and is equipped with an angular scale from 0—40° in both directions.

THIS PAGE INTENTIONALLY
LEFT BLANK

Steering Geometry and Wheel Position

General Description

The correct functioning of all parts of the front axle and steering, and proper wheel position, are the main factors which govern the smooth riding qualities of the car. The following points are important.

1 - Correct play of all parts of the front axle and steering and freedom of movement, resulting from regular lubrication.

2 - Correct setting of steering gear.

3 - Correct running of front wheels i. e. correctly adjusted wheel bearings, and no excessive unbalance and inadmissible wheel run-out.

4 - Correct adjustment of the track of the front axle and the front wheels.

5 - Correct tire pressure, and even tire wear.

6 - Correctly functioning shock absorbers.

Front Wheel Alignment

The front wheels must be steered and guided in a certain manner when running. In accordance with the springing and steering geometry, the following factors influence car steering control:

1 - Camber of front wheels.

2 - King pin inclination, due to the inward tilt of the top of the king pin.

3 - Caster, formed by inclining the top of the king pin towards the rear of the vehicle as viewed from the side.

4 - Caster of the axle assembly, formed by the angle between the line through the axle tubes and the vertical plane perpendicular to the centre-line of the car.

5 - Toe-in with the front wheels in straight-ahead position.

This angular relationship between the front wheels, the front wheel attaching parts, and the frame insure easy steering and riding qualities.

If one angle is altered by wear or distortion of certain front axle components, the springing and steering geometry is disturbed, even if all the other angles are correct. Misaligned front wheels due to mal-adjustment, excessive wear or bending through a collision may result in:

1 - Excessive tire wear.

2 - Hard steering.

3 - Car wander or car pulling to one side.

4 - Poor road holding, especially on wet or icy roads, also poor cornering.

5 - Front wheel shimmy.

It is, therefore, of utmost importance to check the specified values for camber, toe-in and caster at regular intervals. Apart from adequate mechanical skill, it is essential to posses a good knowledge of the theory, if the testing and adjustment are to be carried out correctly.

Camber and King Pin Inclination

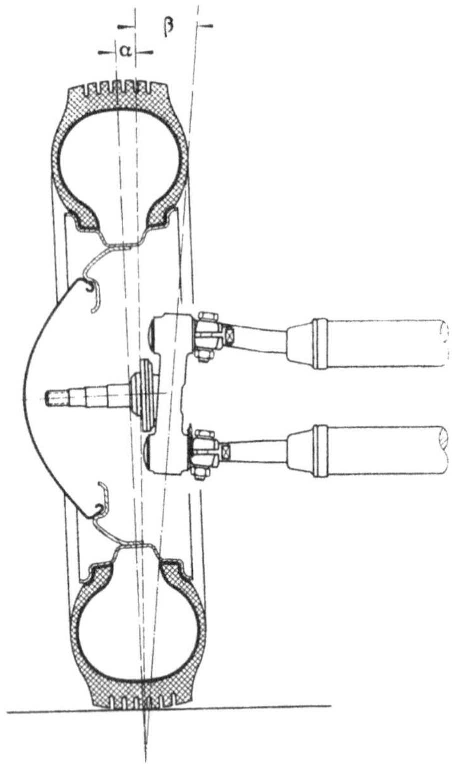

Angle α = Camber
Angle β = King pin inclination

Camber is the amount the wheel inclines away from the vertical at the top as viewed from the front of the vehicle. It is formed by the line through the stub axle not being at right angles to the line through the king pin. The angle formed by these lines is 90°+ approximately 5°. Of this amount 4° 20′ is the angle of the king pin inclination, and 40′ the camber angle.

The king pin inclination can be defined as the amount in degrees that the top of the king pin is inclined away from the vertical towards the center of the vehicle.

Camber and king pin inclination insure that there is a definite, favourable distance between point of contact of the tire with the ground and the point of intersection of the center-line of the king pin and the ground. Apart from reducing the shock imparted by the road to the steering, it also minimizes tire wear when the car is conering.

The king pin inclination as well as the caster, cause the front wheel camber to change in a certain respect when the wheels are turned.

Caster

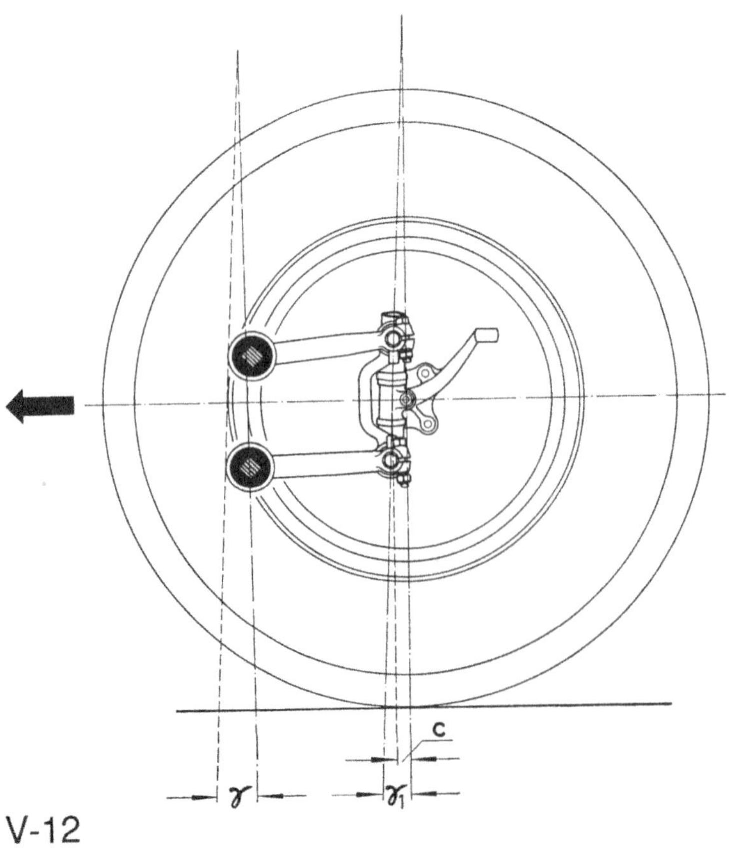

The caster is formed by

1 - the caster angle of the front axle beam,

2 - the amount the top of the king pin inclines away from the vertical, towards the center of the vehicle,

3 - the offset between king pin and stub axle.

Angle γ = Caster angle of front axle tubes.

Angle γ_1 = Caster angle of king pin.

Distance c = Displacement between king pin and wheel spindle.

The offset means that the turning point of the wheel is somewhat behind a line drawn through the king pin i. e., the wheels are trailing.

The king pin inclination also increases the stability of the front wheels because, when the vehicle is cornering, the side on the inside of the curve is slightly lowered, while that on the outside is slightly raised. The resistance to this motion returns the wheels to the straight-ahead position. Due to the caster angle, the camber of the inner wheel rounding a corner is greater, and of the outer wheel smaller as opposed to the camber angle in the straight-ahead position.

Insufficient caster increases the tendency of the car to give in to side forces (potholes, uneven road surface, side wind).

Too much caster makes steering difficult as increased force must be applied to turn the steering wheel to counteract the tendency of the wheels to remain in the straight-ahead position.

Toe-in and Toe-out

The camber and the rolling resistance tend to force the front wheels outward at the front. This is possible depending on the amount of play present in the front wheel bearings, wheel suspension and track rod joints. This tendency is compensated by the toe-in, in other words, the front wheels are adjusted to point inward at the front to insure that they roll parallel when the car is moving forwards. The circles on which the four wheels turn when rounding a curve should have a common center point which lies on a line extended from the rear axle.

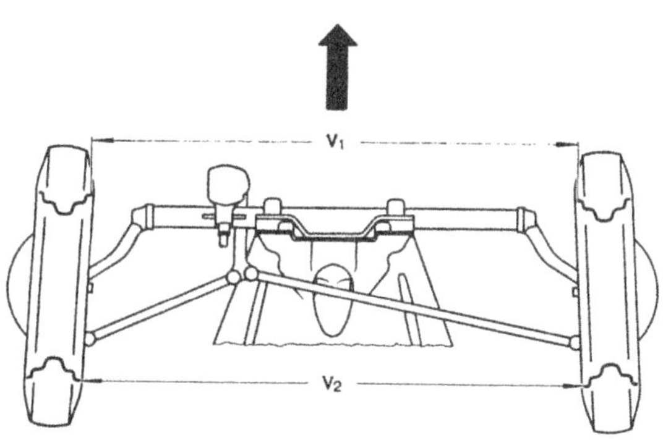

$v_1 < v_2$ = Toe-in

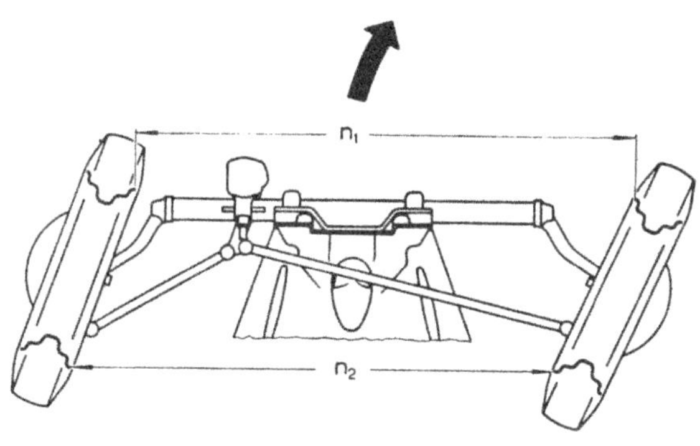

$n_1 > n_2$ = Toe-out

The steering is so constructed that these conditions are fulfilled. When cornering, the inner wheel is turned at a sharper angle and, therefore, describes a smaller circle than the outer wheel. When turning the wheel from the straight ahead position toe-in is reduced and eventually changes to toe-out.

If the steering is incorrectly adjusted or a steering arm is bent, scuffing occurs which leads to excessive tire wear.

THIS PAGE INTENTIONALLY
LEFT BLANK

Testing and Adjusting Position of Wheels and Toe-in

General Description

To test and adjust the position of the front wheels and their toe-in, there are a variety of alignment gages of different makes available, with which camber, caster, toe-in and difference in wheel angularity at 20° lock can be measured. A track gage is required for the front axle. Axle alignment gages are used for dimensional surveys. Further details regarding suitable gages for Volkswagen cars are obtainable from the Technical Service Department.

Camber and Caster can also be easily checked with the aid of a protractor, as described below.

Preliminary Checks

1 - Check tire pressures.

1—2 Occupants	front: 1.1 kg/sq. cm (16 lbs./sq. in.) rear: 1.4 kg/sq. cm (20 lbs./sq. in.)
3—4 Occupants	front: 1.2 kg/sq. cm (16 lbs./sq. in.) rear: 1.6 kg/sq. cm (23 lbs./sq. in.)

2 - Lift front end of car and check front wheels for run-out.

Permissible lateral and radial run-out: max. 1.5 mm (0.06").

3 - Check wheel bearing play; adjust if necessary.

4 - Lower the car and place it on a level surface.

5 - The horizontal position can be checked by taking two measurements with the protractor VW 245 a or a spirit level:

a - Place the gauge on both door openings. Take the mean value.

b - Place the gauge on the frame tunnel in front of the fork.

When measuring the vehicle longitudinally the inclination of the vehicle towards the rear must be considered. This angle is 30'. In the event of excessive deviations check the spring plate adjustment.

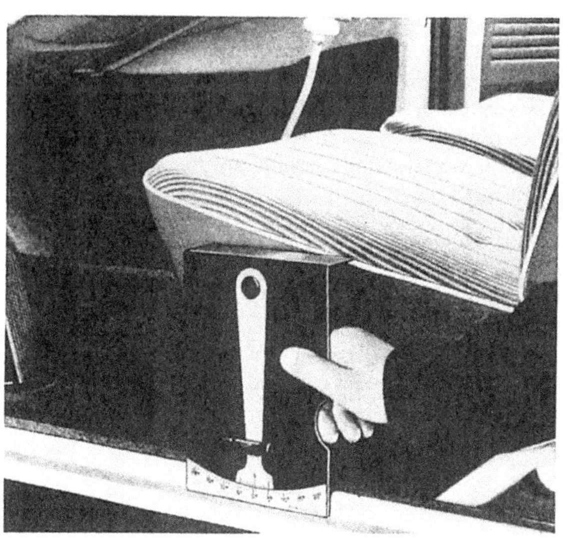

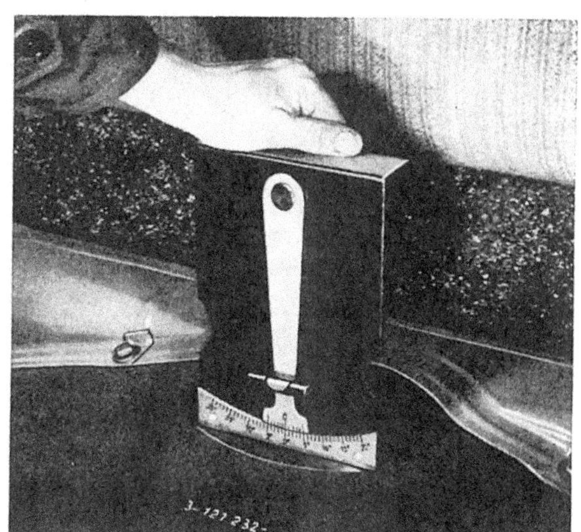

Checking Camber (0° 40′ ± 30′)

The camber of the front wheels is checked by means of the protractor VW 245 a with the car in an unloaded condition. The front wheels should exactly be in the straight-ahead position.

The camber angle is not adjustable.

Deviations from the specified value may be the result of wear or distortion of front axle components. If deviations are found, proceed as follows:

1 - Remove torsion arm link and stub axle (steering knuckle).

2 - Check torsion arm link pin bushes and shims for wear, renew if necessary.

3 - Check king pins and bushes for wear, renew if necessary.

4 - Check stub axle (steering knuckle) with gage VW 258 a.

5 - Check torsion arm link with gage VW 259 and a depth gage.

6 - Check displacement of torsion arms, If the displacement is in excess of the specified limit, the torsion arms may be twisted and should be checked on test plate VW 282 b.

7 - When installing torsion arm link, note proper arrangement of shims. Correction of the camber by an unauthorized arrangement of the shims is inadmissible as this may cause the torsion arm link pins to become jammed, which would result in premature wear.

Important
It is important for the camber to be as equal as possible on the two sides.

Checking Caster (2° ± 15′)

To check the caster, only the tilt of the front axle tubes is measured at both ends of the tubes by means of the protractor VW 245 a.

Deviations from the caster angle specified above may be due to the front axle tubes or the frame head being distorted. The adjustment of the rear torsion bars also influences the caster angle of the front wheels and should in this connection be checked.

Checking Toe-in (2—4 mm = .08"—.16")

It is advisable to check the front axle track when the vehicle is unladen

toe-in: 2—4 mm (0.08"—0.16")

with permissible total weight,

toe-in: 1—3 mm (0.04"—0.12")

The first step in checking toe-in is to make sure that there is no run-out of the front wheels and that wheel bearings, torsion arm link pins and tie rod joints are not excessively worn.

1 - Place the car facing forwards and bring the front wheels into the straight-ahead position.

2 - Force the two front wheels apart at the front to take up any play in the wheel suspension and tie (track) rods.

3 - Put the alignment gage in front of the wheels in such a position that the gage pins are against the wheel rims, roughly level with the center of rotation of the wheels.

4 - Set the gage dial to zero and remove the gage.

5 - Roll car forward by half a wheel turn and place alignment gage behind the wheels so that the gage pins bear equally against the wheel rims.

6 - Read off the measurement. It ought to be between 2 and 4 mm (.08" and .16").

Adjusting Toe-in

1 - Bend up locking plates at both ends of adjustable tie rod and slacken the nut.

2 - Adjust toe-in by turning the tie rod until the correct wheel position is reached.

3 - Tighten and secure the nuts. Check toe-in.

Measuring Difference in Wheel Angularity

It is necessary to measure the difference in wheel angularity when checking the steering parallelogram for correct operation. The measurement is read with the wheels on a 20° lock to right and left. In each case the wheel on the outside of the curve will be inclined at an angle less than 20°. The difference in wheel angularity is the difference between 20° and the inclination of the outer wheel, and is always measured on the outer wheel.

The difference in wheel angularity for all models of VW Passenger Cars is:

$2° \pm 30'$.

The difference in wheel angularity can only be reliably measured with an axle alignment gage suitable for use with Volkswagen cars. Further information may be found under the heading "Vehicle Measurements".

THIS PAGE INTENTIONALLY
LEFT BLANK

Vehicle Measurement

Correct wheel alignment is of paramount importance to the road and curve holding of the car. Inaccurate wheel alignment, or inadmissible track values, not only have an adverse influence on road holding ability and cornering safety but are also prone to cause abnormal tire wear.

If the car should act strangely on the road, show signs of tire wear, or it is thought that the wheel alignment has been affected by accident damage or other operational influences, the car should be checked while conducting a dimensional survey as regards steering geometry and wheel alignment and, if necessary, readjusted.

On cars which have been damaged in an accident it may be necessary to check whether the front axle is perpendicular to the longitudinal axis of the car, at the same time checking the front axle assembly for distortion. By measuring the wheelbase on both sides (2400 ± 4 mm/94.5 ± 0.16″) it is possible to discover any serious deviation by the axles. If the frame is obviously damaged, it is advisable to inspect for torn metalwork or parted welds, for safety's sake. A damaged frame head or front axle must be replaced. Deviation in the wheelbase on either side can possibly be corrected by adjusting the position of the rear wheels.

A proper check of the above cases nearly always demands a full survey of the car because inadmissible deviations from the prescribed values for the wheel position or improper functioning of one part of the suspension of one axle may adversely affect one wheel of the other axle. The survey can only be carried out on the Volkswagen with the aid of a Wheel Alignment Gage which operates independent of the car and possesses sufficient accuracy. Further details of Wheel Alignment Service Equipment suitable for VWs can be obtained from the Technical Service Department.

The prescribed values for a car survey are listed on the card P 6a and at the end of this section. The values are applicable only under the following conditions.

a - The tires are at the prescribed pressure,

b - The car is unladen,

c - The spring plates are correctly set.

For a survey of the car cards are available on which the limit values are printed, and on which the measurements can be entered. The manner in which the alignment equipment functions, and the procedure to be adopted, are given in the instructions issued by the manufacturer of the particular equipment. The Volkswagen must be measured at the following points:

1 - Toe-in with wheels not forced inward.

2 - Toe-in with wheels forced inward at the rear.

3 - Camber of wheels in straight-ahead position.

4 - Difference in wheel angularity with wheels turned 20° to the right. Turn the steering to the left until the right-hand wheel is at 20° to the straight-ahead position. Measure difference in wheel angularity on left-hand wheel.

5 - Camber with wheels turned 20° to right.

Measure camber on right-hand wheel. Turn steering further to the right till left-hand wheel is also turned 20°. Measure camber on left-hand wheel.

6 - Difference in wheel angularity with wheels turned 20° to the left. Turn the steering to the left until the left-hand wheel is at 20° to tha straight-ahead position. Measure difference in wheel angularity on right-hand wheel.

7 - Camber with wheels turned 20° to left.

Measure camber on left-hand wheel. Turn steering further to the left till right-hand wheel is also turned 20°. Measure camber on right-hand wheel.

8 - Measure position and angle of camber of rear wheels.

For further details see chapter H, section 9.

Re 1 and 2: The front wheels are correctly adjusted if toe-in (or out) is 10′ ± 10′ with wheels forced inward in the straight-ahead position, i. e. about 0 ± 1 mm. In order to set the toe-in absolutely correctly it is important to know the amount of the play in the wheel bearings and suspension, and in the tie (track) rod ends, so that the wheels can toe-out when running. The play is determined by the difference between measurements 1 and 2.

Re 3: The values should be as equal as possible on both sides.

Re 4 and 6: If the measurement of the difference in wheel angularity indicates inadmissible deviation, first examine the steering linkage [i. e. drop arm, tie (track) rods, stub axle steering arms] for distortion, and carefully check the position of the steering gear (inclination and distance from center of front axle assembly).

Re 5 and 7: The front wheels then posses the desired caster with the steering turned, if the camber of the wheel on the inside of the curve is greater than that in the straight-ahead position, while that of the outside wheel is less. The permissible difference between the minimum and maximum angle of camber for one wheel on right and left lock is 2° 15′ ± 40′.

Specifications for Vehicle Measurement

1	With 15″ tires 10 angular minutes correspond to	1.2 mm (0.05″)
2	Toe-in with wheels forced inward at the rear with car unladen	10′ ± 10′
3	Max. permissible difference in the toe-in before and after having forced the wheels inward	max. 25′
4	Amount of pressure needed to force front wheels inward	10 ± 2 kg (22.0 ± 4.4 lbs.)
5	Difference in wheel angularity with inside wheel turned 20° to left and right	1° 30′—2° 30′
6	Camber of front wheels in straight-ahead position with car unladen Max. permissible difference between both wheels	0° 40′ ± 30′ 30′
7	Difference in camber of a wheel turned 20° to the left and right	2° 15′ ± 40′
8	Caster angle of axle tubes	2° ± 15′
9	Track of rear wheels with correct adjustment of spring plates and car unladen	— 5′ ± 15′
10	Max. permissible departure of rear wheel track from perpendicularity to the fore and aft axis of the car Evaluate positions of rear wheels by means of the diagram on wheel alignment chart	10′
11	Camber of rear wheels with correct adjustment of spring plates and car unladen Spring plate adjustment: 16° 30′ + 50′ Max. permissible difference between both wheels	3° ± 30′ 20′

Meaning of the signs: + = toe-in, — = toe-out

THIS PAGE INTENTIONALLY
LEFT BLANK

Description of Roller Steering

(From August 1961)

General

In the roller type steering gear the adjustable steering worm engages the steering roller by means of a roller mounted on needle bearings. The worm runs in two ball bearings in the steering case and the steering roller shaft is mounted in bronze bushes in the steering gear housing and in the housing cover.

The steering worm is adjusted by a shim underneath the upper ball bearing. The meshing depth of the roller shaft and the resultant free of play contact between roller and worm is adjusted by an adjusting washer in the steering gear housing cover.

The main advantage of this type of steering gear is that the rotary movement of the worm is transmitted to the roller shaft by rolling friction instead of sliding friction. This makes the steering lighter in operation and gives very precise steering control. The degree of wear is also reduced considerably.

One feature of roller type steering is that when the steering is turned more than a quarter of a turn with the vehicle stationary, a steadily increasing amount of play is noticed. The play is necessary because the steering gear is subject to the maximum amount of stress in the central position and there must be some means of adjustment available. When the vehicle is moving, the self centering force at the wheels is such that the roller rests against one of flanks of the steering worm and the play is not noticed.

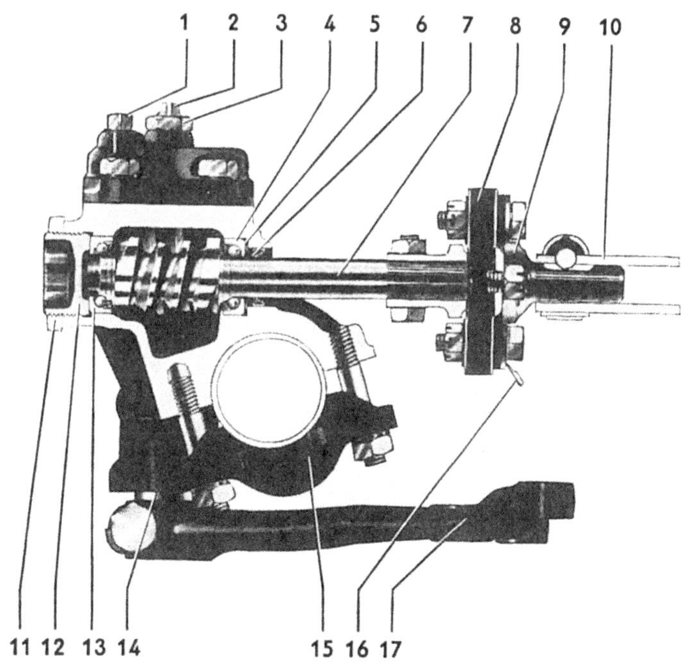

1 - Oil filler plug
2 - Adjusting screw for roller shaft
3 - Lock nut
4 - Upper worm bearing
5 - Shim for worm adjustment
6 - Oil seal
7 - Steering worm
8 - Coupling disc
9 - Coupling upper flange
10 - Steering column
11 - Lock nut
12 - Worm adjusting screw
13 - Lower worm bearing
14 - Roller shaft
15 - Mounting clamp
16 - Ground connection terminal
17 - Drop arm
18 - Roller
19 - Roller needle bearings
20 - Roller shaft pin

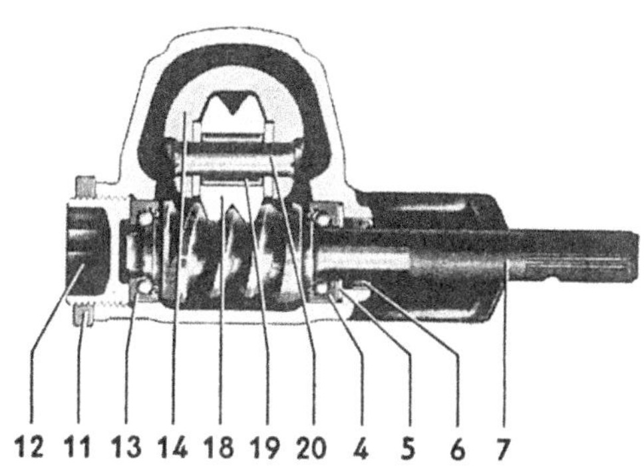

Lubrication

The steering gear must only be lubricated with Hypoid oil SAE 90 and not with grease or other oils. The oil capacity is 0.16 liters (0.32 U. S. pints; 0.26 Imp. pints).

The oil level should be checked regularly as specified in the lubrication chart. The oil level should be up to the lower edge of the threaded hole for the oil filler plug. The filler hole is accessible after removing a cover plate in the reinforcement panel.

Maintenance

The steering gear adjustment should be checked at the intervals specified in the maintenance chart. Timely and accurate adjustment of the steering gear has an important bearing on the service life of the steering and the driving characteristics of the vehicle.

Apart from the removal and installation of the tie rods the section V-9 "Steering Linkage" remains valid for the roller type steering.

Removing and Installing Tie Rods

Removal

1 - Lift vehicle and take front wheels off.

2 - Bend up lock plates and remove the tie rod end nuts.

3 - Bend up lock plates and remove nuts at drop arm.

4 - Mark left side of tie rods (left hand thread).

5 - Press tie rods out with special tool VW 266 h.

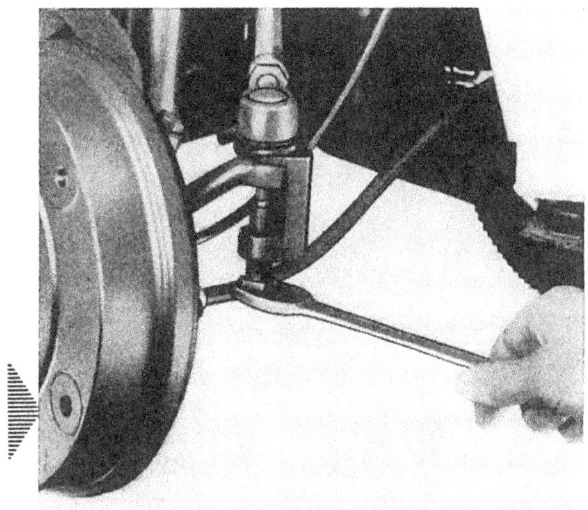

> **Important**
> Do not damage rubber seals when removing tie rod ends. Take care not to squeeze grease out of the seals when working on the steering or carrying out maintenance. The service life of the joints is only satisfactory when they are packed with the correct amount of grease (about 3.5 gr.)

Installation

1 - Check tie rods for damage. **Bent tie rods must be replaced, not straightened.**

2 - Check tie rod ends for wear and tightness. If there is any play or if the pin cannot be moved by hand the tie rod end must be replaced. The thread on the pin must be undamaged.

3 - Check rubber seals for wear and damage and replace where necessary.

4 - Install both rods so that the marked ends are on the left.

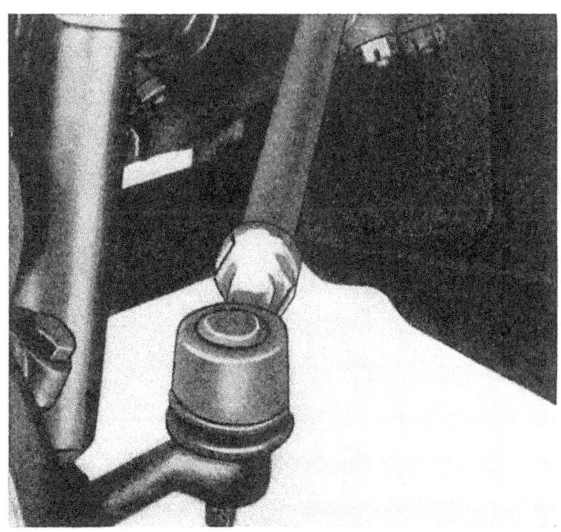

5 - Both ends on each rod must be tilted as far as possible in one direction to front or rear so that they align properly.

6 - Tighten tie rod end nuts to 16—18 ft. lbs. in this position and lock.

7 - Adjust toe-in.

THIS PAGE INTENTIONALLY
LEFT BLANK

The removal and installation of the steering gear is identical with the procedure adopted with the sector type steering. The maintenance-free tie-rods are pressed off with the remover VW 266h.

Disassembly and Assembly of Steering Gear

Disassembly

1 - Remove plug and drain oil.

2 - The assembly and test jig VW 280 consists of a tube and a measuring head. Mount steering gear on tube so that it is quite firm. The measuring head must be removed first.

3 - Bend back lock plate of drop arm pinch bolt and remove pinch bolt.

4 - Pull the drop arm off roller shaft.

5 - Remove the lock nut of the roller shaft adjusting screw.

6 - Remove the four screws of the steering gear case cover.

7 - Turn the adjusting screw to detach the steering gear case cover from the roller shaft and remove cover.

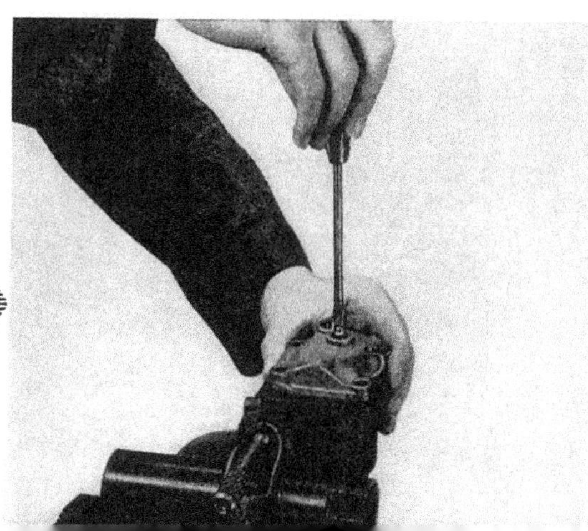

8 - Slide the oil seal protective sleeve VW 649 (local manufacture) fully home over the roller shaft splines.

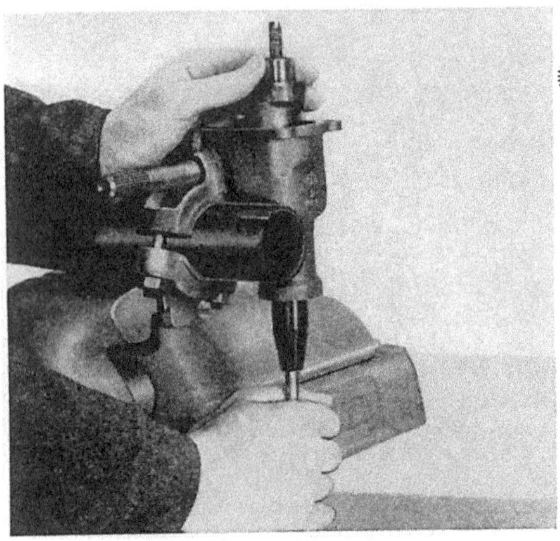

9 - Turn worm to central position and press roller shaft out upwards with a drift.

10 - Remove the circlip for the roller shaft adjusting screw.

11 - Take the adjusting screw and shim out of shaft.

Note:

The roller shaft cannot be dismantled further.

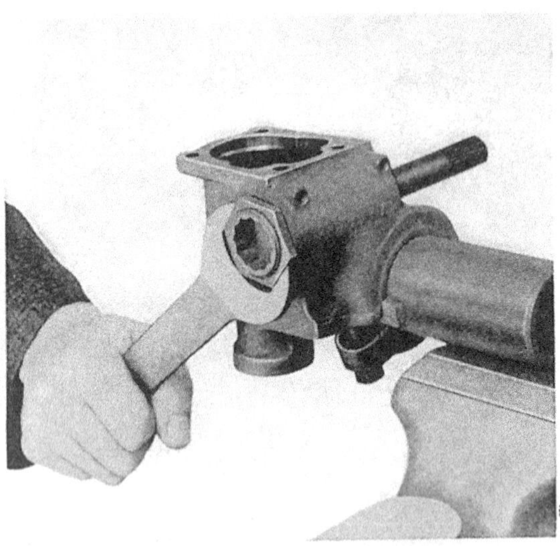

12 - Loosen the lock nut of the steering worm adjusting screw with open-end wrench VW 277.

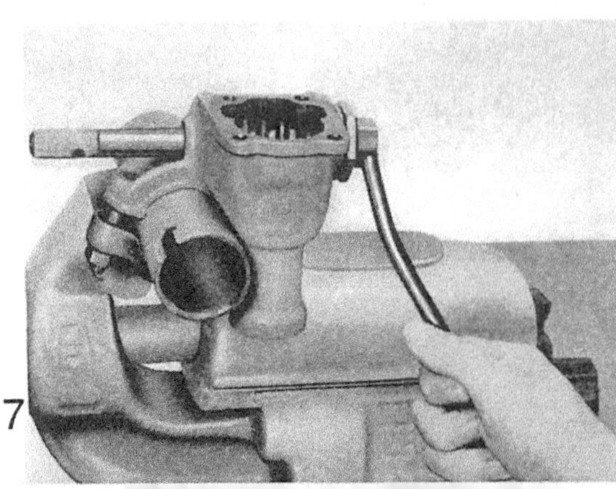

13 - Remove adjusting screw with wrench VW 278a.

14 - Knock worm and lower bearing out by taping lightly with a rubber hammer.

15 - Press oil seal, shim and upper bearing out inwards with 21.5 mm tube VW 423.

16 - Examine all parts for wear and damage and replace as required.

Assembly

On assembly, the worm and roller must be adjusted so that the steering can be turned approximately the same amount in each direction before play appears.

Proceed as follows:

1 - Insert worm and upper bearing into the housing with a shim of medium thickness (approximately 0.35 mm/0.12″)

Note:

The worm spindle oil seal is installed after the steering gear has been assembled and adjusted.

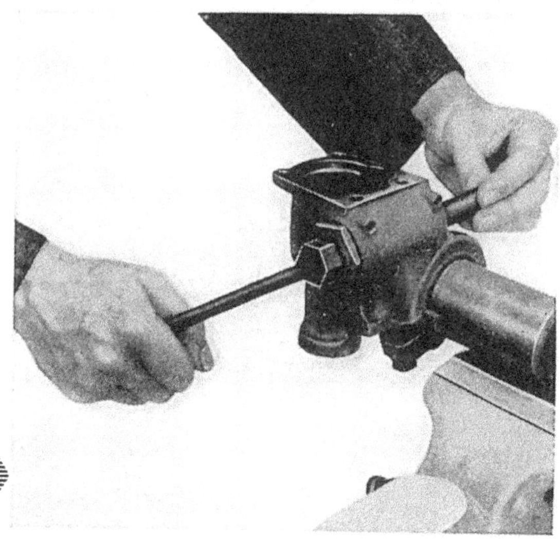

2 - Install lower bearing.

3 - Install worm adjusting screw with sealing compound.

4 - Screw adjusting screw in and tighten lightly.

5 - Loosen screw and retighten until worm runs roughly when turned.

A torque of 1.30—2.17 in. lbs. should be sufficient to turn the worm (1.74—2.17 in. lbs. with oil seal).

6 - Tighten adjusting screw lock nut.

7 - Check that the roller axial play does not exceed 0.04 mm (0.0015″).

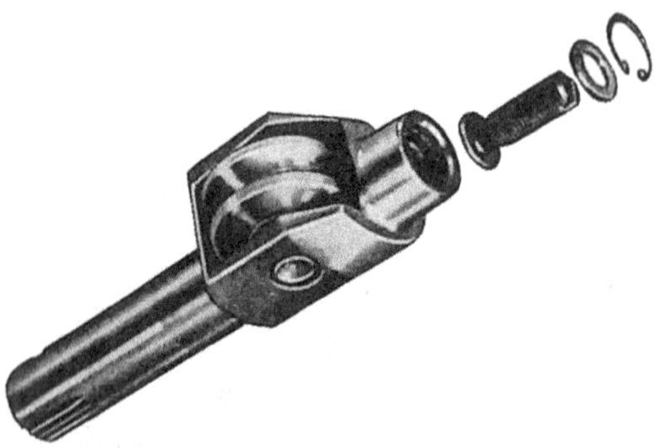

8 - Assemble roller shaft.

a - Check adjusting screw. If the thread or the contact surface is damaged, the screw must be replaced.

b - Place adjusting screw and shim in the bore in the roller shaft. The shim should be selected so that when in the bore it can just be turned with thumb and forefinger without rock. The shims are supplied in thicknesses from 2.0 mm to 2.55 mm in 0.05 mm stages.

The adjusting screw must be coated with oil when installing and not, under any circumstances, with grease.

c - Install circlip and check play.

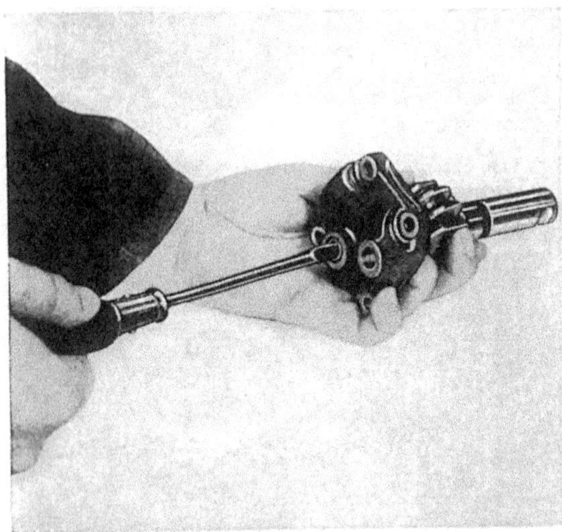

Important

The circlip must engage in the groove properly all the way round.

9 - Screw adjusting screw with roller shaft into housing cover as far as it will go.

10 - Insert oil seal protective sleeve VW 439 (local manufacture) into housing.

11 - Install the roller shaft so that the roller is at right angles to the worm.

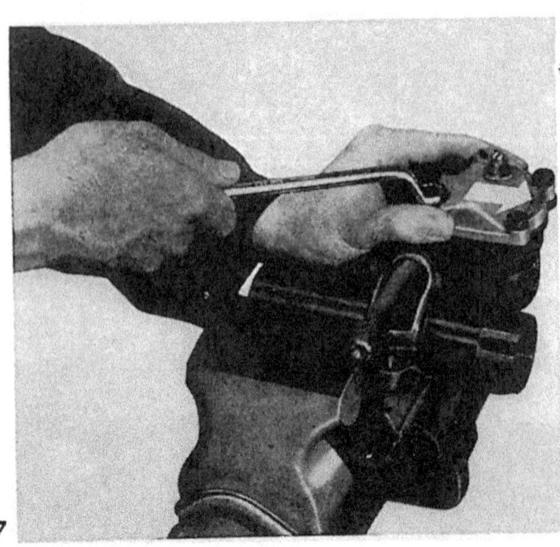

12 - Tighten housing cover screws to 14—18 ft. lbs. While doing this push the cover away from the worm so that it is firmly against the side of the housing opposite to worm. This prevents the cover from moving and causing premature play in the central position.

13 - Push drop arm on to steering roller shaft until the upper edge intersects the bevelled end of the shaft. Tighten pinch bolt to 50 ft. lbs.

14 - Check operation of steering gear by moving the drop arm back and forth several times.

Adjustment (Steering gear removed)

1 - Insert the measuring head of the test jig VW 280 into the assembly tube and clamp in position. Ensure that the feeler plate on the head contacts the worm spindle.

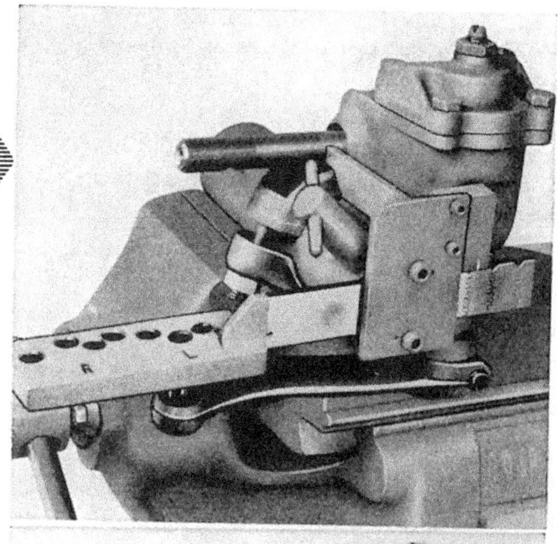

2 - Slide the setting plate until the appropriate drop arm number is under the mark.

L = Left hand steering

341 415 371 Type 3: Karmann Ghia Coupé,

141 415 371 A Type 1: Karmann Ghia Coupé, 2-seater Convertible

311 415 371 Type 3: Sedan, Variant

113 415 371 Type 1: (except Standard): Sedan, 4-seater Convertible

R = Right hand steering

342 415 371 Type 3: Karmann Ghia Coupé,

142 415 371 A Type 1: Karmann Ghia Coupé, 2-seater Convertible

312 415 371 Type 3: Sedan, Variant

114 415 371 Type 1: (except Standard): Sedan, 4-seater Convertible

3 - Move drop arm into central position.

a - Align right-hand hole in drop arm roughly under the appropriate hole in the setting plate.

b - Insert pilot into correct hole in plate.

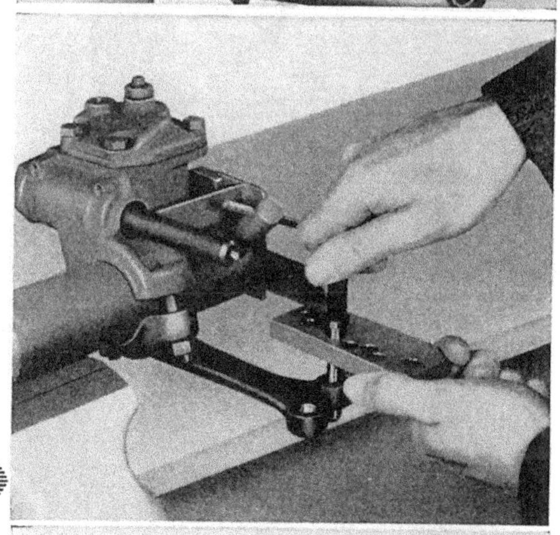

c - Move drop arm and plate to and fro slightly until the pilot drops into hole in arm under its own weight.

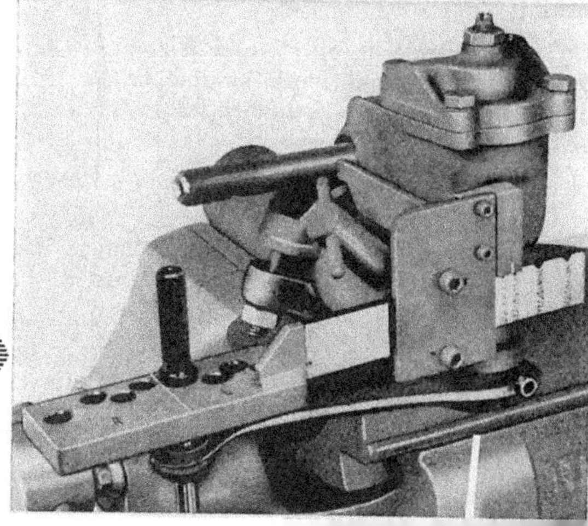

4 - Insert the scale from the adjusting device VW 279 into the slot in the drop arm and clamp in position.

5 - Place the pointer from VW 279 device on the boss on the steering case and set it at zero.

6 - Pull pilot out.

7 - Turn drop arm 11° to left or right.

> **Important**
>
> The 11° ± 2° angle is only valid if the steering roller shaft and worm spindle or one of these parts has been replaced, otherwise the steering is set at 5°.

8 - Screw the roller shaft adjusting screw in until no play can be detected. Move the drop arm back and forth while doing this and hold the worm spindle with the other hand. The steering flange should be installed on the worm spindle to facilitate holding.

9 - Tighten adjusting screw lock nut to 16 to 18 ft. lbs.

10 - Check free of play range on the other lock. It should be possible to turn the worm through the central position with a torque of 7.8 to 10.4 in. lbs. There should be no play at the drop arm at an angle of 11° ± 2°. If there is play at this angle, the steering gear must be dismantled again and the setting of the worm to the roller rectified by inserting a different shim.

a - No-play movement to left side of scale greater than 11° ± 2°.
Install thicker shim.

b - No-play movement to left-hand side of scale less than 11° ± 2°.
Install thinner shim.

The adjustment should be repeated until the desired degree of accuracy is obtained. Shims are available in thicknesses of 0.20 mm to 0.50 mm in 0.05 mm steps.

11 - When adjustment is finished, press the worm spindle oil seal in with the 21.5 mm tube VW 423.

12 - Remove four steering case cover securing screws, coat them with sealing compound, screw them in again and tighten to 14 to 18 ft. lbs.

13 - Mark the central position of the steering again.

Note:

From Chassis No. 4010995, the center position of the steering gear is indicated by a mark on the worm spindle (A) after the steering gear has been adjusted in production. When the wheels are in the straight-ahead position, this mark must align with the casting seam on the housing. The mark is protected with a coat of clear varnish.

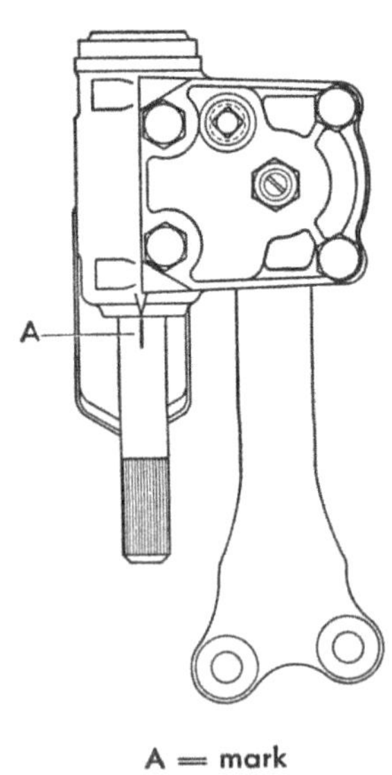

A = mark

The steering worms supplied as spares do not have this mark. When installing a new worm, the center position must be set with the test jig VW 280. This position must then be marked by pushing the marking ring (Part No. 311415213) on to the worm spindle.

Note:

The steering gear — Part No. 113315061 — is supplied as a spare part without a drop arm and without a marked center position. Before installation, the appropriate from arm must be fitted and the center position located with test jig VW 280 and marked with the marking ring.

Note:

1 - From 24th February 1964, Chassis No. 6147942, the radius of the steering roller stops in the steering box (Part No. 311415105) was altered from 3 mm to 2.5 mm. The housings are marked with the figures "14" moulded into the casting.

When the steering gear housing or the steering roller shaft are replaced during a repair, the clearance of the roller shaft in the housing must be checked with a feeler gauge. This also applies when service installing a housing of the modified type. If the shoulders of the roller shaft contact the stops or if the clearance is less than 0.3 mm, the stops should be reworked with a milling tool. The clearance should not exceed 0.5 mm so that the steering roller can contact the stops fully. The metal cuttings should be removed carefully.

When present stocks are exhausted, only the modified steering gear housing will be supplied.

2 - From 30th November 1964, Chassis No. 115327044, the steering gear (Part No. 113415061) is lubricated with transmission grease (160 cc) instead of SAE 90 hypoid oil. As a result of this change, the steering housing cover has been modified (new Part No. 311415129 A). The oil filler screw has been replaced by two holes which are sealed with plastic plugs (Part No. 311415133).

The operation "Steering gear: check oil level" in the lubrication chart is no longer necessary on this type of steering gear.

The previous type of steering gear can be converted to the grease-filled type by fitting a new cover. When existing stocks are run down, only the new type cover (Part No. 311415129 A) will be supplied.

Checking and Adjusting

(Steering installed)

The roller steering should be checked as follows during maintenance services and adjusted, if necessary:

A - Checking

1 - Vehicle on the ground and the front wheels in straight-ahead position.

2 - Move the steering wheel at the outer end of the spoke lightly to and fro with your finger until resistance is felt in both directions. This range at the center position is determined by the steering adjustment as well as by the tie rods and steering coupling. Measured at the steering wheel circumference this range should be 25 mm/1".

B - Adjusting

If there is excessive play at the center position there are three points which can be the cause. These points should be checked and adjusted in the following sequence.

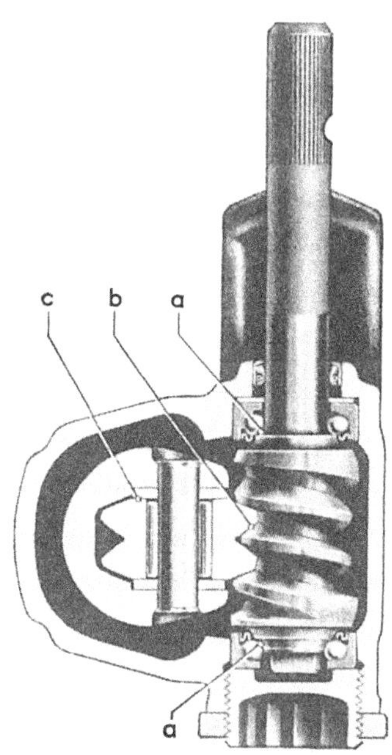

a - Axial play of steering worm.
b - Play between steering roller and steering worm.
c - Axial play of steering roller.

a - **Axial play of steering worm**

The axial play of the worm is determined by turning the worm to and fro at the steering coupling. To do this, the vehicle must be raised. If there is play, it can be eliminated as follows:

1 - Turn steering wheel to left or right.

2 - Loosen the lock nut of the steering worm adjusting screw with 41 mm wrench VW 277.

3 - Turn the steering worm to and fro at the steering coupling and at the same time tighten the adjusting screw with the wrench VW 278a until play is no longer detectable. The VW 278 wrench can also be used if it is modified as follows:

a - Shorten hexagon to 20 mm/.78″.

b - Extend handle by 100 mm/4″ (weld on a 4″ piece of a round bar 10 mm dia./.04″).

c - Bend handle as shown in drawing.

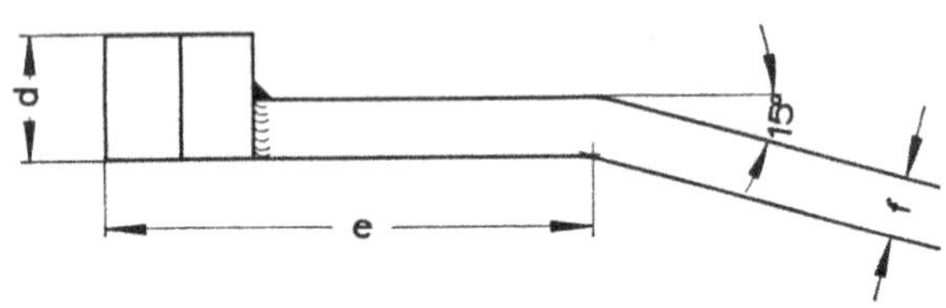

d = 20 mm/0.78″ e = 80 mm/3.14″ f = 10 mm dia./.04″

4 - Hold adjusting screw and tighten lock nut.

5 - Turn steering worm from lock to lock. When turning there should be no unusually tight positions noticeable. If there are any, the adjusting screw is too tight; the adjustment must then be corrected.

6 - If the excessive play at the center position is still present, the roller must be adjusted to the steering worm as described under b.

b - **Play between steering roller and steering worm**

At the center position the steering roller must have no play at the steering worm if the steering movements are to be transmitted exactly to the wheels. The vehicle must be on the ground if the steering roller is to be adjusted correctly.

1 - Turn steering wheel 90° to left or right.

2 - Loosen roller shaft adjusting screw lock nut.

3 - Loosen the adjusting screw approximately one turn.

4 - Tighten the adjusting screw until you feel the roller contact the steering worm.

5 - Hold adjusting screw and tighten the lock nut to 2.2 mkg (16 ft. lbs.) max. 2.5 mkg (18 ft. lbs.).

6 - Checking: Move the steering wheel at the outer end of the spoke lightly to and fro with your finger until resistance is felt in both directions. If the range of play is greater than 25 mm (1″), the roller shaft must be removed and checked as described under c.

7 - If the play is less than 25 mm (1″) on this lock, the adjustment must be checked on the other side at a lock of 90°. If the range of play on that side is greater than 1″, the roller shaft adjusting screw must be tightened until you feel the roller contact the steering worm.

If the range of play cannot be adjusted to the prescribed dimension, the roller shaft must be removed and checked as described under —c—.

> **Important**
>
> In any case the adjustment must be checked on **both** sides at a lock of 90°.

8 - Check toe-in, correct if necessary.

9 - Roadtest the vehicle. If the steering does not self-center automatically to approximately 45° of the center position after negotiating a corner at between 15—20 kph (9—12 mph.), the steering roller is too tightly adjusted. The adjustment must be carried out again as otherwise the steering worm and roller can become damaged.

c - **Axial play of steering roller**

If the prescribed range at the center position cannot be obtained by adjusting the steering worm and roller shaft, the axial play of the steering roller must be checked. This operation cannot be carried out during maintenance services.

1 - Remove and disassemble steering gear.

2 - Check axial play of steering roller, 0.04 mm (0.0016″) is the maximum play. Check the play between roller and washer with a feeler gauge. If the feeler gauge can be inserted between roller and washer, the steering roller shaft must be replaced.

3 - Assemble, adjust amd install steerin gear.

4 - Check toe-in, correct if necessary.

5 - Road test the vehicle.

Testing and Adjusting Toe-in

General

With the exception of toe-in adjustment, the sections V-12 to V-14 — Steering Geometry, Wheel Position and Vehicle Measurement remain valid for the roller type steering.

The toe-in of the front wheels should be between 2—4 mm (.08—.16") with the vehicle unladen, and between 1—3 mm (.04—.12") with permissible total weight.

The following data must be correct when measuring the toe-in:

1 - Tire pressures front 17 lbs./sq. ins.
rear 22 lbs./sq. ins.

2 - Even wear on tires.

3 - Spring plates correctly adjusted.

4 - No excessive run out at rims.

5 - No play in steering linkage and torsion arm link pins.

6 - Correctly adjusted front wheel bearings.

7 - Correctly adjusted steering.

8 - The test surface must be level and horizontal.

Checking

1 - Position wheels in straight-ahead position.

2 - Place alignment gauge in front of wheels.

3 - Position the gauge pins against both wheel rims approximately level with the center of the wheels.

4 - Set the gauge dial to zero and mark the point of contact with chalk.

5 - Roll the vehicle forward half a turn of the wheels.

6 - Locate gauge behind wheels and place gauge pins on the marked spots.

7 - Read measurement.

Toe-In Adjustment

The adjustment of the toe-in is one of the most important items of vehicle alignment. Incorrect toe-in causes excessive tire wear in every case and should, therefore, be checked and adjusted most carefully.

The toe-in is adjusted by turning the tie-rod tubes. The steering must be in the central position according to the marks on the worm spindle when doing this. Both tubes must be lengthened or shortened the same amount to avoid altering the central position.

Note:

If the central position of the roller steering is not watched carefully when making alterations to the toe-in, the roller can run out of the steering worm on full lock and jam in the corner of the steering gear case.

1 - Bend lock plates up.

2 - Loosen securing nuts.

3 - Cut lock plates and take off.

4 - Knock tapered rings off tie-rod tubes.

5 - Install new lock plates.

6 - Correct toe-in by turning both tie-rod tubes in the same direction.

a - Turning the rods to the front increases the toe-in.

b - Turning the rods to the rear decreases the toe-in.

7 - After adjustment, slide the tapered rings on to tie-rod tubes.

8 - Turn both ends of each rod in the same direction as far as they will go so that they are properly aligned.

9 - Tighten lock nuts in this position.

10 - Secure tie-rod end nuts by bending the lock plates up alternately.

11 - Check the central position again at the steering gear with the wheels straight ahead.

12 - Check toe-in and road test car.

Important

If the steering wheel spoke is at an angle when driving straight ahead, the steering center position must be rectified. According to which side the spoke is inclined, one tie-rod must be shortened and the other lengthened by the same amount so that the toe-in is not altered. Under no circumstances should the position of the spoke be altered by repositioning the steering wheel on the column.

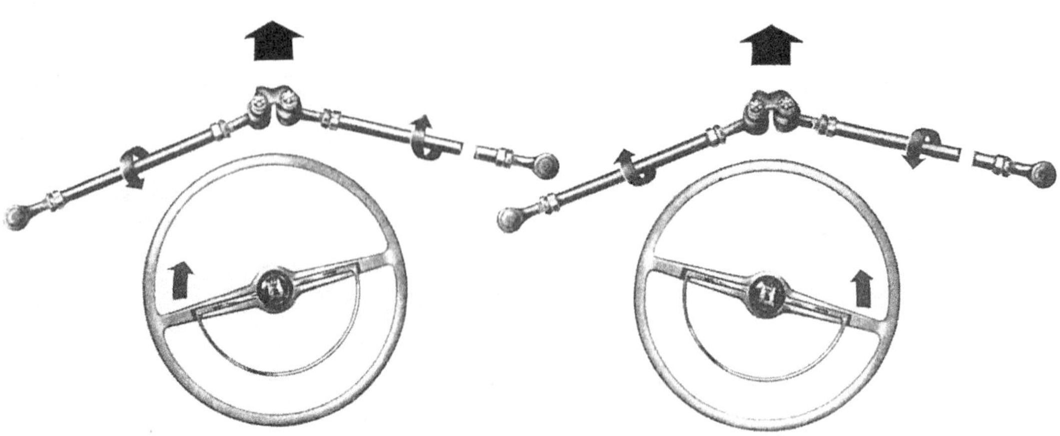

Note:

Since the introduction of the roller-type steering it has been noted that in isolated cases the vehicles tend to pull to the right.

This trouble is usually caused by the steering damper. The slight pressure exerted by the damper is only noticeable if the steering wheel is released when driving on a level road. If the vehicle does not run on a straight course when the damper has been removed, the following factors may be responsible:

1 - Uneven tire pressures. The vehicle will pull to the side on which the pressure is low.

2 - Uneven camber. The vehicle will pull to one side when the difference between the camber angles on left and right sides is more than specified in the test charts.

3 - Unevenly worn tires. The vehicle will pull to the side on which the tires are worn most. Interchange wheels at front or rear according to wear.

4 - Uneven load distribution.

5 - Incorrect toe-in. The toe-in should be adjusted as described on pages V-18, 1/2.

6 - Toe-in, wheel alignment and camber difference of rear axle must not exceed the tolerances specified in the test charts.

THIS PAGE INTENTIONALLY
LEFT BLANK

1 - VW Special Service Tools

VW 113	Two Open-end Wrenches, 27 mm
VW 150	Offset Handle
VW 156	Key, 8 mm
VW 202	Extractor
VW 202 c	Extractor Ring
VW 202 d	Extractor Hooks
VW 202 i	Thrust Pad
VW 202 s	Extractor Hooks
VW 217	Facing Cutter
VW 224	King Pin Bush Reamer
VW 244	Driving Sleeve
VW 254 a	Protractor
VW 256 a	Front Axle Tube Alignment Gauge
VW 258 a	Stub Axle Gauge
VW 259	Torsion Arm Link Gauge
VW 266 f	Tie Rod End Remover
VW 266 h	Tie Rod Remover
VW 270 a	Torsion Arm Offset Gauge
VW 273 a	Torsion Arm Bush Drift
VW 274 a	Torsion Arm Bush Reamer
VW 277	Open-end wrench 41 mm
VW 278	Hexagon wrench
VW 279	Adjusting Device for Roller Steering
VW 282 b	Torsion Arm Test Plate
VW 309	Holder
VW 400	Repair Press 15 t
VW 401	Thrust Plate
VW 407	Punch
VW 408	Punch
VW 410	Punch
VW 411	Punch
VW 412	Thrust Disc
VW 415	Tube 75 mm dia.
VW 417	Tube, 30/31.9 mm dia.
VW 418	Tube, 32.5 mm dia.
VW 421	Tube, 28 mm dia.
VW 422	Tube (split)
VW 423	Tube, 21.5 mm dia.
VW 431	Thrust Pad 16.5/28 mm dia.
VW 432	Arbor, 50 mm dia.
VW 433	Thrust Pad
VW 434	Arbor
VW 436	Guide Pin (Tapered)
VW 438	Guide Pin (Cylindrical)
VW 446	Tube
VW 447	Thrust Plate

2 - VW Workshop Equipment for Local Manufacture

VW 603/1	Vehicle trolley
VW 606	Front axle jack
VW 610	Front axle cradle for trolley jack
VW 633	Trestle
VW 637	Inner and outer hub cap removal tool
VW 638	Torsion Arm Bush Drift
VW 643	Stand
VW 648	Mounting Tube for Steering Gear
VW 649	Guide Sleeve for Roller Shaft
VW 668	Checking device for mounting angle of front axle torsion bars

3 - Normal Hand Tools

Screwdriver, 6 mm
Combination pliers
Water pump wrench
Cold chisel
Prick punch (center punch)
Mechanic's hammer, 300 grams
Mechanic's hammer, 500 grams
Rubber mallet, 85 × 50 mm
Allen wrench, 8 mm
Triangular scraper
Flat file, 180 mm in length
Open-end wrench, 14 mm
Open-end wrench, 17 mm
Open-end wrench, 22 mm
Box wrench, 14 mm
Box wrench, 17 mm
Box wrench, 19 mm
Wire brush
Oil-can
Can for derusting fluid
Grease container
Scratch awl
Set of feeler gages, 0.1—0.5 mm
Micrometer caliper, 0—25 mm
Micrometer caliper, 25—50 mm
Micrometer caliper, 50—75 mm
Caliper square, 200 mm in length, measuring $^1/_{10}$ mm
Tap M 10
Tap M 12 × 1.5
Tap M 16 × 1.5
Tap M 18 × 1.5
Die stock, size 2
Die M 10
Die M 12 × 1.5
Die M 18 × 1.5
Tap wrench, size 1, adjustable
Tap wrench, size 2, adjustable
Torque wrench, 0—8 mkg (58 ft. lbs.)
Drill, 8.5 mm
Drill, 10.0 mm

Drill, 10.5 mm
Drill, 12.0 mm
Inspection lamp with cable and plug
Electric drill

4 - Supplementary Workshop Equipment

Special Cutter for Brake Drum Tapped Inserts
Track (tread) Gauge
Wheel Alignment Service Equipment
Mobile Pit Jack
Turning and Grinding Machine for Brake Drums
Run-out tester (radial and lateral run-out of wheels)

THIS PAGE INTENTIONALLY
LEFT BLANK

Contents: H

Important — Do not re-use any fasteners that are worn or deformed in normal use. Many fasteners are designed to be used only once and become unreliable and may fail when used a second time. This includes, but is not limited to, nuts, bolts, washers, self-locking nuts or bolts, circlips, cotter pins. Always replace these fasteners with new parts.

THIS PAGE INTENTIONALLY
LEFT BLANK

Description of Rear Axle

General Description

Transmission, rear axle, and engine are in unit in the rear of the vehicle. The rear axle is of the swing half axle type. The rubber-mounted transmission case is secured to the frame at three points. It incorporates the transmission, final drive and differential.

Transmission Case

One-piece, die-cast transmission case of tubular construction with integral clutch housing and engine mounting flange at one end and gear carrier and gearshift housing attached at other end. Two final drive covers with ball-shaped surfaces for mounting rear axle tubes.

Transmission

The transmission has four forward speeds and one reverse. The forward gears are fitted with baulk type synchromesh units and the gearwheels are in constant mesh to ensure quiet operation.

Gear ratios:

1st	3.80 : 1
2nd	2.06 : 1
3rd	1.32 : 1
Top	0.89 : 1
Reverse	3.88 : 1

Note:

From August 1966, **Chassis No. 117 000 001** the ratio of the third gear of all vehicles was altered to suit traffic conditions.

The new gear wheels of the third gear, Part No. 113 311 275 A, with a ratio of 29 : 23 teeth = 1.26 : 1 can be installed in older transmissions without difficulty.

Gear Shifting

The gears are shifted with a floor-mounted lever via a shift rod which runs through the central frame tunnel. When the forward gears are engaged, the synchromesh units which are situated on the pinion for 1st and 2nd gears and on the drive shaft for 3rd and 4th gears, come into operation. The synchromesh units consist mainly of a sliding sleeve, the synchronizer rings for the various gears and the clutch hubs for the gearwheels. During the shifting process, the end faces of the shifting plates which are attached to the operating sleeve by springs press the synchronizer ring on to the conical face of the gear wheel, so that a friction drive is created. Due to the difference in speed, the synchronizer ring is turned until stopped by the shifting plates. The synchronizing pressure is then increased via the bevelled surface of the sleeve and the synchronizer ring teeth until the speeds of the gearwheel and sleeve are identical. The sleeve is then able to turn the ring back from the locked position and pass into engagement with the synchromesh teeth on the gearwheel.

In order to ensure that the baulk synchromesh mechanism operates properly, it is essential that the clutch disengages completely. For this reason the adjustment of the free-play at the clutch pedal must be carried out very carefully. Careless declutching or a dragging clutch plate (due to damaged clutch linings or distortion of plate when installing engine) create excessive synchronizing pressures so that the synchronizer rings wear rapidly. If the clutch plate is locked completely as can be caused by a piece of clutch lining breaking off, the gears cannot be shifted at all even though the transmission is in order, because the mechanism cannot synchronize and the synchronizer rings remain in the blocked position. Only factory approved clutch linings should be used.

Every time an engine is installed, check that the clutch is freeing properly by declutching and engaging reverse gear with the engine running.

Further information is given in the section on the "Clutch".

Final Drive

Power is transmitted through a helically-cut drive pinion and ring gear, provided with differential bevel gears, via two swing axles to the rear wheels. Silent operation and long service life of the final drive mechanism are only ensured by carefully adjusting drive pinion and ring gear.

Gear ratio: 4.375 : 1

Note:

From August 1966, **Chassis No. 117 000 002** the Type 1 models with 1.5 liter engines have a rear axle ratio of 33 : 8 teeth = 4.125 : 1. By this means, the transmission is adapted to the higher engine output and vehicle performance. The ring gear and pinion as well as the differential of this transmission have been taken from the Type 3.

The differential is fitted to compensate for the difference in wheel travel as the car makes a turn, thus maintaining an equal drive to the wheels.

Rear Suspension

The rear wheels are sprung independently. Road shocks are transmitted to the left or right torsion bar by a trailing spring plate on either side of the rear axle. Screwed to the spring plate and the axle shaft bearing housing is a progressive rubber bumper which becomes active in the upper deflection range.

The torsion bars are anchored in the center of the cross tube in a splined adaptor.

The torsion bars have different numbers of splines at inner and outer ends so that the suspension can be adjusted properly. Hydraulic double-acting shock absorbers dampen road shocks and prevent excessive rebound.

Lubrication

The transmission gears and the differential are combined in the transmission case and are both lubricated with the same hypoid oil. The oil capacity is 3.0 liters for the first filling and 2.5 liters at oil changes.

The oil should be changed at intervals indicated in the Lubrication Chart, using oil of the proper specification. The used oil should be drained by simultaneously removing the two magnetic oil drain plugs, while the oil is still warm.

The magnetic oil drain plugs should be cleaned carefully and the transmission refilled with 2.5 liters of oil.

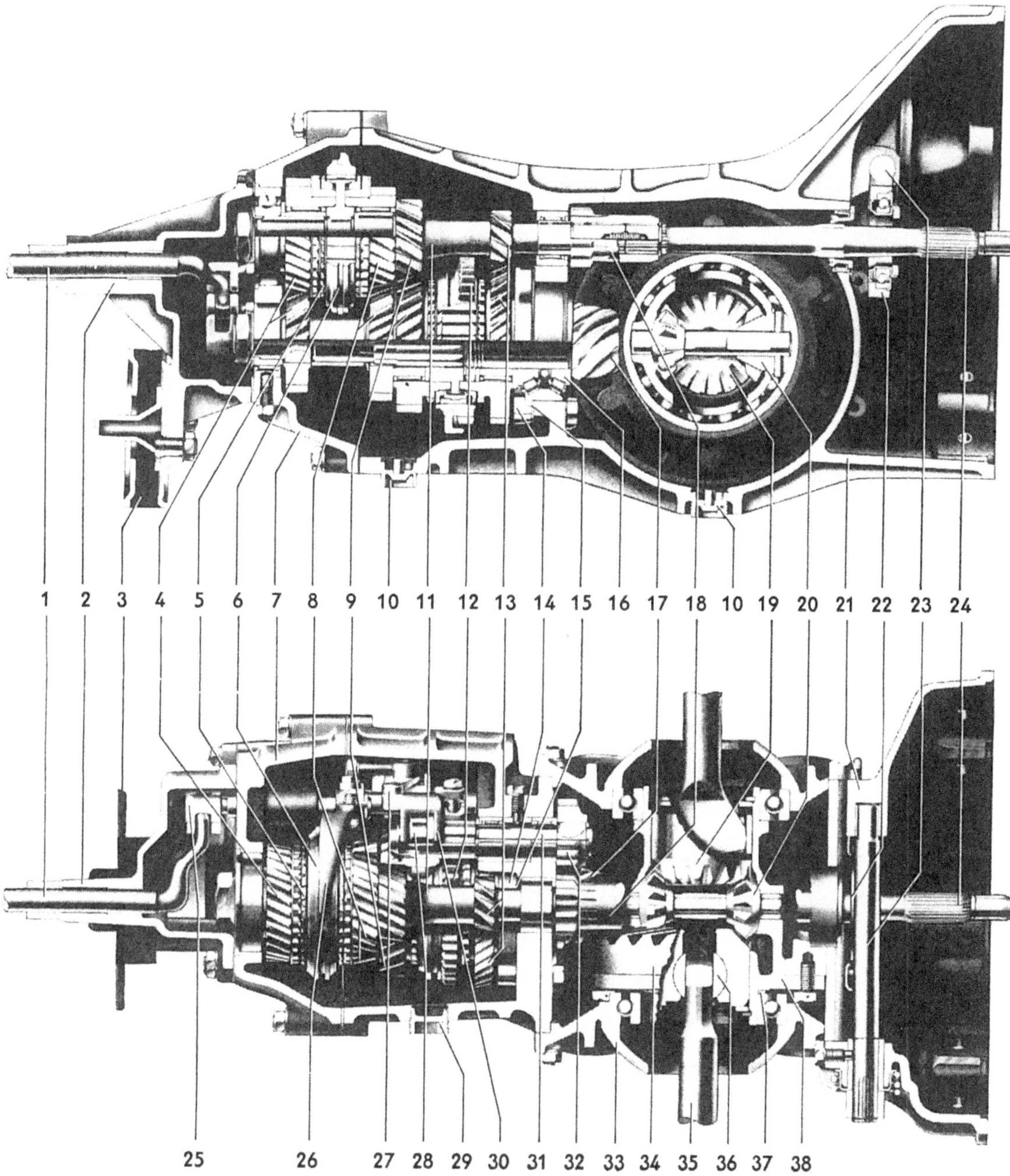

Volkswagen — Rear Axle and Transmission

(Sectional View)

1 - Transmission shift lever
2 - Gearshift housing
3 - Front bonded rubber mounting
4 - 4th gear train
5 - Synchronizer stop ring for 4th gear
6 - Clutch gear for 3rd and 4th gear
7 - Gear carrier
8 - 3rd gear train
9 - 2nd gear train
10 - Oil drain plugs, magnetic
11 - Main drive shaft, front
12 - Clutch gear for 1st and 2nd gear and reverse gear
13 - 1st gear train
14 - Double taper roller bearing retainer
15 - Pinion shims
16 - Double taper roller bearing
17 - Drive pinion
18 - Reverse gear on main drive shaft
19 - Differential side gear
20 - Differential pinion
21 - Transmission case
22 - Clutch release bearing
23 - Clutch operating shaft
24 - Main drive shaft, rear
25 - Selector shaft for 3rd and 4th gear
26 - Selector fork for 3rd and 4th gear
27 - Reverse lever for reverse gear
28 - Reverse sliding gear
29 - Oil filter plug
30 - Selector fork for reverse gear
31 - Reverse gear shaft
32 - Reverse drive gear
33 - Final drive cover
34 - Ring gear
35 - Rear axle shaft
36 - Fulcrum plate
37 - Shim
38 - Differential housing

THIS PAGE INTENTIONALLY
LEFT BLANK

Removing and Installing Rear Axle

General Description

If it is intended to disassemble the rear axle upon its removal from the vehicle, loosen the axle shaft nuts before lifting the vehicle.

Removal

1 - Disconnect battery ground (earth) strap.

2 - Lift vehicle, support it on trestles and remove engine.

3 - Remove rear wheels.

4 - Disconnect the brake hoses at the rear.

5 - Detach brake cables at hand brake. Remove hand brake and withdraw cables from conduits.

6 - Loosen dust sleeves.

7 - Remove lower shock absorber mounting bolts.

8 - Mark the position of the spring plate in relation to the axle shaft bearing housing by means of a chisel on the spring plate dead in line with the groove in the axle shaft bearing housing.

Note:

It is not necessary to make the marks if a new rear axle is installed, or if a front rubber cushion, frame or spring plate is replaced since the rear wheels have to be re-adjusted in any case. Make the marks on top of the spring plates as opposed to the grooves already on the side of the spring plates which are intended to facilitate rear wheel adjustment without the aid of a rear wheel alignment gage.

9 - Remove bolts at rear axle shaft bearing housing.

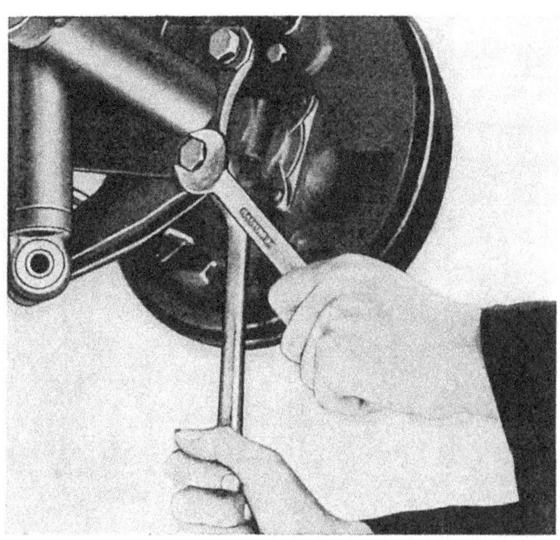

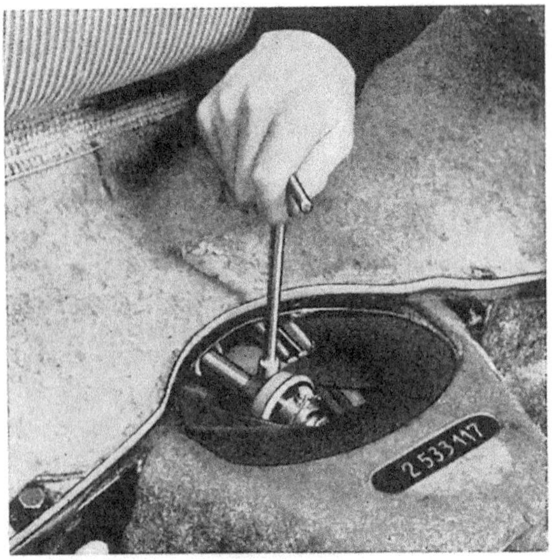

10 - Disconnect clutch cable from clutch operating shaft lever, slide off the rubber boot and withdraw cable and sleeve from the bracket on the left final drive cover.

11 - Unhook accelerator cable from retainer on gear carrier.

12 - Disconnect cable from terminals 30 and 50 at starting motor.

13 - Remove frame end inspection cover under rear seat. Remove the rear screw of the shifting rod coupling, using T-Wrench VW 114, and move the gearshift lever to withdraw the coupling from the transmission shift rod.

14 - Remove the nuts at the front rubber cushion of the transmission case.

15 - Push trolley jack under vehicle and clamp rear axle cradle VW 609 to rear axle.

16 - Remove the two bolts at transmission carrier by means of a 27 mm box wrench.

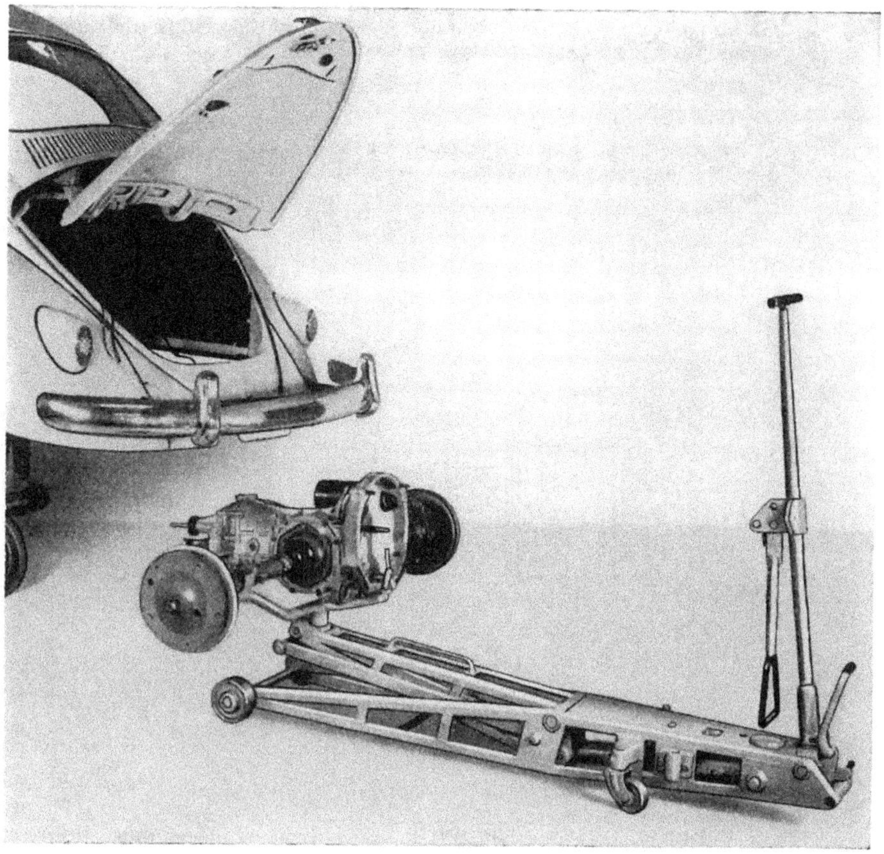

17 - Withdraw the rear axle towards the rear with the trolley jack.

Installation

When installing, the following points should be observed:

1 - Use trolley jack with cradle VW 609 to reposition rear axle in vehicle.

2 - Grease the two transmission carrier mounting bolts well and tighten them.

3 - Tighten nuts of mounting plate at the front of the transmission case.

Important

When installing a new rear axle, the nuts attaching the transmission carrier to the rear rubber mountings should be loosened and only tightened after the nuts of the front rubber mounting have been tightened. This procedure is also applicable when the transmission carrier has been removed. The order of tightening the nuts is to avoid distortion and premature wear of the rubber mountings.

Note:

When placing the rear axle on the ground, make sure that the main drive shaft does not become damaged or bent. Use wooden support with hole for main drive shaft.

4 - Make sure that the points of the coupling screws are correctly bedded in their recesses. Secure screws individually with a piece of wire.

Note:

1 - Incorrect installation may render it difficult or even impossible to shift down to 2nd or 1st gears. Under certain conditions, there is a possibility of reverse gear being selected when engaging 2nd gear, which could lead to serious transmission damage if force is used.

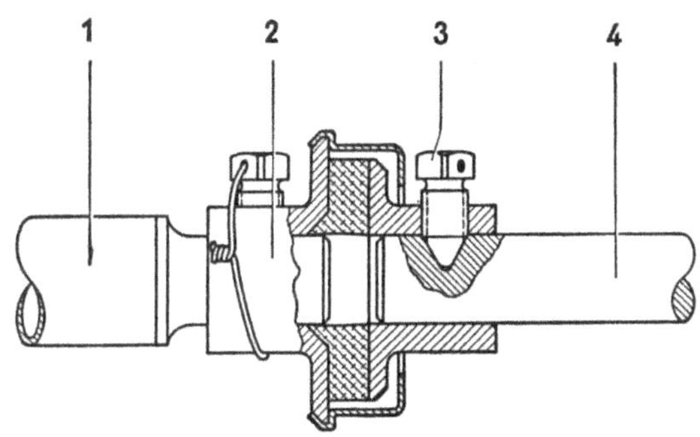

1 - Shifting rod
2 - Coupling
3 - Screw
4 - Transmission shift rod

2 - From November 1963, Chassis No. 5 911 561 (Rear Axle No. 6 344 500) a modified gearshift rod coupling is installed which prevents the gears from jumping out of engagement even when driving on bad roads. The coupling consists of a sheet metal housing which has a rubber guide ring on each side. The coupling is connected to the gearshift rod, which has also been modified, by a pin which passes through the guide rings. The guide rings have about 2 mm (.08″) axial play in the housing so that the vibration from the gear lever and gearshift rod which is caused by the movement of the engine-transmission unit cannot be transmitted directly to

the transmission. As a result the engagement of the gears is no longer affected unfavourably.

At the same time the access opening for the gearshift rod coupling has been widened 6 mm.

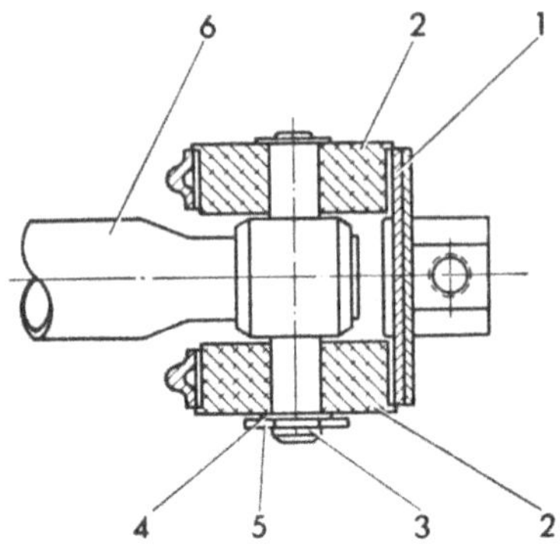

Part	Part No.
1 - Housing for gearshift rod coupling	311 711 211
2 - Guide ring	311 711 233
3 - Pin for coupling	311 711 209
4 - Washer	311 711 227
5 - Lock washer	311 711 229
6 - Gearshift rod	311 711 155 B Type 3
Gearshift rod	111 711 155 D Type 1

Service Installation

The modified coupling can be installed — without widening the access hole — from Chassis No. 929 746 if the gearshift rod is replaced at the same time. From Chassis No. 3 140 046 to 5 309 049, the new gearshift rod guide bush (Part No. 111 701 259 A) must be installed as well.

5 - Hook accelerator cable into retainer on gear carrier.

6 - Adjust rear wheels and tighten spring plate mounting bolts to 11 mkg (80 ft. lbs.).

7 - Securely tighten lower shock absorber mounting bolts.

8 - Check splines in hub of brake drum. Renew brake drum if splines are worn or damaged. Coat splines lightly with graphite grease.

9 - After the engine has been installed, adjust clutch pedal free-play to between 10 and 20 mm (.4 and .8").

10 - Tighten rear axle shaft nuts with a torque wrench to 30 mkg (271 ft. lbs.). If the cotter pin cannot be inserted turn on to the next hole.

11 - Bleed and adjust brake system.

Disassembly and Assembly of Rear Axle

It is recommended that the following sequence of operations be adopted for the disassembly and assembly of the rear axle:

1 - Remove starting motor

2 - Attach rear axle with fixture VW 307 to stand.

3 - Remove magnetic drain plugs and drain oil.

4 - Remove axle shaft nuts and withdraw brake drums.

5 - Remove wheel brake parts and back plates.

6 - Remove rear axle tubes and shafts.

7 - Remove gearshift housing.

8 - Remove differential.

9 - Remove rear main drive shaft.

10 - Remove transmission.

11 - Remove reverse drive gear and reverse gear shaft.

12 - Remove main drive shaft and drive pinion from gear carrier.

13 - Remove selector shafts and detent balls and springs.

Assembly

Observe the hints on the following pages when assembling.

THIS PAGE INTENTIONALLY
LEFT BLANK

Renewing Oil Seal or Rear Wheel Bearing

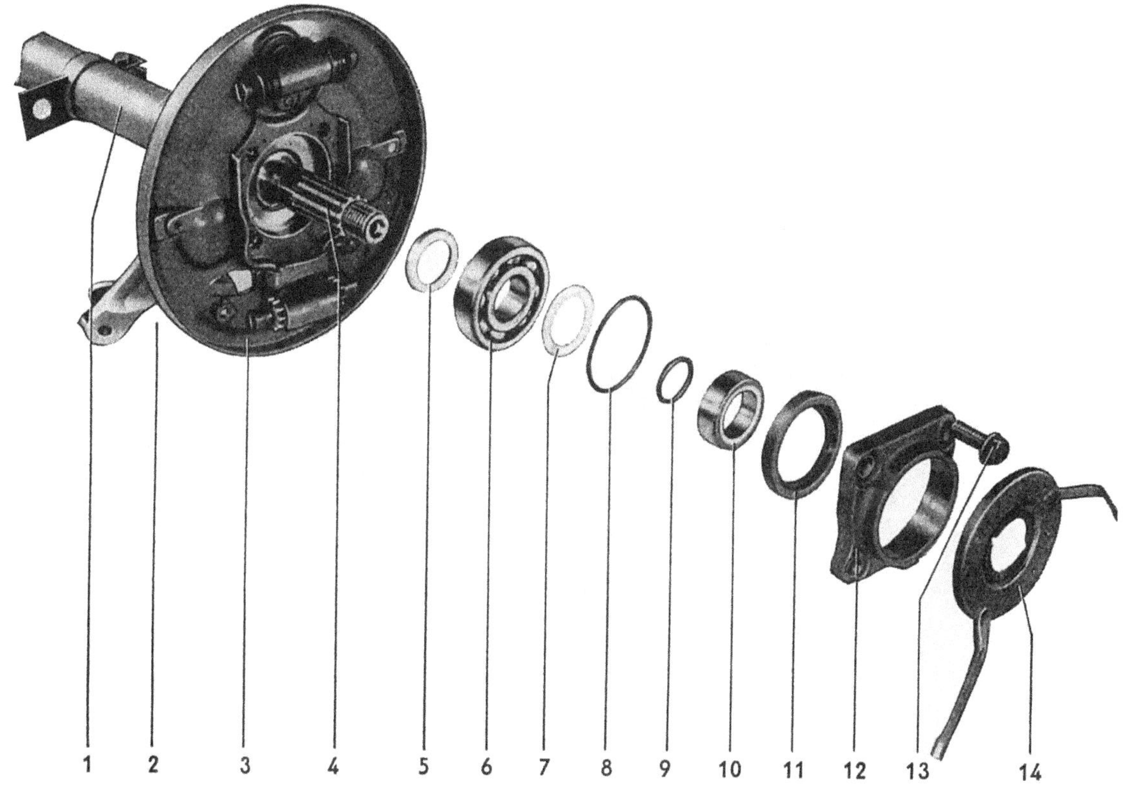

1 - Rear axle tube	5 - Inner spacer	9 - Gasket	
2 - Bearing housing	6 - Ball bearing	10 - Outer spacer	
3 - Brake back plate	7 - Washer	11 - Oil seal	13 - Cover retaining screw
4 - Axle shaft	8 - Gasket	12 - Cover	14 - Oil deflector

Removal

1 - Take off axle shaft nut and remove brake drum using extractor VW 202 in conjunction with VW 202c and 202d.

2 - Remove cover retaining screws and take off cover and oil seal.

3 - Remove brake back plate.

4 - Take off outer spacer, gasket between spacer and ball bearing, oil deflector and cover sealing ring.

5 - Remove rear wheel bearing, using extractor VW 241 a in connection with thrust pad VW 202 k and take off inner spacer.

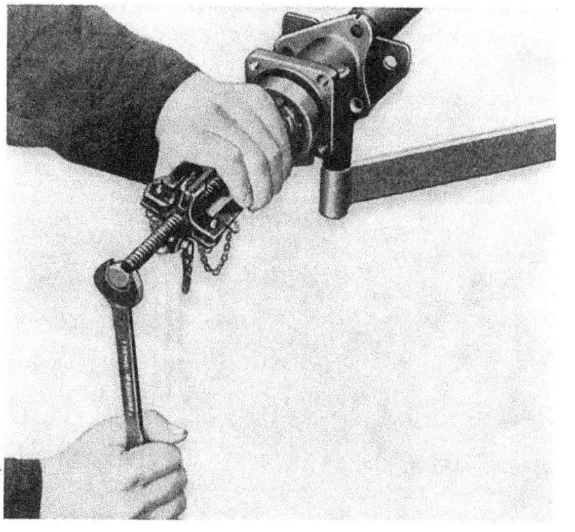

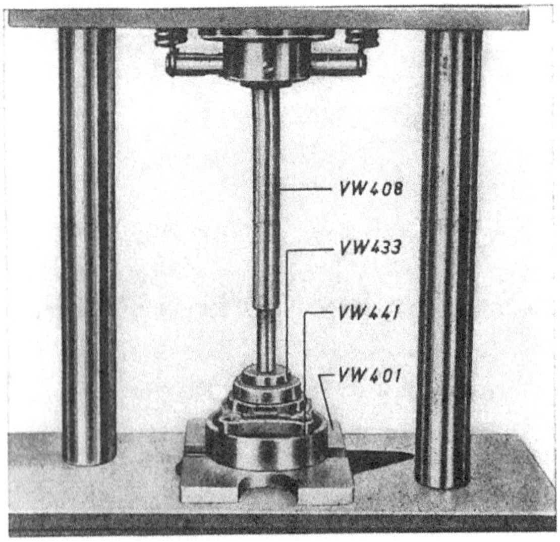

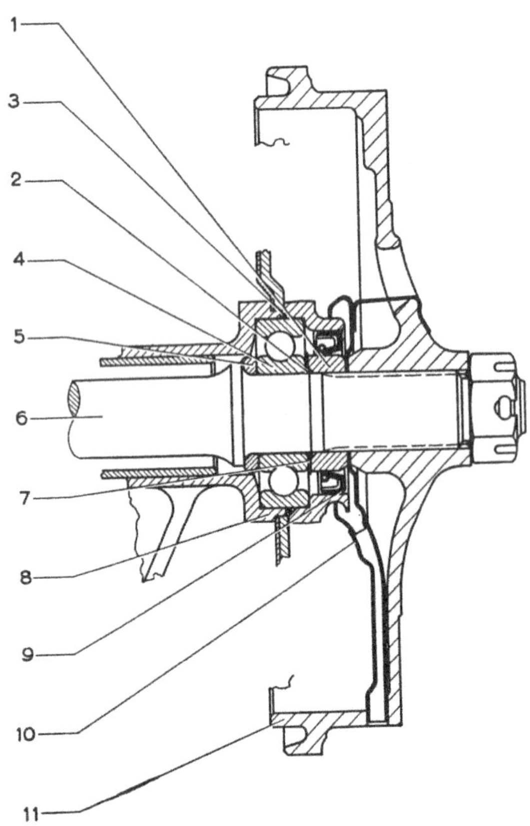

Installation

When installing, the following points should be observed:

1 - Examine condition of ball bearing, renew if worn or damaged.

2 - Renew the two gaskets.

Note:

From Chassis No. 4388450 (Rear Axle No. 4530936), the thickness of the inner spacer for the rear wheel bearings (New Part No. 111501281 A) was altered from 5.9—6.1 mm (0.232—0.240″) to 6.45—6.65 mm (0.252—0.260″).

Important

The modified spacer can only be installed on Types 1 and 3. The former spacer (Part No. 111501281) remains available for the Transporter.

3 - Examine oil seal. If any signs of damage or unevenness are visible on the lip, the oil seal should be renewed. Remove the seal using the Repair Press in conjunction with VW 401, VW 408, VW 433 and VW 441. Lightly coat the new oil seal with oil and press it into the cover conjunction with VW 401, VW 408 and VW 442.

4 - The outer spacer must neither be scored, cracked nor show signs of rust.
To avoid damage to the oil seal lip by friction, the spacer should be lightly coated with oil. Make sure that all components to be installed are absolutely clean.

5 - Replace cover so that the oil drip nose points downwards.

6 - Clean the oil deflector before reinstalling it. Make sure that the oil drip tube is bearing tightly against the brake drum to prevent it fouling the brake shoes.

7 - Check splines in brake drum hub. Replace brake drum if splines are worn.

8 - Tighten rear axle shaft nut with a torque wrench to 30 mkg (217 ft. lbs.) and secure it with a cotter pin.

Note:

The axle shaft nuts should be tightened when the vehicle is standing on its wheels again after the rear axle has been installed. On axles which are sent in for repair as assemblies, the nuts must be tightened to the proper torque and cotter pinned before axle is taken out of repair stand.

9 - Check transmission oil level, top up if necessary.

10 - Bleed and adjust brakes.

1 - Outer spacer
2 - Gasket
3 - Gasket
4 - Ball bearing
5 - Inner spacer
6 - Axle shaft
7 - Washer
8 - Bearing housing
9 - Oil seal
10 - Oil deflector
11 - Brake drum

Renewing Oil Seal or Rear Wheel Bearing

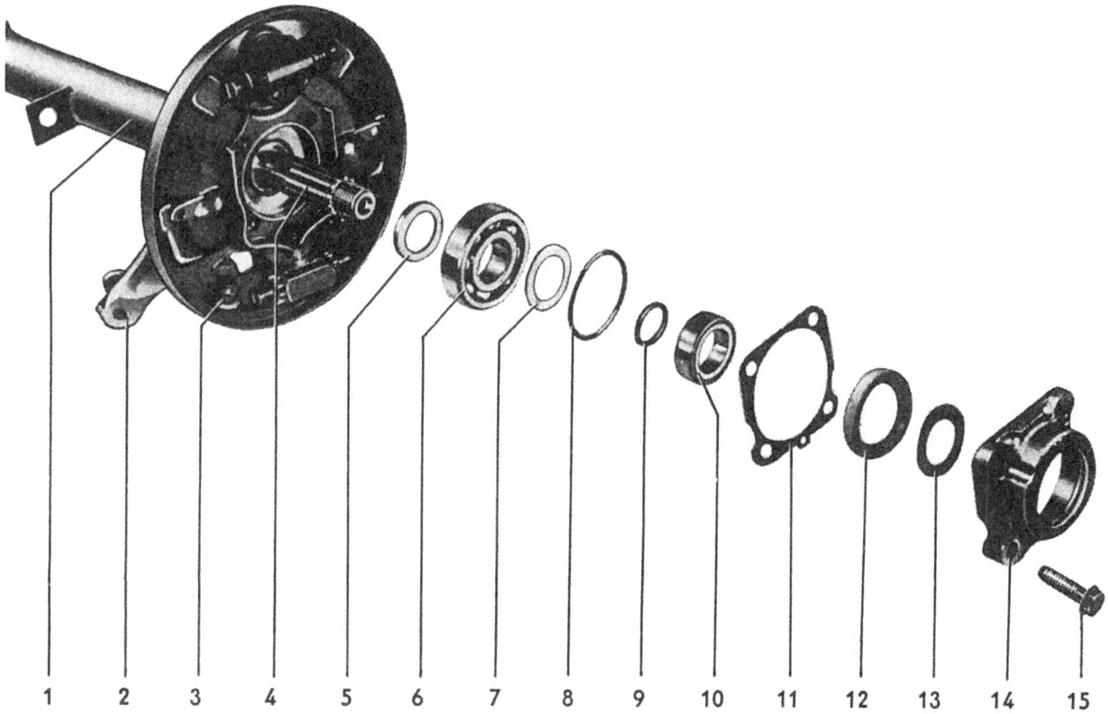

1 - Rear axle tube
2 - Bearing housing
3 - Brake back plate
4 - Axle shaft
5 - Inner spacer
6 - Ball bearing
7 - Washer
8 - Sealing ring
9 - Sealing ring
10 - Outer spacer
11 - Paper gasket
12 - Oil seal
13 - Oil deflector
14 - Cover
15 - Cover retaining screw

Removal

1 - Take off axle shaft nut and remove brake drum.

2 - Remove cover retaining screws and take off cover and oil seal.

3 - Remove brake back plate and paper gasket.

4 - Take off outer spacer, gasket between spacer and ball bearing, washer, and cover gasket.

5 - Remove rear wheel bearing and take off inner spacer.

Note:

When selecting the puller, bear in mind that from **March 1966** ball bearings with only **seven** balls instead of the usual eight have been installed intermittently. These bearings are marked with "Fafnir" on the outer race.

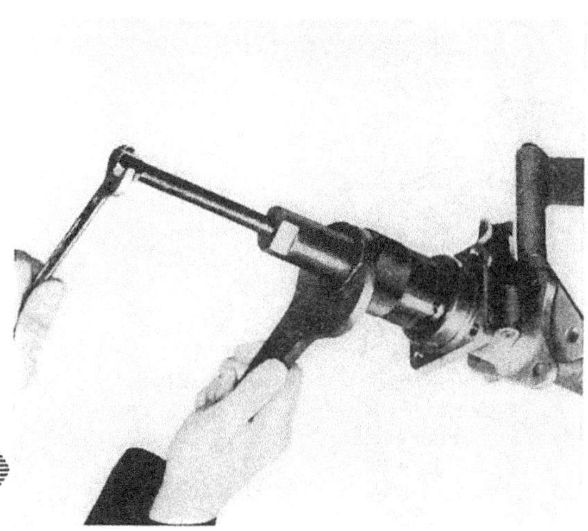

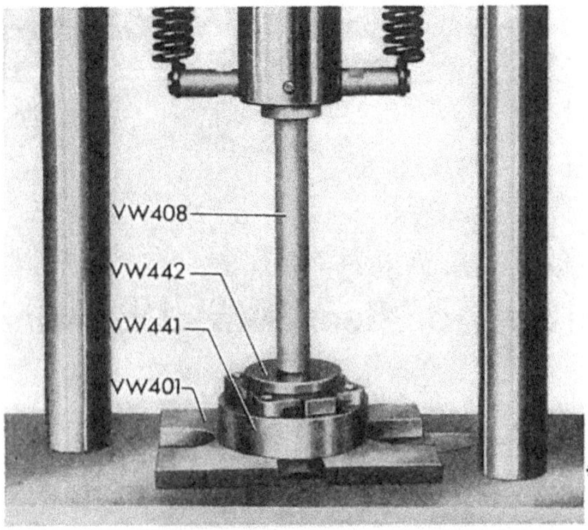

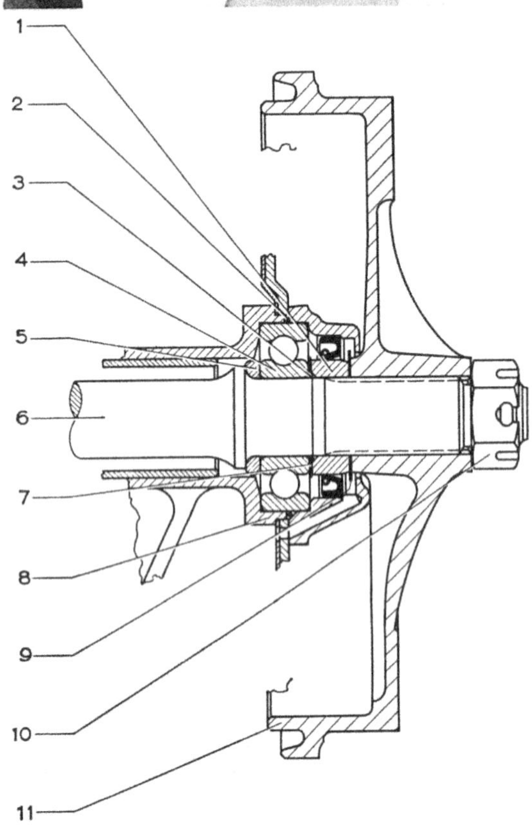

1 - Outer spacer
2 - Sealing ring
3 - Sealing ring
4 - Ball bearing
5 - Inner spacer
6 - Axle shaft
7 - Washer
8 - Bearing housing
9 - Oil seal
10 - Nut
11 - Brake drum

Installation

When installing, the following points should be noted:

1 - Examine condition of ball bearing, renew if worn or damaged.

2 - Renew the two gaskets.

3 - Examine oil seal. If any signs of damage or roughness are visible on the lip, the oil seal should be renewed. Remove the seal using the repair press in conjunction with VW 401, VW 408, and VW 447h. Coat the new oil seal lightly with oil and press it into the cover in conjunction with VW401, VW408, VW441 and VW442.

4 - The outer spacer must neither be scored, cracked nor show signs of rust.

To avoid damage to the oil seal lip by friction, the spacer should be lightly coated with oil. Make sure that all components to be installed are absolutely clean.

5 - Clean the oil drilling.

6 - Replace cover.

7 - Check splines in brake drum hub. Replace brake drum if splines are worn.

Note:

Since August 1965, from Chassis No. 116000001 all Type 1 Models have been fitted with brake drums the hubs of which have stiffening ribs. On the rear drums the measurement from wheel disc contact surface to drum hub contact surface on the rear wheel bearing has been increased slightly. This change has altered the rear track from 1288 to 1300 mm and will give more clearance in the wheel housing for snow chains.

The new rear drums can be service installed in vehicles from Chassis No. 1673411 without difficulty.

8 - Tighten rear axle shaft nut to a torque of 216 lb.ft (30 mkg) and secure it with a cotter pin.

Note:

It is best to tighten the axle shaft nuts when the vehicle is standing on the ground. On axles supplied as assemblies for repair, however, the nuts **must** be tightened fully and cotter pinned while the axle is in the repair stand.

9 - Check transmission oil level, and top up if necessary. The oil should be up to the edge of the filler hole.

10 - Bleed and adjust brakes.

Removing and Installing Rear Axle Tube and Shaft

Removal

1 - Withdraw brake drum. Remove bearing cover, brake back plate, and pull off rear wheel ball bearing.

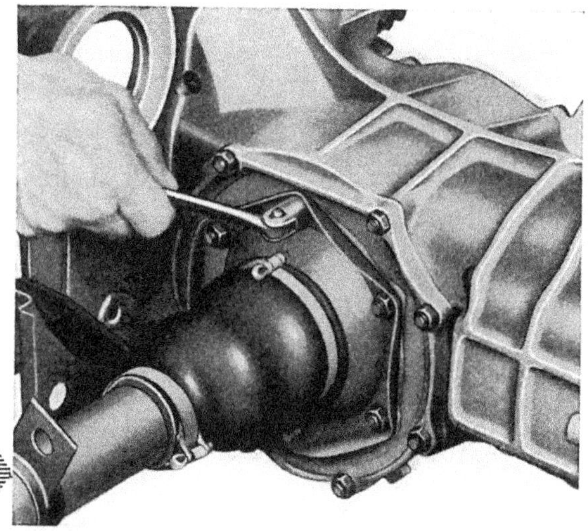

2 - Remove nuts at axle tube retainer.

3 - Withdraw rear axle tube and retainer and take off gasket and plastic packing.

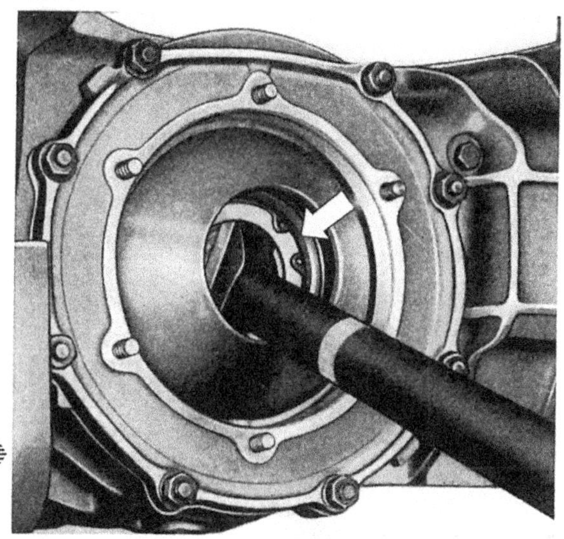

4 - Remove differential side gear lock ring.

5 - Remove differential side gear thrust washer and withdraw axle shaft.

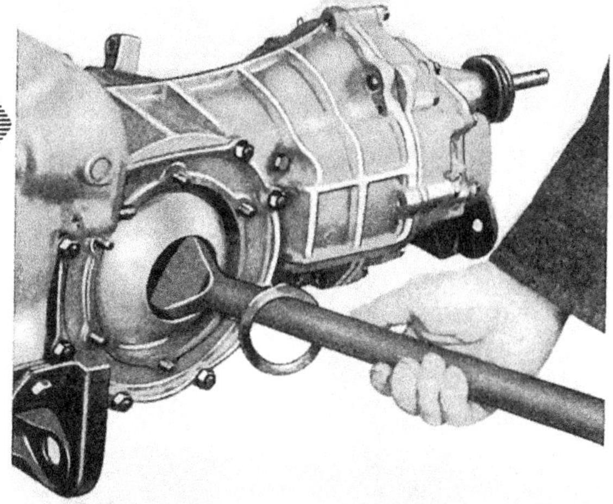

6 - Remove differential side gear and fulcrum plates from differential housing.

7 - Drive out lock pin in axle shaft bearing housing.

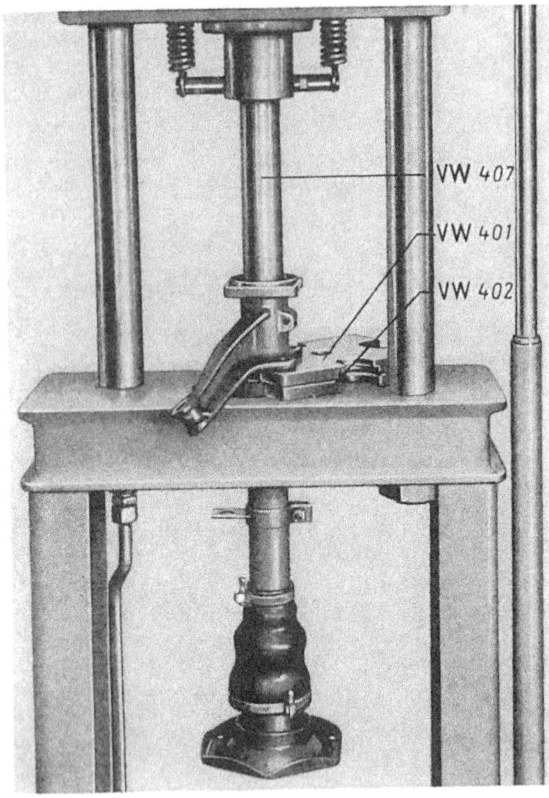

8 - Loosen dust sleeve.

9 - Remove axle tube from bearing housing on VW Repair Press in conjunction with VW 401, 402 and VW 407.

Note:

The bearing housing is liable to suffer damage when removed in an unskilled manner. Damaged bearing housings should not be re-used.

10 - Withdraw dust sleeve from axle tube and remove axle tube retainer.

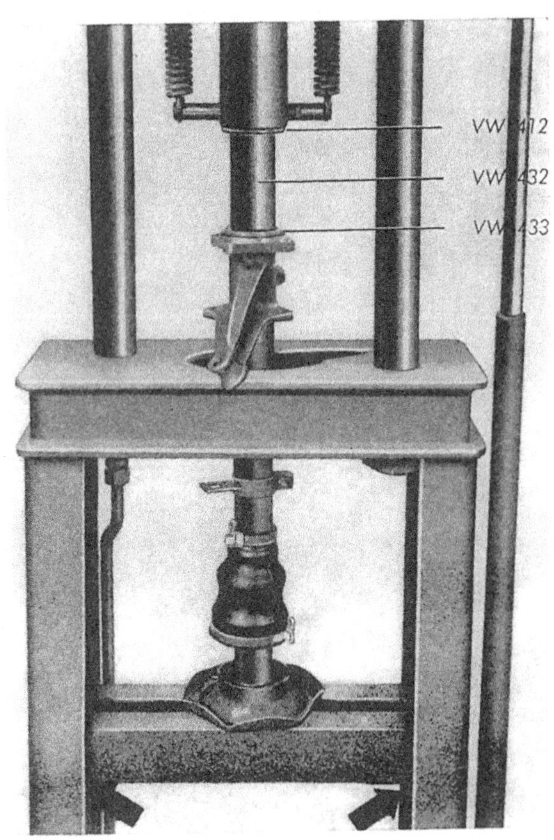

Installation

When installing, the following points should be observed:

1 - Clean axle tube retainer and axle tube retainer seat on final drive cover.

2 - Check dust sleeve for damage, replace if necessary.

3 - Before pressing the bearing housing into place, throughly clean and oil all seating surfaces. Press it into place by using VW Repair Press in conjunction with VW 412, 432 and 433. The cross member of the press should be raised by placing a 90 mm (3.5″) spacer under both sides.

4 - Check axle shaft, differential side gear, and thrust washer for damage and wear. Replace, if necessary, paying attention to the following points:

The fitting clearance between the flat end of the rear axle shaft (measured across the ball-shaped sides) and the inner diameter of the differential side gear is between 0.03 mm and 0.1 mm (.0012 and .004″).

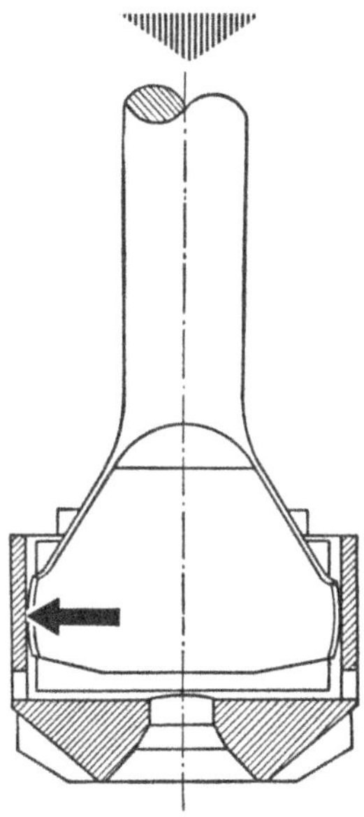

Rear axle shafts and differential side gears fall into three tolerance groups and should be mated accordingly:

Paint mark	Side gear inner diameter	Axle shaft outer diameter
blue	59.97–60.00 mm (2.3610″–2.3622″)	59.90–59.94 mm (2.3582″–2.3598″)
pink	60.01–60.04 mm (2.3625″–2.3638″)	59.95–59.97 mm (2.3602″–2.3610″)
green	60.05–60.07 mm (2.3642″–2.3650″)	59.98–60.00 mm (2.3614″–2.3622″)

Only side gears and axle shafts marked blue and pink are supplied as spares. The paint mark of the side gear is found on the recessed face in the form of a dot. The axle shaft is marked by a ring painted 150 mm (6″) from the flat end.

Excessive clearance can lead to rear axle noise.

Note:

The SP sets "52E" and "53E" have been discontinued since May 1964. In future, the side gears and pinions will be supplied as separate parts.

They must still be matched according to their tolerances.

Color	Side gear Inside dia.	Axle shaft Outside dia.
yellow	59.93–59.97 mm	59.87–59.90 mm
blue	59.98–60.00 mm	59.91–59.94 mm
pink	60.01–60.04 mm	59.95–59.97 mm

Note:

Grinding noises may occasionally be heard on new vehicles when moving off in first gear or in reverse gear. These noises originate in the rear axle between the differential side gears, fulcrum plates and the flanges of the axle shafts and continue until the parts are adequately broken in.

The noise does not indicate any mechanical fault and usually disappears during the break-in period of its own accord. There is no justification for disassembling the axle to locate the noise.

5 - Check rear axle shaft for run-out at ball bearing seat, the permissible run-out being 0.05 mm (.002″). If the run-out is found to be in excess of this limit, straighten rear axle shaft in VW Repair Press in conjunction with VW 405 and VW 406 in cold condition.

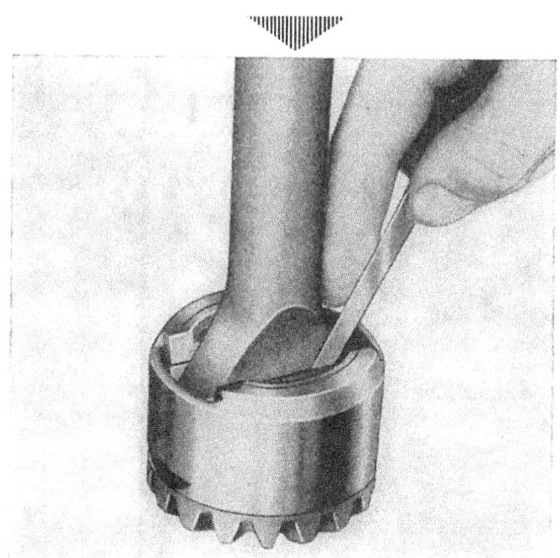

6 - Check fitting clearance of rear axle shaft/ fulcrum plates / differential side gear (0.035 to 0.244 mm/.0014—.0096″). In the case of excessive clearance install oversize fulcrum plates (provided with a groove on their faces) or replace worn parts.

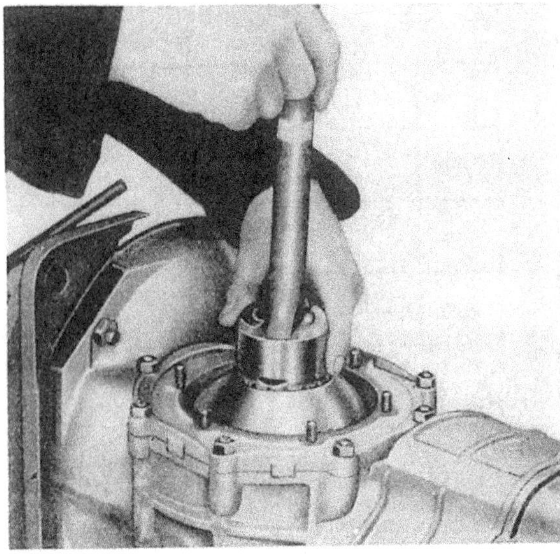

7 - Install differential side gear, axle shaft and thrust washer in differential housing and insert lock ring.

8 - The rear axle tube should be fitted without end play by selecting an axle tube retainer gasket of appropriate thickness. The maximum end play of 0.2 mm (.008″) must not be exceeded. Tighten nuts of axle tube retainer to a torque of 2 mkg (14 ft. lbs.).

9 - Do not tighten dust sleeve clips before the rear axle is installed to prevent the sleeves from becoming twisted and eventually damaged.

Renewing Rear Axle Dust Sleeve

(Rear Axle Installed)

General

To avoid removal or disassembly of the rear axle when renewing a damaged dust sleeve, a split type dust sleeve is available.

Removal

1 - Remove both retaining clips.

2 - Cut the damaged dust sleeve and remove it.

3 - Clean axle tube and axle tube retainer.

Installation

1 - Lightly coal joining faces of the split dust sleeve with sealing compound (VW Sealing Compound D 1 a).

2 - The joining faces of the split dust sleeve should point horizontally towards the rear.

Important

The dust sleeve screws and retaining clips should not be overtightened. The tightening should be carried out with the rear axle in the loaded condition. Take care the dust sleeve is not distorted or strained.

3 - Tighten dust sleeve screws. Install and tighten retaining clips (9 mm/.35″ in width).

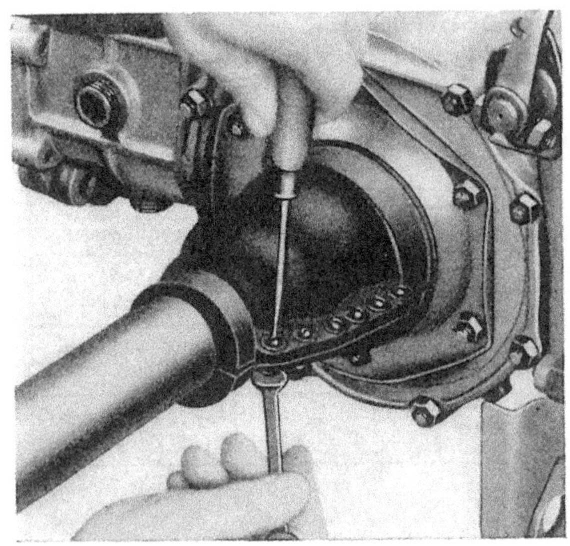

Note:

From January 1961, Chassis No. 3700001 (Rear Axle No. 3787854) the outer diameter of the axle tube retainer neck was increased from 85 mm (3.3″) to 89 mm (3.5″).

Two split type dust sleeves with different inner diameters on the axle tube retainer side are available for repairs:

Part No. 111 501 153 A,
inner diameter 85 ± 0.5 mm (3.3″ ± .02″)

Part No. 111 501 153 B,
inner diameter 89 ± 0.5 mm (3.5″ ± .02″)

The closed type of dust sleeve — Part No. 111 501 151 — is used in current production for both types of axle tube retainer.

Note:

Make sure that the correct dust sleeve is always used for all repair and sealing operations.

Removal and Installation of Transmission Carrier

Removal

1 - Remove the transmission carrier attaching screws and take off transmission carrier.

2 - Remove the nuts that attach the rubber mountings to the transmission case and take off mountings.

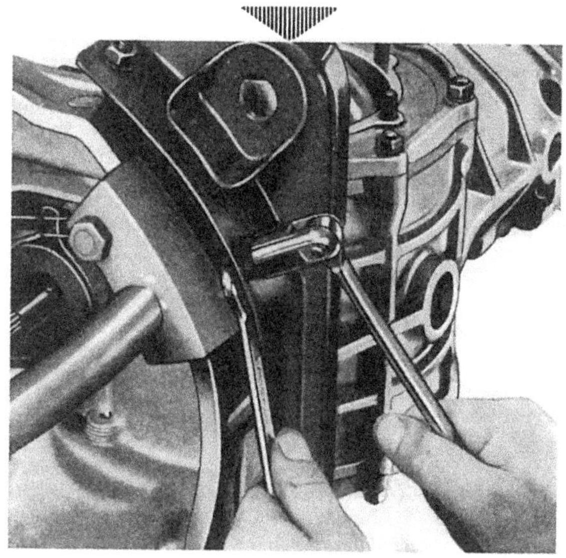

Installation

When installing, the following points should be observed:

1 - Inspect rubber mountings for damage, replace if necessary.

2 - Tighten transmission carrier attaching screws but not before the transmission is installed and the screws of the front rubber mounting are tightened.

Note:

From Chassis No. 4 477 631 (Rear Axle No. 4 611 625), the Shore hardness of the bonded rubber mounting was charged from sh 65 to sh 70. The mountings can be identified by the figure "70" embossed on the rubber. Only the new harder mountings will be supplied as spares. The part numbers remain unchanged.

Removing and Installing Gearshift Housing

Removal

1 - Remove the nuts that attach the rubber mounting to the gearshift housing and take off rubber mounting.

2 - Remove the nuts of the gearshift housing.

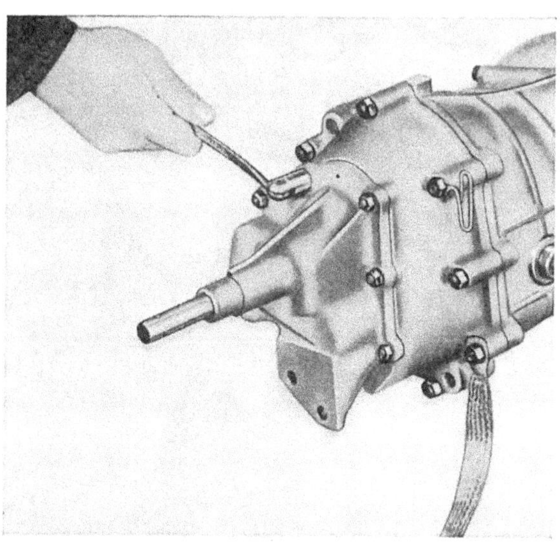

3 - Take off gearshift housing and transmission shift lever.

4 - Remove gasket and clean joining faces.

Installation

When installing, the following points should be observed:

1 - Inspect transmission shift lever bushes, replace if necessary. Worn bushes must be replaced by using the VW repair press in conjunction with VW 401, VW 412 and VW 439.

The bushes are to be reamed up to

a = 15 mm ϕ E7 =

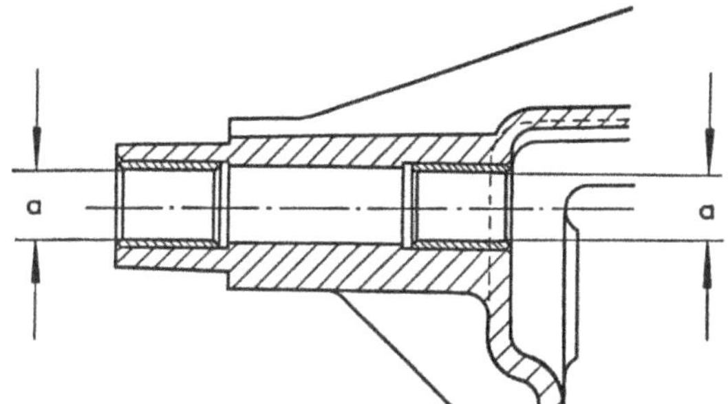

$$a = 15\ \varnothing\ E7\ \frac{15.050\ mm}{15.032\ mm} = \frac{.5925''}{.5918''}$$

2 - Make sure that the selector shafts are in neutral position when attaching the gearshift housing.

3 - Tighten nuts with a torque wrench to 2 mkg (14 ft. lbs.).

4 - Check rubber mounting for damage, replace if necessary.

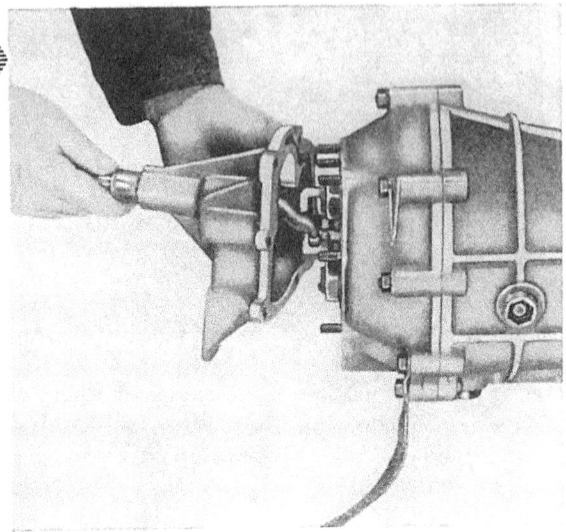

> **Important**
>
> If the rubber mounting is replaced, the rear wheels have to be re-adjusted.

Note:

1 - The softer engine-transmission mountings have, in some instances, caused excessive shift lever vibration to occur. This is particularly evident when the hardness of the bonded rubber mountings is at or below the lower tolerance limit and can cause the gears to jump out in districts where the roads are very bad.

The complaint can be rectified as follows:

a - Replace the front mounting Part Number 111 301 265 by the front mounting, Part. No. 211 301 265 used on the Transporter.

b - Replace the rear mounting. Part Number 111 301 263 B by the rear mounting, Part No. 111 301 263 used on the Transporter.

The nuts securing the mountings should be properly tightened as otherwise the holes in the gearshift housing will be enlarged. Gearshift housing with enlarged holes should be replaced.

If a hard knock is heard from the transmission when the clutch is engaged quickly in 1st or reverse gear, the neck of the gearshift housing is probably not central in the hole in the frame. In such cases, the holes for securing the mounting to the frame should be filed out as required.

2 - From Chassis No. 4 173 001 (Rear Axle No. 4 288 201) the previous front bonded rubber mounting Part No. 111 301 265 has been replaced by a progressive mounting Part No. 311 301 265. The progressive action is achieved by means of a lug which engages in the modified support on the rear cross tube.

The mounting is now secured to the gearshift housing with two M 10 studs instead of the former M 8 studs. The part number of the modified gearshift housing is 113 301 205 D.

Service Installation

As the support on the rear cross tube was also modified, the progressive effect cannot be attained by subsequently installing the new bonded rubber mounting. The former mounting — Part No. 111 301 265 — will remain available.

The previous type gearshift housing will be discontinued when stocks are exhausted. When the modified housing is installed subsequently, the new mounting must also be installed.

3 - From Chassis No. 4 477 631 (Rear Axle No. 4 611 625), the Shore hardness of the bonded rubber mounting was changed from sh 53 to sh 60. The mountings can be identified by the figure "60" embossed on the rubber.

Only the new harder mountings will be supplied, as spares. The part numbers remain unchanged.

Note:

4 - From May 1963, Chassis No. 5 540 290 (Rear Axle No. 5 893 615) the gearshift housing is manufactured without the internal rib. The lower web of the breather compartment has also been modified. The new Part. No. is 113 301 205 E. This will prevent the loss of oil from the breather hole in the gearshift housing which occurred occasionally on vehicles from November 1962 (Type 1 from Chassis No. 5 093 461).

Note:

As a temporary measure the internal rib was machined away on Type 1 vehicles from Chassis No. 5 447 350 (Rear Axle No. 5 845 236).

Service modification:

1 - Remove gearshift housing.

2 - Remove internal rib (A) in the housing.

3 - Mill the lower web of the breather compartment (B) away down to the base of the compartment for a width of 9 mm. Ensure that the transfer from the base of the compartment to the inner space is rounded off properly.

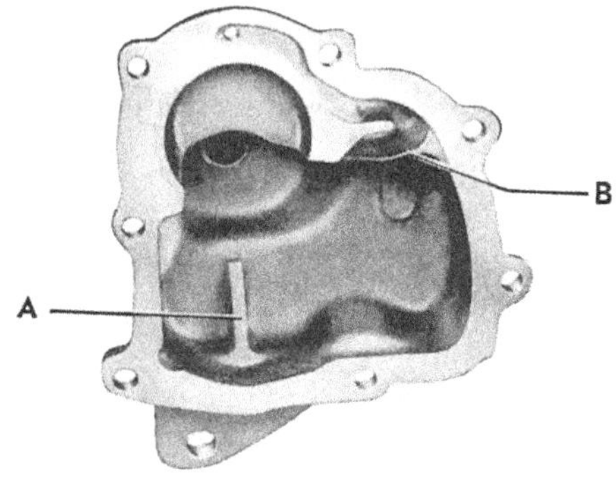

Former housing

Modified housing

Removal

1 - Remove gearshift housing.

2 - Pry off lock plates for drive pinion and main drive shaft nuts.

3 - Lock transmission by engaging reverse and third or fourth gears.

4 - Remove drive pinion and main drive shaft nuts and take off lock plates. The lock plates must not be re-used.

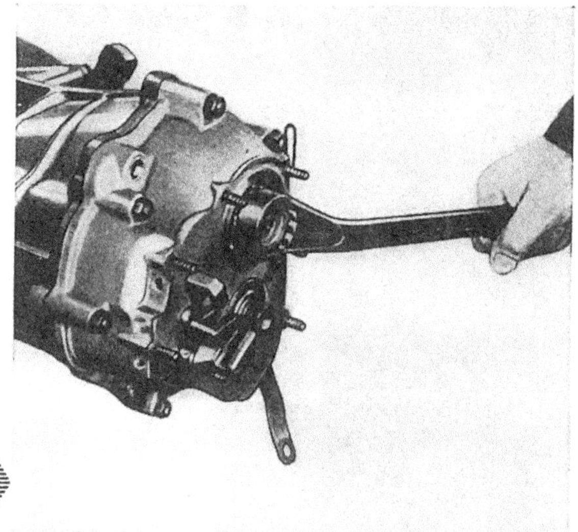

5 - Remove gear carrier stud nuts, take off ground strap and retainer for accelerator cable.

6 - Turn the transmission case so that the left-hand final drive cover faces upwards.

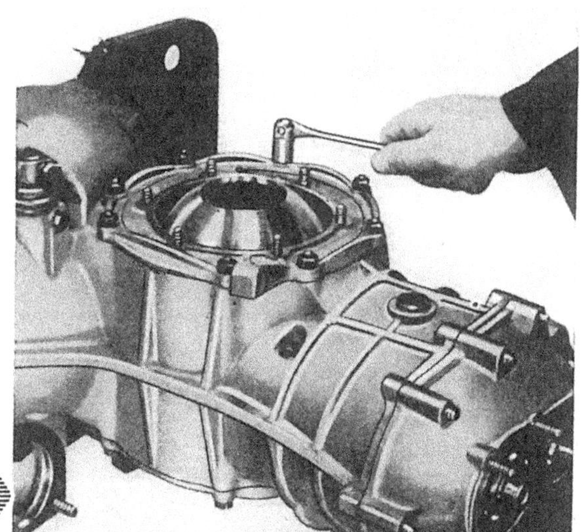

7 - Remove stud nuts of left-hand final drive cover.

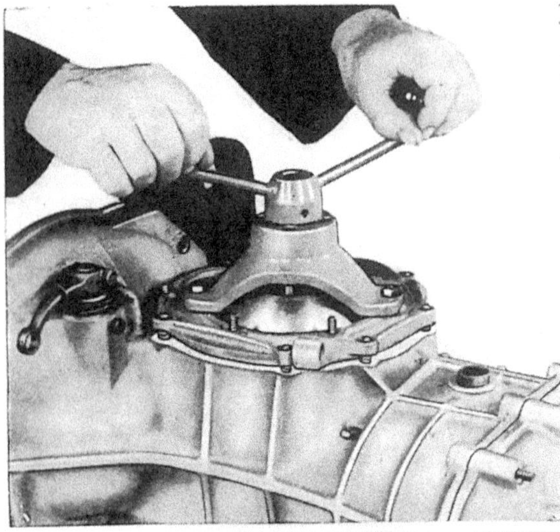

8 - Remove final drive cover by means of the Device VW 297. The thrust plate is positioned on the differential housing flange and the spindle is attached to two of the axle tube retainer studs.

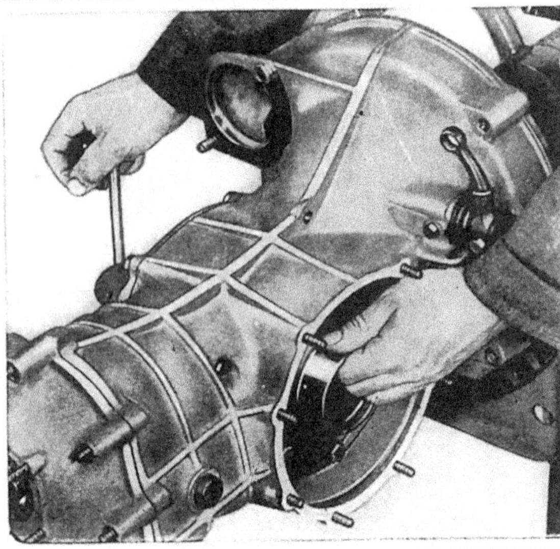

9 - Position the spindle and thrust plate on the right-hand final drive cover and press out the differential.

Important

When removing the differential, take a note of the thicknesses and arrangement of the differential shims to facilitate reassembly.

10 - Loosen retaining ring for reverse gear on main drive shaft. Slide reverse gear rearward and screw main drive shaft apart.

11 - Take off reverse gear and retaining ring and withdraw rear main shaft towards the rear, taking care not to damage the oil seal.

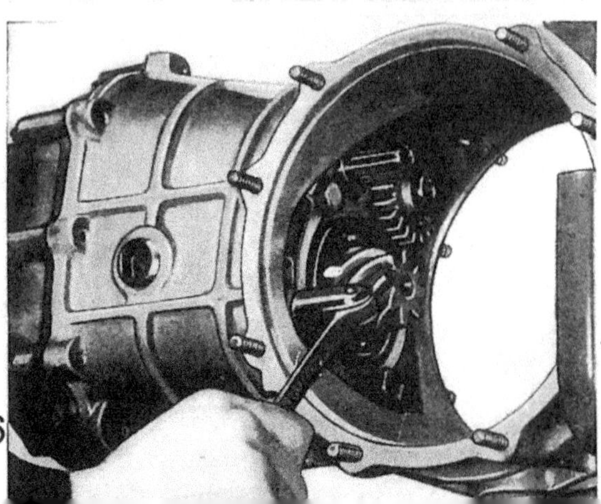

12 - Remove right-hand final drive cover.

13 - Pry up lock plates of screws attaching drive pinion ball bearing retainer and remove screws.

14 - Push transmission out of case by means of removal tool VW 296.

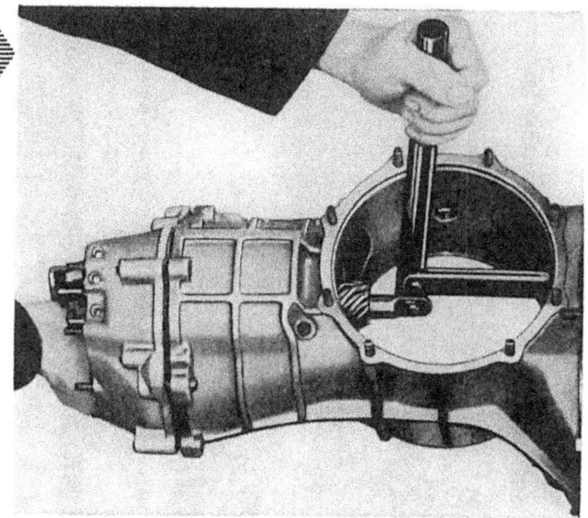

Important

In order to facilitate the assembly, it is advisable to note the thicknesses of the pinion shims.

15 - Remove snap ring and reverse drive gear from reverse gear shaft.

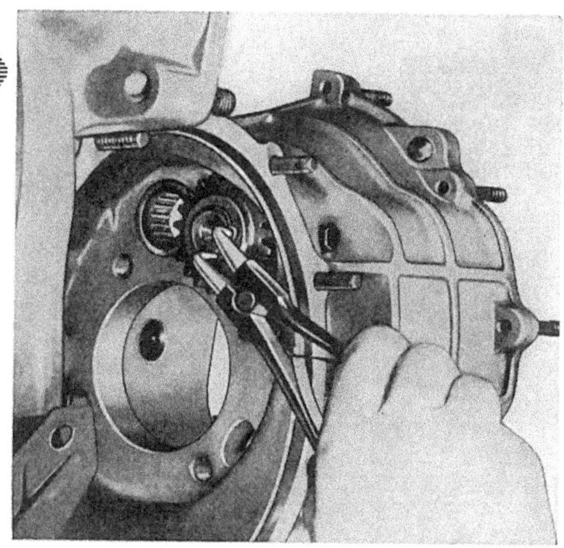

16 - Remove woodruff key and withdraw reverse gear shaft and thrust washer out of the transmission case.

17 - Remove screw which secures the spacer sleeve for the reverse gear shaft needle bearings.

18 - Drive out needle bearings for reverse gear shaft and spacer sleeve by means of the drift VW 295.

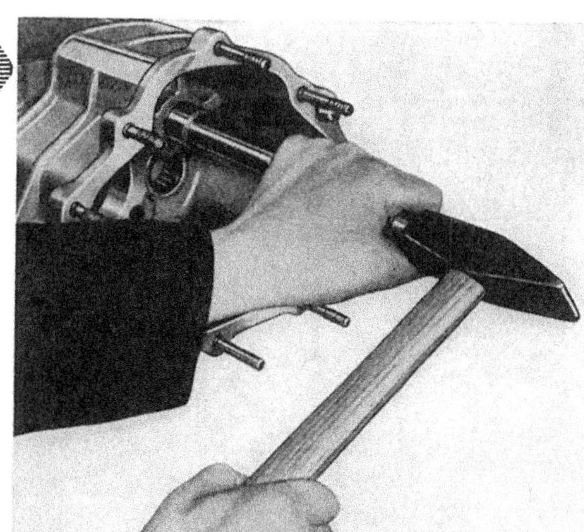

19 - Remove screw which secures the needle bearing on the main drive shaft.

20 - Drive out the main drive shaft needle bearing with the drift VW 295 in conjunction with VW 295 a.

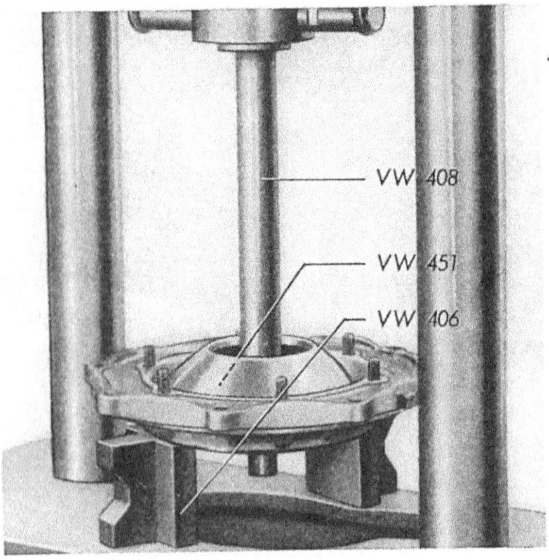

21 - Remove ball bearing from left and right final drive covers by means of VW Repair Press in conjunction with VW 406 (use two pieces), VW 408 and VW 451.

Note:

It can happen that the two ball bearings come out when removing the differential. In this case the bearings can be removed from the differential housing by using the extractor VW 202 in conjunction with the extractor hooks VW 202 a, the thrust plate of device VW 297 and the thrust pad VW 202 k.

22 - Remove clutch release bearing and operating shaft.

Installation

When installing, the following points should be observed:

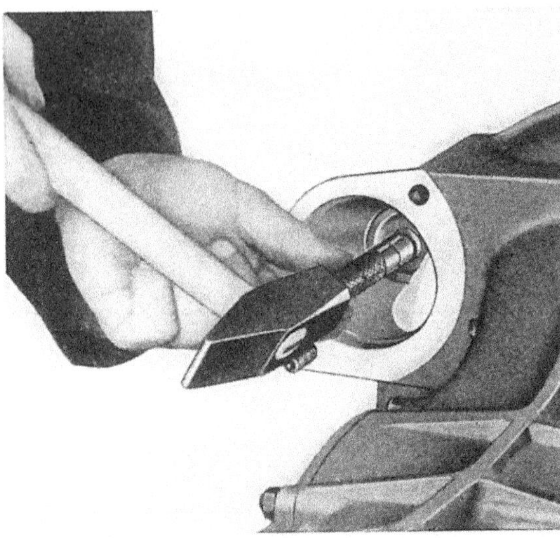

1 - Clean and inspect transmission case and final drive covers for damage. Replace damaged parts.

2 - The starting motor armature bush should be inspected for wear by means of the gauge VW 246. If worn, the bush should be replaced by means of the drift VW 222.

Note:

Replacing the starting motor armatures bush with the engine in the vehicle:

Removal: the extractor VW 228a.
Installation: the drift VW 222.

3 - Check clutch operating shaft bushes for wear, replace if necessary.

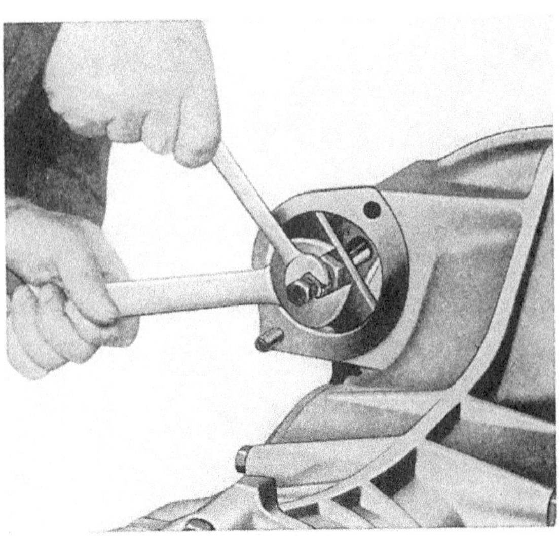

Important

The clutch operating shaft should be checked at every transmission repair for free movement, and special grease VW—051 applied, if necessary.

4 - Check all bearings before installation, and renew if necessary.

5 - Insert needle bearings for reverse gear shaft and spacer sleeve by means of the drift VW 295 and secure.

6 - Install main drive shaft needle bearing with the drift VW 295 in conjunction with VW 295 a and secure.

7 - Install reverse gear shaft with thrust washer and drive gear. Do not forget the woodruff key. Check snap ring for proper tension.

8 - Place drive pinion shims over ball bearing and screw two studs (approx. 100 mm long/4″) into ball bearing retainer. This prevents the retaining ring from turning when installing the transmission.

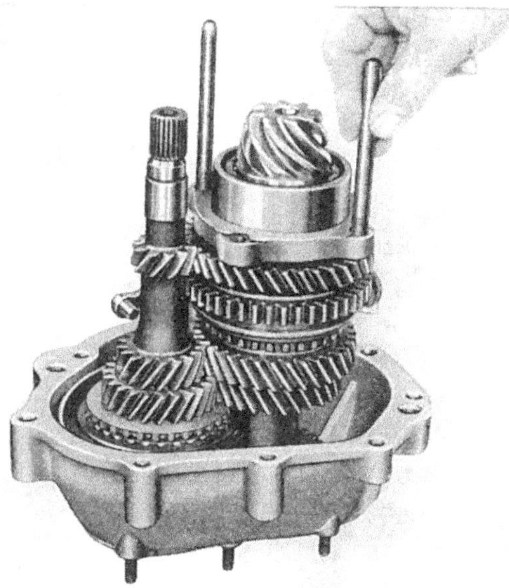

9 - Push the reverse selector fork and sliding gear on to reverse lever and engage reverse gear.

Note:

On very steep inclines it was occasionally found that the reverse gear jumped out of engagement. To eliminate this trouble, the width of 3 of the splines in the reverse sliding gear, spaced at 120°, was reduced by 0.2 to 0.3 mm from February 1965, Chassis No. 115 400 207 so that more pressure was placed on the other teeth.

Only the modified sliding gears (Part No. 113 311 531 A) will be supplied as replacement parts. If the sliding gear is replaced because of jumping out of gear, the 1st and 2nd gear operating sleeve (Part No. 311 311 255) should be replaced at the same time.

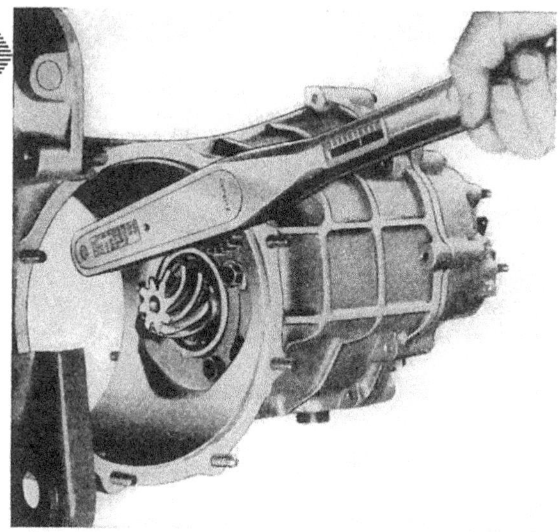

10 - Insert transmission into transmission case. Use a rubber hammer to position the pinion correctly in the bearing seat. Use a new gear carrier gasket.

11 - Tighten ball bearing retainer screws to 5 mkg (36 ft. lbs.). Use new lock plates.

Important

Only use screws of the quality specification "10 K".

12 - Apply oil to the lip of the oil seal before installing the rear half of the main drive shaft. Screw both halves of the drive shaft together. Back them off until the splines for the reverse gear are in line. The halves of the main drive shaft must not be screwed tightly together. Make sure that the pretension of the reverse gear snap ring is correct.

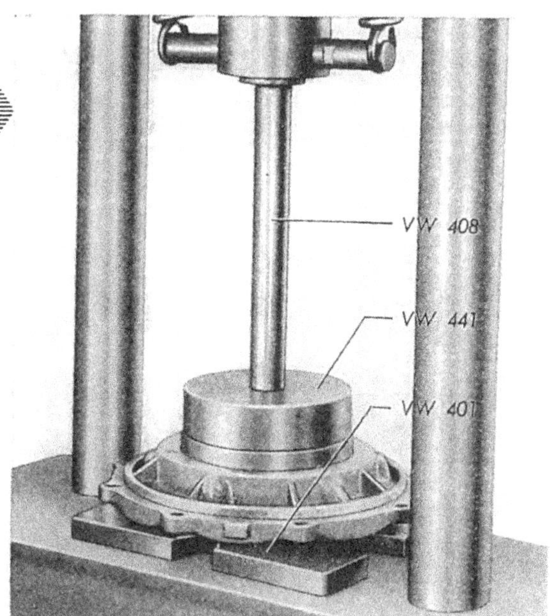

13 - Press ball bearings into left and right final drive covers on the VW repair press in conjunction with VW 401, VW 408 and VW 441.

14 - Check differential and recondition it if necessary.

15 - Install right final drive cover and use a new gasket. Tighten nuts to 2.5 mkg (18 ft. lbs.).

Note:

a - The final drive covers can settle slightly when the vehicle is first used. The prescribed tightening of the rear axle nuts after 500 km (300 miles) should also be carried out after repairs so as to prevent leaks.

b - From Chassis No. 4297692 the material for the studs and nuts of the final drive covers has been improved and the tightening torque altered to 3 mkg (22 ft. lbs.).

Only transmission cases — Part No. 113301051 C with the new studs will be supplied as spares. Until further notice these cases and those installed in production will be marked with a yellow paint spot near the Part No.

When carrying out repairs on the former rear axle, the studs should be replaced by the new type (Part No. N 144004) with the aid of two locked nuts, and hexagon nuts — N 110085 — then installed.

16 - Install differential in transmission case. Make sure that the shims are inserted correctly.

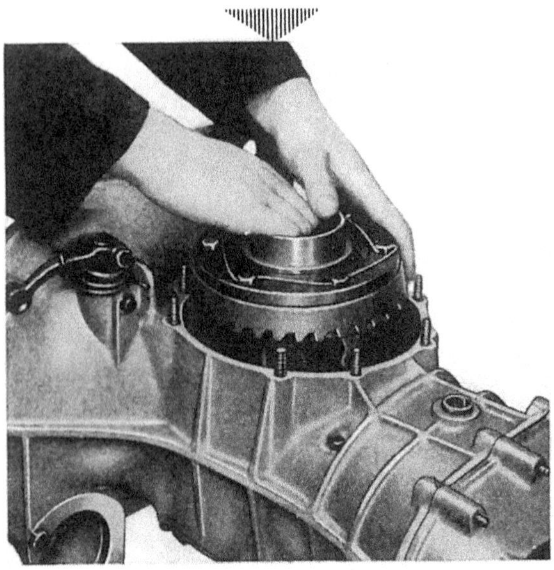

17 - Tighten gear carrier stud nuts to 2 mkg (14 ft. lbs.).

Note:

From April 1965, Chassis No. 115685587 the final drive covers, new Part No. 113301179 B/180 B, were modified. The paper gasket has been discontinued and it is replaced with a rubber seal, Part No. 113301185 A. The transmission case, new Part No. 113301051 F, is being machined deeper to accommodate the new final drive cover.

The clutch cable eye on the left final drive cover has been relocated, and a straight clutch operating lever, new Part No. 131141719, is now being installed. The spherical nut for adjusting the clutch cable has been replaced with a wing nut, Part No. 131721349.

18 - Block the transmission by engaging both the reverse and third or fourth gears.

19 - Tighten main drive shaft nut with a torque wrench and 32 mm socket to 12 mkg (87 ft. lbs.). Back off again, and tighten to 6 mkg (43 ft. lbs.). Secure with lock plate.

20 - Tighten drive pinion nut to a torque of 6 mkg (43 ft. lbs.) and secure with lock plate.

21 - Make sure that the three selector shafts are in neutral position when attaching the gear shift housing.

Service installation

The modified final drive covers cannot be installed in transmission cases of the former design. It is possible, however, to install final drive covers of former design with paper gaskets in new transmission cases. In such a case, ensure that the bent clutch operating lever, Part No. 111141719 A, and the spherical nut are also installed.

Gear Carrier Disassembly and Assembly

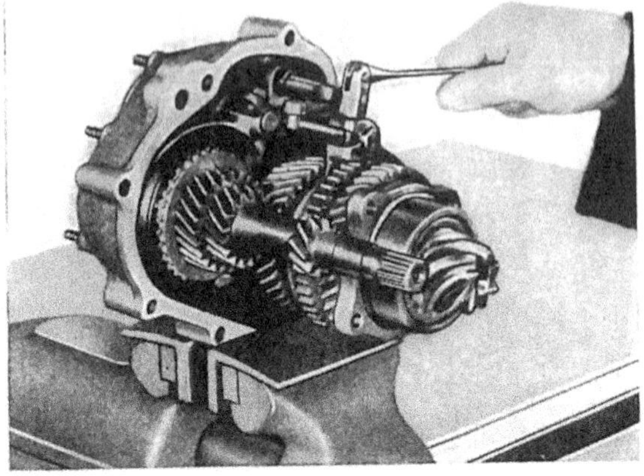

Disassembly

1 - Remove reverse selector fork including reverse sliding gear from reverse lever.

2 - Remove shims from drive pinion ball bearing. Note the thickness of the shims.

3 - Place gear carrier in vice equipped with aluminum clamps. Loosen locking screws of the first-and-second and third-and-fourth selector forks, and remove selector fork for first and second gears.

4 - Fully withdraw the selector shaft for third and fourth gears out of the selector fork.

5 - Place rubber band around the operating sleeve or first and second gear, and main drive shaft.

6 - Place gear carrier into VW452. Use VW Repair Press in conjunction with VW412 and VW434 to remove transmission from gear carrier. The force should be applied to the main drive shaft. The drive pinion should be carefully guided to avoid tilting which could lead to damage of the gear or needle bearing in the gear carrier.

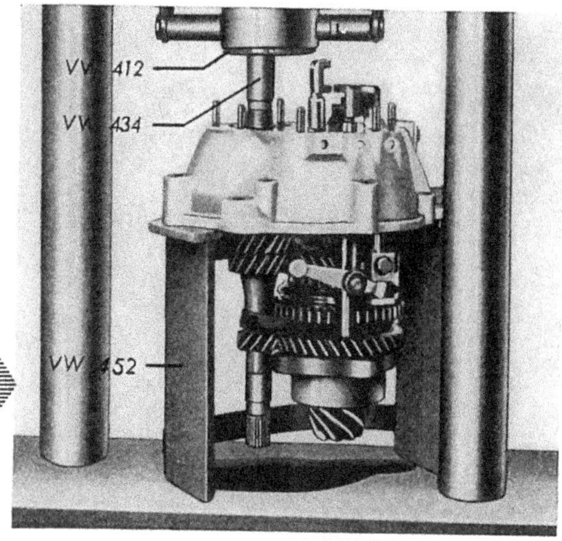

7 - Remove screw that secures drive pinion needle bearing and press out needle bearing.

8 - Press out ball bearing for main drive shaft on the VW Repair Press in conjunction with VW 401, VW 408, VW 433 and VW 441.

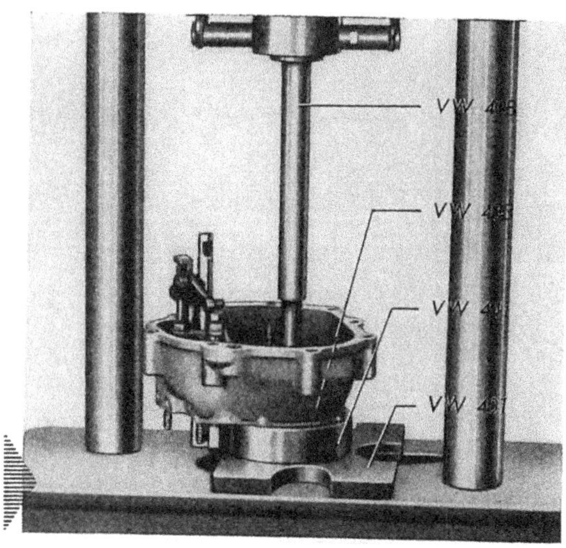

9 - Place gear carrier in vice (use aluminum clamps) and remove screw on reverse lever guide.

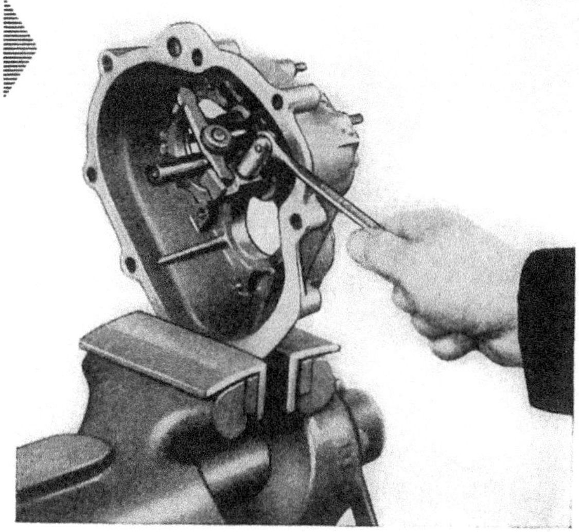

10 - Withdraw reverse gear selector shaft and remove reverse lever guide.

11 - Withdraw selector shaft for first and second gears and remove reverse lever from the support.

12 - Remove selector shaft for third and fourth gears.

13 - Take out plungers and detent balls.

14 - Remove springs with a small screwdriver.

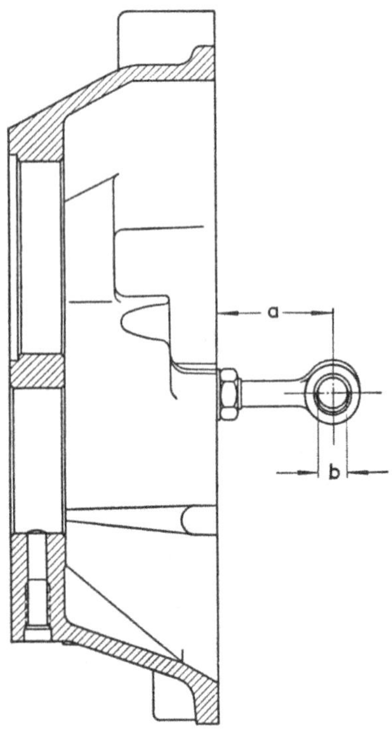

a = 38.6 ± 0.4 mm (1.520 ± 0.016″) b = 10 mm (0.400″)

Note:

The reverse lever support does not usually need to be removed. If, however, it is necessary to do so, the dimensions given to the left should be observed. Check the reverse lever for correct positioning when adjusting selector forks in gearshift test appliance VW 294. The lock nut should be tightened to 3.5 mkg (25 ft. lbs.).

Assembly

When assembling, the following points should be observed:

1 - Check detent springs, replace if necessary. Free length of detent springs: 25 mm (1″); wear limit 23 mm (0.9″). The force applied to overcome detent ball grooves on the selector shafts should be approximately 15—20 kg (33—44 lbs.). A test should be carried out with the drive pinion and main drive shaft removed, if difficult gear shifting is experienced.

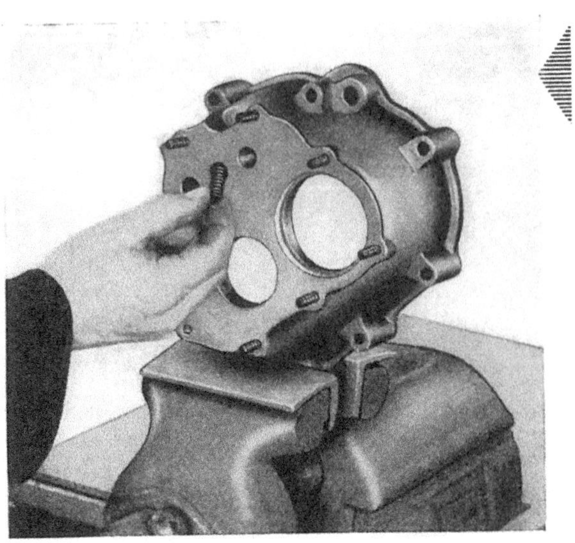

2 - The detent springs should be inserted through the holes for the selector shafts. Since the top halves of the detent spring bores are without bushes, the detent springs for 1st and 2nd and reverse gears can be more easily installed by inserting them into the top halves first.

3 - Install reverse selector shaft including reverse lever and reverse lever guide.

4 - Install selector shafts for first and second and for third and fourth gears taking care not to omit the two interlock plungers. Check for proper interlocking by engaging a gear. The selector shaft next to the one used must be locked. When engaging first or second gears, the two other selector shafts should be locked

5 - Check needle bearing for drive pinion and ball bearing for main drive shaft, replace as necessary. Secure drive pinion needle bearing after it has been installed in the gear carrier.

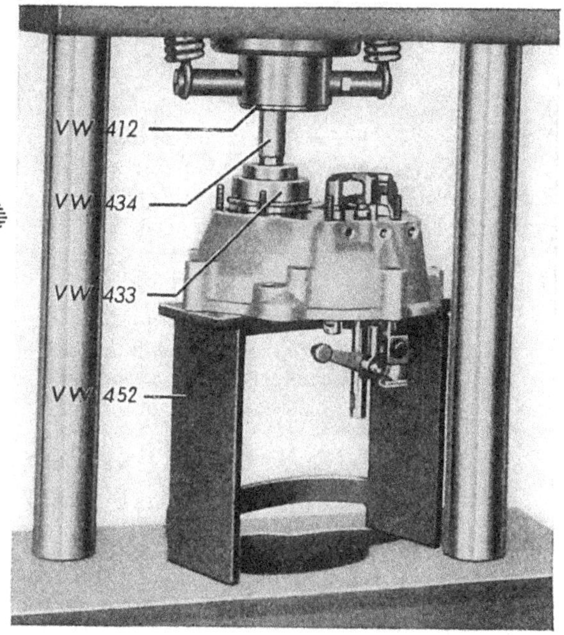

6 - Place gear carrier on support VW 452 and press main drive shaft ball bearing into position by means of VW repair press in conjunction with VW 412, VW 433, and VW 434.

7 - Check selector forks for wear. The clearance between selector forks and operating sleeves should be 0.1—0.3 mm (0.004—0.012"). Replace worn parts.

8 - Check main drive shaft and drive pinion. Recondition if necessary.

9 - Press transmission into gear carrier on the VW repair press in conjunction with VW 401, VW 412, and VW 422. The selector fork for the 3rd and 4th gears should be positioned in the operating sleeve beforehand. When pressing, the drive pinion should be lifted slightly and care taken that the selector fork for 3rd and 4th gears does not become jammed on the selector shaft. Fully insert the selector shaft into the fork beforehand.

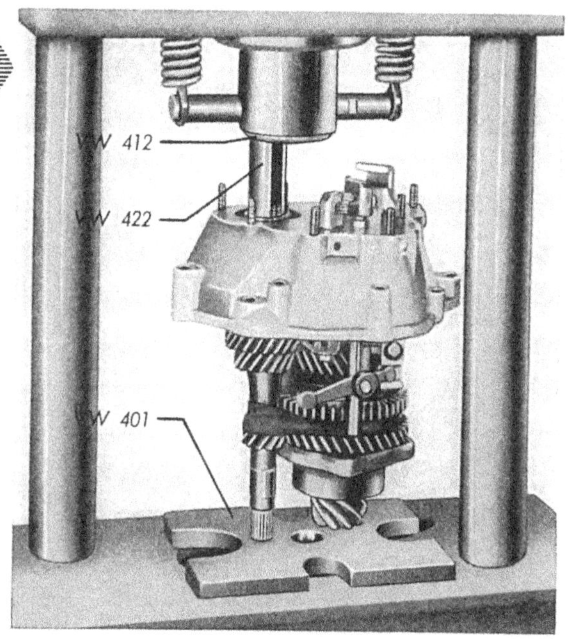

Note:

When pressing the transmission into position, it is recommended that the drive pinion and main drive shaft be held together by placing a rubber band round the operating sleeve for the 1st and 2nd gears and the main drive shaft.

10 - Install 1st-and-2nd selector fork.

11 - Attach reverse gear selector fork with reverse sliding gear on to reverse lever.

12 - Adjust selector forks.

Adjustment of Selector Forks

The adjustment of the selector forks can only be carried out properly with the help of the gear shift test appliance VW 294. Since the adjustment of the 1st-and-2nd and reverse selector forks alters in accordance with the adjustment of the drive pinion, the drive pinion adjustment must be carried out beforehand. Apart from that, the nuts for the drive pinion and main drive shaft must be tightened beforehand to the prescribed torque.

1 - Place transmission with drive pinion shims and gasket for gear carrier on gear shift test appliance VW 294 and attach gear carrier with 4 screws.

2 - Tighten drive pinion ball bearing retainer with two screws diagonally opposed to 5 mkg (36 ft. lbs.).

3 - Push crank of the test appliance onto splines of main drive shaft so that the main drive shaft is locked by the crank handle. Engage 1st or 2nd gears.

4 - Tighten main drive shaft nut with a torque wrench to 12 mkg (87 ft. lbs.). Loosen nut and then retighten to 6 mkg (43 ft. lbs.) and secure it.

5 - Tighten drive pinion nut to a torque of 6 mkg (43 ft. lbs.) and secure it.

7 - Set selector forks for 1st/2nd and 3rd/4th gears so that the forks more freely in the operating sleeve not only in neutral position, but also when the different gears are engaged.

8 - Set reverse gear selector fork so that the reverse sliding gear is in the middle between operating sleeve and 2nd gear of main drive shaft with the 2nd gear engaged and engages properly in the reverse gear on the drive pinion when the reverse gear is engaged.

9 - The locking screws of selector forks should be tightened to a torque of 2.5 mkg (18 ft. lbs.), using a T-handle torque wrench in conjunction with an 11 mm socket. Tighten the reverse lever guide screw to 2 mkg (14 ft. lbs.).

6 - Attach gearshift housing and shifting handle. By attaching the gearshift housing, a proper seating of the main drive shaft ball bearing in the gear carrier is ensured.

10 - Remove gearshift housing and take out transmission.

Replacing Main Drive Shaft Oil Seal

(With Rear Axle Either Installed or Removed)

Removal

1 - Remove engine.

2 - Remove clutch release bearing.

3 - Remove the damaged oil seal carefully from transmission case with the removal tool VW 681.

Installation

When installing, the following points should be observed:

1 - Lightly coat the exterior of the oil seal with sealing compound. Oil main drive shaft and oil seal lip.

2 - Slide oil seal on main drive shaft and drive it in position by means of the Driving Sleeve VW 244 b.

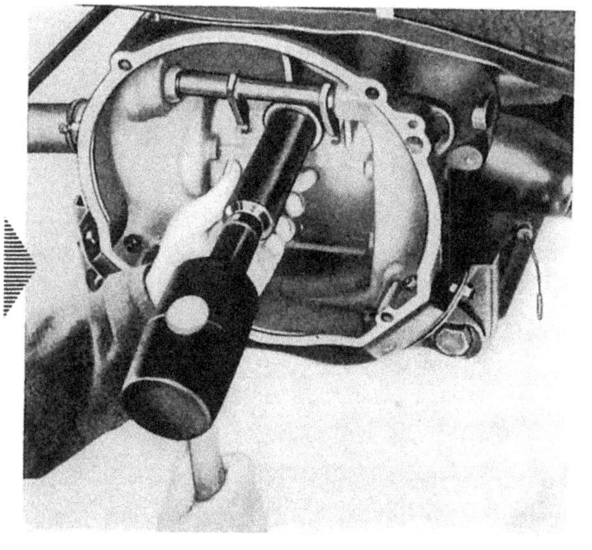

> **Important**
> Carefully slide the oil seal on the shaft to avoid the spring around the lip coming out of place.

THIS PAGE INTENTIONALLY
LEFT BLANK

Reconditioning the Drive Pinion

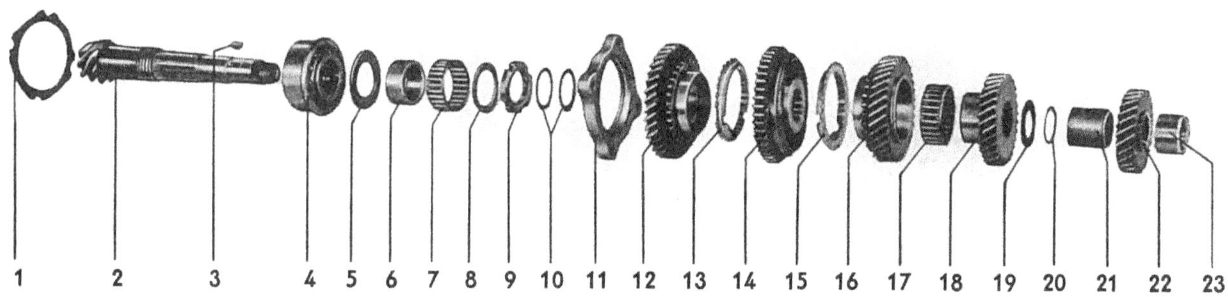

1 - Shim
2 - Drive Pinion
3 - Woodruff key for 4th gear
4 - Ball bearing
5 - Thrust washer for 1st gear
6 - Needle bearing inner race (1st gear)
7 - Needle cage (1st gear)
8 - Thrust washer for needle bearing (1st gear)
9 - Round nut
10 - Shims, end play 1st gear
11 - Ball bearing retainer
12 - 1st gear
13 - Synchronizer stop ring (1st gear)
14 - Clutch gear for 1st and 2nd gears, and reverse gear
15 - Synchronizer stop ring (2nd gear)
16 - 2nd gear
17 - Needle cage (2nd gear)
18 - 3rd gear
19 - Concave washer
20 - Shims for concave washer.
21 - Spacer sleeve
22 - 4th gear
23 - Inner race, needle bearing in gear carrier

Disassembly

1 - Press out inner race of needle bearing and 4th gear by means of VW Repair Press in conjunction with VW 401 and 408 and remove woodruff key for 4th gear.

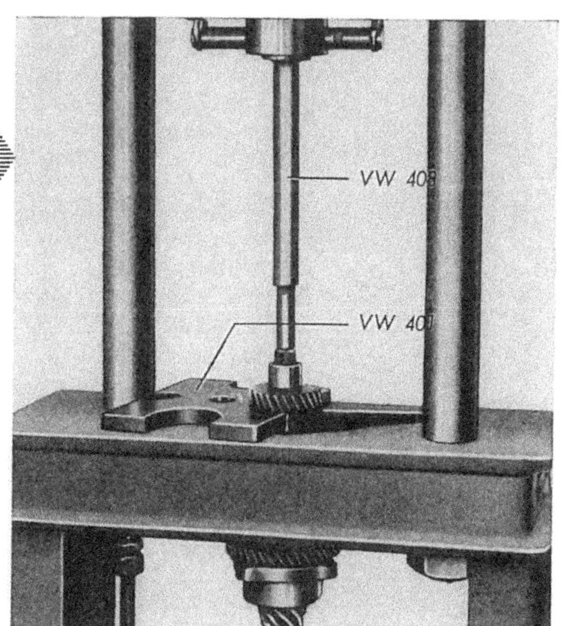

2 - Take off spacer sleeve, concave washer shims and concave washer.

3 - Remove 2nd and 3rd gears including needle cage and synchronizer stop ring for 2nd gear.

4 - Remove clutch gear for 1st and 2nd gears complete with springs, shifting plates, and operating sleeve and disassemble.

5 - Remove synchronizer stop ring, 1st gear and ball bearing retainer.

6 - Remove the shims for 1st gear.

7 - Remove round nut with a 32 mm box wrench in conjunction with tool VW 293.

8 - Remove thrust washer and needle cage for 1st gear.

9 - Remove inner race of needle bearing, thrust washer for 1st gear and ball bearing by means of VW Repair Press in conjuction with VW 401, 409 and 449 f.

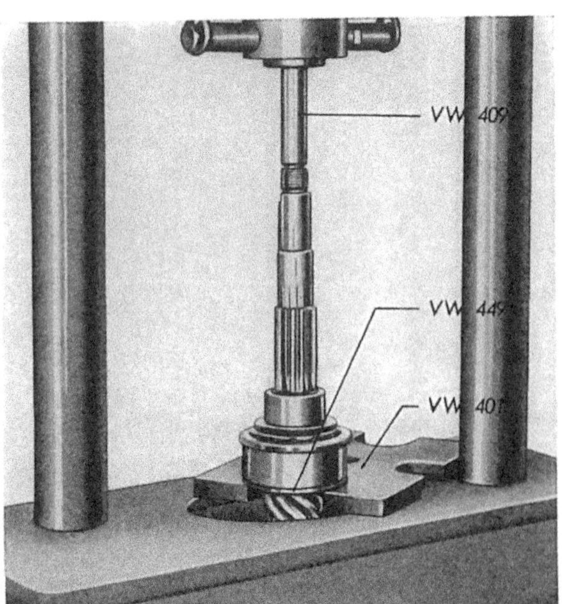

Inspection

1 - Inspect drive pinion for wear and damage, if necessary replace drive pinion and ring gear (crown wheel) as a pair. (Note matching number on pinion and ring gear).

2 - Check condition of ball bearing and needle bearing; replace if necessary.

> **Important**
>
> If the drive pinion or ball bearing are replaced the drive pinion and ring gear have to be re-adjusted.

3 - Check gears for wear and damage; replace if necessary.

> **Important**
>
> Whenever a damaged gear is replaced, the mating gear should also be replaced. Worn or damaged 1st and 2nd gears necessitate a replacement of the front main drive shaft.

4 - Check all synchronizer components for wear:

a - Clean the internal cone surface of the stop rings with a wire brush.

b - Check the clearance between the stop ring face and the clutch teeth of the corresponding gear (normal a = 1.1 mm/.043"). If the wear limit is found to be reached (a = 0.6 mm/.024"), the stop ring should be replaced. Premature wear of the stop rings is attributable to the clutch being in a bad condition or the driver incorrectly operating the clutch.

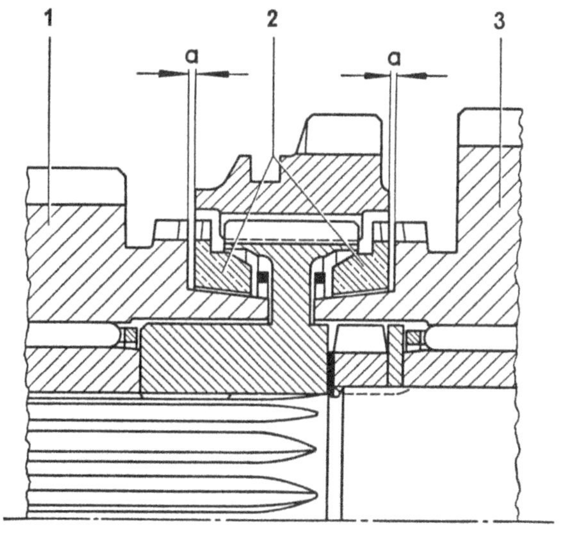

1 - 2nd gear
2 - Synchronizer stop rings
3 - Ast gear

c - If a gear will not engage, although the clutch is fully released, it may be due to the teeth of the stop ring being too much out of line with the splines of the operating sleeve. This condition is caused by the slots in the stop ring being worn.

d - Replace worn parts.

5 - Check condition of thrust washers and shims; replace if necessary.

Assembly

When assembling, the following points should be observed:

A - Ball bearing

1 - Before being pressed into position, the inner races of the ball bearing and the needle bearing inner race for the 1st gear should be heated in an oil bath to approx. 90° C (194° F).

2 - Slide one of the ball bearing inner races on to the drive pinion.

3 - Slide the ball bearing onto drive pinion and then slide the second inner race on in such a manner that the bearing numbers indented in both races (305309, 505921 or 2—51) are exactly opposite to one another.

> **Important**
> Noisy bearing operation may be caused if the races are not correctly positioned.

4 - Slide 1st gear thrust washer and needle bearing inner race onto the drive pinion.

5 - Press home all parts on the Repair Press in conjunction with VW 401, VW 411 and VW 430.

6 - Tighten round nut with a torque wrench to 12 mkg (87 ft. lbs.) in conjunction with VW 293.

7 - Complete assembly of pinion as described at points 7 to 15 under B.

B - Taper roller bearing

From September 1964, Chassis No. 115 085 239 the ball bearing on the drive pinion was replaced by a double taper roller bearing. When installed, the bearing should have a certain pre-load which can only be checked properly with a torque gauge. Note the following when installing:

1 - Heat the inner rings of the taper roller bearing and the 1st gear needle bearing to about 100°C.

Note:

Two types of double taper roller bearing are being used. One bearing has a spacer ring between the two inner rings and the other one has slightly wider inner rings which contact each other.

2 - Install double taper roller bearing, thrust washer for 1st gear and inner needle bearing ring on the pinion.

3 - Cool pinion to room temperature in benzin or kerosene then press all parts fully home with VW 401, VW 411 and VW 430, using a pressure of about 3 tons.

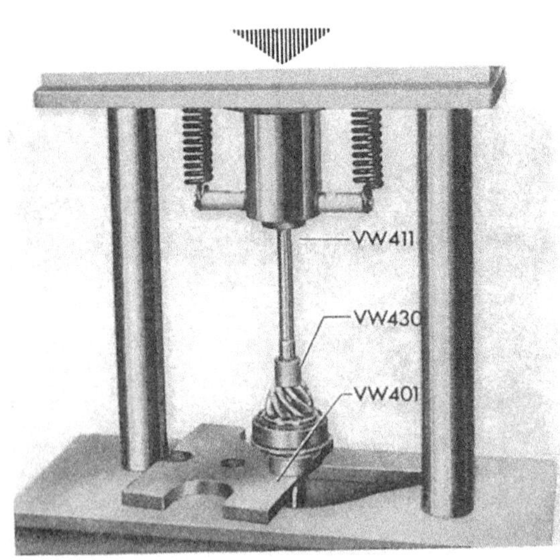

4 - Fit needle cage and thrust washer for 1st gear and tighten round nut to 15—20 mkg (108—144 ft. lbs.).

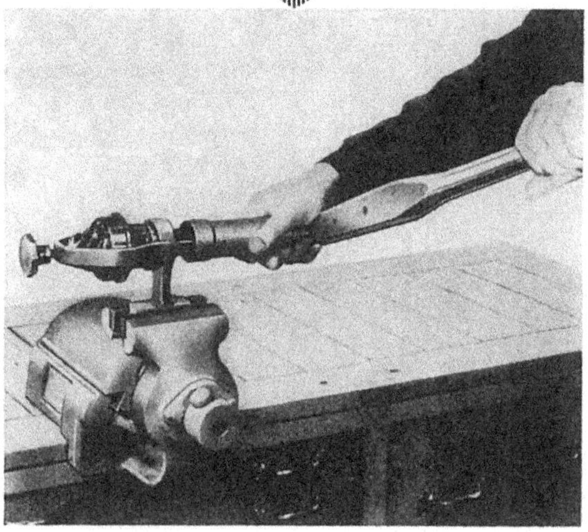

5 - Install pinion in transmission housing and tighten retainer screws to 5 mkg (36 ft. lbs.).

6 - Place socket from special tool VW 293 on the pinion.

Fit torque gauge and check torque. Turn the pinion about 15 to 20 revolutions in each direction first and then read the torque value while turning steadily.

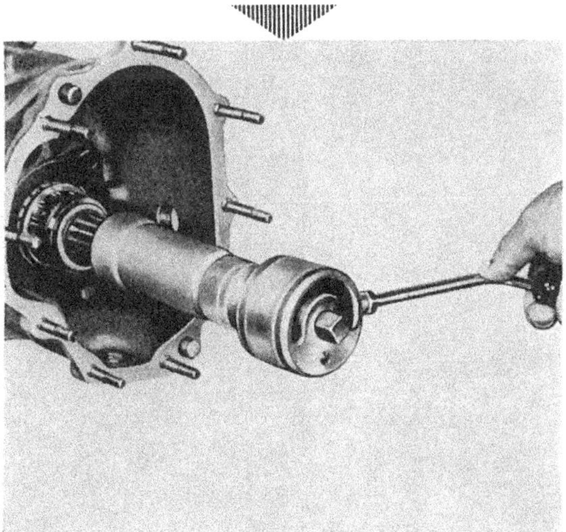

Important

It is essential that the torque is checked. When fitting new bearings, it should be 6—21 cmkg (5—18 in. lbs.). When the bearings have been run in, which covers bearings from vehicles which have done more than about 500 km (300 miles), the torque should be between 2 and 7 cmkg (1.7 and 6.1 in lbs.). On no account may the torque be 0 or the bearing have any end play.

If the torque is not in the specified ranges, it can only be corrected by using another bearing, housing or pinion.

7 - Install shims for 1st gear. The end play (0.10 to 0.25 mm/.004—.010″) should be checked between thrust washer and 1st gear after the clutch gear for 1st and 2nd gears has been installed, and corrected if found necessary.

Shims of 0.10, 0.15, 0.20, 0.25, 0.30 and 0.40 mm are available for selective fitting.

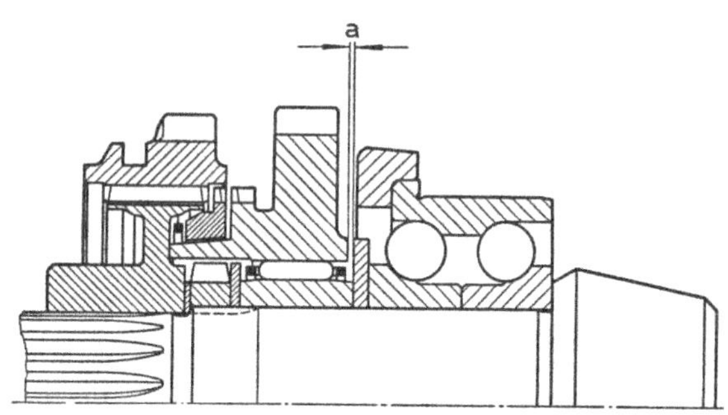

a = 0.10—0.25 mm (.004″—.010″)

Note:

From April 1964, Chassis No. 6284358 the 1st speed gear wheel on the pinion was widened from 11.8 mm (.464″) to 13.55 mm (.532″) and the oil grooves in the gear wheel discontinued. The thrust washer between the 1st speed gear wheel and the bearing now has two oil grooves on each side.

At the same time, the teeth of the 1st speed gear wheel on the main drive shaft have been widened 2 mm (.078").

The screws for the pinion bearing retainer have been shortened from 35 mm to 34.6 mm to prevent them from rubbing on the 1st speed gear wheel.

Part Nos. of the new parts:

1st speed gear wheel on pinion	113311251 A
Drive shaft, front	113311101 A
Thrust washer between 1st speed gear wheel and ball bearing	113311221 A
Hexagon head screw	113311215

Service installation

Note the following when installing individual parts of the modified version:

1 - The drive shaft, front — 113311101 A — and the 1st speed gear wheel on the pinion —113311251 A — can be installed individually when necessary.

2 - The previous type of thrust washer must **not** be installed together with the wider 1st speed gear wheel without oil grooves.

3 - Only the shorter bearing retainer screws — 113311215 — may be installed together with the wider 1st speed gear wheel on the pinion. If necessary, the shanks of the previous screws must be shortened to 34.6 mm (1.35").

Note:

1 - From November 1961, Chassis No. 4274401 (Rear Axle No. 4371905) the synchro clutch gear was turned out as far as the center web to give better centering of the synchronizer rings in the clutch gear. The Part No. of the clutch gear for 1st and 2nd gears (113311243) was not changed.

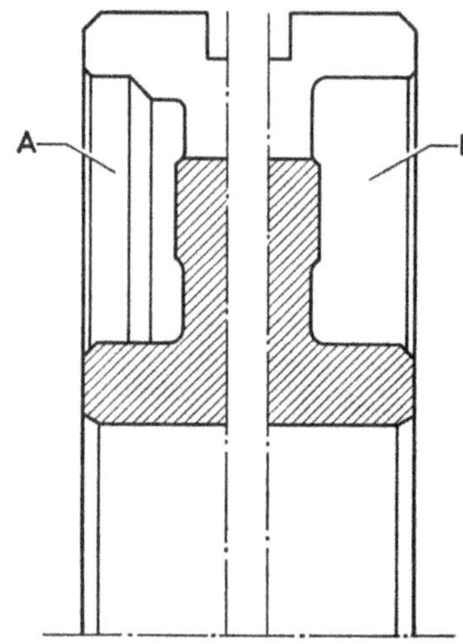

A = old version
B = **new** version

2 - The width of the shoulder on the synchronizer ring (a) was increased from 3 to 5 mm from March 1962, Chassis No. 4572833 (Rear Axle No. 4763480). The Part No. of the modified synchronizer ring is 113311247 A/295 A. Rings of the former type — Part Nos. 113311247/295 — will remain available.

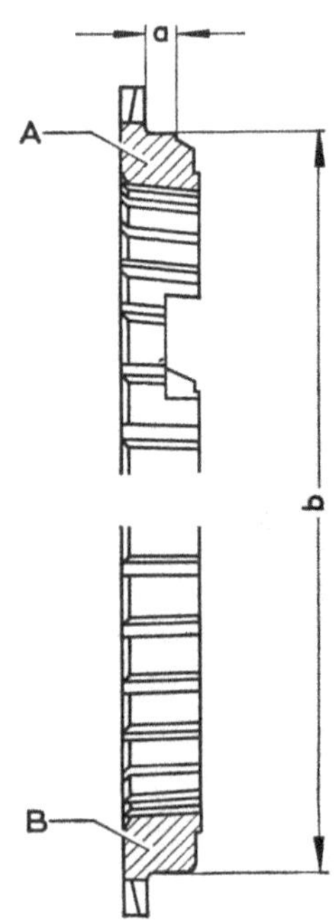

A = old version
B = **new** version

8 - Place 1st speed stop ring on the cone surface of the gear. The synchronizer stop rings for the 1st and 2nd gears are not interchangeable.

9 - Assemble synchro unit for 1st and 2nd gears. Slide the operating sleeve on the clutch gear so that its shifting plate slots are in line with the slots in the clutch gear. Put the shifting plates in position, and install the two snap rings offset to one another. Make sure that the ends of each ring engage behind the shifting plates.

Important

It should be noted that the modified synchronizer rings (B) should not be installed together with the old clutch gear (A) as this could cause seizure of the rings.

The following may be installed together:

1 - Synchronizer clutch gear (B)
Part No. 113 311 243
Synchronizer rings (B)
Part No. 113 311 247 A/295 A

2 - Synchronizer clutch gear (B)
Part No. 113 311 243
Synchronizer rings (A)
Part No. 113 311 247/295

3 - Synchronizer clutch gear (A)
Part No. 113 311 243
Synchronizer rings (A)
Part No. 113 311 247/295

The nominal dia. of the synchronizer ring (b = 70.2—0.15/2.762—.006") should be as near the upper limit as possible. The end play for the 1st speed gearwheel should be 0.1 mm (.004").

Note:

The synchronizer clutch gear — Part No. 113 311 307 — will no longer be supplied as a spare part. Existing stocks should only be used on Type 1 vehicles.

10 - Slide the assembled synchro unit on to the drive pinion. The longer hub portion should be toward the face of the drive pinion splines. Turn the 1st speed stop ring until the shifting plates engage with the slots.

11 - Adjust the concave washer with the device VW 299 to give the prescribed spring travel of 0.17 ± 0.01 mm (.007 ± .0004").

12 - The 4th speed gear and needle bearing inner race should be heated in an oil bath to 90° C (194° F) before being pressed into position.

13 - Insert woodruff key for 4th gear into drive pinion.

14 - Slide the 4th gear on to the drive pinion with its wide shoulder facing the spacer sleeve.

15 - Press 4th speed gear and needle bearing inner race fully home by means of VW Repair Press in conjunction with VW 401, VW 412 and VW 422.

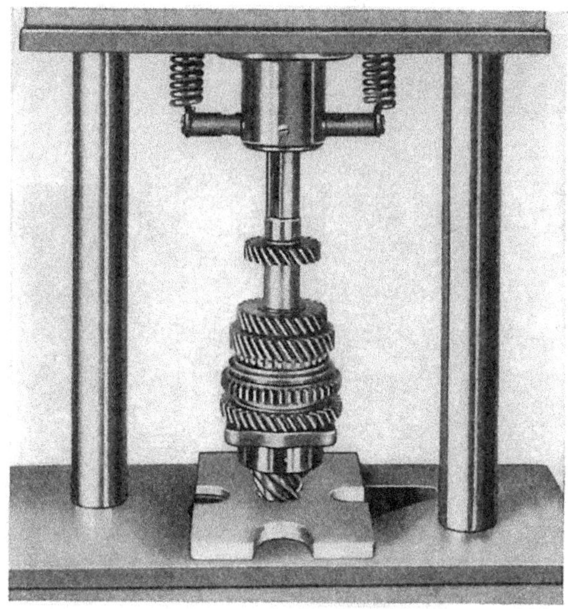

Adjustment of Concave Washer

General

The drive pinion concave washer must be adjusted to give a spring travel of 0.17 ± 0.01 mm (.007 ± .0004"). The concave washer exerts a pressure of approximately 100 kg (220 lbs.) on the 3rd gear and the clutch gear for 1st and 2nd gear and also reduces self-oscillation of these parts which are fitted to the drive pinion with a minimum backlash of 0.05 mm (.002").

The silent running of the rear axle will be adversely affected if the clutch gear and 3rd gear are tightly seated and no backlash is present as a result of the concave washer being too tightly adjusted. If the maximum spring travel is exceeded the 2nd gear can tend to jump out.

A - Measuring Procedure

1 - Assemble the drive pinion up to and including the 3rd gear.

Important

The clutch gear for 1st and 2nd gear and the 3rd gear should not be tightly seated on the drive pinion. Check the parts by hand for backlash.

2 - Use the Repair Press in conjunction with VW 401, VW 412 and VW 415 to ensure that the clutch gear for 1st and 2nd gears and the 3rd gear are correctly positioned. The end play of the 1st gear 0.10—0.25 mm (.004 to .010") should be rechecked and corrected if necessary.

3 - Slide the measuring sleeve of the adjusting device VW 299 on to the drive pinion as far as the shoulder for the 4th gear and set the dial indicator to zero. The dial indicator tracer pin must be replaced by an extension piece 28 mm long.

4 - Slide the drive pinion spacer sleeve on to the measuring pin of the adjusting device.

5 - Remove the measuring sleeve from the drive pinion. Slide it on to the measuring pin of the adjusting device and position it firmly on the spacer sleeve.

The dial indicator reading represents the amount by which the spacer sleeve is shorter than the distance from the shoulder on the pinion for the 4th gear to the 3rd gear.

B - Calculating the shim thickness

When the drive pinion is finally assembled the 4th gear is pressed on to the pinion as far as the shoulder. The dial indicator reading must be compensated by shims whilst taking the concave washer thickness (constant 1.04 mm/.041") and the prescribed spring travel 0.17 mm (.007") into consideration.

Example:

Measurement	1.84 mm
— Thickness of concave washer	1.04 mm
— Spring travel	0.17 mm
Thickness of shims	0.63 mm

The thicknesses of the shims available are as follows: 0.15; 0.2; 0.25; 0.3; 0.4; 0.6; 0.8; 1.0 and 1.2 mm. The shims must be measured carefully with a micrometer.

Finally the shim thickness is checked by placing the shims under the spacer sleeve on the measuring pin of the adjusting device. Press the measuring sleeve down firmly. The dial indicator reading must not exceed 1.21 ± 0.01 mm/.048 ± .0004" (thickness of concave washer + spring travel).

Reconditioning Main Drive Shaft

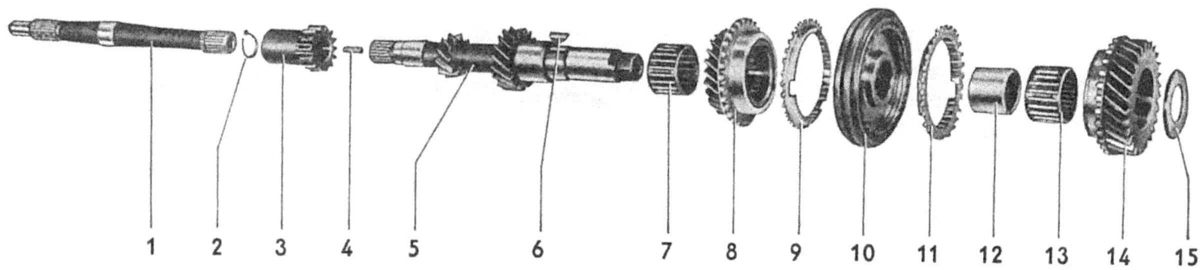

1 - Main drive shaft rear half
2 - Snap ring for reverse gear
3 - Reverse gear on drive shaft
4 - Stud
5 - Main drive shaft front half
6 - Woodruff key for clutch gear
7 - Needle cage (3rd gear)
8 - 3rd gear
9 - Synchronizer stop ring (3rd gear)
10 - Clutch gear (3rd and 4th speeds)
11 - Synchronizer stop ring (4th gear)
12 - Needle bearing inner race (4th gear)
13 - Needle cage (4th gear)
14 - 4th gear
15 - Thrust washer (4th gear)

Disassembly

1 - Remove thrust washer, 4th gear, needle cage and stop ring.

2 - Remove 4th speed needle bearing inner race, clutch gear for 3rd and 4th speeds and 3rd gear on the VW Repair Press in conjunction with VW 401, VW 411 and VW 435.

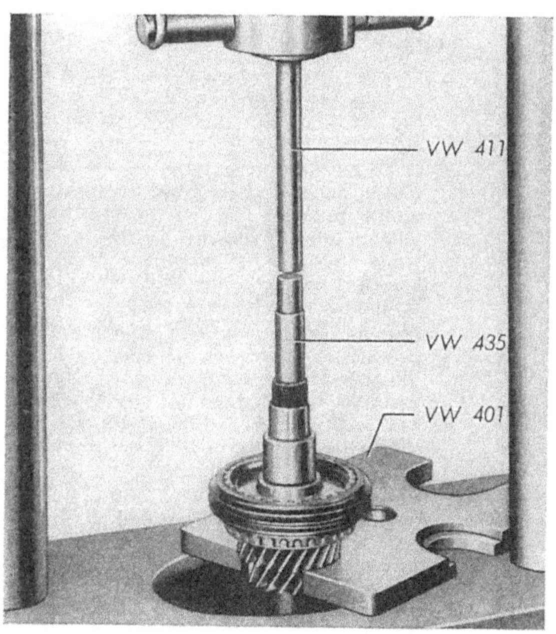

3 - Remove needle cage for 3rd gear.

4 - Disassemble synchro unit for 3rd and 4th gears.

Inspection

A - Check main drive shaft, front, for wear:

1 - Check the splines of reverse gear on drive shaft for wear.

2 - Check needle bearing contact surfaces for wear.

3 - Check 1st and 2nd gears for damage and wear.

4 - Place front main drive shaft between two points and check at contact surface of 3rd gear needle bearing for run-out. Permissible run-out: Max. 0.015 mm (.0006").

Important

In the case of undue wear or excessive runout the front main drive shaft should be replaced. Since the gear wheels may only be replaced in pairs, the gear wheels for the 1st and 2nd speeds on the drive pinion must be simultaneously replaced.

5 - Check needle bearings and 3rd and 4th gears for damage and wear. Replace gear trains if necessary.

6 - Check all synchronizer components for wear:

a - Clean the internal cone surface of the stop rings with a wire brush.

b - Check the clearance between the stop ring face and the clutch teeth of the corresponding gear with a feeler gauge (normal a = 1.1 mm/.043"). If the wear limit has been reached (a = 0.6 mm/.024"), the stop rings should be replaced. Premature wear of the stop rings indicates poor clutch condition or incorrect operation of clutch.

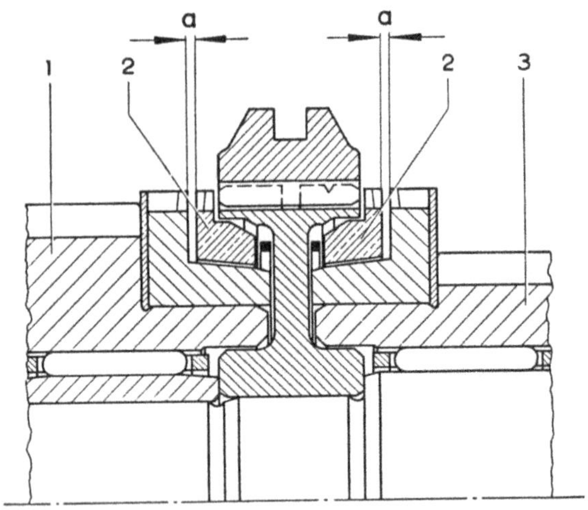

1 - 4th gear
2 - Synchronizer stop rings
3 - 3rd gear

c - If a gear will not engage, although the clutch is fully released, it may be due to the teeth of the stop ring being too much out of line with the splines of the operating sleeve. This condition is caused by the slots in the stop ring being worn.

7 - Check 4th gear thrust washer for wear, replace if necessary.

B - Rear Main Drive Shaft:

1 - Check main drive shaft pilot for wear.

2 - Check splines for wear and damage.

3 - Check oil seal seating surface for scores and signs of wear.

4 - Check reverse gear on main drive shaft for wear and damage, replace if necessary.

Assembly

When assembling, the following points should be observed:

1 - Assemble synchro unit for 3rd and 4th gears.

a - Clutch gears and operating sleeves must only be replaced in pairs.

b - The 1 mm deep groove in the operating sleeve must be towards 4th gear on assembly. The clutch gear has a chamfer on the 3th gear side only to facilitate installation on the drive shaft.

c - The assembly position of the clutch gear/operating sleeve is marked with an etched line-A. The parts must not be assembled with the marks out of line.

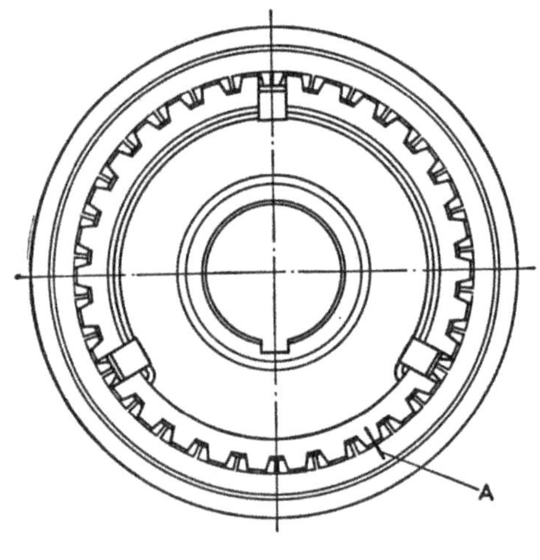

A - Etched line

d - Put the shifting plates in position and install the two snap rings offset to one another. Make sure that the ends of each ring engage behind the shifting plates.

Note:

a - To prevent the 3rd gear jumping out when pulling, the driving flanks of the teeth on the 3rd gear side of the clutch gear are set back slightly.

b - If the 3rd gear jumps out even though the clutch gear is assembled correctly, the clutch gear for the 3rd and 4th speeds with operating sleeve and the 3rd gear train have to be replaced. Adjust the selector forks carefully. Worn selector forks must also be replaced.

c - Due to an unfavourable combination of production tolerances, the end play of the 4th speed gear wheel can sometimes exceed the specified limit of 0.10 to 0.25 mm (.004—.010"). In certain cases this can cause the 4th gear to jump out when pulling.
The only satisfactory way rectify this is to replace the 3rd and 4th speed clutch gear with operating sleeve, both 4th speed gears and the selector fork for 3rd and 4th gear.

Note:

1 - The selector fork should be adjusted carefully.

2 - Only gear wheels which are also relieved on the driving flanks of the synchro teeth should be installed. This can be recognised by the fact that 3×7 teeth are noticeably thinner than the remaining 3×3. On gear wheels which are only relieved on the coasting flanks the difference is not so noticeable.

Important

To avoid confusion, only 4th gear wheel pairs on which the gear for the drive pinion has the larger 38 mm dia. shoulder on one side, should be installed.

Gear wheel pairs on which the gear for the drive pinion has shoulders of the same size on both sides may only be used up together with the operating sleeves with retaining shoulders.

3 - The end play of the 4th speed gear wheel on the drive shaft should be kept as near the lower limit as possible.

d - From November 1961, Chassis No. 4 274 401 (Rear Axle No. 4 371 905) the clutch gear is turned out as far as the center web to improve the centering of the synchronizing rings. The new Part No. of the clutch gear with operating sleeve for 3rd and 4th gears is 311 311 307 A.

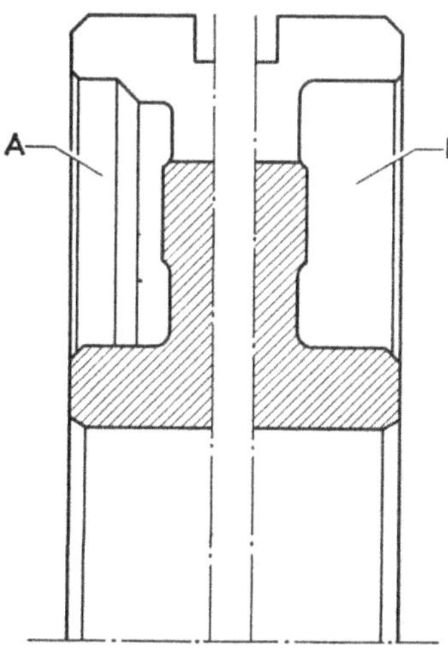

A = old version
B = **new** version

e - The shoulder on the synchronizer ring (a) was increased in width from 3 to 5 mm with effect from March 1962, Chassis No.

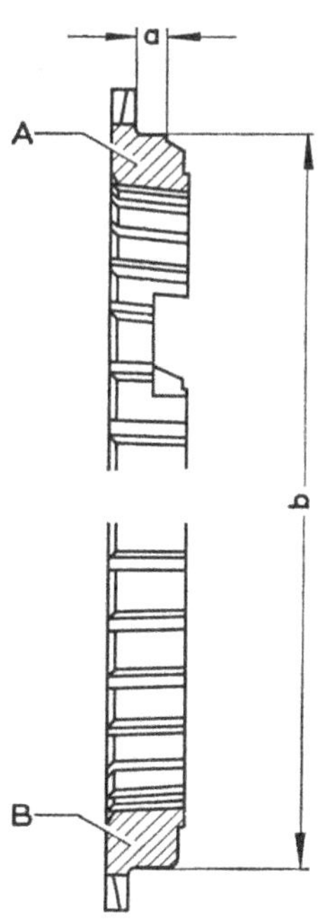

A = old version
B = **new** version

4 572 833 (Rear Axle No. 4 763 480). The Part No. of the modified ring is 113 311 295 A. Rings of the former type — 113 311 295 — will remain available.

Important

Note that the modified synchronizer ring (B) should not be installed together with the old type clutch gear (A) as this could cause seizure of the synchronizer rings.

The following can be installed together:

1 - Clutch gear (B) — Part No. 311 311 307 A
Synchro ring (B) — Part No. 113 311 295 A

2 - Clutch gear (B) — Part No. 311 311 307 A
Synchro ring (A) — Part No. 113 311 295

3 - Clutch gear (A) — Part No. 113 311 307
Synchro ring (A) — Part No. 113 311 295

The nominal diameter of the synchronizer ring (b = 70.2—0.15) should be as near the upper limit as possible. The end play for the 1st speed gearwheel should be 0.1 mm.

Note:

The clutch gear — Part No. 113 311 307 — will no longer be supplied as a spare part. Existing stocks should only be used on Type 1 vehicles.

2 - Insert woodruff key for clutch gear in the main drive shaft and place 3rd gear synchronizer stop ring on the cone of the gear.

3 - Press clutch gear for 3rd and 4th gears into position on the VW repair press in conjunction with VW 401, VW 412 and VW 419. The identification figure 4 on the clutch gear must be towards the 4th gear. The 3rd gear is lifted slightly and turned until the stop ring engages in the shifting plates.

4 - Press 4th gear needle bearing inner race into position on the VW repair press in conjunction with VW 401, VW 412 and VW 419.

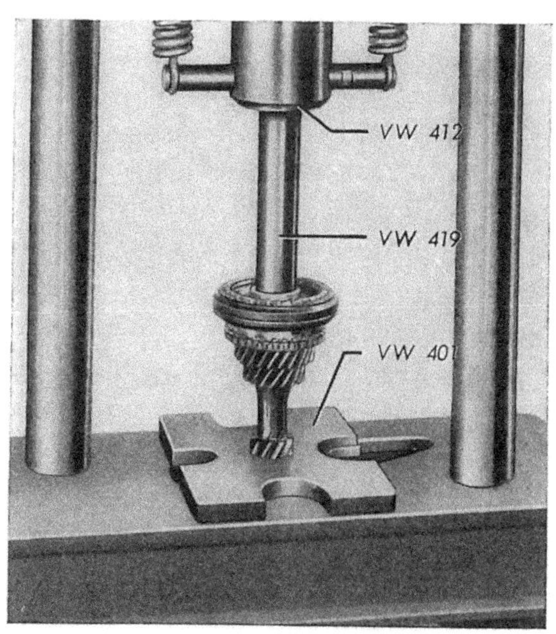

Disassembly

1 - Place the differential in holding fixture VW 664/1.

2 - Remove lock wire and ring gear (crown wheel) attaching screws.

3 - Lift off ring gear.

4 - Drive out differential pinion shaft with a drift after removing the pin, and remove differential pinions.

Assembly

1 - Check concave differential pinion contact surfaces in differential housing for damage and wear. If necessary, replace differential housing.

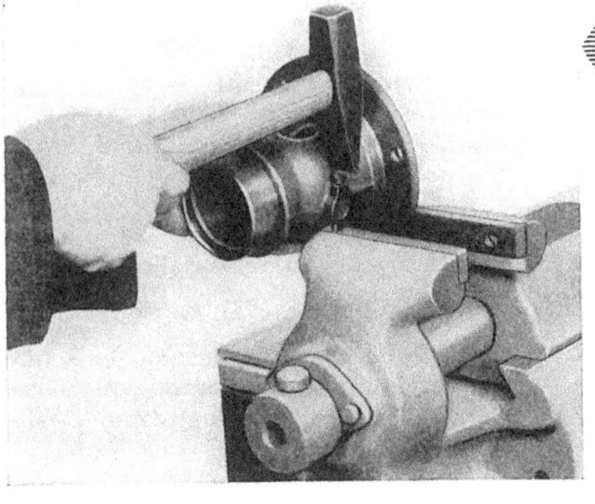

2 - The lock pin for the differential pinion shaft should be peened.

3 - Check ring gear for wear and damage; if necessary replace drive pinion and ring gear as a pair (note matching number on pinion and ring gear).

> **Important**
>
> A replacement of drive pinion and ring gear or differential housing necessitates a readjustment of the transmission.

4 - Keep the contact surfaces of the differential housing and ring gear absolutely clean to ensure uniform backlash.

5 - Tighten ring gear attaching screws to a torque of 6 mkg (43 ft. lbs.).

6 - Insert the wire so that it imposes a clockwise tension on the attaching screws when intertwining its ends.

Note:

From Chassis No. 3 192 507, the wrench size of the ring gear retaining screws has been altered from 15 to 17 mm. This has given the screws a larger contact surface so that the spring washers, which were installed in addition to the lock wire, could be discontinued.

Instead of the recesses in the differential housing flange for the screw heads, the thickness of the complete flange has been reduced. The shim contact area on the ring gear side is no longer undercut.

The part numbers are unchanged.

Adjustment of Drive Pinion and Ring Gear (Crown Wheel)

General

Silent operation and minimum wear of the final drive depend on the proper adjustment of drive pinion and ring gear. For this reason, the drive pinion and ring gears are machined in pairs during production, and inspections carried out with special test appliances ensure correct tooth contact and silent meshing in both turning directions. Silent operation is obtained by adjusting the pinion endwise with the ring gear lifted sufficiently out of the fully engaged position (without backlash) to ensure that the backlash is within the prescribed tolerance of 0.17—0.25 mm (0.0067″—0.0098″). The tolerance discrepancy r from the standard fitting dimension of the drive pinion R is measured and marked on the pinion face. Each gear set is given a matching number and replacements must be made in pairs.

Up to 1965

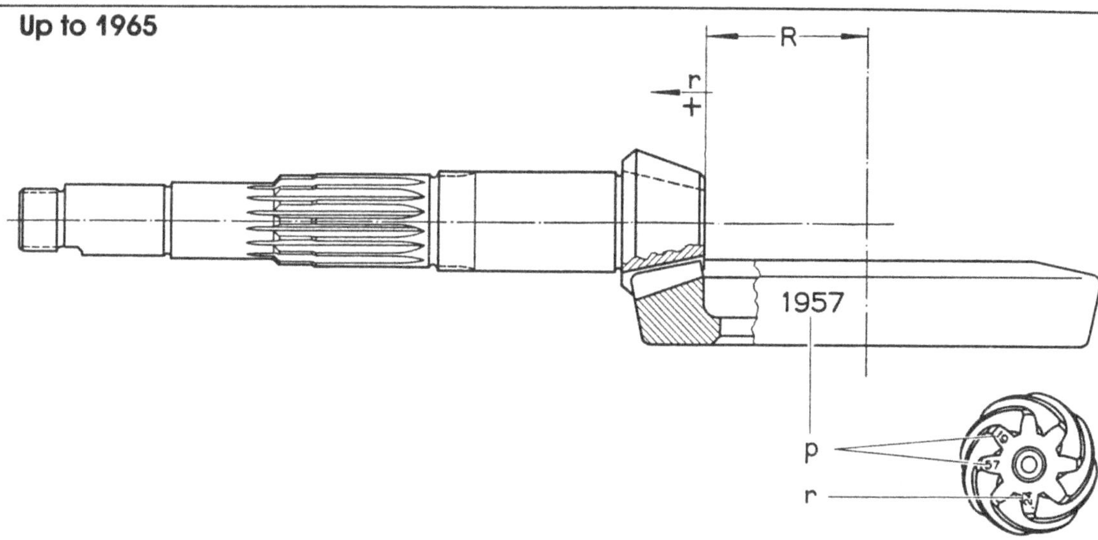

R - Standard fitting dimension/ring gear center line to drive pinion face;
Gleason toothing — No. of teeth 8 : 35 .... 58.70 mm (2.3110″)

r - Tolerance discrepancy from R (given in hundredths of a millimeter)

P - Matching number of gear set

1965

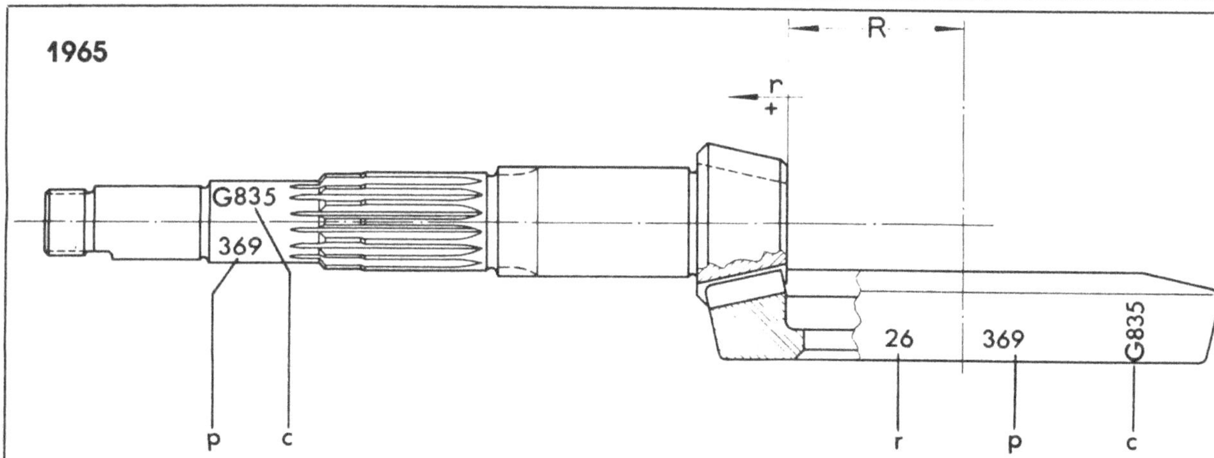

R - Fitting dimension/ring gear center line to pinion face = 58.70 mm (2.3110″)

r - Deviation from R (given in hundredths of a millimeter)

p - Matching number of gear set

c - Type of toothing (G = Gleason, K = Klingelnberg) and the number of teeth (8 : 35)

It is usually only necessary to re-adjust the ring gear and drive pinion when carrying out repairs to the rear axle if parts have had to be replaced which directly affect the adjustment. It is sufficient to re-adjust the ring gear if the differential housing, a final drive cover or a differential bearing have been replaced. The ring gear and drive pinion must be re-adjusted if the transmission case, the gear set itself, ot the drive pinion ball bearing have been replaced.

The object of the ring gear and drive pinion adjustment is to obtain the same silent operation as determined in the factory by test appliances. To ensure this, the pinion must first of all be adjusted by installing shims between the ball bearing and the contact surface at the transmission case to ensure that the distance from the ring gear center line to the drive pinion face coincides with the fitting dimension R + r determined in the factory. Then the ring gear is adjusted to give the prescribed backlash and the thickness of the shims for the differential housing determined. It is important, however, to note, that both of the final drive covers must be installed with a pre-load of 0.14 mm (0.0055″). After determining the thickness of the shims, a pre-load of 0.07 mm (0.0028″) must be taken into consideration on both sides.

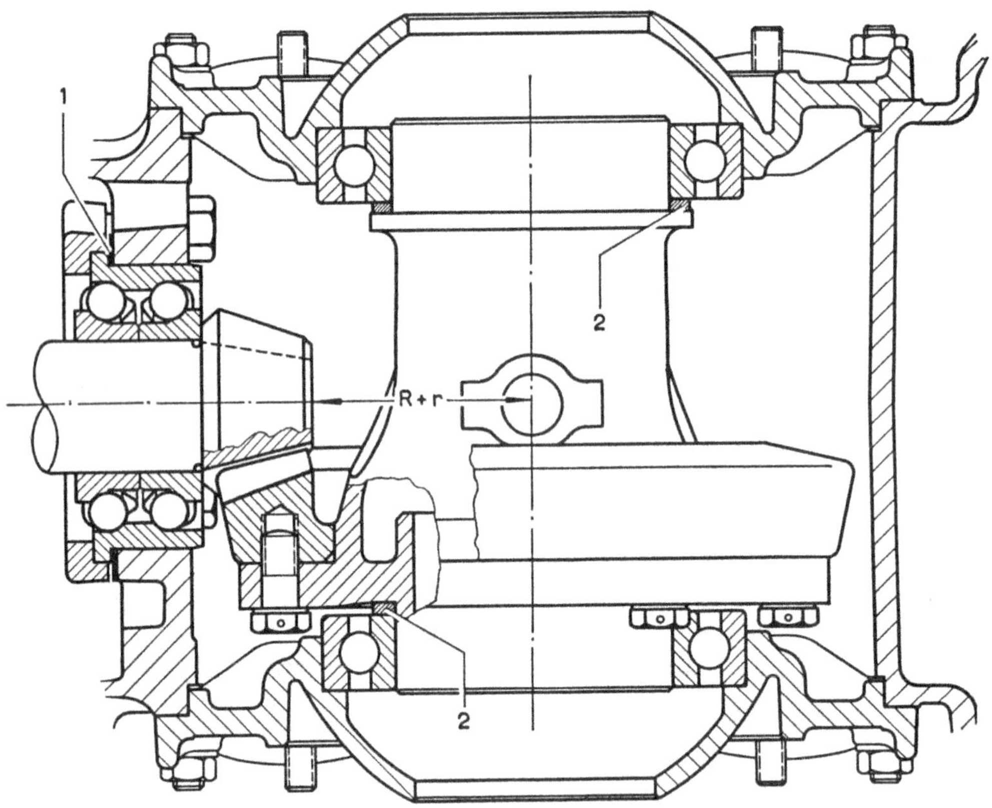

1 - Pinion shims
2 - Shims S_1 (ring gear side) and S_2 for the differential

Care and cleanliness during all assembly work and measuring operations are an absolute necessity to obtain corredt results.

A - Drive pinion adjustment and measuring the depth of the transmission case between the two differential ball bearings

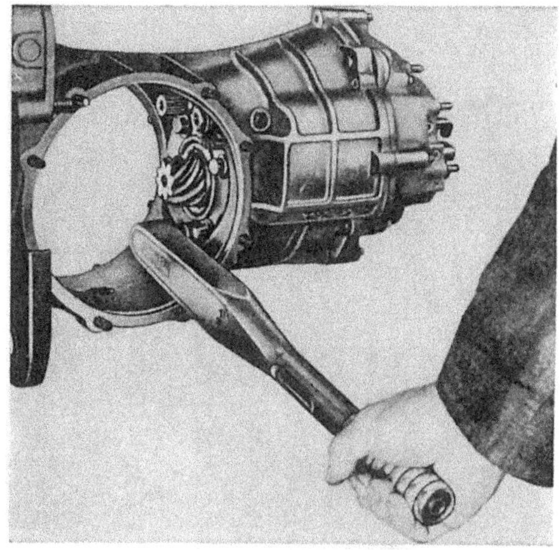

1 - Install the transmission into the transmission case without pinion shims.

2 - Tighten the 4 ball bearing retainer bolts correctly to 5 mkg (36 ft. lbs.). Use washers.

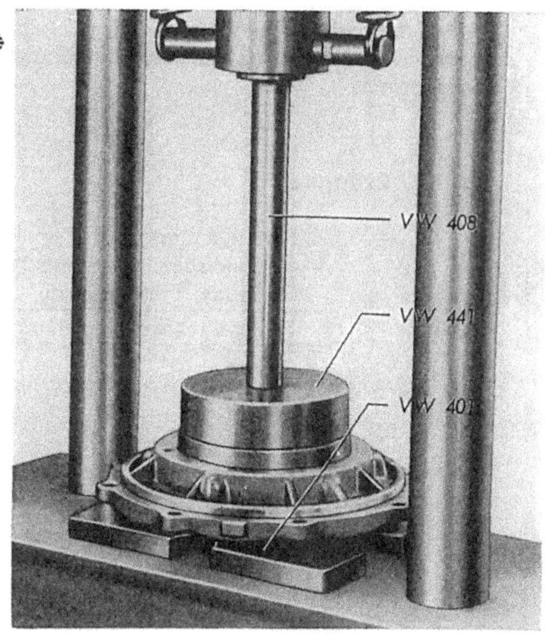

3 - Make sure that the bearings in the final drive covers are correctly seated by using the Repair Press in conjunction with VW 401, 408 and 441.

4 - install the right final drive cover with gasket and tighten it.

Important

Since the depth of the transmission case has still to be determined, on no account use a hammer when installing the final drive covers as this practice would cause the bearings in the covers to work loose. Tighten the 8 nuts evenly to ensure that the final drive cover seats correctly in the transmission case.

5 - Position the mandrel VW 289 d on the adjusting plate. Insert the dial indicator with a 5 mm tracer pin into the mandrel and set the dial indicator to zero with a pre-load of 1.0 mm (0.039"). Make sure that the adjusting plate and the contact surface of the mandrel are perfectly clean.

6 - Insert the mandrel into the transmission case, install the left final drive cover with gasket and tighten it.

7 - View the dial indicator through the hole in the mandrel, allow the tracer pin to carefully contact the pinion face and rotate the mandrel until the dial indicator gives the maximum reading.

Taking the pre-load of the dial indicator into consideration, the dial reading is subtracted from the nominal dimension of the mandrel with adjusting plate and gives the fitting dimension of the drive pinion without shims.

Example

	Nominal dimension of mandrel with adjusting plate	58.70 mm
—	Dial indicator reading (1.0 mm pre-load of dial indicator taken into consideration)	0.56 mm
	Fitting dimension of pinion without shims	58.14 mm

The correct fitting dimension of the pinion is based on the standard fitting dimension R + the tolerance discrepancy r. The difference between these two values denotes the thickness of the shims required:

Example

	Standard fitting dimension of pinion R	58.70 mm
+	Tolerance discrepancy r	0.24 mm
		58.94 mm
—	Fitting dimension of pinion without shims	58.14 mm
	Thickness of shims required	0.80 mm

The shims must be carefully measured at various points with a micrometer. It is good practice to choose the shims carefully so as to get as close as possible to the thickness required. On no account must the thickness of the shims differ by more than ± 0.02 mm (0.0008″) from the calculated thickness.

The thicknesses of the shims available are as follows: 0.1; 0.15; 0.2; 0.3; 0.4; 0.5; 0.6; 0.7; 0.8; 0.9; 1.0 and 1.2 mm. The tolerance of the shims according to thickness is between 0.01 and 0.03 mm (0.0004″ and 0.0012″).

8 - Attach the dial indicator bracket of device VW 297 to one of the studs for the axle tube retainer. Place the dial indicator in the bracket and set to zero.

9 - Turn the transmission case 180° so that the mandrel falls of its own accord and presses against the bearing in the left final drive cover. The dial indicator reading is added to the length of the mandrel and gives the transmission case depth J which is needed for the ring gear adjustment.

Example:

Nominal length of mandrel	107.88 mm
+ Dial indicator reading	1.82 mm
Depth of transmission case J ...	109.70 mm

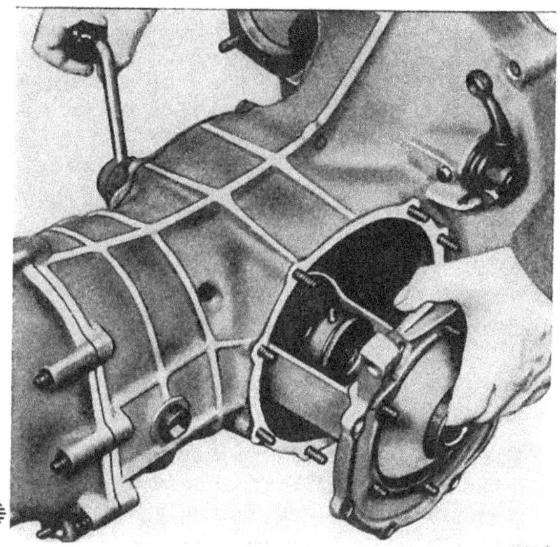

10 - Turn the transmission case to the horizontal position, remove the dial indicator bracket and take off the nuts on the left final drive cover.

11 - Place the spindle of the Device VW 297 on the right final drive cover and tighten it.

12 - Press off the left final drive cover with the spindle and remove the mandrel from the transmission case.

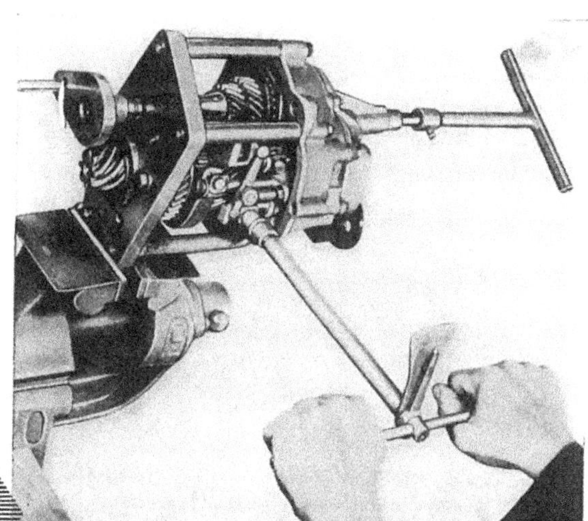

13 - Take off the right final drive cover, remove the ball bearing retainer and press transmission out of transmission case.

14 - Slide the shims on to the ball bearing. Adjust the selector forks with the gearshift test appliance VW 294 and finally install the transmission in the transmission case.

B - Ring Gear Adjustment

1 - Insert the dial indicator with a 28 mm tracer pin into the device VW 287a. Place the adjusting plate on the polished surface for the differential and set the dial indicator to zero.

2 - Raise the tracer pin and insert the differential into the device.

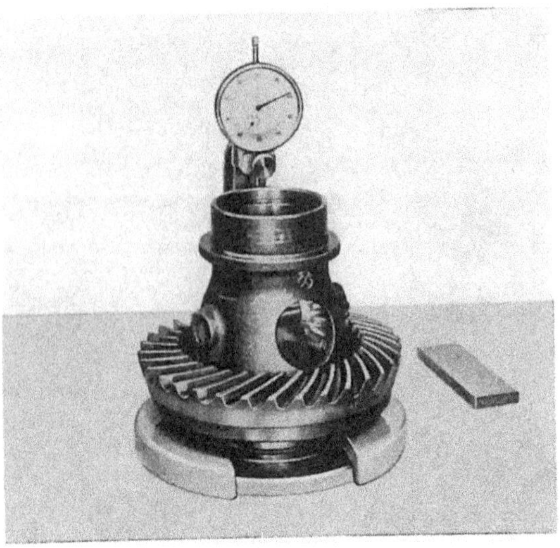

The dial indicator reading is added to the nominal dimension of the adjusting plate gives the length of the differential housing L:

Example:

Nominal dimension of adjusting plate	102.51 mm
+ Dial indicator reading	0.39 mm
Length of differential housing ..	102.90 mm

Note:

If the device VW 287a is not available, the length of the differential housing can be measured with a caliper square. It is best to use a magnifying glass to ensure correct readings.

3 - Install the differential in the transmission case with the device VW 297.

a - Place the differential between the two thrust plates of the device. Insert the retaining bolts from the ring gear side and tighten the nuts.

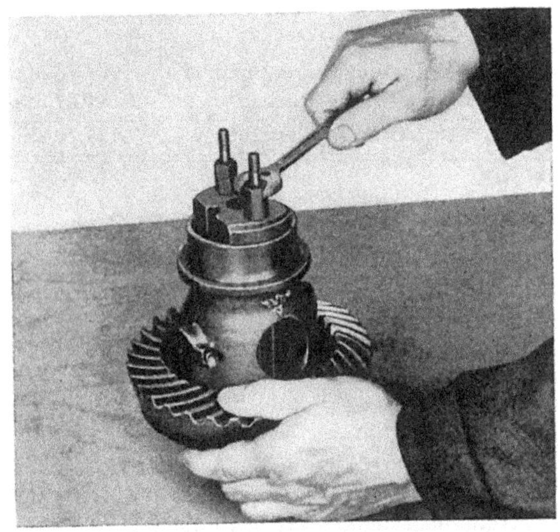

b - Install the right final drive cover with gasket and install the differential in the transmission case.

c - Insert the gauge ring VW 298 on the ring gear side and install the left final drive cover with gasket. The two final drive covers must be tightened correctly on all the studs to 2.5 mkg (18 ft. lbs.).

d - Position the spindle on the left final drive cover. Fully tighten the spindle attaching nuts.

e - Attach the drive pinion retaining bracket to the gear carrier.

f - Turn the transmission case 180°. Firmly attach the bracket of the dial indicator for measuring the ring gear backlash to the clamping bolts of the two thrust plates with two of the axle tube retainer nuts.

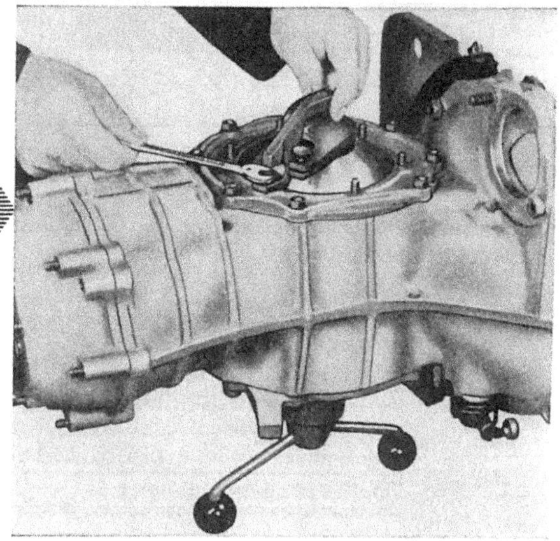

g - Attach the bracket of the dial indicator for the differential axial adjustment.

h - Insert the dial indicators.

4 - Pull the differential fully home with the spindle into the bearing in the left final drive cover. Back off the spindle and set the dial indicator for the differential axial adjustment to 2.8 mm (thickness of the gauge ring).

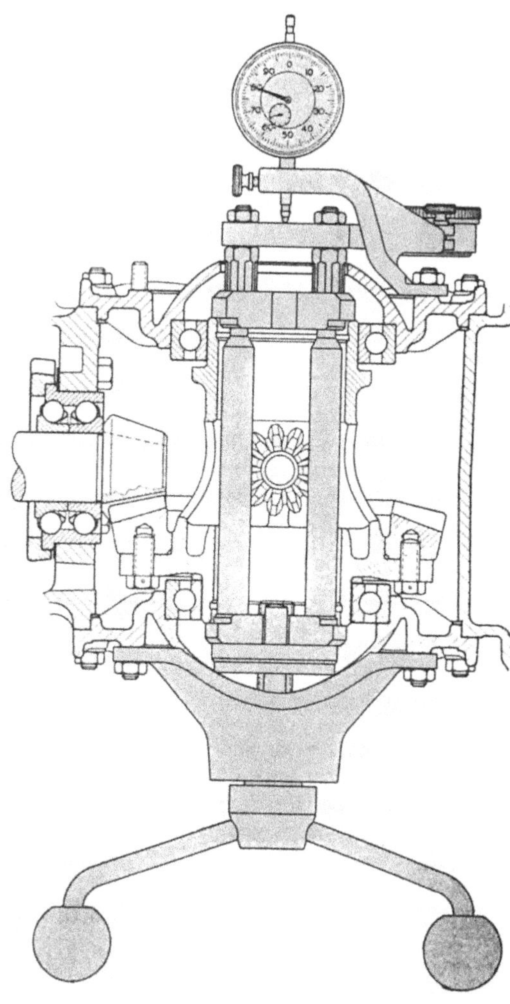

Important

To avoid measuring errors as a result of axial play in the differential bearings it is important when carrying out the following measurements that the ring gear side of the differential is at the bottom of the transmission case. Moreover, make sure when setting the dial indicator to 2.8 mm that the bearing in the left final drive cover is correctly seated. Tighten the spindle again, back it off and check the dial indicator setting.

5 - Rotate the differential in both directions as far as possible. Place a 32 mm socket on the pinion nut and lock the pinion so that the differential is approximately half way between the two stop positions.

6 - Measure the ring gear backlash.

7 - Adjust the differential carefully towards the drive pinion with the spindle until a backlash of between 0.20 and 0.22 mm (0.0079" and 0.0087") is obtained.

Important

The spindle must be backed off each time when measuring the backlash. The axial movement of the differential will then correspond with the dial indicator reading.

The thickness of the shim on the ring gear side, S_1, can be read direct from the dial indicator for the differential axial adjustment.

Example:

Commencing position (reading) of dial indicator (thickness of gauge ring inserted)	2.80 mm
Axial movement up to prescribed backlash	0.35 mm
Final reading of dial indicator (S_1)	3.15 mm

The shim S_2 is calculated from the depth of the transmission case between the two differential bearings J, the length of the differential housing L, and S_1:

Example:

J	109.70
— L	102.90
— S_1	3.15
S_2	3.65

Finally the pre-load on both sides is considered with which the two final drive covers have to be installed. The pre-load for both final drive covers amounts to 0.14 mm so that 0.07 has still to be added on each side to the final shim thickness. The final shim thicknesses are given in the following example:

$$S_1 = 3.15 + 0.07 = 3.22 \text{ mm}$$

$$S_2 = 3.65 + 0.07 = 3.72 \text{ mm}$$

Measure the shims at various points with a micrometer. The shims are available in thicknesses ranging from 2.8 to 4.0 mm in units of 0.1 mm. A washer 0.25 mm thick provides for adjustments in units of 0.05 mm. Since the actual shim thickness may differ up to 0.03 mm from the thickness marked, make sure you choose those shims which get as near as possible to the shim thickness calculated.

8 - Take off the dial indicator brackets and press off the left final drive cover. Place the spindle on the right final drive cover and press out the differential.

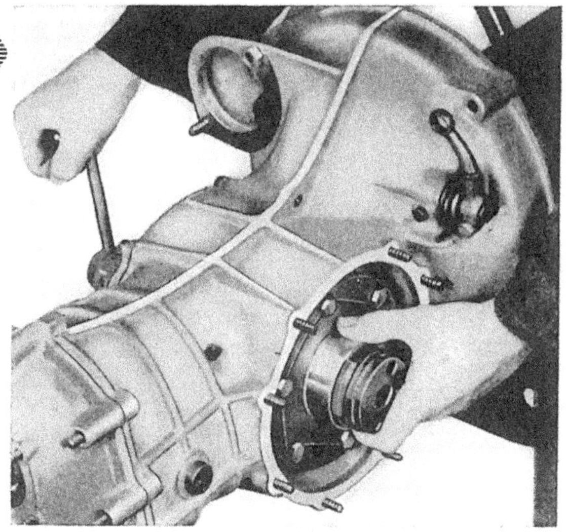

9 - Install rear main drive shaft.

10 - Install the shims S_1 and S_2 with their chamfered edges facing the differential. The washers are positioned between the shim and ball bearing.

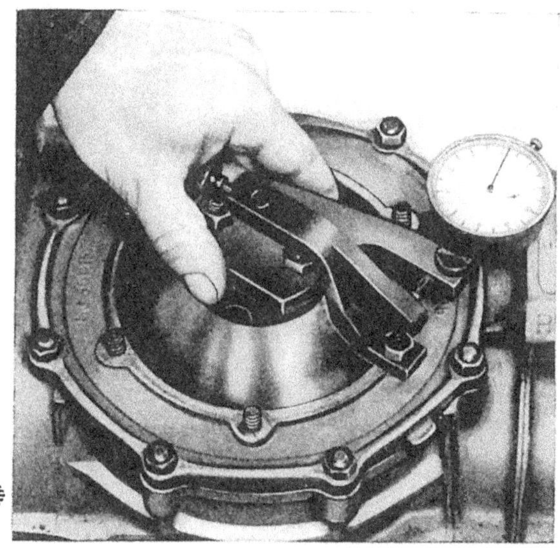

11 - Tighten the final drive covers as prescribed to 2.5 mkg (18 ft. lbs.).

Note—1965

- Tighten the final drive covers as prescribed to 3 mkg (22 ft. lbs.).

12 - Check the ring gear backlash again in various positions. The backlash tolerance is between 0.17 and 0.25 mm (0.007—0.010″) and the measurements taken must not differ by more than 0.05 mm (0.002″).

13 - Remove the dial indicator brackets.

14 - Take off retaining bolts and remove the thrust plates through the openings for the rear axle shafts.

THIS PAGE INTENTIONALLY
LEFT BLANK

Gearshift Lever Removal and Installation

Removal

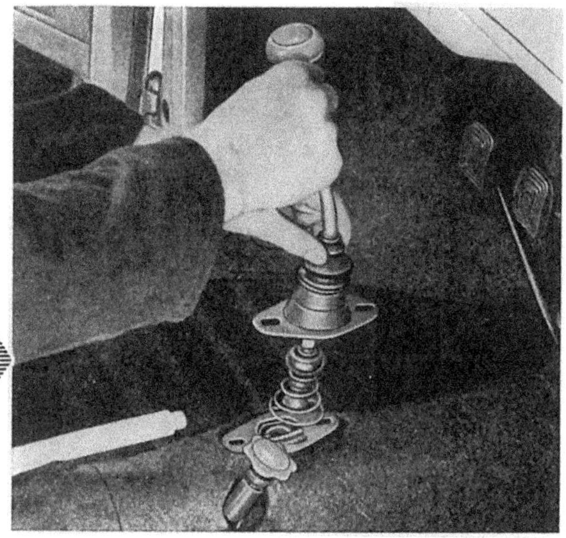

1 - Remove front floor mat.

2 - Remove screws that attach the gearshift lever ball housing to the frame tunnel.

3 - Take off gearshift lever, ball housing, rubber boot, and spring as a unit. The spring should be turned for removal to clear the pin.

4 - Take off stop plate.

5 - Clean all components.

Installation

When installing, the following points should be observed:

1 - Check gearshift lever collar, stop plate, and gearshift lever ball socket in shifting rod for wear. Replace worn parts.

2 - Make sure the gearshift lever locating pin is secure. Check spring-loaded steel ball for tension. Replace if necessary.

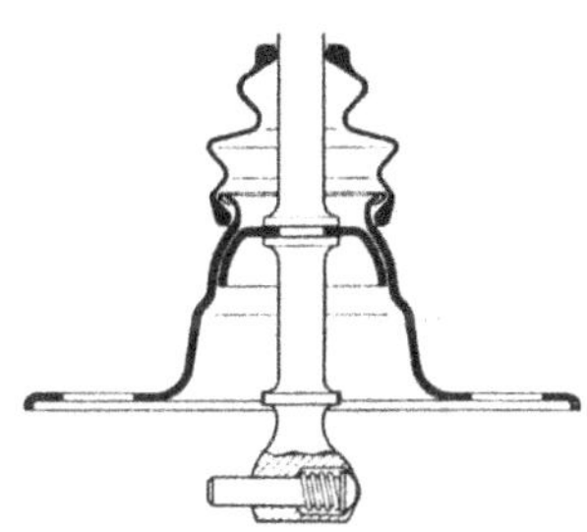

Note:

From Chassis No. 4 581 537, the pin soldered into the gearshift lever was replaced by a spring-loaded, sliding round-head bolt as on the Type 3. The bolt is peened and must not jam.

Only the modified gearshift lever will be supplied as a spare. The Part Nos. — 111 711 121 D or 211 711 121 A — have not been altered.

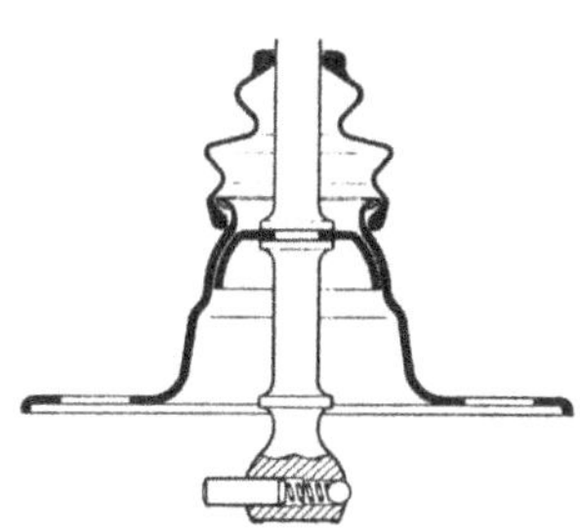

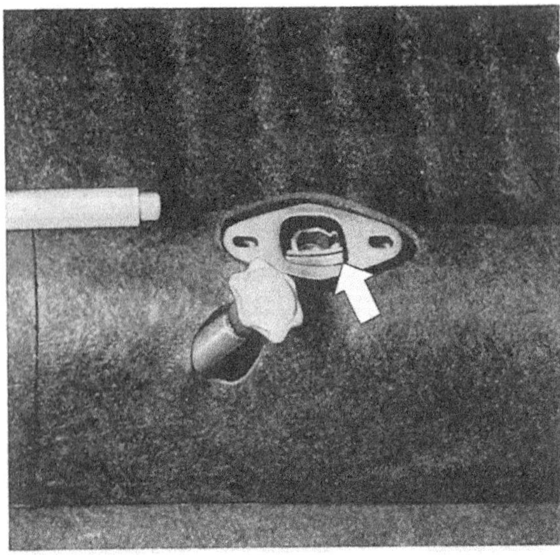

3 - When installing stop plate, make sure that the turned-up edge is on the right-hand side.

Note:

From September 1963, Chassis No. 5 749 573 all Type 1 models are being fitted with the Type 2 shift lever stop plate (Part No. 211 711 149).

The rearward angled recess on the raised front edge of the web (reverse gear stop) prevents the shift lever spring from becoming jammed between the ball housing and the stop plate on installation.

The shift lever stop plate — Part No. 111 711 149 — will be discontinued when stocks are exhausted.

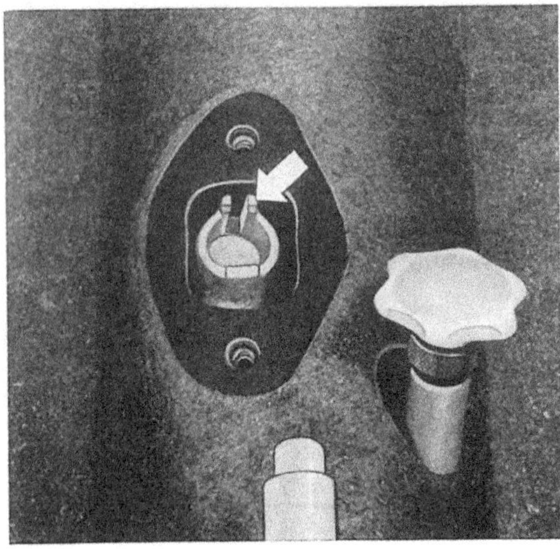

4 - Grease all moving parts with universal grease.

5 - The gearshift lever ball housing must occupy a position which ensures that the gearshift lever lower (straight) portion stands vertical in neutral position and the gearshift lever locating pin engages in the slot provided in the ball socket.

The stop plate must be embedded in the hollow flange of the ball housing.

6 - Check condition of gearshift lever rubber boot, replace if damaged.

7 - Check position of gearshift lever by engaging the gears. Correct if necessary.

Note:

1 - In isolated instances it may be found that 2nd or 1st gears are difficult or impossible to engage when changing down. Under certain conditions, there is a possibility of reverse gear being selected when attempting to change into 2nd gear, which could lead to serious transmission damage if force is used.

The trouble can be eliminated as follows:

a - Loosen gearshift lever ball housing securing screws, move the housing hard over to the left at right angles to direction of travel and tighten securing screws again.

If, in exceptional cases, it is necessary to offset the ball housing further to the left, the elongated holes should be lengthened 1 mm (.040").

b - On vehicles up to Chassis No. 3 228 000, check that the two securing screw holes in the gearshift rod coupling are properly aligned. If necessary, replace coupling.

2 - On new vehicles it may occasionally be found that one of the gears is difficult to engage or does not synchronize properly.

Trouble of this nature can be caused by the combination of unfavourable tolerances at the conical surfaces of gearwheel and synchronizer ring. It generally only appears during the first 100 miles and ceases as the mating parts become run in.

The transmission should not be disassembled if this shifting difficulty is experienced. It is usually sufficient to shift the gears frequently during the initial period. If a gear cannot be engaged, the shift lever should be returned to neutral and a further attempt made after de-clutching again. Force should not be used to engage the gears as this may cause the synchronizer rings to seize on the gears or give rise to other damage.

3 - From Chassis No. 4 289 952 the gearshift lever is tapered and has a smaller knob.

Part Nos. of the modified parts:

Gearshift lever:	111 711 121 D	or 141 711 121 C
Knob:	111/113 711 141 A	or 141 711 141 A

The new gearshift lever can be installed in all vehicles with the cranked lever from Chassis No. 1 929 746.

Observe the hints on page H-6, 2 also.

Removal

1 - Remove front floor mat and gearshift lever.

2 - Remove rear seat.

3 - Remove inspection cover on frame tunnel.

4 - Remove lock wire at the coupling screw and release front coupling screw.

5 - Open the front hood and remove spare wheel.

6 - Remove front bumper.

7 - Remove frame head cover.

8 - Withdraw shift rod from the coupling, using a pair of combination pliers, and push it toward the frame head.

9 - Completely withdraw shift rod through the openings in the body.

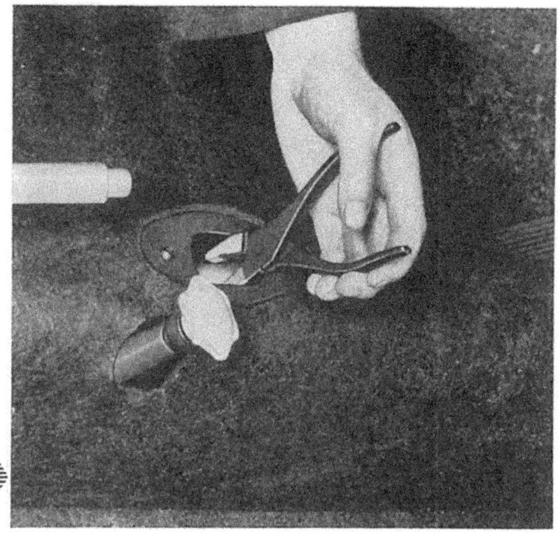

Note:

1 - The plastic bush in the frame tunnel behind the opening for the gearshift lever does not usually need to be replaced. If a replacement is necessary, however, proceed as follows:

a - Remove the bush from the shift rod guide with a pair of combination pliers.

b - Insert the new bush through the shift lever opening, and starting from the slot, press it fully into the shift rod guide.

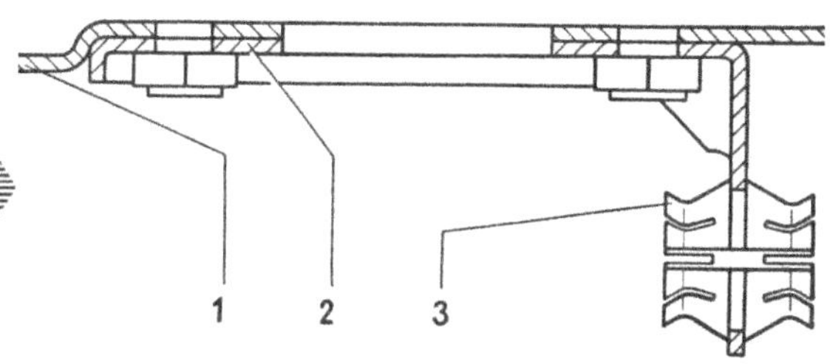

1 - Frame tunnel
2 - Shift rod guide
3 - Plastic bush

2 - From Chassis No. 4 027 316 (August 3rd, 1961), the synthetic bush (Part No. 111 701 259) in the gearshift rod guide in the frame was reduced in diameter at the front by means of a wire ring (Part No. 111 701 263). This increases the friction between gearshift rod and bush.

Service Installation

If shift lever vibration occurs on previous vehicles. The wire ring for the gearshift rod guide bush can be service installed in vehicles from Chassis No. 3 140 046 (June 1960).

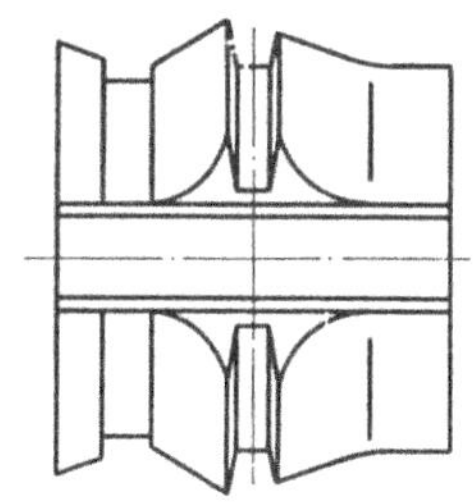

new

3 - From February 1963, Chassis No. 5 309 050 the sleeve for the shift rod guide — new Part No. 111 701 259 A — is made of a different material.

The new sleeve gives a constant and more uniform frictional effect. The shift rod noises which occur occasionally will also be eliminated by the new sleeve. Universal grease is used as a lubricant.

The new sleeve can be distinguished from the previous version by the asymmetrical shape.

Service Installation

The new sleeve can be installed from Chassis No. 3 140 046.

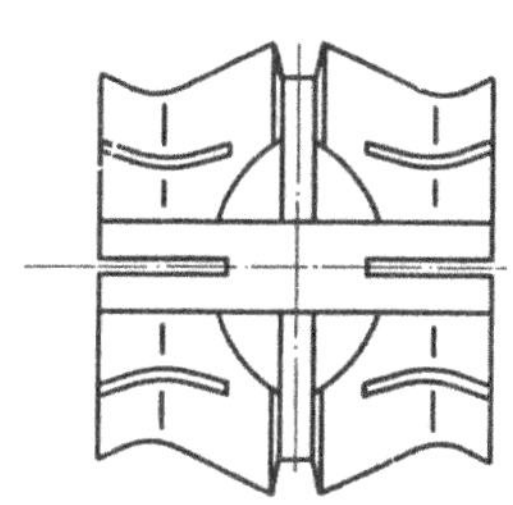

old

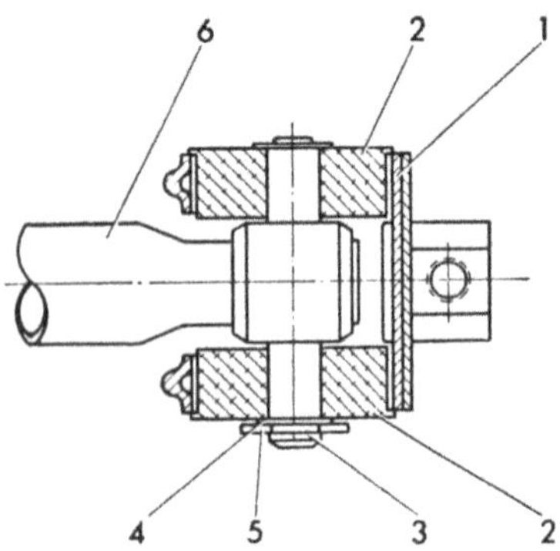

Part	Part No.
1 - Housing for gearshift rod coupling	311 711 211
2 - Guide ring	311 711 233
3 - Pin for coupling	311 711 209
4 - Washer	311 711 227
5 - Lock washer	311 711 229
6 - Gearshift rod	111 711 155 D

Installation

When installing, the following points should be observed:

1 - Check shifting rod for wear and twist. Replace if necessary.

2 - Coat the entire shift rod with universal grease.

3 - Insert the shift rod through the openings in the body into the shift rod guide in the tunnel.

4 - Make sure that the gearshift rod coupling screw is correctly seated in the gearshift rod and secure the screw with locking wire.

Note:

From November 1963, Chassis No. 5 911 561 (Rear Axle No. 6 334 500) a modified gearshift rod coupling is installed which prevents the gears from jumping out of engagement even when the vehicle is driven on bad roads. The coupling consists of a sheet metal housing which has a rubber guide ring on both sides. The coupling is connected to the gearshift rod, which has also been modified, by a pin which passes through the guide rings. The guide rings have approximately 2 mm (.08″) axial play in the housing so that vibration of the gear lever and gearshift rod which is caused by the movement of the engine-transmission unit is not transmitted directly to the transmission. Consequently the engagement of the gears will not be unfavourably affected.

At the same time the access opening for the gearshift rod coupling was widened 6 mm.

Service Installation

The modified coupling can be installed without widening the access opening — from Chassis No. 929 746 — if the gearshift rod is replaced at the same time. From Chassis No. 3 140 046 to 5 309 049, the new gearshift rod guide bush (Part No. 111 701 259 A) should be installed as well.

General Description

The rear wheels are sprung independently. The inner ends of the two torsion bars are anchored in the center of the frame cross member by a splined tube which is welded in position. The outer ends of the torsion bars, which are also provided with splines, carry the spring plates (radius arms) the hubs of which are rubber-cushioned. The rear axle tubes are attached to the rear ends of the spring plates. The suspension is adjustable by means of the splines. Screwed to the spring plate and the axle shaft bearing housing is a progressive rubber bumper which becomes active in the upper deflection range. Double-acting telescopic shock absorbers prevent excessive rebo' nd.

Removing and Installing Torsion Bar

Removal

1 - Loosen rear wheel mounting bolts.

2 - Support car on trestles in a horizontal position and remove rear wheel.

3 - Disconnect handbrake cables at handbrake lever and withdraw them slightly towards the rear. Remove handbrake lever.

4 - Mark the position of the spring plate in relation to the rear axle shaft bearing housing by means of a chisel on the spring plate dead in line with the groove in the rear axle shaft bearing housing.

5 - Remove lower shock absorber mounting bolts.

6 - Remove bolts at axle shaft bearing housing.

7 - Pull rear axle towards the rear until it clears the spring plate.

8 - Remove screws that attach spring plate hub cover and take off cover.

9 - Withdraw spring plate and both rubber cushions.

10 - Remove about five of the foremost fender screws.

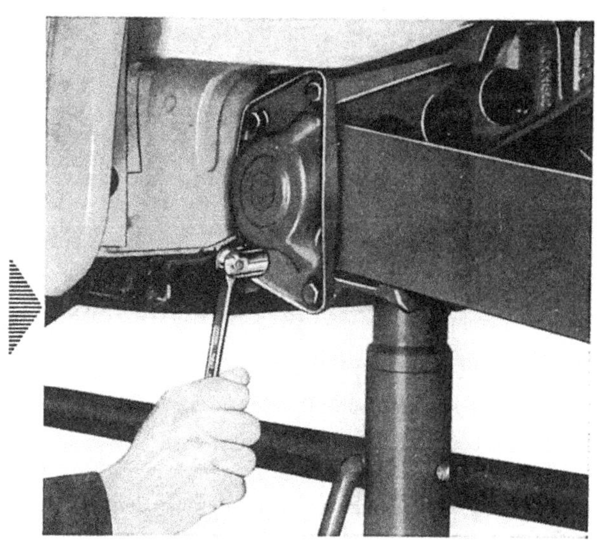

11 - Pull the fender aside and withdraw torsion bar from frame cross tube.

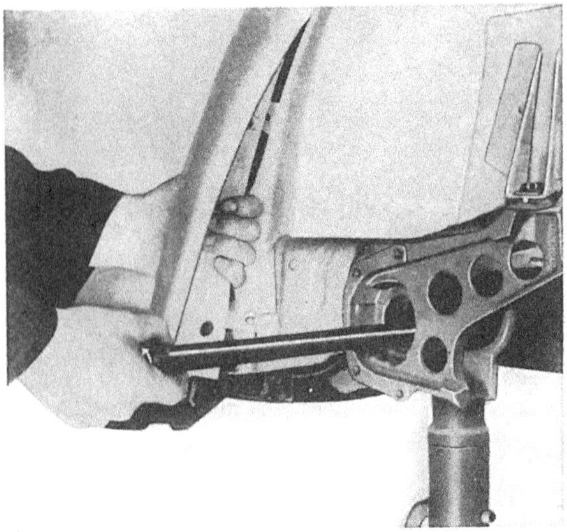

Note:

In the case of a broken torsion bar, push the broken end from the splined center anchor with a steel rod after removing the opposite torsion bar.

Installation

When installing, the following points should be observed:

1 - Inspect torsion bar splined ends and paint for damage. Make sure there are no signs of rust replace if necessary.

2 - Grease splines of torsion bar.

3 - Install torsion bar and spring plate and adjust them.

> **Important**
>
> The torsion bars are prestressed. Left and right bars are not interchangeable. They can be distinguished by an arrow on the outside face showing the direction of torque.

4 - Coat rubber bushes with flaked graphite when installing.

Note:

1 - From Chassis No. 4325735, the mounting of the spring plates in the rubber bushes was altered. Type 1 is now being provided with the eccentric rubber cushion previously used for Type 3 only so that the spring plate is centrically mounted when loaded. When installing the eccentric rubber bush make sure that the wider edge of the bush is facing upward. The wider edge of the new rubber bush has been marked "oben" to distinguish it from the previous type one.

Part Nos. of modified rubber bushes:

311 511 245 — rubber bush, outer left and inner right

311 511 246 — rubber bush, inner left and outer right

The rubber bushes can be service installed in vehicles from Chassis No. 2528668. **Both** bushes on one side must be replaced by the eccentric type.

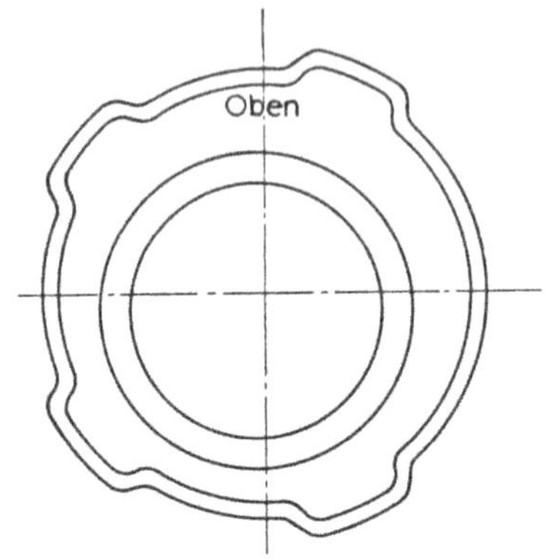

new

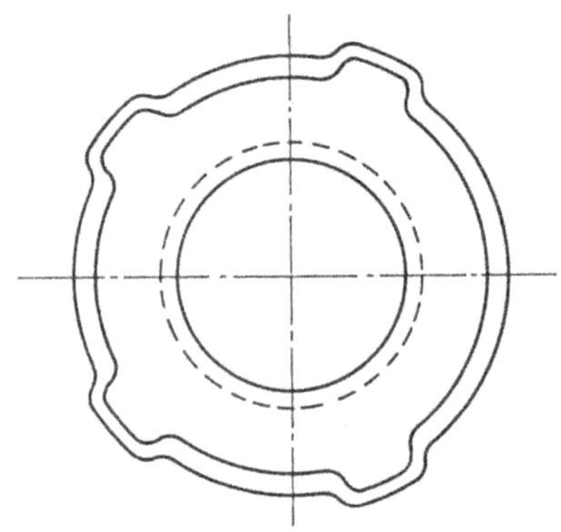

formerly

2 - From Chassis No. 5112045, the spring plate rubber bush was modified. The modified bush (Part No. 311 511 245/246 as before) will locate the spring plate more positively. The new bushes are not marked specially but they differ slightly in shape from the previous type.

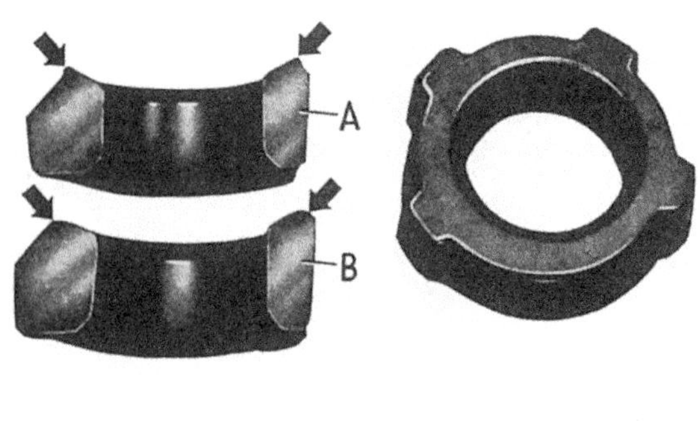

A = Previous type
B = New type

Service Installation

The modified bush can be installed in older vehicles.

Subsequent installation will help to prevent noises occurring at the spring plate hub cover.

5 - Lift the spring plate with the spring plate tensioner VW 655 until its lower edge is higher than the lower stop in the cross tube flange.

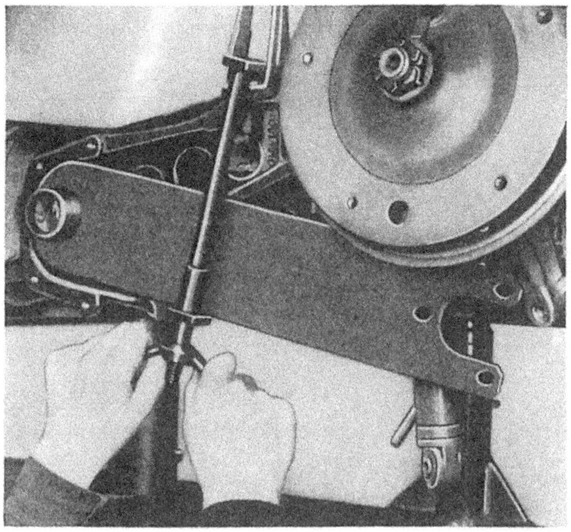

6 - Press the spring plate home by using the spring plate installing tool VW 656.

Note:

a - To facilitate the installation of the hub cover and to avoid damage to the threads, it is advisable to screw two tapered guide pins (approx. 45 mm/1.8" long) into two of the tapped holes, diagonally opposite. The cover is then pushed over the two guide pins, allowing two of the cover bolts to be screwed in. Afterwards the two guide pins are removed and the two other bolts screwed in.

b - From Chassis No. 3672390 (Feb. 13th, 1961), the ring-shaped recess in the cover for the spring plate hub was modified and increased in diameter from 38 to 40 mm (1.496—1.574"). This will prevent the spring plate hub from coming into contact with the cover and creating noise.

Note:

The Part No. of the cover remains unchanged — 111 511 225 B.

7 - Install and tighten the spring plate hub cover.

8 - Clean mating faces between spring plate and axle bearing housing.

9 - When bolting axle bearing housing and spring plate together, ensure that the mark made on the spring plate when removing the torsion bar is dead in line with the groove in the axle bearing housing. Tighten axle bearing housing mounting bolts to a torque of 10—12 mkg (72—87 ft. lbs.).

10 - Install handbrake lever, adjust handbrake.

Spring Plate Adjustment

General description

It is a point of major importance that there be no difference in the inclinations of both spring plates if perfect riding qualities under all conditions of load are to be ensured. When adjusting one spring plate, the inclination of the other should be checked and adjusted if necessary.

An exact adjustment of the rear suspension is effected by measuring the inclination of the spring plate with the frame in the horizontal position and spring plate in an unloaded condition.

The inclination of the unloaded spring plate should be:

16° 30′ + 50′

Note:

Due to the various modifications which have taken place during recent years, particularly the introduction of the fresh air heating, the rear axle loading has increased.

From March 1964, Chassis No. 6200001, the setting angle was increased to 17° 30′ + 50′.

From Chassis No. 2528668 (August 1959), the spring plates can be subsequently set to 17° 30′ + 50′.

If this measure does not give the desired results on vehicles which are operated **mainly with full load,** the setting can be increased to a maximum of 19°. This is, however, the limit and must not be exceeded.

Adjustment

1 - Check horizontal position of the vehicle by placing protractor VW 245a on door opening.

2 - Insert inner end of torsion bar in the center anchor.

> **Important**
>
> The torsion bars are prestressed. Left and right bars are not interchangeable. They can be distinguished by an arrow on the outside face showing the direction of torque.

3 - Press spring plate on outer end of torsion bar.

4 - Place protractor VW 245a on the unloaded spring plate. To obtain a correct reading, the half axle must be supported.

5 - Adjust the protractor until the bubble is in the center position.

If the protractor indicates a departure from the prescribed inclination, the adjustment is to be corrected. There are more splines at the outer end of the torsion bar than at the inner end, so that exact adjustment is possible.

Inner end:	40 splines
Outer end:	44 splines

If the inner end of the bar is turned by one spline, the adjustment is altered by 9°. If the spring plate is displaced by one spline, the adjustment is altered by 8° 10′. Thus, the inclination of the spring plate can be corrected by 0° 50′.

The rear wheels are sprung independently. The inner ends of the two torsion bars are anchored in the center of the frame cross member by a splined tube which is welded in position. The outer ends of the torsion bars, which are also splined, carry the spring plates the hubs of which are rubber cushioned. The rear axle tubes are attached to the rear ends of the spring plates and the holes in the spring plates are slotted so that the wheels can be set properly. The suspension can be adjusted by means of the splines at the ends of the torsion bars.

Screwed to the spring plate and the axle shaft bearing housing is a progressive rubber bumper which becomes active in the upper deflection range. Double-acting telescopic shock absorbers prevent excessive rebound.

Removing and Installing Spring Plates and Torsion Bars

Removal

1 - Loosen rear wheel mounting bolts.

2 - Support car on trestles in a horizontal position and remove rear wheels.

3 - Disconnect hand brake cables at hand brake lever and withdraw them slightly towards the rear.

4 - Mark the position of the spring plate in relation to the rear axle shaft bearing housing by making a chisel mark on the spring plate exactly in line with the groove in the rear axle shaft bearing housing.

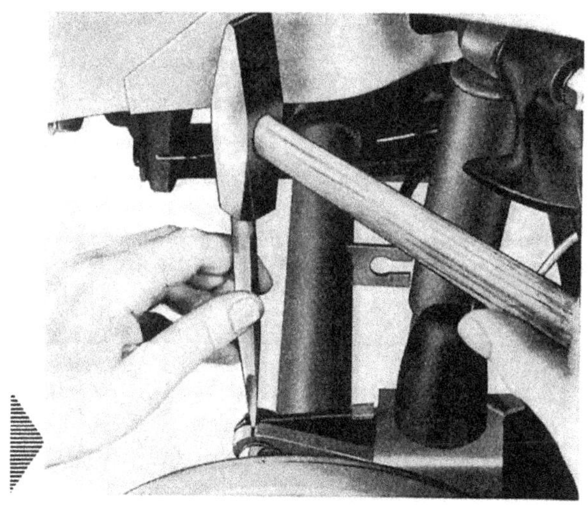

5 - Remove lower shock absorber mounting bolts.

6 - Remove bolts at axle shaft bearing housing.

7 - Pull rear axle out of the spring plates.

8 - Remove screws that attach spring plate hub cover and take off cover.

9 - Release spring plate tension and pull plate off torsion bar.

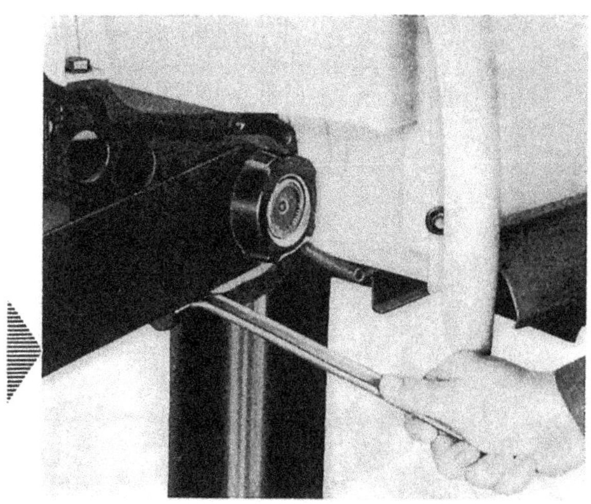

10 - Remove about five of the foremost fender screws.

11 - Pull the fender aside and withdraw torsion bar from frame cross tube.

> **Important**
>
> The protective paint on the torsion bar must not be damaged on any account. Even slight damage can lead to the formation of corrosion and this, in time, causes fatigue fractures.

Note:

To remove a broken torsion bar, push the broken end from the splined center anchor with a steel rod after removing the opposite torsion bar.

Installation

When installing, the following points should be observed:

1 - Inspect torsion bar splined ends and paint for damage. Make sure there are no signs of rust, replace if necessary.

2 - Grease splines of torsion bar.

3 - Install torsion bar and spring plate and adjust them.

> **Important**
>
> The torsion bars are prestressed in manufacture. Left and right bars must not be interchanged. They can be distinguished by the letters R (right) and L (left) on the outer end face.

Diameter of torsion bars:

a - On vehicles without equalizer spring
Sedan and Coupé .... 22 mm

b - On vehicles with equalizer spring
Sedan and Coupé .... 21 mm

4 - Coat rubber bushes with flaked graphite when installing. Ensure that the thicker part of the bush which is marked "oben" is always at the top.

5 - Fit cover and tighten all four screws uniformly. If necessary, use two longer screws to pull the cover down until the proper screws can be installed without difficulty.

6 - Lift the spring plate with the spring plate tensioner VW 655 until its lower edge is higher than the lower stop in the cross tube flange.

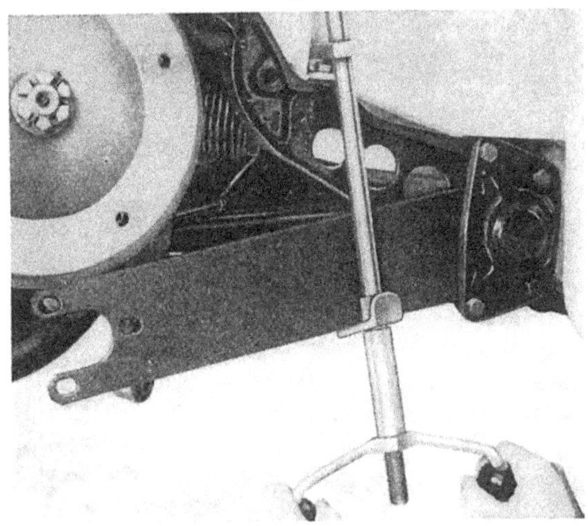

7 - Install and tighten the spring plate hub cover.

8 - Clean mating faces between spring plate and axle bearing housing.

9 - When bolting axle bearing housing and spring plate together, ensure that the mark made on the spring plate when removing the torsion bar is exactly in line with the groove in the axle bearing housing.

> **Important**
>
> If a new spring plate has been fitted the rear wheels must be re-aligned. If an optical axle alignment gauge is not available the track and wheel alignment cannot be set accurately. In this case the wheels should be set so that the marks in the **sides** of the spring plates are aligned with the marks in the axle shaft bearing housings. When using an optical gauge, the parts are installed with marks aligned and the alignment checked with the instrument and rectified where necessary. Movement of the bearing housing 1 mm alters the track of the wheels 8′. Further details on the optical alignment check are given in Technical Recommendation V 2.

10 - Tighten axle bearing housing mounting bolts to a torque of 80 lb. ft. (11 mkg).

11 - Connect cables to hand brake again and adjust hand brake.

Setting the Spring Plates

Proper wheel alignment and adequate spring travel under all conditions of load which are essential to good roadholding can only be obtained if the angles at which both spring plates are set are exactly the same. When one side has been set, the angle on the other should also be checked and rectified as necessary.

Note:

The rear wheel camber angle and track values have been revised to take into account the settling which occurs with new torsion bars. The figures given in this table are, therefore, only valid for vehicles which have run **at least 300 miles (500 km).** Different readings will be obtained if the angles are checked earlier.

The permissible minimum camber angle is for guidance only and the camber should not be below this angle even after a considerable mileage.

The new values should only be used on vehicles manufactured **before August 1966** if the torsion bars have to be reset during a repair. **Both torsion bars** should always be reset on vehicles which have been in use for some time, to avoid upsetting the driving characteristics with two different camber angles.

The torsion bars have different numbers of splines on the inner and outer ends so that this angle can be set accurately.

Inner end 40 splines

Outer end 44 splines

Turning the inner end of the torsion bar one spline, alters the angle by 9° and turning the spring plate one spline gives an alteration in angle of 8° 10′. This means that the smallest possible alteration in spring plate angle is 50′.

1 - Insert the torsion bar into the inner splines in cross tube.

> **Important**
>
> The torsion bars are prestressed in manufacture. Right and left hand bars must not be interchanged. The bars are marked on the outer ends with an L or an R.

2 - Fit the spring plate.

3 - Place the VW 261 protractor in the door opening.

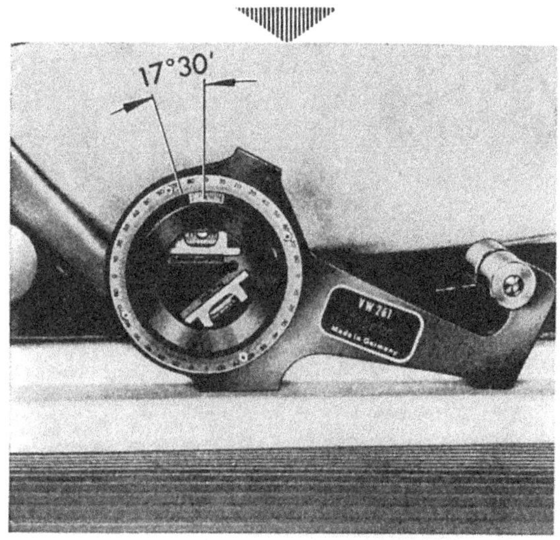

4 - Adjust the protractor so that the bubble in the spirit level marked "Achskörper/Winkel" is in the center position.

5 - Turn the level carrier from this position by the amount **specified for the vehicle involved.**

6 - Place the protractor on the untensioned spring plate and eliminate the play in the plate mounting by lifting the plate lightly in the working direction.

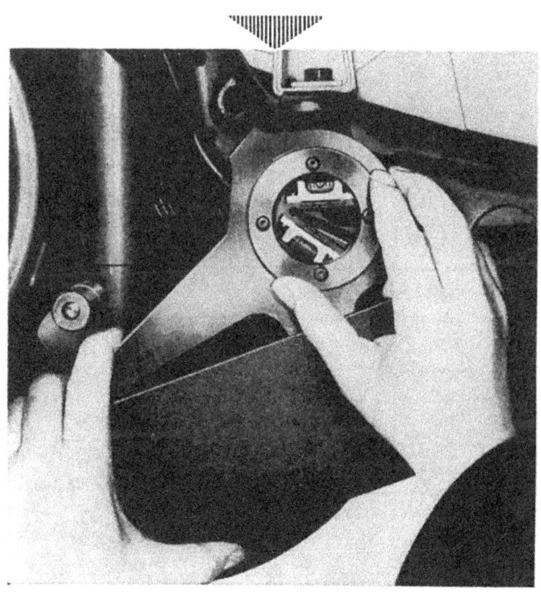

7 - Rectify the spring plate setting as necessary:

A - A bubble deviation of one graduation from the center position equals a setting error of 50′. This means that bar and spring plate must each be turned one spline in opposite directions to correct the error.

The spring plate must be turned in the direction in which the level carrier would have to be turned to bring the bubble into the center position.

B - Larger deviations which cannot be read on the scale are established by finding the angle by which the level must be turned to bring the bubble to the center position. The number of splines which the torsion bar and spring have to be turned, in this case, in opposite directions, is then found by dividing this angle by 50′ (smallest possible alteration).

When making this calculation, the remainder which is usually left over should be rounded off up to a full 50′ (alteration of bar and spring plate by one spline) if the initial angle of the spring plate is too flat (example 1). If the angle is too acute, the remainder should be ignored (example 2). This ensures, in each case, that the setting remains within the tolerance of 50′.

Examples:

1 - Protractor reading in door opening 40′
Setting angle for spring plates (+18° 30′) . . . . 19° 10′
Measured spring plate angle . . . . 17° 50′
Difference between 19° 10′ and 17° 50′ . . . . 1° 20′
1° 20′ = 80′ divided by 50′ = 1 (remainder 30′).

As the initial angle of the spring plate was too flat in this case, the 30′ remainder is rounded off to 50′ so that the torsion bar and spring plate must be moved 2 splines in opposite directions.

2 - Protractor reading in door angle . . 30′
Setting angle for spring plates (+18° 30′) . . . . 19°
Measured spring plate angle . . . . 21° 20′
Difference between 19° and 21° 20′ 2° 20′
2° 20′ = 140′ divided by 50′ = 2 (remainder 40′).

As the initial angle of the spring plate was too acute in this case, the 40′ remainder is ignored.

Model	Introduction from	to	Chassis Numbers from	to	Spring plate setting angle	Camber angle after 300 miles	Permissible minimum camber angle	Total track	Permissible deviation in wheel alignment
A - Without equalizer spring									
all (except 147 and 111 112, 115, 116 with M 86)	Aug. 59	July 66	2 528 668	1161021298	**17° 30′+50′**				
147	April 65		145 395 753		**18° 30′+50′**	2° 30′ ± 1°	0°	N 5′ ± 10′	max. 10′
111, 112, 115 116 with M 86	Jan. 67		117 483 306		18° 30′+50′				
B - With equalizer spring									
111—118	Aug. 66	—	117 000 001	—		1° ± 1°	−1°		
141—144 151—152	Aug. 66	—	147 000 003 157 000 002	— —	20° +50′	15′ ± 1°	−1° 30′	N 5′ ± 10′	max. 10′

The underlined values have not changed N = Toe-out

Removing and Installing Equalizer Spring

The equalizer spring is a torsion rod transversely mounted under the luggage compartment floor and connected to the two axle tubes by levers and operating rods. As the equalizer spring works in conjunction with the main springing (torsion bars) a progressive springing is attained when both rear wheels are deflected. In addition, the equalizer spring together with the specially adapted main springing causes the dynamic wheel load of the outer wheels to be indirectly distributed when cornering, thus making the cornering characteristics of the vehicle more neutral.

Removal

1 - Disconnect battery ground strap.

2 - Loosen wheel mounting bolts.

3 - Raise vehicle and remove wheels.

4 - Remove nuts on operating rods.

5 - Remove rubber caps and unscrew operating rod hexagon nuts on the levers.

6 - Remove operating rods and rubbers buffers.

7 - Unscrew hexagon nuts for supports and remove and rubber bushes.

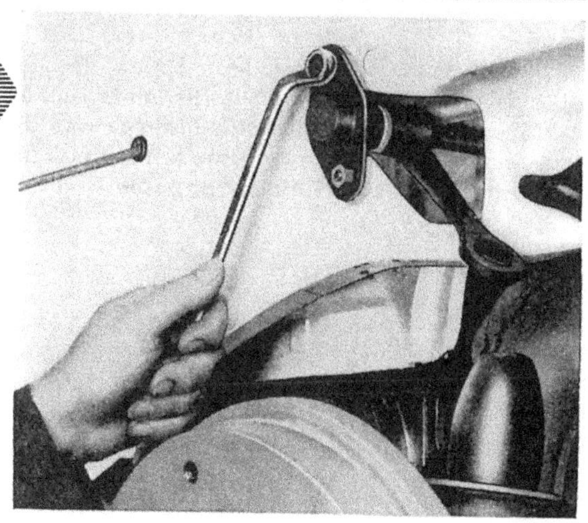

8 - Withdraw equalizer spring and levers towards right.

9 - Slacken the lock nuts and socket head bolts securing the levers on the equalizer spring a few turns until the levers can be removed.

10 - Check the operating rod guide rings and replace if necessary. They can be removed with a screwdriver.

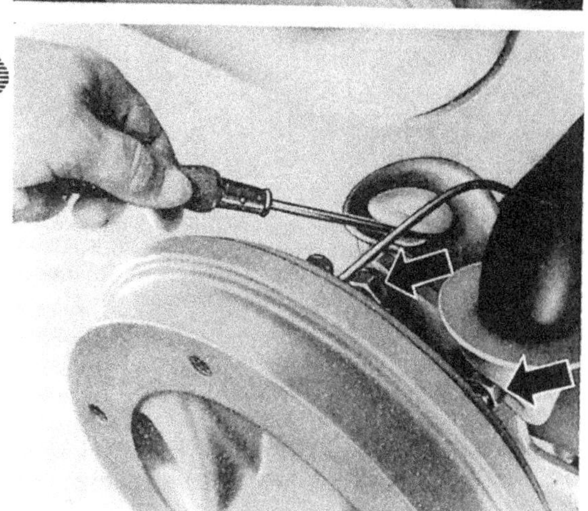

11 - Unscrew two hexagon nuts on each side and take left and right guides off axle tube flange.

12 - Check equalizer spring, rubber bushes, damping rings, guide tubes and rubber stops and replace them if necessary.

Installation

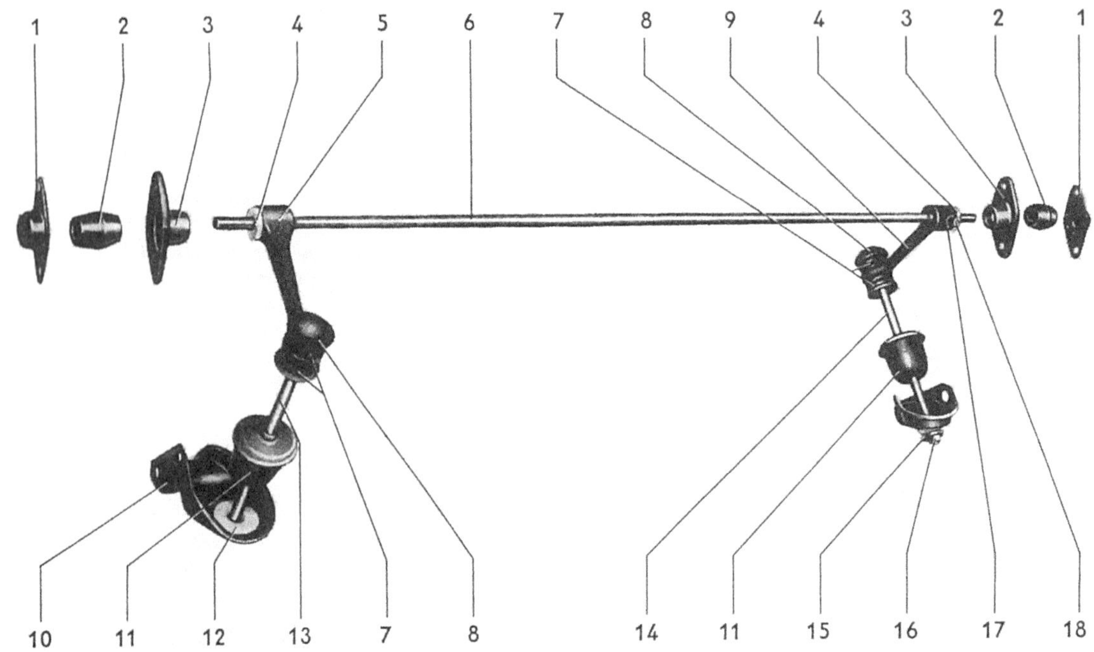

1 - Outer support
2 - Rubber bush
3 - Inner support
4 - Hard rubber washer
5 - Left lever
6 - Equalizer spring rod
7 - Damping ring
8 - Protective cap
9 - Right lever
10 - Left guide
11 - Rubber stop
12 - Guide ring
13 - Left operating rod
14 - Right operating rod
15 - Washer
16 - Self-locking nut
17 - Lock nut
18 - Clamping screw

1 - Push lever on to equalizer spring. The left lever is marked with an L. The left lever must point downwards towards the rear and the clamping screw towards the front. Install the right lever symmetrically opposite.

2 - Tighten clamping screw and lock it.

3 - Install equalizer spring, together with lever, from the right.

Note:

The two supports are installed symmetrically opposite so that the inclined position of the equalizer spring to the lateral parts is compensated for. Install cover oppropriately.

4 - Install supports and rubber bushes. Place a hard rubber washer between lever and support.

5 - Install operating rods. The long operating rod is installed on the left-hand side. Place a damping ring above and below the levers. Install protective caps.

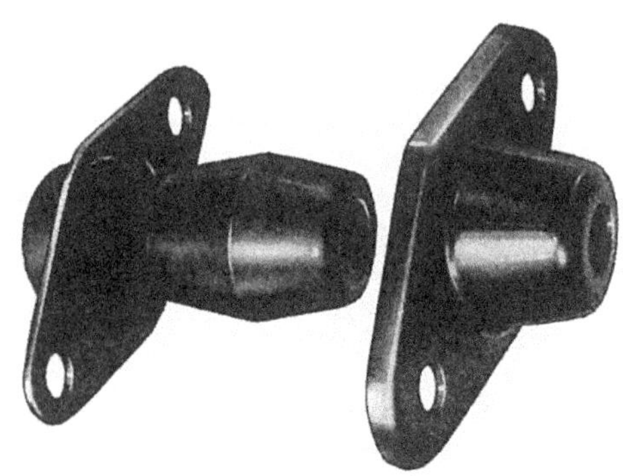

Note:

The operating rods should be fitted to the levers first to ensure that the damping rings fit properly. The levers must not rest in the guides.

6 - Insert operating rods into guides. Install washers and tighten nuts.

7 - Install all other parts.

Removing and Installing Shock Absorbers

General

Correctly operating rear shock absorbers and properly adjusted spring plates are of great importance to perfect suspension and road-holding of the car The compression and re-bound actions of the shock absorbers correspond to the springing characteristics of the car. The shock absorber prevents the spring plate from striking the lower stop at the cross tube flange during re-bound.

The shock absorbers require no maintenance and have an adequate fluid reservoir which compensates smaller losses. A more accurate check of the efficiency of the shock absorbers can be made with special testing appliances. Shock absorber action may be roughly checked by bouncing each corner of the car in turn or by driving the car over a very uneven road. Checking by compressing the removed shock absorber by hand will only give indication whether or not there is a resistance, the degree of efficiency of the compression and re-bound strokes cannot be determined by this method.

Removal

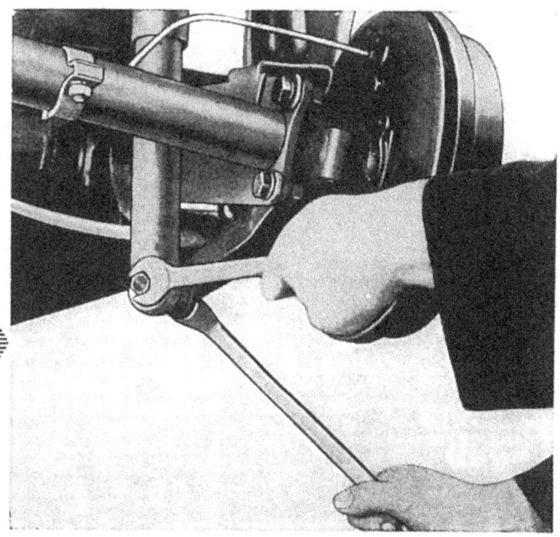

1 - Lift the vehicle.

2 - Remove the two nuts of shock absorber mounting bolts.

3 - Take off shock absorber.

Installation

When installing, the following points should be observed:

1 - Test shock absorber and, if necessary, replace.

2 - Inspect sleeves and rubber bushes for wear. Replace if necessary. Worn bushes should be removed by means of the VW Repair Press in conjunction with VW 401, VW 410, VW 421, VW 438.

Replace damaged rubber bushes.

Install new sleeves with the VW Repair Press in conjunction with VW 401, VW 410, VW 421, and VW 436.

Rear Wheel Track and Rear Wheel Alignment

General Description

The rear wheel position is of paramount importance to the roadworthiness and cornering stability of the vehicle and to the tire life.

If the vehicle should act strangely on the road, show signs of tire wear, or it is thought that the wheel alignment has been affected by accident damage, the vehicle should be checked as regards steering geometry and wheel alignment and, if necessary, readjusted.

In order to check the vehicle exactly, it is in most cases necessary to measure the vehicle completely. Excessive departures from the prescribed wheel alignment values or a faulty suspension can adversely affect a wheel of the other axle. The measurements of the Volkswagen, especially of the rear wheel alignment, can only be taken by means of an axle tester which is sufficiently exact and operates independent of the vehicle. Upon request, the Technical Service Department will give information as regards other axle testers which are suitable for the Volkswagen.

The nominal values for the Volkswagen are given in both the specification chart P 6 a and section V of the Workshop Manual. The values are valid only for tests carried out under the following conditions:

a - Correct tire pressures

b - Vehicle unladen

c - Correct adjustment of spring plates.

Particulars concerning the operation of these axle testers are contained in the instruction manuals published by the manufacturers. The following points are of importance when checking the rear axle of the Volkswagen:

1 - Rear axle tread (track)

2 - Position of rear wheels

3 - Camber angles

Rear Wheel Tread

The following readings are valid with correct spring plate adjustment and unladen:

Adjustment:	toe-in 1.0 mm — toe-out 2.5 mm
Specification chart:	—5′ ± 15′

The values for the rear wheel position represent the tread. Plus values for both wheels, or minus values should be added. If the values differ, i. e., (+/—), subtract the smaller value from the other.

The tread can only be altered by correcting the wheel position. Even if the wheel track is correct, the rear wheel positions may be incorrect. For this reason the measurement of the wheel positions is of great importance.

Rear Wheel Alignment

The values for the rear wheel alignment must meet the following requirements:

1 - The rear wheel tread must be in accordance with the specified nominal values.

2 - The departure of the wheel alignment from the longitudinal vehicle center line must not be more than 10′.

Note:

Any departure in the rear wheel alignment from the vehicle centerline may be caused by one wheel having toe-in (+) and the other toe-out (—), or one wheel being straight-ahead and the other not, for example:

left + 10′ (toe-in)	right — 20′ (toe-out)
left — 25′ (toe-out)	right 0′

Both rear wheels should have approximately the same toe-out values, for example:

left — 5′ (toe-out)	right — 5′ (toe-out)
left 0	right — 5′ (toe-out)

The holes provided in the spring plate for attaching it to the axle bearing housing are elongated. To make it possible to carry out wheel alignment sufficiently accurately without the aid of a rear wheel alignment gauge a groove is provided on the side of the spring plate and on the axle bearing housing above the top mounting hole. When adjusting the wheels without an alignment gauge, these two grooves must be dead in line.

With new vehicles in unloaded condition the track values of the rear axle is — 5′ ± 10′ which is achieved in the factory by special equipment and ensures precise alignment. When carrying out repairs such as rear axle removal or loosening the spring plate at the axle bearing housing, the exact position of the rear wheels has to be marked to ensure proper wheel alignment when reinstalling. This can best be done by making a mark on the top of the spring plate dead in line with the groove in the axle bearing housing.

A re-adjustment is necessary if a new rear axle is installed, or if a frame, spring plate or front rubber cushion is replaced. Workshops not possessing an optical rear wheel alignment gauge are not in a position to carry out a precise alignment. In this case, the groove in the side of the spring plate must be dead in line with the one in the axle bearing housing when the rear axle is installed. In order to obtain a precise adjustment the vehicle is measured completely, after the installation with the grooves dead in line, and the wheel position corrected if necessary. Moving the axle bearing housing forward alters the position in the direction of toe-in, and backward in the direction of toe-out. 1 mm alters the alignment approximately 8′.

Camber of Rear Wheels

The camber angle of the rear wheels with the vehicle in unloaded condition and spring plates adjusted correctly is 3° ± 30′. Small deviations on both wheels do not influence the road-holding. It is, however, important that the wheels have the same camber angle. The maximum permissible deviation between the readings of both sides is 20′.

Excessive deviations in the camber readings are to be corrected as follows:

Symptom	Cause	Remedy
Camber angle too large but equal on both wheels	Spring plates wrongly adjusted	Adjust spring plates
Camber angle too small, but equal on both wheels	Spring plates have settled	Reset spring plates to proper position
Camber angle uneven on both wheels	a - Spring plates are unevenly adjusted b - The friction indices of the rubber cushions in the spring plates are uneven and too large	a - Correct spring plate adjustment b - Apply graphite when installing rubber cushions

THIS PAGE INTENTIONALLY
LEFT BLANK

1 - VW Special Service Tools

VW 112	Special Wrench 36 mm with Guide Plate
VW 114	T-Wrench 8 mm Square Socket
VW 172	Allen Key for Oil Filler Plugs
VW 202	Extractor
VW 202 a	Extractor Hooks
VW 202 c	Extractor Ring
VW 202 d	Extractor Hooks
VW 202 k	Thrust Pad
VW 222	Starter Shaft Bush Pilot Drift
VW 228 a	Starter Shaft Bush Extractor
VW 241 a	Ball Bearing Extractor
VW 244 b	Driving Sleeve
VW 245 a	Protractor
VW 246	Not-Go Plug Gauge
VW 287 a	Differential Housing Gauge
VW 289 d	Transmission Case Mandrel
VW 293	Drive Pinion Round Nut Removing and Installing Tool
VW 294	Gear Shift Test Appliance
VW 295	Needle Bearing Drift
VW 295 a	Additional Tool for VW 295
VW 296	Removal Tool for Transmission Gear Assy
VW 297	Adjusting Device for Concave Washer
VW 299	Adjusting Device for Ring Gear
VW 298	Gauge Ring
VW 307	Fixture
VW 400	Repair Press 15 t
VW 401	Thrust Plate
VW 402	Thrust Plate
VW 405	Punch with V-Block
VW 406	V-Blocks (two)
VW 407	Punch
VW 408	Punch
VW 409	Punch
VW 410	Punch
VW 411	Punch
VW 412	Punch
VW 415	Tube, 75 mm dia.
VW 419	Tube, 32 mm dia.
VW 421	Tube, 28 mm dia.
VW 422	Tube, Slotted
VW 430	Thrust Pad (Bronze)
VW 432	Thrust Pad 50 mm dia.
VW 433	Thrust Pad
VW 434	Thrust Pad
VW 435	Thrust Pad
VW 436	Guide Pin (Tapered)
VW 438	Guide Pin (Cylindrical)
VW 439	Guide Pin (Shouldered)

VW 441	Base Block
VW 442	Thrust Pad
VW 449 f	Toothed Thrust Ring
VW 451	Thrust Pad
VW 452	Gear Carrier Support

2 - VW Workshop Equipment for Local Manufacture

VW 605	Gantry Crane
VW 609	Rear Axle Cradle for Trolley Jack
VW 633	Threstle
VW 643	Stand
VW 655	Torsion Bar Tensioner
VW 656	Spring Ptate Installing Tool
VW 664/1	Differential Housing Holding Fixture
VW 681	Oil Seal Remival Tool

3 - Normal Hand Tools

Combination Pliers
Circlip Pliers, Noses angled at 90°
Circlip Pliers, Long Nosed
Cold Chisel
Prick Punch (center punch)
Pin Punch 4 mm
Mechanic's Hammer, 300 grams
Mechanic's Hammer, 500 grams
Rubber Hammer 85 x 50 mm
Triangular Scraper
Flat Scraper
Flat File, 180 mm long
Half-Round File, 180 mm long
T Handle for Sockets with Reduction Pieces
Socket, 10 mm
Socket, 11 mm
Socket, 13 mm
Socket, 14 mm
Socket, 15 mm
Socket, 17 mm
Socket, 19 mm
Socket, 32 mm
Socket, 36 mm
Open-End Wrench, 14 mm
Open-End Wrench, 17 mm
Open-End Wrench, 19 mm
Box Wrench, 13 mm
Box Wrench, 14 mm
Box Wrench, 15 mm
Box Wrench, 17 mm
Box Wrench, 19 mm
Box Wrench, 27 mm
Box Wrench, 32 mm
Twist Drill 5 mm
Twist Drill 7.5 mm
Twist Drill 12 mm
Tap M 6
Tap M 7

Tap M 8
Tap M 10
Tap M 12 x 1.5
Tap Wrench, adjustable, size 1
Tap Wrench, adjustable, size 2
Die M 6
Die M 7
Die M 8
Die M 10
Die M 12 x 1.5
Die Stock, size 1
Die Stock, size 2
Torque Wrench, 2—30 mkg (14—217 ft. lbs.)
Scratch Awl
Dial Indicators, readings of 1/100 mm
Set of Feeler Gauges 0.1—0.5 mm
Micrometer Caliper 0—25 mm
Caliper Square, 300 mm in length, measuring 1/10 mm
Electric Drill
Inspection Lamp with Cable and Plug
Wire Brush
Oil Can
Can for Derusting Fluid
Grease Container

4 - Supplementary Workshop Equipment

Hydraulic or Mechanical Trolley Jack
Run-out Testing Device
Optical Wheel Alignment Gage
Heli-Coil Assortment for Volkswagens

THIS PAGE INTENTIONALLY
LEFT BLANK

Contents: B

Hydraulic Brakes

Wheels and Tires

Disc Brakes

Important — Friction materials such as brake or clutch discs may contain asbestos fibers. Do not create dust by grinding, sanding, or by cleaning with compressed air. Avoid breathing asbestos fibers and asbestos dust. Breathing asbestos can cause serious diseases such as asbestosis or cancer, and may result in death.

THIS PAGE INTENTIONALLY
LEFT BLANK

Description of Hydraulic Brakes

General

The function of the hydraulic brake is based on Pascal's law:

> "Pressure applied to a fluid enclosed in a vessel is transmitted uniformly to every part of the vessel"

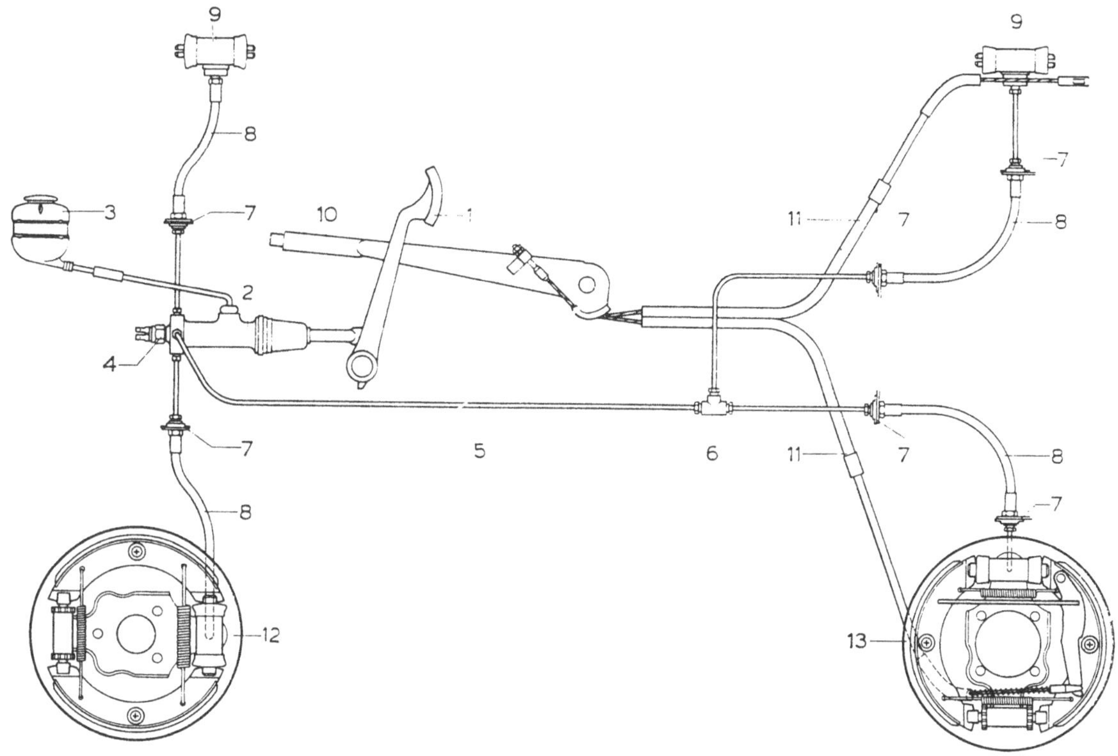

Diagrammatic view of complete hydraulic brake system

1 - Brake pedal
2 - Master cylinder
3 - Fluid reservoir
4 - Stop light switch
5 - Brake line
6 - Three-way connection
7 - Brake hose bracket
8 - Brake hose
9 - Wheel cylinder
10 - Hand brake lever
11 - Cable and conduit tube
12 - Front wheel brake
13 - Rear wheel brake

The brake system consists of:

The master cylinder, in which the hydraulic pressure is generated;

The fluid reservoir, which maintains the constant volume of the brake fluid;

The wheel cylinders, which force the brake shoes against the drums;

The hydraulic lines, connecting master cylinder to wheel cylinders, hoses being used at moving parts.

The master cylinder is provided with one piston and each wheel cylinder with two pistons. Rubber cups prevent leakage.

Operation

When the brake pedal is depressed, the piston in the master cylinder operates and forces the brake fluid through the lines and hoses to the wheel cylinders. The fluid enters the wheel cylinders between the pistons and forces them outwards. The piston movement moves the brake shoes outwards into contact with the brake drums.

Increasing the force on the brake pedal increases the pressure in the wheel cylinders and this in turn exerts a greater effort at the brake shoes.

When the pedal is released, the return springs bring the shoes back to the rest position. This return movement of the shoes forces the pistons in the wheel cylinders back to their starting positions and the brake fluid flows back to the master cylinder.

Hand Brake

The mechanical hand brake is operated by pulling up the horizontally placed hand brake lever, which is of the ratchet-and-pawl type. Both cables are anchored direct to the lever and are guided in tubes to the rear wheels.

Note:

It is possible for the braking force to vary slightly between the two wheels of an axle. If the braking effort of the right and left wheels on an axle is found to be unequal during an official vehicle inspection, the permissible variation must be established when making the assessment. For the hydraulic and mechanical brake systems of the VW Passenger Car and Transporter the figure is 15%.

When the hydraulic system is properly maintained and the brake drums and linings are fairly evenly worn, the variation will usually be slight. With the mechanical brakes, the braking effort can be kept below the specified value if the cables and brakes are adjusted correctly.

Master Cylinder

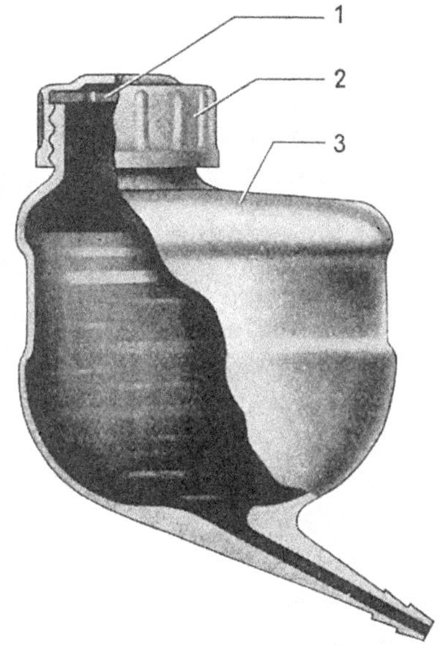

1 - Gasket
2 - Screw cap
3 - Reservoir
4 - Brake light switch
5 - Master cylinder body
6 - Check valve
7 - Spring
8 - Plug
9 - Washer for plug
10 - Compensating port
11 - Main cup
12 - Washer
13 - Intake port
14 - Piston
15 - Secondary cup
16 - Stop washer
17 - Lock ring
18 - Push rod
19 - Boot

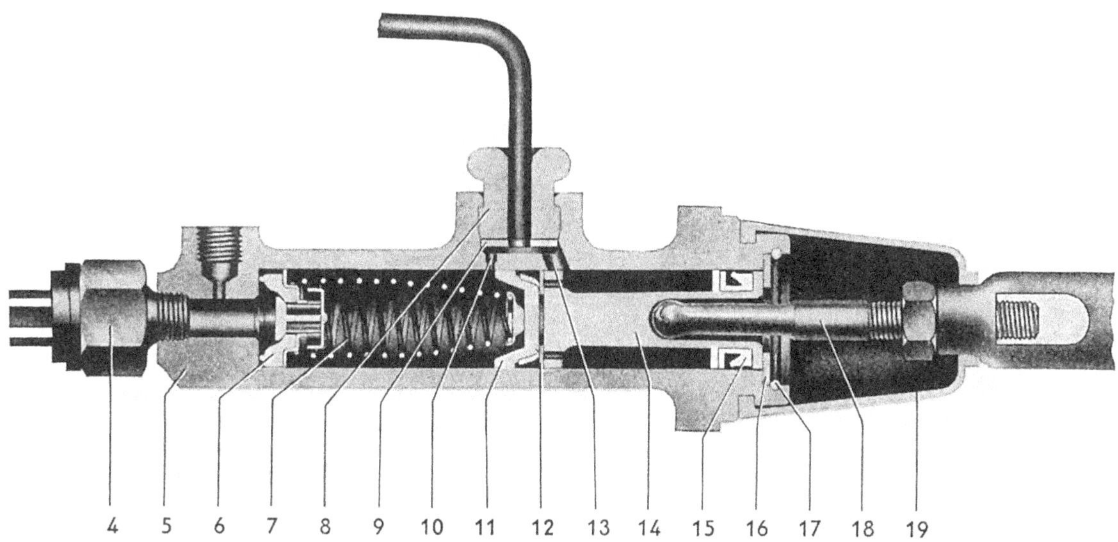

Internal dimensions of master cylinder: Diameter = .687 in. (17.46 mm)
Stroke = 1.299 in. (33 mm)

The piston in the master cylinder is operated by a push rod which is connected to the brake pedal. When the pedal is depressed, the pressure exerted on the brake fluid by the piston is transmitted uniformly through the system to the wheel cylinders.

Fluid Reservoir

The vent hole in the cap on the reservoir must always be clear so that fluid can flow into the master cylinder.

When topping up the fluid in the reservoir, take care that any dirt present on the cap or near the opening is removed carefully so that it does not fall into the reservoir and get into the brake system. The reservoir must always be filled up to about .6—.8 in. (15—20 mm) below the cap.

To facilitate topping-up and inspection, the reservoir is mounted on the body in the front luggage compartment behind the spare wheel and separated from the master cylinder. It is connected to the master cylinder by a pipe.

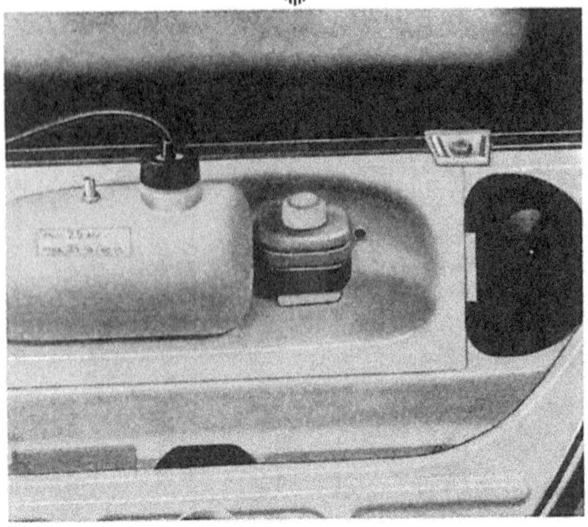

Important

Never put mineral oil into the brake reservoir. Use only genuine VW brake fluid. Do not spill brake fluid on the paintwork.

When carrying out the Delivery Inspection or other service inspections, the complete braking system must be checked for leaks. This includes also the checking of the connection between the brake fluid reservoir and the master cylinder. When doing this, check the fluid reservoir securing nuts for tightness and for leaks, and check also the sealing plug in the master cylinder. It is not sufficient to merely top-up the brake fluid reservoir.

Note:

Since May 1966, from **Chassis No. 116872032,** all Type 1 Models have been fitted with a new fluid reservoir with a screw connection. Also the lines between the fluid reservoir and the brake master cylinder as well as the connecting hose for the line between reservoir and brake master cylinder have been altered.

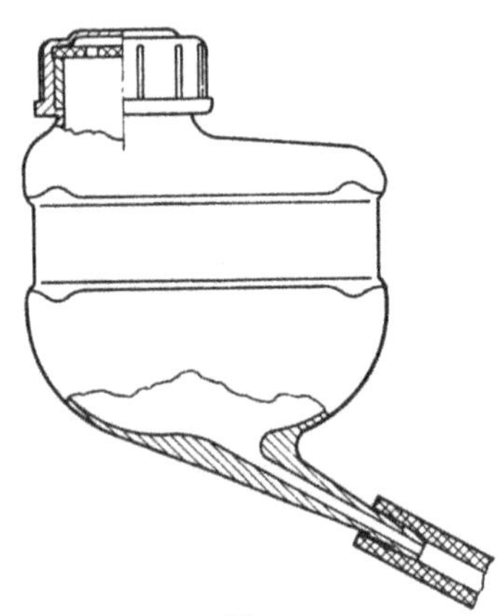

old

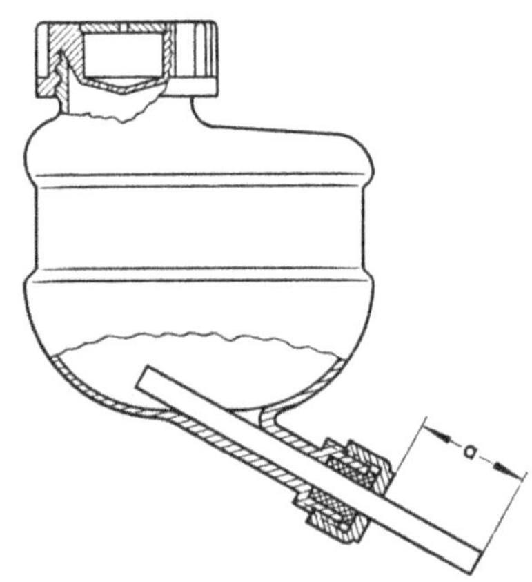

new
a = .9 in. (23 mm)

In August 1966, from **Chassis No. 117000001,** the cross section of the connecting lines from fluid reservoir to master cylinder was enlarged. (Outside diameter/thickness of the line were previously 4.75 × 0.72 mm and are now 6.00 × 0.7 mm). In this connection, also the fluid reservoir and the plug for the brake line were modified.

Vehicles with dual circuit brake systems also have 6 mm brake lines.

Service installation of the modified brake lines in older vehicles is not being contemplated.

Important new part nos.

		Line diameter	
		4.75 mm	6.0 mm
Fluid reservoir		113 611 301 G	113 611 301 J
Line between master cylinder and fluid reservoir	LH drive	113 611 805 C	113 611 805 E
	RH drive	114 611 805 B	114 611 805 D
Screw cap		131 611 349 A	131 611 349 A
Securing nut for line		131 611 341	131 611 341
Plug for master cylinder		113 611 817	131 611 817*
Connecting hose for line		113 611 801 A	131 611 801

* The plug for the 6 mm line has sharp edges on the upper shoulder to distinguish it from the plug for the 4.75 mm line.

Note:

Model		from Chassis No.
11	Single circuit disc brakes	117 469 710
11	Single circuit drum brakes	117 488 486

In January 1967, the connecting line from fluid reservoir to master cylinder (with exception of short portion in reservoir) was increased in size on all Type 1 Models (old outer dia./wall thickness: 6.00×0.7 mm; new: 8.00×0.7 mm). In addition, the line is no longer fitted directly into the master cylinder but is connected via an 8 mm hose and an elbow. The connection between the 6 mm line and the 8 mm one is made with a shouldered hose (conical).

In the future only 4.75 mm and 8.00 mm diameter connecting lines will be supplied as replacement parts. When parts are required for vehicles with 6 mm dia. lines, convert to 8 mm dia. lines as described here.

Main Part Nos.:

Connecting hose, front (6 mm and 8 mm dia. ends)		113 611 813
Line between front and rear hoses	LHD	113 611 805 G
	RHD	114 611 805 E
Connecting hose, rear		211 611 801
Elbow 105° (black)		113 611 153
Plug for master cylinder		311 611 817
SP set (for conversion to 8 mm line)	LHD	113 698 995
	RHD	114 698 995

When fitting the 8 mm dia. lines, note the following points:

Brake fluid damages paint and should therefore be sucked out of the reservoir before starting work or caught in a suitable container when the lines are pulled out of the master cylinder.

Moisten the ends of the lines with brake fluid before fitting the hoses.

1 - Remove fuel tank.

2 - Pull line between master cylinder and reservoir off at master cylinder and catch fluid.

3 - Pull line and hose off the line in the reservoir.

4 - Take plug and washer out of master cylinder.

5 - Insert plug (Part No. 311 611 817) into master cylinder correctly and **without** washer.

6 - Fit small end of front connecting hose (113 611 815) on to pipe in reservoir.

7 - Fit rear hose (211 611 801) on brake line (8 mm dia.) 113 611 805 G for LHD; 114 611 805 E for RHD.

8 - Push 105° elbow (113 611 153) into rear connecting hose.

9 - Push line into front connecting hose.

10 - Press elbow into master cylinder.

11 - Secure line to frame with clip.

12 - Install fuel tank.

13 - Fill with **new** brake fluid.

14 - Bleed entire system as specified.

Layout of brake lines

LHD

RHD

On vehicles with single circuit brakes, from Chassis No. 117 000 001 (8/66) to Chassis No. 117 349 408, the fluid reservoir must also be raised. This entails the following additional work:

1 - Remove container for windshield washer.

2 - Remove brake fluid reservoir retaining strap.

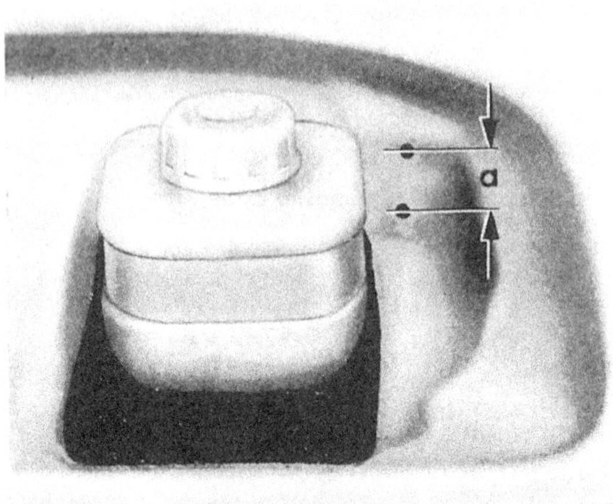

3 - Drill a clearance hole for tapping screw B 3.9×13 (N 139 612) on the second side of the strap with a 4.2 mm drill.

4 - Mark off hole in reinforcement plate as shown in illustration and drill a 2.7 mm hole for existing B 3.5×13 tapping screw. a = 15 mm.

5 - Secure strap on one side and mark off and drill a 3.1 mm hole for B 3.9×13 tapping screw on the other side.

6 - Stick packing for reservoir on to reinforcement plate.

7 - Secure reservoir with strap.

8 - Install windshield washer container.

Brake Fluid Compensation

The fluid in the brake system is subject to external and internal influences which can cause a change in volume. A temperature rise and the consequent expansion of the fluid can cause an excess of fluid. A drop in temperature, on the other hand, creates an increased demand for fluid which must be fed into the system.

Compensation Port

To cope with this fluid movement, the master cylinder is equipped with an automatic fluid regulation system. In the cylinder wall just in front of the primary cup is a compensation port which provides a connection between the fluid reservoir and the cylinder pressure chamber so that the excess fluid caused by expansion can flow from the pressure chamber into the reservoir. Vice versa, when a fluid shortage exists, fluid can flow from the reservoir back into the master cylinder via the compensation port. As it is the task of the compensation port to maintain a state of harmony in the fluid system it will be appreciated that any interference with this task will affect the functioning of the brakes.

Important — The compensation port must always be free when the system is at rest.

If the compensation port becomes blocked it will prevent the fluid from flowing back to the reservoir when it heats up (continuous braking on hills). The fluid will then expand toward the wheel cylinders where it will overcome the return spring pressure and cause the brakes to bind continuously (brake light will remain on).

Operating Rod

In order to ensure that the compensation port is not covered by the primary cup when the brakes are not in use, the push rod must be set so that there is a clearance of .04 in. (1 mm) between tip of rod and bottom of hole in piston.

Important

The clearance must always be set by moving the brake pedal stop.

Clearance S of .04 in. (1 mm) in the master cylinder can be measured also at the brake pedal plate.

S = .04 in. (1 mm) is equal to a pedal movement x of .2—.28 in. (5—7 mm).

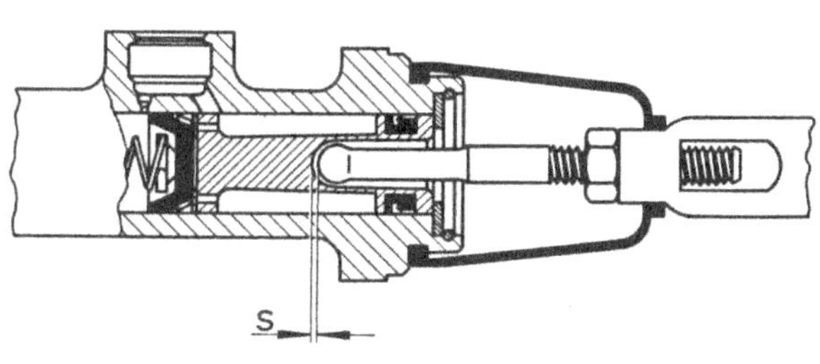

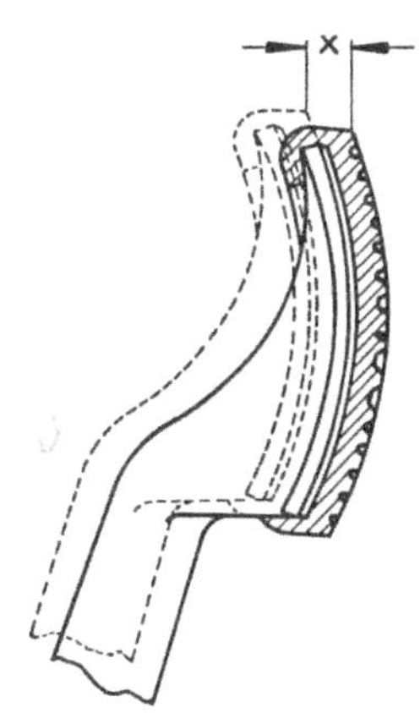

An alteration of the length of the push rod (dimension a) set in production is not permissible and must not be carried out even at a customer's request (e. g. alteration of the pedal plane toward the front or rear). If a push rod has to be replaced, the old push rod must be measured and the length transferred to the new push rod. The measurement is taken from the push rod tip to the center of the attachment hole.

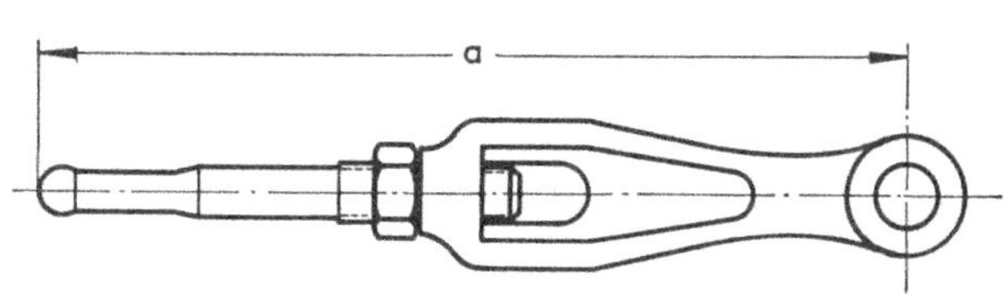

Check Valve

A further part in the self-filling master cylinder is the double-acting check valve which is fitted to control the fluid movement already described. If a vacuum occurs in the brake system, the check valve reacts to the slightest change and permits the necessary quantity of fluid to flow from the reservoir through the cylinder pressure chamber into the system. The valve is also lifted from its seat by excessive pressure in the system so that the surplus fluid can return to the reservoir.

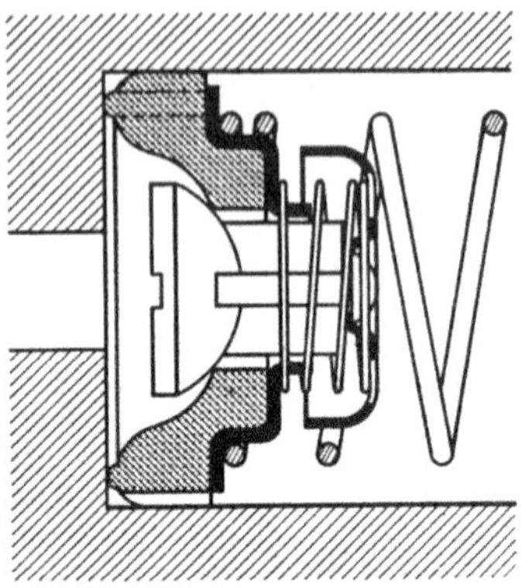

Rest position

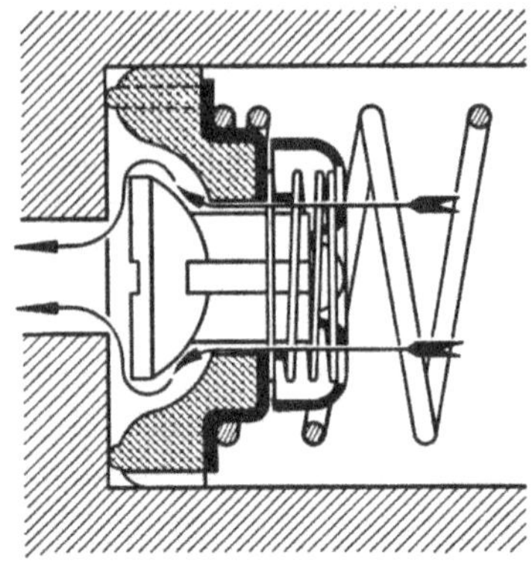

Braking position

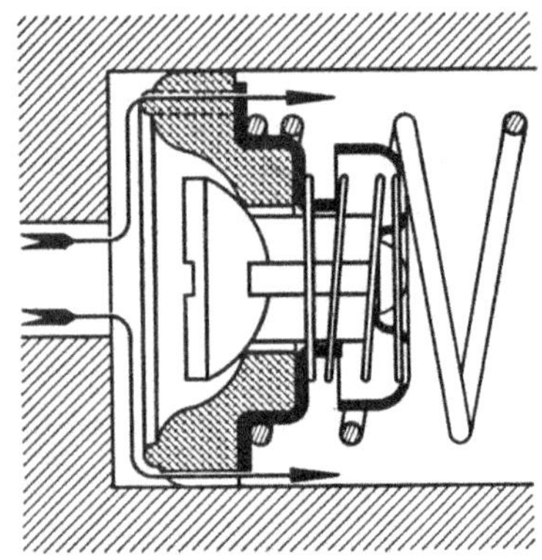

Return position

When braking, the fluid flows past the check valve into the brake lines and returns freely when the pedal is released. The check valve is fitted with a small spring which maintains a slight pressure in the system. This ensures that the system is always completely full and the force exerted at the brake pedal is transferred to the complete brake system without delay.

Special Check Valve

Contrary to the drum brakes, the disc brakes must be completely without pressure when they are not in operation. The special check valve can be distinguished from the conventional check valve by the hole in the piston head.

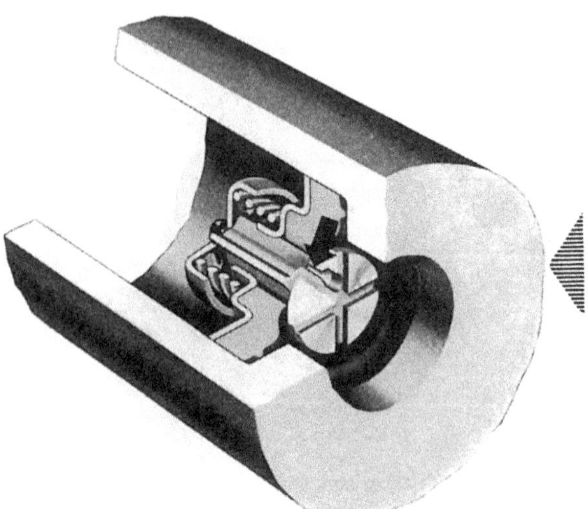

This so-called pressure relief port arrow enables the pressure to be completely relieved after braking and ensures that when the brake pedal is depressed quickly several times in sucession, brake fluid can be "pumped" by the master cylinder piston (e. g., filling and bleeding the brake system).

Important

As the master cylinder with special check valve cannot be distinguished outwardly from a master cylinder with a normal check valve, the former is marked with a blue sticker bearing the words "**Anschluß mit Spezial-Bodenventil**".

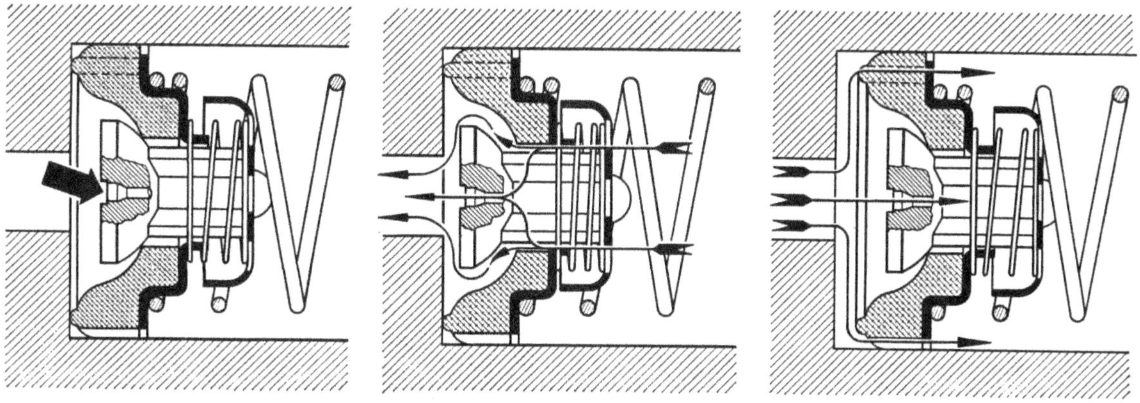

Rest position Braking position Return position

Rest position: When the brakes are in their rest position, the whole brake system is without pressure. This is attained by the pressure relief port (arrow).

Braking position: When the master cylinder piston moves forward, the hydraulic pressure in the master cylinder increases and moves the check valve from its rubber seating. The brake fluid flows through the pressure relief port and past the lifted check valve into the system.

Return position: On releasing the foot brake, the pressure in the brake lines is higher than that in the master cylinder so that the complete check valve is lifted from its seating. The brake fluid flows past the rubber seating and through the pressure relief port back into the master cylinder. If the pressure in the brake lines is lower than the force of the check valve spring, the check valve moves back to its original position. The pressure is then completely equalized via the pressure relief port.

Primary Cup

To prevent air from being drawn in when the pedal is released, an annular space is provided round the piston behind the primary cup. The brake fluid can flow from this space into the brake line side via the drillings in the piston, the cup washer and the grooves in the primary cup when the piston returns rapidly. If a depression is created in the master cylinder by the rapid movement of the piston, it will always be filled by fluid from behind the primary cup and never by air.

The outer end of the annular fluid chamber round the piston is sealed by the secondary cup. The coordinated operation of the by-pass port, double-acting check valve and the primary cup ensures completely automatic filling of the system and prevents the ingress of air which would be detrimental to the brake system efficiency.

Important

Assembled master and wheel cylinders should not be stored for longer than 1 year. If a longer period of storage is unavoidable, the parts must be disassembled and examined. The rubber parts should not be exposed to daylight when stored for prolonged periods.

Removing and Installing Master Cylinder

The master cylinder should be removed for cleaning and inspection of the component parts. The reservoir should also be removed when it requires cleaning.

Removal

1 - Remove master cylinder plug with line.

2 - Disconnect brake light cables from switch.

3 - Remove brake lines and plug them with the bleeder valve dust cap.

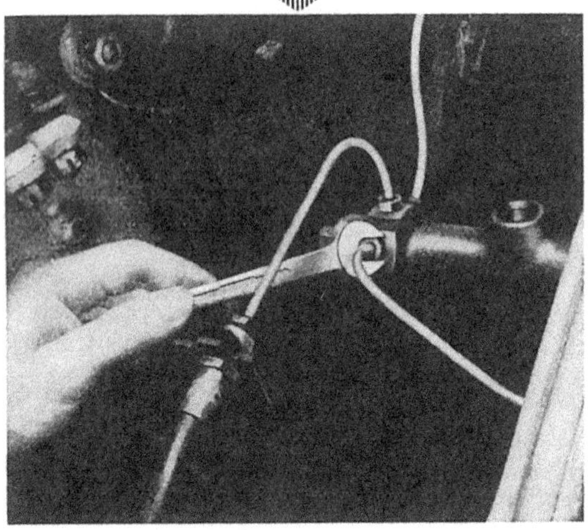

4 - Bend up lock plate of pin in master cylinder operating rod and remove pin.

5 - Loosen brake pedal stop bracket and take push rod out.

6 - Remove master cylinder securing screws and take cylinder out to the front.

Installation

Note the following when installing:

1 - Place distance piece for master cylinder in front cross member.

2 - Install master cylinder complete with push rod. Ensure that boot is located properly.

Important
Always fit a new lock plate on the push rod pin.

3 - Adjust pedal by moving the stop plate until there is a clearance of .04 in. (1 mm) between tip of push rod and recess in piston.

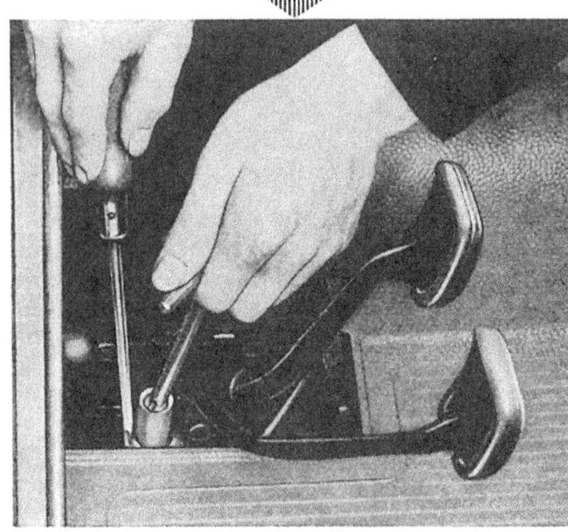

4 - Check clutch pedal free play.

5 - Connect brake lines and brake light cables.

6 - Check that vent hole in fluid reservoir cap is clear.

7 - Fill reservoir with fresh brake fluid.

8 - Bleed brake system.

9 - Check brakes on road test. Check operation of brake lights.

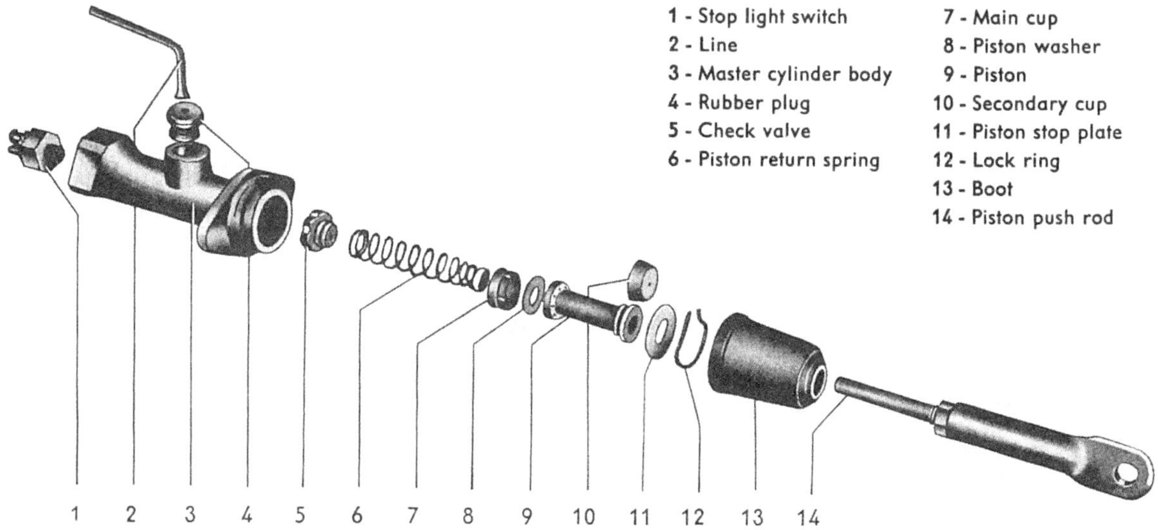

1 - Stop light switch
2 - Line
3 - Master cylinder body
4 - Rubber plug
5 - Check valve
6 - Piston return spring
7 - Main cup
8 - Piston washer
9 - Piston
10 - Secondary cup
11 - Piston stop plate
12 - Lock ring
13 - Boot
14 - Piston push rod

Reconditioning Master Cylinder

Disassembly

1 - Remove the rubber boot.

2 - Remove the lock ring.

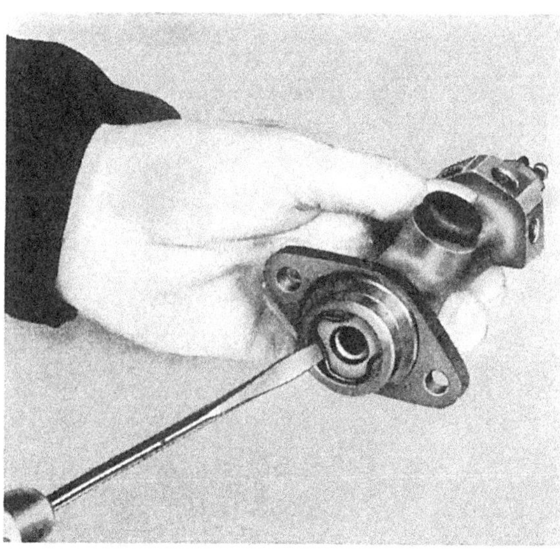

3 - Take out piston stop plate and piston.

4 - Remove piston washer, main cup, return spring and check valve.

5 - Screw off stop light switch.

Assembly

When assembling the following points should be observed:

1 - Clean all parts in denatured alcohol or Genuine VW Brake Fluid. Fuel, paraffin, mineral oil, etc., destroy the rubber parts.

2 - Examine all parts for wear. Make sure the intake and by-pass ports are open and free from burrs. Check the piston fit in the cylinder bore.

3 - Always renew the two rubber cups.

4 - Install check valve, return spring and return spring seat.

5 - Install piston with Genuine VW Brake Cylinder Paste.

6 - See that the lock ring is bedding correctly in its recess.

Note:

The master and wheel cylinders of the hydraulic brake system are manufactured by two different firms and consequently there are two different identification marks i.e., "Ate" or "S". The parts are identical in appearance, and can only be recognized by the identification marks.

Some of the component parts of the master cylinder marked "S" are different to those in the pattern marked "Ate".

Interchangeable	Part No.	Non-interchangeable	Part No.
Stop light switch	113 945 515	Valve seat ring	113 611 121
Master cylinder housing	113 611 021	Check valve	113 611 127
Sealing plug	113 611 817	Check valve (modified)	113 611 127 B
Stop washer	113 611 179	Piston return spring	113 611 143
Lock ring	113 611 185	Main cup	113 611 155
Boot	113 611 195	Washer	113 611 167
		Piston — master cylinder	113 611 165
		Secondary cup	113 611 173

Important

Spare parts are available only for the master cylinder marked "Ate" and all parts of these cylinders can be renewed individually. When it becomes necessary to replace one of the non-interchangeable parts of a master cylinder marked "S" **all** the non-interchangeable components must be replaced together.

Note:

1 - Checking the brake system for leaks is one of the tasks to be carried out during the maintenance service. On vehicles from Chassis No. 3 444 340 to 3 511 064, thejoint between the reservoir neck and the rubber insert should be inspected with particular care.

2 - Intermittently from Chassis No. 3 507 791 to 3 522 064 and continuously from Chassis No. 3 511 065 the line between master cylinder and reservoir (Part No. 113 611 805 B or 114 611 805 A, unchanged) was divided into two parts which are connected by a flexible hose (Part No. 113 611 801). The shorter part of the line (Part No. 113 611 317 B) is inserted into the rubber insert in the reservoir to a depth of 25 mm.

The flexible connection between the two parts of the line reduces the possibility of leakage.

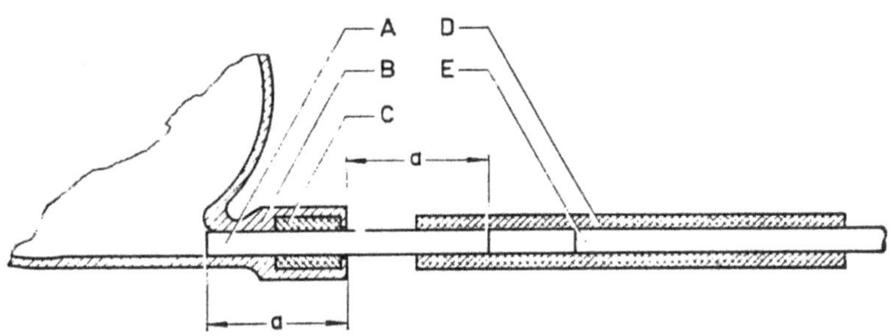

a - 25 mm/1"

A - Line, 50 mm/2" long

B - Reservoir

C - Rubber insert

D - Hose 113 611 801

E - Line between reservoir and master cylinder

3 - If the reservoirs on vehicles from Chassis No. 3 444 340 to 3 511 064 are found to be leaking between the neck and the rubber insert, proceed as follows:

a - Remove the line between master cylinder and reservoir.

b - Cut 75 mm/3" off at the front end. Shorten this piece by 25 mm and clean up rough edges.

c - Insert the 50 mm long piece into the reservoir neck to a depth of 25 mm.

d - Connect the two parts with the hose (Part No. 113 611 801) and install the line.

e - Bleed the brake system.

f - If leakage still occurs, the reservoir (Part No. 113 611 301 B) should be replaced.

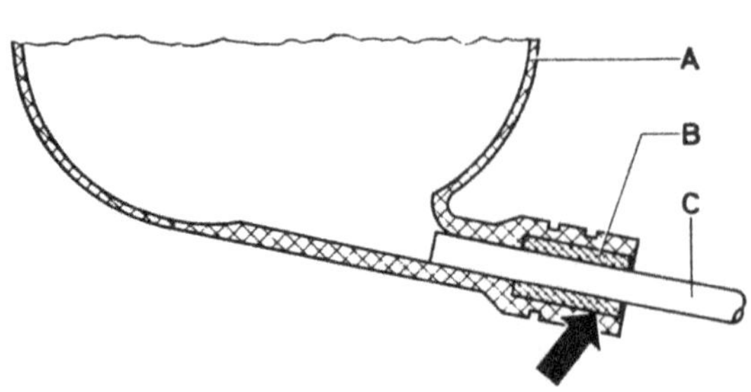

A - Reservoir

B - Rubber insert

C - Line between reservoir and master cylinder

On right hand drive vehicles it is only necessary to cut a 50 mm length off the line and connect it with the hose. The line should also be inserted into the reservoir to a depth of 25 mm.

Brake Hoses

General

Brake hoses connect the wheel cylinders to the brake lines which are clipped to the frame.

When installing a brake hose, care should be taken to see that there are no sharp bends at the connection. The brake hoses must not be strained in any steering position nor under springing action of the wheels. The hose should not be allowed to chafe at the chassis or body with the car in motion.

To prevent the front brake hoses from sagging too far, they should be installed as follows:

After the brake hose is attached to the brake wheel cylinder, twist the free end of the hose 90° — or max. 180°. The twisting should be done in a way which makes the tube bend toward the **front** of the car.

This method of fitting must always be employed when installing the front brake hoses on the De Luxe and Convertible models, taking care that the front wheels are completely free of load.

The nature and the extent of the damage done to brake hoses indicates in every case that excessive force has been used. The cause of the damage is usually carelessness during installation.

In order to avoid such damage in the future and to maintain vehicle safety, the above instructions should be noted carefully when fitting brake hoses.

Important

Brake hoses must not be painted nor be left in contact with grease, oil, gasoline or kerosene for prolonged periods. Traces of grease which get onto the hoses when greasing the vehicle should be removed immediately.

The brake system must be bled every time a line or hose has been disconnected. Top up as necessary, using Genuine VW Brake Fluid.

Furthermore, ensure that the hose to cylinder and line to cylinder connections are tightened to the correct torque 1.5—2.0 mkg (10—14 ft. lbs.).

Renewing Brake Hose

Removal

1 - Remove wheel.

2 - Loosen union nut at hose bracket and remove the hose retainer.

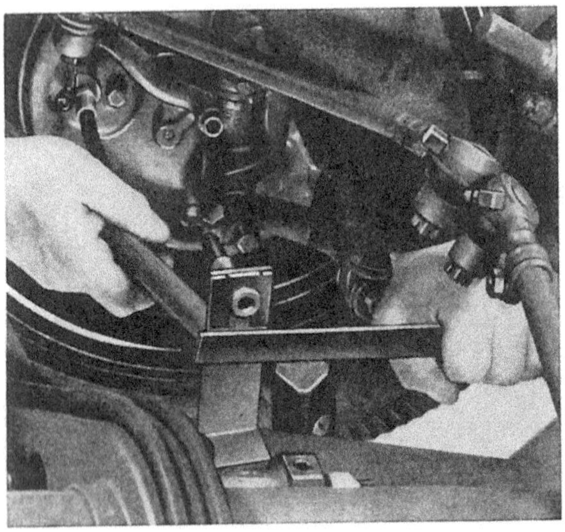

3 - Withdraw brake hose from bracket.

4 - Disconnect brake hose from wheel cylinder.

Installation

When installing the following points should be observed:

1 - Install the brake hose as prescribed.

2 - Check for proper position of brake hose by turning steering from lock to lock.

3 - Bleed the system. Do not forget to re-install the dust cap on the bleeder valve.

Brake Lines

General

The rest of the brake lines consists of steel tubes of 4.75 x 0.72 mm. No trouble should arise provided the tubing is at all times tightly clipped to the chassis to prevent vibration. The tubes are tested to withstand pressure far in excess of that developed in braking.

Tube Connections

The tube ends are double flared for added protection against splitting and leakage. They are forced against the beveled faces in the unions when tightening the union nuts, ensuring strong and tight joints.

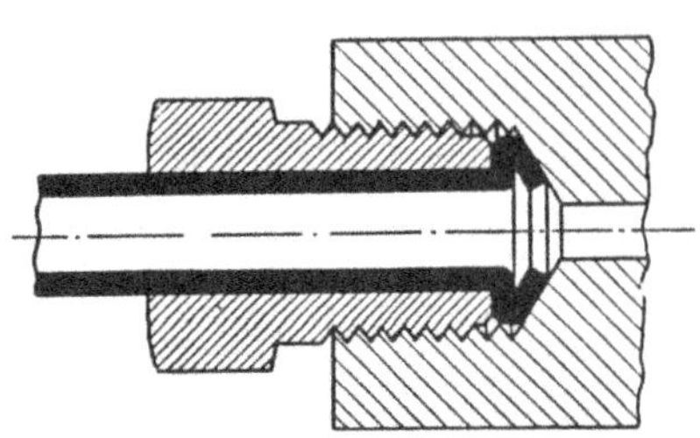

Moisten the flared ends with a few drops of brake fluid prior to tightening the nuts.

> **Important**
>
> When checking the brake system for leaks at routine services, all brake lines should be inspected for signs of corrosion or other damage, i. e. flying stones, twisting. This check should also cover the brake line along the frame tunnel as this portion is exposed to dampness.

Note:

The sealing compound which coats the brake lines situated alongside the frame tunnel should be renewed when repairs have been carried out. This compound can be either sprayed or applied by brush. It may be thinned by means of benzine so as to facilitate spraying.

In the course of routine maintenance the brake system should be checked regularly for leaks and general condition. Workshops should, therefore, inform their customers of any apparent damage to ensure safety in operation.

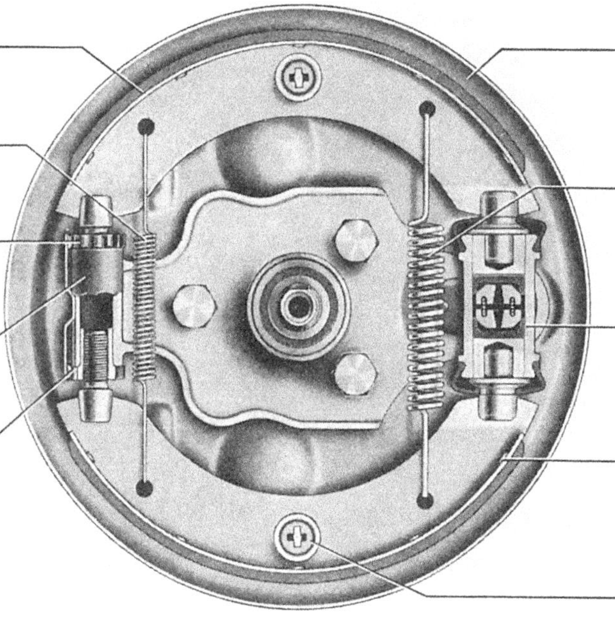

1 - Brake back plate
2 - Secondary (trailing) brake shoe
3 - Primary (leading) brake shoe
4 - Hold-down spring and spring seat
5 - Wheel cylinder
6 - Anchor block
7 - Adjusting nut
8 - Adjusting screw
9 - Return spring
10 - Return spring

General

The front wheel brake assembly is shown in the above illustration.

The brake shoes are self-centralising and rest freely in the slots of the piston push rods and the adjusting screws, thereby reducing the tendency of the brake to bind to a minimum.

The two shoe hold-down springs assure a constant contact of the shoes with the bosses on the back plate. When the pedal returns to the "off" position, the brake shoes are pulled to the released position by two return springs.

The brake shoes are adjustable by screws and nuts at the anchor block.

Important

Before attempting to pull the brake drums off, turn the shoe adjusting screws right back. Otherwise the shoes may jam in the brake drum if worn and break the wheel cylinder at the mounting flange.

Note:

The brake linings can be checked during the maintenance service by means of the inspection hole in the brake drum. If the linings are found to be excessively worn, i. e., down to about 2.5 mm (.1"), they should be replaced.

Renewing Front Brake Shoes

Removal

1 - Remove front wheel and brake drum.

2 - Remove shoe hold-down spring seats, springs and pins.

3 - Unhook brake shoe return springs.

4 - Take off brake shoes.

Installation

When installing, the following points should be observed:

1 - When replacing brake shoes, take care the linings are of the same type to ensure uniform braking.

2 - Correctly install brake shoes: the slots in the brake shoe webs must be towards the wheel cylinder.

3 - Connect return springs so that they do not make contact with other components of the wheel brake.

4 - Before installing the brake drum, make sure the oil seal is in perfect condition.

5 - Clean brake drum hub and ball bearing and apply universal grease VW — A 052.

6 - Adjust front wheel bearing as prescribed.

7 - Adjust the brakes and bleed the system. Do not forget to re-install the bleeder valve dust cap.

Lining Replacement

General

Brake shoe lining replacement should be made only in axle sets (both front wheels or both rear wheels) to ensure uniform braking. It is also of great importance to use the same type of brake lining.

Oil-soaked brake linings should be replaced. It is no use washing the linings in fuel or other grease solvent, as the oil in the lining will again appear on the surfaces due to the heat developed when braking.

Relining Brake Shoes

1 - Remove brake shoes.

2 - Carefully remove the lining rivets to avoid damage and distortion to the brake shoe.

3 - Clean and inspect the brake shoe. Remove any burr from rivet holes.

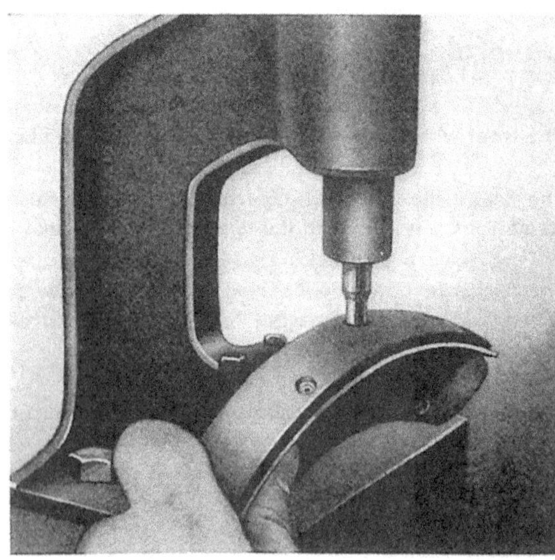

4 - The center rivets should be set first, then work towards both ends of the lining. It is very important that a snug fit exists between the lining and the shoe, as otherwise brake noises and premature failure of the brakes would be the result.

Drive in rivet perpendicularly to avoid tension on the brake linings.

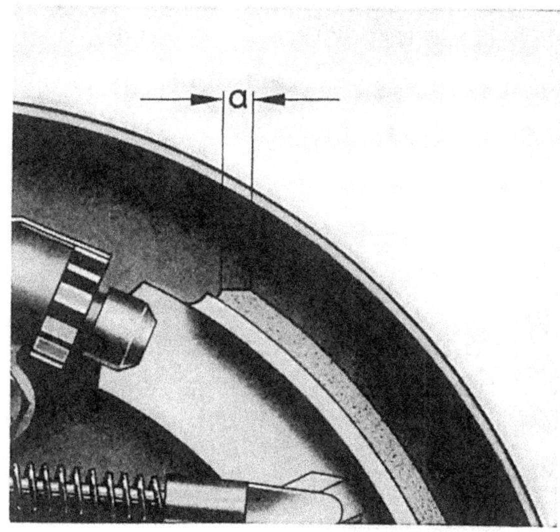

a = approx. 5 mm/0.2″

5 - Both ends of the lining should be chamfered (5 mm/0.2″). Round off sharp edges of friction surface.

> **Important**
> Only use rivets supplied by the VW Factory. Never use aluminum rivets.

Removing and Installing Front Brake Back Plate

Removal

1 - Remove front wheel.

2 - Remove cotter pin securing speedometer drive cable. Remove grease cap.

3 - Withdraw brake drum.

4 - Disconnect brake hose and plug it up by means of a piece of wood.

5 - Remove brake shoes.

6 - Remove wheel cylinder.

7 - Take off adjusting screws and nuts.

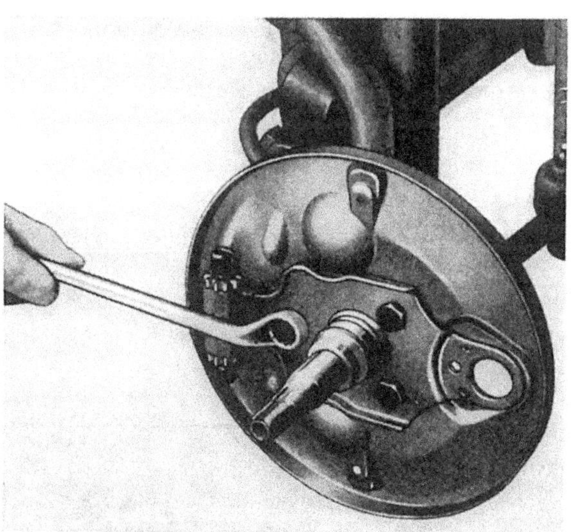

8 - Remove the three back plate mounting bolts.

9 - Take off back plate.

Installation

When installing, the following points should be observed:

1 - Thoroughly clean contact surfaces between back plate and stub axle (steering knuckle).

2 - Tighten back plate screws to the following torque:

Bolts of quality specification 8 G
4 to 4.5 mkg (29 to 33 ft. lbs.)

Bolts of quality specification 10 K
5 mkg (36 ft. lbs.)

Secure each bolt with a spring washer so that bolts do not work loose.

3 - Check pretension of the leaf spring on the anchor block. Bend weak springs.

4 - Check if the adjusting screws and nuts can be turned easily. Apply Universal Grease VW — A 070 (graphitic high melting point grease).

5 - Before installing the brake drum, make sure the oil seal is in perfect condition.

6 - Clean brake drum hub and ball bearing and apply Universal Grease VW — A 052.

7 - Adjust front wheel bearing as prescribed.

8 - Adjust the brakes and bleed the system. Do not forget to re-install the bleeder valve dust cap.

Front Wheel Cylinder

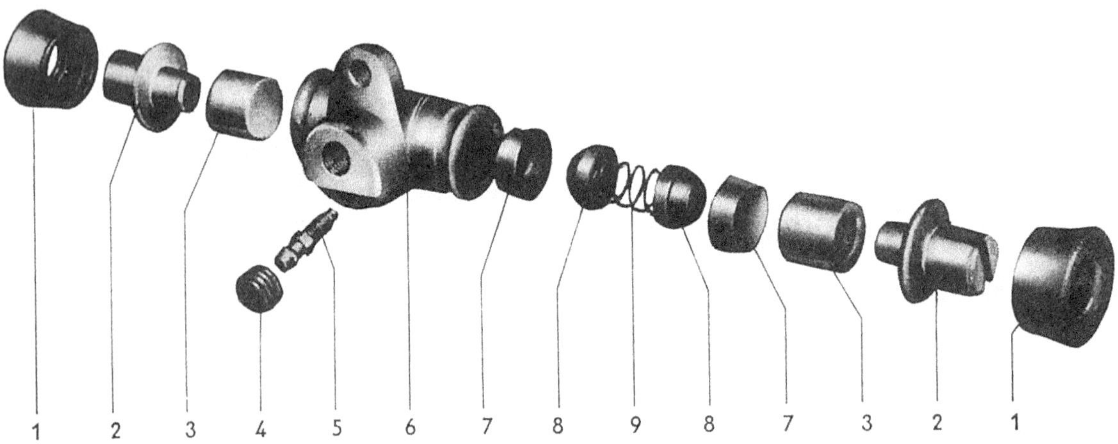

1 - Boot	4 - Dust cap	7 - Cup
2 - Push rod	5 - Bleeder valve	8 - Cup expander
3 - Piston	6 - Cylinder body	9 - Return spring

Description

The wheel cylinders transmit the pressure generated in the master cylinder to the brake shoes.

The wheel cylinder is attached to the brake back plate. The cast cylinder body contains two cups and two pistons. Between the cups there is a return spring and two cup expanders. The return spring and the fillers press the cups against the cylinder wall. When depressing the brake pedal, the fluid displaced by the master cylinder forces the cups and pistons outward, expanding the brake shoes against the drum via the two push rods.

The open ends of the cylinder are fitted with rubber boots to prevent ingress of dirt.

With the brake system at rest, the wheel cylinder is practically filled up, that is, there is no space for the fluid. This insures a perfect bleeding.

Between the pistons is an opening for the bleeder valve, which is turned to the open position when bleeding the system.

Important Note when replacing a wheel cylinder: —

Wheel cylinder, front	22.20 mm (0.874") dia.
Wheel cylinder, rear	19.00 mm (0.75") dia.

Front Wheel Cylinder Removal and Installation

Removal

1 - Remove front wheel and brake drum.

2 - Disconnect the hose and plug it up with a piece of wood.

3 - Remove shoe hold-down spring seats, springs and pins.

4 - Unhook brake shoe return springs.

5 - Remove brake shoes.

6 - Remove wheel cylinder attaching screw and take off wheel cylinder.

Installation

When installing, the following points should be observed:

1 - Install wheel cylinder. When renewing the wheel cylinder, attention should be paid to the size.

2 - Correctly install brake shoes; the slots in the webs must be towards the wheel cylinder.

3 - Before installing the brake drum, make sure the oil seal is in perfect condition.

4 - Clean brake drum hub and ball bearing and apply Universal Grease.

5 - Adjust front wheel bearing as prescribed.

6 - Adjust the brakes and bleed the system. Do not forget to re-install the bleeder valve.

Reconditioning Front Wheel Cylinders

Disassembly

1 - Remove wheel cylinder.

2 - Remove both boots.

3 - Take off push rods, pistons, cups, cup expander and return spring.

4 - Screw out bleeder valve.

Assembly

When installing, the following points should be observed:

1 - Clean all parts only in denatured alcohol or brake fluid.

2 - Examine all parts for signs of wear. Check the piston fit in the cylinder bore.

3 - Renew both cups (note diameter of wheel cylinder).

4 - Install pistons with Genuine VW Brake Cylinder Paste.

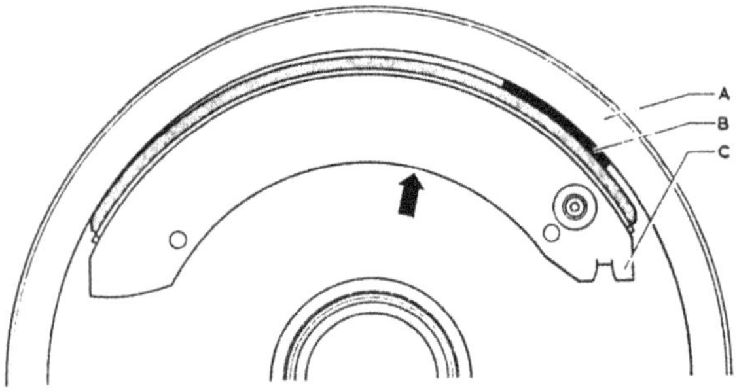

A - Brake drum
B - Metal strip
C - Brake shoe

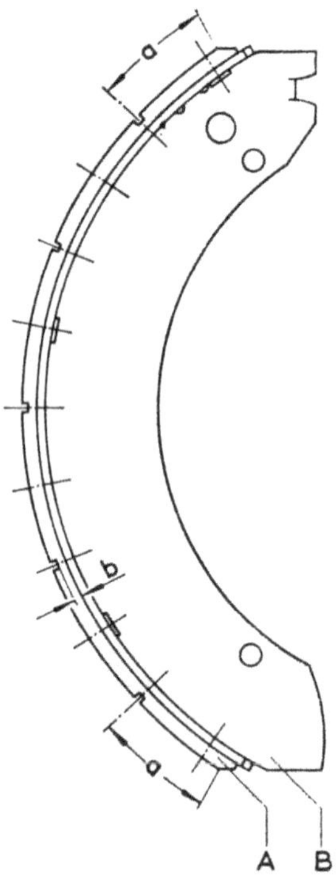

A - Brake lining   a - approx. 35 mm/1.375″
B - Brake shoe   b - 1.5—2 mm/0.060—0.080″

Note:

Isolated instances of brake squeaking have been noted on the Passenger Car, and, to a lesser extent, on the Transporter. It is caused mainly by vibration between the brake shoes and linings and the brake drums. The squeaking is particularly noticeable in the early morning before the thin film of rust on the drums, caused by the moisture in the air, has worn off.

An additional cause of squeaking is an accumulation of road dust and worn-off lining material on the linings themselves. An accumulation of this type is encouraged when the brakes are nursed too much and the vehicle always braked very gently or when the driver makes a practice of changing down through the gears to stop the vehicle instead of using the footbrake.

Remedy for squeaking brakes

1 - Remove the drums concerned, mark and remove the brake shoes. Place the shoes in the drum individually and check that they fit properly. If necessary, bend the shoe with a hydraulic press at the place where it does not contact the drum. This can be done by using an old drum which has been cut in half and a piece of metal strip to reduce the shoe radius.

2 - Check the brake back plates for misalignment and distortion and replace if necessary. The brake shoes must contact the shoulders in the back plate and the linings should be at right angles to the back plate.

3 - Check that the spring Part No. 113 609 283 has sufficient tension (free length 21 mm/0.826″, loaded length 13.5 mm/0.531″, with a load of 11.3 kg/25 lbs. ± 10%).

4 - Clean the linings with a wire brush.

5 - Cut five grooves across the brake linings between the rivets. The first groove must be at least 30 mm from the end of the linings. The cuts should not reduce the thickness of the lining material (b) to less than 1.5 to 2 mm.

1 - Brake back plate
2 - Secondary (trailing) brake shoe
3 - Primary (leading) brake shoe
4 - Hold-down spring and spring seat
5 - Wheel cylinder
6 - Anchor block
7 - Adjusting nut
8 - Adjusting screw
9 - Return spring
10 - Return spring
11 - Brake shoe lever
12 - Operating link

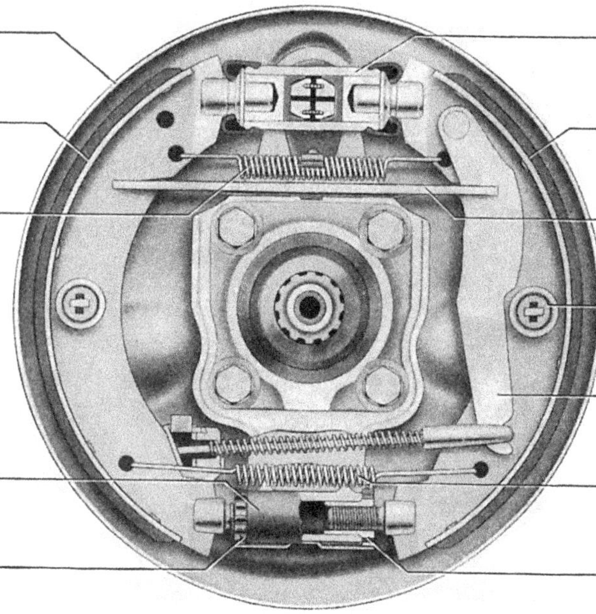

General

The front wheel assembly is shown in the above illustration.

The brake shoes are self-centralising and rest freely in the slots of the piston push rods and the adjusting screws, thereby reducing the tendency of the brake to bind to a minimum.

The two shoe hold-down springs assure constant contact of the shoes with the bosses on the back plate. When the pedal returns to the "off" position, the brake shoes are pulled to the released position by two return springs.

The brake shoes are adjustable by screws and nuts at the anchor block.

Important

There is a risk of accident when loosening and tightening the rear axle shaft nuts with the vehicle raised on a lift. It is advisable, for safety's sake, to carry out these operations before lifting or after lowering the vehicle.

Before attempting to pull the brake drums off, turn the shoe adjusting screws right back. Otherwise the shoes may jam in groove worm in drum and break the wheel cylinder at the mounting flange.

Note:

The brake linings can be checked by means of the inspection hole in the brake drum. If the linings are found to be excessively worn, i. e., down to about 2.5 mm (.1"), they should be replaced.

Renewing Rear Brake Shoes

Removal

1 - Remove rear wheel, brake drum, and oil deflector.

2 - Remove shoe hold-down spring seats, springs, and pins.

3 - Unhook lower brake shoe return spring.

4 - Disconnect brake cable.

5 - Remove brake shoes, return spring, brake shoe lever, operating link, and clip.

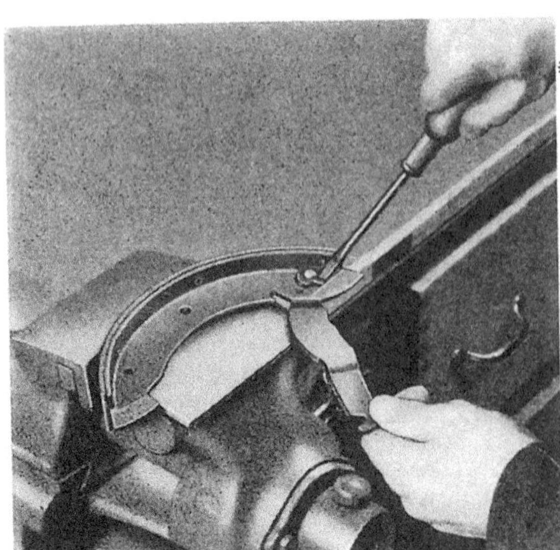

6 - Detach brake lever from brake shoe by removing the circlip from the anchor pin.

Installation

When installing, the following points should be observed:

1 - When replacing brake shoes, take care the linings are of the same type to insure an equal braking.

2 - Correctly install brake shoes, return spring, brake shoe levers, operating link, and clip. The slots in the brake shoe webs must be towards the wheel cylinder. Connect brake cable.

3 - Connect return springs so that they do not make contact with other components of the wheel brake.

4 - When re-installing brake drum, note proper position of oil deflector.

5 - Tighten rear axle shaft nut to 30 mkg (217 ft. lbs.).

6 - Bleed the system. Adjust foot and hand brakes. Do not forget to re-install the bleeder valve dust cap.

The brake linings can be checked by means of the elongated hole in the brake drum. If the linings on inspection are found to be excessively worn, i. e., down to about 2.5 mm (.1''), they should be replaced.

Removing and Installing Rear Brake Back Plate

Removal

1 - Remove rear wheel.

2 - Withdraw brake drum.

3 - Disconnect brake line and plug it up by means of a piece of wood.

4 - Remove brake shoes, brake lever, operating link, and clip. Unhook brake cable.

5 - Remove wheel cylinder.

6 - Remove brake cable retainer from back plate.

7 - Take off adjusting screws and nuts.

8 - Remove the four bolts of the bearing housing cover and take off back plate.

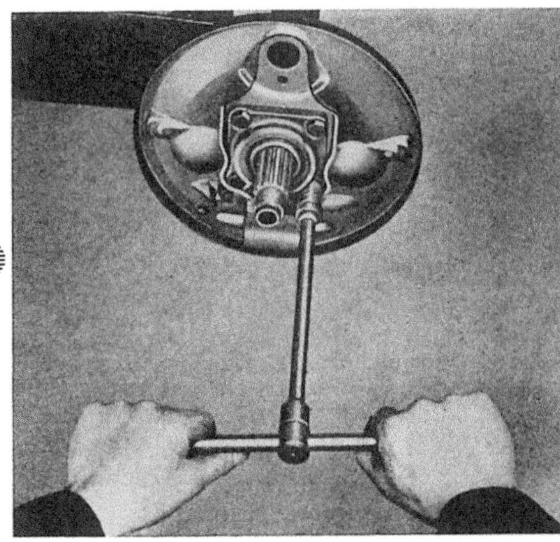

Installation

Installation is a reversal of the removal procedure, but the following points should be observed:

1 - Thoroughly clean contact surfaces of back plate.

2 - Before replacing the bearing housing cover, check condition of oil seal and see that it is correctly bedding in its recess. Renew the two gaskets for bearing housing cover and spacer. The oil drop nose of the cover must point downward. Tighten back plate mounting bolts to the following torque:

Bolts of quality specification 8 G
4 to 4.5 mkg (29 to 33 ft. lbs.)

Bolts of quality specification 10 K
5.5 to 6.5 mkg (40 to 47 ft. lbs.)

3 - Check pretension of the leaf spring on the anchor block. Bend weak springs.

4 - Check if the adjusting screws and nuts can be turned easily. Apply Universal Grease VW — A 070 (graphitic high melting point grease).

5 - Correctly install brake shoes; the slots in the webs must be toward the wheel cylinder.

6 - When re-installing brake drum, note proper position of oil deflector.

7 - Tighten rear axle shaft nut to a torque of 30 mkg (217 ft. lbs.) and secure with cotter pin.

8 - Bleed the system. Adjust foot and hand brakes. Do not forget to re-install the bleeder valve dust cap.

Rear Wheel Cylinder Removal and Installation

Removal

1 - Remove rear wheel, brake drum, and oil deflector.

2 - Disconnect brake line and plug it up by means of a piece of wood.

3 - Remove shoe hold-down seats, springs, and pins.

4 - Unhook brake shoe return springs.

5 - Remove brake shoes, brake shoe levers, operating link, and clip. Disconnect brake cable.

6 - Remove one wheel cylinder attaching screw and take off wheel cylinder.

Installation

When installing the following points should be observed:

1 - Install wheel cylinder. When renewing the wheel cylinder, attention should be paid to the size.

2 - Correctly install brake shoes, brake shoe levers, operating link, and clip. The slots in the brake shoe webs must be towards the wheel cylinder. Connect brake cable.

3 - When re-installing brake drum, note proper position of oil deflector.

4 - Tighten rear axle shaft nut to a torque of 30 mkg (217 ft. lbs.).

- Bleed the system. Adjust foot and hand brakes. Do not forget to re-install the bleeder valve dust cap.

Reconditioning Rear Wheel Cylinder

Disassembly

1 - Remove wheel cylinder.

2 - Remove both boots.

3 - Take off push rods, pistons, cups, cup expanders

3 - Take off push rods, pistons, cups, cup expanders, and return spring.

4 - Screw off bleeder valve.

Assembly

When assembling, the following points should be observed:

1 - Clean all parts only in denatured alcohol or brake fluid.

2 - Examine all parts for signs of wear. Check the piston fit in the cylinder bore.

3 - Renew both cups (note diameter of wheel cylinder).

4 - Install piston with Genuine VW Brake Cylinder Paste.

THIS PAGE INTENTIONALLY
LEFT BLANK

Removal and Installation of Hand Brake Lever

Removal

1 - Remove front seats and take rear floor covering out.

2 - Remove lever cover.

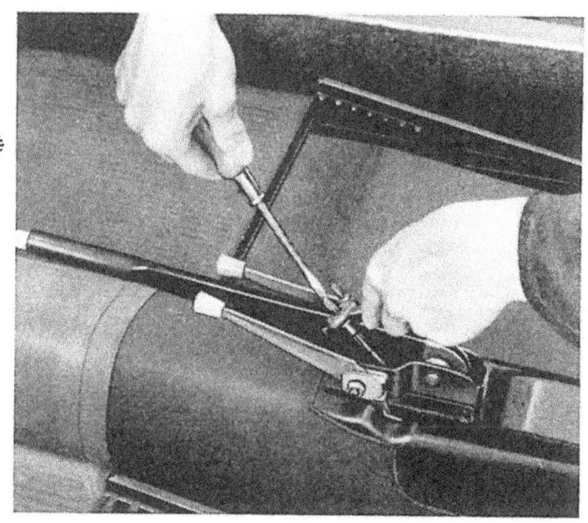

3 - Remove lock and adjusting nuts from cables.

4 - Remove circlip from lever pin and take pin out.

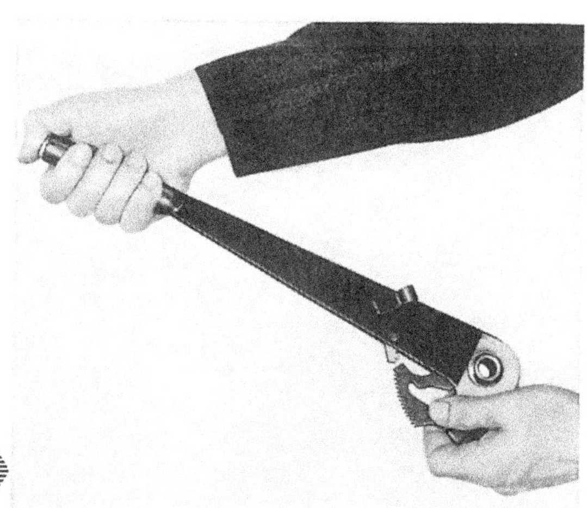

5 - Press lever to the rear until it can be lifted out with the ratchet segment. Do not press the release button when doing this.

6 - Press button and take ratchet segment out.

Installation

Note the following before installing:

1 - Disassemble lever. Clean pawl rod, press button, pawl spring and ratchet segment. Grease all parts and assemble.

2 - Insert ratchet segment so that the recess fits over the tube in the lever (A) and the teeth engage in the pawl. Ensure that the rounded end of the pawl is positioned correctly (B).

3 - Insert lever from above, guiding the threaded ends of the cables to the sides.

Important

The slot in the ratchet segment must engage the frame edge when the lever is inserted (C).

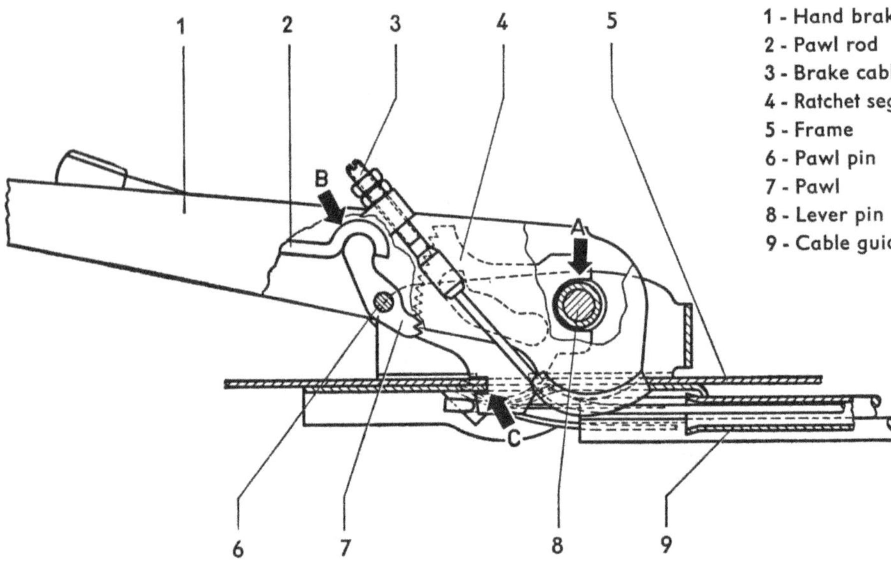

1 - Hand brake lever
2 - Pawl rod
3 - Brake cable
4 - Ratchet segment
5 - Frame
6 - Pawl pin
7 - Pawl
8 - Lever pin
9 - Cable guide tube

4 - Grease lever pin and install. Do not forget the circlip.

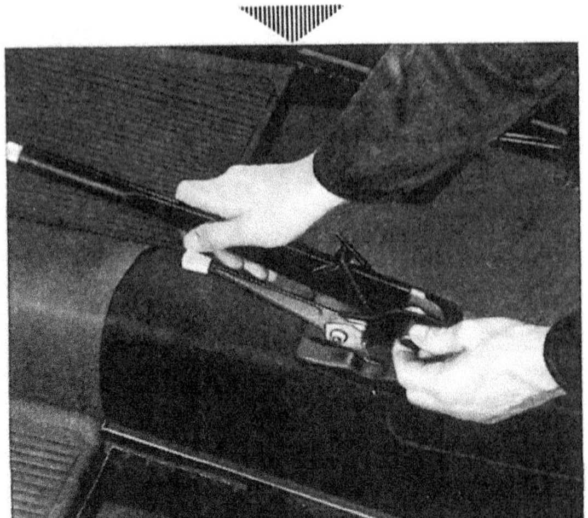

5 - Attach cables and adjust hand brake.

6 - Secure the cables with the locknuts.

7 - Fit hand brake lever cover.

8 - Install floor covering and front seats.

9 - Check hand brake on the road.

Removal and Installation of Brake Cables

Removal

1 - Remove hand brake.

2 - Remove rear wheel and brake drum.

3 - Remove brake shoes.

4 - Detach cable clip from back plate.

5 - Pull cable with guide hose out of back plate and cable guide tube.

Installation

Note following when installing:

1 - Clean cable and guide tube.

2 - When fitting new cables, check length.

3 - Lubricate cables well with universal grease.

4 - Adjust hand brake.

5 - Check hand brake on the road.

Note:

From February 1965, **Chassis No. 115400109,** all Type 1 models were fitted with a hand brake cable equalizer.

All work on the hand brake remains unchanged. When adjusting the hand brake, ensure that the equalizer plate is horizontal when the brake is applied.

The cable equalizer cannot be installed in older vehicles. Spare parts of the previous type will remain available.

Main parts:

Hand brake lever	113711305A
Equalizer plate	111711331
Brake cable	113609721G
Adjusting nut	111711349
Heating lever	111711711A/712
Cover for brake lever	111863339A

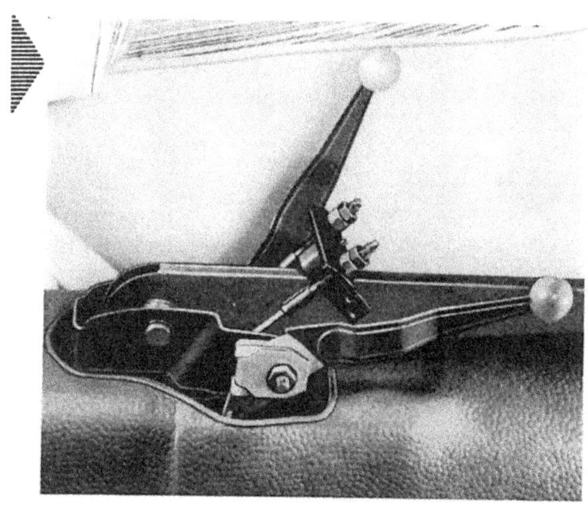

THIS PAGE INTENTIONALLY
LEFT BLANK

Filling and Bleeding

General

The hydraulic system must be topped up and bled whenever a fluid line has been disconnected. Air in the braking system seriously impairs braking efficiency resulting in soft, spongy pedal action. Air is eliminated by adding brake fluid and bleeding the system.

Brake Fluid

Only Genuine VW Brake Fluid should be used for the hydraulic system of the Volkswagen. It ensures a correct and reliable function of the brakes under all climatic conditions. The Genuine VW Brake Fluid is composed of ingredients which will not derange nor deform the structure or surface of the brake components. Should there be any difficulty in obtaining this brake fluid, Lockheed Brake Fluid may be used in lieu.

Brake Cylinder Paste

The sealing components of the hydraulic system must not be lubricated with mineral oil or grease. The Genuine VW Brake Cylinder Paste has been introduced to obtain a proper lubrication of the pistons and cylinders This lubricant does not affect the sealing components of the system and greatly improves the sliding action of the pistons and cups.

Disassemble the cylinder and carefully clean them in denatured alcohol. When completely dry, the pistons and the cylinder wall should be coated with the brake cylinder paste. Whenever a cylinder has been disassembled, the brake cylinder paste should be applied to the aforementioned parts.

Bleeding the Brake

a - The bleeding can best be carried out by two men. Proceed as follows:

1 - Remove bleeder valve dust cap.

2 - Attach one end of the bleeder drain hose to the valve.

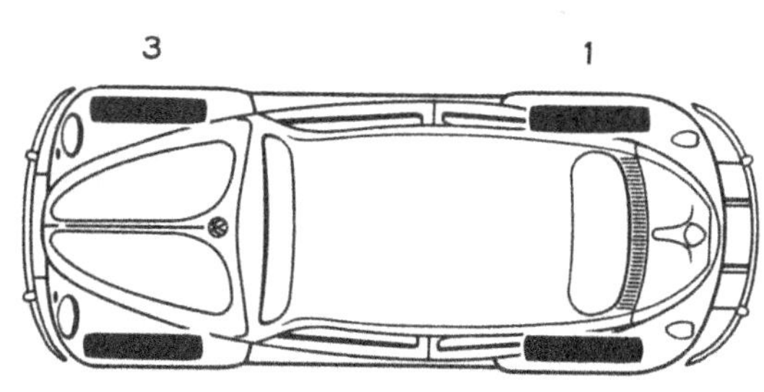

3 - Allow the free end of the hose to be submerged in a clean glass container partially filled with hydraulic fluid. The end of the drain hose should, if possible, be above the level of the bleeder valve.

4 - Slacken the bleeder valve about one turn, using a 7 mm wrench.

5 - Depress the brake pedal quickly, allowing it to return without assistance, and continue to pump in this manner until the fluid runs out of the bleeder hose in a continuous stream, without air bubbles.

It is extremely important that the master cylinder reservoir be full of fluid and that it be refilled before being completely exhausted.

6 - The brake pedal should be kept in the fully depressed position, until the bleeder valve is closed.

7 - Tighten bleeder valve and replace dust cap.

8 - Repeat the operation on each of the remaining wheel cylinders, If necessary, top up brake fluid in reservoir after the complete system has been bled.

b - A combined filling and bleeding operation, which may be carried out by one man, is described below:

This operation requires a fill-bleeder tank, these are manufactured by a number of firms.

Before starting to bleed, make sure the fill-bleeder tank is sufficiently filled with brake fluid. Then fill the tank with compressed air until the dial indicates 4 to 5 kg./sq. cm. (57 to 71 lbs./sq. in.).

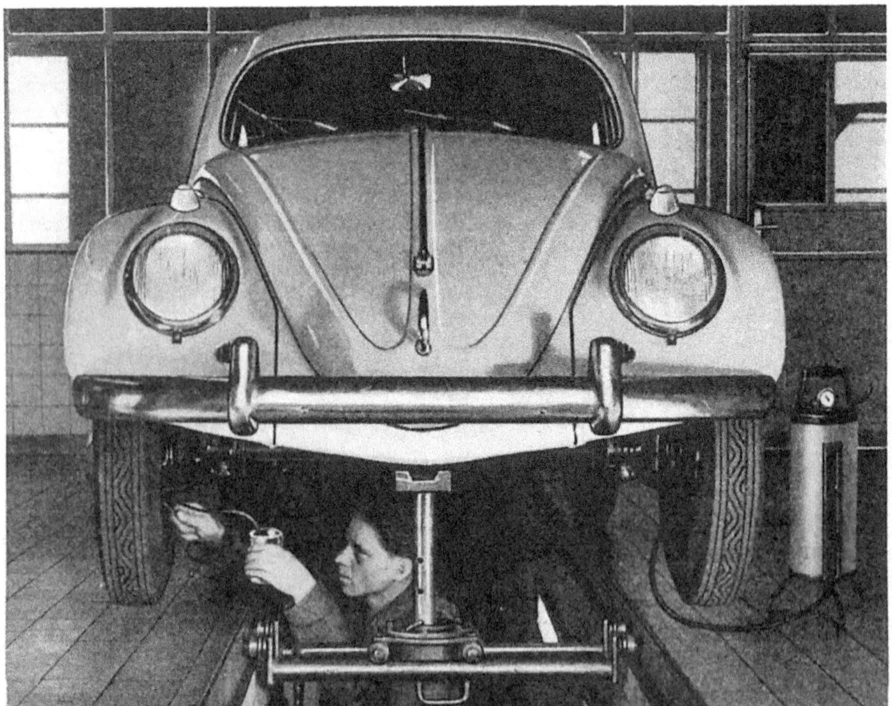

1 - Completely depress brake pedal and fix it in this position.

2 - Remove bleeder valve dust caps of all four wheel cylinders.

3 - Attach free end of tank hose to bleeder valve of left-hand front wheel and slacken bleeder valve about one turn using a 7 mm wrench.

4 - Attach one end of the bleeder drain hose to the bleeder valve of the right-hand rear wheel cylinder and allow the free end of the hose to be submerged in a clean glass container partially filled with hydraulic fluid. Slacken bleeder valve about one turn, using a 7 mm wrench.

5 - Open the shut-off valve in the tank hose and allow the fluid to be bled into the glass container until bubbles cease to appear in the container and the fluid stream is solid.

6 - Tighten bleeder valves.

7 - Close the shut-off valve in the tank hose.

8 - Repeat the operation on the left-hand rear wheel and right-hand front wheel as outlined under 4 to 7.

9 - Replace all bleeder valve dust caps.

10 - Release brake pedal.

11 - If necessary, top up brake fluid in reservoir after the complete system has been bled.

Also refer to manufacturer's instructions for the various filling appliances.

It is not advisible to reuse the brake fluid pumped out when bleeding as it may be contaminated or dirty. Before the fluid is reused it should be filtered thoroughly through filter-paper.

Flushing the system

To flush the system, only use brake fluid, never fuel or mineral oil. Even the use of alcohol is not recommended, as the flushing agent cannot be completely removed. When refilling the system, the remaining alcohol will mix with the brake fluid, leading to vapour locks at higher temperatures of the system.

Adjusting Brakes

General

The clearance between the brake shoes and drums increases as the vehicle is used, due to lining wear. The hydraulic brake system automatically takes up this play as far as the volume of the master cylinder permits. However, when the distance the pedal moves before the brakes are applied becomes excessive, the brake shoes must be individually adjusted.

Before adjusting the brakes, check the front wheel bearing play.

Adjustment

1 - Raise the car and release hand brake.

2 - Prior to adjusting, completely depress the brake pedal several times, to centralize the brake shoes in the drums.

3 - Turn the wheel to be adjusted forward until the hole in the brake drum is in line with one of the adjusting nuts.

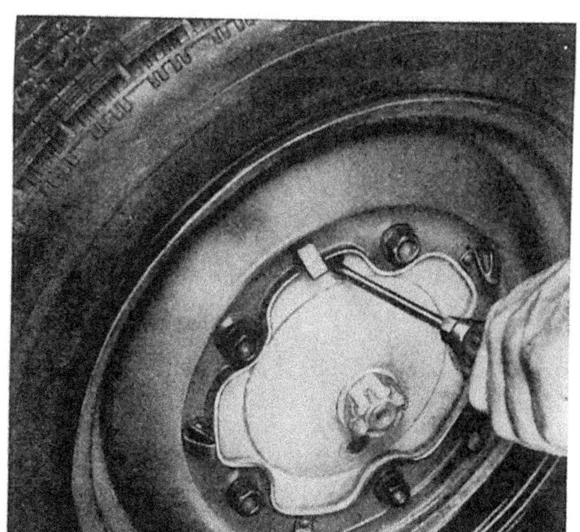

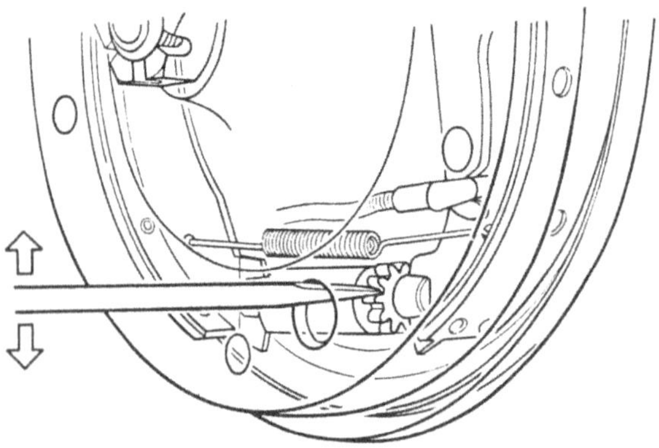

4 - Insert a screwdriver through the hole and turn the adjusting nut, using screwdriver as a lever, until a light drag is noted when wheel is turned by hand. Back off adjusting nut 3 to 4 teeth to allow the wheel to turn freely.

5 - Repeat the procedure on the other adjusting nut. Note the opposite turning direction of the two nuts.

6 - Repeat the above operations on the other wheels.

7 - Road-test the car.

Adjusting Hand Brake

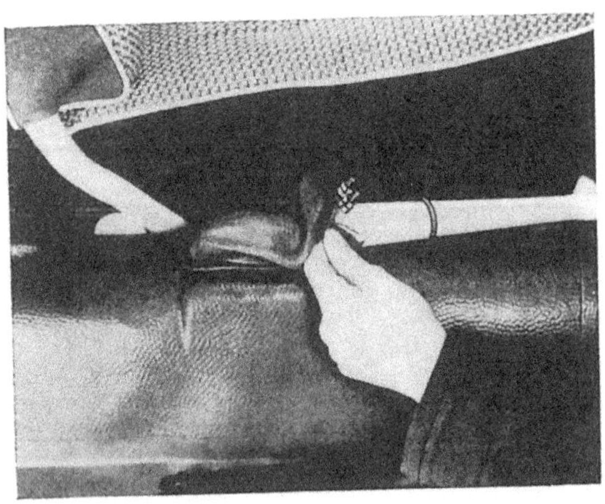

Adjustment

1 - Raise the car at the rear.

2 - Slide off the rubber ring, and fold back the hand brake lever rubber boot until the brake cable adjusting nuts are free.

3 - Back off lock nuts and tighten adjusting nuts to a degree which will still allow the rear wheels to turn freely when the hand brake lever is in the released position.

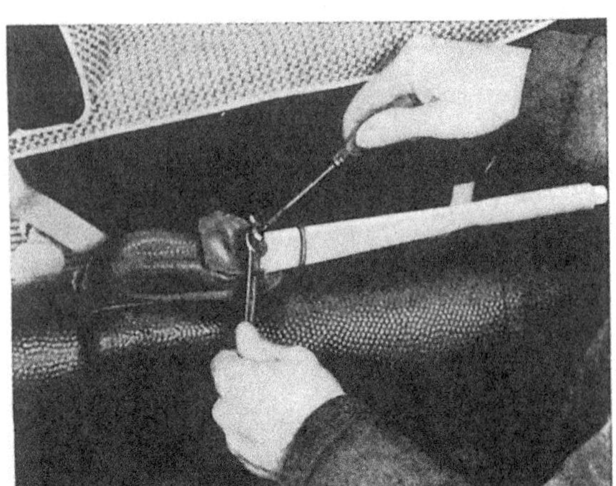

4 - Pull the hand brake lever up two notches, and make sure both rear wheels have the same braking effect. At the fourth notch it should be impossible to turn the wheels by hand.

5 - Securely tighten lock nuts and fasten hand brake lever rubber boot.

Note:

From Chassis No. 4 036 536 the protective sleeves on the hand brake cables are maintenance-free and do not need to be lubricated. The Part Nos. of these cables are 113 609 721 F and 211 609 701 E. Service installation in older vehicles is possible.

Turning Brake Drums, Fitting New Linings

Turning Brake Drums

Brake drums which are out-of-round or are worn or scored can often be made usable by turning.

1 - Turn out internal diameter of drum to measurement "a". After turning, the wall thickness must not be less than measurement "b". The braking surface must not taper more than 0.1 mm. The maximum permissible axial run-out is 0.25 mm. Measurement "c" is the wear limit.

2 - Brake shoes for reconditioned drums should be fitted with oversize linings as these linings are ground to match the turned out drums.

Size	Brake drum diameter	
	front	rear
Original	230.1+0.2 mm (9.059+.008")	230.0+0.2 mm (9.055+.008")
a	231.1+0.2 mm (9.098+.003")	231.0+0.2 mm (9.094+.008")
b	4 mm	4 mm
c	231.5 mm (9.103")	231.5 mm (9.103")

Relining Brake Shoes

When renewing brake shoe linings, the linings on both front or rear wheels should be renewed together to ensure uniform braking. Linings of the same quality should be used on both sides for the same reason. Oil-soaked linings should be renewed. It is no use washing the linings in fuel or similar grease solvents as the oil in the linings will reappear when the brakes warm up.

Relining Brake Shoes

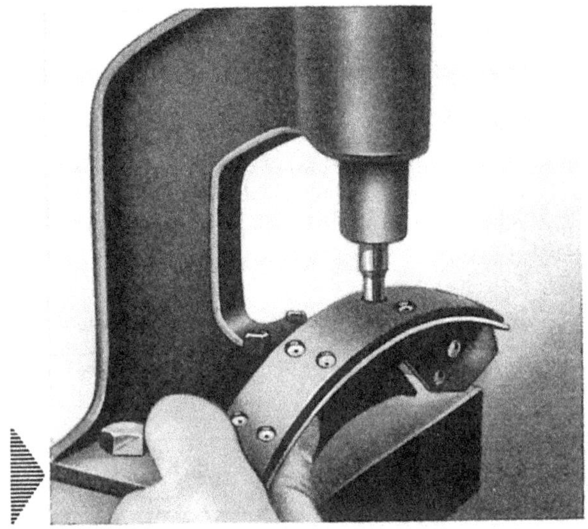

1 - Remove brake shoes.

2 - Remove linings carefully to avoid damage and distortion to shoes.

3 - Clean brake shoes and remove burr from rivet holes.

4 - Rivet new linings from the center outwards.

The linings must not overhang at the sides and must contact the shoe properly all over as otherwise noises will occur and the braking efficiency will suffer.

Press rivets in vertically to avoid tension in linings.

Important
Note the different brake lining widths.

Front lining width = 40 mm (1.57″)
Rear lining width = 30 mm (1.18″)

Brake System Trouble Shooting

Symptom	Cause	Remedy
Pedal travel increased to such an extent that pedal contacts floor.	Linings worn.	Adjust brake shoes. Do not adjust at pedal cluster.
Pedal can be depressed a long way and feels spongy.	a - Air in system. b - Insufficient fluid in reservoir.	a - Bleed system. b - Top fluid up and then bleed system.
Pedal can be depressed a long way without brake action even though shoes have been adjusted and system bled.	a - Check valve in master cylinder defective. b - Valve seat dirty.	a - Replace parts in master cylinder. b - Clean valve seat, fit new valve if necessary.
Braking effort only obtained after pumping pedal several times despite fitting new check valve.	a - Air in system. b - Spring weak.	a - Bleed system. b - Replace parts in master cylinder.
Braking effort decreases and pedal goes to floor after adjusting shoes.	a - Brake lines leaking. b - Damaged or defective cup in master or wheel cylinders.	a - Tighten connections or fit new lines and hoses. b - Fit new cup. If in master cylinder, fit new internal parts.
Brakes overheat.	a - Compensation port in master cylinder blocked. b - Insufficient clearance between brake pedal rod and master cylinder piston. c - Return spings weak. d - Rubber parts swollen due to use of unsuitable brake fluid.	a - Clean master cylinder. b - Adjust pedal clearance. c - Fit new return springs. d - Drain fluid, remove all rubber parts and flush system well with VW or Lockheed brake fluid. Replace all parts in master cylinder.

Symptom	Cause	Remedy
Poor braking effort despite high pedal pressure.	a - Linings oiled-up. b - Excessive fading after continuous use due to unsuitable linings.	a - Clean drums, fit new seals if necessary. b - Fit new linings.
Brakes bind while vehicle is in motion.	a - Compensation port blocked, possibly by swollen cup. b - Use of unsuitable fluid. c - Incorrect brake pedal stop setting.	a - Clear port with a piece of 0.7 mm wire. Remove burr. Fit new cups. b - Flush system well, and then fill system with VW or Lockheed fluid. c - Check brake pedal stop and rectify clearance so that compensation port is clear with brakes in rest position.
Brakes chatter and tend to grab.	a - Linings worn, rivets proud. b - Drums out of round.	a - Fit new linings or new rivets. b - Turn drums out or fit new.
Brakes squeak.	a - Unsuitable linings. b - Loose rivets. Badly fitting linings. c - Brakes dirty. d - Back plates distorted. e - Return springs weak.	a - Fit new linings. Use Genuine VW parts. b - Fit new rivets, or possibly reline. c - Clean brakes. d - Check back plates for distortion and fit new parts if necessary. e - Fit new return springs.
Brakes uneven in operation.	a - Oil or grease on linings. b - Poor lining contact pattern due to brake shoe distortion. c - Brake shoes too tight in slots in adjusting screws or wheel cylinder pistons. d - Different types of linings on one axle. e - Incorrect tyre pressures or unevenly worn tires.	a - Fit new seals or repair wheel cylinders. Fit new linings on both wheels on one axle to ensure equal coefficient of friction. b - Shape shoes so that there is 0.2 mm clearance at lining ends. c - Free shoes off. d - Fit new linings. e - *Check and rectify pressures or* replace worn tires.

Symptom	Cause	Remedy
Brakes uneven in operation.	a - Oil or grease on linings.	a - Fit new seals or repair wheel cylinders. Fit new linings on both wheels on one axle to ensure equal coefficient of friction.
	b - Poor lining contact pattern due to brake shoe distortion.	b - Shape shoes so that there is .008 in. (0.2 mm) clearance at lining ends.
	c - Brake shoes too tight in slots in adjusting screws or wheel cylinder pistons.	c - Free shoes off.
	d - Different types of linings on one axle.	d - Fit new linings.
	e - Incorrect tire pressure or unevenly worn tires.	e - Check and rectify pressures or replace worn tires.
	f - Drums out-of round or scored.	f - Turn drums out or fit new drums.
	g - Drums distorted by unevenly tightened wheel bolts.	g - Tighten bolts to 72 lb. ft. (10 mkg).
	h - Contact area on wheel disc out of true and causing distortion in drum.	h - Measure drum with wheel disc bolted on and try to reduce ovality to a minimum by mounting wheel in different positions. This can sometimes be achieved by interchanging the wheels. Otherwise fit a new wheel disc.
	i - Brake shoes not contacting back plate. Brake shoes or back plates distorted.	i - Remove shoes and straighten or renew. Fit new back plates.
	k - Pistons tight in wheel cylinders.	k - Free pistons off.
	l - Dirt in brake lines or hoses.	l - Clean lines. Replace defective. hoses.
Brakes pulsating.	a - Front axle attaching bolts unevenly tightened or too loose.	a - Back off attaching bolts and retighten to prescribed torque of 36 lb. ft. (5 mkg).
	b - Brake drum distorted, caused by unevenly tightening wheel bolts.	b - Tighten wheel bolts to 72 lb. ft. (10 mkg).

Symptom	Cause	Remedy
Brakes pulsating.	c - Seating surface of wheel disc on brake drum not flat, thus causing distortion of drum.	c - Check brake drum, while still bolted to wheel disc, with the "appliance for checking brake drum ovality, VW 759". If the ovality is more than .004 in. (0.1 mm) relocate the wheel disc to the drum to try to attain less brake drum distortion. In some circumstances this can be achieved by interchanging the wheel discs. If the ovality cannot be kept below the maximum permissible limit of .004 in. (0.1 mm), replace a wheel disc. If a suitable lathe is available, use it to check the brake drum for ovality, while still bolted to the wheel disc, in conjunction with a dial gauge. If necessary, turn out brake drum while still bolted to wheel disc.
	d - Brake shoe webs too tight in slots in adjusting screws or wheel cylinder pistons.	d - Smoothen shoe web ends and contact surfaces of slots in adjusting screws and wheel brake cylinder pistons with crocus cloth and lubricate with high melting point graphite grease on installation.
	e - Poor lining contact pattern due to brake shoe distortion.	e - Re-shape brake shoes so that there is a .008 in. (0.2 mm) clearance at the lining ends as shown in the illustration.

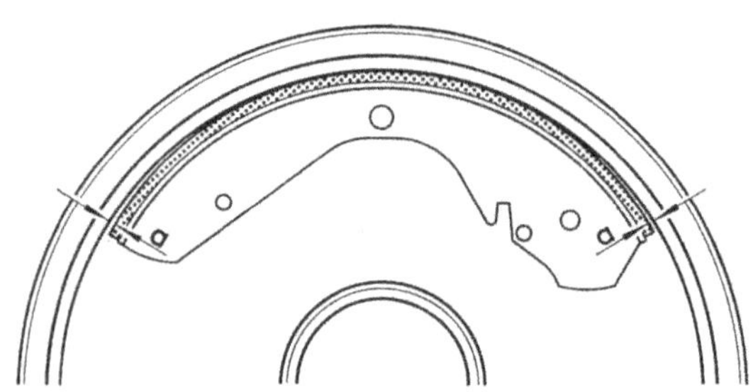

Important

Do not try to remedy pulsating by installing the recessed cups for Type 3 vehicles only (SP 136) in the wheel brake cylinders.

Storage and Installation of Brake Cylinders, Rubber Parts and Brake Hoses

(up to 1965)

New brake master and wheel cylinders in assembled condition should be stored in a dustfree, cool and dry atmosphere and never in excess of 6 months. Separate rubber parts, such as cups, boots, check valves, should not be stored for a longer period than 12 months. As to brake hoses, the storage is recommended not to exceed a period of 18 months. On no account should brake components be exposed to the sun or permitted to come into contact with oil, grease, fuel or water.

If assembled master and wheel cylinders are to be installed after a storage period of over 6 months, they should be disassembled and cleaned in denatured alcohol. The components are then dried and coated with Genuine VW Brake Cylinder Paste (Part No. B 1) prior to reassembly.

All parts should also be checked as to their serviceability. Parts showing signs of corrosion should be discarded. This applies also to hardened or damaged rubber components, no matter how irrelevant the damage seems to be, and to pistons revealing too tight a fit in the cylinder. Care should be taken to prevent dirt or other foreign matter from entering the brake system, during assembly or by reusing contaminated brake fluid.

All parts packages containing master and wheel cylinder units or components will henceforward be marked with the date of packing to permit an exact control of the storage periods.

Storage of Brake Parts

(1965)

All brake parts should be stored in a dust-free, cool, dry place and not longer than **two years.** They must also be protected from oil, grease, gasoline and water. Rubber parts must also be protected additionally from the sun.

Assembled master cylinders and wheel cylinders as well as disc brake calipers which have been stored for more than two years, must be disassembled, cleaned in methylated spirits, dried and, prior to reassembly, be lubricated with original VW brake cylinder paste. At the same time all the parts should be checked for serviceability. Parts showing signs of corrosion must not be used. Rubber parts which have become brittle or are only slightly damaged and also pistons which are tight must be replaced. The bores of unpacked parts must be closed with rubber plugs.

To keep a check on storage times, the cartons of replacement master cylinders and wheel cylinders as well as the disc brake calipers are marked with the packing date.

Note:

When installing disc brake calipers which have been stored a long time it is possible for the protective wax layer on the piston to be pushed away by the inherent tension of the rubber seal, the pistons "stick" and the friction pads are then not automatically adjusted (running clearance too large). In such cases, remove the friction pads and push the pistons forward a little. Then push the piston back to its original position with the piston retaining pliers and reinstall the friction pads. Drepress the brake pedal several times while the vehicle is stationary.

THIS PAGE INTENTIONALLY
LEFT BLANK

General

Drop-center rim type wheels are used on the Volkswagen.

The rim size is: 4 J x 15

The tire size is: 5.60—15 tubeless

Proper wheel and tire maintenance is essential to economical and safe automobile operation.

The maintenance includes:

1 - Properly tightened wheel mounting bolts.

2 - Maintaining correct tire pressures.

3 - Inspecting tires for damage and signs of wear.

4 - Inspection of rims as regards damage, especially to flange and shoulder.

5 - Interchanging the wheels if tires are wearing unevenly.

6 - Wheel balancing.

Note:

From Chassis No. 5677119, the VW sign on the chrome-plated wheel caps is no longer painted black.

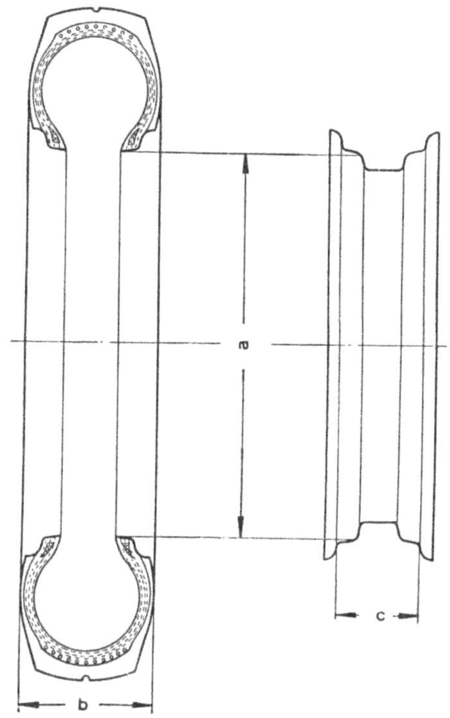

a = 15″ b = 5.60″ c = 4″

Changing the Wheels

Removal

1 - Set the hand brake.

2 - Insert jack into the square tube below the sill panel. The jack should stand on solid ground.

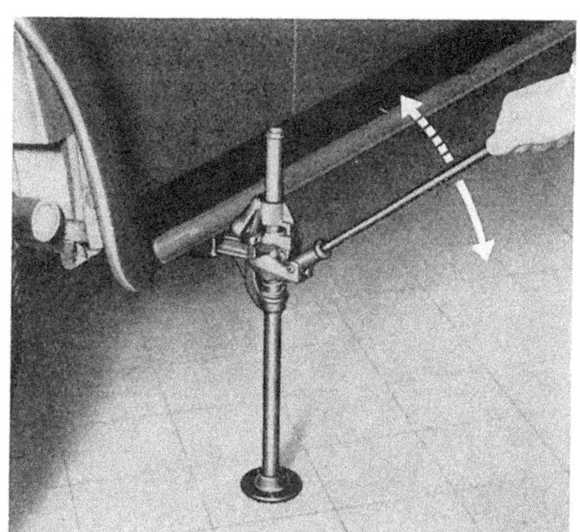

3 - Remove the wheel cap with the cap removing tool VW 636 (drawing for local manufacture) or a similar tool.

4 - Loosen wheel bolts with the hexagon socket wrench.

5 - Jack up the car.

6 - Remove wheel mounting bolts and take off wheel.

Installation

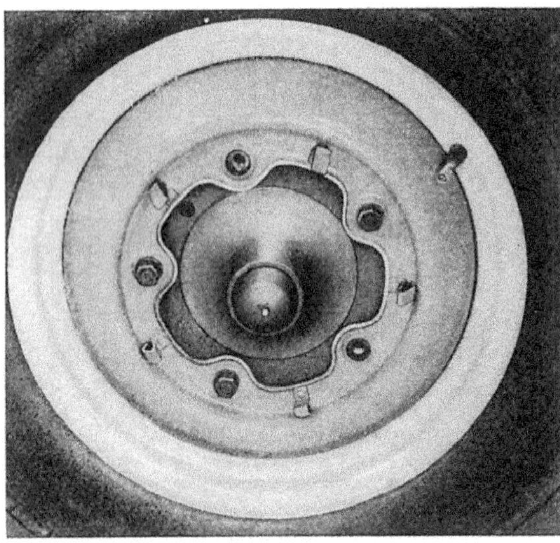

Installation takes place in the reverse order to the instruction on removal, but the following points should be observed:

1 - Screw in the wheel mounting bolts until the countersunk heads are concentric with the corresponding recesses.

2 - Lower the car.

3 - Tighten all bolts diametrically opposite in turn to a torque of 10 mkg (72 ft. lbs.). Oversize bolts M 14 are to be tightened to 12 mkg (86 ft. lbs.).

Replacing Broken Wheel Cap Retaining Springs

1 - Remove wheel.

2 - Remove broken spring and rivet, using a flat chisel. Drive out the remaining piece of the rivet.

3 - Clamp the dolly in a vise and position a button-head rivet (4.5 x 7 mm) and the retaining clip as shown to the right.

4 - Place wheel horizontally (outside down) over the dolly so that rivet enters the corresponding hole in the wheel.

5 - Rivet the retaining spring in place.

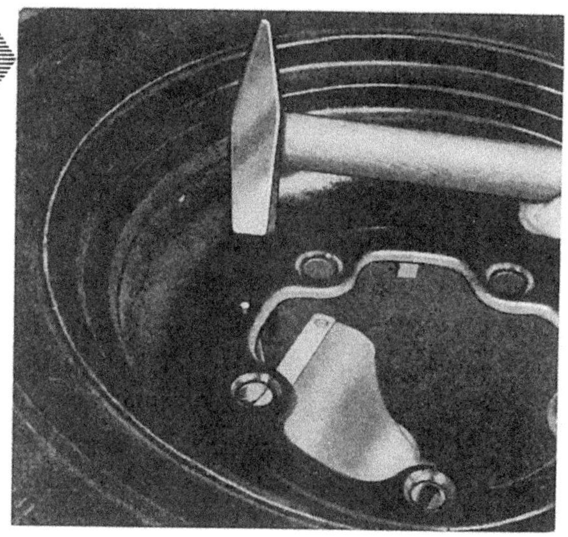

6 - Paint retaining spring and rivet on both sides.

7 - Position the wheel and tighten the mounting bolts.

Damaged Wheels

Slightly damaged rims, especially rim flanges can be repaired without difficulty. If more severe damage is present, e. g., after an accident, the serviceability of the wheels should depend on the result of a thorough test. Wheels that are not likely to meet the following requirements after repair should be discarded.

Tire must fit and seal properly.

The welding seam must be airtight.

The maximum permissible lateral and radial run-out (1.5 mm, 0.006″) must not be exceeded.

In any case, the wheel assembly must be balanced both statically and dynamically after a repair has been carried out. Distorted rims must not be straightened.

THIS PAGE INTENTIONALLY
LEFT BLANK

General

The condition of the tires greatly influences the driving characteristics of the vehicle and the road safety. It is therefore essential to ensure that adequate attention is given to the tires at all times.

A reasonable amount of care will help to keep tire wear to a minimum and increase the service life. Balanced wheels and tires also reduce tire and wheel bearing wear and improve driving characteristics.

Incorrect pressures, driving habits, faulty wheel alignment, condition of road surface and excessive unbalance can all cause abnormal and rapid wear.

Overloading should be avoided and the tires should be protected from intense sunshine and contact with fuel and oil as far as possible.

Note:

To comply with a unification of standards for Europe, in future **all** tires must be given a PR designation. Previously a designation was only necessary from a carcass strength of 6 PR.

PR means Ply Rating and is the international designation for the carcass strength.

In addition, new maximal permissible loads for individual tire sizes have been decided upon so that for the Micro Van (Model 147) tires with stronger carcasses must now be used.

For the Type 1, therefore, tires bearing the following designation are required:

Models 11 and 15	5.60—15 4 PR tubeless
Model 14/1200 and 1300	5.60—15 4 PR tubeless
Model 14/1500	5.60 S 15 4 PR tubeless
Model 147	5.60—15 6 PR tubeless

As only tires conforming to the new standards will be manufactured in future, the aforementioned designations apply also to older vehicles with immediate effect.

> All tires manufactured to previous standards and bearing no PR designation must only be used as 4 PR tires.

Tire Size: 5.60—15

Dimensions:

Outside diameter	**651 ± 6 mm**
Operating width	**max. 154 mm**
Effective static radius	**304 ± 3 mm**
Effective dynamic radius	**309 ± 3 mm**

Inflation Pressures:

With 1—2 occupants:

front 1.1 kg/cm² (16 psi) rear 1.7 kg/cm² (24 psi)

Fully loaded:

front 1.2 kg/cm² (17 psi) rear 1.8 kg/cm² (26 psi)

For continuous driving increase all tire pressures by 0.2 kg/cm² (3 psi).

To keep tire wear to a minimum and get the best possible ride with these tires it is essential that the pressures bechecked regularly and kept accurate.

> **Important**
>
> **Driving with the pressures too low affects the security of the tires and impairs the safety of the vehicle.**

Tire Markings

To permit easier identification of the two types of tires, all standard tires 5.60—15 with tubes have been given a new marking. As from February 1958, they will be marked with a green dot instead of the red one that used to indicate the lightest point of the tire where the valve should be when mounting. All tubeless tires are still marked with a dot.

Consequently, tires marked with green dots must always be mounted with tubes.

During a transitional period, standard tires 5.60—15 with tubes marked with either red or green dots will be stored together with the tubeless tires marked with red dots and we should, therefore, like to advise you to watch for the lettering "Tubeless" or "Schlauchlos" to avoid any confusion.

Tubeless Tires

Tubeless tires can be changed with simple tools. Two long tire irons with carefully rounded edges are required, also a valve installing tool and a bead breaker to break the tire bead away from its seat in the rim.

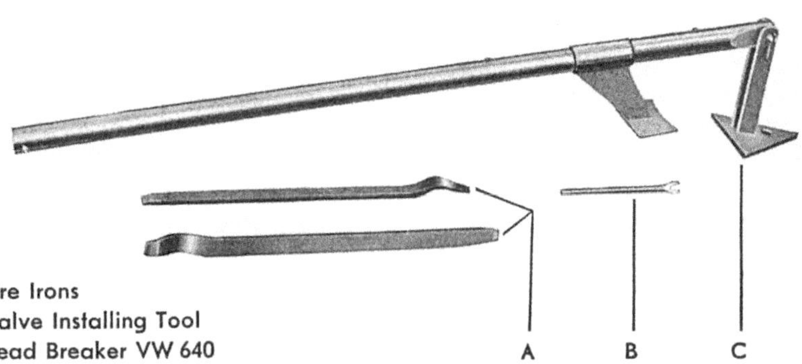

A - Tire Irons
B - Valve Installing Tool
C - Bead Breaker VW 640

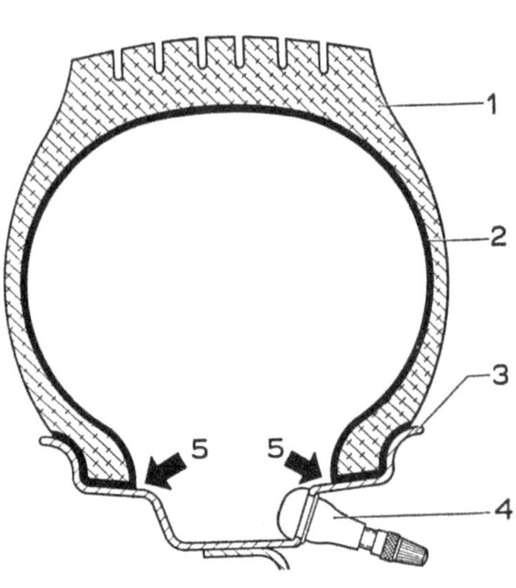

1 - Tubeless tire
2 - Air retaining and sealing lining
3 - Rim flange
4 - Valve
5 - Rim sealing surface

Apart from these tools, there are a number of tire tool sets from various manufacturers that have been tested for their suitability for the Volkswagen. Further details may be seen from the manual "Workshop Equipment".

When working on tubeless tires it is of paramount importance to make sure that the air retaining and sealing lining does not suffer damage.

The following instructions should be heeded when changing tubeless tires.

The rims 4 J x 1.5 suitable for tubeless tires have a rectangular recess at the hole for the valve. They are stamped with a "T" between two of the wheel bolt holes for better identification.

A rubber base valve (43 GS) according to DIN 7780 is used as standard equipment for tubeless tires.

Removing Tire

1 - Screw off valve cap and valve core.

2 - Break the tire bead loose from its seat in the rim. This is best accomplished with the Bead Breaker VW 640.

3 - Using tire irons, pry the bead over the edge of the rim.

4 - Check the air-retaining and sealing lining inside the tire for damage and bruises. Inspect the exterior of the tire for embedded flints and stones, cuts, grease, and for signs of wear.

5 - Check valve rubber base for cracks and breakage.

> **Important**
>
> To guard against damaging the sealing lining on the bead, only use long tire irons having carefully rounded edges.

Mounting the Tire

Mounting the tire is, in general, accomplished by reversing the dismounting procedure.

1 - Check the rim for damage. Never mount tubeless tires on rims having damaged, distorted or deeply scratched shoulders or flanges.

2 - Remove any dirt from rim shoulders and flanges, use a wire brush, if necessary.

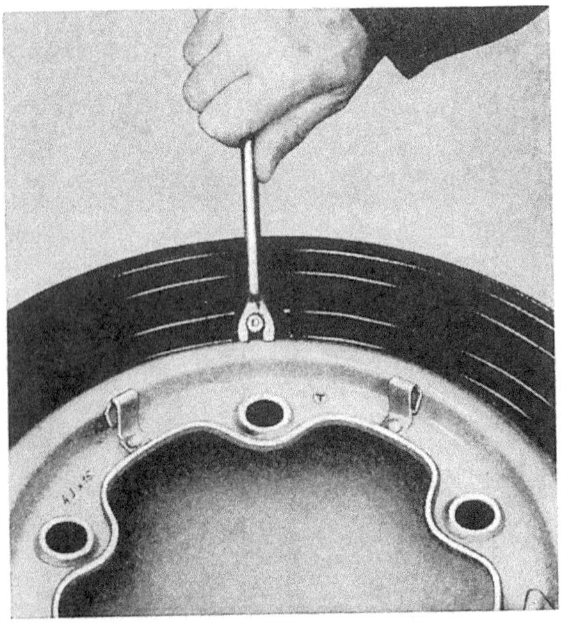

3 - Pull the rubber base valve into position, using the Valve Installing Tool for tubeless tires.

4 - Mount tire on the rim with the red dot on the sidewall of the tire located at the valve. Be careful not to damage the tire beads.

> **Important**
>
> The diameter of the tire beads is smaller than the rim shoulders so as to provide a better seal between tire and rim. Moreover, the rim shoulders are not horizontal but inclined upward. The tire beads must, therefore, have a certain degree of preload when positioned on the rim shoulders.

5 - Remove the valve core.

6 - Allow the full rush of air to overinflate the tire to at least 4 kg/cm² (57 lbs./sq. in.). This initial rush of air will push the tire beads against the rim flange and provide a quick seal.

Note:

If the sidewalls of the tubeless tire are pressed together because of improper storing, a bead expander (being a spring steel band placed round the circumference of the uninflated tire) should be applied to spring the beads against the rim flange for proper sealing.

Tubeless tires should always be stored in an upright position.

7 - Screw in the valve core and deflate the tire to the correct pressure.

8 - Check air seal of tire and valve by immersion in a water tank.

> **Important**
>
> All tubeless tires are uniformly inflated to 1.7—1.8 kg/cm² (24.2—25.6 lbs./sq. in.) at the factory. Be sure they are brought to the specified pressures before delivery of the car.

Valve Installing Tool

(Local Manufacture)

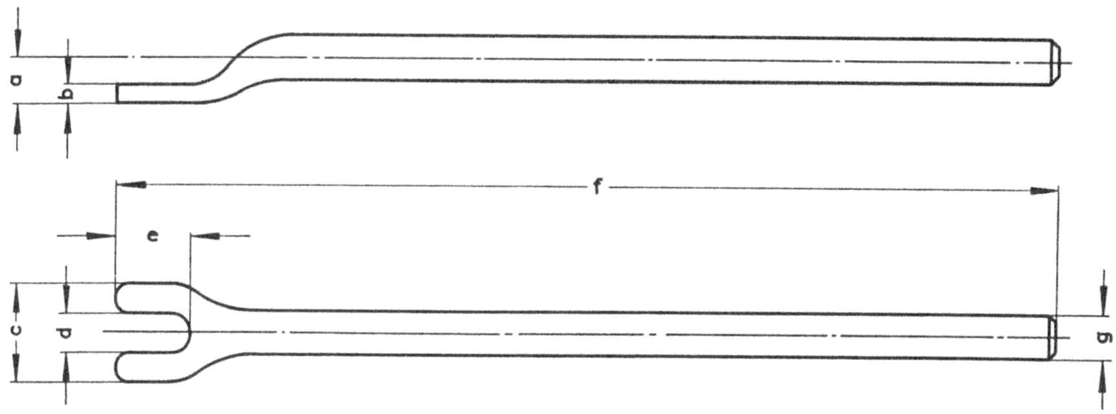

a = 9 mm (0.35″)
b = 3.5 mm (0.14″)
c = 20 mm (0.78″)
d = 8 mm (0.31″)
e = 15 mm (0.59″)
f = 195 mm (7.68″)
g = 9 mm dia. (0.35″)

Tire Inspection

The tires should be frequently checked but always during Maintenance Service for correct inflation and undue signs of wear as well as for cracks, grease, cuts, and bruises. Flints and stones that are embedded in the tire should be removed.

Tire inflation

Correct inflation is one of the most important factors for long tire life and perfect riding qualities of the VW Passenger Car. Tires in ordinary use should be checked at least once each week with an approved tire gage.

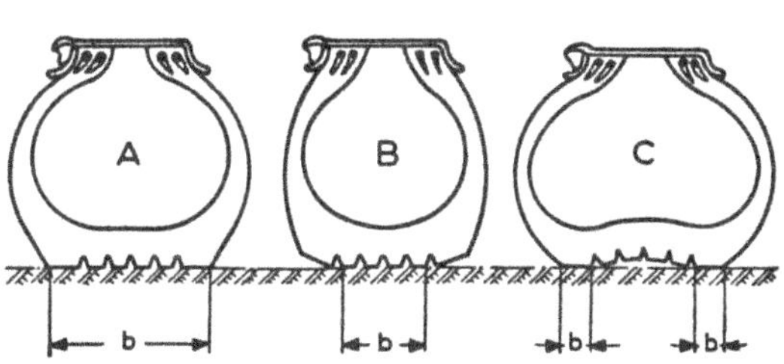

A = Proper inflation
B = Overinflation
C = Underinflation
b = Road contact

Tire pressure should be checked only when tires are cold. If the pressure is checked when the tires are hot i. e., after high speed operation, the pressure should not be reduced as otherwise it will be too low after the tires have cooled down.

Note:

Care should be taken when using a simple tire gage for a prolonged period. After a certain time of use, such tire gages are subject to variations. Although the readings obtained often depart only slightly from the real pressure in the tube, the detrimental over or underinflation may lead to abnormal tire wear. Therefore, it is absolutely necessary to regularly check the tire gage for accuracy.

The tightness of the valve cores can easily be checked by moistening a finger and placing it on the valve opening. Renew leaky valve cores.

Abnormal tire wear

Following are the causes for abnormal tire wear:

- Underinflation or overinflation
- Poor driving habits
- Overloading the vehicle
- Bad road conditions
- Improper wheel adjustments

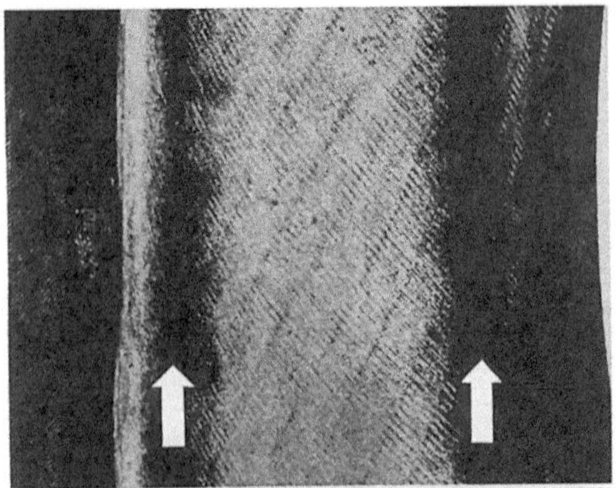

Underinflation

An underinflated tire will be subjected to excessive flexing of the side walls and resultant heat.

Underinflation can be determined by wear at the sides of the tread.

Two black stripes will first form inside the tire casing.

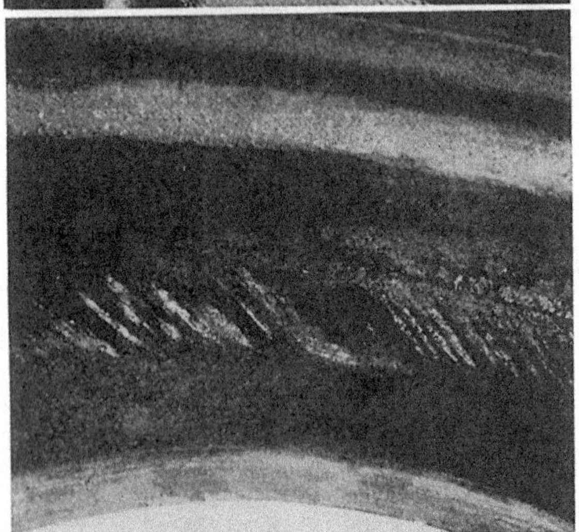

They will lead to damage of the fabric structure.

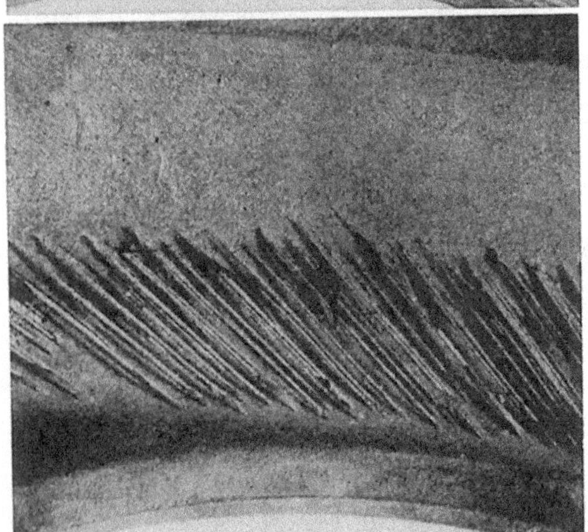

If the tires are further operated with insufficient air pressure, the plies in the side walls will separate so that the tires are ruined.

The underinflated tire contacts the road along the sides, where the tread is subjected to greater wear than in the center.

Overinflation

Overinflated tires cannot absorb road shocks and abnormal wear will occur at the center of the tread, making the tires susceptible to bruises and breaks.

Driving habit

The average speeds of automotive vehicles has gone up considerably in the course of the past decade. The tire wear increases rapidly with the speed. The chart indicates how much more rapidly tires wear at high than at low speed.

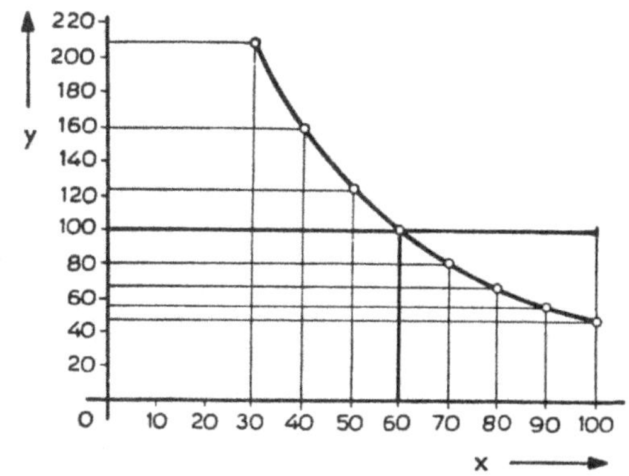

High-speed operation causes much more rapid tire wear because of the high temperature and greater amount of scuffing and rapid flexing to which the tires are subjected. Fast acceleration, sudden severe brake application, turning corners and rounding curves too fast or sharply will additionally contribute to increased and uneven tire wear.

Influence of speed on tire life
x = Speed
y = Relative tire life
The average value for tire life is indicated at the speed of 60 k.p.h. = 100 per cent on the chart.

The reasons for increased tire wear when applying the brakes are clear, but it should be borne in mind that unequally or improperly adjusted brakes or differences in the quality of the brake shoe linings may lead to abnormal wear of some tires. Eccentric brake drums may also be the cause for unequal tire wear.

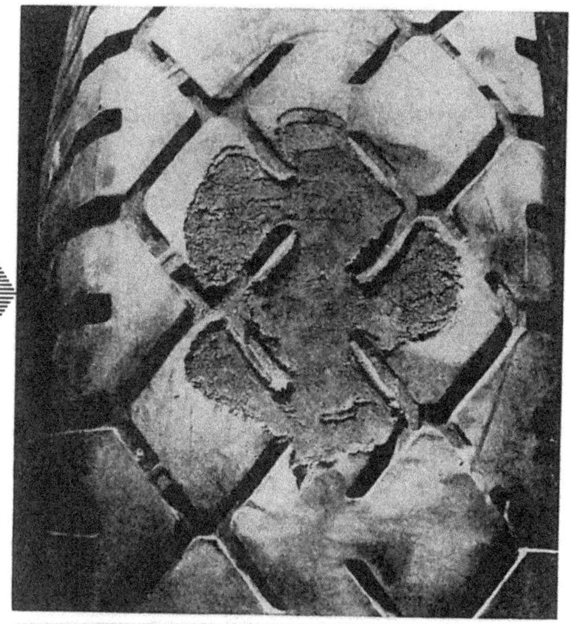

Overloading the car

Tires interpose a cushion between the road and the car wheels. The compressibility of the air in the tube allows the tires to absorb shocks resulting from irregularities in the road. It is the interrelation of air pressure, air volume and load that determines the correct tire size. Each tire size has to cope with a certain continuous load at a given inflation pressure. A temporary overloading is common in the operation of a motor vehicle; a fact which has been taken into account in designing the tires. Continuous overloading, however, has a detrimental effect on the tire structure.

This wear results in a saw-tooth effect with one end of each tread block worn more than the other.

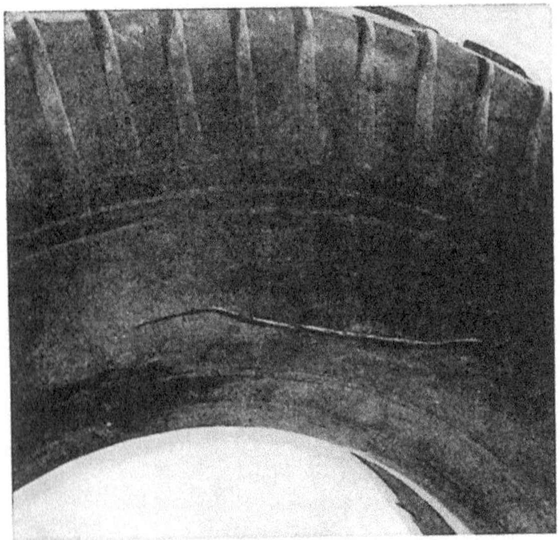

With the tire continuously overloaded, the fabric structure inside the tire will crack. After some time of operation, the cracks appear outside in the tire walls.

Road Surface

The tire life largely depends on the condition of the surface. To contribute to safe driving, antiskid material is now used to cover the roads. This provides frictional contact between the wheels and the road so that good traction is secured. This will make the tires subject to increased wear.

On steeply cambered roads (high crowned roads), the tires have to resist a side force which the driver attempts to compensate by applying a certain force on the steering wheel toward the center of the road so that the tires drag at an angle to the direction of vehicle movement.

Even a slight departure from the correct toe-in adjustment is very detrimental to the tire life under such conditions. If the toe-in is excessive, the tire toward the nearer curbstone wears more rapidly. If the toe-in adjustment is too low, the wheel toward the center of the road wears faster. This wear is usually distinguished by side wear and feather edges of rubber that appear on one side of the tread design.

A faster wear on steeply cambered (high crowned) roads must be taken into consideration, even if the toe-in adjustment is correct, but such wear will not be confined to one tire.

Misaligned Wheels

There are also several types of wear which can be attributed to wheel misalignment. In each case of abnormal tire wear, it is strongly recommended to check the as follows:

Front axle track
Track with steering at full lock
Position and track of the rear wheels
Position of both axles in relation to one another
Wheel base on both sides
Camber
Spring plate adjustment
Shock absorber operation

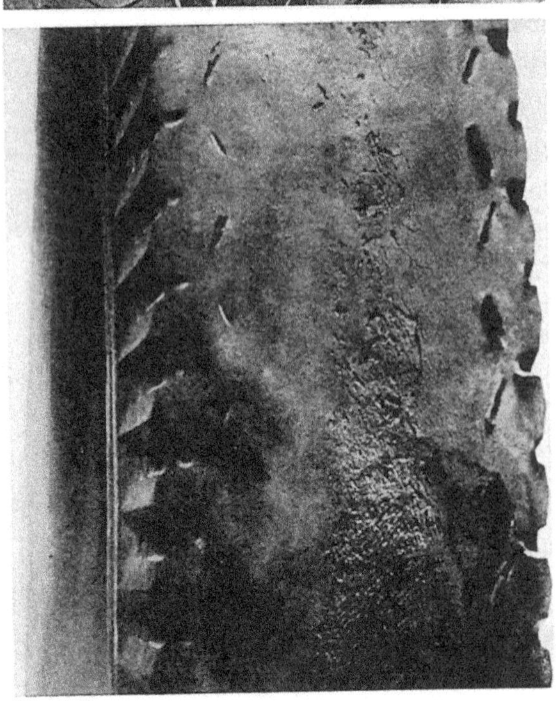

Misaligned wheels cause part of the tread rubber to be scraped off; this is indicated by feather edging of the tread and side wear.

A chassis that has been involved in an accident should in all cases be checked for proper alignment to assure a perfect track of the wheels.

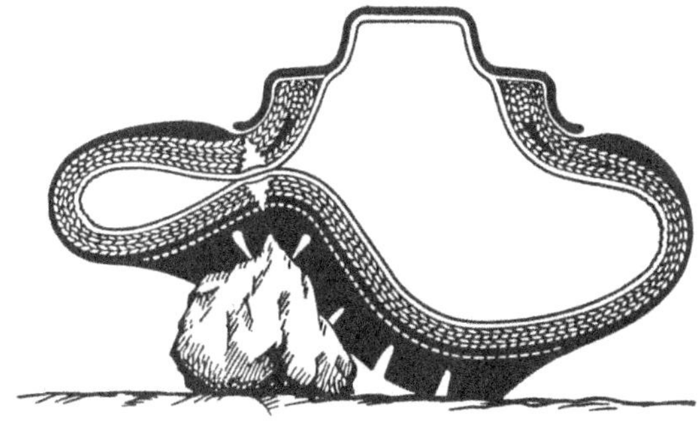

Cracks and Breaks

Sudden and violent shocks resulting from stones or gutters, etc., may cause cracks or breaks in the fabric structure of the tire casing. The illustration on the right indicates how such damage occurs.

The damage so caused may be of various natures.

Single diagonal break

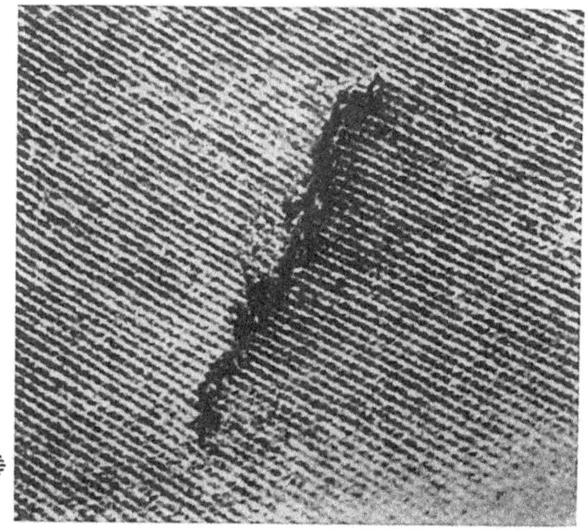

Double thrust break

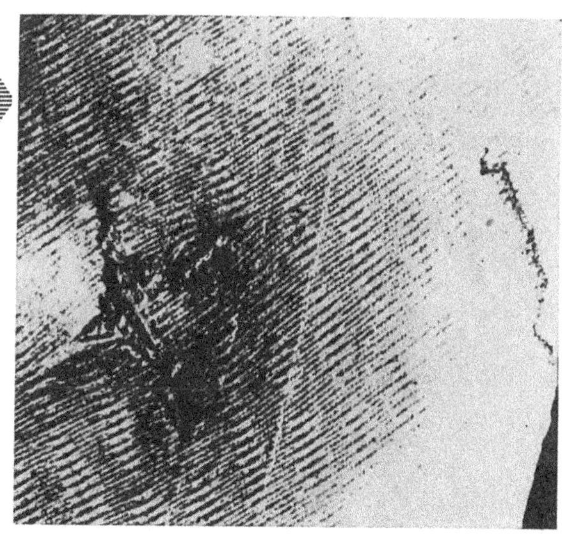

This kind of damage is generally difficult to locate outside. It is, therefore, absolutely necessary to carry out a thorough inspection at the inside of the tire.

It is often the case that only some of the plies are broken and that the damage expands when the trouble had not been located within a reasonable time.

Grease, Foreign Matter, and Cuts

Grease or oil soaked dirt should be removed from the tire tread and the spots cleaned by means of fuel. Never use paraffin.

Foreign matter embedded in the tire should be removed as soon as possible. Small damage to the tread material is of minor importance. More severe damage to the tread material or even to the tire casing should be eliminated by vulcanizing to avoid the tire becoming completely ruined by ingress of moisture.

After the flints, nails, etc., have been removed from the tire, it is good practice to fill up the holes with suitable compound.

Service Diagnosis of Abnormal Tire Wear

Type of Wear	Cause
Wear at the sides of the tread.	Underinflation.
Wear at the center of the tread.	Overinflation.
Spotty or irregular wear (gouges and waves).	Wheel assembly static and dynamic unbalance. Lateral wheel run-out. Excessive play in wheel bearings or at king pin.
Lightly worn spots at the center of the tread (cups).	Wheel and tire static unbalance. Radial wheel run-out.
Flat spots at the center of the tread.	Violent brake application. Brake drum out of round. Check brakes
Heel and toe wear (saw-tooth effect with one end of each tread block worn more than the other) leading to cracks in the fabric structure which become visible outside in the long run.	Typical for overloading. Check inside of tire casing for cracks.
(a) Side wear. (b) Feather edge of rubber on one side of the tread blocks. (c) Rounding and roughening of the tread at the outside shoulder of the tire.	(a) Incorrect camber. Continual driving on steeply cambered (high-crowned) roads. (b) Wheels toe-in or toe-out excessively. If the rear wheels are concerned, check adjustment of spring plate and efficiency of shock absorbers. (c) Caused by high speed driving on curves, called cornering wear.

Tire Rotation

(Interchanging)

To equalize wear of all tires it is recommended to rotate the wheels periodically every 5000 km (3000 miles). The spare wheel should be included in the rotation in order to obtain additional tire mileage. The tires should be rotated as shown below.

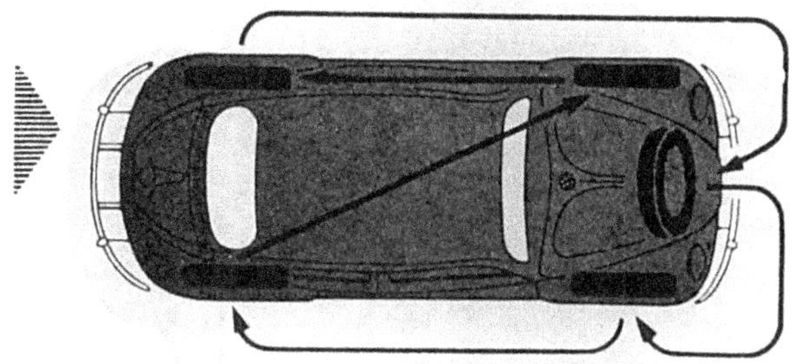

Not only is there a difference in the amount and type of wear between front and rear position, but also between the left and right side of the car.

As a precautionary measure, you should employ the better tires on the front axle, if already unevenly worn.

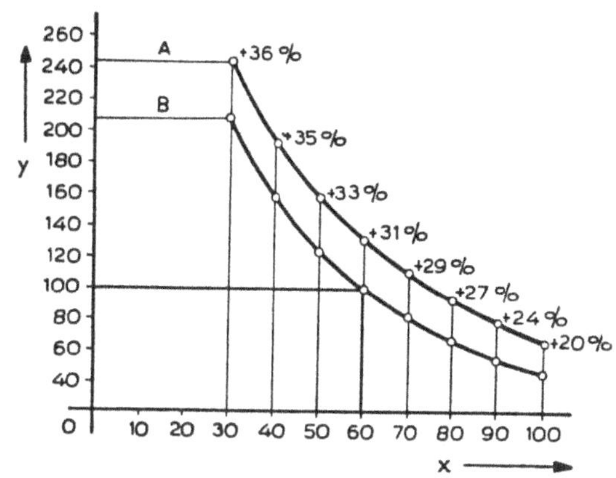

Seasonal influence on tire life (outer temperature)
A = Winter B = Summer
x = Speed of car (k. p. h.)
y = Relative tire life

New Tires

It is good practice to renew the tires in autumn as tire wear is generally higher in summer than in the cold season. At the same time it should be taken into consideration that the treads of the new tires provide good traction during the cold season when bad road condition such as snow, ice and thaw prevail.

Wheel and Tire Balancing

General

The wheel and tire assembly balance must be maintained to insure good riding qualities and best steering performance of the car. A statically unbalanced condition of a wheel and tire assembly is indicated by an up and down hopping or pounding action. The wheel wobbles or shimmies if it is dynamically out of balance. The faster a car the more important is a perfect wheel and tire balance.

It is, in general, not necessary to balance new wheels and tires. They are statically balanced during the course of production. Any remaining unbalance is so negligible that it does not affect the driving qualities of the Volkswagen. The wall of the tire is marked by either a single or double dots. The valve is supposed to be situated at this location during mounting so as to counteract the remaining unbalance.

Wheel balance is the first item to ckeck after the tire has been repaired, retreaded or recapped. Should it not be possible to have repaired tires balanced both statically and dynamically, the car owner should be advised to use such wheels on the rear axle.

Under these circumstances fast driving should be avoided.

Influence of unbalance on tire wear

A = Balanced wheel and tire after 14,200 km (8,800 Miles)
B = Unbalanced wheel and tire after 11,500 km (7,100 Miles)

Wheel balance is the equal distribution of the weight of the wheel, tire, brake drum and hub around the axis of rotation. The complete wheel assembly should be balanced both statically and dynamically.

A wheel assembly may be considered in static balance when it will remain static in any position to which it is revolved on its axis.

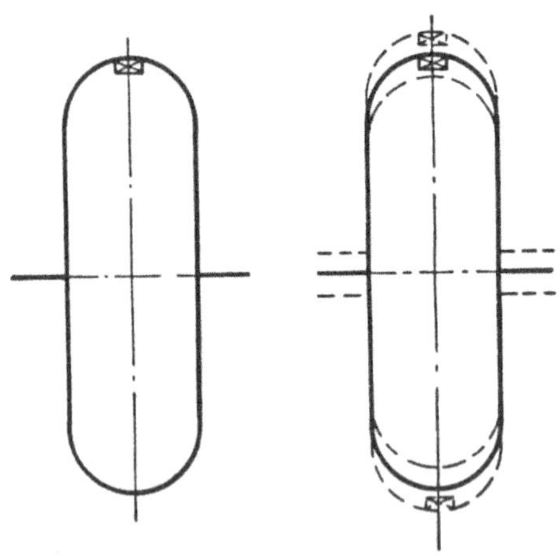

A wheel which is not statically balanced causes the tire to bounce at each revolution. The effect of the static unbalance increases with the speed of the car.

Perfect static balance does not assure that the wheel assembly is in complete balance. A dynamically out-of-balance wheel does not have even weight distribution in a plane vertical to the wheel axle.

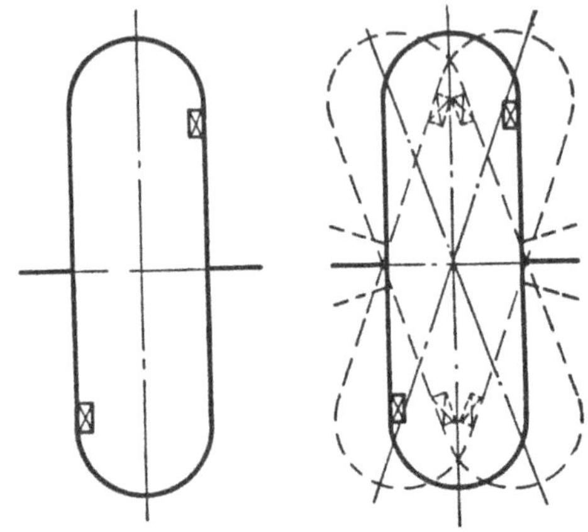

Dynamic unbalance causes the wheels to vibrate rapidly from side to side in their effort to revolve in a straight vertical plane. The greater the speed the more pronounced the condition becomes and steering ability is affected accordingly. This produces excessive wear of the tires, wheel bearings, king pins and steering connections. Prior to balancing the wheel and tire assembly, the wheel rim should be checked for lateral and radial run-out (maximum permissible: 1.5 mm/0.06").

Static balancing can be carried out on a locally-made balancing fixture. Dynamic balancing requires special equipment to determine the value and location of weights in order to correct the condition without disturbing the static balance. Several types of wheel balancer are in use, all operating in a similar manner. The wheel balancer should be used as instructed by the manufacturer.

Wheels which have been balanced should be re-balanced when the wheels are interchanged (i. e. at Maintenance Services).

If the wheels are removed from the vehicle and balanced on a wheel balancer, it is necessary to re-balance them again after 10,000 km (6,000 miles) as tire wear may result in the wheels becoming out of balance. Moreover, if a damaged tire has been repaired it is necessary to re-balance the wheel again.

Details concerning wheel balancers tested by the Technical Service Department are contained in the manual "Workshop Equipment".

The balancing weights are made of lead. Only weights with small retaining tongues should be used.

Bearing of Weather Conditions on Tire Life

Among the factors which determine the tire mileage are also the weather conditions.

Heat is the worst enemy of the tire. Under the influence of high outside temperatures, the operating temperature of the tire increases rapidly — especially during a long non-stop trip — so that the structure of the tire is liable to alter. Such alteration in the structure of the tire causes abnormal wear if the tire is continually subjected to an excessively high thermal load.

Therefore, it is recommended that attention is paid to the heat development of the tires on long trips at high speed when high outside temperature prevails. It pays to stop from time to time and let the tires cool down in the shade. Tire pressures vary with temperature and the pressure increases when tires are hot. That is why the pressure should be checked only when tires are cold, that is before operation.

Tire wear is less when the road surface is wet, as the water acts as a lubricant, reducing the frictional contact.

Anti-Skid Qualities of Tires

General

A wet or ice-covered road surface reduces the frictional contact between the wheels and the road, but this condition will also reduce the traction of the car. The tendency of the car to skid is influenced by

a - Road surface material

b - Weather and season

c - Tire tread

Road Surface Material

Roads of concrete, granite paving-stones, or asphalt mixed with basalt chippings combat the tendency of the car to skid even when they are wet. Asphalt roads are very slippery and dangerous when wet, especially if the surface is not fluted. Care should also be exercised when driving on wet basalt or wood block surfaces. Basalt can be recognized by its dark-blue colour.

Weather and Season

Drizzle, ice, and snow increase the tendency of the car to skid. A road surface which is dry in the open may be wet under the trees. Fallen leaves are a cause of skidding. One of the greatest hazards is a wet road which may freeze over.

Tire Tread

The tire tread design is of great importance to the steering stability of the car under bad road and weather conditions. Where the tread has worn smooth, it is sometimes possible to retread the tire in order to restore the traction and anti-skid properties.

Tires

I New marking

From October 1952, all VW Passenger Cars were fitted with standard tires 5.60—15 with inner tubes but from the middle of July 1956 we have been using exclusively tubeless tires of the same size. On the outer side wall these tires are marked "Tubeless" or "Schlauchlos".

To permit easier identification of the two tire types, all standard tires 5.60—15 with inner tubes have been given a new marking. From February 1958, they are marked with a green dot instead of with the red dot that used to indicate the lightest point of the tire where the valve should be when mounting. All tubeless tires are still marked with a red dot.

Consequently, tires marked with green dots must always be mounted with inner tubes.

During a transition period, standard tires 5.60—15 with inner tubes marked with either red or green dots will be stored together with the tubeless tires marked with red dots and we should, therefore, like to advise you to watch for the lettering "Tubeless" or "Schlauchlos" to prevent mix-ups.

II Tire pressure

Tubeless tires keep the pressure much longer than tires with inner tubes and customers as well as workshops are beginning to see the advantage. Here and there, the opinion seems to prevail that the tire pressure no longer requires the same degree of observation and attention as was essential previously. This opinion is mistaken and quite dangerous.

The tubeless tire reacts to insufficient pressure just as unfavorably as a tire with inner tube. It will become excessively hot, diminishing road holding qualities and even causing the car to skid.

In the case of tires with inner tubes, a certain loss in pressure had come to be taken for granted and the pressure was checked and corrected regularly.

With tubeless tires, a noticeable loss in tire pressure is, however, an indication that something is wrong with either the wheel, the tire, or the valve. If, after a loss in tire pressure has been noticed repeatedly, no leaks can be detected in the outer cover (including the bead), the following causes may come into consideration:

a - leaky rim (replace rim),

b - leaky valve (replace valve or valve core),

c - foreign matter or unevenness between the tire and its seat in the rim (clean or straighten the tire bead and the tire seat in the rim),

d - inadequate tire seat in the rim in the case of exceptional border-line tolerances (replace disc wheel and outer tire).

In the last-mentioned case (d), disc wheel and tire should not be used again until we have had an opportunity to examine them with a view to determining the cause of the pressure loss (inland only). On principle, it may be said that to merely adjust the pressure and to let the customer continue to operate the vehicle is not good practice.

Use of Snow Chains

On the Volkswagen 1200, only very thin chains which do not stand clear of the tire tread more than 13 mm (including connecting link) can be used. They should be of the radial chain type to give good lateral stability.

Snow chains can only be fitted to the rear wheels.

In Germany, chains which meet these conditions can be obtained from accessory dealers or from the manufacturers. The chain manufacturers are responible for the operation and quality of the chains offered by accessory shops.

Export countries can obtain suitable chains from our Parts Department under the following numbers:

111 016 901 and 111 016 901 A.

Influence of Weather Conditions on Tire Life

Among the factors which determine the tire mileage are the weather conditions.

Heat is the worst enemy of the tire. Under the influence of high outside temperatures, heavy loading and low pressure, the operating temperature of the tire increases rapidly — especially during a long spell of fast driving — so that the structure of the tire is liable to alter. Such an alteration in the structure of the tire causes abnormal wear if the tire is continually subjected to an excessively high thermal load.

Therefore, it is advisable to pay attention to the heat development of the tires on long trips at high speed when high outside temperatures prevail. Tire pressures vary with temperature and the pressure increases when tires are hot. That is why the pressure should be checked only when tires are cold, that is before starting a trip.

Tire wear is less when the road surface is wet, as the water acts as a lubricant, reducing the frictional contact.

These factors are responsible for the increased tire wear in summer.

Anti-Skid Qualities of Tires

A wet or ice-covered road surface reduces the frictional contact between the wheels and the road, and affects the roadholding ability of the vehicle. The tendency of the car to skid is influenced by

a - Road surface material

b - Weather and season

c - Tire tread

Road surface material

Roads of concrete, granite stones, or asphalt mixed with basalt chippings combat the tendency of the car to skid even when they are wet. Asphalt roads are very slippery and dangerous when wet, especially if the surface is not fluted. Care should also be exercised when driving on wet basalt or wood block surfaces. Basalt can be recognized by its dark-blue color.

Weather and season

Fog, fine rain and wet snow increase the tendency of the car to skid. Wet leaves, dirt from agricultural vehicles, snow and ice are also potential causes of skidding.

A long spell of rain, however, tends to wash the grease film from the road surface.

Tire tread

The tire tread design is of great importance to the steering stability of the car under bad road and weather conditions.

Use of Snow Chains

Only thin chains which do not stand clear of the tire tread and inner side wall more than .5 in./13 mm (including connecting link) can be used on the Volkswagen. They should be of the radial chain type to give good lateral stability.

Chains can only be fitted on the rear wheels.

Suitable snow chains for all Type 1 Models are obtainable from our Parts Department.

Radial Ply Tires

The radial ply tire, or braced tread tire as it sometimes called, differs from the normal type in that between the tread and the inner carcass there is a radial layer of rigid textile or steel cord which reinforces the surface of the tire which contacts the road and holds the cross section more or less constant. Because of their design, radial ply tires have certain features which make them appear superior to the conventional tire in some ways, but these also bring certain disadvantages.

The advantages radial ply tires offer are longer service life and better cornering characteristics when driving fast and in a sporty manner. On wet or snow-covered surfaces radial ply tires provide better grip and directional stability than standard tires.

The following disadvantages have been found: Most radial ply tires are considerably noisier on the Volkswagen at medium speeds and there is more vibration from cobbled roads than is the case when using normal tires. Moreover, the vehicle reacts more directly to steering corrections and it would appear that the changes in the behavior of the vehicle on fast cornering call for special study of the particular properties of tires of this sort. The less favorable points should be mentioned only because they may be found troublesome to some drivers and we do not wish to have the causes sought on the Volkswagen.

There are no radial ply tires specially suitable for the Volkswagen. Steel corded tires are particularly durable. They have the lowest rolling resistance. When driving at low speeds they are the noisiest — at least, on the Volkswagen. These disadvantages (road noise and body drone on cobbled roads) are also present on radial ply tires with textile bracing, but not to the same extent. The advantages of long service life and low rolling resistance are not as predominant on this type as on the metal corded tire. With the semi-braced tire the attempt has been made to obtain a mean between the advantages and disadvantages. As in all branches of technology the tire industry has had to resort to compromise solutions, since advantages in one direction involve disadvantages in another.

The increased durability of the radial ply tire is shown to be of considerable advantage if a vehicle is regularly driven at high speeds on motorways. There is no doubt that the roadholding of the vehicle can be improved still further by the braced tire. In consequence our opinion is certainly favorable towards these tires. As already mentioned, the criterium is their hard running at low speeds on poor roads. In view of their design we consider that radial ply tires are essentially tires for fast driving. In addition, their benefits are only appreciated on good roads. A person who only seldom covers long distances and who lives in an area where road surfaces are poor is not likely to get much pleasure from braced tires on his Volkswagen. In such cases it is also feared that the hardness of the tires with the resultant vibration would not do any good to the chassis, body, suspension and shock absorbers in the long run.

Volkswagen owners who are considering going over to radial ply tires are advised to first take a trial run in a Volkswagen fitted with this type of tire to enable them to assess the vehicle's behavior and roadholding. When a person realizes that not all the properties of radial ply tires are positive and is not troubled by the negative characteristics, there is no reason to advise against changing to radial ply tires.

Radial ply tires must always be fitted to all four wheels.

The sizes for the Type 1 are:

155—15 or 5.60—15 X

All radial ply tires recommended for the Volkswagen are with tubes and can be used on the normal rims.

At the present stage of development, radial ply tires of a tubeless type are not suitable regarding rim seal on the Volkswagen. All makes of radial ply tires we have tested can separate from the rims of our vehicles under extreme cornering conditions, with the vehicle fully loaded and with the tires at the correct inflation pressures. The result of this is sudden loss of air which could prove most dangerous.

To obtain good road holding with radial ply tires it is important that they be adequately and equally inflated. For the 5.60—15 X tires, the pressures are the same as for the normal tires. For 155—15 radial ply tires the inflation pressures should be 1.5 psi (0.1 kg/cm²) higher than those prescribed for the normal tire.

THIS PAGE INTENTIONALLY
LEFT BLANK

1 - Splash shield 2 - Brake disc 3 - Brake caliper

The front brakes are of the disc type, the main parts being the brake disc and the brake caliper which contains the hydraulically operated components of the brake system. A splash shield, attached to the steering knuckle, protects the brake disc inner surface from dirt, stone damage or similar damage. The outer side of the brake disc in protected from damage by the wheel. Viewed in driving direction, the brake caliper is positioned behind the wheel axis.

The brake caliper consists of two housings, the inner housing and the outer one, these housings being positioned one each side of the brake disc. The caliper is secured to the steering knuckle by two bolts. Four bolts hold the two caliper housings firmly together. A cylinder is machined into each caliper housing and each cylinder contains a piston and a rubber seal. The rubber seal, of square cross section, is positioned in an annular groove in each of the cylinders and prevents fluid leakage past the piston as well as the entry of moisture and dirt. Cylinder, piston and rubber seal are also protected against moisture and dirt by a rubber boot which is held in the recess at the front end of the cylinder by a spring ring and against the piston skirt by the inherent tension of the boot.

To prevent the pistons from rotating when braking, they are provided with retaining plates which are pushed into the piston crowns and held in the caliper grooves.

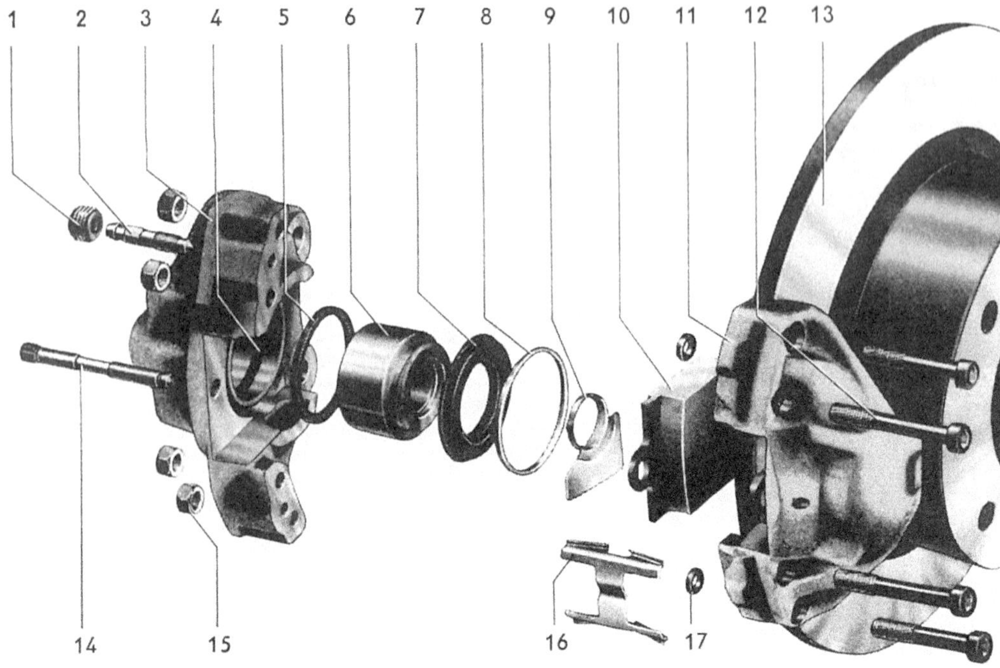

1 - Bleeder valve dust cap
2 - Bleeder valve
3 - Brake caliper inner housing
4 - Groove for rubber seal
5 - Rubber seal
6 - Brake caliper piston
7 - Rubber boot
8 - Spring ring
9 - Piston retaining plate
10 - Friction pad
11 - Brake caliper outer housing
12 - Caliper housing securing bolt
13 - Brake disc
14 - Friction pad retaining pin
15 - Nut
16 - Spreader spring
17 - Fluid channel "O" ring

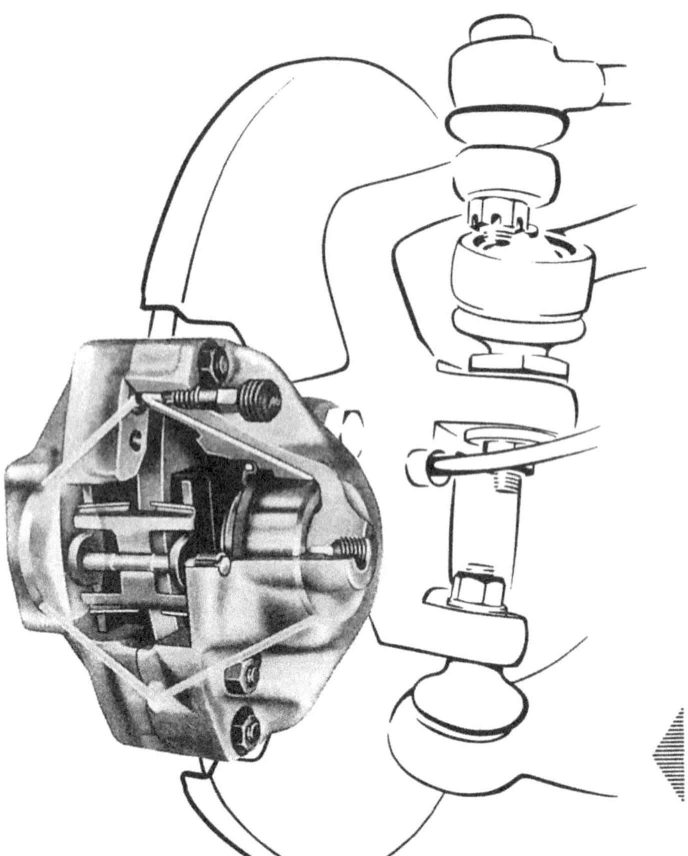

The friction pads slide axially in grooves in the housings, between the pistons and the brake disc. The pads are held in a radial direction by a retaining pin which, in turn, is held in the brake caliper by a split clamping bush. The spreader spring, which presses the friction pads against the pistons and thus eliminates any vibrations and rattling noises, is positioned under the friction pad retaining pin. The brake caliper inner housings are each provided with a threaded fitting for the connection to the hydraulic system and the inner housings also carry the bleeder valves. The cylinders of the brake caliper are connected to each other by internal fluid channels.

Disc Brake Operation

When the brake pedal is depressed, hydraulic pressure from the master cylinder piston is transferred to the brake caliper pistons. The pistons move toward each other and press the friction pads, after eliminating the clearance between pads and disc, against the friction surfaces of the disc with equal pressure on both sides. The rubber seals which bear on the skirts of the pistons are then deflected in the direction of the piston movement.

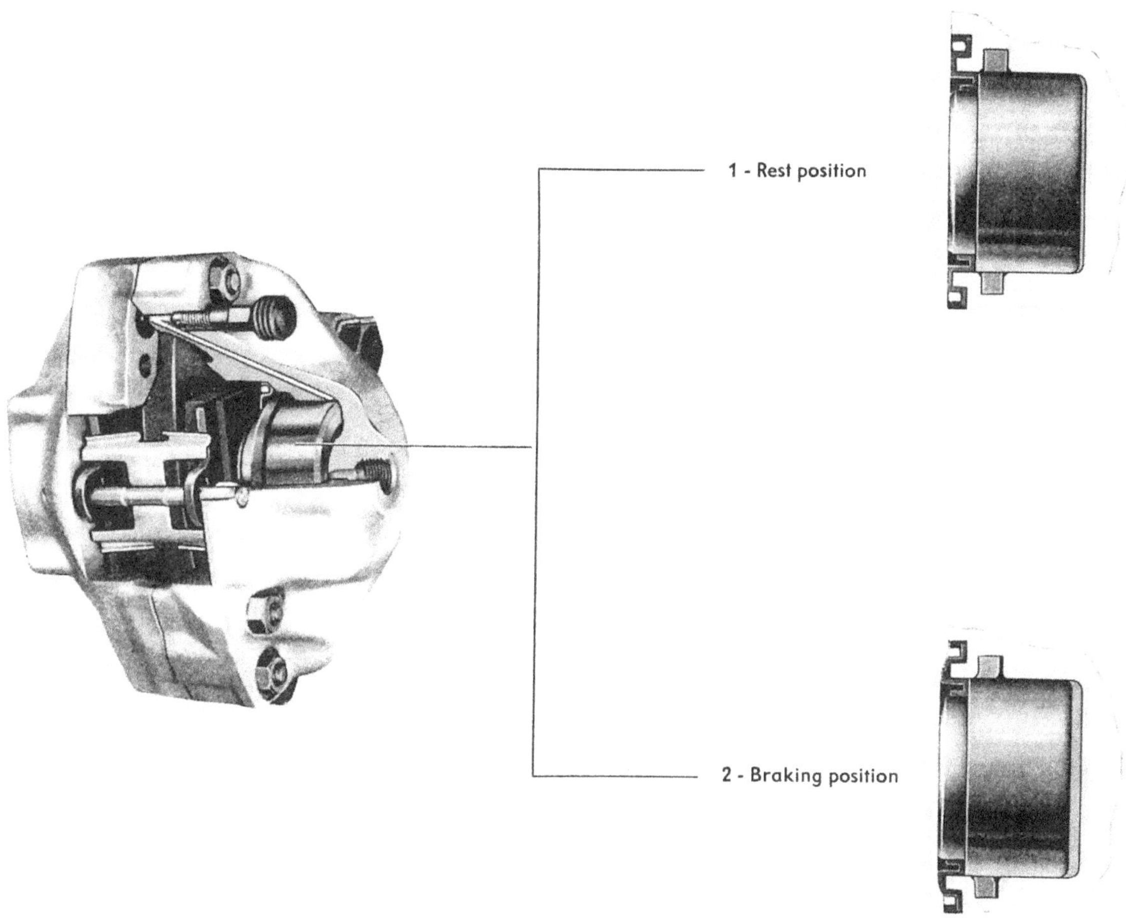

On releasing the brake pedal, the spring forces the master cylinder piston back to its original position and the complete system is relieved of pressure due to the pressure relief port in the check valve or in the master cylinder. Simultaneously, the pistons in the brake caliper are retracted by the rubber seals resuming their normal condition. The friction pads, which are pressed against the pistons by the spreader springs, move away from the brake disc, thus allowing the disc to rotate freely again. The amount of clearance between the friction pads and the brake disc when the brakes are off depends upon the elasticity of the rubber seal. The clearance is approximately .0006 in. (.15 mm). This clearance does not increase as the friction pads wear, as the pistons, when they have to cover a distance toward the brake disc larger than the lateral deflection of the rubber seals, slip through the rubber seals. The friction pads adjust themselves automatically according to the amount of wear.

Friction Pads

The friction pads consist of a pad of friction material and a metal plate to which the friction material is bonded. A layer of sound deadening plastic is cemented to the back of the metal plate, thus keeping noises down to a minimum. The friction material is provided with a groove .08 in. (2 mm) wide and .3 in. (8 mm) deep running in a radial direction which enables a water or oil film forming on the brake disc to be immediately broken. By this means, good braking in bad weather is attained.

Checking Friction Pads

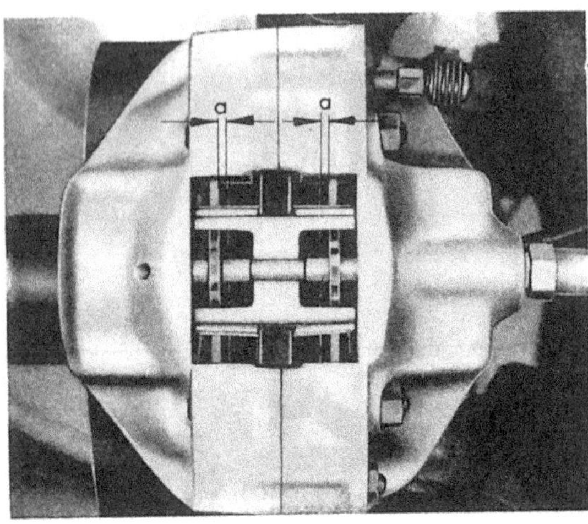

The friction pads must be checked for wear during the regular maintenance inspections every 6,000 miles (10,000 km). Friction pads which have worn down to a thickness a = .08 in. (2 mm) must be replaced with new ones.

Important
All four friction pads must always be renewed together. It is not permissible to renew single friction pads or both pads of only one wheel. In addition, the spreader springs of both calipers must also be renewed with the friction pads. All necessary parts are contained in the genuine VW repair kit SP 419.

Removing and Installing Friction Pads

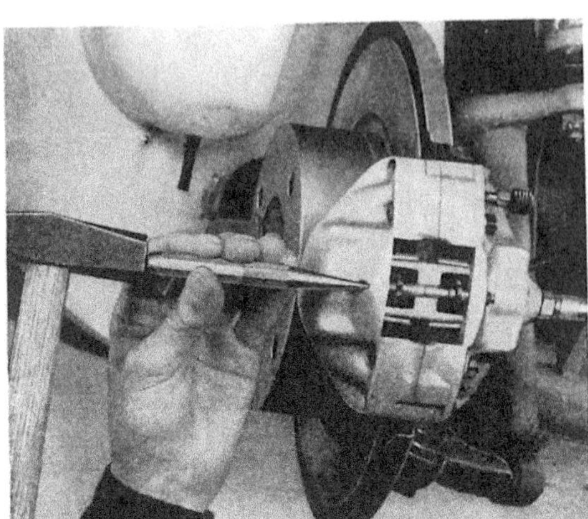

Removal

1 - Remove front wheel.

2 - Using a punch, drive out friction pad retaining pin.

3 - Remove friction pad spreader spring.

4 - Pull friction pads out of brake caliper, using the extractor hooks.

> **Important**
> If the friction pads are to be reused, they and the appropriate brake caliper housings must be marked as it is not permissible to change the pads from the outside to the inside and vice versa or from the right to the left wheel.

Installation

On installation, note the following:

1 - Oily friction pads, those with deep cracks or those which have become detached from the metal plate must be renewed.

In this instance, too, all four friction pads must be renewed.

2 - Push both pistons into their end positions with the piston retaining pliers.

3 - Clean seating and sliding surfaces of friction pads in brake caliper. To do this, remove the piston retaining plates.

> **Important**
> When doing this, the brake fluid behind the piston in the cylinder is forced back into the fluid reservoir. To prevent the reservoir overflowing, some of the fluid must be removed from the reservoir with a container which is used **only** for brake fluid.
> Brake fluid is **poisonous** and must not be siphoned off with a hose.

> **Important**
> Use only methylated spirits for cleaning. Never use mineral oil based solvents. Sharp edged tools, too, must not be used.

After cleaning, blow out brake caliper with compressed air.

4 - Check rubber boot (arrow) for damage. Hardened, brittle or cracked boots must be renewed. To replace the rubber boot, the brake caliper must be removed.

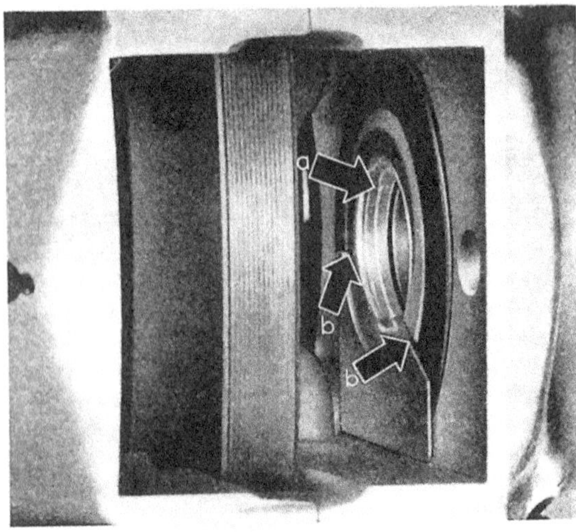

5 - Correctly insert piston retaining plate. The circular part of the plate (arrow a) must be firmly pressed into the piston crown. In addition, the plate must lie below the relieved part of the piston (arrow b). When the piston retaining plate is correctly installed, the 20° position of the piston is correct.

Corroded or damaged piston retaining plates must be renewed.

6 - Correct position of piston with the piston rotating pliers if necessary.

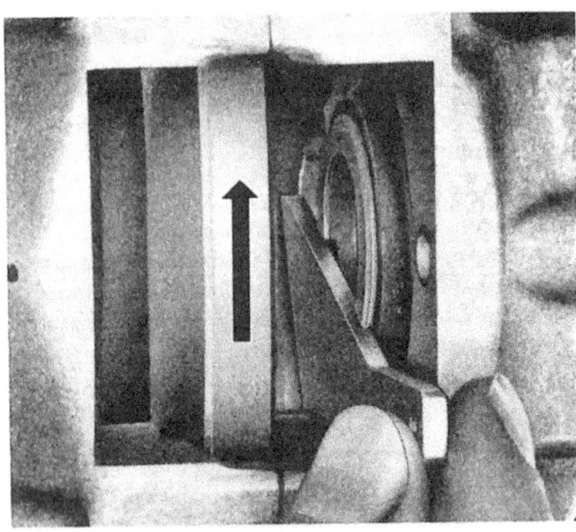

7 - Check position of piston again accurately, using the piston setting gauge. The gauge must always be held against the lower guide surface in the brake caliper i. e., counter-clockwise to the brake disc rotation (arrow) when driving the vehicle forward.

8 - Check brake disc for wear.

9 - Insert friction pads into brake caliper.

Important

Install already used friction pads according to the markings made on removal.

The friction pads must be free enough to move to and fro in the brake caliper.

10 - Correctly install **new** friction pad spreader spring.

11 - Insert friction pad retaining pin into brake caliper.

> **Important**
> The retaining pin must not be driven in with a punch smaller in diameter than the pin, otherwise the front shoulder can easily be sheared off by the split clamping bush.
> It is advisable to drive in the retaining pins, therefore, with a hammer only and no additional tools.

Renew corroded or damaged retaining pins.

Important

12 - **Depress brake pedal several times while vehicle is stationary, to enable the pistons and friction pads to take up their correct positions to the brake disc.**

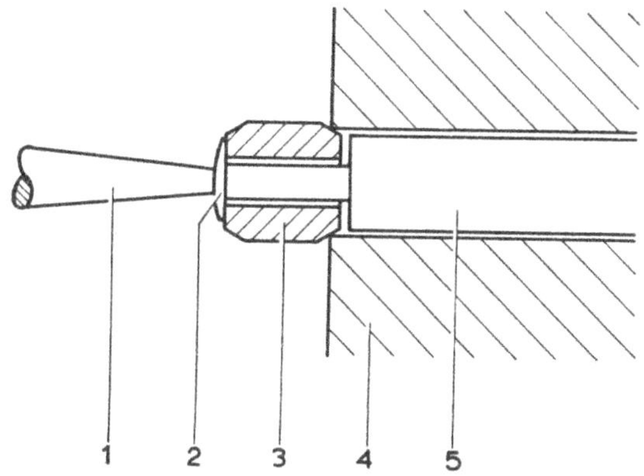

1 - Punch
2 - Endangered retaining pin shoulder
3 - Split clamping bush
4 - Housing
5 - Retaining pin

13 - Check brakes on road test.

Brake Caliper

The brake caliper may be removed only after it has cooled down to normal room temperature. The brake caliper may be disassembled only if the joint between the two housings becomes leaky and the fluid channel "0" rings have to be renewed. It is not permissible to disassemble the brake caliper for other repair work.

> **Important**
> The brake caliper must always be installed with the bleeder valve in the inner housing **at the top.**

Removing and Installing Brake Caliper

Removal

1 - Remove front wheel.

2 - Remove brake hose and close brake line with bleeder valve dust cap.

3 - Bend up brake caliper lock plate on both bolts.

4 - Unscrew brake caliper attaching bolts.

5 - Remove brake caliper.

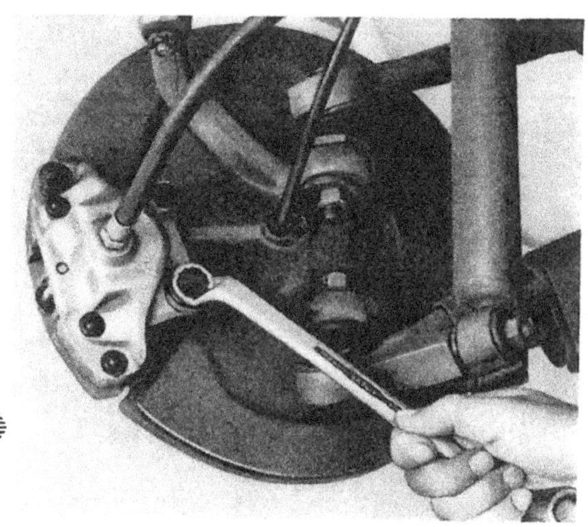

Installation

On installation, note the following:

1 - Clean mating surfaces of brake caliper and steering knuckle.

2 - Tighten brake caliper attaching bolts with a torque wrench.

Bolt grade: 10 k
Torque: 43 lb. ft. (6 mkg)

The attaching bolts and the lock plate **must be renewed.**

3 - Lock the bolts.

4 - Bleed brake system. Do not forget dust caps on the bleeder valves.

5 - Check brakes on road test.

Repairing Brake Caliper

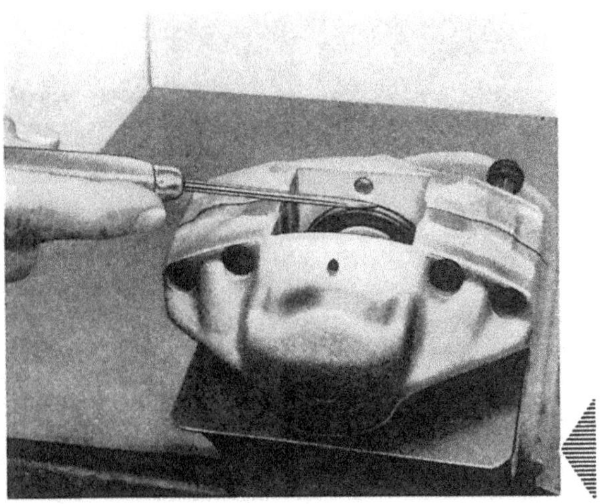

Disassembly

1 - Remove friction pads.

2 - Remove brake caliper and remove excessive dirt.

3 - Hold brake caliper by the flange in a vise.
Vise clamps must be used.

4 - Remove piston retaining plates.

5 - Using a screwdriver, pry out rubber boot spring ring.

When doing this, do not damage rubber boot.

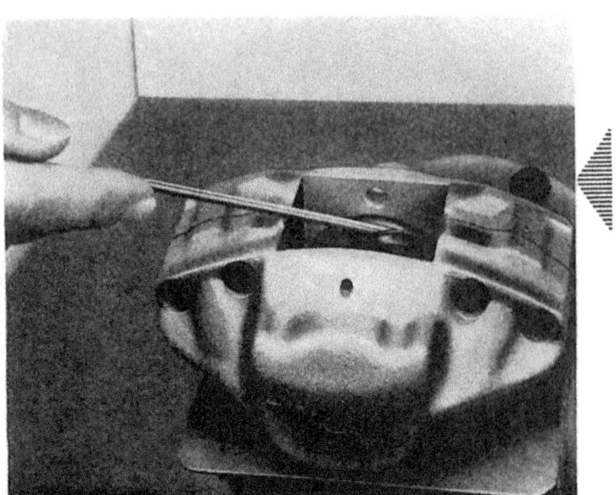

6 - Remove rubber boot, using a plastic or rubber rod.

Do not use sharp-edged tools.

7 - Remove one piston from the brake caliper with compressed air. To do this, the second piston must be held by the piston retaining pliers. In addition, a .2—.4 in. (5—10 mm) thick hardwood or rubber block must be inserted into the housing grooves so that the piston is not damaged.

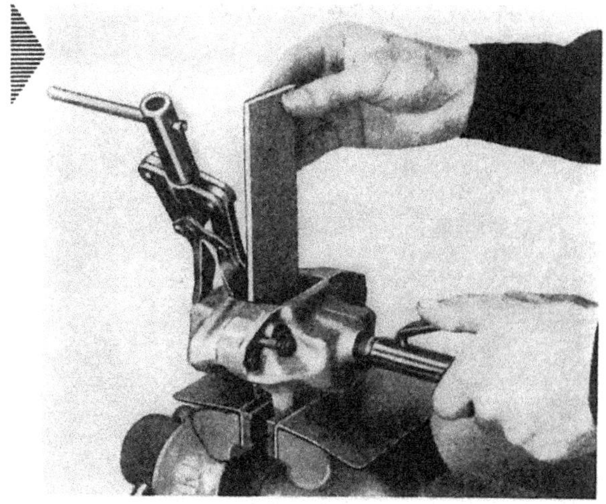

Important

The cylinders can only be repaired one at a time, as, with one piston removed, no pressure can be built up in the brake caliper.

If a piston cannot be completely removed from the cylinder immediately, it must be pushed back into its original position again with the piston retaining pliers.

8 - Remove rubber seal with a plastic or rubber rod.

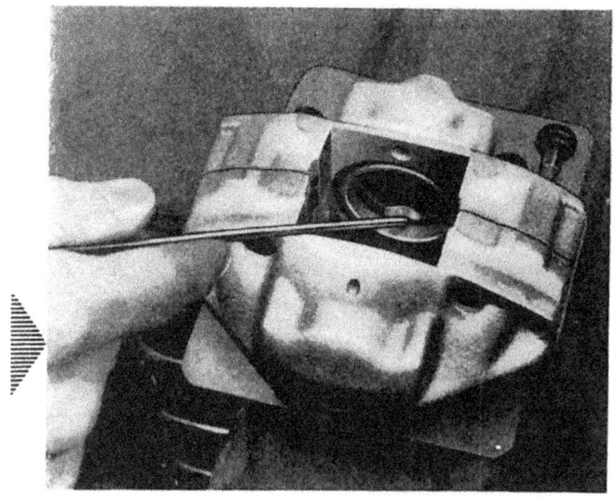

Assembly

On assembly, note the following:

1 - All parts must be cleaned in methylated spirits or brake fluid only.

2 - Check parts for wear. If a cylinder is damaged, the complete brake caliper must be replaced with a new one.

Important

The rubber seal, rubber boot, spring ring and piston retaining plate must be replaced with new ones each time the brake caliper is repaired. All the necessary parts are contained in the genuine VW repair kit SP 423.

3 - Thinly coat piston and new rubber seal with genuine VW brake cylinder paste and install them.

When pressing the piston into the cylinder with the piston retaining pliers, the piston couldt ilt. For this reason, always use the piston installing clamp.

4 - Install new rubber boot and new spring ring.

5 - Check 20° position of piston with the checking gauge and correct with piston rotating pliers if necessary.

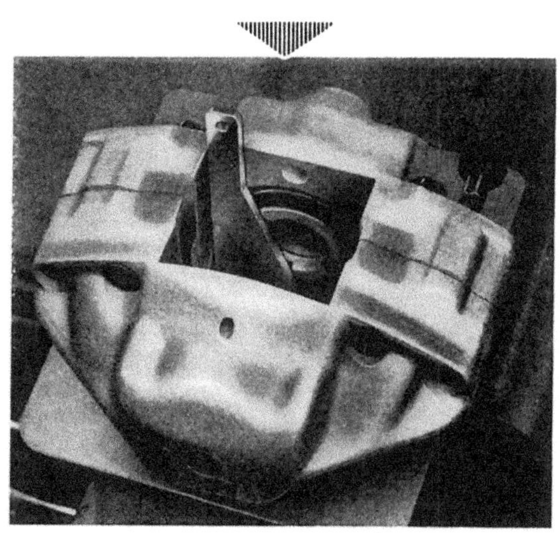

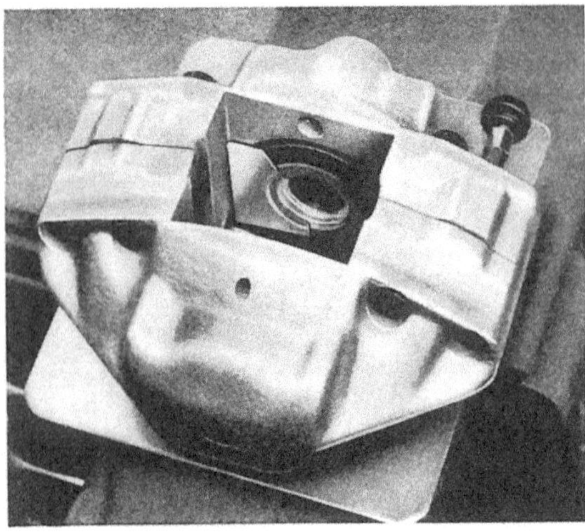

6 - Correctly install piston retaining plate.

7 - Remove the second piston and repeat the operations as for the first piston.

Assembling and Disassembling Brake Caliper

The brake caliper may be disassembled only if the joint between the two housings becomes leaky and the fluid channel "0" rings have to be renewed.

Disassembly

1 - Remove brake caliper housing attaching bolts.

2 - Remove outer housing.

Assembly

On assembly, note the following:

1 - Renew both fluid channel "0" rings.

2 - Use new bolts and nuts.

> **Important**
>
> Pay attention to the various bolt lengths. The shorter bolts are on the outside.

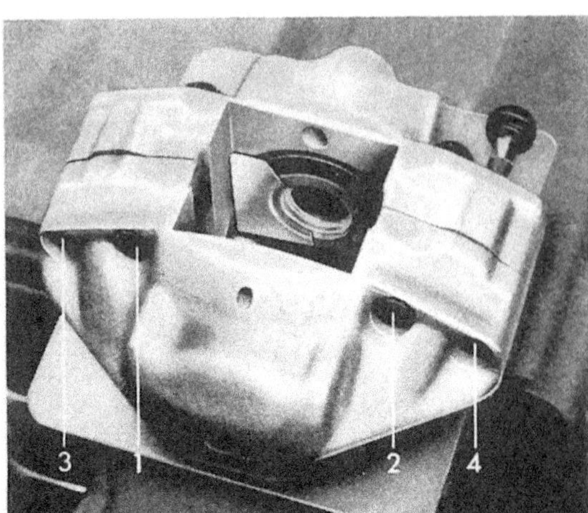

3 - Align housings with each other.

4 - Tighten bolts in the precribed sequence first to a torque of **7 lb. ft. (1 mkg).**

5 - Check the position of the housings to each other.

6 - Tighten the bolts in the same sequence to a torque of **14 lb. ft. (2 mkg).**

Brake Disc

The brake discs should be checked for wear each time repairs are carried out. Worn, scored or cracked brake discs must be renewed. Brake discs, the thickness of which is only .315 in. (8 mm), must also be renewed.

Reconditioning Brake Disc

The brake disc can be reworked only down to a thickness of .335 in. (8.5 mm), noting the following:

The brake disc must be reworked equally on both sides.

The maximum permissible thickness tolerance is .0008 in. (.02 mm) which means that the thickness of the disc may not vary more than this amount when measured at several locations.

Thickness of new brake disc		**.374 in. (9.5 mm) — .004 in. (0.1 mm)**
Minimum thickness of brake disc after reworking		**.335 in. (8.5 mm)**
Wear limit of brake disc thickness		**.315 in. (8 mm)**
Brake disc thickness tolerance	**max.**	**.0008 in. (.02 mm)**
Runout of installed brake disc	**max.**	**.008 in. (.2 mm)**

Note:

Depending on climatic conditions and how long and where the vehicle is left standing, a thin film of rust can form on the brake discs.

For this reason, the brake discs must be checked for signs of corrosion during the pre-delivery inspection. If corroded brake discs are found, they must be cleaned before the vehicle is handed over to the customer.

To smoothen installed brake discs, polishing blocks have been made and are obtainable from the Parts Department. They have a **red** mark on the metal plate to avoid confusion with the friction pads.

Part No. 113 615 259

Instructions for use:

Install polishing blocks in place of the friction pads. According to the amount of corrosion, brake the vehicle several times, with a medium pedal pressure, from a speed of 30 mph. (50 kph) to 20 mph. (30 kph). Make short pauses between each braking to enable the polishing blocks to cool down.

The braking efficiency is not affected in any way with polishing blocks installed. Nevertheless, these runs should be made either on your premises or in a quiet street. Emergency braking or heavy application of the brakes should be avoided so as not to wear the polishing blocks out too quickly.

Prior to reinstalling the original friction pads, carefully remove the dust particles from the brake caliper recesses. Clean polishing blocks with a wire brush.

Note:

To prevent the brake discs from rusting when vehicles are stored for a prolonged period, the discs can be treated with a special preservative. This solution can be obtained in spray cans from our Parts Department under the part number G 5.

Application:

1 - Clean rusty brake discs.
  If rust is very light, the discs can be cleaned by applying the brakes a few times.
  If discs are badly rusted, they can be cleaned with the polishing block (see above).
2 - Lift front of vehicle.
3 - Take both front wheels off.
4 - Remove friction pads on both sides.
5 - Spray outer surface of disc with preservative.
6 - Spray inner surface of disc through the slot in the caliper housing.
  Note: The disc must be turned while spraying.
7 - Spray second disc in same way.
8 - Allow preservative to dry for at least 15 minutes.
9 - Install friction pads.
10 - Fit both front wheels and tighten wheel bolts to **94 lb. ft. (13 mkg).**

Whenever possible, the preservative should be applied at the location where the vehicle is to be stored. If this work is carried out in the workshop, allow a drying time of at least 1 hour at room temperature. On the way to the final location, try to use only the handbrake so that the preservative solution is not rubbed off. If the footbrake has to be used, the discs must be given a fresh coat of preservative.

Removing and Installing Brake Disc

Removal

1 - Remove front wheel.

2 - Detach brake caliper from steering knuckle and hang it on the tie rod with a wire hook.

3 - Remove wheel bearing clamp nut and remove brake disc.

Installation

On installation, note the following:

1 - Check brake disc splash shield for damage.

2 - Adjust front wheel bearing play as prescribed. (See section V).

3 - Install brake caliper.

The attaching bolts and the lock plate must be renewed.

Tighten attaching bolts to a torque of 43 lb. ft. (6 mkg).

Checking Brake Disc Runout

1 - Remove front wheel.

2 - Remove friction pads.

3 - Adjust front wheel bearing play as prescribed.

4 - Insert measuring appliance in brake caliper grooves and tighten with wing nut.

5 - Install dial gauge.

6 - Turn brake disc slowly by hand.

Important

When doing this, the brake disc must be neither tilted nor pushed in an axial direction as this will give a false reading.

The maximum permissible runout is **.008 in. (.2 mm).**

7 - Brake discs, the runout of which is greater, must be renewed.

8 - Remove dial gauge.

9 - Loosen wing nut and remove measuring appliance.

10 - Install friction pads.

11 - Check brakes on road test.

Removal

1 - Remove front wheel.

2 - Detach brake caliper from steering knuckle and hang it on tie rod with a wire hook.

3 - Remove brake disc.

4 - Unscrew splash shield attaching bolts and remove splash shield.

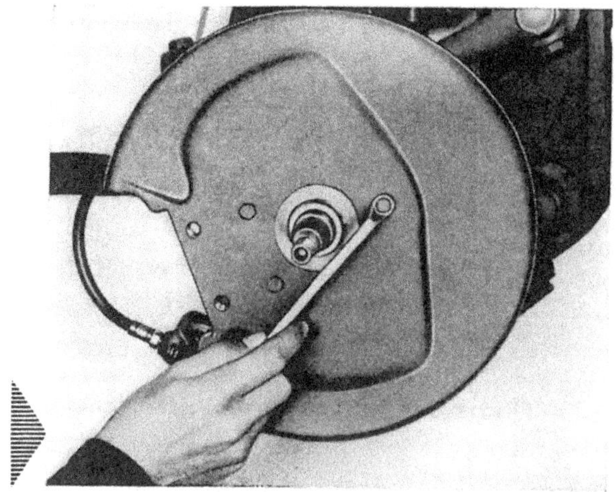

Installation

On installation, note the following:

1 - Clean mating surfaces of splash shield and steering knuckle.

2 - Install splash shield. Renew damaged splash shields.

3 - Check brake disc for wear.

4 - Adjust front wheel bearing play as prescribed.

5 - Install brake caliper. The attaching bolts and the lock plate must be renewed.

Tighten attaching bolts to a torque of 43 lb. ft. (6 mkg).

6 - Check brakes on road test.

Bleeding the Brakes

Two mechanics are required for the bleeding operation.

1 - Place the fluid container on the reservoir so that the reservoir is kept full during the bleeding process.

2 - Take dust cap off bleeder valve on wheel cylinder. Start bleeding at the wheel cylinder farthest from the master cylinder.

3 - Place bleeder hose on wheel cylinder nipple.

4 - Depress and release brake pedal quickly and repeatedly until a resistance shows that pressure has built up in the system.

5 - Hold brake pedal in its lowest position.

6 - Back off bleeder valve about one turn.

7 - Close bleeder valve as soon as the pressure has been released from the system i. e., when no more fluid flows from the bleeder hose.

Important

The brake pedal must not be released until the bleeder valve has been closed.

8 - Repeat the operation on the same wheel until no more air bubbles are visible in the sight tube in the bleeder hose.

9 - Remove bleeder hose and install dust cap.

10 - Repeat operation on each wheel in turn.

Adjusting the Brakes

Front brakes

The front wheel disc brakes do not require adjusting.

Rear brakes

The clearance between the brake shoes and drums increases as the vehicle is used, due to lining wear. When the distance the pedal moves before the brakes are applied becomes excessive, the brake shoes must be adjusted individually.

Contents: S

Lubrication

Maintenance

THIS PAGE INTENTIONALLY
LEFT BLANK

Every attempt is being made to increase the life, reliability and economy of the Volkswagen still further by design, construction and method of production. In spite of this fact, however. maintenance and lubrication remain an indispensable necessity. For this reason the Volkswagenwerk has laid down lubrication instructions which, if strictly followed, will substantially help to maintain the value of the car and its driving qualities.

The Service Booklets contain vouchers listing all lubrication operations to be carried out at specified intervals. A survey of the periodical lubrication services and the lubricants to be used at the various lubrication points is contained in the Lubrication Chart.

It is importance to use lubricants of well-known brands which are in accordance with the specifications laid down by the Volkswagenwerk. These specifications are known to all the major mineral oil firms and make it possible to produce the lubricants to the specifications and to check the products when necessary. In view of the world-wide distribution of the Volkswagen it is obvious that these specifications cover all possible climatic conditions from arctic cold to tropical heat.

By "branded lubricants" we mean the products of well-known mineral oil firms whose names alone are a guarantee for adequate and constant quality. The fact that the Volkswagen requires no special lubricants, makes it possible for VW owners and also VW workshops to use the available commercial lubricants of well-known firms all over the world. On the other hand, this also does away with the necessity of recommending lubricants by name to customers and VW workshops.

A variety of modern tools and equipment are available for carrying out the Lubrication Service as efficiently and economically as possible.

Important — Greases, lubricants and other automotive chemicals contain toxic substances, many of which are absorbed directly through the skin. Read manufacturer's instructions and warnings carefully. Use hand and eye protection. Avoid direct skin contact.

THIS PAGE INTENTIONALLY
LEFT BLANK

Lubricants and Lubricant Specifications

The ever-increasing demand for improved performance and riding qualities has greatly influenced the development of automobile lubricants.

The requirements to be met by engine lubricants vary with the operating conditions. It is very difficult to gauge the exact effect the different operating conditions have on the lubricant. The oil of engines operating at high r.p.m. or under full load for a long period becomes very hot, but this may also be the case with engines operating at low r.p.m. and under part load when atmospheric temperatures are high.

Under the influence of high operating temperatures and oxygen from the air the oil tends to oxidize. This process thickens the oil and may form a varnish-like film on the top portion of the pistons, in the ring grooves and on the valve stems.

Incomplete combustion is often the result of an overrich mixture, or the engine being seldom or never subjected to full load or mainly operating below normal working temperature. These conditions will cause the formation of sludge, resin, acid and tar. The unburnt fuel runs down the cylinder walls, washing away the oil film in cylinders and pistons, and finds its way into the crankcase. This results in dilution of the oil and in the frictional surface of the pistons not being properly lubricated.

The basic requirement of an engine oil is that its lubricating qualities are not affected by extreme operating temperatures. It should resist any chemical reaction and provide efficient protection against corrosion.

Due to changes in temperature every oil tends to alter its viscosity. Increasing temperature causes the oil to thin and consequently influences the adhesion and "body" of the oil film. Decreasing temperature causes the oil to thicken and increases the resistance of the oil to moving parts. It is, therefore, necessary to use an oil, the viscosity of which does not change materially with varying temperatures. When the engine is cold, the oil should be thin enough to facilitate starting and to flow immediately to all lubrication points.

The viscosity is similar to the internal friction of an oil and is distinguished by numbers according to the SAE system (Society of Automotive Engineers). Examples of these numbers are, SAE 30, SAE 20, SAE 10 W. High numbers signify thick oils, low numbers thin oils. The viscosity of an oil has no bearing on its lubricating properties.

VW Lubricant Specifications

The composition and properties of lubricants suitable for the Volkswagen are specified by the factory and laid down in the "S" group Workshop Bulletins. The bulletins also contain information on lubrication points, capacities and viscosities.

Engine Oils

A well-known brand of HD oil should be used for both new and reconditioned VW engines: the viscosity should be selected in accordance with the outside temperature as follows:

8/66

VW Specifications as laid down in Workshop Bulletins	Viscosity	Outside Temperature
No. S-1	SAE 10 W*	Only for districts where the average temperature is below —15° C (5° F)
No. S-2	SAE 20/W 20	in winter
No. S-3	SAE 30	for use in warm weather
No. S-5	SAE 5 W*	below —25° C (—13° F)

SAE 5 W oil should only be used in countries where the average temperature is lower than —25° C (—13° F).

* Do not drive at high speeds for long periods when using these oils if the outside temperature is above 0° C (32° F) or —15° C respectively.

The recommendation of HD oils is solely based on the good results obtained with a view to their lubricating qualities. HD oils contain additives which improve oxidation stability, bearing corrosion preventive properties and the detergent-dispersant characteristics which tend to hold in suspension contaminants which would normally deposit on engine parts. These contaminants will drain out with the oil at the periodical oil changes.

Years of experience with millions of VW engines all over the world have shown clearly that normal HD oils of the specified viscosities are adequate for the VW engine under all operating conditions. They have proved satisfactory from the trouble-free running point of view and also from the consumption angle. This still applies today.

There is, therefore no justification at all for the use of the considerably more expensive multigrade oils in VW engines. This observation is founded on the basis of economy which, for the customer, has always been one of the main reasons for buying a Volkswagen.

In countries where the API classification (American Petroleum Institute) is used, the HD oils suitable for the VW engine are labelled "For Service MS".

Transmission Oils

The transmission and differential gears are combined in the transmission case and both lubricated with SAE 90 hypoid oil. The thinner SAE 80 oil should only be used in countries with arctic climates. In these conditions it can be left in all the year. The thin oil facilitates gear shifting when the gearbox is cold.

Hypoid oils contain, amongst other things, EP (Extreme pressure) additives.

When using hypoid oils in the VW transmission, note the following points:

1 - Do not store transmissions filled with hypoid oil for long periods.

2 - If the vehicle is laid up for a few months during the winter, the hypoid oil should be drained off and the transmission filled with anti-corrosion oil.

3 - Check oil seals for leaks.

Grease

All moving parts of the chassis and body are subject to a certain amount of wear. In order to ensure constant operating and driving safety, whilst at the same time keeping repair costs down to a minimum, it is of great importance to have the vehicle lubricated at regular intervals as laid down in the lubrication chart. Depending of the kind of stress imposed on the individual parts, different greases should be used which conform with the VW Lubricant Specifications.

Workshop Bulletin	VW Description	General Description
S-8	Universal Grease	High Pressure Grease
S-9	Transmission Grease for steering box	Liquid Grease
S-10	Special Grease	High Melting Point Graphite Grease
S-11	Lithium Grease	Multi-purpose Grease
—	Special Lubricants	Lubricants with a Molybdenum Disulphide base

High pressure greases, mostly saponified by calcium, must be cold resistant and water-repellent. The melting point must be above 110° C (230° F).

Transmission grease is saponified with sodium and contains a large percentage of oil. This grease has a low consistency and is very viscous. The drip point should be above 140° C (284° F).

The special grease mentioned in Workshop Bulletin No. S-10 is a grease saponified by sodium and containing approximately 10% graphite. The melting point should be at least 170° C (338° F)

The multi-purpose greases which are recommended by well-known mineral oil firms correspond generally with the lubricant specifications for lithium grease. These have a high melting point (at least 165° C) and should be water-repellent and cold resistant. Moreover, for arctic climates below —25° C (—13° F), the grease must have sufficient cold resistant properties down to —35° C.

The special lubricants mentioned here are of the molybdenum disulphide type, which are manufactured in the form of both oil and paste, and are used in current production in the factory.

Other Lubricants

VW engines which are stored for prolonged periods must be protected against corrosion. Various well-known anti-corrosion oils are available.

Anti-corrosion agents are also required for the underside of the vehicle. There are various types of underseal available for this purpose including one based on a wax solution. This type forms a layer of wax when sprayed on but needs further treatment from time to time to maintain maximum protection.

Undersealing agents with a rubber basis give protection for an almost unlimited period. This coating is impact and friction-proof and thus also provides protection against stone damage and drumming noises.

THIS PAGE INTENTIONALLY
LEFT BLANK

Workshop Bulletin No. S-6

Lubricant specifications
September 1962

Hypoid Transmission oil	SAE 90	Type: Transmission oil with EP additives

Specific gravity	g/cm³/+20° C	< 0.950
Viscosity	° E/+100° C ° E/—17.8° C	> 2.5 < 10000
Pressure resistance test with "Four Ball Apparatus (VKA)"	kg	> 400

Wear test with "Four Ball Apparatus (VKA)"	1^h/15 kg/mm dia.	< 0.3
Water	% permissible	0.0
Ash (oxide ash)	% permissible	> 0.5

Corrosion Test: a - Steel b - Copper Weight loss after wiping with filter paper and normal benzine	48^h/100° C/2% H_2O mg/test strip (50×5×1 mm)	 No rust deposit, discoloration permissible. Black discoloration or film permissible. < 10
Niemann Test: a - Sudden rise into high wear range b - Specific wear	 Load stage mg/HPh	 > 12 < 0.20
Reaction to rubber: Swelling	 % permissible	Rubber parts in contact with the above oil (oil seals) must not exceed the value given below after a period of 100^h/130° C > 100

Important: Do not store rear axles fitted with hypoid oil for prolonged periods.

Note: If the vehicle is laid up for a few months, the hypoid oil should be drained off and the transmission filled with anti-corrosion oil.

Type	Lubrication point	Quantity		Remarks
1, all Models except Standard Sedan	Transmission	Capacity Refill	3.0 litres 2.5 litres	all the year (In countries with arctic climates: SAE 80)

Workshop Bulletin No. S-7

Lubricant Specifications

April 1961

Hypoid Transmission oil	SAE 80	Type: Transmission oil with EP additives

Specific gravity	g/cm^3/20° C	< 0.930
Viscosity	° E/+100° C ° E/—17.8° C	> 1.90 < 3000
Pressure resistance test with "Four Ball Apparatus (VKA)"	kg	> 400

Wear test with "Four Ball Apparatus (VKA)"	1^h/15 kg/mm dia.	< 0.30
Water	% permissible	0.0
Ash (oxide ash)	% permissible	< 0.5

Corrosion Test a - Steel b - Copper Weight loss after wiping with filter paper and normal benzine	48^h/100° C/2% H_2O mg/test strip (50×5×1 mm)	 No rust deposit, discoloration permissible. Black discoloration or film permissible. < 10
Niemann Test: a - Sudden rise into high wear range b - Specific wear	 Load stage mg/HPh	 > 12 < 0.20
Reaction to rubber: Swelling	 % permissible	Rubber parts in contact with the above oil (oil seals) must not exceed the value given below after a period of 100^h/130° C. > 100

Important: Do not store rear axles fitted with hypoid oil for prolonged periods.

Note: If the vehicle is laid up for a few months, the hypoid oil should be drained off and the transmission filled with anti-corrosion oil.

Type	Lubrication point	Quantity		Remarks
1, all Models	Transmission	Capacity	3.0 litres	all the year
		Refill	2.5 litres	Only in countries with arctic climates

Workshop Bulletin No. S-8

Lubricant Specifications
December 1962

Universal Grease	Type: High Pressure Grease, Cold-resistent and Water-repellent

Saponified	by calcium
Melting point °C	> 110
VW strainer test 24h/+90° C; 24h/+20° C %; 24h/+90° C percolating; 24h/+20° C	0
Consistency mm WS/+20° C; mm WS/— 5° C; mm WS/—35° C	500— 600; 900—1100; 5000—6000
Consistency after strainer test mm WS/+20° C	600— 800

Hot water test 15 minutes/+90° C	No change
Ash % permissible	< 3.0
Water % permissible	< 0.5
Mech. impurity %	0.0
Neutralization No. mg KOH/g	< 0.3
Storage stability	The product must not separate or harden after storage 6 months at a room temperarute of approx. 20° C (68° F)

Type	Lubrication Points
1, all Models	Front axle tubes, king pins, torsion arm link pins, fuel pump, needle bearing in flywheel gland nut, gear and hand brake levers, clutch cable, accelerator cable, heating control cables, door and hood locks, window regulator.

Workshop Bulletin No. S-9

Lubricant Specifications
August 1964 — 1st Edition

Transmission grease for steering gear	Type: Liquid grease

Color	natural
Saponified	by sodium
Drip point °C	> 140

Consistency mm WS/+20° C; mm WS/—25° C; mm WS/—35° C	150±50; 5000±1000; 10000±1000
VW Strainer test 2×24 hours/90° C % percolation	maximum permissible 50

Type	Lubrication point
1, all Models	Steering box (roller steering only)

Workshop Bulletin No. S-10

Lubricant Specifications
October 1961

Special Grease	Type: High Melting Point Graphite Grease

Saponified		by Sodium
Melting point	° C	> 170
Consistency	mm WS/+20° C	500—600
Graphite	% permissible	10 ± 1
Water	% permissible	< 0.5

Type	Lubrication Points
1, all Models	Hydraulic brakes: Brake shoe adjusting nuts. Linkage in heater junction box

Workshop Bulletin No. S-11

Lubricant Specifications
December 1962

Lithium grease	Type: Multi-purpose grease Water-repellent, coldresistant

Saponified		Lithium
Melting point	° C	> 170
Consistency	mm WS/+20° C mm WS/—35° C	600—800 < 10000
VW strainer test	% percolating % percolating	0
Consistency after strainer test	mm WS/+20° C	< 1000
Oxide ash	% permissible	< 0.50
Water	% permissible	< 0.1

Hot water test 15 minutes/+90° C	No change
Neutralization No. mg KOH/g	< 0.2
Impurity	The grease must be free from all foreign and abrasive particles
Storage stability	The product must not separate or harden after 6 months storage at a temperature of approx. 20° C (68° F)

Type	Lubrication point
1, all Models	Front wheel bearings fiber block on distributor contact, bush for starter pinion

Engine Oils

A well-known brand of HD oil should be used for both new and reconditioned VW engines: the viscosity should be selected in accordance with the outside temperature as follows:

VW Specifications as laid down in Workshop Bulletins	Viscosity	Outside Temperature
No. S/1	SAE 10 W	below 0° C (32° F)
No. S/2	SAE 20 W/20	from 0° C to 30° C (32° F to 86° F)
No. S/3	SAE 30	above 30° C (86° F)
No. S/4	SAE 10 W-30	all the year
No. S/5	SAE 5 W	below —25° C (—13° F)

SAE 5 W oil should only be used in arctic countries when the average temperature is lower than —25° C (—13° F).

Multigrade oils, i. e., oils which have HD properties and cover several SAE viscosity ranges, are also suitable for the VW engine.

The recommendation of HD oils is solely based on the good results obtained with a view to their lubricating qualities. HD oils contain additives which improve oxidation stability, bearing corrosion preventive properties and the detergent-dispersant characteristics which tend to hold in suspension contaminants which would normally deposit on engine parts. These contaminants will drain out with the oil at the periodical oil changes.

In countries where the API classification (American Petroleum Institute) is applied, the HD oils suitable for the VW engine are labelled "For Service MS".

Transmission Oils

The transmission and differential gears are combined in the transmission case and both lubricated with SAE 90 hypoid oil.The thinner SAE 80 oil should only be used in countries with arctic climates. In these conditions it can be left in all the year. The thin oil facilitates gear shifting when the gearbox is cold.

Hypoid oils contain, amonget other things, EP (Extreme pressure) additives.

When the transmission is stored for a prolonged period the hypoid oil can cause corrosion, and may lead to a premature hardening of the main drive shaft oil seal. The following points should, therefore be observed:

1 - Transmissions which are to be stored for a prolonged period of time should not be filled with hypoid oil.

2 - If the vehicle is laid up for a few months, the hypoid oil should be drained off and the transmission filled with anti-corrosion oil.

3 - Check oil seals for leaks.

Grease

All moving parts of the chassis and body are subject to a certain amount of wear. In order to ensure constant operating and driving safety, whilst at the same time keeping repair costs down to a minimum, it is of great importance to have the vehicle lubricated at regular intervals as laid down in the lubrication chart. Depending on the kind of stress imposed on the individual parts, different greases should be used which conform with the VW Lubricant Specifications.

Workshop Bulletin	VW Description	General Description
S-8	Universal Grease	High Pressure Grease
S-9	Transmission Grease for steering box	Liquid Grease
S-10	Special Grease	High Melting Point Graphite Grease
S-11	Lithium Grease	Multi-purpose Grease
—	Special Lubricants	Additives with a Molybdenum Disulphide base

High pressure greases, mostly saponified by calcium, must be cold resistant and water-repellent. The melting point must be above 110° C (230° F).

Transmission grease is saponified with sodium and contains a large percentage of oil. This grease has a low consistency and is very viscous. The drip point should be above 140° C (284° F).

The special grease mentioned in Workshop Bulletin No. S-10 is a grease saponified by sodium and containing approximately 10% graphite. The melting point should be at least 170° C (338° F).

The multi-purpose greases which are recommended by well-known mineral oil firms correspond generally with the lubricant specifications for lithium grease. These have a high melting point (at least 165° C) and should be water-repellent and cold resistant. Moreover, for arctic climates below —25° C (—13° F), the grease must have sufficient cold resistant properties down to —35° C.

The high quality greases mentioned here are of the molybdenum disulphide type, which is manufactured in the form of both oil and paste, and are used in current production in the factory.

Other Lubricants

VW engines which are stored for prolonged periods must be protected against corrosion. Various well-known anti-corrosion oils are available.

Anti-corrosion agents are also required for the underside of the vehicle. There are various types of underseal available for this purpose including one based on a wax solution. This type forms a layer of wax when sprayed on but needs further treatment from time to time to maintain maximum protection.

Under sealing agents with a rubber basis give protection for an almost unlimited period. This coating is impact and friction-proof and thus also provides protection against stone damage and drumming noises.

Lubrication Chart

The lubrication chart summarizes the operations which should be carried out at regular intervals. Each lubrication service is covered by a voucher in the Service Booklet. When carrying out the operations listed on the reverse side of the vouchers, the following hints should be observed.

Engine Oil Changes

The oil should only be drained when the engine is warm. The oil is drained by removing the plug in the oil strainer cover. The oil is changed at 500 km and then at 5000 km (300 and 3000 miles) and then regularly every 5000 km (3000 miles).

Note:
For the VW 1200 A as produced from August 1965 and the 1300, the new oil change intervals on page S/3-9 are valid.

Since the oil strainer collects a large amount of deposits it should be removed and cleaned thoroughly at every oil change. When installing, both gaskets and the copper washers on the cap nuts should always be renewed.

Before refilling the engine with oil, replace the drain plug and tighten to a torque of 3 mkg (22 ft. lbs.).

8/66

Important

The quantity of oil used for the initial filling and the quantity to be used at oil changes is 2.5 liters (5.3 US pints, 4.4 Imp. pints) for all VW engines.

The viscosity grades for the initial filling are as follows:

from October to March	SAE 20/W 20
from April to September	SAE 30

Due to slight dimensional variations, it can sometimes occur that after refilling the engine with the pro per amount of oil the oil level is slightly above the top mark on the dipstick. The oil circulation and the lubrication of the engine are not affected if the oil level is a few millimeters above the top mark, or between the two marks after the vehicle has covered a few hundred miles.

When changing the oil, the same type, and if possible the same brand of oil should be used. It is, therefore, advisable to attach a tag to the engine which states the brand, type and viscosity of the oil used.

The use of different types of oils is not recommended. Oils of the same type, but of different brands should, if possible, not be mixed. Oil of the same type and brand, but of different viscosities may be mixed if required during seasonal changes.

The oils "For Service MS" (HD oils), should be changed every 5000 km (3000 miles). In the case of vehicles which are operated for short distances and in city traffic only, it is good practice, especially in winter, to carry out an additional oil change, say every 2500 km (1500 miles). However, if only a few hundred miles are covered per month under these conditions, it is advisable to change the oil every 6—8 weeks. The recommended intervals for oil changes are based on extensive experience, and take such factors as operating conditions and operational safety into consideration.

In arctic climates where temperatures are below —25° C (—13° F), it is advisable to change the oil every 1250 km (750 miles). It is advisable to change the oil more often in the winter on a vehicle which is only covering a few hundred miles a month. The oil strainer should be cleaned each time.

In the case of vehicles which cover less than 10000 km (6000 miles) a year, at least two oil changes should be carried out during the year.

As the oil change intervals have been increased to 5000 km (3000 miles), the oil level should be checked regularly every 2500 km (1500 miles) at the lubrication services. Due to detergent characteristics of HD oils, it is not necessary to flush the engine when oil changes are carried out.

In very dusty climates the air cleaner should be serviced more frequently.

Air Cleaner

The oil bath air cleaner should be checked every 5000 km as specified in the lubrication chart and the lower part cleaned if necessary. In very dusty conditions the cleaner must be checked more frequently, even daily if necessary. If the check reveals that there is only 4—5 mm (.16—.2") of oil above the sludge layer, the lower part should be carefully cleaned and filled to the mark with fresh SAE 20 engine oil.

If the oil level is more than 4 mm below the mark but there is very little dirt in the cleaner, fresh oil can be added to bring the level up to the mark.

The top part does not normally need cleaning with gasoline or other solvents. However, if the filter insert has become so dirty due to overdue cleaning or oil shortage that the air inlet holes on the underside are partly blocked, the encrusted dirt should be removed, preferably with a chip of wood.

Note:

The upper part of the air cleaner should always be laid down with the opening downwards.

Note:

From August 1965, the oil bath air cleaner is only checked every 10000 km (6000 miles).

Transmission Oil

The transmission oil should be changed every 50000 km (30000 miles) whilst the oil is still warm. Both drain plugs should be removed for this purpose. The first oil change is at 500 km (300 miles) and the next change at 5000 km.

Note:

From August 1965, the second transmission oil change at 5000 km has been deleted.

At every oil change, both magnetic oil drain plugs must be cleaned thoroughly with a brush, gasoline and compressed air. This regular cleaning is essential as the magnets can only hold a limited amount of material.

The initial filling quantity of the fully synchronised transmission is 3.0 liters (5.3 Imp. pints, 6.3 U.S. pints). The refill requirement is 2.5 liters (4.4 Imp. pints, 5.3 U.S. pints).

The oil should be checked as laid down in the lubrication chart and oil added if found necessary. The oil level should be up to the bottom edge of the filler hole.

Various amounts of oil will remain in the half axles depending whether the oil change is carried out with the rear axle horizontal or in an unloaded position. After being filled with the proper amount of oil, the oil level in the transmission case can vary slightly. This however, does not affect the lubrication of the transmission and rear axle. The oil change can be carried out with the rear axle in either of the above-mentioned positions.

It should be noted that in some cases the oil fills very slowly. If the oil is poured in too quickly it may overflow and lead one to believe that the correct level has already been reached, whereas in reality, only about 1.0—1.5 liters (2.1—3.2 U.S. pints, 1.8—2.6 Imp. pints) have been put in. It is great of importance the life and silent operation of the rear axle that the correct amount of oil is used to when refilling. It is, therefore, good practice to wait a few minutes after having poured in 1.5 liters and then to put in the rest.

SAE 90 transmission oil can generally be used in the winter and need not be replaced by another grade of oil. In countries with arctic climates it is necessary, however, to use the thinner SAE 80 oil. This oil can be used all the year round.

Steering Gear

The steering gear (roller type only) is filled with a liquid transmission grease. Checking and topping-up of the steering gear is thus unnecessary. The grease need only be changed when the steering has been dismantled. A grease which conforms to the lubricant specifications should be used.

Front Axle

Good springing and steering properties con only be ensured if all moving parts are properly lubricated: Lubrication should be carried out at the following points:

4 Lubrication points on the axle beams,

2 Lubrication points on each torsion arm link for torsion arm link pins and king pins.

Under normal operating conditions it is sufficient to grease the front axle every 2500 km (1500 miles). If, however, the vehicle is used mostly on bad roads, or the total distance covered each year is less than 12000 km (7500 miles), approx. 1000 km (600 miles) per month, it is advisable to additionally grease the torsion arm links every 1250 km (750 miles).

Note:

From August 1965, there are only four grease nipples on the front axle. These are for the torsion arm bearings and should be lubricated every 10000 km (6000 miles). Under unfavourable conditions or if the yearly mileage is less than 12000 km, it is advisable to lubricate the axle every 5000 km (3000 miles).

Important

The greasing should be carried out when the front axle is unladen, i. e. when raised. This ensures that the grease reaches the lubrication points in adequate quantities.

Before lubricating, the grease nipples should be cleaned. Damaged, blocked up, or missing nipples should be replaced. Continue with the greasing until the fresh grease begins to emerge at the edges of the lubrication points.

Important

Grease or oil must not come in contact with either the tires or brake hoses. Even small amounts should be wiped off immediately.

Note:

The maximum pressure premissible when greasing ball joints is

400 kg/sq. cm (5700 lbs./sq. in.)

There is no danger of this pressure being exceeded when using compressed air operated lube units as there is always a pressure relief valve or some other safety device in the system. With the foot operated lube unit however, it is possible through incorrect use, such as jerking on the foot pedal with the full weight of the body, to achieve pressures up to 1000 kg/sq. cm (14200 psi). Pressures of this nature can lead to the parts receiving damage which will not be immediately apparent, but might cause an accident at a later date.

Front Wheel Bearings

The front wheel bearings are packed with grease on assembly. They should be cleaned and packed with lithium grease of the correct specification during the inspection at 50000 km (30000 miles) and each time they are removed. Observe the following points:

1 - Remove all grease from brakes, wheel spindles and the chamber in the brake drum between the bearing seats.

2 - Remove old grease from rings, cages, and the balls. Rinse out the bearings.

3 - Use the specified type of lithium grease when lubricating the bearings. It should be inserted into the cage and between the balls. Bearing seats and rings should be lightly greased. Use only well-known brands of grease.

4 - Pack the space in the brake drum between the bearings with grease.

5 - The hub caps should be free of grease.

Important

Never mix greases of different brands and types as this is liable to affect the lubricating properties and cause damage.

Too much grease is also harmful.

The amount of grease required for each wheel is approximately 50 grams.

6 - Adjust bearings properly.

Door and hood locks, door hinges

Door check rods and front hood lock should be greased lightly. Door and hood hinges should be cleaned first and then oiled. The door hinges should be oiled at each lubrication service.

The door hinges of all Karmann Ghia models require no maintenance and need, therefore, not be lubricated.

Striker plates should be lightly greased on the friction surfaces with molybdenum-disulphide based paste. Excess paste should be cleaned off so as to avoid any stains to clothes when getting in or out of the car.

The door lock cylinders should be lubricated with graphite. The key should be dipped in graphite, inserted into the lock and turned to and fro a few times.

If the door trim panels have been removed during a repair, it is advisable to lubricate the moving parts of the window mechanism with universal grease.

Carburetor linkage

The carburetor linkage must be oiled at the prescribed regular intervals in order to ensure correct operation. A few drops of oil should be applied at the following locations: Choke valve shaft and fast idle cam, throttle valve shaft, accelerator cable swivel pin, connector rod and lever for accelerator pump.

Distributor

There must always be a small amount of grease on the fiber block of the contact arm in order to lubricate the cam lobes on the distributor shaft. Every 5000 km this point must be checked to see if it requires cleaning and fresh grease. Only lithium grease should be used and none of it must be allowed to get on to the points as otherwise trouble will be experienced with the ignition.

Front Seats

The seat runners of the front seats should be lightly greased. Clean the runners and apply the grease to the sliding surfaces.

Gearshift Lever

The gearshift lever is not greased in the course of the lubrication services, but if it has been removed it should be greased, if necessary. All moving parts, such as gearshift lever ball socket in shifting rod, spring and stop plate should be well greased with universal grease. At the same time the plastic bush in the shift rod guide should also be lubricated with universal grease.

Additional Lubrication Operations on VW Convertible and Karmann Ghia Convertible

In addition to the aforementioned lubrication operations which should also be carried out in the same manner and at the same intervals on the Convertible models, the joints of the top linkages should be lubricated as well. To do so, the top should be opened. The joints should be cleaned first, and then lubricated, making sure that oil does not come in contact with the cover.

Special Cold Weather Hints

Cables

In winter the chassis is exposed to the elements and open to the effects of the cold and dampness. In order to avoid the trouble arising out of frozen cables and such like, it is advisable to lubricate the guide tubes for the carburetor, clutch, and heating cables with a cold-resistant grease before the cold weather starts.

The clutch cable adjusting nut situated in the concave socket of the clutch operating lever should be greased with universal grease as required, but at least once a year before the cold season begins.

Frame

During the summer, the underside of the vehicle does not require any special treatment. In winter, however, the underside may tend to rust because the paint cannot resist indefinitely the damaging influence of slush and snow mixed with salt, gravel and sand which is being used for ice and snow thawing in ever increasing quantities from year to year.

Two different types of anti-corrosion agents are available:

1 - **Wax-based underseal**

This type gives protection for a limited period depending on actual conditions and the coating must be repaired or resprayed from time to time. Apart from washing the underside, no further preparatory work is required before applying this compound. Some brands can even be sprayed on when the chassis is still wet.

2 - **Rubber-based underseal**

This type of underseal gives protection for an almost unlimited period. The coating also gives protection against stone damage, has very good adhesion and helps to damp down noise.

The underside of the vehicle must be cleaned very thoroughly and all grease and oil removed before this coating is applied.

If a vehicle with wax underseal is subsequently to be sprayed with a rubber-based underseal, it is not necessary in every instance, to remove the wax film. Teroson underseal from Messrs. Teroson-Werke GmbH, Heidelberg can be sprayed straight on to the wax coated surface. If undersealing compounds from other firms are to be used, the instructions given for preparatory work should be followed closely.

The treatment of Volkswagen vehicles is described in a Technical Recommendation of group S from December 1963.

Lubrication Chart

(from August 1964)

At km/miles 500 km 300 miles	2500 km 1500 miles	5000 km 3000 miles	No.	Lubrication points	Every
			1	Engine: Check oil level, top up if necessary	**2500 km 1500 miles**
			2	Lubricate torsion arms	
			3	Lubricate king pins and torsion arm link pins	
			4	Oil door hinges	
			5	Engine: Change oil	**5000 km 3000 miles**
			6	Clean oil strainer	
			7	Check air cleaner, clean lower part if necessary	
			8	Transmission: Check oil level, top up if necessary	
			9	Oil carburetor linkage	
			10	Lubricate door and hood locks	
			11	Transmission: Change oil, clean magnetic oil drain plugs	**50 000 km 30 000 miles**

Lubricants

Lubricant	Lubrication points	Specifications			
		Temperature	°C	°F	
Engine oil (HD oil for spark ignition engines)	Engine, oil bath air cleaner, carburetor controls, door hinges	above	30	86	SAE 30
		from up to	0 30	32 86	SAE 20 W/20
		below	0	32	SAE 10 W
		below	—25	—13	SAE 5 W
Hypoid gear oil	Transmission case	SAE 90 all the year*			
Transmission grease	Steering gear	Liquid grease			
Universal grease	Torsion arms, king pins, torsion arm link pins, gearshift lever, door and hood locks	**Cold resistant water-repellent grease**			
Lithium grease	Front wheel bearings, breaker arm fiber block	**Multi-purpose grease**			

***) SAE 80 all the year in countries with arctic climates**

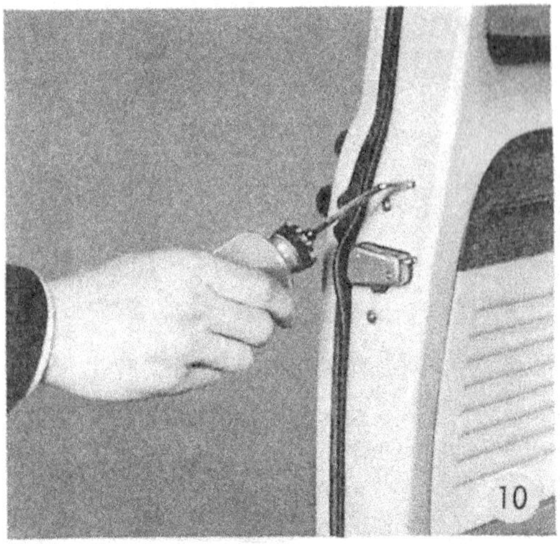
10
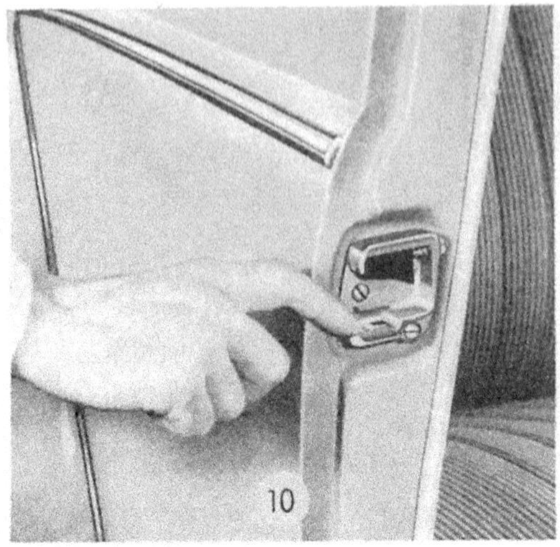
10

3
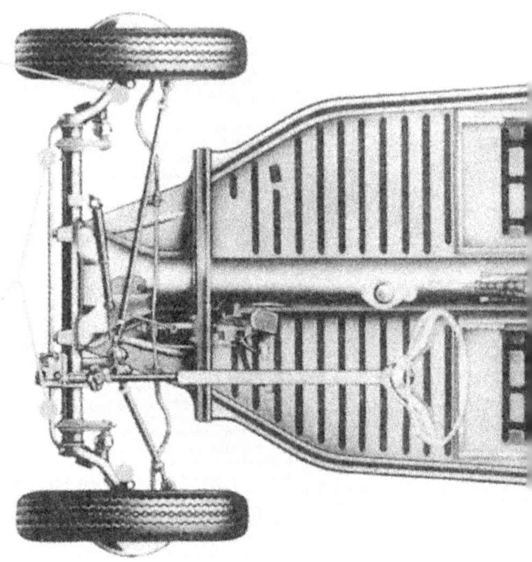

2
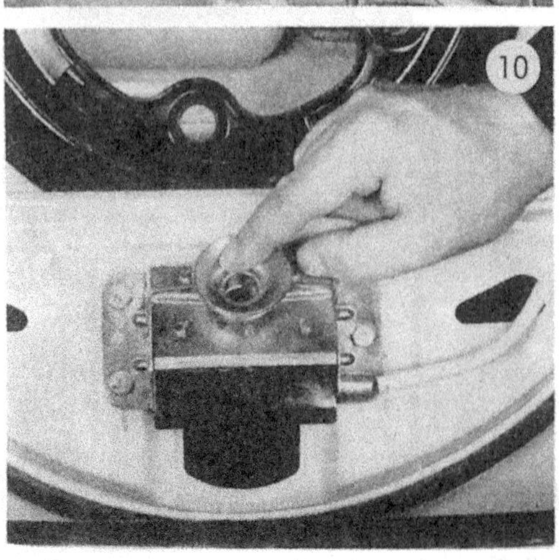
10
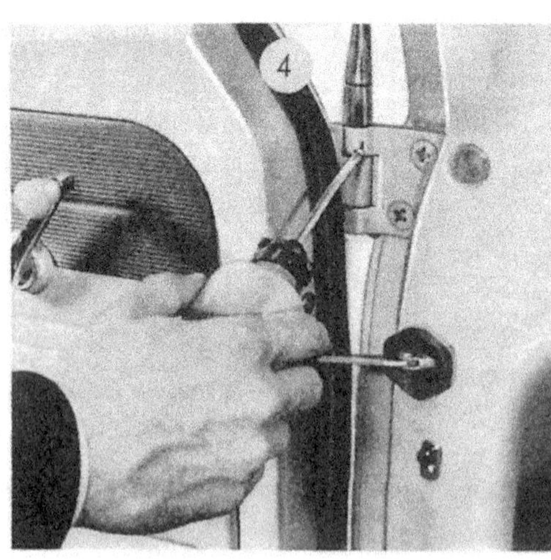
4

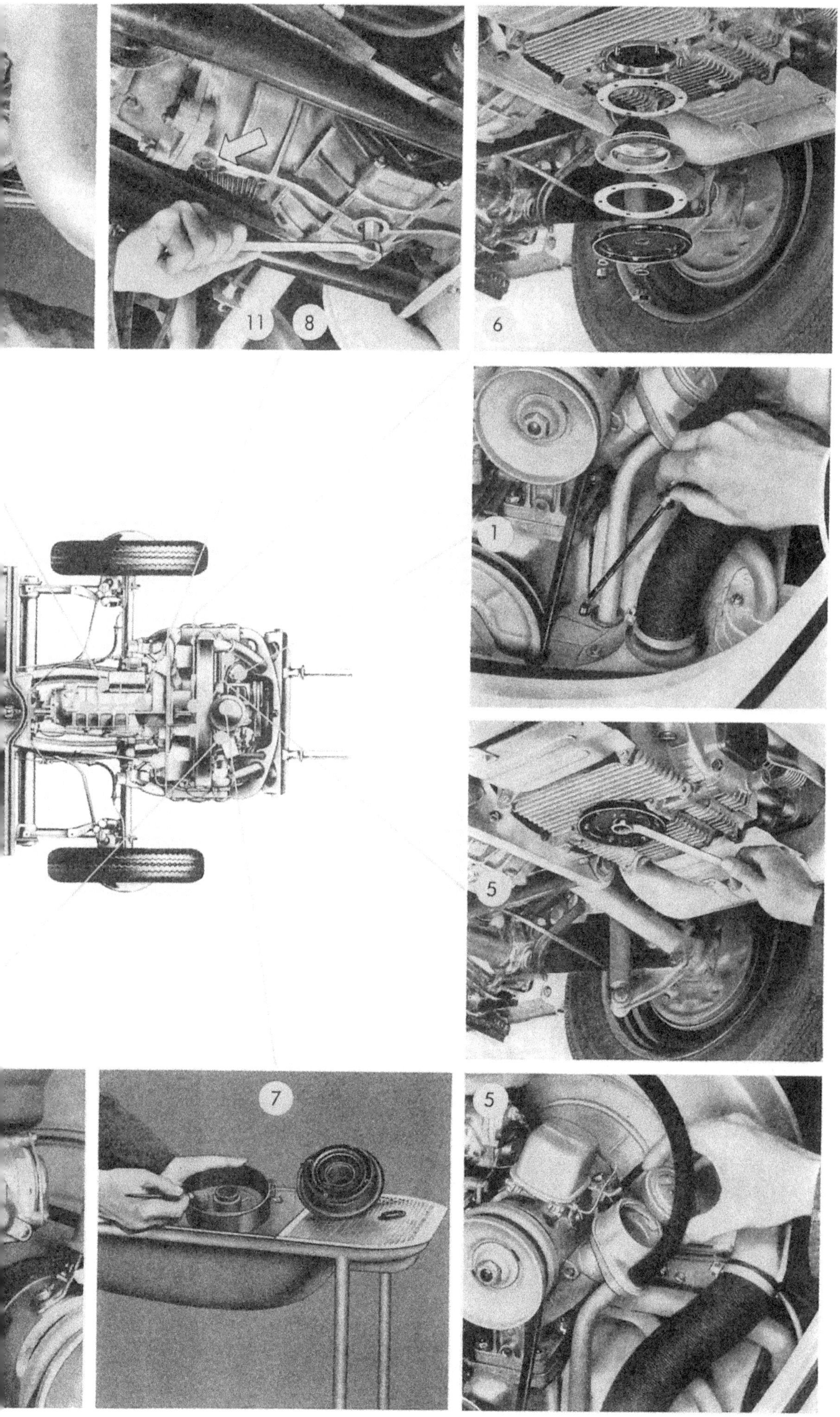
11
8
6
1
5
7
5

THIS PAGE INTENTIONALLY
LEFT BLANK

Lubricant Additives

Various upper cylinder lubricants and special lubricating agents are offered for use on the Volkswagen. Sometimes a certain kind of additive is needed, for instance under very severe operating conditions where the normal type of lubricant is not sufficient.

It is generally agreed that the ordinary petrol engine, and especially the VW engine does not require any additives to obtain its performance and life whilst maintaining the normal average fuel consumption. Particular cases where the pressure and temperature is very severe do not arise in the case of the VW engine. As a matter of fact experience has shown that the oils we recommend are suitable for the VW engine under any operating conditions.

Tests carried out so far with various additives have shown that in the case of a properly maintained VW engine, neither the performance nor the fuel consumption has been increased in such a way as to justify the additional costs incurred when using the additives regularly. For this reason, as far as the VW engine is concerned, any type of additive, be it mixed with the fuel or engine oil, is superfluous.

The question as to whether additives can be used with HD lubricants has not yet been clarified to the extent that their use with HD oils can be recommended without hesitation. For that particular reason it is stated in the Instruction Manual for the Volkswagen that no additives of any kind should be mixed with HD oils.

Bypass Oil Filter for the VW Engine

There are no objections to the subsequent installation of bypass oil filters which have been tested by the Volkswagenwerk. However, these filters are only of advantage when the vehicle is operating in extremely dusty regions. Under such adverse conditions the use of a bypass oil filter has its advantages as far as the life of the engine is concerned.

The best protection which can be afforded the engine when operating in dusty regions is the proper care of the oil bath air cleaner. The engine can be protected from excessive wear by frequently checking the oil level in the air cleaner and cleaning it. If the oil level has dropped it should be topped, but only as far as the oil level mark whilst at the same time making sure that the gaskets fit properly. The oil bath air cleaner must be cleaned in any case when the oil is not visible above the layer of dust in the bowl of the cleaner itself. If necessary, an oil bath air cleaner connected in series to a cyclone filter as shown in the Technical Bulletin K/7, can be subsequently installed.

It is, however, rather difficult to state what extent the oil change periods can be increased when a bypass filter is being used. For certain reasons, it is of great importance that during the guarantee period, the oil change intervals are observed. It is not advisable to extent the oil change intervals during the cold season.

The lubricating properties depend not only on the cleanliness of the oil as regards foreign matter, but are also influenced by chemical reactions. Any particular affects on the lubricating properties of an oil are of greater importance during the winter and are not counteracted by the use of a filter. Therefore, the oil should be changed at the prescribed intervals.

Tool Rack foor Tool Trolley VW 678/1-3

The tools required during the lubrication and maintenance service can be accomodated on a tool rack. By this means the mechanic has all his tools ready to hand.

The tool rack can only be used in conjunction with a tool trolley. It can be quickly and easily installed.

A description and the necessary drawings for the local manufacture of the tool rack appear in the 6th supplement of the publication "Local Manufacture of Workshop Equipment".

Maintenance Chart

500 km 300 miles	5000 km 3000 miles	Operation	Every
■		Check nuts and bolts on chassis, body, engine rear axle, front axle, and steering for security	
■		Check rear axle and engine for leaks	
■		Check tire pressures and security of wheel mounting bolts	
■	■	Check fan belt	
■	■	Clean fuel pump filter	
■	■	Clean breaker points, lubricate distributor, check contact breaker gap and timing	
■	■	Check valve clearance	
	■	Check spark plugs and compression	
	■	Check exhaust system for damage. Check rubber valve for crankcase ventilation	
■	■	Check clutch pedal free-play	**5000 km 3000 miles**
■	■	Check torsion arm link pins, tie rod dust seals, security of tie rods and steering damper, toe-in	
■	■	Check steering adjustment	
	■	Check tires for wear and damage, tire pressures	
■	■	Check brake system for leaks and damage. Check brake fluid level and adjustment of hand and foot brakes	
	■	Check thickness of brake linings	
■	■	Check battery, operation of electrical system, headlight adjustment	
■	■	Road test: Check foot and hand brake operation, heating, idling adjustment	
		Clean, grease and adjust front wheel bearings	**50000 km 30000 miles**

Note: On vehicles with automatic clutch the following operations have to be carried out every 25000 km (15000 miles):
1 - Clean control valve filter
2 - Clean and adjust gearshift lever contacts

THIS PAGE INTENTIONALLY
LEFT BLANK

Contents: T

THIS PAGE INTENTIONALLY
LEFT BLANK

Engine

Design	4-cycle, air-cooled, internal-combustion engine in unit with transmission and differential in rear of car
Number of cylinders	4
Arrangement of cylinders	Two pairs, horizontally opposed (flat four)
Bore	77 mm (3.031")
Stroke	64 mm (2.520")
Piston displacement	1192 cc (72.74 cu. ins.)
Compression ratio	7.0:1
Maximum performance (DIN)	34 bhp at 3600 rpm.
(SAE, with air cleaner)	41.5 bhp at 3900 rpm.
Maximum torque (DIN)	8.4 mkg (61 ft. lbs. at 2400 rpm.)
(SAE, with air cleaner)	9.0 mkg (65 ft. lbs. at 2400 rpm.)
Total weight, dry	Approx. 108 kg (238 lbs.)
Crankcase	Two pairs, horizontally opposed (flat four) through the centre lines of both the main bearings and the camshaft bearings
Cylinders	Individual cylinders cast from special grey cast iron, finned
Cylinder heads	Cast in pairs, of aluminium alloy, finned for air cooling
Valve seat inserts	Shrunk in position, made of sintered steel alloy
Valve guides	Shrunk in position, made of bronze
Spark plug thread	Cut in the cylinder head
Crankshaft	High quality steel stamping, four plain bearings
Main bearings 1, 3 and 4	Sleeve-type bearings of aluminium alloy
Main bearing 2 (center)	Split shell bearing of aluminium alloy
Main bearings 1–3	55 mm diameter
Main bearing 4	40 mm diameter
Flywheel	Steel forging, with integral starter gear ring
Connecting rods	H section steel forgings
Connecting rod bearings	3-layer, steel-backed
Piston pin bearings	Pressed-in bronze bushings
Pistons	Of aluminium alloy with steel inserts
Pistons pins	Fully floating, secured by circlips
Piston rings	2 compression rings 1 oil control ring
Valve actuating mechanism	1 camshaft situated below crankshaft, valves operated via push rods, cam followers and rocker arms
Camshaft	Of grey cast iron, runs in three bearings machined direct in crankcase
Camshaft drive	By helical gears from the crankshaft
Valves	1 intake valve and 1 exhaust valve for each cylinder
Exhaust valve	Heat-resistant seating surface
Arrangement	Overhead
Clearance: Intake	0.10 mm (.004") } up to a max. oil temperature of 50° C (122° F)
Exhaust	0.10 mm (.004") } up to a max. oil temperature of 50° C (122° F)
On engines **without** sticker	
Intake	0.20 mm (.008")
Exhaust	0.30 mm (.012")
Valve springs	1 spring per valve

Valve timing with a valve clearance of 1 mm (.04")

Intake opens	6° before TDC.
Intake closes	35° 30′ after BDC.
Exhaust opens	42° 30′ before BDC.
Exhaust closes	3° after TDC.
Cooling system	Air cooling by fan on generator armature shaft
Fan drive	From crankshaft by V-belt
Ratio: crankshaft/generator	approx. 1.8 : 1
Cooling air intake	Thermostat-controlled
Amount of cooling air	500 liters (18 cubic feet) per second at 3600 engine rpm.
Lubrication	Pressure feed lubrication by gear pump
Oil cooling	Oil cooler situated in cooling air stream
Oil pressure control	By warning light
Ignition	Battery ignition
Ignition coil	Bosch TE 6 B 4 / VW 111 905 105 F } intermittently
Ignition distributor	Bosch 111 905 205 M / VW 211 905 205 L } intermittently
Spark timing	10° before TDC.
Firing order	1–4–3–2
Spark advance	By vacuum control
Breaker point gap	0.4 mm (.016")
Spark plugs	14 mm thread, heat range 175 Bosch W 175 T 1 Beru 175/14 Champion L 87 Y or plugs with similar values from other manufacturers
Spark plug gap	0.6–0.7 mm (.024" to .028")

Clutch

Design	Single disc, dry, K 10 (Fichtel and Sachs)
Pedal free-play	10 to 20 mm (.4" to .8")
Total lining area	268 sq. cm (41.6 sq. ins.)

Fuel System

Carburetor	Downdraft, SOLEX 28 PICT-1, with accelerator pump and automatic choke
Venturi	22.5 mm diameter integral
Main jet	122.5
Air correction jet:	
Sedan and Convertible	130 y (with emulsion tube)
Karmann Ghia Models	145 y (with emulsion tube)
Pilot jet	55 g
Pilot jet air bleed	2.0 mm diameter
Pump jet	0.5
Power fuel jet	1.0
Float needle valve	1.5 mm diameter
Float weight	5.7 grams (.20 oz.), plastic material
Pump feed	approx. 1.1–1.4 cc per stroke
Air cleaner	Oil-bath type with pre-heater tube
Fuel pump	Diaphragm type, mechanically operated } at 3400 rpm.
Feed pressure	max. 0.2 kg/sq. cm (2.8 psi) } at 3400 rpm.
Fuel delivery via float needle valve 1.5	min. 24 l/h (400 cc per minute) } at 3400 rpm.
Fuel level indication	Gauge with mechanical sender unit
Fuel filter	Fuel filter in fuel pump

Transmission and Rear Axle

Design	Four speed, combined with differential in one housing
Transmission	Four forward speeds, 1 reverse, 1–4 gear in constant mesh, gears helically-cut, provided with cone-type synchro-devices
Gear ratios	First 3.80 : 1
	Second 2.06 : 1
	Third 1.32 : 1
	Fourth 0.89 : 1
	Reverse 3.88 : 1
Gear control	Remote, through linkage with central ball-type shift lever on frame tunnel
Final drive	Spiral bevel drive, swinging half axles
Ratio	4.375 : 1

Chassis

Frame	Flat platform reinforced by central tubular backbone with wide head for mounting the front axle and forked at the rear to accommodate the power unit
Front axle	with stabilizer
Wheel suspension: Front	Independent, two cranked trailing links on either side
Rear	Independent, swinging half axles, one trailing link on either side
Springs: Front	Two square laminated torsion bars of 8 steel leaves each, passing through front axle beams
Rear	One round torsion bar on each side
Adjustment of rear torsion bars, unloaded	17° 30′ + 50′ (inclination of the spring plate)
Shock absorbers: front and rear	Double acting, telescopic
Steering	Worm and roller with divided tie rod and steering damper
Overall ratio	14.34 (Karmann Ghia Models 14.7)
Turns of steering wheel, lock to lock	2.6
Angle of wheels at full steering lock:	
Inside wheel	34° ± 2°
Outside wheel	28° — 1°
Wheel alignment (unladen)	
a) Front wheels	
Toe-in, wheels not pressed	+ 2 to + 4 mm (.08 to .16″) [1]
Camber, wheels straight-ahead	0° 40′ ± 30′
Axle beam angle	2° ± 15′
King pin inclination	4° 20′ [2]
b) Rear wheels [3]:	
Total track angle	—5′ ± 15′ [1]
Camber	3° ± 30′
Wheels	Steel disc wheels with drop-center rims 4 J × 15
Tires	5.60—15, tubeless
Rolling radius	309 mm (12.17″)
Tire pressures:	
1–2 occupants	Front 1.1 kg/sq. cm (16 psi.) Rear 1.7 kg/sq. cm (24 psi.)
3–5 occupants	Front 1.2 kg/sq. cm (17 psi.) Rear 1.8 kg/sq. cm (26 psi.)
	For long periods of high speed driving, increase tire pressures by 0.2 kg/sq. cm (3 psi.)
Brakes:	
Foot brake	Hydraulic, acting on all wheels
Hand brake	Mechanical, acting on rear wheels
Effective lining area	approx. 620 sq. cm (96 sq. in.)
Hand brake lever	Centrally mounted between front seats
Lubrication system	Single lubrication points

1) + = toe-in, – = toe-out
2) at maximum permissible weight
3) with spring plates set correctly

Body

	VW Sedan, VW Convertible
Design	Two-door, all-steel body with sloping front hood and smooth, evenly sloping rear end, bolted to the platform-type frame
Fenders and sill panels	Bolted in position, replaceable
Doors:	
Width	950 mm (37.4")
Opening angle	Approx. 80°
Windows:	
Windshield	Slightly curved, with clear vision area
Door windows	Vent wings with clamp, vertical wind-down type windows
Rear quarters windows	Fixed in position. **Convertible:** Drop-type
Rear window	Curved
Type of glass	Safety glass
Hoods:	
Front	Rear-hinged, held in open position by springs Pin type lock with cable control **Convertible:** Front hood pull knob lockable by means of door key
Rear	Top-hinged, with press button lock
Seats:	
Seating capacity	4–5, **Convertible:** 4
Front	Adjustable seats with hinged backs which can be set to 3 positions
Rear	Seat bench with hinged back which can be folded down to extend luggage compartment
Instrument panel:	
Instrument cluster	Combining speedometer with mileage recorder and variable speedometer light, built-in warning lights for flashing indicators, generator charge, headlights, and oil pressure, fuel gauge with variable lighting on right
Flashing indicator operation	Self-cancelling lever on the steering column
Ignition switch	Combined steering/ignition lock with starting switch and non-repeat lock
Glove compartment	Equipped with lid with press button lock on passenger side **Convertible:** lockable
Miscellaneous	Space for loudspeaker on left of speedometer, space for radio between fuel gauge and glove box. Ashtray. Grab handle for passenger
Interior trim:	
Floor	Covered with rubber mats
Frame tunnel	Rubber-covered
Front panel, front side panels, lower body side members	Covered with haircord carpet
Doors and side panels	Covered with synthetic material
Roof	Perforated leatherette headlining
Miscellaneous	Foot support for front passenger. Kick boards under rear seat. Rear view mirror, two padded sun visors which can be swivelled to the side Door pocket on driver's side, arm rest on passenger's door 2 safety belt mountings for each seat
Convertible additional	Door pocket also on passenger's side. Stone guards on the rear fenders
Heating	Hot air from the engine. Outlets in front foot well, under rear seat and two defroster vents on the windshield
Control	Right-hand lever on tunnel
Luggage compartment, rear	Approx. 120 liter/4.3 cu. ft. behind rear seat 280 liters/10 cu. ft. more with backrest down
front	Approximately 140 liters/4.94 cu. ft. under the front hood
Miscellaneous:	
Bumpers	At front and rear, each equipped with two overriders
Spare wheel	Theft-proof under front hood
Fuel tank	Under front hood
Tools and accessories	Under front hood
Convertible	Wheel trim rings

Sliding roof .... Steel, stays in any position when open
Make .... Golde
Roof opening:
Clear length .... 375 mm (14.75″)
Clear width .... 665 mm (26.18″)

Karmann Ghia Coupé and Convertible

Design .... Two-door, all-steel body, of stream-lined pontoon shape bolted to a full width platform-type frame
Fenders .... Welded to body
Doors .... Half-door, no upper frame, opening angle limited by hold-open mechanism
Width .... 1000 mm (39.4″)
Windows:
Windshield .... Curved, inclined by approx. 50°
Door windows .... Vertical wind-down type without frame
Rear quarter windows .... Hinged (Coupé only)
Rear window .... Curved, wrap-around (Coupé only)
Glass .... Safety glass (except for rear window of Convertible)
Hoods:
Front .... Released by pulling a knob under instrument panel, provided with safety catch, held in open position
Rear .... Released by a knob on the vertical panel below rear seat, hood incorporates air intake slots
Seats:
Front .... Adjustable single seats with hinged backs. Back rest rake adjustable to 3 positions
Rear .... Emergency or children's seat bench, back can be folded forward to increase luggage space
Instrument panel:
Instrument cluster .... Combining speedometer, mileage recorder and variable speedometer light, built-in warning lights for flashing indicators, generator charge, headlight high beam and oil pressure
Clock .... Electric. At right of instrument cluster, variable lighting
Fuel gauge .... Between instrument cluster and clock
Ignition switch .... Combined gearshift/ignition lock with starting switch and non-repeat lock
Flashing indicator .... Flashing indicator lever on steering column, self-cancelling. Headlamp flasher button in indicator lever
Glove compartment .... Equipped with lid with press button lock on passenger side

Convertible: lockable

Miscellaneous .... Space for radio on right of clock, ashtray further to left. Grab handle for passenger. Loudspeaker space under instrument panel
Interior trim:
Floor .... Rubber matting
Frame tunnel, front panel, front side panels and lower side members ..... Lined with haircord carpet
Doors and side panels .... Trimmed with synthetic material
Roof .... Trimmed with perforated leatherette
Miscellaneous .... Door pockets and arm rests with grip recess on both doors, two padded swivelling sun visors, rear view mirror with interior light, grab handle for passenger. Two safety belt mounting points for each seat
Heating and ventilation:
Heating .... By warm air from engine. Outlets at foot level, under rear seat and defroster vents at the windshield and rear window. Adjustable by means of a lever on frame tunnel
Ventilation: .... Fresh air enters apertures in front panel and is directed to defroster vents
Control .... Either fresh or warm air or both
Miscellaneous:
Bumpers .... Wrap-around at front and rear, each equipped with two over-riders
Spare wheel .... Theft-proof under front hood
Fuel tank .... Under front hood
Tools and accessories .... Under front hood
Wheels .... with slotted trim rings

Electrical System

	All Models
Electrical system	6 volts, with voltage regulator
Battery	6 volts, 66 ampere hours
Generator	Bosch 111 903 021 G VW 113 903 021 C } intermittently
Regulator	Bosch RS/TAA 180/6/A 4 VW 113 903 801 C } intermittently
Ratio, crankshaft — generator	Approx. 1 : 1.8
Cut-in rpm.	Between 1660 and 1950 generator rpm.
Starting motor	Bosch EEF 0.5/6 L 1 VW 113 911 021 A

	Sedan and Convertible
Lighting System:	
Two headlamps	Asymmetrical low beams, combined with parking lights, adjustable
Rim inner diameter	180 mm (7.09″)
Headlamp bulbs	45/40 watts
Parking light bulbs	4 watts
Two stop/tail/reflector units	on rear fenders
Stop lamp, Tail lamp } twin filament bulbs	18/5 watts each
Turn indicators, rear	18 watts
One license plate lamp	In center of engine hood, also serving as engine compartment light
One ball type bulb	10 watts
One interior light	In left-hand roof side member with built-in switch
Interior light bulb	10 watts

	Karmann Ghia Models
Lighting:	
Two headlights	With asymmetrical low beams and built-in parking lights, adjustable
Reflector dia.	170 (6.7″)
Head light bulbs	45/40 watts
Parking light bulbs	4 watts
Two stop/tail/reflector/turn indicator units	In rear fenders
Tail lights	5 watts
Stop lights	18 watts
Turn indicator lights	18 watts
1 licence plate light	In the centre of the engine hood
2 bulbs	5 watts
1 interior light	With switch, above inside mirror
Bulb	10 watts

	All Models
Turn indicators	Flashing type. Self-cancelling lever on the steering column
Turn indicator light, front	18 watts
All warning light bulbs	1.2 watts
Speedometer lighting	Rheostat-controlled
Bulb	1.2 watts
Fuel gauge	Mechanical, with cable from sender in tank
Bulb	1.2 watts
Windshield wipers	Electric, with two arms and automatic-parking
Windshield washer	Pneumatically operated with press button in windshield wiper switch. Transparent water container under the front hood
Capacity	Approximately 1 quart
Fuses	8 point fuse box below the instrument panel to the right of the steering column

	Karmann Ghia Models
Clock	Electric
Dial lighting	Indirect
1 bulb	1.2 watts
Headlamp flasher	Push button mounted in flashing indicator lever

Dimensions and Weights

	Sedan and Convertible	Karmann Ghia Models
Wheel base	2400 mm (94.5 ins)	2400 mm (94.5 ins)
Track: front	1305 mm (51.4 ins)	1305 mm (51.4 ins)
rear	1288 mm (50.7 ins)	1288 mm (50.7 ins)
Length, without bumper guards	4070 mm (160 ins)	4140 mm (163 ins)
with bumper guards	4080 mm (160.7 ins)	4140 mm (163 ins)
Width	1540 mm (60.5 ins)	1634 mm (64.3 ins)
Height (unladen)	1500 mm (59 ins)	1330 mm (52.4 ins)
Smallest ground clearance with the car fully laden	152 mm (6 ins)	152 mm (6 ins)
Angle of approach	27° 40′	24° 10′
Angle of departure	12° 30′	12° 30′
Smallest turning circle (between walls)	approx. 11.0 m (36 ft.)	approx. 11.25 m (37 ft.)
(between kerbs)	approx. 10.5 m (34 ft. 6 ins)	approx. 10.5 m (34 ft. 6 ins)

	Sedan*)		4 Seater Convertible		Karmann Ghia Models	
	kg	lbs	kg	lbs	kg	lbs
Kerb weight	760	1675	810	1764	820	1808
Maximum load	380	838	360	793	320	705
Permissible total weight	1140	2513	1170	2579	1140	2513
Max. front axle load	480	1058	480	1058	480	1058
Max. rear axle load	700	1543	700	1543	700	1543

*) Weights apply also with steel sliding roof

Capacities

Fuel tank	40 liters (8.8 Imp. gals.; 10.5 U. S. gals.), including 5 liters reserve (1.3 U. S. gals., 1.1 Imp. gals.)
Crankcase	Capacity: 2.5 liters (4.4 Imp. pints; 5.3 U. S. pints) Oil change quantity: 2.5 liters (4.4 Imp. pints; 5.3 U. S. pints)
Transmission case and differential	Capacity: 3.0 liters (5.3 Imp. pints; 6.3 U. S. pints) Oil change quantity: 2.5 liters (4.4 Imp. pints; 5.3 U. S. pints)
Brakes	0.25 liter (0.44 Imp. pint; 0.53 U. S. pint)
Oil bath air cleaner	Refill to indicated level (approx. 0.25 liter/0.44 Imp. pint; 0.53 U. S. pint).

Performance

	Sedan and Convertible	Karmann Ghia Models
Maximum and cruising speed	approx. 115 kph (72 mph)	120 kph (75 mph)
at an engine speed	approx. 3870 rpm.	4040 rpm.
Mean piston speed at maximum speed	approx. 8.26 m/sec (1626 ft. min)	8.62 m/sec (1697 ft. min)
Speeds at 3600 engine rpm.		
1st gear	25 kph (15 mph)	25 kph (15 mph)
2nd gear	46 kph (30 mph)	46 kph (30 mph)
3rd gear	72 kph (45 mph)	72 kph (45 mph)
4th gear	107 kph (66 mph)	107 kph (66 mph)
Reverse	25 kph (15 mph)	25 kph (15 mph)
Mean piston speed	7.68 m/s 1512 ft./min.	7.68 m/s 1512 ft. min.

Climbing ability with two occupants on normal roads:

		Sedan	VW Convertible and Karmann Ghia Models
1st gear	approx. %	43.5	39
2nd gear	approx. %	22.5	20.5
3rd gear	approx. %	13.5	12
4th gear	approx. %	7.5	6.5
Acceleration from 0 to 80 kph (0 to 50 mph) through the gears	approx. seconds	18	18

Fuel Consumption

Fuel consumption	Metric: 7.5 liters per 100 km, according to DIN 70030 (U. S.: 31.5 mpg; Imp. 37.5 mpg)
Fuel rating	86 Octane (Res. F 1)
Oil consumption	Between 0.3 and 1.0 liter per 1000 km (1—3.4 U. S. pints/1000 miles, .8—2.8 Imp. pints/1000 miles)

Technical Data

(from August 1960)

Engine

Weight, dry	approx. 110 kg (242 lbs.)
Valve timing with a clearance of 1 mm (.040″)	from Engine No. 6864207
Intake opens	6° before TDC.
Intake closes	35° 30′ after BDC.
Exhaust opens	42° 30′ before BDC.
Exhaust closes	3° after TDC.
Ignition	
Distributor (from Chassis No. 4057000)	Bosch ZV/PAU 4 R 5 mk, VW 113905205 B } intermittently
Spark plugs	Bosch W 175 T 1 Beru 175/14 Champion L 85 or plugs with similar values from other manufacturers

Fuel System

Fuel reserve indication (from Chassis No. 4010995)	Fuel gauge with mechanical sender unit and cable drive

Chassis

Steering (from Chassis No. 4010995)	Worm and roller type with divided tie-rod and steering damper
Overall ratio	14.34 : 1 (Karmann Ghia Models 14.7)
Turns of steering wheel from lock to lock	2.6
Maximum wheel angle	
inner wheel	34°+2°
outer wheel	28°—1°
Front wheel settings:	
a) With vehicle fully loaded	
Toe-in	1—3 mm (.04—.12″)
Camber	0°
King pin inclination	4° 20′
Axle tube angle	2° ± 15′
b) With vehicle unladen	
Toe-in	2—4 mm (.08—.16″)
Camber	0° 40′ ±30′

Electrical System

Combined tail, brake and turn indicator lamps	on rear fenders
Brake light	18 Watt } Twin-filament 18/5 Watt bulb
Tail light	5 Watt } Twin-filament 18/5 Watt bulb
Turn indicator light, rear	18 Watt

Capacities

Fuel tank	40 liters, with 5 liters reserve
Crankcase	2.5 liters, oil change 2.5 liters
Transmission and rear axle drive	3.0 liters, oil change 2.5 liters
Steering	
Sector type	0.125 liter
Roller type	0.16 liter
Brakes	0.25 liter
Oil bath air cleaner	Fill to mark (approx. 0.25 liter)

Fuel Consumption

Fuel type .... 86 Octane (Res. F 1)

Alterations from August 1963

Fuel System

Fuel delivery	Mechanical diaphragm type pump	
Delivery pressure	Maximum 0.2 kg/sq. cm (2.8 psi)	at 3600 rpm.
Delivery capacity via 1.5 float needle valve	Minimum 24 liters per hour (400 cc per minute)	at 3600 rpm.

[1]) from Chassis No. 5578122

Accelerator pump
(from Chassis No. 5220713) .... 1.1—1.4 cc per stroke

Chassis

Tire pressure, rear .... 1.7 kg/cm² (24 psi)
(with full load)

Weights

Weights in kg (lbs.)	Modell 11	Modell 14	Modell 15
Unladen weight	760 (1675)	820 (1808)	810 (1785)
Permissible load	380 (838)	320 (705)	360 (794)
Permissible total weight	1140 (2513)	1140 (2513)	1170 (2579)
Permissible front axle load	480 (1058)	480 (1058)	480 (1058)
Permissible rear axle load	700 (1543)	700 (1543)	700 (1543)

Tolerances and Wear Limits

(from August 1960)

General

The term WEAR LIMIT means that parts which approach, or have already reached, the limit given should not be re-used when carrying out an overhaul. When deciding the wear limit of pistons and cylinders, due consideration should also be given to the oil consumption of the respective engine.

		Tolerance Limits (new parts)	Wear Limit
Engine (1192 cc = 72.74 cu. in.; 34 bhp, 41.5 SAE bhp)			
1 - Cylinder seating depth in cylinder head		13.7—13.8 mm (.538—.542″)	
2 - Cylinder	out of round	0.01 mm (.0004″)	
3 - Piston/cylinder	clearance	0.036—0.054 mm (.0014—.0021″)	0.20 mm (.008″)
4 - a) Upper compression ring	side clearance	0.065—0.092 mm (.0026—.0036″)	0.12 mm (.005″)
b) Lower compression ring	side clearance	0.045—0.072 mm (.0018—.0028″)	0.10 mm (.004″)
5 - Oil scraper ring	side clearance	0.025—0.052 mm (.0010—.0020″)	0.10 mm (.004″)
6 - Compression rings	gap	0.30—0.45 mm (.012—.018″)	0.90 mm (.035″)
Oil scraper ring	gap	0.25—0.40 mm (.010—.016″)	0.95 mm (.037″)
7 - Weight tolerance of pistons in one engine		max. 5 g (.18 oz.)	max. 10 g*) (.35 oz.)
8 - Weight tolerance of connecting rods in one engine		max. 5 g (.18 oz.)	max. 10 g*) (.35 oz.)
9 - Piston pin/con rod bush	clearance	0.003—0.016 mm (.0001—.0006″)	0.04 mm (.002″)
10 - Connecting rod bearing	clearance	0.018—0.076 mm (.0007—.0030″)	0.15 mm (.006″)
	end play	0.1—0.4 mm (.004—.016″)	0.70 mm (.028″)
11 - Crankshaft main bearing (taking into account the preload by the crankcase)			
a) Bearings 1 to 3	clearance	0.036—0.098 mm (.0010—.0039″)	0.18 mm (.0072″)
b) Bearing 4	clearance	0.047—0.102 mm (.0019—.0040″)	0.19 mm (.0075″)
c) Steel-backed bearings 1—3 (for cold climates only)	clearance	0.028—0.087 mm (.0011—(.0034″)	0.17 mm (.0066″)
12 - Crankshaft at 2nd and 4th main bearing journals (1st and 3rd bearing journals on V-blocks)	run-out		0.03 mm (.001″)
13 - Crankshaft/main bearing 1	end play	0.06 —0.12 mm (.0023—.0047″)	0.15 mm (.006″)

*) when repairing

		Tolerance Limits (new parts)	Wear Limit
14 - Crankshaft	out of balance	max. 8 cmg (.11 oz. in.)	
15 - Main bearing journal	out of round		0.03 mm (.0012″)
16 - Crank pin	out of round		0.03 mm (.0012″)
17 - Crankcase bore for crankshaft			
a) Bearings 1 to 3	diameter	65.000—65.019 mm (2.5591—2.5598″)	65.030 mm (2.560″)
b) Bearing 4	diameter	50.000—50.025 mm (1.9685—1.9695″)	50.040 mm (1.970″)
c) Seat for oil seal	diameter	90.000—90.054 mm (3.5433—3.5453″)	
18 - Fan pulley	radial run-out	max. 0.8 mm (.031″)	
	lateral run-out	max. 0.3 mm (.012″)	
19 - Crankcase bore for camshaft	diameter	25.020—25.041 mm (.98504—.98587″)	
20 - Camshaft	clearance	0.020—0.054 mm (.0008—.0021″)	0.12 mm (.0047″)
	end play	0.060—0.114 mm (.0024—.0045″)	0.14 mm (.0055″)
up to engine No. 5067817	end play	0.030—0.084 mm (.0012—.0033″)	0.10 mm (.004″)
(between centers)	run-out	0.02 mm (.0008″)	0.04 mm (.0016″)
21 - Camshaft timing gear	backlash	0.000—0.052 mm (.0000—.0020″)	
22 - Flywheel	lateral run-out	max. 0.30 mm (.012″)	
	out of balance	max. 5 cmg (.07 oz. in.)	
Shoulder below gear ring	outer diameter	69.9—70.1 mm (2.75—2.76″)	69.40 mm (2.731″)
Remachining width of teeth			max. 2.0 mm (.008″)
23 - Valve stem: intake	diameter	7.95—7.94 mm (.3130—.3126″)	7.90 mm (.3110″)
exhaust	diameter	7.92—7.91 mm (.3118—.3114″)	7.87 mm (.3098″)
	out of round	0.01 mm (.0004″)	
24 - Valve head: intake	diameter	31.5 mm (1.24″)	
exhaust	diameter	30.0 mm (1.18″)	
25 - Valve guides: intake and exhaust	inner diameter	8.000—8.020 mm (.3150—.3157″)	8.060 mm (.3173″)
26 - Valve guide/valve stem: intake	clearance	0.050—0.080 (.0020—.0031″)	0.16 mm (.006″)
exhaust	clearance	0.080—0.11 mm (.0031—.0043″)	0.16 mm (.006″)
27 - Valve seating face:			
a) intake	width	1.3—1.6 mm (.051—.063″)	
b) exhaust	width	1.7—2.0 mm (.067—.079″)	
c) valve seating face	run-out	0.015 mm (.0006″)	0.02 mm (.0007″)
28 - Valve springs loaded length 33.4 mm (1.31″)	load	43.8±3 kg (96 ± 7 lbs.)	
up to engine No. 6850939: loaded length 34.3 mm (1.35″)	load	46.3 ± 2.3 kg (102 ± 5 lbs.)	

		Tolerances (new parts)	Wear Limit
29 – Valve clearance up to a max. oil temperature of 50° C (122° F):			
Intake and exhaust	adjustment	0.10 mm (.004")	
Engines **without** sticker:			
Intake	adjustment	0.20 mm (.008")	
Exhaust	adjustment	0.30 mm (.012")	
30 – a) Rocker arm	inside diameter	18.00–18.02 mm (.7086–.7093")	18.04 mm (.710")
b) Rocker arm shaft	diameter	17.98–17.97 mm (.7077–.7073")	17.95 mm (.7066")
31 – a) Crankcase bore for cam follower..	diameter	19.00–19.02 mm (.7480–.7485")	19.05 mm (.7499")
b) Cam follower	diameter	18.98–18.96 mm (.7472–.7464")	18.93 mm (.7452")
32 – Compression pressure (To be checked with the throttle open, engine warm, all spark plugs removed, pressure gauge in spark plug seat and the engine turned over by the starter motor) VW 1200	pressure	9.0–7.0 kg/sq. cm (128–100 psi)	6 kg/sq. cm (85 psi)
VW 1300	pressure	9.5–7.5 kg/sq. cm (135–107 psi)	
Pressure variation between cylinders ..	max.	1 kg/sq. cm (14 psi)	
33 – Oil pump: Gears/body with gasket (cover removed)	end play	0.07–0.18 mm (.0027–.0070")	0.20 mm (.008")
Gears/body (gasket removed)	end play		0.10 mm (.004")
Gears	backlash	0.03–0.08 mm (.0012–.0031")	0.15 mm (.006")
34 – Oil pressure (Only for SAE 10 W-30 oil) at oil temperature of 70° C (158° F):			
a) At 550 rpm. (idling)		min. 0.5 kg/sq. cm (7 psi)	
b) At 2500 engine rpm.		min. 2.0 kg/sq. cm (28 psi)	
35 – Spring for oil pressure relief valve: Length compressed: 23.6 mm	load	7.75 kg (17 lbs.)	
36 – Oil pressure contact opens	pressure	0.15–0.45 kg/sq. cm (2.1–6.4 psi)	
37 – Thermostat: After a water bath of 65 to 70° C (149 to 158° F)	pressure unit length	min. 46 mm (1.81")	

Clutch

a – VW 1200

1 – Clutch driven plate	lateral run-out	max. 0.8 mm (.03")	
2 – Clutch thrust spring:			
Length loaded 28.3 mm (1.11")	load, new	58.5–63.5 kg (128–140 lbs.)	
	load, settled	51.0–56.0 kg (112–124 lbs.)	

		Tolerances (new parts)	Wear Limit
From March 1965:			
a – Dark brown spring	load, new	62.5–65.5 kg (138–144 lbs.)	
	load, settled	55.0–58.0 kg (121–128 lbs.)	
Light brown spring	load, new	59.5–62.5 kg (131–138 lbs.)	
	load, settled	52.0–55.0 kg (115–121 lbs.)	
3 – Clutch pedal free-play		10–20 mm (.4–.8″)	
4 – Clutch pressure plate	run-out		0.10 mm (.004″)
5 – Clutch release plate	run-out	0.40 mm (.016″)	
6 – Clutch assembly	unbalance	max. 15 cmg (.21 oz. in.)	
7 – Distance from flywheel to release plate		26.8–27.2 mm (1.054–1.070″)	
8 – Total pressure		320–345 kg (705–760 lbs.)	
up to Chassis No. 115 318 170		315–350 kg (694–771 lbs.)	
b – VW 1300			
with the following deviations:			
2 – Clutch thrust springs:			
Length loaded 29.2 mm (1.15″)			
a – Dark blue spring	load, new	62.0–66.0 kg (137–146 lbs.)	
	load, settled	54.5–58.5 kg (120–129 lbs.)	
b – Light blue spring	load, new	60.0–64.0 kg (132–141 lbs.)	
	load, settled	52.5–56.5 kg (116–125 lbs.)	
8 – Total pressure		345–370 kg (760–816 lbs.)	
Automatic Clutch (Saxomat)			
(1200 only, with 41.5 SAE bhp engine)			
1 – Clutch plate	run-out	max. 0.8 mm (.03″)	
2 – Clutch lining	outside diameter	160±1 mm (6.299±.040″)	
	inside diameter	110±1 mm (4.330±.040″)	
	thickness	2.75±1 mm (.108±.040″)	
3 – Clutch pressure springs:			
a) outer, compressed 21.5 mm	load	26–28 kg (57–61.5 lbs.)	
b) inner, compressed 19.5 mm	load	11–12 kg (24–26 lbs.)	
4 – Gearshift lever	contact clearance	0.25 mm (.010″)	

		Tolerances (new parts)	Wear Limit
Front Axle			
a – VW 1200 De Luxe Model (up to July 1965)			
1 – Torsion arm	twist	max. 0.5 mm (.020″)	
2 – Torsion arm/plastic bush (higher limit preferable as bush swells)	clearance	0.20–0.27 mm (.008–0.11″)	0.40 mm (.016″)
3 – Torsion arm link pin/sintered iron bush	clearance	0.04–0.09 mm (.0016–.0035″)	0.22 mm (.0086″)
4 – Torsion arm link pin	diameter	17.94–17.91 mm (.7063–.7052″)	17.80 mm (.7008″)
5 – King pin/bush	clearance	0.02–0.05 mm (.0008–.002″)	0.08 mm (.003″)
Steering knuckle/torsion arm link	preload	0.00–0.04 mm (.0000–.0016″)	
6 – Steering drop arm/steering gear case..	end play	0.4–1.0 mm (.016–.039″)	
7 – Steering gear			
a) Roller shaft/seat for bush	diameter	23.99–23.98 mm (.9445–.9440″)	
b) Bush for roller shaft	inside diameter	24.02–24.00 mm (.9456–.9448″)	
b – VW 1300 and VW 1200 Sedan			
(from August 1965)			
1 – Lower ball joints	clearance	max. 0.5 mm (.02″)	2.0 mm (.08″)
2 – Metal bushes for torsion arm, in axle tube	inside diameter	37.06–37.16 mm (1.4591–1.4630″)	37.38 mm (1.4717″)
3 – Steering knuckle/stub axle	distortion	± 30′ △ 0.3 mm [1] (.0118″)	
4 – Axle beam			
a – Upper needle bearing seat	diameter	45.97–45.99 mm (1.8098–1.8106″)	
Needle bearing to suit	diameter	46 mm (1.8110″)	
Oversize seat	diameter	46.17–46.19 mm (1.8177–1.8185″)	
Needle bearing to suit	diameter	46.2 mm (1.8189″)	
b – Lower needle bearing seat	diameter	49.97–49.99 mm (1.9673–1.9681″)	
Needle bearing to suit	diameter	50 mm (1.9685″)	
Oversize seat	diameter	50.17–50.19 mm (1.9752–1.9760″)	
Needle bearing to suit	diameter	50.2 mm (1.9764″)	

[1]) Measured with VW 258 k

		Tolerances (new parts)	Wear Limit
Rear Axle and Transmission			
1 – Main drive shaft/needle bearing in gland nut	clearance	0.12–0.19 mm (.0047–.0075")	0.25 mm (.0098")
2 – Main drive shaft front, contact surface for needle bearing of the 3rd gear	run-out	max. 0.02 mm (.0008")	
3 – Bushes for gearshift housing	inside diameter	15.05–15.03 mm (.5924–.5916")	15.25 mm (.6004")
4 – Transmission shift lever	diameter	15.00–14.96 mm (.5900–.5889")	14.75 mm (.5807")
5 – Preload of final drive covers on the differential ball bearings		0.14 mm (.0055")	
6 – Rear axle shaft:			
a) Flat end/fulcrum plates/differential side gear (4 parts)	clearance	0.04–0.24 mm (.002–.009")	0.25 mm (.010")
b) Flat end/differential side gear (measured across the convex sides)	clearance	0.03–0.10 mm (.0012–.0039")	0.20 mm (.008")
c) Between centers, measured at bearing seats	run-out	max. 0.05 mm (.002")	
7 – Plastic packing/transmission case/rear axle tube/axle tube retainer	clearance	0.25–0.35 mm (.010–.014")	0.40 mm (.016")
8 – Starter shaft bush....................	inside diameter	12.55–12.57 mm (.4941–.4949")	12.65 mm (.4980")
9 – Starter shaft/bush	clearance	0.09–0.14 mm (.0035–.0055")	0.25 mm (.010")
10 – Gear for 1st speed	end play	0.10–0.25 mm [1]) (.0039–.0098")	
11 – Gear for 4th speed	end play	0.10–0.25 mm (.0039–.0098")	
12 – Selector shaft / operating sleeves for 1st and 2nd, 3rd and 4th gears	end play	0.10–0.30 mm (.004–.012")	
13 – Synchronizer stop rings/gears (measured between clutch teeth faces)	clearance	1.1 mm (.043")	min. 0.6 mm (.024")
Brakes and Wheels			
1 – Brake master cylinder................	diameter	17.46 mm (.687")	
	stroke	33 mm (1.299")	
2 – Wheel cylinder: Front	diameter	22.20 mm (.874")	
Rear	diameter	19.05 mm (.750")	

[1]) Try to obtain lower clearance (0.10 mm/.0039")

		Tolerances (new parts)	Wear Limit
3 – Brake drums:			
Front	inside diameter	230.1 + 0.2 mm (9.059 + .008")	231.5 mm (9.114")
Rear	inside diameter	230.0 + 0.2 mm (9.055 + .008")	231.5 mm (9.114")
Turning sizes for 0.5 mm oversize linings:			
Front	inside diameter	231.1 + 0.2 mm (9.089 + .008")	231.5 mm (9.114")
Rear	inside diameter	231.0 + 0.2 mm (9.094 + .008")	231.5 mm (9.114")
Front and rear	wall thickness		4.0 mm (.16")
	out-of-round	max. 0.1 mm (.004")	
	taper	max. 0.1 mm (.004")	
	lateral run-out	max. 0.25 mm (.010")	
Measured at friction surface	radial run-out	max. 0.15 mm (.006")	
4 – Brake linings:			
Front	width	40 mm (1.6")	
Rear	width	30 mm (1.2")	
Front and rear	thickness	4.0–3.8 mm (.16–.15")	2.5 mm (.09")
Oversize	thickness	4.5–4.3 mm (.18–.17")	2.5 mm (.09")
Effective lining area		620 cm² (96 sq. ins.)	
5 – Wheel	radial run-out	max. 1.5 mm (.06")	
	lateral run-out	max. 1.5 mm (.06")	

THIS PAGE INTENTIONALLY
LEFT BLANK

Contents: E

THIS PAGE INTENTIONALLY
LEFT BLANK

Description and Wiring Diagrams

Generator

The six-volt electrical system is operated by a generator with a nominal output of 180 W at 2500 r.p.m. The generator output is controlled by a regulator. A generator warning light is located in the speedometer dial on the bottom, left-hand side.

Starting Motor

The solenoid starting motor of the overrunning-clutch type produces 0.5 HP. It is operated by the ignition key. When the ignition key is turned, it actuates a solenoid switch which connects the motor to the battery, causing the pinion to engage.

Battery

The six-volt battery consists of three cells and has y capacity of 66 amp. hours at a twenty-hour discharge rate. The battery is located on the right-hand side of the frame under the rear seat on the Sedan und VW Convertible and is securely held in position by a metal strap. The negative terminal of the battery is grounded. On the Karmann Ghia Models the battery is located in the engine compartment, on the right hand side.

Ignition

The ignition system operates from the battery with a coil and distributor, the spark timing being advanced by vacuum. It is switched on by operating the combined ignition and starting key.

Lighting System

The built-in headlamps incorporate filaments for the high beam, asymmetrical low beam, and parking light. The reflectors are horizontally and vertically adjustable. The main lighting switch is of the multi-stage push-pull type which also controls the instrument lights and their intensity. The selection of the high or low beam is done by means of a foot-controlled switch on the left-hand side of the clutch pedal. The high beam is indicated by a blue warning light in the upper half of the speedometer dial.

The rear lighting consists of two tail lights on the fenders and a license plate light on the rear hood. The two stop lights are combined with the flashing indicator and tail lights in bezels on the rear fender. The stop light is operated by a switch at the brake master cylinder.

An oil pressure switch, in connection with a green warning light in the speedometer dial serves as a control of oil circulation.

The interior light in the left-hand roof side member above the door pillar is controlled by a switch incorporated in the interior light fitting. The interior light is automatically switched on by the door contact switches when opening the doors. A manual switch at the interior light allows the light to be switched off with the doors open, and on when the doors are closed.

Electrical Accessories

The horn is operated by a half ring in the steering wheel.

A self-cancelling indicator switch on the left of the steering column below the steering wheel operates the flashing indicator lamps on the front fenders via a flasher relay. At the same time, the stop lamp filament is used to give the flasher signal at the rear. Two arrows in the speedometer dial light up when the indicators are switched on.

The windshield wiper motor, which operates two blades, is operated by a turn switch on the instrument panel. The blades are self-parking.

The fuse box under the instrument panel to the right of the steering column contains the fuses for: High and low beam, park, stop and tail lamps as well as the flashing indicators, interior lamp, windshield wiper and horn.

The windshield washer system is operated by a suction pump which is combined with the windshield wiper switch. A twin jet is installed in the center of the cowl in front of the windshield. The transparent water container, with a capacity of approx. 1 liter, is located behind the spare wheel in the front luggage compartment.

Instruments

Speedometer and mileage recorder are driven by the left-hand front wheel through a cable.

Electrical Cables

All parts of the electrical system are equipped with tab connectors. The cable terminals are pushed on to the connections of the various electrical items. The terminals are crimped on to the cables ends to ensure good contact.

Important

To avoid short circuits, the negative cable should be disconnected from the battery whenever an operation on the electrical system requires the removal of cables.

Note:

Repairs on the electrical system are generally limited to the removal of defective or worn components and reconditioning of the wiring. To avoid excessive tension or a voltage drop, it is important to use cables of the same cross section when renewing (see key to wiring diagram).

Repairs on BOSCH parts should, if possible, be carried out by BOSCH Service Stations. The electrical system service is fully described on the following pages.

Fuses

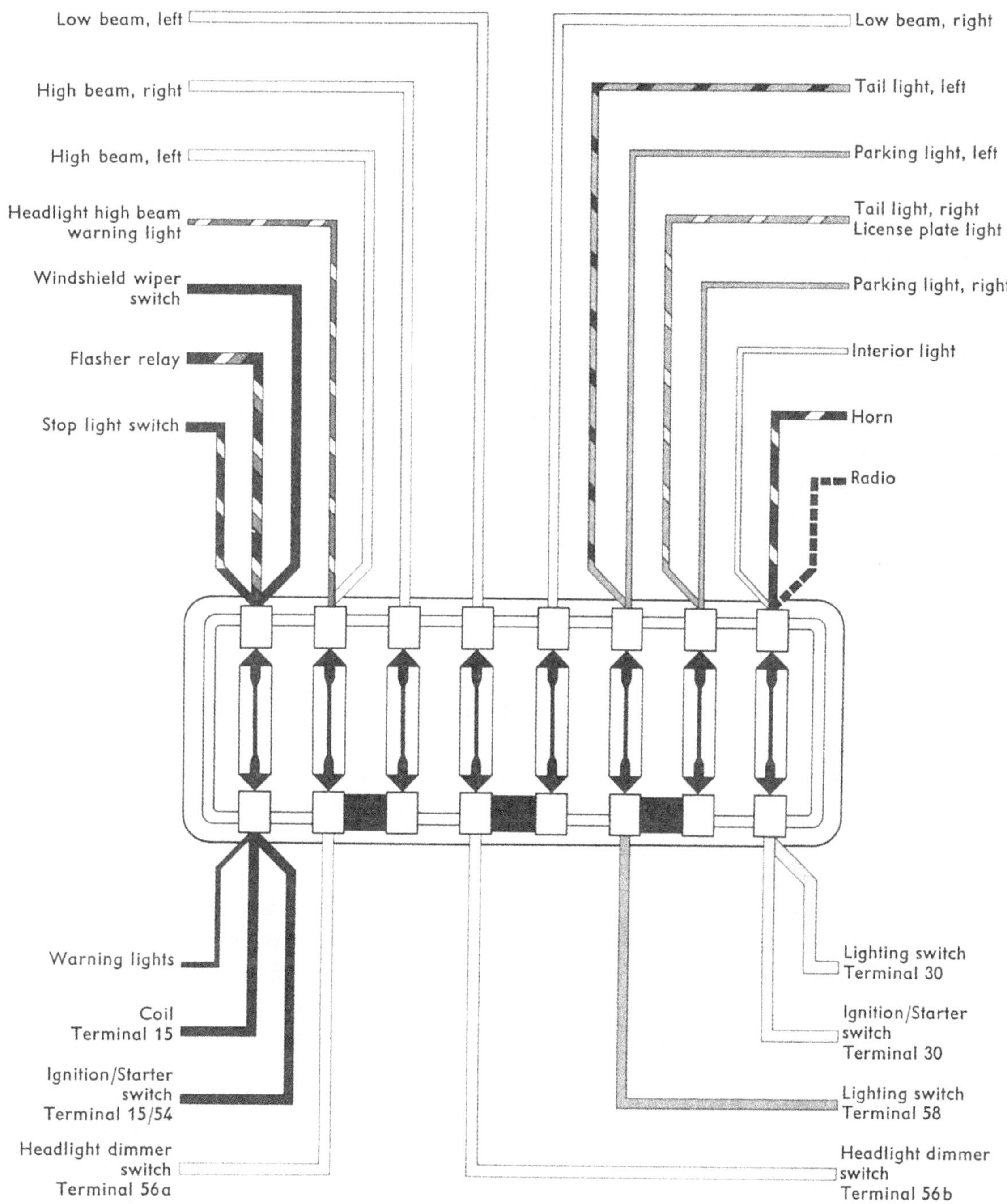

Fuse box below instrument panel

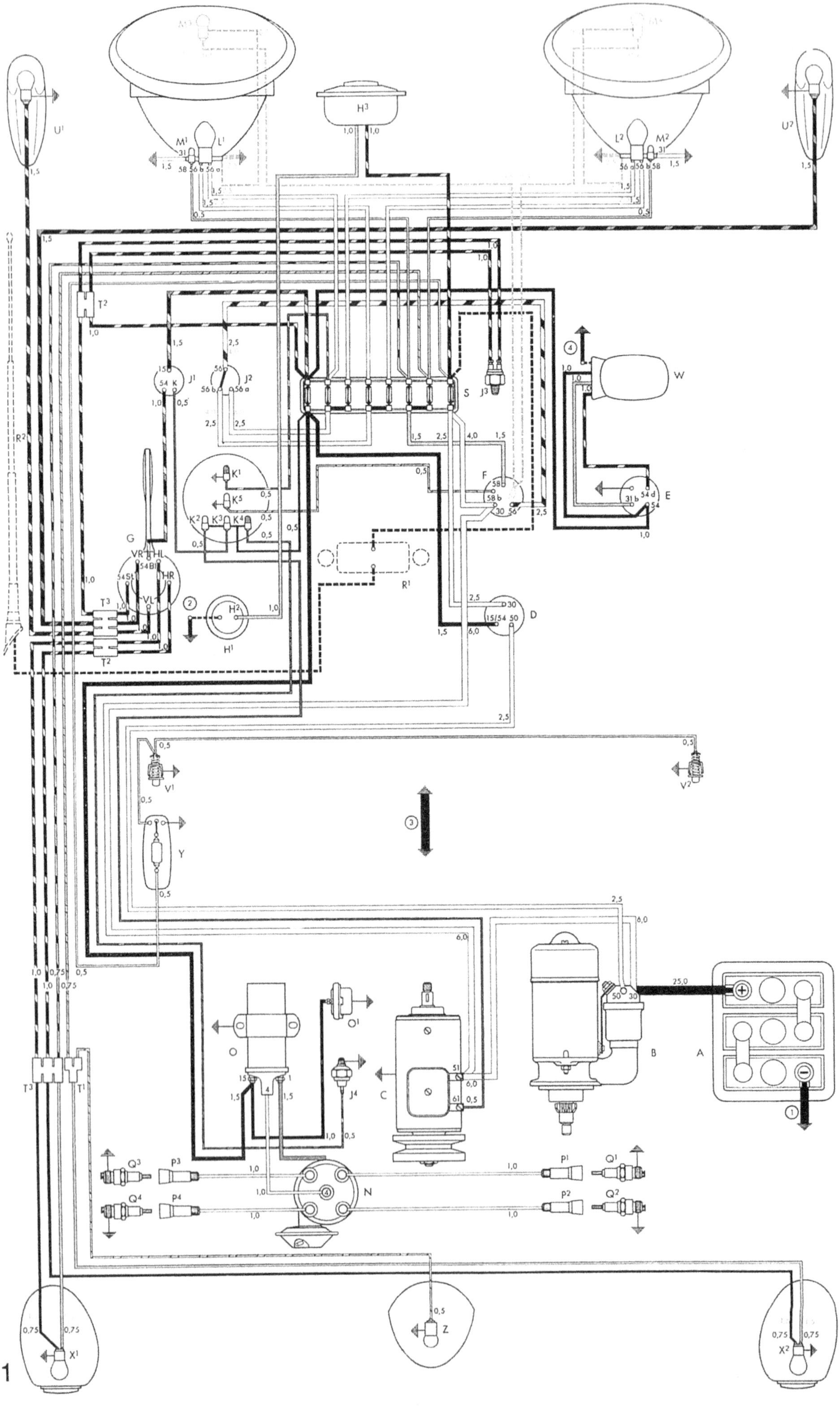

U1
M1
L1
H3
L2
M2
U2
T2
R2
J1
J2
S
J3
W
K1
K5
K2
K3
K4
F
E
G
R1
T3
H2
H1
D
V1
V2
Y
O1
O
J4
C
B
A
T1
Q3
P3
Q4
P4
N
P1
Q1
P2
Q2
X1
Z
X2

Key to Wiring Diagram · Sedan and Convertible

From August 1960

A - Battery
B - Starting Motor
C - Generator
D - Ignition Starter Switch
E - Windshield Wiper Switch
F - Light Switch and Instrument Panel Lighting
G - Flashing Indicator Switch (self-cancelling)
H^1 - Horn Half-ring
H^2 - Steering column connection
H^3 - Horn
J^1 - Flasher relay
J^2 - Dimmer Switch
J^3 - Stop Light Switch
J^4 - Oil Pressure Switch
K^1 - High Beam Indicator Light
K^2 - Generator Control Light
K^3 - Flashing Indicator Control Light
K^4 - Oil Pressure Warning Light
K^5 - Speedometer Light
L^1 - Bifilament Bulb for Headlight, Left or Sealed-Beam insert, left
L^2 - Bifilament Bulb for Headlight, Right or Sealed-Beam insert, right
M^1 - Parking Light, Left
M^2 - Parking Light, Right
M^3 - Parking light, Sealed-Beam insert, left
M^4 - Parking light, Sealed-Beam insert, right
N - Ignition Distributor
O - Ignition Coil
O^1 - Automatic Choke on Carburetor
P^1 - Spark Plug Connector for Cylinder 1
P^2 - Spark Plug Connector for Cylinder 2
P^3 - Spark Plug Connector for Cylinder 3
P^4 - Spark Plug Connector for Cylinder 4
Q^1 - Spark Plug for Cylinder 1
Q^2 - Spark Plug for Cylinder 2
Q^3 - Spark Plug for Cylinder 3
Q^4 - Spark Plug for Cylinder 4
R^1 - Radio
R^2 - Antenna
S - Fuse box (eight fuses)
T^1 - Connector
T^2 - Connector, double
T^3 - Connector, triple
U^1 - Flasher Light, Left
U^2 - Flasher Light, Right
V^1 - Door Contact Switch, Left
V^2 - Door Contact Switch, Right
V^3 - Roof Switch, Convertible
W - Windshield Wiper Motor (3 connections)
X^1 - Indicator, Stop and Tail Light, Left
X^2 - Indicator, Stop and Tail Light, Right
Y - Interior Light
Z - License Plate Light

① - Battery Ground Strap
② - Steering Column Flange Ground Strap
③ - Ground Strap between Transmission and Frame
④ - Ground Strap between Windshield Wiper Motor and Body

Black dotted line = Service Installation
Grey or brown dotted line = USA Version
Grey cross section data = USA Version

Fuses and Wiring Diagram

Type 1 — Sedan and Convertible

(From August 1961) Chassis No. 4 010 995

Fuses

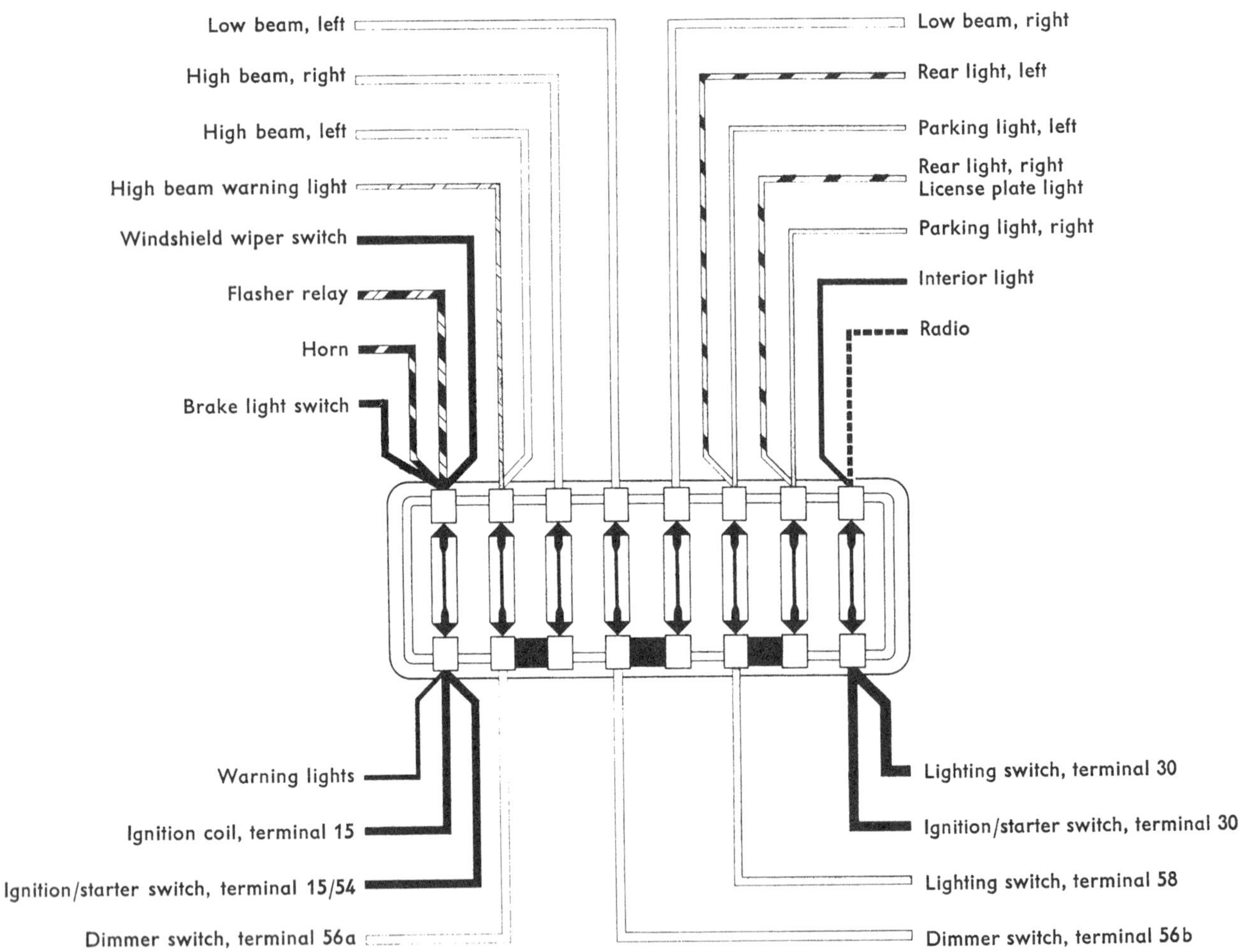

Fuse box below instrument panel

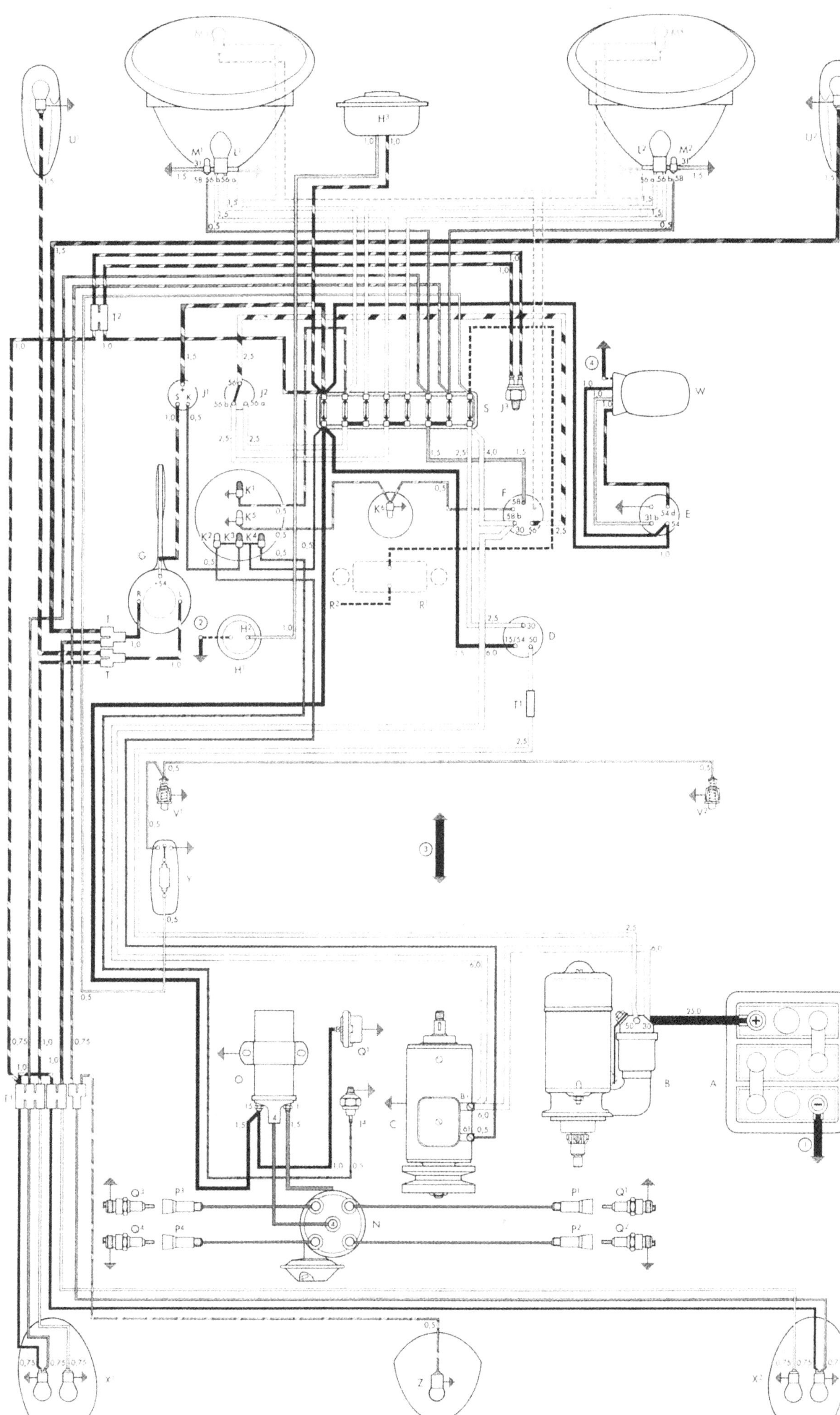
U1
M1
L1
H3
L2
M2
U2
T2
J1
J2
S
J3
W
K1
K5
K6
F
E
K2
K3
K4
G
R2
R1
T
H2
H1
D
T1
V1
V2
Y
Q1
O
C
B
A
J4
P3
Q3
P1
P4
Q4
N
P2
Q2
X1
Z
X2

Key to Wiring Diagram — Type 1 Sedan and Convertible

(From August 1961) Chassis No. 4 010 995

A - Battery

B - Starting Motor

C - Generator

D - Ignition/Starter Switch

E - Windshield Wiper Switch

F - Light Switch and Instrument Panel Lighting

G - Flashing Indicator Switch (self-cancelling)

H^1 - Horn Half-ring

H^2 - Steering Column Connection

H^3 - Horn

J^1 - Flasher Relay

J^2 - Dimmer Switch

J^3 - Stop Light Switch

J^4 - Oil Pressure Switch

K^1 - High Beam Indicator Light

K^2 - Generator Control Light

K^3 - Flashing Indicator Control Light

K^4 - Oil Pressure Warning Light

K^5 - Speedometer Light

K^6 - Fuel Gauge Light

L^1 - Bifilament Bulb for Headlight, Left or Sealed-Beam Insert, Left

L^2 - Bifilament Bulb for Headlight, Right or Sealed-Beam Insert, Right

M^1 - Parking Light, Left

M^2 - Parking Light, Right

M^3 - Parking Light, Sealed-Beam Insert, Left

M^4 - Parking Light, Sealed-Beam Insert, Right

N - Ignition Distributor

O - Ignition Coil

O^1 - Automatic Choke on Carburetor

P^1 - Spark Plug Connector for Cylinder 1

P^2 - Spark Plug Connector for Cylinder 2

P^3 - Spark Plug Connector for Cylinder 3

P^4 - Spark Plug Connector for Cylinder 4

Q^1 - Spark Plug for Cylinder 1

Q^2 - Spark Plug for Cylinder 2

Q^3 - Spark Plug for Cylinder 3

Q^4 - Spark Plug for Cylinder 4

R^1 - Radio

R^2 - Aerial Connection

S - Fuse Box (eight fuses)

T - Cable Connector

T^1 - Connector, single

T^2 - Connector, double

T^3 - Connector, triple

U^1 - Flasher Light, Left

U^2 - Flasher Light, Right

V^1 - Door Contact Switch, Left

V^2 - Door Contact Switch, Right

W - Windshield Wiper Motor (3 connections)

X^1 - Indicator, Stop and Tail Light, Left

X^2 - Indicator, Stop and Tail Light, Right

Y - Interior Light

Z - License Plate Light

① - Battery Ground Strap

② - Horn ring to Steering Column Flange Ground Strap

③ - Ground Strap between Transmission and Frame

④ - Ground Strap between Windshield Wiper Motor and Body

Black dotted line = Service Installation

Grey or brown dotted line = USA Version

Grey cross section data = USA Version

Wiring Diagram and Fuses

Karmann Ghia Models

Fuses

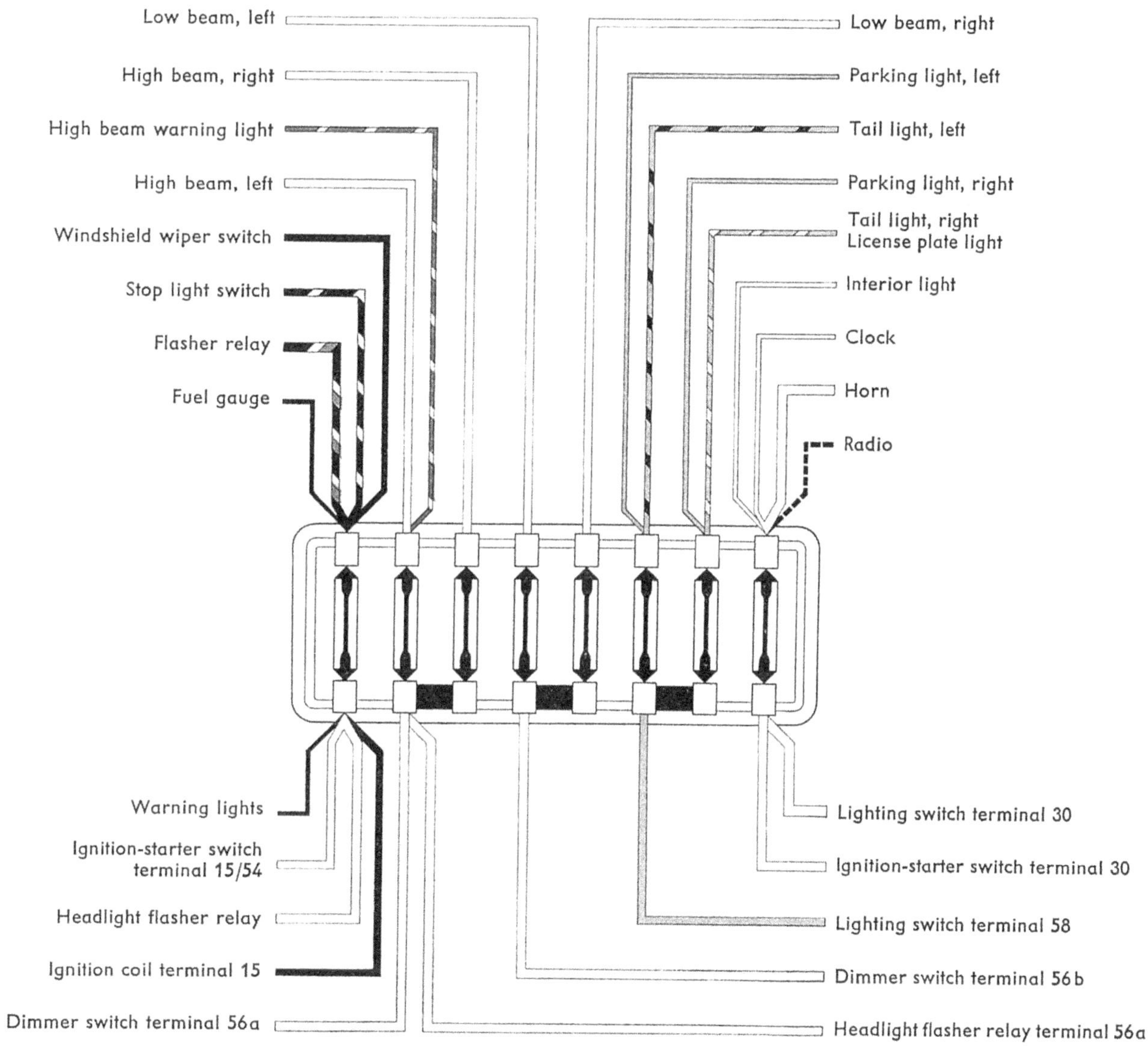

Fuse box below instrument panel

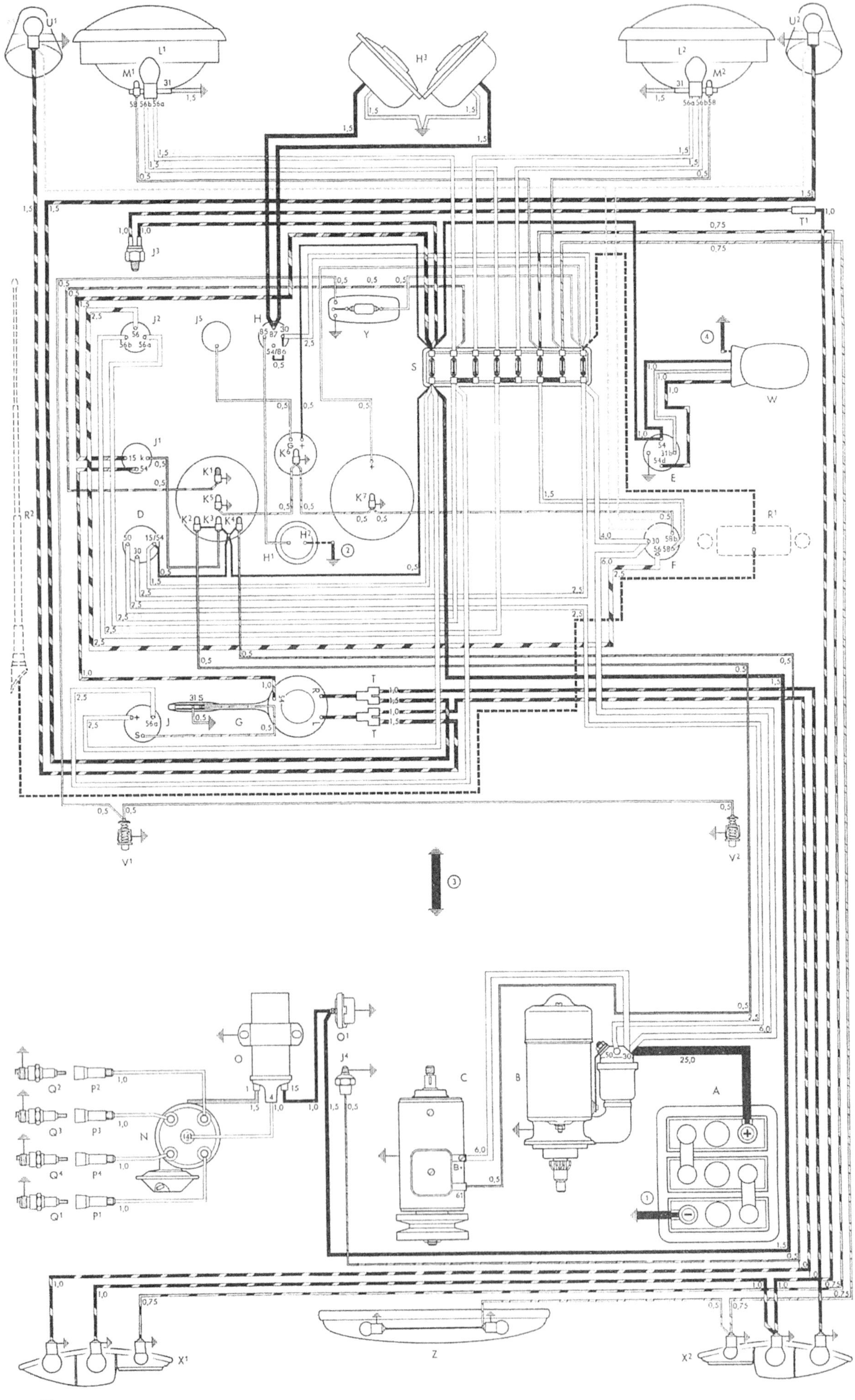

U¹
L¹
M¹
H³
L²
M²
U²
J³
T¹
J²
J⁵
H
Y
S
W
J¹
K¹
K⁵
K⁶
K⁷
D
K²
K³
K⁴
H¹
H²
E
R¹
F
R²
T
J
G
V¹
V²
O
O¹
J⁴
C
B
A
N
Q¹
Q²
Q³
Q⁴
P¹
P²
P³
P⁴
X¹
Z
X²

Key to Wiring Diagramm - Karmann Ghia

(Coupé and Convertible) from August 1960

A - Battery
B - Starter
C - Generator
D - Ignition-starter switch
E - Windshield wiper switch
F - Lighting switch and instrument panel light control
G - Indicator switch with headlight flasher
H - Twin horn relay
H^1 - Horn ring
H^2 - Steering column connection
H^3 - Twin horn
J - Headlight flasher relay
J^1 - Indicator flasher relay
J^2 - Dimmer switch
J^3 - Stop light switch
J^4 - Oil pressure switch
J^5 - Fuel gauge sender unit
K^1 - High beam warning light
K^2 - Generator control light
K^3 - Indicator control light
K^4 - Oil pressure control light
K^5 - Speedometer light
K^6 - Fuel gauge light
K^7 - Clock light
L^1 - Bifilament bulb for headlight, left or Sealed-Beam insert, left
L^2 - Bifilament bulb for headlight, right or Sealed-Beam insert, right
M^1 - Parking light, left
M^2 - Parking light, right
M^3 - Parking light, left (USA version)
M^4 - Parking light, right (USA version)

N - Ignition distributor
O - Ignition coil
O^1 - Automatic choke
P^1 - Spark plug connector for No. 1 cylinder
P^2 - Spark plug connector for No. 2 cylinder
P^3 - Spark plug connector for No. 3 cylinder
P^4 - Spark plug connector for No. 4 cylinder
Q^1 - Spark plug for No. 1 cylinder
Q^2 - Spark plug for No. 2 cylinder
Q^3 - Spark plug for No. 3 cylinder
Q^4 - Spark plug for No. 4 cylinder
R^1 - Radio
R^2 - Antenna
S - Fuse box
T - Cable connector
T^1 - Cable connector, single
U^1 - Front indicator light, left
U^2 - Front indicator light, right
V^1 - Door contact switch, left
V^2 - Door contact switch, right
W - Windshield wiper motor (3 connections)
X^1 - Indicator, stop and tail light, left
X^2 - Indicator, stop and tail light, right
Y - Interior light
Z - License plate light

① - Battery ground strap
② - Steering column flange ground strap
③ - Transmission to frame ground strap
④ - Windshield wiper motor to body ground strap

Black dotted line = Service installation
Grey dotted line = USA version
Grey cross section data = USA version

Karmann Ghia 1200

(from August 1964)

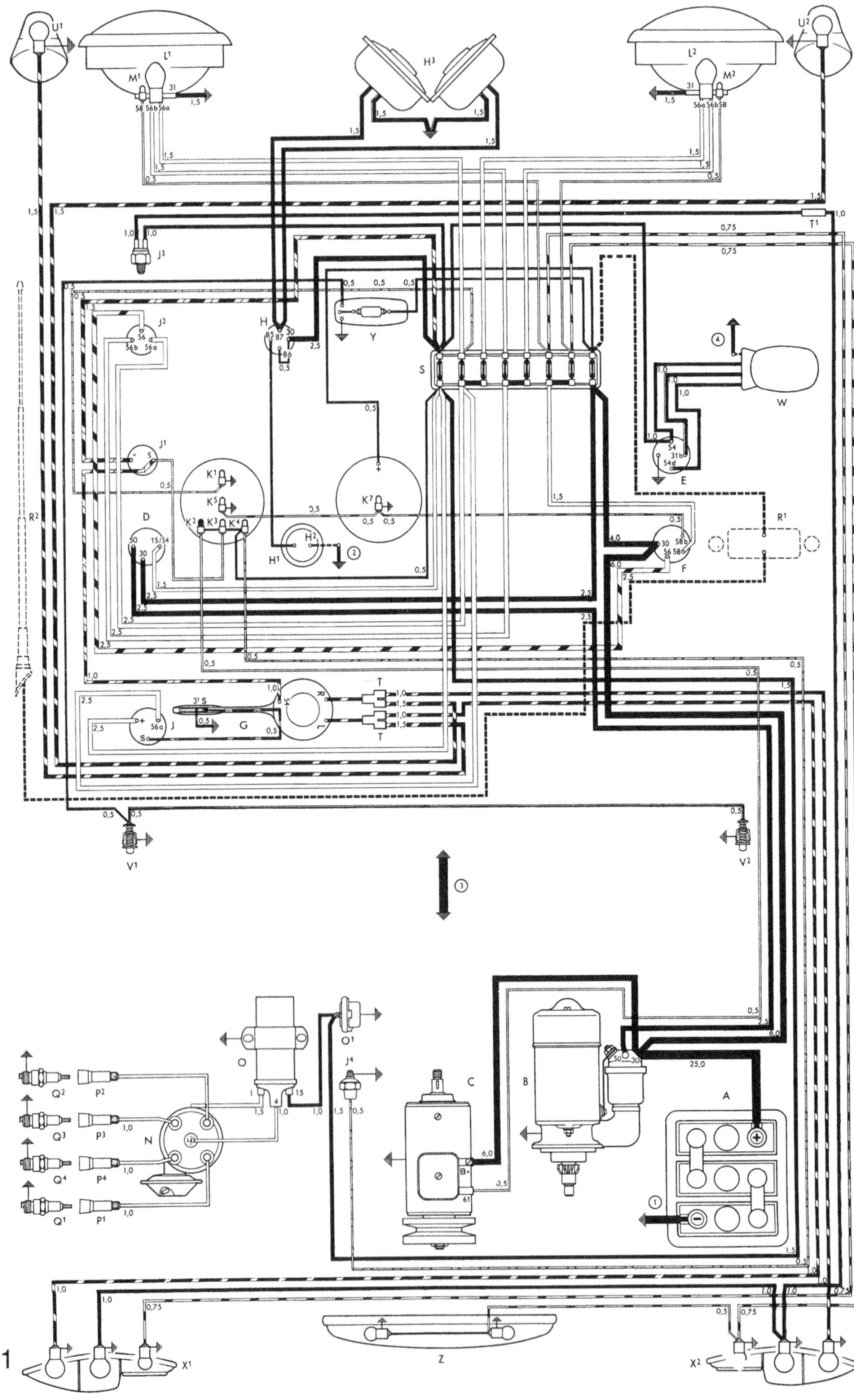

Key to Wiring Diagram

A - Battery
B - Starter
C - Generator
D - Ignition-starter switch
E - Windshield wiper switch
F - Lighting switch
G - Indicator switch with headlight flasher
H - Twin horn relay
H^1 - Horn ring
H^2 - Steering column connection
H^3 - Twin horn
J - Headlight flasher relay
J^4 - Indicator flasher relay
J^2 - Dimmer switch
J^3 - Stop light switch
J^4 - Oil pressure switch
K^1 - High beam warning light
K^2 - Generator control light
K^3 - Indicator control light
K^4 - Oil pressure control light
K^5 - Speedometer light
K^6 - Fuel gauge light
K^7 - Clock light
L^1 - Bifilament bulb for headlight, left
or Sealed-Beam insert, left
L^2 - Bifilament bulb for headlight, right
or Sealed-Beam insert, right
M^1 - Parking light, left
M^2 - Parking light, right
M^3 - Parking light, left (USA version)
M^4 - Parking light, right (USA version)

N - Ignition distributor
O - Ignition coil
O^1 - Automatic choke
P^1 - Spark plug connector for No. 1 cylinder
P^2 - Spark plug connector for No. 2 cylinder
P^3 - Spark plug connector for No. 3 cylinder
P^4 - Spark plug connector for No. 4 cylinder
Q^1 - Spark plug for No. 1 cylinder
Q^2 - Spark plug for No. 2 cylinder
Q^3 - Spark plug for No. 3 cylinder
Q^4 - Spark plug for No. 4 cylinder
R^1 - Radio
R^2 - Antenna
S - Fuse box (8 fuses)
T - Cable connector
T^1 - Cable connector, single
U^1 - Front indicator light, left
U^2 - Front indicator light, right
V^1 - Door contact switch, left
V^2 - Door contact switch, right
W - Windshield wiper motor (3 connections)
X^1 - Indicator, stop and tail light, left
X^2 - Indicator, stop and tail light, right
Y - Interior light
Z - License plate light

① - Battery ground strap
② - Steering column flange ground strap
③ - Transmission to frame ground strap
④ - Windshield wiper motor to body ground strap

Black dotted line = Service Installation
Grey or brown dotted line = USA version
Grey cross section data = USA version

THIS PAGE INTENTIONALLY
LEFT BLANK

Generator

Make: BOSCH LJ/REG 180/6/2500 L3

General Description

The output of the generator is controlled by the regulator BOSCH RS/TAA 180/6/4.

Specifications

Nominal voltage:	6 volts
Nominal output:	180 watts
Nominal speed:	2500 r.p.m.
Cut-in speed:	1660—1950 r.p.m.

The output of 180 watts at 2500 r.p.m. corresponds to 1385 r.p.m. of the crankshaft 48 kph/30 mph in the fourth speed. The generator is attached to its support by means of a strap. The armature is supported by a single-thrust ball bearing at each end. The armature shaft carries an adjustable pulley on the commutator end and the fan on the other end.

The generator regulator consists of two independent units encased in a sealed, moisture-proof and dirt-proof box mounted on the generator. The function of the cutout relay or circuit breaker is to close the circuit between the generator and the battery when the generator is producing current and to open this circuit so that the battery cannot discharge back through the generator when the generator slows or stops. The current limiting regulator and voltage regulator is designed to prevent the output and voltage of the generator from exceeding the predetermined safe maximum. It is important only to use the regulator which corresponds to the nominal output of the generator. A regulator designed to suit a higher output would lead to overcharging while a regulator for a lower output would cause the generator to produce insufficient current for the electric circuit.

Connections:

Terminal B+ (51) to terminal 30 at the starting motor (battery lead).

Terminal 61 to generator warning lamp.

Maintenance

The ball bearings of the generator are packed with BOSCH High Melting Point Grease and require no attention under normal conditions. Lubrication is, in general, only carried out when the generator is undergoing an overhaul. Never use ordinary grease. The brush wear should be inspected. Worn brushes should be replaced if trouble is experienced or when carrying out repairs on the generator.

Generator Warning Lamp

General Description

A red lamp is connected to a cable between terminals 15 and 61 of the regulator to provide an indication when the ignition is switched on. This lamp lights up when the engine is stationary or running slowly and goes out when the generator commences to charge.

The lamp also provides a control over the fan belt and the cooling fan. If the belt breaks, the generator and fan cease to revolve and the lamp lights up.

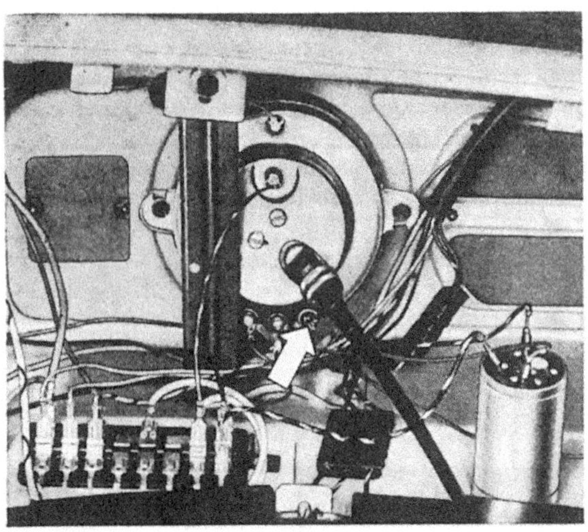

Bulb Type:

J 6 V 1.2 W DIN 72601

Bulb Replacement

The lamp is accessible from behind the speedometer.

1 - Remove socket and bulb from the speedometer.

2 - Press bulb lightly into its socket, turn it to the left and pull it out.

3 - Install new bulb in reverse order.

Testing Generator and Regulator

The fact that the generator warning lamp goes out after starting and when the engine picks up speed is not in itself a definite indication of a proper regulator setting and adequate charging of the battery. The generator can first be tested in situ. A good electric current and voltage testing instrument or a moving coil-type voltmeter (calibrated 0—30 volts) and a moving coil-type ammeter (calibrated 50—0—50 amperes) are required.

1 - Disconnect cable from terminal B+ (51) at the regulator. Connect the positive lead of the voltmeter to terminal B+ (51) at the regulator and ground the negative lead.

2 - Start engine. Increase the engine speed gradually from idle (approx. 500 rpm.) to 1895—2220 rpm., which corresponds to a generator speed of 3500—4000 rpm. The hand should jump from 0 to 6—7 volts at increased idling speed and should then constantly register between 7.4 and 8.1 volts (exactly 8.1 volts at a room temperature of 20° C = 68° F), provided that the regulator is correctly set.

3 - When stopping the engine, watch carefully for the flicking back of the voltmeter hand, which indicates that the cutout relay is operating correctly.

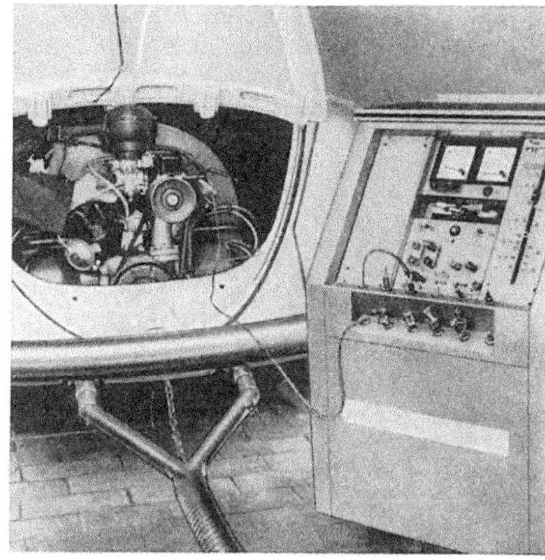

Testing Generator without Regulator

(Quick check)

1 - Disconnect both cables from regulator.

2 - Connect cable F (DF) on the generator to ground (D—).

3 - Connect voltmeter positive terminal to + (D+) cable on generator and negative terminal to ground (D—).

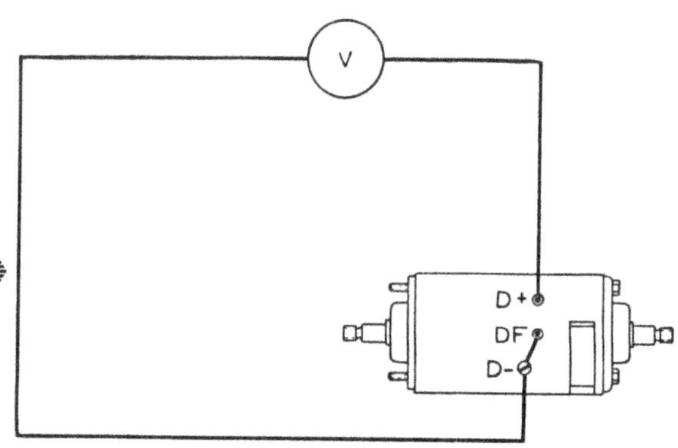

4 - Run generator at the following speeds for a short time and check the voltage produced.

Rpm	Voltage
1500	approx. 6 V
3000	approx. 18 V

5 - If the generator produces no voltage or too low a voltage it should be removed and checked.

Note:

Under certain weather conditions a graphite coating can form on the commutator. This can be removed with fine polishing cloth without removing the generator.

Important

This test should only be carried out for a few seconds as otherwise the field windings will burn out.

Testing Charging Current

Even if the regulator adjustment is correct, the battery may not be receiving sufficient charge.

To check the charging current proceed as follows:

1 - Disconnect battery ground cable.

2 - Disconnect cable from regulator terminal B+ and connect a suitable ammeter (50—0—50 A range) between this cable and regulator B+ terminal.

3 - Connect battery ground cable again.

4 - Run engine at high speed and switch on normal current consuming equipment.

5 - If the ammeter shows no current output from the generator, the regulator is defective and must be replaced.

6 - Decrease the engine speed slowly to idling level. The ammeter needle should drop past the zero mark on the instrument into the negative range and thus indicate that the current is flowing back from battery to generator. Just before the engine reaches idling speed the regulator cut-out switch should operate and the ammeter reading return to zero. The regulator adjustment cannot be judged from the strength of the charging current as this current is governed by the state of charge of the battery.

Testing Current Regulator

To check the current regulator independently of the state of charge the battery you require:

a - a moving coil voltmeter calibrated 0—30 volts.

b - a moving coil ammeter calibrated 50—0—50 amperes.

c - a sliding resistance (R), suitable for a load of 50 amperes.

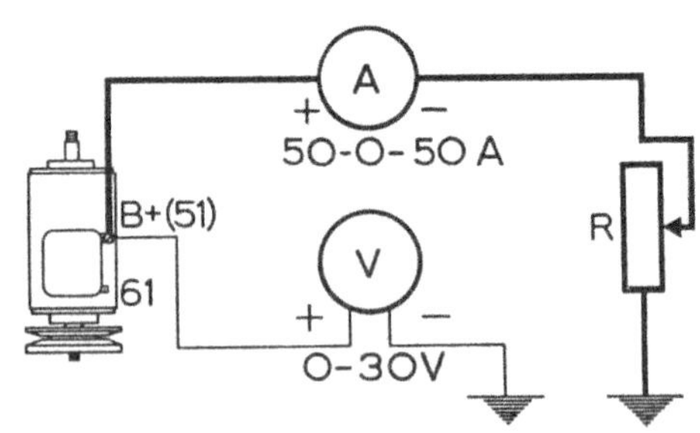

The check can be carried out with the generator in the vehicle or on the test bench.

1 - Disconnect cable from terminal B+ (51) at the regulator.

2 - Connect sliding resistance in series with the ammeter between terminal B+ (51) at the regulator and ground.

3 - Connect voltmeter positive terminal to B+ (51) terminal on the regulator and the negative terminal to generator ground.

> **Important**
>
> Use only short cables with a minimum cross section of 6 mm². Check for good ground and terminal connections to avoid high resistances and false readings.

4 - Start engine. With a generator speed of 4000 r.p.m. and the resistance set to give a load of 45 Amps, the voltage should be at least 6—7.15.

If the readings are not within these limits, the regulator should be replaced. Any re-setting or repairing of the regulator should be referred to a special workshop which is equipped with the testing appliances required. Any interference with the regulator system, such as cleaning or re-machining the contact points, may completely alter the electrical adjustments, and cause great damage to the electrical system.

> **Important**
>
> Cable connections on the generator and regulator must only be connected or disconnected with the engine stationary and cable B+ (51) disconnected as a short circuit will render the regulator unserviceable. Interchanging the cables on + (D+) and F (DF) terminals will also damage the regulator.

Removing and Installing Regulator

Removal

1 - Disconnect cables from terminals B+ (51) and 61 at the regulator.

2 - Remove the slotted screws attaching the regulator to the generator and take off regulator.

3 - Disconnect the two cables from the terminals marked + (D+) and F (DF) at the bottom of the regulator.

Installation

This is a reversal of the above procedure, but the following points should be noted:

1 - The thicker cable coming from the positive brush must be connected to the terminal + (D+) at the bottom of the regulator.

2 - The thin cable from the field windings must be connected to the F (DF) terminal at the bottom of the regulator. If replacing the regulator does not give the desired results, then generator is defective.

Checking Brushes and Commutator

1 - Examine brushes for wear. If they no longer portrude from the holders they are worn out and should be replaced. By new brushes of the same type.

2 - If the commutator is dirty or oily, clean with a cloth dampened with fuel.

If the commutator is worn, rough or burned, it should be remachined.

3 - Replacement of springs or reconditioning the commutator calls for a removal of the generator.

Removing and Installing Generator

(Engine installed)

Removal

1 - Disconnect cables from regulator.

2 - Remove carburetor.

3 - Take off fan belt.

4 - Disconnect generator mounting strap.

5 - Remove throttle ring and the screws on both sides of the fan housing. Lift housing a few inches.

6 - Remove four screws on fan cover and take off generator and fan.

Installation

When installing, the following points should be observed:

1 - When installing the fan housing, make sure that it seats properly against the cylinder cove plates all around.

2 - Note concentric position of throttle ring.

Disassembling and Assembling Generator

Disassembly

1 - Take off fan pulley and spacer washers.

2 - Remove fan nut by means of the wrench VW 112 and take off fan and spacers, fan hub and woodruff key. The fan should be held in position by a second mechanic.

3 - Remove fan cover and reinforcement flange.

4 - Remove regulator.

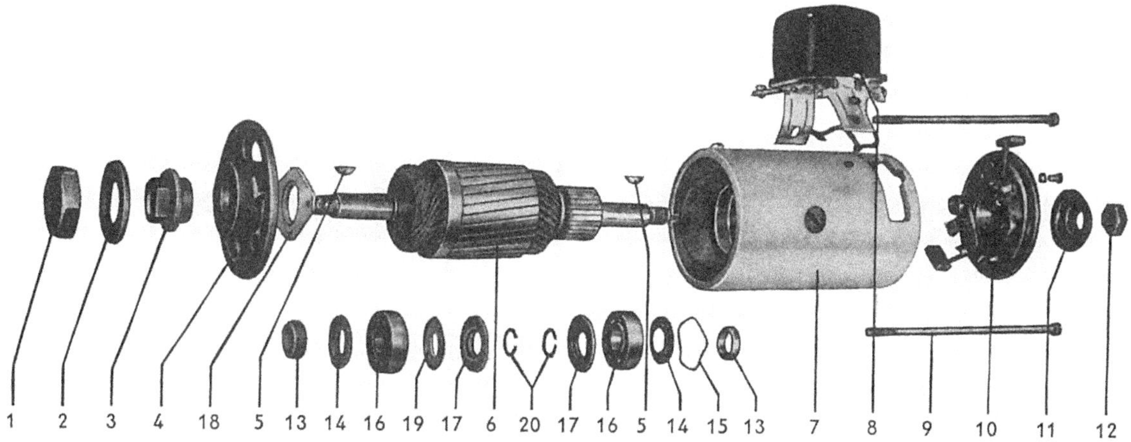

1 - Fan Nut
2 - Carrier Plate
3 - Fan Hub
4 - End Plate
5 - Woodruff Key
6 - Armature
7 - Housing and Field Assembly
8 - Regulator
9 - Housing screws
10 - Brush Holder End Plate
11 - Spacer Washer
12 - Pulley Nut
13 - Spacer Ring
14 - Oil Slinger
15 - Spring Ring
16 - Ball Bearing
17 - Oil Slinger
18 - Flange
19 - Cover Washer
20 - Circlip

5 - Disconnect field coil terminal from brush holder of the positive brush.

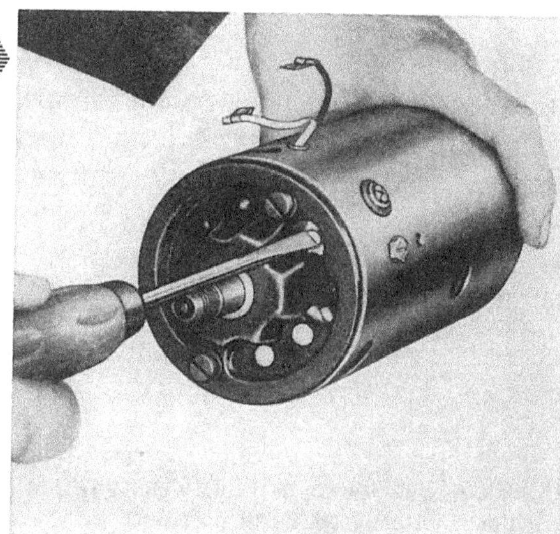

6 - Screw out the two housing screws.

7 - Lift up brushes and take off commutator end plate. Take off ball bearing spring ring.

8 - Withdraw end plate and armature as a unit from the side of the frame.

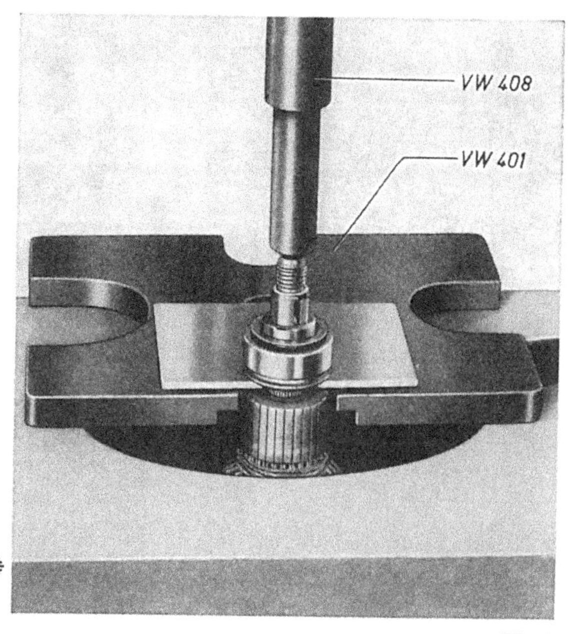

9 - Press the ball bearing on the commutator side, including oil slingers, and spacer ring off the armature with the Repair Press VW 400 in conjunction with VW 401 and VW 408.

A plate of 2.5 mm (.1") thickness, machined to the dimensions given below should be placed under the bearing.

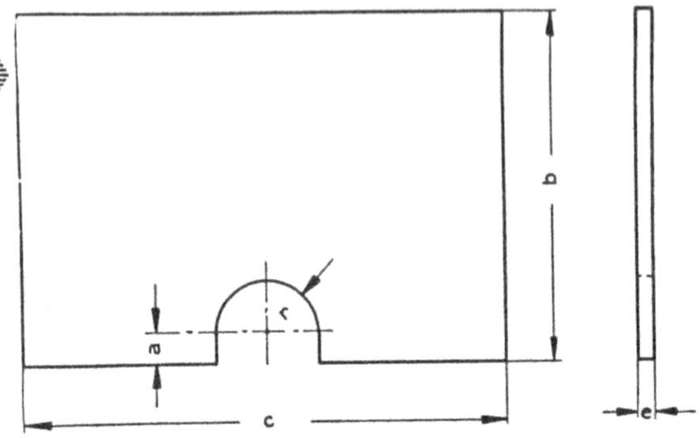

a = 6 mm (.24") c = 100 mm (4")
b = 70 mm (2.8") e = 2.5 mm (.1")
r = 11 mm (.43")

10 - Remove end plate on the fan end by means of Repair Press VW 400 in conjunction with VW 401 and VW 408.

11 - Remove ball bearing oil slinger from armature and take spacer ring out of end plate.

12 - Unscrew flange and take bearing, cover washer and oil slinger out ot end plate.

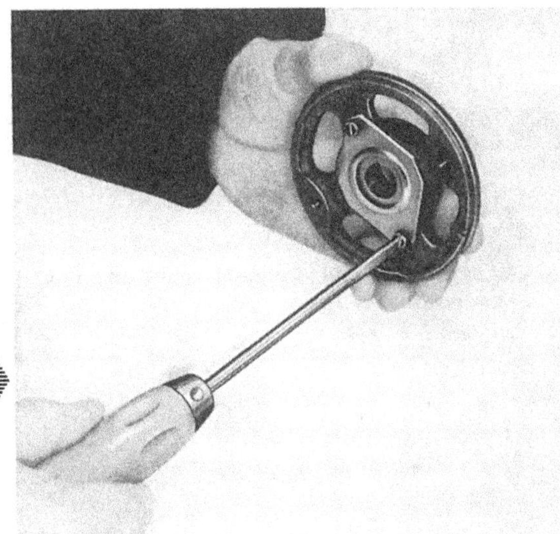

13 - The two retainer rings can be left on the armature shaft.

14 - If necessary, remove field coils with the special appliance designed for the purpose.

Inspection

1 - Inspect generator parts for wear, open circuit, short circuit and ground, replace as necessary.

2 - Thoroughly rinse the bearings in fuel and re-pack them with BOSCH High Melting Point Grease.

3 - Check commutator for signs of wear and bur-ned spots and overhaul if necessary.

Test

In a large number of cases the armature does not give visible evidence of trouble. The armature is tested for open circuits, short circuits and ground.

1 - Open circuits in the armature are usually readily apparent, since this condition causes burned spots between the commutator segments. Open circuits can also be determined by a sensitive growler, which is, however, rarely available in workshops.

2 - A shorted winding can only be tested be means of a growler. Place the armature on the growler, turn the armature slowly and hold a thin steel strip or hacksaw blade over it, as shown in the illustration. Short circuits in the armature cause the steel strip or hacksaw blade to vibrate against the core when it is held above the slot containing the shorted winding.

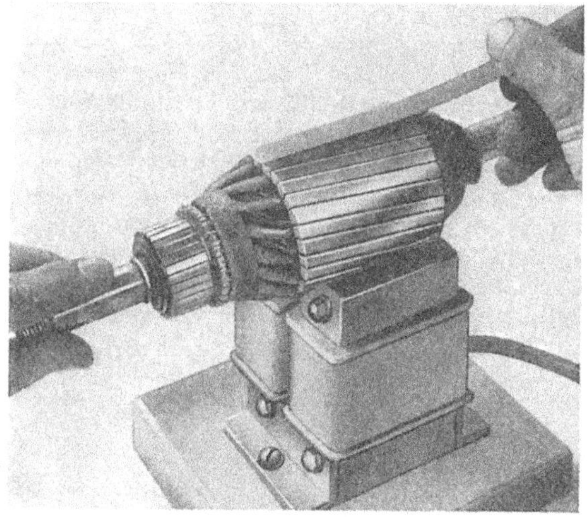

With another type growler the armature is turned, while the mechanic moves a feeler along the armature core. A short circuit in the windings is indicated by a growling noise in the earphones due to alternating current generated in the coil of the feeler by induction.

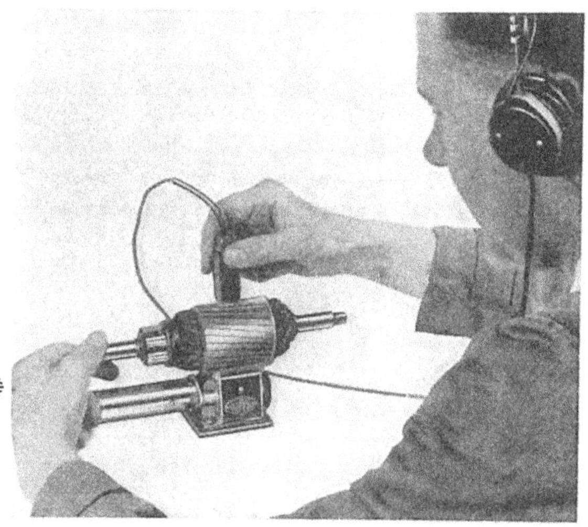

3 - The armature is grounded when the armature core contacts the windings or when carbon dust has entered the windings.

A test is carried out with a normal 220 volts test lamp (not a glow lamp) with test points held on commutator and armature core.

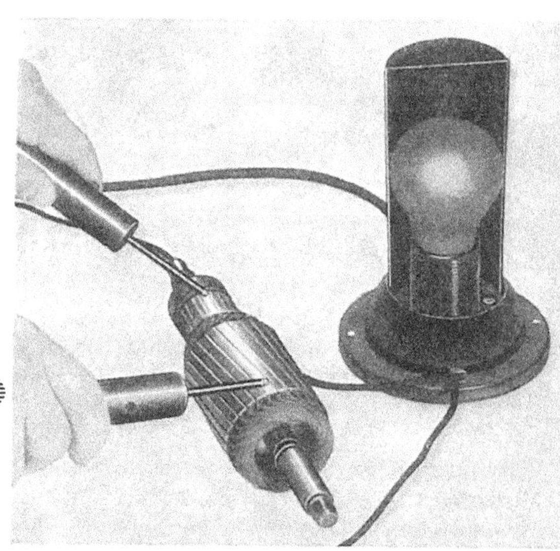

Commutator Repair

The commutator consists of copper segments which are insulated from each other. If the commutator is out of round, burnt or ridged by the brushes, it must be turned down and polished. The armature should be clamped at the bearing surfaces for turning and not held between the lathe centers. The normal diameter of the commutator is 33 mm and it must not be turned below 30.5 mm. The permissible commutator run-out is 0.02 mm. The insulation between the segments is undercut with a special commutator saw until it is approximately 0.3—0.4 mm below the surface of the commutator. Large workshops are advised to use a commutator milling machine for this purpose. Take care that no chippings are left between the segments as this could cause a short circuit in the armature windings.

Testing Field Coils

The two field coils are tested for open circuits, short circuits and ground.

Test

1 - Test each field coil individually for open circuits by connecting their ends with a normal 220-volt test lamp or a battery in series with a 6-volt test lamp.

2 - Short circuits in the field windings can be checked by connecting an ohmmeter to the ends of each coil and comparing the readings.

Should such an instrument not be available, connect a 6-volt battery in series with an ammeter to the coil ends and compare the current draw of the two coils. If the current draw of one coil is higher (difference more than 0.5 amperes), there is a short circuit in the windings.

3 - Test the field coils for ground by connecting a normal 220-volt test lamp with the end of one field coil and the generator housing.

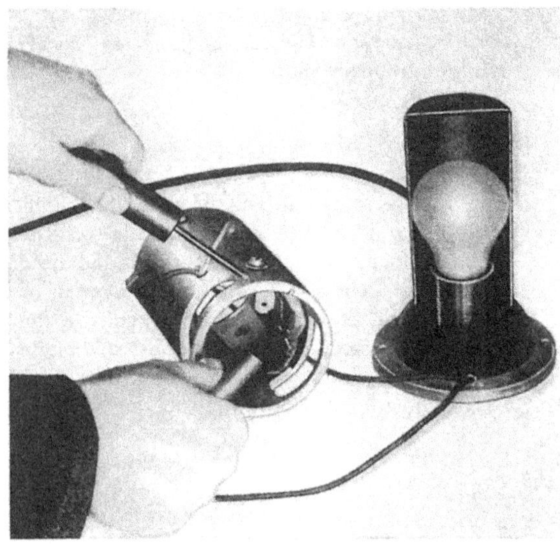

4 - Test for ground as at 3 with the regulator installed but without the + (D+) and F (DF) cables connected.

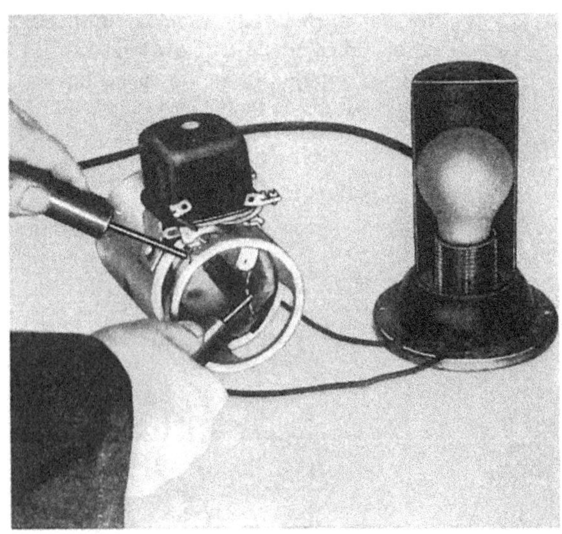

Important

When installing a regulator or subsequently installing a suppression condenser, check the length of the securing screws. If these screws are too long they can ground the field coils.

Assembly

To assemble, reverse the removal procedure while observing the following points:

1 - Install the oil slinger and ball bearing in the fan side end plate, place the cover washer on the bearing and secure the flange to the end plate.

2 - Place an oil slinger on the armature shaft (fan side) and press the end plate on with the VW Repair Press 400, with VW 401 and VW 409.

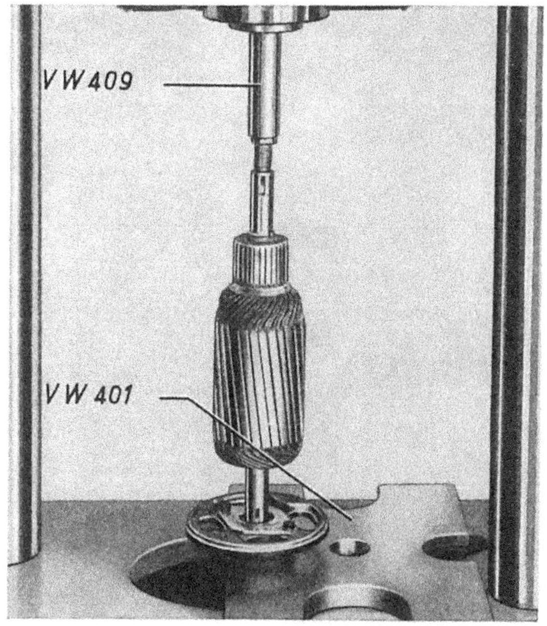

3 - Press both oil slingers and bearing on the commutator side of the armature shaft with Repair Press VW 400 in conjunction with VW 401, VW 412 and VW 421.

4 - Insert fan side end plate and armature into the housing so that the nose engages in the groove.

5 - Place spring ring in commutator end plate and insert the end plate in the housing so that the nose engages in the groove.

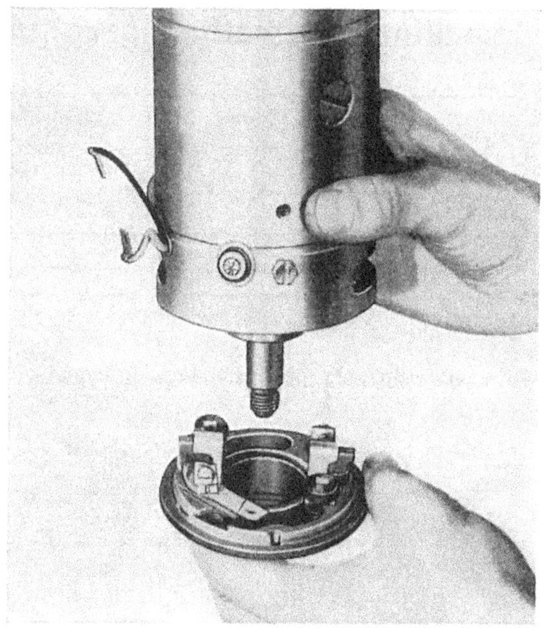

6 - Connect field coil lead to brush holder of positive brush.

7 - Tighten the two housing screws.

8 - Install the two spacers on the armature shaft with the Repair Press VW 400 in conjunction with VW 401, VW 412 and VW 421.

9 - Press brushes against the commutator and make sure the brush springs are properly seated.

10 - Connections of regulator:

light colored cable to terminal + (D+)
dark colored cable to terminal F (DF)

11 - Place spacer washers for fan in position so that the installed fan does not foul fan housing or the fan cover.

12 - Tighten fan nut with torque wrench and a 36 mm socket to a torque of between 5.5 and 6.5 mkg (40—47 ft. lbs.).

Note:

From 4th December 1963, Chassis No. **5 967 385**, Engine No. **8 204 702**, a modified Bosch generator with the designation 0 101 212 003* (VW Part No. 111 903 021 F) is installed. The connecting cables to the voltage regulator pass through holes in the housing. The test data for the modified generator is the same as for the Bosch LI/REG 180/6/2500 L 3 type.

* **Note:**

Messrs. Bosch have discontinued the previous type of part designation. We have, therefore, quoted the Bosch Hollerith number here in addition to the VW part number. In future, all Bosch parts will be marked with both designations in manufacture.

The new generator is fitted with a modified Bosch Type RS/TAA 180/6/A 4 regulator. The former Bosch RS/TAA/180/6/4 regulator (Part No. 113 903 801 B) cannot be combined with the new generator.

Repair instructions:

On the modified generator the cables on the carbon brushes are welded to the brush holders (A). When new brushes have to be fitted, the cable should be cut off close to the brush holder.

The 111 903 515 A carbon brushes can be used as replacements. The cable terminal can be attached to the tab of the brush holder (B) with a screw.

Checking Generator Operation and Output

Cut-in speed *) rpm	Cut-in voltage V	Return current A	No-load regulating V	Regulating voltage under load V	Load current A
1660—1650	6,2—6,8	3—7	7,4—8,1	6,4—7,3	34

*) generator rpm.

All values apply at a generator housing temperature of 20° C, 68° F

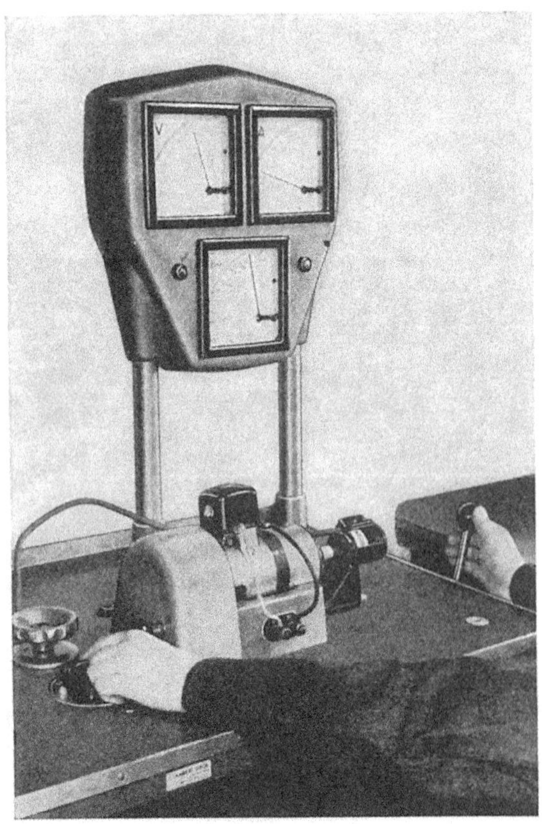

The generator can be accurately tested on a suitable test bench. In order to judge the condition and the output capacity of the generator the following readings are required:

1 - No-load voltage.

2 - Charging current.

3 - Cut-in rpm.

4 - Cut-in voltage.

5 - Output.

6 - Check regulator contacts for contact resistance.

Generator Trouble Shooting

The red generator lamp lights up when the ignition is switched on and should go out when the engine has been started and the speed increases.

Symptom	Cause	Remedy
Generator lamp does not light up with ignition switched on	a - Battery discharged b - Battery defective c - Bulb burned out d - Corroded or loose battery terminals e - Loose connections or broken cables f - Ignition starting switch defective g - Generator brushes not making contact with commutator	a - Charge battery b - Renew battery c - Renew bulb d - Clean or tighten terminals respectively e - Tighten or repair cables respectively f - Renew ignition starting switch g - Free the brushes or renew them. If necessary renew the brush springs
Generator lamp does not go out or flickers when engine is accelerated	a - Fan belt loose or faulty b - Regulator faulty c - Charging cable loose or defective d - Generator faulty e - Commutator graphited	a - Adjust belt tension or renew belt b - Renew regulator c - Check cables and connections d - Check generator e - Clean commutator with fine polishing cloth
Generator lamp goes out only at high speed	a - Generator faulty b - Regulator faulty	a - Check generator b - Renew regulator
Generator lamp remains on with the ignition switched off	a - Regulator contact points sticking (burned)	a - Renew regulator

Charging control lamp glows when driving

If the red charging control lamp glows when the vehicle is in motion but the generator and voltage regulator are in order, the cause can only be an excessively high voltage drop in the connections between regulator terminal B+ and the charging control lamp. This voltage drop can appear at the following points in the electrical system:

Regulator, terminal B+

Lighting switch, terminal 30

Fuse box, terminal 30 (Fuse 1 lower)

Ignition/starter switch, terminal 30

Ignition/starter switch, terminal 15/54

Fuse box, terminal 15/54 (Fuse 8 lower)

All these connections should be checked as follows:

1 - Start engine and run it at a fast idling speed.

2 - Switch on headlamps, windscreen wipers and turn indicators.

3 - Strip insulation at both ends of piece of 2.5 mm² cable about 13 feet long and connect it to terminal B+ on regulator.

4 - Contact all the connections listed here with the free end of the cable, one after the other and have an assistant watch the charging control lamp.

5 - If the charging control lamp ceases to glow when one of the connections is touched with the cable, the fault has been found.

Between this connection and the one tested before it, is the voltage drop which is causing the control lamp to glow.

The voltage drop can be caused by loose connections, poor contacts in switches or damaged cables. Damaged parts should be replaced and loose or corroded connections repaired.

General Description

The voltage-controlled generator is equipped with a regulator of the type VW 113 903 801 C.

Only the operations which differ from those on the Bosch generator are described here.

Data

		Nominal speed:	2400 r.p.m.
Nominal voltage:	6 volts	Cut-in speed:	1350—1600 r.p.m.
Nominal output:	180 watts	Zero watt speed:	1500 r.p.m.

Maintenance

The ball bearings of the generator are packed with high melting point grease and require no attention under normal conditions. Re-packing with Bosch high melting point grease is, in general, only necessary when the generator is undergoing an overhaul. Never use ordinary grease.

Examine brushes for wear and make sure they are free in their guides. Worn brushes must be replaced. The regulator requires no maintenance.

Regulator

The regulator contains no replaceable wearing parts and cannot be repaired in a VW Workshop; the regulator cap should, therefore, not be removed. Any warranty claims will be rendered void should it be found that the cap was removed. The regulator should, if necessary be tested in connection with a generator which is known to be in good condition.

Generator Disassembly and Assembly

Disassembly

1 - Remove fan pulley, and spacer washers.

2 - Remove fan nut with wrench VW 112 and take off fan, spacer washers, fan hub, and key.

3 - Remove fan cover and reinforcement flange.

4 - Remove regulator.

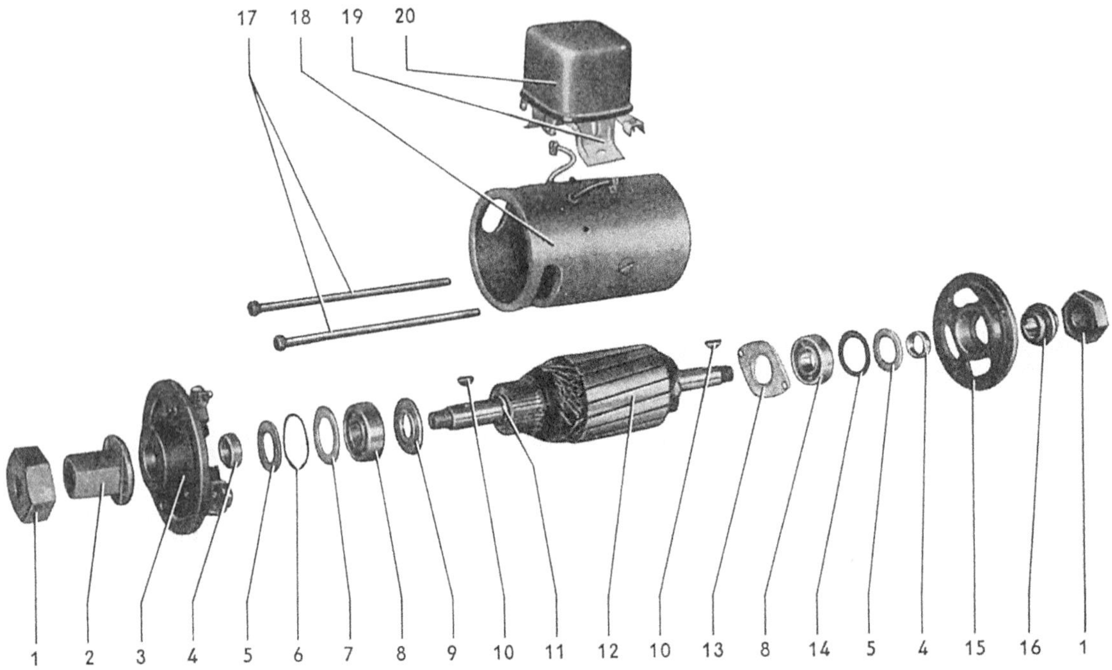

1 - Nut
2 - Pulley hub
3 - Brush holder end plate
4 - Spacer ring
5 - Felt washer
6 - Thrust ring
7 - Retainer
8 - Ball bearing
9 - Guard
10 - Key
11 - Spacer
12 - Armature
13 - Bearing retainer
14 - Thrust ring
15 - End plate
16 - Fan hub
17 - Housing screws
18 - Housing and field assembly
19 - Slotted screw
20 - Regulatur

5 - Disconnect field coil lead from brush holder of positive brush.

6 - Lift up the brushes and fix them in the raised position by means of the brush springs.

7 - Remove the two housing screws.

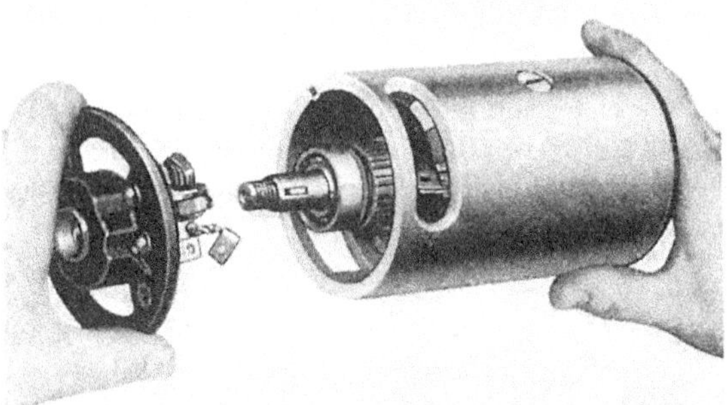

8 - Pull off brush holder end plate. Take off thrust ring, felt washer, retainer, and spacer.

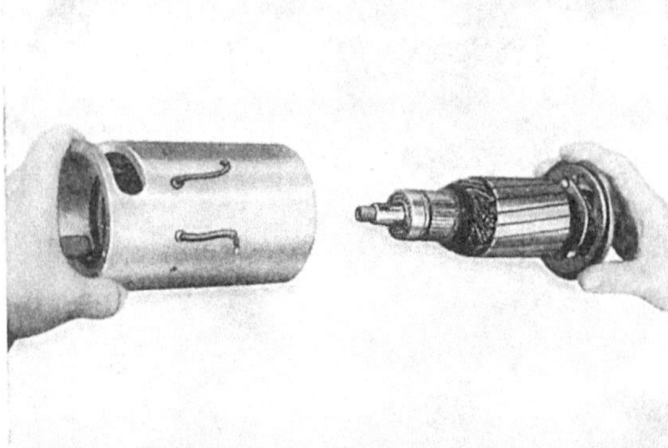

9 - Withdraw end plate and armature as a unit from the fan side of the housing.

10 - Remove ball bearing and guard on the commutator side by means of the Repair Press VW 400 in conjunction with VW 401 and VW 408.

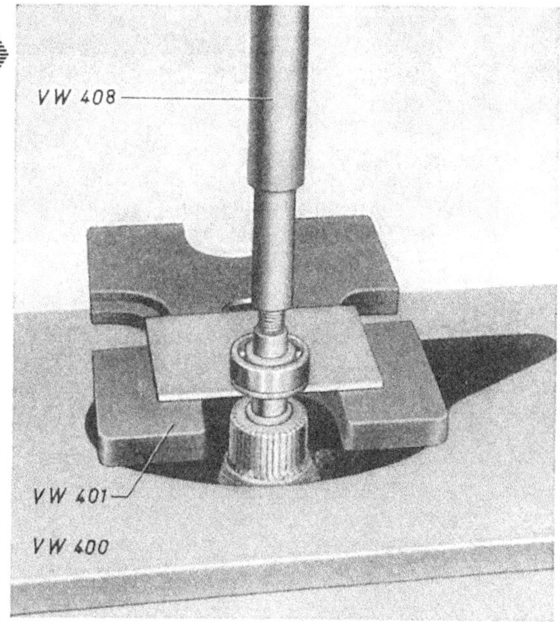

A plate of 2.5 mm (.1″) thickness, machined to the dimensions given below, should be placed under the bearing.

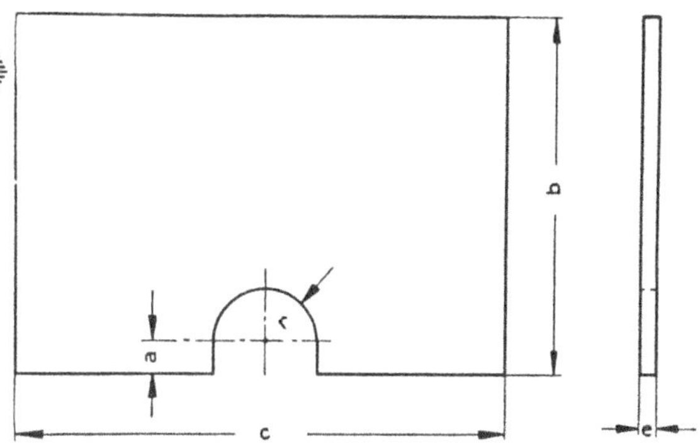

a = 6 mm (.24″) c = 100 mm (4″)
b = 70 mm (2.8″) e = 2.5 mm (.1″)
r = 11 mm (.43″)

11 - Take off the spacer ring.

12 - Remove the two bearing retainer screws and withdraw the end plate.

13 - Take off felt washer, thrust ring, and spacer.

14 - Remove ball bearing and retainer on Repair Press VW 400 as on commutator side.

15 - If necessary, remove field coils by means of a specially designed appliance after unsoldering the terminals.

Inspection

1 - Inspect generator parts for wear, open circuit, short circuit, and ground.

2 - If the commutator is out of round, rough due to burnt spots, or if it is grooved, it should be turned down in a lathe and polished to obtain a smooth surface. When doing this, special attention should be paid to the correct centering of the armature in the lathe. The run-out of the commutator must never exceed 0.02 mm (.0008″) to assure a perfect operation of the generator. The diameter of the commutator must not be reduced below 32 mm (1.26″).

3 - Test the brush spring tension with the brushes in position by means of a spring scale. The correct tension in 450 ± 25 grammes (16 ozs. ± .88 oz.). Check that the spring tension is neither above nor below the specified value.

4 - Thoroughly rinse the bearings in fuel and re-pack them with BOSCH High Melting Point Grease.

Assembly

To assemble, reverse the preceding operations, while observing the following points:

1 - Assemble the end plate of the fan side by placing felt washer in its recess, inserting thrust ring and bearing in end plate bore and screwing the retainer in position.

2 - Install assembled end plate on armature shaft with Repair Press VW 400 in conjunction with VW 401 and VW 409. Slide spacer ring into position.

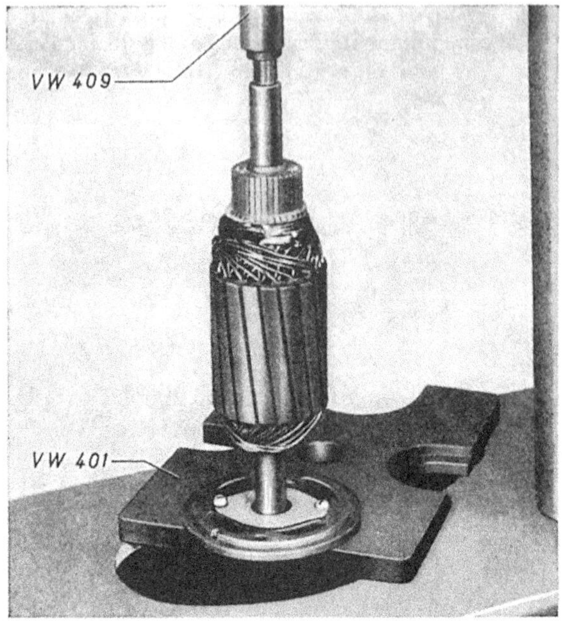

3 - Install guard and ball bearing on the commutator side with VW Repair Press VW 400 in conjunction with VW 401, VW 412 and VW 421.

4 - Before installing an overhauled generator its polarity must be matched to the battery. This is done by montoring it for a short time by connecting as follows:

Terminal F (DF) of generator to ground (D—)

Battery positive to terminal + (D+)

Battery negative to ground (D—)

Important
If the cables on terminals + (D+) and D— are interchanged, even for a short time, the generator polarity will be reversed and this will, in turn, damage the regulator.

5 - Connections of regulator.

Thicker cable to terminal 61,
Thinner cable to terminal DF.

6 - Place spacer washers for fan in position so that the installed fan does not foul the fan housing.

7 - Tighten fan nut with Torque Wrench and socket VW 163 to a torque of between 5.5 and 6.5 mkg (40—70 ft. lbs.).

Checking Generator Operation and Output

No-load voltage	Charging current	Cut-in r.p.m. *)	Cut-in Voltage	Rated output
7.4—8.1 V	max. 45 A at 6.0—7.2 V	1350—1660	5.9—6.6	180 W at 6 V and 2500 r.p.m.

*) generator r.p.m.

All measurements apply at a generator housing temperature of 20° C, 68° F.

Starting Motor

Type: BOSCH EED 0.5/6 L 49

General Description

The VW Passenger Car engine is started by a solenoid motor of the overrunning-clutch type, producing 0.5 HP. The starting motor draws a heavy current to give maximum torque. The drive end of the armature shaft is supported in a bush which is recessed in the transmission case. The starting motor is operated by turning the ignition key further, causing the solenoid starter switch to shift the pinion (9 teeth) into mesh with the flywheel ring gear (109 teeth). The final movement of the solenoid closes the copper contacts, switches on main current and turns the starting motor. As soon as the engine fires, the magnetic switch no longer draws current when the key is released and the pinion is returned by spring tension. If the starting motor is not immediately switched off when the engine starts, the overrunning clutch prevents the armature from being turned and protects the parts against damage.

A non-repeat lock in the ignition switch prevents the starter switch being operated when the engine is running and causing the starter pinion to strike the flywheel starter ring. Before the operation of the starter can be repeated, the ignition must be switched off by turning the key fully to the left.

Maintenance

The armature bearings should only be lubricated when overhauling the starting motor. When removing the engine, the armature bush should be inspected for wear (VW Gauge 246) and renewed if badly worn. The bush is to be filled with Special Grease prior to re-fitting the starting motor.

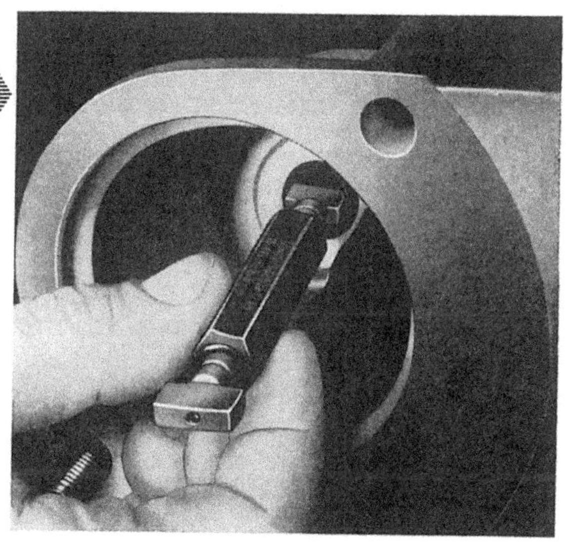

During repair, or if trouble is experienced take off starting motor end cap, inspect brushes for wear and make sure that they slide freely in their guides. Renew worn brushes and weak brush springs. If the commutator is rough and pitted, or shows burned spots, the starting motor must be overhauled. If dirty or oily the commutator can be cleaned by means of a clean cloth dampened with fuel and wrapped around a piece of wood.

Removing and Installing Starting Motor

Removal

1 - Disconnect negative lead from battery.

2 - Disconnect battery cable and cables to generator and ignition switch at terminal 30 of the starting motor.

3 - Disconnect control cable (to ignition/starting switch) at terminal 50 of the starting motor.

4 - Remove bolt and nuts that attach starting motor to transmission case.

5 - Withdraw starting motor.

Installation

When installing, the following points should be observed:

1 - Lubricate starter shaft bush with Special Grease.

2 - Apply Genuine VW Sealing Compound D 1 a between intermediate bracket and transmission case.

3 - Make sure that terminals are clean and tight.

Connections:

1 - Battery lead (positive terminal).
2 - Terminal 51 at generator.
3 - Terminal 50 ignition/starting switch.

Removing and Installing Solenoid Switch

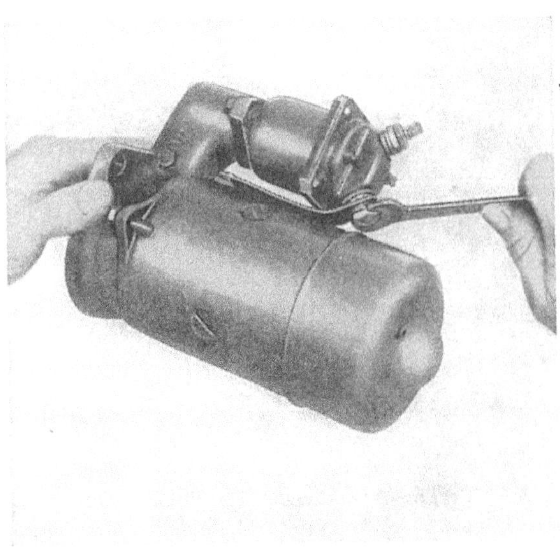

Removal

1 - Release connector strip from solenoid switch.

2 - Remove screws that attach solenoid switch to intermediate bracket.

3 - Remove shift lever bolt.

4 - Withdraw solenoid switch by overcoming the spring load.

Defective solenoid switches must be replaced. Never alter setting of the switch. Care should be taken that the distance between flange and center of the mounting bolt corresponds exactly with that of the old part.

Installation

When installing, the following points should be observed:

1 - The contacts of the solenoid switch may have become twisted by using too much force when tightening the nuts, which occasionally leads to damage of the insulation. Test them for ground with a 220-volt test lamp.

2 - To facilitate inserting solenoid switch, pull out drive pinion to bring the shift lever forked end as far as possible toward the solenoid switch clevis.

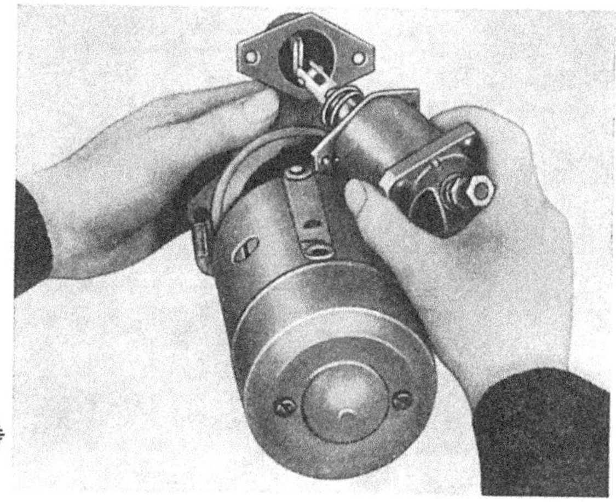

Inspecting Brushes and Commutator

1 - Remove starting motor end cap.

2 - Inspect brushes for wear and make sure that they slide freely in the guides of the brush holders. If the brushes are so deep in the holders that the flexible connector touches the holder edge, they are worn out and must be replaced by new ones of the same type. Also replace brushes which are oil-saturated or have loose flexible connectors. When replacing brushes, see that the flexible connector is free to avoid sticking of the brushes during operation.

3 - If the commutator is oily or dirty, clean with a cloth dampened with fuel and wrapped around a piece of wood.

Take care that no dirt or fuel enters the bearing.

4 - If the commutator surface is rough and pitted, or shows burned spots, the starting motor should be overhauled.

Disassembling and Assembling Starting Motor

Disassembly

1 - Detach connector from solenoid switch, remove end cap and lift up brushes.

2 - Clamp armature shaft in a vise (use soft jaws) and remove nut at the commutator end of the starting motor.

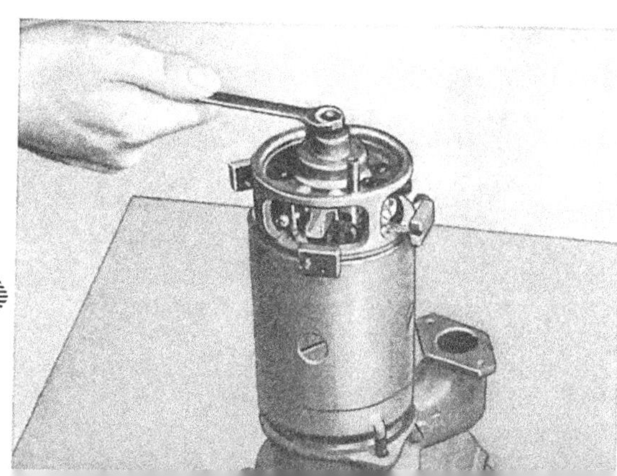

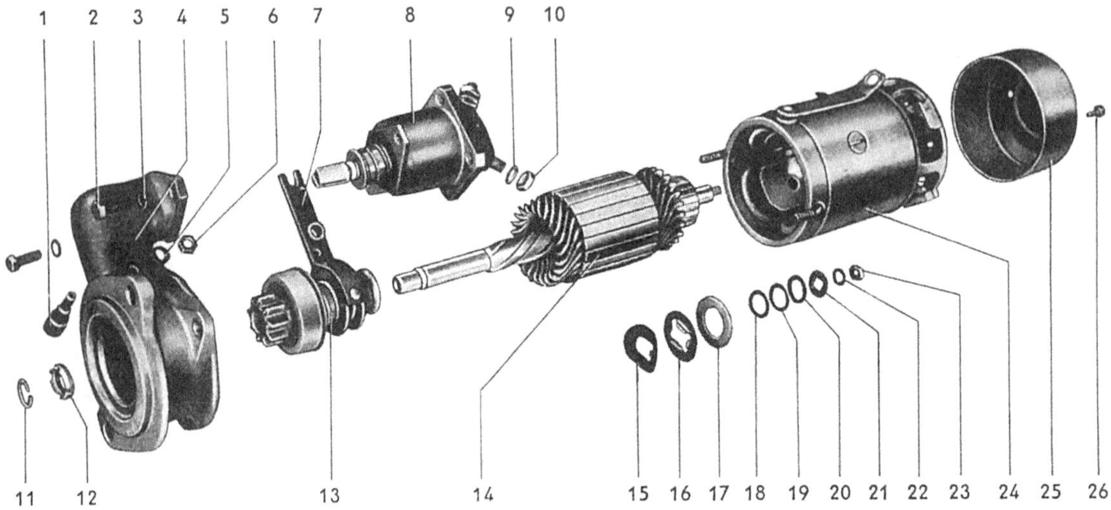

1 - Shift Lever Pivot Screw
2 - Bolt
3 - Washer
4 - Intermediate Bracket
5 - Washer
6 - Nut
7 - Shift Lever
8 - Solenoid Switch
9 - Spacer
10 - Nut
11 - Circlip
12 - Pinion Stop Ring
13 - Drive Pinion Assy
14 - Armature
15 - Lockwasher
16 - Retainer Plate
17 - Insulating Washer
18 - Thrust washer
19, 20 - Shims
21 - Cone Washer
22 - Lockwasher
23 - Nut
24 - Housing and Field Assy
25 - End Cap
26 - Slotted Screw

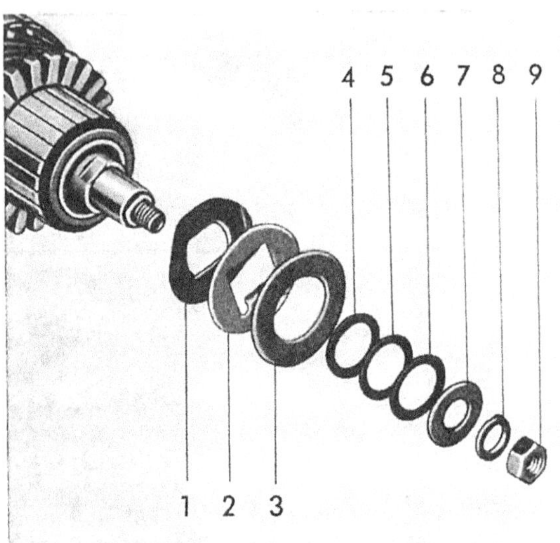

3 - Remove nuts of intermediate bracket hook studs and pull out intermediate bracket with armature. Note arrangement of washers for proper reassembly.

Inside:
1 - Lockwasher
2 - Retainer Plate
3 - Insulating Washer

Outside:
4 - Thrust washer
5, 6 - Shims
7 - Cone Washer
8 - Lockwasher
9 - Nut

4 - Place armature in a vise with the commutator end pointing downwards and drive back pinion stop ring by means of a hollow punch.

5 - Remove circlip and pull off stop ring. Remove any burr which has developed at the circlip groove.

6 - Withdraw armature from intermediate bracket and remove drive pinion.

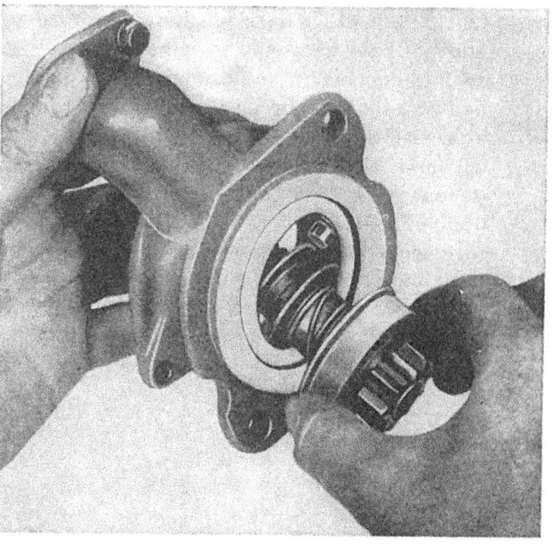

7 - Remove shift lever from intermediate bracket.

8 - Defective drive pinions should be replaced.

Testing Armature

In a large number of cases the armature does not give visible evidence of trouble. The armature is tested for open circuits, short circuits and ground.

1 - Open circuits in the armature are usually readily apparent, since this condition causes burned spots between the commutator segments. Check soldered commutator riser-bar connections.

2 - The armature is tested for short circuits on the growler. Place the armature on the growler and slowly revolve it while holding a thin steel strip or hack-saw blade above the armature core. Short circuits in the armature cause the steel strip or hack-saw blade to vibrate against the core when it is held above the slot containing the shorted winding.

See also pages 7-8 in Section E-2.

3 - The armature is grounded when the armature core comes into contact with the winding or when carbon dust has entered the windings (direct and indirect ground). The armature is tested electrically for ground by placing one test point of a normal 220 volt test lamp on the armature core and the other on the commutator.

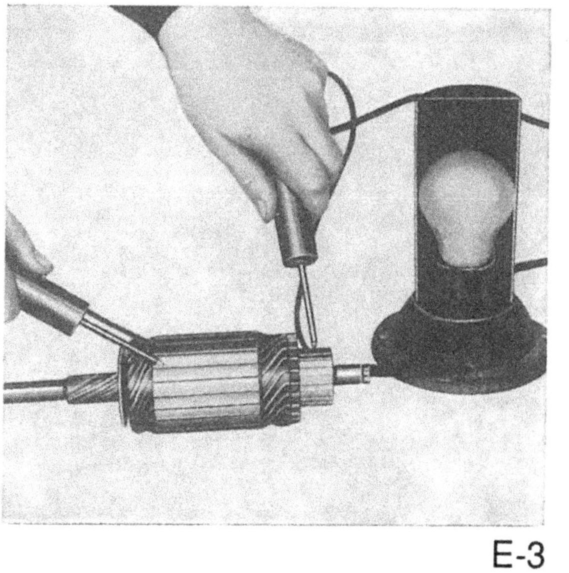

4 - The commutator consists of copper segments which are individually insulated. If the commutator is out of round and burned, or if is grooved, it should be turned in a lathe to obtain a smooth surface. The armature should

be clamped at the bearing surfaces for turning and not held between the lathe centers. The normal diameter of the commutator is 36 mm and it must not be turned down below 33.5 mm. The permissible run-out at the commutator is 0.03 mm. The insulation between the segments is undercut with a commutator saw until it is about 0.3—0.4 mm (.012"—.016") below the surface. Large workshops are advised to use a commutator miller for this operation. Make sure there are no metal chips between the segments, as these may lead to short circuits.

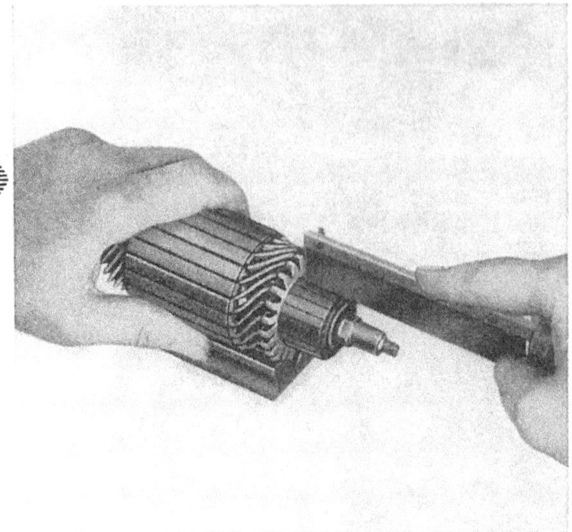

Test

The two field coils are tested for open circuits, short circuits and ground.

1 - The field coils can be tested for an open circuit by connecting a 6-volt battery in series with a test lamp and placing the points at the ends of each coil.

2 - If the outer insulation of the field coils is found to be in order, there is rarely a short circuit in the windings. To determine short circuits is generally beyond the scope of a workshop, as this requires special testing appliances.

3 - Test for a grounded field with points of a 220-volt test lamp between one coil end and housing.

4 - Make sure that the electrical connections between the two field coils are in order as the full starting current passes through the field coils.

Assembly

The components are washed in fuel and then dried with compressed air. The compo-bushing of the commutator bearing is only cleaned as far as it is accessible. The drive pinion should only be washed in fuel if it has become oily and does not engage in cold weather. If necessary, renew crankshaft oil seal or main drive shaft oil seal.

When assembling, the following points should be observed:

1 - When replacing bush in commutator bearing, the new bush must be placed in a hot oil bath prior to assembly.

2 - Bearing points, armature brake, drive pinion seat, and shift lever should be greased with Universal Grease.

3 - The stop ring should be peened when the circlip has been installed.

4 - The armature shaft end play must be within 0.1—0.3 mm (.004"—.012") and can be adjusted by fitting shims.

5 - Do not forget to install end cap rubber seal, replace if damage is apparent.

6 - To avoid starting trouble caused by water entering the starting motor, the following points should be sealed with VW Sealing Compound D 1 a:

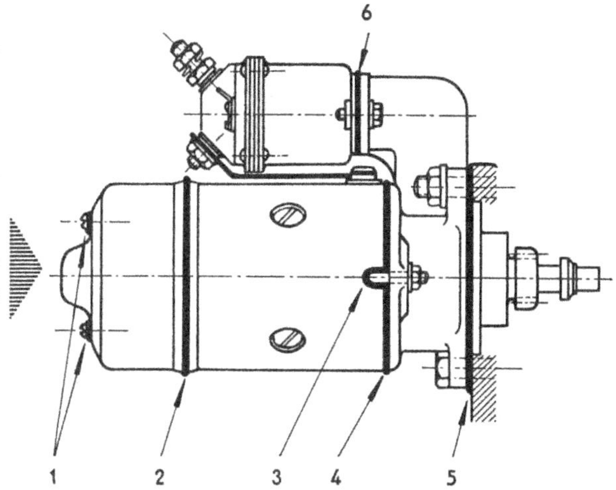

1 - Holes for slotted screws in end cap.

2 - Rubber seal between housing and end cap.

3 - Holes in frame for hook studs of intermediate bracket.

4 - Joining faces of housing and intermediate bracket.

5 - Joining faces of transmission case and intermediate bracket.

6 - Joining faces between solenoid switch and intermediate bracket.

Testing Starting Motor Operation and Output

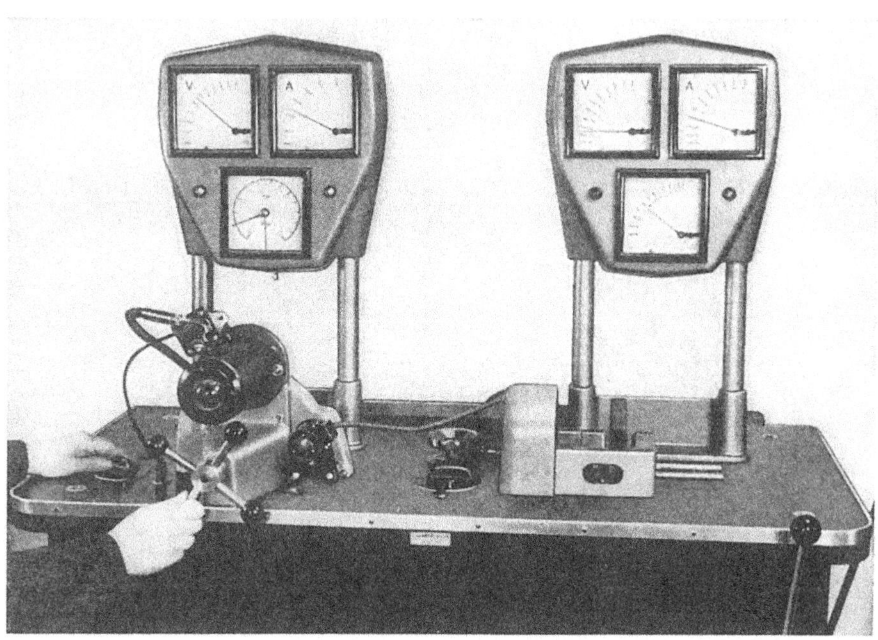

The operation of the starting motor can be checked and its output measured on a suitably equipped test bench. The following readings provide adequate information regarding the condition of the starting motor.

1 - Battery voltage.

2 - No-load current and speed.

3 - Current draw, torque and voltage drop of battery when loaded (speed of starting motor approx. 1000 rpm.).

4 - Current, torque and voltage drop in the event of short circuit (load starting motor till it stops).

5 - Check engagement of pinion under load.

Important

If these tests are carried out with a car battery, check first that the battery is in good condition and fully charged. The readings depend on its condition.

A reconditioned starter with new brushes will not give accurate readings as the commutator and brushes take some time to bed-in properly.

Test Readings

No-load test			Load test			Stall torque test				Minimum pull-in voltage for solenoid V
Current A	Voltage V	Speed rpm x	Current A	Voltage V	Torque ft. lbs.	Speed rpm x	Current A	Voltage V	Torque ft. lbs.	
60—80	5.5	5300—7300	260	4.5	4	1000—1300	450—520	3.5	8.6	4

The values apply to 135 Ah battery or two 77 Ah batteries in parallel and a temperature of 20° C (68° F)

Deviations from the above values should not exceed ±10%.

x = Speed of starting motor A = Current in Amps V = Voltage

Starting Motor Trouble Shooting

Symptom	Cause	Remedy
Starter does not operate with the ignition key in the starting position	Switch on the lamps when testing:	
	a - Lights do not burn. Loose cables or poor ground connection. Battery run down	a - Check battery cables and connection. Test voltage of battery, charge if necessary
	b - Lights go out when moving ignition key to starting position. Insufficient current due to loose connections or corroded terminals	b - Clean battery terminals and cable clamps, clean and tighten connections between battery, starting motor and ground
	c - Lights go dim when moving ignition key to starting position. Battery run down	c - Charge battery
	d - Lights stay bright when moving ignition key to starting position. Make a jumper contact between terminals 30 and 50 at starting motor. If the starting motor operates, there is an open circuit in cable 50 to ignition-starting switch, or in cable 30 to lighting switch, or the ignition-starting switch is defective	d - Eliminate open circuits, replace defective parts
	e - Lights stay bright and solenoid switch operates. Disconnect battery cable from terminal 30 at starting motor and connect it to terminal stud of connecter (contact blade). If the starting motor operates, the contacts of the solenoid switch are worn or dirty	e - Replace solenoid switch
Starting motor does not operate when battery cable is directly connected with terminal stud of connector (contact blade)	a - Brushes sticking	a - Clean brushes and guides of brush holders
	b - Brushes worn	b - Replace brushes
	c - Weak spring tension. Brushes do not make contact	c - Replace springs
	d - Commutator dirty	d - Clean commutator
	e - Commutator rough, pitted, or burned	e - Overhaul starting motor
	f - Armature or field coils defective	f - Overhaul starting motor
Sluggish or slow action of the starting motor	a - Battery run down	a - Charge battery
	b - Insufficient current flow due to loose or corroded connections	b - Clean battery terminals and cable clamps, tighten connections
	c - Brushes sticking	c - Clean brushes and guides of brush holders
	d - Brushes worn	d - Replace brushes
	e - Commutator dirty	e - Clean commutator
	f - Commutator, rough pitted, or burned	f - Overhaul starting motor
	g - Armature or field coils defective	g - Overhaul starting motor
Starting motor is heard to operate, but cranks engine erratically or not at all	a - Drive pinion defective	a - Replace drive pinion
	b - Flywheel gear ring defective	b - Replace flywheel or remachine gear ring
Drive pinion does not move out of mesh	a - Drive pinion or armature shaft dirty or damaged	a - Overhaul starting motor
	b - Solenoid switch defective	b - Replace solenoid switch

THIS PAGE INTENTIONALLY
LEFT BLANK

General Description

The Bosch starting motor EEF 0.5/6 L 1 differs from the EED type in the internal construction and the modified armature brake which is combined with the drive pinion.

The removal and installation, the checking of armature, fields coil, commutator and brushes is carried out as on the EED 0.5/6 L 49 type.

When testing the starter on a test stand, the values given for the EED type can be used.

Removing and Installing Solenoid

Removal

1 - Release connector strip from solenoid.

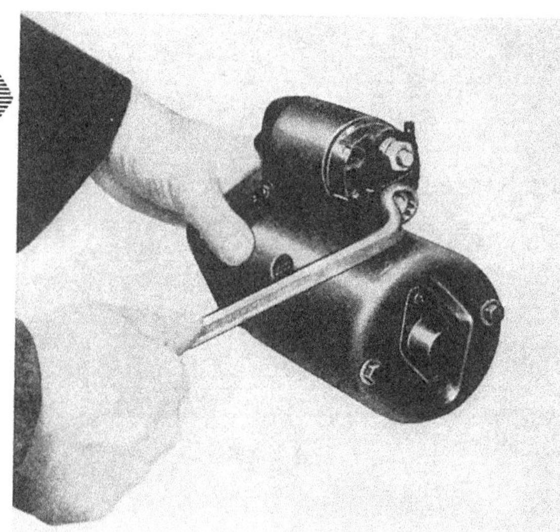

2 - Remove two screws securing solenoid to intermediate bracket.

3 - Lift the pull rod up out of the shift lever and remove solenoid.

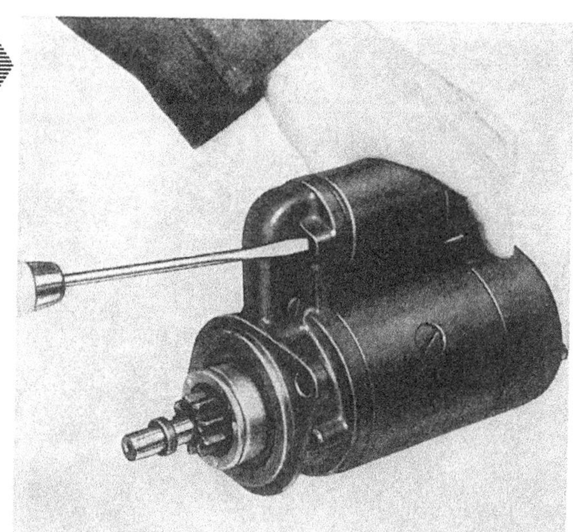

Defective solenoids should be replaced. The setting of the solenoid must not be altered. When installing a new solenoid, the distance from the flange to the pull rod eye must correspond with that of the old part.

Installation

The following points should be noted:

1 - Check that the rubber seal on the intermediate bracket is correctly seated.

2 - Coat the outer edge of the solenoid face with a thin layer of VW Sealing Compound D 15.

3 - To facilitate the solenoid insertion, pull the drive pinion and shift lever fork out as far as possible.

Checking Brushes and Commutator

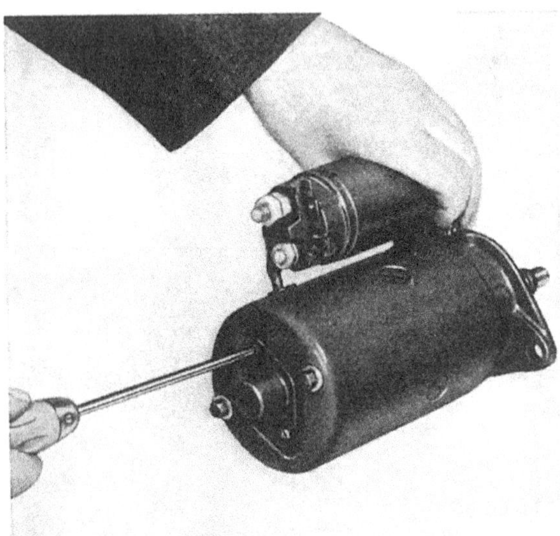

1 - Remove end cap.

2 - Remove lockring and steel shims from the drive shaft.

3 - Remove housing screws and take off commutator end-plate.

4 - Lift the four brushes out of their guides.

5 - Pull the brush holder off the armature shaft. Unsolder two brushes on the field coil connections and two brushes on the brush holder and solder on new brushes.

6 - If the commutator is dirty or oily, it can be cleaned with a clean, fuel soaked rag wrapped round a piece of wood.

7 - If the surface of the commutator is grooved or burned, the starter should be overhauled.

The diameter of the commutator must not be reduced to less than 34.5 mm (1.35").

Assembly takes place in the reverse order, taking care that the rubber seal for the solenoid cable is properly seated.

Disassembling and Assembling Starter

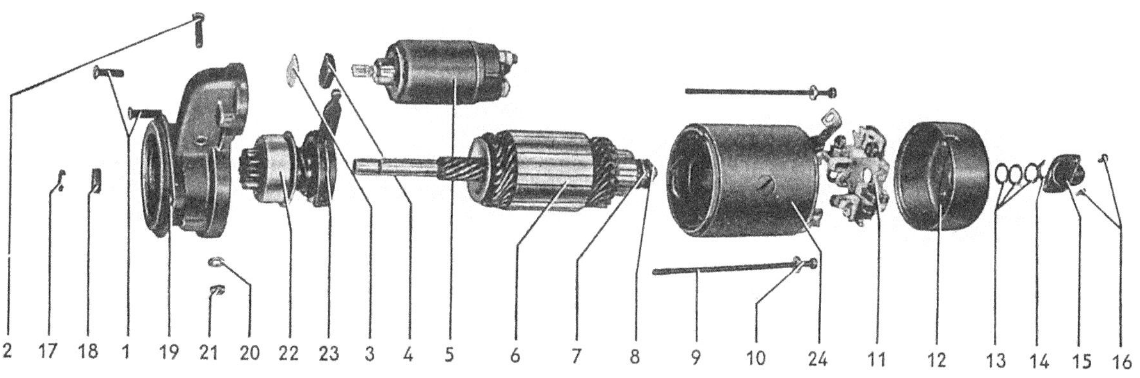

1 - Securing screws
2 - Bearing bolt
3 - Support washer
4 - Rubber seal
5 - Solenoid
6 - Armature
7 - Steel washer
8 - Synthetic washer
9 - Housing screws
10 - Washer
11 - Brush holder
12 - End plate
13 - Shims
14 - Lockwasher
15 - End cap
16 - Screws
17 - Circlip
18 - Stop ring
19 - Intermediate bracket
20 - Spring washer
21 - Nut
22 - Drive pinion
23 - Shift lever
24 - Housing

Disassembly

1 - Remove solenoid.

2 - Remove end cap.

3 - Remove lockring and shims from armature shaft.

4 - Remove housing screws and take off commutator endplate.

5 - Lift the four brushes out of their guides.

6 - Remove brush holder from armature shaft.

7 - Pull starter housing off intermediate bracket.

8 - Hold the armature with the commutator downwards and knock the pinion stop ring back with a suitable punch.

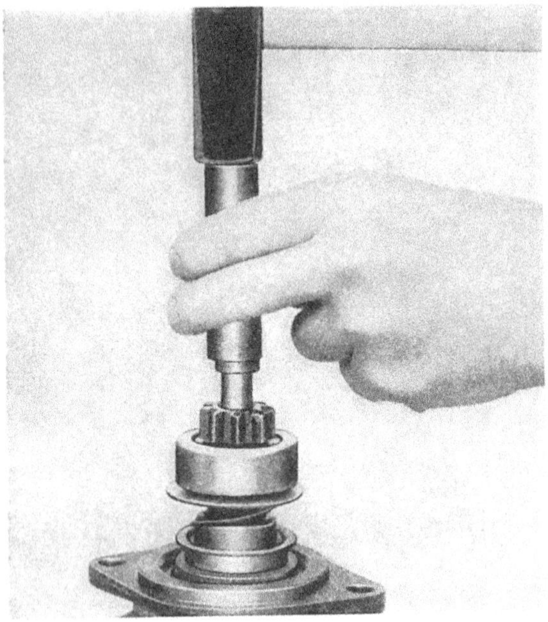

9 - Pull circlip and stopring off. Remove any burr existing on the circlip groove.

10 - Take the armature out of the intermediate bracket and remove drive pinion.

11 - Screw the shift lever bolt out and remove the lever.

If the drive pinion is defective it should be replaced. Individual parts are not supplied.

Note:

From Chassis No. 4 753 580, the meshing force of the pinion on the Bosch EEF starter was increased to ensure that the pinion engages properly even under unfavourable conditions such as when a pinion tooth is opposite a flywheel tooth.

The new pinion can be installed on previous starters and is marked on the face of the freewheel with the Bosch designation.

New: ALGT L 3 Z (Part No. 113 911 335 A)
Formerly: ALGT L 1 Z (Part No. 113 911 335)

The armature and field coils are checked as on the EED type starter.

Assembly

The parts should be washed in fuel and dried with compressed air. The commutator bush should only be cleaned externally. The drive pinion should be washed in fuel if it is very oily and does not engage properly in cold weather. It may be necessary to renew the crankshaft and main drive shaft oil seals.

The assembly takes place in the reverse order, noting the following points:

1 - When replacing the bush in the commutator end plate, soak the new bush in hot oil before installation.

2 - Lubricate bearing surfaces, shift lever and drive pinion thread with universal grease.

3 - Assemble the armature with intermediate bracket and drive pinion.

4 - When the circlip has been installed, the stop ring must be pushed over it. The stop ring must be free to turn on the armature shaft.

5 - Check that the support washer and the seal between the starter housing and the intermediate bracket are correctly seated.

6 - Insert the seal for the solenoid connection strip carefully. Bolt the housing and intermediate bracket together whilst lifting the bracket slightly.

7 - The armature end-play should be 0.1—0.3 mm and is adjusted by the installation of suitable shims.

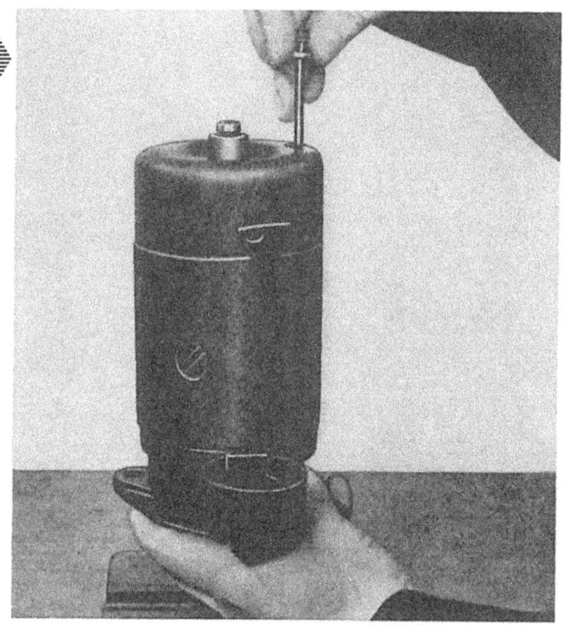

8 - To prevent starter damage due to the ingress of water, the following places should be sealed with VW Sealing Compound D 1 A:

a - Holes for end cap screws.

b - Joining surface between end cap and end plate.

c - Holes for the housing screws.

d - Joint between starter housing and end plate.

e - Joint between starter housing and intermediate bracket.

f - Holes for solenoid screws.

g - The surface between the solenoid and end plate should be sealed with VW Plastic Sealing Compound.

THIS PAGE INTENTIONALLY
LEFT BLANK

Starting Motor

Type: VW 113 911 021 A

General Description

The VW starting motor is of the solenoid shift type with an overrunning clutch drive. Described below are the main features:

a - When at rest, the overrunning clutch is secured on the armature shaft by detent balls.

b - When the pinion has been shifted into mesh with the flywheel, it is held in that position by means of the detent balls until the ignition key is released. This prevents the pinion from being pulled back out of mesh when the engine fires irregularly.

c - The spring-loaded pinion shifting linkage insures a correct meshing even when flywheel and pinion teeth are opposite each other.

d - The two windings in the solenoid consist of a pull-in winding and a hold-in winding. They work together to pull the core in, meshing the pinion and closing the starting motor circuit. As the switch contacts close, the pull-in winding is shorted out as the magnetic power of the current saving hold-in winding is sufficient to hold the core in.

e - The armature shaft runs in self-lubricating bearings which require no service attention. For the removal and installation of the starting motor and checking armature, field coils, commutator and brushes refer to the instructions given for the BOSCH starting motor.

Checking Starting Motor

When checking the starter function and output on a test stand, the values given for the BOSCH EED 0.5/6 L 49 can be taken.

Permissible armature end play: 0.1—0.3 mm (.004"—.012").

Checking Overrunning Clutch Drive

a - The solenoid switch should shift the pinion 6 mm (.24"). If the movement is below 6 mm the shifting linkage may be worn or loose, or the armature end play excessive. Either replace the linkage or reduce end play by inserting shims respectively.

b - If, when holding the armature, the pinion cannot be turned anticlockwise, this indicates that the overrunning clutch is tight on the armature shaft. Replace damaged or worn parts. The overrunning clutch drive can only be replaced as a unit. If the pinion does not completely spring back to the rest position, it may be due to the solenoid core or shifting linkage sticking, or the pressure spring not engaging in the overrunning clutch.

Removal and Installation of Solenoid Housing

Removal

1 - Remove connector strip from solenoid.

2 - Remove the two 7 mm nuts and withdraw housing and insulating disc.

3 - Remove 14 mm nut at field coil lead.

4 - Take off profiled rubber seal.

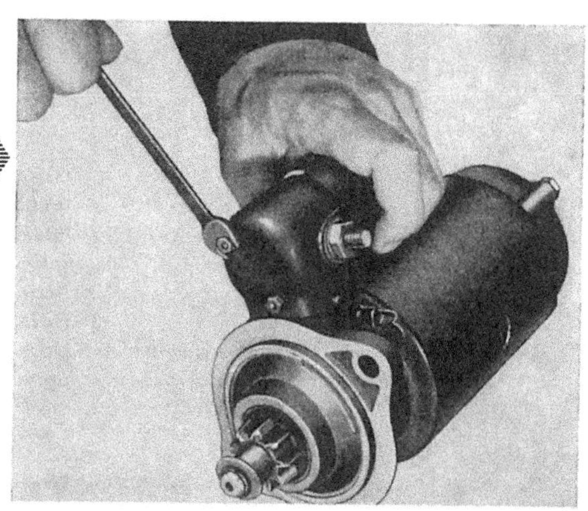

Testing Solenoid

1 - Check current draw of pull-in winding and hold-in winding with 6-volt battery:

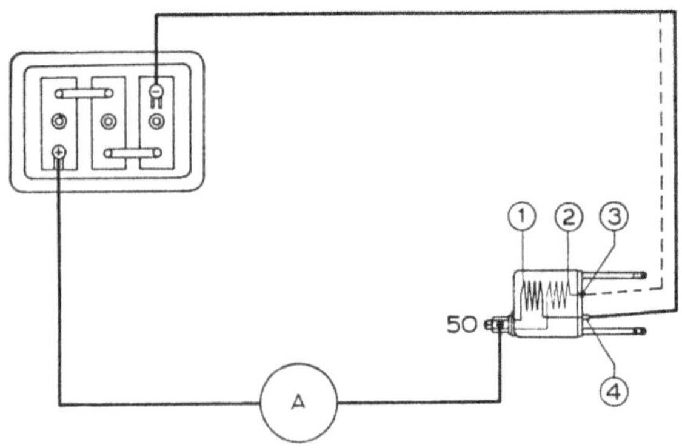

1 - Pull-in winding
2 - Hold-in winding
3 - Cover plate
4 - Terminal

a - Insert ammeter in lead from positive battery terminal to terminal 50.

b - Checking pull-in winding: Connect negative battery terminal with pull-in winding terminal.
Current draw: 35 to 40 amperes.

c - Checking hold-in winding: Connect negative battery terminal with cover plate.
Current draw: 10 to 12 amperes.

2 - If defective, the whole solenoid housing should be replaced.

Installation

This is a reversal of the preceding operations. When installing the solenoid housing, make sure there is a perfect contact at the terminal and that the profiled rubber seal is seated properly.

Disassembly and Assembly of Starting Motor

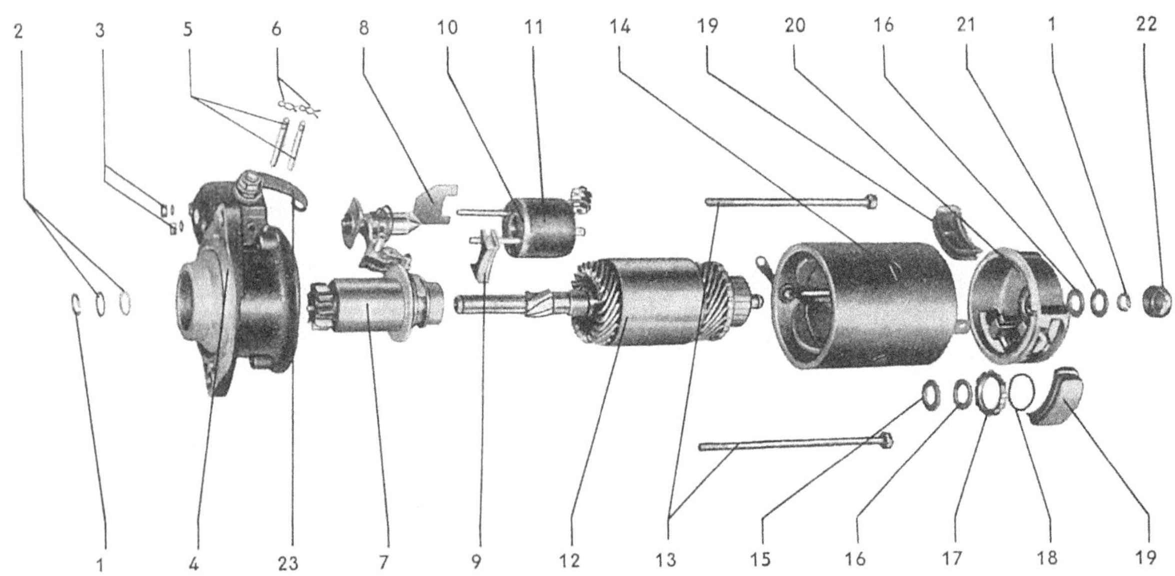

1 - Circlip
2 - Cup washer
3 - Hex. nuts and lockwashers
4 - Intermediate bracket
5 - Pivot pins
6 - Spring clips
7 - Overrunning clutch drive with linkage and solenoid core
8 - Insulating plate
9 - Profiled rubber seal
10 - Insulating disc
11 - Solenoid housing
12 - Armature
13 - Housing screws
14 - Housing and field assembly
15 - Steel washer
16 - Bronze washer
17 - Brake washer
18 - Trust ring
19 - Brush inspection cover
20 - Commutator end plate
21 - Steel washer
22 - Cap
23 - Connecting strip

Disassembly

1 - Remove starting motor.

2 - Remove solenoid housing.

3 - Disconnect connecting strip at end plate.

4 - Remove cap from commutator end.

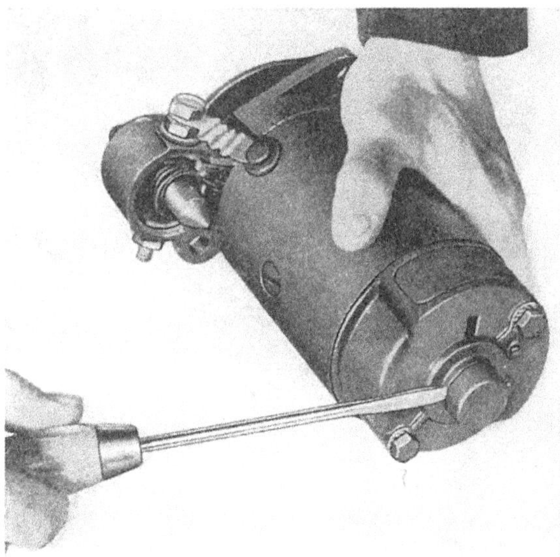

5 - Remove circlip with pliers VW 161 a, take off steel washer and bronze washer. Remove any burr that has developed.

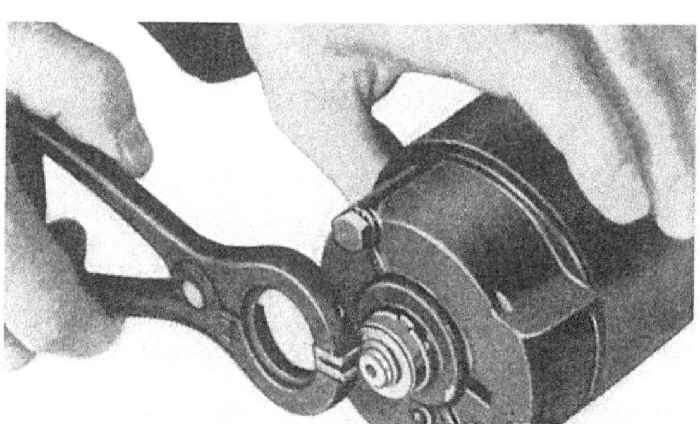

6 - Remove the two brush inspection covers, lift up the brushes, and fix them in position.

7 - Remove the two housing screws.

8 - Withdraw intermediate bracket and armature from the housing as a unit. Note position and number of shims for proper reassembly.

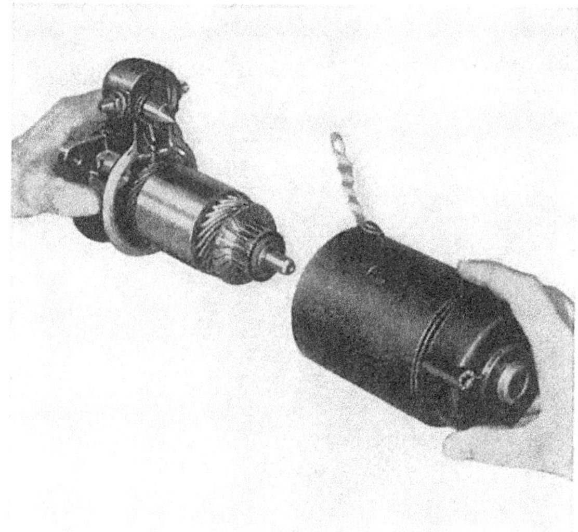

9 - Disconnect field coil lead from brush holder of positive brush and take off commutator end plate. Remove brake washer and thrust ring.

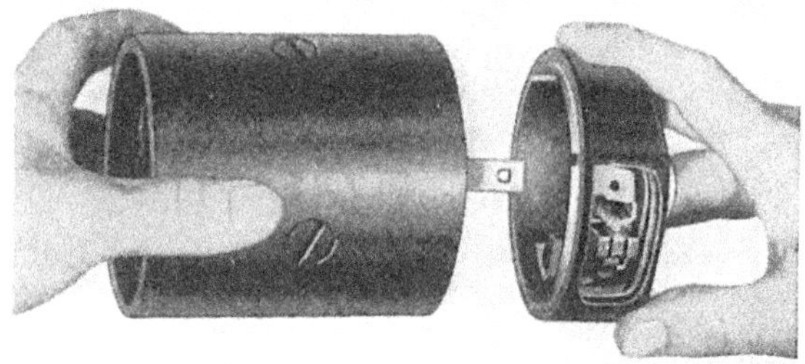

10 - Remove the two pivot spring clips and push out the pivot pins.

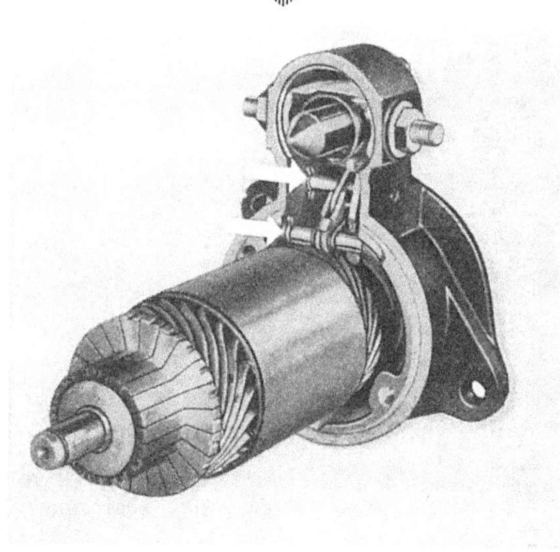

11 - Pull out insulating plate and turn contact plate of solenoid core 90 degrees.

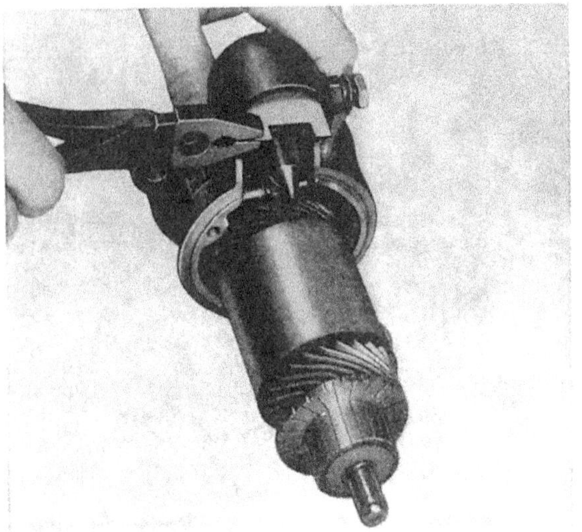

12 - Withdraw armature, shifting linkage and solenoid core as a unit from the intermediate bracket.

13 - Take off circlip and cup washers from the armature drive end.

14 - Withdraw shift collar approx. 3—6 mm (.12"—.24") from the overrunning clutch and remove the whole drive assembly by turning it clockwise and jerking slightly.

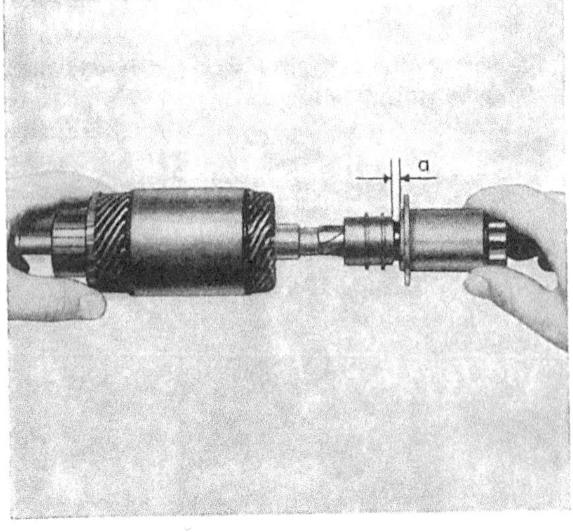

a = 3—6 mm (.12"—.24")

15 - Remove shift collar and the five steel balls from the overrunning clutch. The overrunning clutch cannot be disassembled.

16 - If necessary, remove the two terminals 30 together with insulation and contact blade from intermediate bracket.

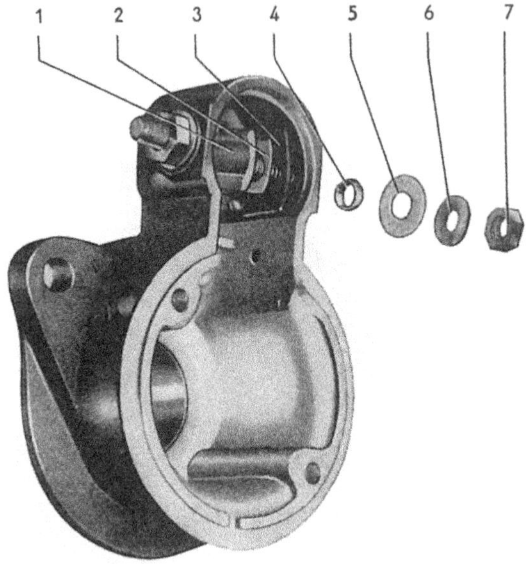

1 - Terminal
2 - Contact blade
3 - Insulating plate
4 - Insulating bush
5 - Insulating washer
6 - Lockwasher
7 - Nut M 8×1

Testing

1 - Clean parts in fuel, as found necessary. The self-lubricating bearing bushes must not be cleaned with fuel.

2 - Check brushes for wear, replace if necessary.

3 - Examine commutator surface for wear, pitting, roughness and burned spots. If found necessary, turn commutator down in a lathe. The diameter of the commutator must nit be reduced to below 33 mm. Carbon dust can be removed from the commutator with a cloth dampened with fuel. If the solder at the commutator riser bars has melted and been thrown out, resolder the connections.

4 - Inspect armature and field coils for open circuit, short circuit and ground. If necessary, replace armature or frame and field assembly.

5 - Examine contact surfaces of solenoid core contact plate for burned spots and remachine or replace contact plate as necessary.

Assembly

Assembly takes place in the reverse order, noting the following points:

1 - To facilitate assembly, apply high melting point grease to the steel balls when inserting them in the corresponding holes.

Pushing the overrunning clutch into position with slight jerks will insure that the inner spring passes over the shoulder of the pinion and engages into a groove. The engaging of the spring can only take place with the overrunning clutch at rest position. The overrunning clutch should be pulled forward to check whether the spring is engaged.

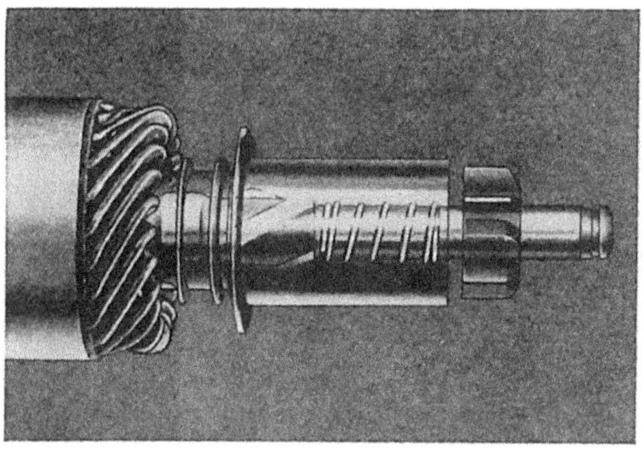

2 - Install the circlip by means of a tapered pin and a sleeve to be made locally in accordance with the dimensions given below. Pin and sleeve must be of the Morse taper 1.

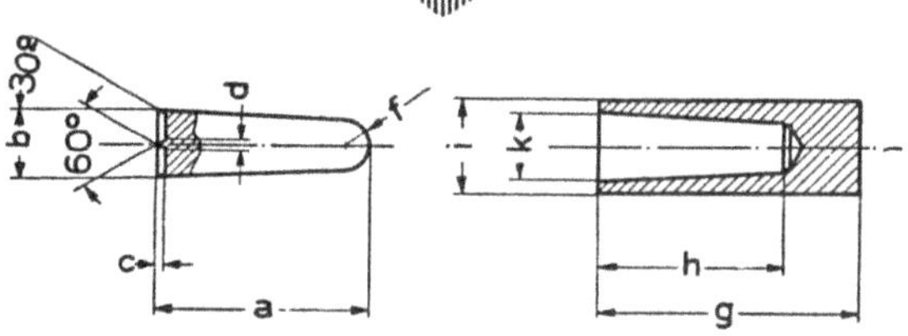

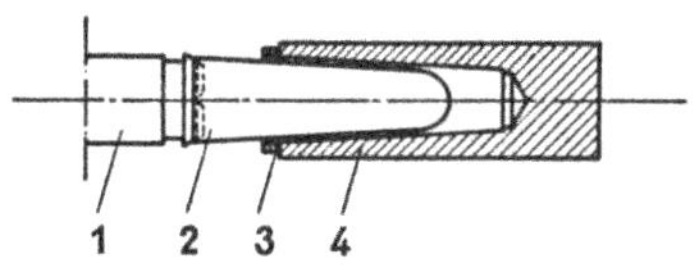

a = 34 mm (1.34″) f = ball radius
b = 11.8 mm (.46″) dia. g = 50 mm (2″)
c = 1.2 mm (.05″) h = 40 mm (1.6″)
d = 3 mm (.12″) dia. i = 16 mm (.6″) dia.
k = 12.065 mm (.47″) dia.

1 - Armature shaft 3 - Circlip
2 - Tapered pin 4 - Sleeve

3 - Place shifting linkage and solenoid core on the shift collar and insert the drive end of the armature in the intermediate bracket.

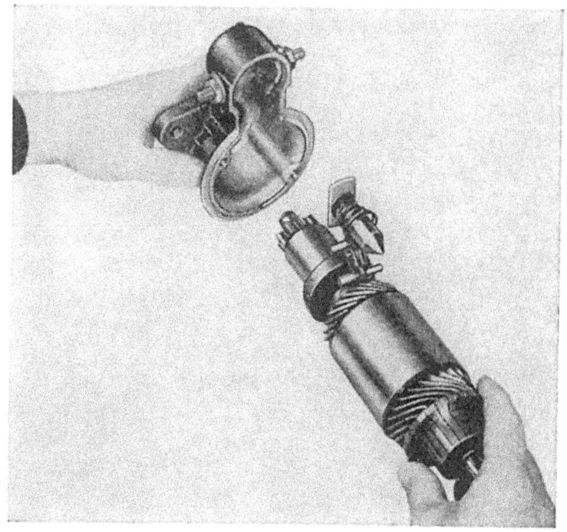

4 - Turn contact plate 90 degrees and insert the insulating plate up to the stop.

5 - Insert the two pivot pins and secure them with the spring clips.

6 - First place thrust ring and then brake washer. in the commutator end plate.

7 - Place housing and field assembly against the commutator end plate so that the nose of the frame enters the groove in the commutator end plate. Connect field coil lead to positive brush holder.

8 - Insert armature with overrun clutch and end plate into the housing and tighten the two housing bolts.

9 - Make sure the washers and shims occupy their original position.

10 - Check for tight seat of brush inspection covers and end cap. If necessary, replace parts to avoid ingress of water.

11 - When installing the solenoid core housing be sure that the profiled rubber makes a perfect seal between starting motor housing and solenoid core housing.

THIS PAGE INTENTIONALLY
LEFT BLANK

Battery

Type: 6 volts, 66 amp.-hours

General Description

The battery stores the energy produced by the generator and supplies current for operating the starting motor, lights, and other electrical accessories. If consists of three cells, each of which has a positive lead peroxide plate group (PbO_2) and a negative lead plate group (Pb). The electrolyte is made up of about 40 per cent sulphuric acid and about 60 per cent water ($H_2SO_4 + H_2O$) having a specific gravity pf 1.285 = 32° Bé (Baumé).

The cells are united in the battery case of acid-proof insulating material and connected electrically by lead sell connectors. To avoid confusion, the positive battery terminal post is made thicker than the negative terminal post.

Cell Voltage

The average voltage of each cell is 2 volts. It increases to about 2.5—2.7 volts when the battery is being charged and decreases to between 2.1 and 2.0 volts soon after the charging current has been switched off. The battery is discharged when the cell voltage has dropped to 1.75—1.8 volts under no-loaded conditions.

Battery Rating

The discharging rate of the battery amounts to 66 ampere-hours. The 20-hour rate represents the amount of current a battery can deliver for 20 hours at a temperature of 20° C (70° F). Thus, a battery of 66 ampere-hours can deliver a current of 3.3 amperes for 20 hours at a temperature of 20° C (70° F).

Discharging process

The lead (negative plate) and lead peroxide (positive plate) change to lead sulphate ($PbSO_4$) during the discharge process. Thus the specific gravity of the diluted sulphuric acid decreases.

Charging process

The direct current from the generator causes the lead sulphate to change back to spongy lead in the negative plates and to brown lead peroxide in the positive plates, and the sulphuric acid reappears in the electrolyte, increasing the specific gravity of the electrolyte. If the charging process continues, the charging current electrolyses the water of the diluted sulphuric acid and generates hydrogen and oxygen. The battery commences to boil.

Battery Maintenance

Ready starting of the engine depends upon perfect condition of the battery. It should, therefore, be inspected at regular intervals.

An unused battery can discharge itself within 6 weeks to such an extent that it requires recharging. This applies to new as well as used batteries.

Batteries which are not charged in good time will sutphate. This reduces its current absorption and capacity. It is necessary, therefore, to check an unused battery at least every 4 weeks and recharge it if required. This check is particularly important on new vehicles with long delivery times.

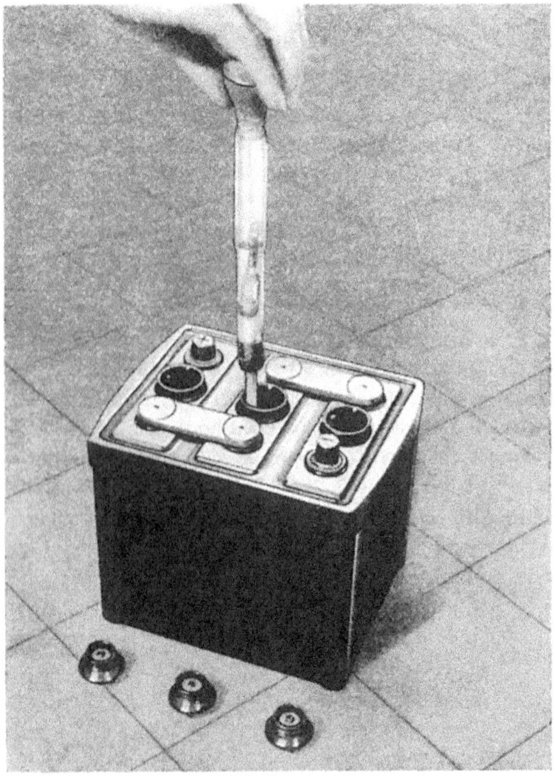

Hydrometer Test

The state of charge of the battery may be checked by means of a battery hydrometer. The specific gravity of the battery liquid will increase with the charging of the battery. Tested with the hydrometer, the gravity can be read from the scale of a float.

Battery fully discharged	18° Bé	= spec. gravity 1.142
Battery semi-charged	27° Bé	= spec. gravity 1.230
Battery fully charged	32° Bé	= spec. gravity 1.285

Electrolyte Level

During operation, the electrolyte level in the battery drops due to evaporation of the water. Only distilled water should be added, as other water contains chemical properties which are harmful to the electrolyte. The electrolyte level should be approximately 5 mm (.20″) above the plates. If there is a level indicator, adjust the level according to it.

To facilitate accurate checking and topping up all batteries are equipped with level indicators. There are two different systems depending on the make of battery supplied.

a - Level indicator (bar)

With this level indicator the electrolyte should cover the level indicator bar (Arrow).

b - Level indicator (insert)

With the insert the electrolyte should be up to the lower edge of the basket-type part of the insert.

Note:

During prolonged periods of day time operation or when fully charged, batteries may tend to boil over if the electrolyte level is too high.

Never add acid, unless it is known that acid has been spilled from the battery. Check specific gravity afterwards and compensate if necessary.

Voltage Test

The high rate discharge test is carried out with an individual cell tester consisting of a low-reading voltmeter and a heavy resistance of between 80—100 amperes.

The two prods of this instrument are placed across the terminal posts of each battery cell in turn. The voltage of each cell should not fall below 1.6 volts while taking the reading (10 to 15 seconds). Otherwise the cell is discharged or defective. The normal voltage is 2 volts. The difference in the voltage of the individual cells should not be in excess of 0,2 volt.

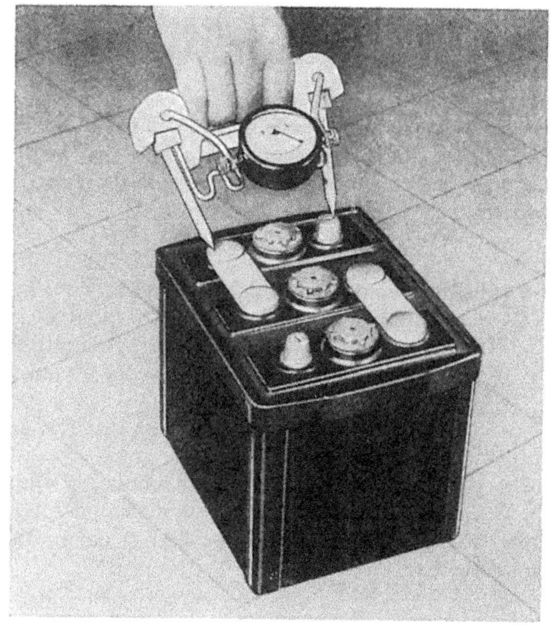

How to Maintain Battery

Because of the high stress imposed on the battery when starting, the average service life of a battery is about two years. When operating the starter continuously the battery is heavily strained since a current up to 250 amperes is drawn from the battery.

The battery must be held firmly in its mounting. Terminal posts and cable clamps should be kept free from corrosion to prevent excessive electrical resistance. They should be cleaned with a clean rag, or, in severe cases, with some battery pole cleaner. Then coat the posts and cable clamps with petroleum jelly to prevent corrosion. Cable clamps which are difficult to remove from the terminals posts due to corrosion should be removed with a special tool.

Important

Since some splashing of battery electrolyte is natural as the battery is being charged, traces of electrolyte should be cleaned off with a common baking-soda solution to avoid damage to the fabric and metal parts.

If the car is left stationary for a long period, the battery must be recharged at intervals of 6—8 weeks.

The charging current depends on the battery and should, with a battery capacity of 66 amp.-hours, not exceed the rate of 6.6 amperes. Thus, the charging requires about 10 hours and more at a lower charging rate. The fully charged condition is reached when the cell voltage has increased to approx. 2.5 to 2.7 volts, the battery is gassing freely and there is no further rise in voltage for three hours. This should be checked at intervals of 1 hour.

Before charging, the vent plugs should be removed.

Recharging Battery

It is good practice to remove the battery at intervals of three or four months and discharge it down to a cell voltage of 1.8 volts before recharging it. Batteries are subject to a self-discharge of 1 per cent per day with the battery in good condition.

High-Rate charging

A high-rate charging with a fast charger should only be carried out in isolated cases. This charging can only be performed on batteries that are already in operation and in good condition. New batteries should be dealt with as follows.

Charging New Batteries

New batteries are generally received in an unfilled condition. If no instructions are attached by the manufacturers, proceed to charge for service as follows:

1 - Remove vent plugs and fill cells with chemically pure battery acid diluted with distilled water to a specific gravity of 1.285 at 20° C (70° F). The level should be approx. 15 mm (.6") above the plates including separators.

2 - Let the battery stand for 5 or 6 hours to allow the plates become saturated with electrolyte. The electrolyte level drops slightly during this period.

3 - Add electrolyte to restore correct level.

4 - Charge battery at a rate of 5 amperes until the voltage of each cell is between 2.5 and 2.7 volts and all cells gas freely.

5 - Check temperature of electrolyte regularly. If in excess of 40° C (105° F) reduce charging current.

6 - After charging, check specific gravity (1.285 = 32° Baumé) and adjust it if necessary. Should it be necessary to add acid or distilled water the battery must be recharged for a short time to ensure a good mixing of the fluid.

7 - The vent plugs should not be replaced for at least two hours after charging. Wash off splashed electrolyte and then dry battery.

Cold Weather Operations

It is true that the conductivity and degree of viscosity of the electrolyte largely depends on temperature conditions. Extreme cold considerably reduces battery output. At an electrolyte temperature of —15° C (5° F), the output amounts to only 50 per cent of that at a temperature of +20° C (70° F).

The higher the gravity of the electrolyte, the lower its temperature must be before it will freeze. The battery must, therefore, be kept in a sufficiently charged condition to prevent if from freezing. A frozen battery gives no current but can usually be restored by thawing out and recharging.

Spec. Gravity	Freezing Temperature
1,285	—65° C (—85° F)
1,18	—22.5° C (—8° F)
1,14	—13° C (+9° F)

The increased stress imposed on the battery at low temperatures by the starting motor, necessitates a more frequent inspection. In winter, it is recommended that the battery be removed at 4-week intervals for recharging and checking specific gravity and electrolyte level.

Batteries which are not used for prolonged periods tend to sulphate. If the cell tester shows large differences between the voltages of the individual cells or if the battery starts to gas as soon as charging commences or very shortly afterwards (typical sign of sulphated plates), the charging current should be kept as low as possible (max. 3 A) because **high charging currents can damage sulphated batteries.** Batteries in which the acid has thickened due to the cold should be thawed out in a warm room before charging.

Important Smoking and open flames are not permitted in the battery charging room. It is advisable not to have precision tools and instrument in such rooms.

Note:

From 30th April 1964, Chassis No. **6 317 836** (Models 111 and 118) and from 11th May 1964, Chassis No. **6 337 511,** the clip on the battery securing strap was moved to the top to facilitate the removal and installation of the battery cover during maintenance checks, etc. It is now no longer necessary to remove the right-hand kick board.

The new strap can be installed in all vehicles from Chassis No. **1-0 929 746** (August 1955).

General Description

The ignition equipment consists of the battery, switch, ignition coil, ignition distributor with vacuum advance mechanism, spark plugs and wiring. The 6 Volt low tension current, supplied by the battery, is converted to high tension current by the ignition system and fed to each of the four plugs in the correct sequence and at the proper time.

Ignition Coil

Type: BOSCH TE 6 B 4

General Description

The ignition coil consists of an iron core around which is a primary circuit made up of some windings of heavy wire and a secondary circuit made up of numerous windings of fine wire. The process in the ignition coil is similar to that of a transformer. As the contact points separate in the distributor, the flow of current from the battery through the primary winding of the coil is interrupted and the magnetic field collapses. The collapsing magnetic field induces high-voltage surge that is conducted through the distributor rotor and cap to a spark plug. A condenser is connected in parallel with the contact breaker to increase the surge in the circuit and prevent the formation of sparks at the points so that the primary current drops quickly.

Note:

From 17th February 1964, Chassis No. **6 135 252**, Engine No. **8 388 270**, a new smaller Bosch coil is fitted. In addition to the Bosch coil TE 6 B 4 and the VW 111 905 105 F coil. The designation of the new coil is 111 905 105 J and this is also the VW part number.

All three coils can be interchanged.

Maintenance

The coil insulating cap must be kept clean and dry to prevent short circuits and tracking across the cap.

Inspection

To test the ignition coil for serviceability, the length of the spark produced by it should be measured. This can be done on a test bench or on the engine.

If, after inspection, the supply of current and the distributor are considered in good order, disconnect lead 4 at the distributor cap and hold it about 10 mm (.4") from the crankcase.

If no spark occurs while cranking the engine, the ignition coil must be replaced, assuming that the condenser is known to be in order.

Connections:

Terminal 15 to ignition/starting switch
Terminal 1 to distributor (contact breaker)
Terminal 4 to distributor cap (high-tension lead)

Distributor

Type: BOSCH ZV/PAU 4 R 2 mk

General Description

The distributor serves the purpose of distributing the high voltage current to the four spark plugs in proper sequence and time. Ignition timing is controlled by a vacuum spark-advance mechanism.

Maintenance

In the course of time, the slight burning which takes place at the points creates a small crater and a point. This does not normally affect the operation of the system but if trouble is experienced,the points should be removed and cleaned up with an oilstone. Badly burned points should be replaced.

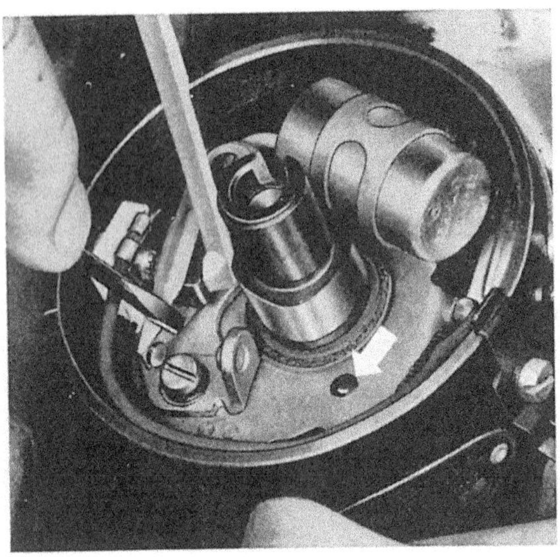

Blow out the inside of the distributor carefully. The fiber block on the breaker arm must be lubricated with multi-purpose grease (Lithium grease). To avoid the risk of grease getting onto the points, a very small quantity should be used and this pressed into the corner between fiber block and breaker arm.

The rotor and the four segments of the distributor cap are subjected to a certain amount of wear because of the sparking which occurs during operation. Trouble may occur if the insulating material of the distributor cap or the rotor is cracked. The cap must be kept clean and dry inside and outside to avoid high-tension leakage and tracking. When mounting the cap, make sure that the spring-loaded brush for the rotor is installed.

Lubricate the felt ring in the contact breaker base-plate with one drop of engine oil every 5000 km.

Take care that no oil gets on to the breaker points.

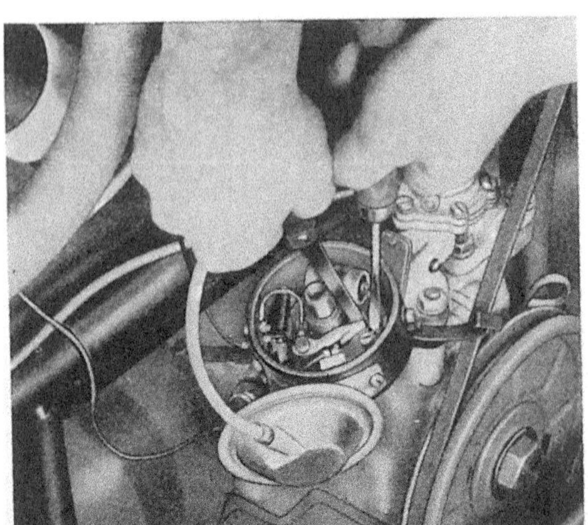

Adjusting Breaker Points

To adjust the breaker points, proceed as follows:

1 - Remove distributor cap and rotor.

2 - Crank the engine until fiber block on the breaker arm rests on the highest point of the cam lobe.

3 - Loosen the lock screw of the fixed breaker point.

4 - Insert screwdriver between the two small pins on the contact breaker plate and in the slot at the end of the fixed point. Turn the screwdriver until the clearance 0.4 mm (0.016") is obtained.

5 - Tighten lock screw.

Important

Do not allow oil or grease to come into contact with the points (on the feeler gauge blades, for instance) as this will cause premature point burning or misfiring. When the points have been adjusted, the ignition timing must be re-set as an alteration of 0.1 mm/ 0.004″ in the points gap alters the ignition timing about 3°.

The points can only open and close at the proper times if there is no excessive radial play in the distributor shaft bearings.

Replacing breaker points

If the course of time, a certain amount of natural wear takes place at the points due to burning. When adjustment is no longer possible and when the points are badly burned they should be replaced.

1 - Condenser
2 - Oil drilling
3 - Return spring
4 - Ground connection
5 - Pull rod
6 - Low tension cable
7 - Breaker arm spring
8 - Insulator
9 - Hexagon screw
10 - Condenser cable
11 - Breaker arm
12 - Breaker point
13 - Retaining screw
14 - Pins and adjusting slot

1 - Remove distributor cap and rotor.

2 - Loosen screw on breaker contact insulator.

3 - Remove low tension cable and lift out breaker arm and spring.

4 - Remove breaker point screw.

5 - Lift contact breaker point out and loosen cable for condenser on insulator.

Replacement takes place in the reverse order, taking care that the washers for the breaker arm are in the correct order.

6 - Adjust contact breaker gap and firing point.

Ignition Timing

Before timing the ignition to 10° before top dead center, the breaker point gap must be checked.

Important

There are two marks on the crankshaft pulley:

a - 10° before T. D. C. (right-hand mark in direction of rotation.

b - 7.5° before T. D. C. (left-hand mark in direction of rotation)

The ignition timing is set to 10° before T.D.C.

If the engine tends to "pink", as it does when low octane are used, the ignition timing should be adjusted to 7.5 before T. D. C. (left-hand mark).

Note:

On pulleys which have only one mark at 7.5° before T. D. C., a second mark should be made with a triangular file 4 mm/0.16″ to the right of the first.

1 - Crank the engine until the right-hand mark on the crankshaft pulley lines up with the vertical crankcase joint and the distributor rotor arm is in the position for firing on No 1 cylinder (see mark on rim of distributor base).

2 - Loosen clamp screw of distributor retainer.

3 - Connect one lead of 6-volt test lamp to terminal 1 of ignition coil and the other to ground.

4 - Switch on ignition.

5 - Rotate the distributor body clockwise until the contact points are closed and then slowly counter-clockwise until the breaker points just begin to open and the test lamp lights up.

6 - Tighten clamp screw of distributor retainer.

7 - Reinstall rotor and distributor cap.

The ignition is correctly timed for all four cylinders, if the lamp lights up when the right hand mark on the pulley is exactly in its highest or lowest position (in line with the crankcase jointing faces) while slowly cranking the engine.

> **Important**
>
> The adjustment of ignition firing point on a cold engine must always be done with a 6 Volt test lamp. A stroboscopic lamp should not be used as it will alter the entire setting range. If it is necessary, in exceptional cases, to adjust the timing with a warm engine, the following points should be noted:

Adjusting ignition timing with engine warm

The ignition timing should be adjusted to 10° before T.D.C. (right-hand mark on pulley) with the engine cold or oil temperature not exceeding 50° C/122° F.

1 - The cooling down of the engine can be accelerated if a special engine cooling fan is available.

2 - The engine temperature is taken by inserting a special thermometer into the crankcase via the dipstick hole. The accuracy of the instrument should be checked at regulator intervals.

3 - The ignition timing can be adjusted when the oil temperature is below 50° C. Above this point the oil temperature changes too quickly to permit accurate adjustment.

> **Important**
>
> If the engine tends to pink, as it does with low octane fuels, the ignition timing should be adjusted to 7.5° before T.D.C. (left-hand mark).

The spark advance is controlled over the whole range of engine speed by means of vacuum. The vacuum is taken from the lower part of the carburetor and operates a diaphragm in the vacuum unit. A pull rod transfers the movement of the diaphragm to the contact breaker plate and turns it against the direction of rotation of the distributor shaft. A return spring is installed to pull the breaker plate back to its basic position when the vacuum ceases.

Checking vacuum spark advance

A special control scale is secured to the crankcase with a bracket and shows the ignition timing adjustment and the operation of the vacuum advance mechanism by means of a stroboscopic lamp.

Application

1 - Secure control scale on the generator support flange.

2 - Turn the crankshaft until the mark in the pulley aligns with the crankcase joint.

3 - Make a 2—3 mm wide paint or chalk mark on the pulley corresponding with the 20° mark on the scale.

4 - Connect the test lamp to the plug cable of No. 1 cylinder in accordance with the manufacturer's instructions.

5 - Remove vacuum pipe from carburetor and seal with a plug.

6 - Start the engine and allow it to idle. The paint or chalk mark which shows up white in the flashes from the test lamp should be aligned with the 20° mark on the scale and should not move even if the engine speed is increased.

7 - Re-connect the vacuum pipe to the carburetor and run the engine at a fast idling speed, about 1200 to 1300 rpm. The vacuum unit should move the breaker plate to the fully advanced position at this speed and the white mark should move across the scale from 15° to 21° to the left.

Important

The advance curve can only be accurately checked on a distributor test stand with the distributor removed.

8 - If the amount of advance is found to be less than 15° following checks should be made:

a - Check ease of movement of contact breaker plate, if necessary, disassemble distributor and free plate off.

b - Check vacuum pipe and unit for leakage and replace parts as required.

Checking the spark advance with an ignition tester

If an ignition tester with a vacuum gauge and a vacuum adjuster is available, the spark advance can be checked with the distributor on the engine. The measuring range of the gauge must be laid out from 0 to 80 mm Hg so that accurate readings can be taken.

Checking the spark advance on an ignition tester is carried out as follows:

1 - Connect vacuum adjuster.

a - Pull carburetor vacuum pipe off and connect it to connection "U" of the vacuum adjuster.

b - Connect hose from connection "V" of vacuum adjuster to the carburetor.

c - Connect long hose to connection "J" of the adjuster and to the vacuum gauge.

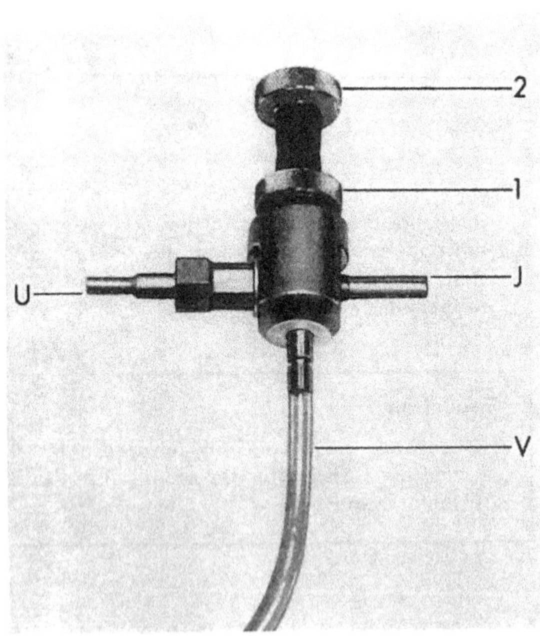

2 - Attach control scale.

3 - Turn engine until the right-hand mark on the pulley aligns with the crankcase joint.

4 - Make a 2 to 3 mm wide paint or chalk mark on the pulley opposite the 20° mark on the control scale.

5 - Connect the ignition tester according to manufacturers instructions.

6 - Start engine and increase the speed until the gauge indicates maximum vacuum. If no reading is shown, turn the knurled screw (2) out slightly.

7 - Turn knurled screw (1) out until the gauge needle is at zero.

8 - Turn the screw (1) in slowly and direct the stroboscopic lamp flashes on to the graduated scale so that the spark advance can be read off the scale at the given vacuum values.

mm Hg	Crankshaft degrees
15—25	Advance begins
26—36	10
42—50	15—21

If these values are not obtained, the tension of the breaker plate return spring in the distributor can be adjusted with a special wrench.

Messrs. Robert Bosch GmbH. Stuttgart supply a wrench for this purpose under the designation EFZV 28.

From May 1961, the shape of the setting pin on the Bosch distributor was altered for production reasons. Wrench EFZV 31 must be used for this distributor.

If the values are obtained at a lower vacuum reading, the spring tension should be increased by turning the eccentric pin in the direction shown by the arrow. If the values are obtained at higher vacuum readings, the tension should be reduced by turning the pin in the opposite direction.

Checking vacuum unit for leakage

The vacuum unit can be checked for leakage on the ignition tester.

1 - Turn the knurled screw (1) in to the stop.

2 - Start engine and increase speed until the gauge indicates maximum reading.

3 - Tighten screw (2).

4 - Stop engine and watch gauge. The vacuum must remain constant for approximately one minute, otherwise the vacuum unit should be replaced.

Vacuum unit removal and installation

1 - Remove vacuum hose from unit.

2 - Remove distributor cap and rotor.

3 - Disconnect low tension cable at coil.

4 - To facilitate the removal and installation of the vacuum unit, the distributor can be taken out.

5 - Take breaker plate return spring off.

6 - Remove the screws and take vacuum unit and pull rod off. After installing in the reverse order, the ignition timing should be reset. The vacuum unit cannot be repaired or adjusted.

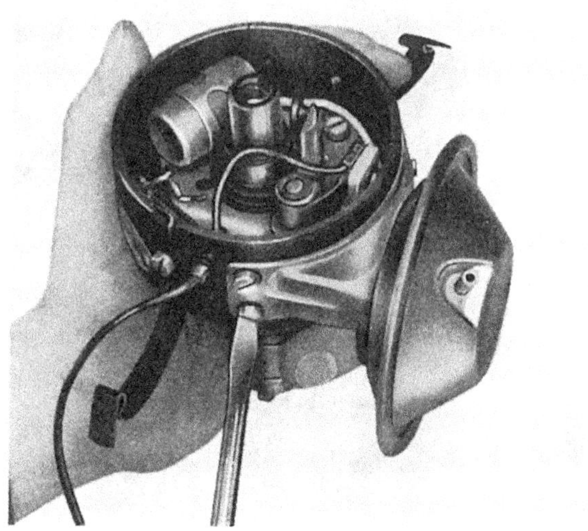

Note:

	30 hp (36 SAE hp)	**34 hp (40 SAE hp)**
From Chassis No.	**4 245 398** (approx)	**4 245 398** (approx)
From Engine No.	3 928 696 (approx)	6 219 139 (approx)

the material for the diaphragm in the vacuum units on the Bosch distributor ZV PAU 4 R 5 has been improved to make it more resistant to damage by the fuel.

Leaking vacuum units can cause poor engine output and acceleration, "flat spots" and high fuel consumption. They should be checked as follows:

1 - Pull vacuum hose off at the distributor and remove distributor cap.

2 - Turn the contact breaker plate to the left as far as it will go and close the small pipe in the vacuum unit with a finger.

3 - The breaker plate will be pulled back about 3 mm by the return spring and should then remain in this position for about one minute. If it moves further to the right with the small pipe sealed, the vacuum unit is leaking and must be replaced. The removal and installation of the vacuum unit is described in the various Workshop Manuals.

All vacuum units with diaphragms of improved material which are supplied as spares (Part No. 113 905 271 CH or 113 905 271) have a black paint spot above the connection pipe.

From January 1962 the paint marking will be discontinued and replaced by the production stamp "1 T", "2 T", "3 T" and so on.

1 = Figure for month (January)
T = Letter for year (1962)

From November 1961, all distributors with improved vacuum units are marked on the distributor housing below the low tension cable with the production stamp "11 S", "12 S", "1 T" and so on.

Vacuum Hose

To prevent fuel from entering the vacuum unit a vacuum pipe with a loop has been installed in place of the vacuum hose from **Chassis No. 4 423 336** (Engine No. 6 411 578). The vacuum pipe is connected to the distributor and vacuum unit pipes by means of two 40 mm lengths of hose.

The looped pipe (Part No. 111 129 489) can be used for service installation in previous vehicles. The existing vacuum hose should be cut about 40 mm from the carburetor (measurement "a") and the pipe inserted so that the loop points vertically upwards.

The looped pipe (Part No. 111 129 489) can also be used as a spare to replace the vacuum pipe shown in above picture.

Note:

From Chassis No. 4 981 020 (Engine No. 3 945 210), the vacuum hose (Part No. 113 129 491) between the vacuum pipe on the carburetor and the distributor vacuum unit on Models 111, 112, 115 and 116 only, was lengthened from 310 to 375 mm (12.2 to 14.5″). This will ensure that there is no tension on the pipe and thus prevent it from being pulled down.

> **Important**
>
> Ensure that the loop in the pipe on the carburetor points upwards so that it prevents fuel from entering the vacuum unit. It may be necessary to replace the previous hose with the 375 mm long version.

Checking contact breaker plate rock

1 - Remove distributor cap and rotor.

2 - Loosen hexagon screw on breaker contact insulator and take low tension cable off.

3 - Locate adaptor VW 292 with dial gauge on the edge of the distributor housing so that the dial gauge pin contacts the insulator (1) on the breaker plate.

4 - Press the dial gauge adaptor firmly against he edge of the distributor with the right hand.

5 - Press the breaker arm (2) and the condenser (3) with the left hand and rock the breaker plate up and down.

If the breaker plate rock is more than 0.15 mm (.006″), the distributor must be removed and the breaker plate replaced.

Note:

To reduce the tendency of the breaker plate to rock as the ignition advances, the breaker contacts, the condenser and the low tension connection have been rearranged.

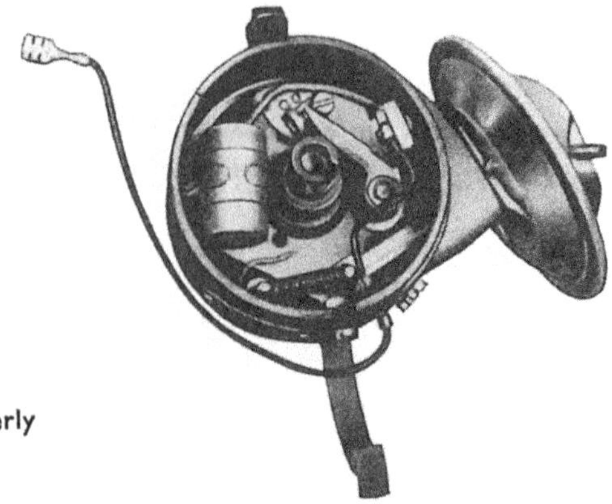

formerly

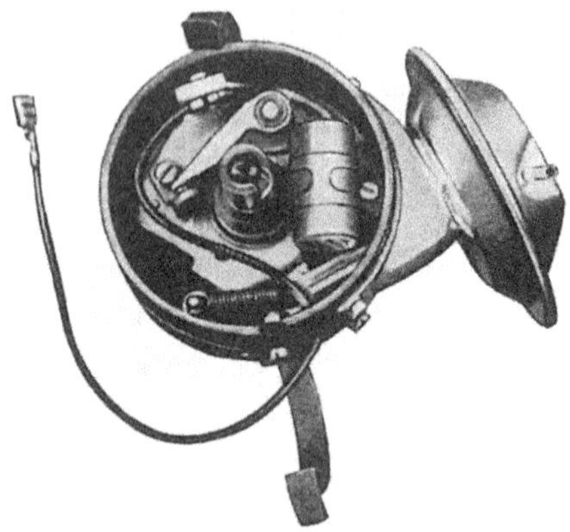

new

	30 hp (36 SAE hp)	34 hp (40 SAE hp)
From Chassis No.	**4 057 000**	**4 057 000**
From Engine No.	3 924 900	6 024 500

The Bosch distributor now has a different type designation but the part number remains the same.

Formerly: Bosch distributor ZV PAU 4 R 2 mk
New: Bosch distributor ZV PAU 4 R 5 mk

These distributors are interchangeable.

The contact breaker plates (Part No. 113 905 227 A as before) are also interchangeable.

Note:

From 5th February 1964, Chassis No. **6 100 163**, Engine No. **8 350 310**, a new Bosch distributor ZV/JU 4 R 3 is installed instead of the Bosch ZV/PAU 4 R 5. This distributor is slightly smaller than the ZV/PAU 4 R 5 model and the condenser is mounted outside on the aluminium housing.

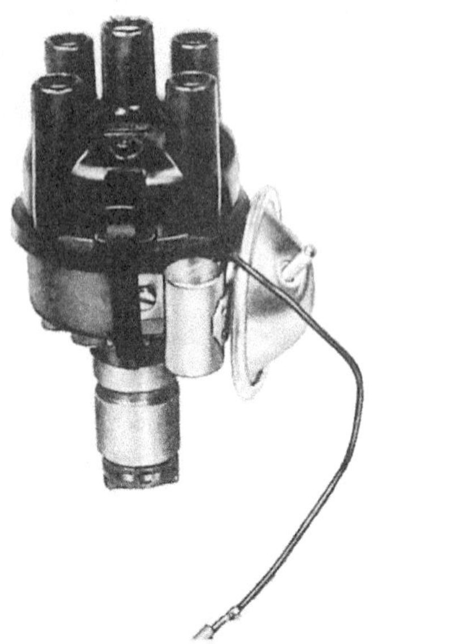

Adjusting data:

Firing point setting: 10° before TDC. (right-hand mark on pulley)

Dwell angle: 47°—53°.

Vacuum advance:

mm Hg	crankshaft degrees
18,5—24,5	Advance begins
29—38	10°
50	16°—22° (end of advance)

From about May 1963, the engines are fitted with a modified 28 PICT carburetor which has a jet in the vacuum connection. The ZV/JU 4 R 3 distributor must only be installed together with the modified 28 PICT carburetor or the 28 PICT-1 carburetor.

Condenser

The condenser has a considerable influence on the attaining of the high voltage necessary for the ignition system. It also reduces the spark formation when the points are breaking and prevents them from being burnt prematurely.

A defective condenser causes badly burned points, a poor spark with starting difficulty or a complete loss of ignition spark.

Only condensers of the prescribed type should be used as replacements, as condensers with incorrect capacities will adversely affect the service life of the points.

Testing

With a condenser tester it is possible to check the condenser for internal breakdown, insulation loss and capacity. If a tester is not available, proceed as follows:

1 - Remove distributor cap.

2 - Turn crankshaft until the points are fully open.

3 - Disconnect cable 1 at the ignition coil.

4 - Connect one lead of a 6 Volt test lamp to terminal 1 on the ignition coil and the other lead to the distributor cable.

5 - Switch on the ignition. If the test lamp lights up, the condenser is grounded and should be replaced.

6 - Remove test lamp and re-connect cable 1 from the distributor to the ignition coil.

7 - Remove the main high tension cable (No. 4) from the distributor cap and hold it about 10 mm (1/4") from a suitable ground.

8 - Turn the engine over with the ignition on. If the spark does not jump at the correct distance, repeat the test with a satisfactory condenser.

THIS PAGE INTENTIONALLY
LEFT BLANK

Condenser removal and installation

1 - Remove distributor cap and rotor.

2 - Remove breaker points.

3 - Disconnect condenser cable at breaker point insulator.

4 - Remove securing screw and take out condenser.

Installation takes place in the reverse order. Breaker point gap and ignition timing must be re-adjusted.

Distributor Removal and Installation

1 - Remove vacuum pipe at the distributor.

2 - Disconnect cable 1 at the ignition coil.

3 - Take off distributor cap.

4 - Remove the distributor retaining bracket screw on the crankcase.

5 - Lift the distributor out.

Installation takes places in the reverse order, noting the following points:

1 - Position No. 1 cylinder at firing point. The slot in the drive pinion must then be at right angles to the longitudinal axis of the engine and offset towards the crankshaft pulley and the mark on the pulley aligned with the joint in the crankcase.

2 - When inserting the distributor, turn the distributor shaft until the rotor points to the mark for No. 1 cylinder on the distributor housing and the connecting part of the shaft engages in the drive shaft slot after being turned slightly back and forth.

3 - Adjust ignition timing.

Distributor Disassembly and Assembly

Disassembly

1 - Take off distributor cap and remove distributor.

2 - Remove rotor.

3 - Remove breaker points.

4 - Remove condenser.

5 - Remove low tension cable with rubber sleeve.

6 - Detach breaker plate ground cable on housing and take off retaining spring.

7 - Remove vacuum unit return spring.

8 - Remove three screws and take vacuum unit off.

9 - Remove spring from driving dog.

10 - Knock driving dog pin out.

11 - Take driving dog off and remove shaft.

12 - Remove contact breaker plate and take off plastic washer.

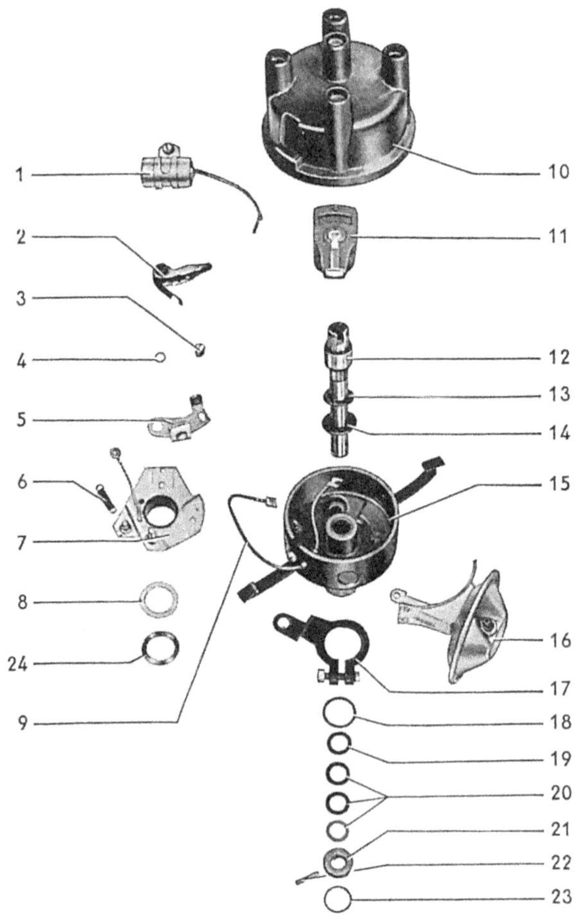

1 - Condenser
2 - Contact breaker arm
3 - Securing screw with flat and spring washers
4 - Insulating washer
5 - Contact breaker point
6 - Return spring
7 - Breaker plate with ground cable
8 - Plastic washer
9 - Low tension cable
10 - Distributor cap
11 - Rotor
12 - Distributor shaft
13 - Steel washer
14 - Fiber washer
15 - Distributor housing
16 - Vacuum advance unit
17 - Clip
18 - Sealing ring
19 - Fiber washer
20 - Shim
21 - Driving dog
22 - Pin
23 - Locking spring
24 - Shim

Checking

1 - Wash all parts in benzine, with the exception of the composition bush in the distributor housing.

2 - Check points for wear and replace if necessary.

3 - Check shaft bearing areas for wear. If radial play is excessive replace complete distributor. Adjust axial play with shims.

4 - Replace the breaker plate if the plastic bushes are worn.

5 - Renew sealing ring for distributor shaft.

Assembly

Assembly takes place in the reverse order, with special attention to the following points:

1 - Oil distributor shaft and felt ring for breaker plate.

2 - Check correct order and quantity of steel and fiber washers on the shaft. Compensate for axial play.

3 - Turn the securing slot for the rotor towards the No. 1 cylinder mark on the distributor housing. Place the fiber and steel washers on the shaft end and install the driving dog so that the lugs are offset towards the mark on the housing. Insert the pin and install the lock-ring in the groove on the driving dog.

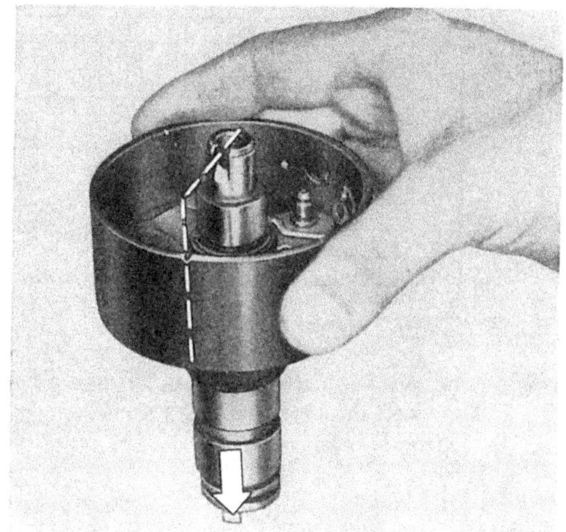

4 - The retaining spring with the guide piece for the distributor cap should be on the same side as the breaker plate ground connection.

5 - Lubricate the fiber block with universal grease.

Checking the Distributor on a Test Stand

With a distributor test stand the ignition timing (cam offset), the vacuum advance, the dwell angle and the breaker arm spring tension can be checked.

The best sequence for carrying out the check is as follows:

1 - Check the ignition timing of the individual cylinders (cam offset).

2 - Check the dwell angle (51°—55°).

3 - Check the spring tension with a special stroboscope.

4 - Check the vacuum unit for leakage. A vacuum of 100 mm Hg should remain constant for approximately 1 minute with the connections closed.

5 - Check the vacuum advance curve.

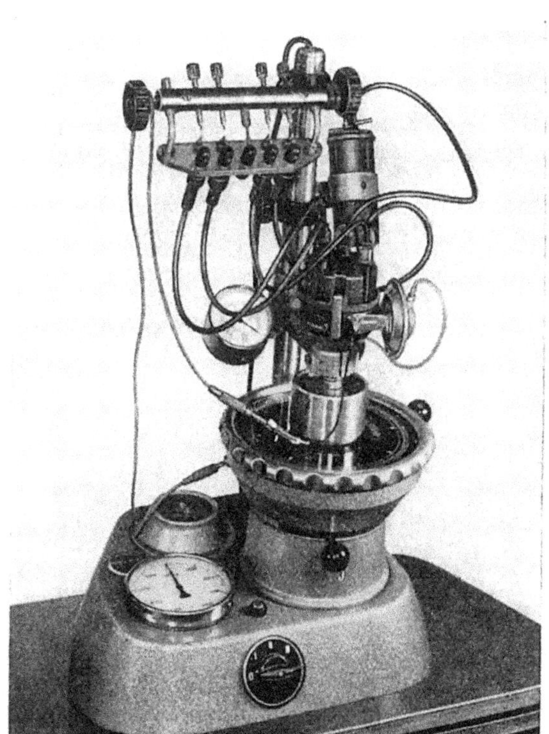

Vacuum Advance Curve for the Bosch ZV PAU 4 R 2 and 4 R 5 Distributors

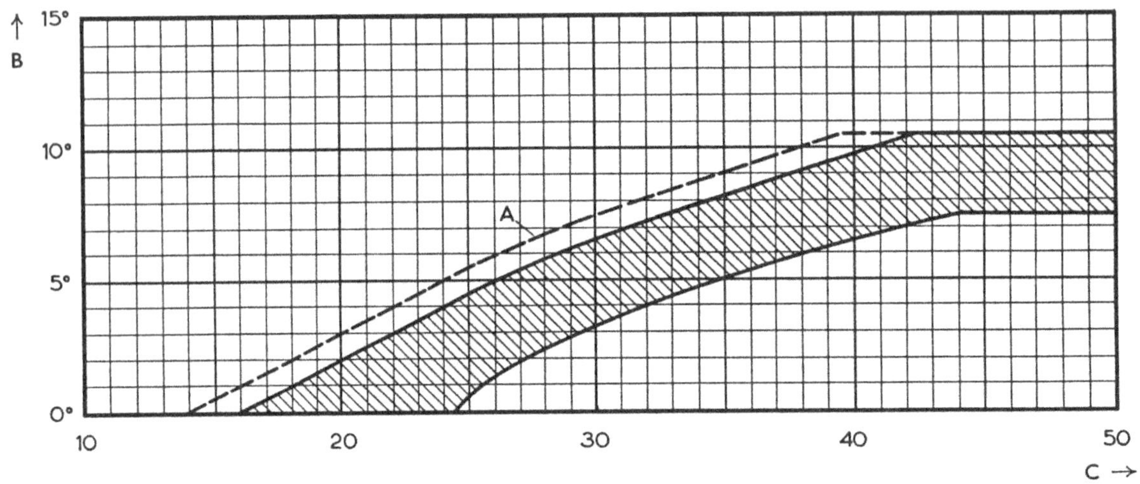

A - Wear limit when measuring with falling vacuum
B - Advance in distributor shaft degrees
C - Vacuum in mm Hg

If the readings show excessive variation from the permissible tolerance, the following tests should be made:

a - Check the freedom of movement of the breaker plate, if necessary disassemble the distributor and free the breaker plate off.

b - Check the vacuum unit for freedom of movement, replace if necessary. (The diaphragm becomes hard and brittle after being in use for a long time.)

c - Check return spring and adjust tension with special wrench Bosch EFZV 28 or EFZV 31 if necessary.

Note:

From Chassis No. 4 710 048 (Engine No. 6 744 903 to 6 750 577), 840 34 hp engines were fitted with BOSCH ZV PAU 4 R 7 distributors (not supplied as a spare). The differences between this distributor and the ZV PAU 4 R 5 type are that the former has a stronger return spring and a jet in the connecting pipe for the vacuum unit. These parts are not available as spares. The vacuum advance curve has also altered slightly.

The BOSCH ZV PAU 4 R 7 distributor may only be installed together with the standard modified carburetor SOLEX 28 PICT (Part No. 113 129 023 E/141 129 023 D). On this carburetor the vacuum pipe is extended to below the center of the fuel discharge arm in the mixing tube. All SOLEX 28 PICT carburetors from number 3 239 081 and 6 824 776 are fitted with the long vacuum pipe. The modified carburetor can be installed in all vehicles with 34 hp engines.

If trouble is experienced with the ZV PAU 4 R 7 distributor, the complete distributor should be replaced.

Ignition Coil

Type: VW 111 905 105 F

The instructions for the maintenance and testing of this coil are the same as those given for the Bosch coil.

Ignition Distributor

Type: VW 113 905 205 B
Type: VW 113 905 205 H

General

The spark advance over the whole engine speed range is controlled by the vacuum advance unit. A ball joint connects the vacuum unit pull rod to the contact breaker plate which is mounted off-center on a steel leaf spring.

The removal and installation of the VW distributor, the adjustment of the ignition timing, condenser testing and vacuum spark advance are the same as with the Bosch distributor.

Maintenance

The maintenance of the VW distributor is also the same as with the Bosch. The lubrication of the felt ring in the contact breaker plate is deleted.

Adjusting contact breaker points

When adjusting breaker points, proceed as follows:

1 - Remove distributor cap and rotor.

2 - Turn the distributor shaft by moving the crankshaft until the points are fully open.

3 - Loosen lock screw of breaker point.

4 - Insert screwdriver between slot in breaker plate and the breaker fixed point. Move the screwdriver to and fro until the clearance 0.4 mm (0.016") is obtained.

5 - Tighten lock screw.

6 - Check clearance on all four cam lobes.

Important

The contact breaker plate must not be moved from the basic position when adjusting the points as otherwise the point gap will be altered.

Replacing breaker points

1 - Remove distributor cap and rotor.

2 - Remove fixed point lock screw.

3 - Loosen breaker arm spring securing screw.

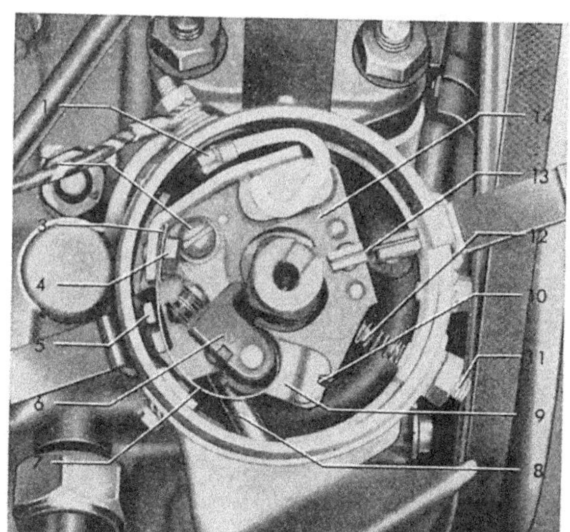

1 - Primary connection with cable
2 - Securing screw
3 - Leaf spring
4 - Hexagon head screw
5 - Stop bracket
6 - Breaker arm
7 - Breaker arm spring
8 - Pull rod
9 - Fixed point
10 - Adjusting slot
11 - Threaded rod
12 - Spring
13 - Leaf spring for breaker plate
14 - Breaker plate

4 - Lift breaker arm out.

5 - Remove fixed point with insulator.

Installation takes place in the reverse order.

6 - Adjust point gap and ignition timing.

Condenser removal and installation

1 - Unscrew nut from terminal 1 on the distributor and take off low-tension cable and washers.

2 - Remove condenser securing screw and take condenser out.

Installation takes place in the reverse order. Only condenser of the prescribed type should be used as replacements, as condensers with incorrect capacities will adversely affect the service life of the points.

Distributor Disassembly and Assembly

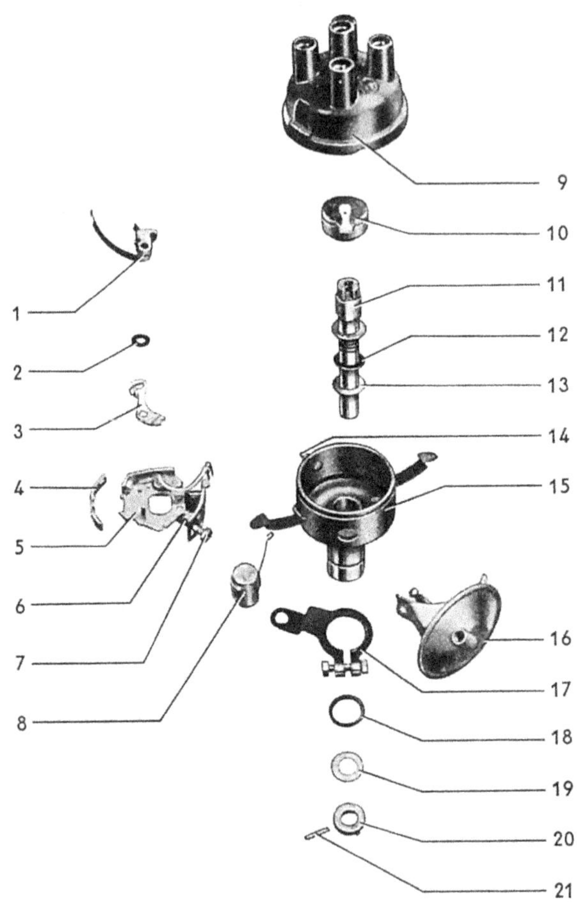

1 - Breaker arm
2 - Insulating washer
3 - Fixed breaker point
4 - Stop bracket
5 - Breaker plate
6 - Spring
7 - Threaded rod with nut and lock washer
8 - Condenser
9 - Distributor cap
10 - Rotor

11 - Distributor shaft
12 - Steel washer
13 - Plastic washer
14 - Low-tension cable
15 - Distributor housing
16 - Vacuum advance unit
17 - Clamp bracket
18 - Sealing ring
19 - Plastic washer
20 - Driving dog
21 - Pin

Disassembly

1 - Remove distributor cap and rotor.

2 - Unscrew nut from terminal 1 and take off low-tension cable and washers. Pull out connecting cable to breaker plate.

3 - Remove condenser.

4 - Remove contacts.

5 - Push the vacuum unit pull rod out of the ball joint by inserting a suitably bent wire hook in the hole provided in the pull rod.

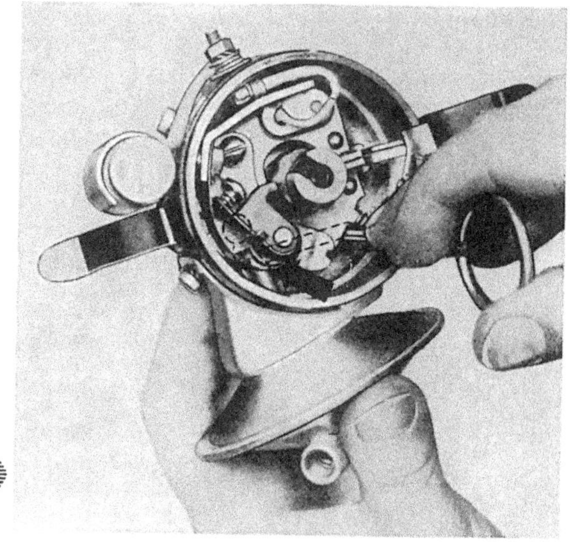

6 - Remove nut with threaded pin and vacuum unit securing screw.

7 - Remove vacuum unit.

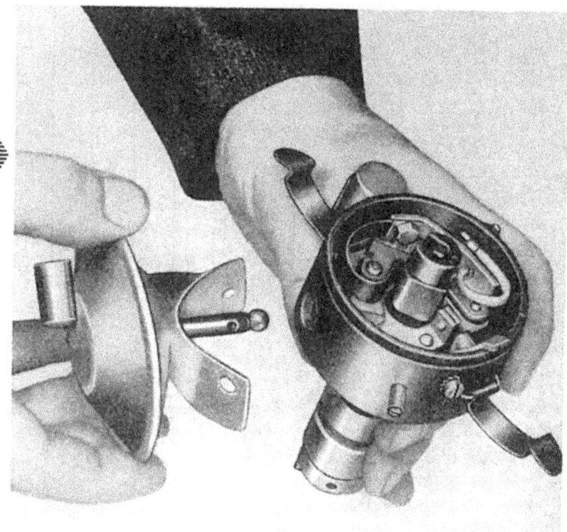

8 - Remove screw holding breaker plate retaining bracket. Take breaker plate and stop bracket out of distributor housing.

9 - Pull low-tension cable out.

10 - Knock out driving dog pin and take off driving dog and fiber washer.

11 - Remove distributor shaft with steel and fiber washers.

12 - Take rubber ring and clamp arm off distributor base.

Checking

1 - Wash all parts with benzine except the self-lubricating bush in the distributor housing.

2 - Examine contact breaker points for wear and renew if necessary.

3 - Examine distributor shaft bearing areas for wear. If radial play is excessive, replace distributor complete. Axial play is eliminated with shims.

4 - On the breaker plate, check the steel leaf spring, the insulation of the low tension connection and the engagement of the spark advance pull rod in the ball joint. If necessary, replace breaker plate.

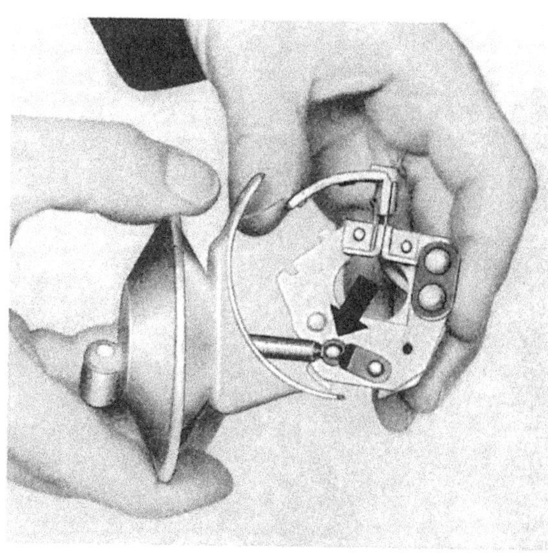

5 - Replace distributor shaft sealing ring.

Assembly

The assembly takes place in the reverse order with attention to the following points:

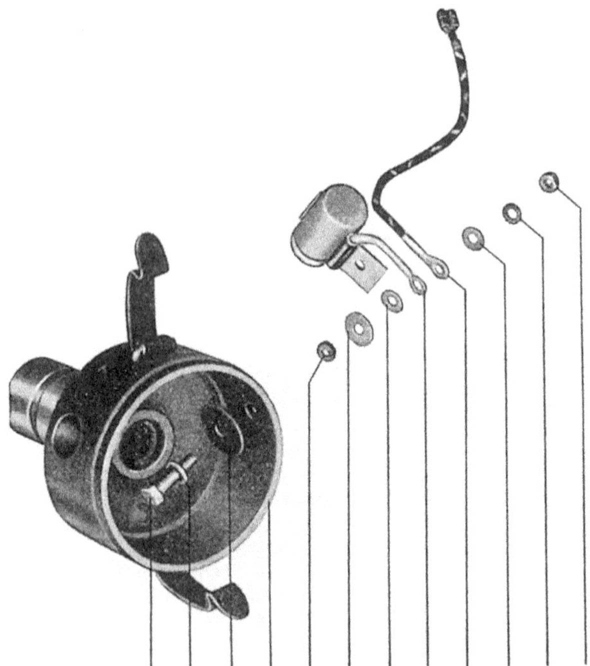

1 2 3 4 5 6 2 7 8 2 9 10

1 - Hexagon head screw
2 - Washer
3 - Insulating washer
4 - Distributor housing
5 - Insulating bush
6 - Insulating washer
7 - Condenser cable
8 - Low-tension cable
9 - Lock washer
10 - Nut

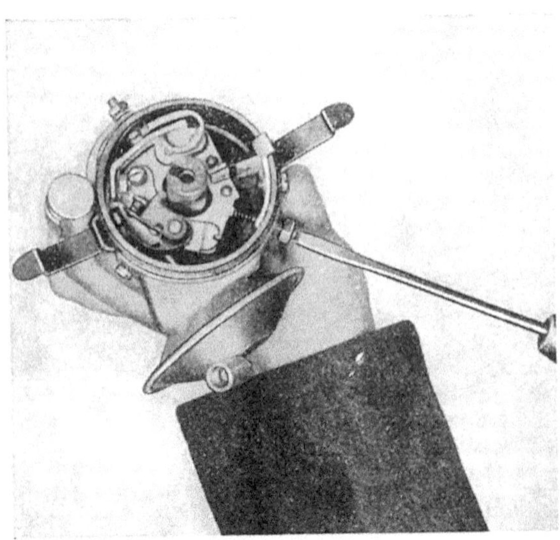

1 - Oil the distributor shaft and fill the space in the housing between the bushes with Bosch Special Grease FT 1 v 8.

2 - Insert the shaft with steel and fibre washers into the distributor housing.

3 - Turn the securing slot for the rotor to the No. 1 cylinder mark on the edge of the distributor housing. Place the fibre washer on the bottom of the shaft. Install the driving dog so that the lugs are offset towards the mark on the housing edge. When the pin has been inserted, peen both ends of the hole.

4 - Check that the washers on the primary connection screw are in the correct order to ensure proper insulation.

5 - Lubricate the ball socket underneath the breaker plate with Universal Grease.

6 - Place the stop bracket and breaker plate in the distributor housing and secure with a screw. Attach vacuum advance unit to the housing.

7 - Press the vacuum advance unit pull rod into the ball socket in the breaker plate from underneath with a suitably shaped wire hook.

8 - Install the condenser and breaker points and adjust breaker point gap.

Important

When securing the vacuum advance unit, the threaded rod must only be screwed in to the stop on the spring. The adjustment of the spring influences the spark advance curve of the distributor. Every time the distributor is overhauled the advance curve must be readjusted on a distributor test stand by altering the spring tension.

Replacing the vacuum advance unit

1 - Press the pull rod down out of the breaker plate by inserting a suitably bent wire hook (2.8 mm dia.) in the hole provided in the pull rod.

2 - Loosen not on threaded rod, remove the securing screw and take the unit off.

The vacuum advance unit cannot be repaired or adjusted.

Assembly takes place in the reverse order after the pull rod ball joint has been lubricated with Universal Grease.

> **Important**
>
> The adjustment of the threaded rod and the spring must not be altered as otherwise the distributor advance curve will be affected.

Checking the Distributor on a Test Stand

With a distributor test stand, the ignition timing (cam offset), the vacuum advance, the dwell angle and the breaker arm spring tension can be checked.

The best sequence for carrying out the check is as follows:

1 - Check the ignition timing of the individual cylinders (cam offset).

2 - Check the dwell angle (48 to 52°).

3 - Check the spring tension with a special stroboscope.

4 - Check the vacuum unit for leakage. With connection closed, a vacuum of 100 mm Hg should remain constant for a period of 1 minute.

5 - Check the spark advance curve against the diagram below.

If the readings show excessive variation from the permissible tolerance, the following tests should be made:

a - Check the setting of the spring for the breaker plate. If the curve commences to rise too soon, the threaded rod should be screwed in further to increase the spring tension. If the curve rises too late, the spring tension should be reduced by screwing the rod out.

b - Check the ease of operation of the vacuum unit and replace if necessary. (The diaphragm can become hard and brittle after being in use for a long time.)

Vacuum Advance Curve for the VW Distributor

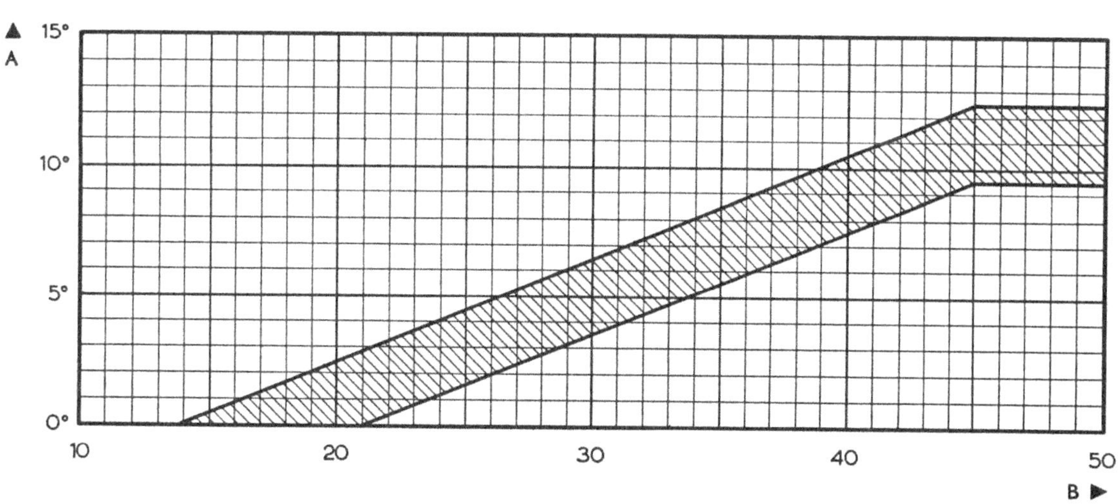

A - Advance in distributor shaft degrees   B - Vacuum in mm Hg

Spark Plugs

General

Types:	1192 ccm
Bosch	W 175 T 1;
Beru	175/14
and plugs of other makes with similar values	

Spark plug thread = 14 mm
Spark plug gap "a" = 0.6 to 0.7 mm (.023"—.027")

The electric spark jumps the gap between the two electrodes to ignite the fuel-air mixture in the combustion chamber. Starting, idling, acceleration and maximum performance greatly depend on the right choice of spark plug to suit engine operating conditions. The heat range of the correct plug can be obtained from the above table.

Maintenance and Inspection

Spark plugs should be cleaned and the gaps checked at 5000 km (3000 miles) intervals.

To an experienced and skilled mechanic, the appearance of the spark plug gives evidence of the adjustment and engine operating condition. The following rules are generally applicable:

Electrodes and insulator

medium brown — good carburetor setting and spark plug working properly.
black — mixture too rich;
light grey — mixture too lean;
oiled up — failure of spark plug or leaking piston.

In the case of fuel containing tetra-ethyl-lead (anti-knock fuel), the insulator will show a grey color, provided the engine is correctly adjusted.

During operation, the plug gap increases due to natural burning. If the gap has increased too much, the plug may fail to operate. The gap is checked by means of a gauge and adjusted by bending the ground (outer) electrode to the correct value.

A number of testing devices for spark plugs are in use, where the spark must occur under pressure (6—8 kg sq. cm = 85—114 lbs./sq. in.) and can be watched through an inspection hole.

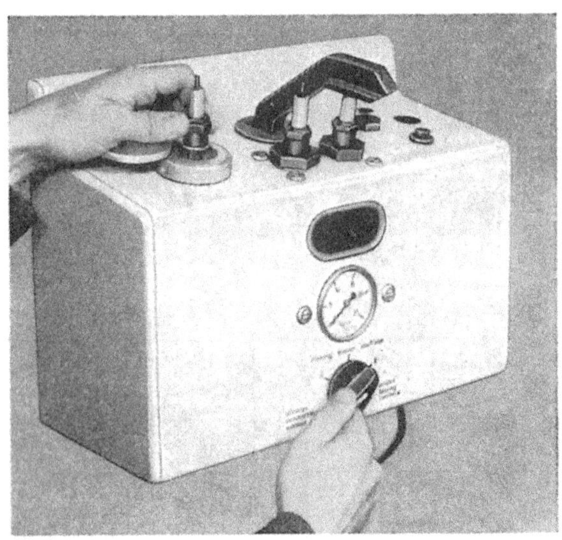

It is recommended that spark plugs are replaced about every 15 000 km (9000 miles). Dirty spark plugs are cleaned with a wire brush and a chip of wood. Oiled-up spark plugs can be made serviceable by burning out. The insulator must be kept clean and dry to avoid short circuits and high-tension leakage. A sand-blast type cleaner should be used for cleaning spark plugs in large workshops.

Starting from cold at very low temperatures will be facilitated if the spark plug gaps are reduced from .7 mm to about .5 mm (.020").

Suppressed Ignition Cables

To prevent radio and television interference, resistor type ignition cables which have a graphite coated synthetic core in place of the normal copper core, are installed. To distinguish them from the previous black non-suppressed type these cables are red. Only the spark plug connectors (Part No. 111 905 445 A) and the cable connectors (Part No. 111 905 427 A) may be used for securing these cables.

These connectors have gland screws which retain the graphite coated synthetic core better and are, therefore, only suitable for the suppressed cables.

Important

The connectors must be screwed to the cable ends in such a manner that the contact between the synthetic core and the connector is perfect. In order to ensure that the synthetic core is not pushed out of the cable by the gland screw when screwing on the connector, the cable must be squeezed together with a pair of pliers at the place where the connector fits. Faulty contact at this point is invariably caused by incorrect assembly and results, in most cases, in the cable being burned through.

If trouble occurs, replace the individual ignition cables.

When installing a VHF Radio set the following measures are necessary to ensure interference-free reception;

a - Install a suppressed distributor rotor.

b - Install four suppression adaptors in the spark plug connectors. These adaptors have a thread at one end to screw into the spark plug connector and the normal clip at the other end for pushing onto the spark plug.

Further suppression of the vehicle by means of condensers is not affected by the above measures.

Note:

All engines built up to the 10th October 1960 were intermittently equipped with light-red (two shades) and dark-red colored resistor type ignition cables.

The **light-red** colored cables have proved to be of inferior quality in various respects. Even when pulled or bent slightly during installation the graphited synthetic core tended to break and create very high resistances within the cables. The spark jumping across the break in the core eventually damaged the red outer coating of the cable.

From the 20th October 1960 (**Chassis No. 3 390 251,** Engine No. 5 242 646) the light-red cable has been replaced by an improved **orange-red** type. This cable is of the same quality as the dark-red type which has been used since the resistor type cable was introduced and has recently been designated "KMN F 25".

The cables can be checked for internal breakage with an ohmmeter when no external damage is visible.

The cables should be removed complete with the connectors and coupled to the ohmmeter so that a good contact is ensured at both ends. During the test the cable should be moved about and stretched slightly. If the reading does not remain constant and the needle flicks up to a considerably higher resistance value, the synthetic core is broken and the cable should be replaced.

(up to 1965)

The readings should not be above or below the following resistance values:

Ignition cable	Minimum value in kΩ at 20° C	Maximum value in kΩ
Main cable	4	130
No. 1 cylinder cable	15	550
No. 2 cylinder cable	15	500
No. 3 cylinder cable	10	350
No. 4 cylinder cable	10	350

From Chassis No. 3 960 131 (Engine No. 5 870 247) only blue resistor type ignition cables (Part No. unchanged N 18 387 1) are being installed in all engines. One manufacturer identifies the cables with the letters "KMN".

The blue cables have the same characteristics and resistance values as the red type used previously and can be service installed in all engines.

Note:

From 1 st July 1961, the ignition systems of all vehicles in Germany must be suppressed in accordance with local Traffic Regulations. Please observe current regulations in this respect when carrying out repairs on engine and electrical units.

From Chassis No. 3 192 507 Volkswagens with the 34 hp (40 SAE hp) engine and all 34 hp exchange engines are suppressed in production.

Additional suppression equipement must be installed when fitting radio sets.

Note:

To prevent water from entering the connection between spark plug connector and ignition cable, the connectors are fitted with plastic caps (Part No. 311 905 433) **from Chassis No. 5 199 980** (Engine No. 7 336 420).

The caps can be subsequently installed in previous engines.

(1965)

The readings should not be above or below the following resistance values:

Ignition cable	Minimum value in kΩ at 20° C	Maximum value in kΩ
Main cable	4	19
No. 1 cylinder cable	18	86.5
No. 2 cylinder cable	17	82
No. 3 cylinder cable	11.5	54
No. 4 cylinder cable	11	52

When installing a VHF radio, the following additional measures are necessary in order to ensure interference free reception.

a - Install a suppressed distributor rotor.

b - Install four suppression adaptors in the spark plug connectors. These adaptors have a thread at one end to screw into the spark plug connector and the normal clip at the other end for pushing onto the spark plug.

Further suppression of the vehicle by means of condensers is not affected by the above measure.

Note:

From 5th March 1965, Chassis No. **115 594 027**, Engine No. **9 285 001** (34 bhp engine) or Chassis No. **115 594 028**, Engine No. **3 994 721** (30 bhp engine), the resistor type ignition cables were replaced by cables with a copper core. In order to conform with the suppression regulations which exist in various countries, the spark plug connectors are fitted with resistances (1 Kilo ohm).

Suppressed distributors are also fitted.

The copper ignition cables may only be installed together with the plug connectors with resistances and a suppressed rotor arm.

Part	Part No.
Distributor, suppressed	Bosch 111 905 205 N (as before)
1 Set of ignition cables, complete	SP 210 B
Spark plug connector, suppressed (1 Kilo ohm)	113 035 255 A

Spare parts for the resistor type ignition cables used formerly will remain available. It is not necessary to subsequently convert earlier vehicles.

Important
Additional suppression equipment is also required on vehicles with copper core cables and suppressed rotor arms when a radio set is fitted.

THIS PAGE INTENTIONALLY
LEFT BLANK

Headlights

General

The headlights are flush mounted in the fenders and combine the high beam, asymmetrical low beam and the parking light. A twin-filament bulb for the high and low beam and the parking light bulb are held in position in the reflector by a cap. The headlight beams can be aimed by altering the setting of the reflector.

Bulb types:

Twin-filament bulb A 6 V 45/40 W DIN 72601

Parking light bulb HL 6 V 4 W DIN 72601

Bulb replacement

1 - Loosen the slotted screw at the bottom of the headlight rim.

2 - Take out headlight unit.

3 - Turn the cap to the left and take bulb out of reflector. Pull the connector off the bulb holder.

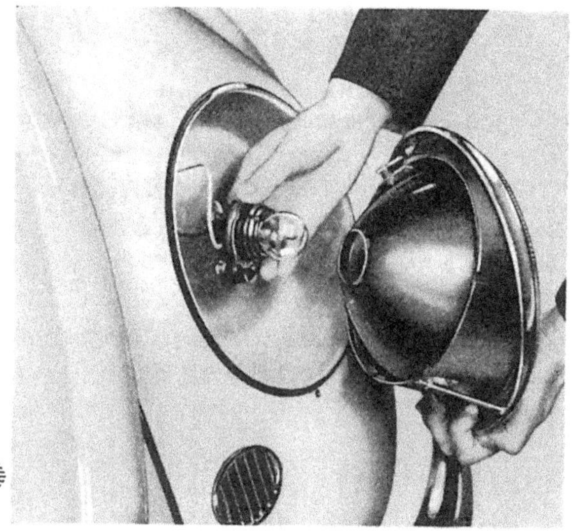

4 - Replace bulb. Hold the bulb with a clean cloth to avoid touching it with the bare hand.

When installing, note the following points.

1 - The lug in the bulb holder must engage in the notch provided in the reflector.

2 - The contact strip must press firmly on the base of the parking light bulb.

Headlight lens replacement

1 - Remove headlight unit.

2 - Turn the cap to the left and take bulb out of reflector. Take out the parking light bulb.

3 - Lift the lens retaining springs out of the rim with a screwdriver.

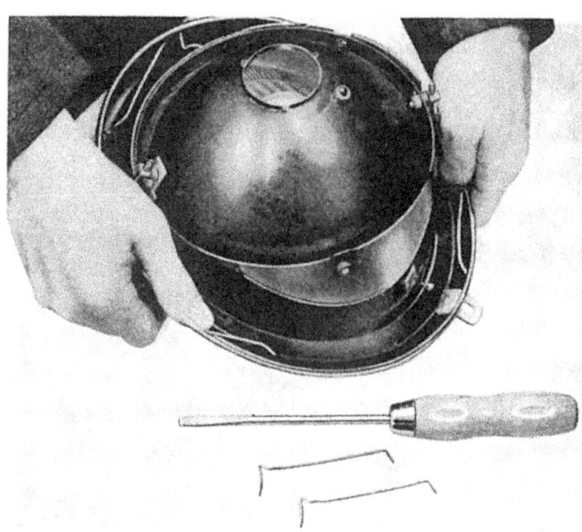

4 - Clean up the peened threads at the ends of the two adjusting screws and remove the screws.

5 - Take out the retaining ring wirh reflector, lens and sealing ring.

Important

When replacing the lens, take care not to touch the reflector as otherwise the sensitive mirror finish will be soiled or damaged and the reflecting properties affected.

6 - Lay the sealing ring on the new lens and place the lens in the rim with the "TOP" mark correctly positioned.

7 - Place the retaining ring and reflector in position and check that the sealing ring is properly seated between rim and retaining ring.

8 - Install the adjusting screws and peen the ends.

9 - Place retaining springs in position.

10 - Install headlight unit, taking care that the bulb and contact strip are properly seated.

11 - Adjust headlights.

Note:

The headlight adjusting screws are threaded right up to the screw head. The stop for the 5 mm long spring is formed by a threaded washer which is squeezed together by the manufacturers. To replace a headlight rim or adjusting screw, proceed as follows:

1 - Cut open the threaded washer and remove adjusting screw and spring.

2 - Insert the new screw (56 mm long, Part No. 111 941 141 C or 96 mm long, Part No. 111 941 141 D) into the rim.

3 - Install the spring and compress it to a length of 3.5 to 4 mm (0.14—0.16") by screwing on a 4 mm nut which is locked with a second nut.

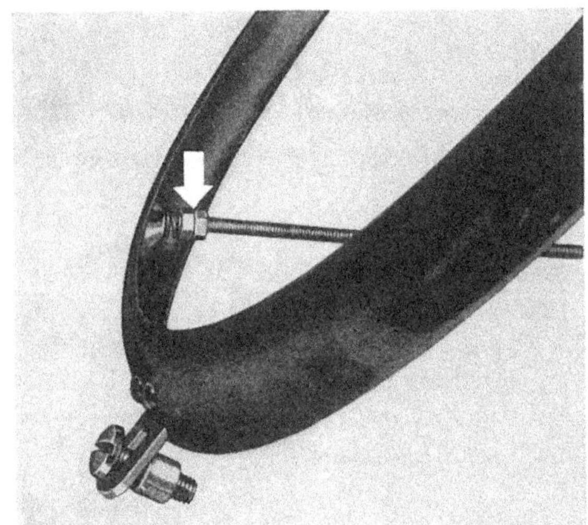

Aiming the Headlights

The headlights can be aimed on an adjustable vertical surface. The test surface should be light colored and provided with a line for the headlight center and a parting mark for the light-dark border line. A suitable screen is shown in "Local Manufacture of Workshop Equipment" under VW 635/1.

When using headlight aiming instruments, the manufacturers instructions should be followed. The settings obtained with such instruments must conform to the legal requirements.

The following instructions must be noted before and during the headlight aiming operation.

1 - The vehicle must be on a level surface, 5 m (16.5") from the test screen.

2 - The test surface must be vertical and the line parallel to the ground.

3 - The tires pressures must be correct.

4 - The vehicle should be rolled back and forth a few metres to allow the suspension to settle.

5 - The headlights should be adjusted separately with the second light covered up.

6 - The headlights should be aimed horizontally and vertically when switched to low beam.

7 - The rear seat must be loaded with one person or a weight of 70 kg (154 lbs.).

Vertical adjustment

The headlights should be aimed vertically so that the light-dark border is horizontally on the adjusting line to the left of the cross and slopes upward to the right of the cross.

Note:

With the vehicle 5 m from the screen or wall the adjusting line should be 5 cm (2″) — or 1% — below the headlight center line.

Horizontal adjustment

The headlights should be aimed horizontally so that the angle on the light-dark border line is exactly on the center of the cross.

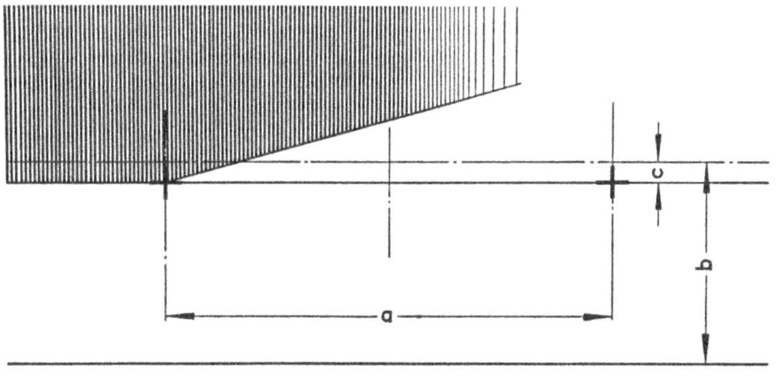

	Sedan and Convertible	Karmann Ghia models
a =	1004 mm (39.5″)	1240 mm (48.0)

b = the height of the headlight center from the floor.
c = 1% of the distance from vehicle to test surface, that is 5 cm at a distance of 5 m.

Adjustment	Bosch and Hella
Vertical	Turning upper screw To the right — lowers To the left — raises
Horizontal	Turning lower screw To the right — moves beam to left To the left — moves beam to right

Light switch removal and installation

1 - Disconnect battery ground cable.

2 - Remove instrument panel protective cover in front luggage compartment.

3 - Unscrew operating knob.

4 - Unscrew retaining ring with special wrench VW 674 and remove switch.

Installation takes place in the reverse order, taking care that the cables are connected correctly.

Dimmer switch removal and installation

1 - Disconnect battery ground cable.

2 - Remove left front partition lining and rubber covering.

3 - Remove securing screws and take out switch.

4 - Pull off three cable connections.

Installation takes place in the reverse order, taking care that the cables are connected correctly. Cement lining and rubber cover into position.

Headlight Indicator Light

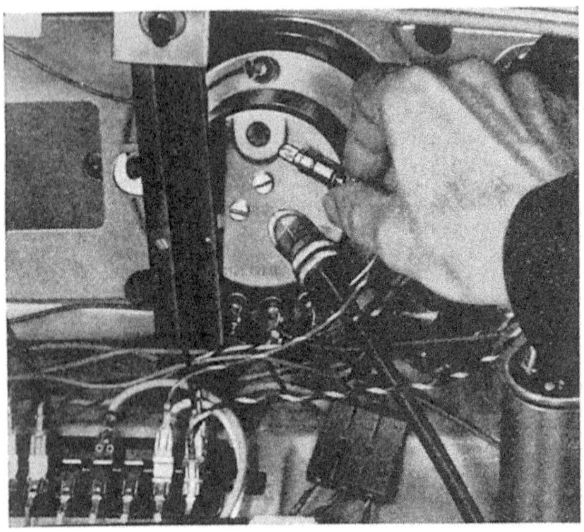

General Description

When the high beam is switched on a blue control lamp in the instrument panel lights up. The lamp is connected with the cable to terminal 56a (high beam) of the left headlamp.

Bulb type

J 6 V 1.2 W DIN 72601

Bulb Replacement

The bulb is accessible after the front hood has been opened and the lining in front of the instrument panel turned down.

1 - Pull out socket and bulb.

2 - Press bulb lightly into its socket, turn it slightly and pull it out.

License Plate Light

General Description

The license plate light housing is fitted to the rear hood and is accessible by opening the rear hood.

Bulb Type

License Plate light bulb G 6 V 5 W DIN 72601

Bulb Replacement

1 - Open rear hood halfway.

2 - Loosen both screws on light housing and take out lens.

3 - Replace bulb.

To make sure that the bulb is tightly seated and makes good contact, check the tension of the contact springs and clean them if dirt is present. Check housing gasket and renew it if found necessary.

License plate light removal and installation

1 - Open rear hood halfway.

2 - Remove three nuts inside the rear hood.

3 - Remove light with gasket and take off cable.

When installing, check that the gasket is properly seated.

Note:

From Chassis No. 5 888 185 (Model 113, 114) and Chassis No. 5 852 989 (Model 151, 152) the license plate lamp has a wider housing. The new lamp cannot be installed in vehicles built before July 1963 because the rear hood has been modified.

Stop Light Switch

The stop light switch for the hydraulic system is located on the brake master cylinder. It is not adjustable.

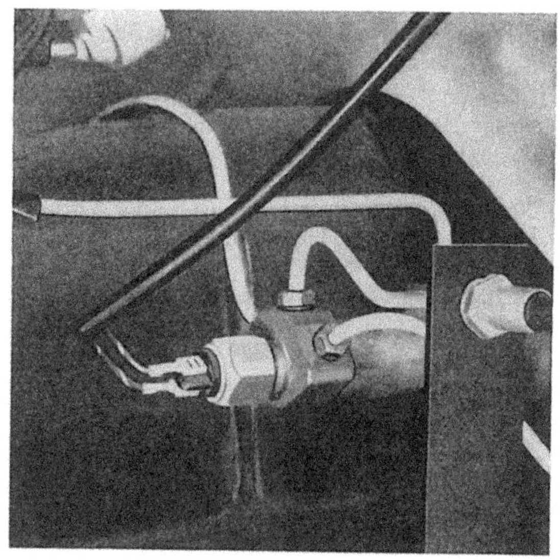

Stop light switch removal and installation

1 - Lift front of vehicle.

2 - Remove left front wheel.

3 - Pull cables from switch.

4 - Screw the switch out.

5 - When the new switch has been installed the hydraulic system must be bled to remove air.

Interior Light

General

The interior light which is located in the left hand roof side member over the door pillar can be controlled by a tumbler switch in the housing. The switch enables the light to be switched on or off with the doors open or closed.

Bulb type:

K 6 V 10 W DIN 72601

Bulb replacement

1 - Grip the light at both sides and pull it out of the hole in the roof member.

2 - Replace the bulb, taking care that it is seated properly and making good contact.

Door contact switch

The De Luxe and Convertible models are equipped with door contact switches which are connected in parallel to the switch on the instrument panel and operate the light when a door is opened.

Door contact switch removal and installation

1 - Remove securing screw, take out switch and disconnect cable.

Oil Pressure Switch

General

The oil pressure switch is installed in the oil pressure line between the oil pump and the oil cooler. When the engine is stationary the contact on the diaphragm is held closed by spring pressure. When the ignition is switched on the battery current flows from terminal 15 of the ignition via the green warning lamp and the oil pressure switch to ground. The green warning lamp lights up.

When the engine is running, the oil pressure actuates the diaphragm, opens the contact and the lamp goes out.

Removal

1 - Disconnect cable at oil pressure switch.

2 - Remove oil pressure switch with wrench VW 159a.

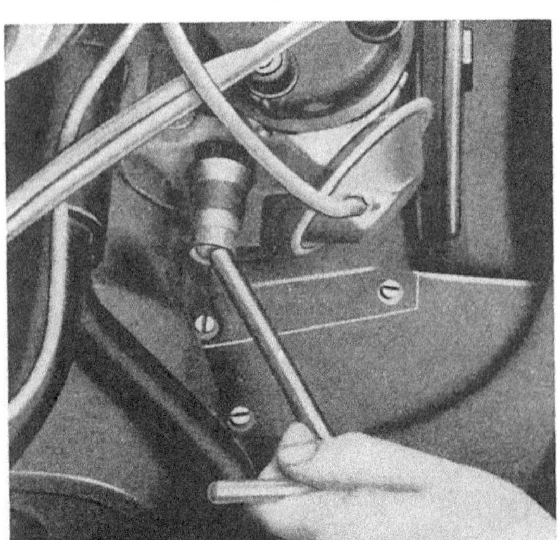

Installation

Sealing is effected by the tapered thread. To avoid damaging the thread do not use undue force when tightening the switch.

After installation, the switch should be checked in connection with the green warning light in the speedometer dial.

Testing Oil Pressure Switch

The oil pressure switch can be tested on the engine, using a simple testing device with a gauge (drawing VW 662/2 for local manufacture) and a test lamp, after the engine has attained operating temperature.

1 - Screw oil pressure switch into the testing device.

2 - Screw testing device into seat for oil pressure switch and connect one lead of the test lamp to the oil pressure switch and the other lead to terminal 15 at ignition coil. The test lamp should then light up. If the lamp does not light up, the switch is defective and must be replaced.

3 - Start engine. The lamp must go out when the gauge indicates pressure above 0.3—0.6 kg/sq. cm (4.3—8.5 lbs./sq. in.). At a lower pressure, the contact should remain closed and the lamp light up. When the engine is cold, the lamp should go out at normal idling speed, and when the engine is warm, it should go out as the speed increases.

If the lamp goes out too late, the switch should be replaced.

4 - Stop engine. A slight delay may occur before the lamp lights up, as the oil pressure decreases slowly.

The switch cannot be repaired.

Note:

From Chassis No. 3 924 800 (Engine No. 5 843 201) the operating pressure of the oil pressure switch (Part No. unchanged 113 919 081) was changed from 0.3—0.6 kg/sq. cm to 0.15—0.45 kg/sq. cm (2.1—6.4 lbs/sq. in.). This will prevent the oil pressure warning lamp from lighting up at or just above normal idling speed as it did occasionally.

The operating pressure is stamped in the upper surface of the switch and cannot be altered. The new switch can be installed in all engines from Engine No. 1 090 762 (January 1947). The cable connector will have to be converted to the push-on type.

Oil Pressure Warning Lamp

General

The green oil pressure warning lamp is located between terminal 15 at ignition switch and oil pressure switch terminal. The warning lamp lights up when the ignition is switched on and goes out when the engine starts.

Bulb type:

J 6 V 1.2 W DIN 72601

Note:

If the lamp lights up continuously whilst driving, the chances are that the oil circulation has been interrupted, which means that the lubrication of the engine has ceased.

If the lamp lights up occasionally it does not indicate trouble, as long as it goes out again as the speed increases. At low outside temperatures the lamp lights up in general only at idling speed, but at high outside temperatures the lamp may also light up at low speed in the various gears, or when changing gears, if the engine oil is of a very low viscosity.

The normal engine oils in use nowadays are of a fairly low viscosity. This facilitates the starting of a cold engine by reducing the frictional losses and has the further advantage that at a comparatively low pressure the oil circulates quickly to all bearing surfaces and gives favourable lubricating and cooling conditions. If the warning lamp only goes out at high speeds when the engine is warm, the oil pressure switch should be checked.

Replacing Bulb

The lamp is accessible after the front hood has been opened and the lining in front of the instrument panel removed.

1 - Pull out socket and bulb from speedometer.

2 - Press the bulb down lightly into the holder, turn it slightly and pull it out.

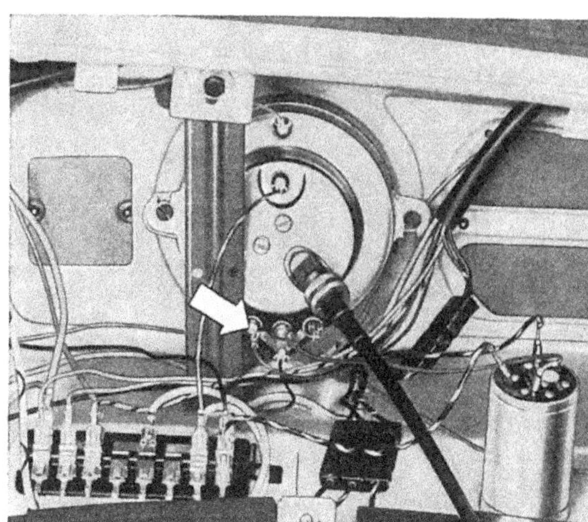

Stop and Tail Lights

Description

Each rear fender carries a combined stop/indicator and tail lamp with reflector and a twin filament bulb.

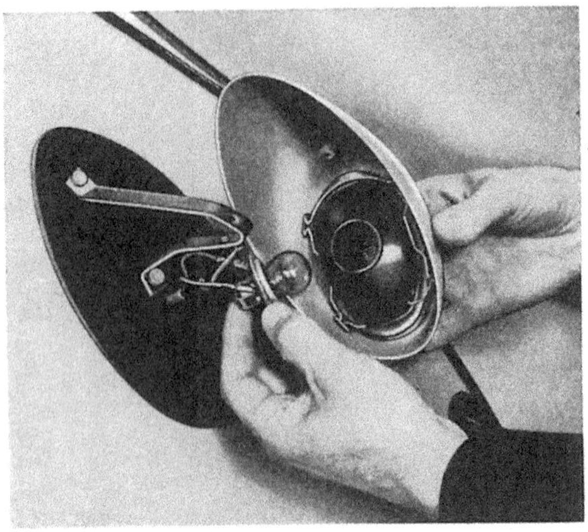

Bulb Type:

Twin filament bulb S 6 V 18/5 W DIN 72601

Bulb replacement

1 - Take off the bezel after removing the Phillips screw.

2 - Withdraw the bulb holder from the reflector.

3 - Replace bulb. When inserting the holder, make sure that the lug engages properly in the slot in the reflector.

4 - Be sure the rubber gasket between bezel and fender is in good condition, replace if necessary.

5 - When placing the bezel in position, ensure that the rubber gasket is correctly seated.

6 - The bezel mounting bracket can be taken off after removing the two nuts below the fender.

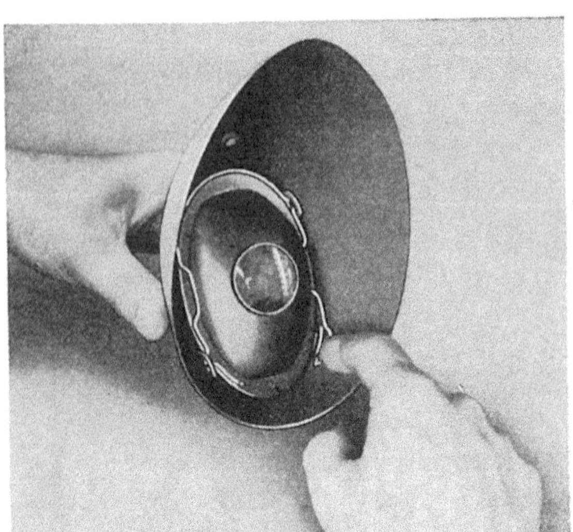

Reflector Replacement

1 - Take off the bezel after removing the Phillips screw.

2 - Withdraw the bulb holder from the reflector.

3 - Remove reflector spring clip by pressing it out of the retaining tabs. Take off reflector and glass.

4 - Install new reflector.

5 - When placing the bezel in position, be sure the rubber ring between the reflector and the housing is properly seated.

Two-compartment Rear Lights

From Chassis No. 3 856 472 (Inland only) and **from Chassis No. 4 010 995** (Export) all Volkswagen De Luxe and Standard models are being equipped with modified two-compartment stop/tail/ flashing turn indicator lights. The special versions installed for certain export countries differ only slightly from the standard type described here.

A - Flashing turn indicator (yellow) Bulb type: R 6 V 18 W DIN 72601
B - Stop/tail light (red) Bulb type: S 6 V 18/5 W Din 72601
C - Reflector (red)

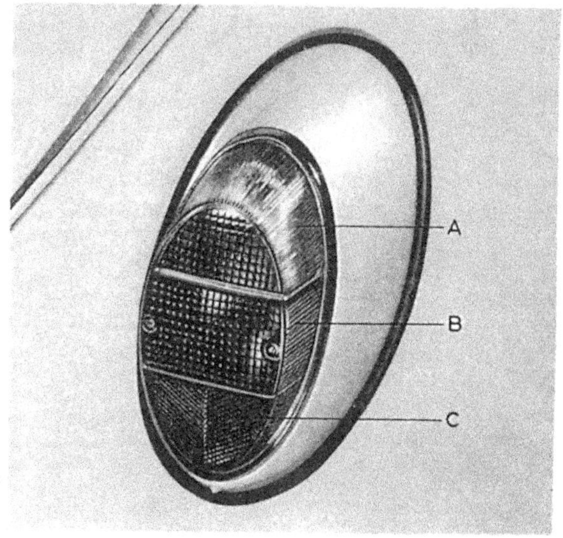

The new rear lights have three connections. The stop light filaments of the stop/tail bulbs are connected directly to the stop light switch and not via the indicator switch as they were formerly. The twin-circuit indicator switch used previously has been replaced by a single circuit switch.

The rear fenders have been modified to suit the new lights.

The most important spare parts are:

1	Indicator switch (Standard model only)	111 953 513 B
1	Indicator switch (De Luxe model only)	141 953 517 C
1	Stop/tail/indicator light, left	111 945 095 F
1	Stop/tail/indicator light, right	111 945 096 F
1	Stop/tail/indicator light, left (USA)	111 945 095 G
1	Stop/tail/indicator light, right (USA)	111 945 096 G
2	Gaskets between rear light and fender	111 945 191 D
2	Grommets	311 971 913
2	R 6 V 18 W indicator bulbs	N 17 731 1
2	S 6 V 18/5 W stop/tail bulbs	N 17 737 1
1	Rear fender, left	111 821 305 G
1	Rear fender, right	111 821 306 G
1	Rear fender, left (bumper guards)	111 821 305 H
1	Rear fender, right (bumper guards)	111 821 306 H

Bulb replacement

1 - Remove two screws and take lens off.

2 - Press bulb lightly into holder, turn it to the left and take out.

Do not interchange the bulbs when installing.

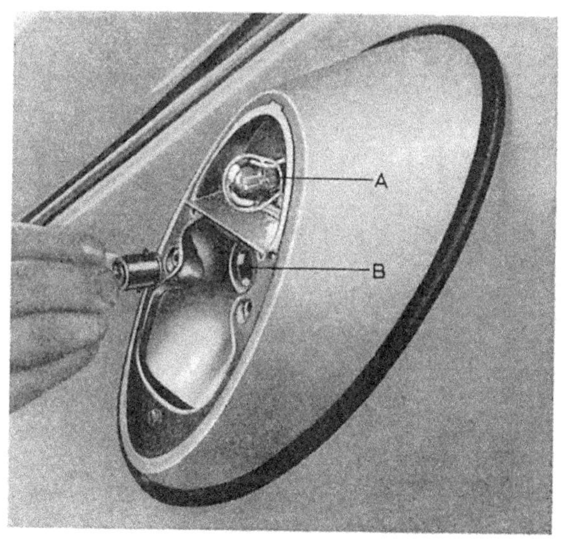

A - Single filament 18 W
B - Twin filament 18/5 W

When inserting the twin filament bulb, the retaining pin nearest to the bulb glass must point downwards.

The installation of the modified rear lights in previous vehicles is described in the 2nd supplement to Technical Bulletin E - 12.

THIS PAGE INTENTIONALLY
LEFT BLANK

Sealed Beam Headlights

General Description

The cars delivered to countries such as, USA, Canada, and Guam are equipped with Sealed Beam units. The sockets for the parking light bulbs are screwed to the headlamp shells. Replacing the front glass and aiming the headlights is done in the same way as detailed for the standard headlights.

Replacing Sealed Beam Unit and Parking Light Bulb

Removal

1 - Loosen front rim retaining screw and pull out complete headlight unit.

2 - Pull the headlight connector off the sealed beam unit.

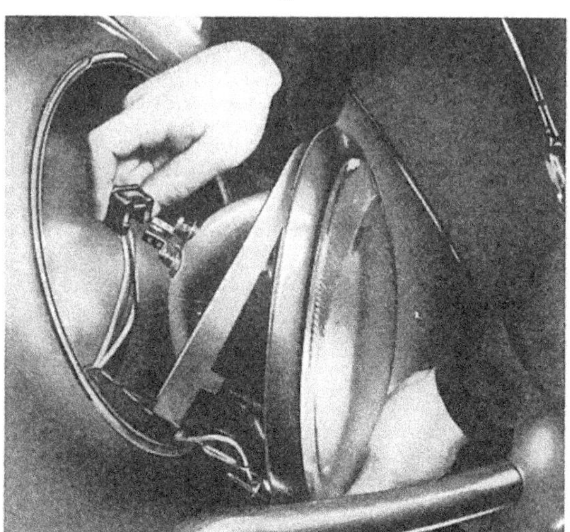

3 - Disconnect the two cables from the parking light bulb socket.

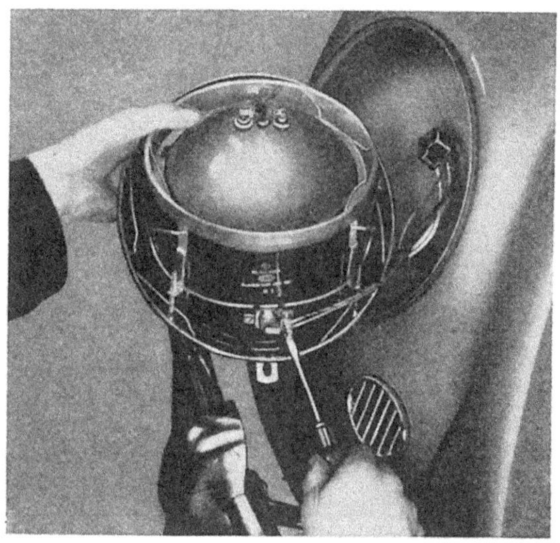

4 - Screw off the parking light bulb socket.

5 - Remove the five spring clips from the sealed beam unit retaining ring.

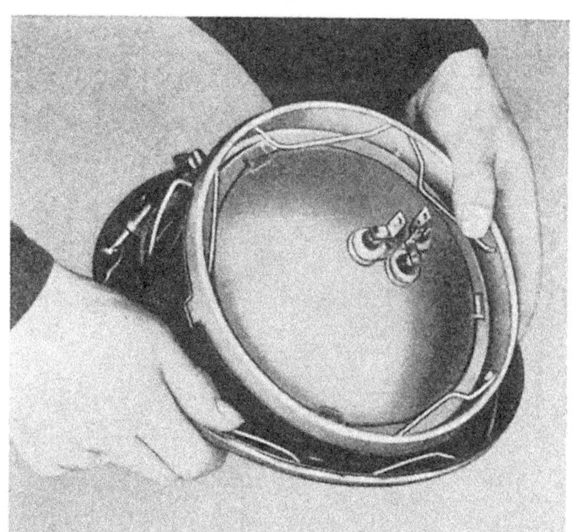

6 - Withdraw retaining ring and sealed beam unit.

Installation

The following points should be noted when installing the headlight:

1 - The tabs and slots of the retaining ring ensure a correct positioning of the sealed beam unit.

2 - Be sure the cables are properly connected.

3 - Be sure the rubber seal between front rim and fender is properly seated.

4 - Recheck headlight adjustment.

Headlight glass replacement

1 - Remove headlight unit.

2 - Detach connector and parking light cable.

3 - Lift glass retaining springs out of the rim with a screwdriver.

4 - Unscrew headlight adjusting screws as far as possible and detach spring clip.

5 - Lift the retaining ring and sealed beam unit until the headlight glass and gasket can be taken out.

6 - Place gasket on the new glass and insert the glass in the rim so that the VW sign is upright when installed and the arrow points downwards.

7 - Locate retaining ring and sealed beam unit and check that the gasket between rim and retaining ring is properly seated.

8 - Insert retaining springs.

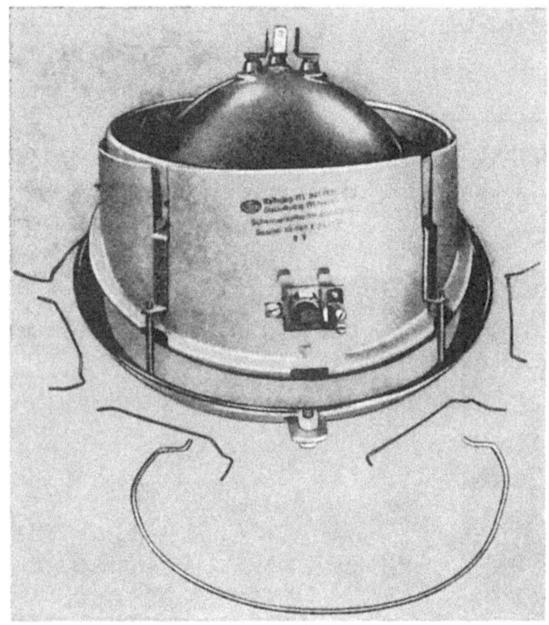

Headlight Adjustment

The headlights should be adjusted by means of the two screws to conform to the Lighting Regulations of the particular country concerned.

Adjustment	
Vertical	left screw turning to right — raises beam turning to left — lowers beam
Horizontal	right screw turning to right — moves beam to left turning to left — moves beam to right

Right or left screw seen in direction of travel.

Note:

From 29th November 1963, Chassis No. **5 943 202,** and owing to changes in the regulations in various export countries, Sealed-Beam headlamps with a modified adjusting mechanism were installed.

Adjustment

A = Vertical aim	in direction of arrow — raises against arrow —lowers
B = Horizontal aim	in direction of arrow — to left against arrow — to right

The new headlamps can be installed instead of the former type.

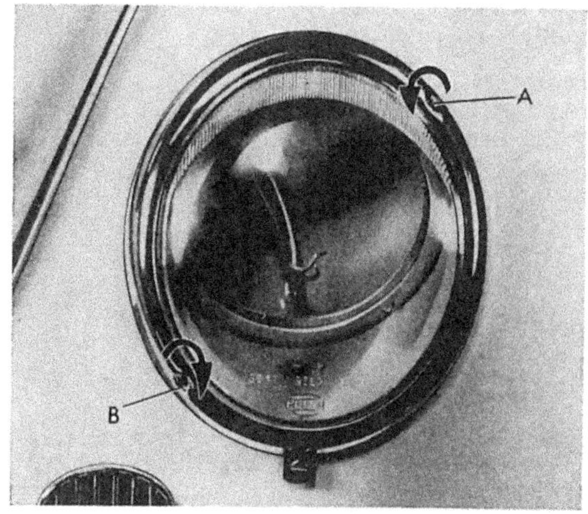

THIS PAGE INTENTIONALLY
LEFT BLANK

Headlights

General Description

The flush fitting headlights in the front fenders combine high beams, asymmetrical low beams, and parking lights. The twin filament bulb for the high and low beams and the parking light bulb are held in position in the reflector by a cap.

Provision is made for the headlight beams to be aimed vertically and horizontally.

Bulb types:

Twin filament bulb A 6 V 45/40 W DIN 72601

Parking light bulb HL 6 V 4 W DIN 72601

Bulb replacement

1 - Remove the screw in the center below the headlight and take off the trim ring.

2 - Remove the securing screw at the lower right and take out the headlight unit.

3 - Turn the cap to the left and take bulb out of reflector. Pull connector off bulb base and replace bulb.

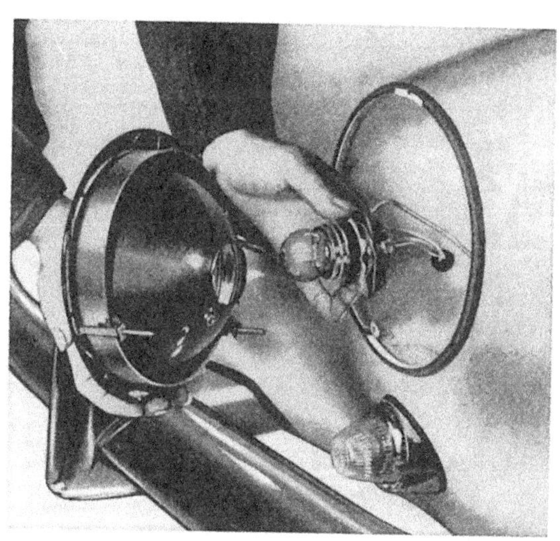

Note the following points when installing:

1 - The notch in the bulb holder must engage in the slot provided in the reflector.

2 - The contact strip must press firmly on the base of the parking light bulb.

Headlight Adjustment

The headlights are adjusted after the trim ring has been removed.

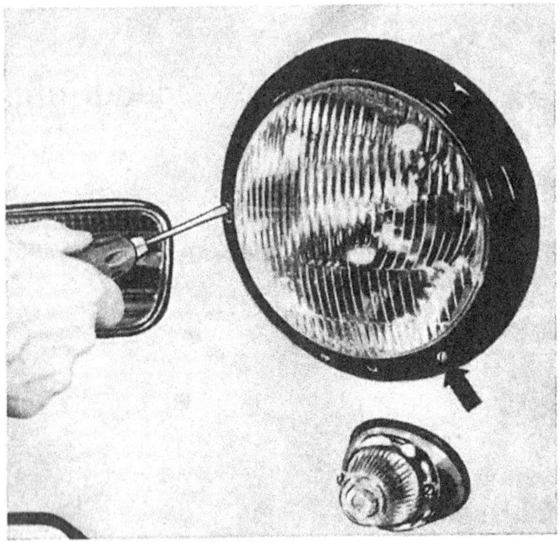

The adjustment is the same as for the VW Passenger Car, see Section E 6, pages 2 and 3.

Adjustment	Hella
Vertical	lower left screw turning to right — raises beam turning to left — lowers beam
Horizontal	upper right screw turning to right — moves beam to left turning to left — moves beam to right

Right or left screw in direction of travel.

Sealed Beam Headlights

Vehicles for export to certain countries are equipped with Sealed-Beam units.

The parking light is combined with the indicator light. A twin filament bulb S 6 V 18/5 W DIN 72601 carries out the functions of indicator and parking light.

Headlight adjustment

The headlights should be adjusted by means of the three adjusting screws to conform to the Lighting Regulations of the particular country concerned.

Adjustment	
Vertical (When adjusting, the two screws must be turned simultaneously in opposite directions)	upper left screw turning to right — lowers beam turning to left — raises beam lower left screw turning to right — raises beam turning to left — lowers beam
Horizontal	upper right screw turning to right — moves beam to left turning to left — moves beam to right

Sealed Beam unit replacement

When the trim ring has been removed and the headlight unit taken out, the sealed beam unit can be replaced as on the VW Sedan.

Note:

From 14th November 1963, Chassis No. **5 930 400**, Sealed-Beam headlights with a different adjusting mechanism were installed.

Adjustment

A = Vertical aim	turning to right — raises turning to left — lowers
B = Horizontal aim	turning to right — moves beam to right turning to left — moves beam to left

General

The combined stop, indicator and tail lights with the reflectors are flush mounted in the rear fender.

Bulb types:

1 - **Indicator** **R 6 V 18 W DIN 72601**

2 - **Stop** **R 6 V 18 W DIN 72601**

3 - **Tail** **G 6 V 5 W DIN 72601**

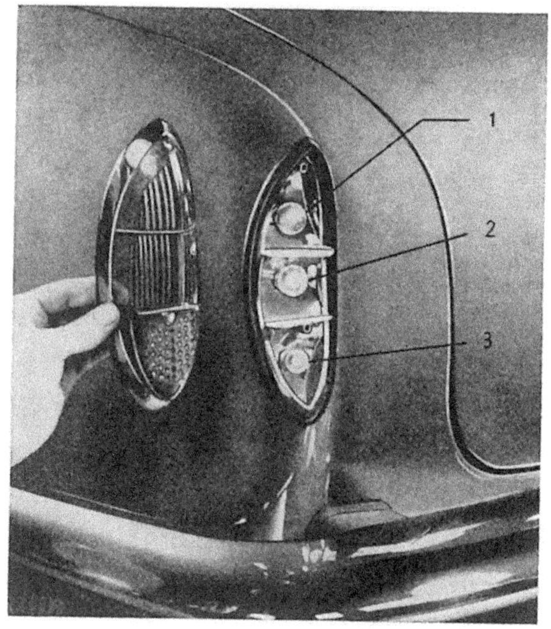

Bulb replacement

1 - Remove two Phillips screws and take off chrome rim with lens.

2 - Replace bulbs. Check that they are properly seated and making good contact.

3 - When installing, ensure that the chrome frame seats properly and that the lens are sealed correctly on the fender.

Installation takes place in the reverse order. Take care that the gaskets are properly seated.

Stop, indicator and tail light housing replacement

1 - Remove two Phillips screws and take off chrome rim with lens.

2 - Remove sound absorbing material from left or right of engine compartment after bending down the securing clips and taking out the air cleaner insert and the battery.

3 - Disconnect cables.

4 - Loosen two wing screws.

5 - Remove housing with gasket to the rear.

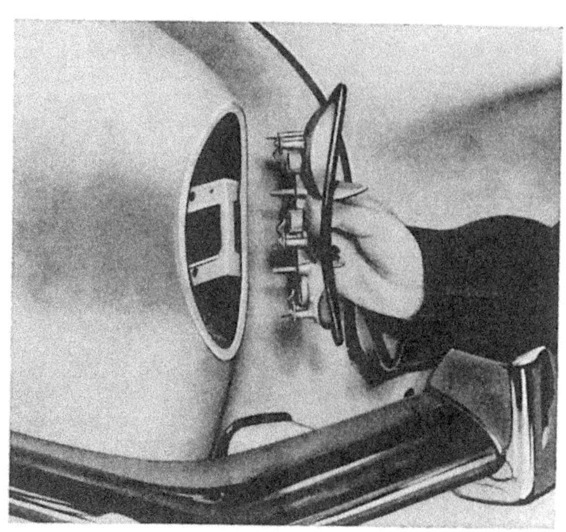

License Plate Light

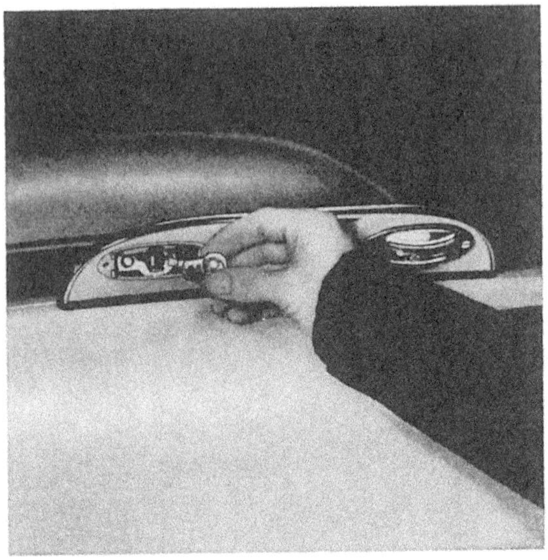

The two license plate lights are located in the handle of the rear hood.

Bulb type:

License plate — G 6 V 5 W DIN 72601

Bulb replacement:

1 - Open rear hood.

2 - Remove two securing screws and take out glass insert.

3 - Replace bulbs.

When installing, check that the groove in the glass is located over the lug in the housing.

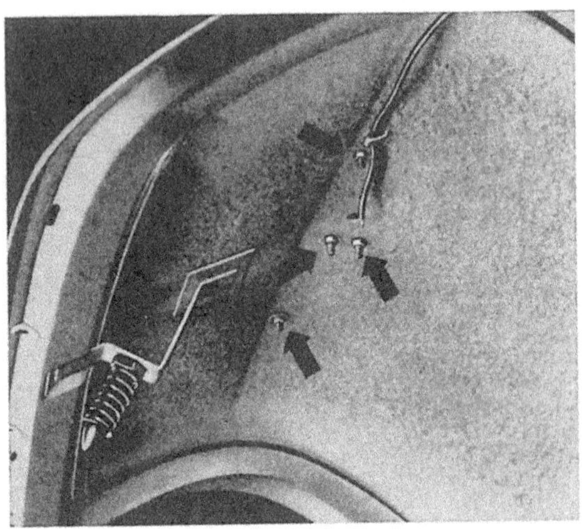

License plate light housing removal and installation

1 - Open rear hood and bend up cable clips.

2 - Remove four securing screws and take off hood handle with gasket.

3 - Disconnect cable.

4 - Unscrew glass insert and take lamp housing out of handle.

5 - Unsolder cable from old light and solder to new housing.

Installation takes place in the reverse order.

Interior Light

The interior light is automatically operated by opening or closing either of the doors. A manual three-position switch incorporated in the light fitting allows the light to be turned off and on independently of the door contacts.

Position of switch:

Upper	— on
Intermediate	— off
Lower	— door contacts

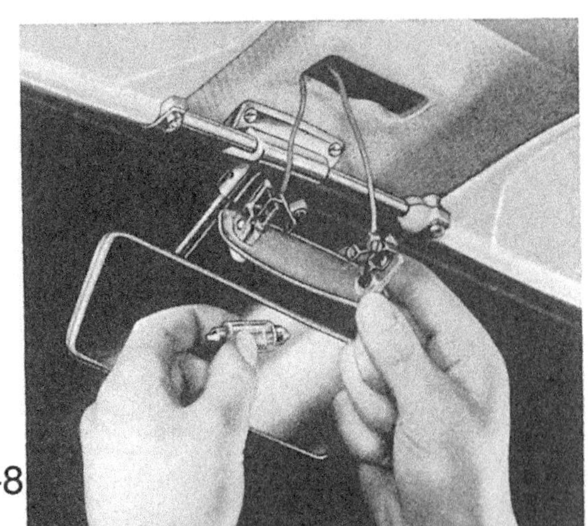

Bulb Type:

K 6 V 10 W DIN 72601

Bulb replacement

1 - Pull out interior light by gripping the two sides of the rim with both hands.

2 - Replace the bulb.

Light Fitting Removal

Pull out interior light and disconnect both cables.

When installing, make sure the cables are properly connected.

Note:

From 3rd August 1964, Chassis No. **145 000 001**, the interior light was modified and is now located in the mirror bracket.

The new interior light cannot be service installed in earlier vehicles.

Door Contact Switch

Each door is provided with a contact switch, by means of which the interior light is automatically turned on with the light tumbler switch in the lower position.

Door Contact Switch Replacement

1 - Pull door contact switch out of the door hinge pillar.

2 - Disconnect the cable.

THIS PAGE INTENTIONALLY
LEFT BLANK

Windshield Wiper

General

The windshield wiper motor and the two spindles with linkage are mounted on a common frame which is accessible when the front hood has been raised. The windshield wiper motor is controlled by a switch on the instrument panel to the right of the steering wheel. The same switch operates the suction and pressure pump of the windshield washer.

When the windshield wiper motor is switched off, the wiper arms swing automatically to the right and come to a standstill at the correct point.

A double contact in the wiper motor, actuated by an eccentric plate on the wiper shaft, continues to supply current to the motor after the wiper switch has been switched off until the contact is opened by the eccentric plate. At this moment the flow of current is cut off and simultaneously short circuited by the second contact. The motor is stopped quickly by this short circuit and remains in the end position. Movement of the motor from the end position can only be achieved by operating the wiper switch.

Important

The automatic end position switching works correctly as long as the movement of the wiper arms is unrestricted. If the wiper arms should stop outside the end switching area, the current supply to the motor will continue even though the wiper system has been switched off. When this occurs the armature windings can burn in a few minutes and render the motor useless.

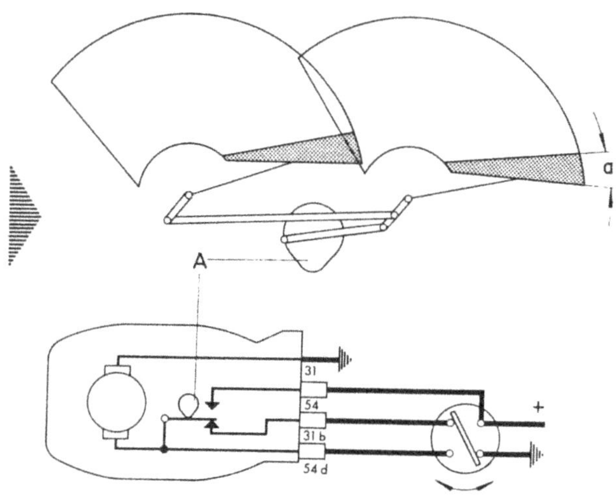

A - Eccentric plate a - End position

Observation of the following measures will help to prevent damage occurring to the wiper motor:

1 - Blades which have frozen to the windshield should be freed before operating the wiper switch. If this is not done, the blades may move slightly when switched on and then stop again. Should this occur at the switched on position, the current supply to the motor will continue despite the fact that the switch has been moved to off.

2 - When snow is falling, take care that it does not build up on the windshield and stop the blades reaching the end position.

3 - When the windshield is dry and voltage low (almost flat battery) the blades can stop in the center of their arc. It is essential to move the wiper arms to the end position.

Note:

Wiper motors from Messrs. SWF and Bosch are intermittently installed. The Bosch motor can be recognised by the firm's trade-mark and the fact that the cover is secured to the housing with a screw and not with a clip like that on the SWF motor.

The terminals for the electrical connections are arranged in the same order on both products and the motors are interchangeable.

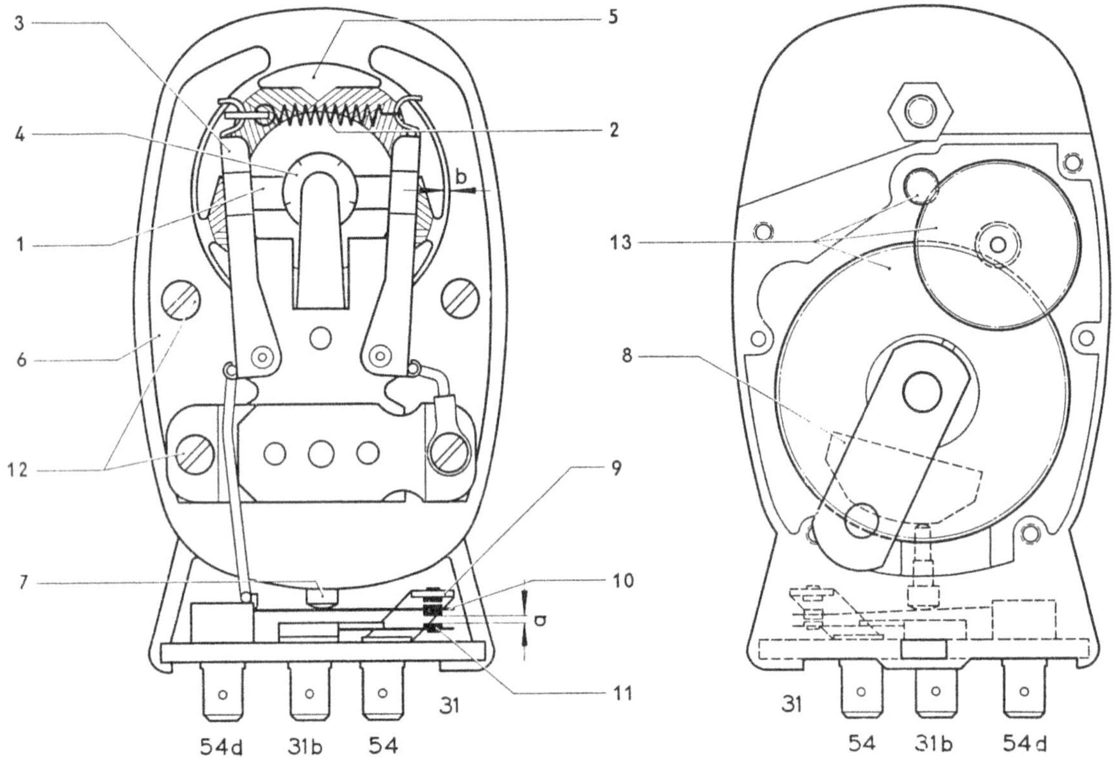

1 - Brush
2 - Spring
3 - Brush holder
4 - Commutator
5 - Armature
6 - Pole shoe
7 - Cylinder bolt
8 - Eccentric plate
9 - Contact
10 - Contact spring
11 - Contact
12 - Attaching screws
13 - Gears

a = 0.8 mm (.031")
b = Clearance

Maintenance

When repairs are being carried out on the windshield wiper system, the linkage, the spindles and the wiper arm joints should be oiled. The blade contact, uniformity of stroke to each side and the end position switching should be checked at each Maintenance Service.

Removal and installation of windshield wiper frame with motor

1 - Disconnect battery ground cable.

2 - Loosen clamp screws in wiper arm brackets and remove arms.

3 - Remove both wiper bearing hexagon nuts with washers and outer bearing seals.

4 - Remove instrument panel cover in front luggage compartment.

5 - Disconnect cable from wiper motor.

6 - Remove screw securing wiper frame to body.

7 - Take out frame complete with motor and linkage.

Installation takes place in the reverse order, noting the following points:

1 - Move the wiper frame by means of the elongated hole so that the wiper spindles are vertical to the windshield.

2 - Check that the seals and washers are in the correct order.

3 - Check that the ground strap on windshield frame securing screw is making good contact.

Removal and installation of wiper motor

1 - Remove windshield wiper frame complete with motor and linkage.

2 - Remove lock washer and spring washer from the driving shaft and disconnect driving link.

3 - Loosen wiper shaft securing nut, remove one motor securing nut and take the motor off the frame.

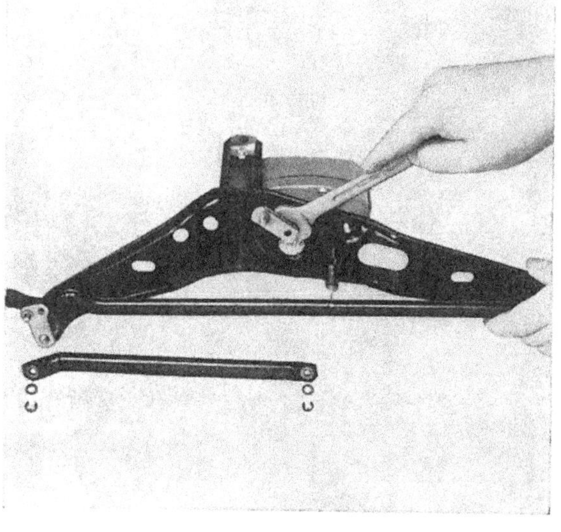

Installation takes place in the reverse order.

Windshield wiper bearing replacement

1 - Remove windshield wiper frame with motor.

2 - Unhook spring between frame and connecting rod.

3 - Remove lock and spring washers at the bearings. Take off driving link and connecting rod.

4 - Remove inner seal and washer.

5 - Unscrew retaining nut and take out wiper bearings.

Installation takes place in the reverse order, noting the following points:

1 - The pressed lug on the wiper frame must engage in the groove in the wiper bearing.

2 - Check the plastic bushes in the linkage for wear. If necessary, replace the links complete.

3 - The hollow side of the links must face towards the frame and the angled end of the driving link towards the right hand wiper bearing.

4 - The inner bearing seal should be positioned so that the shoulder of the rubber moulding faces towards the wiper arm when installed.

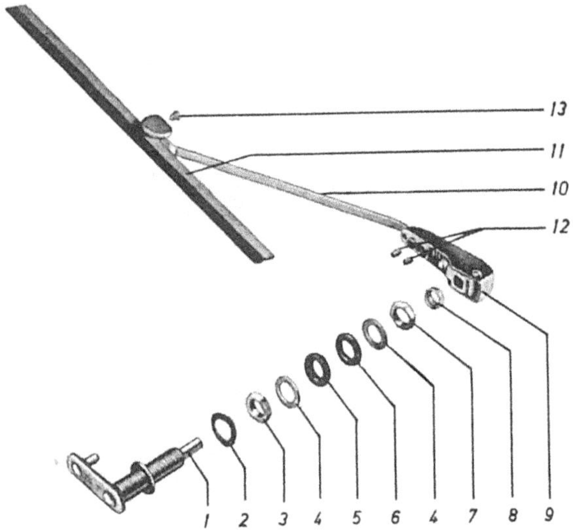

1 - Wiper shaft with crank
2 - Spring washer
3 - Brass nut
4 - Washer
5 - Inner bearing seal
6 - Outer bearing seal
7 - Nut
8 - Wiper shaft seal
9 - Wiper bracket
10 - Wiper arm
11 - Wiper blade
12 - Grub screws
13 - Fillister head screw

Replacement of windshield wiper motor carbon brushes

1 - Remove windshield wiper frame with motor.

2 - Take motor off frame.

3 - Take cover off after removing screw or clip.

4 - Unhook brush holder spring.

5 - Swing brush holders outwards.

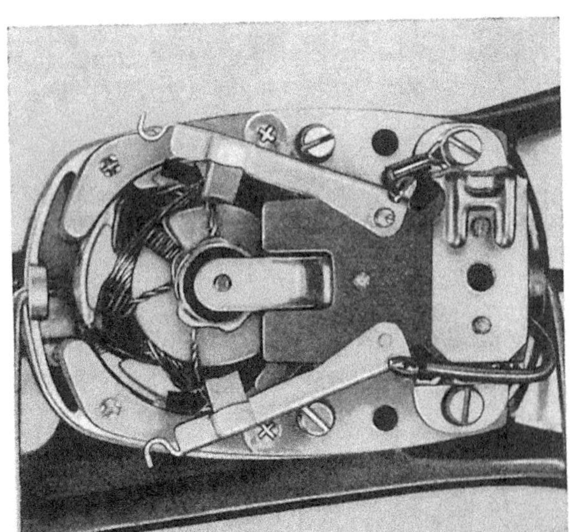

6 - Pull brushes out with a pair of long-nosed pliers and insert new brushes. Take care that the brushes are tight in the holders and that the ends are properly seated.

Important

When replacing the carbon brushes, the commutator should be checked and the slots between the segments cleared of carbon dust. The brush surfaces should be free of oil or grease. The commutator can usually be cleaned with a benzine moistened cloth. Fine polishing cloth should only be used if burn marks are found.

7 - Replace brush holder spring and cover. When installing the cover on the Bosch motor, do not overtighten the securing screw. The cover must not bend inwards round the screw head.

If more extensive repairs are necessary on the motor, an exchange part should be fitted.

Wiper switch removal and installation

1 - Wiper motor "OFF"
2 - Wiper motor "ON"
3 - Windshield washer pump operation

1 - Disconnect battery ground cable.

2 - Screw switch knob off.

3 - Remove instrument panel cover in front luggage compartment.

4 - Unscrew switch retaining ring on instrument panel with Special Wrench VW 674 and take switch out.

5 - Disconnect four cables and the pipe for the windshield washer pump.

Installation takes place in the reverse order.

Contact set for windshield wiper motor with automatic end position switching

The automatic end position switching of the "Bosch" and "SWF" permanent magnet windshield wiper motors is controlled by a double contact.

If trouble is experienced the complete contact set can be replaced when the motor has been removed.

Bosch contact set

1 - Remove four screws and take off drive housing cover with drive.

2 - Remove one screw and take off protective cover.

3 - Unsolder cable to contact set at brush holder.

4 - Remove contact set.

5 - Install new contacts and ensure that the connections 31 b and 54d are towards the drive housing cover.

6 - Solder cable to brush holder.

7 - Install drive housing cover with drive.

8 - The correct wiper motor switch (Part No. 111 955 517 A, for example) must be used to check the operation of the end position switching. The motor is connected as shown below.

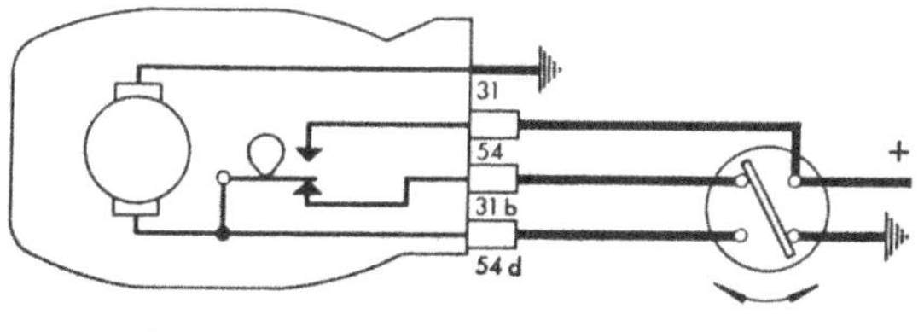

A - Windshield wiper motor
B - Windshield wiper switch

SWF contact set

When this contact set is replaced, the contacts must be adjusted after installation. The following operations are necessary:

1 - Take protective cover off wiper motor, unhook retainer.

2 - Remove switch housing cover.

3 - Remove defective contacts after unsoldering cable in brush holder.

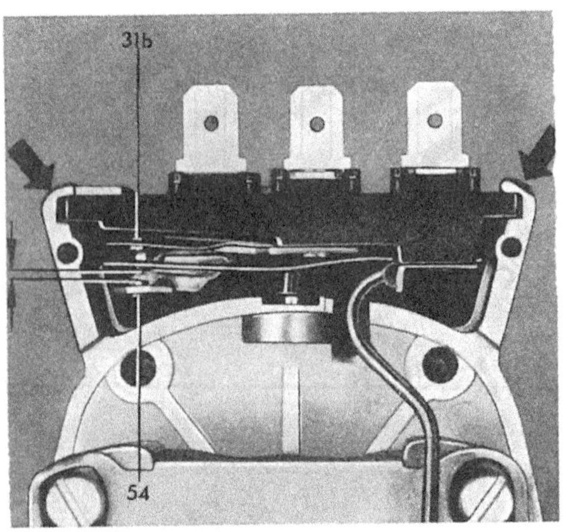

4 - Solder cable of new contact set to brush holder.

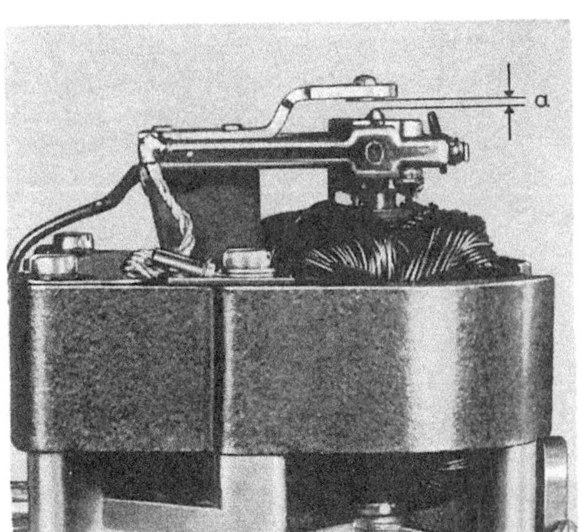

5 - Insert contacts carefully, taking care that the long contact spring does not catch on the switch cam and become distorted. If necessary, the motor armature must be turned until the cam is fully withdrawn.

6 - Peen the corners of the grooves carefully (arrow).

7 - Turn the armature until the cam presses the center contact against contact 31 b. The center contact should be depressed about 0.1 mm (.004″) in this position. Rectify as necessary.

8 - Bend contact 54 back until the switch travel "a" is 0.6—0.8 mm (.024″—.031″). When the center contact has been released by the cam it must bear firmly on contact 54.

9 - The end position switching should be tested as on the Bosch wiper motor.

Note:

When carrying out repairs on wiper motors, ensure that the clearance "a" between the thrust point on the armature shaft and the thrust surface does not exceed 0.1 mm (.004″). If necessary, rectify by bending the thrust bracket.

Windshield Wiper Motor Trouble Shooting

Symptoms	Cause	Remedy
Windshield wiper motor operates too slowly, cuts out or comes to a standstill	a - Brushes worn	a - Replace brushes
	b - Brush tension spring too weak or annealed	b - Replace tension spring
	c - Brush levers not free on their pivots	c - Free the brush levers
	d - Commutator dirty	d - Clean the commutator
	e - Moving joints of windshield wiper linkages devoid of grease or jammed	e - Thoroughly lubricate all moving joints with Universal Grease, eliminate jamming
	f - Battery voltage too low	f - Charge battery, check cables and connections
Windshield wiper motor continues to run after manual switch is turned off or does not return the blades to parking position	a - Contacts in housing damaged	a - Replace contacts
	b - Contact spring bent	b - Replace contacts
	c - Contact bracket (insulation plate) broken	c - Replace contacts
	d - Spring contact terminal 31 b does not make contact with spring contact terminal 54d	d - Adjust contact gap to (0.8 mm, 0.031″) or replace contacts
	e - Contacts dirty	e - Clean contacts
	f - Windshield wiper motor cannot be switched off	f - Screw button back slightly, bend contacts
	g - Poor connection from terminal 31 b via wiper switch to ground	g - Check connection, replace switch if necessary
Squeaking noise when motor in action and in some cases slow operation of motor or burnt armature	a - Moving joints of windshield wiper linkages devoid of grease	a - Thoroughly grease all moving joints with Universal Grease
	b - Point of armature spindle rests against stop of brush holder	b - Bend stop into shape
	c - Motor cover not correctly positioned on housing	c - Seal cover properly
Windshield wiper motor inoperative	a - Armature burnt by short circuits	a - Replace the motor or armature
	b - "On" position of switch not correct No connection between terminal 1 and 2, Faulty connection from terminal 1 to terminal 54 of motor	b - Replace switch Repair faulty connection (loose contacts)
	c - Same as under "Windshield wiper motor operates too slowly, cuts out or comes to a standstill"	

Windshield Wiper Blades

Good wiper blades and plenty of water are essential to ensure a perfectly clear wiper field on the windshield.

During long dry periods the wiper blades become clogged with tar splashes and dead insects. In this condition the elastic, friction-proof rubber lips of the wiper blades cannot fulfil their function which is to remove the water film from the windshield as completely as possible.

To clean the blades thoroughly it is necessary to take them off and scrub them with a hard nylon brush and methylated spirits or a strong detergent solution. Care must be taken to avoid the rubber blades themselves being damaged. If, when examined with an ordinary magnifying glass, the blades show cracks or signs of perishing and the edges are no longer sharp, they should be renewed to prevent the windshield from being scratched.

Important

Replacing or cleaning the blades alone is often not enough to effect a complete remedy. The slightest trace of silicon, which is present in many paint and chrome polishes, furniture polishes and even hand creams, is sufficient to cause streaks and clouding within the rage of the wiper blades when it rains.

The following agents can be used to clean the silicon from the windshield:

1 - "1 Z" paste from Messrs. Werner Sauer & Co., Bensberg.

3 - "Rex" paste from Messrs. Rex-Autopflege, Mainz.

3 - "Silicontra" paste from Messrs. Anton Fatka, Lüneburg.

3 - "Carnu-Lackreiniger" from Messrs. Johnson's Detergents, Hamburg.

5 - "Sidol" from Messrs. Siegel & Co., Köln. Applied as a paste.

6 - A paste consisting of 2 weight-parts of Vienna lime and 3 volume-parts of water. A few drops of liquid ammonia will further improve the cleaning efficiency, but will also give it a strong smell. This type of paste should not come into contact with the paintwork. Rub the paste on the windshield, allow it to dry and then wipe it off.

7 - Clean windshield with benzine and then treat it with an acid solution consisting of 1 part muriatic acid and 9 parts water. Afterward, rinse carefully with clear water.

The preparations listed at 1 to 5 have no detrimental affect on oven and air dried synthetic enamels provided that they are not allowed to work on the finish for a long time in strong sunlight. After cleaning the windshield all traces of paste should be removed thoroughly.

The preparations listed at 1 to 4 have a strong affect on cellulose enamels. As the silicon removers are used only on the outside of the body and cellulose enamels are invariably used for paint repairs on internal surfaces it is not considered that any special precautions are required.

Important

When cleaners containing silicon are used, the brushes, sponges, leathers and rags used to clean the paintwork should not be used for the windshield.

When spraying the paintwork with preservatives containing silicon, the windshield and windows should be covered with masks of paper or other material.

Trouble	Cause	Remedy
1 - Smearing	a - Blade dirty	a - Clean blade with a hard nylon brush and a soap solution or methylated spirits
	b - Frayed blade lips, rubber damaged or worn out	b - Fit new blades
	c - Old blades, surface cracked	c - Fit new blades

Trouble	Cause	Remedy
2 - Traces of water left in blade range forms small beads	Glass soiled by paint polish, oil or diesel exhaust deposits	Clean windshield with a clean cloth and a grease-oil-silicone remover

Trouble	Cause	Remedy
3 - Blade misses parts of windshield	a - Blade torn out of retainer	a - Fit blade back into retainer carefully
	b - Blade not contacting glass uniformly due to distorted springs or retainer	b - Fit new blade. This defect is often caused by careless fitting of blades
	c - Pressure exerted by wiper insufficient	c - Oil arm linkage and spring lightly or fit new arm

Trouble	Cause	Remedy
4 - Blade wipes well on one side and badly on the other, judders	a - Blade distorted to one side, no longer "flips"	a - Clean blade with a hard nylon brush and a soap solution or methylated spirits or fit new blade
	b - Wiper arm distorted, blade not vertical on the windshield	b - Twist arm carefully until it is vertical

Note:

These photographs were made on a Type 3 vehicle but the troubles are identical on the Type 1.

Important

The wiper blades are subject to a certain amount of natural wear even if given the best of care. It is advisable, in the interest of road safety, to fit new blades about every 12 months or earlier if necessary.

Replacing wiper blades

Loosen fillister head screw and take blade off. When the new blade has been fitted, check the wiping range with a wet windshield.

To adjust the stroke of the blade, fold the bracket forward, loosen the hexagon retaining screw and adjust the bracket to give a uniform wiping area on the wet windshield. The blades must not strike the windshield frame.

General

The windshield washer, in conjunction with the wipers helps to keep the windshield clean when on the move.

The water container, with a capacity of approximately one litre (1 quart) is located in the front luggage compartment behind the spare wheel. A suction pump which is combined with the windshield wiper switch draws water from the container via a non-return valve and forces it via a second non-return valve to a twin jet in front of the windshield.

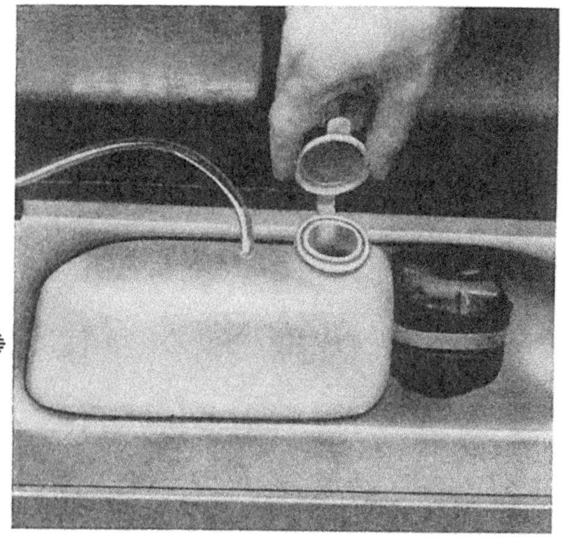

> **Important**
>
> The container can be refilled after the spare wheel has been removed. It can be removed for cleaning by carefully levering the upper and then the lower plastic retaining lugs out of the reinforcement plate with a screwdriver.
>
> It is an advantage to mix methylated spirits with the water in the container in proportions of 1 to 3. This assists the removal of insects from the windshield in the summer and prevents the liquid from freezing in the winter, down to —12° C (10° F).

Spray jet removal and installation

To remove the jet and rubber seal, press it upwards out of the hole in the cowl panel after removing the instrument panel cover. When installing, check that the rubber seal is seated properly. The two spray holes can be set with a piece of fine wire so that the jets of water strike the windshield uniformly.

Non-return valve removal and installation

1 - Remove instrument panel cover.

2 - Detach three pipes and take valve out.

When installing the valve, check that the arrow is pointing towards the jet.

Pneumatic Windshield Washer

From Chassis No. 4 010 995 all Volkswagen models are equipped with a pneumatic windshield washer.

	Sedan and Convertible	**Karmann Ghia Models**
Container Mounting bracket	111 955 945 A —	113 955 941 A 113 955 949
Wiper switch with valve Hose	111 955 517 N 18 053 1	

The container has a capacity of about 1 liter (1 quart) and is located in the front luggage compartment behind the spare wheel on the Sedan and the four-seater Convertible. On the Karmann Ghia models it is mounted on the spare wheel.

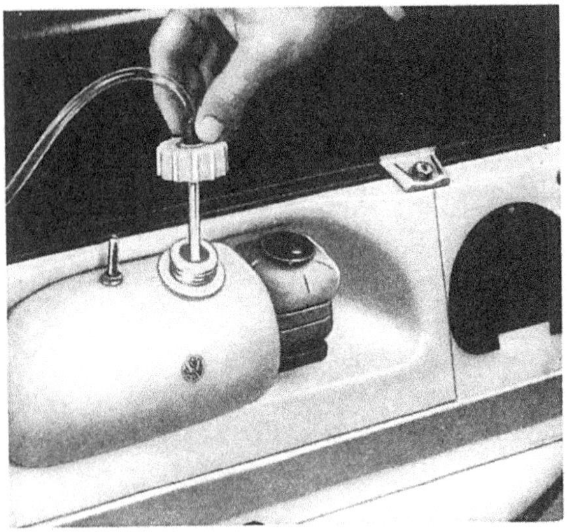

Sedan and Convertible

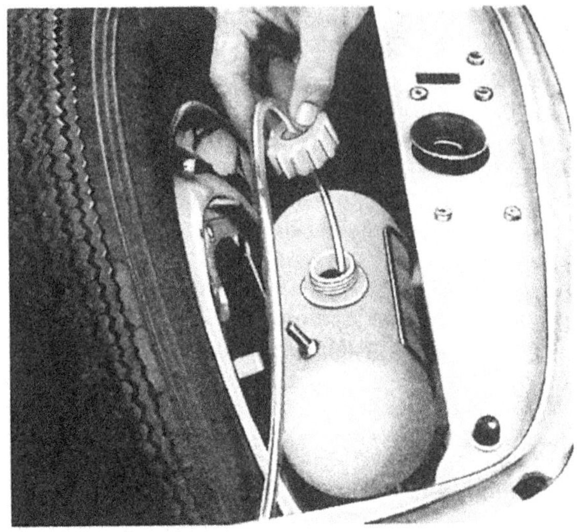

Karmann Ghia Models

The container has a valve and can be filled with compressed air like a tire. The pressure must not exceed 2.5 kg sq./cm (35 lbs. sq./in.) which will suffice to completely empty the container. The washer is operated by means of a valve in the windshield wiper switch.

The pressure in the container should be checked and rectified as necessary every time fuel is put in the tank. When the cap has been removed, the container should only be filled with water up to the lower end of the small tube to ensure that the air cushion is sufficiently large.

It is advisable to mix methylated spirits with the water in the container in the ratio of 1:3. This will remove insects from the windshield better in the summer and prevent the water from freezing down to —12° C (10° F) in the winter.

Note:

As an accessory, a pressure hose with valve (Part No. 113 955 971) can be connected between the spare wheel and the container. This has the following advantages:

1 - The larger volume of air in the tire enables the container to be filled with water several times without having to check the air pressure.

2 - The tire and container can be filled with air more easily via the valve insert (1) in the pressure hose.

3 - A spring-loaded valve (2) in the hose closes the connection between tire and container when the pressure in the tire falls below 1.3 kg sq./cm (18 lbs. sq./in.).

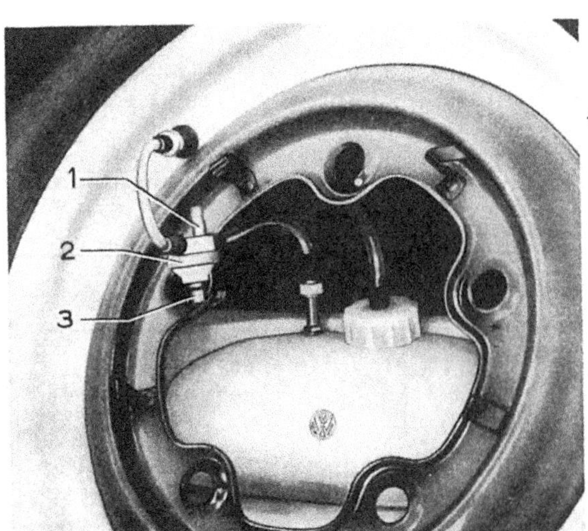

When installing or removing the pressure hose or filling the container with water, the small valve (3) must be closed. The hose is attached by screwing the connection marked "RES.RAD" on to the tire valve. The tire should be inflated to 2.5 kg sq./cm (35 lbs. sq./in.) every time fuel is put in the tank.

The modified windshield washer system can be installed in previous vehicles and is described in Technical Bulletin Z-13.

Note:

Checking windshield washer container for leakage
In order to enable all workshops to check these parts properly, a test cap with which the neck of the container to be checked can be sealed, has been developed.

Method:

1 - Remove container and empty.

2 - Screw test cap on tightly.

3 - Apply an air pressure of 2.5 kg/sq. cm (35 psi) to the container, hold it under water and check for bubles.

4 - If no sign of leakage is found, the defect must be sought in the pipes or in the valve of the washer system.

Note:

A burr (arrow) on the sealing washer of the container pipe can cause leakage. This burr can be removed easily with a small fine file (key or contact file).

Electrical Horn

General

An electro-magnet in the horn produces vibration in connection with a breaker mechanism. The vibration is transmitted to the diaphragm which produces the sound. A condenser or short circuit ring reduces the tension of the spark occurring between the points of the breaker mechanism to avoid damage by burning. The horn is operated by a half-ring on the steering wheel hub, which closes the circuit with ground (negative terminal of battery). Current is drawn through the steering column tube and three contact pins in the steering wheel hub.

Maintenance

Care should be taken that the horn bracket is not damaged and the horn does not touch the body, as the vibration will otherwise be affected. Damage may result from worn or dirty breaker points, ingress of water, and defective condenser.

Removing and Installing Electric Horn

1 - Remove bolt at horn bracket under left-hand front fender.

2 - Disconnect cable and remove horn.

When installing the horn, make sure that it does not contact the body.

Note:

The adjusting screw on the horn should not be turned. Careless adjustment of this screw can seriously damage the horn contact points.

If trouble is experienced with the horn, always check for defects in the cable from fuse box to horn or from the horn via the horn ring or horn button to ground. Corroded or loose connections often affect the operation of the horn.

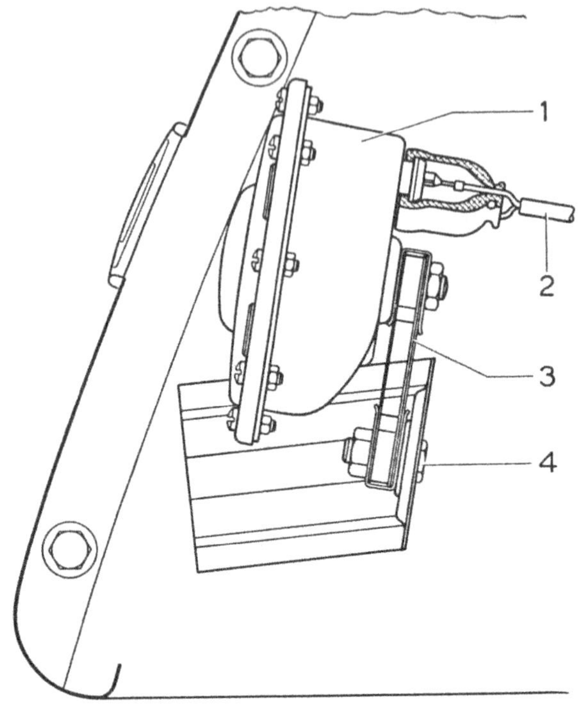

1 - Horn
2 - Cable
3 - Bracket
4 - Bolt

Horn Circuit

General

The steering column tube serves as the current feed and the steering column as the ground connection for the steering wheel hub. The steering column tube is insulated from the body by the rubber mountings in the front partition and under the instrument panel. The steering column is insulated from the column tube at the ball bearing, the spring and the brass washer by a plastic sheet.

When the horn half ring is actuated, the current flows via the column tube, ball bearing, spring and brass washer to the upper part of the steering wheel hub. The spring loaded contact pins which connect the upper and lower parts of the steering wheel hub when the horn half ring is actuated, provide the contact between the steering column (ground) and the current carrying parts.

Particular attention should be paid to the ground connection between the steering column and steering.

Note:

From Chassis No. 4 010 095 the horn connection in the fuse boy and the ground return cable have been altered and the horn can now only be operated when the ignition is switched on. The first fuse on the left side of the fuse box has been increased from 8 to 16 amps.

The insulating sleeve (Part No. 113 415 527) and the insulating washer (Part No. 113 415 533) for the steering column have been deleted. A ground cable attached to the horn ring which is insulated from the steering wheel hub, passes through the hollow steering column and is connected to the steering gear by a terminal on the coupling disc. When the horn ring is operated, the current flows from horn or the horn relay through the steering column tube, steering column bearing, steering column, steering wheel hub upper part, horn ring and horn ground cable to the steering gear and then via the front axle to ground on the frame.

The following parts have been altered.

1	Horn ring	113 951 531 D
3	Insulating bushes	113 951 547
1	Contact washer	113 951 563
3	Spring for contact washer	113 951 555 A
1	Clamp	113 951 559
1	Cheese head screw AM 4×5	N 10 701 1
1	Steering wheel	113 415 651 E
3	Insulating washers for horn ring	113 951 539
1	Steering column with lock shells for roller steering	113 415 505 E
1	Steering column for VW sector steering (new cable routing) — overseas only —	113 415 515 C
1	Cancelling ring	113 415 661 A

Spare parts of the previous pattern will remain available for older vehicles.

Note:

On some vehicles the steering column tube contacts a rib in the luggage compartment. The vibration caused by the motion of the vehicle rubs away the paint at the points of contact so that the column tube grounds and the horn operates.

When this trouble is experienced on a vehicle, a piece of plastic material should be cemented between steering column tube and luggage compartment.

THIS PAGE INTENTIONALLY
LEFT BLANK

Horn ring removal and installation

1 - Remove horn fuse.

2 - Remove horn ring cap.

3 - Remove three screws and take ring off.

4 - Take out the three contact pins, check and replace if necessary.

Installation takes place in the reverse order.

Fuse Box

General

The fuse box which has a transparent cover, is located below the instrument panel near the steering column tube.

Fuse replacement

1 - Remove transparent cover.

2 - Take out fuse.

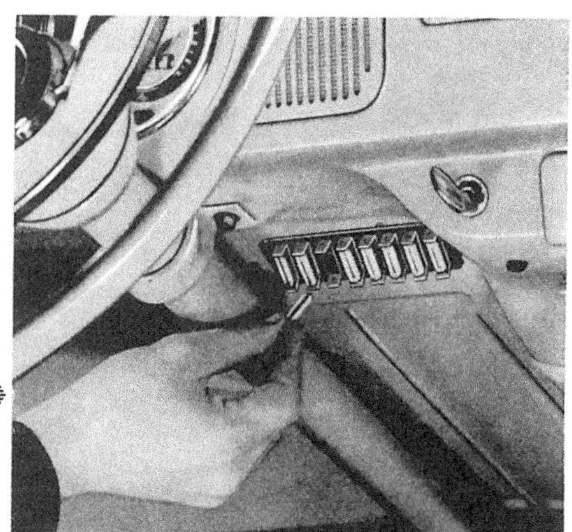

When a fuse burns out it is not sufficient to merely replace it with a new one. The cause of the short-circuit or overload must be established.

The patching up of fuses with wire or silver paper is not permitted as this can cause serious damage in other parts of the system. It is advisable to always have a few spare fuses (8 Amp.) handy on the vehicle.

Fuse box removal and installation

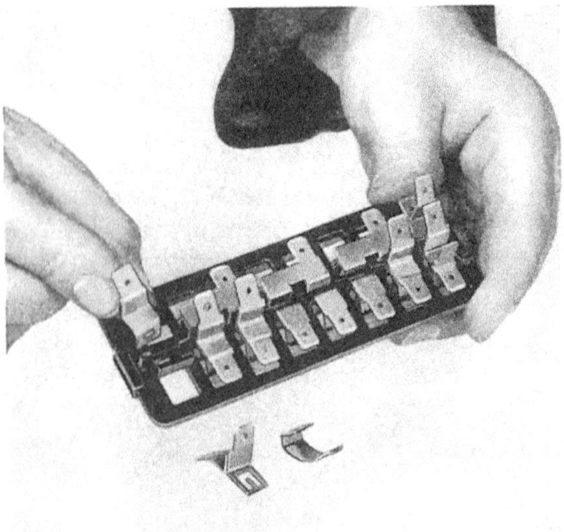

1 - Remove instrument panel protective cover.

2 - Disconnect battery ground cable.

3 - Pull all connections off the fuse box.

4 - Press retaining spring out and remove fuse box.

5 - The individual terminal and fuse retaining springs can be replaced when the fuse box is removed.

Note:

It is also possible to replace terminals and springs when the fuse box is in situ.

Installation takes place in the reverse order, taking care that the cables are connected properly.

Flashing Indicator System

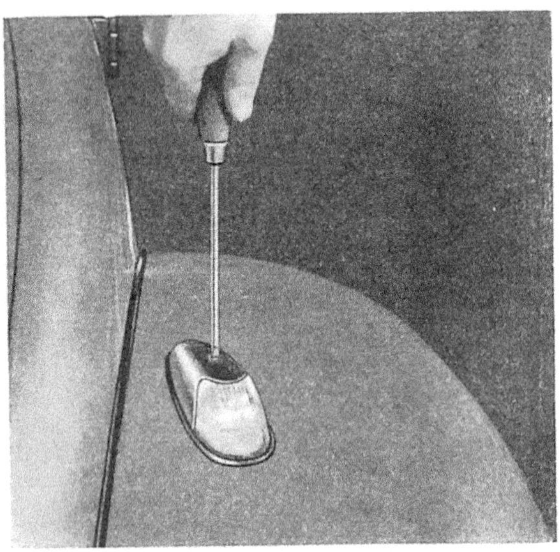

General

The front direction indicator lamps are positioned on the front fenders above the head lights. The rear direction indicators are housed in the combined stop and tail light units, utilizing the stop light filament of the twin filament lamps. The direction indicator lamps are operated by a self-cancelling switch below the steering wheel. The flashing is controlled by a flasher unit.

Bulb Type:

Front R 6 V 18 W DIN 72601
Rear S 6 V 18/5 W DIN 72601

Bulb Replacement (front):

1 - Take off screw holding lens to bulb holder.

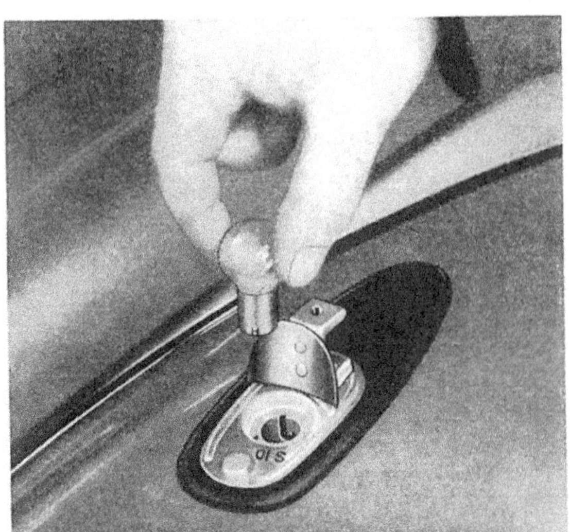

2 - Lift off lens.

3 - Replace bulb, make sure new bulb seats firmly and makes good contact.

4 - Be sure the rubber seal between bulb holder and fender is in good condition, replace if necessary.

5 - When installing the lens, make sure the bulb holder is properly seated in the channel provided in the packing.

Bulb Replacement (rear)

To replace the direction indicator rear bulb, follow the instructions given under "Stop and Tail Lights".

Front Indicator Lamp Assembly and Disassembly

Disassembly

1 - Take out screw holding lens to bulb holder.

2 - Take out lens of direction indicator lamp.

3 - Screw off both bulb holder attaching nuts below fender.

4 - Pull up bulb holder with bulb and disconnect cable.

5 - Pull out cable.

6 - Loosen cable sleeve from side panel, pull up packing and take it out.

Assembly

This is a reversal of the preceding operations, but the following points should be noted:

1 - Do not damage cable sleeve when pushing it into side panel.

2 - Do not over-tighten bulb holder attaching nuts.

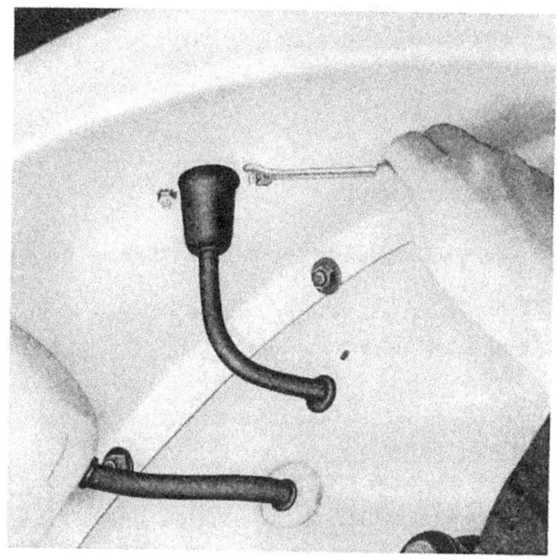

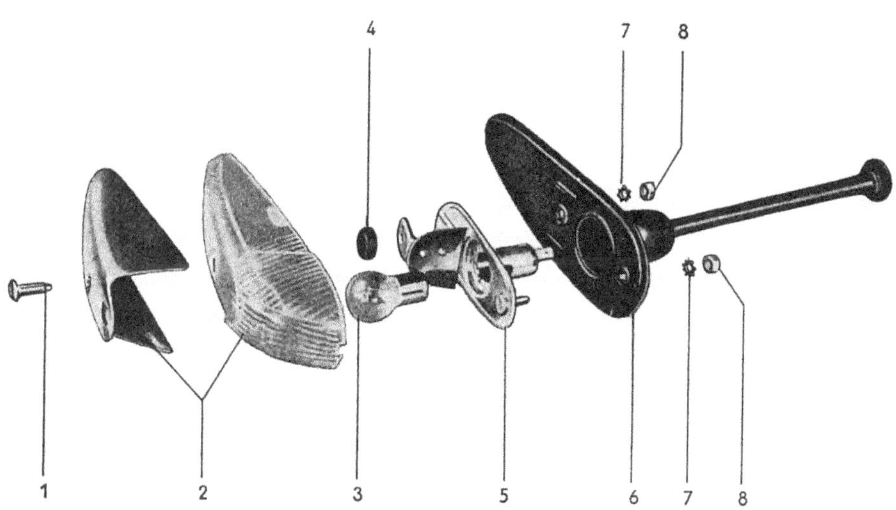

1 - Screw
2 - Lens
3 - Bulb
4 - Sealing ring
5 - Bulb holder
6 - Packing
7 - Lock washer
8 - Attaching nuts

Note:

From 31st October 1963, Chassis No. **5 888 185** (Model 111—118) and 31st October 1963, Chassis No. **5 852 937** (Model 151 and 152) the shape of the front indicators was changed. The front fenders were also altered. The new turn indicator lamps cannot be installed in earlier vehicles without altering certain other parts.

Indicator warning light

Bulb type:

Warning light J 6 V 1.2 W DIN 72601

Bulb replacement

1 - Remove instrument panel protective cover.

2 - Pull holder with bulb out of retainer on the back of the speedometer.

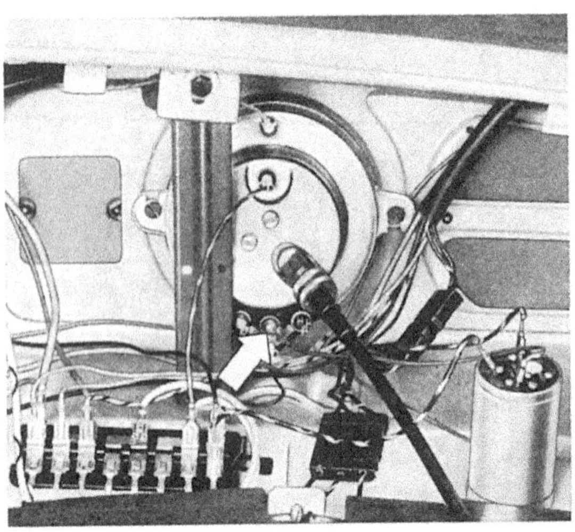

3 - Press bulb lightly into the holder, turn and pull out.

Indicator switch replacement

The self-cancelling indicator switch is secured to the steering column tube below the steering wheel with a clamp screw. The automatic cancelling is carried out by a cancelling ring which is screwed to the steering wheel hub.

Removal

1 - Disconnect battery ground cable.

2 - Remove horn half ring and take out contact pins.

3 - Loosen steering wheel nut with ring wrench and take wheel off.

4 - Remove instrument panel cover in front luggage compartment.

5 - Disconnect one cable at flasher relay terminal 5 and 5 cables (3 and 2) on the two cable connectors. Pull cable with protective sleeve out of the front partition.

6 - Loosen switch securing screw and take switch off column tube.

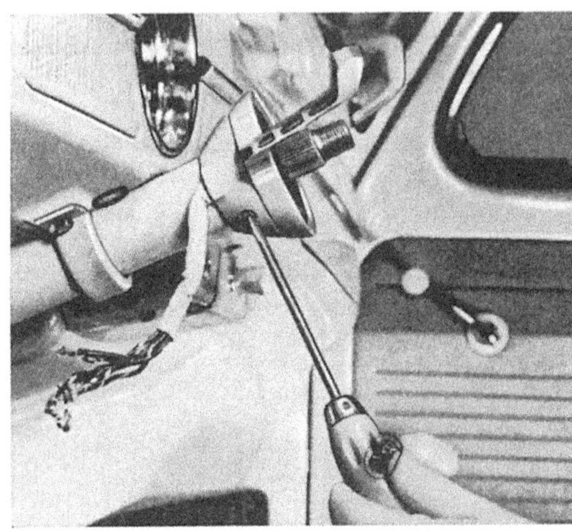

Installation takes place in the reverse order, noting the following points:

Note:

The indicator switches are supplied with the cables soldered on.

1 - Place the switch on the column against the stop, with the lever pointing to the left and clamp into position.

2 - Feed the cable connections with insulating sleeve through the front partition into the front luggage compartment, taking care that the grommet is properly seated.

3 - Check that the brass washer is located correctly, The recess must point to the right when the front wheels are in the straight-ahead position.

4 - With the wheels in this position, install the steering wheel so that the tongue on the cancelling ring which actuates the automatic cancelling engages in the brass washer. The spokes of the steering wheel must be horizontal and the recess in the spokes for the horn half ring towards the driver's seat.

Important

The indicator switch must not be switched on when installing the steering wheel as otherwise the operating cam will be damaged and the switch rendered unserviceable.

5 - Insert spring washer and tighten steering wheel nut to a torque of 5.0 to 6.0 mkg (35—42 ft. lbs.).

6 - With the steering wheel installed and the switch properly secured, adjust the clearance between the switch and wheel hub to 2.0 + 0.5 mm (.080 + .020") by moving the steering column tube up or down.

7 - Check that the switch cancels properly on right and left turns and, if necessary, turn the switch slightly.

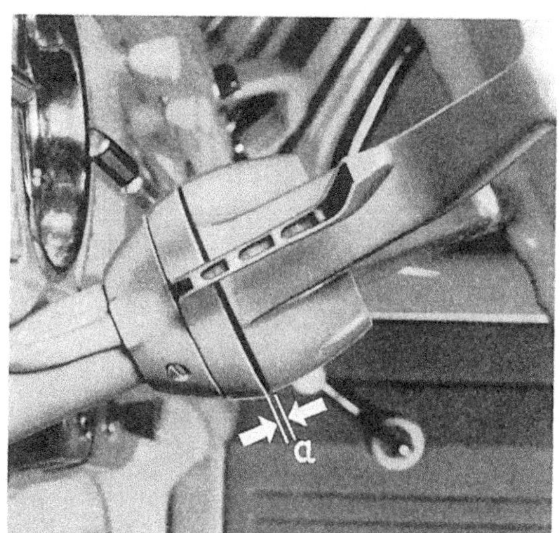

a = 2 + 0.5 mm

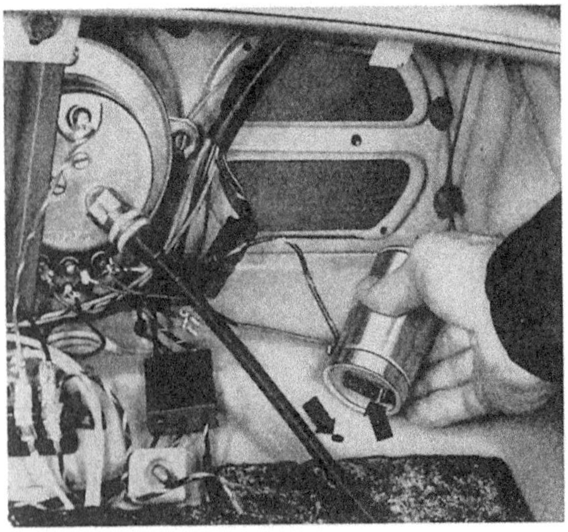

Flasher relay replacement

To avoid short circuits, remove the battery ground cable before commencing work.

The flasher relay is mounted at the back of the instrument panel and can be reached when the panel protective cover is removed.

1 - Remove three cables.

2 - Press the relay out of the retainer. The relay is held by a riveted leaf spring which engages in a hole in the front partition.

On installation, check that the cables are correctly connected.

Note:

A poor ground connection between the flasher relay and the body can put the indicator system out of action. This fact is often overlooked and the flasher relay replaced even though there is really nothing wrong with it.

When carrying out repairs on the indicator system, the ground connection between relay and body should always be carefully checked. After removing the instrument panel lining, turn the relay a few times about 10°. The sharp edges underneath the relay will then cut through the paint and ensure a good ground connection to the body.

If the spring clip has become bent so that it no longer fits tightly, it should be reshaped to fit properly when the relay is removed. If the flasher relay still does not work correctly it should be replaced.

Flasher relays are often damaged internally by short circuits occurring externally. When carrying out repairs on the indicator system, ensure that the ignition is switched off and the indicator lever is in the central position. The cables from the flasher relay and indicator switch to the direction indicators should also be checked carefully as a short circuit in these cables can damage the flasher relay.

If the indicator warning lamp only lights up **once** when the indicator switch is operated this means that one of the indicator bulbs is defective or the ground or bulb circuit is broken. The other indicator bulb on the same side then flashes considerably faster than usual.

Note:

In order to improve the ground connection between the flasher unit and the body, the flasher mounting was modified from Chassis No. 4 846 836.

Instead of the spring clip previously used the new version of the flasher unit (Part No. 111 953 185) is provided with a riveted-on tapping screw for attaching it to the body.

A metal bracket to which the flasher unit is screwed is welded to the new steering column support (Part No. 111 415 621 C).

Note:

From Chassis No. 5 436 196, the flasher relay was modified internally and the terminal "K" (warning lamp) discontinued. On the new relay, the cable to the indicator warning lamp is connected to terminal "S" (indicator switch) on the relay together with the cable to the switch.

The previous type of relay can be replaced by the new relay and the cable to the warning lamp connected to terminal "S" on the relay with the aid of an adaptor.

When fitting a radio with a separate vibrator and final stage amplifier the flasher unit must be screwed to the support bracket of the vibrator. To do this the flashing unit bracket on the steering column support must be bent upwards.

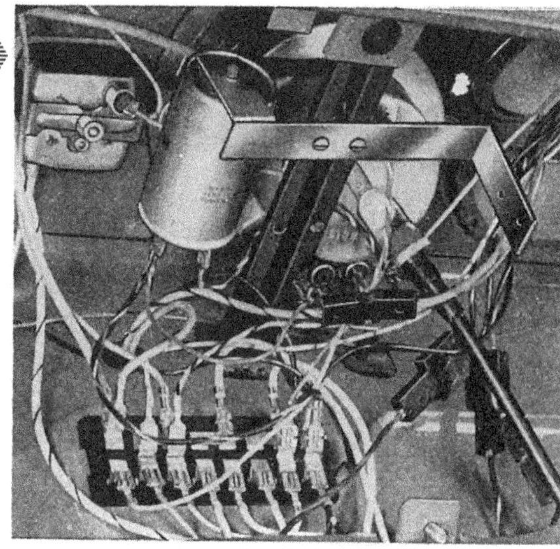

Steering/Ignition Lock

From Chassis No. 4 010 995 a steering/ignition lock (Part No. SP 201 C) can be installed at the factory as an optional extra (M-No. 55). When the key is withdrawn at the "Halt" position, a bolt engages the lock shells on the steering column and locks the steering. At the "Garage" position the ignition is switched off and the key can be withdrawn without locking the steering.

Type A (36 mm/1.4" dia.)
— Part No. 111 905 865 A —

Type B (28 mm/1.1" dia.)
— Part No. 111 905 865 B —

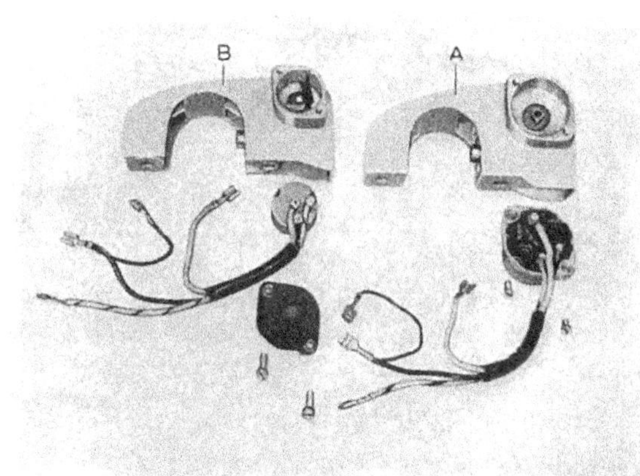

The ignition/starter switch with non-repeat lock is combined with the steering lock. The ignition is switched on at the "Fahrt" position and the starter is operated when the key is turned to the "Start" position. The key returns to the "Fahrt" position automatically. A mechanical non-repeat lock in the ignition/starter switch prevents the starter pinion from striking the starter ring on the flywheel when the engine is running. Before the starter can be operated again, the ignition key must be turned to the left to the "Garage" position which switches the ignition off.

Note:

From Chassis No. 5 012 112, the red cable between terminal 50 on steering/ignition lock and the cable connector was shortened 40 mm/1.5" and the corresponding red cable in the main harness lengthened by the same amount.

If the cables are improperly routed, the end of the flat connector projecting out of the cable connector can touch the steering support and cause a short circuit when starting.

All vehicles between Chassis No. 4 010 995 (August 1961) and Chassis No. 5 012 112 (1st October 1962) fitted with a steering lock in production should be checked when inspections or repairs are being made, to see if there is a danger of a short circuit in the red cable 50.

As an additional safety measure, the door and steering locks have different lock cylinders and different keys. The door lock cylinders do not, therefore, need to be altered when the steering lock is replaced.

The ignition/starter switch in the steering lock can be replaced.

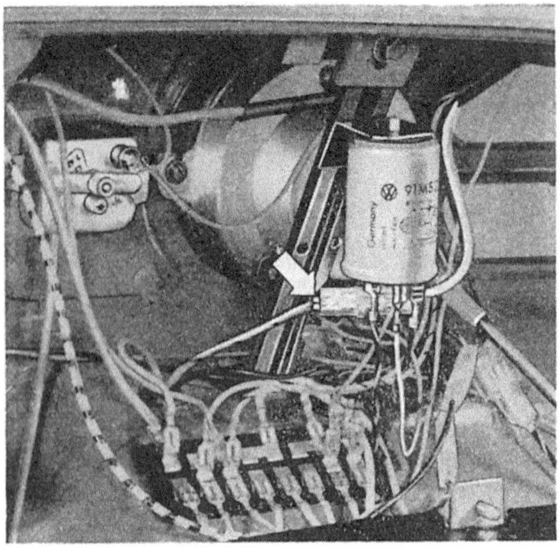

Short circuit danger

If there is, a piece of insulating sleeving about 6 cm (2") long should be pushed over the cable connector so that the ends of the flat connectors are covered.

Note:

When service installing a steering/ignition lock on vehicles with **Chassis Nos. from 2 533 000 to 4 010 995** the following points should be noted:

1 - Transfer horn cable in fuse box from terminal 30 to terminal 15.

This will prevent the horn from operating if a second key contacts the steering column tube when the ignition key is inserted into the lock.

2 - Replace the 8 Amp. fuse at terminal 15 by a 16 Amp fuse (Part No. 111 937 095).

These measures should also be carried out on all vehicles which have been subsequently fitted with a steering lock.

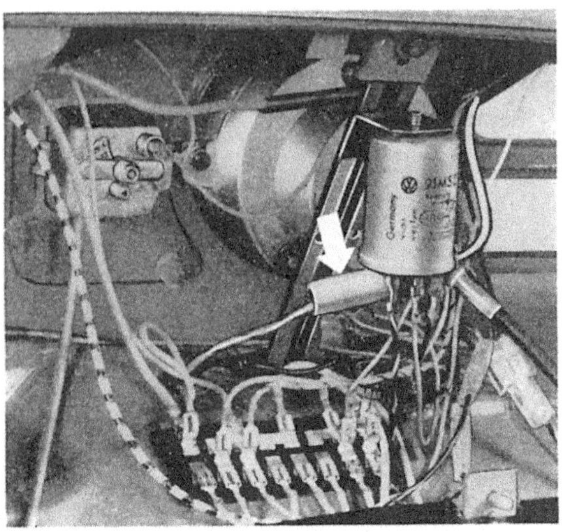

Connectcr insulated

Replacing ignition/starter switch

1 - Disconnect battery ground cable.

2 - Remove protective cover behind instrument panel.

3 - Disconnect one cable at cable connector, one cable at the speedometer and two cables at the fuse box and pull all the cables through into the body.

4 - Remove two securing screws and take out ignition/starter switch. The steering lock does not need to be removed.

5 - Install in the reverse order and ensure that the lug on the switch engages in the groove in the steering lock.

The steering lock is secured to the instrument panel with two special screws (67 mm long — Part No. 111 905 881 A and 36 mm long — Part No. 111 905 881). When the lock is being installed, these screws are tightened until the heads break off. To remove the steering lock, the screws must be drilled 15 mm (.6") deep with a 5.5 mm (.216") dia. drill and then removed with a stud extractor.

Removal of broken ignition keys

If excessive force is used, the ignition key can break off in the ignition switch. The broken-off part of the key in the switch can be removed as follows:

1 - The ignition must be switched off. If this is not the case, the lock cylinder must be turned to the left to the stop with a small screwdriver.

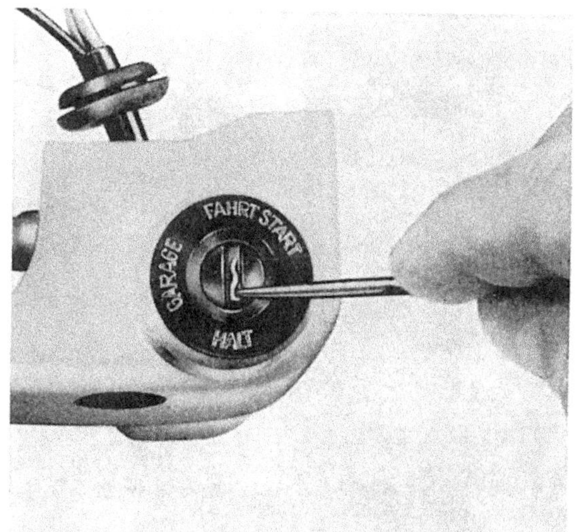

2 - Insert a suitably ground scriber into the lock carefully and press it against the broken-off part of the key sideways.

Pull the key web out by levering with the scriber.

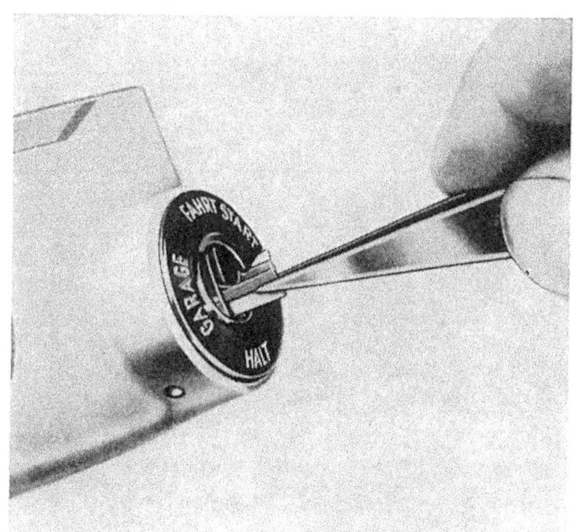

Note:

Take care not to press the key web deeper into the lock cylinder. If one of the tumblers has moved across behind the web, there is no point in continuing to try to extract the broken part in this manner. The ignition/starter switch should be replaced in this case.

3 - Pull the broken portion out with a pair of pliers or strong tweezers.

THIS PAGE INTENTIONALLY
LEFT BLANK

Horn

General Description

A matched pair of horns is located in the spare wheel compartment behind an opening that inclines downward. A rubber boot encloses the horns and provides a seal. The horns are operated by a half ring on the steering wheel hub via a relay.

Maintenance

It is important to see that the flexible mounting parts of the horns and the rubber boot are free from damage.

Horn Removal and Installation

1 - Remove the horn fuse from the fuse box.

2 - Disconnect cable 87 at the relay.

3 - Loosen rubber boot in front panel.

4 - Lift spare wheel out and disconnect cable at horns.

5 - Remove securing screws and take out horns with rubber boot.

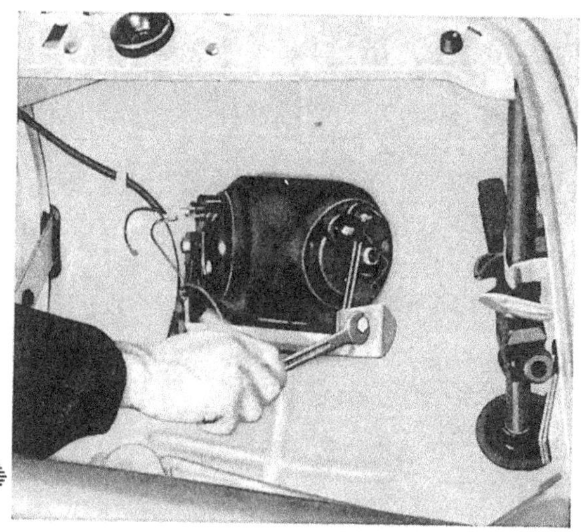

6 - Detach horns from rubber boot.

When installing, make sure there is a good ground connection and that the rubber boot is properly positioned. The horns must not contact the body.

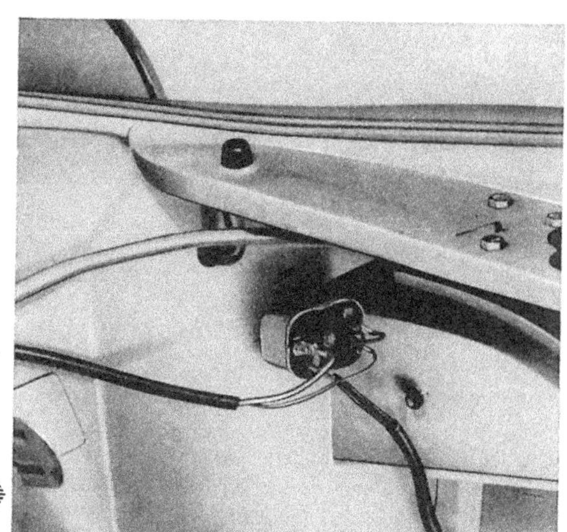

Replacing Horn Relay

1 - Remove the fuse from the fuse box.

2 - Disconnect the four cables from the relay.

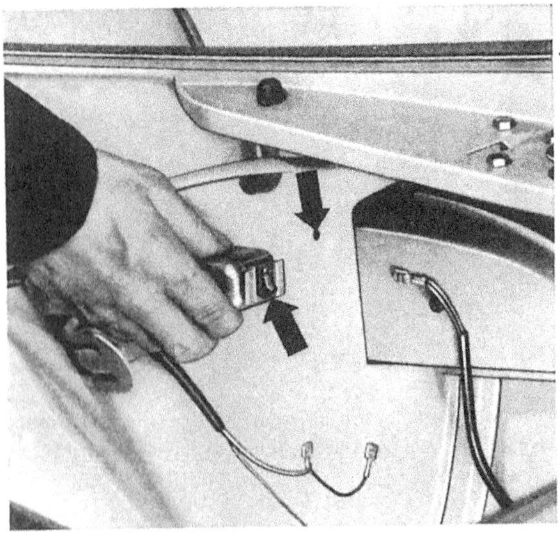

3 - Press the relay out of the hole in the side panel.

When installing, make sure the cables are properly connected.

Horn half ring removal and installation

1 - Remove horn fuse from fuse box.

2 - Remove cover.

3 - Remove three screws and take horn half ring off.

4 - Remove two contact pins, check, and replace if necessary.

Installation takes place in the reverse order.

Headlight flasher

General

The Karmann Ghia models are equipped with a headlight flasher. The flasher relay is located near the indicator relay behind the protective cover in the front luggage compartment. *The headlight flasher button* is in the indicator lever. When the ignition is switched on and the button pressed, current flows via the headlight flasher relay to the high beam filaments of the headlights and causes them to light up.

Flasher contact replacement

1 - Disconnect battery ground cable.

2 - Remove indicator switch.

3 - Detach cable ends from both flasher cables.

4 - Tie cable ends together with thin wire and pull cables out of the insulating sleeve.

5 - Remove two screws and take headlight flasher contact and spring out of the indicator switch lever.

6 - Secure both cables of the new contact to the wire in the insulating sleeve and pull the new cables through.

7 - Attach new cable ends to the flasher cables.

8 - Secure the flasher contact and spring to the indicator switch lever with two screws.

9 - Install indicator switch and connect cables correctly.

Headlamp flasher relay replacement

1 - Disconnect battery ground cable.

2 - Remove protective lining from front luggage compartment.

3 - Detach three cables from relay.

4 - Press the relay out of the securing hole and insert new relay.

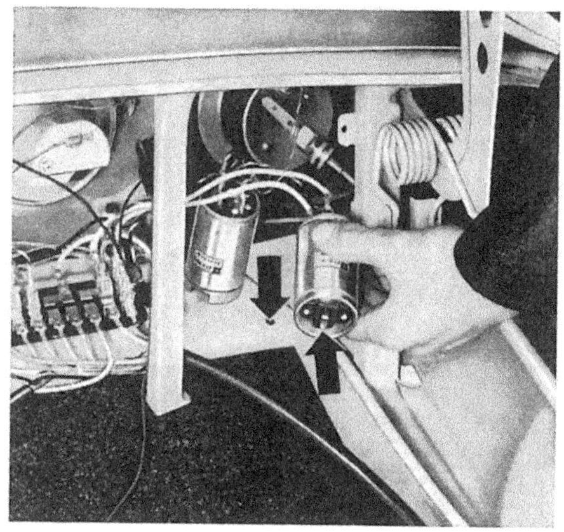

5 - Take care that the cables are connected correctly.

Fuse box

General

The fuse box with transparent cover is located below the instrument panel near the steering column tube.

The fuse box and fuses are replaced as on the VW Sedan.

Windshield Wiper

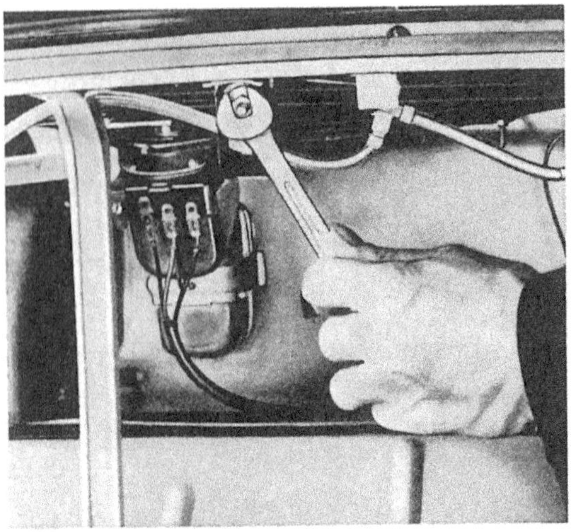

General

The windshield wiper motor with the two spindles and linkage are mounted on a common frame which is accessible when the front hood has been raised.

The removal, installation and repair of the windshield wiper system is carried out as on the VW Sedan.

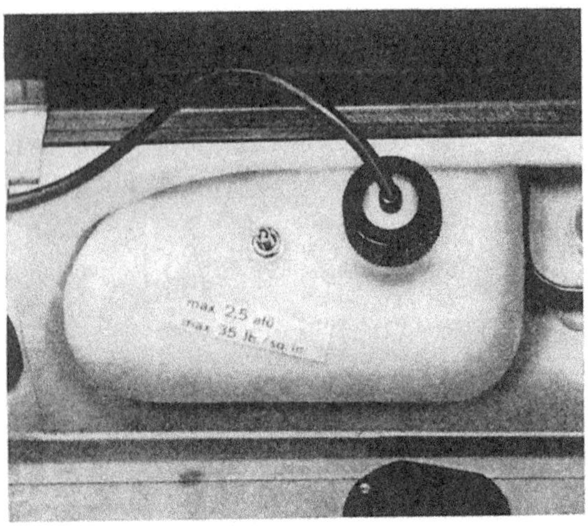

Note:

From 19th October 1963, Chassis No. **5 851 619**, the Karmann Ghia 1200 models are fitted with the same water container as is used on the Sedan. This container can be installed in vehicles from Chassis No. 3 192 507 without difficulty. It may be necessary to bend the spare wheel retaining brackets on these vehicles slightly so that the wheel does not touch the container.

The former type of windshield washer container (Part No. 113 955 941 A) and the support bracket (Part No. 113 955 949) will remain available.

Flasher Indicator System

General

The front indicator lights are mounted on the fenders below the headlights. The rear lights have three compartments for stop, tail and indicator lights. The flashes are controlled by a flasher relay and the self-cancelling switch is located below the steering wheel.

Bulb types:

Front: R 6 V 18 W DIN 72601
Rear: R 6 V 18 W DIN 72601

Bulb replacement (front)

1 - *Remove two securing screws.*

2 - Take off bezel with lens.

3 - Replace bulb.

When installing, check that the gaskets between bezel and lens and between lens and bulb holder are properly seated.

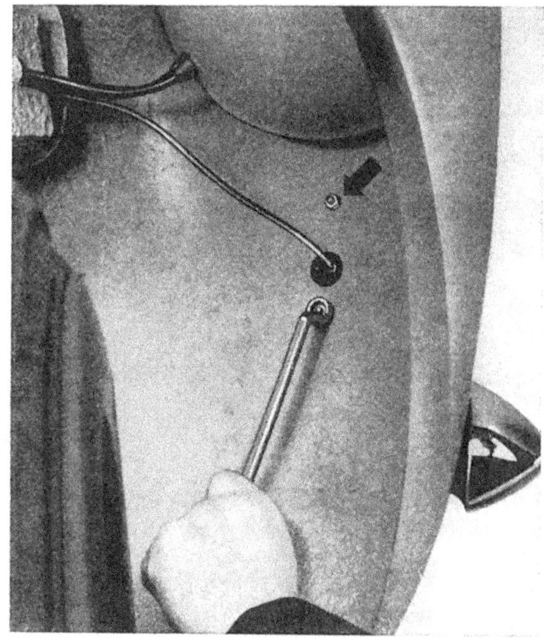

Bulb holder replacement

1 - Remove two screws and take off bezel with lens.

2 - Unscrew two nuts under the fender and remove bulb holder.

3 - Disconnect cable.

When installing, check that the gasket between bulb holder and fender is sealing properly.

Bulb replacement (rear)

The replacement of the rear indicator bulbs is carried out as described in the section on stop and tail lights.

Note:

From 1st November 1963, Chassis No. **5 953 217**, the Karmann Ghia models are fitted with a new bulb holder and the same lens as is used in the front indicator of the Volkswagen 1500 N.

The new turn indicator cannot be installed in vehicles manufactured before November 1963 without altering certain other parts.

Flasher relay replacement

To avoid short circuits, remove the battery ground cable before commencing work.

The flasher relay is mounted at the back of the instrument panel and can be reached when the panel protective cover is removed.

1 - Remove three cables.

2 - Press the relay out of the retainer. The relay is held by a riveted leaf spring which engages in a hole in the front partition.

On installation, check that the cables are correctly connected.

Indicator switch removal and installation

The removal and installation of the indicator switch is the same as on the VW Sedan with the exception of the cable connections.

Remove the lining in the luggage compartment and disconnect one cable each from headlight flasher relay and indicator flasher relay (terminal S), one cable from each of the two connectors and the ground cable from the speedometer mounting.

Gearshift Lever Lock with Ignition/Starter Switch

From Chassis No. 3 933 247 a gearshift lever lock with ignition/starter switch and non-repeat lock can be installed at the factory as an optional extra (M-No. 59). The Part Nos. are 113 016 051 A and — for vehicles with Saxomat — 113 016 051 B.

The gearshift lever can be locked in neutral or with reverse gear engaged to prevent the vehicle from being stolen.

When necessary, the ignition starter switch (Part No. 113 016 063 A) can be replaced:

1 - Disconnect battery ground cable.

2 - Remove front floor mat and take foot rest out.

3 - Lift frame tunnel covering as required.

4 - Disconnect three cables at gearshift lever lock.

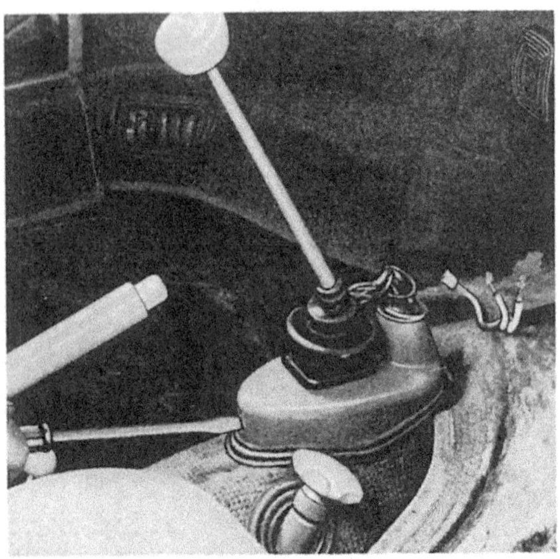

5 - Remove securing screw, push lock upper part forward and take off.

6 - Remove ignition/starter switch.

Install in the reverse order and ensure that cables are connected properly.

The service installation of the gearshift lever lock is described in the Technical Bulletin Z - 6.

Speedometer and Drive Cable

General Description

Speedometer and mileage recorder are driven by a cable from the left-hand front wheel. The speedometer operates on the eddy current generator principle. A ringshaped magnet is rotated by the speedometer drive shaft. An armature is located within an aluminium cup, which is free to rotate without touching the magnet.

The magnetic eddy-currents produced by induction when the armature revolves, cause the cup to rotate together with the speedometer needle shaft which is firmly attached to it. A fine spring is fitted to the upper end of this shaft as a brake, and counteracts the rotations of the cup and shaft, permitting only a predetermined movement of the speedometer needle. Under the braking effect of the spring, the deflection of the needle becomes greater as the speed of the magnet increases with the speed of the car.

The speedometer unit incorporates a mileage recorder which is driven through a triple worm mechanism. Five numeral rolls show the mileage covered by the vehicle.

The speedometer cable consists of several strands of wire and is protected by a synthetic housing. If a speedometer cable breaks, it is worth while having it repaired at a VDO Service Workshop, as long as the outer housing is intact.

Note:

From Chassis No. 3 862 145 all speedometers marked in km have a red line on the scale at 50 km (Part No. unchanged). This will facilitate driving in various countries which have a local speed limit of 50 km per hour.

To bring the warning lights into line with the international regulations, all speedometers on vehicles from **Chassis No. 4 010 995** have a green flashing indicator warning light. At the same time, the range from 0—10 km has been deleted. The indicator scale now begins between 10 and 15 km.

Removing and Installing Speedometer

Removal

1 - Remove indicator bulbs and instrument light bulb from the speedometer.

2 - Pull out cable at the speedometer after releasing the knurled nut.

3 - Remove the two slotted screws that attach the speedometer to the instrument panel.

4 - Remove speedometer from instrument panel.

Installation

Install the speedometer in the reverse order. Before tightening the slotted screws, make sure that the speedometer is correctly located and the figures on the face are upright.

Note:

No attempt should be made to alter the mileage indicated or interfere with the mechanism. When replacing or repairing the speedometer, the mileage figure must be the same as that indicated by the speedometer on its removal.

When a replacement becomes necessary, the supplier should be given the mileage which was indicated by the old speedometer.

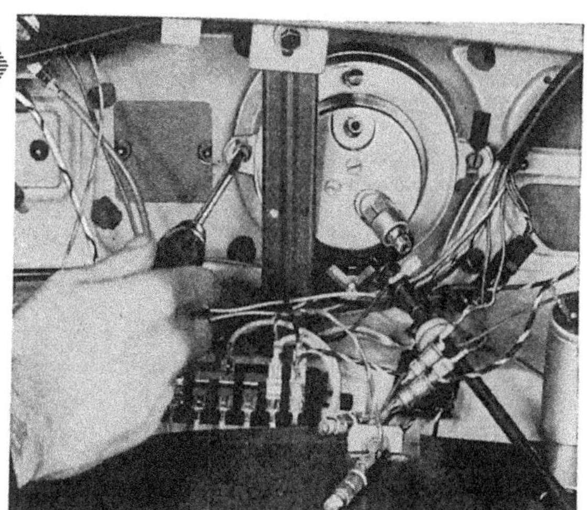

Removing and Installing Speedometer Cable

Removal

1 - Loosen knurled nut at speedometer.

2 - Remove outer hub cap from left-hand front wheel.

3 - Remove cotter pin in square end of speedometer cable at hub cap.

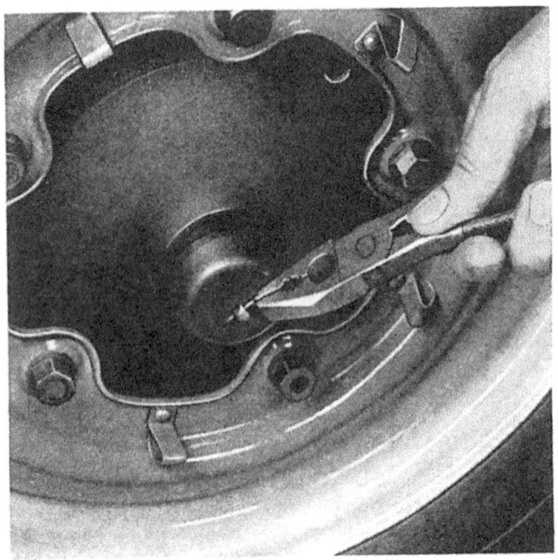

4 - Pull cable out of the stub axle.

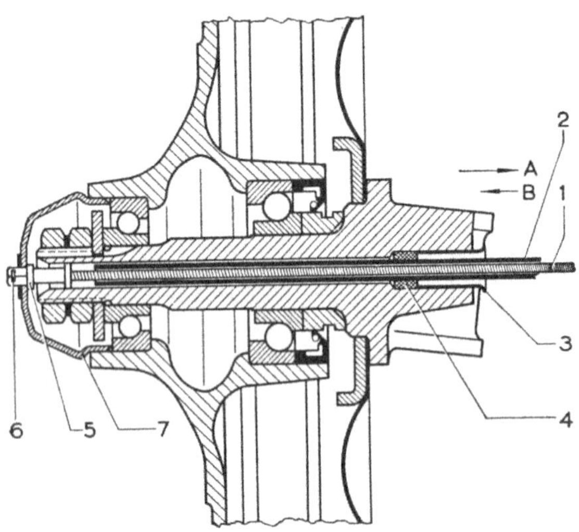

1 - Speedometer cable
2 - Synthetic housing
3 - Metal sleeve
4 - Rubber sleeve
5 - Cable drive end
6 - Cotter pin
7 - Cap with square hole

5 - Pull cable out the guide channel and grommet on the body.

Installation

A new speedometer cable should be installed in the following sequence to avoid damage to speedometer or cable:

1 - Do not kink or pull the cable when passing it through the grommet and locating it in the guide channel.

2 - Insert the square end into the speedometer and tighten the union nut.

3 - Insert the cable into the steering knuckle with a new rubber sleeve (4) and drive in the metal sleeve. Do not use any lubricant when installing these parts.

4 - Cotter-pin the square cable drive at the hub cap.

5 - On left-hand drive vehicles the speedometer cable must be behind and on right-hand drive vehicles in front of the left tie-rod, looking in direction of travel.

6 - As the synthetic casing varies slightly in length under the influence of extreme temperature changes, it must be located in a certain position in the steering knuckle on installation. This is done by pulling the casing (2) out of the steering knuckle in direction "A" until resistance is felt and then pushing it back approximately 6 to 7 mm (.23—.27") into the steering knuckle in direction "B".

Important

When carrying out repairs or adjustments on the left front wheel, take care that the cable is not disturbed when installing the hub cap. Otherwise the cable must be centralised as described under point 6.

If the cable makes a ticking noise or the speedometer needle oscillates, the synthetic casing should be centralised as described under point 6 and the routing of the cable checked carefully to ensure that there are no sharp bends or kinks.

Make sure the rubber sleeve is properly seated in the stub axle. The function of this sleeve is to seal the stub axle unit against splash water, which may lead to bearing trouble and, in winter, freezing of the speedometer cable.

Only cold-resistant and water repellent grease should be used as lubricant. Oil or normal grease is not suitable.

Note:

From Chassis No. 4 482 110 all speedometer cables have a metal casing with a plastic sheathing. As opposed to the previous versions used the plastic sheathing is now sprayed on to the metal casing.

It is now no longer necessary to centralise the plastic casing as was formerly the case when installing a new cable.

The Part Nos. of the new speedometer cables are:

	Left-hand drive	Right-hand drive
Model 11:	111 957 801 F	112 957 801 F
Model 14:	141 957 801 B	142 957 801 B

Note:

From 13th April 1964, Chassis No. **6 281 950** to Chassis No. **6 351 950**, the fixing of the speedo cable in the steering knuckle was modified on the vehicles listed above. The previous metal and rubber sleeves have been replaced by a modified rubber sleeve. This simplifies installation and improves sealing between cable and steering knuckle.

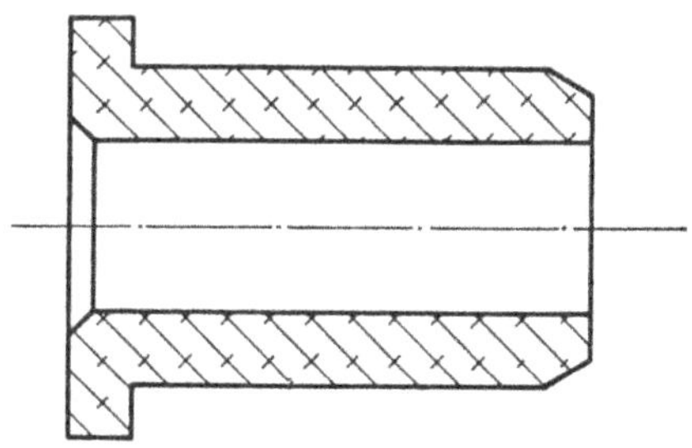

THIS PAGE INTENTIONALLY
LEFT BLANK

Clock

General Description

The electrically driven clock is situated to the right of the speedometer. To reset the clock, push in the knob in the center of the dial and turn to correct the time. The clock light is automatically switched on with the exterior lights. It is rheostat-controlled in conjunction with the instrument light.

Removal

1 - Remove front luggage compartment trim panel.

2 - Pull out the two bulbs and disconnect the clock supply cable.

3 - Remove the two clock attaching screws.

4 - Take the clock out of the instrument panel.

Installation

To install, reverse the preceding operations and make sure the clock is properly positioned before tightening the attaching screws.

Fuel Gauge

General

A fuel gauge is installed in the Karmann Ghia models. This fuel gauge consists of a gauge located in the instrument panel between speedometer and clock and a tank unit in the fuel tank.

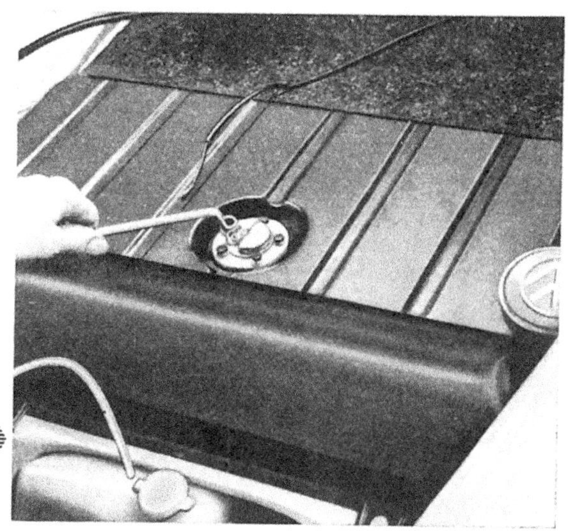

Disassembly of tank unit

1 - Disconnect cable.

2 - Unscrew mounting screws.

3 - Lift out tank unit and cork gasket from tank.

Disassembly of fuel gauge

1 - Take out lining of front luggage compartment.

2 - Disconnect electrical cables and pull out instrument lamps.

3 - Unscrew both nuts and take off mounting ring.

4 - Take instrument out of instrument panel from the passenger compartment.

Assembly

1 - When installing the tank unit, make sure cork gasket is in good condition and has a ground lead attached to it.

2 - The contact spot of the ground lead must be scraped clean and should not be corroded.

3 - Tighten mounting screws evenly.

Main Cable Harness

General

The main cable harness on the VW Sedan is housed in the left hand roof member. It can be removed and installed as follows:

Removal:

1 - Disconnect battery ground cable.

2 - Removal sound absorbing lining from left of engine compartment.

3 - Pull sound absorbing material out of roof member from the engine compartment.

4 - Disconnect electrical connections at rear.

5 - Remove instrument panel cover and luggage compartment lining.

6 - Tape cable ends into a smooth harness.

7 - Attach a thin wire about 4 m long to the longest cable at the front.

8 - Pull the harness out to the rear.

Installation

The wire is now threaded through the roof member in place of the harness. The new harness is attached to the wire in the same manner as the old one was and pulled through the roof member.

1 - Connect all cables.

2 - Check that the electrical system functions properly.

3 - Replace the sound absorbing material in the roof member from the engine compartment.

On the VW Convertible the main harness is encased in a tube inside the lower right side member. The tube runs from the front partition just behind the glove compartment to the rear quarter panel level with the muffler.

The harness can be removed and installed in the same manner as on the VW Sedan.

Reduction of voltage drop

When carrying out repairs on the electrical system or removing and installing parts in the system, take care that good ground connections are made. All metallic parts should be cleaned down to the bare metal before installation.

Push-on and Solderless Cable Terminals

All parts of the electrical system are equipped with tab connectors. The cable terminals are pushed onto the various connections in the system. The terminals are firmly fixed on the cables by means of a crimped solderless connection.

When repairing the electrical system, a suitable repair kit must be used. Special kits are supplied by the following firms:

Deutsche AMP-GmbH, Düsseldorf-Benrath, Bonner Strasse 203

Burndy, Elektro-GmbH, Köln, Burgunderstr. 35/37

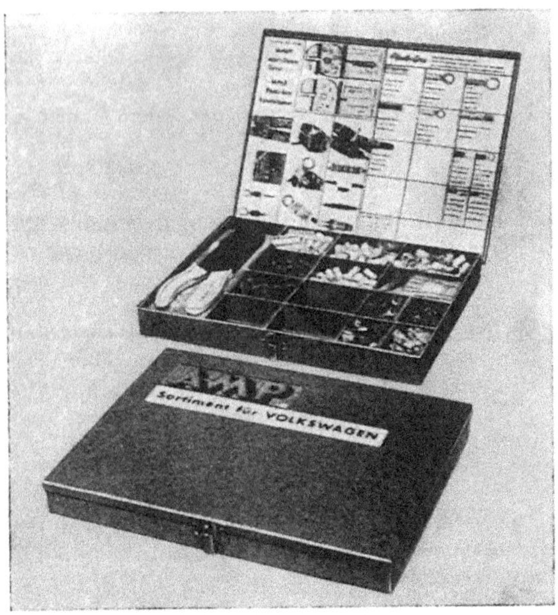

The kit consists of a special pair of pliers, open and closed terminals, flat terminals, cable connectors, flatterminal connectors, butt and parallel connectors.

These parts are required to enable all connections to be made without the use of solder. The insulation material used is oil-, fuel-, and acid-proof. All the necessary cable connecting operations can be carried out with the special pliers which are designed to suit all normal cable sizes. The notches in the pliers are marked with the same colors as the appropriate connector which also corresponds to the various cable sizes. The jaws of the pliers are provided with holes for cutting off screws of various diameters.

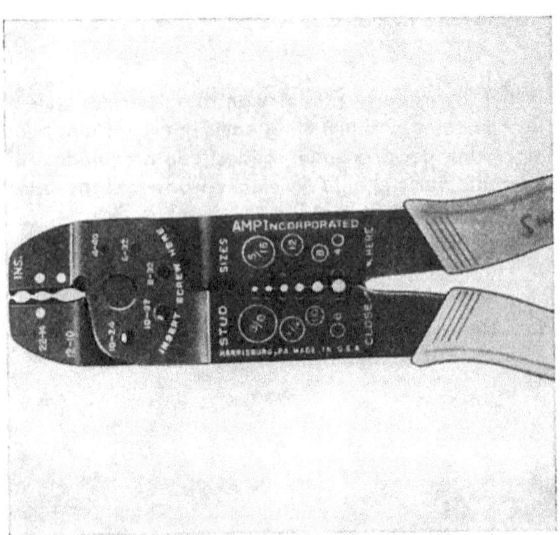

Installation of Cable Terminals and Connectors

1 - Flat terminal

Remove cable insulation. Push the terminal onto the cable until the wire is visible at the end of the sleeve and crimp the terminal with the appropriately colored notch in the pliers.

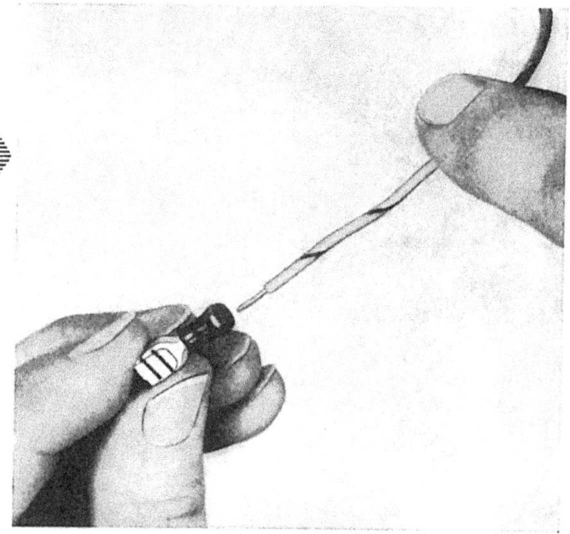

Note:

The directions inside the lid of the repair outfit should be followed carefully when making the connections.

2 - Open or closed terminal

As described under 1.

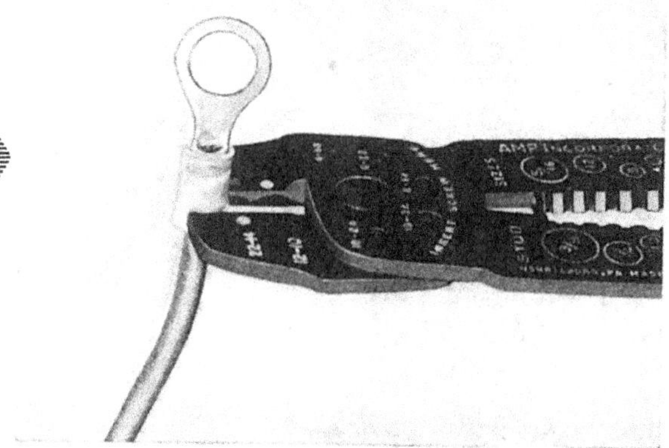

3 - Parallel connector

Remove the insulation from the ends of the two cables to be joined, insert the ends into the connector and crimp connector with the special pliers. The butt connector is installed in a similar manner.

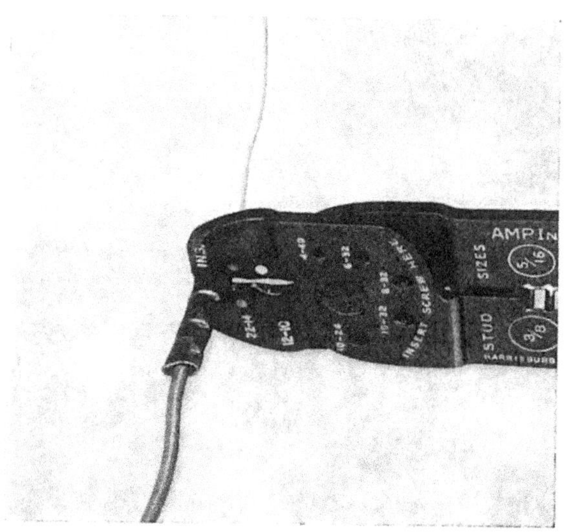

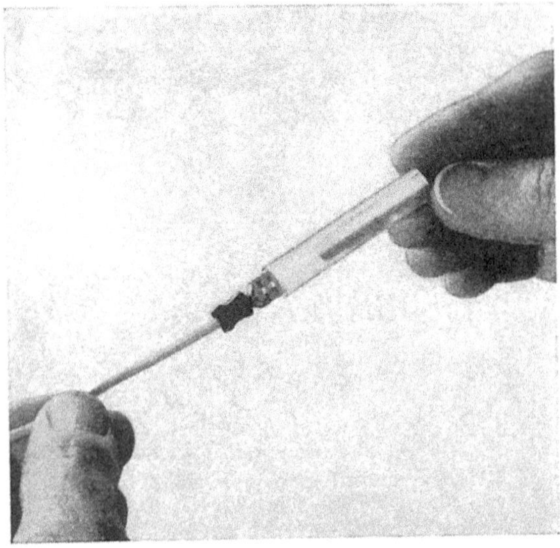

4 - Flat terminal connector

Install flat terminals on the ends of the two cables to be joined — as described under 1. Push the flat terminals into the connector.

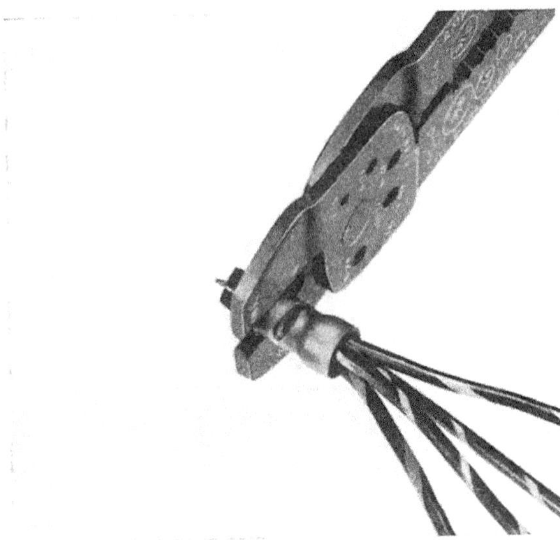

5 - End connector

Remove cable insulation. Insert the bare ends of the cables into an end connector and crimp with the special pliers.

6 - Cable adaptor

A cable adaptor is provided to enable two cables to be attached to one tab in the fuse box. The adaptors are fitted as follows:

Pull the cable with flat terminal off the tab in the fuse box and push the adaptor over the tab which is then free. The normal cable is then attached to one tab of the adaptor. The other side is then available for a further connection.

If the cable adaptor connections are inadequate a cable connector must be installed on a cable attached to one of the fuse box terminals. Further connections can then be made at the cable connector.

Main Cable Harness

The main cable harness in Karmann Ghia models is encased in a tube inside the lower right side member.

It can be removed or installed as follows:

Removal

1 - Disconnect battery ground cable.

2 - Take out left and right sound absorber in engine compartment.

3 - Disconnect electrical connections of main cable harness in rear.

4 - Pull a thin wire along with cable from left tail light to the rear of the sound absorber on the engine compartment rear plate. With this wire the new cable will later be pulled in.

5 - Take off right quarter panel lining.

6 - Unscrew cover plate at right hinge pillar.

7 - Take out lining of luggage compartment and remove glove compartment.

8 - Disconnect electrical connections of main cable harness at the rear of instrument panel.

9 - Bend cable ends back and tape into a smooth harness.

10 - Attach thin wire approximately 4 m long to the longest cable in front.

11 - Pull out main cable harness towards the rear.

Installation

The wire is now in the tube instead of the cable harness. The new cable harness can be fastened to it and pulled into the tube in the same manner as the old one was pulled out.

Make sure that the grommets stay in position when pulling the cable in. If necessary correct their positions.

1 - Connect all cables.

2 - Check functioning of electrical system.

Reduction of voltage drop

When carrying out repairs on the electrical sytem or removing and installing parts in the system, take care that good ground connections are made. All metallic parts should be cleaned down to the bare metal before installation.

THIS PAGE INTENTIONALLY
LEFT BLANK

SERVICE

Special Instructions

Volkswagens with 12 Volt Systems

From November 1962 a limited number of vehicles will be available with a 12 Volt electrical system as an optional extra. These vehicles will be marked with a transfer consisting of a red electrical flash and the wording "12 Volt". The transfer is located inside the glove compartment lid.

Important

When working on the electrical system, always check the voltage.

The following list includes all the parts of the electrical system which are converted to 12 Volt operation:

Bosch regulator RS/VA 450 M 12 A 5	113 903 803
Bosch generator LJ/GEG 450 M 12 3700 FL	113 903 031
Bosch circuit relay*)	SH/SE 31/2[1])
Bosch starter EEF 0.8/12 L 1	113 911 023
Bosch ignition coil TK 12 A 16	122 905 115 B
Flasher unit	113 953 183
Windshield wiper motor, fully suppressed	113 955 113 A
Horn	113 951 113 A
Cover with bi-metal spring and heater element	113 129 193
Battery 12 Volt, 44 Ah*)	

Factory installed suppression parts[1])

Distributor rotor, suppressed	113 905 225 A
Suppression condenser for generator*)	Bosch EMKO 21 Z 1 Z Beru SK 210 RL
Suppression condenser for ignition coil, terminal 15	113 035 267 B
Suppression condenser for regulator, terminal 61	113 035 265 B
Suppression condenser for regulator, terminal B+/51	113 035 266

*) Not available as a Genuine VW spare part

[1]) Only for service installation of a two-way radio system

Bulbs*)

Headlamp	A 12 V 45/40 W
Sealed-Beam unit (only for certain countries)	SAE 60 12 (Westinghouse)
License plate light	G 12 V 5 W
Interior light	K 12 V 10 W
Instrument panel and warning light	J 12 V 2 W
Parking light	HL 12 V 4 W
Turn indicator light	R 12 V 18 W
Twin-filament for tail light	S 12 V 18/5 W

*) Not available as Genuine VW spare parts

Radio interference suppression

Interference can be caused to radio reception in a vehicle by the ball bearing on the steering column. The interference appears when the steering wheel is turned with the vehicle stationary or in motion. On assembly the bearing is packed with a contact grease which has good conducting properties. In the course of time, a high resistance can appear at this bearing and this causes radio interference. When this trouble occurs, the bearing should be removed and a new one fitted.

Test readings and settings

1 - Checking generator and regulator

The generator and regulator are tested as described for the 6 Volt equipment. The test values are given below.

Bosch generator Type LJ/GEG 450 M 12/3700 FL

Regulator RS/VA 450 M 12 A 5

Nominal output		450 Watts
Nominal output speed	cold	3550 rpm
	warm	3800 rpm
Regulating voltage	no-load	13.6—14.4 V
	under load	12.8—13.6 V
Return current		3—9 Amps
Cut-in voltage		12.4—13.1 V

2 - Checking starter

The 12 Volt starter is also tested as described for the 6 Volt equipment. The test values are as follows:

Bosch starter Type AL/EEF 0.8/12 L 1

	Current (A)	Voltage (V)	Torque (ft. lbs.)	Speed (rpm)
No-load test	38— 45	12	—	6400—7900
Load test	165—200	9	4.3	1100—1400
Stall torque test	250—285	6.0	6.5—8.2	—

Note:
These values are for a 135 Ah battery at 20° C
The rpm. figures are based on the starter shaft
Tolerance ±10%

Normal commutator diameter $35.3^{-0,2}$ mm
Minimum commutator diameter 31.0 mm
Maximum run-out at commutator 0.03 mm

Solenoid switch

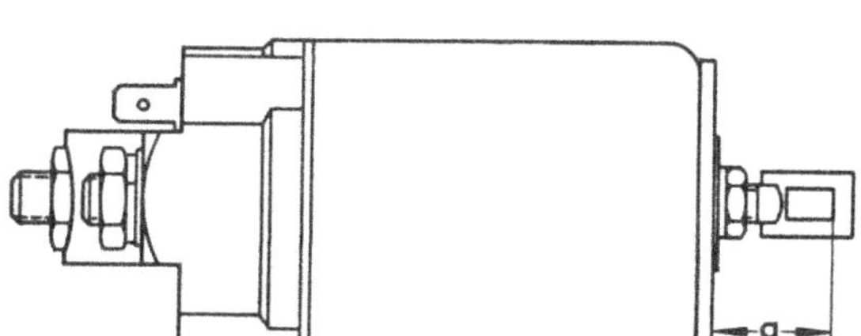

Minimum solenoid pull-in voltage 8 V

Measurement "a" (see drawing) when pulled-in 19±0.1 mm.

THIS PAGE INTENTIONALLY
LEFT BLANK

1 - VW Special Service Tools

VW 112	Special wrench
VW 159 a	Oil Pressure Switch Wrench
VW 161 a	Circlip pliers
VW 246	Plug Gauge
VW 400	Repair press
VW 401	Thrust plate
VW 408	Mandrel
VW 409	Mandrel
VW 412	Mandrel
VW 421	Thrust tube 28 mm dia.

2 - VW Workshop Equipment for Local Manufacture

VW 662/2	Oil Pressure Switch Tester
VW 674	Special wrench
VW 635/1	Adjusting screen

3 - Normal Hand Tools

Electrician's Screwdriver, 3 mm
Screwdriver, 6 mm
Combination pliers
Side-Cutting pliers
Center punch
Drift 2.5 mm
Mechanic's hammer, 300 grams
Triangular scraper
Flat file, 180 mm in length
Round file, 180 mm in length
Half-round file, 180 mm in length
Phillips screwdriver, 3 mm
Open-end wrench, 6 mm
Open-end wrench, 7 mm
Open-end wrench, 9 mm
Open-end wrench, 10 mm
Open-end wrench, 13 mm
Open-end wrench, 14 mm
Open-end wrench, 17 mm
Box wrench, 9 mm
Box wrench, 13 mm
Box wrench, 14 mm
Box wrench, 17 mm
Spark plug wrench
Socket 27 mm
Socket 36 mm
Wire brush
Oil can
Can for derusting fluid
Grease container
Scratch awl
Test lamp, 6 volts
Contact file
Feeler gauge, 0.1—0.5 mm
Caliper square, 200 mm in length, measuring 1/10 mm
Inspection lamp with cable and plug

4 - Other Workshop Equipment

Electric hand drill
Hydrometer
Voltmeter
Spark plug cleaner and tester
Distributor test stand
Headlight aiming and testing device
Ammeter and voltmeter
Ignition timing light (Stroboscope lamp)
Generator- and Starting motor test stand
Multi-purpose electric tester
Electric soldering iron
Repair kit for solderless connectors
Ignition timing scale with bracket No. 959/4080 S from Messrs. Matra.

Installation of a retractable aerial

From 1st December 1964, Chassis No. **115 481 187**, the joint in the front side panel level with the groove above the trim moulding was shortened.

The front heater hoses have been turned 90° so that the seams on the hoses point in the direction of motion.

Both these measures will facilitate the installation of a retractable aerial.

It may be difficult to install an aerial of this type on vehicles which are not modified in this way. The trouble can be remedied by inserting a half-round file through the aerial hole and filing the joint in the side panel away until the aerial tube passes freely.

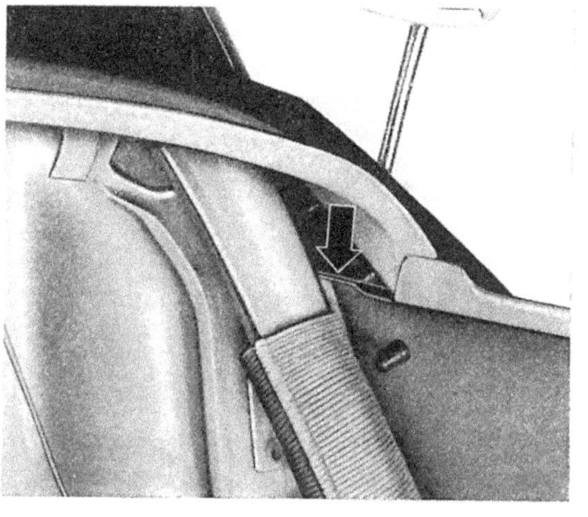

THIS PAGE INTENTIONALLY
LEFT BLANK

Contents: R

THIS PAGE INTENTIONALLY
LEFT BLANK

The frame consists of several sheet metal pressings welded together. The two floor plates are welded to the sides of the frame tunnel (central tube type frame), and the frame head for the front axle attachment is welded to the front of it. The frame fork supports the engine and transmission. A cross tube, which accommodates the torsion bars and carries the spring plate supports, is welded to the front of the fork where it joins the tunnel.

The tunnel, the main supporting member of the frame, consists of the high, curved upper section and a flat bottom plate reinforced by ribs. Inside the tunnel are the fuel line, the guide tubes for the hand brake clutch, carburetor and heater control cables as well as the gear lever attachment. In addition, provision is made on the frame tunnel for the pedal cluster, the gear lever, the hand brake and the heater control. Welded-in reinforcements behind the hand brake lever mounting bracket serve as attachments for the safety belts.

The platform consists of two pieces of ribbed sheet steel welded into position between the front cross member and the frame end plate. Both floor plates are provided with front seat runners. A socket for the car jack is spot welded to each side underneath the platform. A guide tube for the control cables of the rear foot level heating is attached to each side of the frame tunnel. The battery mounting parts are welded to the rear of the right floor plate and the accelerator pedal attachment is spot welded to the front of the left floor plate. The front cross member, the channel-shaped edges of the floor plates and the frame end plate provide the contact surfaces for the body.

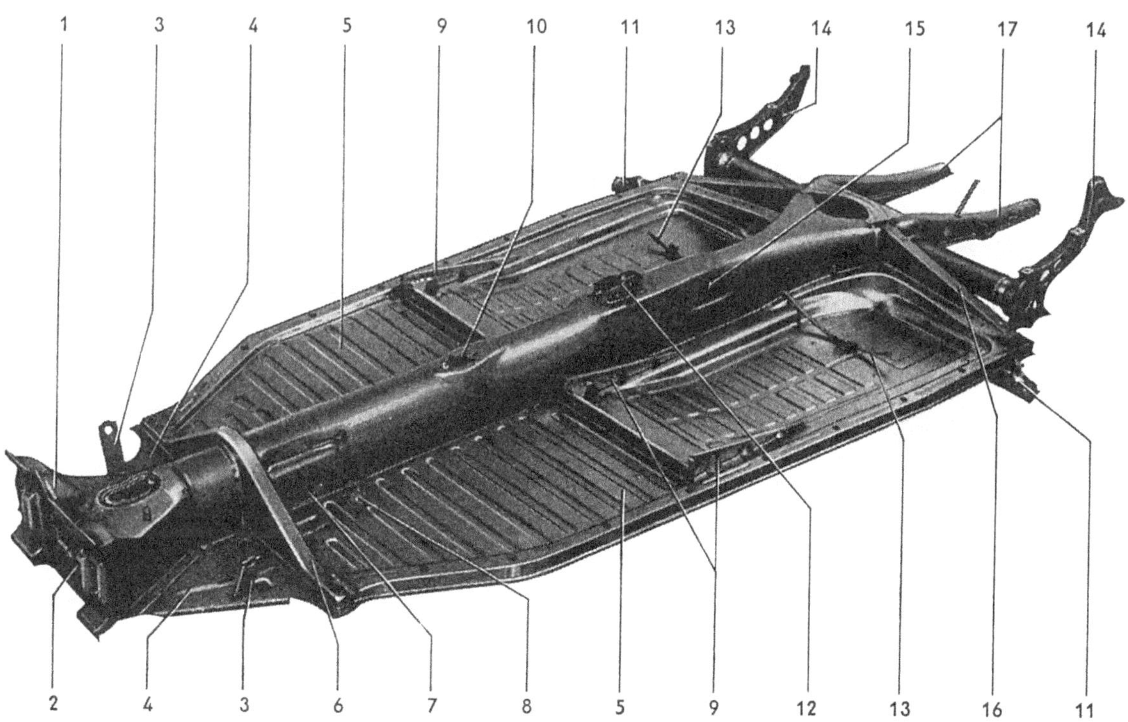

1 - Frame head — upper part
2 - Frame head — front plate
3 - Brake hose retainer
4 - Frame head — lower part
5 - Floor plate
6 - Front cross member
7 - Hole for pedal cluster
8 - Accelerator pedal attachment
9 - Seat runners
10 - Hole for gear lever
11 - Jack sockets
12 - Hand brake lever and heater control mounting bracket
13 - Guide tube for rear foot level heating control cable
14 - Spring plate supports
15 - Safety belt attachments
16 - Frame end plate
17 - Frame fork

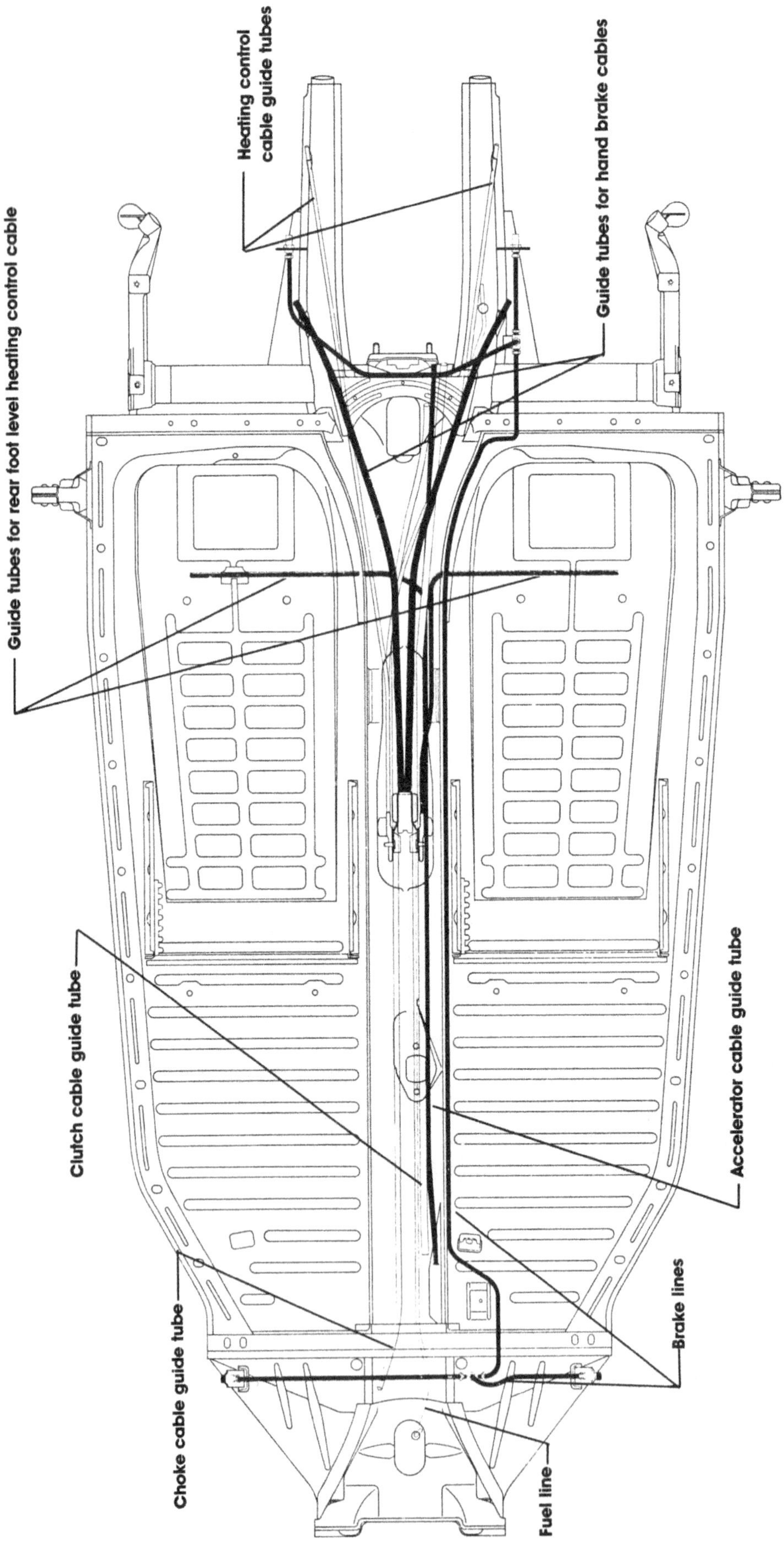
Heating control cable guide tubes
Guide tubes for hand brake cables
Guide tubes for rear foot level heating control cable
Clutch cable guide tube
Accelerator cable guide tube
Brake lines
Choke cable guide tube
Fuel line

Modifications to Frame

From February 1965, Chassis No. 115 543 024 (Model 11), Chassis No. 145 540 882 (Model 14), and Chassis No. 155 540 891 (Model 15), all frames (Part No. unchanged) are provided with small raised points on the body contact surface (arrows) in order to ensure uniform spacing between body and frame.

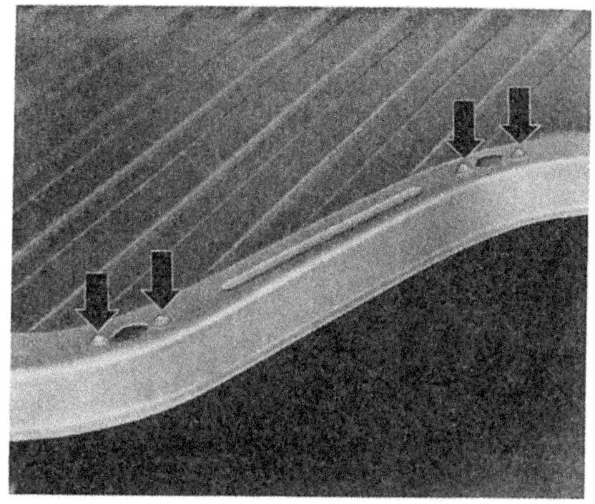

At the same time, the welding nuts in the body were replaced by threaded plates.

These two modifications will improve sealing between body and frame. Furthermore, the tightening of the body bolts which is specified in the 500 km maintenance check can be discontinued from the Chassis No. given above.

Note:

From August 1965, Chassis No. 116 000 001, all Models are fitted with a modified frame. The frame head has been altered to suit the new axle.

The modified parts can only be installed in earlier vehicles in conjunction with the new front axle.

Note:

From January 1966, Chassis No. 116 538 284, all frames are fitted with a shorter clutch cable guide tube because the shank of the cable has been lengthened. The part number of the frame remains the same.

Note:

From August 1966, Chassis No. 117 000 001, the frame tunnel lower part of all Models was provided with a depression and a central drain hole in front of the frame fork. This hole is sealed against entry of moisture by a rubber valve, Part Number 111 701 309. Water that collects can drain off through a slot.

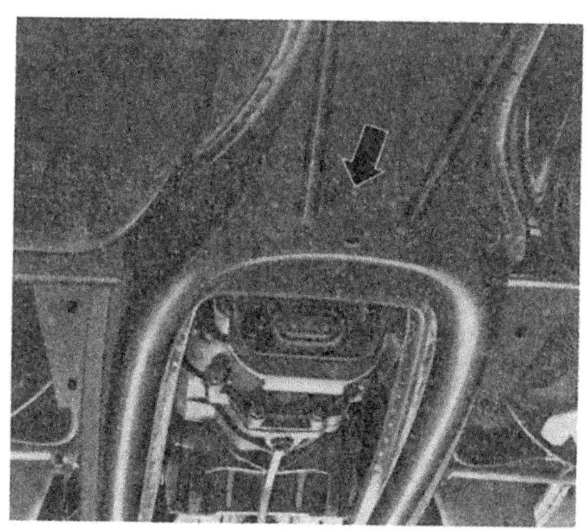

Note:

If leaks on the frame tunnel occur on older vehicles, the rubber valve can be service installed. Note the following:

Make a .55 in (14 mm) diameter hole in the frame tunnel lower part in front of the frame fork. Deburr and paint the edge of the hole. Burn a notch into the shoulder of the rubber valve at right angles to the slot with a red-hot welding rod, so that the water that has collected can run into the valve. The rubber valve must be installed with the notch facing the driving direction.

a = 86 mm (3.4 in.) b = 20 mm (.8 in.)

Note:

In August 1966, from Chassis No. 117 000 001, because of modifications to the braking system, the brake hose brackets on the frame head lower part were positioned 86 mm (3.4 in.) farther outward and 20 mm (.8 in.) toward the rear.

When replacing a frame, therefore, note the following:

a - Installing a frame of old design on a vehicle manufactured after August 1966, if these frames are still in stock.

Detach brake hose bracket from frame head lower part and reweld bracket on farther outwards and towards the rear.

b - Installing a frame of new design on a vehicle manufactured before August 1966.

Detach brake hose bracket from frame head lower part and reweld bracket on 86 mm (3.4 in.) farther inward and 20 mm (.8 in.) farther toward the front.

Frame Removal and Installation

The following sequence should be followed when removing or installing the frame.

Detailed instructions regarding removal and installation of individual parts are given in the appropriate sections of the manual.

Removal

1 - Remove both front seats, rear seat bench and backrest.

2 - Remove floor covering.

3 - Remove battery.

4 - Place car on trestles.

5 - Remove wheels.

6 - Detach fuel hose and pull it off fuel line in frame tunnel.

Note:

On the Sedan, the choke cable must also be removed.

7 - Pull connecting hose off brake fluid reservoir and also off brake line between brake fluid reservoir and brake master cylinder.

8 - Detach stop light wiring from stop light switch.

9 - Pull speedometer cable out of steering knuckle of left front wheel.

10 - Detach steering column from steering gear.

11 - Remove left and right hot air tube between engine and body.

12 - Detach cable 30 and 50 from starter.

13 - Detach cable 51 and 61 on generator, cable 15 on ignition coil and automatic choke and cable on oil pressure switch.

14 - Remove attaching bolts between body and chassis (see A 11/2).

15 - Lift body off chassis.

16 - Remove engine.

17 - Remove rear axle together with brake cables.

18 - Remove rear suspension.

19 - Remove front axle.

20 - Remove pedal cluster and hand brake system.

21 - Remove brake master cylinder, brake lines and T-piece.

22 - Remove control cables.

23 - Remove gear shift rod.

Installation

When installing, observe the following points:

1 - Check frame for distortion and damage (see also page R 5/1—3).

2 - Clean guide tubes and check that they are clear.

3 - Install removed units and parts correctly.

4 - Install new sealing strips between body and frame.

5 - Place body on to chassis. Tighten attaching bolts to prescribed torque. (See also section A-11, pages 3 and 4).

6 - Bleed and adjust brakes.

7 - Check front wheel alignment.

8 - Check electrical system.

9 - Road test car.

THIS PAGE INTENTIONALLY
LEFT BLANK

Pedal Cluster – L. H. Drive

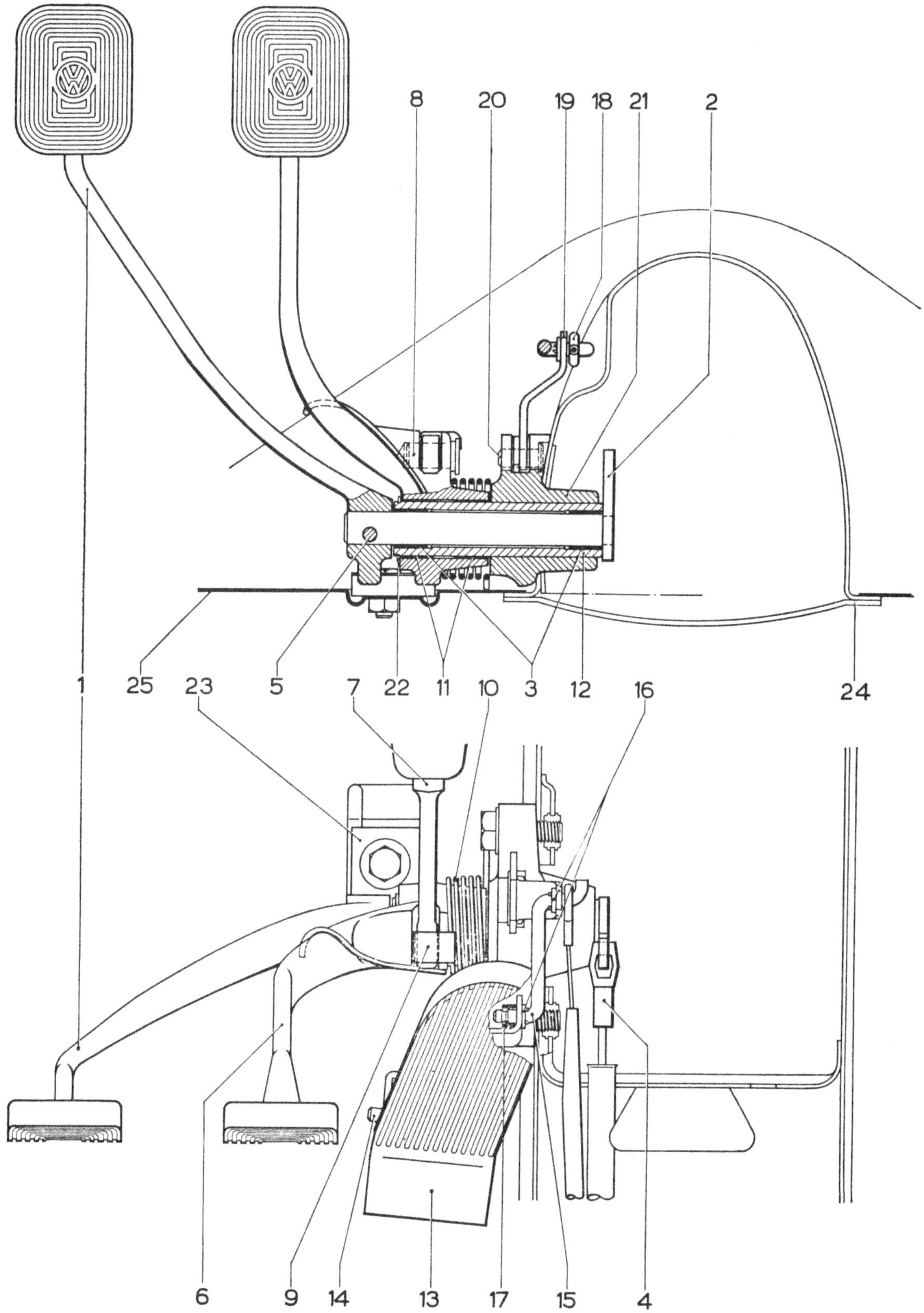

1 - Clutch pedal
2 - Clutch pedal shaft
3 - Clutch pedal shaft bushing
4 - Clutch cable
5 - Locating pin
6 - Brake pedal
7 - Brake master cylinder push rod
8 - Bolt for push rod
9 - Lock plate
10 - Brake pedal return spring
11 - Brake pedal bushing
12 - Mounting tube
13 - Accelerator pedal
14 - Mounting bolt
15 - Push rod
16 - Washers
17 - Accelerator pedal push rod spring
18 - Accelerator cable
19 - Accelerator cable connecting lever
20 - Bolt
21 - Pedal cluster mounting
22 - Circlip
23 - Stop plate
24 - Frame tunnel
25 - Floor plate

Removing and Installing Pedal Cluster

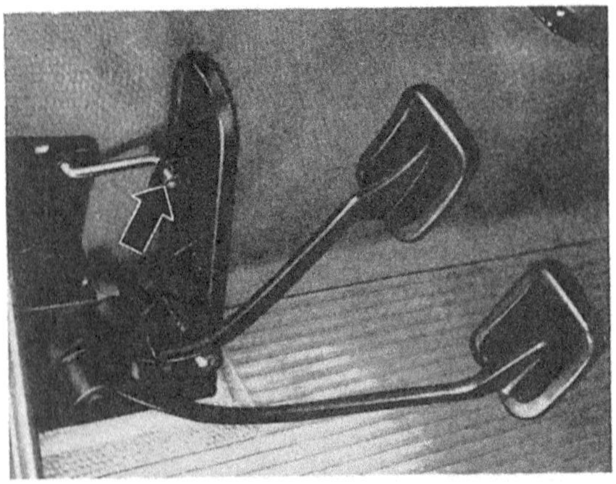

Removal

1 - Place car on trestles.

2 - Remove driver's seat.

3 - Detach clutch cable from lever on clutch pedal shaft.

4 - Remove spring (arrow) from accelerator push rod.

Detach push rod from accelerator pedal.

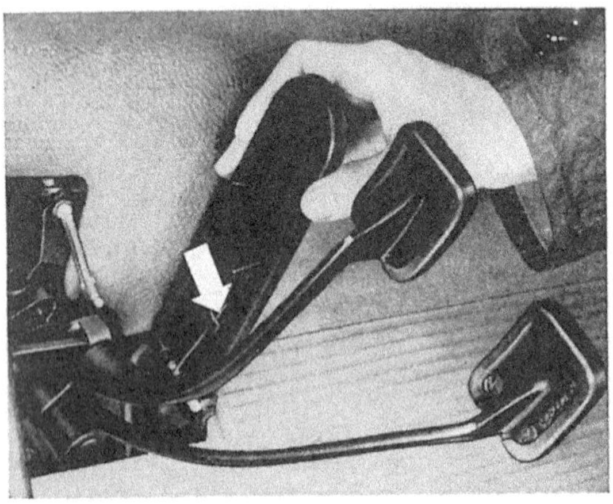

5 - Detach return spring from rear of accelerator pedal (arrow).

6 Using pliers, pull accelerator pedal mounting bolt out of moulding.

Remove accelerator pedal.

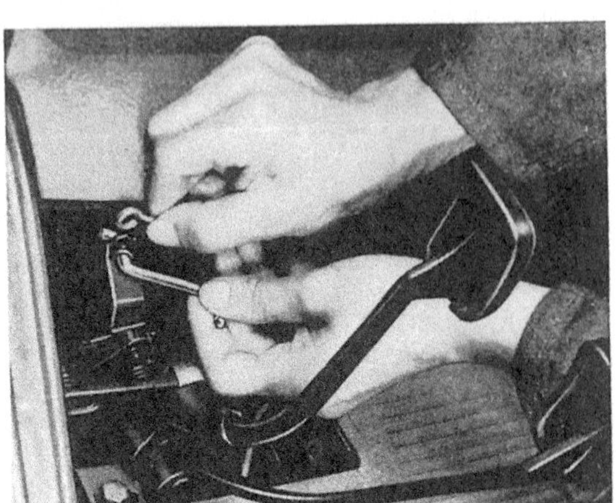

7 - Detach accelerator pedal push rod from accelerator cable and from connecting lever.

8 - Bend up lock plate of brake master cylinder push rod attaching bolt.

9 - Pull out bolt and remove brake master cylinder push rod.

10 - Remove brake and clutch pedal stop. When doing this, hold clutch pedal vertical to prevent clutch cable being detached.

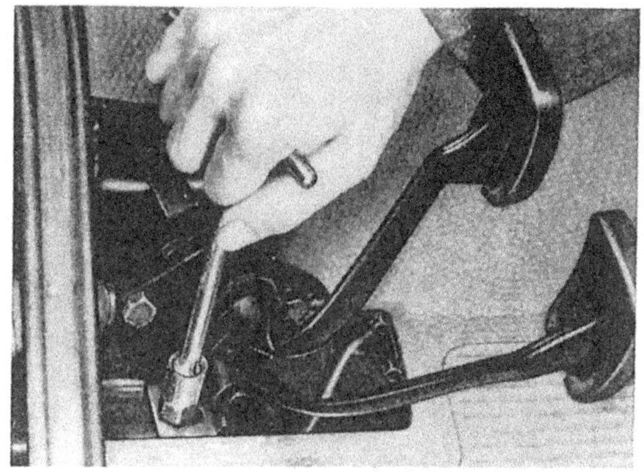

11 - Unscrew pedal mounting attaching bolts, remove pedal cluster and detach clutch cable.

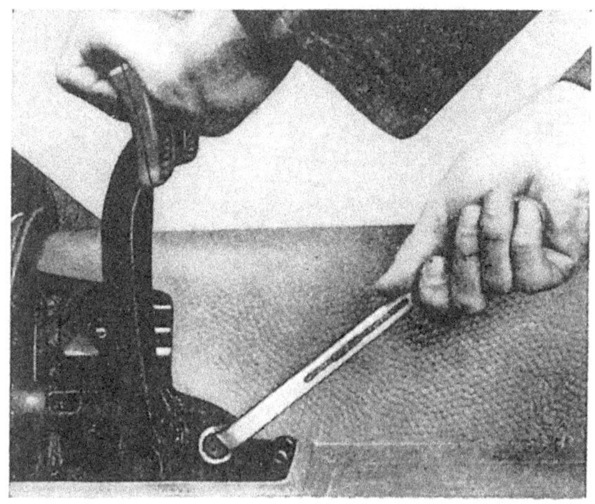

Installation

On installation, note the following:

1 - Clean pedal cluster and check for free operation as well as for excessive wear.

2 - After attaching the clutch cable, hold the clutch pedal vertical as otherwise the clutch cable will detach itself. If necessary, a second mechanic can keep the cable taut while the pedal cluster is being installed.

3 - Ensure that the brake pedal return spring is correctly installed.

4 - Install brake master cylinder push rod.

A new lock plate for the push rod attaching bolt **must always be used.**

5 - Set brake pedal by adjusting the stop so that there is a clearance of approximately .04 in. (1 mm) between the push rod and the piston of the master cylinder.

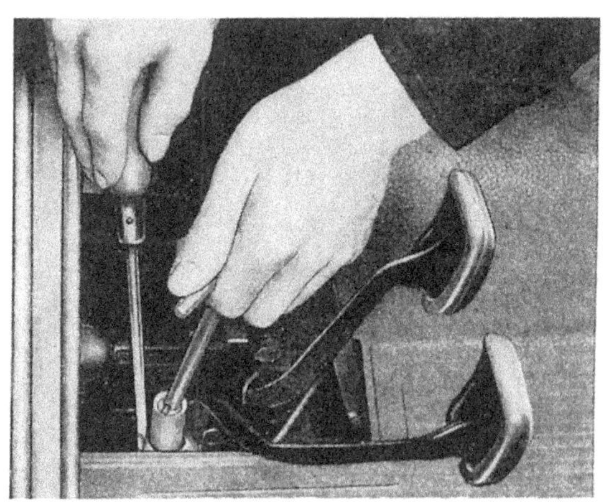

Important

The length of the brake master cylinder push rod must not be altered.

6 - When installing the accelerator pedal mounting bolt, ensure that the accelerator pedal return spring is located in the annular groove in the center of the bolt.

7 - Ensure that the accelerator pedal push rod spring is correctly installed.

8 - Adjust clutch pedal free play as specified.

9 - Road test car.

THIS PAGE INTENTIONALLY
LEFT BLANK

Disassembly and Assembly of Pedal Cluster

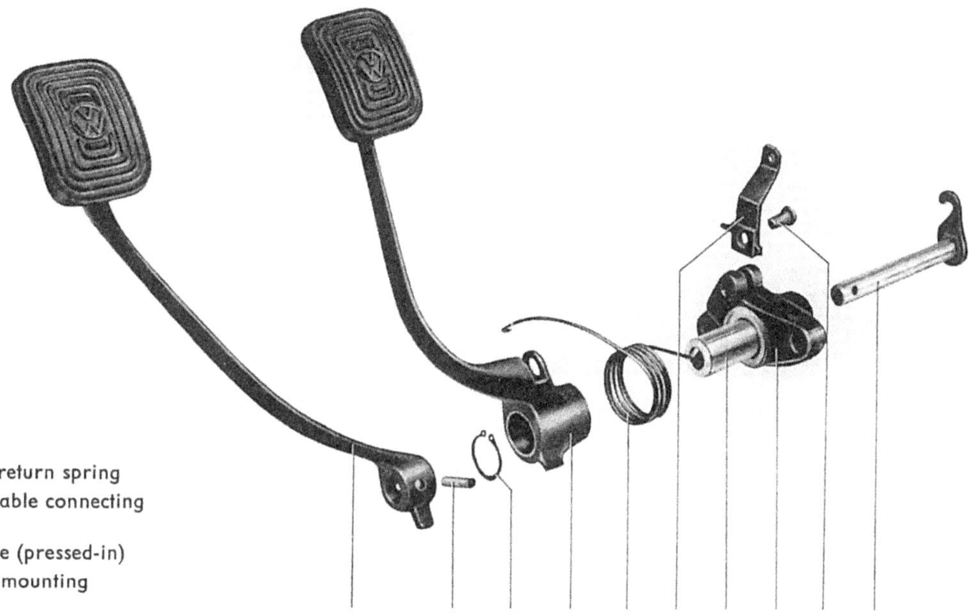

1 - Clutch pedal
2 - Locating pin
3 - Circlip
4 - Brake pedal
5 - Brake pedal return spring
6 - Accelerator cable connecting lever
7 - Mounting tube (pressed-in)
8 - Pedal cluster mounting
9 - Bolt
10 - Clutch pedal shaft

Disassembly

1 - Press accelerator cable connecting lever attaching bolt out of pedal cluster mounting and remove connecting lever.

2 - Drive locating pin out of clutch pedal and clutch pedal shaft.

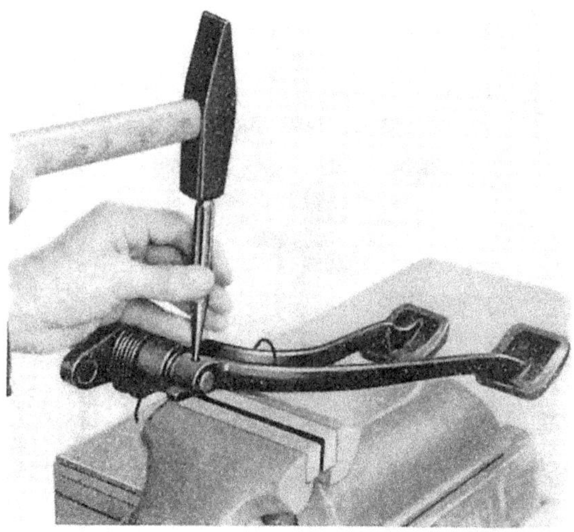

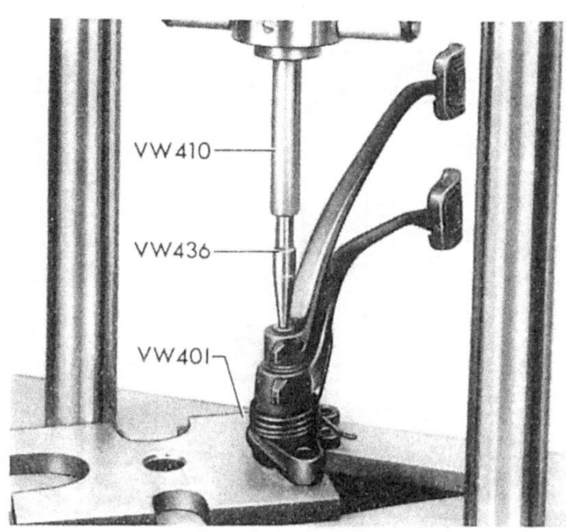

3 - Press clutch pedal shaft out of clutch pedal on a repair press, in conjunction with VW 410, VW 436 and VW 401.

4 - Pull clutch pedal shaft out of pedal cluster mounting.

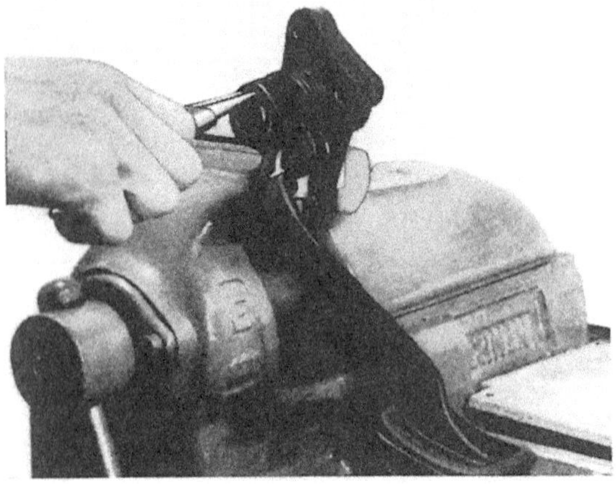

5 - Remove circlip from mounting tube, using circlip pliers.

6 - Remove brake pedal and return spring from mounting tube by turning them by hand.

Assembly

On assembly, note the following:

1 - Clean individual parts and check for wear. Replace excessively worn parts with new ones.

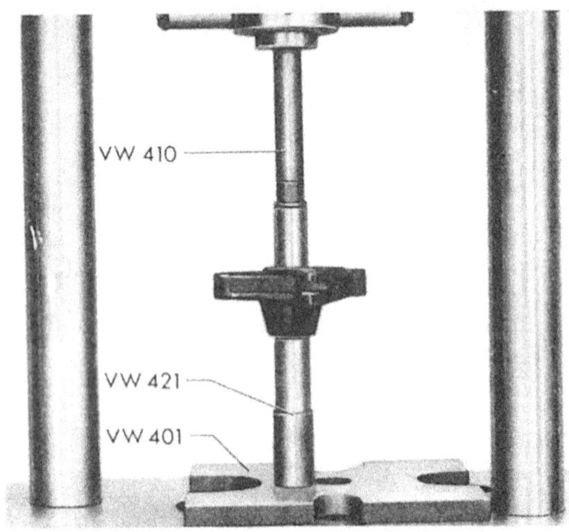

2 - If necessary, drive clutch pedal shaft bushings out of mounting tube. Press in new bushings on a repair press, in conjunction with VW 410, VW 421 and VW 401.

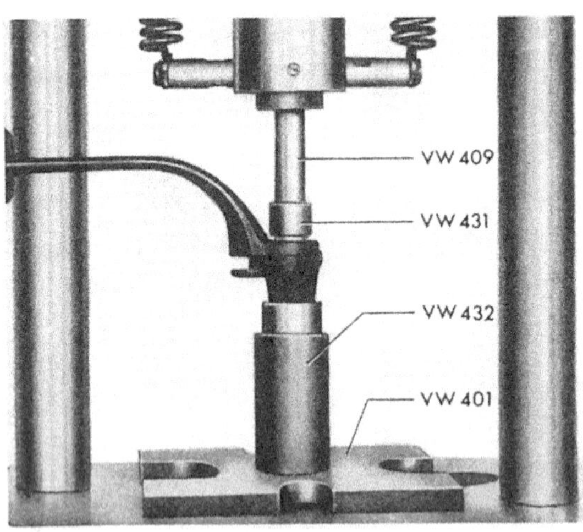

3 - If necessary, drive bushings out of brake pedal. Press in new bushing on a repair press, in conjunction with VW 409, VW 431, VW 432 and VW 401.

4 - Prior to assembling the pedal cluster, apply universal grease to all bearing surfaces.

5 - Push brake pedal on to mounting tube, ensuring that the foot brake return spring is correctly positioned.

6 - Do not forget the brake pedal circlip.

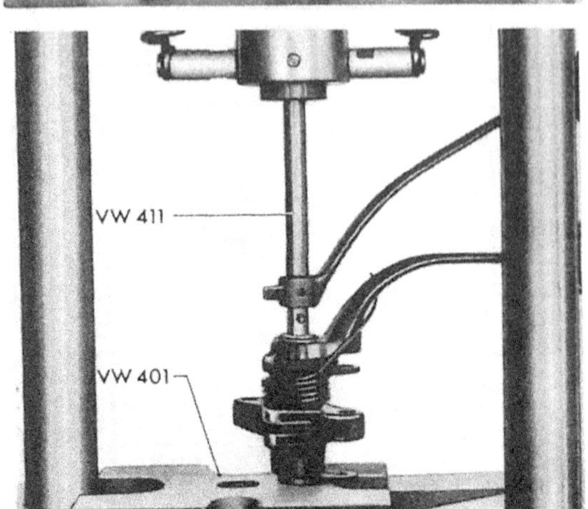

7 - Press clutch pedal on to clutch pedal shaft on a repair press, in conjunction with VW 411 and VW 401.

8 - Use a new locating pin for the clutch pedal.

9 - Install accelerator pedal correctly.

Pedal Cluster – R. H. Drive

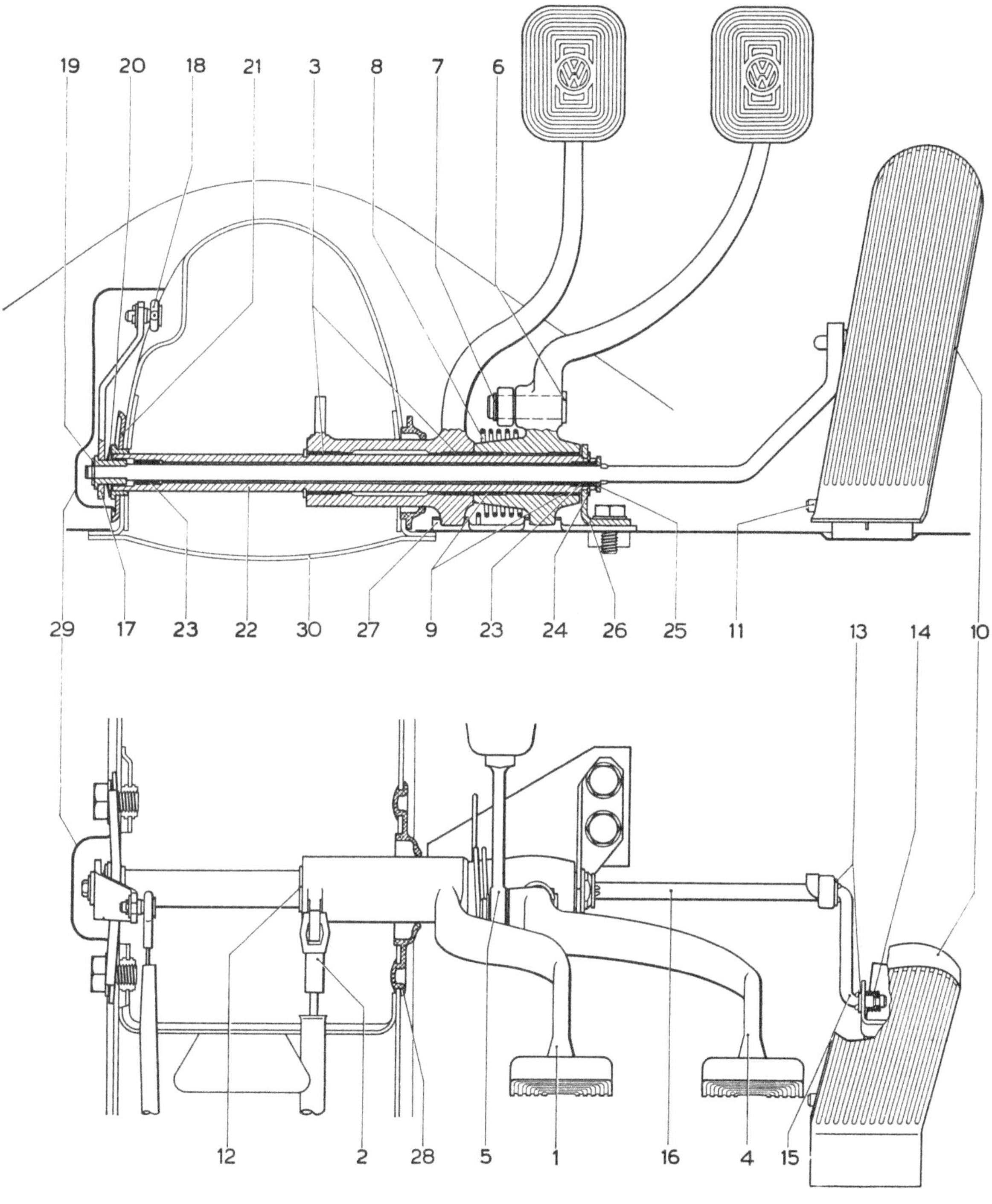

1 - Clutch pedal
2 - Clutch cable
3 - Bushing
4 - Brake pedal
5 - Operating rod
6 - Bolt
7 - Circlip
8 - Return spring
9 - Bushing
10 - Accelerator pedal
11 - Bolt
12 - Circlip
13 - Washer
14 - Compression spring
15 - Pushrod
16 - Accelerator pedal shaft
17 - Accelerator cable connecting lever
18 - Accelerator cable
19 - Circlip
20 - Concave washer
21 - Bearing
22 - Mounting tube
23 - Bushing
24 - Circlip
25 - Washer
26 - Pedal cluster bracket
27 - Stop plate
28 - Rubber seal
29 - Cover plate
30 - Frame tunnel

Removal and Installation of Pedal Cluster

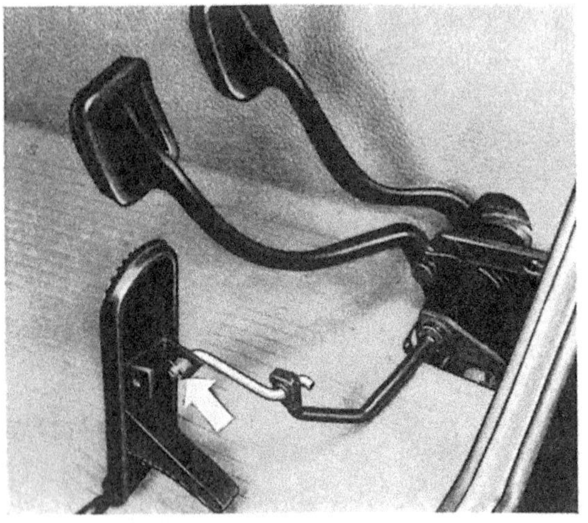

Removal

1 - Place car on trestles.

2 - Remove front seats.

3 - Detach clutch cable from lever on clutch pedal shaft.

4 - Remove spring (arrow) from accelerator push rod.

Detach push rod from accelerator pedal.

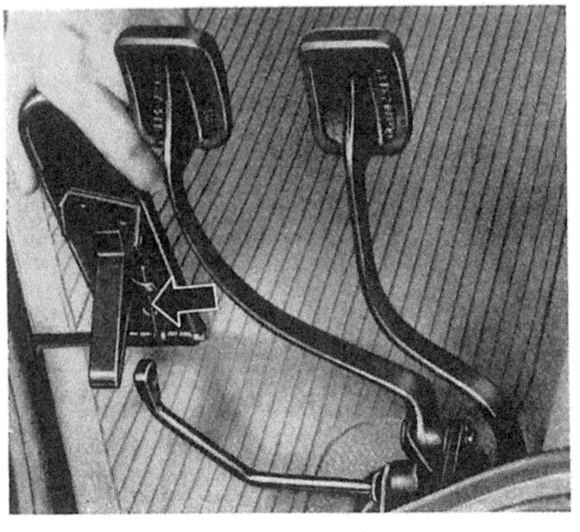

5 - Remove push rod from accelerator pedal shaft.

6 - Detach return spring from rear of accelerator pedal (arrow).

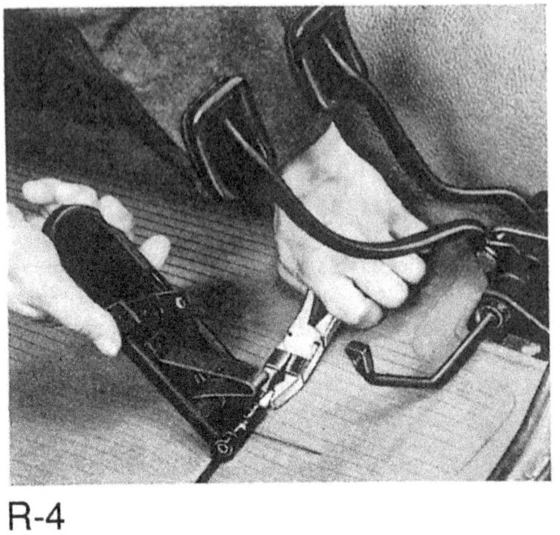

7 - Using pliers, pull accelerator pedal mounting bolt out of mounting.

Remove accelerator pedal.

8 - Using circlip pliers, remove circlip from brake master cylinder push rod attaching bolt.

9 - Pull out bolt and remove brake master cylinder push rod.

10 - Remove accelerator cable cover plate.

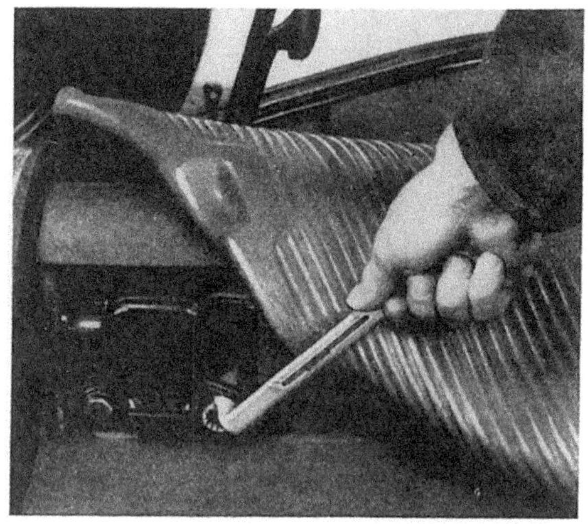

11 - Detach accelerator cable from connecting lever.

12 - Using circlip pliers, remove accelerator cable connecting lever circlip from accelerator pedal shaft.

Remove accelerator cable connecting lever and concave washer from accelerator pedal shaft.

13 - Remove mounting tube bearing.

14 - Unscrew pedal cluster bracket attaching bolts, holding the clutch pedal vertical to prevent the clutch cable from being detached.

15 - Remove pedal cluster and detach clutch cable.

Installation

On installation, note the following:

1 - Clean pedal cluster and check for free operation as well as for excessive wear.

2 - After attaching the clutch cable, hold the clutch pedal vertical as otherwise the clutch cable will detach itself. If necessary, a second mechanic can keep the cable taut while the pedal cluster is being installed.

3 - Ensure that the brake pedal return spring is correctly installed.

A new circlip must be used for the brake master cylinder push rod attaching bolt.

4 - Tighten the pedal cluster bracket attaching bolts after installation of the accelerator pedal and the accelerator cable cover plate.

5 - Insert the mounting tube bearing so that the stop (arrow) for the accelerator pedal is above and behind the accelerator pedal shaft axis.

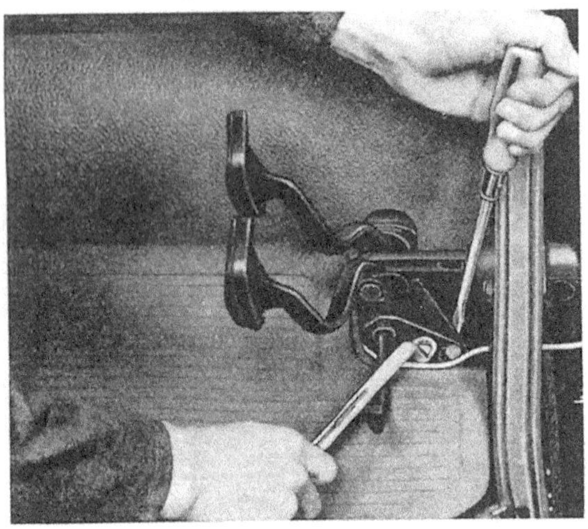

6 - Set brake pedal by adjusting the stop so that there is a clearance of approximately .04 in. (1 mm) between the push rod and the piston of the master cylinder.

Important

The length of the brake master cylinder push rod must not be altered.

7 - When installing the accelerator pedal mounting bolt, ensure that the accelerator pedal return spring is located in the annular groove in the center of the bolt.

8 - Ensure that the accelerator push rod spring is correctly installed.

9 - Adjust clutch pedal free play as specified.

10 - Road test car.

Disassembly and Assembly of Pedal Cluster

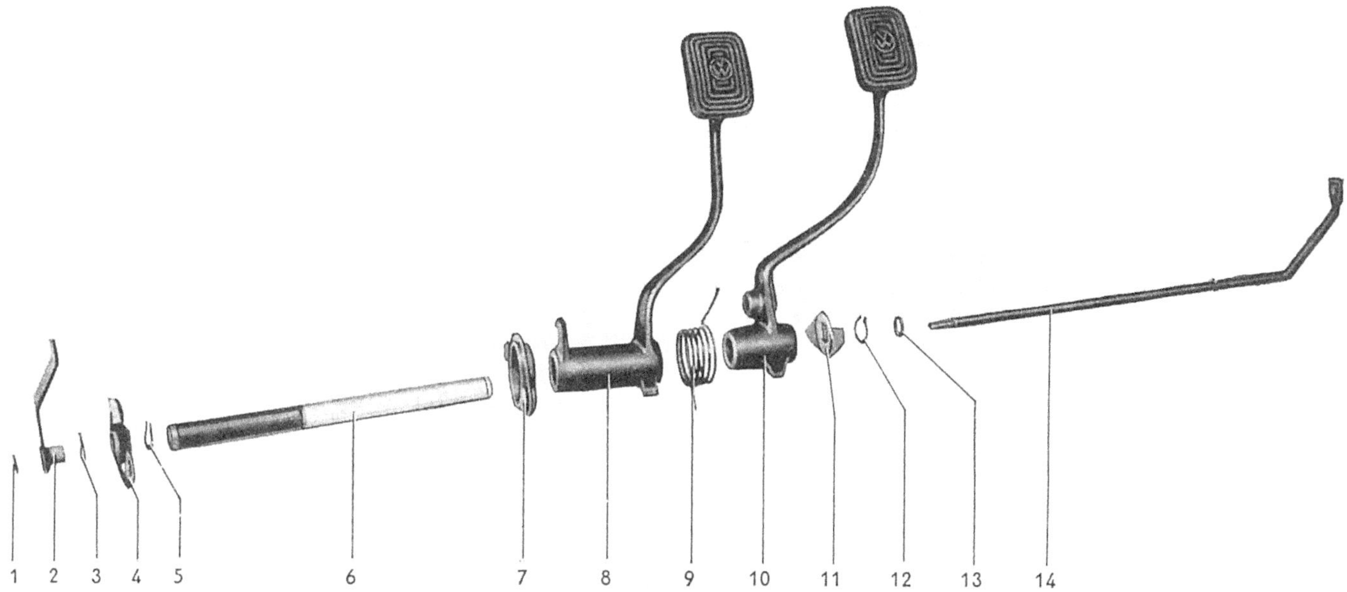

1 - Circlip
2 - Accelerator cable connecting lever
3 - Concave washer for accelerator pedal shaft
4 - Mounting tube bearing
5 - Circlip
6 - Pedal cluster mounting tube
7 - Pedal cluster seal
8 - Clutch pedal
9 - Brake pedal return spring
10 - Brake pedal
11 - Pedal cluster bracket
12 - Circlip
13 - Washer
14 - Accelerator pedal shaft

Disassembly

1 - Pull accelerator pedal shaft out of pedal cluster mounting tube.

2 - Using circlip pliers, remove circlip from pedal cluster mounting tube.

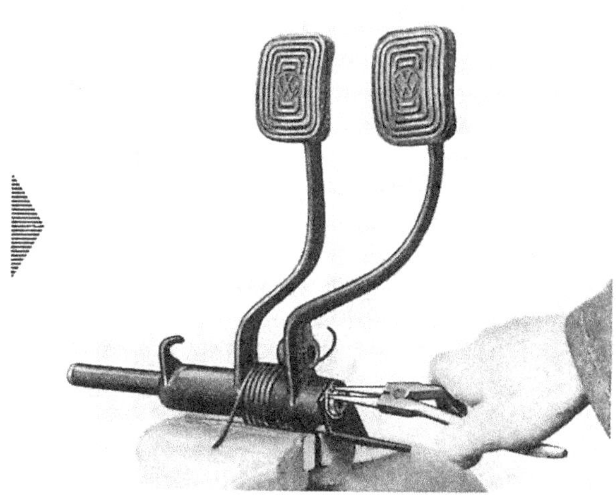

3 - Remove pedal cluster bracket, brake pedal, brake pedal return spring and clutch pedal from pedal cluster mounting tube by hand.

Assembly

On assembly, note the following:

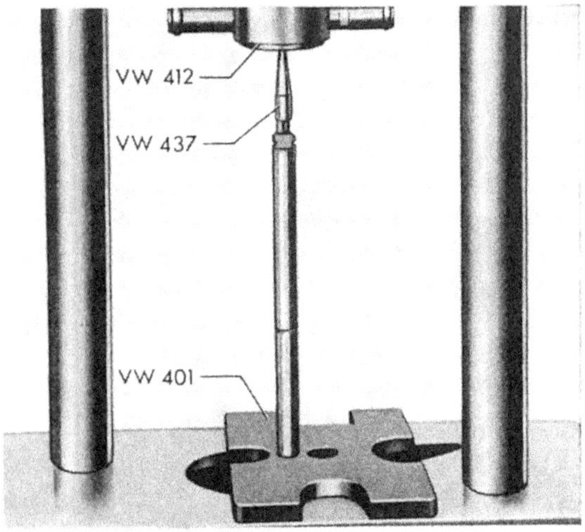

1 - Clean individual parts and check for wear. Replace excessively worn parts with new ones.

2 - If necessary, drill accelerator pedal shaft bushings out of pedal cluster mounting tube. Press in new bushings on a repair press, in conjunction with **VW 412**, **VW 437** and **VW 401**.

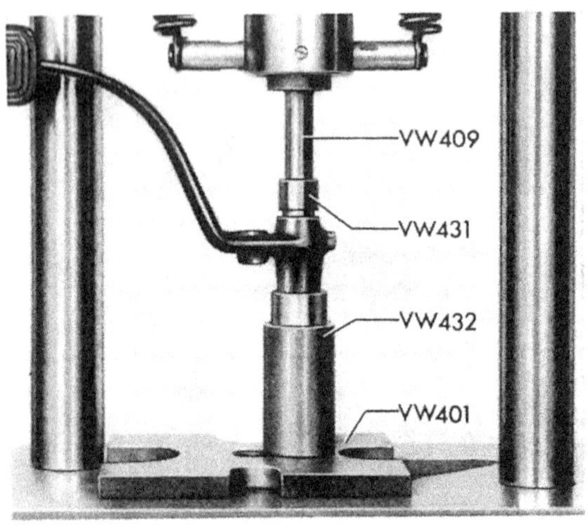

3 - If necessary, drive bushings out of brake pedal. Press in new bushings on a repair press, in conjunction with **VW 409**, **VW 431**, **VW 432** and **VW 401**.

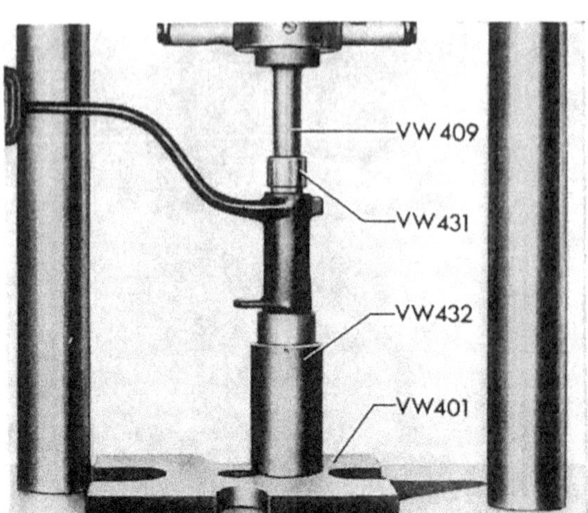

4 - If necessary, drive bushings out of clutch pedal. Press in new bushings on a repair press, in conjunction with **VW 409**, **VW 431**, **VW 432** and **VW 401**.

5 - Prior to assembling the pedal cluster, apply universal grease to all bearing surfaces.

6 - Ensure that the brake pedal return spring is correctly installed.

7 - Do not forget the washer for the accelerator pedal shaft.

Checking Frame

The frame can be checked while installed, using the frame checking gauge as described in the Workshop Equipment Manual on pages R-1 and R-2.
If the frame has been removed, it can also be checked on the repair jig VW 1025.

1 - Stands
2 - Main beam
3 - Slide and locating plate for frame head front plate
4 - Lever for slide
5 - Clamps
6 - Positioning pins
7 - Floor plate locating pins
8 - Slides and locators for spring plate support
9 - Levers for slides
10 - Supports for cross tube
11 - Locating bolts for rear axle attachment

Checking Frame

1 - Prior to placing the frame on to the repair jig VW 1025, clean all locating points.

2 - Check frame in the following sequence:

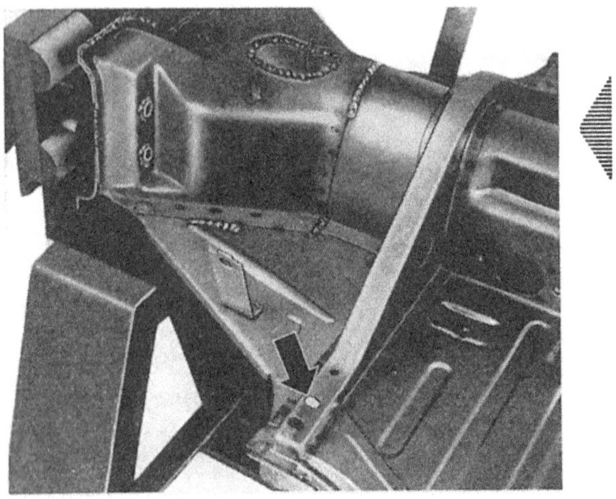

Place frame on to the floor plate locating pins (arrow). Should difficulties arise, align the spacer sleeves in the front cross member with the elongated holes. If the frame still cannot be located, a rubber hammer can be used when the elongated are only slightly out of alignment with the pins. If they are badly out of line, the frame is damaged.

Push both slides together with locators against the spring plate support. If a locating bolt cannot be inserted, the rear cross tube is distorted and a new frame must be installed.

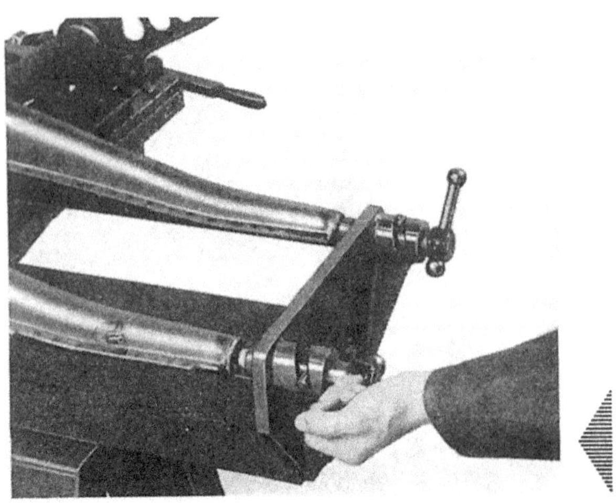

Insert rear axle attachment centering bolts.

The ends of the frame fork may be slightly aligned with each other laterally but not vertically.

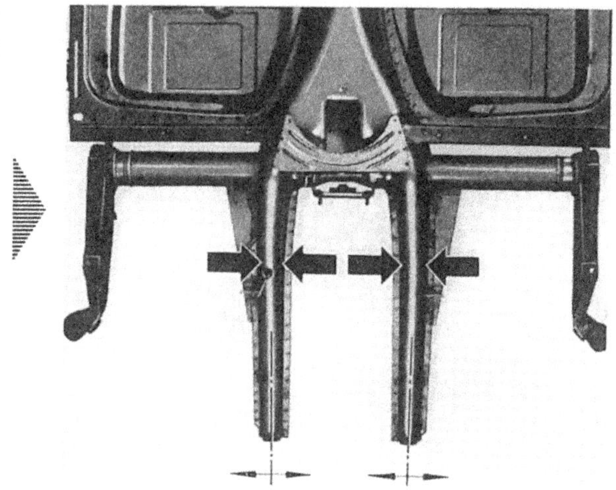

Deviations up to ± .2 in. (5 mm) can be corrected by stretching a welding flange (arrows).

If the frame fork is badly distorted or damaged, a new frame must be installed.

Push slide and locating plate against frame head front plate.

The clearance between the cylindrical pins and the frame head holes is equivalent to the permissible vertical and lateral deviation of the frame head.

The permissible longitudinal deviation from the basic dimension is ± .16 in. (4 mm). The difference between right and left (measured at the outside edge of the frame head front plate) must not exceed .08 in. (2 mm).

The basic dimension is set by the positioning pins.

A floor plate or the frame head or the complete frame must be replaced according to the extent and type of frame damage.

Important

If the frame tunnel is distorted or torn, the frame must be replaced with a new one.

THIS PAGE INTENTIONALLY
LEFT BLANK

Welding and cutting work requires a high degree of skill and only experienced welders should be entrusted with this type of work.

Before starting frame repairs, blow out the frame tunnel and fuel lines well with compressed air to clear all traces of gasoline fumes.

Frame Head Replacement

Prior to replacing the frame head, the frame must always be accurately checked on the repair jig VW 1025 (see page R-5/1—3).

Removing the Frame Head

When cutting, the torch must always be held so that there is no possibility of damaging or overheating the frame tunnel.

1 - Remove sealing compound from frame head.

2 - Cut the frame head front plate along the outer edge of the frame tunnel with a torch.

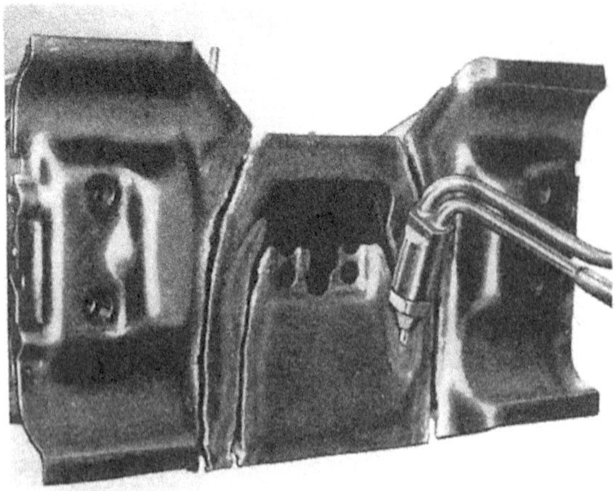

3 - Cut the frame head front plate along the inner edge of the frame tunnel with a torch.

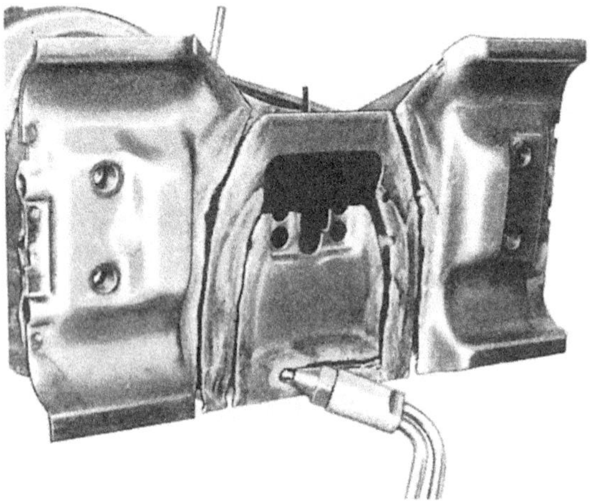

4 - Cut out the frame head front plate center piece along the inner edge of the frame head lower part.

5 - With a torch, make a cut along the center of the frame head lower part over a length of dimension a = 6.7 in. (170 mm) and from this point make two cuts, each at approx. 30°, towards the outside up to the frame tunnel.

6 - Make two cuts (arrows) in the remaining frame head front plate center piece with a torch.

Important
When doing this, *hold the torch in such a* way that the frame tunnel cannot be damaged.

7 - Cut the frame head upper part on both sides along the frame tunnel with a torch.

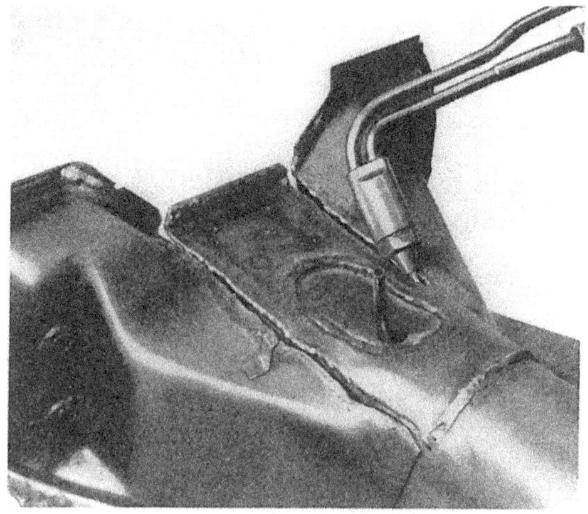

8 - Cut the frame head lower part on both sides along the welding flange of the front cross member and the frame tunnel to the frame head upper part, using a torch.

Loosen the welding seam on frame head upper part and frame head (arrow) with a torch, proceeding as follows:

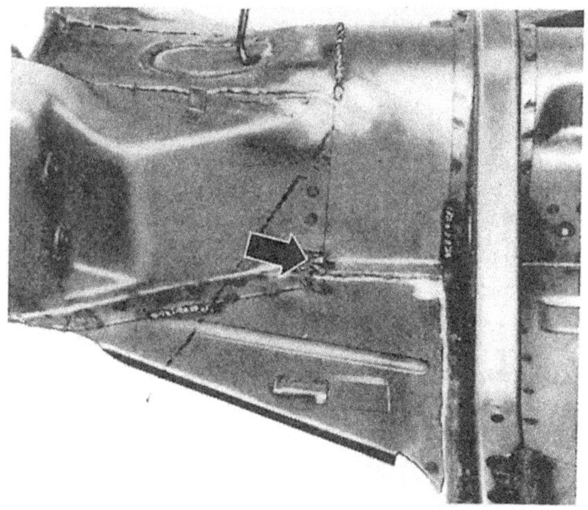

Hold the torch so that only the upper sheet metal or welding material is heated and is blown away by the pressure of the oxygen. The connection has only to be weakened and the sheet metal underneath must on no account be damaged.

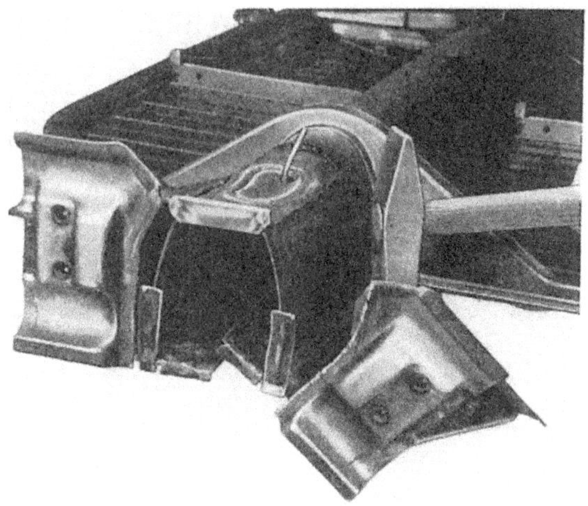

9 - Separate frame head halves from frame tunnel by applying heavy hammer blows.

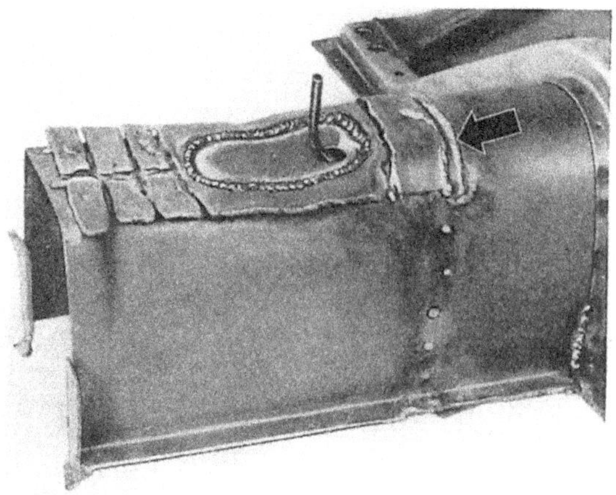

10 - Cut the rest of the frame head upper part into strips, using a torch. Detach welding seam (arrow) from frame tunnel, proceeding as outlined at position 8, par. 3.

11 - Pull remants of frame head upper part off frame tunnel with pincers.

12 - Press fuel line through opening into frame tunnel interior.

13 - Detach welding seam (arrows) from frame tunnel with a torch, proceeding as outlined at position 8, par. 3.

14 - Pull frame head upper part remnants off frame tunnel with pincers.

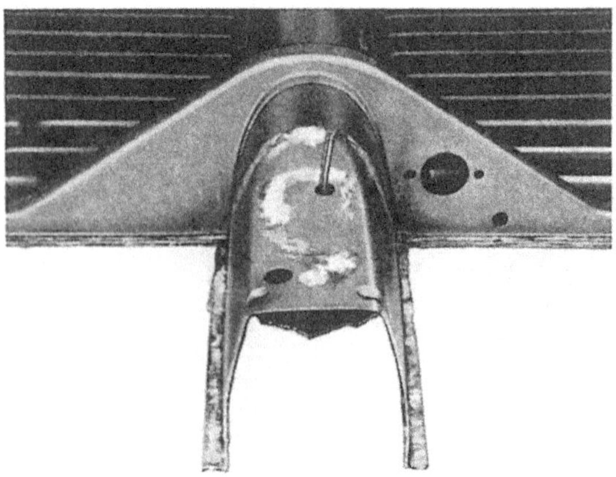

15 - Cut off frame head front plate remnants with a torch.

16 - Grind all contact surfaces, welding and parting edges clean.

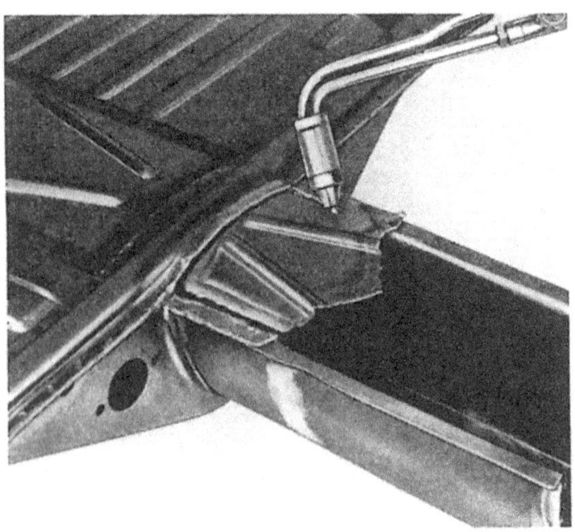

17 - Turn frame over.

18 - With a torch, cut out center piece of frame head lower part along the inner edge of the welding flanges of the frame tunnel and of the front cross member.

19 - Cut frame head lower part remnants into strips with a torch.

20 - Pull frame head lower part remnants off frame tunnel with pincers.

21 - Straighten all contact surfaces and welding flanges and grind them clean.

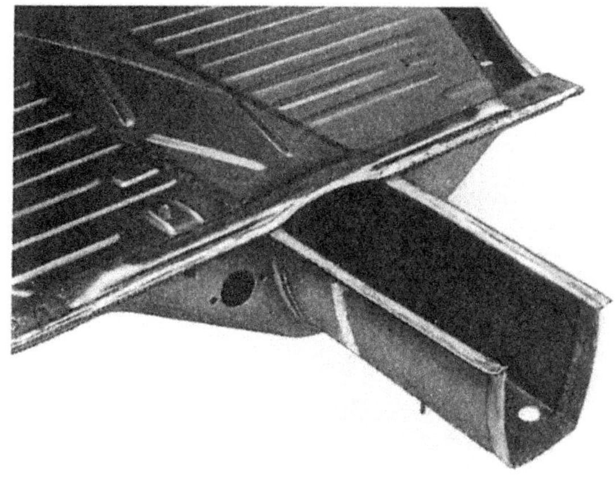

Installing Frame Tunnel

Before starting welding operations, lower the front end of the frame so that any scraps of metal etc. can fall out.

> **Important**
>
> Do not bump the frame on the ground as otherwise the frame tunnel may be damaged.
>
> Check that the pipes in the tunnel are secure.

1 - Degrease frame head parts.

2 - Weld tapped sleeves of frame head front plate.

3 - Place frame on to repair jig VW 1025 and clamp it at the rear attaching points.

4 - Place frame head lower part on to repair jig VW 1025.

5 - Place frame head front plate on to repair jig VW 1025. Push slide and locating plate against front plate and insert positioning pin. Hold front plate firmly in the clamps.

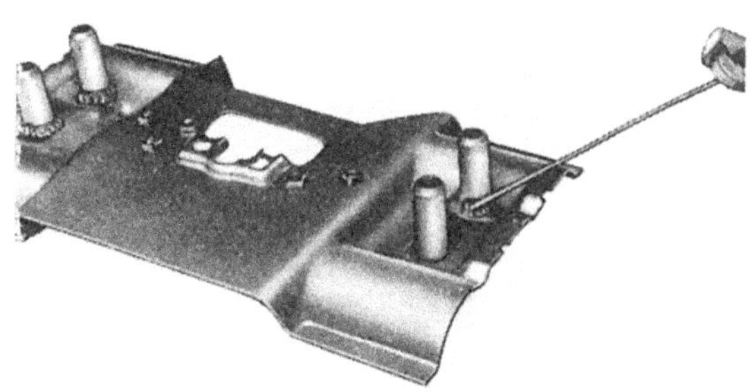

> **Important**
>
> All welding should be done electrically, using electrodes of .08 in. (2 mm) dia. and a tensile strength of 40 to 50 kg per sq. mm. at a current of 40—60 Amps.
>
> Final operations outlined at positions 17, 18 and 24 can also be carried out with electrodes of .12—.14 in. (3—3.5 mm) dia.

6 - Weld frame head front plate to frame tunnel at 3 locations over a distance of .6 in. (15 mm) on each side.

7 - Align frame head lower part to frame center and tack to front cross member on both sides.

8 - Lift the front of the frame head lower part and tack to the frame tunnel welding flange on both sides.

9 - Weld frame head lower part to front cross member on both sides over a length of .4 in. (10 mm) every 2 in. (50 mm).

10 - Weld frame head lower part to the frame tunnel welding flange on both sides. Carry out the welding only in the area which is afterwards covered by the frame head upper part.

11 - Open the clamps, place frame head upper part on to repair jig VW 1025 and close clamps.

12 - Tack frame head upper part to center of frame tunnel and also to frame head lower part at the bottom on both sides.

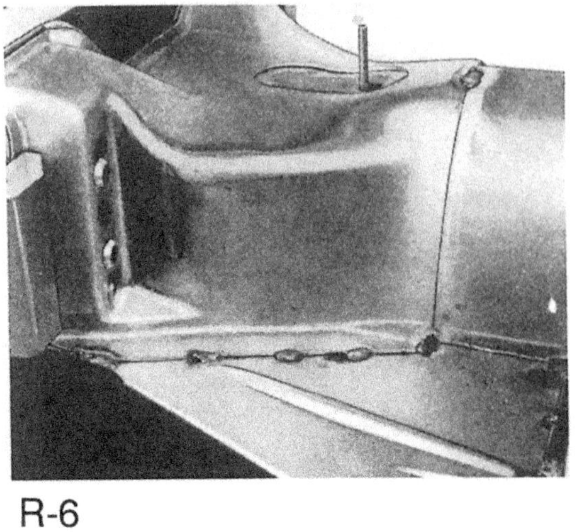

13 - Weld frame head upper part to frame head lower part on both sides over a length of .4 in. (10 mm) every 2 in. (50 mm).

14 - Weld frame head upper part to frame tunnel over whole length of join.

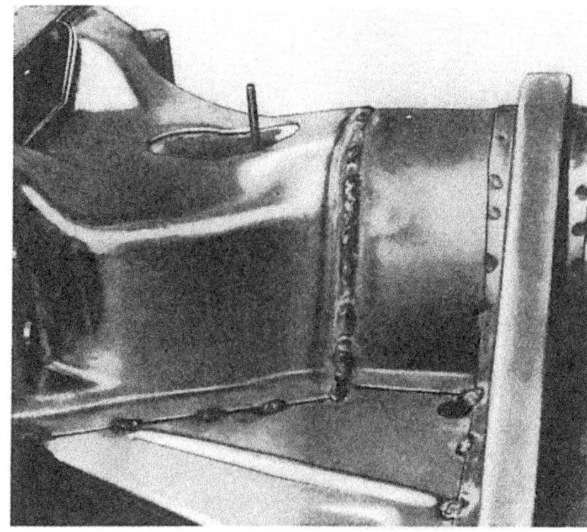

15 - Weld tapped sleeves of frame head front plate to both sides of frame head upper part.

16 - Weld frame head front plate to upper edge of frame head upper part over whole length of join.

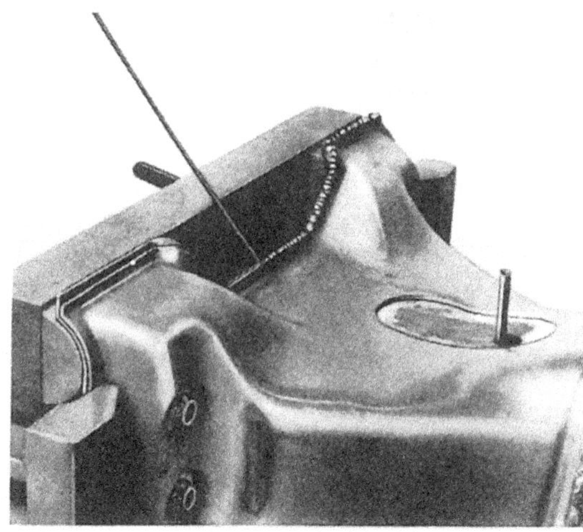

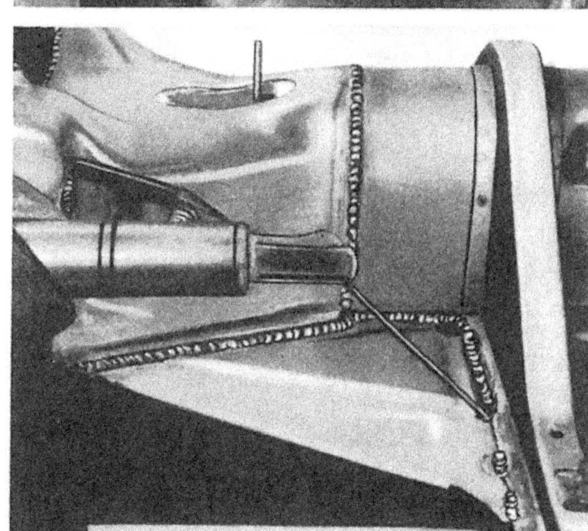

17 - Weld frame head lower part to frame head upper part, frame tunnel and front cross member on both sides over whole length of join.

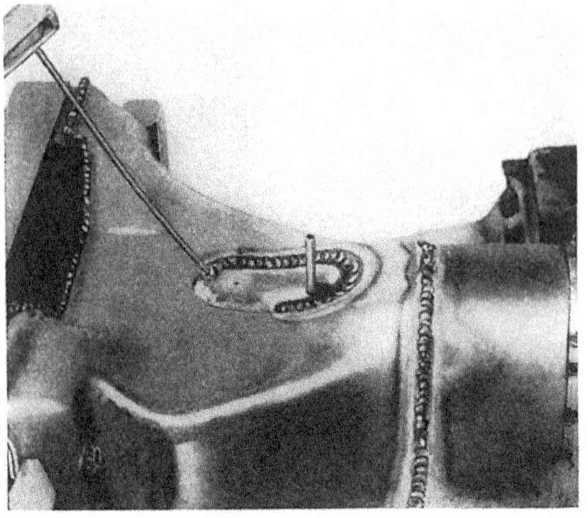

18 - Weld frame head upper part to center of frame tunnel.

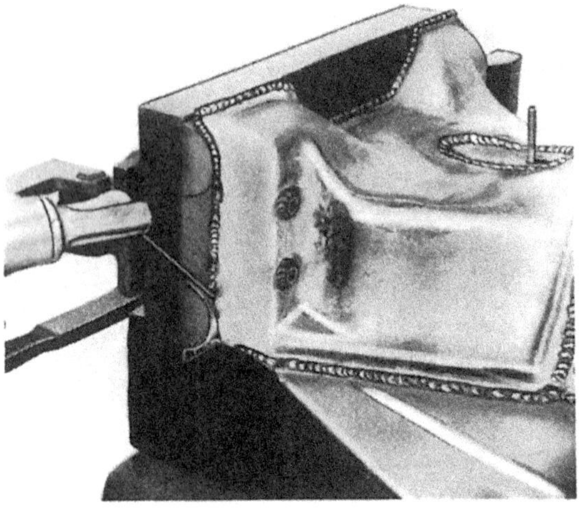

19 - Open clamps of repair jig VW **1025** and weld frame tunnel front plate to both sides of frame head upper part over whole length of join.

20 - Weld brake hose bracket to frame head lower part.

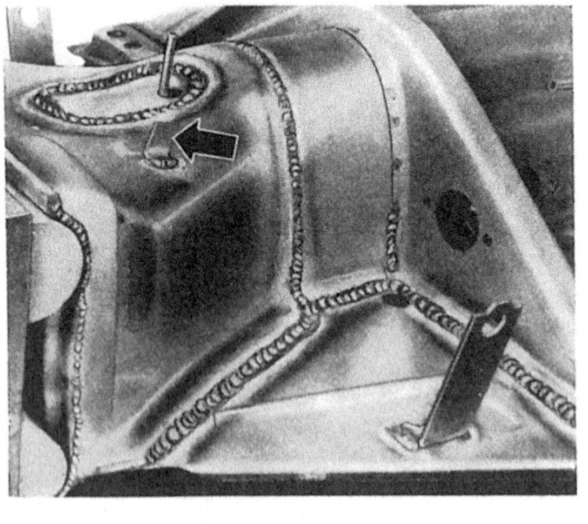

21 - Weld sheet metal clamping tongue for brake line to frame head upper part (arrow).

Important

Leave frame in repair jig VW 1025 for at least 5 minutes to prevent it distorting when cooling down.

22 - Take frame out of repair jig VW **1025**.

Turn frame over.

23 - Weld frame head front plate to frame head lower part over whole length of join.

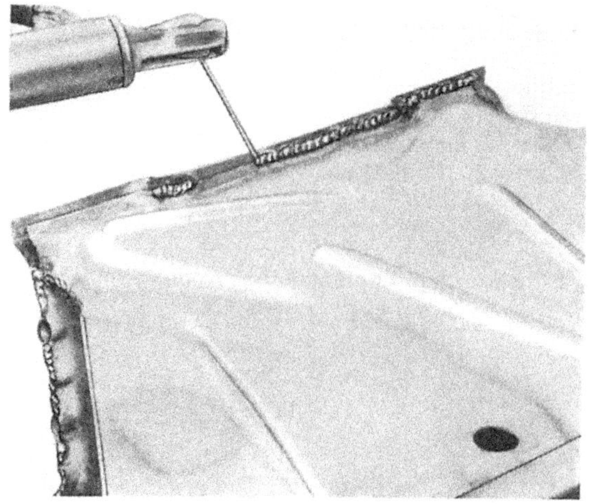

24 - Weld frame head lower part to front cross member over whole length of join.

25 - Seal frame head at all welded joints with genuine VW sealing compound D 17.

26 - Paint frame head.

Replacement of Floor Plate

Before a floor plate is replaced, the frame should be checked accurately in the repair jig VW 1025 (see page R-5/1—3). If it is found that the frame head must also be replaced, a new frame must be fitted.

If the left-hand floor plate has to be replaced, remove the relocated brake line on the frame tunnel.

Cutting Out Floor Plate

The floor plate can be separated from the frame easily and cleanly, utilizing a compressed air hammer as described in the Workshop Equipment Manual, page A-38.

Two chisels are required for carrying out the following operations:

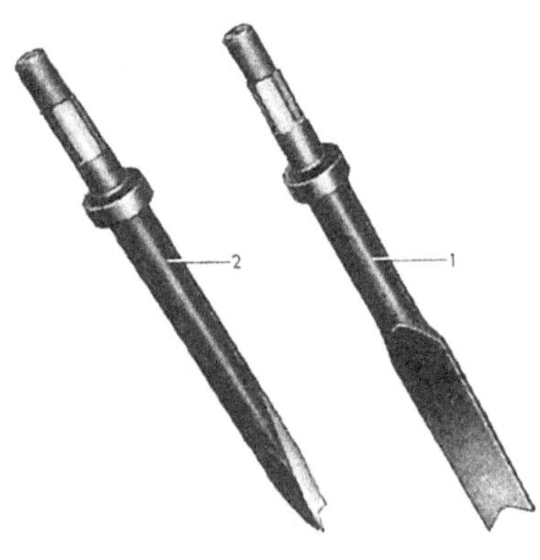

Type 1 - Chisel for cutting out sheet metal parts with even cut.

Type 2 - Parting chisel for spot welds.

All work carried out with the compressed air hammer and described below can also be done with a parting disc or a torch.

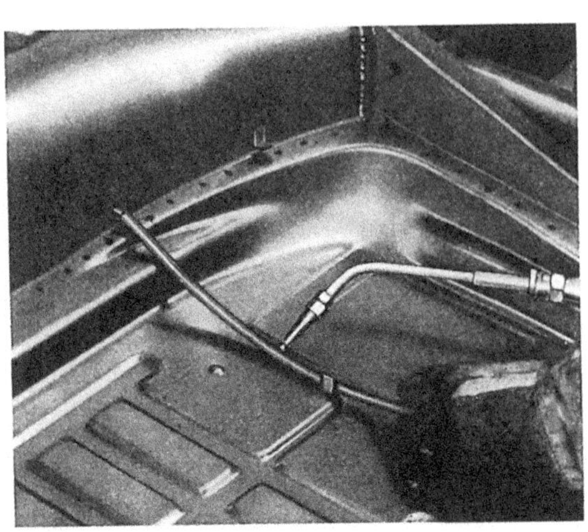

1 - Remove insulation material and as much sealing compound as possible.

2 - Using a torch, heat solder holding rear foot level heating control cable guide tube.

Remove guide tube.

3 - Using a torch, cut floor plate along welding flange of floor end plate over a length of 4 in. (100 mm). The welding flange of the floor end plate must not be damaged.

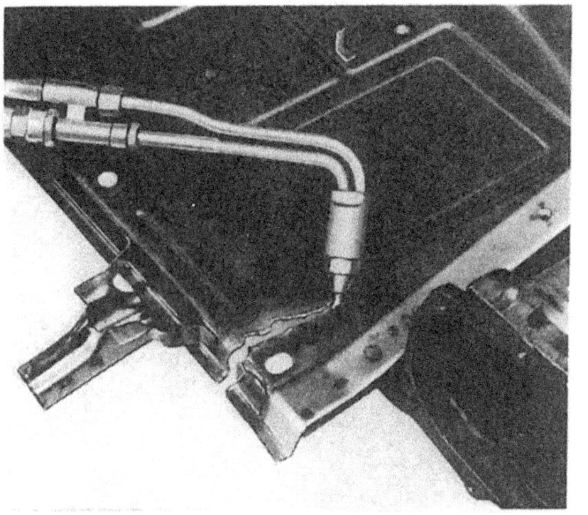

4 - Using a torch, cut floor plate along welding flange of front cross member over a length of 4 in. (100 mm). The moulded edge (arrow) and the welding flange of the front cross member must not be damaged.

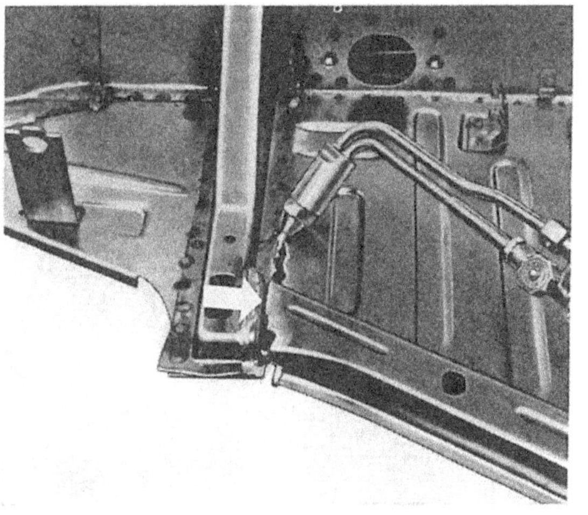

5 - Turn frame over.

6 - Cut floor plate along welding flange of front cross member, using a compressed air hammer and a type 1 chisel.

7 - Cut floor plate along welding flange of floor end plate, using a compressed air hammer and a type 1 chisel.

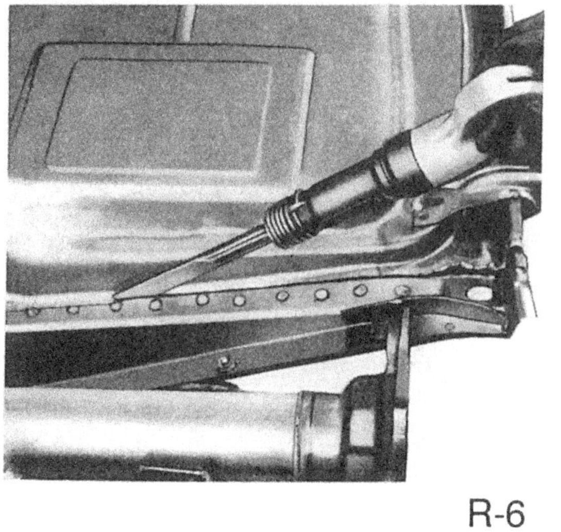

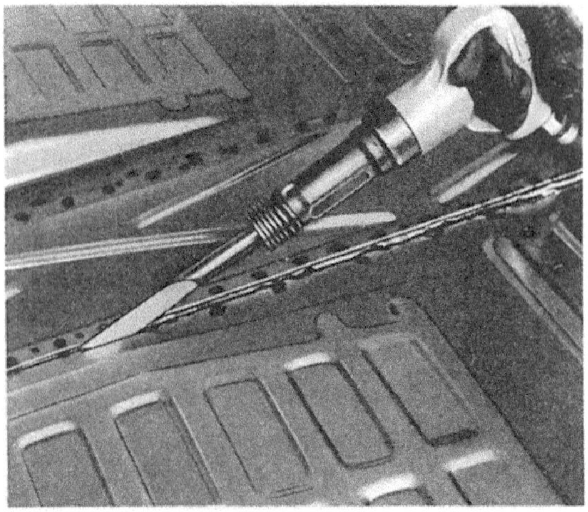

8 - Cut floor plate along welding flange of frame tunnel, using a compressed air hammer and a type 1 chisel.

9 - Smoothen and clean all welding flanges with a hand grinder.

10 - Turn frame over.

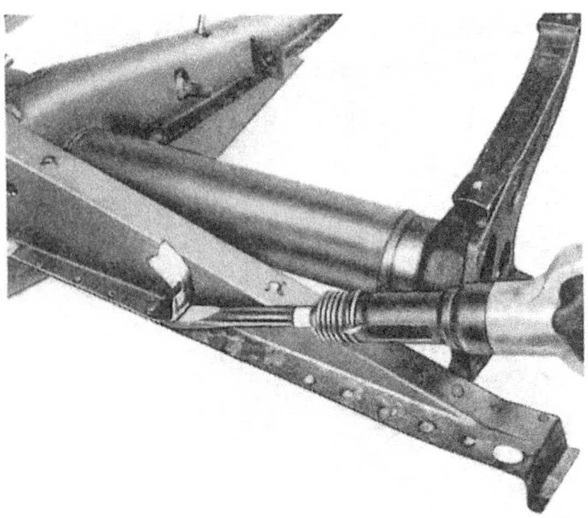

11 - Remove floor plate remnants from welding flange of floor end plate, using a compressed air hammer and a type 2 chisel.

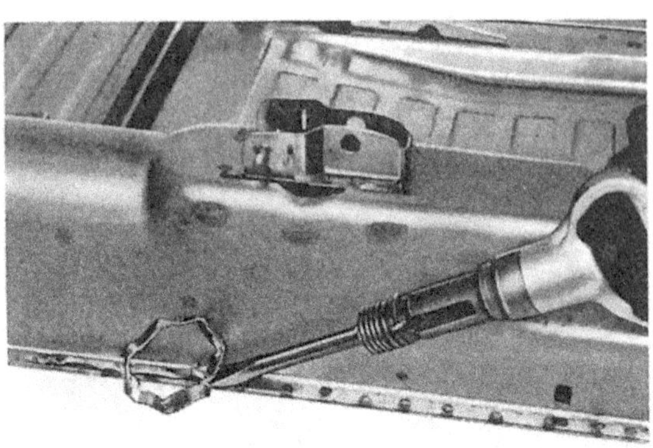

12 - Remove floor plate remnants from welding flange of frame tunnel, using a compressed air hammer and a type 2 chisel.

13 - Remove floor plate remnants from welding flange of front cross member, using a compressed air hammer and a type 2 chisel.

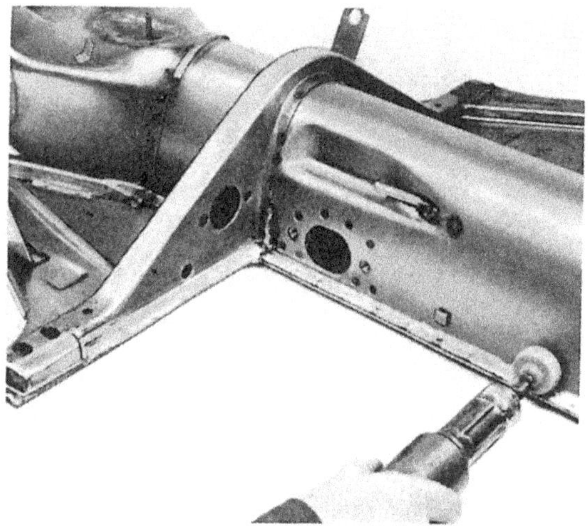

14 - Straighten all contact surfaces and welding flanges, then smoothen and clean them with a hand grinder.

Installing Floor Plate

1 - Degrease floor plate to be installed.

2 - Place floor plate into frame and clamp to floor end plate and to front cross member.

3 - Spot weld floor plate to frame tunnel at five locations, the welds being at 4 in. (100 mm) intervals and starting at the floor end plate.

4 - Spot weld floor plate to frame tunnel at five locations, the welds being at 4 in. (100 mm) intervals and starting at the front cross member.

5 - Spot weld floor plate to front cross member at 4 in. (100 mm) intervals.

6 - Spot weld floor plate to floor end plate at 4 in. (100 mm) intervals.

7 - Spot weld floor plate to frame tunnel at 4 in. (100 mm) intervals over whole length of tunnel.

8 - Prepare floor plate between the tacked locations on the welding flange for spot welding.

9 - Spot weld floor plate to front cross member, frame tunnel and floor end plate at .8 in. (20 mm) intervals.

10 - Electrically weld lateral part of floor plate to front cross member.

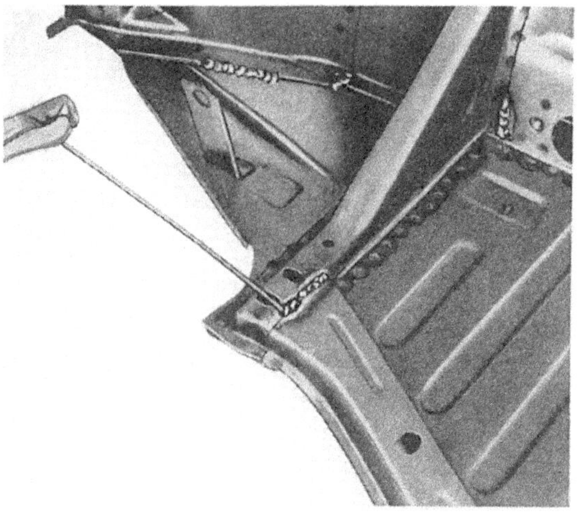

11 - Solder guide tube for rear foot level heating control cable to floor plate.

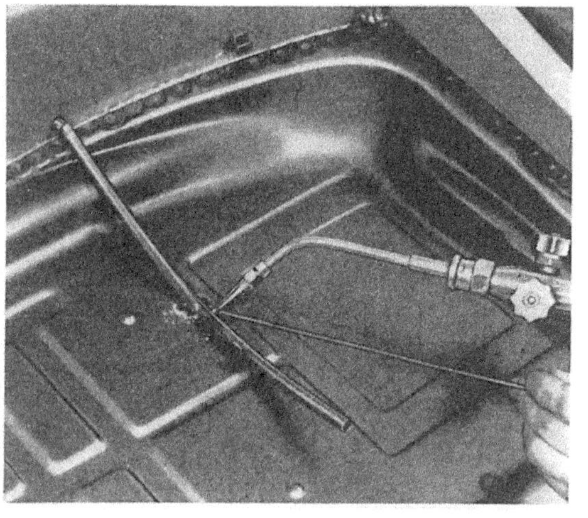

12 - Turn frame over.

13 - Electrically weld floor plate to spot welded flange of front cross member, frame tunnel and floor end plate over a length of .4 in. (10 mm) every 4 in. (100 mm).

14 - Seal all welds with genuine VW sealing compound D 17.

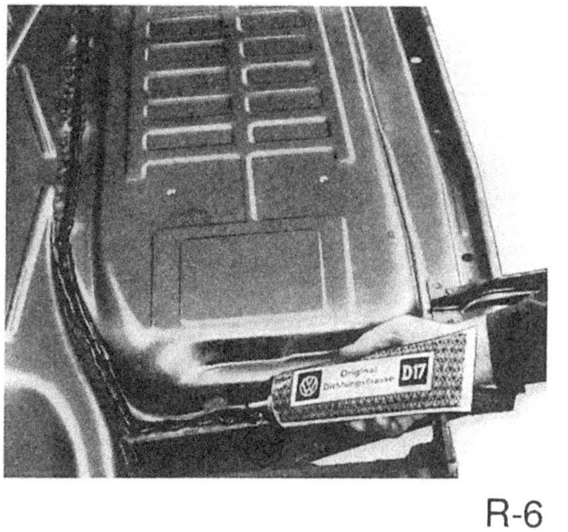

15 - Turn frame over and seal this side.

16 - Paint floor plate.

17 - Install insulation material.

THIS PAGE INTENTIONALLY
LEFT BLANK

3/66

Clutch Cable Guide Tube

If a clutch cable guide tube becomes detached at any point, it can be reattached as follows.

Cutting and welding work requires a high degree of skill. This work can, therefore, be carried out only by workshops who employ skilled welders.

Repair instructions:

1 - Remove seats, rear seat cushion and floor mats.

2 - Pull front and rear hoses off the fuel line. Blow out fuel line and frame tunnel to remove any combustible gases.

3 - Remove clutch cable and boot from guide tube.

4 - Determine where the guide tube has become detached.

A - Rear attaching point loose

1 - Remove cover plate for gearshift rod coupling. Weld guide tube to frame.

B - Center and front attaching points loose

1 - Cover windows and interior trim to prevent damage by flying sparks.

2 - Detach fiber board on right side and bend back to left side.

3 - Cut two sides of a triangle shape, 2 in. (50 mm) long, into the frame tunnel at distances a—c or b—c with a cutting disc.

Dimensions a - 7.87 in. (200 mm), b - 3.15 in. (80 mm), c - 5.9 in. (150 mm)

4 - Bend the resulting triangle back.

5 - Insert the welding electrode through the opening and make a weld.

Note:

The welding electrode must be placed on the frame insert or the clip and then drawn on to the guide tube. The guide tube must not get too hot or burn through as otherwise the clutch cable will chafe through at this location. Then the frame would have to be replaced. The guide tube is brazed to the clip in the center of the frame tunnel. In some circumstances it can be reattached at this location by brazing.

6 - Make sure the welds are firm and check by pushing a clutch cable through the guide tube.

7 - Bend the cut out triangle into position and weld it electrically to the frame tunnel.

Important

All welding work should be carried out with electrodes leaving very little slag (.1 in./2.5 mm diameter).

8 - Grind and paint welding seams.

9 - Install clutch cable with sufficient grease.

10 - Carry out other operations in reverse sequence.

C - All attaching points loose

1 - Move guide tube to its original position so that the welds on the guide tube coincide with the welds of the attaching points.

2 - Weld guide tube first at the front, then at the center and at the rear.

Note:

These instructions supersede those given in section R 12-1, of the Workshop Manual for VW Passenger Cars 1958, Volume 2.

Transmission Mounting Plate

When a vehicle is crashed into from the rear, the transmission mounting plate can, in unfavorable cases, be bent or become partly detached from the cross tube.

As, for construction reasons, the mounting plate must be installed accurately when replacing it, — an offset or an incline to the vertical plane lead to gearshift difficulties and transmission defects, — a fixture has been designed which facilitates the installation of a new mounting plate.

Repair work with this fixture can be carried out on all Models from Chassis No. 3 192 507.

The fixture will shortly appear in the publication "Workshop Equipment" and can be manufactured locally.

Repair instructions:

1 - Raise vehicle.

2 - Remove engine and rear axle.

3 - Ensure that the brake line from the T-piece to the right rear wheel is not on top of the frame fork. If necessary, remove brake line.

4 - Separate mounting plate from cross tube with a torch, then remove welding seam remnants.

5 - Screw new mounting plate to fixture.

6 - Place fixture on to cross tube and insert the aligning pins into the holes of the spring plate supports.

7 - Align fixture so that the distance (dimension a) between the adjusting screw and the inside face of the frame fork is equal on left and right.

8 - Tack mounting plate to the cross tube.

> **Important**
> All welding must be done electrically.

9 - Check whether the distance between adjusting screw and frame fork has altered.

10 - Weld mounting plate to cross tube. Welding runs must be made inside and outside the plate.

11 - Remove slag from welding runs. Paint mounting plate and cross tube.

12 - Carry out the remaining operations in reverse sequence.

THIS PAGE INTENTIONALLY
LEFT BLANK

Replacing the Fuel Line

The fuel line is held in the tunnel by three retainers which also hold the conduit tubes for the heating cable, the choke cable and, in the Standard Model, the rear brake cable. At the rear below the opening in the frame fork it is also clamped to the left-hand heating cable conduit tube.

There are two alternative methods of replacing the fuel line.

A - The damaged line is replaced; for this the body has to be removed.

B - The damaged line is not removed; a new line is simply run along the outside of the tunnel.

A - Replacing the fuel line

1 - Put the car on trestles.

2 - Remove the wheels.

3 - Remove the seats and floor mats.

4 - Remove the battery.

5 - Remove the gear shift lever, spring and stop plate.

6 - Remove body.

7 - After removing the locking pin take off the knob of the heating cable.

8 - Remove the tunnel carpets.

9 - Remove the brake push bar and push-rod of the mechanical brake.

10 - Remove hand-brake lever and ratchet assembly.

11 - Take off the frame fork inspection cover.

12 - Remove gearshift rod.

13 - Pull gearshift rod out forwards.

14 - Remove pedal cluster.

15 - Detach fuel hoses and rubber grommets from fuel line.

16 - Use a small torch to cut through the fuel line by the hole for the pedal cluster. The torch should be held at an acute angle to the pipe to prevent drops forming at the rear end of the line.

Caution!
Before using the torch blow through the tunnel and fuel line with compressed air.

17 - Pull out the front end of the cut fuel line.

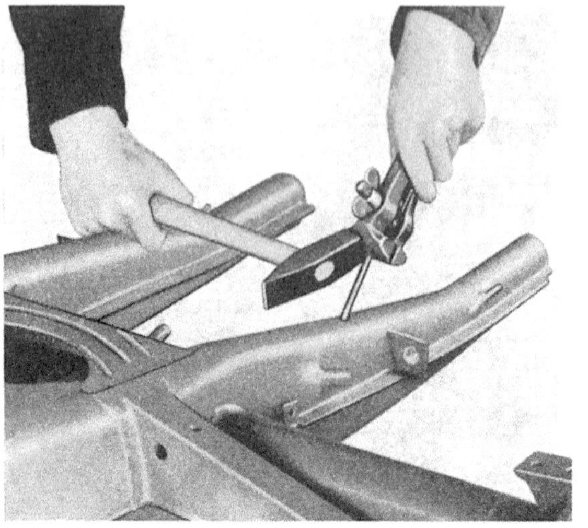

18 - Clamp the rear end of the fuel line in a hand vice where it emerges from the fork, and drive the line out by tapping the vice with a hammer.

Installation

When installing, the following points should be observed:

1 - Check that the new line is quite free of blockage.

2 - Insert the new line through the opening in the front plate of the frame head as far as the first retainer on the frame insert for the pedal cluster in the tunnel.

With Standard Model frames the line must be guided under the clip in the frame head for the return spring of the brake push bar.

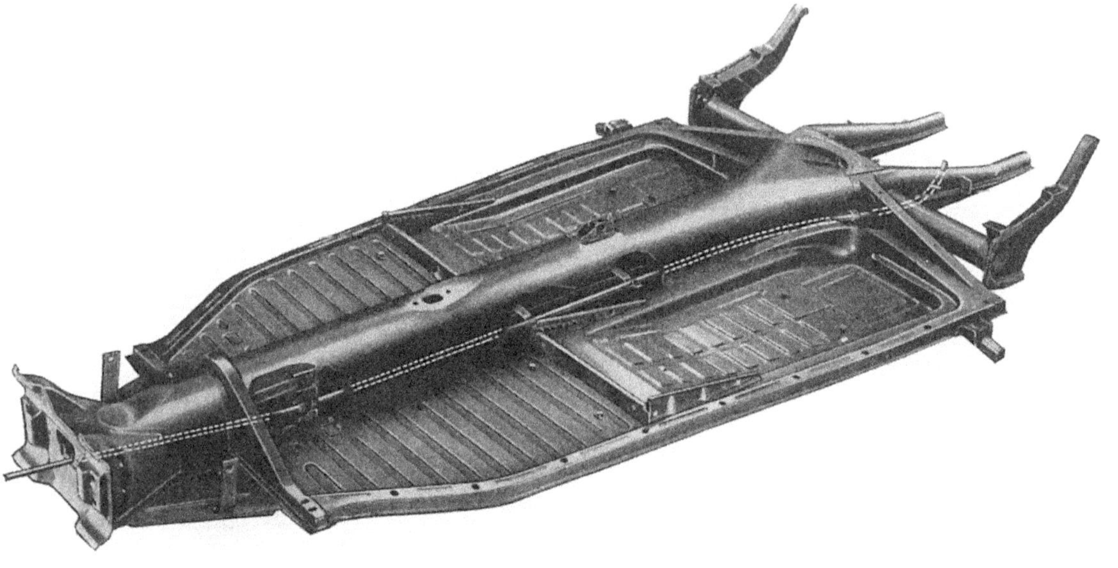

3 - By lightly tapping the end of the fuel pipe work it through the three retainers in the tunnel until it is just visible through the hole in the frame fork. To insure that the line runs straight it is advisable for an assistant mechanic to watch its progress down the tunnel, using the openings for the pedal cluster, the gearshift assembly and the hand brake lever.

4 - Insert a steel wire 4 mm (1.6″) in diameter through the opening in the fork provided for the fuel line and through the clip which holds the fuel line and the left-hand conduit of the heating cable, and let it enter the rear end of the fuel line.

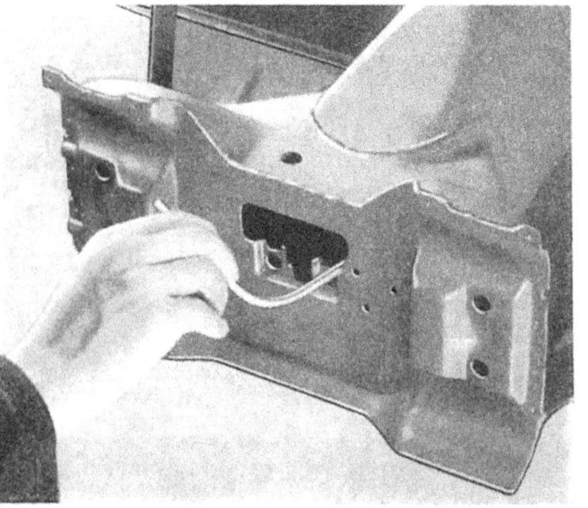

5 - Tap the fuel line still further with great care until it projects about 40 mm (1.5″) out of the frame fork.

6 - Bend about 85 to 90 mm (3.35—3.54″) of the front end of the fuel line, projecting from the frame head, upwards at an angle of 80° to the horizontal.

7 - Clamp the end of the line projecting from the frame fork in a hand vice. To prevent the vice from squeezing the line leave the steel wire in position.

8 - Using a hammer tap the rear end of the line backwards until the bent front end can be inserted in the hole provided in the frame head.

On Standard Model frames the line in the frame head must pass under the right hand rear (in driving direction) brake cable conduit.

9 - With a hacksaw cut off the ends of the fuel line until it projects about 50 mm (2″) at the front and about 80 mm (3″) at the rear, remove burrs inside and outside.

10 - Slide on the rubber grommets, blow through with compressed air and check that the line is airtight.

11 - When assembly is finished adjust the brakes and road-test the car.

B - Fitting an additional fuel line

1 - Put the car on trestles.

2 - Remove right hand front wheel and left hand rear wheel.

3 - Remove seats and floor mats.

4 - After driving out the locking pin remove the heating knob.

5 - Remove the tunnel carpets.

6 - Drill a hole 15 mm (.6″) in diameter through the front cross member, and another through the left-hand frame end plate.

7 - Make six clips from steel sheet 0.8 mm (1/32″) thick and 10 mm (3/8″) wide, then weld them as shown on the sketch below to the frame head, the right hand floor plate and the frame fork.

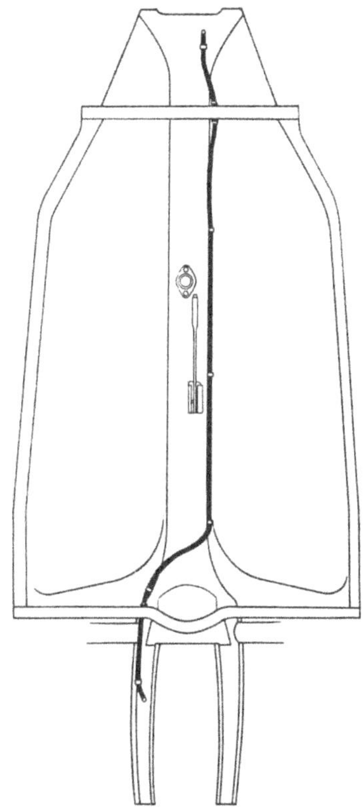

8 - Slide two rubber grommets over the ends of the line, bend it cold and place it in position.

Run the fuel line over the tunnel just before the frame fork, taking care that it cannot be damaged by the back seat.

9 - Fit a third grommet from outside in the hole through the front cross member.

10 - Blow through the fuel line with compressed air and fit the fuel hoses.

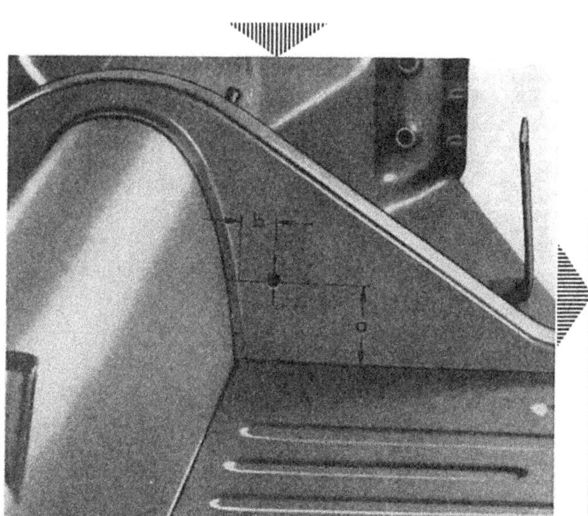

a - 60 mm (2.36″) b - 40 mm (1.57″)

a - 60 mm (2.36″) b - 15 mm (0.59″)

THIS PAGE INTENTIONALLY
LEFT BLANK

Replacing the Heating Control Conduit

It is only necessary to replace the heating control sleeve when its thread is damaged.

Sequence of Operations

1 - Put the car on trestles.

2 - Take out the front seats.

3 - Remove the gearshift lever with spring and stop plate.

4 - Remove the heating controls and cable.

5 - Remove the tunnel carpets.

6 - Cut through the sleeve about 20 mm (.8″) below the upper edge. Smooth the cut surface with a file and chamfer.

7 - Cut a new sleeve to the same length, smooth the surface and chamfer.

8 - With the aid of a drift, to be made locally in accordance with the sketch below, place the new sleeve on the remaining end of the tube.

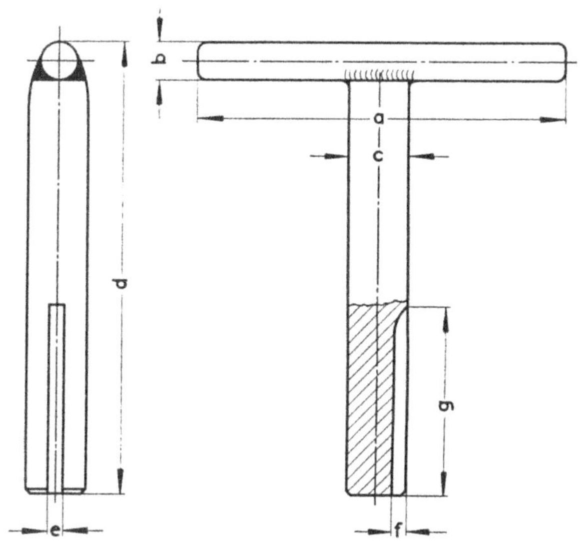

a - 105 mm (4.15″)
b - 10 mm (.4″)
c - 15.8 mm (.62″)
d - 120 mm (4.7″)
e - 4 mm (.16″)
f - 3 mm (.12″)
g - 50 mm (2″)

9 - Join the sleeve to the tube all round by acetylene welding and smooth the weld seam.

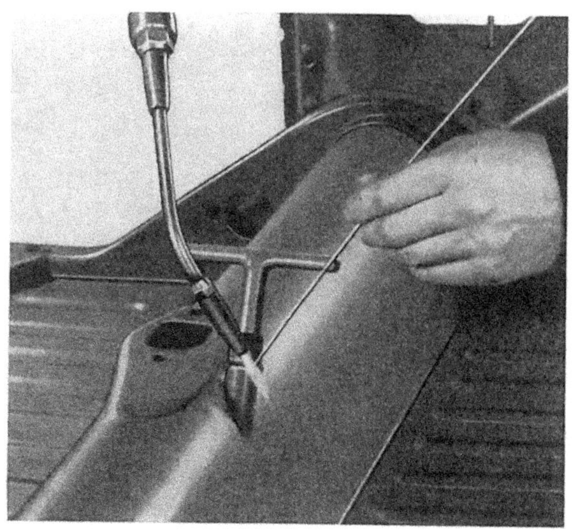

10 - Repaint the sleeve and frame tunnel.

THIS PAGE INTENTIONALLY
LEFT BLANK

Replacing the Shift Rod Guide

The shift rod guide is secured in the tunnel with six spot welds. If the guide has to be replaced a new type guide should be installed.

Note:

From Chassis No. 3 140 046 the shift rod guide with leaf spring was replaced by a guide with a plastic bush.

Removal

1 - Put the car on trestles.

2 - Take out the front seats.

3 - Remove the gearshift lever with spring and stop plate.

4 - After driving out the locking pin remove the heating knob.

5 - Remove the tunnel carpets.

6 - Remove the push bars of the mechanical brake.

7 - Remove the shifting rod.

8 - Drill out the spot welds with a 4 mm (5/32") drill.

If the spots are not clearly visible on the tunnel, rub the area down with fine emery cloth first to remove the paint.

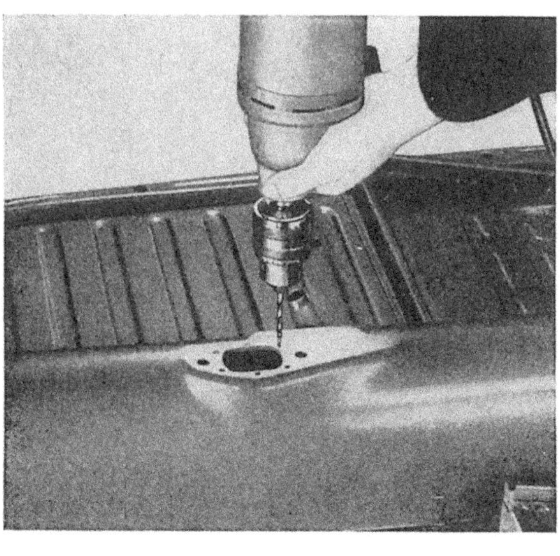

9 - Detach the shift rod guide from the tunnel with a cold chisel.

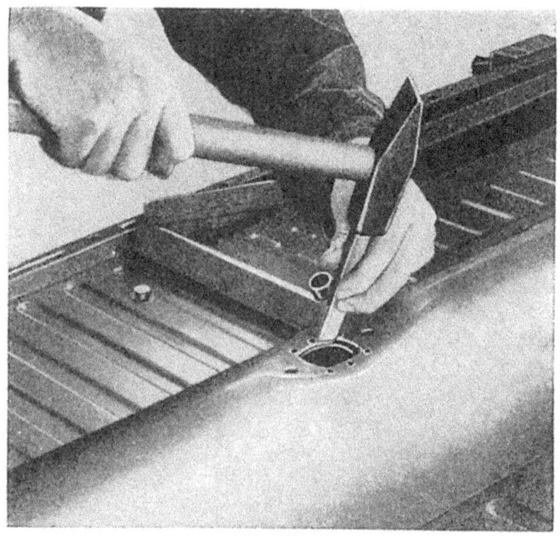

10 - Remove the guide as far as possible through the hole in the tunnel, finally cutting it to pieces with a torch until it can be removed altogether.

Important

Before starting work with the torch blow through the tunnel and fuel line with compressed air.

Installation

When installing he following points should be observed:

1 - Smooth the inner surface of the tunnel in the area where the shift rod guide is fitted, to ensure that the new guide fits properly to the tunnel.

2 - Pass a piece of wire through the guide hole and push it forward to the frame head cover opening. Attach new guide to wire and pull it into the tunnel.

3 - Pull guide up into position, locate correctly and secure with two hexagon head M 8 × 20 screws.

4 - Weld up the holes drilled in the tunnel. This also welds the guide to the tunnel.

5 - Trim the welds with a hand grinder.

6 - Repaint the tunnel.

7 - Apply Universal Grease to the gear shift rod before reinstalling.

Clutch cable conduit

In order to prevent the clutch cable conduit in the frame tunnel from working loose even under unfavourable operating conditions, the following modifications were introduced from Chassis No. 2115600:

1 - The end of the clutch cable conduit is now bell-mouthed instead of obliquly cut and is shortened in front by 5 mm (0.2″). This modification allows a better welding of the conduit tube to the insert for the pedal cluster.

2 - The intermediate conduit tube retainer in the tunnel has been modified. The clutch cable conduit is positioned 10—15 mm (0.4″—0.6″) further towards the centerline of the frame tunnel and is welded to the retainer at full width.

3 - The opening for the accelerator, choke and clutch cable conduit at the frame tunnel rear end has been reshaped at the left hand top corner to conform with the contour of the clutch cable conduit thus allowing a more efficient welding.

Note

There is no need for the frame to be replaced if the clutch cable conduit has worked loose at one of the aforementioned points.

In order to reweld the conduit to the insert for the pedal cluster the body has to be taken off and the pedal cluster removed. The welding can then be carried out with a suitably bent welding torch through the hole for the pedal cluster and can be observed from the frame head.

The conduit can be rewelded to the intermediate retainer in the tunnel through the opening for the handbrake after the handbrake lever and the shifting rod have been removed without removing the body.

The rear end of the conduit is accessible by removing the rear seat cushion and the inspection cover.

Caution

The frame tunnel and the fuel line are to be carefully blown out with compressed air prior to starting welding jobs to avoid an explosion of fuel vapor.

THIS PAGE INTENTIONALLY
LEFT BLANK

1 - VW Special Service Tools

VW 400	Repair Press 51
VW 401	Thrust plate
VW 409	Punch
VW 410	Punch
VW 411	Punch
VW 412	Thrust Disc
VW 415	Tube 75 mm dia.
VW 416	Tube 38 mm dia.
VW 419	Tube 32 mm dia.
VW 421	Tube 28 mm dia.
VW 431	Thrust pad 16.5/28 mm dia.
VW 432	Arbor 50 mm dia.
VW 433	Thrust pad
VW 435	Arbor
VW 437	Guide Pin (Tapered)

2 - VW Workshop Equipment for Local Manufacture

VW 670	Frame Head Assembly Jig
VW 671	Frame Alignment Gauge
VW 672	Frame Alignment Gauge

3 - Normal Hand Tools

Screwdriver 6 mm
Screwdriver 8 mm
Combination pliers
Water pump pliers
Side cutting and pipe — grip pliers
Handvice
Flat chisel
Cape chisel
Center punches
Punch 2 mm
Punch 4 mm
Fitter's hammer 300 g
Fitter's hammer 500 g
Triangular scraper
Flat scraper
Flat file, 180 mm long
Round file, 180 mm long
Half round file 180 mm long

Open-end Wrench 14 mm
Open-end Wrench 17 mm
Box Wrench 14 mm
Box Wrench 17 mm
Wire brush
Oil can
Can for derusting fluid
Grease tin
Scriber
Caliper rule 300 mm long, with 1/50 vernier
Tap M 12 x 1.5
Tap wrench, size 2, adjustable
Die plate M 12 x 1.5
Die stock, size 2
Twist drill 4.0 mm
Twist drill, 10.0 mm
Twist drill, 12.0 mm
Inspection lamp with cable and plug
Electric drill

4 - Miscellaneous Workshop Equipment

Frame repair jig VW 1025
Spot welding equipment
Spot-welding gun, extension 650 mm (26″)
Arc welding equipment
Surface grinder
Acetylene or oxygen supply with
Torches of various sizes

Contents: A

THIS PAGE INTENTIONALLY
LEFT BLANK

Painting with Genuine VW Paints of the LKL and L Groups

General

All Volkswagen vehicles are painted on a four coat system using synthetic resin priming and finishing paints. The primer is oven dried at 170° C for 12—16 minutes and the finish coats at 130° C for 25 minutes.

The advantages of this paint build-up and this paint quality lie in the high gloss and weather-resistant properties, impact resistance and absence of fading as well as the insensitivity to normal cleaning agents such as benzine, kerosene, mild paint and enamel solvents and so on.

How these properties are achieved during paint repairs and why this is necessary, is described in the following sections.

In recent years, highway authorities in all countries where the roads are covered with snow and ice at certain times of the year, are using gritting or thawing agents containing salt in steadily increasing quantities. It has been found that only properly applied synthetic resin paints can resist the increased stresses continuously. However, not only in countries with winter climates but also in coastal areas and everywhere where the roads are treated with saltbased agents all the year or for certain periods, the paintwork of all vehicles is exposed to severe chemical influences.

As a general rule, cellulose or cellulose based materials tend to be severely affected by the action of the salt, particularly when the paint surface is damaged by stone impact or scratches.

Even a finish coat of air-drying synthetic resin cannot prevent the underneath layers of cellulose primer being attacked by the salt solution and the paint undermined in a very short time. This results in paint peeling and severe corrosion. It is stressed, therefore, that first-class paintwork can only be obtained by the use of LKL group primer and finish paints (pure synthetic resin type).

Cellulose-based primer and finish paints should only be used where the paint is not exposed directly to the weather or to stone damage. Even here it is safer and better to use a pure synthetic resin build-up.

Spare parts (complete bodies, fenders, hoods and other body parts) are given a coat of synthetic resin dip primer with a water thinnable basis at the factory. This coat forms a high grade primer base for the paint layers which follow. The winter of 1962/63 in particular, showed that this primer has a higher resistance to salt solutions than any conventional synthetic resin primer. It must of course be assumed that any damage incurred during transport or storage, especially marks which go down to the bare metal, are given proper preparatory treatment. (See page A-1/6.)

The common practice of applying the finish coats straight on to the dip primer regardless of primer condition can never produce durable paintwork as the maximum paint layer thickness achieved in this manner is about 30—35 my (mm). This is why our vehicles are painted by the four coat method which gives layer thicknesses of 80—110 my (mm) and this guarantees good resistance to weathering. The primer used for the vehicles is also used for the spare parts.

The paints to be used for repair work are:

a - air drying synthetic resin paint of the LKL group (referred to as "LKL paint" in following text)

for limited usage:

b - cellulose based paint of the L group (referred to as "L paint" in following text).

1 - If applied properly and in accordance with the specified build-up, the LKL paint gives the same paint finish and the same quality as is obtained at the factory. The main difference when compared with the factory paint is in drying times. LKL finish paint is dust-free after 3—4 hours air drying, depending on the air temperature and is tack-free after about 12 hours. In a drying room or oven with an object temperature of 80° C the same degree of dryness is reached after 30 minutes.

2 - An optically satisfactory paint repair can only be obtained when the color shade of the paint used for patching up is exactly the same as that of the existing body paint.

Note:

It is almost impossible to get absolute matching of paint shades between different deliveries and between individual paint types (oven drying to air drying). It is, therefore, not always possible to avoid tinting or toning of the patch-up paint.

Further elaboration of these circumstances, such as the influence of the spraying process, natural ageing of original paint, the impossibility of measuring exact shades and so on would take too long here.

3 - The L group covers cellulose combination paints which have a very short drying time and good polishing properties but also have completely different ageing, gloss and fading characteristics to the LKL paint. This means that L paints are only suitable for protected areas and areas which are not readily visible. Furthermore, L paints must not be applied to areas which come into contact with sliding roof covers, of PVC material as this will set up a flow of softener between paint and roof material and the paint will remain soft and smeary.

Genuine VW Mixing Paints are available for LKL and L paints so that it is quite easy to match the Genuine VW paint when necessary.

The following priming materials are available:

a - To improve adhesion and to neutralise the surface in place of the phosphating and thus prevent the formation of blisters at a later date

Genuine VW Wash Primer L 143 and
Genuine VW Additive L 144

should be used.

Note:

The wash primer is **not** a cellulose priming material but a primer-neutraliser agent which should always be used when the paint has been sanded through to the metal.

b - For large areas and complete resprays in conjunction with wash primer L 143/L 144

Genuine VW Primer LKL 140

should be used to even out the sheet metal surface and as a general quality coating for the complete respray.

c - In conjunction with L 143/L 144 to give good binding and LKL 140 primer as quality coating

Genuine VW Rapid Filler LKL 145

should be used when painting spare parts and for complete resprays when the old paint has been removed. It then acts as an intermediate layer and gives the paint a good depth.

d - For paint repairs on protected areas, mainly body interior and trim parts

Genuine VW Filler L 141

should be used together with L 143 / L 144 primer when there are bare metal areas, to give smooth blending into the original paint.

e - To even out small depressions, limited to small areas, but not on bare metal

Genuine VW Universal Filler, L 146 (grey)

should be used.

Basic Notes

1 - The spraying booths and drying rooms must be absolutely clean and free of dust in order to ensure good paint finishes. It is essential to keep the room free of drafts which cause excessive air movement, to prevent people walking through the room unnecessarily and to install an effective air extraction system. The floor should be kept damp and the room temperature at 18—25° C when spraying is taking place. The compressed air should be free of oil and water and constant in pressure. The spray gun jet size should be 1 mm. In addition to the normal items of equipment, a bank of infra-red drying lamps on a mobile support should be available. The compressor should have a capacity of 400 liters a minute at 6 kg/cm² (14 cu.ft at 85 psi).

2 - The success of the painting operation depends largely on the cleaning of the body or part when it is being prepared for painting. All traces of sanding deposits should be washed off with plenty of water. To avoid sanding water faults, the water should not be dried off in an oven. The best method is to dry the surface thoroughly with a leather and then blow all joints and seams off with high pressure compressed air. This removes all dust, dirt and water completely and prevents the paint finish from being spoiled by moisture and dirt reappearing at these places.

3 - To ensure that the color matches correctly, the paint and any tinting paint used must be very thoroughly stirred before use. There must not be the slightest trace of solid matter on the bottom of the container. The paint must only be thinned with the correct thinner in every case (see also page A-1/9).

4 - A paint cup with a 4 mm drain hole and a stop watch are required to mix the paint to the correct spraying viscosity which has an important bearing on the success of the paint job, particularly when using LKL paints.

5 - A pressure paint container with a capacity of 5 liters should always be used when completely respraying a body. The paint pressure is 1.2—1.4 kg/cm² (17—20 psi) for all materials to be sprayed (primer and finish coats). This allows the actual spraying to be carried out in the shortest possible time which is a decisive factor in obtaining a flawless finish.

Further advantages are:

a - The spraying can be completed without interruption.

b - The flow of paint to the spray jet remains constant and permits rapid uniform work so that the paint layers are of equal thickness and free of texture.

c - It is possible to employ two painters as the pressure tank is fitted with connections for two guns. The cup type guns should be reserved for partial spray jobs and repair work.

6 - Cellulose paint spray tends to burn on contact with synthetic resin paint spray. As it is not always possible to avoid spraying both materials alternately in the same booth, it is essential that the booth, filter system and air extraction ducts are cleaned thoroughly at least every weekend. Accumulations of paint deposits should be avoided if the parts concerned are not kept constantly wet with water (depending on system).

7 - LKL paints should be dried at a temperature of 80° C as this enables the existing drying plant to be used to maximum capacity. At this temperature the part is hardened to the point where it can safely be handled in 30 minutes.

Note:

When drying paint repairs at temperatures of about 60 to 80° C, it may be found that flats form on the contact surfaces of the tires under certain conditions. These flats then either remain in the tire or take a long time to disappear. To prevent this from happening, note the following points:

Increase tire pressures to 3—4 kg/cm² (42—56 psi) before painting vehicle.

Do not continue drying process longer than absolutely necessary.

Do not dry at more than 80° C.

Cover the tires up in drying rooms with direct infra-red heaters.

After painting and when the tires have cooled down, the pressures must be reduced to normal.

The following table shows the relationship between drying time and drying temperature, assuming the same degree of hardness in each case:

Drying temperature in °C	Drying time
80	30 minutes
70	60 minutes
60	80 minutes
50	2 hours
40	4 hours
20	approx. 10—12 hours

Note:

If the times for a given temperature are exceeded for any reason it will not normally affect the finish. In the case of very light colors, however, the drying times in the 70—80° C range should not be exceeded as otherwise a slight yellowing is liable to appear.

8 - When drying a paint job on a complete vehicle, the temperature must never rise above 90° C as otherwise the various rubber and plastic parts as well as the adhesive and sealing materials used will be damaged.

The following parts are particularly susceptible to damage in the drying room and should be removed beforehand:

Speedometer
Clock } when fitted
Variode regulator }
Outside mirror
Sun visors
Plastic number and international registration plates
Turn indicator and rear light lenses

Various plastic accessories are also not sufficiently heatproof and should be removed to be on the safe side.

9 - The scratch resistance of the paint can be improved after the specified time in the drying room or the air drying time by treating the painted areas with Genuine VW Preservative L 190. The preservative is applied thinly with cotton wool and allowed to dry for 20 minutes. The surface is then polished with cotton wool or a soft cloth until iridescent colors can no longer be seen when looking across the surface at an angle.

10 - LKL paints can only be polished properly when they are thoroughly dry. If oven-dried, this will take 30—40 minutes at 80° C plus 12 hours air-drying or 48 hours if dried solely by air. If the paint coat is faulty due to excessive dust or incorrect application it is useless to attempt to polish out the defect immediately. If it is not possible to wait until the paint is completely dry, the area must be resprayed after rubbing lightly with very fine wet sanding paper. This coat must be as thin as possible.

Paint Application

1 - When using LKL paints to repair damaged parts or areas, the complete part or the complete area must be painted. Only sharply defined areas such as the door lower panel, front side panel and complete roof panel etc. should be sprayed and the surrounding parts masked with tape and paper.

This type of repair does not produce patches with fairly sharp borders and the slightly larger outlay on paint and masking paper is offset by the saving of polishing time and material.

If drying rooms are not available, the drying time of the LKL paint can be reduced considerably by use of mobile banks of infra-red lamps. It is important to position the lamps at such a distance from the paint surface that the temperature of the parts to be dried cannot exceed 70° C.

2 - Skilled and experienced painters can touch up spots with LKL paints satisfactorily, particularly on large areas such as the roof.

The damaged area is first feather-edged well with No. 360 sanding paper. If a large area is rubbed down to the bare metal while doing this, a thin coat of Wash Primer L 143/L 144 should be applied and allowed to air-dry for 15—20 minutes before spraying on a coat of VW Surfacer L 141. If filling is required, it should be applied when the wash primer has dried using VW Filler L 142.

The drying time for the filler is 15—20 minutes for each layer of filler applied. The filler is rubbed down with No. 360 paper and given a coat of L 141 surfacer. When the sprayed-on surfacer has dried for 30—40 minutes, rub down carefully with No. 600 paper and soapy water, covering an area about the width of a hand larger than the damaged spot or the area sprayed with surfacer. The area is then thoroughly dried and the spot carefully sprayed with LKL paint. The sanded area should not be covered completely with paint but should be sprayed so that a ring about one inch wide remains visible round the edge. Immediately afterwards, the border area is given a mist coat with a mixture of one part paint and two parts thinner.

When the paint has dried thoroughly — an infra-red lamp must be used after repairs of this nature — the area is polished in with VW Polishing Paste L 180.5 or treated with VW Polishing Liquid L 170.

3 - The method described under point 2 cannot be used for metallic finish paints. In such cases, use the method detailed under point 1. Owing to the changing properties of metallic finish paints, it is for more difficult to get good results with these paints than with normal paints. The reason is that even a different thickness of coat or a difference in the color of the primer coats gives a different changing effect and thus a different shade of color. For partial paint repairs, all the skill of the painter is required to ensure that the tinting of the paint and the spraying of the test plate are successful in giving the color effect of the original paintwork. To obtain the same undercoats, the area for painting or the test plate should be sprayed first with VW Filler, grey LKL 145. This coat should be applied at a viscosity of 21—22 seconds in the 4 mm cup (DIN). To avoid floating in the paint layer, a retatively dry spray should be used. This is achieved by increasing the distance between spray gun and surface and should change over to the finest mist spray as the thickness of the paint coat increases. The LKL 161 thinner which was specially developed for metallic finishes, should be used.

Example

A. Painting prime coated spare parts
Bodies
and body parts

1 - **Sanding the primer**

a - Wet sand with No. 360 paper. Paint damage incurred in transport and the resultant rust spots should be feather-edged or sanded off completely.

b - Clean the part with running water (hose brush) and dry with a leather. Blow off with clean compressed air and, if a body is involved, dry in a drying room.

2 - **Neutralizing**

a - Spray all bare metal areas lightly with VW wash primer L 143. If the primer was removed extensively by the sanding and sanding marks are through to the metal, the entire area should be sprayed with wash primer L 143.

Note:

The wash primer is mixed in the proportion of

2 parts Wash Primer L 143
1 part Additive L 144

and then thinned with L 160 thinner to a spraying viscosity of 15—16 seconds in a 4 mm cup (DIN).

Important!
As this solution only remains usable for a maximum of 48 hours, only mix enough for one day at a time.

b - Air dry for 15—20 minutes. The wash primer should not be sanded as the layer is very thin.

3 - **Priming and/or filling**

a - The items are then given 1 1/2 cross coats of VW Rapid Filler LKL 145 inside and out.

b - Data:

Thinner:	Genuine VW Thinner L 160
Viscosity:	18—20 seconds at 20° C in 4 mm cup (DIN)
Spraying pressure:	4—5 kg/cm² (57—71 psi)
Jet size:	1 mm
Drying time:	2—2 1/2 hours air drying or 40 minutes at 60—75° C or 30 minutes at 80—85° C in a drying room.

4 - Sanding the filler

a - Wet sand with No. 400 paper. A satisfactory paint finish depends on texture and surface sanding.

b - Clean with running water (hose brush) and dry with leather. Blow off with clean compressed air and, in the case of bodies, dry in drying room.

5 - Painting

a - Spray parts inside and out — in the case of bodies include wheel housings and floor — with 1 $^1/_2$ to 2 cross coats of VW Synthetic Resin LKL paint to give good coverage, free of texture and runs.

b - Data:

Thinner:	VW Thinner LKL 160 (for metallic finishes LKL 161)
Viscosity:	approx. 22—24 seconds at 20° C in 4 mm cup (DIN)
Spraying pressure:	5—6 kg/cm² (71—85 psi)
Jet size:	1 mm
Coats:	1 $^1/_2$ to 2 cross coat (heavier application can cause sagging)
Drying time:	10—12 hours air drying or 30 minutes at 80° C in the drying room

Note:

The above example is not valid when climatic conditions are unfavourable (mainly high humidity and salt sprinkling during the winter months). In these cases, a four coat paint system should be used as follows:

a - Neutralizing — see point 2.

b - Intermediate primer as follows:

Spray parts inside and out, including wheel housings on bodies, with 1 $^1/_2$ cross coats of VW Spray Primer LKL 140.

Data:

Thinner:	VW Thinner L 160
Viscosity:	20—22 seconds at 20° C in a 4 mm cup (DIN)
Spraying pressure:	5—6 kg/cm² (71—85 psi)
Jet size:	1 mm
Drying time:	2 hours air drying or 30 minutes at 70° to 80° C in drying room

c - Wet sand with No. 360 paper, clean and dry off.

d - Apply primer — see point 3.

e - The next operations are as at points 4 and 5.

B. Complete respray without removing original paint

Note:

This can only be done if the old finish is free of damage such as hair lines, checkering, blistering or under-surface rust.

1 - Examine condition of old paint closely. Damaged or badly bonded areas of paint should be removed completely and feather-edged well.

2 - To remove all the wax used to clean the vehicle, clean the complete body well with benzine and sand with No. 360 or 400 paper until the old finish is no longer shiny.

3 - If cleaning and rubbing down can be carried out without exposing bare metal even at the edges, the finish coat can be applied as at para A, point 5 as soon as the water has been dried off.

Otherwise, the spots which have been rubbed through must be sprayed lightly with wash primer L 143/L 144 and the complete body then sprayed with 1—1$^1/_2$ cross coats of filler L 145 as at para A, point 4. If the rapid filler viscosity is set properly and the spraying carried out well enough to ensure a good filler coat, the wet sanding process can be replaced by dry sanding (trimming) with damp No. 400 paper.

Note:

When painting a prime-coated new vehicle, the prime-coated surfaces must be cleaned and prepared by wet sanding. The sanding should be continued until all traces of dirt and weathering due to transport and storage (chalking) are completely removed. After drying off the water, proceed with painting as at points 3 to 5 of example A.

If the primer is sanded through to the metal in small spots, possibly to remove edge rust, these spots must be treated with wash primer L 143 — as at para A, point 2 — before applying the L 145 filler.

C. Complete respray with removal of old paint

1 - Remove old paint down to the metal with chemical remover or by sanding.

2 - The following solutions are suitable for removing old paint and for derusting and neutralizing the surface at the same time:

a - Remover from Messrs. Glasurit-Werke, Hiltrup (Westf.), (Designation: Glasso Paint Remover 219/4) or from Messrs. Nobel & Co., Chemical Works, Hamburg-Eidelstedt-Ost, (Designation: Nobel Universal Paint Remover 1825 fl.) or from Messrs. Agep & Co., Horrem, Köln (Designation: Horral Paint Remover E 137).

When using chemical paint removers, ensure that all traces of paint, remover, rust, acid and water are removed thoroughly so that perfectly clean sheet metal is ready for painting.

b - Derust and neutralize with a combined degreasing and derusting agent, "Deoxydine Perfect" from Messrs. Collardin, Köln-Braunsfeld, or "Antox Special" from Messrs. Metallgesellschaft, Frankfurt/Main. Both solutions are supplied ready for use and should be applied to the metal uniformly with a brush or sponge. The solution should be allowed to work for 5—10 minutes until the rust is removed completely. The action of the solution can be accelerated on heavy rust deposits by rubbing the surface with steel wool. Ensure that the derusting solution does not dry on the metal. If there is a danger of this happening, another coat of solution should be applied.

The solution is removed with a cloth, brush or sponge soaked in benzine or ethylglycol. The surface is then dried with a clean, dry cloth and compressed air, paying particular attention to the corners and flanged seams.

After degreasing and derusting, the body must be primed as soon as possible to prevent the formation of surface rust again.

3 - Painting takes place as follows:

VW Wash Primer L 143/L 144
VW Spray Primer LKL 140
VW Surfacer LKL 145
VW Paint LKL . . .

The data for this work is the same as given in example A.

If filler is required, L 146 should be applied to the LKL 140 spray primer coat. Thick layers are built up applying several thin layers and allowing 15—20 minutes for each layer to dry thoroughly.

Note:

It is advisable to fill partly used tins of L 146 filler (grey) with water before closing them again. This will prevent the formation of skin on the surface of the filler while it is standing.

Paints, polishes and preservatives

All materials required for the painting operations are obtainable from the Parts Department of the Volkswagenwerk.

The following are available:

a - Priming materials	**Correct thinner**
VW Wash Primer + Additive L 143/L 144 VW Spray Primer, red-brown LKL 140 VW Rapid Filler, grey LKL 145 VW Surfacer, grey L 141	VW Thinner L 160
VW Universal Filler, grey L 146	
VW Patching Spray Primer red-brown L 140 VW Filler Paste grey-green L 142	VW Thinner L 160

Note:

Paintwork which is built up solely on priming materials with a cellulose basis does not possess the same degree of quality as paintwork carried out completely with synthetic resin materials, particularly in respect of resistance to weathering. The last two priming materials should, therefore, only be used for small area patch-up jobs.

b - Paints

1 - VW Synthetic Resin paints of the LKL group in all standard production colors	VW Thinner LKL 160
2 - VW Synthetic Resin paints of the LKL group in all standard production metallic effect colors	VW Thinner LKL 161
3 - VW Combination paints of the L group in standard colors	VW Thinner L 160

All the finishing paints listed are supplied in 1 kg tins. Tinting colors (VW Mixing Paints) must be ordered separately from the Parts Department. They are also supplied in 1 kg tins. The following tinting colors are available:

green	LKM 10	violet	LKM 33	wine red	LKM 55
dark green	LKM 11	black	LKM 40	yellow	LKM 60
yellow green	LKM 12	bright red	LKM 50	ochre	LKM 61
medium blue	LKM 30	ruby	LKM 51	yellow	LKM 62
dark blue	LKM 31	orange	LKM 53	red-brown	LKM 70
blue	LKM 32	red	LKM 54	white	LKM 80
				aluminium	LKM 90

The tinting colors with the designation letters LKM are used to tone VW synthetic resin paints (Group LKL). Tinting colors with the designation LM are available (colors as above) for toning VW combination paints of the L group.

c - Polishes and preservatives

VW Liquid Polish L 170
VW Polishing Paste L 180.5
VW Preservative L 190.

Note:

The L 170 and L 190 solutions are also available in 1/2 kg tins which are designated L 170.5 and L 190.5.

Care and cleaning of textile materials

Normally the interior trim should be cleaned with a vacuum cleaner or brushed with a soft brush. Grease and oil spots can be removed with spot remover. The spot remover must not be poured directly on to the material as this is bound to leave marks. The best way is to moisten a clean white cloth with the spot remover and rub the spot with the cloth, working with a circular movement inwards towards the center of the mark.

Care and cleaning of plastic materials with smooth or grained surfaces

Plastic materials have a dust-repellent surface and require no special preservative.

Normal dirt can be removed as follows:

a) With a soap solution made of water and any normal washing powder.

b) With a solution made of water and any normal plastic cleaner.

A soft brush will facilitate removal of dirt from grained surfaces.

Always deal with bad stains as soon as they are noticed. Suitable agents are given in the following table. Note that the various cleaners, particularly benzine, methylated spirits and thinner, should never be poured directly on to the surface to be cleaned as they will then penetrate into the seams and padding. These liquids should be used on a clean cloth. They should not be allowed to work on the surface of the material for too long as otherwise the dirt-repellent film will be destroyed.

After cleaning, the plastic material must always be rubbed dry with a soft cloth, especially in the seams.

Stains	Removal of	
	new stains	old stains
Oil or grease	Remove with dry, soft cloth, turning this several times. Do not make the stain larger by rubbing. Any discoloration on the surface can be dabbed off with a rag moistened with benzine. Rub dry with a clean, soft cloth	Moisten a clean, soft cloth with benzine or spirit, wipe carefully and then rub dry. Turn the cloth several times in order not to spread the stain over a wider area
Shoe polish	The same applies as with oil or grease Methylated spirits, and turpentine may also be used as well as benzine	
Artificial resin, cellulose or oil paints	Remove with dry, soft cloth as with oil and grease. Remove the remaining stains with a rag moistened in water or with a piece of rubber	Moisten a cloth with cellulose thinner, turpentine or benzine, wipe carefully and then rub until dry. Cellulose thinners for synthetic resin and cellulose paint stains and turpentine or benzine for oil paint stains
Blood	Moisten a cloth in cold or lukewarm water and dab the stain off. Do not make it larger by rubbing	
Rust	Moisten a soft rag with an acid solution (1 part of hydrochloric acid to 10 parts of water), dab rust spots carefully but do not spread them by rubbing. The solution must not soak into crevices, corners or seams, as otherwise more rust will be created. After this treatment wipe off with a rag moistened with clean water to remove all traces of the solution. The rags used should be destroyed	

Cleaning Air-permeable Plastic Upholstery Material

Upholstery material of air-permeable plastic should be cleaned dry with a soft cloth or brush wherever possible. Liquid cleaners must not be used because they will penetrate through the grooves in the plastic into the seat padding.

If dirt cannot be removed with a cloth or a brush, commercial dry foam cleaners should be used. As these cleaners also contain a certain amount of liquid, they should not be applied to the surface to be cleaned in large quantities. After cleaning, the material should be wiped with a dry cloth.

It is advisable to brush the material in the direction of the grooves when it is completely dry to remove all traces of the cleaning solution from the grooves.

To remove spots, the agents listed above can be used. The precautions to be exercised when using these agents on plastic materials with smooth or grained surfaces should also be noted when cleaning air-permeable material.

1 - Cleaning plastic roof materials

a - Remove normal dirt with a brush and luke-warm water.

b - If warm water alone is not sufficient, use a solution of pure, mild soap (soap flakes). A cleaning solution of water and a normal PVC or plastic cleaner is also suitable, using a hard brush to get the best results. Afterwards rinse with clear water until all traces of soap are removed from roof and from vehicle paintwork, particularly between roof and paint.

c - Stains caused by tar, soot, oil, grease, paint, etc. which cannot be removed with water or a soap solution, should be removed with an organic solvent such as trichlorethylene or benzine. Apply the solvent with a moistened cloth only, do not pour on to the surface, allow to work for short time then wipe off, wash surface with a soap solution and rinse with clear water. Organic solvents tend to harden or crack the PVC material if left in contact too long. These solvents must not get on to the paintwork or into vehicle interior.

2 - Cleaning the cloth sun roof

The appearance and life of the sun roof depend to a large extent on the care and treatment it receives. The roof must never be opened when wet. When dusty, the top should be beaten lightly and then brushed in line with the pile with a soft brush as the sharp particles of dust will damage the material and cause friction marks and other damage.

Never use benzine, benzol spot remover or other solvents to remove spots as these liquids will attack the rubber layer in the roof and endanger the waterproofing and life of the material.

The top should not be washed with the car shampoos which are often used nowadays as some of these agents affect the impregnation of the material and can cause leakage even after one application.

Dirt which cannot be removed by just brushing can often be removed by rubbing the dry top with a soft eraser and then brushing. In this manner, too frequent washing with soap solutions and the consequent premature bleaching of the material can be avoided.

The roof should only be washed when very dirty. Remove loose dust first by beating lightly or brushing. Only good quality soap powder will ensure that the roof is cleaned properly with the minimum influence on the life of the actual material. Mix two dessert spoonsful of soap powder in a gallon of clear water without any chemical additives or other agents.

Moisten the top with clear water and apply the solution with a soft brush in one direction only. Then rinse the top with clear water and brush it at the same time. If necessary, wash the roof a second time. The final rinsing should be continued until all traces of soap are removed and the water runs off quite clear. The top must then be left closed until dry.

The soap suds should then be washed off the vehicle paintwork and the paint dried with a leather.

3 - Re-impregnation of the fabric sun roof

The following waterproofing preparations are recommended:

a - "Happich-Viktoria-Impraegnierung", made by Gebr. Happich GmbH., Wuppertal-Elberfeld, Germany.

b - "Impraegnol M-Extra", made by Pfersee GmbH., Augsburg, Germany.

c - "Primenit VS", made by Farbwerke Hoechst, Hoechst/Main, Germany, to be used in conjunction with "Ramasit", made by Bad. Anilin- u. Sodafabriken, Ludwigshafen, Germany.

Please comply with the manufacturer's instructions which are delivered together with the product.

4 - Dyeing of the fabric sun roof

Fabric sun roofs which have become bleached or faded through frequent washing or the influence of the weather, may be dyed. Sometimes, however, stripes and paint spots in the fabric may remain more or less visible even after dyeing.

The textile dyes of

Messrs. Artekobin-Gesellschaft Gerhard & Co., Oberntorwall 14, Bielefeld, Germany,

are suitable for dyeing the textile sun roof. The sun roof has to be closed during treatment. Newly dyed sun roofs need not be impregnated again as the dyes contain an impregnating agent.

3 - **Replacing larger body parts**

The complete side, front or rear end of a body can be renewed without involving dimensional errors by using body sill templates that can be additionally screwed to the trolley.

The body repair trolley has rubber wheels for easy transport.

The trolley is delivered with

2 stanchions	2 locking pins
2 stanchion brackets	2 turnbuckles

The additional parts listed below can also be ordered from the firm Karmann:

2 body sill templates
for VW Sedan and Convertible

2 body sill templates
for Karmann Ghia models

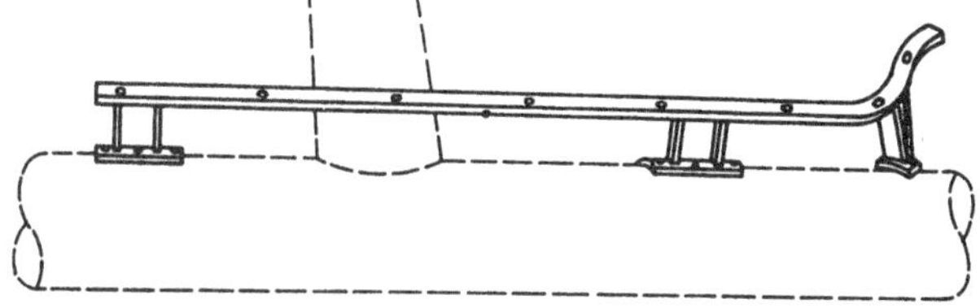

1 insert for head light opening
(Karmann Ghia models)

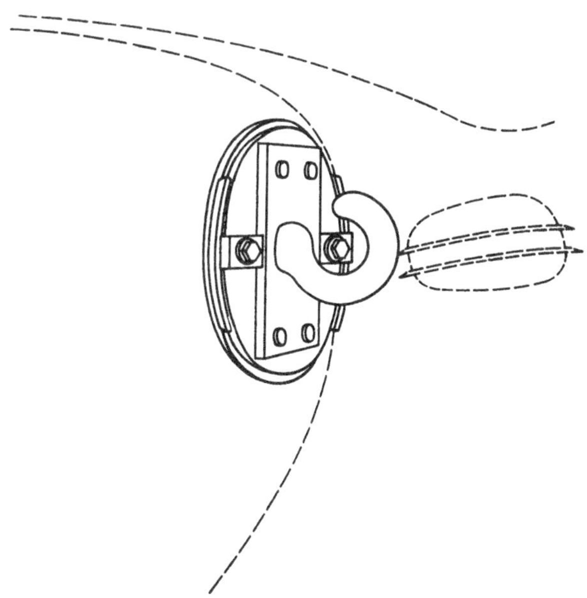

1 Attachment for front and rear
bumper brackets

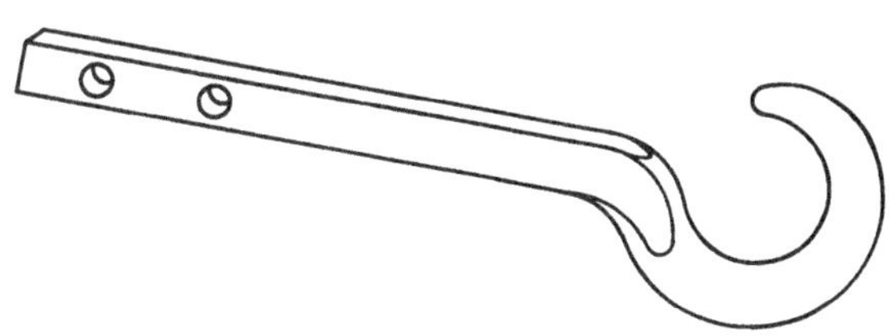

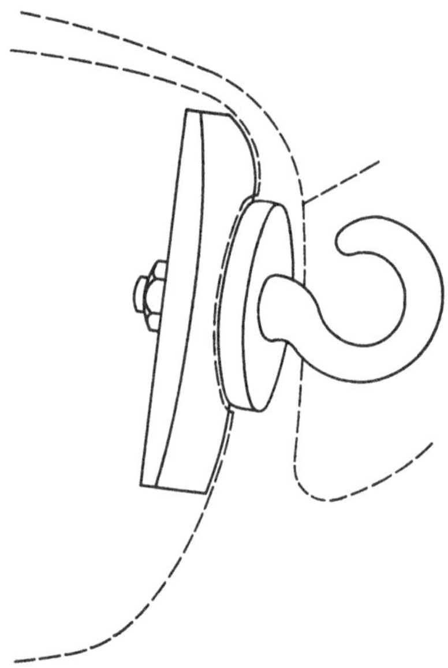

1 insert for tail light
(Karmann Ghia models)

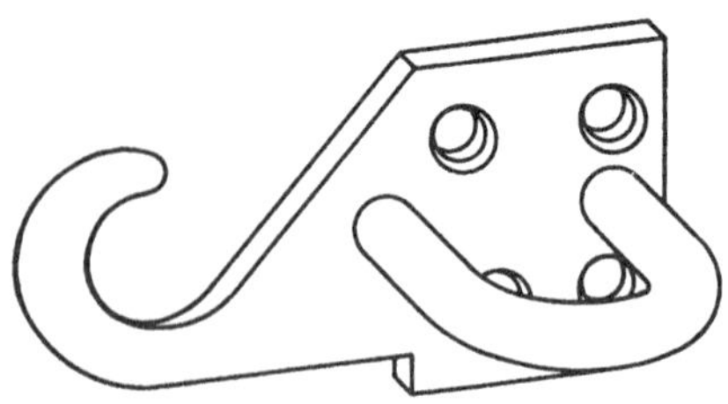

1 Attachment for hinge pillar —
upper hinge

Prices and details of supply can be had on request from the firm of W. Karmann GmbH.

Adhesives and Sealing Compounds

Part No.	Description	Use	Solvent	Packed
G 2	High Quality Graphite	Door and lid locks to prevent freezing in winter	Wipe off with a clean rag	250 gram tins
D 10	Genuine VW Window Cement	Sealing window glass when installing	Benzine (Glass to be cleaned with alcohol)	1 kg
D 10/5	Genuine VW Window Cement	Sealing window glass when installing	Benzine (Glass to be cleaned with alcohol)	375 gram tube
D 11	Plastic Adhesive	For sliding roof PVC covers and leatherette	Terocal thinner D	1 kg
D 12	Genuine VW Adhesive	Fixing insulating material, upholstery cloth and carpets. Fixing rubber to metal (e. g. rubberstrip to body at the hinged window)	Benzine	1 kg
D 14	Plastic Sealing Compound	Weather resistant seal between wood, metal, and plastic parts; especially for platform protection battens of Pick-up	Benzine	1 kg
D 17	Genuine VW Sealing Compound	Between body and frame, filling weld seams and insulating brake lines	Benzine	750 g
D 19	Plastic Sealing Tape	Sealing front fender (rear edge) to front side panel on Type 3. Eliminating leaks		20 m rolls
D 20	Joint Sealer	For coating visible joints and folded edges	Benzine	750 gram tins

THIS PAGE INTENTIONALLY
LEFT BLANK

Care and Cleaning of Artificial Leather Covers on PVC Basis

Artificial leather covers have a very efficient dust-repellent surface. No special preservatives are needed.

With ordinary dirt the following may be used for cleaning:

a - soap suds, made of soft water and a commercial detergent.

b - cleaning solutions made of water and an ordinary plastic cleaner.

A soft brush will facilitate removal of dirt from grained surfaces.

Always deal with dirt of all kinds as soon as you notice it. Cleaning agents especially suited for removing certain stains can be found in the table below. Be careful not to pour benzine, spirit, or thinners onto the area to be cleaned but rub it on with a moist rag in order to prevent it from penetrating into the seams or upholstery. It is advisable not to allow the liquid for cleaning to work on the surface for any length of time as the transparent, dirt-repellent finish of the artificial leather might suffer.

After **each** cleaning, the artificial leather should be rubbed dry with a soft rag, especially at the seams.

Stains	Removal of	
	new stains	old stains
Oil or grease	Remove with dry, soft cloth, turning this several times. Do not make the stain larger by rubbing. Any discoloration on the surface can be dabbed off with a rag moistened with benzine. Rub dry with a clean, soft cloth	Moisten a clean, soft cloth with benzine or spirit, wipe carefully and then rub dry. Turn the cloth several times in order not to spread the stain over a wider area
Shoe polish	The same applies as with oil or grease Besides benzine or spirit, turpentine may also be used	
Artificial resin, cellulose or oil paints	Remove with dry, soft cloth as with oil and grease. Remove the remaining stains with a rag moistened in water or with a piece of rubber	Moisten a cloth with cellulose thinner, turpentine or benzine, wipe carefully and then rub until dry. Cellulose thinners for synthetic resin and cellulose paint stains and turpentine or benzine for oil paint stains
Blood	Moisten a cloth in cold or lukewarm water and dab the stain off. Do not make it larger by rubbing	
Rust	Moisten a soft rag with an acid solution (1 part of hydrochloric acid to 10 parts of water), dab rust spots carefully but do not spread them by rubbing. The solution must not soak into crevices, corners or seams, as otherwise more rust will be created. After this treatment wipe off with a rag moistened with clean water to remove all traces of the solution. The rags used should be destroyed	

Care and Cleaning of Roof Covers

1 - Cleaning the plastic (PVC) sun roof

a - Remove normal dirt by washing with luke-warm water and a brush.

b - If warm water fails to affect satisfactory cleaning, use a solution of pure, mild soap suds and then rinse with clear water until all soap has been washed off. A solution of water and a normal PVC or leatherette cleaner is also suitable for this purpuse. Good results can be obtained with a hard brush. Make sure the soap is washed completely off the paintwork of the vehicle, too, especially from between roof cover and paintwork.

c - Spots caused by tar, soot, oil, grease or paint which cannot be cleaned off with water or soap solution, should be removed with gasoline. Do not pour the detergent on the spot to be removed but apply it with a damp cloth. Let it act for a short time only, then wipe if off and treat the spot first with soap solution and then with clear water.

Stain removers containing chlorine or thinners should not be used as these attack the plastic cover and the paintwork.

2 - Cleaning of the fabric sun roof

a - To clean the top, beat or brush it lightly with a soft brush to remove the surface dirt and dust. Small spots can often be removed with a soft india-rubber.

b - If the top is very dirty it must be washed. Only clear water without chemical or other additives should be used for this purpose. Alkali-free soap flakes should be beaten into suds in a bucket of lukewarm water and the soap suds applied to the top which has been moistened with clear water beforehand. The suds are then brushed in with a soft brush by rubbing in one direction only.

The top is then rinsed with clear water and brushed at the same time. If necessary, the soaping process should be repeated. The top must be rinsed until the last trace of soap is removed and the water runs off quite clear. It is advisable to re-impregnate the top with a solution of 1 part aluminium acetate and 5 parts water after washing. This solution is applied evenly to the top while it is still wet. After this treatment the top must remain closed until properly dry.

c - The vehicle paintwork should be rinsed with clear water to remove all traces of soap and impregnating solution and then dried with a leather.

Benzine, benzol, spot removers and other solvents must not be used for removing stains as they attack the rubber in the material, cause leaks and shorten the service life of the top.

3 - Re-impregnation of the fabric sun roof

The following waterproofing preparations are recommended as well as the aluminium acetate already mentioned:

a - "Happich-Viktoria-Impraegnierung", made by Gebr. Happich GmbH., Wuppertal-Elberfeld, Germany.

b - "Impraegnol M-Extra", made by Pfersee GmbH., Augsburg, Germany;

c - "Primenit VS", made by Farbwerke Hoechst, Hoechst/Main, Germany, to be used in conjunction with "Ramasit", made by Bad. Anilin- u. Sodafabriken, Ludwigshafen, Germany.

Please comply with the manufacturer's instructions which are delivered together with the product.

4 - **Dyeing of the fabric sun roof**

Fabric sun roofs which have become bleached or faded through frequent washing or the influence of the weather, may be byed. Sometimes, however, stripes and paint spots in the fabric may remain more or less visible even after dyeing.

The textile dyes of

Messrs. Artekobin-Gesellschaft Gerhard & Co., Oberntorwall 14, Bielefeld, Germany,

are suitable for dyeing the textile sun roof. The sun roof has to be closed during treatment. Newly dyed sun roofs need not be impregnated again as the dyes already contain an impregnating agent.

Cleaning the windshield

Some liquid or paste preservatives contain silicon for greater efficiency. If such preservatives come into contact with the windshield, the silicon contents will cause streaks and clouding within the range of the wiper blade when it rains, thus impairing visibility and driving safety.

As not all usual cleaning agents and solvents are suitable for removing such traces of silicon from the windshield, we recommend the following agents:

1 - "1 Z" paste of the firm Werner Sauer & Co., Bensberg.

2 - "Rex" paste of the firm Rex-Autopflege, Mainz.
Important! This paste should not come into contact with the paintwork.

3 - "Silicontra" paste of the firm Anton Fatka, Lüneburg.
Important! This paste should not come into contact with the paintwork.

4 - "Carnu-Lackreiniger" of the firm Jonson's Detergents, Hamburg.
Important! This paste affects the paintwork. Wipe off any traces immediately.

5 - "Sidol" of the firm Siegel & Co., Köln. For this purpose it is easiest to apply it as a paste, spread it over the windshield, let it dry and then wipe it off. "Sidol" should not come into contact with the paintwork.

6 - A paste consisting of 2 weight-parts of Vienna lime and 3 volume-parts of water. A few drops of liquid ammonia will further improve the cleaning efficiency, but will also give it a strong smell. This type of paste should not come into contact with the paintwork. It is applied as described under paragraph 5.

7 - Clean windshield with benzine and then treat it with an acid solution, which consists of 1 part hydrochloric acid and 9 parts water. Afterward, rinse carefully with clear water.

The wiper blades should be cleaned at the same time. Sometimes, however, new blades may have to be installed to keep the windshield clear.

In this connection, we should like to draw your attention to the following points:

1 - When using any preservatives containing silicon, the brushes, sponges, chamois and rags used to clean the paintwork should not be used for the windshield.

2 - Whenever the paintwork is to be sprayed with preservatives containing silicon, be sure to cover windshield and windows with cardboard or suitable material first.

The Genuine VW Preservative L 190 does not contain silicon.

Removing and Installing Heater Flap Cable

General

The heater flap cables are fixed together near the operating lever by a piece of tube. It is, therefore, not possible to replace one cable.

Removal

1 - Loosen nut on clamping bush pin.

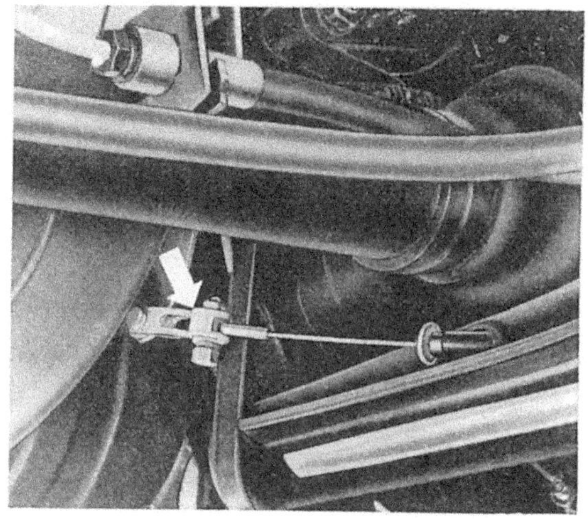

Note:

To avoid breaking or bending the heater flap cables, hold the pin with a 9 mm wrench and loosen nut with a 10 mm wrench.

11/65

Note:

On 5 November 1965, from Chassis No. 116 314 087 (Engine No. F 0 295 592) the attachment of the heater flap cable to the heater flap lever on the heat exchanger was simplified. Similar to the accelerator cable attachment, the new attachment consists of the link, Part No. 111 255 309 B, the swivel pin, Part No. 311 129 777 and the screw N 10 205 2.

The new link can be service installed on **all** previously manufactured engines (also on those with heat control boxes). For this purpose the SP Set of clamping parts for the heater flap cable with link, SP 117 A, is available.

SP Set SP 117 will still be supplied for the previous attachment of the heater flap cable.

2 - Pull cable ends out of clamping bushes.

3 - Pull sealing plugs out of guide tubes and slide them over the cables.

4 - Remove hand brake cover and unscrew nut from right-hand heating operating lever.

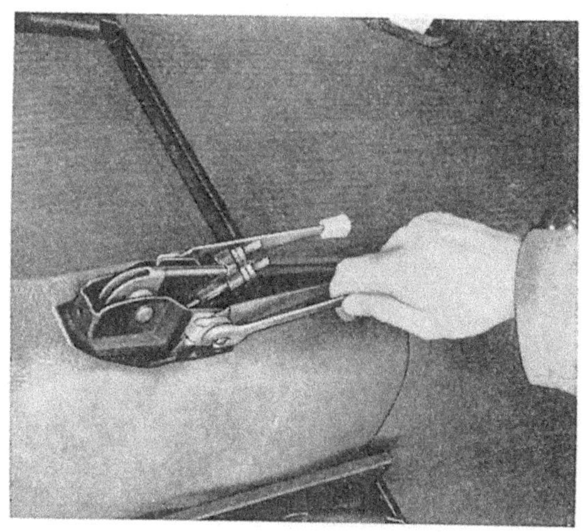

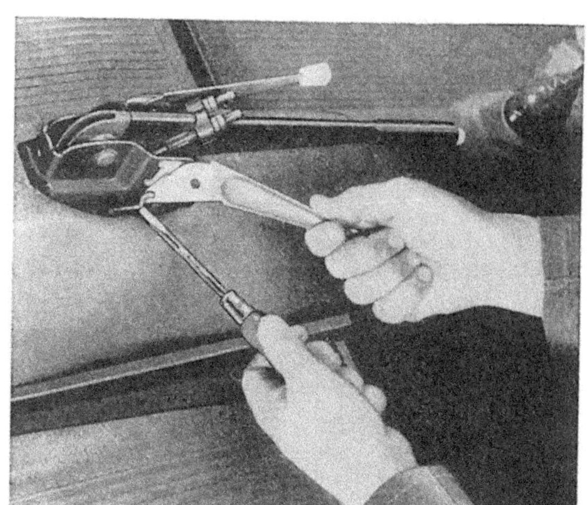

5 - Take lever and friction discs off mounting, disconnect cables and pull them out upward.

3 - Hook lever into the hook-shaped end of the cable and attach lever to mounting with friction discs.

4 - Seal guide tubes carefully with the plugs.

5 - Clamp cables to heater flaps so that the flaps open and close fully. Check heating by opening and closing flaps a few times.

Installation

1 - Check all parts and replace if worn or damaged. Coat cables lightly with universal grease.

2 - Push cables into the tubes.

Note:

The nut on the mounting can be loosened or tightened to alter the effort required to move the lever.

Removing and Installing Cables for Rear Footwell Heating

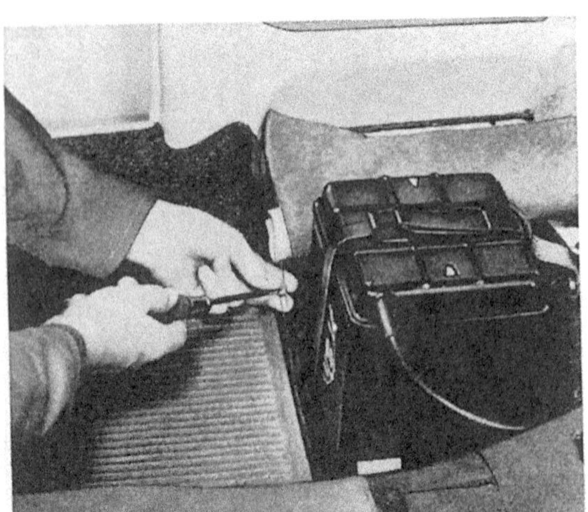

The heater flap cables are joined at the lever by a piece of tube. It is, therefore, not possible to replace one cable only.

Removal

1 - Remove kick plates.

2 - Loosen screws at each clamp.

3 - Pull cable ends out of clamp.

4 - Take off handbrake lever cover and remove nut from left heater lever.

5 - Take lever and friction discs off mounting, disconnect cables and pull them out upwards.

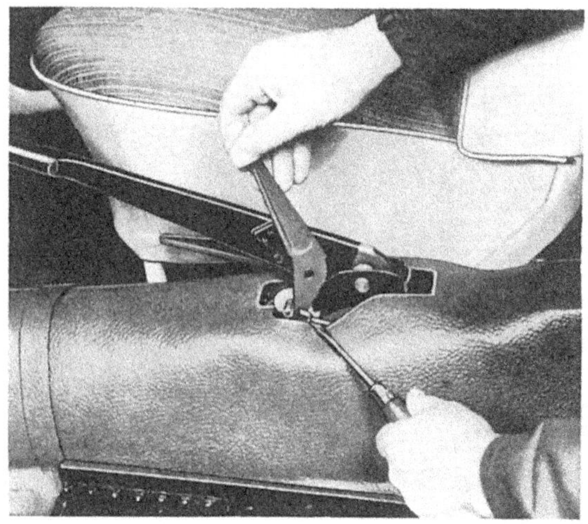

Installation

1 - Check all parts and replace if worn. Coat cables lightly with universal grease.

2 - Push cables into guide tubes.

> **Important**
>
> The longest cable belongs in the lower guide tube.

3 - Hook lever into hook-shaped end of cables and attach lever to mounting with friction washers.

4 - Clamp cables to the flaps so that the flaps open and close fully. Check operation of footwell heating by opening and closing flaps several times.

Note

The nut on the mounting can be loosened or tightened to alter the effort required to move the levers.

5 - Install kick boards. Ensure that the rubber rings contact the warm air outlets properly.

Removing and Installing Rear Footwell Heater Flaps

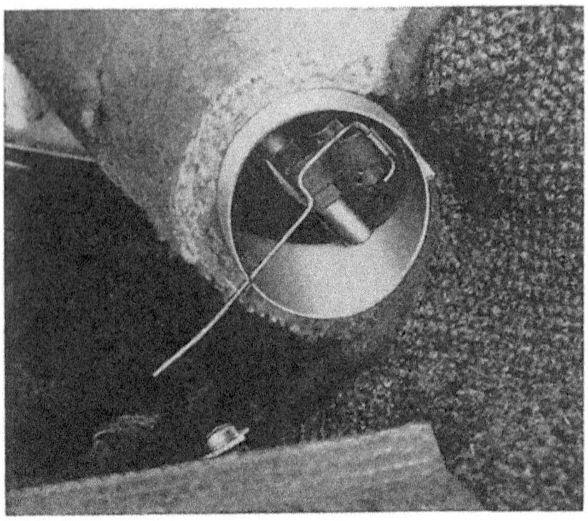

Removal

1 - Remove kick boards.

2 - Loosen clamps at cable ends and pull flaps off pivots in warm air outlet pipe.

Installation

1 - Check that the spring clips are tight in the flaps and replace if necessary.

2 - Press flap on to pivot. Move to and fro to see if the spring clip is gripping the pivot properly and that the short cable can move freely.

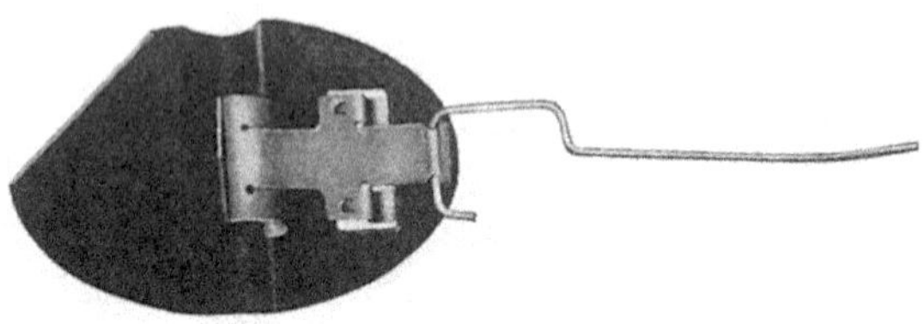

3 - Connect cable to flap. Check function of footwell heating by moving the lever to and fro a few times. If necessary, rectify cable length after loosening clamp.

4 - Install kick boards. Ensure that the rubber rings contact the warm air outlets properly.

Removing and Installing Front Footwell Heater Slides

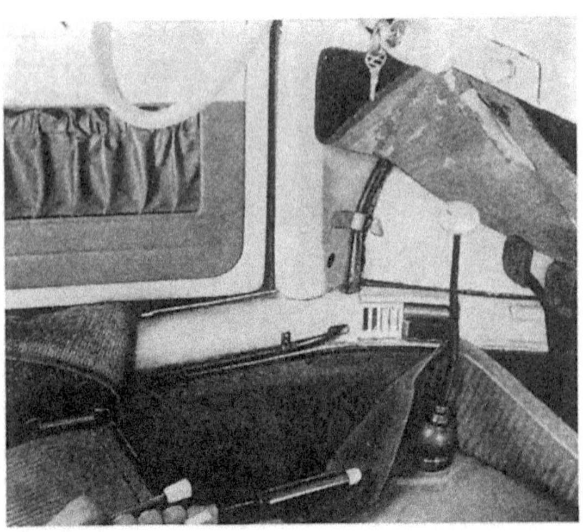

Removal

1 - Detach front lining carefully near the warm air outlets and fold it back.

2 - Pull tensioning wire out of side member.

3 - Pull slide forward out of the upper groove.

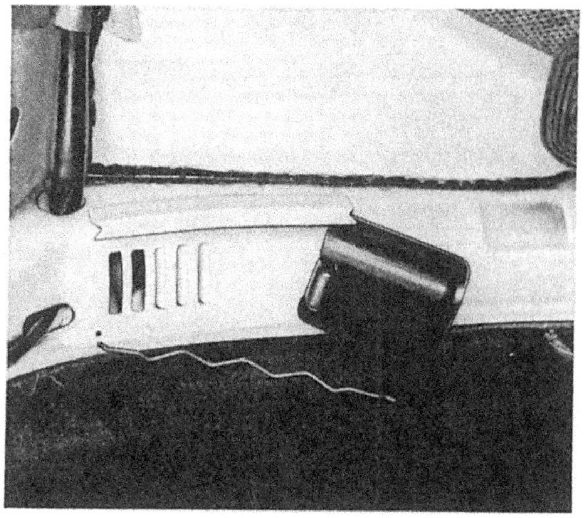

Installation

1 - Check slide and retaining wire, and replace if necessary.

2 - Check upper groove on side member for damage and straighten as necessary.

3 - Insert slide and fit retaining wire.

4 - Check ease of movement of slide by moving it to and fro a few times. Straighten upper groove if necessary.

5 - Coat front panel lining lightly with D 12 adhesive and stick it down again. Ensure that lining material fits properly round the outlet.

Removing and Installing Front Heater Hose

Removal

1 - Pull heater hose downward off the defroster vent at the windshield.

2 - Bend up the clip on the upper side of the hinge pillar.

3 - Pull hose carefully off the adaptor on the side member and take it out.

Note:

If the adaptor on the side member has to be replaced, it is advisable to remove the front hood so that the fit of the adaptor in the side member can be checked properly.

Installation

1 - When installing the heater hose, ensure that the small diameter is upward.

2 - Finally, run the engine, close the front and rear heater outlets and check that the adaptor and heater hose are not leaking.

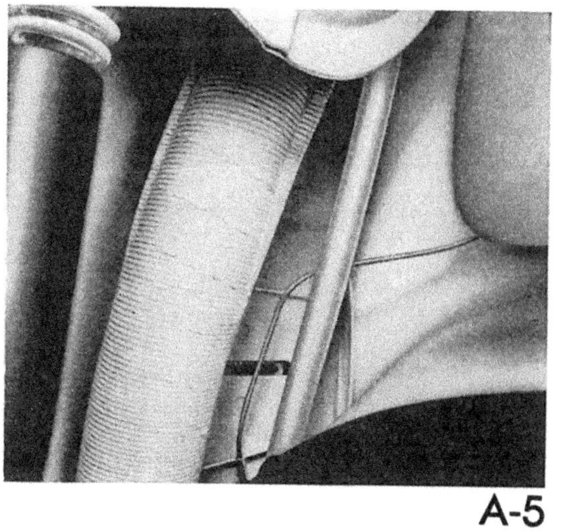

Note:

Since 2 August 1965, Chassis No. 116 000 002, all vehicles have been fitted with an additional defroster vent in the center of the instrument panel. This will improve de-misting of the windshield at this point under adverse conditions.

For service installation the following parts are required:

Quantity	Description	Part Number
1	Defroster vent, center	111 255 473
1	Cap	111 255 483
2	Hose with branch pipe	111 255 463

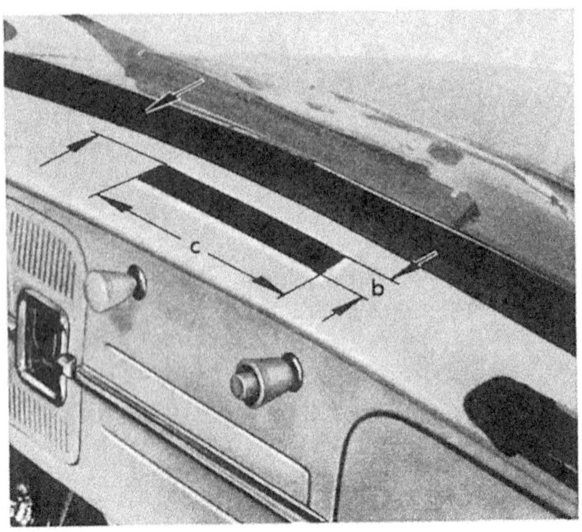

a = .6 in. (15 mm)
b = .43 in. (11 mm)
c = 5.9 in. (150 mm)

A - Preparations:

1 - Disconnect battery and remove the following parts:

a - Luggage compartment lining and cover in front of instrument panel.

b - Windshield wiper motor with frame and linkage.

c - Glove box.

d - Light and wiper switches.

2 - Make a hole for center vent in the instrument panel and paint the edges of the hole.

B - Installation

1 - Shorten the existing hoses so that the branch pipe with the hoses for the side vents can be inserted.

> **Important**
> The existing hoses should not be removed to do this as it may not be possible to install them again properly.

2 - Press cap into hole in instrument panel.

3 - Press center vent into cap from underneath.

4 - Route connecting hoses to center vent close to cowl panel and fit them to vent and branch pipes.

5 - Attach connecting hoses to the body near the cowl panel with adhesive tape to prevent rattling noises.

6 - Reinstall all parts and check operation of heating system.

Rear Defroster Vent Flexible Pipe Removal and Installation

Removal and Installation

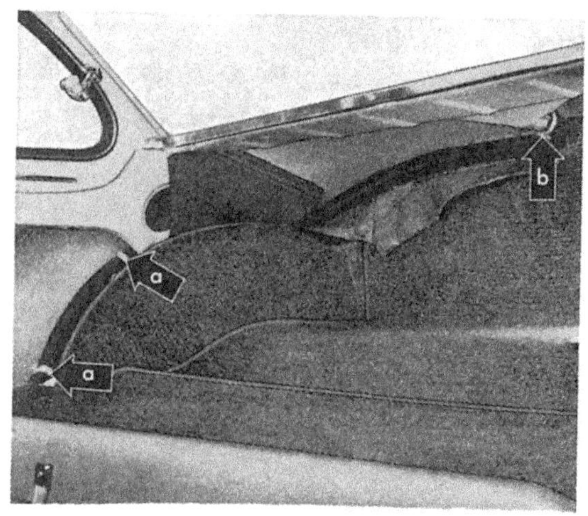

1 - Remove right quarter trim panel and carefully detach haircord carpet near flexible pipe.

2 - Loosen clamp (b) on defroster vent and junction pipe, bend up two clips (a) and remove flexible pipe.

3 - Prior to installation, check condition of flexible pipe and install new one if necessary.

4 - Install flexible pipe, secure with clamps and carefully bend clips down.

5 - Install quarter trim panel. Thinly coat haircord carpet with universal adhesive and cement in position.

THIS PAGE INTENTIONALLY
LEFT BLANK

Contents: A—Sedan

Assembly Work

Replacement of Body Parts

Body—General
1 through 9
(see table of contents preceding section A-1)
Body—Convertible
35 through 53
(see table of contents preceding section A-35)
Body—Karmann Ghia Coupe
60 through 79
(see table of contents preceding section A-60)
Body—Karmann Ghia Convertible
80 through 89
(see table of contents preceding section A-80)

THIS PAGE INTENTIONALLY
LEFT BLANK

General

The two-door all-steel body is bolted to the platform-type frame and sealed with U-section rubber sealing strips. The air resistance is exceptionally low due to the curved front hood (bonnet), the inclined windshield, the convex roof panel and the stepless, evenly sloping rear end. The body is divided into luggage compartment, passenger compartment and engine compartment by partitions. Air is allowed to enter the engine space through a series of slots below the rear window.

The body comprises the following parts, which are welded together:

a - Front panel with reinforcement plate, instrument panel and door-hinge pillars;

b - Front side panels;

c - Floor of rear luggage compartment;

d - Rear quarter panels, inside and outside, with roof members;

e - Side members

f - Roof

g - Front and rear aprons.

All fenders (mudguards) are detachable and can easily be replaced. Sill panels between the front and rear fenders add to the practical appearance of the car.

Doors

The 950 mm (37.4″) wide doors are attached to the body front pillars by two hinges. The door locks are operated from both inside and outside by handles. The door on the driver's side is locked from the outside, whereas the right-hand door is locked by means of the inside handle. The door fit in the body can be corrected by adjustable rubber buffers and door lock striker plates. The door check strap assembly allows the door to be opened 70°. A weatherstrip around the door prevents ingress of dust and water.

Hoods (Bonnets)

The front hood, which covers the fuel tank, spare wheel, tools and front luggage compartment, is locked, and is operated by a pull-knob situated below the instrument panel to the left of the steering column. The front hood is kept in the open position by means of two stays. The rear hood lock is manually operated by turning the T-handle. A spring holds the hood in the open position.

Windows

The undivided windshield inclines at 32°. It offers an exceptionally wide field of vision. The max. clear width is 1035 mm (40.7″) and the max. clear height 330 mm (13.0″). Defroster vents at the bottom edges of the windshield frame prevent condensation or icing.

The window regulators require 3¼ turns for full travel of window. Vent wings in door windows permit draughtless control of fresh air; the spring-loaded pivot mechanism holds the wings in any position required. The wing locks are operated by handles and released by push-buttons.

The rear quarter panel window panes are fixed. The rectangular rear view window has a slant of 53°. The maximum width of the pane is 810 mm (31.9"), maximum height 360 mm (14.2").

All glass panes are of heat-treated safety glass. The windshield also possesses a specially treated zone which permits clear vision even when the pane is destroyed. For certain countries windshields are made of laminated glass.

Front Seats

The separate adjustable seats of the De Luxe Sedan are held in runners which rise slightly towards the front. To facilitate adjustment springs are provided which assist the forward movement of the seats. The seats of the Standard Sedan are held by clamps and wing nuts.

The backs of the seats in the De Luxe Sedan can be fixed in three different positions.

The seats have a tubular steel frame and an interior of interlinked coil springs.

Springs are used for the backs. On the De Luxe Sedan rubber hair padding is used with extra bulges being provided at the sides of the seat backs. In the Standard Sedan the padding is of quilted cotton wool.

Rear Seats

The back rest of the bench-type rear seat can be folded down, and is held in the upright position by a rubber check strap. The interior of the seat and back rest consists of springs.

Interior Trim

De Luxe Sedan

The floor is covered with rubber mats. The lower part of the front partition is lined with rubber, likewise the frame tunnel. The upper part of the front partition, the front quarter panels, the side members and the rear luggage compartment are all lined with haircord carpet. The roof, roof side members, and the trim panel around the rear view window are covered with cloth. The door pillars are lined with leatherette.

The door trim panels and rear quarter trim panels are covered with two-tone plastic material and finished at the top with a polished decorative strip. On the driver's side the door has a pocket with elastic through the top edge. On the passenger's side an arm-rest with decorative strip is provided. In the right-hand rear quarter trim panel there is an ash-tray which can be tipped up to empty.

The handbrake lever is sealed at the frame tunnel by a rubber boot.

The rear-view mirror on a ball-and-socket mounting, with the sun vizor for the driver, is attached to the middle of the windshield frame.

The door pillars are also equipped with coat hooks and assist straps.

Standard Sedan

The floor is covered with rubber mats. The lower part of the front partition is lined with rubber. Between the front partition and the gearshift lever, and between the front and rear seats, the frame tunnel is lined with rubber.

The upper part of the front partition, the front quarter panels, and the side members are lined with haircord carpet. Above the rear wheel housing the rear luggage compartment is lined with haircord carpet, while the floor of the compartment is covered with fiberboard. The roof is lined with cloth down to the upper rim of the rear-view window.

The door and rear quarter trim panels are covered with plain plastic material.

The handbrake lever is sealed at the frame tunnel by a rubber boot.

The rear-view mirror on a ball-and-socket mounting is attached to the middle of the windshield frame.

Seat Upholstery

The cushions and back-rests of the front and rear seats are covered with cloth upholstery in both the De Luxe and Standard Sedans, the sides of the cushions and back rests, as well as the rear portion of the back rests are covered with leatherette. The back of the rear seat is covered with a sheet of fiberboard reinforced by metal straps.

For certain countries the seats are lined with leatherette throughout.

Sound Deadening and Insulation

The open spaces on the left and right between the rear of the rear luggage compartment and the rear quarter trim panels are sealed off from the engine compartment by padding. The rear openings in the roof side members are plugged with foam rubber. Between the engine compartment and the rear of the luggage compartment and the wheel housings are panels of sound absorbent material. The floor of the rear luggage compartment is coated with cork-felt on the inside. The frame fork and the floor beneath the rear seat, and in the De Luxe Sedan the frame tunnel up to the front partition, are covered with bituminized felt. The front luggage compartment is lined with cork-felt. The outer panels of the doors thereof, and the rear quarter panels are insulated with a layer of absorbent material cemented to the inside.

Heating

Heated air, which is taken from the air flow warmed up by the engine, is guided via sound deadeners situated under rear seat through flexible metal pipes and the bottom side members (sills) of the body into the interior of the car through outlets at foot level in the front compartment and two defroster vents at the windshield.

Luggage Accommodation

Luggage space is provided behind the rear seat, which has its backrest hinged to facilitate loading. Another luggage compartment is under the front hood.

Exterior Trim

Standard and De Luxe Sedans

The sill panels are rubber-covered. The width of front and rear bumpers is 90 mm (3.54″). The hight of the overriders amounts to 190 mm (7.5″). Headlamp rims are chrome-plated. The VW sign is indented in the hub cap.

Standard Sedan

Bumpers, overriders, hub-caps, vent wing frames, outside mirror, door, window and hood handles, the grille for the electric horn, loudspeaker and ashtray are painted.

De Luxe Sedan

The mouldings fitted to the front hood, instrument panel, along the waist-line, the doors, outer edge of the sill panels, the window frames and the grilles in the front fenders are all of highly-polished lightmetal.

Bumpers, overriders, hub-caps, vent wing frames, outer mirror, door, window and hood handles are chromium-plated. The additional bumper guards provided for export are also chromium-plated.

The front hood exhibits a VW sign and the emblem of the town of Wolfsburg, where the Volkswagen is manufactured.

Sliding Roof

The sliding roof, make Golde, has a clear length of 690 mm (27.2″) and a clear width of 730 mm (28.7″) when fully opened. It can be locked in any open position desired.

THIS PAGE INTENTIONALLY
LEFT BLANK

License Plate Installation

As license plate sizes vary considerably from country to country, the following instructions are only general recommendations and local regulations should be observed when attaching plates. It is advisable to avoid drilling the holes through a letter or figure, including the space within the figure "0", as this will affect the readability of the plate.

Fitting Front License Plate to Sedan and Convertible

1 - Drill two 5.3 mm holes in the plate, outside letter and figure groups if possible, and mark off and drill front bumper.

2 - Clean holes up and paint on the inside to prevent rust formation.

3 - Secure plate with two M 5×15 screws (Part No. N 10 815 1), spring washers, nuts and rubber washers (Part No. 211 955 193) as shown in sketch (Fig. 1).

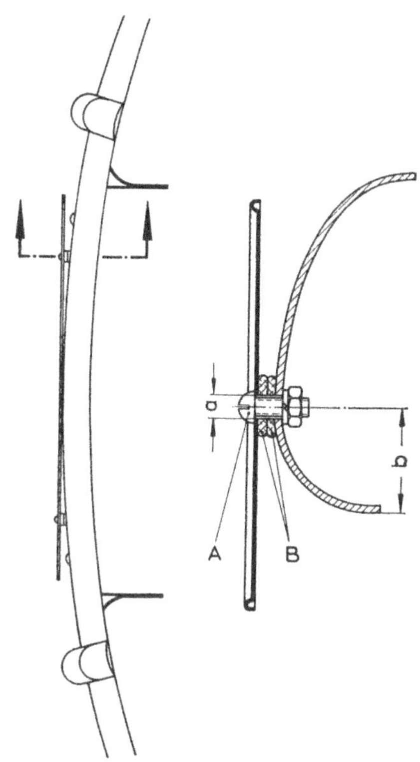

A - Screw
B - Rubber washer

a - 5,3 mm ∅
b - 28 mm

Fitting Rear License Plate to Sedan and Convertible up to Chassis No. 1 600 439

1 - Drill two 5.3 mm holes in plate and mark off and drill rear hood. The distance between license plate lamp and upper edge of plate should be as shown in sketch (Fig. 2).

2 - Clean holes up and paint on the inside to prevent rust formation.

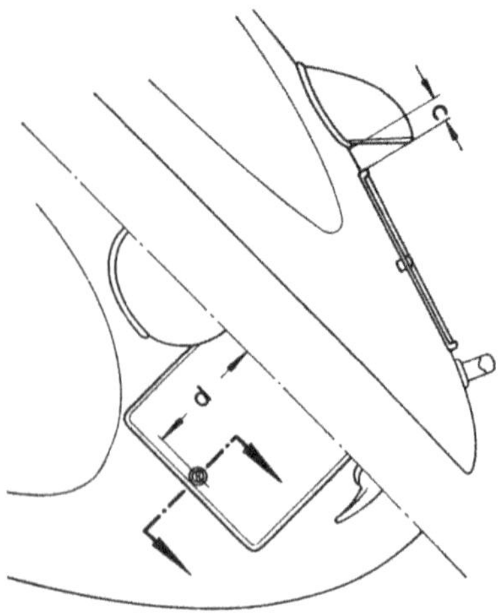

3 - Secure plate with two AM 5×25 screws (Part No. N 10 723 1), washers, spring washers, nuts and two spacers (Part No. 111 012 299).

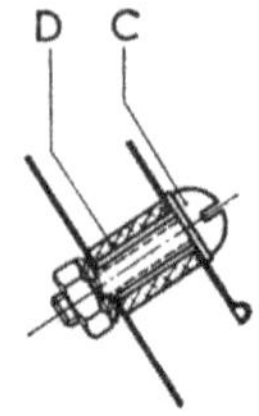

C - Screw
D - Spacer

c - 25 mm
d - 125 mm

Fitting Rear License Plate to Sedan and Convertible from Chassis No. 1 600 440

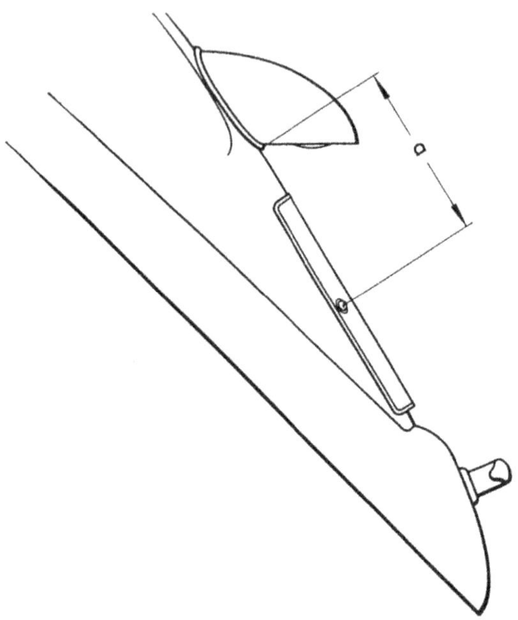

The plate should be fitted so that the distance between license plate lamp and center of plate is 145 mm as shown at "a" in sketch (Fig. 3).

On vehicles with reinforced bumpers, the measurement "a" should be 140 mm, to ensure that the plate is properly illuminated.

a = 145 mm without bumper guards
b = 140 mm with bumper guards

Fitting Rear License Plate to Karmann Ghia models

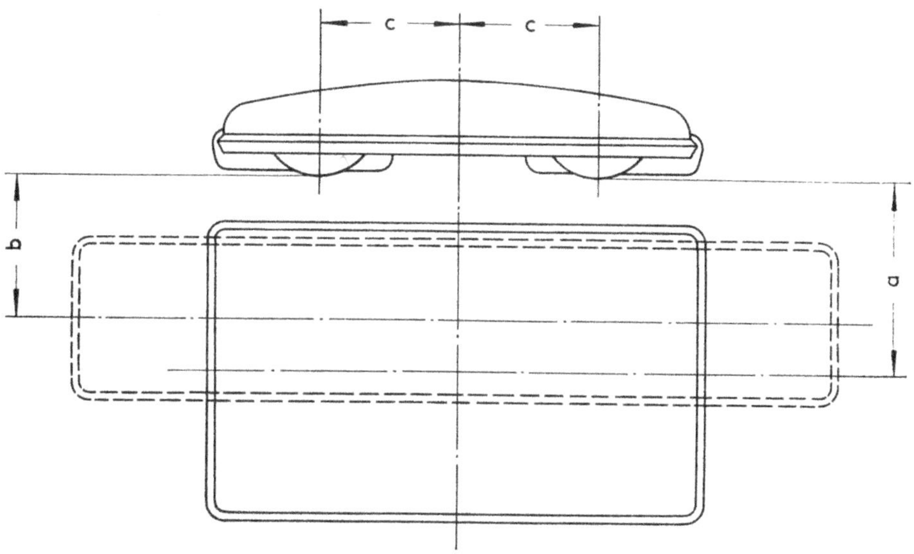

a = 130 mm without bumper guards
b = 95 mm with bumper guards
c = 95 mm

On the Karmann Ghia models with reinforced bumpers the single line type of license plate is most suitable at the rear. In countries where only the square type of plate is used, the plate should be mounted so that it is easily readable at all times.

The front plate can generally be installed as on the Sedan.

Note:

International registration plates should also be fitted in accordance with local regulations.

THIS PAGE INTENTIONALLY
LEFT BLANK

Body Removal and Installation

Removal

To facilitate the removal and installation of the body, it is recommended to adopt the following sequence of operation:

1 - Place car on trestles and remove rear wheel.

2 - Remove front and rear seats.

3 - Disconnect battery.

4 - Close fuel tap, detach fuel pipe from fuel tap and remove cotter pin between operating rod and fuel tap.

5 - Remove fuel tank.

6 - Disconnect choke control cable from instrument panel and withdraw it from the body.

7 - Detach the hose between the reservoir and master brake cylinder from the pipe to the latter.

8 - Disconnect cables from stoplight switch. On the Standard sedan this involves removing the cover of the frame head.

9 - Disconnect speedometer drive shaft from left-hand front wheel hub.

10 - Detach steering column from steering gear and remove complete with steering wheel. Pull off heating pipes after having released the clips on the heating junction boxes.

11 - Detach cables 30 and 50 from starting motor.

12 - Disconnect cables 51 and 61 from the generator and cable 15 from the ignition coil. Also detach cable from oil pressure control switch.

13 - Remove the 2 bolts (metric 17 mm) on each side of the front crossmember.

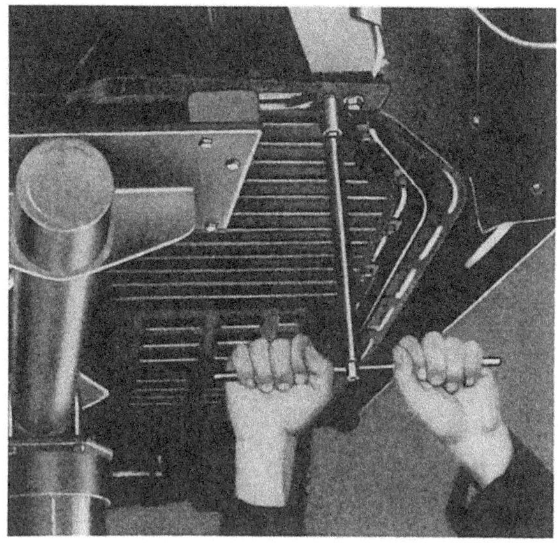

14 - Remove the 18 body bolts (metric 14 mm) from the side members.

15 - Remove the bolts located at the rear quarter panel reinforcement, one on each side.

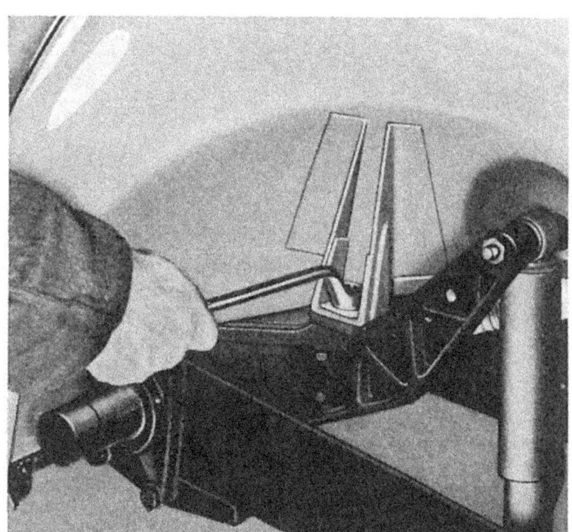

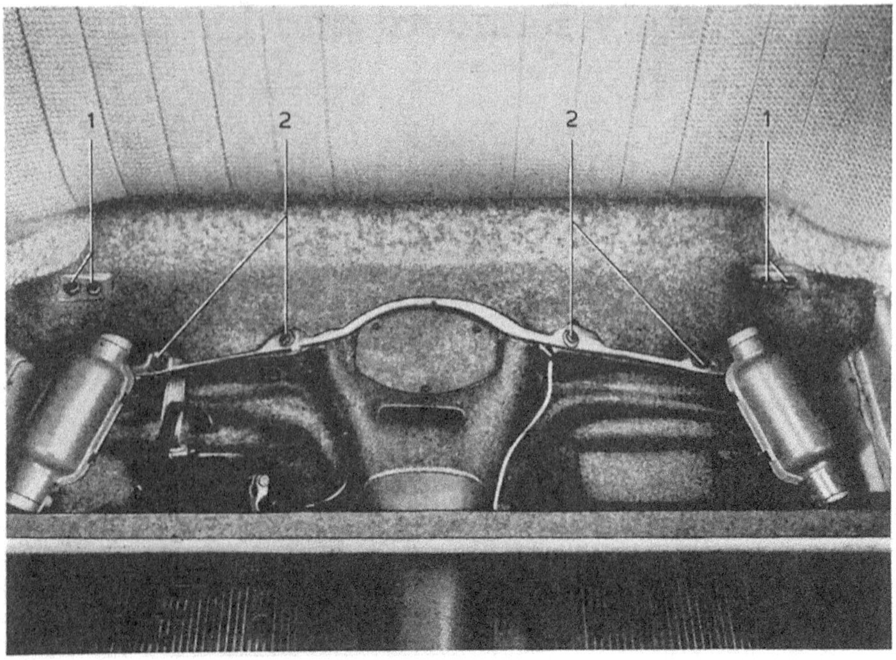

1 - Body attachment at rear cross tube

2 - Body attachment at rear cross-member

16 - Remove the 4 bolts (metric 17 mm) that attach the body to the rear cross tube, and the 4 bolts (metric 14 mm) from the rear cross members.

17 - Remove the two bolts (metric 17 mm) that attach the body to the front axle.

18 - Remove the front quarter panel lining and the covering of the side members on left and right, which are also cemented to the floor plate, if necessary.

Note:

If it is intended to lift off the body by means of the gantry crane VW 605 (for local manufacture), reinstall the wheels and lower the car to the floor.

19 - Attach the hooks of the gantry crane (VW 605 for local manufacture) to the fenders in such a manner that the felt pad inside the hook is not displaced. Lift the body off the chassis and withdraw the chassis towards the rear.

Installation

Before replacing the body, attention should be paid to the points below:

1 - Renew weather strip between body and frame. The weather strip is supplied in lengths of 6 m (20 ft.) and should be cut to size. Attach it to the frame with tacks and join at the front and rear corners with a stapler.

2 - With an 8-mm hollow punch and wood backing, punch holes in the weather strip coinciding with the holes in the floor plates.

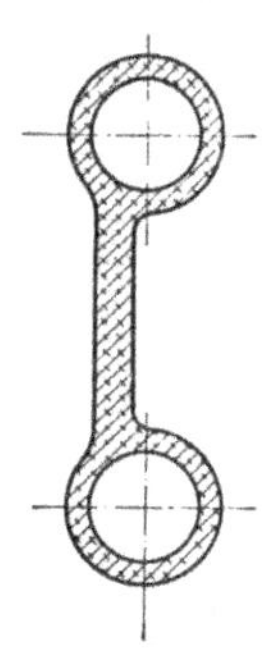

modified seal

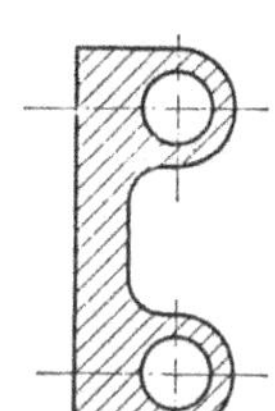

old seal

Note:

From August 1963, Chassis No. 5 677 119, a modified seal is installed between frame and body. The new seal — Part No. 111/141 701 605 A — can be installed in older vehicles. Seals of the former type should be used up.

3 - Above the slots in the front cross members punch holes in the weatherstrip, using a 10-mm hollow punch.

4 - Fill the depression in the weatherstrip, and particularly at the front and rear corners, with Genuine VW Sealing Compound D 9.

Note:

As an additional seal, a rubber packing is inserted in the depression of the weatherstrip between the frame and body in the region of the frame end plate above the rear cross-tube. This became a standard feature from March 1st, 1957 with Chassis No. 1 461 126. Owing to certain modifications to the body and frame this additional packing was omitted as from March 3rd, 1958 with Chassis No. 1 862 104. It is still available as a spare part and can be subsequently installed.

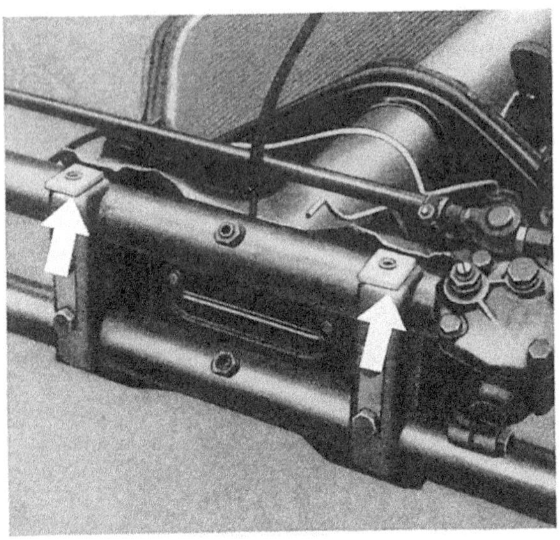

5 - Place rubber packings on front axle and rear cross tube.

6 - To guide the body into place, it is recommended that studs be screwed into the holes at rear cross tube. Remove these studs after the body is replaced.

7 - Replace the body. Take care that the weatherstrip does not become displaced sideways. If necessary attach it at points where it might slip, using adhesive or additional tacks.

8 - After replacing the body, tighten up all 17 mm bolts first, then the 14 mm bolts, on alternate sides in the sequence indicated in the sketch below.

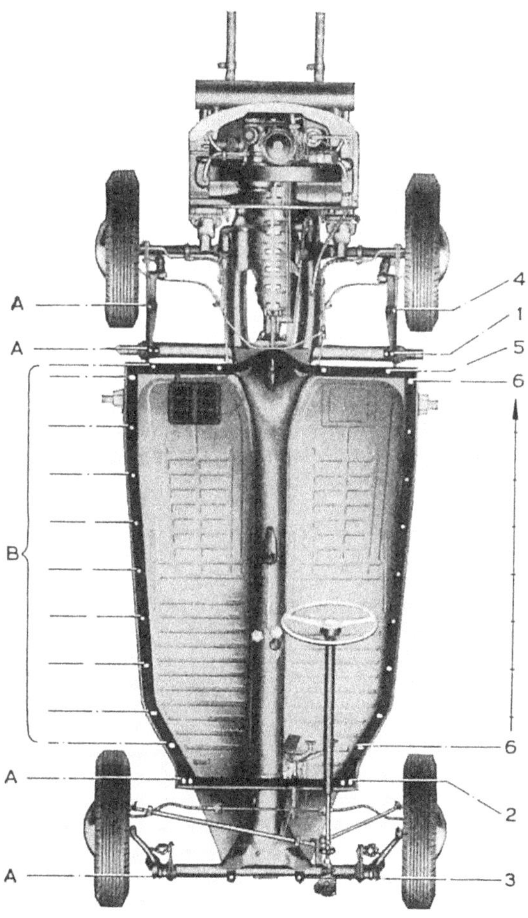

A = 17 mm bolts
B = 14 mm bolts

9 - The 17 mm bolts to be inserted through the front cross-member are screwed into a plate provided with two tapped holes. These plates are freely mounted in metal guides which are welded to the body.

10 - If a gap exists between the packings on the front axle and the reinforcement plate, additional packing pieces should be inserted. The gap should never exceed 3 mm (.12").

11 - Tighten all body mounting bolts to the following torques:

M 8 screws — } 1,5—2,0 mkg
M 10 screws — } (11—14 ft. lbs)

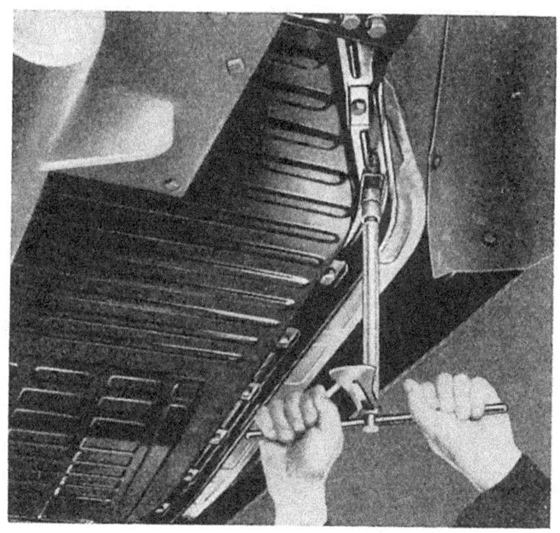

The body seal must be watched carefully when tightening the screws. If it commences to distort, do not tighten the screws further. When a vehicle is in the workshop for repair or inspection, the tightness of the body securing should be checked, using the following torques:

M 10 and M 8 body screws — 1.0—1.5 mkg (7—11 ft. lbs).

Important

Leakage between frame and body cannot be eliminated by tightening the body screws. In most cases this will merely result in the seal being damaged by over-tightening.

12 - When fastening the body to the front axle, ensure that the washers, metal plates, etc. are put on in the correct order.

1 - Mounting bolt
2 - Lockring
3 - Washer
4 - Metal plate
5 - Rubber packing

Note:

From 1st August 1960, Chassis No. 3 192 507 the threaded bushes in the front axle beam into which the body securing bolts are screwed, have been raised 7 mm (.275"). This has led to the following modifications:

1 - The securing bolts are now located below, instead of in, the front reinforcement plate.

2 - When installing a new axle beam in an old vehicle, a second rubber packing piece — Part No. 111 899 115 — must be placed under each of the upper rubber packing pieces — Part No. 111 899 123.

3 - When a new body is installed on a vehicle with an old front axle, a 7 mm thick spacer washer — Part No. 111 899 109 A — must be fitted on each side in addition to the two new packing pieces — Part No. 111 899 115 A — and — 111 899 123 A.

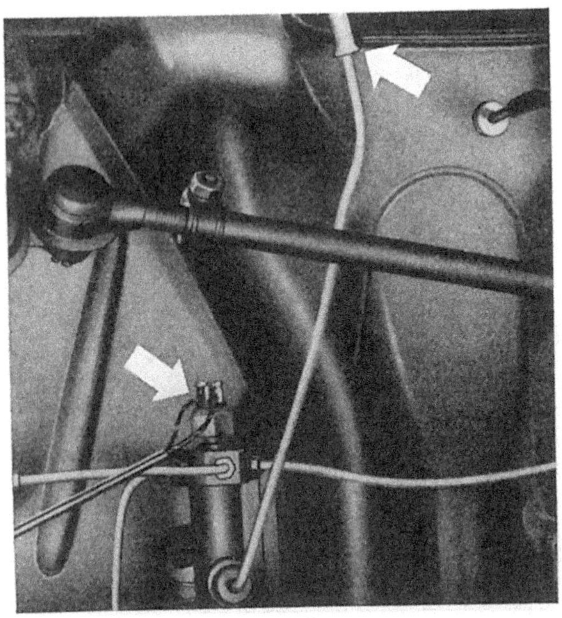

13 - Replace the connection from the reservoir on the pipe to the brake master cylinder and connect the cable to the stoplight switch.

14 - Push the lower lip of the rubber sealing strip for the engine compartment down into position, using a suitable piece of metal. Check that the strip is seating properly.

15 - Replace all parts that were removed, connect the leads to the battery and open the fuel tap.

16 - Bleed the brakes.

Front Fender Removal and Installation

Removal

1 - Place front end of car on a trestle, remove front wheel and spare wheel.

2 - Remove headlamp and remove cable from headlamp housing support.

3 - Remove retaining screws of headlamp housing support.

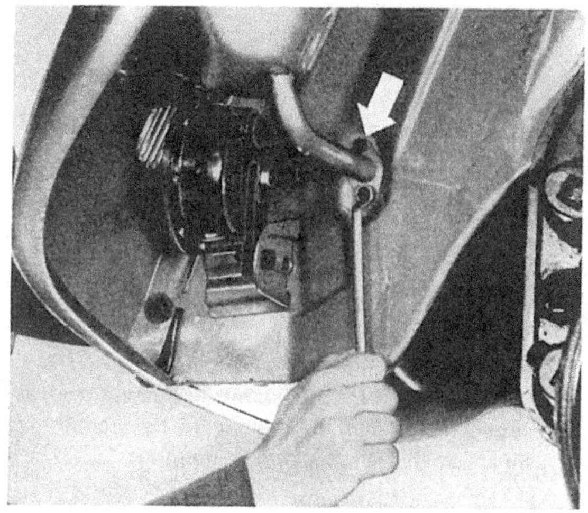

4 - Remove the nut and bolt between fender and sill panel.

Note:

As from April 14th, 1958 commencing with Chassis No. 1 904 235 the 14 mm bolts and nuts for fixing the sill panels were replaced by 13 mm bolts (Part No. N 10 047 2) with nuts (Part No. N 11 131 2).

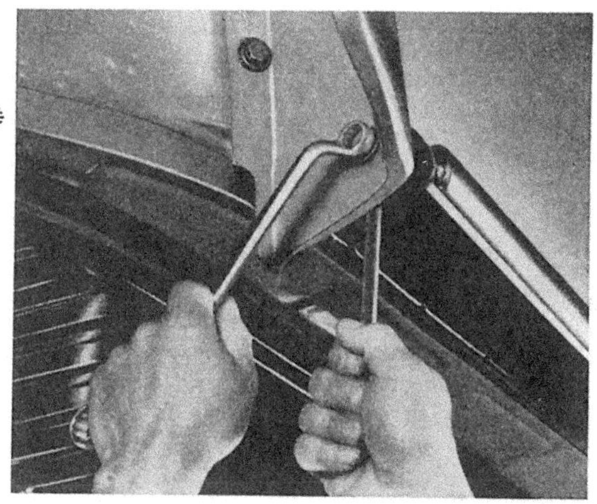

5 - Remove the fender mounting bolts.

Note:

As from April 14th, 1958 commencing with Chassis No. 1 904 235 the 14 mm bolts for fixing the fenders were replaced by 13 mm bolts (Part No. 111 821 145). The lock ring and washer used with the former bolts were replaced at the same time by a lock-washer.

Bolts, lock rings and washers of the old type can be exchanged for bolts and lockwashers of the new type. Both types are still available as spares.

6 - Take off fender and beading.

Installation

When installing the fender observe the following points:

1 - Check condition of fender beading, replace if necessary. Note correct position of beading when installing the fender.

2 - If necessary, recut the threads of the holes for the fender mounting bolts. Grease the bolt.

3 - If necessary, replace rubber washer between fender and sill panel.

4 - Aim the headlights.

5 - Tighten wheel mounting bolts to a torque of between 9 and 11 mkg (65 and 80 ft. lbs.).

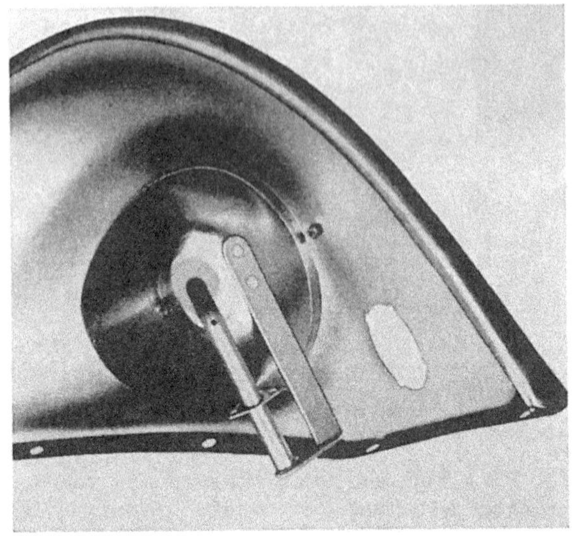

As from February 1958 the fenders are supplied without headlamp housing support (Part No. 111 821 131) and flange for the support (Part No. 111 821 129). Therefore when installing new fenders the support must first be welded to the headlamp housing. For this there are two methods:

Welding on the support:

1 - Without a gage.

2 - With a gage.

1 - Without a gage

a - Attach the fender to the vehicle with 3 bolts.

b - After sliding the flange on the support, place the latter against the hole in the headlamp housing and attach the flange to the front quarter panel with 2 bolts.

c - Weld the support to the headlamp housing at two points.

d - Remove the fender with flange from the vehicle.

e - Finish welding the support to the headlamp housing.

2 - With a gage

a - Screw the appropriate gage to the headlamp housing, according to whether right or left fender.

b - Slide the flange on the headlamp housing support.

c - Push the end of the support over the boss of the gage and place the other end against the hole in the headlamp housing.

d - Weld the support to the headlamp housing. Check for proper welding seam.

e - Remove the gage after the seam has cooled down.

To make the gage

The welding gages are made in accordance with the sketch shown below:

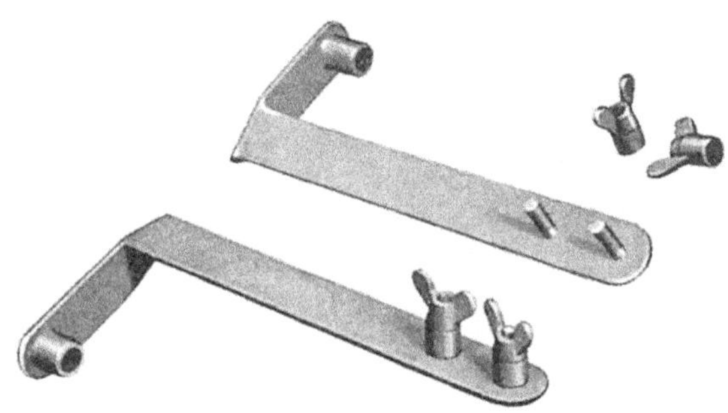

A - 2 flat bars 4 × 25 × 200 mm.

B - 2 angles of flat bar 4 × 25 × 85 mm.

C - 4 wing-nuts M 6.

D - 2 bosses of bar or pipe 15 mm dia. × 15 mm.

E - 4 spacers 12 mm dia. × 8 mm × 10 mm.

F - 4 bolts M 5 × 25 with tapered head 7 mm dia.

To make the gauge

The welding gauges are made in accordance with the sketch shown below:

A - 2 flat bars 4 × 25 × 200 mm.

B - 2 angles of flat bar 4 × 25 × 85 mm.

C - 4 wing-nuts M 6.

D - 2 pieces of bar or pipe 15 mm dia. × 15 mm.

E - 4 spacers 12 mm dia. × 8 mm × 10 mm.

F - 4 bolts M 5 × 25 with tapered head 7 mm dia.

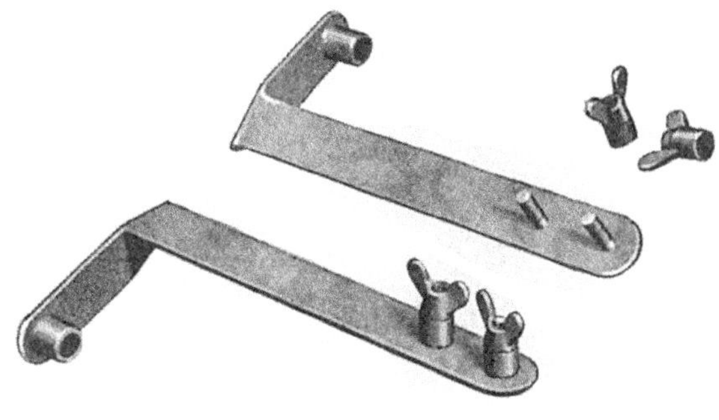

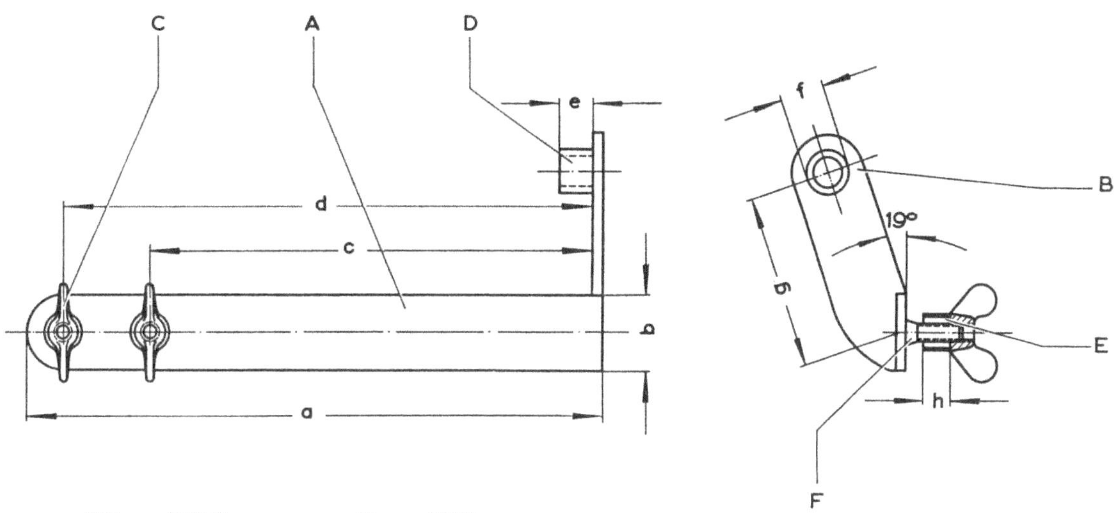

a — 200 mm (7.87″)
b — 25 mm (.98″)
c — 153 mm (6.02″)
d — 183 mm (7.20″)
e — 15 mm (.59″)
f — 15 mm (.59″) dia.
g — 59 mm (2.32″)
h — 10 mm (.39″)

The sketch shows the gauge for the right-hand fender. For the left-hand fender the gauge should be angled to the opposite side.

Note:

From 31st October 1963, Chassis No. 5 888 185, for Models 111—118 and from 1st November 1963, Chassis No. 5 852 837, for Models 151—152, the hole pattern in the front fenders (Part No. 111 821 021 G/113 821 022 G) was altered to suit the new location of the turn indicators.

To simplify parts provisioning of front fenders, the fenders without holes (111 821 021 D and 113 821 022 D) can be stocked instead of 111 821 021 F/G and 113 821 022 F/8.

The holes can then be made to suit the turn indicators with the aid of a template.

Note

The center elongated securing holes in the fender has been reduced in size. To avoid difficulty when fitting the fender, it is advisable to attach the fender at this point first.

THIS PAGE INTENTIONALLY
LEFT BLANK

Removal and Installation of Sill Panels

Removal

1 - Remove screws securing sill panel to front and rear fenders.

2 - Loosen screws securing panel to side member.

3 - Take sill panel off upwards.

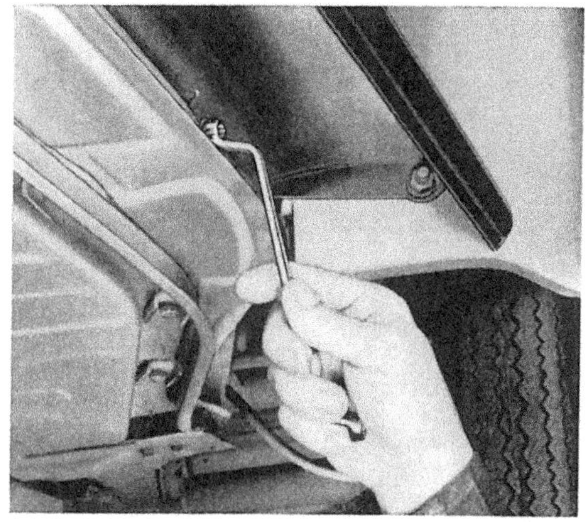

Installation

1 - Install sill panel so that the slots engage behind the washers on the screws.

2 - Place rubber washers between sill panel and fender and fit nuts loosely on screws.

3 - Tighten screws on side member first and then tighten the nuts and screws at the ends.

Fitting New Sill Panel Lining

Removal

1 - Remove sill panel.

2 - Press trim moulding out after bending up the retaining tabs.

3 - Detach lining from securing flange on sill panel and pull it out of the outer fold.

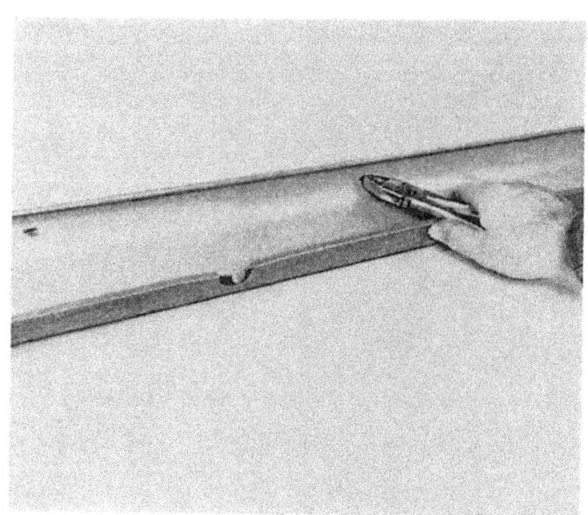

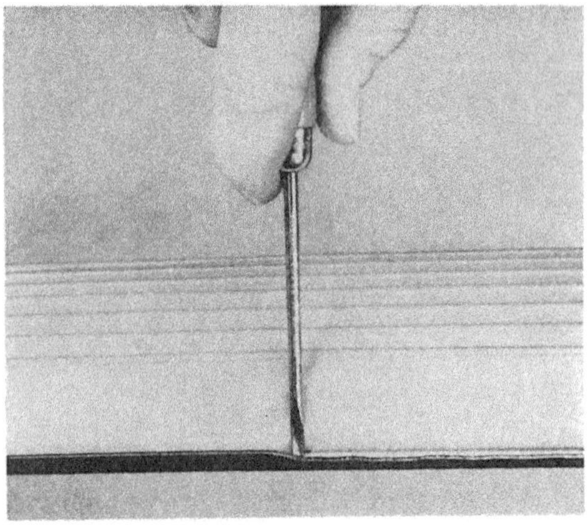

Installation

1 - Before fitting the lining, remove all dirt and rust from sill panel and paint it.

2 - Open the fold on the sill panel slightly, insert lining and tap fold down again.

3 - Secure the rib on the lining under the retaining flange.

4 - Install sill panel.

THIS PAGE INTENTIONALLY
LEFT BLANK

Rear Fender Removal and Installation

Removal

1 - Place rear end of car on a trestle and remove rear wheel.

2 - Remove tail and stop light casing, rubber seal and lampholder from fender, and disconnect cable. Withdraw the cable from the fender.

3 - Unscrew the bracket for the stop and tail light and remove it with rubber gasket.

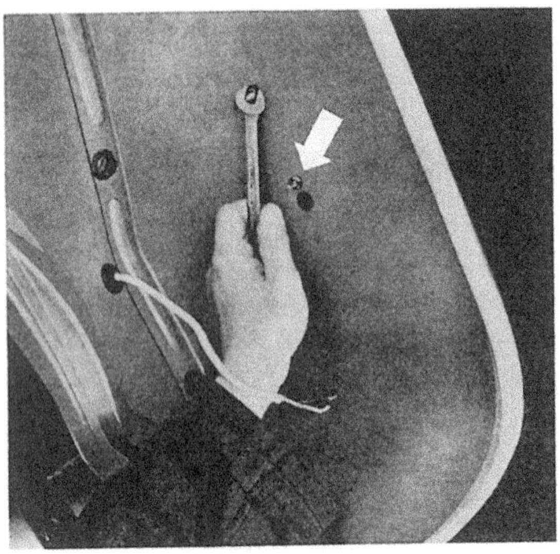

4 - Remove the rear bumper with brackets, and take off the rubber grommet.

5 - Remove the fender mounting screws and the bolt between fender and sill panel.

Note:

From 14th April 1958 and Chassis No. 1 904 235 the former 14 mm bolts used for mounting the fender were replaced by 13 mm bolts (Part No. 111 821 145). The spring and flat washers of the old type were replaced by a lockwasher. The mounting bolt for the sill panel (Part No. N 10 047 2) with nut (Part No. N 11 131 2) is now also 13 mm.

Bolts, nuts, spring and flat washers of the old type can be replaced by the new type. Both types are available as spares.

6 - Take off fender and beading.

Installation

1 - Check condition of fender beading, replace if necessary. Note correct position of beading when installing the fender.

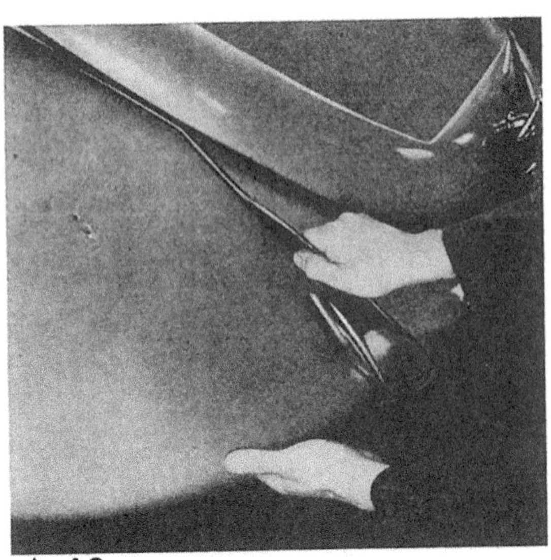

2 - If necessary, recut the threads of the holes for the fender mounting screws. Grease the screws.

3 - If necessary, replace rubber washer between fender and sill panel.

4 - Press bumper bracket grommet into the slot in the fender. Insert bumper bracket from the rear and screw it into place, ensuring that the grommet is seated properly.

5 - Note position of rubber seal between lamp casing and fender, renew rubber seal if necessary.

6 - Tighten wheel mounting bolts to a torque of 10 mkg (72 ft. lbs.).

Front Hood Removal and Installation
(up to 1962)

Removal

To simplify removal and installation of the front hood it is advisable for the job to be performed by two mechanics. The hood can then be properly supported, avoiding damage to the front cowl panel.

Should this not be possible, place a suitable piece of woollen or plastic material on the cowl before commencing.

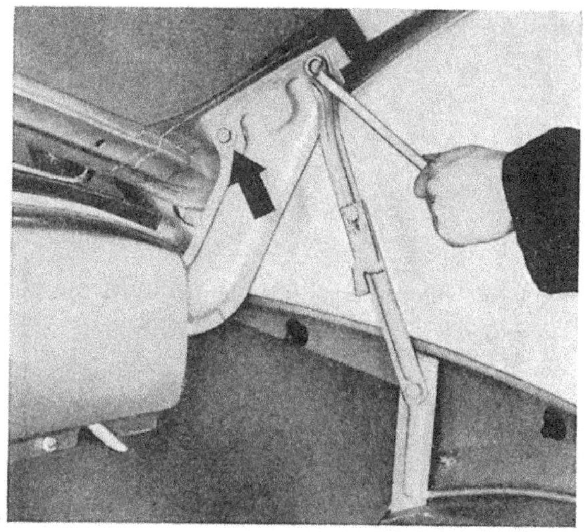

1 - Open the hood and lock the collapsible bracket.

2 - Remove the two bolts from the hinges on either side of the hood and take off the hood.

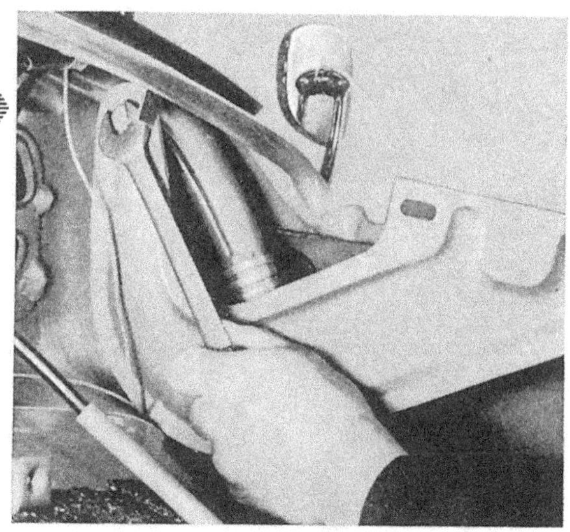

3 - If necessary, screw out the two mounting bolts of the hood hinges, first removing the panel protecting the instrument panel.

4 - If necessary, dismantle the hood handle and lock assembly. Remove the moulding, emblem, VW sign and sealing strip.

Installation

1 - Check condition of weather seal. If necessary, replace it.

It is important that, after removing the weather-strip, the retaining channel be bent apart slightly with a broad-bladed screwdriver. After inserting the new strip force it into position using a suitably shaped block of wood.

2 - Loosely attach the hood to the hinges and adjust hood in the slots until a perfect hood alignment is obtained. Tighten the hinge bolts.

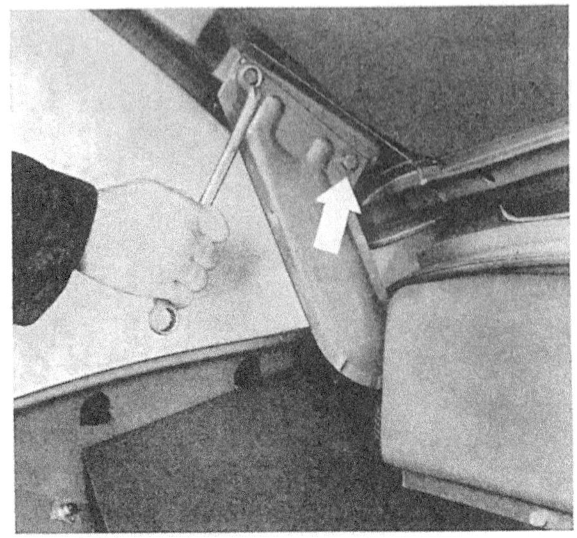

In individual cases it may be necessary to slightly lift the top cowl panel, in order to obtain a good fit between the cowl panel and the hood. The best way to do this is to use the lever jack VW 735 (for local manufacture).

3 - Check the lock assembly for proper functioning by closing and opening the hood several times. If necessary, adjust position of the lock assembly by moving it in its slots.

Front Hood Removal and Installation

(1962 and later)

Removal

To avoid damage to the cowl panel, this operation should be carried out by two mechanics. The cowl panel should also be covered up.

1 - Open hood.

2 - Remove hinge screws and take hood off.

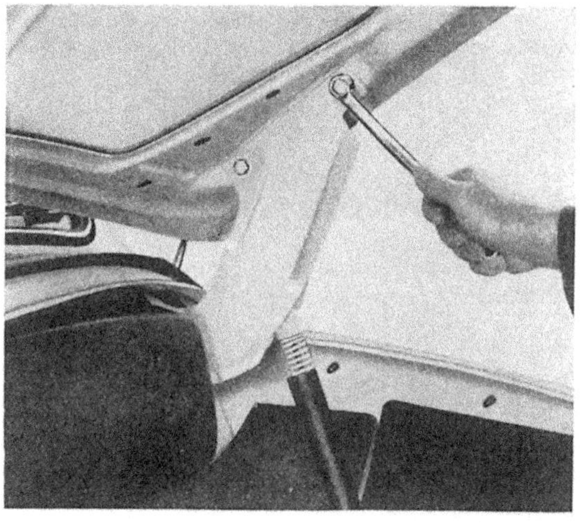

Installation

1 - Before fitting the front hood, check that the seal is in good condition and located properly. If necessary, a new seal should be used.

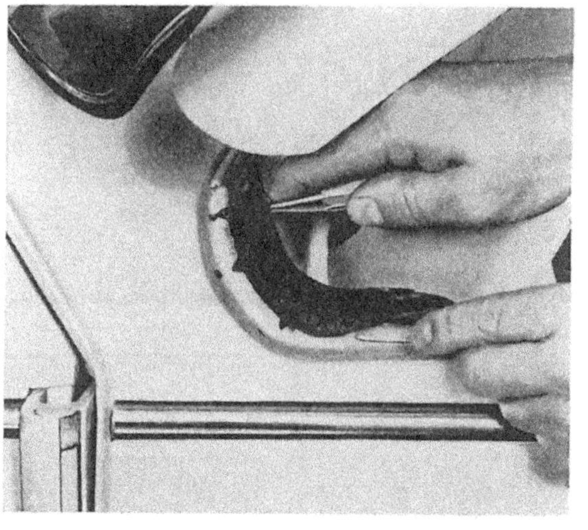

2 - Attach hood loosely and move it in the slots until it contacts uniformly all round and seals properly. The fully tighten the screws.

3 - Check lock by opening and closing hood several times. If necessary, move lock in the slots.

Note:

In some cases it may be necessary to shape the cowl panel to match the curvature of the hood. This can be done by lifting the cowl panel with the lever shown here.

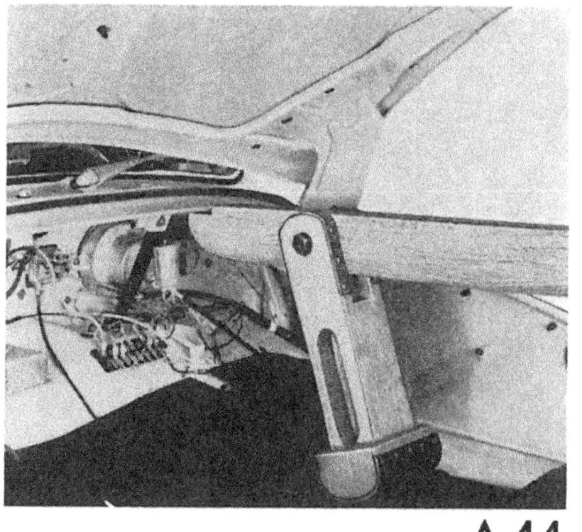

Support Spring Removal and Installation

(1962 and later)

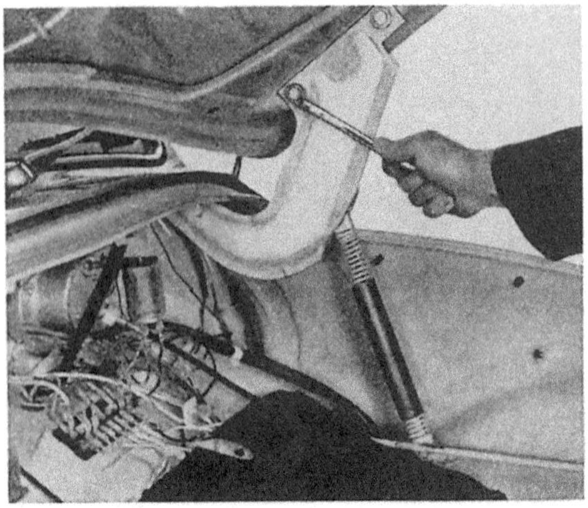

Removal

1 - Remove luggage compartment lining.

2 - Unscrew knurled nuts and take off lining in front of instrument panel.

3 - Remove hood.

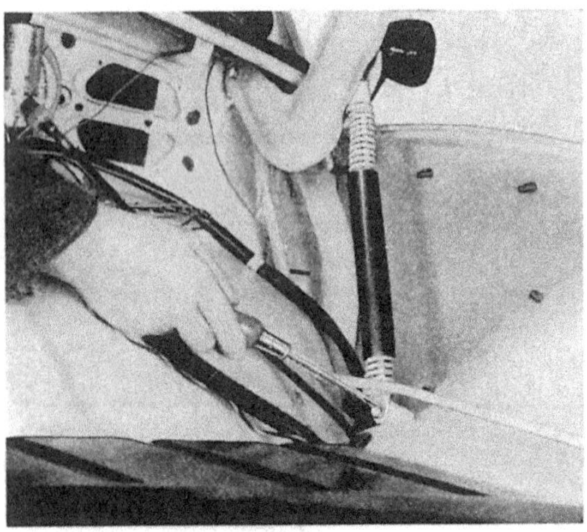

4 - Remove one screw on hinge and the guide bar securing pin after pressing out the lock washer.

5 - Take hinge and support spring off.

> **Important**
>
> If the hood support spring is to be replaced, the riveted pin on the hinge must be ground off and punched out first. When doing this, the spring, which is under considerable pressure, must be pressed against the lower spring plate by a second mechanic.

Installation

1 - Check guide bar, support spring, spring plate and upper guide piece and fit new parts where necessary.

2 - Rivet support to hinge and secure with hexagon screw and pin.

3 - Open and close hood several times to check operation of support.

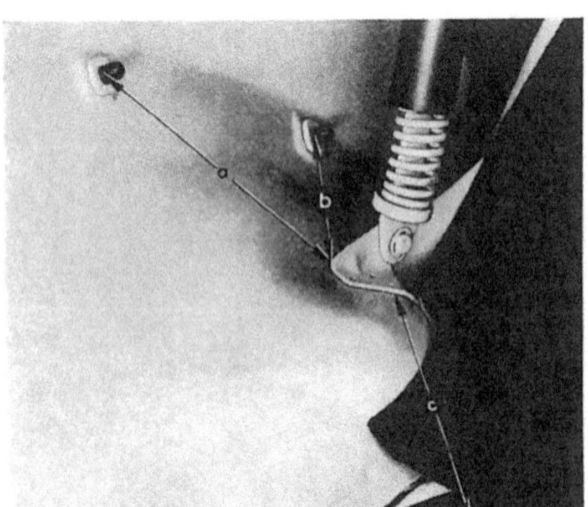

a = approx. 150 mm (5.9")
b = approx. 100 mm (3.9")
c = approx. 200 mm (7.8")

Note:

From 15th June 1965, Chassis No. **115 928 504**, the front hood supports on the 1200 A — Part No. 111 823 315 — were replaced by the spring loaded type as used on the De Luxe Models.

The spring-loaded support can be service installed by riveting the support to the hinge at one end and spot welding the bracket — Part No. 113 809 205 A/206 A — to the side panel as shown in picture. It is not necessary to replace the hinge to do this.

Supports of the previous type will remain available.

Note:

To prevent water entering at the front hood when driving at high speeds in heavy rain, a one-piece weatherstrip was installed from Chassis No. 3 862 489 (4th May 1961).

The new weatherstrip can be service installed in older vehicles.

The weatherstrip (Part No. 111 823 731) is secured in the corners of the cowl panel by means of three lugs at each side. The 4.7 mm (.185") holes for these lugs are drilled with the aid of a template.

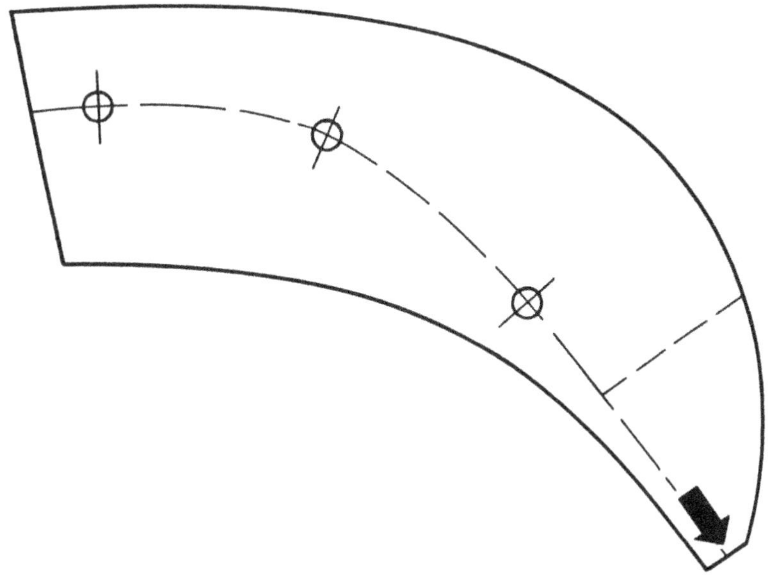

The dotted lines on the template should be aligned with the angled edges of the cowl panel. The end indicated by the arrow should be lined up with the tapering end of the roof where it joins the front side panel.

The weatherstrip Part No. 111 823 705 will be discontinued when existing stocks are exhausted but the weatherstrip Part No. 111 823 711 will remain available.

Note:

If the cowl panel has been drilled with the old template for the weatherstrip Part No. 111 823 733/734 which was introduced (intermittently on 12th October 1960 at Chassis No. 3 364 161, these holes should be filled with solder before drilling the holes with the new template.

Note:

The front hood support was altered from 31st July 1961, Chassis No. 4 010 995. The hood is now held in the open position by two springloaded supports so that it only has to be pushed down when closing.

The supports are riveted to the hood hinges at the top and mounted in brackets welded to the left and right front side panels at the bottom.

Each support consists of the following parts:

Guide rod for spring	113 823 319
Spring seat, lower	113 823 325
Guide piece, upper	113 823 329
Pin	113 823 335
Support spring	113 823 337
8×40 washer for hood hinge	113 823 339
Protective tube for spring	113 823 355

The modified supports cannot be installed in previous vehicles.

Note:

The handle on the front hood was modified from 11th February 1961, Chassis No. 3 666 506. To prevent paint damage, plastic packings are now installed between the handle ends and the hood. The Part Nos. of the packings are: upper 111 823 581 B, lower 111 823 579.

The modified handle, Part No. 111/113 823 565 with the plastic packings can be installed on older vehicles without difficulty. The previous type handles will be discontinued when the existing stocks are exhausted.

Note:

From 1st August 1960 and Chassis No. 3 192 507 the lock cable — Part No. 113 823 531 (VW Convertible) for opening the front hood is being installed on the right of the steering column for vehicles with right-hand steering.

At the lock on the front panel — Part No. 152 823 509 — the cable is on the right-hand side.

Service installation in right-hand drive vehicles is possible.

Front Hood Lock Removal and Installation

From 2nd September 1958 and Chassis No. 2 071 106 the front hood lock was changed. The new lock carrier with pin and lock on the apron can only be exchanged as a unit. Installation is the same for both types.

Removal

1 - Open the front hood.

2 - Remove the two retaining screws of the lock carrier and take off handle and lock bolt assembly.

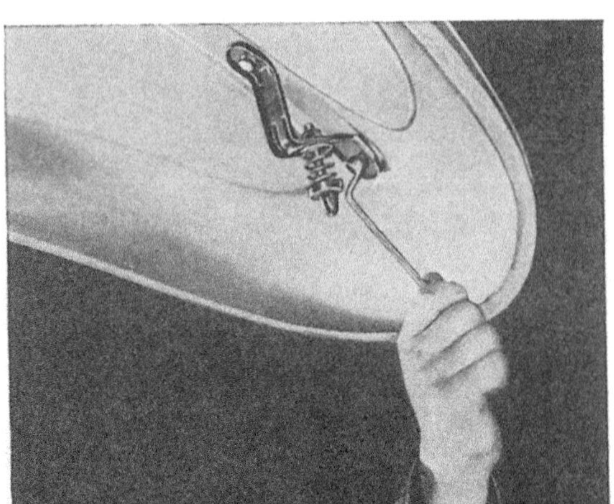

3 - Remove the mounting bolts and nuts.

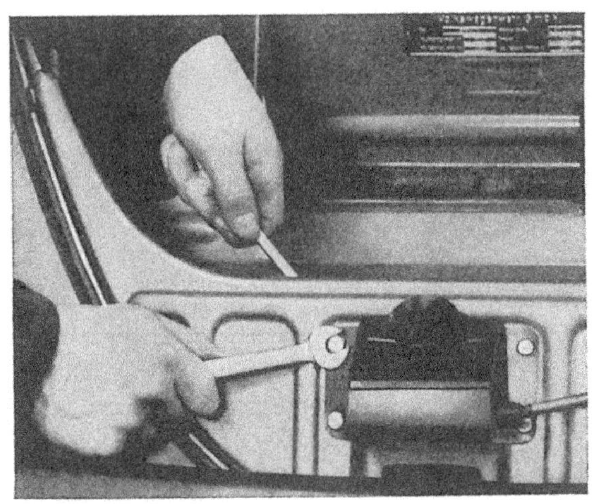

4 - Remove the cover plate of the lock assembly downwards and remove the cable after undoing the clamping screws. Remove the lock assembly.

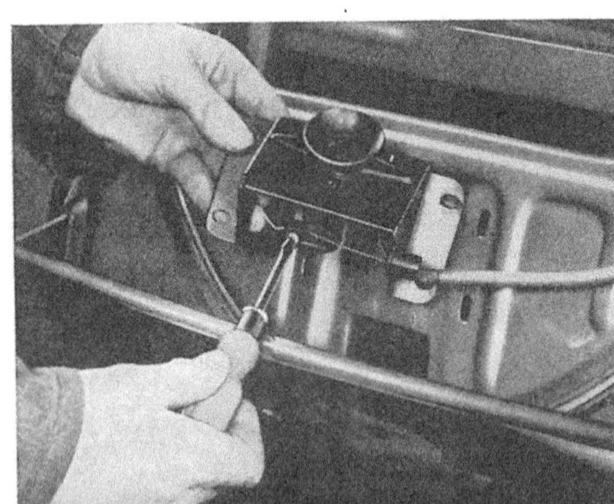

5 - Pull out the hood lock cable from the guide tube.

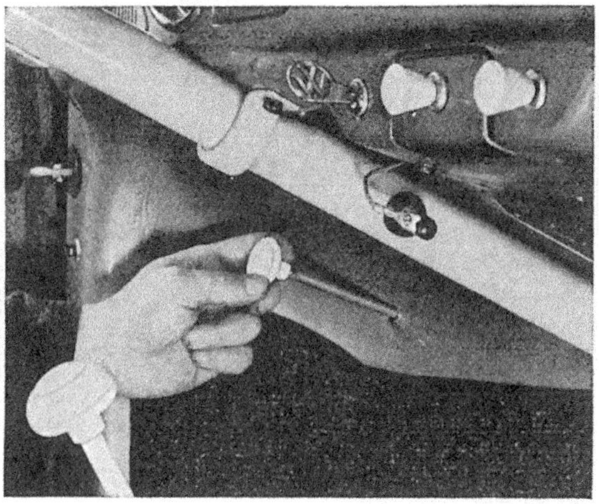

Installation

1 - Grease the hood lock cable and insert it into the guide tube.

2 - The hood lock is so designed that, if the cable should break, the latch which retains the bolt on the hood bracket when the hood is closed springs back automatically. The bolt is then free and the hood can be opened.

When attaching the lock cable, the crank under the latch plate, to which the cable clamping screw is fixed, must be turned **against the spring** until the latch projects into the bolt opening. In this position insert the cable into the lock and clamp it up with the clamping screw. Bend back excess cable.

3 - Install the latch plate assembly and cover plate.

4 - Install hood handle and lock bolt assembly.

5 - Check length of lock bolt and proper position of latch plate assembly by opening and closing the hood several times. If necessary, correct the length of the lock bolt at the adjusting nuts and adjust latch plate assembly by moving it in its slots.

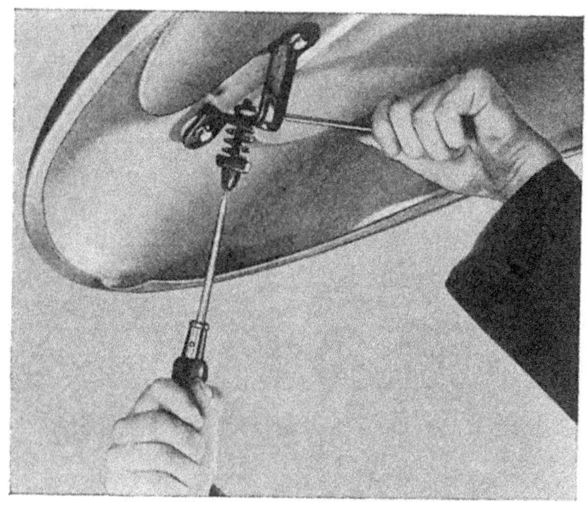

6 - Check adjustment of hood lock cable. If necessary, adjust the cable after removing the cover plate. Bend back excess cable.

7 - Grease lock bolt.

THIS PAGE INTENTIONALLY
LEFT BLANK

Removal

1 - Open the rear hood.

2 - Remove the air cleaner.

3 - Bend up the clips of the license plate light cable and pull the cable out of the clamping plate on the hood.

4 - Loosen the two hinge bolts on each side.

5 - Pull the hood upwards and outwards, thereby unhooking the counterbalancing spring from the hood bracket.

6 - If necessary, remove the hinge carriers on each side, which are attached to the roof by 3 bolts, or remove the hood lock.

Installation

1 - Check the condition of the weatherstrip, replace if necessary. In doing so, the retaining flange should be bent up slightly with a broad-bladed screwdriver. After inserting the new strip turn the flange down again, using a suitably shaped block of wood.

2 - After hooking in the springs, bolt the hood loosely into position, then adjust by moving in the slots until the hood fits uniformly and is water-tight down both sides. Finally tighten up the fixing bolts.

Note:

The rear hood (Part No. 111 827 025, installed up to Chassis No. 1 - 397 022 is no longer available as spare part. Replacements should be made by using the rear hood, Part No. 111 827 025 D (from Chassis No. 1 - 397 023 to 1 - 929 746).

As the stop light was accommodated in the license plate light housing up to October 1952, the tail lights must be modified on all of these vehicles when installing a later rear hood. This is being done by installing the combined tail and stop lights, Part No. 111 945 095 or 111 945 096 respectively, used from Chassis No. 1 - 929 746 onwards.

Apart from a new hood and the license plate light, the following parts are additionally required:

1 tail and stop light, left	111 945 095
1 tail and stop light, right	111 945 096
2 rubber seals for stop and tail light	111 945 191 B
1 cable for tail light, left	111 971 165 A
1 cable for tail light, right	111 971 166 A
2 bulbs	N 17 736 1

Modification

1 - Remove both tail lights incl. cables.

2 - Provide the right and left rear fenders with three holes each as specified on the drawing.

3 - Run new cables and connect to the fuse box in the engine compartment as shown on the wiring diagrams.

4 - Install and connect tail and stop lights.

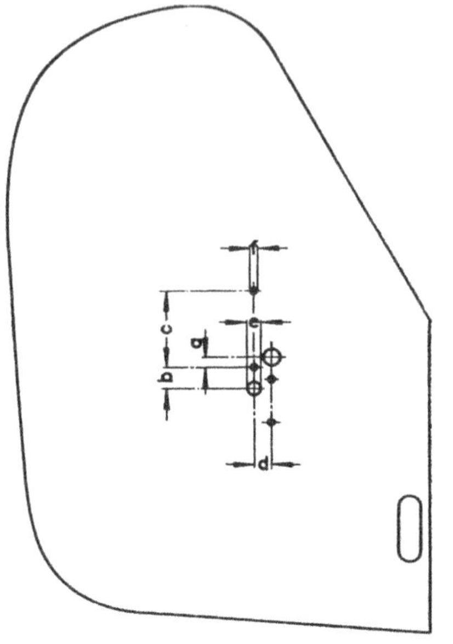

a - 8 mm (0.315")
b - 15 mm (0.591")
c - 73 mm (2.874")
d - 16 mm (0.630")
e - 12 dia. (0.472")
f - 5.5 dia. (0.217")

New Wiring Diagram

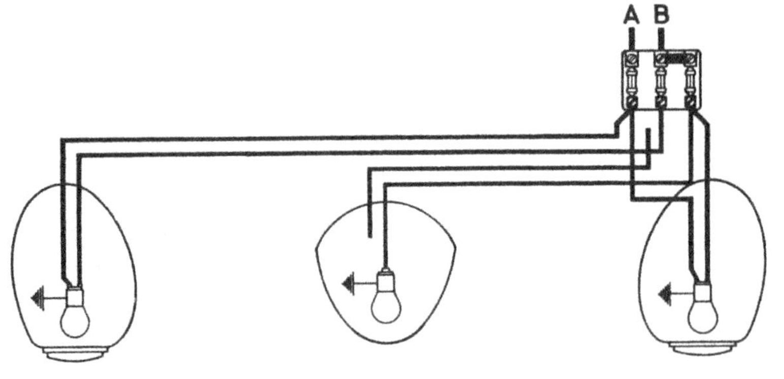

Earlier Wiring Diagram

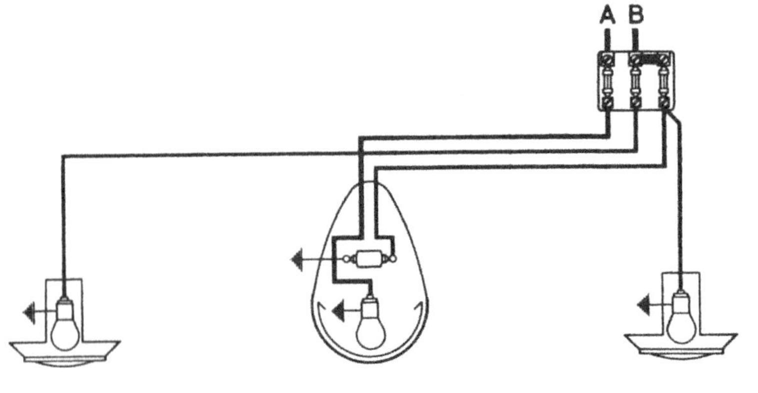

A - to stop light switch
B - to light switch

Note:

From August 1963, the rib in the rear hood was altered and the holes for the license plate screws moved on the De Luxe Sedan and the Convertible to suit the wider license plate light.

The new rear hood — Part No. 113 827 025 and 151 827 025 D — can be service installed if the new license plate light is fitted at the same time. The previous type rear hoods — Part. No. 111 827 025 E and 151 827 025 C — will remain available as spare parts.

Rear Hood Lock Removal and Installation

Removal

1 - Lift the rear hood.

2 - Remove retaining nut from inside of hood.

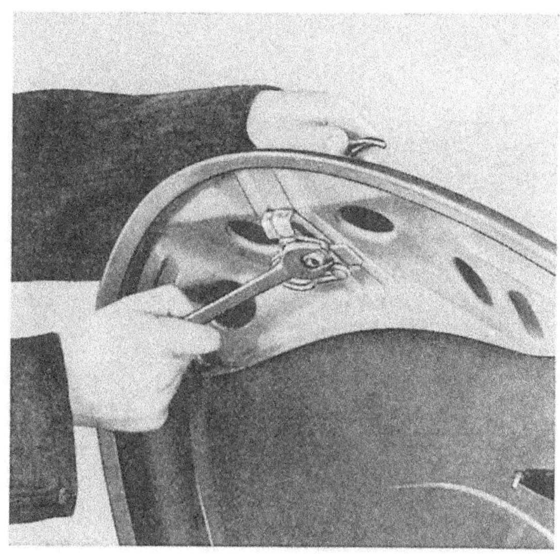

3 - Hold the lock and pull out handle and escutcheon.

Installation

1 - Check condition of gasket and replace as required.

2 - Position hood lock in the guide on the inside of the hood and insert complete door handle from the top.

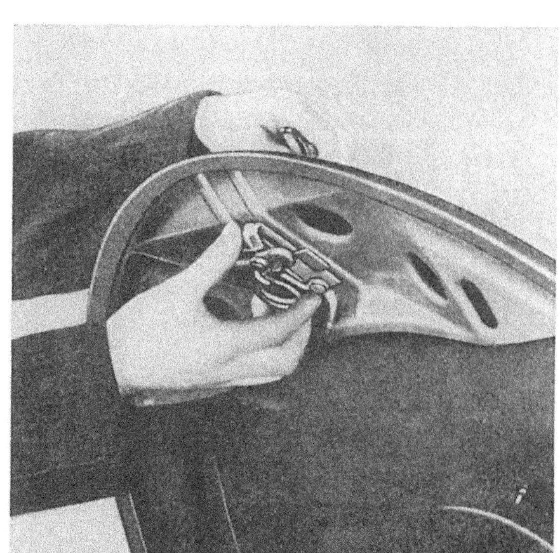

3 - Tighten nut so far that the lock still works easily.

Note:

From 3rd August 1964, Chassis No. 115 000 001, the "T" handle rear hood lock was replaced by a press button type lock. The lock carrier has been discontinued.

To make the lid close softly, a spring with a toggle link was fitted. To prevent this slightly longer spring from touching the oil bath air cleaner, it is hooked in at the top of the support bracket. At the same time, the stops on the hood hinges have been increased in height from 7 mm to 12 mm (.275 to .472").

The following parts have been altered:

Part	Part No.
Hood, rear	111 827 025 H
Hood, rear	113 827 025 A
Hood, rear	151 827 025 E
Apron, rear	111 813 301 H
Hood lock, lower part	113 827 503
Hood lock, lower part (lockable)	113 827 503 A
Packing for knob	113 827 517
Lock cylinder with key	111 827 573 B
Hood lock, upper part	111/113 827 507
Striker plate for lock	111 827 513
Spring for hood	111 827 329

Hoods of the previous type will be discontinued, but all other parts remain available.

Service installation

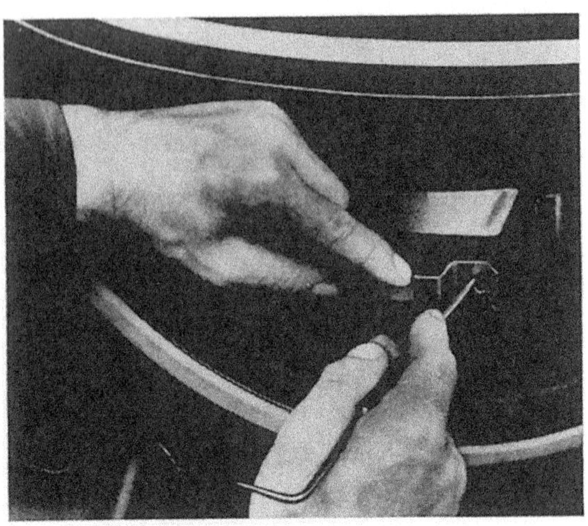

1 - When a new hood is installed in an older vehicle, the complete press button lock and the striker plate must also be replaced.

The striker plate is attached over the old plate on the rear apron with two hexagon N 10 216 2 screws. When marking off the two holes, place the new plate on the old one so that the upper edges are in line. Mark off holes, center punch and drill 6.2 mm (.24") holes.

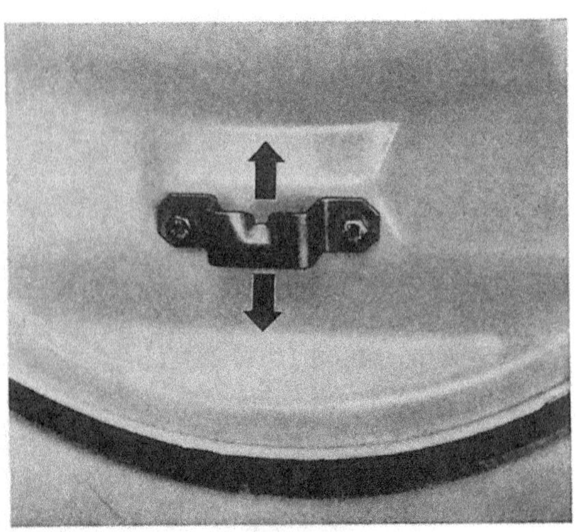

If the hood does not open or close properly, the striker plate should be moved slightly.

2 - The spring with toggle link can only be service installed *together with the new hood. Ensure that* the spring is hooked into the upper recess in the support.

Front Bumper Removal and Installation

Removal

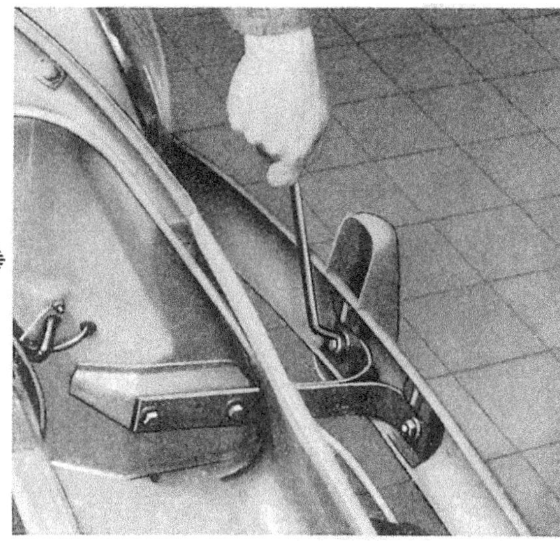

1 - Lift front hood and remove the spare wheel.

2 - Remove bumper bracket bolts on both sides and remove the bumper, reinforcement plates and overriders.

3 - Detach the left and right-hand bumper brackets and pull them out of the front apron.

4 - Take off bumper bracket grommets.

Installation

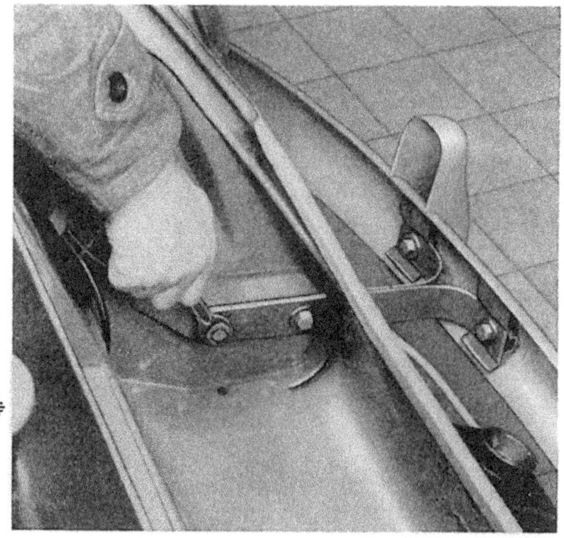

1 - Check condition of bumper bracket grommets and replace as necessary. Position grommets in the front apron.

Note:

From April 1963, Chassis No. 5 419 817, the bumper seals are made of plastic instead of rubber. Only the new seal — Part No. 111 707 197 A — will be supplied as a spare part. This seal is suitable for front and rear bumpers. The former seals — Part No. 111 707 197/347 — have been discontinued.

2 - Install the bumper and brackets.

3 - Ensure uniform clearance between bumper and fenders.

Rear Bumper Removal and Installation

Removal

1 - Remove bumper bolts and nuts and take off bumper and the four reinforcement plates.

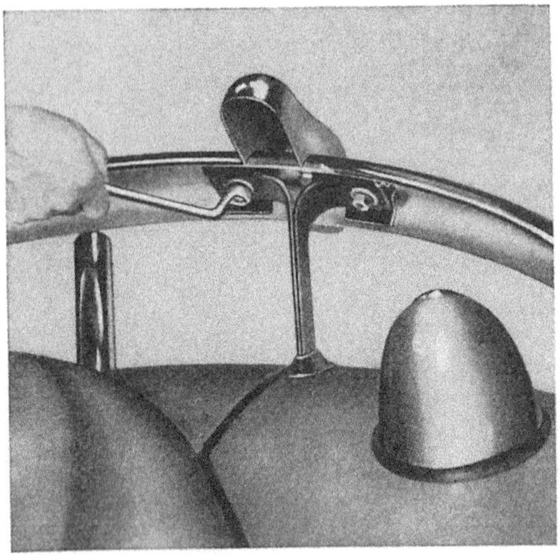

2 - Remove left and right overriders.

3 - Remove bumper brackets on both sides and withdraw through the fender.

4 - Remove bumper bracket grommets.

Installation

1 - Check condition of bumper bracket grommets and replace as necessary. Position grommets in rear fenders.

2 - Install the bumper and brackets.

3 - Ensure uniform clearance between bumper and fenders.

Front Bumper Removal and Installation

(USA Version)

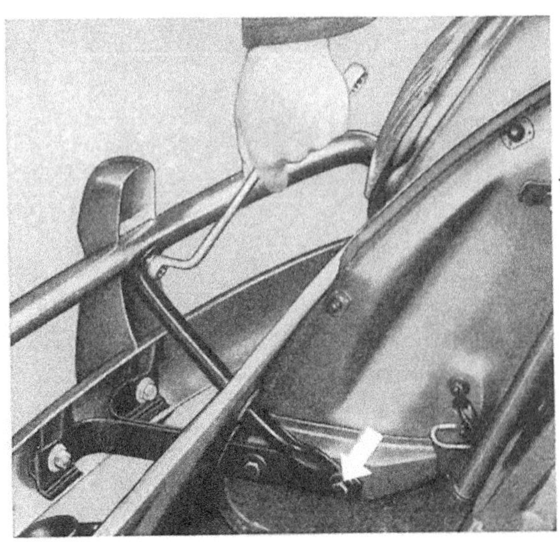

Removal

1 - Remove the two bolts from the overrider supports left and right.

2 - Detach the bumper brackets left and right.

3 - Remove the bumper with extra bow, overriders and supports from the front apron.

4 - Remove the two mounting bolts left and right, and take off the extra bow.

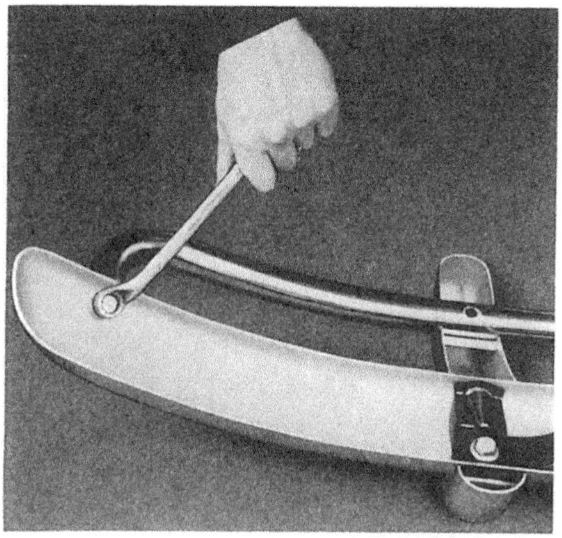

5 - Unscrew the bumper brackets and remove the overriders complete with reinforcement plates.

6 - Take off rubber grommets for bumper brackets and supports from apron.

Installation

1 - Check condition of bumper bracket grommets, replace if necessary. Position grommets in holes provided.

2 - Reassemble bumper with extra bow, overriders, reinforcement plates and brackets.

3 - Insert bumper brackets in the openings provided.

4 - Insert the overrider support and bolt to the body with the bumper.

5 - Bolt the supports to the overriders.

Rear Bumper Removal and Installation

(USA Version)

Removal

1 - Remove the two mounting bolts of the overriders left and right.

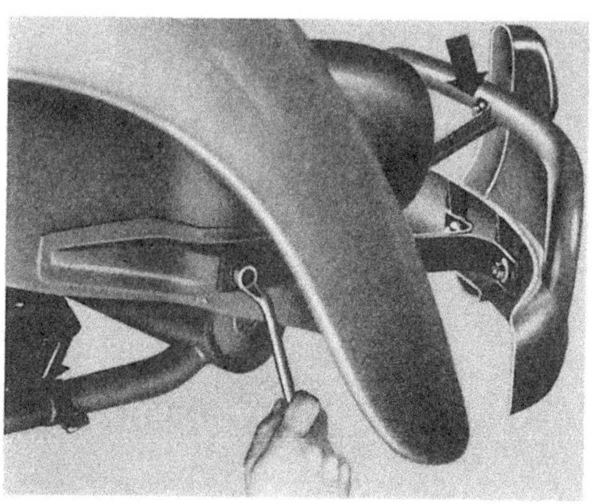

2 - Remove the bumper brackets on either side.

3 - Remove the bumper with the two extra bows, overriders and brackets from the rear fender.

4 - Remove the two mounting bolts and detach the extra bows left and right.

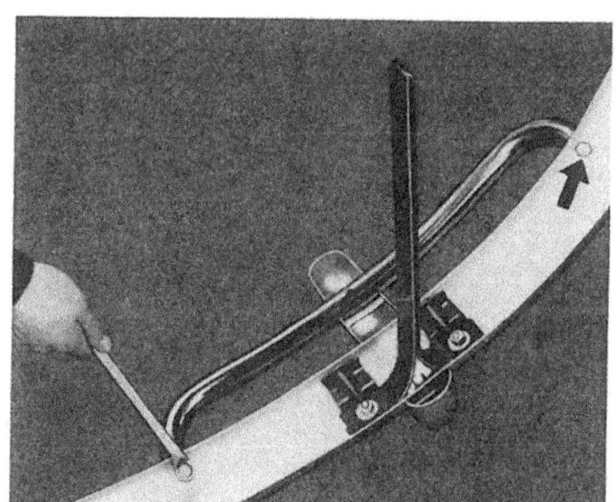

5 - Detach the bumper brackets with overriders and reinforcement plates.

6 - Remove the grommets for the bumper brackets and the overrider supports from the rear fenders.

Installation

1 - Check condition of the grommets, replace if necessary. Insert grommets in the holes provided.

2 - Reassemble bumper with extra bows, overriders, reinforcement plates and brackets.

3 - Insert the bumper brackets in the openings provided.

4 - Insert the overrider supports and bolt to the body with the bumper.

5 - Bolt the supports to the overriders.

Door Removal and Installation

Removal

1 - Remove the check rod after withdrawing the pin from the door hinge pillar.

2 - Remove the rubber plugs from the door hinge to gain access to the inner hinge screws.

3 - First loosen the four Phillips screws of the door hinges with a hammer and metal-headed screwdriver.

Remove the countersunk screws with a Phillips screwdriver. Hold the door while doing so.

4 - Lift off the door.

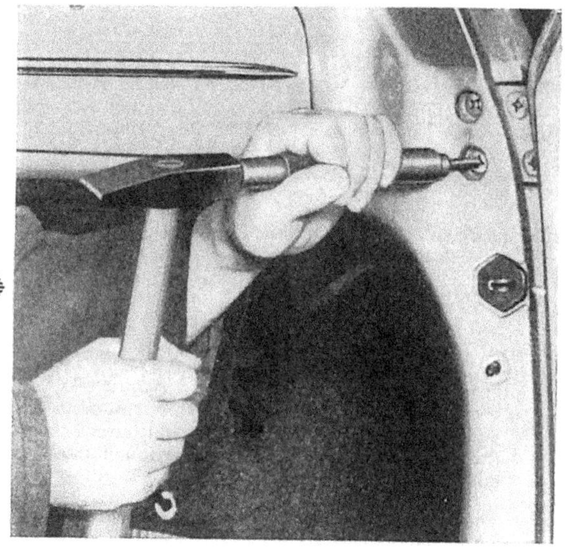

Note:

From Chassis No. 4 671 926, the door hinges are attached to the hinge pillar with three screws instead of four.

The clearance holes in the hinge pillars have been reduced to three but the hinges still have four holes. The part number of the instrument panel has not been changed. The modified instrument panel can be installed in previous vehicles without difficulty.

Installation

1 - Install the door and fit it into the body opening so that it will open and close freely and presses uniformly against the weather strip all round. To check this it is advisable to remove the striker plate.

The hinges are bolted to adjustable threaded plates, which allows the door to be correctly adjusted and aligned with the exterior of the body.

2 - Check condition of door weather strips, replace as necessary. Cement the new weather strips into place, using Genuine VW Compound D 12.

3 - Adjust the striker plate on the body until the door is flush with the rear quarter panel and closes easily (see Adjustment of Striker Plate).

4 - Oil the door hinges, making sure the oil slots are clean.

The contact surfaces of the latch housing and the striker plate and wedge should be greased lightly with Vaseline or molybdenum disulphide paste.

The lock cylinders must be lubricated with powdered graphite only. Under no circumstances may oil or grease be used. It is recommended that the key is dipped in the graphite powder and then turned back and forth several times in the lock.

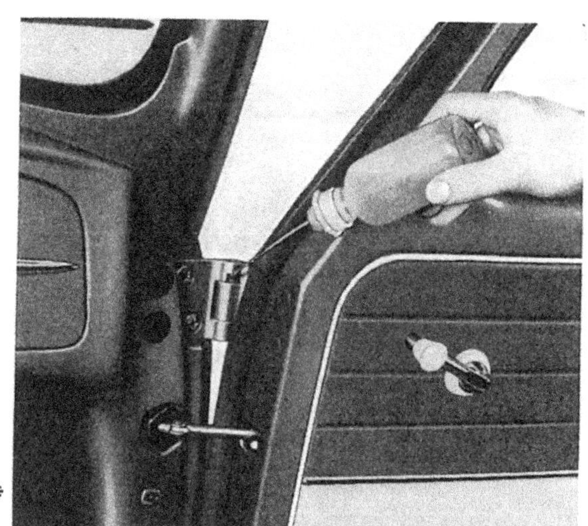

Note:

To prevent noises occurring at the door check rod, a spring — Part No. 111 837 275 A — was installed under tension between the check rod mounting and the rod from 1st August 1960 and Chassis No. 3 192 507.

Note:

From Chassis No. 4 010 995 (31 st July 1961), the doors are equipped with modified check rods — Part No. 111 837 249 B — which hold the door firmly in the fully opened position.

Note:

Only the new doors — Part Nos. 111 831 051 G/052 E, 151 831 051 D/052 D, 113 831 051 G/052 E will be supplied for vehicles from Chassis No. 1 929 746 (August 1955). When a complete door has to be replaced on these vehicles, the modified door check rod must also be installed.

The modified check rod cannot be installed in doors of the previous type.

Note:

a - From Chassis No. 4 060 506 (23rd August 1961), a grooved pin (Part No. N 12 810 1) was fitted in the door check rod. On older vehicles which have check rods without these pins, it may be found that the doors close too easily.

The force required to close the door should be 2.5+1.5 kg (5.5+3.3 lbs.). This can be measured by hooking a spring balance into the oil hole for the door lock.

If the spring balance indicates less then 2.5 kg, rectify as follows:

1 - Remove check rod.
2 - Fit a pin into the rod as shown.
3 - Install the check rod.
4 - Open and close door several times.
5 - Check force required to close with the spring balance. If this is still below 2.5+1.5 kg, install a new check rod with pin.

b - From March 1963, the check rod cover on the Sedan (from Chassis No. 4 377 334 and on the Convertible (from Chassis No. 5 393 245) was provided with an additional sealing lip to prevent drafts at the front edge of the door.

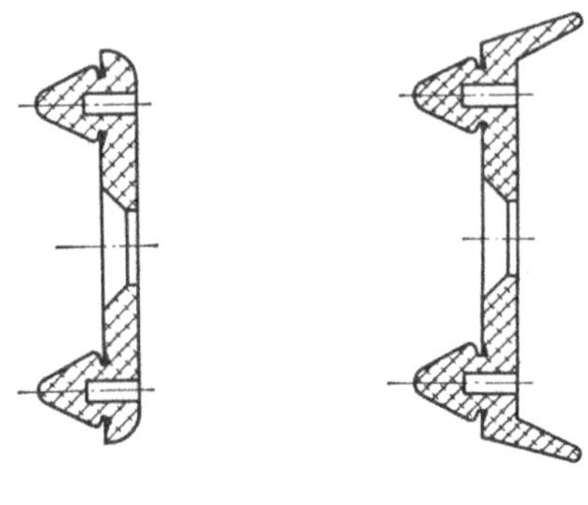

Note:

From Chassis No. 4 357 893, the door hinges are no longer oiled in production, to avoid damage to new vehicles in transit due to dripping oil. The hinge pins are treated with molybdenum disulphide before assembly.

> **Important**
>
> With immediate effect, all door hinges should be oiled during the delivery inspection, before the vehicle is handed over to the customer.

Note:

From Chassis No. 5 264 811, the door hinges are oiled after painting. The hinge pins are still treated with molybdenum disulphide before assembly.

> **Important**
>
> All hinges must still be oiled during the delivery inspection.

Door Window Removal and Installation

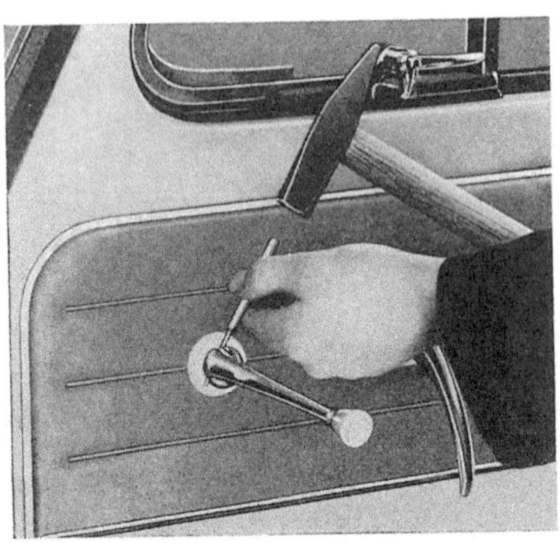

Removal

1 - Press the escutcheons for the window regulator and inside handle against the door trim panel, exposing the tapered pins. Knock out the pins with a punch, remove handle, window regulator and escutcheons.

2 - Pull out the door trim retaining springs and remove the door trim panel. This should be done with care to avoid damaging the paintwork.

3 - Remove the escutcheon *springs and rubber* buffers from the regulator and door handle shaft.

4 - Remove the 4 regulator attaching screws (A).

The arm-rest on the passenger door is provided with a support which can be removed after taking out the screws.

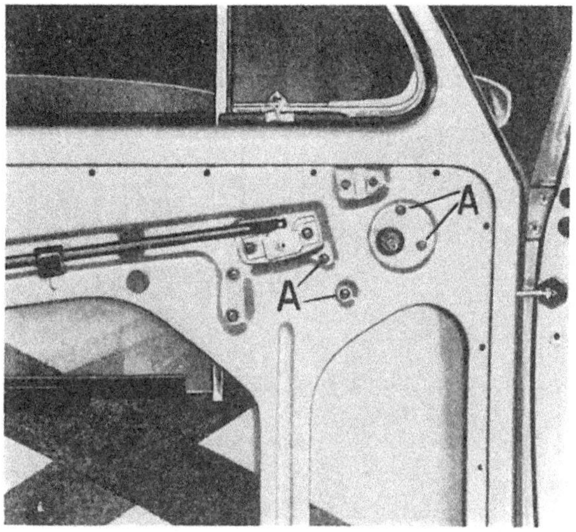

5 - Push the regulator inwards and pull downwards. Remove the linkage with ball-joint from the bottom retainer channel of the window and take the regulator out of the door.

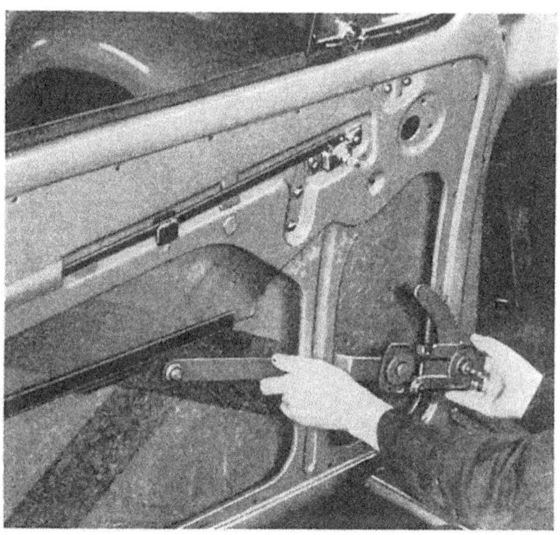

Note:

From Chassis No. 4 672 922, the window lifter is located in the window lifter channel with a spring (Part No. 111 837 507) instead of the roller.
The modified window lifter (Part No. 111 837 501 E / 502 E) can be installed in older doors without difficulty. Stocks of the former lifter (Part No. 111 837 501 D / 502 D) should be used up.

Note:

From Chassis No. 4 420 885, the following modifications were made to the doors:

a - The **rubber seal for the glass** in the window lift channel was altered. The present seal — Part No. 311 837 565 A — has been extended so that it forms a water drain channel.

b - The **window lift channel** — Part No. 111 837 571 D/ 572 D (new) — was lengthened. Water which has entered at the glass is guided by the lift channel and the modified seal to the side window guide channels where it can run down.

c - The **window winder** was fitted with an additional guide plate. The guide plate prevents the rubber seal from being bent over. The window winder with guide plate — Part No. 111 837 501 E/502 E — can be service installed in previous vehicles.

d - **Service installation**
The rubber seal and the window lift channel can only be installed in previous vehicles together with the modified window winder.

6 - Remove the two fixing bolts for the front glass-run channel.

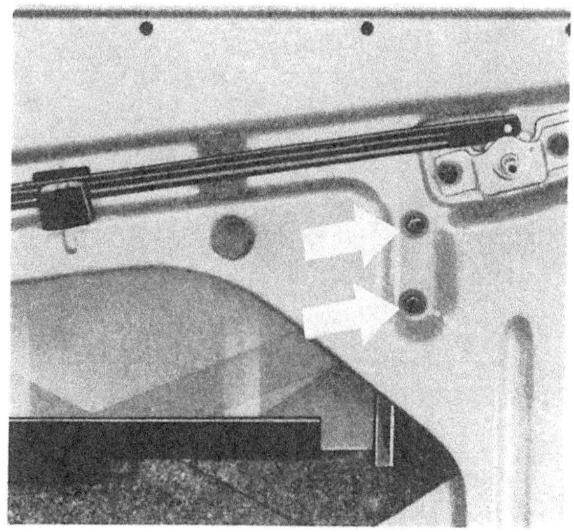

7 - Pull the window down. In doing so, push the front glass-run channel slightly forwards so that the glass can be removed from the channel. Tilt the glass slightly and take it out of the door.

8 - If necessary, remove the fixing screws at the top and rear of the window opening and at the bottom of the rear glass-run channel, and lift the rear channel out upwards.

The upper and side fixing screws of the rear glass-run channel of the De Luxe Sedan also hold the decorative frame with the rubber weatherstrip for the door window.

9 - Slightly raise the rear glass-run channel, unhook it from its upper mounting and pull it out downwards.

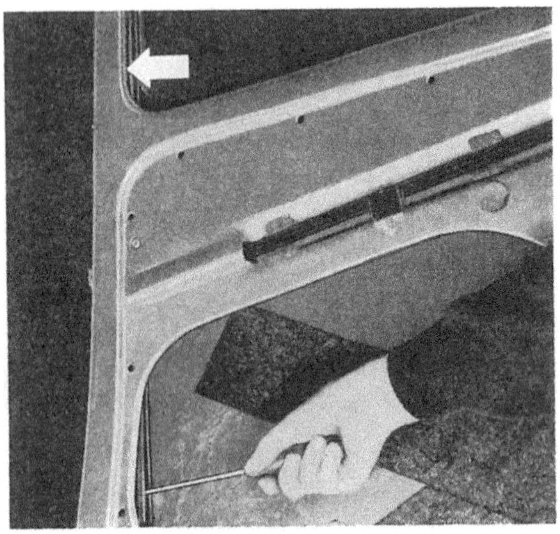

Note:

To prevent the glass-run channels from squeaking the method of fixing the rear channel was changed from 20th January, 1958 and Chassis No. 1 804 134.

The rear glass-run channel of the new type (Part No. 111 837 419 A) now has a spring clip (Part No. 111 837 421) which is fixed in the top of the window opening in the door instead of a sheet-metal lug. This ensures that the channel is held more firmly than before.

Rear glass-run channels of the old and new types are **not** interchangeable.

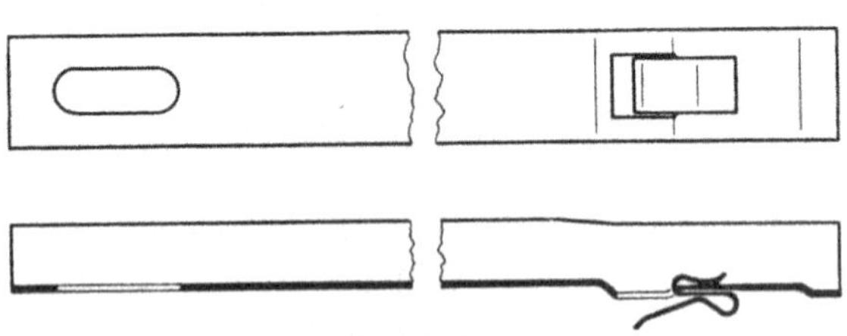

New design

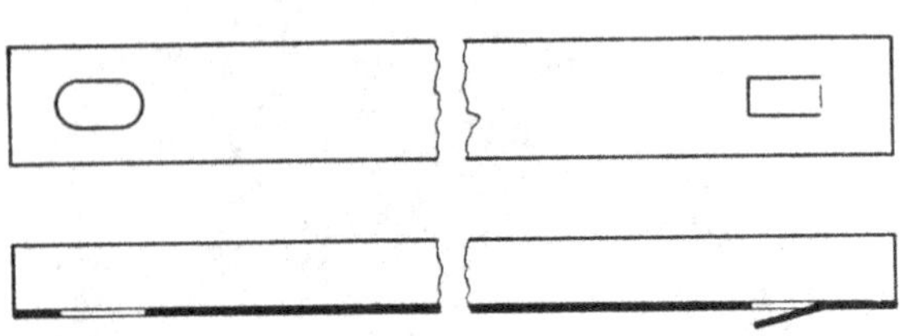

Old design

From 30th December 1959 and Chassis No. 2 799 685 the rear window run channel was shortened by 80 mm (3.15"). The window run channel is held in the run channel retainer by a clip (Part No. 111 837 443). The run channel retainer is now secured to the door by a fillister head screw (Part No. N 141301) with a spring washer (Part No. N 12 225 1).

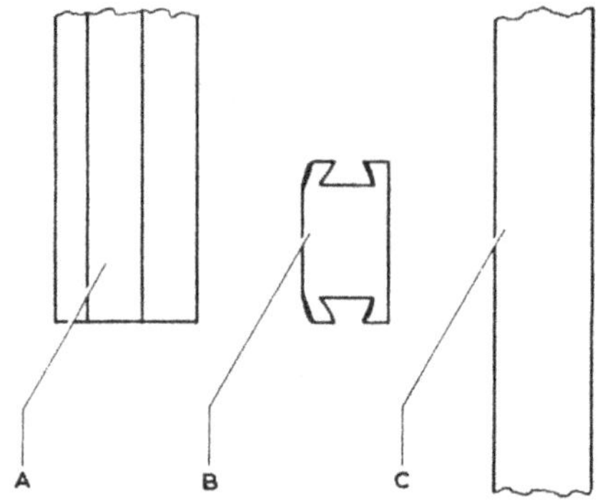

A = Window run channel, rear
B = Clip
C = Run channel retainer

10 - If the regulator bar at the bottom of the window or its rubber gasket is damaged, replacement is possible.

The bottom bar is tapped off carefully, using a rubber hammer and a suitable block of wood. Reassembly is carried out in a similar manner.

11 - When the window-pane is removed, the moulding with rubber strip on the outside of the door, and the rubber strip on the inside of the door, can be replaced.

To simplify removal of the moulding for the window it is advisable to remove the clips holding this moulding by compressing them and pushing them out.

The weather strip for the window compartment opening, which is fitted to the inside of the door with wire clips, can be detached by hand or with the aid of a screwdriver.

Note:

From 19th March 1959 and Chassis No. 2 342 782 the sealing strip riveted to the door window garnish molding (Part No. 113 853 321/322) was replaced by a modified strip (Part No. 113 853 321 A/322 A). The new sealing strip is longer and has a wider lip to increase the contact area. The lip is reinforced by longitudinal ribs. The end towards the door lock pillar follows the radius of the door window opening. The angled edge of the door outer panel to which the garnish molding and the sealing strip are attached is now nearly parallel to the door window glass. The new type sealing strip can be installed *on earlier cars. When current stock is* exhausted only new type sealing strips will be supplied.

12 - When removing the front glass-run channel, the two upper fixing screws must be removed.

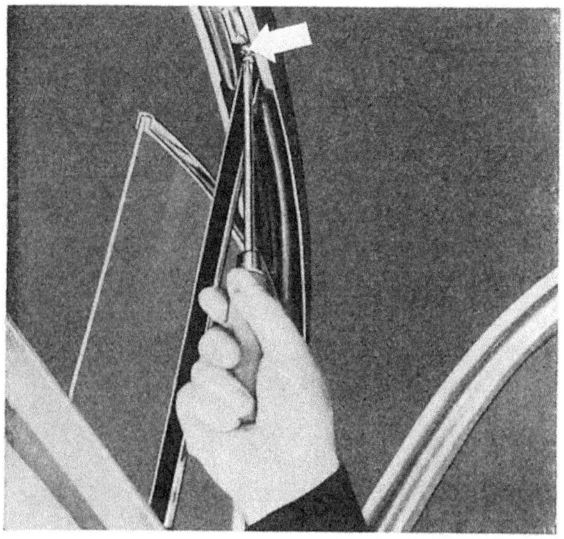

The front channel and the rear weatherstrip of the vent, which is incorporated in the channel, can also be replaced.

Installation

When installing, the following points should be observed:

1 - All glass-run channels and retainers should be inserted and fastened in position. The rear channel is supplied in a straight piece ready-cut. When installing, it has to be bent to shape and screwed in the window opening with the moulding. The holes for the fixing screws must first be drilled in new parts.

2 - Grease all moving parts of the regulator mechanism with Universal Grease.

3 - Push the window-pane about two thirds of the way into the glass-run channels, screw up the front retainer to the pane with a light pressure and install the regulator.

4 - Check that the window slides easily. If necessary, the pressure of the front retainer must be adjusted to suit.

5 - Ensure that the damper strip in the door is firm and that the drainage hole in the bottom of the door is not blocked.

6 - When replacing the interior trim panel, the impregnated paper lining must be placed over the metalwork at the bottom of the door to allow any water which may have entered to drain off.

7 - On the passenger's side the arm rest at the back of the interior trim panel must be hooked into the bracket provided on the inner rib of the door.

Note:

1 - From Chassis No. 4 420 885, the following modifications were made to the doors:

a - The lower and side **holes** for the door trim panel securing clips are now sealed with rubber caps — Part No. 113 857 219 A — which prevent water which has entered between the glass and the window slot weatherstrip from getting into the interior of the vehicle.

The rubber caps can be service installed.

b - The **waterproof paper** for the door trim panel has been altered. The present paper is cemented to the door inner panel with Genuine VW Universal Adhesive D 12.

Water which has entered at the glass is guided down to the drain holes in the bottom of the door and can no longer penetrate into the vehicle interior via the trim panel.

The waterproof paper can also be service installed. Ensure that it is stuck smoothly to the door panel without creases.

2 - From August 1963, Chassis No. 5 677 119, the sealing of the door was altered:

a - The beading and the oiled paper on the door trim panel have been discontinued and a seal — Part No. 111 867 109 — cemented to the inside of the panel. This seal seals the door inner panel to the trim panel and water which has entered the window slot drains down to the holes in the bottom of the door.

b - The guide plate on the window winder — Part No. 111 837 501 D/502 D — has been discontinued and the additional lip on the winder lifter channel seal — New Part No. 111 837 565 A.

When installing a lifter channel seal or a window winder, ensure that the window winder with guide plate is always fitted with the seal — Part No. 111 837 565 and the window winder without guide plate with the seal — Part No. 111 837 565 A. Otherwise, when installing a new window winder in an older vehicle, the existing sealing lip must be cut off.

8 - Install rubber inserts, escutcheon springs and trim panel. The large diameter of the spring must towards the trim panel.

9 - Reinstall the regulator and door handles in their prescribed positions.

10 - Check that all parts function properly by operating several times.

Vent Wing Removal and Installation

Removal

1 - Remove inner door handle and window regulator handle.

2 - Take off door trim panel, escutcheon springs, and rubber inserts.

3 - Remove the two screws holding the vent wing clamp in place.

4 - Pull the weather strip of the vent wing out as far as the upper pivot point, then remove the fixing screws of the pivot with a Phillips screwdriver.

5 - Partially open the vent wing and lift it out.

6 - Take off vent wing clamp after removing the fixing screws from inside the door.

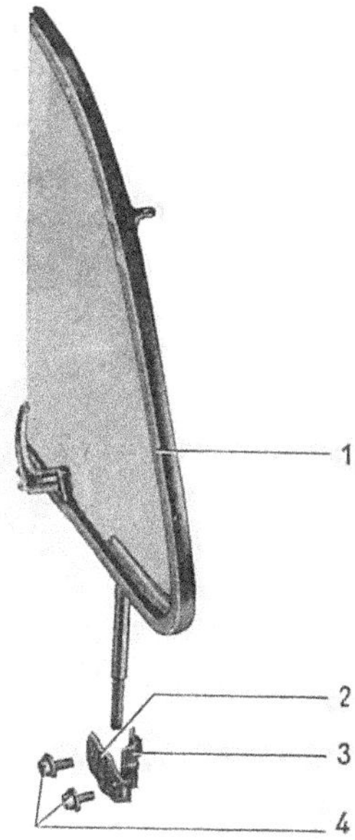

1 - Vent wing
2 - Clamp, upper part
3 - Clamp, lower part
4 - Attaching screws

7 - Remove vent wing lock plate, if necessary.

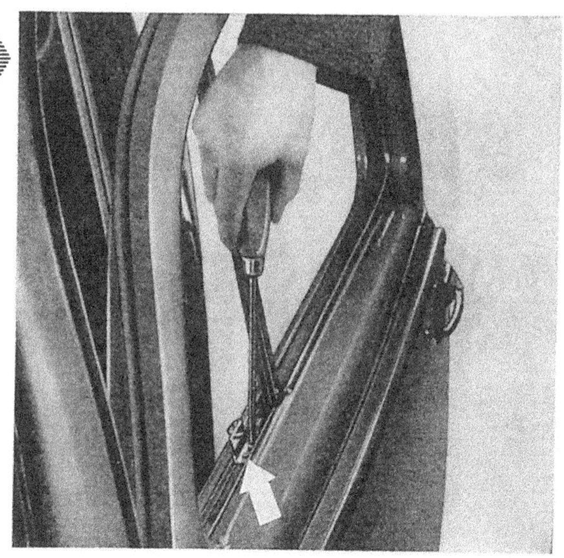

Note:

From Chassis No. 1 852 263 the following modification was made to the door window on the De Luxe Sedan, to prevent the ingress of water between handle, retainer and weatherstrip of the vent wing. The angle and the radii of the vent wing retainer (Part No. 111 837 635 A) have been modified so as to ensure a snug fit of the rubber lip at this point.

8 - If only the vent wing frame has to be renewed, it is sufficient if the hollow rivet in the upper pivot is ground down on its underside and knocked out with a punch.

Installation

1 - Check condition of weather strip and replace if necessary.

2 - Place the vent wing in the door window frame.

3 - Hook clamp parts together and hold the assembly against the inner door panel from inside the door shell — the joint of the clamp must be toward the door hinges — tighten the clamp screws.

4 - Rivet the vent wing to the upper pivot bracket, if necessary.

5 - The amount of frictional resistance can be adjusted by tightening or loosening the rear screw.

6 - Reassemble the door, taking care that the regulator and door handles are in their correct positions.

Door Lock Removal and Installation

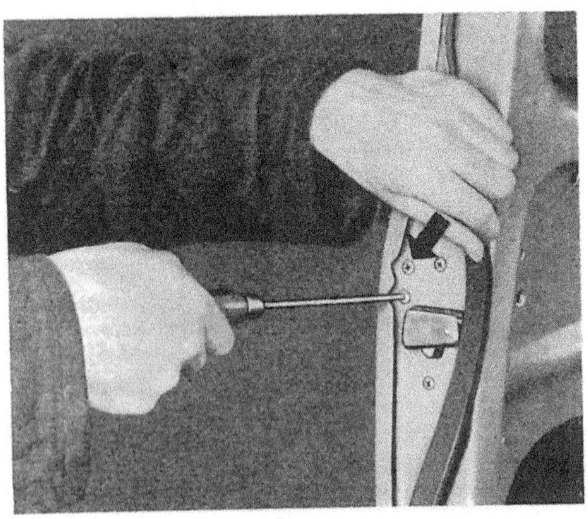

Removal

1 - Remove inner door handle and window regulator handle.

2 - Remove door trim panel.

3 - Remove window regulator and lift out glass.

4 - Remove the rear glass-run channel and retainer.

5 - Pull the door weatherstrip out of its retaining flange slightly in the region of the lock and remove the two fixing screws of the outer handle.

6 - Take off the outer door handle.

7 - Remove the three door lock attaching screws.

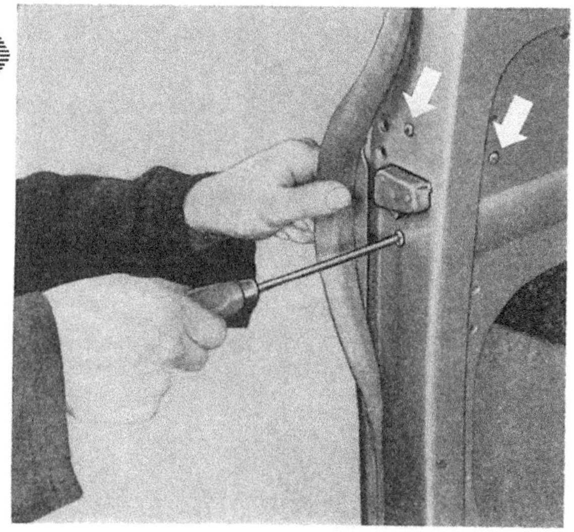

8 - Remove the two attaching screws of the remote control assembly.

9 - Pull lock and remote control rod towards the hinge, twist the lock 180° and pull it out downwards with the remote control rod.

10 - Remove the lockwasher from the remote control rod with a screwdriver.

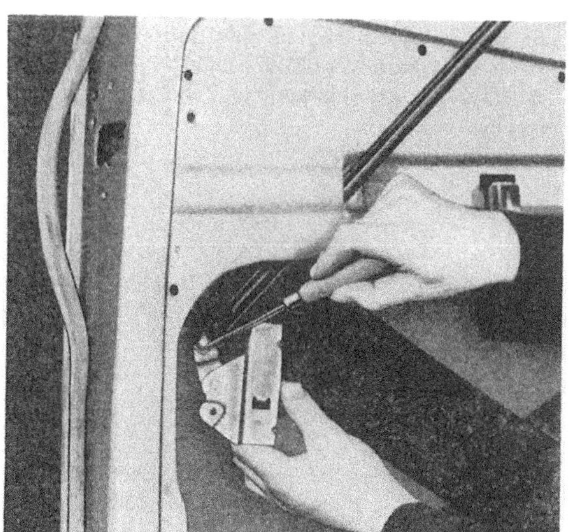

11 - Take off the lock and remove the remote control assembly with rod.

Installation

1 - Check all parts for wear, replace if necessary, lubricate all moving parts.

2 - If necessary, the remote control assembly or rod can be renewed separately after removing the rivets.

3 - Check that the felt pads and rubber sleeve fitted on the inside of the door, to prevent the control rod from rattling, are properly positioned.

4 - When reassembling the remote control rod and the lock, a springwasher should be placed underneath the lockwasher.

5 - The remote control assembly has slots to facilitate adjustment in either direction. When the lock is mounted, the remote control assembly should be pushed towards the hinge until light resistance is felt and then tightened up.

6 - Before reassembling the door, check that the lock is functioning correctly by operating the handle. The lock must fit properly in the striker plate; if this is not so, the remote control assembly must be readjusted.

7 - Reassemble the door, ensuring that the regulator and inner door handles are correctly positioned.

Door Handles, Locks and Lock Cylinders

From 6th August 1959 and Chassis No. 2 533 000 the previous pull type outer door handle was replaced by a push-button operated rigid type (Part Nos. 111 837 205 B/206 B and 113 837 205 B/ 206 B). The new type is installed by inserting the end of the handle into the hole at the front portion of the grip recess and by attaching the other end to the lock. When removing, take out Phillips head screws on lock side, push handle forward and unhook.

To facilitate easy closing of the door, the door lock and the striker plate have been modified. On the inside the door lock has been provided with an oil deflector plate and with an oil hole above the lock to facilitate regular lubrication.

New door locks (Part Nos. 111 837 015 B/016 B or 112 837 015 A/016 A) and new striker plates (Part Nos. 111 837 035 B/036 B) may only be installed as a unit. The new striker plates can be distinguished by the Part No. indented on the rear side. Parts of previous type are still available as spares.

Due to modifications to door lock and handle, the front portion of the grip recess is now provided with an opening and with an oil hole above the door lock (see Workshop Bulletin A-36).

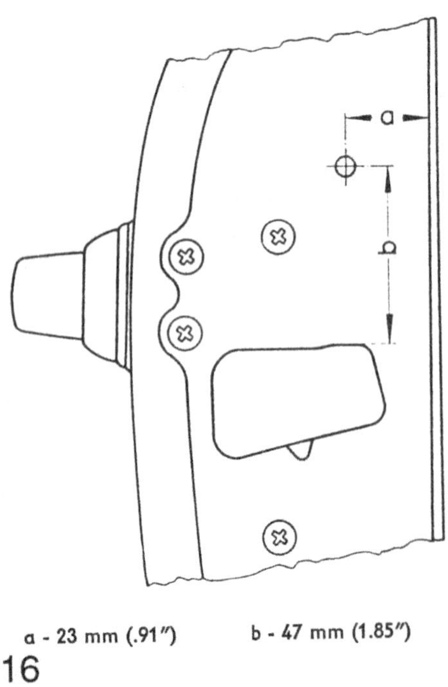

a - 23 mm (.91″) b - 47 mm (1.85″)

To simplify stock-keeping only doors of previous make (Part Nos. 111 831 051 C/052 C and 113 831 051 C/052 C) i.e. without the 2 holes, are being supplied.

When it is necessary to install a new door, the oil hole must be made as shown in the sketch.

As the opening for the hook of the handle requires reinforcement, a plate (Part No. 111 837 247) is to be installed.

When installing a new door, the hole for the handle retaining hook must be made to the measurements in the sketch below.

The sketch can also be used to manufacture a template.

a - 13 mm (.51″) c - 120 mm (4.27″) e - 8 mm (.31″)
b - 50 mm (1.97″) d - 16 mm (.63″)

Previous vehicles can also be modified in the above way.

Note:

If difficulty is experienced when locking or unlocking the door, and the trouble originates from dampness and low temperature or production defects in the lock cylinder (Part No. 111 837 217 B), proceed as follows:

When frozen:

Dampness in the lock cylinder often causes the locks to freeze in the winter.

Experience has shown that a large percentage of this trouble is due to the water which gets into the lock cylinder when the vehicle is being washed. Before washing a vehicle at temperatures below freezing point, it is essential to cover the key hole with adhesive tape or with one of the small plastic caps which are available from accessory dealers.

Freezing of the lock cylinders can be prevented by squirting a solution consisting of

40% Methanol or methylated spirits
50% Glycerine
10% Anti-corrosion oil

through the key hole.

The same solution can also be used to thaw out frozen locks. It is, however, advisable to remove the lock cylinder when it has been thawed out and to dry it properly. Before installation, the lock cylinder should be coated with grease externally. A few drops of the anti-freeze solution should be put inside the cylinder.

Other difficulties:

1 - Remove lock cylinder, wash in benzine, dry with compressed air and dip in anti-freeze solution or sewing machine oil.

2 - Check operation of lock cylinder and install a new cylinder if necessary. Restore key uniformity.

3 - The door handle (Part No. 113 837 205 B) should not be replaced.

Note:

The door handle and lock cylinder were modified from 1st August 1960 and Chassis No. 3 192 507.

1 - The key slot in the lock cylinder — Part No. 111 837 217 C — is now horizontal and has a small taper to facilitate the insertion of the key.

2 - An internal flap closes the key slot when the key is withdrawn.

3 - The press button is secured by a slide when the door is locked. The slide is constructed so that the operation of the lock is neutralised if the press button is pushed in by force. This provides a double security measure.

4 - Three new key profiles with the designations SC, SU and SV have been introduced at the same time.

The Part Nos. of the modified door handle are

111 837 205 C and 113 837 205 C.

The new door handle with lock cylinder can be service installed in vehicles of previous pattern if the ignition/starter switch is replaced at the same time to maintain the lock uniformity.

The door handle — Part No. 111 837 205 B and 113 837 205 B — and the lock cylinder — Part No. 111 837 217 B — will remain available as spare parts.

The lock cylinders must be lubricated with powdered graphite only. Under no circumstances may oil or grease be used. It is recommended that the key is dipped in the graphite powder and then turned back and forth several times in the lock.

Note:

From Chassis No. 3 682 335 (20th February 1961) the door weatherstrip was modified in the area round the striker plate. The lip on the new weatherstrip is about 4 mm wider than the lip on the former weatherstrip. See measurements "a" and "b".

This modification ensures better sealing between door and lock pillar at this point.

The modified weatherstrip, Part No. 111 837 721/722 can be installed in all previous vehicles without difficulty. The former type weatherstrip will be discontinued when existing stocks are exhausted.

Note:

1 - If the gap between the door and the lock pillar is wider than 4 mm/.16″, the door position must be altered by bending the hinges. This can be done by inserting a hardwood wedge between the upper hinge and closing the door firmly. The lower hinge is then moved by holding a wooden block against the hinge and tapping it lightly with a hammer until the door gap is a uniform 4 mm.

2 - If the door weatherstrip on vehicles produced before 1st February 1961 does not seal correctly near the lock, a strip of beading (Part No. Kö 107) should be inserted between the door and the weatherstrip. This ensures that the lip of the weatherstrip contacts the lock pillar properly.

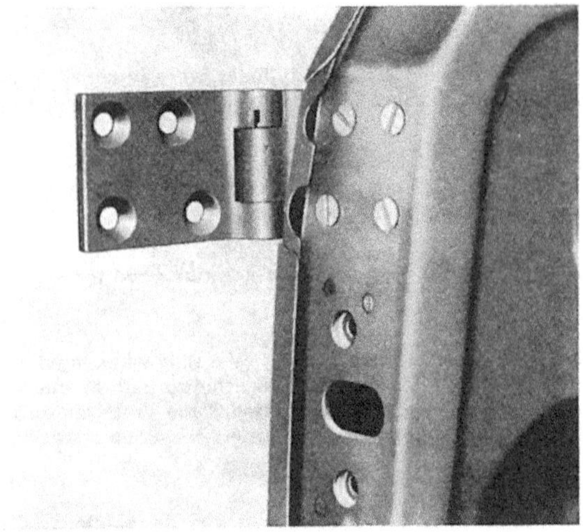

Note:

Loose door hinge rivets are usually very difficult to tighten properly. An 8 mm countersunk head screw with an extra thin head was taken up in the parts program in November 1963 to replace loose rivets. The Part No. of the screw is N 10 997 1.

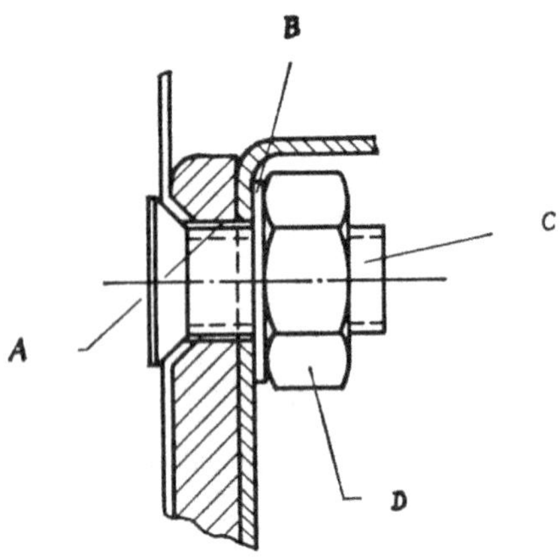

After drilling out the rivet with an 8.4 mm drill, the hinges are attached with screw (N 10 997 1), lock washer (N 12 108 2) and nut (N 11 008 3).

A — 8.4 mm hole
B — Lock washer
C — Screw
D — Nut

Door Lock for Right-Hand Drive Cars

RHD Sedans and Convertibles are provided with a door lock on the driver's side as standard equipment. If it is intended to install this kind of lock the two locks in the car must be exchanged for two new ones.

The following points should be observed:

1 - Remove both door locks inclusive of pull rods, remote control locks and exterior door handles.

2 - Remove rivets for pull rods at the remote control locks and take off pull rods.

3 - Rivet new remote control locks to the old pull rods. Install the old spring washer between the pull rod and remote control lock.

4 - Install new locks with remote control locks and pull rods.

5 - Interchange both outer door handles, so that the lockable door handle is on the right.

Adjustment of the Striker Plate

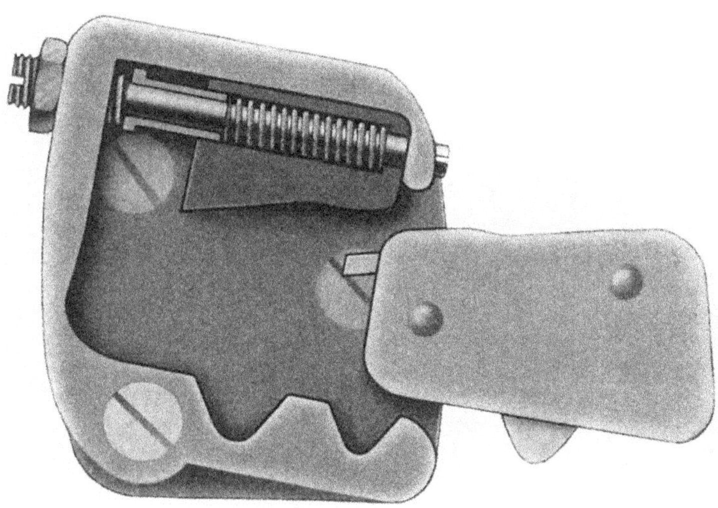

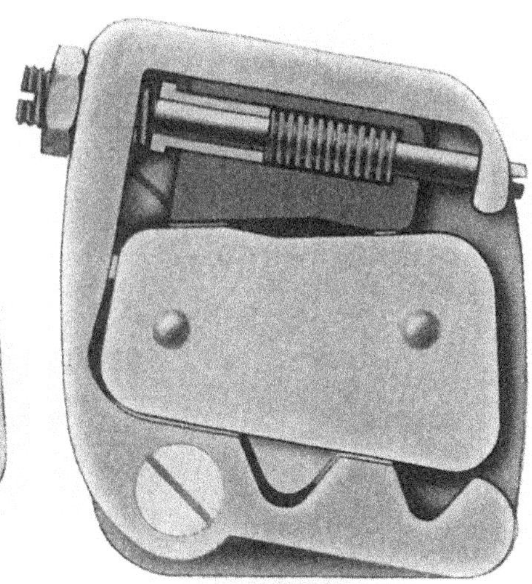

Function

Projecting from the latch housing are the latch bolt on the underside and the safety catch on the end.

When the door is closed, the latch bolt first jumps over the outer notch (safety notch) and engages in the second, inner notch.

At the same moment the safety catch strikes the inner surface of the striker plate and is pushed back into the housing. This locks the latch and prevents the door from flying open.

To prevent the housing from rattling due to the vibration caused when the car is running, the striker plate is equipped with an adjustable plastic wedge. When the door is shut this wedge presses the latch bolt against the right-hand flank of the inner notch, and the housing against the bearing surface of the striker plate. To be fully effective though, it is important for the door, striker plate and wedge to be correctly installed.

Basic Adjustment

1 - Remove the striker plate.

2 - Check the fit of the door in the opening. It is properly fitted when

a - the gap between the top of the door and edge of the roof, and between the edge of the door and the side panel of the body are approximately equal,

b - the door is flush with the side panel of the body,

c - the waistline mouldings of the door and body are in line,

d - the door does not rub, either at the top or bottom,

e - the weatherstrip is uniformly compressed and bears uniformly all round the door.

If necessary, adjust the door. This may involve loosening the hinges and readjusting the position of the door.

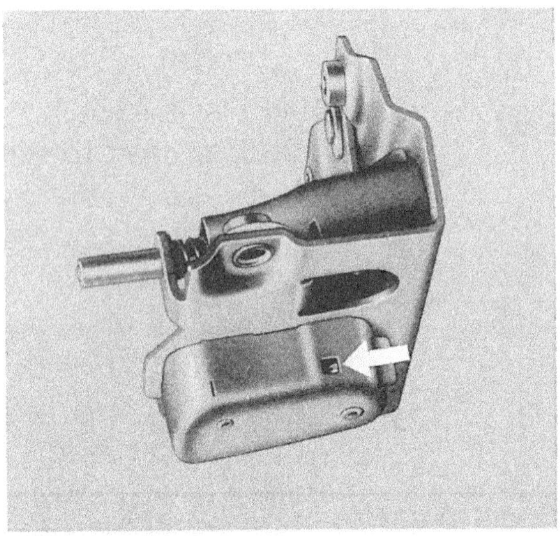

3 - Check that the lock is functioning correctly. The following points should be observed:

a - The top and bottom surface of the latch housing must be perfectly flat. The openings for the latch bolt and the safety catch must not have worn edges.

b - When the door handle is pulled the latch bolt should retract completely into the housing, otherwise it may prove difficult to close or open the door.

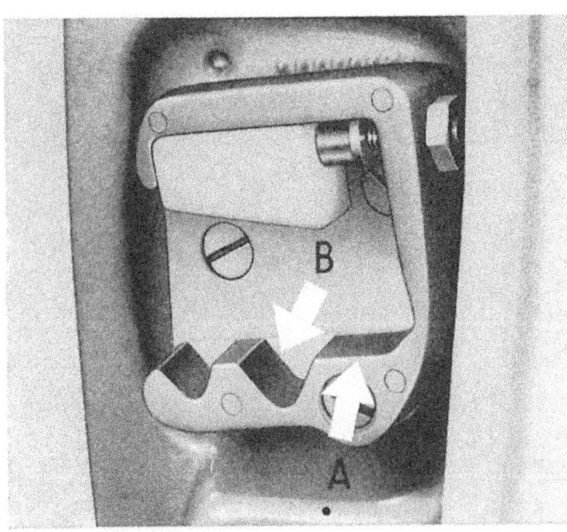

4 - Inspect the striker plate. If the bearing surfaces for the latch housing (A) and the notch (B) are badly worn, the striker plate should be renewed.

If the plastic wedge is scored or worn due to the latch housing being damaged, renew the wedge.

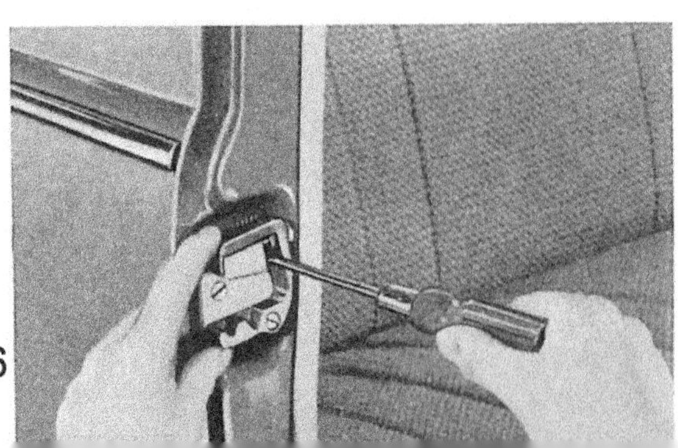

5 - The striker plate, which was removed to check the door fit, can now be replaced.

The striker plate is fastened to a movable tapped plate, allowing it to be adjusted vertically and horizontally.

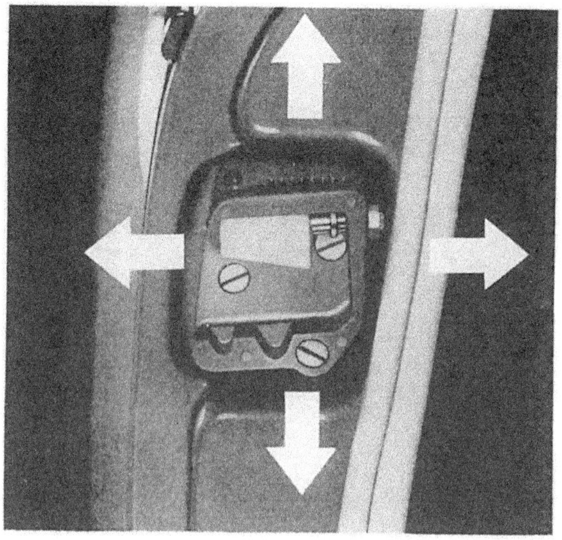

> **Important!**
>
> The adjustment of the striker plate influencas the fit of the door in the opening.

6 - Loosen the lock-nut and turn the adjusting screw to the right until the shoulder strikes the body of the striker plate. In this position the plastic wedge has greater freedom of movement.

7 - Adjust the striker plate.

a - **Lateral adjustment**
Adjust the striker plate sideways until the door and rear quarter panel are flush with one another.

b - **Vertical adjustment**
When making the vertical adjustment the clearance between the latch housing and the striker plate must be less at the bottom (y) than at the top (x).

When the door is closed, the latch should strike the bearing edge of the striker plate and be

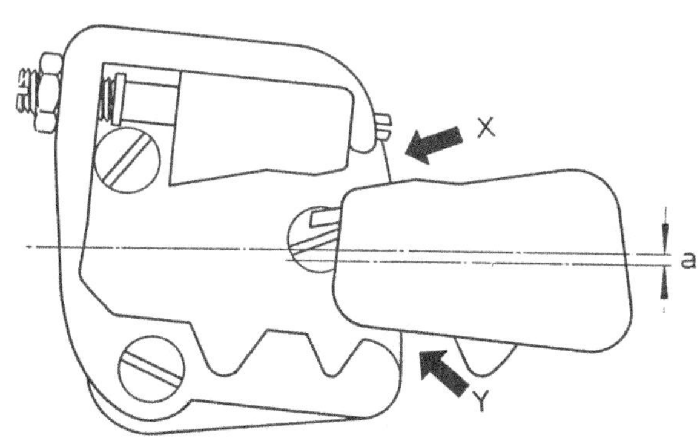

lifted by about 2 mm (0.08″) (a). The height of the striker plate is correct if the weatherstrip is slightly displaced with the door nearly shut.

8 - Having completed the adjustment of the striker plate, check the bearing of the latch on the striker plate by opening and closing the door several times.

When correctly adjusted, the bearing surfaces of the latch housing must bear evenly on the surfaces of the striker plate.

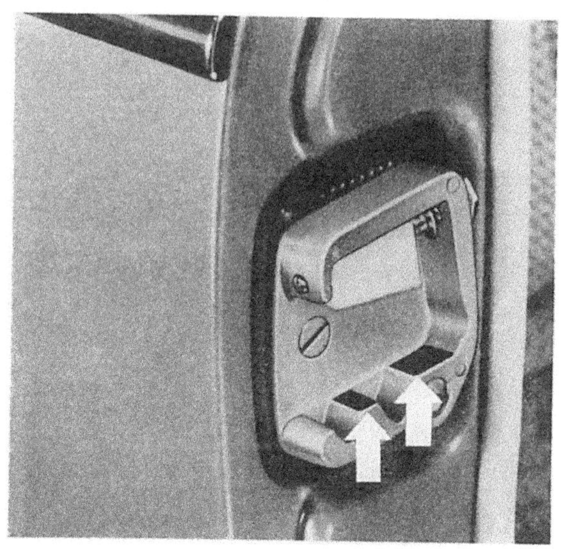

Otherwise the striker plate must be slightly tilted.

9 - Tighten up the securing screws of the striker plate.

10 - To adjust the wedge

This new type striker plate is to prevent the wedge from deflecting when the body is subject to torsional stress and road shocks. An adjusting screw allows the stop to be positioned so that the wedge cannot move out of place when the door is closed.

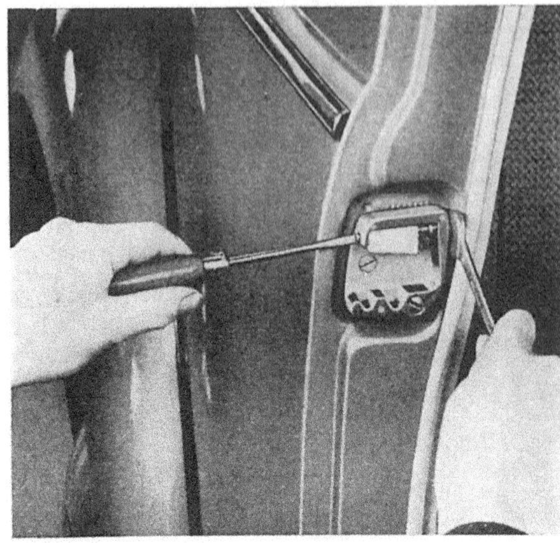

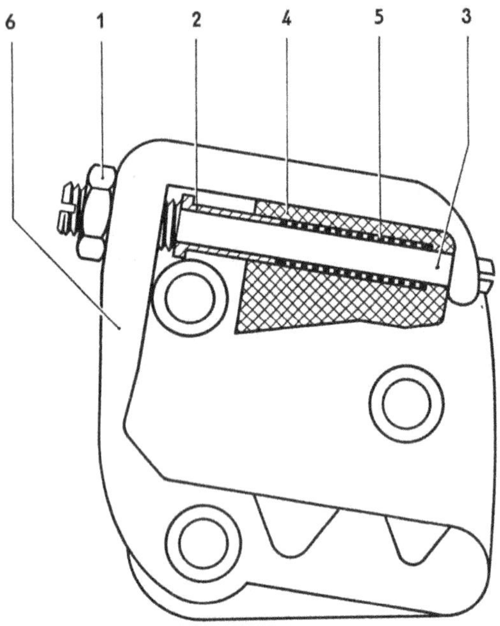

1 - Lock-nut
2 - Stop
3 - Adjusting screw
4 - Plastic wedge
5 - Spring
6 - Body of striker plate

Adjust as follows:

a - Loosen the lock nut (1) of the adjusting screw (3), while holding the adjusting screw with a screwdriver.

b - Turn the adjusting screw anti-clockwise until the stop (2) is far enough toward the wedge. This position of the wedge (4) is reached when the resistance (closing pressure) is felt to increase when opening the closed door.

If, when opening the door, the resistance is felt to be excessive, or if the door springs back when trying to close it, reduce the pressure of the wedge by screwing the adjusting screw further inward.

c - Hold the adjusting screw with a screwdriver, while tightening the lock nut.

d - New striker plates may somewhat settle in the initial period, necessitating a readjustment after a certain mileage. To readjust, follow the instructions given under a-, b-, c- of this bulletin.

e - Spread a thin film of vaseline on the contact surfaces for the door latch and for the wedge.

Keys and Locks

Since January 1954 the door locks and the ignition/starting switch can be operated with one key. If a lockable handle is subsequently installed on the rear hood the lock can be adapted to the tumbler combination used on the door lock and ignition/starting switch.

Whenever exchanging a lock cylinder for a new one it is possible to restore the uniform tumbler combination.
As it would be too complex to stock the entire range of locks our Parts Department will supply disassembled lock cylinders as SP sets.

When replacing a door handle or a lock cylinder, the tumblers can be arranged in the desired combination in accordance with the code list, which our Spare Parts Department will supply. If replacement of the ignition/starting switch becomes necessary the tumbler combination of the door and rear hood locks, if provided, should be adapted to that of the new ignition/starting switch. It is not possible to remove the lock cylinder of ignition/starting switches.

Our Parts Department no longer supplies locks with pre-set tumbler combinations. In future the desired assembly of lock cylinders is to be carried out by VW Distributors or General Importers.

Orders for code lists and the corresponding SP sets should be placed with the Parts Department.

Door Check Rod Removal and Installation

Removal

1 - Remove regulator and inner door handle and interior trim panel.

2 - Remove circlip and pin.

3 - Turn the check rod 90° and pull it with the rubber buffer between the inner and outer door panels through the slot in the leather washer.

4 - If necessary, remove the leather washer from its recess in the door.

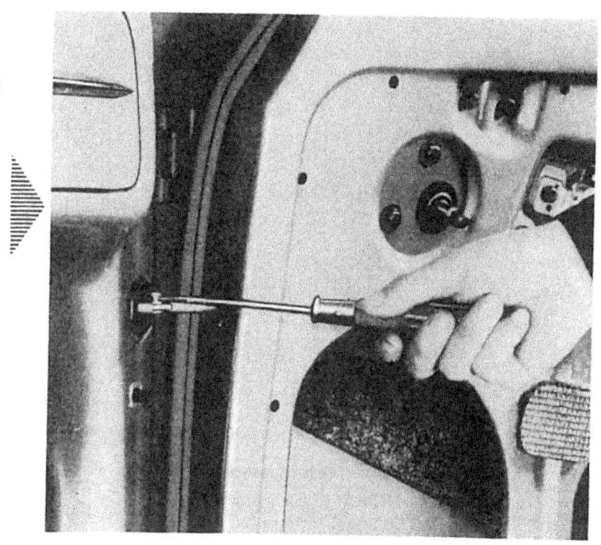

Installation

1 - Reinsert the leather washer, position the rubber buffer with the end washer and push the check rod outwards through the leather washer.

1 - Pivot pin
2 - Circlip
3 - Check rod
4 - Leather washer
5 - Rubber buffer
6 - Washer

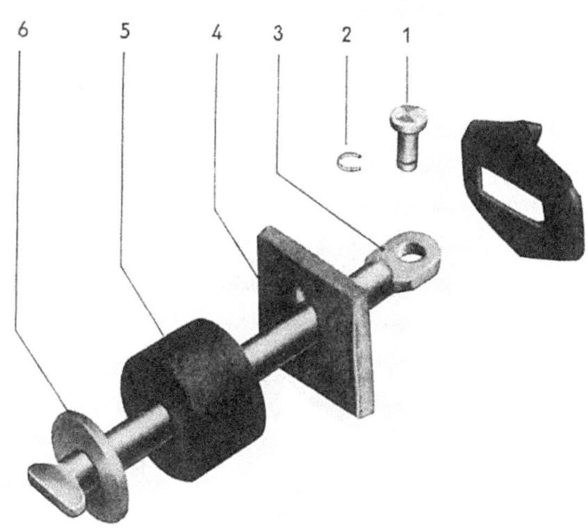

2 - Lightly oil the check rod joint. A drop of oil on the rod will prevent the leather packing from squeaking.

Windshield Removal and Installation

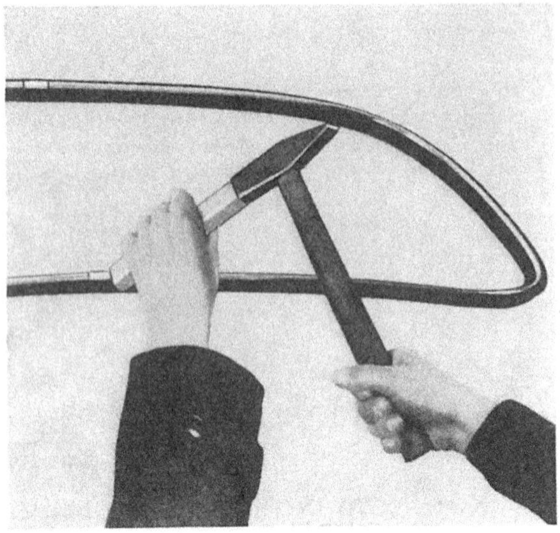

Removal

1 - Remove the windshield wiper arms.

2 - Remove windshield and windshield weather strip by pushing outward, beginning at one of the upper corners.

De Luxe only

3 - Drive the upper and lower sleeves from the garnish moulding ends, using a piece of wood, and exposing the joints.

4 - Remove the garnish moulding halves from the weather strip.

Standard and De Luxe

5 - Remove the weather strip from the windshield.

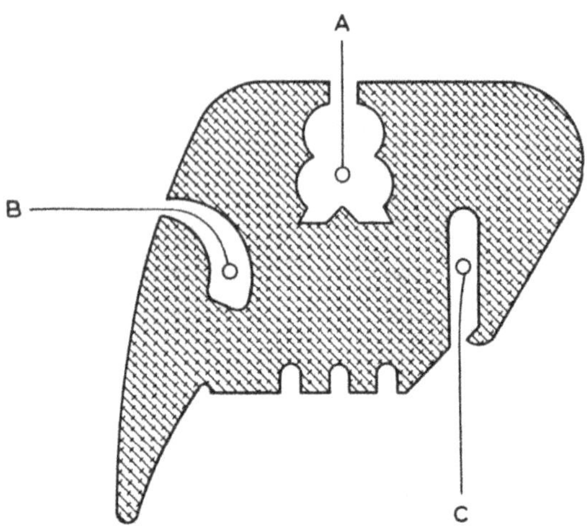

Section of windshield weather strip
A - Windshield channel
B - Garnish moulding channel
C - Windshield frame channel

Installation

1 - The windshields are marked in the bottom left or right-hand corner. When installing, the following points should be observed:

Non-laminated Safety glass

The marking must always be visible on the driver's side.

Laminated glass

The marking must always be legible from outside.

2 - Remove all traces of old sealing compound from the windshield frame, using benzine.

3 - Examine condition of weather strip, replace if necessary.

4 - Place weather strip around windshield edge. The ends of the weather strips should meet in the center of the upper windshield side.

Installing Garnish Moulding (De Luxe)

4 - Lay a length of cable (preferably an old electric cable with insulation, outside diameter about 2—3 mm = 0.08"—0.12") in the channel for the garnish moulding.

This is done by means of a piece of tube, one end of which should be flattened. Insert the flat end into the channel and move it completely around the windshield as shown below. The cable, passing through the tube, is thus embedded in the channel as the tube is moved. The two ends of the cable should meet in the middle of the upper or lower edge of the windshield.

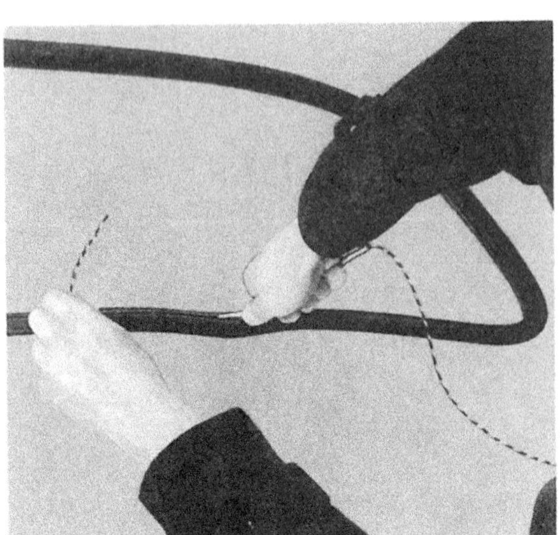

5 - First push one of the two moulding halves completely into place and then the other. Starting at the point where the cable ends meet, slowly pull out the cable while pushing the garnish moulding into place by hand until the operation is complete.

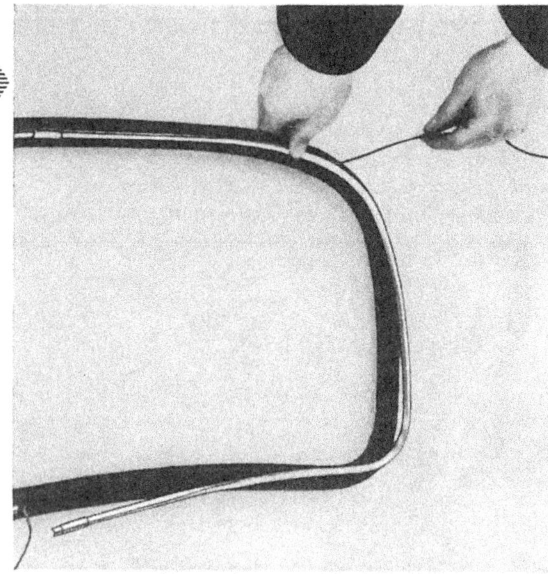

6 - Install the two garnish moulding sleeves.

Windshield Installation

7 - Fit a cable around the outer channel of the strip. The ends should cross in the center of the bottom edge of the windshield.

8 - Coat outside of windshield frame with VW Window Cement D 10 all way round.

9 - Position the glass against the windshield frame from outside. The cable ends must hang inside the body.

10 - The installation of the glass requires two persons.

A helper inside the body pulls at one end of the cable parallel to the glass and around the frame to force the lip of the weather strip over the frame.

During this operation, the glass must be tapped into place from outside the body, following the pull of the cable.

11 - Remove excess sealing compound with benzine. Test the windshield for leakage and clean the glass with spirit.

Important

If panes of laminated glass are fitted instead of nonlaminated, they must only be pressed into the window frame by even gentle pressure with both hands, while the second man inside the car pulls the lip of the rubber over the frame with the cable. A rubber mallet must not be used with laminated glass owing to the danger of it shattering.

Removal and Installation of Side Window and Rear Window

These operations are nearly the same as those outlined under "Windshield Removal and Installation". The garnish mouldings of the side windows are in one piece; that of the rear window in two pieces.

Removal and Installation of Vent Wing Glass

(Vent Wing Removed)

Removal

1 - Place vent wing in appliance VW 737 so that the glass frame ends come to rest against the shoulders of the support (see arrows).

2 - Place glass clamp on the spindle and turn the spindle toward the glass as far as it will go.

3 - Place rubber packings between glass and clamp, one on each side. Insert the pressure plate so that the wing screw ends come to rest in the two recesses.

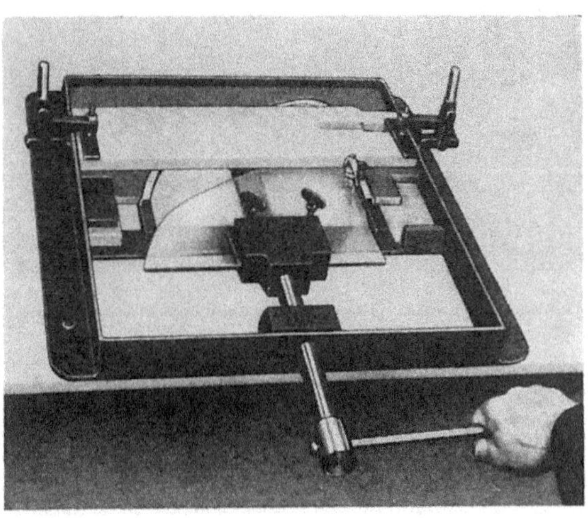

4 - Fully tighten the two wing screws by hand.

5 - Place the wooden thrust plate on the vent wing and tighten it in position by means of the two clamps.

6 - Withdraw the vent wing glass by turning the spindle anticlockwise.

7 - Loosen the wing screws and remove the glass.

8 - Release the clamps, take off thrust plate and vent wing frame.

Installation

1 - Remove all traces of old cement at the vent wing handle.

2 - Coat the frame channel at the handle with VW Cement D 10.

3 - Coat both sides of the weather strip with powdered talc.

4 - Clamp vent wing pivot pin in a vise and place the weather strip in the frame channel as shown below.

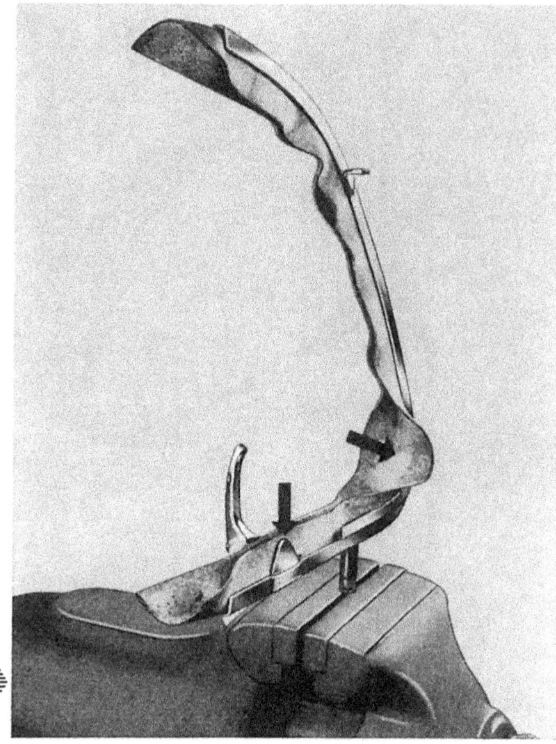

Note:

Special care should be taken that the weatherstrip is not pushed into the channel at the point indicated by the arrow but, as illustrated, is perfectly flat, thus avoiding the formation of folds when pressing in the glass. Make sure that the two triangular plates at the vent wing handle are fully covered by the weather strip.

5 - Slide in the glass so that the weather strip overlaps the glass on either side. When doing this, support the frame with one hand, to prevent dislocation.

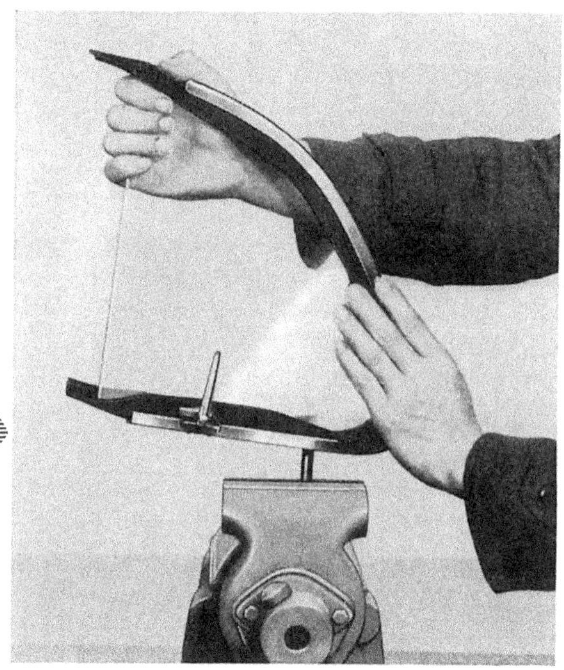

6 - Place vent wing in appliance VW 737.

7 - Place the wooden thrust plate on the vent wing and tighten it in position by means of the two clamps.

8 - Place the wooden thrust channel against the glass and press in the glass by turning the spindle clockwise.

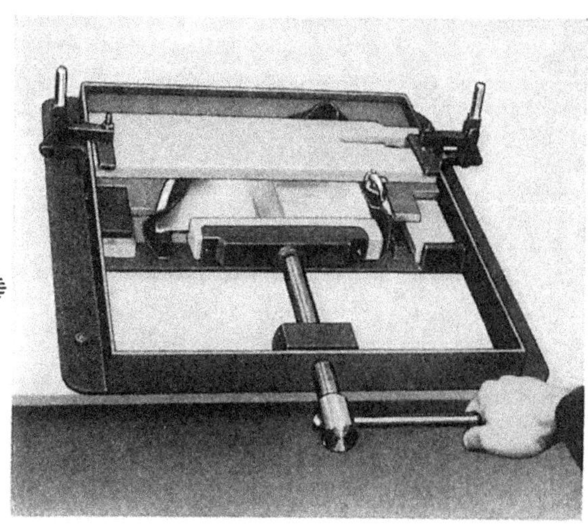

9 - Back off the spindle and remove thrust channel. Release the clamps and take off thrust plate and vent wing.

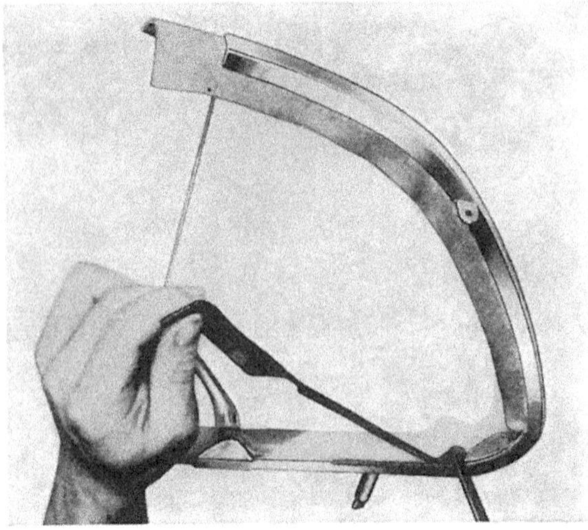

10 - Cut off the surplus rubber. Care should be taken to insure that the cut edge is below 90 degress to allow the water to drain off.

11 - Check for proper seal by spraying water against the vent wing.

Note:

The two adaptors should be reversed when removing and installing the glass of the opposite vent wing.

Door Removal and Installation

Removal

1 - Pull check strap out of bracket on pillar after taking out retaining pin.

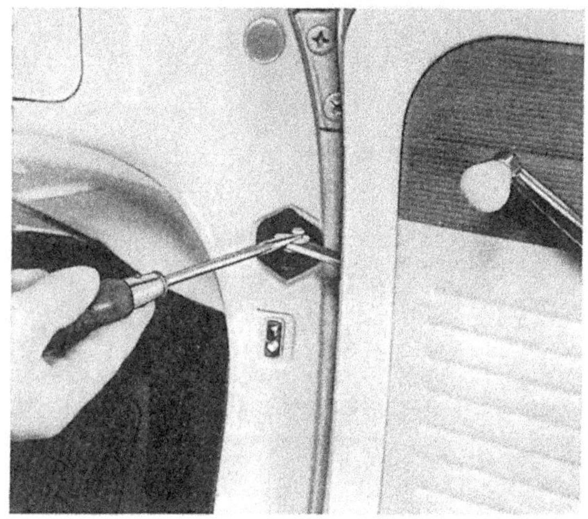

2 - Remove hinge screw cover plugs.

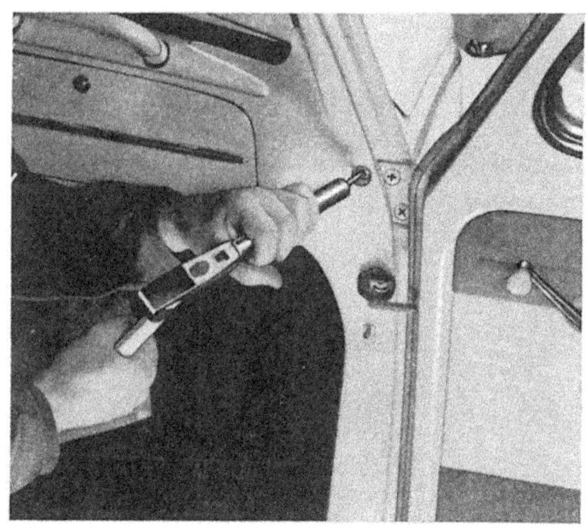

3 - Loosen the hinge screws first with a punch screwdriver and then remove them with a Phillips screwdriver.

4 - Take door with hinges sideways out of hinge pillar.

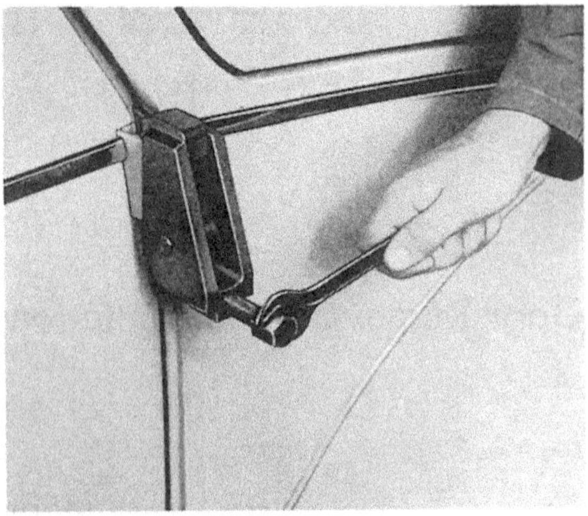

Note

If the same door is to be fitted again, it is advisable to press out the hinge pins. It is then not necessary to adjust the door afterwards.

Installation

1 - Check door weatherstrip and fit new seal if necessary. Stick new seal on with universal adhesive D 12.

2 - Attach door and fit it in the door opening so that the seal makes even contact all round and the door can be opened and closed without jamming. The striker plate must be removed when doing this.

The hinges are screwed to movable threaded plates. This makes it possible to adjust the door so that it aligns with the outer panels properly.

3 - Fit striker plate again and adjust it so that the door aligns with the rear quarter panel and closes easily.

Detailed instructions are given in the section on "Striker Plate Adjustment".

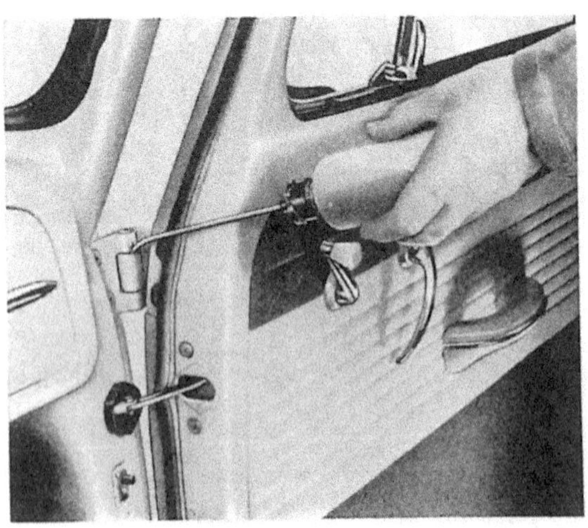

4 - Oil the hinges and ensure that the oil slots are clean.

5 - Coat the contact surfaces of the latch housing on the striker plate and on the wedge lightly with petroleum jelly or molybdenum-disulphide paste.

Important

The lock cylinder in the lockable door handle must be lubricated only with graphite powder and never with oil or grease. It is advisable to dip the key in the graphite powder and then turn it back and forth in the lock a few times.

Removal

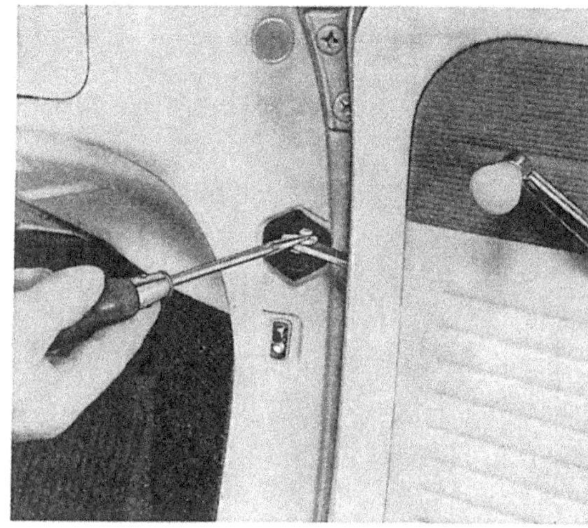

1 - Press retaining ring off check strap pin and take pin out.

2 - Press escutcheons for window crank and inner handle against the trim panel and knock out the retaining pins with a punch. Take off crank, handle and escutcheons.

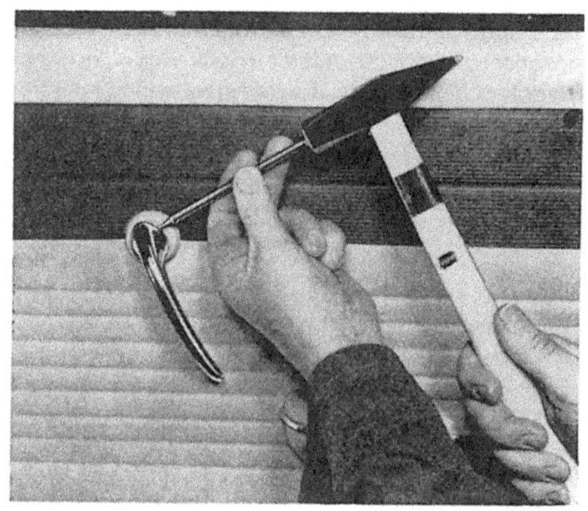

3 - Take trim panel off, taking care not to damage the panel or the paintwork.

> **Important**
>
> On the passenger's side the trim is hooked on the armrest support on the door inner panel with the arm rest retaining plate. Pull trim panel away slightly, then lift it clear.

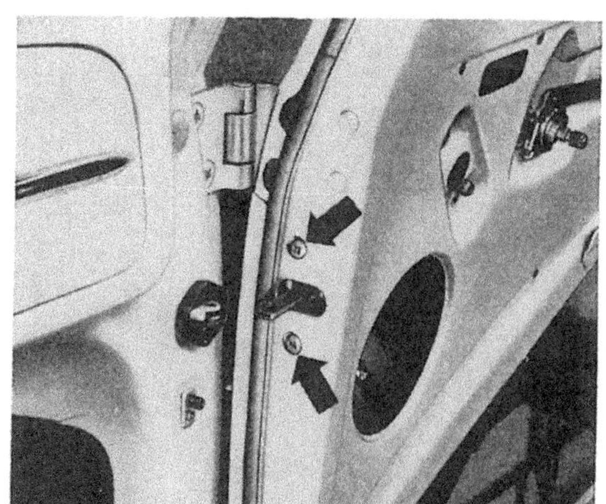

4 - Remove two screws on hinge side of the door and take check rod out between inner and outer door panels.

Installation

1 - Check cover for check rod and renew if necessary.

2 - Place cover in body.

3 - Attach check rod to door.

4 - Insert pin again, press ring on and oil check rod joint lightly.

5 - Open and close door several times to check operation of check rod.

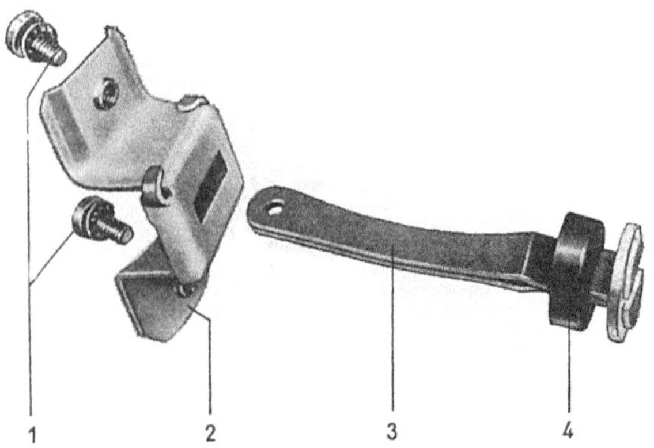

1 - Screws with lock washers
2 - Anchor
3 - Check rod
4 - Buffer

Removing and Installing Door Handle and Lock Cylinder

The door on the driver's side can be locked from outside. When a lock cylinder is replaced, it is not possible to order a certain tumbler combination. It is however, possible to restore key uniformity again by using tumblers from a special assortment which contains all the parts necessary.

With the aid of a code list it is possible to re-arrange the sequence of the tumblers in the cylinder so that a new lock cylinder can be operated with the existing key.

12/65

Note:

During the cold season there is a danger of door locks freezing-up due to condensate or the entry of water when washing the vehicle.

For this reason it is important to cover the keyhole with adhesive tape or a cap, obtainable as an accessory, prior to washing the vehicle.

To thaw lock cylinders that are already frozen-up, an anti-freeze solvent consisting of

40% methylated spirits
50% glycerine
and 10% anti-corrosion oil

can be usea, this being squirted into the keyhole. We recommend you to remove the lock cylinder and dry it after thawing it out. Apply universal grease to the outside of the lock cylinder prior to installation.

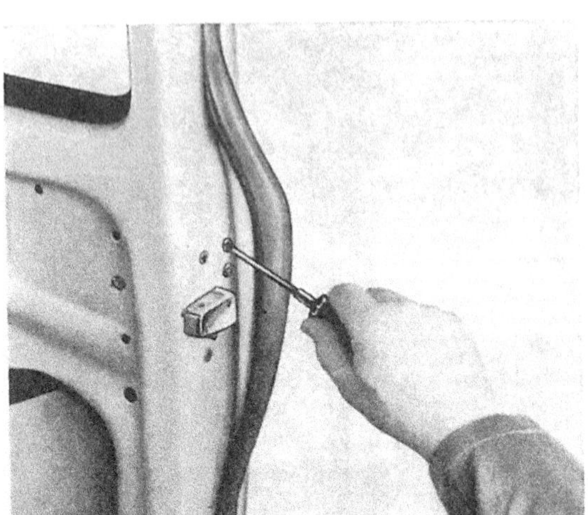

Removal

1 - Loosen the weatherstrip slightly round the door lock and remove the two screws holding the outer handle.

2 - Slide handle towards the hinge side and take it off.

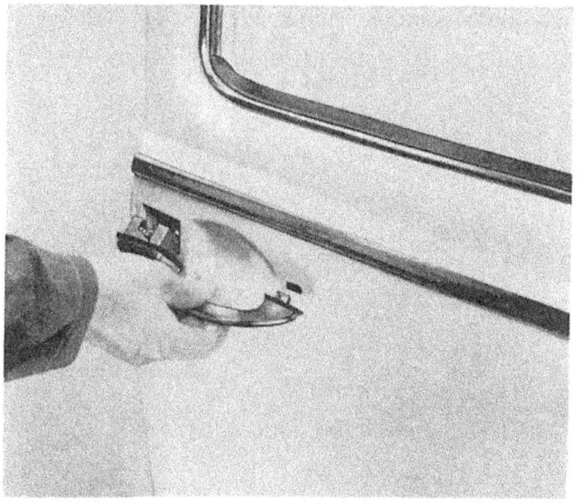

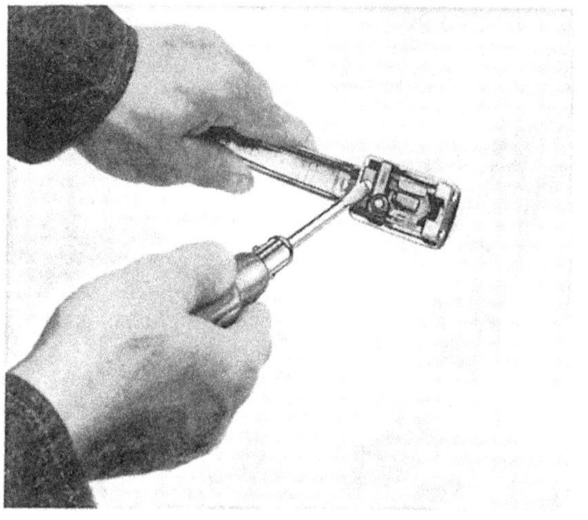

3 - Remove screw from button guide. Take off guide with spiral spring.

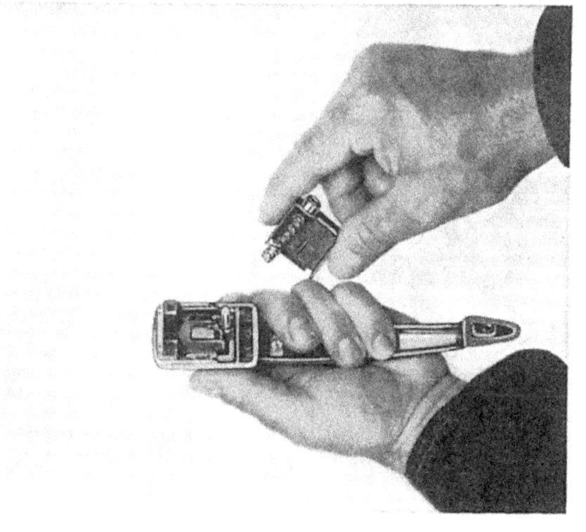

4 - Take press button with lock cylinder out of handle.

5 - Remove grubscrew at side of button.

6 - Push cylinder retainer sideways out of the guide.

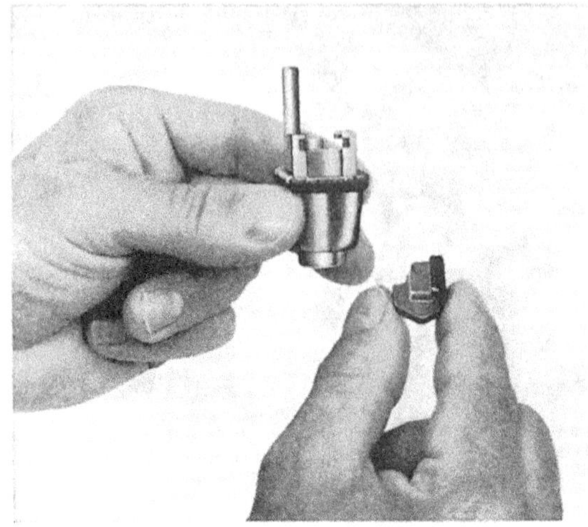

7 - Take cylinder out of press button.

8 - Remove tumblers with springs after removing the burred over material at the ends of the holes for the tumbler springs.

Installation

Before installation, the outside of the lock cylinder should be coated lightly with universal grease. The inside must be lubricated only with graphite powder and never with oil or grease. It is advisable to dip the key in graphite powder and then turn it to and fro in the lock several times.

1 - Clean all parts in benzine and dry with compressed air. Replace any damaged tumblers or springs.

2 - Insert tumblers with springs into cylinder in correct order and burr the holes.

3 - Place cylinder in press button.

4 - Slide retainer into guide and secure lock cylinder with grubscrew.

5 - Insert key and check operation of lock. It may be necessary to rearrange the tumblers.

6 - Insert button into door *handle and fit handle* to door.

Door Lock Removal and Installation

Removal

1 - Press escutcheons for window crank and inner handle against the trim panel and knock out the retaining pins with a punch. Take off crank, handle and escutcheons.

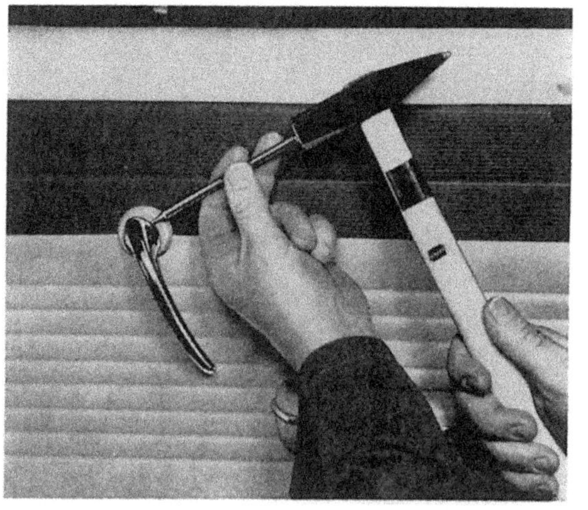

2 - Take trim panel off, taking care not to damage the panel or the paintwork.

> **Important**
>
> On the passenger's side the trim is hooked on the arm rest support on the door inner panel with the arm rest retaining plate. Pull trim panel away slightly, then lift it clear.

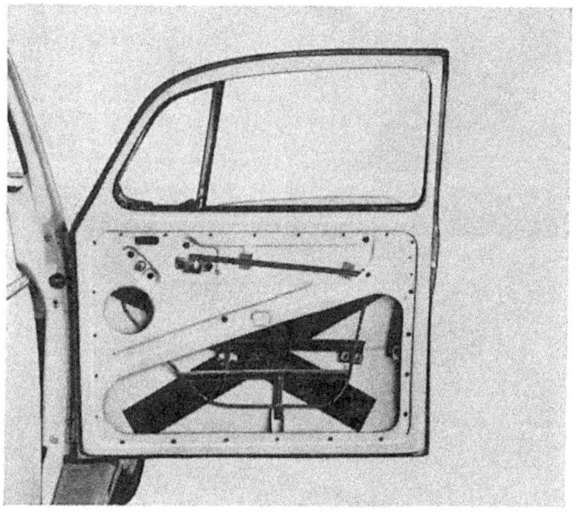

3 - Remove springs and rubber plugs from the spindles of the crank and inside handle.

4 - Remove four screws from the window lift channel, pull window down and slide it forward.

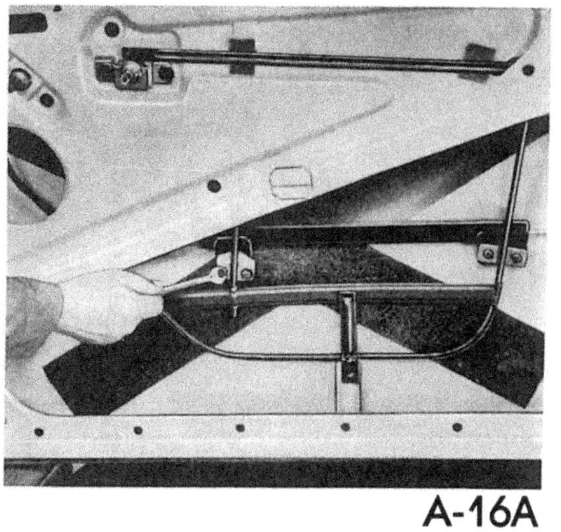

5 - Detach window lifter at lower mounting point.

6 - Detach rear window guide in window slot. Press guide channel up slightly in order to release the upper clip from the door inner panel. Remove guide channel downwards.

Note

The removal of the guide channels can be facilitated by carefully inserting a wooden wedge into the window slot to press the seals apart.

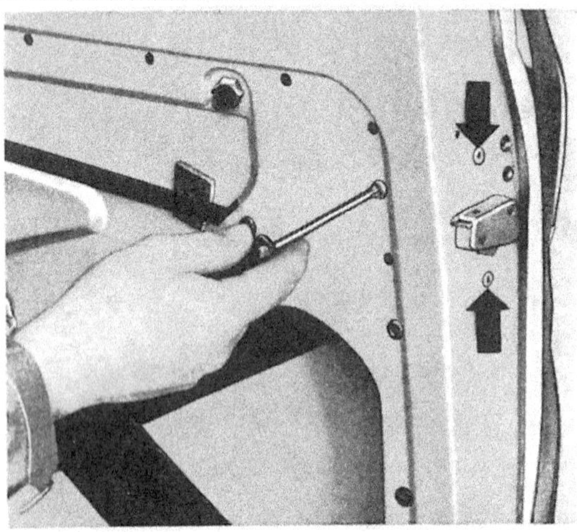

7 - Remove outer handle screws, slide handle towards hinge and take it off.

8 - Remove three lock securing screws.

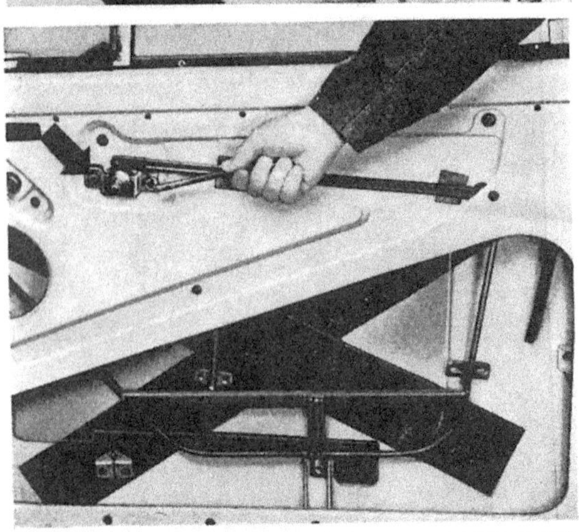

9 - Remove two screws securing remote control lock.

10 - Press lock and pull rod towards hinge and take out downwards with the remote control rod.

11 - Press pull rod lock ring off with a screwdriver.

12 - Take lock off and take out pull rod with remote control lock upwards.

Installation

1 - Check all parts for wear and renew as necessary. Lubricate moving parts. If required unrivet pull rod or remote control lock and replace separately.

2 - Check security of felt pad on door inner panel. Apply fresh adhesive if necessary.

3 - Insert spring washer between pull rod and lock and press lock ring on.

4 - Install door lock. If the lock jams when placing it in the door, press either the pin or the latch into the housing with a screwdriver. Take great care when doing this to avoid making marks in the outer door panel.

5 - Install remote control lock. Slide it towards the hinges in the slots until a slight resistance is felt.

Important

Before assembling the door further, operate the press button in the handle to check that the lock opens properly and that the latch is withdrawn fully into the latch housing. If this is not case, the remote control lock must be adjusted.

6 - Assemble door completely. Check window lifter and note position of window crank and inner handle.

Note:

From 3rd August 1965, Chassis No. **115 000 001**, both doors on the Convertible can be locked and unlocked from inside and out. The vehicle can be locked by merely moving the inner lever forward and pressing the button in the outer handle as the door is closed. If the door should close accidentally after the inner lever has been moved forward it will unlock automatically.

The following parts have been altered:

Part	No.
Door handle	113 837 205 F
Door lock, left	151 837 015
Door lock, right	151 837 016
Remote control, left	111 837 021 D
Remote control, right	111 837 022 C
Pull rod for remote control	111 837 193 B
Rivet for pull rod	111 837 199

The service installation of the modified door locks in earlier vehicles is possible if all parts are replaced at the same time. When doing, this, ensure that the control hole in the door inner panel is in line with that in the door lock.

The remote control lock must be pushed towards the hinge side of the door until the pull rod is tensioned slightly.

If the old pull rod is to be used again, it must be removed from the old remote control lock and riveted on to the new one with a special rivet (Part No. 111 837 199).

Door Window Removal and Installation

Removal

1 - Press escutcheons for window crank and inner handle against the trim panel and knock out the retaining pins with a punch. Take off crank, handle and escutcheons.

2 - Take trim panel off, taking care not to damage the panel or the paintwork.

Important

On the passenger's side the trim is hooked on the arm rest support on the door inner panel with the arm rest retaining plate. Pull trim panel away slightly, then lift it clear.

3 - Press retaining ring off check strap pin and take pin out.

4 - Remove four screws from the window lifter channel.

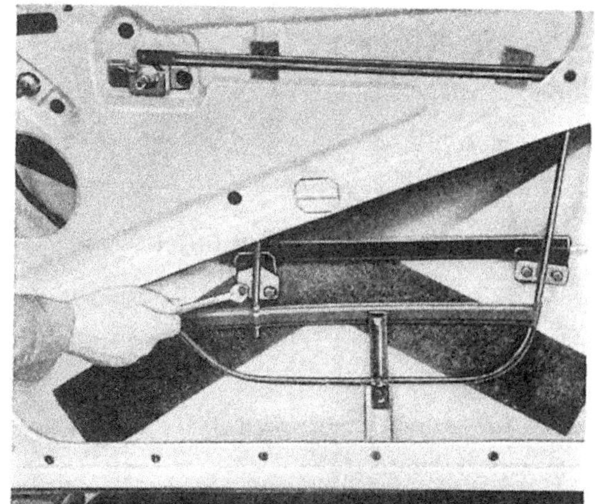

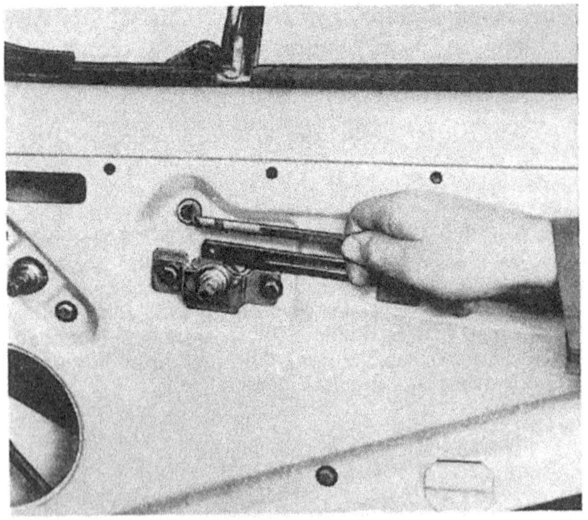

5 - Push window upwards. Remove five screws in window lifter and one screw for vent wing.

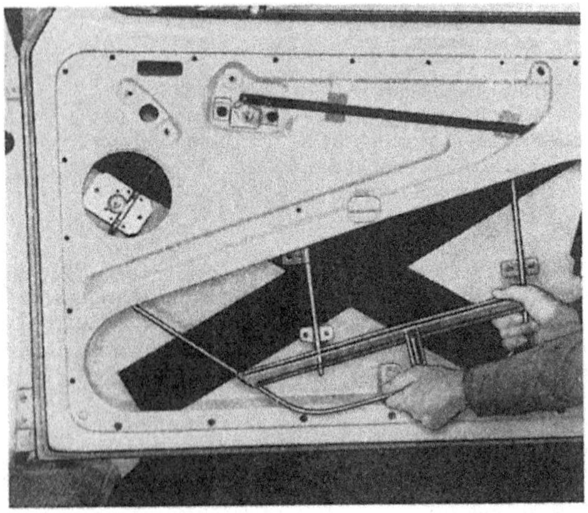

6 - Press window lifter towards outer panel and take it out downwards.

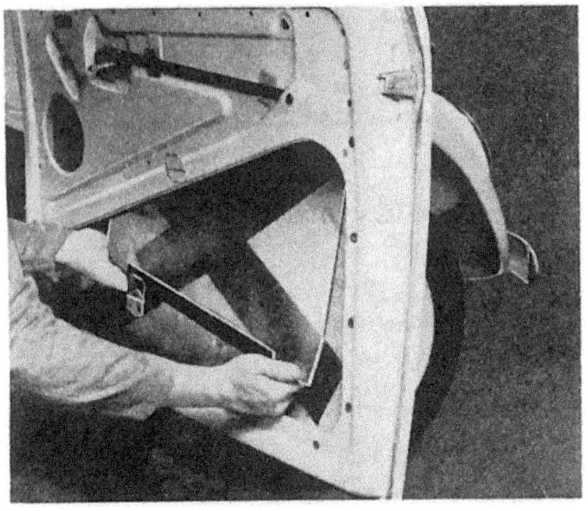

7 - Pull window glass down, tilt it slightly and take it out of door.

8 - If necessary, remove trim frame with weather-strip. It is advisable to press the trim frame clips out from the rear.

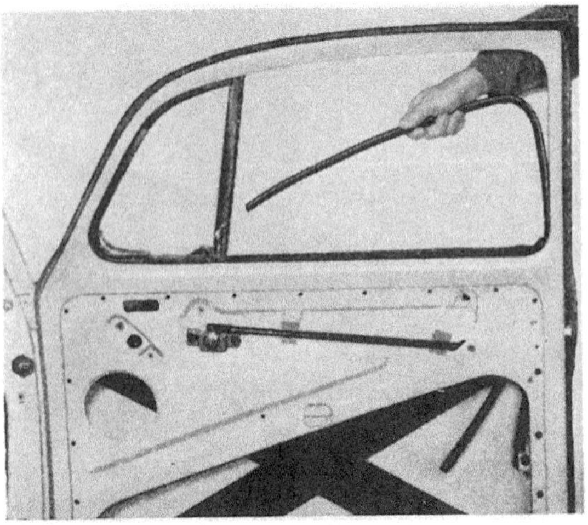

9 - Remove window slot seal on inner door panel.

10 - If necessary, pull the window run channel out of the door slot upwards. The window guide

channel can be taken out downwards after removing one screw from the mounting.

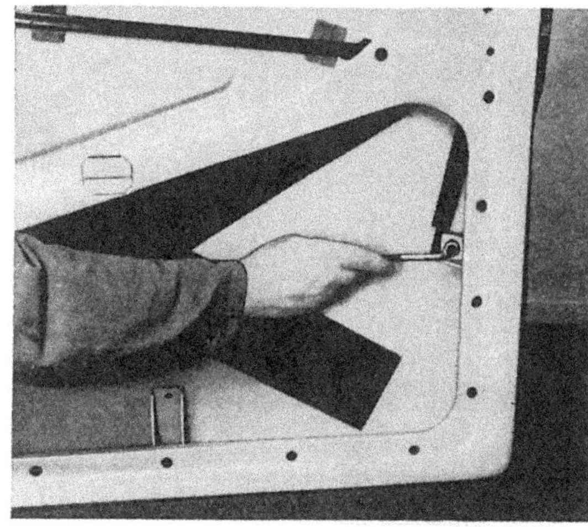

11 - Remove Phillips screw in upper window frame and hexagon head screw near the remote control lock on the inner panel and then take out front window guide channel with vent wing frame.

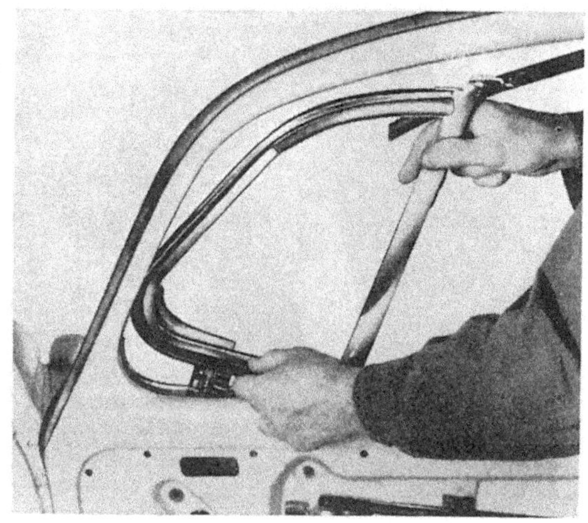

Note:

The front run channel and the rear seal for the vent wing which are fitted in this guide channel and in the vent wing frame can also be replaced separately if necessary.

Installation

1 - Check guide channel, run channel and retaining clips and renew as necessary. The rear run channeling is supplied in straight pieces, cut to size. Before fitting, it must be bent to shape and then pressed into the clips in the door frame and the rear guide channel.

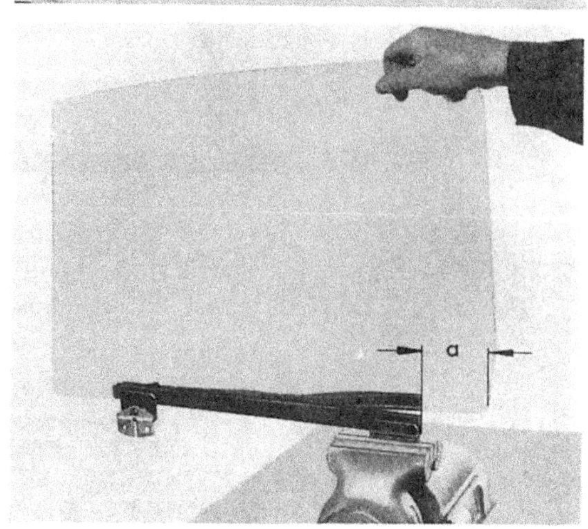

> **Important**
>
> If the lifter channel was taken off the glass it must be installed again at a distance of about 80 mm from the straight guide edge.

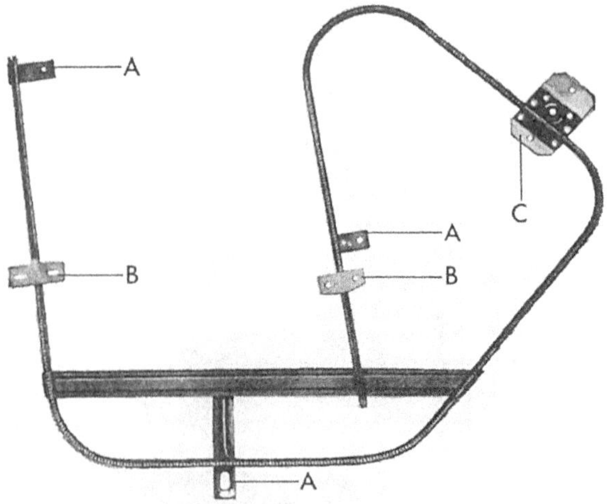

2 - Coat lifter cable lightly with universal grease.

3 - Check that the damping strips on the door are secure and that the drain holes at bottom of door are clear.

4 - Install run channel with vent wing frame and secure it.

5 - Insert window into run channel from below and slide it upwards.

A - Mounting point — Window lifter/inner panel
B - Mounting point — Window lifter channel
C - Cable drive

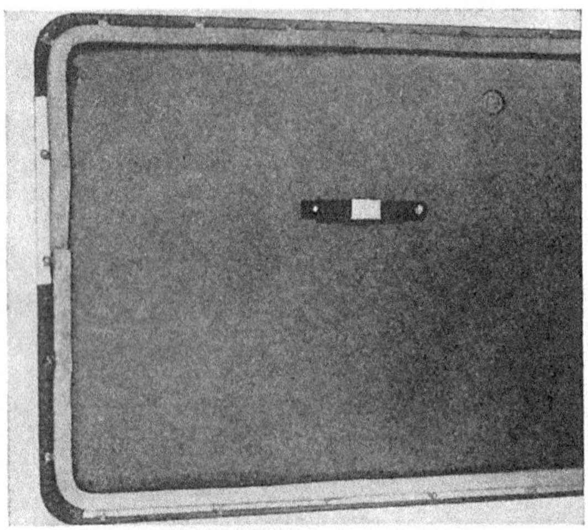

6 - Install window lifter and bolt it to door inner panel.

> **Important**
>
> If the lifter cannot be pushed through the front window guide washer near the remote control lock, the vent wing must be lifted slightly.

7 - Slide window down and attach lifter channel loosely to window lifter on each side. Open and close window a few times then tighten screws in window lifter channel.

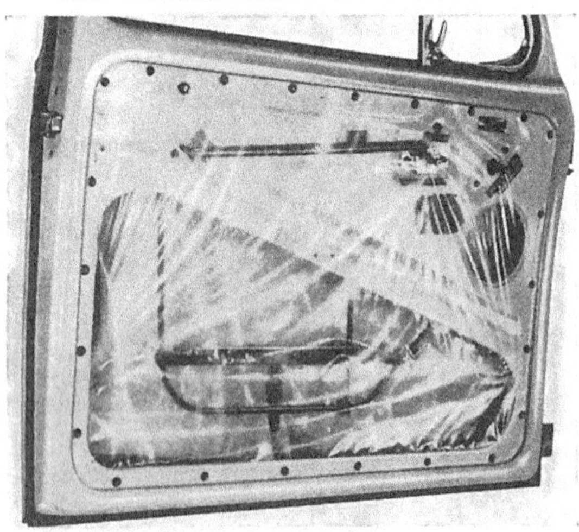

8 - Before fitting the trim panel, check that the foam rubber strip is secure. If not, stick it on again properly.

On the passenger's side the trim panel must be hooked on to the armrest support with the retaining plate for the armrest.

Note:

From 16th August 1965, Chassis No. **116 073 288**, a plastic sheet was cemented to the door inner panel to provide better sealing against the entry of water between the door inner panel and the door trim panel. The plastic foam seal in the door trim panel has been discontinued. The plastic sheet is now obtainable under part number 111 867 147 B.

The rubber seals for the clips must be pressed into the holes in the door inner panel **after** attaching the plastic sheet.

Important

Only universal adhesive D 12 guarantees a good attachment of plastic sheet to door inner panel.

9 - Fit rubber buffers, springs and trim panel. The large end of the conical springs must be towards the trim panel.

10 - Install window crank and inner handle at the correct angles.

11 - Check all parts by operating several times.

Removal

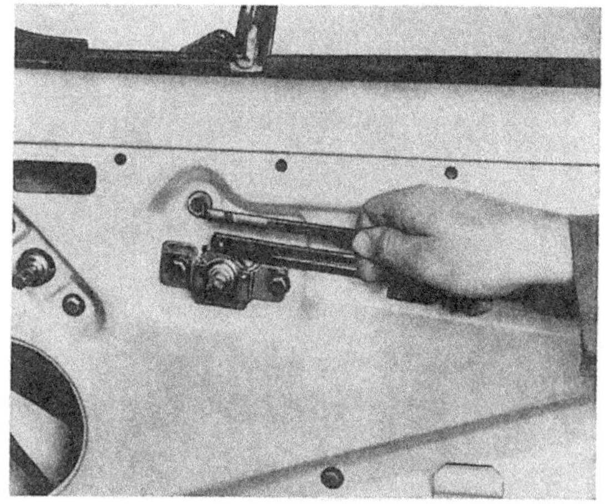

1 - Remove door trim panel.

2 - Remove screw securing the front window guide channel.

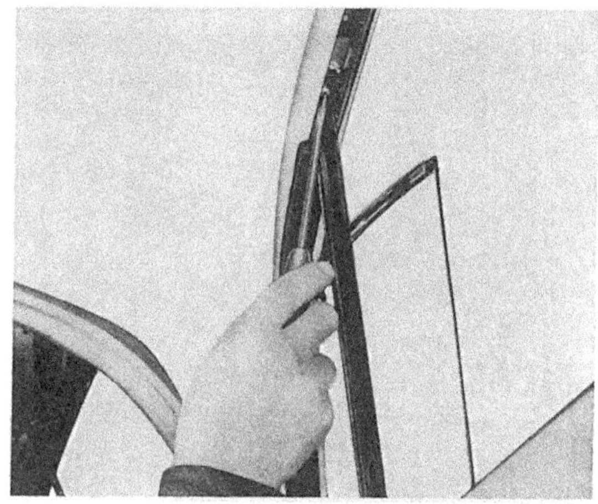

3 - Pull rear run channel away slightly near the front window guide channel and remove screw in upper door window frame.

4 - Incline vent wing with frame inward at the top.

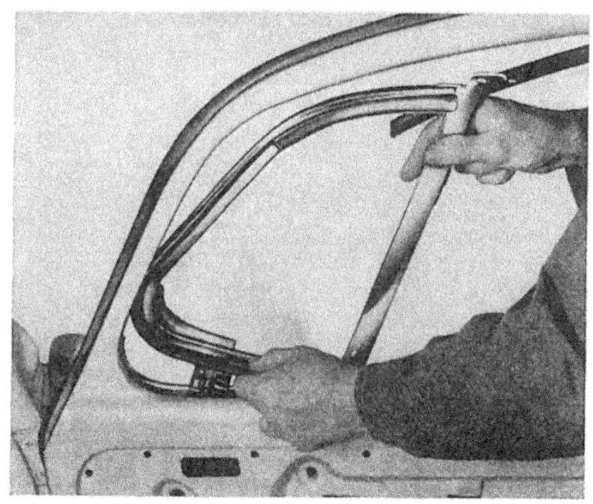

5 - Pull vent wing and frame and guide channel carefully out of the door slot.

6 - Remove vent wing clamp.

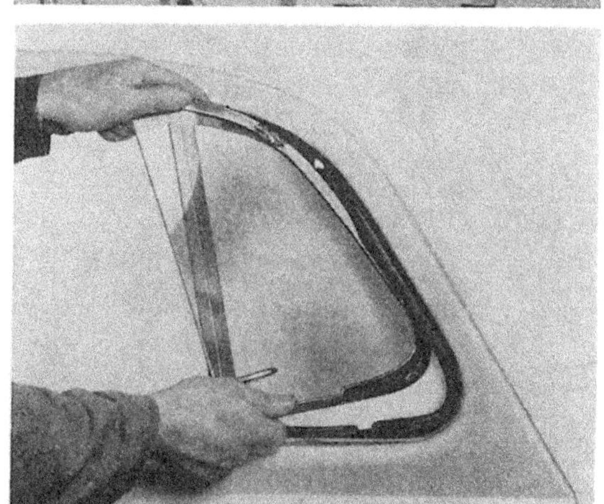

Note:

If the vent wing frame or glass has to be replaced it is only necessary to grind the underside of the rivet in the upper mounting off slightly and then to knock the rivet out with a punch. The spindle at the bottom can then be pulled out of the clamp without difficulty.

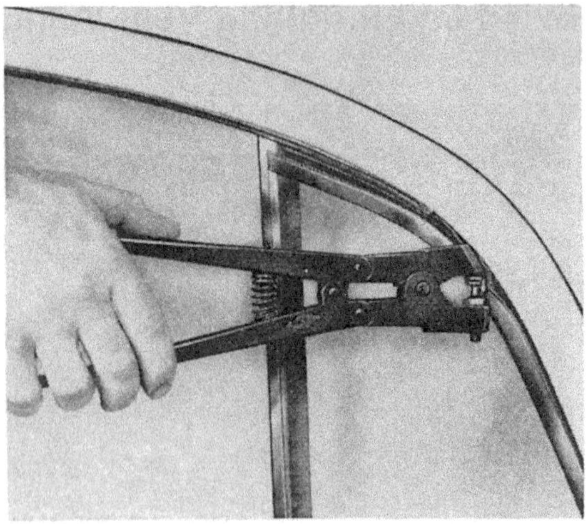

Installation

1 - Check vent wing seal and replace if necessary.

2 - Fit a new rivet at the upper mounting, using the riveting pliers.

3 - Insert vent wing into window slot from above, slide it forward and secure the frame.

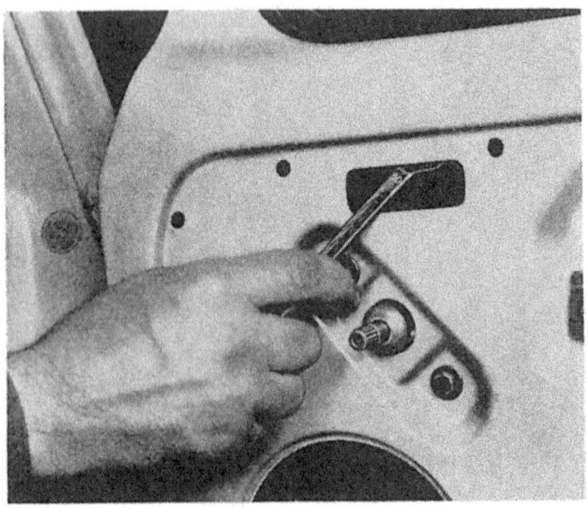

4 - Adjust clamp by loosening or tightening the securing screw. This screw is accessible through a hole in the inner door panel.

5 - Assemble door and ensure that the window crank and inner handle are positioned correctly.

Note:

The vent wing can jam in its frame and, therefore, become difficult to open.

There are various reasons for the vent wing being difficult to open, these difficulties can be eliminated as follows:

1 - **Vent wing jams at location A:**

Remedy:

Pull lower weatherstrip out of vent wing frame up to pivot. Place plastic or wooden wedge on to exposed frame and drive frame deeper into window aperture in door by applying light strokes with a rubber hammer.

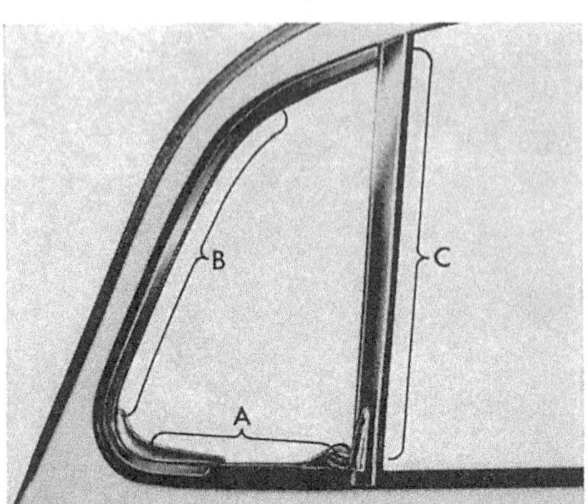

2 - **Vent wing jams at location B:**

Remedy:

Open door window fully and remove door trim panel. Back-off vent wing frame and front window guide channel attaching screws. Press vent wing frame and window guide channel to the rear up to the stop and retighten attaching screws.

3 - **Vent window glass at location C does not run parallel with front window guide channel:**

Remedy:

Open window so that it forms a right angle and carefully align it upward or downward.

On completion of work, thinly coat vent wing weatherstrip with glycerine or talcum powder. **Do not use mineral oils.** If necessary ease off the fastener with a few drops of oil.

Removal

1 - Take vent wing out.

2 - Pull glass and seal out of frame using the special appliance VW 737 (see section "Vent Wing Glass Removal and Installation").

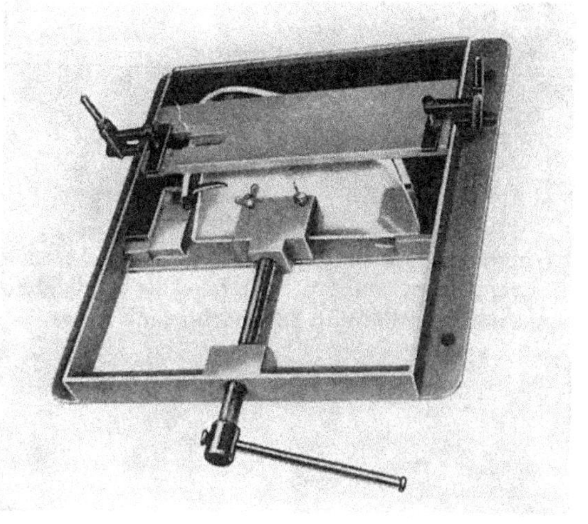

3 - Drill rivets out and take fastener off.

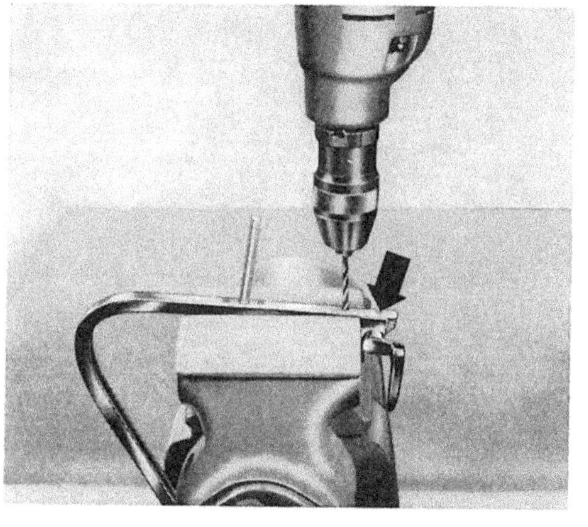

Installation

1 - Check frame, straighten if necessary and remove any burr round the holes.

2 - Rivet fastener on and file off surplus rivet material.

3 - Insert glass with a new seal. A coat of D 10 window sealing compound should be applied between frame, seal and fastener in the area near the fastener.

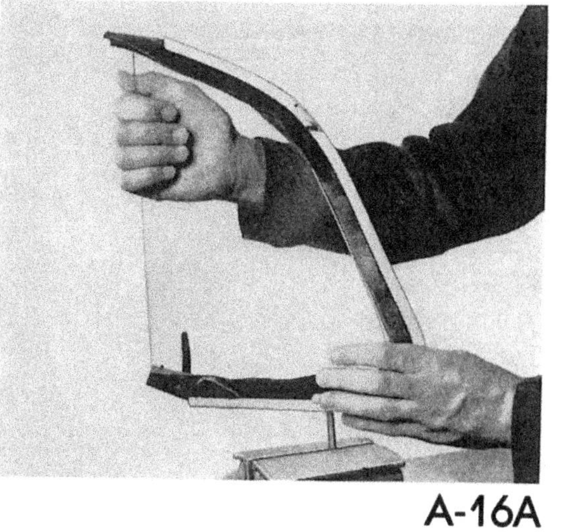

4 - Install vent wing again.

Striker Plate Adjustment

Function

Projecting from the latch housing are the latch on the underside and safety slide at the end.

When the door is closed, the latch springs over the safety groove and then engages in the somewhat deeper locking groove. At the same time the safety slide contacts the inside of the striker plate and is pushed back into the latch housing so that the slide holds the latch and prevents the door from springing open.

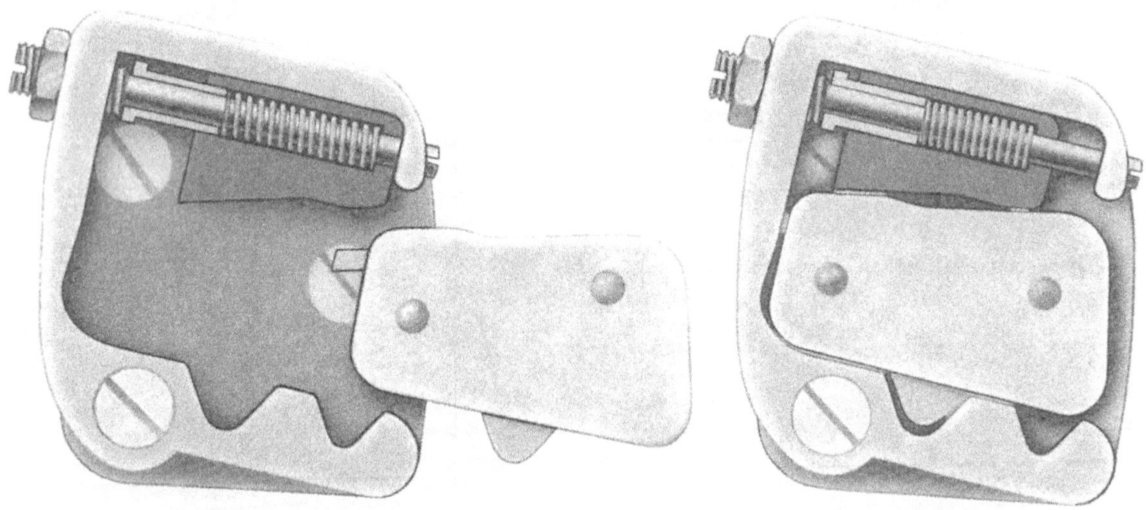

To prevent the latch housing moving back and forth due to vibration when the vehicle is in motion, the striker plate is fitted with an adjustable plastic wedge. This wedge presses the door latch against the flank of the locking groove when the door is closed and also presses the latch housing against the contact surface of the striker plate. This is, of course, assuming that the door, striker plate and wedge are all adjusted properly.

Basic Adjustment

1 - Remove striker plate.

2 - Check the fit of the door in the body opening.

The fit is right when

a - the gap between top edge of door and roof and between rear edge and quarter panel is more or less uniform, b - the door and quarter panel are in line,

c - the waistlines of door and body are in line,

d - the door does not rub at top or bottom,

e - the door weatherstrip contacts and is compressed evenly all round.

If necessary, the door must be adjusted. This may involve loosening the hinges and realigning the door in the frame.

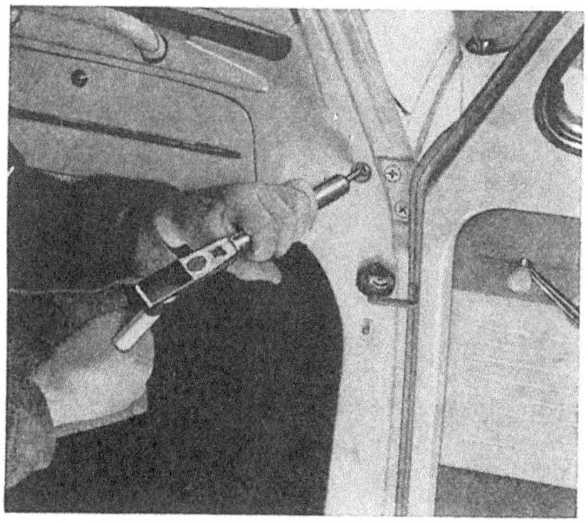

3 - Check that the lock works properly. Note the following points:

a - The top and bottom surfaces of the lock housing must be perfectly smooth. The holes for the safety slide and the latch must not be burred at the edges.

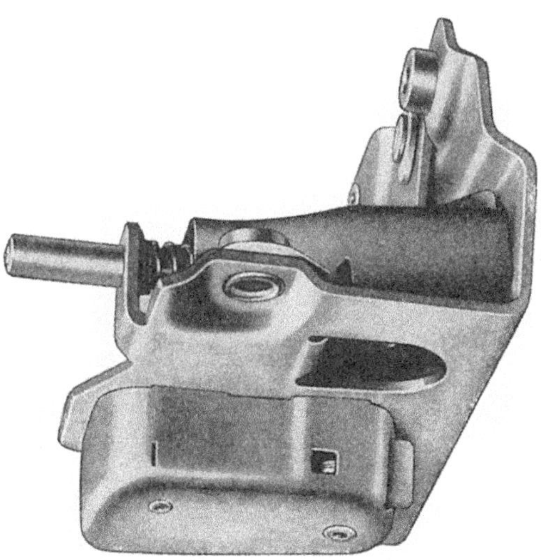

b - The latch must be withdrawn right into the housing when the button is pressed. If this is not to the door may be difficult to open.

4 - Check the striker plate. If the contact surfaces for the latch housing (a) and the groove for the latch (b) are badly worn, a new striker plate should be fitted.

If the plastic wedge is marked on the underside due to roughness on the latch housing or if there are other signs of wear, the wedge should be renewed.

5 - Install striker plate but do not tighten the screws fully.

6 - Loosen the locknut on the wedge adjusting screw and turn the screw to the right until the stop bush contacts the striker plate housing. This gives the wedge a maximum amount of free movement.

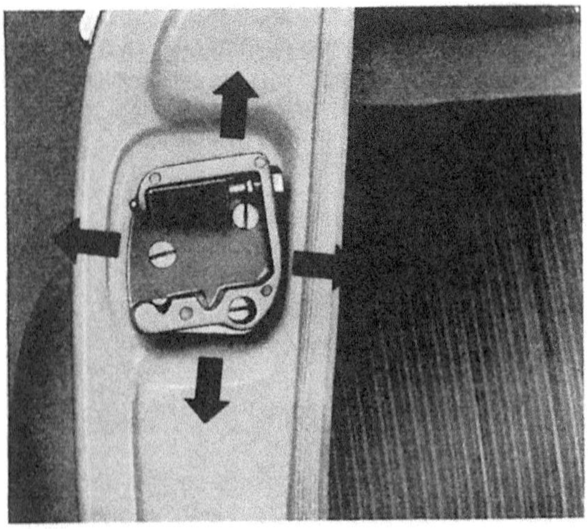

7 - Adjust striker plate.

a - **Lateral adjustment**

Adjust the plate sideways so that the door and quarter panel are in line.

b - **Vertical adjustment**

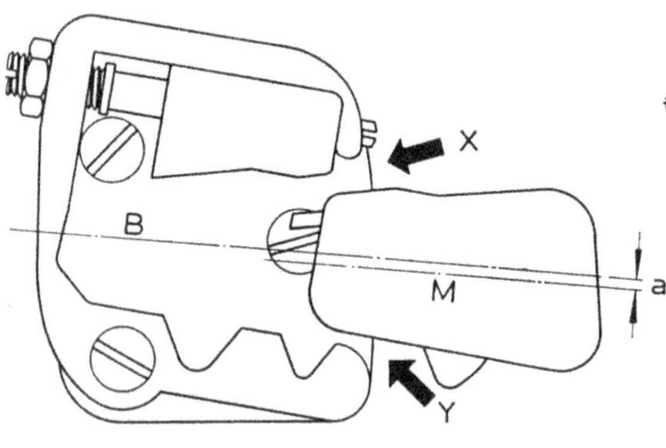

When adjusting vertically, ensure that the space between the latch housing (M) and the striker plate is smaller at the bottom (Y) than at the top (X).

8/65

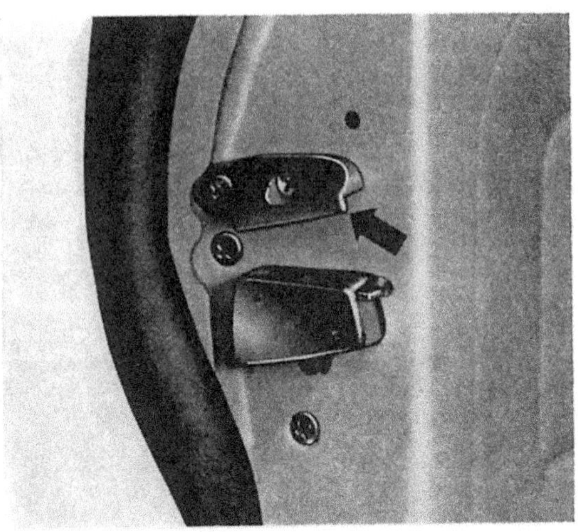

Note:

Since 2 August 1965, Chassis No. 116 000 002, all vehicles have a locking plate on the door end face. The plate engages a shoulder cast on the striker plate — Part No. unchanged — when the door is closed.

Note:

The locking plates can be service installed without difficulty. It is merely necessary to replace the upper lock securing screw by a 5 mm longer version — Part No. N 14 267 5. On vehicles which have not already got striker plates with this shoulder, the striker plates must also be replaced.

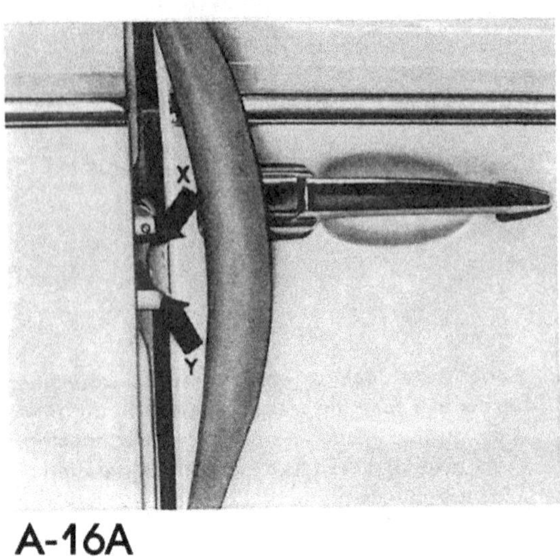

When the door is closed, the latch housing must contact the wedge and be lifted about 2 mm (a).

The height of the striker plate can be observed if the weatherstrip on the door is pressed to the side slightly.

8 - After carrying out the adjustment, open and close the door several times and check the contact pattern on the striker plate.

When adjustment is correct, the contact areas of the latch housing on the strike plates should show a uniform contact pattern.

9 - Tighten striker plate screws fully.

If necessary, the striker plate should be turned slightly.

10 - Adjust the plastic wedge. The movement of the wedge and thus the movement of the latch under the influence of vehicle vibration is limited by an adjustable stop. A screw with a bush is adjusted so that the plastic wedge is more or less fixed when the door is closed.

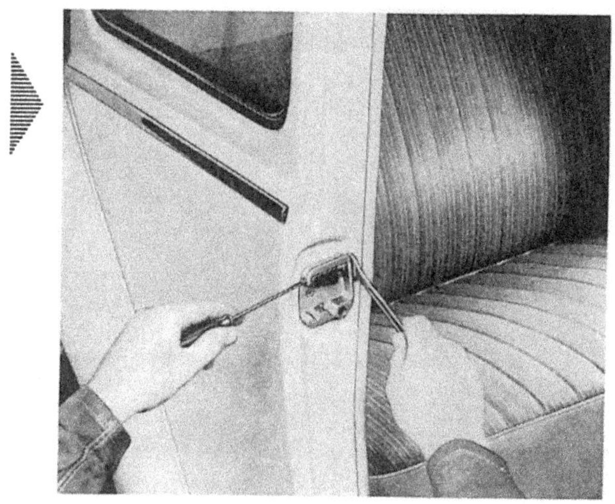

The adjustment is carried out as follows:

a - Loosen locknut (1) on the adjusting screw (3). Hold the screw with a screwdriver.

b - Move the stop bush (2) towards the wedge (4) by turning the screw to the left until the resistance (locking pressure) increases noticeably when opening the door.

The resistance can be felt better if the door is opened with the inner handle.

If the resistance when opening becomes excessive or if the door springs back when slammed to, the screw must be turned to the right until the resistance is correct.

c - Hold screw with a screwdriver and tighten locknut.

The stops on new striker plates tend to bed in slightly so that it may be necessary to adjust the wedge again after a short time in use. Adjust as at points a, b and c.

d - Coat the latch housing contact surfaces on the striker plate lightly with petroleum jelly or a molybdenum disulphide paste.

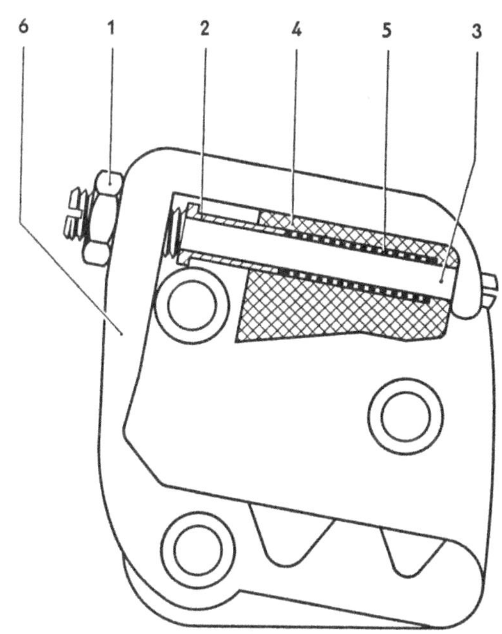

1 - Locknut
2 - Bush
3 - Adjusting screw
4 - Wedge
5 - Spring
6 - Striker plate housing

Windshield Removal and Installation

Removal

1 - Starting at left and right upper corners, push the windshield and weatherstrip out of body.

2 - Push the upper and lower trim moulding sleeves to one side with a piece of wood until the moulding ends are free.

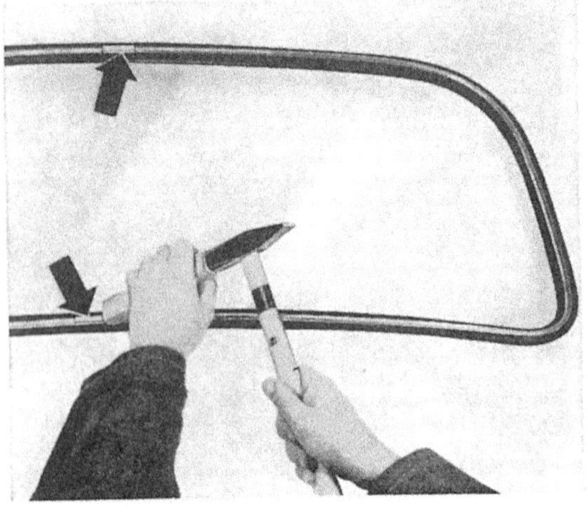

3 - Pull the trim mouldings out of the weatherstrip.

4 - Take weatherstrip off glass.

Installation

All windshields have a mark in the left or right lower corner. Note the following when installing:

a - With toughened glass windshields, the mark should be readable from the driver's side.

b - With laminated glass windshields, the mark must be readable from outside.

1 - Remove all traces of old cement from the body and windshield frame and patch up paint damage.

2 - Check condition of weatherstrip and fit new if necessary.

3 - Check trim mouldings, straighten or replace as necessary.

4 - Fit weatherstrip on to glass. The joint should be in the center at the top.

Fitting trim moulding

1 - Insert a length of cable, preferably a piece of 2—3 mm dia. electrical cable with insulation, into the groove for the trim moulding. This can be done by passing the cable through a piece of tube which is flattened at one end slightly. Place the flat end of the tube in the groove and pass it right round the windshield so that the cable is laid in the groove in the seal. The ends of the cable should meet in the center of the lower edge of the windshield.

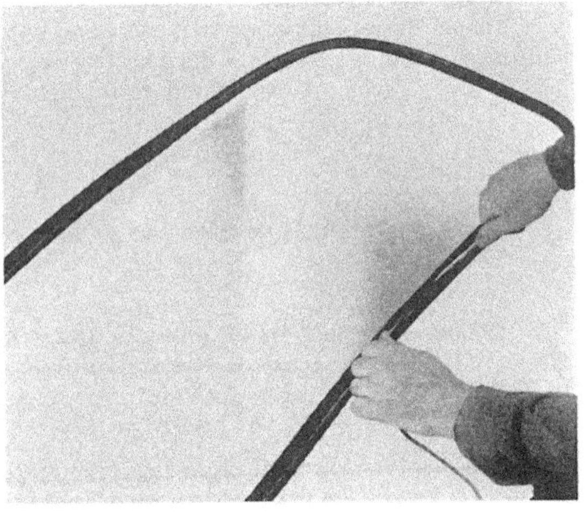

2 - Starting at the cable ends, place the trim moulding in the weatherstrip. Pull the cable out slowly and follow round pressing the moulding into the groove.

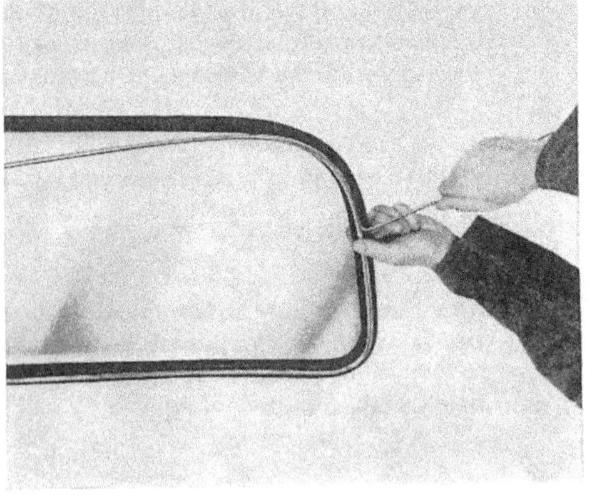

3 - Push the sleeves over the moulding ends.

Fitting windshield

1 - Insert cable into the outer slot in the weatherstrip so that the ends meet in the center of the lower windshield edge.

2 - Powder the weatherstrip with talcum or coat it lightly with glycerine.

3 - Lay the glass in the frame from outside with the cable ends hanging down inside.

4 - Two mechanics are required to install the windshield. Starting at one end of the cable, one mechanic pulls the lip of the weatherstrip over the frame edge, keeping the cable at a right angle to the inside surface of the glass. The second mechanic strikes the glass from outside with his hand, following the pull of the cable and seating the windshield evenly all round.

5 - Test for leakage and then clean glass with methylated spirit.

Important

If a laminated glass windshield is being installed instead of the toughened glass type, pressure must be applied evenly with both hands to seat the glass while the second mechanic lifts the lip over the frame edge with the cable. Rubber hammers must not be used owing to the danger of cracking the glass.

Side and Rear Window Removal and Installation

These windows are removed and installed in accordance with the instructions for the windshield. The trim mouldings of the side windows are in one piece and those for the rear window are in two pieces. Instructions on the replacement of the drop window are given in a special section.

Vent Wing Glass Removal and Installtion

(Vent Wing Removed)

Removal

1 - Place the vent wing in the VW 737 appliance so that the frame edges are properly in contact with the mounting.

2 - Place the clamp on the spindle and screw spindle forwards as far as it will go.

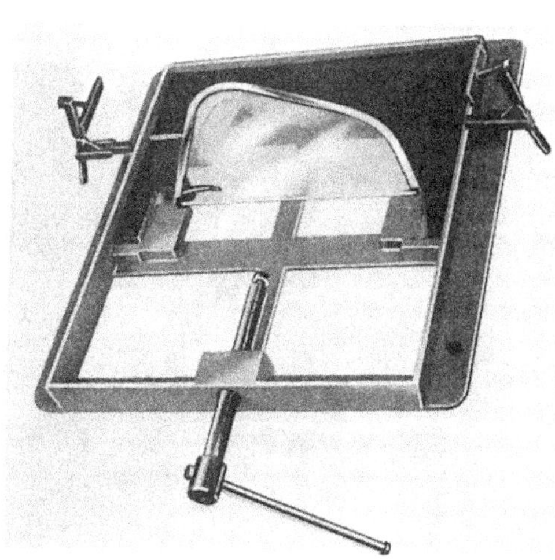

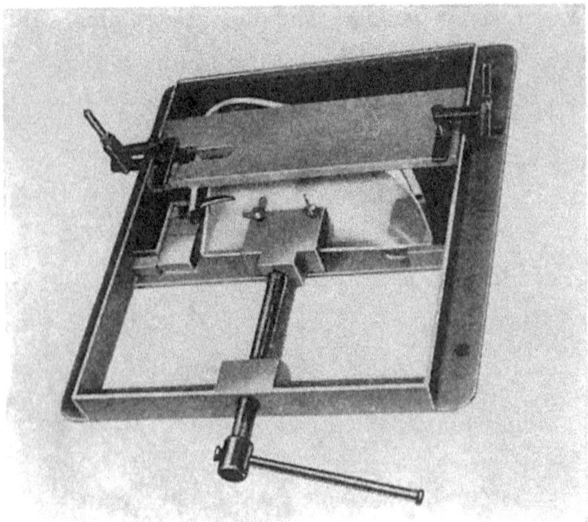

3 - Install rubber packings above and below the glass and lay the clamping plate on the upper packing with the countersunk depressions in line with the ends of the wing bolts.

4 - Tighten the wing bolts by hand.

5 - Place pressure plate on vent wing and secure with clamps.

6 - Turn spindle back to withdraw the glass.

7 - Release wing bolts and take glass out of device.

8 - Release clamps, take pressure plate off and lift frame out of device.

Installation

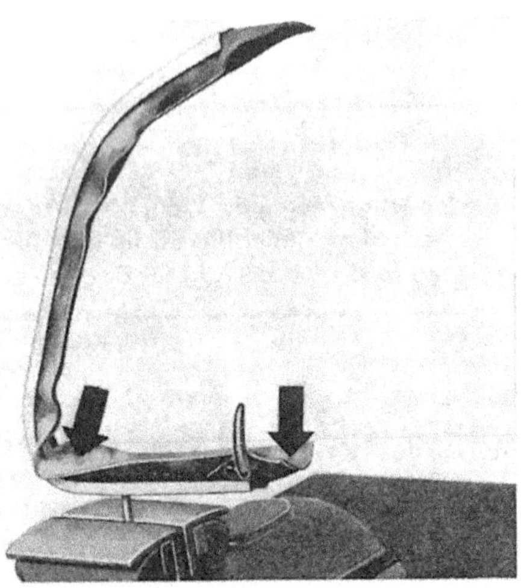

1 - Remove all traces of old adhesive from frame and round fastener.

2 - Coat inside of frame near the fastener with window cement D 10.

3 - Coat both sides of the new weatherstrip with talcum powder and rub it well in.

4 - Clamp the frame in a vise and insert the weatherstrip into the frame as shown in illustration.

Note

Ensure that the weatherstrip is not pushed into the frame channel at the points indicated by the arrows but remains flat as shown in illustration thus avoiding the formation of creases when the glass is pressed in. Check that the two fastener shoulders are adequately covered by the weatherstrip.

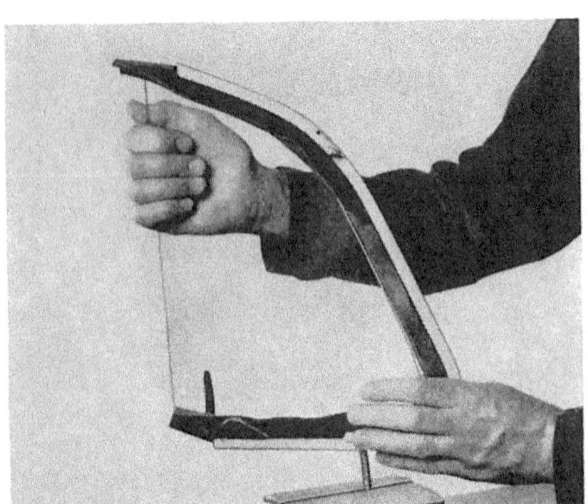

5 - Press the glass in by hand until the weatherstrip overlaps the glass on both sides. When doing this, press the frame with the other hand to prevent distortion.

6 - Place vent wing and adaptor in the appliance.

7 - Install retaining plate and secure with clamps.

8 - Place the thrust plate on the glass and press glass in by turning the spindle.

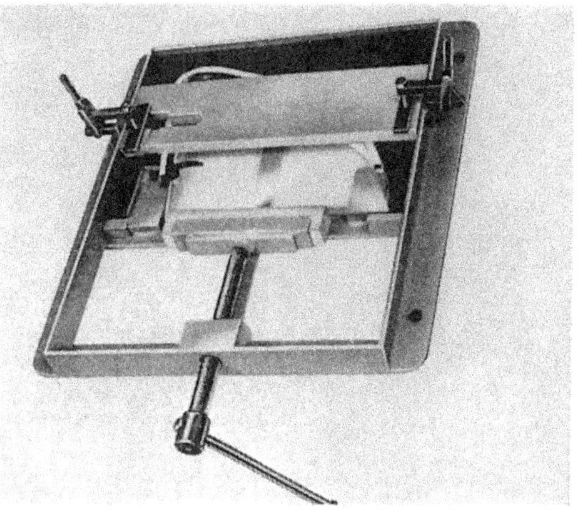

9 - Turn spindle back slightly, take thrust plate off and release clamps. Take off pressure plate and vent wing.

10 - Cut off surplus rubber with an angled knife. Take care that the cut edge is angled so that water can run over the edge and drain off.

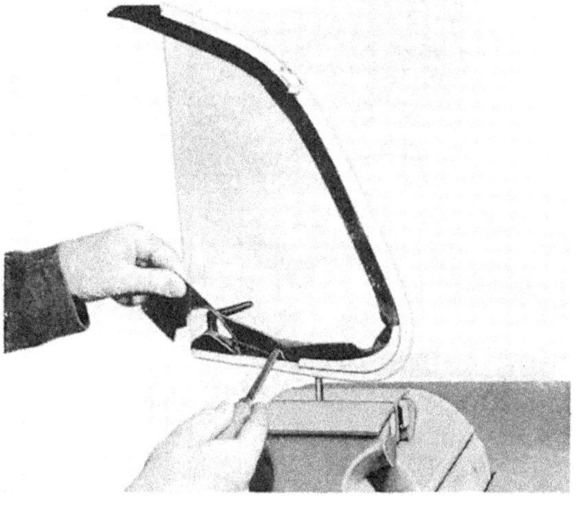

11 - Check for leakage by spraying with water and then clean glass with methylated spirit.

Important

When removing and installing the glass in the left or right vent wings, the two adaptors should be turned 180° in the appliance as required.

THIS PAGE INTENTIONALLY
LEFT BLANK

Front Seat Removal and Installation

Removal

De Luxe Sedan

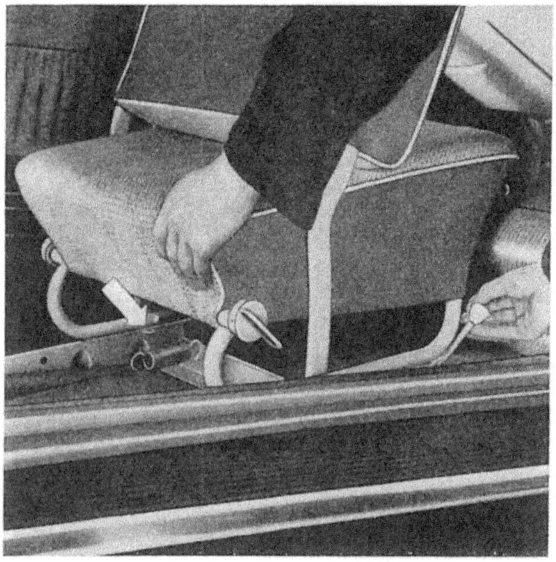

1 - Lift up locking mechanism on the right hand side of the seat and slide the seat forward until the retaining spring on the left side of the seat and the seat frame can be disengaged. Then push the seat further forward and take it out of the guide rails.

2 - Take out the seat.

Note:

When interchanging the front seats the adjusting mechanism must also be taken out and reinstalled at the proper side. The adjusting cams with wings must always point outward.

Grease the seat runners with VW Universal Grease A-052 prior to installing the seat.

Standard Sedan

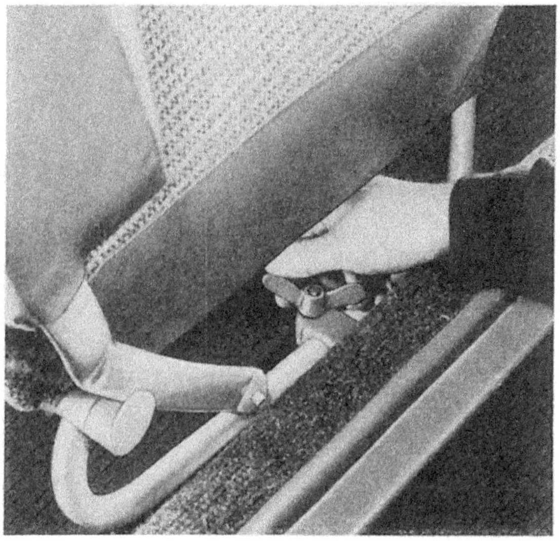

1 - Remove the two wing nuts and take off the seat clamps.

2 - Remove the seat.

Removal and Installation of the adjusting cams

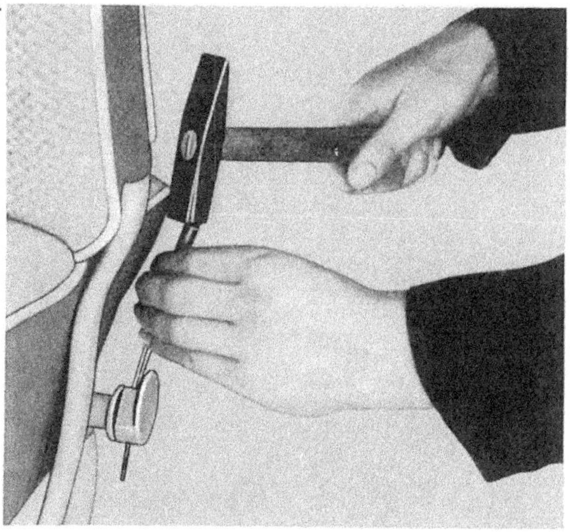

1 - Remove seat.

2 - Remove the pin from the left or right hand adjusting cam by means of a punch.

3 - Remove cam.

4 - Remove the opposed cam with connecting tube from the guide tube.

Note:
The adjusting cams of the front seats are not interchangeable.

Installation is a reversal of the preceding operations. Prior to installation the cams should be lightly greased on the friction side to the guide tube with Genuine VW Universal grease A - 052.

Note:
The stop plates for the front seat back rests on the Standard Sedan are also secured by a pin at the rear cross tube.

Rear Seat Removal and Installation

Removal

1 - Push the seat cushion slightly backward, raise it and take it out from its support.

2 - Remove seat cushion diagonally.

3 - To remove the seat cushion support raise the felt lining at the two ends of the rail and remove the two hex. screws.

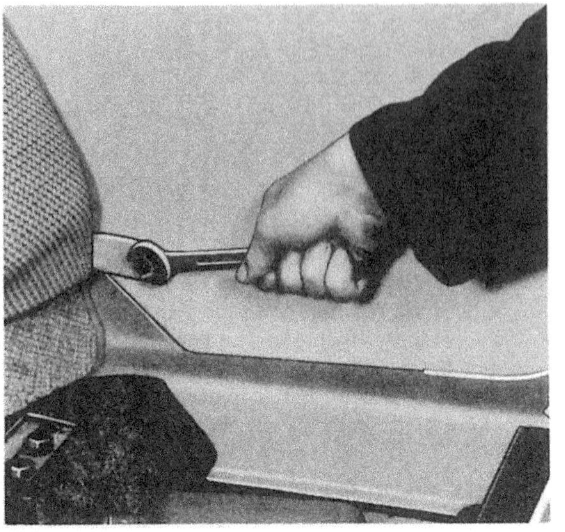

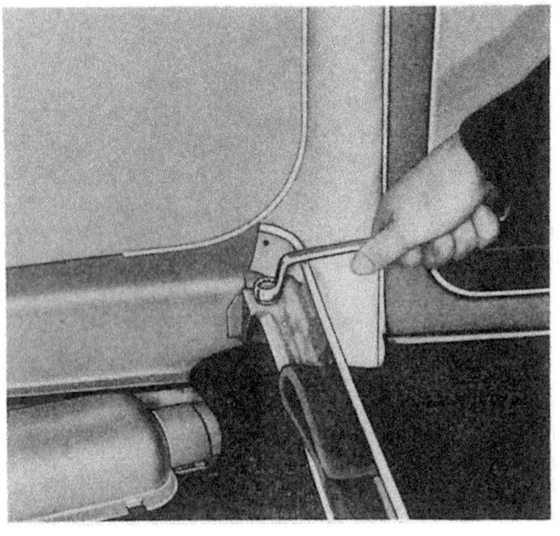

4 - Unhook the loop at the right hand back of the rear backrest.

5 - Remove the two backrest securing screws (hex. head) and take out backrest to front.

Installation

1 - Check condition of rubber strips on the support brackets, re-new if necessary.

2 - Check condition of the felt lining and the three hard felt pads and the inside of the support bracket and replace if necessary. The three pads prevent the seat cover from fraying on the lower seat frame support bracket.

3 - When installing the rear seat cushion it must be pushed back until the lower seat frame slots in behind the edge of the support bracket. Push seat cover down until it comes to rest on its support. Force will damage the seat covers.

When the stocks of spring interiors (Part No. 111 881 705 B) for the front seat back rests are used up, only the spring interiors (Part No. 111 881 705 C) will be supplied. These parts are, however, 17 mm/ ,67" too long for the seat frames on vehicles up to Chassis No. 2 528 668 (August 1959).

On installation they should be altered as follows:

The four ribs in the lower frame of the spring interior should be reshaped in the vice with a flat hammer until the ribs are approximately 17 mm shorter and the spring interior can be attached to the back rest frame.

Before fitting the material check that the reworked places are covered with padding.

Note:

From Chassis No. 4 057 923 (21st August 1961), the range of seat adjustment on the De Luxe models has been increased from 100 mm (3.9") to 120 mm (4.7"). The extra movement is towards the rear.

At the same time, the cams on the back rest adjusters have been modified so that the back rests can be set further forward and backward than they could formerly. This alteration also applies to the drivers seat of the Standard model.

These alterations will enable the front seats to be adjusted to give the desired seating position for persons of all sizes.

Note:

From Chassis No. 3 192 507 (1st August 1960), a seat with an adjustable backrest is being production installed on the driver's side on the Standard model. The seat — Part No. 111 881 033 — has the same adjusting arrangement as on the De Luxe model.

The complete seat can be service installed in vehicles of previous pattern.

Note:

On right-hand drive vehicles, adjusting cams — Part No. 113 881 164 A and 113 881 167 A must be used.

Note:

To prevent the front seat back rest from coming off the pins on the seat frame, it is secured with a cap nut (Part No. N 11 062 3) on each side from Chassis No. 3 771 255 (28th March 1961).

The pins are now longer and are threaded. The retaining nuts can only be service installed by replacing the seat frame (Part No. 113 881 107 C as before).

THIS PAGE INTENTIONALLY
LEFT BLANK

A. Front Seats

Removal

1 - Remove seats and take backrests off.

2 - Bend up the ten tabs on the seat frame and release the tensioning wire of the seat material.

3 - Take cover and padding off.

4 - Bend up the ten retaining lugs on the seat frame and take spring core off.

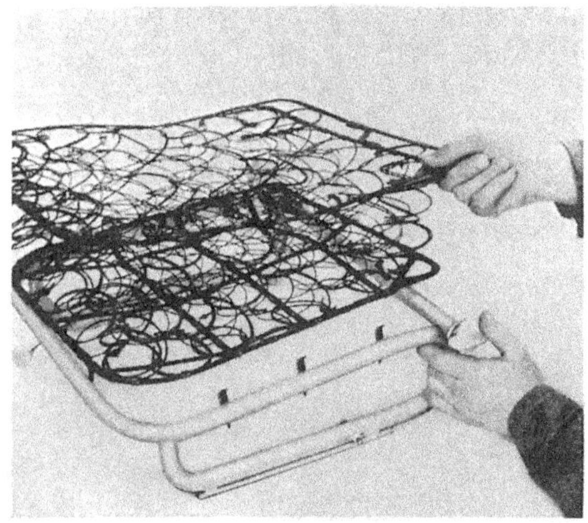

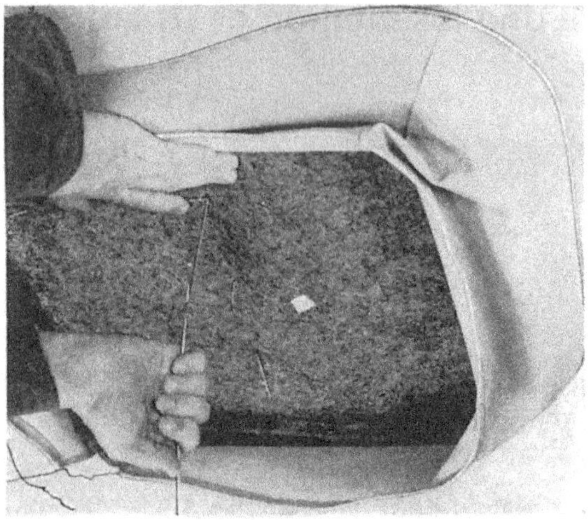

Installation

1 - Replace unserviceable parts, straighten seat frame if necessary and remove burr from runners. Check the seat catches, oil lightly and repair paint damage.

2 - Place spring interior on frame and bend back the lugs to secure it. Install the padding.

3 - Cover the outer contours with additional wadding and fit cover over spring interior and padding.

4 - Join cover and padding with three tackstitches. This makes it look better and prevents the material from slipping on the padding.

5 - Pull seat cover material over the tabs on the seat frame and bend the tabs over.

6 - Pull the ends of the tensioning wire tight, twist them together and tuck them under the spring interior.

7 - Install backrest.

8 - Install seat in vehicle.

B. Front seat backrest

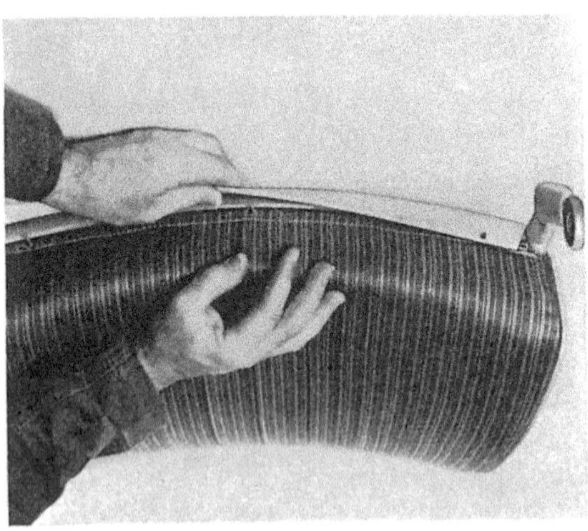

Removal

1 - Remove seat and take backrest off.

2 - Bend up three tabs on backrest frame, unhook cover and pull out two tensioning wires.

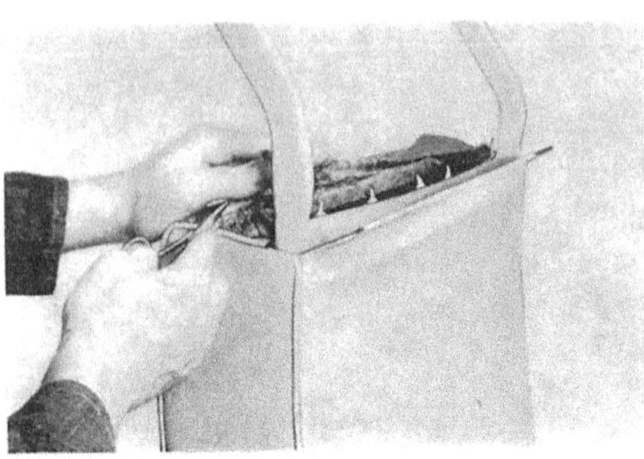

3 - Part the tack stitching on each side and pull material off the padding.

4 - Take padding off spring core.

5 - Remove four screws, bend up retaining lugs and take spring core with cardboard cover off the side tubes.

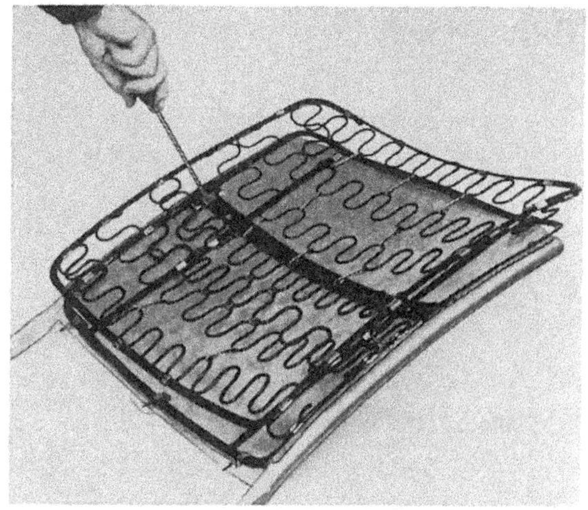

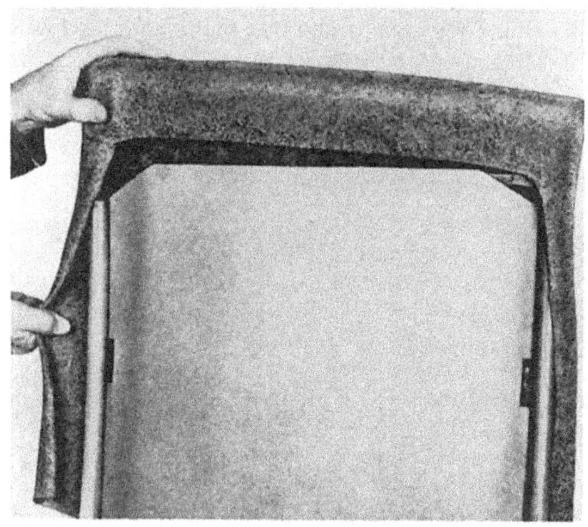

Installation

1 - Replace unserviceable upholstery, straighten backrest tubes if necessary and touch up paint damage at bottom of tubes.

2 - Attach spring interior and cardboard to frame and bend tabs up.

3 - Place padding over spring core. Cover outer contours with additional wadding.

4 - Pull cover over backrest.

> **Important**
>
> To prevent the padding from slipping down or moving sideways, press the backrest together slightly and pull the cover material down carefully.

5 - Push tensioning wires into cover.

6 - Hook front part into retaining lugs and then the rear part and bend the lugs down.

7 - Tack the material near the tubes.

8 - Fit backrest to seat.

9 - Fit seat in vehicle.

C. Rear Seat

Removal

1 - Remove rear seat.

2 - Release tensioning wire.

3 - Bend up 16 clips and free them from the lower frame of the spring core.

4 - Take seat cover and padding off spring core.

Installation

1 - Replace unserviceable upholstery parts or spring cores, straighten spring core frame if necessary.

2 - Place padding on spring core. Pad up the contours with additional wadding.

3 - Pull cover over spring core and fit it to the shape of the spring core lower part by twisting the tensioning wires together.

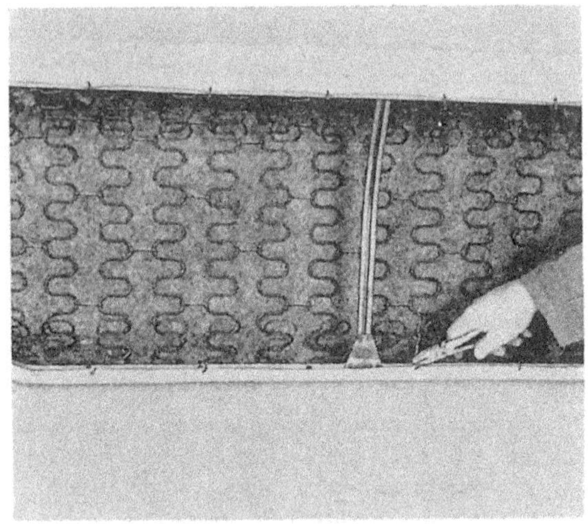

4 - Secure cover to frame of spring core with clips.

Note

These clips are not supplied as spare parts. Where necessary the seat cover should be attached to the spring core with thin welding wire.

5 - Install seat.

D. Rear Seat Backrest

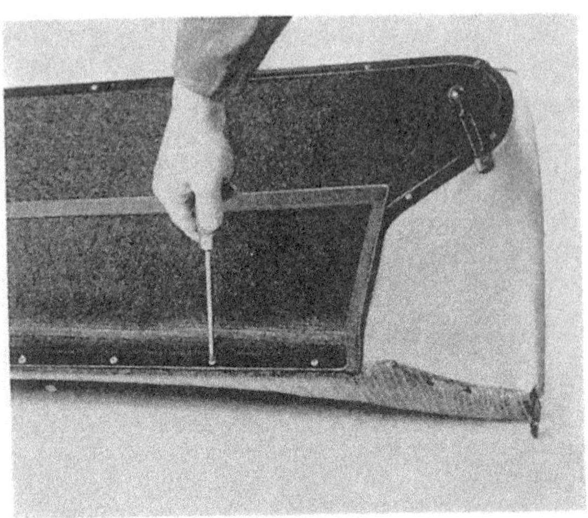

Removal

1 - Remove backrest.

2 - Remove eight screws and take off retaining strip with strip of haircord carpet.

3 - Remove screw and take retaining loop off.

4 - Remove 13 screws and take off rear part with protective strip.

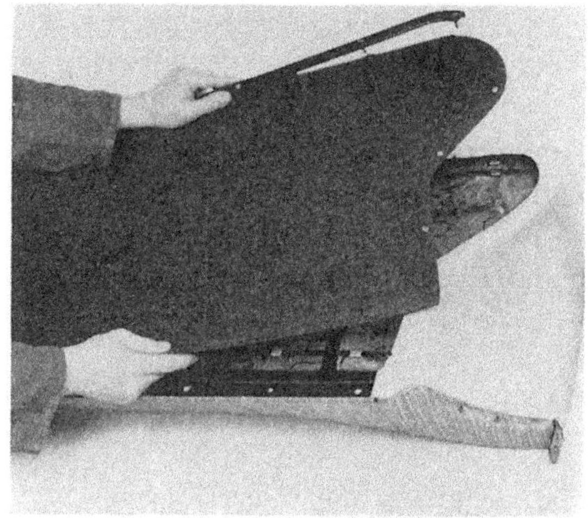

5 - Loosen tensioning wire, bend up tabs and detach cover from backrest frame at both lower corners.

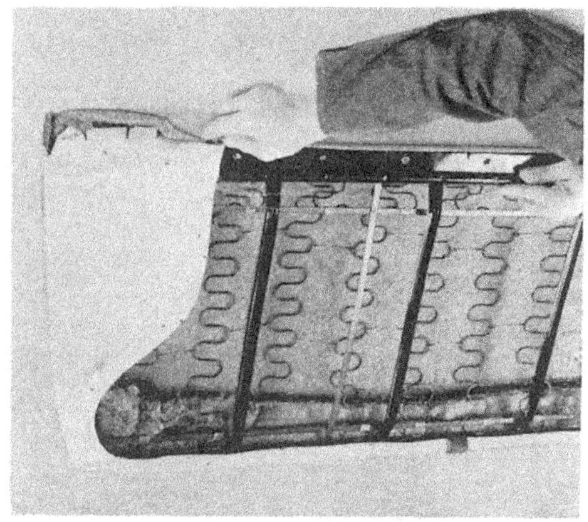

6 - Detach cover from the retaining tabs in upper part of frame. The retaining strap should be taken off the center tab beforehand.

7 - Take off cover with padding.

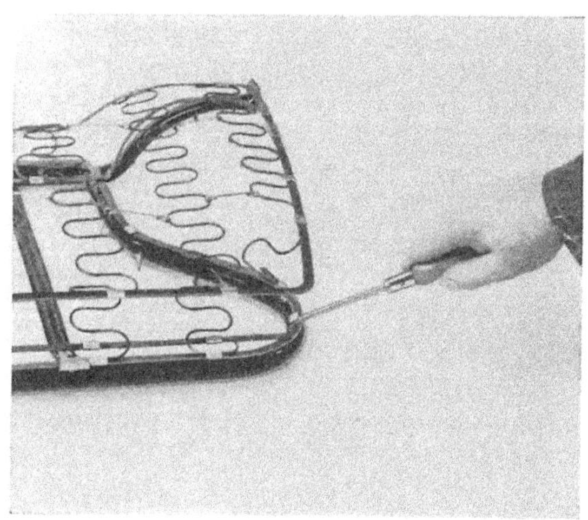

8 - Bend up 12 tabs on backrest frame and take spring core off.

Installation

1 - Replace unserviceable upholstery parts, straighten frame securing and protective strips as required and paint.

2 - Secure spring core to frame by bending down the tabs.

3 - Place padding on spring core and pad out contours with additional wadding.

4 - Pull cover over spring cover and hook it onto the retaining tabs, tighten tensioning wire and bend down tabs.

Note

The retaining strap for the backrest should be hooked to the center tab first.

5 - Fit rear part with protective strip, carpet and retaining loop and install backrest.

THIS PAGE INTENTIONALLY
LEFT BLANK

Removal and Installation of Front Hood Moulding, Emblem and VW Sign

Removal

1 - Remove the moulding clip gaskets.

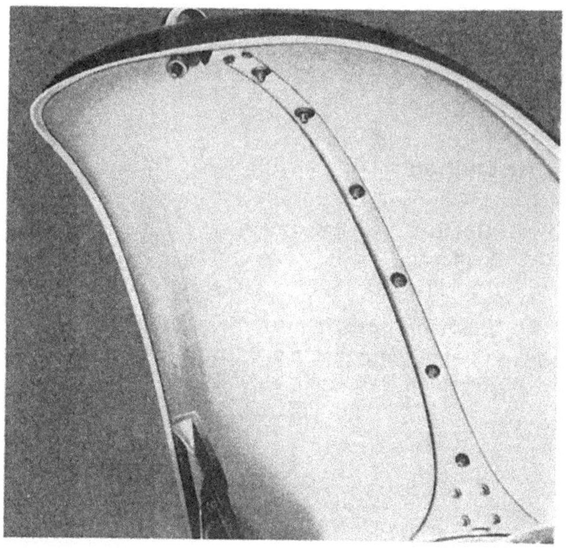

2 - Push the clips through the holes from inside the hood and take off the moulding.

Note:

From Chassis No. 5 010 448, the following alterations were made on the front hood:

The coat-of-arms on the front hood has been discontinued and the VW sign redesigned.

The modified VW sign (Part No. 113 853 601 B) cannot be installed on hoods of the previous pattern. The old VW sign (Part No. 113 853 605 A) will remain available. The coat-of-arms (Part No. 113 853 621 B) will also remain available as a spare part.

The front hood (Part No. 113 823 031 B as before) and the trim moulding have also been altered. The modified trim moulding (Part No. 113 853 505 F) cannot be installed on hoods of the previous pattern. The old trim moulding (Part No. 113 853 505 C) will remain available.

When installing the modified hood on earlier vehicles, the new trim moulding (Part No. 113 853 605 F) and VW sign (Part No. 113 853 601 B) must be fitted.

3 - Bend up the emblem retaining tongues and take off gaskets, emblem and base plate.

4 - Bend up the three VW sign retaining tongues, pull off the gaskets and remove VW sign.

Installation

1 - Replace bent or damaged moulding clips. Replace VW sign and emblem if retaining tongues have broken off.

2 - Cement the VW sign inside the hood using VW Compound D 15.

3 - Place the emblem in position with the rubber gasket inside the hood and bend down the retaining tongues.

4 - Push the gaskets over the moulding clips and cement them into place, using Genuine VW Compound D 12.

Moulding on Cowl Side Panel

Removal

1 - Raise the front hood, remove the front luggage compartment lining and pull off the moulding clip gaskets.

2 - Push the clips through the holes from the inside and take off the moulding.

Installation

1 - Bent or damaged moulding clips should be replaced.

2 - Push the gaskets over the moulding clips and cement them into place, using Genuine VW Compound D 12.

Door Moulding

Removal

1 - Remove regulator and inner door handles.

2 - Remove door trim panel.

3 - Pull off moulding clip gaskets.

4 - Push out moulding clips through the holes from inside and take off the moulding.

Installation

1 - Bent or damaged moulding clips should be replaced.

2 - Push the gaskets over the moulding clips and cement them into place inside the door, using Genuine VW Compound D 12.

Rear Quarter Panel Moulding

Removal

1 - Remove rear quarter trim panel.

2 - Pull off moulding clip gaskets.

3 - Push the clips through the holes from inside and take off the moulding.

Installation

1 - Damaged or bent moulding clips should be replaced.

2 - Push the gaskets over the moulding clips and cement them into place inside the panel, using Genuine VW Compound D 12.

Outside Rear View Mirror

Every Passenger Car for the home market is provided with an outside rear view mirror as standard equipment. To avoid damage during transport the mirror is placed in the glove compartment.

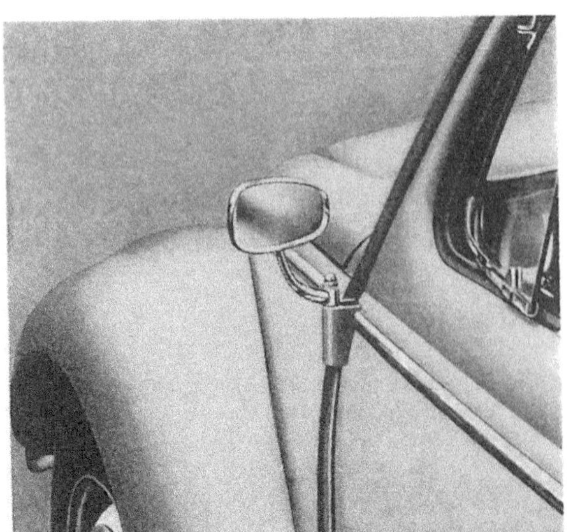

Prior to delivery, the mirror should be mounted on the left-hand door hinge pin so that the mirror rod is approx. at right angles to the longitudinal car axis. The cap nut should never be tightened to a torque of more than 1 mkg (7 ft. lbs.).

The following SP Sets are available for subsequent installation:

VW Sedan SP 96; VW Convertible SP 97

It is recommended that the hinge pin for the mirror be inserted with a hollow drift or a piece of suitable tube to prevent damage to the thread.

Left-hand drive cars for export are not provided with an outside rear view mirror as standard equipment, but the upper door hinge of the left-hand door has a hinge pin with cap nutto allow service installation o fthe mirror. Right-hand drive cars continue to be provided with the usual hinge pins.

Note:

From Chassis No. 2 149 710 the outside mirror (Part No. 113/114 857 513 A) was enlarged and is held in position by means of a tapered locking sleeve (Part No. 111 857 519) and a union nut (Part No. 113 831 423 C) in the correspondingly modified hinge pin (Part No. 111 831 423 C), which is also available in oversizes (111 831 423 D and E).

The mirror glass is no longer exchangeable.

When installing the new mirror, the following should be observed:

1 - Insert tapered locking sleeve into hole provided in the hinge pin top end.

2 - Screw union nut on hinge pin with two turns.

3 - Insert mirror mounting pin fully through the hole of the union nut into the hinge pin.

4 - Tighten union nut (16 mm) to a torque not exceeding 1 mkg (7 ft. lbs.).

Only cars for the German market are provided with an outside rear view mirror as standard equipment. On request, export cars will be delivered with an outside rear view mirror as an optional extra (M 158).

The following SP sets are available for service installation

VW Sedan and Convertible

Left-hand drive

SP set 96 B consisting of:
1 No. 113 857 513 A mirror
1 No. 113 857 517 A union nut
1 No. 111 857 519 tapered locking sleeve

Right-hand drive

SP set 99 A consisting of:
1 No. 111 831 423 C hinge pin
1 No. 114 857 513 A mirror
1 No. 113 857 517 A union nut
1 No. 111 857 519 tapered locking sleeve

In order to prevent damage to the hinge pin thread, the pin should be driven into position with a suitable drift inserted in the tapped hole.

Left-hand drive export cars are not provided with an outside rear view mirror as standard equipment, but the upper door hinge of the left-hand door has a hinge pin with cap nut (Part No. 111 831 433) to allow for an easy installation of the mirror on request.

Right-hand drive cars will be provided with the usual hinge pins.

The earlier type mirrors should be used up. The mounting parts will remain available as spares.

Note:

From Chassis No. 3 550 952 (2nd January 1961), the method of securing the VW sign on the front hood was changed.

The four hooks have been replaced by three grooved pins (B).

Plastic thimbles (A) Part No. 113 853 615, which are pressed into the holes in the hood, are provided to seal the modified sign — Part No. 113 853 601 A.

The modified sign cannot be installed in previous vehicles nor can the plastic thimbles be used for sealing the old VW sign.

Note:

Only front hoods with holes for the new sign will be supplied. (Part No. unchanged).

THIS PAGE INTENTIONALLY
LEFT BLANK

Removal and Installation of Luggage Compartment Lining and Sound-Absorbing Pad

Removal

1 - Remove rear seat and backrest.

2 - **De Luxe Sedan**

Remove the luggage compartment lining from the sound absorbing pad, starting at the bottom. At the upper edge the carpeting is folded in between the roof and the rear panel of the luggage compartment, and held with a cord. The cord must therefore be pulled out when removing the lining.

Standard Sedan

The Standard Sedan has a cardboard lining instead of carpet. It is also held between the roof and back panel by a cord. At the front the cardboard is held by two sheetmetal tacks. In order to remove the lining, the tacks and cord must first be removed. Since the cardboard is stuck in position, care must be taken when removing or it will tear.

3 - Remove the sound-absorbing pad from floor of luggage compartment. The pad is cemented in place and will be damaged if removed with insufficient care.

Installation

1 - Before cementing the sound-absorbing pad into place, thoroughly remove all traces of the old pad and cement. The use of benzine as a solvent is recommended.

2 - Cement the sound-absorbing pad into place, using Genuine VW Compound D 12.

3 - Cement the carpet or cardboard lining to the sound-absorbing pad and push the upper edge between the roof and rear panel of the luggage compartment. Finally push the cord, doubled, into the gap, using the headlining tool VW 736.

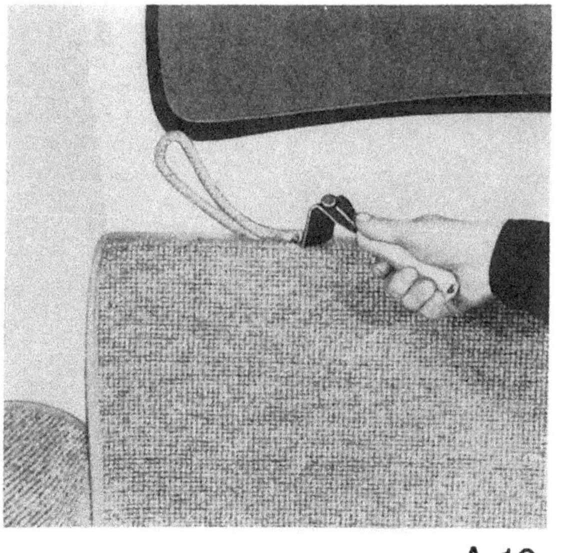

Rear Wheel Housing Carpet

1 - Remove rear seat and backrest.

2 - Carefully pull off the carpet on the appropriate side.

3 - Lift the lining of the luggage compartment floor and pull out the rear of the wheel housing lining.

4 - Before cementing the carpet into place, thoroughly remove all traces of the old cement. The use of benzine as a solvent is recommended.

5 - Cement the carpet into place, using Genuine VW Compound D 12.

6 - The carpeting must be tucked under the luggage compartment floor lining and the side panel trim.

Rear Quarter Trim Panel Removal and Installation

1 - Remove the rear seat and backrest.

2 - Prise off the rear quarter trim panel from the rear quarter inner panel.

3 - The moulding on the De Luxe Sedan can be removed after bending up the clips.

4 - Check that the sound absorbent strips inside the body are properly in position, if necessary cement them on again.

5 - Make sure the rubber gaskets of the mouldings are correctly in position.

6 - Remove damaged or broken clips.

7 - When reinstalling the rear quarter trim panel, make sure that the oiled paper at the bottom overlaps the inner panel to allow the water to drain.

Headlining Removal and Installation

Sedan without sliding roof

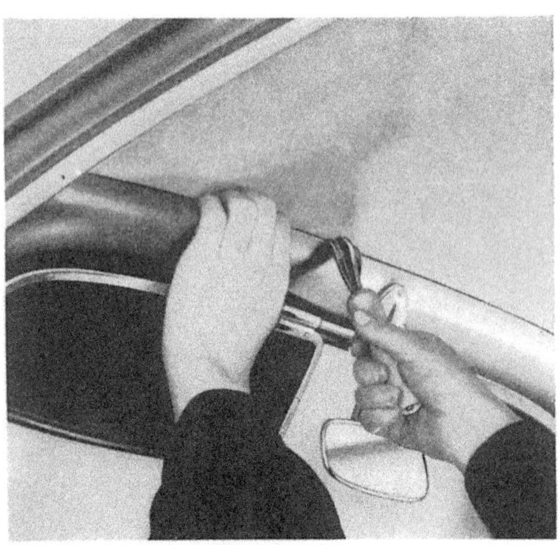

Removal

1 - Lower sun visor.

2 - Insert the headlining tool VW 736 (local manufacture) carefully between the front roof member and the headlining.

3 - Move the tool along the front roof member, lifting out the headlining toward the front.

4 - Reinsert the tool behind the corner of one door hinge pillar and move it up to the first bow. Lift the bow with one hand and continue under it with the tool. Continue to the rear end of the headlining.

Standard Model

When installing a sun visor on Standard Sedan on either the driver's or the passenger's side, the following should be observed:

1 - Remove the rear view mirror holder.

2 - Remove visor plugs.

3 - Push padded sun visor into holes.

4 - Insert mounting plate on the rear of the mirror holder.

5 - Tighten mirror attaching screws.

From 1st August 1960 and Chassis No. 3 192 507 a second sun visor — Part No. 113 857 552 A — and a grab handle for the front seat passenger — Part No. 151 857 641 B — are being production installed. Service installation in vehicles of previous pattern is possible.

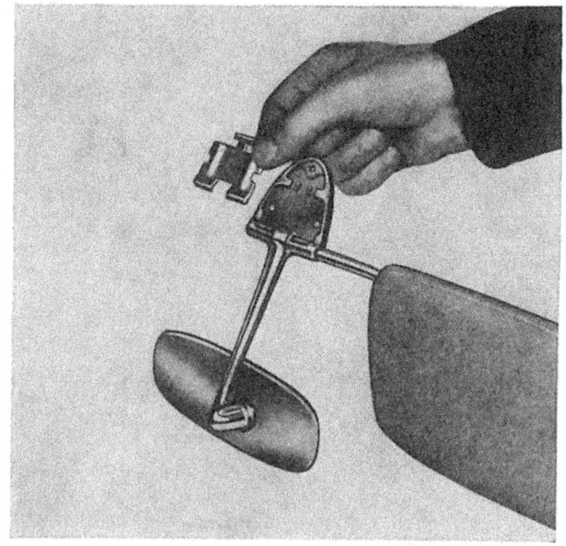

Headlining Removal and Installation

Sedan without sliding roof

Note:

From Chassis No. 4 846 836, (De Luxe Sedan) and 5 580 432 (1200 Sedan) a plastic headlining is installed. This headlining can be service installed in vehicles from Chassis No. 1 600 440.

When repairing or renewing plastic interior trim materials such as the headlining, for example, only the plastic adhesive D 11 should be used. Otherwise the materials may be discolored.

Previous type headlinings will remain available.

Note:

From August 1963, Chassis No. 5 677 119, the plastic headlining of the De Luxe Sedan was modified so that it now covers the entire width of the body interior. As a result, the side roof pillar lining has been discontinued with the exception of the part between quarter window and rear window. This part is now supplied separately as rear roof pillar lining under Part No. 117 867 519/520.

The modified headlining — Part No. 113/117 867 501 G — can be installed in previous vehicles without difficulty. When necessary, the rear roof pillar lining must also be installed.

Removal

1 - Lower sun visor.

2 - Insert the headlining tool VW 736 (local manufacture) carefully between the front roof member and the headlining.

3 - Move the tool along the front roof member, lifting out the headlining toward the front.

4 - Reinsert the tool behind the corner of one door hinge pillar and move it up to the first bow. Lift the bow with one hand and continue under it with the tool. Continue to the rear end of the headlining.

5 - Insert the tool behind the corner of the other door hinge pillar and then repeat the procedure specified in para 4.

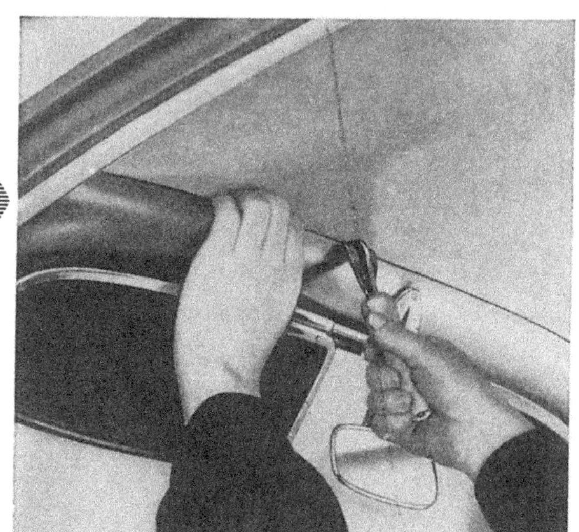

6 - Similarly insert the tool at the rear and completely remove the headlining.

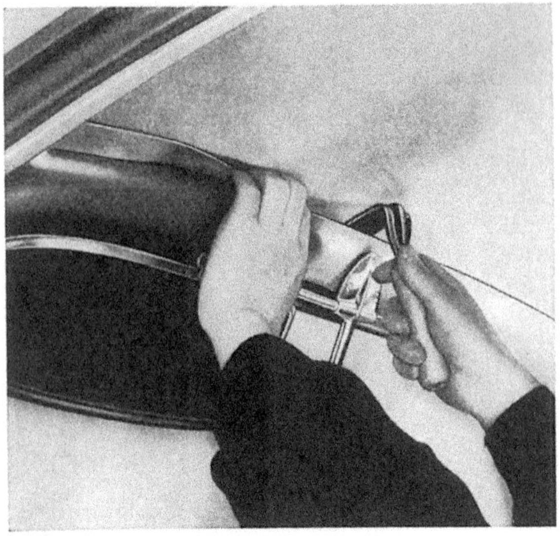

Installation

The following points should be observed:

1 - Check that the roof insulation is firmly in position, if not refix it with Genuine VW Universal Cement D 12.

2 - Insert the headlining, beginning at the rear. Place tool VW 736 between the cardboard strip and headlining cloth and press down the cardboard strip until it is held back by the roof member flange.

3 - In each case the rounded edges of the bows should be on the same side as the main cable harness.

Sedan with sliding roof

Removal

1 - Open the roof slightly. Detach the roof covering from the ends of the guide bows and fold back after removing the two screws for the locking bow.

2 - Having removed the countersunk screws remove the rear guide rails and take out the front bow with the middle of the roof covering.

3 - Bend the retaining rail for the headlining up at the front and sides with a broad-bladed screwdriver.

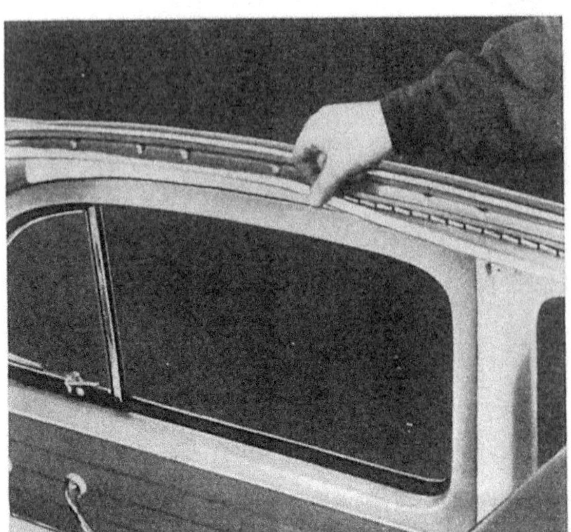

4 - Pull the headlining out of the retaining rail at the sides and front.

5 - Remove the hooks for the rear backrest, also the pads on which the backrest is supported.

6 - Pull off the floor lining of the luggage compartment and the wheel-house lining as far as necessary. Remove the rear quarter trim panels.

7 - Slit the seams between the lining of the roof side member and the centre door pillar.

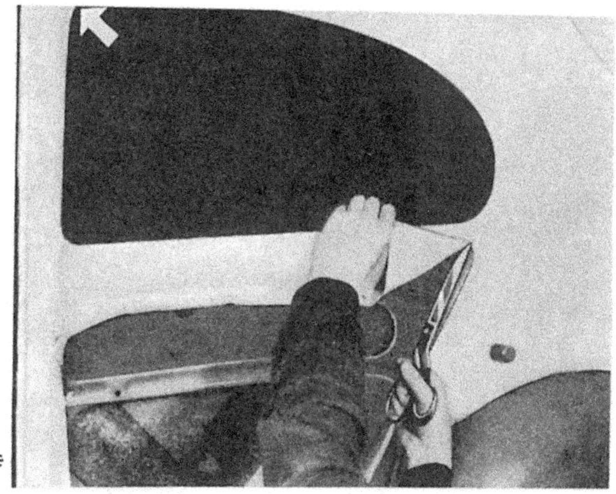

8 - Bend down the retaining strip of the roof side member with a broad-bladed screwdriver and pull out the side-member lining.

Remove the side-member lining, beginning at the front, finishing with the lining round the rear quarter windows.

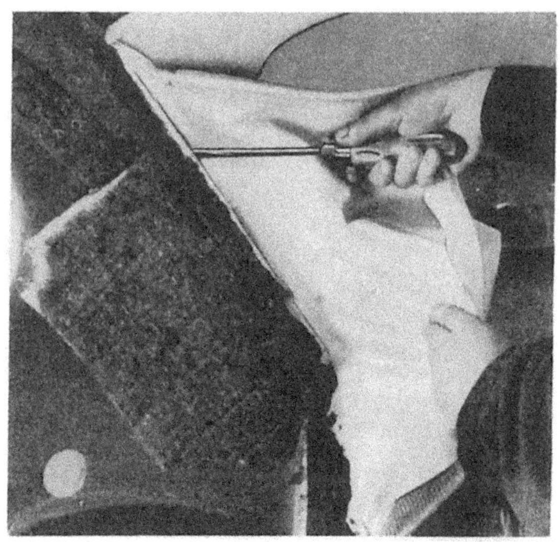

9 - The side-member lining is held to the roof by a retaining strip which is accessible when the padding is removed and the lining lifted up. Bend up the strip with a broad-bladed screwdriver and remove the lining.

10 - Pull off the centre door pillar lining from the member below the rear quarter window. The lining is held at the pillar by a retaining strip which is accessible when the lining is folded back after removing the padding. Bend open the strip with a broad-bladed screwdriver and remove the lining.

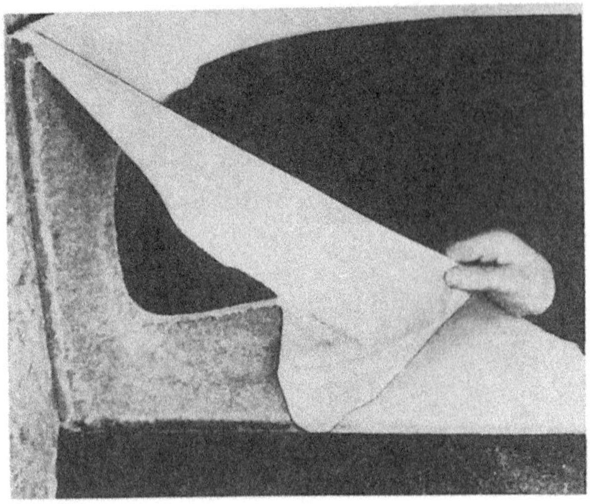

11 - Carefully remove the lining surrounding the rear window.

Installation

1 - Check that the felt pads on the left and right of the rear window are firm, if not cement them in position.

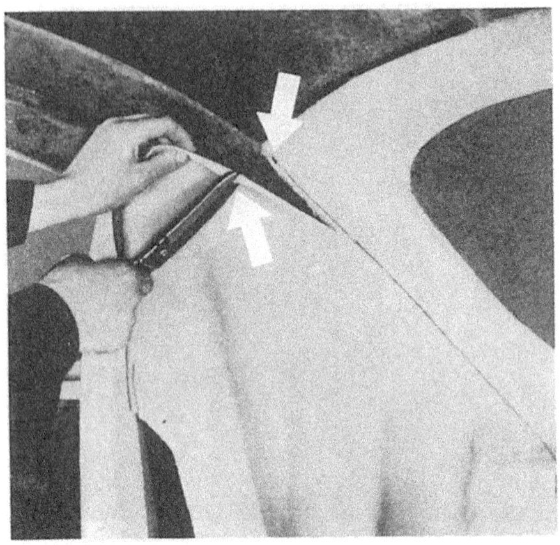

2 - Apply a thin film of VW Compound D 12 to the felt pads and the rear window frame. Fit the lining carefully, beginning at the top, then the sides and finishing at the bottom, avoiding wrinkles. Cut out the opening for the rear window and trim off until no more than about 10 mm (0.4") overlaps the frame all round.

Nick the overlapping material, especially at the curves and attach to the window frame pointing outwards.

3 - Attach the new side-member lining to the rear retaining strip on the roof, at first at the bottom.

4 - Cut a nick of about 10 mm (0.4") in the side-member lining at the upper end of the retaining strip.

This nick is necessary, so as to allow for the kink between the straight retaining strip and the curved side member. If it is not done the material will wrinkle when fixed in position.

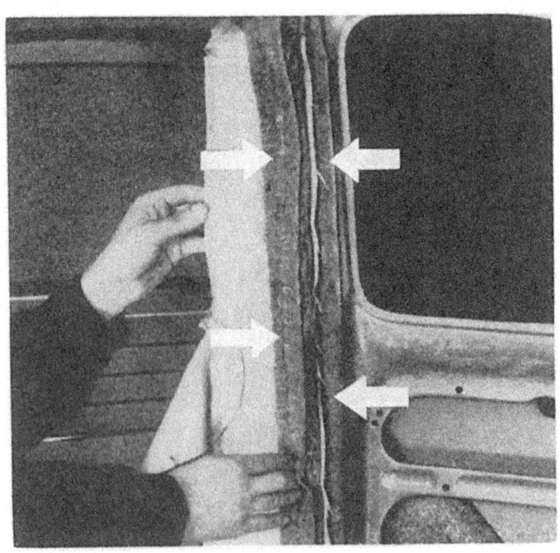

5 - Insert the material at the top and fasten by turning over the retaining strip.

6 - Fasten the lining to the centre door pillar, first at the top by turning over the retaining strip, then pull tight down. Then insert the lining in the retaining strip as far as the seam and turn over the retaining strip. Apply a coat of Universal Adhesive D 12 to the door pillar and stick a double thickness layer of upholstery padding to the full length of the door pillar. Then cement another single layer of padding over the previous layer.

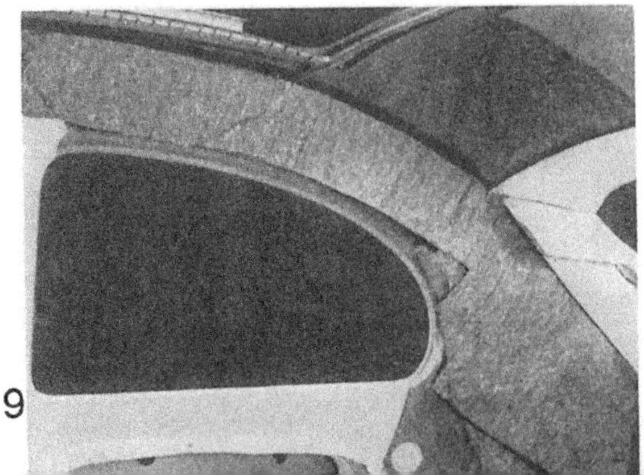

7 - Cement the lining material, without wrinkling it, to the pillar and to the member beneath the rear quarter window.

8 - Apply VW Compound D 12 lightly to the roof side member and rear quarter panel as far as the centre pillar, as required. Affix a double thickness of padding in the joint between the side member and the side panel; and at the rear of the side panel a large piece of padding folded double.

11 - Carefully remove the lining surrounding the rear window.

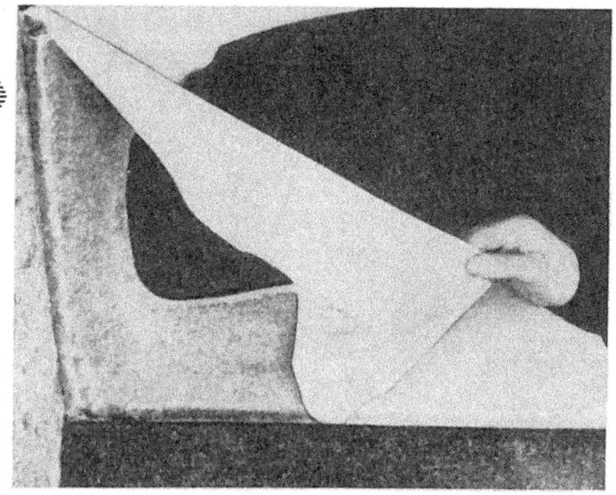

Installation

1 - Check that the felt pads on the left and right of the rear window are secure, if not cement them in position.

2 - Apply a thin film of VW Compound D 12 to the felt pads and the rear window frame. Fit the lining carefully, beginning at the top, then the sides and finishing at the bottom, avoiding wrinkles. Cut out the opening for the rear window and trim off until no more than about 10 mm (.4") overlaps the frame all round.

Nick the overlapping material, especially at the curves and attach to the window frame pointing outwards.

3 - Attach the new side-member lining to the rear retaining strip on the roof at the bottom first.

4 - Cut a nick of about 10 mm (.4") in the side-member lining at the upper end of the retaining strip.

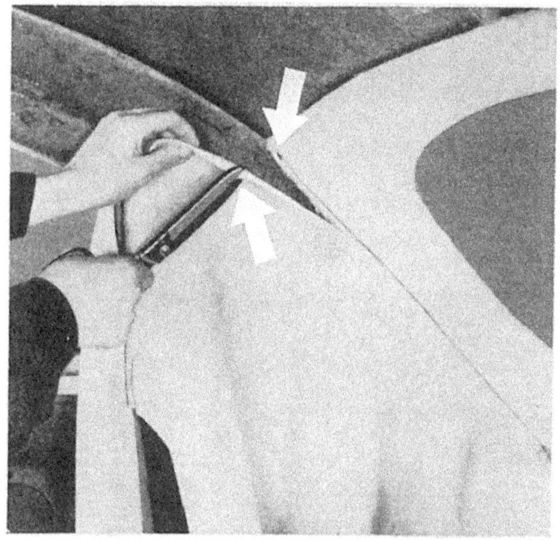

This nick is necessary, so as to allow for the kink between the straight retaining strip and the curved side member. If it is not done the material will wrinkle when fixed in position.

5 - Insert the material at the top and fasten by turning over the retaining strip.

6 - Fasten the lining to the centre door pillar, first at the top by turning over the retaining strip, then pull down tight. Then insert the lining in the retaining strip as far as the seam and turn over the retaining strip. Apply a coat of Universal Adhesive D 12 to the door pillar and stick a double thickness layer of upholstery padding to the full length of the door pillar. Then cement another single layer of padding over the previous layer.

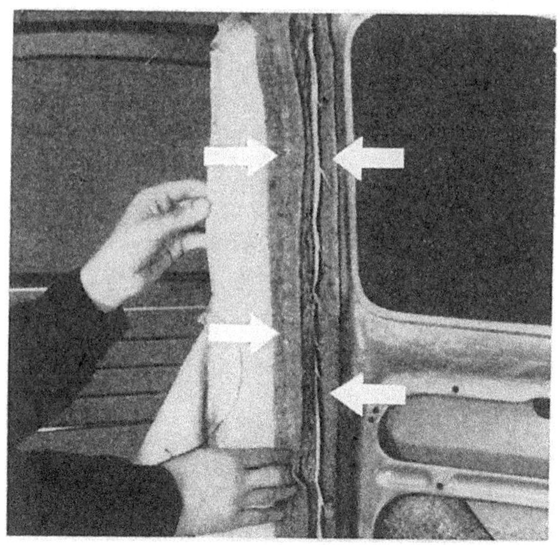

7 - Cement the lining material, without creases, to the pillar and to the member beneath the rear quarter window.

8 - Apply VW Compound D 12 lightly to the roof side member and rear quarter panel as far as the cente pillar, as required. Affix a double thickness of padding in the joint between the roof member and the side panel, and at the rear of the side panel.

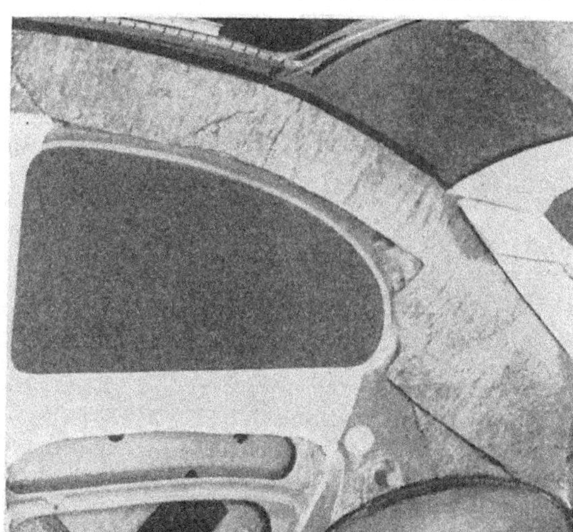

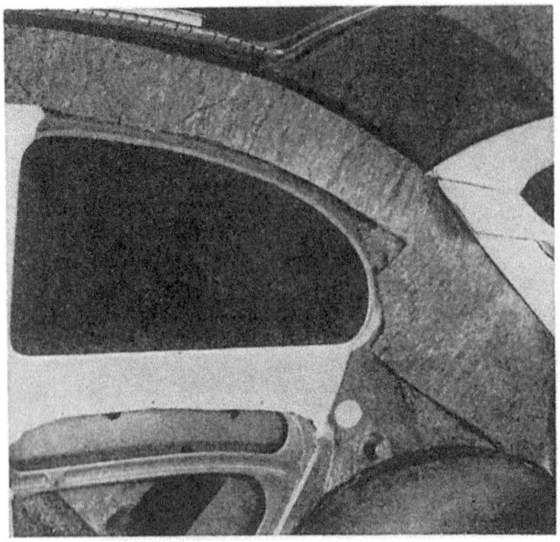

9 - Having stuck the padding strips into place, apply a layer of cement and stick a further layer of padding over the entire width of the side member and over the first piece of padding at the rear.

Note:

Leakage at the joints between the rear apron and the rear quarter panels will cause a slight oil odor inside the vehicle when the heating is in operation.

Remedy:

The joints should be sealed with D 15 Sealing Compound at the points indicated by the arrows.

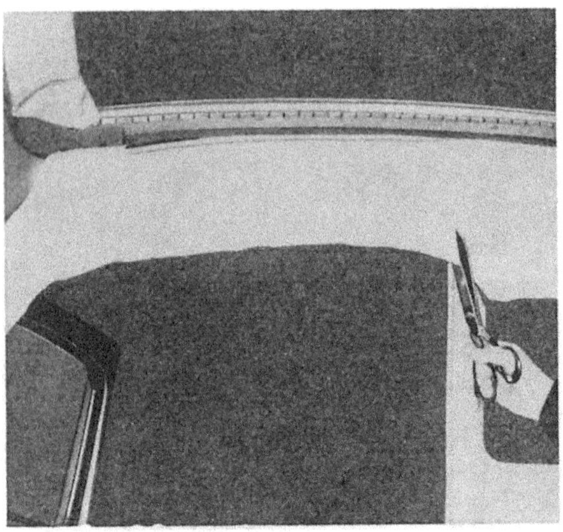

10 - Pull the side-member lining forwards and stick it to the side member without wrinkling. At the front end fold in the lining material, nicking it behind the front reinforcement at the corner in Sedans with sliding roof. At the lock pillar the material must also be nicked as far as the retaining strip.

11 - When fixing the material at the rear do not forget the hole which has to be cut for the rear seat backrest support.

12 - Fasten the material to the front retaining strip in the side member above the door and stick it to the rear quarter window frame. In the latter case the 10 mm (.4") overlap must be nicked, especially at the radii, then turned outwards.

13 - Sew the seams between the roof-member lining and the lock pillar lining.

14 - Replace all parts which were removed.

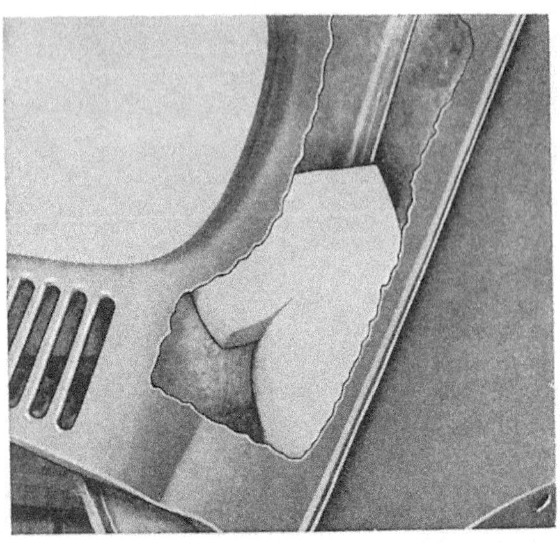

Note:

The interior trim of De Luxe Sedans (from Chassis No. 1 600 440) may become dirty due to dust accumulating on the side panels, lock pillars, roof members and headlining as a result of a faulty seal at the engine compartment partition. Furthermore, engine noise and odors will penetrate into the body interior. To affect a remedy the following alterations have been made:

1 - The sponge-like packings, Part No. 111 867 835B, are positioned in such a way as to ensure a more efficient sealing of the roof members from the engine side. If the packings are placed too far forward, the longitudinal slits in the roof members will be uncovered, allowing dust in the drawn-in air to settle on the roof members.

Subsequent remedy:

Withdraw packings from roof member with a piece of bent wire until they are seated properly in the opening. As found necessary, cement roof member lining properly in position.

At the same time, care should be taken that the openings between the side panels and the engine compartment partition are fully closed by the cushions provided at these points.

2 - Sealing compound is applied to the joint between door lock pillar and roof member at the body assembly line.

Subsequent remedy:

Partly loosen roof member lining and seal the joint between pillar and roof member by applying Sealing Compound D 15. Carefully cement loosened lining into place.

3 - A hole for a trim panel clip in the inner side panel has been omitted at the curved edge towards the door lock pillar.

Subsequent remedy:

Seal the hole with either adhesive tape or a patch of plastic material. Carefully cement any loosened lining above the hole in position.

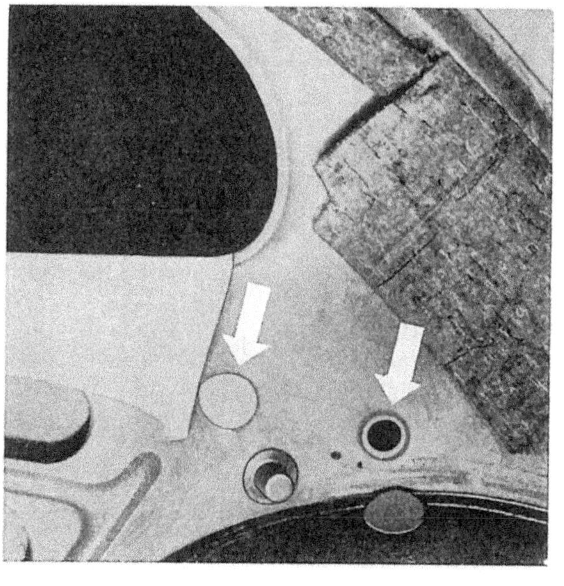

4 - From February 1956, the two holes in the inner side panels on the Sedan De Luxe are covered by leatherette disc. In a number of cases, these discs have been found to get loose.

On the Standard Model, rubber plugs, Part No. 111 853 953, are used instead of the discs.

Subsequent remedy:

Loosen lining of both side panels as far as necessary and re-cement leatherette discs in position.

5 - Leakage at the joints between the rear apron and the rear side panels can give rise to slight oil odors inside the vehicle when the heating is on.

Remedy:

Seal the openings with Plastic Sealing Compound D 15.

Body Side Member Carpet Removal and Installation

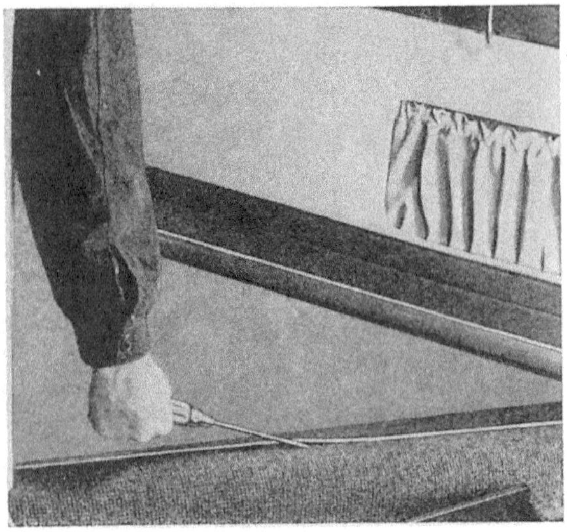

1 - Pry up the carpet retaining flange with a broad screwdriver.

2 - Remove the tacks and pull off the carpet.

3 - Prior to cementing the carpet into place, carefully remove all traces of the old cement. The use of benzine as a solvent is recommended.

4 - Cement the carpet into place, using VW Compound D 12.

5 - Bend down the carpet retaining flange.

6 - Secure at the front end of the carpet with a tack.

Front Panel Carpet Removal and Installation

The three front panel carpets are tacked in position at the upper corners. When these tacks have been removed it is an easy matter to strip off the carpets which are cemented in position. For removal of the right and left carpets, the tacks at the front end of the side member carpets should also be removed.

When installing, observe the following points:

1 - Prior to cementing the carpets into place, carefully remove all traces of the old cement. The use of benzine as solvent is recommended.

2 - Cement the carpets into place, using VW Compound D 12. Be sure the carpets make perfect contact with the front panel, especially round the hot air openings.

3 - Install tacks at the upper corners and the side members.

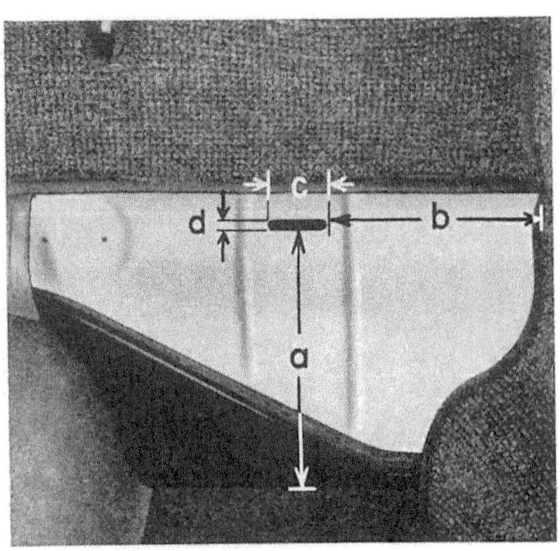

Note:

From 6th August 1959 and Chassis No. 2 533 000 the De Luxe Sedan is equipped in standard production with a foot support (Part No. 113/114 863 691 or 151/152 863 691) on the passenger's side. This foot support is hooked into the front panel and is covered by the floor mat which has been lengthened accordingly.

The front panel lining has been altered as well. The rubber lining under the carpet now covers the front panel on the driver's side only. The previous type lining will still be available as a spare. When installing the new type lining in vehicles of earlier make it should be cut to the required size or an opening should be provided for the hook on the foot support.

a - 205 mm (8.1")
b - 145 mm (5.7")
c - 40 mm (1.6")
d - 7 mm (.28")

The foot support can be service installed in vehicles of earlier make by using the new rubber floor mat (Part No. 113 863 712 B or 114 863 711 B). The opening for the hook in the front panel should be provided.

Note:

From 6th August 1959 and Chassis No. 2 533 000 the De Luxe Sedan floor had to be altered as a result of installation of a foot support for the passenger and kick boards.

A 2-section floor mat is now installed instead of the 5-section floor covering hitherto used for floor plate and frame tunnel. The front floor mats, the frame tunnel covering up to the gear lever and the mats for the rear leg room and rear portion of the frame tunnel are now of one continuous piece. Both coverings (Part No. 113/114 863 703 and 113 863 717) meet in the region between hand brake lever and gear lever and are no longer stuck down.

The new 2-section floor covering can only be installed as a unit in previous vehicles.

The previous floor and frame tunnel covering are still available as spares.

Frame Tunnel Lining Removal and Installation

1 - Remove the rear seat cushion.

2 - Remove rear seat support bracket.

3 - Take out the rubber mats.

4 - Remove heating control knob.

5 - Pull the rubber lining off the frame tunnel.

On the Standard Sedan pull the rubber lining sections off at the front and back of the tunnel.

6 - If necessary remove the sound-deadening material.

7 - Check the rubber boot of the handbrake lever, replace if necessary.

8 - Cement the lining into place, using VW Compound D 12.

Note:

From Chassis No. 4 846 836, the insulation of the floor plates was improved.

The bitumen insulation which is cemented to the floor plates and frame tunnel has been reinforced by additional insulation material.

When existing stocks have been used up, only the thicker insulation material will be supplied. The new material can be installed in earlier vehicles by replacing the kick boards or cutting the insulation material to fit.

The foot support has also been modified. The new foot support (Part No. 113/114 863 691 A) can be installed in earlier vehicles. The previous type of foot support will be discontinued when present stocks are exhausted.

The rubber mats have been altered to suit the thicker bitumen insulation. The modified front rubber mats (Part No. 113/114 863 703 A) and the modified rear rubber mats (Part No. 113/114 863 717 B) can be installed in earlier vehicles without difficulty as long as the colors match. The previous type rubber mats will remain available apart from the colors graphite and olive.

Note:

From 6th August 1959 and Chassis No. 2 533 000 the De Luxe Sedan is equipped with leatherette-covered kick boards (Part Nos. 113 863 375/376; 151 863 375/376) which cover the openings underneath the rear seat. The kick boards are held in position at the top by the rear seat support, and are attached to the floor plate by means of spotwelded brackets. The rear floor mats are flush with the kick boards.

The kick boards may be service installed in previous vehicles. For better support on the left side, it is advisable to cut the rubber mat in front of the kick board and to attach a metal bracket (10 mm high) to the floor plate. On the right side the battery clamping strap acts as a stop.

From 1st August 1960 and Chassis No. 3 192 507 the interior trim on the De Luxe Model is painted to match the color of the vehicle.

Glove Compartment Removal and Installation

1 - Raise the front hood and remove the instrument panel protective cover.

2 - Remove the retaining screw and take off the glove compartment.

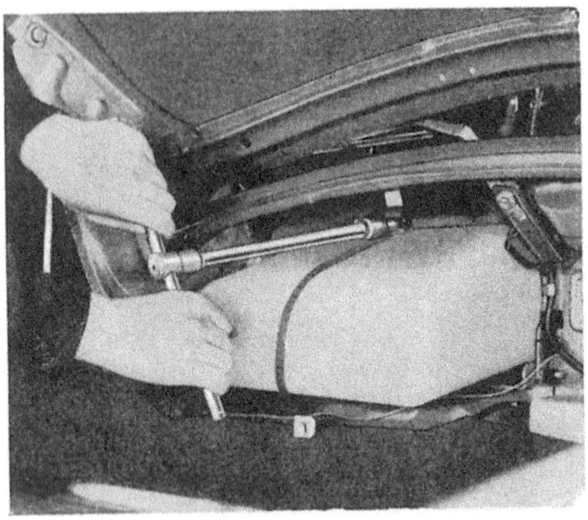

3 - Remove the four screws at the lid check straps and take off the lid.

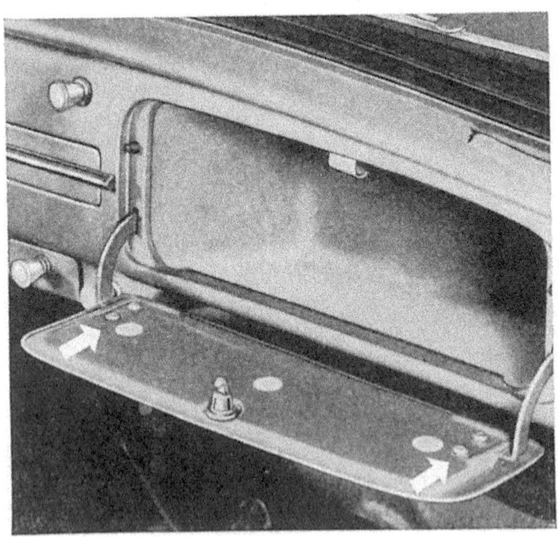

4 - To remove the lid check straps the spring must first be removed and the heads of the two rivets removed.

5 - Remove the two rubber plugs from the glove compartment frame.

6 - Screw out the lid lock, using the circlip pliers VW 122 b or a suitable tool of local manufacture.

7 - When installing the glove compartment and tightening the retaining strap, take care that the glove compartment is correctly positioned against the instrument panel. Poor contact between the glove compartment and the instrument panel should be eliminated by means of adhesive tape.

8 - Check condition of rubber plugs. Replace if necessary.

9 - Check the lid alignment by opening and closing the lid repeatedly.

10 - Apply a few drops of oil to the lid lock.

11 - If the lid is difficult to open or close, the catch in the compartment should be bent slightly.

Note:

From Chassis No. 4 010 995 (31st July 1961), mountings for safety belts for driver and passengers are installed.

De Luxe and Standard Model:

For the front seats:
On lock pillars and frame tunnel.

For the rear seats:
On quarter panels and floor of the rear luggage compartment below the back rest.

The mounting holes are sealed with plastic plugs (a) which are removed when the belts are fitted.

The service installation of safety belts in previous vehicles is described in Technical Bulletin Z - 14.

De Luxe 1200 Model with Steel Sliding Roof

Removal

1 - Remove sliding roof.

2 - Remove assist straps, coat hooks and interior light.

3 - Remove quarter windows.

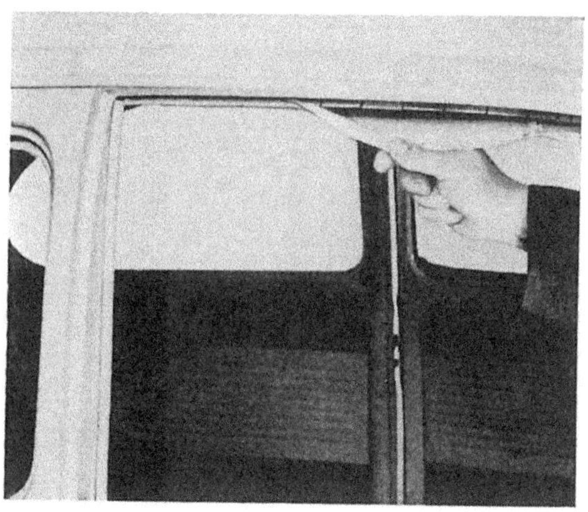

4 - Bend up the retaining flanges over the doors with a broad-bladed screwdriver and pull headlining out carefully.

5 - Detach headlining carefully round the sliding roof opening.

6 - Pull headlining off body and side lining around the quarter windows.

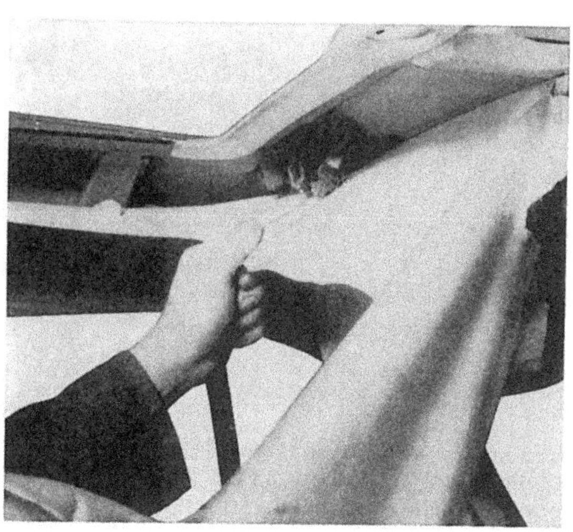

7 - Insert headlining tool between roof member and headlining round the windshield and rear window, pass it along the roof frame and lift the lining out to the front.

Note

Remove wadding between roof and roof member near the front water drain hoses so that the lining can be lifted out.

8 - Lift stretcher rods with springs out of roof member and take headlining off.

Before installing the headlining, ensure that the padding of the roof members and the damping material on the roof panel is secure.

Installation

1 - Fit headlining in around the rear window.

2 - Insert rear stretcher rod into roof member.

3 - Hook retaining springs for center and front stretcher rods into roof member and then press the rods into the springs.

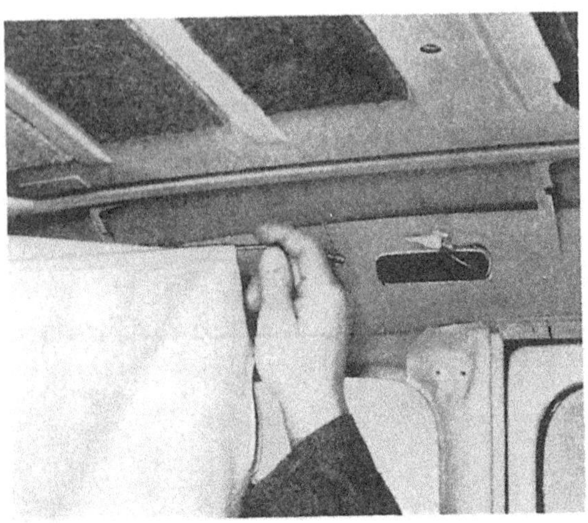

4 - Fit lining into inner windshield frame. In the area of the recess for the crank, the lining must be stuck on with D 11 adhesive.

5 - Press wadding in between roof, member and lining near the front drain hoses.

6 - Coat lining lightly with D 11 adhesive round the quarter windows, doors and sliding roof opening and press into position.

7 - Bend the retaining flange over the door opening down carefully.

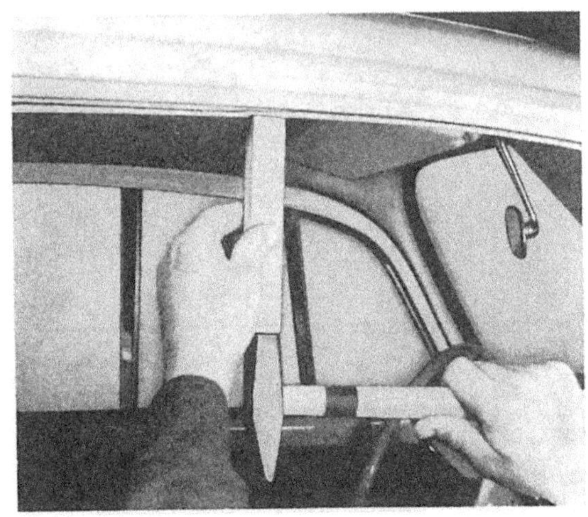

8 - Install sliding roof and interior trim.

Volkswagen de Luxe 1200 Sedan

The removal and installation of the headlining is the same as on the vehicle with the steel sliding roof. Note that installation should start from the rear window and that the stretcher rods should be checked for parallelism before pressing the front headliner cardboard strip into the windshield frame.

Top Cover Removal and Installation

Removal

1 - Open the roof about 1/3 of the way.

2 - Remove the end covering from the guide and locking bows. When doing this take care that the protective caps are not pulled off the ends of the locking bows.

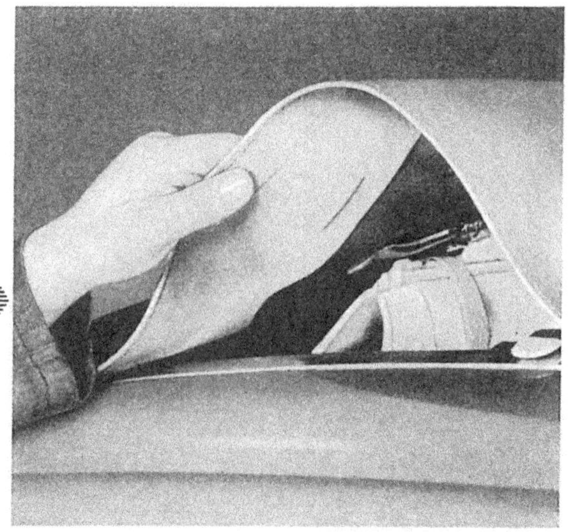

3 - Remove the two screws which attach the top cover bow to the locking bow.

Lift the top bow out of its fastenings and fold back the top cover.

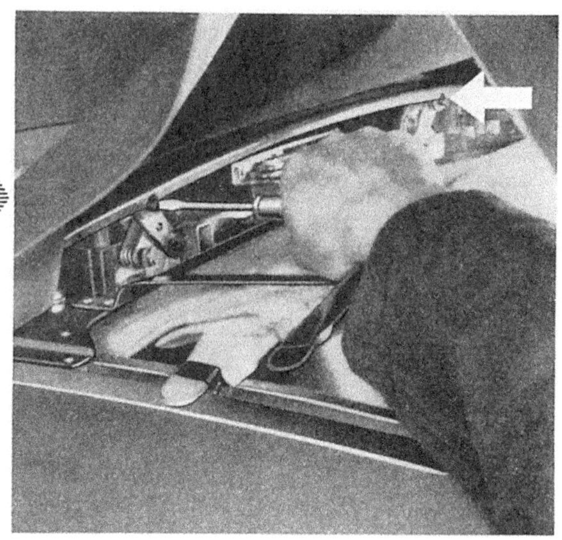

4 - Detach the rear headlining with concealed fasteners.

This work is best done with a suitable tool, in order to prevent damage to the headlining.

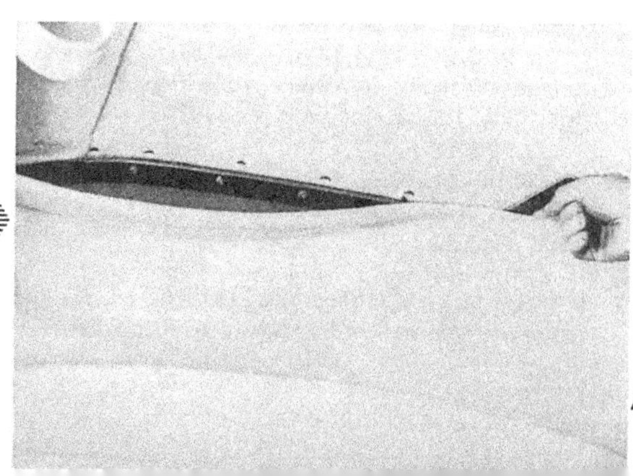

5 - Remove the nuts holding the tension bow and detach the lower part.

6 - Remove the top cover and tension bow.

7 - If necessary, unrivet the top cover from the locking bow and pull the tension bow out from the back of the cover. Check the position of the rubber strip under the top cover bow.

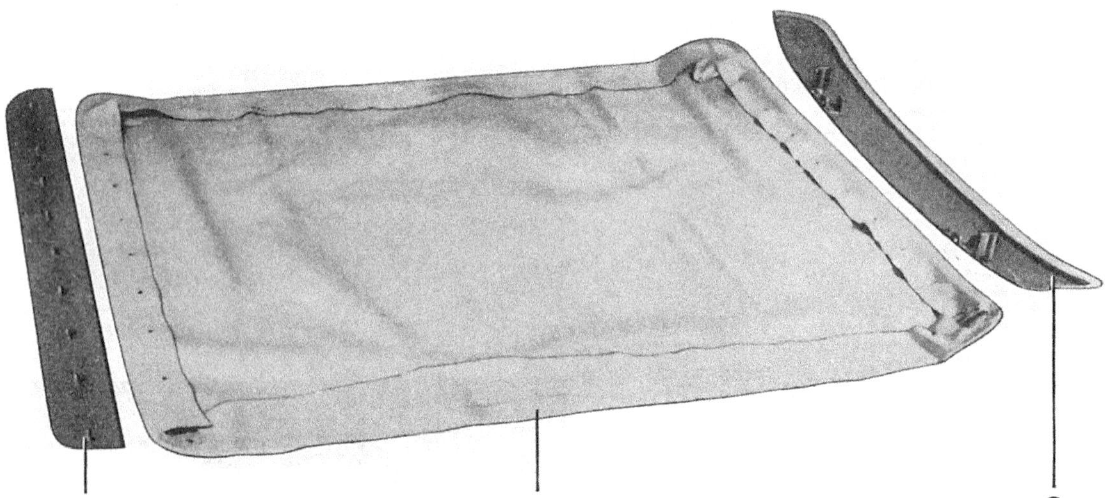

A - Tension bow
B - Top cover
C - Top cover bow

Installation

1 - Rivet the new top cover to the top cover bow using the holes provided, and taking care to avoid wrinkles.

2 - Apply some vaseline to the side guide rails and sliding roof lock.

3 - When using new screws for attaching the top cover to the locking bow, check them for free movement in the locking bow before installation. Coat the screws with grease.

4 - Check the position of the protective covers at the ends of the bows, if necessary cement with fresh adhesive.

5 - If necessary, add a drop of oil to the axles of all rollers. Wipe any old grease out of the guide rails, if necessary using benzine. Then apply a thin film of new grease to the rails.

6 - Refit the top cover. Screw up the nuts holding the tension bow finger-tight.

7 - Close the roof and check the tension of the top cover. If necessary, slacken the nuts and adjust the tension bow until the cover is adequately and uniformly tight.

When the tension is correct, tighten the nuts up fully.

8 - Reattach the rear headlining.

9 - Test the roof by opening and closing several times.

Lock and Stretcher Mechanism Removal and Installation

Removal

1 - Open the sliding roof slightly.

2 - Remove the top cover from the ends of the bows. Unscrew the top cover bow and fold the top cover back.

3 - Press the escutcheon plate against the locking bow and drive out the pin. Take off handle, escutcheon plate, spring and rubber insert.

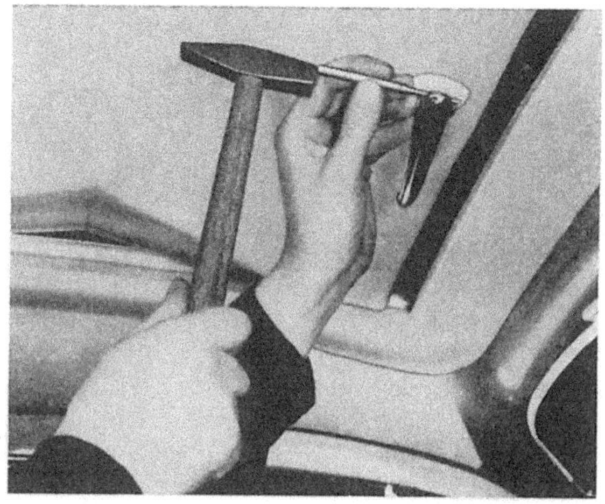

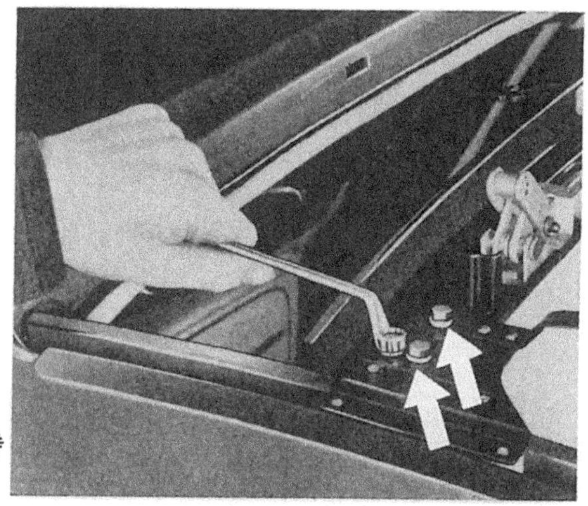

4 - Remove the three countersunk screws from each locking bow guide plate.

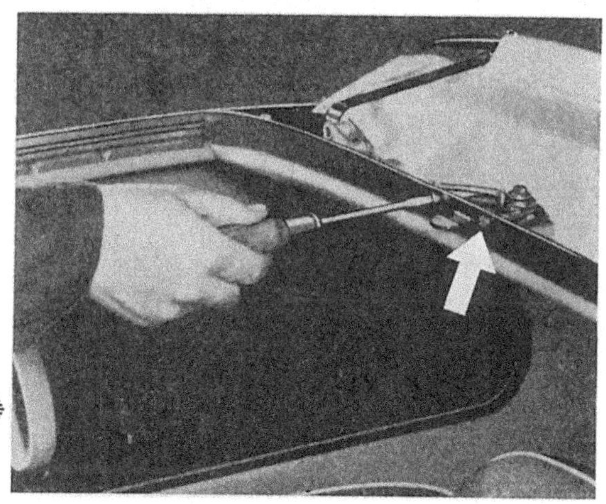

5 - Remove the two cylindrical screws at the lock.

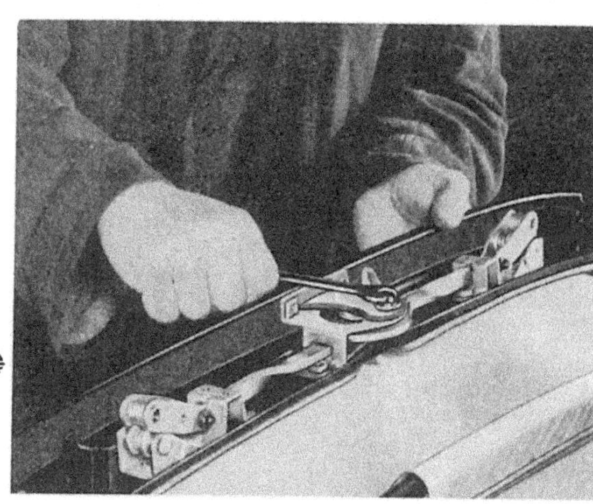

6 - Unhook the lock spring and remove the screw for the locking hook. Lift the hook off the tapered stud and push into the lock.

7 - Turn the locking bar over after angling slightly and remove the upper part of the escutcheon plate and spring.

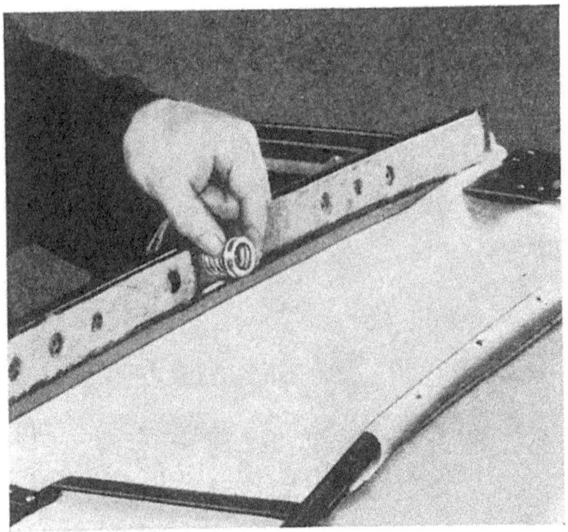

8 - Remove the eight countersunk Phillips head screws and remove the lock and locking bow.

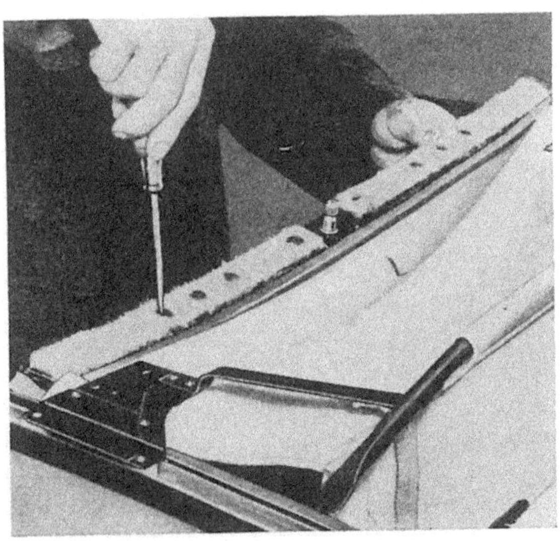

Installation

1 - Lightly grease all moving parts of the lock and locking bow with Vaseline.

Note:

From Chassis No. 4 846 836, the sliding roof handle is provided with a hinge and is flatter. The modified handle — Part No. 115 875 219 — cannot be service installed in previous vehicles.

The handle — Part No. 115 875 221 A — will still remain available.

2 - Having screwed on the lock, check the felt strips and rubber seal, replacing or recementing where necessary.

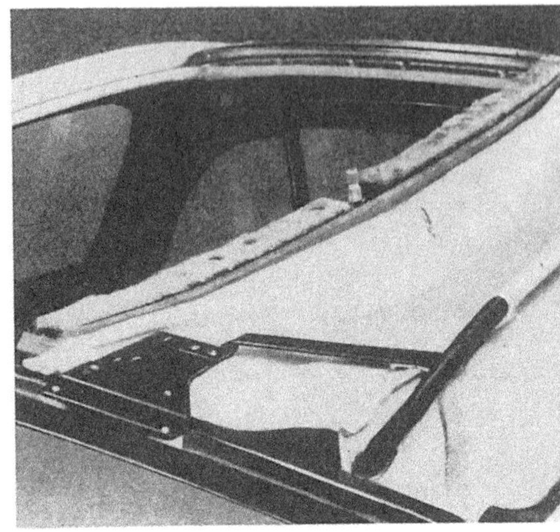

3 - When the locking bar is turned over, insert the upper part of the escutcheon plate and spring for the handle.

4 - Tighten the fixing screw of the locking hook only so far as allows the hook to move freely.

5 - If the locking hook jams in the slot when closing the sliding roof, loosen the screws at each locking bow guide plate. Shift the guide plates in their slots until the sliding roof can be opened and closed without difficulty.

6 - Before inserting the bolts which hold the guide bow to the top cover bow, check that they are free in their fixings.

7 - Fit the handle with escutcheon plate and rubber insert so that when the roof is closed the handle is parallel to the bow and pointing towards the right-hand door.

8 - After assembly, check locking mechanism by opening and closing the sliding roof several times.

Removal

1 - Open the roof slightly.

2 - Detach the roof cover from flat bow ends. Remove the two roof cover bow anchor bolts from roof locking bow and fold back the roof cover.

3 - Remove the countersunk screws which hold the rear guide rails, and take off the rails.

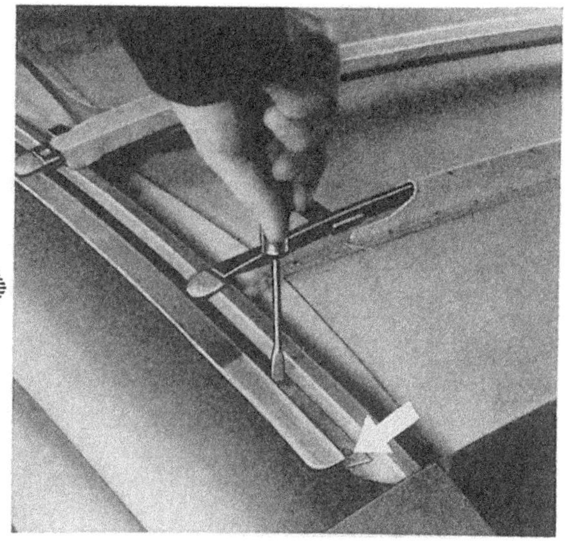

4 - Push locking bow and headlining back and lift locking bow out of the roof opening. Remove the rollers for the guide bow.

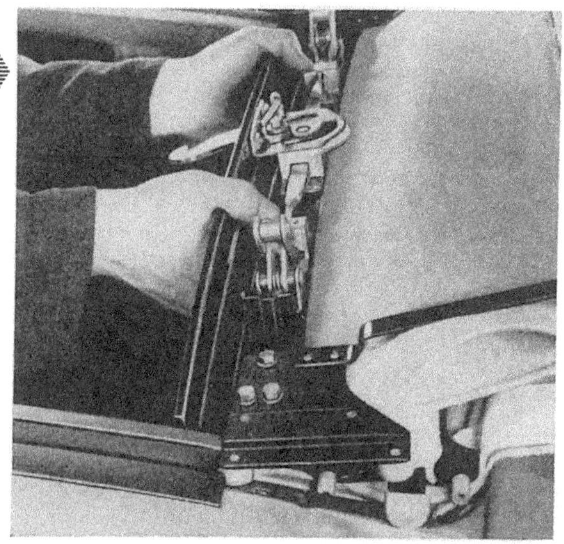

5 - Remove the countersunk screws which hold the front guide rails, and take off the rails.

Installation

1 - Wash out any old grease from the guide rails with benzine.

2 - After installation of front guide rails, slide the locking bow with top cover forward.

3 - When installing rear guide rails, make sure that they are flush with the front guide rails.

Note:

From April 1963, Chassis No. 5 488 894, the runners for the sliding roof are made in one piece. Formerly a long runner — Part No. 115 875 709 C/710 C — and a short one — Part No. 115 875 725 C/726 C — were fitted on each side. Only the short runners will remain available as spare parts.

The one-piece runners — Part No. 115 875 709 D/710 D — can be installed in previous vehicles. There are now two screws less on each side. The nuts for these screws should be fixed with adhesive to stop rattling noises at the runners.

> **Important**
>
> When installing the sliding roof, fit one runner first then insert sliding roof into this runner and install the second one-piece runner.

4 - Lightly grease the guide rails.

5 - After assembly, check locking mechanism by opening and closing the sliding roof several times.

Headlining and Locking Bow Removal and Installation

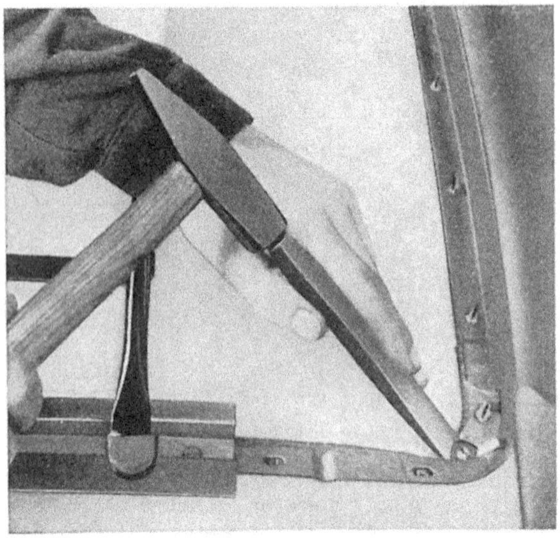

Removal

1 - Open the roof.

2 - Remove locking handle, escutcheon plate, spring, and rubber insert.

3 - Detach the roof cover from flat bow ends. Remove the two cover bow anchor bolts from roof locking bow and fold back the roof cover.

4 - Remove the rear guide rails after removing the two fixing screws in each case.

5 - Remove the two rivets, one in each rear corner of the roof opening.

6 - Detach the rear headlining with concealed fasteners, and the center headlining which is cemented to the rear of the opening for the sliding roof.

7 - Push the locking bow with top cover out towards the back. Remove the locking rollers of the guide bow.

8 - Unrivet the headlining retaining strip and detach the headlining from the tension bow.

9 - Unrivet the struts of the rear flat bow.

Pull the flat bows at front and back out of the headlining.

10 - Turn over the headlining and remove the screws which hold it to the guide bow.

11 - Pull off the rubber sealing strip from the locking bow.

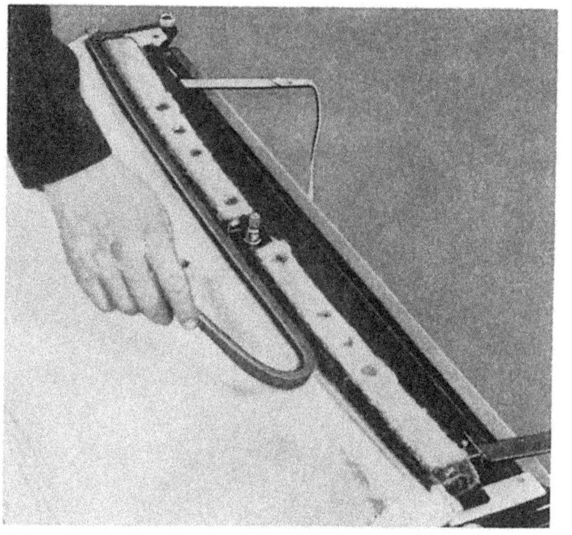

12 - Pry up the headlining retaining strip on the locking bow and pull the headlining away from the locking bow.

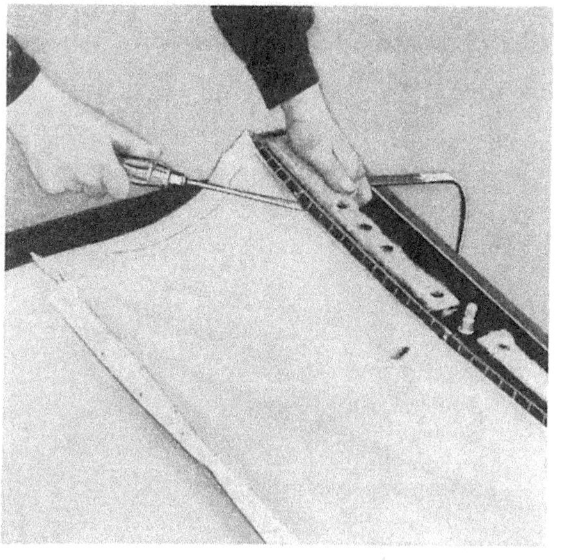

Installation

Proceed in the reverse sequence, observing the following points:

1 - Before clamping the headlining in the retaining strip of the locking bar, check that the material overlaps uniformly on either side. Then turn the retaining strip hard over.

2 - Stick the sponge-rubber strip to the retaining strip and to the sides of the locking bow.

3 - Cut the opening for the shaft of the lock in the material. Place the spring and the upper part of the escutcheon plate on the shaft and turn the locking bow over. Finally rivet the headlining to the corners of the locking bow.

4 - Insert the tension bow in the loops of the top cover and rivet to the struts.

5 - Wash any adhesive remaining on the rear of the roof opening, using benzine.

6 - Re-install the headlining.

7 - Close the sliding roof and cement the rear part to the rear of the roof opening under light tension, avoiding wrinkles.

8 - Nick the rear corners of the headlining and rivet to the roof using two leather washers.

9 - Re-install the top cover.

10 - Check the operation of the roof by opening and closing several times.

11 - Screw the headlining to the guide bow from below at each end.

Trouble Shooting on Sliding Roof

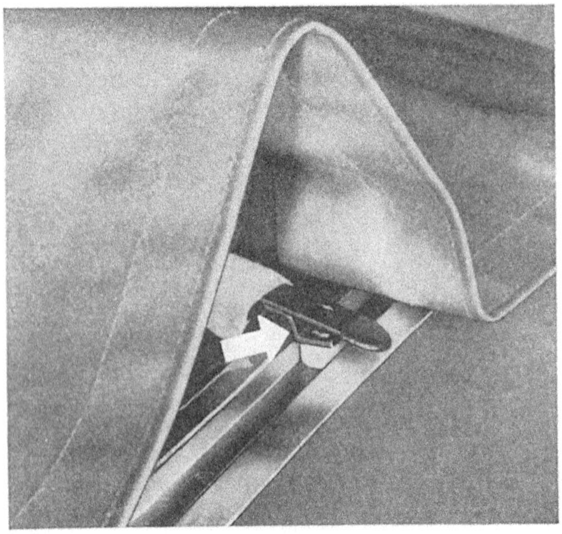

Rattle

a - Eliminate vertical play by bending the roller bracket until the two lower rollers come to bear against the guide rail and no play is perceptible.

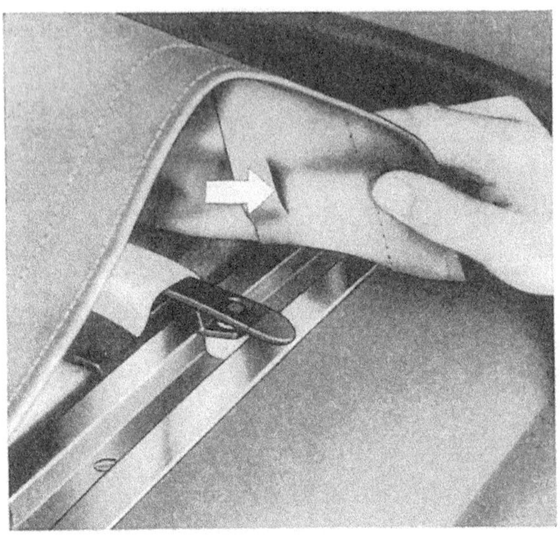

b - The sliding roof bows are inserted into pockets of the now glued cover flaps. This makes rattling of the bows at the guide rails impossible and the cover folds properly when the roof is pushed backward.

> **Attention!**
> Only the special Plastic Adhesive D 11 may be used for the top cover.

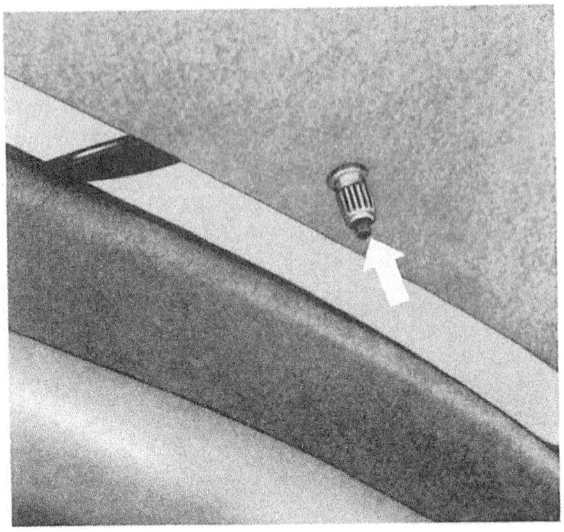

c - In the case of a rattling handle, make sure the rubber insert is not missing. If necessary, interpose a felt washer between bow and spring.

Stiffness

If undue force is required to turn the handle to the locked position, lessen the tension of the roof cover at the tension bow.

Leakage

a - Leakage at the front edge of the sunshine roof indicates that the tension of the top cover is excessive or that the rubber sealing strip under the top cover bow is out of place. Undue tension when closing the roof lifts off the top cover bow and the roof ceases to be water-tight. This trouble may be overcome by loosening the top cover at the tension bow or by cementing the rubber sealing strip in afresh. In some cases it may be necessary to straighten the locking bow first if it has become bent due to the use of force when closing the roof.

THIS PAGE INTENTIONALLY
LEFT BLANK

The sliding roof is a Golde Model with a clear roof opening of 14.75 × 26.2".

It is operated with a crank. The crank handle is flush mounted in a recess between the sun visors and is folded out to open the roof which can be fixed in any position automatically.

Sliding roof panel removal and installation

Removal

1 - Open roof halfway.

2 - Remove trim panel at front by pulling securing clipsout.

3 - Push trim panel to rear as far as it will go.

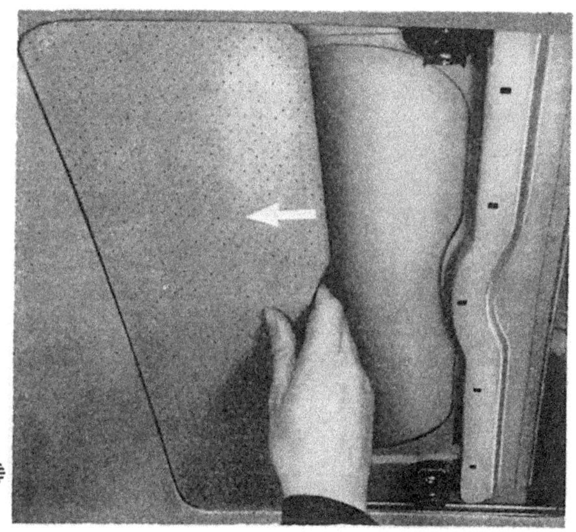

4 - Close roof until it is only open about two inches.

5 - Unscrew both front guides.

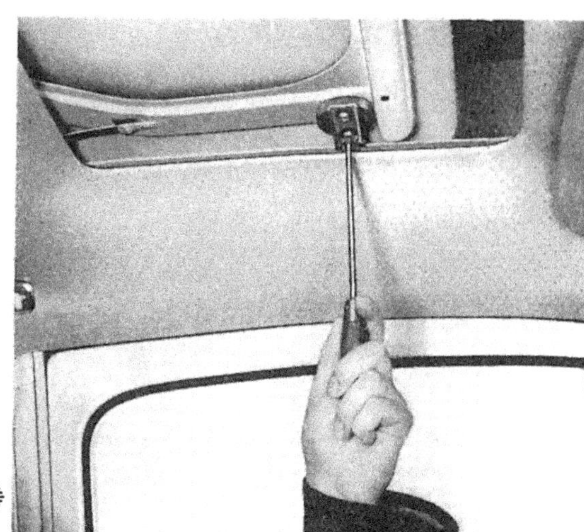

6 - Close roof. Unhook the leaf springs on the rear guides at left and right and turn them to the front.

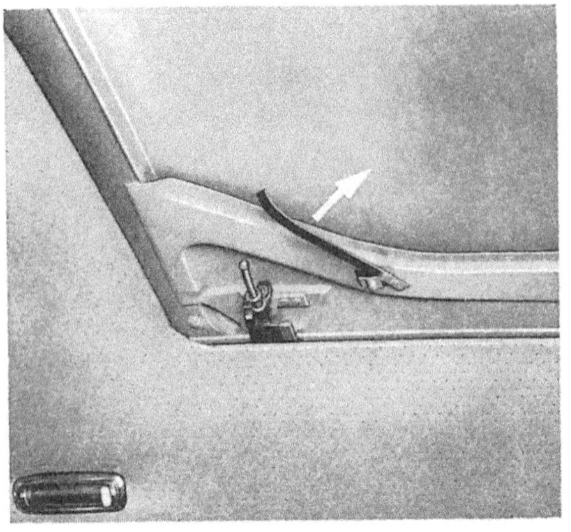

7 - Pull the rear guide brackets out of the supports at left and right.

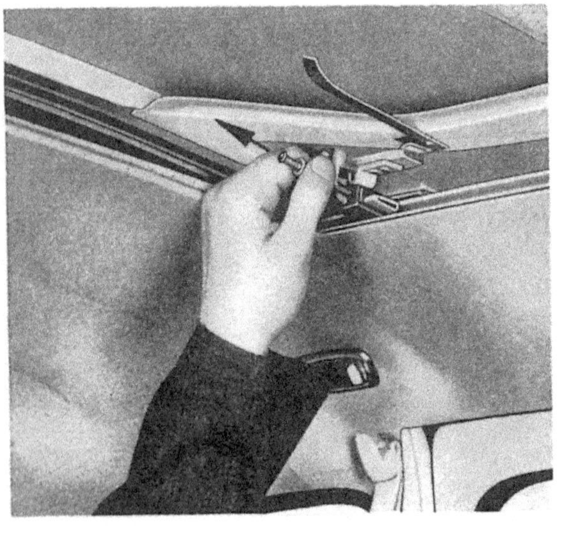

8 - Lift sliding roof panel out of the opening, taking care to lift it evenly and to avoid damaging the seal.

Installation

1 - Press the trim lining down and insert the roof panel into the rear of the opening at an angle.

2 - Slide panel to the rear and lower front end slowly. Take care not to damage the seal.

3 - Pull roof panel fully to the front.

4 - Insert rear guide brackets into the supports at left and right.

5 - Install front guides at left and right.

6 - Open roof halfway and pull trim lining to the front.

7 - Attach trim lining by pressing the clips into position.

Trim lining removal and installation

The trim lining for the sliding roof panel can only be taken out of the opening when one of the side runners has been removed

When installing, ensure that the trim lining is inserted together with the side runner.

Runner and cable removal and installation

Removal

1 - Remove sliding roof panel.

2 - Remove four screws in the central cable guide housing and take upper part off.

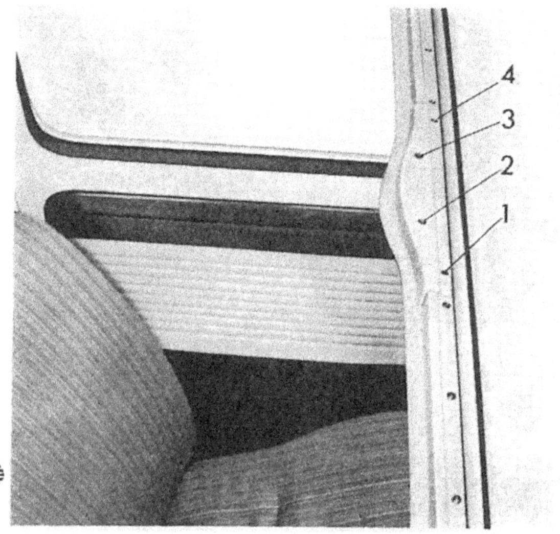

3 - Take tensioning plate out of housing.

4 - Remove eight screws from the upper part of front cable channels and side runners on both sides.

5 - Take off upper parts of front cable channels.

6 - Take side runners with cables and trim out of the opening to the front.

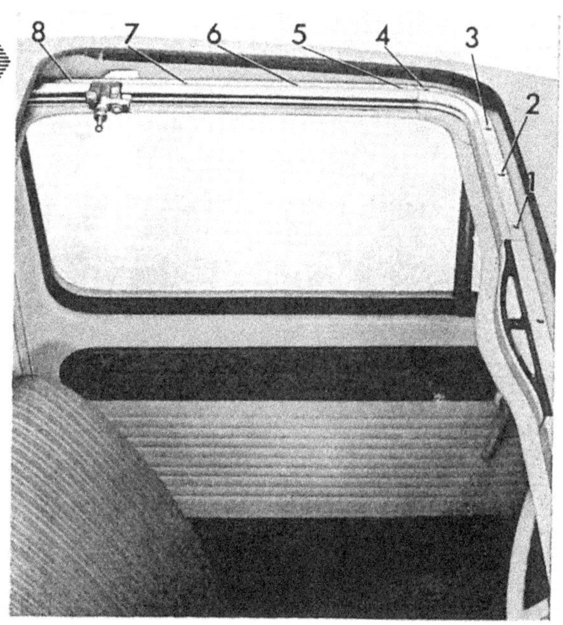

> **Important**
>
> This should be done very carefully to avoid damaging the paint on the roof.

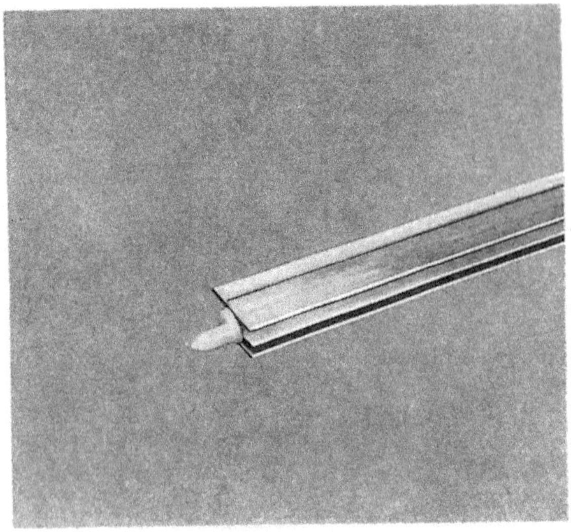

Installation

Note the following points when installing:

1 - When inserting the side runners, the front retainers which are made of plastic material must be guided into the supports.

If this is not done the runners can move sideways when the roof panel is opened and closed and the panel will not slide properly.

2 - Before securing the side runners, insert the roof panel trim lining into the lower groove and push it to the rear as far as it will go.

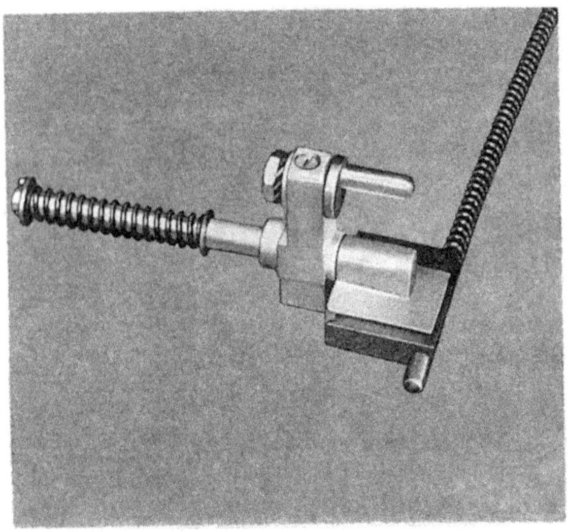

3 - Check the cables for wear, distortion and damage before installation.

Damaged parts should be renewed. If only one cable is damaged it is advisable to fit two new cables as this ensures that the movement of the roof panel is accurate and parallel.

Check that the new cables are not shorter than the cables taken out.

4 - The cables may only be lubricated with molybdenum-disulphide grease.

5 - Install cables. The cables should cross over at the center cable guide housing.

6 - Install guide housing upper part.

7 - Install sliding roof panel and adjust cables.

Cable adjustment

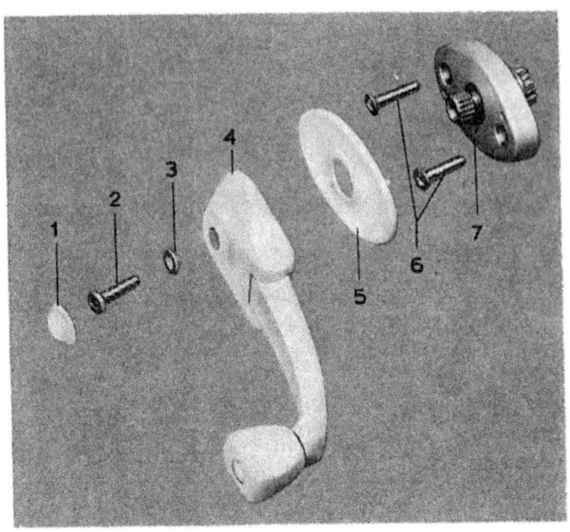

1 - Close sliding roof.

2 - Remove crank after taking off the small cap.

1 - Cap
2 - Screw
3 - Washer
4 - Crank
5 - Escutcheon
6 - Drive gear screws
7 - Drive gear

3 - Remove escutcheon.

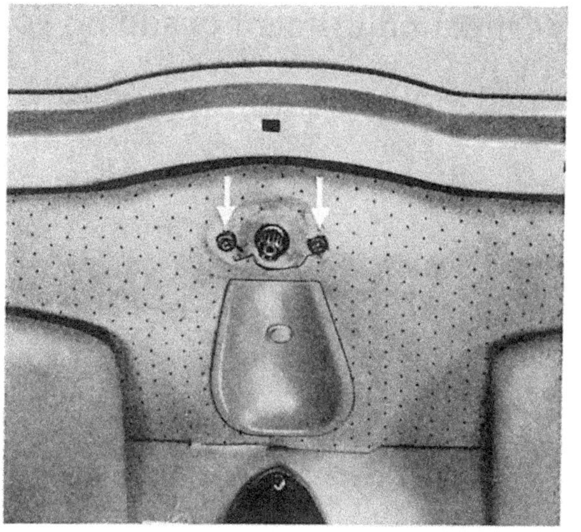

4 - Unscrew the two cable drive gear securing screws about six turns.

5 - Pull the drive gear down until the gear no longer engages in the cables.

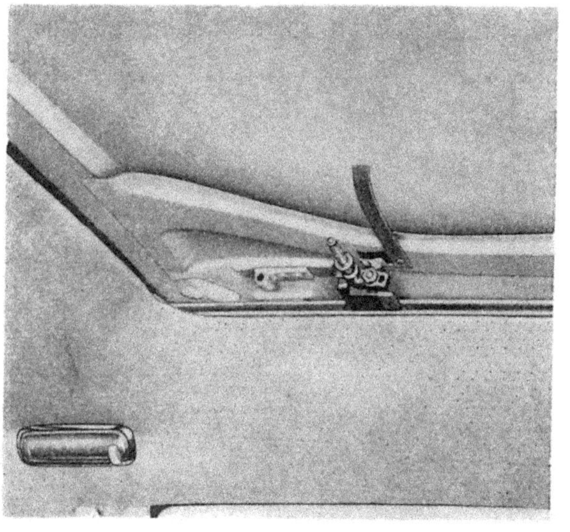

6 - Pull the two rear guides with the cables up to the sliding roof supports.

7 - Insert the rear guide brackets into the supports in the vertical position.

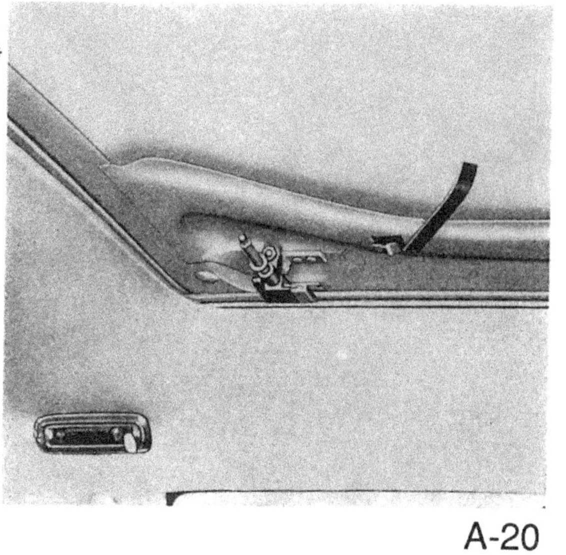

8 - Turn the drive gear shaft to the right as far as it will go (maximum 12 turns).

9 - Press cable drive gear up and secure. The gear must engage both cables.

10 - Install escutcheon and crank.

11 - Check cable adjustment by opening and closing roof.

Vertical adjustment of sliding roof panel

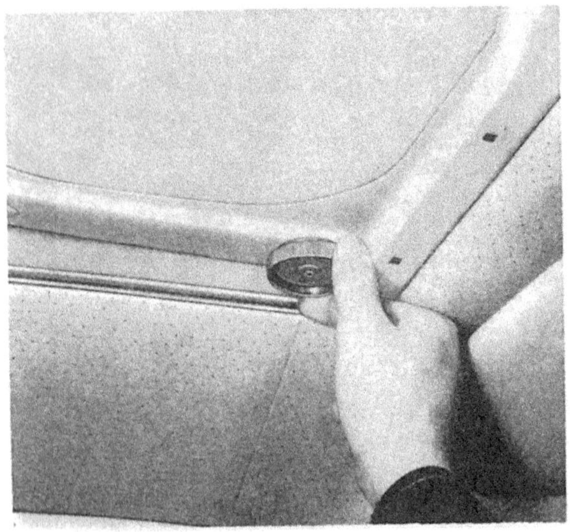

Front:

1 - Open roof halfway, detach trim lining and push to rear.

2 - Close roof.

3 - Loosen screws in front guides.

4 - Adjust sliding roof panel to correct height by turning the knurled nuts at left and right.

5 - Tighten front guide screws.

6 - Pull trim lining forward and attach to panel.

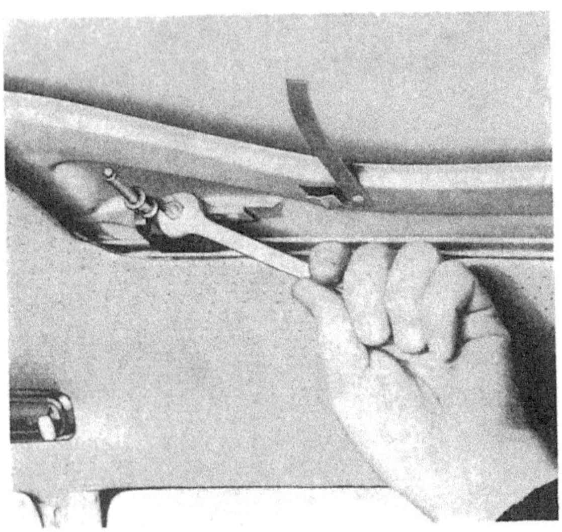

Rear:

1 - Open roof halfway, detach trim lining and push to rear.

2 - Close roof until rear guide brackets are just about to be lifted.

3 - Loosen nuts on upper pins.

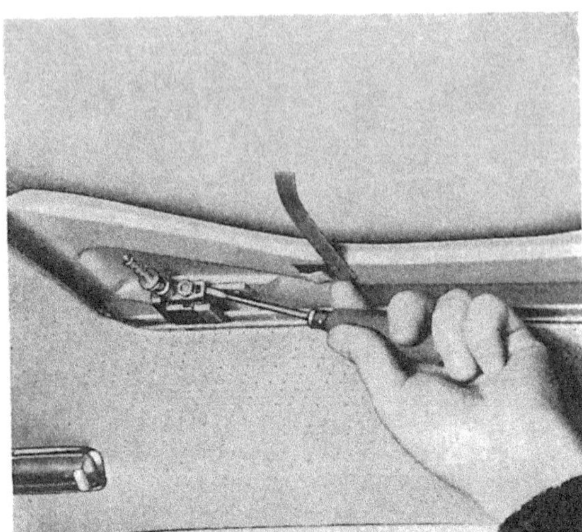

4 - Adjust the position of the bracket pin to the correct height with the screw on the bracket.

5 - Tighten nuts.

6 - Check that roof slides parallel and adjust cables if necessary.

7 - Pull trim lining forward and attach to panel.

Water drain channels

Water which enters the roof opening runs along the water channels in the sliding roof frame to the drain holes.

It then runs through the front and rear roof members to the fenders and side panels and drains away.

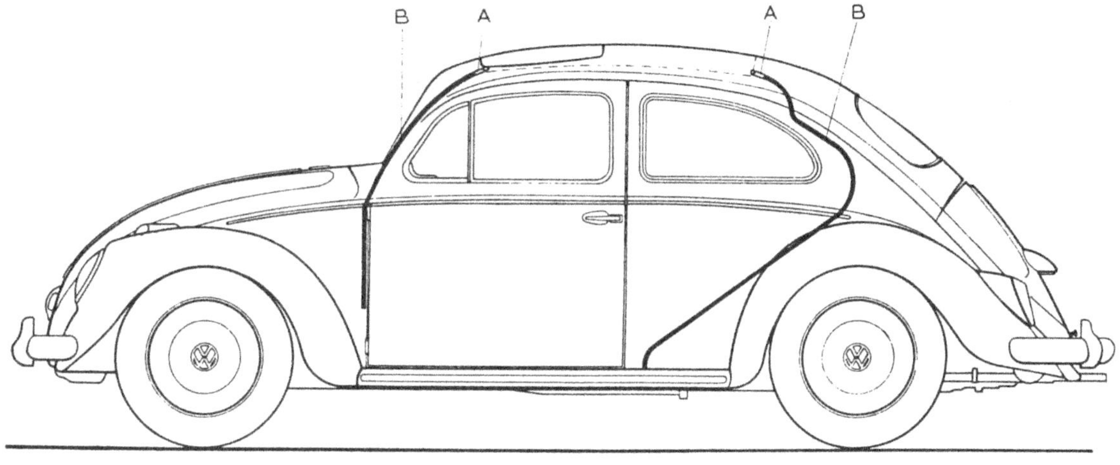

A = Welded water drain pipes
B = Drain hoses

Blocked drain hoses can be cleaned from underneath with compressed air or a flexible steel wire.

Note:

The water drain hoses must be routed so that they are not kinked.

Sliding roof panel sealing

1 - The long seal (Velvet beading) is cemented to the front and sides of the roof opening and the short one is cemented to the rear edge of the roof panel.

2 - Ensure that the short seal is cemented round to the ends of the rear panel radii. The long seal must be fitted so that there is no space between the two seals when the roof is closed.

3 - An additional rubber seal is fitted immediately behind the short seal at the rear of the roof panel.

4 - The rubber seal can only be renewed with the roof panel removed.

Sliding roof trouble shooting

1 - Sliding roof panel lifts on one side.

Cause:

a - Lifter is not running on the ramp.

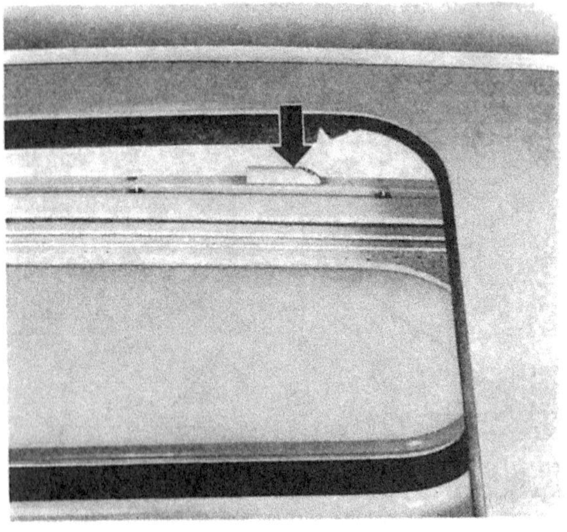

b - Cables or drive gear damaged.

c - Brackets set too low.

Remedy:

a - The ramps in the sliding roof water channels must be properly aligned with the lifter (angled metal lug of the rear roof panel support).

Set ramp so that the contact point of the lifter rests on the center of the ramp. The ramps must be so located longitudinally that at the moment the front panel edge touches the front seal, the brackets are at an angle of 45°. The lifter contact point can be altered by tapping the ramps forward or backward slightly.

b - Renew cables or drive gear.

c - Adjust brackets properly.

2 - Sliding roof panel does not run parallel.

Cause:

a - Adjustment incorrect.

b - Cables or drive gear damaged.

Remedy:

a - Adjust roof panel (cables and height).

b - Renew cables or drive gear.

THIS PAGE INTENTIONALLY
LEFT BLANK

Note:

From 16th June 1965, Chassis No. **115 929 751**, the channel lower parts for the left and right cable, previously made of plastic, were replaced with parts made from stainless steel.

The cable channel upper parts, the cable guide housing together with tensioning plate and cover as well as the left and right guide rails have also been altered. The cross section of the cable channel upper parts are adapted to the cable channel lower parts of stainless steel. In addition, they have been provided with a lug for location in the guide rails.

Plastic packing strips, part number 313 877 385, are obtainable for insertion between roof and cable channel lower parts.

Service Installation

a - The modified parts can be service installed **completely** in vehicles manufactured earlier.

b - If **only** the cable channel lower part has to be replaced on earlier vehicles, use parts 117 877 363 B/364 B.

They can be distinguished from the production parts by their shorter, conically ended tubes which provide a better lead-in to the guide rails. The guide rails will still be supplied as spare parts.

c - A new cable guide housing, part number 313 877 369 A, can be service installed on earlier vehicles only together with the reinforcement, part number 313 877 373, which replaces the tensioning plate.

Both types of guide housing cover can be combined with old and new guide housings as well as with the tensioning plate or the reinforcement.

New part numbers:

Part	old	new
Cable channel lower part, left/right	117 877 363/364	117 877 363 A/364 A
Cable channel upper part, left/right	117 877 365/366	117 877 365 A/366 A
Cable guide housing	313 877 369	313 877 369 A
Guide housing tensioning plate	313 877 371	—
Guide housing reinforcement	—	313 877 373
Guide housing cover	313 877 375	313 877 375 B
Guide rail, left/right	117 877 355/356	117 877 355 A/356 A
Cable channel lower part, left/right for service installation on earlier vehicles	—	117 877 363 B/364 B
Packing strip	—	313 877 385

Notes:

I - On 3 August 1964, from Chassis No. 115 000 001, the headlining for the steel sliding roof, Part No. 117 877 505 A was altered.

1 - The headlining for the sliding roof is now provided with a seam at the rear into which a tensioning wire is inserted.

2 - The frame now has five tabs which are used to secure the headlining.

3 - On the roof frame, the headlining has been extended about .8 in. (20 mm) at the rear so that it now reaches to the front groove.

These measures make the headlining on the steel sliding roof and on the roof frame more secure.

Important

The following headlinings are available for the steel sliding roof under the same Part No. 117 877 505 A:

A - In **silver-beige** for frames of the previous type (without securing tabs).

B - In **cloud-white** with seam and tensioning wire for frames with tabs.

Note:

To firmly reattach detached headlinings, proceed as follows:

A - **Sliding roof lining**

1 - Remove sliding roof.

2 - Pull lining off completely.

3 - Coat frame and lining evenly with PVC adhesive D 11.

4 - Allow adhesive to dry for 20—30 seconds then place lining on frame and press firmly.

5 - Cut a strip of leatherette (.8 in./20 mm wide) and stick it over the lining at the rear.

B - **Headlining in roof frame**

1 - Remove sliding roof panel.

2 - Pull headlining off completely in the roof frame area.

3 - Coat frame and lining evenly with PVC adhesive D 11 and allow to dry for 20—30 seconds.

4 - Tension headlining slightly, fold into roof frame and press firmly.

5 - Powder sticky parts with talcum to prevent the sliding roof lining sticking to the headlining when the roof is closed.

II - On 25 September 1964, from Chassis No. 115 148 174, slotted plastic sliding pieces (313 877 275) were fitted on the sliding roof and secured with steel clips (313 877 277). These slides support the roof lining against the roof frame and thus eliminate the dirty marks which sometimes appeared after opening and closing the roof.

The plastic slides can be fitted subsequently without difficulty:

Note:

Dirty sliding roof linings can be cleaned satisfactorily with the R 3 leatherette cleaner.

Contrary to the details given in this Workshop Manual the lining for the steel sliding roof can be removed without taking the sliding roof out of the runners:

1 - Open roof halfway.

2 - Detach the front part of the lining from the sliding roof: This is done by inserting a screwdriver carefully between the lining and the roof.

3 - Pull clips out of roof panel by levering with screwdriver.

4 - Open roof fully.

5 - Slide roof lining right forward.

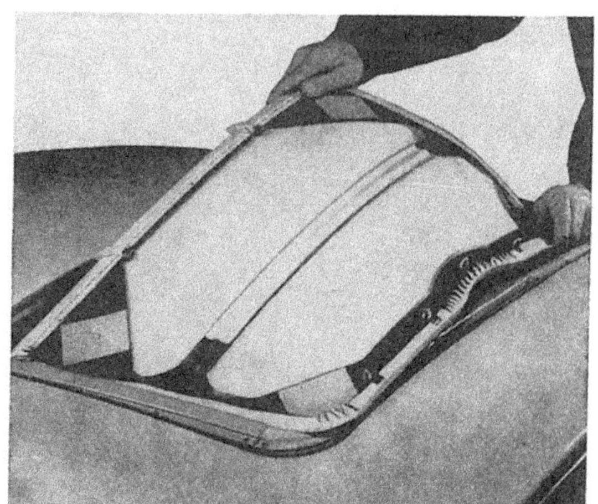

A-20

6 - Bend the lining upward and take it out of the runners diagonally.

7 - Place slides on the sliding roof lining with the curved surface toward the headlining. The two slides must be located 100 mm to each side of the centerline.

8 - Only if the slides are not secure should they be fixed with steel clips.

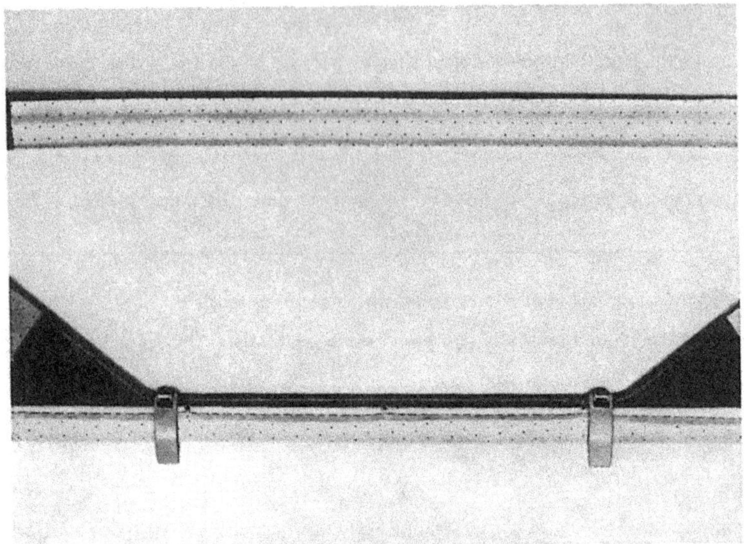

Installation takes place in the reverse order.

Note the following:

1 - Insert sliding roof lining into the lower groove of the runners.

2 - Straighten roof lining frame as necessary before pressing the clips into the roof panel.

5/66

III - Since 25. February 1966, Chassis No. 116 680 470 the previously used special sealing compound has been replaced with an elastic rubber seal to attain better sealing between the casing, into which the roof panel slides, and the roof.

Primed **bodies and roofs without fittings** obtained from the Parts Department must be subsequently sealed at the roof casing. To do this, the rubber seal, Part No. 317 877 297, must be evenly and firmly pressed into the channel-shaped edge round the roof casing. The rubber seal must not be pulled taut, otherwise it will contract in time and cause drafts.

Drafts inside the Passenger Car can generally be eliminated with simple materials by the following measures:

1 - Glove compartment

Cut two rubber strips with a 7×7 mm (.275″) cross section and 470 mm (18.5″) and 370 mm (14.5″) in length out of weatherstrip (Part No. 211 831 711 Transporter door lower weatherstrip). Remove glove compartment and cement the strips round the glove compartment opening above and below the hinges so that the edge of the compartment, when installed, rests on the rubber strip. Seal the hinge openings with pieces of felt 1—2 mm thick, slotted to receive the hinge.

2 - Upper choke cable hole in luggage compartment

Install grommet (Part No. 111 711 535). Cut grommet through to the center, grease lightly and insert in the hole with the cut at right angles to the direction of travel.

3 - Gap between door and lock pillar

Cement two cardboard strips 200×8×2 mm (7.8″×.314″×.078″) under the door weatherstrips as shown in sketch. The center of the cardboard strips should be level with the lock. If the gap between door and lock pillar cannot be sealed by inserting the cardboard strips the door should be moved to the rear as required by adjusting the hinges.

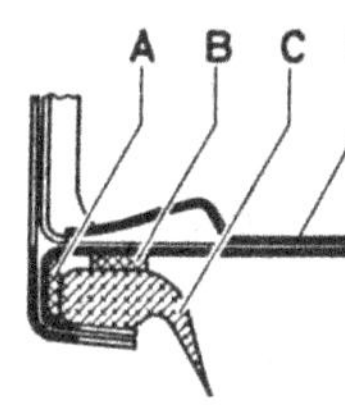

Left door cross section (at lock level)
A and B = Cardboard strips
C = Door weatherstrip
D = Door

4 - Check rod opening in door

Cut a piece of foam rubber (Part No. 111 867 835 B) as shown in sketch. Remove door check rod pin, push the foam rubber over the check rod and cement it over the opening in the door.

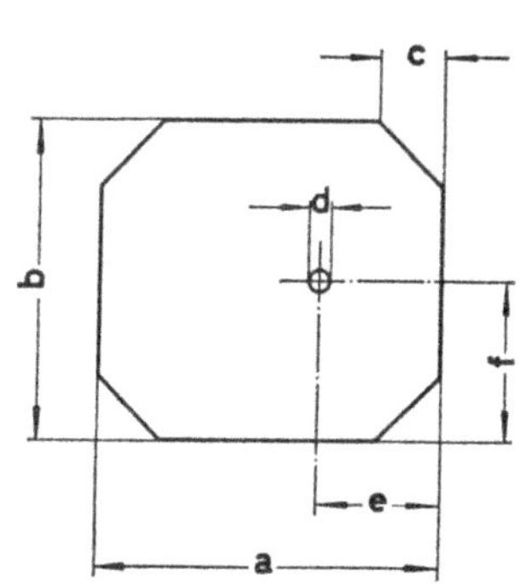

a = 55 mm (2.165″) b = 50 mm (1.986″)
c = 10 mm (.383)×45° d = 3 mm (.118″) dia.
e = 20 mm (.787″) f = 25 mm (.984″)

5 - Gap at door upper frame

Cement 2 mm thick cardboard strips under the door weatherstrip.

Note:

From Chassis No. 4 010 995 (31st July 1961), the warm air outlets at foot-level in the front of the body are equipped with slides. When the outlets are closed, all the warm air is directed to the windshield.

Slide for left outlet	113 255 961 A
Slide for right outlet	113 255 962 A
Tensioning wire	113 255 965
Channel	113 801 119

The slides can be service installed in all Volkswagens from Chassis No. 1 385 974 (January 1957).

There is now a warm air outlet in each of the kick boards. The warm air is taken from the right-hand connecting pipe between side member and heater pipe and directed to the outlets through a hose.

Adaptor, rear	113 255 511
Hose for rear foot well heating	113 255 515 A
Kick board **new**	113 863 376 A
previously	113 863 376
new	151 863 376 A
previously	151 863 376

The rear foot well heating can be installed in all Volkswagens from Chassis No. 3 060 711 (May 1960).

Note:

From Chassis No. 4 513 596, the tensioning wires for the slides on the heater outlets in the front foot well were altered. The modified tensioning wire is retained by two 3 mm dia. holes.

Service installation

The modified wire cannot be service installed in previous vehicles.

The wire — Part No. 113 255 965 — will remain available as a spare part.

Note:

From Chassis No. 4 723 425, the plastic support cage in the heater hose was replaced by a wire mesh hose.

Only heater hoses with wire mesh hoses (Part No. 111 255 355 as before) will be supplied as spares.

Note:

From Chassis No. 5 199 980, the rear foot well heating was modified so that it can be controlled with regulating levers.

The outlets in the kick boards — Part No. 113/151 863 376 B — have been moved. The modified kick boards — Part No. 113/151 863 376 C — have round openings for the heater outlets.

Branch pipes — Part No. 113 255 429 A — which end in the kick boards are installed under the rear seat in place of the connecting pipes — Part No. 111 255 431. The branch pipes are covered with insulating material — Part No. 113 255 485 B/486 B.

Note:

The modified heating system for the rear foot well can be installed in previous vehicles.

Complete replacement bodies are only supplied with the branch pipes installed. When fitting these bodies on previous chassis, the foot well heating can be installed at the same time. In this case, the kick boards — Part No. 113/151 863 376 C — should be used. The fitted branch pipes can, however, be replaced by the connecting pipes — Part No. 111 255 431.

Note:

From Chassis No. 5 199 980, a split plastic tube with a metal sheet (Part No. 113 255 423/424) attached to it was installed on the heater pipes between the heater muffler and the body to improve heat insulation. The plastic tube is held together by a spring clip (Part No. 111 255 545).

Replacement of Body Sections

General

The description of the replacement of individual sections of the body is given in the form of instructions for the procedure to be adopted. A reasonable knowledge of bodywork repairs is taken for granted. The work should, where possible, be carried out by a skilled man, well versed in bodywork repairs.

Unless oxy-acetylene welding is expressly stipulated, the various sections refer to this method in addition to the more usual electric spot welding.

It should, however, be stressed that spot welding with an electrode holder or push-type welding gun is preferable to oxy-acetylene welding.

Lack of knowledge of the welding process or of the material can result in changes in the structure producing appreciable stresses in the material when welded with oxy-acetylene. These later become apparent as fractures in the metalwork or as noise caused by the stresses.

Every effort should therefore be made to use spot welding for the particular job when the bodywork has to be repaired.

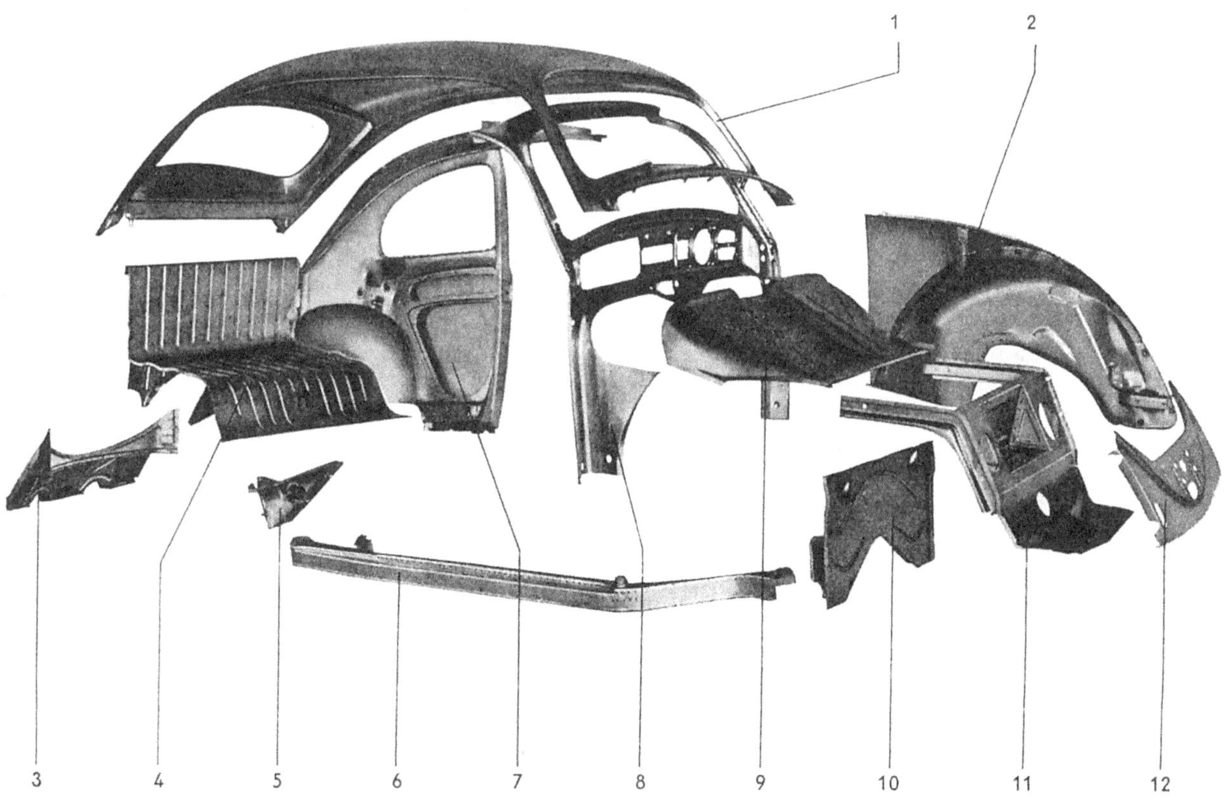

1 - Roof

2 - Front quarter panel

3 - Tail plate

4 - Luggage compartment floor-plate

5 - Rear cross-member

6 - Side member

7 - Rear quarter panel

8 - Instrument panel

9 - Front luggage compartment

10 - Front cross panel

11 - Lower reinforcement plate

12 - Front apron

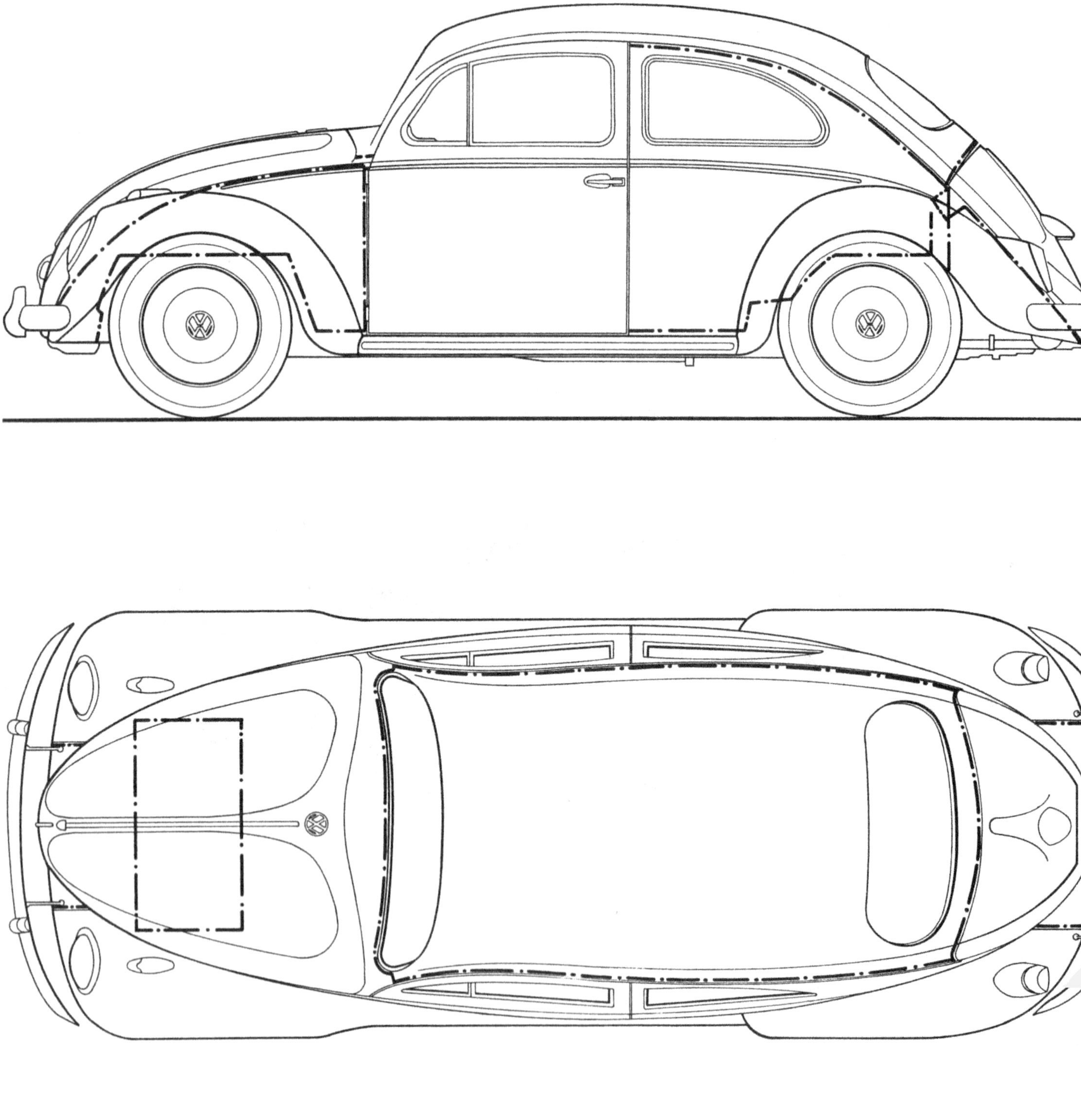

A-22

THIS PAGE INTENTIONALLY
LEFT BLANK

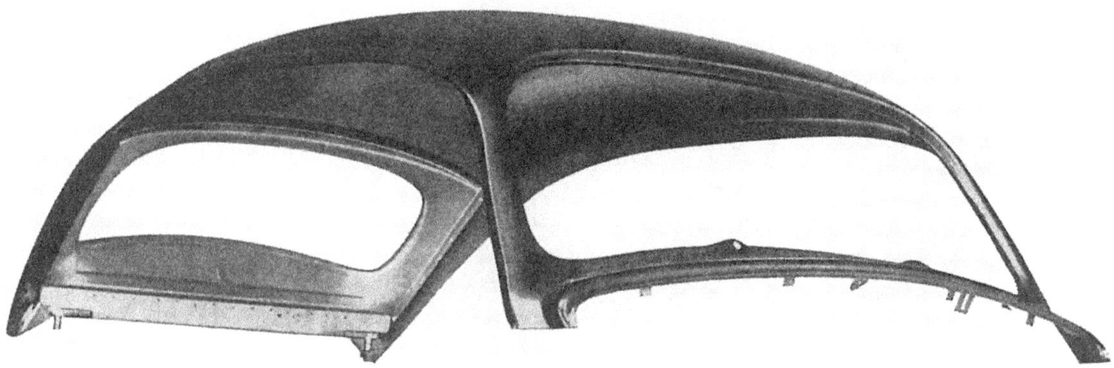

To Replace a Roof

Preparation

1 - Disconnect the battery.

2 - Remove engine, sound-absorbent lining and pads.

3 - Remove all seats and mats.

4 - Remove windshield, rear-view and rear quarter windows.

5 - Lower the door windows and cover the vent wings with paper.

6 - Remove all interior trim, including rear-view mirror and sun visor.

7 - Remove steering wheel.

8 - Remove front and rear hoods, including supports.

9 - Disconnect and remove main cable harness with direction indicators.

10 - Remove cardboard lining of front luggage compartment.

11 - Remove windshield wiper system.

12 - Remove fuel tank.

13 - Remove glove compartment.

14 - Strip off half the sound-absorbent lining from the top of the rear of the rear luggage compartment.

15 - Remove the rubber weatherstrip for the front hood (from the upper cowl). Pull out the side weatherstrips of the front hood, down to about halfway.

16 - Pull the side weatherstrips of the rear hood out of the retaining flange, down to about halfway.

17 - Remove the mouldings from the two front side panels.

Body repairs

1 - Mask off the rear quarter panels with damp asbestos wool below the gutter.

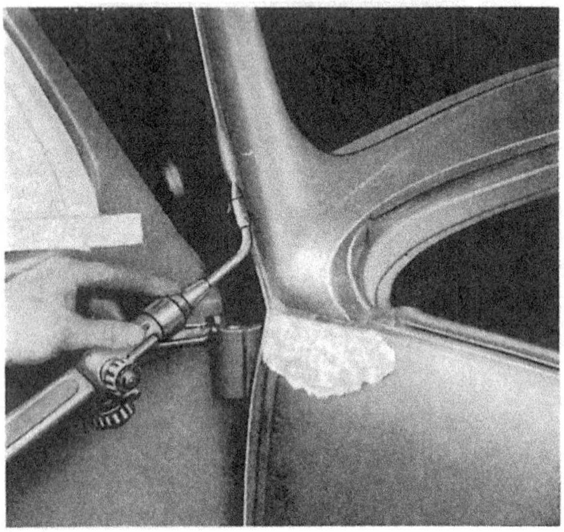

2 - Mask off the two front side panels below the upper cowl with damp asbestos wool.

3 - Heat the gutter to a dull red with a small torch and gentle flame.

This heating is necessary to prevent the gutter from cracking when it is curled up later.

4 - Remove the damp asbestos from the rear quarter panels.

5 - Prize open the joints between the front side panels and upper cowl and curl over with a suitable tool for a distance of about 6 cm.

This is necessary because the edge curling tool used with the gutter on the right-hand side of the car has to be applied at this point. On the left-hand side of the car the gutter is curled over from the rear to the front. The joint between the left-hand front side panel and the cowl has also to be prized open for removal of the tool.

6 - Curl up the edge of the gutter on the left and right of the roof, using a gutter parting tool (VW 734 for local manufacture).

7 - Cut away the rear roof lining plate (drip catcher) about 15 mm (0.6") above and below (dotted line) the welded seam with the rear luggage compartment floor, using an oxy-acetylene torch.

8 - Having masked with damp asbestos wool, the spot-welded joint between the rear end of the roof and the rear side panel left and right should be cut through with an oxy-acetylene torch about 20 mm (0.8") above the joint.

Do **not** cut through the flanged edge of the roof side member.

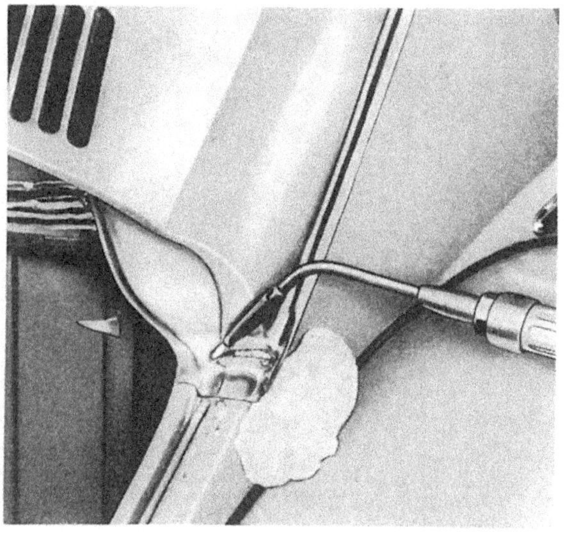

9 - Split the welded joints between the side roof members and the roof lining plate open with a cold chisel.

10 - Cut away the front cowl above its joints left and right with the side panels.

The flanged edge of the instrument panel must **not** be cut through.

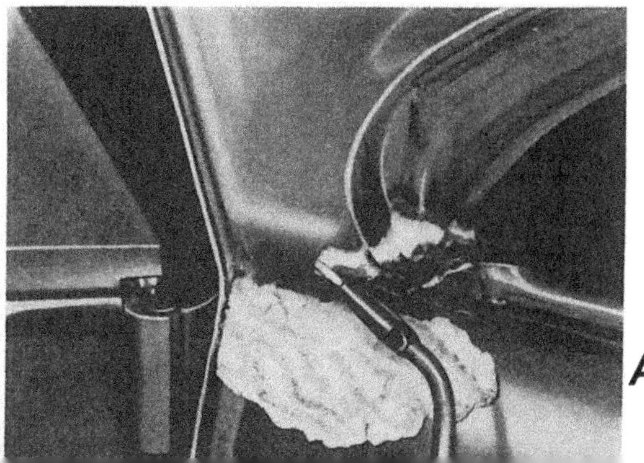

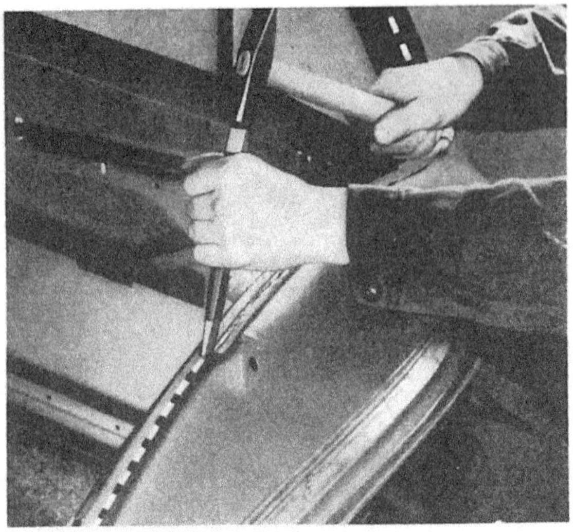

11 - Split open the joint between the roof and instrument panel (window frame) on the outside all round with a cold chisel.

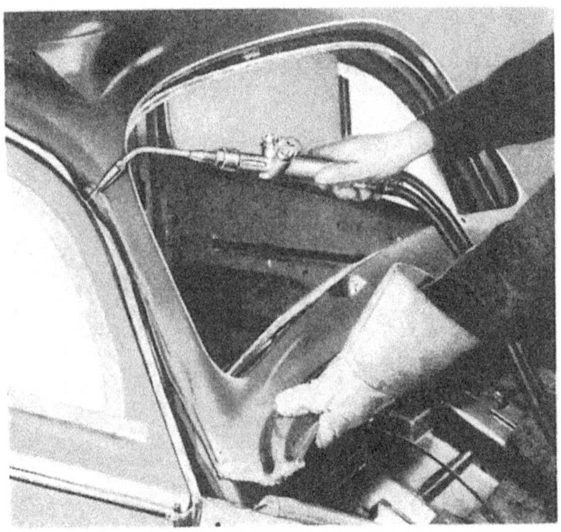

12 - Lift the roof slightly. Heat the welds at the front of gutter left and right, finally breaking them through by lifting the roof sharply.

If these spot welds were to be split open with a chisel or similar tool, there would be a risk of the gutter being damaged.

Attention!

Protective gloves should worn when removing or refixing the roof.

13 - Take off the roof.

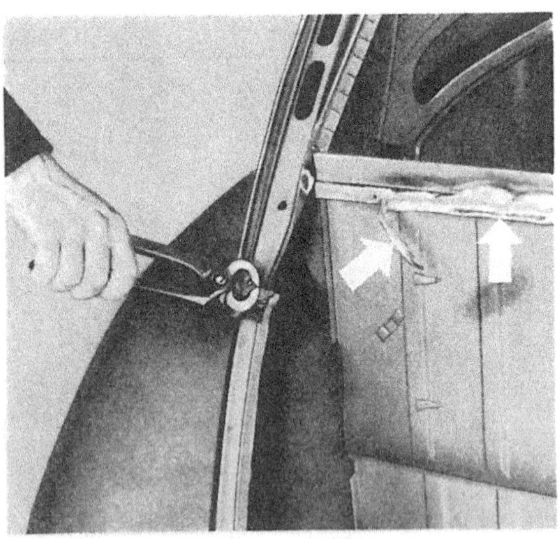

14 - Scraps of metal remaining attached to the rear quarter panels and the luggage compartment floor should be removed with pincers.

Note:

When removing such scraps the metal at the welded spots must not be torn away but sheared off by twisting the pincers. Otherwise the spot weld will be torn out too and will leave a hole at that point.

15 - Pull away the metal strips at the instrument panel (window frame), also with pincers.

Metal scraps remaining attached to the front side panels should be ground, milled or filed away, never pulled or chiselled. Otherwise deformation will result.

16 - Grind or file all spot welds smooth. At the seams and flanged edges the metalwork should be straightened and any dents pressed back into shape.

17 - All points on the replacement roof and body should be ground clean.

If a spot welding unit is not available, holes should be bored 4 mm (0.16") diameter at intervals of 150 mm (5.91") in the underside of the flanged edge of both roof side members. These holes then have to be filled in with filler metal when the new roof is placed in position, thus welding the roof to the body. At the joint between the front cowl and the front side panel apply spot welding paste. (See Special Hints, page A 29-1.)

18 - Place the new roof in position.

19 - Fix clamps to the windshield opening, the front side panels and roof side members.

Note:

If insufficient clamps are available, fix them first to the window frame and to the left and right sides of the cowl. When the front part has been welded together, shift the clamps further back and continue welding there.

20 - Weld the joint between the cowl and front side panels on both sides to the door lock pillar, for a length of about 20 mm (0.8"), with oxy-acetylene.

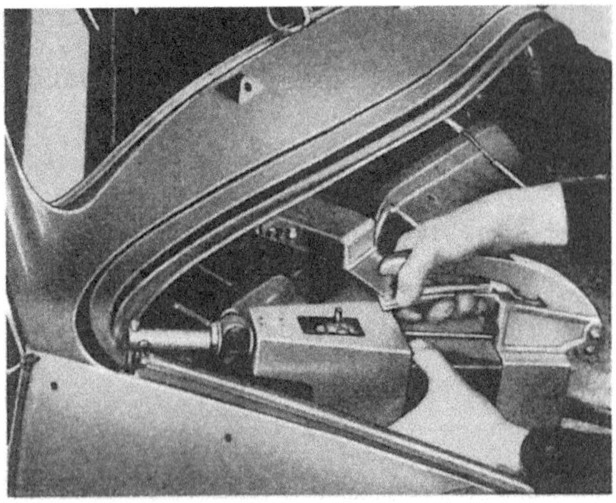

21 - Spot weld the cowl and front side panels together.

If oxy-acetylene welding is employed, the area must be masked with damp asbestos wool.

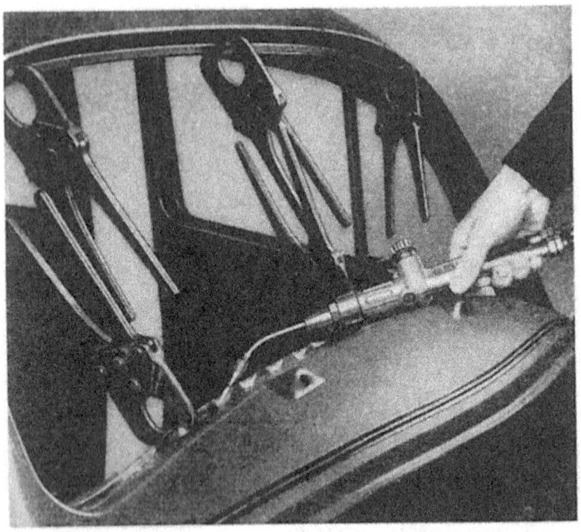

22 - Weld the instrument panel to the roof in the windshield opening.

a - **Electric**

Spots at intervals of about 20 mm (0.8″).

b - **Oxy-acetylene**

Fuse the edges of the sheets together for a length of 15 mm (0.6″) at intervals of about 20 mm (0.8″).

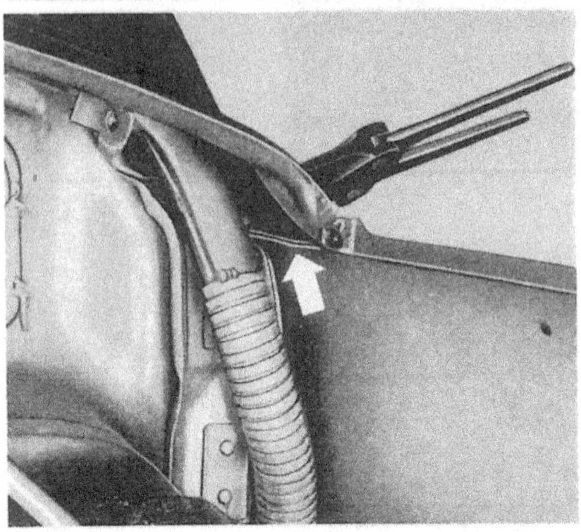

23 - Weld the cowl and front side panels together along the entire length from the inside.

In this region the side panels should first be masked with damp asbestos on the outside.

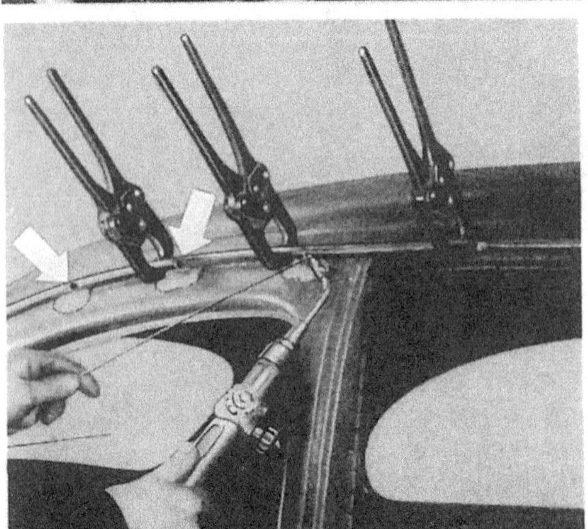

24 - The roof has now to be welded with the side member from front to back.

a - **Electric**

Spots at intervals of about 100 mm (4″).

b - **Oxy-acetylene**

Holes drilled in the underside of the flanged edge should be welded with filler metal, the welds being ground down before the gutter is turned down.

Mask the side panels in this area with damp asbestos.

25 - Spot weld the rear ends of the roof to the right and left rear quarter panels. If only gas welding equipment is available, two 4 mm (.16") holes should be drilled in the roof and drip catcher, and welded to the side panel by adding metal.

The joints between the ends of the roof and the side panels should always be welded inside and out with gas.

Mask the side panels in the area with adequate damp asbestos wool.

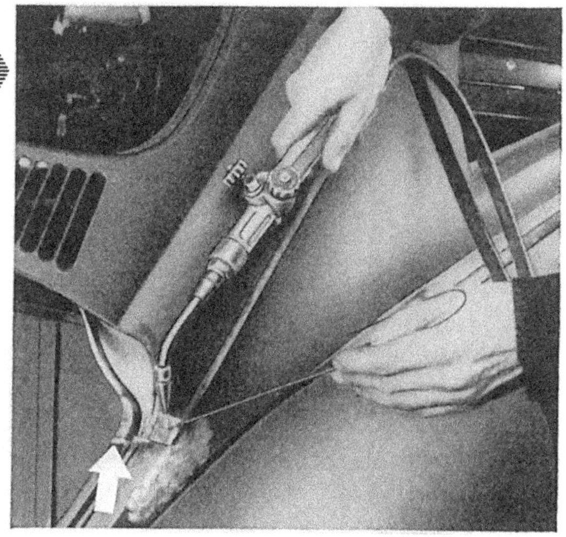

26 - Weld the inside roof lining (drain channel) with the rear of the luggage compartment (see arrow).

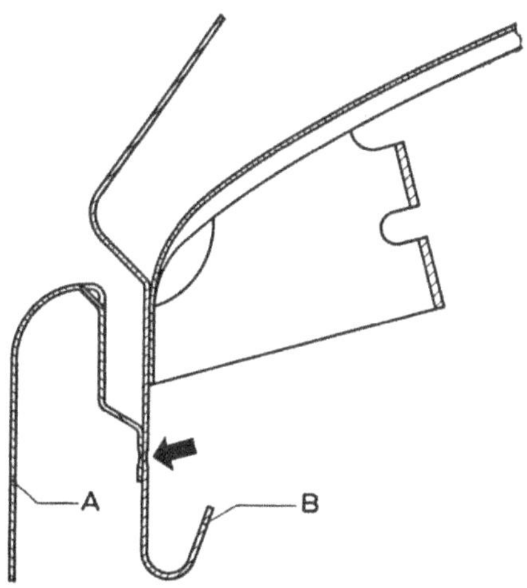

A = Rear of luggage compartment
B = Roof (drain channel)

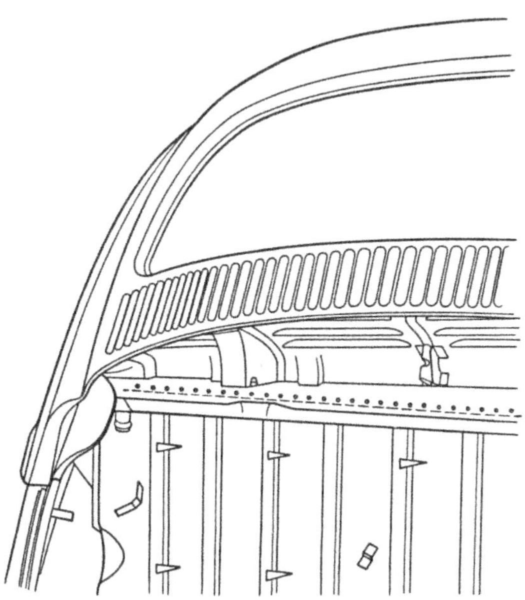

a - **Electric**

Spot weld the two parts at intervals of 20 mm (.8").

b - **Gas**

Drill 4 mm (.16") holes in the roof lining at intervals of about 40 mm (1.16") and weld it to the luggage compartment rear panel.

---- = Luggage compartment rear panel
• • • • = Holes

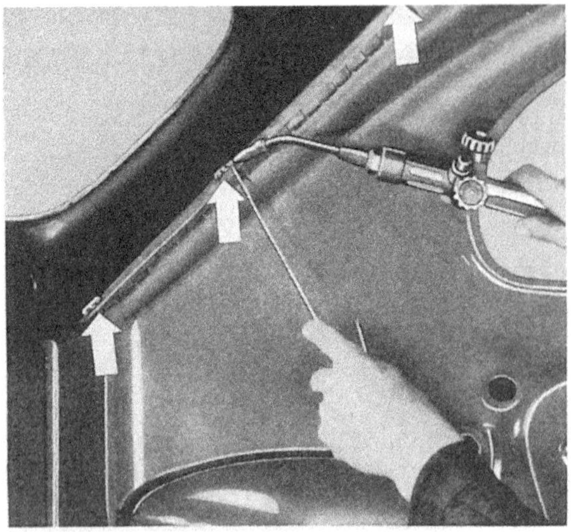

27 - Gas weld the roof lining to the side members at three points on the left and right.

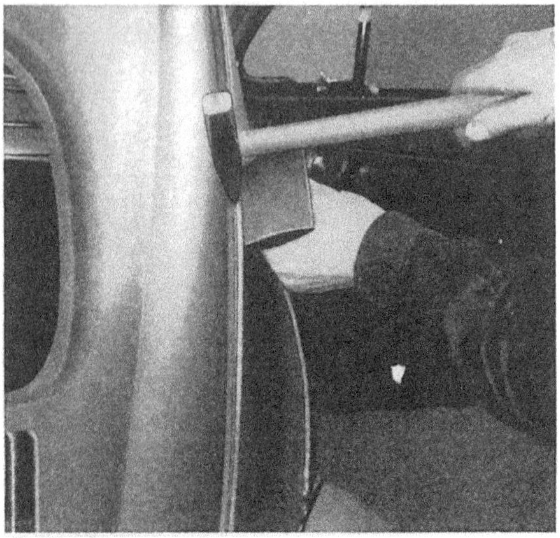

28 - Curl over the gutter with a hammer and iron. The most convenient method is to begin at the door lock pillar and work in both directions. Straighten out afterwards.

29 - Parts which have been drilled or welded should be ground smooth and the vehicle prepared for painting.

Note:

From 6th August 1959 and Chassis No. 2 533 000 the water drain channel in the engine compartment has been altered.

The water collected in the channel is no longer disposed off through the two pipes to the lower edge of the hood; it now flows out at the corners of the hood. This method renders the two pipes previously used superfluous.

When existing stocks have been used up, only roofs with the modified water channels will be supplied.

The water pipes are still available as spares.

Note:

When the present stocks are exhausted, only roofs No. 111 817 021 A and 115 817 021 C (with sliding roof) will be supplied. When installing these roofs in vehicles which have the oval or divided rear window, the upper part of the instrument panel must also be replaced. This is done as follows:

1 - Remove the damaged roof.

2 - Separate the instrument panel upper part as shown.

3 - Cut the upper part from the new instrument panel (Part No. 113 805 105 F) and weld in place.

4 - Place roof in position, tack and then weld to the body.

25 - Spot weld the rear ends of the roof to the right and left rear quarter panels. If only oxy-acetylene welding equipment is available, two 4 mm (0.16″) holes should be bored in the roof and drip catcher, and welded to the side panel with filler metal.

The joints between the ends of the roof and the side panels should always be welded inside and out with oxy-acetylene.

Mask the side panels in the area with adequate damp asbestos wool.

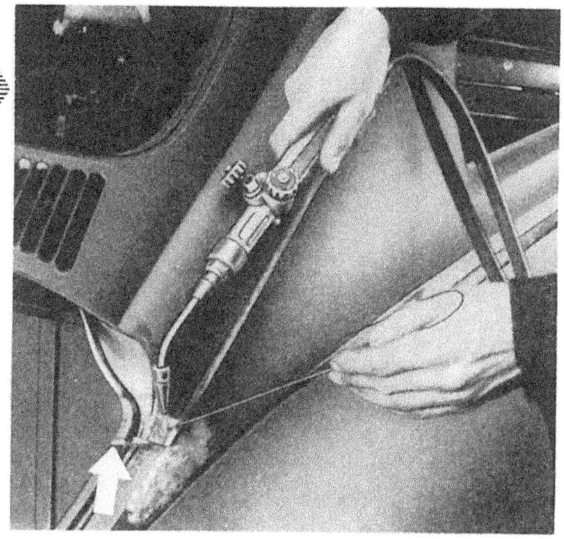

26 - Weld the inside roof lining (drip catcher) with the rear of the luggage compartment (see arrow).

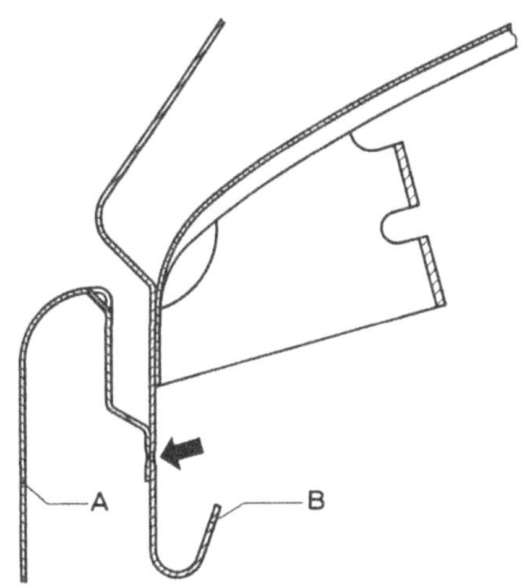

A = Rear of luggage compartment
B = Roof (drip catcher)

a - **Electric**

Spot weld the two parts at intervals of 20 mm (0.8″).

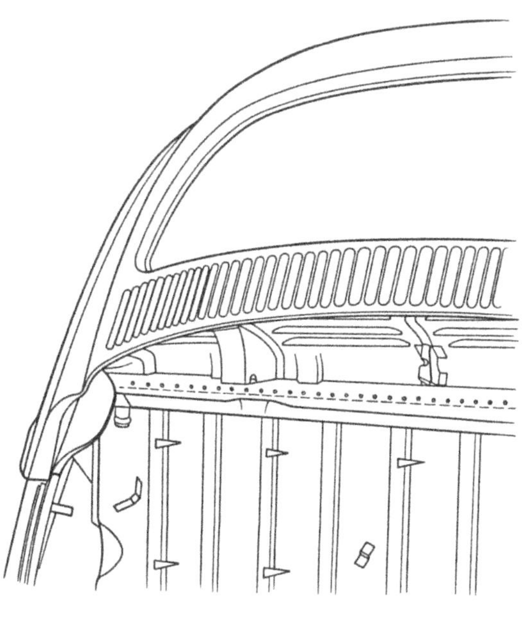

b - **Oxy-acetylene**

Bore 4 mm (0.16″) holes in the roof lining at intervals of about 40 mm (1.6″) and weld through them with the luggage compartment floor.

---- = Lower edge of luggage compartment floor
• • • • = Holes

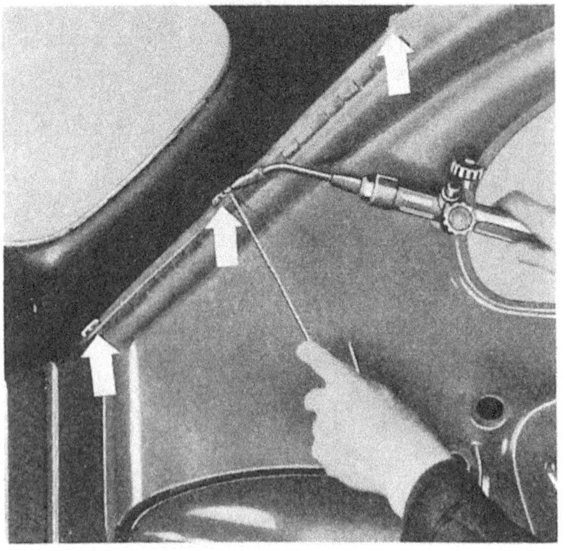

27 - Oxy-acetylene weld the roof lining to the side members at three points on the left and right.

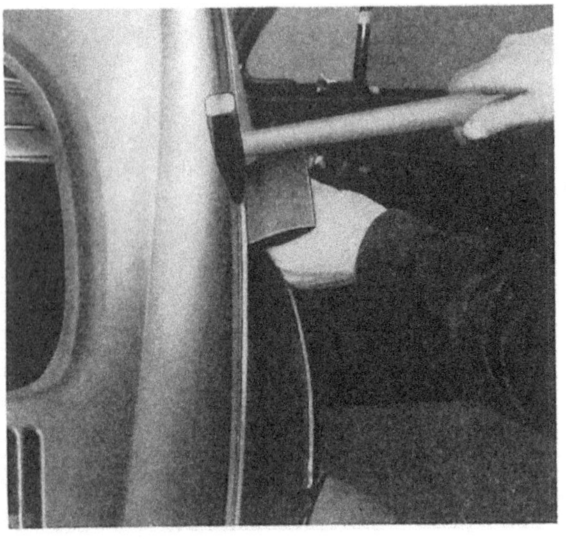

28 - Curl over the gutter with a hammer and iron. The most convenient method is to begin at the door lock pillar and work in both directions. Straighten out afterwards.

29 - Points which have been drilled or welded should be ground smooth, if necessary, and the vehicle prepared for painting.

Replacement of a Front Side Panel

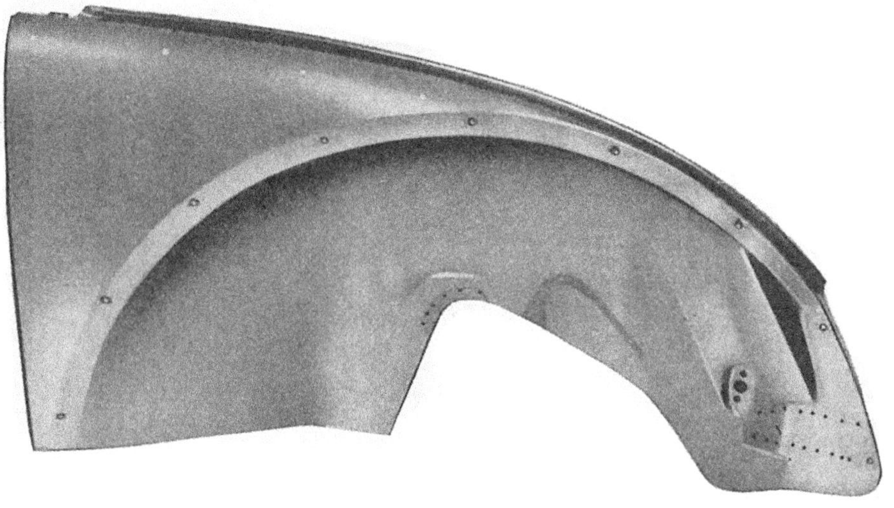

Note:

From April 1959 the front side panels (Part. No. 111 809 021/022) are supplied with a lateral fuel tank support (Part No. 111 805 571/572). The Part No. of the front side panels remains unchanged. When replacing a front side panel, the lateral fuel tank support should also be removed.

General

The front side panel is supplied with the following parts welded on:

a - Sheet-metal nuts for attachment of the fender bolts.

b - Supports for the bumper brackets.

c - Retaining flange for the weatherstrip.

d - On the left-hand panel: a support for the horn and one for the hood bracket.

The support for the car jack and the support, supplied as an extra, for the second horn on the right-hand panel have to be welded on subsequently.

All parts listed above are available as individual spares.

Preparation

1 - Disconnect battery.

2 - Remove front hood.

3 - Pull weatherstrip out of its retaining flange, also on cowl, as far as necessary.

4 - Remove fuel tank.

5 - Remove door, fender, sill panel, bumper and wheel on the affected side.

6 - For work on the right-hand side, remove the glove compartment.

7 - Remove windshield.

8 - Remove floor carpeting as far as necessary.

9 - Remove sealing compound at the welds.

10 - Remove cardboard and sound-deadening material from the front luggage compartment.

11 - Disconnect the wiring for the headlights and horn as far as necessary, and run it in the front luggage compartment.

Body repairs

1 - Heat the gutter down the hinge pillar to a dull red with a small torch and gentle flame.

This is necessary to prevent the gutter from cracking when curled up later.

2 - Curl up the gutter for a distance of about 20 mm (0.8") above the joint between the cowl and the front side panel, using the edge curling tool (VW 734 for local manufacture).

a = ~ 20 mm

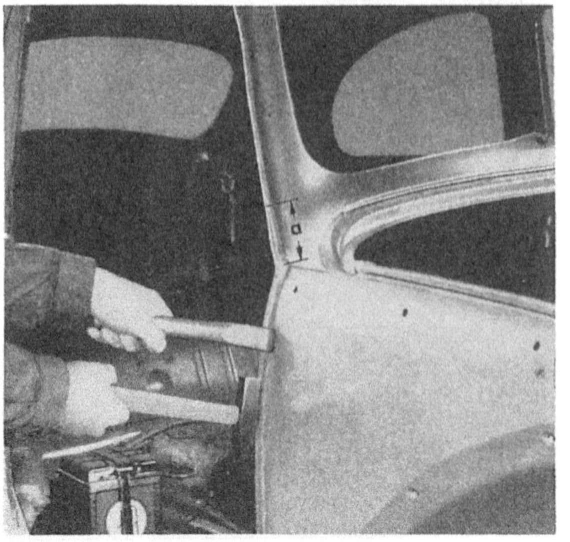

3 - With a body chisel VW 732 (for local manufacture) or with an oxy-acetylene burner, cut away the front side panel along the lines indicated.

Note:

The cut should, where possible, be made about 10—15 mm (0.4—0.6") below the spot-welded joint, otherwise the connecting pieces may be damaged.

4 - Take off the front side panel.

> **Attention!**
> For removal of the side panel wear protective gloves.

5 - Remove any scraps of metal with pliers. Grind down all spot welds carefully. Straighten the connecting pieces, rub down and prepare for installation of the new panel.

Note:

When removing such metal, do not tear but shear away by twisting the pliers. Otherwise, the spot weld will be torn out and will leave a hole at that point.

6 - Grind all points clean on the new side panel at which it is to be spot welded.

7 - If it is only possible to weld with oxy-acetylene, bore 4 mm (0.16") holes at intervals of about 50 mm (2") in the gutter of the hinge pillar.

a = 50 mm

8 - Install the new side panel, beginning at the top.

9 - Adjust the side panel to fit the hinge pillar and cowl, clamp up and prepare for welding.

10 - Spot weld the side panel to the hinge pillar. If only oxy-acetylene is available, drill holes and weld with filler-metal.

11 - Oxy-acetylene weld the joint between the cowl and side panel for a length of about 20 mm (0.8") (a) and spot weld at the front corner of the cowl (b).

12 - Oxy-acetylene weld the joint between side panel and cowl all the way along the inside.

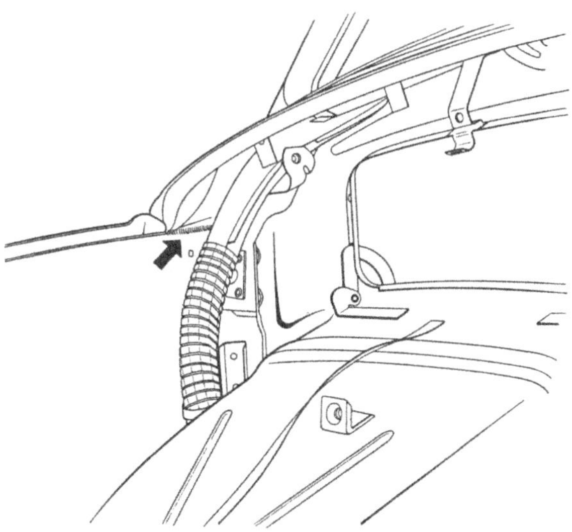

13 - Grind the spot welds down the gutter of the hinge pillar and curl the edge over.

14 - Insert the door and check that it fits in the opening. Adjust where necessary.

15 - Weld the side panel to the front apron.

16 - Spot weld the side panel to the front cross panel reinforcement plate and apron in the front luggage compartment, with spots at intervals of 20 mm (0.8").

If no spot welding equipment is available, use beads 15 mm (0.6") long at intervals of about 20 mm (0.8").

17 - Oxy-acetylene weld the upper and lower joints between the side panel and the apron.

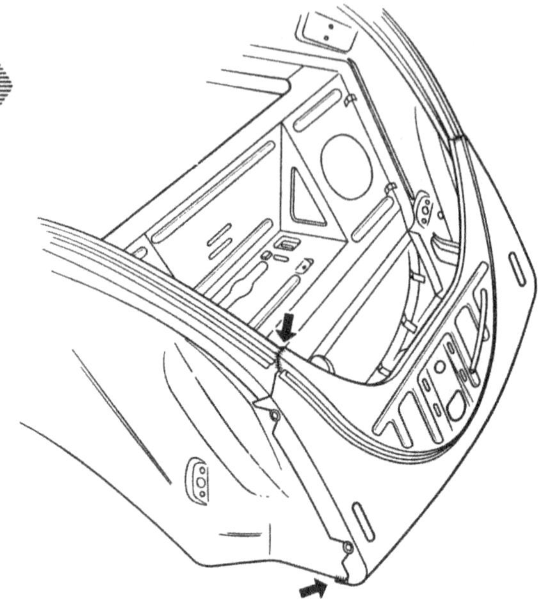

18 - Clamp the side panel to the front cross panel and spot weld it to the lower side member. If only oxy-acetylene welding is possible, use beads 15 mm (0.6") long at intervals of about 20 mm (0.8").

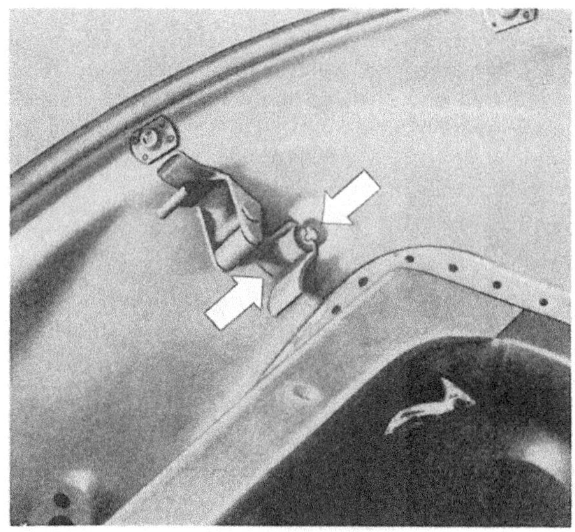

19 - Spot weld the side panel to the instrument panel and front cross panel around the footrest.

20 - Weld in the car-jack support.

21 - Points which have been drilled or welded should be ground smooth, where necessary.

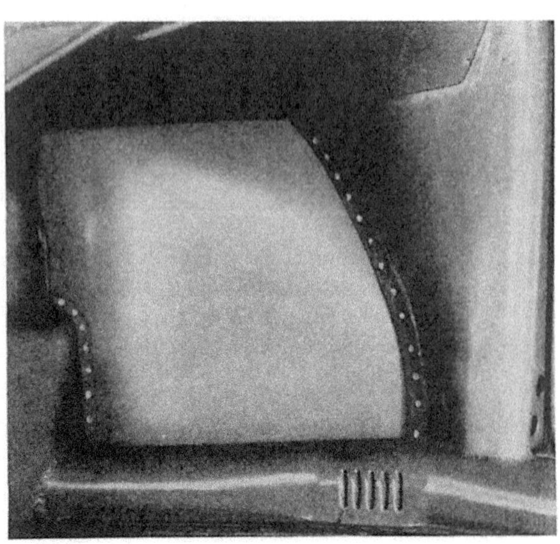

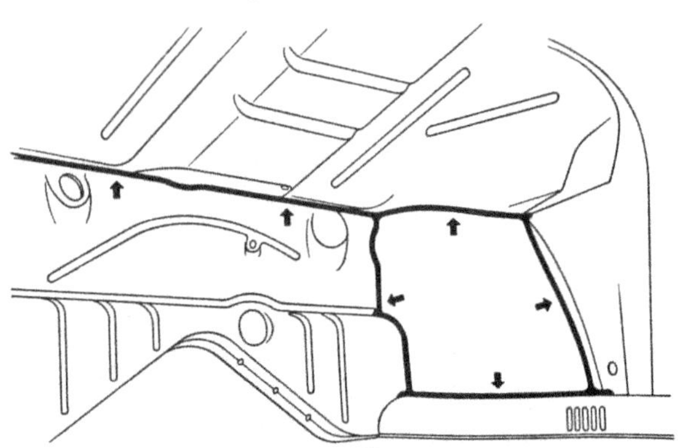

22 - All joints between the front side panel around the footrest and the rest of the body should be amply coated with Sealing Compound D 9.

23 - Prepare the vehicle for the necessary painting work.

Note:

I - From Chassis No. 3 192 507 (1 st August 1960) the front panels (111 809 021 B/022 B) have a recess in the upper part. These front side panels cannot be installed in older vehicles. The former type panels (111 809 021 022) will remain available. The shape and method of fitting the front fenders has not been changed.

II - From February 1964, the cable holes in the side panel were altered. Part No. 111 809 021 B/022 B and 113 809 021 F/022 F as before. The position of the previous hole is marked with a center punch.

When service installing a new panel, the existing hole should be plugged with a rubber plug — Part No. N 20 025 1 — and a 12 mm hole made at the center punched mark.

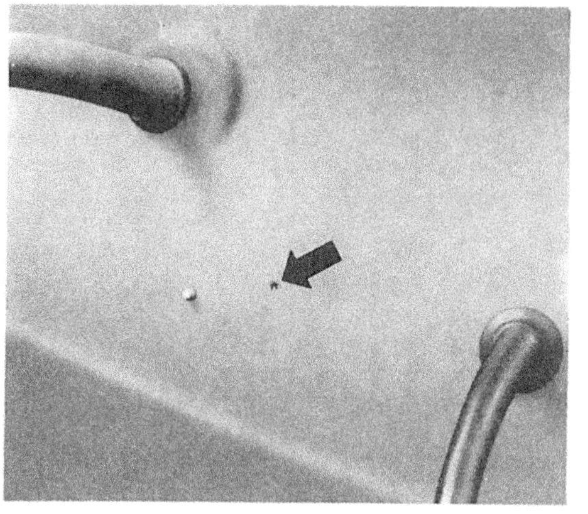

Note:

From August 1964, all Type 1 vehicles are supplied with a new jack (Part No. 111 011 031 C) which has separate links for raising and lowering.

The lowering of the vehicle can now be controlled better with the bottom link.

The tube of this jack is only 20 mm in diameter and the bracket on the side panel (new Part No. 111 809 371 B) has been altered to suit.

Note

As only side panels with the new bracket (Part No. as before) will be supplied when stocks of the former panel are exhausted, the bracket must be modified as necessary to suit the old type jack:

1 - Bend bracket so that jacks of the previous type with the 22 mm tube can be fitted.

2 - Grind 2 mm off the edge of the bracket so that the clamping strap can engage properly.

THIS PAGE INTENTIONALLY
LEFT BLANK

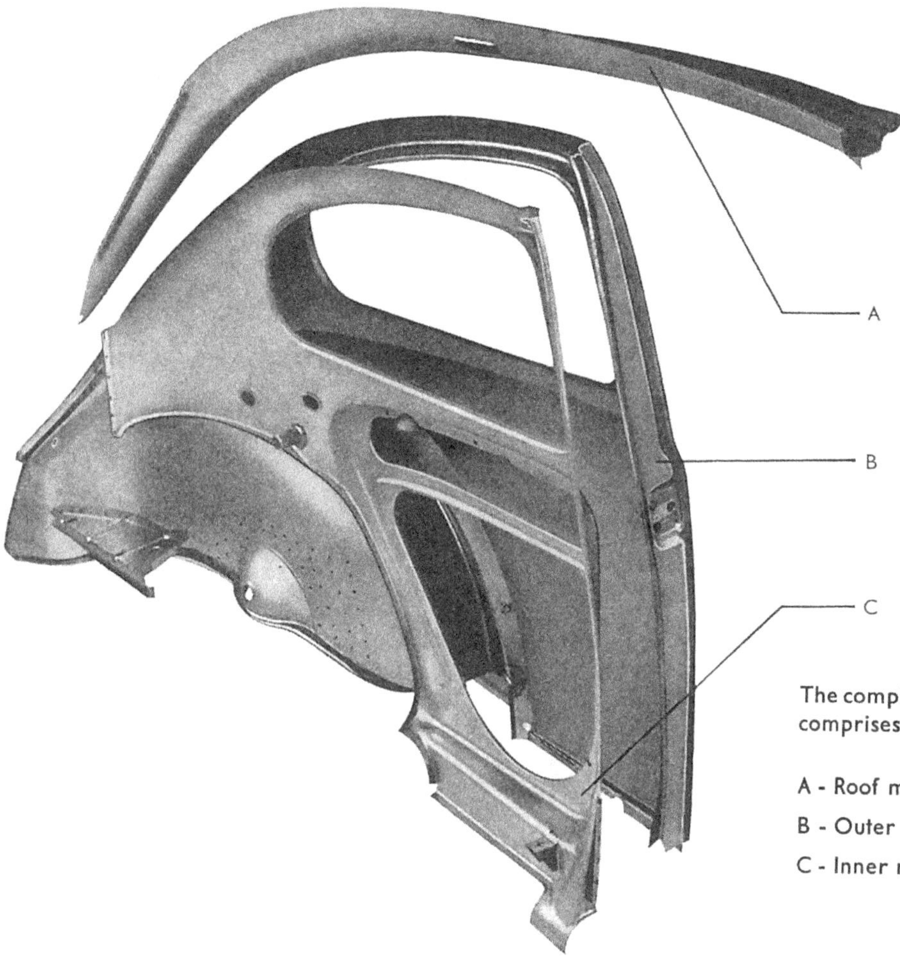

The complete rear quarter panel comprises the following parts:

A - Roof member

B - Outer rear quarter panel

C - Inner rear quarter panel

These parts are available as spares, either separately or assembled.

Note:

From Chassis No. 2 154 170 the rear quarter panel has been modified in the following respects:

1 - The inside rear quarter panels (Part No. 113 809 061 B / 062 B) have been recessed in the top portion to give a better stop for the rear seat back rest. One rubber stop (Part No. 111 885 553 B) on either side of the recessed portion is attached by means of a screw (Part No. N 13 961 1). The spot-welded brackets with rubber buffers previously used have been discarded.

2 - The hook (Part No. 111 885 561) that keeps the rear seat back rest in position is attached with a new type screw (Part No. N 14 093 1). The previously used threaded plate in the rear quarter panel has been discontinued.

3 - The inside rear quarter panels are provided with recesses to do away with the spot-welded brackets previously used. The support (Part No. 111 813 103 A) has been lengthened.

Service Installation

Vehicles from Chassis No. 1-0 929 746 (August 4th, 1955) may be serviced — equipped with the modified inside rear quarter panels if the necessity arises or a complete rear quarter panel with roof member (Part No. unchanged) may be installed as a unit or in part. The following should be observed:

a - Brackets or recesses have been adopted to each other as regards dimension. A modification of the original version is, therefore, not required.

b - The new type, lengthened cushion support (Part No. 111 813 103 A) for the rear seat may be service-installed in vehicles of previous make. The holes for the attaching screws should be staggered.

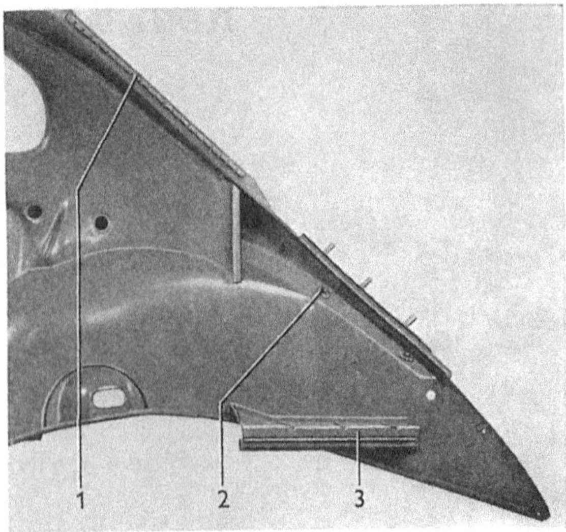

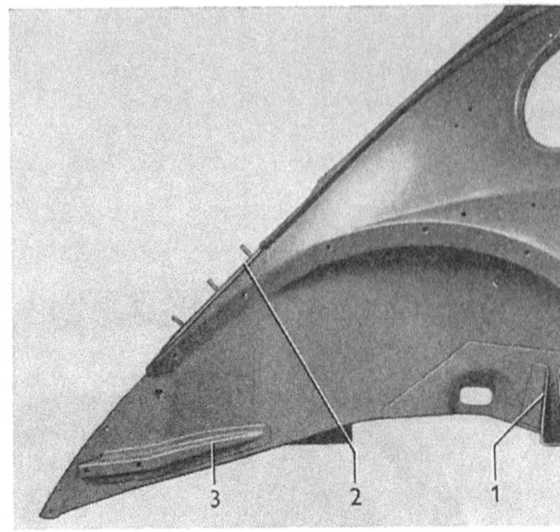

The following parts available as spares are welded to the rear quarter panel:

Inside:

1 - Retaining strips to hold the inside trim of the De Luxe Sedan.

2 - Sheet-metal nuts for attaching the fender bolts.

3 - An engine cover plate with retaining strip for rubber gasket.

Outside:

1 - Support for the rear side panel.

2 - Retaining strip for weatherstrip.

3 - Support for rear bumper.

The support for the rear bumper and the support for the rear side panel can be welded in with the aid of the templates (local manufacture) VW 705 and 706, respectively.

When replacing or carrying out repairs to the rear quarter panel observe the following points:

A - Complete replacement of the outer panel, inner panel and roof member (welded together) is necessary when the roof is undamaged and not being replaced.

B - Individual replacement of the outer or inner panels or roof member is possible when the roof is removed.

C - Pieces of the outer or inner panels can be replaced by cutting out the damaged area and butt welding new pieces of the same shape, if the dents cannot be removed in the normal way, or this is uneconomical.

A - Replacement of a complete rear quarter panel with roof member

Preparation

1 - Disconnect battery.

2 - Remove engine, sound-deadening pads in the engine compartment and roof member.

3 - Remove all seats and floor covering.

4 - Remove the windshield, rear-view window and the affected rear quarter window.

5 - Lower the window of the door on the affected side and cover with paper.

6 - Remove the lock striker plate.

7 - Remove all inside trim from the affected side, including rear-view mirror and sun vizor. Other parts of the inside trim should be removed or suitably covered.

8 - If the left-hand quarter panel is to be replaced, the main cable harness must be removed with the direction indicator.

9 - Remove rear bumper, fender and sill panel.

10 - Remove the garnish moulding with the gasket for the fixing clips from the affected side.

11 - Remove the rear hood.

12 - Pull the weatherstrip for the rear hood partly out of its retaining strip.

13 - Remove sealing compound from the welded seams of the panel being replaced.

Body repairs

1 - Curl up the gutter as far as the joint between the roof side member and instrument panel with the edge curling tool (VW 734 — local manufacture).

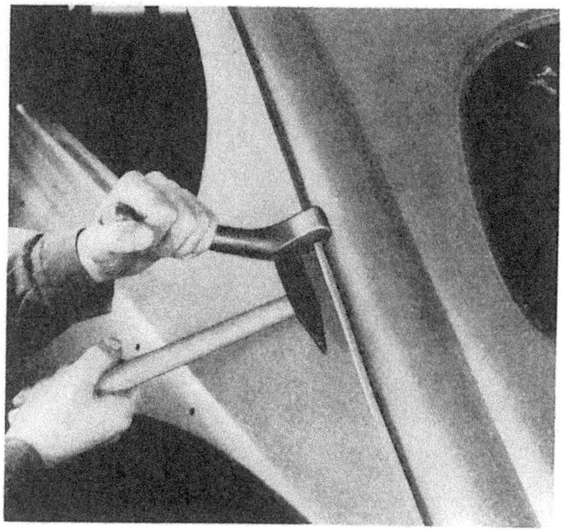

2 - Chisel the joint open between the roof side member and instrument panel.

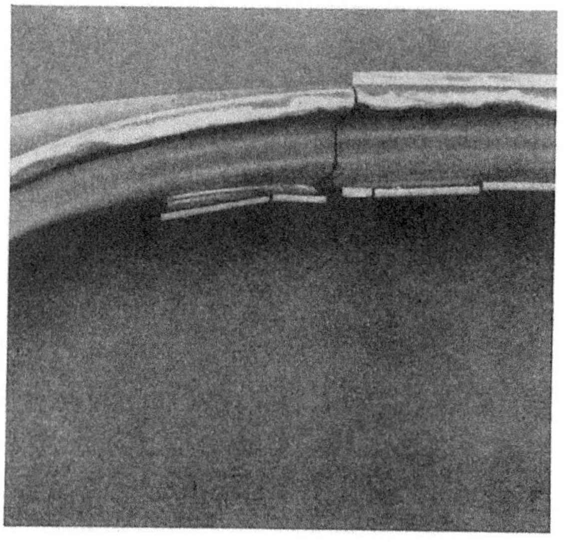

3 - Cut away the rear quarter panel with a chisel or oxy-acetylene torch at the following points:

Outside:

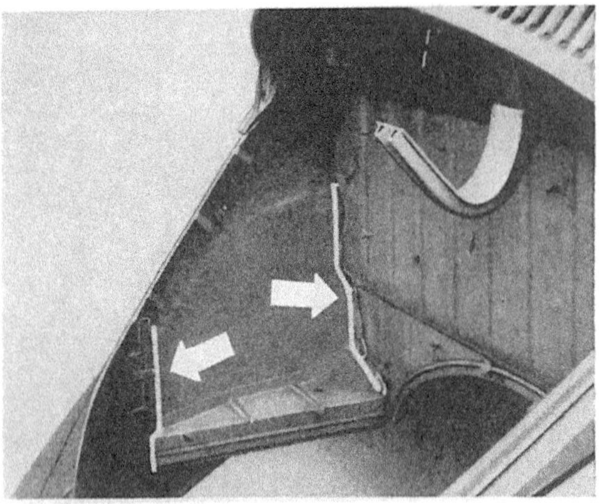

Inside (Engine compartment):

Inside (Passenger compartment):

4 - Remove rear wheel and shock absorber.

5 - Remove bolt of the body support with lock ring, washer and rubber gasket.

6 - It is advisable to begin cutting away the rear quarter panel at the tail-plate.

Cut away 10—15 mm (0.4"—0.6") above and below the spotwelded seams or the connecting plates will be damaged.

Note:

To simplify the task, the tail-plate can be chiselled away from the rear quarter panel. This renders subsequent tearing away of metal scraps from the connecting plate of the tail-plate unnecessary.

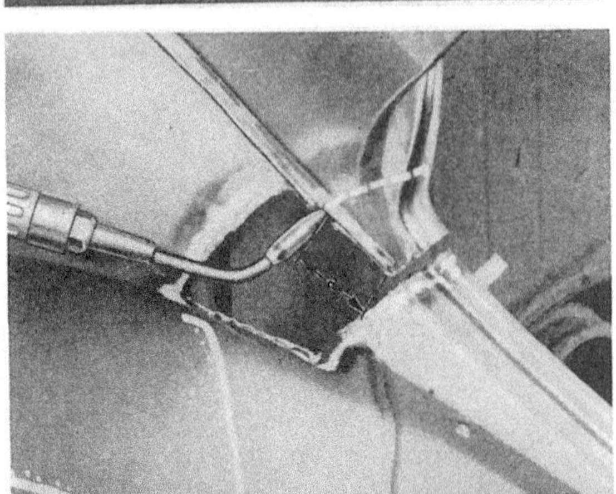

7 - Cut the rear quarter panel away from the end of the roof, about 50 mm (2") above the joint between the two parts. For this it is necessary to cut through one of the windows since the side panel must be cut away **underneath** the rear end of the roof.

The end of the roof must not be damaged.

a = ~ 50 mm (2")

8 - Cut away the quarter panel on the inside.

9 - Cut away the door lock pillar from the lower side member with a chisel.

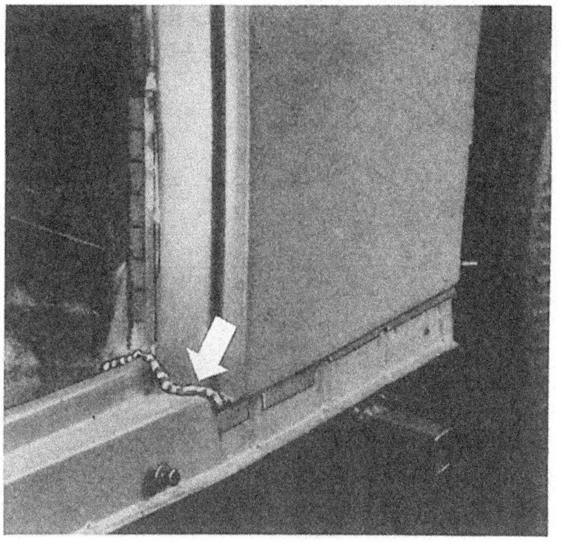

10 - Cut away the quarter panel from outside.

11 - Cut through the roof side member, either with a torch or hacksaw, about 40 mm (1.6") before the joint between the roof member and instrument panel.

The roof must not be damaged in doing so.

a = ~ 40 mm (1.6")

12 - Heat up the spot welds along the gutter with a torch and break the quarter panel away by bending to and fro.

Attention!

Wear protective gloves when removing or installing the rear side section.

13 - Remove with pincers any scraps of metal remaining from the quarter panel. Grind all spot welds smooth. Straighten connecting plates, rub down and prepare for installation of the new quarter panel.

Note:

When removing such metal, do not tear but shear away by twisting the pincers. Otherwise the spot welds will be torn out and leave a hole at that point.

14 - Grind all points at which the new side section is to be welded. If it is only possible to weld with oxy-acetylene, bore 4 mm (0.16") holes at intervals of about 150 mm in the underside of the flanged edge of the roof member.

15 - Place the new quarter panel in position and support it on the lower side member first. Then pull the top under the roof until the gutter engages with the bead in the upper side member.

Note:

To simplify installation, the lower recess for the studs in the quarter panel can be cut through and twisted up.

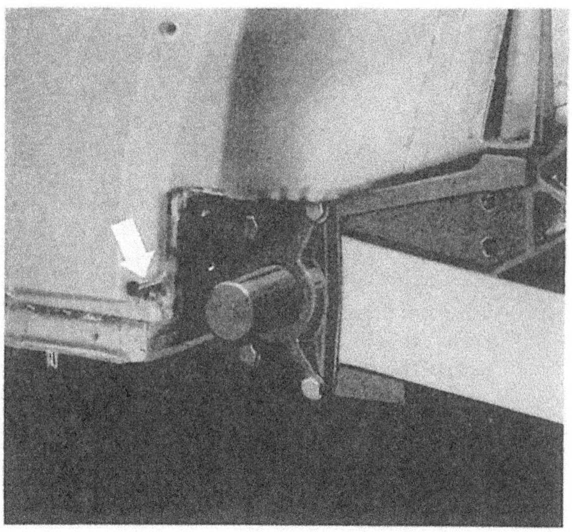

16 - Fit the quarter panel, align and prepare for welding.

a - Attach clamps to the roof side member.

b - Bolt the quarter panel to the frame as specified at the support.

c - Bolt the quarter panel to the fender fixing stud, after placing a nut and washer underneath.

d - Bolt the tail-plate to the quarter panel.

e - Apply two C clamps so that the quarter panel lines up uniformly with the lower side member and floor of the luggage compartment.

f - If necessary, apply an extra C clamp at the bottom of the lock pillar to press the quarter panel against the lower side member.

17 - Check that the door fits in the opening. If necessary, readjust the rear quarter panel.

18 - Lock pillar (outer panel) and then inner panel should now be welded at three points by oxy-acetylene to the lower side member. Spot weld the inner panel.

Note:

The inner and outer panels of a new quarter panel are not welded together at the bottom, thus simplifying installation and alignment with the lower side member.

19 - Weld the quarter panel to the lower side member outside.

a - **Electric**
Spot welds at intervals of 15 mm (0.6").

b - **Oxy-acetylene**
Beads 15 mm (0.6") long at intervals of about 20 mm (0.8").

20 - Weld the tail-plate with the quarter panel on the outside. With oxy-acetylene welding use beads 15 mm (0.6") long at intervals of 20 mm (0.8").

The joint between the tail-plate and the quarter panel should be welded with oxy-acetylene across the top.

21 - Inside the engine compartment weld the quarter panel to the tail-plate and the floor ofthe luggage compartment.

a - **Electric**
Spot welds at intervals of 15 mm (0.6").

b - **Oxy-acetylene**

Beads 15 mm (0.6") long at intervals of 20 mm (0.8").

22 - In the passenger compartment weld the quarter panel to the lower side member and floor of the luggage compartment.

a - **Electric**

Spot welds at intervals of 15 mm (0.6").

b - **Oxy-acetylene**

Beads 15 mm (0.6") long at intervals of 20 mm (0.8").

23 - With oxy-acetylene weld the roof lining to the roof side member at three points, and the roof member to the quarter panel at the top of the luggage compartment partition.

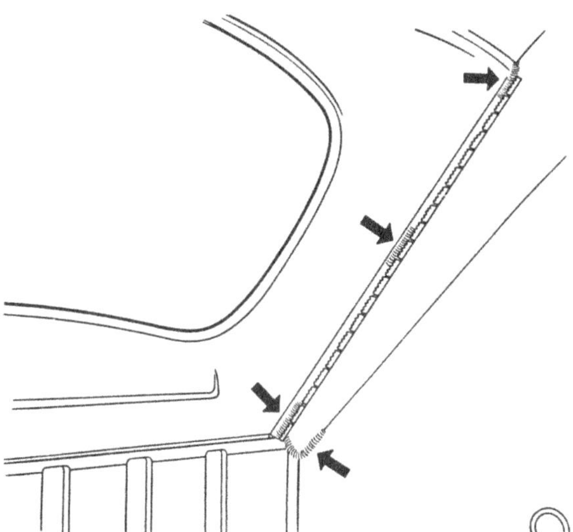

24 - Oxy-acetylene weld the lock pillar with the roof side member.

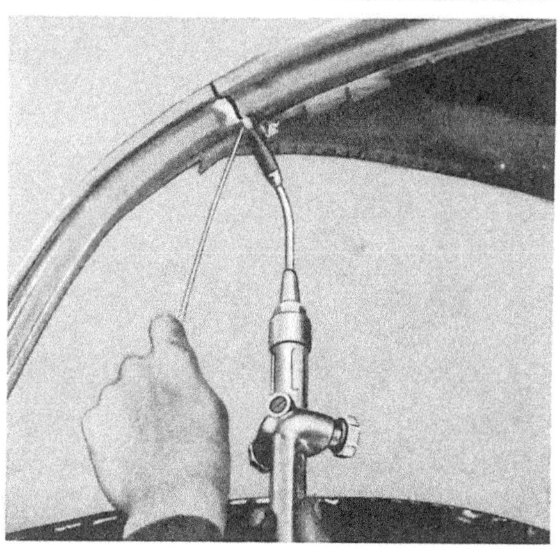

25 - Weld the joint between the roof member and instrument panel with oxy-acetylene.

On the Standard Model in particular this joint must be ground down with care as the roof side member is not covered.

26 - Weld the roof to the roof member at front and back.

a - **Electric**

Spot welds at intervals of about 100 mm (4").

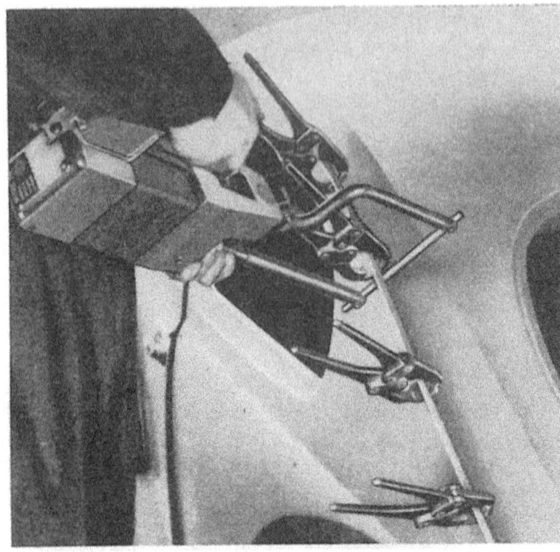

b - **Oxy-acetylene**

Weld holes on underside of gutter to roof using filler metal. Before curling over, grind or file welds smooth.

27 - Curl over the gutter with hammer and iron. It is best to work in both directions beginning at the lock pillar. Finally straighten out the gutter.

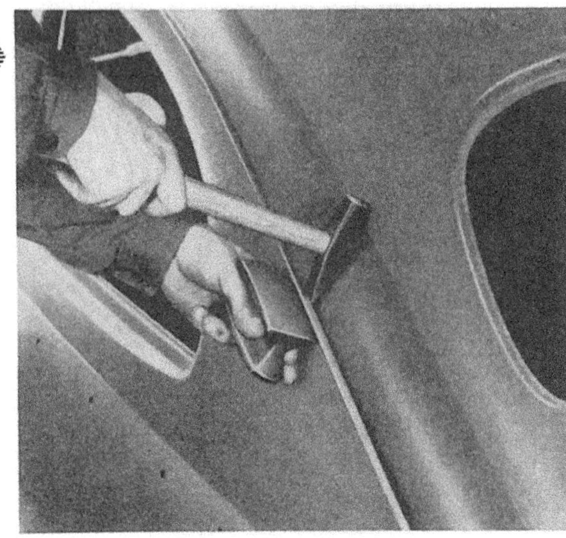

28 - Grind smooth all welds and treated parts.

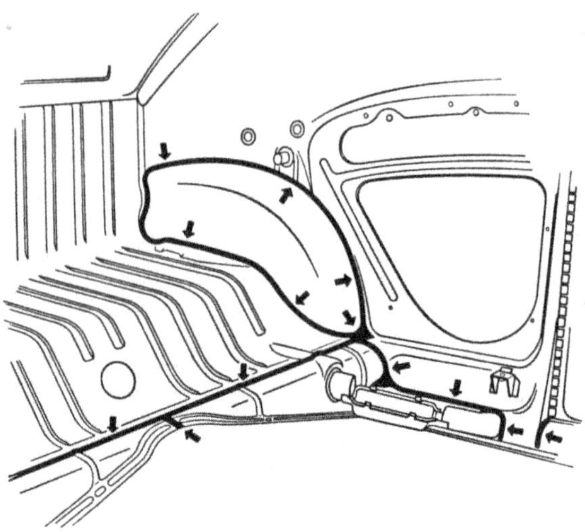

29 - Fill joints at the wheel house and lower side member with Sealing Compound D 9.

30 - Prepare the vehicle for necessary painting.

B - Replacement of individual parts of the rear quarter panel or roof side member

Individual replacement of the outer or inner panels, or the roof member, is only possible when the roof is removed. The necessary body repairs are carried out in accordance with the appropriate part of the previous section "Replacement of a complete rear quarter panel with roof member".

C - Replacement of parts of the inner or outer rear quarter panel

If the lower part of a side section is so badly damaged that it cannot be restored to its original shape, or to do so would be uneconomical, yet replacement of the entire quarter panel appears unnecessary, it is advisable to replace part of the outer panel.

1 - Remove the damaged outer panel at the points marked.

2 - Chisel open the three welded joints at the bottom of the lock pillar.

3 - Chisel the lock pillar away from the lower side member.

4 - Curl up the flanged edge between inner and outer panel.

5 - After removal, strip metal scraps with pincers.

6 - Grind smooth all cut edges and weld spots.

7 - Fit new outer panel and cut to shape. Take care that a 20 mm (0.8″) wide overlap is allowed for welding to the wheel house.

8 - The points on the quarter panel and on the new piece where the two are to be spot welded should be ground smooth.

9 - Place the new panel in position and clamp up.

····· hacksaw

- - - chisel

——— with chisel or oxy-acetylene torch.

10 - Check that the door fits properly in its opening. If not, readjust the new panel.

11 - Spot weld the quarter panel inside and outside the window frame at intervals of 20 mm (0.8″), if oxy-acetylene is used, with 15 mm (0.6″) beads at the same interval.

12 - Weld the quarter panel to the outside of the lower side member, either with spots or 15 mm (0.6″) beads.

13 - Weld the lock pillar to the lower side member with oxy-acetylene.

14 - Weld the new portion of the quarter panel at its edges after masking with damp asbestos wool, using oxy-acetylene.

15 - Weld the inner and outer panels to three points at the bottom of the lock pillar with oxy-acetylene; at the top with spot welds. Curl over and trim the flanged edge with hammer and iron.

16 - Grind down the metalwork, particularly weld spots, and prepare the vehicle for the necessary painting.

Reinforcement Plate

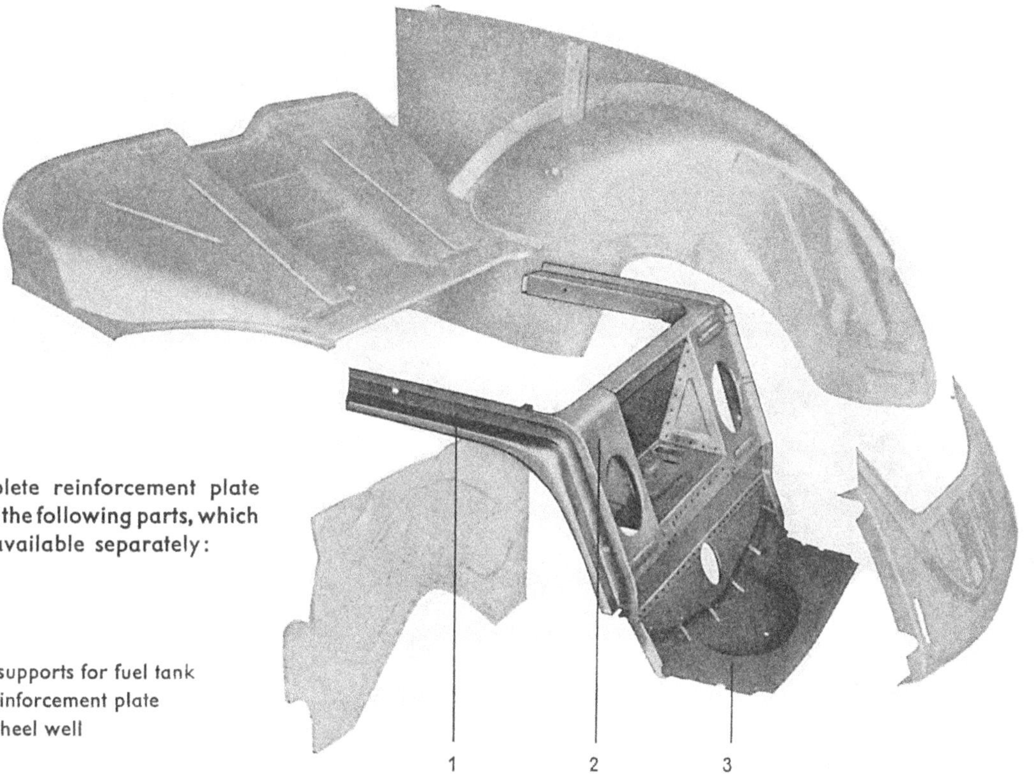

The complete reinforcement plate comprises the following parts, which are also available separately:

1 - Lateral supports for fuel tank
2 - Front reinforcement plate
3 - Spare wheel well

Note:

Since April 1959 the reinforcement plate for the front apron (Part No. 111 805 505) is available under Part No. 111 805 505 B, however without the 2 lateral fuel tank supports. These supports continue to be available as spares. The front apron is supplied under Part No. 111 805 583 without the cable holder.

Preparation

1 - Disconnect battery.

2 - Remove spare wheel and car jack.

3 - Remove front bumper and brackets, and both front fenders.

4 - Remove fuel tank.

5 - On De Luxe Sedan remove equalizing cylinder of hydraulic brakes.

6 - Remove the two hex. bolts attaching the body to the front axle.

7 - Disconnect the wiring to the headlights and horn, and lay them in the luggage compartment as far as necessary.

8 - Remove identification plate.

Note:

If, on a car whose reinforcement plate has to be renewed, the front side panels must also be renewed or straightened, the latter work must be completed before replacing the reinforcement plate.

Body repairs

1 - Cut away front apron.

2 - Cut away the reinforcement plate on either side about 10 mm (0.4") away from the connecting plates.

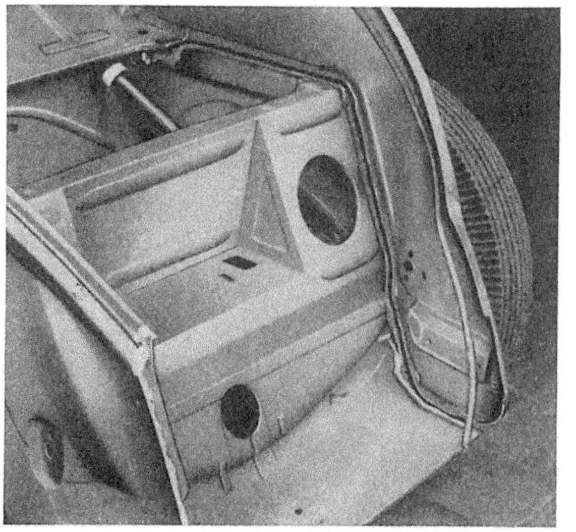

3 - Remove reinforcement plate.

4 - Strip off any remaining scraps of metal with pliers.

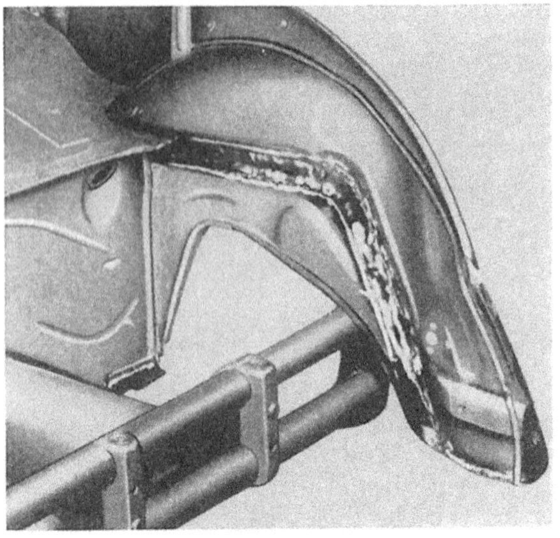

Note:

When removing such scraps the metal at the welded spots must not be torn away but sheared off by twisting the pliers. Otherwise the spot weld will be torn out too and will leave a hole at that point.

5 - File or grind down all welded spots. If necessary, straighten connecting plates and dented places.

6 - Grind all points clean at which the new reinforcement plate is to be welded.

7 - Insert new reinforcement plate, adjust and align. Apply clamps in the spare wheel well, bolt the lateral supports for the fuel tank to the luggage compartment and the reinforcement plate to the front axle.

8 - First spot weld the reinforcement plate in the sparewheel well and at the fuel tank supports, from the top. If it is only possible to weld with oxy-acetylene, use 15 mm (0.6″) beads at intervals of about 20 mm (0.8″).

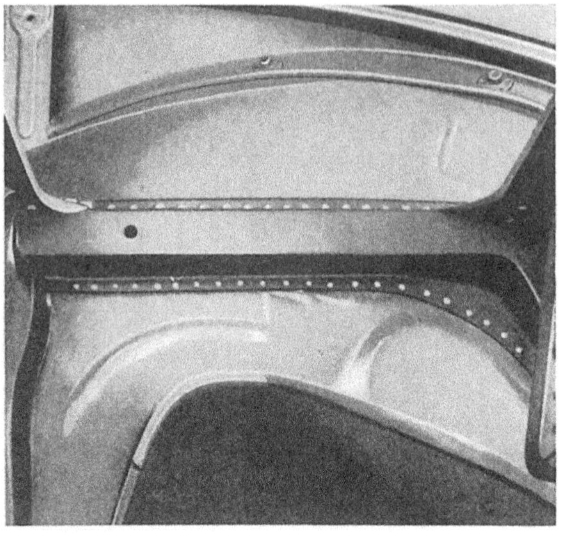

9 - Then weld the reinforcement plate and the fuel tank supports to the side sections from underneath.

10 - Insert the front apron and weld to the side panels.

11 - Grind down the metalwork, particularly weld spots, where necessary.

12 - Stamp the body number in the space provided on the new reinforcement plate.

13 - Prepare vehicle for necessary painting.

Attention!

After painting, attach the identification plate to the space provided on the reinforcement plate.

Scuttle with Instrument Panel

In the event of accidental damage to the bodywork including the instrument panel, it is necessary to insert a complete new scuttle with instrument panel, in addition to any other parts which may have to be renewed.

The complete scuttle comprises the following parts which are also available as spares:

instrument panel	=	blue
front luggage compartment	=	red
front partition	=	yellow

Note:

From April 1959 the scuttle with instrument panel is only available in the individual parts specified.

The following parts, available separately, are welded to the scuttle.

Instrument panel

a - Hinge-pillar reinforcements left and right with tapped plates for door hinges, held loosely in hinge straps.

b - Hook-type hinges with rivet studs for the lid of the glove compartment.

c - Welded to the top of the right and left hinge-pillar lining is a pivot for the hinge of the front hood.

Luggage compartment

d - On the De Luxe Sedan there are angle brackets left and right for fixing the cover plate in front of the instrument panel.

Preparation

Note:

From past experience it is usual for at least one front side panel to be renewed when accidental damage affects the scuttle. Therefore the following instructions will be based on the renewal of a complete scuttle when the reinforcement plate and right-hand side panel are removed.

1 - Take the complete body off the frame and place it on a framework suitable for use as a template.

2 - Remove both doors and front fenders.

3 - Remove all accessories fitted to the instrument panel including windshield. Remove linings at sides, front partition, and main cable harness with lighting wiring.

4 - Remove sealing compound from welded joints in the same area.

Body repairs

If the upper cowl and the upper part of the windshield frame are undamaged, it is recommended that, for the sake of simplicity, only the lower part of the instrument panel be replaced. This assumption was taken as the basis of the following instructions.

1 - Curl up the gutter on the side panel which has not to be renewed, using the deflanging tool VW 734 (local manufacture).

2 - Curl up the roof gutter to about 20 mm (0.8″) above defroster vents at the windshield on both sides.

3 - Saw through the inner side members of the windshield at the upper edge of the defroster vents. But take care not to damage the outer members, i. e. the roof.

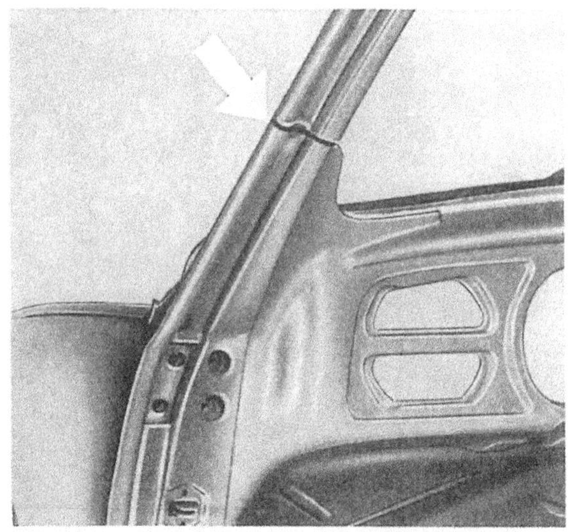

4 - Strip off the gutter with the side panel in position, but not beyond the cut above the defroster vent.

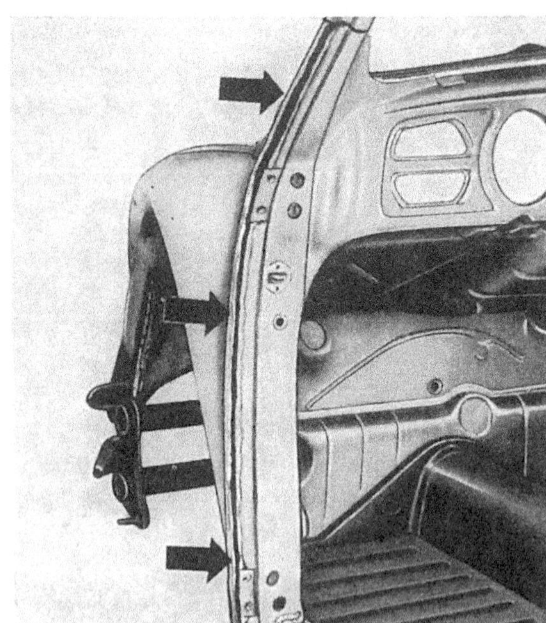

But take care that the side panel and heating pipe are not damaged.

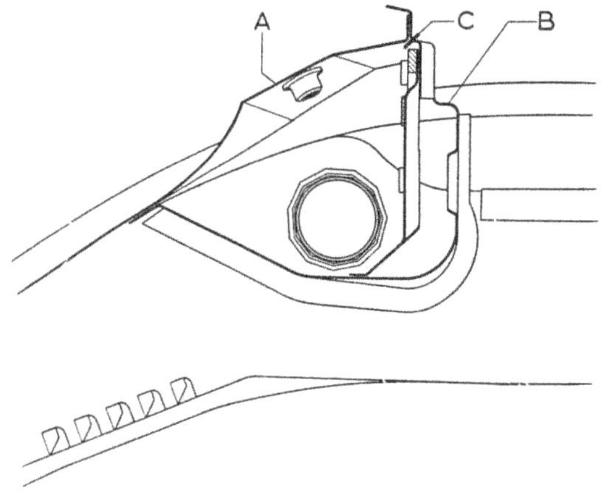

A - Front side panel
B - Instrument panel
C - Cut here

5 - With the side panel removed, the gutter should be stripped from the cowl to the cut above the warm-air outlet.

The cowl must not be damaged in the process.

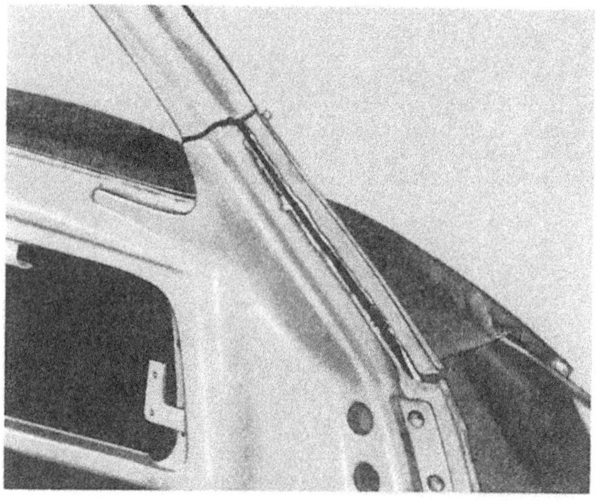

6 - Chisel open the window frame as far as the heating nozzles.

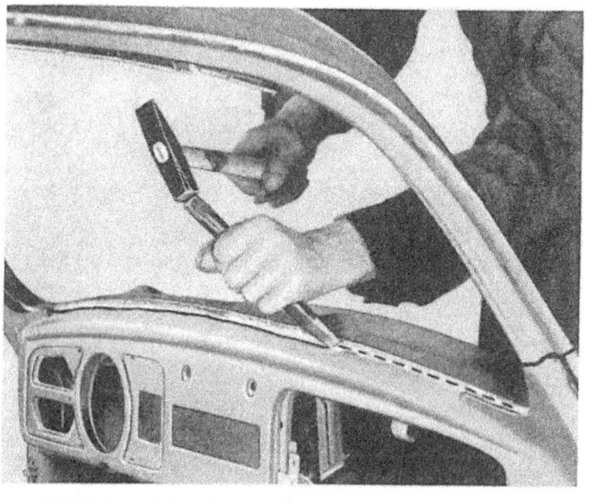

7 - Detach the frontal unit from the side section and lower side members around the front footrest.

When cutting the hinge pillars away from the side members be sure not to damage the heating pipes.

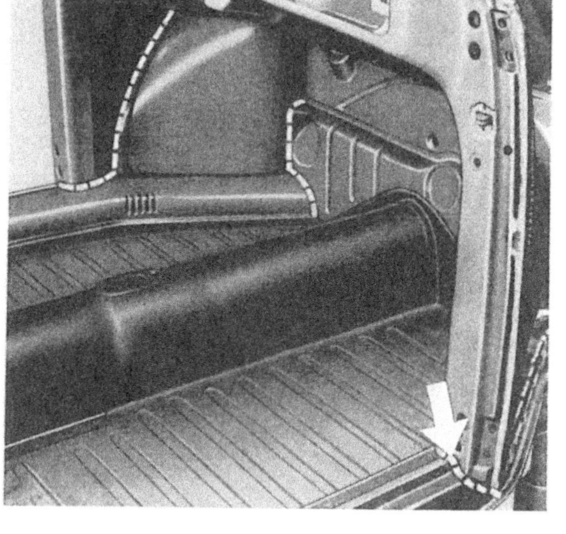

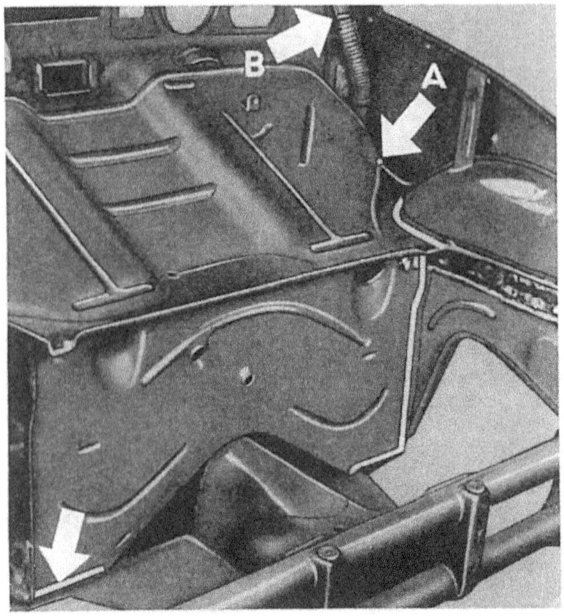

8 - Detach the scuttle from the side panel and side members in the front luggage compartment.

Remove the conduit for the front hood cable (A).

Remove the heating nozzles at the top of the hoses (B).

9 - To simplify removal of the complete scuttle it is advisable to cut openings in the left and right sides of the instrument panel in the vicinity of the heating nozzles. The entire scuttle can then be lifted and pulled off over the side members.

10 - Remove scuttle.

Attention!

Protective gloves should be worn when removing or inserting the scuttle.

11 - Remove any scraps of metal with pincers.

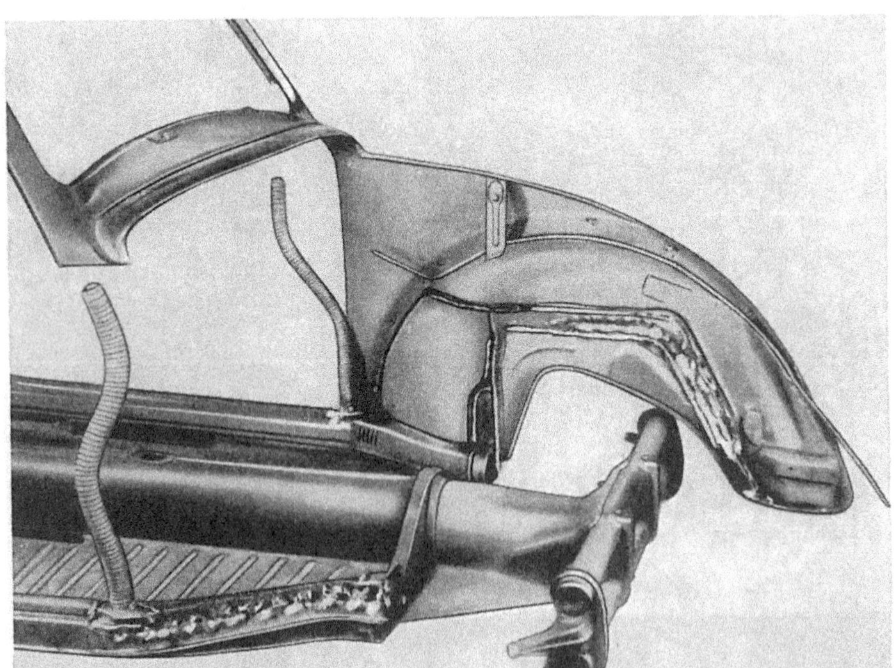

Note:

When removing such scraps, the metal at the welded spots must not be torn away but sheared off by twisting the pincers. Otherwise the spot welds will be torn out too and will leave a hole at that point.

12 - Grind or file all points which are to be welded. Where necessary straighten connecting plates and dents.

13 - Grind all points on the new scuttle which are to be welded.

14 - Cut off the side members of the windshield frame of the new unit above the warm-air outlet.

15 - Insert the new scuttle with instrument panel, adjust and align.

16 - Apply clamps to the windshield frame and side section.

17 - Fix the dimension of the door opening at the lower side member.

a = 945 mm (37.20")

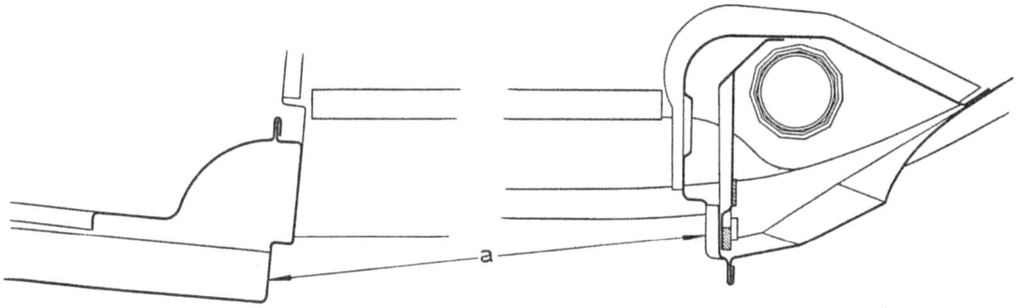

a = 945 mm

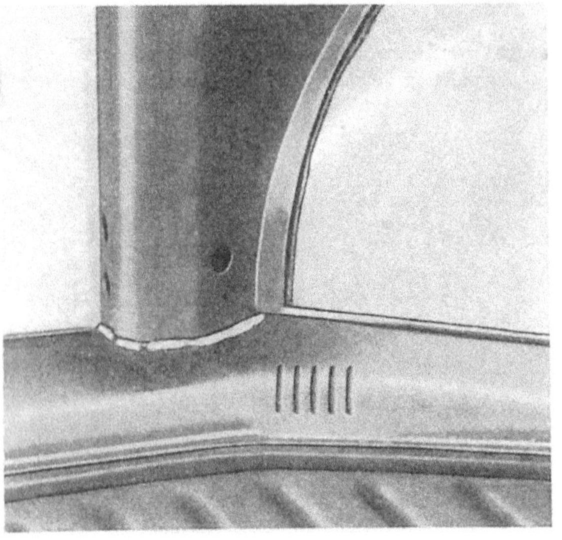

18 - Oxy-acetylene weld the hinge pillars to the side members.

19 - Spot weld the scuttle to the front side panels.

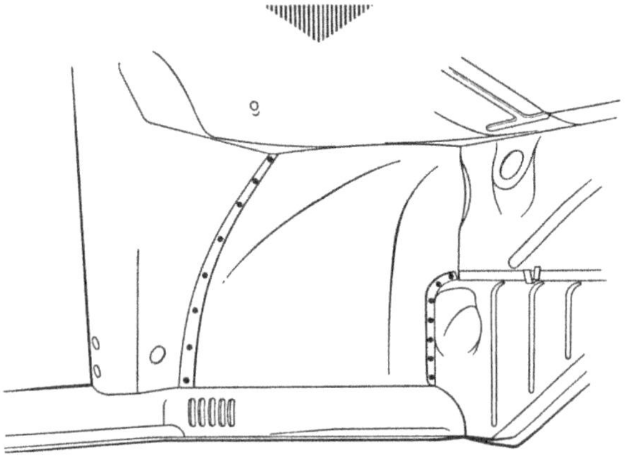

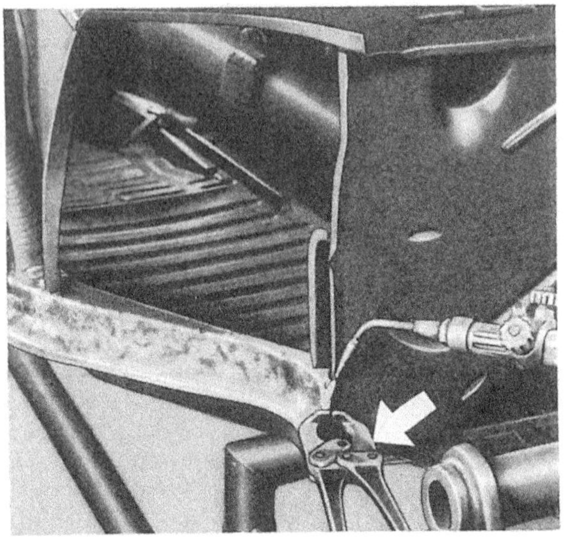

20 - Having clamped the front end, oxy-acetylene weld the front cross panel to the side member on the underside.

To weld the underside, the body must be slightly lifted from the framework.

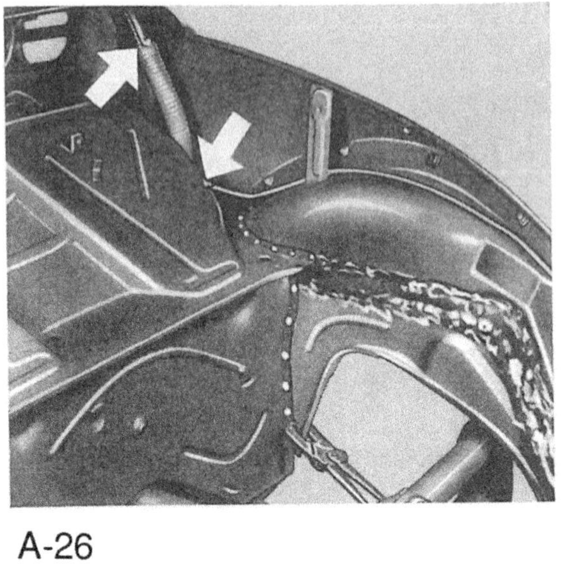

21 - Weld the scuttle to the front side panel in the luggage compartment.

Weld the heating nozzles and pipes to one another with a short seam.

Weld in the conduit for the front hood cable.

22 - Weld the instrument panel to the roof in the windshield frame. The joint on the inner side member of the frame must at all times be butt-welded by oxy-acetylene.

23 - Spot weld the instrument panel to the side section, with oxy-acetylene beads 15 mm (0.6") long at intervals of about 20 mm (0.8").

24 - Grind down spot welds and turn over the gutter of the old side panel with hammer and iron.

25 - Install the new side panel on the opposite side. Insert the reinforcement plate and front apron and weld them to the side panels.

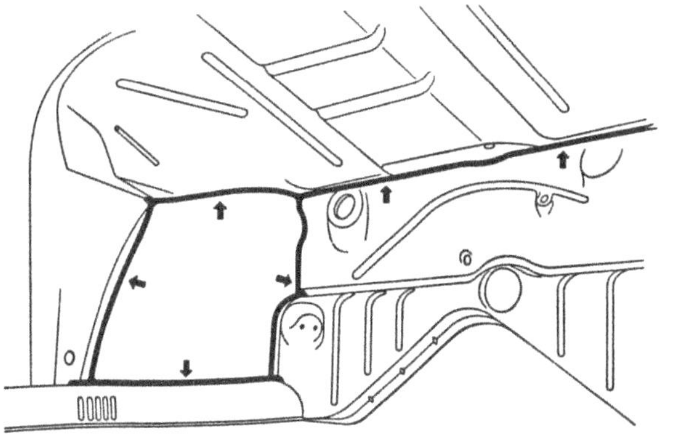

26 - Grind down points which have been welded or otherwise treated.

27 - Coat joints at wheelhouse and side members with Sealing Compound D 9.

28 - Prepare vehicle for the necessary painting.

Note:

To simplify stock keeping, the cowl and instrument panel unit (Part Nos. 113 805 021 C and 151 805 025 A) for Sedans and Convertibles from October 1952 (Chassis No. 1 - 397 023) up to August 1955 (Chassis No. 1 - 929 745) will be discarded as soon as they have run out of stock.

Henceforward, only one cowl and instrument panel unit will be delivered for Sedans (Part No. 113 805 021 D) and for Convertibles (Part No. 151 805 025 B) which can be used for all VW Passenger Cars from October 1952 up to August 1957. The same applies to the corresponding right-hand drive parts.

The only difference between the two versions lies in the mounting of the door check rods.

As the doors for VW Passenger Cars produced in the period from October 1952 up to August 1955 can only be used with the corresponding door check rods, the following modification should be carried out on the hinge pillars prior to installing the new instrument panel.

1 - Remove the two rivets from the door check rod mounting plate on the new instrument panel and take off mounting plate.

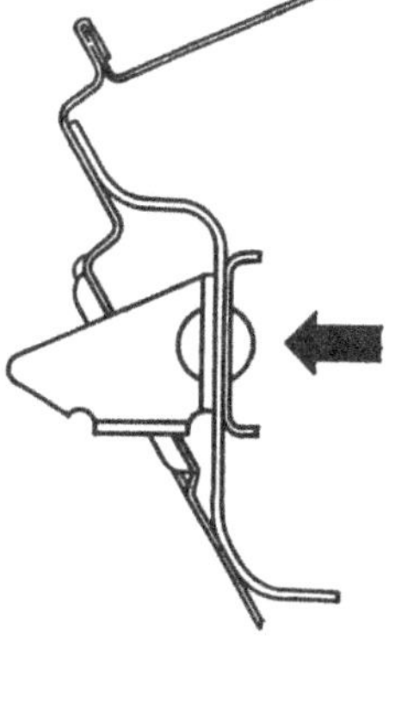

2 - Carefully cut out the recess for the mounting plate in the hinge pillar, as indicated by the dotted line.

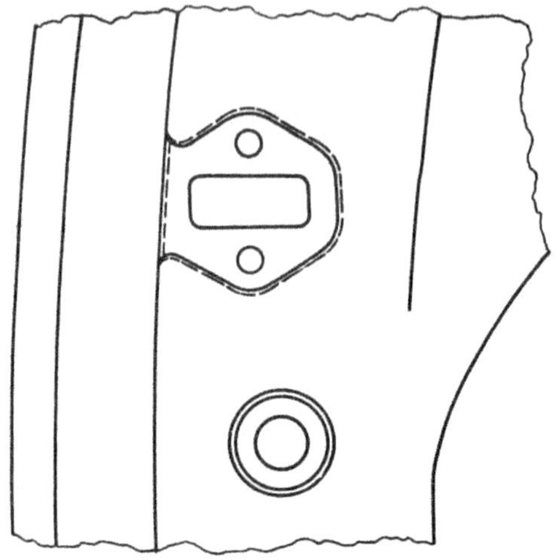

3 - Provide the inner side of the hinge pillar with two slotted holes according to the drawing to the right.

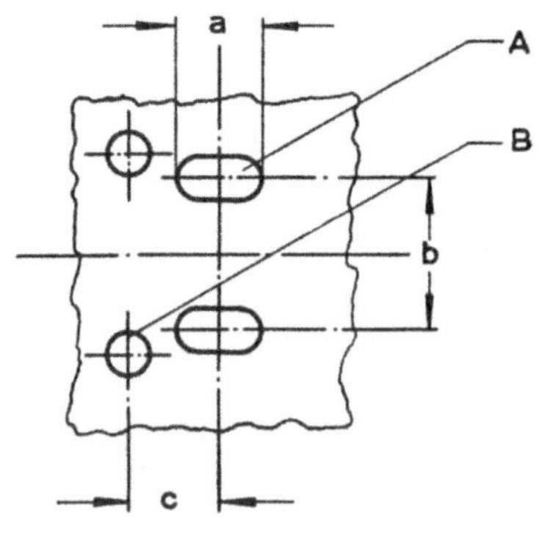

a - 12 mm (.472″) b - 21 mm (.827″) c - 13 mm (.512″)

A - New slotted holes
B - Holes for the rivets

4 - Remove the tapped plate for the door check rod bracket from the old instrument panel and attach it to the new one with a sheet metal strap.

Proceed in the same order on the other hinge pillar.

Replacement of the Apron

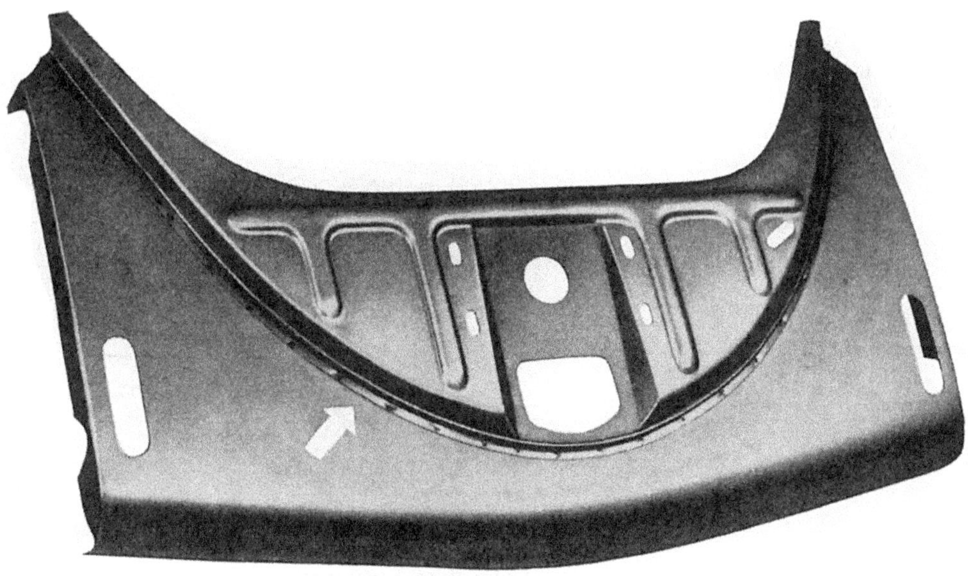

The retaining strip for the weatherstrip, which is welded to the apron, is also obtainable as a spare.

Preparation

1 - Disconnect battery.

2 - Remove spare wheel, car-jack and fuel tank.

3 - Remove front bumper with bracket and rubber grommets.

4 - Unscrew both front fenders.

5 - Remove horn.

6 - Disconnect cable harness for headlights and horn and lay in luggage compartment.

7 - Remove hood lock with operating cable.

1 - Cut away the apron at both sides. Remove the hood cable conduit from the apron.

When cutting away the apron take care **not** to damage the side panels

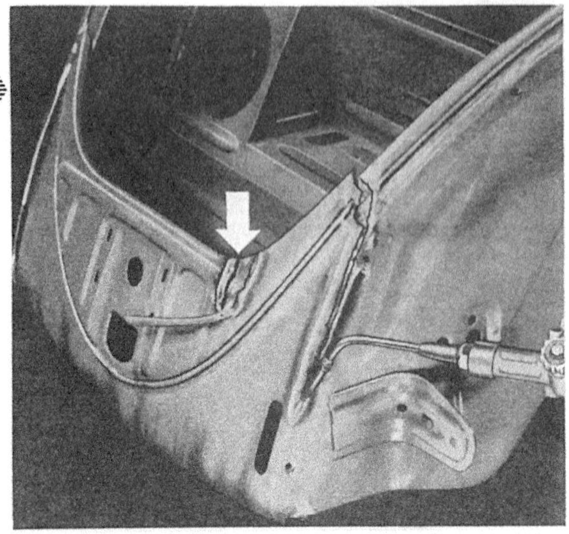

2 - Cut away the apron in the spare-wheel well about 10 mm (0.4″) in front of the joint with the well bottom.

3 - Remove apron.

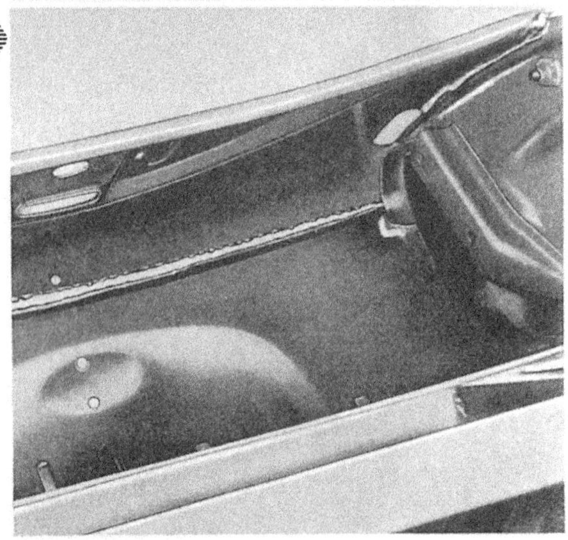

4 - Remove any scraps of metal with pincers. Grind down all weld spots. If necessary straighten connection plates, rub down and prepare for installation.

Note:
When removing such scraps, the metal at the welded spots must not be torn away but sheared off by twisting the pincers. Otherwise the spot welds will be torn out too and will leave a hole at that point.

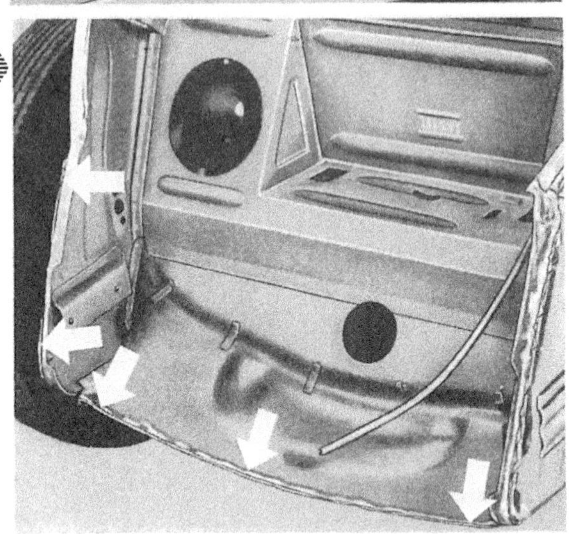

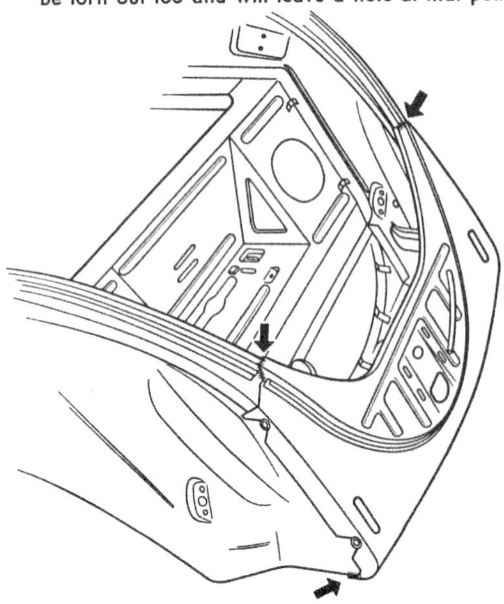

5 - Grind down all points on the new apron which are to be welded. Insert the new apron and bolt to the side panels.

6 - Weld the *top and bottom at either side of the* apron to the respective side panels, using oxy-acetylene.

7 - Spot weld the apron to the two side panels.

If only oxy-acetylene welding is possible, use beads 15 mm long at intervals of about 20 mm.

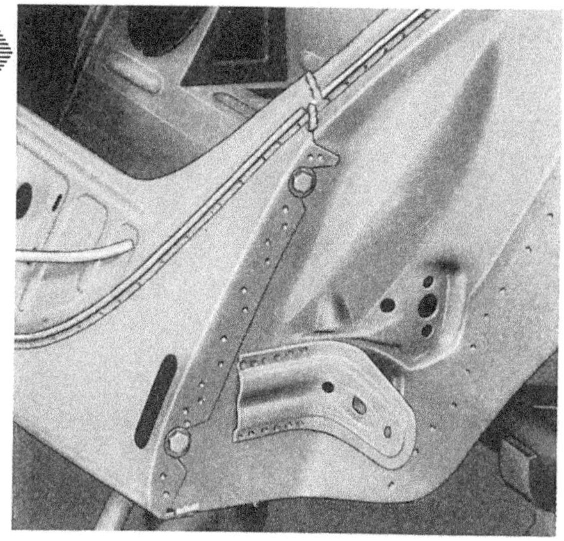

8 - Oxy-acetylene weld the conduit for the hood cable.

9 - Spot weld the spare-wheel well bottom with the reinforcement plate.

With oxy-acetylene welding use beads 15 mm (0.6″) long at intervals of about 20 mm (0.8″).

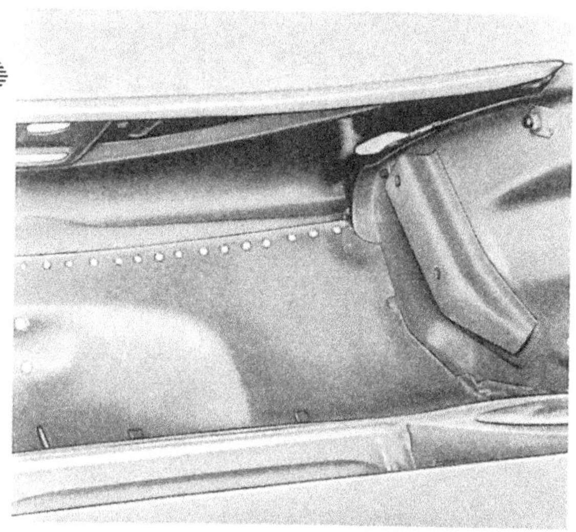

10 - Grind smooth parts which have been welded or treated otherwise and prepare the vehicle for necessary painting.

THIS PAGE INTENTIONALLY
LEFT BLANK

Replacement of the Tail-Plate

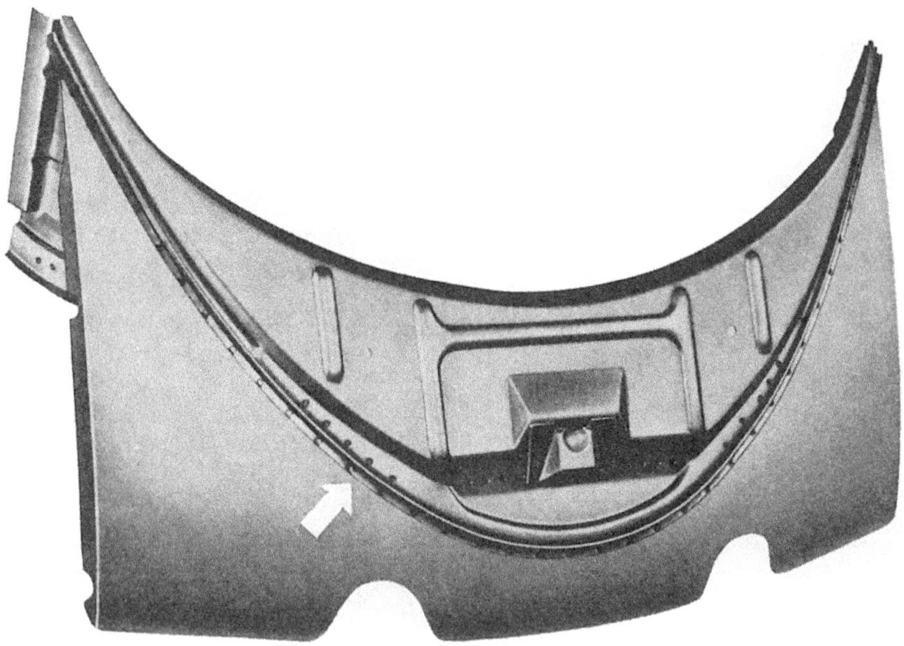

The retaining strip for the weatherstrip, which is welded to the tail-plate, is also obtainable as a spare.

Preparation

1 - Remove engine.

2 - Remove rear bumper with brackets and rubber grommets.

3 - Unbolt both rear fenders.

4 - Pull out the sound-absorbent lining at the side of the engine compartment, and remove the drain pipe. Pull out the weatherstrip of the rear hood.

Body repairs

1 - Cut away the tail-plate along the joint with the rear quarter panel and at the side exactly along the edge.

2 - Cut away the tail-plate in the engine compartment (arrow).

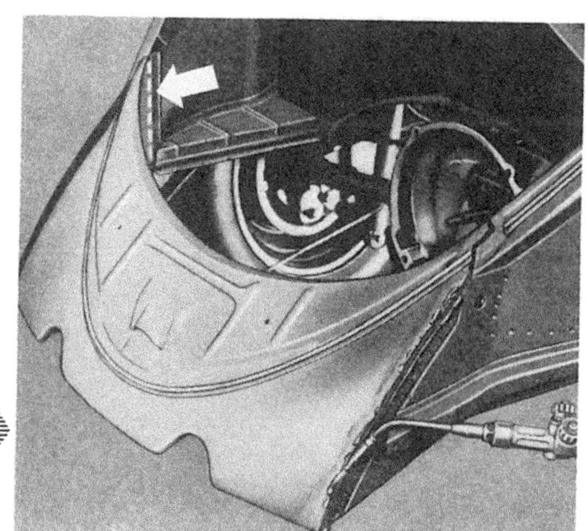

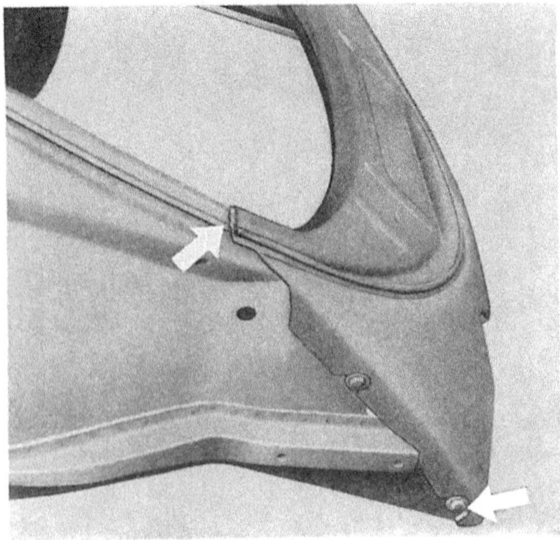

3 - Remove tail-plate and strip off any scraps of metal with pincers. Grind down all weld spots. Straighten connecting plates if necessary, rub down and prepare for installation.

Note:

When removing such scraps, the metal at the welded spots must not be torn away but sheared off by twisting the pincers. Otherwise the spot welds will be torn out too and will leave a hole at that point.

4 - Grind smooth all points on the new tail-plate which are to be welded.

5 - Insert the new tail-plate from above and bolt to the rear quarter panels.

6 - Weld the tail-plate with oxy-acetylene to the rear quarter panels at the top and bottom.

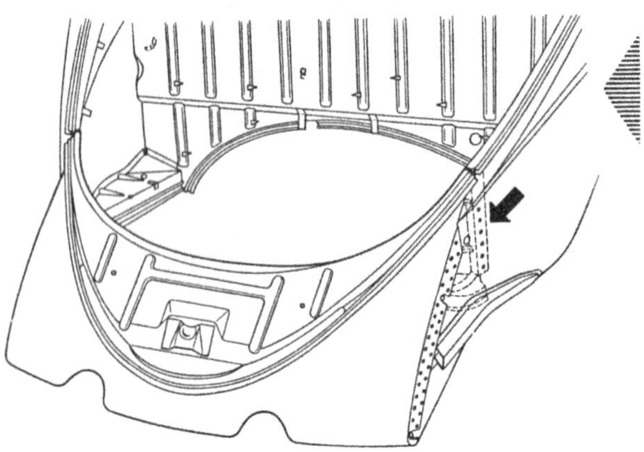

7 - Spot weld the tail-plate to the rear quarter panels inside and outside.

If only oxy-acetylene welding is possible, use beads 15 mm (0.6") long at intervals of about 20 mm (0.8").

8 - Points which have been welded or treated otherwise should be ground smooth and the vehicle prepared for the necessary painting.

THIS PAGE INTENTIONALLY
LEFT BLANK

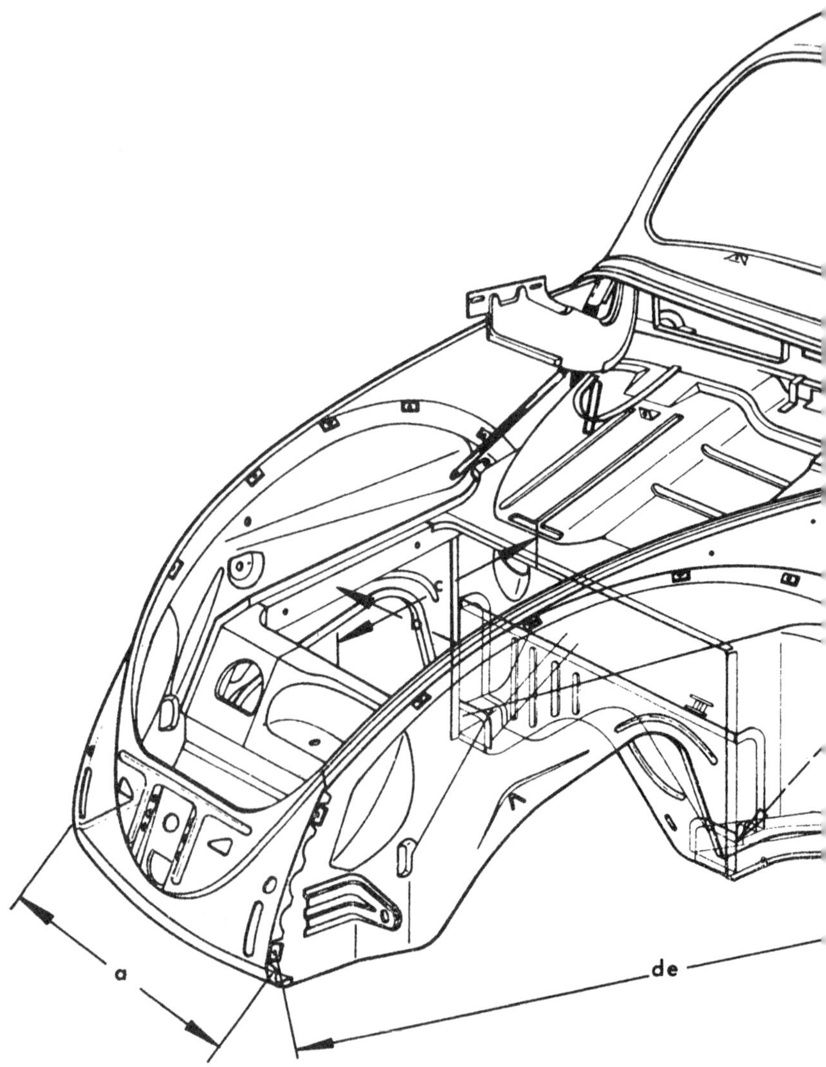

Basic Dimensions for Body Repairs

Dimen-sion	Location	mm	Remarks
a	Width of front apron at first fender securing nut	600	
b	Width of tank supports, side to side	693	Measured inside tank su
c	Width of tank supports, fore and aft	368	Measured inside tank su
d	Distance between first and last fender securing nuts on front side panel	1118	**Left-hand side of vehi** measured at hole centers
e	Distance between first and last fender securing nuts on front side panel	1120	**Right-hand side of veh** measured at hole centers
f	Distance from lower welding seam on hinge pillar to upper corner roof member/quarter panel	1394	**Left-hand side of vehic** agonal measurement of d
g	Distance from lower welding seam on hinge pillar to upper corner roof member/quarter panel	1394	**Right-hand side of v** agonal measurement of d
h	Distance between hinge pillar and quarter panel	935	**Left-hand side of ve** door opening at striker
j	Distance between hinge pillar and quarter panel	937	**Right-hand side of ve** door opening at striker

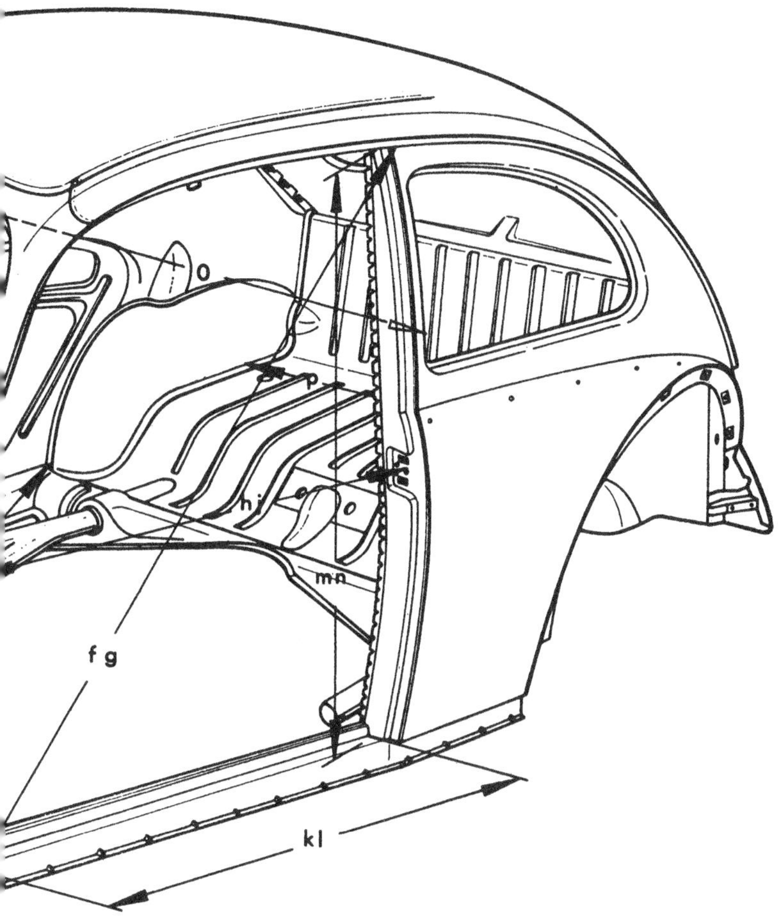

Location	mm	Remarks
:e between hinge pillar and quarter panel	924	**Left-hand side of vehicle,** door opening at side member level
.e between hinge pillar and quarter panel	923	**Right-hand side of vehicle,** door opening at side member level
:e between roof and side members at quarter panel	1049	**Left-hand side of vehicle**
e between roof and side members at quarter panel	1049	**Right-hand side of vehicle**
of body interior between spot-welded flanges of · panel window openings level with safety belt ages	1023	
of luggage compartment floor between wheel js along spot-welded seam	939	Measured at holes in luggage compartment floor
e between front corner side members/cross panel ır corner side member/wheel housing	1989	Diagonal measurement of interior, **left front - right rear,** measured on side member
:e between front corner side member/cross panel ır corner side member/wheel housing	1985	Diagonal measurement of interior, **right front - left rear,** measured on side member

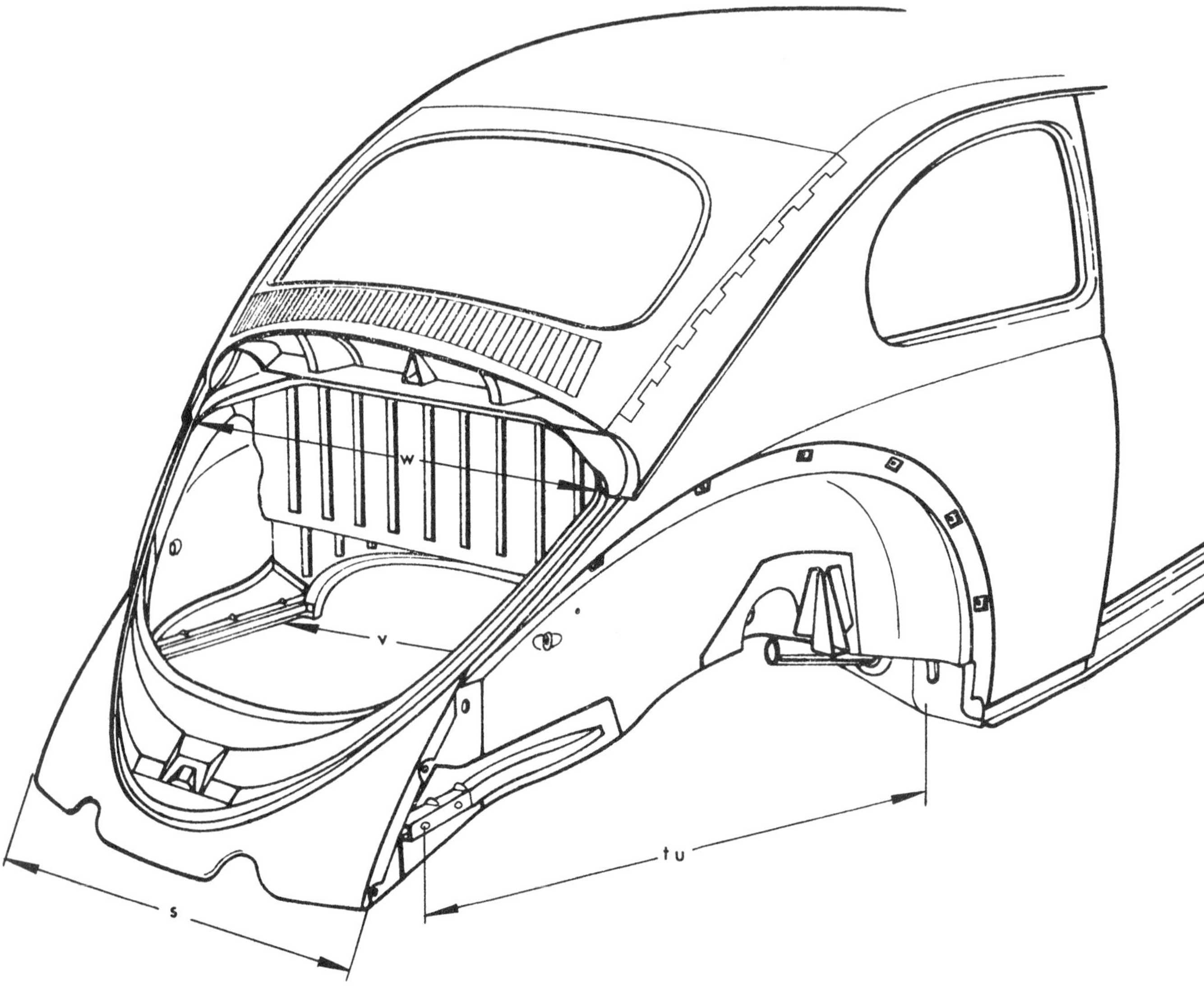

Basic Dimensions for Body Repairs

Dimension	Location	mm	Remarks
s	Width of rear apron at lowest fender securing nut	748	
t	Distance between cross member and rear hole for bumper bracket screw	1133	**Left-hand side of vehicle,** measured to center of hole
u	Distance between cross member and rear hole for bumper bracket screw	1135	**Right-hand side of vehicle,** measured to center of hole
v	Distance between right and left engine cover plates on front engine cover plate	748	
w	Distance between right and left side panels in engine compartment on upper radius of hood opening	909	

Spot Welding Pastes and Paint

A - Spot welding paste

In order to obtain water-tight joints, the parts of metal sheets which are to be welded are given a coating of sealing paste which is not affected by welding.

This standard manufacturing procedure should also be adopted for bodywork repairs.

The paste can be obtained direct from the manufacturers:

Manufacturer	Designation
Bonaval-Werk, Bonn, Germany Brühler Straße 2—20	Spot welding paste 59852 or 60506
Teroson-Werke GmbH, Heidelberg, Hans-Bunte-Straße 4, Germany	Spot welding paste 2257

The paste is applied before welding commences. Before and after welding the paste should not run. At the joints it should form a film which adheres firmly, prevents corrosion and renders the joint water-tight. It must also remain unaffected by degreasing agents and subsequent painting.

Paste which burns during spot welding should cease to do so as soon as the heat is removed.

The paste should be employed where water-tight seams are stipulated and where it is impossible or difficult to apply sealing compound after welding.

B - Spot welding paint

Spot welding paint is used to protect hollow parts against corrosion, which cannot be painted after welding due to inaccessibility.

Manufacturer	Designation
Teroson-Werke GmbH, Heidelberg, Hans-Bunte-Straße 4, Germany	Spot welding paint 2273

Before welding, spray or paint all components with spot welding paint.

General

The use of the above products does not simplify the welding process. It may be necessary to step up the welding current, depending on the thickness of the coat applied.

Superfluous material must, however, be removed before the part is given a coat of paint which must stand up to the conditions stipulated for the outside finish. Neither welding paste nor paint can be considered as a substitute for primer to which a top coat can be applied.

Note:

From Chassis No. 2 533 000 (6th August 1959), the heating efficiency has been improved by enlarging the flexible front heating pipes from 32 mm to 38 mm (1.26"—1.50") and increasing the size of the openings at the side members.

The service installation of new type flexible heating pipes (Part No. 111 255 455 C) in vehicles of previous make can only be carried out by using front heating tube adapters (Part No. 111 255 467). When installing new side members with larger outlets for the front heating pipes in previous vehicles, i.e. vehicles with earlier type heating pipes, the adapters must be used as well.

Heating pipes of previous design will also be available in the future.

When existing stocks have been used up, only new type side members will be supplied.

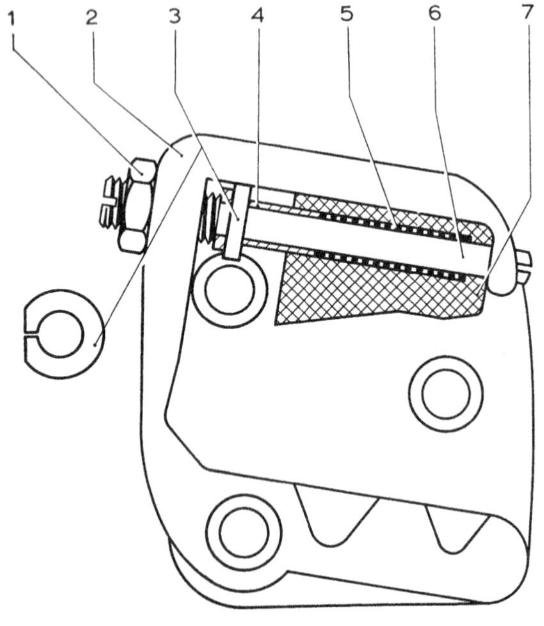

Note:

To eliminate noises in the door lock, a rubber washer (Part No. 151 837 177) was installed between the shoulder on the adjusting screw stop sleeve and the plastic wedge from Chassis No. 3 340 032 (30th September 1960).

1 - Lock nut
2 - Striker plate housing
3 - Rubber washer
4 - Stop sleeve
5 - Spring
6 - Adjusting screw
7 - Plastic wedge

When installing the rubber washer in older vehicles, cut a slit in it and push it over the adjusting screw.

The washer can be installed in the De Luxe and Standard models.

This section deals only with the assembly and body repair operations which differ from those on the Sedan.

Contents: A—Convertible

Assembly Work

Replacement of Body Parts

THIS PAGE INTENTIONALLY
LEFT BLANK

Apart from the top and minor points described here, the construction, shape and trim of the Convertible is identical to that of the Sedan.

To compensate for the slightly reduced strength even when the top is closed the body of the Convertible is reinforced at the following places:

a - In the front luggage compartment by means of cowl reinforcement plates welded to the upper rear part of the side panel, the instrument panel and the luggage compartment panel at each side.

b - At the windshield frame by the extended and channel-shaped warm air jets.

c - In the front compartment by means of lower cowl reinforcement plates spot welded to the hinge pillar, the side member, the front partition and the luggage compartment panel on each side.

d - Between the front side panel and the frame side member by means of corner plates welded on each side.

e - Between the quarter panel and the frame side member by means of corner and reinforcement plates welded on each side.

f - In the engine compartment by reinforcement plates welded to the luggage compartment rear panel and the quarter panel on each side.

g - In the engine compartment by a reinforcement plate for the rear body bow which is welded to the luggage compartment rear panel, the rear hood hinge carrier plate and the stiffener plates on each side.

h - Between the quarter panels by a welded-in cross member with corner plates which also acts as a support for the rear seat.

i - At the side members by reinforcement channels welded underneath.

Top

The top is supported by the top linkage which is composed of two side metal roof frames and two hinged linkages joined together by wooden and tubular bows which give cross support. The whole assembly is supported by two main hinges mounted in brackets bolted to special pillars in the body.

The outer cover is made of PVC material or special waterproof top material which is rubberised underneath. A rubberised hair pad, sewn into a linen sheet and secured to the top linkage, pads out and shapes the top in addition to providing insulation against heat and cold. The contours of the top are, where necessary, evened out by additional wadding. The underside of the top linkage is covered by a headlining of perforated leatherette.

In the folded position the top is retained by two spring catches.

Doors and Windows

Instead of a complete top frame the doors are fitted with a chrome-plated frame for the vent wing which also serves as the front guide channel for the door window. The door windows can be fully lowered and have a chrome-plated frame on the top and rear edges. The window are guided by sliding shoes on adjustable brackets located on the lock side of the door.

The quarter windows are of the drop type with chrome-plated frames and can also be fully lowered.

The inner and outer window slot weatherstrips are held by chrome-plated mouldings and clips.

Hoods

Cooling air for the engine enters through a series of horizontal slots in the rear hood.

Water which passes through the slots is diverted by a plate inside the hood to two drain pipes and runs away outside the weatherstrip at the lower edge of the hood.

The knob for the front hood cable is fitted with a lock which can be locked by the door key.

The glove compartment lock has a separate key.

Seat Trim

The wearing surfaces of all seats are of air-permeable leatherette and the side and back parts are of leatherette.

Additional Trim

1 - A door pocket on the passenger's side

2 - Protective corners on the rear fenders.

3 - Trim rings for the wheel rims.

Care of Top

The protective boot should always be fitted when the top is open to stop the top from flapping in the wind which will cause tearing and friction marks and spoil the appearance of top and headlining.

Top catches which do not hold the top tightly should be replaced or straightened. If the top is not held firmly, this will also cause friction marks.

Care of the PVC Top

The special PVC top cover on the Convertible does not require any particular care. It is, however, important to clean the top promptly and regularly. When very dirty, the top can be cleaned with a soap powder solution or one of the normal PVC cleaners. A hard brush will help to remove dirt from the grained surface of the material but care must be taken at the rear to avoid scratching the paint with the bristles. The top should be rinsed thoroughly with clear water after washing.

After driving for a long time on dusty roads, the top should be wiped clean before being lowered otherwise the sharp dust particles will damage the surface of the PVC material and cause friction marks.

Spots in the top material must never be removed with paint thinner, chlorine based spot removers or similar solutions as this will damage the material.

Stubborn spots can be removed by wiping with a cloth moistened with white spirit (benzine) and then washing thoroughly with a lukewarm soap powder solution.

Care of Textile Top

The appearance and life of the top depends on the care and treatment it receives. The top must never be lowered when wet. When dusty, it should be beaten lightly and brushed with a soft brush in line with the lay of the material, as the sharp particles of dust damage the material and cause friction marks and other damage.

Never use benzine, benzole, spot removers or other solvents to remove spots as these fluids will destroy the rubber layer in the top cover and cause leaks and shorten the life of the material.

The top should not be washed with the automobile shampoos used frequently nowadays. Some of these compounds attack the impregnation of the material and can cause leakage even after one application.

Marks which cannot be removed by just brushing can often be eliminated by rubbing the top, when dry, with a soft eraser and then brushing with a soft brush. This treatment avoids too frequent washing with soap solutions and the premature bleaching of the top material which this causes.

The top should only be washed when very dirty. Remove all loose dust first by beating lightly and brushing. Only good quality soap flakes or powder will give the best cleaning results with the least possible effect on the material. Dissolve 2 dessert spoonfuls of soap compound in a gallon of clear water which is free of any chemical additives. Moisten the top with clear water and apply the solution with a soft brush, rubbing in one direction only. Then rinse the top with clear water and brush at the same time. If necessary, scrub with the soap solution again. The final rinsing should be continued until all traces of soap are removed and the water runs off quite clear. The top must be left up until dry. The soap suds should then be washed off the vehicle paintwork with clear water and the surface dried with a leather.

Lubrication

A few drops of oil should be applied to the hinge points of the top linkage when required. Wipe the dirt and dust off the joints before oiling.

Door Window Removal and Installation

(up to 1965)

Removal

1 - Press window regulator handle and inside door handle escutcheons against the door trim panel until the dowel pin is visible. Knock out the pin with a punch and remove window regulator and door handles and escutcheons.

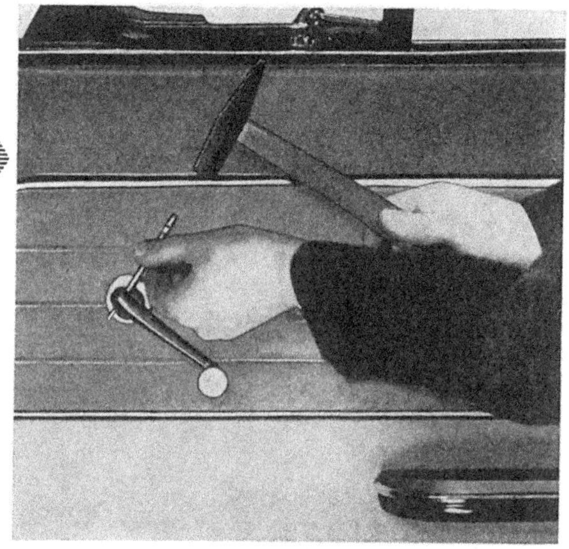

2 - Remove the door trim panel by pulling the clips out of the door. Care must be taken to avoid damaging the paintwork.

When the door trim panel clips on the passenger's side have been pulled out of the holes in the door, the panel must be lifted until the arm rest retainer on the back of the panel is hooked out of the arm rest support bracket.

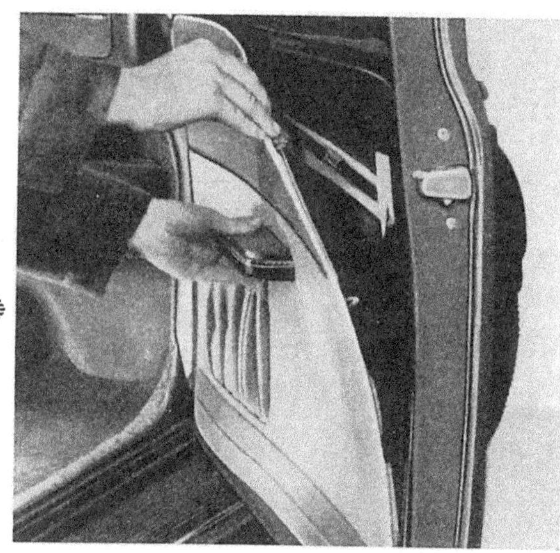

3 - Remove escutcheon springs and the rubber inserts from the window regulator and door handle shafts.

4 - Remove the upper securing screw and take off the arm rest support bracket on the passenger's side.

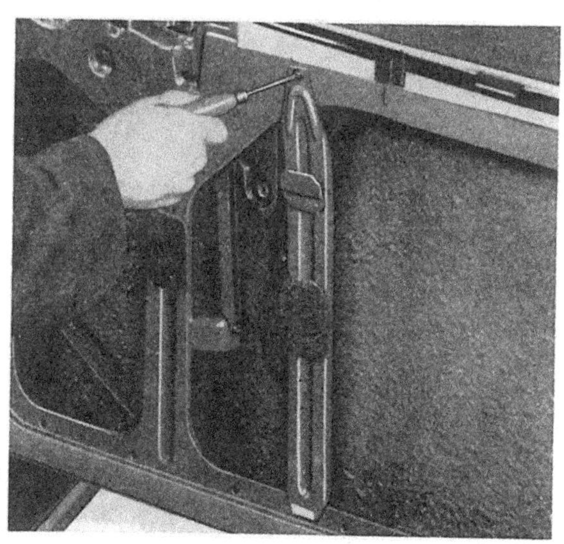

5 - Remove the lower screw from the front window run channel.

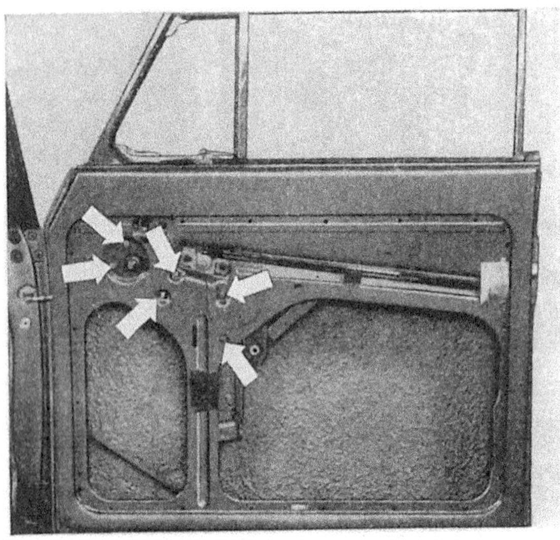

6 - Remove the six screws retaining the window regulator.

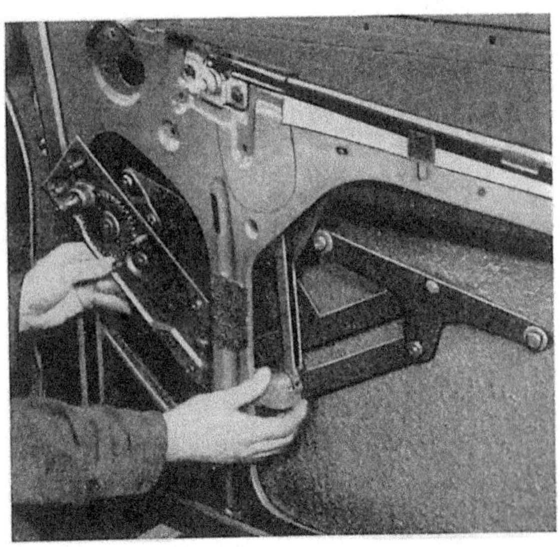

7 - Press the window regulator inwards and push the window lifting linkage towards the lock until the two ball heads in the window lifting channel are free.

Lower the window regulator and take it out of the door.

8 - Loosen the top retaining screw of the rear window run channel.

9 - Lift the window up and out of the door.

10 - If necessary, remove the window rear run channel after taking out the upper and lower screws.

The screws securing the window rear run channel are screwed into captive nuts which can slide sideways. This enables the window to be adjusted by moving the run channel as required.

11 - The window front run channel is cemented in position and can be replaced when the vent wing frame has been removed (see "Vent Wing removal and installation").

12 - The inner and outer window slot weatherstrips are secured by the trim moulding clips. To remove or renew the weatherstrips and trim mouldings lift the clips out with a screwdriver.

13 - Damaged door window frames with weatherstrips and window lift channels with rubber inserts can be replaced.

The window frames and lift channels can be removed by tapping gently with a hammer and a suitable block of wood. They are replaced in a similar manner, taking care that the rubber strips are seated properly.

Installation

The following points must be observed when installing the door window:

1 - Lubricate all the moving parts of the window regulator and linkage, including the rear guide channel, with VW Universal Grease - A 052 before installation.

2 - Insert upper and lower screws retaining the window rear run channel and leave loose so that the channel can be moved as necessary.

3 - When inserting the window in the run channel take care that the window slot weatherstrip is not damaged.

4 - Install window regulator.

5 - Tighten window run channel screws.

6 - Check the window for ease of operation. If operation is stiff, loosen the run channel screws and move the channels sideways as required. Tighten screws again.

7 - The door window must be adjusted sideways so that it aligns with the drop window from front to rear and rests evenly on the drop quarter window weatherstrip.

The side adjustment is carried out by loosening the upper screw securing the window rear run channel.

8 - If the window tilts when being raised, loosen the adjusting nut on the regulator linkage.

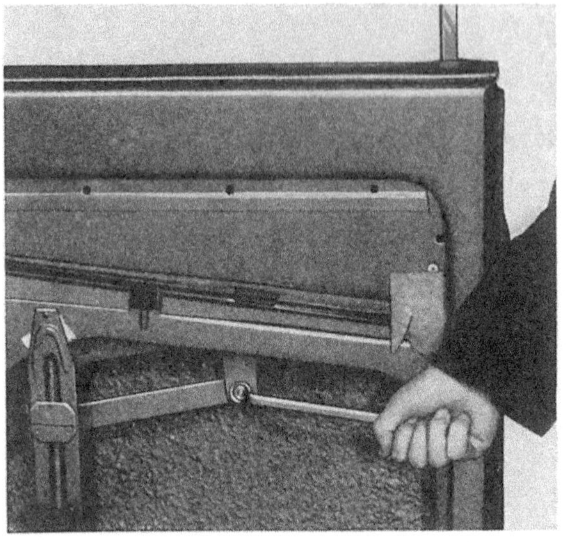

Push the window towards the front run channel and tighten the nut again.

9 - The height adjustment of the window is controlled by a stop on the window regulator which rests against an adjusting screw behind the door inner panel (arrow).

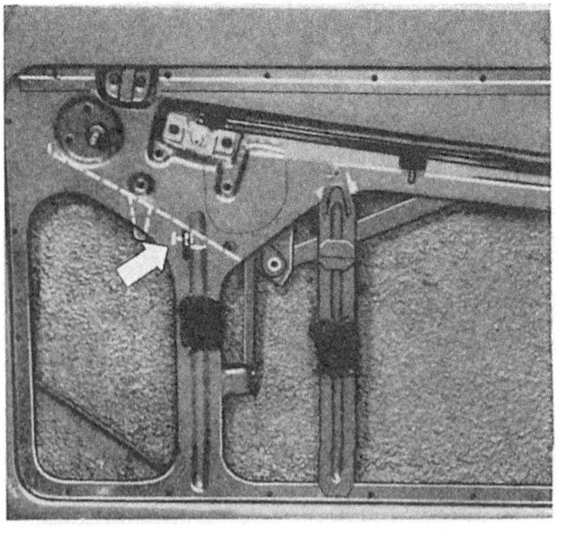

The height is adjusted by turning the adjusting screw after loosening the lock nut which is accessible through the slot in the door inner panel.

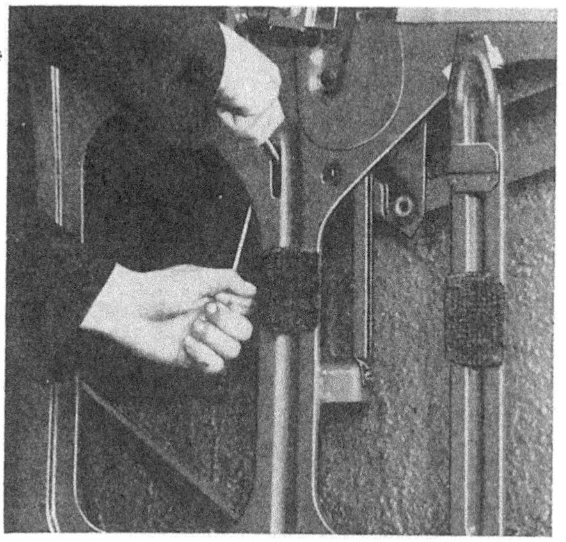

The door window must be in line with the top edge of the vent wing.

10 - Before installing the door trim panel check that the anti-rattle strip and felt pads are secure and correctly located.

11 - Install rubber inserts, escutcheon springs and door trim panel. The large diameter of the springs must face the trim panel.

12 - Install door inner handle and window regulator handle in the correct positions.

13 - Check that all parts are working properly by operating several times.

THIS PAGE INTENTIONALLY
LEFT BLANK

Door Window Removal and Installation

Removal

1 - Remove vent wing with frame.

2 - Remove three hexagon head screws for window lifter.

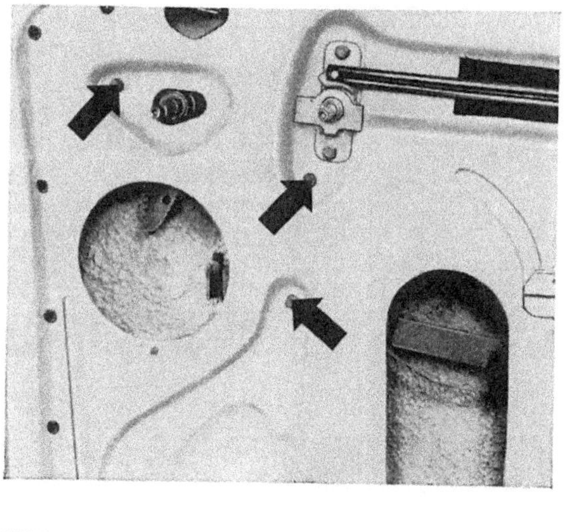

3 - Press window lifter inwards and slide it towards door lock until the two plastic guide pieces can be pressed through the recesses in the window lift channel.

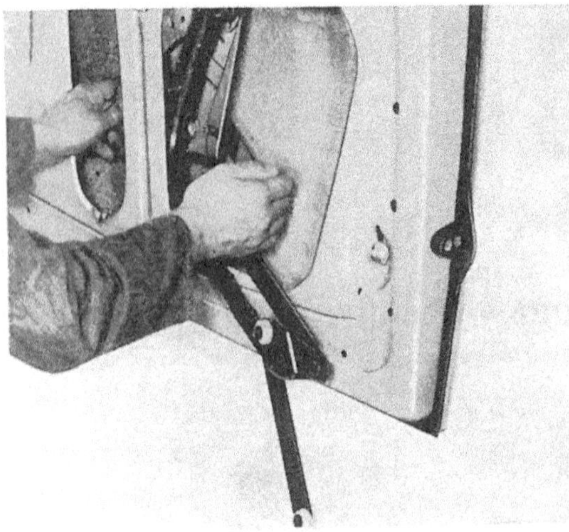

4 - Take window lifter out of door downwards.

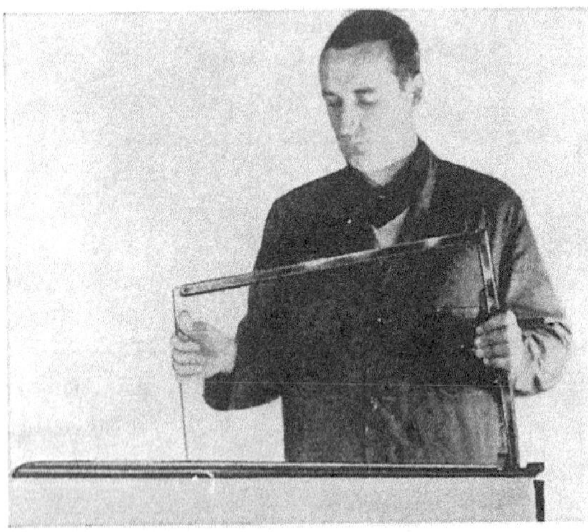

5 - Take window glass out upwards.

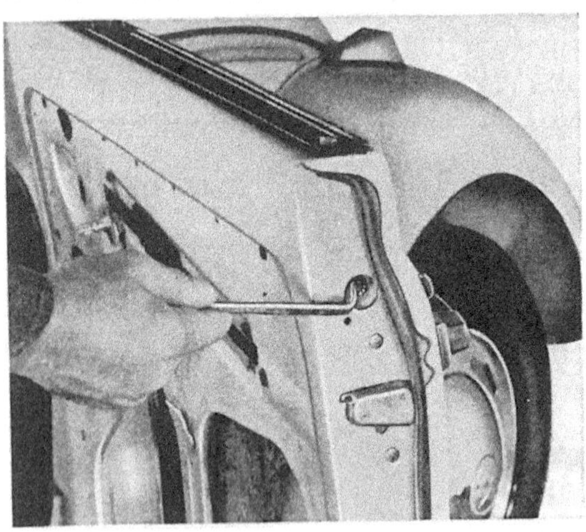

6 - Take out upper screw securing the rear window run channel after pressing out the cover plug.

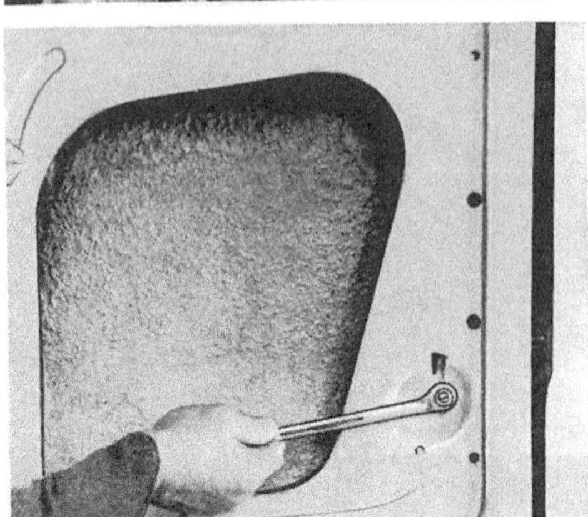

7 - Remove nut and threaded pin of rear window run channel.

8 - If necessary, remove trim moulding with riveted-on outer rubber seal and inner seal in window slot.

a - Detach inner seal from groove and pull off at bottom of metal strip in window slot.

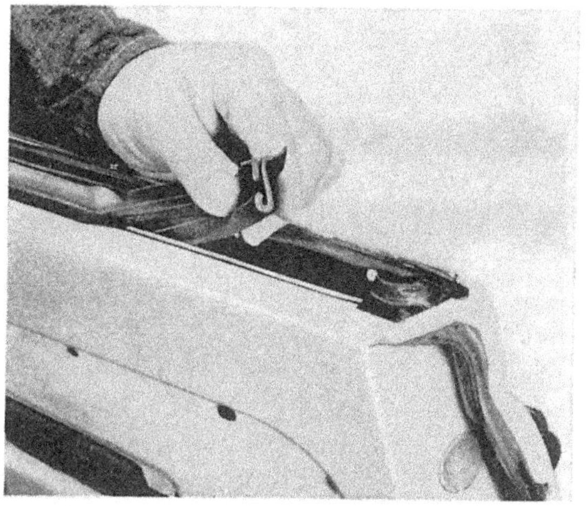

b - Remove two screws in metal strip in window slot and take out trim strip with outer seal.

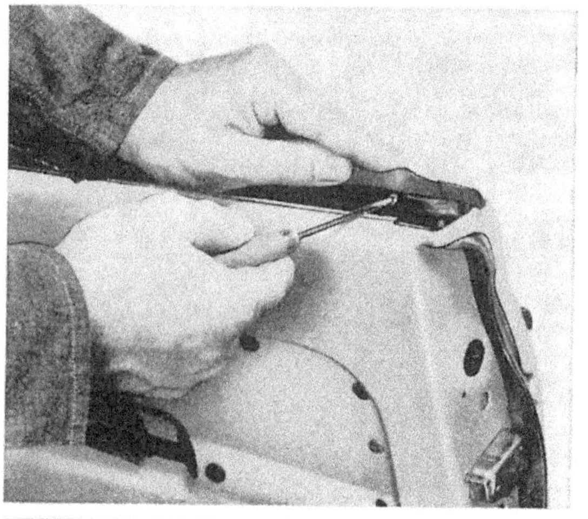

c - Remove rubber end piece from window slot.

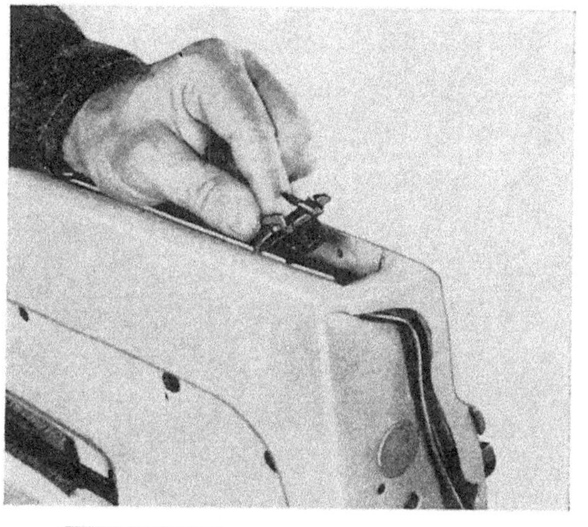

Installation

If the window glass is to be replaced, the window lift channel — A — and the window frame — B — can be removed with a rubber hammer and a wooden wedge. The channel is fitted on the new glass in the same way.

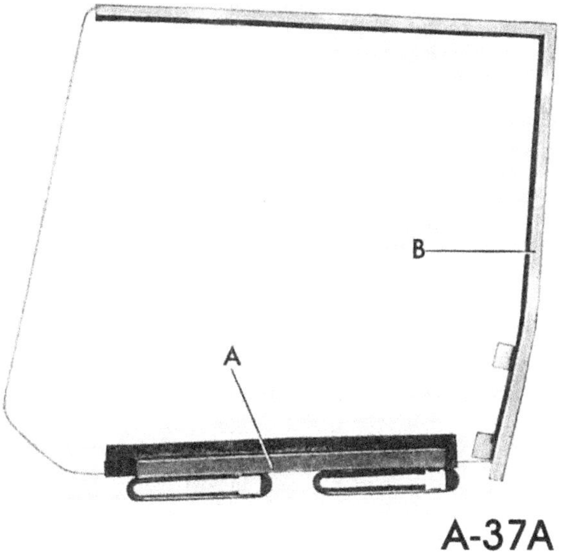

Note:

The front window channel is stuck in and can only be replaced with the vent wing removed.

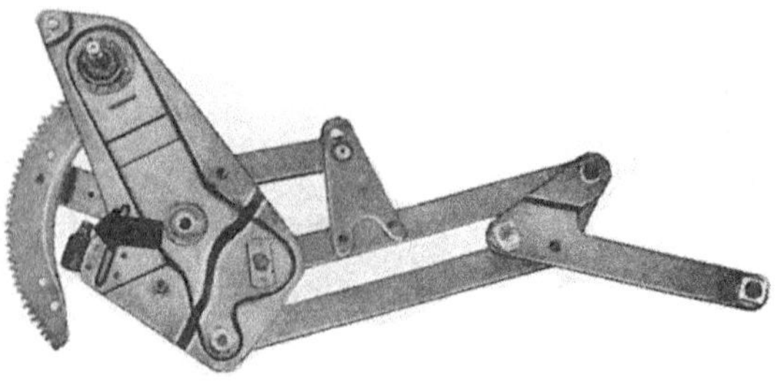

1 - Check the trim moulding with outer seal, inner seal in window slot, window lifter channel with rubber seal and window lifter and fit new parts where necessary.

2 - Coat all moving parts of window lifter and the rear window run channel lightly with grease before installing.

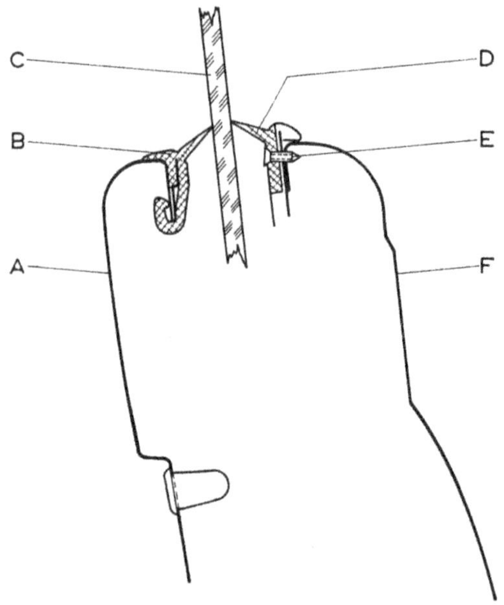

Note:

Loosen adjusting screw (arrow).

3 - Attach rear window run channel at top and bottom.

A - Door inner panel
B - Inner seal
C - Glass
D - Outer seal with trim moulding
E - Tapping screw
F - Door outer panel

4 - Insert inner and outer rubber seals in window slot.

a - Press outer seal in window slot and secure with metal strip.

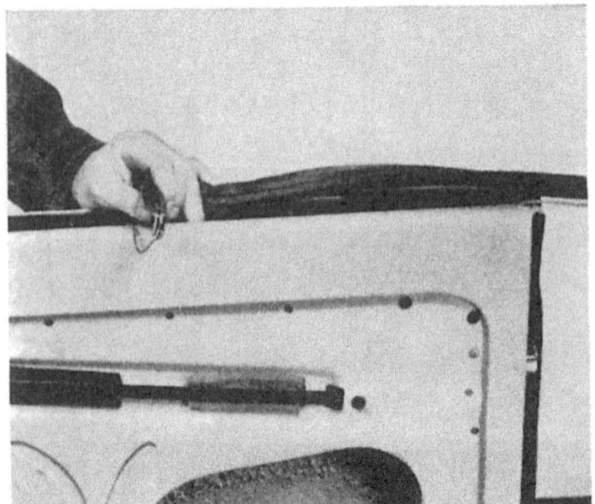

b - Press the wedge-shaped part of the inner seal into the groove in the window slot first and then lift the lower lip over the metal strip in the window slot.

This operation can be facilitated by laying a piece of cable in the lip of the seal.

The seal is pressed into the window slot with one hand while the other hand keeps the cable taut.

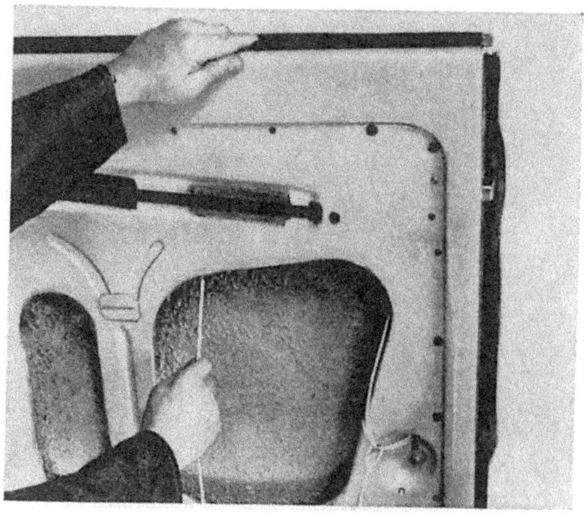

c - Stick rubber end piece in window slot.

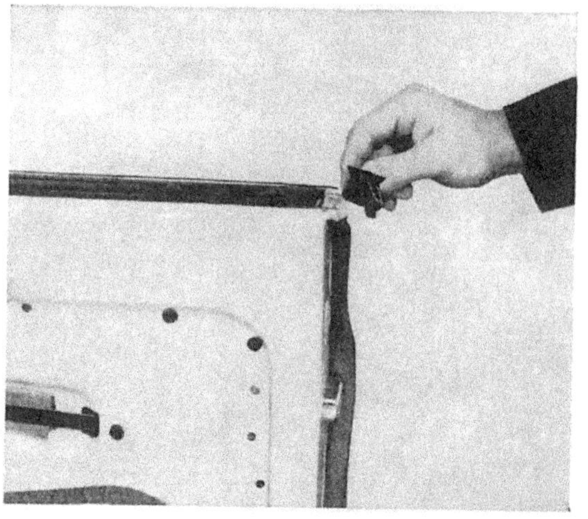

5 - Slide glass into window slot. Take care not to damage the rubber seals in the slot.

6 - Install window lifter and attach loosely.

7 - Slide vent wing frame into window slot at an angle and screw threaded pin into bottom of window run channel.

8 - Insert window glass into front run channel. Attach vent wing frame in window slot and window run channel to door inner panel. Close door.

9 - Secure rear window run channel.

10 - Tighten window lifter screws.

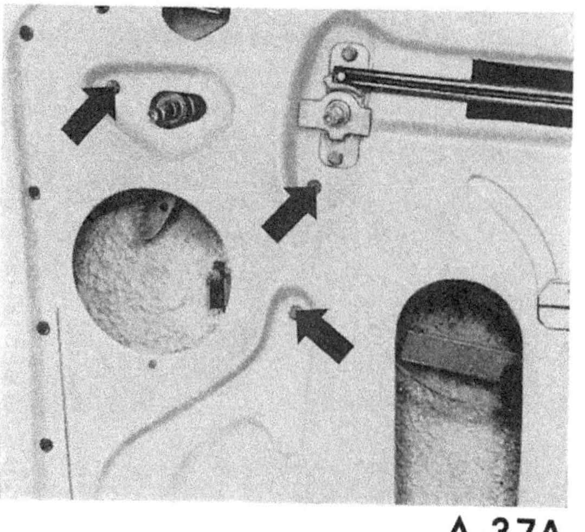

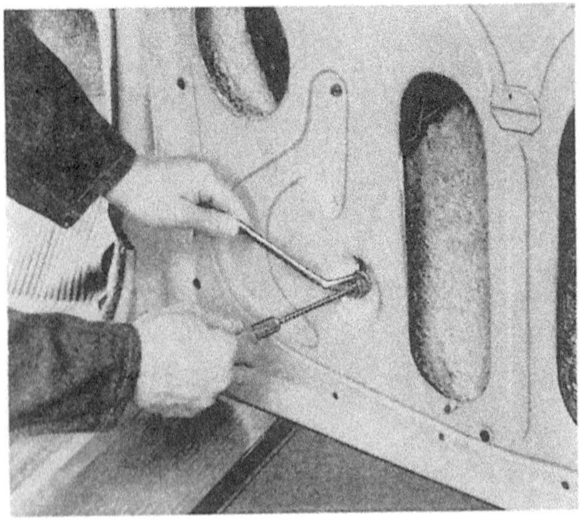

11 - Adjust door window sideways so that it aligns with the drop window longitudinally and the door window frame contacts the drop window seal uniformly.

Note:

Adjustment is made by loosening the upper securing screw and screwing the lower threaded pin of the rear run channel in or out.

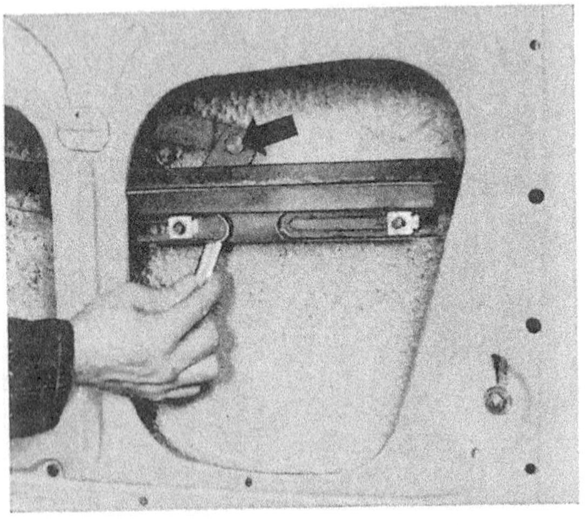

12 - If the window glass tilts when the window is raised, the adjusting nut on the crank linkage should be loosened — arrow — the window pushed towards the front or rear window run channel and the nut tightened again.

13 - Adjust height of door window so that it aligns with the upper edge of the vent wing.

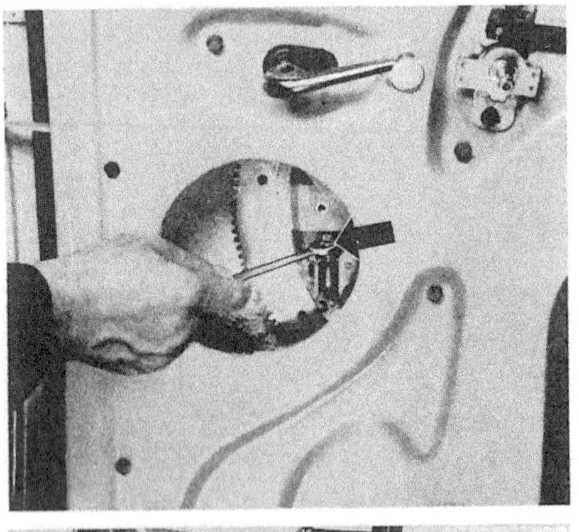

Note:

The height of the window is limited by a stop on the window lifter which contacts a locked adjusting screw (arrow). The lock nut is loosened and the screw turned to make the adjustment.

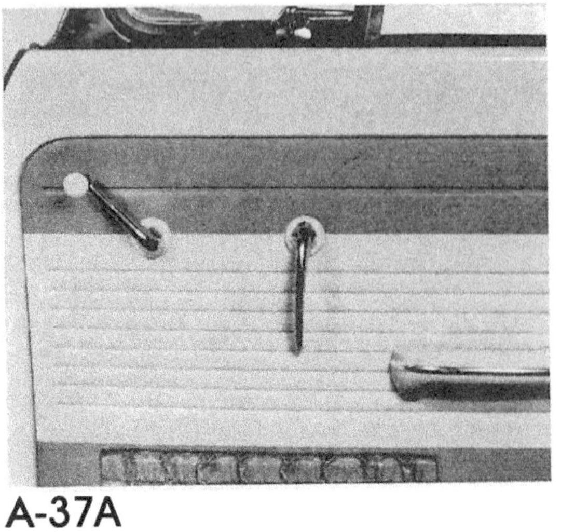

14 - Open and close door window to check ease of movement of window lifter. If necessary, the run channel screws should be loosened slightly and the run channels moved carefully. Tighten screws again afterwards.

15 - Assemble door completely and check that all parts are working.

Vent Wing and Frame Removal and Installation

(up to 1965)

Removal

1 - Remove window regulator handle and door inner handle.

2 - Remove trim panel and take the rubber inserts and escutcheon springs off the regulator and door handle shafts.

3 - Remove the two securing screws and take the vent wing clamp out of the door.

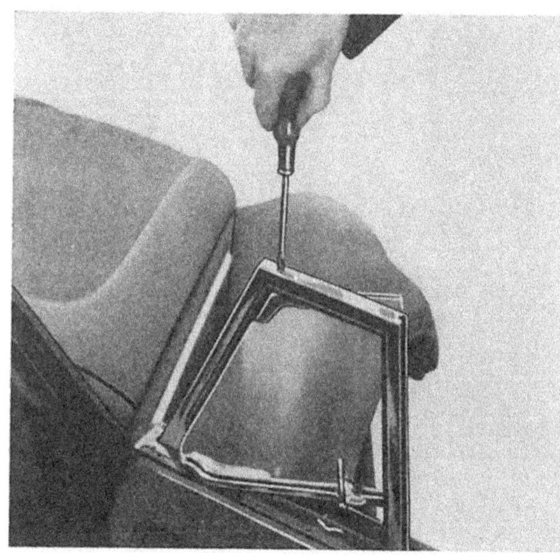

4 - Screw out the vent wing upper pivot.

5 - Tilt window outwards slightly and lift out.

6 - To remove the vent wing frame, which also serves as front run channel for the door window it is necessary to remove the door window and the trim moulding with weatherstrip.

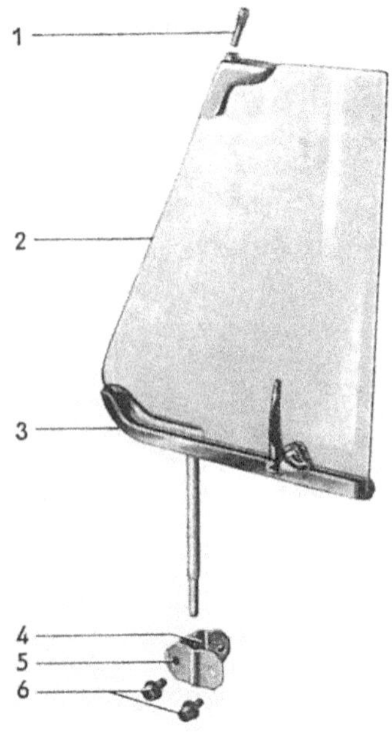

1 - Threaded pin
2 - Vent Wing Glass
3 - Vent Wing Fitting
4 - Clamp-bottom part
5 - Clamp-top part
6 - Securing Screws

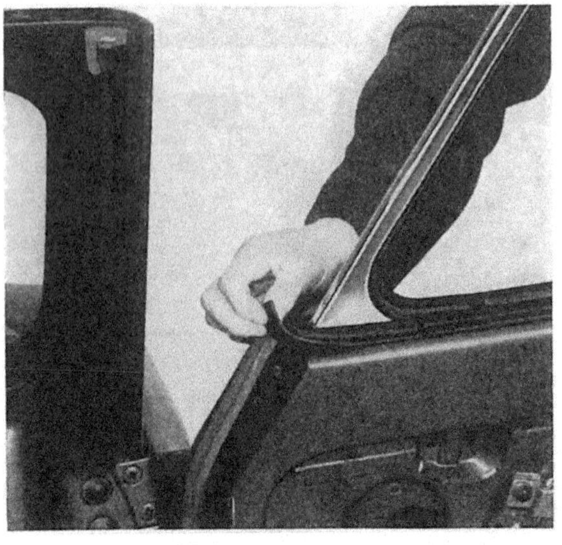

7 - Remove end piece of window slot weatherstrip.

8 - Remove the two countersunk screws above the door top hinge.

9 - Lift the weatherstrip and remove the vent wing frame screw in the window slot.

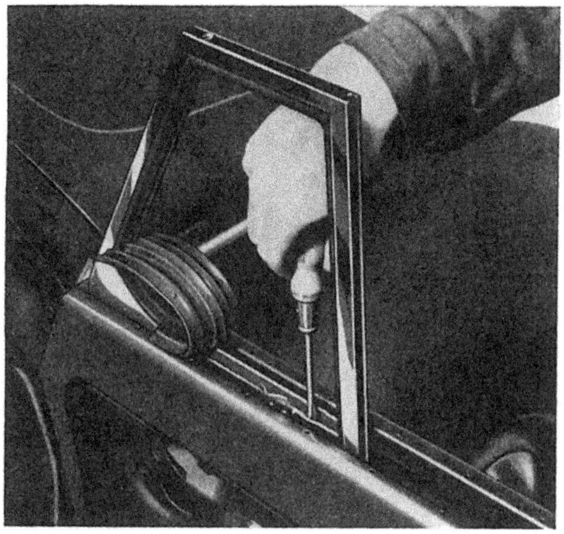

10 - Lift the vent wing frame out.

Installation

1 - Check vent wing and window slot weatherstrips and renew if necessary.

2 - Install the vent wing frame and cement the window slot weatherstrip end piece into position with Genuine VW Adhesive D 12.

3 - Check that vent wing weatherstrip is properly seated and install window.

Hook vent wing clamp parts together, locate in the door from below and secure. The clamp joint should face towards the door hinges.

4 - Adjust the vent wing clamp by tightening or loosening the rear securing screw.

5 - Assemble the door again. Check the height adjustment and ease of operation of door window.

6 - The vent wing frame must press firmly against the sealing lip of the windshield frame weatherstrip on the inside. The distance between the weatherstrip and the vent wing frame must also be uniform over the full length.

If necessary, correct the clearance by inserting suitably shaped strips of rubber.

THIS PAGE INTENTIONALLY
LEFT BLANK

Vent Wing Removal and Installation

Removal

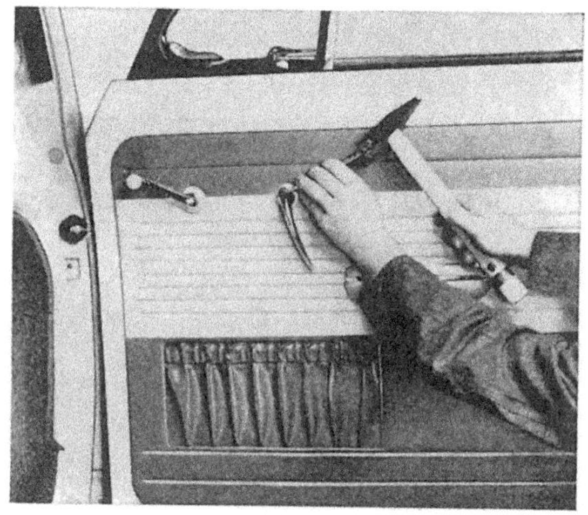

1 - Press window crank and inner door handle escutcheons against door trim panel and knock out pins with a punch. Take crank, door handle and escutcheons off.

Important

On the passengers side the trim panel is hooked on to the armrest support on the door inner panel with the armrest retaining plate.

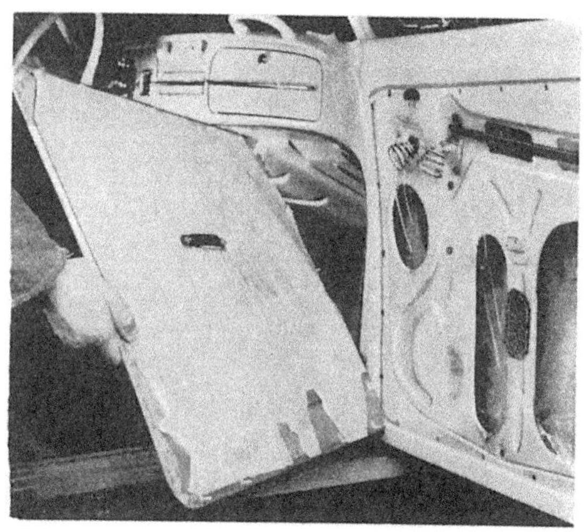

2 - Take door trim panel off door inner panel. Take care to avoid damage to paintwork.

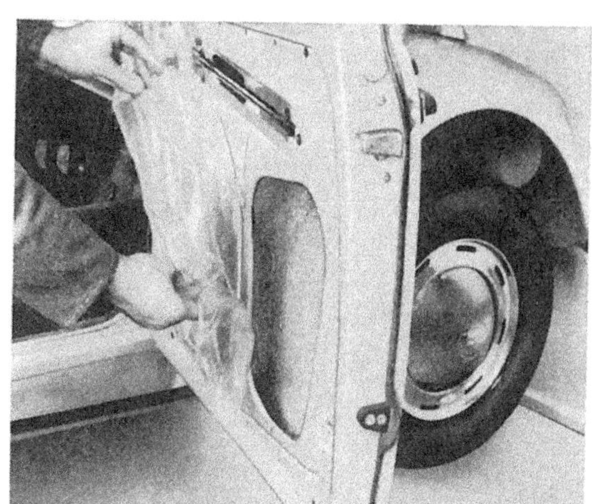

3 - Take springs and rubber inserts off the door handle and window crank spindles.

4 - Take PVC sheet off door.

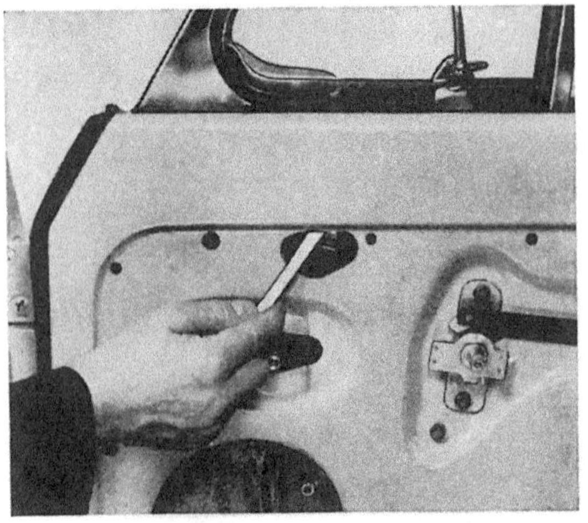

5 - Loosen vent wing clamp by slackening securing screw.

6 - Remove threaded pin for upper vent wing mounting.

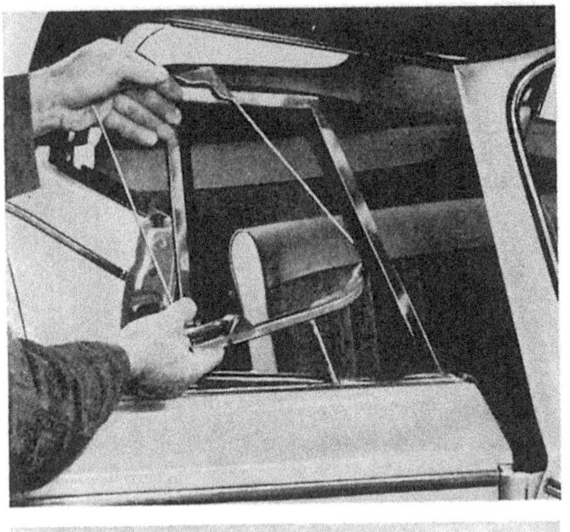

7 - Tilt vent wing outwards slightly and lift it out upwards.

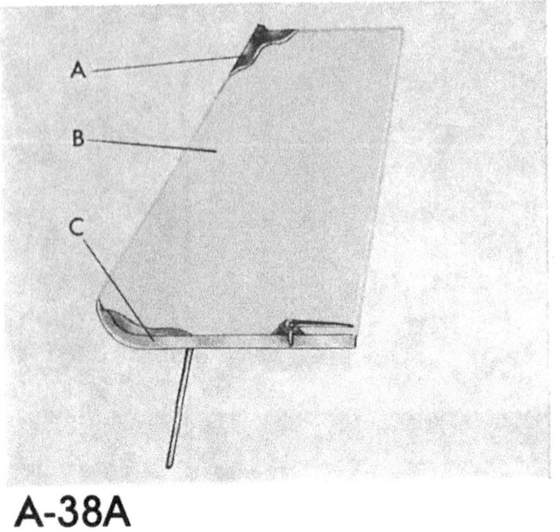

Installation

If the vent wing glass or frame are to be replaced, the upper and lower fittings can be taken off the glass with a rubber hammer and a suitable wooden wedge. The fittings are installed on the new glass in the same manner.

A = Upper vent wing fitting
B = Vent wing glass
C = Lower fitting with fastener and spindle

1 - Check both vent wing seals and replace where necessary.

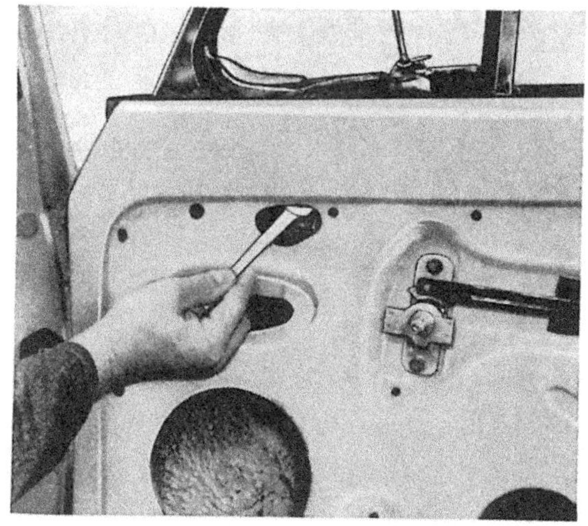

2 - Place vent wing in window slot from above and attach it to frame with threaded pin.

3 - Adjust vent wing clamp by loosening or tightening the securing screw. This screw is accessible through a special hole in the door inner panel.

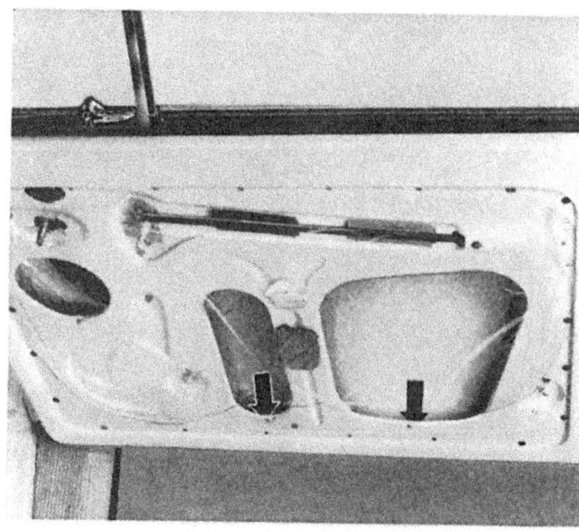

4 - Coat door inner panel lightly with D 12 adhesive and stick the PVC sheet in position. Ensure that the lower edge of the sheet is inside the window slot.

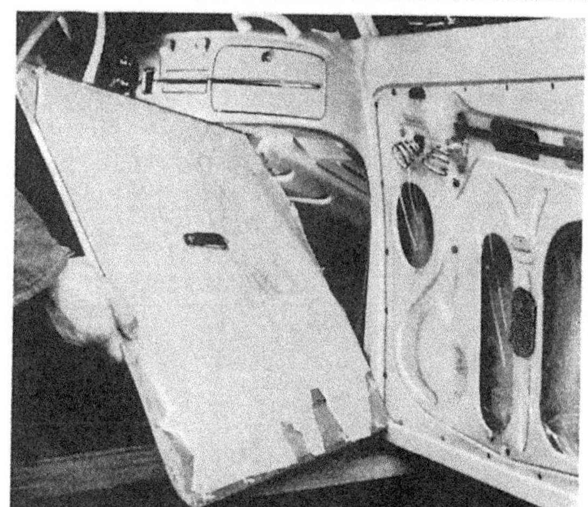

5 - Install rubber inserts, springs and trim panels. The large end of the springs must be towards the trim panel.

Note:

On the passengers side, the trim panel should be hooked on to the armrest support on the door inner panel with the armrest retainer.

6 - Install door inner handle and window crank in the correct positions.

Vent Wing Frame Removal and Installation

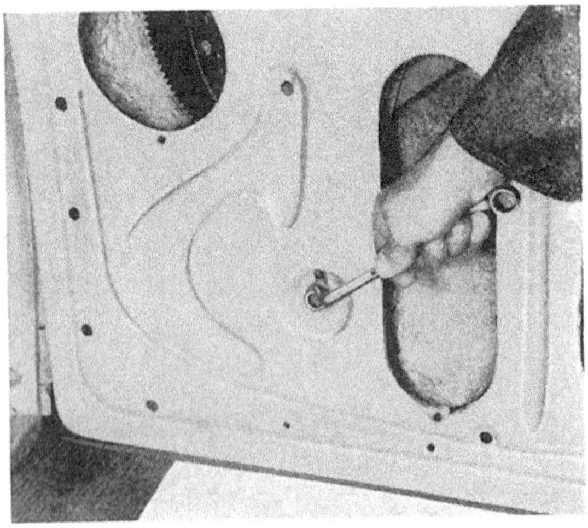

Removal

1 - Remove door trim panels and pull PVC sheet off inner panel.

2 - Take hexagon nut off threaded pin for front window run channel and remove lock and flat washers.

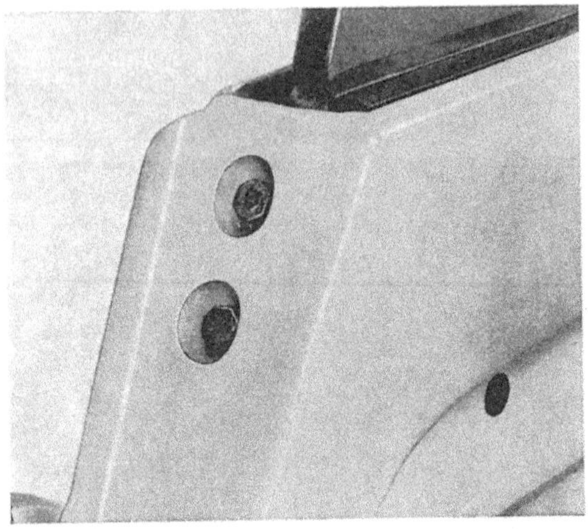

3 - Pull door weatherstrip off slightly near upper hinge and press cover plugs out.

4 - Remove two screws in vent wing frame.

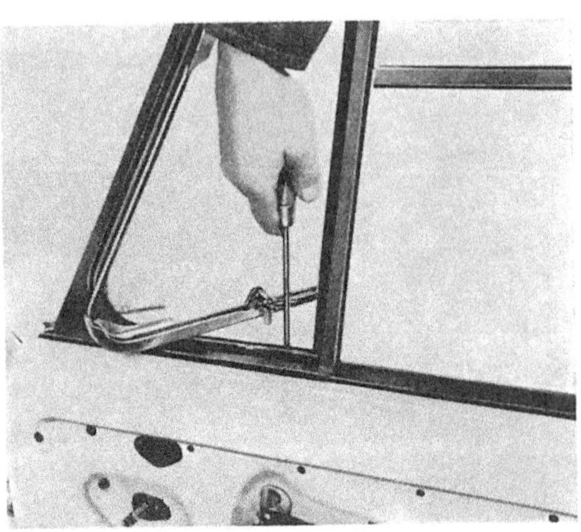

5 - Remove Phillips screw securing vent wing frame in window slot.

6 - Screw threaded pin out of front run channel.

7 - Lift frame out upwards.

Installation

1 - Check rubber seals for vent wing frame and window slot and replace as necessary.

2 - Check retaining bracket and upper part of vent wing clamp and replace as necessary.

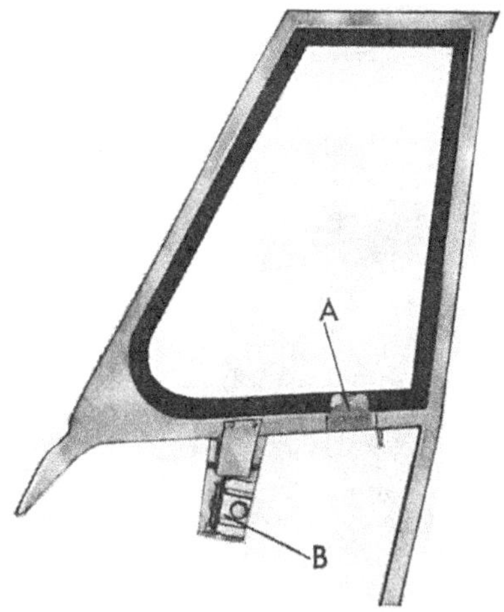

A - Retaining bracket
B - Upper part of clamp

3 - Slide vent wing frame into window slot at an angle and screw threaded pin into bottom of window run channel.

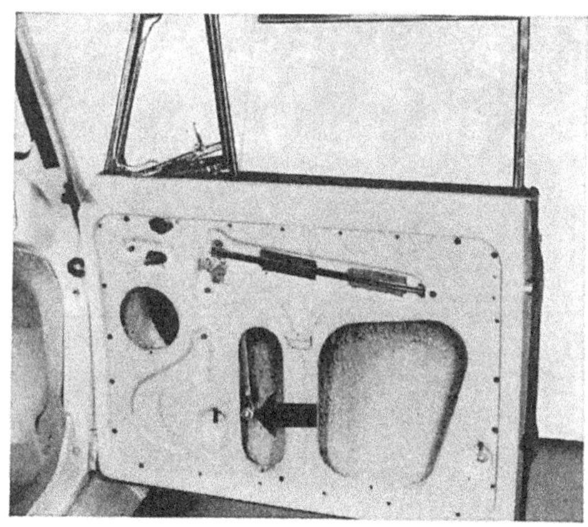

4 - Insert window glass into front run channel. Attach vent wing frame in window slot and window run channel to door inner panel.

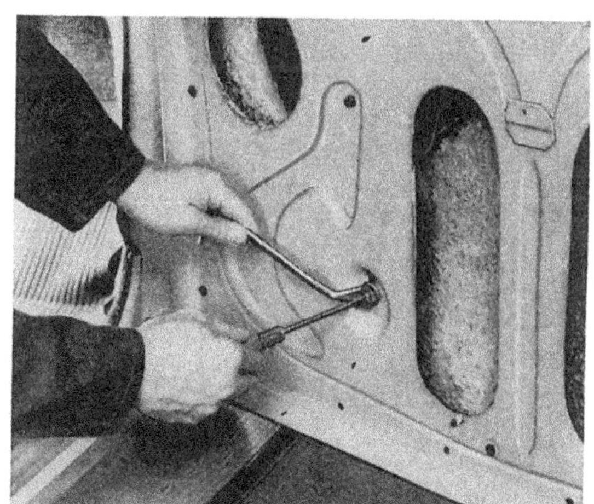

5 - Open and close door window to check ease of movement of window lifter and position of window glass. If necessary loosen the threaded pin in the door inner panel and the securing screw in the window slot and move the frame and front run channel. Tighten screws afterwards.

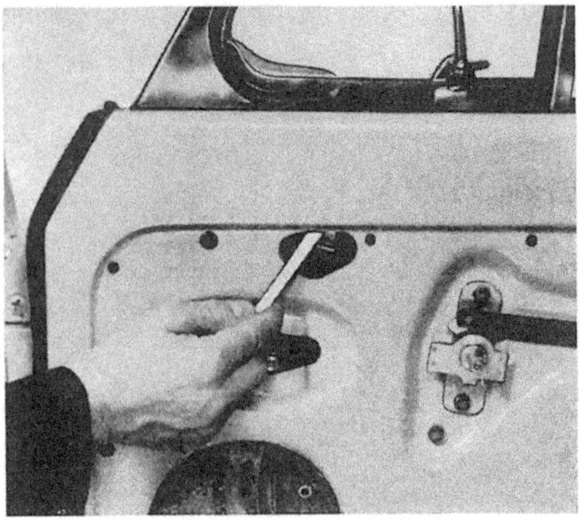

6 - Adjust the vent wing clamp by loosening or tightening the securing screw.

7 - Assembly door again and ensure that the window crank and door inner handle are positioned correctly.

8 - Close door and check location of vent wing frame on the rubber seal of the windshield frame.

Note:

The inner edge of the vent wing frame must press firmly and uniformly on the lip of the seal on the windshield frame. If it does not, the rubber seal must be replaced.

Drop Quarter Window Removal and Installation

(up to 1965)

Removal

1 - Open the top.

2 - Remove rear seat and back rest.

3 - Press the handle escutcheon against the trim panel until the dowel pin is visible. Drive the pin out with a suitable punch and take off handle.

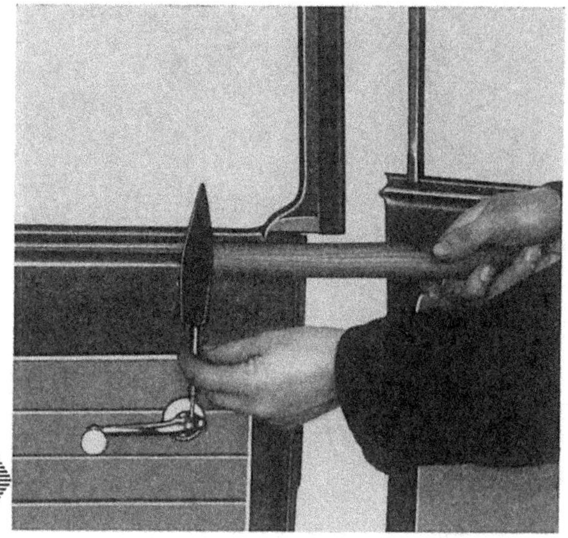

4 - Remove the trim panel by lifting the clips out of the quarter panel. Care must be taken to avoid damaging the paintwork.

5 - Remove four window regulator screws.

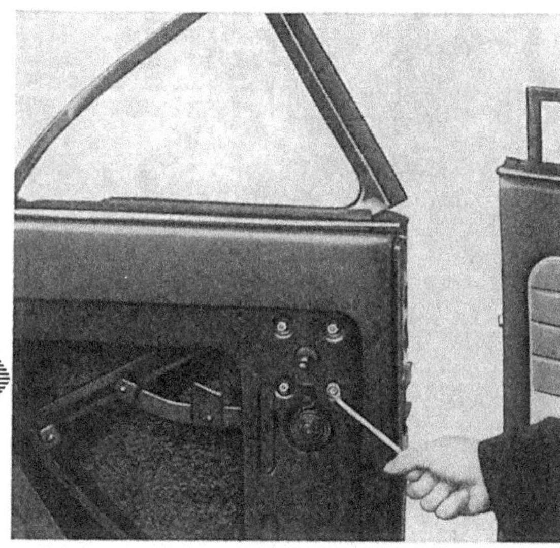

6 - Push the regulator in towards the outer panel. If only the window regulator is to be renewed, bend the window guide channel up slightly so that the ball head is free and take the regulator out.

If the window and the lifting linkage is to be removed it is not necessary to remove the regulator mechanism separately.

7 - Take out the rubber plugs in the lock pillar above the striker plate and remove the screws securing the bearing for the drop window.

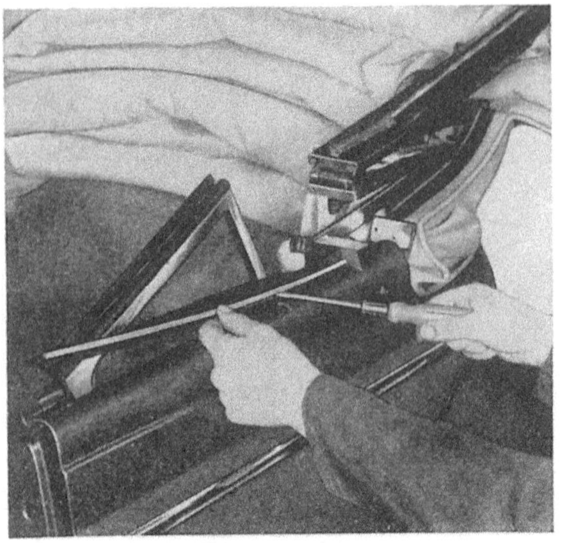

8 - To facilitate the removal and installation of the drop window and avoid damage to the window slot weatherstrips it is advisable to remove the weatherstrips and trim mouldings by lifting out the clips.

9 - Lift the window out of the guide channel and lower towards the wheel housing and over the inner panel at the same time. Take window out from below.

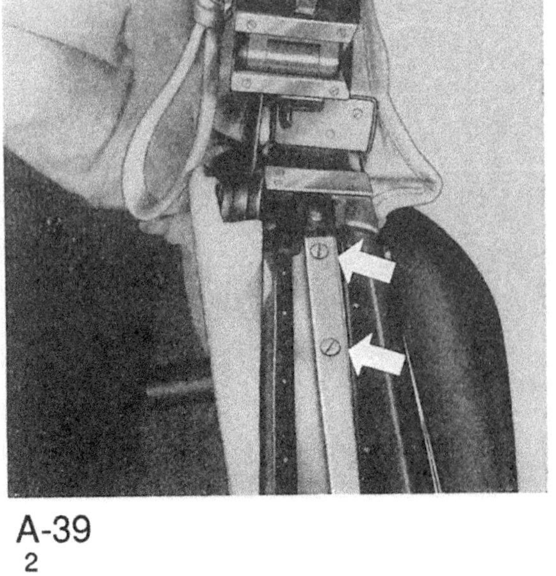

10 - If necessary, the guide channel in the window slot can also be removed. To do this, remove the two rear screws on the main bow pillar.

and at the front on the quarter panel inner part.

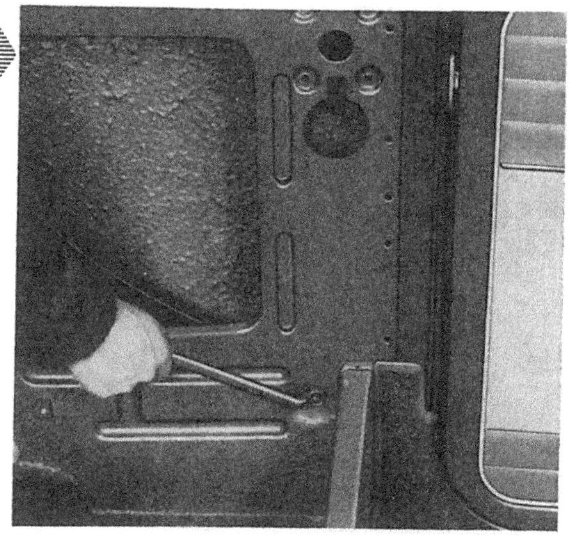

11 - If the window lift channel, the drop window or the weatherstrips belonging to them are damaged, they can be replaced.

Remove the window frame and lifting channel screws.

Take off the lifting channel by tapping carefully with a rubber hammer and suitable wooden block and remove the window glass from the frame. The replacement takes place in a similar manner.

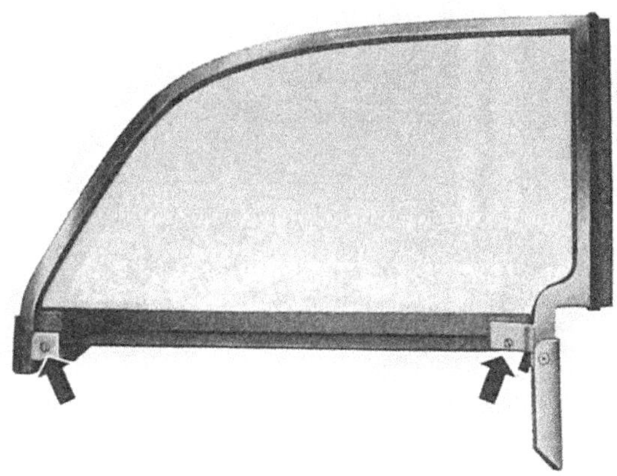

Installation

1 - Clean window regulator and lubricate moving parts with Universal Grease VW A 052.

2 - Install guide channel and drop window again.

3 - Check that window opens and closes properly and that the weatherstrip is correctly seated.

4 - The top edge of the drop window must be level with the door window.

5 - If a gap exists between drop and door windows when the drop window is fully raised, note the following:

A stop plate is screwed to the window lifting arm.

A - Lifting arm

B - Stop plate

The stop plate can be moved to the left or right when the securing screw is loosened.

The stop plate rests against the top rear threaded pin.

If the stop is moved towards the ball head the window can be raised more, if moved downwards the height is reduced.

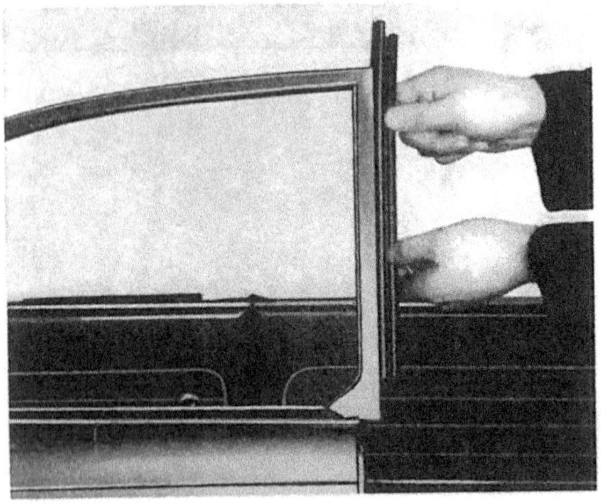

This operation is carried out with the window installed. To adjust the stop plate lower the window slightly.

The adjustment of the drop window is governed by the clearance between the door and drop windows.

6 - If necessary, the front weatherstrip on the drop window can be pulled out and replaced.

When a new weatherstrip is installed, about 6—8 mm should be cut off the sealing lip at the upper end.

This measure is necessary to ensure that the operation of the drop window is not affected by the sealing lip of the roof frame weatherstrip. Check, therefore, that the drop window can be wound up fully when the roof is closed.

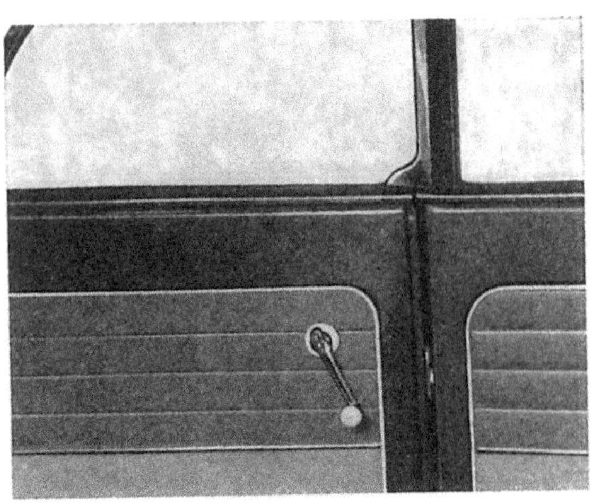

7 - Install handle in the proper position.

Drop Window Removal and Installation

Removal

1 - Open top

2 - Remove rear seat and backrest.

3 - Press window crank escutcheon against trim panel and knock out pin then visible with a pinch. Take off crank and escutcheon.

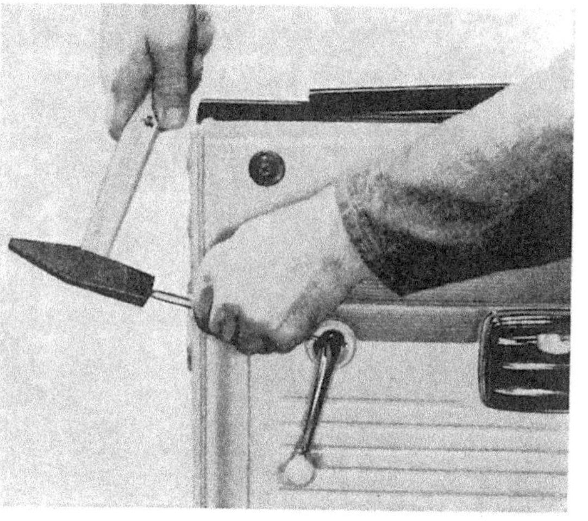

4 - Take trim panel off quarter panel, taking care not to damage panel and paintwork.

5 - Pull PVC sheet off inner panel.

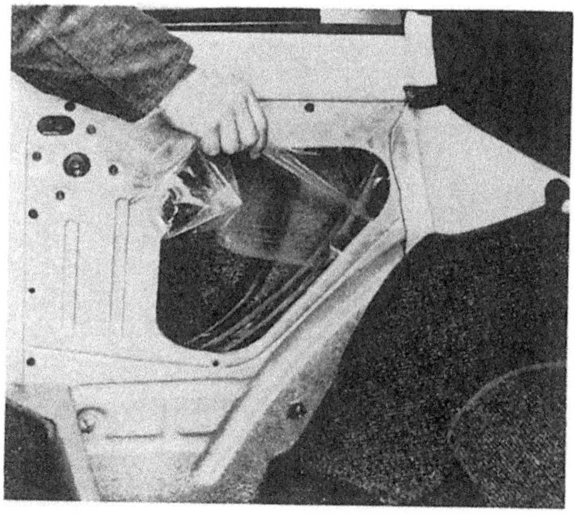

6 - Remove four screws from window lifter.

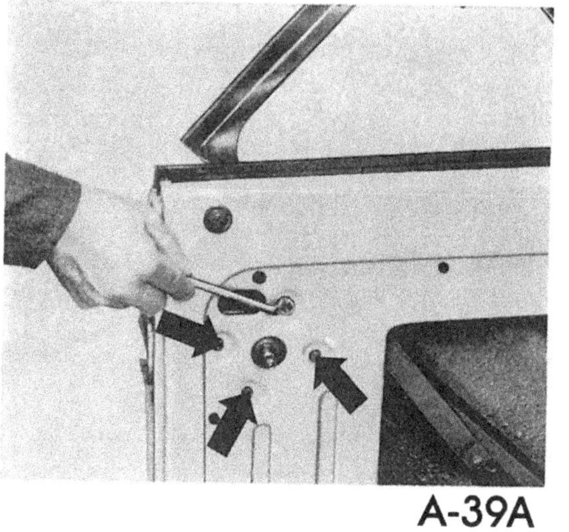

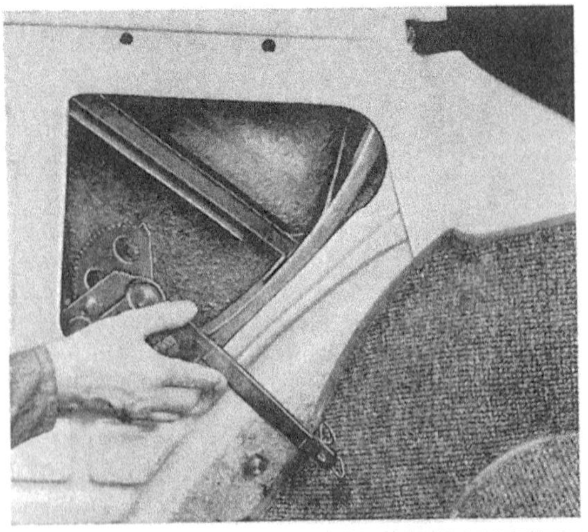

7 - Press window lifter inwards and to the rear until the sliding spring can be taken out of the window lift channel.

8 - Press out two cover plugs in the lock pillar above the striker plate and remove drop window mounting screws.

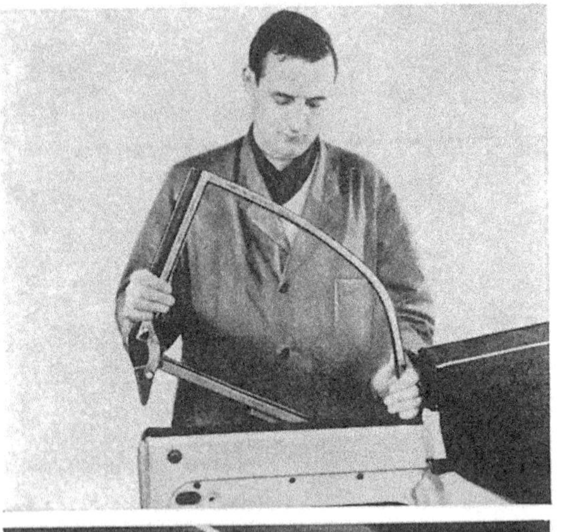

9 - Take drop window out upwards.

10 - If necessary, remove the drop window run channel in window slot after taking out:

a - Remove rubber end piece from top hinge support.

b - Two screws on the top hinge pillar.

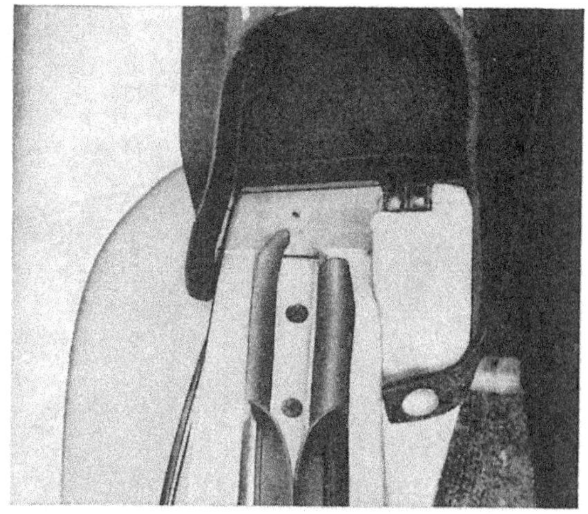

c - A hexagon nut at bottom of quarter panel inner part.

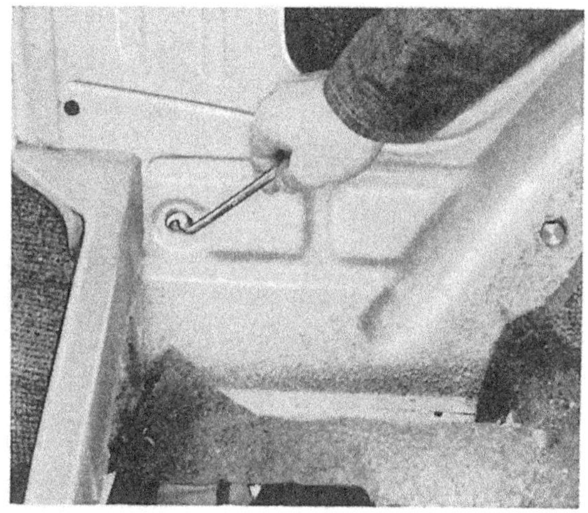

d - Tilt run channel inwards slightly and pull it out.

11 - If necessary, take window slot rubber seal off.

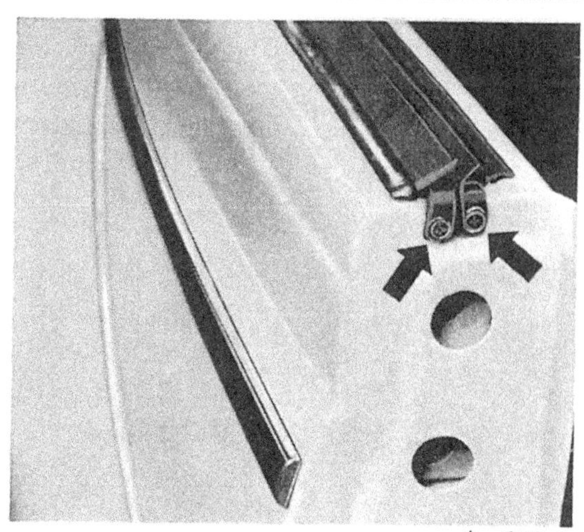

a - Take two screws out of front face of lock pillar and remove front seal.

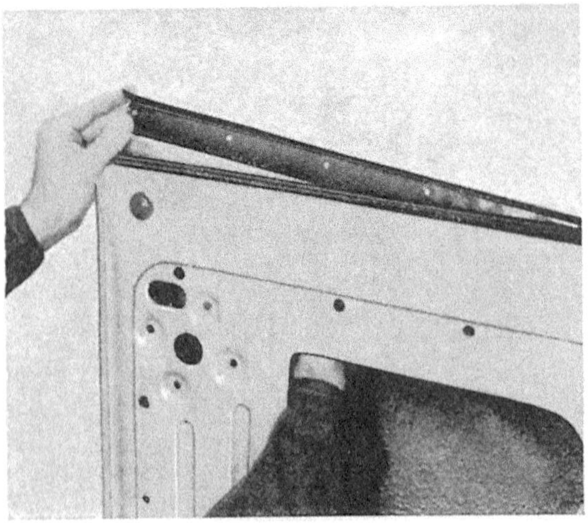

b - Detach outer rubber seal from metal strip in window slot and press upwards against the trim moulding at the same time.

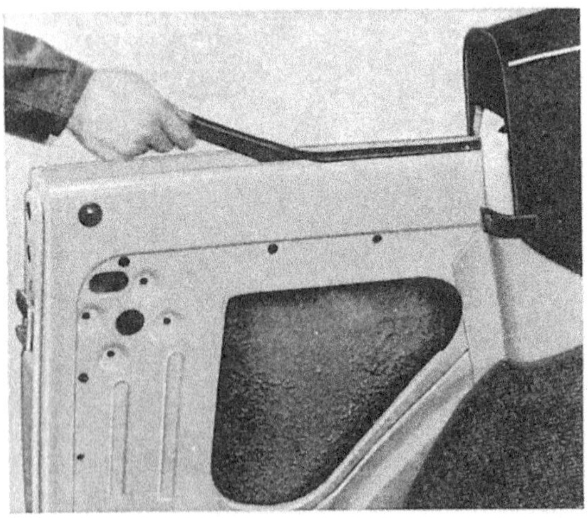

c - Detach inner seal from groove at the top and pull it out of metal strip in window slot.

Installation

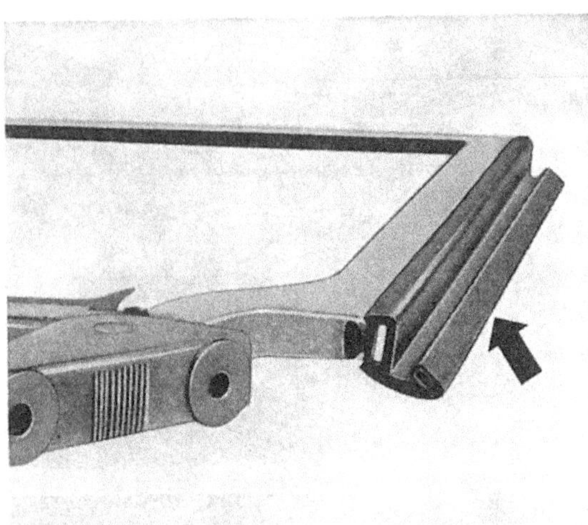

1 - Check window lift channel, window frame and rubber seals and fit new where necessary.

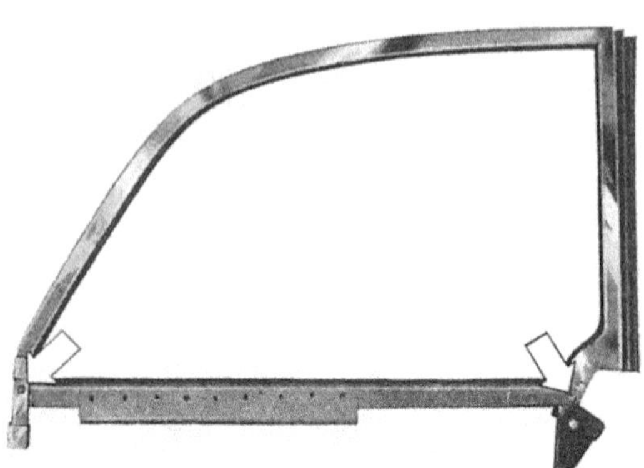

Note:

To remove the glass, take out two screws in frame and remove the window lift channel.

2 - Clean window lifter and lubricate moving parts lightly with universal grease.

Note:

Loosen adjusting screw (arrow).

3 - Secure window run channel in window slot and on inner panel.

4 - Install rubber seals in window slot. The fitting of the inner and outer rubber seals can be facilitated by laying a piece of cable in the lower tip and lifting the lip over the metal strip in the slot:

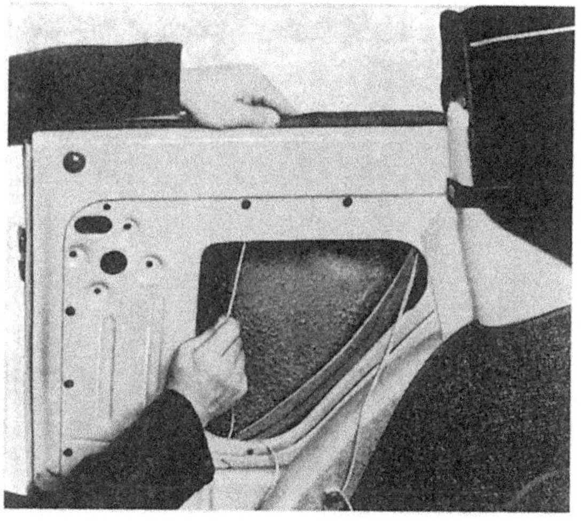

a - Press the upper wedge-shaped part of the inner seal into the groove in the window slot first and then lift the lower lip over the metal strip by pulling one end of the cable.

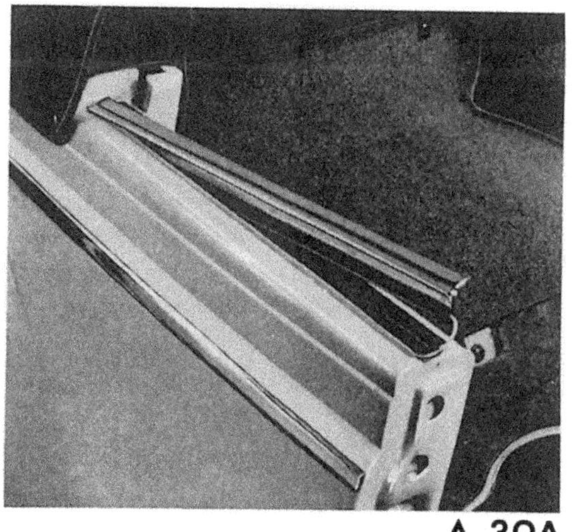

b - To install the outer seal press the trim moulding against the top hinge pillar first then lift the lower lip over the metal strip by pulling the cable.

c - Fit rubber end piece on top hinge support.

d - Fit front seal on the end face of the lock pillar.

5 - Insert drop window into slot from above, taking care not to damage the rubber seals.

6 - Bolt drop window mounting to the lock pillar.

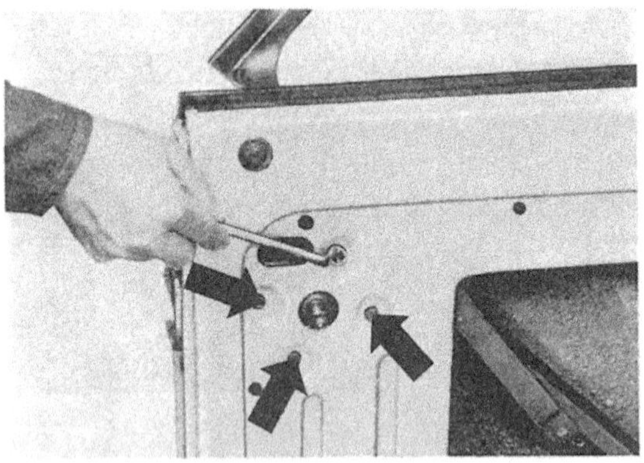

7 - Fit window lifter channel first and then secure it to quarter panel.

8 - Open and close drop window to check window lifter for ease of movement. If necessary, loosen nuts of window lifter and move lifter as required. Tighten nuts again afterwards.

9 - Adjust height of drop window so that it aligns with the top edge of the door window.

The height of the drop window is limited by a stop on the window lifter which contacts an adjusting screw (arrow). After loosening lock nut, the screw can be turned to adjust drop window.

10 - Press both plugs into lock pillar above striker plate and seal lower run channel mounting with plastic sealing compound.

11 - Coat inner panel lightly with D 12 universal adhesive and stick PVC sheet on.

12 - Install rubber inserts, springs and side trim panels. The large end of the springs must be towards the trim panel.

13 - Install window crank at correct angle.

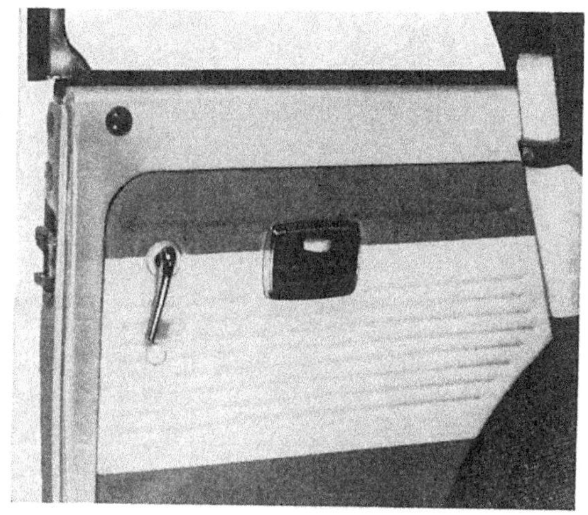

14 - Install rear seat and backrest.

15 - Check all parts by operating several times.

THIS PAGE INTENTIONALLY
LEFT BLANK

Replacing Header

1 - Remove one chromed screw from outer side of left and right roof side members.

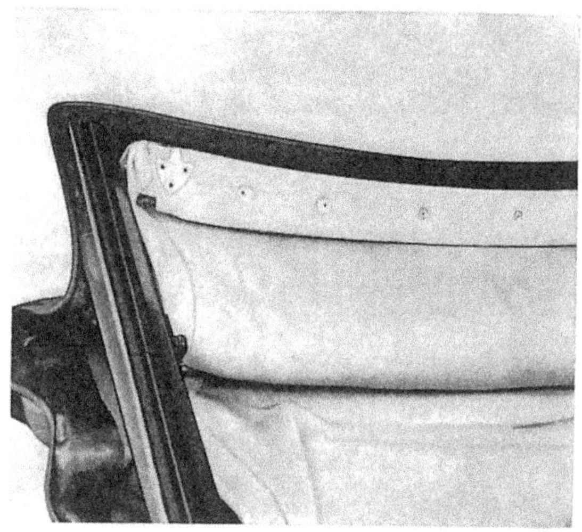

2 - Remove one Phillips screw and washer from inner side of left and right roof side members. Unscrew Phillips screws from lateral parts of top headlining.

3 - Unscrew top handles and fasteners from header.

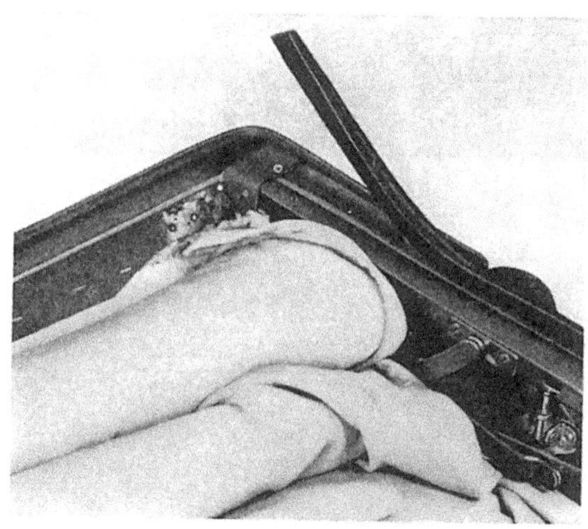

4 - Unscrew two Phillips screws from header and remove leatherette covered molding.

5 - Detach weatherstrips over a length of 15.75" (400 mm) from both roof side members, near the header.

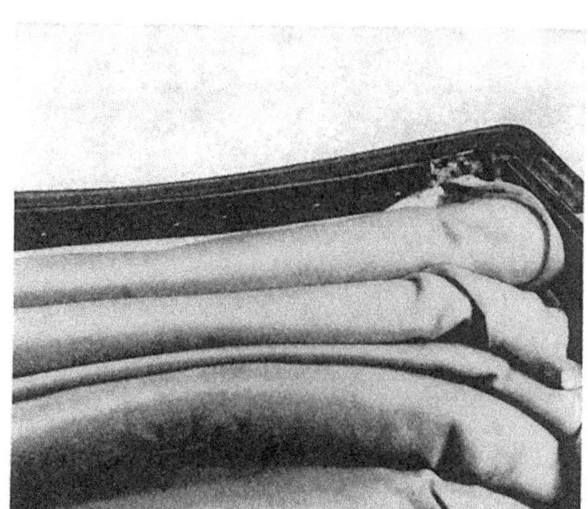

6 - Pull staples for additional attachment of top pocket out of roof side members.

Note:

Exercise extreme care to avoid tearing the top cover.

7 - Detach top pocket from header and fold top cover back.

Note:

On older vehicles, an additional top cover strip may have to be pulled off the weatherstrip.

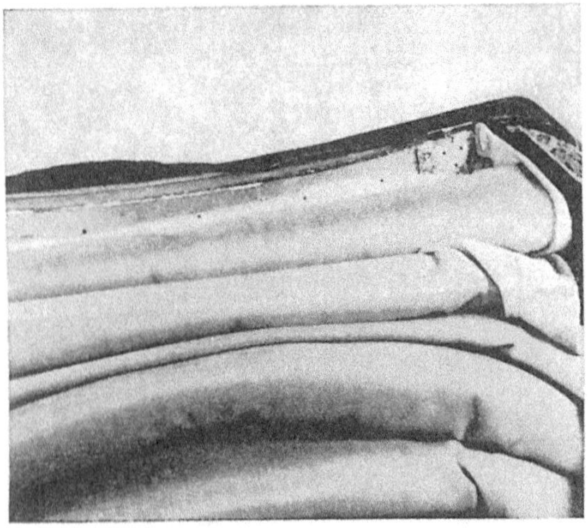

8 - Carefully detach lining strips from top of header.

9 - Detach PVC top cover material strips and roof part of linen sheet from top of header.

10 - Detach top headlining from top of header.

Note:

Exercise extreme care to avoid damaging the top headlining.

11 - Pry plastic wood out of header holes and remove two screws on left and right. Hold one nut (arrow) on outside of header.

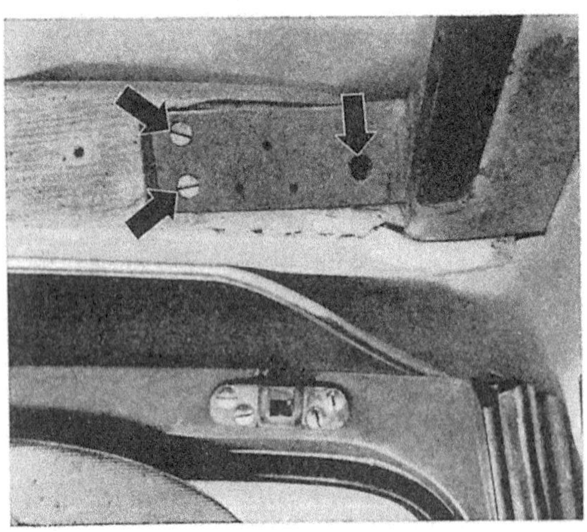

12 - Unscrew attaching screws from roof side member brackets and remove header.

Installation

From Chassis Number 1 600 440 headers without holes for attaching screws will be supplied for all Convertibles. This makes it necessary for the position of the holes for the attaching screws to be marked first and for the holes to be drilled afterwards.

Prior to installing a new header, clean any paint, rust or cement off the contact surfaces of the top side members and repaint them after first treating them with wash primer.

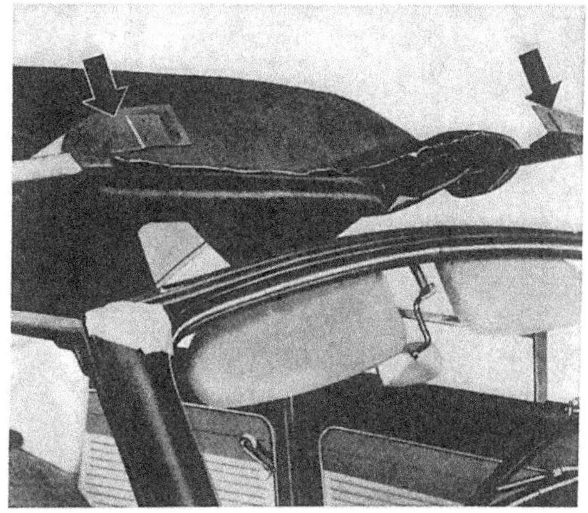

1 - Place header on windshield frame and push header back to roof side member brackets.

2 - Align roof side members with outer edges of header and mark a hole on each side from inside for the outer attaching screws.

Note:

Because of their location, the upper holes cannot be reached with a scriber. Therefore, blow powdered graphite into the welded nut of the left and right roof side member brackets to mark the position of the hole.

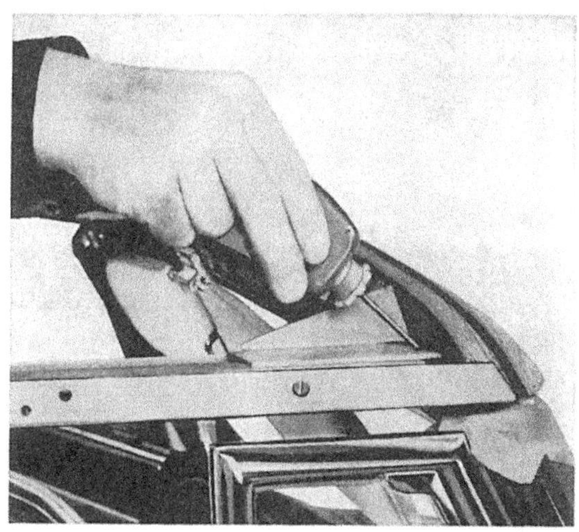

3 - Remove header and drill .14" (3.5 mm) pilot holes.

> **Important**
>
> The holes must be drilled at right angles to the contact surface of the header and the roof side member brackets.

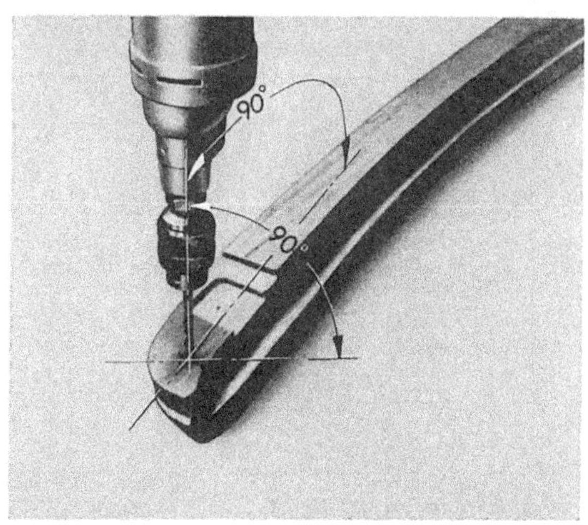

4 - Enlarge holes to .32" (8.2 mm) and countersink at the front so that the heads of the attaching screws are completely below the surface of the wood.

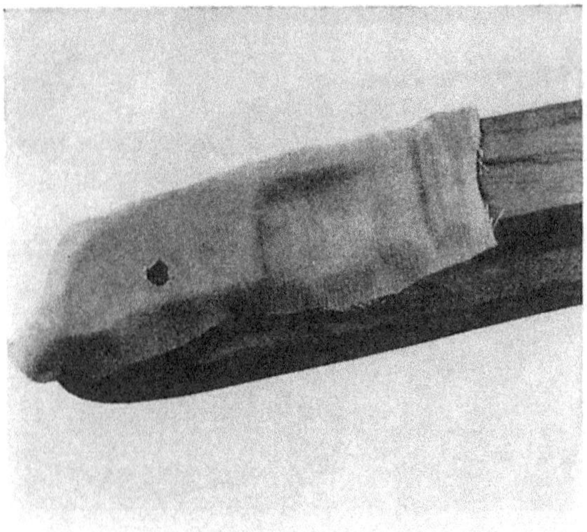

5 - Cement top cover material to header to avoid squeaking noises on contact surfaces of roof side members.

6 - Place header on windshield frame and secure it to roof side members on both sides with one hexagon head bolt M 8×35 each.

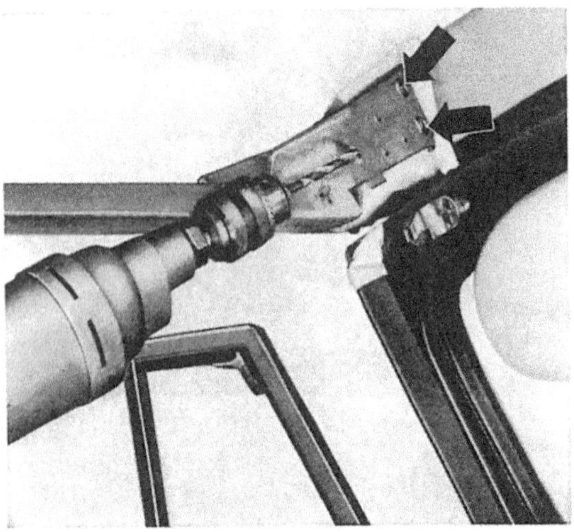

7 - Drill three .14" (3.5 mm) holes on both sides, through roof side member bracket holes into header. Then enlarge these holes to .32" (8.2 mm) and countersink them from the front so that the heads of the attaching screws are completely below the surface of the wood.

8 - Insert other attaching screws into header and check that header is properly seated by opening and closing top frame several times. Correct if necessary.

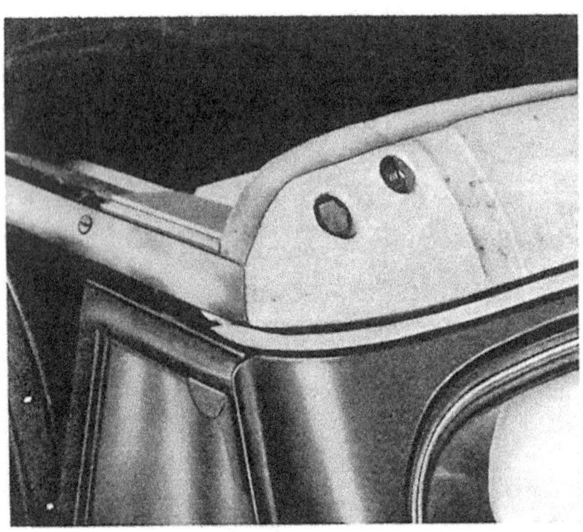

Note:

Using a rasp, file the rounded contours so that they blend with the roof side members.

9 - Check tightness of header attaching screws, then fill countersunk holes in header with plastic wood.

A - One-piece Rear Body Bow

It the rear body bow is being removed only because a part of the body, such as the rear panel for example, has to be replaced, proceed very carefully and without using force in order to avoid damaging the rear body bow.

Small cracks caused by inserting staples too closely, mainly in the vicinity of the webbing straps, can be eliminated by using wood cement. The bow is made up of five separate parts and if any of the joints are loose, they should be glued again.

If a rotten or broken rear body bow is being replaced, use a wood chisel to cut the bow into small pieces which can then be removed easily.

Removal

1 - Detach luggage compartment lining as far as necessary.

2 - Take trim molding off.

Note:

This must be done very carefully to avoid damaging the top cover.

3 - Detach top cover from rear body bow carefully. Lift the staples out of rear body bow by inserting a suitably ground screwdriver between top cover and bow and then levering.

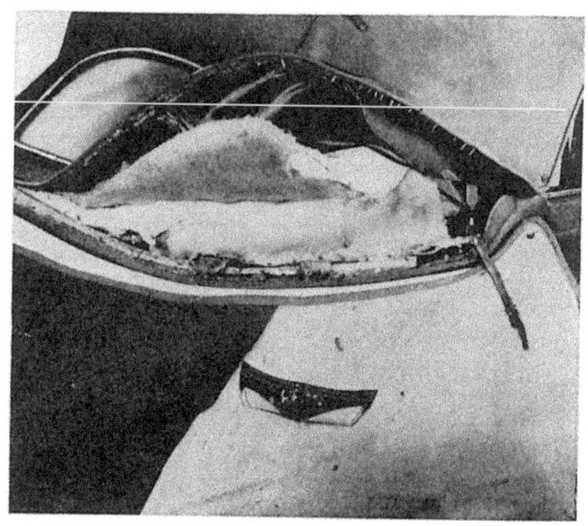

4 - Carefully detach rear window linen and side padding from rear body bow.

5 - Pull out the staples securing the webbing straps to the bow, using a nail-puller.

6 - Pull headlining off rear body bow, and pull out the tacks securing the rubber bands on each side. Tape up the detached part of the headlining near the main bow.

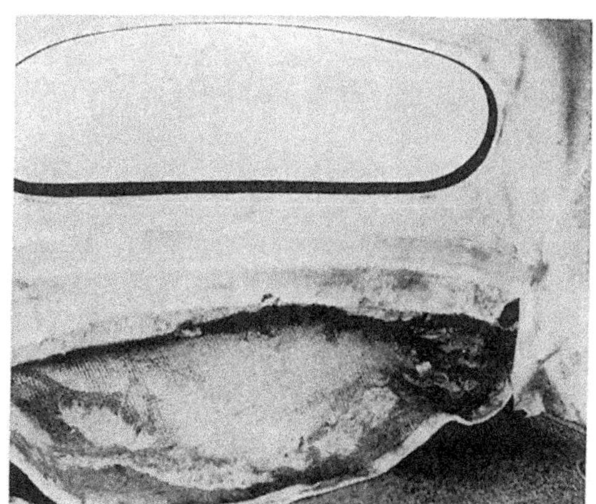

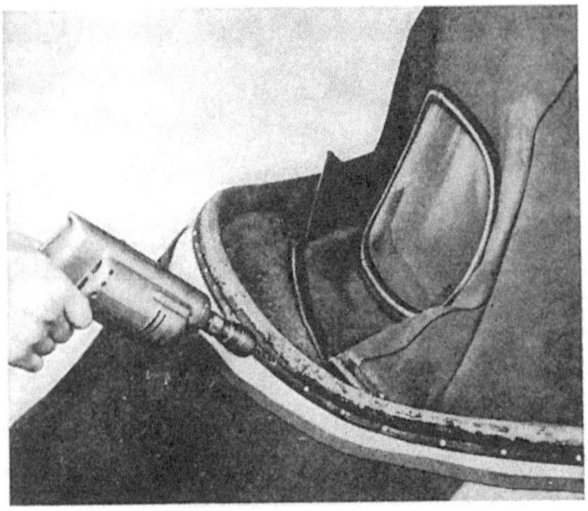

7 - Rear body bow up to August 1964:

Drill the heads off the barbed nails in the rear body panel.

Note:

If these nails are pulled out with a nail-puller, the rear panel will be badly damaged: As it is not possible to knock out any dents at this point, they must be filled with solder.

8 - Fold detached parts of top cover and corner padding forwards and attach them near main bow with adhesive tape.

9 - Pull nine barbed nails out of bow securing brackets.

10 - Rear body bow from August 1964 to November 1966:

Remove six nuts from screws holding bow and take screws out.

11 - Pull three barbed nails out of bow securing brackets.

12 - Lift bow uniformly at front ends and take it out of rear panel upwards.

Important

If this is not done, the rear panel may be badly damaged because the body radius is smaller than that of the rear body bow.

Installation

One-piece rear body bows supplied as replacement parts are without holes so that they can be used for bodies with nine brackets and barbed nails as well as for the type with three brackets and screws. To prevent noises, stick a piece of top material on the surfaces of the bow which contact the body sheet metal.

A - Rear body bow up to August 1964:

1 - Place bow between rear panel and brackets, starting from the center.

2 - Fit wedges at each end of bow to force it towards the center so that the radius makes contact all round.

3 - Knock bow into position by tapping it lightly with a mallet. Any loss in tension caused by the bow settling can be corrected by driving the two wedges down further.

4 - Shape the nine brackets to fit bow by tapping them with a hammer.

5 - Secure bow to rear panel with barbed nails.

6 - Secure bow to brackets with barbed nails.

7 - Pull both wedges out upwards.

B - Rear body bow from August 1964 to November 1965:

1 - Place bow between rear body panel and brackets, starting from the center.

2 - Fit wedges at each end of bow to force it towards the center so that the radius makes contact all round.

3 - Knock bow into position by tapping it lightly with a mallet. Any loss in tension caused by the bow settling can be corrected by driving the two wedges down further.

4 - Mark position of holes for securing screws on bow through the holes in the rear inner panel and the hinge support panel.

5 - Take rear body bow out.

6 - Drill six .24 in. (6.2 mm) holes for the securing screws and countersink them from above .28 in. (7 mm) deep with a .5 in. (12 mm) drill.

7 - Install bow again, insert wedges and shape the brackets to fit.

8 - Secure bow with the screws and nuts and bow to brackets with barbed nails.

9 - Cover countersunk holes with adhesive tape.

10 - Pull wedges out upwards.

B - Three-piece Rear Body Bow

Removal

1 - Take seats out and detach luggage compartment lining as far as necessary.

2 - Pull headlining off rear body bow and pull out tacks securing rubber bands on each side. Tape up the detached parts of the headlining near the main bow.

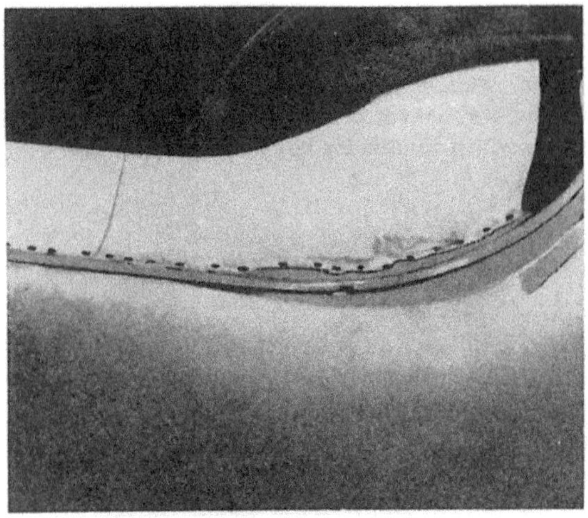

3 - Remove hexagon nuts from the threaded ends of tensioning wire.

4 - Remove screws securing the trim molding and take trim molding off.

5 - Pull top cover and wire out of "U" channel.

6 - Carefully detach rear window linen and side padding from rear body bow.

7 - Pull out the staples securing the webbing straps to the bow, using a nail-puller.

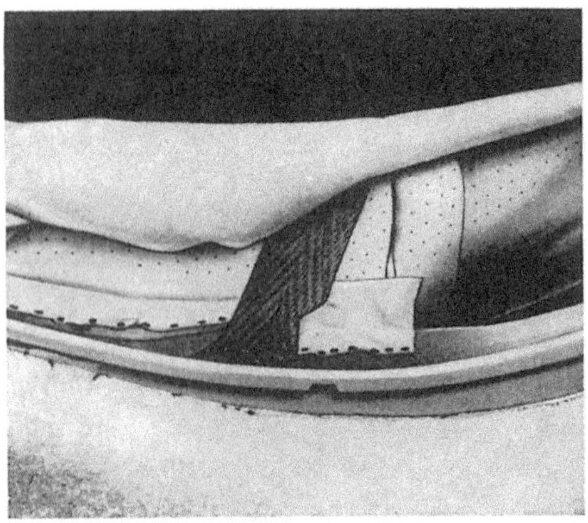

8 - Fold detached parts of top cover and corner padding forwards and attach them near the main bow with adhesive tape.

9 - Pull out the three nails attaching the bow to the brackets.

10 - Remove two screws from each section of the rear body bow.

11 - Take individual parts of bow out upwards.

Installation

The rear body bow is supplied in three separate parts and it should be noted that the side parts are not interchangeable.

Rear Body Bow

Left side part

Center part

Right side part

To prevent noises, stick strips of top material on the surfaces of the bow which contact the sheet metal.

1 - Place the three parts of the bow between rear panel and securing brackets.

2 - Bolt the bow parts to the body.

Note:

Ensure that the ends of the bow segments align properly.

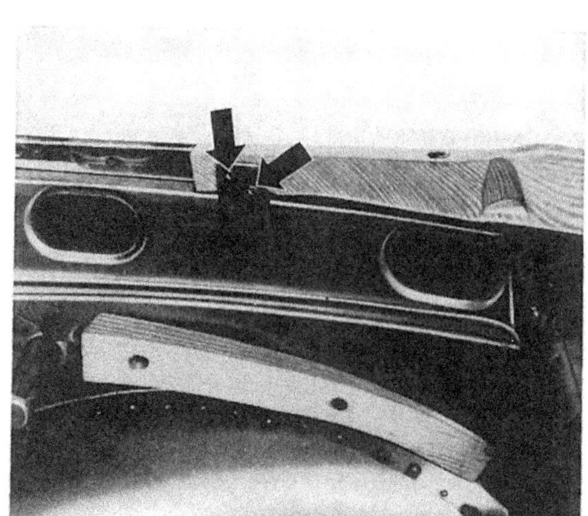

3 - Align the brackets and secure the bow segments by driving in the barbed nails.

The individual parts of the three-piece bow can be replaced without difficulty. Small cracks caused by inserting the staples too closely together, mainly in the vicinity of the webbing straps, can be repaired with wood cement.

The remaining operations such as fitting the headlining and padding as well as attaching the webbing straps and the top cover to the rear body bow are the same as decribed in section A 40 — "Top Assembly".

It is advisable to fit a new beading and two-piece trim molding on the nailed top cover and a new trim molding in the case of the top cover with tensioning wire.

To avoid corrosion, attach the webbing straps and top cover with brass tacks or stainless steel staples.

A strip of plastic sealing compound must be inserted between the rear panel and the **one-piece** rear body bow.

Replacing Left or Right Roof Framet

Top frames manufactured **up to** August 1964 and **after** August 1964 are different. For this reason, bows and roof side members **cannot** be interchanged.

In addition, the guide bars on the top frames manufactured after August 1964 are attached to the roof side members by rivets instead of bolts.

Top frame up to August 1964

Top frame from August 1964

Removal

1 - Unscrew roof side member weatherstrip.

2 - Unscrew handles and top fasteners together with molding from top header.

3 - Unscrew chromed screw from roof side member and detach top cover, padding and headlining from top header.

4 - After bending up sheet metal tab in top side member, remove spring and tensioning cable for top cover.

5 - Carefully detach top cover and headlining near inner and outer wooden facings on main bow.

6 - Unscrew outer wooden facing on main bow.

7 - Fold top cover and padding as far back as possible and hold it in position with tape.

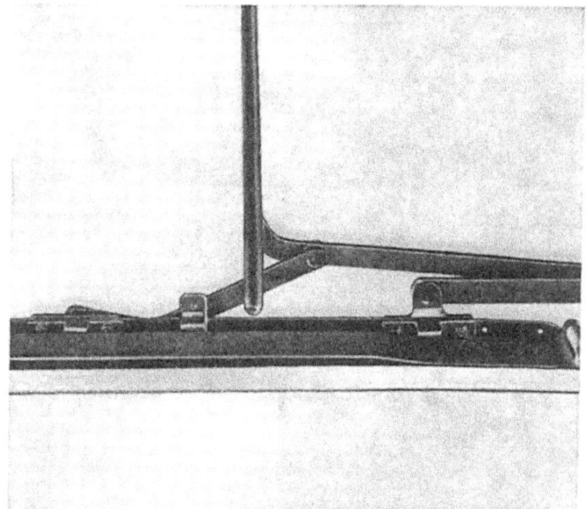

8 - Detach roof side member guide bar and main bow guide bar from side member as follows:

a - Up to August 1964: remove two hexagon head bolts and their nuts from each of the front and rear brackets on roof side member, then push out guide bars.

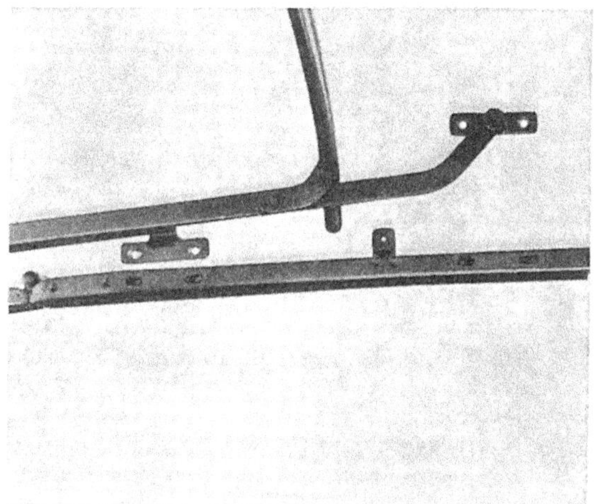

Note:

To allow the tapped plate to be adjusted in the elongated holes of the roof side member, separate a welding run. It is not necessary to weld the tapped plate to the roof side member, after adjusting the top frame, as long as the hexagon bolts are firmly tightened.

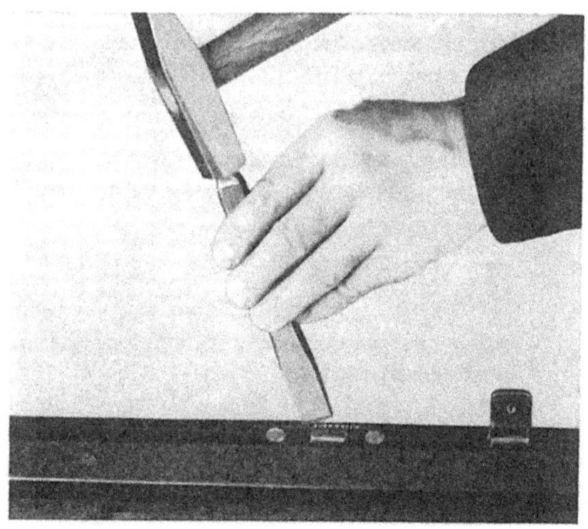

Important

The shims on the rivets must be reinstalled in the same positions.

When driving out the rivets, use a heavy hammer as a support.

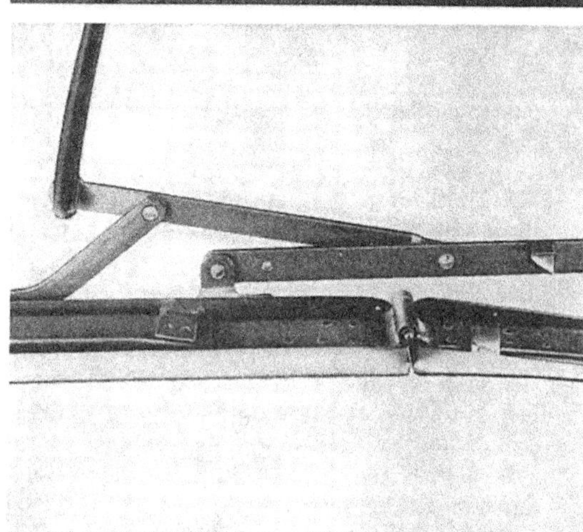

b - From August 1964: detach front guide bar/ metal bow guide bar connection and rear pivoting point of roof side member guide bar by drilling out the rivets and push out guide bars.

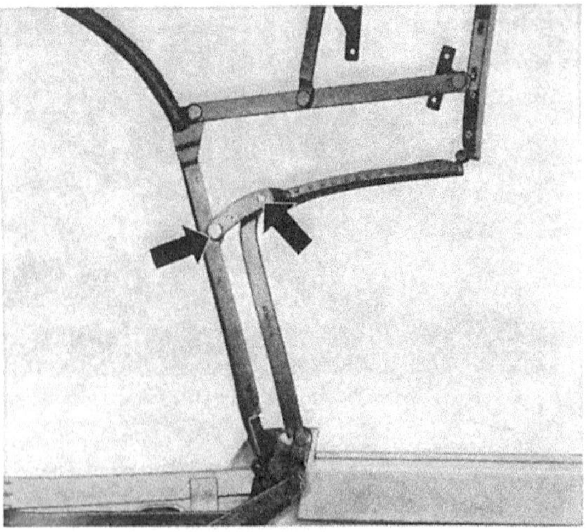

9 - Detach main bow from roof side member by drilling out the rivets. Drive the drilled rivets out of main bow with a punch.

10 - Remove screws connecting header to roof side member and remove roof side member.

Installation

A - **Roof side members up to** August 1964

1 - Insert rivets, with the correct shims, into roof side member at main bow and peen them.

2 - Bolt top header to roof side member. Do not forget plastic wood.

3 - Bolt front and rear brackets reasonably tightly in the elongated holes in roof side member.

Note:

When carrying out this operation, the top frame must remain closed.

4 - Set height of tubular bow to that of top header. The height can be adjusted by moving the bracket in the elongated holes.

5 - Install top lock and fasten top on both sides.

6 - Check distance of drop quarter window from roof side member. The height of the roof side member can be altered by moving rear bracket.

7 - Tighten bracket attaching screws. Carefully fold top back and compare distance of support pillar and roof side member hinges on one side with those of the other side. The distance must be equal. If necessary, correct this at rear bracket. (See also A 41-17/18.)

8 - Check operation of top by opening and closing it several times.

9 - Check position of windows to weatherstrips. If necessary reposition weatherstrips or adjust window glass.

B - **Roof side members from** August 1964

1 - Insert rivets, with the correct shims, into main bow and peen them.

2 - Bolt top header to roof side member.

3 - Insert front guide bar rivets, with the correct shims, and rivets for rear pivot point/roof side member, then peen them.

4 - Carefully fold top back and compare distances of support pillar and roof side member hinges with those of the other side. The distances must be equal. If necessary, correct these at the rear bracket. (See also A 41-17/18.)

5 - Completely assemble top and check operation by opening and closing it several times.

6 - Check position of windows to weatherstrips. If necessary, reposition weatherstrips or adjust window glass.

Replacing Top Headlining

(with top cover installed)

Removal

1 - Remove rear window and, if necessary, interior light.

2 - Detach luggage compartment floor trim panel and left and right wheel housing trim panels at the sides.

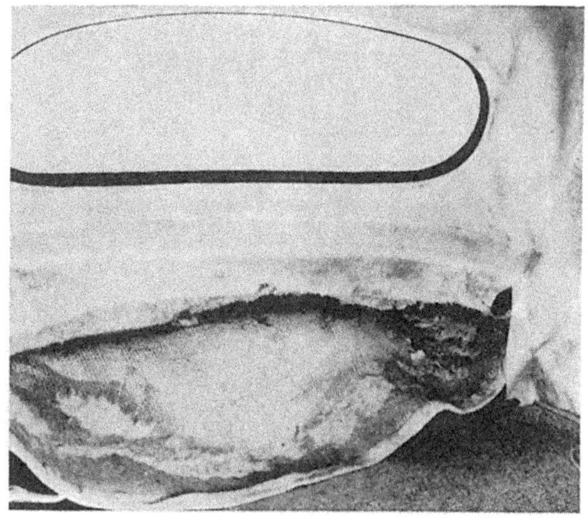

3 - Detach and remove top headlining from rear body bow, rear window frame and laterally at the main bow.

4 - Starting at rear bow detach headlining from support strips.

5 - Remove top handles, fasteners and molding from header.

6 - Detach left and right headlining side panels from roof side members.

7 - Detach headlining at the header.

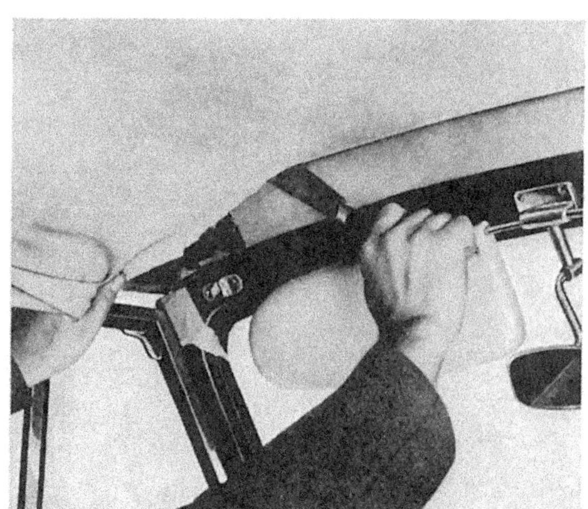

Note:

When doing this, exercise extreme care to avoid damage to the linen sheet beneath it.

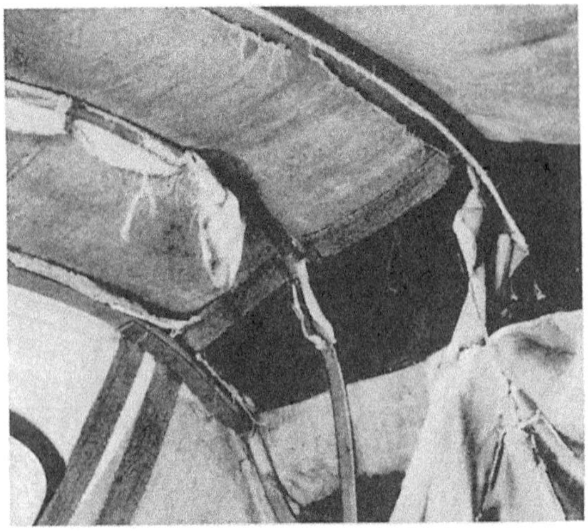

8 - Remove headlining support strips from the bows.

9 - Remove rear wooden facings if their covering material is worn or water stained. In this case, recover side facings, support pillar wedges and outer wooden facings on main bows.

Installation

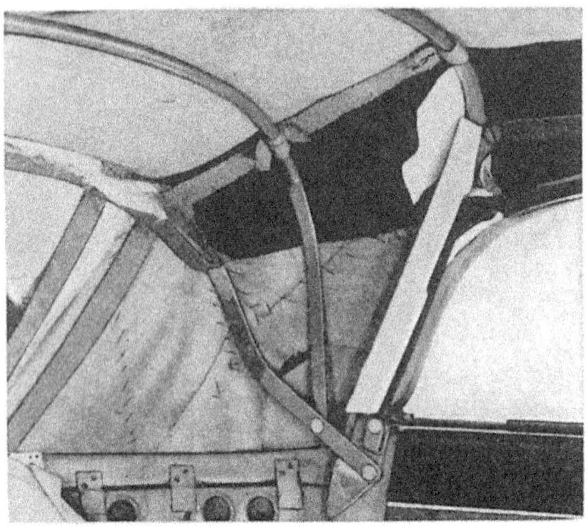

1 - Attach left and right inner wooden facings to main bow with three Phillips screws each.

2 - Fold excess leatherette of facings inside round the main bow and cement it in position under slight tension. Staple leatherette at outer edges and cut off excess material.

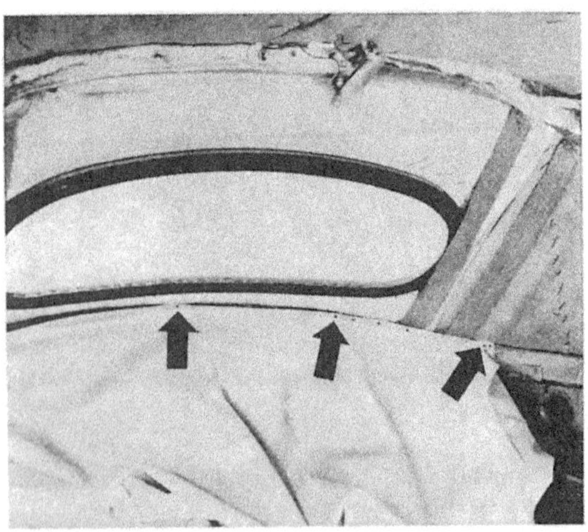

3 - Tack support strip of headlining exactly in the center and then laterally, under slight tension, to rear body bow.

4 - Check position of headlining near rear body bow and correct if necessary. Then staple support strip to rear body bow, without creases.

5 - Tack headlining support strip to rear bow, according to the center marking.

6 - Check that headlining is correctly positioned, then staple support strip under tension and without creases to rear bow.

7 - Secure interior light bracket to rear bow.

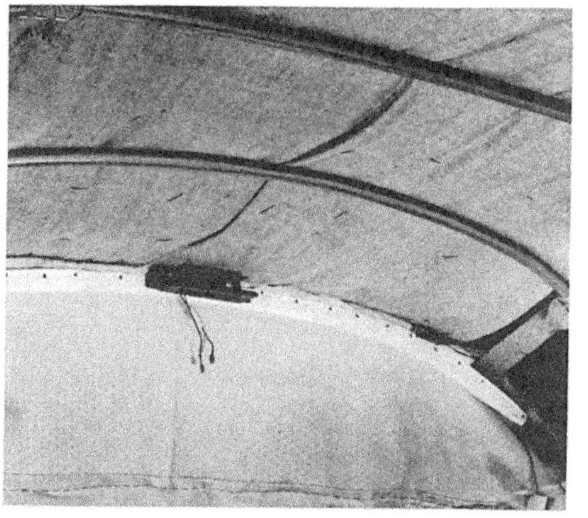

8 - Thinly coat rear tubular bow with universal adhesive D 12 and, according to the center marking, stick support strip to tubular bow. Ensure that the support strip is pulled up high enough so that its seam contacts underside of tubular bow.

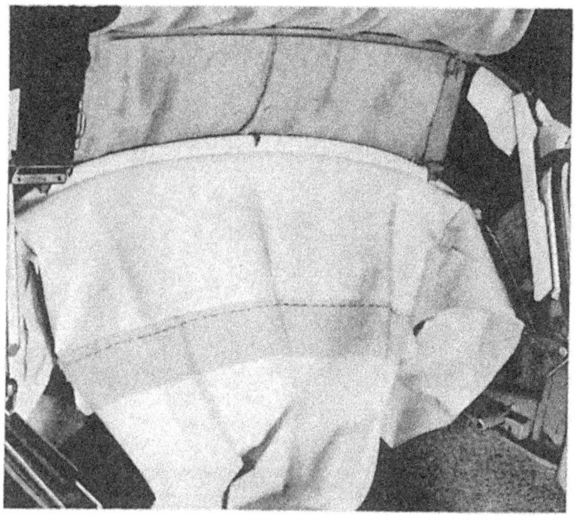

9 - Additionally sew outside of support strip to tubular bow over a length of .8″ (20 mm).

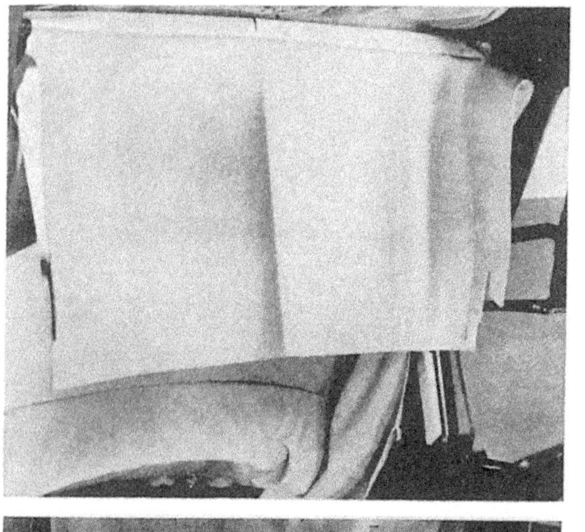

10 - Thinly coat main bow and front tubular bow with universal adhesive D 12 and, according to the center marking on the support strips, pull headlining upwards until the seam contacts the underside of the bows.

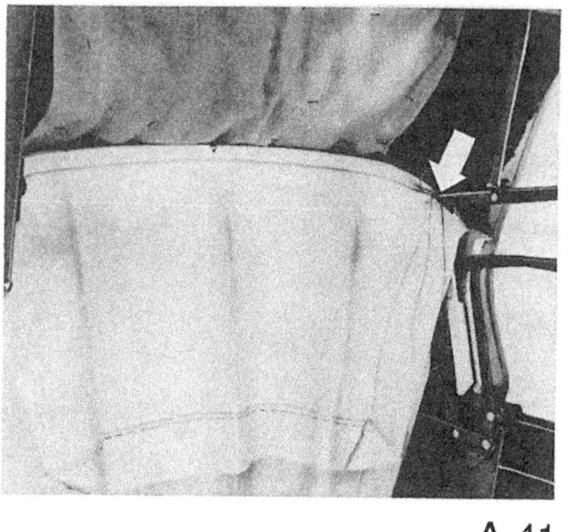

11 - Additionally sew support strips on outside to the bows, over a length of .8″ (20 mm).

12 - Attach headlining to header. To do this, proceed as follows:

a - Cut a piece of cardboard .24" × 45.3" (6 × 1150 mm).

b - Cut out two aluminum strips .8" × 4.7" (20 × 120mm) and drill two .14" (3.5 mm) holes in them. (As material, use a weatherstrips rail).

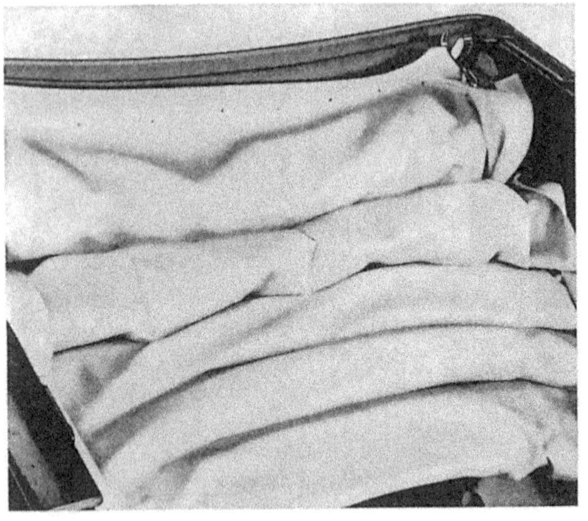

c - Cut out strip of headlining material 3.2" × 63" (80 × 1600 mm). This strip is used for covering the gap between molding and top headlining.

d - Install top fasteners.

e - Tack headlining to header, leaving enough gap between front tubular bow and header, then carefully close top. Check top headlining tension. If necessary, make any corrections at this stage.

Important
Ensure that the seam of the front support strip runs parallel to the front tubular bow.

f - Open top and unscrew top fasteners.

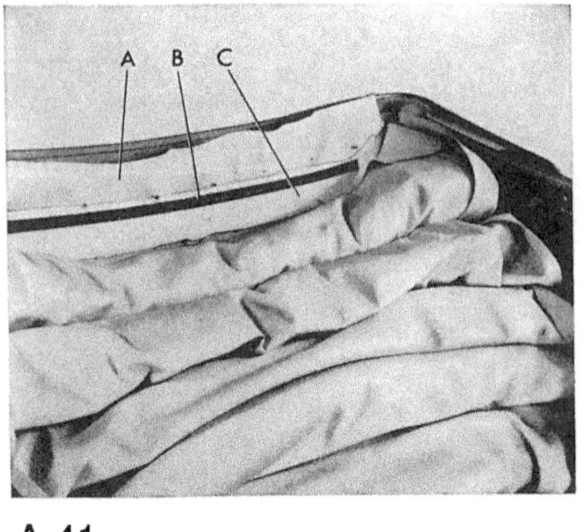

g - Staple cardboard strip, together with material strip on left-hand side, to header so that it fits evenly to top of header.

A - Top headlining
B - Cardboard strip
C - Leatherette strip

h - At the side near the roof side member, cement material strip to roof side member attaching bracket, then secure it additionally to roof side member with an aluminum strip and screws.

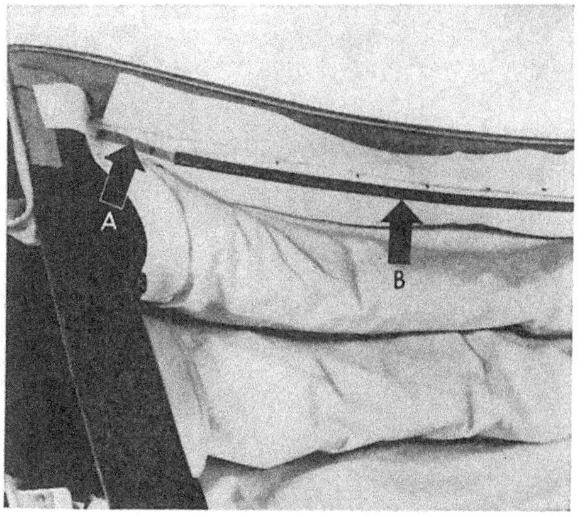

A - Aluminum strip
B - Cardboard strip

i - Pull tacks out of top header and cleanly cut off overhanging strip of headlining at cardboard or aluminum strip.

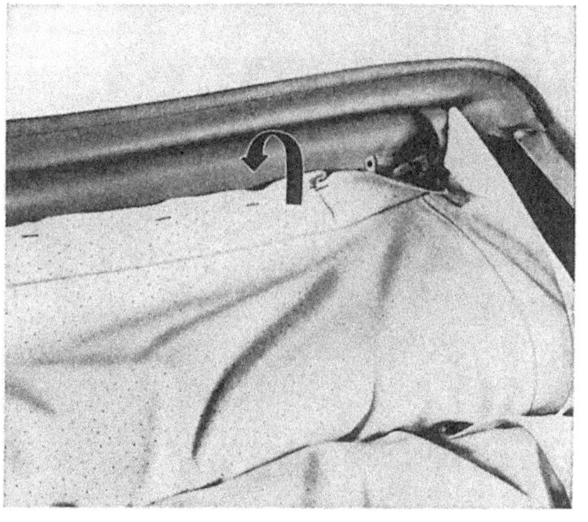

j - Thinly coat material strip with universal cement D 12 and cement to header. Attach it additionally with several staples.

13 - Secure molding to header with two Phillips screws. Close top carefully and check position of molding near headlining. Correct, if necessary.

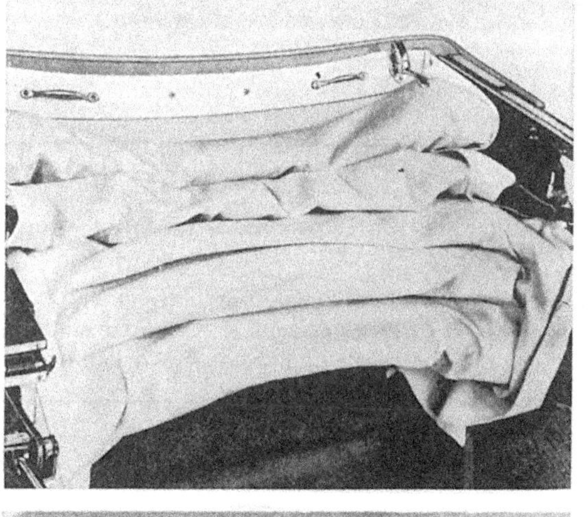

14 - Attach handles and top fasteners to header and close top.

15 - Secure lateral parts of headlining under slight tension to roof side member. Do not forget washers.

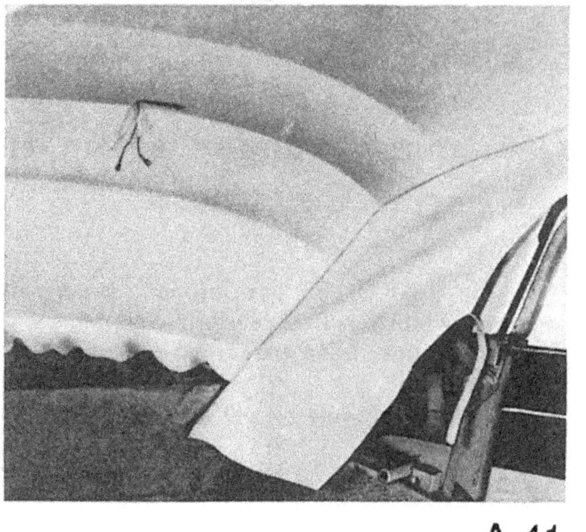

16 - Carefully cut out headlining round interior light bracket.

17 - Attach headlining with two large-headed tacks to inner wooden facings of main bow, cement without creases near luggage compartment and additionally staple at rear body bow.

18 - Tension headlining, cement to left and right wooden facing on wheel housings and additionally staple it.

19 - Cut out opening for rear window and insert window.

20 - Connect interior light, then attach press studs for the boot, backrest securing hook and, if necessary, cover plates for safety belt anchorages.

21 - Check rear window for leaks by spraying the rear part of the top with an even jet of water. If the rear window leaks, remove it and seal it as outlined in section A-42.

Important

When work is completed, check that the top opens and closes properly and if the fasteners engage correctly.

New tops always have a certain amount of tension. It is not necessary to slacken this tension in any way. If, after installing a PVC top cover, there are still creases in the material, these can be easily eliminated by the application of warmth (fan or lamp).

Subsequent Fitting of Rear Window when Top Cover has Shrunk Excessively

If the top cover has not been properly looked after or if it has been affected by industrial fall-out, the top material of fabric or PVC can shrink excessively. Due to the resulting expansion of the wooden frame holding the rear window, leaks are sure to appear near the rear window.

1 - Remove rear window and weatherstrip.

2 - Place rear window and the cleaned weatherstrip loosely in rear window frame and push them upwards.

The gap appearing between rear window frame and rear window weatherstrip shows where a correction by inserting cardboard strips must be made.

3 - Cut an appropriate cardboard strip and shape the ends to make a good joint with the contours of rear window frame.

4 - Detach staples near lower part of top cover and carefully remove top cover. Cement cardboard strip to rear window frame and additionally secure it with a few staples.

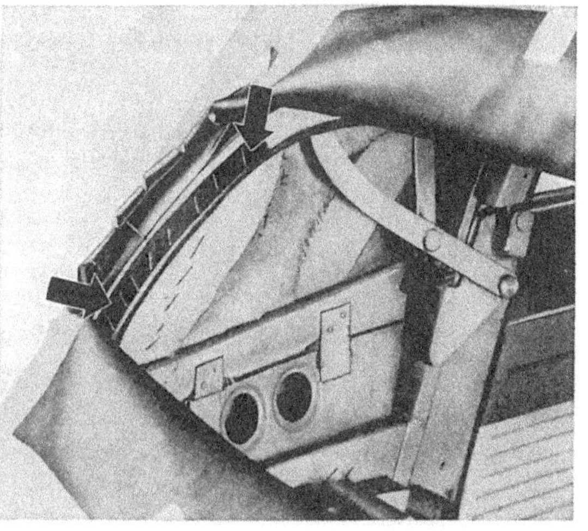

5 - Unfasten top and open it somewhat.

6 - Pull top cover over cardboard strip and staple it, free of creases, to rear window frame.

7 - Thinly coat stapled surface of wooden frame with glazing cement D 10.

8 - Place rear window and weatherstrip into window frame. The ends of the cable used for assembly must hang down inside.

Note:

On installation, a second helper must counterhold the wooden frame from car interior, as otherwise the pressure applied to install the window from outside can tear the linen material off the rear window frame.

9 - Carefully remove excessive sealing compound with benzine, clean window with white spirits and check for leaks.

Corrections to Roof Side Members and Bows

Guide tongues of top locks strike bottom part of retaining lugs.

If the guide tongues of the top locks strike the retaining lugs when the top is closed, check whether the roof side members are located properly on the window frames and if the gap between the quarter panel and the doors widens towards the top.

If the roof side members are resting on the window frames, the top frame has probably been damaged by being opened in a careless manner. In this case, straighten the side members. See Section "Straightening the Roof Side Members".

If the door gaps widen towards the top, unscrew the body securing bolts at the spring plates. Place hard rubber packing pieces .1" (3 mm) thick under the body support brackets on both sides. Reinstall body bolts and tighten them. If the door gaps are still not the correct width, thicker packing pieces must be installed or the existing ones must be ground down.

If the roof side members are not resting on the window frames or the door gaps do not widen, bend the guide tongues towards the front until they enter correctly.

Guide tongues of top locks strike the windshield frame.

If the guide tongues of the top locks strike the windshield frame when the top cover and top headlining are properly tensioned, shims should be inserted under the upper parts of the locks until the guide tongues enter correctly.

Guide tongues of top locks slide to the side of the retaining lugs.

If the guide tongues slide to the side of the retaining lugs, lay the top loosely on to the windshield frame (do not fasten the top).

Press the top frame, without applying excessive force, in the required direction until the guide tongues enter the retaining lugs properly.

Straightening the Roof Side Members

The distance between the rear drop windows and the roof side members is .5" (13 mm). If this distance cannot be attained, carry out the following corrections:

If the roof side members are too flat, press them outwards and if the bend is too pronounced, press them inwards. For carrying out this work, use either a drift and a heavy hammer or hydraulic tools.

If a roof side member sags near the hinge, the two metal covers must be removed, provided with supports and be reinstalled. If the bend is too weak it is advisable to reduce the thickness of the metal covers.

Straightening the Bows

The bows must run in one plane. A sight check can ascertain whether the bows are bent laterally or whether they are uneven near the radii. The sight check is carried out from the front to determine the position of the bows to each other, mainly near the lateral radii, and also from the rear of the vehicle.

According to the amount of correction work to be carried out, press at one location or, if space is limited, use a counterweight (heavy hammer or mallet) and apply blows with a rubber hammer.

If the top frame is bent laterally, the linkage has only to be pressed in the required direction.

Convertible Top
(Early)

The top is supported by the top linkage which is composed of two side metal roof frames and two hinged linkages joined together by wooden and tubular bows which give cross support. The whole assembly is supported by two main hinges mounted in brackets bolted to special pillars in the body.

The outer cover is made of special waterproof top material which is rubberised underneath. A rubberised hair pad, sewn into a linen sheet and secured to the top linkage, pads out and shapes the top in addition to providing insulation against heat and cold. The contours of the top are, where necessary, evened out by additional wadding. The underside of the top linkage is covered by a headlining of loosely woven material.

In the folded position the top is retained by two spring catches.

Lubrication

A few drops of oil should be applied to the hinge points of the top linkage when required. Wipe the dirt and dust off the joints before oiling.

Saddlers Tools

(for work on Convertible top)

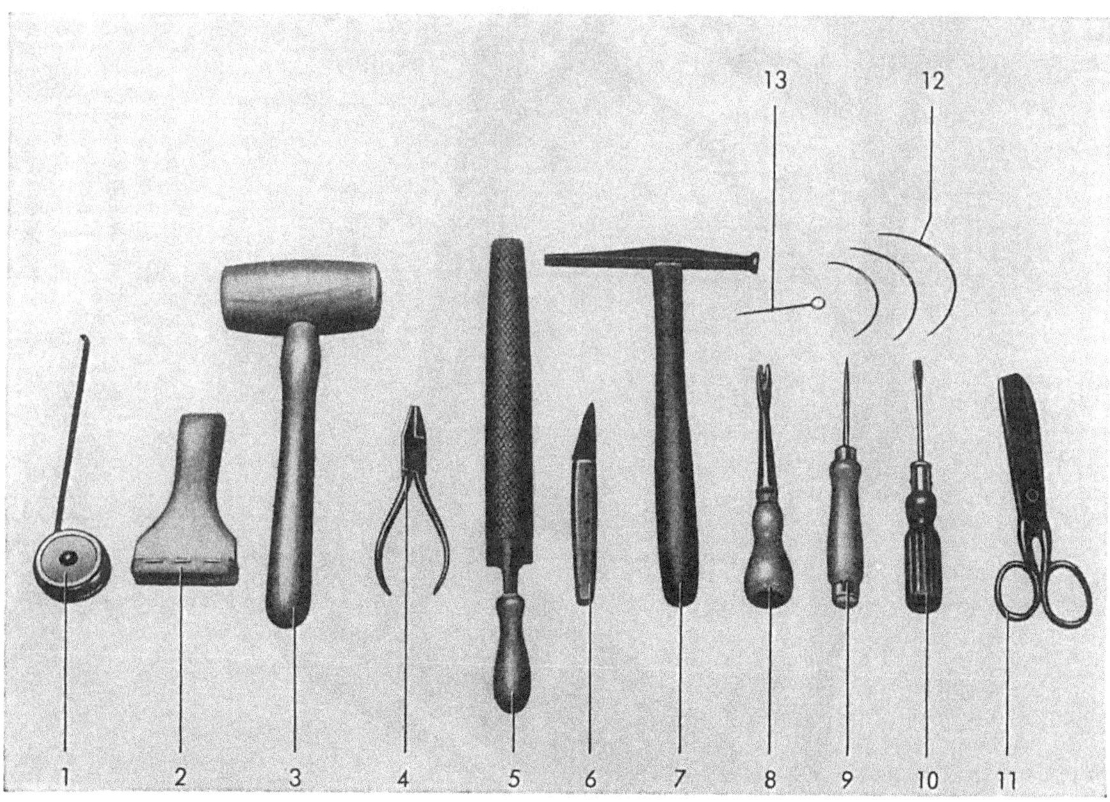

1 - Steel tape
2 - Moulding inserter
3 - Mallet
4 - Flat nose pliers
5 - Rasp
6 - Upholsterer's knife
7 - Upholsterer's hammer
8 - Nail claw
9 - Awl
10 - Screwdriver
11 - Scissors
12 - Upholsterer's needles (different sizes)
13 - Pins

Top Frame

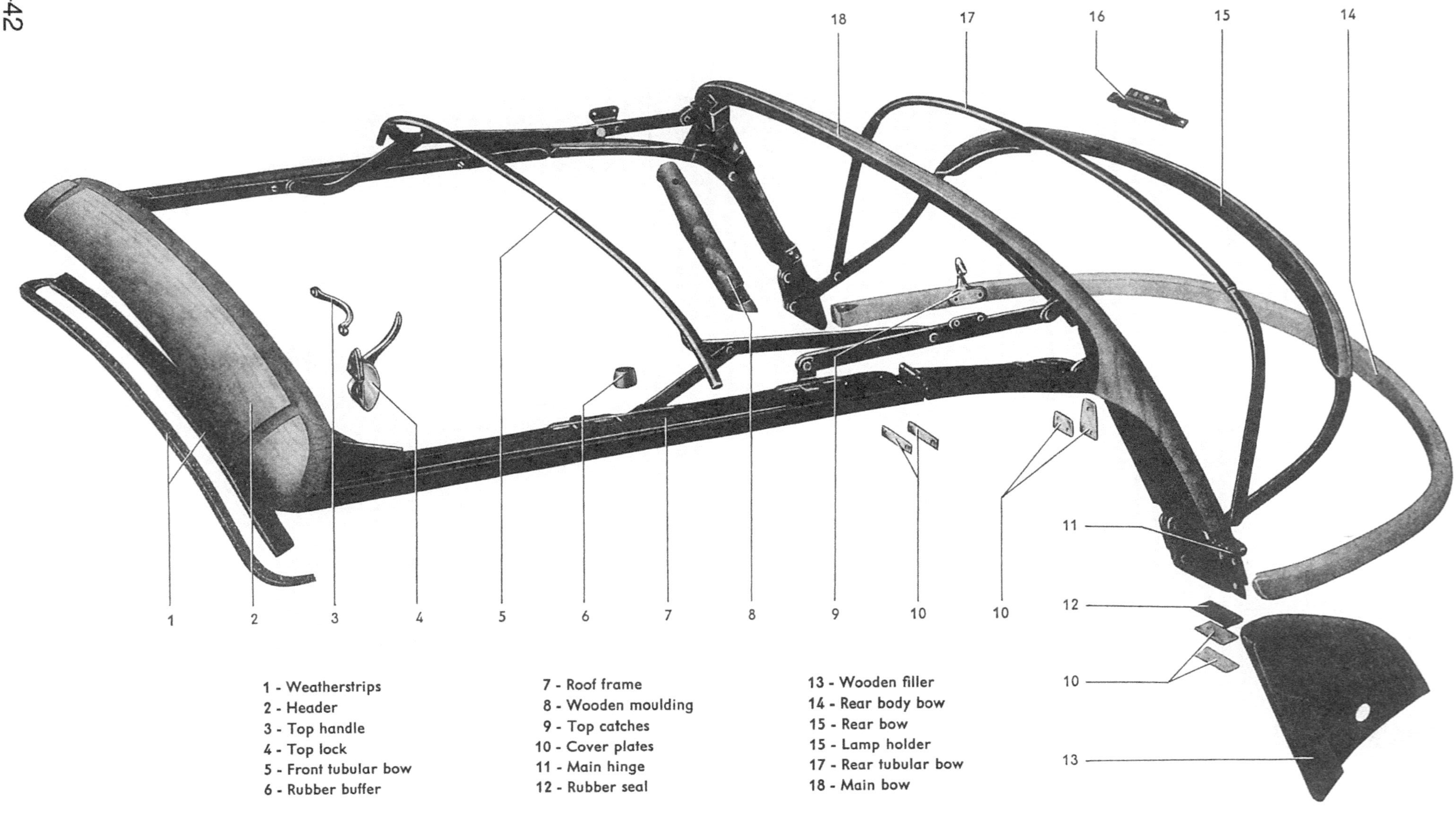

1 - Weatherstrips
2 - Header
3 - Top handle
4 - Top lock
5 - Front tubular bow
6 - Rubber buffer

7 - Roof frame
8 - Wooden moulding
9 - Top catches
10 - Cover plates
11 - Main hinge
12 - Rubber seal

13 - Wooden filler
14 - Rear body bow
15 - Rear bow
15 - Lamp holder
17 - Rear tubular bow
18 - Main bow

Top Assembly

The description of the Convertible top assembly serves as a guide for the sequence of operations. The work should, where possible, be carried out by a skilled man who is familiar with top construction. If two men can be employed, the task of fitting the top cover will be made considerably easier. The assembly of individual parts of the top described in some paragraphs can naturally be carried out separately.

The tools required are listed in the "Body General" section — under the heading "Workshop Equipment".

Note:

From Chassis No. 3 492 689 (2nd January 1961) synthetic material, instead of cloth, is used intermittently for the VW Convertible top.

These tops are now welded instead of being sewn.

Installation of Top Frame

1 - Before installing the top frame, remove the seats, quarter trim panels and the carpets from the luggage compartment and both rear wheel housings. Wind the drop windows down.

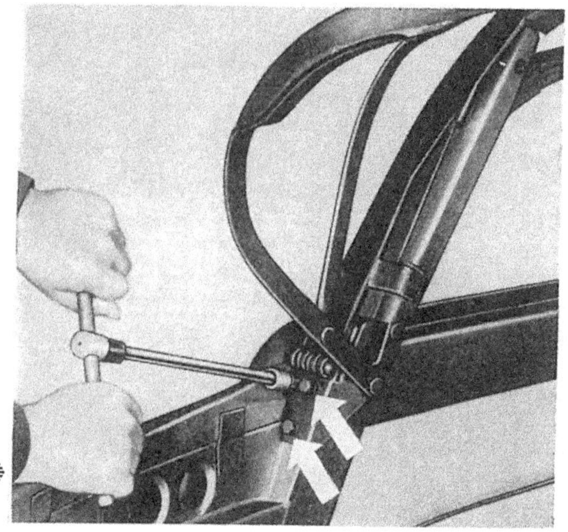

2 - Place top frame in position and insert the three screws in the tapped plate of the box pillars on each side but do not tighten them.

3 - Push the top frame sideways until the wooden part of the main bow projects about 2 mm inwards at left and right as otherwise leaks will appear in this area. If this measurement cannot be obtained by adjustment the wood must be rasped off the appropriate amount.

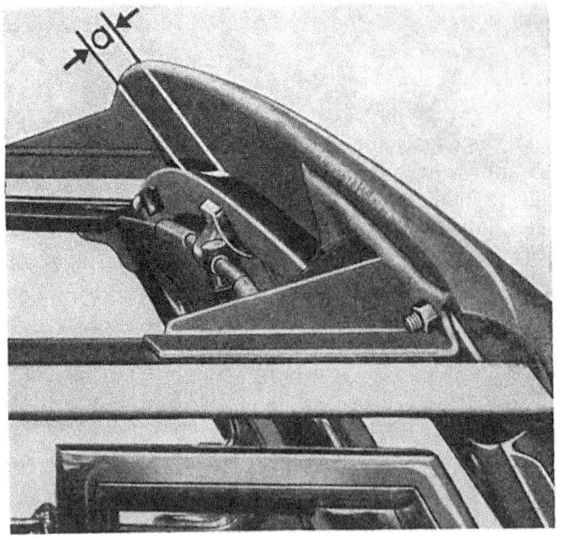

4 - Check that the header contacts the windshield frame properly. The rear surface of the header must be about 10 mm in front of the inner edge of the windshield frame as the frame curves more after the cover has been fitted and tensioned and pulls the header back.

a = 10 mm (.4″)

5 - Remove the lamp-holder from the rear bow and the wooden mouldings from the main bow.

6 - Lay the top frame back.

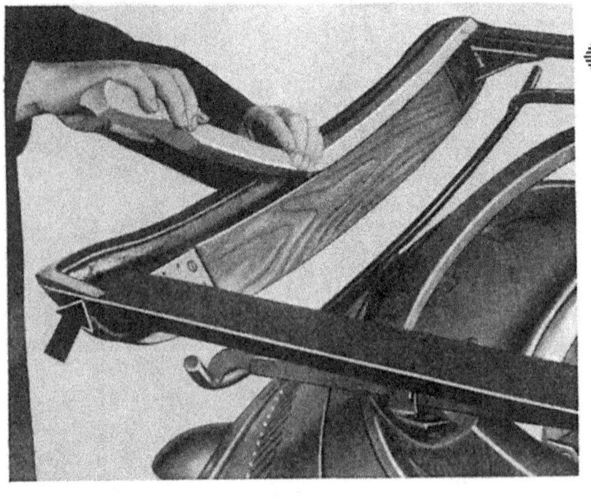

7 - Coat the underside of the header with adhesive. Cement a strip of foam rubber (10x16x1350 mm) Part No. 151 871 349 to the front edge and a strip (18 x 38 x 1250 mm) Part No. 151 871 345 to the underside of the header in the recess provided. Both strips should be tapered at the ends slightly to give a smooth contour.

8 - Cut a piece of top cover material and a piece of headlining roughly 1250 x 150 mm (49.2″x 6″) in size and of the same color as the top and headlining for the vehicle.

Note:

A suitable piece of leatherette — similar to that used in current production from Chassis No. 2 336 614 — can be used instead of the headlining material.

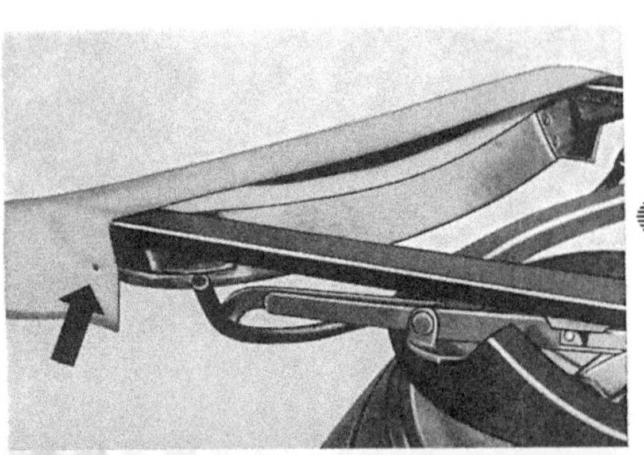

9 - Tack the top cover strip to the ends of the *header under tension.*

Ensure that the top cover strip extends about 30 mm over the ends of the roof frame.

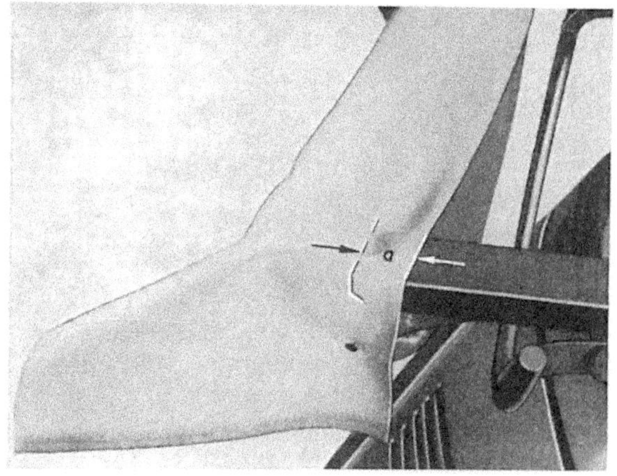

a = 30 mm (1.2")

10 - Cement the top cover strip to both roof frames.

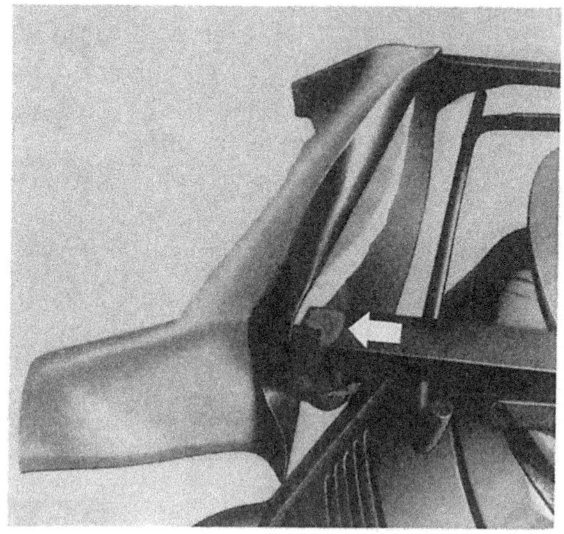

11 - Tack the top cover strip inside the header to within 50 mm of the rear edge. Cut the material suitable at the corners and at the holes for the the top locks.

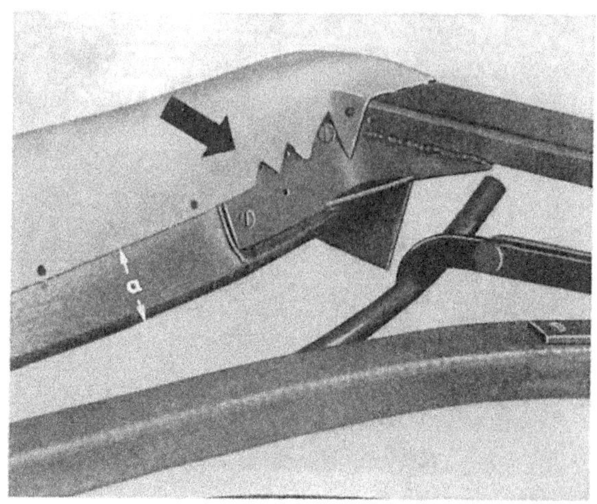

a = 50 mm (2")

12 - Tack the headlining on the inside of the header over the top cover material. Check that the back of the material is outside so that when it is folded over later on the right side will be visible.

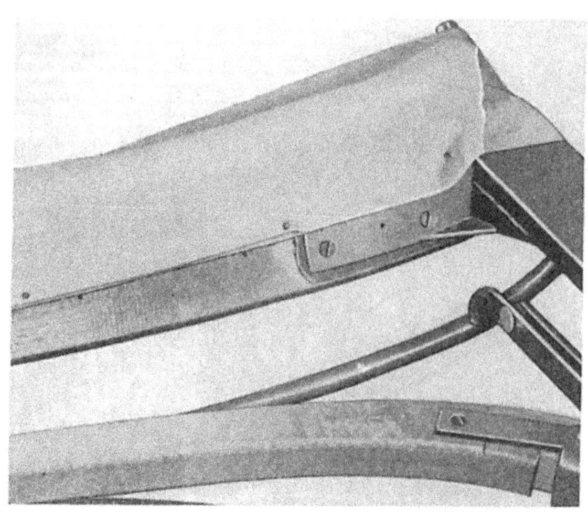

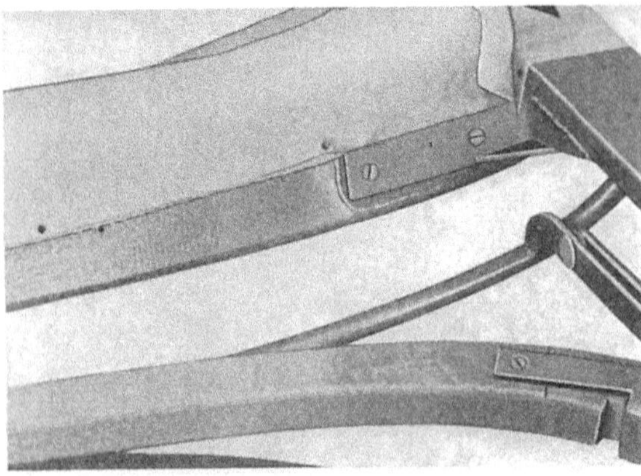

13 - Fold the headlining ends over and cement them to the material strip.

14 - Cut a strip of cardboard for the header to the following dimensions.

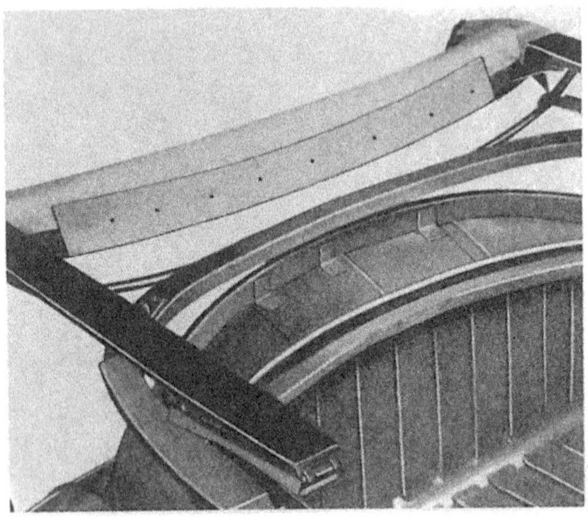

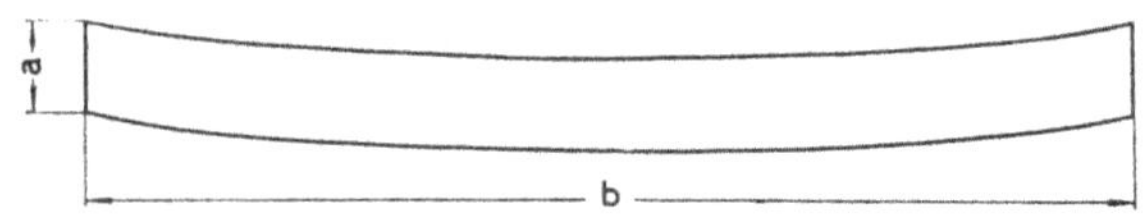

a = 75 mm (3")
a = 860 (33.8")
Approximately 3 mm (0.12") thick

15 - Nail the cardboard strip to the inside of the header so that the radius almost cuts the upper edge of the header.

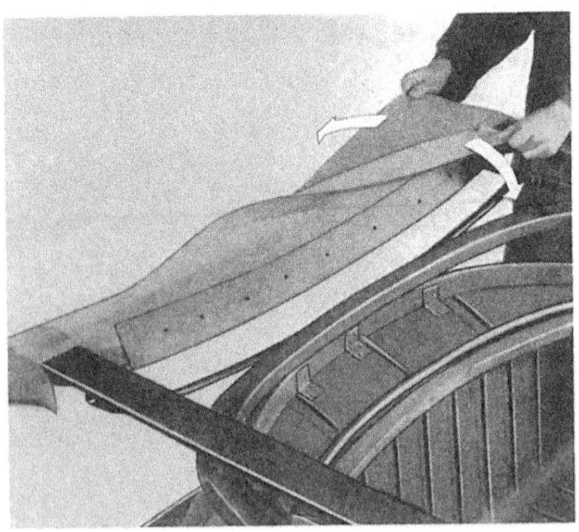

16 - Coat the inside and the lower part of the front of the header lightly with adhesive.

17 - Fold the headlining inwards and the top cover material outwards and cement in position without creases.

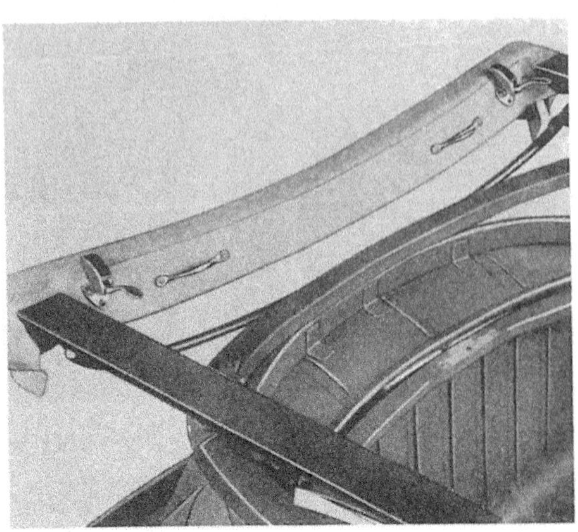

18 - Nail the headling in the recess on the top of the header, free of creases, and cut off surplus material.

19 - Screw the top locks and the handle to the header at the places marked by the holes.

20 - Close the top and nail the top cover strip to the header under side tension, about 25 mm (1") from the lower edge. To ensure uniform tension it is advisable to nail from the center outwards.

At both ends of the header the top material should be laid in small pleats to fit the curves better. Do not cut the material as this will cause leaks.

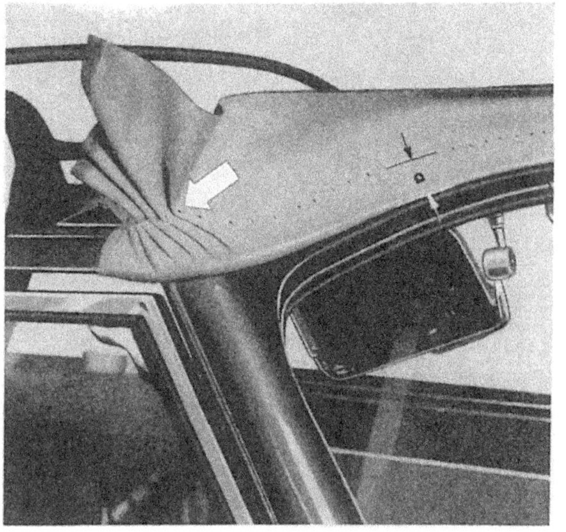

21 - Cut off the surplus material above the line of tacks.

a = 25 mm (1")

Note:

12 mm brass pins, Part No. 151 871 471
14 mm tacks, Part No. 151 871 473

Important

To avoid rust formation use only brass pins and tacks to secure the top cover and the webbing strips.

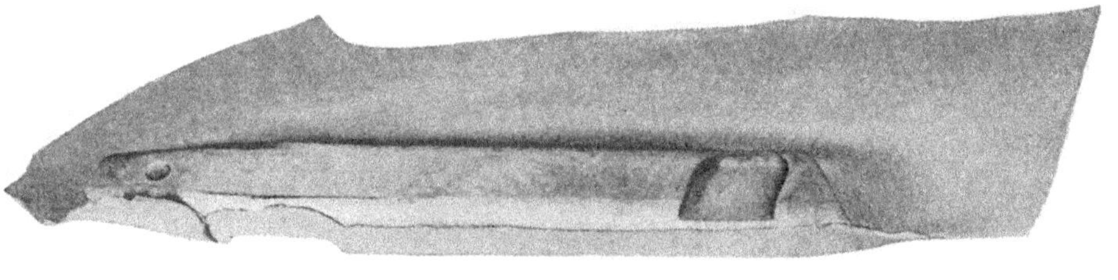

22 - Cement a piece of headlining material, 500 x 200 mm (19.6" x 7.8") of the same color as that on the vehicle, to the outside of the wooden mouldings.

23 - Lay the top halfway back.

24 - Cement a piece of headlining material, roughly 600 x 300 mm (23.5" x 11.8") to the sides of the wooden part of the main bow up as far as the curve on the inside. Cut out the openings for the brackets for the wooden mouldings.

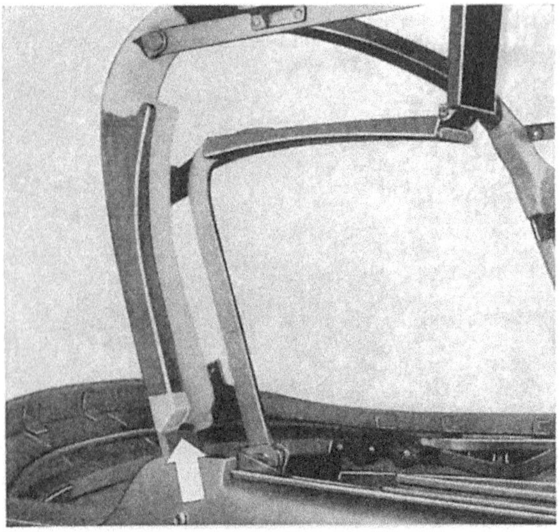

25 - Cement a piece of leatherette under the corners of the main bow and secure with a nail. This leatherette patch prevents the covering material from absorbing moisture.

26 - Install the wooden mouldings on the main bow. Lay the surplus material round the lower part of the main bow on the inside and cement and tack into position under slight tension. Cut off surplus material.

27 - Tack two small strips of leather on the front part of the rear body bow at each side to give a smooth contour between the body and the body bow.

28 - Close the top frame and secure the locks. Set the rear bow and the rear tubular bow according to the measurements below.

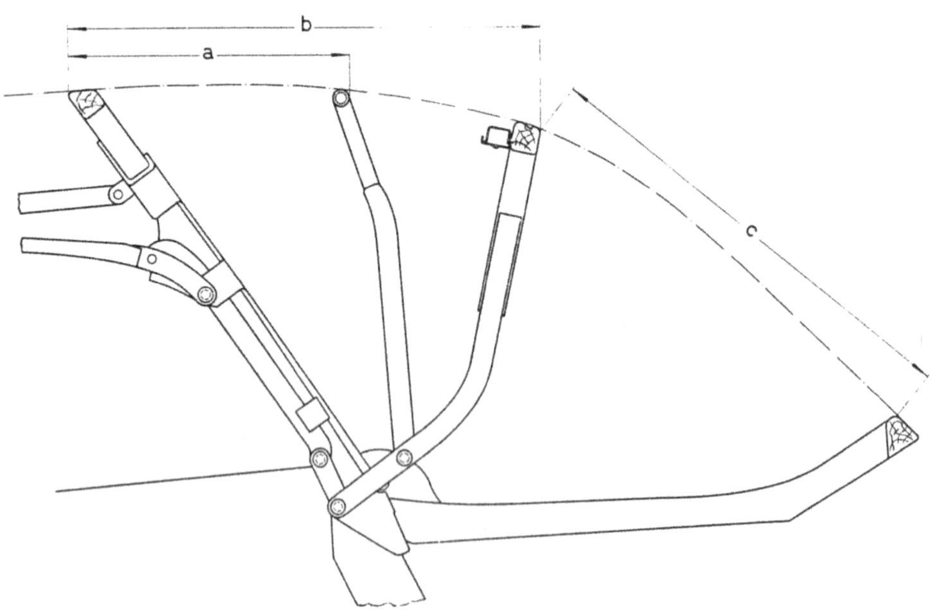

a = 320 mm (12.5")
b = 575 mm (22.6")
c = 515 mm (22.3")

Important

These measurements should be strictly adhered to as they govern the satisfactory padding and folding of the top.

29 - Nail the webbing strips to the body, rear and main bows under slight tension to the measurements given.

a = 390 mm (15.3") (from center of rear body bow)
b = 120 mm (4.7")
c = 250 mm (9.8")

Webbing tacks should be used to secure the webbing strips to the rear body bow.

The two webbing strips should be nailed to the sides of the rear bow as otherwise the trim moulding which is installed later will be too high at the ends and stand clear.

Nail the webbing strips to the wooden part of the main bow.

30 - Attach the webbing strips to the rear tubular bow by sewing on two loops of webbing material.

31 - Sew the interior lamp cable to the rear webbing strip between the body bow and rear bow and insert it through the hole provided in the rear bow.

Headlining Installation

The spare headlinings are supplied complete with the support strips, which are used for attaching it to the individual bows, sewn on.

To ensure satisfactory headlining fitting the material is attached to the various bows in the following order with the roof closed:

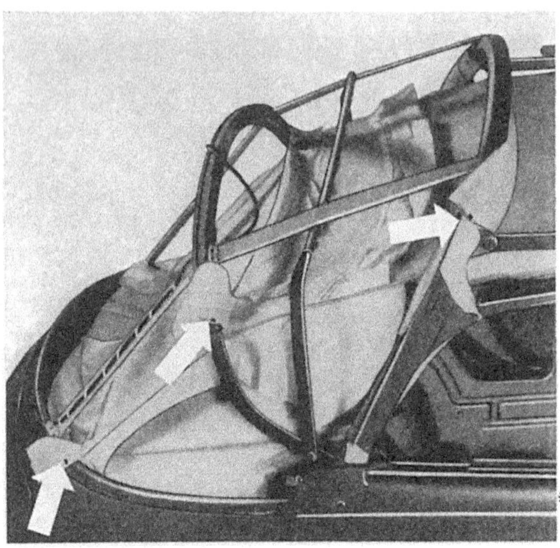

a - Main bow
b - Rear bow
c - Rear body bow
d - Header
e - Wooden mouldings
f - Rear tubular bow
g - Front tubular bow

Further details of the operation are as follows:

1 - First tack the headlining under slight tension to the ends of the header, the main, rear and body bows.

2 - Coat the front and top edges of the main bow with adhesive.

3 - Pull the headlining up by the appropriate support strip until the seam in the lining contacts the underside of the main bow. Cement the support strip to the bow and secure on the top with one tack each side. Cut off surplus material.

4 - Secure the headlining support strip to the rear bow in the same manner.

5 - Nail the support strip to the top edge of the rear body bow. Cut opening for the webbing strips.

6 - Nail the headlining into the recess on top of the header under uniform tension. Cut off surplus material.

7 - Attach the headlining to the rear edge of the header with one tack each side. This ensures that the longitudinal seams of the headlining are uniformly tensioned.

Make a wedge-shaped cut at the front of the two seams to give a smooth contour.

8 - Tack the headlining to the two wooden mouldings inside the body.

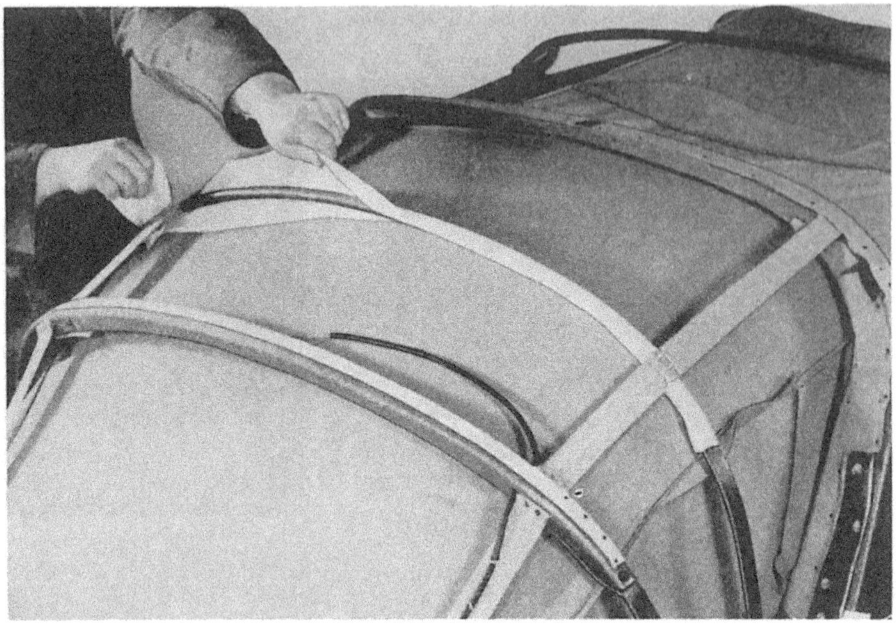

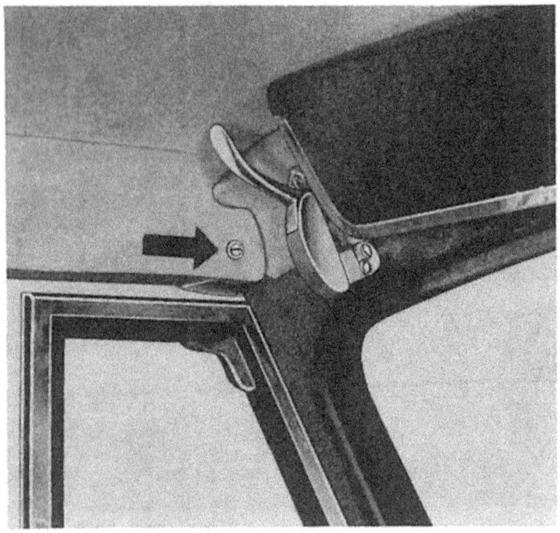

9 - Coat the rear tubular bow with adhesive, pull the support strip up to the seam and cement round the tube. Cut off surplus material.

10 - Secure the headlining to the roof frame inside by inserting one screw each side into the holes provided.

11 - Cement the headlining support strip to the front tubular bow.

Take care that the headlining material hangs down roughly 10 mm below the roof frame on each side.

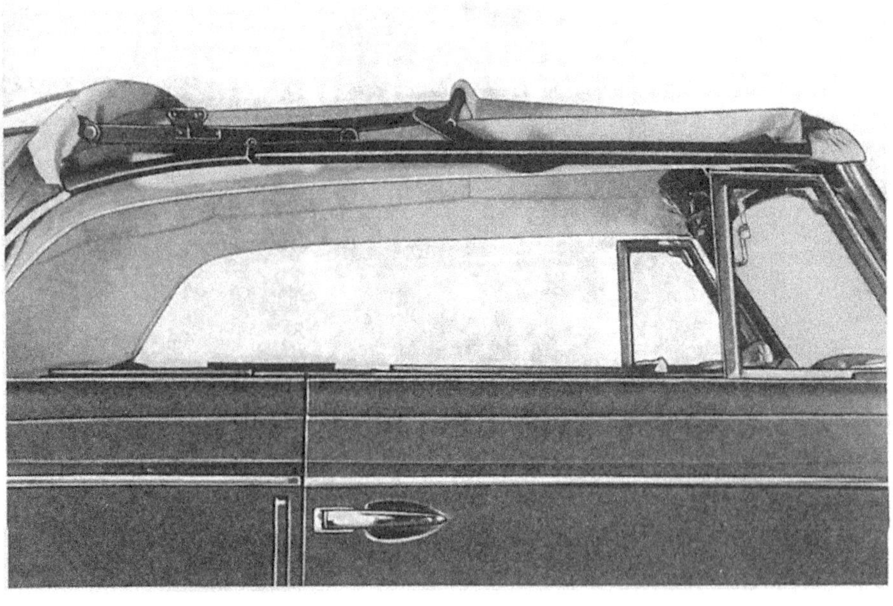

Sew the outer edges of the headlining.

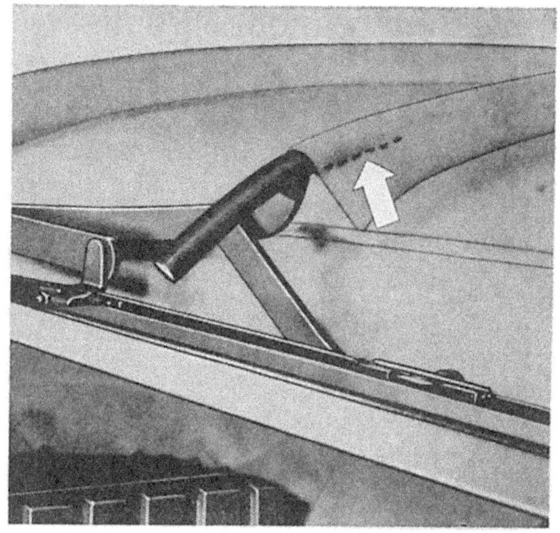

When these operations are completed, check the headlining from inside to ensure that the lining is taut, free of folds and that the seams of the support strips are straight. If necessary, correct at the appropriate bows.

12 - Replace the lamp holder on the rear bow and insert screws into the holes provided. Cut a hole in the headlining for the interior lamp and cement the lining to the lamp holder.

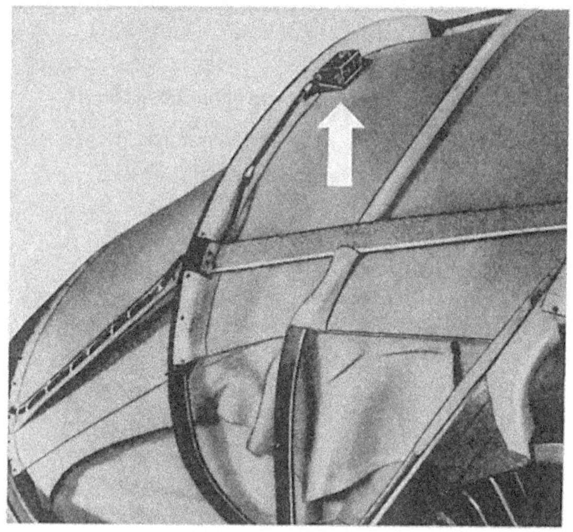

13 - Nail a length of 4 mm diameter cord to the main bow on each side, level with the headlining side seams. Loop once round the rear tubular bow and nail, under tension, to the rear bow.

14 - Nail a piece of headlining material on the curve of the main bow at each side.

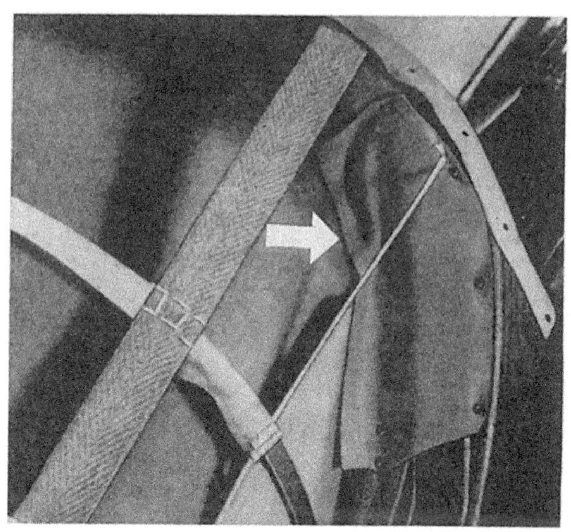

15 - Sew the headlining side seams to the cord under tension.

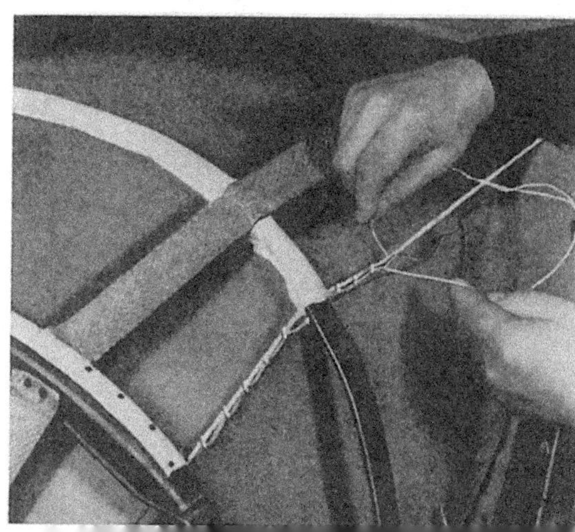

Finally, check from inside that the headlining side seams are straight. The lining must be re-tensioned if folds are present.

Installing Rear Window Frame

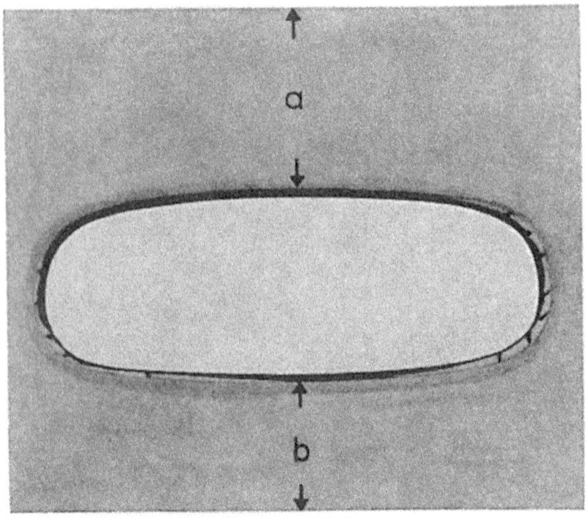

From Chassis No. 4 846 836:

a — 190 mm (7.48″)	a — 155 mm (6.1″)
b — 80 mm (3.15″)	b — 90 mm (3.5″)

The top must remain closed when the rear window frame is being installed.

Note:

From Chassis No. 4 846 836, the rear window was increased in size by 30%. The top frame was also modified.

The enlarged rear window — Part No. 151 845 501 B — can only be service installed in older vehicles if the complete top frame is replaced.

The previous rear window — 151 845 501 A — will remain available.

The top cover — Part No. unchanged 151 871 035 B — is suitable for both the previous and modified rear window.

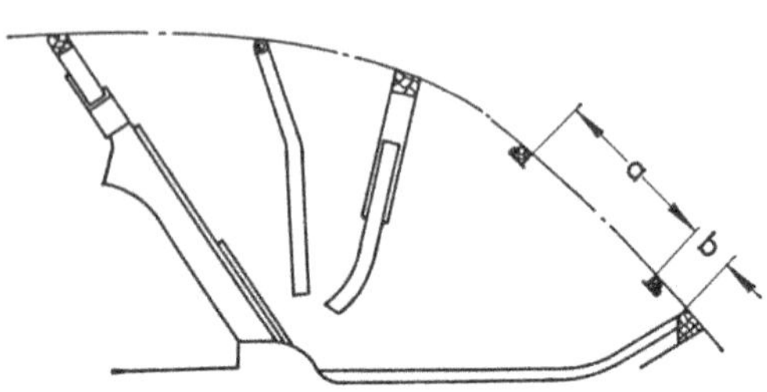

From Chassis No. 4 846 836:

a — 255 mm (10″)	a — 295 mm (11.6″)
b — 80 mm (3.15″)	b — 90 mm (3.5″)

1 - Cement a piece of coarse lining material (roughly 650×850 mm / 25.5″×33.5″) to the window frame so that the width is 250 mm at the top and 200 mm at the bottom.

2 - Cut the material round inside the window frame so as to leave a 15 mm wide border. Cement this border to the inner edge of the window frame and secure with tacks.

3 - Mark off the center line of the rear body bow, the rear bow and also the center of the window frame on the lining material.

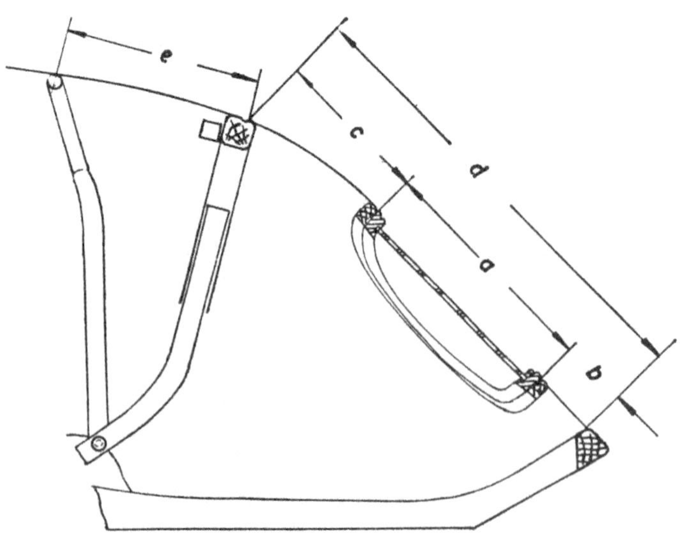

Dimension 1 up to Chassis No. 1 600 439
2 from Chassis No. 1 600 439 to Chassis No. 4 846 835
3 from Chassis No. 4 846 836

Dimension a - 1 - 192±2 mm
2 - 255±2 mm
3 - 295±2 mm

Dimension b - 1 - 120±2 mm
2 - 80±2 mm
3 - 90±2 mm

Dimension c - 1 - 165±2 mm
2 - 190±2 mm
3 - 155±2 mm

Dimension d - 1 - 480±5 mm
2 - 525±5 mm
3 - 540+5 mm

Dimension e - 1 - 240+3 mm
2 - 226+3 mm
3 - 213+3 mm

4 - Nail the material with the window frame to the recess in the rear bow and on the rear body bow under tension and in line with the center marks.

5 - Nail two additional webbing strips to the rear bow and on the rear body bow under tension so that the rear window frame rests on the curve without affecting the size of the window opening.

Important
Measure the window frame to check that it is horizontally central and correct if necessary.

6 - Fold the lower edge of the material over and nail to the rear body bow again to ensure adequate strength. Cut off surplus material.

7 - Cut the headlining in the window frame, leaving a 15 mm wide border which is cemented to the frame without folds.

Installing Linen Sheet

First cut five pieces of linen material to the following dimensions:

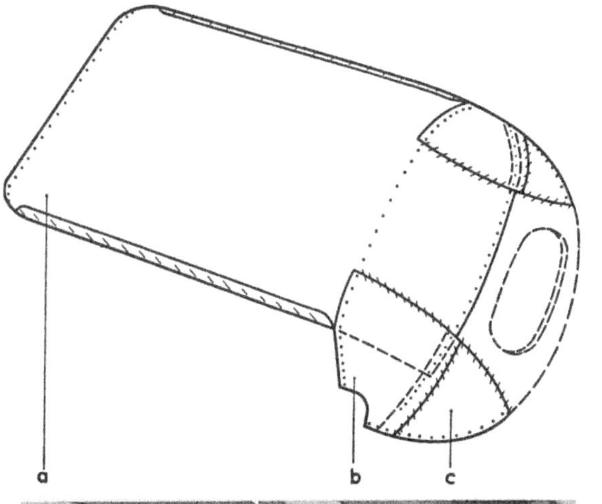

= nailed
= hand sewn
= seamed

a - 1370×1700 mm (54" ×67") 1
b - 450× 650 mm (17,7"×25,5") 2
c - 450× 480 mm (17.7"×18.9") 2

1 - Nail the two side parts 'c' on the rear body bow and in the recess in the rear bow. Cut off surplus material.

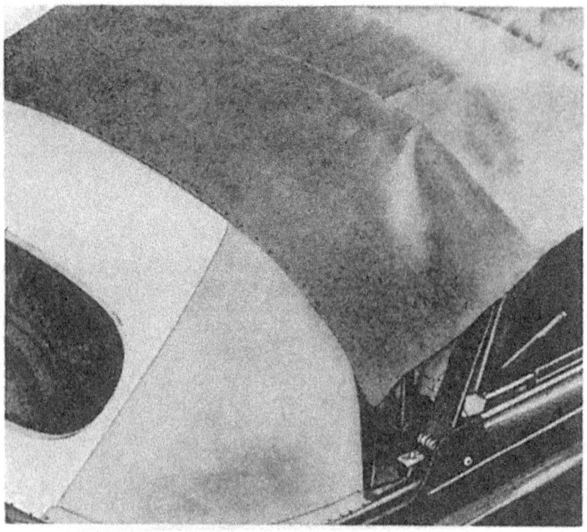

2 - Nail the roof part 'a' under side tension in the recess in the rear bow.

3 - Tack the roof part 'a' to the header without folds and nail it to the main bow.

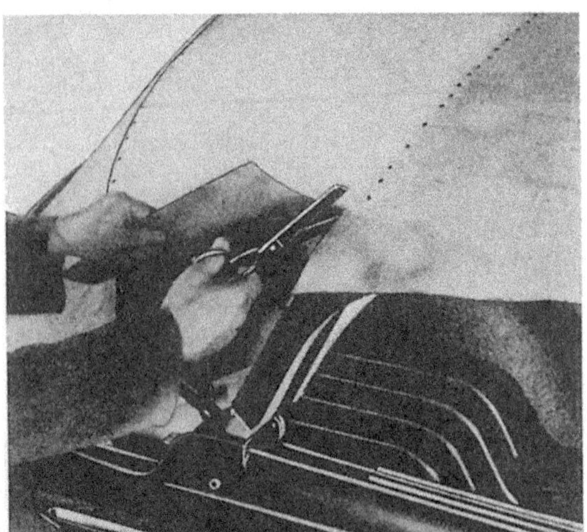

4 - Make a small cut in the roof part 'a' in front of the rear bow and behind the main bow.

5 - Fold the edges of the roof linen up to the roof frame and nail to the rear edge of the main bow. This fold is sewn to the rubberised hair padding later on.

6 - Nail the two side parts 'b' to the main and rear bows and to the appropriate place on the rear body bow under tension and free of folds.

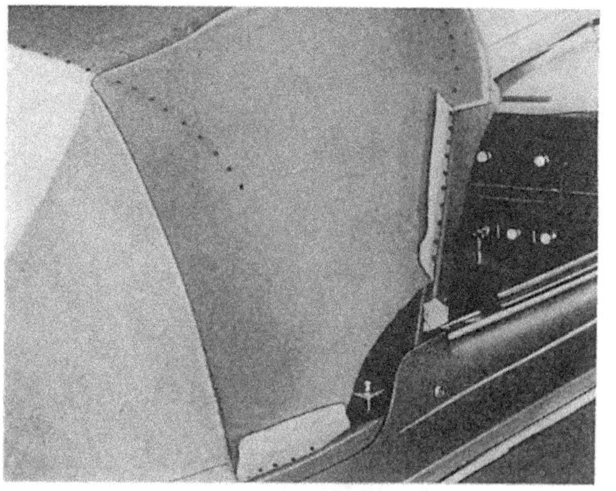

7 - Cut a half circle of approximately 150 mm (5.9") radius out of the side parts 'b' near the main hinges.

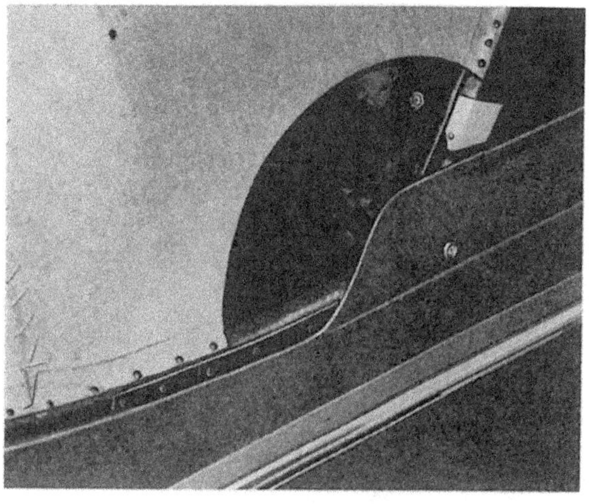

8 - Sew the side parts 'c' to the rear window frame material and to the side parts 'b' under tension. Sew the side parts 'b' to the roof frame 'a' as well.

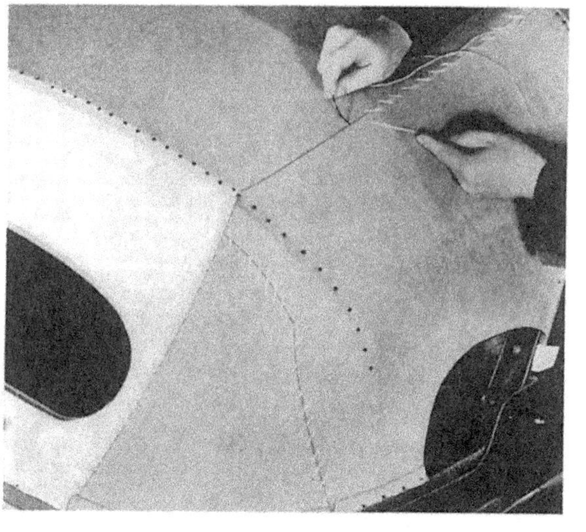

9 - Loop stitch the curves on both sides in the area between the rear bows.

Note:

Loop stitches are large, loose stitches made with a thick thread under which the padding material is pushed. They are intended to hold the loose padding in position.

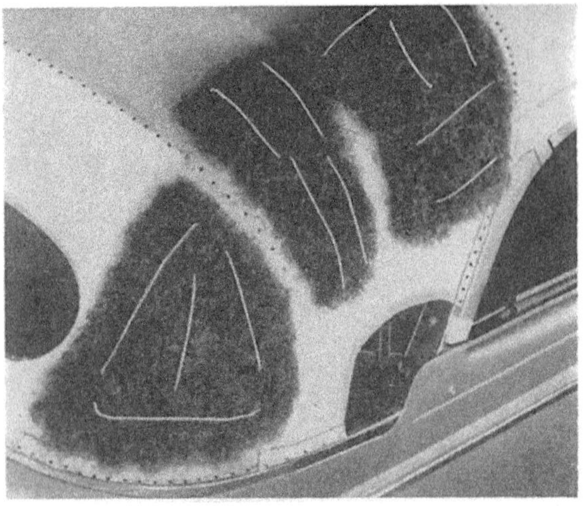

10 - Push the padding material — horse hair if possible — under the loop stitches in uniform thickness to level out the depression between the bows and sew it again.

> **Important**
>
> The padding controls the contour of the top at the rear. If the padding is too thick or too thin the top will be uneven.

Installing Rubberised Hair Padding

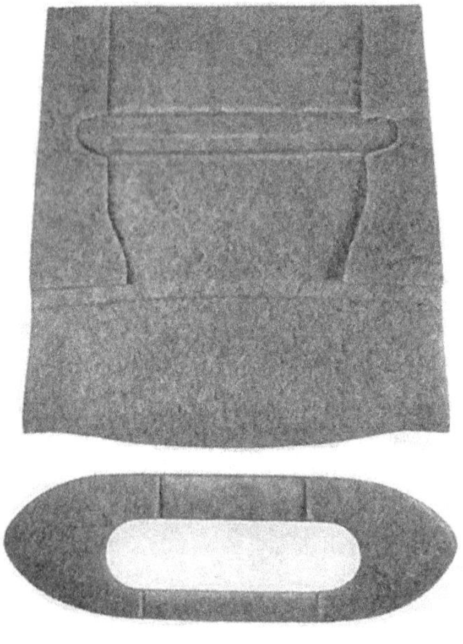

The rubberised hair padding upholsters and shapes the Convertible Top and also provides good insulation. It consists of two parts — the rear part and the top part — but is supplied under one part number.

1 - Pluck the rubberised hair pads out slightly at the edges to give a smoother contour at the roof sides.

2 - Lay both parts in position on the roof.

3 - Pull the *rubberised hair from the rear bow to* the rear body bow at the rear window and secure to the linen sheet with large stitches.

4 - Sew the roof part to the linen sheet with large cross stitches, leaving a small gap at the rear bow.

Pull the roof part almost up to the header rear edge and sew to the linen sheet with small, firm stitches.

5 - Sew the rubberised hair pad to the linen sheet with rows of large stitches 100 mm apart across the roof.

6 - Sew the outer sides to the fold in the linen sheet with small firm stitches.

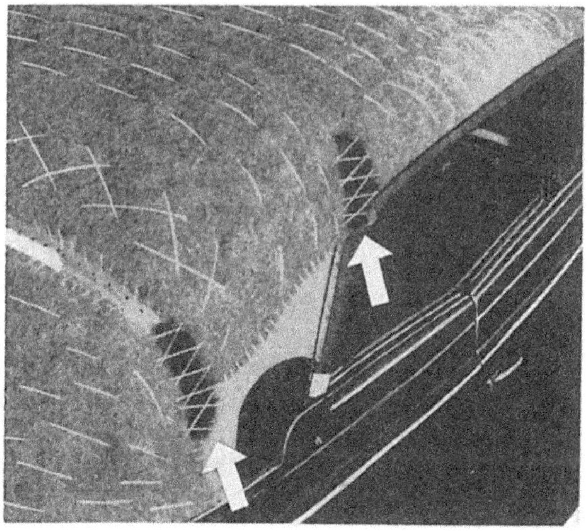

7 - Level out the rubberised hair pad with padding material at the places indicated to ensure a smooth contour.

8 - Cover both rear corners of the top with pieces of coarse lining material (550 × 400mm / 21.5" × 15.75") and stitch them firmly to the linen material at the sides and to the rubberised hair padding. Nail the material in the recess on the rear bow and to the rear body bow.

Installing Top Cover

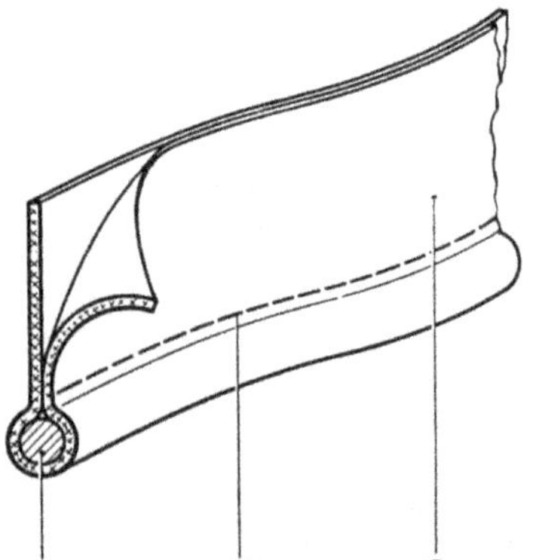

A - Hard cord 4 mm dia.
B - Machine seam
C - Top material 50×1850 mm (2″×72.8″)

The top must remain closed when installing the top cover.

A piece of lining material is sewn at both sides under the front part of the top cover. The linen sheet, with rubberised hair sewn to it, is detached at the front and pulled through between the lining and top cover materials.

1 - Make a beading strip 50×1850 mm as shown in the drawing.

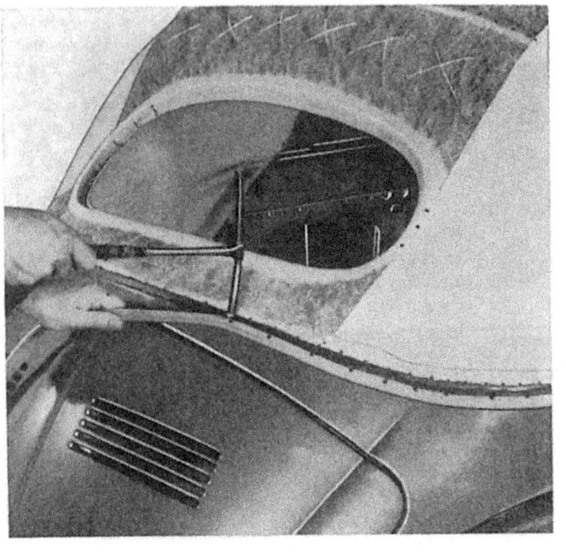

The beading is included with the top covers supplied as spares.

2 - Coat the rear body bow with sealing compound and nail the beading on under side tension so that it covers the nails securing the rear body bow to the body.

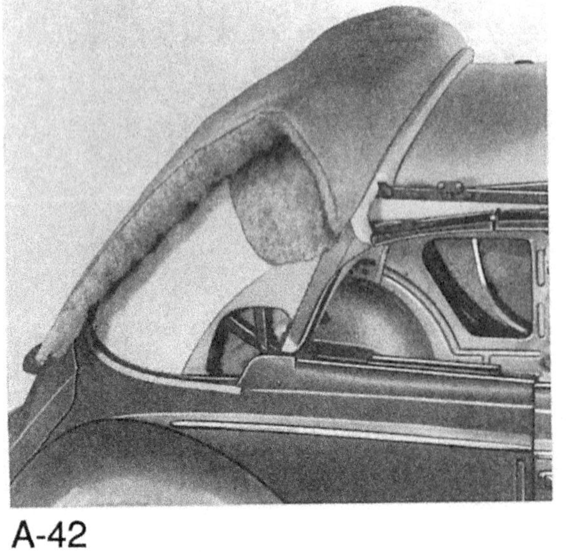

3 - Detach the linen sheet with the rubberised hair padding where it was tacked to the header and lay it back.

Important

Do not loosen the nails securing the linen sheet to the main bow.

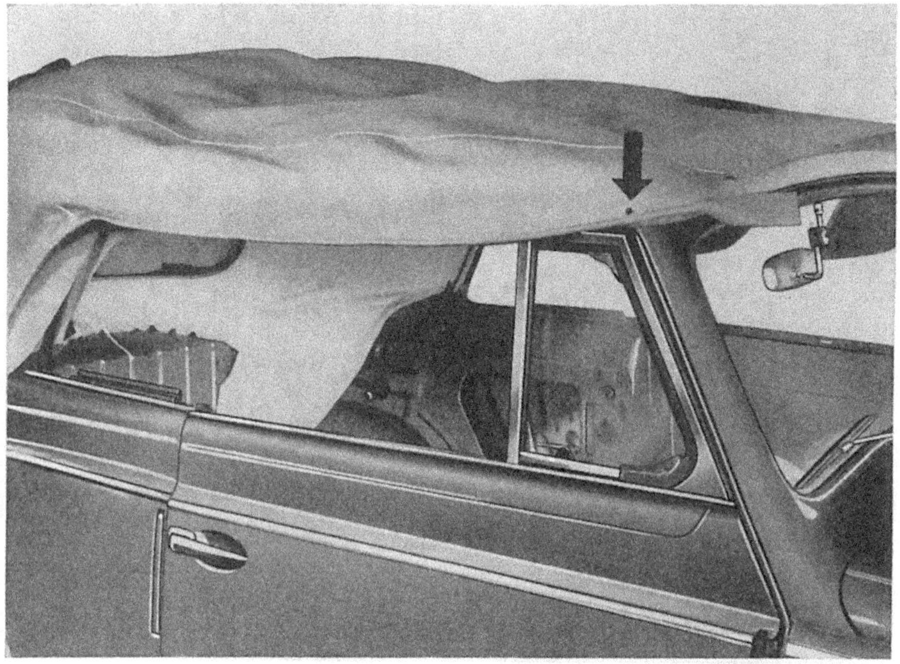

4 - Lay the top cover on the top and tack to the header with one nail each side to make the fitting easier.

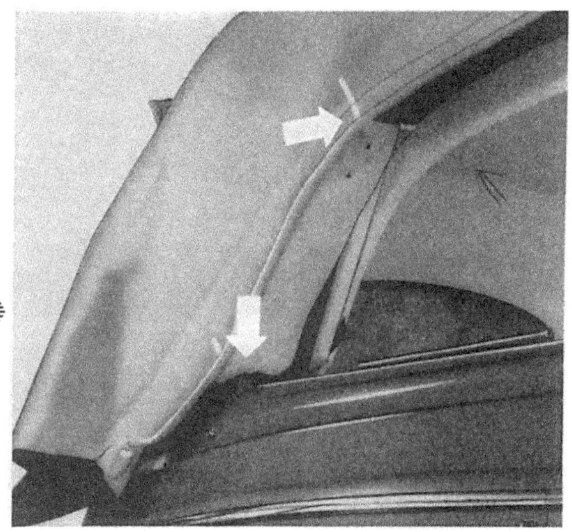

5 - The top cover is marked on both sides near the main hinges with two white lines which serve as locating points when fitting the top. Pull the cover on the right hand side first until the lower mark is in line with the bottom edge of the wooden part of the main bow. The upper mark must be in line with the joint between the roof frame and the main bow.

6 - Tack the top cover flap with two nails to the inside of the main bow on the right hand side.

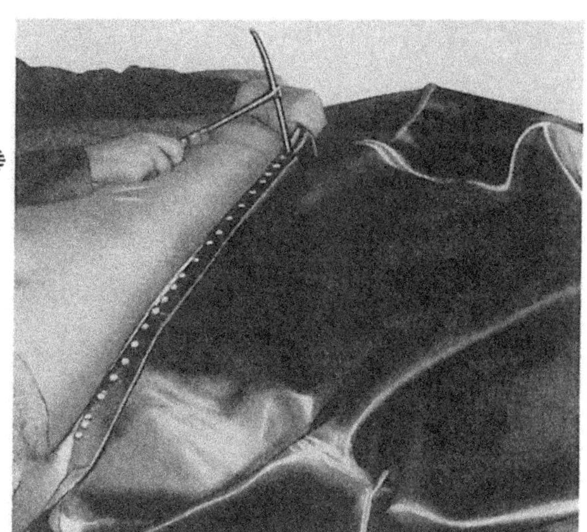

7 - Fold the rear part of the top cover forwards and nail the lining material to the main bow, from the right hand side to the center, without creases.

8 - Detach the top cover flaps on the right hand side where they were tacked to the main bow and repeat the operation on the left hand side.

Watch the marks on the top cover!

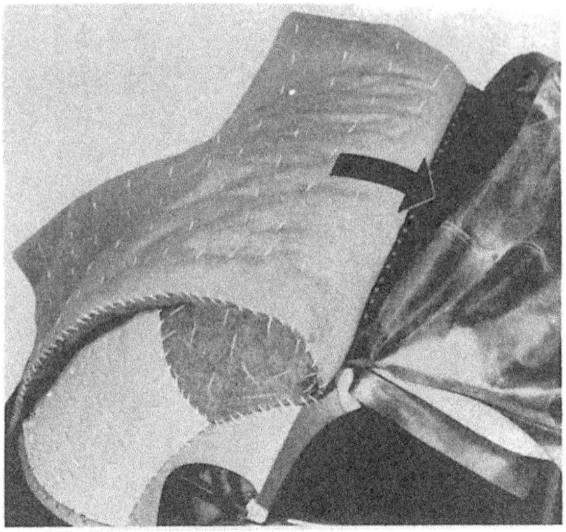

9 - Loosen the top cover flaps on the left hand side where they were tacked to the main bow. Powder the rubberised side of the top cover lightly with talc and push the rubberised hair padding forward as shown by the arrow.

Remove the two tacks on the header before doing this.

10 - Lay the top cover material to the rear again and tack the material strip to the main bow under tension.

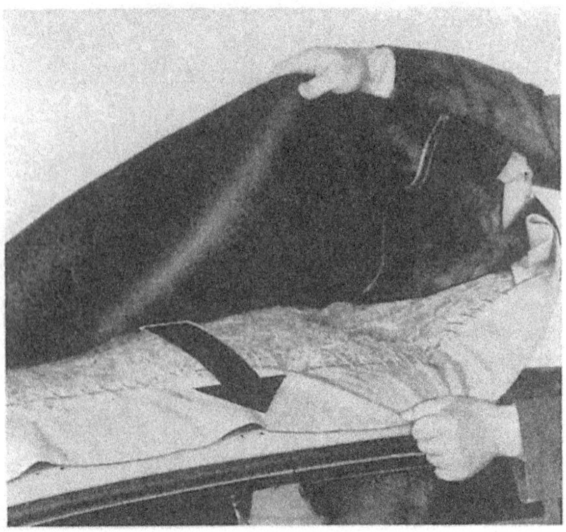

11 - Pull the rubberised hair padding hard towards the header.

12 - Pull the top cover towards the rear. The cut in the top cover must be exactly over the rear bow.

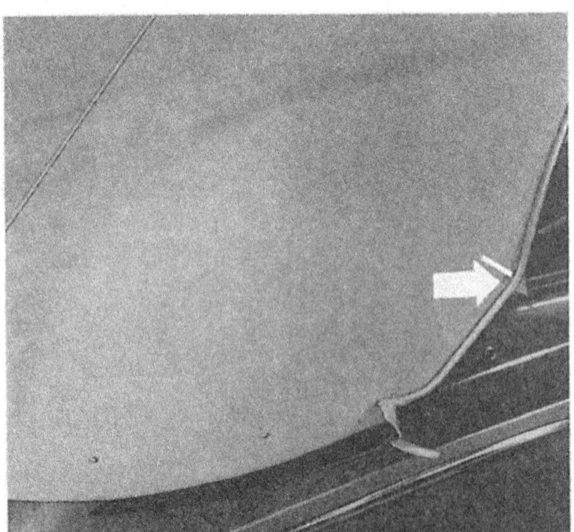

13 - Pull the top cover at both sides and secure with a nail inserted into the hole where the chrome-plated screw is placed later (see arrow).

Tack the top cover to the rear body bow from the longitudinal seams to the main bow. For the time being do not tack the cover to the rear body bow in the center underneath the window.

Important

The longitudinal seams of the top cover must be straight.

14 - Pull the top cover to the rear and tack with two nails to the rear body bow.

To avoid tearing the material, drive the two nails through the seams.

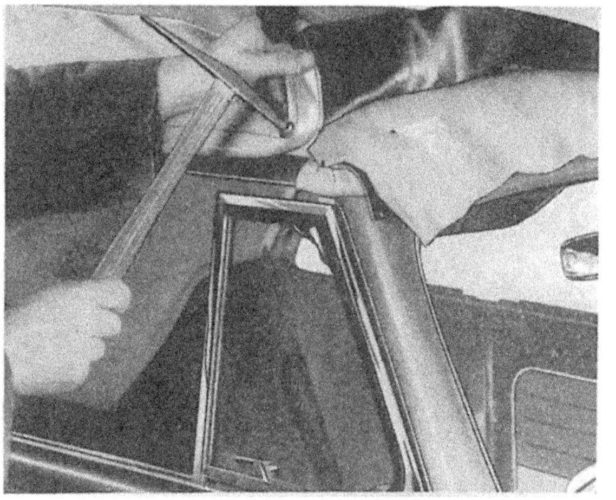

15 - Pull the top cover hard towards the front and tack the linen sheet and cover to the ends of the header.

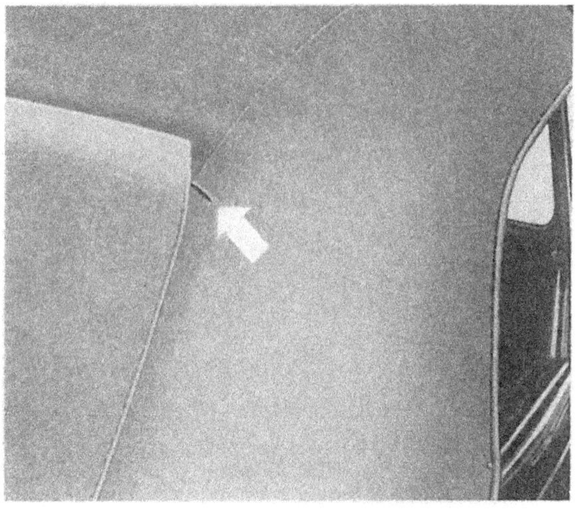

16 - Extend the cut in the top cover at the rear bow about 30 to 40 mm (1.2" to 1.6") on each side. Make sure, however, that the trim moulding, which is installed later, overlaps the cut by 10 mm (0.4") at each end. This cut is essential to avoid ridges or creases forming in the top cover at this point.

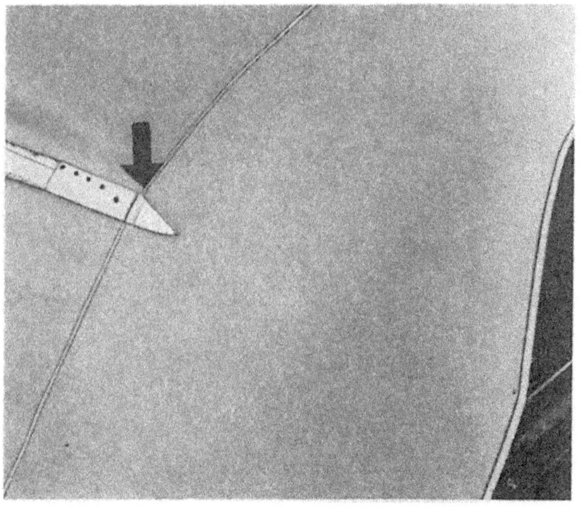

17 - Chisel 3 mm grooves in the rear bow where the seams come. This measure is only necessary on new roof frames and rear bows and must be carried out in order to lower the seams and prevent the trim moulding, which is nailed on later, from standing proud.

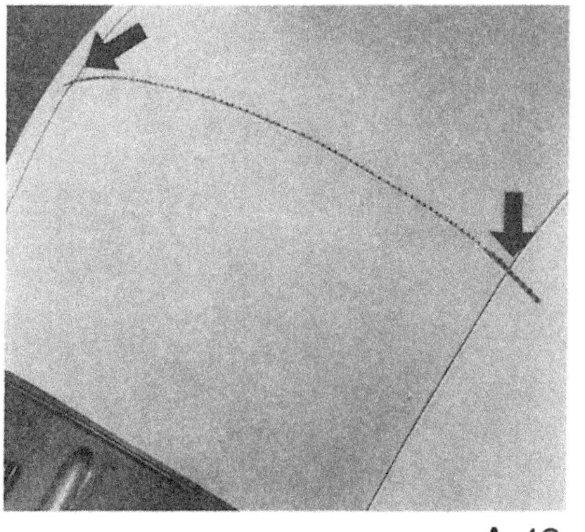

18 - Nail the top cover in the recess on the rear bow evenly and free of creases. Attach the lower part first and then nail the upper part over it so that the water running off the roof will not enter the cut. Cut off surplus material.

19 - Coat both ends of the row of nails with rubber solution, particularly at the seams. Take care, however, that the coated area is not wider than the trim moulding which is nailed on later.

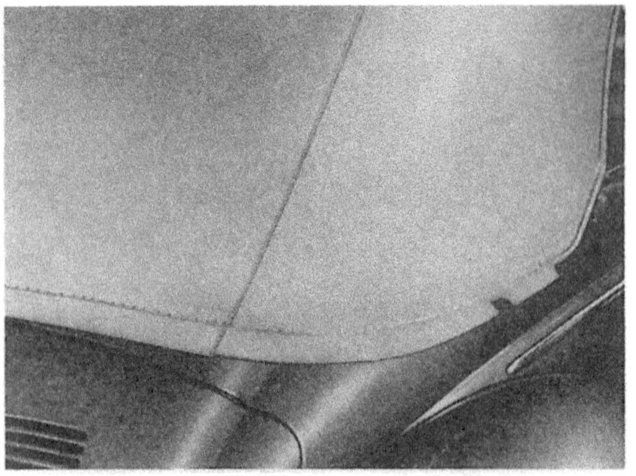

20 - Nail the top cover to the rear body bow under even tension and free of creases, taking care that the row of nails is at a uniform distance from the beading.

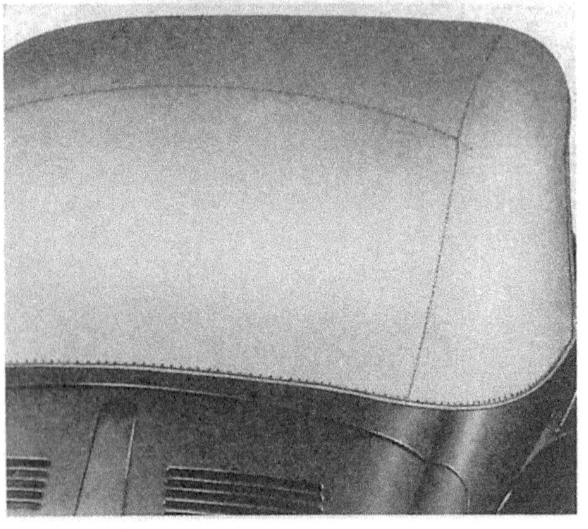

21 - Cut surplus material off cleanly so that top cover and beading overlap properly.

22 - Nail the linen sheet with rubberised hair padding and the lining material of the top cover in the recess on the header under tension and free of creases. Cut off surplus material.

Important

The lining material and linen sheet must be secured flush in the recess on the header as otherwise a ridge will be formed at the rear edge of the header.

23 - Cement a wide piece of sealing tape into the recess on the header and secure with tacks. This is intended to smooth out the contour between the edge of the recess and the rounded part of the header.

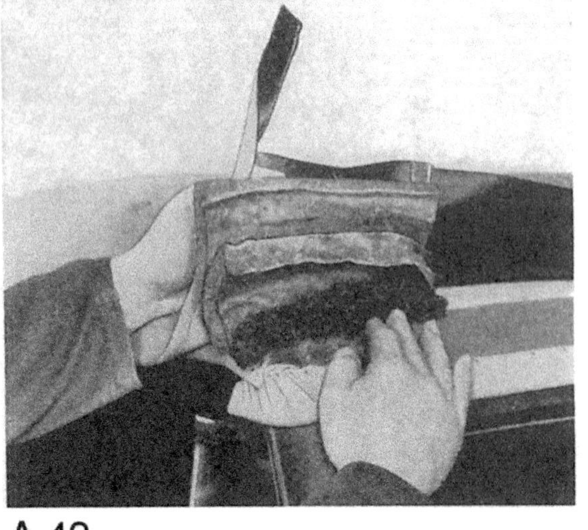

24 - Pad the top cover out behind the header at each corner with a small amount of horsehair and a small piece of wadding about 150 × 200 m (6″ × 7.8″).

Important

This padding controls the contour of the header and if it is too thick or too thin it will cause bumps and folds.

25 - Tack the top cover to the header with three tacks each side.

26 - Open the top about 180 mm (7").

27 - Nail the tacked-on top cover to the main bow at both sides so that the top material and the headlining material join exactly and the beading projects over the outer edge of the main bow. Cut off surplus material.

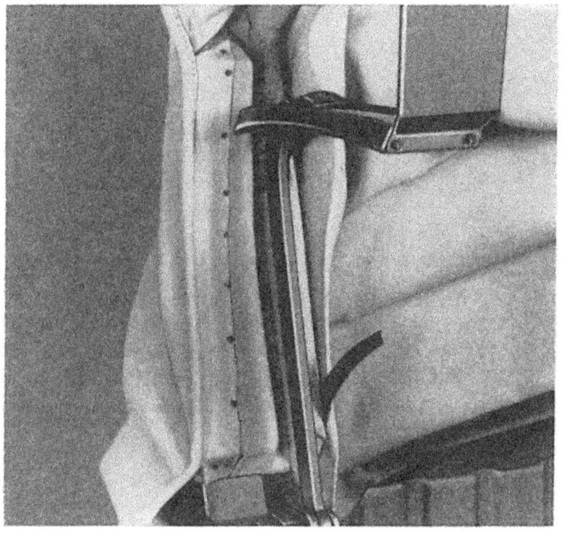

28 - Tack the top cover under tension to the header at the side seams with two tacks each side.

29 - Close the top. Wind up the windows and check the clearance at the window frames when opening the door.

The side parts of the top cover must be as low as possible without contacting the window frames. Check the side tension of the cover as slackness here can cause fluttering noises. If necessary, the top cover should be detached at the header and re-tensioned with the top open.

30 - Close the top again and make small cuts into the top cover at left and right where the beading is sewn on.

31 - Lay the beading strips round the corners of the header and nail under tension (arrow).

32 - Starting at the corners and working to the center, nail the top cover to the header, 20 mm from the covered lower edge, evenly and free of folds.

a = 20 mm (0.8")

33 - Cut the surplus material off cleanly below the row of tacks.

34 - Cut the top cover out inside the rear window frame leaving a good overlap. Make a series of small cuts in the corners.

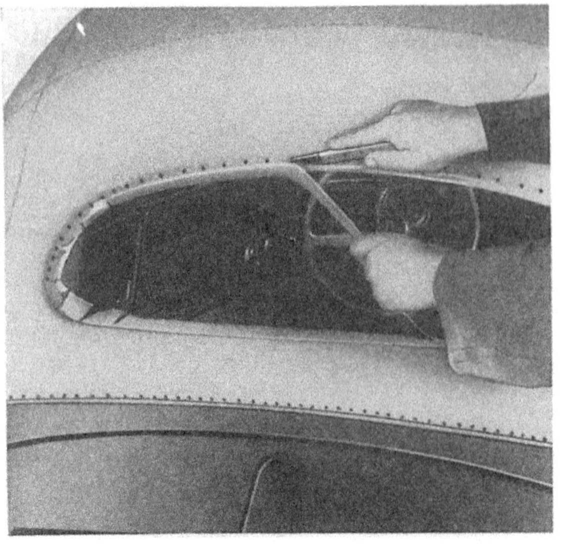

35 - Cement and nail the top cover to the rear window frame. Take care that the corners are nailed first and then the straight sides.

36 - Cut surplus material off cleanly round the window frame.

37 - Check all nailed areas for folds and uniform tension and correct as and where necessary by re-tensioning or re-nailing.

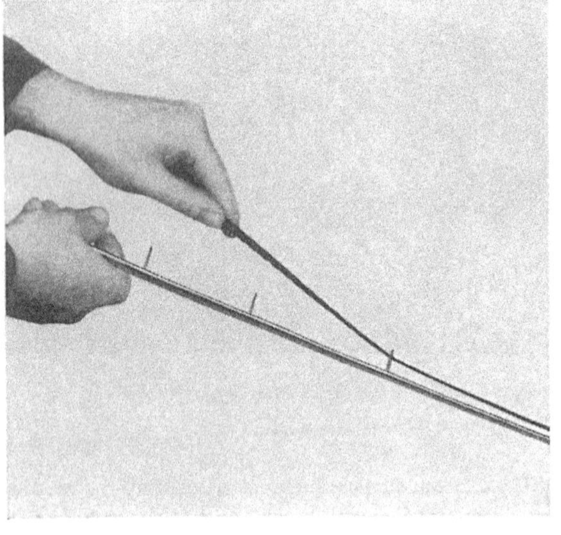

38 - Nail strips of sealing band over the rows of nails on the header and rear body bow.

This sealing band is a natural rubber strip of the same width as the trim moulding and is obtainable under Part No. 151 871 435. The band has a light-coloured protective strip on one side which must be removed before the band is fitted.

39 - To facilitate the fitting of the sealing band and trim moulding on the rear bow it is recommended that the band is placed on the moulding first.

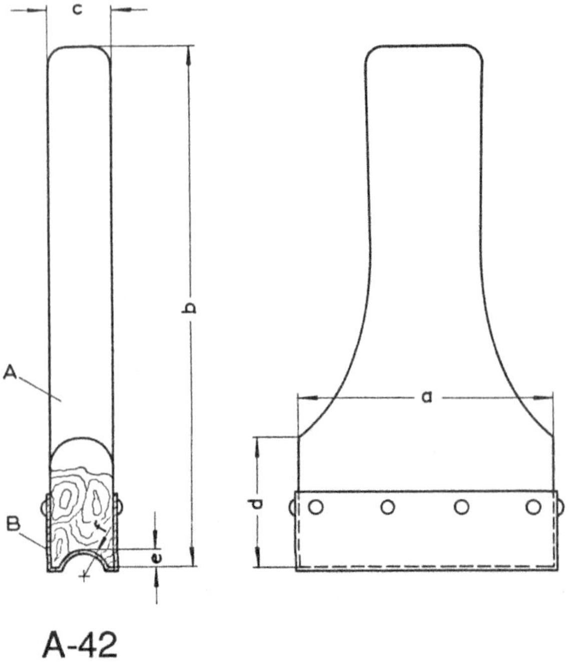

Important

The trim moulding must be installed with the special wooden block to ensure proper fitting and avoid damage to the moulding.

A - Special block
B - Leather cover (approximately 1 mm thick)
a = 70 mm (2.75″)
b = 140 mm (5.5″)
c = 17 mm (0.66″)
d = 35 mm (1.35″)
e = 5 mm (0.2″)
f = 7 mm (0.27″)

A-42

40 - Nail trim moulding.

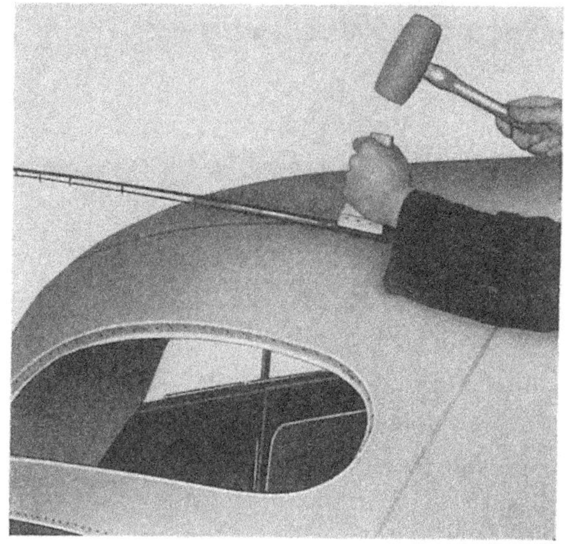

To avoid difficulties — such as overhanging at the ends and similar faults — it is advisable to determine the exact location of the trim moulding by fitting, measuring or marking with chalk. Then start by nailing at one side and work across to the other.

The trim moulding at the rear body bow must fit closely against the beading.

41 - Insert the chrome-plated screws and special washers and coat the screw heads with paint of the same colour as the top material to avoid rust formation.

42 - Place the assembled rear window into the frame without sealing compound and check for fit. Then coat the nailed surface in the wooden frame lightly with sealing compound and insert the window. To avoid soiling the cover material do not coat the weatherstrip with compound.

When pressing the window into position one mechanic must support the frame from inside the vehicle to prevent the linen material from tearing.

Securing Headlining

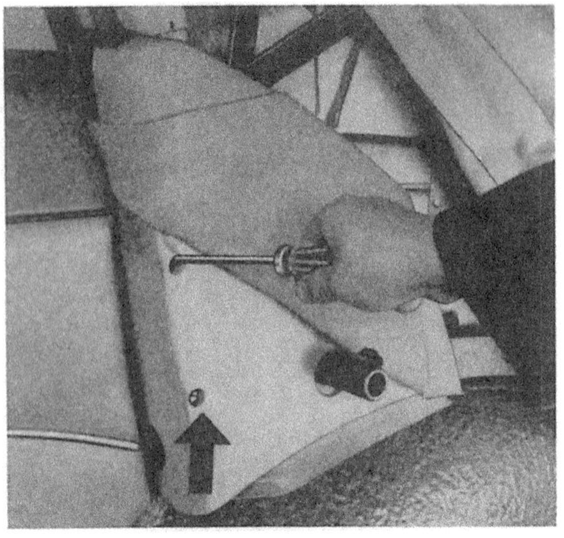

1 - Secure the material covered filler blocks for the main bow pillars with screws in the holes provided on each side.

2 - Cement the detached material to the blocks and to the sheet metal under the quarter trim panels. Secure the side trim panels again.

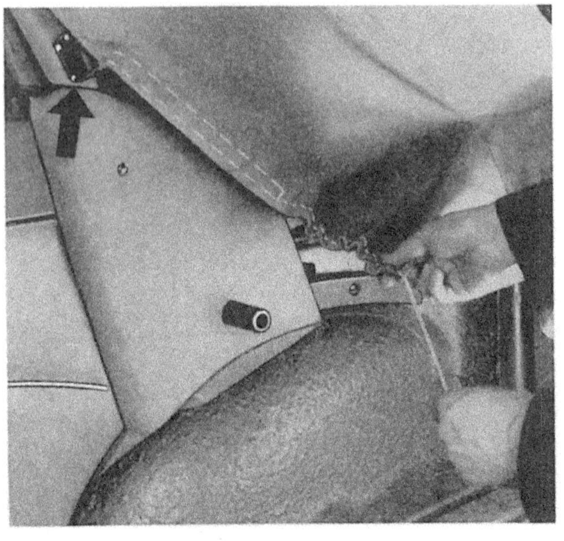

3 - Nail the rubber bands to the wooden mouldings and tension them.

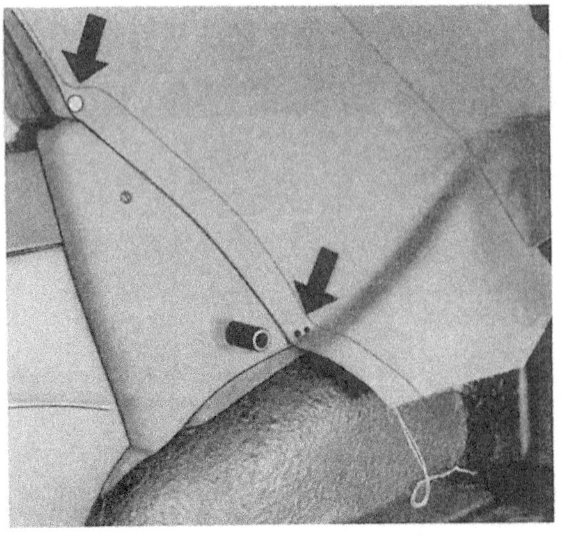

4 - Secure the headlining to the wooden mouldings with one cloth covered nail on each side and nail the lining with the rubber band under tension to the bottom edges of the filler-blocks.

5 - Connect interior lamp and secure it to the roof bow.

6 - Cement carpets for luggage compartment and wheel housing into position.

7 - Secure the assist straps with brackets on each side in the holes provided in the roof frame.

8 - Screw the three weatherstrips with the channels to the roof frame on each side so that when the window is raised it comes to rest on the inside of the rubber lip.

When installing the weatherstrips on a new top frame, holes must be drilled to correspond with those in the channel under the rubber lip.

The weatherstrips are nailed to the main bow.

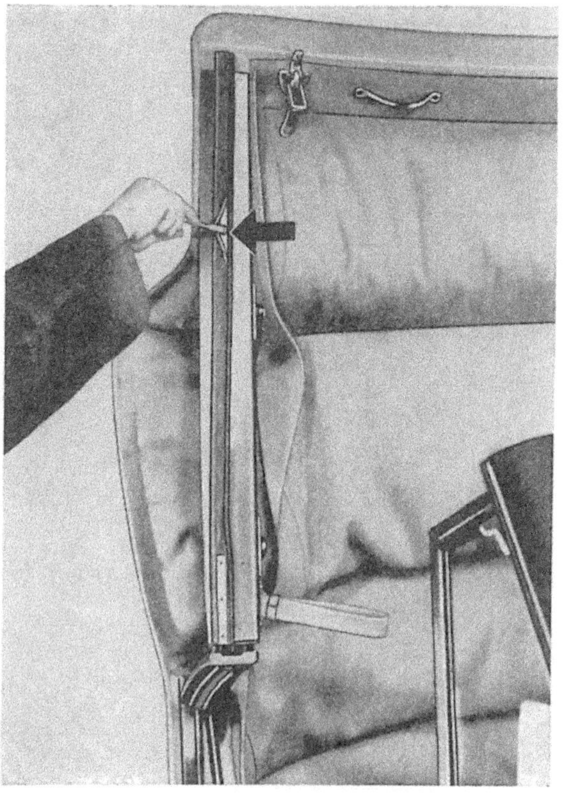

Note:

From Chassis No. 2 232 780 the weatherstrips on the windshield frame (Part No. 151 871 353), on the roof side rails (Part No. 151 871 357/361) and on the main bow (Part No. 151 871 369) have an additional sealing lip. The new rubber lip presses against the window frame when the window is raised and thereby improves the sealing effect. The weatherstrips (Part No. 151 871 357 A) can be subsequently installed without any further alteration. Weatherstrips of the earlier version should be used up on vehicles up to Chassis No. 2 232 779.

To eliminate noises caused by friction between window frames and weatherstrips it is advisable to apply brake cylinder paste (Part No. B 1) to the rubber lips. Wipe the weatherstrips afterwards with a clean dry cloth.

Previous version New Version

9 - Fit the six chrome-plated cover plates to the main bow and roof frame on both sides.

A sealing rubber is installed under the cover plates at the bottom of the main bow on both sides.

On new top frames some of the holes have to be drilled first.

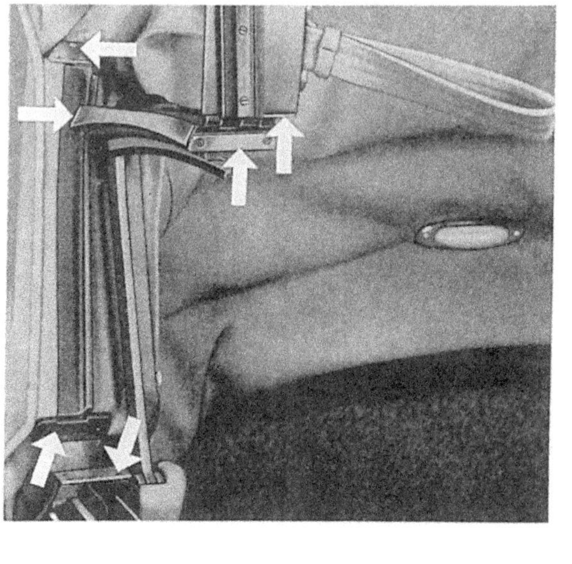

Note:

As the top frame is inclined to bend excessively when a new top has been fitted the clearance between the window and the roof frame can become too large. In these cases it is possible to lower the frame by placing packing under the cover plates for the front hinge near the assist straps. When the frame is too low (older vehicles) the roof linkage can be raised by grinding the cover plates down.

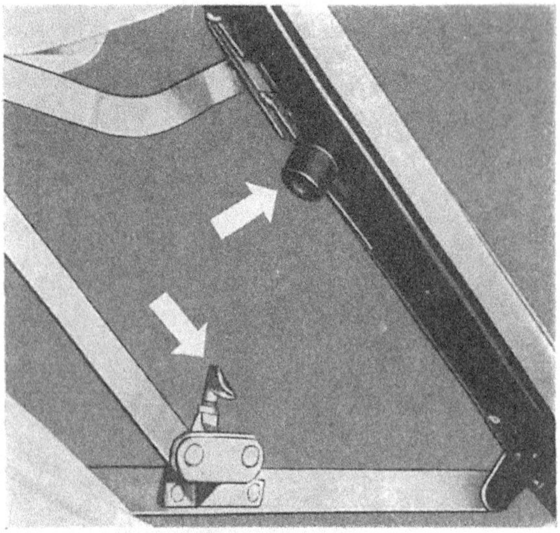

10 - Secure the top catches and rubber buffers in the holes provided.

When the top is completely installed, check that it opens and closes properly.

Brush the top cover with a soft brush and spray with clear water. Leave the top closed until it is completely dry.

New tops always have a certain amount of tension. It is not necessary to slacken this tension in any way as the cover will stretch after it has been in use for some time until it can be closed easily.

Top Assembly
(Late)

The description of the Convertible top assembly serves as a guide for the sequence of operations. The work should, where possible, be carried out by a man who is familiar with this type of work. If two men can be employed, the task of fitting the top cover will be made considerably easier. The assembly of individual parts of the top described in some paragraphs can naturally be carried out separately.

Note:

All photographs in the sections: "Fitting Headlining" and "Fitting Top Cover" show parts of PVC material.

Work on headlinings and top covers of cloth is carried out in the same way as described here.

The tools required are listed in the Workshop Equipment section.

Preparation

1 - Take seats out.

2 - Detach quarter panel trim at rear on both sides.

3 - Pull luggage compartment floor carpet off at the sides.

A-42A

4 - Pull carpet material off the sides of the wheel housings.

5 - Wind the drop windows down on both sides.

6 - Cover all parts of the body in the working area.

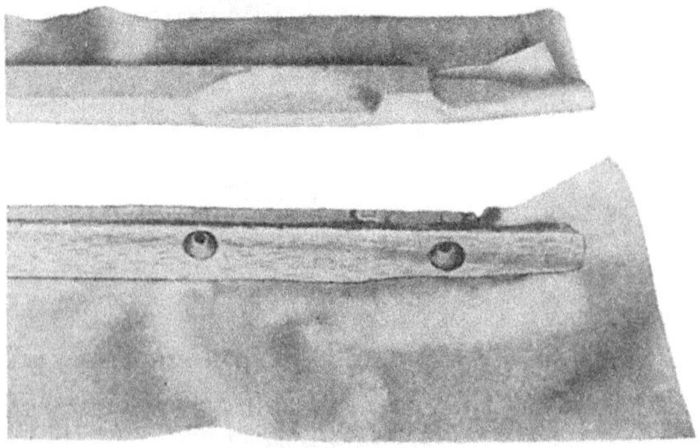

7 - Cement a strip of leatherette to each inner wooden moulding.

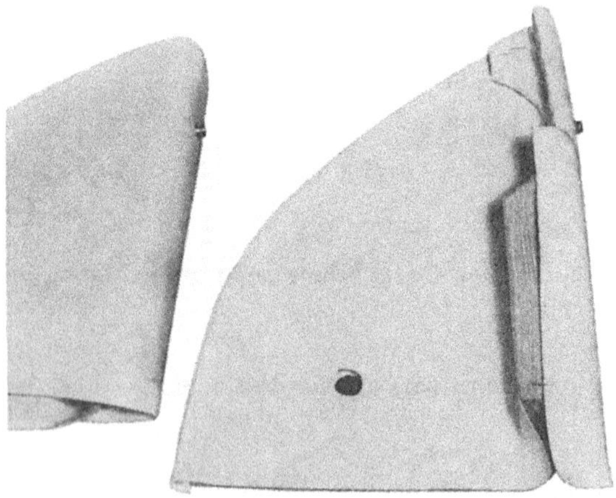

8 - Cement strips of leatherette to the support blocks for the main bow pillars on each side.

Fitting top frame and straps

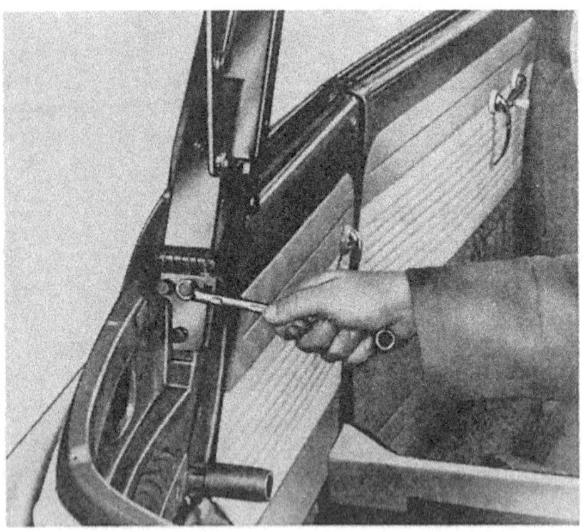

1 - Place top frame in position and insert the three screws securing main hinge to support pillar on each side but do not fully tighten the screws.

Note:

Install a 5 mm thick distance piece between main hinge and support pillar on each side so that there is sufficient space to fit the rubber seals and cover plates later.

2 - Push the top frame onwards until the outer wooden moulding on the main bow is about 2 mm in from side of body on each side.

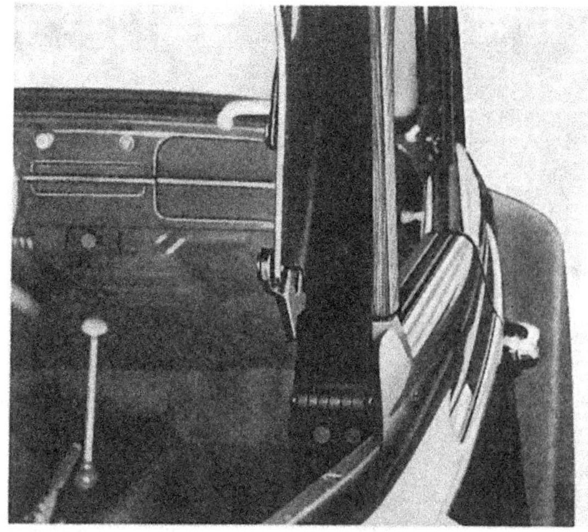

Note:

If this measurement cannot be obtained by adjustment, rework the wooden moulding with a rasp. If this is not done, there will be leaks at this point later on.

3 - Check the position of the header in the windshield frame. The rear edge of the header must be about 6—8 mm in front of the rear edge of the windshield frame.

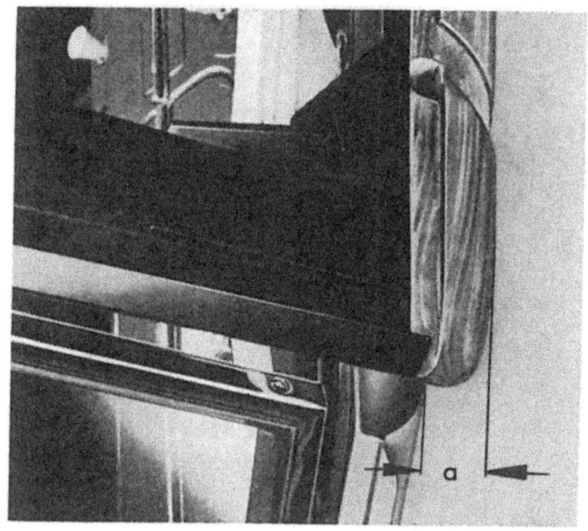

Note:

This position must be obtained because the top linkage curves slightly after the top cover has been fitted and tensioned and this pulls the header back.

The position is corrected on the main bow directly above the hinges on the top support pillars.

4 - Fill two holes on each side of header with plastic wood.

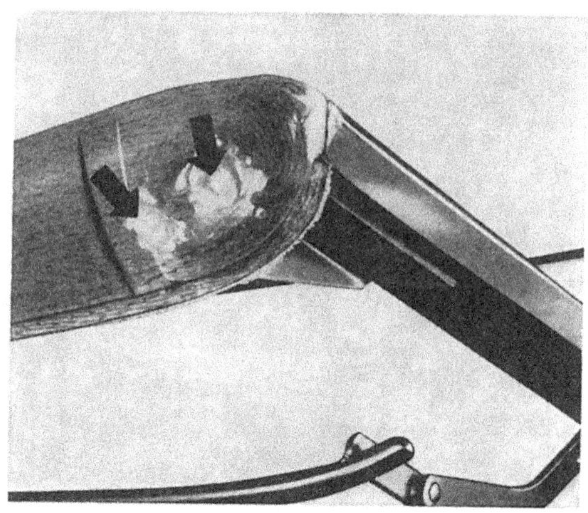

5 - Coat the top and bottom surfaces of the header with D 12 universal adhesive and fit two strips of leatherette (1250 × 100 mm). The material should be cut at the corners so that it fits better. Cut off surplus material.

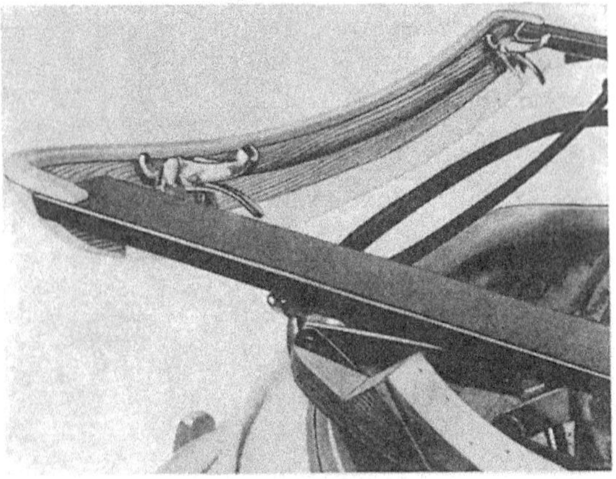

6 - Lay the top frame back.

7 - Fit the top fasteners.

8 - Coat the lower edge of the header with D 12 adhesive and stick the sealing rubber (1280 × 16 × 10 mm) to the front edge of the header. The rubber strip should be cut off at the ends so that it is flush with the header.

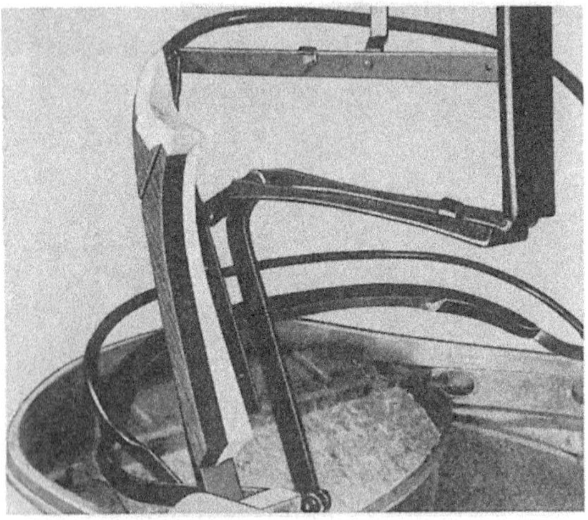

9 - Cement strips of leatherette (460 × 120 mm) to the inside of the wooden moulding for the main bow on each side. Cut out the holes for the retaining brackets for the wooden mouldings. Excess material should be cut off on the outside and stuck round the main bow near the top linkage.

10 - Attach the outer wooden mouldings to the main bows with three Phillips screws each.

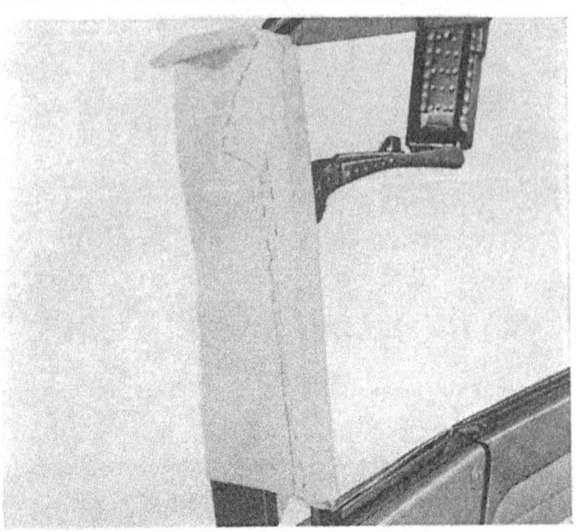

11 - Lay the surplus leatherette round the bow and cement it into position under slight tension. Tack the material at the outer edges and cut off the surplus.

12 - Close the frame and secure it with the fasteners. Set the rear bow and the rear tubular bow to the following measurements.

a = 304 mm (12")
b = 530+5 mm (20.8+.2")

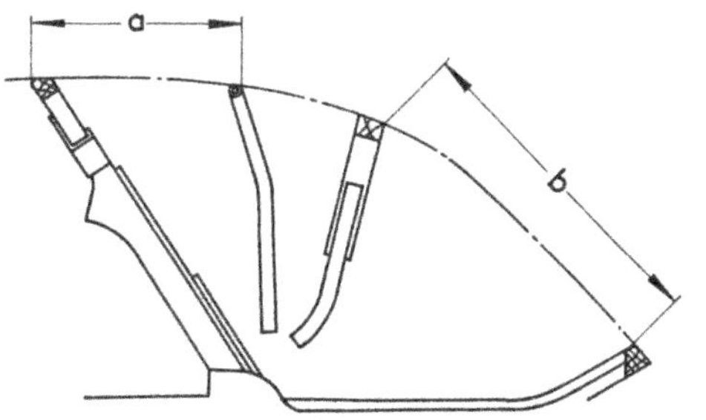

Important

These measurements should be strictly adhered to as they govern the satisfactory padding of the top and the folding of the top when opened.

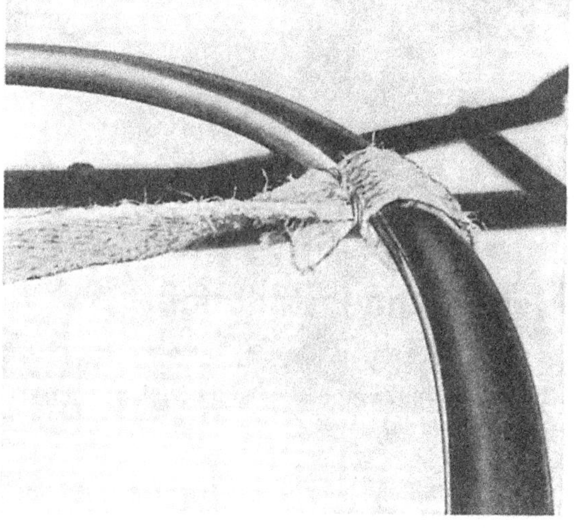

13 - Lay the webbing strips which support the rear window round the main bow at the sides and sew into position.

The spacing to the roof frame should be 310 mm (12.2").

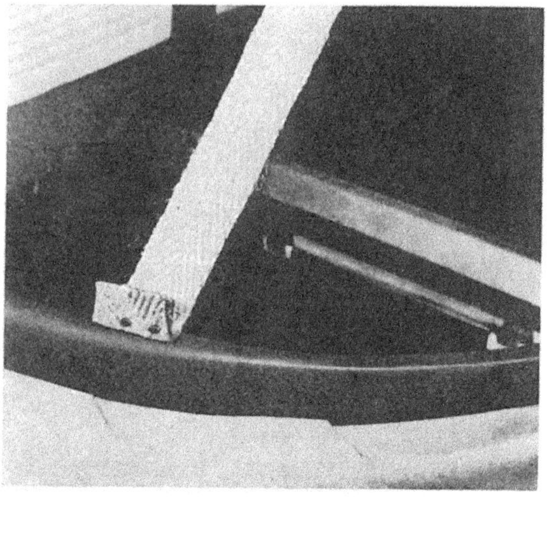

14 - Secure webbing strips **on** the rear body bow with webbing tacks.

The spacing from the center of the rear bow should be 450 mm (17.7").

Important

The webbing strips should be secured with brass tacks to avoid rust formation.

15 - Tacking webbing strips to the rear bow **in the recesses.**

If the webbing is not attached in the recesses, the trim moulding which is installed later will be too high at the ends and will stand clear.

16 - Attach the webbing strips to the rear bow with a loop on each side. The loops are made of webbing material and sewn to the strips to support the rear window.

Headlining Installation

The headlinings supplied as spare parts are complete with the support strips which are used to attach the lining to the individual bows.

To ensure that the headlining fits properly it is essential that it is attached to the various bows in the following order with the roof closed:

1 - Main bow

2 - Rear bow

3 - Rear body bow

4 - Header

5 - Outer wooden mouldings on main bow

6 - Rear tubular bow

7 - Front tubular bow

Important

Do not move the webbing straps on the tubular bows. Check measurements again.

1 - Coat all bows with D 12 universal adhesive where the support strips of the headlining fit.

2 - Tack the headlining to the outer ends of the main and rear bows and the rear body bow under light tension.

a - 450 mm (17.7″) center of rear body bow
b - 130 mm (5.1″)
c - 310 mm (12.2″)

A-42A

3 - Tension material lightly and tack it to the header at each side near the seams.

Note:

To prevent tearing, the nails should be driven through the side seams.

4 - Tack the material to the inner wooden mouldings on each side under light tension.

5 - Pull the headlining up near the main bow with the support strips until the seam in the lining contacts the underside of the bow. Lay strip round bow and press firmly into place. Cut off excess material.

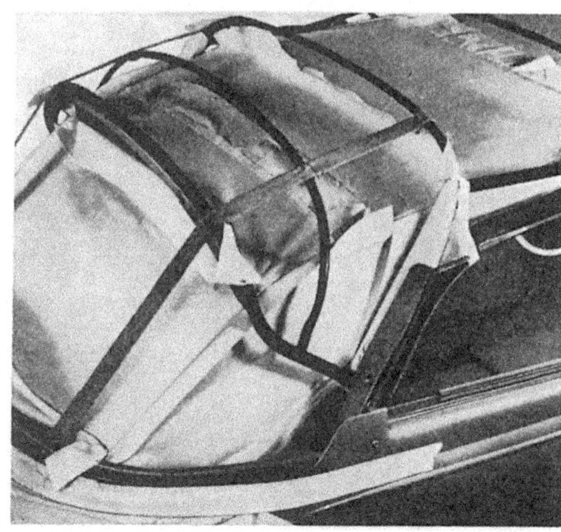

6 - Secure the headlining to the rear bow in the same manner.

7 - Tack the support strip to the top edge of the rear body bow. Cut openings for the webbing strips.

8 - Tension the headlining uniformly and tack it between the side seams into the flattened part of the header.

9 - Pull the left and right seams of the lining further forward as necessary and secure in the grooves at the corners of the header with two tacks on each side. This re-tensioning ensures that the longitudinal seams of the headlining are tensioned uniformly.

Important

It is essential to ensure that the seam of the front support strip is parallel to the front tubular bow.

10 - Cut off surplus leatherette on the flattened surface of the header. Both headlining seams must run forward in a wedge shape on the top of the header.

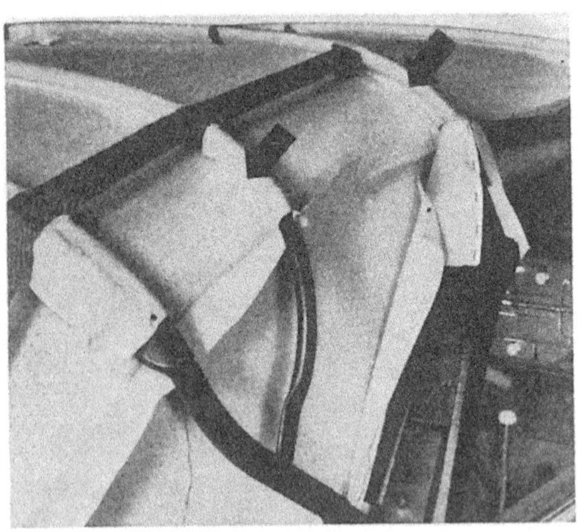

11 - Pull the support strip for the rear tubular bow up until the headlining almost contacts the bow and then wrap the strip round the tube.

12 - Repeat this operation at the front tubular bow.

Important

The headlining should hang down about 10 mm below the two roof frames on the inside.

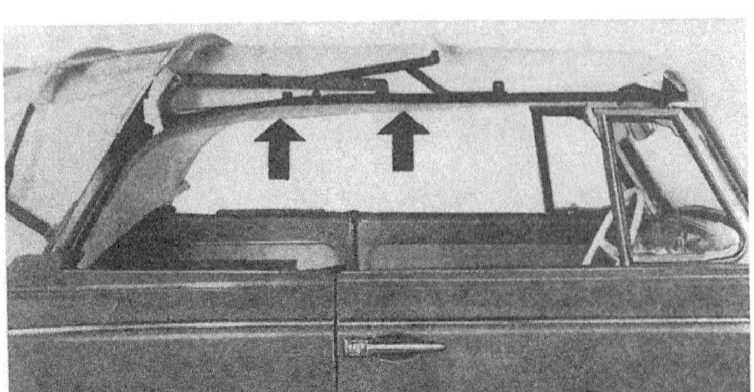

13 - If necessary, pull the headlining down slightly. Cut off surplus support strip material.

14 - Check the headlining from inside to ensure that the lining is taut, free of folds and that the seams of the support strips are parallel to the bows. If necessary, correct at the individual bows.

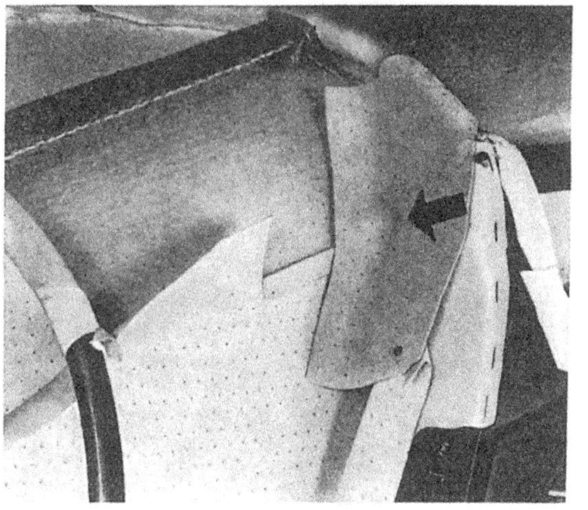

15 - Sew the outer edges of the support strips on the tubular bows for about 20 mm (.8") and tack the outer edges of the strips to the main and rear bows.

16 - Tack strips of leatherette 250 × 100 mm (9.8" × 4") to the rounded parts of the main bows. This leatherette covers the space between headlining and main bow.

17 - Nail a length of 4 mm (.2") diameter cord to the main bow on each side, level with side seams in the headlining. Loop the cord round the rear tubular bow once and nail it, under tension, to the rear bow.

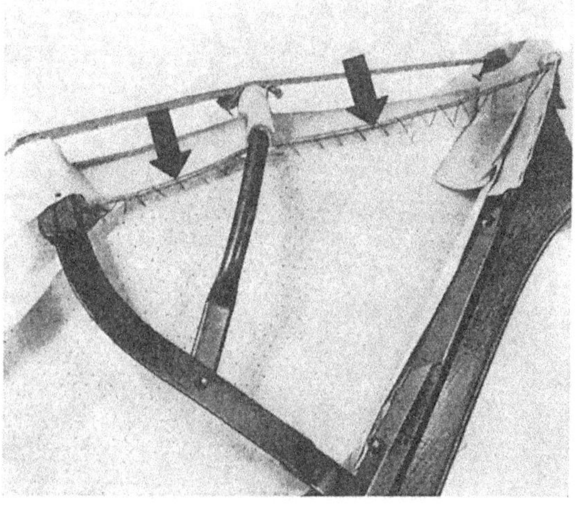

18 - Sew the headlining side seams to the cord under tension.

Note:

Check from inside that the headlining side seams are straight. If there are folds in the material, it must be detached and sewn on to the cord again.

19 - Nail two webbing strips to side of rear bow and **on** rear body bow to give additional support for the corner padding. The outer edges of the webbing strips must align with the left and right grooves of the rear panel.

Installing Rear Window Frame

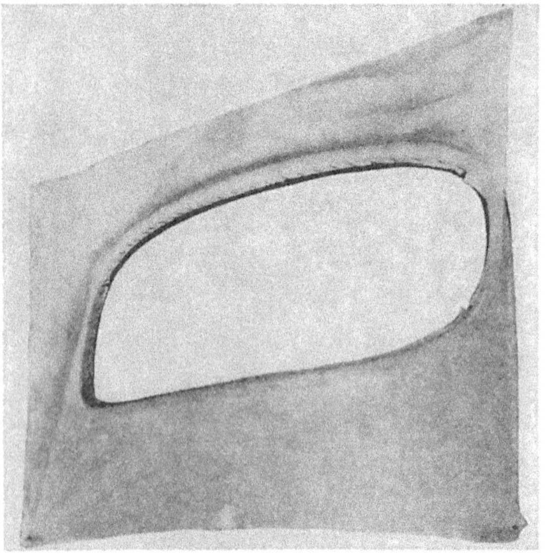

1 - Stick a piece of nettle cloth 37.4" × 25.6" (950 × 650 mm) to the frame so that it overlaps about 8" (200 mm) below and 10" (250 mm) above the frame.

2 - Cut the material round inside the frame so as to leave a .6" (15 mm) wide border. Stick this border to the inside of the frame and secure it with tacks.

3 - Mark the center line of the rear body bow and the rear bow.

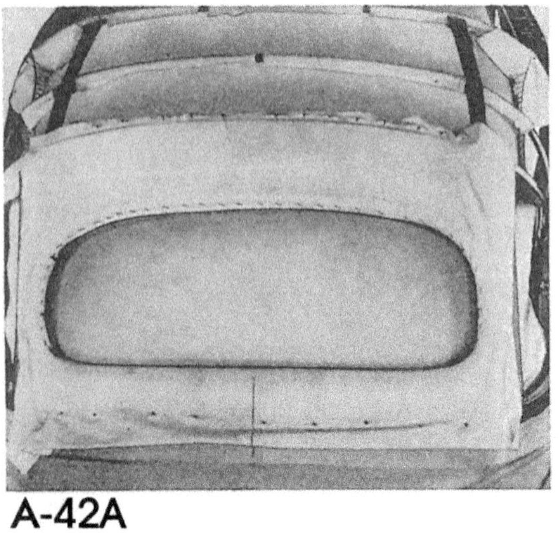

4 - Mark the center of the window frame on the nettle cloth.

5 - Tack the nettle cloth with frame to the body bow and rear bow with the marks aligned. The measurement "a" (see drawing) must be 11.6" (295 mm).

6 - Tension the nettle cloth and nail it into the recess on the rear bow and on to the body bow. The measurement "b" must not exceed 3.54"±.08" (90±2 mm) under any circumstances.

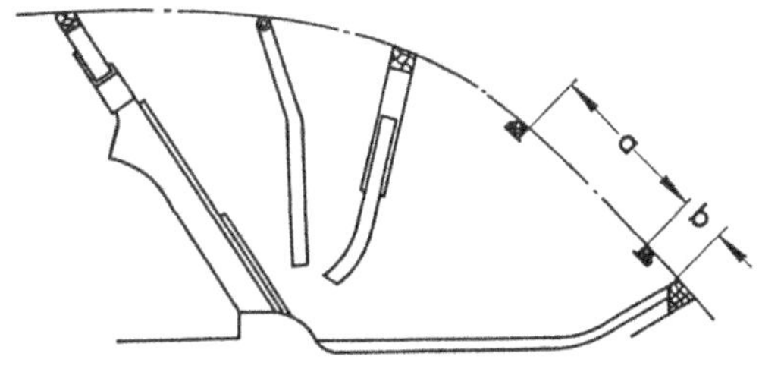

Important

Check position of rear window frame and rectify as necessary.

Dimension	1 up to Chassis No. 1 600 439 2 from Chassis No. 1 600 440 to Chassis No. 4 846 835 3 from Chassis No. 4 846 836
Dimension a	1 - 192±2 mm 2 - 255±2 mm 3 - 295±2 mm
Dimension b	1 - 120±2 mm 2 - 80±2 mm 3 - 90±2 mm
Dimension c	1 - 165±2 mm 2 - 190±2 mm 3 - 155±2 mm
Dimension d	1 - 480+5 mm 2 - 525+5 mm 3 - 540+5 mm
Dimension e	1 - 240+3 mm 2 - 226+3 mm 3 - 213+3 mm

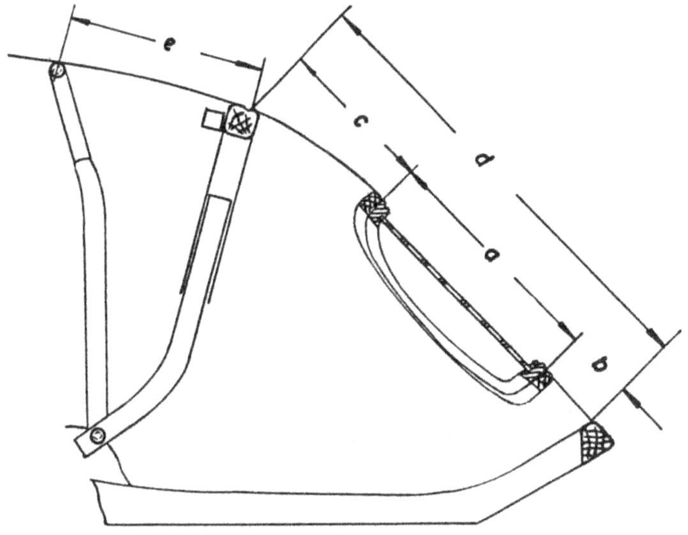

The dimensions at the rear bow are measured up to the folded edge

7 - Fold overlapping ends of nettle cloth and support strip over at the body bow and nail again. Excess material should be cut off.

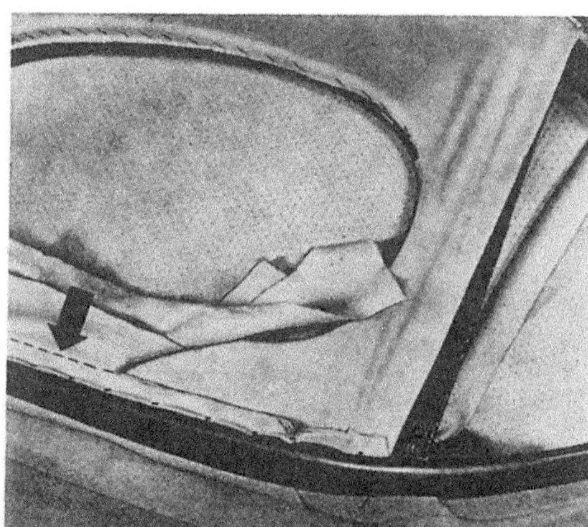

Note:

The double nailing increases the security of the nettle cloth for the window frame and the headlining support strip.

Fitting Linen Sheet

Two different types of cloth are used:

a - a coarse, tightly woven, light-colored linen

b - a slightly thinner, black lining material

The linen material forms a base for the rubberised hair. The lining material is used to enclose the rubberised hair padding.

The linen and lining material is supplied in separate parts which consist of a large roof piece and two rear panels which are each made up of a piece of linen and a piece of lining material.

The roof part of the linen sheet is made up of a piece of linen material, two strips of PVC top material and two strips of lining material.

The strips of PVC material sewn on each side of the linen sheet are used solely to prevent the linen sheet becoming visible at the sides when the roof is folded back.

The two strips of black lining material hold the rubberised hair padding together and give it additional support.

Furthermore, four pieces of linen are required for the rear side panels and two pieces of nettle cloth for the side panels. All the parts for the linen sheet are contained in a set which is supplied complete as a spare part.

1 - Nail the two side parts of linen **on** the body bow and in the recess in the rear bow under tension. The parts should be stuck at the sides. Cut surplus material off.

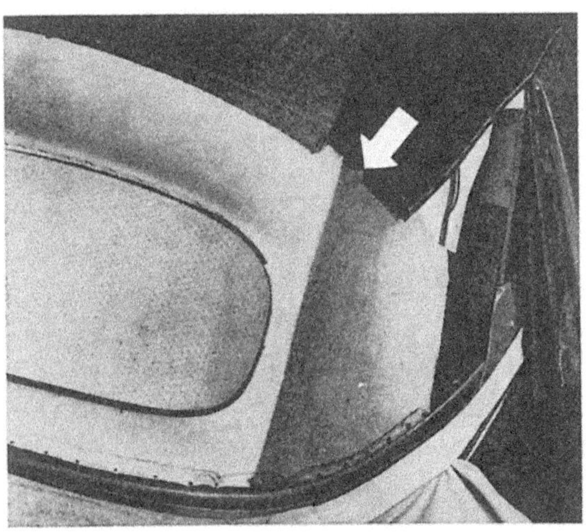

2 - Place the roof part with the sewn in PVC and lining strips on the roof frame so that the cuts in the PVC material are pointing to the rear (arrow).

Note:

At the rear the side seams must be parted up to the main bow.

3 - Tension roof part with side strips of PVC sideways and nail it in the recess in the rear bow.

Note:

Make wedge-shaped cuts to the rear in both seams.

4 - Lift top frame about 8″ (200 mm).

5 - Pull the *roof part forward until it is taut and* nail it to the header without folds. Part the side seams slightly near the header.

6 - Close the roof again.

7 - Nail roof part and PVC side strips carefully in the recess in the header. Cut off surplus material. The seams should be cut at an angle (white arrow).

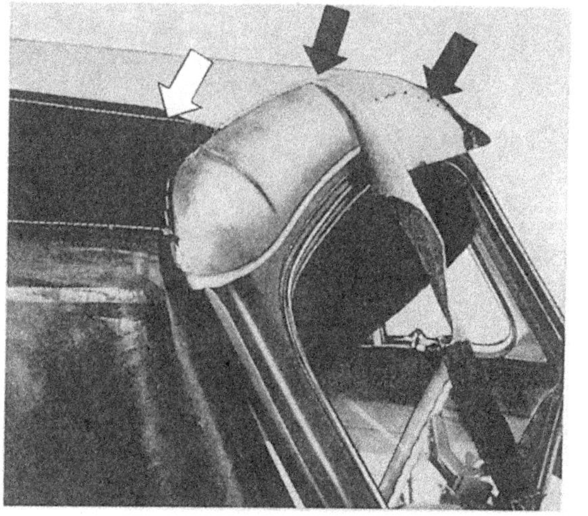

8 - Hook the spring with tensioning cable on to the lugs in the roof frame and pull the cable sideways out of the frame.

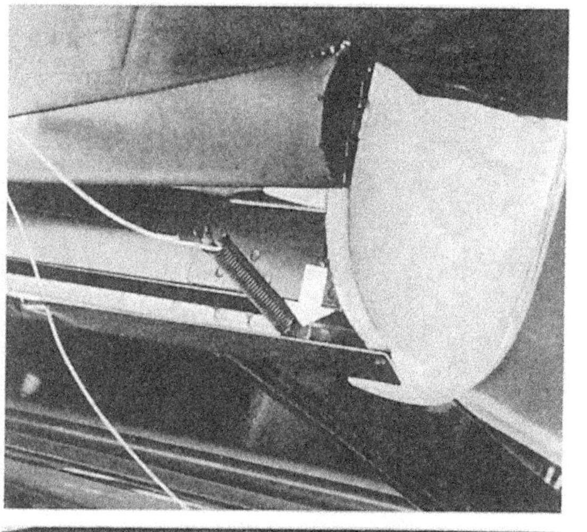

9 - Nail the front part of the linen side panels to the main and rear bows and to the body bow, under tension and free of folds.

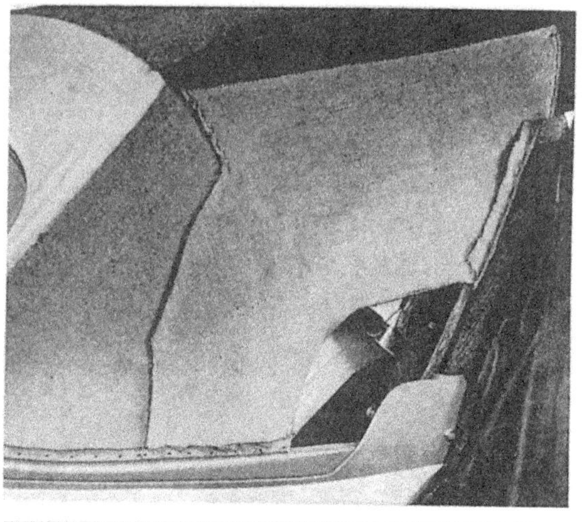

10 - Cut a half circle of approximately 150 mm radius out of the linen front side parts near the main hinge.

11 - Sew the rear linen side part to the nettle cloth of the rear window frame and to the front linen side part under tension.

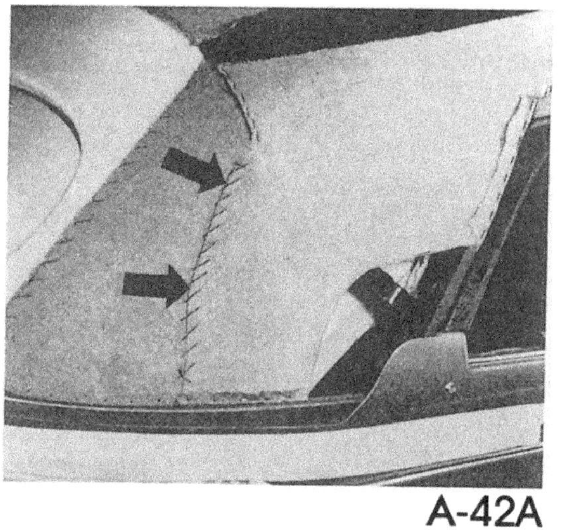

12 - Sew the front linen side part to the roof part under tension.

Note:

Small, tight stitches fairly close together should be used when sewing these materials. If the parts of the linen sheet are sewn together loosely, the shape of the top will suffer, particularly at the rear.

Installing Rubberised Hair Padding

The rubberised hair padding shapes the top of the Convertible and also provides good insulation. It consists of the rear part and the roof part but is supplied under one part number.

1 - Lay the roof part of the padding on the roof frame with the smooth surface uppermost and the thin part towards the header.

2 - Secure the roof part of the padding to the rear bow on each side.

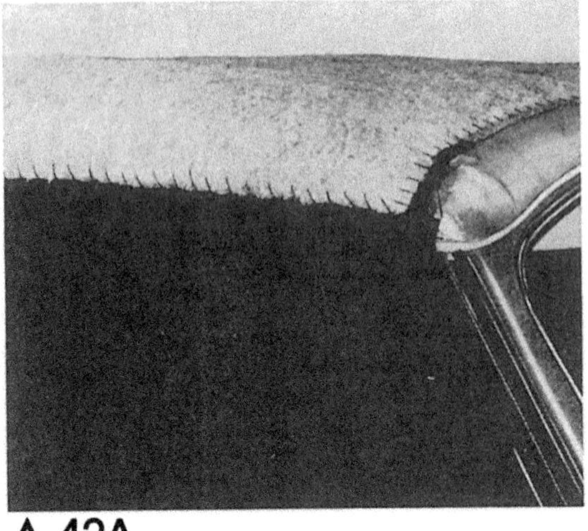

3 - Pull the padding gently towards the header and secure it.

4 - Pull the padding up to the rear edge of the header and sew it to the linen sheet there with small firm stitches.

5 - Sew the outer edges of the roof padding to the seam of the PVC strips with small firm stitches. Cut the excess padding off at an angle at the sides and pluck the material out slightly to taper it off.

6 - Sew the roof padding to the linen sheet at the rear bow with small firm stitches.

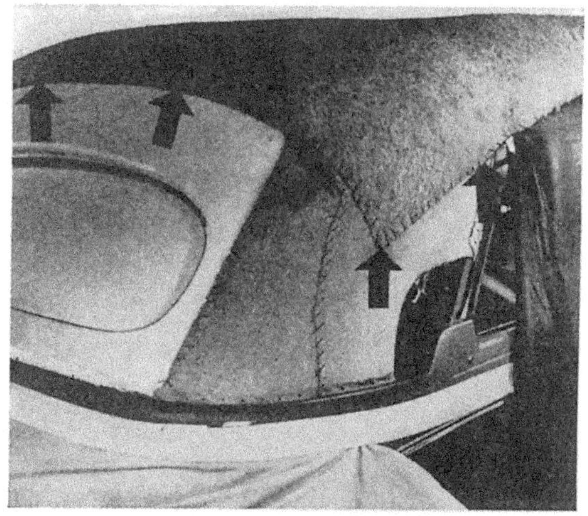

7 - Sew the roof padding to the linen sheet with rows of large stitches about 100 mm apart across the roof.

8 - Sew the roof padding to the linen sheet near the rear bow and the header, using large cross stitches.

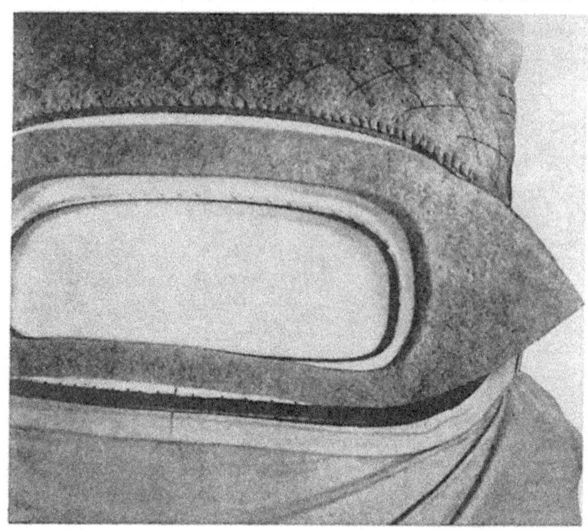

9 - Lay the rear part of the padding on the roof with the wide part upwards, stretch it from rear bow to body bow and secure it.

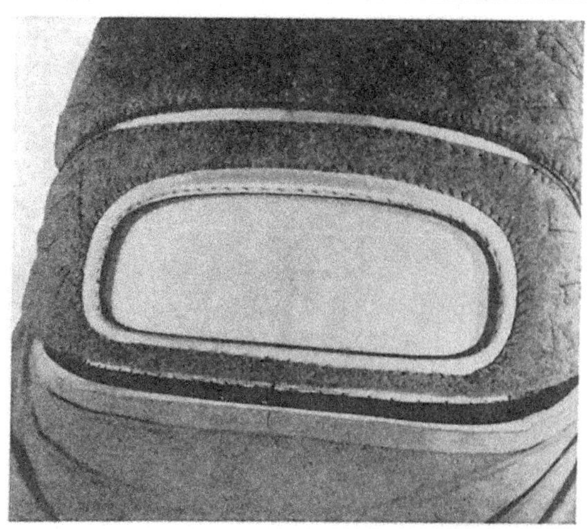

10 - Sew the rear part of the padding at the body bow, to the nettle cloth round the rear window and to the side parts of linen sheet, using small firm stitches.

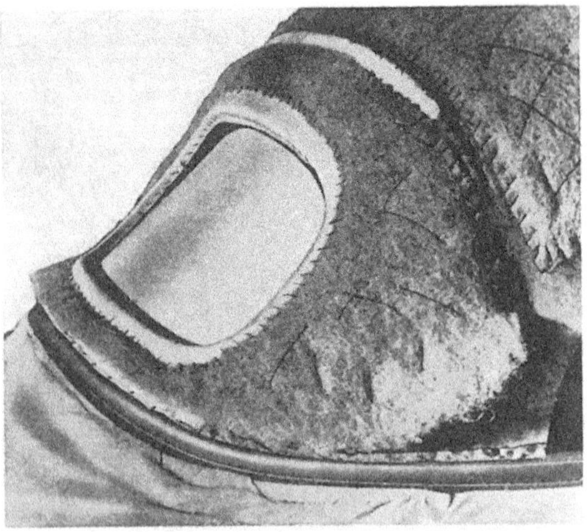

11 - Sew the rear part of padding to the nettle cloth and to the rear side linen panels with large cross stitches.

Important
The rear part of the padding is not sewn near the body bow.

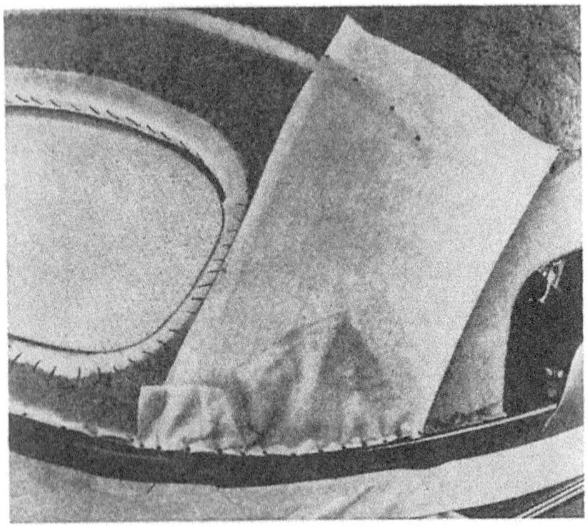

12 - Place pieces of nettle cloth (550 × 350 mm) over the corner parts of the padding on each side and tack them in the recess in the rear bow and **on** the upper edge of the body bow.

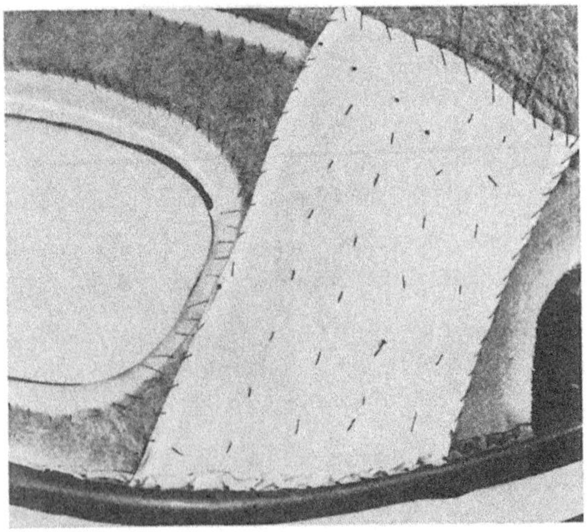

13 - Sew the nettle cloth strips to the linen side parts and the rubberised hair padding at the sides with small firm stitches. Cut off surplus material.

14 - Sew the nettle cloth strips again with several rows of large stitches.

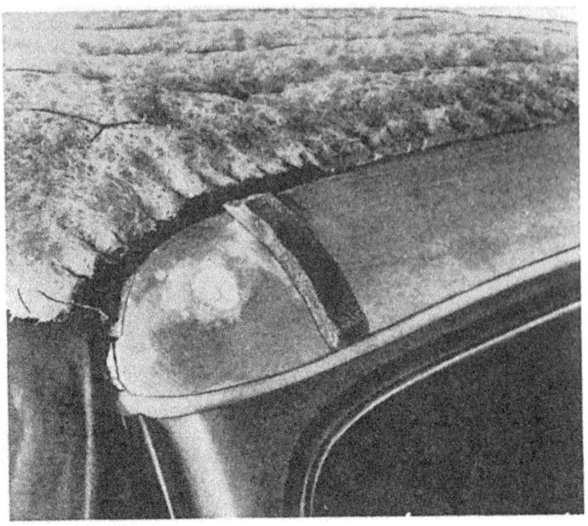

15 - Coat the header lightly with D 12 universal adhesive and pad it out as follows:

a - Cement a thin strip of foam rubber in the recess for the top seams.

b - Cement a strip of foam rubber on each corner of the header.

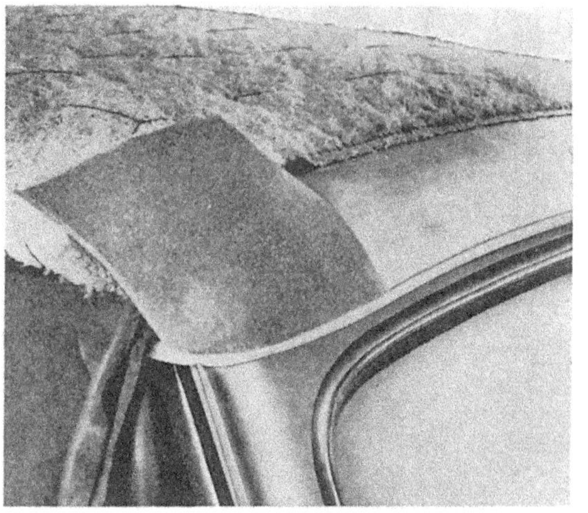

c - Cement a strip of foam rubber to the main bow.

16 - Lay the strips of lining material on the roof padding over one another and tack them into the recess on the rear bow free of creases.

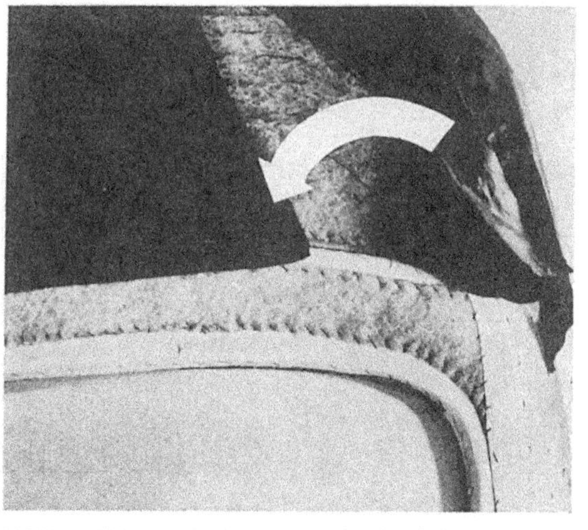

17 - Open top slightly, pull the strips of lining material forwards firmly and tack to the header.

18 - Lift top further, pull the strips of lining material taut round the header and cement them underneath in the recess.

19 - Close top and check position of lining material strips on the header, particularly at the rounded parts. Rectify position as necessary.

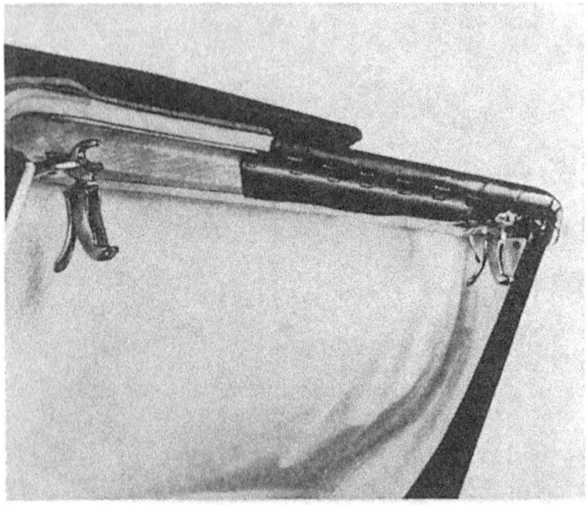

20 - Open top and nail the lining material strips to the header from underneath, starting at one corner.

21 - Cut off surplus lining material at the row of tacks on the rear bow and at the header.

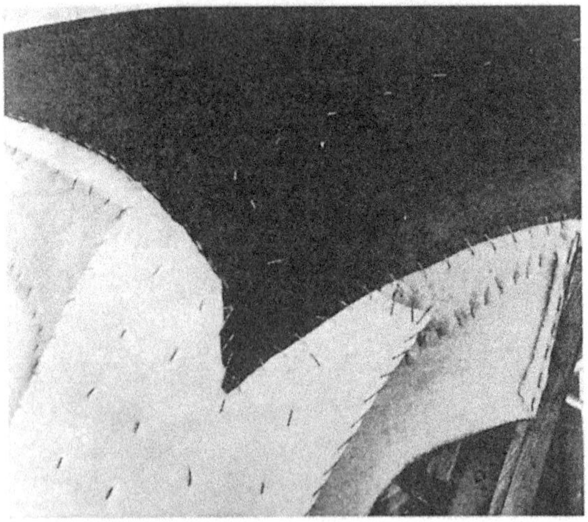

22 - Close top again and sew the lining strips to the padding all round.

23 - Pull the lining strips which are laid one over the other together in the center and sew with small stitches.

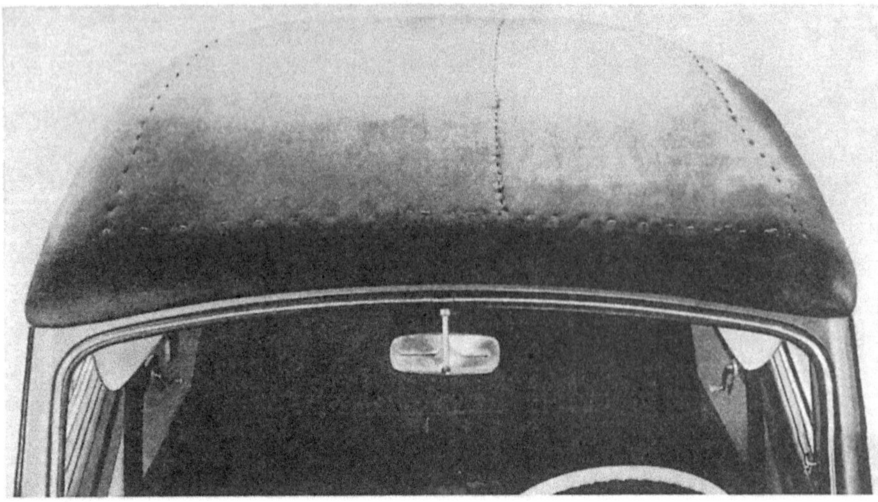

A-42A

Fitting Top Cover

1 - Nail the body bow beading on. The round part of the beading must be located 3 mm below the edge of the rear panel. At the sides, roughly level with the front edges of the coarse linen cloth, leave the beading loose for the time being.

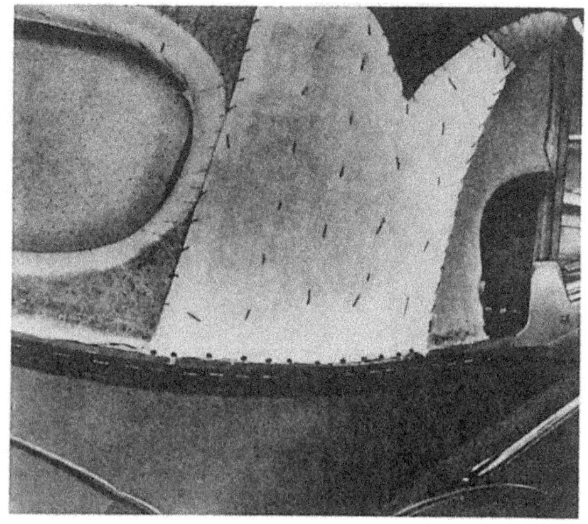

2 - Cement a strip of foam rubber (250 × 100 × 5 mm) on each side the rear bow.

3 - Lay the top cover on the closed top.

4 - Position the side seams of the top cover in the two locating points on the rear bow so that the rear part of the cover is under the center part.

Note:

If a new rear bow has been fitted, the locating points (recesses about 15 × 15 mm) must be made about 5 mm deep in the wood with a chisel.

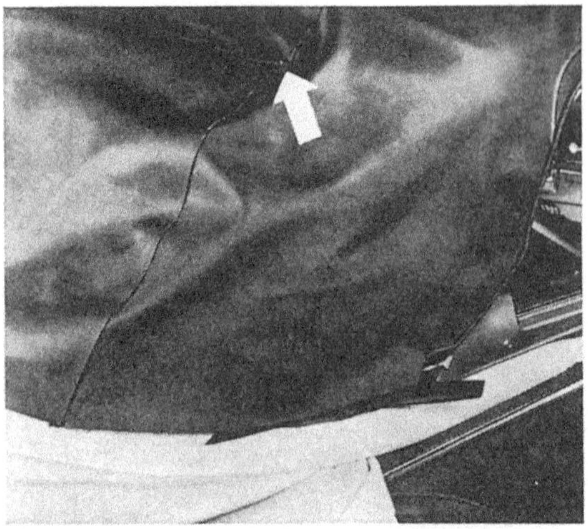

5 - Nail the top cover in the two locating points (recesses) on the rear bow with two brass tacks each side, noting the following measurements carefully:

a - Distance between locating points on rear bow 35.8 in. (910 mm).

b - Distance from locating point to lower edge of outer wooden molding on main bow 17.3 in. (440 mm).

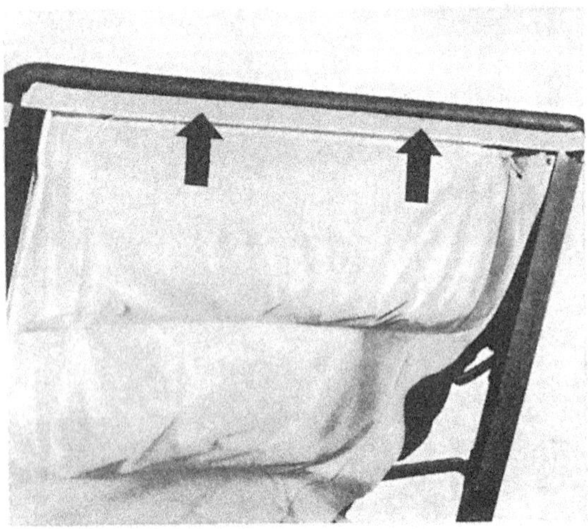

6 - Open roof and remove both fasteners.

7 - Cement header sealing rubber (110 × 30 × 18 mm) in the depression provided in the header. The sealing rubber must be angled at the ends so that a smooth joint is obtained.

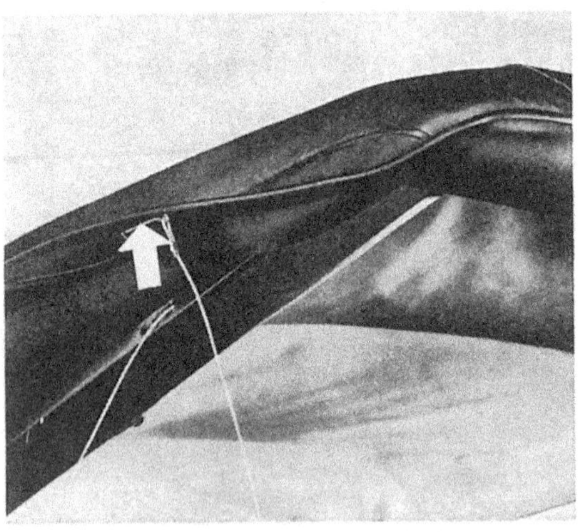

8 - Pull pocket in top cover over header.

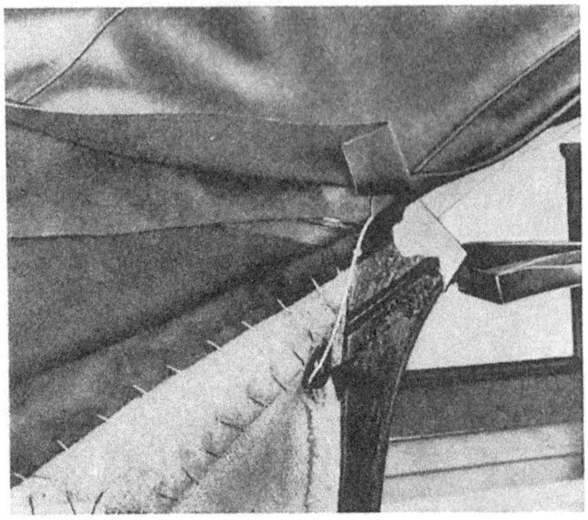

9 - Pull two tensioning wires into top cover from front to rear and secure them to the rear bows on each side. The wires can be pulled through with a piece of welding wire with a hook at the end.

10 - Attack the assist straps and coat hooks to the roof frame.

11 - Secure the side securing strips of the top cover to the inside of the outer wooden mouldings on the main bow with a few nails, starting at the top.

Note:

The top cover is marked on both sides near the main hinges with two white lines which serve as locating points when fitting the top.

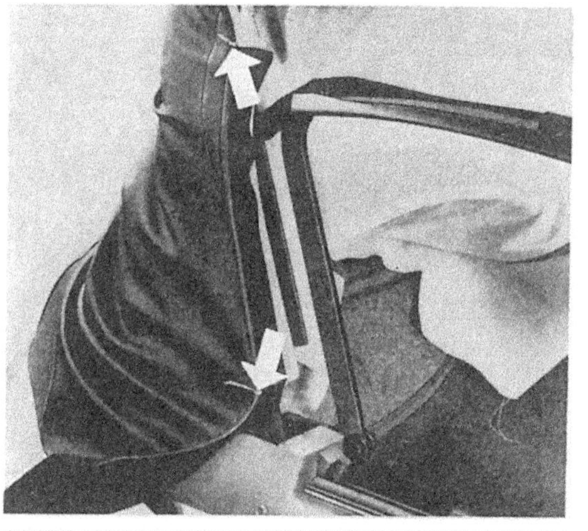

Pull the cover on the right side first until the upper mark with the sewn-on flap is in line with the joint between roof frame and main bow. Then pull the cover down firmly until the lower mark is in line with bottom edge of the outer wooden moulding on the main bow.

Secure top cover on left side in the same way.

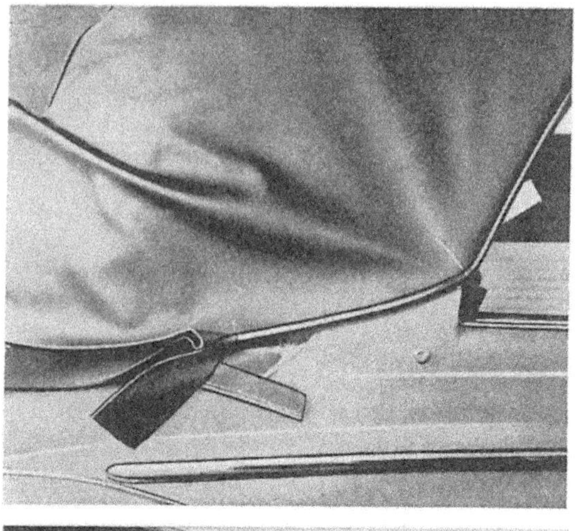

12 - Tension the top cover sideways and secure to the ends of the main bow wooden mouldings with one nail on each side. These two nails should be knocked in at the points where the chrome-plated screws are inserted later.

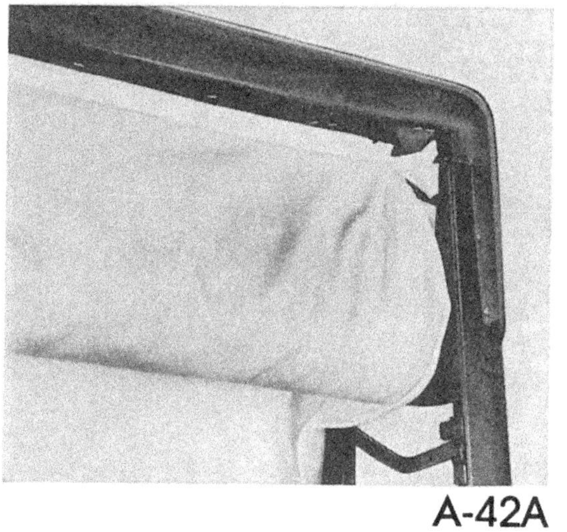

13 - Press the top to the closed position and check if the pocket in the top fits properly on the header, particularly at the corners.

14 - Open roof and stick top pocket to the underside of the header and the two side frames with D 12 universal adhesive. The pocket should also be tacked at the corners.

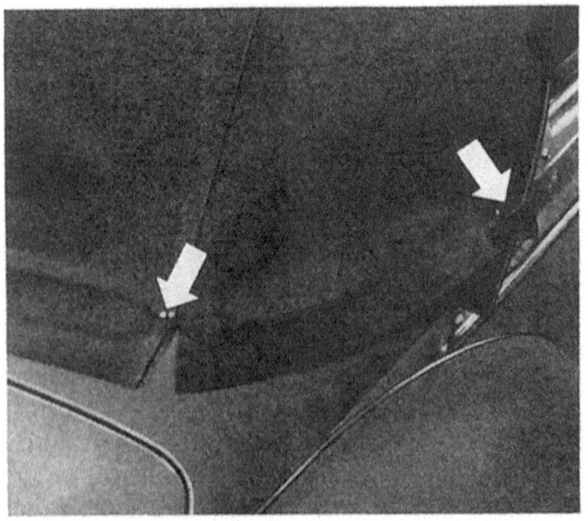

15 - Attach fasteners to header and close roof.

16 - Starting at the rear bow, pull the top cover down at the sides and tack it to the body bow with two nails each side.

17 - Pull cover to rear and nail the center part to the body bow with two nails.

Important
To avoid tearing the material, drive the two nails through the seams.

18 - Lay the body bow beading in the top cover on both sides and nail it to the body bow.

19 - Tension the fitting strips for the cover and nail them to the body bow on each side.

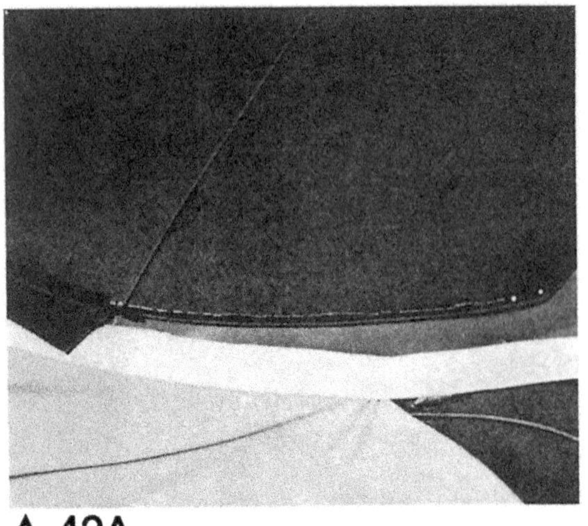

20 - Tack the sides of the top cover, from the two side seams towards the main bow, to the body bow under tension. For the time being, the top cover is not nailed to the body bow under the rear window.

Note:

Place a small strip of top cover material between top cover and body bow near the side seams. This gives a smoother surface for the trim moulding which is nailed on later.

21 - Tension top cover uniformly and nail it to the recess in the rear bow free of creases. Cut off surplus strip of center part.

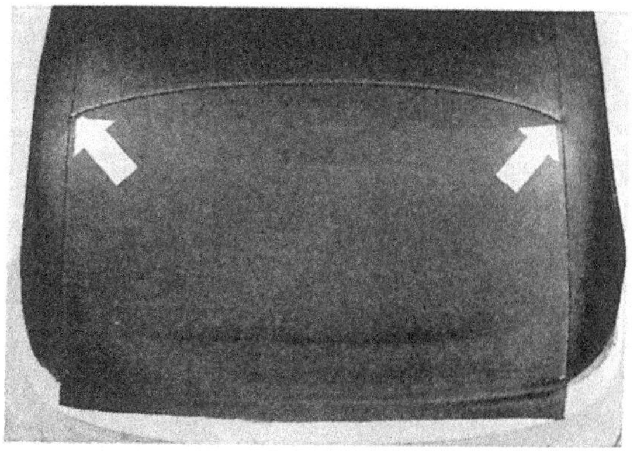

Important

Nail the rear part first and then nail the center part over it so that water does not run off into the joint.

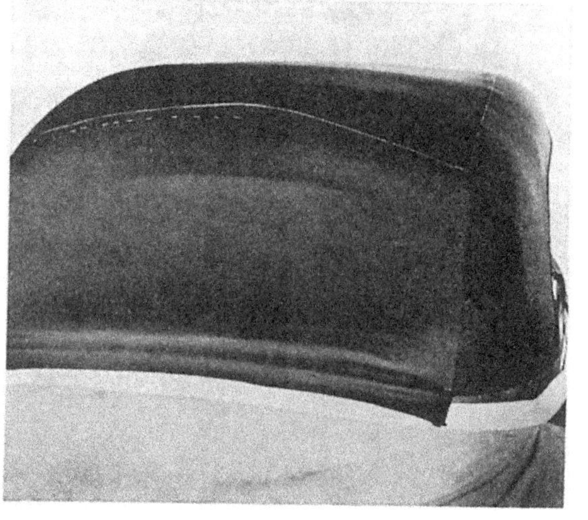

22 - Pull top cover down taut and nail it to the body bow under tension and free of creases.

Important

The space between the beading for the body bow and the row of nails must be uniform.

23 - Cut surplus material off cleanly with a knife so that the top cover material and the beading overlap properly.

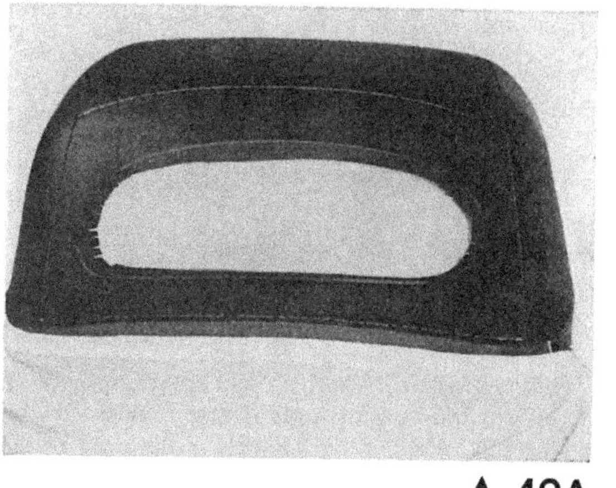

24 - Cut the top cover out inside the rear window frame and make a series of small cuts in the corners.

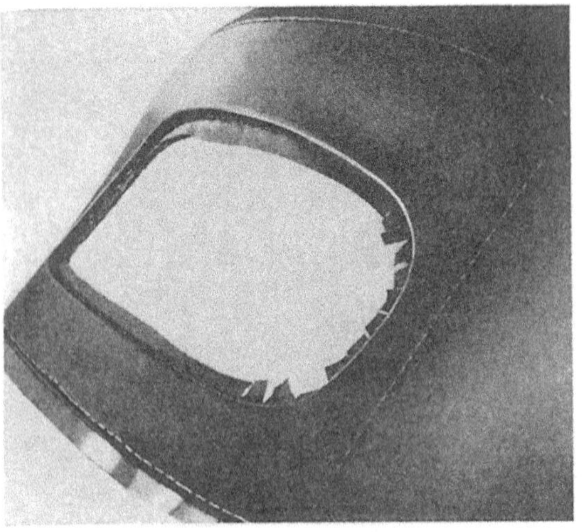

25 - Nail the top cover to the rear window frame in the corners first and then the straight parts.

26 - Cut surplus material off round inside the frame.

27 - Cut out the headlining inside the rear window frame and make small cuts in the corners.

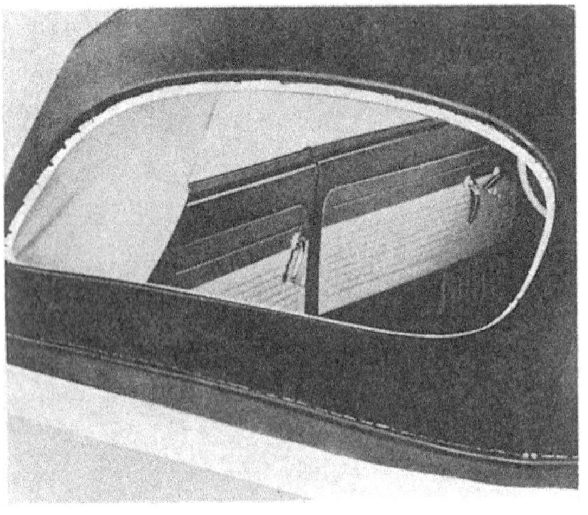

28 - Nail the headlining to the window frame in the corners first and then the straight parts.

29 - Cut off surplus material in the window frame.

Fitting Trim Moulding

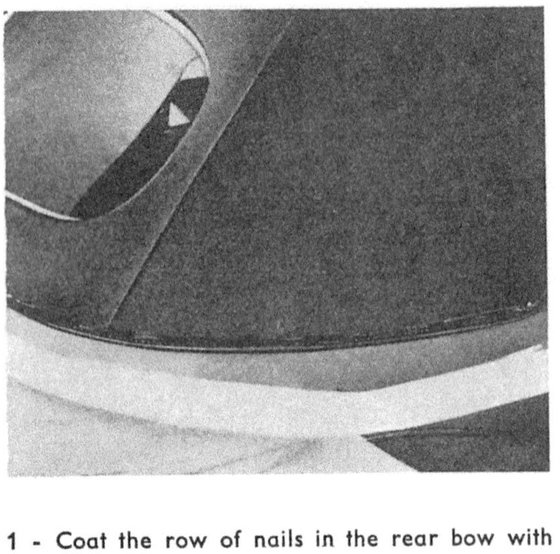

1 - Coat the row of nails in the rear bow with rubber solution particularly in the area of the seams at the ends. Take care that the coated area is not wider than the trim moulding to be nailed on.

2 - Nail a strip of sealing tape over the row of nails on the rear bow.

This sealing tape is a natural rubber strip of the same width as the trim moulding. The light colored protective strip must be removed before the tape is fitted.

3 - Fit a strip of sealing tape on the trim molding.

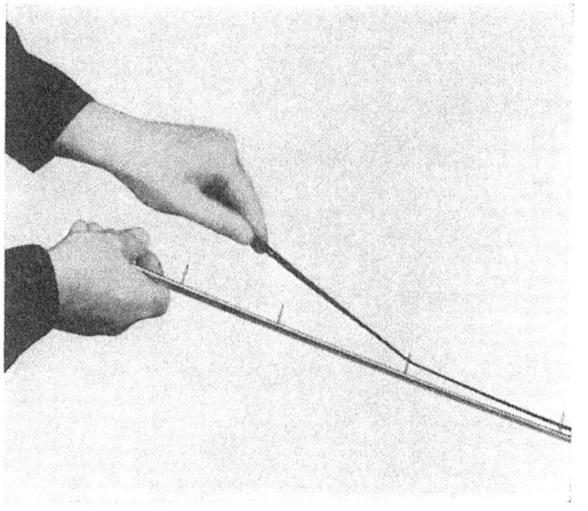

4 - Starting at one side, nail the trim molding on.

Important

A special wooden block must be used when fitting the molding to avoid damage and ensure that molding fits properly.

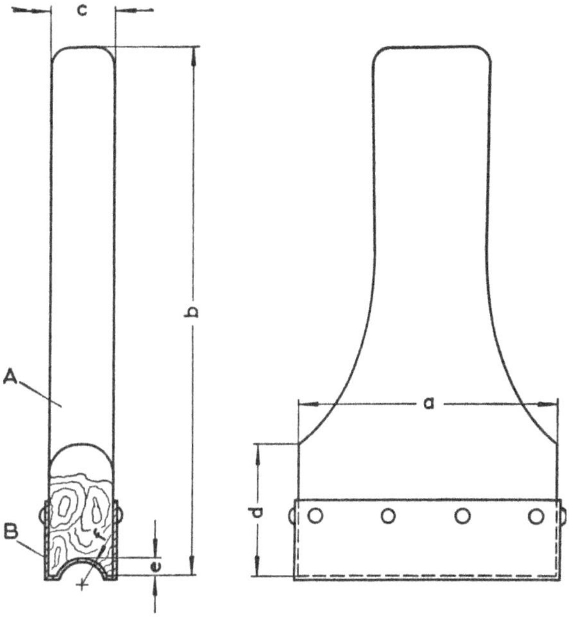

A - Special block
B - Leather cover (approximately 1 mm thick)

a = 2.75 in. (70 mm)
b = 5.5 in. (140 mm)
c = .66 in. (17 mm)
d = 1.37 in. (35 mm)
e = .2 in. (5 mm)
f = .27 in. (7 mm)

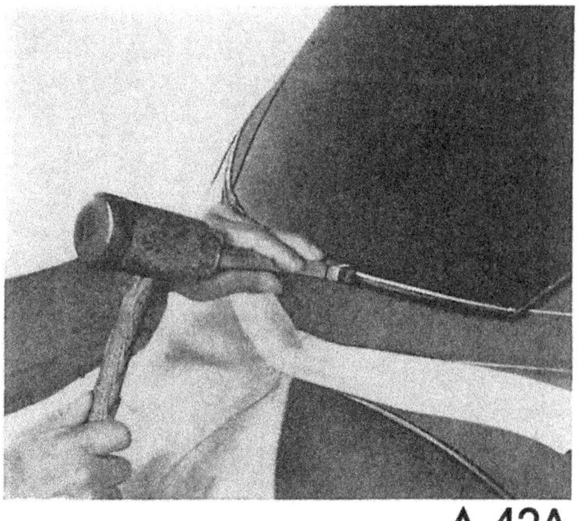

5 - Nail on the two-piece molding for the rear body bow, starting in the middle, and slide clip over joint. Ensure that the trim molding goes right round to the beading at the sides.

A-42A

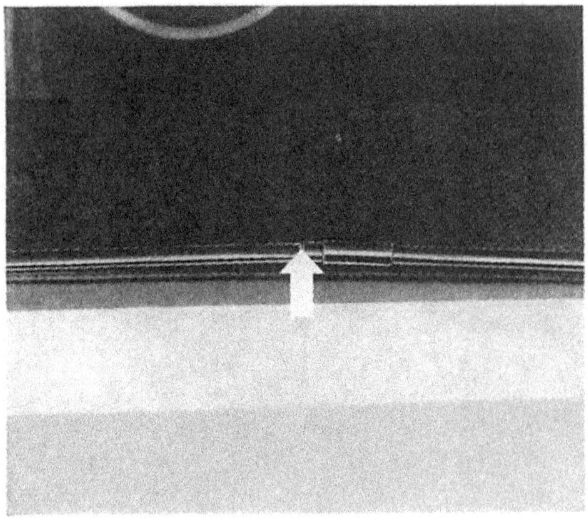

Note:

To avoid difficulties, such as overhanging at the ends, it is advisable to determine the exact location of the trim molding by fitting, measuring or marking with chalk. If necessary, the ends of the molding should be ground off at the center joint.

The repair sequence for the fitting of headlining and top padding has only changed very slightly with the introduction of the new tensioning wire method of securing the top cover. The following differences should be noted:

1 - The rear body bow consists of three separate segments which can be replaced without difficulty.

2 - Only one webbing strip is used on each side of the rear window and these strips are nailed to the **side** of the rear body bow.

The following measurements are important:

a - Center of rear panel to webbing strip = 16.5 in. (420 mm)

b - Main bow to rear tubular bow = 11 in. (280 mm)

c - Main bow over tubular bow to rear bow = 18.1 in. (460 mm)

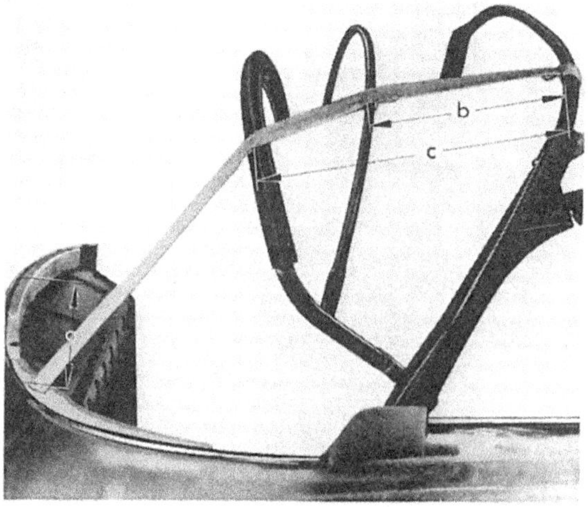

Note:

There are small depressions on each side of the rear bow to accomodate the webbing strips. For this reason, no measurements are given.

This top cover is no longer nailed to the rear bow because the center part of the cover is in one piece and runs right down to the rear beading which holds the tensioning wire.

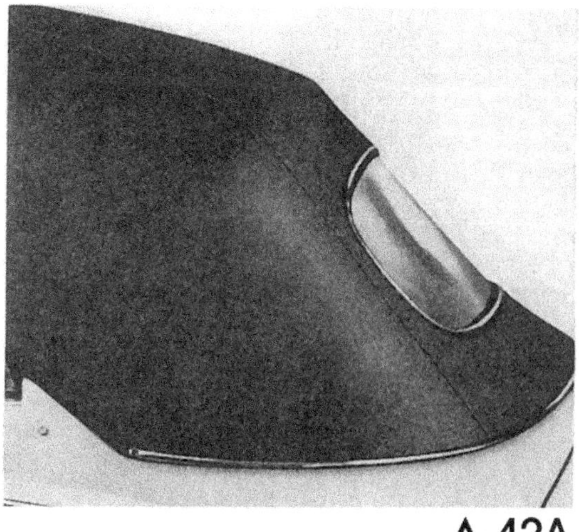

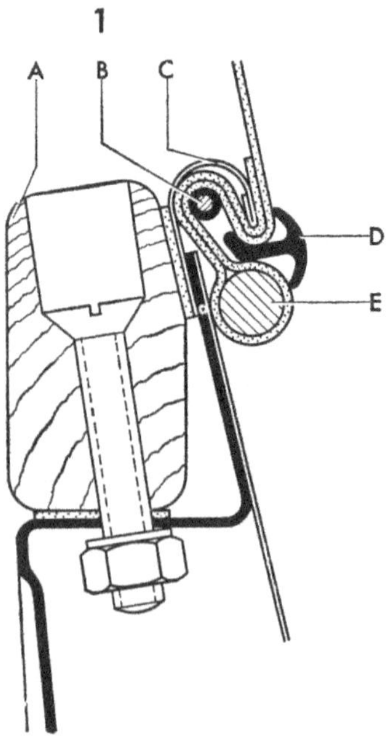

1 - Cross section at side

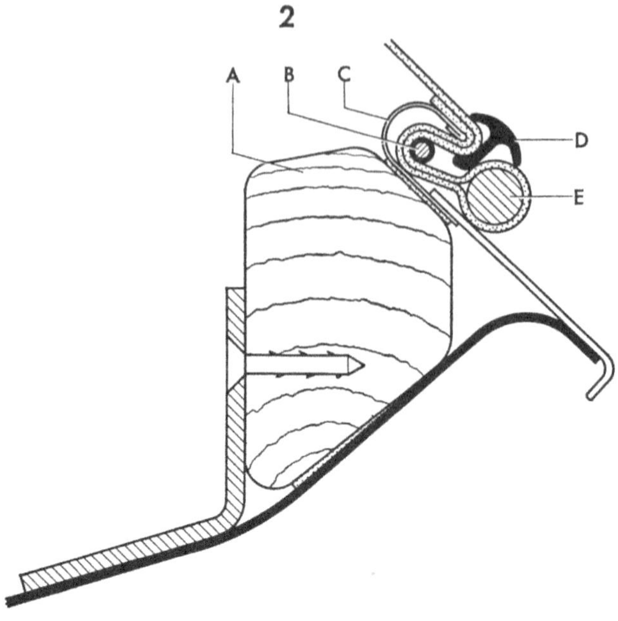

2 - Cross section in center

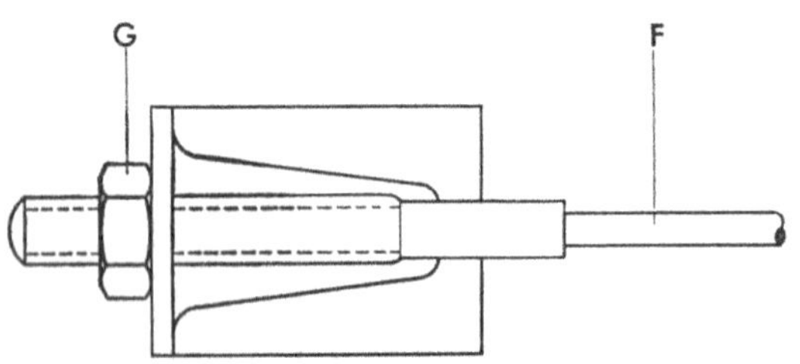

Tensioning bracket

A - Rear body bow
B - Tensioning wire
C - 'U' channel
D - Trim molding
E - Beading (sewn in)
F - Tensioning wire
G - Hexagon nut

Fitting Top Cover

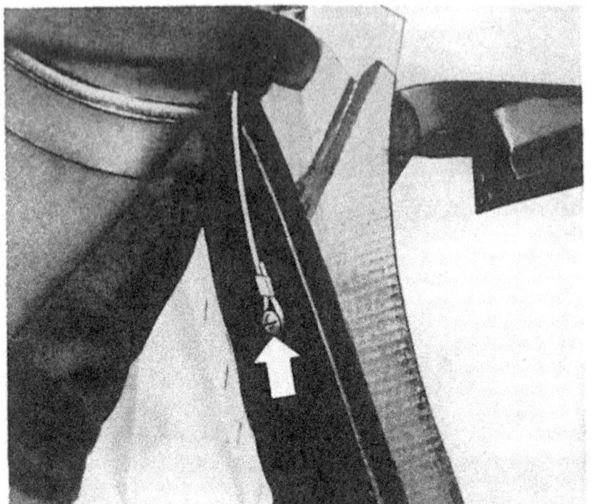

A-42A

1 - Lay the top cover on the closed top.

2 - Pull pocket in top cover over header.

3 - Pull the two front tensioning wires into top cover from front to rear and secure them to the main hinges.

4 - Secure the top cover and sewn-on material strips to the front face of main bow by inserting one tack at the lower edge on each side.

5 - Press top down and check if the pocket in the top cover fits properly on the header, particularly at the corners.

6 - Open top slightly and stick top pocket to the header and the sides of roof frame with D 12 universal adhesive. The corners of the pocket should also be secured with one brass tack on each side.

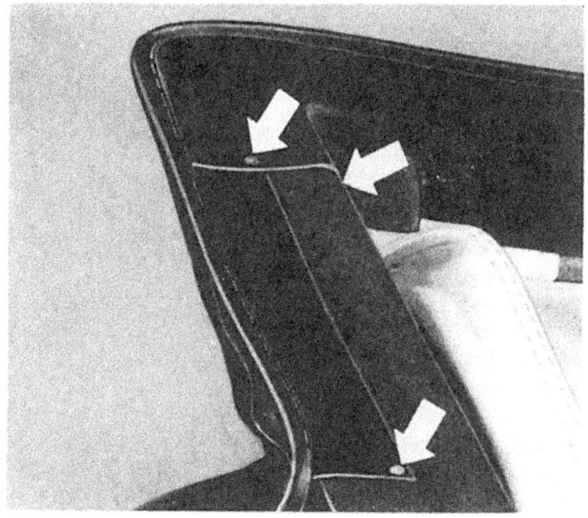

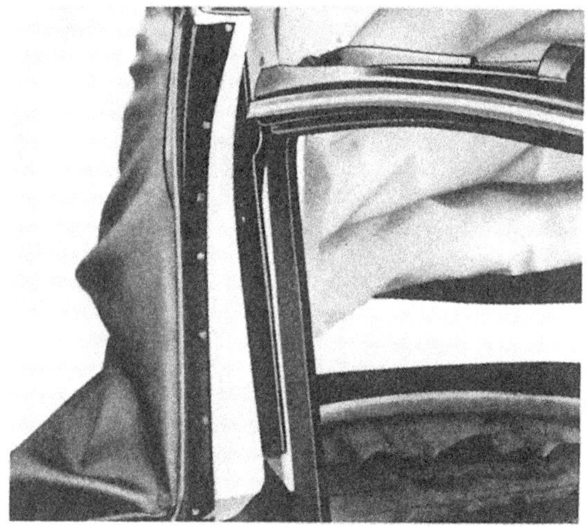

7 - Tack the edges of top cover to the front face of *main bow* provisionally with a few tacks and working from bottom to top.

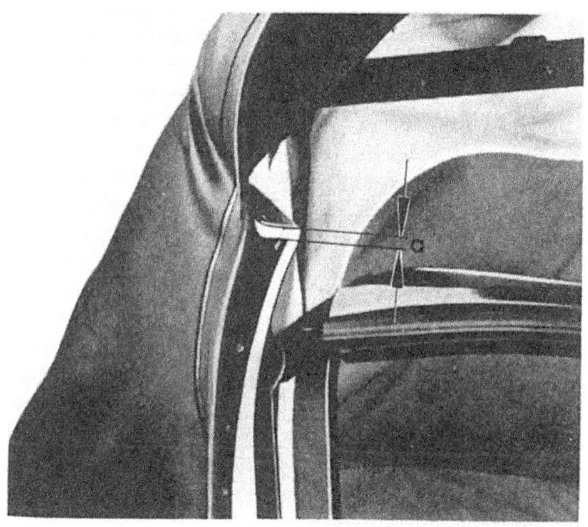

Important
Dimension "a" must be accurate.

a = .24 in. (6 mm)

8 - Close top and check that the top cover fits smoothly on the wooden moldings on the main bow. If necessary, the fit must be rectified straight away.

THIS PAGE INTENTIONALLY
LEFT BLANK

14 - Pull side parts of headlining forward and attach them to the roof frame with one chrome-plated Phillips screw and washer on each side.

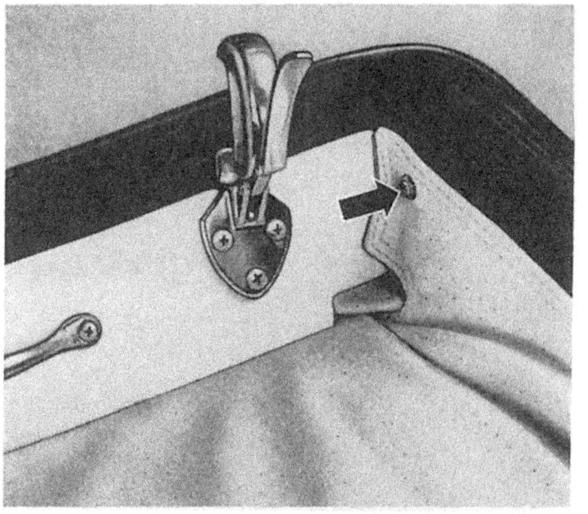

15 - Open top and secure top cover to rear body panel and quarter panels with the tensioning cable.

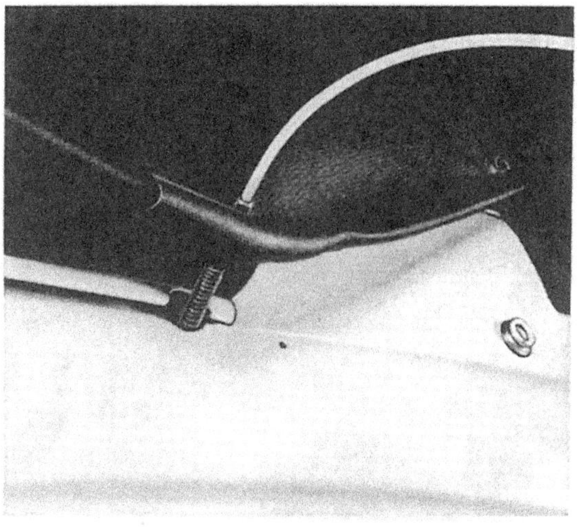

a - Pass the two threaded ends of the cable through the holes in the top cover and side panels into the body and fit them into the tensioning brackets.

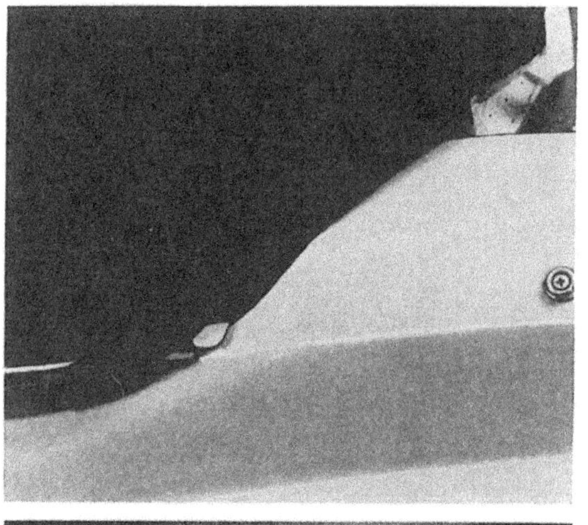

Note:

When a new top cover is being fitted, the holes at the tips of the small "pockets" on each side must be enlarged with a suitable tool (punch).

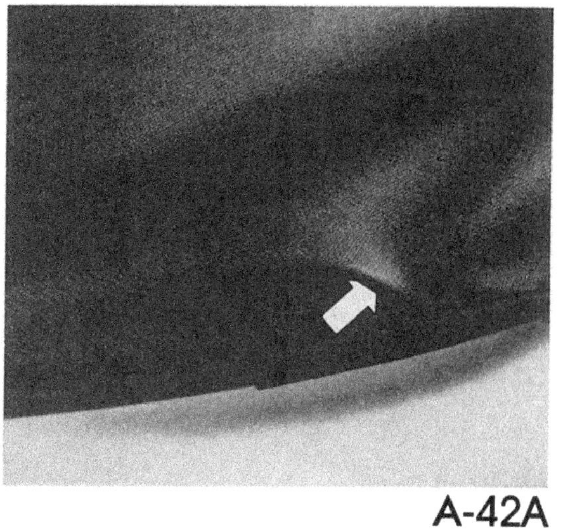

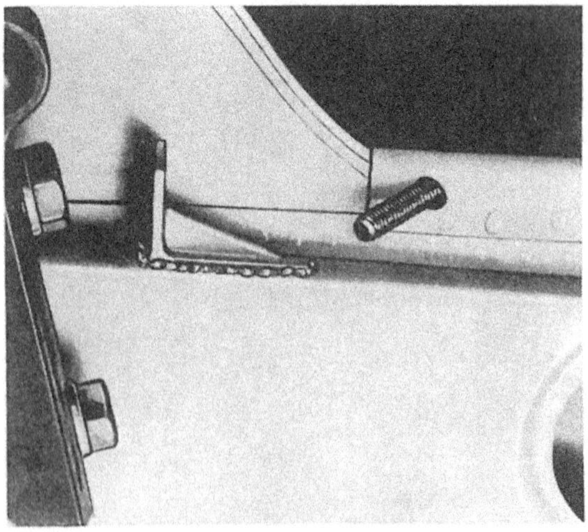

b - Screw hexagon nuts on a few threads.

c - Pull the top cover beading down so far that it is level with the "U" channel all along. The reinforcement sewn underneath the top cover must not form creases when the top cover is pulled down.

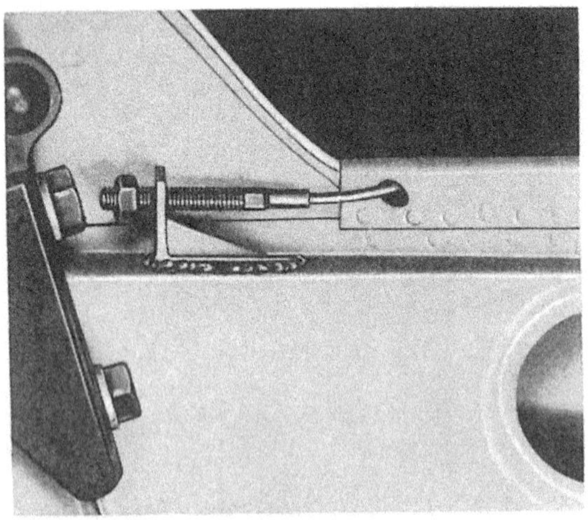

Two persons are required for the remaining operations:

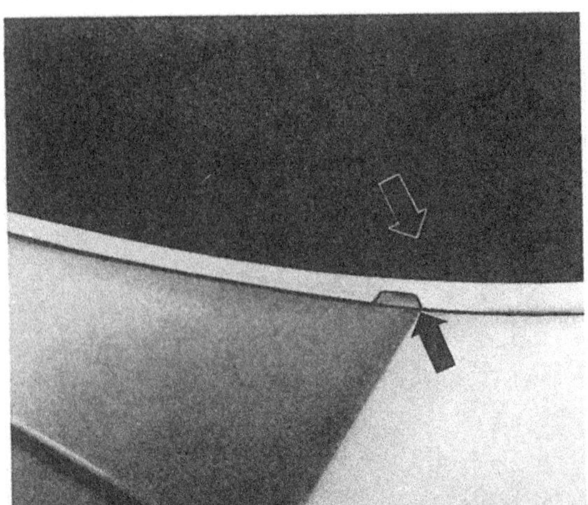

Important

The two longitudinal seams in the top cover must be in line with the cut-outs in the "U" channel.

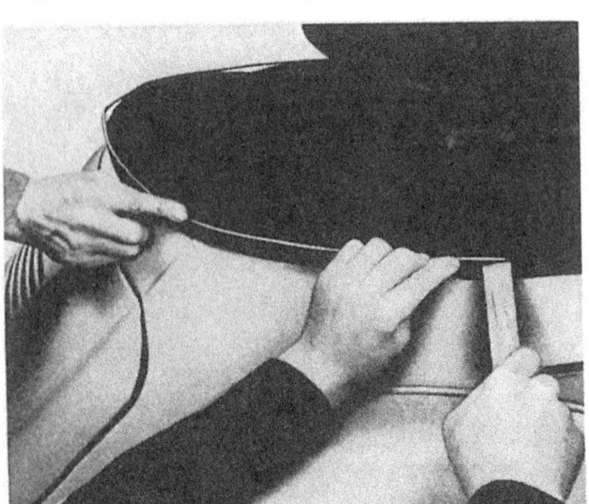

d - Starting on one side, press the top cover and tensioning wire into the "U" channel.

Important

The hole in the side of the top cover must be aligned with the cut-out in the "U" channel and the hole in the side panel.

e - While one person presses the beading and tensioning wire under the "U" channel at the side, the second person holds the side seam in the top cover in line with the cut-out in the "U" channel. At the same time, he keeps the tensioning wire taut.

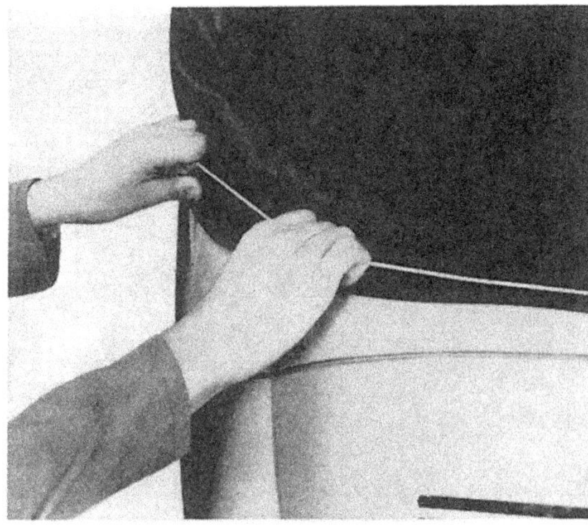

f - Press the top cover into the "U" channel in the rear panel in the same manner. Ensure that the seam in the top cover is also in line with the cut-out here.

Important

To facilitate fitting of the top cover, the tensioning wire outside the top cover must be pulled taut at the sides so that it reaches properly from side to side. This also makes it easier to press the top cover into the "U" channel.

g - Tighten the two hexagon nuts alternately. Following the pull of the wire as it tightens, the beading on the top cover must be driven into the "U" channel all round with a special wooden block.

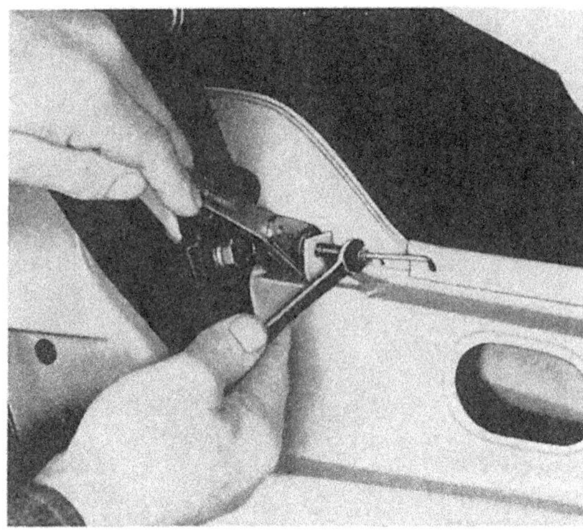

Note:

To prevent the wire from twisting, the threaded ends of the wire must be held with an open-end wrench when the nuts are being tightened.

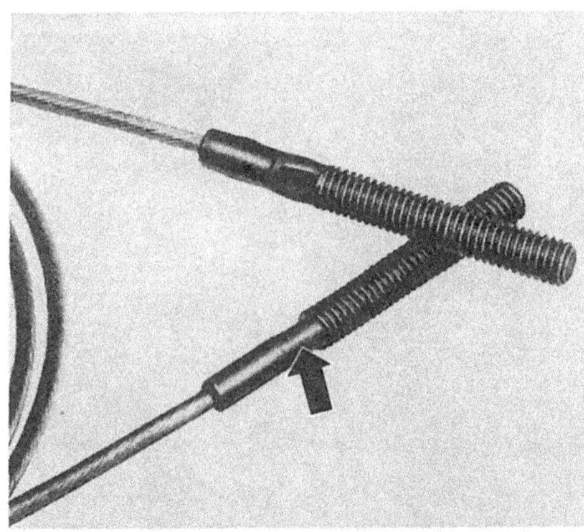

16 - Before the wire is finally tightened, the top should be closed and the beading positioned properly with the special block:

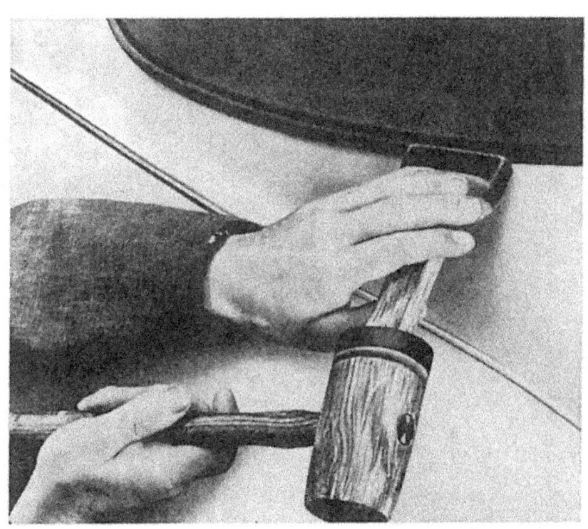

A-42A

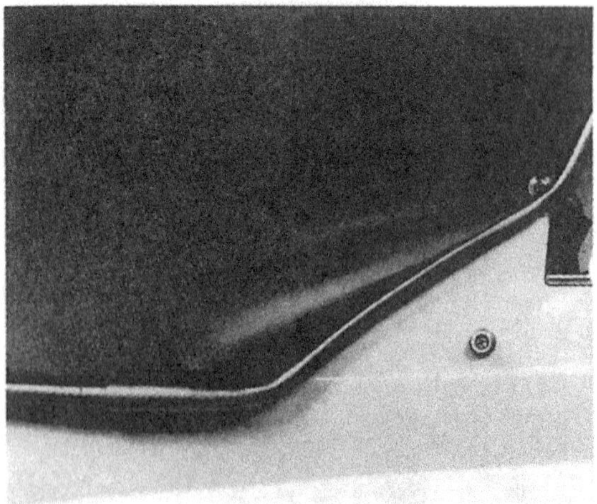

a - The wire must not be visible at any point.

b - The beading must contact the "U" channel all along without forming creases.

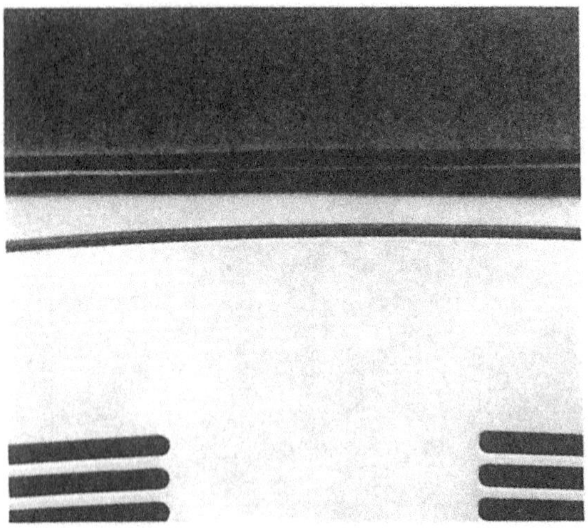

17 - Fit the trim molding as follows:

a - Mark the centers of the top cover and the trim molding.

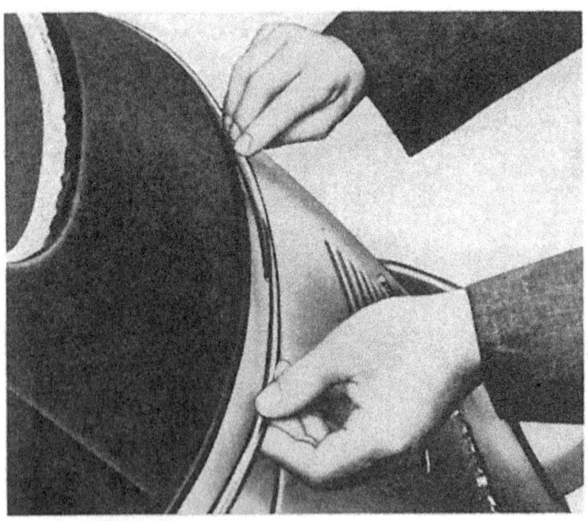

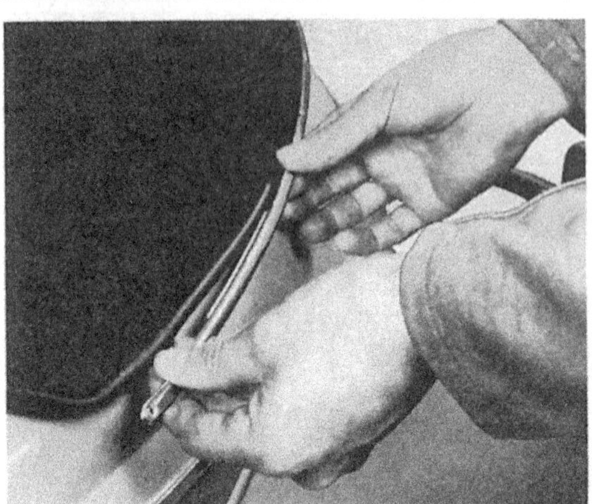

b - Starting in the center and working to one side first, press the trim molding into the gap between top cover and "U" channel. In the curves from rear to side panels, the special block should be used to ensure that the trim molding fits properly.

c - Hold trim molding firmly and drill a hole for the securing screw in the side panel. Attach trim molding loosely.

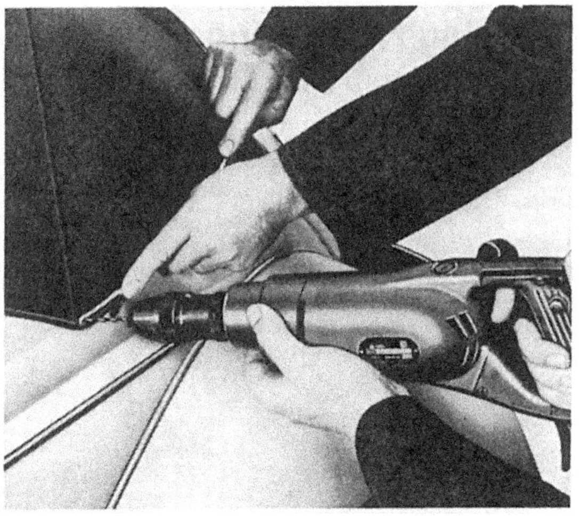

d - **Open top,** pull top cover down at end of beading, press trim molding over beading and tighten securing screw.

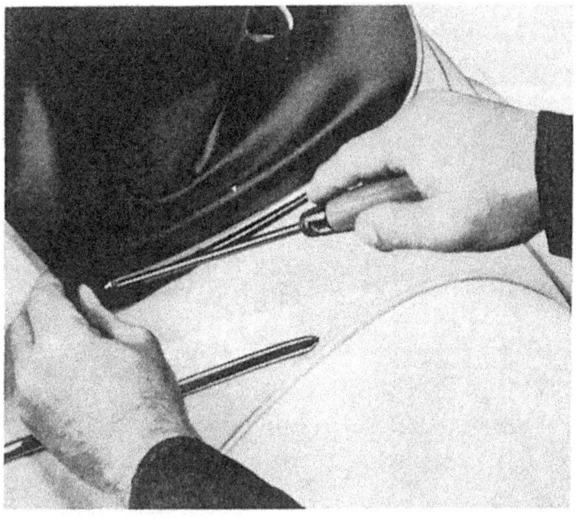

e - Close top, check position of beading and trim molding. If rectification is necessary, the top must be opened first.

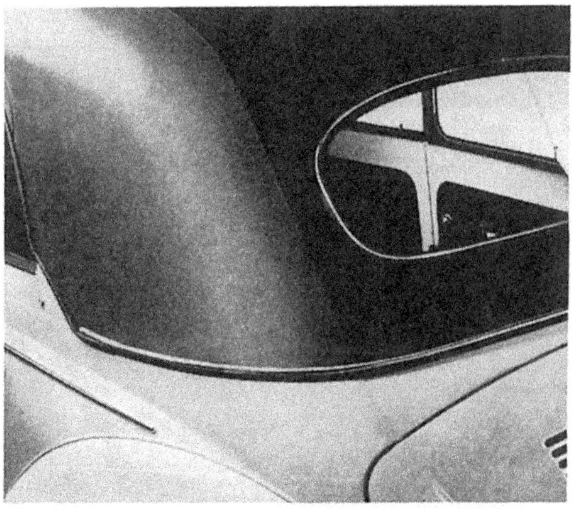

18 - Open doors and check that the door window frames are not rubbing on top.

Note:

The door window frames must not come into contact with top material but the parts must be as low as possible at this point. Corrections can be made by adding or removing padding.

Securing Headlining

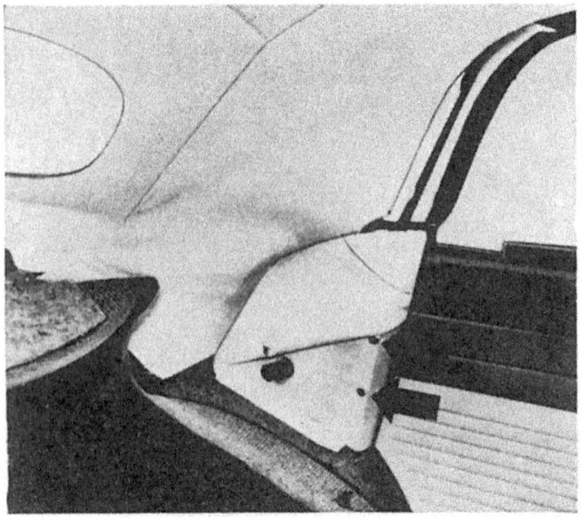

1 - Secure the blocks for the main bow pillars, using the holes provided.

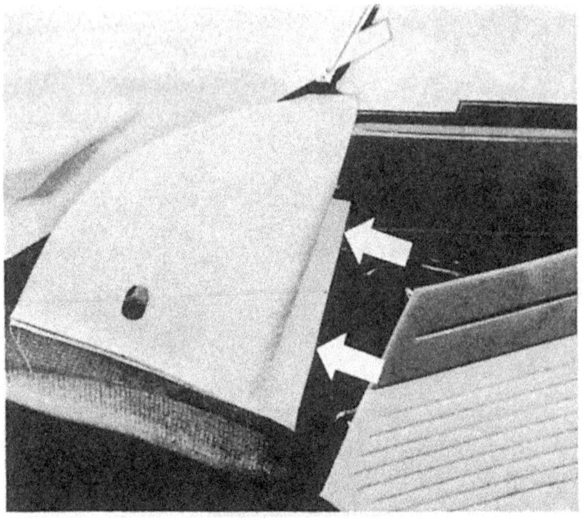

2 - Stick the excess leatherette to the blocks and to the metal under the quarter panel trims. Secure quarter panel trims again afterwards.

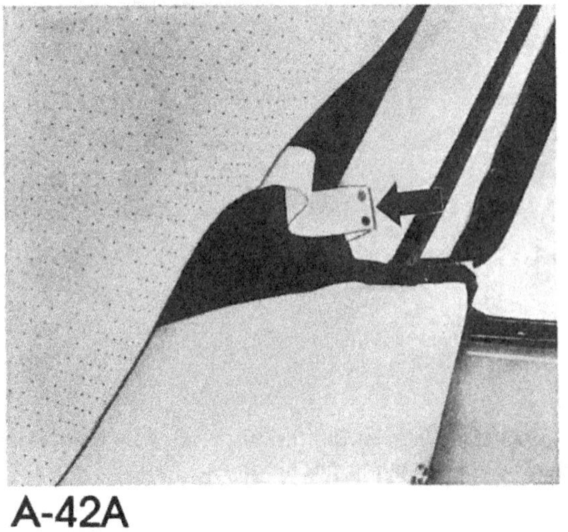

3 - Nail the rubber bands to the inner wooden moldings on the main bow.

4 - Nail the headlining and tensioned rubber bands to the lower side of the blocks.

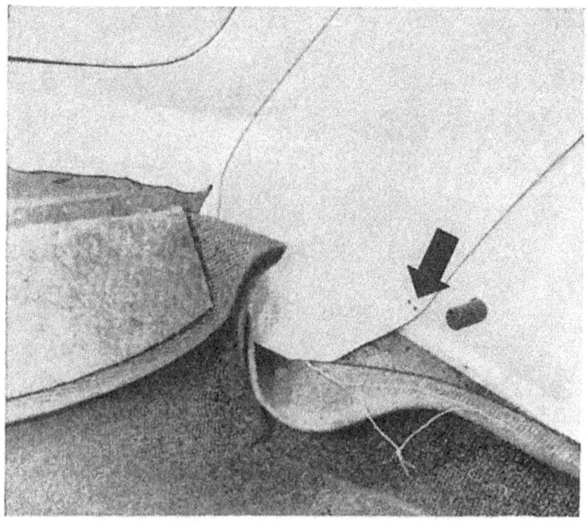

5 - Nail the headlining and tensioned rubber bands to the inner wooden moldings on the main bow.

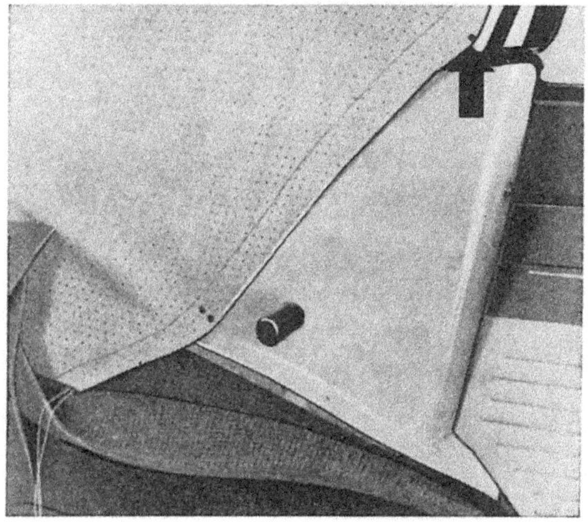

6 - Nail headlining to inner wooden moldings on main bow with two large headed tacks on each side.

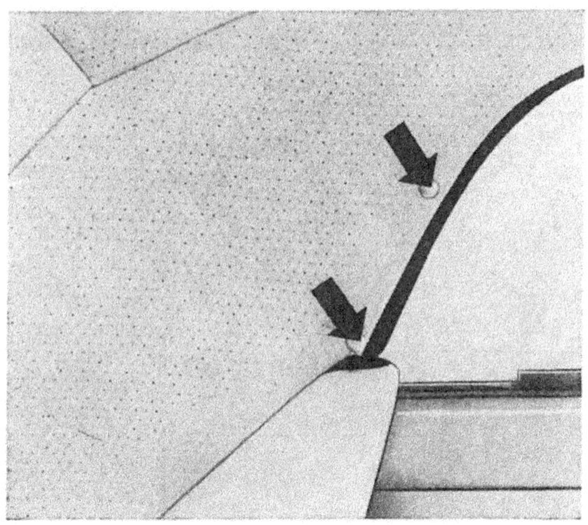

7 - Stick headlining in position round the luggage compartment and nail to rear body bow as well.

8 - Stick the detached luggage compartment lining and the carpet for the wheel housings on again.

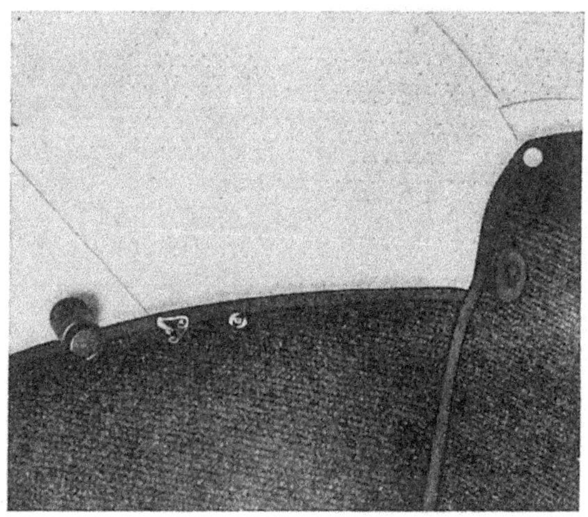

9 - Fit the press studs for the boot, the backrest securing hook and the bushes and cover plates for the safety belt anchorages.

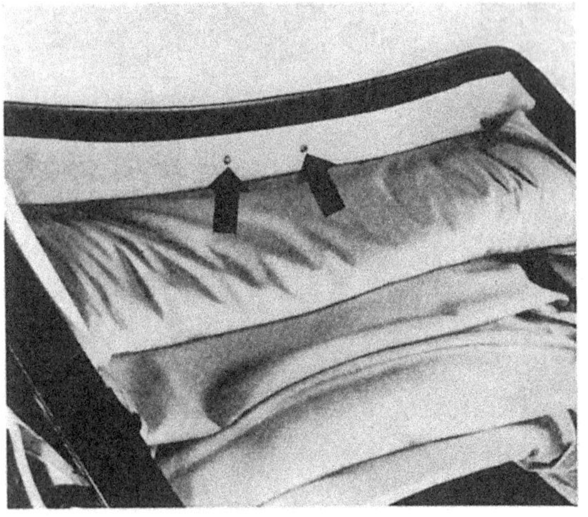

10 - Open top and take off the front fasteners.

11 - Stick the leatherette covered molding on the inside of the top header and secure additionally with two screws.

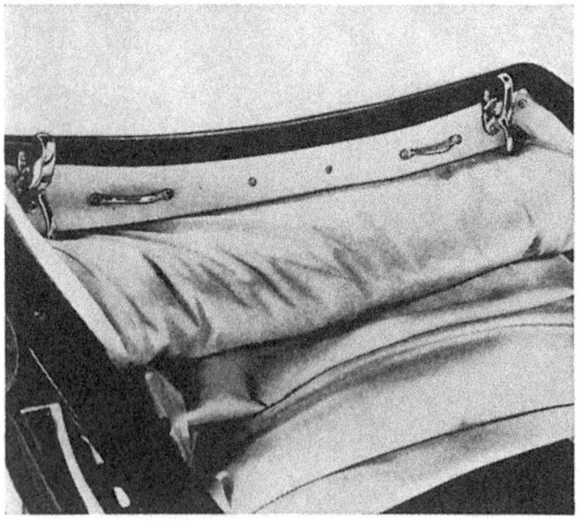

12 - Attach the handles and fasteners to the header at the places marked with a center punch.

13 - Close roof, pull side parts of headlining to the front and attach to left and right roof frame with screws.

14 - Secure top cover to front roof frame at each side and to the main bow with chrome plated screws.

Important

When work is completed, check if the top opens and closes properly and if the fasteners engage correctly.

New tops always have a certain amount of tension. It is not necessary to slacken this tension in any way as the cover will stretch after it has been in use for some time and can then be closed easily.

1 - Attach the three seals and the metal strips to the roof frame on each side so that when the window is raised it contacts the inner side of the rubber lip.

Note:

The metal strips are under the lip of the seal. When a new top frame has been fitted, holes must be drilled in the frame to suit the holes in the metal strips.

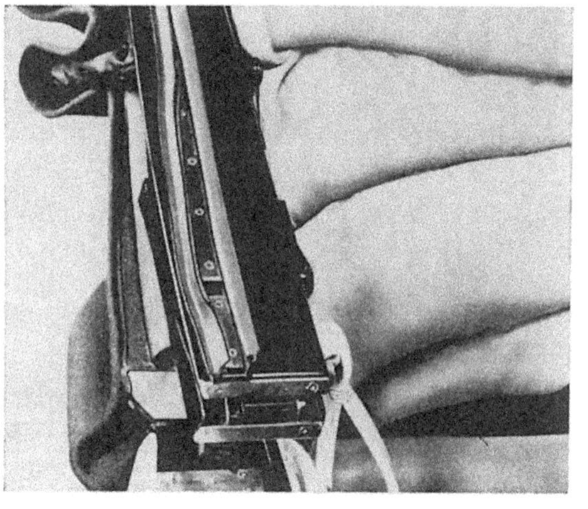

2 - Nail seals and metal strips to the main bow on each side.

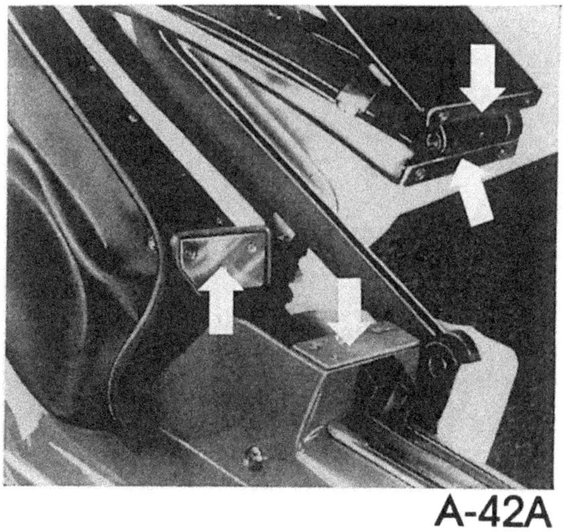

3 - Fit the four chrome-plated cover plates to the end faces of main bow and roof frame on each side.

A-42A

Note:

A rubber seal is installed under each of the cover plates for the main bow.

4 - Fit the clips to the roof frame on each side.

Installing Rear Window

The rear window is installed in accordance with the instructions on the fitting of the windshield on the Sedan with the exception that D 10 Window Sealer is used between seal and top to improve sealing.

1 - Coat nail surface in wooden frame lightly with D 10 Window Sealer.

2 - Place window in frame with seal fitted. The ends of the cable used to facilitate installation should hang down inside vehicle.

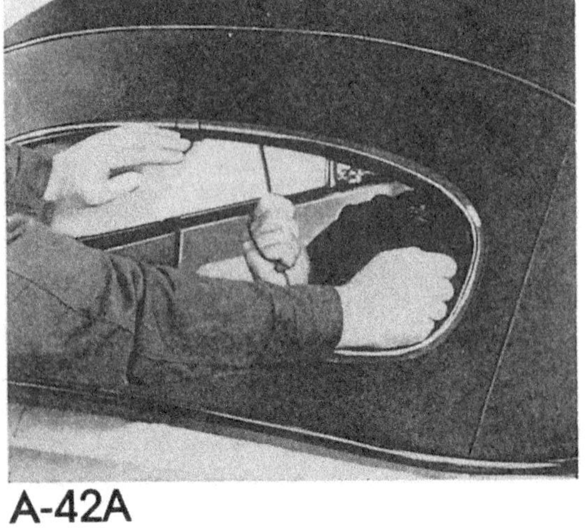

Note:

When fitting the window, another mechanic must hold the wooden frame from inside as otherwise the pressure exerted from outside to insert the window is liable to tear the linen material of the window frame.

3 - Clean off excess sealing compound with benzine, clean glass with methylated spirits and test for leakage.

Adjustments on Convertible Top

A - Guide tongues of both top locks strike the bottom part of the retaining lugs.

a - Windshield frame
b - Retaining lug
c - Guide tongue

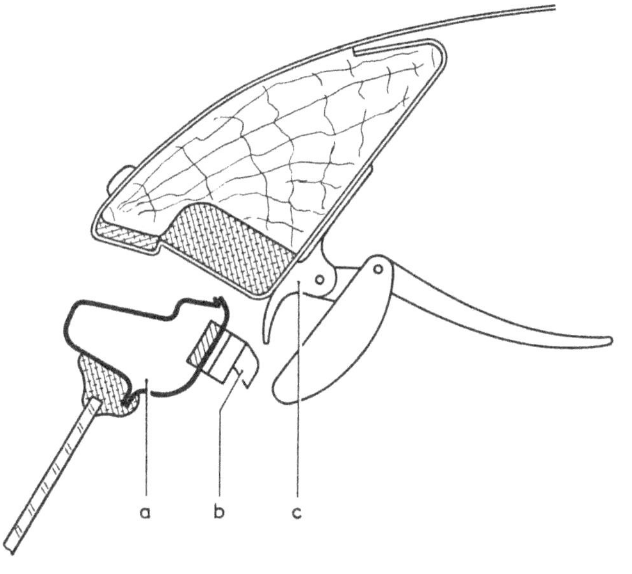

If the guide tongues of the top locks strike the retaining lugs when the top is closed, check whether the roof side members are located properly on the window frames and if the gap between the quarter panel and the doors widens towards the top.

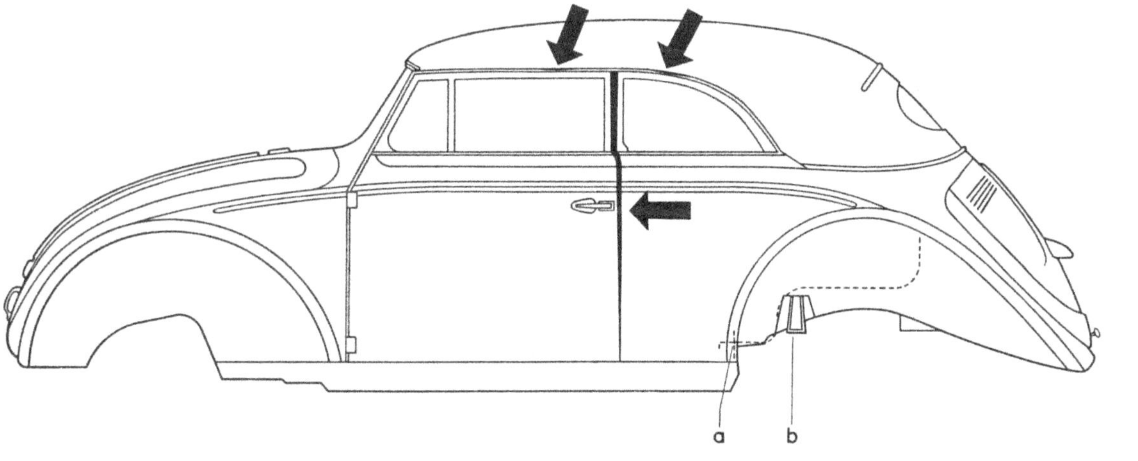

If the roof side members are resting on the window frames, the top frame has probably been damaged by being opened in a careless manner. In this case straighten the side members cold as required.

If the door gaps widen towards the top, loosen the body securing screws and remove them at points a and b. Place hard rubber packing pieces about 3 mm thick under the body support brackets on both sides. Tighten the body bolts again to 3 mkg (22 ft. lb.) torque. If the door gaps are still not the correct width, thicker packings must be used.

Should the roof side member location and the door gap both be in order, the guide tongues of the top locks can be bent slightly to the front or small leather pieces inserted under the bottom part of the locks until the guide tongues enter correctly.

B - Guide tongues of both top locks strike the windshield frame

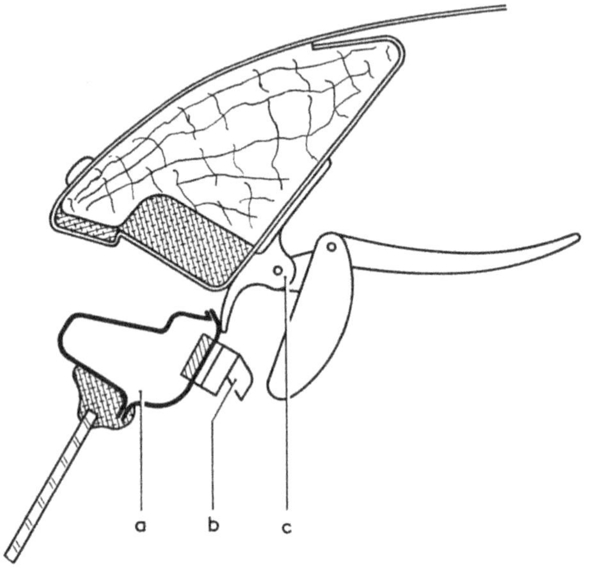

a - Windshield frame

b - Retaining lug

c - Guide tongue

If the guide tongues of the top locks strike the windshield frame when the top cover is properly tensioned, small leather pieces should be inserted under the locks until the guide tongues enter correctly. Otherwise — particularly on older vehicles — it is necessary to re-tension the top cover on the header.

C - Guide tongues of both top locks slide to the side of the retaining lugs.

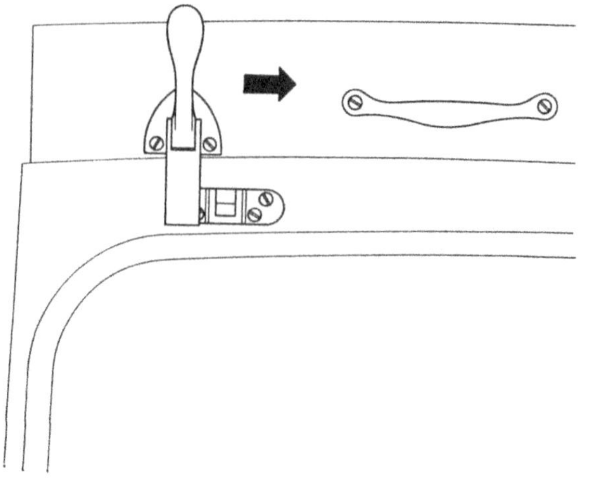

If the guide tongues slide to the side of the retaining lugs, leave the top resting on the windshield frame and pull the frame in the required direction until the normal position is reached and the guide tongues will enter properly.

Leakage at the Rear Window

If the rear window leaks, water stains will be visible in the headlining below the rear window. This leakage is attributable to the top cover having shrunk lengthwide due to improper treatment of the top material.

As the top is already tensioned to a certain degree when being attached to the window frame, any excessive shrinkage will cause the top material to pull the window frame apart.

To eliminate the leaks proceed as follows:

1 - Remove rear window.

2 - Check the internal size of the rear window frame.

Up to Chassis No. 1 600 439
a = 192 ± 2 mm (7.5″ ± 0.08″)

From Chassis No. 1 600 440
a = 225 ± 2 mm (10″ ± 0.08″)

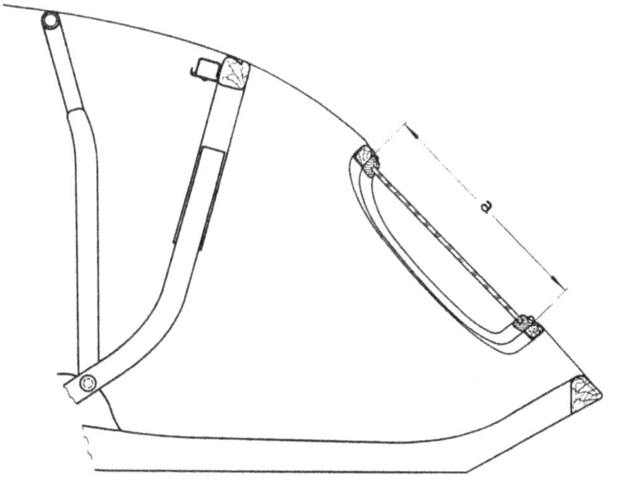

3 - If these dimensions are exceeded, detach top cover from lower edge of rear window frame, insert a cardboard strip of appropriate thickness and tack the top coverinto position again.

4 - Before installing the rear window assembly again, place the window in the frame without sealing compound and check fit.

5 - Apply sealing compound lightly to the inside of the frame (along the tacks) and install the rear window. To avoid soiling the top, do not apply sealing compound to the weatherstrip.

6 - Carry out a water test by spraying the rear part of the top with an even jet of water and check the window for leakage.

7 - Water stains on the headlining can be removed with a solution of one part ammonia and three parts water.

If the headlining is dirty the places which have been treated will naturally be lighter in colour. In such cases it is advisable, therefore, to wipe the entire headlining with a well soaked cloth after the water stains have been removed.

Leakage at Top Cover Seams

If the top cover leaks, water stains will be visible at the following places:

1 - at the front near the top locks

2 - at both sides of the interior lamp

3 - at both sides of the rear window

4 - at the lowest point of the rear body bow

5 - rust formation on the cloth covered nails in the headlining

To locate the leak exactly it is necessary to carry out a water test. To do this, loosen the headlining as far as possible at the rear and spray the whole top with an even jet of water until the ingress of water is observed.

The top can only be effectively sealed if the thread itself is in good condition. If the thread is damaged or rotted to such an extent that it is no longer tight the only remedy is to replace the cover. Re-sewing is not recommended as experience has proved that the stitches cannot be spaced to conform to the original holes when sewing with a machine. Apart from this the seam holes would be enlarged too much by the tension of the new thread.

The leaks can be eliminated by the use of the seam protecting solution "Happich 7303" which is obtainable from Gebr. Happich GmbH., Wuppertal-Elberfeld, Neuenteich 72.

The solution is applied underneath all the top cover seams including those at the side beading and rubbed in vigorously until small drops can be seen in the stitch holes from outside. The solution must not come into contact with the outer surface of the top cover as it cannot be removed.

The following operations are necessary to enable the solution to be rubbed into the front seams

1 - Lift the trim moulding at both ends of the header for about a third of its length.

2 - Remove the screws securing the top cover to the roof frame at the sides.

3 - Loosen the top cover at both corners of the header until access is gained to the seams.

4 - Open the top until it is free of tension.

5 - Rub the solution into all the seams which can be reached.

6 - Secure the top cover to the header and the roof frame again. More detailed information about work on the top can be found under the section "Convertible-Top".

The following operations are necessary when sealing the rear seams

1 - Remove rear body bow trim moulding completely.

2 - Remove the screws securing the top cover to the main bow at the sides.

3 - Remove drop window weatherstrip on the main bow and detach cover from the main bow.

4 - Detach top cover from the rear body bow to approximately 10 cm (4") behind the seams.

5 - Open the top half way and rub seam protective solution into the seams.

6 - Attach the top cover to the bows again. More detailed information about work on the top can be found under the section "Convertible-Top".

The water stains in the headlining caused by the entry of water can be removed with a solution of one part ammonia and three parts water. The solution is applied to the dry headlining with a well soaked cloth and rubbed evenly into the places concerned.

If the headlining is dirty, the places which have been treated will naturally be lighter in colour. In such cases it is advisable, therefore, to wipe the entire headlining with a well soaked cloth.

Padded Sun Visors

From Chassis No. 2 252 685 the Convertible was fitted with padded sun visors instead of the dark-coloured plastic type.

The new visors can be installed in place of the previous type without any further alteration and are obtainable separately under the Part No. 151 857 551 A.

The new rear view mirror with spindles (less visors) is supplied under Part No. 151 857 523 A.

When service installing the new sun visors on the previous rear view mirror, observe the following:

1 - Remove mirror with sun visors.

2 - Tap the previous sun visors off the spindles carefully with a piece of wood to avoid damage.

3 - Shorten the left hand spindle by 30 mm (1.2") and the right hand by 80 mm (3.15").

4 - Push the padded sun visors on to the spindles as far as possible and tighten clamp screws.

Eliminating Fluttering Noises in the VW Convertible Top

From 29th March 1960 and Chassis No. 2 967 166 a wire cable was inserted into the hollow part of both top cover side seams along the roof frame as far as the main bow to eliminate fluttering noises.

The wire cables, with tension springs, are attached to hooks welded to the roof frame behind the header and with one round-head 5×25 screw on each side at the rear of the main bow.

The tension springs are located inside the roof frame at the front behind the header.

The part number of the top cover with the cables inserted is 151 871 035 C and the tension springs are under Part No. 151 871 953.

Further instructions on the subsequent installation in older vehicles are contained in Technical Bulletin A-20.

Header

Note:

From April 1964, the headers — Part Nos. 151 871 189 A/B/C — were discontinued and replaced by a new header (151 871 189 D) without holes for the securing screws. This header can be used for all Convertibles from Chassis No. 1 600 440.

The complete top frame with cover — Part No. 151 871 021/021 A/025/025 C — will remain available.

General

This section describes only the replacement of body parts where the operations differ from those on the Sedan.

The description serves as a guide for the sequence of operations. A reasonable knowledge of body repair is essential and the work should, wherever possible, be carried out by a skilled man well versed in this type of work.

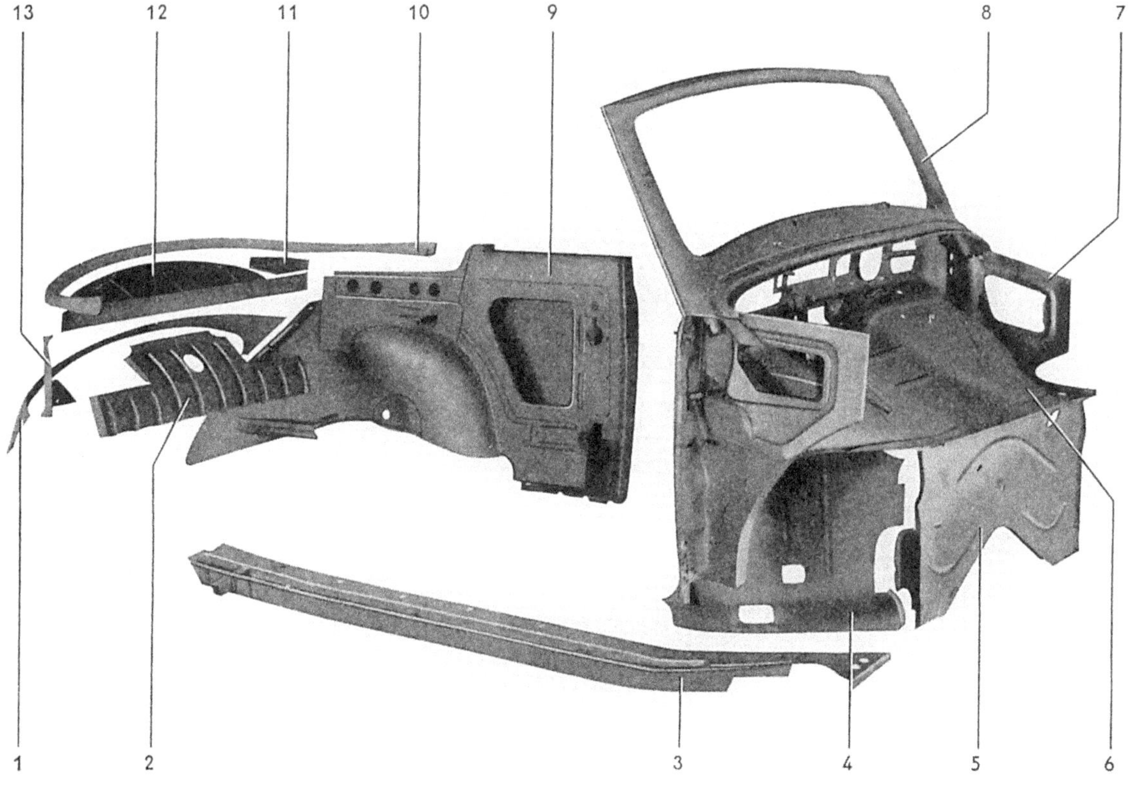

1 - Rear panel
2 - Reinforcement plate for rear body bow
3 - Side member reinforcement
4 - Lower cowl reinforcement panel
5 - Front partition
6 - Luggage compartment panel
7 - Upper cowl reinforcement panel
8 - Windshield frame
9 - Quarter panel
10 - Rear body bow
11 - Rear corner plate
12 - Hood hinge carrier plate
13 - Upper quarter panel reinforcement

Wherever gas welding is not expressly stipulated it is mentioned in the individual sections in addition to the more generally used spot welding.

It must be emphasised however, that spot welding with spot welding tongs or a push type welding gun is preferable to gas welding.

Gas welding calls for great care during repair operations as lack of skill and material knowledge can cause structural alterations and stresses in the thinly walled panels. This in turn leads to cracks in the metal and stress noises in the body.

For this reason, an endeavour should be made to use spot welding wherever possible.

When the body has been removed for repair it should be placed on the frame specially designed for this purpose. The frame is described in detail, along with other gauges and jigs, under the heading "Workshop Equipment" in "Body General" section.

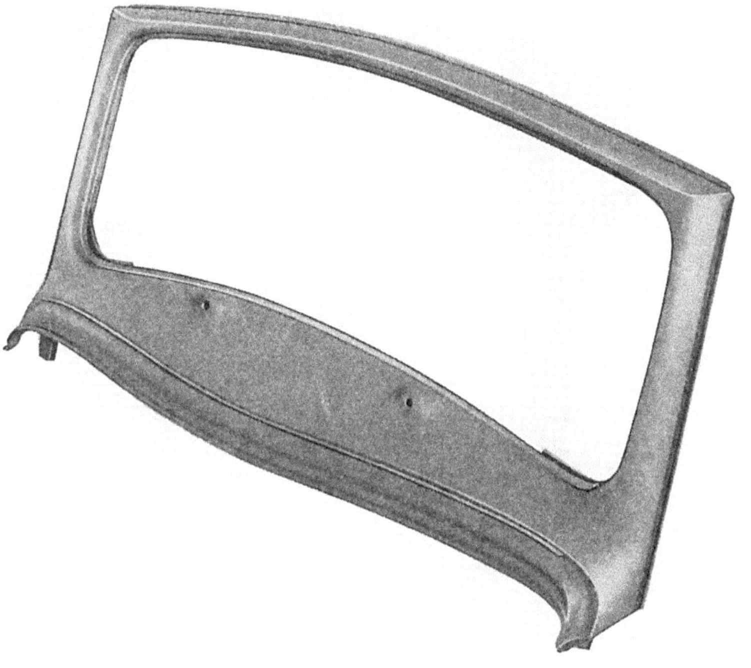

General

The spare windshield frames are supplied as a complete section consisting of the following parts:

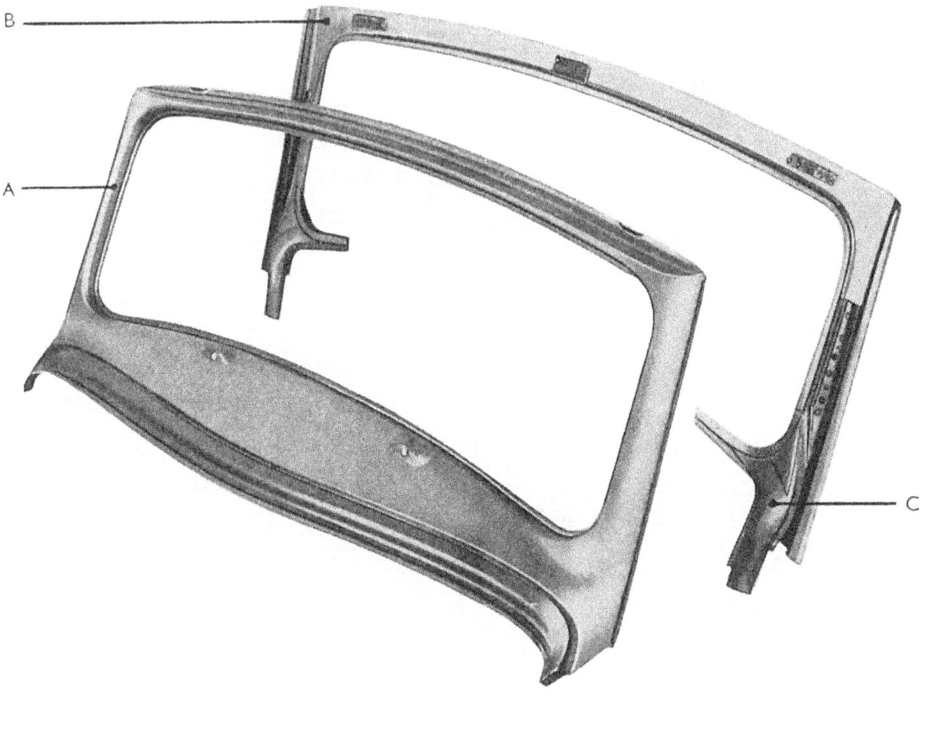

A - Outer panel

B - Inner panel

C - Warm air jets

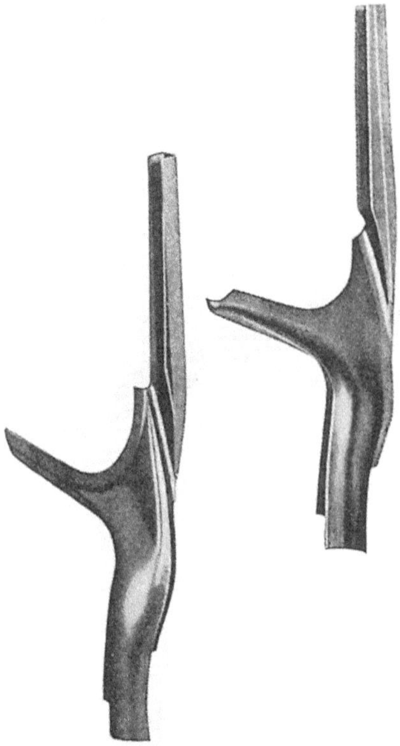

The warm air jets are made of two parts joined together and extended upwards by channel sections which also serve as stiffeners for the windshield frame.

All the parts are welded together and the sides of the windshield frame are folded into a flanged edge.

When necessary, the complete windshield frame must be replaced. The cutting of the side pillars and welding on of a suitably shaped upper frame is not recommended, despite the use of the special jig, as the warm air jet stiffeners cannot be welded into the pillars.

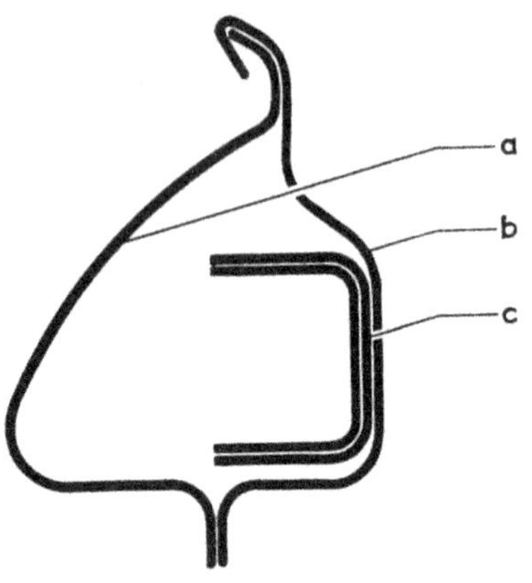

a - outer panel
b - inner panel
c - warm air jet extension

Windshield Frame Replacement

Preparation

1 - Disconnect battery.

2 - Remove front hood.

3 - Remove cardboard and sound insulation material from front luggage compartment.

4 - Remove all instrument panel parts

5 - Remove windshield wiper mechanism.

6 - Disconnect main cable harness as far as necessary and cover up.

7 - Remove fuel tank.

8 - Take out front seats and mats.

9 - Remove windshield glass.

10 - Remove vent wing weatherstrip from windshield frame.

11 - Remove steering wheel.

12 - Remove both doors.

13 - Remove front side panel trim moulding.

14 - Fold top back and cover up to avoid soiling.

1 - Heat the hinge pillars above the holes for hinge securing screws with the welding torch and scrape the holder off with a suitable spatula.

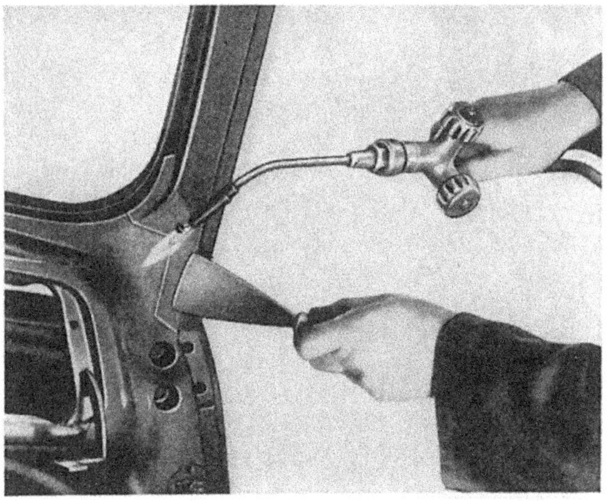

2 - Chisel open the joint between the windshield frame and the instrument panel on the outside and along to the top of the warm air jets in the side pillars.

3 - Chisel the windshield frame on the outside above the cowl/side panel joint and up the pillar to the top of the warm air jets, saw off above the jets.

4 - Take off windshield frame.

5 - Loosen the spot welded cowl reinforcement plates on the side panel and bend away slightly. Do not damage the side panel during this operation.

6 - Chisel the cowl reinforcement plate off the warm air jet without damaging the plate.

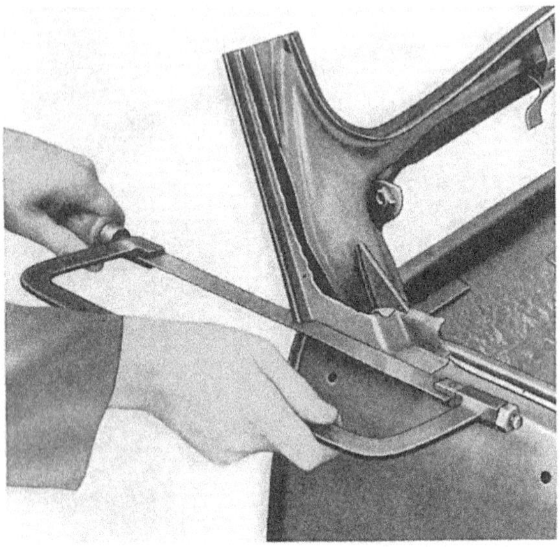

7 - To avoid damage to the side panel, the remaining metal parts of the windshield frame should be sawn off above the cowl/side panel joint.

Part the warm air jets at the top of the heater pipes.

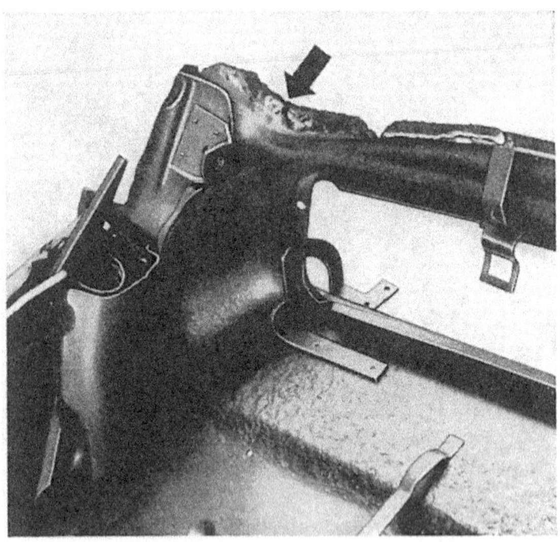

8 - Remove the metal edges on the instrument panel with a pair of pliers.

Note:

Remove these edges by twisting with the pliers and not by pulling as otherwise the spot welds will be pulled out and leave holes.

9 - Grind or file all spot welds clean. Straighten all cut edges and hammer out dents.

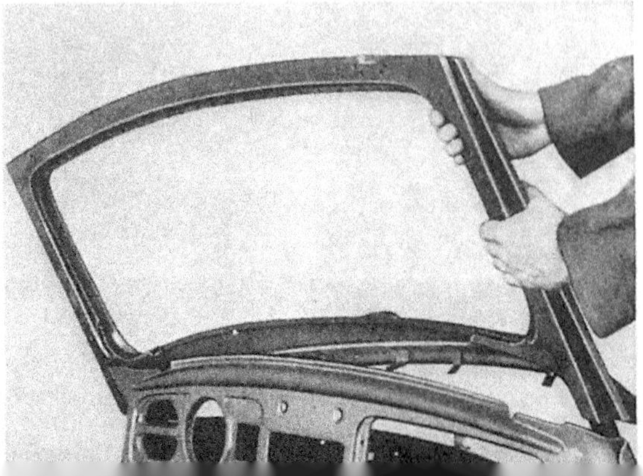

10 - Grind all spot welds on the new windshield frame clean.

11 - Place new frame in position.

12 - Fit body clamps in windshield glass opening on the front side panel.

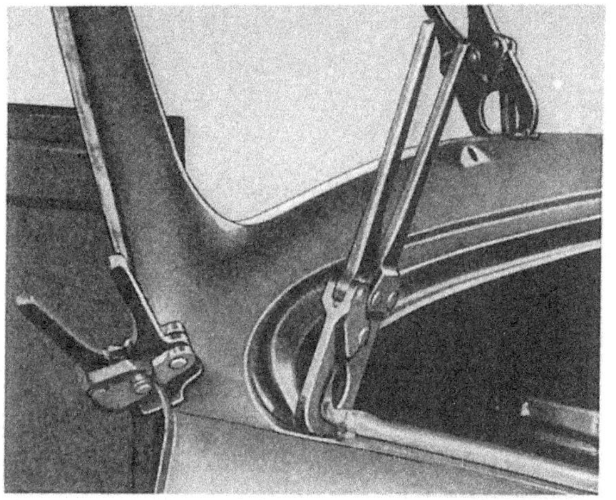

13 - Install both doors to locate the windshield frame

Note:

If repairs are being carried out frequently on the Convertible it is advisable to make the gauge VW 707 (Local manufacture). With the help of this gauge the exact position of the windshield frame can be checked by two methods.

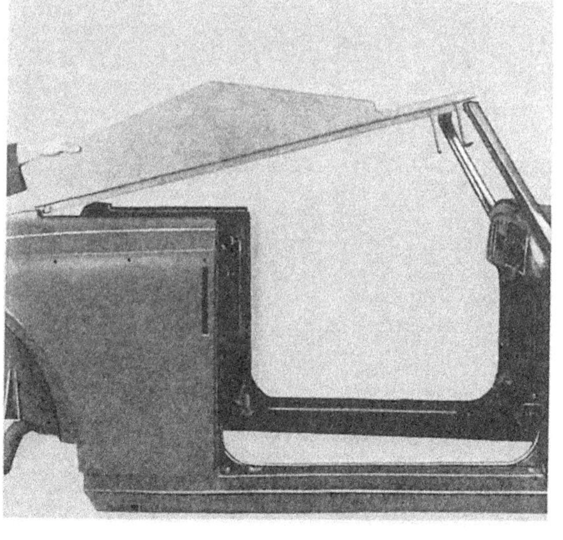

14 - Gas weld the windshield frame to the side panel at front and rear.

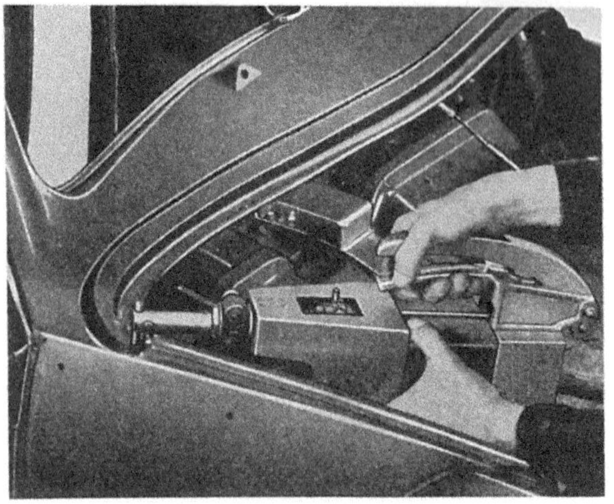

15 - Spot weld the windshield frame to the front side panel.

16 - Finally check the position of the windshield frame again.

The clearance between the windshield frame and the vent wing frame must be uniform over the full length. The doors must also fit properly in the door opening.

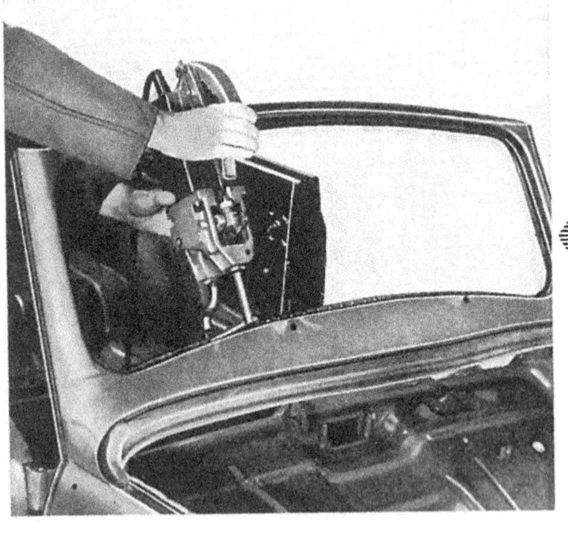

17 - Weld the instrument panel and windshield frame in the frame opening.

a - **electric**

Spot weld the parts at intervals of 20 mm.

b - **gas**

Weld with 15 mm runs at intervals of 20 mm

18 - *Gas weld the joint between* the windshield frame and the instrument panel on the inside. At the same time weld the warm air jets to the instrument panel.

19 - Gas weld the windshield frame and side panel on the inside. Bend the reinforcement plate away slightly to facilitate this operation. Tack weld the warm air pipes to the jets with the torch.

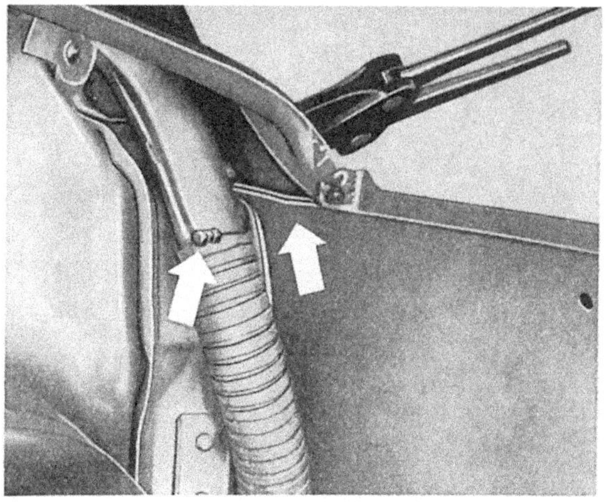

20 - Gas weld the cowl reinforcement plates to the warm air jets and spot weld them to the side panel.

21 - Remove the doors again.

22 - Braze the joint between the windshield frame and the side panel.

23 - Fill the joint between the windshield frame and the instrument panel with solder until a smooth contour is obtained.

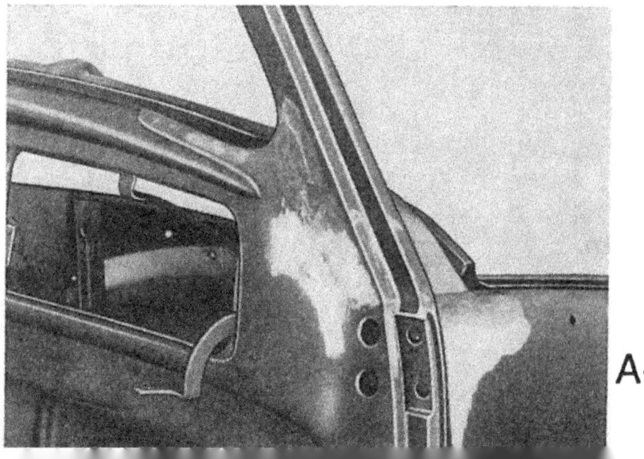

24 - Grind all reworked and welded places smooth and prepare the vehicle for painting.

THIS PAGE INTENTIONALLY
LEFT BLANK

Scuttle with Instrument Panel

General

Where accident damage to the front part of the body extends to the instrument panel, it is advisable to install a complete new scuttle with instrument panel in addition to the replacement of other parts.

The complete scuttle consists of the following parts which are also available separately:

Blue	= windshield frame	Yellow	= front partition
Black	= instrument panel	Green	= upper cowl reinforcement plate
Red	= front luggage compartment		

The following parts, which are obtainable as spares, are welded to the scuttle:

Instrument panel

a - Hinge pillar reinforcements and movable tapped plates for door hinges held by retaining bands.

b - Hinges with rivet studs for glove compartment lid.

c - Bearings for the front hood hinges which are spot welded to the top of the hinge pillars.

Luggage Compartment

d - Securing brackets for the cover plate in front of the instrument panel.

On account of the reduced stability when the top is open the scuttle and instrument panel is reinforced by the following parts:

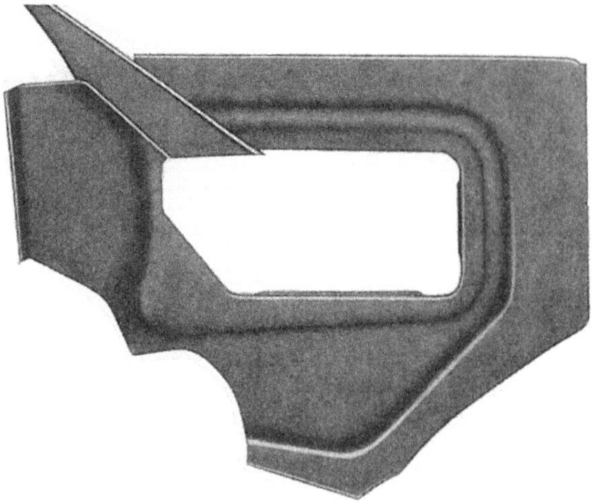

1 - Upper cowl reinforcement plates at both sides.

The upper cowl reinforcement plates are gas welded to the warm air jets, the hinge pillar reinforcement plates and the luggage compartment.

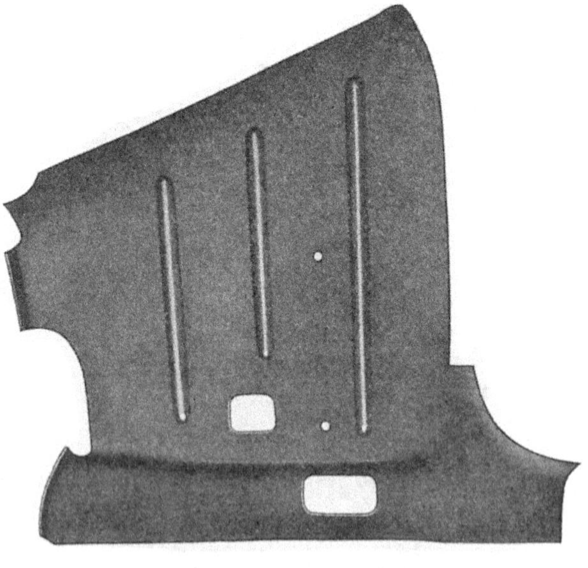

2 - Lower cowl reinforcement plates at both sides.

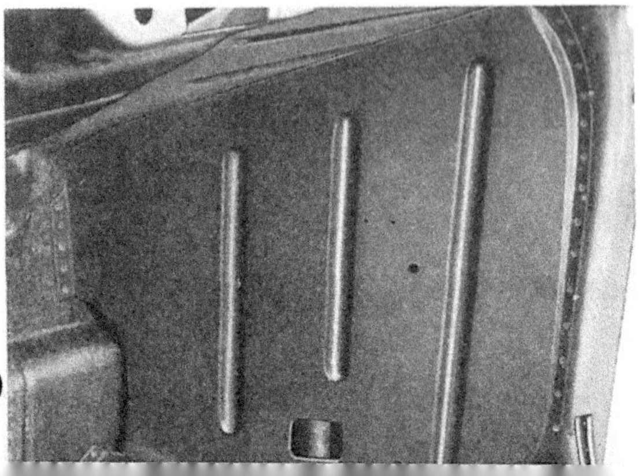

The lower cowl reinforcement plates are spot welded to the side members, the front partition, the luggage compartment and the hinge pillars in the lower compartment.

Small corner plates, also obtainable separately are welded to the lower rear corners of the lower cowl reinforment plate.

These corner plates are gas welded to the hinge pillars and the frame side members.

Note:

Similar plates are welded to the lock pillars and the frame side members for reinforcement purposes.

Replacement of Scuttle with Instrument Panel

Preparation

1 - Take body off the chassis and mount it on the repair frame or a frame suitable for use as a jig.

Note:

The following operations were carried out on the repair frame specially designed for this purpose. This repair frame is described in detail under the heading "Workshop Equipment" in the "Body-General" section.

2 - Remove both doors, the front hood and the front fenders.

3 - Remove all parts fitted to the instrument panel, including windshield glass, front lower compartment lining and main cable harness with lamp wiring.

4 - Remove sealing compound from all weld seams in the front lower compartment.

Body Repair Work

The following repair operation was carried out on a Convertible which had turned over and received serious damage to the windshield frame, instrument panel and luggage compartment. The side panel and other parts of the front showed no signs of damage worth mentioning so it was only necessary to replace the scuttle with instrument panel.

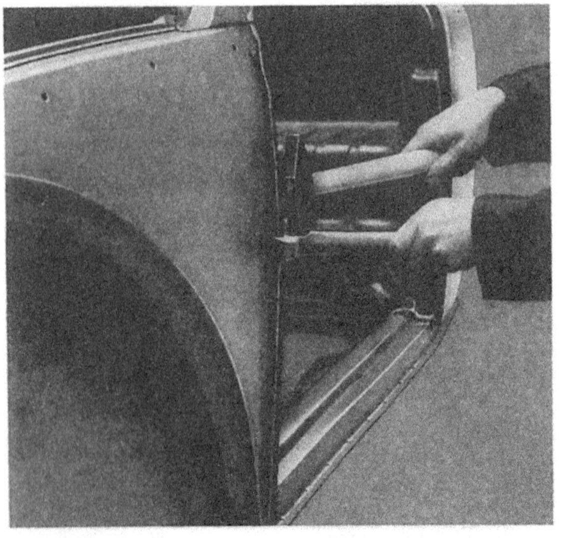

1 - Fold back the edges of the side panel and instrument panel joint with the rain channel parting tool (Local manufacture VW 734).

2 - Cut the folded joint of the instrument panel from the frame side member to the windshield frame.

Take care not to damage the side panel or the warm air pipes.

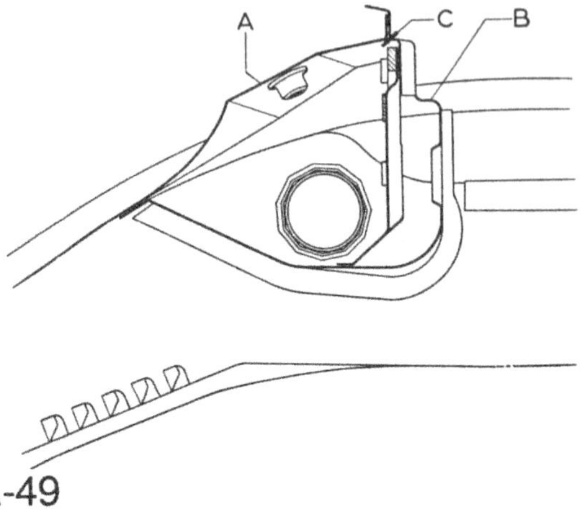

A - Front side panel

B - Instrument panel

C - Cutting line

3 - Cut out the lower cowl reinforcement plate on both sides. Take care not to damage the side panel, the side member or the main cable harness tube (arrow) on the right hand side.

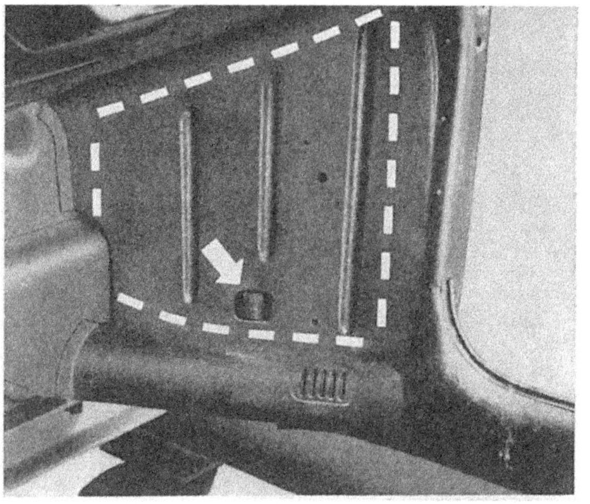

4 - Remove the remaining metal edges of the reinforcement plate by twisting with a pair of pliers and not by pulling, as otherwise the spot welds will be pulled out and leave holes in the parts which are to be used again.

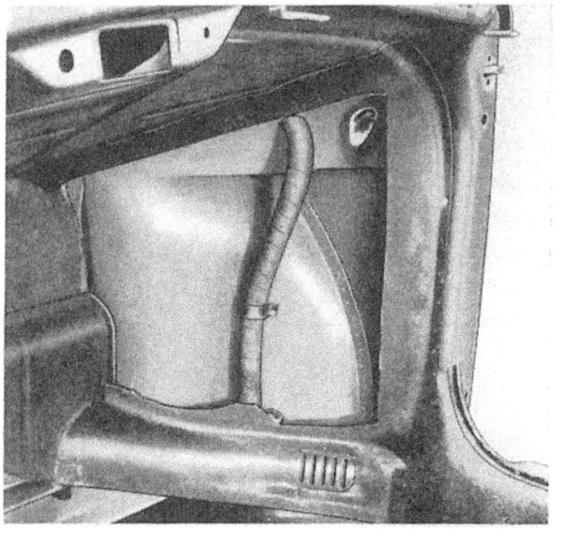

5 - Part the scuttle from the side panel and side member in the front lower compartment.

When cutting the hinge pillars take care not to damage the warm air pipes.

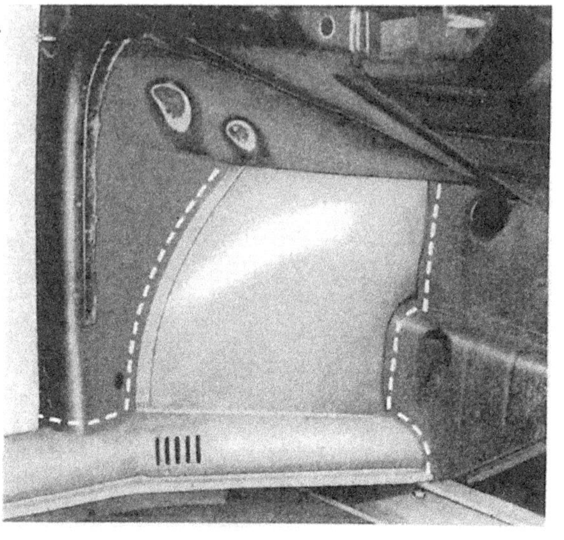

6 - Part the upper cowl reinforcement plate above the cowl/side panel joint.

7 - Part the scuttle from the side panel and the side member in the luggage compartment.

Cut the guide tube for the hood cable out of the luggage compartment.

Cut the warm air jets off at the top of the warm air pipes.

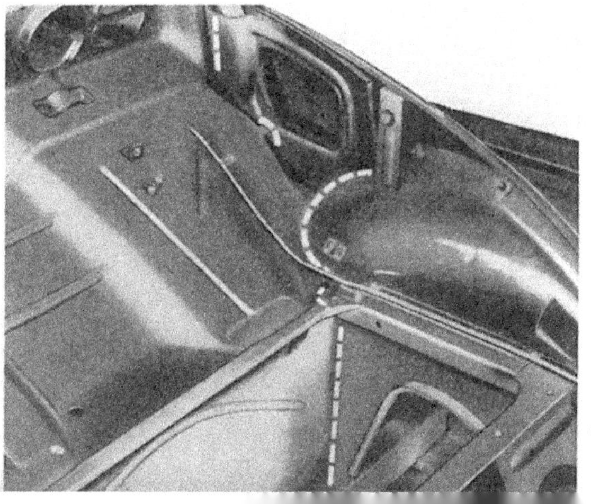

Loosen the upper reinforcement plate at the side panels and bend back.

Do not damage the side panels during this operation.

8 - Take out the scuttle.

9 - Remove remaining metal edges with a pair of pliers.

Note:

Remove the metal edges by twisting with the pliers and not by pulling as otherwise the spot welds will be pulled out and leave holes.

Attention

Protective gloves should be worn when removing and installing the scuttle.

10 - Grind or file all spot welds clean. Straighten jointing flanges as required and hammer out dents.

a = 945 mm (37.2")

11 - Grind all spot welds on the new scuttle and on the body clean.

12 - Insert new scuttle with instrument panel in the vehicle and locate correctly.

13 - Place body clamp on the side panels.

14 - Check the measurement of the door opening on the side member.

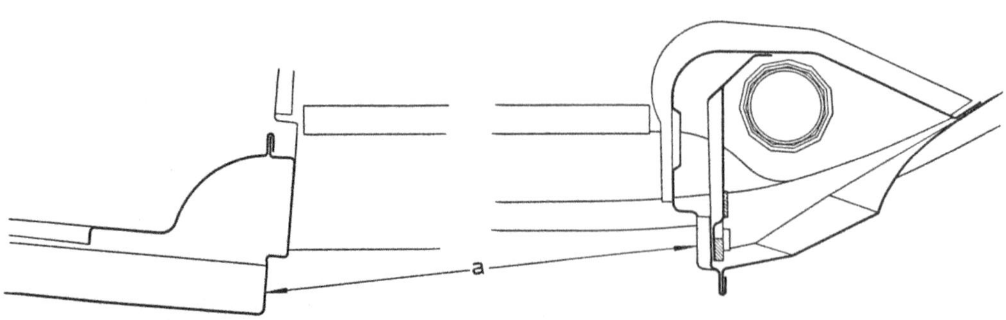

15 - Bolt the luggage compartment to the fuel tank support brackets.

16 - Gas weld the hinge pillars and the front partition to the side members.

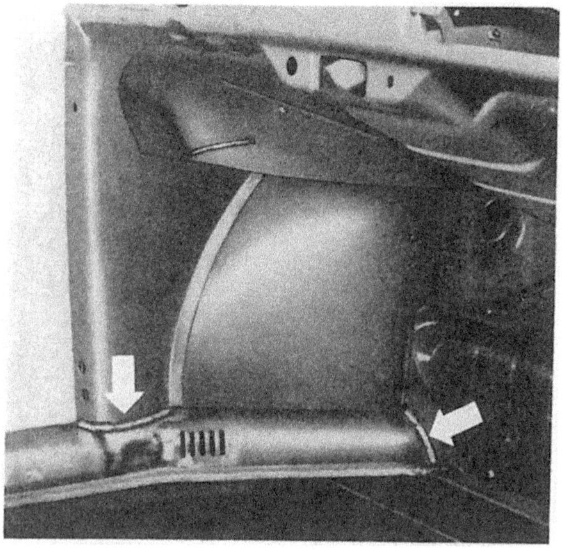

17 - Spot weld the hinge pillars and the front partition to the side panels or gas weld with runs of 15 mm length at intervals of 20 mm.

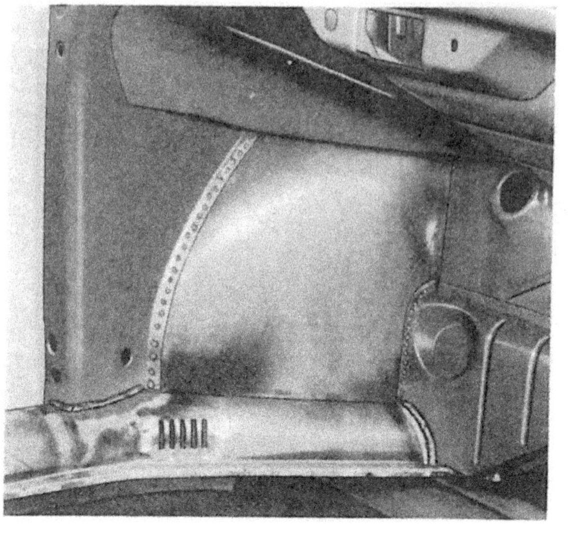

18 - Weld the luggage compartment and the front partition to the side panels.

Weld warm air jet and pipes with a small run of gas weld.

Gas weld the guide tube for the front hood cable to the luggage compartment.

Lift the body and gas weld the front partition to the side members (arrow).

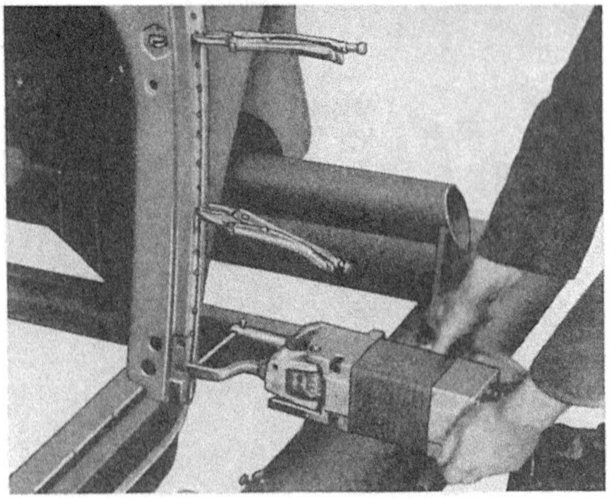

19 - Spot weld the hinge pillars and the side panels at the flanged edges.

20 - Grind spot welds smooth and fold the flanged edge over with a hammer and dolly.

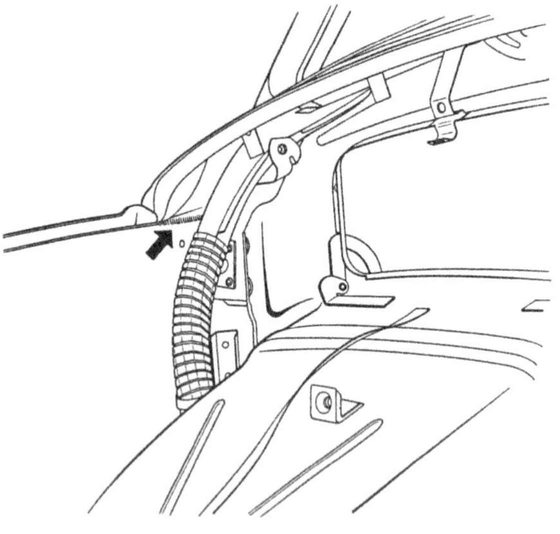

21 - Gas weld the windshield frame cowl panel and the side panels on the inside.

22 - Braze the joint between the windshield frame cowl panel and the side panels on the outside.

23 - Cut the hole for the cable harness tube in the right hand side of the luggage compartment panel and insert the tube.

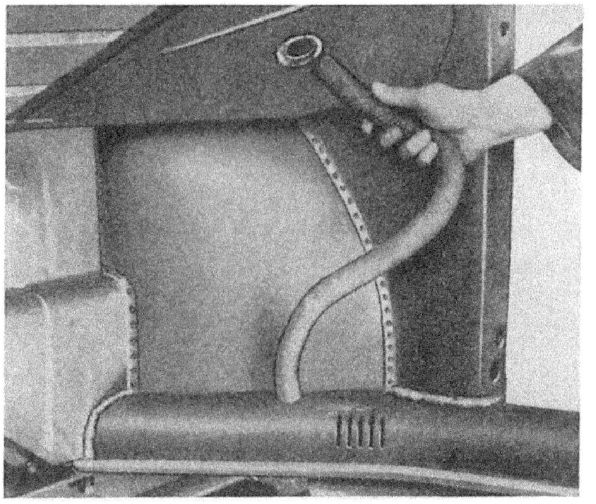

24 - Spot weld the upper cowl reinforcement plates to the side panels.

25 - Gas weld the corner plates to the hinge pillars and side members.

This operation is only necessary when the corner plates are not welded to the lower cowl reinforcement plate.

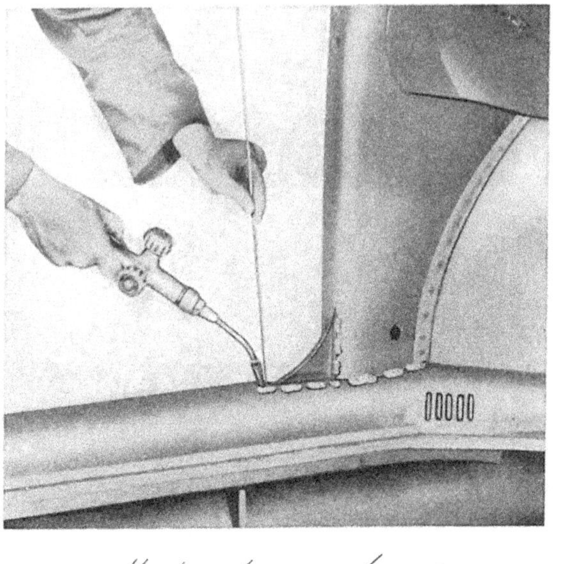

26 - Coat the seams on the wheel housing, the luggage compartment and the side members adequately with Sealing Compound D 9.

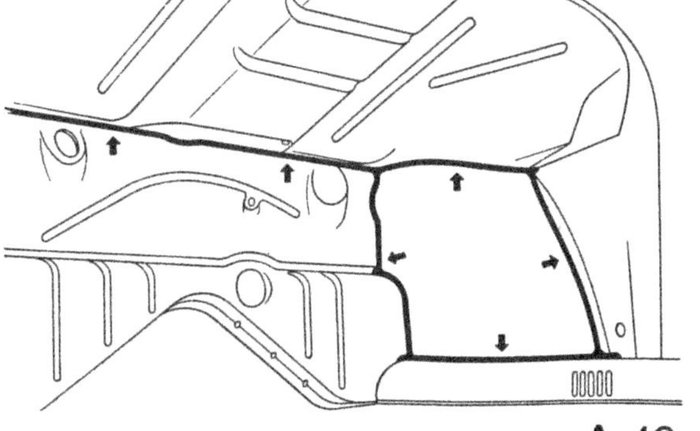

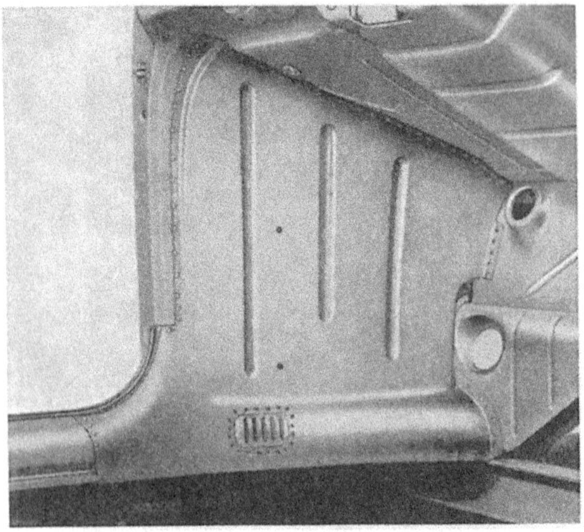

27 - Insert the lower cowl reinforcement plate, position correctly and spot weld.

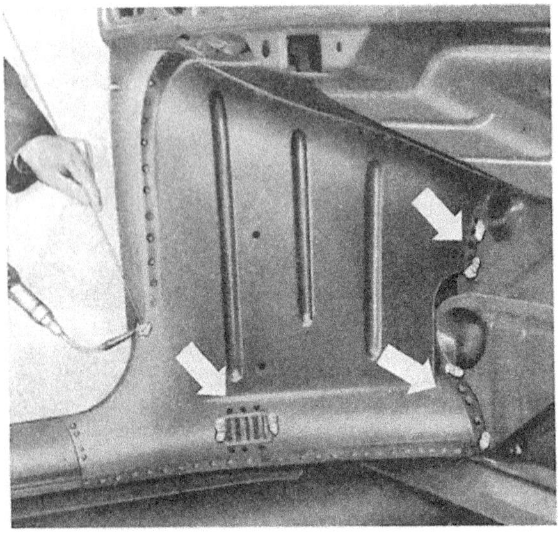

28 - Gas weld the lower reinforcement plate at the places indicated.

29 - Spot weld the carpet channel to the corner plates and the side member.

30 - Grind all reworked and welded places smooth.

31 - Prepare vehicle for painting.

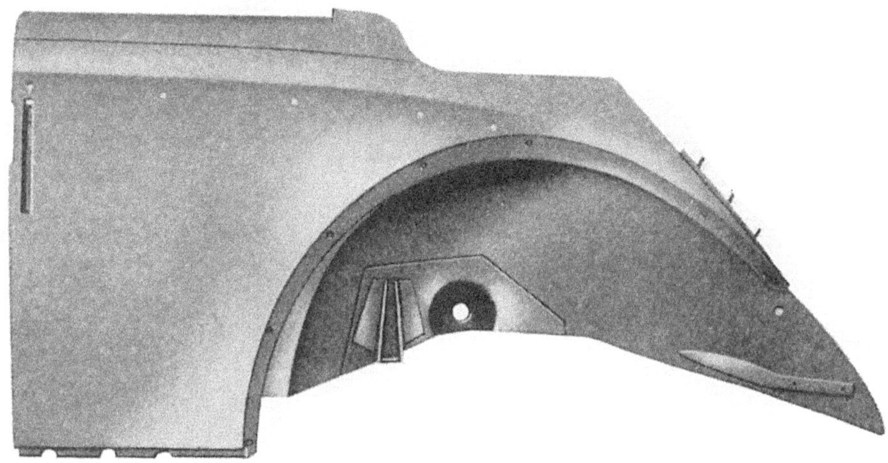

General

The spare quarter panels are supplied complete — outer and inner panels welded together — or the outer panel only.

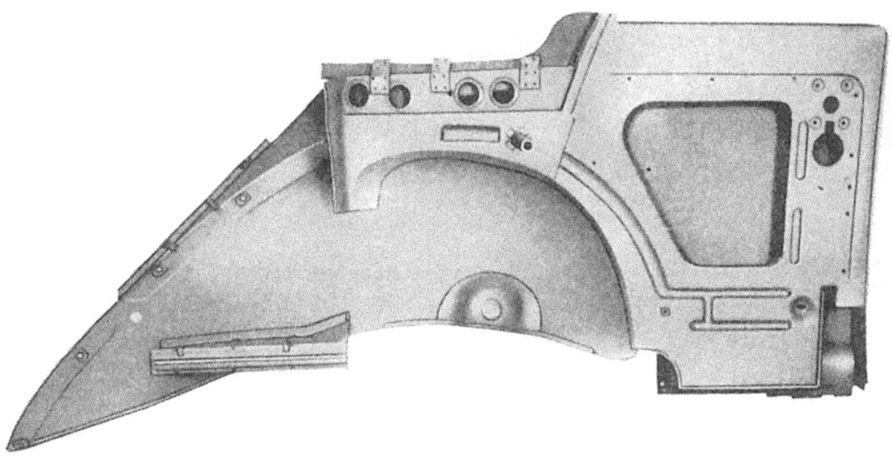

Outer and inner panels

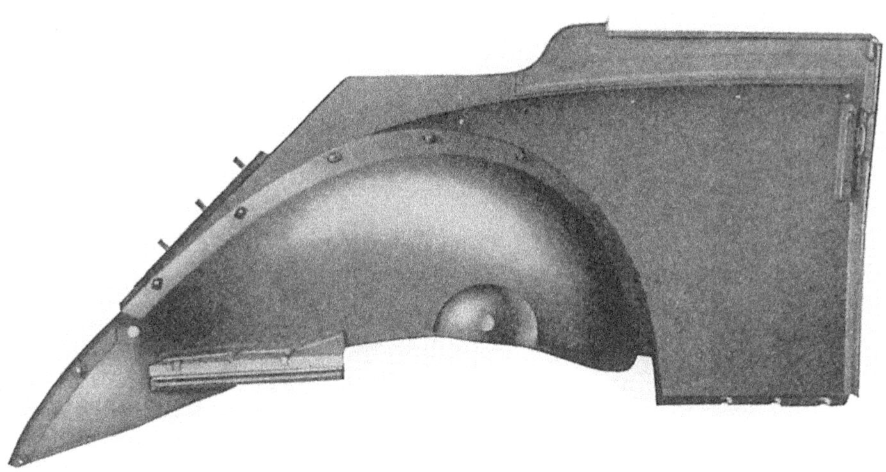

Outer panel

The following parts, which are obtainable separately, are welded to the quarter panel:

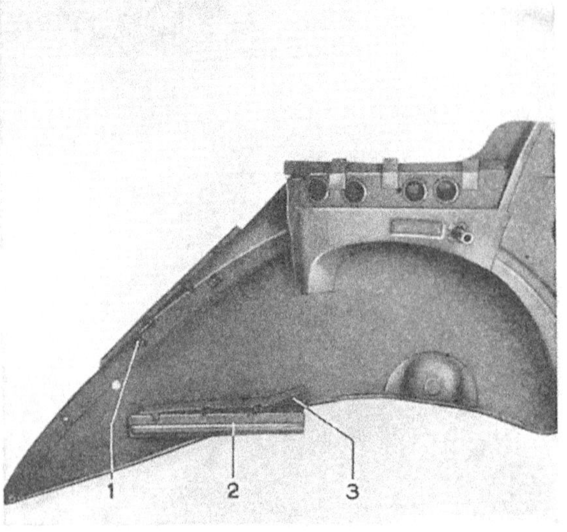

Inside

1 - Captive nuts for fender securing screws.

2 - One channel for the engine cover plate weatherstrip.

3 - One engine side cover plate.

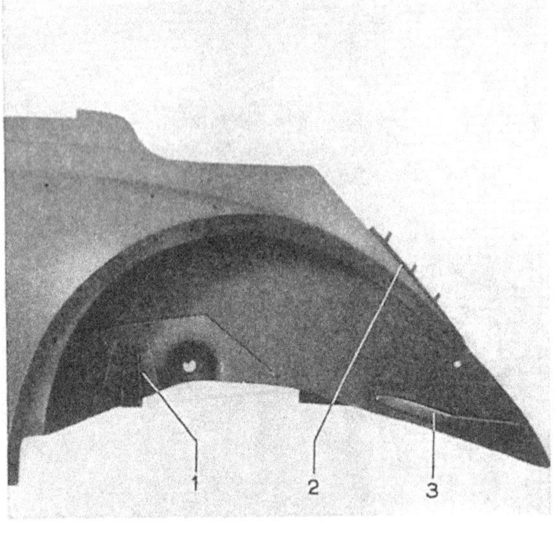

Outside

1 - One quarter panel bracket.

2 - Channel for weatherstrip.

3 - One retainer for the bumper bracket.

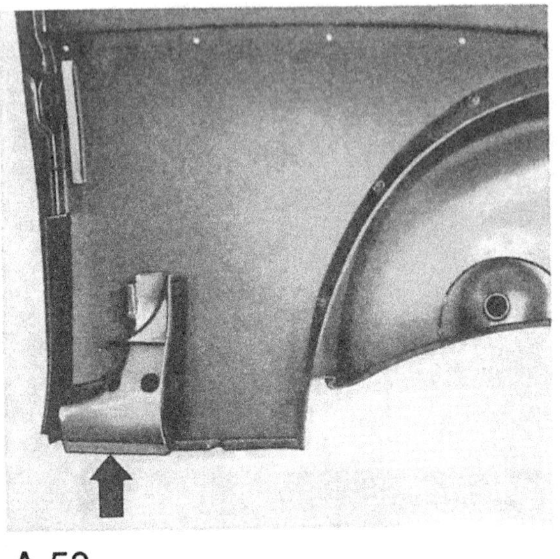

The bumper bracket retainer and quarter panel support can be welded to the quarter panel with the local manufacture gauges, VW 705 "Gauge for rear bumper bracket retainer" and VW 706 "Gauge for quarter panel bracket".

A lock pillar reinforcement plate and a corner plate for the side member are attached to the lock pillar on the inside of the quarter panel with one spot weld. These two parts, *the installation of which* is described separately, must be removed before commencing work on the body.

The following operations are described separately:

A - Replacement of a complete quarter panel (inner and outer panels).

B - Replacement of a part of the outer panel.

A further operation which is not dealt with separately, is the replacement of the end section of a quarter panel. The necessary operations can be extracted from part 'B'.

Note:

Spare quarter panel end sections are only available in limited quantities.

A - Replacement of a complete quarter panel

Preparation

1 - Remove body from chassis and place it on the repair frame or a frame suitable for use as a jig.

Note:

The following operations were carried out on the repair frame specially designed for the purpose. This repair frame is described in detail under the heading "Workshop Equipment" in the "Body-General" section.

2 - Detach the Convertible top at the rear body bow and at the main hinges and remove.

3 - Remove both doors and rear fenders.

4 - Remove all interior trim including the drop window and sound insulating material.

5 - Disconnect and remove main cable harness.

6 - Cover the instrument panel and windshield to avoid damage.

7 - Remove rear hood and all weatherstrips in the area of the quarter panel to be removed.

8 - Remove all sealing compound from the weld seams on the appropriate quarter panel.

9 - Take out the sound insulating material behind the reinforcement plates in the engine compartment on the appropriate side.

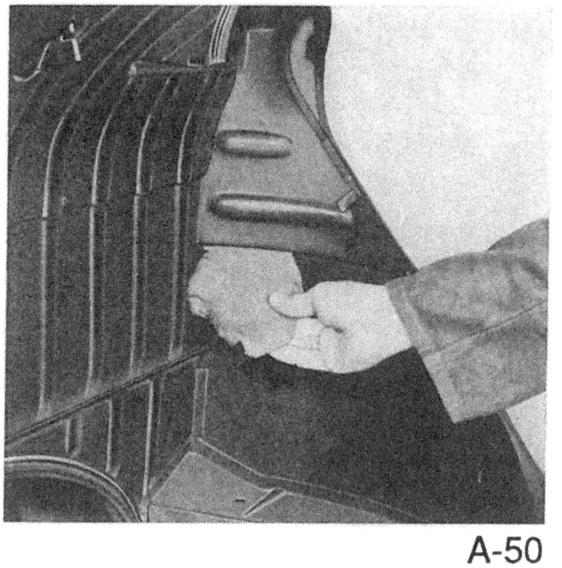

Body Repair Work

1 - Bend back the rear body bow retaining plates after chiseling off the securing screws.

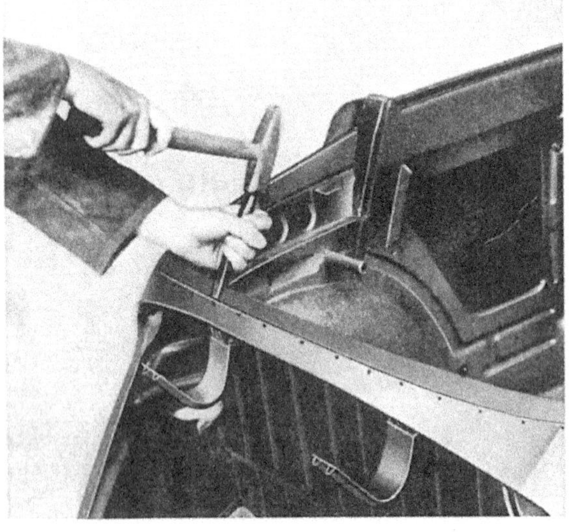

2 - Chisel off the screws securing the rear body bow to the rear panel.

3 - Remove rear body bow.

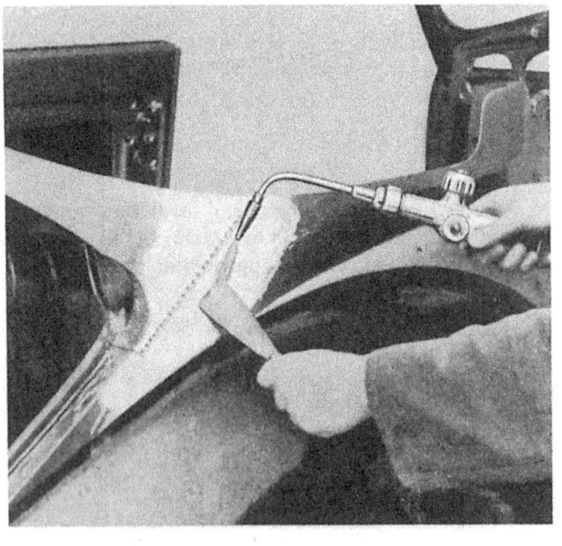

4 - Heat the solder at the joint between the quarter panel and the rear panel with the welding torch and scrape off with a suitable spatula.

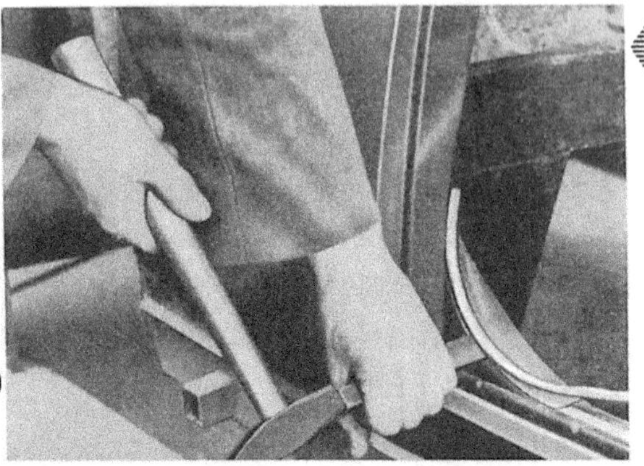

5 - Chisel the carpet channel off at the rear corner plate.

6 - Cut the quarter panel at the following places with a body parting chisel (Local manufacture VW 732) or with a gas cutter:

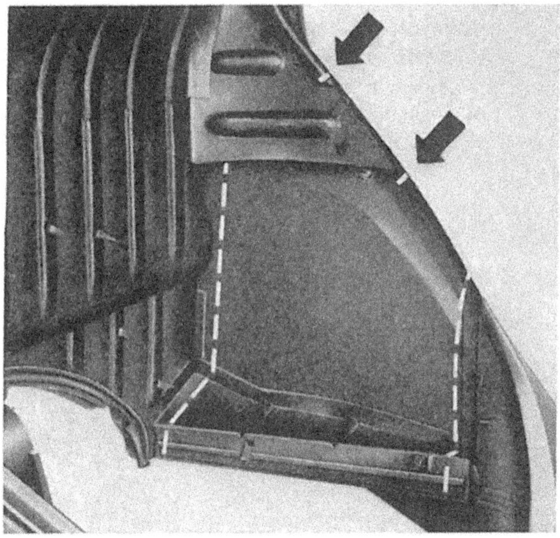

Inside (engine compartment)

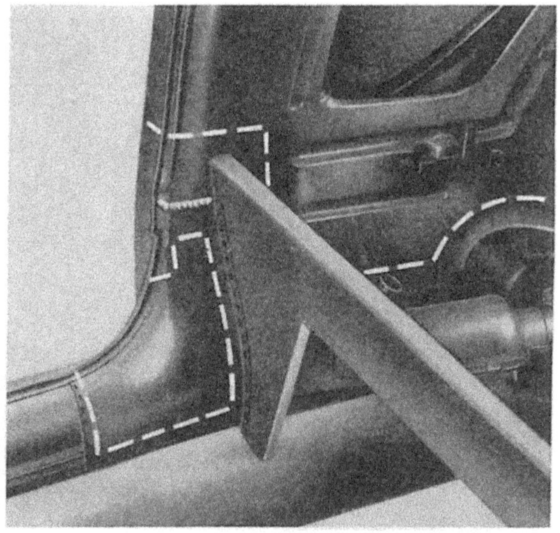

Inside (lock pillar)

Inside (passenger compartment)

Outside (front part)

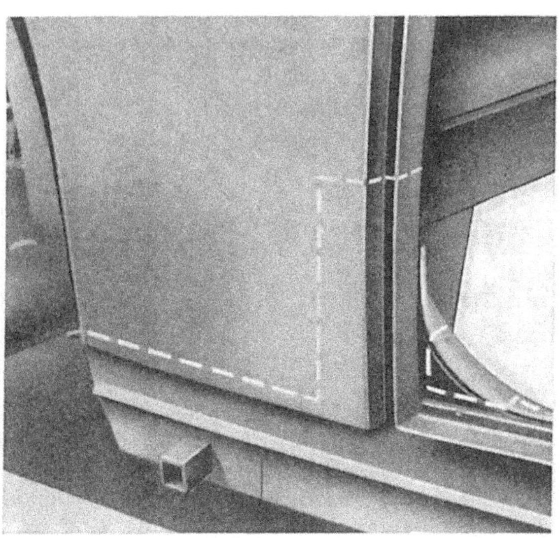

Outside (rear part)

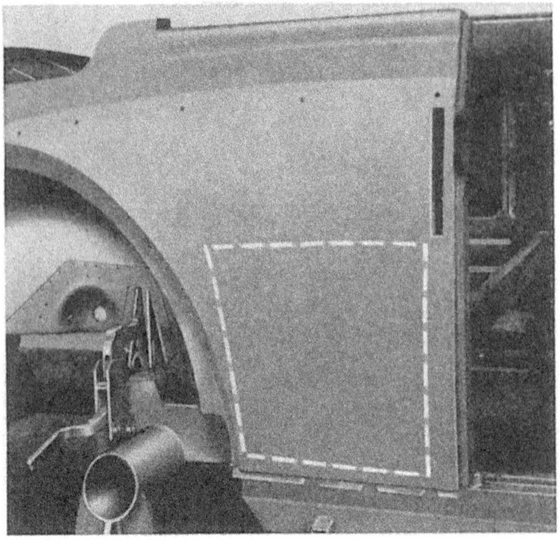

To avoid damaging the cross member (seat support) between the outer and inner panels it is advisable to cut a window in the outer panel first.

When the metal is removed the cross member is visible and it is possible to cut the quarter panel properly.

7 - It is best to commence at the tail plate when cutting out the quarter panel.

The cut must be made 10—15 mm above or below the spot welded seam or joint in each case as otherwise the joining flanges will be damaged.

Important
Protective gloves should be worn when removing and installing the quarter panel.

Note:

To facilitate the operation the tail plate can be separated from the quarter panel by chiseling off thes pot welds. This does away with the task of tearing the metal edges off the joining flanges of the tail plate.

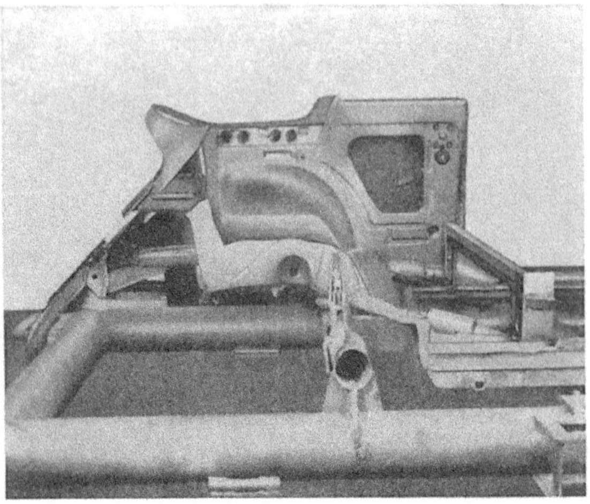

8 - Take out the quarter panel.

9 - Remove the remaining metal edges of the quarter panel with a pair of pliers. Grind all spot welds smooth, *straighten joining flanges* and grind smooth and prepare for installation of the new part.

Note:

Remove the metal edges by twisting the pliers and not by pulling as otherwise the spot welds will be torn out and leave holes.

10 - Before fitting the quarter panel, check that the reinforcement for the cross member (seat support) is not damaged or distorted, and rectify if necessary.

11 - Cement the felt strips to the side extensions of the luggage compartment rear wall.

12 - Grind all spot welds on the new quarter panel clean.

13 - Position new quarter panel and secure to the body with clamps.

Note:

To make the fitting easier the hole in the quarter panel for the stud can be cut open and bent up.

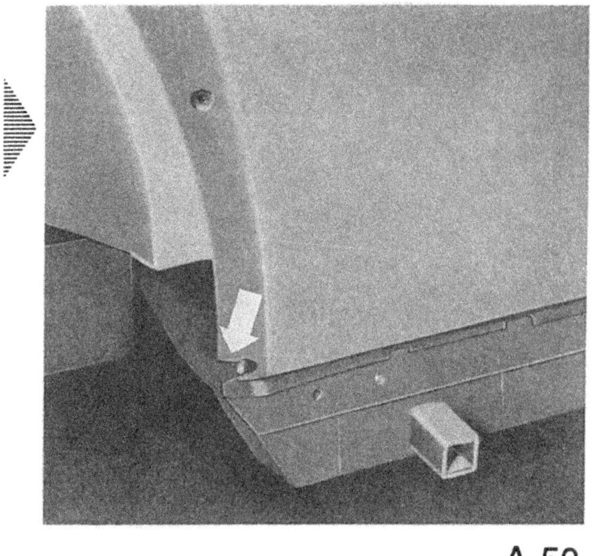

14 - When the quarter panel has been positioned it must be fitted exactly. The "Gauge for Convertible Body" VW 707 (Local manufacture) is suitable for this purpose.

15 - Check the measurement of the door opening on the side member.

a = 945 mm (37.2")

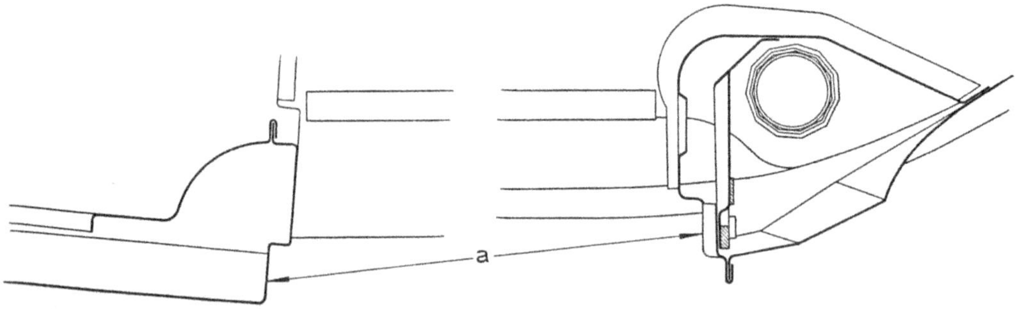

16 - Install the door to assist location of the quarter panel. The door gap must be of uniform width over the full length.

17 - Gas weld the lock pillar (outer panel) to the side member and the reinforcement (A). Gas weld the inner panel to the top of the cross member (seat support) (B) and to the reinforcement (C).

18 - Weld the side panel to the side member.

a - **electric**

Spot weld at 15 mm intervals.

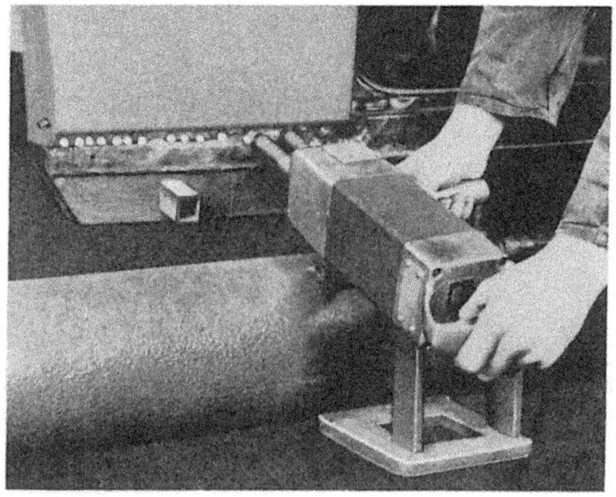

b - **gas**

With runs of 15 mm at 20 mm intervals.

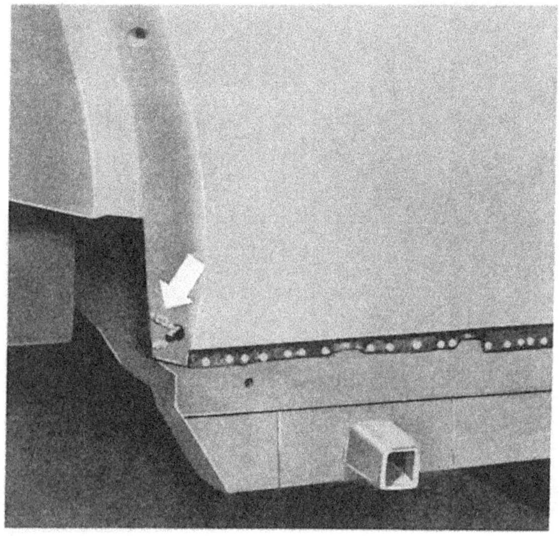

19 - Weld the stud hole which was cut open.

20 - Bolt the quarter panel to the tail plate and gas weld the top and bottom joints.

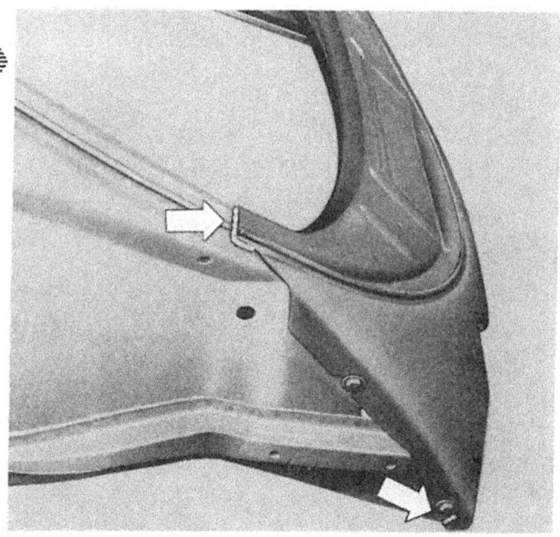

21 - Spot weld the tail plate to the quarter panel outside. If gas welding is used, weld with 15 mm runs at intervals of 20 mm.

22 - Weld the quarter panel in the engine compartment to the tail plate, the luggage compartment and the reinforcement for the quarter panel.

a - **electric**

Spot weld at 15 mm intervals.

b - **gas**

With runs of 15 mm at 20 mm intervals.

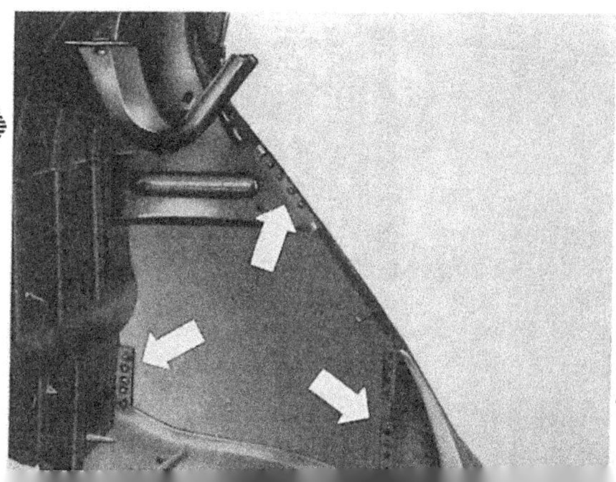

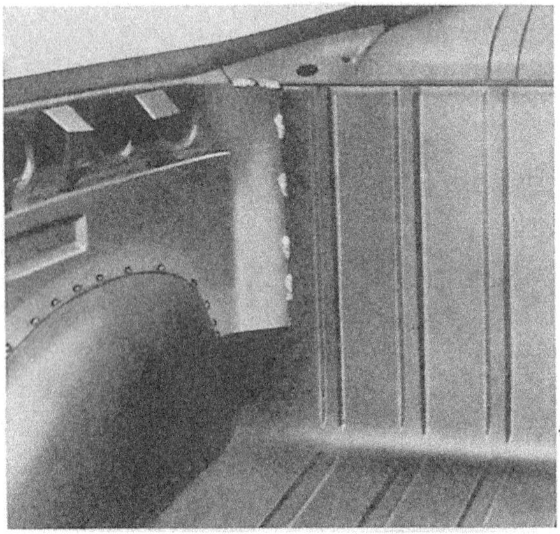

23 - Spot weld the quarter panel inside the body to the side member and the luggage compartment rear walls.

24 - Gas weld the quarter panel to the luggage compartment rear wall and the blanking plates for the air passage with 15 mm runs.

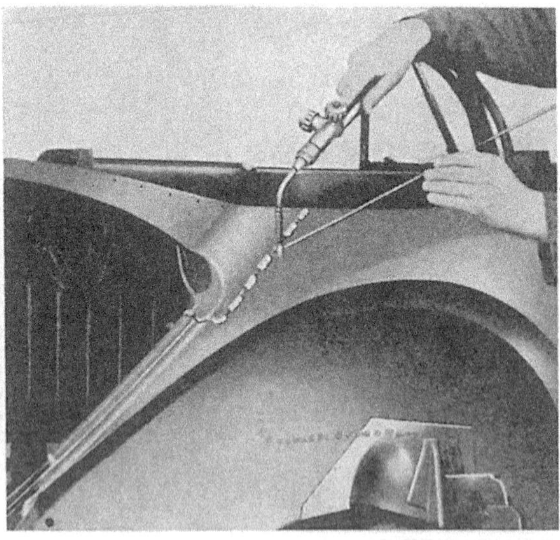

25 - Gas weld the quarter panel to the rear panel with 15 mm runs. Take care that the panels are free of stress.

26 - Before welding the corner plate and reinforcement panel to the lock pillar and the side member, coat the seams in this area adequately with spot welding paste (see: Special Hints, page A - 53/1).

27 - Gas weld the corner plate to the side member and the lock pillar.

28 - The reinforcement panel has two joining flanges and when it has been welded into position the outer flange (arrow) must be folded over.

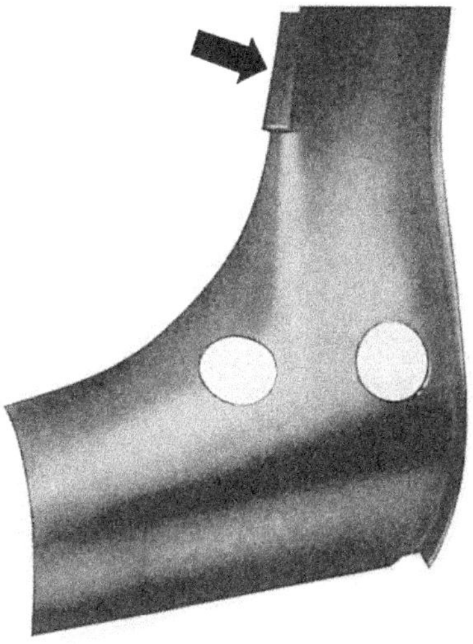

29 - Insert the reinforcement panel and spot weld to the cross member (seat support), the side member and the corner plate. Gas weld it to the side panel and to the side member at the front.

Fold outer flange over.

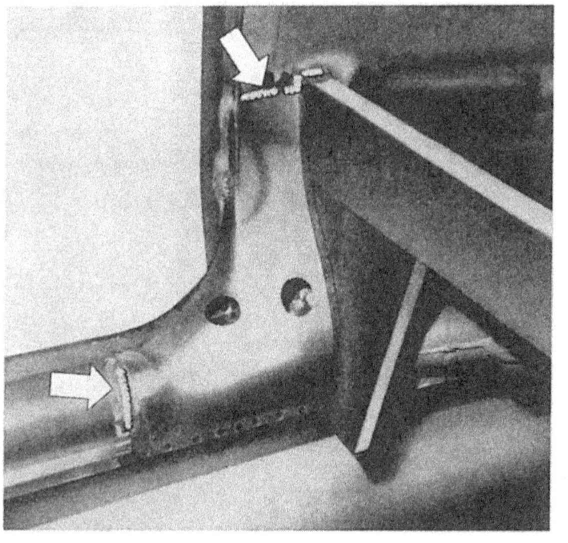

30 - Spot weld the carpet channel to the side member and the reinforcement panel.

31 - Cement a strip of material to the lower half of the rear body bow before installation.

This is essential to prevent the bow from rubbing on the body and causing noises.

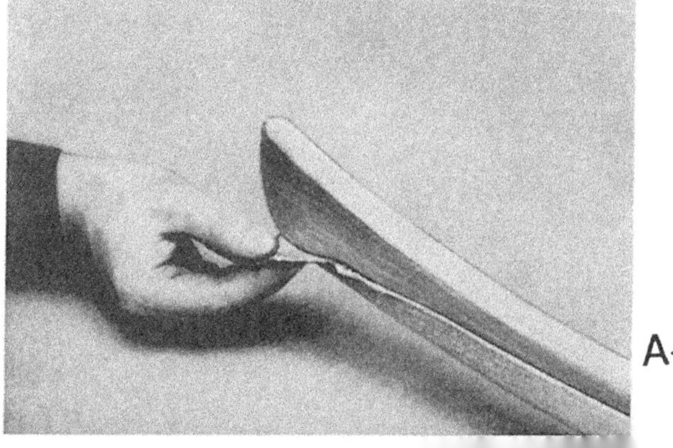

32 - Install the rear body bow, locate properly and secure to the retaining plates inside with wood screws.

The rear body bow may sometimes be slightly too long, if so, shorten by an equal amount at each end.

33 - Secure the rear body bow to the body on the outside with wood screws. When a new quarter panel has been fitted, the holes for the screws must be drilled first.

34 - Coat the joint between the rear panel and quarter panel with solder to ensure a smooth contour.

35 - Coat all seams adequately with Sealing Compound D 9.

36 - Grind all reworked and welded places smooth and prepare the vehicle for painting.

B - Replacement of a part of the Quarter Panel

If the quarter panel lower section has been so badly damaged that it is impossible or uneconomical to straighten it, but replacement of the complete panel does not seem justified, the renewal of a part of the panel is recommended. Proceed as follows:

1 - Before the body repair work commences all the operations given under "Preparation" in part "A" (Replacement of a complete quarter panel) must be carried out.

2 - Cut the outer panel at the places indicated, at an adequate distance from the side member and wheel housing.

During this operation take care that the hinge box pillar (see *marking) is not damaged.*

3 - Cut the outer panel at the hinge pillar.

4 - Remove the cut out portion and the remaining metal edges (marked area).

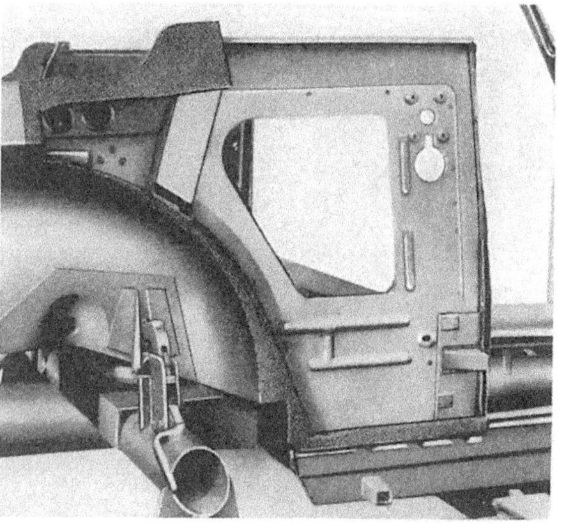

Note:

Remove these metal edges at spot welded places by twisting with a pair of pliers and not by pulling as otherwise the spot welds will be pulled out and leave holes.

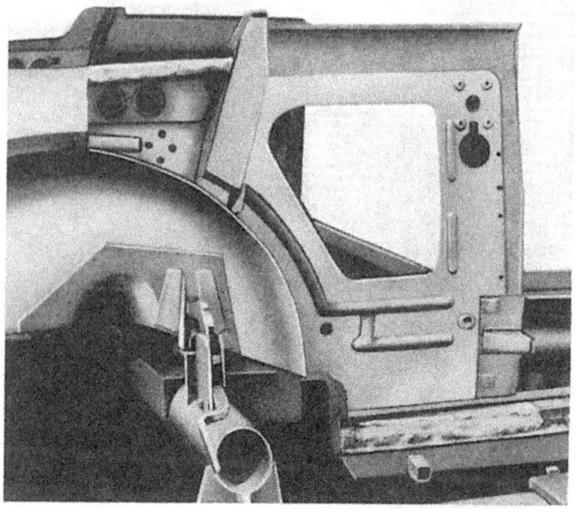

5 - Grind cut edges and spot welds smooth.

Check the condition of the reinforcement for the cross member (seat support) and the folded edge of the inner panel.

6 - Cut the new outer panel to shape and check for fit. Leave a 20 mm overlap at the wheel housing for welding purposes.

7 - Grind the appropriate places on the new panel clean ready for spot welding.

8 - Install the new outer panel section and secure with clamps.

9 - Check the position of the door in the door opening.

10 - Tack the new outer panel at various places with the welding torch.

11 - Spot weld the outer panel to the side member outside or gas weld with runs of 15 mm length.

12 - Gas weld the lock pillar to the side member.

13 - Pack the edges of the cut with wet asbestos and gas weld the panels together. Take care that the panels are free of stress.

14 - Fold the edges of the lock pillar over with a hammer and dolly, straighten where necessary.

15 - Weld the corner plate and reinforcement into position.

16 - Grind the reworked and welded places smooth and prepare the vehicle for painting.

Replacing the "U" channel

The "U" channels are shaped in a special machine, using a certain amount of heat. For this reason, only that part of a new "U" channel which corresponds to the cut-out portion of the old channel, can be used. When carrying out repairs, therefore, try to replace only that part which must be renewed. This can be done by cutting a suitable segment out of the "U" channel supplied as a replacement part.

Do not try to reshape a part of the channel to fit where it does not belong as this is more or less impossible due to the variation in the curvature of the channel.

1 - Hold "U" channel against body side or rear panel, mark off approximate length and cut channel to size.

2 - Secure "U" channel on side or rear panel with clamps.

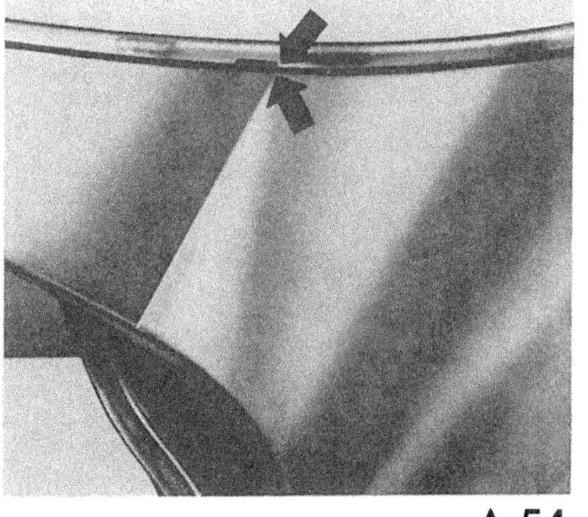

Important

The cut-out in the channel must be aligned with the groove in body rear panel as shown in illustration. The outer edge of the channel must also be level with the edge of body panel all round.

3 - Spot weld the "U" channel to body at 20 mm intervals.

Note:

The electrode set shown here can be used for this purpose. A copper plate (A) should be placed under the outer electrode in order to prevent it from marking the body panel.

Otherwise the spot weld marks must be filled with solder.

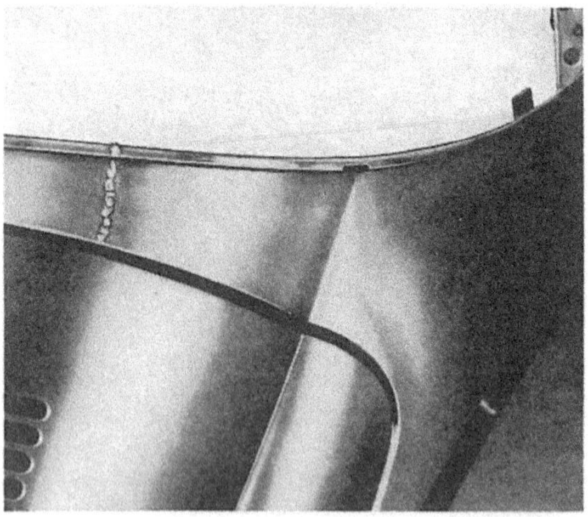

4 - Gas weld the individual parts of the channel together.

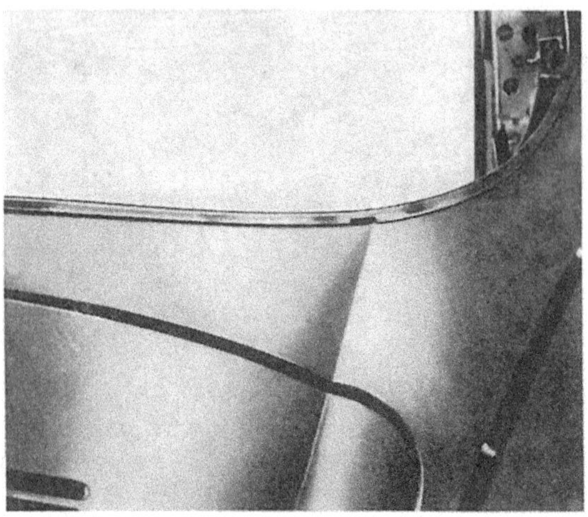

5 - Clean up the welds and make sure that the joints are smooth inside and out.

Important
The top cover will rub through quickly if any roughness is left at these points.

6 - Drill a 6 mm hole near the cut-out in the side of the "U" channel for the tensioning wire.

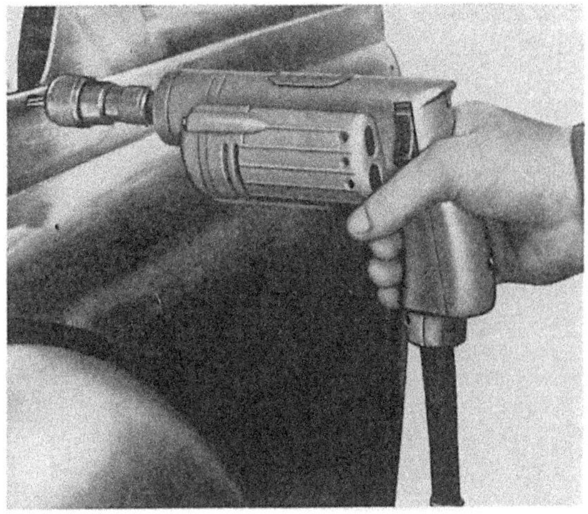

7 - Insert a punch through the hole and twist it towards the tensioning bracket.

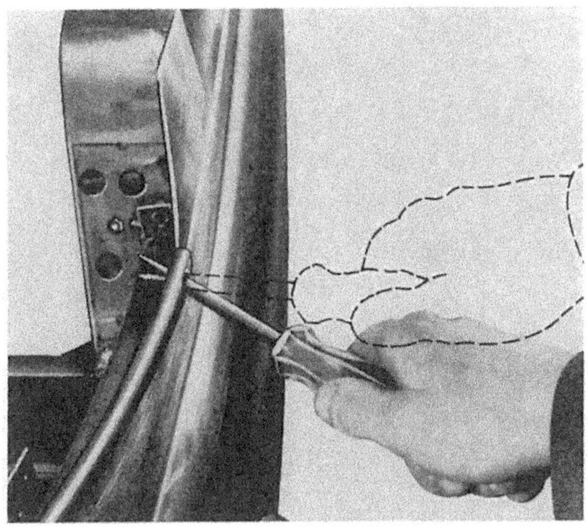

Note:

This aligns the edges of the hole so that the tensioning wire does not rub on the side panel.

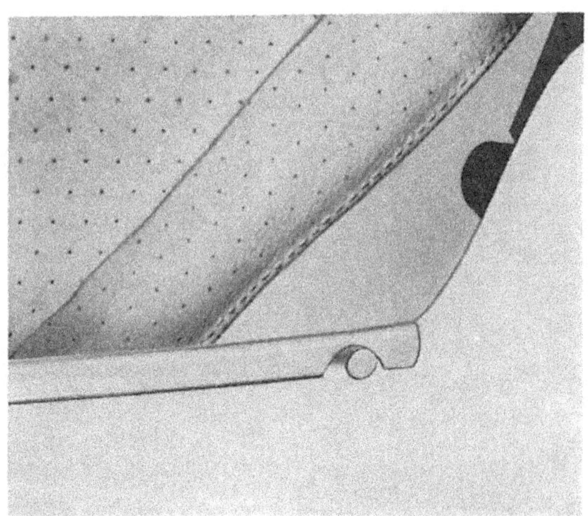

8 - Prepare the various parts for painting.

THIS PAGE INTENTIONALLY
LEFT BLANK

Spot Welding Pastes and Paint

A - Spot welding paste

In order to obtain water-tight joints, the parts of metal sheets which are to be welded are given a coating of sealing paste which is not affected by welding.

This standard manufacturing procedure should also be adopted for bodywork repairs.

The paste can be obtained direct from the manufacturers:

Manufacturer	Designation
Bonaval-Werk, Bonn, Germany Bruehler Strasse 2—20	Spot welding paste 59 852 or 60 506
Teroson-Werke GmbH., Heidelberg, Hans-Bunte-Strasse 4, Germany	Spot welding paste 2257

The paste is applied before welding commences. Before and after welding the paste should not run. At the joints it should form a film which adheres firmly, prevents corrosion and renders the joint water-tight. It must also remain unaffected by degreasing agents and subsequent painting.

Paste which burns during spot welding should cease to do so as soon as the heat is removed.

The paste should only be employed where water-tight seams are stipulated and where it is impossible or difficult to apply sealing compound after welding.

B - Spot welding paint

Spot welding paint is used to prevent corrosion in hollow parts, which cannot be painted after welding due to inaccessibility.

Manufacturer	Designation
Teroson-Werke GmbH., Heidelberg Hans-Bunte-Strasse 4, Germany	Spot welding paint 2273

Before welding, spray or paint all components with spot welding paint.

General

The use of the above products does not simplify the welding process. It may be necessary to step up the welding current, depending on the thickness of the coat applied.

Superfluous material must, however, be removed before the part is given a coat of paint which must stand up to the conditions stipulated for the outside finish. Neither welding paste nor paint can be considered as a substitute for primer to which a top coat can be applied.

THIS PAGE INTENTIONALLY
LEFT BLANK

Contents: A Body—Karmann Ghia Coupe

Assembly Work

Replacement of Body Parts

THIS PAGE INTENTIONALLY
LEFT BLANK

Description of Body

Karmann Ghia Coupé

General

The two-door, all-steel, open bottomed body of streamlined pontoon shape is bolted to the widened platform frame and sealed with U-section rubber sealing strips. The strong side members under the doors combine with the platform frame and tunnel to form a bend and twist proof foundation. The fenders are welded to the outer body panels.

The body is divided into front luggage compartment, passenger compartment and rear engine compartment by partitions. The engine compartment is ventilated by means of a series of slots in the rear hood. The area of these cooling air slots must not be restricted by the installation of accessories.

The body is comprised of the following parts, which are welded together:

a - Front section outer panel, inner wheel housings, reinforcement plate, instrument panel and hinge pillars.

b - Rear luggage compartment floor plate

c - Rear quarter panels with wheel housings

d - Side members

e - Roof

f - Front and rear panels

Doors

The 1000 mm (39.3") wide half doors have no top frames and are attached to the front body pillars by concealed, adjustable hinges. When the vertically curved window panes are in the closed position they bear directly on the body weatherstrip.

The doors are lockable and have press button controls in the rigid outer handles and buttons on the inside. The door opening angle is controlled by a check strap with an intermediate stop. A strap on the driver's side and an arm rest with grip recess on the passenger's side are provided for pulling the doors from inside. The door is sealed in the door frame by means of an all-round, moulded foam rubber strip.

The hood of the front compartment, which contains the fuel tank, spare wheel and tools, and the engine compartment hood have locks which are controlled from the vehicle interior by Bowden cables. The control knob for the front hood is located underneath the instrument panel on the left hand side. After the front hood has been released the safety catch near the lock must be pushed to the rear before the hood can be opened. The rear hood is released by means of a knob situated on the left side of the emergency seat reinforcement panel. The number plate lamp is shaped to form a handle for lifting the hood which is held in the open position by springs at the hinges.

Windows

The windshield is curved and inclined at approx. 50°. The maximum clear width is 1200 mm (47.275″) and the max. clear height 425 mm (16.73″). The rear window extends round the sides of the roof contour and is inclined at 60°. The maximum clear width is 1100 mm (43.33″) and the height 430 mm (16.93″).

The windows can be fully lowered by rotary handles and are sealed at the fixed side windows and the roof by a rubber weatherstrip.

All windows are of heat-treated safety glass. The windshield has a specially treated zone which remains clear even when the glass is damaged. For certain countries windshields of laminated glass are installed.

Front Seats

The wide, adjustable, individual seats are held in runners which rise slightly towards the front. The seat backs can be adjusted to three different angles. The seat frames are of tubular steel and the spring interiors are of interlocked coil springs. Seats and seat backs are padded with rubberised hair.

Rear Seats

The seat and back rest of the rear bench are upholstered and serve as an emergency seat. The backrest can be folded down and the space used as an additional luggage compartment.

Instrument Panel

In front of the driver, to the left of the steering column, is the speedometer which reads up to 140 k.p.h. (87 MPH) with the warning lights let into the dial. In the center is the fuel gauge and on the right the electric clock. The center portion of the instrument panel is prepared for the installation of a radio set and loud speaker. On the right side is an ashtray and a glove box with lid. The combined ignition/starter switch and the choke control are on the left near the speedometer, the light switch and the windshield wiper switch are on the right by the clock. The top part of the instrument panel is covered with padded, non-reflecting material.

Interior Trim

The door and side trim panels are covered with leatherette. The seats are finished in leatherette or cloth according to choice. The floor, front partition, frame tunnel and side members are covered with rubber matting. The floor is also coated with an additional sound absorbing material. The luggage compartment and the luggage space inside the body is lined with haircord carpet. Both doors are fitted with pockets. The rear view mirror on a ball and socket mounting and two padded sun visors are attached in the center of the roof. A grab handle is provided on the instrument panel in front of the passenger.

Heating and Ventilation **(without Fresh Air Heating)**

Air warmed by the engine is taken via flexible metal pipes, sound dampers and frame side members to the outlets at foot level in the front compartment and at the windshield. Fresh air for ventilation enters at openings in the front panel and passes via ducts arranged on each side to the defroster vents. The fresh air flow is regulated by valves in the ducts. The valves are operated by regulator levers — one for each side — located under the instrument panel at the left side. The warm and cold air can be controlled separately or mixed in the desired quantities.

Luggage Space

Access to the rear luggage space with a capacity of 6.4 cu.ft., is obtained by folding the emergency seat back rest down. Under the front hood is another luggage compartment of 2.5 cu. ft. capacity. If the emergency seat back rest is left folded down the rear luggage space is increased by a further 7.1 cu.ft.

Exterior Trim

Chrome-plated, wrap-round bumpers with over-riders.

Chrome-plated trims on headlamps, direction indicator lamps, tail and number plate lamps and windows.

Trim strips on body and door lower edges, on the window slots and on the fresh air intake grilles in the front panel.

Chrome-plated door handles, exhaust tail pipes, wheel rims and ornamental hub caps.

Heating and Ventilation **(with Fresh Air Heating)**

Fresh air is drawn in by the fan on the engine and heated in heat exchangers. In each of the front parts of the heat exchangers is a shaft with a double-acting valve. According to the position of this valve, the outlet leading to the car interior or that leading to the atmosphere is closed. When operating the right lever between the front seats, the outlet leading to the atmosphere is closed against the tension of the spring by control cables and the outlet leading to the car interior is opened. The hot air passes partly or wholly, according to position of valve, through sound deadening hoses then through tubes in the frame side members to the outlets on the left and right at foot level in the front passenger compartment and to the defroster vents at the windshield. The rear compartment is heated by air coming from two outlets below the emergency seat and from a defroster vent below the rear window.

The outlets at foot level in the front passenger compartment can be closed by flaps, while the outlets at foot level in the rear compartment can be partly or fully closed by flaps operated by control cables connected to the left lever.

Fresh air for ventilation enters through two openings in the front panel, each covered by a grille, and travels via ducts to the defroster outlets. At the end of each duct is a mixing tube with a regulating valve, each valve being operated separately by two levers on the left-hand side under the instrument panel.

Warm and cold air can be regulated separately or can be mixed.

THIS PAGE INTENTIONALLY
LEFT BLANK

Front Hood Removal and Installation

Removal

The removal and installation of the front hood is simplified if the job is carried out by two mechanics. The hood can then be properly supported and damage to the front cowl panel avoided.

If this is not possible, a suitable piece of woollen or plastic material should be placed over the cowl before work commences.

1 - Open hood.

2 - Remove two bolts from the hinges on each side of the hood and take off the hood.

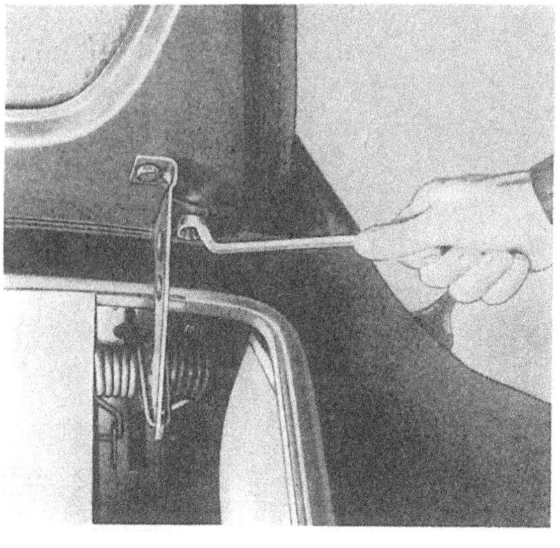

3 - If necessary, unhook the spring and after removing the fiberboard luggage compartment lining unscrew the hinge securing bolts and remove hinges.

4 - If necessary, remove the lock bolt and safety hook from the hood.

Installation

1 - Check condition of weatherstrip and replace if necessary.

When renewing the front hood weatherstrip remove all traces of old cement with benzine and secure the new weatherstrip in position with Genuine VW Universal Adhesive D 12. Take care that holes in the weatherstrip line up with the holes in the front panel.

2 - Bolt the hood loosely to the hinges so that it can be moved in the elongated holes to obtain proper seating and sealing all round. Then fully tighten the bolts.

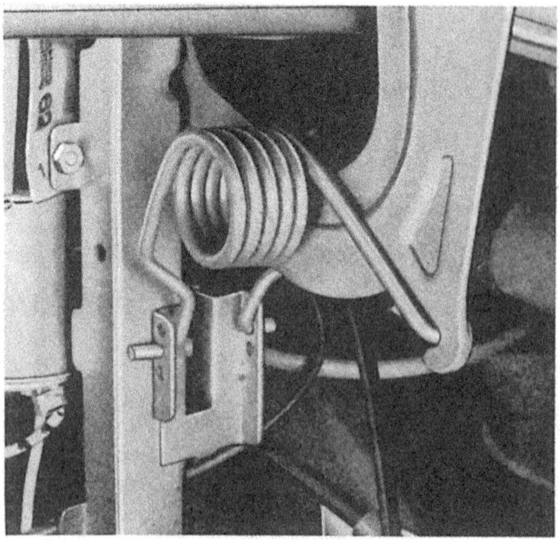

3 - Attach the springs to the hinges so that the hood remains in the open position.

4 - Open and close the hood several times to check lock operation. If necessary adjust the lock in the elongated holes.

Front Hood Lock Removal and Installation

Removal

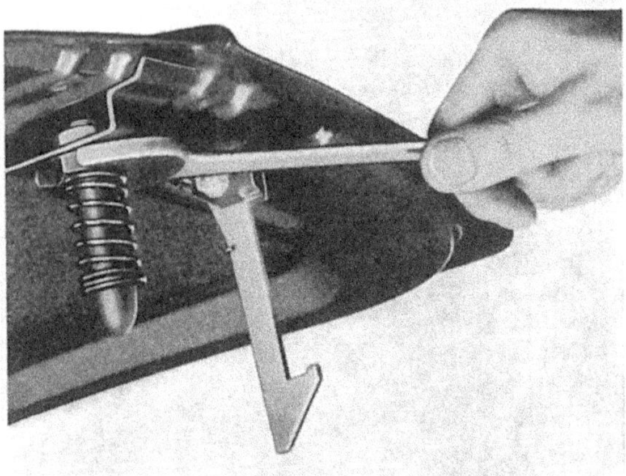

1 - Open front hood.

2 - Remove lock bolt and safety hook.

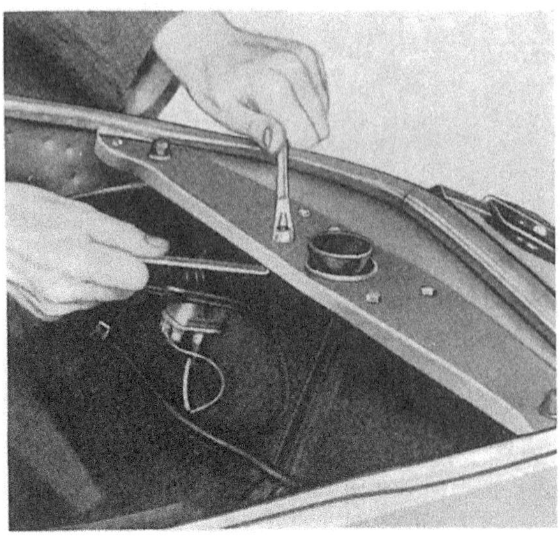

3 - Unscrew five retaining nuts and bolts.

4 - Loosen lock cable clamp screw and withdraw cable from catch. Remove lock.

5 - Pull lock cable out of guide tube.

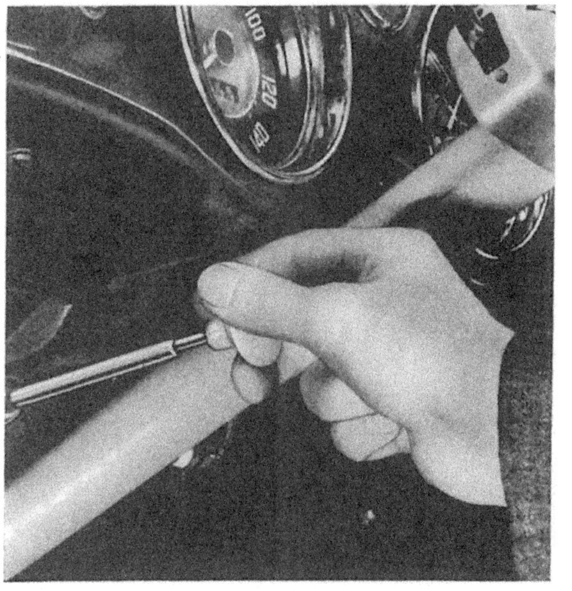

Installation

1 - Grease hood lock cable and insert into guide tube.

2 - The lock is so constructed that if the lock cable breaks the latch plate, which retains the lock bolt when the hood is closed, springs back automatically. The bolt is then free and the hood can be opened after pushing back the safety hook.

When attaching the lock cable the arm under the lock which carries the cable clamp screw must be turned against the spring tension until the latch plate projects into the opening for the lock bolt. In this position the lock cable should be inserted into the lock and secured to the arm by tightening the clamp screw. Bend surplus cable back.

3 - Secure hood lock.

4 - Install lock bolt and safety hook.

5 - Open and close the hood several times to check length of lock bolt and position of lock.

If necessary, adjust the position of the lock by moving it in the elongated holes, and the lock bolt length by loosening the lock nuts and turning the bolt in or out.

6 - Check the lock cable adjustment. If required adjust cable. Bend surplus cable back again.

7 - Grease lock bolt.

Rear Hood Removal and Installation

Removal

The removal and installation of the rear hood is simplified if the job is carried out by two mechanics. The hood can then be properly supported and damage to the rear upper panel avoided.

If this is not possible, a suitable piece of woollen or plastic material should be placed over the rear upper panel before work commences.

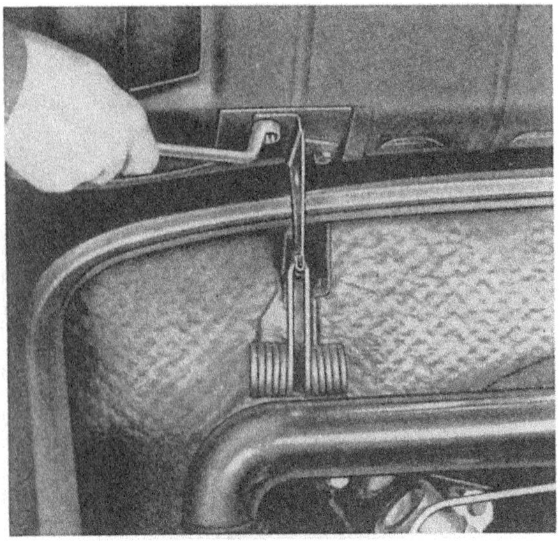

1 - Open rear hood.

2 - Unclip the license plate lamp cable and pull it out of the clamping plate on the hood.

3 - Remove two bolts from the hinges on each side of the hood.

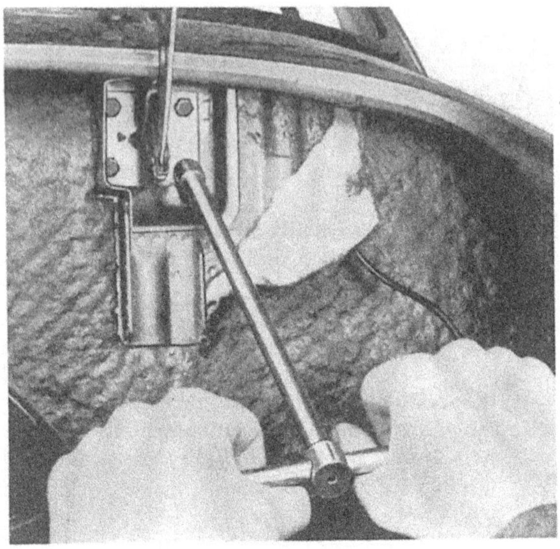

4 - Remove the hood.

5 - If necessary, remove the hinges after lifting the sound damping material, or remove the hood lock.

Installation

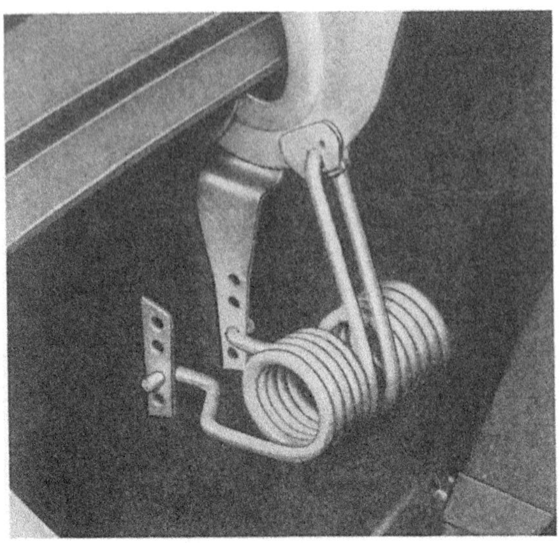

1 - Check condition of weatherstrip and replace if necessary. When renewing the rear hood weatherstrip remove all traces of old cement with benzine and secure the new weatherstrip in position with Genuine VW Universal Adhesive D 12.

2 - Bolt the hood loosely to the hinges so that it can be moved in the elongated holes to obtain proper seating and sealing all round. Then fully tighten the bolts.

3 - Attach the springs to the hinges so that the hood remains in the open position.

4 - Open and close the hood several times to check lock operation. If necessary, adjust the lock in the elongated holes.

Rear Hood Lock Removal and Installation

Removal

1 - Open rear hood.

2 - Unscrew lock bolt.

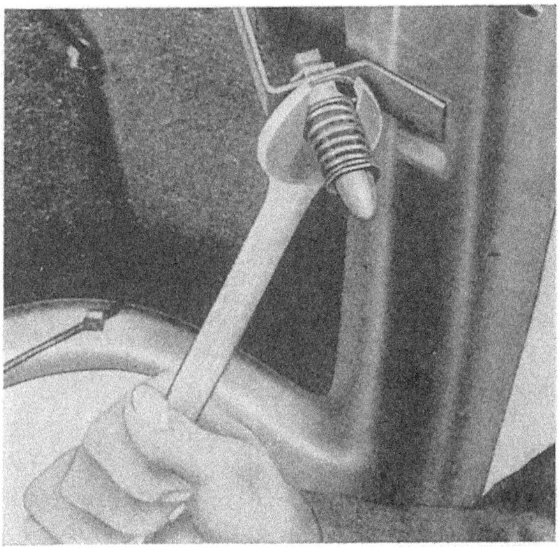

3 - Remove four bolts.

4 - Loosen cable clamp screw and remove cable from catch. Remove lock.

5 - Pull lock cable out of guide tube.

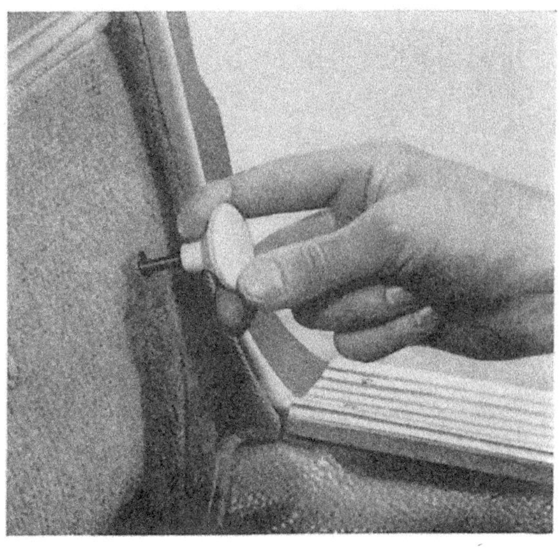

Installation

1 - Grease lock cable and insert into guide tube.

2 - The lock is so constructed that if the lock cable breaks, the latch plate, which retains the lock bolt when the hood is closed, springs back automatically. The bolt is then free and the hood can be opened.

When attaching the lock cable the arm under the lock which carries the cable clamp screw must be turned against the spring tension until the latch plate projects into the opening for the lock bolt. In this position the lock cable should be inserted into the lock and secured to the arm by tightening the clamp screw. Bend surplus cable back.

3 - Secure hood lock.

4 - Install lock bolt.

5 - Open and close the hood several times to check length of lock bolt and position of lock.

If necessary, adjust the position of the lock by moving it in the elongated holes, and the lock bolt length by loosening the lock nuts and turning the bolt in or out.

6 - Check the lock cable adjustment. If required, adjust cable. Bend surplus cable back.

7 - Grease lock bolt.

Front Bumper Removal and Installation

Removal

1 - Lift front hood and remove spare wheel.

2 - Remove bumper brackets bolts on each side.

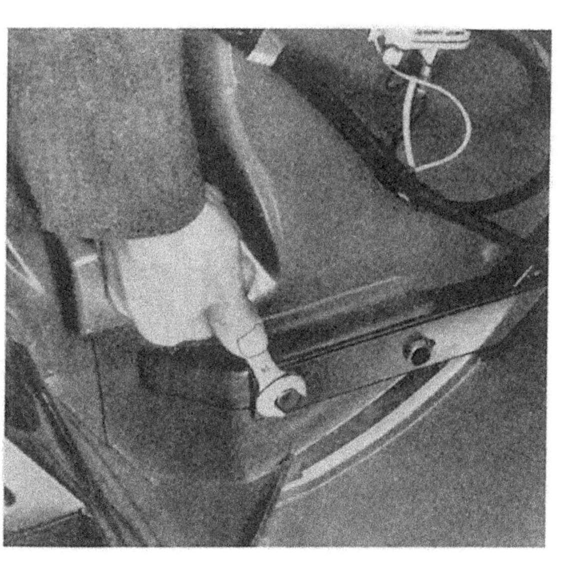

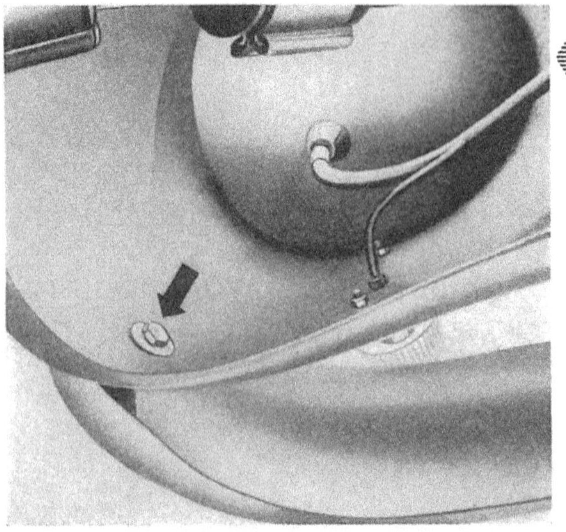

3 - Remove retaining bolts from both outer parts.

4 - Pull bumper complete with overriders and brackets, out of the front panel.

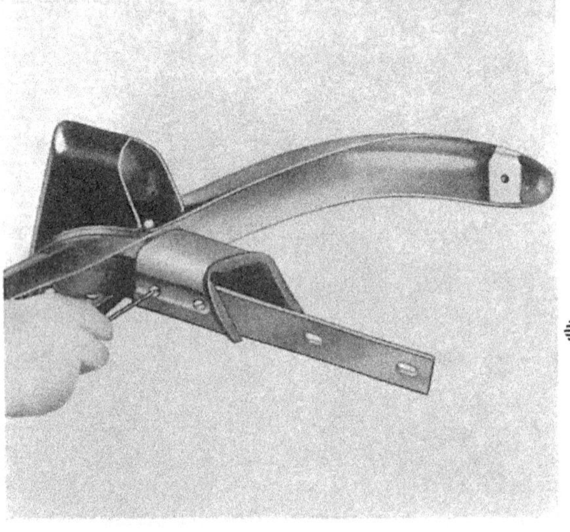

5 - Remove bumper bracket covers.

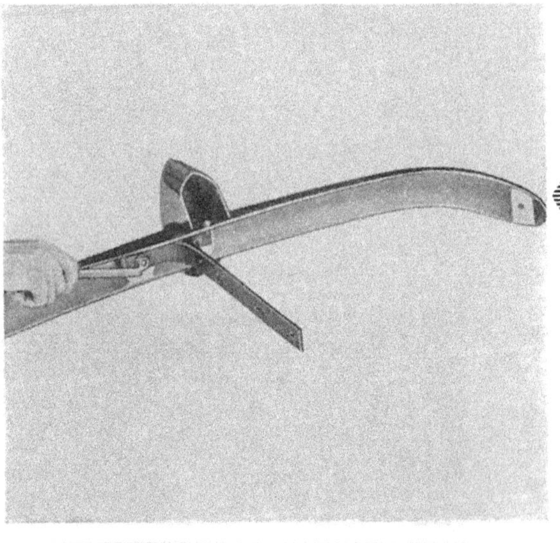

6 - Unscrew the two bolts at each side and remove brackets, spacers and overriders.

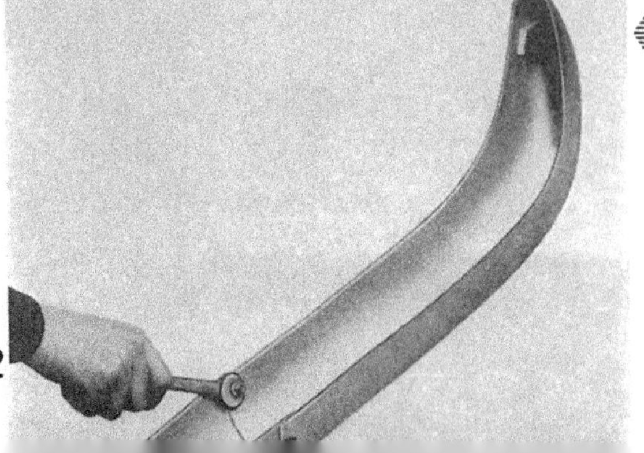

7 - Unscrew two bolts at each side and detach outer parts from bumper center part.

8 - Remove bumper bracket grommets.

Installation

1 - Check condition of seals for overriders, bumper brackets and bumper bracket covers and renew if necessary.

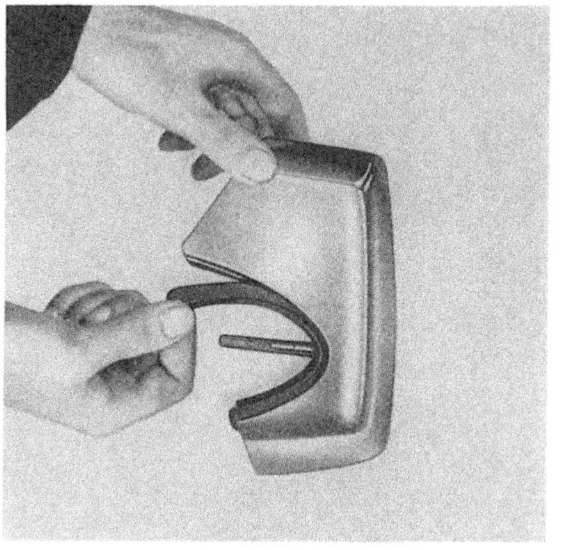

2 - Bolt outer parts to bumper center part.

3 - Bolt brackets with bumper to body.

4 - Replace both outer part bolts and check that clearance between bumper and body is uniform.

5 - Replace bumper bracket covers so that the rubber seal contacts body all round.

Rear Bumper Removal and Installation

Removal

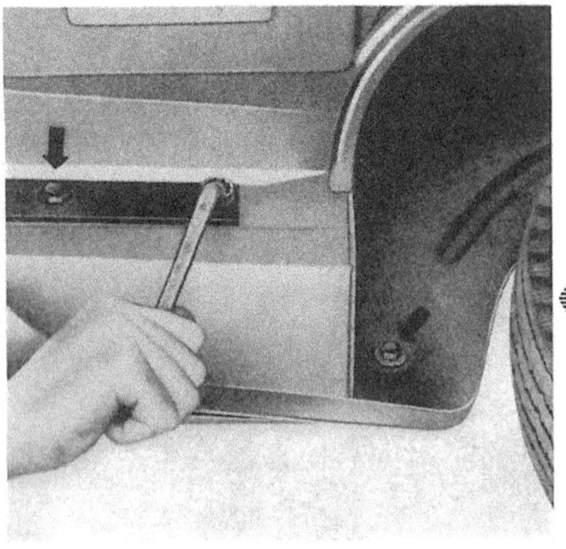

1 - Remove two bumper bracket bolts and one outer part securing bolt on each side.

2 - Pull bumper, complete with overriders and brackets out of the rear panel.

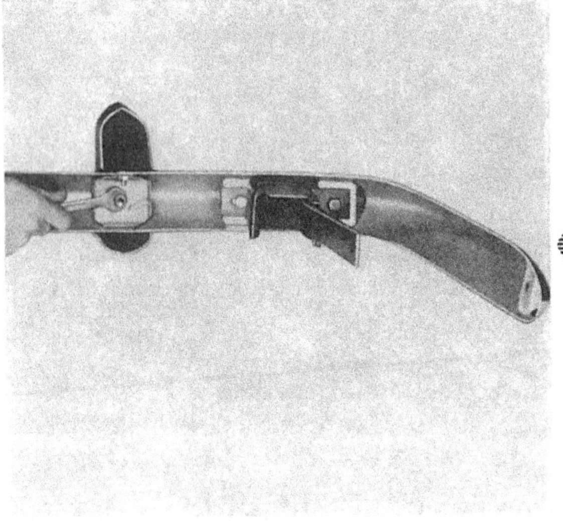

3 - Remove one overrider bolt on each side and detach overriders and spacers.

4 - Remove bumper bracket covers.

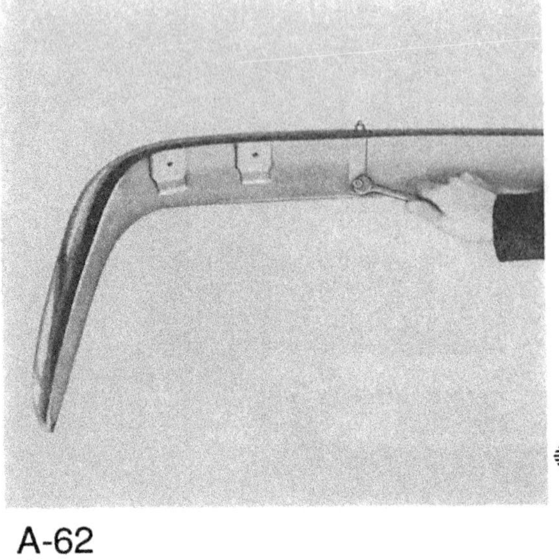

5 - Detach brackets from bumper.

6 - Unscrew two bolts at each side and detach outer parts from bumper center part.

Installation

1 - Check condition of seals, renew if necessary and replace.

2 - Install bumper with brackets.

3 - Secure outer parts and check that clearance between bumper and body is uniform.

4 - Replace bumper bracket covers so that the rubber seal contacts body all round.

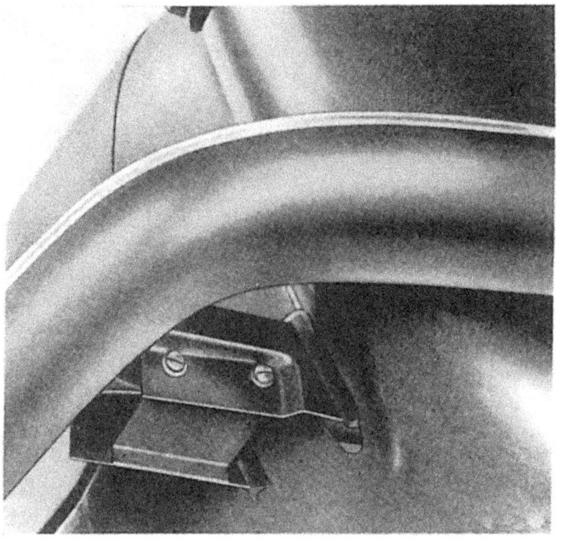

Front Bumper Removal and Installation

(USA Version)

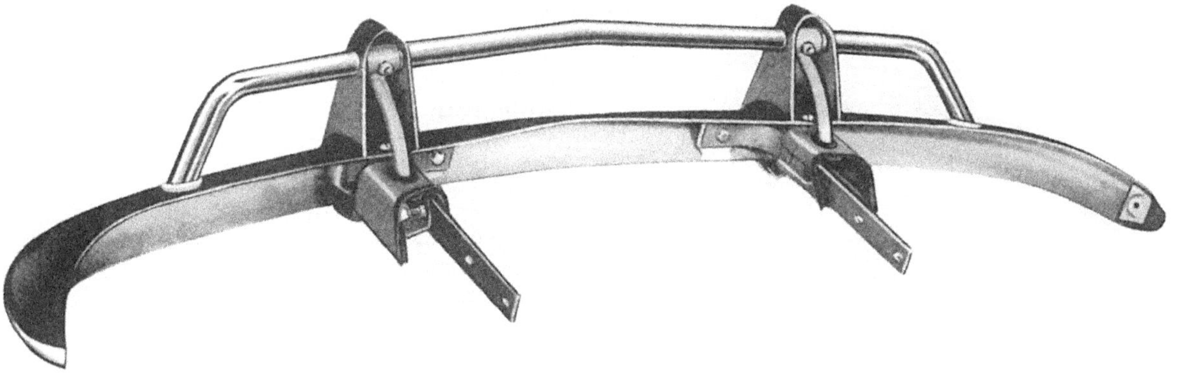

Removal

1 - Remove bolts from both brackets.

2 - Remove bolts from both outer parts.

3 - Pull bumper, complete with bows, overriders and brackets out of the front panel.

4 - Remove two overrider support nuts on each side.

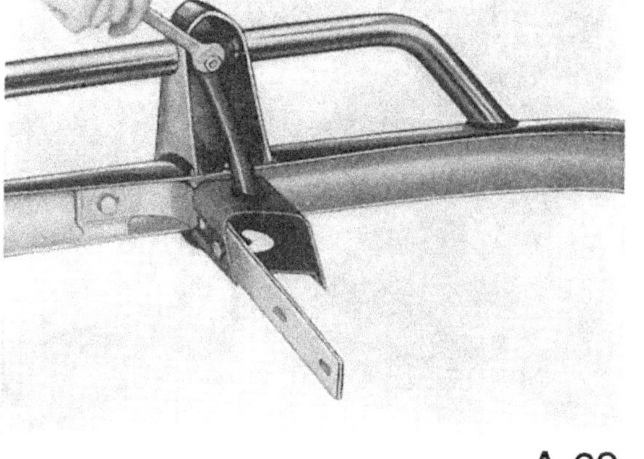

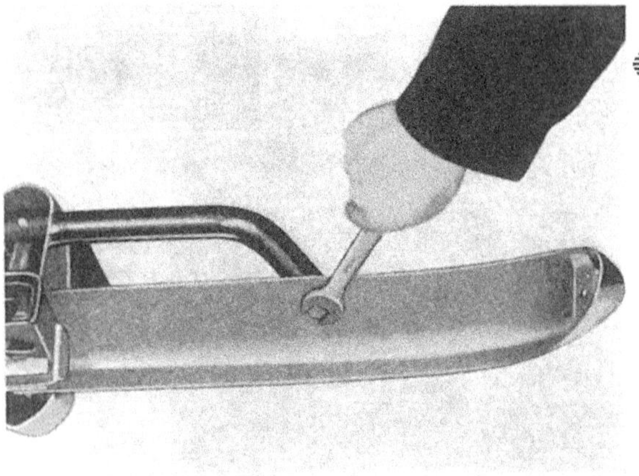

5 - Detach bow after removing the bolts on both sides.

6 - Remove bumper bracket covers.

Unscrew two bolts at each side of bumper and remove brackets, spacers and overriders.

7 - Unscrew two bolts at each side and detach outer parts from bumper center part.

8 - Remove bracket grommets from front panel.

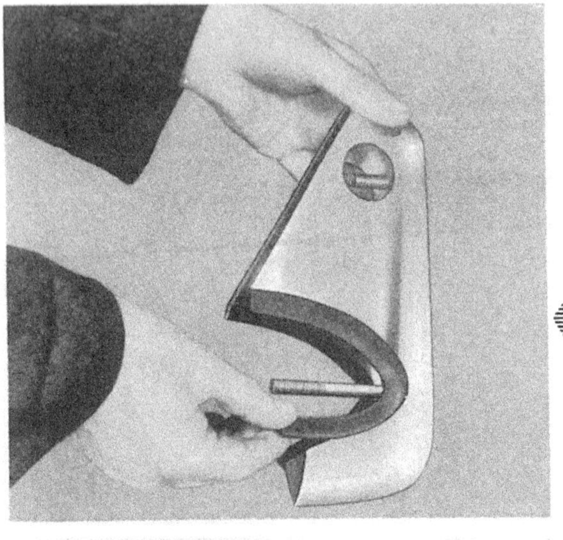

Installation

1 - Check condition of seals, renew if necessary and replace.

2 - Bolt outer parts to bumper center part.

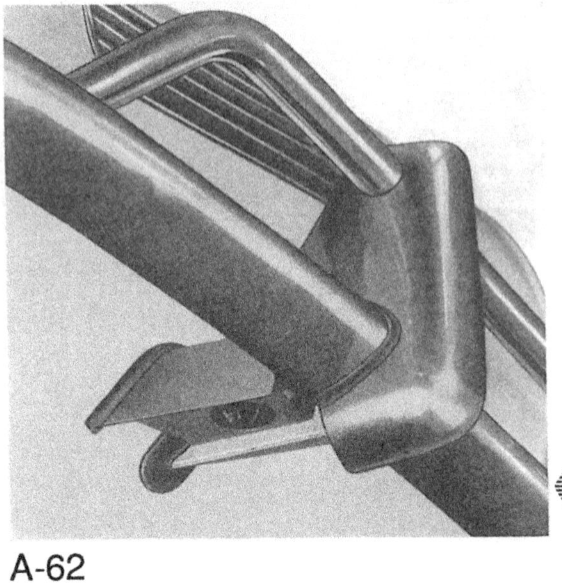

3 - Bolt bumper, bows, overriders, spacers and brackets together.

4 - Bolt bumper with brackets to body.

5 - Secure outer parts and check that clearance between bumper and body is uniform.

6 - Replace bumper bracket covers so that the rubber seal contacts body all round.

Rear Bumper Removal and Installation
(USA Version)

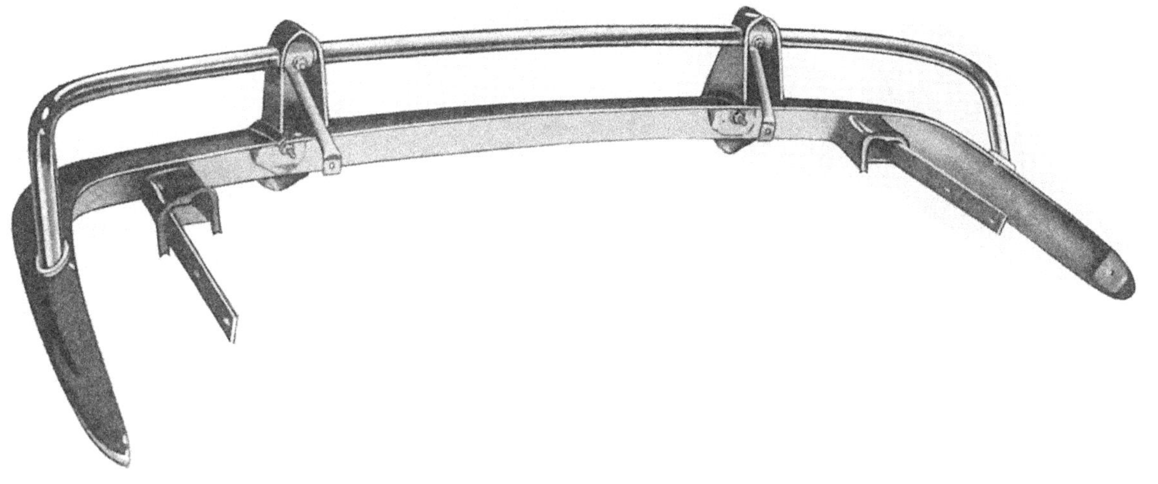

Removal

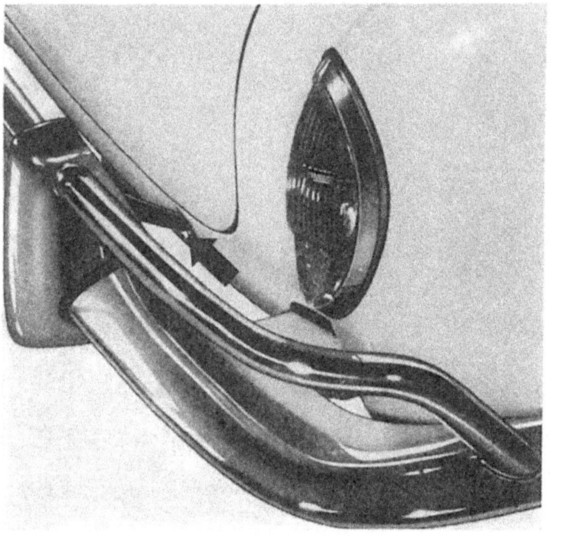

1 - Remove two overrider support bolts on each side.

2 - Remove two bracket bolts and one outer part securing bolt on each side.

3 - Remove bumper, complete with bows, overriders and bumper brackets.

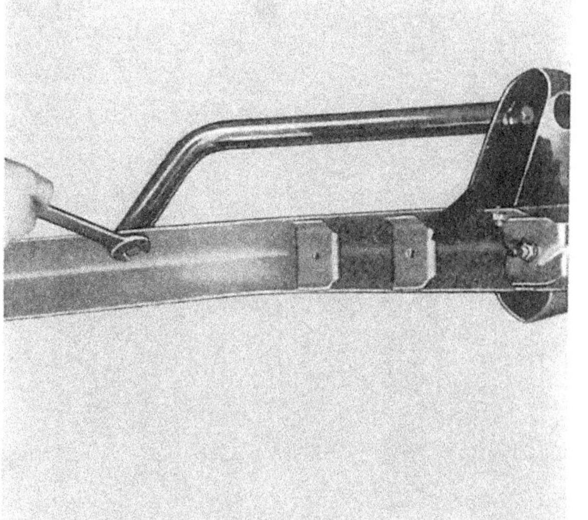

4 - Detach bow after removing the bolts on each side.

5 - Unscrew two bolts at each side and detach outer parts from bumper center part.

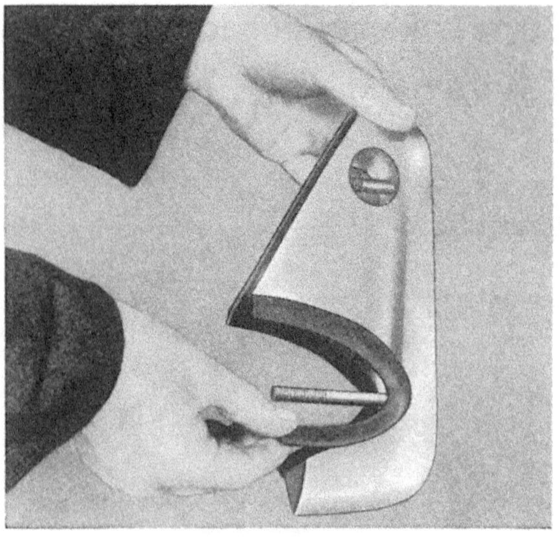

Installation

1 - Check condition of seals, renew if necessary and replace.

2 - Bolt bumper, bows, overriders, spacers and bumper brackets together.

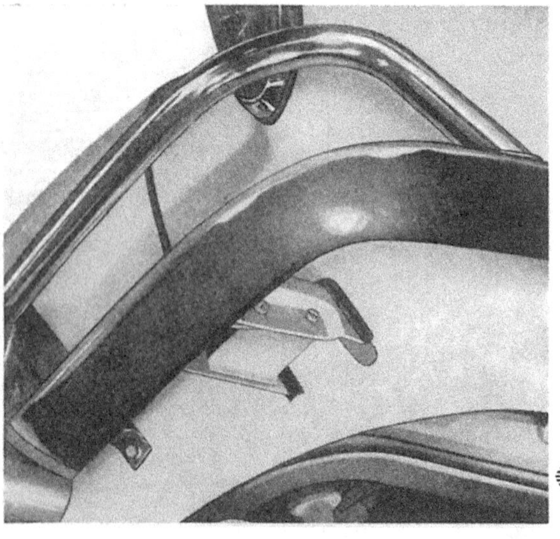

3 - Pass brackets through holes provided and bolt them to the body. Secure both outer parts and check that the clearance between bumper and body is uniform.

4 - Bolt overrider supports to body.

5 - Replace bumper bracket covers so that the rubber seal contacts the body all round.

Door Removal and Installation

Removal

1 - Pull out door check strap after removing retaining pin on hinge pillar.

2 - Loosen four Phillips screws on each hinge with a punch screwdriver, then unscrew them with a Phillips screwdriver.

3 - Remove door, together with hinges, to one side out of the hinge pillar.

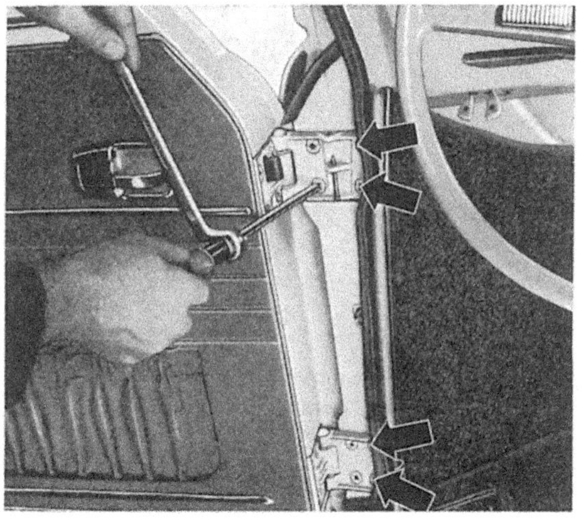

Installation

1 - Check condition of door weatherstrip and install new one if necessary. Cement weatherstrip in position with genuine VW universal adhesive D 12.

2 - Install door and position it in the opening so that weatherstrip seals all round and door can be easily opened and closed. For this purpose, remove striker plate.

The door hinges are bolted to moveable threaded plates to facilitate correct adjustment and alignment of door to outer panel.

3 - Install striker plate and adjust it so that door and quarter panel are in alignment and the door can be closed easily. Detailed instructions are contained in the section "Striker Plate Adjustment".

4 - Thinly coat mating surfaces of latch, striker plate and wedge with vaseline or usual molybdenum disulfide paste.

Important

The door lock cylinder of the lockable door handle must be lubricated only with powdered graphite and never with oil or grease. The key should be dipped in graphite powder, inserted in the lock and turned back and forth several times.

Striker Plate Adjustment

To avoid the latch moving back and forth due to vibration when the vehicle is being operated, the striker plate is fitted with a sprung, plastic wedge. This wedge is not adjustable. To check whether the latch still fits snugly in the striker plate at the wedge, remove striker plate.

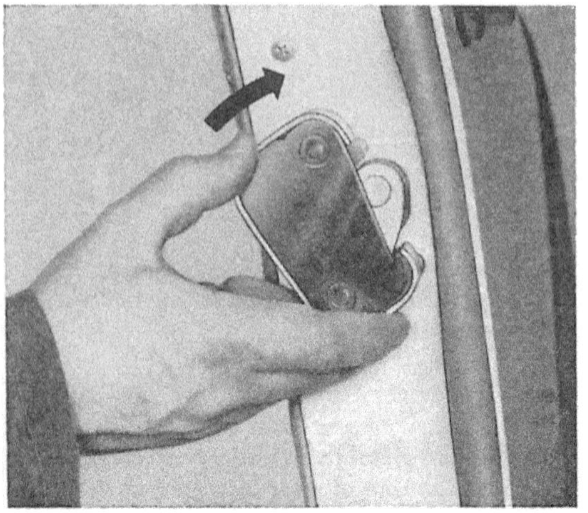

With latch completely depressed — locking position — insert striker plate into latch, starting at the bottom.

Afterwards rotate the striker plate farther.

If, in this position, clearance can be determined by moving striker plate up and down, striker plate must be renewed.

After removing striker plate, check position of door in door opening as follows:

1 - Tightness of hinge attaching screws.

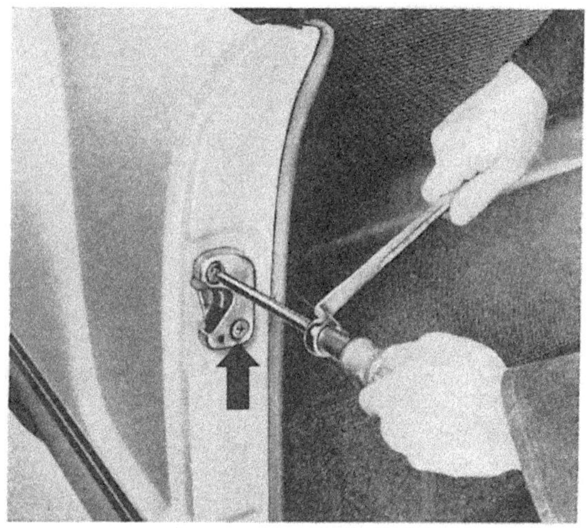

2 - Alignment of door and front side panel.

3 - Distortion in the door.

4 - Uniform gap between door and hinge pillar.

After installing striker plate, check position of door in door opening as follows:

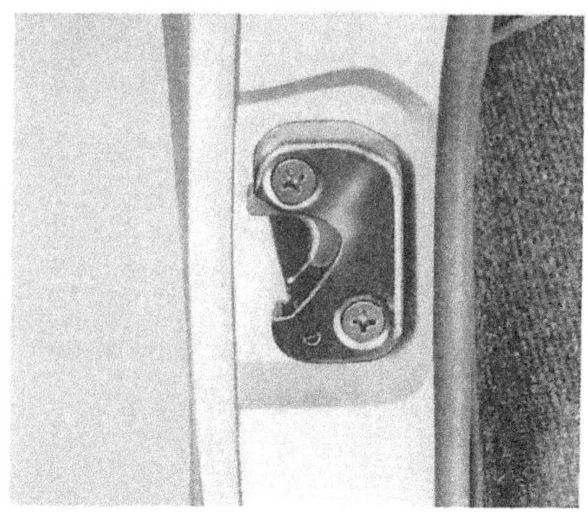

1 - Alignment of door and quarter panel.

2 - Uniform gap between door and lock pillar.

If necessary, the following work has to be carried out.

With striker plate removed — ref. pos. 2:

Loosen door hinges only at the hinge pillar. Move door appropriately inwards or outwards and tighten hinge Phillips screws.

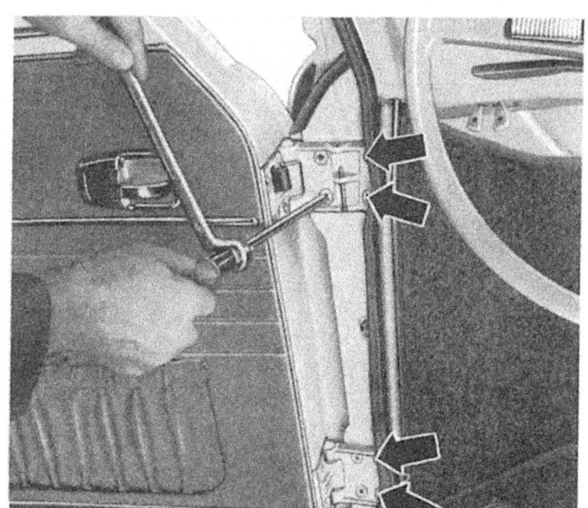

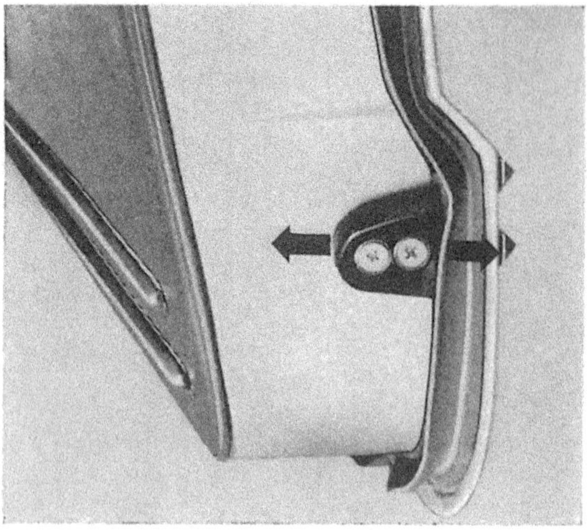

Ref. pos. 3:

Readjust door buffer and remove material by grinding if necessary.

Ref. pos. 4:

If the gap between door and hinge pillar is not uniform over the whole length, the door must be moved the required amount by adjusting the hinges.

With striker plate installed — ref. pos. 1:

The striker plate is correctly installed if

a - the door is in alignment with the quarter panel,

b - there is no noticeable clearance when pulling firmly on the door handle,

c - there is a noticeable clearance of approx. .04 in. (1 mm) when pushing firmly against the door handle,

d - the door can be opened from inside and out without excessive effort.

An incorrect adjustment of the striker plate can be corrected as follows:

a - If the door is difficult to close, the push button is hard to operate. The reason for this is that the striker plate is inclined too far inwards at the top.

Remedy: Incline striker plate outwards.

b - If the door does not remain in closed position when slamming to, but springs back to the safety position, the striker plate is inclined outwards. The door can be easily opened with the push button.

Remedy: Incline striker plate inwards.

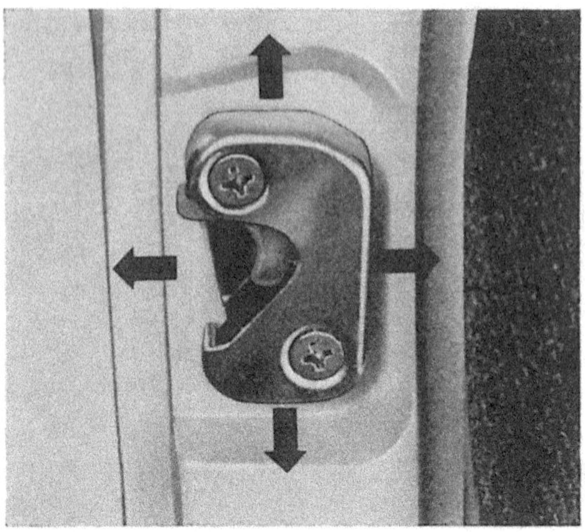

c - If the striker plate has been set too high, the door is difficult to open with the push button. When the door is opened, it does not run parallel out of the opening but drops down.

Remedy: Move striker plate downwards.

d - If the striker plate has been set too low, the door engages only in the safety position when slamming it to. It springs out of the closed position.

Remedy: Move striker plate upwards.

Ref. pos. 2:

The gap between door and lock pillar should be approx. .16 in. (4 mm). If this gap widens towards the top instead of being uniform over the whole length, loosen the body mounting screws at point A and B.

Lift body at A and insert an appropriately thick packing piece under the quarter panel rear support at the spring plate support. Tighten the body mounting screws to the correct torque.

If the door gap is still not correct, install a further packing piece. After completing this work, check position of door windows in door openings. If necessary, adjust door windows.

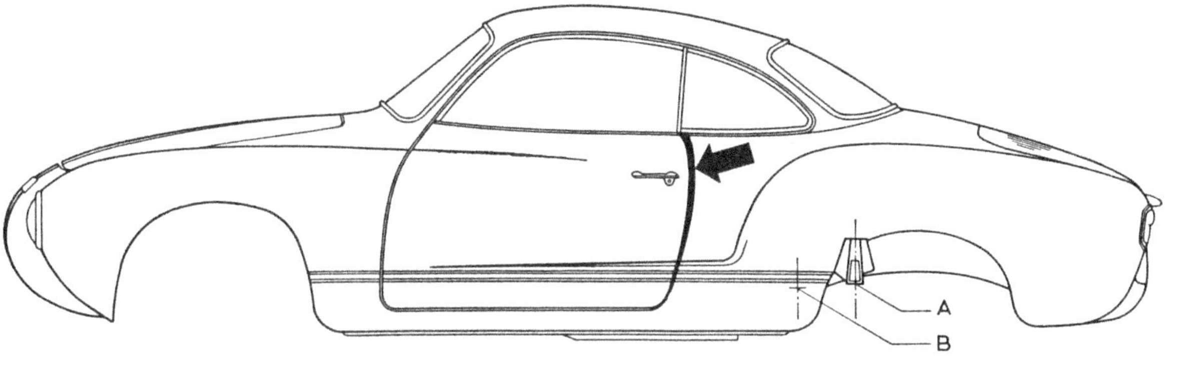

Door Check Strap Removal and Installation

Removal

1 - Remove circlip from check strap retaining pin and, with a punch, drive out retaining pin.

Note:

To avoid damaging door, support it carefully prior to driving out pin.

2 - Loosen eight Phillips screws in upper hinge, using a punch screwdriver, and unscrew them with a Phillips screwdriver.

3 - Remove door hinge at an angle from hole in door.

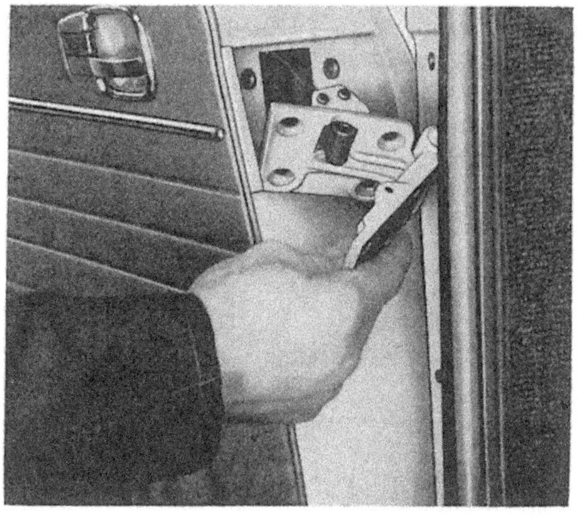

4 - Remove circlips from retaining pins and, with a punch, drive out retaining pins.

5 - Remove door check strap and rollers.

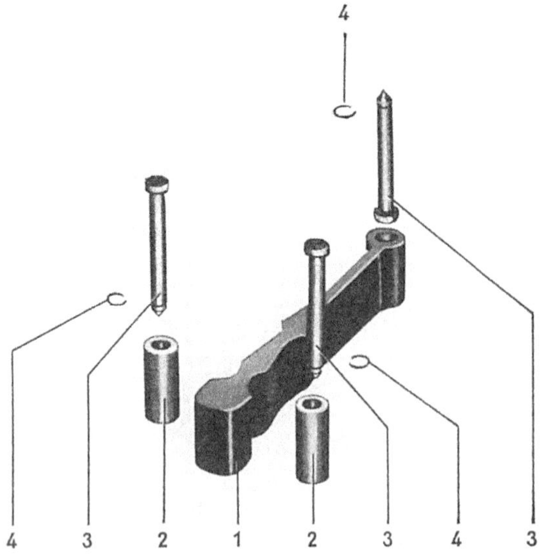

1 - Door check strap
2 - Rollers
3 - Retaining pins
4 - Circlips

Installation

1 - Check condition of all parts and install new ones if necessary.

2 - Insert door check strap and secure rollers to hinge with the retaining pins. Do not forget circlips.

> **Important**
> The door check strap must be free from grease or oil.

3 - Reinstall hinge and secure door check strap with retaining pin. Do not forget circlip.

Door Trim Panel Removal and Installation

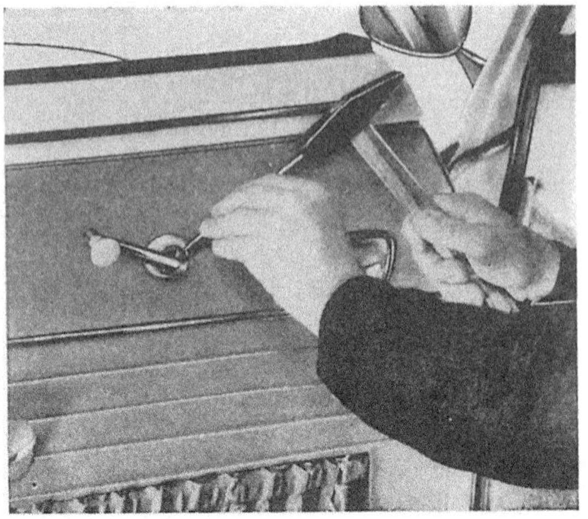

Removal

1 - Press window regulator handle escutcheon against door trim pad and drive out dowel pin with a punch. Remove handle and escutcheon.

2 - Remove door lock inner control finger plate.

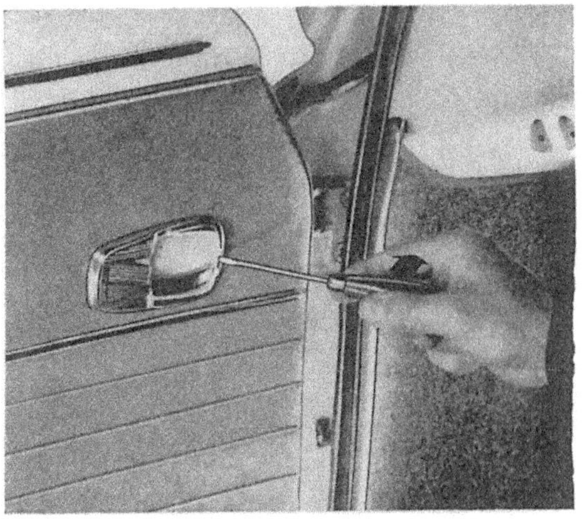

3 - Remove two Phillips screws and take off inner control trim frame.

4 - Remove door trim panel carefully, ensuring that it and the paintwork are not damaged.

5 - Remove conical spring and rubber insert from window regulator handle shaft.

6 - Remove plastic sheet.

Installation

1 - Thinly coat inside surface of door with universal adhesive D 12 and attach plastic sheet. Ensure that the sheet is pulled into bottom of window aperture.

2 - Install rubber insert, conical spring and door trim panel. The larger diameter of the spring must face the trim panel.

Note:

Hang door trim panel in position by hooking the arm rest retainer strip over bracket of arm rest support on guide channel, ensuring that foam rubber pad for inner control does not move.

3 - Install window regulator handle in correct position.

4 - Check function of all parts by operating them several times.

Door Handle and Lock Cylinder Removal and Installation

No pre-set tumbler combinations are supplied for replacing a lock cylinder. It is possible, however, to restore the key uniformity. For this purpose an assortment of all the necessary tumblers is obtainable so that individual tumblers can be used when required.

With the aid of the code list, the tumblers in the lock cylinder can be altered so that the newly installed lock cylinder for the door can be operated with the existing key.

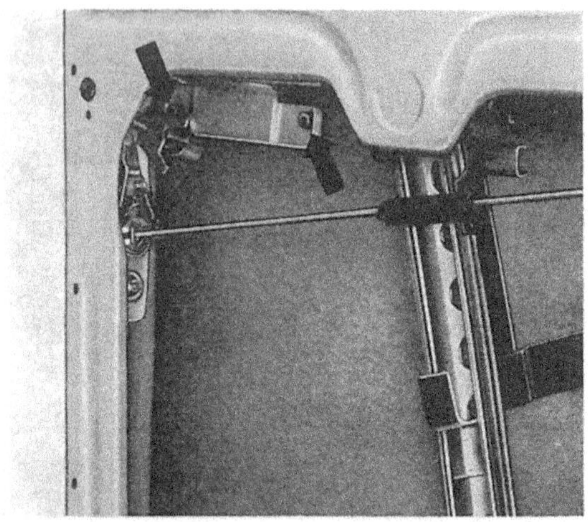

Removal

1 - After removing door trim panel, unscrew two hex. head screws on outer door handle.

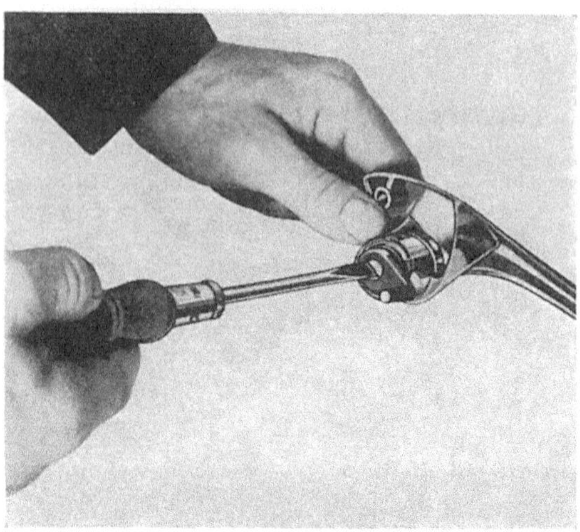

2 - Remove outer door handle.

3 - Remove countersunk head screw from lock cylinder. Remove lock cylinder operating cam and return spring.

4 - Pry circlip out of groove in door handle. Remove push button guide, push button return spring and push button together with lock cylinder.

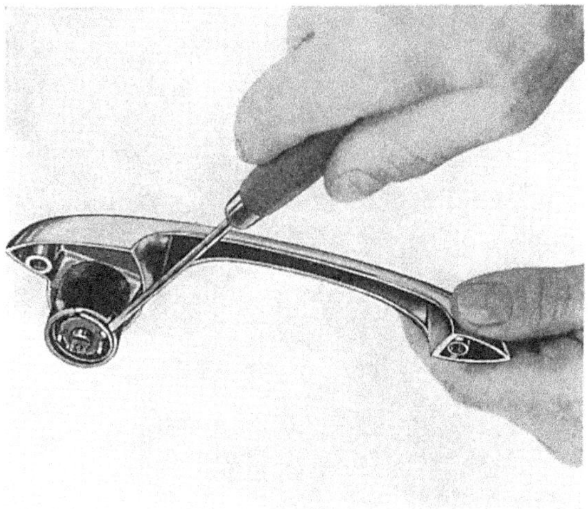

5 - Unscrew threaded pin on push button and remove lock cylinder from push button.

6 - Take out tumblers and springs, after removing the peened material on outside of tumbler spring hole.

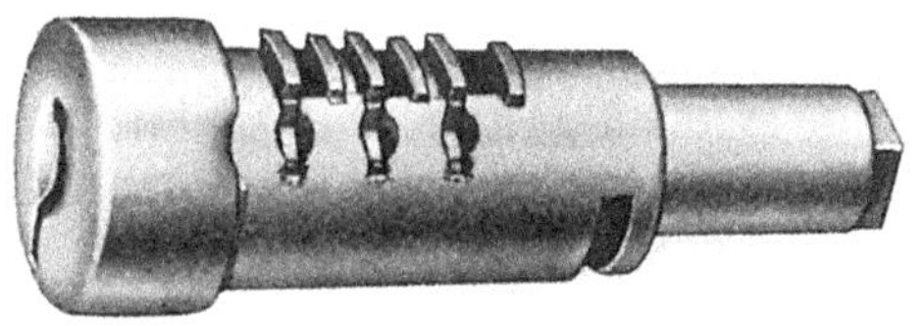

Installation

Prior to installation, thinly coat the outer surface of the lock cylinder with universal grease. The interior of the lock cylinder must only be lubricated with powdered graphite. Never use oil or grease. The key should be dipped in graphite powder, inserted in the lock and turned back and forth several times.

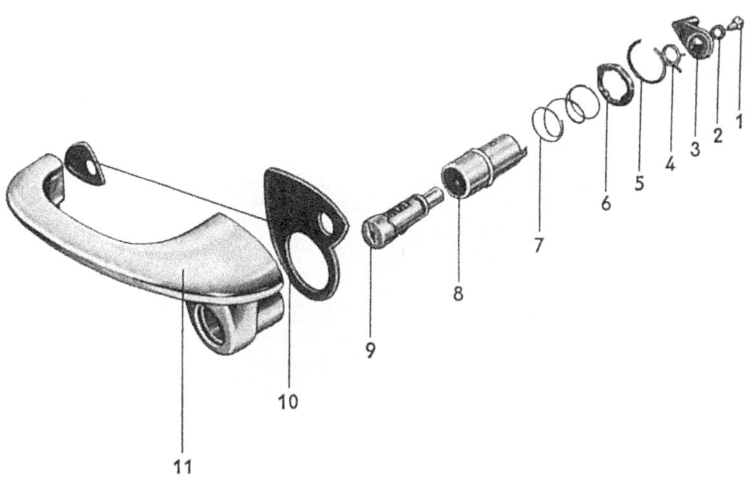

1 - Countersunk head screw
2 - Lockwasher
3 - Operating cam
4 - Return spring
5 - Circlip
6 - Push button guide
7 - Return spring
8 - Push button
9 - Lock cylinder
10 - Seals
11 - Door handle

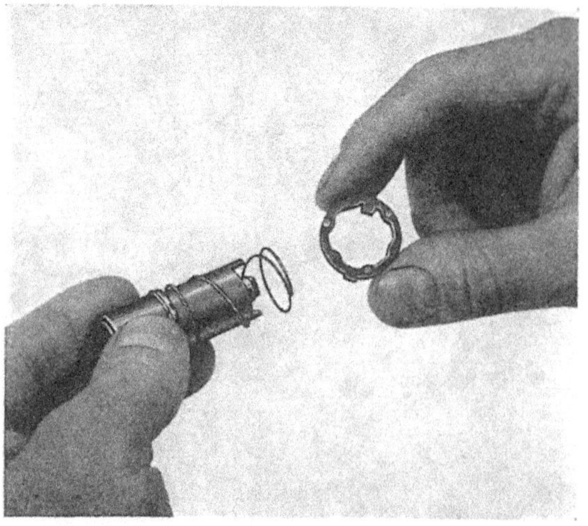

1 - Clean all parts in benzine and blow out with compressed air. Replace damaged tumblers or springs.

2 - Insert tumblers in their correct order, together with the springs, into the lock cylinder. Peen material in vicinity of spring holes.

3 - Place lock cylinder into push button and secure with threaded pin. Secure threaded pin by centerpunching.

4 - Insert push button, return spring and guide into door handle and secure with circlip.

Important

Ensure that the inner projection of the guide is located in the groove of the push button and the outer projection in the groove of the door handle.

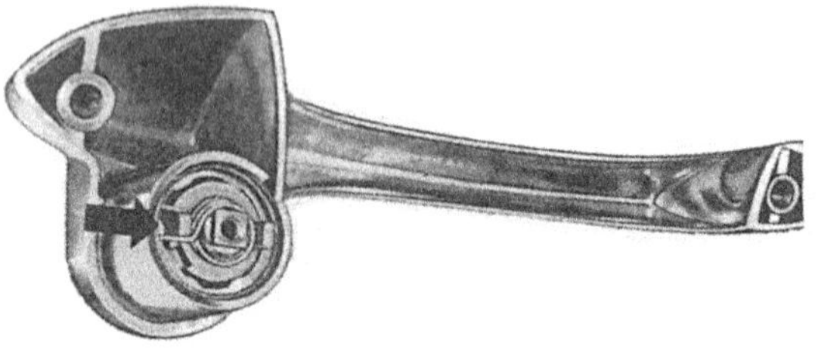

5 - Insert lock cylinder return spring. The ends of the spring must bear against a stop on the push button.

6 - Push operating cam on to square part of lock cylinder and install lockwasher and countersunk head screw.

Important

The pin on the operating cam must always face driving direction.

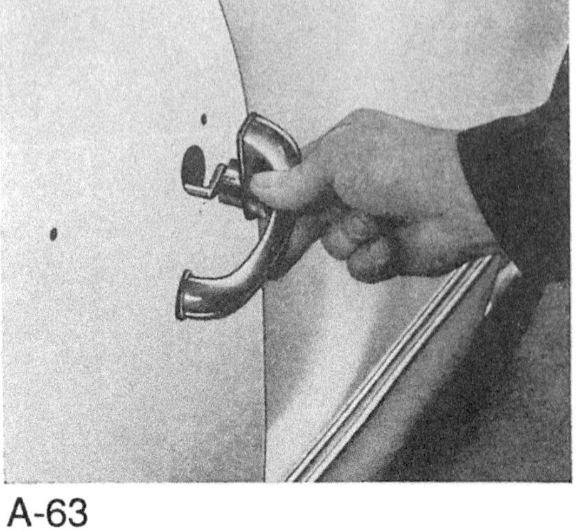

7 - Insert key and check operation of lock cylinder. If necessary, invert the tumblers.

8 - Attach door handle and seals to door.

Removal

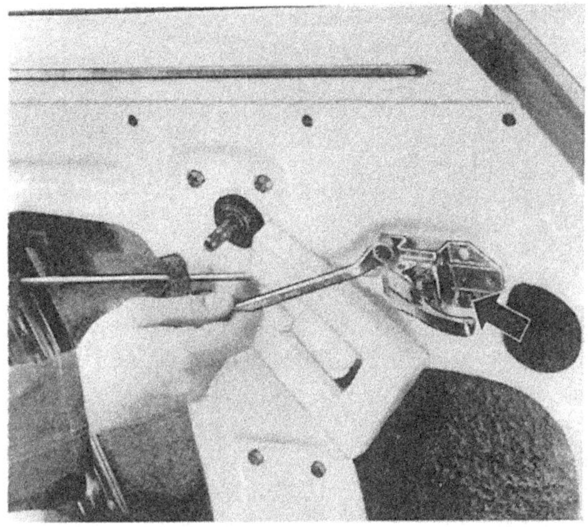

1 - Unscrew two hex. head screws on door lock inner control.

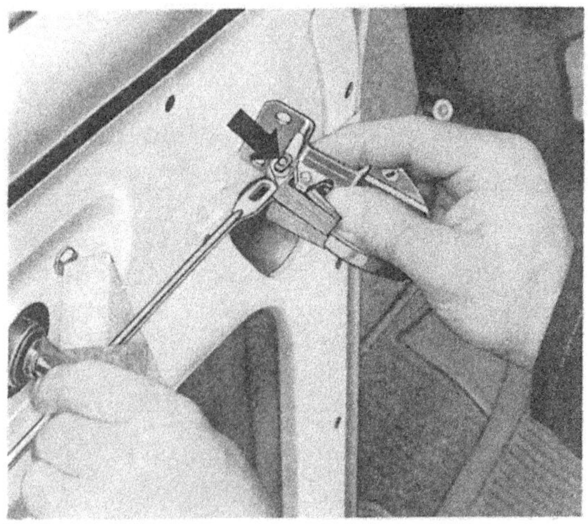

2 - Detach operating lever pull rod from inner control.

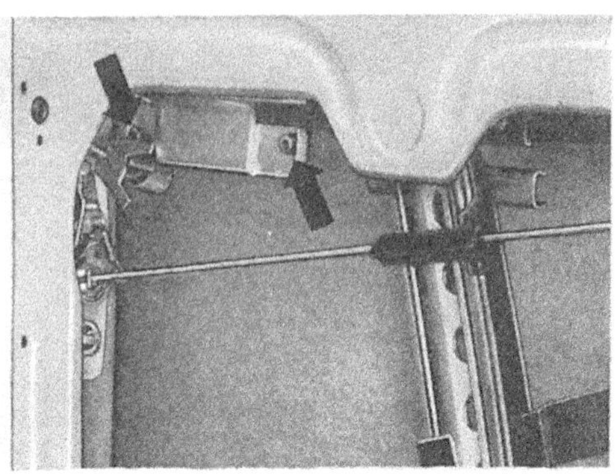

3 - Unscrew two hex. head screws on outer door handle and remove handle.

4 - Move lock latch to vertical position and unscrew three door lock Phillips screws.

5 - Pull door lock downwards out of door.

Installation

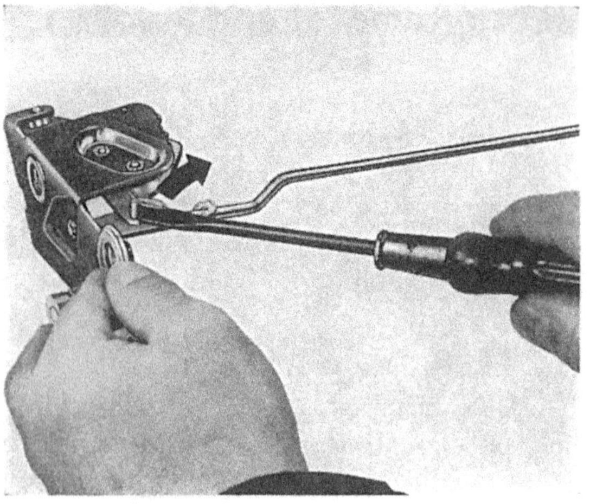

1 - Check all parts for wear and install new ones if necessary. Lubricate moving parts. If necessary, detach pull rod spring from door lock and replace pull rod or inner control separately.

2 - Install door lock, ensuring that lock latch is vertical.

3 - Install inner control, pushing it in the elongated holes in door hinge direction until a slight resistance is felt.

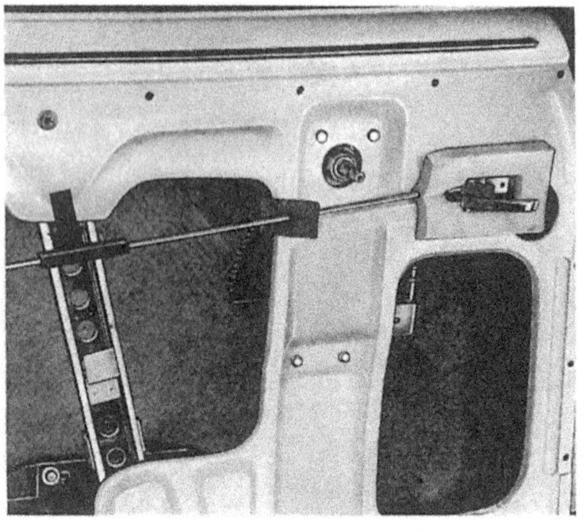

> **Important**
>
> Prior to further assembly of the door parts, operate the opening and locking lever to check whether the lock opens, closes and locks correctly. If this is not the case, the length of the pull rod must be corrected by moving the inner control.

4 - Check whether piece of felt and foam rubber pad on door inner surface are correctly positioned. If necessary, reattach them.

5 - Check whether anti-rattle rubber on pull rod is correctly positioned and replace if necessary.

6 - Completely *assemble door. Check operation* of door window regulator and whether window regulator handle is correctly positioned.

Removal

1 - Remove door trim panel and pull off plastic sheet.

2 - Unscrew four hex. head screws for window lift channel and push window downwards.

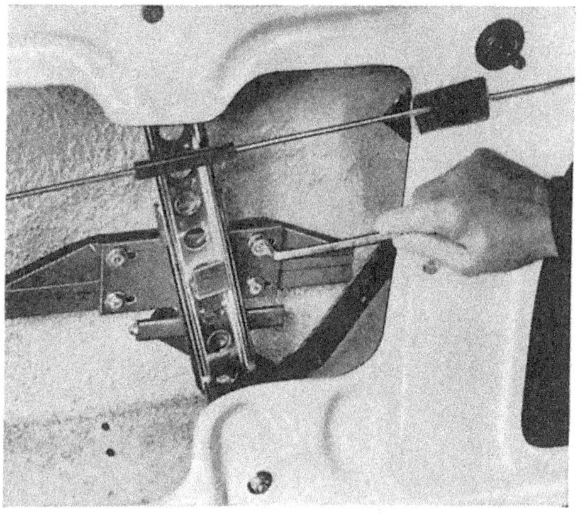

3 - Carefully remove outer trim moulding and window aperture rubber seal together with clips from door, to avoid damage to paintwork.

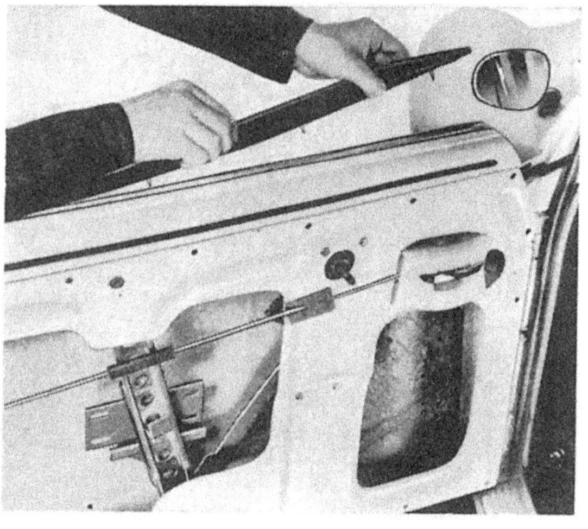

4 - Using a screwdriver, carefully pry off window inner guide strip press studs ensuring that paintwork is not damaged, then remove inner guide strip.

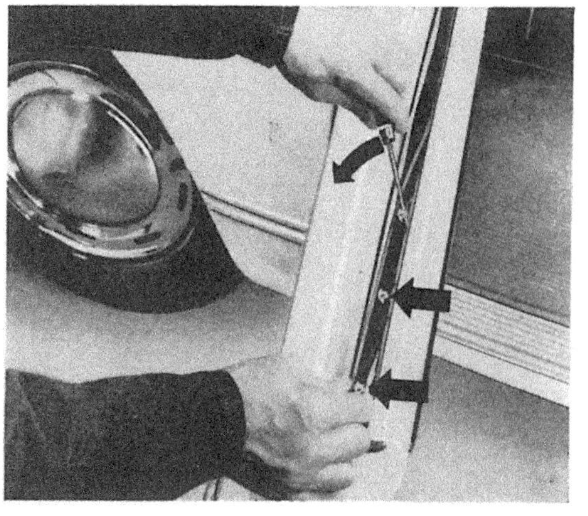

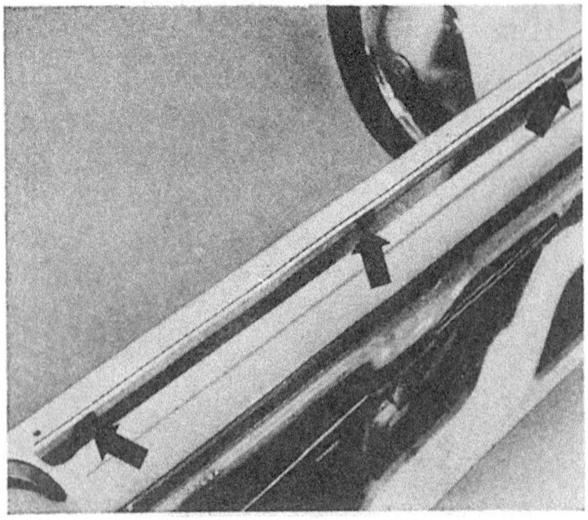

5 - Pry out seven window glass guide plugs.

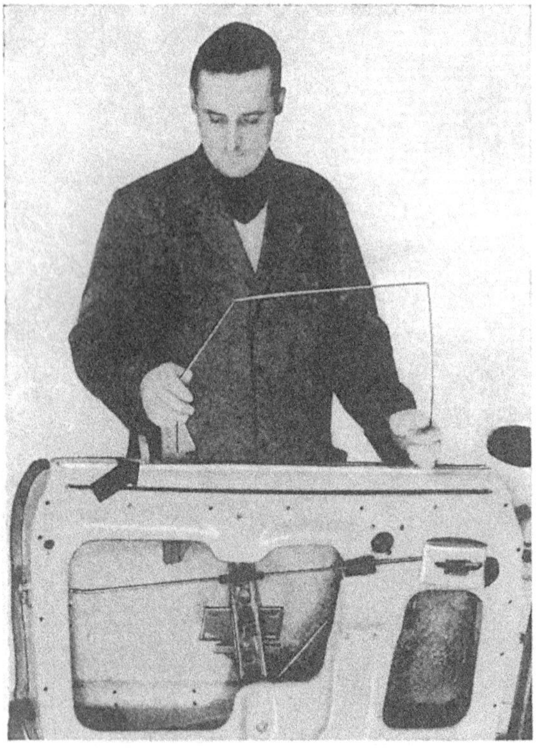

6 - Turn window 90° and remove from above.

Note:

If the window lift channel jams in the window aperture, press the sheet metal panels apart with a wooden block.

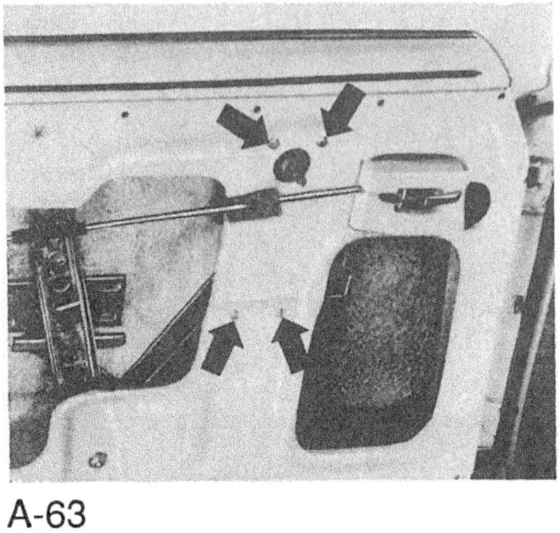

7 - Unscrew four window regulator hex. head screws, press regulator against door outer panel and push downwards.

8 - Pull window regulator linkage and spring (arrow) out of window lift channel slide to one side, then remove window regulator.

9 - Unscrew one Phillips screw and one hexagon nut from guide channel threaded pin and remove guide channel and window lift channel slide.

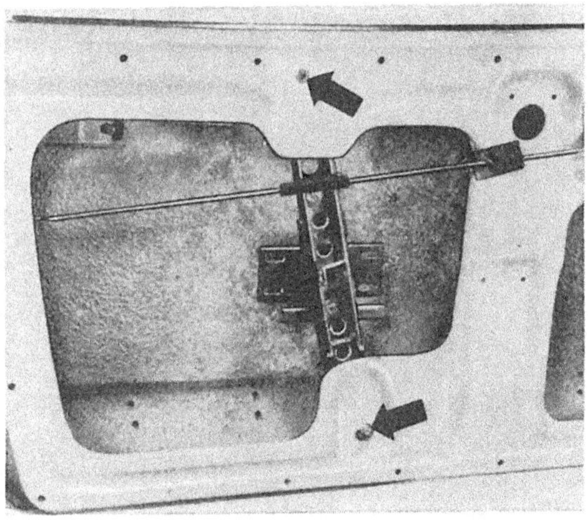

Installation

If the window glass has to be replaced, the window lift channel can be detached with a rubber hammer and an appropriate wooden block. The lift channel can be installed in the same way.

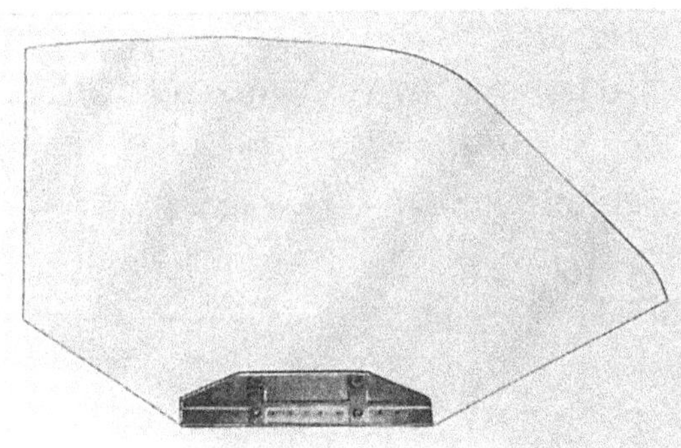

6/66

Note:

Only trim molding, Part No. 141 853 333 C, will be available for vehicles manufactured up to March 1960 (Chassis No. 2 960 113).

Fitting instructions:

A - **New molding / old weatherstrip**
Increase diameter of molding attaching holes in door outer panel to .16 in. (8 mm). New holes to suit the holes for the retaining clips in the molding must be made in the weatherstrip.

B - **New molding / new weatherstrip**
Increase diameter of molding retaining clip holes in door outer panel to .16 in. (8 mm). Alter position of holes in new weatherstrip as the retaining clips must be moved.

1 - Check condition of trim molding and window slot rubber seal, window inner guide strip, window glass guide plugs, window lift channel and seal as well as window regulator and install new parts if necessary.

Check clearance between runners of window lift channel slide and guide channel. Eliminate excessive clarance by pressing runners together.

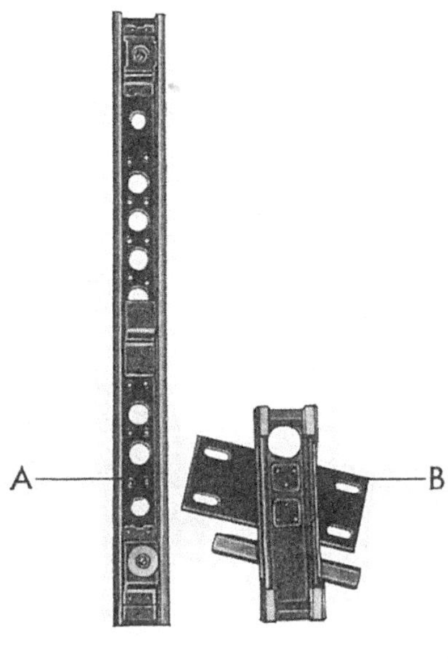

A - Guide channel
B - Window lift channel slide

2 - Grease all moving parts of window regulator sparingly prior to installation.

Note:

Loosen adjusting screw and locknut (arrow).

3 - Thinly coat sliding surfaces of guide channel with grease prior to installation.

4 - Install guide channel and window regulator.

5 - Insert window glass into slot from above and push downward.

6 - Install window glass guide plugs, outer trim molding and window slot rubber seal as well as inner channel.

7 - Attach window lift channel to slide.

8 - Adjust door window glass.

Detailed instructions are contained in the section "Adjusting Door Window Glass".

9 - Install plastic sheet and door trim panel, ensuring that window regulator handle is in correct position.

10 - Check function of all parts by operating them several times.

Adjusting Door Window Glass

Prerequisite for correct window glass adjustment is that the door fits correctly in the door opening.

Vertical adjustment

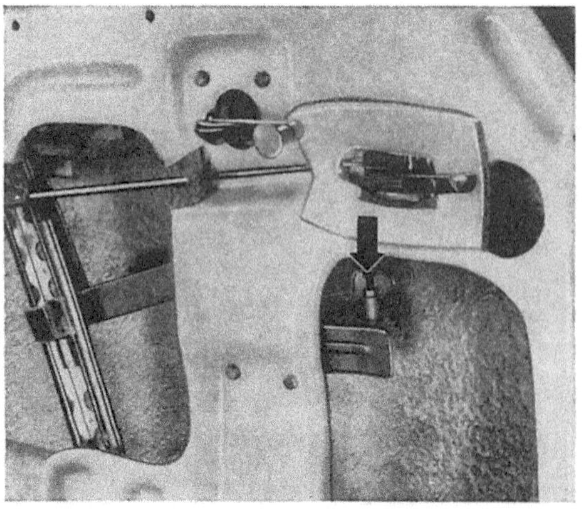

1 - Back-off locknut of adjusting screw on window regulator.

2 - Adjust window vertically by turning adjusting screw.

Counterlockwise — lowers window

Clockwise — raises window

3 - Check for correct vertical adjustment of window glass with door closed. Set adjusting screw by tightening locknut.

Longitudinal Adjustment

1 - Check that the moulding and weatherstrip between door and quarter window are properly positioned and secure. If necessary, tighten moulding screws.

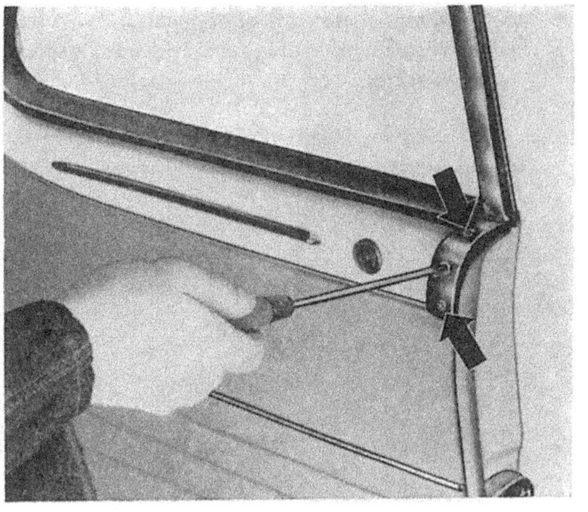

Note:

If the Phillips screws in the door panel do not hold, use screws of larger diameter.

2 - Check clearance between runners of window lift channel slide and guide channel. Excessive clearance can also be eliminated with parts installed, by pressing runners together.

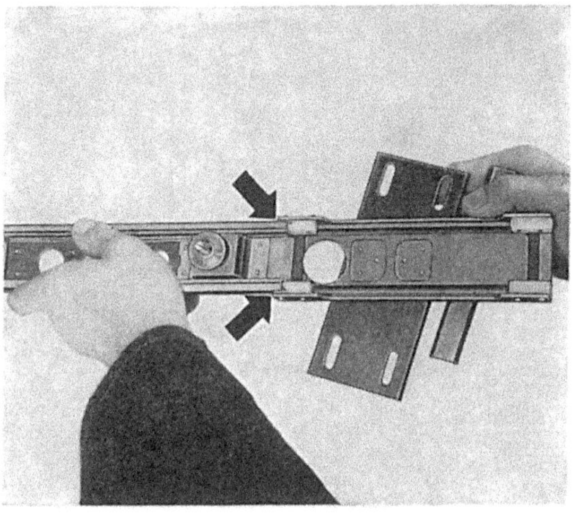

3 - Ensure that clearance between door window glass and outer lip of door and quarter window weatherstrip is uniform.

A - Door window glass
B - Weatherstrip
C - Moulding between door and quarter window
D - Quarter window

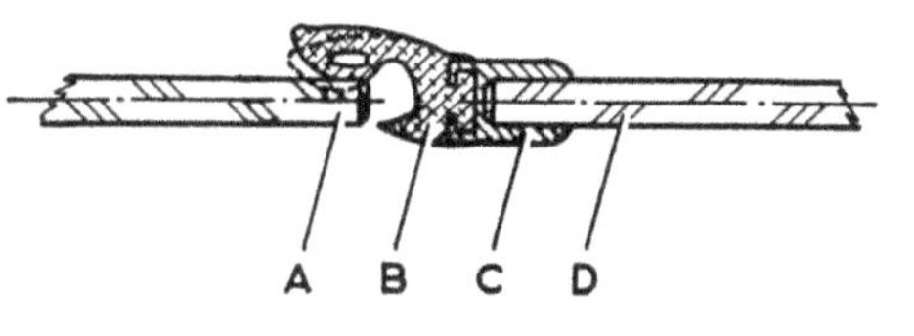

4 - Back-off four hex. head bolts on window lift channel.

5 - Move window glass in window lift channel towards front or rear.

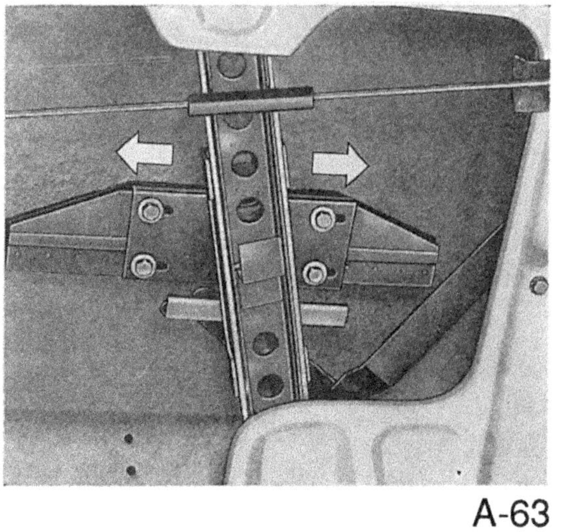

6 - Tighten hex. head bolts.

Side Adjustment

1 - Check door window for side rock. If necessary, bend sheet metal of door panel, to which window glass guide plugs are attached, towards window glass.

2 - Back off one Phillips screw and one hex. nut on guide channel threaded pin.

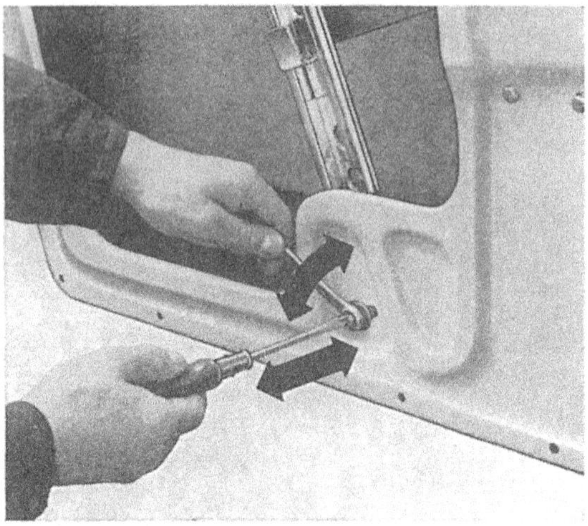

3 - Correct position of window glass by screwing the guide channel threaded pin in or out.

The location for correct position of window glass is given by the weatherstrip on roof and quarter window.

4 - Tighten Phillips screw and hex. nut. When locking hex. nut, hold threaded pin with screwdriver.

Guide Plug Adjustment

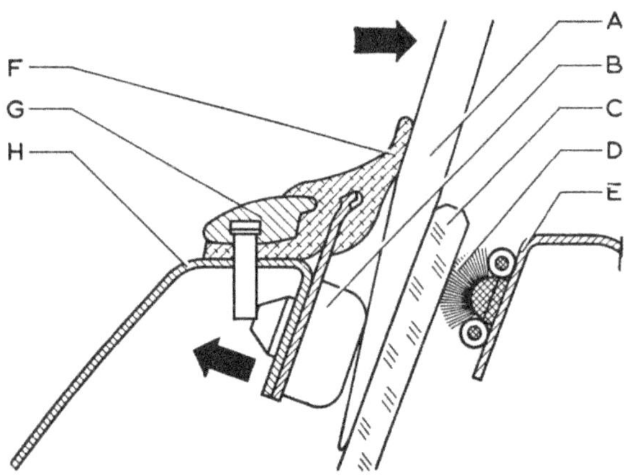

The movement of the glass in the door window aperture is additionally controlled by plastic guide plugs. If these plugs press unevenly or too heavily on the glass they will cause squeaking noises when the window is raised or lowered. Excessive pressure can also cause scratches if dirt lodges between plug and glass.

To avoid such damage, the plugs should be set uniformly. If the plug pressure is too great, it can be reduced by inserting a thin, wooden wedge between plug and glass and pressing the plug outwards.

A - Wooden wedge
B - Plastic plug
C - Window glass
D - Inner guide strip
E - Door inner panel
F - Weatherstrip
G - Moulding
H - Door outer panel

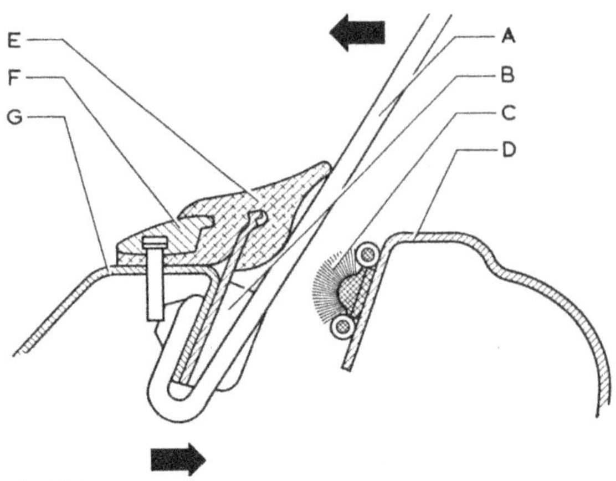

If a guide plug does not contact the glass, the support should be pulled in at the appropriate location. This can be done with a piece of .8×.12 in. (20×3 mm) strip steel bent to form a hook at one end.

If the center guide plug has to be adjusted, the window glass must be removed.

A - Bending hook
B - Plastic guide plug
C - Inner guide strip
D - Door inner panel
E - Weatherstrip
F - Moulding
G - Door outer panel

Removal

1 - Remove windshield wiper arms.

2 - Starting in one of the top corners, push windshield and weatherstrip outward.

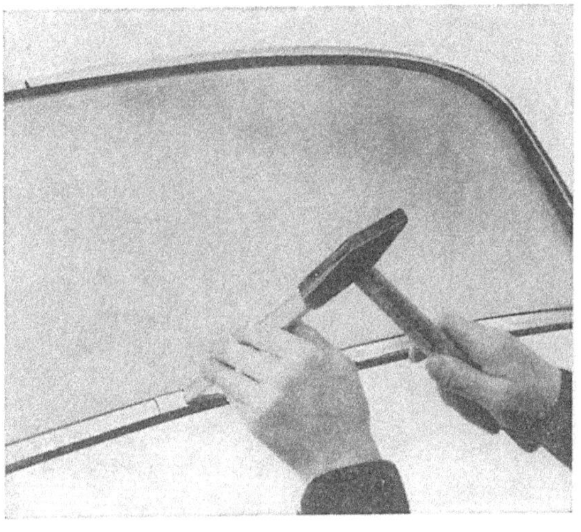

3 - Slide upper and lower trim molding sleeves to one side with a piece of wood so that the molding ends are free.

4 - Carefully remove molding halvesfrom weatherstrip.

5 - Remove weatherstrip from windshield.

Installation

1 - Remove traces of old cement from windshield opening in body with benzine and touch up damage to paintwork.

2 - Check condition of weatherstrip and fit a new one if necessary.

3 - Check molding halves, straighten or replace if necessary.

4 - Fit weatherstrip to glass. The ends of the weatherstrip should meet in the center at the top.

Fitting molding

1 - Insert a length of cable, preferably a piece of .08—.12 in. (2—3 mm) diameter electric cable with insulation, into groove for molding. This can be done by passing the cable through a piece of tube flattened slightly at one end. Place the flat end of the tube in the groove and pass it right round the windshield so that the cable lies in the groove in the weatherstrip. The ends of the cable should meet in the center of the lower edge of the windshield.

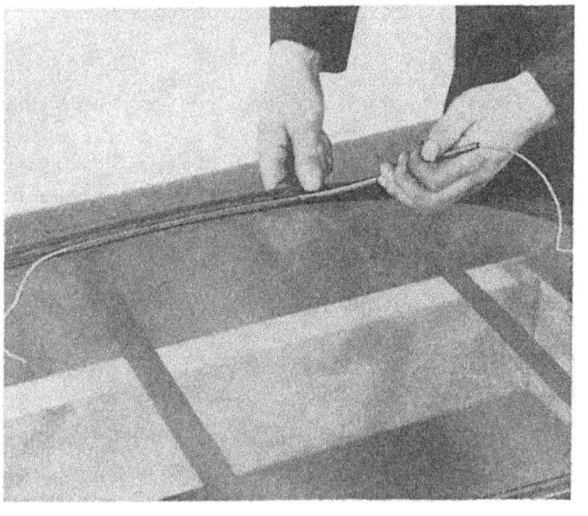

2 - Starting at cable ends, place molding in weatherstrip. Pull cable out slowly and follow round, pressing molding into groove.

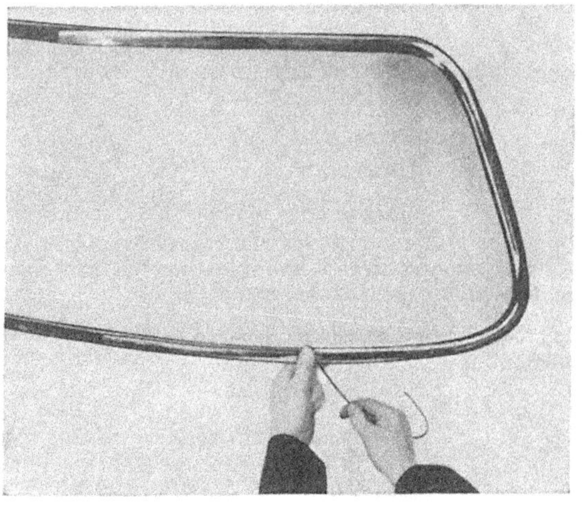

3 - Push both sleeves over molding ends.

Fitting windshield

1 - Insert cable into outer slot in weatherstrip so that cable ends meet in the center of the windshield lower edge.

2 - Powder weatherstrip with talcum or thinly coat it with glycerine. Coat inside of sealing lip, which bears against body outer panel after installation of windshield, with genuine VW glazing cement D 10.

3 - Check that paint drain holes are well sealed and replace rubber plugs if necessary.

4 - Place glass into frame from outside.

The cable ends must hang down inside.

5 - Two mechanics are required to install the windshield. Starting at one end of the cable, one mechanic pulls the lip of the weatherstrip over the frame edge, keeping the cable at right angles to inside surface of the glass, while the second mechanic strikes the glass from outside with his hand, following the pull of the cable, and seats the windshield evenly all round.

6 - Remove excess sealing compound from windshield and windshield frame with benzine. Check windshield for leaks, especially at the radii, and clean it with methylated spirits.

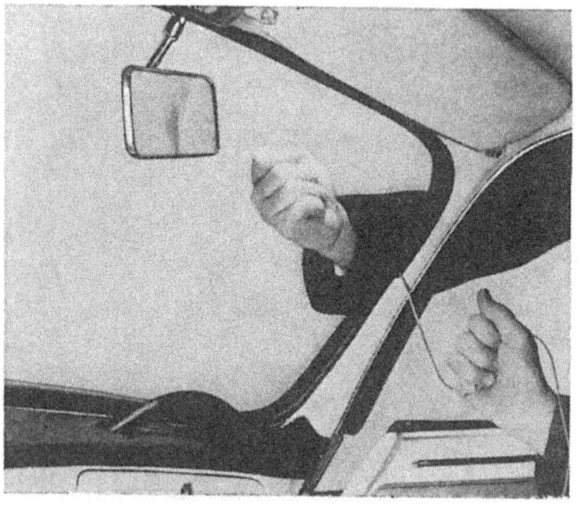

Important

If a laminated glass windshield is being installed instead of the toughened glass type, pressure must be applied evenly with both hands to seat the glass while the second mechanic lifts the lip over the frame edge with the cable. Rubber hammers must not be used owing to the danger of cracking the glass.

Removing and Installing Rear Window

The rear window is removed and installed in the same way as the windshield.

Check that paint drain holes are well sealed and replace rubber plugs if necessary.

2/66

Note:

On 6 July 1965, from Chassis No. 145 908 542 (Model 141/142) and on 13 January 1966, Chassis No. 146 530 703 (Model 143/144), the weatherstrip, trim frame and sleeve for the windshield and the rear window were modified to attain an even better sealing, especially of laminated glass.

As the new trim frame is of T-shaped cross section instead of the previous U-shape, previous and new parts cannot be interchanged.

Description	New Part No.
Weatherstrip for windshield	143 845 121 A
Left half of trim frame for windshield	143 853 305 A
Right half of trim frame for windshield	143 853 306 A
Sleeve for trim frame	143 853 309 A
Weatherstrip for rear window	143 845 521 A
Left half of trim frame for rear window	143 853 355 A
Right half of trim frame for rear window	143 853 356 A
Sleeve for trim frame	143 853 309 A

THIS PAGE INTENTIONALLY
LEFT BLANK

Quarter Window Removal and Installation

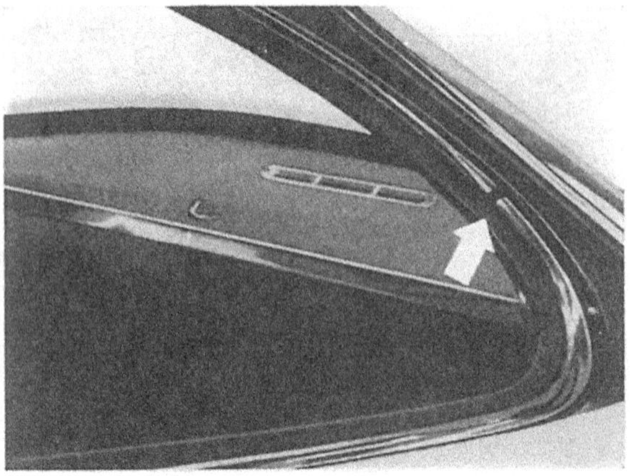

Removal

1 - Detach weatherstrip and trim moulding for door window at the top.

2 - Pull quarter window out forwards. Take weatherstrip and channel off the glass.

3 - Push moulding sleeve to one side with a piece of wood so that moulding ends are free.

4 - Push trim moulding forwards and take off.

5 - Take off weatherstrip and unscrew retaining channel.

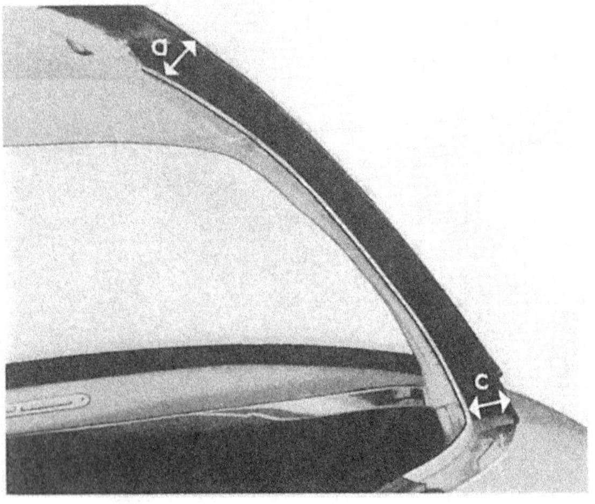

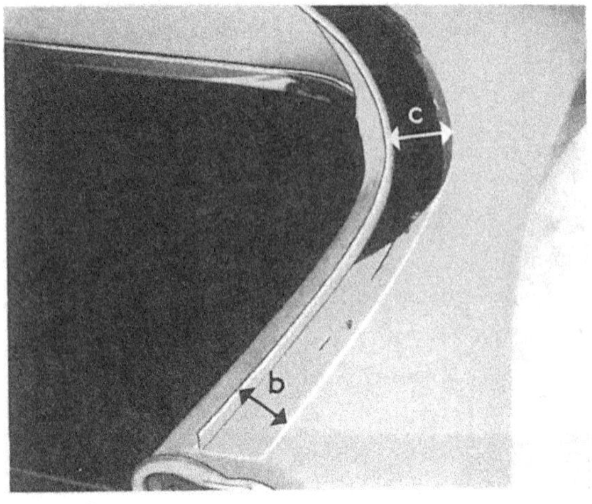

Installation

1 - Install retaining channel, observing the measurements given as far as possible.

a = 30 mm (1.18")
b = 34 mm (1.34")
c = 38 mm (1.50")

2 - Coat retaining channel on the outside with plastic sealer D 15.

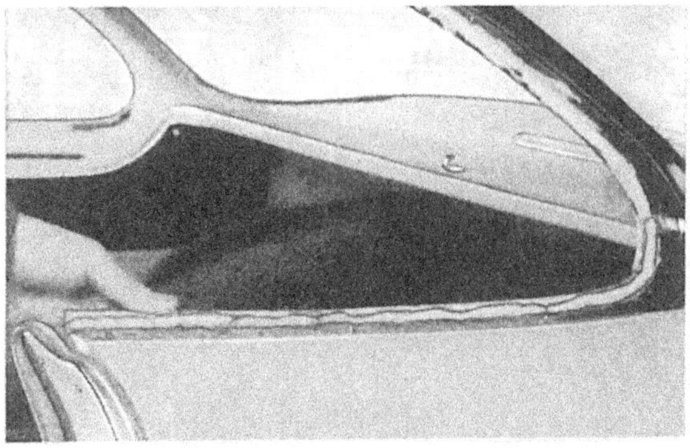

3 - Insert trim moulding and slide sleeve over the ends.

4 - Place weatherstrip in the quarter window opening.

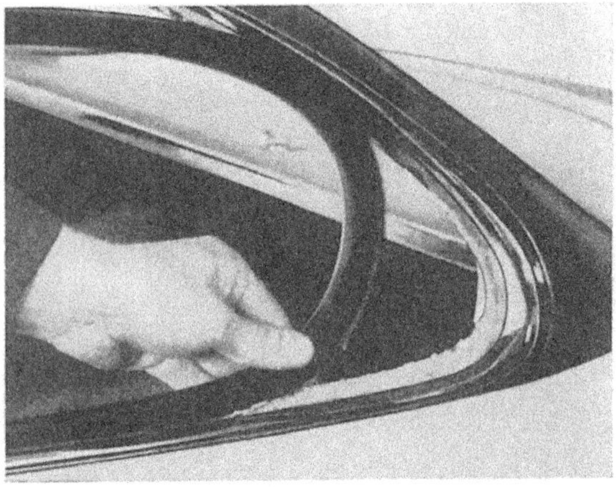

5 - Insert quarter window glass in the weatherstrip and lift the lip over the edge of the glass with a wooden wedge.

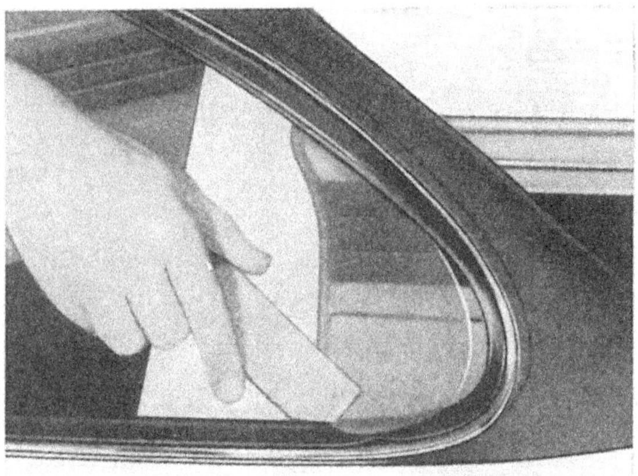

6 - With a screw gun and a suitable nozzle force VW Window Cement D 10 between the quarter window glass and the weatherstrip.

7 - Tap the glass fully into the quarter window opening with a rubber hammer.

8 - Place a strip of insulating tape on the straight edge of the window.

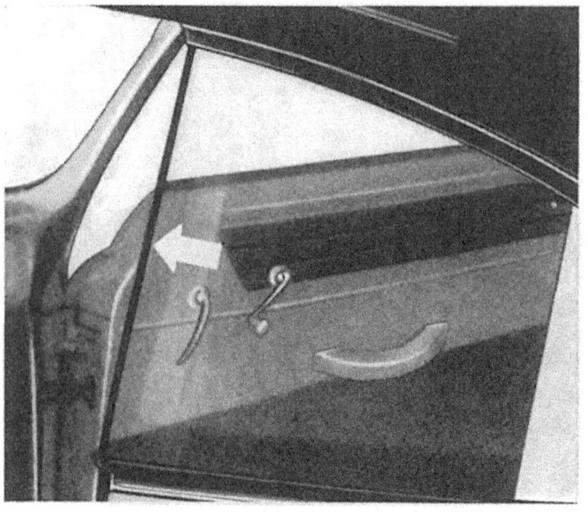

9 - Install channel with weatherstrip. Screw door window trim moulding to the roof frame.

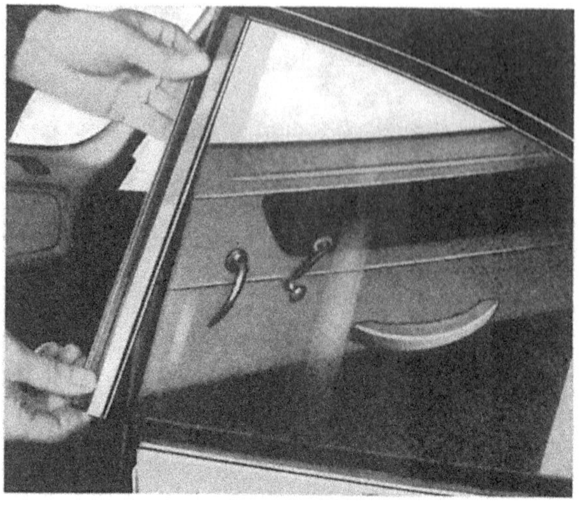

Quarter Window Removal and Installation

(Hinged)

Removal

1 - Open window.

2 - Unscrew the three toggle retaining screws inside the body.

3 - Pull weatherstrip out of the channel at the door side.

10 - With a pointed punch, drive quarter window ledge trim molding toward front until the three slots in the molding guide are accessible.

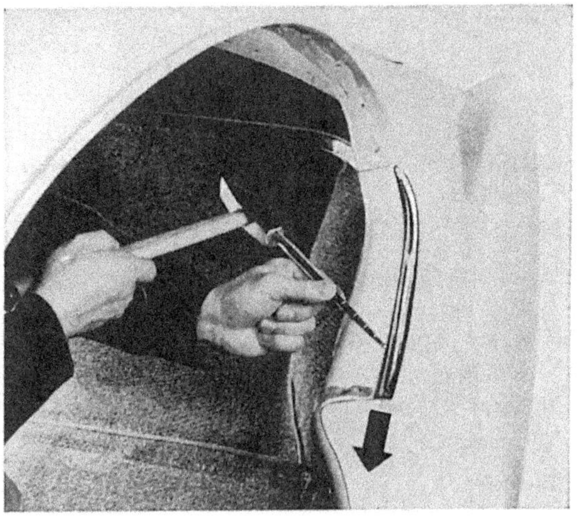

Note:

If necessary, drill a hole approximately .2 in. (5 mm) deep in the trim molding to support punch.

11 - Remove trim molding by lifting it up.

Installation

1 - Detach remnants of old sealing strip from window opening in body and from quarter window ledge trim molding. Use benzine if necessary.

2 - Check channel between quarter window and door window as well as trim molding and weatherstrips and replace if necessary.

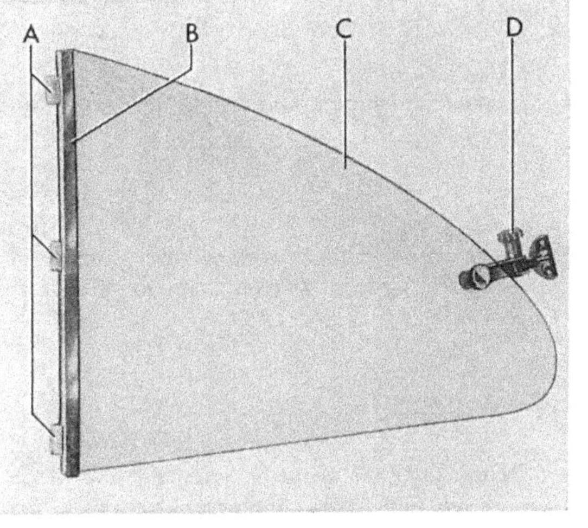

A - Connecting pieces
B - Quarter window glass channel
C - Window glass
D - Window fastener

3 - Cement sealing strip under quarter window trim frame and quarter window ledge trim molding.

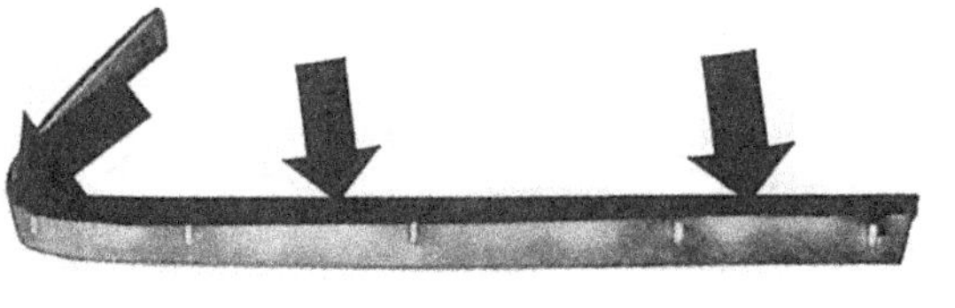

4 - Check condition of three countersunk head screws for holding quarter window ledge trim molding and use new ones if necessary.

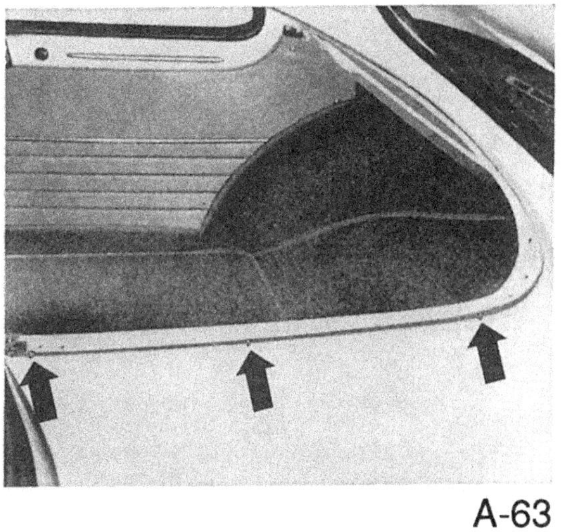

5 - Attach quarter window trim frame in window opening.

6 - Press quarter window weatherstrip into window trim frame.

Installation

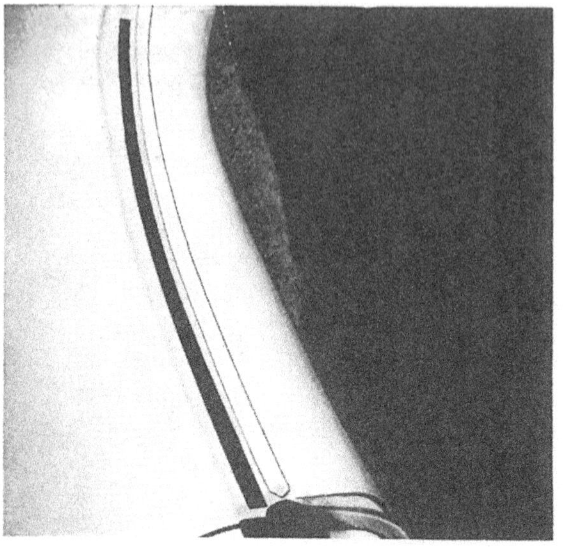

1 - Remove all traces of the old, hard sealer from the body.

2 - Place a piece of sealing strip under the trim moulding and a length of sealing band under the trim frame.

3 - Secure trim moulding on the quarter window shoulder.

4 - Secure trim frame in quarter window opening.

5 - Check condition of weatherstrip, renew if necessary. Insert weatherstrip in the trim frame.

6 - Secure channel between quarter and door windows with three countersunk screws at the bottom and to the roof frame with one hexagon head screw.

7 - Check condition of connection pieces and rubber hinge. Renew if necessary.

8 - Lay rubber hinge round the channel for the quarter window glass.

9 - Place quarter window in the frame and secure with three countersunk screws.

10 - Screw toggle to body and check the contact between glass and weatherstrip.

11 - Insert weatherstrip between quarter and door window into the channel.

Note:

Since March 1966, the Parts Department have no longer been supplying the quarter window, Part No. 143 845 301 A / 302 A, and the quarter window channel, Part No. 143 847 311 / 312 separately. In place of these, only quarter windows with cemented-on channels, Part No. 143 847 041 / 042, are available.

Note:

Quarter windows still in stock can be used after carrying out the following:

If, prior to cementing, there is too large a gap between window and channel, other parts from stock must be paired. Only if both parts fit together properly can they be firmly cemented with **WEVO two component metal cement.**

This cement is not obtainable from our Parts Department. If necessary it can be obtained directly from the manufacturer, Messrs. WEVO-Gesellschaft, 7 Stuttgart, Postfach, Germany. Exact instructions are enclosed with the metal cement.

Front Seat Removal and Installation

Removal and Installation

1 - Lift up the locking mechanism on the right hand side of the seat and push the seat forward out of the guide rails on the floor.

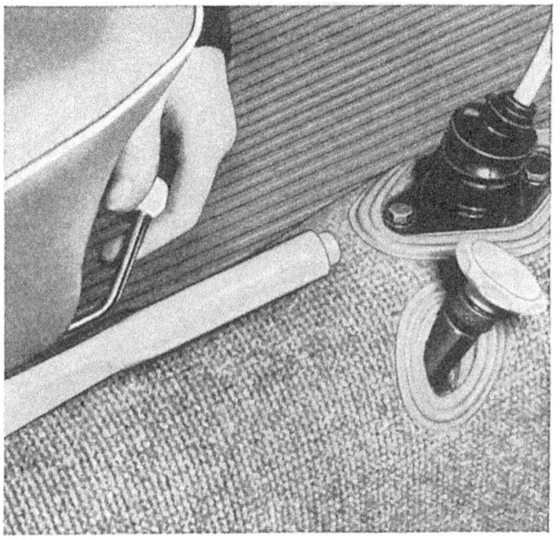

2 - Take the seat out.

Grease the guide rails with VW Universal Grease before installing the seat.

Adjusting Cam Removal and Installation

1 - Remove seat.

2 - Knock the pin out of the left or right hand adjusting cam with a punch.

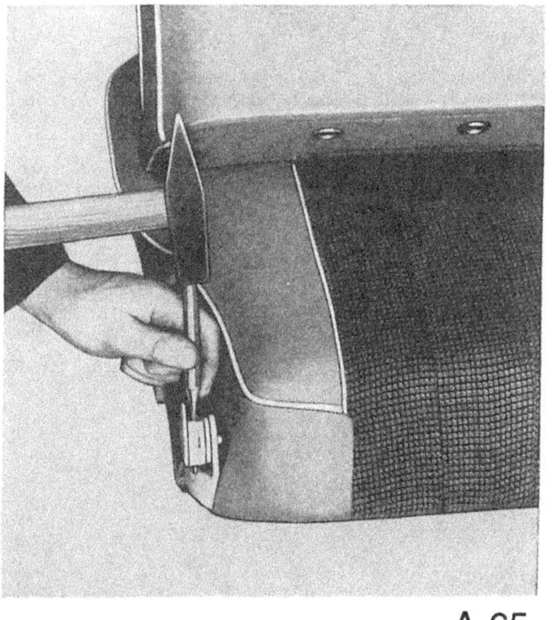

3 - Take cam off.

Note:

The adjusting cams must not be interchanged.

Before reassembly the cams should be lightly greased with VW Universal Grease at the face which contacts the guide tube.

Emergency Seat Removal and Installation

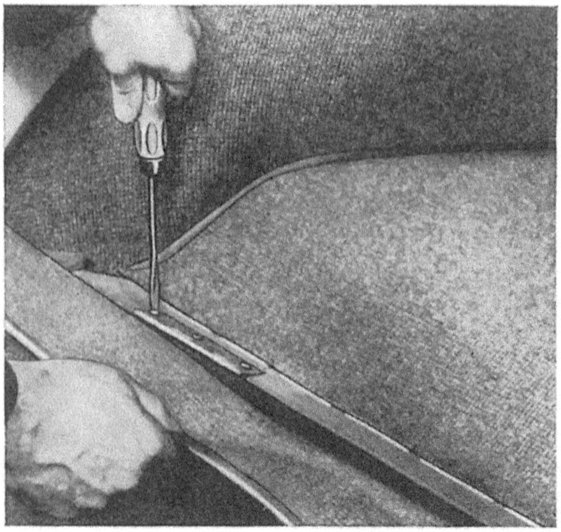

Removal

1 - Lift emergency seat and take it out of the support channels.

2 - Fold emergency seat back rest down. Lift luggage compartment lining on the reverse side of the back rest slightly and unscrew four hinge screws on each side.

3 - Take out back rest.

Installation

1 - Check back rest retaining strap for damage and renew if necessary.

2 - Screw back rest to hinges and cement lifted part of luggage compartment lining into position with VW Universal Cement D 12.

Note:

From Chassis No. 3 192 507 (1st August 1960), the spring interiors and frames of the front seats and back rests were modified to improve the seating position.

1 - Front seats

The modified spring interiors (Part No. 141 881 305 B / 306 B) and frames (Part No. 141 881 107 B/108 B) cannot be installed in the previous pattern seats.

The upholstered padding has been altered to suit the shape of the spring interiors and frames. The new padding (Part No. 141 881 375 C/376 C) cannot be installed in the previous front seats.

The new seat complete (Part No. 141 881 031 B/032 B) can, however, be installed in all vehicles manufactured previously.

2 - Front seat backrests

The modified spring interiors (Part No. 141 881 705 B) back rest frames (Part No. 141 881 511 B/512 B) and padding (Part No. 141 881 775 B) cannot be installed in backrest of the previous pattern.

The new backrests complete (Part No. 141 881 041 B/ 042 B) will not fit front seats of the previous pattern.

Ventilation Control Removal and Installation

Removal

1 - Disconnect battery.

2 - Position both levers of the ventilation control so that they can be pushed through the openings in the instrument panel. Unscrew one cap nut (arrow).

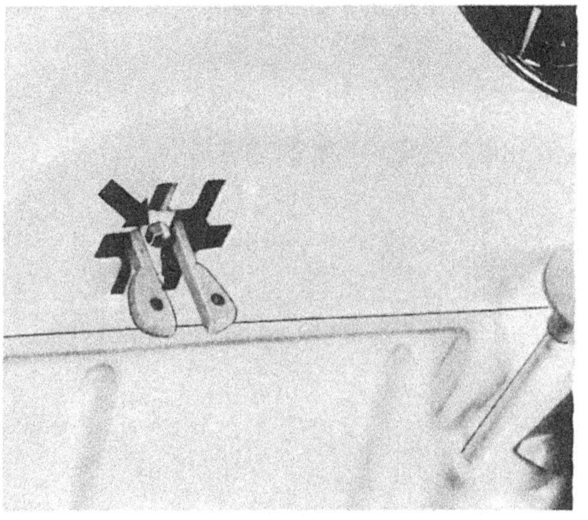

3 - Open front hood and remove luggage compartment cardboard lining.

4 - Pry circlip off left and right lever, detach ventilation control cables from levers and remove ventilation control.

Installation

1 - Check rubber seal, circlips and levers of ventilation control and replace if necessary.

2 - Carefully insert ventilation control into instrument panel and secure.

3 - Connect left and right cable to levers and secure with circlips.

4 - Install luggage compartment cardboard lining and connect battery.

Mixing Tube Removal and Installation

Removal

1 - Disconnect battery, open front hood and remove luggage compartment cardboard lining.

2 - Loosen hex. head screw on throttle valve shaft and detach ventilation control cable.

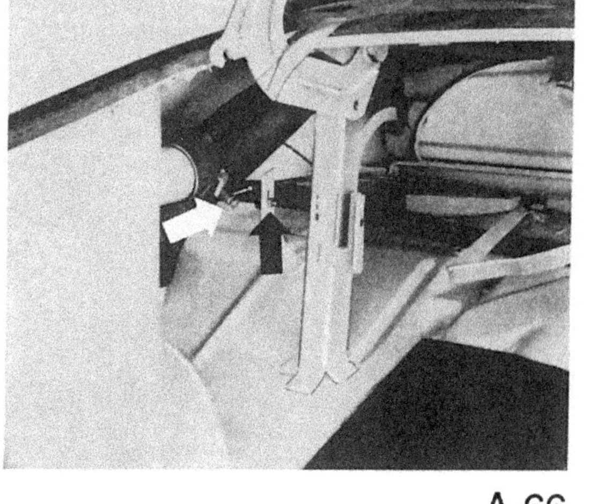

White arrow: Cable to throttle valve attachment.

Black arrow: Guide tube to bracket attachment.

3 - Press three rubber connections off mixing tube and remove mixing tube.

Installation

1 - Check condition of mixing tube, throttle valve and shaft as well as rubber connections and use new parts if necessary. The pivoting locations of the throttle valve can be lubricated with a few drops of oil.

2 - Insert mixing tube and attach it to defroster vent, warm and fresh air outlets with rubber connections.

3 - Insert ventilation control cable into throttle valve shaft and secure with hex. nuts.

Note:

To adjust, hold both levers in the "off" position and tighten the hex. nuts in both throttle valves when the ventilation control valves are fully closed.

4 - Connect battery and check operation of ventilation on a road test. If necessary, readjust cables.

Fresh Air Inlet Trim Frame Removal and Installation

Removal

1 - Remove three retaining screws.

2 - Take trim frame out of front panel opening.

3 - Remove three screws and take grille and flyscreen off.

Installation

1 - Check condition of trim frame, rubber seal, grille and screen and renew parts as required.

2 - Place flyscreen and grille in the fresh air opening and secure with three screws.

3 - Place rubber seal round trim frame.

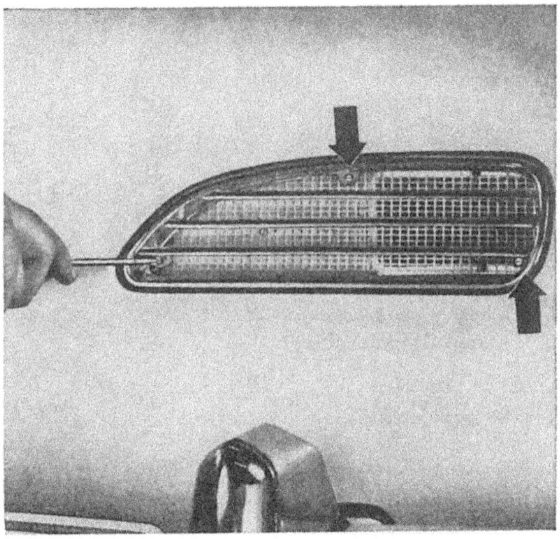

Insert trim frame in fresh air opening and secure with three chrome-plated screws.

Front Side Panel Trim Mouldings

Removal and Installation

1 - Push moulding clips out of the holes in the front side panel from inside and remove moulding.

2 - Renew bent or damaged moulding clips.

3 - Seal moulding clips inside the front side panel with plastic sealer D 15.

Door Trim Mouldings

Removal

1 - Remove window regulator handle and inner door handle.

2 - Remove door trim panel.

3 - Remove plastic sheet.

4 - Remove plastic sealer from the moulding clips.

5 - Push the moulding clips out of the holes from inside the door and take moulding off.

Installation

1 - Renew bent or damaged moulding clips.

2 - Seal moulding clips inside the door with plastic sealer D 15.

Quarter Panel Trim Mouldings

Removal

1 - Remove side trim panel.

2 - Remove plastic sealer from moulding clips.

3 - Push moulding clips out of the holes from inside the rear quarter panel and remove moulding.

Installation

1 - Renew bent or damaged moulding clips.

2 - Seal the clips with plastic sealer D 15.

Outside Rear View Mirror

Removal

1 - Unscrew two hexagon head screws from cover plate on hinge pillar.

2 - Remove cover plate and rubber seal.

3 - Unscrew two hexagon head screws on the inside of the front side panel and remove mirror and packing.

Installation

1 - Bolt mirror, with packing and reinforcement plate, to side panel.

2 - Check condition of cover plate rubber seal and renew if necessary.

3 - Secure cover plate to hinge pillar with two hexagon head screws.

VW and "Karmann Ghia" Sign Removal and Installation

Removal and Installation

1 - Pull speed nuts off and remove signs.

2 - Check speed nuts and signs and renew if necessary.

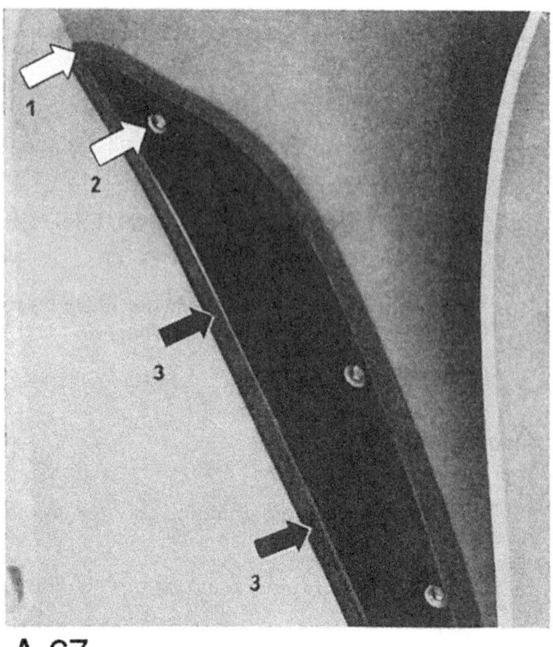

Note:

From chassis No. 3 412 252 (22nd November 1961), the upper corner (1) of the hinge pillar cover plate (Part No. 143 805 377/378) under the front side panel, is welded up. This seals the cover plate to the hinge pillar better.

Furthermore, an additional mounting has been provided on the hinge pillar so that the cover plate can be more rigidly fixed by a third securing screw (2). This will ensure uniform pressure on the rubber seal.

When installing an external mirror or a radio aerial, the hinge pillar cover plate must be taken out. Ensure that the openings at the upper edge and inner side of the plate (3) are filled with Plastic Sealing Compound D 14 and sealed with Genuine VW Sealing Compound D 17 when the plate is re-installed.

Note:

From Chassis No. 4 764 158, a "VOLKSWAGEN" sign (Part No. 361 853 687) is fitted to the rear hood.

Removal and Installation of Luggage Compartment Lining

Removal

1 - Fold emergency seat back rest down.

2 - Remove luggage compartment lining.

3 - Remove sound-absorbing felt from luggage compartment.

Installation

1 - Before cementing the sound-absorbing material into position, thoroughly remove all traces of old cement and felt. Benzine is suitable as a solvent.

2 - Coat the floor with VW Universal Adhesive D 12 and cement the sound-absorbing felt smoothly into position.

3 - Cement the carpet on to the felt.

Headlining Removal and Installation

Removal

1 - Remove windshield, quarter window and rear window.

2 - Remove door window weatherstrip trim moulding and quarter window trim frame.

3 - Remove interior rear view mirror and coat hooks.

4 - Dismantle interior lamp.

5 - Pull headlining off. The lining is cemented into position and must be pulled carefully to avoid damage.

6 - Unhook stretcher bars and take lining out.

Installation

1 - Check that the sound-absorbing felt on the roof is firmly in position, and refix with VW Universal Cement D 12 if necessary.

2 - Clean remains of cement from the roof side members, windshield and rear window frames. Benzine is suitable as a solvent.

3 - Hook stretcher bars in so that headlining is lightly tensioned and free from creases.

4 - Coat roof side members, windshield and rear window frames lightly with VW Universal Adhesive D 12 and cement headlining into position without creases.

5 - Reinstall all parts removed.

Front Roof Pillar Lining Removal and Installation

1 - Remove windshield.

2 - Remove door window weatherstrip trim moulding.

3 - Loosen headlining slightly and pull roof pillar lining off.

4 - Thoroughly clean remains of cement from the roof pillar before refixing lining. Benzine is suitable as a solvent.

5 - Cement roof pillar lining in position with VW Universal Adhesive D 12, without creases.

6 - Reinstall all parts removed.

Rear Roof Pillar Lining Removal and Installation

1 - Remove rear and quarter windows.

2 - Remove quarter window trim frame.

3 - Loosen headlining slightly and pull roof pillar lining off.

4 - Thoroughly remove remains of cement from roof pillar before refixing lining. Benzine is suitable as a solvent.

5 - Cement door pillar lining in position with VW Universal Adhesive D 12, without creases.

6 - Reinstall all parts removed.

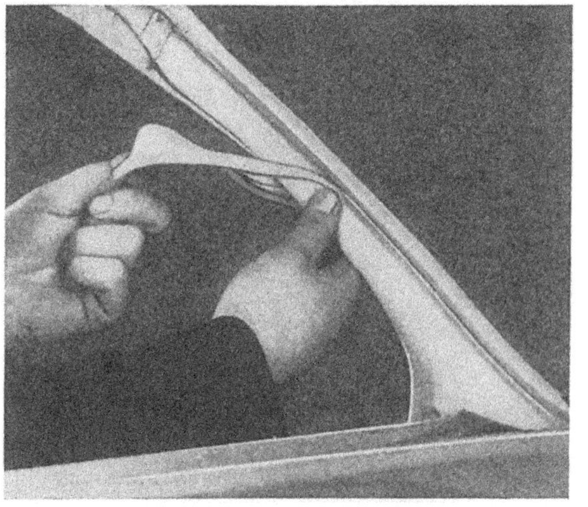

Quarter Panel Lining Removal and Installation

1 - Fold emergency seat back rest down.

2 - Unscrew two countersunk tapping screws.

3 - Pull quarter panel lining and clips from the inner panel.

4 - The quarter panel lining trim moulding can be removed after bending up the retaining lugs.

5 - Renew broken clips.

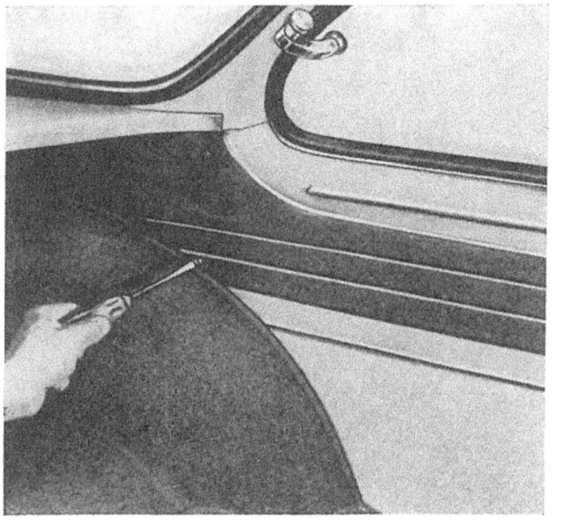

Front Panel Lining Removal and Installation

The right and left front panel linings can be lifted after the retaining channel lip has been bent up. Pull the remainder of the lining off. Unscrew the retaining channel.

The following points must be observed during installation:

1 - Thoroughly remove the remains of cement from the inside of the body before refixing the lining. Benzine is suitable as a solvent.

2 - Install retaining channel.

3 - Cement front panel lining into place with VW Universal Adhesive D 12. Take care that the lining makes good contact with the body all round and is correctly located round the warm air outlets.

4 - Bend the lip of the retaining channel down again.

Body Side Member Carpet Removal and Installation

1 - Remove eight screws and take off scuff plate.

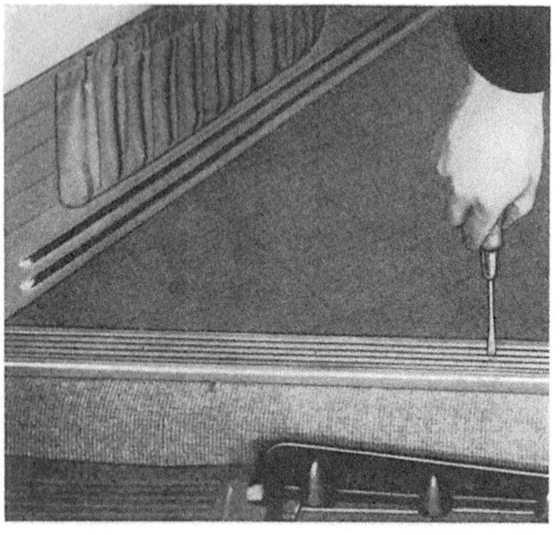

2 - Remove carpet nail and pull carpet off.

3 - Thoroughly clean the remains of cement and sealing compound from the body side members before refixing the new carpet. Benzine is suitable as a solvent.

4 - Cement carpet in position with VW Universal Adhesive D 12.

5 - Place a strip of sealing band D 16 under the scuff plate and reinstall plate.

6 - Secure the rear end of the carpet with a carpet nail.

Glove Compartment Removal and Installation

1 - Open front hood and remove luggage compartment fiber board lining.

2 - Unscrew retaining strap screw and take out glove compartment.

3 - Remove two hinge screws on glove compartment lid and take lid off.

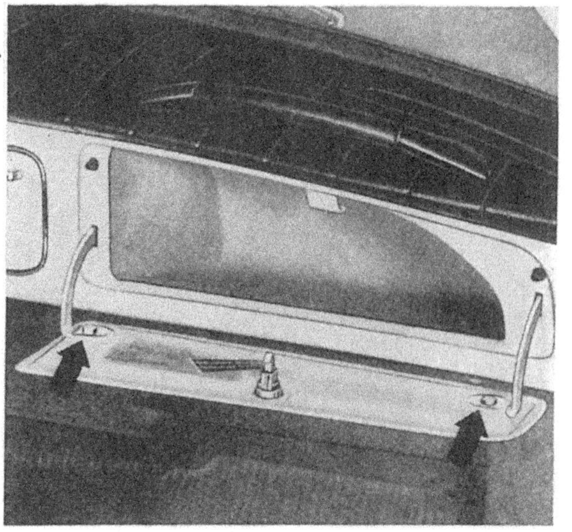

4 - To detach the hinges it is necessary to remove the compartment lid spring and the two riveted pins.

5 - The two rubber plugs for the lid can be pulled out.

6 - The lid lock can be removed with circlip pliers VW 122 b or a suitable tool of local manufacture.

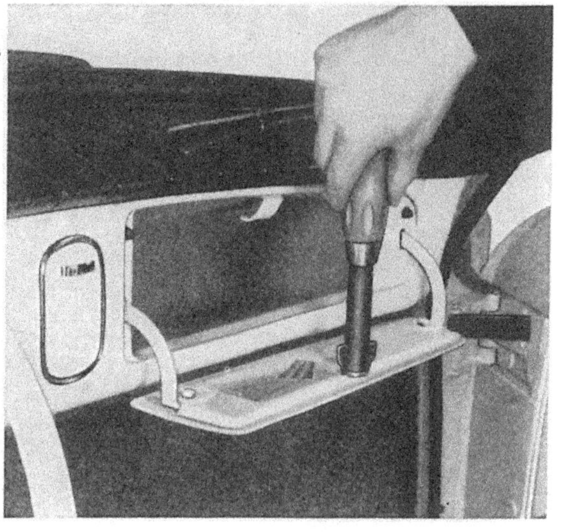

7 - When installing the glove compartment and tightening the retaining strap, take care that the glove compartment is correctly seated against the instrument panel. If contact between glove compartment and instrument panel is poor, rectify with adhesive tape.

8 - Check condition of rubber plugs for lid and renew if necessary.

9 - Check lid alignment by opening and closing several times.

10 - Apply a few drops of oil to the lid lock.

11 - If the lid is difficult to open or close bend the sheet metal catch slightly.

Package Shelf Covering Removal and Installation

Removal

1 - Remove rear window.

2 - Remove emergency seat back rest hooks and retainer plate for warm air outlet.

3 - Remove package shelf trim moulding.

4 - Carefully pull off the package shelf covering.

The cardboard is cemented on and can be damaged if insufficient care is taken when pulling off.

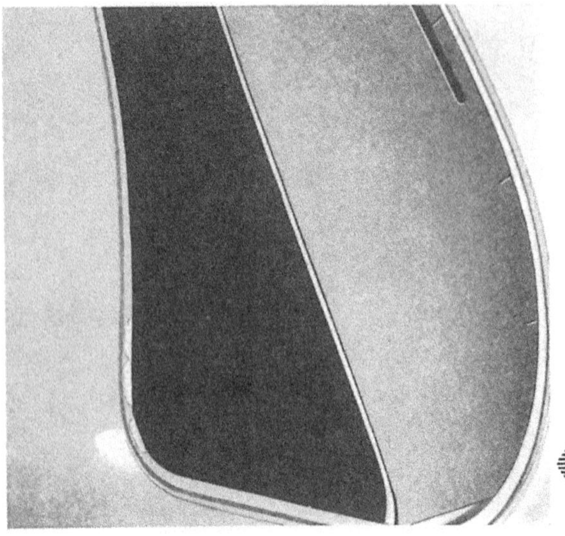

Installation

1 - Thoroughly remove all cardboard and cement remains from the package shelf before refixing the cardboard. Benzine is suitable as a solvent.

2 - Coat package shelf with VW Universal Adhesive D 12 and cement cardboard smoothly into position.

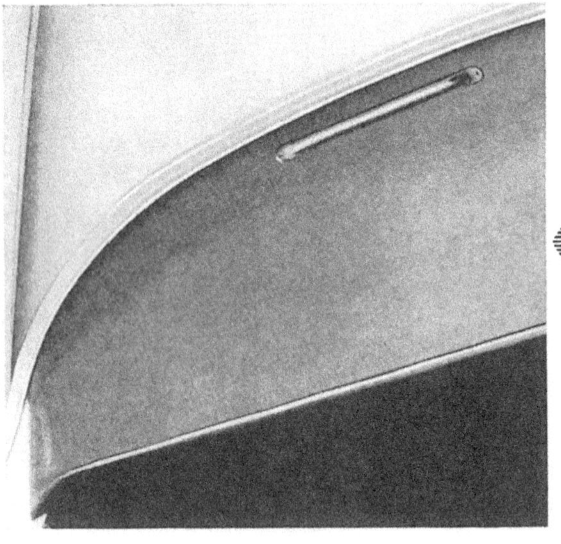

3 - Coat cardboard with adhesive and cement covering into position smoothly and without creases. Cut out opening for the warm air outlet.

4 - Reinstall all parts removed.

Instrument Panel Cover Removal and Installation

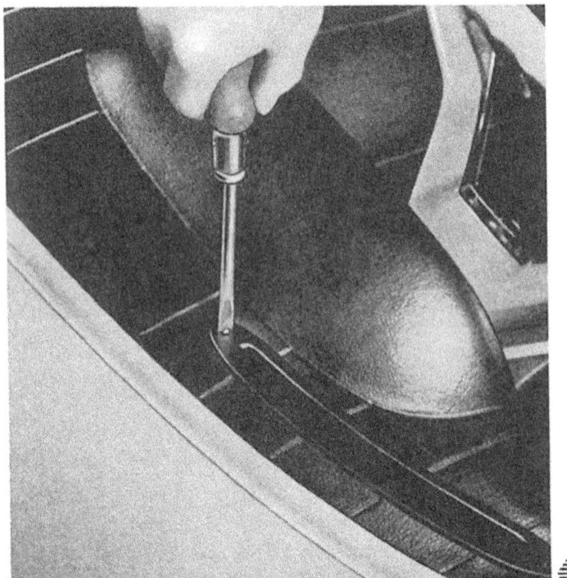

Removal

1 - Remove windshield.

2 - Remove fiberboard lining from luggage compartment.

3 - Remove glove compartment.

4 - Remove passenger's grab handle.

5 - Remove loud speaker opening insert.

6 - Remove two screws from each defroster vent retainer plate.

7 - Unscrew retaining strip. Loosen roof pillar lining slightly and pull off instrument panel cover.

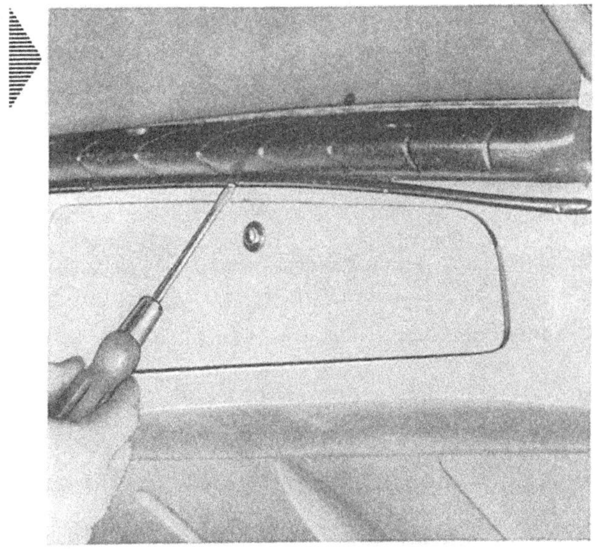

Installation

1 - Clean the cement remains thoroughly from the instrument panel with fine sand paper.

2 - Coat the contact area of instrument panel evenly with BOSTIK 675 or 475. Allow the adhesive applied to the instrument panel 5 or 6 minutes to dry off. The application of the adhesive to the instrument panel cover must be carried out so that the drying time of 2 to 3 minutes is not exceeded. These drying times must be strictly adhered to in order to ensure good adhesion. Apply two coats of adhesive around the defroster vents.

3 - Place instrument panel cover in position, press down and smooth evenly from the loud speaker outwards to both sides, taking care that the various openings line up correctly. The adhesion must be particularly good at the outer edges.

It is essential to press the instrument panel cover down evenly and thoroughly, as BOSTIK is a contact adhesive which bonds only under pressure.

4 - Place rubber spacers under passenger's grab handle with the thicker spacer in the recess nearest the center of the vehicle. The rounded edges of the spacers must be at the bottom, facing the panel cover (see arrow).

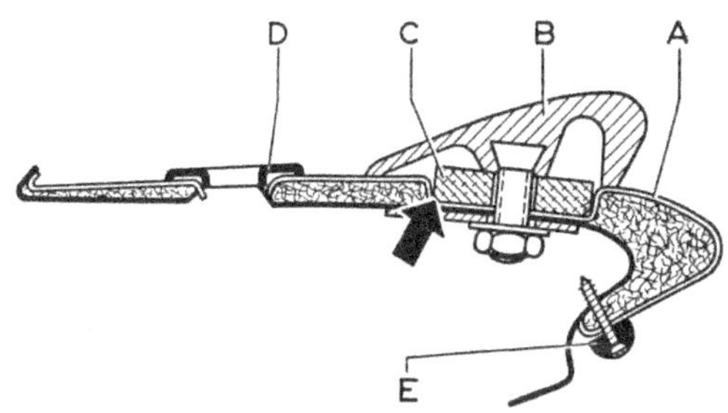

A - Instrument panel cover
B - Passenger's grip
C - Spacer
D - Retaining plate
E - Retaining strip

5 - Install defroster vent retainer plates and retaining strip.

6 - Paint all screw heads with matt black lacquer.

7 - Reinstall all parts removed.

Note:

Only BOSTIK 675 or 475 should be used for fixing the instrument panel cover. This adhesive can be obtained from the following firms:

Australia: Boston Blacking Chemical Company of Australia (Pty.) Ltd., Epping Road, Thomastown, Victoria.

Canada: Boston Blacking Chemical Company of Canada Ltd., 2610, Bennet Avenue, Maisonneuve, Montreal P.Q.

Great Britain: Boston Blacking Chemical Co. Ltd., Ulverscroft, Leicester.

South Africa: Boston Blacking Chemical South Africa (Pty.) Ltd., Newark Street, Port Elizabeth.

U.S.A.: Boston Blacking Chemical Co., 784 Memorial Drive, Cambridge 39, Mass.

A supply of BOSCODUR A is included in every consignment of adhesive. One part of BOSCODUR A must be mixed with nine parts of BOSTIK 675 or 475 shortly before application. The mixture must be used within 8 hours of being prepared otherwise it will solidify.

THIS PAGE INTENTIONALLY
LEFT BLANK

Door Window Weatherstrip Trim Moulding Removal and Installation

Removal

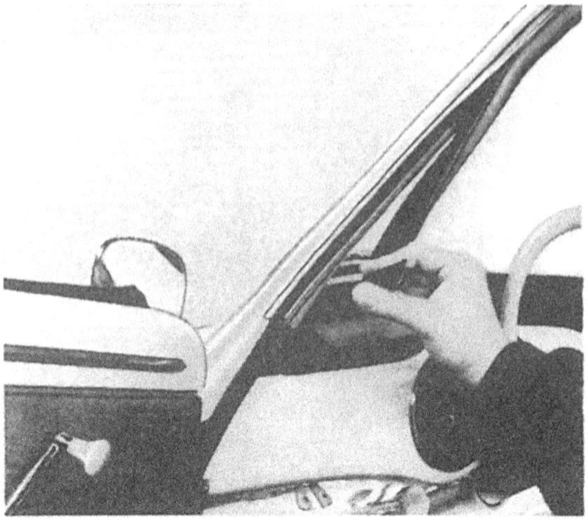

1 - Pull door window upper weatherstrip out of trim moulding channel.

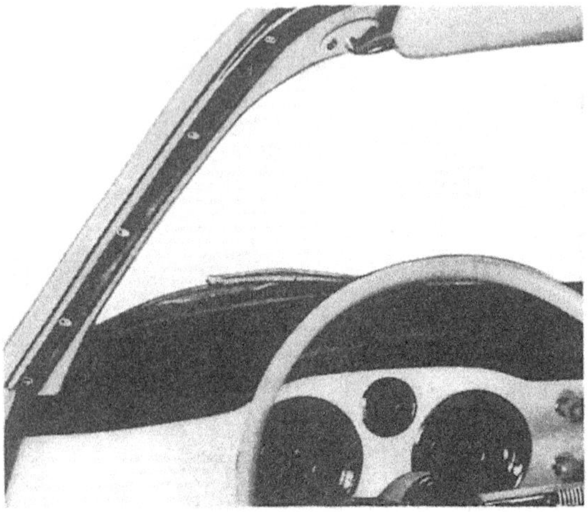

2 - Unscrew Phillips tapping screws.

3 - Remove trim moulding.

4 - Remove remains of old sealing strip from door opening or from trim moulding. If necessary, use benzine.

Installation

1 - Check condition of trim moulding, weatherstrip and Phillips tapping screws and use new parts if necessary.

2 - Cement sealing strip to outer side of trim moulding.

3 - Screw trim moulding to roof frame and adjust it in elongated holes so that the door window glass seals well against the weatherstrip.

Heavy springing action when the steering is locked hard over can cause the front wheels to rub at the sharp edge of the upper portion of the wheel arch. This can be rectified as follows:

Place a suitable dolly on the inner edge of the wheel arch. The edge should be turned up over a length of 400 mm (16") with a rubber hammer as shown on the sketch.

a = 400 mm (16")

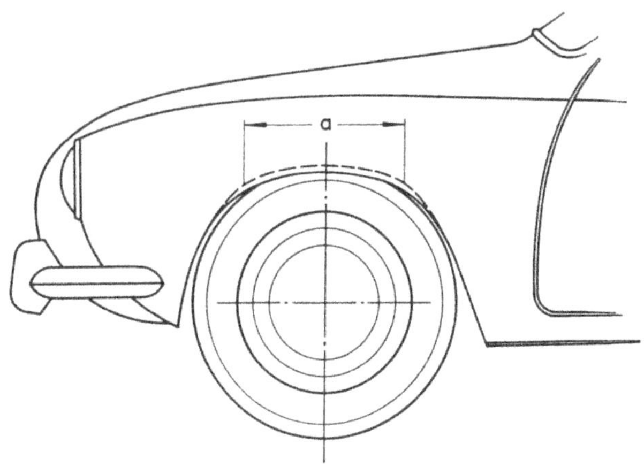

Note:

From 1st August 1960 and Chassis No. 3 192 507 the front wheel housings, the luggage compartment panel and the front reinforcement plate were modified to suit the shape of the new fuel tank.

1 - The wheel housings — Part No. 141 805 071 B/072 B — now have recesses in the upper portions. They can be installed in older pattern vehicles and the former parts will be discontinued when present stocks are exhausted.

2 - The fuel tank support surfaces in the luggage compartment panel — Part No. 143 805 405 A — are angled. As the modified luggage compartment panel cannot be installed in older pattern vehicles, the former parts will remain available.

3 - The front reinforcement plate — Part No. 141 805 503 A — is now located deeper in the body and has a larger recess to accomodate the water and brake fluid containers. As the new parts cannot be installed in older pattern vehicles, the former parts will remain available.

Leaks at quarter window

Leaks at the quarter window can be located and eliminated as follows:

There are five places at which leakage can occur and they should each be sprayed in turn with a weak jet of water.

1 - Spray the bottom of the channel between quarter and door window for about 5 minutes. If no water appears on the inside of the window, there is no leakage at this point.

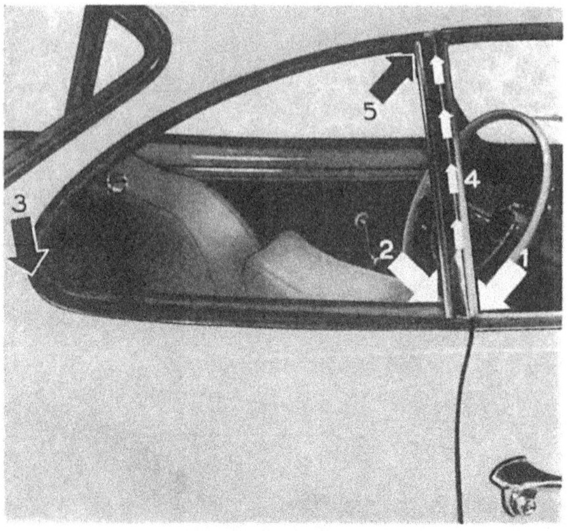

Take care that the water does not spread to the next point as it is then difficult to locate the leak exactly.

Remedy if leaking:

Seal the angle weatherstrip with Window Cement D 10.

2 - Spray the bottom end of the quarter window weatherstrip.

Remedy if leaking

Remove window, fill the joint between weatherstrip and channel with Plastic Sealer D 15 and coat with Sealing Compound D 17. Shorten or tuck-in the upper end of the beading on the lock pillar so that its ends about 5 mm/.2" short of the inside of the channel and cannot absorb moisture at this point.

3 - Spray the lower window weatherstrip from the rear end of the trim moulding.

Remedy if leaking

Remove window, weatherstrip and retaining channel. Clean off old sealer and, if necessary, trim the headlining material so that it does not extend beyond the retaining channel screw holes and absorb moisture.

Apply a 5 mm layer of Plastic Sealer D 15 to the outside of the row of screw holes. Coat the underside of the retaining channel with adhesive, install it and tap it lightly into position with a block of wood.

4 - Spray the joint between the vertical weatherstrip and the channel between quarter and door window from bottom to top.

Remedy if leaking

Install the vertical weatherstrip with Window Cement D 10.

5 - Spray the joint between the door window weatherstrip on the roof frame and the vertical weatherstrip between quarter and door windows.

Remedy if leaking

Check the diagonal joint between the door window weatherstrip and the vertical weatherstrip. Fill the joint between channel and door window weatherstrip with plastic filler. Join the upper end of the vertical weatherstrip to the door window weatherstrip with adhesive D 12.

Note:

From Chassis No. 3 273 002 (27th September 1960), a rubber packing (Part No. 143 847 347) was installed at the upper end of the channel between the door and quarter window to eliminate noises in the roof frame.

To install the packing in older vehicles, proceed as follows:

1 - Pull the rear end of the upper door window weatherstrip out of the trim moulding for about 20 cm/8".

2 - Remove the countersunk screw (B) securing the channel between quarter and door window to the roof frame.

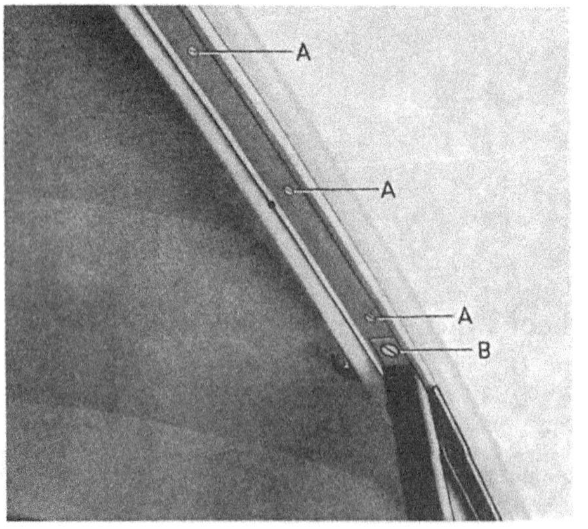

3 - Remove the three rear screws (A) for the upper door window weatherstrip and bend the trim moulding down.

4 - Bend the channel between quarter and door window forward as shown by the lower arrow and slide the rubber packing (C) into the trim moulding for the quarter window (see upper arrow).

5 - Coat the quarter window with D 12 Adhesive and bend the channel between quarter and door window back against the quarter window.

6 - Replace the countersunk screw securing the channel to the roof frame.

7 - Bend the trim moulding upwards against the roof frame, replace the screws and press weatherstrip into position.

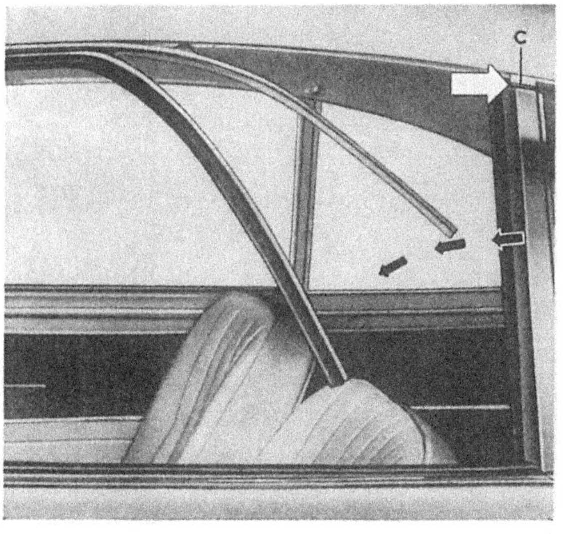

Note:

1 - Heating modifications

From Chassis No. 5 188 470, the rear foot well heating was modified so that it can be controlled with regulating levers.

Branch pipes — Part No. 141 255 431/432 — with attached heater outlets — Part No. 141 255 515 — which end in the reinforcement panel, are installed in place of the connecting pipes. The branch pipes are insulated.

The foot well heating can be service installed in previous vehicles.

All bodies supplied as spares are fitted with branch pipes. When installing these bodies on previous chassis the foot well heating can be installed at the same time. In this case, clearance holes should be made in the reinforcement panel. The connections on the branch pipes can, however, be welded up.

2 - Door modifications

From August 1963, Chassis No. 5 718 489, the inner lock controls were replaced by the same type as is used on the Type 3.

The following parts have been modified:

Designation	New Part Number
1 - Door	143 831 051 E/052 E
2 - Operating lever	311 837 018/020
3 - Pull rod	141 837 193 A/194 A
4 - Door lock	141 837 015 B/016 B
5 - Striker plate	141 837 295 A/296 A
6 - Door handle	141 837 205 A/206 A
7 - Lock pillar	141/143 837 395 B/396 B

Striker plate

A - old

B - new

Door lock

A - old

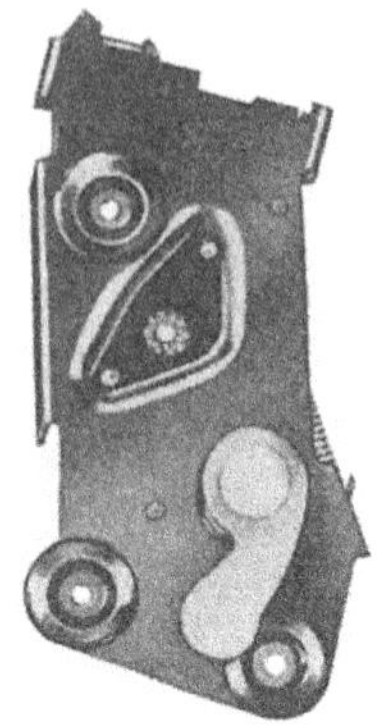

B - new

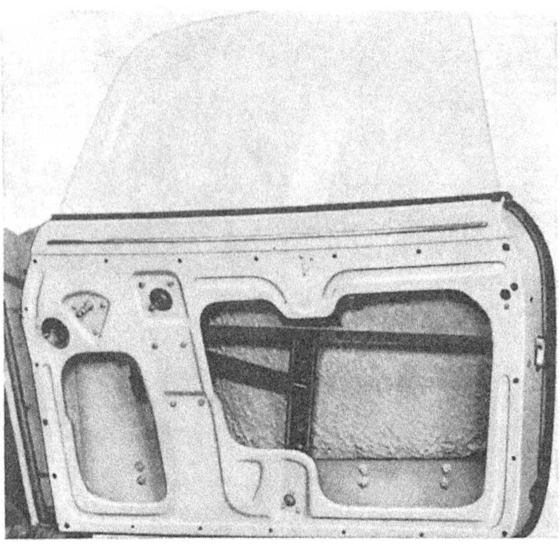

A - previous pull rod

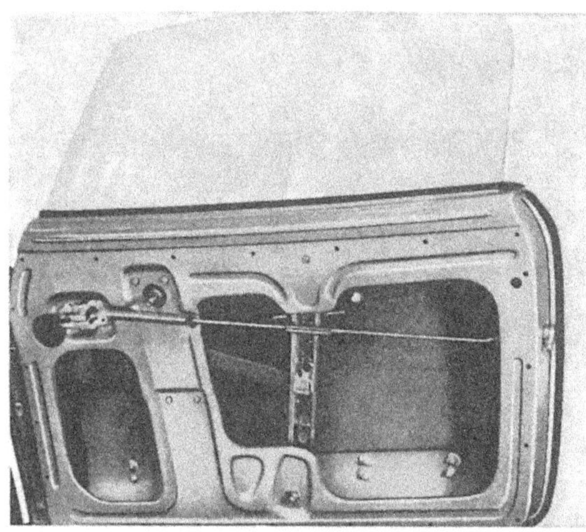

B - new pull rod with modified operating lever

The modified parts cannot be installed in previous vehicles.
The former parts will still be supplied as spares.

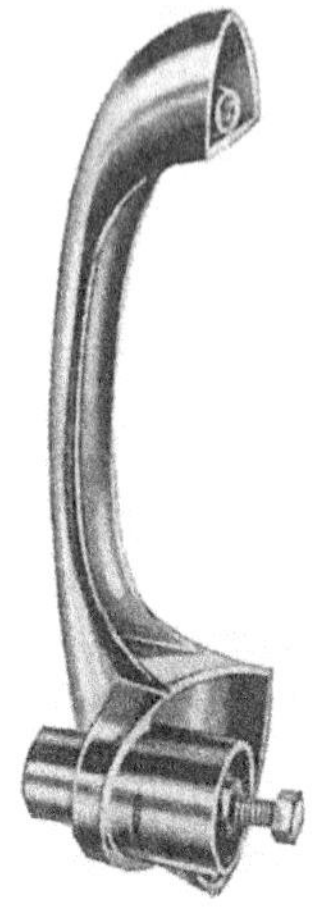

A - previous door handle

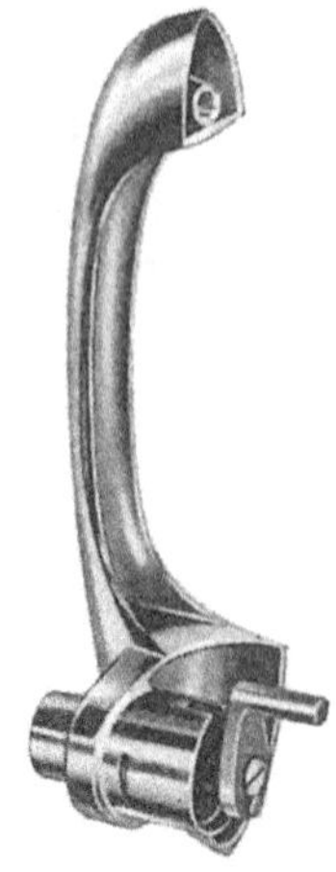

B - new door handle

General

The description of the replacement of individual body sections serves as a guide for the sequence of operations. A reasonable knowledge of body repair is essential and the work should wherever possible, be carried out by a skilled man, well versed in body repair.

As opposed to the Sedan or VW Convertible, the body sheet metal of both Karmann Ghia Models must often be butt welded. This work in particular calls for great care during repairs, as lack of skill and material knowledge can cause structural alteration, stress and distortion in the thinly walled paneld which is impossible or very difficult to eliminate.

Where gas welding is not expressly stipulated it is mentioned in the individual sections in addition to the more generally used spot welding.

It must be emphasised however, that spot welding with spot welding tongs or a push type welding gun is preferable to gas welding.

For various operations, especially the replacement of large body parts, the body must be taken off. The use of the specially designed repair frame is recommended as it guarantees the body measurements being kept accurate when replacing body parts and enables any distortion to be detected and eliminated. The frame can be used for all Passenger Car Models and is described in detail, with other gauges and jigs, under the heading "Workshop Equipment" in "Body General" section.

Before commencing to cut out any sheet metal parts it is essential to plan the individual cuts to be made.

The interchangeability of the other body panels can be seen in the drawing.

When cutting damaged body panels out, make sure that the cutting line does not follow the weld seam but is roughly 10 mm above or below it. The remaining metal is removed when the new part has been matched up.

Identification	Description of Part
A	Front panel
B	Front hood
C	Front side panel, left Front side panel, right
D	Cowl panel
E	Roof
F	Quarter panel, left Quarter panel, right
G	Rear top panel
H	Rear hood
I	Door bare, left Door bare, right
K	Sill panel, left Sill panel, right
L	Rear panel

THIS PAGE INTENTIONALLY
LEFT BLANK

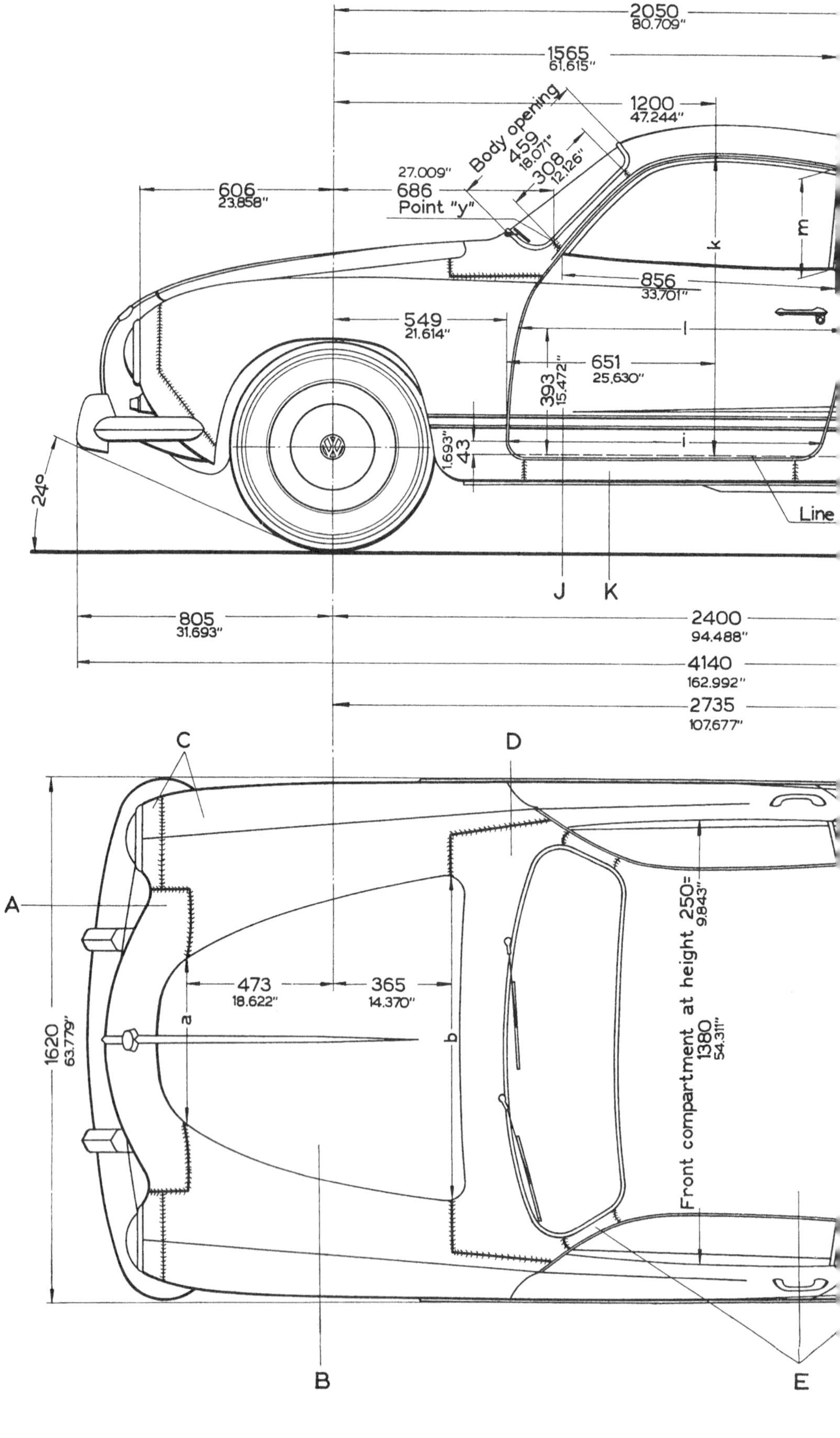

2050
80.709"
1565
61.615"
1200
47.244"
Body opening
459
18.071"
308
12.126"
27.009"
686
Point "y"
606
23.858"
k
m
856
33.701"
549
21.614"
l
651
25.630"
393
15.472"
1.693"
43
i
Line
24°
J
K
805
31.693"
2400
94.488"
4140
162.992"
2735
107.677"
C
D
A
1620
63.779"
473
18.622"
365
14.370"
a
b
Front compartment at height 250=
9.843"
1380
54.311"
B
E

A-72

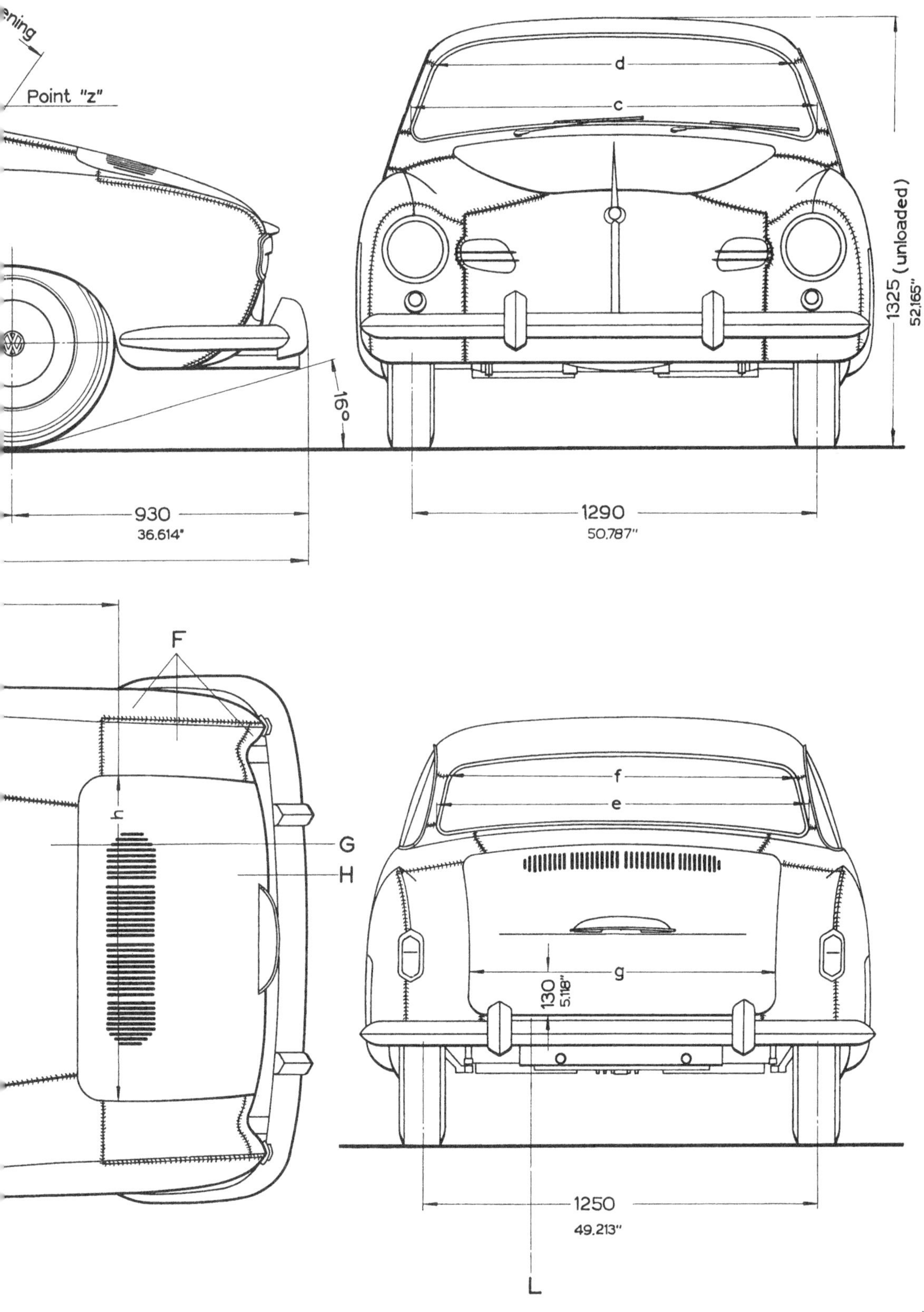
Point "z"
d
c
1325 (unloaded)
52.165"
16°
930
36.614"
1290
50.787"
F
G
H
h
f
e
g
130
5.118"
1250
49.213"
L

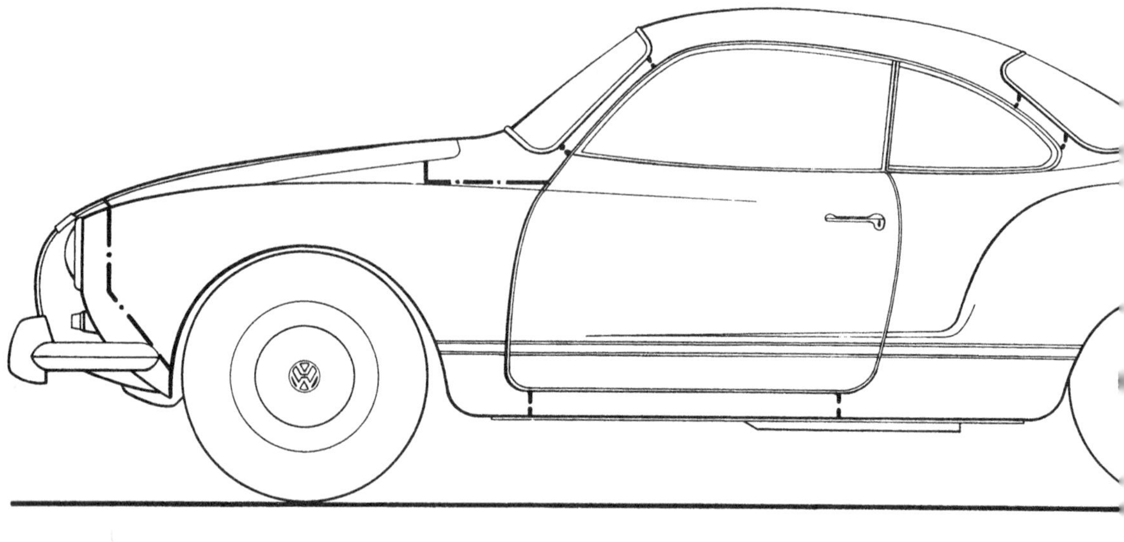

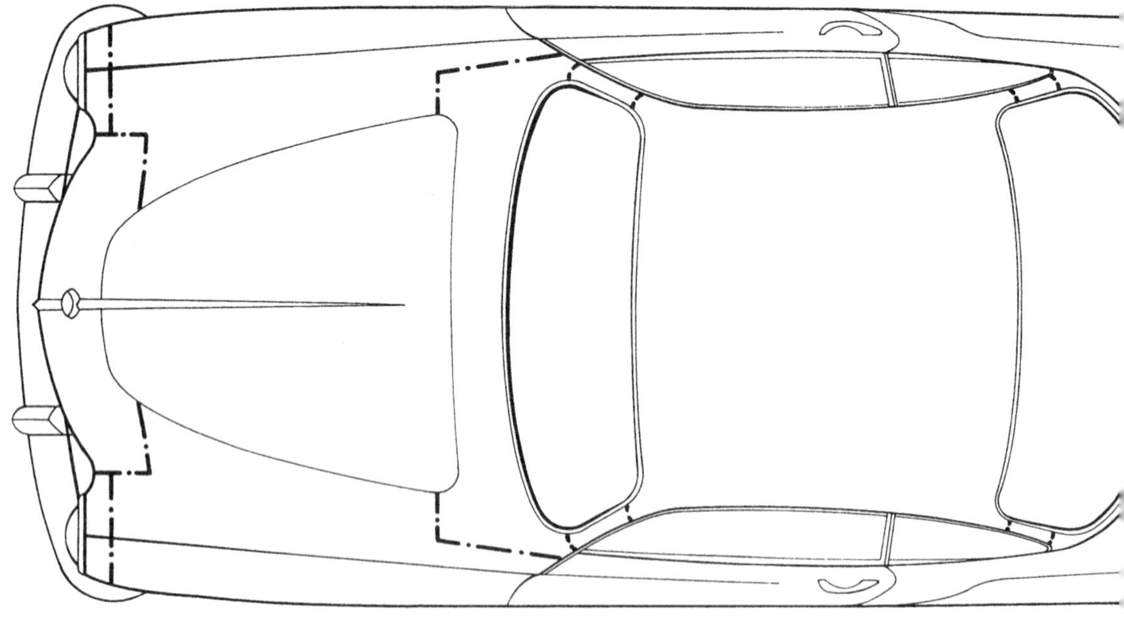

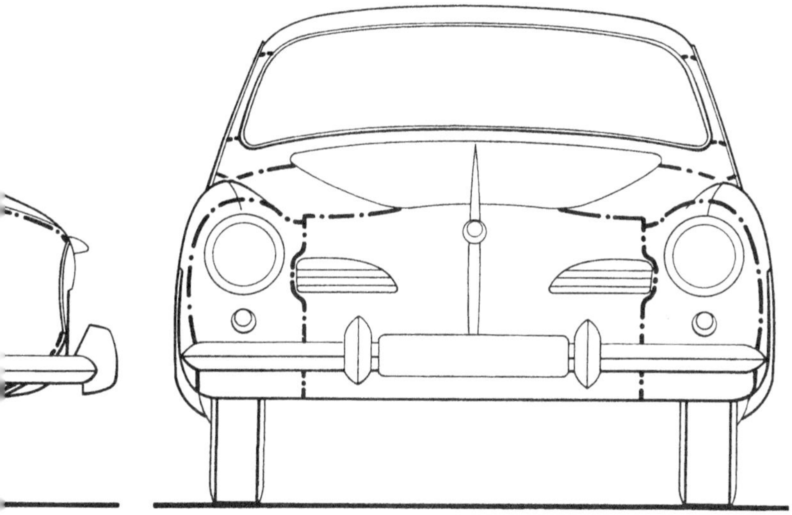

THIS PAGE INTENTIONALLY
LEFT BLANK

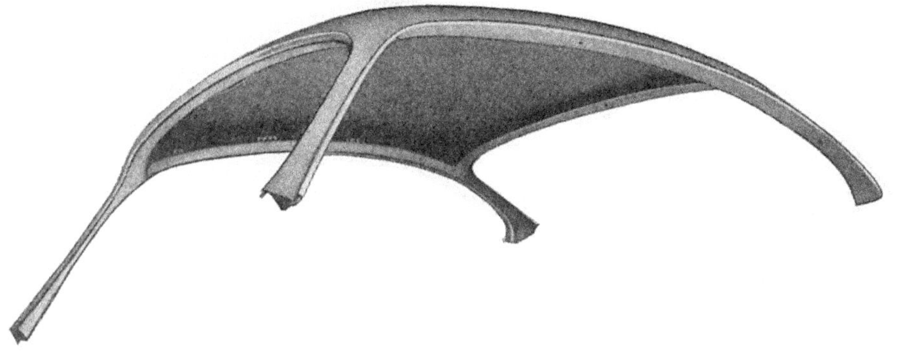

The roof panel is supplied complete with roof pillars. The roof pillars can be replaced separately if required and the operation is described at the end of the section.

Four perforated angle plates are welded to each side member inside the roof. The stretcher bars for the headlining are hooked into the holes in these plates.

When replacing the roof or a roof pillar the use of the specially designed gauges is essential in every case. Details of body gauges and jigs are given under the heading "Workshop Equipment" in the "Body General" section.

Roof Replacement

Preparation

1 - Disconnect the battery.

2 - Remove all seats and mats or cover adequately.

3 - Remove windshield, rear window and both quarter windows.

4 - Lower the door windows and cover the window slots.

5 - Remove all interior trim including rear view mirror and sun visors. Cover the rear luggage compartment.

6 - Remove steering wheel.

7 - Remove windshield wiper fittings. Cover or remove the instrument panel parts as necessary.

8 - If no other body repairs are to be carried out, adequately cover all body parts within the area of the operation.

Body Repair Work

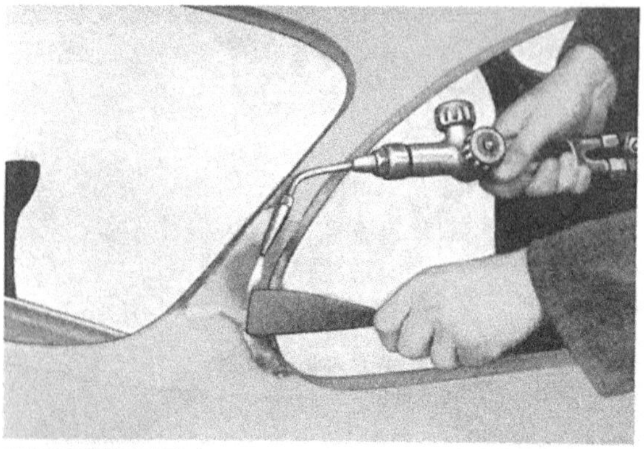

1 - Heat the solder in the area of the lower roof pillar welds with a welding torch and remove with a suitable spatula.

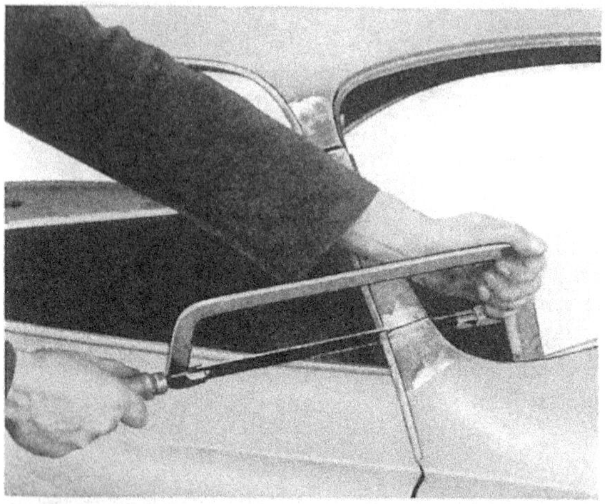

2 - Saw the roof pillars off about 50 mm above the weld seams.

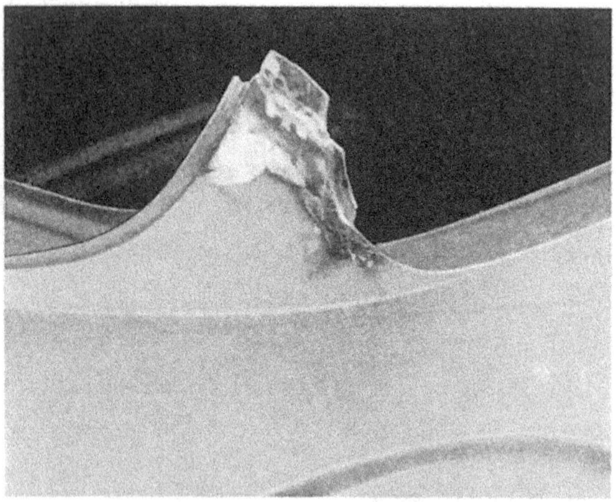

3 - Remove the remaining metal down to the welds. Clean up the welds by grinding or filing. If necessary remove dents and straighten the sheet metal.

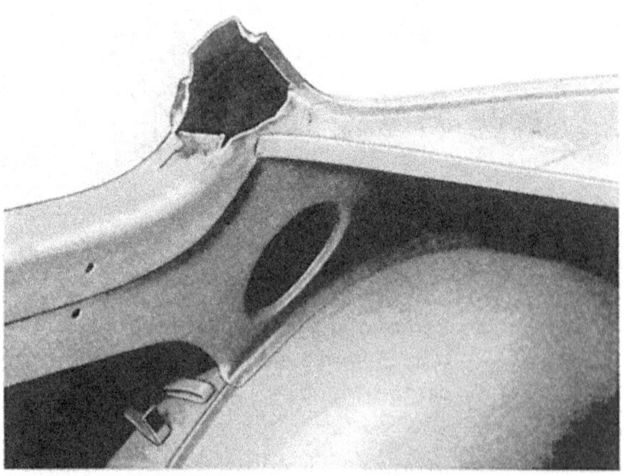

4 - Take care that the inner panels on the rear roof pillars, which serve as overlaps, are not distorted.

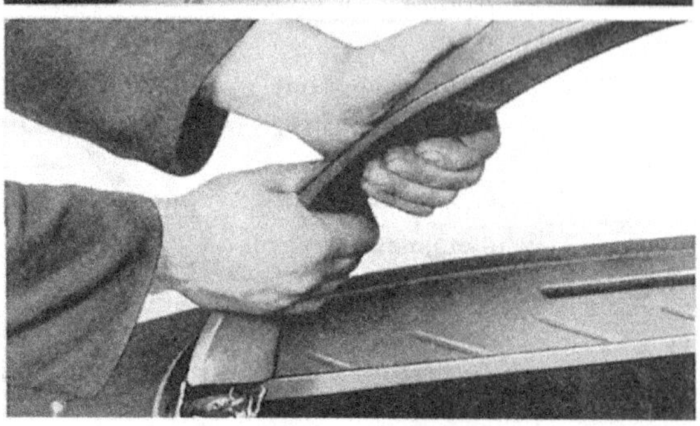

5 - Place roof in position.

6 - Check that the inner panels on the rear roof pillars are overlapping.

7 - Install window gauges and fit the roof.

8 - Tack-weld the roof at the front first.

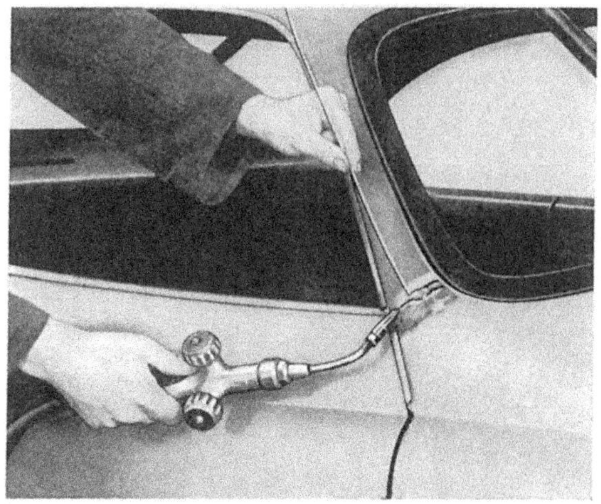

9 - Tack-weld the rear roof pillars with the welding torch with the window gauges in position.

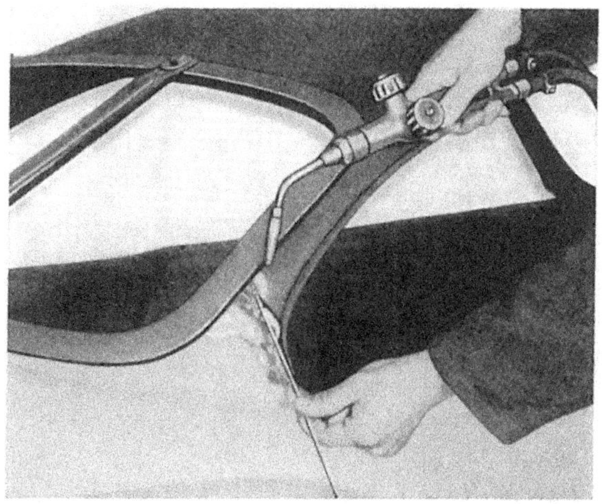

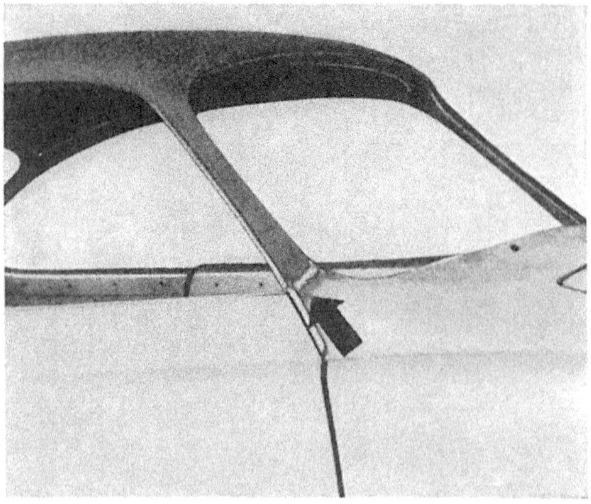

Gas weld the roof pillars inside and out. Insert the window gauge occasionally to check that the position of the roof is correct.

10 - Grind the weld seams down as far as possible. Fill the welded and reworked areas with solder until a smooth contour is obtained.

11 - Grind the joints smooth and prepare vehicle for painting.

Roof Pillar Replacement

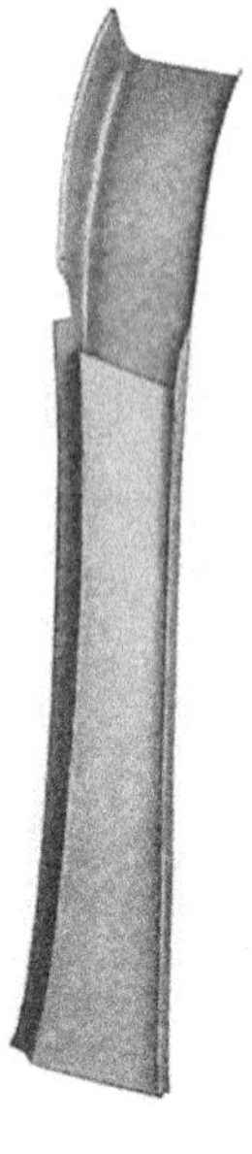

Front roof pillar

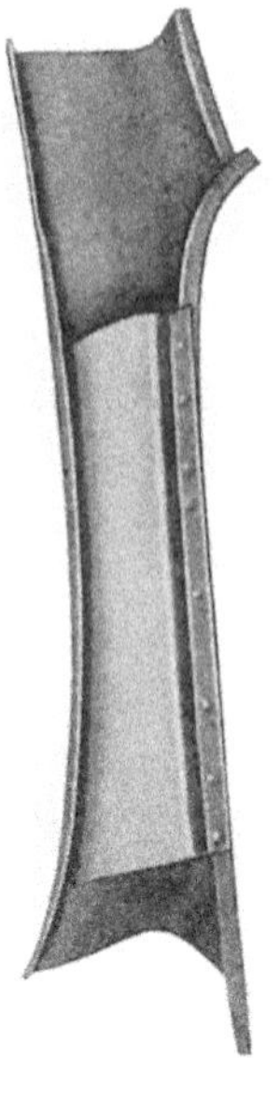

Rear roof pillar

Front and rear roof pillars are supplied as spares.

When replacing one or more roof pillars, the use of the specially designed gauges is essential.

The preparatory work is roughly the same as that required when replacing the roof. All parts within the vicinity of the pillar being repaired must be removed or covered up as necessary.

1 - Heat the solder on the appropriate pillar in the area of the upper and lower weld seams with a welding torch and remove with a suitable spatula.

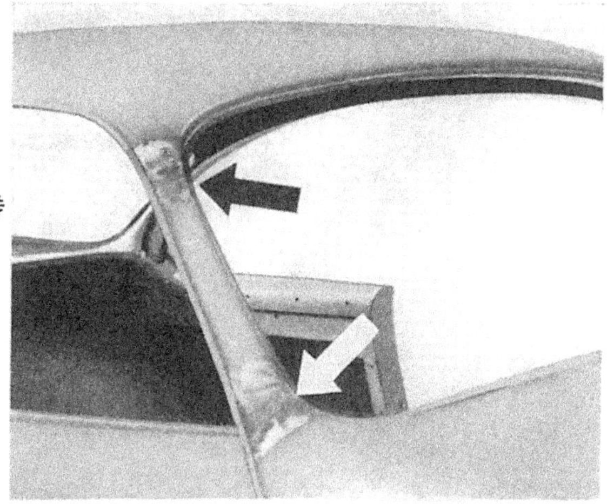

2 - Saw the roof pillar off 50 mm above the lower and below the upper weld seams.

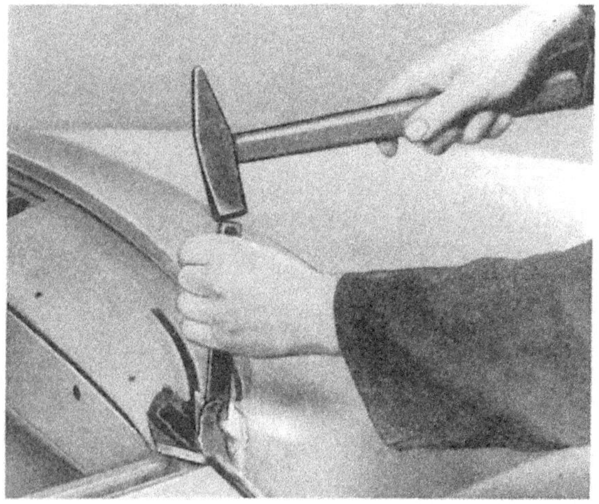

3 - If the upper part of the roof pillar concerned is not damaged it is recommended that the work be simplified by merely pushing the bottom of the pillar over the bottom joining flange and butt-welding the top joint. In this case it is only necessary to remove the remaining metal on the body.

4 - Grind or file the top weld area and edges of the cut clean. If necessary remove dents and straighten sheet metal.

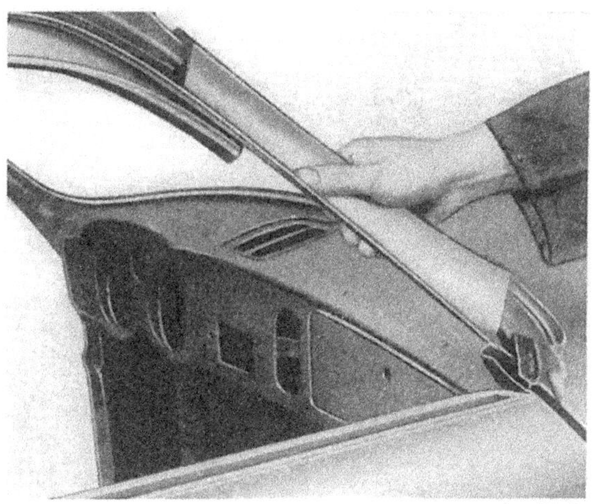

5 - Push roof pillar over the bottom joining flange.

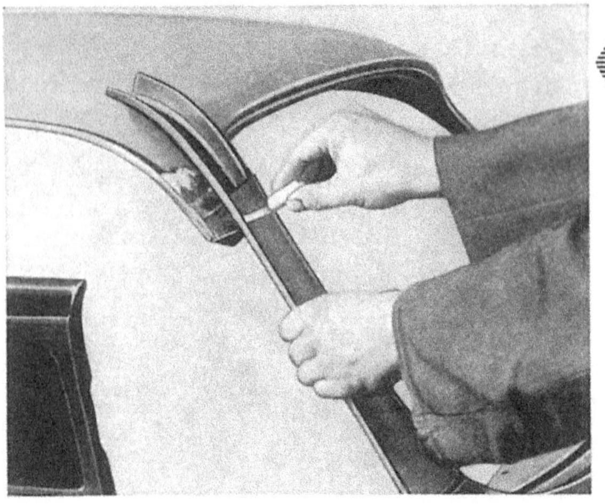

6 - Fit pillar to the roof and mark off for cutting.

7 - Cut roof pillar as marked.

8 - Insert window gauge.

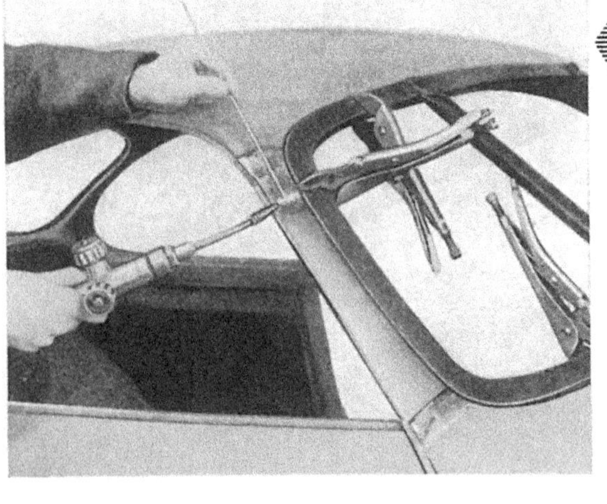

9 - Push roof pillar over the joining flange on the body, clamp it to the window gauge with body clamps and tack with gas welding.

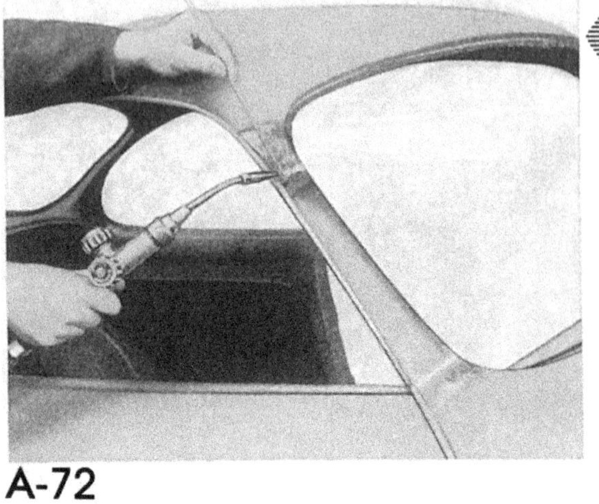

10 - Take window gauge out and gas weld roof pillar.

11 - Grind the weld seams down where necessary. Fill the welded and reworked areas with solder until a smooth contour is obtained.

12 - Grind the joints smooth and prepare vehicle for painting.

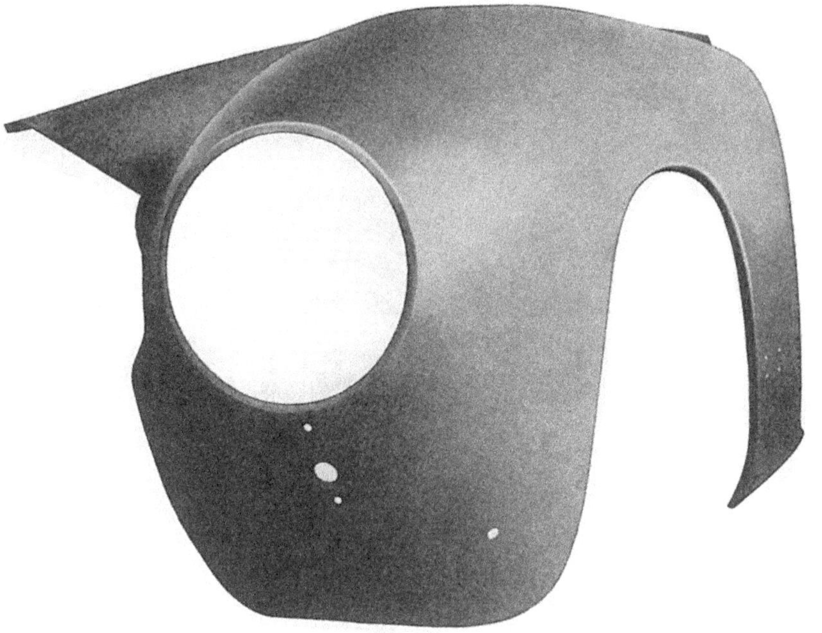

The front side panel is composed of two parts, the side panel proper and the so-called front panel. As a spare part the front side panel is supplied with the front panel welded to it.

The replacement of the front panel can only be carried out with the specially designed gauge (see section "Body General", Workshop Equipment).

Preparation

1 - Disconnect the battery.

2 - Remove door and front hood.

3 - Remove front wheel.

4 - Remove bumper bar.

5 - Remove hinge pillar cover plate in front wheel housing.

6 - Remove front hood weatherstrip.

The following repair operation was carried out on a Karmann Ghia Convertible, the front side panel and hinge pillar of which was so badly damaged that part replacement or straightening was not possible.

1 - With a sheet metal chisel (Local Manufacture Drawing VW 732) cut the side panel out at the places shown.

Top

Side (Hinge pillar)

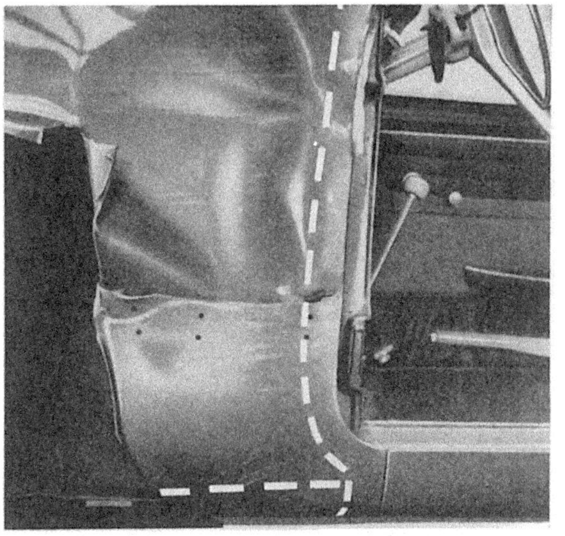

Front

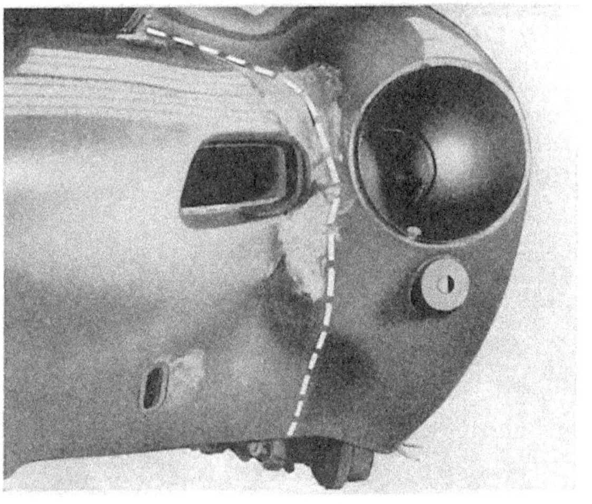

2 - It is best to commence cutting at the hinge pillar.

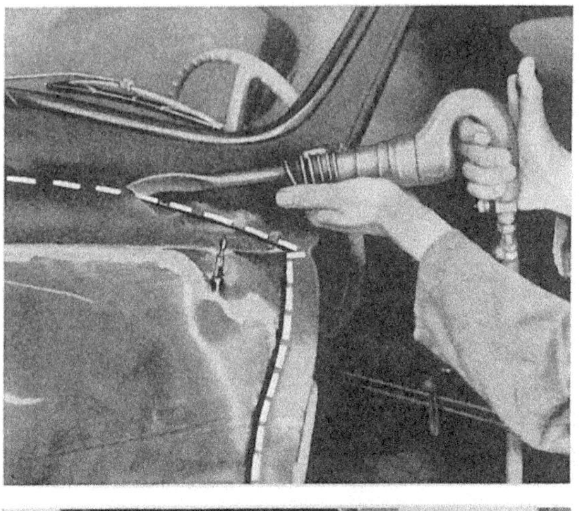

3 - Remove cut out side panel.

Attention

Protective gloves should be worn when removing or installing the side panel.

4 - Where damage to the side panel makes replacement necessary it is generally also necessary to replace the hinge pillar gusset plate. (The gusset plate is a separate part.) The parting cut to remove the gusset plate should be made as shown in the instructions (see picture).

5 - A complete hinge pillar replacement is not advisable for repair technical reasons if the roof is not being replaced. The top of the hinge pillar fits into the roof pillar and it is recommended therefore, that only parts of the hinge pillar are replaced. To do this, it is necessary to saw the hinge pillar off at the bottom of the roof pillar in such a manner that the new hinge pillar section, cut to fit, can be butt-welded into position.

6 - Remove remaining pieces of the old gusset plate and grind spot welding marks flat.

7 - Coat all spot welding areas with spot welding paint or paste (see Special Hints, page A 76-1). Fit new gusset plate and clamp in position with body clamps.

Spot weld gusset plate with a push type spot welder.

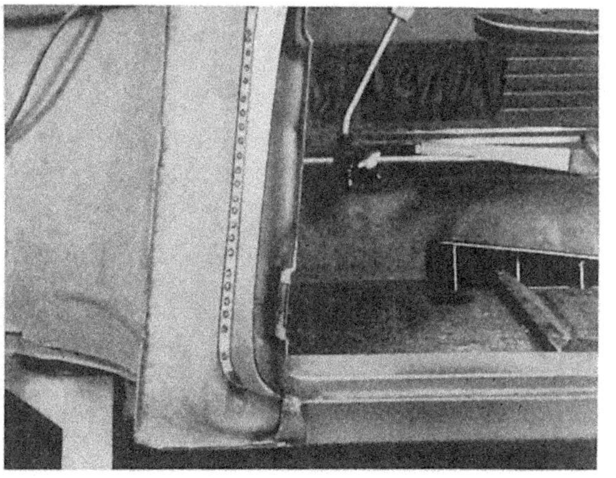

Gas weld top

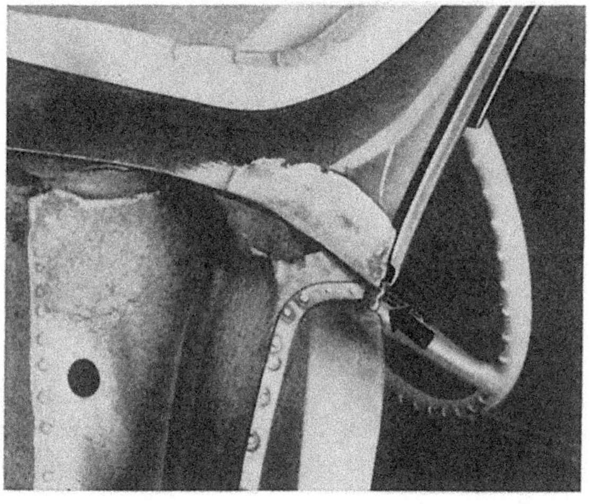

and bottom joints

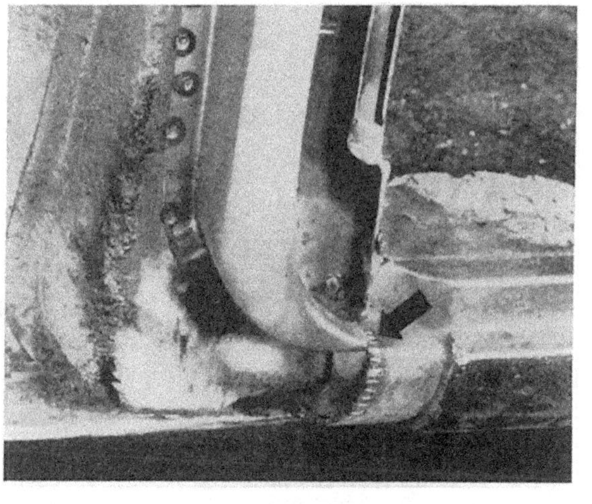

8 - Spot weld brackets for cover plate.

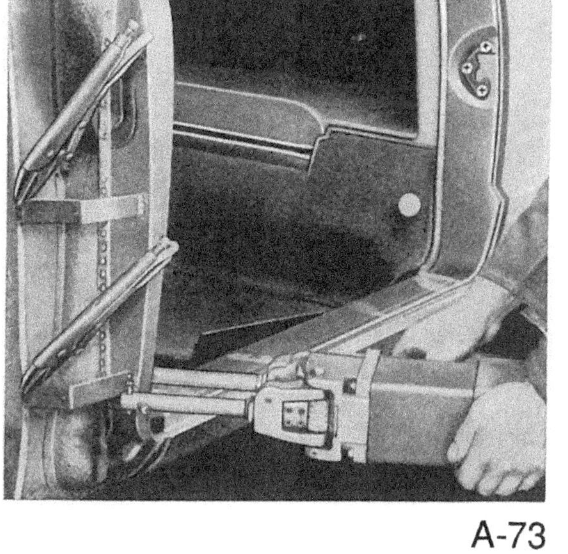

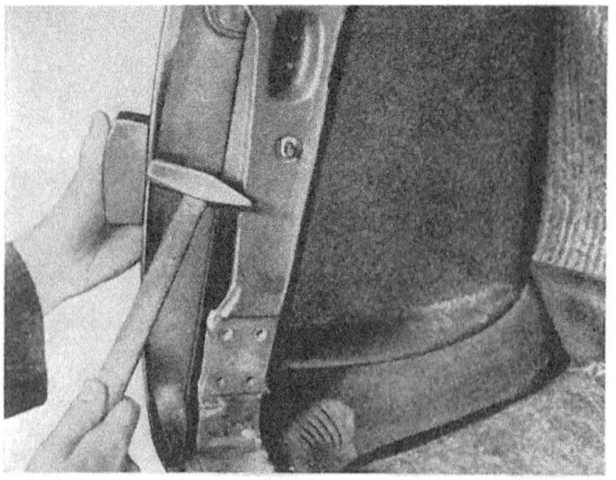

9 - Fold the gusset plate outer edge over.

10 - Before proceeding any further it is essential to install the door and check that the clearance between door and gusset plate is uniform over the whole length. The clearance should be approximately 8—10 mm.

This comparatively large clearance is necessary in order to leave sufficient space for the folded edge of the side panel which is to be installed.

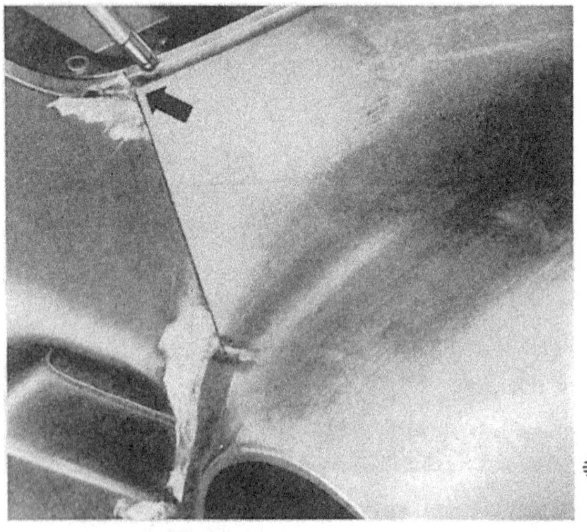

11 - Remove door again.

12 - Install new side panel and check fit. Location at front and rear is given by the weld seams (see arrow).

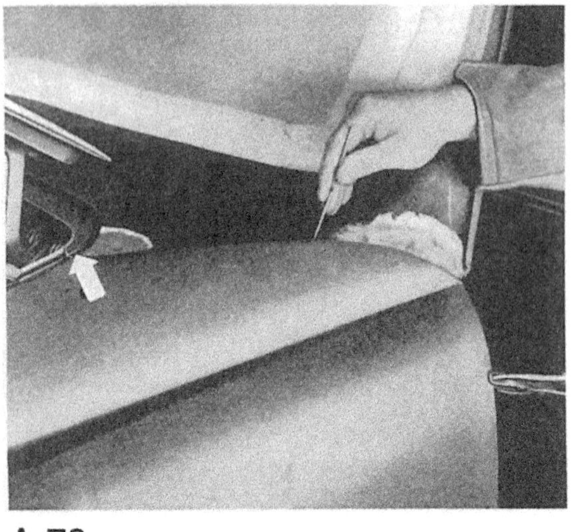

Mark off new side panel on the body.

Remove side panel and cut overhanging edges on body back to the marked line. Install side panel again, position and prepare for welding.

13 - Tack weld side panel to sill panel.

Tack weld inner and outer ends of joint at windshield frame.

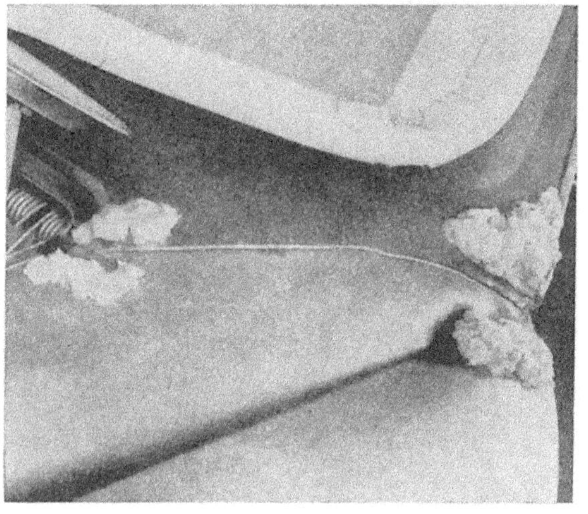

Tack weld at front panel.

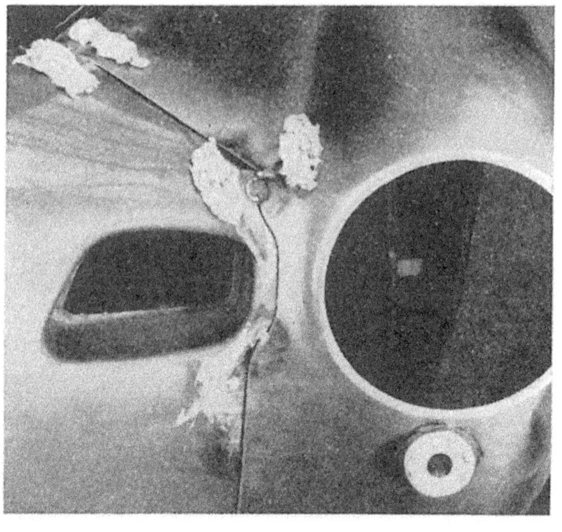

14 - Spot weld to body along the front hood opening.

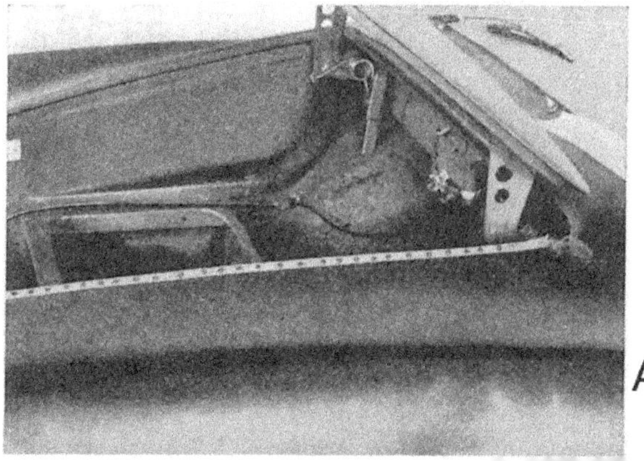

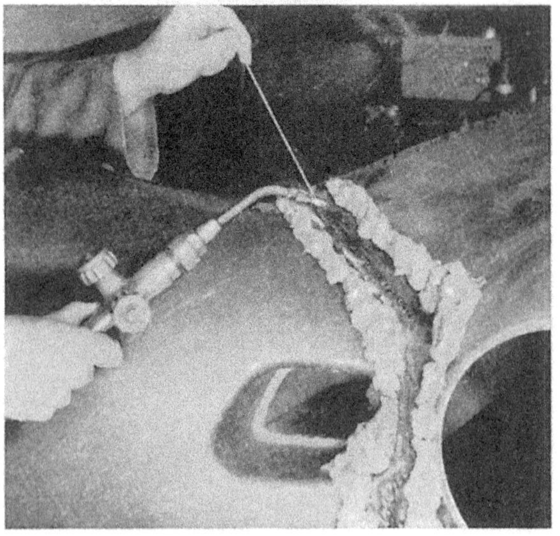

15 - Butt weld the new side panel to the body with gas welding.

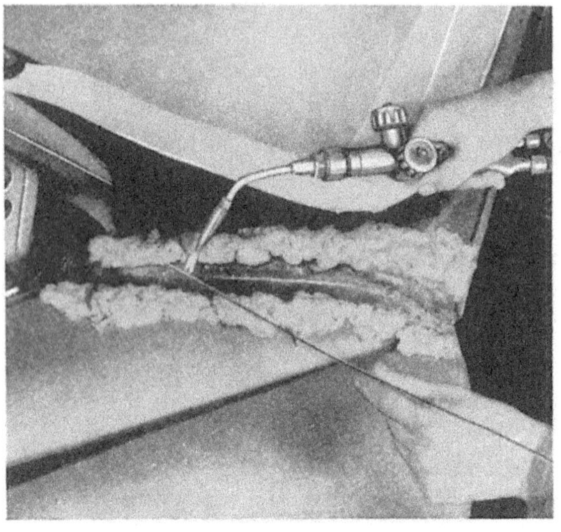

Note:

Pack the windshield frame, front and side panels adequately with asbestos along the areas to be welded.

16 - Install door again and check fit.

If necessary, the edge of the front side panel can be corrected with the Door Frame Tool VW 740 (Local Manufacture).

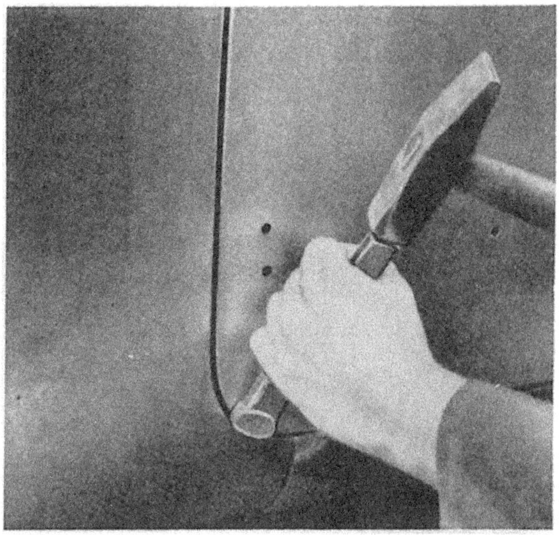

To pull the side panel outwards slightly in the door opening the bending bar VW 741 (Local Manufacture) can be used.

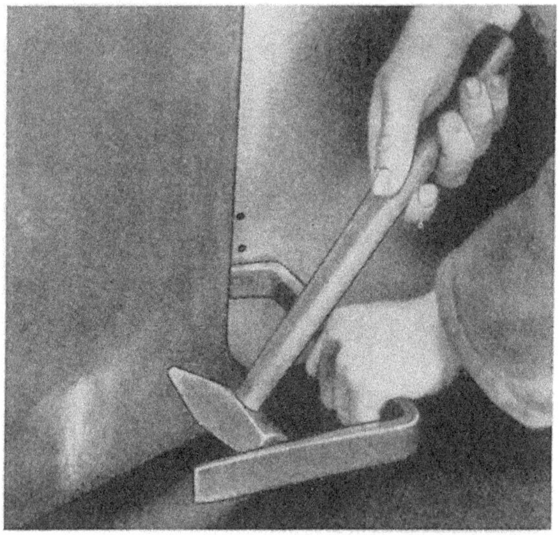

17 - Finally remove the door again, fold the outer edge of the side panel over and weld it

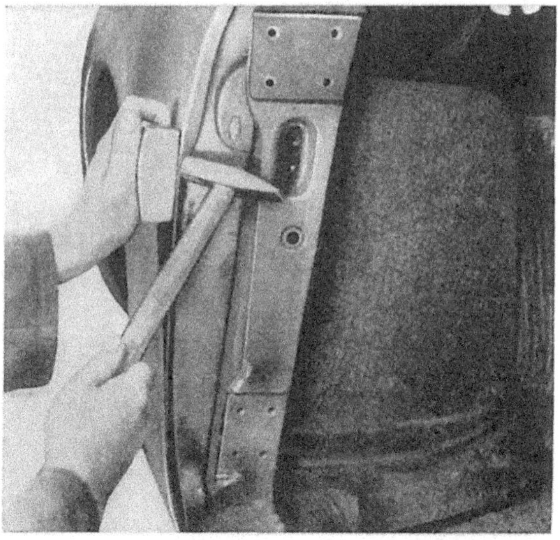

to the gusset plate with a push type spot welder.

18 - Weld the lower edge of the side panel to the hinge pillar.

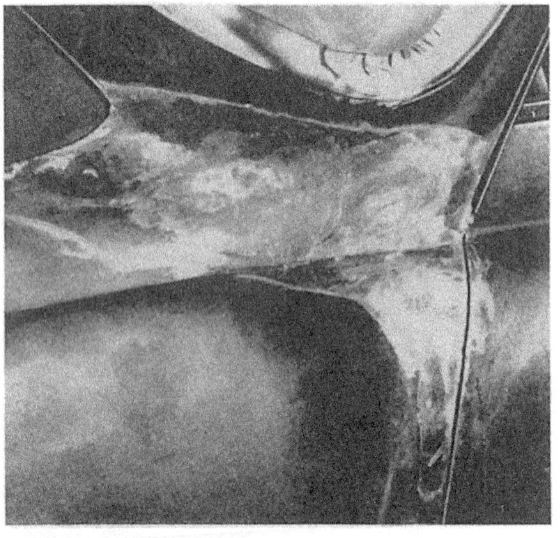

19 - Smooth all spot welds and gas welded joints as far as possible and coat with solder.

20 - When this work is completed the head lamp housing is inserted. It is essential to ensure that the housing is located exactly in the center before spot welding.

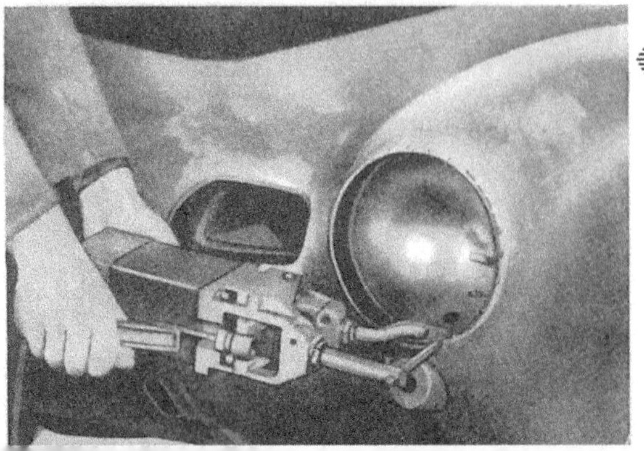

21 - Grind all reworked areas smooth and prepare vehicle for painting.

Body Repair Work

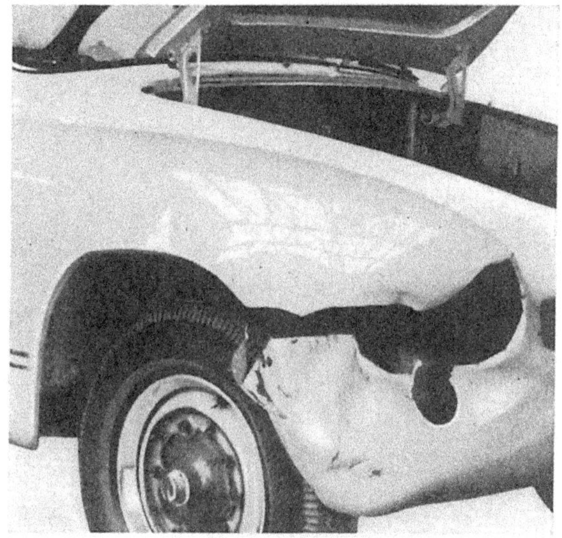

The following repair operation was carried out on a Karmann Ghia Coupé where the damage to the right front side panel was such that a partial replacement was possible.

1 - Cut out the damaged section with a sheet metal chisel (Local Manufacture Drawing VW 732) at the places shown.

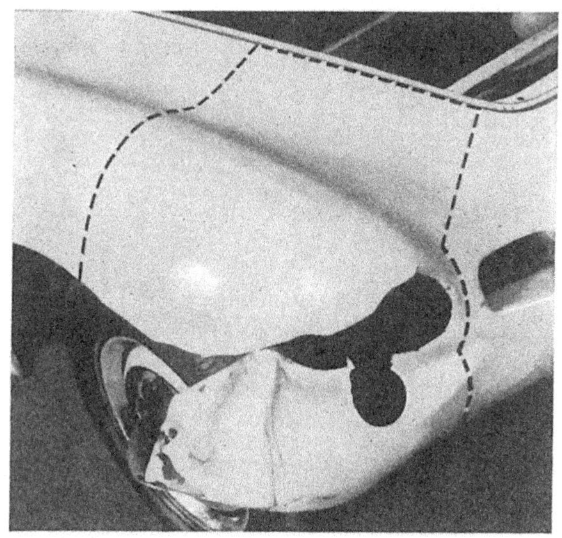

2 - Remove damaged section.

Attention

Protective gloves should be worn when removing and later when installing the section.

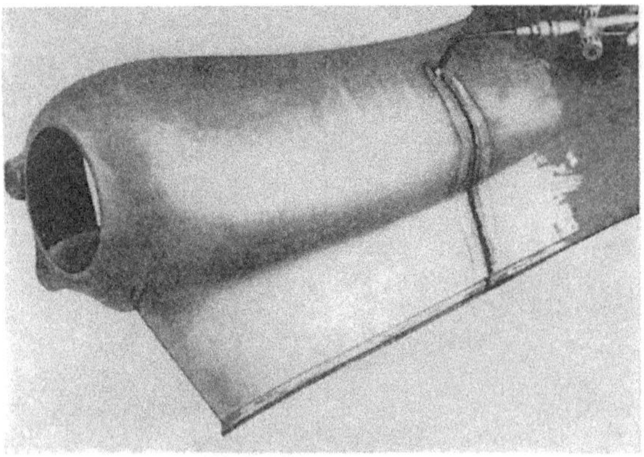

3 - Cut a section from a new side panel slightly larger than opening in the old panel.

Straighten the edges of the new section with tin snips.

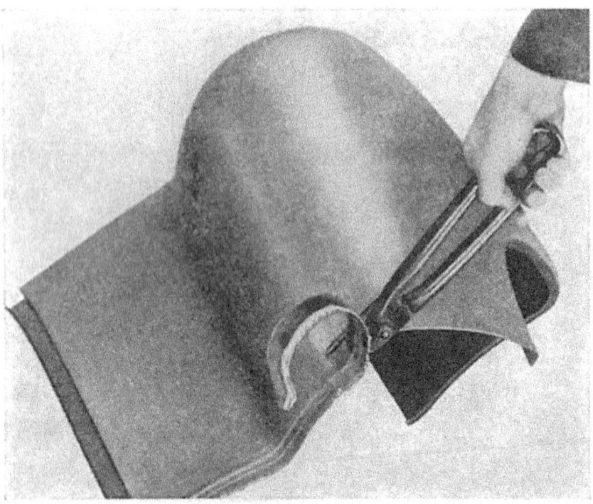

4 - Clamp the new section in position and mark off on the body.

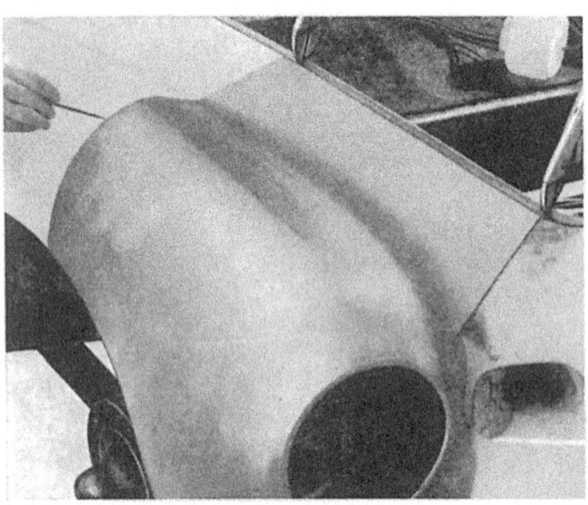

Remove new section and cut the overhanging edge on the body back to the line marked. Install section again, position and prepare for welding.

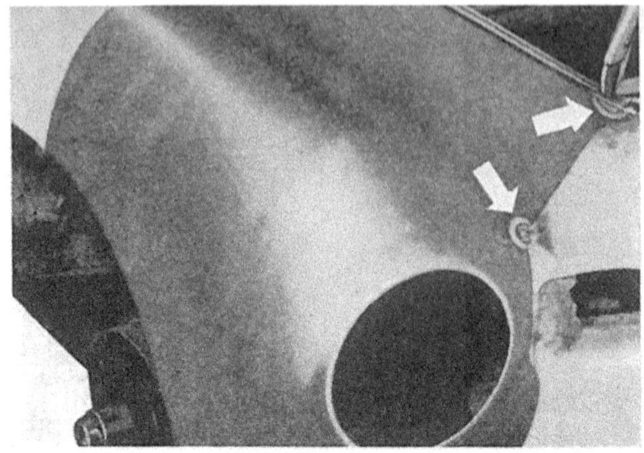

5 - Tack weld the section to the body (see arrows).

6 - Pack the side and front panels with asbestos along the joints to be welded.

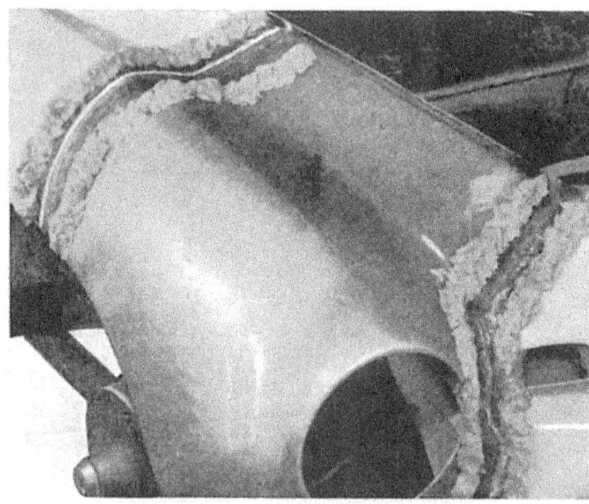

7 - Butt-weld the new section to the body.

8 - Smooth all weld seams as far as possible and coat with solder. Spot weld the headlamp housing and grind smooth. Prepare vehicle for painting.

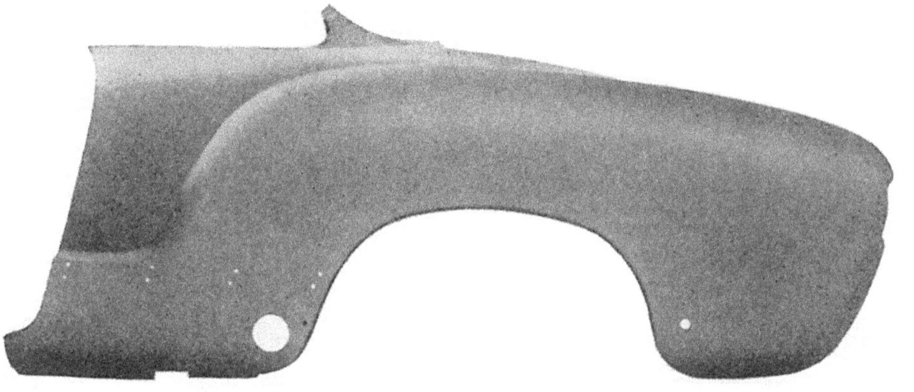

The quarter panel is pressed out of a single sheet and is supplied as a spare part in this form.

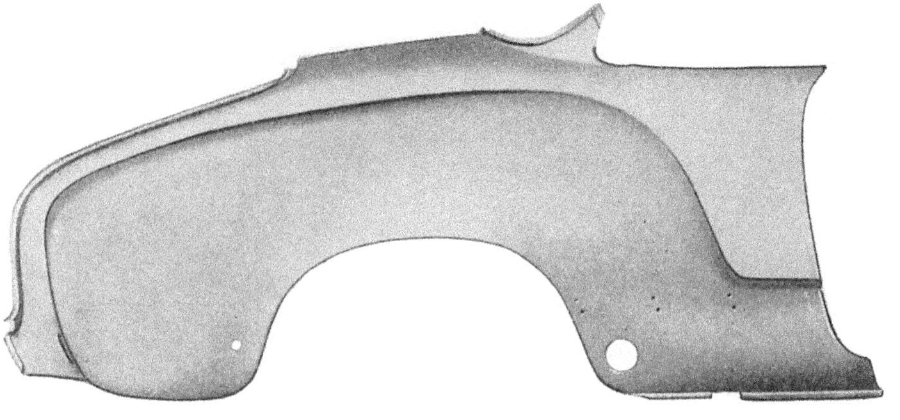

When installing the quarter panel in the Karmann Ghia Coupé it should be spot welded to the lock pillar which is available as a separate part.

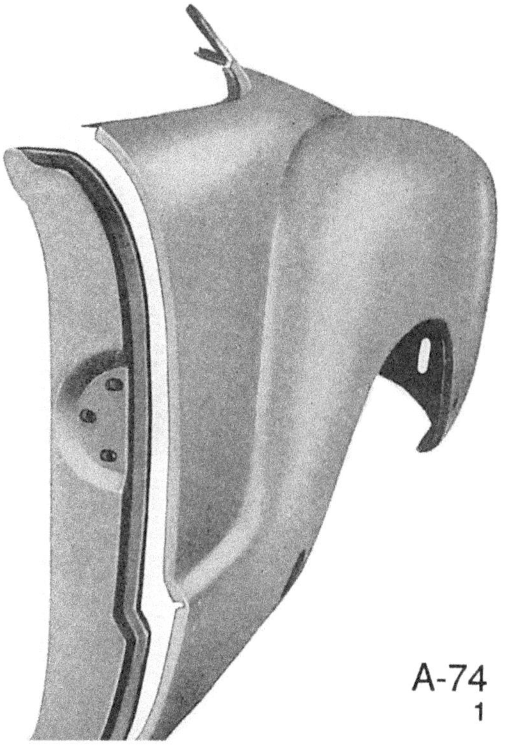

The inner quarter panel is also supplied separately and is attached in the quarter window area.

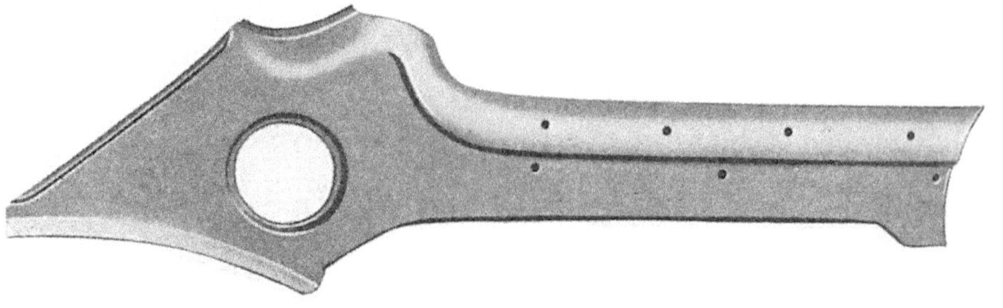

It is welded to the inner wheel housing, the quarter panel, the lock pillar and the roof pillar.

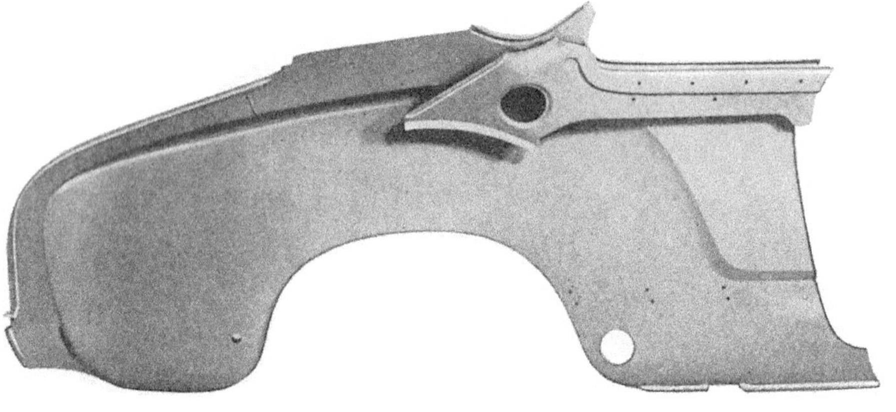

In the lower area the quarter panel is spot welded to the side member, the outer wheel housing and the rear panel. Outer and inner wheel housings are supplied separately and can be renewed when the quarter panel is being replaced.

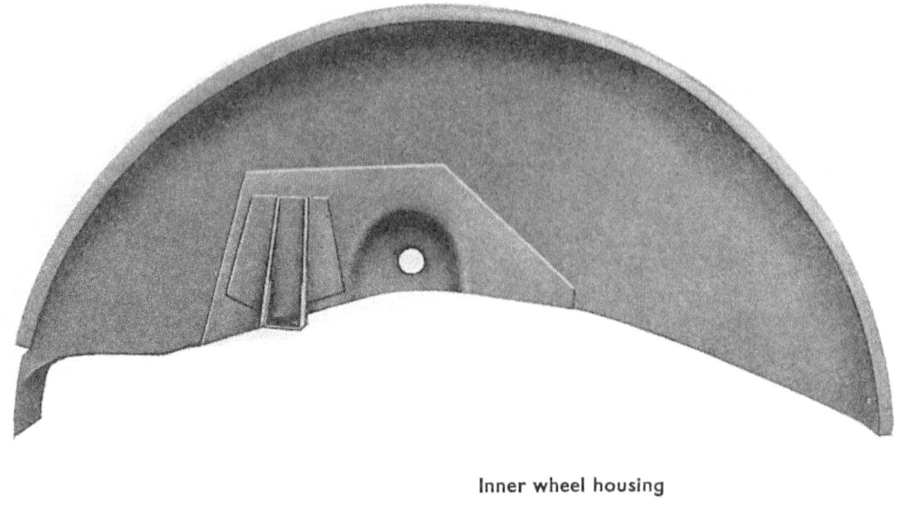

Inner wheel housing

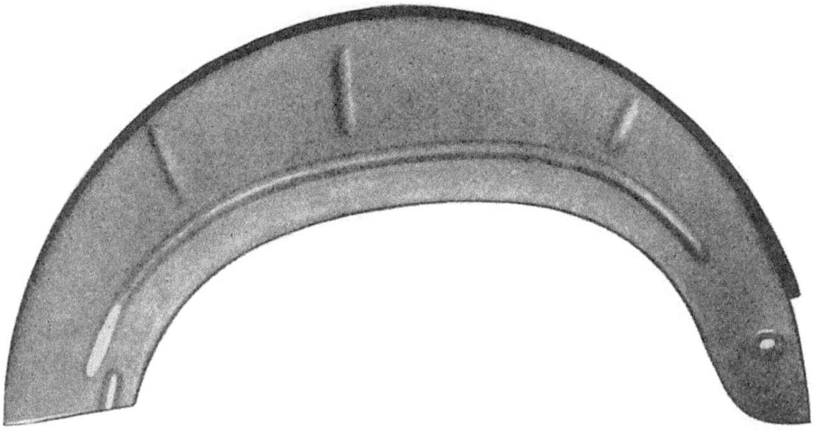

Outer wheel housing

The quarter panel is welded to the body and to the parts mentioned above by means of the so-called joining flanges. Butt welding is only necessary at the rear top panel.

The quarter panel can also be partially replaced if its condition is considered suitable. The damaged section should be cut out and replaced by a section cut to correspond and butt welded in position.

The following repair operation was carried out on a Karmann Ghia Coupé where the left quarter panel and left hand door were so badly damaged that a partial replacement or straightening was not possible. The outer wheel housing was also badly dented and distorted. This operation, therefore, covers the replacement of the quarter panel, the outer wheel housing and the door.

Preparation

1 - Disconnect the battery.

2 - Remove door and rear hood.

3 - Remove striker plate.

4 - Remove rear and quarter windows.

5 - Remove seats and floor mats.

6 - Remove interior trim where necessary or cover adequately.

7 - Remove rear hood weatherstrip or partly withdraw from channel.

8 - Remove rear lamp and place the cable in the engine compartment so that it will not be damaged.

9 - Remove quarter window weatherstrip channel.

10 - Remove rear bumper.

11 - Remove sealing compound from weld seams where necessary.

12 - Remove sound insulating lining and packing from the engine compartment on the appropriate side.

Body Repair Work

1 - Cut the quarter panel out with a sheet metal chisel (Local Manufacture drawing VW 732) at the places shown.

The use of a chisel for this operation is recommended in preference to a welding torch owing to the risk of the cut edges not being straight and the interior body parts being damaged.

2 - The cut line must run roughly 20 mm from the weld seam to avoid damage to the joining flanges or the inner body panels. At the top half of the wheel opening the spacing should be 250 mm to prevent the outer wheel housing being damaged.

To enable the cutting line between quarter panel and rear top panel to be marked the position of the weld seam must be located. The weld seam is visible on the rear window frame and in the channel for the rear hood weatherstrip.

a = 250 mm (9.8″)

Attention

Protective gloves should be worn when removing and later when installing the quarter panel.

3 - Remove the section which has been cut out.

4 - The weld seams and joints between quarter panel and window frame (quarter window) are coated with solder to give a smooth finish. Before removing the sheet metal edges it is necessary to heat this solder with a welding torch and scape it off with a suitable scraper or wire brush. The seams are then easier to see and the metal edges can be removed better.

5 - The operation is facilitated if the vehicle is placed on blocks and the rear wheel taken off, after the large sections of the side panel have been removed.

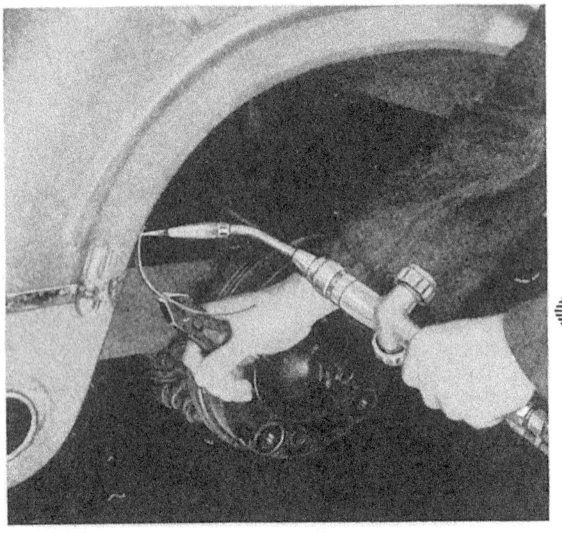

6 - Remove the remaining edges of the quarter panel and grind spot welds clean. Straighten joining faces where required, grind smooth and prepare for installation.

Note:

Remove the metal edges by twisting with a pair of pliers and not by pulling as otherwise the spot welds will be pulled out and leave holes. In some instances it may be advisable to heat the spot welds with the welding torch and then pull the metal strips off with the pliers.

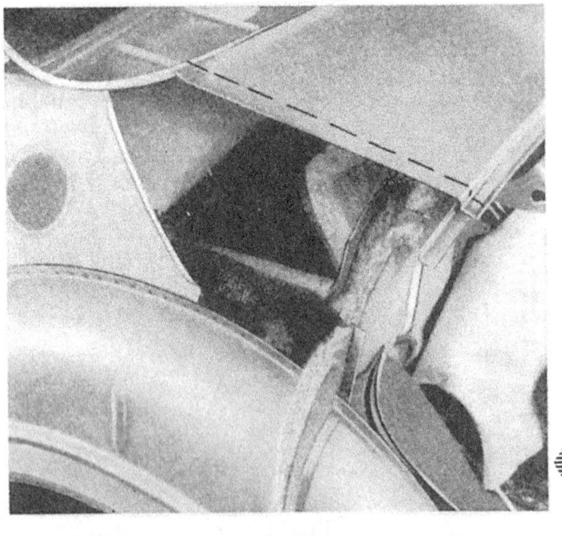

7 - At the top rear panel cut the metal off cleanly about 10 mm from the weld seam (see dotted line).

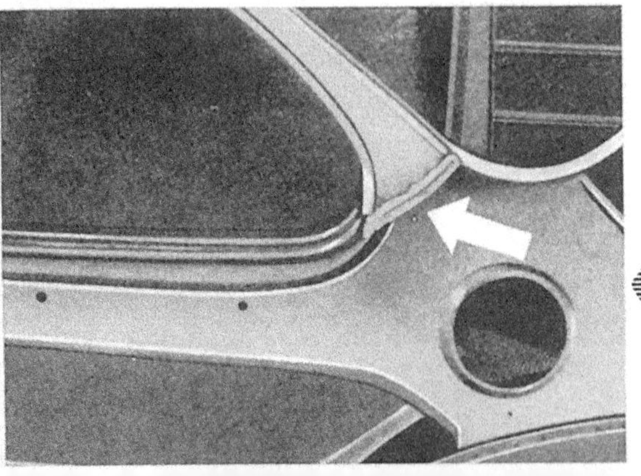

8 - The roof pillar should be cut off about 5 mm under the rain channel (see arrow) and not at the weld seam.

9 - Remove the metal edges from the joining flanges at the back in the same manner as above and straighten the flange if required.

10 - Cut the outer wheel housing off near the joining flange with a sheet metal chisel.

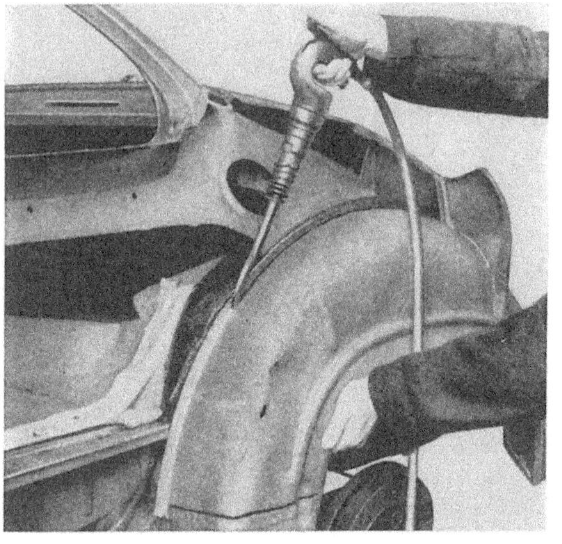

11 - Remove metal edges from the spot welded seam of the inner wheel housing. Grind spot welds smooth and prepare for assembly.

12 - Check new outer wheel housing for fit and rectify if necessary.

13 - Seal the area to be spot welded with a coating of spot welding paste. Details of the manufacturers and usage of spot welding paint and paste are given under "Special Instructions" on page A 76/1.

14 - Place outer wheel housing in position and clamp to body.

15 - Spot weld outer housing to inner housing.

16 - The lower corners between outer and inner wheel housing should be gas welded at front

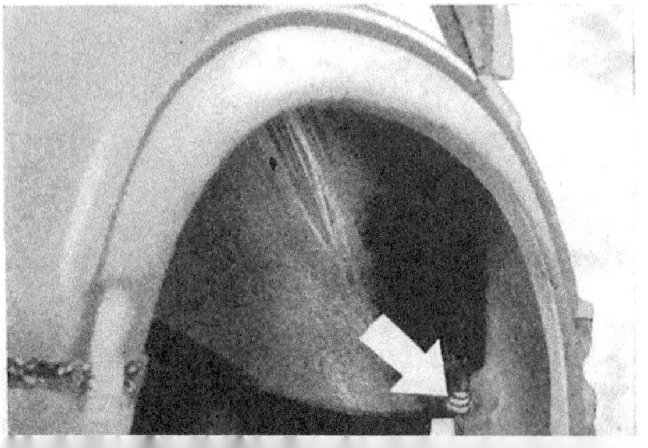

and rear

17 - Clamp the extension to the welded part of the luggage compartment rear panel onto the body.

Spot weld the extension to the wheel housing and cement a felt strip of suitable width to the top edge for sealing.

18 - Install door and check fit. Verify that the door gap at the hinge pillar is the same width over the whole length. At the same time the top edge of the door at the lock pillar must be correctly aligned with the quarter panel (Window frame top edge). The door gap at the lock pillar is aligned by the application of solder at a later stage when the quarter panel has been installed.

19 - Prepare quarter panel for installation. Grind spot welds smooth.

20 - Gas weld the quarter panel to the top rear panel.

The gas welding calls for sound knowledge and experience in butt welding. Unskilled welding can cause the formation of waves and distortion which will prove very difficult or even impossible to eliminate. It must be emphasized, however, that good gas welding is equivalent to the original condition.

To ensure good welding the joining edges of the quarter panel and the top rear panel must be perfectly straight and parallel over the whole length when the panels are installed.

21 - All spot welds on the wheel housing which have been ground smooth should be coated with spot welding paste to prevent corrosion (see Special Instructions — page A 76/1).

22 - Install side panel, position correctly and clamp.

23 - Insert window gauge. Further details of the window gauge can be found in "Body General" section under the heading "Workshop Equipment".

24 - Tack the quarter panel to the top rear panel at front and rear with the welding torch.

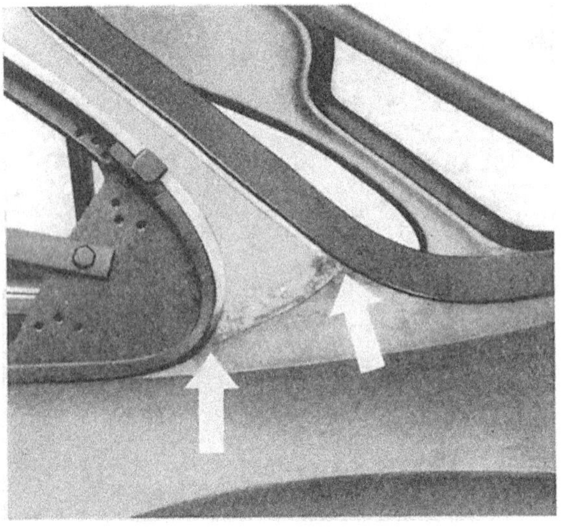

25 - Tack the quarter panel to the roof pillar at the left and right with the welding torch.

The space between the quarter window gauge and the quarter panel is filled with solder later on.

26 - Tack the quarter panel to the lock pillar at the top and to the lock pillar and sill panel at the bottom with the welding torch.

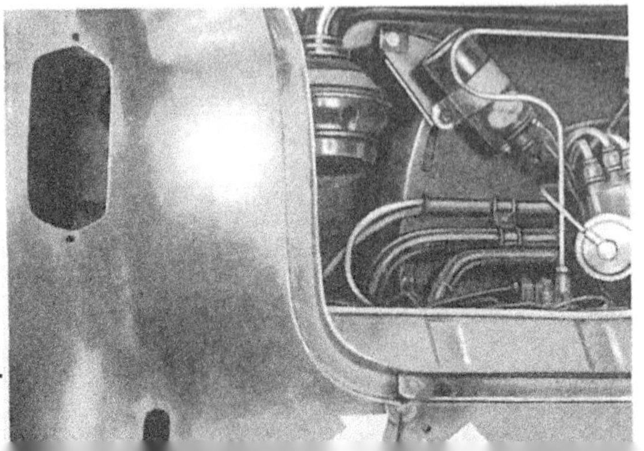

27 - Tack the quarter panel to the rear panel at top and bottom with the welding torch.

28 - Remove the window gauge and spot weld the quarter window to the lock pillar and to the quarter window frame.

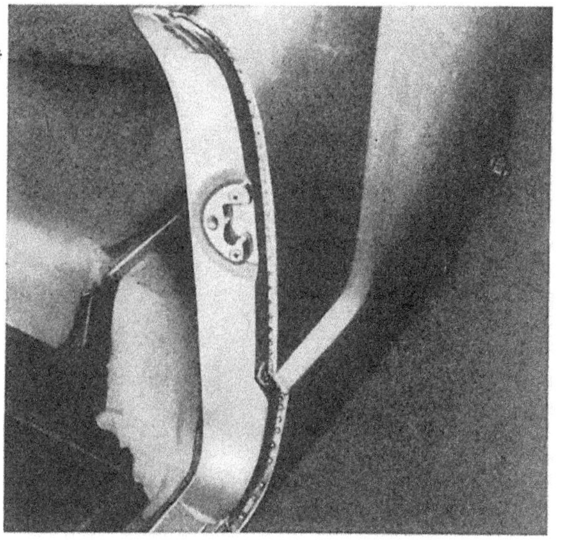

29 - Spot weld quarter panel to the side member.

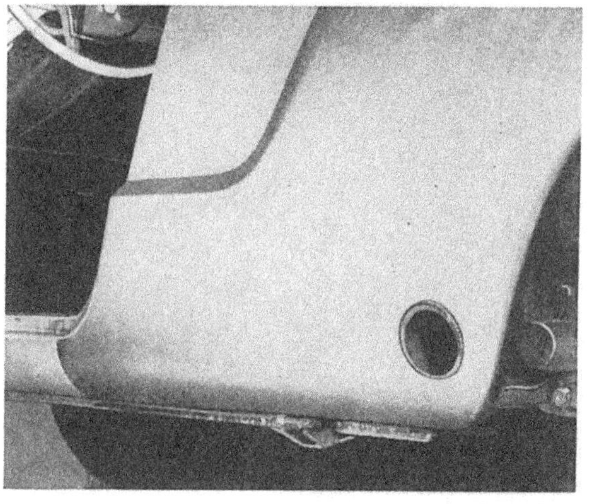

30 - Spot weld the quarter panel near and below the opening for the bumper bar bracket and also to the outer wheel housing. Gas weld the quarter panel to the rear panel in the channel for the rear hood weatherstrip.

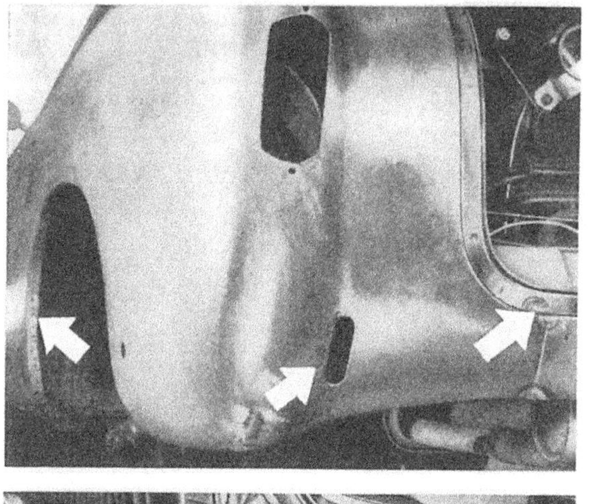

31 - Pack roof pillar and quarter panel adequately with asbestos along the area to be welded and gas weld.

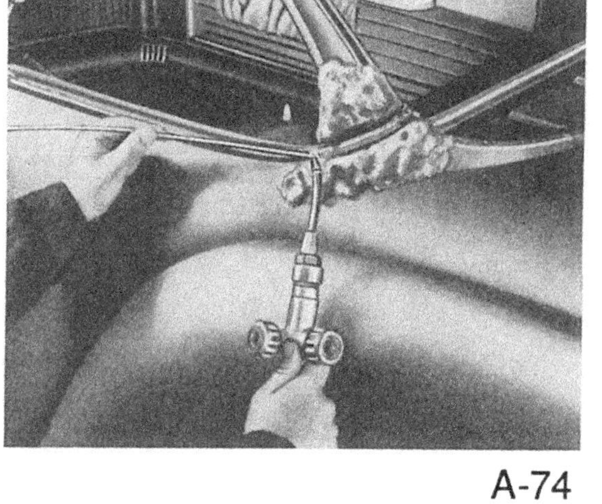

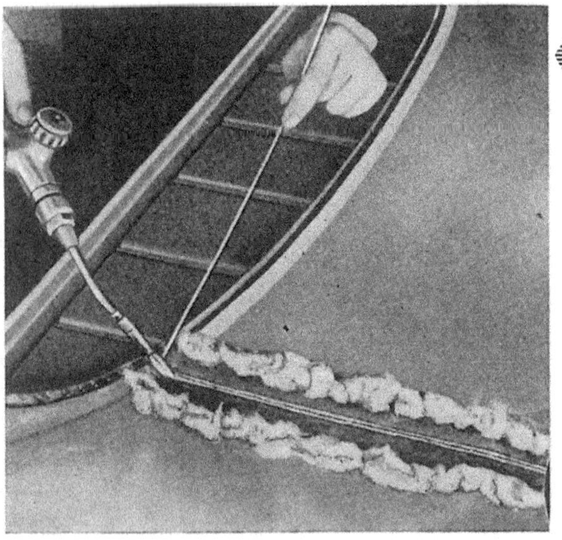

32 - Gas weld the quarter panel and top rear panel in the same manner with special attention to the following: Weld the panels first with short welds of 10 mm length spaced at intervals of 20 mm to avoid distortion of the sheet metal. When this operation is completed the joint can be finish welded over the entire length.

Note:

When tacking the panels together take constant care to ensure that no waviness or distortion appears. If this should happen it must be removed immediately, before the weld seams have cooled. This is the only way to ensure a perfect gas welded joint between the panels.

33 - Grind all spot and gas welded seams smooth.

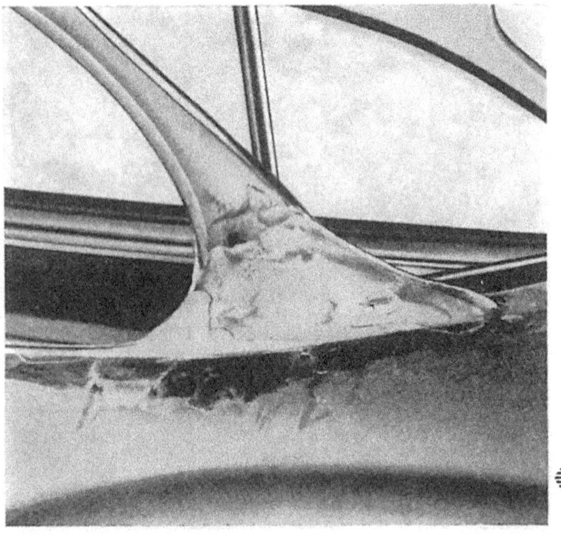

34 - The uneveness caused by the welding on top rear panel, roof pillar, quarter window frame and the lock pillar is levelled out with solder. The sheet metal in the appropriate areas must be smooth and free from all traces of welding metal. The tinning compound, consisting of soldering powder and soldering solution is then applied. The compound must be warmed after application and the layer rubbed with a clean cloth. This wipes all the deposits off and leaves a thin layer of tin which forms a foundation for the soldering which follows.

35 - Apply the solder, and with a steady even heat spread it as required.

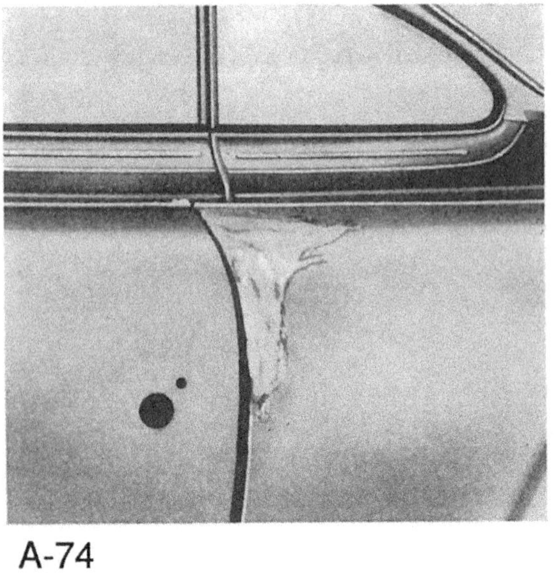

Take care that the door gap at the lock pillar is of uniform width over the full length. The picture shows the upper part of the door gap reduced by solder and below it, the original width.

36 - Grind all reworked places smooth and prepare the vehicle for painting.

Spot Welding Paste and Paint

A - Spot welding paste

In order to obtain water-tight joints, the parts of metal panels which are to be welded are given a coating of sealing paste which is not affected by welding.

This standard manufacturing procedure should also be adopted for bodywork repairs.

The paste can be obtained direct from the manufacturers:

Manufacturer	Designation
Bonaval-Werk, Bonn, Germany, Brühler Straße 2—20	Spot welding paste 59 852 or 60 506
Teroson-Werke G.m.b.H., Heidelberg, Hans-Bunte-Straße 4, Germany	Spot welding paste 2257

The paste is applied before welding commences. Before and after welding the paste should not run. At the joints it should form a film which adheres firmly, prevents corrosion and renders the joint water-tight. It must also remain unaffected by degreasing agents and subsequent painting.

Paste which burns during spot welding should cease to do so as soon as the heat is removed.

The paste should only be employed where water-tight seams are stipulated and where it is impossible or difficult to apply sealing compound after welding.

B - Spot welding paint

Spot welding paint is used to prevent corrosion in hollow parts which cannot be painted after welding due to inaccessibility.

Manufacturer	Designation
Teroson-Werke G.m.b.H., Heidelberg, Hans-Bunte-Straße 4, Germany	Spot welding paint 2273

Before welding, spray or paint all components with spot welding paint.

General

The use of the above products does not simplify the welding process. It may be necessary to step up the welding current, depending on the thickness of the coat applied.

Superfluous material must, however, be removed before the part is given a coat of paint which must stand up to the conditions stipulated for the outside finish. Neither welding paste nor paint can be considered as a substitute for primer to which a top coat can be applied.

Templates for Karmann Ghia Coupé

The following body templates are available either in a set or individually for the Karmann Ghia Coupé. They are supplied by Messrs. Wilhelm Karmann GmbH., Osnabrück, Martinistraße 59.

Template for windshield

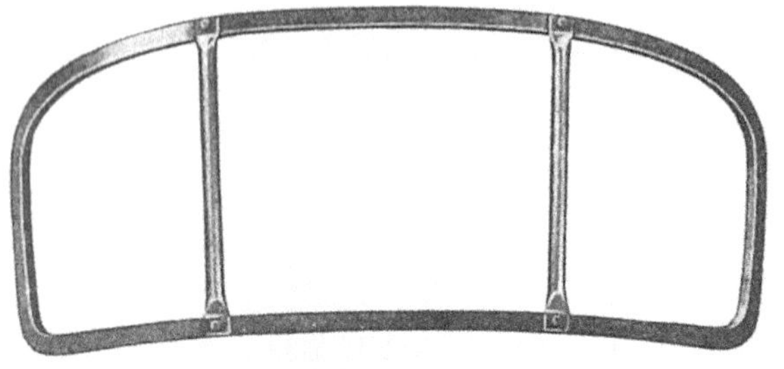

Template for rear view window

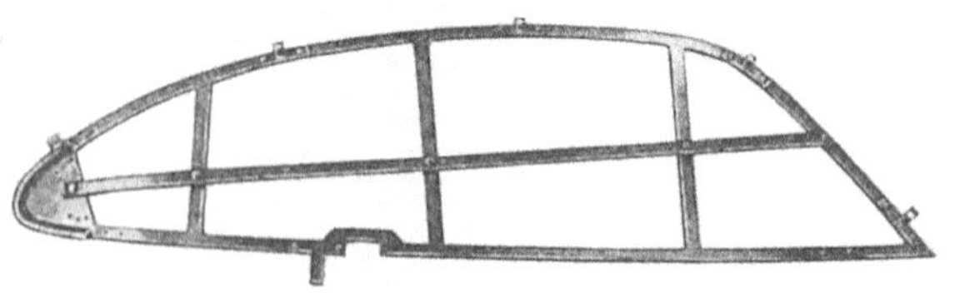

Template for quarter window, right

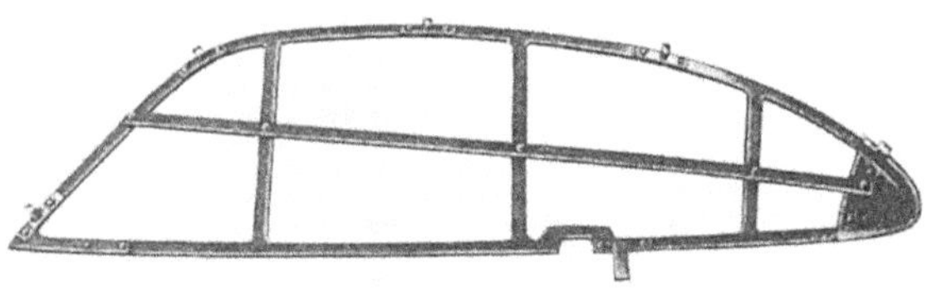

Template for quarter window, left

These templates are essential when replacing the roof, individual roof side members, rear quarter panels, or front and rear portions, in so far as any of the windows are affected by these operations. Moreover, the window openings can be checked for proper measurement by means of these sturdy templates which are resistant to distortion.

All questions as regards supply and prices should be addresed to the above-mentioned firm.

Body Repair Trolley for Passenger Car Bodies

Messrs. Wilhelm Karmann GmbH. produce a body repair trolley which can be used for the bodies of all VW 1200 and Ghia models.

The rigid tubular construction of the body repair trolley with additional reinforcement is distortion-resistant even when subjected to considerable stress. Its range of application is as follows:

1 - Measuring the body

Mounting and measuring points (A, B, C, and D), as well as welded on contact gauges (E and F) for the body sills make it possible to check the body accurately for proper alignment.

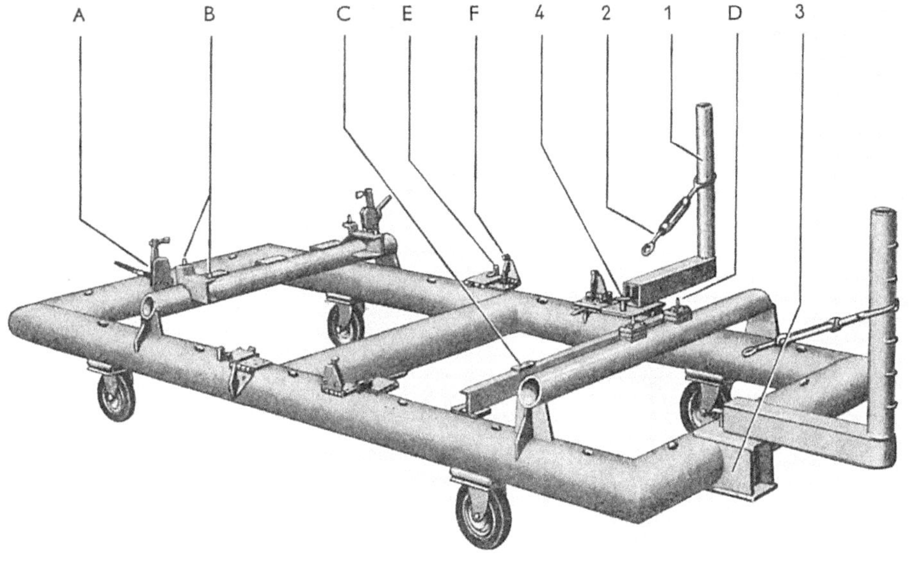

2 - Straightening of body

Stanchions (1) and turnbuckles (2) which can be attached to the trolley frame at various points by means of locking pins (4) inserted at the stanchion brackets (3) allow panel beating operations and straightening to be carried out on the body in transverse and longitudinal directions in conjunction with a number of adaptors. As the body is bolted to the trolley and also held by clamps it cannot move out of position.

For example, on Ghia models, damaged front and rear side panels can be pulled out to the proper body contour and repaired in this condition. Hinge pillars can be straightened by means of the turnbuckles (2) on the stanchions (1). Dented front and rear ends can be straightened with said tools where appropriate.

3 - Replacing larger body parts

The complete side, front or rear end of a body can be renewed without involving dimensional errors by using body sill templates that can be bolted to the trolley.

The body repair trolley has rubber wheels for easy transport.

The trolley is delivered with

2 stanchions	2 locking pins
2 stanchion brackets	2 turnbuckles

The additional parts listed below can also be ordered from the firm Karmann:

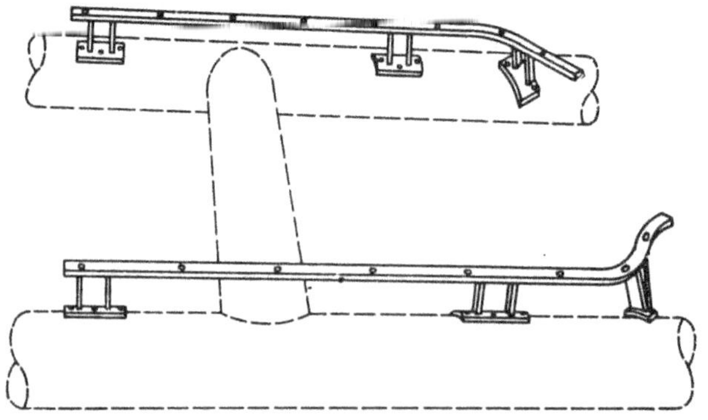

2 body sill templates
for VW Sedan and Convertible

2 body sill templates
for Karmann Ghia models

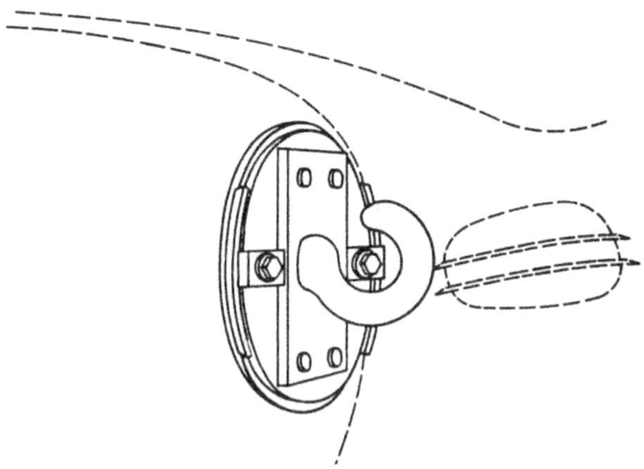

1 insert for headlight opening
(Karmann Ghia models)

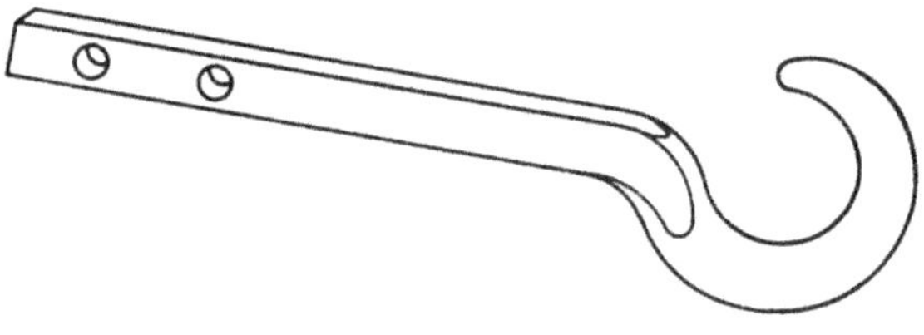

1 attachment for front and rear
bumper brackets

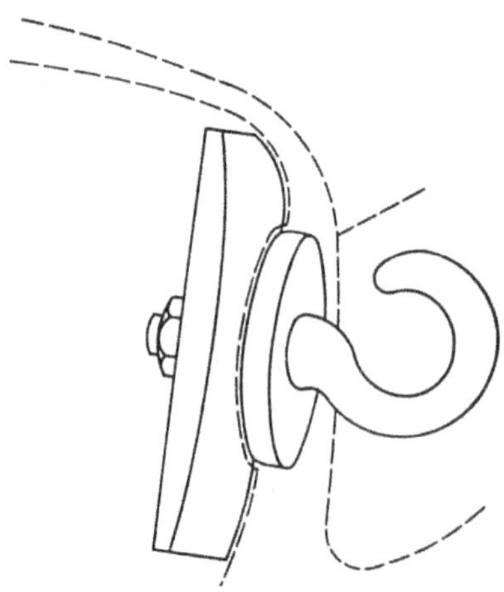

1 insert for tail light
(Karmann Ghia models)

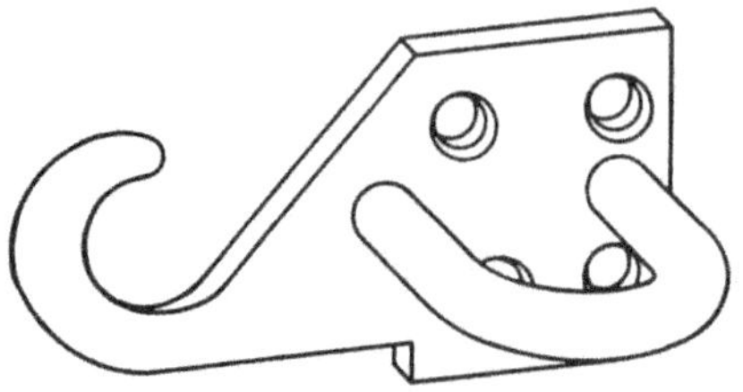

1 attachement for hinge pillar —
upper hinge

Contents: A Body—Karmann Ghia Convertible

Assembly Work

Replacement of Body Parts

THIS PAGE INTENTIONALLY
LEFT BLANK

General

Apart from the top and minor points described here, the construction, shape and trim of the Karmann Ghia Convertible is identical to that of the Coupé.

To compensate for the slightly reduced stability **when the top is open** the body of the Karmann Ghia Convertible is reinforced at the following places:

a - at the hinge pillar upper sections by means of stiffeners welded into the windscreen frame;

b - at the side members by welded reinforcements;

c - at the lock pillars by the main hinge brackets and their supports;

d - at the quarter panels by welded-on reinforcement plates;

e - at the upper part of top compartment and quarter panels by welded-in sheet metal stiffeners.

Variations from the Karmann Ghia Coupé

1 - The rear window is 730 × 320 mm and made of flexible plastic material.

2 - No quarter windows.

3 - The interior lamp is under the instrument panel.

THIS PAGE INTENTIONALLY
LEFT BLANK

Door and Striker Plate Adjustment

When the door and striker plate have been installed the position of the door in the door opening should be checked with the top open as follows:

1 - Alignment of door and side panel

2 - Alignment of door and quarter panel

3 - Uniform clearance between door and lock pillar

If these positions are not correct proceed as follows:

Re 1 - Remove striker plate and loosen door hinges at the hinge pillar only. Move the door in or out as required and re-tighten hinge screws.

Re 2 - Install striker plate and adjust. The striker plate is properly adjusted if:

a - The door is correctly aligned with the quarter panel;

b - No play can be felt between lock and striker plate when forcing the door handle in and out. To align the door and quarter panel move the striker plate in or out.

Play will exist between lock and striker plate if the upper part of the latch does not rest against the striker plate.

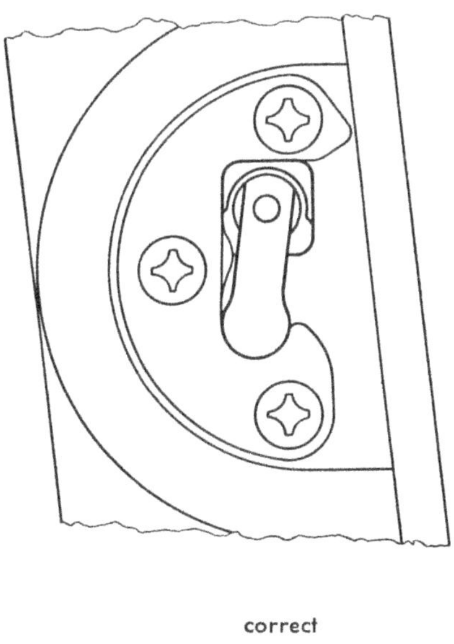

correct

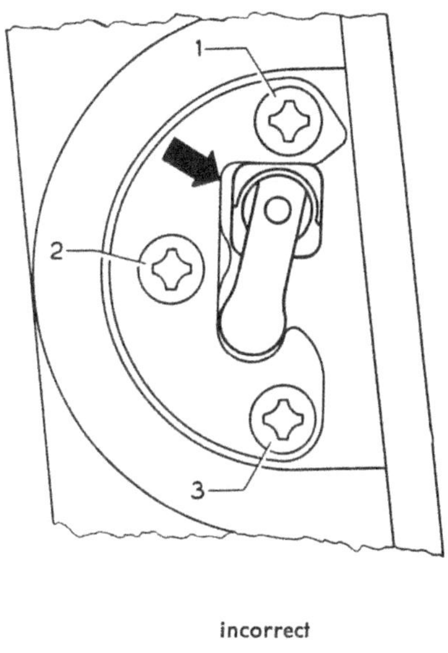

incorrect

In this case loosen screws 1 and 2 slightly and move the striker plate out at the top. Tighten screws and check play.

If the door springs back when slammed to, loosen screws 1 and 2 again and move the striker in slightly at the top.

Re 3 - When the gap between door and lock pillar widens towards the top instead of being uniform over the full length, loosen the body securing screws and remove them at points A and B.

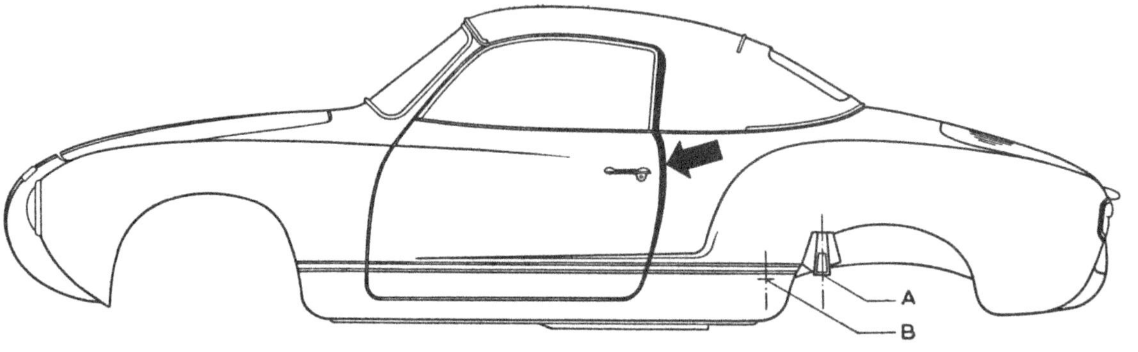

Insert 3 mm thick hard rubber packings left and right at the points indicated. Re-tighten the screws which were loosened to a torque of 3 mkg. If the door gap is still not correct, insert thicker packings.

Adjustment of Door Window Glass

One essential requirement for door glass adjustment is the correct fit of the door in the door opening.

Vertical Adjustment

1 - After removing the window regulator handle, inner door handle, door trim panel and plastic sheet, loosen adjusting screw lock nut.

2 - Regulate the door glass by turning the adjusting screw to left or right.

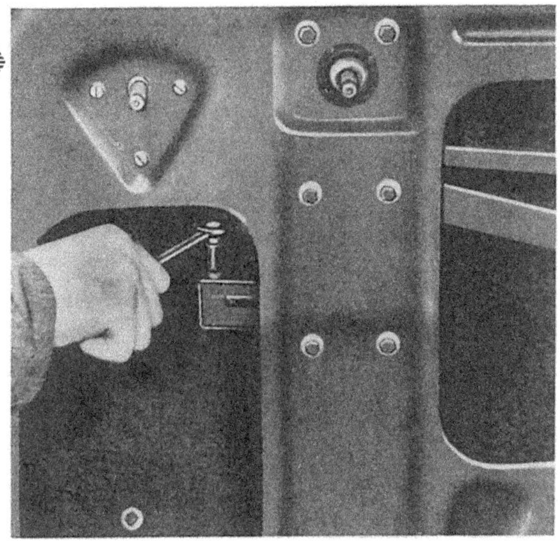

Turning to left — lowers glass
Turning to right — raises glass

3 - With top and door closed, check correct vertical adjustment of door glass.

4 - Secure adjusting screw by tightening lock nut.

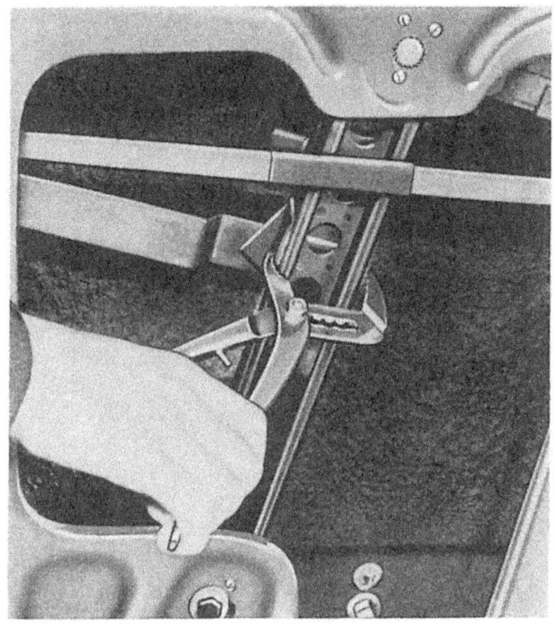

Longitudinal Adjustment

1 - Check play between window lift channel runner and the guide rail. If necessary reduce the clearance by squeezing the runner together with a suitable pair of pliers (Waterpump pliers).

2 - Loosen four window lift channel screws.

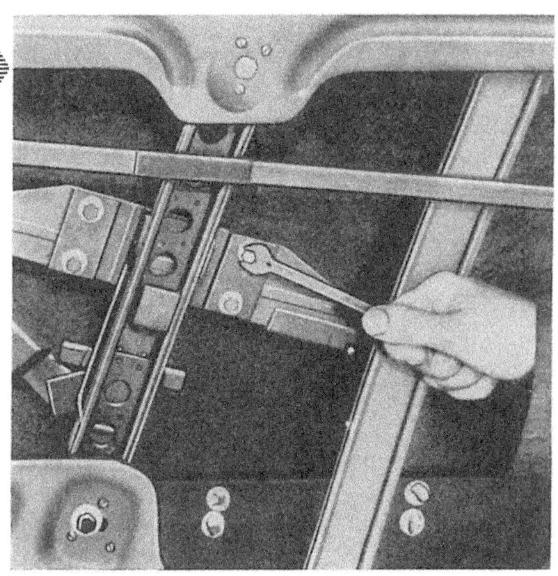

3 - Move door glass with lift channel to the front or rear as required.

4 - Tighten screws.

Side Adjustment

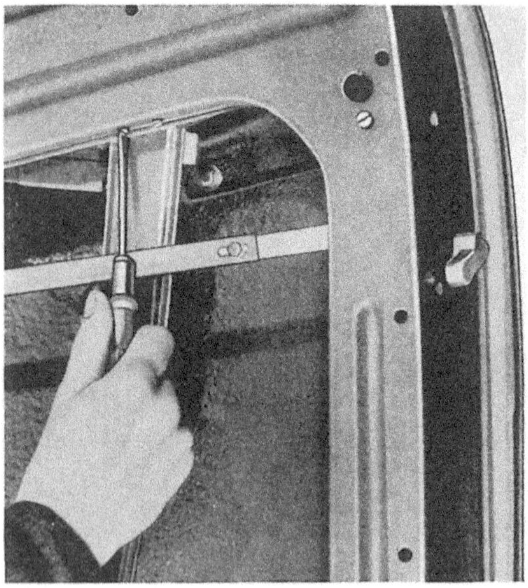

1 - Check door glass for side rock. If necessary loosen the top screws of the window roller bracket and with the window in the fully raised position press the bracket hard against the roller and tighten screws.

2 - Loosen hexagon and countersunk head screws on guide rail.

3 - Correct the window glass position by screwing the 3 threaded pins at top and bottom of guide rail in or out.

The location for the adjustment of the door glass is given by the main roof frame weatherstrip at the top and the front and rear weatherstrips at the sides.

4 - Tighten hexagon and countersunk head screws.

Convertible Top

(Early)

General

The top is supported by the top linkage which is composed of two side metal roof frames and two hinged linkages joined together by wooden and tubular bows which give cross support. The whole assembly is supported by two main hinges mounted in brackets bolted to special pillars in the body.

The outer cover is made of special waterproof top material which is rubberised underneath. A rubberised hair pad, sewn into a linen sheet and secured to the top linkage, pads out and shapes the top in addition to providing insulation against heat and cold. The contours of the top are, where necessary, evened out by additional wadding. The underside of the top linkage is covered by a headlining of loosely woven material.

In the folded position the top is retained by two spring catches.

Lubrication

A few drops of oil should be applied to the hinge joints of the top linkage when required. Wipe the dirt and dust off the joints before oiling.

Care of the Roof

The proper treatment and care of the top has a large influence on its appearance and service life.

The top must always be perfectly dry before being lowered. When very dusty the top should be beaten out lightly and brushed thoroughly with a soft brush in line with the lay of the material, as the sharp dust particles can damage the material and cause friction marks.

Friction marks can also appear when the catches do not hold the opened top firmly in position. In this case the catches must be adjusted by loosening the lock nuts, screwing the catches further into their retainers and tightening the lock nuts again.

Never use fuel, spot removers or other solvents for removing spots as these fluids will destroy the rubber layer in the top cover, causing leaks and shortening the life of the top. Light surface marks can often be removed by rubbing gently with a crust of white bread or a soft white rubber (eraser).

Top Frame

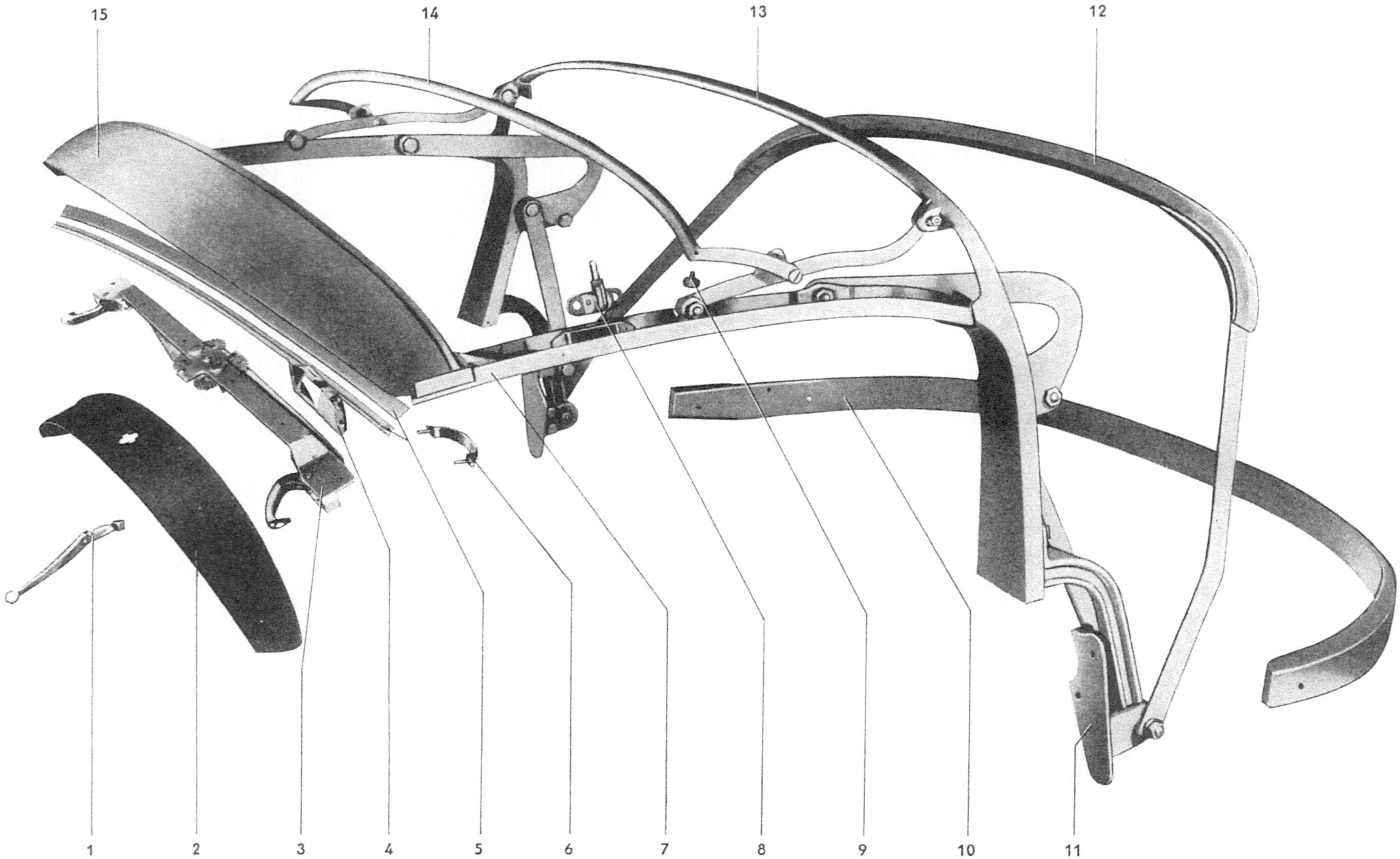

1 - Handle for Top Lock
2 - Cover Plate for Header
3 - Top Lock
4 - Cover Plate for Lock Hooks
5 - Weatherstrip

6 - Top Handle
7 - Roof Frame
8 - Top Retaining Catches
9 - Plug — Intermediate Bow
10 - Rear Body Bow

11 - Main Hinge
12 - Rear Bow
13 - Main Bow
14 - Intermediate Bow
15 - Header

The top should be washed only when it is exceptionally dirty. Only clear water, without any chemical solutions or additives should be used. Beat the top out lightly and brush thoroughly before washing. Add soap flakes to lukewarm water and beat into suds. Moisten the top with clear water, apply the thick suds and scrub with a soft brush, rubbing in one direction only. Rinse the top with clear water and if necessary, repeat the scrubbing with the soap suds. The final rinse must be continued until all traces of soap are removed and the water runs off quite clear. Be sure the top is thoroughly dry before lowering.

Washing with soap flakes should not be carried out more than once every six months.

After washing the top, rinse all traces of soap suds from the vehicle finish and dry off.

Top Assembly

The description of the Convertible top assembly serves as a guide for the sequence of operations. The work should, where possible, be carried out by a skilled man who is familiar with top construction. If two men can be employed, the task of fitting the top cover will be made considerably easier.

The assembly of individual parts of the top described in some paragraphs can naturally be carried out separately.

The tools required are listed in the "Body General" section — under the heading "Workshop Equipment".

In order to avoid paint damage it is recommended that the front and rear parts of the body are covered with sheets.

The sheet at the front should be secured about halfway up the windshield with adhesive tape so that any nails which fall down will not get underneath the sheet.

The corners of the windshield frame should also be covered with adhesive tape to prevent paint damage occurring if the pliers should slip when tensioning the top cover.

Installation of Top Frame

Note:

From 11th April 1964, Chassis No. 6 264 621, the lower part of the main bow and the main hinge are sheet metal pressings and not castings as they were formerly. This improves the elasticity and rigidity considerably.

At the same time other parts of the top frame have been changed. The part number is 141 871 025 as before.

The modified top frame, with and without cover, can be installed in earlier vehicles without difficulty. The following parts, however, are not interchangeable.

Designation	Part No. old	Part No. new
Main bow	141 871 801	141 871 801 A
Main bow, lower part, left/right	141 871 815/816	141 871 815 A/816 A
Stud on main bow for lower part	—	141 871 817
Guide bar, lower, left/right	141 871 831 A	141 871 831 A/832
Main hinge, left/right	141 871 851/852	141 871 851 A/852 A
Bolt for main hinge	141 871 855	141 871 855 A

Note

When a complete top frame or a main bow is replaced, the clearance between the lower part of the main bow and the main hinge support bracket may be reduced due to a combination of tolerances. This can also occur on the previous version. In this case, the edge of the support bracket (141 809 417/418) which is angled to 90° normally, should be folded completely over.

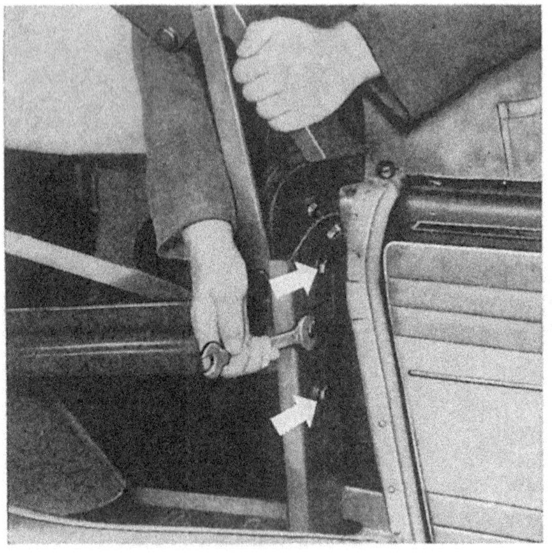

1 - Before commencing work on the top assembly remove front seats and rear quarter trim panels. Wind the door windows down.

2 - Place top frame in position and bolt the main hinges to the screwed plates on the support pillars. Pull the main bow outwards as far as possible.

3 - Extend the linkage to the front and secure the header to the windshield frame with the top lock. In this — normally closed — condition the clearance between the lower edge of the main bow and the lock pillar should be 6 mm. If this is not so, loosen the frame and move up or down as required.

a — 6 mm (0.236″)

4 - In the closed position the clearance between the uncovered header and the windshield frame should be:

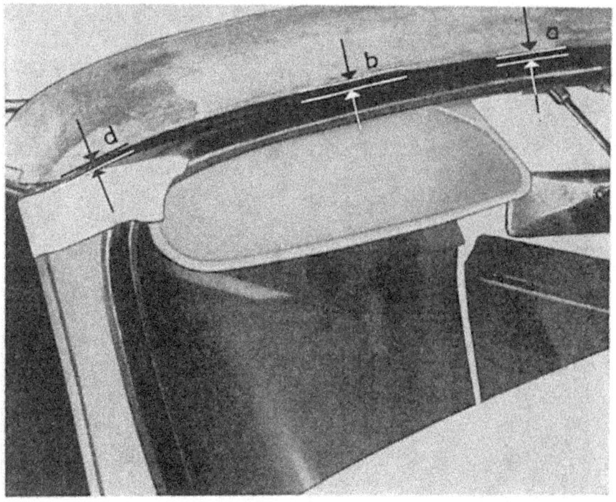

a - in the center 7 mm (0.275")
b - in the area of the locking hooks 5 mm (0.196")
c - tapered to 2 mm (0.078") at the ends

This clearance is essential in order to leave sufficient space for the rubber weatherstrip and top cover.

Note:

If the clearance is too small the front moulded rubber seal of the header can lift or press out the windshield weatherstrip at the top and cause leaks to appear.

Insufficient clearance between header and windshield frame can also create excessive tension at the top locks and lead to distortion of the windshield frame.

In these cases it is possible to rasp the header off slightly.

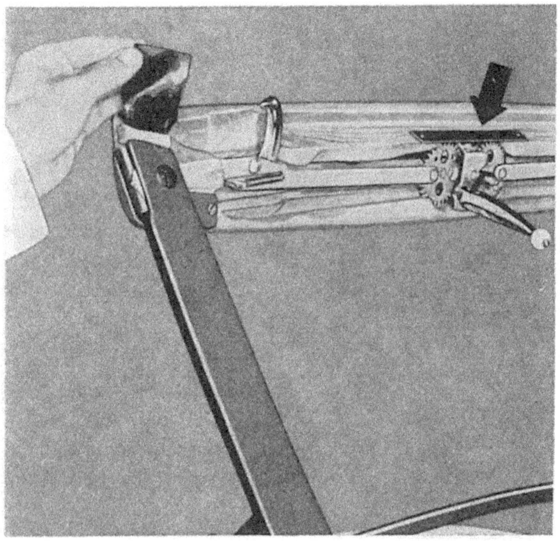

5 - Tack a 1 mm thick cardboard strip, 200 mm (7.8") long and 15 mm (0.6") wide under the header in the area of the lock mechanism (Arrow).

Cement two pieces of foam rubber underneath the header at the frame ends and finally cement two suitably shaped pieces of top cover material over roof frame, foam rubber pieces and header.

This preparatory work is necessary in order to achieve a uniform surface between roof frame and header and so avoid leaks at this point.

6 - Make a groove at each end of the header with a rasp. These grooves accommodate the top cover seams which are nailed on later.

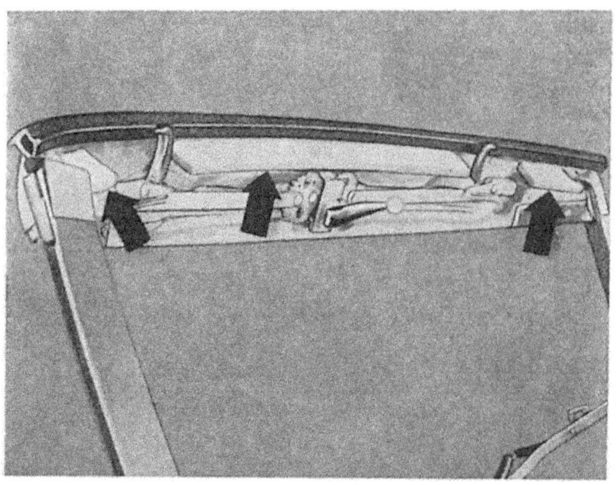

7 - Coat the underside of the header with adhesive and cement suitably cut lengths of foam rubber strip (Part No. 141 871 605) in between and to the left and right of the locking hooks (Arrows).

Cement the weatherstrip (Part No. 141 871 607) to the front edge of the header. The overhanging ends should be cut off exactly at the joint between roof frame and header.

8 - Cut a piece of headlining material roughly 100 mm × 1250 mm (3.9″ × 49.2″) and a piece of top material 180 mm × 1400 mm (7″ × 55″) of colors to match the top and headlining of the vehicle.

9 - Coat the header rear edge up to the recess on the top with adhesive and cement the headlining strip into position. Cut off the surplus material round the locking hooks and at the edge of the recess.

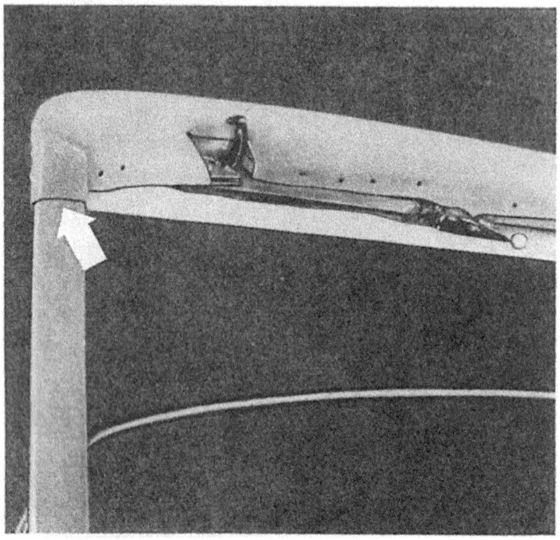

10 - Tack top material strip underneath the header and cut the strip out to fit round the locking hooks. Coat the roof frame ends with adhesive and cement the strip into position (Arrow).

> **Important**
>
> To avoid rust formation use only brass pins and tacks to secure the top cover and the webbing strips.

Note:

12 mm Brass pins Part No. 151 871 471
14 mm Tacks Part No. 151 871 473

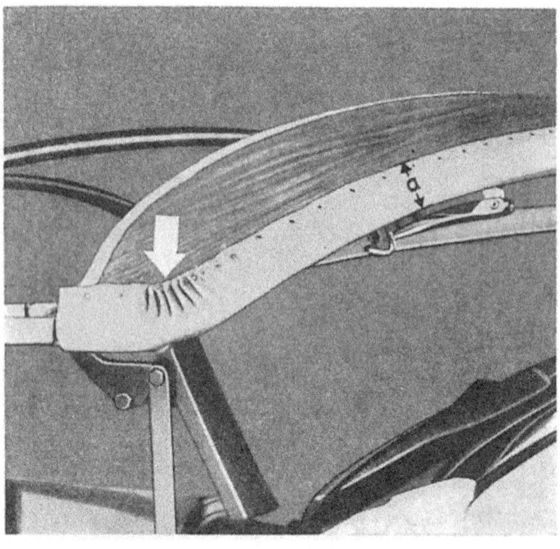

a = 40 mm (1.574″)

11 - Lightly tension the top cover strip and tack to the top of the header. Close the top and nail the strip at intervals of 40 mm whilst keeping a slight tension on the material. To ensure uniform tension it is advisable to nail evenly in both directions from the center outwards.

To ensure a better fit at the ends of the header lay the material in small folds round the curves and then tack. Do not cut the material or make the folds too large as this may cause leaks.

Trim the surplus material off above the line of tacks.

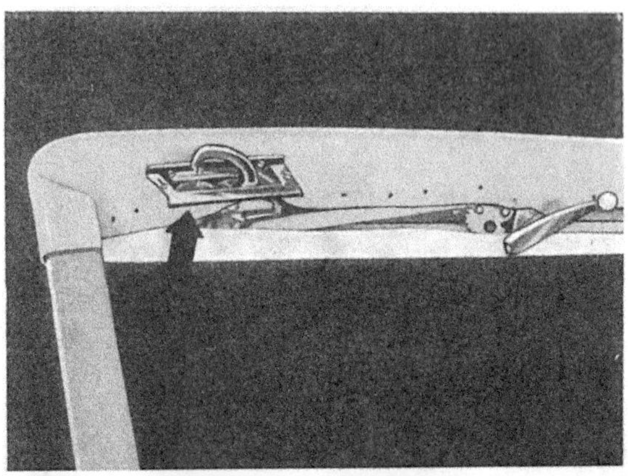

12 - Screw the lock hook cover plate into position, after checking that the hooks have sufficient working clearance. It is advisable therefore, *to tighten the side screws first and finally* the front ones.

13 - Coat the intermediate bow rubber plugs with adhesive and insert them in the appropriate holes.

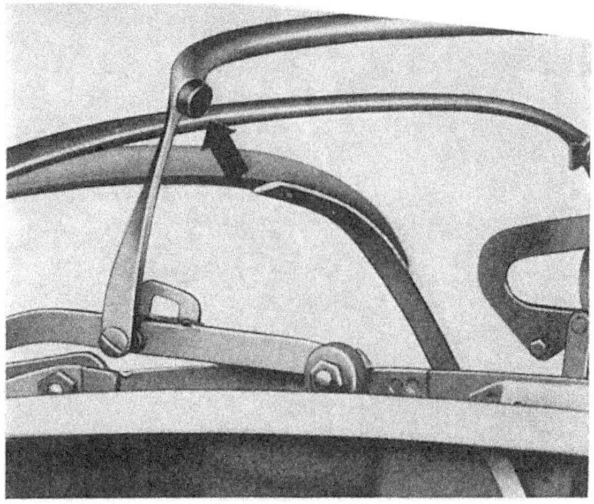

Note:

The plugs are intended to stop rattling noises. Check that they are securely seated.

14 - Lay the top frame forward and secure with the lock. Adjust rear bow to measurements.

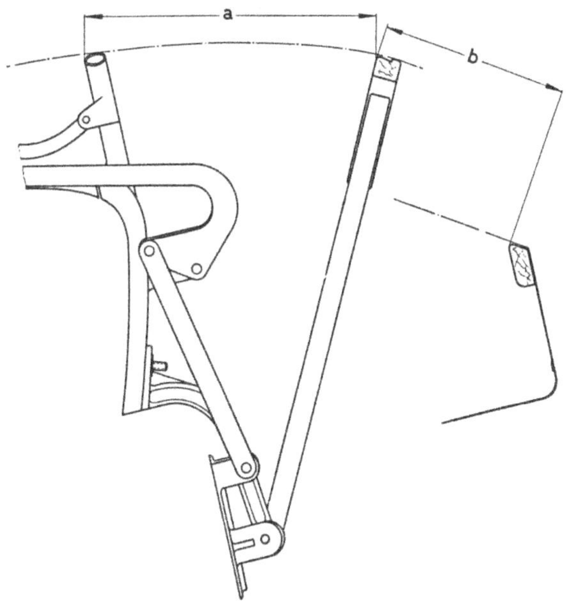

> **Important**
>
> These measurements should be strictly adhered to as they govern the satisfactory padding and folding of the top.

a = 350 mm ± 5 mm (13.78 ± 0.2″)
b = 469 mm ± 5 mm (18.5 ± 0.2″)

15 - Cut the webbing strips to size and tack them to the rear body bow 420 mm to left and right of the center line.

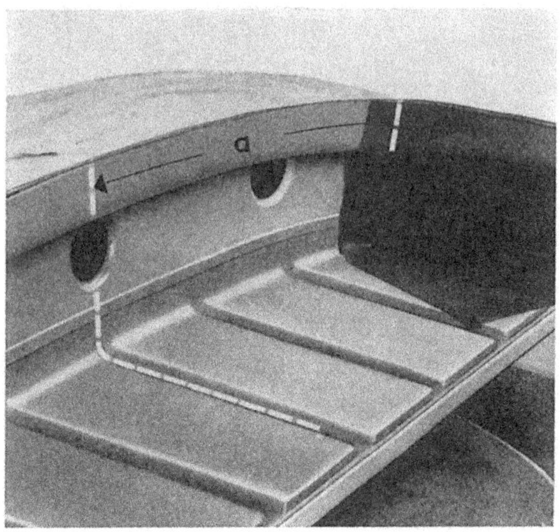

a = 420 mm (16.53″)

16 - The main bow has two grooves on the top to take the webbing strips. The grooves should be coated at top and bottom with adhesive.

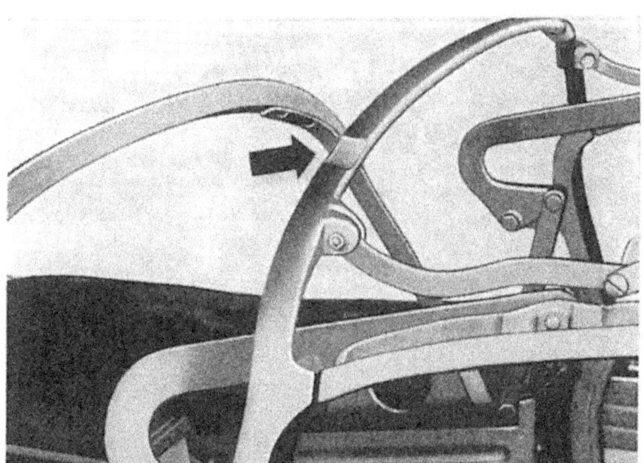

17 - Pull the webbing strips hard to the front, cement them into the grooves and sew securely.

a = 420 mm (16.53")

18 - Mark the width of the webbing strips on the rear bow. Lay the bow back slightly and chisel or rasp out the places marked to a depth equal to the thickness of the webbing strips so that the strips do not stand proud.

19 - Position rear bow to the correct measurement and tack on the webbing strips.

Headlining Installation

The headlining — without rear portion — is supplied as a spare part completely finished with the support strips which are used to secure it to the roof bows. The rear portion with the rear window is sewn to the main headlining after the top cover has been installed.

To ensure that the headlining fits properly it should be secured to the individual bows in the following order with the **roof closed**:

a - Main bow
b - Rear bow
c - Header
d - Intermediate bow

It is secured to the rear body bow at a later stage.

The operation is carried out as follows:

1 - Coat the top and front edges of the main bow with adhesive.

2 - Pull the headlining up by the appropriate support strip until the seam in the lining contacts the underside of the main bow. Cement the lining in position, cut out the openings for the webbing strips and cut off the surplus headlining material.

3 - Pull headlining to the front, tension slightly and secure to the header with one tack each side. To avoid the possibility of tearing the material it is advisable to insert the tacks in the longitudinal seams (Arrow).

4 - Coat rear and top edges of rear bow up to the recess with adhesive. Pull supporting strip up to seam and cement to rear bow under tension. Insert a tack each side for additional security. Cut off surplus headlining material.

5 - Tension the headlining uniformly and tack to the recessed surface on the top of the header.

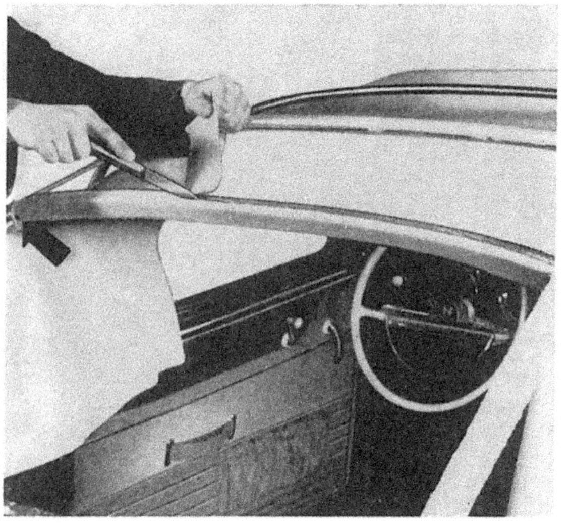

6 - Coat the top of the intermediate bow with adhesive. Pull the support strip up to the seam and cement round the bow. Cut off surplus material.

7 - After completion of these operations check from inside that the headlining is taut, free from folds and that the seams of the supporting strips are straight. If necessary, correct on the appropriate bows.

The surplus headlining material at the header can now be cut off and the ends of the supporting strips on the tubular bows sewn.

a = 90 mm (3.54")

8 - Cut 20 mm (0.78") wide openings in the supporting strips 90 mm (3.54") to the left and right of the webbing strips.

Insert two 550 mm (21.6") long rubber bands through these holes and sew them together.

These rubber bands pull the intermediate bow back when the roof is opened and thus ensure proper folding.

9 - A 2.10 m (82.6") length of beading, in the same color as the headlining, is supplied with the spare headlining.

This beading is tacked to the lower rear edge of the rear bow between the webbing strips so that the bead protrudes slightly **below** the rear bow.

The ends, which should be of equal length, remain hanging down for the time being.

10 - A cardboard strip 1 mm thick, 15 mm (0.6") wide and of suitable length is tacked over the beading and holds it straight and firm against the rear bow.

Linen Sheet and Rubberised Hair Padding Installation

General

The rubberised hair pad upholsters and shapes the Convertible top and also provides good insulation. It consists of three parts, the roof portion (a) and the two rear portions (b).

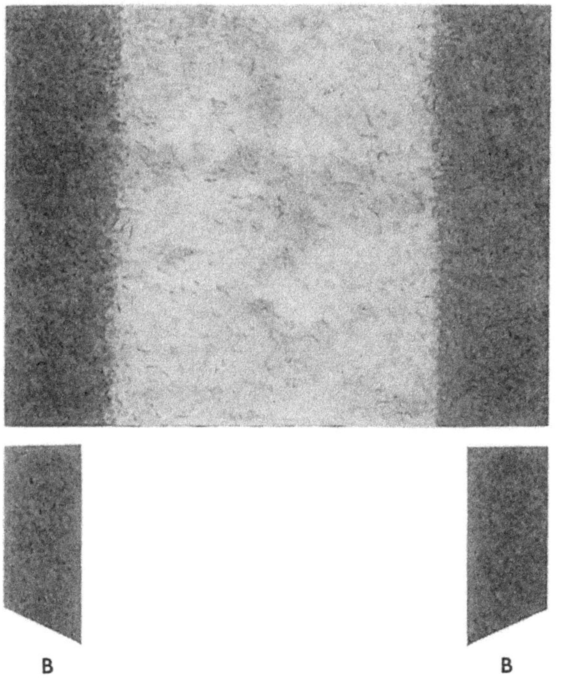

The rubberised hair padding is sewn in and onto the appropriately shaped pieces of linen sheet. The linen sheets are then secured to the roof frame.

Two different types of material are used:

a - a thick, light-colored linen material

b - a thinner, black lining material, known to the trade as "Nessel or Molton Cloth".

The linen material serves as the foundation to which the rubberised hair pads are sewn and the lining material is used to enclose the pads.

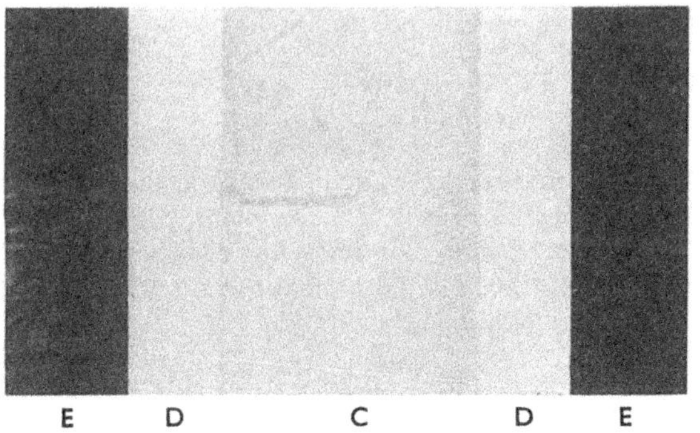

The linen and lining material is supplied in three separate parts in the same manner as the rubberised hair padding, namely the large roof portion and the two rear portions which each consist of one piece of linen and one of lining material. The roof portion comprises one linen sheet (C), two top material pieces (D) and two lining material pieces (E).

The two pieces of top material sewn to left and right of the linen sheet are for appearances sake only and merely prevent the linen material from showing at the sides when the roof is laid back.

All the rubberised hair parts and linen material pieces are combined in an installation set and supplied as a spare part under the designation "Top Padding" (Part No 141 871 057).

1 - Tack the two black lining material pieces to the rear bow and rear body bow to the correct measurements. Cut the material out at the webbing strips. Do not cut off the surplus lining material on the inside.

a - 180 mm (7″)
b - 200 mm (7.8″)

2 - Tack the linen material pieces to the rear bow and rear body bow in the same manner. Fold the edges of the linen and lining material back at the outside. Sew the linen sheet securely to the webbing strips.

3 - Lay the rubberised hair rear portions in position and sew to the linen sheet at top and sides with small firm stitches. The pads must not cover the rear bow or be sewn on at the bottom near the body bow.

4 - Fold the surplus lining material over the rubberised hair padding, tack to the rear and body bows and sew the outer edges. Finally sew padding and lining material together with large stitches. Trim off surplus material at top and bottom.

5 - Lay the roof portion of the linen material in position, taking care that the black, rubberised side of the top cover material faces upwards.

6 - Tack the linen sheet to the rear bow first, pull firmly to the front and tack to the header.

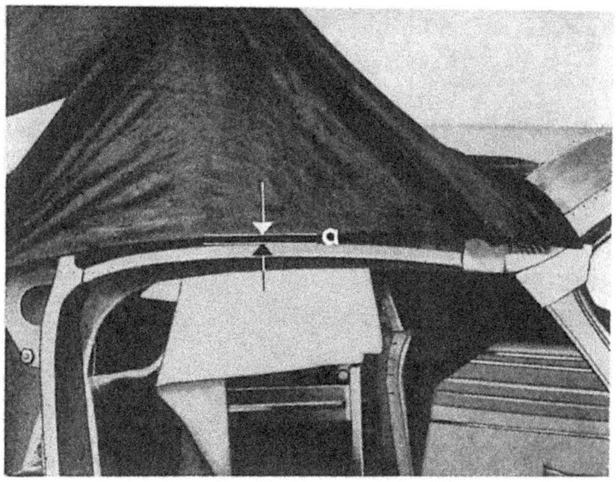

When tacking, take care that the outer sides of the top cover strip are roughly 10 mm above the roof frame (measured in the center).

a - 10 mm (0.4")

Correct if necessary.

7 - Tack the linen sheet in the recess on the rear bow and to the header. To ensure uniform tension in the cross direction as well it is advisable to tack from both sides towards the center.

As the linen material is always cut fuller it is necessary to cut wedge shaped pieces out of the material at the header and rear bow as the formation of folds will cause ridges to appear.

Cut off surplus linen material (see dotted line).

Near the individual bows cut off the surplus material of the longitudinal seams as far as possible without damaging the stitch threads. This will stop the formation of ridges which will later be visible on the top cover.

8 - Securely sew the cuts in the linen material and loop stitch the top cover strips in the area between rear intermediate bows.

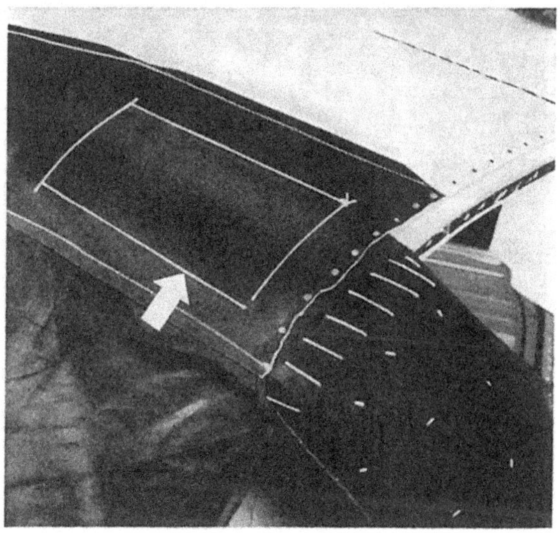

Note:

Loop stitches are large, loose stitches made with a thick thread under which the padding material is pushed. They are intended to hold the loose padding material in position.

9 - Push the padding material — horse hair if possible — under the loop stitches in uniform thickness to level out the depression between the bows.

> **Important!**
>
> The padding controls the contour of the top at the rear. If the padding is too thick or too thin the top will be uneven.

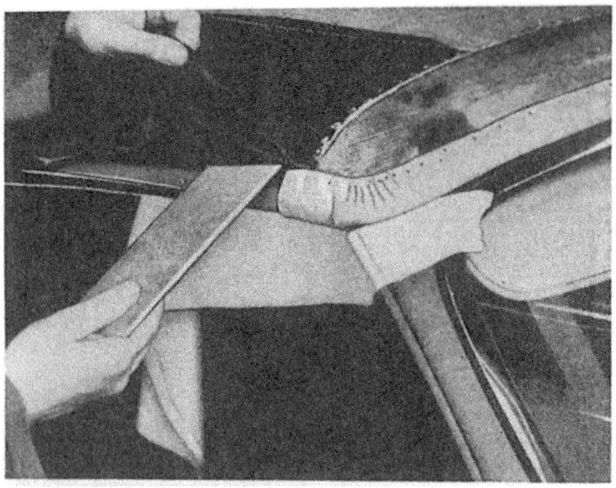

10 - Before sewing the rubberised hair into position insert a piece of cardboard 100 mm wide and of suitable length between the top cover and the headlining at the header.

This cardboard, which is removed when the operation is completed, prevents the headlining being caught when the rubberised hair is sewn to the linen sheet.

11 - Place the roof portion of the rubberised hair padding in position. The thickened edges should be on the left and right outer sides.

Sew the rubberised hair padding firmly to the linen sheet close to the rear bow.

Pull the padding almost up to the rear edge of the header and sew to the linen sheet. Finally sew the padding to the linen sheet longitudinally with rows of large stitches 100 mm apart.

Sew the side outer edges firmly to the top cover strips.

12 - To level out the space between the header and the rubberised hair padding, tack a 200 mm (7.8″) wide layer of upholstering wadding which has been thinned out at the edges, to the header.

Cut the corners of the wadding to suit the roof shape and tack high enough to avoid the appearance of leaks. The top cover material *tacked to the header must butt up against the* wadding.

13 - Lay one flap of the black lining material over the rubberised hair padding and tack in the recess in the rear bow. At the header end tack the lining over the top material strip. Treat the other flap in the same manner.

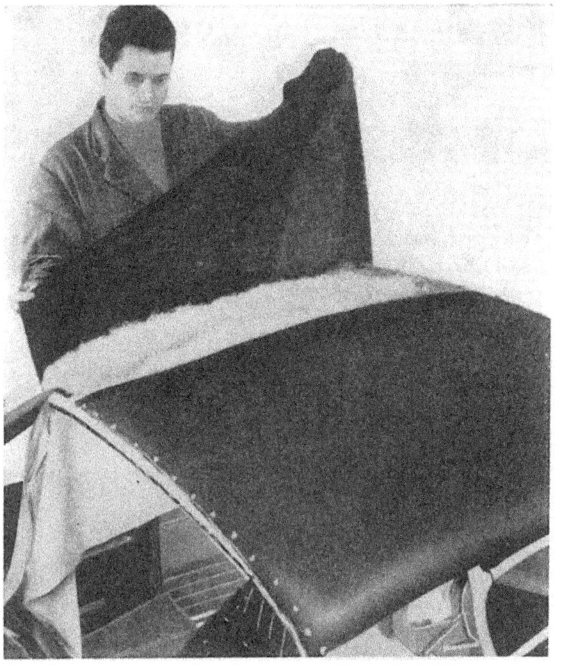

14 - Cut off surplus lining material and sew the flap joint together with small firm stitches.

15 - To give the padding additional support at the front it is advisable to sew the lining and padding together about 100 mm (4") from the header.

16 - Remove the cardboard strip which was inserted before fitting the padding.

Installing the Top Cover

The top cover is supplied in various colors and the choice must be stated when placing the order. The beading for the body bow is supplied with the top cover in the matching color.

The rear window (Part No. 141 845 501) is complete with top cover material and the rear portion of the headlining. The rear portion of the headlining is sewn to the main headlining after the top cover has been completely fitted.

Apart from a few operations the fitting of the top cover is carried out with the top closed.

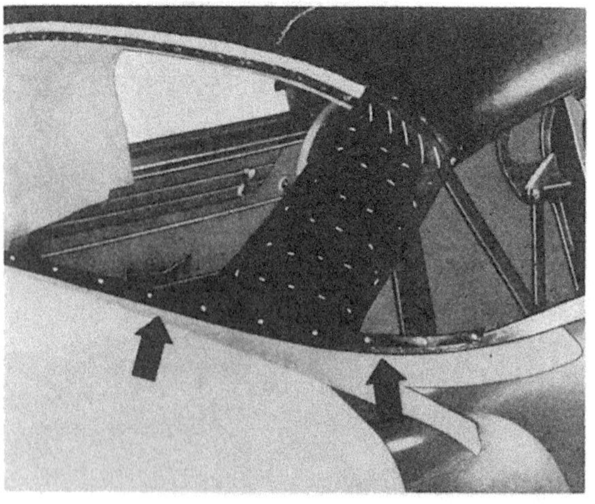

1 - Tack a filler strip to the body bow to reduce the comparatively large clearance between the bow and the body panel. A suitably shaped piece of top cover material can be used for this purpose.

2 - Coat the filler strip with sealing compound.

3 - Tack the beading to the body bow under side tension so that the bead rests in the body groove against the bow.

For the time being do not tack the beading at the sides, roughly from the ends of the rubberised padding.

4 - Attach the headlining to the main bow on each side with two chrome-plated screws and protective washers.

5 - Attach the ends of the headlining beading directly under the webbing bands at the center of the bow width with two tacks each and to the bottom panel of the top compartment with a hollow rivet.

The bead must face towards the center of the vehicle and not be damaged in any way.

6 - Tack the headlining to the beading at the rear. The sewing takes place when the top cover fitting is completed.

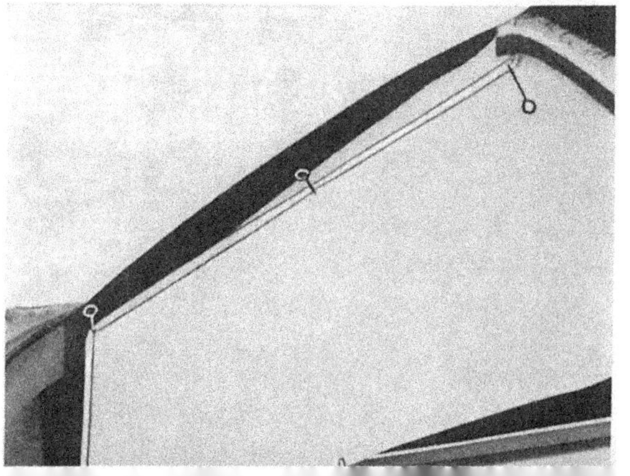

7 - Place top cover in position and secure to th lower ends of the main bow with a chrome plated screw and protective washer on eac side.

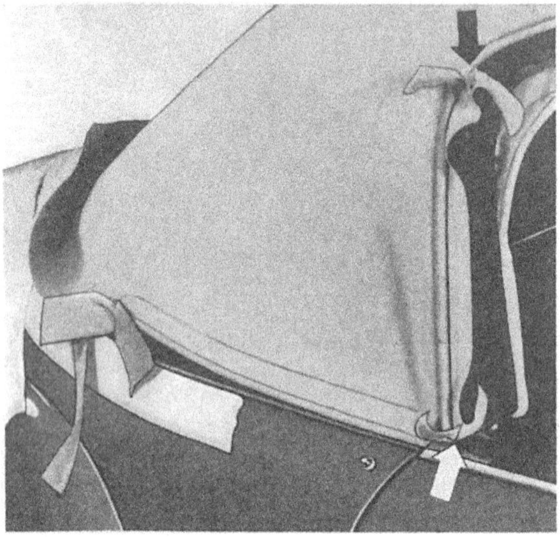

8 - Part the surplus top cover material at the side of the main bow for about 50 mm at top an bottom.

Note:

Parting in this instance means separating the rubbe layer from the material layer of the top cover t reduce the thickness of the material, thus avoidin the creation of noticeable ridges.

To facilitate this operation and avoid damaging th material it is advisable to slightly dampen the are to be parted with turpentine. Otherwise a shar saddler's or parting knife must be used.

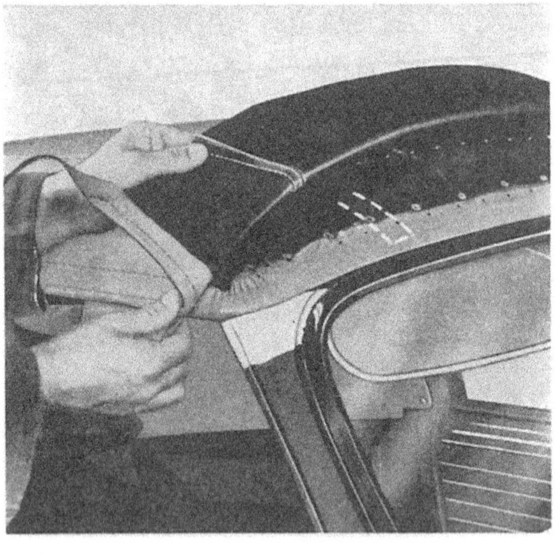

9 - Cut the rubber layer of the parted material o up to the main bow edge at top and bottom.

10 - To prevent the longitudinal seams in the to cover standing proud of the surface it i necessary to cut the lining and wadding out a the appropriate places and chisel out th header slightly.

To locate the exact position for the cuts, la the top cover beads on the lower edge of th top frame and mark off the seams.

11 - *Chisel a groove in the header about 2 mm dee* and extending down to the front weatherstri

12 - Apply sealing tape to the joint between the tacked-on top cover strip and the lining material, particularly at the corners. This serves as an additional seal against the possible ingress of water.

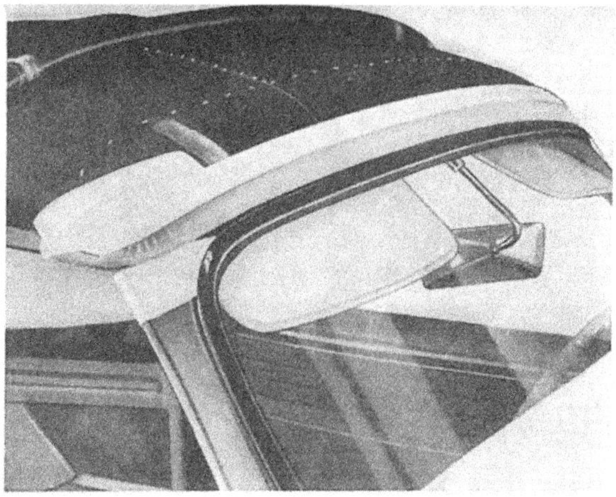

13 - Attach the top cover to the body bow under tension, with three tacks each side.

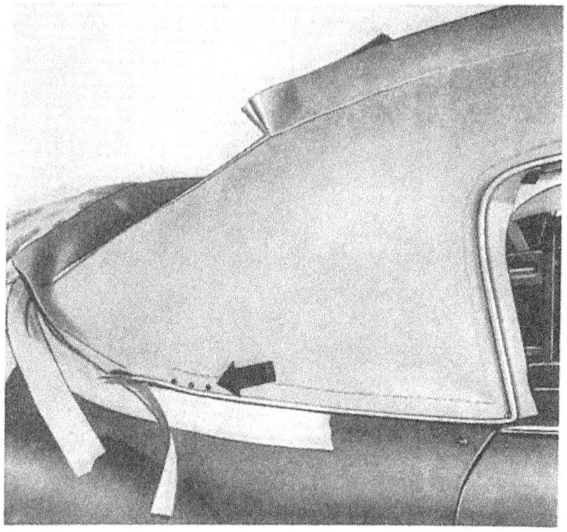

14 - Pull the top cover tightly to the rear and attach to the body bow with two tacks.

Drive the tacks through the seams to avoid tearing the material.

15 - Pull the top cover forward and tack to header at top and sides under tension and free of folds. The top tacks should also be driven through the seams.

When tensioning the top cover take care that the cut in material over the rear window rests exactly in the recess in the rear bow. Correct if necessary.

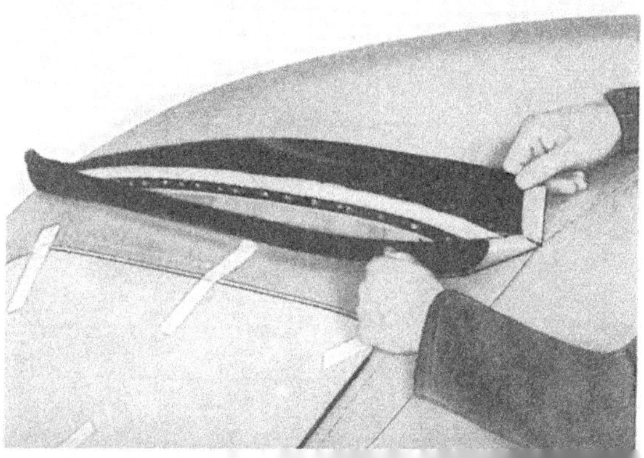

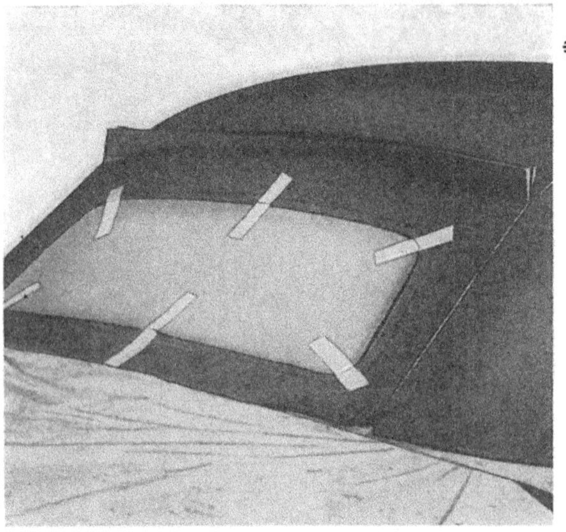

16 - Pull the top cover taut and tack to rear bow and to body bow.

Check that the rear window is horizontally central and correct if necessary by loosening pulling in the desired direction and retacking.

17 - Under even tension and free of folds tack the top cover in the recess of the rear bow. Tack the lower portion first and then the upper portion over it so that the water running of the roof will not enter the cut. Cut off surplus material.

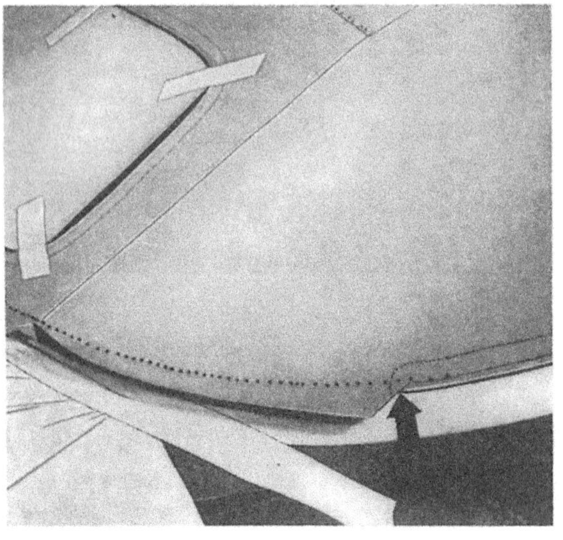

18 - Coat both ends of the row of tacks with rubber solution — particularly at the seams — taking care that the coated surface is not wider than the trim moulding which is nailed on later.

19 - Cut off the ends of the body beading at the point where the top cover beading ends (see arrow). Part the surplus ends of the top cover beading up to the seam, cut off rubber layer and fold the material part round the body beading to form a smooth joint.

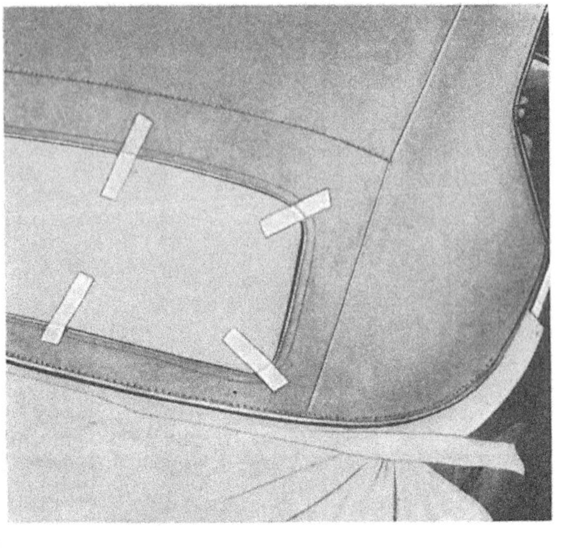

Finally tack top cover and body beading evenly round the top cover seams.

20 - Under tension and free from folds tack the top cover between the longitudinal seams — i. e. *under the rear window — to the body bow.* Cut off the surplus material taking care not to damage the body beading.

21 - Loosen the top cover tacks from the header on one side and at the same time pull the cover forward. Secure the top cover by inserting a chrome-plated screw with washer into the hole provided in the roof frame, taking care that the top cover beading covers the lower edge of the roof frame.

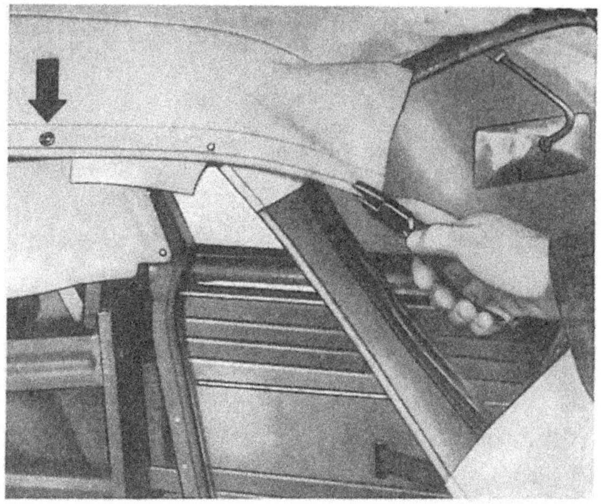

This operation can be carried out best with the assistance of a second saddler.

Treat the other side of the top cover in the same manner.

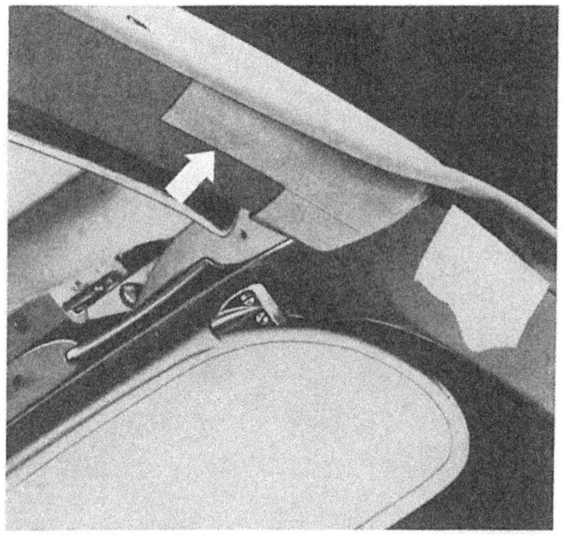

Important

The top must not be opened while tensioning.

22 - Cement the surplus top cover strips to the front roof frame over half its width on both sides.

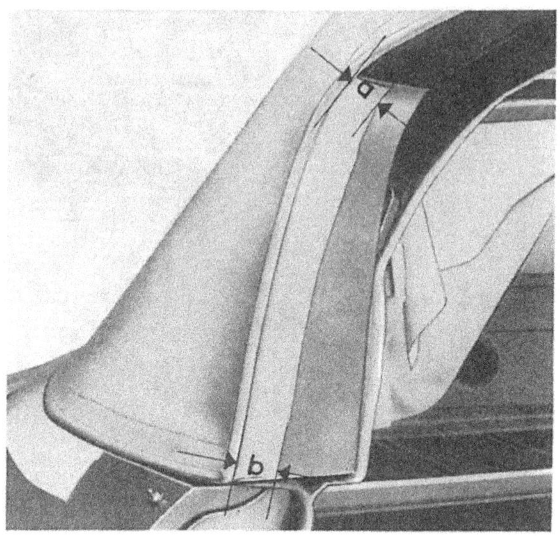

23 - Cement the surplus top cover strips to the main bow on both sides. The strips must not be more than 30 mm wide at the top and 25 mm at the bottom. Cut off any remaining material.

a - 30 mm (1.18")
b - 25 mm (0.98")

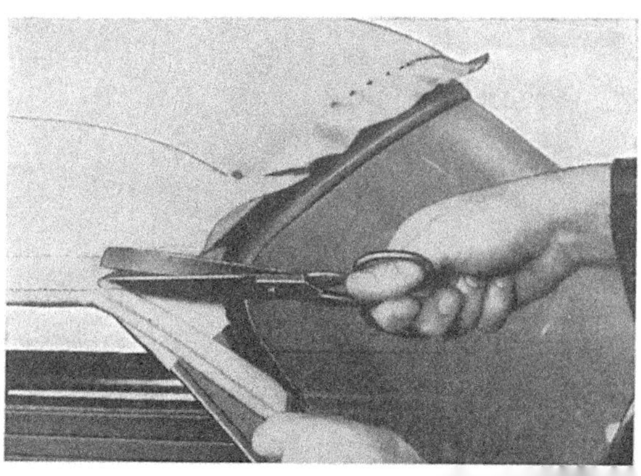

24 - Cut into the top cover beading far enough to allow a proper fit at the ends of the header.

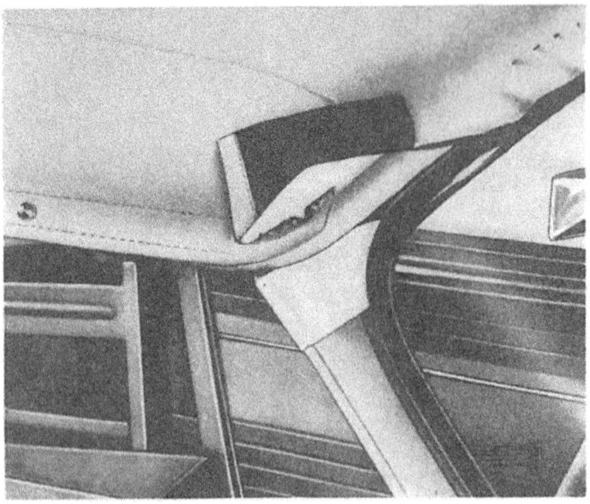

25 - Lay the beading round the curved ends of the header and tack so that it tapers off into groove in the header. Coat the row of tacks with rubber solution to improve sealing.

26 - If necessary, even out the corners of the header with small amounts of wadding and tack the top cover, without folds, approximately 30 mm above the covered lower edge of the header.

27 - Cut the surplus material off straight and cleanly below the row of tacks.

a - 30 mm (1.18")

28 - Coat the tack row with rubber solution, particularly at the corners, taking care that the coated surface is not wider than the trim moulding which is nailed on later.

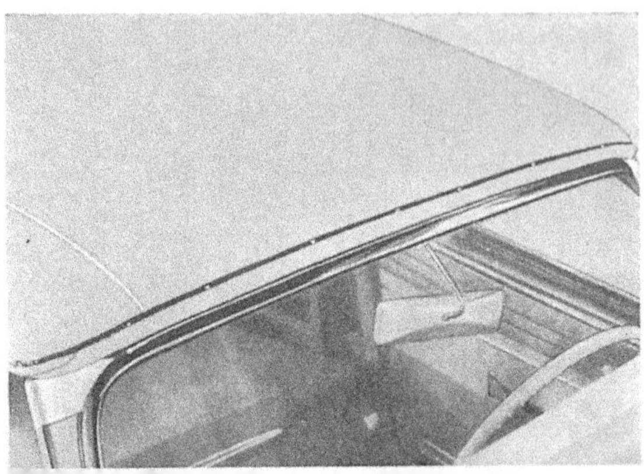

29 - Tack the trim moulding sealing band over the tack rows on the header and the body bow.

This sealing band is a natural rubber strip of the same width as the trim moulding which is obtainable under Part No. 151 871 435. The band has a *light-colored* protective strip on one side which must be removed before the band is fitted.

30 - To facilitate the correct placing of the band and the trim moulding for the rear bow it is recommended that the band is placed on the moulding first.

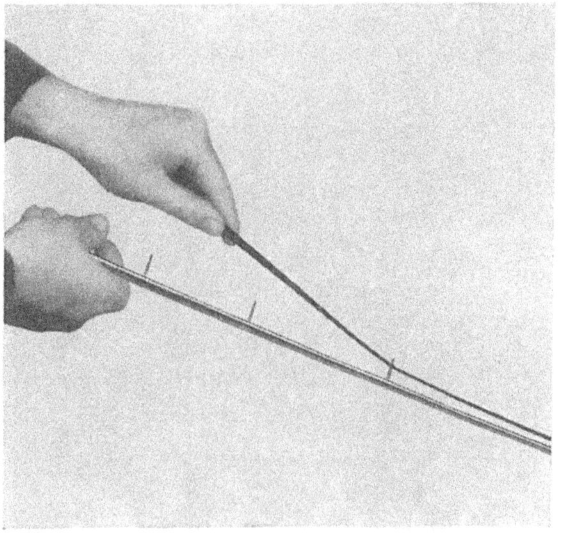

Important

The trim moulding must be installed with the special wooden block to ensure proper fitting and avoid damage to the moulding.

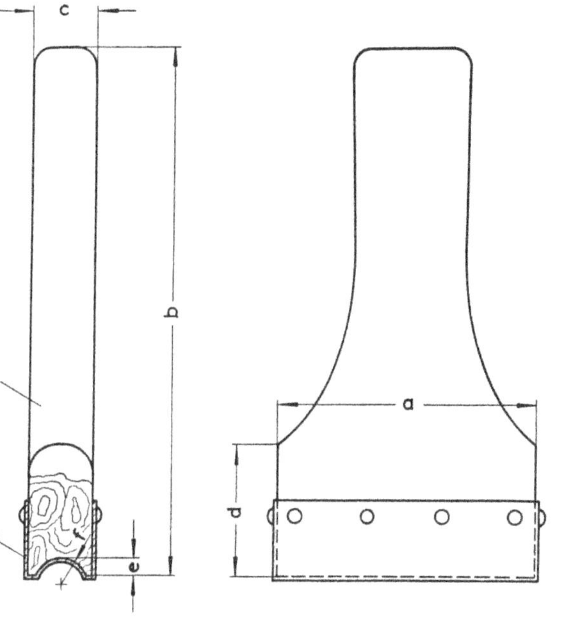

A - Special block
B - Leather cover (approximately 1 mm thick)

a - 70 mm (2.75")
b - 140 mm (5.5")
c - 17 mm (0.66")
d - 35 mm (1.37")
e - 5 mm (0.2")
f - 7 mm (0.27")

31 - Nail trim moulding in position.

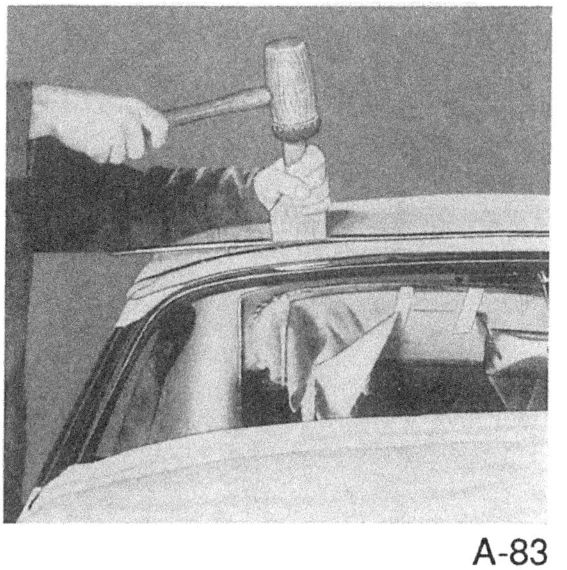

To avoid difficulties — such as overhanging at the ends and similar faults — it is advisable to first determine the exact location of the trim moulding by fitting, measuring or marking with chalk. Then start by nailing at one side and work across to the other.

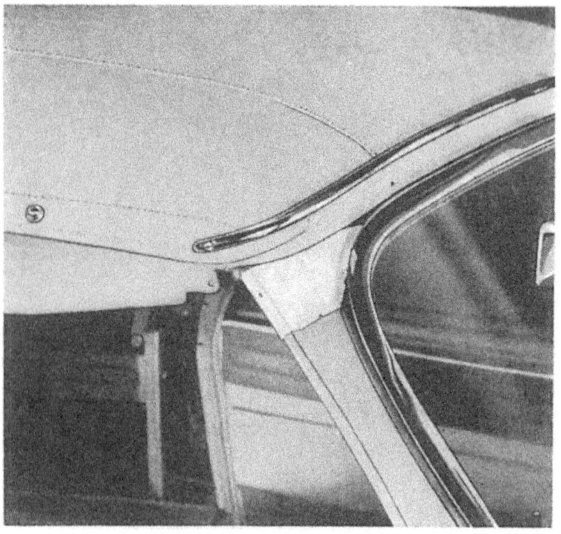

When installing the trim moulding take care that the corners are properly located. Secure the ends of the trim moulding with the special screws (Part No. N 13 864 1).

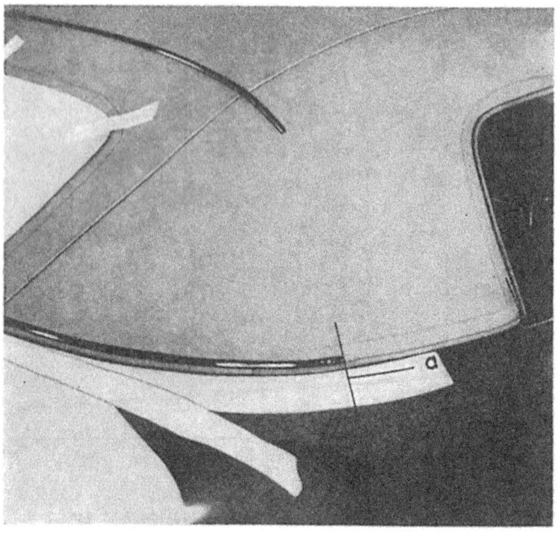

Before nailing the trim moulding for the body bow on, check that the distance between the ends of the moulding and the lock pillar is 360 mm.

a - 360 mm (14.2")

Secure the ends of the trim moulding for the rear body bow to the body with the countersunk tapping screws (Part No. 11 456 1).

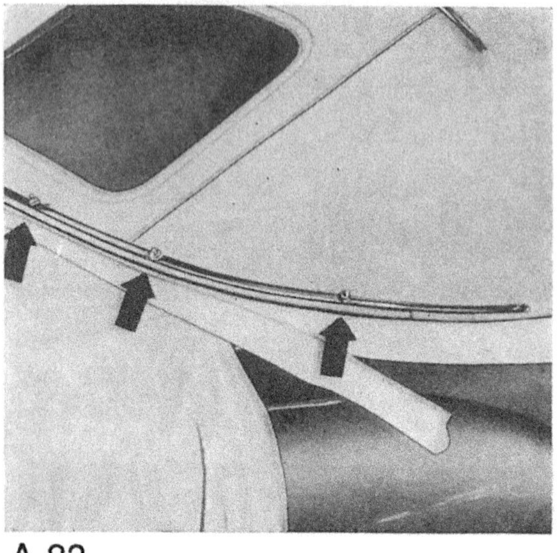

32 - Screw the bottom parts of the press buttons for the top boot and the necessary bases for the buttons into the holes provided in the trim moulding.

33 - Attach the leatherette covered plate for the header to the inside of the header with the special screws (Part No. N 12 611 1) and the protective washers (Part No. 141 871 619).

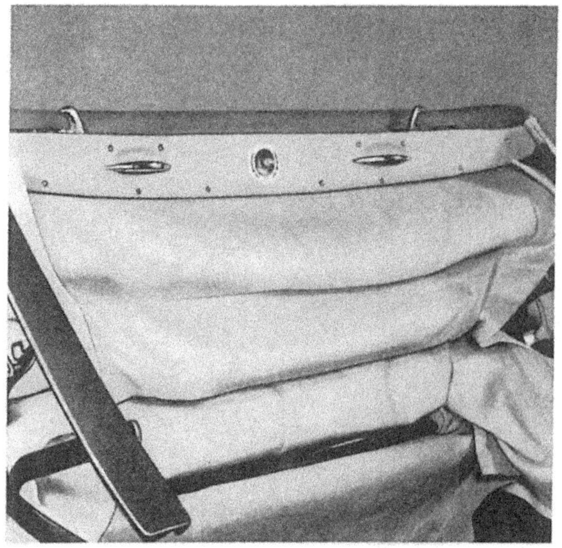

34 - Push the top lock handle onto the shaft and secure with the cap nut (Part No. N 11 143 1). Take care that the handle points to the left when the top is closed. Finally attach the headlining to the plate by inserting a counter-sunk screw and special washer in the holes provided at left and right.

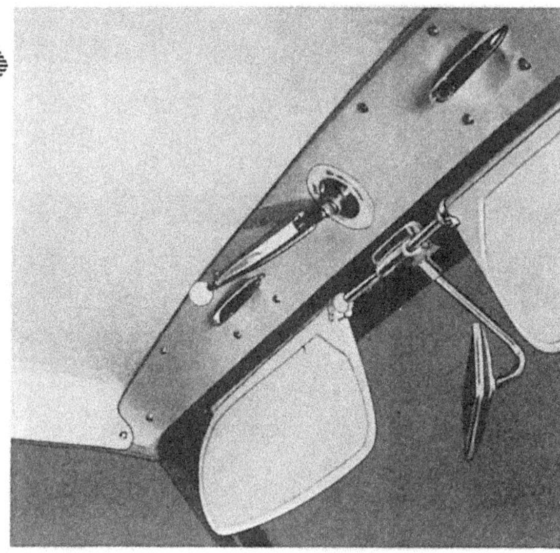

The screw points must not touch the roof frame as this will result in noises in the header, if necessary bend the ends of the header plate up away from the roof frame.

35 - Open and close the top several times to check the lock.

36 - When all the top installation work is finished the roof frame can be adjusted.

An adjuster screw and lock nut is fitted to each of the main hinge pillars for this purpose.

As already pointed out during the roof frame fitting operation there should be a clearance of 6 mm (0.236") between the lower edge of the main bow and the quarter panel.

Bearing this measurement in mind, loosen the lock nut and turn the screw in until it contacts the main bow without exerting any pressure.

Tighten the lock nut.

Securing Headlining

As already stated the rear portions of the headlining and the top cover are sewn to the rear window.

The headlining rear portion is not sewn to the main part until the top cover is completely fitted.

1 - Tension the rear portion slightly and pin it to the headlining and the top beading.

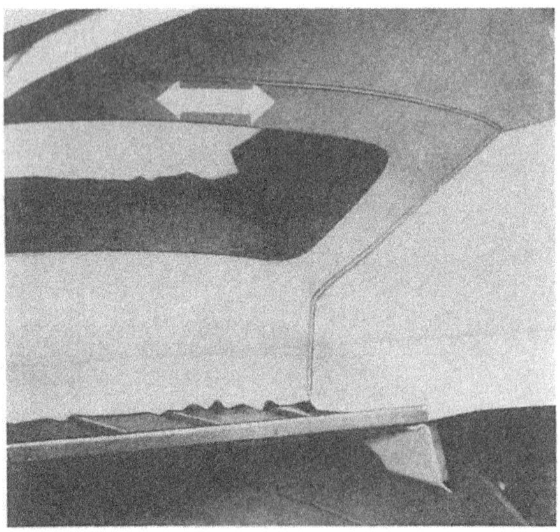

2 - Sew the rear portion, top beading and headlining together with small invisible stitches.

It is advisable to start in the center and work outwards to avoid creases.

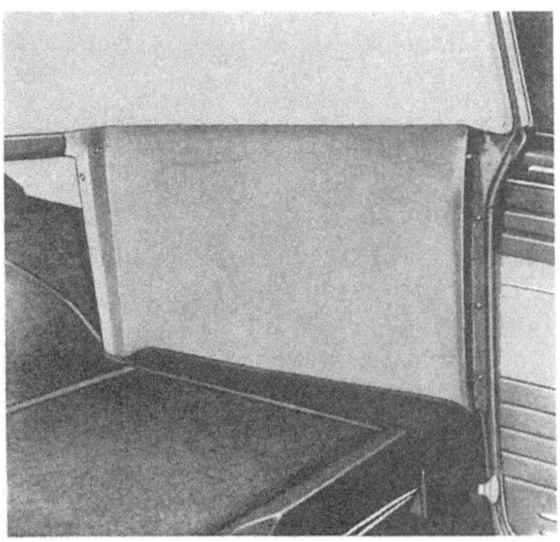

3 - Cement the surplus rear portion of the headlining to the top compartment panel.

4 - Secure the rear quarter trim panels.

5 - Coat the bottom panel of the top compartment with adhesive on the top, front and halfway underneath.

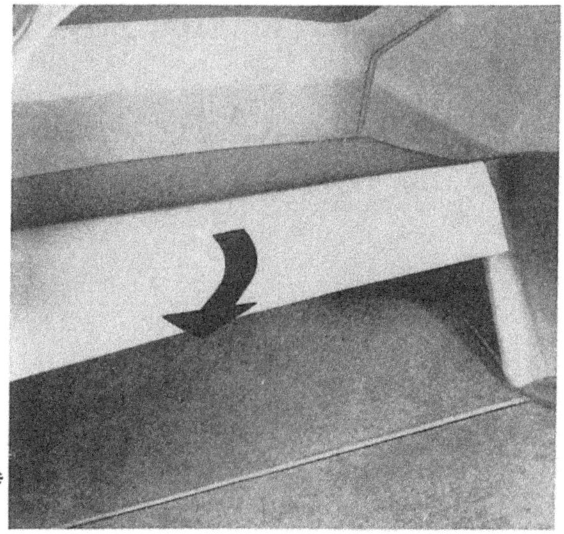

6 - Cement the leatherette covered cardboard sheet onto the top compartment bottom panel.

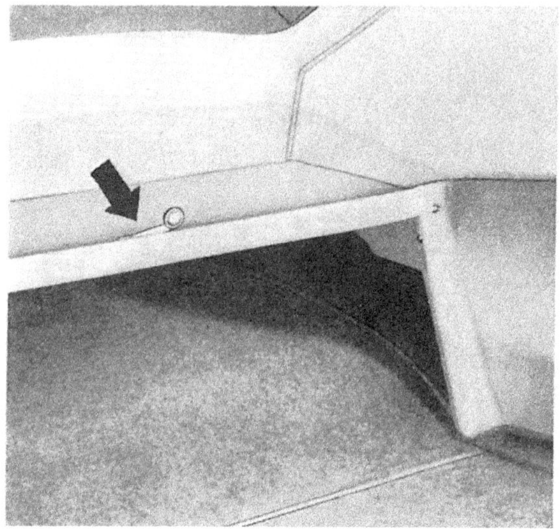

7 - Secure the retaining strap for the emergency seat.

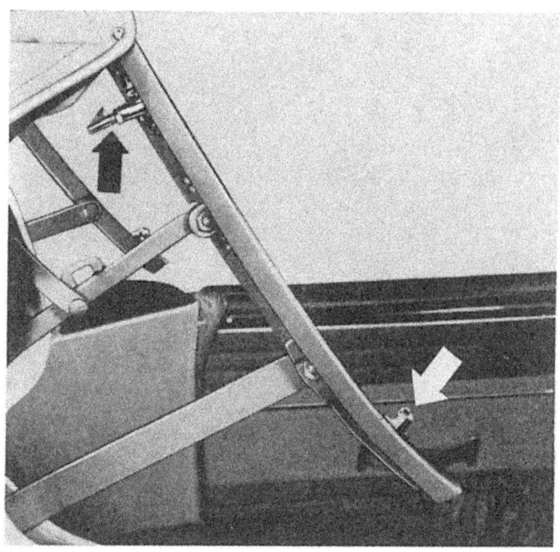

8 - Open the top and screw the top catches and coat hooks into the holes provided in the top frame.

Installing Lock Pillar Trim Plate

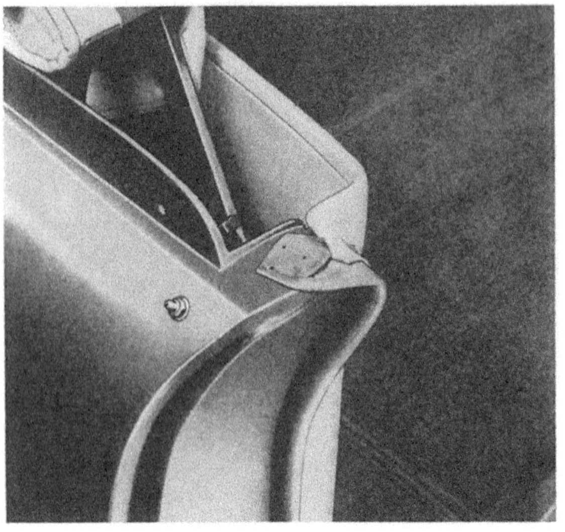

Before installing the door window weatherstrips, the seals (Part Nos. 141 853 367/368) and the trim plates (Part Nos. 141 853 365 A/366 A) must be fixed to the lock pillars.

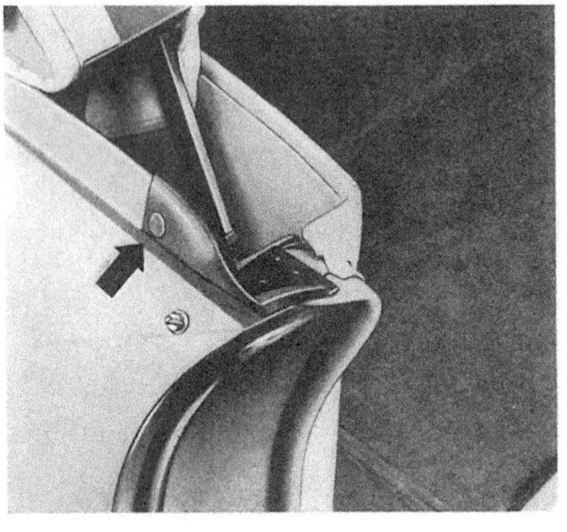

1 - Secure the sealing rubbers to the lock pillars with one hollow rivet each (Part No. 141 853 369).

2 - Attach the trim plates with three chrome-plated screws each (Part No. N 11 322 1).

Installing Door Window Weatherstrips

The door window weatherstrips, under Part Nos. 141 871 925 A/926 A and 141 871 923, are attached to the roof frame, the main bow and the windshield frame.

The weatherstrips are supplied longer than finally required to permit proper fitting and cutting to size. They are attached to the roof frame by metal strips which fit inside the rubber moulding and are secured to the roof frame with tapping screws.

From Chassis No. 2105320 softer weatherstrips were installed. These weatherstrips conform to the window shape better, make window adjustment easier and stop friction noises between the window glass and the rubber lips.

The new weatherstrips, which are supplied under the previous part number, can be subsequently installed without any alterations.

To stop friction noises between weatherstrip and window glass, apply ATE brake cylinder paste to the rubber lips. Wipe the weatherstrips afterwards with a clean cloth.

To stop noises with the softer weatherstrips, roughen the lips with rough emery cloth only as this rubber is self-lubricating and the normal chemical solutions will increase the noise.

1 - Before screwing the weatherstrips into position it is necessary to locate them exactly.

To do this, wind up the window glass and place the weatherstrip in position where it has to be screwed on. Take care that the weatherstrip makes good contact on the door window.

2 - Mark the position of the weatherstrip on the inside of the vehicle with a pencil. Do not use a scriber or similar tool.

3 - Cut the weatherstrip to size. It is advisable to be on the safe side and leave the weatherstrip slightly too long.

4 - To ensure good sealing it is recommended that a strip of sealing band (Part No. 151 871 435) be cemented under the weatherstrip.

5 - Drill holes to correspond with the holes in the metal strips and secure the weatherstrip in position.

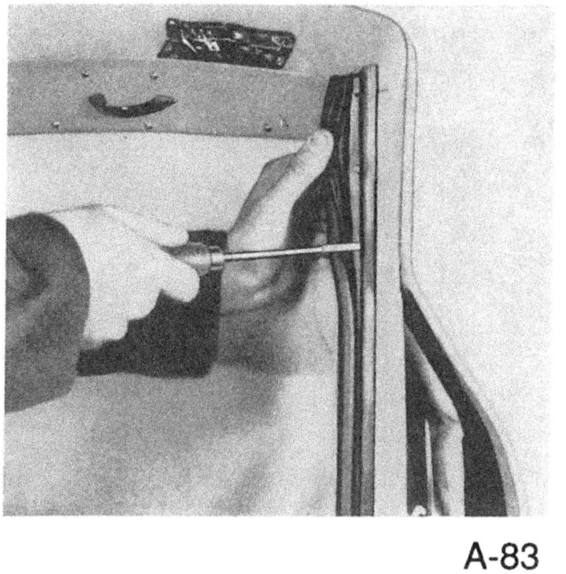

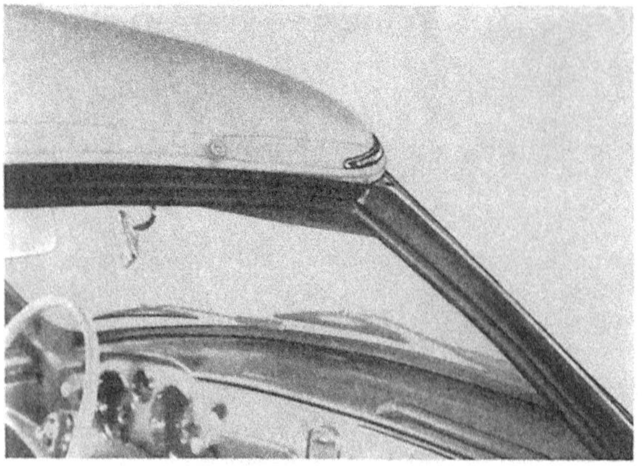

6 - Cut the weatherstrip to the necessary length. Pay particular attention to the corners, the weatherstrips must fit closely together at this point.

Top Assembly
(Late)

The description of the top assembly serves as a guide for the sequence of operations. The work must be carried out by a skilled man who is familiar with the top assembly. By employing two men for this work, the task of fitting the top cover will be considerably simplified.

The assembly of individual parts of the top described in the following sections can naturally be carried out separately.

Note:

All photographs in the sections "Fitting Headlining" and "Fitting Top Cover" show parts made of plastic material.

Work on cloth headlining and top cover is carried out in the same way as described here.

All tools required for the top assembly are listed in the section "Workshop Equipment".

Important

Brass tacks were not used for tacking the padding parts and the top. A compressed air tacker was used, the magazine of which was filled with non-rusting clips.

Top Frame Installation

Preparation

1 - Cover all body parts in the working area.

2 - Remove front seats and lower rear seat backrest.

3 - Pull haircord carpet off luggage compartment floor.

4 - Pull haircord carpet off wheel housings.

5 - Pull cardboard off package shelf on rear panel.

6 - Remove quarter trim panel on both sides.

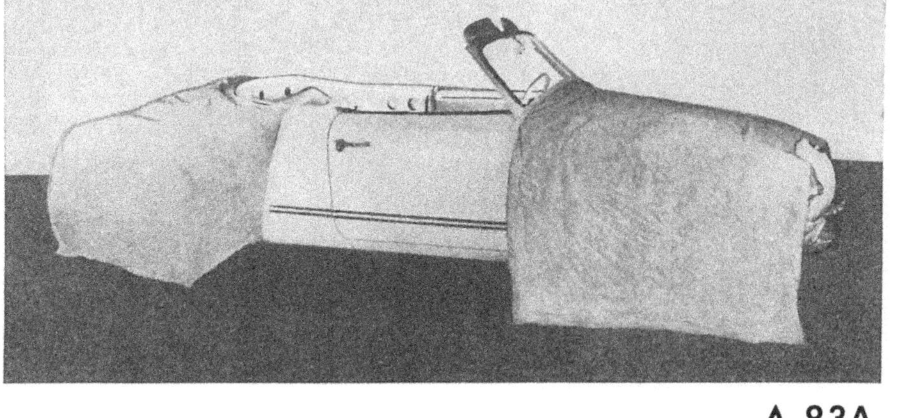

A-83A

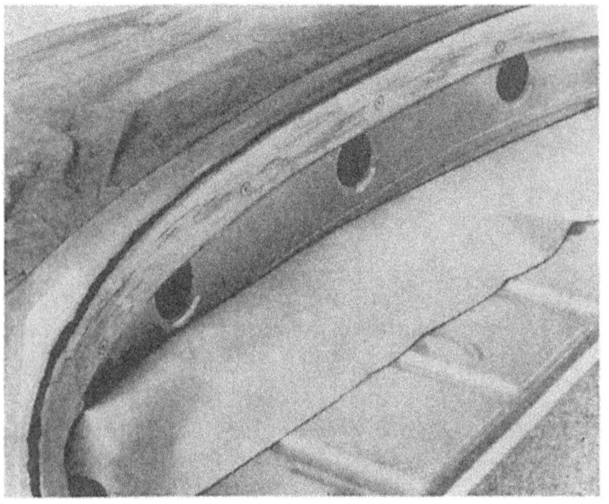

7 - Detach trim material from rear panel near rear body bow.

8 - Wind both door windows down.

9 - Press rubber sealing beading (Terostat) between body and rear body bow.

Fitting Top Frame and Straps

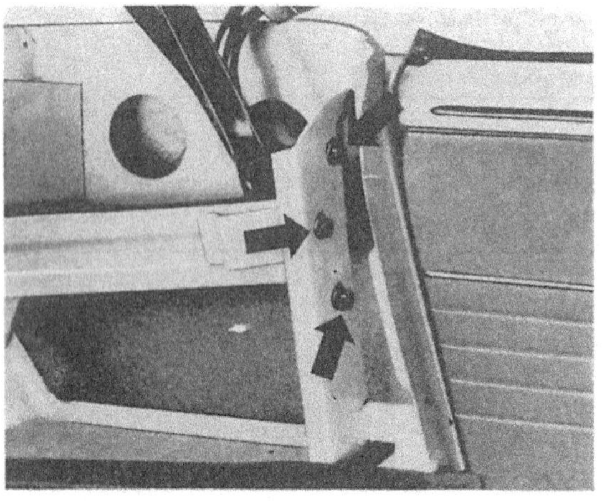

1 - Place top frame in position and loosely attach main hinges to the tapped plates of the left and right support pillars with three hexagon head bolts on each side. Do not forget washers and circlips.

Note:

From 11th May 1965, Chassis No. **145 806 650**, the screws connecting the roof frame to the upper guide bar and the roof frame to the main guide bar were replaced with rivets. As a result of this, the left and right roof frames and the upper guide bars have been modified.

New Part Numbers:

Part	Old Part No.	New Part No.
Roof frame, left/right	141 871 701/702 (deleted)	141 871 701 A/702 A
Upper guide bar, left/right	141 871 723/724	141 871 723 A/724 A
Top frame upper guide bar bolt	141 871 751	—
Washer (also for rivet)	151 871 185	—
Top frame upper guide bar rivet	—	141 871 751 A
Main guide bar bolt	141 871 825	—
Washer (also for rivet)	151 871 185	—
Main guide bar rivet	—	141 871 825 A

Repair note:

The roof frames, Part No. 141 871 701/702, will be discontinued when stocks are exhausted. If a new roof frame has to be installed on an older vehicle, this can be done only by installing also an upper guide bar of present manufacture which has to be riveted to the roof frame using rivet, Part No. 141 871 751 A.

a = .24 in. (6 mm)

2 - Place a .24 in. (6 mm) thick distance piece between main bow and left and right support pillars.

3 - Push sides of top frame outwards until main bow is about .08" (2 mm) from side of body on each side. Tighten main hinges.

a = .08" (2 mm)

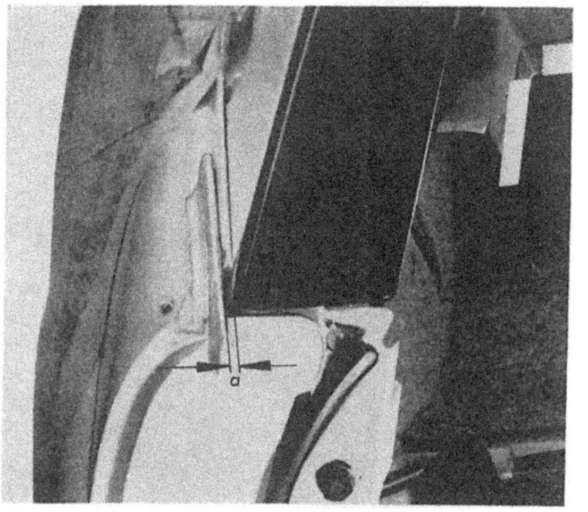

4 - In the closed position, the clearance between uncovered header and windshield frame should be:

a - in the center .3" (7 mm).

b - near the fasteners .2" (5 mm).

c - at the ends tapering to .08" (2 mm).

Note

These distances must be adhered to otherwise there will not be sufficient space for the rubber seals and the top cover to be installed later.

If the space is too small, the front rubber seal of the header can lift or press out the windshield weatherstrip at the top, thus causing leaks.

Insufficient space between header and windshield frame can also cause excessive tension at the top fasteners and lead to distortion of the windshield frame.

If the necessary space is not available, rework the header with a rasp.

5 - Tack a .04" (1 mm) thick cardboard strip 7.9 × .6" (200 × 15 mm) under the header, near the top lock.

6 - Apply universal adhesive D 12 to top and rear of header and attach leatherette strip 49.2 × 4" (1250 × 100 mm). The leatherette strip should be cut at the corners so that it fits better. Cut off surplus material.

7 - Apply universal adhesive to underside of header and cement rubber seal 48.4×.6×.4" (1230×16×10 mm) to front edge of header.

Note:

From end of rubber seal make a recess 2.4" (60 mm) long and cement seal to side of top frame on each side. To improve sealing, additionally cement a strip of rubber between header and roof frame on each side (arrow).

8 - Lay the webbing strips, which support the rear window at the sides, round the main bow and sew in position.

Note:

On the upper side of the main bow there are two recesses for accommodating the webbing strips. Prior to sewing the webbing strips in position, thinly apply universal adhesive D 12 to these recesses to safely avoid the webbing strips slipping.

9 - Secure webbing strips **on** rear body bow with tacks.

(Dimension a = 16.5" (420 mm) from car center)

Important

Secure webbing strips only with brass tacks to avoid rust formation.

10 - Close and lock frame. Set corner bows to the following measurements:

a = 13.8" (350 mm) ± .2" (5 mm)
b = 18.5" (469 mm) ± .2" (5 mm)

Important

These measurements must be strictly adhered to as they govern the satisfactory padding of the top and the folding of the top when opened.

11 - Tack webbing strips to rear bow **in the recesses.**

Note:

If the webbing strips are not attached in the recesses, the trim moulding which is installed later will be too high at the ends and will stand clear.

Headlining Installation

The headlining is supplied as a spare part, complete with the support strips for attaching it to the individual bows.

The rear part is not sewn to the headlining but is attached to the top cover together with the rear window. It covers the lateral parts of the actual headlining after the top cover has been fitted, while it is tacked to the rear bow at the top and to the rear body bow below.

To ensure that the headlining fits properly, it is essential that it is attached to the individual bows, the rear body bow and the header in the following sequence, with top frame closed but not locked:

1 - Main bow
2 - Header
3 - Rear bow
4 - Intermediate bow

1 - Apply universal adhesive D 12 to front and top surface of main bow.

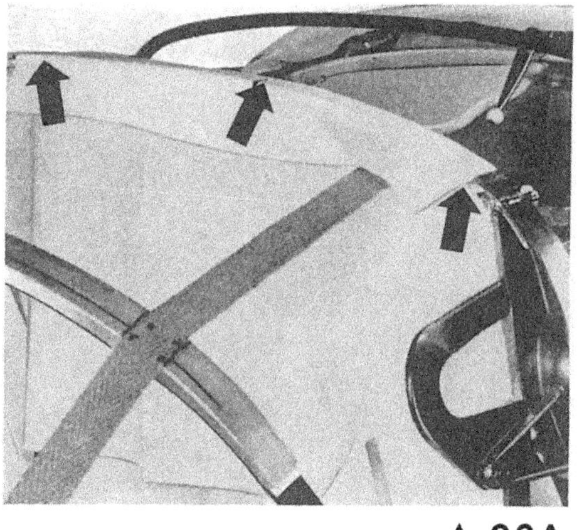

2 - Pull headlining up near main bow by support strips until seam contacts underside of bow. Lay strip round bow and press firmly into place. Cut opening for webbing then cut off excess material.

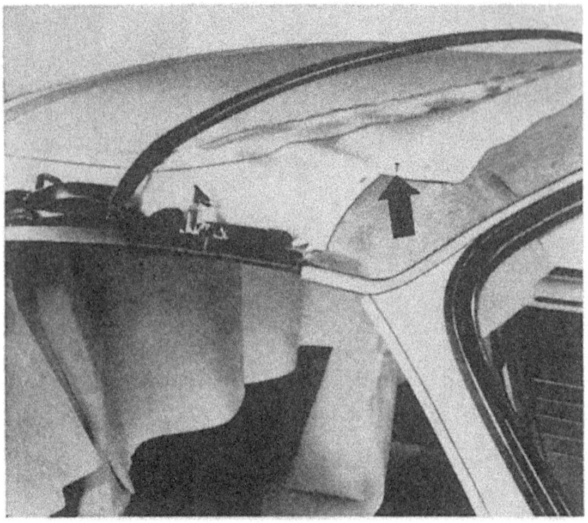

3 - Slightly tension headlining and tack it near the longitudinal seams to both sides of header. To prevent tearing, drive tacks through side seams.

4 - Apply universal adhesive D 12 to rear and upper surfaces of rear bow as far as recess, tension headlining towards rear, lay round bow and press firmly in position. Tack headlining in bow recess and cut off excess material.

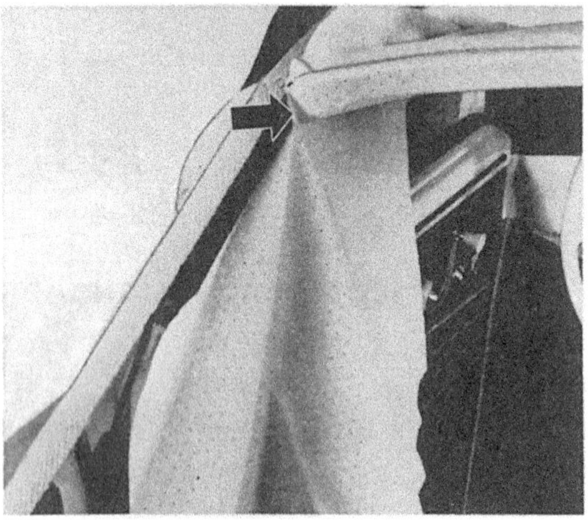

Important

The side seam of the headlining must line up exactly with the inside of the webbing.

5 - Tension headlining towards front and tack to top of header between side seams, keeping headlining under equal tension. It is essential to ensure that seam of front support strip is parallel with intermediate bow.

6 - Apply universal adhesive D 12 to intermediate bow, then pull support strip up until seam contacts underside of bow. Lay support strip round bow and press strip in place. Cut off excess material.

Additionally sew the outside of the support strips at the intermediate and main bows.

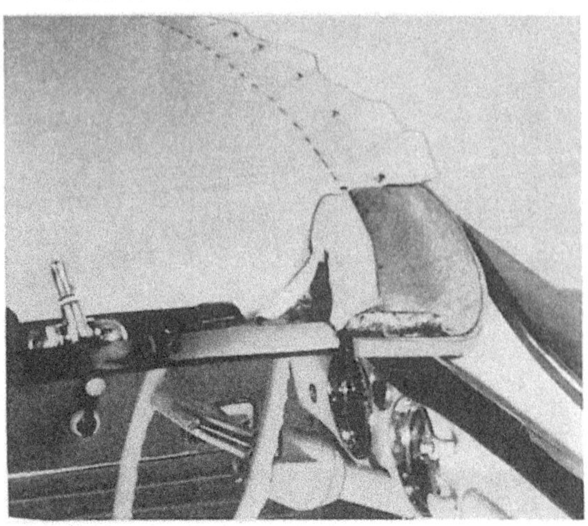

7 - Carefully staple headlining to header on the flattened surface.

8 - Pull left and right headlining seams farther forward as necessary and secure at corners of header. This re-tensioning ensures that longitudinal seams of headlining are uniformly tensioned.

7 - *Cut off surplus leatherette on flattened surface* of header.

10 - Place a piece of doubled nylon cord round main bow at outside on both sides and tack it to rear bow. Sew longitudinal seam of headlining between bows to the nylon cord under tension.

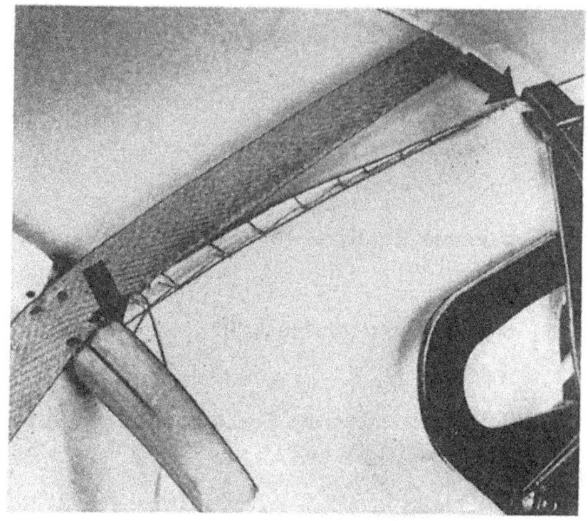

11 - Check from inside that headlining is taut and free from folds and that seams of support strips run parallel with bows. If necessary, correct at appropriate bows.

12 - Cut .8" (20 mm) wide openings in support strips of main and intermediate bows.

Dimension a = 3.5" (90 mm)

13 - Pull a rubber strip 21.7" × .6" (550 × 15 mm) through these openings on left and right and stitch them together. Additionally stitch rubber strips at main bow.

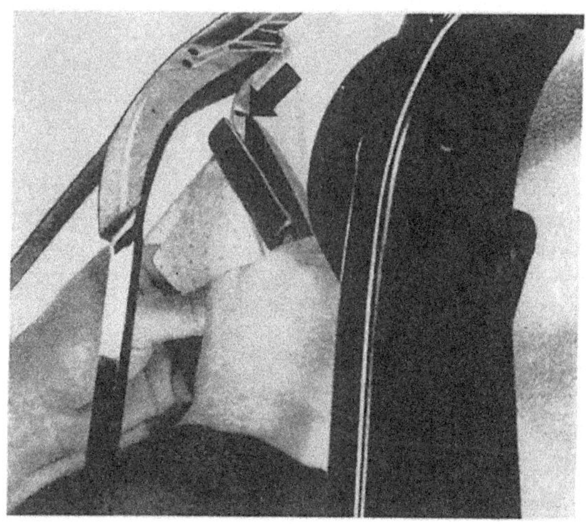

These rubber strips pull the intermediate bow towards the rear when opening the top, thus ensuring that the top cover folds correctly.

14 - Cut lateral parts of headlining hanging down at rear bow, as far as the rear bow.

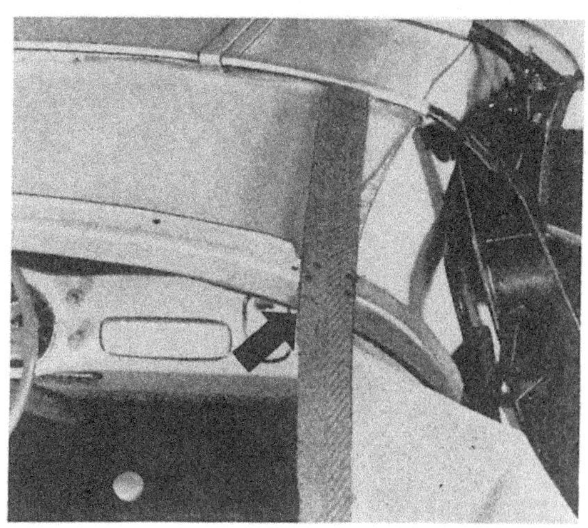

Fitting Linen Sheet

Two different types of cloth are used:

a - a coarse, tightly woven, light-coloured linen,

b - a slightly thinner, black lining material.

The linen material forms a base for the rubberised hair. The lining material is used only to enclose the rubberised hair padding.

The linen and lining material is supplied in separate parts which consist of a large roof piece and two rear panels each made up of a piece of linen and a piece of lining material.

The roof part of the linen sheet is made up of a piece of linen material A, two strips of plastic top material B and two strips of lining material C sewn together.

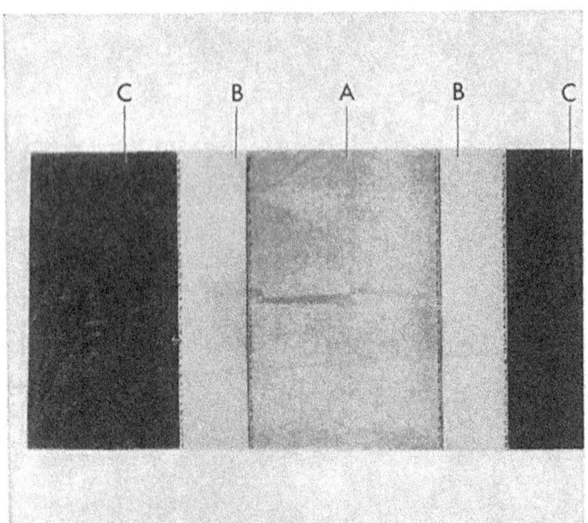

A - Linen material
B - Plastic top material
C - Lining material

The strips of plastic material sewn to each side of the coarse linen sheet are used solely to prevent the linen sheet becoming visible at the sides when the roof is opened.

The two strips of black lining material hold the rubberised hair padding together and give it additional support.

In addition, two pieces of coarse linen are required for the rear side panels and two pieces of nettle cloth for the two side panels. All the parts for the linen sheet are contained in a set which is supplied complete as a spare part.

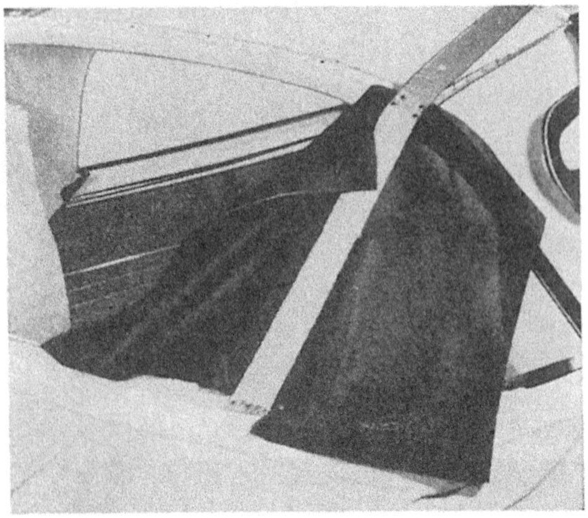

1 - Tack a piece of black lining material each side to rear bow and rear body bow.

Important

Cut the pieces of linen at the webbings. The overhanging lining strip on the inside must, under no circumstances, be cut off.

2 - Tack a rear side panel of coarse linen to rear bow and rear body bow on each side.

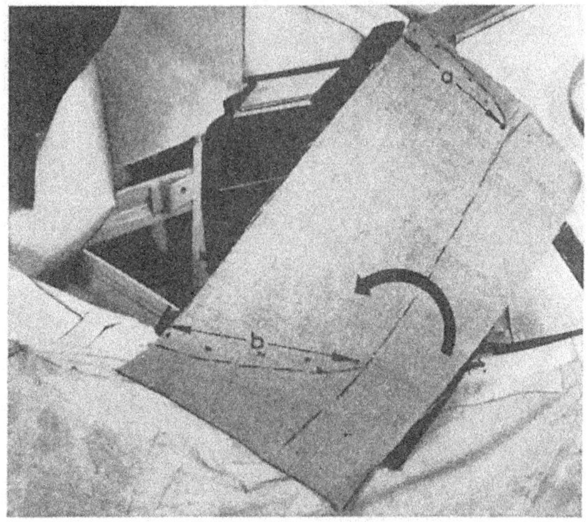

a = 7.1″ (180 mm)   b = 7.9″ (200 mm)

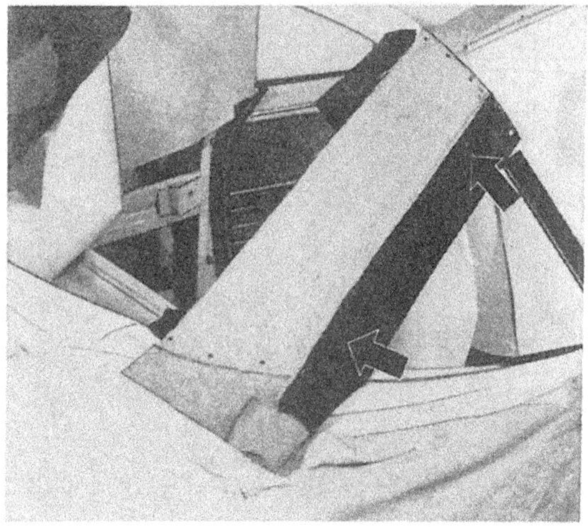

3 - Fold up the course linen and black lining material at the sides and tack in position. Cut off excess material at rear bow.

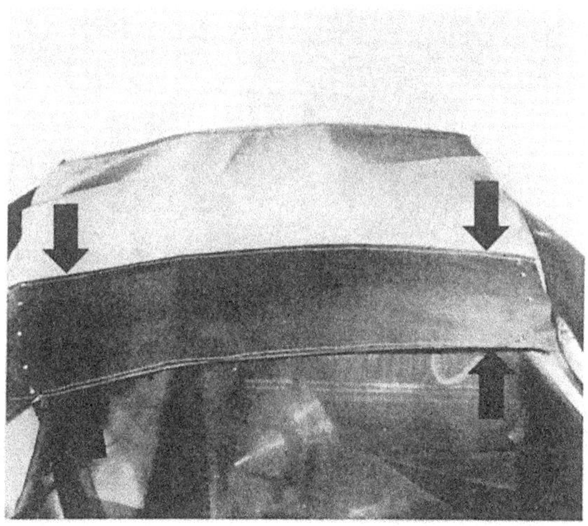

4 - Lay roof part, together with sewn-on top and lining material strips, on to top frame with the wider end of left and right plastic strips facing rearwards.

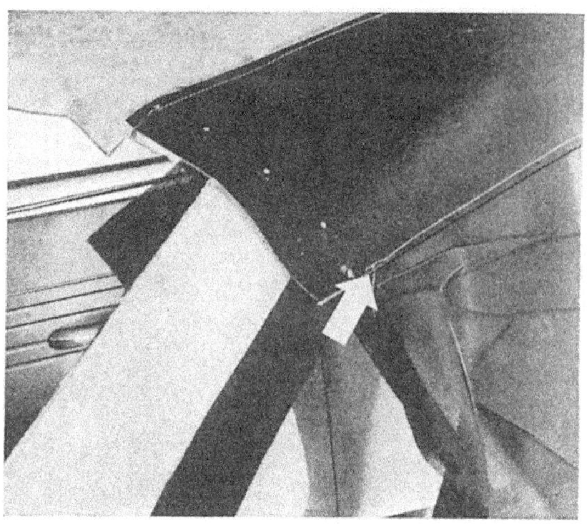

Note:

At the rear, the side seams must be parted up to the main bow.

5 - Tension roof part and side strips of plastic top material and carefully staple them, working inwards from outer edge, in recess in rear bow. Cut off excess material.

Note:

Either cut a wedge-shaped piece out of the additional material in the center or make a cut in it, fold it and staple it to the rear bow.

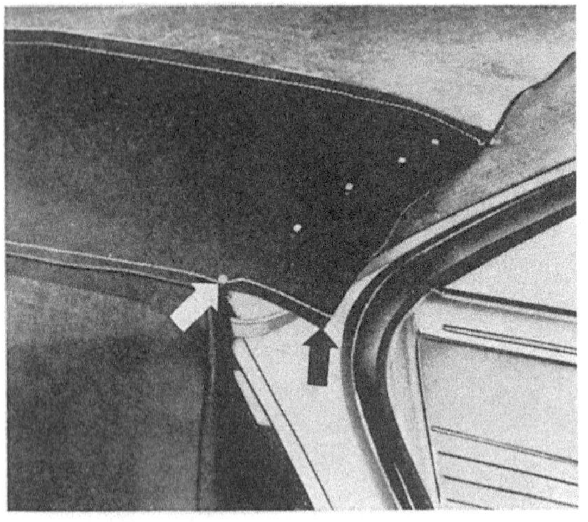

6 - Pull roof part and side strips forward until they are taut and tack to header without folds. Part side seams slightly near header.

7 - Make wedge-shaped cuts towards rear in both seams and near the bow.

8 - Carefully staple roof part and side strips of plastic top material in recess in header.

9 - Cut off excess material.

Installing Rubberised Hair Padding

The rubberised hair padding shapes the top of the Convertible and also provides good insulation. It consists of a roof part and two rear parts but is supplied under one part number.

1 - Place a rubberised hair rear pad on the linen material and lining material on each side, then insert it into the folded up lining and coarse linen material.

2 - Sew rubberised hair rear pads to webbing and side panels of coarse linen, using small, tight stitches. Tack them in the center with several large stitches.

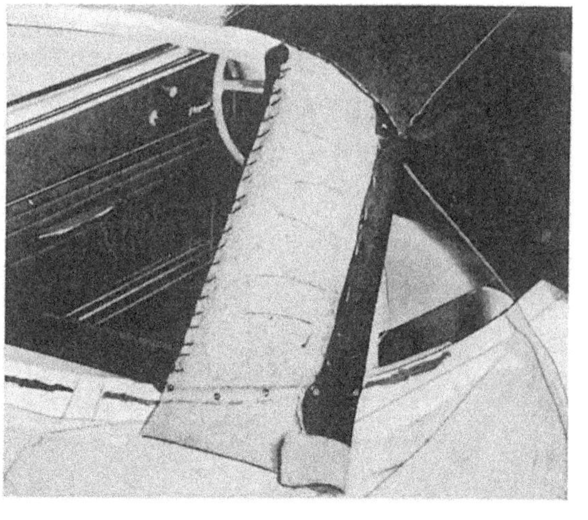

Note:

The rubberised hair pads must not cover the rear bow and be sewn on at the bottom near the rear body bow.

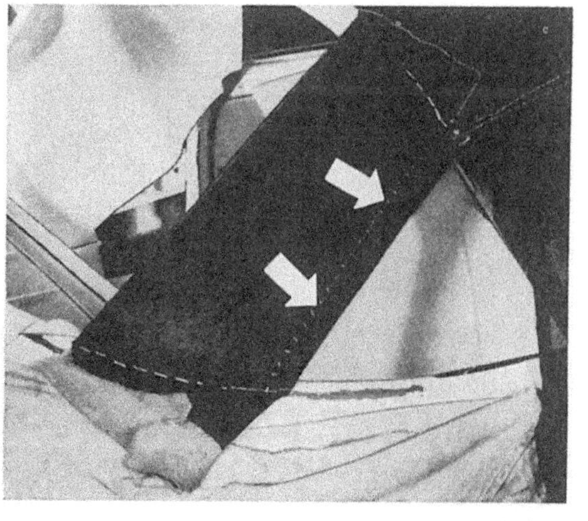

3 - Fold lining material hanging down inside over left and right rear rubberised hair pads and staple to rear bow and rear body bow. Sew lining material at outer edges.

4 - Sew lining material to rubberised hair pads with large stitches and cut off surplus lining at top and bottom.

5 - Lay roof part of rubberised hair padding on to top frame with smooth surface facing upwards and thickened edges on left and right outer sides facing downwards. Cut a strip 27.6×2" (700×50 mm) out of rubberised hair padding near the indentation for rear bow on left and right.

Note:

Place additional padding between rear and main bows under rubberised hair.

6 - Secure roof padding to rear bow on each side.

7 - Pull roof padding gently towards header and secure it.

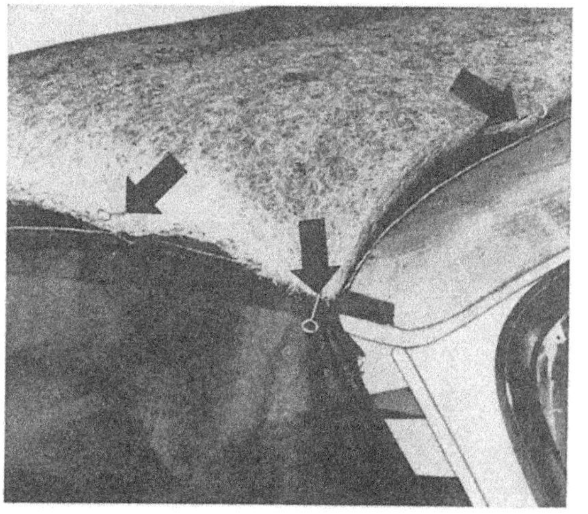

8 - Pull roof padding to rear edge of header and sew to linen sheet with small, tight stitches.

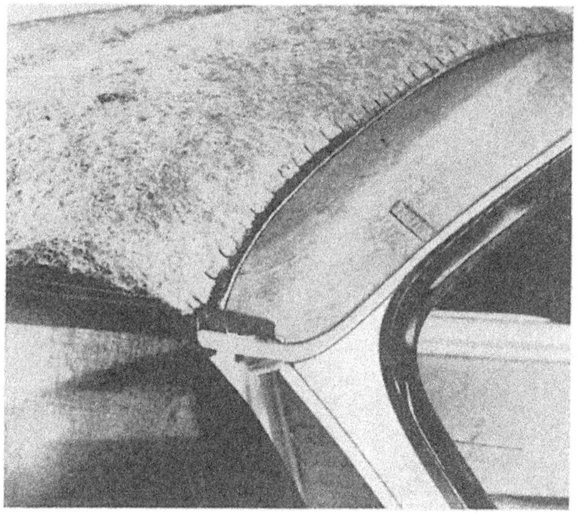

9 - Sew outer edges of roof padding to seam of plastic strips with small, tight stitches. Cut excess padding off at an angle at seam and pluck material out slightly to taper it off.

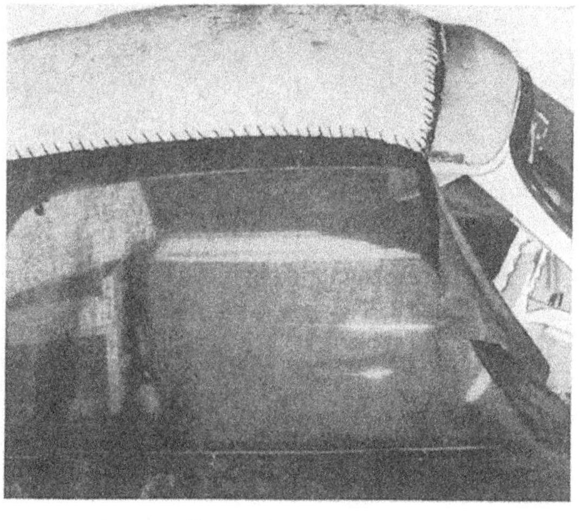

10 - Sew roof padding to linen sheet at rear bow with small, tight stitches.

11 - Sew roof padding to linen sheet with rows of large stitches across the roof.

12 - Thinly coat header with universal adhesive D 12 and pad it out with .2" (5 mm) thick foam rubber as follows:

a - Cement a narrow strip of foam rubber in recesses for top seam.

Note:

The strip of foam rubber does not have to be installed if a fabric top cover is used.

b - Cement a narrow strip of foam rubber at outer edges in recess transverse to driving direction.

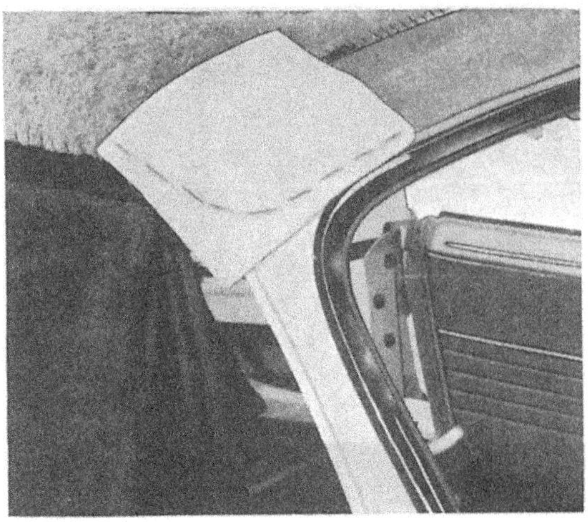

c - Cement a square piece of foam rubber to each corner so that it fits header contours.

d - Cement a 5.5″ (140 mm) wide strip of foam rubber over whole length of header, ensuring that it also covers front part of rubberised hair.

Note:

When installing a fabric top cover, wool padding glued on both sides can be used instead of the foam rubber.

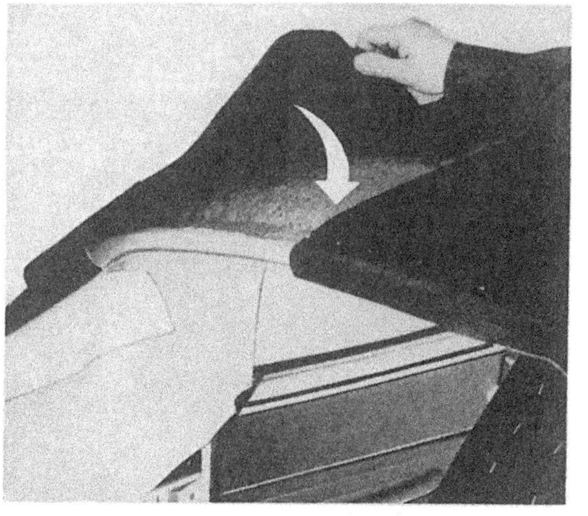

13 - Lay roof padding lining material strips one over the other and tack them free of creases in recess in rear bow.

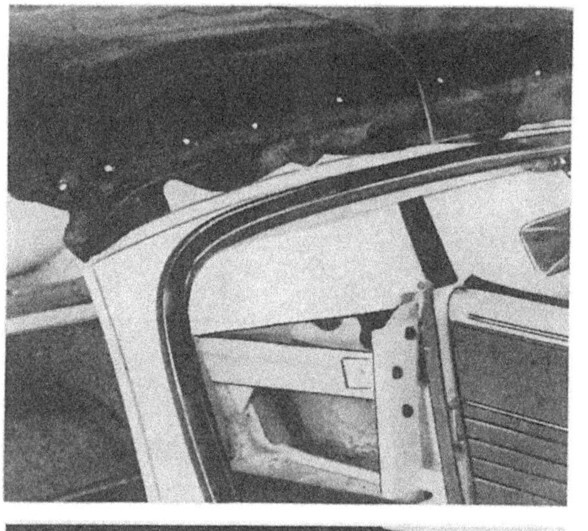

14 - Slightly open top, pull strips of folded lining material firmly forward and tack them to header.

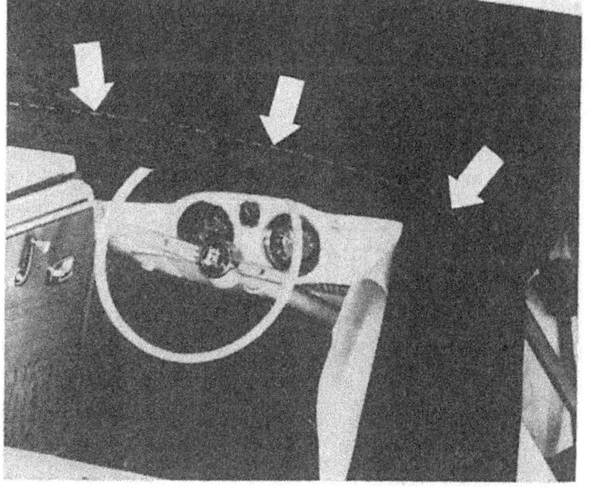

15 - Carefully staple folded lining material strips in recess in rear bow.

16 - Open top farther, pull lining material strips taut round header and cement them underneath in recess.

17 - Close top and lock it. If necessary, correct the position of the lining material on header particularly at rounded side parts.

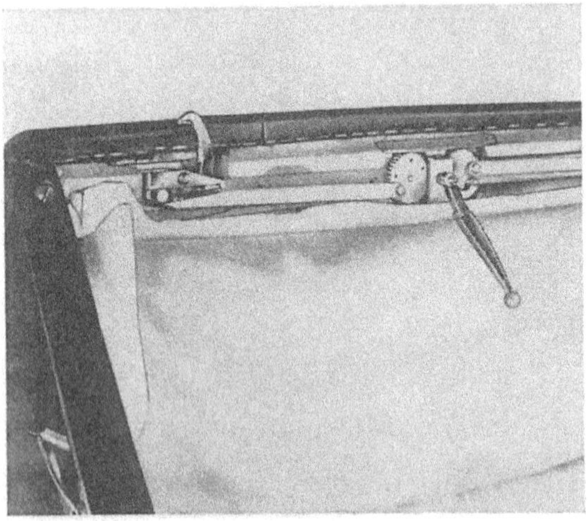

18 - Open top and staple folded lining material strips to header from underneath, starting at one corner.

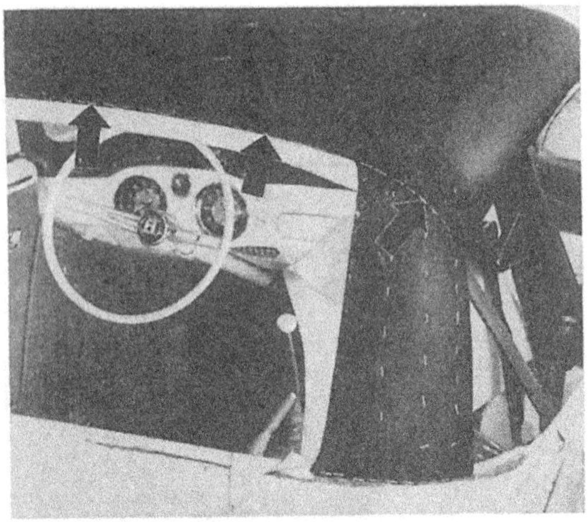

19 - Cut off surplus lining at row of staples on rear bow and at header.

20 - Close top and sew lining strips to padding all round.

21 - Pull lining strips, which are laid over each other, together in the center and sew with small stitches.

The flexible rear window is sewn to the top cover and the rear part of the headlining. This leatherette is not sewn to the already installed headlining but is tacked, together with the top cover, only to the rear bow and rear body bow.

1 - To compensate for the thickness of the sheet metal, staple a narrow strip of top cover to the rear body bow between the rear padded parts. Use rubber tape (Terostat) near the rear padded parts.

2 - Tack beading to rear body bow. The bead must rest in body groove below rear body bow. Do not yet tack beading at the sides near sewn rear part of rubberised hair padding.

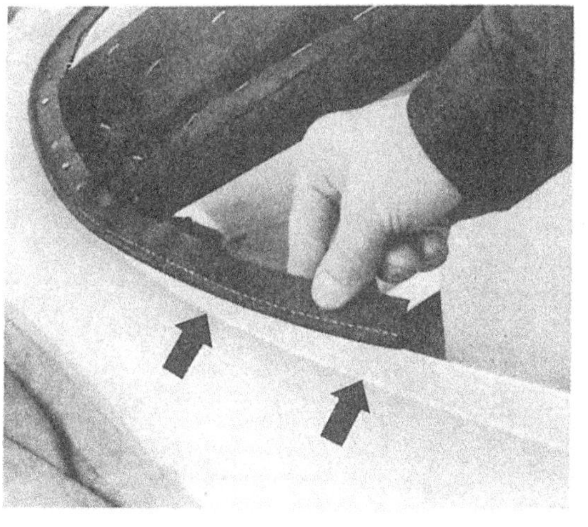

3 - Staple rear panel covering to rear body bow and cut off surplus material.

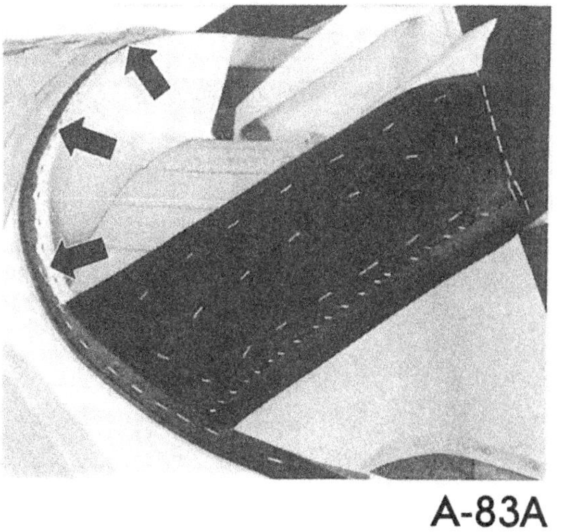

Note:

The leatherette must be cut off in front of the bead.

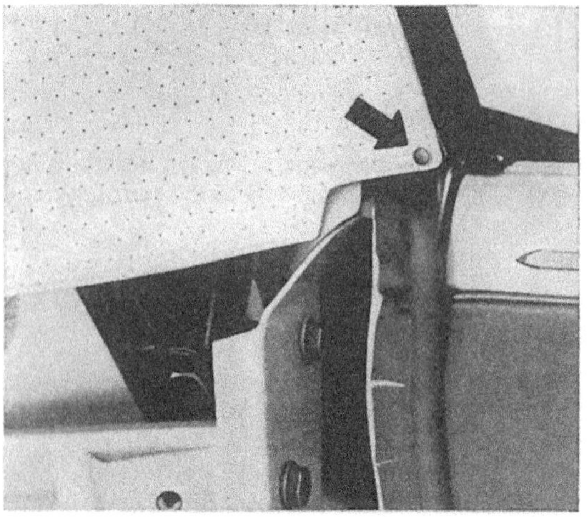

4 - Tack headlining side parts to main bow at the bottom where the chrome-plated screw is inserted later.

5 - Pull headlining side parts taut towards rear and tack under tension to rear bow and to rear padded parts.

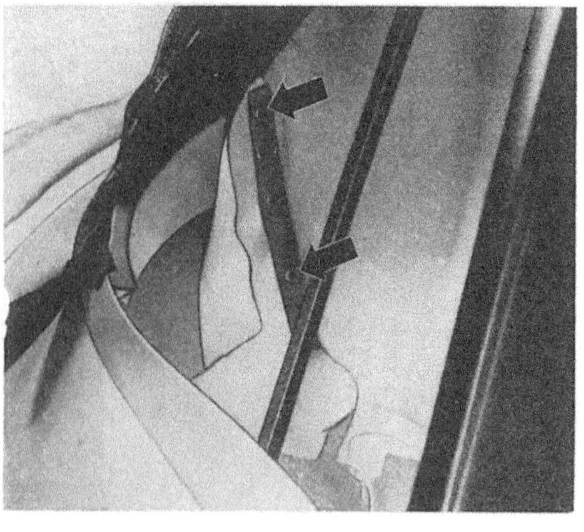

6 - Tack headlining side parts near rear panel sides under tension to inside of rear body bow and to left and right wooden wedge.

Important

Do not forget cardboard layer.

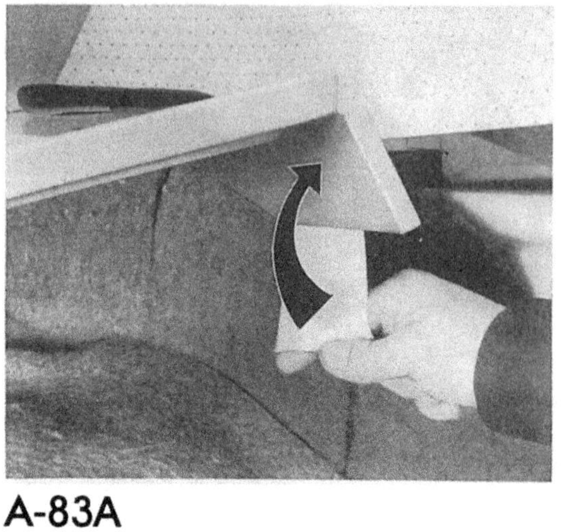

7 - Cement surplus leatherette of headlining side parts to top compartment lower part:

a - to top compartment side facing downwards.

b - to the sides on top of top compartment lower part.

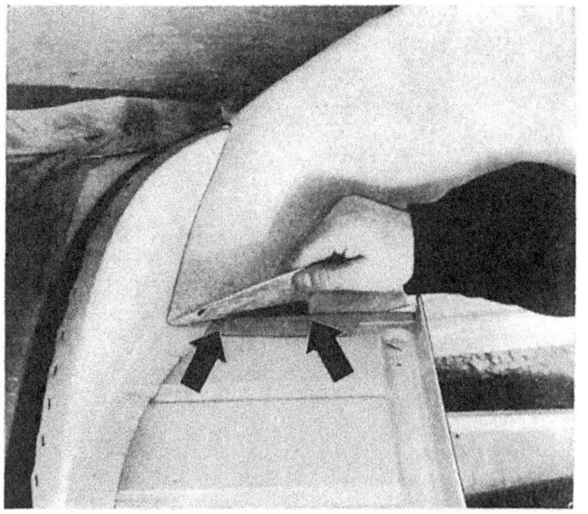

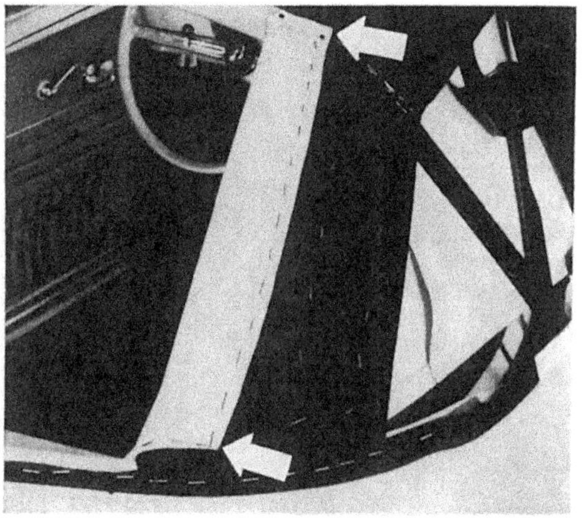

8 - Pull headlining side parts taut round rear padding and carefully sew together. Additionally sew headlining with small, hidden stitches in left and right corners of rear bow.

9 - Cement foam rubber strip to top on both sides near rear bow.

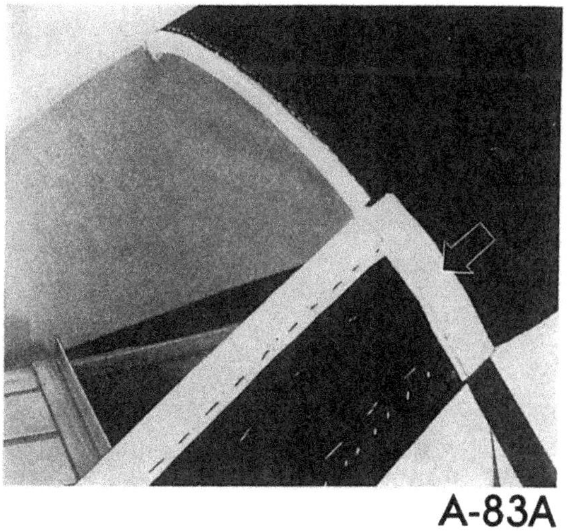

Note:

If new a top cover is being installed on old padding, proceed as follows:

a - Make loop stitches in lining strips near rear and intermediate bows.

Loop stitches are large, loose stitches made with a thick thread, under which the padding material is pushed. They are also intended to hold the loose padding material in position.

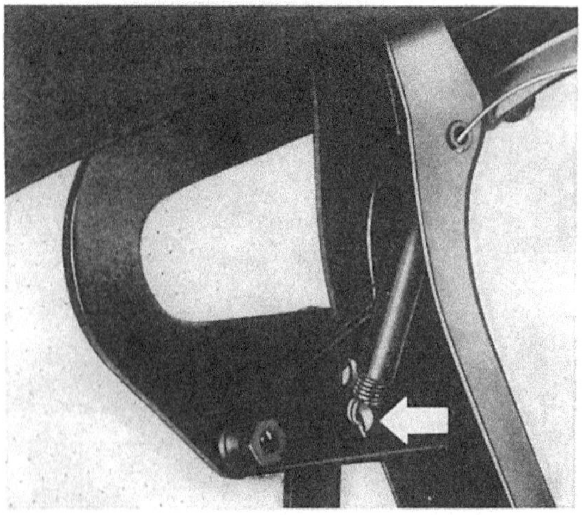

b - Push padding material, horse hair if possible, under the loop stitches in uniform thickness to equalize the depression between both bows.

The padding shapes the contours at the rear of the top. If the padding is too thick or too thin, this will cause a visible unevenness in the top.

10 - Attach spring and tensioning wire to hook on main bow on both sides and secure by bending hook down.

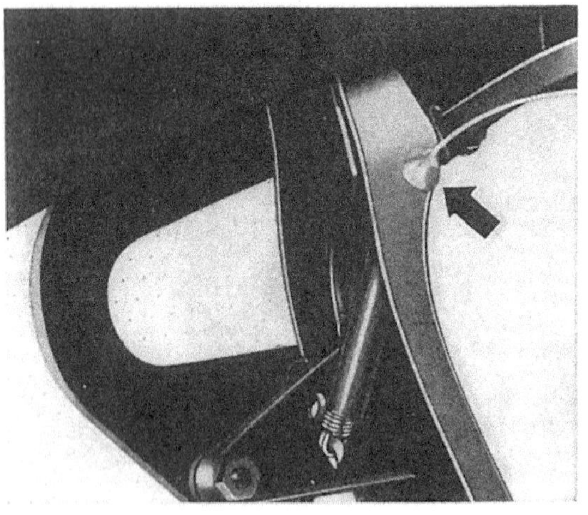

11 - Pull tensioning wires to one side out of main bow, then close openings with sealing compound (Terostat).

12 - Place top cover on to closed top frame.

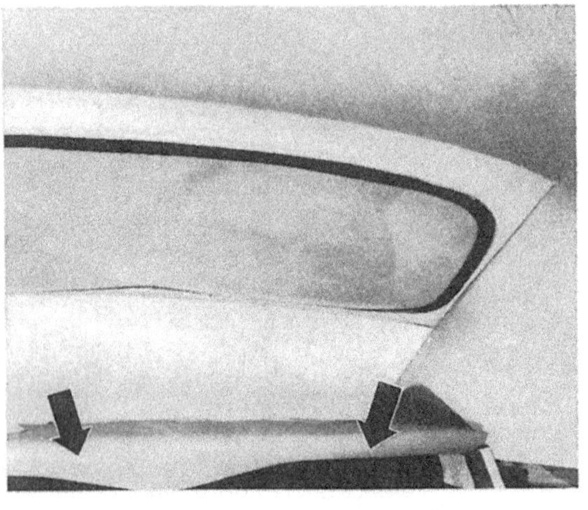

Important

The top material strip attached to the outside of the top cover and serving to protect the rear window must be removed only after completion of the top installation. The inner protective strip must be attached to the top compartment lower part.

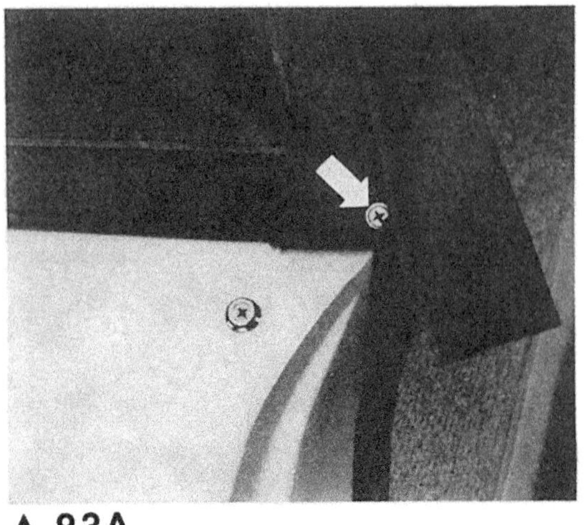

13 - Tension top cover on both sides at main bow and secure it with a chrome-plated screw each side. Do not forget protective washers.

Note:

Each side of the top cover is marked at the top near the main bow with a white line and below with a hole for the chrome-plated screw. These markings serve as a guide when installing the top cover. First pull top cover on to right side until the upper line lines up with the joint between roof frame and main bow. Then pull cover tightly downwards until the hole coincides with the lower edge of the main bow.

Attach top cover to left side in the same way.

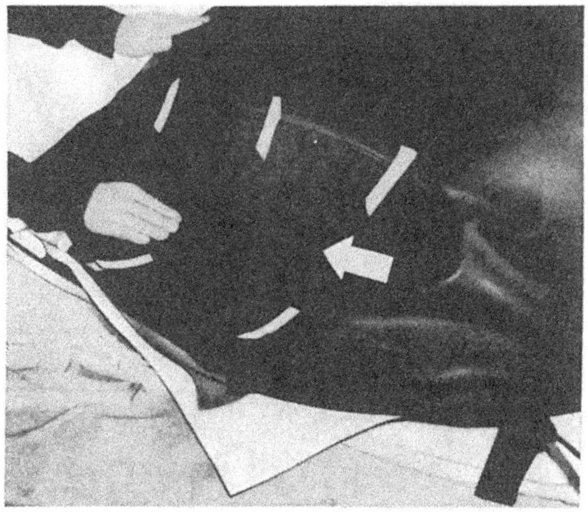

14 - Open top ensuring that the flexible rear window does not bend. Damage can be avoided by applying slight pressure to the window.

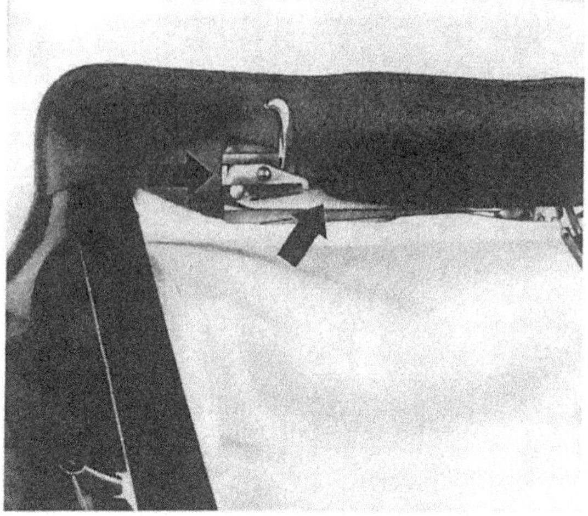

15 - Pull top cover pocket over header and cut it near the two fasteners.

16 - Push top into closed position and check whether top cover pocket is correctly positioned, especially near the outer rounded contours. Further installation of the top cover is carried out with top closed.

Important

When closing the top the first time, do not use force as otherwise the sides of the top cover at the roof frame can tear.

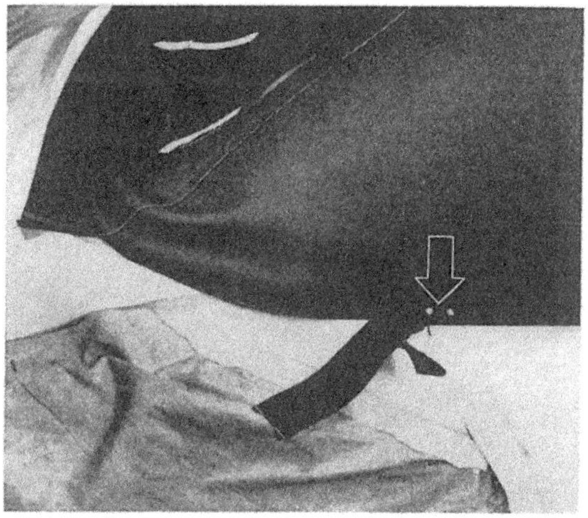

17 - Beginning at rear bow, pull top cover and sewn-on rear part of headlining down at the sides and secure under tension to rear body bow with two nails each side.

The beading all round top cover must not protrude outwards near main bow, otherwise the door windows will bear against it.

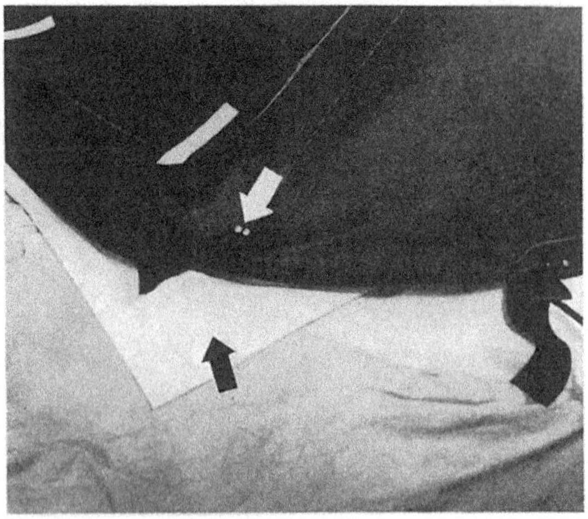

18 - Pull top cover taut towards rear and secure under tension in center of rear body bow with two nails each side.

To prevent top cover tearing at these locations, drive the nails only through the seams.

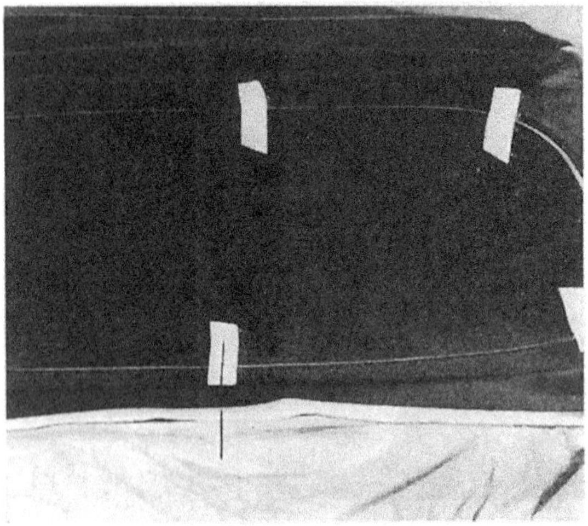

19 - Check that rear window is horizontal and centrally positioned, and correct if necessary. To correct, detach top cover material somewhat at appropriate location, pull in the required direction and re-attach it.

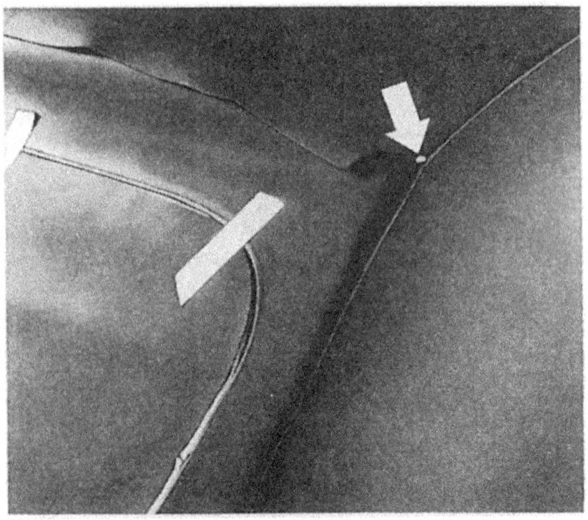

20 - Tack top cover to rear bow with two brass pins each side, ensuring that dimensions a and b are carefully maintained.

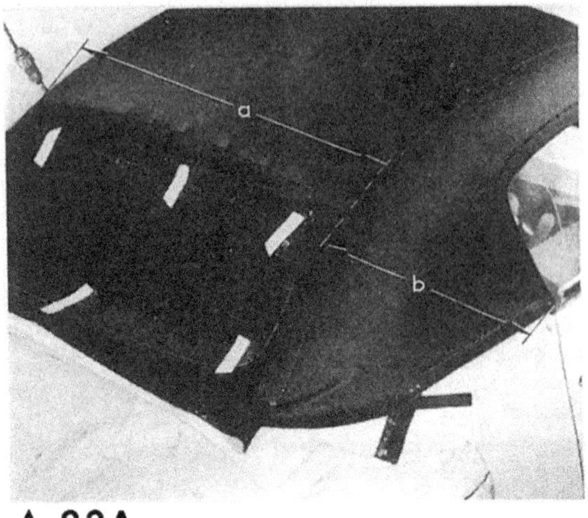

a - Distance from attaching location to attaching location of rear bow = 35.8" (910 mm)

b - Distance from one attaching location of rear bow to lower edge of main bow = 17.3" (440 mm)

Important

The rear part of the top cover must be under the center part.

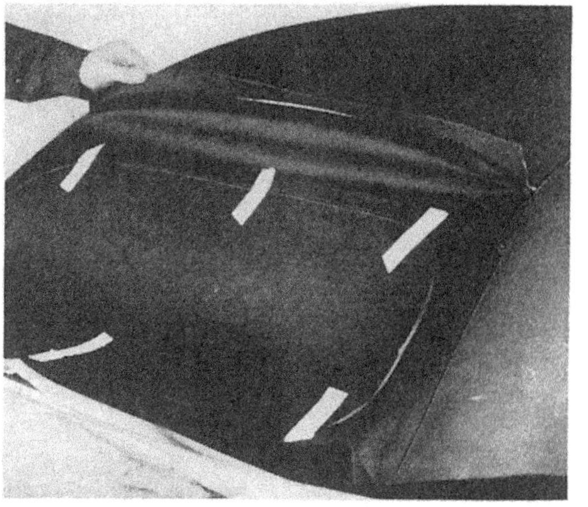

21 - Open top, ensuring that flexible rear window does not bend. This can be avoided by applying slight pressure to the window.

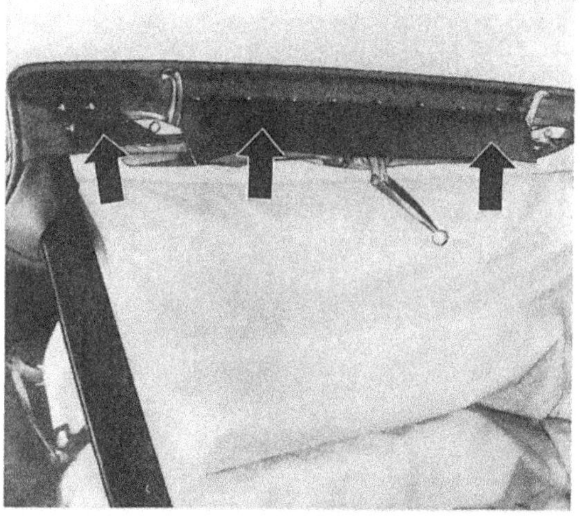

22 - Tack top cover pocket to underside of header with several nails.

23 - Close and fasten top. Check again carefully that top cover is tensioned all over and that no folds are visible. The same applies to headlining near rear window.

If necessary, carry out any necessary corrections at the locations marked by arrows as these will not be possible later.

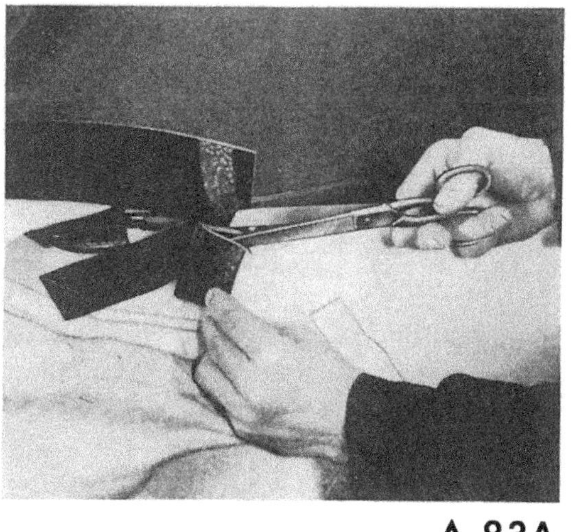

24 - Join ends of beading on rear body bow, which are not yet attached, to sewn-on beading on top cover:

a - Pull support strip taut in downward direction, tack it and cut off near rear body bow.

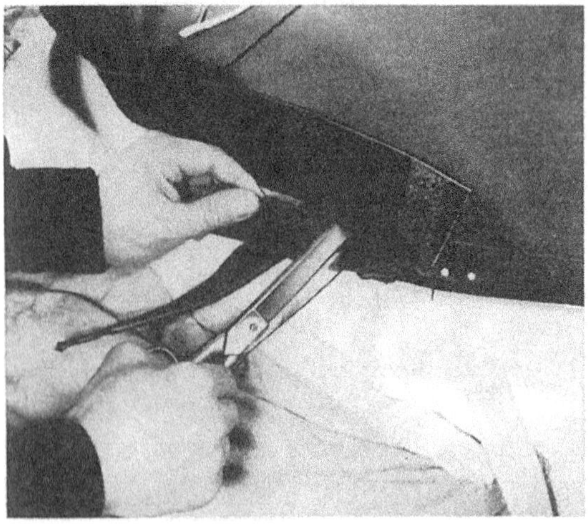

b - Cut off lateral support strip level with top beading, ensuring that the length is equal on both sides.

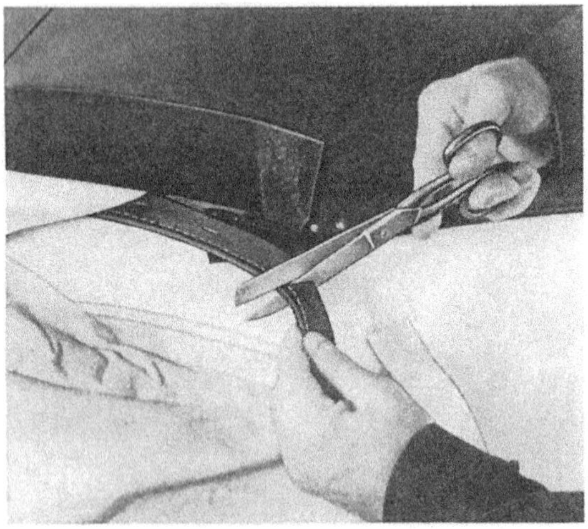

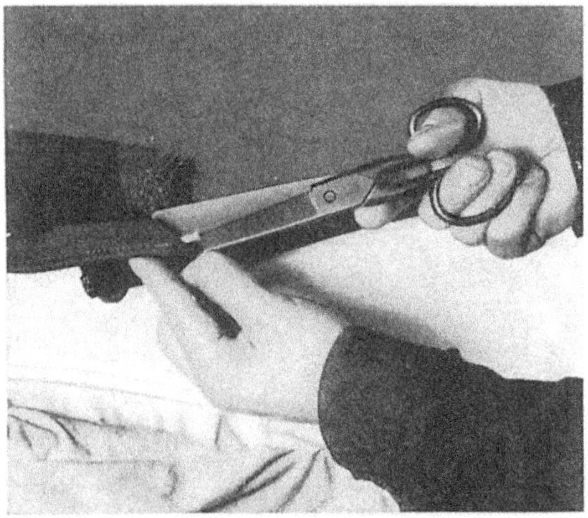

c - Leave beading on rear body bow 1.6" (40 mm) longer than the cut off beading of top cover, cut end at an angle and separate over a length of .8" (20 mm).

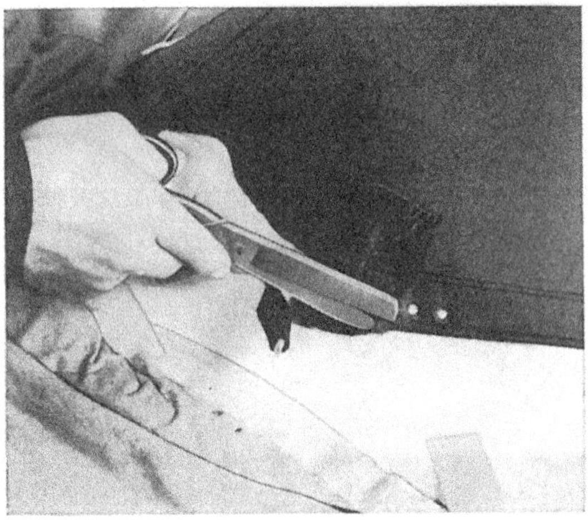

d - Make a cut in beading on top cover, then press cut end outwards.

A-83A

e - Push free end of cord in beading tacked to rear body bow into cut beading on top cover. Secure beading to rear body bow.

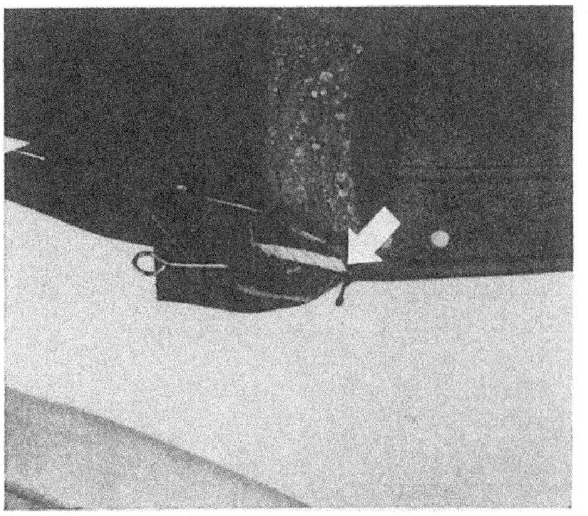

f - Pull the cut and pushed out end of top beading taut towards rear, lay round beading on rear body bow and staple in position.

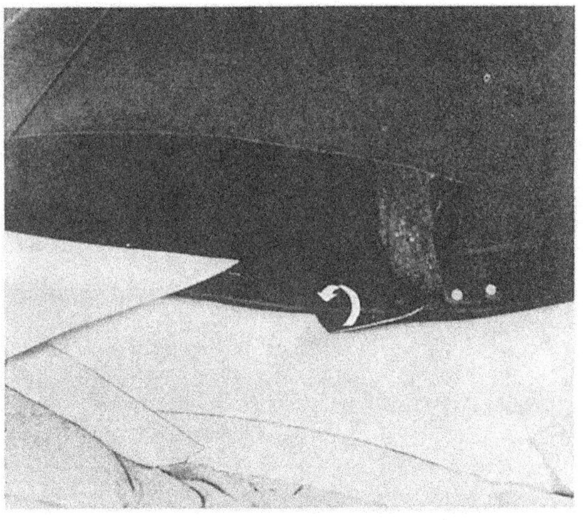

g - Pull outer support strip of headlining tightly downwards and staple to rear body bow together with beading.

Note:

To avoid leakage, the top material must overlap the headlining.

A - Beading
B - Headlining

25 - Pull top cover tightly downwards and staple under tension to rear body bow, starting from the longitudinal seam and working towards main bow.

Do not yet secure top cover below rear window to rear body bow.

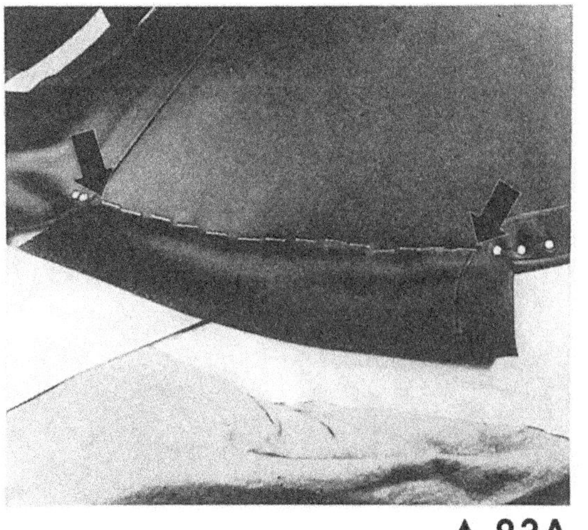

A-83A

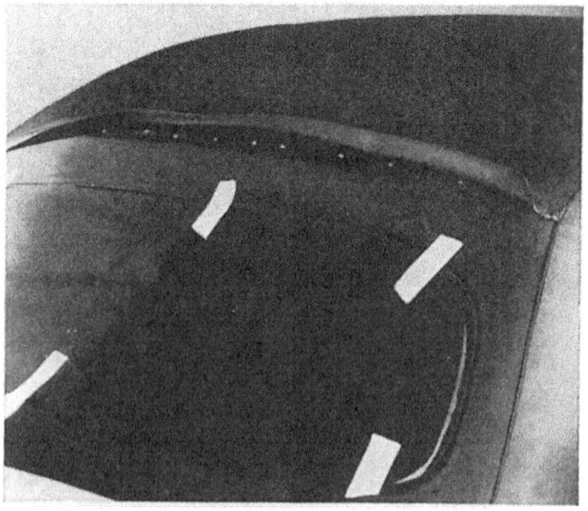

26 - Tack rear part of top cover, together with sewn-on rear part of headlining, without creases and under even tension to rear bow.

Insert tacks as near front edge of bow as possible so that part already perforated is not touched by tacks when finally securing it.

27 - From interior, carefully check position of headlining. If necessary, remove tacks and eliminate creases by retensioning.

Important

By measuring, check that rear window is horizontal and centrally positioned. If necessary, correct by detaching top cover at the appropriate location, pulling in required direction and retacking.

28 - Tack rear part of top cover, together with sewn-on rear part of headling, without creases and under even tension to rear body bow.

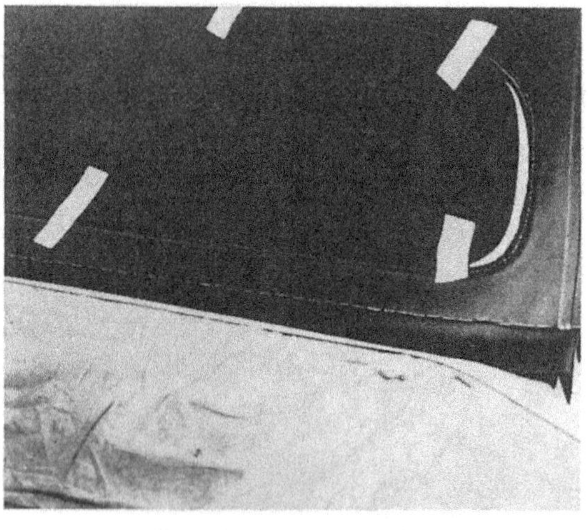

Note:

To simplify this, remove protective cover from rear window.

29 - From interior, check again that headlining is correctly positioned, and correct if necessary.

30 - Staple top cover between both longitudinal seams (below rear window) without creases and under tension to rear body bow.

The row of staples must be at an even distance from rear body bow beading edge.

31 - Staple top cover, together with sewn-on rear part of headlining, without creases and under even tension in rear bow recess. First staple the lower part, then staple the upper part over it to prevent water draining off roof from entering the cut. Carefully cut off excess material.

32 - Cut off excess material on rear bow cleanly, so that top material and beading edge overlap.

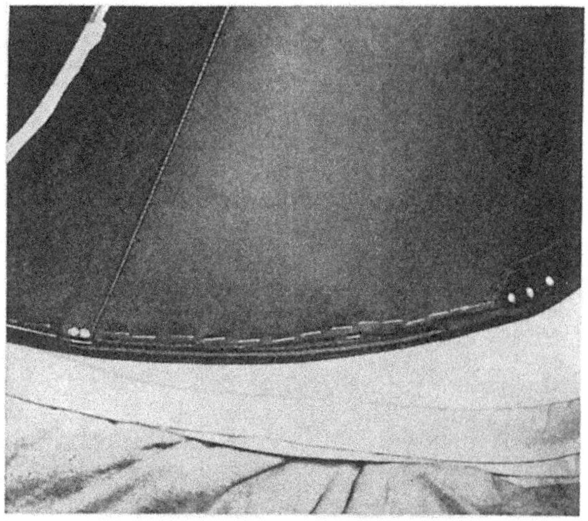

33 - Make a hole in each side of top cover for chrome-plated screws.

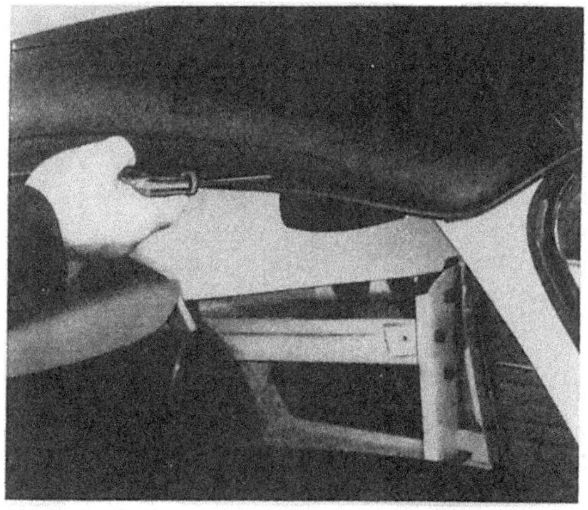

Important The top must be fastened and the header pocket correctly positioned on header.

34 - Open top and pull tensioning wires into top on both sides.

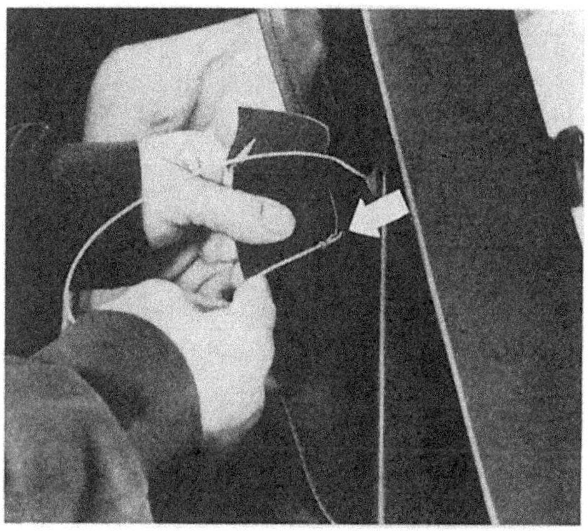

Note:

The insertion of the tensioning wires is simplified by using a .12" (3 mm) dia. welding wire, to the bent end of which the loop of the tensioning wire is attached.

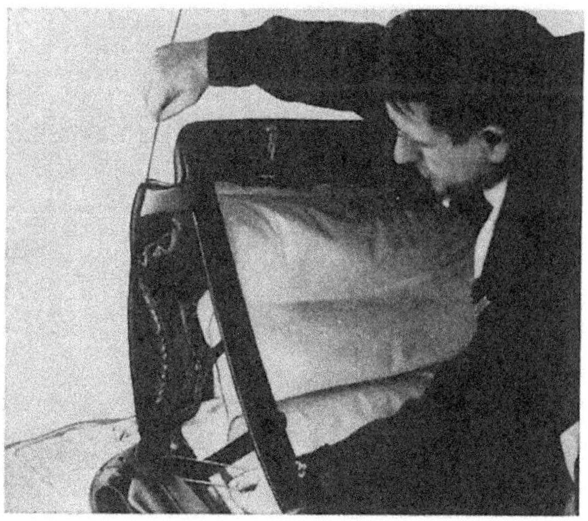

The welding wire must be inserted into the top cover from above and pulled out on the inside near the bottom of the main bow.

35 - Push chrome-plated screw, with washer installed, into top cover, attach loop of tensioning wire and screw into roof frame.

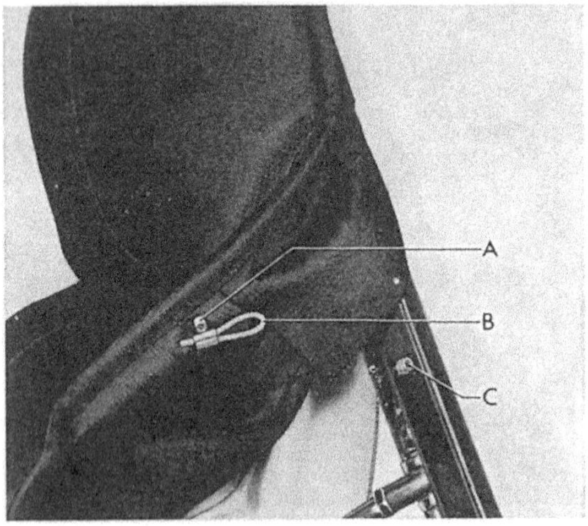

A - Chrome-plated screw
B - Tensioning wire loop
C - Hole in roof frame

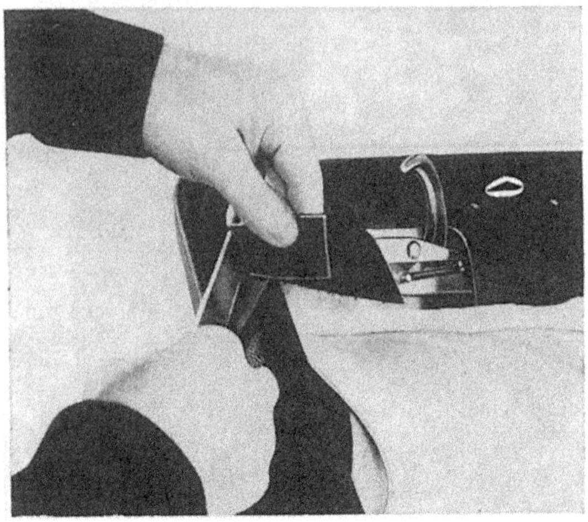

36 - Open top. Cut a piece of material, equal in width to the weatherstrip installed later, out of underside of header pocket. Cement header pocket to left and right roof frame.

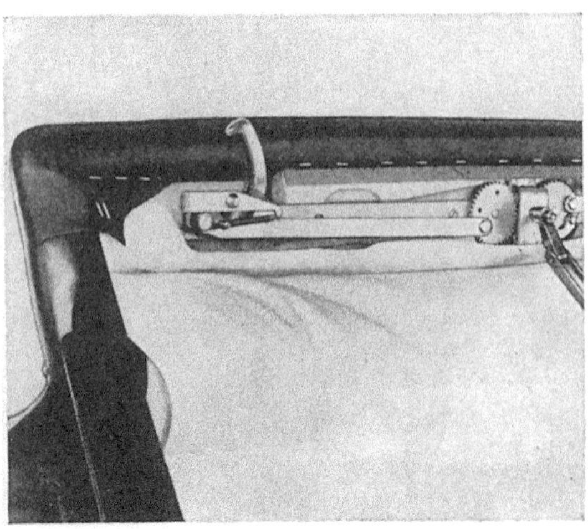

37 - Carefully staple tacked header pocket to header, remove tacks and cut off excess material.

38 - Staple a strip of top material between top fasteners, with outside facing header.

For this purpose the excess material cut off header pocket can be used.

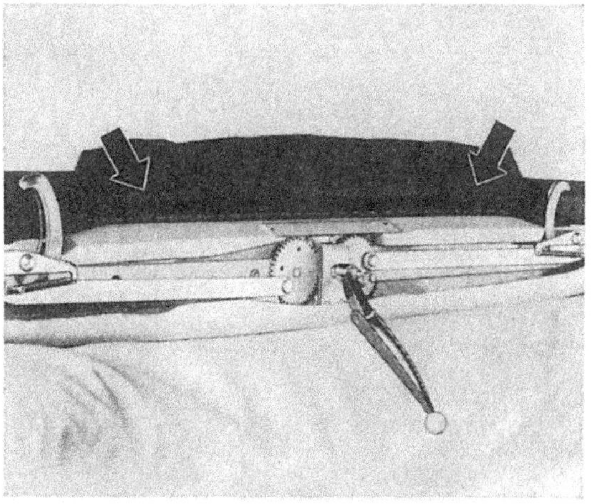

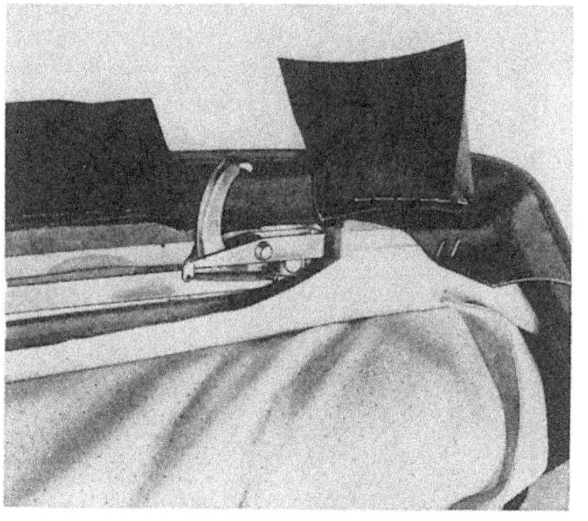

39 - Attach a piece of top material 4.7×5.9" (120×150 mm) near each corner by stapling, the narrower side facing upwards. The material must be folded at the outside.

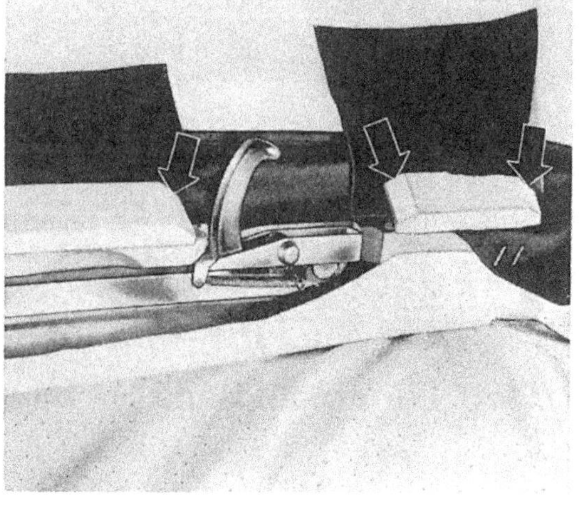

40 - Cement three-piece rubber seal, 13.8×.6" (350×16 mm), in recess provided in header. Chamfer ends of seal facing top fasteners.

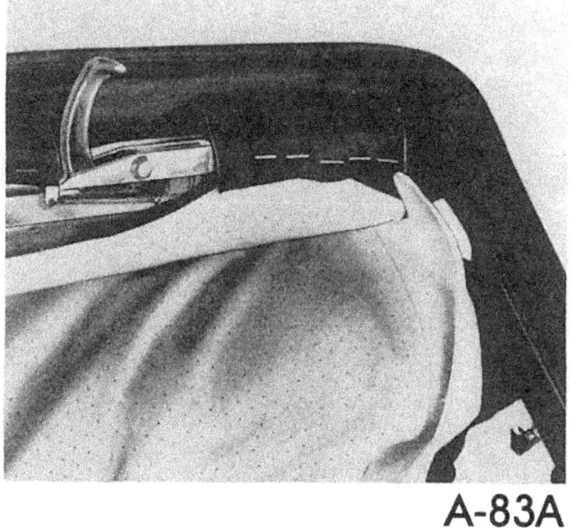

41 - Pull top material strip tightly over header rubber seals and staple, without creasing, to header.

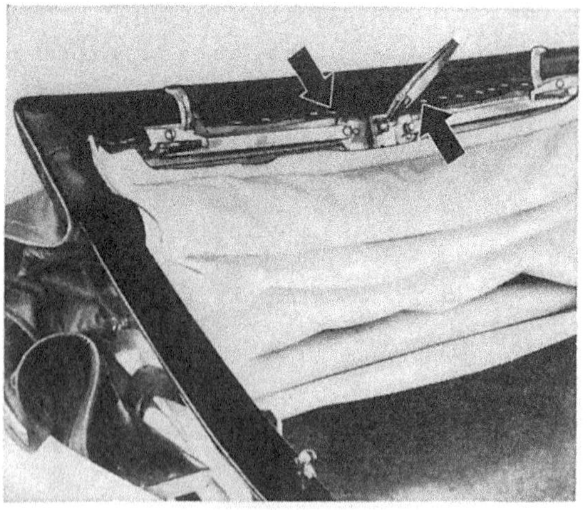

Cement top material strip and header rubber seal to locking mechanism as the header wood has been hollowed out here.

42 - Close and lock roof without applying force.

Important
The beading of the header pocket must not be trapped underneath.

43 - Cement loose top material strips on both sides to main bow. The strips must be 1". 2". (30 mm) wide at the top and 1 in. (25 mm) wide at the bottom. The strips must be cut off at the top and bottom, level with the joint of the main bow. Cut off excess material.

a = 1.2" (30 mm)
b = 1" (25 mm)

Fitting Trim Mouldings

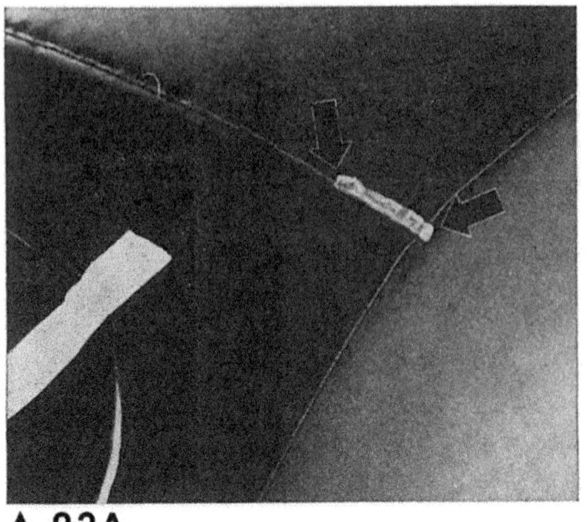

1 - Apply plastic sealing compound to staples in rear bow, at both ends *near seams. The sealing* compound must not be wider than the moulding to be installed.

2 - Staple a strip of sealing tape over row of staples on rear bow.

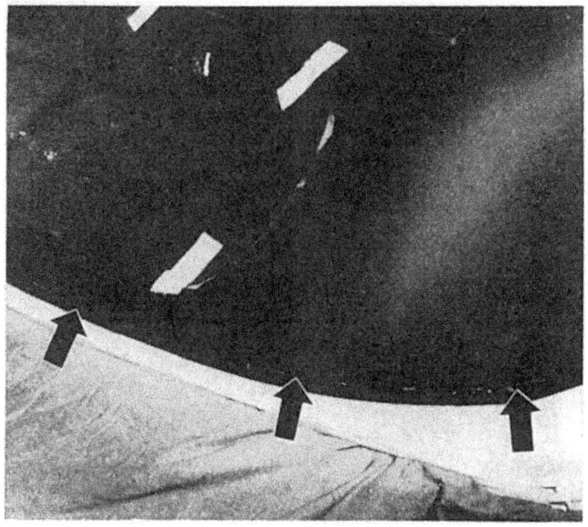

Note:

This sealing tape is a natural rubber strip of the same width as the trim moulding. The light coloured protective strip must be removed before installation.

3 - Fit a strip of sealing tape to trim moulding.

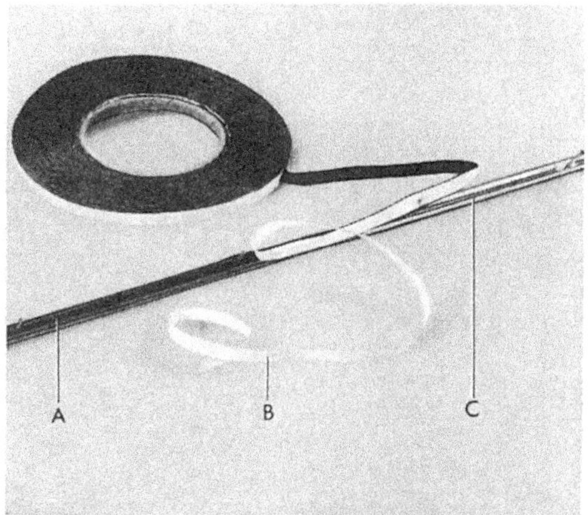

A - Natural rubber strip
B - Protective strip
C - Trim moulding

4 - Mark position of trim moulding on rear bow with saddler's chalk.

5 - Nail on rear bow trim moulding, starting at one side and working to the other.

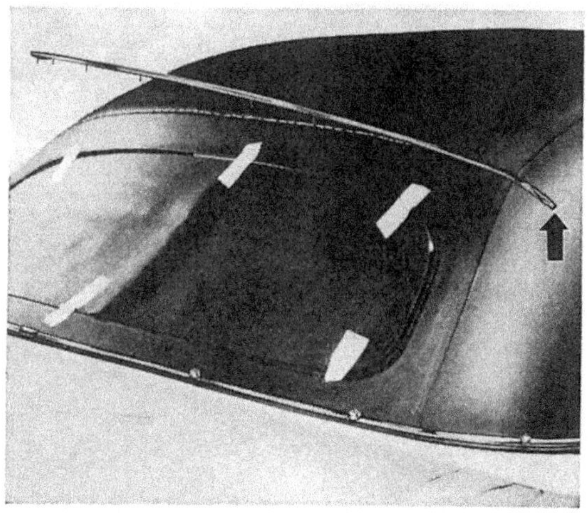

Important

A special wooden tool must be used when fitting the moulding, to avoid damage and ensure that moulding fits properly.

6 - Nail two-piece moulding to rear body bow and slide clip over joint.

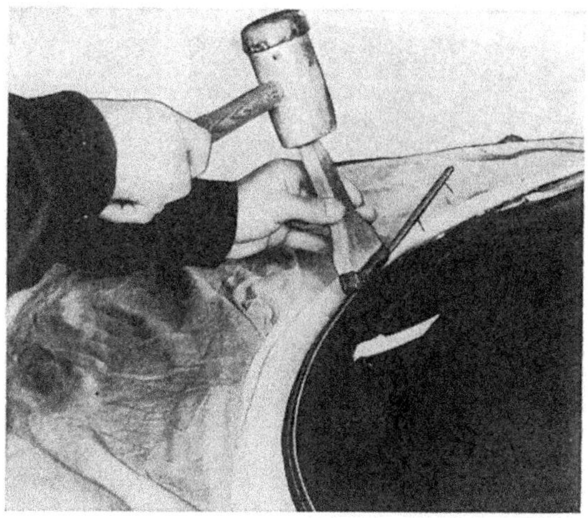

Note:

The distance between trim moulding and lock pillar is 14.2" (360 mm). To avoid difficulties such as overhanging at the ends, it is advisable to determine the exact location of the trim moulding by fitting, measuring or marking with saddler's chalk. If necessary, grind off moulding ends at center joint.

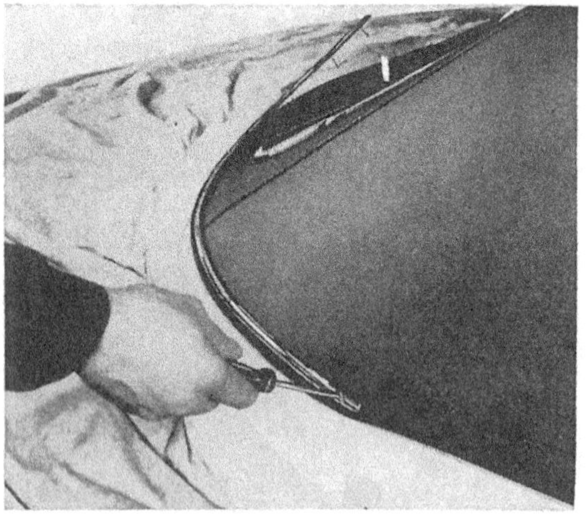

Nail on one trim moulding half, starting at outside.

a - Drive in outer nail, drill a hole through trim moulding and into rear body bow, then secure moulding with a Phillips screw.

b - Bend moulding round rear body bow and nail in position, using special wooden tool.

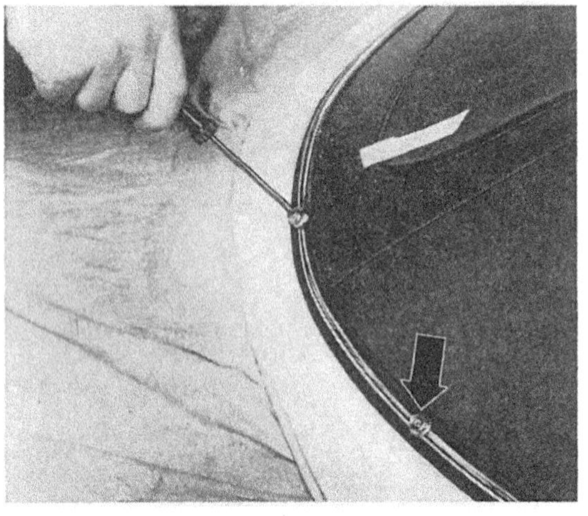

c - Starting at center, nail other moulding half **on to** rear body bow and additionally secure outside with a Phillips screw.

7 - Screw press stud lower parts, with supports in position, into holes in trim moulding.

Securing Headlining

1 - Open top and unscrew top lock handle.

2 - Screw top handles to header plate.

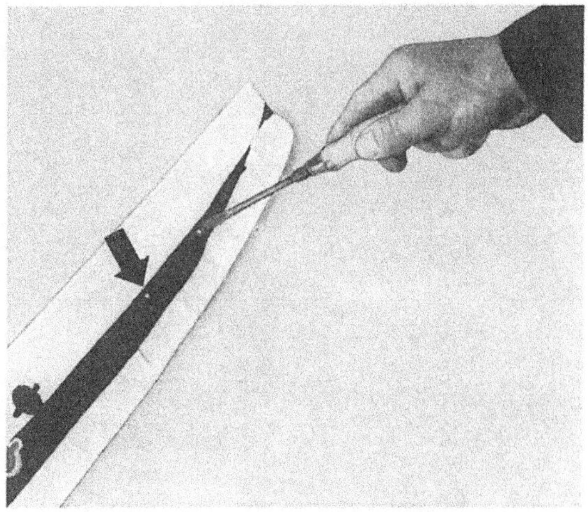

3 - Secure fastener cover plate to cut-outs in header with four Phillips screws each.

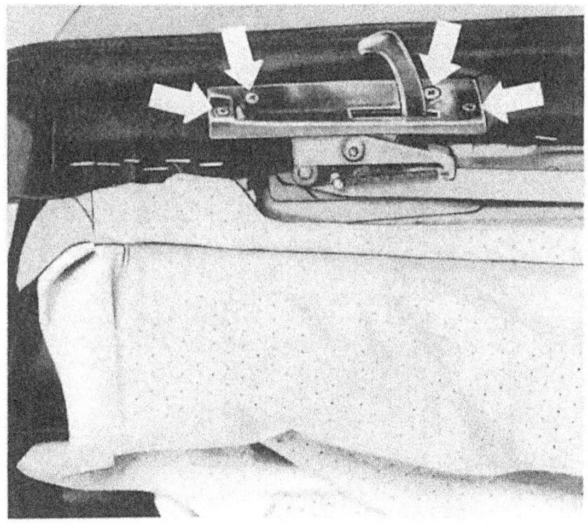

Note:

If a new top header or top lock has to be installed, check for free operation of the fasteners in the cover plate cutouts, prior to installation of the cover plates, by opening and closing top lock.

4 - Attach header plate, by means of chrome-plated top lock handle escutcheon, to inside of top header.

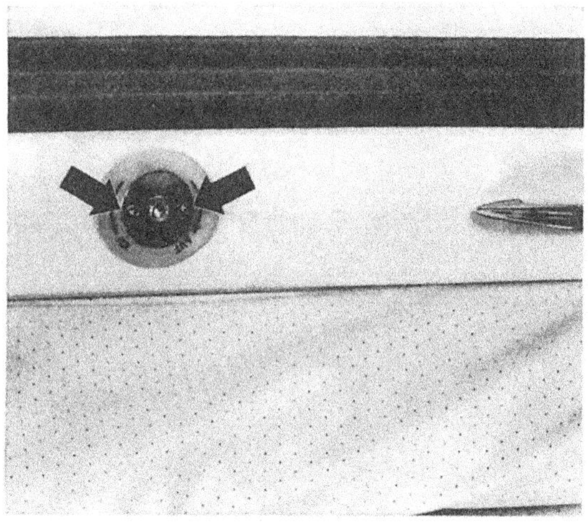

5 - Attach header plate to top header with twelve Phillips screws.

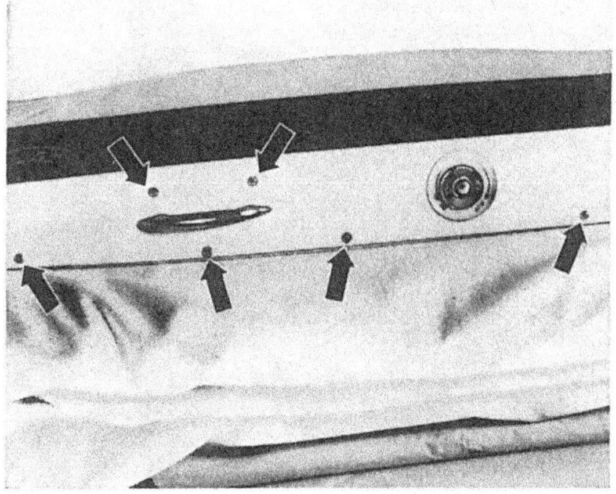

Note:

If a new header plate has to be installed, the holes for the attaching screws must be drilled first.

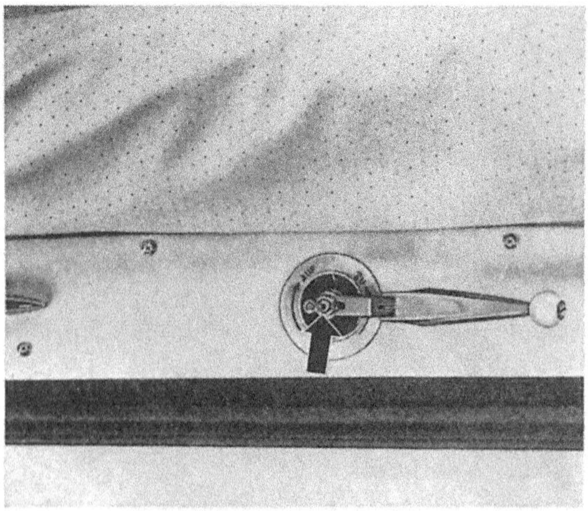

6 - Push top lock handle on to shaft and secure with cap nut.

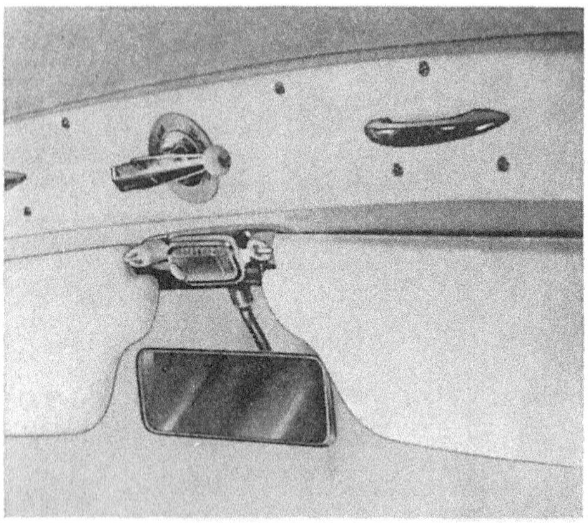

Note:

Ensure that the knob of the folded handle faces right when top is closed.

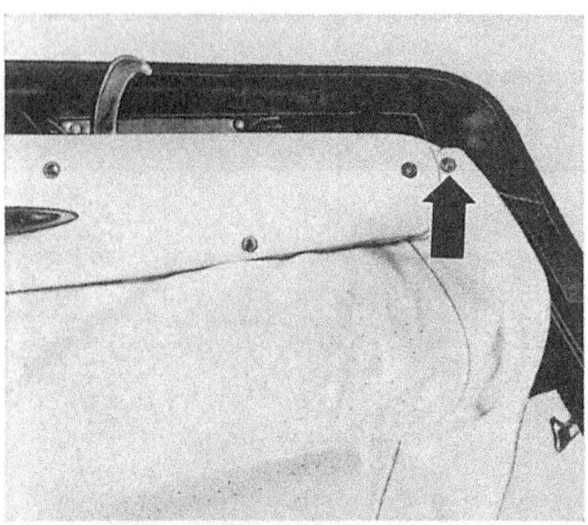

7 - Pull side parts of headlining taut towards front and screw them to top header plate. The screws must not contact roof frame as otherwise noises can occur at the header. If necessary, bend **ends** of leatherette covered top header plate slightly away from roof frame.

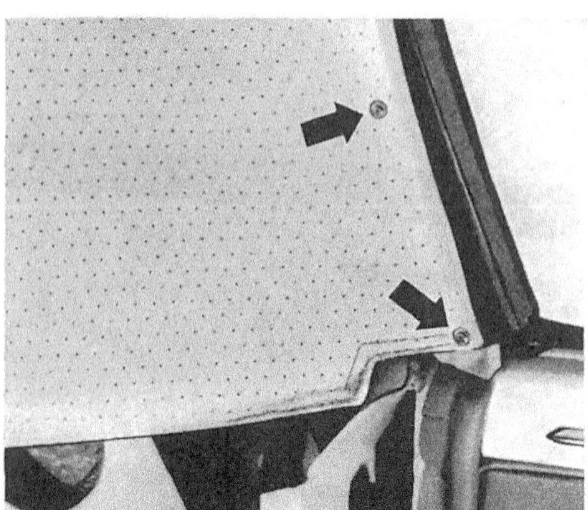

8 - Close top.

9 - Attach headlining to main bow on both sides by inserting two chrome-plated Phillips tapping screws in holes provided. *Do not forget* washers.

10 - Apply universal adhesive D 12 to top compartment lower part.

11 - Attach haircord carpet to wheel housings and a piece of leatherette (arrow), the same colour as the quarter trim panel pads, in both corners of top compartment lower part.

12 - Secure leatherette covered quarter panel trim pads to lock pillars with three Phillips screws each (white arrows) and to top compartment lower part with two Phillips screws each (black arrows).

13 - Cement protruding leatherette strip of quarter panel trim pad to top compartment lower part and reattach haircord carpet for emergency seat support. Insert safety belt mounting point cover bushings and escutcheons.

Note:

If the quarter panel trim pad has been recovered, install a press stud support and a fastener for protective boot on each side.

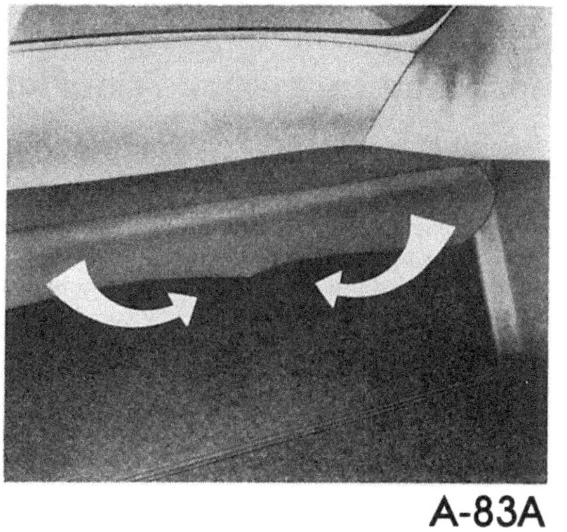

14 - Cement leatherette covered cardboard to top compartment lower part, then pull hanging leatherette strip taut and cement it to underside of top compartment lower part.

15 - Cement haircord carpet to luggage compartment floor.

16 - Install emergency seat backrest retaining strap.

Important

After completion of installation, check whether the top can be easily opened and closed and whether the fasteners position themselves correctly in the sockets on the windshield frame.

There is always a certain amount of tension in a new top. It is not necessary to carry out any form of correction, as, after a certain period of time, the top stretches so that it can be easily closed.

Installing Lock Pillar Trim Plate and Weatherstrips

The weatherstrips are of ample length so that they can be fitted and cut to length. They are attached to the roof frame by metal strips which fit inside the rubber moulding and are secured by Phillips tapping screws. In the case of a new top frame, the holes must first be drilled in the roof frame so that they coincide with the holes in the metal strips.

Before the individual weatherstrips can be secured, they must be accurately fitted. For this purpose, the correctly adjusted windows must be wound up and the weatherstrips be inserted in the position in which they will be secured later. Special attention must be paid to the contact of the lip on the door window glass.

1 - Cement rubber seal to lock pillars and secure with tubular rivets.

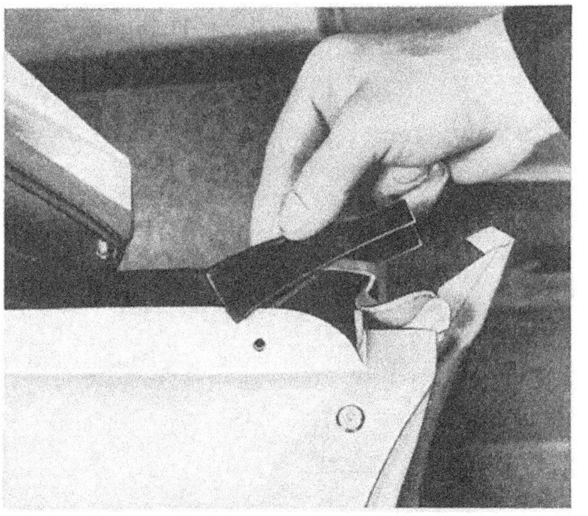

Note:

Two-piece rivets are used, these being pressed together with a pair of combination pliers. Place pieces of leatherette between the jaws of the pliers to avoid scratching the rivets.

2 - Pull end of beading on lock pillar tightly upwards and staple it. Pull strip of quarter panel trim pad material tightly upwards, place it over beading and staple it.

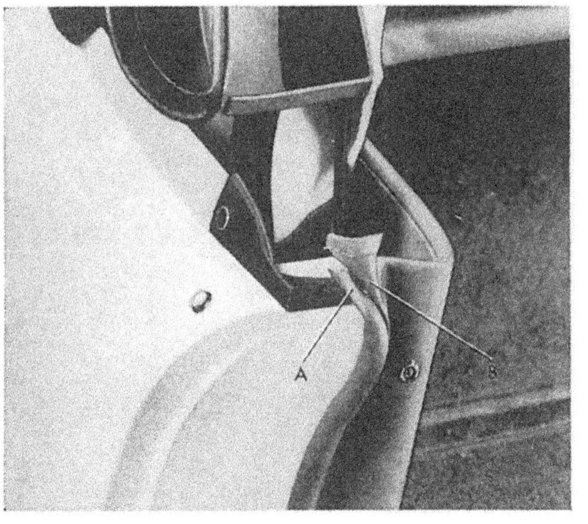

A - Beading on lock pillar
B - Leatherette of quarter panel trim pad

3 - Attach lock pillar trim plates with three chrome-plated Phillips tapping screws at the top and two at the bottom on each side.

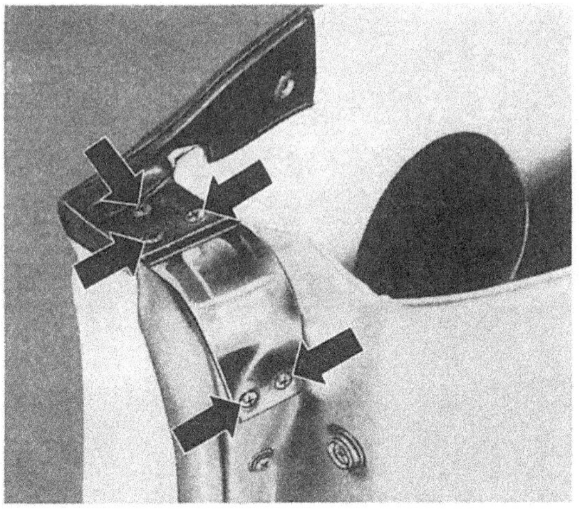

4 - Mark position of weatherstrip on windshield frame, along wound up window glass, with a pencil.

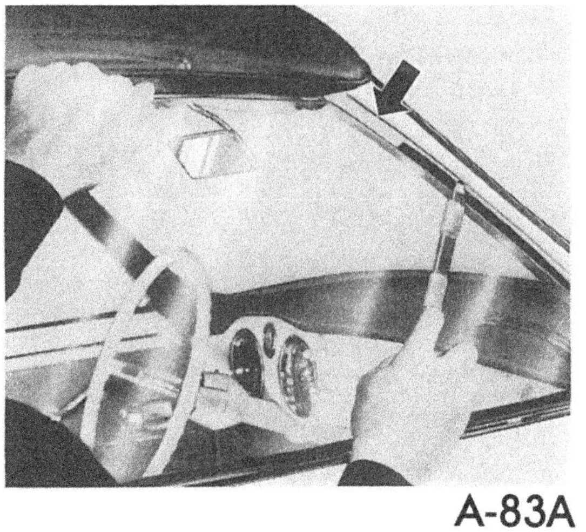

Important
Do not use a scriber or similar sharp tool.

5 - Cut weatherstrips to approximate length, leaving them slightly longer than necessary.

6 - Mark and drill screw holes so that they coincide with holes in metal strips.

Important

The weatherstrips must be attached .2" (5 mm) farther outwards at the top and .08" (2 mm) farther outwards at the bottom so that they contact the window glass firmly.

It is recommended that the weatherstrip be secured to roof frame .2" (5 mm) offset towards outside from marking.

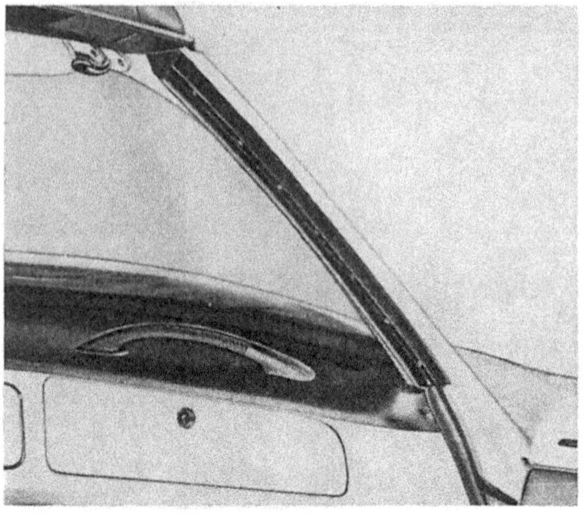

7 - Attach weatherstrips together with metal strips.

As additional sealing it is recommended that the sealing tape be attached to weatherstrip underside facing outwards.

Leaky Top Cover Seams

When the seams of the Convertible top show signs of leakage, carry out a test by spraying the top evenly and lightly with water until the location of the leak is discovered.

The top seams, including those round the rear window, can only be effectively sealed if the thread itself is in good condition. If the thread is damaged or rotted to such an extent that it is no longer tight the only remedy is to replace the cover. Re-sewing is not recommended as experience has proved that the stitches cannot be spaced to conform to the original holes when sewing with a machine. Apart from this the seam holes would be enlarged too much by the tension of the new thread.

The leaks can be eliminated by the use of the seam protecting solution "Happich 7303" which is obtainable from Gebr. Happich GmbH, Wuppertal-Elberfeld, Neuenteich 72. The solution is applied underneath all the top seams, including the edges, and rubbed in vigorously until small drops can be seen in the stitch holes from outside. The solution must not come into contact with the outer surface of the top cover as it cannot be removed.

The following operations are necessary to enable the solution to be rubbed into the front seams.

1 - Lift the header trim moulding at each side about a third of its length.

2 - Remove the screws securing the top cover to the roof frame at the sides.

3 - Loosen the top cover at both corners of the header until access is gained to the seams.

4 - Open the top until it is free of tension.

5 - Rub the solution well into the seams.

6 - Secure the top cover to the header and roof frame again.

To treat the rear seams it is merely necessary to open the top about half-way. The seams can then be reached from the side, between the roof frame and the top cover.

The water stains in the headlining caused by the leakage can be removed by rubbing them evenly with a cloth well soaked in a solution of 1 part ammonia and 3 parts water. The headlining must be perfectly dry before the solution is applied.

If the headlining is dirty, the places which have been treated will be lighter in colour. It is advisable, therefore, to wipe the entire headlining with a well soaked cloth after the water stains have been removed.

Leakage at Rear Window

If leaks are found at the rear window of the Convertible, first check the condition of the thread securing the rear window in the top cover. Should the thread be damaged or rotted to such an extent that it is no longer tight, the only remedy is to replace the top cover. Re-sewing is not recommended as experience has proved that the stitches cannot be spaced to conform to the original holes when sewing with a machine. Apart from this the seam holes would be enlarged too much by the tension of the new thread.

Carry out a test by spraying the area lightly with water. If water penetrates round the window even though the seams appear sound, seal the part between window and top cover with the sealing compound "Epple 55". This compound can be obtained direct from Chemischen Fabrik Epple and Co., Stuttgart-W., Seidenstr. 57.

Waterstains on the headlining under the rear window do not always indicate that the window is leaking. In most cases these stains are caused by condensed water running off the window and soaking into the headlining. A remedy in these cases is possible only if the window is wiped dry when condensation forms. The water stains in the headlining can be removed by rubbing them evenly with a cloth well soaked in a solution of 1 part ammonia and 3 parts water. The material must be perfectly dry before the solution is applied.

If the headlining is dirty, the places which have been treated will be lighter in colour. It is advisable, therefore, to wipe the entire headlining with a well soaked cloth after the stains have been removed.

Noises at the Header

From Chassis No. 2125345 onwards anti-squeak washers (Part No. 141 871 505) and anti-squeak sleeves (Part No. 141 871 509) were installed in the corner fittings of the Convertible to eliminate noises between the header fittings and the roof frame caused by the stressing of the top frame.

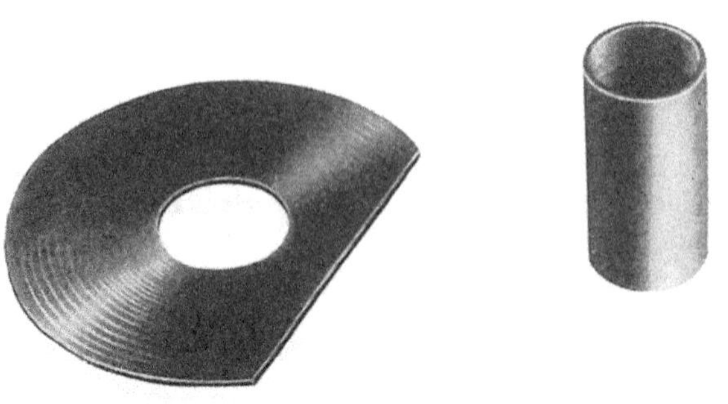

If these noises occur on older Convertible models the plastic parts can be service installed. The following operations are necessary:

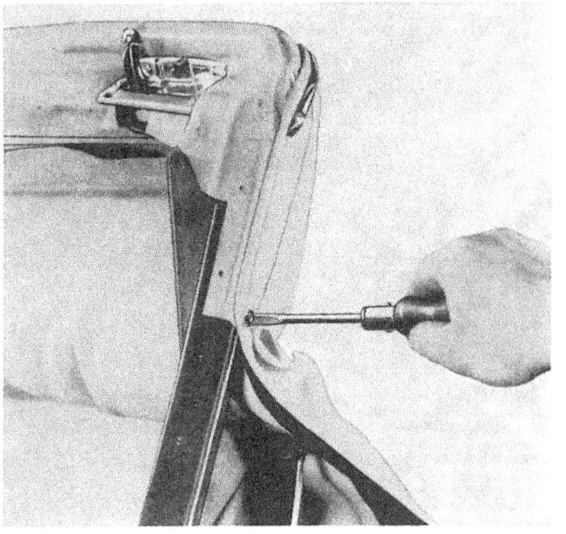

1 - Open top and fold it back. Remove top lock handle and header cover plate.

2 - Loosen both door window weatherstrips on the top frame for about half their length.

3 - Remove the screws securing the top cover to the frame at the sides.

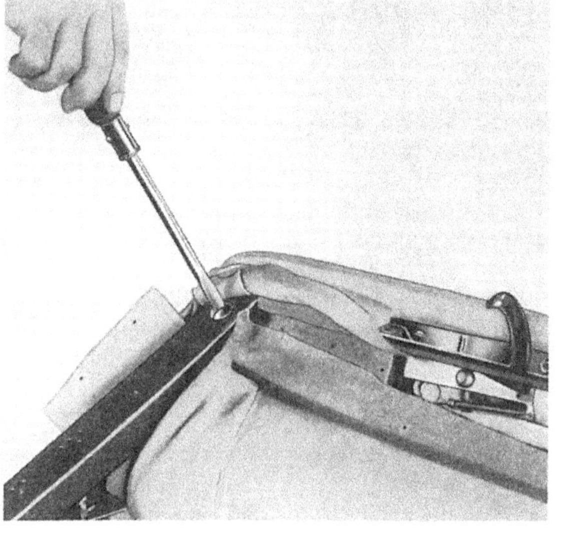

4 - Loosen the top cover at both corners where it is nailed to the header.

5 - Lift the top cover at both sides where it is cemented to the frame until the screws securing the header are exposed.

6 - Remove the header securing screws taking care that the brass washers under the screws are also withdrawn from the holes.

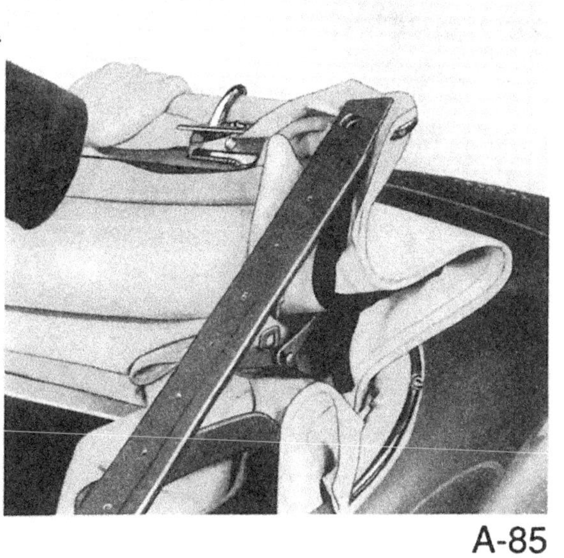

7 - Push header out of the top frame.

8 - Remove scraps of old material and cement remains from the header corner fittings.

9 - Place the sleeves over the rear ends of the corner fittings.

Coat the ends of the corner fittings lightly with adhesive to stop the sleeves slipping off when the header is installed. (Allow adhesive to dry well as the synthetic material does not stick easily.)

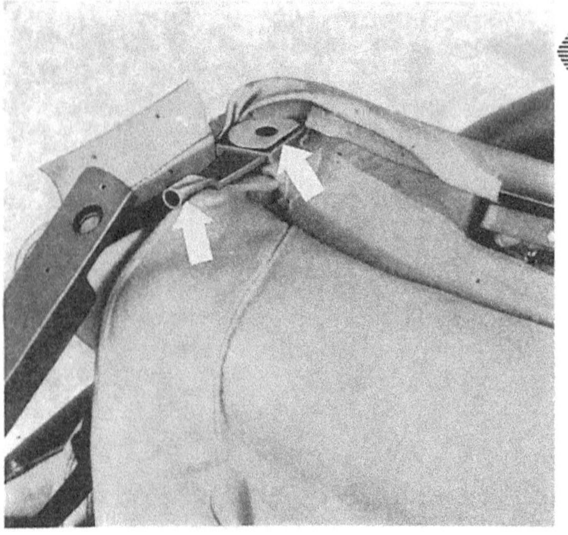

10 - Cement the anti-squeak washers over the holes for the header securing screws so that the straight edge faces the vehicle interior. Place a shake-proof washer in the hole in the header fitting.

11 - Locate the header in the roof frame again and insert screws. Take care that the brass washers are correctly in position. While the screws are being tightened move the appropriate side of the roof frame back and forth so that the screws seat properly. The screws should be tightened to a torque of 4 mkg (29 ft. lbs.).

12 - Place the top cover in position again and secure with cement, nails and screws. Install the header cover plate and the lock handle.

Padded Sun Visors

From Chassis No. 2252455 the Convertible has been fitted with padded sun visors instead of the dark-colored plastic material type.

The new type visors can be installed in place of the previous type without any further alteration and are obtainable separately under the Part Nos. 141 857 551 B/552 B.

The new rear view mirror with spindles (less visors) is supplied under Part No. 141 857 523 B.

When service installing the new type sun visors on the previous rear view mirror with holder observe the following:

1 - Remove securing screws and remove previous sun visors with the two end supports (Part No. 141 857 559).

2 - Push the padded sun visors onto the spindles as far as possible and tighten the clamp screws.

The end supports are no longer necessary.

Noises at the Header Cover Plate

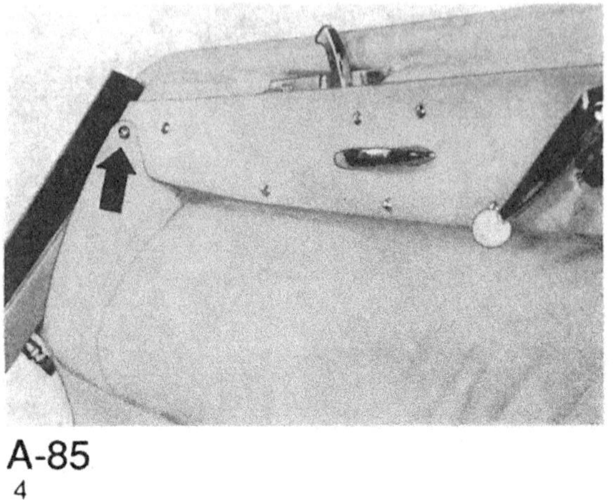

Noises occasionally occur at the corners of the header on the Convertible which can be traced to the following source.

The two headlining securing screws at the ends of the header cover plate contact the metal roof frame when the vehicle vibrates.

A remedy can be affected by shortening the screws and slightly bending the ends of the header cover plate so that the screws cannot touch the metal frame of the roof. To avoid soiling or damaging the cover plate when bending it, wrap a piece of cloth round the jaws of the pliers.

Leakage at the Top Header

When leakage occurs at the header of the Convertible, the cause can be as follows:

a - Insufficient or uneven contact between the header and the windshield frame, particularly at the corners.

b - Incorrect nailing of the top cover to the header under the front trim moulding.

The leaks can be located and eliminated as follows:

Re a - Light spots on the header contact surface usually indicate the entry of water at this point.

Check that the contact is even by inserting strips of paper between the header and the windshield frame. At the points where the paper strips are not firmly pressed, the header must be sealed by installing additional foam rubber strips of appropriate thickness. First of all check whether the original weatherstrip is deformed or misplaced. Should this be the case, install a new weatherstrip. The most suitable foam rubber strip for sealing is the type used under the roof corners (Part No. 141 871 605).

To replace the weatherstrip, the handle for the top lock and the header cover plate must be removed. If the nails on the inside of the header are removed the front top cover strip can be lifted enough to allow the weatherstrip to be renewed or additional pieces inserted. Nail the top cover in position and carry out a water test. If the leaks have been eliminated, install header cover plate and handle.

Re b - The main top cover is nailed under the front trim moulding together with the strip of material for the header. At the corners of the header the strip of material is nailed in small pleats to follow the curve. If the pleats are too large they act as canals through which the water is forced under the main top cover when the vehicle is moving at high speeds, thus wetting the padding and header. If wadding is used to even out the pleats it must not be visible on the row of nails between main top cover and the header cover as the wadding will soak up moisture.

This type of leakage shows as damp spots in the front part of the headlining and can be eliminated as follows:

Remove:

1 - Handle for top lock

2 - Header cover plate

3 - Header trim moulding

4 - Sealing band under the trim moulding.

Loosen the top cover at the roof bow corners as far as necessary and fold up. Renew damp wadding and rubberised hair and dry header. If the header is cracked or the glue is faulty the header must be renewed.

When nailing the header strip in position make the corner pleats smaller before nailing. Coat the upper ends of the pleats well with rubber solution and cover with the widest possible strip of linen. Finally, nail the main top cover to the header again, under tension and free of creases with brass tacks only (Part No. 151 871 471). The top beading must seat firmly and evenly particularly at the corners of the header. Coat the row of nails on the header with rubber solution. Tack a strip of sealing band (Part No. 151 871 435) over the nails and install the trim moulding.

Protective Cloth for Rear Window

From Chassis No. 2120400 a soft protective cloth (1050 mm × 650 mm) was supplied with the Convertible. This cloth should be placed over the rear window when the roof is open to avoid the window being scratched.

A folding card was also included with the cloth, giving more detailed instructions as follows:

"Please note the following when opening the top:

When you have unlocked the top and lifted it slightly, place the protective cloth over the rear window. The top can then be folded back, smoothing out the folds between the bows with the back of the hand.

Press the top down until the catches engage. Secure the top boot on the inside first and then outside."

This text was included in the Instruction Manual at the same time.

Windshield Frame

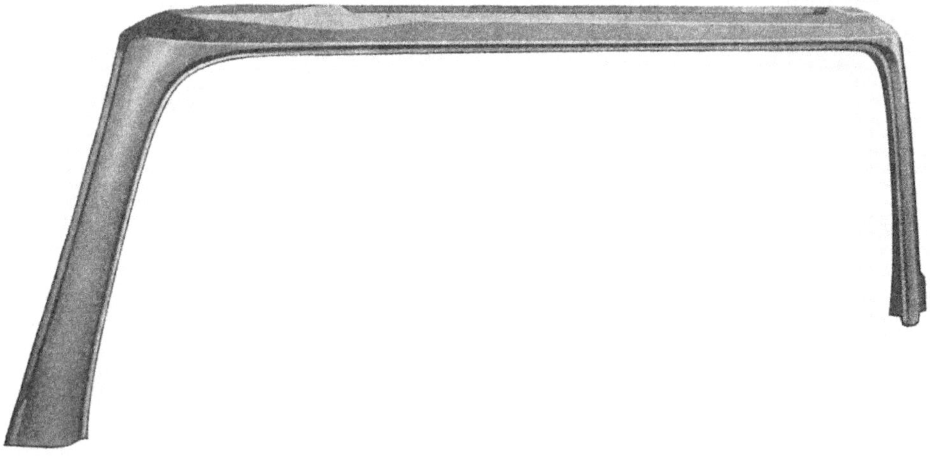

The windshield frame is supplied as a complete item and consists of:

a - Inner panel

b - Outer panel

c - The recesses and rollers for the locking hooks

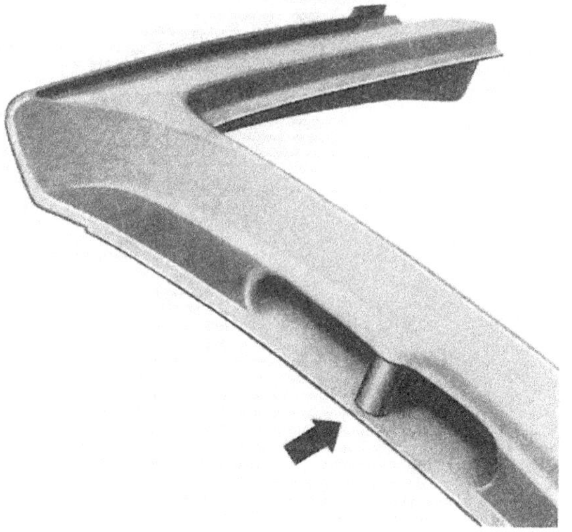

All the parts are welded together. The outer edges of the windshield frame are folded together.

If necessary, replace the complete frame and weld into position with the assistance of the special jig. The use of the jig is strongly recommended as otherwise difficulty can appear when installing the windshield or when closing the top. Further details of other jigs and appliances can be found in "Body General" section under "Workshop Equipment".

Replacing the Windshield Frame

Preparation

1 - Disconnect battery.

2 - Remove or cover the instrument panel as far as necessary.

3 - Remove windshield wiper mechanism.

4 - Remove front seats and floor mats.

5 - Remove windshield glass.

6 - Unscrew the door window weatherstrip at the windshield frame.

7 - Remove steering wheel.

8 - If no other body repair work is being carried out, remove or adequately cover both doors.

9 - Fold top back and cover up to avoid *soiling.*

Body Repair Work

1 - Heat the layer of solder on the weld seam with a gas torch and scrape off with a suitable spatula.

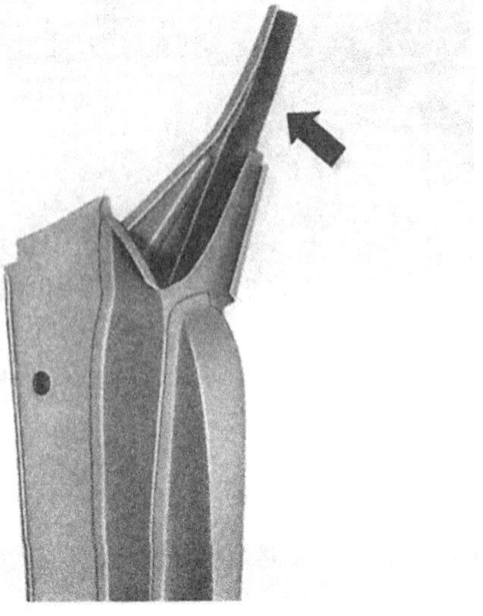

2 - As already known the hinge pillars of the Convertible have welded-in reinforcement channels at the top which are also welded into the windshield frame.

These reinforcement channels must be used again as it is impossible or extremely difficult to weld them into the hinge pillar in situ. For this reason *the channels are not supplied* separately as spare parts.

3 - Saw the pillar off about 200 mm above the weld seam.

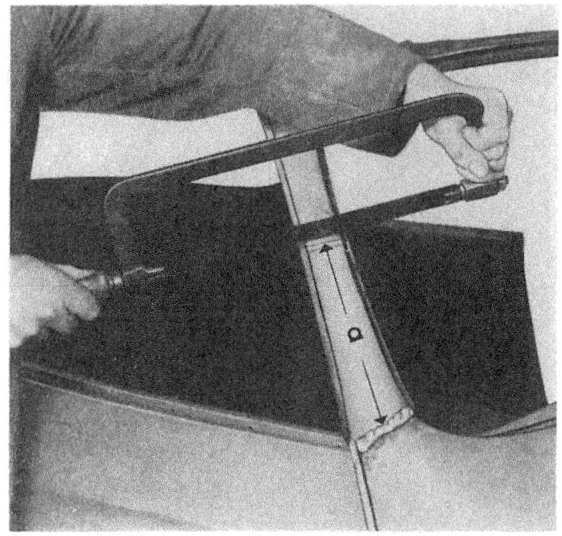

a - 200 mm (7.8")

4 - Bend up the folded edges of the windshield frame and chisel down the sides of the remaining windshield pillar parts. The reinforcement channels must not be damaged during this operation.

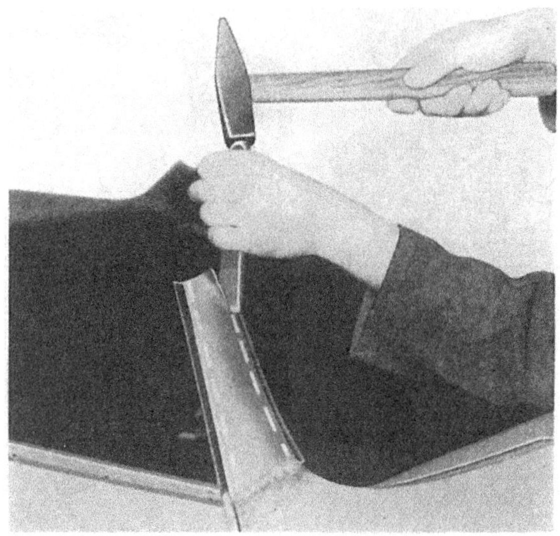

5 - Remove metal scraps down to the weld seam and grind or file the weld seams clean. If necessary straighten the reinforcement channel and beat out any dents.

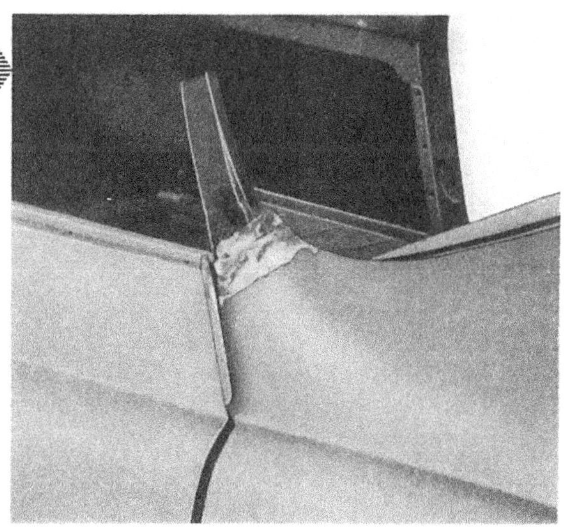

6 - Position the new windshield frame, insert the special jig for the windshield glass and secure with clamps.

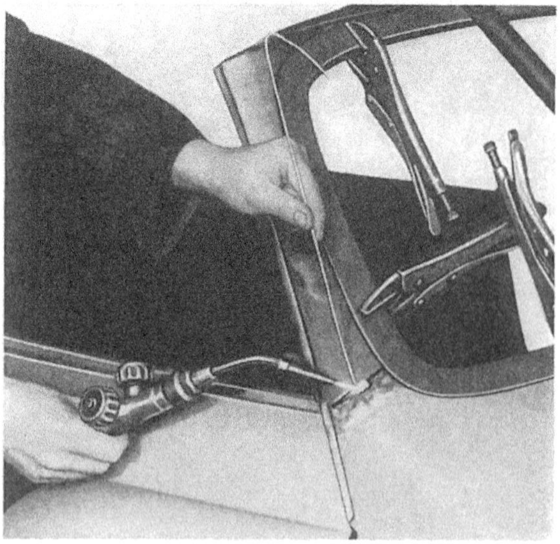

7 - Tack weld the windshield frame to the body.

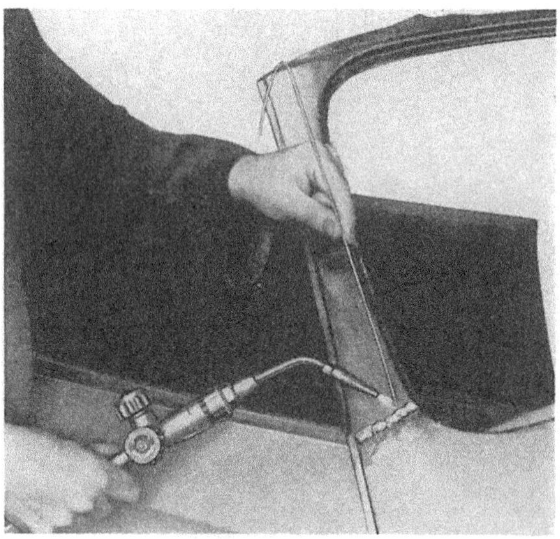

8 - Remove the windshield jig and gas weld the frame.

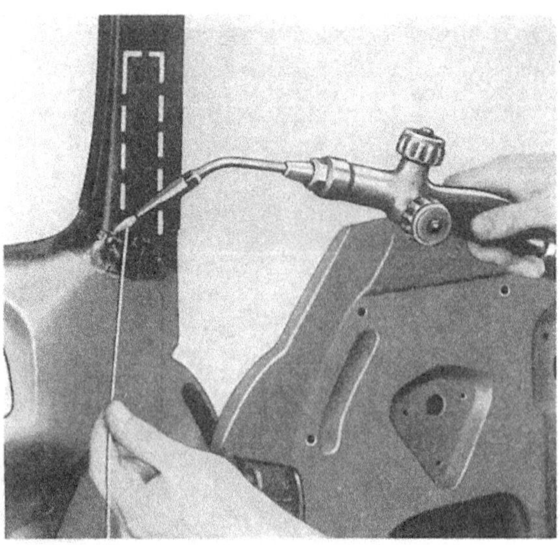

9 - At the bottom of each windshield pillar are three holes which enable the windshield frame to be gas welded to the reinforcement channels.

The dotted line indicates the position of the channel.

10 - Insert the windshield jig again and check the position of the frame. Correct if necessary.

11 - Grind the weld seams off as far as possible and coat all welded and repaired places with solder to ensure a uniform contour.

12 - Grind all reworked places smooth and prepare the vehicle for painting.

Spot Welding Pastes and Paint

A - **Spot welding paste**

In order to obtain water-tight joints, the parts of metal sheets which are to be welded are given a coating of sealing paste which is not affected by welding.

This standard manufacturing procedure should also be adopted for bodywork repairs.

The paste can be obtained direct from the manufacturers:

Manufacturer	Designation
Bonaval-Werk, Bonn, Germany, Brühler Straße 2—20	Spot welding paste 59 852 or 60 506
Teroson-Werke G.m.b.H., Heidelberg, Hans-Bunte-Straße 4, Germany	Spot welding paste 2257

The paste is applied before welding commences. Before and after welding the paste should not run. At the joints it should form a film which adheres firmly, prevents corrosion and renders the joint water-tight. It must also remain unaffected by degreasing agents and subsequent painting.

Paste which burns during spot welding should cease to do so as soon as the heat is removed.

The paste should only be employed where water-tight seams are stipulated and where it is impossible or difficult to apply sealing compound after welding.

B - **Spot welding paint**

Spot welding paint is used to prevent corrosion in hollow parts, which cannot be painted after welding due to inaccessibility.

Manufacturer	Designation
Teroson-Werke G.m.b.H., Heidelberg, Hans-Bunte-Straße 4	Spot welding paint 2273

Before welding, spray or paint all components with spot welding paint.

General

The use of the above products does not simplify the welding process. It may be necessary to step up the welding current, depending on the thickness of the coat applied.

Superfluous material must, however, be removed before the part is given a coat of paint which must stand up to the conditions stipulated for the outside finish. Neither welding paste nor paint can be considered as a substitute for primer to which a top coat can be applied.